PRIMER ON CEREBROVASCULAR DISEASES

SECOND EDITION

PRIMER ON CEREBROVASCULAR DISEASES

SECOND EDITION

Edited by

LOUIS R. CAPLAN
Department of Neurology
Beth Israel Deaconess Medical Center
Harvard Medical School
Boston, MA, United States

JOSÉ BILLER
Department of Neurology
Loyola University Chicago
Stritch School of Medicine
Maywood, IL, United States

MEGAN C. LEARY
Department of Neurology
Lehigh Valley Hospital and Health Network
Allentown, PA, United States
and
Morsani College of Medicine
University of South Florida
Tampa, FL, United States

ENG H. LO
Departments of Neurology and Radiology
Massachusetts General Hospital
Harvard Medical School
Boston, MA, United States

AJITH J. THOMAS
Division of Neurosurgery
Beth Israel Deaconess Medical Center
Harvard Medical School
Boston, MA, United States

MIDORI YENARI
Department of Neurology
University of California, San Francisco
San Francisco Veterans Affairs Medical School
San Francisco, CA, United States

JOHN H. ZHANG
Departments of Anesthesiology and Neurosurgery
Loma Linda University School of Medicine
Loma Linda, CA, United States

ACADEMIC PRESS

An imprint of Elsevier
elsevier.com

ELSEVIER

Academic Press is an imprint of Elsevier
125 London Wall, London EC2Y 5AS, United Kingdom
525 B Street, Suite 1800, San Diego, CA 92101-4495, United States
50 Hampshire Street, 5th Floor, Cambridge, MA 02139, United States
The Boulevard, Langford Lane, Kidlington, Oxford OX5 1GB, United Kingdom

Cover: Cover design by Lauren J. Lo. Images kindly provided by Dr. José Biller, Dr. Thomas P. Davis, Dr. Joe Herndon, Dr. Dong-Eog Kim, Ms. Caroline Sodja (NRC), and Dr. Danica Stanimirovic. Left panel: immunostaining of ZO-1 (red) and DAPI (blue) in cerebral endothelial cell cultures; Middle panel: angiogram of arteriovenous malformation; Right panel: infarct frequency map derived from a database of 400 diffusion-weighted MRI scans from patients with acute middle cerebral artery strokes; Background: IBA-positive (red) peri-vascular microglia and MHCII-positive (green) cerebral endothelium.

Notices
Knowledge and best practice in this field are constantly changing. As new research and experience broaden our understanding, changes in research methods, professional practices, or medical treatment may become necessary.

Practitioners and researchers must always rely on their own experience and knowledge in evaluating and using any information, methods, compounds, or experiments described herein. In using such information or methods they should be mindful of their own safety and the safety of others, including parties for whom they have a professional responsibility.

To the fullest extent of the law, neither the Publisher nor the authors, contributors, or editors, assume any liability for any injury and/or damage to persons or property as a matter of products liability, negligence or otherwise, or from any use or operation of any methods, products, instructions, or ideas contained in the material herein.

Library of Congress Cataloging-in-Publication Data
A catalog record for this book is available from the Library of Congress

British Library Cataloguing-in-Publication Data
A catalogue record for this book is available from the British Library

ISBN: 978-0-12-803058-5

For information on all Academic Press publications visit our
website at https://www.elsevier.com/books-and-journals

  Working together
to grow libraries in
developing countries

www.elsevier.com • www.bookaid.org

Publisher: Mara Conner
Acquisition Editor: Melanie Tucker
Editorial Project Manager: Kristi Anderson
Production Project Manager: Julia Haynes
Designer: Matt Limbert

Typeset by TNQ Books and Journals

Contents

SECTION III: NEUROPROTECTION

SECTION IV: MOLECULAR MECHANISMS

II

CLINICAL CHAPTERS

SECTION VI: CLINICAL ASPECTS: MEDICAL AND SURGICAL

List of Contributors

J.A. Abbatemarco   Lehigh Valley Hospital and Health Network, Allentown, PA, United States

R.J. Adams   Medical University of South Carolina, Charleston, SC, United States

D.L. Adkins   Medical University of South Carolina, Charleston, SC, United States

Y. Akamatsu   University of California, San Francisco and the San Francisco Veterans Affairs Medical Center, San Francisco, CA, United States; Tohoku University Graduate School of Medicine, Sendai, Japan

O. Akyol   Loma Linda University School of Medicine, Loma Linda, CA, United States

A.V. Alexandrov   The University of Tennessee Health Science Center, Memphis, TN, United States

I. Alim   Burke Medical Research Institute, White Plains, NY, United States; Weill Medical College of Cornell University, New York, NY, United States

A.M. Alkhachroum   University Hospitals Case Medical Center, Neurological Institute, Cleveland, OH, United States

S. Amin-Hanjani   University of Illinois at Chicago, Chicago, IL, United States

A.V. Andjelkovic   University of Michigan, Ann Arbor, MI, United States

J. Anrather   Weill Cornell Medical College, New York, NY, United States

R. Applegate II   Loma Linda University School of Medicine, Loma Linda, CA, United States

K. Arai   Harvard Medical School, Boston, MA, United States

C. Ayata   Harvard Medical School, Boston, MA, United States

M.A. Aziz-Sultan   Brigham and Women's Hospital, Harvard Medical School, Boston, MA, United States

I. Ballesteros   Universidad Complutense, Madrid, Spain; Memorial Sloan-Kettering Cancer Center, New York, NY, United States

B. Bar   Loyola University Chicago, Stritch School of Medicine, Maywood, IL, United States

F.C. Barone   SUNY Downstate Medical Center, New York, NY, United States

D.L. Barrow   Emory University School of Medicine, Atlanta, GA, United States

M.K. Başkaya   University of Wisconsin–Madison, Madison, WI, United States

K. Bateman   University of Cape Town, Cape Town, South Africa

N.G. Bazan   Louisiana State University Health New Orleans, New Orleans, LA, United States

J.S. Beecher   UT Southwestern Medical Center, Dallas, TX, United States

A. Beer-Furlan   Wexner Medical Center, The Ohio State University, Columbus, OH, United States

L. Belayev   Louisiana State University Health New Orleans, New Orleans, LA, United States

P. Bhattacharya   Saint Joseph Mercy Oakland, Pontiac, MI, United States

R. Bhole   University of Tennessee Health Science Center, Memphis, TN, United States

J. Biller   Loyola University Chicago, Stritch School of Medicine, Maywood, IL, United States

V. Biousse   Emory University School of Medicine, Atlanta, GA, United States

C.V. Borlongan   University of South Florida Morsani College of Medicine, Tampa, FL, United States

M.J.R.J. Bouts   University Medical Center Utrecht, Utrecht, The Netherlands; Massachusetts General Hospital, Charlestown, MA, United States; Leiden University, Leiden, The Netherlands; Leiden University Medical Center, Leiden, The Netherlands

R.L. Brey   University of Texas Health Science Center at San Antonio, San Antonio, TX, United States

R. Bronstein   Stony Brook University, Stony Brook, NY, United States

A. Bryer   University of Cape Town, Cape Town, South Africa

K.R. Bulsara   Yale University School of Medicine/Yale New Haven Hospital, New Haven, CT, United States

A. Can   Harvard Medical School, Boston, MA, United States

P. Canhão   University of Lisbon, Lisbon, Portugal

L.R. Caplan   Harvard University, Beth Israel Deaconess Medical Center, Boston, MA, United States

S.T. Carmichael   University of California Los Angeles, Los Angeles, CA, United States

R. Carrau   Wexner Medical Center, The Ohio State University, Columbus, OH, United States

J. Castaldo   Lehigh Valley Hospital and Health Network, Allentown, PA, United States

L. Catanese   Harvard University, Beth Israel Deaconess Medical Center, Boston, MA, United States

H. Chabriat   Centre de référence pour les maladies rares des vaisseaux du cerveau et de l'œil (CERVCO), DHU-NeuroVasc and INSERM U1161, Université Denis Diderot, Paris, France

S. Chaturvedi   University of Miami Miller School of Medicine, Miami, FL, United States

N. Chaudhary   University of Michigan, Ann Arbor, MI, United States

Jieli Chen   Henry Ford Hospital, Detroit, MI, United States

S. Chen   Zhejiang University, Hangzhou, Zhejiang, China

Jun Chen   University of Pittsburgh, Pittsburgh, PA, United States; Fudan University, Shanghai, China

D.W. Choi   State University of New York at Stony Brook, Stony Brook, NY, United States; Korea Institute of Science and Technology, Seoul, South Korea

B. Choi   Massachusetts General Hospital and Harvard Medical School, Boston, MA, United States

M. Chopp   Henry Ford Hospital, Detroit, MI, United States; Oakland University, Rochester, MI, United States

D.Y. Chung   Harvard Medical School, Boston, MA, United States

C.-P. Chung   Taipei Veterans General Hospital, National Yang Ming University, Taipei, Taiwan

M.J. Cipolla   University of Vermont, Burlington, VT, United States

F. Colbourne   University of Alberta, Edmonton, AB, Canada

Q. Colburn   University of South Florida Morsani College of Medicine, Tampa, FL, United States

B.J. Cord   Yale University School of Medicine/Yale New Haven Hospital, New Haven, CT, United States

B.M. Coull   The University of Arizona College of Medicine, Tucson, AZ, United States

M.I. Cuartero   Universidad Complutense, Madrid, Spain; Instituto de Investigación Hospital 12 de Octubre (i+12), Madrid, Spain

J.L. Cummings   Cleveland Clinic Las Vegas, NV, United States; Cleveland Clinic Lerner College of Medicine of Case Western Reserve University, Cleveland, OH, United States

R.M. Dafer   Rush University Medical Center, Chicago, IL, United States

T. Dalkara   Hacettepe University, Ankara, Turkey

B. Daou   Thomas Jefferson University and Jefferson Hospital for Neuroscience, Philadelphia, PA, United States

K.R. Dave   University of Miami, Miami, Florida, United States

T.P. Davis   University of Arizona, Tucson, AZ, United States

M. De Georgia   University Hospitals Case Medical Center, Neurological Institute, Cleveland, OH, United States

T.M. De Silva   The University of Iowa Carver College of Medicine, Iowa City, IA, United States; Monash University, Clayton, VIC, Australia

A. Dharap   JFK Medical Center, Edison, NJ, United States

M.R. Di Tullio   Columbia University, New York, NY, United States

W.D. Dietrich   University of Miami Miller School of Medicine, Miami, FL, United States

R.M. Dijkhuizen   University Medical Center Utrecht, Utrecht, The Netherlands

B.H. Dobkin   University of California Los Angeles, Los Angeles, CA, United States

R. Du   Harvard Medical School, Boston, MA, United States

A.F. Ducruet   University of Pittsburgh, Pittsburgh, PA, United States

K.R. Duncan   Lehigh Valley Hospital and Health Network, Allentown, PA, United States

L. Edvinsson   Lund University Hospital, Lund, Sweden

M.J. Edwards   St Georges University of London, London, United Kingdom

E. Egemen   Koç University Hospital, Istanbul, Turkey

M. El-Hunjul   Lehigh Valley Hospital and Health Network, Allentown, PA, United States

M. Emanuele   Loyola University Chicago, Stritch School of Medicine, Maywood, IL, United States

N. Emanuele   Hines VA Medical Center, Hines, IL, United States

M.K. Erdman   Los Angeles County Hospital and USC Medical Center, Los Angeles, CA, United States

A. Ergul   Augusta University, Augusta, GA, United States

S.C. Fagan   University of Georgia College of Pharmacy, Augusta, GA, United States

F.M. Faraci   The University of Iowa Carver College of Medicine, Iowa City, IA, United States

C. Federau   Stanford University, Stanford, CA, United States

J.M. Ferro   University of Lisbon, Lisbon, Portugal

M. Fisher   University of California, Irvine, Irvine, CA, United States

K.D. Flemming   Mayo Clinic, Rochester, MN, United States

C. Foerch   Goethe University, Frankfurt am Main, Germany

R.S. Freitas   Louisiana State University Health New Orleans, New Orleans, LA, United States

R.M. Friedlander   University of Pittsburgh, Pittsburgh, PA, United States

T. Gaberel   Harvard Medical School, Boston, MA, United States

C. Gakuba   Harvard Medical School, Boston, MA, United States

R.G. Giffard   Stanford University School of Medicine, Stanford, CA, United States

M.P. Goldberg   UT Southwestern Medical Center, Dallas, TX, United States

R.G. González   Harvard Medical School, Boston, MA, United States

S. Gopinath   Baylor College of Medicine, Houston, TX, United States

P.B. Gorelick   Mercy Health Hauenstein Neurosciences, Grand Rapids, MI, United States; Michigan State University College of Human Medicine, East Lansing, MI, United States

C. Goshgarian Mercy Health Hauenstein Neurosciences, Grand Rapids, MI, United States

D.A. Greenberg Buck Institute for Research on Aging, Novato, CA, United States

C.J. Griessenauer Harvard Medical School, Boston MA, United States

K.A. Groshans Walter Reed National Military Medical Center, Bethesda, MD, United States

R. Gupta Wellstar Health System, Marietta, GA, United States

R.A. Hachem Wexner Medical Center, The Ohio State University, Columbus, OH, United States

Z.A. Hage University of Illinois at Chicago, Chicago, IL, United States

E.D. Hall University of Kentucky College of Medicine, Lexington, KY, United States

E. Hamel McGill University, Montréal, QC, Canada

Q. Hao The Johns Hopkins University School of Medicine, Baltimore, MD, United States

A.S. Haqqani National Research Council of Canada, Ottawa, ON, Canada

R. Hariman Medical College of Wisconsin, Milwaukee, WI, United States

D. Hasan University of Iowa Hospital and Clinics, Iowa City, IA, United States

D.C. Haussen Emory University School of Medicine, Atlanta, GA, United States

L. He Vanderbilt University, Nashville, TN, United States

D.M. Heiferman Loyola University Chicago, Stritch School of Medicine, Maywood, IL, United States

J.M. Herndon University of Arizona, Tucson, AZ, United States

W.M. Ho Loma Linda University School of Medicine, Loma Linda, CA, United States

S. Hoffmann Charité – Universitätsmedizin Berlin, Berlin, Germany

B.M. Howard Emory University School of Medicine, Atlanta, GA, United States

B.R. Hu Shock Trauma and Anesthesiology Research Center, University of Maryland School of Medicine, Baltimore, MD, United States

J.D. Huber West Virginia University, Morgantown, WV, United States

B. Huisa University of California San Diego, San Diego, CA, United States

P.D. Hurn University of Michigan, School of Nursing, Ann Arbor, MI, United States

J.J. Iliff Oregon Health & Science University, Portland, OR, United States; University of Rochester Medical Center, Rochester, NY, United States

P. Jabbour Thomas Jefferson University and Jefferson Hospital for Neuroscience, Philadelphia, PA, United States

A.O. Jamshidi Wexner Medical Center, The Ohio State University, Columbus, OH, United States

B. Jankowitz University of Pittsburgh Medical Center, Pittsburgh, PA, United States

G.C. Jickling University of California at Davis, Sacramento, CA, United States

M. Johansen The Johns Hopkins University School of Medicine, Baltimore, MD, United States

T.G. Jovin University of Pittsburgh Medical Center, Pittsburgh, PA, United States

S.S. Karuppagounder Burke Medical Research Institute, White Plains, NY, United States; Weill Medical College of Cornell University, New York, NY, United states

E.M. Kasper Harvard Medical School, Boston, MA, United States

R.F. Keep University of Michigan, Ann Arbor, MI, United States

H.-H. Kim Harvard Medical School, Boston, MA, United States

D.E. Kim Dongguk University, Goyang, Republic of Korea

J.S. Kim Asan Medical Center, University of Ulsan, Seoul, South Korea

J.Y. Kim University of California, San Francisco and the San Francisco Veterans Affairs Medical Center, San Francisco, CA, United States

A.C. Klahr University of Alberta, Edmonton, AB, Canada

M.J. Koch Massachusetts General Hospital and Harvard Medical School, Boston, MA, United States

M. Kole Henry Ford Health System, Detroit, MI, United States

S.M. Koleilat The University of Arizona College of Medicine, Tucson, AZ, United States

A. Kozan University of Wisconsin–Madison, Madison, WI, United States

S. Kuroda University of Toyama, Toyama, Japan; Hokkaido University Graduate School of Medicine, Sapporo, Japan

C. Lamy Paris Descartes University, Paris, France

G. Lanzino Mayo Clinic, Rochester, MN, United States

A.G. Larsen Harvard Medical School, Boston, MA, United States

Y. Laviv Harvard Medical School, Boston, MA, United States

M.T. Lawton University of California, San Francisco, San Francisco, CA, United States

M.C. Leary Lehigh Valley Hospital and Health Network, Allentown, PA, United States; University of South Florida, Tampa, FL, United States

E.C. Leira University of Iowa, Iowa City, IA, United States

L. Li Stanford University School of Medicine, Stanford, CA, United States

Q. Li The Johns Hopkins University School of Medicine, Baltimore, MD, United States

D.S. Liebeskind University of California, Los Angeles, Los Angeles, CA, United States

L. Lin Harvard Medical School, Boston, MA, United States; Wenzhou Medical University Wenzhou, People's Republic of China

V.A. Lioutas Beth Israel Deaconess Medical Center, Boston, MA, United States

T. Lippert University of South Florida Morsani College of Medicine, Tampa, FL, United States

R. Liu University of North Texas Health Science Center, Fort Worth, TX, United States

J. Liu University of California, San Francisco and the San Francisco Veterans Affairs Medical Center, San Francisco, CA, United States

C.L. Liu Shock Trauma and Anesthesiology Research Center, University of Maryland School of Medicine, Baltimore, MD, United States

I. Lizasoain Universidad Complutense, Madrid, Spain; Instituto de Investigación Hospital 12 de Octubre (i+12), Madrid, Spain

E.H. Lo Harvard Medical School, Boston, MA, United States

C.M. Loftus Loyola University Chicago, Stritch School of Medicine, Maywood, IL, United States

A.F. Logsdon West Virginia University, Morgantown, WV, United States

B.P. Lucke-Wold West Virginia University, Morgantown, WV, United States

S. Madhavan University of Illinois Chicago, Chicago, IL, United States

V. Madhugiri Stanford University School of Medicine, Stanford, CA, United States

K. Malhotra University of California, Los Angeles, Los Angeles, CA, United States

W.J. Manning Harvard Medical School, Boston, MA, United States

S.J. Marcell Louisiana State University Health New Orleans, New Orleans, LA, United States

J.-L. Mas Paris Descartes University, Paris, France

K. Masamoto University of Electro-Communications, Chofu, Tokyo, Japan

C. Matute Achucarro Basque Center for Neuroscience, Zamudio, Spain; CIBERNED, Madrid, Spain; Universidad del País Vasco-UPV/EHU, Leioa, Spain

L.D. McCullough The University of Texas Health Science Center at Houston, Houston, TX, United States

M.M. McDowell University of Pittsburgh, Pittsburgh, PA, United States

M. Mehdiratta Trillium Health Partners, Mississauga, ON, Canada; University of Toronto, Toronto, ON, Canada

D. Mehta Lehigh Valley Hospital and Health Network, Allentown, PA, United States

A. Meisel Charité – Universitätsmedizin Berlin, Berlin, Germany

J. Messegee University of New Mexico, Albuquerque, NM, United States

B. Miller University Hospitals Case Medical Center, Neurological Institute, Cleveland, OH, United States

S. Mirza UT Southwestern Medical Center, Dallas, TX, United States

J.M. Modak Beth Israel Deaconess Medical Center, Boston, MA, United States

M.A. Moro Universidad Complutense, Madrid, Spain; Instituto de Investigación Hospital 12 de Octubre (i+12), Madrid, Spain

M.A. Nagel University of Colorado School of Medicine, Aurora, CO, United States

S. Namura Morehouse School of Medicine, Atlanta, GA, United States

M. Nedergaard University of Rochester Medical Center, Rochester, NY, United States; University of Copenhagen, Copenhagen, Denmark

D.W. Newell Seattle Neuroscience Institute, Seattle, WA, United States

N.J. Newman Emory University School of Medicine, Atlanta, GA, United States

K.L. Ng University of California Los Angeles, Los Angeles, CA, United States

D. Nguyen University of California San Diego, San Diego, CA, United States

H. Nguyen University of South Florida Morsani College of Medicine, Tampa, FL, United States

G. Nielsen UCL Institute of Neurology, London, United Kingdom

Y. Nishijima University of California, San Francisco and the San Francisco Veterans Affairs Medical Center, San Francisco, CA, United States; Tohoku University Graduate School of Medicine, Sendai, Japan

N. Nishimura Cornell University, Ithaca, NY, United States

R.G. Nogueira Emory University School of Medicine, Atlanta, GA, United States

C.S. Ogilvy Harvard Medical School, Boston, MA, United States

D.B. Orbach Boston Children's Hospital/Harvard Medical School, Boston, MA, United States

A.P. Ostendorf Ohio State College of Medicine, Columbus, OH, United States

B. Otto Wexner Medical Center, The Ohio State University, Columbus, OH, United States

A. Ozpinar University of Pittsburgh Medical Center, Pittsburgh, PA, United States

D.M. Panczykowski University of Pittsburgh Medical Center, Pittsburgh, PA, United States

A.B. Patel Massachusetts General Hospital and Harvard Medical School, Boston, MA, United States

Y. Perez Trillium Health Partners, Mississauga, ON, Canada; University of Toronto, Toronto, ON, Canada

M.A. Perez-Pinzon University of Miami, Miami, Florida, United States

C. Potey University of British Columbia, Vancouver, BC, Canada

J.M. Pradillo Universidad Complutense, Madrid, Spain; Instituto de Investigación Hospital 12 de Octubre (i+12), Madrid, Spain

D.M. Prevedello Wexner Medical Center, The Ohio State University, Columbus, OH, United States

K. Rajamani Wayne State University School of Medicine, Detroit, MI, United States

L. Rangel-Castilla University at Buffalo, State University of New York, Buffalo, NY, United States

N.M. Rao David Geffen School of Medicine at University of California, Los Angeles, Los Angeles, CA, United States

R.R. Ratan Burke Medical Research Institute, White Plains, NY, United States; Weill Medical College of Cornell University, New York, NY, United States

A.P. Raval University of Miami, Miami, Florida, United States

G.D. Reddy Baylor College of Medicine, Houston, TX, United States

C. Reis Loma Linda University Medical Center, Loma Linda, CA, United States

E.S. Roach Ohio State College of Medicine, Columbus, OH, United States

P.T. Ronaldson University of Arizona, Tucson, AZ, United States

C.L. Rosen West Virginia University, Morgantown, WV, United States

G.A. Rosenberg The University of New Mexico, Albuquerque, NM, United States

W.C. Rutledge University of California, San Francisco, San Francisco, CA, United States

R. Sabzwari Loyola University Medical Center and Edward Hines Jr. Veteran Administration Hospital, Hines, IL, United States

G. Salzano Northeastern University, Boston, MA, United States

P.A. Santucci Loyola University Chicago, Stritch School of Medicine, Maywood, IL, United States

J.L. Saver David Geffen School of Medicine at University of California, Los Angeles, Los Angeles, CA, United States

T. Schallert The University of Texas at Austin, Austin, TX, United States

M.L. Schermerhorn Beth Israel Deaconess Medical Center, Boston, MA, United States

M.J. Schneck Loyola University Chicago, Stritch School of Medicine, Maywood, IL, United States

A.P. See Brigham and Women's Hospital; Boston Children's Hospital/Harvard Medical School, Boston, MA, United States

H.J. Shakir University at Buffalo, State University of New York, Buffalo, NY, United States

F.R. Sharp University of California at Davis, Sacramento, CA, United States

F. Shuja Beth Israel Deaconess Medical Center, Boston, MA, United States

A.H. Siddiqui University at Buffalo, State University of New York, Buffalo, NY, United States

M.A. Silva Harvard Medical School, Boston, MA, United States

A.B. Singhal Massachusetts General Hospital and Harvard Medical School, Boston, MA, United States

K. Sivakumar Lehigh Valley Hospital and Health Network, Allentown, PA, United States

D.H. Slade Loyola University Medical Center and Edward Hines Jr. Veteran Administration Hospital, Hines, IL, United States

E.R. Smith Harvard Medical School, Boston, MA, United States

F. Sohrabji Texas A&M Health Science Center, Bryan, TX, United States

I. Solaroglu Koç University, Istanbul, Turkey

S.K. Sriraman Northeastern University, Boston, MA, United States

B. Stamova University of California at Davis, Sacramento, CA, United States

D.B. Stanimirovic National Research Council of Canada, Ottawa, ON, Canada

C.J. Stapleton Massachusetts General Hospital and Harvard Medical School, Boston, MA, United States

C.M. Stary Stanford University School of Medicine, Stanford, CA, United States

G.K. Steinberg Stanford University School of Medicine, Stanford, CA, United States

C. Stephen Massachusetts General Hospital, Boston, MA, United States

R.A. Stetler University of Pittsburgh, Pittsburgh, PA, United States; Fudan University, Shanghai, China

J. Stone University of Edinburgh, Edinburgh, United Kingdom

R. Sumbria Keck Graduate Institute, Claremont, CA, United States; University of California, Irvine, Irvine, CA, United States

R. Sweis Loyola University Chicago, Stritch School of Medicine, Maywood, IL, United States

R. Tahir Henry Ford Health System, Detroit, MI, United States

R. Tarawneh Cleveland Clinic, Cleveland, OH, United States; Cleveland Clinic Lerner College of Medicine of Case Western Reserve University, Cleveland, OH, United States

J. Tarsia Ochsner Health Systems, New Orleans, LA, United States

R. Tehrani Loyola University Chicago, Stritch School of Medicine, Maywood, IL, United States

M.K. Teo   Stanford University School of Medicine, Stanford, CA, United States

F.D. Testai   University of Illinois at Chicago, Chicago, IL, United States

A.S. Thrane   University of Rochester Medical Center, Rochester, NY, United States; Haukeland University Hospital, Bergen, Norway

M.K. Tobin   University of Illinois at Chicago, Chicago, IL, United States

M.E. Tome   University of Arizona, Tucson, AZ, United States

M.A. Topcuoglu   Massachusetts General Hospital and Harvard Medical School, Boston, MA, United States; Hacettepe University Hospitals, Ankara, Turkey

C.H. Topel   University of Texas Health Science Center at San Antonio, San Antonio, TX, United States

V. Torchilin   Northeastern University, Boston, MA, United States; King Abdulaziz University, Jeddah, Saudi Arabia

R.J. Traystman   University of Colorado Denver, Aurora, CO, United States

S.E. Tsirka   Stony Brook University, Stony Brook, NY, United States

Y. Turan   University of Wisconsin–Madison, Madison, WI, United States

M. Tymianski   Krembil Research Institute, Toronto, ON, Canada; University of Toronto, Toronto, ON, Canada; University Health Network, Toronto, ON, Canada

K. van Leyen   Massachusetts General Hospital, Charlestown, MA, United States; Harvard Medical School, Boston, MA, United States

P. Varade   Lehigh Valley Hospital and Health Network, Allentown, PA, United States; University of South Florida, Tampa, FL, United States

J.S. Veluz   St. Mary Medical Center, Langhorne, PA, United States

R. Vemuganti   University of Wisconsin, Madison, WI, United States

P. Venkat   Henry Ford Hospital, Detroit, MI, United States; Oakland University, Rochester, MI, United States

Z.S. Vexler   University of California, San Francisco, San Francisco, CA, United States

C.M. Vial   Sutter Health/Palo Alto Medical Foundation, Palo Alto, CA, United States

H.V. Vinters   David Geffen School of Medicine at University of California, Los Angeles, Los Angeles, CA, United States

M.R. Vosko   Kepler Universitätsklinikum, Linz, Austria

C. Waeber   University College Cork, Cork, Ireland

B.P. Walcott   University of California, San Francisco, San Francisco, CA, United States

J. Wang   The Johns Hopkins University School of Medicine, Baltimore, MD, United States

X. Wang   Harvard Medical School, Boston, MA, United States

Y.T. Wang   University of British Columbia, Vancouver, BC, Canada

Z.Z. Wei   Emory University School of Medicine and Atlanta Veterans Affairs Medical Center, Decatur, GA, United States

L. Wei   Emory University School of Medicine, Atlanta, GA, United States

B.G. Welch   UT Southwestern Medical Center, Dallas, TX, United States

H.R. Winn   Mount Sinai Medical School, New York, NY, United States; University of Iowa, Iowa City, IA, United States

M. Wintermark   Stanford University, Stanford, CA, United States

R.J. Wityk   The Johns Hopkins University School of Medicine, Baltimore, MD, United States

O. Wu   Massachusetts General Hospital, Charlestown, MA, United States

K.C. Wu   Brigham and Women's Hospital, Boston, Harvard Medical School, MA, United States

G. Xi   University of Michigan, Ann Arbor, MI, United States

H.A. Yacoub   Lehigh Valley Hospital and Health Network, Allentown, PA, United States

A. Yakhkind   Brown University, Providence, RI, United States

Y. Yamamoto   Kyoto Katsura Hospital, Kyoto, Japan

S.-H. Yang   University of North Texas Health Science Center, Fort Worth, TX, United States

M. Yenari   University of California, San Francisco and the San Francisco Veterans Affairs Medical Center, San Francisco, CA, United States

K. Yigitkanli   Polatli Government Hospital, Ankara, Turkey

H. Yonas   University of New Mexico, Albuquerque, NM, United States

Z. Yu   Harvard Medical School, Boston, MA, United States

S.L. Zettervall   Beth Israel Deaconess Medical Center, Boston, MA, United States

J. Zhang   Loma Linda University Medical Center; Loma Linda University School of Medicine, Loma Linda, CA, United States

W. Zhang   University of Pittsburgh, Pittsburgh, PA, United States; Fudan University, Shanghai, China

H. Zhao   Stanford University, Stanford, CA, United States

Introduction

Twenty years have passed since the first edition of *Primer on Cerebrovascular Diseases* was published.[1] The book sought to reduce the growing gap in the cerebrovascular field between physicians and surgeons who actively treated patients and researchers who worked in basic and clinical research. The term "translational medicine" was first being discussed at that time. All agreed that the best way to ensure progress was intimate communication and cooperation between clinicians and researchers. Clinicians needed to have some sense about what was happening and forthcoming from the laboratory and researchers needed to know what were the most important targets to help patient care at the bedside and in the clinic. Dr. Arthur Kornberg, who received the Nobel Prize in Physiology and Medicine in 1959 for his work on DNA, commented in his autobiography[2] that the single most important year in his training was his clinical internship. That exposure provided targets for needed advancement for his entire career, which was spent in various basic research laboratories.

During the past two decades since publication of the first edition, the clinical-research gap has probably widened. Clinicians and surgeons have become even more specialized, each dealing with more restricted situations, technology, compounds, and conditions. Basic researchers have had to become even more competitive for grants. Many work in very specialized areas. I have been at Princeton Cerebrovascular Disease Conference meetings in which all attendees are instructed to sit through all sessions—researchers listening to clinical topics and clinicians taking in research discussions. My sense was that these did not work well. Clinical and research topics were too focused; researchers lacked the clinical background to place the discussions into perspective and clinicians were at sea in the biochemical and technical details of the basic discussions. A few days was too short a time for the education needed.

This second edition of the Primer is aimed directly at providing a clinical-research interface, a repository of information that is basic, concise, simply written, and easily understood for individuals who are unfamiliar with a particular topic. Unlike a short meeting, a volume (hard copy or e-book) can serve as a frequently perused source of information that can bridge a large educational–informational gap. This edition has expanded with more editors and more topics. Editors have carefully selected authors who are working within their topics. They are instructed to make their chapters concise and easily understood. The volume has been thoroughly edited to ensure simplicity and completeness. I hope that it will help reduce the gap and aid progress in translational research and in the clinical care of future stroke patients.

Louis R. Caplan, MD
Boston, Massachusetts
November 2016

1. Caplan LR, Siesjo BK, Weir B, Welch KM , Reis DJ. *Primer on Cerebrovascular Diseases*, San Diego, Academic press, 1997
2. Kornberg A. *For the love of enzymes. The odyssey of a biochemist.* Harvard U press, Cambridge, 1989

BASIC SCIENCES

ANATOMY AND PHYSIOLOGY

SECTION

ANATOMY AND PHYSIOLOGY

1

Cerebrovascular Anatomy and Hemodynamics

R.J. Traystman

University of Colorado Denver, Aurora, CO, United States

INTRODUCTION

The adult human brain represents about 2% of total body weight, but receives nearly 15% of total resting cardiac output. Under normal conditions, the brain is highly perfused and is extremely sensitive to any change or interruption in its blood supply. If the brain's circulation is completely obstructed, loss of consciousness occurs within seconds and irreversible pathological changes occur within minutes. For example, in cardiac arrest, the extent of injury of the central nervous system (CNS) is the critical factor that determines the degree of recovery. It is, therefore, not surprising that the physiological mechanisms that regulate cerebral circulation are designed to ensure the constancy of cerebral blood flow (CBF) over a broad range of internal and external conditions. This may even occur at the expense of adequate blood flow to other organs.

ANATOMICAL CONSIDERATIONS

Arterial System

The brain of essentially all mammalian species is supplied with blood from several major sources, that is, the internal and external carotid, vertebral, and spinal anterior arteries. However, the relative importance of these channels in any species is unclear. Although the internal carotid artery leads directly to the brain, in some species this vessel is unimportant, and it may be the external carotid that carries the major proportion of blood reaching the brain. In humans, the anterior three-fifths of the cerebrum, except for parts of the occipital and temporal lobes, are supplied by the carotid arteries. The posterior two-fifths of the cerebrum, the cerebellum, and brain stem are supplied by the vertebral–basilar system. The carotid and vertebral arteries unite at the base of the brain to form the circle of Willis (Fig. 1.1 [1]). This vascular ring then gives rise to three pairs of arteries, the anterior, middle, and posterior cerebral arteries, which cover the external surface of the corresponding regions of the cerebral cortex. These arteries divide into progressively smaller arteries, penetrating brain tissue and supplying blood to specific regions. Branches of the vertebral and basilar arteries form the blood supply for the cerebellum and brain stem. While there is some variability among individuals, the internal carotids and the vertebral–basilar system generally contribute equally to the circle of Willis. Even though the internal carotid and basilar arteries converge (forming the circle of Willis), blood from the two tributaries normally does not mix completely because blood pressure in each arterial tributary is almost equal. Angiography and/or dye injections indicate that blood from the various tributaries is ultimately distributed to relatively specific and delineated brain regions. Under normal conditions, vertebral–basilar arterial blood is mainly distributed to tissues in the posterior fossa while the internal carotids supply the remainder of the brain. In addition, there is relatively little bilateral crossing, again due to the similarity in blood pressure. Normally the circle of Willis functions primarily as an anterior–posterior shunt than as a side-to-side shunt. However, under pathological conditions, especially those that involve focal obstructions in arterial feeders to the circle, the balance of pressures may be altered and the circle of Willis can then serve either as an anterior–posterior or as a side-to-side shunt.

In addition, there are a number of arterial anastomotic vessels on each side of the head between the intracranial and extracranial circulations. These include: (1) a connection between the vertebral and occipital arteries; (2) a communication between the ascending pharyngeal and internal carotid arteries; (3) the middle meningeal artery branching off from the internal maxillary artery and connecting with the internal carotid artery, (4) the anastomotic

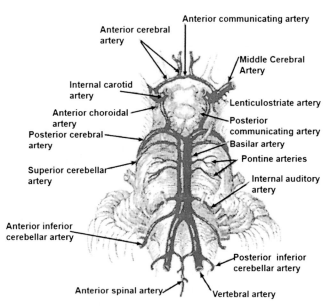

FIGURE 1.1 Major cerebral arteries and the circle of Willis. *Modified from Chusid JG. Correlative neuroanatomy and functional neurology. 14th ed. Los Altos, CA: Lange Medical Publishers; 1970.*

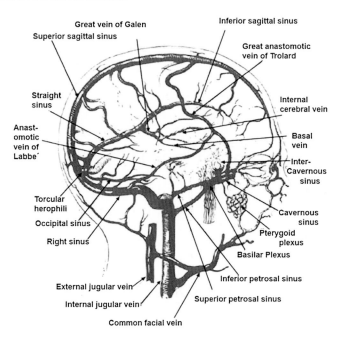

FIGURE 1.2 The brain venous system. *Modified from Shenkin HA, Harmel MH, Kety SS. Dynamic anatomy of the cerebral circulation. Arch Neurol Psychiat 1948;60:240–52.*

artery between the internal maxillary and internal carotid arteries, (5) pathways between the external and internal ophthalmic arteries, (6) anastomosis between the external and internal ethmoidal arteries, (7) collaterals between the vertebral and the omocervical arteries, and (8) connections between the spinal anterior and vertebral arteries. In some species the external carotid system branches into a complicated network of arteries, the rete mirable, prior to its entrance to the circle of Willis. This rete system has been proposed to be involved in a heat-exchange countercurrent mechanism, which acts to lower the temperature of the blood entering the brain.

As the major arteries leave the circle of Willis they reduce their diameter to become arterioles and pial vessels. Pial arteries then plunge at a 90 degree angle into brain parenchyma. There is much evidence that there is a close relationship between pial vessels and the leptomeninges. These vessels, as they enter the parenchyma, are invested with a leptomeningeal sheath and are surrounded by a cerebrospinal fluid (CSF)–containing space. It should be mentioned that most studies of cerebral vessels using methods of staining and light microscopy have not shown differences between brain vessels and vessels in other organs.

Venous System

While the brain's arterial system is complicated, the configuration of the brain's venous system is even more complex and provides many opportunities for mixing of blood draining various brain regions (Fig. 1.2 [2]). Blood is drained from the brain via two primary sets of veins:

the external group and the deep or internal group. These drain into the dural sinuses and then the internal jugular veins. The external venous system is divided into the superior, middle, inferior, and occipital cerebral veins, which drain the outer portion of the cerebral hemispheres. The superior cerebral vein drains the cortex and underlying white matter above the corpus callosum. Several veins on each side merge to form three large trunks, which enter the superior sagittal sinus or straight sinus. The most prominent superior cerebral vein is the great anastomotic vein of Trolard connecting the superior sagittal sinus with the Sylvian vein. The internal cerebral or deep veins include a variety of small transcerebral veins draining the bulk of white matter from the anterior and middle group of the brain. This system eventually drains through the great vein of Galen and the straight sinus. The cerebellum is drained primarily by two sets of veins. The inferior cerebellar veins are larger and end in the transverse superior petrosal and occipital sinuses. The superior cerebellar veins are smaller and empty in part into transverse and superior petrosal sinuses and in part into the great vein of Galen and the straight sinus. The brain stem is drained by veins terminating in the inferior and transverse petrosal sinuses. Veins from all parts of the brain drain into many sinuses situated between two layers of dura, that is, superior sagittal sinus, inferior sagittal sinus, occipital sinus, superior petrosal sinus, cavernous sinus, and transverse sinus. Extensive intervenous collateral anastomoses exist between the two main venous draining systems and with the extracranial venous draining system.

Capillary System

The brain contains a rich network of capillaries; however, the density of capillaries within the CNS is less than that in the heart, kidney, and muscle. Gray matter contains about 2½times as many capillaries as does white matter; that is, cerebral cortex has about 1000 capillaries/mm^3 and white matter about 300 capillaries/mm^3. It has been proposed that capillary density is correlated with the number of synapses in a particular brain region. The high density of capillaries in the cervical sympathetic ganglion, which contains synapses, compared to that in the trigeminal ganglion, which lacks them, demonstrates this point. Oxygen consumption may be the link between capillary density and synaptic frequency. Support for this hypothesis comes from work demonstrating that glucose utilization of gray matter is greater than that of white matter by a factor similar to the ratio of capillary densities for the two tissues. Exposure of experimental animals to hypoxia for a long period leads to an increase in capillary density. Thus, a lack of oxygen must be either a direct or indirect stimulus to capillary growth; however, the precise mechanism responsible for this increased vascularity remains unknown. Cerebral capillary density also varies with age and it has been demonstrated that capillary density at birth is about 30% of that in the adult and is even lower in the premature infant.

Blood–Brain Barrier and Capillary Permeability

The concept of a blood–brain barrier (BBB) arose from the work of a number of investigators who demonstrated that certain dyes and pharmacologically active compounds did not enter the brain but could enter most other organs. The barrier had to be a vascular one because the same substances would readily enter the brain when injected directly into the CSF. It was subsequently shown that the dyes that could not penetrate the brain from the vascular side were bound to plasma proteins, so that the barrier was actually dye–protein complexes. Morphological localization of the barrier to circulating protein has been shown using horseradish peroxidase, a protein that could be localized by electron microscopy [3].

The BBB separates two of the major compartments of the CNS, the brain and CSF, from the third compartment, the blood. The sites of the barrier are the interfaces between the blood and these two compartments: the choroid plexus, the blood vessels of the brain and subarachnoid space, and the arachnoid membrane. All BBB sites are characterized by cells connected by tight junctions, which restrict intercellular diffusion. These cells are represented by endothelia of blood vessels, epithelia of the choroid plexus, and cells of the arachnoid layer. When cells are connected via tight junctions, they act as if they were one single layer of cells and solute exchange occurs transcellularly. These cells thus determine the solubility and transport functions of the entire layer of cells. Lipid-soluble substances penetrate easily and equilibrate between brain and blood quickly. There is only minimal transport by pinocytotic vessels at the barrier site, and this in addition to the tight junctions limits protein transport into the extracellular fluid.

While passage of more permeable substances (sugars and amino acids) into brain is determined by both CBF and permeability characteristics, the permeability of ions and large molecules depends largely on the characteristics of the BBB membrane rather than blood flow. Nonelectrolytes of small molecular weight penetrate faster than their lipid solubilities and diffusion coefficients would predict. Thus, the presence of water channels across the barrier is likely. The BBB may not be equally permeable in all areas of the brain. For example, area postrema, choroid plexus, hypophysis, pineal, and areas in the hypothalamus have no BBB at all. In these areas cerebral capillaries have fenestrations, and there are a large number of pinocytotic vesicles in the endothelial cells. Besides these areas, within brain there are moderate differences in barrier permeability in different regions. The entry of most solutes into gray matter is about three to four times faster than into white matter. This may be correlated with a similar difference in the length of capillaries per unit volume of gray compared to white matter.

Breakdown of the BBB can be caused by mechanisms that either alters the tension in the walls of small vessels or damage the vessel wall in other ways, that is, chemically or by radiation. Several investigators have shown that inhalation of a high concentration of CO_2 (20%) increases the penetration of labeled proteins into the brain. The effect of CO_2 on BBB permeability is reversible. Repeated seizure activity also gives rise to extreme cerebral vasodilation, and again the barrier may be opened. As with CO_2 administration BBB breakdown is enhanced by elevated blood pressure. In a classic article, Rapoport et al. [4] characterized the properties of osmotic opening of the barrier. The degree of opening was determined by the amount of extravasation of dye (Evan's blue) as hyperosmolar solutions of different concentrations were applied to the surface of the cerebral cortex. The conclusions of their study were that barrier opening was reversible and that the degree of opening increased with increased osmolality. They suggested that hyperosmolar solutions shrink endothelial cells and open the tight junctions. However, it has also been suggested that hyperosmolar solutions increase vesicular transport in endothelial cells.

The structural integrity of cerebral endothelium is dependent on metabolism, and endothelia resist hypoxia and ischemia much longer than do other cells of the brain. It has been shown that barrier impermeability to dye is retained for up to 12h after an animal is killed. However, histochemical and biochemical studies have shown that cerebral cortical neurons do not survive O_2 deprivation for more than 3–8 min. Following occlusion of the blood supply to a region of brain, cerebral endothelial cells swell or flatten, but endothelium continuity is undisturbed. Vascular disruption develops after several hours and is preceded by irreversible metabolic and cytological changes in cells of the surrounding parenchyma. Thus, hypoxia and ischemia are not potent causes of BBB breakdown. A large number of other miscellaneous insults may increase barrier permeability. These include physical, chemical, infective, allergic, and neoplastic processes, and they generally result in barrier opening by sensitizing or damaging directly cerebral blood vessels.

OVERVIEW OF CEREBRAL HEMODYNAMICS

The brain's function and survival is dependent on providing it with oxygen and energy-producing substrates. In order to fulfill its needs, this complex neuronal system uses a significant portion of total body blood flow (about 15–20% of cardiac output) and consumes roughly a quarter of resting total body oxygen consumption. This section briefly reviews the hemodynamic principles which determine CBF and cerebrovascular resistance (CVR) and emphasizes mechanical control mechanisms within the macrocirculation of the brain.

General Hemodynamic Principles

Hemodynamics can be defined as the physical factors that govern blood flow. The physical laws formulated to describe laminar flow of fluids through nondistensible tubes are helpful in understanding in vivo cerebrovascular macrohemodynamics. These are the same physical factors that govern the flow of any fluid, and are based on a fundamental law of physics, namely Ohm's law, which states that current equals the voltage difference divided by resistance. In relating Ohm's law to fluid flow, the voltage difference is the pressure difference (ΔP; sometimes called driving pressure, perfusion pressure, or pressure gradient), the resistance is the resistance to flow (R) offered by the blood vessel and its interactions with flowing blood, and the current is the blood flow. This hemodynamic relationship can be summarized by:

$$Q = \frac{\Delta P}{R} = \frac{(P_A - P_V)}{R} \qquad (1.1)$$

For the flow of blood in a blood vessel, ΔP is the pressure difference between any two points along a given length of the vessel. When describing the flow of blood for an organ, the pressure difference is generally considered to be the difference between arterial pressure (P_A) and venous pressure (P_V). For example, in brain, cerebral perfusion pressure (CPP), the difference between arterial inflow and downstream outflow pressure (ΔP), is used as the relevant driving pressure for CBF. CPP is the difference between intraarterial pressure and the pressure in the thin-walled cerebral veins, collapsible at the point of entry into the venous sinuses. Cerebral venous pressure changes in parallel with intracranial pressure (ICP), and is normally 2–5 mmHg higher than ICP. Therefore, the driving pressure in brain is calculated as the difference between mean arterial blood pressure (MAP) and cerebral venous pressure or ICP, whichever is higher. Resistance is determined principally by vessel radius and can be calculated from Eq. (1.1) to estimate total CVR or resistance of any vascular segment of interest in which flow and upstream and downstream pressure gradients are known.

Under ideal laminar flow conditions, in which vascular resistance is independent of flow and pressure, the relationship between pressure, flow, and resistance can be depicted as shown in Fig. 1.3. Because flow and resistance are reciprocally related, an increase in resistance decreases flow at any given ΔP. Also, at any given flow along a blood vessel an increase in resistance increases ΔP. Changes in resistance are the primary means by which blood flow is regulated within organs because control mechanisms in the body generally maintain arterial and venous blood pressures within a narrow range. However, changes in perfusion pressure, will affect flow. The above relationship also indicates that there is a linear and proportionate relationship between flow and perfusion pressure. This linear relationship, however, is not followed when pathological conditions lead to turbulent

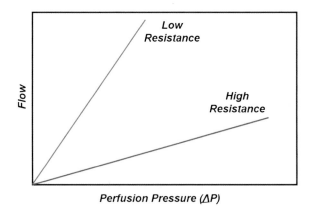

FIGURE 1.3 Pressure, flow, and resistance relationship.

flow, because turbulence decreases flow at any given perfusion pressure. Furthermore, the pulsatile nature of flow in large arteries alters this relationship so that greater pressures are required for a given flow. In other words, pulsatility, like turbulence, increases resistance to flow.

There are three primary factors that determine the resistance to blood flow within a single vessel: vessel diameter (or radius), vessel length, and blood viscosity—the most important, quantitatively and physiologically, being vessel diameter. The reason for this is that vessel diameter changes because of contraction and relaxation of vascular smooth muscle in the wall of the blood vessel. Furthermore, as discussed in the following, very small changes in vessel diameter lead to large changes in resistance. Vessel length does not change significantly and blood viscosity normally stays within a small range, except when hematocrit changes. The flow of fluids through blood vessels (tubes) can be described by Poiseuille's law (Fig. 1.4),

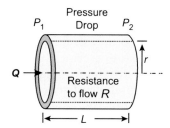

FIGURE 1.4 Depiction of Poiseuille's law.

$$Q = \frac{\Delta P \cdot r^4}{\eta \cdot L} \qquad (1.2)$$

which shows that the major determinants of CBF (Q) are perfusion pressure (ΔP), blood viscosity (η), vessel radius (r), and vessel length (L).

Blood vessel resistance (R) is directly proportional to the length (L) of the vessel and the viscosity (η) of the blood, and inversely proportional to the radius of the vessel to the fourth power (r):

$$R = \frac{\eta \cdot L}{r^4} \qquad (1.3)$$

Thus, a vessel having twice the length of another vessel (and each having the same radius) will have twice the resistance to flow. Similarly, if the viscosity of the blood increases twofold, the resistance to flow will increase twofold. In contrast, an increase in radius will reduce resistance.

Furthermore, the change in radius alters resistance to the fourth power of the change in radius (Fig. 1.5). For example, a twofold increase in radius decreases resistance by 16-fold! Therefore, vessel resistance is

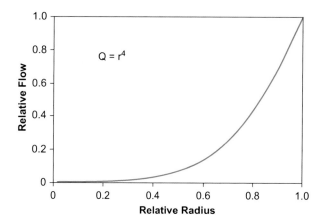

FIGURE 1.5 Relationship between flow and vessel radius to the fourth power (assumes constant ΔP, L, η, and laminar flow).

exquisitely sensitive to changes in radius of the vessel. It is by this mechanism that CVR can change rapidly, and dramatically alter regional or global CBF during normal or pathophysiological conditions. The stability of CBF (cerebral autoregulation) is achieved via the fine tuning of vessel diameter in response to fluctuations in perfusion pressure. The major cerebral and pial arteries are located outside the brain parenchyma and are surrounded anatomically by displaceable CSF, which maximizes the arteries' ability to widely change diameter and geometry. Intracerebral vessels, however, are constrained by extravascular tissue and structural elements.

Blood viscosity is the thickness and stickiness of blood, and it is a direct measure of the resistance of blood to flow through vessels. The primary determinants of blood viscosity are hematocrit, red cell deformability, red blood cell aggregation, and plasma viscosity. Viscosity (internal frictional resistance) of the transported fluid is often overlooked in predicting CBF because direct measurement is uncommon. Importantly, viscosity varies directly with hematocrit and any increase in the aggregation state of blood cellular components. Because blood viscosity also varies inversely with shear rate (i.e., is a non-Newtonian fluid), Poiseuille's law does not precisely describe the relationship between CBF and viscosity, particularly in the microcirculation [5]. Shear rate, a parameter that describes the velocity gradient of laminar blood flow through a vessel, is inversely proportional to vessel radius. Therefore, for a given blood velocity, shear rates are greater in small vessels than in larger vessels (300–500 s^{-1} in arterioles vs. 50 s^{-1} in aorta) [6], and apparent viscosity is thus lower in the microcirculation. The latter effect is the well-known Fahraeus–Lindqvist effect and predicts that viscosity is reduced by about 5% at 300 μm diameters and by 50% at 20 μm diameters relative to large bore tubes [6]. There has been some controversy concerning the effects of blood viscosity on CBF. Some investigators have shown that changes in viscosity alters CBF [7], whereas others have found that changes in viscosity are counteracted by compensatory vascular responses of the cerebral microcirculation, resulting in a well-maintained CBF [8]. It is likely that an increased viscosity is compensated by vasodilation to keep CBF unchanged. Conversely, CBF may be decreased by an increased viscosity when the vasodilatory capacity of the vessels is exhausted.

Segmental Vascular Resistance

The principal regulation and major source of vascular resistance within the cerebral circulation was always thought to be at the arteriolar level, as predicted by Poiseuille's law. However, data from experiments in which the pressure gradient was measured directly across different segments of the vascular bed indicate that the large extracranial vessels (internal carotid and vertebral arteries) and intracranial pial vessels contribute roughly 50% of total CVR (Table 1.1 [9]). The unusually prominent vasoregulatory role of large cerebral arteries may serve to equalize flow during focal neuronal activity and dampen pressure fluctuations in downstream vascular beds, for example, mitigating changes in microvascular pressure during systemic hypertension or hypotension.

Pressure–Volume Relationships

The vascular network of the brain is contained within a rigid bony structure and membranous dura

TABLE 1.1   Distribution of Segmental Vascular Resistance in Cerebral Circulation

Species	Segment	Percentage of Total
Dog	Aorta to 250- to 400- µm pial arteries	10
	Aorta to circle of Willis	22–31
Rhesus monkey	Aorta to 250- to 400- µm pial arteries	13
Cat	Aorta to 200- to 455- µm pial arteries	39
	Aorta to 250- to 455- µm pial arteries	10
	200- to 455- µm pial arteries to 25- to 40- µm	39
	Aorta to 30- to 40- µm pial arterioles	17
	Circle of Willis to 150- to 200- µm pial arteries	26
	Aorta to 150- to 250- µm pial arteries	51

Permission from Heistad DD, Kontos HA. Cerebral circulation. In: Berne RM, Sperelakis N, editors. Handbook of physiology: the cardiovascular system III. Bethesda American Physiological Society; 1978. p. 137–82.

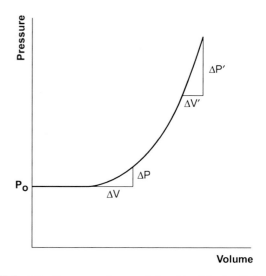

FIGURE 1.6 Pressure–volume relationship (idealized) in brain. An initial increase in volume of the cranial space (ΔV) produces an increase in ICP (ΔP). As the curve steepens (in intracranial hypertension), subsequent increases in volume generate exponentially larger increases in ICP.

with a fixed volume and a limited potential for external, vented release of intracranial contents. Therefore, extravascular pressure is critically important in cerebrovascular hemodynamics because arterial transmural pressure (difference between intraarterial and extravascular pressures) is equivalent to ICP. Under normal conditions, ICP is less than 15 mmHg and reflects the volume of three compartments: brain parenchyma (1200–1600 mL in the adult human), extracellular/CSF (100–150 mL) and intravascular blood (100–150 mL). Because the intracranial vault is fixed in volume, increases in the size of one compartment must be compensated by removal of an equivalent amount of another compartment, or ICP will increase. This pressure–volume relationship is exponential (Fig. 1.6), and the point at which perfusion-compromising ICP elevation occurs is dependent on brain elastance and potential displacement of intracranial contents (CSF displacement).

Cerebral Blood Volume

Intracranial blood volume is determined by two factors, CBF and capacitance vessel diameter (small veins and venules). Cerebral blood volume (CBV) increases with vasodilation and will decrease with vasoconstriction. Although CBF frequently changes in the same direction as CBV, these variables are inversely related under some normal (autoregulation) or pathological conditions. Further, CBV is not equally distributed throughout brain; blood volume per unit weight is greater in gray than in white matter with further variation among the various nuclei. Average CBV in humans is 3–4 mL/100 g tissue. Pathology that affects either CBF or cerebral venous capacitance may modulate CBV with subsequent effects on ICP. More quantitatively, the central volume principle [10] relates the volume that intravascular blood occupies within brain (CBV in mL) and the volume of blood that moves through the brain per unit time (CBF in mL/min):

$$Q = CBV/t \qquad (1.4)$$

Change in vascular diameter will directly affect CBV but not necessarily CBF when mean transit time (t) is simultaneously altered. For example, although CBV is increased during vasodilation, CBF may not change if blood flow velocity is correspondingly reduced. Surplus CBV accumulates primarily within cerebral veins, known to receive sympathetic innervation and to respond to sympathetic stimulation, and within capillaries to a smaller degree. Normally, increases in CBV can be physiologically controlled by two maneuvers: (1) increased blood outflow of the extracranial venous circulation and (2) restricted inflow via constriction of the major feeding arteries.

Pressure–Flow Relationships

The ability of CBF to remain constant despite changes in blood pressure (i.e., CPP) is referred to as cerebral autoregulation [11,12]. Autoregulation has been well demonstrated in both animals and humans. It has also been shown to be impaired or completely abolished during hypoxemia or hypercapnia. In healthy brain, arterial diameter increases or decreases within 30 s to 2 min in order to actively control CBF and maintain flow constant over a range of CPP (about between 60 and 160 mmHg) (Fig. 1.7). Within the

autoregulatory range, CBF remains constant via cerebral vasoconstriction when CPP increases or vasodilation when CPP decreases. Below the autoregulatory range (60 mm Hg-"lower limit") a decrease in CPP decreases CBF. Above the "upper limit" (160 mmHg) an increase in CPP increases CBF. Individuals with chronic hypotension demonstrate a leftward shift of the autoregulation curve, while chronic hypertension shifts the curve to the right. In neonates or infants, there is some question regarding the upper and lower limits of autoregulation because MAP in this age group is below 60 mmHg. Systolic pressures of from 41 to 70 mmHg have been reported in normal newborns and these values are close to the lower limit of autoregulation in adults. Furthermore, MAP above 60 mmHg does not appear to be common until the end of the first year of life. In addition, in the premature infant, arterial blood pressure has been reported to be in the range of 64/39 mmHg. Thus, it appears that the neonate and infant exhibit autoregulation; however, the critical CPP may be around 40 mmHg instead of 60 mmHg as in adults. Autoregulation may be abnormal or lacking in injured brain.

Autoregulatory limits have been extensively studied [11] and values vary in animals and humans. The upper and lower limit of autoregulation has been defined by numerous criteria: the CPP where the first significant change in CBF occurs; the pressure at which no further change in CVR is observed; or by direct observation to determine the pressure at which maximum vessel diameter is reached. While concern about CPP generally focuses on hypotension, it has been shown that with hypertension there is an "escape" (upper limit) from autoregulation such that CBF varies directly with CPP. The level of hypertension above which there is escape from autoregulation is in the range of 150–170 mmHg. It is unclear if this escape occurs in fetuses or newborns, or at which pressure levels. Furthermore, the mechanism accounting for autoregulation remains in debate despite many years of study, with data available to support both the myogenic and metabolic hypothesis [13,14].

CONCLUSION

Application of general hemodynamic principles and an appreciation for the complicated anatomical vascular system in brain, offers an initial understanding of CBF, mechanical control of cerebral macro- and microcirculation, and the basis for understanding CBF in cerebrovascular disease. Pressure–volume and pressure–flow relationships have been well characterized in brain, although the mechanisms controlling these relationships remain of interest and controversial.

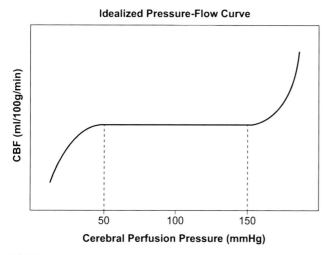

Idealized Pressure-Flow Curve

CBF (ml/100g/min)

50   100   150

Cerebral Perfusion Pressure (mmHg)

FIGURE 1.7 Idealized autoregulatory curve. The lower and upper limits or "breakpoints" are indicated by dotted lines, beyond which CBF varies as a function of CPP.

References

[1] Chusid JG. Correlative neuroanatomy and functional neurology. 14th ed. Los Altos (CA): Lange Medical Publishers; 1970.

[2] Shenkin HA, Harmel MH, Kety SS. Dynamic anatomy of the cerebral circulation. Arch Neurol Psychiatry 1948;60:240–52.

[3] Reese TS, Karnovsky MJ. Fine structural localization of a blood–brain barrier to exogenous peroxidase. J Cell Biol 1967;34:207–17.

[4] Rapoport SI, Hori M, Klatzo I. Testing a hypothesis for osmotic opening of the blood–brain barrier. Am J Physiol 1972;223:323–31.

[5] Hurn PD, Traystman RJ, Shoukas AA, Jones MD. Pial microvascular hemodynamics in anemia. Am J Physiol 1993;264:H2131–5.

[6] Milnor WR. Hemodynamics. 2nd ed. Baltimore: Williams and Wilkins; 1989. p. 11–57.

[7] Massik J, Tang YL, Hudak ML, Koehler RC, Traystman RJ, Jones Jr MD. Effect of hematocrit on cerebral blood flow with induced polycythemia. J Appl Physiol 1987;62:1090–6.

[8] Hudak ML, Jones Jr MD, Popal AS, Koehler RC, Traystman RJ, Zeger S. Hemodilution causes size-dependent constriction of pial arterioles in the cat. Am J Physiol 1989;257:H912–7.

[9] Heistad DD, Kontos HA. Cerebral circulation. In: Berne RM, Sperelakis N, editors. Handbook of physiology: the cardiovascular system III. Bethesda American Physiological Society; 1978. p. 137–82.

[10] Meier P, Zierler KL. On the theory of the indicator-dilution method for measurement of blood flow and volume. J Appl Physiol 1954;6:731–44.

[11] Traystman RJ. Microcirculation of the brain. In: Mortillaro NA, editor. The physiology and pharmacology of the microcirculation. San Diego: Academic Press; 1983. p. 237–98.

[12] Rapela CE, Green HD. Autoregulation of canine cerebral blood flow. Circ Res 1964;15:205–11.

[13] McPherson RW, Koehler RC, Traystman RJ. Effect of jugular venous pressure on cerebral autoregulation in dogs. Am J Physiol 1988;255:H1516–24.

[14] Paulson OB, Strandgaard S, Edvinsson L. Cerebral autoregulation. Cerebrovasc Brain Metab Rev 1990;2:161–92.

CHAPTER

2

Cerebral Microcirculation

T. Dalkara

Hacettepe University, Ankara, Turkey

INTRODUCTION

Research on microcirculation has led to an unprecedented progress with help of amazing developments in imaging technology. It is now possible to image capillaries with high resolution down to the depths of cortex, measure O_2 saturation and tension in them, track fast erythrocyte movements, and study Ca^{2+} signaling, vascular diameter changes, and neurovascular coupling in wild-type as well transgenic animals expressing a large variety of reporter or actuator proteins. Not unexpectedly, it emerges that microvasculature is a complex system evolved to optimally deliver O_2, glucose, and other nutrients to match the local need of the surrounding tissue. Its function is regulated by even more complex mechanisms. This progress has been complemented with discovery of surprisingly important roles of microcirculation in brain diseases. The present review will briefly outline some of these developments primarily in the context of stroke and microvascular dysfunction.

ANATOMY AND PHYSIOLOGY OF MICROCIRCULATION

Several arteries penetrate into the brain tissue after originating from the pial arterial/arteriolar network covering the brain surface. The penetrating arteries branch into arterioles and then to a network of capillaries. CNS capillaries are composed of a single layer of endothelium surrounded by a continuous basement membrane and are covered by astrocyte end feet over the abluminal wall (Fig. 2.1). Endothelial cells are connected by tight junction proteins and form the blood–brain barrier

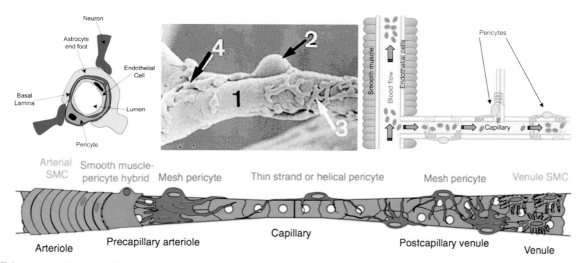

FIGURE 2.1   (Left) Schematic illustrating a capillary. Pericytes are located outside the endothelial cells and are separated from them and the parenchyma by a layer of basal lamina. In the parenchyma, astrocyte end feet and neuronal terminals are closely associated with the capillary. (Middle) Scanning micrograph of a vascular cast of a cortical capillary (1) with a pericyte-like structure (2) having primary and secondary processes (3) distributed around the vascular cast and the capillary branching points (4). (Right) Potential blood flow control sites in cerebral vasculature: arteriolar smooth muscle and pericytes placed especially on first-order capillaries and branching points. (Lower row) Schematic showing the continuum of mural cell types along the cerebral vasculature. *SMC*, smooth muscle cell. *Reproduced with permission from (left) Hamilton NB, Attwell D, Hall CN. Pericyte-mediated regulation of capillary diameter: a component of neurovascular coupling in health and disease. Front Neuroenergetics 2010;2; (middle) Rodriguez-Baeza A, et al. Anatomical Rec 1998;252:176–84; (right) Peppiatt CM, et al. Nature 2006;443:700–4; (lower row) Hartmann DA, Underly RG, Grant RI, Watson AN, Lindner V, Shih AY. Pericyte structure and distribution in the cerebral cortex revealed by high-resolution imaging of transgenic mice. Neurophotonics 2015;2:041402.*

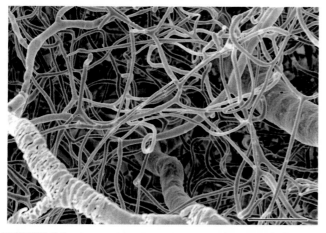

FIGURE 2.2   **Corrosion cast scanning electron micrograph of the capillary plexus in chinchilla auditory cortex.** Several interconnected capillary loops between a precapillary arteriole (red) and postcapillary venule (purple) are illustrated in different colors. Bar = 50 μm. *Reproduced with permission from Harrison RV, et al. Cereb Cortex 2002;12:225–33.*

(BBB), limiting and regulating the access of solutes to the CNS. Smooth muscle cells on arterioles are replaced by pericytes positioned between the two layers of basement membrane on capillaries [1,2]. The capillary density varies to match the regional metabolic demand and is correlated with synaptic density; hence, it is higher in the cortex than in the subcortical areas. It is estimated that the mean capillary size in human cortex is 6.5 μm in diameter and 53.0 μm in length. CNS capillaries form interconnected loops to adequately supply the tissue by passive O_2 diffusion (Fig. 2.2). Capillaries are not found

around large vessels (Pfeifer space); this area is directly oxygenated by diffusion of O_2 from the large vessel (Krogh tissue cylinder) [3]. Studies in the intact mouse brain show that, during baseline activity, arterioles provide 50% of the total extracted O_2, whereas the majority of the remaining O_2 is extracted from the first few capillary branches, which is followed by recruitment of high-branching-order capillaries during activation [3]. It should be noted that capillary segments are not linearly aligned but form loops that bring low- and high-branching-order segments in proximity such that their supply territories can interact [3] (Fig. 2.2).

The pressure gradient between the arterial and venous sides drives the microcirculatory flow. The venous outflow pressure is in equilibrium with the intracranial pressure. Accordingly, microcirculatory flow and tissue oxygenation can be adversely affected by high intracranial pressure as well as reduced arterial pressure. All cerebral capillaries are perfused by red blood cells (RBCs) and plasma under physiological conditions. The RBC velocity in capillaries is highly variable (around 0.2–4.4 mm/s). Random variations in RBC velocity including occasional stalls are observed in experimental animals under resting conditions. Capillary transit times of RBCs therefore significantly vary between neighboring branches of the network. A high capillary transit time heterogeneity (CTH) decreases the efficiency of O_2 transfer from RBCs [4]. The variability in transit times is reduced on neuronal activation, allowing more O_2 to be extracted. Although debated, it appears so that, to match the very focal demand by a small group of nearby cells, the cerebral microvascular blood flow requires a final step of regulation after the arterioles, which serve a larger cohort of cells [1,4]. In addition to the enhanced perfusion in dilated arterioles during neuronal activity, pericyte relaxation as well as the vasodilatory stimulus provided by ATP released from shear-stressed RBCs may reduce the microvascular resistance and promote the RBC transit. Also, the capillary endothelial surface is covered by a 0.5-μm-thick glycocalyx that facilitates the passage of blood cells and, when damaged, disrupts capillary flow.

Pericytes express several vasoactive receptors, suggesting that they have the capacity to respond to neurotransmitters and vasoactive mediators released from nearby neurons and astrocytes. Indeed, in vivo studies with two-photon microscopy have shown that cortical capillaries dilated about 1 s before arterioles during sensory stimulation in mice under anesthesia, supporting the view that capillaries may play a direct role in flow regulation [5]. Capillary endothelia, pericytes communicating with them, and astrocyte end-feet surrounding capillaries can transmit dilatory signals to the upstream arteriole through gap junctions between them to promote arteriolar dilation for providing the blood volume increase in microcirculation [2].

NEUROVASCULAR UNIT AND COUPLING

An increase in neuronal activity leads to an enhanced blood flow to the active brain area (functional hyperemia). In addition to the vasodilation caused by adenosine and lactate produced during metabolic activity [6], the coupling of the blood flow with activity is orchestrated by the neurovascular unit, which is composed of the endothelia, pericytes, astrocyte end-feet, and terminals of the vasoregulatory nerves at the microcirculatory level [2,6] (Fig. 2.1).

Local interneurons or projections of the subcortical and brainstem nuclei terminate in the vicinity of microvessels or on astrocyte end feet around the vessels without making direct synaptic contacts [7]. The subcortical projections promote more global flow changes (e.g., during attention), whereas excitatory or inhibitory interneurons integrate local neuronal activity and differentially contribute to neurovascular coupling by releasing their distinctive mediators. For example, activity-induced Ca^{2+} increase in neuronal NO synthase expressing GABAergic interneurons causes NO synthesis, which rapidly diffuses across plasma membranes and relaxes smooth muscle cells and pericytes on vasculature.

Astrocytes may also play a role in neurovascular coupling by monitoring the neuronal activity with their perisynaptic end feet and translating this information as the vasoactive signals released from their perivascular end feet [6]. Moreover, Ca^{2+} increase in perivascular end feet may release K^+ through the large-conductance Ca^{2+}-activated K^+ channels, which initially induces vasodilation but then vasoconstriction with increasing concentrations of K^+. However, the in vivo significance of these astrocytic mechanisms is currently being evaluated. Of note, relative contributions of the above-outlined pathways to the neurovascular coupling vary in different brain regions and also with the developmental stage, species, and activation pattern (quick, short lasting vs. sustained vasodilatory responses) and, most likely, with the segment of the vascular tree (e.g., the proximal penetrating arteries vs. distal capillary bed).

Interest in pericytes has started to disclose several important microvascular functions mediated by these long-neglected cells on the microvessel wall. Pericytes are located on precapillary arterioles, capillaries, and postcapillary venules [8] (Fig. 2.1). Unlike smooth muscle cells, pericytes have a prominent soma, protruding out from the vascular wall (bump on a log appearance) and are embedded within two layers of basement membrane [9]. Processes that originate from pericyte soma extend along and branch around the microvessels. Processes are more circumferential at the arteriole end of the microvascular bed and at downstream bifurcating points, more longitudinal in the middle of a capillary, and have a stellate morphology at the venule end of the microcirculation (Fig. 2.1). In accordance with a contractile, flow-regulating phenotype, pericytes with circumferential processes express smooth muscle α-actin. In vitro as well as in vivo studies have shown that microvascular pericytes contract or dilate in response to vasoactive mediators applied or to sensory stimulation [1,5]. Such flow regulation with fine spatial resolution at the microcirculatory level may be essential for the brain and retina. However, it should be noted that not all microvascular pericytes are contractile and the ratio of the contractile ones may vary with the species, tissue, and developmental stage, as well as along the arteriovenous axis [8,9].

A close communication between the endothelia, pericytes, and astrocytes is required for development and functioning of the neurovascular unit as well as the BBB. The microvascular pericyte density is the highest in the retina and CNS compared with peripheral vessels in line with their role in formation and maintenance of the blood–brain/retina barrier [9]. The surface area of the vascular wall covered by pericytes and the number of pericytes per endothelial cell are correlated with the permeability of capillaries. Accordingly, pericyte dysfunction as well as deficiency leads to a leaky BBB [9].

MICROCIRCULATION AFTER STROKE AND THROMBOLYSIS

The microvascular injury inflicted by ischemia/reperfusion plays a critical role in determining tissue survival after recanalization by inducing microcirculatory clogging (no reflow) and disrupting the BBB integrity [10]. Evidence from clinical trials show that recanalization does not always lead to reperfusion of the ischemic tissue in patients with stroke treated with tissue plasminogen activator or interventional methods [11]. Since current imaging techniques assess perfusion by measuring the plasma transit through the microcirculation, the current average rate of 26% likely underestimates the rate of incomplete reperfusion in the setting of satisfactory recanalization [11] because the plasma continues to flow at the periphery of the microvessel wall bypassing the RBCs entrapped within the narrowed lumen [10].

Starting 1h after middle cerebral artery (MCA) occlusion in the mouse, some of the microvessels show constrictions (Fig. 2.3). Narrowed lumina are filled with entrapped erythrocytes, leukocytes, and fibrin–platelet deposits. Leukocytes also populate the postcapillary venules and adhere to their wall to enter the CNS. Reducing microvascular clogging by inhibiting oxygen/nitrogen radical production, leukocyte adherence, platelet activation, or fibrin–platelet interactions have been shown to restore microcirculation and improve stroke outcome in animal models. BBB-impermeable agents such as the endothelial NO synthase inhibitor (L-N-5-(1-iminoethyl)-ornithine) or adenosine (continuously released from circulating nanoparticles) reduces the microvascular clogging and infarct volume, suggesting that restoring microvascular patency alone can improve stroke outcome without direct parenchymal neuroprotection [10,12]. These observations support the view that the outcomes of recanalization therapies can be improved by promoting microcirculatory reflow and preventing BBB leakiness, hence, hemorrhagic conversion and vasogenic edema. They also point to the critical but partly neglected importance of the microcirculation in neuroprotection [13].

Narrowing of the microvessel lumen was generally attributed to external compression by swollen astrocyte end feet and to endothelial cell edema. Pericytes on microvessels were reported to play an important role in incomplete microcirculatory reperfusion as some of them contracted during ischemia and remained contracted despite reopening of the occluded artery [10]. Even subtle decreases in microvessel radius caused by pericyte contractions can cause RBC entrapments because capillary luminal size hardly allows RBC passage [1,10]. Entrapped RBCs may also impede flow of other blood cells and promote platelet and fibrin aggregation. The contracted pericytes die in rigor along the course of ischemia, which may contribute to loss of the BBB integrity [5].

MICROVASCULAR DYSFUNCTION AND DEMENTIA

In light of the discoveries at the microcirculatory level, the vascular hypothesis of dementia has been rekindled [9,14]. Pathological examination of brain specimens of patients with Alzheimer disease has revealed a reduced microvascular density and several morphological alterations in capillaries. β-Amyloid accumulation on capillaries as well as within degenerating pericytes has also been reported. Based on these findings, the vascular hypothesis of Alzheimer disease posits that microvascular dysfunction may secondarily cause degeneration of nerve endings of the subcortical projections innervating the cortex and then retrograde death of subcortical neurons (e.g., cholinergic) [2,9,14]. Of note, several cerebrovascular risk factors such as aging, hypertension, and diabetes can impair microcirculation over the course of years.

CAPILLARY TRANSIT TIME HETEROGENEITY AS A MEASURE OF CAPILLARY DYSFUNCTION

Modeling studies imply that capillary dysfunction can cause tissue hypoxia/ischemia without the presence of a proximal occlusion limiting the blood supply [4,15]. These models predict that an increase in blood

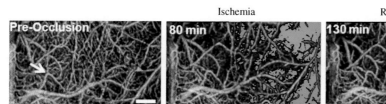

FIGURE 2.3 **Incomplete microcirculatory reperfusion after 90 min of middle cerebral artery (MCA) occlusion.** The *green area* depicts the lack of blood flow in the area of the MCA 80 min after occlusion and 130 min after reopening of the MCA as detected by optical microangiography. The image in the right is the optical microangiography image taken at 50 min of ischemia overlaid on the 24-h infarct analysis by histological staining as the area of pallor. Scale bar = 500 μm. *Reproduced with permission from Dziennis S, et al. Sci Rep 2015;5:10051.*

flow can maintain tissue oxygenation when the capillary dysfunction (increased CTH) is mild. However, when CTH is further increased due to variable transit times through dysfunctional capillaries and accelerated transit from patent ones, neurovascular coupling mechanisms reduce the incoming blood flow to prevent functional shunting through patent capillaries because too rapid transit of RBCs drastically reduces O_2 extraction. In other words, to maintain tissue normoxia, neurovascular mechanisms reduce the blood flow rather than increasing it. Emerging evidence from asymptomatic apolipoprotein E (APOE) ε4 carriers supports these predictions; resting and activity-related CBF levels were found to be elevated in 19- to 28-year-old young carriers, whereas a long-term follow-up study of asymptomatic APOE ε4 carriers illustrated the transition from hyper- to hypoperfusion. The hypothesis that a high CTH could be a source of symptomatic tissue hypoxia has the potential to account for several other neurological conditions such as transient ischemic attacks without embolism, prolonged oligemia accompanying cortical spreading depression, stroke-like attacks in Cerebral Autosomal-Dominant Arteriopathy with Subcortical Infarcts and Leukoencephalopathy (CADASIL), and hyperperfusion after recanalization [15].

References

[1] Hamilton NB, Attwell D, Hall CN. Pericyte-mediated regulation of capillary diameter: a component of neurovascular coupling in health and disease. Front Neuroenergetics 2010;2.

[2] Iadecola C. Neurovascular regulation in the normal brain and in Alzheimer's disease. Nat Rev Neurosci 2004;5:347–60.

[3] Sakadzic S, Mandeville ET, Gagnon L, et al. Large arteriolar component of oxygen delivery implies a safe margin of oxygen supply to cerebral tissue. Nat Commun 2014;5:5734.

[4] Jespersen SN, Ostergaard L. The roles of cerebral blood flow, capillary transit time heterogeneity, and oxygen tension in brain oxygenation and metabolism. J Cereb Blood Flow Metab 2012;32:264–77.

[5] Hall CN, Reynell C, Gesslein B, et al. Capillary pericytes regulate cerebral blood flow in health and disease. Nature 2014;508(7494):55–60.

[6] Attwell D, Buchan AM, Charpak S, Lauritzen M, Macvicar BA, Newman EA. Glial and neuronal control of brain blood flow. Nature 2010;468(7321):232–43.

[7] Cauli B, Hamel E. Revisiting the role of neurons in neurovascular coupling. Front Neuroenergetics 2010;2:9.

[8] Hartmann DA, Underly RG, Grant RI, Watson AN, Lindner V, Shih AY. Pericyte structure and distribution in the cerebral cortex revealed by high-resolution imaging of transgenic mice. Neurophotonics 2015;2:041402.

[9] Winkler EA, Bell RD, Zlokovic BV. Central nervous system pericytes in health and disease. Nat Neurosci 2011;14:1398–405.

[10] Yemisci M, Gursoy-Ozdemir Y, Vural A, Can A, Topalkara K, Dalkara T. Pericyte contraction induced by oxidative-nitrative stress impairs capillary reflow despite successful opening of an occluded cerebral artery. Nat Med 2009;15:1031–7.

[11] Dalkara T, Arsava EM. Can restoring incomplete microcirculatory reperfusion improve stroke outcome after thrombolysis? J Cereb Blood Flow Metab 2012;32:2091–9.

[12] Gaudin A, Yemisci M, Eroglu H, et al. Squalenoyl adenosine nanoparticles provide neuroprotection after stroke and spinal cord injury. Nat Nanotechnol 2014;9(12):1054–62.

[13] Gursoy-Ozdemir Y, Yemisci M, Dalkara T. Microvascular protection is essential for successful neuroprotection in stroke. J Neurochem 2012;123(Suppl. 2):2–11.

[14] Iadecola C. The overlap between neurodegenerative and vascular factors in the pathogenesis of dementia. Acta Neuropathol 2010;120:287–96.

[15] Ostergaard L, Jespersen SN, Engedahl T, et al. Capillary dysfunction: its detection and causative role in dementias and stroke. Curr Neurol Neurosci Rep 2015;15(6):37.

CHAPTER

3

The Glymphatic System and Brain Interstitial Fluid Homeostasis

J.J. Iliff[1,2], A.S. Thrane[2,3], M. Nedergaard[2,4]

[1]Oregon Health & Science University, Portland, OR, United States; [2]University of Rochester Medical Center, Rochester, NY, United States; [3]Haukeland University Hospital, Bergen, Norway; [4]University of Copenhagen, Copenhagen, Denmark

INTRODUCTION

The maintenance of the interstitial compartment is a basic element of an organ's function that is of utmost importance in the brain, given neural cells' exquisite sensitivity to changes in their extracellular environment. In peripheral tissues, interstitial fluid (ISF) is formed from the filtration of plasma across the permeable capillary endothelium. Although a portion of the ISF is reabsorbed into postcapillary venules, much of the ISF and virtually all extracellular proteins are collected into primary lymphatic vessels, which return lymph fluid and proteins to the blood circulation. In the central nervous system (CNS), the presence of the blood–brain barrier (BBB), which restricts the filtration of water and proteins from the plasma, and the apparent absence of lymphatic vessels from brain tissues require that the "lymphatic" function of interstitial homeostasis be subserved by an alternative mechanism. In the CNS, the cerebrospinal fluid (CSF) circulation supports this function, serving as a sink and a crossroad for extracellular proteins and metabolites that cannot readily cross the BBB [1].

The exchange of ISF and CSF is a physiological process in the CNS that likely affects many aspects of brain function, including waste clearance, lipid metabolism, growth factor and neurohormone distribution, and immune surveillance. Dysfunction of these processes appears to contribute to the edema formation after cerebral ischemia and traumatic brain injury (TBI), to the accumulation of aggregated proteins in neurodegenerative conditions such as Alzheimer disease (AD), and to neuroinflammatory diseases such as multiple sclerosis [2].

INTERSTITIAL SOLUTE CLEARANCE

Under physiological conditions, the brain extracellular space comprises approximately 14–23% of the overall brain volume. ISF is formed from a combination of water crossing the BBB, water produced through cellular metabolism, and fluid from the CSF compartments. The relative contributions of these processes to ISF production is not clearly known, and may vary by brain region and physiological state. Interstitial solutes move through brain tissue both by the process of diffusion and bulk flow, with the relative contribution of each process being determined by the physical and chemical properties of the solute and the properties of the extracellular environment [3].

Diffusion and the Blood–Brain Barrier

Many interstitial solutes, including nonpolar molecules and those with specific BBB efflux transporters are readily cleared to the blood stream across the BBB. The process of diffusion, which is the movement of molecules down their concentration gradients driven by thermal motion, is very efficient on the microscopic scales of the brain microcirculation. Hence, the clearance of nonpolar molecules such as CO_2 and BBB efflux transporter substrates such as the drug verapamil (a substrate for P-glycoprotein) is diffusion limited [4]. Diffusion-limited clearance kinetics will be strongly influenced by molecular weight and the dimensions and nature of the extracellular space, with larger solutes moving more slowly than smaller, and with more rapid diffusion as the extracellular space becomes larger and less tortuous [3]. Beyond clearance of nonpolar molecules and those with specific transporters at the BBB, diffusion also dominates the exchange of ISF and CSF in tissue closest to internal and external CSF compartments, including periventricular and subpial brain tissue (Fig. 3.1).

Bulk Flow and the Cerebrospinal Fluid Sink

The CSF compartment, either within the ventricles or the subarachnoid space, serves as a sink for interstitial solutes that cannot be cleared across the BBB. This process was initially thought to be driven solely by diffusion. However, as the inverse relationship between molecular mass and the rate of diffusion dictates, the exchange between most brain tissue and the nearest CSF compartment of larger molecules would be prohibitively slow. This becomes increasingly apparent as molecular masses approach macromolecular dimensions, such as the serum protein albumin (65 kD), which requires approximately 100 h to diffuse 1 cm within brain tissue. When inert solutes spanning more than two orders of magnitude of molecular mass that are not cleared across the BBB are injected into the brain, their clearance kinetics are virtually identical. For example, Groothuis et al.

demonstrated that sucrose [molecular weight (MW) 342 Da] and dextan-70 (MW 70,000 Da) are cleared from the rat brain with a half-life of 2.75 and 2.96 h, respectively [4]. This suggests that the clearance of interstitial solutes from the brain depends on bulk or convective flow of ISF, which was estimated by Cserr and colleagues in rats to be 0.11–0.29 $\mu L \cdot g^{-1} \cdot min^{-1}$, a value that is in line with the rate of lymph flow in peripheral organs [2].

Although these results have shown that bulk flow of ISF supports the clearance of interstitial solutes, several studies suggest that bulk flow is a feature of specific anatomical elements of brain tissue, rather than the wider ISF compartment. Tracers injected into brain tissue spread most rapidly along white matter tracks and along perivascular spaces surrounding cerebral blood vessels, but have a slower rate of spread through the bulk interstitium. This suggests that interstitial solute clearance may be driven by the combined actions of diffusion and bulk flow, with diffusion governing the microscopic movement of solutes between the interstitium and local perivascular spaces or white matter tracks, and bulk flow governing the macroscopic flux through brain tissue to distant CSF compartments (Fig. 3.1).

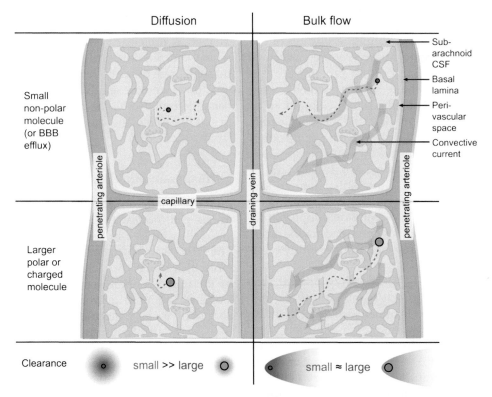

FIGURE 3.1 **The contribution of diffusion and convection to the clearance of interstitial solutes form the brain parenchyma.** Diffusion (*left column*) is a key driver of the clearance of small nonpolar molecules and those with specific blood–brain barrier (BBB) efflux transporters across the local BBB. The rate of diffusion is strongly influenced by molecular mass, with small molecules diffusing more rapidly than larger molecules. Bulk flow of brain interstitial fluid (*right column*) supports the clearance of solutes across long distances. Because interstitial fluid moves along with associated solutes, molecules are cleared by bulk flow at the same rate, independent of molecular size. *CSF*, cerebrospinal fluid.

PERIVASCULAR BULK FLOW AND THE GLYMPHATIC SYSTEM

Perivascular spaces surrounding cerebral blood vessels have long been known to facilitate the exchange of CSF and ISF. Although this process was generally held to occur slowly along the vasculature, a series of studies carried out in dogs and cats in the mid- to late 1980s suggested that the interaction of CSF and ISF along perivascular spaces was both rapid and polarized along the arterial and venous sides of the circulation, with CSF entering the brain along perivascular spaces surrounding penetrating cerebral arteries and ISF being cleared from the brain along perivascular spaces surrounding cerebral veins [1,2]. More recent studies have substantiated these findings employing dynamic imaging approaches in living animals rather than analysis of fixed or frozen tissues. In mice, in vivo two-photon microscopy demonstrated that fluorescent tracers injected into the subarachnoid CSF at the cisterna magna moved rapidly into and through the brain parenchyma along perivascular spaces surrounding cerebral arteries [5,6] (Fig. 3.2). Fluorescent tracers injected into the brain interstitium were in turn cleared along specific anatomical pathways including perivascular spaces surrounding large-caliber draining veins that drain to extraparenchymal venous sinuses, such as the internal cerebral veins that form the origin of the straight sinus. These findings have been confirmed both in rats and with brain-wide CSF tracer imaging using dynamic contrast-enhanced MRI. As detailed later, because perivascular bulk flow along this pathway depends on glial water transport, and because it assumes the lymphatic function of interstitial solute clearance, this brain-wide perivascular network has been termed the "glymphatic" system [6].

Anatomical Basis

Brain-wide imaging studies demonstrate that the CSF moving into the brain along perivascular spaces surrounding cerebral arteries originates within the basal cisterns along the conduit arteries of the circle of Willis, moves distally over the cortical surface along resistance vessels such as the middle cerebral arteries, and then enters the brain parenchyma along perivascular spaces surrounding penetrating cerebral arteries (Fig. 3.2A–C) [2,6]. Electron microscopy studies of the human leptomeningeal vasculature show that a layer of the pia mater invests leptomeningeal arteries and veins, forming a perivascular space, the Virchow–Robin space, that surrounds the elastic lamina of the arterial wall [1,2]. The pial investment of the leptomeningeal arteries follows the vessels as they penetrate the parenchyma, becoming fenestrated and discontinuous with increasing depth from the brain surface. In contrast, the pial investment of the leptomeningeal veins reflects back upon the pia mater overlying the glia limitans, leaving perivascular spaces surrounding cerebral veins open to the subpial space. Although the anatomical pathway that permits CSF from the basal cisterns to enter into leptomeningeal perivascular spaces has not yet been defined, it seems likely that fenestrations in the pia mater ensheathing vessels at the base of the brain allow CSF entry into the proximal perivascular network. CSF is then propelled rapidly through these perivascular spaces by arterial pulsations or by hydrostatic pressure gradients between different cisternal CSF compartments to enter the brain parenchyma along penetrating arteries.

Studies with fluorescent tracers demonstrate that CSF entering the brain along the Virchow–Robin space moves readily along the perivascular spaces surrounding the arteriolar wall, eventually reaching the basal lamina of cerebral capillaries [6]. This demonstrates that the Virchow–Robin space, the perivascular spaces, and the vascular basement membranes are connected and that CSF from cisternal compartments is rapidly transported along these pathways (Fig. 3.2D–G). Fluorescent CSF tracers <70 kD in size exchange quickly between perivascular spaces and the surrounding interstitium, whereas high-molecular-weight tracers (500–2000 kD) remain largely trapped within the perivascular spaces. This suggests that the 20–80 nm extracellular clefts between overlapping perivascular astrocytic end feet, which completely ensheath the cerebral vasculature, restrict the free movement of solutes and cells between perivascular spaces and the wider brain interstitium.

Role of Astrocytes

In addition to physically bounding perivascular spaces with end-foot processes, astrocytes also support both perivascular CSF influx and interstitial solute clearance through the activity of the astroglial water channel aquaporin-4 (AQP4). AQP4 is a trans-membrane water channel expressed in astrocytes throughout the brain and in ependymal cells lining the cerebral ventricles. In astrocytes, AQP4 is localized primarily to the perivascular end-foot process, with up to 50% of the end-foot surface facing the cerebral vasculature being occupied by square arrays of AQP4 (Fig. 3.3A–D) [1,2]. Perivascular CSF influx is dramatically reduced in *Aqp4*-null mice, as is the clearance of interstitial solutes [6]. A computational study provides a biophysical basis for the role of AQP4 in supporting perivascular bulk flow and the clearance of interstitial solutes [7]. In this model, water moves rapidly through the intracellular astrocytic network that bridges perivascular spaces surrounding cerebral arteries and veins. The ready exchange of water from the intracellular astrocytic network and extracellular water across the vast surface area of the nonperivascular

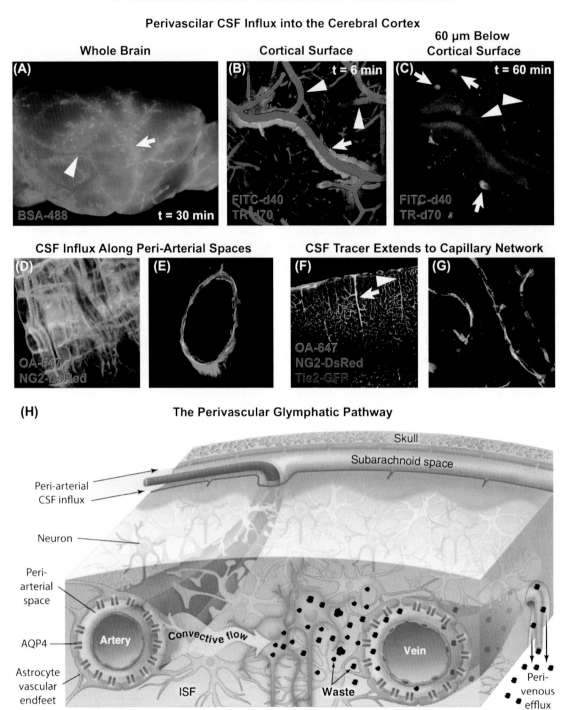

FIGURE 3.2 **The glymphatic pathway supports the exchange of cerebrospinal fluid (CSF) and interstitial fluid (ISF) along perivascular pathways.** CSF tracer moves through the subarachnoid space along perivascular pathways primarily surrounding cerebral surface arteries (*arrows*), but not veins (*arrowheads*) (A). (B, C) Dynamic two-photon imaging shows that CSF tracers enter the brain parenchyma within perivascular spaces along penetrating arterioles, then exchange with surrounding interstitial fluid. Confocal imaging shows that CSF tracer moves along basement membranes in the wall of penetrating arteries (D, E), and reaches the basal lamina surrounding the cerebral microcirculation (F, G). (H) Schematic diagram showing overview of the perivascular glymphatic pathway. *AQP4*, aquaporin-4; *BSA-488*, bovine serum albumin–conjugated Alexa488 (65 kD, CSF tracer); *FITC-d40*, FITC-conjugated dextran (40 kD, CSF tracer); *NG2-DsRed*, red fluorescent protein driven under the NG2 promoter (vascular smooth muscle cell and pericyte marker); *OA-647*, ovalbumin-conjugated Alexa647 (45 kD, CSF tracer); *Tie2-GFP*, green fluorescent protein driven under the Tie2 promoter (vascular endothelium marker); *TR-d70*, Texas Red-conjugated dextran (70 kD, blood tracer). *Reprinted from Illiff JJ, Wang M, Liao Y, et al. A paravascular pathway facilitates CSF flow through the brain parenchyma and the clearance of interstitial solutes, including amyloid beta. Sci Transl Med 2012;4(147):147ra111; Nedergaard Science 2013;340(6140):1529–30 with permission.*

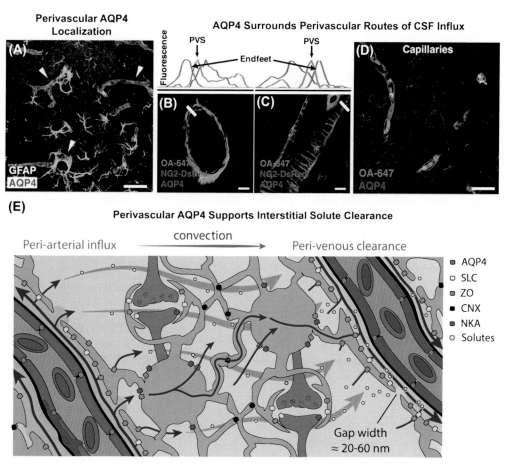

FIGURE 3.3 **Localization of Aquaporin-4 (AQP4) to perivascular astroglial end feet supports interstitial solute clearance along the glymphatic pathways.** (A) AQP4 is expressed in astrocytes and is localized primarily to perivascular astrocytic end feet surrounding the cerebral vasculature. (B, C) Perivascular cerebrospinal fluid (CSF) influx pathways are surrounded by AQP4-bearing perivascular end feet. (D) CSF follows perivascular spaces to reach terminal capillaries, which are surrounded by astroglial AQP4. (E) Schematic showing the role of astroglial water transport in supporting perivascular bulk flow and interstitial solute clearance. *CNX*, connexin; *GFAP*, glial fibrillary acidic protein; *NG2-DsRed*, red fluorescent protein driven under the NG2 promoter (vascular smooth muscle cell and pericyte marker); *NKA*, Na+-K+-ATPase; *OA-647*, ovalbumin-conjugated Alexa647 (45kD, CSF tracer); *PVS*, perivascular space; *SLC*, solute cotransporter proteins; *ZO*, zonula occludens. *Reprinted and modified from Illliff JJ, Wang M, Liao Y, et al. A paravascular pathway facilitates CSF flow through the brain parenchyma and the clearance of interstitial solutes, including amyloid beta. Sci Transl Med 2012;4(147):147ra111; Thrane AS, Rangroo Thrane V, Nedergaard M. Drowning stars: reassessing the role of astrocytes in brain edema. Trends Neurosci 2014;37(11):620–628 with permission.*

astroglial process helps to couple intracellular water flux to solute clearance through the interstitial compartment (Fig. 3.3E).

Sleep–Wake Regulation

Glymphatic pathway function including perivascular influx of CSF through and the clearance of interstitial solutes from the brain is a primary feature of the sleeping, rather than the waking, brain [5]. Compared with the naturally sleeping brain, the influx of fluorescent subarachnoid CSF tracers into the mouse cortex was reduced by more than 90% in the waking brain. The clearance of interstitial tracers was similarly reduced by a factor of two in the waking versus the naturally sleeping brain. Sleep–wake changes in the glymphatic pathway function appear to be underpinned by changes in extracellular volume, as

electrophysiological recordings demonstrated that a 65% increase in the extracellular volume fraction occurred between the sleeping and waking brain. Increasing the extracellular volume facilitates more rapid diffusion of interstitial solutes, supporting access to perivascular bulk flow pathways underlying efflux from the brain parenchyma. Similarly high glymphatic pathway function was observed under anesthesia with ketamine and xylazine, an anesthetic regimen associated with slow wave activity common to stage 3 and 4 non–rapid eye movement sleep. Sleep–wake changes in extracellular volume and glymphatic function appear to be regulated in part by cortical noradrenergic tone underlying arousal state, as local inhibition of cortical noradrenergic signaling with pharmacological antagonists increased the extracellular volume fraction and improved glymphatic pathway function in the waking state.

TERMINAL EFFLUX ROUTES

Perivascular pathways throughout the brain, including those surrounding large-caliber veins that drain into dural sinuses, facilitate the efficient clearance of interstitial solutes to the CSF. A fraction of subarachnoid CSF enters the brain along perivascular spaces surrounding penetrating cerebral arteries, indicating that some portion of the interstitial solutes recirculate through the brain parenchyma along the glymphatic pathway. The remaining CSF and its associated solutes are cleared from the cranium by bulk reabsorption with the CSF. Thus, the CSF compartment is a critical crossroad linking solute and fluid movement into and out of the brain interstitium with movement into and out of the blood stream. Although CSF reabsorption pathways are a topic covered in another chapter of this textbook and are not the central focus of the present chapter, it is important to articulate the functional connections between perivascular CSF–ISF exchange and the routes of terminal CSF efflux from the cranium, including the classically described arachnoid villi and perineural pathways, in addition to the recently described lymphatic vessels that are associated with dural sinuses and the dural vasculature.

Classical descriptions of CSF reabsorption focus on the role of arachnoid villi, valve-like structures within the walls of dural sinuses through which CSF enters the blood stream directly, as well as perineural sheathes, which surround cranial and spinal nerves and permit efflux of CSF and associated solutes to extracranial tissues and peripheral lymphatic vessels draining the head and neck [1,2]. The relative contribution of these classical CSF efflux pathways to the clearance of solutes originating within the brain interstitium is not clear at present, and may differ by species, and developmental and physiological state. One important distinction between these two pathways, however, is that transit via the arachnoid villi provides efflux directly to the blood stream, whereas reabsorption along perineural pathways involves clearance first along cranial lymphatic vessels to the deep cervical lymph nodes. This distinction may have important implications for the function of peripheral immune surveillance within the CNS and the surrounding CSF compartments.

The absence of conventional lymphatic vessels from the CNS has been a central feature of our understanding of BBB and CSF physiology for more than a century, dictating that the lymphatic functions of interstitial protein and fluid homeostasis and peripheral immune surveillance are accomplished within the brain in a manner distinct from that of other organs. However, a network of apparently classical lymphatic vessels associated with the dural sinuses and the dural vasculature were described in mice [8]. These vessels expressed histological markers of lymphatic endothelial cells, were associated with a large number of peripheral T lymphocytes and antigen presenting cells, and provided an efflux pathway for tracers injected both into the CSF and the brain parenchyma to be drained to the cervical lymphatics via the deep cervical lymph nodes. Although these were termed a "cerebral lymphatic system," it is notable that these lymphatic vessels do not appear to extend into brain tissue, but instead are associated with the outer layers of the meninges. For this reason, it appears likely that the perivascular glymphatic pathway, the CSF compartment, and the dural lymphatic vasculature are three elements of a single integrated system that supports the lymphatic functions of interstitial waste clearance and immune surveillance behind the curtain of the BBB and while maintaining the relative "immune privilege" of the CNS parenchyma.

PHYSIOLOGICAL ROLES OF PERIVASCULAR EXCHANGE AND CEREBRAL LYMPHATICS

In the periphery, the lymphatic vasculature makes important contributions to basic aspects of organ function, including ISF, lipid and protein homeostasis, and trafficking of lymphocytes and antigen presenting cells to lymph nodes. Within the brain, it appears that similar functions are supported by the combined activity of the perivascular glymphatic system, the CSF circulation, and sinus-associated lymphatic vessels [2,8].

Experimental studies defining the function of the glymphatic system and sinus-associated lymphatic vessels have focused on the clearance of relatively inert exogenous tracer molecules, including fluorescently or radiolabeled mannitol, inulin, dextrans, and polyethylene glycols, or proteins such as albumin or ovalbumin. Biologically active proteins such as amyloid β (Aβ) and tau, which are released into the brain extracellular space during neural activity, are also cleared along perivascular pathways [1,2]. Deletion of the *Aqp4* gene slows the clearance of soluble Aβ by 55% [6], whereas Aβ is cleared twice as quickly from the sleeping brain compared with the waking brain [5]. Sampling of mouse ISF by in vivo microdialysis and human CSF by serial lumbar CSF sampling demonstrates that levels of interstitial Aβ increase during waking, and decline with sleep, whereas this drop in Aβ during sleep is prevented by sleep deprivation [9]. These findings suggest that the clearance of potentially toxic metabolites, such as Aβ, from the brain interstitium by the glymphatic system may be one of the mechanisms underlying the restorative function of sleep.

A study describing the presence of classical lymphatic vessels associated with dural sinuses also reported that under quiescent conditions, T lymphocytes and antigen presenting cells were strongly associated with these vessels [8]. Tracers injected both into the brain parenchyma and into the CSF drained through these lymphatic structures to the deep cervical lymph nodes, whereas ligation of the afferent drainage of the deep cervical lymph nodes caused the distension of sinus-associated lymphatic vessels and the accumulation of peripheral immune cells within the dura. These findings suggest that sinus-associated lymphatic vessels are a key site for peripheral immune surveillance of the CNS. Intriguingly, the choroid plexus is a key point of entry for peripheral immune cells into the CSF space. This suggests that drainage of interstitial solutes and antigens from the brain parenchyma along the perivascular glymphatic pathway to the CSF may permit interactions with immune cells originating in the choroid plexus, transiting the CSF compartments, and exiting along sinus-associated lymphatic vessels, permitting peripheral immune surveillance from the edge of the CNS, without compromising the relative immune privilege of the CNS.

PATHOLOGICAL ROLES OF PERIVASCULAR CEREBROSPINAL FLUID–INTERSTITIAL FLUID EXCHANGE

Cerebral Edema

Brain edema represents a potentially fatal buildup of excess fluid within the confines of the rigid skull. Edema can accumulate acutely, following, for instance, TBI or stroke, but it can also develop more chronically in the context of brain tumors or metastases. When excess edema fluid builds up predominantly inside cells as a consequence of impaired energy metabolism, it is normally referred to as cytotoxic edema. In the context of cerebrovascular disease, the traditional view is that cytotoxic edema results from net inward movement of water and salts across the BBB into ischemic and infarcted tissue. Vasogenic edema, on the other hand, usually develops days after a brain infarct, and is traditionally thought to involve a breakdown of the BBB, with consequent exudation of plasma proteins and extravasation of leukocytes. However, this traditional understanding of brain edema leaves several explanatory gaps. In the context of focal ischemia, brain edema principally accumulates in the better-perfused "penumbra," rather than in the infarct core, where cellular metabolism is most severely affected. This ischemic penumbra becomes edematous within hours of arterial occlusion and before opening of the BBB in the setting of focal ischemia. Moreover, imaging studies indicate that brain edema is more prone to buildup in the interstitial rather than in the intracellular compartment, likely due to the robust volume regulation of astroglia in vivo. Finally, in the context of vasogenic

edema, salt and water exudation, rather than the osmotic effects of protein leakage, are best correlated with the degree of edema. It is therefore possible that redistribution of solutes into the infarct core would set up osmotic gradients that favor excess glymphatic influx of CSF into the still perfused penumbra. This acute influx of excess glymphatic fluid would depend on astroglial water-channel AQP4, consistent with previous experimental data on cytotoxic brain edema. A failure of the energy-deprived ischemic tissue to maintain ionic gradients and normal perivascular anatomy might also hinder glymphatic ISF clearance and contribute to the development of interstitial and cytotoxic edema (Fig. 3.4) [10]. In addition, during vasogenic edema one would expect the exudation of high-molecular-weight serum proteins and leukocytes to occlude perivascular pathways, creating "perivasculitis" that could further impair glymphatic ISF clearance and contribute to the edema. In summary, it therefore seems likely that glymphatic activity and brain edema are interconnected, but the exact details of how glymphatic pathway impairment might contribute to edema in the context of stroke and other brain pathologies remain to be determined.

Aging and Neurodegeneration

A hallmark of neurodegenerative diseases is the age-related accumulation of fibrillary protein aggregates, such as extracellular senile plaques comprising Aβ and intracellular neurofibrillary tangles comprising hyperphosphorylated tau in AD. Experimental studies carried out in rodents demonstrate that interstitial Aβ and tau are cleared from the brain interstitium during sleep along perivascular spaces surrounding large-caliber draining veins. In the aging rodent brain, glymphatic pathway function is impaired, including a slowing of perivascular CSF–ISF exchange and interstitial Aβ clearance [1,2]. In this setting, impaired glymphatic pathway function was closely associated with the loss of perivascular AQP4 localization, suggesting that mislocalization of astroglial AQP4 may be one of the features that renders the aging brain vulnerable to protein misaggregation and subsequent neurodegeneration. It is possible that similar mechanisms may underlie the aggregation of other proteins in other neurodegenerative conditions, such as α-synuclein in Parkinson disease or Lewy body dementia. Sleep disruption is also a common feature of aging [9], and may further contribute to the impairment of effective Aβ clearance along the glymphatic pathway. In two clinical studies focusing on subjects with mild cognitive impairment (before diagnosis with AD), worsening sleep quality or sleep duration were associated with Aβ plaque deposition evaluated either with CSF biomarker measurement or amyloid positron emission tomography imaging.

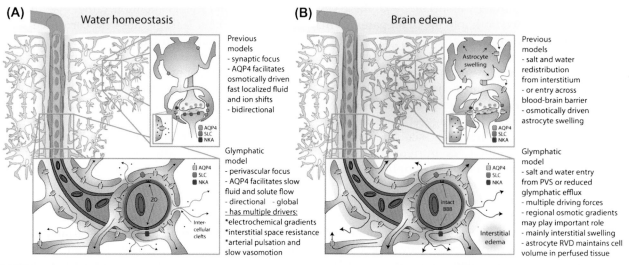

FIGURE 3.4 Proposed role of glymphatic pathway function in the development of cerebral edema after ischemic brain injury. Distinctions between previous models of salt and water homeostasis (*top*) and the glymphatic model (*bottom*) in the physiological setting (A) and in the setting of cerebral edema (B). In the glymphatic model, hydrostatic and electrochemical gradients drive fluid movement along perivascular spaces, whereas astroglial AQP4 supports fluid exchange along these pathways by permitting rapid water movement through the intracellular glial syncytium. In the setting of edema, perivascular salt and water entry combined with slowing glymphatic efflux results in interstitial swelling and inflammation within the perivascular space ("perivasculitis"). *AQP4,* aquaporin-4; *NKA,* Na$^+$-K$^+$-ATPase; *OA-647,* ovalbumin-conjugated Alexa647 (45 kD, CSF tracer); *PVS,* perivascular space; *RVD,* regulatory volume decrease; *SLC,* solute cotransporter proteins. *Reprinted and modified from Thrane et al. Trends Neurosci 2015;38(6):333–5. http://dx.doi.org/10.1016/j.tins.2015.04.009. Epub 2015 May 22 with permission.*

Chronic Glymphatic Pathway Impairment and Posttraumatic Neurodegeneration

TBI, including mild TBI (known commonly as concussion), is a risk factor for the development of early-onset dementia and AD. Among contact sport athletes and armed forces service members exposed to repeated mild TBI, a progressive tauopathy, chronic traumatic encephalopathy is characterized by perivascular deposits of tau aggregates initially in superficial layers of the cerebral cortex. The factors that make the posttraumatic brain vulnerable to neurodegeneration in the decades following injury are not clear. Experimental studies carried out in rodents demonstrate that after TBI, perivascular AQP4 localization and glymphatic pathway function is chronically impaired [1,2]. When glymphatic pathway function is impaired by *Aqp4* gene deletion, tau phosphorylation and neurocognitive deficits are worsened after moderate to severe TBI. This suggests that chronic impairment of glymphatic pathway function in the posttraumatic brain may set the stage for aberrant tau aggregation and neurodegeneration.

CONCLUSION

ISF and protein homeostasis and immune surveillance are two basic physiological functions that in the periphery are served by the lymphatic system. In the CNS, which is sheltered by the BBB, these vital functions must be accomplished through an alternative process. Emerging research suggests that in the brain these functions are supported through the combined activity of the perivascular glymphatic pathway, the CSF circulation, and the recently described sinus-associated lymphatic vasculature. Because the physiological mechanisms underlying the cerebral glymphatic and dural lymphatic systems remain largely unexplored, we are only beginning to appreciate the role that their dysfunction may play in the development of neurovascular, neurodegenerative, and neuroinflammatory conditions. Although the studies that have defined glymphatic pathway function have focused primarily on the pathogenic events underlying the development of neurodegenerative diseases such as AD, it appears likely that the impairment of ISF clearance in the presence of ongoing perivascular CSF influx may be a key driver of the development of cerebral edema after stroke or TBI. Similarly, the development of perivasculitis, with BBB dysfunction and perivascular immune cell infiltration, may underlie many of the consequences of vasogenic edema associated with ischemic brain injury.

References

[1] Jessen NA, Munk AS, Lundgaard I, Nedergaard M. The glymphatic system: a beginner's guide. Neurochem Res 2015;40(12):2583–99.

[2] Simon MJ, Iliff JJ. Regulation of cerebrospinal fluid (CSF) flow in neurodegenerative, neurovascular and neuroinflammatory disease. Biochim Biophys Acta 2015;1862(3):442–51.

[3] Sykova E, Nicholson C. Diffusion in brain extracellular space. Physiol Rev 2008;88(4):1277–340.

[4] Groothuis DR, Vavra MW, Schlageter KE, et al. Efflux of drugs and solutes from brain: the interactive roles of diffusional transcapillary transport, bulk flow and capillary transporters. J Cereb Blood Flow Metab 2007;27(1):43–56.

[5] Xie L, Kang H, Xu Q, et al. Sleep drives metabolite clearance from the adult brain. Science 2013;342(6156):373–7.

[6] Iliff JJ, Wang M, Liao Y, et al. A paravascular pathway facilitates CSF flow through the brain parenchyma and the clearance of interstitial solutes, including amyloid beta. Sci Transl Med 2012;4(147):147ra111.

[7] Asgari M, de Zelicourt D, Kurtcuoglu V. How astrocyte networks may contribute to cerebral metabolite clearance. Sci Rep 2015;5:15024.

[8] Iliff JJ, Goldman SA, Nedergaard M. Implications of the discovery of brain lymphatic pathways. Lancet Neurol 2015;14(10):977–9.

[9] Lim MM, Gerstner JR, Holtzman DM. The sleep-wake cycle and Alzheimer's disease: what do we know? Neurodegener Dis Manage 2014;4(5):351–62.

[10] Thrane AS, Rangroo Thrane V, Nedergaard M. Drowning stars: reassessing the role of astrocytes in brain edema. Trends Neurosci 2014;37(11):620–8.

CHAPTER

4

Cerebrospinal Fluid: Formation, Absorption, Markers, and Relationship to Blood–Brain Barrier

G.A. Rosenberg

The University of New Mexico, Albuquerque, NM, United States

INTRODUCTION

Cerebrospinal fluid (CSF) reflects pathology in the brain, and is essential in diagnosis of many neurological diseases, including those due to inflammation, infection, and immunological processes. CSF indicates pathological processes because of the continuity of the brain interstitial fluid (ISF) with the CSF across the ependymal lining of the ventricles. CSF and ISF are actively secreted by energy-requiring mechanisms: mainly epithelial cells make CSF in the choroid plexus, whereas capillaries and cellular metabolism form ISF. Both processes involve the sodium–potassium ATPase electrolyte pumps. ISF is thought to drain into the CSF spaces by movement along white matter tracts and perivascular spaces. When inflammatory cells cross the capillaries or are activated within brain, the brain edema that results causes brain swelling with raised intracranial pressure. CSF and ISF are continuously produced, resulting in about 500 mL per day, which must be removed or hydrocephalus will result. This dynamic system is routinely sampled for diagnostic and therapeutic reasons, making it essential for the clinician to understand the underlying molecular physiology.

An important concept that was identified by early investigators was the limited capacity of the brain to swell due to the dura and the bones of the skull, which contain the brain tissue, blood, and CSF/ISF; increase in any of these compartments raises intracranial pressure, which if high enough can lead to herniation of the compressed tissue and death. The skull protects the vulnerable brain, composed of 80% water, from minor injury, but severe blows to the head result in movement of the brain tissues against the skull, creating damage to brain cells that can lead acutely to swelling with cell death and over longer periods to neurodegeneration.

Besides the strong external layers protecting vulnerable brain tissues, there are multiple, more delicate internal layers separating brain tissue from the systemic circulation and protecting the internal milieu from being exposed to blood cells and circulating proteins by preventing the blood with toxic electrolyte and protein levels from mixing with the CSF/ISF. These multiple layers prevent cellular dysfunction by maintaining the stable internal milieu needed for normal brain cell function.

The first layer of protection is provided by the endothelial cells, which are joined together by tight junctions [1]. In addition to the physical barrier, there are molecular and enzymatic barriers, including carrier molecules that move substances in and out of the brain, delivering nutrients and removing toxins; enzymes in the endothelial cells degrade unwanted substances and prevent them from entering the brain. Another important interface between the blood and the CSF/ISF occurs at the choroid plexuses. Finally, another site of the blood–brain barrier BBB is the arachnoid villi cells in the subarachnoid space, which transfer CSF/ISF into the blood. All of the interfaces have a common property of being joined with tight junctions. The brain lacks a true lymphatic system to drain metabolic products and deliver nutrients. The CSF and ISF provide the lymphatic function, and have been referred to as the third circulation [2]. Thus this complex series of interfaces, acting as the brain's lymphatic circulation, moves molecules between cells and along perivascular spaces.

Because of the unique interfaces formed at critical sites in the brain, immunological reactions are limited in brain tissue, creating a site of "immunological privilege" [3]. The tight junctions at each interface prevent the levels of electrolytes in the brain from fluctuating widely with changes in the systemic circulation, which would occur during strenuous exercise. The tight junctions prevent the entrance of large protein molecules into the brain, which results in 40 mg% of albumin in CSF normally with a corresponding level of 4 g in the blood. They also block entry of circulating blood cells.

Sampling of the CSF is a common clinical procedure that is relatively safe and inexpensive compared with other procedures, making it a cost-effective addition to the diagnostic workup. The aim of this chapter is to describe the physiology and biochemistry of the CSF and the blood–brain interfaces, particularly as they relate to stroke [4].

BLOOD–BRAIN INTERFACES

Elaborate protective mechanisms are found in the brain mainly to separate the blood from brain cells. Release of blood into the brain as occurs with the rupture of an aneurysm produces a devastating effect. Each of the sites where blood comes into contact with brain fluids has a specialized epithelial-like layer of cells. The major interface is formed by the endothelial

cells, which have specialized proteins between the cells that seal them together, forming tight junctions that prevent proteins and cells from passing, and maintains a high electrical resistance. In addition to the endothelial interface, there are similar tight junction proteins found at the choroid plexus and the arachnoid granulations (Table 4.1). Tight junctions are composed of membrane proteins, occludin, claudins, and junctional adhesion molecules. Other proteins that are important in the maintenance of the tight junction cytoplasmic scaffolding proteins include zonula occludens, actin cytoskeleton, and associated proteins, such as protein kinases, small GTPases, and heterotrimeric G-proteins.

Tight junctions restrict passage of all but lipid-soluble molecules. Albumin, for example, which is 4–5 g/dl in the blood, is normally less than 50 mg/dl in the CSF. An important function of the BBB is to maintain differences in the concentrations between the blood and the brain for many electrolytes (Table 4.2), and it prevents the wide fluctuations in blood levels of certain electrolytes related to food intake, exercise, and other factors.

TABLE 4.1 Three Interfaces Between Blood and Brain That Form the Sites of the Blood–Brain Barrier

Interface	Site	Anatomical Correlate
1. Blood–brain	Cerebral blood vessels	Tight junctions between endothelial cells; transport functions; basal lamina
2. Blood–cerebrospinal fluid (CSF)	Choroid plexuses	Tight junctions between epithelial cells lining the choroid plexus
3. CSF–blood	Arachnoid granulations and arachnoid layer of the meninges	Tight junctions between arachnoid cells; a one-way valve-like action of the arachnoid granulations

TABLE 4.2 Comparative Concentrations of Substances Between Cerebrospinal Fluid (CSF) and Blood

Substance	CSF Concentration	Blood Concentration
ELECTROLYTES (MEQ/L)		
Sodium	138	138
Potassium	2.8	4.5
Chloride	19	102
Bicarbonate	22	24
Calcium	2.1	4.8
PROTEINS (MG/DL)		
Total protein	35	7000
Albumin	16	3700
IgG	1.2	1000

BLOOD–CSF INTERFACE

Choroid plexuses are the major source of CSF. Epithelial cells line the apical surface of the choroid plexuses (Fig. 4.1). Choroid plexuses float freely in the lateral and fourth ventricles. The rate of production of CSF in animals and in man has been measured experimentally, by placing needles to infuse fluids into the lateral ventricles and draining needles into the cisterna magna. Using ventriculocisternal perfusion and the dye and isotope dilution methods, a substance is introduced into the perfusion solution, and the amount of its dilution can be used to calculate the rate of formation of new CSF. In man, the rate of CSF formation is about 0.3 mL/min or about 500 mL/day, which results in the turnover of the CSF several times during the day [5].

Sodium–potassium ATPase pumps on the outer surface of the epithelial cells provide the ionic imbalance that results in a chemiosmotic energy for CSF secretion with the water flowing along the osmotic gradient established by the removal of three sodium ions for two potassium ions returned to the cell. Carbonic anhydrase is also important in fluid formation, but the mechanism is uncertain. Osmotic agents, such as mannitol, reduce production of CSF. Inhibition of carbonic anhydrase by acetazolamide also reduces CSF production. The production of CSF

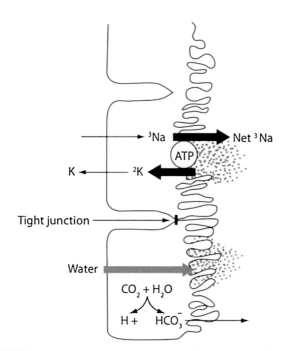

FIGURE 4.1 A schematic diagram of choroid plexus epithelial cells. Fenestrated capillary is shown in the stroma, which allows protein to escape. The epithelial cells have tight junctions between the apical surfaces, which also have the microvilli. An ATPase pump is on the apical surface. Carbonic anhydrase converts carbon dioxide to bicarbonate, which is removed into the cerebrospinal fluid and contributes protons to the sodium–proton exchange pump. Several other exchange pumps are shown.

is insensitive to increased pressure, and continues even when the CSF pressures are dangerously high.

Epithelial cells on the choroid plexus surface have tight junctions joining their apical surfaces. Beneath the epithelial cells is a stroma containing the blood vessels, which lack tight junctions. In this situation, there is no BBB but rather porous, fenestrated blood vessels in the stroma of the choroid plexus; tight junctions at the epithelial cell apical surfaces, providing the barrier at this interface, block larger molecules that leave the fenestrated blood vessels. The ependymal cells that form the surface of the choroid plexus are the site of the tight junctions, which is not the case in other ependymal covered regions that provide no barrier function, allowing fluid and proteins to pass from the CSF into the brain.

ISF production contributes to the CSF; estimates range from 30% to 60%, depending on the species. Most of the ISF formation is thought to be by the capillaries with a small contribution from metabolic water. In cats and rabbits, ISF is estimated to contribute 30% of the CSF, whereas in nonhuman primates some estimates are as high as 60% in an experiment where the choroid plexuses were removed [6]. Cerebral endothelial cells have an ATPase pump on the abluminal surface of the capillary, which is important in ISF formation. ISF moves through the 15–20% of brain that comprises the extracellular space and along perivascular pathways, combining with the CSF in the ventricles. Animal studies showed that substances injected into brain tissue appeared in the cervical lymphatics by draining along the olfactory nerves. The lymphatic route drains significant amounts of ISF in rats, cats, and rabbits, but the significance in humans is uncertain. Drainage of ISF into the cervical lymph nodes could allow brain antigens to enter the systemic circulation, causing an immunological response. The perivascular pathway conducts fluid transport within the brain. Astrocytes and capillaries are involved in the movement of fluid, and there appears to be a role for aquaporin, but the exact mechanism is uncertain [7]. Clearance of ISF via the perivascular route is altered in sleep [8]. The relevance of these studies done in mice to humans remains to be established.

The fluid formed by the cerebral blood vessels joins that formed by the choroid plexus in the cerebral ventricles, beginning the circulation of the CSF. From the ventricles the CSF exits into the cisterna magna through the foramina of Luschka and Magendie. Then, the CSF moves over the cerebral convexities. Arachnoid granulations absorb CSF into the blood. Cerebrospinal fluid percolates into the lumbar sac, where some is absorbed, before flowing up over the cerebral hemispheres.

Hydrocephalus results when the outflow of CSF from the cerebral ventricles is obstructed. This life-threatening situation requires surgical intervention to insert a ventriculoperitoneal shunt. When the obstruction is at the level of the arachnoid granulations, resistance to absorption results in nonobstructed hydrocephalus. If the fluid is removed transependymally, normal pressure can occur. Selection of patients with normal pressure hydrocephalus for shunting is confounded in patients with white matter disease secondary to hypertension since the results of magnetic resonance imaging (MRI) can be confused in those two conditions [9].

CSF–BLOOD INTERFACE

Absorption of CSF occurs across the arachnoid villi by a valve-like mechanism. Electron microscopic images of the arachnoid granulations show a series of channel-like structures [10]. The channels behave as one-way valves, allowing CSF to drain into the blood, but preventing blood from entering the CSF. Absorption of the CSF is pressure dependent. The resistance to the outflow of CSF determines the CSF pressure. When the resistance to absorption is increased, the CSF pressure is raised. This occurs, for example, in bacterial meningitis or subarachnoid hemorrhage, where the white or red blood cells clog up the valve-like channels in the arachnoid granulations. Thus both the rate of CSF production and the resistance to CSF absorption contribute to the CSF pressure measured at a lumbar puncture.

CSF PRESSURE

CSF pressure can be measured during a lumbar puncture performed to collect CSF. Pressure measurements need to be done with the patient in the lateral recumbent position shortly after the needle is placed in the lumbar sac and, if possible, before fluid is removed; otherwise loss of fluid will interfere with an accurate measurement of pressure. When the patient is in a sitting position, the pressure cannot be measured.

Cerebral veins are the main contributors to the CSF pressure because the thin-walled veins transmit pressure from the vascular system to the CSF (Fig. 4.2A). Thicker-walled arteries with muscular layers exert less of an effect because the vessel walls create an equal and opposite force, preventing the transmission of pressure to the CSF. Venous pressure is lower than arterial, reflecting the pressure of the blood as it drains into the heart. Arteries contribute less to the CSF pressure than the veins. The correct location of the spinal needle in the thecal sac can be verified by observations of fluctuations in CSF pressure related to respiration; the expansion of the chest wall reduces the venous pressure and vice versa. The balance between production of CSF and absorption determines the CSF pressure; increased resistance to absorption results in an increase in CSF pressure (Fig. 4.2B).

Diseases that can cause an increase in venous pressure in the heart can lead to increased intracranial pressure, headaches, and rarely papilledema. Increases in

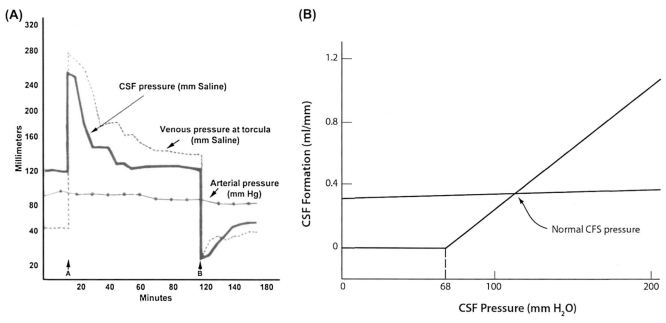

FIGURE 4.2   (A) Graph of influence of arterial and venous pressures on measurement of cerebrospinal fluid (CSF) pressure. Increase in CSF pressure is shown to be driven by the venous pressure, whereas there is no influence from arterial pressure. (B) Absorption of CSF is shown to begin at a CSF pressure of 68 mm H_2O, and to gradually increase as a function of absorption at the arachnoid villi. Formation of CSF is shown as a constant. The CSF pressure is a function of the rates of absorption and formation (*arrow*). *Composite drawing modified from Rosenberg GA. Molecular physiology and metabolism of the nervous system. New York, USA: Oxford University Press; 2012.*

arterial blood pressure affect CSF pressure only when drastic changes in blood pressure occur. For example, malignant hypertension in eclampsia dilates cerebral capillaries, leading to brain edema and increased CSF pressure. Benign intracranial hypertension is a poorly understood syndrome where the CSF pressure is markedly elevated.

There are a number of causes of increased intracranial pressure. Failure to absorb CSF can lead to hydrocephalus, which in the elderly can have normal pressure probably due to the opening of other absorption routes such as the transependymal absorption. In young obese women an idiopathic increase in intracranial pressure can occur [11]. In addition to unknown causes it can be due to medications, such as vitamin A and antibiotics, and to occlusion of the venous sinuses draining blood from the brain. Increased intracranial pressure can produce sixth nerve palsies as a remote effect because of the long course of the sixth nerve and the sharp bend over the clivus. Headache is the most common symptom, and many patients have visual obscurations. Papilledema is the major finding; the vision is preserved until the swollen optic disc encroaches on the macula. It is possible to follow the severity of the papilledema by serial measurements of the size of the blind spot created by the optic disc. Treatment with acetazolamide or steroids is helpful. Occasionally patients need a lumboperitoneal shunt to drain excess fluid. When vision is threatened, the optic nerve sheath fenestration can be done surgically. When the venous sinuses are blocked, there is some evidence that placement of a stent to open the sinus can be

helpful [12]. However, there are no controlled trials to assess this approach, and some of the improvement may have occurred spontaneously.

BLOOD–BRAIN BARRIER INTERFACE

The large surface areas of the cerebral capillaries make the BBB the most important of the interfaces between the blood and brain tissue (Fig. 4.3). Cerebral capillaries are unique and differ from systemic capillaries (Table 4.3). They have endothelial cells with continuous tight junctions. Sodium–potassium ATPase pumps are found on the abluminal side of the vessel. Glucose and amino acids undergo carrier-mediated transport. Enzymes are present in the vessel that degrade substances and prevent them from crossing the capillary. An extracellular matrix forms a basal lamina around the capillaries and arterioles (Fig. 4.4).

Brain tissue requires a constant supply of glucose and amino acids for normal metabolism. Cerebral capillaries have glucose transporter molecules that participate in the carrier-mediated transport across the capillary [13]. Abnormalities in the glucose transporter lead to reduced brain glucose and cellular damage. Monoamine oxidase, both A and B types, deaminate and inactivate biogenic amines, preventing them from entering the brain. Other enzymes identified in brain capillaries include choline acetyltransferase, which is important in choline metabolism, and γ-glutamyl transpeptidase, which is used as a marker for blood vessels.

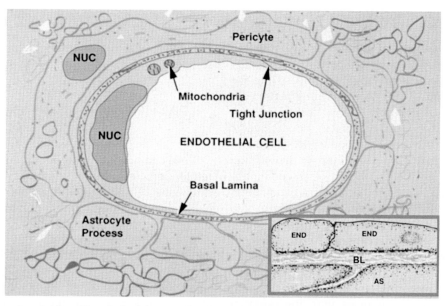

FIGURE 4.3 Drawing of cerebral capillary derived from an electron micrograph. The tight junctions are shown. Increased numbers of mitochondria are shown. Pericytes and astrocytes are shown around the blood vessels. Insert: Basal lamina of cerebral capillary is shown in this electron micrograph tracing. Endothelial cells are separated by basal lamina from astrocytes and pericytes. *AS*, astrocyte; *BL*, basal lamina; *END*, endothelial cell; *NUC*, nucleus.

TABLE 4.3 Unique Features of Cerebral Capillaries Important in the Blood–Brain Barrier

1. Tight junctions (zonula occludens)
2. Sodium–potassium ATPase
3. Carriers for glucose and amino acids
4. Proteolytic enzymes
5. Basal lamina

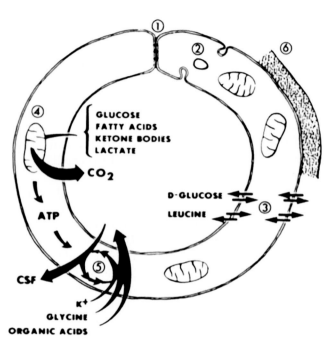

FIGURE 4.4 Cerebral capillary with it main functions. (1) Tight junctions, (2) few pinocytotic vesicles, (3) transport of glucose and amino acids, (4) large number of mitochondria, (5) secretion of cerebrospinal fluid by ATP-dependent pumping, and (6) basal lamina [4].

The glucose transporter is found in higher concentration in the cerebral capillaries than in most systemic vessels. At low blood glucose concentrations the carrier is unsaturated and helps movement of glucose into the brain along with diffusion. At high serum glucose concentrations the carrier molecule becomes saturated, and any additional transport is by diffusion alone. Specific carrier molecules in the cerebral blood vessel also transport amino acids into the brain. Competitive inhibition of a carrier causes reduced transport of essential substances. For example, an amino acid carrier transports L-Dopa, which is used to treat Parkinson disease, into brain. When a large protein meal is digested, other amino acids compete for the carrier sites, and less of the drug is absorbed. Another example is found in patients with an enzyme deficiency that leads to accumulation of excessive amounts of phenylalanine in the blood. Phenylalanine competes with the essential amino acid, tryptophan, reducing its absorption into the brain. Reduction of levels of phenylalanine in the blood through dietary restrictions can prevent permanent brain damage.

Basal lamina surrounds the capillary. It is composed mainly of type IV collagen and laminin along with lesser amounts of fibronectin and heparan sulfate. The function of the basal lamina in brain is uncertain. In other organs it plays an important physiological role in control of permeability. For example, type IV collagen provides a layer of structural support around the blood vessel, heparan sulfate is a charge barrier, and laminin and fibronectin bind hormones and growth factors.

Proteolytic enzymes participate in the modulation of BBB permeability. These enzymes are important in injury and repair, complicating attempts to block their actions with drugs. The major proteases involved in the BBB are the matrix-degrading metalloproteinases, which are neutral proteases that break down the extracellular matrix in many pathological processes, including those involving brain, but they are also important in angiogenesis and neurogenesis during tissue repair after injury. Several brain cells make matrix metalloproteinases and other proteases in response to injury [14].

Several types of molecules can cross the BBB. Lipids dissolve in the membrane and equilibrate rapidly in brain with levels in the blood. Substances that are not lipid soluble enter the brain more slowly. The ability of a substance to cross the cerebral capillary determines its therapeutic potential. For example, heroin is a chemically modified morphine molecule that more rapidly enters the brain. Penicillin slowly crosses the BBB, whereas the newer generations of antibiotics, such as cephalosporin, more easily enters the brain. Drugs used to treat cancer, such as methotrexate, only slowly enter the brain. Injection into the thecal sac can be used to increase the amount of the agent reaching brain tissue. Increasing the ability of a drug to cross the BBB is a major area of research especially for drug companies.

PATHOLOGICAL CHANGES IN THE CAPILLARY

Cells can survive many insults if the cell's membranes remain intact. During an injury, when membranes break down the cells die. Less dramatic forms of cell death occur with the orderly involution of a cell during apoptosis and with autophagy [15]. Another consequence of damage to the cell is brain edema: when the cells swell and the membrane is intact, there is cytotoxic edema; damage to the blood vessels results in vasogenic edema. Capillaries affected by the injury have increased permeability, which results in the movement of solutes and water into the brain. Brain cells do not necessarily die of brain edema alone. Vasogenic edema spreads through white matter in the extracellular space. Either cytotoxic or vasogenic edema can contribute to the tissue damage and interfere with recovery.

Opening of the BBB can be determined radiologically by computed tomography (CT) or MRI, using the injection of an appropriate contrast agent into the blood. Contrast-enhanced CT and MRI show leakage of a contrast agent into the brain in tumors, inflammation, and injury. For example, serial MRI in patients with multiple sclerosis has shown that the blood vessel is damaged before the formation of permanent changes in the myelin and blood vessel inflammation contributes to the demyelinating process [16].

Bacterial meningitis causes opening of the BBB in and around the subarachnoid space. The blood vessels in the subarachnoid space penetrate into the cortex along the Virchow–Robin spaces. As the vessels enter the cortex they have glial limitans around them, but deeper in the cortex, they have the basal lamina. Inflammatory cells derived from the meningeal vessels can enter the Virchow–Robin spaces and cause inflammation of the cortical surface. Antibiotics cause the bacterial cell walls to lyse with the release of toxic bacterial cell wall products, including lipopolysaccharide. Cytokines are formed, along with free radicals and proteases, which result in secondary damage to the blood vessel. Thus the antibiotic can control the infection, but the secondary damage can produce delayed vasogenic edema. In childhood meningitis treatment, steroids are given along with antibiotics to reduce the secondary damage from the inflammatory response, but use of steroids has not been adequately tested in adults with meningitis.

Cerebral ischemia causes opening of the BBB. The factors involved in the opening of the BBB are complex and include cytokines, free radicals, and proteases. Permanent occlusion of a blood vessel damages the cerebral capillaries in the ischemic tissue. Animal studies have shown that when the tissue undergoes reperfusion, BBB injury occurs more rapidly and is more severe. Reintroduction of oxygenated blood into the damaged tissue allows free radicals to form and neutrophils to enter the tissue, worsening the ischemic damage. The critical factor is to reduce the time before reperfusion to reduce the subsequent reperfusion injury.

CONCLUSION

The interfaces between blood and brain preserve the integrity of the neuronal microenvironment. Choroid plexuses and cerebral capillaries form CSF and ISF, which circulate through the cerebral ventricles and the extracellular spaces, leaving the brain mainly via the arachnoid granulations. CSF functions as the brain's lymph, delivering nutrients and removing metabolic products. At each interface between blood and brain, anatomic or enzymatic processes are present that block movement of substances that are not lipid soluble. Carrier-mediated processes transport nutrients essential to brain cell function, such as glucose and amino acids, into the central nervous system. The BBB is frequently opened during injury, allowing brain edema to form, causing damage to brain cells. Many cytotoxic factors, including free radicals, cytokines, and proteases, participate in disruption of the BBB. Understanding the interaction of multiple factors in these processes will lead to more effective therapeutic agents.

References

[1] Hawkins BT, Davis TP. The blood–brain barrier/neurovascular unit in health and disease. Pharmacol Rev 2005;57(2):173–85.

[2] Cushing H. The third circulation. London: Oxford University Press; 1925.

[3] Yong VW, Rivest S. Taking advantage of the systemic immune system to cure brain diseases. Neuron 2009;64(1):55–60.

[4] Rosenberg GA. Molecular physiology and metabolism of the nervous system. New York, USA: Oxford University Press; 2012.

[5] Cserr HF. Physiology of the choroid plexus. Physiol Rev 1971;51:273–311.

[6] Milhorat TH, Hammock MK, Fenstermacher JD, et al. Cerebrospinal fluid production by the choroid plexus and brain. Science 1971;173:330–2.

[7] Iliff JJ, Wang M, Zeppenfeld DM, et al. Cerebral arterial pulsation drives paravascular CSF-interstitial fluid exchange in the murine brain. J Neurosci 2013;33(46):18190–9.

[8] Xie L, Kang H, Xu Q, et al. Sleep drives metabolite clearance from the adult brain. Science 2013;342(6156):373–7.

[9] Brean A, Fredo HL, Sollid S, et al. Five-year incidence of surgery for idiopathic normal pressure hydrocephalus in Norway. Acta Neurol Scand 2009;120(5):314–6.

[10] Upton ML, Weller RO. The morphology of cerebrospinal fluid drainage pathways in human arachnoid granulations. J Neurosurg 1985;63:867–75.

[11] Bastin ME, Sinha S, Farrall AJ, et al. Diffuse brain oedema in idiopathic intracranial hypertension: a quantitative magnetic resonance imaging study. J Neurol Neurosurg Psychiatry 2003;74(12):1693–6.

[12] Higgins JN, Cousins C, Owler BK, et al. Idiopathic intracranial hypertension: 12 cases treated by venous sinus stenting. J Neurol Neurosurg Psychiatry 2003;74(12):1662–6.

[13] Spector R. Nutrient transport systems in brain: 40 years of progress. J Neurochem 2009;111(2):315–20. doi:JNC6326 10.1111/j.1471-4159.2009.06326.x, [Published Online First: Epub Date] [pii].

[14] Candelario-Jalil E, Yang Y, Rosenberg GA. Diverse roles of matrix metalloproteinases and tissue inhibitors of metalloproteinases in neuroinflammation and cerebral ischemia. Neuroscience 2009;158(3):983–94. S0306-4522(08)00895-6 10.1016/j.neuroscience.2008.06.025, [Published Online First: Epub Date] [pii].

[15] Iadecola C, Anrather J. The immunology of stroke: from mechanisms to translation. Nat Med 2011;17(7):796–808. http://dx.doi.org/10.1038/nm.2399. [Published Online First: Epub Date].

[16] Filippi M, Rocca MA, Barkhof F, et al. Association between pathological and MRI findings in multiple sclerosis. Lancet Neurol 2012;11(4):349–60. http://dx.doi.org/10.1016/S1474-4422(12)70003-0. [Published Online First: Epub Date].

CHAPTER

5

Anatomy of Cerebral Veins and Dural Sinuses

E. Egemen[1], I. Solaroglu[2]

[1]Koç University Hospital, Istanbul, Turkey; [2]Koç University, Istanbul, Turkey

INTRODUCTION

Venous infarction of brain comprises only 1% of all strokes [1]. Many predisposing conditions such as dehydration, coagulopathies, pregnancy, trauma, surgical interventions, inherited collagen disorders, and autoimmune vascular diseases may result in cerebral vein thrombosis. Fortunately not all of veins lead to severe complications. However, in case of clinical manifestations developed, diagnosis must be done immediately in order to investigate and treat possible reasons. This chapter aims to present anatomical configuration of cerebral venous system regarding possible significant origin of "stroke."

DURAL VENOUS SINUSES

The dural sinuses are large endothelial-lined trabeculated channels that collect cerebral blood from the superficial, deep, and posterior fossa and drain into the internal jugular vein at the level of the jugular bulb (Fig. 5.1) [2]. These sinuses, which lie between the superficial (periosteal) and deep (meningeal) layers of the dura mater, also excrete cerebrospinal fluid (CSF) via arachnoid granulations (i.e., Pacchionian granulations) that emerge from the subarachnoid space [3]. These arachnoid granulations (villi) are commonly found around the superior sagittal and transverse sinuses [1]. Furthermore, cavernous sinuses (CS), which are paired dural venous sinuses, contain cranial nerves, arteries, and veins.

Superior Sagittal Sinus

The superior sagittal sinus (SSS), which is the longest dural sinus, lies along the superior edge of the falx cerebri, which is attached to the crista galli at the interhemispheric space just underneath the cranial vault. The SSS originates from the anterior part of the frontal lobe at the foramen caecum and drains into the torcular herophili [4]. Contact with the SSS leads to the development of an impression on the frontal and parietal bone.

The SSS enlarges posteriorly due to tributaries from cortical veins and arachnoid granulations. Concurrently, emissary veins carry diploic blood into the SSS. Thus, the posterior portion of the SSS is more visible in the venous phase of digital subtraction angiography (DSA) and magnetic resonance venography (MRV). The radiological appearance of the SSS is curvilinear with an enlarged line in the sagittal view and a reverse triangular shape in the coronal view. A rudimentary view of the anterior one-third portion of the SSS has been well characterized [5].

Inferior Sagittal Sinus

The inferior sagittal sinus (ISS) originates from the inferior edge of the anterior one-third portion of the falx cerebri and lies within the interhemispheric spaces. This relatively small sinus collects anterior pericallosal veins [3]. The ISS has a curvilinear shape like the SSS. The ISS joins with the great cerebral vein (i.e., vein of Galen) at the level of the falcotentorial junction, which both drain into the straight sinus [5].

Straight Sinus

The straight sinus (SS) originates from the falcotentorial junction via the union of the great cerebral vein and the ISS. The SS receives veins from the falx cerebri, tentorium cerebelli, and adjacent brain parenchyma. The SS drains into the torcular herophili together with the SSS [3]. As a variation, the SS may also drain into the transverse sinus; this variation tends to occur more frequently on the left side [1].

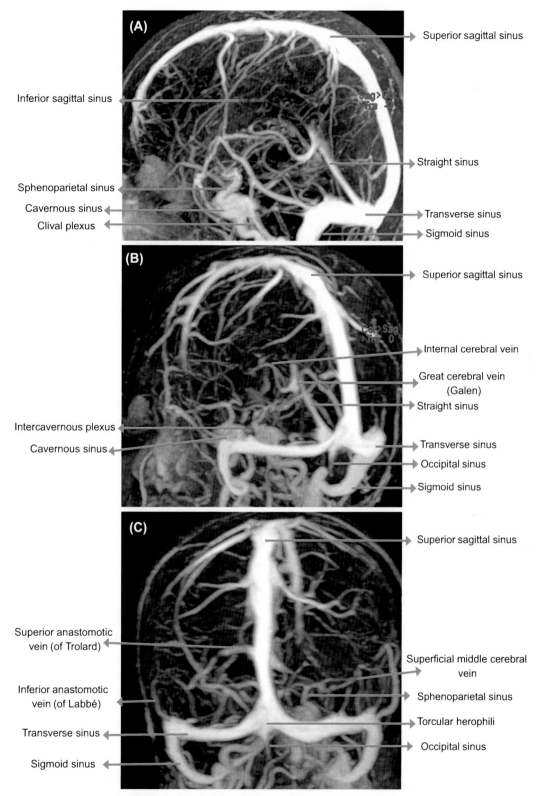

FIGURE 5.1 Magnetic resonance venography of brain with lateral (A), oblique (B), and anteroposterior (C) views. Superior sagittal, straight, and occipital sinuses join at the point of torcular herophili. Then they drain into transverse, sigmoid sinuses and internal jugular vein orderly. Superficial middle cerebral vein makes anastomosis to superior sagittal sinus via vein of Trolard, to transverse sinus via vein of Labbé, and to cavernous sinus via sphenoparietal sinus. Intercavernous plexus and clival plexus also make connection between two cavernous sinuses.

Transverse (Lateral) Sinuses

The transverse (lateral) sinuses (TS) originate from the torcular herophili and drain into the sigmoid sinuses. The TS are located at the posterior edge of the tentorium cerebelli. Like the SSS, the TS also absorb CSF via arachnoid granulations. The TS are mostly asymmetric; the right TS tends to be larger [3]. Radiological view of hypoplastic TS or absence of TS should differentiate from TS thrombosis.

Torcular Herophili (Confluens Sinuum)

The torcular herophili, which are also known as the confluence of sinuses (confluens sinuum), represents the crossroads of the SSS, ISS, TS, and the occipital sinus, when present [3]. Anatomical localization of the torcular herophili is varied. The torcular herophili appear as an asymmetric pouch in the radiological view [5].

Sigmoid Sinuses

Sigmoid sinuses are so named due to their S-shaped curves extending from the lateral edge of the tentorium cerebelli to the jugular bulb. Asymmetry can be seen in accordance with asymmetry in TS [3].

Cavernous Sinuses

These complex large sinuses are about 2 cm long and 1 cm wide including trabeculations and contain important vessels and cranial nerves. Anatomically placed laterally to the sella turcica both CS are connected by intercavernous venous plexus at anterior and posterior edge of sella turcica and also by clival venous plexus [6].

The main tributaries of the CS are the superior and inferior ophthalmic veins and the sphenoparietal sinuses. Thus, the CS collect blood from not only the inferior parts of the frontal and parietal lobe but also the orbital cavity [4]. Sphenoparietal sinuses are anastomotic vein, which are placed on lesser sphenoid wing and make connection between CS and superficial middle cerebral vein (Sylvian vein) and also receive veins of temporal pole. The CS also receive tributaries from the skull base via the superior and inferior petrosal sinuses [3].

Cavernous segment of internal carotid artery enters in CS and runs posteriorly to superior (posterior ascending segment), then turns anteriorly in middle part of CS (horizontal segment) and before exit from CS runs to superior (anterior ascending segment) [6].

Oculomotor nerve (CNIII), trochlear nerve (CNIV), and ophthalmic divisions of trigeminal nerve (CNVI) attach to lateral wall of CS and leave intracranial space via superior orbital fissure. Maxillary divisions of trigeminal nerve (CNV2) also attach to the lateral wall of CS, however, exit extracranial space via foramen rotundum. Abducens nerve (CNVI) is in middle part of CS, which is placed laterally to the ICA [6].

Superior and Inferior Petrosal Sinuses

The superior petrosal sinuses (SPS) are located between the petrous part of the temporal bone and anterolateral edge of the tentorium cerebelli. The SPS connect the CS to sigmoid sinuses and less frequently to TS. The inferior petrosal sinuses (IPS) lie on the petro-occipital fissure and drain the CS into the jugular bulb via the clival venous plexus [3].

Occipital Sinus

The occipital sinus, which is the smallest dural venous sinus, runs along the inner surface of the occipital bone. The occipital sinus is attached to the posterior margin of the falx cerebelli and receives tributaries from the margins of the foramen magnum. It may anastomosis with the sigmoid sinuses and posterior internal vertebral plexus that drain into the torcular herophili. The occipital sinus is an important vascular structure during posterior fossa surgery. Variations in the occipital sinus, such as double or oblique occipital sinuses or the absence of the occipital sinus, are observed in rare cases [7].

CEREBRAL VEINS

Cerebral veins accompany arteries in the subarachnoid space. Unlike veins in other parts of the body, cerebral veins do not have valves. Thus, bidirectional flow is possible in cerebral veins. The walls of cerebral veins are thin and vulnerable as they do not contain muscle tissue. Cerebral veins are subdivided into three groups according to their anatomical location [2]:

1. superficial venous system (external);
2. deep venous system (internal);
3. veins of the brain stem and posterior fossa.

Superficial Supratentorial Cortical Veins (External)

Superficial supratentorial cortical veins are located on the surface of the brain and categorized into three main groups according to their drainage [3].

- *Superior Cortical Veins*: These veins are also known as the superior sagittal group because they drain into the SSS [1]. The most prominent vein of this group is the superior anastomotic vein, which is known as the vein of Trolard. This anastomotic vein connects the superficial middle cerebral vein (SMCV) to the SSS. In addition, there are 8–12 unnamed smaller veins in the upper part of the hemispheric convexity [2].

- *Middle Cortical Veins*: These veins are also known as the sphenoidal group because they drain into the CS via the sphenoparietal sinuses. The dominant vein in this group is the SMCV, which is also known as the Sylvian vein due to its location in the Sylvian fissure (i.e., the lateral cerebral fissure). Middle cortical veins receive tributaries from the inferior part of the frontal lobe, superior temporal gyrus, and parietal opercula. The vein of Labbé (Fig. 5.2), which is another anastomotic vein of the SMCV, receives tributaries from the posterior and inferior temporal lobe and adjacent parietal lobe and drains primarily into the TS and rarely into the sigmoid sinuses [2].
- *Inferior Cortical Veins*: The major vein of this group is the deep middle cerebral vein (DMCV), which receives tributaries from the inferior frontal lobes and temporal lobes such as the insula, basal ganglia, and parahippocampal gyrus. The basal vein of Rosenthal (BVR), which is considered part of the deep venous system, is an anastomotic vein of the DMCV [2,3].

Deep Supratentorial Cortical Veins (Internal)

Deep structures of the cerebral hemispheres, including the basal ganglia, corpus callosum, thalamus, and posterior part of the limbic system, are drained by the deep venous system, which has two major components: internal cerebral vein (ICV) and BVR [1].

Numerous smaller medullary veins emerge from the subcortical area, directly cross the white matter, and drain into the subependymal veins. Septal veins and thalamostriate veins are the most prominent vascular structures of subependymal veins. Septal veins are localized to the frontal horn and course posteriorly toward the septum pellucidum. Thalamostriate veins are anatomically located medially to the caudate nucleus and thalamus [3]. These two veins meet near the foramen of Monro which is composed of ICVs; this point of intersection is called the "venous angle." The ICVs then course between the roof of the third ventricle and the fornices. Choroid veins, which lie along the floor of the lateral ventricle, also drain into ICVs (Fig. 5.3) [4].

The BVR originates from the intersection of the anterior cerebral vein, DMCV, and striate vein [2,4]. The great cerebral vein (i.e., the vein of Galen, VofG) originates from the intersection of two ICVs and the BVR. VofG is a 2-cm long, U-shaped midline vein that courses under the splenium of the corpus callosum in the quadrigeminal cistern [4]. The VofG and ISS intersect to form straight sinuses [3].

Brain Stem and Posterior Fossa Veins

The brain stem and posterior fossa veins are categorized into three subgroups according to their drainage system [2,3]:

- *Superior (Galenic) Group*: This group includes precentral cerebellar (PCV), superior vermian (SVV), and anterior pontomesencephalic veins (APMV). The PCV courses between the lingual and central lobule

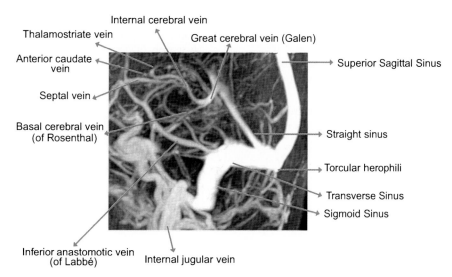

FIGURE 5.2 Cerebral veins of deep circulation and posterior fossa are shown in lateral view MR venography. Dominant structure of this system is great cerebral vein (of Galen), which is formed by joining of two internal cerebral veins, and basal cerebral vein of Rosenthal. Then it drains into straight sinus together with inferior sagittal sinus.

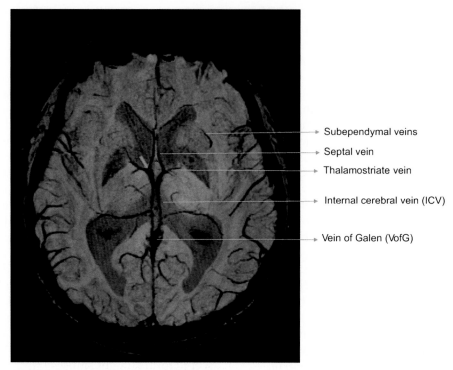

FIGURE 5.3 Axial imaging of susceptibility-weighted (SW) magnetic resonance shows septal vein and thalamostriate vein which meet near the foramen of Monro and are composed of internal cerebral vein. This point of intersection is called the "venous angle" (*orange arrow*). The great cerebral vein (of Galen) originates from the intersection of two internal cerebral veins and basal vein of Rosenthal.

of vermis. The superior vermian vein originates from the top of the vermis, courses through the culmen, and drains into the PCV. The APMV includes many smaller veins that cover the cerebral peduncles and anterior surface of the pons. All veins in this group drain into the VofG.

- *Anterior (Petrosal) Group*: The petrosal vein (PV) is the dominant vascular structure of this group, which is important during cerebellopontine angle cistern surgery. Tributaries from the brain stem and cerebellum are observed as a "petrosal star" on DSA or computed tomography venography (CTV). The PV forms an anastomosis with the lateral mesencephalic vein and SPS [2].
- *Posterior (Tentorial) Group*: The inferior vermian veins lie under the vermis and receive tributaries from the inferior part of the cerebellum.

References

[1] Kılıç T, Akakın A. Anatomy of cerebral veins and sinuses. Front Neurol Neurosci 2008;23:4–15.

[2] Osborn AG. Veins and Venous Sinuses. In: Harnsberger HR, Osborn AG, Macdonald AJ, Ross JS, Moore KR, Salzman KL, Wiggins RH, Davidson HC, Hamilton BE, Carrasco CR, editors. Utah: Amirsys Publishing; 2011. p. I 334–87.

[3] Rhoton Jr AL. The cerebral veins. Neurosurgery 2002;51(4 Suppl.):159–205.

[4] Uddin MA, Haq TU, Rafique MZ. Cerebral venous system anatomy. J Pak Med Assoc 2006;56(11):516–9.

[5] Ayanzen RH, Bird CR, Keller PJ, McCully FJ, Theobald MR, Heiserman JE. Cerebral MR venography: normal anatomy and potential diagnostic pitfalls. AJNR Am J Neuroradiol 2000;21(1):74–8.

[6] Yasuda A, Campero A, Martins C, Rhoton Jr AL, de Oliveira E, Ribas GC. Microsurgical anatomy and approaches to the cavernous sinus. Neurosurgery 2008;62(6 Suppl 3):1240–63.

[7] Tubbs RS, Bosmia AN, Shoja MM, Loukas M, Curé JK. Cohen – Gadol AA. The oblique occipital sinus: a review of anatomy and imaging characteristics. Surg Radiol Anat 2011;33(9):747–9.

CHAPTER

6

Cerebral Vasa Vasorum

W.M. Ho[1], C. Reis[2], O. Akyol[1], J. Zhang[1,2]

[1]Loma Linda University School of Medicine, Loma Linda, CA, United States; [2]Loma Linda University Medical Center, Loma Linda, CA, United States

INTRODUCTION

Vasa vasorum (VV) are defined literally as vessels of vessels, and are predominantly observed in large vessels with an important role under pathological conditions. Research on noncerebral VV has been established for over a century, including cardiac, pulmonary, aortic, and portal vein VV. Intracranial vessels were misrepresented to be devoid of VV, and increasing evidence for their existence widens the field of cerebral VV research. Constantly improving technology and modern radiologic methods reveal new aspects of intra- and extracranial cerebral VV, and provide further elucidation on how they act in neurovascular disease. However, cerebral VV remain a highly discussed topic with contradicting and discrepant theories [1].

STRUCTURE, FUNCTION, AND LOCALIZATION

VV are considered microvessels within the wall of a host vessel, forming a microvascular network predominantly in large arteries. Their main functions are supplying oxygen and nutrients to the adventitia and the outer media and eliminating waste, while the inner vascular layers are nourished by intraluminar blood diffusion.

Cerebral arteries compared to systemic vessels have a thinner tunica media and adventitia. Instead of an external elastic lamina, there are only few elastic fibers, and the internal elastic lamina is fenestrated [1]. Further, tiny openings of 1–3 μm diameter were observed on the surface of intracranial vessel walls, which connects the tunica media with the cerebrospinal fluid (CSF). Zervas et al. described that these small channels enable CSF to get in contact with deeper layers of the vessel walls, building an adventitial network called the rete vasorum [1–3]. These differences compared to systemic vessels enable intracranial arteries to be nurtured by CSF diffusion, therefore decreasing the demand for supplemental nourishment by VV [1,2]. Animal studies in canine, feline, and rodents confirmed the lack of VV in cerebral vessels, the same as in an examination of intracranial arteries of neonates and children without cerebrovascular diseases [2]. This suggests that intracranial VV do not exist at birth [1]. However, small capillary-like vessels of 10–20 μm in diameter were detected in adult patients, consistent with VV [3]. They were predominantly localized in the proximal part of the internal carotid arteries, the basilar, and vertebral arteries [2–4]. Cerebral VV were reported in even further distal arteries in two studies, but not in the distal part of the middle and anterior cerebral artery. The reason for the discrepancy might be including vessels with smaller diameter and different patient selection in rather small cohort studies. Connolly et al. suggested an association of VV with the enlargement of vessel and intima thickness, using highly sensitive and specific immunohistochemical staining of Factor VIII [3]. In a similar study by Aydin et al., no VV were seen in arteries with a tunica media of less than 250 μm thickness, disregarding the size of systemic arteries. Furthermore, cerebral VV were observed for a distance of 1–1.5 cm after dural penetration following the proximal intracranial internal carotid artery and vertebral artery [4]. Regarding the layers in the vessel wall, cerebral VV are localized almost exclusively in the tunica adventitia, but rarely in the media [3,4](Figs. 6.1 and 6.2).

Primer on Cerebrovascular Diseases, Second Edition
http://dx.doi.org/10.1016/B978-0-12-803058-5.00006-0

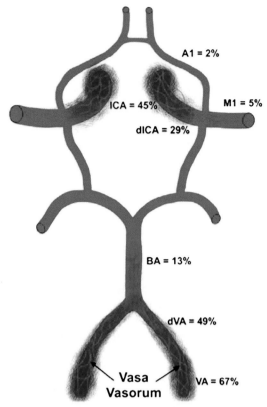

FIGURE 6.1 Localization and incidence of cerebral vasa vasorum (VV). Percentage numbers are derived from the light microscopical investigation by Takaba and colleagues of 50 patients with intracranial diseases, such as aneurysm, intracerebral hemorrhage, stroke, brain tumor, or trauma [4]. Cerebral VV most frequently occurred in the vertebral artery (VA), distal segment of the vertebral artery 1 cm proximal to the formation of the basilar artery (dVA), internal carotid artery (ICA), and distal segment of internal carotid artery to origin of posterior communicating artery (dICA). They are less likely to occur in the basilar artery (BA), proximal M1-part of the middle cerebral artery, and proximal A1-part of the anterior cerebral artery.

CEREBRAL VASA VASORUM NEOVASCULARIZATION

All published data so far agree in terms of a direct correlation between VV incidence and cerebrovascular disease, although whether cerebral VV development is the cause or the response remains poorly understood. Most pathophysiological research on VV neovascularization has been on cardiovascular and pulmonary arteries. However, few experimental investigations on cerebral VV assume similar angiogenic pathways involved in cerebral neovascularization as in other VV [5,6]. There is evidence that VV neovascularization is associated with cardiovascular risk factors and both initiation as well as progression of atherosclerosis. Hypertension and intimal thickening as a response to atherosclerosis leads to insufficient oxygen and nutrient diffusion. Thicker intima and plaque growth becomes a diffusion obstacle for intraluminal nourishment,

thereby contributing to a hypoxic response and vascular cell damage. As a compensatory mechanism to oxygen and nutritional demand, VV neovascularization is induced, and preexisting VV grow deeper toward the vessel lumen to support the supply of the inner layers. Underlying pathophysiological signaling pathways involve hypoxic inducible factor (HIF) as a key regulator of atherosclerosis and angiogenic response, although multiple pathological triggers are assumed to be involved. HIF is known to activate vascular endothelial growth factor (VEGF) and E26 transformation-specific (Ets) factor, both are important enhancers of hypoxia-induced angiogenesis and VV development. VEGF is strongly correlated with neovascularization and angiogenic sprouting under physiological and pathophysiological circumstances. In a hypertensive rat model, elevated expression of VEGF and HIF coincided with VV growth in the aorta. Ets is a transcription factor that regulates gene expression of matrix metalloproteinases (MMPs), and therefore contributes to extracellular matrix degradation and migration of vascular endothelial cells supporting VV neovascularization. Additionally, Ets upregulates the angiogenic factors, hepatocyte growth factor, and VEGF. Besides VEGF and Ets, multiple other mechanisms are involved in VV neovascularization with either synergistic or independent effects on angiogenesis, including fibroblast growth factors (FGF) and epidermal growth factor [7,8].

Interestingly, vascular occlusion seems to be a stronger catalyst than atherosclerosis alone for promoting both intra- and extravascular neovascularization [2]. A major contributor to VV neovascularization is initiation and perpetuation via vascular inflammation caused by intraluminar monocyte adhesion, lipid oxidation, and increased release of cytokines or growth factors. Consequently, neovessels are most likely present where chronic inflammatory cells, such as macrophages and lymphocytes, infiltrate the vessel wall. Novel publications support the hypothesis of vascular inflammation initiated in the tunica adventitia and its advance toward the intima. As mentioned before, lipid oxidation and reactive oxygen species cause vascular inflammation, hence VV formation is suggested to be lipid dependent. For instance, oxidized lipids 15-Deoxy-δ-12 and 14-prostaglandin J2 promote neovascularization by activating PPAR-γ, which upregulates VEGF expression in vascular smooth muscle cells. Furthermore, cholesterol modulates lipid rafts in upregulating VEGF signaling-dependent angiogenesis. Supporting the hypothesis, that lipid-dependent vascular inflammation originates in the adventitia, perivascular adipose tissue is assumed to support angiogenesis, especially in advanced stages of atherosclerosis, when neovessels might emit inflammatory mediators to the vascular network of the adipose tissue [7,8].

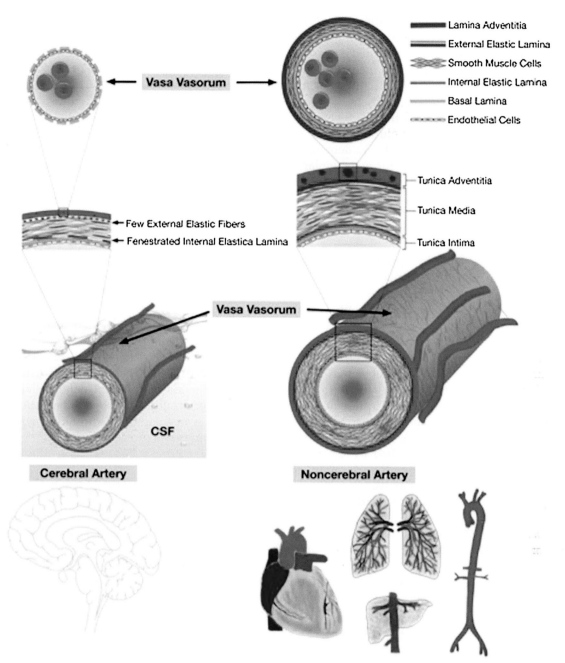

FIGURE 6.2 Vasa vasorum (VV) of the cerebral versus the noncerebral type, such as portal vein VV, cardiac VV, aorta VV, and pulmonary VV. Structural difference of the vascular walls in cerebral arteries leads to distinct expression of VV due to different demands of supplementary nourishment, as intracranial vessels are surrounded by nutrient-rich cerebrospinal fluid (CSF). Cerebral VV are capillary-like, while the structure of the other VV may be completely consistent with small arteries due to their larger lumina.

MOLECULAR IMAGING AND DETECTION

Imaging and detection of VV in general is a challenging task for radiologic methods to distinguish and visualize, because of their delicate anatomical nature and potential overlay by larger vessels due to topical proximity. Especially cerebral VV require high-resolution imaging techniques to be sensitive enough for offering possibility of detailed in-depth evaluation to estimate the degree of neovascularization.

Recent advances in neuroimaging modalities enable direct and indirect detection of cerebral VV in greater detail. Molecular CT imaging presupposes iodinated nanoparticles for increased X-ray absorption at the focused anatomic region. The novel method of high-resolution cone beam CT (CB–CT) could demonstrate in 18 carotid stent patients that contrast enhancement in the tunica adventitia correlates to vulnerable plaque. Additionally, cerebral VV neovascularization covering carotid plaque could be distinguished from calcified plaque localized in the intima and media [9]. Other promising sensitive techniques for detecting neovascularization are contrast-enhanced magnetic resonance imaging (MRI) and ultrasound (US). Micro-computed tomography (CT) with a spatial resolution affords high spatial-resolution images of cerebral VV, demonstrating wall enhancement as an index for neovascularization. Although, this technique offers excellent imaging results, it is restricted as an ex vivo imaging modality [1].

Contrast agent–enhanced ultrasonography (CEUS) has promising VV diagnostic capability using second-generation ultrasound contrast agents. These are molecular agents detectable by specific contrast software, encapsulated in 1–5 μm diameter gas microbubbles, which are stabilized by a lipid or protein shell. Microbubbles enable real-time imaging of vascular blood flow in vessels and cerebral VV, therefore offering an estimation of neovascularization. The adventitial VV network in cerebral vessels can be visualized using this method. In CEUS, microbubbles in atherosclerotic plaque indicate the presence of intraplaque VV. Therefore, contrast enhancement is presented to be higher in patients with atherosclerosis due to VV density and cross-sectional thickness of VV [7]. However, every ultrasound imaging method is limited by the experience of the examiner, restricted depth penetration, and motion artifacts from arterial pulsations.

VV are correlated with a positive feedback loop of inflammation and angiogenesis. Therefore, molecular imaging of cerebral VV is mainly based on the identification of inflammation and angiogenesis. Further development of conjugated microbubbles for binding specific ligands in VV offers a new perspective in experimental molecular imaging, like targeting endothelial cell adhesion molecules and fibrin. In a 2011 study in an atherosclerosis rat model, microbubbles coupled to the dual endothelin-1/VEGF end-signal peptide receptor (DEspR) were suggested to increase the detection rate of neovascularization. DEspR-targeted contrast-enhanced ultrasound micro-imaging could reveal a correlation between an increased level of DEspR expression in carotid artery lesions, the degree of luminal endothelial pathology, and the density of VV neovascularization [5].

High tesla MRI is probably considered the most feasible technique to identify cerebral VV in the clinical setting. Dynamic contrast–enhanced perfusion MRI is most specific for demonstrating cerebral microcirculation characteristics, illustrating its distribution, and angiogenic activity. For molecular imaging, MRI is used with gadolinium or super paramagnetic iron oxide compounds, which increases the T1 signal or decreases the T2 signal. Gadolinium chelates contribute in enhancing markers for neovascularization and inflammation, therefore contrast enhancement in the outer border of extra- and intracranial carotid arteries offers a quantitative assessment of cerebral VV. MRI-derived parameters for angiogenic activity include the endothelial transfer coefficient (Ktrans). Ktrans monitors the effect of antiangiogenic molecules, and quantifies the transfer of contrast agent from plasma to adventitia, which is regulated by VV presence. Consequently, Ktrans of gadolinium enhancement in the carotid adventitial plaques is a suitable quantitative parameter for cerebral VV density measurement, according to the theory, that more VV provide additional endothelial area for contrast agent to diffuse and enhanced adventitial thickness indicate inflammation and neovascularization. The correlation between the total neovascularization area and Ktrans has been confirmed by histological measurements. There is also a strong connection of increased adventitial enhancement and the recent occurrence of an ipsilateral cerebrovascular ischemic event. The advancement of αvβ3 integrin-targeted gadolinium chelates in animal studies gained new aspects for molecular imaging using MRI. αvβ3 integrin-targeted gadolinium nanoparticles offer more accurate localization of neovascularization, since integrin αvβ3 is histologically colocalized with angiogenic molecules in vessel proliferation [1,7].

CLINICAL SIGNIFICANCE IN DISEASE

As mentioned in the foregoing, VV presence most likely coincides with cerebrovascular diseases. Studies have shown that cerebral VV are more frequent and of larger diameter in elder patients (>40 years old) with severe atherosclerosis [2,4] or in thrombotic occluded vessels. Under the latter circumstances, cerebral VV might even develop in childhood, implicating that vascular occlusion is the superior trigger for inducing neovascularization. The fact that cerebral VV develop with time and are absent in neonates, contradicts the theory of VV initiating intracranial atherosclerosis [2]. Extracranial carotid plaques are established as a strong indicator for coronary atherosclerosis, yet not as sensitive for intracranial involvement. Yet, a strong correlation has been reported between recent ipsilateral cerebrovascular ischemia and cerebral VV occurrence, characterized by increased adventitial enhancement in black-blood MRI. MRI is currently the most frequently used method for assessing vulnerable plaque associated with VV, although a study using high-resolution CB–CT showed equal sensitivity. Yet, CB–CT has been shown to offer better feasibility and to be superior in distinguishing cerebral VV as a marker for plaque vulnerability [9]. The 2015 contrast-enhanced ultrasound study in healthy patients investigated carotid

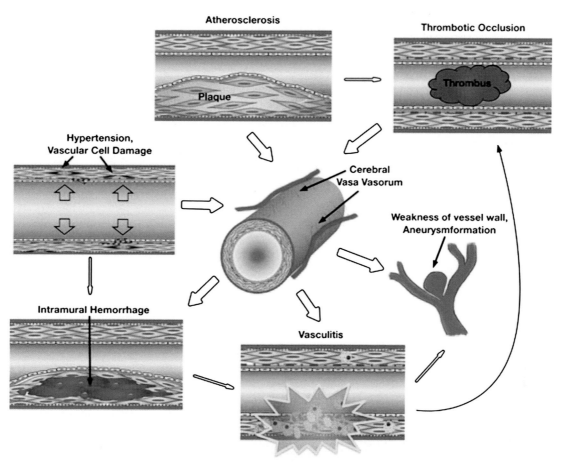

FIGURE 6.3 Overview of neurovascular diseases, involving cerebral VV. The exact underlying pathophysiological mechanism, and whether cerebral VV development is causative or reactive, is not clearly understood. Cerebral VV develop with age, therefore suggesting not to be the initial indicator for atherosclerosis, although the incidence increases with severe atherosclerosis, thrombotic occlusion, hypertension, and cerebral aneurysms, involving intramural hemorrhage and vasculitis.

adventitial VV, concluding that these VV provide a promising marker for atheromatosis detection and progress monitoring [10]. Hypertension leads to reduced blood flow through compressed VV and therefore medial hypoxia, followed by vascular cell damage and necrosis [1]. In rare cases of cerebral VV reaching from the tunica adventitia to the media, Takaba et al. observed an association with complicated intramural hemorrhage and vessel dissection at all times, suggesting the underlying mechanism to be the fragility of medial VV [4]. This phenomenon is similar to intraplaque hemorrhage and progression caused by disruption and leakiness of incompetent fragile VV. Furthermore, cerebral VV have been reported frequently in association with cerebral aneurysms, especially of larger saccular type, of fusiform type, and giant aneurysms. The occurrence of cerebral VV is suspected to be correlated with the enlargement in size and higher risk of aneurysm rupture. Inflammation, intramural hemorrhage, and occluded cerebral VV are factors contributing to the weakening of the vascular wall structures, and therefore permitting aneurysm growth (Fig. 6.3). Additionally, inflammatory response and release of growth hormones are induced by vascular cell damage, as well as fragile VV causing repeated microbleeding. Besides aneurysms, vasculitides appear more frequently in extracranial vessels with higher VV densities, than intracranial arteries. Consequently, VV are assumed to have less impact on the development of intracranial vasculitides [1].

As more clarification continues to be revealed regarding cerebral VV, the more their clinical role in the pathophysiology of neurovascular diseases gains significance.

References

[1] Portanova A, Hakakian N, Mikulis DJ, Virmani R, Abdalla WMA, Wasserman BA. Intracranial vasa vasorum: insights and implications for imaging. Radiology 2013;267(3):667–79. http://dx.doi.org/10.1148/radiol.13112310.

[2] Aydin F. Do human intracranial arteries lack vasa vasorum? A comparative immunohistochemical study of intracranial and systemic arteries. Acta Neuropathol 1998;96(1):22–8. http://dx.doi.org/10.1007/s004010050856.

[3] Connolly ES, Huang J, Goldman JE, Holtzman RNN. Immunohistochemical detection of intracranial vasa vasorum: a human autopsy study. Neurosurgery 1996;38(4):789–93. http://dx.doi.org/10.1227/00006123-199604000-00031.

[4] Takaba M, Endo S, Kurimoto M, Kuwayama N, Nishijima M, Takaku A. Vasa vasorum of the intracranial arteries. Acta Neurochir 1998;140(5):411–6. http://dx.doi.org/10.1007/s007010050118.

[5] Decano JL, Moran AM, Ruiz-Opazo N, Herrera VLM. Molecular imaging of vasa vasorum neovascularization via DEspR-targeted contrast-enhanced ultrasound micro-imaging in transgenic atherosclerosis rat model. Mol Imaging Biol 2011;13(6):1096–106. http://dx.doi.org/10.1007/s11307-010-0444-4.

[6] Langheinrich AC, Michniewicz A, Bohle RM, Ritman EL. Vasa vasorum neovascularization and lesion distribution among different vascular beds in ApoE-/-/LDL-/- double knockout mice. Atherosclerosis 2007;191(1):73–81. http://dx.doi.org/10.1016/j.atherosclerosis.2006.05.021.

[7] Xu J, Lu X, Shi G-P. Vasa vasorum in atherosclerosis and clinical significance. Int J Mol Sci 2015;16(5):11574–608. http://dx.doi.org/10.3390/ijms160511574.

[8] Mulligan-Kehoe MJ. The vasa vasorum in diseased and nondiseased arteries. AJP: Heart Circ Physiol 2010;298(2):H295–305. http://dx.doi.org/10.1152/ajpheart.00884.2009.

[9] Tanabe J, Tanaka M, Kadooka K, Hadeishi H. Efficacy of high-resolution cone-beam CT in the evaluation of carotid atheromatous plaque. J Neurointerv Surg January 2015. http://dx.doi.org/10.1136/neurintsurg-2014-011584.

[10] Arcidiacono MV, Rubinat E, Borras M, Betriu A, Trujillano J, Vidal T, et al. Left carotid adventitial vasa vasorum signal correlates directly with age and with left carotid intima-media thickness in individuals without atheromatous risk factors. Cardiovasc Ultrasound 2015;13:20. http://dx.doi.org/10.1186/s12947-015-0014-7.

CHAPTER

7

Cerebral Vascular Muscle

T.M. De Silva[1,2], F.M. Faraci[1]

[1]The University of Iowa Carver College of Medicine, Iowa City, IA, United States; [2]Monash University, Clayton, VIC, Australia

INTRODUCTION

Cerebral blood flow (CBF) is controlled predominantly by the level of arterial pressure (perfusion pressure) and the diameter of resistance vessels in brain. The moment-to-moment regulation of cerebral arterial and arteriolar diameter, and thus CBF, is primarily the function of vascular muscle. Vascular muscle receives, integrates, and responds to mechanical forces as well as signals from endothelium, neurons, astrocytes, and other cell types. Intrinsic and extrinsic factors regulate the amount of tone that specific segments of the vasculature generate. These factors include contractile forces such as myogenic tone and myogenic reactivity along with molecules and pathways that produce

vasodilation including nitric oxide (NO), reactive oxygen species (ROS), and activation of potassium ion channels. As such, a defect in vascular muscle may disrupt normal regulation of CBF and can have dire consequences for the brain. For example, defects in vascular muscle impact CBF and cognitive function in cerebral autosomal dominant arteriopathy with subcortical infarcts and leukoencephalopathy (CADASIL), the most common genetic cause of small vessel disease known. In this chapter, we summarize recent advances regarding the regulation of vascular tone in cerebral arteries and the microcirculation. As part of this overview and as proof of principle, we briefly discuss our current understanding of the clinical features and pathobiology of CADASIL.

Primer on Cerebrovascular Diseases, Second Edition
http://dx.doi.org/10.1016/B978-0-12-803058-5.00007-2

MYOGENIC TONE: ROLE OF TRP CHANNELS

One inherent feature of vascular muscle is its ability to dynamically respond to changes in intraluminal or transmural pressure. Myogenic responses involve constriction of arteries and arterioles when intraluminal pressures rises and dilation as pressure drops. This characteristic of smooth muscle in resistance vessels is a major contributor to what is known as autoregulation, where CBF remains relatively stable over a wide range of perfusion pressures [1]. However, despite extensive study, the mechanism by which changes in pressure is sensed by vascular muscle and translated into a change in vessel diameter is not entirely understood. Key aspects remain unclear including the identity of the mechanosensor as well as the precise signaling cascade that links these events. G-protein-coupled receptors, ion channels, cytoskeletal elements, and extracellular matrix components have all been suggested to have mechanosensor properties [2]. Which of these protein(s) are of greatest importance is still debated. However, it is also possible that the mechanosensor and signal transduction pathway mediating the myogenic response varies between vascular beds, along the vascular tree, as well as in health versus disease.

In an attempt to better define these mechanisms, studies have begun to unravel the molecular details by which increased intravascular pressure translates to vasoconstriction (Fig. 7.1). Transient receptor potential (TRP) channels are a family of nonselective cation channels that have recently become a focus of effort in regard to regulation of vascular function. There are 28 identified members of the TRP family grouped into six subfamilies based on sequence homology. The six TRP subfamilies are designated canonical (TRPC), vanilloid (TRPV), melestatin (TRPM), ankyrin (TRPA), mucolipin (TRPML), and polycystin (TRPP) [3]. As of 2016, seven have been identified in cerebral vascular muscle (TRPC1, TRPC3, TRPC5, TRPC6, TRPV4, TRPM4, and TRPP2). The majority of these channels have been implicated in the regulation of vasoconstriction, particularly myogenic tone, with the exception of TRPV4, which has been shown to mediate vasodilation (see the following paragraphs). Three TRP channels are thought to contribute to myogenic vasoconstriction, TRPC6, TRPM4, and TRPP2 [3]. The other described roles of TRP channels in regulating vascular muscle function are summarized in the Table 7.1.

A 2015 study proposed a pressure-sensing signaling system that is dependent on TRPC6 and TRPM4. TRPC6 channel activation, which may occur via a phospholipase Cγ1 (PLCγ1)-dependent pathway or by direct mechanical activation, results in calcium influx and triggers further calcium release from the sarcoplasmic reticulum. The localized calcium event (calcium spark) activates TRPM4 channels resulting in membrane depolarization, the opening of voltage-dependent calcium channels (VDCCs) and vasoconstriction [3]. In addition, but separate from TRPC6 and TRPM4 channel activation, TRPP2 channels have also been implicated in the generation of myogenic tone. Knockdown of TRPP2 channels significantly attenuated myogenic tone in isolated cerebral arteries [3].

Ultimately, the increase in intracellular calcium activates calcium/calmodulin and myosin light chain kinase, resulting in increased phosphorylation of myosin light chain and cell contraction. Calcium sensitization is also an important component of the myogenic response and involves activation of Rho kinase (ROCK). Inhibitors of ROCK dilate pressurized cerebral arteries and arterioles with myogenic tone. Two isoforms of ROCK are expressed in vascular cells (ROCK1 and ROCK2). We found recently that a selective ROCK2 inhibitor dilates pressurized cerebral parenchymal arterioles to the same extent as a nonselective inhibitor of both ROCK isoforms [7], suggesting ROCK2 has the greater importance in brain arterioles in relation to myogenic tone.

In addition to activating TRPM4 channels, calcium sparks activate large-conductance calcium-activated K (BK) channels. Opening of BK channels hyperpolarizes the cell, resulting in vasodilation. Thus, modulation of the myogenic response by BK channels prevents excessive vasoconstriction in response to rises in blood pressure or other stimuli.

Apart from TRP channels, the calcium-activated chloride channel, TMEM16A, has been implicated in the generation of myogenic tone. Activation of chloride channels causes chloride efflux, membrane depolarization, activation of VDCC, and contraction of vascular muscle. Inhibition of TMEM16A with specific antibodies or knockdown with siRNA significantly reduces development of myogenic tone in cerebral arteries.

As discussed in the preceding paragraphs cerebral arteries and particularly cerebral arterioles display a robust myogenic response. However, there are uncertainties surrounding the identity of the cellular proteins and pathways that mediate these responses. Further investigation is needed to determine which ion channels and/or other molecules are of most importance in both health and disease.

MECHANISMS UNDERLYING VASODILATION

Nitric Oxide Signaling

NO is a small, uncharged gaseous molecule that readily crosses cellular membranes. However, due to its short half-life and reactivity with select molecules

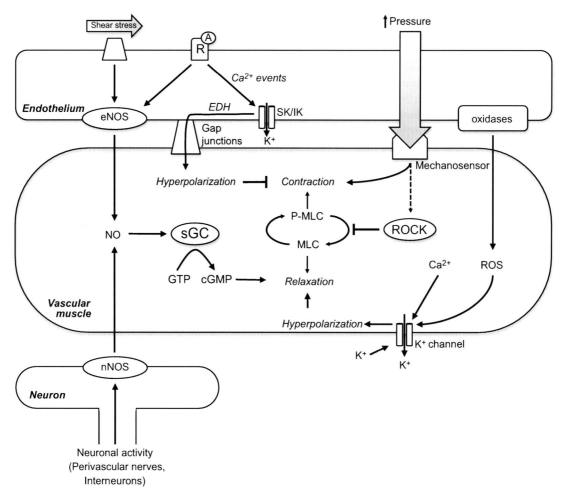

FIGURE 7.1 Schematic diagram depicting some of the major mechanisms regulating contraction and relaxation of vascular muscle. Increased intraluminal pressure activates mechanosensor(s) on vascular muscle, which promotes cell contraction. The contractile state of vascular muscle is determined by the ratio of phosphorylated to unphosphorylated myosin light chain (MLC). ROCK (perhaps ROCK2 in particular) plays a central role in the myogenic response by increasing calcium sensitivity, although the mechanism by which it is activated is unclear. NOS enzymes generate nitric oxide (NO) basally and their activity can be increased by shear stress, receptor-mediated signaling (eNOS), or neuronal activity (nNOS). NO activates soluble guanylate cyclase (sGC) generating cyclic guanine monophosphate (cGMP), resulting in relaxation of vascular muscle. Other mechanisms of relaxation include activation of various K$^+$ channels, which hyperpolarizes the cell and thus, cause relaxation. Numerous molecules can activate K$^+$ channels, including reactive oxygen species (ROS), low concentrations of extracellular K$^+$, and increased intracellular calcium. Activation of small and intermediate conductance K$^+$ channels (SK/IK) in endothelium generates a hyperpolarization current (EDH) that is transmitted to vascular muscle via gap junctions, opposing contraction. *A*, agonist; *EDH*, endothelium-dependent hyperpolarization; *R*, receptor; *ROCK*, Rho kinase.

TABLE 7.1 TRP Channels That Mediate Constriction or Dilation of Cerebral Vascular Muscle

TRP	Function/Vessel
TRPC1	Receptor-operated calcium entry
TRPC3	GPCR-mediated constriction
TRPC5	Store-operated calcium entry
TRPC6	GPCR-mediated constriction, myogenic tone
TRPM4	Depolarization, myogenic tone
TRPP2	Myogenic tone
TRPV4	Calcium influx, vasodilation

See Ref. [3] for detailed review.

(see the following paragraphs), the paracrine effects of NO are spatially limited. Nonetheless, NO has profound effects on vascular function, which are generally protective [4]. NO inhibits inflammation and platelet aggregation and plays an important role is preventing pathological changes in vascular structure. NO is also an important dilator of vascular muscle. Normally, NO is synthesized in endothelium or neurons by endothelial (eNOS) or neuronal NO synthase (nNOS), respectively. Both eNOS and nNOS generate relatively low levels of NO. Basal generation of NO is sufficient to dilate the cerebral vasculature as evidenced by the immediate vasoconstriction that occurs following application of NOS inhibitors. eNOS activity can be further stimulated by increased shear stress on the endothelium- or receptor-mediated agonists including neurotransmitters and metabolic factors. A subpopulation of neurons uses NO as a signaling molecule that also increases local CBF in response to cellular activation, thus contributing to neurovascular coupling. Inducible NOS (iNOS) can be expressed during disease (e.g., ischemia) and can generate pathological concentrations of NO. All NOS enzymes generate NO from the amino acid L-arginine and require the cofactors tetrahydrobiopterin, nicotinamide adenine dinucleotide phosphate (NADPH), flavin adenine dinucleotide, and flavin mononucleotide.

Upon entering the vascular muscle, NO binds to and activates its receptor, the heme site on soluble guanylate cyclase (sGC), which then synthesizes cyclic 3',5'-guanosine monophosphate (cGMP) from guanine triphosphate (GTP). cGMP activates downstream targets, predominantly cGMP-dependent protein kinase I [4]. The resultant decrease in intracellular calcium levels causes relaxation of vascular muscle. This sGC-cGMP-dependent process mediates the vast majority of the vascular response to NO in brain. Pharmacological agents that donate NO and activate this pathway, such as nitroglycerin and sodium nitroprusside, are used therapeutically or experimentally to activate sGC and induce vasodilation.

In cerebral arteries and arterioles, NO-sGC-cGMP is a robust vasodilator mechanism. However, NO is also susceptible to oxidative stress when present. Virtually all forms of vascular disease are associated with an increase in oxidative stress, which tips the redox balance of the tissue to one where oxidants, such as superoxide, are abundant. Superoxide reacts with NO extremely effectively, not only reducing the bioavailability of NO but also generating peroxynitrite, a highly damaging oxidant. As a result, the ability of NO to reach and initiate dilator mechanisms in vascular muscle is reduced, thus compromising perfusion. Numerous disease states including hypertension, diabetes, ischemia–reperfusion injury and hypercholesterolemia are associated with oxidative stress and reduced vasodilator responses [4].

K+ Channels

K+ channels play a fundamental role in maintaining proper membrane potential in vascular muscle and as such are key regulators of vascular tone [5]. Five subtypes of K+ channels—BK, inward rectifier K+ (K_{ir}) channel, ATP-sensitive K+ (K_{ATP}) channel, voltage-sensitive K+ (K_V) and two-pore domain or tandem pore channels have been identified in cerebral arteries. Activation of K+ channels results in K+ efflux, membrane hyperpolarization, and vessel dilation. Their role in vascular muscle function is briefly discussed.

BK channels play an important role in modulating myogenic constriction in cerebral arteries. However, there appear to be differences in the relative importance of BK channels in different vascular segments. Interestingly, while BK channels are expressed in parenchymal arterioles, they may be less active basally. Inhibition of BK channels in cerebral arteries results in constriction, whereas parenchymal arterioles exhibit little response to BK channel inhibition. BK channels are activated by elevations in intracellular calcium in the form of calcium sparks as well as vasodilators that elevate intracellular levels of cGMP or cyclic 3',5'-adenosine monophosphate (cAMP). BK channels are also activated by metabolites of arachidonic acid, reduced pH and ROS. Hypertension, diabetes, and genetic interference with the transcription factor peroxisome proliferator-activated receptor γ (PPARγ) in vascular muscle disrupts proper BK channel function.

Increasing the extracellular concentration of K+ above 3 mmol/L and up to about 20 mmol/L activates K_{ir} channels. At concentrations above 20 mmol/L, K+ depolarizes the cell and causes constriction. Barium is the only known K_{ir} channel inhibitor and blocks K+-induced dilation up to 20 mmol/L but has no effect at higher concentrations of K+. K_{ir} channel activation decreases normal intracellular calcium oscillations, resulting in hyperpolarization and dilation. Recent evidence indicates that genetic interference with PPARγ in vascular muscle severely impairs K_{ir} channel function. Because they are highly sensitive to small changes in K+, it has been suggested that K_{ir} channels may be an important mediator of neurovascular coupling.

K_{ATP} channels may serve as a sensor of the metabolic state of the cell. K_{ATP} channels are thought to be in the closed, inactive state under normal conditions. These channels are sensitive to decreases in intracellular levels of ATP (or changes to the ATP:ADP ratio) or mild hypercapnia (reductions in pH). Additionally, K_{ATP} channels are activated by vasodilators that activate the cAMP-dependent protein kinase pathway and inhibited by vasoconstrictors that activate protein kinase C (PKC). ROS may also activate K_{ATP} channels in the microcirculation. K_{ATP} channel function in vascular muscle is impaired by genetic interference with PPARγ. Interestingly, K_{ATP} channel function is significantly augmented in hypertensive rats following subarachnoid hemorrhage.

K_V channels also oppose or limit contraction of vascular muscle as they are voltage dependent. Application of K_V channel inhibitors constricts pressurized cerebral arteries suggesting these channels are active basally. It is thought that K_V channels may play an important role in maintaining membrane potential and oppose myogenic constriction in a similar manner to BK channels. K_V channels are inhibited by activation of PKC and potentially by increased extracellular concentrations of glucose. As discussed in the following section, vascular muscle cells from mice harboring a mutation that causes CADASIL have increased K_V channels at the cell membrane, resulting in reduced myogenic tone.

Two-pore domain K^+ channels are also expressed in vascular muscle in brain. In contrast to the subtypes outlined in the preceding paragraphs, much less is known regarding mechanisms and stimuli that control expression and activity of this subgroup of channels.

Reactive Oxygen Species

ROS have emerged as important regulators of vascular tone in the cerebral circulation [4]. While ROS can constrict cerebral arteries under some conditions, the majority of evidence suggests that physiological levels of ROS are vasodilators. However, during disease, ROS can have deleterious effects, especially on the bioavailability of NO (see earlier). Major sources of ROS include NADPH oxidases, cyclooxygenases, and mitochondria.

Both endogenous and exogenous ROS, via direct application or local generation using ROS-generating systems, have been shown to dilate large and small cerebral arteries. Endogenous ROS may be generated by application of enzyme substrates such as NADH/NADPH (NADPH oxidase) or arachidonic acid (cyclooxygenase). Dilation in response to ROS is generally mediated via activation of K^+ channels, either K_{ATP} or BK channels.

Although less common, constriction of cerebral arteries to ROS has been described and may involve activation of PKC and stimulation of calcium entry into vascular muscle via L-type calcium channels.

TRPV4

TRPV4 channels facilitate calcium influx in vascular muscle but surprisingly, mediate vasodilation [3]. Activation of TRPV4 channels elicit calcium-induced calcium release in the form of calcium sparks from sarcoplasmic reticulum stores, thus mediating responses to 11,12-epoxyeicosatrienoic acid (11,12-EET). The calcium sparks activate BK channels causing hyperpolarization of vascular muscle and dilation.

Another described role for TRPV4 is the modulation of angiotensin II (Ang II)-induced vasoconstriction. In isolated vascular muscle, Ang II-dependent activation of angiotensin-type receptors increased TRPV4 sparklet activity. In intact cerebral vessels, Ang II-induced constriction was offset by TRPV4 activity. As in the isolated cells, TRPV4-dependent calcium sparks activated BK channels to modulate Ang II-induced vasoconstriction.

CADASIL: A VASCULAR MUSCLE DEFECT THAT REDUCES CEREBRAL BLOOD FLOW AND IMPAIRS COGNITIVE FUNCTION

Due to the importance of proper coupling of CBF to neuronal activity, defects in vascular muscle that compromise its ability to participate in this integrated multicellular response as well as other fundamental elements of CBF regulation may have profound effects on cellular integrity and cognitive function [6,8]. CADASIL is a proof of principle example of a defect in vascular muscle that severely affects brain function. CADASIL is the most common known genetic cause of small vessel disease and is caused by mutations in the *NOTCH3* gene. As of 2016 there are no specific treatments for CADASIL or small vessel disease in general.

Magnetic resonance imaging techniques reveal the presence of lacunar infarcts and microbleeds early in the progression of small vessel disease. Histological analysis of human brain tissue shows profound thickening and fibrosis of white matter and deep gray matter arterioles. The presence of granular osmophilic material around the cell membrane of vascular muscle are thought to be composed of the extracellular domain of the NOTCH3 receptor and other extracellular proteins [9]. The most common symptoms that are observed in CADASIL patients are migraine with aura, subcortical ischemic events, mood disturbances, apathy, and cognitive impairment [9]. While the incidence of these symptoms can vary, the vast majority of patients will have ischemic events and develop dementia in later life.

Studies reported in 2015 on transgenic mice that capture many aspects of the human condition provide important insight into disease pathology. Mice that express a mutated form of *Notch3* reveal that CADASIL is associated with hypoperfusion at baseline, a loss of myogenic responsiveness, disruption of neurovascular coupling, vascular remodeling, and reduced capillary density. Studies have also revealed myelin degradation and increased presence of vacuoles in axons. A channelopathy involving increased expression of K_V1 channels in vascular muscle is thought to account for the decreased myogenic tone observed during this disease [10].

SUMMARY

In this chapter, we have summarized some of the key properties of cerebral vascular muscle. We have highlighted some important mechanisms that regulate contraction and relaxation of vascular muscle. The integration and response to these various stimuli is a key determinant of vascular tone and thus overall

CBF. As discussed, CADASIL is an example of how a defect in vascular muscle function can adversely affect the vasculature and as a result compromise brain function.

Acknowledgments

Work summarized in this chapter was supported by research grants from the National Institute of Health (HL-62984 and HL-113863), the Department of Veteran's Affair's (BX001399), the Fondation Leducq (Transatlantic Network of Excellence), and the National Health and Medical Research Council of Australia (1053786).

References

[1] Cipolla MJ. The cerebral circulation. In: Granger DN, Granger J, editors. Integrated systems physiology: from molecule to function. 2nd ed. 2016, p. 1–69, San Rafael: Morgan & Claypool Life Sciences.

[2] Hill-Eubanks DC, et al. Vascular TRP channels: performing under pressure and going with the flow. Physiology 2014;29(5):343–60.

[3] Earley S, Brayden JE. Transient receptor potential channels in the vasculature. Physiol Rev 2015;95(2):645–90.

[4] Faraci FM. Protecting against vascular disease in brain. Am J Physiol Heart Circ Physiol 2011;300(5):H1566–82.

[5] Dunn KM, Nelson MT. Potassium channels and neurovascular coupling. Circ J 2010;74(4):608–16.

[6] Iadecola C. The pathobiology of vascular dementia. Neuron 2013;80(4):844–66.

[7] De Silva TM, et al. Heterogenous impact of ROCK2 on carotid and cerebrovascular function. Hypertension 2016;68:809–17.

[8] De Silva TM, Faraci FM. Microvascular dysfunction and cognitive impairment. Cell Mol Neurosci 2016;36:241–58.

[9] Chabriat H, et al. CADASIL. Lancet Neurol 2009;8(7):643–53.

[10] Dabertrand F, et al. Potassium channelopathy-like defect underlies early-stage cerebrovascular dysfunction in a genetic model of small vessel disease. Proc Natl Acad Sci 2015;112(7):E796–805.

CHAPTER

8

Endothelium

R. Sumbria[1,2], M. Fisher[2]

[1]Keck Graduate Institute, Claremont, CA, United States; [2]University of California, Irvine, Irvine, CA, United States

INTRODUCTION

The vascular endothelium is a simple monolayer of cells that line the capillary lumen and is optimally placed at the interface of the blood circulation and vessel wall. Once considered to be merely a "cellophane wrapper" of the vessel wall, it is now well known that the endothelium is an active endocrine organ responsible for regulating vascular tone, blood coagulation and thrombosis, nutrient delivery, cellular adhesion, vascular smooth muscle cell proliferation, and inflammation, both in the brain and the periphery. The vascular endothelium in brain is unique since brain capillary endothelial cells lack fenestrations, are cemented together by extremely "tight junctions," have minimal pinocytotic activity, and express enzymes capable of degradation of a number of molecules [1]. The endothelial cells of the brain capillary are the anatomical site of the blood–brain barrier (BBB), a diffusional barrier formed by a complex network of the endothelial cells, pericytes, astrocytes, neurons, and the basement membrane that lines the brain capillary wall (Fig. 8.1).

Besides the structural peculiarities of the brain vascular endothelium, certain functional characteristics of the brain vascular endothelium also contribute to brain-specific

Primer on Cerebrovascular Diseases, Second Edition
http://dx.doi.org/10.1016/B978-0-12-803058-5.00008-4

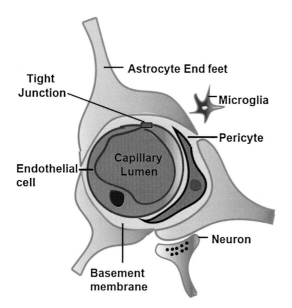

FIGURE 8.1   Brain capillary endothelium and the network of cells that form the specialized blood–brain barrier at the interface of the brain and the peripheral circulation.

thrombosis and hemostasis [2]. Antithrombotic molecule thrombomodulin and tissue plasminogen activator (tPA) are constitutively downregulated, whereas plasminogen activator inhibitor-1 (PAI-1) is upregulated, favoring an overall prothrombotic-antifibrinolytic environment at the brain capillary endothelium. Furthermore, vascular endothelial homeostasis is maintained by endothelium-derived vasoactive factors including nitric oxide (NO), prostaglandins, endothelium-derived hyperpolarizing factor (EDHF), endothelin-1, prostanoids, and angiotensin II.

Endothelial dysfunction is associated with ischemic stroke and other cerebrovascular diseases, and is characterized by altered vasodilation, inflammatory response, oxidative stress and vascular proliferation [3]. In general, endothelial dysfunction implicates loss of any of the several functions of the endothelium; however, the most commonly observed dysfunction is related to loss of NO activity. The mechanisms involved in the pathogenesis of endothelial dysfunction are multifactorial and interwoven. For example, oxidative stress and reactive oxygen species (ROS) can disrupt the balance of NO, causing inflammation and damage the brain endothelium. This chapter will focus on the role of the cerebral endothelium, mechanisms involved in endothelial dysfunction, and strategies that can be used to protect brain endothelial function.

ENDOTHELIUM AND VASCULAR TONE

One of the main functions of the endothelium is to regulate vascular relaxation and contraction to maintain vascular tone by releasing endothelium-derived factors. Among the endothelium-derived factors, NO is central to the maintenance of vascular homeostasis. NO is produced from L-arginine by endothelial NO synthase (eNOS) and this gaseous signaling molecule diffuses to the vascular smooth muscle cells to activate guanylate cyclase. The resultant increase in cyclic guanosine monophosphate (cGMP) causes vasodilation and subsequent relaxation. The brain endothelium is exposed to NO produced by three different NOS enzymes: eNOS and nNOS (neuronal nitric oxide synthase) that are constitutively expressed, and iNOS (inducible nitric oxide synthase) that is induced under active inflammation. Besides NO, endothelium-derived hyperpolarizing factor is an important vasodilator in the cerebral circulation especially in small cerebral arteries during conditions of diminished NO. To maintain vasomotion, the endothelial cells also release powerful vasoconstrictors known as endothelium-derived contracting factors including endothelin-1, prostanoids including thromboxane A_2 and prostaglandin H_2, and angiotensin II [3].

ENDOTHELIUM AND THROMBOSIS

The endothelium plays a crucial role in maintaining the balance between the pro- and anticoagulation systems of the vasculature. The endothelium exhibits heterogeneity in the expression of a number of antithrombotic factors, which assures adequate hemostasis in different tissues determined by local needs. Such tissue-specific thrombosis and hemostasis is particularly true for the cerebral endothelium. Thrombomodulin, a surface endothelial integral membrane protein, serves as a cofactor in the thrombin-induced activation of protein C anticoagulation pathway. Studies show low brain endothelial microvascular thrombomodulin expression. Furthermore, there is restricted expression of tPA, a serine-protease important for fibrin-mediated fibrinolysis, by brain microvascular endothelium. Moreover, studies show enhanced expression of PAI-1 and low expression of tissue factor pathway inhibitor by brain capillary endothelium. Overall, brain capillary endothelium exhibits low expression of most anticoagulant factors, thus shifting the balance toward a more prothrombotic state, which may be specifically important to prevent cerebral hemorrhages [2].

ENDOTHELIUM AND OXIDATIVE STRESS

The brain utilizes 20% of the total oxygen consumed by the body, thus making it very susceptible to oxidative stress. The main enzymes involved in the formation of superoxide anion (O_2^-) are nicotinamide-adenine dinucleotide phosphate oxidase (NADPH oxidase) and xanthine oxidase. This superoxide anion is then either

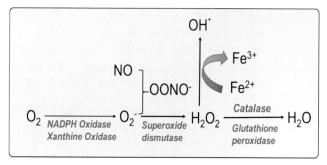

FIGURE 8.2   Enzymatic pathways that result in oxidative stress.

that facilitate leukocyte rolling. Leukocytes then firmly attach to the activated endothelium, a process that is mediated by immunoglobulins including intercellular adhesion molecule-1, vascular cell adhesion molecule-1, and locally produced inflammatory mediators such as tumor necrosis factor-α, interleukin (IL)-1β, and IL-6, C-reactive protein, as well as ROS. This facilitates leukocyte adherence to the endothelium, allowing peripheral inflammatory cells to infiltrate the brain tissue, where these inflammatory cells can further release proinflammatory mediators [5].

spontaneously converted to hydrogen peroxide (H_2O_2) in a process that is enzymatically accelerated by superoxide dismutase (SOD) or can be converted to the highly reactive peroxynitrite ($ONOO^-$) in the presence of NO. The reaction of O_2^- with NO is three times faster than dismutation of O_2^- by SOD. This is one of the main mechanisms of NO inactivation associated with cerebrovascular diseases. $ONOO^-$ can induce poly(ADP-ribose) polymerase activation and vascular dysfunction, and has been implicated in cerebrovascular dysfunction. H_2O_2 can stimulate vasodilator and vasoconstrictor response in cerebral blood vessels. Although not a free radical, H_2O_2 can easily cross the cell membrane and react with cellular metals such as Fe^{++} to give rise to the toxic hydroxyl radical ($\cdot OH$; half-life 10^{-10} s) via the Fenton reaction or via the iron-catalyzed Haber–Weiss reaction; this $\cdot OH$ may impair cerebrovascular function. Alternately, the H_2O_2 that is formed can be removed by reactions catalyzed by glutathione peroxidase (GTX) or catalase (CAT) (Fig. 8.2).

Processes such as cerebral ischemia-reperfusion (I/R) result in an influx of oxygen and thus a surge of ROS that overwhelms the cells' stores of ROS scavenging enzymes. Poor CAT activity and only moderate amounts of SOD and GTX in the brain render the endogenous protective mechanisms insufficient to contain ROS as they are formed. Release of ROS can trigger a number of downstream pathways that can impact cerebral endothelial function including enzymatic conversion of arachidonic acid to prostanoids via cyclooxygenases, formation of mitochondrial permeability transition pore, metabolism of hypoxanthine and xanthine, lipid peroxidation, activation of matrix metalloproteinases, increased leukocyte endothelial adhesion, and alteration of proinflammatory gene expression [4].

ENDOTHELIUM AND INFLAMMATION

Active inflammation has been implicated in cerebrovascular dysfunction and is mediated by a complex mechanism. Briefly, immediately upon an inflammatory insult, activated endothelial cells release stored von Willebrand Factor and adhesion molecule P-selectin

ENDOTHELIUM AND THE RHOA/RHO-KINASE PATHWAY

Evidence shows that rhoA, a small G-protein (guanine nucleotide-binding protein), and its downstream effector rho-kinase, both play an important role in regulating basal vascular tone and mediating responses to vasoconstrictor agonists including angiotensin-II, endothelin-1, and oxyhemoglobin. An increase in the rho-kinase pathway activity in vascular cells has been implicated in cerebrovascular dysfunction associated with stroke, chronic hypertension, and aging. Cerebral endothelial permeability induced by monocyte chemoattractant protein-1 is prevented by a pharmacological inhibitor of rhoA/rho-kinase, Y-27632, suggesting the role of this pathway in maintaining cerebral endothelial function. Furthermore, inhibition of the rhoA/rho-kinase pathway by fasudil and Y-27632 improves cerebral blood flow (CBF) transiently following ischemia reperfusion. This increase in CBF associated with rhoA/rho-kinase pathway inhibition is suggested to be mediated by increased NO bioavailability. Moreover, age-related increase in cerebral endothelial rhoA/rho-kinase activity may be one of the mechanisms leading to endothelial dysfunction in aging. The rhoA/rho-kinase pathway is also implicated in cerebral vasospasm after subarachnoid hemorrhage (SAH). Rho-kinase activity is increased in the basilary artery after SAH and this increase can be prevented by Y-27632. Overall, studies indicate that cerebral endothelial rhoA/rho-kinase activity is also augmented in cerebrovascular disease and is associated with endothelial dysfunction in the cerebral circulation [6].

ENDOTHELIUM AND ANGIOGENESIS

Angiogenesis is a process of sprouting of new blood vessels from the preexisting ones and is one of the main functions of the endothelium. Brain angiogenesis is a controlled process that is regulated by endothelial-derived growth factors that bind to tyrosine kinase receptors on the endothelium. Angiogenesis is enhanced before birth in the embryonic neuroectoderm, and occurs in the adult brain under pathological conditions such as ischemic stroke and glioblastomas. The main growth factor involved in brain angiogenesis is vascular endothelial growth factor (VEGF). VEGF binds to its receptors (VEGFR-1 and -2) with high affinity to induce new blood vessel formation and endothelial cell proliferation. Besides VEGF, angiopoietin is also involved in the development of the vascular system. Angiopoietin binds to Tie-1 and -2 (tyrosine kinase with immunoglobulin and epidermal growth factor homology domains), the only other receptors apart from VEGFR that are endothelial-specific receptor tyrosine kinases. The angiopoietin-Tie system is important for endothelial cell proliferation, migration, and survival during angiogenesis [7].

THERAPEUTIC STRATEGIES TO MAINTAIN CEREBRAL ENDOTHELIAL FUNCTION

One of the most promising classes of drugs with beneficial vascular effects include 3-hydroxy-3-methyl-glutaryl-coenzyme A (HMG-CoA) reductase inhibitors (statins). Apart from their cholesterol-lowering effects, statins increase vascular NO availability and thus improve endothelial function. Statins prevent isoprenoid intermediate formation thereby inactivating rhoA, causing an increase in eNOS and NO (Fig. 8.3). Suppression of isoprenylation also results in reduction of NADPH oxidase. Thus, endothelial protective effects of statins may also involve reduction in oxidative stress mediated by NADPH oxidase [3].

Targeting ROS formation by administering antioxidants is another approach to provide endothelial protection (Fig. 8.3). However, antioxidants have shown limited success in clinical trials of vascular diseases. Studies examining the therapeutic effects of an O_2^- scavenger (edaravone) and $ONOO^-$ scavenger (ebselen) have shown some clinical benefits in reducing oxidative stress and stroke outcomes; however, the protective effects of SOD mimetics have not been clinically validated [8]. In addition, rho-kinase inhibition increases eNOS activity, NO bioavailability, and reduces BBB permeability and exerts antiinflammatory effects. This represents a promising strategy to prevent cerebral endothelial dysfunction, and rho-kinase inhibitors such as fasudil may be clinically useful for diseases of the cerebral endothelium [6] (Fig. 8.3).

Another class of drugs that exerts endothelial protection is the phosphodiesterase (PDE) inhibitors (Fig. 8.3). PDEs are enzymes that degrade cellular cAMP and cGMP thereby regulating the levels of these second messengers. Studies have shown that an increase in intracellular cAMP levels induces brain endothelial cell barrier properties and offers endothelial protection. Furthermore, an increase in cGMP levels enhances the effects of NO. One of the most extensively studied PDE inhibitors is cilostazol, a selective PDE3 inhibitor. In experimental models, cilostazol reduces endothelial dysfunction, ROS formation, and inflammation. Dipyridamole is a nonselective PDE inhibitor (acting on both PDE3 and PDE5) and increases both cAMP and cGMP levels. Vascular protection by dipyridamole is mediated by free radical scavenging and antiinflammatory effects of this PDE inhibitor. Thus, PDE inhibitors represent another class of drugs that exert endothelial protection and hold promise for disorders of cerebral endothelial function [9].

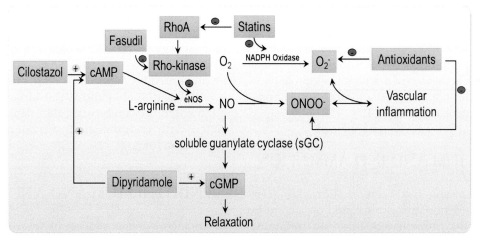

FIGURE 8.3   Important mediators of vascular endothelial function/dysfunction (*highlighted in green*) and important therapeutic strategies to maintain endothelial function (*highlighted in orange*).

Funding Statement

Supported by NIH NS20989 (MF) and Alzheimer's Association NIRG-15-361188 (RS).

References

[1] Abbott NJ, et al. Structure and function of the blood–brain barrier. Neurobiol Dis 2010;37(1):13–25.

[2] Fisher MJ. Brain regulation of thrombosis and hemostasis: from theory to practice. Stroke 2013;44(11):3275–85.

[3] Miller AA, Budzyn K, Sobey CG. Vascular dysfunction in cerebrovascular disease: mechanisms and therapeutic intervention. Clin Sci (Lond) 2010;119(1):1–17.

[4] Pun PB, Lu J, Moochhala S. Involvement of ROS in BBB dysfunction. Free Radic Res 2009;43(4):348–64.

[5] Madden JA. Role of the vascular endothelium and plaque in acute ischemic stroke. Neurology 2012;79(13 Suppl. 1):S58–62.

[6] Chrissobolis S, Sobey CG. Recent evidence for an involvement of rho-kinase in cerebral vascular disease. Stroke 2006;37(8):2174–80.

[7] Plate KH. Mechanisms of angiogenesis in the brain. J Neuropathol Exp Neurol 1999;58(4):313–20.

[8] Allen CL, Bayraktutan U. Oxidative stress and its role in the pathogenesis of ischaemic stroke. Int J Stroke 2009;4(6):461–70.

[9] Liu S, et al. Phosphodiesterase inhibitor modulation of brain microvascular endothelial cell barrier properties. J Neurol Sci 2012;320(1–2):45–51.

CHAPTER

9

Development and Maintenance of the Blood–Brain Barrier

J.M. Herndon, M.E. Tome, T.P. Davis

University of Arizona, Tucson, AZ, United States

INTRODUCTION

The blood–brain barrier (BBB) serves to limit the exposure of the brain parenchyma to foreign substances. Early experiments illustrate the "barrier" function present in the neurovasculature, as tracers such as peroxidase injected systemically are confined to vessel lumens or in a small number of pinocytic vesicles and do not reach the brain parenchyma [1]. Endothelial cells, pericytes, astrocytes, neurons, and microglia comprise the "neurovascular unit (NVU)," a term used to encapsulate those cell types critical to the maintenance of the BBB. Endothelial cells that surround the capillary lumen comprise a physical and biochemical barrier between the blood and brain. The physical barrier is maintained by tight junctions (TJs) between endothelial cells that restrict the diffusion of substances across the paracellular space between endothelial cells. The biochemical barrier comprises a host of transporters, including members of the ATP-binding cassette (ABC) family of transporters, and solute carrier (SLC) family of transporters, as well as intracellular enzymes that all serve to restrict the transcellular passage of substances and aid in their efflux from the brain.

Chemical compounds and microorganisms constantly gain access to the bloodstream and threaten the tightly regulated homeostatic environment the brain attempts to maintain. A wide variety of chemicals and pathogens encountered in the environment are capable of causing severe neurotoxicity and decreased quality of life if they reach the brain parenchyma. However, the CNS also has high energy demands, so nutrients (e.g., glucose and amino acids) in the blood must be able to gain access to the brain across the BBB. Therefore the development and maintenance of the BBB is critical to ensure proper functioning of the CNS, by simultaneously excluding toxicants and allowing the passage of molecules that nourish the CNS.

Primer on Cerebrovascular Diseases, Second Edition
http://dx.doi.org/10.1016/B978-0-12-803058-5.00009-6

Maintenance of the BBB requires complex regulation that is carried out by a number of intracellular signaling cascades as well as cross talk between cell types of the NVU. It is important to understand how the BBB functions to develop strategies that might be utilized in the treatment of a variety of CNS disorders. The BBB presents a significant clinical challenge, because not only does the BBB exclude harmful chemicals, but it also excludes candidate drugs that might otherwise be used to treat diseases and disorders of the CNS. This chapter briefly summarizes the current understanding of the development and maintenance of the BBB.

CELL TYPES OF THE NEUROVASCULAR UNIT

Endothelial Cells

Endothelial cells comprise the core of the BBB, as this is the cell type in direct contact with the blood. Endothelial cells are held together via adherens junctions and seal tightly through TJs (Fig. 9.1). The resulting tubular, single-cell-thick, "cobblestone" arrangement of endothelial cells comprises the arterioles, venules, and capillaries of the CNS. These cells house the proteins that maintain a physical and biochemical barrier between blood and brain. Paracellular transit of polar compounds between endothelial cells is prevented by TJs, which also serve to maintain endothelial cell polarity by limiting lateral diffusion of lipids and membrane proteins, and provide an intracellular scaffold for signaling complexes [2].

TJs between endothelial cells in the BBB primarily comprise occludin, claudins (particularly claudin-3 and -5), and (junctional adhesion molecules A, B, C) JAM-A, -B, and -C. Extracellular strands of TJ proteins from one endothelial cell adhere to the extracellular strands of TJ proteins on a neighboring endothelial cell. In electron microscopy images, the TJs bring two neighboring cells into such close association that the outer leaflets of the plasma membranes appear to fuse. Molecules such as transforming growth factor β (TGF-β) and glucocorticoids initiate signaling cascades that affect the expression and function of TJ proteins, and posttranslational modifications of the TJ proteins (e.g., phosphorylation, palmitoylation) further regulate TJ properties [3]. Furthermore, the trafficking of TJ proteins from intracellular compartments and insertion of such complexes into the endothelial membrane is another means by which barrier properties can be modulated [4]. The intracellular portion of TJ proteins form complexes with zonula occludens (ZO) proteins that, in turn, bind actin and thereby anchor the TJ complex to the cytoskeleton of the endothelial cell as well as provide a means

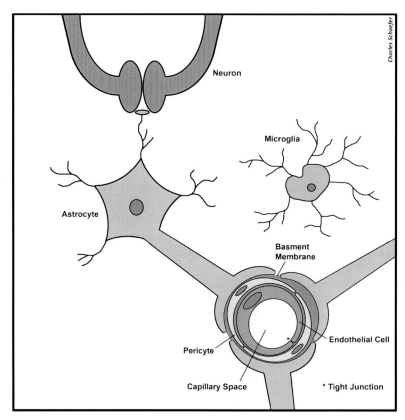

FIGURE 9.1　**Cell types of the neurovascular unit (NVU).** A schematic depicting components of the NVU at the level of a cerebral capillary.

to relay TJ status to intracellular compartments via ZO proteins or other signaling molecules [5].

Transporters located in the luminal and/or abluminal membranes of endothelial cells facilitate efflux of many potential neurotoxicants into the blood or influx of nutrients from the blood. ABC transporters are typically efflux transporters that couple the hydrolysis of ATP to the movement of substrates against their concentration gradients. P-glycoprotein, breast cancer resistance protein, and members of the multidrug resistance protein family are examples of ABC transporters. These transporters efflux a vast number of substrates. Some, like chlorpyrifos or ivermectin, would be toxic if they crossed the BBB; others are drugs that require access to the CNS (e.g., morphine, indinavir, taxol, prazosin) for efficacy. SLC transporters are also present at the BBB. These transporters have a wide variety of substrates that contain both toxicants (e.g., aflatoxinB1, ethidium bromide) and drugs (e.g., paclitacel, acyclovir, ketoprofen). The expression and trafficking of these transport proteins at the endothelial membrane are regulated by a multitude of signaling pathways and are capable of being altered in pain or disease states [6,7].

Finally, intracellular enzymes such as the cytochrome P450 family members, monoamine oxidases, glutathione-S-transferases, and catechol-O-methyltransferase act upon compounds that gain entry into the endothelial cell from the vessel lumen to inactivate or increase the clearance of these compounds [8]. Drugs such as caffeine (CYP1B1), dextromethorphan (CYP46A1), and risperidone (CYP2D6) are examples of compounds that are metabolized by such enzymes. The capacity of these enzymes to affect CNS bioavailability of certain compounds is appreciated, but not yet exhaustively studied.

Pericytes

Pericytes are fibroblast-like cells with extensive cytoplasmic processes that wrap around endothelial cells in arterioles, capillaries, and venules (Figs. 9.1 and 9.2). Pericytes cover between 22% and 99% of the endothelial cell surface. Pericytes are surrounded by a basement membrane, but contact the endothelial cells with "peg and socket" contacts through holes in the basement membrane. They also communicate with endothelial cells via paracrine signaling. Pericytes are required for normal development and function of the CNS vasculature; absence of pericytes is embryonic lethal [9].

During embryogenesis, secretion of TGF-β causes the differentiation of platelet-derived growth factor receptor (PDGFR)-positive pericyte progenitors from mesodermal precursors at the vessel plexus. Secretion of signals, including platelet-derived growth factor (PDGF),

by endothelial cells, attracts pericyte precursors and pericytes. Pericytes surround the immature vessels and contribute to vessel maturation [10]. Pericytes continue to differentiate based on signals in situ; however, they remain pluripotent and are able to differentiate into other cell types under conditions where alternate signaling pathways are triggered.

Pericytes contribute to the regulation of blood–brain barrier permeability. The percentage of endothelial cell coverage by pericytes is inversely proportional to the permeability of the BBB. Pericytes regulate TJ formation and influence transendothelial vesicular transport [10]. Signals from pericytes regulate the expression of genes in the endothelial cells that influence BBB permeability. Pericytes also have contractile fibers. Contraction of these fibers constricts blood flow through the capillaries and can lead to increased BBB permeability.

Advances in our knowledge of pericyte biology suggest that pericytes likely have additional functions in the NVU. Lack of specific pericyte markers, lethality of pericyte knockout mice, and the tight integration of pericyte function with that of the endothelial cells has limited our ability to fully understand the contribution of pericytes to the NVU. An understanding of the function of pericytes in disease-specific processes and their role in drug delivery across the BBB will be critical for the treatment of pathologies with a CNS component.

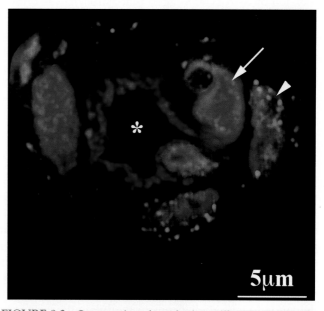

FIGURE 9.2 **Cross-section of a rat brain capillary.** An immunofluorescence image of a capillary cross-section obtained from a rat brain microvessel isolate. The sample was stained for P-glycoprotein (green) and glucose transporter 1 (red), with a nuclear counterstain (DAPI (4′,6-diamidino-2-phenylindole)/blue). The *white arrow* highlights an endothelial nucleus and the *arrowhead* highlights a pericyte nucleus. The asterisk marks the capillary lumen. This image was masked and processed for brightness, contrast, and RGB levels.

Astrocyte

Astrocytes form the bridge that connects neuronal signaling to the CNS vasculature. Anatomically, astrocytes have specialized processes called astrocyte end feet that extend from the astrocyte cell body and attach to the basement membrane that surrounds the endothelial cells and pericytes (Fig. 9.1). Estimates of the relative basement membrane surface area covered by astrocyte end feet range from 80% to 99% of the CNS microvasculature. Processes also extend from the astrocyte cell bodies toward the neurons so that bidirectional signaling between neurons and the vasculature can occur. This bidirectional signaling coordinates blood flow with neural activity.

Astrocytes contribute to induction and maintenance of the blood–brain barrier through paracrine interactions with the pericytes and endothelial cells. Astrocytes secrete classes of factors with either barrier-promoting or barrier-disrupting effects depending on signals received from neurons and/or endothelial cells. Paracrine factors from the astrocytes also contribute to regulation of blood flow [9]. One of the major regulatory functions of astrocytes in the NVU occurs during inflammation. Astrocytes, through regulation of matrix metalloproteinase, control breakdown of the basement membrane that allows movement of immune cells from the circulation to the brain [11].

The complex signaling pathways to which astrocytes respond combined with the plethora of paracrine factors secreted by astrocytes suggest that astrocytes are critical for the regulation of the BBB. However, it also indicates that defining the role of astrocytes in a specific physiological or pathological process is difficult. Targeted drug delivery across the BBB will require an understanding of disease-specific astrocyte function and accommodation of the barrier properties modulated by astrocyte signaling.

Neurons

Neurons are typically not in direct contact with endothelial cells, but projections from neurons are often located near vessels, perivascular astrocytes, and pericytes. Neurons release a wide variety of effectors that have the potential to regulate various aspects of other cells in the NVU and, therefore, the BBB. Neurons require nutrients and oxygen that is carried to the brain via the vasculature, and must communicate their energy demands to endothelial cells to modulate the exchange of molecules across the BBB. The perivascular location of neurons allows for paracrine signaling between neurons and other cells of the NVU.

Activity at neuronal synapses ultimately causes vasodilation to supply neurons with nutrients and oxygen to meet energy demands. This process is mediated by Ca^{2+}

increases in astrocytes and neurons. Ca^{2+} waves propagate between astrocytes via purinergic signaling and result in the production of vasoactive molecules such as prostaglandins and epoxyeicosatrienoic acids derived from arachidonic acid [10]. Certain interneurons may also release vasoactive substances such as nitric oxide and prostaglandin that signal for vasodilation of vessels in a paracrine fashion. This neurovascular coupling is crucial to proper functioning of neurons, and disruption of this coupling could facilitate the extravasation of blood components such as proteins and inflammatory cells that would promote neurodegeneration.

Microglia

Microglia are the resident macrophage of the CNS. These specialized cells derive from yolk sac primitive macrophages and are distinct from bone marrow–derived monocytes and their lineage. Microglia are ubiquitous across brain regions and each individual microglial cell surveys a defined and relatively small territory, often in close proximity to CNS capillaries (Fig. 9.3). In response

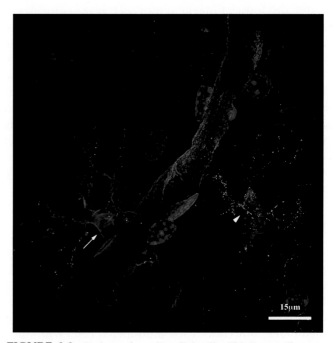

FIGURE 9.3 **Perivascular microglial cells.** This image illustrates the proximity of microglial cells to a cerebral capillary in the adult rat hindbrain. A 30-μm rat brain section was stained for the microglial marker Iba1 (red), and major histocompatibility complex (MHC) II (green), which is upregulated in activated microglia but also stains endothelial cells. Nuclei were stained with (DAPI) 4′,6-diamidino-2-phenylindole (blue). The *white arrow* highlights a surveillance microglia and the *arrowhead* highlights an activated microglia that has increased expression of MHC II. Note the capillary between the two microglia. This image is a maximum intensity projection of a 10-μm segment of the brain slice and was processed for brightness, contrast, and RGB levels.

to chemical insult, infection, or necrosis, microglia adopt a response phenotype in which these cells defend the brain by mechanisms that might include proliferation, phagocytizing foreign factors, or the release of proinflammatory cytokines and antimicrobial factors.

Microglia play vital roles in the development of the CNS. These cells participate in the pruning and remodeling of neurons during embryonic development. Microglia also participate in the shaping of CNS vasculature during embryonic development, as mice lacking the macrophage-critical transcription factor PU.1 exhibited reduced vascular intersections in the subventricular vascular plexus [12].

Microglia modulate a number of neuronal processes via the release of molecules that signal to neighboring cells in a paracrine fashion. Microglia elicit responses such as increasing excitatory postsynaptic current frequency via paracrine signaling. Finally, activated microglia can increase permeability of the BBB via the release of nitric oxide and interleukin-1β [9].

DEVELOPMENT

Vascularization of the embryonic CNS begins with formation of the perineural vascular plexus (PNVP) around the neural tube. The PNVP comprises endothelial cells that arise from the mesoderm germ layer of the embryo, whereas the neural tube, which will ultimately become the brain, arises from the ectoderm and is initially avascular [13]. In the mouse, with a total gestational period of about 20 days, vascular sprouts from the PNVP begin to invade the neural tube around embryonic day 9.5 [14]. Elaborate branching of these vascular sprouts takes place thereafter and the vessels anastomose around the ventricle of the neural tube. In these developing vessels, endothelial "tip cells" serve to guide the vascular sprout toward angiogenic stimuli and trailing "stalk cells" proliferate to extend the developing vessel [13]. Tightly regulated sprouting and pruning events continue throughout embryonic development and the resulting vascular network will eventually comprise the postnatal CNS vascular system.

Molecular signaling that governs the formation of both the CNS vascular network and the BBB is complex and comprises a large host of ligands, receptors, and downstream effectors. Crucial roles for a number of molecules facilitating angiogenesis in the developing CNS have been described. Vascular endothelial growth factor (VEGF) released by cells of the neuroectoderm act on VEGF receptors (VEGFR) of endothelial cells (most notably VEGFR2) to stimulate angiogenesis via activation of downstream VEGF effectors [9]. VEGF signaling in embryonic angiogenesis is crucial

as heterozygous $vegf^{+/-}$ mice are embryonic lethal [9]. Other regulators of angiogenesis include angiopoietin-1 and -2 released from pericytes, which act on their receptors, including endothelial Tie2. Tie2 stimulation facilitates endothelial stabilization and barrier formation [15].

Some endogenous factors, such as Wnt family members, facilitate angiogenesis specifically in the CNS. A number of Wnt family members, including Wnt1, 3, 3a, 4, 7a, and 7b were identified in areas of the developing brain and spinal cord [13]. In the canonical Wnt/β-catenin pathway, stimulation of Frizzled receptors leads to stabilization of β-catenin within endothelial cells, allowing for translocation of β-catenin to the nucleus where it facilitates transcription of genes involved in proangiogenic processes. Genetic deletion of β-catenin in all endothelial cells of the mouse proves embryonic lethal, but vascular abnormalities are only observed in the CNS and are absent in nonneural tissues [9]. The list of molecules implicated in CNS angiogenesis is growing and also includes neuropilins, integrins, GPR124, and TGF-β.

Historically, it was suspected that development of the BBB was an astrocyte-dependent process, but emerging evidence suggests that a functional barrier is present before astrocytes are generated, relatively quickly after the initial vascular sprouts from the PNVP invade the neural tube. In mice at embryonic day 13.5, a small tracer dye permeated the vessels of the brain, but by embryonic day 15.5, systemically injected dye was confined to the vascular lumen and completely absent from the surrounding neural tissue, illustrating the functional barrier present in the developing vascular network [16].

Critical to the functioning of the BBB are TJs between endothelial cells that restrict the paracellular passage of solutes. TJ proteins such as claudin-1, -3, and -5, occludin, and JAM-A have all been identified in relative abundance in endothelial cells of the developing CNS [3]. A number of proteins regulate the expression of these TJ proteins, including sonic hedgehog, Wnt3a/7a/7b, and Norrin [13]. Relatively few blood-borne solutes pass the developing BBB via the paracellular route (through TJs), however. The majority of solutes pass through the BBB via a transcellular route, whereby molecules are carried through endothelial cells in vesicles [16].

Vesicular transcytosis is active in CNS vasculature earlier in embryogenesis, and becomes less prevalent later in development [17]. One regulator of this vesicular transcytosis pathway is the major facilitator super family domain containing 2a (Mfsd2a). This molecule was selectively found in blood vessels of the CNS and genetic deletion of Mfsd2a in mice caused a significant increase in vesicular transcytosis in endothelial cells without disrupting TJ complexes [16]. Interestingly, pericytes regulate the expression of Mfsd2a, which illustrates the importance of these cells regarding BBB integrity. Pericytes are initially recruited to endothelial cells via stimulation of PDGFR-β located on pericytes by PDGF released from endothelial cells. Genetic ablation of *pdgf* or *pdgfrβ* is embryonic lethal, but mutant mice with reduced pericyte density exhibit increased transcellular permeability across the BBB without disrupting TJ formation, illustrating the importance of this cell type in barriergenesis [18].

CONCLUSION

The brain is an extremely dynamic organ and requires complex regulation. Development and maintenance of the BBB ensures that the brain parenchyma receives only necessary factors from the bloodstream. Cerebral vasculature originates via angiogenesis and barriergenesis occurs concurrently in this vasculature. A functional barrier (BBB) is present relatively early in embryonic development. The importance of the BBB is illustrated by the fact that deletion of any number of critical molecular players in angiogenesis/barriergenesis, such as VEGF, Wnt7b, GPR124, and sonic hedgehog, results in embryonic lethality [9].

Disorders such as Alzheimer disease, epilepsy, and stroke are characterized by BBB disruption and increased CNS vascular permeability. Increased permeability may be manifest as any number of processes, including paracellular leak via TJ disruption, altered expression of drug efflux transporters and metabolizing enzymes, or extravasation of proteins and leukocytes. The opening of the BBB may be mediated by a number of factors including matrix metalloproteases, reactive oxygen species, and proinflammatory cytokines [9]. Increased permeability of the BBB is often deleterious to the brain parenchyma and further exacerbates disease pathogenesis.

Facets of BBB development and maintenance discussed in this chapter comprise a brief summary of the complex regulation of this essential barrier. Understanding pathways involved in this regulation would allow for more precise strategies for drug delivery across the BBB, as well as improved treatment of CNS disorders.

References

[1] Reese TS, Karnovsky MJ. Fine structural localization of a blood–brain barrier to exogenous peroxidase. J Cell Biol 1967;34(1):207–17.

[2] Luissint AC, Artus C, Glacial F, Ganeshamoorthy K, Couraud PO. Tight junctions at the blood–brain barrier: physiological architecture and disease-associated dysregulation. Fluids Barriers CNS 2012;9(1):23.

[3] Haseloff RF, Dithmer S, Winkler L, Wolburg H, Blasig IE. Transmembrane proteins of the tight junctions at the blood–brain barrier: structural and functional aspects. Semin Cell Dev Biol 2015;38:16–25.

[4] McCaffrey G, Staatz WD, Quigley CA, Nametz N, Seelbach MJ, Campos CR, et al. Tight junctions contain oligomeric protein assembly critical for maintaining blood–brain barrier integrity in vivo. J Neurochem 2007;103(6):2540–55.

[5] Hawkins BT, Davis TP. The blood–brain barrier/neurovascular unit in health and disease. Pharmacol Rev 2005;57(2):173–85.

[6] Miller DS. Regulation of P-glycoprotein and other ABC drug transporters at the blood–brain barrier. Trends Pharmacol Sci 2010;31(6):246–54.

[7] Tome ME, Schaefer CP, Jacobs LM, Zhang Y, Herndon JM, Matty FO, et al. Identification of P-glycoprotein co-fractionating proteins and specific binding partners in rat brain microvessels. J Neurochem 2015;134(2):200–10.

[8] Agundez JA, Jimenez-Jimenez FJ, Alonso-Navarro H, Garcia-Martin E. Drug and xenobiotic biotransformation in the blood–brain barrier: a neglected issue. Front Cell Neurosci 2014;8:335.

[9] Obermeier B, Daneman R, Ransohoff RM. Development, maintenance and disruption of the blood–brain barrier. Nat Med 2013;19(12):1584–96.

[10] Dalkara T, Alarcon-Martinez L. Cerebral microvascular pericytes and neurogliovascular signaling in health and disease. Brain Res 2015;1623:3–17.

[11] Combes V, Guillemin GJ, Chan-Ling T, Hunt NH, Grau GE. The crossroads of neuroinflammation in infectious diseases: endothelial cells and astrocytes. Trends Parasitol 2012;28(8):311–9.

[12] Fantin A, Vieira JM, Gestri G, Denti L, Schwarz Q, Prykhozhij S, et al. Tissue macrophages act as cellular chaperones for vascular anastomosis downstream of VEGF-mediated endothelial tip cell induction. Blood 2010;116(5):829–40.

[13] Engelhardt B, Liebner S. Novel insights into the development and maintenance of the blood–brain barrier. Cell Tissue Res 2014;355(3):687–99.

[14] Tata M, Ruhrberg C, Fantin A. Vascularisation of the central nervous system. Mech Dev 2015;138(Pt 1):26–36.

[15] Winkler EA, Bell RD, Zlokovic BV. Central nervous system pericytes in health and disease. Nat Neurosci 2011;14(11):1398–405.

[16] Hagan N, Ben-Zvi A. The molecular, cellular, and morphological components of blood–brain barrier development during embryogenesis. Semin Cell Dev Biol 2015;38:7–15.

[17] Saunders NR, Liddelow SA, Dziegielewska KM. Barrier mechanisms in the developing brain. Front Pharmacol 2012;3:46.

[18] Armulik A, Genove G, Mae M, Nisancioglu MH, Wallgard E, Niaudet C, et al. Pericytes regulate the blood–brain barrier. Nature 2010;468(7323):557–61.

CHAPTER

10

Cerebral Autoregulation

S.-H. Yang, R. Liu

University of North Texas Health Science Center, Fort Worth, TX, United States

INTRODUCTION: FROM STATIC TO DYNAMIC

The human brain constitutes only 2% of the body weight, but receives 15% of cardiac output, accounts for almost 20% of the total oxygen consumption, and consumes approximately 25% of total body glucose utilization. The human brain is by far the most expensive organ in term of energy expenditure in the whole body. Maintenance and restoration of transmembrane resting potential dissipated by postsynaptic and action potential and neurotransmitters recycling represent the main energetic cost at the brain. In addition, the brain has very limited energy storage and relies on coincidental increase of cerebral blood flow to comply the local brain activity. Mammalian brain is characterized by high metabolic activity with fine regulatory mechanisms, defined as cerebral autoregulation, to ensure adequate energy substrates supply in register with neuronal activity. Cerebral autoregulation enables the cerebral blood flow to remain relatively constant during variations of arterial pressure. It is critical for normal brain function, given the high metabolic demand and low energy reserve of the brain tissue.

The term autoregulation for the cerebral circulation was first introduced by Lassen in 1959. Before then, it was generally believed that the cerebral perfusion passively followed the changes of arterial blood pressure and that the cerebral vasculature did not possess any significant capacity for intrinsic control of the vascular tone. Lassen summarized the previous quantitative studies and concluded that the cerebral blood flow is independent of the changes of mean arterial blood pressure within a wide range of mean arterial blood pressure. He further suggested that the autoregulation was governed by the metabolic demands of the cerebral tissues [6].

It is important to note that the earlier studies of cerebral autoregulation relied on pharmacological interventions to establish a sustained period of hypotension and hypertension and, thus, described the response of cerebral blood flow, averaged over a longer period of time, to long-term changes in blood pressure with gradual onset. By plotting these changes in cerebral and mean arterial blood pressure, it was found that cerebral blood flow remains constant over a wide range of mean arterial blood pressure, defined as the plateau phase. In normotensive adults, the lower and upper arterial blood pressure limits of the cerebral autoregulation have been determined as about 50–60 and 150–160 mmHg, respectively [8]. There is much evidence to support the contention that cerebral autoregulation does not maintain constant perfusion through a mean arterial blood pressure range of 60–150 mmHg as is so often cited in the literature [14]. Within the limitation, a slight slope is frequently observed with a small change of cerebral blood flow from 80% to 120% of the baseline [13]. Nonetheless, within the boundaries, cerebral autoregulation is effective. This adaptation not only enables instant cerebral blood flow regulation in register to the neuronal function, but also protects the brain against fatal consequences of hypoxia and energy deficit. In addition, it is noted that the upper and lower boundaries of the plateau phase are not fixed but can be modulated by a plethora of factors such as sympathetic nervous activity, the renin–angiotensin system, and many diseases.

The classic cerebral autoregulation was subsequently substantiated by many brain circulation researchers and became the key dogma of the brain circulation. However, it is important to note that the classic methods permit sampling of regional cerebral blood flow at intervals of minutes. Thus, the classic, or static, cerebral autoregulatory responses might be the consequence of autoregulation but not the process of cerebral

Primer on Cerebrovascular Diseases, Second Edition
http://dx.doi.org/10.1016/B978-0-12-803058-5.00010-2

autoregulation itself. Modern techniques such as transcranial laser Doppler ultrasonography enable to measure cerebral autoregulatory responses with a much high temporal resolution with the assumption that cerebral blood flow is proportional to blood velocity. Dynamic studies of cerebral autoregulation quantify the fast modifications in cerebral blood flow velocity in a major cerebral artery in relation to rapid alterations in blood pressure within the upper and lower limits of the static cerebral autoregulation. This approach allows differentiation of the cerebral autoregulation response to fluctuations in beat-to-beat blood pressure of different magnitudes and durations. The dynamic studies have identified events of reduced blood flow in the period needed for cerebral blood flow to return to baseline after hypotension [13]. Despite the distinction of long-term steady-state and transient changes in arterial pressure between static and dynamic cerebral autoregulation, the two may represent the same phenomenon of cerebral circulation.

MECHANISMS UNDERLYING CEREBRAL AUTOREGULATION

The mechanisms of cerebral autoregulation remain poorly understood, especially in human. It is clear that cerebral autoregulation is multifactorial phenomenon of the cerebral circulation, including myogenic, autonomic, and metabolic mechanisms (Fig. 10.1).

The Myogenic Mechanism

Myogenic response is the intrinsic property of vascular smooth muscle to respond to changes in intravascular pressure. The innate myogenic activity is crucial for autoregulation of blood flow for normal hemodynamic function and maintaining vascular resistance. The myogenic activity arises from arteries and arterioles denuded of endothelium and autonomic nerve control. Nonetheless, endothelium produces several vasoactive mediators, including nitric oxide, prostacyclin, and endothelium-derived hyperpolarizing factor that play critical roles in vascular tone and cerebral autoregulation.

The Autonomic Mechanism

The cerebrovascular bed is well innervated by both sympathetic and parasympathetic nerve fibers. However, the role of the autonomic nerve system in the regulation of cerebral blood flow remains controversial. Studies have shown that cerebral autoregulation is preserved in sympathetically and parasympathetically denervated animals, arguing against a major

autonomic contribution to cerebral autoregulation. Recently, modeling of the dynamic arterial blood pressure and cerebral blood flow velocity relationship has provided us better insight into the mechanisms of cerebral autoregulation. Dynamic cerebral autoregulation has been found to be altered markedly after blockage of ganglion, suggesting the involvement of autonomic control in beat-to-beat autoregulation. Similarly, sit-to-stand maneuver causes an immediate decrease in cerebral blood flow velocity. Thus, there is increasing evidence supporting the involvement of autonomic control in the cerebral autoregulation [11]. Nonetheless, the action of autonomic regulation may be dominated by myogenic and metabolic action under normal physiological condition.

The Metabolic Mechanism

In addition to the extreme high energy consumption, brain energy supply and expenditure are tightly coupled by neurovascular and neurometabolic mechanisms. As the brain has very limited energy storage, local brain activity has to be complied with a coincidental increase of cerebral blood flow, referred as neurovascular coupling. Despite the extensive knowledge of neurovascular coupling, the underlying mechanisms have not been delineated.

Brain vasculature has many unique structural and functional features that are distinct from the peripheral characterized by the intimate relationships between endothelium, pericyte, astrocyte, and neuron, termed together as the neurovascular unit [3,5]. Neurovascular unit is manifested in an intimate anatomical and metabolic relationship between cellular and vascular components at the central nerve system, which provide a robust coupling of neuronal activation to autoregulation. Since the mid-2000s, our knowledge in neuroenergetics has been rapidly evolving from the "neurocentric" view to an integrated picture of neuron–astrocyte coupling. Anatomically, the neurovascular coupling has to be orchestrated by the synergistic action of neurovascular unit. Therefore, cerebral autoregulation is likely mediated through the concerted action of numerous vasoactive agents derived from the neurovascular unit components [4,9].

Under normal physiological condition, adult human brain almost exclusively relies on aerobic respiration for energy metabolism, encompassing glycolysis, citric acid cycle, and oxidative phosphorylation, consuming oxygen and producing ATP and carbon dioxide. Interestingly, oxygen seems to play little role in cerebral autoregulation. On the other hand, brain perfusion is highly sensitive to changes in partial pressure of arterial carbon dioxide ($PaCO_2$). The high vascular sensitivity to carbon dioxide is unique to cerebrovasculature and manifests from carotid arteries to pial arterioles and parenchymal vessels. Carbon dioxide has a profound dilatory action in cerebral arteries and arterioles resulting in an increase of cerebral blood flow. Correspondingly, increase of $PaCO_2$ exerts suppressive effect on brain activity [7]. Therefore, carbon dioxide may play a reciprocal action on both autoregulation and brain activity to maintain the energy supply and expenditure homeostasis.

CEREBRAL AUTOREGULATION IN HYPERTENSION AND NEUROLOGICAL DISEASES

Hypertension is associated with cardiovascular hypertrophy and increased sympathetic activity, both of which might cause cerebral autoregulation impairment. Consistently, chronic hypertension has been found to reduce cerebral blood flow and increase cerebrovascular resistance [10]. In addition, hypertension has long been known to be associated with impaired neurovascular regulation of the cerebral circulation and associated neurodegeneration and cognition dysfunction. It is still unclear whether hypertension itself impairs cerebral autoregulation directly or the angiotensin signaling is the culprit for the neurovascular dysfunction.

Given the crucial role of cerebral autoregulation in the maintenance of normal brain function, we might expect that cerebral autoregulation will be significantly impaired in the elders and patients of neurological diseases. Indeed, impaired cerebral autoregulation has been found in the transgenic mouse models of Alzheimer's disease and aged mice [1,12]. As of 2016, research on cerebral autoregulation in human Alzheimer's disease and elders has only recently been undertaken. Surprisingly, there is no evidence support for severe impairment of cerebral autoregulation in either the elders or Alzheimer's disease patients [1,13]. The unexpected and controversial findings might be due to the methodological limitations of cerebral autoregulation. Further efforts on the physiopathology of cerebral autoregulation are needed before we come to understand its complex involvement in both aging and neurological diseases.

Despite our increasing knowledge of cerebral autoregulation, harnessing cerebral autoregulation as a basis for therapy remains mostly conceptual rather than practical. Nonetheless, there is increasing evidence that multimodality monitoring of cerebral autoregulation might aid prognostication and help identify optimal cerebral perfusion pressure level in patients of ischemic stroke and traumatic brain injury [2]. In the future, such monitoring may facilitate patients-specific control of cerebral autoregulation and, hence, improve outcome of different neurological diseases.

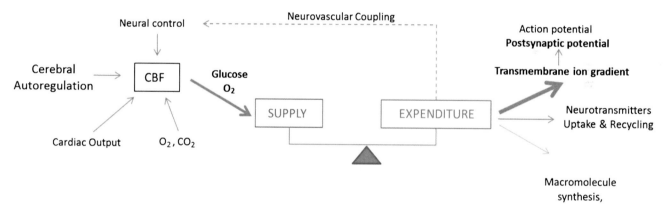

FIGURE 10.1 Schematic Diagram illustrating potential mechanisms underlying cerebral autoregulation. Cerebral autoregulation is multifactorial phenomenon of the cerebral circulation mediated by myogenic, autonomic, and metabolic mechanisms.

References

[1] Claassen JA, Zhang R. Cerebral autoregulation in Alzheimer's disease. J Cereb Blood Flow Metab 2011;31:1572–7.

[2] Czosnyka M, Miller C. Monitoring of cerebral autoregulation. Neurocrit Care 2014;21(Suppl 2):S95–102.

[3] Del Zoppo GJ. Toward the neurovascular unit. A journey in clinical translation: 2012 Thomas Willis Lecture. Stroke 2013;44:263–9.

[4] Girouard H, Iadecola C. Neurovascular coupling in the normal brain and in hypertension, stroke, and Alzheimer disease. J Appl Physiol 2006;100:328–35 (1985).

[5] Iadecola C. Neurovascular regulation in the normal brain and in Alzheimer's disease. Nat Rev Neurosci 2004;5:347–60.

[6] Lassen NA. Cerebral blood flow and oxygen consumption in man. Physiol Rev 1959;39:183–238.

[7] Lin AL, Fox PT, Hardies J, Duong TQ, Gao JH. Nonlinear coupling between cerebral blood flow, oxygen consumption, and ATP production in human visual cortex. Proc Natl Acad Sci USA 2010;107:8446–51.

[8] Paulson OB, Strandgaard S, Edvinsson L. Cerebral autoregulation. Cerebrovasc Brain Metab Rev 1990;2:161–92.

[9] Petzold GC, Murthy VN. Role of astrocytes in neurovascular coupling. Neuron 2011;71:782–97.

[10] Sanders RD, Degos V, Young WL. Cerebral perfusion under pressure: is the autoregulatory 'plateau' a level playing field for all? Anaesthesia 2011;66:968–72.

[11] ter Laan M, van Dijk JM, Elting JW, Staal MJ, Absalom AR. Sympathetic regulation of cerebral blood flow in humans: a review. Br J Anaesth 2013;111:361–7.

[12] Toth P, Tucsek Z, Sosnowska D, Gautam T, Mitschelen M, Tarantini S, et al. Age-related autoregulatory dysfunction and cerebromicrovascular injury in mice with angiotensin II-induced hypertension. J Cereb Blood Flow Metab 2013;33:1732–42.

[13] van Beek AH, Claassen JA, Rikkert MG, Jansen RW. Cerebral autoregulation: an overview of current concepts and methodology with special focus on the elderly. J Cereb Blood Flow Metab 2008;28:1071–85.

[14] Willie CK, Tzeng YC, Fisher JA, Ainslie PN. Integrative regulation of human brain blood flow. J Physiol 2014;592:841–59.

CHAPTER

11

Cerebral Blood Flow Regulation (Carbon Dioxide, Oxygen, and Nitric Oxide)

R.J. Traystman

University of Colorado Denver, Aurora, CO, United States

INTRODUCTION

The effects of carbon dioxide (CO_2), oxygen (O_2), and nitric oxide (NO) on the cerebrovasculature are the most pronounced, easily demonstrated, and reproduced phenomena observed in the cerebral circulation. Studies in man and animals, using many different techniques, have shown that CO_2, O_2, and NO exert a profound influence on cerebral blood flow (CBF). Cerebral vasodilation to hypercapnia, hypoxia, and NO, and vasoconstriction to hypocapnia, hyperoxia, and NO inhibitors are universal findings in mammals, regardless of age or sex. Thus, these gases are considered to be fundamental regulators of CBF. Here I briefly review the effects of CO_2, O_2, and NO on the cerebral vasculature and the potential mechanisms of action which account for these effects.

PHYSIOLOGICAL RESPONSES OF CO_2

An increase in arterial CO_2 tension ($PaCO_2$) produces perhaps the most marked and consistent cerebral vasodilation of any known agent. In man, 5% CO_2 inhalation increases CBF by about 50%, and 7% CO_2 by 100% [1]. It had been proposed that the CBF response to alterations in $PaCO_2$ was a threshold phenomenon; however,

it was subsequently shown that this response is a continuous one [2] (Fig. 11.1). Furthermore, the CBF/$PaCO_2$ relationship can be described by an S-shaped curve. There also appears to be a maximal increase in CBF with hypercapnia. When $PaCO_2$ is altered from 4 to over 400 mmHg, a maximal increase in CBF occurs at about 150 mmHg. On the other hand, reducing $PaCO_2$ from about 45 to 25 mmHg reduces CBF by about 35%. These alterations in CBF with hyper- or hypocapnia are reversible.

Despite all cerebral vessels respond to changes in CO_2, hypercapnia dilates smaller cerebral arterioles more than larger ones, but the hypocapnic vasoconstriction effect is size independent. While the effects of hypercapnia are reversed when $PaCO_2$ is reduced, animals exposed to prolonged increases in $PaCO_2$ increase CBF initially; however, after many hours or days of exposure, CBF returns toward baseline despite continued elevated $PaCO_2$. Men exposed to high altitude for 3–5 days have higher CBF than those at sea level, and prolonged hypercapnia reduces CBF responsiveness to acute alterations in $PaCO_2$. Prolonged hypocapnia also alters CBF responsiveness to acute changes in $PaCO_2$. During prolonged hypocapnia CBF initially decreases but later increases toward baseline despite the continued lowered $PaCO_2$.

Primer on Cerebrovascular Diseases, Second Edition
http://dx.doi.org/10.1016/B978-0-12-803058-5.00011-4

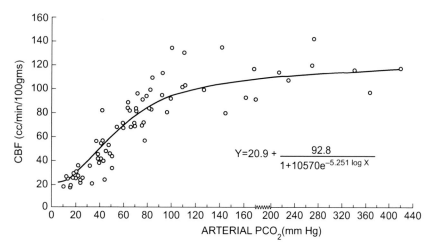

FIGURE 11.1 Curve and its equation describing relationship between cerebral blood flow (CBF) and arterial PCO_2, individual data points for each of eight monkeys. *From Reivich M. Arterial PCO2 and cerebral hemodynamics. Am J Physiol 1964;206:25–35.*

Anesthetized animals exposed to concentrations of CO_2 respond to a lesser degree than conscious animals. This may, at least in part, be explained by the depressive effect of anesthetics on brain metabolism and CO_2 production. This effect may be due to effects of metabolism on tissue levels of PCO_2 because a reduction in O_2 consumption would likely decrease the amount of CO_2 that is generated and diffuses from brain tissue to the vessel wall. The question of whether alterations in $PaCO_2$ change CBF equally in all brain regions is controversial. Some investigators show no differences in CO_2 reactivity among brain regions and that blood flow to the hemispheres, brainstem, cerebellum, and medulla is altered by the same percentage per mmHg change in $PaCO_2$. Others found that gray and white matter blood flow increased with hypercapnia; however, the white matter increase was less than the gray matter increase. Other brain regional areas such as the posterior pituitary and choroid plexus demonstrate minimal increases in flow with hypercapnia.

There appears to be a developmental difference in the CBF response to CO_2, although in all age groups, CBF increases with increasing $PaCO_2$. In both fetus and newborn, gray matter blood flow increases at $PaCO_2$ greater than 40 mmHg, but changes little at lower $PaCO_2$ levels. It has also been demonstrated that the change in CBF/$PaCO_2$ is higher in the newborn than in the fetus, and this suggests that the cerebrovascular response to CO_2 may not be completely developed at birth. This depressed CO_2 response in the fetus may be correlated with a difference in cerebral O_2 consumption (cerebral metabolic rate of O_2, $CMRO_2$). However, when CBF responses are normalized for $CMRO_2$, the increase in CBF is greatest in newborns, smaller in adults, and even smaller in fetuses. The reactivity of cerebral vessels in mid-gestational fetuses (sheep, 93 days) versus near-term fetuses (sheep, 133 days) is interesting. CBF and $CMRO_2$ increase threefold between

93 and 133 days of gestation. The CBF response to hypercapnia is greater at 133 days in mL/min/100 g of flow, but not as a percentage of baselines or as a ratio of CBF/$CMRO_2$. Thus, CO_2 reactivity appears normal relative to metabolism by 93 days gestation. Old age may also affect the responses to CO_2, and a decreased CBF responsiveness has been observed with increasing age in humans.

MECHANISMS OF ACTION OF CO_2

Several mechanisms have been proposed to account for the effects of CO_2 on the cerebrovasculature: extracellular fluid $[H^+]$, prostaglandins, NO, and neural pathways.

Extracellular Fluid $[H^+]$ pH Hypothesis

The main mechanism of the potent effect of CO_2 on CBF is a local action on cerebral arteries mediated by extracellular fluid $[H^+]$ [3,4]. Marked changes in $PaCO_2$ and bicarbonate ion concentration of cerebrospinal fluid (CSF) do not affect pial arteriolar caliber unless a change in pH occurs. This demonstrates that molecular CO_2 and bicarbonate ion have no vasoactivity and that it is the $[H^+]$ which is the important vasoactive agent. The cerebral vasodilation produced by hypercapnia can be completely counteracted by a change in extravascular $PaCO_2$ of the same magnitude but in the opposite direction. This indicates that local effects of CO_2 can explain the alterations in vascular caliber produced by $PaCO_2$ changes. The pH hypothesis regulation of the cerebrovasculature was originally described more than 50 years ago [5] and states that the actions of CO_2 are mediated by direct effects of $[H^+]$ on cerebrovascular smooth muscle. The $[H^+]$ in the area of vascular muscle depends on bicarbonate concentration and PCO_2 of the extracellular fluid at that site. In turn, extracellular fluid PCO_2 depends on both $PaCO_2$ and PCO_2 in CSF. Since the blood–brain barrier (BBB) is impermeable to bicarbonate and $[H^+]$, but freely permeable to CO_2, when $PaCO_2$ increases, molecular CO_2 diffuses across the barrier to increase local PCO_2 of vascular muscle, reduces extracellular fluid pH, and produces vasodilation. The reverse occurs when $PaCO_2$ is decreased. This local nature of CO_2 control by $[H^+]$ has been verified using ventriculocisternal perfusion techniques. Alteration of bicarbonate concentration in one lateral ventricle lowered caudate nucleus blood flow when bicarbonate increased and suppressed the increased flow when $PaCO_2$ was elevated compared to contralateral caudate blood flow [6].

Prostaglandins

Prostaglandins may be mediators of the CBF CO_2 response. Vasodilator prostanoids are important in vasodilation to hypercapnia in some species (gerbil, mice, rat, and baboon), but not in others (rabbits and cats). That prostanoids are important in hypercapnia comes from the observation that indomethacin, a cyclooxygenase inhibitor, decreases the CBF response to CO_2 inhalation in baboons [7]. Others have shown a complete abolition of the CBF response to hypercapnia with indomethacin and with no alteration in $CMRO_2$. In premature infants, indomethacin blunts the cerebrovascular response to hypercapnia, but while indomethacin affects CBF in man, administration of aspirin and indomethacin does not decrease control of CBF or attenuate the increase in CBF with hypercapnia. Other more specific cyclooxygenase inhibitors (AHR-5850-sodium amfenac) do not alter diameter of pial arterioles during normo- or hypercapnia. There may be interaction between the prostanoid system and NO production, with prostacyclin facilitating the release of NO. Thus, in species in which prostanoids act as mediators of hypercapnic vasodilation, inhibition of NO synthase may impair the cerebrovascular response to hypercapnia. The cerebral circulatory response to CO_2 may be gender specific, and it has been demonstrated that this response is altered more by indomethacin in women than in men. The response of cerebral vessels to CO_2 is universal among species; thus, it is curious that the prostaglandin mechanism of hypercapnic vasodilation is species dependent. For a response so prevalent, the mechanism of action is likely to be similar across species.

Nitric Oxide

NO is an important messenger involved in a wide variety of biological processes including regulation of the cerebral circulation. It plays a role in the maintenance of resting cerebrovascular tone and perhaps in evoked vasodilation. Since NO is a diffusible, short lived, highly reactive molecule, its effects have usually been inferred from studies of NO synthase (NOS) activity or inhibitors of this enzyme. Therefore, the importance of NO in the mechanism of hypercapnic cerebrovasodilation is somewhat unclear. However, a large number of studies have found that NOS inhibitors attenuate the increase in CBF with hypercapnia [8] by 35–95%. Because cerebral vessels remain responsive to other vasodilator stimuli (papaverine, nitroprusside, hypotension, and hypoxia) after NOS inhibition, the absent or reduced CBF response to hypercapnia is not due to nonspecific reduction of cerebral vascular responsivity. However, the studies are limited because cerebral vascular resistance was not calculated and to determine whether cerebral vessels truly vasodilated this must be known. This is important because blood pressure increases considerably following administration of NOS inhibitors, and CBF is decreased. On the other hand, other investigators have found little or no attenuation of cerebral vasodilation to hypercapnia following NOS inhibition [9]. Other data indicate that NO may play a small role in cerebral vasodilation to hypercapnia at moderate $PaCO_2$ levels (~50 mmHg) but not at higher levels (70 mmHg). Recent data in early gestation (93 days) and near-term gestation (133 days) sheep fetuses demonstrate that NOS inhibition does not alter cerebrovascular reactivity to CO_2.

The precise factors which account for these discrepant findings may involve species differences, methodological differences, dose of NOS inhibitor and consequent inhibition, timing of NOS inhibition relative to hypercapnia onset, anesthetic, degree of hypercapnia, and the failure to calculate cerebrovascular resistance which truly defines vasodilation or vasoconstriction. The fact is that in all species studied, hypercapnia leads to cerebral vasodilation and an increase in CBF, and NOS inhibition does not completely ablate CO_2 reactivity. Considering that there are region-specific responses to CO_2 within the brain, this likely means that there is more than one mechanism that accounts for the CO_2-mediated vasodilation. At best it would appear that the role of NO in the mechanism of the cerebrovascular response to CO_2 is as a modulator. There is no doubt that the major mechanism is increased perivascular $[H^+]$ during hypercapnia which reduces extracellular fluid pH and relaxes cerebral vascular smooth muscle. Other additional overlapping mechanisms involving NO or prostanoids are likely to involve reduced extracellular fluid pH. It is possible that increased extracellular fluid $[H^+]$ increases NOS activity or prostanoid production and/or release. It is also possible that there are multiple mechanisms accounting for the effects of CO_2 on the cerebrovasculature.

Neural Pathways

While this mechanism is understudied, the available literature is conflicting. Years ago it was suggested that hypercapnia stimulates cholinergic vasodilator reflex pathways [10]. Possibly, CO_2 exerts its effects on cerebral vessels via remote neural sites such as arterial chemoreceptors or brainstem vasomotor centers. In fact, the CBF response to CO_2 may be abolished or attenuated by a number of interventions: atropine administration, α-adrenergic blockade, arterial chemoreceptor denervation, vagal section, section of the seventh nerve, and certain brainstem lesions. There is also impressive evidence arguing against a neural role in the regulation of cerebral vessels by CO_2. Atropine, α-adrenergic blockade, section of the seventh, ninth, and tenth nerves, and arterial chemoreceptor denervation do not alter the CBF response to CO_2 [11,12]. This potential mechanism of action is controversial, and there is no convincing evidence of neural involvement in cerebrovascular responses to CO_2.

CONCLUSION

Hypercapnia profoundly increases CBF, and hypocapnia decreases CBF. Although there may be several factors that can influence hyper- and hypocapnic CBF responses, the major mechanism is related to the $[H^+]$ of extracellular fluid. This mechanism appears to occur across species but could work in conjunction with other mechanisms such as prostanoids, NO, and neurogenic components.

PHYSIOLOGICAL RESPONSES OF O_2

A tremendous amount of information concerning the effects of alteration in arterial oxygen tension (PaO_2) on the cerebral circulation has been reported. The relatively sparse capillarity and high $CMRO_2$ of brain indicate that the brain relies on a continuous supply of O_2. It is generally agreed that if PaO_2 is lowered sufficiently, CBF will increase (Fig. 11.2) [13]. The increase in CBF during hypoxemia has been observed in different animal species including man, with different anesthetics and different CBF techniques, regardless of accompanying alterations in PCO_2. Inhalation of low O_2 mixtures results in an increase in pial vessel diameter whether or not $PaCO_2$ is controlled.

Many investigators have dealt with whether there is a threshold PaO_2 for alterations in CBF. A general threshold number has been determined to be around 50 mmHg below which CBF increases markedly (Fig. 11.2) [13]. In these experiments, the animals (dogs) were ventilated and normocapnic. CBF began to increase as PaO_2 approached 50 mmHg and, at 30 mmHg, CBF increased to 220%. Others have reported an increase in CBF at PaO_2 as high as 85 mmHg. Thus, it is a most consistent finding that if PaO_2 is reduced, an increase in CBF occurs. The increased CBF during hypoxemia maintains a normal $CMRO_2$ up to a limit. $CMRO_2$ is maintained constant even when PaO_2 is reduced to 30–40 mmHg (an O_2 content of around 8–10 vol % or less). Brain tissue concentrations of ATP, ADP, and AMP have also been shown to remain unchanged at lower levels of PaO_2 [14]. The consistent findings of a maintenance of cerebral energy production with severe hypoxemia have led to the conclusion that the functional symptoms accompanying hypoxemia are not due to energy failure but depend on other metabolic perturbations, and that powerful homeostatic mechanisms come into play to prevent energy failure. This compensatory response is predominantly the increase in CBF.

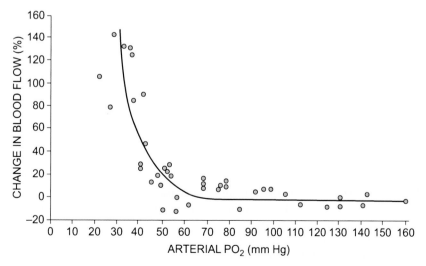

FIGURE 11.2 Effect of alterations in arterial PO_2 on cortical blood flow in dogs. *From MacDowell DG. Interrelationships between blood oxygen tensions and cerebral blood flow. In: Payne JP, Hill DW, editors. Oxygen measurements in blood and tissues. London: Churchill ltd; 1966. p. 205–19.*

MECHANISMS OF ACTION OF O_2

Direct Effects of O_2

While it is clear that hypoxemia produces cerebral vasodilation and increased CBF, the precise mechanism by which hypoxemia produces this vasodilation is not. Hypotheses to explain this mechanism include direct effects of O_2, chemical or metabolic mediators, and neurogenic and NO theories. Little evidence exists concerning the direct effects of O_2 on cerebral vessels. However, there is some evidence that O_2 may act directly on the smooth muscle of cerebral vessels with high PaO_2 resulting in vasoconstriction and low PaO_2 leading to vasodilation. The dependence of the contractile response to PO_2 is explained if one assumes that O_2 plays a metabolic role within the mitochondria of smooth muscle cells. It has also been suggested that receptors exist in smooth muscle which are sensitive to PO_2 and work as chemoreceptors.

Chemical or Metabolic Mechanism

The mechanism of cerebral vasodilation with hypoxemia may be mediated chemically by extracellular acidosis secondary to cerebral lactate production. Reducing PaO_2 to less than 50 mmHg increases CBF and the concentration of intracellular and extracellular cerebral lactate. Thus, cerebral metabolic acidosis could affect cerebral vascular smooth muscle by altering pH within the cell. Cerebral vasodilation correlates well with cerebral cortical acidosis and it is possible that hypoxemia exerts its effects on cerebral vessels secondary to the formation of parenchymal lactate from anaerobic glycolysis. However, others refute this finding and have demonstrated that during the initial, rapid, nonsteady state increases in CBF during hypoxemia, there is only a slight increase in lactate or none at all. Also, this increase in CBF leads to a reduction in tissue PCO_2 and a subsequent increase in pH. Thus, the increased CBF must be related to some aspect of cellular metabolism less sluggish than lactate formation.

The relationship between organ blood flow and the metabolism of that organ is a very old physiological issue (dating back to 1870s), and a close relationship between CBF and the concentration of metabolic by-products in the interstitial fluid was proposed in 1880s. One such metabolic by-product, adenosine, has been proposed to be the mechanism by which metabolic demands of the brain are transformed into the stimulus to increase CBF in hypoxemia [15]. Hypoxemia increases brain adenosine levels rapidly (within 2–3 s) and to extremely high levels. Coupled with the fact that adenosine is a strong dilator of pial arterioles when applied to the perivascular space and that adenosine causes cerebral vasodilation, this supports the potential role of adenosine as a chemical link between metabolism and CBF during hypoxemia. Other metabolic substrates such as oxygenases may also play a role in hypoxic vasodilation since oxygenase inhibitors attenuate cerebral vasodilation with hypoxemia [16]. The precise nature and location of these oxygenases is unclear, but it has been suggested that these "receptors" for hypoxemia are located close to the CSF. The idea of an O_2 sensor is not new and it has been proposed to exist in cerebral parenchymal tissue, or in CSF areas. These O_2 receptors could participate in a neural feedback loop originating within cerebral tissue to produce vasodilation with hypoxemia. In addition to the already mentioned adenosine and oxygenases, other vasoactive mediators of blood flow are bradykinin, histamine, prostaglandin, and serotonin.

Neurogenic Mechanism

During mid-1960s it was thought that neurogenic mechanisms were responsible for the vasodilator response to hypoxemia. It was suggested that the carotid chemoreceptors acting through neurogenic mechanisms were responsible for virtually all of the cerebral vasodilation with hypoxemia. It was shown that carotid chemoreceptor and baroreceptor denervation abolished the cerebral vascular response to hypoxemia; however, this was subsequently shown not to be the case [17]. In addition it was demonstrated that the cerebral vasodilation to hypoxemia was not different from that induced by elevating carboxyhemoglobin concentration, so that the arterial O_2 content was reduced equally with both types of hypoxemia. With carbon monoxide hypoxia, O_2 content is decreased but PaO_2 is unchanged, thus providing no stimulus to the chemoreceptors. The aortic chemoreceptors, glossopharyngeal and vagus nerves, are also not involved in the mechanism for the increased CBF with hypoxemia. Finally, it is possible that central brainstem mechanisms are involved with cerebral hypoxic vasodilation and the importance of the pons has been demonstrated.

Nitric Oxide

As opposed to the moderate role of NO in hypercapnic vasodilation, its role in hypoxic vasodilation is much less clear, and if there is a role for NO it is much less robust. Initial studies suggested that vasodilation and increased CBF with hypoxemia were not dependent on NO, however, more recent studies indicate that NO may play a larger role in severe hypoxia and in newborn animals. The determination of whether NO is or is not involved with the CBF response to hypoxemia has been made using NO inhibitors and comparing the hypoxemic CBF response before and after the inhibitor. Most of these studies indicate that the effect of NO is minimal, although there are studies that show a mild effect. The reasons for the discrepancy in these data, as with hypercapnia, may involve species differences, methodological differences, age of animals, dose of NOS inhibitor and consequent inhibition, timing of NOS inhibition relative to hypoxemic onset, anesthetic, degree of hypoxemia, and the failure to calculate cerebrovascular resistance.

CONCLUSION

Hypoxemia profoundly increases CBF at PaO_2 of 50 mmHg or less, and hyperoxia slightly decreases CBF. Although there may be several factors that can influence both hypoxemic and hyperoxic CBF responses, it is likely that the major mechanism is related to the alterations in brain metabolism, or chemical aspects such as lactacidosis and adenosine. This mechanism appears to occur across species but could work in conjunction with other mechanisms such as prostanoids, NO, and neurogenic components.

PHYSIOLOGICAL RESPONSES OF NITRIC OXIDE

Biology of NO

More than 35 years ago, it was recognized that there was a factor released by endothelium that relaxed vascular smooth muscle and resulted in vasodilation [18]. This factor, "endothelium-derived relaxing factor," EDRF, was subsequently identified as NO [19]. Since that time there have been volumes written concerning the physiology and pharmacology of NO in mediating many physiological functions, and its role in the pathophysiology of a variety of disorders particularly those dealing with the regulation of blood flow and inflammation [20]. NO is an inorganic, uncharged gas that easily crosses biological membranes, including the BBB. NO is synthesized together with L-citrulline by NOS from the precursor L-arginine in the presence of O_2 and cofactors, including NADPH, tetrahydrobiopterin (BH4), heme, FAD, FMN, and calmodulin. It is deactivated rapidly via oxidative pathways to nitrite or nitrate and has a short half-life of only a few seconds. It is also scavenged by superoxide-generating agents such as pyrogallol, hydroquinone, oxyhemoglobin, and others. There are also conditions in which superoxide and NO are generated to form peroxynitrite, another free radical that has biological actions. Thus, overproduction of NO can be neurotoxic via the production of peroxynitrite and superoxide which can bind directly to DNA, changing its structure and causing cell injury and enhancement of apoptosis.

Isoforms of NOS

There are at least three isoforms of NOS: type I NOS from neurons (nNOS constitutive), which is Ca^{2+} dependent; type II NOS from macrophages, astrocytes, and glia (iNOS inducible), which is inducible by inflammatory substances such as cytokines and endotoxins and generates high levels of NO; and type III NOS from endothelium (eNOS-constitutive), which is Ca^{2+} dependent, and under physiological conditions generates low levels of NO. The three forms of NOS have been cloned and appear to be products of different genes that have about 55% amino acid identity. NOS in brain and endothelium are quite similar, with the major difference being that NOS in brain is cytosolic and NOS in endothelium is mainly a membrane-associated protein. Two major strategies are used to study the normal and pathophysiological function of these NOS isoforms. One is pharmacological and the other is genetic. The pharmacological strategy is based on several NOS inhibitors which have been used to demonstrate the functional roles of endogenous NO and are as follows: NG-monomethyl-L-arginine (L-NMMA), NG-nitro-L-arginine (L-NA), L-NA methyl ester (L-NAME), and asymmetric dimethylarginine (ADMA) as nonselective inhibitors; 7-nitroindazol (7-NI), as a relatively selective inhibitor of nNOS; and aminoguanidine, N6-iminoethyl-L-lysine, and Wl1400 as selective inhibitors of iNOS. The genetic approach is based on the development of mutant mice lacking expression of the genes for nNOS, iNOS, and eNOS. There exists knockout (KO) mice for each of the NOS isoforms and much work has been performed with each of these NOS-KO mice to determine the physiological and pathophysiological characteristics in these animals and their involvement with a variety of disease processes.

NO and Cerebrovascular Physiology

Endothelial NOS

eNOS plays an important role in the regulation of CBF and in the regulation of the vasculature throughout the body. eNOS results in cerebral vasodilation, a reduction in vascular resistance, a reduced blood pressure, platelet aggregation and adhesion inhibition, leukocyte adhesion and migration inhibition, and a reduction of smooth muscle proliferation. It also has an important role in vascular remodeling and angiogenesis. Using pharmacological inhibitors of eNOS and eNOS KO mice, many investigators have demonstrated the role of eNOS in the control of cerebral circulation under a variety of conditions, even to control basal CBF levels. NOS inhibitors lead to vasoconstriction and decreased CBF and NO release leads to vasodilation and increased CBF. Vasodilation occurs when NO, released from endothelium, stimulates soluble guanylate cyclase in smooth muscle which results in vasorelaxation. Endothelial NO production is important for other aspects of vascular function such as intimal proliferation which is a response to arterial injury. eNOS KO mice develop more neointimal proliferation after vascular injury than do wild-type mice. What is curious and interesting is that in both KO and wild-type mice there is less intimal proliferation in female animals compared with male animals, indicating a possible role for estrogen in this response.

The role of NO in the cerebral vascular response to CO_2 and O_2 was previously discussed. Cerebral autoregulation, the maintenance of CBF despite changes in perfusion pressure, is another basic physiological characteristic of the cerebral circulation. Since autoregulation involves cerebral vasodilation and vasoconstriction, NO is thought to be involved with this mechanism, and most studies have shown that NO is involved with the mechanism of cerebral autoregulation. While much work has been performed in this area, as with the role of NO in the mechanism of CO_2 and O_2, the precise mechanism involving NO in autoregulation is unclear and somewhat controversial.

Inducible NOS

The inducible form of NOS (iNOS) has been described in many cell types: macrophages, astrocytes, microglia, leukocytes, and endothelial cells. iNOS is upregulated as a response to immunological stimuli such as endotoxin and cytokines. Many cytokines induce iNOS: interferon gamma, tumor necrosis factor, interleukin-1, 1B, 6, and so on. Lipopolysaccharide also induces iNOS. These agents can produce cerebral vasodilation which is dependent on the formation of NO and may be mediated by NOS. iNOS produces large amounts of NO which can damage or even destroy cells. Cerebral ischemia and other types of brain injury (cardiac arrest and traumatic brain injury) result in a marked inflammatory reaction. This inflammation activates a variety of cells; iNOS is upregulated and NO is produced in high, even toxic amounts, can then combine with superoxide to produce peroxynitrite, and results in injury. While this mechanistic view prevails, it is a simplified view, and it is likely that iNOS generated NO, like eNOS NO has several roles to play, and the balance between neurotoxicity and neuroprotection depends on many factors. iNOS-derived NO also promotes lipid peroxidation, DNA damage, and BBB breakdown, all of which have prominent roles in brain ischemic injury.

Neuronal NOS

nNOS is a constitutive enzyme expressed in brain, peripheral nervous system, and skeletal muscles. nNOS is expressed in neuronal cell bodies, and NO derived from nNOS is an important neurotransmitter associated with neuronal plasticity, memory formation, regulation of CBF, transmission of pain signals, and neurotransmitter release. The excitotoxic transmitter glutamate increases cellular Ca^{2+} with ischemia and Ca^{2+} is required for nNOS activity. Ca^{2+} mediates calmodulin binding to nNOS. Since nNOS is produced from excitotoxicity, nNOS plays a crucial role in ischemic injury. Thus nNOS NO mediates synaptic plasticity and neuronal signaling, but promotes neurotoxicity following ischemic damage, whereas NO produced by eNOS is protective. Neurogenesis also appears to be decreased by nNOS, while iNOS seems to stimulate it.

CONCLUSION

NO is an endogenous signaling agent involved in many physiological and pathophysiological processes. It is an important mediator in the regulation of CBF in the resting state, acting as a key mediator on the pathways responsible for maintaining resting CBF and perfusion in physiological situations, such as hypoxia, hypercapnia, and autoregulation. Disruption of NO synthesis and metabolism underlies many pathophysiological processes that occurs following brain injury. NO is also involved as a neuroprotective and neurotoxic agent following cerebral ischemia. This contradiction can be explained by the type of NOS, the cell type producing the NOS, the amount produced, and whether NO undergoes further oxidation. NO has wide-ranging, complex effects in brain and it is important to consider both protective and detrimental effects of NO while inhibiting the particular NOS for therapeutic effect against cerebral ischemia.

References

[1] Kety SS, Schmidt CF. The effects of altered arterial tensions of carbon dioxide and oxygen on cerebral blood flow and cerebral oxygen consumption of normal young men. J Clin Invest 1948;27:484–92.
[2] Reivich M. Arterial PCO2 and cerebral hemodynamics. Am J Physiol 1964;206:25–35.
[3] Kontos HA, Raper AJ, Patterson JL. Analysis of vasoactivity of local pH1. PCO2 and bicarbonate on pial vessels. Stroke 1977;8:358–60.
[4] Kontos HA, Wei EP, Raper AJ, Patterson JL. Local mechanisms of CO2 action on cat pial arterioles. Stroke 1977;8:226–9.
[5] Gotoh F, Tazaki Y, Meyer JS. Transport of gases through brain and their extravascular vasomotor action. Exp Neural 1961;4:48–58.
[6] Koehler RC, Traystman RJ. Bicarbonate ion modulation of cerebral blood flow during hypoxia and hypercapnia. Am J Physiol 1982;243:H33–40.

[7] Pickard JD, MacKenzie ET. Inhibition of prostaglandin synthesis and the response of baboon cerebral circulation to carbon dioxide. Nature 1973;245:187–8.

[8] Iadecola C. Does nitric oxide mediate the increases in cerebral blood flow elicited by hypercapnia? Proc Natl Acad Sci USA 1992;89:3913–6.

[9] McPherson RW, Kirsch JR, Ghaly RF, Traystman RJ. Effect of nitric oxide synthase inhibition on the cerebral vascular response to hypercapnia in primates. Stroke 1995;26:682–7.

[10] Shalit MN, Reinmuth OM, Shimojyo S, Scheinberg P. Carbon dioxide and cerebral circulatory control. III. The effects of brain stem lesions. Arch Neurol 1967;17:342–53.

[11] Bates DB, Chir B, Sundt Jr TM. The relevance of peripheral baroreceptors and chemoreceptors to regulation of cerebral blood flow in the cat. Circ Res 1976;38:488–93.

[12] Hoff JT, MacKenzie ET, Harper AM. Responses of the cerebral circulation to hypercapnia and hypoxia after seventh cranial nerve transection in baboons. Circ Res 1977;40:258–62.

[13] MacDowell DG. Interrelationships between blood oxygen tensions and cerebral blood flow. In: Payne JP, Hill DW, editors. Oxygen Measurements in Blood and Tissues. London: Churchill ltd; 1966. p. 205–19.

[14] Siesjo BK. Hypoxia. In: Siesjo BK, editor. Brain Energy Metabolism. New York: John Wiley and Sons; 1978. p. 398–452.

[15] Berne RM, Rubio R, Curnish RR. Release of adenosine from ischemic brain. Effect on cerebral vascular resistance and incorporation into cerebral adenine nucleotides. Circ Res 1974;35:262–71.

[16] Traystman RJ. Regulation of Cerebral Blood Flow. In: Vanhoutte P, Leusen I, editors. Vasodilation. New York: Raven; 1981. p. 39–48.

[17] Traystman RJ, Fitzgerald RS. Cerebrovascular response to hypoxia in baroreceptor and chemoreceptor-denervated dogs, 1981 baroreceptor and chemoreceptor-denervated dogs. Am J Physiol 1981;241:H724–31.

[18] Furchgott RF, Zawadzki JV. The obligatory role of endothelial cells in the relaxation of arterial smooth muscle by acetylcholine. Nature 1980;288:373–6.

[19] Ignarro LJ, Buga GM, Wood KS, Byrns RE, Chadhuri G. Endothelium-derived relaxing factor produced and released from artery and vein is nitric oxide. Proc Natl Acad Sci USA 1987;84:9265–9.

[20] Iadecola C. Bright and dark sides of nitric oxide in ischemic brain injury. Trends Neurosci 1997;20:132–9.

CHAPTER

12

CBF–Metabolism Coupling

P. Venkat[1,2], M. Chopp[1,2], J. Chen[1]

[1]Henry Ford Hospital, Detroit, MI, United States; [2]Oakland University, Rochester, MI, United States

INTRODUCTION

The brain requires oxygen and glucose to meet its metabolic demands, and cerebral blood flow (CBF) is its supply channel. The brain has high energy requirements but limited storage capacity, which means persistent CBF is critical for its proper functioning and prevention of damage and death. Therefore, in spite of constituting only about 2% of total body weight, the brain is easily the most perfused organ with almost 15–20% of the total cardiac output directed as CBF. Moreover, a process called cerebral autoregulation thrives to maintain adequate CBF at a constant rate. The arteries supplying the brain namely the internal carotid arteries and vertebral arteries that merge to form the basilar artery arrange themselves into the "circle of Willis" creating collaterals in the cerebral circulation. This is a defense mechanism against CBF drop such that if an artery supplying the circle is blocked, blood flow from the other blood vessels is able to sustain cerebral circulation.

The demand–supply relationship between CBF and cerebral metabolism is tightly coupled and brain regions are either hypoperfused or hyperperfused depending on metabolic needs. A sudden decrease in CBF (either temporary or permanent) due to the occlusion of a cerebral artery is called cerebral ischemia and leads to ischemic stroke, neurological deficits, tissue damage, and even death. An excess of blood flow results in hyperemia in which the intracranial pressure may increase and evoke

Primer on Cerebrovascular Diseases, Second Edition
http://dx.doi.org/10.1016/B978-0-12-803058-5.00012-6

tissue compression and damage. Therefore, to maintain brain homeostasis, local neuronal activity and subsequent changes in CBF are tightly coupled and termed neurovascular coupling. The neurovascular unit is a conceptual model encompassing the anatomical and metabolic interactions among the neurons, vascular components (endothelial cells, pericytes, vascular smooth muscle cells), and glial cells (astrocytes and microglia) in the brain.

CBF MEASUREMENTS

CBF–metabolism coupling is typically studied in two levels: at the whole-brain level and regionally depending on brain activation/stimulation. Advances in imaging techniques such as PET (positron emission tomography), MRI (magnetic resonance imaging), fMRI (functional MRI), NIRSI (near-infrared spectroscopic imaging), and optogenetics have fostered the understanding of CBF–metabolism coupling. There is clearly a close relationship between neural activity in the brain and local CBF which is captured by imaging techniques such as PET, fMRI, and NIRSI. Since hemoglobin, the oxygen-carrying component of blood releases more oxygen to activated neurons compared to inactive neurons, several imaging techniques rely on the differences in magnetic susceptibility (fMRI) and absorption coefficients (NIRSI, optical intrinsic signal imaging) between oxyhemoglobin and deoxyhemoglobin to image CBF variations in response to neural activity. Changes in CBF can be additionally imaged by laser Doppler technique, laser speckle contrast imaging, two-photon microscopy and optical coherence tomography among others. High spatial (differentiating local and global effects) and temporal (rapid sampling to reconstruct time course of dynamic processes) resolution two-dimensional and three-dimensional in vivo optical imaging is now available and is being used to study neurovascular coupling.

CBF AND METABOLISM COUPLING

To study the coupling between CBF and metabolism, a metabolic parameter called CMRO2 is widely analyzed. A linear relationship between fractional changes in CBF and cerebral CMRO2 coupling upon activation and deactivation has been found, under conditions of a high (~60%) baseline oxygen extraction fraction [1]. PET studies have reported that a disproportionately large change (~30–50% increase) in CBF is required to support a relatively small change (~5%) in the O_2 metabolic rate [2]. While this was initially interpreted as uncoupling between flow and oxidative metabolism meaning that increase in CBF during neural stimulation serves a need other than oxidative metabolism, it was later explained as a result of a tight

nonlinear coupling of flow and oxidative metabolism [3]. The rate of oxygen delivery to tissue will increase much more slowly than the increase in regional CBF, because oxygen extraction fraction decreases with increasing regional perfusion [3]. Hence, the large CBF increase supplies a small increase in the oxygen delivery rate to the tissue and increases glucose oxidation [3].

NEURONAL COUPLING TO CBF AND METABOLISM

Neural activity induces mediators that generate hemodynamic responses causing vasodilation and increased CBF. Chemical mediators of neurovascular coupling are of immense research interest and may function as post hoc mechanisms or occur in anticipation or in parallel with neural activity. Typically, products of neural and glial metabolism with vasodilator properties have been studied such as adenosine, nitric oxide (NO), ions like hydrogen (H^+), potassium (K^+), calcium (Ca^{2+}), and lactate. However, most of these biochemical mediators may not have sufficiently high spatial and temporal resolution to function independently, and it is likely that there is an interrelationship between several of these mediators and their pathways to effectively maintain neurovascular coupling.

CBF changes induced by functional activation is a useful surrogate for neural activity and is widely used by functional neuroimaging studies which assume that the CBF response is a linear-time invariant (LTI) transform of the underlying neural activity. Activity-induced CBF increase is dependent on synaptic excitation, and it has been suggested that active as well as passive postsynaptic mechanisms involving ionic fluxes may induce regional CBF increase while the nature of the neuronal circuitry stimulated determines the relative contribution from either mechanisms [4]. Neurons, glial cells, and vascular cells through different pathways can produce NO which is a powerful vasodilator. NO is also released by the endothelial cells that line the cerebral vessels. In the hippocampus, use of direct and simultaneous in vivo measurements of NO and CBF changes revealed that neurovascular coupling is mediated by NO by diffusion between active glutamatergic neurons and blood vessels [5]. During brain activation and metabolism, lactate produced may mediate functional hyperemia via increasing H^+ concentration and producing vasodilation [6].

ROLE OF ASTROCYTES

Astrocytes are a type of glial cells that are almost twice as abundant as neurons. They are at a strategic advantage to detect synaptic activation and couple it with glucose uptake in that (1) astrocytic end feet form a continuous layer surrounding cerebral blood vessels, (2) the processes of many vasoactive interneurons synapse onto astrocytes rather than directly onto blood vessels, (3) astrocytic processes ensheath synaptic contacts and neurotransmitter receptors, and (4) neurotransmitter receptors uptake sites are expressed on astrocytes. While the exact molecular mechanisms and messengers that relay neural activity to blood vessels are not understood, recent studies suggest a novel role for astrocytes in communicating neuronal activity levels to blood vessels and coordinating energy demand with oxygen and glucose supply. For instance, it has been suggested that upon neural activation, astrocytes can shunt vasoactive mediators like potassium ions from the synapse to astrocytic end feet and thereby increase the perivascular potassium ion concentration [7].

The two key hypotheses of astrocytes mediation include (1) the metabolic hypothesis, in which changes in neuronal activity drive changes in energy metabolism, and astrocytes can induce vasodilation or vasoconstriction by calcium-dependent release of vasoactive substances to alter CBF [8] and (2) the neurogenic hypothesis, in which changes in neuronal activity drive CBF and energy metabolism by inducing vasodilation or vasoconstriction by feed-forward mechanisms releasing neurotransmitter and neuropeptide molecules related to neuronal signaling [8]. Astrocytes can mediate the synthesis and release of vasoactive gliotransmitters in response to neurotransmitters and neuropeptides.

To study the role of astrocytes in neurovascular coupling, recently genetically modified mice such that cortical astrocytes express a light-sensitive cation channel, which can be transcranially activated with a blue laser has been developed [9]. When activated with a brief photostimulation, a transient, fast, and widespread CBF increase was observed which may be due to the diffusion of astrocyte-derived vasoactive signals involving potassium ion signaling [9]. Neuroglial metabolic coupling is thought to regulate brain metabolism. Glial cells absorb glutamate from the synaptic cleft and the byproduct of glial glycolysis which is lactate is consumed by neurons and metabolized. However, since the brain is usually supplied with much greater glucose than its requirement during rest and activation, lactate as a major secondary fuel is required in critical circumstances such as hypoglycemia or strenuous exercise that raise glucose demand and lactate production. Lactate produced by astrocytes is either locally oxidized, or selectively diffused to astrocytic end feet to increase CBF. In vivo and in vitro data suggest that cortical astrocytes maintain a steady-state reservoir of lactate that is immediately mobilized via a small rise in potassium channel during neuronal activation [10]. Lactate release was regulated by a positive feedback loop [10].

UNCOUPLING BETWEEN CBF AND METABOLISM

Neurovascular uncoupling may result from aging and from several neurological disorders including ischemic stroke, traumatic brain injury, epilepsy, dementia, hypertension, diabetes mellitus, and glioma. The consequences of such uncoupling may include compromise of the blood–brain barrier (BBB) which in turn triggers an inflammatory cascade, mitochondrial dysfunction and oxidative stress, neuronal death, and brain tissue atrophy. The BBB is formed by microvascular endothelial cells connected by tight junctions, a thick basement membrane and astrocytic end feet. The BBB is highly selective in its permeability and forms a biochemical barrier protecting the brain from invasion of neurotoxins. Endothelial cells can regulate vascular tone by releasing potent vasoactive factors. Vascular tone is determined by the contractile activity of smooth muscle cells that line the vessel walls, and largely affects the resistance to blood flow through the circulation. Cerebral endothelial cells mediate vasodilation and vasoconstriction by releasing NO, prostacyclin, endothelium-derived hyperpolarizing factors, and activation of potassium ion channels which can induce vessel relaxation. Under neurological disease states, BBB dysfunction and endothelial damage decrease vasodilation induced by the endothelium and via the release of endothelin which can induce vascular contraction.

During ischemic stroke, BBB disruption occurs acutely and triggers an inflammatory cascade of events which can worsen brain damage. Particularly in diabetes, stroke exacerbates BBB leakage, white matter damage, vascular damage, and inflammatory responses contributing to increased mortality and poor long-term functional recovery. Exercise raises metabolic demand and the brain receives higher CBF from an elevated cardiac output. However, in type 2 diabetes, due to decreased vasodilation capacity, patients suffer from a diminished ability to increase cardiac output when responding to the increased oxygenation demands during exercise often resulting in the perception of exertion.

Neurovascular uncoupling can also result from other conditions such as aging, hypertension, diabetes, traumatic brain injury, and dementias. Hypertension decreases CBF with greater severity of hypertension-inducing greater CBF reduction and oxygen metabolism impairment. In the elderly, such hypertension-induced CBF decrease can induce white matter lesions and cause vascular dementia. Aging impairs dynamic CBF regulation and leads to cognitive dysfunction. Aging hinders mediation of neurovascular coupling by (1) NO due to endothelial dysfunction and (2) glutamate due to astrocytic dysfunction, the end result being cognitive decline. Hence, neurovascular uncoupling plays a key role in the pathophysiology of several diseases and developing therapeutic strategies that maintain or restore neurovascular coupling is of great interest.

References

[1] Hoge RD, Atkinson J, Gill B, Crelier GR, Marrett S, Pike GB. Linear coupling between cerebral blood flow and oxygen consumption in activated human cortex. Proc Natl Acad Sci USA 1999;96(16):9403–8.

[2] Fox PT, Raichle ME. Focal physiological uncoupling of cerebral blood flow and oxidative metabolism during somatosensory stimulation in human subjects. Proc Natl Acad Sci USA 1986;83(4):1140–4.

[3] Buxton RB, Frank LR. A model for the coupling between cerebral blood flow and oxygen metabolism during neural stimulation. J Cereb Blood Flow Metab 1997;17(1):64–72.

[4] Mathiesen C, Caesar K, Akgoren N, Lauritzen M. Modification of activity-dependent increases of cerebral blood flow by excitatory synaptic activity and spikes in rat cerebellar cortex. J Physiol 1998;512(Pt 2):555–66.

[5] Lourenco CF, Santos RM, Barbosa RM, Cadenas E, Radi R, Laranjinha J. Neurovascular coupling in hippocampus is mediated via diffusion by neuronal-derived nitric oxide. Free Radic Biol Med 2014;73:421–9.

[6] Girouard H, Iadecola C. Neurovascular coupling in the normal brain and in hypertension, stroke, and Alzheimer disease. J Appl Physiol (1985) 2006;100(1):328–35.

[7] Paulson O, Newman E. Does the release of potassium from astrocyte endfeet regulate cerebral blood flow? Science 1987;237(4817):896–8.

[8] Lindauer U, Leithner C, Kaasch H, et al. Neurovascular coupling in rat brain operates independent of hemoglobin deoxygenation. J Cereb Blood Flow Metab 2010;30(4):757–68.

[9] Masamoto K, Unekawa M, Watanabe T, et al. Unveiling astrocytic control of cerebral blood flow with optogenetics. Sci Rep 2015;5:11455.

[10] Sotelo-Hitschfeld T, Niemeyer MI, Machler P, et al. Channel-mediated lactate release by K(+)-stimulated astrocytes. J Neurosci 2015;35(10):4168–78.

CHAPTER

13

Perivascular Neurotransmitter Regulation of Cerebral Blood Flow

L. Edvinsson

Lund University Hospital, Lund, Sweden

INTRODUCTION

The brain circulation is generally believed to be controlled by (1) chemical factors as perivascular pH; (2) autoregulation, a response to changes in systemic blood pressure; and (3) intrinsic mechanisms within the brain via neurovascular units and the microvasculature. However, for decades it has been known that the major cerebral arteries and arterioles are supplied by perivascular nerves with origin in the sympathetic, parasympathetic, and sensory ganglia [2]. With start during the 1970s, development of new histochemical methods allowed the demonstration of neurotransmitters in nerve fibers, followed by detailed analysis of their possible roles in regulation of cerebral blood flow (CBF).

The early demonstration of autonomic nerves (noradrenaline, NA, and acetylcholine, ACh), receptors, and effects on CBF provided novel input. A fabulous development in peptide chemistry followed, providing isolation of gut peptides like vasoactive intestinal polypeptide (VIP). Technology made it possible to produce quantities of peptides that could be used in functional tests, localized with immunohistochemistry and quantified with radioimmunoassay (RIA). This allowed for a paradigm shift with the first demonstration of perivascular VIP-containing nerve fibers in the wall of cerebral arteries [7]. The initial study led to a decade when a new perivascular neuropeptide was demonstrated almost every year [4] followed by attempts to understand their respective role in regulation of CBF [3].

Nature of the Perivascular Nerves Innervating Intracranial Blood Vessels

It is well recognized that the cerebral arteries have a dense network of perivascular nerves mainly localized around the major arteries of the circle of Willis and pial vessels on the surface of the brain. These intracranial blood vessels are supplied with nerve fibers that emanate from cell bodies in ganglia belonging to the sympathetic, parasympathetic, and sensory nervous systems (Fig. 13.1) [4,5]. In addition, intracerebral arterioles and microvessels are

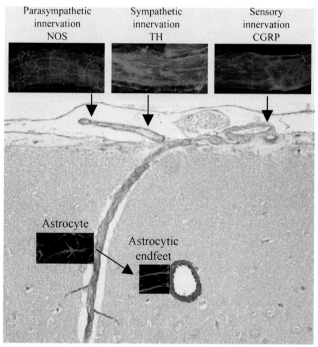

FIGURE 13.1 **Schematic overview of nerve supply to the cerebral circulation.** The nervous supply of the large cerebral arteries belonging to the circle of Willis and the vessels on the surface of the brain and proximal in some penetrating arterioles emanate in the cranial ganglia. The sympathetic fibers store NA (here shown by tyrosine hydroxylase immunohistochemistry, TH) and NPY, originate in the superior cervical ganglion. The parasympathetic fibers contain VIP, PACAP, NOS, and ACh, and originate in the sphenopalatine and otic ganglia. The trigeminal ganglion is the source of intracranial sensory nerves, which contain CGRP, SP/NKA, PACAP, and NOS. Arterioles or microvessels located in the brain parenchyma are supplied with "intrinsic" nerve pathways that have an origin in the CNS. There is evidence that cortical microvessels receive NA, 5-HT, ACh, GABAergic, and NOS afferents from either subcortical neurons from the locus coeruleus, raphe nucleus (shown as brain stem neuron in green), basal forebrain, or local cortical interneurons. The latter is illustrated by NOS-containing interneurons in cortex of rat and man. The "neurovascular unit" depicts yet another intricate mode of control where astrocytic end feet may communicate signaling in the brain to the microcirculation.

mainly surrounded by astrocytes but may also be innervated by nerve fibers that originate within the brain itself and thereby representing an intrinsic nerve supply [2,6].

Sympathetic Nervous System

The sympathetic nerves that supply the cerebral blood vessels arise mainly from the ipsilateral superior cervical ganglion [9], while some nerve fibers that supply the vertebral and basilar arteries originate from the inferior cervical ganglion and the stellate ganglion. The activation of these fibers results in vasoconstriction, modulation of cerebrovascular autoregulation, reduction of intracranial pressure, and a decrease of cerebral blood volume and cerebrospinal fluid production [2]. The neural responses are mainly mediated by NA and neuropeptide Y (NPY); at least 40–50% of the NA-positive cells in the superior cervical ganglion contain NPY.

The neurotransmitter content in the nerve cell bodies is influenced by various factors. For example, activation may increase catecholamine synthesis and NPY mRNA, while denervation results in depletion of NA and NPY. Furthermore, there are age-dependent changes in sympathetic neurons; for instance, in old rats there is a selective loss of NPY with a concomitant increase of nerve fibers containing VIP and calcitonin gene-related peptide (CGRP) around cerebral blood vessels. In man there is a significant reduction with age of NPY, VIP, substance P (SP), and CGRP in cerebral arteries.

Electron microscopic and functional studies have revealed that NA, NPY, and adenosine triphosphate (ATP) are costored in large dense-cored vesicles. Stimulation of the sympathetic nerves results in the release of these transmitters; the stimulus intensity determines the relative contribution of NA and NPY. At resting conditions, little NPY is released, and hence sympathetic vasoconstriction is largely due to activation of adrenoceptors and purinergic receptors whereas in situations of high sympathetic activity, the contribution of NPY becomes prominent.

It has been suggested that the small arterioles on the cortical surface may be supplied by NA-containing fibers emanating from an intracerebral source, for example, the locus coeruleus (LC) and/or the hypothalamus. However, this hypothesis has limited support. On the other hand, studies showing that destruction of the LC induces a reduction in the number of NA-containing nerve fibers in intracerebral vessels seems better founded [2]. Moreover, central stimulation of NA neurons in the hypothalamus is associated with an increase in hypothalamic blood flow which is unaffected by superior cervical ganglionectomy or by the β-adrenoceptor antagonist propranolol. It has been suggested that such a pathway may be involved in coupling neuronal activity to local blood flow regulation [8].

Parasympathetic Nervous System

Early studies revealed that the "classical" transmitter in parasympathetic nerves ACh and the cell bodies of the sphenopalatine and otic ganglia contain acetylcholinesterase (AChE). Cerebral blood vessels were found to contain perivascular nerves that display AChE activity and later choline acetyltransferase (ChAT). At the ultrastructural level, varicosities that contain numerous small agranular vesicles (40–60 nm in diameter) and that remain after sympathectomy were presumed to represent cholinergic parasympathetic nerve terminals. These varicosities frequently occur in close apposition to large dense-cored vesicles in the neuroeffector area, thus suggesting that parasympathetic nerves have the potential to interact with sympathetic nerve terminals near cerebrovascular smooth muscle [2]. In several species, ACh induces constriction of isolated cerebral arteries when deprived of the endothelium, while transmural nerve stimulation predominantly induces relaxation in the same preparations. The neurogenic vasodilatation in these preparations is not blocked by atropine, thus noncholinergic. It was suggested that additional substances are released together with ACh to mediate dilatation. Several neuronal messengers, which induce cerebral neurogenic vasodilatation, were suggested. Among these VIP, pituitary adenylate cyclase–activating peptide (PACAP), and nitric oxide (NO) received particular attention. Interestingly, all three seem to mediate the vasodilator responses in vivo as demonstrated by CBF measurements, although in some cases in human isolated blood vessels the potency of PACAP seems limited. In fact, NO might be the last link in cholinergic transmission. Another possibility would be that ACh mainly acts prejunctionally to inhibit neurotransmitter release from autonomic nerves. The vast majority of parasympathetic nerve fibers to cerebral blood vessels originate from the sphenopalatine and the otic ganglia.

Peptides of the VIP family, such as peptide histidine isoleucine (PHI), and its human form PHM (peptide histidine methionine) are seen in nerve fibers that supply cerebral blood vessels [2]. In most species, VIP-containing nerves are most abundant in the circle of Willis and in the major cerebral arteries (Fig. 13.1). The density of the nerve plexus is highest in the carotid system and diminishes in caudal direction. In humans, the VIP-immunoreactive nerve supply is sparse in both cerebral arteries and veins, while sphenopalatine ganglia show a rich supply of cell bodies containing VIP, PACAP, and NO and express mRNA for NPY Y_1 and VIP_1 receptors. A recent study examined the expression of the parasympathetic transmitters and their receptors in the human and rat sphenopalatine ganglion (SPG). VIP- and PACAP-containing neurons and fibers were seen in the SPG. NOS-positive neurons co-localized with the peptides. Common VIP and PACAP receptors (VPAC1 and VPAC2), and PACAP receptor (PAC1) proteins were identified by western blot.

PACAP is a vasoactive peptide that mediates dilatation and displays 68% homology to porcine VIP. PACAP is about 1000 times more potent than VIP in stimulating adenylate cyclase activity in cultured rat anterior pituitary cells. PACAP immunoreactivity and PACAP mRNA have been found in the sphenopalatine and otic ganglia. Perivascular nerve fibers that contain PACAP immunoreactivity can be seen in cerebral blood vessels. The majority of the PACAP-immunoreactive nerve fibers express VIP/NOS immunoreactivity as verified by tracing, denervation, and colocalization experiments. VIP/PACAP receptors have been found in human and rodent cerebral vessels and in the SPG.

NO is a highly labile molecule and its cellular localization has largely been attained by immunocytochemistry for nitric oxide synthase (NOS). There is evidence that NO is not only a candidate for the endothelium-derived relaxing factor in the endothelium, but also acts as a neurotransmitter. NO is a nonconventional transmitter, since it appears to be released by diffusion rather than exocytosis upon formation, is not stored in vesicles, and its action is not dependent on the presence of conventional membrane-associated receptors. There is a rich supply of NOS-immunoreactive nerve fibers around cranial blood vessels from several species, including humans. In the human circle of Willis, NOS-containing fibers are relatively sparse and mainly detected in posterior arteries [2]. In rodent and bovine cerebral blood vessels NOS-immunoreactive nerve fibers contain both VIP and AChE, and are assumed to represent parasympathetic nerves originating mainly from the SPG. It was early suggested that NO is a primary postjunctional messenger. ACh rather acts to limit the release of NO via a prejunctional cholinergic receptor.

SENSORY NERVOUS SYSTEM

Most cranial sensory fibers derive from the trigeminal ganglion (TG) and contain several signaling molecules [4]. Based on early sequencing of cDNA for SP, three different SP precursors (α-, β-, and α-preprotachykinin A) have been predicted. δ-Preprotachykinin produces SP only, whereas the proteolytic cleavage of β- or δ-preprotachykinin produces SP and a second tachykinin, neurokinin A (NKA). The cerebrovascular distribution of NKA resembles that of SP. Coexistence of SP and NKA in cell bodies of sensory ganglia and in perivascular nerve fibers have been demonstrated. NKA-immunoreactive nerve fibers have a similar distribution to that of SP-containing nerves and are also depleted after capsaicin treatment.

CGRP-immunoreactive nerve fibers supply the major cerebral arteries and pial arterioles on the cortical surface of several species including humans. While cerebral arteries of laboratory animals receive a dense supply of CGRP fibers, human cerebral blood vessels contain only a sparse network. In perivascular nerve fibers CGRP is often co-localized with SP/NKA. The first studies to show SP-containing nerve fibers in the dura mater, meningeal arteries, and in the cerebral blood vessels were followed by findings that SP receptor blockers were effective against neurogenic inflammation. However, several clinical studies with different SP receptor blockers showed that they were without effect on migraine. In addition, there are indications that SP does not take part in vascular nociception in humans.

In human TG, CGRP-immunoreactive neurons occur in high numbers (50% of all neuronal cells), whereas SP-immunoreactive neurons are less numerous (18%). This agrees well with observations in cat and rat in which the relation of CGRP to SP is about 3:1. In situ hybridization has revealed that 40% of all nerve cell bodies contain CGRP and CGRP mRNA. CGRP and SP are potent vasodilators in vivo and in vitro, the former being 10–1000 times more potent. Several studies have suggested that SP is involved in plasma extravasation from postcapillary venules in the dura mater during primary headache attacks. While neurokinin receptor antagonists are potent inhibitors of neurogenic inflammation, clinical studies have shown that these blockers do not have any significant effect in acute migraine attacks. Furthermore, while CGRP is released during the headache phase of a migraine attack, SP is not. This view is supported by intravital microscopy studies demonstrating that vasodilatation during perivascular stimulation of the middle meningeal artery (MMA) in vivo were blocked by a CGRP receptor antagonist, but remained unaffected by neurokinin agonists or antagonists.

Immunohistochemistry has revealed a minor expression of PACAP not only in parasympathetic but also in sensory ganglia which has led to the suggestion that PACAP may act as a neuromodulator in the sensory systems. There is a moderate supply of PACAP-immunoreactive nerve fibers in the cat cerebral circulation. In rats, the majority of the PACAP-containing fibers around cerebral blood vessels seem to derive from the SPG. In human TG, PACAP-containing cell bodies are more numerous than in laboratory animals, amounting to 15–20% of the neuronal population. Double immunostaining has revealed that PACAP colocalizes with CGRP in some cell bodies in the TG, putatively the small- and medium-sized neurons. PACAP dilates cerebral arteries and can increase CBF. Activation of the trigeminovascular system results in release of CGRP and a minor corelease of PACAP into the jugular vein of cats, a model used in studies of migraine. It is possible that this peptide may to some degree participate in antidromic vasodilatation following activation of the trigeminovascular reflex.

Intracerebral Innervation

It has been an established fact for decades that brain metabolism produces substances (e.g., H^+, CO_2, K^+, Ca^{2+}, and adenosine) that mediate local changes in CBF that accompany neuronal activity. However, these experiments were by and large performed at steady-state conditions and in whole brain or large parts of it. Consequently, this view did not seem to account fully for the adaptive responses in vasomotor activity in the microvascular bed. There has been a long quest whether intracerebral neurons can directly control the intracerebral microcirculation, and there are some data to support this hypothesis [8]. One proposal is that regional changes in brain perfusion are controlled directly by neurons and glial cells located within the brain parenchyma and in close apposition to the microvasculature (Fig. 13.1). For instance, stimulation of specific brain regions such as the cerebellar fastigial nucleus, the basal forebrain, or the brain stem raphe nucleus elicits changes in CBF in specific brain areas [2,6]. These changes in perfusion occur independently of those in glucose metabolism, thus implying that neuronal pathways can exert direct effects on the microcirculation. Furthermore, there is a population of neurons whose activity is related to spontaneous waves of CBF in the cerebral cortex, and is suspected to transduce neuronal signals into vasomotor responses. Together, these observations suggest that the neuronal control of the microvascular bed, which is achieved in concert with other mechanisms (vasoactive metabolic substances, ionic gradients and intrinsic endothelial or myogenic responses within the vessel wall), is a key determinant in the spatial and temporal adaptation of local perfusion to cellular activity. This implies that brain neurons send projection fibers to microvessels in target regions and that resistance microarterioles and possibly capillaries have the ability to modify their diameters and consequently local blood flow in response to changes in the level of brain neurotransmitters and/or neuromodulators [1].

Electron microscopy was employed to study the innervation of intracerebral arterioles and capillaries in which axon terminals abutted their abluminal walls [3]. Adrenergic vesicles remained after bilateral cervical sympathectomy, which leads to the hypothesis that intracerebral vessels receive an adrenergic innervation of central origin. The central innervation of intraparenchymal arterioles seems to be located primarily at branching sites—a strategic location for the control of local blood flow. There is now morphological evidence that nerve fibers of central origin associated with brain intraparenchymal blood vessels belong to the noradrenergic system. Studies have confirmed that the LC is the exclusive source of cortical perivascular noradrenergic nerve terminals, and ultrastructural analysis has emphasized the frequent association of these fibers not only with capillaries, but also with microarterioles.

Perivascular application of dopamine in cortical brain slices has been shown to cause vasoconstriction in about 50% of the microvessels studied [2]. The authors also documented the presence of dopaminergic fibers that were closely associated with intracortical microvessels, such as capillaries, microarterioles, and penetrating arteries. In contrast to the relatively minor effect on local perfusion exerted by central noradrenergic neurons, stimulation of the brain stem raphe nuclei (the source of serotonergic nerve fibers throughout the brain) or the ascending serotonergic pathways results in vascular responses in projection areas, such as the cerebral cortex, corresponding primarily to vasoconstriction. Histochemical examination has shown intimate associations between serotonergic neuronal processes and intraparenchymal vessels of the raphe nuclei. This innervation of local microvessels appears to embrace all vascular elements—arterioles, capillaries, and venules. At the ultrastructural level, perivascular nerves containing serotonin (5-hydroxytryptamine; 5-HT) labeled for the 5-HT-synthesizing enzyme tryptophan hydroxylase were associated with capillaries and microarterioles of all sizes, including penetrating arteries. Therefore, several findings support the role of neurally produced substances in the control of microvascular tone and, thereby, local CBF [8] including: (1) the presence of frequent and strategically located neurovascular appositions, their region-selective distribution and perivascular proximity in the regions known to modify their local perfusion in response to the stimulation of specific neuronal populations and (2) the exceptional positioning of cortical interneurons.

Another way of connecting the CNS with the cerebral vasculature is within the so-called neurovascular unit as illustrated in Fig. 13.1 [1]. It is suggested that neuronal activation in the CNS links to cerebral microvessels via astrocytes and glutaminergic signaling. In this system astrocytes are strategically localized to sense glutaminergic synaptic activity over a large area via activation of metabotropic glutamate receptors and subsequent calcium signaling. The astrocyte foot processes can signal to vascular smooth muscle cells and change vascular tone by prostaglandin pathways and by astrocytic and smooth muscle potassium channels. In this microenvironment in the brain non-glutaminergic transmitters released from neurons (e.g., NO, cyclooxygenase-2 metabolites, and VIP) might modulate the neurovascular signaling within the neurovascular unit (Fig. 13.1) [6]. It is of course pivotal to understand how this signaling is integrated in regulation of the microcirculation in the brain in different conditions.

References

[1] Attwell D, Buchan AM, Charpak S, Lauritzen M, Macvicar BA, Newman EA. Glial and neuronal control of brain blood flow. Nature 2010;468:232–43.
[2] Edvinsson L, Krause DN. Cerebral blood flow and metabolism. Philadelphia: Lippincott Williams Wilkins; 2002.
[3] Edvinsson L, MacKenzie ET, McCulloch J. Cerebral blood flow and metabolism. New York: Raven Press; 1993.
[4] Edvinsson L, Uddman R. Neurobiology in primary headaches. Brain Res Brain Res Rev 2005;48:438–56.
[5] Gulbenkian S, Uddman R, Edvinsson L. Neuronal messengers in the human cerebral circulation. Peptides 2001;22:995–1007.
[6] Hamel E. Perivascular nerves and the regulation of cerebrovascular tone. J Appl Physiol 2006;100:1059–64.
[7] Larsson LI, Edvinsson L, Fahrenkrug J, Håkanson R, Owman C, Schaffalitzky de Muckadell O, et al. Immunohistochemical localization of a vasodilatory peptide (VIP) in cerebrovascular nerves. Brain Res 1976;113:400–4.
[8] Lou HC, Edvinsson L, MacKenzie ET. The concept of coupling blood flow to brain function: revision required? Ann Neurol 1987;22:289–97.
[9] Nielsen KC, Owman C. Adrenergic innervation of pail arteries related to the circle of Willis in the cat. Brain Res 1967;6:773–6.

CHAPTER

14

Adenosine and Its Receptors Update: Influence on Cerebral Blood Flow (CBF)

H.R. Winn[1,2]

[1]Mount Sinai Medical School, New York, NY, United States; [2]University of Iowa, Iowa City, IA, United States

INTRODUCTION

This chapter updates the data supporting the hypothesis that adenosine (Ado) is a metabolic regulator of cerebral blood flow (CBF). A metabolic regulator is a substance whose concentration reflects and is linked to cellular metabolism. If a factor is to be considered as a metabolic regulator of blood flow, it must fulfill certain criteria (see Table 14.1) [1].

ADENOSINE BACKGROUND

Physiological Activities and Biochemistry of Ado in Brain: Ado is a purine nucleoside and in brain, it has two main physiological activities and two main biochemical sources. Both the physiological activities and biochemical sources may be interrelated (Fig. 14.1):

1. *Modulation of neuronal activity* [2]: In general, Ado has a depressant effect on neural activity. The source of Ado related to this activity is primarily S-adenosyl-homocysteine (subarachnoid hemorrhage, SAH, Fig. 14.1) which in turn is derived from a metabolic cycle that is active in the catabolism of neuronal transmitters.
2. *Regulation of CBF* [1]: Ado is a potent vasodilator in many organs including the brain. The primary source of Ado involved in flow regulation is via the breakdown (ATP→ADP→AMP→AdO) of adenine nucleotides. The initial step is extrusion of intracellular ATP and subsequent extracellular breakdown to AMP and then conversion to Ado by 5′nucleotidase (Fig. 14.1).

Cellular Source of Ado in Brain: Potential sites for the production of Ado in the brain are the cerebrovascular endothelial (CVE) cells, neurons, and/or astrocytes:

1. *CVE cells*: Unlike the coronary circulation, the CVE cells do not appear to be a source for Ado production in brain.
2. *Neurons*: Neurons are capable of producing Ado, but it is unclear whether neurons are the cellular source for Ado involved in flow regulation. Ado produced at the synaptic cleft would have to transverse a considerable distance through a hostile environment to arrive at the abluminal surface of the arteriolar vascular smooth muscle (VSM). Ado receptors (see the following) on the abluminal surface of the VSM are the target through which Ado evokes relaxation of the VSM and dilatation of arterioles.
3. *Astrocytes*: Astrocytes, being located between the neuron and the vasculature, are anatomically well positioned to integrate the collective and individual status of the neurons [3]. Previous studies revealed that astrocytes produce Ado (or ATP) [4]. In 2008,

TABLE 14.1   Criteria for a Metabolic Regulator

1. Must be a vasoactive agent.
2. The affected organ must be capable of producing the proposed regulator and in appropriate vasoeffective concentrations.
3. Metabolic activity that increases (or decreases) blood flow must correlate with an increase (or decrease) of the regulator.
4. Temporal production of the regulator must correlate with changes in metabolism and blood flow. Thus, there should exist rapid mechanisms for production and catabolism of the regulator.
5. Specific blockade (or enhancement) of the metabolic regulator must diminish (or increase) the expected vascular response.

Primer on Cerebrovascular Diseases, Second Edition
http://dx.doi.org/10.1016/B978-0-12-803058-5.00014-X

we demonstrated quantitatively that astrocytes in culture produced Ado in response to hypoxia in concentration which are in the vasoactive range (Fig. 14.2A) and in a temporal profile similar to the changes in CBF observed during hypoxia (Fig. 14.2B, inset) [5]. In addition, we found that Ado was derived by the extracellular catabolism of adenine nucleotides (ATP→ADP→AMP→Ado, see Fig. 14.1) and that hyperoxia (95% O_2) evoked a decrease in Ado. The latter information may be important in assessing previous data derived from hippocampal slices which have been routinely cultured and studied with 95% O_2.

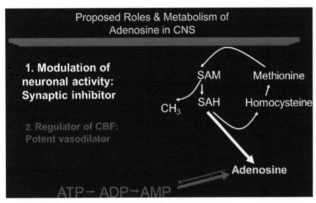

FIGURE 14.1 Overview of metabolic pathway of adenosine.

Ado *receptors*: In brain, Ado's actions are mediated through four receptors as defined by cloning: A1, A2a, A2b, and A3 (Table 14.2) [6].

1. *A1 receptor (A1R)*: Studies in isolated cerebral arterioles indicate that the A1R is not directly involved in vasoactivity [7]. Thus, the actions of Ado on the A1R occur within the parenchyma (glia and neurons) of the brain. The A1R act both presynaptically to inhibit neurotransmitter release [6] and postsynaptically to inhibit neuronal firing. A1R acts to reduce central nervous system activity, decrease cAMP and inhibit the release of excitatory amino acids. In addition, Ado can cause the release of nitric oxide (NO) from cultured astrocytes by means of AdoR (A1 > A2). A1R are located ubiquitously in brain with high concentrations in the hippocampus. Recent studies note the important role of the A1R, for example, in neuroprotection, aging, Parkinson's disease (with A2a), and tremor.

2. *A2 receptor (A2R)* family is subdivided into A2a (high affinity) and A2b (low affinity) [6]. The A2R is primarily responsible for Ado-induced vasodilatation. However, there is some dispute in the literature as to which of the A2 receptor (i.e., A2a or A2b) is primarily responsible for cerebral vasodilatation.

   A2a receptor (A2aR): Pharmacological and knockout (KO) data derived from in vitro (isolated

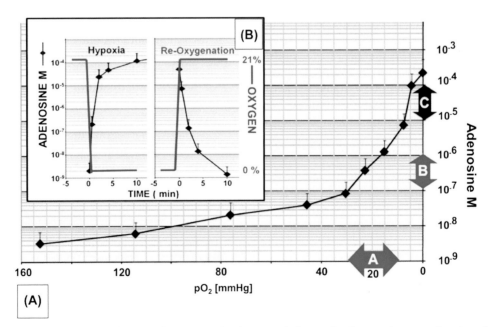

FIGURE 14.2 Adenosine production in glia. (A) Adenosine production to graded, stepwise decrease in oxygen in cultured astrocytes (see Ref. [5] for details; response in separate cultures, n = 18; each point = mean ± SD). As compared to 160 and 30 mm Hg, all values were significantly different (*p* = <.05). *Green arrow* A: pO_2 in brain tissue. *Green arrow* B: Interstitial and CSF Ado concentrations measured during normoxia. *Black arrow* C: Brain and CSF concentrations measured during hypoxia [13]. (B) *Inset*: Temporal changes in adenosine concentration in astrocyte cultures during changes in oxygen. Note the rapidity of the changes both with hypoxia and reoxygenation which is comparable to temporal profile in CBF during hypoxia and reoxygenation [9].

TABLE 14.2 Adenosine Receptors (P$_1$)

Receptor Subtype	Location	Activities	Agonists	Antagonists
A1	Cortex, hippocampus, cerebellum	↓ Adenylate cyclase activity	CHA, NECA, CPA, CCPA	CPX, CPT, XAC theophylline
		↑ K$^+$ flux		
		↓ Ca^{2+} flux		
		↑ IP$_3$		
		↑ Release of excitatory amino acids		
A2a	Striatum, blood vessels, caudate putamen, nucleus accumbens	↑ Adenylate cyclase activity	CGS 21,680, NECA	CSC, KF 17,837 theophylline
		↓ Blood vessel tone		
		↓ Leukocyte–endothelial cell interaction		
		↓ Generation of superoxide anion		
A2b	Hypophyseal pars tuberalis, cortical astrocytes		NECA	None that are selective
A3	Not well documented, hippocampus		APNEA	Varies by species

APNEA, N^6-2-(4-aminophenyl)ethyladenosine; CCPA, 2-chloro-N^6-cyclopentyladenosine; CGS 21,680, 2[p-(2-carbonyl-ethyl)-phenylethylamino]-5'-N-ethylcarboxamidoadenosine; CHA, N^6-cyclohexyladenosine; CPA, N^6-cyclopentyladenosine; CPT, 8-cyclopentyltheophylline; CPX, 8-cyclopentyl-l,3-dipropylxanthine; CSC, 8-(3-chlorostyryl) caffeine; KF 17,837, 1,3-dipropyl-8-(3,4-dimethoxystyryl)-7-methylxanthine; NECA, 5'-N-ethylcarboxamidoadenosine; XAC, xanthine amine congener.

penetrating cerebral arterioles) [7] and in vivo studies [8] indicates that A2aR activation is responsible for the majority (~50–70%) of Ado-induced vasodilatation.

A2b receptor (A2bR): In vitro studies in rat [7] and A2aR KO mice [9,10] only reveal a limited role for an A2bR contribution to Ado vasodilatation. Pharmacological studies of the physiological role of A2bR in vasoregulation are problematic because of the absence of highly specific A2bR antagonists and agonists.

3. *A3 receptor (A3R)*: This receptor does not appear to play a role in the actions of Ado on cerebrovascular tone since vasodilatation evoked by abluminal adenosine is not impaired in the presence of a selective A3 receptor antagonist [7]. This receptor is expressed in cerebral vessels as well as in hippocampus, stratum and on astrocytes. Activation of this receptor reduces ischemic injury in rats.

DATA SUPPORTING ADO'S ROLE IN CEREBRAL BLOOD FLOW REGULATION

A. Topically (i.e., extraluminally) applied, Ado is a dilator of brain resistance vessels (pial and penetrating) in vivo and in vitro and results in increases in CBF [7,8,11]. Ado-induced dilatation occurs through A2 receptors (see Fig. 14.3) located on the abluminal surface of pial and parenchymal arterioles [7,8]. Ado concentrations in CSF and brain extracellular space under control conditions are about 10^{-7}M. This concentration falls on the steepest past of the arteriolar response curve (Fig. 14.3). Therefore, small changes in ado concentration in brain will result in vasodilatation and have a profound effect on CBF since flow is an exponential of radius.

B. Brain tissue produces Ado during ischemia (Fig. 14.4A), hypoxia (Fig. 14.4B), hypotension (Fig. 14.4C), and increased metabolic demand [12–15], as well as during neuronal activation in both in vivo and slice preparations [2].

C. Increases in brain Ado concentration correlate temporally with the changes in cerebrovascular resistance during acute ischemia (Fig. 14.4A), hypoxia (Fig. 14.4A), and seizures (Fig. 14.4D) [12,13,15].

D. As noted previously (see Ado receptors), pharmacological evidence revealed a predominant involvement of the A2aR subtype in eliciting relaxation of VSM [7]. Similar findings were documented in A2aR KO mice [9]. However, residual vasodilatation in A2aR KO mice ([9,10]; and in our earlier pharmacological studies suggested a potential minor role for A2bR in Ado-induced vasodilatation.

E. Alteration in Ado availability and receptor activation exert an expected influence on CBF. For example, during hypoxic hyperemia, CBF is attenuated by Ado receptor blockers in rats (Fig. 14.5) [11] and in A2aR KO mice [9] (Fig. 14.6).

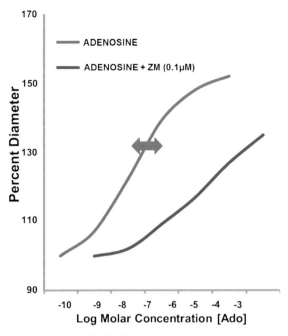

FIGURE 14.3 Dose–response of isolated, perfused small (40–80 μm) penetrating arterioles (*red curve*) to changes in abluminal concentrations of Ado [7,8]. These small arterioles are responsible for a significant component of cerebrovascular resistance and have a large influence in CBF regulation. Note that steepest portion of the dose–response curve (red) occurs in the concentration range of normoxic concentrations of Ado in CSF and brain interstitial (*red arrow* fluid). *Green curve* demonstrates the arteriolar response to Ado in the presence of ZM (1.0 μM) [7]. ZM is primarily an A2aR antagonist. These two curves demonstrate that the majority of the effect of Ado on vasodilation occurs by means of the A2aR.

F. During hypotension which evokes an increase in brain Ado levels [14] (Fig. 14.4C), A2aR KO mice had an attenuated CBF response. Blockade of A2bR in these mice further compromised the autoregulatory response suggesting that both A2aR and A2bR are involved in autoregulation during hypotension [10] (Fig. 14.7).

G. During cortical activation by contralateral sciatic nerve stimulation (SNS), pial arteriolar dilatation can be attenuated by theophylline and caffeine (A1R and A2aR antagonists) and ZM-241385 (an A2aR antagonist) [16,17] (Fig. 14.8). These interventions did not lead to alteration in somatosensory evoked potential (SSEP) or cortical optical imaging suggesting that the primary effect of these Ado A2R agonists was targeted on the vasculature and not the parenchyma. In addition, topical (micromolar) dipyridamole, which inhibits the uptake of Ado into cells and thereby potentiates the extracellular half-life of Ado, exaggerates cortical vasodilatation during SNS. In a similar fashion, pial arteriolar vasodilatation during neuronal activation was potentiated by EHNA (erythro-9-(2-hydroxy-3-nonyl) adenine), a blocker of Ado deaminase (ADA). EHNA would increase in extracellular Ado concentration [16]. In contrast, the A1R antagonist, CPX (8-cyclopentyl-1,3-dipropylxanthine), which acts primarily through the parenchyma (i.e., not via the vasculature), exaggerated neuronal excitation and CBF during SNS [18]. Simultaneous analysis of optical imaging and pial arteriolar diameter demonstrates that the attenuated CBF with theophylline is related to effects on the arterioles and not the parenchyma [18]. Lastly, A2aR KO mice compared to wild-type animals have an attenuated response to SNS (Fig. 14.9). These collected observations and findings support an important role for Ado in the regulation of CBF during neuronal activation.

CONCLUSION

Ado fulfills the criteria for a regulator of CBF. As outlined in the foregoing, during hypoxia, hypotension, and neural activation, a causal relation appears to exist between CBF and changes in Ado concentrations and/or Ado receptor activation. Taken together, these data support the hypothesis that Ado is a regulator of CBF. However, other factors and regulators undoubtedly also regulate CBF and the interactions of these regulators with Ado remain to be defined.

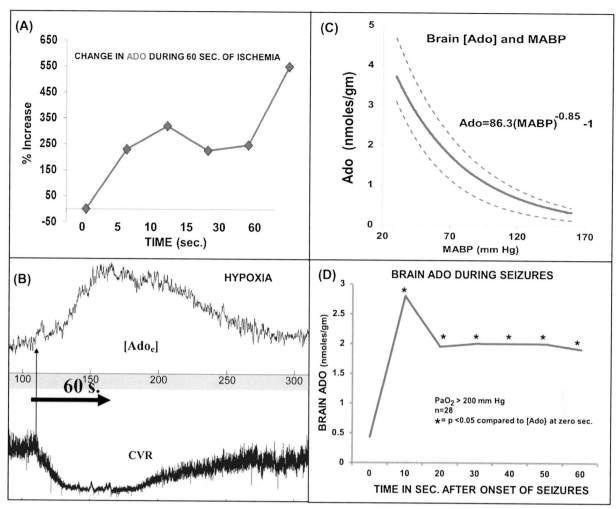

FIGURE 14.4 Changes in brain Ado levels in response to (A) ischemia, (B) hypoxia, (C) hypotension, and (D) increased metabolic demand (seizures). All studies were performed in rats. (A) Ischemia [12]: Rats were subject to acute ischemia by aorta disruption and then brain tissue was obtained by freeze blow technique and analyzed for Ado and other brain metabolites. Ado levels rose more than 2½-fold within 5s of the onset of ischemia. (B) Hypoxia: Changes in brain Ado and cerebral vascular resistance (CVR) during acute (60s=arrow) of hypoxia and reperfusion [13]. Brain Ado was measured by means of Ado-sensitive electrodes during acute (60s—arrow) hypoxia and reperfusion. Changes in brain Ado occurred rapidly during hypoxia and reperfusion. (C) Hypotension [14]: Brain tissue (sampled by freeze blow technique) was obtained during sustained (2min) graded hypotension induced by femoral artery withdrawal. Ado concentrations increased significantly ($p < .05$) as blood pressure decreased from 140 to 70mmHg (i.e., within the autoregulatory blood pressure range). (D) Increased metabolic demand [15]: Seizures were induced by bicuculline and brain tissue was obtained freeze blow technique. Ado levels increased rapidly (<10s) and remained elevated. During the same period, CBF increases many fold.

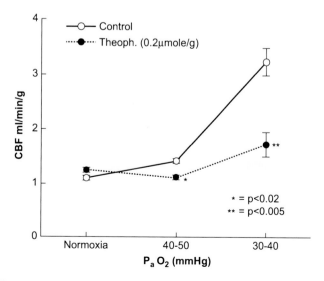

FIGURE 14.5 Effect of theophylline on sustained (2–5min) hypoxia: (○) untreated animals; (●) animals pretreated with theophylline (0.20/μmol/g, ip); P_aO_2, arterial 02 partial pressure. Cerebral blood flow (CBF) was measured by microsphere technique. All values are mean ± SE. *$p < .02$; **$p < .005$ versus untreated animals with comparable P_aO_2. *From Morii S, et al. Role of adenosine in regulation of cerebral blood flow: effects of theophylline during normoxia and hypoxia. Am J Physiol 1987;253(1 Pt 2):H165–75.*

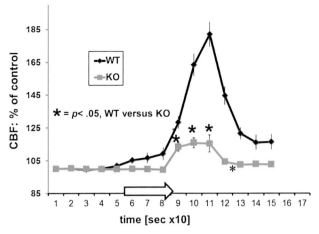

FIGURE 14.6 CBF response to 30 s (*arrow*) of hypoxia in wild-type (WT, *black line*) and A2aR knockout mice (KO, orange) [9]. CBF was measured by laser Doppler. Hypoxic hyperemia was attenuated ($p < .05$) by 85% in the KO mice as compared to the wild-type animals.

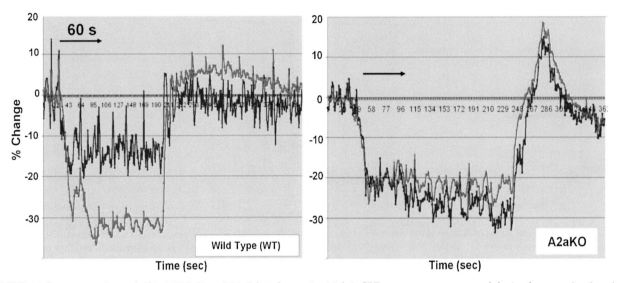

FIGURE 14.7 Autoregulation [10] in WT (left) and A2aR knockout mice (right): CBF response was measured during hypotension (maximum to 30% of resting MABP) induced by sustained (2 min) graded blood withdrawal from the femoral artery. CBF (*blue trace*) and arterial blood pressure (*red trace*, measured from contralateral femoral artery). In WT animals, autoregulation was intact. Note in WT animal (left panel), CBF was better maintained during hypotension and with reperfusion, whereas in the KO animal (right panel), CBF passively followed MABP during both hypotension and reperfusion.

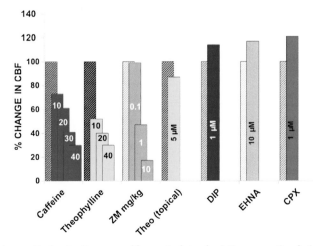

FIGURE 14.8 Pial dilatation during cortical activation caused by contralateral sciatic nerve stimulation (SNS): Influence by adenosine blockade and availability [16,17]. In anesthetized rats, the contralateral sciatic nerve was stimulated while the ipsilateral exposed cortical arterioles where measured through a closed cranial window. Caffeine, theophylline, and ZM (A2a receptor agonist, were administered intravenously or intraperitoneally. Other compounds theophylline; DIP: dypyridomole; EHNA: erythro-9-(2-hydroxy-3-nonyl) adenine) were perfused topically under the cranial window. Ado receptor agonist uniformly and significantly depressed CBF in a dose-related fashion during SNS, whereas compounds that increased extracellular availability (DIP and EHNA) of Ado exaggerated CBF response. CPX, an A1 agonist, increased the blood flow response to SNS, by exaggerating the degree of cortical excitation [18].

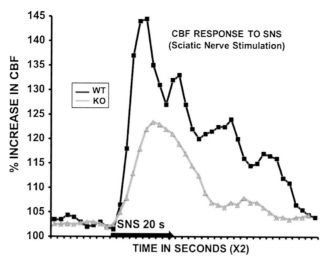

FIGURE 14.9 CBF in wild-type (WT) and A2aR KO (KO) mice during 20 s of sciatic nerve stimulation (SNS). Both the maximal peak (144% vs. 123%) and area under the curve were significantly greater ($p < .05$) in the WT compared to the KO animals.

References

[1] Winn HR, Rubio GR, Berne RM. The role of adenosine in the regulation of cerebral blood flow. J Cereb Blood Flow Metab 1981;1(3):239–44.

[2] Dunwiddie TV, Masino SA. The role and regulation of adenosine in the central nervous system. Annu Rev Neurosci 2001;24:31–55.

[3] Iadecola C, Nedergaard M. Glial regulation of the cerebral microvasculature. Nat Neurosci 2007;10(11):1369–76.

[4] Parkinson FE, et al. Differences between rat primary cortical neurons and astrocytes in purine release evoked by ischemic conditions. Neuropharmacology 2002;43(5):836–46.

[5] Kulik T, et al. Regulation of cerebral vasculature in normal and ischemic brain. Neuropharmacology 2008;55(3):281–8.

[6] Fredholm BB, et al. International Union of Pharmacology. XXV. Nomenclature and classification of adenosine receptors. Pharmacol Rev 2001;53(4):527–52.

[7] Ngai AC, et al. Receptor subtypes mediating adenosine-induced dilation of cerebral arterioles. Am J Physiol Heart Circ Physiol 2001;280(5):H2329–35.

[8] Ngai AC, Winn HR. Effects of adenosine and its analogues on isolated intracerebral arterioles. Extraluminal and intraluminal application. Circ Res 1993;73(3):448–57.

[9] Miekisiak G, et al. Cerebral blood flow response in adenosine 2a receptor knockout mice during transient hypoxic hypoxia. J Cereb Blood Flow Metab 2008;28(10):1656–64.

[10] Kusano Y, et al. Role of adenosine A2 receptors in regulation of cerebral blood flow during induced hypotension. J Cereb Blood Flow Metab 2010;30(4):808–15.

[11] Morii S, et al. Role of adenosine in regulation of cerebral blood flow: effects of theophylline during normoxia and hypoxia. Am J Physiol 1987;253(1 Pt 2):H165–75.

[12] Winn HR, Rubio R, Berne RM. Brain adenosine production in the rat during 60 seconds of ischemia. Circ Res 1979;45(4):486–92.

[13] Winn HR, Rubio R, Berne RM. Brain adenosine concentration during hypoxia in rats. Am J Physiol 1981;241(2):H235–42.

[14] Winn HR, et al. Brain adenosine production in rat during sustained alteration in systemic blood pressure. Am J Physiol 1980;239(5):H636–41.

[15] Winn HR, et al. Changes in brain adenosine during bicuculline-induced seizures in rats. Effects of hypoxia and altered systemic blood pressure. Circ Res 1980;47(4):568–77.

[16] Meno JR, Crum AV, Winn HR. Effect of adenosine receptor blockade on pial arteriolar dilation during sciatic nerve stimulation. Am J Physiol Heart Circ Physiol 2001;281(5):H2018–27.

[17] Meno JR, et al. Effect of caffeine on cerebral blood flow response to somatosensory stimulation. J Cereb Blood Flow Metab 2005;25(6):775–84.

[18] Haglund MM, Meno JR, Hochman DW, Ngai AC, Beb. Correlation of intrinsic optical signal, CBF & evoked potentials during activation of rat somatosensory cortex. J Neurosurg 2008;109(4):654–63.

CHAPTER

15

Cerebrovascular Activity of Peptides Generated by Central Nervous System

C. Gakuba, T. Gaberel

Harvard Medical School, Boston, MA, United States

INTRODUCTION

Peptides and Neuropeptides: Definitions

Peptides are biologically occurring short chains of amino acid monomers linked by amide bonds. Arbitrarily, peptides are distinguished from proteins on the basis of size: they contain approximately 50 or fewer amino acids.

Neuropeptides are peptides used by neurons to communicate with each other [1]. Nerve cells communicate with each other through two mechanisms: (1) fast synaptic transmission through fast-acting neurotransmitter (e.g., glutamate = excitatory; gamma-aminobutyric acid = inhibitory), which achieves effects within 1 ms, and (2) slow synaptic transmission through peptide neurotransmitters that influence the activity of the brain over hundreds of milliseconds to minutes [2].

Some *peptides of cerebral origin* are not *stricto sensu* defined as neuropeptide based on the cell types that release and respond to the molecule:

- Neuropeptides are secreted from neuronal cells and signal to neighboring cells (primarily neurons).
- Peptides of cerebral origin are secreted or act on cells other than neurons, mainly on the cells of the cerebral vasculature, like endothelial cell or smooth muscle cell.

Functions of Peptides and Neuropeptides

Neuropeptides modulate neuronal communication by acting on cell surface receptors. This latter process is mediated through a more complicated sequence of biochemical steps: neuropeptides act mainly on G-protein-coupled receptors, which activate an intracellular cascade of molecular enzymatic events, followed by cellular responses. Subsequently, this response occurs in a time span considerably longer than that of low-molecular-weight fast neurotransmitters [1].

Neuropeptides control the efficacy of fast synaptic transmission by regulating the efficiency of neurotransmitter release from presynaptic terminals and by regulating the efficiency with which fast-acting neurotransmitters produce their effects on postsynaptic receptors [2]. Peptides have several effects on cerebral blood flow, synaptogenesis, glial cells morphology, and gene expression, so subsequently on brain functions like analgesia, food intake, metabolism, reproduction, social behaviors, learning and memory, etc.

ENDOTHELINS

Physiology

Endothelins are 21-amino acid peptides that constrict blood vessels and raise blood pressure. Their name derived from the fact that they were first identified in cultured endothelial cells. Endothelins are the most potent vasoconstrictors known. In a healthy individual, a balance between vasoconstriction and vasodilation is maintained by endothelin and other vasoconstrictors on the one hand, and nitric oxide, prostacyclin, and other vasodilators on the other [3].

There are three isoforms of endothelin: endothelin-1 (ET-1), endothelin-2 (ET-2), and endothelin-3 (ET-3), but ET-1 is the only isoform produced by endothelial cells. There are three main endothelin receptors: ET(A)-receptor, ET(B)-receptor, and ET(C)-receptor [3]. The varying regions of expression of endothelin peptides and receptors implicate its involvement in a wide variety of physiological processes in the body.

Primer on Cerebrovascular Diseases, Second Edition
http://dx.doi.org/10.1016/B978-0-12-803058-5.00015-1

Role in Neurovascular Diseases

Endothelins are implicated in vascular diseases of several organ systems, including the heart, general circulation, and brain. In neurovascular pathology, endothelin dysregulation has mainly an impact in ischemic and hemorrhagic stroke. The pathological condition in which the pathophysiology of endothelins has been the most studied is a particular form of hemorrhagic stroke: aneurysmal subarachnoid hemorrhage (SAH) [4].

In case of intracranial aneurysm rupture, blood raises the subarachnoid space, in which are located the cerebral arteries. The blood degradation products then have several impacts, including inducing arterial vasospasm. Arterial vasospasm is due to the contraction of the smooth muscle cells of the artery's wall. It then decreases the cerebral blood flow, and concurs to the occurrence of delayed cerebral ischemia, which is an ischemic stroke arising usually 5–15 days after the aneurysm rupture. Several data support a key role of ET-1 in the occurrence of arterial vasospasm [3]. The activity of ET-1 is mediated by two receptor subtypes, ET(A)-receptor and ET(B)-receptor. Under physiological conditions the vasoconstrictive effect of ET-1 is mediated by ET(A)-receptor on smooth muscle cells. This vasoconstrictive effect is attenuated by an ET(B)-receptor: its activation induces release of nitric oxide (NO) from endothelial cells, which has a powerful vasodilative effect. In turn, vascular homeostasis is maintained (Fig. 15.1A).

In case of SAH, two phenomena seem to occur, even if it is still debated. First, an increase in expression of ET-1, which in turn activates ET(A)-receptor and ET(B)-receptor [3]. But in the case of SAH, the expression of ET(B)-receptor decreases. So the balance between vasoconstrictive and vasodilative effect is lost, leading to an increase of smooth muscle cell contraction and finally to arterial vasospasm (Fig. 15.1B).

Endothelins as a Therapeutic Target

As ET-1 seems to play a key role in arterial vasospasm after SAH, it has been designated as a promising therapeutic. However, targeting the expression of ET-1 failed to improve arterial vasospasm in clinical trials. This disappointing effect is probably related to the fact that ET-1 acts on its two receptors, which have opposed effects. As cerebral vasospasm seems to be linked to an overstimulation of ET(A)-receptor, tailored approaches remain to be ET(A)-receptor selective antagonists. ET(A)-receptor antagonist clazosentan was the most promising treatment of arterial vasospasm, and in fact it is: clinical trial results confirmed that clazosentan reduced angiographic vasospasm. Unfortunately, clazosentan failed to improve patient's neurological outcome, probably because the mechanisms underlying delayed cerebral ischemia are multiples, and do not only include arterial vasospasm [4].

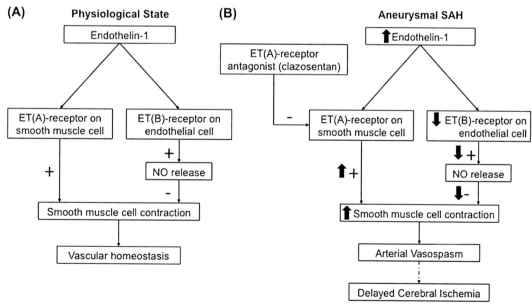

FIGURE 15.1 **Role of endothelin (ET)-1 in cerebral vascular tone, and its impairment after subarachnoid hemorrhage (SAH).** (A) In a healthy individual, ET-1 effects are balanced between vasoconstrictive effect, mediated by ET(A)-receptor, and vasodilative effect through production of NO, mediated by ET(B)-receptor. (B) In case of aneurysmal SAH, two different phenomena occur: an increased expression of ET-1, and a decreased activity of ET(B)-receptor, which in turn imbalance the effect of ET-1 on the vasoconstrictive side. It results in cerebral vasospasm, and contributes to the occurrence of delayed cerebral ischemia. This imbalance can be corrected by the ET(A)-receptor antagonist clozasentan.

THE KALLIKREIN–KININ SYSTEM

Physiology

The kallikrein–kinin system (KKS) is a proteolytic system involved with its receptors in several physiological and pathological phenomena. The KKS is constituted by a family of 15 highly conserved trypsin- or chymotrypsin-like serine proteases operating in many tissues [5]. The components of this complex system were progressively discovered during the last century after isolation from plasma or tissue in animals and humans.

Formation of Kinins

Bradykinin (BK) is the most studied peptidase of the kinins family. It is a nonapeptide constituted by the subsequent peptidic sequence: Arg1-Pro2-Pro3-Gly4-Phe5-Ser6-Pro7-Phe8-Arg9. Two distinct pathways in blood and tissue lead to its formation. The blood pathway involves factor XII after its activation by negative surface (Fig. 15.2). Factor XIIa is able to cleave the prekallikrein into its proteolytically active form, i.e., kallikrein, which in turn hydrolyses high-molecular-weight kininogen and then generates BK. In tissue, a proteolytic cascade also leads to the conversion of prekallikrein into kallikrein. Then tissue kallikrein generates BK from low-molecular-weight kininogen, the main tissue kininogen form. Due to the potent effects of kinins, their formation and degradation are tightly regulated. Circulating or tissue kininases convert BK into potent (such as Des-Arg9BK) or inactive peptides.

Kinin Receptors

The cellular effects of kinins are mediated by two G-protein-coupled receptors: bradykinin receptor 1 (BR1) and bradykinin receptor 2 (BR2). BR2, for which BK has high affinity, is constitutively expressed. In contrast, BR1, for which Des-Arg9BK has preferential affinity, is known as an inductive receptor whose expression is triggered by inflammatory processes (Fig. 15.3).

KKS in the Central Nervous System

The different components of the KKS have been described in the central nervous system [6]. High concentrations of tissue kallikrein have been shown in the hypothalamus, the pituitary, and pineal gland. Detection of BK and BR has also been realized in various regions of rodents' brain and spinal cord. The KKS participates in the genesis of secondary damage following primary insult in brain. Kinins are involved in several physiopathological cascades leading to vasodilatation, edema, or cytotoxicity.

KKS in Acute Neurological Disorders

Traumatic Brain Injury

Since traumatic brain injury is a heterogeneous disease including a large range of lesions, modulation of the KKS that acts on several pathways contributing to brain injury seemed to be an attractive strategy. Experimental data argue in favor of a deleterious role of bradykinin receptors following traumatic brain injury but with contradictory results regarding the specific role of BR1 and BR2 [7,8]. The peptide Bradycor (Deltibant,

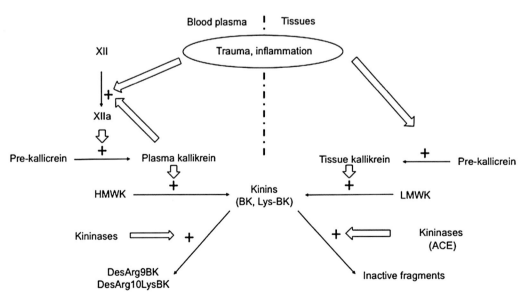

FIGURE 15.2 **Schematic representation of the two main pathways in plasma and tissues for kinins formation and degradation.** Two distinct systems in blood plasma and in tissues lead to the formation of kinins from the proform kininogen. The half-life of kinins is short and once formed they are quickly converted by kininase into active or inactive peptides. *XII*, Factor Hageman; *XIIa*, activated Factor Hageman; *ACE*, angiotensin converting enzyme; *BK*, bradykinin; *LMW*, low molecular weight kininogen; *HMWK*, high molecular weight kininogen.

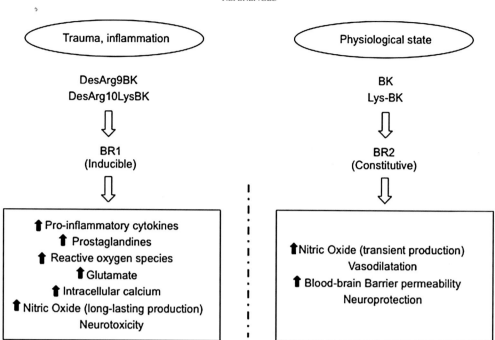

FIGURE 15.3   **Summary of the effects related to the activation of kinin receptors BR1 and BR2.** BR2 is constitutively expressed, whereas BR1 is induced selectively in pathological states. Activation of both BR1 and BR2 triggers inflammatory processes but some effects seem to be more specific for one of the two kinin receptors. *XII*, factor Hageman; *XIIa*, activated factor Hageman; *ACE*, angiotensin converting enzyme; *BK*, bradykinin; *BR*, bradykinin receptor; *HMWK*, high-molecular-weight kininogen; *LMWK*, low-molecular-weight kininogen.

CP-0127) was evaluated in a pilot single-blinded study but without significant effect on intracranial pressure, which was the primary objective [9]. More recently, the safety of a more potent nonpeptide BR2 antagonist, LF 16-0687 (Anatibant), was demonstrated in a double-blind randomized trial including 25 patients with traumatic brain injury [10]. But in regard to the design of this pilot study, no beneficial clinical effects could be shown. Moreover, results of a prematurely stopped phase II trial did not appear to favor Anatibant, although differences observed did not reach statistical significance [11].

Stroke

An interplay between thrombotic and inflammatory pathways, called thromboinflammation, has been described as a key event in the pathophysiology of lesions following acute ischemic stroke [12]. It is noteworthy that KKS is a part of the interface between these two pathways. Logically, strategies aiming at the inhibition of plasma kallikrein have been developed with promising results in preclinical ischemic stroke models [13]. In hemorrhagic stroke, a major prognostic factor is the expansion of hematoma. The modulation of KKS could also be a possible option in this setting since it was shown that hyperglycemia antihemostatic effect is mediated by plasma kallikrein in an intracerebral hemorrhage model [14].

Subarachnoid Hemorrhage

Experimental data evidence the involvement of BR2 in the formation of cerebral edema after SAH suggesting

beneficial effect of the blockade of this pathway [15]. But like in other neurological disorders, clinical relevance of these findings remains to be established. Furthermore, members of the KKS family, such as kallikrein-related peptidase 6, offer the opportunity to develop new biomarkers in brain injuries especially in SAH [16].

References

[1] Merighi A, Salio C, Ferrini F, Lossi L. Neuromodulatory function of neuropeptides in the normal CNS. J Chem Neuroanat 2011;42:276–87.

[2] Greengard P. The neurobiology of slow synaptic transmission. Science [Internet] 2001;294:1024–30. Available from: http://www.ncbi.nlm.nih.gov/pubmed/11691979. [cited 20 March, 2016].

[3] Zimmermann M, Seifert V. Endothelin and subarachnoid hemorrhage: an overview. Neurosurgery [Internet] 1998;43:863–75. discussion 875–6. Available from: http://www.ncbi.nlm.nih.gov/pubmed/9766314. [cited 26 April, 2016].

[4] Macdonald RL. Delayed neurological deterioration after subarachnoid haemorrhage. Nat Rev Neurol [Internet] 2014;10:44–58. Available from: http://www.ncbi.nlm.nih.gov/pubmed/24323051. [cited 26 April, 2016].

[5] Sotiropoulou G, Pampalakis G, Diamandis EP. Functional roles of human Kallikrein-related peptidases. J Biol Chem 2009;284:32989–94.

[6] Walker K, Perkins M, Dray A. Kinins and kinin receptors in the nervous system. Neurochem Int 1995;26:1–16.

[7] Trabold R, Erös C, Zweckberger K, Relton J, Beck H, Nussberger J, et al. The role of bradykinin B(1) and B(2) receptors for secondary brain damage after traumatic brain injury in mice. J Cereb Blood Flow Metab 2010;30:130–9.

[8] Raslan F, Schwarz T, Meuth SG, Austinat M, Bader M, Renné T, et al. Inhibition of bradykinin receptor B1 protects mice from focal

brain injury by reducing blood–brain barrier leakage and inflammation. J Cereb Blood Flow Metab [Internet] 2010;30:1477–86. Available from: http://www.pubmedcentral.nih.gov/articlerender.fcgi?artid=2949241&tool=pmcentrez&rendertype=abstract. [cited 26 April, 2016].

[9] Marmarou A, Nichols J, Burgess J, Newell D, Troha J, Burnham D, et al. Effects of the bradykinin antagonist Bradycor (deltibant, CP-1027) in severe traumatic brain injury: results of a multi-center, randomized, placebo-controlled trial. American Brain Injury Consortium Study Group. J Neurotrauma [Internet] 1999;16:431–44. Available from: http://www.ncbi.nlm.nih.gov/pubmed/10391361.

[10] Marmarou A, Guy M, Murphey L, Roy F, Layani L, Combal J-P, et al. A single dose, three-arm, placebo-controlled, phase I study of the bradykinin B2 receptor antagonist Anatibant (LF16-0687Ms) in patients with severe traumatic brain injury. J Neurotrauma [Internet] 2005;22:1444–55. Available from: http://www.ncbi.nlm.nih.gov/pubmed/16379582. [cited 26 April, 2016].

[11] Shakur H, Andrews P, Asser T, Balica L, Boeriu C, Quintero JDC, et al. The BRAIN TRIAL: a randomised, placebo controlled trial of a Bradykinin B2 receptor antagonist (Anatibant) in patients with traumatic brain injury. Trials [Internet] 2009;10:109. Available from: http://www.pubmedcentral.nih.gov/articlerender.fcgi?artid=2794266&tool=pmcentrez&rendertype=abstract. [cited 20 April, 2016].

[12] De Meyer SF, Denorme F, Langhauser F, Geuss E, Fluri F, Kleinschnitz C. Thromboinflammation in stroke brain damage. Stroke. [Internet] 2016:1165–72. Available from: http://www.ncbi.nlm.nih.gov/pubmed/26786115.

[13] Göb E, Reymann S, Langhauser F, Schuhmann MK, Kraft P, Thielmann I, et al. Blocking of plasma kallikrein ameliorates stroke by reducing thromboinflammation. Ann Neurol [Internet] 2015;77:784–803. Available from: http://www.ncbi.nlm.nih.gov/pubmed/25628066. [cited 26 April, 2016].

[14] Liu J, Gao B-B, Clermont AC, Blair P, Chilcote TJ, Sinha S, et al. Hyperglycemia-induced cerebral hematoma expansion is mediated by plasma kallikrein. Nat Med [Internet] 2011;17:206–10. Available from: http://www.pubmedcentral.nih.gov/articlerender.fcgi?artid=3038677&tool=pmcentrez&rendertype=abstract.

[15] Thal SC, Sporer S, Schmid-Elsaesser R, Plesnila N, Zausinger S. Inhibition of bradykinin B2 receptors before, not after onset of experimental subarachnoid hemorrhage prevents brain edema formation and improves functional outcome. Crit Care Med [Internet] 2009;37:2228–34. Available from: http://www.ncbi.nlm.nih.gov/pubmed/19487935.

[16] Martínez-Morillo E, Diamandis A, Romaschin AD, Diamandis EP. Kallikrein 6 as a serum prognostic marker in patients with aneurysmal subarachnoid hemorrhage. PLoS One [Internet] 2012;7:e45676. Available from: http://www.pubmedcentral.nih.gov/articlerender.fcgi?artid=3458071&tool=pmcentrez&rendertype=abstract. [cited April 20, 2016].

C H A P T E R

16

Eicosanoids in Cerebrovascular Diseases

K. van Leyen[1,2]

[1]Massachusetts General Hospital, Charlestown, MA, United States; [2]Harvard Medical School, Boston, MA, United States

INTRODUCTION

Eicosanoids have long been known to participate in cerebrovascular injury, starting in the early 1970s [1]. Eicosanoids are derivatives of the 20-carbon polyunsaturated fatty acid (PUFA) arachidonic acid (AA); the more generalized term for oxidized versions of PUFAs is oxylipins. Nonetheless, because of the physiological importance of AA-derived prostaglandins and leukotrienes, as well as the hydroxyeicosanoic acids (HETEs), *eicosanoids* is sometimes used as a blanket term for the enzyme-generated oxidation products of PUFAs. Although there are several important oxylipins with shorter or longer chains,

e.g., the 18-carbon linoleic acid derivative 13-HODE and the 22-carbon docosahexaenoic acid protectins, the main focus of this review will be on the eicosanoids as oxidation products of AA, and the enzymes that generate them. A great number of eicosanoids has been identified, and novel functions continue to be discovered. They can be pro- or antiinflammatory, vasoconstrictive or vasodilatory, or can serve as intracellular second messengers. Because of this dizzying array of substances with sometimes overlapping, sometimes counteractive activities, the enzymes leading to the production of eicosanoids have received increased attention. The reasons for this are threefold. One, there is a limited number of enzymes contributing

Primer on Cerebrovascular Diseases, Second Edition
http://dx.doi.org/10.1016/B978-0-12-803058-5.00016-3

to the generation and consumption of eicosanoids; two, these enzymes and their activities are upregulated following experimental stroke and several are implicated in causing neurovascular injury; three, as enzymes they are potentially very promising drug targets.

PRODUCTION OF EICOSANOIDS

AA is not produced in the human body, and thus must be taken up exogenously from various food sources. Once processed, most AA is not present as the free acid, but incorporated into phospholipids, typically in the sn-2 position. These AA-containing phospholipids are especially abundant in the brain, where they make up around 30% of all phospholipids. Under conditions of ischemia, phospholipases A2 (PLA2s) are activated, which liberate AA from the membrane phospholipids. Chief among these is the cytosolic calcium-dependent PLA2, but others may contribute.

With its four double bonds, free AA is highly reactive and prone to oxidation. This is especially relevant in stroke, because oxidative stress is known to be a major injury mechanism.

THE EICOSANOID-PRODUCING ENZYMES

Three classes of enzymes oxidize free arachidonic acid in the brain (see Fig. 16.1): the lipoxygenases, which generate leukotrienes, HETEs, and lipoxins as major oxidation products [2]; the cyclooxygenases I and II, which lead to the production of inflammatory prostaglandins [3]; and cytochromes P450, which generate vasoactive substances including 20-HETE and various protective epoxides, the epoxyeicosatrienoic acids (EETs) [4].

CONTRIBUTIONS TO STROKE INJURY

Due to their ubiquity and wide ranging functions, there have been many surprises as to the role of eicosanoids in stroke pathology. Studying these effects is not always easy, because there is considerable cross talk between eicosanoids and the enzymes they are produced and metabolized by; furthermore their analysis has only recently, through the National Institutes of Health–funded LIPID MAPS program, become more readily accessible. Nonetheless, with these technical advances and the availability of knockout mouse models, we now have a much better understanding of the individual contributions of various eicosanoids to stroke pathobiology.

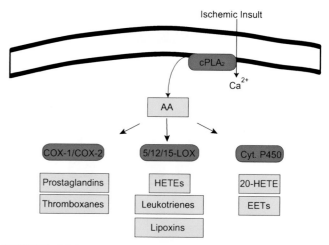

FIGURE 16.1 **Production of major eicosanoids in the brain.** Following an ischemic event, an influx of calcium along with other factors leads to activation of cPLA2, which cleaves phospholipids inside the membrane to liberate arachidonic acid (AA). AA is then oxidized by cyclooxygenases COX-1 and COX-2, lipoxygenases 5-LOX and 12/15-LOX, and proteins of the cytochrome P450 family. The resulting eicosanoids have diverse functions and can be both damaging and protective for the ischemic brain. HETE, hydroxyeicosanoic acid.

Phospholipase A2

Early work documented increases in eicosanoids following ischemic events, related to increased phospholipase A2 activity [1]. This has led to the idea that reducing phospholipase activity might be neuroprotective, and indeed, mice in which the gene encoding cPLA2 is knocked out suffer reduced ischemic injury. It may be possible to exploit this finding therapeutically, but there are several challenges. One lies in the lack of highly specific inhibitors of the enzyme. The second is the early increase in activity detected: it may be difficult in practice to begin stroke treatment early enough to efficiently block the release of AA. Nonetheless, PLA2 inhibition is a concept worth pursuing.

Cyclooxygenases COX-1 and COX-2

The proinflammatory prostaglandins have long been seen as an attractive target for stroke therapy. Although genetic deletion of COX-1 surprisingly increases ischemic injury, COX-2 knockout mice were protected [5]. Unfortunately, however, exploiting this finding therapeutically has turned out to be challenging due to the mixture of positive and negative effects of prostaglandin signaling, leading to an increased focus on individual downstream prostaglandin E_2 receptor (EP), prostaglandin D_2 receptor (DP), and other prostaglandin receptors [6]. For the EP receptors, for example, antagonists of EP1 and EP3 are beneficial, whereas for EP2 and EP4, activation is protective (Fig. 16.2).

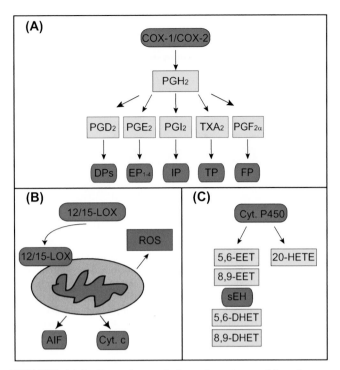

FIGURE 16.2 **Protective and damaging eicosanoid pathways.** (A) Cyclooxygenase: the initially formed prostaglandin H_2 (PGH_2) is further converted to prostaglandins D, E, F, and I series as well as thromboxane (TX) A_2, which each bind to their own class of receptors. In the case of the EP receptors 1–4, they each in turn have different functions, which can be vasodilatory (EP2, EP4) or vasoconstrictive (EP3). *IP*, prostaglandin I_2 receptor; *TP*, thromboxane A_2 receptor; *FP*, prostaglandin $F_{2\alpha}$ receptor. (B) Lipoxygenase: once activated by calcium and/or oxidative stress, 12/15-LOX translocates from cytosol to the mitochondrial membrane, where it causes lipid peroxidation and release of proapoptotic molecules like cytochrome c and apoptosis-inducing factor (AIF), along with increasing levels of reactive oxygen species (ROS). (C) Cytochrome P450: different members of the cytochrome P450 family generate damaging 20-hydroxyeicosanoic acid, as well as beneficial EETs. These are, however, rapidly hydrolyzed by the soluble epoxide hydrolase enzyme.

Lipoxygenases 5-LOX and 12/15-LOX

On the lipoxygenase side, there was an early recognition of the importance of increased levels of leukotrienes and HETEs for ischemic injury [7]. The initial studies targeted eicosanoids produced by 5-LOX, but two developments based on gene knockout studies in the early 2000s led to a shift in focus: first, the finding that 5-LOX knockout mice were subject to the same amount of ischemic injury as their wild-type counterparts [8]; second, gene knockout or inhibitors of 12/15-LOX provided excellent neuroprotection against focal ischemia [9]. Especially this latter finding came with a true paradigm shift, because there is an entirely different mechanism involved. The 12/15-LOX is unique among the lipoxygenases in that it can directly oxidize AA in its membrane-bound form, leading to lipid peroxidation and damage to organelles including mitochondria and membranes of the endoplasmic reticulum. The consequence is an apoptotic type of cell

death that appears to operate mainly through activation of apoptosis-inducing factor AIF, rather than mediated by caspases. Postmortem findings in the brains of patients with stroke support the presence of this mechanism in humans as well. The timing of increased 12/15-LOX activity appears to be suitable for therapeutic intervention. 12-HETE as a readout for 12/15-LOX activity was increased in the brains of mice subjected to focal ischemia at 12 and 24h, whereas immunohistochemistry showed 12/15-LOX protein increased gradually over the first 24h. Significant infarct size reductions in mice were achieved when treatment commenced 4h after onset of ischemia. With recent improvements including the development of novel 12/15-LOX inhibitors with improved selectivity, it may be possible to exploit this window of opportunity.

Cytochrome P450 and Soluble Epoxide Hydrolase

Cytochromes of the P450 family of proteins generate 20-HETE, a potent vasoconstrictor that can be counteracted using inhibitors HET0016 or TS011. The exact role of 20-HETE in ischemic stroke and subarachnoid hemorrhage is still under discussion, but blocking its activity is generally seen as beneficial. Another set of cytochromes P450 produces the EETs, which have mainly protective effects for the vasculature. They are quickly degraded by soluble epoxide hydrolase (sEH), however, to produce the corresponding dihydroeicosatrienoic acids, which are significantly less active. Inhibitors of sEH thus reduce ischemic injury by preserving the protective EETs [10].

CONCLUSION AND OUTLOOK

Although there is a bewildering array of eicosanoids with functions that can be both beneficial and detrimental after stroke, many of the corresponding pathways have been investigated and clarified in recent years. As our understanding of these eicosanoid-related effects increases, we can manipulate these powerful pathways to develop novel treatments for stroke and related cerebrovascular diseases.

References

[1] Bazan Jr NG. Effects of ischemia and electroconvulsive shock on free fatty acid pool in the brain. Biochim Biophys Acta 1970;218(1):1–10.

[2] Kühn H, Banthiya S, van Leyen K. Mammalian lipoxygenases and their biological relevance. Biochim Biophys Acta 2015;1851(4):308–30. http://dx.doi.org/10.1016/j.bbalip.2014.10.002.

[3] Bosetti F. Arachidonic acid metabolism in brain physiology and pathology: lessons from genetically altered mouse models. J Neurochem 2007;102(3):577–86. http://dx.doi.org/10.1111/j.1471-4159.2007.04558.x.

[4] Koerner IP, Zhang W, Cheng J, Parker S, Hurn PD, Alkayed NJ. Soluble epoxide hydrolase: regulation by estrogen and role in the inflammatory response to cerebral ischemia. Front Biosci 2008;13:2833–41.

[5] Iadecola C, Niwa K, Nogawa S, Zhao X, Nagayama M, Araki E, et al. Reduced susceptibility to ischemic brain injury and N-methyl-D-aspartate-mediated neurotoxicity in cyclooxygenase-2-deficient mice. Proc Natl Acad Sci USA 2001;98(3):1294–9. http://dx.doi.org/10.1073/pnas.98.3.1294. pii:98/3/1294.

[6] Iadecola C, Gorelick PB. The Janus face of cyclooxygenase-2 in ischemic stroke: shifting toward downstream targets. Stroke 2005;36(2):182–5. http://dx.doi.org/10.1161/01.STR.0000153797.33611.d8. pii:01.STR.0000153797.33611.d8.

[7] Moskowitz MA, Kiwak KJ, Hekimian K, Levine L. Synthesis of compounds with properties of leukotrienes C4 and D4 in gerbil brains after ischemia and reperfusion. Science 1984;224(4651):886–9.

[8] Kitagawa K, Matsumoto M, Hori M. Cerebral ischemia in 5-lipoxygenase knockout mice. Brain Res 2004;1004(1–2):198–202. http://dx.doi.org/10.1016/j.brainres.2004.01.018. pii:S0006899304000587.

[9] van Leyen K. Lipoxygenase: an emerging target for stroke therapy. CNS Neurol Disord Drug Targets 2013;12(2):191–9.

[10] Huang H, Al-Shabrawey M, Wang MH. Cyclooxygenase- and cytochrome P450-derived eicosanoids in stroke. Prostaglandins Other Lipid Mediat 2016;122:45–53. http://dx.doi.org/10.1016/j.prostaglandins.2015.12.007.

CHAPTER

17

Neurogenesis in Cerebrovascular Disease

D.A. Greenberg

Buck Institute for Research on Aging, Novato, CA, United States

INTRODUCTION

Tissue repair is as old as tissues themselves, as it can be observed in the most primitive multicellular organisms. However, evolution confers tissue complexity and cellular specialization at a price, which includes less effective repair capacity. As the most complex and most specialized tissue, brain is also the most difficult to regenerate after injury, such as that associated with cerebrovascular disease. Nevertheless, mechanisms for brain regeneration and repair exist. One such mechanism involves the manufacture of new cerebral neurons, or neurogenesis, which occurs at select sites in the brain and may contribute to functional recovery from stroke.

ADULT NEUROGENESIS

Developmental neurogenesis associated with brain maturation proceeds with a time course that varies across brain regions. However, neurogenesis continues in the adult brain, most prominently at two sites. These are the subgranular zone (SGZ) of the hippocampal dentate gyrus,

which adds neurons to the adjacent granule cell layer and may thereby participate in memory function, and the subventricular zone (SVZ) along the anterior horns of the lateral ventricles (Fig. 17.1). In rodents, neuroblasts arising in the SVZ migrate along a pathway known as the rostral migratory stream to the olfactory bulb, where they replenish interneurons. A similar pathway exists in humans, but loses prominence after infancy, and populates not only the olfactory bulb but also the prefrontal cortex [1]. Precursor cells from a variety of brain regions, including neocortex, can also generate neurons under certain in vitro conditions, suggesting that a similar phenomenon might also occur in vivo, perhaps in connection with injury or disease.

CEREBROVASCULAR DISEASE AND NEUROGENESIS

Brain injury, like numerous physiological factors (e.g., neurotransmitters, hormones, and growth factors), can enhance adult neurogenesis in excess of basal levels (Fig. 17.2). This has been demonstrated in rodent models of global cerebral ischemia [2], small- [3] and large-vessel

[4] ischemic stroke, and intracerebral hemorrhage [5]. Of these, large-vessel stroke has been studied in most detail. Unilateral occlusion of the middle cerebral artery in rats is accompanied by enhanced proliferation of neuroblasts in both the SGZ and SVZ, beginning within hours of onset. These cells migrate from SVZ, but not SGZ, toward the site of injury. Immature neurons have also been observed adjacent to small blood vessels in the vicinity of cerebral infarcts [6] and intracerebral hematomas [7] from human brain samples obtained at autopsy. However, whether these migrate from distant sites such as SVZ, or arise locally, is uncertain.

FUNCTIONAL SIGNIFICANCE OF NEUROGENESIS IN CEREBROVASCULAR DISEASE

A major issue surrounding neurogenesis induced by cerebrovascular disease is whether it has any functional significance in relation to recovery. This is an important consideration because if neurogenesis improves recovery,

stimulating neurogenesis could provide a new therapeutic approach to stroke, and a large number of clinically available drugs are neurogenic. Testing the functionality of stroke-induced neurogenesis requires selectively inhibiting neurogenesis and measuring the effect on outcome. This has been done in several ways. Ionizing radiation reduces neurogenesis and impairs behavioral recovery following global cerebral ischemia in gerbils [8]. Ablation of neurogenesis by transgenic expression of herpes simplex virus thymidine kinase and administration of ganciclovir worsened outcome from middle cerebral artery occlusion in mice, as shown by an increase in infarct size and exacerbation of sensorimotor behavioral deficits [9]. These findings suggest, albeit indirectly, that neurogenesis helps to promote postischemic recovery in brain, since inhibiting neurogenesis worsens outcome.

MECHANISMS OF NEUROGENESIS-MEDIATED REPAIR

How neurogenesis might enhance structural and functional repair in the ischemic brain is puzzling. The number of neurons lost dwarfs the number that can be produced by this process and only a small fraction of new neuroblasts differentiate into mature functional neurons. This implies that mechanisms other than cell replacement must be involved, and several such mechanisms have been proposed. These include the secretion of trophic factors and antiinflammatory cytokines from neuroblasts [10]. Candidate mediators of neuroblast-induced neuroprotective or immunomodulatory "bystander effects" include bone morphogenic protein, brain-derived neurotrophic factor, ciliary neurotrophic factor, glial cell-derived neurotrophic factor, nerve growth factor, neurotrophin-3, sonic hedgehog, and vascular endothelial growth factor.

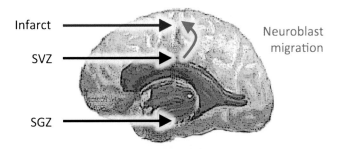

FIGURE 17.1 Anatomical overview of stroke-induced neurogenesis. Cerebral infarction triggers increased proliferation of neuroblasts in the rostral subventricular zone (SVZ) and subgranular zone (SGZ) of the hippocampal dentate gyrus, which is followed by migration of SVZ (but not SGZ) neuroblasts to the infarct site.

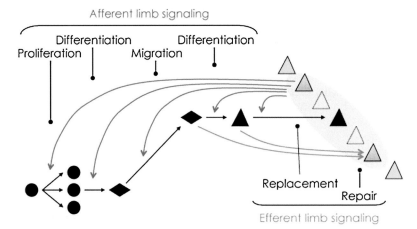

FIGURE 17.2 Stages of stroke-induced neurogenesis. These include an afferent limb (*red arrows*) in which ischemic tissue (*gray shading*) signals to the neurogenic zones to promote proliferation, differentiation, and migration of neuroblasts, and an efferent limb (*black and blue arrows*) in which neuroblasts that migrate to the site of ischemia either replace dead neurons or protect injured neurons from further damage.

CONCLUSION

Neurogenesis occurs, and is likely to contribute to brain repair, after injury from cerebrovascular disease. The two principal sites of adult neurogenesis and its amplification after stroke are the SGZ of the hippocampal dentate gyrus and the rostral SVZ, but neurogenesis may occur at additional sites in injury or disease. Neuroblasts generated in this setting migrate to damaged tissue, where they may differentiate into mature neurons or release neuroprotective trophic factors or antiinflammatory cytokines that protect reversibly injured cells or inhibit detrimental immune processes. The finding that stroke-induced neurogenesis is functional in rodent models implies that it may be a therapeutic target in cerebrovascular disease.

References

[1] Sanai N, Nguyen T, Ihrie RA, Mirzadeh Z, Tsai HH, Wong M, et al. Corridors of migrating neurons in the human brain and their decline during infancy. Nature 2011;478:382–6.

[2] Liu J, Solway K, Messing RO, Sharp FR. Increased neurogenesis in the dentate gyrus after transient global ischemia in gerbils. J Neurosci 1998;18:7768–78.

[3] Gu W, Brannstrom T, Wester P. Cortical neurogenesis in adult rats after reversible photothrombotic stroke. J Cereb Blood Flow Metab 2000;20:1166–73.

[4] Jin K, Minami M, Lan JQ, Mao XO, Batteur S, Simon RP, et al. Neurogenesis in dentate subgranular zone and rostral subventricular zone after focal cerebral ischemia in the rat. Proc Natl Acad Sci USA 2001;98:4710–5.

[5] Masuda T, Isobe Y, Aihara N, Furuyama F, Misumi S, Kim TS, et al. Increase in neurogenesis and neuroblast migration after a small intracerebral hemorrhage in rats. Neurosci Lett 2007;425:114–9.

[6] Jin K, Wang X, Xie L, Mao XO, Zhu W, Wang Y, et al. Evidence for stroke-induced neurogenesis in the human brain. Proc Natl Acad Sci USA 2006;103:13198–202.

[7] Shen J, Xie L, Mao XO, Zhou Y, Zhan R, Greenberg DA, et al. Neurogenesis after primary intracerebral hemorrhage in adult human brain. J Cereb Blood Flow Metab 2008;28:1460–8.

[8] Raber J, Fan Y, Matsumori Y, Liu Z, Weinstein PR, Fike JR, et al. Irradiation attenuates neurogenesis and exacerbates ischemia-induced deficits. Ann Neurol 2004;55:381–9.

[9] Jin K, Wang X, Xie L, Mao X, Greenberg DA. Transgenic ablation of doublecortin-expressing cells suppresses adult neurogenesis and worsens stroke outcome in mice. Proc Natl Acad Sci USA 2010;107:7993–8.

[10] Hermann DM, Peruzzotti-Jametti L, Schlechter J, Bernstock JD, Doeppner TR, Pluchino S. Neural precursor cells in the ischemic brain – integration, cellular crosstalk, and consequences for stroke recovery. Front Cell Neurosci 2014;8:291.

CHAPTER

18

Gliogenesis

K. Arai, E.H. Lo

Harvard Medical School, Boston, MA, United States

INTRODUCTION

Brain pathophysiology is influenced by a dynamic balance between deleterious and beneficial responses to the initial insult [1]. Stroke and brain injury trigger a wide spectrum of neurovascular perturbations, glial activation, and secondary neuroinflammation that may all amplify neuronal cell death cascades. But at the same time, many endogenous neuroprotective responses may also be activated (review by Moskowitz et al. [2]), and these beneficial processes include compensatory gliogenesis.

Gliogenesis (e.g., generation of astrocytes and oligodendrocytes) occurs both in developing and adult brain. During embryogenesis and development, multipotential progenitor cells generate new neurons, astrocytes, and oligodendrocyte lineage cells in germinal zones. In the adult brain, the subventricular zone (SVZ) of the lateral ventricle and the subgranular zone (SGZ) in the dentate gyrus of hippocampus retain multiple stem cells to form the largest germinative areas for new neurons and glial cells. Although the rate of neurogenesis and gliogenesis is much lower in adult brain than in developing brain, adult neurogenesis and gliogenesis may increase after brain injury to repair damaged brain tissue.

Historically, neurons and glial cells (e.g., astrocytes and oligodendrocytes) have been thought to derive from distinct precursor pools, partly because glial cells appear after neurons emerge during development. However, studies have proposed that some populations of glial cells may work as a neural stem cell and give rise to differentiated neurons (and differentiated glial cells, as well). This chapter will first introduce basic processes for neurogenesis and gliogenesis during development. And then, mechanisms of adult gliogenesis under physiological and pathological conditions will be discussed.

RADIAL GLIA GIVES RISE TO NEURONS AND GLIAL CELLS DURING DEVELOPMENT

The term "glial cell" has been used in different ways in the literature. For a long time, glial cells have been considered as differentiated nonneuronal cells that support nerve cells and regulate metabolic activity in the central nervous system. However, the term "glial cell" now refers to both a progenitor population that gives rise to brain cells (e.g., neurons, astrocytes, oligodendrocytes) as well as a differentiated population of parenchymal astrocytes and oligodendrocytes (and sometimes ependymal cells and microglia). In fact, during development, radial glial cells express astrocytic markers and make an astrocyte-like contact with endothelial cells, but at the same time, those cells play a role as neural stem cells to generate differentiated neurons and glial cells. (The term "neural stem cell" is also somewhat confusing, but this chapter follows the definition by Kriegstein and Alvarez-Buylla [3] and uses the term "neural stem cell" to refer to the primary progenitor cells that initiate lineages leading to the formation of differentiated neurons or glial cells.)

During the early embryonic stage of brain development, neuroepithelial cells convert into radial glial cells that are morphologically characterized by the projection of its processes from the ventricular zone to the meningeal zone with apicobasal polarity. Radial glial cells contact the ventricle apically and the meninges, basal lamina, and blood vessels basally. Radial glia may have three major roles: self-proliferation, neurogenesis, and gliogenesis. For self-proliferation, radial glial cells divide symmetrically. But for neurogenesis and gliogenesis, they divide asymmetrically.

Before gliogenesis takes place during embryonic development, radial glia generates neurons directly via asymmetric proliferation or indirectly through intermediate progenitor cells (so-called neurogenic intermediate progenitor cells; nIPCs). During the late embryonic stage, most radial glial cells begin to detach from the apical side and generate astrocytes via direct conversion or through astrocytic intermediate progenitor cells. In a parallel to astrocyte generation, radial glial cells also generate oligodendrogenic intermediate progenitor cells [oIPCs, also known as oligodendrocyte precursor cells (OPCs)], which then differentiate into oligodendrocytes at the end of embryonic development or after birth (Fig. 18.1).

After birth, a subpopulation of radial glial cells that reside at the apical side continues to generate nIPCs and oIPCs for physiological neurogenesis and oligodendrogenesis. However, eventually, some of these neonatal radial glial cells convert into ependymal cells and others into the adult SVZ astrocytes (also known as type B cells) over time. In the adult brain, there are two major subtypes of glial cells that work as a neural stem cell to generate differentiated neurons and glial cells. The first subset comprises adult SVZ astrocytes (type B cells) that contact ependymal cells apically and blood vessels basally. The second subset comprises a subpopulation of astrocytes in SGZ in the dentate gyrus of hippocampus (so-called adult SGZ astrocytes), which may derive from radial glia during embryonic development.

ADULT NEUROGENESIS AND GLIOGENESIS TAKE PLACE IN SVZ OR SGZ IN ADULT BRAIN

As noted, the SVZ in the walls of the lateral ventricle and the SGZ at the interface of the granule cell layer and the hilus of hippocampus are the two major regions that contain adult neural stem cells (Fig. 18.2). The adult SVZ astrocytes (e.g., type B cells) exhibit astrocytic characteristics with typical astrocyte markers, such as glial fibrillary acidic protein (GFAP) and L-glutamate/L-aspartate transporter (GLAST). They are relatively quiescent, but generate actively proliferating type C cells that work as nIPCs or oIPCs (generally referred to OPCs) for neurogenesis or oligodendrogenesis, respectively. nIPCs within type C cells generate type A cells (neuroblasts) that migrate to the olfactory bulb and differentiate into interneurons. On the other hand, oIPCs within type C cells give rise to oligodendrocytes over time. But some of the oIPCs from the adult SVZ astrocytes (type B cells) may reside in an undifferentiated state (i.e., OPCs), which start to proliferate and differentiate into oligodendrocytes as needed, such as under demyelinating conditions. Although the adult SVZ astrocytes generate OPCs, some population of residual OPCs in adult brain may originate from asymmetric proliferation of radial glial cells during the embryonic developmental stage.

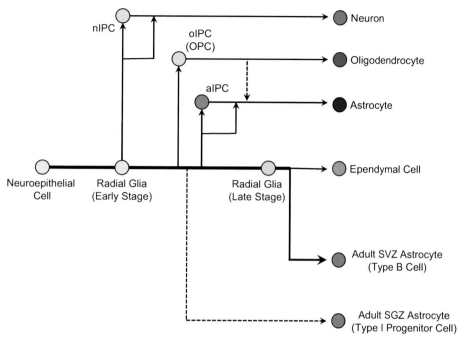

FIGURE 18.1 A basic process for gliogenesis from radial glial cells during development. *aIPC*, astrocytic intermediate progenitor cell; *nIPC*, neurogenic intermediate progenitor cell; *oIPC*, oligodendrogenic intermediate progenitor; *SGZ*, subgranular zone; *SVZ*, subventricular zone.

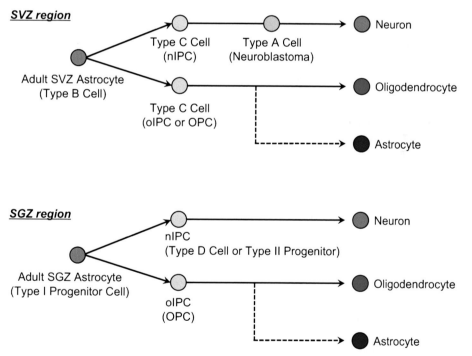

FIGURE 18.2 Models for adult gliogenesis in the subventricular zone (SVZ) and subgranular zone (SGZ) regions. *nIPC*, neurogenic intermediate progenitor cell; *oIPC*, oligodendrogenic intermediate progenitor, *OPC*, oligodendrocyte precursor cell.

Similar to the SVZ, the SGZ also produces new differentiated neurons and glial cells in adult brain. Within these subsets, so-called adult SGZ astrocytes (also known as type I progenitors) function as a neural stem cell for adult neurogenesis and gliogenesis. The adult SGZ astrocytes may not directly generate neurons, but instead generate intermediate progenitor cells (nIPCs, also referred to type D cells or type II progenitors) that eventually differentiate into neurons. Adult oligodendrogenesis from the adult SGZ astrocytes may also occur, but compared with the SVZ, the number of newly generated oligodendrocytes in the SGZ may be very low. Besides functioning as neural stem cells, the adult SGZ astrocytes also exhibit astrocytic characteristics with expressions of astrocyte markers (e.g., GFAP).

Astrocytes are heterogeneous and the underlying mechanism/process for astrocyte generation (e.g., astrogliosis) is complicated. As mentioned, at the end of developmental period, astrocytes arise from transformation of radial glia. In adult brain, the adult SVZ or SGZ astrocytes may continue to generate astrocytes, whereas some residual astrocytes may also divide locally. Although still controversial, OPCs can also give rise to astrocytes (the term "OPC" is somewhat vague, but this chapter uses this term to refer to cell populations that mainly differentiate into oligodendrocytes). When purified from optic nerves or cerebral cortex, these OPCs (so-called oligodendrocyte-type2 astrocyte progenitor cells; O-2A cells) can differentiate into both oligodendrocytes and astrocytes in vitro, depending on culture conditions. Also in vivo, a fate-mapping study confirmed that some OPCs differentiated into protoplasmic astrocytes in gray matter but not into fibrous astrocytes of white matter [4]. In addition, when purified OPCs were transplanted into myelin basic protein (MBP)-depleted conditions, some OPCs were converted to GFAP-positive astrocytes, whereas others differentiated into MBP-expressed mature oligodendrocytes [5]. However, astrocyte generation from OPCs is limited after the postnatal period [6], and therefore, OPC-to-astrocyte conversion in adult brain may depend on the conditions of surrounding environments.

INTRINSIC FACTORS FOR GLIOGENESIS REGULATION

Gliogenesis is tightly regulated by intrinsic factors. Notch signaling is one of the major mediators of cell fate determination in neural and glial lineage cells (see review by Pierfelice et al. [3]). Notch signaling is mediated by the Notch receptors activated by their ligand Jagged on the surface of the neighboring cells. The targets are transcriptional factors Hes (Hes1 and Hes5), which inhibit the neural differentiation from radial glia. The Notch-mediated cell–cell interactions spread to surrounding cells to maintain the undifferentiated conditions in their surroundings, which is known as lateral inhibition. Notch signaling is also an important regulator for GFAP gene expression both genetically and epigenetically. Notch intracellular domain/CBF1 complex upregulates GFAP via direct binding to the GFAP promoter region. There are CpG methylated regions on the GFAP promoter to which Notch signaling-activated nuclear factor I A binds and prevents Dnmt1 from accessing the promoter region. This leads to demethylation of the promoter region and upregulation of GFAP expression.

Phase-dependent activation of Wnt signaling is another critical regulator for neural and glial cell fate (see review by Nusse [7]). Binding of Wnt to its receptor Frizzled inhibits the degradation of β-catenin, which binds the T-cell factor to upregulate downstream genes. The importance of β-catenin in neurogenesis is illustrated by the fact that β-catenin deletion is associated with impaired neurogenesis from immature neuronal progenitor cells. Among genes downstream of Wnt signaling, Ngn1 and Ngn2 are key regulators that are highly expressed transcriptional factors during the neurogenesis phase. Ngn regulates neurogenesis and gliogenesis in a complex way. Ngn strongly promotes neurogenesis to constitute the six layers of brain cortex in an order from deep to superficial layers, and Ngn negatively regulates (janus kinase/signal transducer and activator of transcription) JAK-STAT-mediated demethylation of the GFAP promoter region and inhibits neural stem cell differentiation into astrocytes. Ngn2-knockout mice exhibit prominent reductions in neurogenesis and gliogenesis. Although Wnt signaling upregulates the expression of Ngn during the neurogenesis phase, its effects on chromatin remodeling by histone modification of the promoter region of Ngn in the later phase results in decreased neurogenesis. Wnt-induced chromatin remodeling blunts the sensitivity to Wnt signaling, leading to a shift from neurogenesis to gliogenesis. Since precise mechanisms as to how Ngn regulates neurogenesis/gliogenesis are still mostly unknown, elucidating phase- and region-dependent mechanisms of Ngn signaling in neural stem cell function is warranted for future studies.

Thus far, several key mechanisms have been identified in cell fate decisions for oligodendrocyte lineage cells (see review by Nishiyama et al. [8]). Olig2 is a critical transcriptional regulator, which directs OPCs to become oligodendrocytes during embryogenesis. In fact, under Olig2-deficient conditions, OPCs were directed to become astrocytes, and postnatal hypomyelination was observed in the neocortex and corpus callosum. Epigenetic regulation such as DNA methylation and histone modification associated with chromatin remodeling was also shown to be involved in cell specification for oligodendrocytes. Treatment with histone deacetylase (HDAC) inhibitors or genetic ablation of HDAC1/2 in OPCs resulted in failure of differentiation in vivo. Several groups have confirmed that HDAC activities are required for efficient myelination. For example, HDACs act in conjunction with the transcription factor YY1 to start myelin gene expression by suppressing Id4 and Tcf4. Neuronal activity, microRNAs, and Notch signaling are also well-defined mechanisms for cell fate decisions in oligodendrocyte lineage cells. Working together, the factors and signaling pathways discussed here coordinately regulate oligodendrocyte generation from multipotent stem cells.

GLIOGENESIS AFTER BRAIN INJURY

In adult brain, function and cell number of glial cells are tightly regulated to maintain proper brain function. Although glial cells are generally quiescent in the adult brain, they become activated after brain injury such as stroke. Activated glial cells are now thought to play complex biphasic roles, with both deleterious and beneficial effects depending on the scenarios involved. This section briefly overviews how astrocytes and oligodendrocytes respond under pathological conditions.

In the healthy adult brain, astrocytes play essential roles, including regulating cerebral blood flow, providing energy metabolites to neurons, and maintaining the resting conditions of extracellular ions. In general, astrocytes in adult brain are highly differentiated and quiescent. However, they respond to all sorts of stress and become activated by changing their morphology and protein expression patterns (so-called reactive astrocytes or astrogliosis). During severe brain damage such as stroke or trauma, reactive astrocytes proliferate (e.g., one form of gliogenesis) and form glial scars at the damaged region (see review by Sofroniew [9]). Traditionally, the glial scar was considered to be deleterious, especially for brain repair and remodeling, because they may release deleterious inflammatory factors as well as function as a physical barrier to inhibit axonal extension for synaptogenesis. However, studies propose that the glial scar may work to protect surviving brain tissues from secondary damage, by suppressing inflammation. In addition, reactive astrocytes can secrete prosurvival factors to support brain remodeling after injury. Therefore depending on the context, reactive astrocytes and glial scar would accelerate pathological conditions after brain injury, and at the same time, may also support a compensatory and pro-recovery response.

The oligodendrocyte, which is a major glial cell type in cerebral white matter, contributes to white matter homeostasis by forming myelin sheaths. Myelin impairment associated with loss of oligodendrocytes is well documented in several cerebrovascular diseases. Because injured oligodendrocytes are unable to produce new myelin sheaths and mature oligodendrocytes do not proliferate, oligodendrogenesis (e.g., formation of new oligodendrocytes from residual OPCs) is an essential process for axonal repair after brain injury. Under normal conditions in adult brain, physiological oligodendrogenesis by residual OPCs would occur to maintain white matter homeostasis, whereas the rate for oligodendrocyte generation is very limited. On the other hand, after demyelination after injury, residual OPCs actively proliferate and migrate to the affected areas, attempting to differentiate into mature oligodendrocytes and remyelinate axons (see reviews by Zhang et al. [10] and Maki et al. [11]). The process of oligodendrogenesis is influenced by many intrinsic and extrinsic factors from various types of cells, thus offering a number of pathways for potential therapeutic interventions. However, it should be noted that aging may dampen mechanisms of compensatory oligodendrogenesis [12], and therefore, therapies to boost endogenous repairing response may be more challenging in aged populations.

CONCLUSION

Gliogenesis is most active during development. But even in the adult brain, there remains some plasticity, and adult gliogenesis may be one of the important mechanisms to maintain and regulate brain function. This chapter overviews adult gliogenesis, focusing on generation of astrocytes and oligodendrocytes from their progenitor cells. Although some key intra- and intercellular mechanisms for gliogenesis were discussed, precise and detailed mechanisms that regulate gliogenesis after brain injury are still mostly unknown. Therefore a deeper understanding of signals and substrates for adult gliogenesis under pathological conditions will be required for further development of therapies for cerebrovascular diseases.

References

[1] Lo EH. A new penumbra: transitioning from injury into repair after stroke. Nat Med May 2008;14(5):497–500.
[2] Moskowitz MA, Lo EH, Iadecola C. The science of stroke: mechanisms in search of treatments. Neuron July 29, 2010;67(2):181–98.
[3] Pierfelice T, Alber L, Gaiano N. Notch in the vertebrate nervous system: an old dog with new tricks. Neuron 2011;69(5):840–55. http://dx.doi.org/10.1016/j.neuron.2011.02.031.
[4] Zhu X, Bergles DE, Nishiyama A. NG2 cells generate both oligodendrocytes and gray matter astrocytes. Development January 2008;135(1):145–55.
[5] Windrem MS, Nunes MC, Rashbaum WK, et al. Fetal and adult human oligodendrocyte progenitor cell isolates myelinate the congenitally dysmyelinated brain. Nat Med January 2004;10(1):93–7.
[6] Hill RA, Patel KD, Goncalves CM, Grutzendler J, Nishiyama A. Modulation of oligodendrocyte generation during a critical temporal window after NG2 cell division. Nat Neurosci November 2014;17(11):1518–27.
[7] Nusse R. Wnt signaling and stem cell control. Cell Res May 2008;18(5):523–7.
[8] Nishiyama A, Komitova M, Suzuki R, Zhu X. Polydendrocytes (NG2 cells): multifunctional cells with lineage plasticity. Nat Rev Neurosci January 2009;10(1):9–22.
[9] Sofroniew MV. Molecular dissection of reactive astrogliosis and glial scar formation. Trends Neurosci December 2009; 32(12):638–47.
[10] Zhang R, Chopp M, Zhang ZG. Oligodendrogenesis after cerebral ischemia. Front Cell Neurosci 2013;7:201.
[11] Maki T, Liang AC, Miyamoto N, Lo EH, Arai K. Mechanisms of oligodendrocyte regeneration from ventricular-subventricular zone-derived progenitor cells in white matter diseases. Front Cell Neurosci 2013;7:275.
[12] Miyamoto N, Pham LD, Hayakawa K, et al. Age-related decline in oligodendrogenesis retards white matter repair in mice. Stroke September 2013;44(9):2573–8.

CHAPTER

19

Vascular Remodeling After Cerebral Ischemia

Y. Nishijima[1,2], Y. Akamatsu[1,2], K. Masamoto[3], J. Liu[1]

[1]University of California, San Francisco and the San Francisco Veterans Affairs Medical Center, San Francisco, CA, United States; [2]Tohoku University Graduate School of Medicine, Sendai, Japan; [3]University of Electro-Communications, Chofu, Tokyo, Japan

NEOVASCULARIZATION AFTER CEREBRAL ISCHEMIA AND HYPOXIA

Although the adult brain vascular network becomes quiescent after the completion of brain development, the remodeling of existing blood vessels or de novo vessel formation has been well documented after cerebral ischemia in both humans and laboratory animals as one of the CNS responses to ischemic injury or hypoxia. Because the mechanisms and mediators involved in the above two processes are complex and partially overlap, "neovascularization" has been the commonly recognized term to describe such phenomenon. As in other organs, neovascularization in the brain requires an orchestrated interplay among the immune, endocrine, and vascular systems, which can be categorized into three distinct mechanisms, namely angiogenesis, vasculogenesis, and arteriogenesis.

Arteriogenesis is a crucial process for maintaining bulk blood supply in the event of abrupt obstruction of blood flow like an ischemic stroke, or a chronic adaptation in response to progressive narrowing of the vasculature occurring in human carotid stenoocclusive diseases. Arteriogenesis, enabling blood flowing through the preexisting arterial anastomosis virtually instantaneously after cerebral ischemia, is not only the most immediate neovascularization process but also the most effective flow compensation mechanism in restoring ischemia-induced perfusion deficit and in salvaging potential tissue loss. Unlike arteriogenesis, vascular neogenesis via angiogenesis and vasculogenesis involves cell proliferation and thus requires time; consequently they are unlikely to deliver functional blood flow soon enough to prevent ischemic cell death. It is generally accepted that when capillaries develop in previously avascular tissue, it is called vasculogenesis, a repair process initiated by the recruitment of endothelial progenitor cells (EPCs).

In contrast, angiogenesis often refers to the formation of new capillaries by sprouting from the existing venules.

With respect to the direct functional output of neovascularization processes in restoring blood flow, experimental evidence suggests that the early stage of cerebral blood volume (CBV) increase is predominantly due to the recruitment of collateral flow, which is indispensible in reducing ischemic cell death, whereas the later phase of CBV increase is likely attributed to the emergence of angiogenesis and/or vasculogenesis, which may influence recovery and long-term functional outcome [1,2]. The mechanism and contribution of each neovascularization process following cerebral ischemia is discussed further in this chapter from the perspectives of human and experimental brain ischemia and hypoxia.

ANGIOGENESIS AFTER ISCHEMIA AND HYPOXIA

Ample structural evidence supports the occurrence of angiogenesis and induction of angiogenetic factors following both human stroke and experimental cerebral ischemia or hypoxia [3]. Stroke patients with the higher blood vessel counts in the infarcted brain tissue correlated with longer survival, suggesting that angiogenesis might play a beneficial role in stroke outcome [4]. Angiogenesis is reported to act as a route for infiltrating macrophages that clean up the necrotic debris and thus promote tissue remodeling, adding to the list of benefits. Stroke-induced angiogenesis appears to be transient and restricted to the border of the infarct, suggesting that it is a spatially and temporally self-contained process. However, due to the limited size of small capillaries, angiogenesis is unable to fully restore the function of larger vessels. Besides, factors produced during angiogenesis such as vascular endothelial growth factor (VEGF) and matrix metalloproteinase 9 have the potential to worsen vasogenic edema by increasing blood–brain barrier (BBB) leakage or by increasing the risk of hemorrhagic transformation.

The main stimulus of angiogenesis both in development and following ischemia is hypoxia, activating hypoxia-inducing factor (HIF) and downstream signaling events. HIF led to the increased expression of VEGF and activation of the receptor pathways, which are known as the earliest and the most crucial players in angiogenesis. The other important early regulators of angiogenesis include the Notch–Jagged/Delta-like pathways, whereas later vascular development and lumen formation is controlled by the ephrin–Eph, Wnt, and those involved in axonal guidance, such as the semaphorins, netrins, and ROBO/Slits signaling pathways. The angiopoietin–Tie, platelet-derived growth factor, and transforming growth factor beta families are also known to regulate additional aspects of angiogenesis that may include, but not limited to, vessel maturation. To form mature and functional blood vessels, pericytes and vascular smooth muscle cells are required to stabilize capillaries and control vessel conductance, respectively. To that end, PDGF is most instrumental in recruiting pericytes to cap the newly formed vascular tube and Ang-1 in reducing vascular leakage.

VEGF-A expression is upregulated by hypoxia via the hypoxia response element in the promoter region of the *vegf* gene. The tip cells of growing vessels express a high level of VEGFR-2 and organize a filopodia to migrate toward the area of higher VEGF-A expression, whereas the stalk cells proliferate, adding to the length of the sprouting vessels. Using two-photon longitudinal in vivo imaging technique, a detailed spatiotemporal dynamics of angiogenesis was witnessed in the mouse cerebral cortex after continuous exposure to hypoxia. After 7–14 days of living in the hypoxia chamber, capillary vessels located on average 60 μm away from penetrating arterioles in the cortex began to form sprouts, although red blood cells were still stagnant inside the sprouts at this stage. After 14–21 days of hypoxia, functional blood flow was established once the sprouting vessel made connection with an existing capillary. The maturation of the newly born vessels is evident by the wrapping of vessels with neighboring astrocyte processes, forming a "neuron-glia-vascular" unit [5]. However, emerging evidence suggests that microglia-secreted factors may also interact with VEGF in promoting or guiding vessel sprouting after chronic hypoxia.

In response to mild ischemia such as unilateral common carotid artery occlusion, it appears that vascular remodeling including pial arteries and veins, and capillary dilation in the parenchyma, and collateral growth (Fig. 19.1) are sufficient to restore blood flow without angiogenesis by using repeated in vivo optical imaging of cerebral microvasculature in living mice. Other studies using the permanent middle cerebral artery occlusion (MCAO) mouse models (distal occlusion) also consistently showed various remodeling of the vasculature on the ischemic cortical surface. Whether this remodeling of the vasculature involves capillary angiogenesis triggered by the ischemic insult is not supported by in vivo microscopic imaging studies. As expected, a rapid deterioration of capillary networks in the ischemic cortex was evidenced (Fig. 19.2; [15]). Failure of in vivo detection of angiogenic capillaries in the experimental ischemia

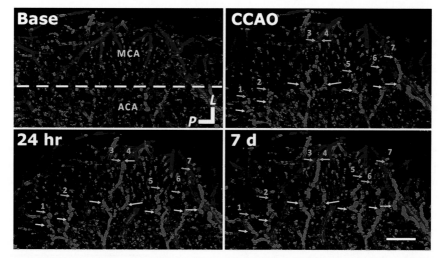

FIGURE 19.1 **Chronic hypoperfusion induces retrograde shift of anastomosis points toward multiple distal MCA branches.** The anatomic orientation of the brain is indicated with *arrows* pointing to the lateral (L) and posterior (P) directions. Immediately following unilateral common carotid artery occlusion (CCAO), a relatively mild ischemic condition compared with distal MCAO, a minor shift of the anastomosis point toward distal MCAs was detected in the normoglycemic *db/+* mice by DOCT, and sustained at 24 h and 7 days after CCAO. Dotted white line marks the divide between MCA and ACA territory at base line. *Yellow and blue arrows* indicated the various anastomosis points as numbered before and after the shift, respectively. The direction of blood flow is color coded, with the blood flowing toward the scanning probe beam coded designated as red, and the opposite direction as green. Scale bar: 1 mm. *MCAO*, middle cerebral artery occlusion; *MCA*, middle cerebral artery; *DOCT*, doppler optical coherent tomography; *ACA*, anterior cerebral artery.

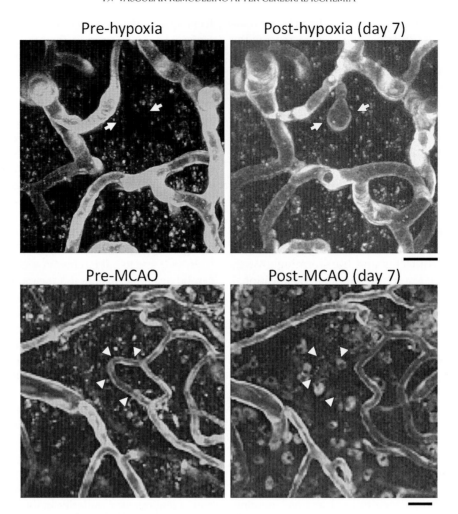

FIGURE 19.2 **Vascular sprouting following exposure to hypoxia as detected by two-photon microscopy.** The vascular images were taken with in vivo two-photon microscopy from the mice expressing GFP in vascular endothelium. The mouse was subjected to chronic cranial window according to "Tomita-Seylaz method" before initiation to hypoxia exposure (8–9% oxygen in normal atmosphere) or permanent MCAO. The same vascular location of the cortex was repeatedly imaged before and after exposure to chronic hypoxia or MCAO. In response to hypoxia, a newly formed vessel appeared (*arrows* in the upper panels) in the cortex 7days after hypoxic induction, and lumen of the capillary increased. *(Images modified from Masamoto K, Takuwa H, Seki C, et al. Microvascular sprouting, extension, and creation of new capillary connections with adaptation of the neighboring astrocytes in adult mouse cortex under chronic hypoxia. J Cereb Blood Flow Metab February 2014;34(2):325–31).* In contrast, induction of MCAO evoked a rapid deterioration of capillary networks (*arrow head* in the lower panels) in the ischemic core as well as peripheral regions. Note that new vessel formation was not detected in this early phase of the cerebral ischemia [15]. Scale bar: 20 μm. *MCAO*, middle cerebral artery occlusion; *GFP*, green fluorescent protein.

models could be due to spatiotemporal heterogeneity of the tissue remodeling after ischemic events that may limit the field of view during the microscopic imaging, or lack of the hypoxia-induced angiogenesis.

Apart from the cerebral cortex in which the vascular remodeling processes were most reported, the thalamus was also known for angiogenesis after middle cerebral artery (MCA) stroke, likely related to local hemodynamic impairment and its synaptic connections to the cortex. Following experimental stroke, the ipsilateral thalamus suffered from temporary hypoperfusion during the first week after MCAO, followed by hyperperfusion chronically for 1–3months. At the end of 3months, angiogenesis was detected as increased blood vessel branching in the ipsilateral thalamus, representing a novel type of remote plasticity that may support the removal of necrotic brain tissue and aid functional recovery [6].

VASCULOGENESIS AND THE ROLE OF ENDOTHELIAL PROGENITOR CELLS

Vasculogenesis is the major mechanism of vessel formation during development, yet continues to occur following brain injury or ischemia [7]. Circulating EPCs are mobilized and recruited to ischemic site and contribute to vessel growth or repair after stroke. Evidence suggests that the increase of circulating EPCs after acute ischemic stroke is associated with good clinical outcome, lending vasculogenesis another viable therapeutic target for stroke therapy.

Angioblasts are a type of endothelial precursor cells derived from the bone marrow, guided by angiogenic signals and migrated to sites where they proliferate, differentiate into endothelial cells, and subsequently develop into blood vessels. Despite the heterogenous nature of EPCs as a cell population, CD34, CD133, and VEGF-R2 are recognized as their common surface markers. Activation and mobilization of EPCs from bone marrow is induced primarily by HIF-a, VEGF, and erythropoietin released during hypoxia and tissue injury. Once circulating in the peripheral blood, stromal cell-derived factor 1 mediates migration and homing of EPCs to vascular endothelium via a CXCR-4-dependent mechanism.

Similar to angiogenesis, the benefit of vasculogenesis after brain ischemia lies not so much in providing the structural resource for vessel growth. Rather, the paracrine hypothesis suggests that the greatest contribution of EPCs in vascular regeneration is by providing a steady source of secreting proteins including growth factors, chemokines, and cytokines that may reduce progressive tissue damage and promote endogenous repair.

ARTERIOGENESIS AND THE REMODELING OF COLLATERAL VESSELS AFTER STROKE OR CEREBRAL HYPOPERFUSION

Arteriogenesis occurs via collateral vessels, a specialized vascular network with a distinct phenotype from the regular arteries, veins, or capillaries. Native collaterals are naturally occurring artery-to-artery or arteriole-to-arteriole anastomoses present in healthy tissue, whereas they undergo outward remodeling to compensate for reduced blood flow in response to vascular occlusive diseases. A number of collateral circuits exist within and between the extracranial and intracranial circulations. The intracranial collateral circuits are directly involved in maintaining the perfusion pressure inside the brain, and they can be further divided into primary (i.e., circle of Willis) or secondary (e.g., ophthalmic and leptomeningeal) collateral pathways [1,8]. Empirical experimental evidence suggests that at least in mice, the extent of native leptomeningeal collateralization is governed by genetic factors and causally related to infarct volume in MCA stroke [9].

Although the native collateral system is determined during the embryonic stages, a reduction in the number and diameter of collaterals in the adult, also known as collateral rarefaction, occurs with aging or disease process. In particular, the presence of vascular risk factors including metabolic syndromes is known to adversely affect the confluence of collateral flow. The collateral status plays an influencing role in altering the risk of hemodynamic stroke. In patients with symptomatic severe internal carotid artery stenosis, the risk at 2 years of stroke or transient ischemic attack is significantly reduced among those with angiographically defined collaterals compared with those without. The collateral status is also an independent predictor of outcome and response to thrombolytic therapies in patients with ischemic stroke. Specifically, among those who received endovascular revascularization therapy, collateral grade determined the recanalization rate and clinical outcome. Furthermore, a good collateral status can also lower the rate of hemorrhagic transformation after thrombolytic and endovascular therapies [8]. Thus, the endovascular triage approach using advanced neuroimaging to ascertain the extent of penumbra and collateral flow status has contributed to the higher success rates seen in some trials of acute stroke intervention [10].

Distinct from angiogenesis, arteriogenesis is triggered by hemodynamic forces such as fluid shear stress (FSS) induced by the pressure gradient during the obstruction or change of blood flow. The collateral flow developed proximal to the occlusion does not require the local expression of HIF-1 or VEGF, although the latter is needed for the formation of collateral vessels during development and during their continuing remodeling after ischemia. The initial increase of FSS results in the activation of endothelium and stimulates a cascade of signaling events including the induction of ion channels and production of nitric oxide. Although the conductance of collateral flow may occur immediately after vessel occlusion, the remodeling of collateral vessels continues to evolve over days to weeks (Fig. 19.1). The additional steps in arteriogenesis involve monocyte invasion and recruitment, activation of inflammatory responses, secretion of growth factors and cytokines, followed by matrix digestion and ultimately proliferation of smooth muscle cells, leading to the outward remodeling of collateral vessels and the increase of collateral flow [1].

THERAPIES TO PROMOTE POSTSTROKE NEOVASCULARIZATION

Collateral Flow Augmentation

Collateral perfusion is widely recognized as a key element that has significant effects on various aspects of clinical practice; hence collaterals are the potential therapeutic targets for acute stroke intervention. Pharmacological approaches to improve collateral circulation include volume expansion, hemodilution, and induced hypertension. Induced mild hypertension may improve the national institute of health stroke score with an acceptable degree of safety if cardiac and neurological status are closely monitored. Device-based approaches to augment collateral flow have also been attempted, including external counterpulsation via air-filled cuffs; partial occlusion of the abdominal aorta via NeuroFlo technology ([safety and efficacy of neuroflo technology in ischemic stroke] SENTIS trial); and transcranial laser therapy via NeuroThera ([necrotizing enterocolitis surgery trial] NEST trial 1&2) [8].

Promoting Vasculogenesis and Angiogenesis With Gene- or Cell-Based Therapies

Therapeutic angiogenesis or vasculogenesis using various kinds of growth factors or bone marrow-derived or local stem/progenitor cells is a new strategy to induce or enhance neovascularization after ischemic diseases. Numerous gene therapy– or cell therapy–based clinical trials aimed at revascularization have been conducted using proangiogenic genes coding for VEGF, fibroblast growth factor (FGF), hepatocyte growth factor (HGF), etc., or cells such as mononuclear cells (MNCs) with enriched EPCs from various sources to patients with peripheral or coronary artery disease. Although data from meta-analysis of initial trials suggest that angiogenic therapies are feasible and relatively safe, further large studies are required to demonstrate true efficacy [7].

Despite the enthusiasm from experimental studies and promise from trials of heart and other ischemic diseases, none of the gene-based therapies has yet been conducted clinical trials in patients with ischemic stroke so far, until pending issues to be resolved in angiogenic therapy in ischemic stroke including concerns regarding potential adverse effects such as progression of vasogenic edema, or hemorrhagic transformations through the process of degradation of the extracellular matrix which is necessary for newly vessel formation [11]. Therapy involving VEGF in particular may increase vascular permeability, thus potentially exacerbating ischemia-induced BBB leakage. To enhance vessel maturation or to prevent angiogenic factor–induced vascular leakage, codelivery of vascular stabilizers (e.g., PDGF, Ang-1) with angiogenic factors or cells has gained considerable acceptance as one of the promising strategies to avoid such adverse effects and should be taken into consideration in future clinical trials.

There is only one on-going angiogenesis-based clinical study using EPCs in patients with acute ischemic stroke (http://ClinicalTrials.gov Identifier: NCT01468064) and this study is not yet completed (estimated study completion date: March 2017). The precise role of angiogenesis-based therapy for ischemic stroke cannot be resolved based on current data;

however, it remains as a promising strategy to promote vascular and functional recovery after ischemic stroke in the future.

Diminished Responses to Native or Therapeutic Neovascularization With Vascular Risk Factors

Patients with metabolic syndromes are associated with poor anatomical collateral status during acute ischemic stroke [12]. Despite the delayed arteriogenic response seen in hypercholesterolemic mice, no clinical evidence supports the negative effect of cholesterol in collateral growth. Similar to the poor stroke outcome commonly observed among patients with hyperglycemia, mice with type 2 diabetes mellitus (T2DM) exhibited impaired retrograde collateral flow compensation from anterior cerebral artery (ACA) during MCA stroke, coincided with larger stroke and more severe neurological impairment [13]. Some evidence suggests that genetic factors contributed to the differences among individuals to form collaterals. For example, Asp298 allele of the endothelial nitric oxide synthase (eNOS) gene is associated with limited collateral development, especially in patients with diabetes mellitus (DM). Differences in genes regulating monocyte function or inflammatory state were also reported in people after heart ischemia, coincided with variation in collateral formation.

Specifically with regards to molecular and cellular mechanisms, clinical and experimental evidence suggests that vascular repair after ischemia is compromised in DM mainly due to the abrogation of proangiogenic pathways, reduced eNOS expression as well as NO production, and the activation of antiangiogenic signals including the overproduction and accumulation of advanced glycation end products, reactive oxygen species, and increased endoplasmic reticulum stress [14]. Thus, the diabetes-associated elevation in oxidative stress may exacerbate angiogenic factors–induced vascular permeability.

The mobilization of BM-derived EPCs depends on gradients of chemokines. T2DM is characterized by an increased M1 and reduced M2 macrophage population, leading to altered inflammatory cytokine profile and angiogenic factors. Since the recruitment of EPCs expressing CXCR4 is regulated by hypoxia and HIF-α, which in turn activates CXCL12, the diminished CXCL12 expression in diabetic mice would potentially impair the recruitment of EPCs. Lastly, DM may also decrease the ability of recruited BM-derived mononuclear cells in differentiating into endothelial cells. Taking all into consideration, the reduced neovascularization capacity in T2DM is further exacerbated by the already dwindling circulating EPC population, apart from the reduced recruitment and differentiation of EPCs, as well as impaired phosphorylation of BM eNOS. Thus, a thorough understanding of the vascular niche in diabetes or other vascular risk factors may overcome the impaired endogenous vascular repair or the blunted treatment effect afforded by revascularization therapies.

Acknowledgments

This work was supported by NIH grant R01 NS071050 (JL), and VA merit award I01RX000655 (JL).

References

[1] Liu J, Wang Y, Akamatsu Y, et al. Vascular remodeling after ischemic stroke: mechanisms and therapeutic potentials. Prog Neurobiol April 2014;115:138–56.

[2] Yanev P, Dijkhuizen RM. In vivo imaging of neurovascular remodeling after stroke. Stroke December 2012;43(12):3436–41.

[3] Arai K, Jin G, Navaratna D, Lo EH. Brain angiogenesis in developmental and pathological processes: neurovascular injury and angiogenic recovery after stroke. FEBS J September 2009;276(17):4644–52.

[4] Krupinski J, Kaluza J, Kumar P, Kumar S, Wang JM. Role of angiogenesis in patients with cerebral ischemic stroke. Stroke September 1994;25(9):1794–8.

[5] Masamoto K, Takuwa H, Seki C, et al. Microvascular sprouting, extension, and creation of new capillary connections with adaptation of the neighboring astrocytes in adult mouse cortex under chronic hypoxia. J Cereb Blood Flow Metab February 2014;34(2):325–31.

[6] Hayward NM, Yanev P, Haapasalo A, et al. Chronic hyperperfusion and angiogenesis follow subacute hypoperfusion in the thalamus of rats with focal cerebral ischemia. J Cereb Blood Flow Metab April 2011;31(4):1119–32.

[7] Silvestre JS, Smadja DM, Levy BI. Postischemic revascularization: from cellular and molecular mechanisms to clinical applications. Physiol Rev October 2013;93(4):1743–802.

[8] Nishijima Y, Akamatsu Y, Weinstein PR, Liu J. Collaterals: implications in cerebral ischemic diseases and therapeutic interventions. Brain Res October 14, 2015;1623:18–29.

[9] Sealock R, Zhang H, Lucitti JL, Moore SM, Faber JE. Congenic fine-mapping identifies a major causal locus for variation in the native collateral circulation and ischemic injury in brain and lower extremity. Circ Res February 14, 2014;114(4):660–71.

[10] Prabhakaran S, Ruff I, Bernstein RA. Acute stroke intervention: a systematic review. JAMA April 14, 2015;313(14):1451–62.

[11] Greenberg DA. Cerebral angiogenesis: a realistic therapy for ischemic disease? Methods Mol Biol 2014;1135:21–4.

[12] Menon BK, O'Brien B, Bivard A, et al. Assessment of leptomeningeal collaterals using dynamic CT angiography in patients with acute ischemic stroke. J Cereb Blood Flow Metab March 2013;33(3):365–71.

[13] Akamatsu Y, Nishijima Y, Lee CC, et al. Impaired leptomeningeal collateral flow contributes to the poor outcome following experimental stroke in the type-II-diabetic mice. J Neurosci March 2015;35(9):3806–14.

[14] Howangyin KY, Silvestre JS. Diabetes mellitus and ischemic diseases: molecular mechanisms of vascular repair dysfunction. Arterioscler Thromb Vasc Biol June 2014;34(6):1126–35.

[15] Masamoto K, Tomita Y, Toriumi H, et al. Repeated longitudinal in vivo imaging of neuro-glio-vascular unit at the peripheral boundary of ischemia in mouse cerebral cortex. Neuroscience June 14, 2012;212:190–200.

PATHOPHYSIOLOGY

An Overview of Atherosclerosis

K. Rajamani[1], M. Fisher[2]

[1]Wayne State University School of Medicine, Detroit, MI, United States; [2]University of California, Irvine, Irvine, CA, United States

INTRODUCTION

Arteriosclerosis is a generic term used for describing hardening and thickening of arteries. Atherosclerosis is the most common form and is responsible for several common clinical manifestations such as stroke, coronary artery diseases, peripheral arterial disease, and aortic aneurysm. It is a major contributor to worldwide morbidity and mortality and in the Western world is responsible for more than half the annual mortality. It is postulated that the socioeconomic impact of this disease will continue to increase.

High plasma levels of low-density lipoprotein (LDL) provide sufficient plasma apolipoprotein B-containing lipoproteins that can cause clinical atherosclerosis, such as in familial hypercholesterolemia. Often, however, symptomatic atherosclerosis develops at LDL levels within or close to the normal range of the population, but in combination with other risk factors that facilitate atherosclerosis or its clinical complications. These other factors include smoking, high blood pressure, diabetes, male gender, and a complex genetic susceptibility to the disease.

CELLULAR AND MOLECULAR MECHANISMS IN ATHEROSCLEROSIS

Atherosclerosis is a disease in which inflammation has been implicated in every step of the process—from the onset, to the slow progression over years and to the final catastrophic plaque rupture, which typically makes it symptomatic. An array of different cellular players and molecular pathways contribute to different steps in atherogenesis. Endothelial cell dysfunction is a key early step resulting in upregulation of leukocyte adhesion molecules on the surface such as P-selectin, E-selectin, vascular cellular adhesion molecule, and intercellular adhesion molecules. Circulating monocytes start the process of adhesion by upregulating receptors on the endothelial surface. Cytokines such as interleukins (IL-1α and IL-1β) are secreted into the plasma by activated monocytes and macrophages, and play an important role in upregulating leukocyte adhesion molecules on the endothelial cells as well cause proliferation of vascular smooth muscle cells and production of other cytokines. These proinflammatory cytokines, chemokines, and growth factors further elaborate this response in the vessel wall. Other external factors such as hypoxia, reactive oxygen species, and even excess nitric oxide help this process onward.

Monocytes enter the intima by a process of diapedesis. Besides monocytes, various other cells such as neutrophils, T cells, and dendritic cells play important roles in mediating the inflammatory process. Subendothelial lipids are oxidated and are taken up by macrophages to form foam cells. Macrophages and the proinflammatory foam cells are by and large the predominant cell types in the fibrous plaque. As these macrophages and foam cells die, the intracellular contents including lipoproteins and cholesterol crystals are released, leading to formation of the plaque's necrotic core. Smooth muscle cells migrate from the tunica media into the intima and are a prominent and important feature of the atherosclerotic plaque. They exist in two distinct phenotypes, the contractile or quiescent form, and the proliferative or active form. Smooth muscle cells importantly also synthesize the collagen strands, fibrin, and proteoglycans that make the fibrous cap in response to various cytokines.

PATHOLOGY

The fibrous plaque or "atheromatous plaque" is the basic pathologic entity of atherosclerosis. The terms arteriosclerosis and atherosclerosis are sometimes used interchangeably. The nomenclature is derived from the Greek word *"atheros"* meaning gruel, a description of the soft lipid core of the atheromatous plaque. Atherosclerosis characteristically affects the large vessels such as the aorta and carotid arteries. The other less common type of arteriosclerosis is Mönckeberg's medial sclerosis that typically affects the medium-sized arteries.

Fatty streaks are the earliest lesions of the atherosclerotic process and are evident very early in life. They appear as flat, pale, yellow, spotty lesions that stain brightly with Sudan IV. Microscopically, they are composed of subintimal collections of macrophages with exclusively intracellular lipid. Macrophage "foam cells" contain massive amounts of cholesterol esters and are a hallmark of both early and late atherosclerotic lesions. Cholesterol accumulation in these cells is mediated primarily by the uptake of modified forms of LDL via so-called scavenger receptors. Some of these lesions can disappear spontaneously, whereas others can progress to atheromatous plaques. Very young adults have been shown to have not only yellow streaks but also raised lesions and early calcified lesions [1].

The fatty streaks often evolve into fibrous plaques, characterized by intimal hyperplasia and changes within the extracellular matrix. These fibrous plaques grow further and can progress toward several possible outcomes depending on microenvironmental factors, e.g., vessel stenosis (as a result of progressive luminal reduction), plaque fibrosis, or mural thrombosis. Occlusive thrombosis can result due to rupture of the fibrous cap. Intraplaque hemorrhage can result from abnormal angiogenesis arising from the vasa vasorum in the adventia, altered vascular permeability, and increasing plaque volume. Inward sprouting of neovessels from the adventitia toward the plaque surface is important for the development of intraplaque hemorrhage in human atheroma. Lipid mediators initiate this process on the luminal side and subsequent stimulation of vascular smooth muscle cell peroxisome proliferator activated receptor-g induces vascular endothelial growth factor expression, causing further sprouting of adventitial vessels toward the vessel lumen. However, neovascularization is highly immature and susceptible to rupture and leakage [2].

The distribution of atherosclerosis in the vascular tree is quite characteristic. Although the endothelial lining is similar in the whole vascular tree, they are exposed to different hemodynamic conditions in different parts. In the aortocranial circulation, atherosclerosis is frequently seen at the origin of the internal carotid arteries, the carotid siphon, the proximal middle cerebral arteries and the proximal basilar arteries. Shear stress and local rheological factors modulate

not only the presence of these plaques but also progression and clinical manifestations of the disease [1].

There are characteristic racial and gender differences in the distribution of atherosclerotic lesions in patients with stroke. In the Warfarin-Aspirin Symptomatic Intracranial Disease (WASID) trial, patients with basilar artery stenosis were older and more likely to have hyperlipidemia, whereas patients with middle cerebral artery stenosis were more likely to be women and African-American [3]. The locations of atherosclerotic lesions are influenced by variations in risk factor profiles and demographic features. For example, in the United States, Caucasians are more than twice as likely as African-Americans to have atherosclerotic occlusive disease of the extracranial carotid arteries. On the other hand, blacks, males, diabetes, and hypercholesterolemia are particularly associated with symptomatic intracranial atherosclerotic disease affecting the middle cerebral artery, carotid siphon, vertebral, and basilar arteries. Worldwide, the most common cause of stroke is estimated to be from intracranial atherosclerotic lesions, whereas in the United States they account for about 8–10% of all ischemic strokes [4]. The reason for association of intracranial atherosclerotic lesions with certain risk factors is unclear. There may be differences based on vascular location in the interaction between systemic risk factors and local hemodynamic and structural factors. The intracranial vessels have been shown to have paucity of the vasa vasorum, which could explain some of these variations and contribute to a more benign form of atherosclerosis in the intracranial vasculature [5].

Another important local factor may be minor anatomical variations that alter the geometry of the blood vessels [6]. Atherosclerosis is common at bifurcation of blood vessels. The normal laminar blood flow is atheroprotective and is disturbed at these locations, such as at carotid artery bifurcations. Subtle variations in the angle of bifurcation and the relative sizes of the parent common carotid artery compared with the daughter internal carotid and external carotid arteries may play an important role in plaque development and progression. These variations result in demonstrable changes in the local blood flow patterns by increased formation of complex flow recirculation patterns, resulting in low shear stress and prolonged exposure of this endothelial surface to the blood and its lipid contents. Moreover, the proximal portion of the plaques (as opposed to more distal ones) has been demonstrated to have increased markers of proteolysis, apoptosis, and plaque rupture. These are all important events that contribute to conversion of an asymptomatic plaque to a symptomatic one.

The Vulnerable Plaque

Traditionally, the risk of acute ischemic stroke and other clinical events from atherosclerosis has been related to the degree of arterial stenosis. It has now been established that other factors are also at play, because high-risk plaques rupture and result in local thrombosis or may cause distal embolism. Vulnerable plaques are histopathologically characterized by active inflammation, large lipid core, thin fibrous cap, intraplaque hemorrhage, and high degree of neovascularization. When the thickness of the intima increases beyond just 0.5 mm, hypoxia induces growth of the vasa vasorum. These neovessels not only sustain the inflammatory process in the plaques but also could cause intraplaque hemorrhages because of poor structural integrity. Moreover, the vasa vasorum invading lipid-rich plaques are highly permeable due to constant exposure to inflammatory factors and leak red blood cells into plaques leading to plentiful red cell markers in unstable plaques. The characteristic pathology of the vulnerable plaque has been studied in great detail in coronary arteries of individuals with acute coronary syndromes. The most common cause of coronary thrombosis is plaque rupture because of a thin fibrous cap associated with a substantial necrotic core, with rare smooth muscle cells and abundant macrophages [7].

Pathophysiologic Mechanisms of Ischemic Stroke With Atherosclerosis

Atherosclerotic disease of the carotid artery is a well-established risk factor of ischemic stroke. In the coronary arteries, plaque ulceration and rupture with in situ thrombosis is the major pathologic event causing myocardial infarction; the role of these events is less certain in the development of ischemic stroke. Systematic evaluation of pathology of atherosclerotic plaques from the Asymptomatic Carotid Artery Surgery (ACAS) trial and the North American Symptomatic Carotid Endarterectomy Trial (NASCET) showed that symptomatic patients with carotid stenosis were more likely to have ulceration and attached thrombus than asymptomatic patients [8]. This helps us to better understand the events that lead to ischemic infarction in the brain.

Ischemic infarcts in patients with atherosclerosis of the internal carotid artery result from two main mechanisms: (1) artery-to-artery embolism and (2) hemodynamic mechanism or hypoperfusion. Thromboembolic phenomena result from rupture of a vulnerable atherosclerotic plaque. The exposure of the thrombogenic lipid core results in thrombus formation with potential for distal embolism by thrombi or calcific material. Embolic infarction of the brain tends to be cortical-based wedge-shaped infarcts. Hemodynamic insufficiency or hypoperfusion can result from severe stenosis or total occlusion (with poor distal collateral circulation) of a proximal artery, typically resulting in watershed distribution infarctions. Watershed territory is that overlap between the distal ends of a circulatory bed of two nonanastomosing arterial systems.

Treatment of Atherosclerosis

Controlling risk factors by a combination of medications, dietary changes, and lifestyle modifications is the cornerstone of therapy for atherosclerosis. Statin medications (hydroxy methyl glutaryl coenzyme A reductase inhibitors) are widely used for reducing elevated LDL cholesterol levels and preventing vascular events including myocardial infarction and ischemic strokes. Statins are also used in patients who have had prior stroke, coronary artery disease, or peripheral vascular disease to reduce risk of recurrence. Most randomized studies that have shown benefit in reducing stroke recurrence have been in patients with prior coronary heart disease. The stroke prevention by aggressive reduction of cholesterol levels study showed significant benefit of high-dose atorvastatin after stroke and transient ischemic attack (relative risk reduction of 20% after 5 years) [9].

Cholesterol lowering can be achieved by lifestyle modifications, including dietary modifications and exercise, statin medications, and nonstatin medications such as fibrates, niacin, and gemfibrozil. Nonstatin medications have not consistently resulted in reduction of stroke risk. A meta-analysis of the various studies showed that lowering total cholesterol by dietary means resulted in modest benefit (but not statistically significant) in stroke risk reduction. Use of statins was the most effective, with significant relative reductions in stroke risk up to 15%; a 4% reduction in risk of stroke was noted with 10% reduction in LDL cholesterol levels [10]. Besides lowering cholesterol levels, statins have other (pleiotropic) effects, including improving endothelial function, antiinflammation, and plaque stabilization. In the JUPITER (justification for the use of statins in prevention, an intervention trial evaluating rosuvastatin) trial, treatment with rosuvastatin resulted in significant reductions in stroke among asymptomatic patents with elevated C-reactive protein (a marker of systemic inflammation) but without increased cholesterol levels [11]. Smoking cessation, control of diabetes and hypertension, weight reduction, healthy dieting, and exercise are also important contributors to reducing stroke risk from atherosclerosis.

Funding Statement

Supported by NIH NS20989.

References

[1] Glass CK, Witztum JL. Atherosclerosis: the road ahead. Cell 2001;104:503–16.
[2] Michel J-B, Martin-Ventura JL, Nicoletti A, Ho-Tin-Noé B. Pathology of human plaque vulnerability, mechanisms and consequences of intraplaque hemorrhages. Atherosclerosis 2014;234:311–9.
[3] Turan TN, Makki AA, Tsappidi S, Cotsonis G, Lynn MJ, Cloft HJ, et al. Risk factors associated with severity and location of intracranial arterial stenosis. Stroke 2010;41(8):1636–40.

[4] Gorelick PB, Wong KS, Bae HJ, Pandey DK. Large artery intracranial occlusive disease, a large world-wide burden but a relatively neglected frontier. Stroke 2008;39:2396–9.

[5] Labadzhyan A, Csiba L, Narula N, Zhou J, Narula J, Fisher M. Histopathologic evaluation of basilar artery atherosclerosis. J Neurol Sci 2011;307:97–9.

[6] Fisher M. Geometry is destiny for carotid atherosclerotic plaques. Nat Rev Neurol 2012;8:127–9.

[7] Virmani R, Burke AP, Farb A, Kolodgie FD. Pathology of the vulnerable plaque. J Am Coll Cardiol 2006;47:C13–8.

[8] Fisher M, Paganini-Hill A, Martin A, Cosgrove M, Toole JF, Barnett HJ, et al. Carotid plaque pathology, thrombosis, ulceration and stroke pathogenesis. Stroke 2005;36:253–7.

[9] Stroke Prevention With Aggressive Reduction of Cholesterol Levels 'SPARCL' Investigators. High dose atorvastatin after stroke or transient ischemic attack. New Engl J Med 2006;355:549–59.

[10] De Caterina R, Scarano M, Marfisi R, Lucisano G, Palma F, Tatasciore A, et al. Cholesterol-lowering interventions and stroke: insights from a meta-analysis of randomized controlled trials. J Am Coll Cardiol 2010;55:198–211.

[11] Ridker P, Danielson E, Fonseca FA, Genest J, Gotto Jr AM, Kastelein JJ, et al. Rosuvastatin to prevent vascular events in men and women with elevated C-reactive protein. New Engl J Med 2008;359:2195–207.

CHAPTER

21

Thrombosis

D. Nguyen[1], B.M. Coull[2]

[1]University of California San Diego, San Diego, CA, United States; [2]The University of Arizona College of Medicine, Tucson, AZ, United States

INTRODUCTION

Thrombosis resulting from various pathologies is the underlying cause of the majority of ischemic strokes. Hemostasis is the complex and highly regulated system that helps to maintain the integrity of blood and the vasculature. Maintenance of blood flow depends upon an intact and fully functional vascular endothelium. Thrombosis is one of the pathological consequences when the hemostatic system is activated within the circulation itself often as a result of disruption or derangements in the vascular endothelium. Although the concept dates back at least 250 years to the time of the Scottish surgeon John Hunter, the current understanding of how thrombotic events happen increasingly links inflammatory mechanisms with thrombosis [1].

PHYSIOLOGY: HOMEOSTATIC SYSTEM

The hemostasis system functions to limit bleeding when blood vessels are injured or broken. Thrombosis is the formation of a clot within a blood vessel or the heart that results in obstruction of blood flow within the vessel or downstream by embolization. With vascular injury there is a rapid multiphase chain-reaction response initiated by platelet activation and fibrin deposition [2]. The thrombotic reactions begin with a series of initiation events that are quickly followed by amplification and propagation of the clotting process. The initiation step can be triggered by mechanical or inflammatory injury to the vascular endothelium that changes the surface into a prothrombotic state or disrupts the endothelium altogether thereby exposing subendothelial prothrombotic molecules including collagen and tissue factor

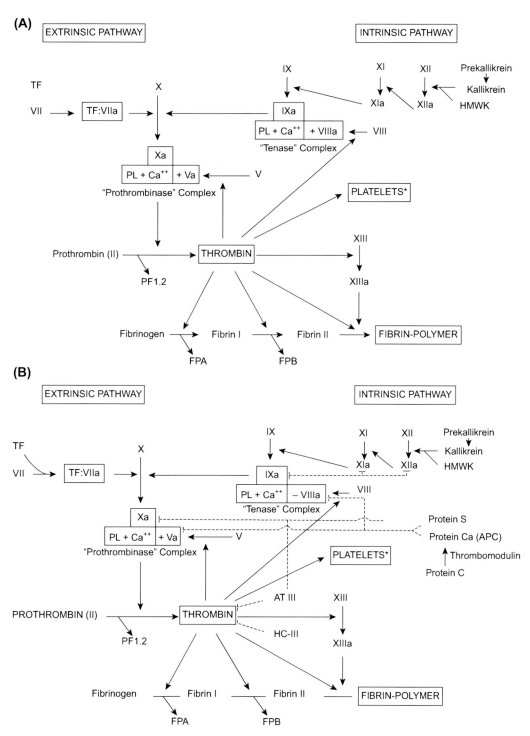

FIGURE 21.1 Schematic diagram of major steps in coagulation. The asterisk (*) indicates that platelets become activated, are a source of phospholipid (PL), and provide receptors for factors V and VIII. *Reprinted with permission by Elsevier [11].*

(TF). These events induce circulating platelets to change shape, express activated surface markers, and adhere to the collagen via von Willebrand factor (vWF) binding. At the location of vascular injury vWF is released from platelets and from stores in endothelial Weibel–Palade bodies. Thrombin is generated in this phase by release from the activated platelets and from the exposure of TF and extrinsic pathway activation as shown in Fig. 21.1.

The amplification phase is instigated by a great increase in the available surface area for the clotting cascade to become active. This process is driven by thrombin generation that further activates surrounding circulating platelets causing them to adhere to the site of injury thereby providing increased membrane surface areas for the clotting cascade assembly. In the propagation phase sometimes described as a traveling wave, thrombin generation is massively increased as a result of a positive feedback effect of thrombin on the activation of coagulation factors V and VIII, principal participants in the "prothrombinase complex" composed of factors Va, Xa, Ca+, and phospholipid that converts prothrombin to thrombin [4]. As outlined in Fig. 21.1 the "contact" or intrinsic pathway plays an important role in both the amplification and especially the propagation phases of thromboses. It bears emphasis that these clotting reactions take place on membrane surfaces that not only include activated platelets but may also include shed cellular microvesicles that have phosphatidylserine exposed on their outer surfaces. Factor XIIIa, which is activated by thrombin, cross-links fibrin and helps to stabilize the clot.

ROLE OF PLATELETS

Platelets contain all the elements needed to initiate the clotting cascade and accordingly they have a major role in maintaining normal hemostasis as well as in the genesis of clinical event such as transient ischemic attack (TIA) and stroke [5]. Platelets remain stable and discoid in shape in the presence of prostacyclin (PGI2) and healthy vascular endothelium. When exposed to subendothelial collagen, high shear stress, thrombin, adenosine diphosphate (ADP), or thromboxane A2, platelets began to change shape and express surface markers that promote adhesion to the vessel wall. This dynamic process is followed by an aggregation reaction. Expressed surface markers on activated platelets include receptors for vWF, ADP, and the IIb/IIIa receptor that avidly binds fibrinogen. Platelet activation also triggers a migration of phosphatidylserine from the inner to the outer leaflet of platelet plasma membrane. This movement is associated with an increase in the activation of prothrombin and factor X, and with the appearance of high-affinity binding sites for factors Va and VIIIa. Migration of anionic phospholipids is also associated with the externalization and shedding of microvesicles containing platelet cytoskeletal proteins and platelet membrane glycoproteins Ib, IIb, IIIa, and IV [3]. Exposure of GPIIb/IIIa receptors on activated platelet surfaces bind fibrin thereby facilitating platelet aggregation. Subsequently, platelets release a group of pro-aggregatory substances from α-granules, including β-thromboglobulin (fl-TG), platelet factor 4 (PF4), vWF, thrombospondin, fibrinogen, and platelet-derived growth factor (PDGF), and release from dense granules (ADP, ATP, serotonin, and Ca^{2+}) which stimulate further platelet aggregation.

ROLE OF VASCULAR ENDOTHELIUM

The vascular endothelium is a distributed organ which in humans has a surface of more than $3000\,m^2$ and in brain approximates the surface area of a tennis court [1, 2]. Vascular endothelium has many important functions including regulation of vascular tone, molecular exchange between blood and tissue compartments, hemostasis and signaling for the immune regulation and inflammation. Brain vascular endothelium is responsible for the blood–brain barrier and is an important regulatory component of the neurovascular unit and in concert with perivascular smooth muscle cells regulates regional perfusion. Many of the functions of the vascular endothelium vary depending upon location within the vasculature and are affected by blood-related shear stress or by subendothelial cellular elements such as components of the neurovascular unit in brain. The intact vascular endothelium expresses or releases a repertoire of molecules and substances that inhibit coagulation or limit clotting once it is triggered by the degradation of thrombin or other activated clotting factors. Some of the principal endothelial-derived coagulation inhibitors as well as prothrombotic molecules are listed in Table 21.1 and include tissue factor pathway inhibitor (TFPI), thrombomodulin, heparin-like proteoglycans, and endothelial

TABLE 21.1   List of Principal Endothelial Derived Coagulation Inhibitors as Well as Prothrombotic Molecules

Anticoagulant	Procoagulant
Thrombin (low concentration)	Thrombin (high concentration)
Release of NO	Release of vWF
Release of PGI2	Release of thrombin
Thrombomodulin	Tissue factor expression
Heparin sulfate	VCAM-1
u-PA and t-PA release	Altered CD39/CD73 expression
C-terminal ADAMTS-18	Endothelin
EPCR	PAI-1
Release of TFPI	ADAMTS-13

ADAMTS-13, A disintegrin and metalloprotease with a thrombospondin type 1 motif member 13; *CD39 and 73*, cluster of differentiation 39 and 73; *C-terminal ADAMTS-18*, C-terminal A disintegrin and metalloprotease with a thrombospondin type 1 motif member 18; *EPCR*, endothelial protein C receptor; *NO*, nitric oxide; *PAI-1*, plasminogen activator inhibitor 1; *PGI2*, prostacycline; *TFPI*, tissue factor pathway inhibitor; *t-PA*, tissue plasminogen activator; *u-PA*, urokinase; *VCAM-1*, vascular cell adhesion molecule 1; *vWF*, von Willebrand factor.

cell protein C receptor. The repertoire of coagulation inhibitors expressed on the healthy vascular endothelium can vary according to different organs in the body or even within the vasculature of an organ such as brain. Anticoagulant substances released by vascular endothelium include prostaglandin I_2 (PG I_2), nitric oxide, and ectonucleotidase CD39/NTPDase1, which hydrolyzes soluble ADP a platelet agonist. In totality the intact healthy vascular endothelium is the most potent regulator of the dynamic hemostatic system. Healthy endothelium releases TFPI into the circulation. Thrombomodulin on the intact endothelial surface binds thrombin and in doing so activates protein C which together with protein S inactivates clotting factors Va and VIIIa. Circulating antithrombin (AT) and heparin cofactor II (HCII) when bound to heparin proteoglycan on the normal vascular endothelial glycocalyx inhibit thrombin and in the case of AT coagulation factors IXa, Xa, XIa, and XIIa. A disintegrin and metalloprotease with a thrombospondin type 1 motif member 13 (ADAMTS-13) that is secreted by hepatic and vascular endothelial cells cleaves large vWF multimers at the endothelial surface and thereby disrupting platelet adhesion.

FIBRINOLYSIS

Fibrinolysis is a process that prevents clots from growing or digests already formed thrombi. Fibrin, which forms the mesh-like substrate of thrombus, is formed when thrombin cleaves fibrinogen molecules. Plasminogen is produced by the liver, circulates in blood, and binds to but cannot degrade the fibrin substrate in thrombi. Precursor single chain urokinase (u-PA) and tissue plasminogen activator (t-PA) are secreted into the circulation in the region of thrombi by activated endothelial cells. Local plasmin converts the u-PA and t-PA to the active form that is able to convert plasminogen within thrombi to plasmin-enabling fibrinolysis. This dynamic process is regulated by substances that block u-PA and t-PA function including plasminogen activator inhibitors (PAI-1 and PAI-2). Plasmin is inactivated in the circulation by alpha 2-antiplasmin and alpha 2-macroglobulin. PAI-1 may have important regulatory nonvascular functions within the brain, and elevated circulating PAI-1 levels have been associated with insulin resistance and are a potential biomarker for increased risk of stroke in some populations [6].

THROMBOTIC MECHANISMS

Specific thrombotic disorders are reviewed elsewhere. Differing patterns of blood flow and related shear stress influence endothelial function and the composition of thrombi formed in various vascular territories reflect these differences. Hemodynamic shear stress is low in the venous system and in a variety of clinical circumstances, including low blood flow, venous stasis, dehydration, congestive heart failure (CHF), inflammatory conditions, cancer, and certain inherited thrombophilias venous thrombi can form. These thrombi tend to be "red clots" containing high amounts of fibrin and large numbers of entrapped erythrocytes [7]. Formation of thrombi in the left atrial appendage in association with atrial fibrillation probably happens by a mechanism similar to venous thrombosis. This pathophysiology also pertains to cerebral vein and sagittal sinus occlusions. In contrast, the arterial circulation is a high-shear flow environment especially at arterial bifurcations, arterial bends, and areas of luminal narrowing and within the smaller arterioles. High hemodynamic shear stress within the arterial circulation is predictive of locations for atherosclerosis, and these areas are a nidus for activation of thrombosis [8]. With disruption of endothelium or rupture of an atheromatous plaque so-called "white thrombi" composed of platelets and fibrin predominate. Such a process could occur within an ulcerated plaque at the bifurcation of the internal carotid artery or a stenotic plaque in the vertebrobasilar system.

Thrombosis within the cerebral small vessels is complex and the mechanistic understanding ischemic white matter changes and small infarctions remains an evolving science [9]. The classic pathological descriptions of arteriolosclerosis, lipohyalinosis, and fibrinoid necrosis of C Miller Fisher from 50 years ago pertain in some but not all examples for cerebral small vessel diseases that produce brain parenchymal changes apparent on magnetic resonance imaging (MRI). An emerging concept is organ-specific regulation of hemostasis [10]. In relative terms the brain has large quantities of tissue factor and very little thrombomodulin and t-PA. Cerebral vascular pericytes and microglia may influence endothelial signaling functions for both hemostasis and inflammation. It is suggested that particularly in the aging brain, the signaling becomes more "thrombogenic" and the BBB and endothelium become leaky thereby allowing various plasma components, including fibrinogen and pro-inflammatory cytokines and related signaling molecules, to enter the perivascular space causing fibrin deposition and a microinflammation that alters physiological regulatory vascular wall functions and produces perivascular infarction even in the absence of complete lumen occlusions [8].

ANTITHROMBOTIC TREATMENT OF ISCHEMIC STROKE

Antiplatelet drugs and anticoagulant medicines are the mainstay of drug treatments for prevention of stroke, whereas fibrinolytics are the principal treatment for most individuals with acute ischemic stroke within the treatment time window. These agents modulate hemostasis by affecting clotting mechanisms in platelets, clotting factors, or for thrombolytics fibrinolysis. The clinical details and guidelines for the use of these drugs are given elsewhere and their mechanism of action is only briefly summarized here.

Antiplatelet agents: Among the antiplatelet agents, acetyl salicylic acid (ASA; aspirin) is very effective in the secondary prevention of cerebral ischemic events. ASA irreversibly inhibits cyclooxygenase (COX)-1 via acetylation of serine 530. This make COX-1 unavailable for arachidonic acid derived from the platelet membrane to produce thromboxane A2, which is an important promoter of platelet aggregation. Clopidogrel is a thienopyridine, class of drug that blocks the $P2Y_{12}$ receptor and thereby inhibits ADP-dependent and thrombin-induced full platelet aggregation response. Dipyridamole has a complex mechanism of action that affects both platelets and the vessel wall via nonreceptor-mediated signaling. Dipyridamole inhibits the activity of adenosine deaminase and phosphodiesterase, which prevents uptake of adenosine into platelets and endothelial cells, leading to the accumulation of cellular and platelet adenosine, adenine nucleotides, and cyclic AMP. This process "stabilizes platelets" and increases PGI_2 release from the vascular wall. Cilostazol is a phosphodiesterase 3 inhibitor that is used clinically in Asia and has a mechanism of action similar to dipyridamole. Triflusal is another antiplatelet agent that is structurally related to ASA. However, it is considered investigational in the United States but available in some European and Latin American countries.

Anticoagulants: Unfractionated heparin is an endogenously produced, collection of glycosaminoglycans which have a variety of anticoagulant, antiinflammatory, and possibly antiangiogenic effects. Both unfractionated heparin and low-molecular-weight (LMW) heparin exert antithrombotic activity by efficiently binding to antithrombin (AT) which converts AT from a slow to a rapid inactivator of thrombin and factor Xa. The enhancement of AT anticoagulant activity by heparins is on the order of 1000- to 4000-fold. Heparins are ineffective in the absence of AT and the vascular glycocalyx. Individuals receiving unfractionated heparin require platelet count monitoring on an approximate 48 h basis.

Vitamin K antagonists such as warfarin are used for primary and secondary stroke prevention in patients with high risk for cardioembolic stroke including individuals with atrial fibrillation or mechanical heart valves. Warfarin inhibits the hepatic carboxylation of factors II, VII, IX, and X and reduces plasma levels of proteins C and S. However, an antithrombotic effect is not achieved until at least 2–3 days of therapy due to the prolonged plasma half-lives of these factors. The possible consequences of insufficient or excessive anticoagulation are extremely serious and often fatal, making it imperative to pursue good control but in some individuals, this can be difficult. A study analyzing the time in therapeutic range (TTR) of INR of 2.0–3.0 showed 59% of the measured INR values were between 2.0 and 3.0, with an overall mean and median TTR of 65 ± 20% and 68% (The median times below and above the therapeutic range were 17% and 10% respectively).

The direct oral anticoagulants (DOACs) are alternative treatments to vitamin K antagonists for primary and secondary stroke prevention in nonvalvular atrial fibrillation. DOACs have also been referred to in the literature as new oral anticoagulants (NOACs) or target-specific oral anticoagulants (TSOACs). Apixaban, edoxaban, and rivaroxaban directly inhibit factor Xa. These DOACs selectively and reversibly block the active site of the molecule factor Xa in both free factor Xa in plasma and factor Xa bounded to the prothrombinase complex. The inhibition of factor Xa leads to a reduction of thrombin generation. Dabigatran inhibits the enzyme thrombin by directly binding to its active binding site. Both the factor Xa inhibitors and direct thrombin inhibitors indirectly block platelet aggregation and reduce thrombus formation. The DOACs have the potential clinical benefit of rapid onset of anticoagulation following initial oral dosing.

Laboratory Testing

The functional integrity of the clotting system can be tested with a series of straightforward clinical tests that are sufficient for most individuals who present with ischemic stroke. The vast majority of individuals with ischemic stroke do not require esoteric testing to detect thrombophilia as such testing is shown to be low yield and wasteful of resources. Routine testing includes a platelet count, and measuring the prothrombin time (PT) and the activated partial thromboplastin time (aPTT). The extrinsic pathway is tested by the PT determination and the intrinsic pathway by the aPTT. Thrombin time (TT) assays measure the conversion of fibrinogen to fibrin and can help detect abnormalities in fibrinogen or the presence of an inhibitor. The fibrinolytic system can be tested with the euglobluin lysis time and measuring D-dimer and fibrin split product levels. A variety of tests are available to test platelet function including the standard bleeding time assay, which is not performed routinely any longer and optical platelet aggregometry. Various bedside platelet function assay devices are available, but they have not gained widespread use. Thrombo-elastography can be used to measure whole blood coagulation, clot integrity, as well as platelet function. This technology is useful in surgical settings that employ cardiac assist devices or the artificial heart.

The DOACs taken at standard therapeutic doses prolong the PT (rivaroxaban and apixaban) and/or the aPTT (dabigatran), but these tests are not reliable for clinical monitoring purposes. Dibigatran can be monitored with the TT, the ecarin clotting time (ECT), or by measuring factor IIa directly. Rivaroxaban and apixaban can be monitored with chromogenic assays for anti-Xa levels. The ECT measures the initial cleavage of prothrombin to meizothrombin an intermediary step in thrombin production that is inhibited by direct thrombin inhibitors. Unfortunately, this test is not available in many clinical laboratories.

Individuals receiving unfractionated heparin who are suspected of experiencing heparin associated thrombocytopenia type 2 should be tested for antibodies against heparin/platelet factor 4 complexes.

References

[1] van Hinsbergh VWM. Endothelium-role in regulation of coagulation and inflammation. Semin Immunopathol 2012;34:93–106.

[2] Yau JW, Teoh H, Verma S. Endothelial control of thrombosis. BMC Cardiovasc Disord 2015;15:130.

[3] Guria K, Guria GT. Spatial aspects of blood coagulation: two decades of the self-sustained traveling wave of thrombin. Thromb Res 2015;135:423–33.

[4] Boilard E, Duchez A-C, Brisson A. The diversity of platelet microparticles. Curr Opin Hematol 2015;22:437–44.

[5] van Roos M-J, Pretorius E. Metabolic syndrome, platelet activation and development of transient ischemic attack or thromboembolic stroke. Thromb Res 2015;135:434–42.

[6] Tjarnlund-Wolf A, Brogren H, Lo EH, Wang X. Plasminogen activator inhibitor-1 and thrombotic cerebrovascular diseases. Stroke 2012;43:2833–9.

[7] Aleman MM, Walton BL, Byrnes JR, Wolberg AS. Fibrinogen and red blood cells in venous thrombosis. Thromb Res 2014;133:S38–40.

[8] Davies PF. Hemodynamic shear stress and the endothelium in cardiovascular pathophysiology. Nat Clin Pract Cardiovasc Med 2009;6:16–26.

[9] Wardlaw JW, Smith C, Dichgans M. Mechanisms underlying sporadic cerebral small vessel disease: insights from neuroimaging. Lancet Neurol 2013;12:483–97.

[10] Fisher MJ. Brain regulation of thrombosis and hemostasis from theory to practice. Stroke 2013;44:3275–85.

[11] Haring H-P, Del Zoppo GJ. In: Welch KMA, Caplan LR, Reis D, Siesjö BK, Weir B, editors. Thrombosis. From Primer on Cerebrovascular Diseases. San Diego: Elsevier/Academic Press; 1997. p. 148–53.

CHAPTER

22

Histopathology of Cerebral Ischemia and Stroke

W.D. Dietrich

University of Miami Miller School of Medicine, Miami, FL, United States

INTRODUCTION

The incidence of focal ischemia or stroke has been increasing over the last two decades with over 16 million new cases of stroke reported worldwide in 2010. Focal ischemic brain injury leads to local neurological deficits and a progression of histopathological changes depending on ischemic severity, location, and duration. In regions of developing infarction, acute neuronal alterations can progress through a series of phases including micro vacuolation, and ischemic and homogenizing cell change

[1]. In other areas adjacent or remote to the evolving focal infarct, a slower progressive neuronal damage termed selective neuronal necrosis, which may have distinct histopathological mechanisms, is also commonly seen in human ischemic tissues. In addition to neuronal changes, alterations in astrocytes, oligodendrocytes, vascular structures, and microglia/macrophage inflammatory responses are important components of the histopathological changes. Because approximately a quarter of all strokes occur in white matter, injured axons and myelin sheaths are commonly observed in pathological specimens [2]. Together, these

Primer on Cerebrovascular Diseases, Second Edition
http://dx.doi.org/10.1016/B978-0-12-803058-5.00022-9

previously described yet complicated pathological responses contribute to the behavioral consequences of the stroke and are important targets for stroke therapies. This review will focus on the histopathological changes that characterize the cellular responses to cerebral infarction and selective neuronal necrosis. For this discussion, events at the tissue and cellular levels resulting from profound alterations in local cerebral blood flow and cerebral metabolism will be emphasized.

CEREBRAL INFARCTION

Focal brain ischemia lasting more than 30–60 min in the majority of cases produces a cerebral infarct. Brain infracts may be a consequence of either embolic or thrombotic processes. Two common sources of embolization to the brain are the heart and internal carotid artery. A thrombotic arterial brain infarct may result from severe atherosclerosis of an artery leading to vascular occlusion and a severe drop in local cerebral blood flow. Common neuropathological characteristics of ischemic infarction, or pannecrosis, include a well-demarcated area of cellular necrosis including neurons, glia, and endothelial cells. This pathology is usually associated with a defined vascular territory of a major cerebral artery or a single vessel. In a fully developed infarct, the gray matter neuron cytoplasm stains intensely with acidophilic dyes (eosinophilic or red neurons) or fails to stain giving the appearance of ghost neurons. In contrast to cytoplasmic changes, the cell nucleus appears dense and pyknotic. The surrounding neuropil adopts a spongy appearance, whereas other cells types such as astrocytes, oligodendrocytes, and endothelial cells may show evidence of early swelling and pale staining due to edema formation. Acute therapeutic interventions including thrombolytic agents or mechanical removal of thrombotic clots are used to restore blood flow and salvage ischemic tissues.

ASTROCYTE AND VASCULAR CHANGES

Cerebral infarction is associated with early changes in astrocytes, oligodendrocytes, microglia, and vascular endothelial cells. Astrocytic changes include early signs of swelling and fragmentation within the evolving infarct. Over time, astrocytes become reactive and go through several patterns of morphological change. Perivascular swelling of astrocytic processes is a common occurrence that might lead in severe cases to cause luminal compression. These structural changes may be associated with the release of inflammatory cytokines, excitatory transmitters such as glutamate leading to a reduced control of ionic homeostasis within the extra cellular space [3]. In chronic stages, reactive astrocytes are observed around infarct borders that may impede axonal growth or cortical circuit plasticity by creating mechanical barriers and producing specific inhibitor factors.

The cerebral vasculature plays an important role in the protection of neuronal cells through regulating levels of local blood flow to brain regions to maintain metabolic needs. In addition to serving as conduits of blood, cerebral microvessels control mechanisms important for cellular inflammation, secondary injury mechanisms, and tissue remodeling [4]. The vascular endothelium helps maintain the blood–brain barrier (BBB), which helps control the exchange of materials between luminal and parenchymal spaces. This barrier is formed by endothelial cell tight junctions and limited endothelial cell pinocytosis. Endothelial cells within the evolving infarct demonstrate light microscopic alterations including swelling and necrosis. During focal ischemia they can lose their cell permeability barrier properties, leading to edema formation and degradation of the vascular basal lamina. As described, microemboli composed of thrombus or other materials originating from remote sites can produce small regions of focal ischemia that also lead to severe endothelial damage and serum protein leakage across the BBB causing perivascular astrocytic swelling and luminal compression. Finally, in situations where the reperfusion of a previously ischemic region occurs, reperfusion injury associated with altered BBB function is observed. The transformation of pale to hemorrhagic infarction may result from reperfusion of an infarct or from the migration of an embolus.

INFLAMMATORY EVENTS

Microglia function as immune cells acting to monitor the microenvironment for changes in signaling, pathogens, and injury [5]. They represent the endogenous inflammatory cells within the central nervous system, exist in different morphological states, and become reactive following periods of cerebral ischemia and stroke [6]. Activation of microglia is an initial event of inflammation in ischemic stroke, which leads to the production of both pro- and antiinflammatory mediators [6]. In response to various signaling molecules, microglia proliferate and migrate to the site of injury. They can progress through various morphological stages characteristic of activation and phagocytosis including ramified, hyperramified, and amoeboid as seen using special stains. Because microglia are an important source of both injurious and reparative products, this "double-edged sword" feature has encouraged research into the search for microglia-targeted therapeutic strategies for both reducing secondary injury processes and promoting reparative processes. The plasticity of inflammatory cells including microglia has been highlighted using M1 and M2 classifications to emphasize their potentially destructive or reparative phenotypes.

Neutrophil infiltration is another early inflammatory event seen after severe cerebral ischemia and is frequently associated with microvessels coursing within the infarcted region. Over time neutrophils can migrate across the vascular endothelial barrier and infiltrate the ischemic tissue. Evidence for intravascular obstruction by polymorphonuclear leukocytes and platelets has been demonstrated [4]. Neutrophils and other white blood cells can also participate in secondary injury mechanisms through the synthesis and release of pro-inflammatory cytokines and oxygen free radicals. Subsequent inflammatory processes including leukocyte adherence and transmigration across the BBB leads to hemodynamic perturbations and destructive secondary injury processes.

WHITE MATTER CHANGES

In addition to the well-described changes that occur in the gray matter after focal ischemia, white matter tracts can also be damaged by a focal ischemic insult producing a loss of axons and other white matter structures [7]. Axonal changes can occur relatively rapidly in white matter regions where severe ischemia is occurring. In other cases, delayed axonal degeneration driven by changes in axon/glial contact and axonal injury metabolism are observed. Researchers have utilized specific strategies for identifying axonal changes including silver staining as well as accumulation of beta amyloid precursor protein. Over time, these axonal changes can lead to altered circuit function and synaptic changes that may participate in the neurological consequences of cerebral ischemia. For these reasons, current research is directed toward determining ways in which to prevent these circuit changes as well as to promote circuit plasticity to promoter improved function in patients with stroke.

Oligodendrocytes provide myelin for CNS axons and conduction fails in myelinated axons with injury to oligodendrocytes or their myelin [2]. Importantly, these myelinating cells have been reported to be highly vulnerable to ischemic insults. Although demyelination can occur after injury, this spontaneous reparative process may not be complete and also result in prolonged conduction deficits. Current research indicates a role for glutamate-mediated oligodendrocyte cell death in white matter through the activation of glutamate receptors on glial cells. Although the pathogenesis of white matter injury is complex, other potentially important strategies for protecting white matter include the use of blockers of the voltage-gated Na^+ channel, the voltage-gated Ca^{2+} channel, or the Na^+–Ca^{2+} exchange.

ISCHEMIC PENUMBRA

Surrounding some infarcted regions are border or transitional zones that also contain populations of irreversibly damaged neurons within an intact neuropil. Currently, there is evidence that an ischemic penumbra exists in animals and humans after the occurrence of focal brain ischemia [8]. In contrast to the infarct central core, glia and endothelial cells are mainly preserved. This margin of the infarct is commonly called the ischemic penumbra and may also show evidence for delayed cell death. Vulnerable neurons within the infract core are generally felt to die mainly by necrosis and associated with robust inflammatory responses. Active areas of research continue to investigate the pathophysiology of the ischemic penumbra for therapeutic interventions. In this regard, repetitive episodes of cortical spreading depolarization have been implicated in the vulnerability of the penumbral tissue. Also, new imaging tools are being developed and tested to determine the occurrence of penumbral regions in individual patients with stroke to clarify salvable tissues for neuroprotective treatments.

SELECTIVE ISCHEMIC CELL DEATH

Selective neuronal damage is characterized by injury to specific populations of neurons that have been identified as vulnerable to periods of cerebral ischemia. In these situations, the neuropil consisting of axons, dendrites, and glial processes is relatively intact by electron microscopic examination. Patterns of selective neuronal vulnerability commonly occur in the CA1 subset of the hippocampus, medium-sized striatal neurons, and cerebellar Purkinje neurons following periods of hypoxia-ischemia. A morphological hallmark of this ischemic cell change is an increased affinity for acid dyes resulting in neuronal acidophilia [1]. Shrunken neural bodies with acidophilia staining and a dark staining pyknotic nucleus are therefore common morphological characteristics of cells dying by selective neuronal damage. Because hematoxylin and eosin staining is generally used to assess neuropathology after cerebral ischemia, the acidophilia is commonly referred to as eosinophilia.

Cellular mechanism underlying selective neuronal cell death may be multifactorial [9]. In contrast to patterns of ischemic cell necrosis, evidence for apoptotic autophagy or other programmed cell death mechanisms has been demonstrated in experimental and clinical conditions related to focal and global ischemia. Apoptotic cells undergo shrinkage, with nucleus collapse and chromatin cleavage resulting in nucleus fragments. In most cases, apoptotic cell death is not generally associated with robust inflammatory responses due to selective death of the apoptotic cells. In these cases, neurons exhibit chromatin concentrated condensation and DNA fragmentation in areas commonly located in the inner border of the infarcted tissue including the penumbra or in other remote brain regions. Neurons in the penumbral regions may also die by apoptosis or other programmed cell death mechanisms. The concept of pyroptotic neuronal cell death has been advanced that is mediated by proinflammatory cell death mechanisms [10]. Pyroptosis, or caspase-1-dependent cell death is triggered by pathological stimuli including stroke and is critical for controlling various pathological stimuli. Data therefore indicate that multiple neuronal cell mechanisms may contribute to the vulnerability of selective neuronal populations and the expansion of an ischemia lesion. Neuroprotective strategies targeting necrotic or programmed cell death are currently being developed and tested.

SUMMARY

The neuropathology of cerebral ischemia and stroke is well described in textbooks and manuscripts. Experimental studies using clinically relevant animal models of focal brain ischemia have replicated many of the histopathological changes observed in human brain tissues after stroke. The magnitude and temporal profile of these cellular changes have been described with the characterization of light microscopic neuronal alterations as well as in other cell types influenced by the ischemic insult. Cellular and molecular mechanisms underlying these vulnerability patterns have now been expanded to include complex cell death mechanism as well as multifactorial secondary injury mechanisms that are emerging targets for therapeutic interventions. The continued clarification of the cellular and molecular mechanisms involved in ischemic cellular vulnerability and damage is important as we attempt to develop and test therapeutic interventions against ischemic brain injury.

References

[1] Auer RN. Histopathology of cerebral ischemia. In: Mohr JP, Choi DW, Grotta JC, Weir B, Wolf PA, editors. Stroke pathophysiology, diagnosis, and management. 4th ed. Churchill Livingstone; 2004. p. 821–8. [Print].

[2] Rosenzweig S, Carmichael ST. The axon-glia unit in white matter stroke: mechanisms of damage and recovery. Brain Res October 14 2015;1623:123–34.

[3] Ding S. Dynamic reactive astrocytes after focal ischemia. Neural Regen Res December 1 2014;9(23):2048–52.

[4] Del Zoppo GJ. Toward the neurovascular unit. A journey in clinical translation: 2012 Thomas Willis Lecture. Stroke January 2013;44(1):263–9.

[5] Ziebell JM, Adelson PD, Lifshitz J. Microglia: dismantling and rebuilding circuits after acute neurological injury. Metab Brain Dis April 2015;30(2):393–400.

[6] Lee Y, Lee SR, Choi SS, Yeo HG, Chang KT, Lee HJ. Therapeutically targeting neuroinflammation and microglia after acute ischemic stroke. Biomed Res Int 2014;2014:297241.

[7] Matute C, Domercq M, Pérez-Samartín A, Ransom BR. Protecting white matter from stroke injury. Stroke April 2013;44(4):1204–11.

[8] Scalzo F, Nour M, Liebeskind DS. Data science of stroke imaging and enlightenment of the penumbra. Front Neurol March 5 2015;6(8).

[9] Cho YS. Perspectives on the therapeutic modulation of an alternative cell death, programmed necrosis. Int J Mol Med June 2014;33(6):1401–6.

[10] Bergsbaken T, Fink SL, Cookson BT. Pyroptosis: host cell death and inflammation. Nat Rev Microbiol Feburary 2009;7(2):99–109.

CHAPTER

23

Histopathology of Intracerebral Hemorrhage

N. Chaudhary, G. Xi

University of Michigan, Ann Arbor, MI, United States

INTRODUCTION

Intracerebral hemorrhage (ICH) is a subtype that is responsible for 10–15% of all strokes with significant morbidity and mortality. An ICH can be primary or secondary [1]. A primary ICH results when a vulnerable small or large blood vessel ruptures without an identifiable reason for structural weakness and blood leaks out under pressure causing considerable damage to surrounding structures. A secondary ICH occurs when an underlying identifiable cause results in parenchymal hemorrhage. The causes include aneurysm, arteriovenous malformation, hemorrhagic transformation of arterial or venous ischemic stroke, trauma, coagulopathy, and hemorrhage from an underlying primary or secondary parenchymal neoplastic lesion. Whatever the cause, based on whether the hemorrhage is from an arterial or a venous pressure head there are consequences of neuronal injury of varying degree.

In the past 20 years there has been considerable technological advancements in improved minimally invasive techniques for the treatment of specific neurovascular disease, however, the mortality rates from ICH at 1 month has remained static at approximately 40% [2]. Treatment of the insult that the brain parenchyma undergoes as a result of a hemorrhage is still by current standards of treatment for ICH largely supportive. These supportive measures include intracranial pressure control, brain edema treatment, and hemodynamic stability maintenance. Unfortunately these treatment strategies, not surprisingly, have not been able to reduce mortality or improve neurological outcome. The human population continues to age around the world and hence it can be extrapolated that the incidence of ICH will continue to increase. Thus it is crucial to improve our understanding of the underlying mechanisms of brain injury to be able to identify more effective therapeutic targets to improve neurological outcome from an ICH. For the purposes of this chapter we will present a detailed analysis of the existing understanding of the mechanisms of injury from ICH, cellular level correlates of severity of neuronal injury, and early and late brain injury.

HISTOPATHOLOGICAL CHANGES

Hematoma Expansion and Mass Effect

The extent of damage is determined by the rapidity of the accumulation of blood and also the size of the hematoma. Both these factors result in mechanical damage to the adjacent brain parenchyma [3]. The hematoma volume and expansion of the hemorrhage are critical factors that determine the extent of neurological injury. Hematoma expansion occurs in about a third of patients with ICH [4]. The rate of expansion of the hematoma also is associated with midline shifts and hence with poor outcomes. Based on these hypotheses of early brain injury resulting in sheer mechanical stress on the brain parenchyma, treatment strategies were developed to prevent hematoma expansion or to evacuate the hematoma to improve neurologic outcome. An ongoing blood pressure lowering trial called "Antihypertensive Treatment of Acute Cerebral Hemorrhage (ATACH II, NCT01176565)" is aimed to test the effects of blood pressure on hematoma enlargement. Early surgical evacuation was also explored in the surgical trial in intracerebral hemorrhage trial but did not show any significant benefit [5]. To test whether using thrombolytic agents may enhance surgical clot removal, minimally invasive surgery plus recombinant tissue-type plasminogen activator (MISTIE) phase III (NCT01827046) is ongoing.

Blood–Brain Barrier Disruption and Brain Edema

Brain edema around the hematoma occurs instantly after hemorrhage. It appears as hypodensity around a

Primer on Cerebrovascular Diseases, Second Edition
http://dx.doi.org/10.1016/B978-0-12-803058-5.00023-0

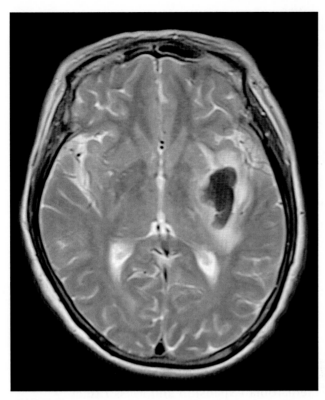

FIGURE 23.1   T2 MRI of a patient with perihematomal edema the first day after intracerebral hemorrhage.

hematoma on computed tomographic scan (Fig. 23.1) and as hyperintensity on T2-weighted or flair magnetic resonance images. Edema develops and peaks several days later, which can cause an increase of intracranial pressure in patients with ICH. Edema formation has an important role in the secondary brain injury following ICH.

There are two major forms of brain edema, vasogenic and cytotoxic. Vasogenic edema results from an increase in permeability of the blood–brain barrier (BBB) and often occurs in white matter. Cytotoxic edema is due to parenchymal cell dysfunction and is most apparent in gray matter. The main form of perihematomal edema is vasogenic. After ICH, the BBB remains intact for several hours. A significant BBB leakage occurs after 8–12 h. Edema development following ICH has been well studied during the past decade. At least three phases of edema are involved in ICH. These include a very early phase (first several hours) involving hydrostatic pressure and clot retraction, a second phase (first 1–2 days) involving the activation of the coagulation cascade and especially thrombin production, and a third phase (several days later) involving red blood cell lysis and iron-induced brain cell toxicity. Activation of the complement system in brain parenchyma also plays an important role in the second and third phases.

Various pathways are involved in edema resolution after ICH. Cerebrospinal fluid is the primary pathway for edema clearance, but brain edema may resolve through the subarachnoid spaces and the vascular system. In addition, uptake of proteins in the extracellular space by neurons and glia also contributes to edema resolution.

Glia Activation and Inflammation

Delving deeper into the evolution of hematoma-mediated injury following early brain injury, there are three different strata of cellular cascades that lead to neuronal damage. Initially there is an immediate inflammatory cascade from blood products in the brain parenchyma. This is mediated by accumulation and activation of brain cells, namely, the microglial cells and astrocytes, which in turn release cytokines and chemokines that then attract the circulating inflammatory cells, namely, the macrophages and T cells. The second wave of cellular cascade is led by complement activation and red blood cell (RBC) lysis, which worsen the existing cell damage. The third wave of cellular cascade is initiated by an ever-increasing volume of cell death that then magnifies the existing inflammatory burden. There are several "danger" signals, which are so-called danger-associated molecular patterns emanating from the apoptotic neuronal cells as a result of the hematoma that reinforces the existing inflammatory cascade manifold via leukocyte induction into the brain.

There are several mechanisms of microglial activation that usually sets in within minutes to hours of the ICH. The RBCs in the hematoma cause phagocytosis of the microglia with expression of proinflammatory markers such as interleukin-1β, tumor necrosis factor-α, inducible nitric oxide synthase, and matrix metalloproteinase-9. On bench side experiments, cultured microglia develop antenna-like extension in response to contact with RBCs and get covered with RBC within hours of exposure along with increased expression of CD36 [6]. Toll-like receptors-4 (TLR4) is expressed in large numbers in microglia stimulated by an ICH. Fibrinogen has been shown to initiate microglial and other downstream cytokine modulation by the TLR4 signaling pathway. Thrombin, another component of the hematoma, amplifies inflammatory ingredients via complement activation by mechanisms that are still not completely understood. Heme, a lysed product of the hemoglobin in RBC also activates microglial cells via TLR4 expression leading to transcription of factor nuclear factor-κB, which in turn upregulates the expression of proinflammatory cytokines.

RBC lysis product–mediated inflammation and cell death ensues at about 24 h from ICH. Hemoglobin has been demonstrated to be toxic in rat ICH model where lysed RBC injected into rat brain resulted in cellular injury. The degradation product of hemoglobin, which is hemin, gets phagocytosed by local microglia and macrophages. Intracellular degradation then involves hemoxygenases, which further metabolize heme to biliverdin and carbon monoxide, releasing cytotoxic iron. When hemoxygenases are inhibited or deleted, the resulting brain injury is reduced in ICH animal models [7]. Similarly deferoxamine also results in reduced ICH-induced brain injury in a rat model of ICH [8]. It has been demonstrated that the inflammatory mediator of the effector pathway of heme in ICH is via the TLR4 receptors. TLRs are a complex transmembrane protein that has three domains (ectodomain, transmembrane complex, and intracellular Toll-interleukin 1 receptor domain). They are the common pathway mediators of innate immunity to foreign pathogens like bacteria and viruses. In the sterile environment of the brain these receptors are active mediators for production of proinflammatory markers activated by several different endogenous ligands including red cell lysis products such as heme, fibrinogen, high-mobility group box 1, heat shock proteins, hyaluronan, oxidized low-density lipoprotein, and amyloid beta [9].

The microglial activation following ICH is initially neuroprotective but the continued stimulation leads to progressive amplification of the inflammatory markers and then leads to apoptotic and necrotic cell death. The temporal variation of microglial activation has been studied in animal models. Typically the peak activity of microglial activation is around 3–7 days correlating with commencement of hematoma resolution (Fig. 23.2) [10]. There are ICH animal models demonstrating peak activity of microglial activation at 3 days and resolution to normal levels by 21 days [6]. The correlation of peak activity of microglial cells with hematoma resolution makes early removal of hematoma in ICH by phagocytosis an enticing therapeutic target.

CD36 is a class B scavenger receptor family and is a key receptor in mediating the phagocytic response of microglial cells. The CD36 expression even in nonphagocytic cells confers phagocytic properties to the cell. Hence CD36 is critical in the phagocytic properties of microglia/macrophages, which can play a significant role in early and effective hematoma reduction by this mechanism. CD36 expression is regulated by an intranuclear receptor super family called peroxisome proliferator-activated receptor (PPARγ) that participates in inflammation. So PPARγ agonist could potentially upregulate CD36 expression, in turn leading to improved phagocytosis by the microglia. PPARγ can favorably upregulate the TLRs, which is linked to CD36 modulation in both inflammation and phagocytosis. In ICH TLR-mediated hematoma resolution mechanisms are not clearly understood but there is evidence of hematoma resolution by day 5 in TLR4 knockout mice suggesting a possible

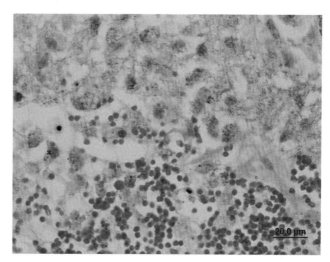

FIGURE 23.2 Erythrophagocytosis by macrophages/microglia. Hematoxylin and eosin–stained section from a pig intracerebral hemorrhage model at day 7. Note the engulfed erythrocytes (*) and the production of hemosiderin (*yellow staining*).

mechanism of upregulation of CD36 by inhibition of the TLR4 signaling pathway. Hence TLR4 pathway modulation is also a potential therapeutic target in ICH.

Brain Cell Death and Brain Atrophy

Cell death that occurs in ICH are necrotic, apoptotic, and autophagic. The apoptotic pathway is complex and has an extrinsic and an intrinsic route leading to cell destruction. The extrinsic pathway employs effectors via death receptors from the tumor necrosis factor family, whereas the intrinsic pathway is activated by cytotoxic stimuli. The common effector pathway from both extrinsic and intrinsic pathways is activation of caspases, which then leads to DNA fragmentation and apoptotic cell death. Necrotic brain cell death is found after ICH. Although there is a good understanding of the apoptotic cell death, there is no clear understanding of the necrotic cell death mechanisms. Necrosis may be caused by mechanical forces during clot formation or chemical toxicity from the hematoma. One key difference in the necrotic cell death mechanism is the early destruction of cell membrane as opposed to apoptotic cell death when the cell membrane integrity is maintained until late stages. Following cell death the cells release a myriad of denatured cellular fragment ligands that can then stimulate inflammatory pathways. Studies also found ICH-induced autophagic cell death.

Brain cell death in the acute phase of ICH results in significant brain tissue loss later. Brain atrophy occurs in patients with ICH (Fig. 23.3). In an aged rat model, ICH caused more than 20% tissue loss in the ipsilateral basal ganglia 2 months later [8]. Brain atrophy with functional outcome has been used for a translational study.

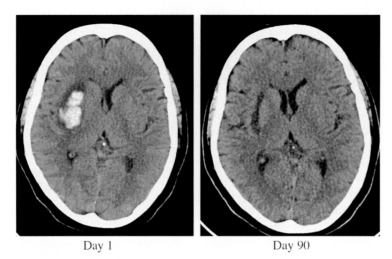

Day 1   Day 90

FIGURE 23.3   Computed tomographic scans showing brain atrophy at day 1 and day 90 after intracerebral hemorrhage.

SUMMARY

There are several phases of ICH-induced cellular injury. The mechanism in the early brain injury phase dictates cell death by sheer mechanical trauma from the volume of the hematoma and the hematoma expansion. The delayed or secondary phase of cellular injury sets in within hours to days mediated by blood components for the clot. The understanding of microglial and astrocytic mechanisms and their role in inducing cell damage has improved significantly. Further studies should focus on clinical translations and apply targeted therapy to improve functional outcome.

References

[1] Mayer SA, Rincon F. Treatment of intracerebral haemorrhage. Lancet Neurol 2005;4:662–72.

[2] van Asch CJ, Luitse MJ, Rinkel GJ, van der Tweel I, Algra A, Klijn CJ. Incidence, case fatality, and functional outcome of intracerebral haemorrhage over time, according to age, sex, and ethnic origin: a systematic review and meta-analysis. Lancet Neurol 2010;9:167–76.

[3] Keep RF, Hua Y, Xi G. Intracerebral haemorrhage: mechanisms of injury and therapeutic targets. Lancet Neurol 2012;11:720–31.

[4] Broderick JP, Brott TG, Tomsick T, Barsan W, Spilker J. Ultra-early evaluation of intracerebral hemorrhage. J Neurosurg 1990;72:195–9.

[5] Mendelow AD, Gregson BA, Fernandes HM, Murray GD, Teasdale GM, Hope DT, et al. Early surgery versus initial conservative treatment in patients with spontaneous supratentorial intracerebral haematomas in the international surgical trial in intracerebral haemorrhage (STICH): a randomised trial. Lancet 2005;365:387–97.

[6] Zhao X, Sun G, Zhang J, Strong R, Song W, Gonzales N, et al. Hematoma resolution as a target for intracerebral hemorrhage treatment: role for peroxisome proliferator-activated receptor gamma in microglia/macrophages. Ann Neurol 2007;61:352–62.

[7] Wang J, Dore S. Heme oxygenase-1 exacerbates early brain injury after intracerebral haemorrhage. Brain 2007;130:1643–52.

[8] Okauchi M, Hua Y, Keep RF, Morgenstern LB, Schallert T, Xi G. Deferoxamine treatment for intracerebral hemorrhage in aged rats: therapeutic time window and optimal duration. Stroke 2010;41:375–82.

[9] Yang QW, Lu FL, Zhou Y, Wang L, Zhong Q, Lin S, et al. HMBG1 mediates ischemia-reperfusion injury by TRIF-adaptor independent toll-like receptor 4 signaling. J Cereb Blood Flow Metab 2011;31:593–605.

[10] Wang J, Rogove AD, Tsirka AE, Tsirka SE. Protective role of tuftsin fragment 1-3 in an animal model of intracerebral hemorrhage. Ann Neurol 2003;54:655–64.

Pathophysiology of Ischemia-Reperfusion Injury and Hemorrhagic Transformation in the Brain

Z. Yu[1], L. Lin[1,2], X. Wang[1]

[1]Harvard Medical School, Boston, MA, United States; [2]Wenzhou Medical University Wenzhou, People's Republic of China

INTRODUCTION

Restoration of blood supply, referred to as "reperfusion," is a desired goal for acute stroke treatment. Spontaneous reperfusion occurs commonly after stroke, in about 50–70% of patients with ischemic stroke. Reperfusion can also be achieved either by thrombolytic therapy using tissue plasminogen activator (tPA) or endovascular therapy, including embolectomy surgery using retrieval devices and thrombus disruption using stents. Since the publication of the first edition of *Primer on Cerebrovascular Diseases* in 1997, there has been considerable progress in the mechanisms study and treatment development for reperfusion injury in stroke. The time window of thrombolysis using tPA has been extended up to 4.5h after stroke onset; meanwhile the embolectomy surgery using stent retrievers has undergone multiple clinical trials in the United States in the past 10 years with significantly beneficial outcomes and is expected to be widely practiced in the near future. However, despite the beneficial effect of oxygen supply brought by reperfusion, rapid reperfusion also has detrimental impact on brain function, the so-called reperfusion injury. This has been documented by both experimental studies using animal stroke models and clinical evidence, such as rat stroke models showing significantly increased infarct volume after reperfusion compared with permanent occlusion. In addition, MRI studies using fluid-attenuated inversion recovery and perfusion-weighted images for human stroke showed that reperfusion could be linked to an early opening of the blood–brain barrier (BBB) and consequently to secondary reperfusion injury and poor outcome [1]. In this chapter, we briefly discuss the pathophysiology and cellular and molecular mechanisms of reperfusion injury and hemorrhagic transformation (HT), and potential therapeutic strategies against these injuries.

CELLULAR AND MOLECULAR MECHANISMS OF ISCHEMIA-REPERFUSION INJURY AND HEMORRHAGIC TRANSFORMATION

To date the deleterious effect of reperfusion in brain function after stroke has been widely recognized and the underlying cellular and molecular mechanisms are being clarified, part of which are learnt from ischemia-reperfusion injuries in other organs such as heart and liver. The mechanisms of reperfusion injury include oxidative stress, leukocyte infiltration, platelet activation, complement activation, and disruption of the BBB, which ultimately lead to edema or HT. HT significantly contributes to the neurological dysfunction and mortality after acute ischemic stroke, and is further worsened by reperfusion caused by either tPA recanalization or embolectomy surgery. In the past few years, the concept of neurovascular unit (NVU) has been widely accepted, in which multiple cell types including endothelial cells, astrocytes, pericytes, oligodendrocytes, microglia, and neurons functionally interact with each other to maintain brain function. As a consequence, in addition to vascular damage, ischemia-reperfusion also causes deleterious effects on these NVU components. Due to the space limit of this chapter, we mainly focus on endovascular damage caused by ischemia-reperfusion.

Oxidative Stress

Oxidative stress results from an imbalance in which the manifestation of reactive oxygen species overwhelms the antioxidant capacity of the cells. The overproduction of reactive oxygen species, mainly peroxides and free radicals, causes damages to all components of the cells, including proteins, DNA, and lipids. Oxidative stress has emerged as one of the mechanisms implicated in the pathogenesis and disease progression of many diseases including stroke. Increased ROS production has been demonstrated in ischemic stroke, both during ischemia and reperfusion [2]. The cerebral ischemia-reperfusion model revealed that oxidative stress mediates BBB dysfunction in mice with superoxide dismutase deficiency. Furthermore, free radical generation and oxidative damage in BBB are main triggers of HT after transient focal cerebral ischemia, which is supported by experimental evidence that free radical scavenger can significantly decrease tPA-induced HT in embolic focal ischemia.

Although a large array of experimental studies have established that oxidative stress is an important mechanism of reperfusion-injury and HT, supportive clinical evidence is scant. A clinical investigation for patients with stroke after tPA thrombolysis showed that the oxidative stress markers including malondialdehyde and myeloperoxidase were already increased at baseline of stroke, whereas no further increases were found for these markers after tPA recanalization, suggesting no relationship between free radical–induced oxidative damage to lipids/proteins and reperfusion injury. However, there exists much limitation in clinical studies; for instance, the peak of oxidative stress might have been missed. The contribution of oxidative stress in reperfusion injury cannot be denied, which warrants further investigation in the future.

Leukocyte Infiltration

Leukocytes play important roles in cerebral reperfusion injury. During reperfusion, activated leukocytes attach to endothelial cells through chemotactic signals, and matrix metalloproteinase and neutrophil-derived oxidants are subsequently produced to break down the BBB. The leukocytes then extravasate from capillaries and infiltrate into brain tissue, releasing proinflammatory cytokines, which eventually result in deterioration of the penumbra [3].

The destructive effect caused by leucocyte infiltration has been validated by numerous animal studies. It was revealed that in rat stroke models neutrophil accumulation at the neuronal injury site occurred earlier and to a greater extent in reperfusion tissue than in tissue with permanent occlusion. In addition, the contribution of leukocyte infiltration in reperfusion injury is also supported by the beneficial effects of neutrophil depletion, in which the animals after transient ischemia showed smaller infarct size when administered with either antineutrophil antiserum or monoclonal antibodies. Furthermore,

leukocyte infiltration is also involved in HT, supported by increased white blood cell count in patients with HT than in those without. The subsequent enhanced leukocyte infiltration may damage microvascular endothelial cells, causing BBB dysfunction and HT.

Leukocyte infiltration is a cascade of processes including leukocyte migration and adhesion to the microvascular endothelial surface, matrix metalloproteinase production for BBB breakdown, leukocyte extravasation into brain tissue, and finally the release of cytokines to brain tissue triggering an inflammatory response. One important regulator of leukocyte adherence to endothelium is endothelial P-selectin, which is upregulated by superoxide free radicals produced during ischemia and reperfusion. P-selectin interacts with its receptor on leukocyte, P-selectin glycoprotein ligand-1, which facilitates the low-affinity "rolling" of leukocytes on the endothelium. Firm adherence of leucocytes to endothelium is mediated by the interaction of leukocyte β_2 integrins CD11a/CD18 and CD11b/CD18 with endothelial intercellular adhesion molecule 1. The subsequent transmigration of leukocytes is regulated by platelet-endothelial cell adhesion molecule-1 along the endothelial cell junction.

Platelet Activation

Increasing evidence supports a role of platelets in the pathogenesis of ischemic-reperfusion injury. Platelets are activated by ischemia-reperfusion and accumulate in vascular beds early after reperfusion. Upon activation, platelets generate reactive oxygen radicals and release proinflammatory factors such as platelet-derived growth factor, arachidonic acid metabolites, thromboxane A2, and platelet factor 4. In addition, activated platelets adhere to microvascular endothelial cells, causing the latter to release mediators for leukocyte chemotaxis and migration, subsequently exacerbating the inflammatory responses [4].

Furthermore, platelets have the potential to modulate leukocyte functional responses, potentially through the interaction between platelet-released fibrinogen and CD11C/CD18. Conversely, activated leukocytes can also alter platelet function. Indeed, accumulating evidence has supported that platelets and neutrophils act synergistically in the pathogenesis of reperfusion injury, largely through cell–cell interactions mediated by P-selectin.

Complement Activation

The complement system as part of the innate immune system comprises a big group of plasma proteins that can be activated by pathogens or other stimuli and subsequently induce inflammatory responses. Previous studies have indicated that the activation of complement system is one of the important mechanisms of reperfusion injury. During reperfusion the complement system can be activated through different pathways, including the antibody-dependent classical pathway, the alternative pathway, or the mannan-binding lectin/mannan-binding lectin-associated serine proteases pathway [5]. As a result, multiple inflammatory mediators will be released, including anaphylatoxins C5a, and the distal complement component C5b-9 [membrane attack complex (MAC)]. C5a can stimulate leucocyte infiltration into damaged tissue, and may also induce the release of proinflammatory factors such as interleukin (IL)-1, IL-6, and tumor necrosis factor-α. MAC forms transmembrane channels that can increase cell membrane permeability, and ultimately disrupt the phospholipid bilayers of cellular membrane, leading to cell lysis and death. Furthermore, MAC plays an essential role in mediating the recruitment of leukocytes to the reperfused tissue, potentially via local induction of IL-8.

Blood–Brain-Barrier Disruption

BBB disruption is a consequence of aforementioned reperfusion-injury mechanisms including oxidative stress, leukocyte infiltration, platelet activation, and complement activation. It has been documented by both animal studies and clinical investigation that BBB disruption occurs during cerebral reperfusion and can lead to vascular edema and HT. Previous studies have reported that BBB disruption occurs after 3 h of reperfusion following transient occlusion, whereas BBB remained intact after 6-h occlusion in a permanent occlusion group. A clinical study involving 144 patients with acute stroke showed that BBB disruption was more common in patients with reperfusion than in those without reperfusion. In addition, in the reperfused group, patients with BBB disruption were more likely to have a poorer clinical outcome than those without disruption [6]. This study established the association of early BBB disruption with HT and poor clinical outcome in humans.

INTERVENTION STRATEGIES FOR ISCHEMIA-REPERFUSION INJURY

With the underlying mechanisms of reperfusion-injury being gradually understood, an increasing number of therapeutic strategies are being developed to limit or rescue ischemia-reperfusion–induced brain injury, targeting different mechanisms of reperfusion injury.

Antioxidant Treatment

Experimental studies have shown that treatments with antioxidants such as angiotensin-converting enzyme inhibitors, iron-chelating compounds, catalase, superoxide dismutase, and vitamin E were able to prevent or attenuate ischemia-reperfusion injury. A clinical trial demonstrated that patients with hemorrhagic shock receiving continuous intravenous injection of recombinant superoxide dismutase for 5 days had significantly improved neurological outcome. However, many human clinical trials of antioxidant therapy to prevent or attenuate ischemia-reperfusion injury have not yielded significant improvements [7]. Thus, despite the large amount of experimental data supporting the role of oxidative stress in ischemia-reperfusion injury, the efficacy of antioxidant therapy is yet to be validated, and further randomized human clinical trials are warranted.

Inhibition of Leukocyte Infiltration

Therapeutic strategies targeting leucocyte-mediated reperfusion injury include inhibition of leucocyte adhesion molecules synthesis, inflammatory factor release, and receptor-mediated leucocyte adhesion to endothelial cells. There exists experimental evidence that this strategy is effective in protecting against reperfusion injury. For example, the inhibitors of leucocyte adhesion molecule synthesis, such as glucocorticoids, gold salts, and D-penicillamine, have been shown to have therapeutic effects against leucocyte-mediated reperfusion injury [8]. In addition, lipoxins are potent inhibitors of leucocyte chemotaxis, adhesion, and transmigration induced by inflammatory factors, suggesting that they are part of innate protective pathways dampening the host inflammatory response. Furthermore, administration of biostable lipoxin analogues attenuated PMN (polymorphonuclear leukocytes)-mediated vascular barrier dysfunction and second-organ injury in several models of ischemia-reperfusion.

Inhibition of Platelet Activation

Platelets can also be a target for therapeutics against ischemia-reperfusion injury. Accumulating studies have shown beneficial effect of platelet depletion in ischemia-reperfusion injury. It has been demonstrated that platelet depletion using filter improved both hepatic and pancreatic function after ischemia-reperfusion injury, probably through reducing lipid peroxidation in cell membrane and the ratio of thromboxane A2 to prostaglandin I2. In addition, antiplatelet agents including dipyridamole and cilostazol improved myocardial function when combined with statin after ischemia-reperfusion [9]. These findings suggest that inhibition of platelet activation might be a potential therapeutic strategy for ischemia-reperfusion injury in the brain as well. However, the potential risk of bleeding caused by antiplatelet agents should be considered for further development of antiplatelet therapy.

Inhibition of Complement Activation

Inhibition of complement activation is also a potential strategy to prevent ischemia-reperfusion injury as demonstrated in multiple experimental models [10]. As an example, C3 convertase is a member of the serine protease family in the complement system. It has been found that an inhibitor of C3 convertase, the soluble complement receptor 1 (CR1), significantly decreased myocardial infarct size and improved myocardial function in a rat model of ischemia-reperfusion. More interestingly, the CR1 short consensus repeats have been shown to protect against cerebral ischemia-reperfusion injury in rats, decreasing cerebral infarct size and improving neurological function. In addition, C5 is another important component of the complement system, which after cleavage can be converted into C5a and C5b-9, two potent inflammatory mediators that increase vascular permeability, leucocyte adhesion and activation, and endothelial activation. Administration of a recombinant antibody against human C5, namely, pexelizumab (Alexion Pharmaceuticals, Inc., Cheshire, CT), has been shown to significantly attenuate complement activation, leucocyte activation, myocardial injury, and acute postoperative mortality in patients undergoing coronary artery bypass grafting surgery. However, to date not many complement inhibitory reagents have been potent enough to enter human clinical trials. More potent complement inhibitory reagents targeting ischemia-reperfusion are yet to be developed in the future.

CONCLUSION

In summary, over the past decade there have been significant advancements in our understanding of molecular and cellular mechanisms of ischemia-reperfusion injury for brain. The major mechanisms of reperfusion injury include oxidative stress, leukocyte infiltration, platelet activation, complement activation, and breakdown of BBB, which ultimately lead to edema or HT. A number of therapeutics studies are ongoing targeting these injury mechanisms, which, however, are still far from achieving clinical success. Further investigations on the mechanisms of reperfusion injury are warranted, which will be helpful for developing effective therapeutics against reperfusion injury in the brain.

References

[1] Warach S, Latour LL. Evidence of reperfusion injury, exacerbated by thrombolytic therapy, in human focal brain ischemia using a novel imaging marker of early blood–brain barrier disruption. Stroke 2004;35:2659–61.

[2] Yamato M, Egashira T, Utsumi H. Application of in vivo ESR spectroscopy to measurement of cerebrovascular ROS generation in stroke. Free Radic Biol Med 2003;35:1619–31.

[3] Pan J, Konstas AA, Bateman B, Ortolano GA, Pile-Spellman J. Reperfusion injury following cerebral ischemia: pathophysiology, MR imaging, and potential therapies. Neuroradiology 2007;49:93–102.

[4] Gawaz M. Role of platelets in coronary thrombosis and reperfusion of ischemic myocardium. Cardiovasc Res 2004;61:498–511.

[5] Collard CD, Lekowski R, Jordan JE, Agah A, Stahl GL. Complement activation following oxidative stress. Mol Immunol 1999;36:941–8.

[6] Latour LL, Kang DW, Ezzeddine MA, Chalela JA, Warach S. Early blood–brain barrier disruption in human focal brain ischemia. Ann Neurol 2004;56:468–77.

[7] Eltzschig HK, Collard CD. Vascular ischaemia and reperfusion injury. Br Med Bull 2004;70:71–86.

[8] Panes J, Perry M, Granger DN. Leukocyte-endothelial cell adhesion: avenues for therapeutic intervention. Br J Pharmacol 1999;126:537–50.

[9] Ye Y, Perez-Polo JR, Birnbaum Y. Protecting against ischemia-reperfusion injury: antiplatelet drugs, statins, and their potential interactions. Ann NY Acad Sci 2010;1207:76–82.

[10] Duehrkop C, Rieben R. Ischemia/reperfusion injury: effect of simultaneous inhibition of plasma cascade systems versus specific complement inhibition. Biochem Pharmacol 2014;88:12–22.

CHAPTER

25

Pathophysiology of Subarachnoid Hemorrhage, Early Brain Injury, and Delayed Cerebral Ischemia

C. Reis[1], W.M. Ho[2], O. Akyol[2], S. Chen[3], R. Applegate II[2], J. Zhang[1,2]

[1]Loma Linda University Medical Center, Loma Linda, CA, United States; [2]Loma Linda University School of Medicine, Loma Linda, CA, United States; [3]Zhejiang University, Hangzhou, Zhejiang, China

INTRODUCTION

Subarachnoid hemorrhage (SAH) is a devastating cerebrovascular disease that occurs after rupture of an intracranial aneurysm, promoting hemorrhage into the subarachnoid space. This leads to impairment of brain perfusion and function, contributing to brain injury after SAH. It has a complex, multisystem, and multifaceted pathogenesis. Intracranial aneurysms may be present in 2–3% of the population with an annual risk of rupture about 0.7–4% [2]. Although SAH counts for only 5% of all strokes, its burden to individuals and society is significant.

The pathophysiological mechanisms of SAH involve early brain injury (EBI) and delayed cerebral ischemia (DCI), including cerebral vasospasm. Immediately after SAH, early brain injury occurs, lasting up to 72 h. Several mechanisms contribute to EBI pathogenesis. These include cell death signaling, inflammatory response, oxidative stress, excitotoxicity, microcirculatory dysfunction, microthrombosis, and cortical spreading depolarization. EBI and DCI are suggested to be linked due to common pathogenic pathways and direct interaction. Both lead to focal neurological and/or cognitive deficits after SAH. Despite advances in experimental research to decrease the effects of SAH, brain injury remains the major cause of death and disability in patients with SAH. There is no sufficient treatment of SAH and its devastating consequences known so far.

ANEURYSM FORMATION

There are three layers in the blood vessel that act to promote integrity and functionality. The outermost layer, tunica externa, comprises connective tissue providing protection for the vessel. The tunica media is composed of smooth muscle cells and elastic tissue that is responsible for autoregulation of cerebral blood flow. Endothelial cells make up the tunica interna, sensing the shear stress due to blood flow over their surface. Neuronal signaling to the surrounding connective tissue and smooth muscle allows the vessel to adapt its diameter according to blood flow.

Aneurysm formation occurs with an initial vascular lesion after interaction of specific biological, physical, and external factors. A tangential force imposed on the vessel wall by blood flow creates aneurysms, or dilations of the vessel wall. This force, called the wall shear stress, along with two other forces, impulse and pressure, encompasses the hemodynamic factors associated with vessel wall degeneration. The endothelium is the first to be damaged by pressure and shear stress from circulating blood. It senses changes in wall stress and adapts the lumen diameter according to the level of wall shear stress to maintain physiology and determine the overall remodeling process. Vascular remodeling influences the progress and growth of the aneurysm. Thus excessive levels of wall shear stress induce focal injury and denude the endothelial barrier leading to intracranial aneurysm formation. Compared with normal intracranial arteries, the aneurysm wall has a thinner tunica media, and lacks organized vascular structure as well as intact internal elastic lamina. As the intracranial aneurysm progresses, the endothelium changes and alters the blood flow. The aneurysm wall subsequently undergoes a process of constant remodeling. With aging and exposure to hemodynamic stress, layers of fibrous tissue develop between the endothelium and the internal elastic lamina, affecting the capacity of the vessel wall to respond [1].

ETIOLOGY OF SAH

There are three morphological forms of cerebral aneurysm related to SAH: saccular, and fusiform aneurysms. The formation and location depends on the cause and type of the aneurysm. The combination of genotype and risk factors determine the morphology and the presentation of aneurysms. Congenital or genetic predisposition may be the initial trigger for the future condition. The most common type of aneurysm, saccular, presents in individuals in their fifth decade of life, resulting in significant morbidity and mortality. Other risk factors associated with SAH include hypertension, smoking, familial history of SAH, alcohol consumption, and drug abuse. Women and adults aged 40–60years are among the highest risk for aneurysm and subsequent SAH. Ehlers–Danlos syndrome, neurofibromatosis type 1, pseudoxanthoma elasticum, and fibromuscular dysplasia are congenital causes of aneurysm formation. These syndromes account for less than 1% of all intracranial aneurysms in the population. Familial intracranial aneurysm represents 7–20% of all cases and rupture at a younger age and smaller size [2].

Inflammation plays a major role in both aneurysm formation and rupture. Proinflammatory activation occurs during vascular remodeling in areas exposed to higher shear stress. This includes activation of endothelial inducible nitric oxide synthetase and downregulation of the endothelial constitutional equivalent (eNOS), induction of matrix metalloproteinases (MMP), such as MMP-2 and MMP-9, and other proinflammatory cytokines [interleukin (IL)-10, IL-1β, IL-6, and tumor necrosis factor (TNF)-α], as well as complement and coagulation cascades. Furthermore, the inflammatory activation of macrophages and T cells contributes to enhanced tissue changes and fibrosis [1]. The extracellular matrix, synthesized mainly by smooth muscle cells, maintains the wall's structural strength and contractile force to maintain cerebral blood flow. This is impaired in aneurysm walls, because of thinned out vascular layers, therefore less functional smooth muscle cells [3]. The complement system is associated with intracranial aneurysms and vascular wall degeneration. Once activated, besides

inducing inflammation by anaphylotoxins, C3a and C5a, it induces activation and degranulation of endothelial cells, mast cells, and macrophages. It therefore indirectly causes smooth muscle cell contraction and capillary permeability [4].

Mechanochemical stress induces atherosclerotic changes in intracranial aneurysms associated with lipid accumulation, the ultimate manifestation of chronic inflammation. In peripheral arteries, early lipid accumulation is accompanied by complement activation and inflammatory cell infiltration. The role of atherosclerosis in the etiology of intracranial aneurysms has been under debate for decades but atherosclerotic modifications are present in over half of intracranial aneurysms.

EARLY BRAIN INJURY

Immediately after aneurysm rupture within minutes of SAH, transient global cerebral ischemia and other pathologies are referred to as EBI. Its symptom onset occurs within the first 2h until 72h after the initial event. The injury after SAH is not restricted to the territory of the ruptured vessel, but includes areas beyond the site of hemorrhage. Signaling cascades are triggered soon after SAH, thereby initiating blood–brain barrier (BBB) disruption, inflammatory responses, cell damage, and oxidative stress.

After aneurysm rupture, the first stimulus is a rapid elevation of intracranial pressure (ICP) reaching approximately the arterial pressure (around 120mmHg) in 1–2min, and then returning to slightly higher than baseline. ICP elevation decreases cerebral perfusion pressure (CPP), consequently reducing cerebral blood flow (CBF), and impairing cerebrovascular autoregulation. A drop in CPP down to almost zero has been reported after SAH, followed by decreased CBF. Although the CPP recovers, a reduced CBF persists indicating acute vasoconstriction within 6h. This causes EBI due to global cerebral ischemia with gradual reperfusion. Early vasoconstriction is reported to be present independently of CPP and ICP changes. Although reduced CPP is not always associated with poor neurological outcome, decreased CBF to less than 40% of baseline in the first hour after SAH predicted 100% mortality [5]. Oxygen and CBF are preserved by the compensatory vascular responses of cerebral autoregulation in a metabolic-, myogenic-, and neurogenic-dependent manner. SAH leads to acute impairment of cerebral autoregulation initiated by hemodynamic mismatch between neurons and vessels, thereby contributing to decreased CBF [5] Fig. 25.1.

Oxidative stress plays an important role in the development of EBI after SAH via production of reactive oxygen species (ROS) radicals. ROS, synthesized early after SAH, includes superoxide anion (O^{2-}), hydroxyl

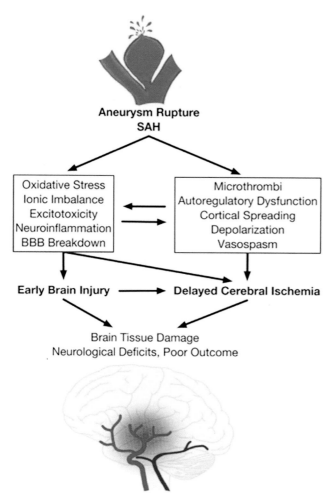

FIGURE 25.1 Overview of pathophysiological mechanism of subarachnoid hemorrhage (SAH). *BBB*, blood–brain barrier.

radical (OH^-), hydrogen peroxide (H_2O_2), NO, and peroxynitrate ($ONOO^-$). After SAH, hypoxia and oxygen shortage disrupts mitochondrial function, and autooxidation of hemoglobin released by lysis of subarachnoid blood occurs. Further contributors to high ROS levels are nitric oxide synthase activity, hypoxic transformation of endothelial xanthine dehydrogenase, lipid peroxidation, and upregulated NADPH oxidase. Mitochondria in damaged cells produce excessive ROS when triggered by elevated levels of Ca^{2+}, Na^+, and ADP. ROS have been observed to damage the neurovascular unit by mediating lipid peroxidation, and protein and DNA breakdown. In addition, ROS lead to inflammation, BBB disruption, and release of vasoconstrictors. Under physiological conditions, a compensatory endogenous antioxidative response is initiated via nuclear factor (erythroid-derived 2)-like 2 (Nrf2) and antioxidant response elements. However, these antioxidants decrease 1h after SAH. This imbalance contributes significantly to EBI development via the major signaling pathways involving the apoptotic cell death

cascade and nuclear factor kappa beta (NF-κB), which is activated by intracellular ROS in upregulating nitric oxide synthase 2.

The main cells contributing to the central nervous system's inflammatory response are microglia and astrocytes. Microglia cells are immunocompetent and phagocytic, whereas astrocytes are able to synthesize and secret factors like cytokines and chemokines after SAH, causing EBI. Other inflammatory mediators are cytokine-secreting leukocytes that are activated by free radicals and extravascular hemoglobin and its lysates. Cytokines strongly associated with EBI are IL-1α, IL-1β, IL-6, IL-8, and TNF-α. IL-1β, IL-6, and Jun N-terminal kinase (JNK)-mediated induction of MMP-9 expression are upregulated after SAH with an early peak at around 6h and a second elevation between 48 and 72h. These are suggested to activate the inflammatory cascade via mitogen-activated protein kinase (MAPK)-extracellular signal-regulated kinase 1/2 (ERK1/2). Two other mediators include high-mobility group box-1, which may have potential in predicting functional outcome after SAH, and NF-κB, a key regulator of inflammatory gene transcription. Chemokines such as monocyte chemotactic protein-1 (MCP-1), chemokine (C–C motif) ligand 5 (CCL5), chemokine (C-X-C motif) ligand 1 (CXCL1), and fibroblast growth factor-2 (FGF2) are able to activate blood-derived inflammatory cells, and their expression has been shown to be upregulated following SAH [5,6].

Adhesion molecules are surface proteins mediating leukocyte–endothelial interaction and inflammation. The number of leukocytes is increased after SAH and is closely related to worse outcome. For instance, activated leukocytes pass through vascular endothelium to brain parenchyma by binding of lymphocyte function associated antigen-1 to intercellular adhesion molecule-1 (ICAM-1). Cell surface proteins detected after SAH include immunoglobulins, integrins, cadherins, and selectins. In the cerebrospinal fluid (CSF) and serum of patients with SAH, the levels of soluble forms of E-selectin, P-selectin, vascular cell adhesion molecule-1 (VCAM-1), and ICAM-1 are increased.

There are three types of cell death mechanisms occurring after SAH, such as apoptosis, necrosis, and autophagy. Apoptosis is a potentially reversible energy-dependent induced cell death used to maintain a homeostasis of a healthy and balanced number of cells. There is a strong association between SAH and apoptotic pathways triggered by neurotoxicity, ischemia, reperfusion injury, and acute vasoconstriction happening in cortical neurons, subcortical neurons, hippocampal neurons, endothelial cells, and vascular cells. The activated intrinsic and extrinsic signaling cascades of apoptosis after SAH are involved in caspase-dependent and caspase-independent death receptor, and mitochondrial response. Related mediators elevated after aneurysm rupture are apoptosis-inducing factor, TNF-α, cytochrome C, P53, and caspases [6]. Neuronal apoptosis has been reported especially in the most exposed and vulnerable region of the basal cortex and hippocampus. A connection to endoplasmic reticulum (ER) stress via C/EBP homologous protein (CHOP), mitogen-activated protein kinases (MAPKs), including ERK1/2, JNK, and p38 seems to be the underlying mechanism [6]. Apoptotic endothelial cell death can lead to BBB disruption, which additionally exposes smooth muscle cells to vasoconstrictor blood components from the subarachnoid space. Necrosis has been evident in the early stage after SAH, triggered by ATP depletion, oxidative stress, and increased intracellular and mitochondrial calcium levels. Autophagy has been detected in neurons within 6h after bleeding in SAH animal models induced by increased cleavage of light chain 3I and beclin 1 expression. Although autophagy has been shown to induce cell death, studies suggest it to be beneficial in maintaining cellular homeostasis and attenuating EBI.

Following SAH, excessive glutamate secretion leads to neurotoxicity and therefore excitotoxicity. Glutamate synthesis happens in activated astrocytes and microglia, especially on the first day of SAH and has an important impact on cell death and EBI. The main underlying pathway downstream of glutamate involves activation of N-Methyl-D-Aspartate receptors causing excessive intracellular Ca^{2+} influx, and subsequently results in apoptosis and necrosis. In addition, Na^+ influx is mediated by ionotropic glutamate receptors causing coincidental cell swelling and edema [7].

Disruption of the homeostasis of ion gradients and ion channels in the brain occurs early after SAH, leading to impaired electrical activity. Impaired voltage-gated K^+ channels, activated neuronal K^+ channels, microthrombosis, and subarachnoid hemoglobin elevate K^+ levels in the serum and CSF. Dysregulated non-L-type Ca^{2+} channels and increased Ca^{2+} release from ER leads to apoptosis of endothelial cells and smooth muscle cells, BBB disruption, and vasoconstriction within 15 min after contact with oxyhemoglobin. Ca^{2+} remains elevated for days after aneurysm rupture. It also impairs muscle relaxation of smooth muscle. A direct correlation was shown in cultured smooth muscle cell between increased intracellular Ca^{2+} concentration and endothelin and oxyhemoglobin, but not bilirubin [5].

Multiple studies with subdural grid electrode strips have proved the existence of cortical spreading depolarization. It is defined by self-propagating waves of neuronal and glial depolarization induced by various noxious compounds. Cortical spreading depolarization leads to EBI and cytotoxic edema through deranged ion gradients and persistent depolarization resulting in ionic imbalance between the intra- and extracellular space, distortion of dendritic structures, neuronal swelling, redistribution of neurotransmitters, slowing of electrical potential, and silencing of electrical activity [6].

The most critical consequence after SAH in terms of EBI is brain edema, and it appears in 6–8% of the patients at the time of hospital admission. Cerebral edema develops quickly after SAH and is considered primarily vasogenic and secondarily cytotoxic-induced edema. Numerous pathophysiological mechanisms cause BBB disruption followed by globalized cerebral edema after SAH [5] (Fig. 25.2).

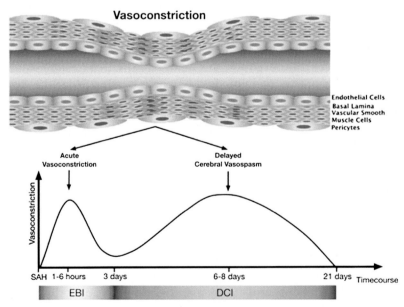

FIGURE 25.2   Time course of subarachnoid hemorrhage–induced vasoconstriction in early brain injury and cerebral vasospasm, demonstrating the points of peak of each one as well as their duration. *DCI*, delayed cerebral ischemia; *EBI*, early brain injury.

DELAYED CEREBRAL ISCHEMIA

For decades, cerebral vasospasm as a consequence of SAH has been considered to be the main and only contributor of DCI.

Cerebral vasospasm has been described as a diffuse and reversible decline in vessel lumen caliber due to arterial smooth muscle constriction after SAH. It is identified by neuroimaging methods such as cerebral angiography, transcranial Doppler, computed tomography (CT), and magnetic resonance imaging. Cerebral vasospasm is detected angiographically in almost 70% of cases, whereas clinically symptomatic vasospasm happens in 30% of patients after aneurysmal SAH. It typically begins 72 h following SAH, peaking at 6–8 days, and terminating 21 days post SAH. The severity of vasospasm is an important clinical risk factor for neurological deterioration and poor outcome.

Vascular dynamics during vasospasm include constriction of the smaller vessels, which induces not only endothelial dysfunction and intramural neuroinflammation, but also smooth muscle constriction of the pial vessels in cerebral vasculature, contributing to symptomatic vasospasm. Endothelin-1 (ET-1) acts on endothelin-A receptors located in vascular smooth muscle cells and causes intracellular Ca^{2+} influx and vasoconstriction by activation of multiple cascades including protein kinase C. Oxyhemoglobin discharge following hemolysis induces ET-1 release, eNOS inhibition, and neuronal apoptosis through stimulating ROS formation, lipid peroxidation, and the MAPK pathway in addition to inhibition of K^+ channel influx. Lowered CSF nitrite levels after SAH have been found during vasospasm. TGF-β, hemoglobin, IL-1, and TNF-α boost ET-1 expression. ET-1 has higher plasma and cerebrospinal fluid levels between days 8 and 14 post hemorrhage with apparent vasospasm.

The (clazosentan to overcome neurological ischemia and infarction occurring after subarachnoid hemorrhage) CONSCIOUS-1 trial demonstrated clazosentan, an ET-1 receptor antagonist can reduce moderate and severe vasospasm after SAH. However, the CONSCIOUS-2 trial failed to show improved outcome, mortality, and vasospasm-related morbidity. Other clinical trials with endothelin receptor antagonists demonstrated that they do not affect neurological outcome, but decrease the incidence of angiographic vasospasm. However, no effect on vasospasm-related cerebral infarction was observed [9].

Vascular remodeling occurs during progression of cerebral vasospasm. SAH induces gene expression of MMP in the arterial wall via p38 MAPK pathways and downstream transcription factors ATF-2 and Elk-1. The balance between tissue inhibitors of metalloproteinases (TIMPs) and MMPs in the pathogenesis of cerebral vasospasm has gained much attention. In the early phase of cerebral vasospasm, upregulation of TIMP-1 supports the restoration

process of the extracellular matrix that lasts during the late phase of cerebral vasospasm. However, there is no direct correlation between MMP-9 and vasospasm [6].

Important molecular changes in endothelial cells are initiated with oxyhemoglobin and bilirubin release in blood. Hemoglobin oxidation products have high affinity for NO and provoke endothelial permeability deterioration and increased levels of both intracellular Ca^{2+} and inositol 1,4,5-triphosphate levels by way of inducing myogenic tone. Bilirubin oxidized products (BOX), detected in the CSF of patients with SAH, are released by oxidation of bilirubin with H_2O_2, superoxide, and NO radicals. CSF levels of BOX are higher in patients with delayed cerebral impulse. Administrating BOX directly to cortical surface vessels in rats produces vasospasm for 30 min up to 24 h and exhibits stress gene induction in subcortical gray matter [4].

Sustained vascular smooth muscle depolarization is one of the pivotal factors during vasospasm. The rise of intracellular Ca^{2+} levels after SAH occurs via voltage-dependent Ca^{2+} channels and neurotransmitter-receptor-operated Ca^{2+} channels. Vascular inflammatory response is related to Ca^{2+} oscillations in cerebral endothelial cells based on ER IP3-receptor-gated Ca^{2+} channels and ER Ca^{2+}-ATPase activity, which also contributes to NF-κB activation, VCAM-1 expression, and vasospasm. Nimodipine, a Ca^{2+} channel blocker, is the only Food and Drug Administration–approved drug in cerebral vasospasm shown to reduce the risk of poor neurological outcome after SAH [10] (Fig. 25.3).

However, the most important triggers for the occurrence of vasospasm and DCI are the amount, density, and prolongation of clot and erythrocyte degradation products in the subarachnoid space. There is a strong association between cerebral vasospasm and the amount of subarachnoid blood demonstrated on CT. Postmortem analysis demonstrated 70% of existing microthrombus in patients with SAH together with smooth muscle proliferation, myofibroblast remodeling, intimal hyperplasia, and collagen deposition in various ratios [8].

Animal SAH models exhibit microthrombi located in both parenchymal arterioles and cerebral cortex from 48 h to 7 days following SAH. The incidence of arteriolar microthrombosis is associated with the strength of vasoconstriction and DCI. Possible explanations for microthrombi formation are due to decreased cortical NO, and increase of P-selectin, fibrinopeptide A, plasminogen activator inhibitor-1, and platelet activated factors such as thromboxane A_2, serotonin, and adenosine diphosphate after SAH. On the other hand, growing evidence proposes that angiographic detection of vasospasm and DCI do not always accompany each other. This inconsistence suggests that cerebral microcirculation mechanisms are diminished by SAH due to microthrombosis and erythrocyte degradation products

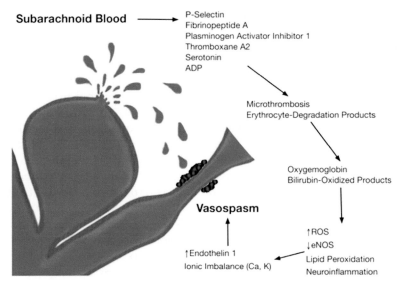

FIGURE 25.3 Overview of subarachnoid hemorrhage mechanisms inducing vasospasm. After subarachnoid hemorrhage a pathophysiological cascade leads to vasospasm. *eNOS*, endothelial nitric oxide synthetase, *ROS*, reactive oxygen species.

precipitating spreading depolarization and inverting the coupling between cortex and cerebral blood flow [9].

Studies have shown that several mechanisms are involved in the pathogenesis of DCI besides cerebral vasospasm. Activated proapoptotic pathways, BBB disruption, microthrombosis, dysfunction in microcirculation, and cortical spreading depolarization play an important role in EBI, yet also contribute significantly to DCI development. Increasing evidence suggest a direct connection between EBI and DCI [7–10].

CONCLUSION

To attain new treatment strategies following SAH for attenuating clinical outcomes after SAH, more knowledge of the molecular correlation between early brain injury, delayed cerebral ischemia, and poor clinical outcome is needed. Although there have been many efforts and achievements in SAH research in the past, SAH pathophysiology remains complex, and many factors are involved. Theories and hypothesis have been established, revised, and renovated over decades. The more we know about the mechanisms of SAH-induced brain injury, the more efficient treatment strategies and targeted drugs can be developed. Understanding all facets of SAH pathophysiology helps also to clarify the challenge and failure in translating experimental SAH results into clinical setting.

References

[1] Tulamo R, Frosen J, Hernesniemi J, Niemela M. Inflammatory changes in the aneurysm wall: a review. J Neurointerv Surg 2010;2(2):120–30.

[2] Frosen J, Tulamo R, Paetau A, et al. Saccular intracranial aneurysm: pathology and mechanisms. Acta Neuropathol 2012;123(6): 773–86.

[3] Frosen J. Smooth muscle cells and the formation, degeneration, and rupture of saccular intracranial aneurysm wall–a review of current pathophysiological knowledge. Transl Stroke Res 2014; 5(3):347–56.

[4] Pradilla G, Chaichana KL, Hoang S, Huang J, Tamargo RJ. Inflammation and cerebral vasospasm after subarachnoid hemorrhage. Neurosurg Clin N Am 2010;21(2):365–79.

[5] Caner B, Hou J, Altay O, Fujii M, Zhang JH. Transition of research focus from vasospasm to early brain injury after subarachnoid hemorrhage. J Neurochem 2012;123(Suppl. 2):12–21.

[6] Chen S, Feng H, Sherchan P, et al. Controversies and evolving new mechanisms in subarachnoid hemorrhage. Prog Neurobiol 2014;115:64–91.

[7] Tso MK, Macdonald RL. Subarachnoid hemorrhage: a review of experimental studies on the microcirculation and the neurovascular unit. Transl Stroke Res 2014;5(2):174–89.

[8] Macdonald RL, Pluta RM, Zhang JH. Cerebral vasospasm after subarachnoid hemorrhage: the emerging revolution. Nat Clin Pract Neurol 2007;3(5):256–63.

[9] Pluta RM, Hansen-Schwartz J, Dreier J, et al. Cerebral vasospasm following subarachnoid hemorrhage: time for a new world of thought. Neurol Res 2009;31(2):151–8.

[10] Wan H, AlHarbi BM, Macdonald RL. Mechanisms, treatment and prevention of cellular injury and death from delayed events after aneurysmal subarachnoid hemorrhage. Expert Opin Pharmacother 2014;15(2):231–43.

CHAPTER

26

Pathophysiology of Ischemic White Matter Injury

S. Mirza, M.P. Goldberg

UT Southwestern Medical Center, Dallas, TX, United States

Cerebral ischemia is a complex injury process that results in damage to both gray matter and white matter. White matter accounts for up to half of the stroke lesion volume in the central nervous system (CNS), with 20% of strokes being purely in the white matter [1–3]. White matter is sensitive to ischemia across all developmental ages: perinatal hypoxia causing periventricular leukomalacia, stroke in adults, and vascular dementia in the elderly. Clinically, ischemic white matter damage brings with it a host of motor, sensory, behavioral, and cognitive consequences due to primary or secondary axonal disruption. Cerebrovascular injury that includes white matter is associated with worse functional outcomes [4]. Magnetic resonance imaging (MRI) shows that the extent of post-stroke disability is related to specific deep white matter damage, and improvement after recanalization therapy is associated with the degree of salvage of these white matter areas [5].

Prior studies of ischemic pathophysiology have focused on injury mechanisms of the neuronal soma and dendrites in gray matter, and have tended to minimize the significance of white matter injury. In part this is due to an inherent deficit of rodent models commonly used to study experimental stroke. About half of the human forebrain is comprised of white matter, while this proportion is only 10–14% in mouse and rat brains [6]. Injury or preservation of cerebral white matter in the rodent would have minimal impact on total infarct volume, a common outcome measure. Compared to gray matter, white matter has lower metabolic rates for glucose and oxygen, and requires relatively less cerebral perfusion; however, this does not provide protection during vascular occlusion. Much of the deep white matter is perfused by direct penetrating arterioles. These areas, lacking collateral supply, are especially vulnerable to focal ischemia (lacunar stroke). Recognition of the prominent role played by the white matter in stroke highlights the importance of identifying distinct molecular mechanisms in white matter ischemia, and provides avenues for new therapeutic exploration.

CELLULAR ANATOMY OF WHITE MATTER

White matter refers to the axonal tracts that interconnect neuronal cell bodies in the brain and spinal cord, and consists of closely packed axons, myelinating oligodendrocytes, astrocytes, and blood vessels. Neuronal cell bodies and dendrites are not present. Cortical projections to and from distant areas such as spinal cord and brain stem form only a small proportion of the CNS white matter tracts. Most of the white matter volume is made up of intracortical connections: short U-fiber bundles between adjacent regions, longer bundles either between contralateral hemispheres (commissural fibers), or different regions (association fibers). CNS white matter varies widely by region in the proportion of myelinated to unmyelinated axons; this variation confers distinct conductance properties and may contribute to regional differences in white matter vulnerability. The relationship between axons and glial cells is rich and complex, both in homeostasis and repair, with multidirectional, harmonized signaling pathways among neurons, glia, and vascular cells [7]. Although there is extensive axon transport of organelles and proteins, the axon does not depend on its neuronal cell body for energy. Axons generate local ATP from metabolic substrates delivered by neighboring glia, and are keenly dependent on glia for function and survival.

Primer on Cerebrovascular Diseases, Second Edition
http://dx.doi.org/10.1016/B978-0-12-803058-5.00026-6

Mechanisms of Injury

It is convenient to consider distinct white matter injury pathways in axons and glial cells, although these pathways occur in parallel and are likely synergistic. The pathophysiology of white matter injury also varies with developmental age. This short review considers established mechanisms involving ionic imbalance on axonal injury, glutamate excitotoxicity on glial cells, and intrinsic axon vulnerability, and briefly examines some other proposed pathophysiological mechanisms.

Sodium and Calcium Accumulation in Axons

Ischemia results in a decline of ATP, which causes a failure of energy-dependent ion transport, mainly Na–K ATPase and Ca-ATPase pumps, resulting in an increase in intracellular Na and extracellular K along their concentration gradients. This triggers membrane depolarization and further opening of voltage-gated channels such as the ATP-dependent K channels or Ca- or Na-dependent K channels. This milieu of elevated extracellular K leads to depolarization, blocked action potentials, uncontrolled neurotransmitter release, and cellular swelling. The shift to anaerobic metabolism results in lactic acid accumulation and potentially toxic reduction in extracellular pH levels.

Calcium-mediated damage is believed to be the "final common pathway" for cellular injury. An increase in intra-axonal Ca can be attributed to three processes: reversal of Na–Ca exchange, influx from voltage-gated Ca channels, and release from intracellular stores. The relative importance of each process differs with age, with experiments suggesting that extracellular calcium entry plays a large role in ischemic damage in young animals, whereas intracellular calcium stores play the pivotal role in this mechanism in older animals leading to possibly differentiated therapeutic targets. The release of intracellular Ca from stores has varied pathways including axonal depolarization, and mitochondria. The end point of increased intracellular Ca with pathological sequelae of vacuolar spaces between axons and myelin and its retraction, swollen mitochondria, and disrupted neurofilaments and tubules is likely to be mediated by auto-digestive enzymes—proteases and lipases—and free radicals.

Glutamate Excitotoxicity

An alternative injury pathway is related to excessive release of the excitatory neurotransmitter glutamate, which may mediate irreversible white matter damage by activating glutamate receptors on oligodendrocytes and their myelin. This represents an extension of the original excitotoxicity hypothesis, which assumed that glutamate-mediated damage was limited to neuronal cell bodies and dendrites. Glutamate is abundant within all cells, and can be released from axons, astrocytes, and oligodendrocytes under conditions of energy depletion.

Intracellular calcium influx is central to the process of oligodendrocyte cell death and myelin destruction due to glutamate receptor activation, setting off a cascade of events including depolarization of the mitochondria, release of free radicals, reactive oxygen species, and pro-apoptotic factors, serving to activate caspases. Acute ischemic injury to the white matter may be mediated by both NMDA (N-methyl-D-aspartate) and AMPA (α-amino-3-hydroxy-5-methyl-4-isoxazolepropionic acid)/kainate classes of ionotropic glutamate receptors [8,9], which are expressed on axons, myelin, and glial cells [10]. In cultured oligodendrocytes, different injury pathways are observed depending on the class of glutamate receptors. Kainate receptor activation leads to apoptosis by activating caspases 3 and 9, whereas AMPA receptor activation stimulates caspase 8, resulting in a different pathway with either necrosis or alternate caspase/polymerase apoptotic end points. In contrast, NMDA receptors appear to be specifically important for Ca signaling within the myelin sheath. Dysregulation of intracellular zinc homeostasis is also implicated in glutamate excitotoxicity.

Terminating glutamate signaling by its uptake in the white matter is crucial to preventing excitotoxicity. Blockade of this uptake, by inhibiting either the glutamate transporter function or its expression in the axons, has been shown to lead to oligodendrocyte loss, resulting in severe demyelination and axonal damage. Hence, dysregulation of this transporter is another mechanism contributing to white matter damage.

Stroke is primarily an age-related disease; hence, it is crucial to note the significant age-dependent findings in white matter excitotoxic damage. Immature and young oligodendrocytes are exceedingly vulnerable to energy depletion, in which it is demonstrated that cell death can be prevented by AMPA/KA receptor antagonists and a Ca–free environment. Blockade of voltage-gated Ca channels or the Na–Ca exchanger is not effective, suggesting that glutamate receptor-mediated Ca entry is sufficient to trigger oligodendrocyte death cascades. These observations highlight the need to focus on age-specific stroke therapies and improved, relevant animal models for research.

Intrinsic Axon Degeneration and Wallerian Degeneration (Slow)

Axons have a unique and intrinsic vulnerability toward self-destruction, which is seen by the spontaneous retrograde and anterograde (Wallerian) degeneration that occurs following axon transection. This distinct, non-apoptotic, auto-destructive axon mechanism is ripe for therapeutic exploration. The discovery of the spontaneous Wallerian degeneration (slow) [Wld(s)] mutation in mice has led to the recognition of possible common mechanisms in axon injury in various conditions such as ischemia, trauma, and other degenerative disorders. This may be relevant to cerebral ischemia, as Wld(s) delays axonal degeneration for hours or days in experimental models of energy depletion [11].

The Wld(s) gene mutation consists of the first 70 amino acids of a ubiquitin factor (Ube4b) and an NAD synthetic enzyme, Nmnat1 (Nicotinamide mononucleotide adenylyltransferase 1). Although normally localized in the nucleus, the Nmnat1 protein and enzyme activity in Wld(s) is aberrantly targeted to axonal components. Wld(s) possibly substitutes for the related NAD synthetic enzyme Nmnat2, which is endogenously expressed in axons but is highly labile and likely disappears shortly after axon transection or injury. Therefore, in Wld(s) axons, prolonged availability of NAD+ or ATP may delay disruption of ionic homeostasis or intra-axonal calcium buffering. Identifying the detailed mechanisms of intrinsic axon degeneration may suggest additional approaches to preserving axons after transient ischemic insults.

Other Injury Pathways

ATP-mediated white matter injury: ATP is released in periods of ischemia and activates purinergic P2X7 receptors on oligodendrocytes, leading to calcium-dependent cell death. Other pathways activated include second messenger systems and enzymes, as well as transcription of nuclear factor NFkB and other such inflammatory gene expressions. ATP-mediated injury may be a target for several potential therapeutic interventions.

Role of oxidative stress in ischemic white matter injury: Ischemia and reperfusion insults result in the formation of free radicals, which contribute to gray and white matter injury through oxidation of lipids, carbohydrates, and proteins of membranes and organelles. Oligodendrocytes and myelin may be especially vulnerable to oxidative injury. Several studies on oxidation targets demonstrate that reduction of free radical damage is associated with a neuroprotective effect (reduced infarct volumes and improved functional outcomes) [3].

Inflammation in ischemic white matter injury: There is a dynamic relationship between the CNS and the immune system after stroke. Following white matter injury, oligodendrocyte precursor cells express MMP-9 (matrix metalloproteinase 9), which results in a disruption of the blood–brain barrier. The influx of innate and adaptive immune cells results in a complex balance of T cells (both CD4 and CD8) enhancing post-ischemic pathology and B cells with potential cytoprotective functions. There may be a cascade of reactive auto-immunity to neuronal antigens such as myelin basic protein, the significance of which is yet to be determined. The role played by inflammatory cells and their cascades remains a vast therapeutic target [7].

Therapeutic Strategies to Protect White Matter From Ischemia

Neuroprotective approaches intended to reduce ischemic damage of neuronal cell bodies alone may not be clinically useful unless there is also preservation of axonal integrity and signal conduction. Table 26.1 provides

TABLE 26.1 Potential Therapeutic Targets for White Matter Protection and Repair

Therapeutic Target	Mechanism of Action
Transmembrane Ion Gradients	
Inhibitors of voltage-gated Na channels	Inhibit intracellular Na influx, limiting reversal of the Na–Ca exchanger
Inhibitors of voltage-gated Ca channels, L-type Ca channels	Inhibit intracellular Ca influx and intracellular Ca release
Inhibitors of the Na–Ca exchanger	Inhibit intracellular Ca influx
Excitotoxicity	
AMPA/KA receptor antagonists	Inhibit glutamate excitotoxicity on oligodendrocytes
NMDA receptor antagonists	Inhibit glutamate excitotoxicity on myelin
Glutamate transporter modulators	Inhibit the Na-Glu transporter, reduce glutamate excitotoxicity in oligodendrocytes
NO synthase inhibitors	Reduce NO synthase in astrocytes
P2X7 receptor blockers	Inhibit toxicity of ATP on microglia and oligodendrocytes
Pannexin-1 blockers	Inhibit ATP release from gap junctions and hemichannels
ATP-degrading enzyme Example: Apyrase	Degrade extracellular ATP, preventing purinergic receptor-mediated toxicity
Oxidative Stress	
Antioxidant pathways	Scavenge free radicals and reactive oxygen species, maintain integrity of membrane phospholipids.
Inflammation	
Anti-inflammatory	Inhibit MMP-9 and BBB leakage, inhibit microglial activation
Axon Degeneration	
NMNAT1 (nicotinamide mononucleotide adenylyltransferase) potentiators	Axonally targeted NMNAT1 inhibits axonal degeneration
P7C3 compounds	P7C3 binds to NAMPT (nicotinamide phosphoribosyltransferase) enzyme, to increase NAD levels
White Matter Repair	
Oligodendrocyte precursor cells (OPC) stimulators Example: Erythropoietin	Stimulate OPCs to proliferate, migrate and differentiate to form mature oligodendrocytes, and restore/repair myelin
OPC apoptosis inhibitors	Inhibit apoptotic pathways in OPCs, increasing oligodendrocyte proliferation
Anti-NOGO antibodies	Block myelin-associated neurite inhibitory factor NOGO, enhance axonal sprouting in white matter
Inosine	Promote axon sprouting

a partial list of potential therapeutic targets for neuroprotection of white matter [3,8]. So far, these interventions remain in the preclinical animal study stage. Future clinical trials of proposed neuroprotective agents should include specific outcome measures, such as MRI diffusion tensor imaging, to assess potential effects on white matter integrity.

Beyond the study of neuroprotection during acute stroke, there are also promising opportunities for improving functional outcomes by preventing delayed axon injury and by repairing damaged white matter tracts. Regeneration of transected axons over long distances (such as corticospinal tracts) does not generally occur in the CNS and may not be a realistic therapeutic goal for disorders such as stroke. In contrast there is extensive spontaneous plasticity of brain connectivity, in development, learning, and after injury. Harnessing therapeutic plasticity to support collateral axon sprouting and functional reconnections offers great promise for functional improvement after stroke. Advances in understanding of glial cell differentiation and development provide approaches to repopulation of oligodendrocytes [12] and replacement of lost myelin after ischemic and other white matter injuries. Understanding the complex and intertwined pathophysiological processes involved in ischemic damage of axons and glial cells is essential for long-term development of effective stroke therapies.

References

[1] Goldberg MP, Ransom BR. New light on white matter. Stroke 2003;34(2):330–2.
[2] Matute C, et al. Protecting white matter from stroke injury. Stroke 2013;44(4):1204–11.
[3] Wang Y, et al. White matter injury in ischemic stroke. Prog Neurobiol 2016;141:45–60.
[4] Wu O, et al. Role of acute lesion topography in initial ischemic stroke severity and long-term functional outcomes. Stroke 2015;46(9):2438–44.
[5] Rosso C, Samson Y. The ischemic penumbra: the location rather than the volume of recovery determines outcome. Curr Opin Neurol 2014;27(1):35–41.
[6] Zhang K, Sejnowski TJ. A universal scaling law between gray matter and white matter of cerebral cortex. Proc Natl Acad Sci USA 2000;97(10):5621–6.
[7] Ortega SB, et al. Stroke induces a rapid adaptive autoimmune response to novel neuronal antigens. Discov Med 2015;19(106):381–92.
[8] Ransom BR, Goldberg MP, Baltan S. 8-Molecular pathophysiology of white matter anoxic-ischemic injury A2-Mohr, J.P. In: Wolf PA, et al., editor. Stroke. 5th ed. Saint Louis: W.B. Saunders; 2011. p. 122–37.
[9] Tekkok SB, Goldberg MP. Ampa/kainate receptor activation mediates hypoxic oligodendrocyte death and axonal injury in cerebral white matter. J Neurosci 2001;21(12):4237–48.
[10] Rosenzweig S, Carmichael ST. The axon–glia unit in white matter stroke: mechanisms of damage and recovery. Brain Res 2015;1623:123–34.
[11] Ransom BR, Waxman SG, Stys PK. Anoxic injury of central myelinated axons: ionic mechanisms and pharmacology. Res Publ Assoc Res Nerv Ment Dis 1993;71:121–51.
[12] McIver SR, et al. Oligodendrocyte degeneration and recovery after focal cerebral ischemia. Neuroscience 2010;169(3):1364–75.

CHAPTER

27

Central Neuroinflammation in Cerebral Ischemia: The Role of Glia

C.M. Stary, L. Li, R.G. Giffard

Stanford University School of Medicine, Stanford, CA, United States

INTRODUCTION

Poststroke neuroinflammation contributes to the delayed phase of neuronal cell death in the penumbra following cerebral ischemia [1]. Therefore antiinflammatory strategies that would seek to inhibit the direct effects of inflammation on neuronal homeostasis have been the focus of numerous neuroprotective strategies, with promising results in animal models of ischemic stroke. However, this approach has failed to translate into any measurable clinical utility, and treatment options for focal ischemic stroke remain largely limited to early reperfusion via thrombolytics or clot retrieval. One explanation for the translational failure of antiinflammatory approaches is that strategies targeting neuronal survival alone may be insufficient to result in any measurable improvement in outcome. Therefore there has been an increased interest in the roles of other cell types in the inflammatory response in the brain following thrombotic stroke. Within minutes of ischemia the cascade of regional neuroinflammatory events that occurs is mediated by the activation of local microglia and astrocytes, specialized glial cells that represent the largest proportion of cells in the mammalian brain. This chapter will discuss the critical role that microglia and astrocytes play in the induction of poststroke neuroinflammation, and their potential as targets in the next generation of therapeutic strategies for thrombotic stroke.

NEUROINFLAMMATORY RESPONSE OF MICROGLIA TO CEREBRAL ISCHEMIA

Microglia share a common myeloid lineage with monocytes and macrophages, and similarly act as the primary form of tissue immune defense. Microglia constitute 10–15% of all cells found within the brain and play an important role in neuronal migration, axonal growth and synaptic remodeling. Microglia are distributed relatively uniformly throughout the brain, although specific microglial populations may exist in different regions, depending on local biochemical features and in response to chemoattractants and inflammatory mediators released by neurons and astrocytes. The primary functions of microglia are: (1) pathogen recognition; (2) phagocytosis of damaged/apoptotic cells, small and inactive synapses, tissue debris, infectious agents, and certain macromolecules; and (3) regulation of T-cell responses and induction of inflammation. Under normal physiological conditions, microglia exist in a resting state, characterized by a ramified morphology with little cytoplasm. Although "resting," these cells are highly dynamic with continuous extension and retraction of processes that actively survey the local microenvironment. Cerebral ischemia induces microglial activation [1], characterized by a change in shape to amoeboid, loss of branching processes and production of lysosomes and phagosomes, and secretion of proinflammatory mediators. Activated amoeboid cells appear abundant in the ischemic core, whereas microglia in the penumbra remain ramified. This active state has been traditionally considered detrimental for neuronal fate, and early drug therapies targeted prevention of microglial activation. However, similar to astrocytes, activated microglia have been observed to exert both injurious and protective effects subsequent to cerebral ischemia [2]. This ambiguity has been partially resolved by the observation that microglia have the ability to transform into a range of activated states with polarization phenotypes, varying between M1 and M2 activation states.

Morphologically, microglia of both M1 and M2 activation states become spherical and retract cell processes; therefore differentiation between the two states is solely

Primer on Cerebrovascular Diseases, Second Edition
http://dx.doi.org/10.1016/B978-0-12-803058-5.00027-8

TABLE 27.1   Markers of M1 and M2 Activation

Activated Microglial Phenotype	Markers			
	Cytokine	Chemokine	Surface Receptor	Metabolic Enzyme
M1	IL-1β, IL-2, IL-6, IL-12, IL-15, IL-17, IL-18, IL-23, IFN-γ, TNF-α	CCL5, CCL8, CCL11, CCL15, CCL19, CCL20, CXCL1, CXCL9, CXCL10, CXCL11, CXCL13	CD16, CD32, CD36, CD68, CD86, MHC-II	iNOS, COX2
M2	IL-4Rα, IL-10, IL-1RA, TGFβ, CD301	CCL14, CCL17, CCL18, CCL22, CCL24, CCL26, CXCL13	CD206, CD163, SR-A1, SR-B1	COX 2, SphK1/2

CCL, chemokine (C-C motif) ligand; *COX2*, cyclooxygenase 2; *CXCL*, chemokine (C-X-C motif) ligand; *IFN-γ*, interferon gamma; *IL*, interleukin; *iNOS*, inducible nitric oxide synthase; *MHC-II*, major histocompatibility complex II; *TGF-β*, transforming growth factor beta; *TNF-α*, tumor necrosis factor alpha.

via antigen expression and cytokine secretion patterns. In M1 or "classical" activation, microglia are characterized by upregulation of proinflammatory surface antigens (Table 27.1). Transformation to the M1 phenotype can be induced in resting microglia by lipopolysaccharides (LPS) or interferon-γ (INF-γ), and it is believed that induction occurs subsequent to chronic inflammation in vivo. Activation of the M2 phenotype by interleukin (IL)-4, IL-10, and/or IL-13 induces phenotypic changes in cell-surface markers that are distinct from those that characterize the M1 activation state. Microglia with M2 phenotype express several distinctive antigens, such as Arg-1 (arginase), heparin-binding lectin Ym-1, CD206, and CD36. The protective role of these cells is thought to be mediated by their high capacity to phagocytize apoptotic and damaged cells (thereby preventing release of their cellular content into the extracellular space) and to phagocytize various potentially injurious molecules that have been already released in the extracellular space. Moreover, the M2 activation state is also characterized by secretion of several antiinflammatory cytokines, including transforming growth factor-β, IL-10, IL-4, IL-13, and insulin-like growth factor (Fig. 27.1). Differentiation into the M2 phenotype can be induced in vivo by pathological events that do not cause abundant necrosis.

Microglial polarization is also mediated by toll-like receptors (TLRs) and complement receptors [3]. TLRs are expressed by myeloid cells in response to activation and can regulate the switch to M1 polarization in two ways: either by inducing nuclear factor kappa-B (NF-κB), which in turn induces proinflammatory cytokines and hypoxia-inducible factor-1α to promote inducible nitric oxide (iNOS) synthesis, or by inducing interferon-regulatory factor-3 (IRF-3) and M2 gene silencing. The induction of IRF-3 by brief activation of TLR4 or TLR9 can also have protective effects. Preconditioning with low doses of either the TLR4 ligand LPS or the TLR9 ligand CpG confers protection against ischemic injury through IRF-3 and IRF-7, an effect mediated by microglia. Complement receptors (including C5aR, C3aR, and C5L2) demonstrate cross

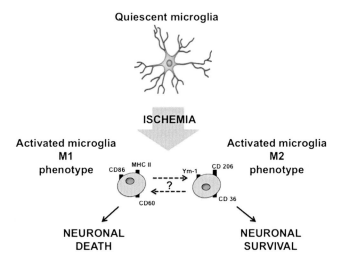

FIGURE 27.1   Microglial activation and polarization states following cerebral ischemia. Activation of microglia varies between M1 and M2 states, which generally have opposing effects on neuronal fate. *MHC*, major histocompatibility complex; *Ym-1*, heparin-binding lectin.

talk with TLRs, resulting in the regulation of innate immune and inflammatory responses in addition to regulation of adaptive immunity. On binding complement receptors, microglia become activated and in turn contribute to the autocrine complement cascade.

Observations suggest the time course of polarization and the relative abundance of the two phenotypes depend on the severity, location, and duration of ischemia and reperfusion [1]. Microglia were previously found to produce proinflammatory cytokines and express M1 markers within the first 24h following the onset of focal ischemia [4]. Subsequently, markers of M2 were observed in the ischemic core at 24h, peaked after 5days, and then declined by 14days after ischemia, whereas the number of cells expressing the M1 phenotype gradually increased after ischemia and peaked after 2weeks [1]. However, the poststroke temporal kinetics of microglial polarization, the mechanisms that determine polarization, the potential for class switching, and even the effect of M1 and M2 activation states on stroke outcome remain elusive and are current areas of active investigation.

NEUROINFLAMMATORY RESPONSE OF ASTROCYTES TO CEREBRAL ISCHEMIA

Astrocytes constitute the most abundant cell type in the mammalian brain and play a critical role in the regulation of function in the normal and injured brain (Fig. 27.2) [5,6]; however, their central role in the neuroinflammatory response to stroke has recently been appreciated. In the uninjured brain, astrocytes maintain functional neuronal homeostasis including scavenging neurotransmitters and metabolic waste following synaptic activity, maintaining ionic and fluid homeostasis, and providing neurotrophic support via release and uptake of trophic factors. Astrocytes contribute to the maintenance of the blood–brain barrier (BBB) as a component of the neurovascular unit (NVU), a dynamic structure also composed of endothelial cells, pericytes, basement membrane, and surrounding neurons.

In response to stroke, astrocytes can undergo morphologic and phenotypic changes that can paradoxically exacerbate injury, termed reactive astrogliosis [7]. Reactive astrocytes are identified by gemistocytic changes, characterized by increases in cytoplasmic mass and branching processes, and increased production of cytoplasmic filaments, most notably glial fibrillary acidic protein (commonly used as an immunohistochemical marker). Reactive astrocytes produce proinflammatory cytokines (IL-1α, IL-1β, and INF-γ), which can contribute to neuronal cell death both directly, by inducing apoptosis, or indirectly, via elevated local production of nitric oxide secondary to activation of iNOS in astrocytes. The cytokine tumor necrosis factor–like weak inducer of apoptosis (TWEAK), a cytokine of the tumor necrosis factor superfamily, and its membrane receptor Fn14, are abundantly expressed in the NVU. Both TWEAK and Fn14 are upregulated following stroke, promoting disruption of the BBB via augmented astrocytic production of matrix metalloproteinases, proteases that contribute to BBB disruption and the development of vasogenic edema. In addition, reactive astrocytes produce chemokines such as monocyte chemoattractant protein-1 [8], resulting in recruitment of circulating immune cells to the site of injury, and these immune cells can contribute to injury. In the acute period following stroke, gap junctions between astrocytes may remain open, allowing proapoptotic factors to spread throughout the syncytium, inducing remote cell death and contributing to the severity of injury. Finally, astrocytes may also inhibit recovery from stroke: activated astrocytes are characterized by hypertrophied processes that form a polarized physical barrier in the form of a glial scar, which both reduces the early spread of injury and inhibits growth of neuronal processes later, by secretion of chondroitin sulfate proteoglycans.

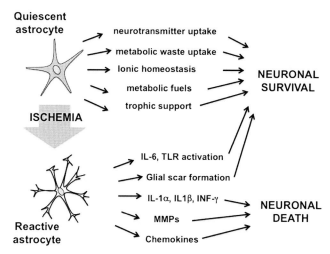

FIGURE 27.2 Astrocytes regulate neuronal homeostasis by a variety of mechanisms. Following cerebral ischemia, astrocytes become activated and may contribute to either neuronal cell death or survival, depending on phenotypic expression patterns. *IL*, interleukin; *INF-γ*, interferon gamma; *MMP*, matrix metalloproteinase; *TLR*, toll-like receptor.

Astrocyte activation prevents propagation of ischemic brain damage, but this may depend in part on the polarization of astrocytes toward specific phenotypes [9]. For example, formation of the glial scar prevents spread of cellular damage and inflammation. Neuroprotective inflammatory mediators are also secreted by activated astrocytes [10]. For example, IL-6, which has both pro- and antiinflammatory properties, can be released from activated astrocytes following transient focal cerebral ischemia. Interestingly, the phenomenon of ischemic preconditioning, whereby a brief, nondamaging period of ischemia results in protection from a subsequent period of more severe, damaging ischemia, depends on induction of TLR signaling. TLRs, originally identified as initiators of innate immunity in response to exogenous microorganisms, have been shown to play a role in the astrocytic response to ischemic injury, whereby ischemic upregulation of TLR4 in astrocytes induces downstream activation of mediators of preconditioning (cyclooxygenase-2, nitric oxide synthase-2) [11]. Astrocyte phenotypes induced by stroke appear to be regulated by surrounding neurons and microglia. Therefore the astrocyte neuroinflammatory response to cerebral ischemia appears to be multifaceted, and broadly antiinflammatory pharmaceutical approaches targeting astrocytes are likely to yield unwanted effects. Elucidation of the mechanisms that induce astrocytes to produce deleterious inflammatory mediators versus protective mediators is an essential next step in the development of a targeted approach to effectively reduce injury and manipulate astrocyte function to improve recovery and regeneration following focal ischemic stroke.

ASTROCYTE–MICROGLIA CROSS TALK FOLLOWING CEREBRAL ISCHEMIA

A key modulator of microglial activation is ATP. Astrocytes are critical regulators of ATP availability to microglia during both normal physiologic function and injury by actively releasing extracellular ATP [7]. Two-photon imaging studies have demonstrated microglial activation following astrocytic release of ATP in response to local injury [7]. Microglia sense levels of extracellular ATP via purinergic receptors, and on activation, microglia can alter the pattern of receptor expression, which in turn determines the microglial functional state. Stimulation of P2Y1 receptors by ATP, released or leaked from injured cells, induces production of proinflammatory cytokines and chemokines by astrocytes via activation of NF-κB [12]. Microglial activation into the ameboid phagocytic phenotype requires a specific receptor expression profile: P2Y12 and A3 receptors cooperate to allow process extension, A2A receptors induce retraction, and P2Y12, P2X4, and A1 receptors interact to induce migration via sensing of the phosphorylation potential (ATP/ADP). They take part in primary damage sensing via chemotaxis along the phosphorylation potential gradient. Early inhibition of ADP sensing by the pharmacologic agent ticagrelor induced protection from ischemia at 48 h and was associated with reduced microglial recruitment to the lesion site, as well as decreased expression of proinflammatory mediators. These observations highlight the importance of astrocytic ATP release in microglial activation and suggest purinergic signaling as an important initial step of the inflammatory response to cerebral ischemia. Elucidation of the relationship between phosphorylation potential sensing and microglial activation, and of the role astrocytes play in M1 and M2 polarization, may provide key insight into future therapeutic strategies for focal cerebral ischemia.

CONCLUSIONS

The inflammatory response to stroke contributes to delayed cell death in the penumbra, and is coordinated by activation of astrocytes and microglia, which demonstrate both independent and integrated responses. Identifying the mechanisms regulating polarization of astrocytes and microglia, investigating the effect of pharmacologically altering microglial polarization, and clarifying the role astrocytes play in microglial activation and polarization are all avenues of investigation that may lead to the development of an effective adjuvant to thrombolysis in stroke therapy.

References

[1] Benakis C, Garcia-Bonilla L, Iadecola C, Anrather J. The role of microglia and myeloid immune cells in acute cerebral ischemia. Front Cell Neurosci 2014;8:461.

[2] Gordon S. Alternative activation of macrophages. Nat Rev Immunol 2003;3:23–35.

[3] Lawrence T, Natoli G. Transcriptional regulation of macrophage polarization: enabling diversity with identity. Nat Rev Immunol 2011;11:750–61.

[4] del Zoppo GJ. Inflammation and the neurovascular unit in the setting of focal cerebral ischemia. Neuroscience 2009;158:972–82.

[5] Nedergaard M, Dirnagl U. Role of glial cells in cerebral ischemia. Glia 2005;50:281–6.

[6] Barreto G, White RE, Ouyang Y, Xu L, Giffard RG. Astrocytes: targets for neuroprotection in stroke. Cent Nerv Syst Agents Med Chem 2011;11:164–73.

[7] Takano T, Oberheim N, Cotrina ML, Nedergaard M. Astrocytes and ischemic injury. Stroke 2009;40:S8–12.

[8] Yepes M. TWEAK and Fn14 in the Neurovascular Unit. Front Immunol 2013;4:367.

[9] Rusnakova V, Honsa P, Dzamba D, Stahlberg A, Kubista M, Anderova M. Heterogeneity of astrocytes: from development to injury - single cell gene expression. PLoS One 2013;8: e69734.

[10] Amantea D, Micieli G, Tassorelli C, Cuartero MI, Ballesteros I, Certo M, et al. Rational modulation of the innate immune system for neuroprotection in ischemic stroke. Front Neurosci 2015;9:147.

[11] Yu S, Wang X, Lei S, Chen X, Liu Y, Zhou Y, et al. Sulfiredoxin-1 protects primary cultured astrocytes from ischemia-induced damage. Neurochem Int 2015;82:19–27.

[12] Kuboyama K, Harada H, Tozaki-Saitoh H, Tsuda M, Ushijima K, Inoue K. Astrocytic P2Y$_1$ receptor is involved in the regulation of cytokine/chemokine transcription and cerebral damage in a rat model of cerebral ischemia. J Cereb Blood Flow Metab 2011;31:1930–41.

CHAPTER

28

Pathophysiology of the Peripheral Immune Response in Acute Ischemic Stroke

J. Anrather

Weill Cornell Medical College, New York, NY, United States

Ischemic stroke triggers an inflammatory response in the affected area, which progresses for days to weeks after the onset of symptoms. The inflammatory reaction involves both tissue resident and peripheral immune cells. Thus the local inflammatory response within the ischemic territory leads to the generation of molecular cues, including cytokines, chemokines and danger-associated molecular patterns (DAMP) setting the stage for the recruitment of peripheral immune cells by weakening the blood–brain barrier (BBB) and by generating chemotactic gradients and activating signals for blood leukocytes. There is evidence that selected aspects of such inflammatory processes contribute to the progression of ischemic brain injury, worsen tissue damage, and exacerbate neurologic deficits. However, research points at a more multifaceted role of immune cells in brain ischemia, where immune cells participate in repair processes during the subacute and chronic stages of brain ischemia. In addition, the interaction of the ischemic brain and the immune system is bidirectional, and although the peripheral immune system is supplying immune cells that participate in the local inflammatory response, neural and humoral signals generated by the ischemic brain modulate the activity of the peripheral immune system leading to immunosuppression and increased risk for nosocomial infections. Understanding these processes and the nature of immune cells involved during deleterious and reparatory phases of postischemic inflammation will be necessary to devise effective therapeutic strategies for human stroke [1]. This chapter will briefly discuss the role of peripheral immune cells participating in the inflammatory response to cerebral ischemia, deliberate potential entry points used by these cells to gain access to the ischemic tissue, and outline the role of the ischemic brain in shaping the peripheral immune response.

CEREBRAL ISCHEMIA AND INFLAMMATION

Infiltration of hematogenous cells into the ischemic territory that persists for days and even weeks after the initial insult is the hallmark of the inflammatory reaction, which parallels activation of brain microglia, astrocytes, and endothelial cells. Cytokines and chemokines are important molecular cues in the inflammatory response to cerebral ischemia. The inflammatory reaction is believed to start at the time of intravascular occlusion due to altered rheology resulting in activation of the endothelium by shear stress and by activation of the coagulation cascade, which leads to the surface expression of adhesion molecules on endothelial cells, the generation of the proinflammatory proteases of the coagulation cascade (thrombin, tissue factor), and release of cytokines and lipid mediators from activated platelets [interleukin (IL)-1 and eicosanoids]. At the same time, cell injury in the ischemic territory generates DAMP including purinergic molecules and high mobility group box 1 protein that signal locally to activate brain resident immune cells but, helped by a deteriorating BBB, might also be released into the blood stream where they are sensed by circulating immune cells or can reach distant lymphoid organs such as spleen and lymph nodes, where they activate local immune cells (Fig. 28.1).

IMMUNE CELLS PARTICIPATING IN ISCHEMIC INJURY AND TISSUE REPAIR

Postischemic inflammation is characterized by activation of brain resident leukocytes, microglia, astroglia, and vascular cells, and the orchestrated recruitment of various blood-borne immune cells which will be discussed in the following text.

Primer on Cerebrovascular Diseases, Second Edition
http://dx.doi.org/10.1016/B978-0-12-803058-5.00028-X

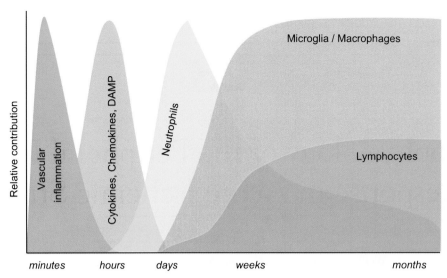

FIGURE 28.1 Temporal profile of the immune response after focal cerebral ischemia in mice. Blood stasis in the affected territory causes shear stress-induced endothelial cell activation resulting in surface expression of adhesion molecules of the immunoglobulin superfamily (vascular cell adhesion molecule 1 (VCAM-1), intercellular adhesion molecule 1 (ICAM-1)) and selectins (P- and E-selectin) resulting in recruitment of leukocytes. At the same time, activation of the coagulation cascade occurs generating proinflammatory proteases (tissue factor and thrombin among others) and release of proinflammatory cytokines and lipid mediators from activated platelets. Signals produced by vascular inflammation propagate to receptive immune cells in the brain parenchyma facilitated by a deteriorating blood–brain barrier. Concurrently, danger-associated molecular patterns (DAMP) released by injured neurons and glia contribute to the inflammatory response that cumulates in the production of chemokines and activator cytokines by microglia, astrocytes, and endothelial cells in the ischemic territory. Leukocytes start infiltrating the ischemic tissue along these chemotactic gradients with neutrophils preceding blood-borne macrophages, whereas lymphocytes are generally observed at later stages of the injury. Microglia are activated early after ischemia and might have overlapping activities with blood-borne macrophages including phagocytosis of dead cells (neurons, neutrophils) and functional polarization into inflammation-resolving macrophages.

Neutrophils

Although intravascular adhesion of neutrophils is a relatively early postischemic event, parenchymal accumulation is generally observed later. Nevertheless, neutrophils are among the first hematogenous immune cells found in the brain after experimental stroke. Although it is not clear whether they enter the brain parenchyma under all circumstances, there is evidence that neutrophils contribute to postischemic inflammation by limiting tissue perfusion due to intravascular clogging, destabilizing the BBB by releasing matrix-metalloproteinases, and by generating reactive oxygen (ROS) and nitrogen (RNS) species. However, a cause-and-effect relationship between the extent of neutrophil trafficking and the severity of ischemic damage has not been firmly established. Attesting to the complex role of neutrophils in cerebral ischemia, "protective" neutrophils, which have undergone functional polarization and express protein markers commonly found on alternatively activated (M2) macrophages, can also be found in the ischemic territory [2].

Microglia, Monocytes/Macrophages, and Dendritic Cells

Historically it has been difficult to separate the relative contribution of resident (microglia) and blood (monocyte)-derived macrophages to ischemic brain injury. Because these cells are not readily distinguishable by morphology or marker gene expression, most results derived from in vivo studies could either implicate microglia or macrophages or both. Advances in multilabel flow cytometry and the availability of bone marrow chimeric animals that express discernable markers (fluorescent proteins or CD45 alloantigen) have allowed to discern microglia and blood-derived myeloid cells. Microglia are activated early after ischemia before extensive neuronal cell death occurs. This early response, characterized by increased exploratory behavior, gives way to macrophage-like transformation within the first 24 h after stroke. Although proinflammatory and cytotoxic activities of microglia have been demonstrated in vitro, eliminating proliferating microglia/macrophages in vivo increased ischemic brain injury and reduced expression of the neurotrophic cytokine insulin-like growth factor 1. In line with their multifaceted role in inflammation, microglia can undergo functional polarization according to the M1/M2 paradigm to acquire a proinflammatory or antiinflammatory/reparatory phenotype, respectively [3]. Similar to microglia, monocyte-derived macrophages can exhibit various phenotypes. Blood monocytes exist as two functionally distinct subpopulations (inflammatory and patrolling) in mice and humans and the relative abundance of these subsets in the blood has been linked to clinical stroke outcome. Initially, infiltrating monocytes are of the "inflammatory" subtype, whereas "patrolling" monocytes are prevalent at later time points. Similar to microglia, monocyte-derived macrophages can undergo functional polarization into M2 macrophages potentially contributing to the resolution of inflammation and tissue repair [4]. However, the precise role and dynamics of monocyte-derived macrophages in postischemic inflammation remains to be defined.

Lymphocytes

T cells are detrimental in the early phase of ischemia and lymphocyte-deficient mice are protected in models of focal ischemia. The mechanism does not involve classical antigen-mediated T-cell activation and the cytotoxic activity might be tied to innate T-cell functions. Accordingly, IL-17 secreting γδT cells, which do not undergo antigen-dependent T-cell activation, have been shown to contribute to ischemic injury. Similarly, natural killer (NK) cells, another innate lymphocyte population, contribute to early ischemic brain injury in an interferon-γ- and perforin-dependent manner. Although effector lymphocytes may contribute to focal ischemic injury, regulatory T cells (Treg) can have a protective effect by downregulating postischemic inflammation. Treg appear in the ischemic tissue after the acute phase and confer neuroprotection by IL-10 secretion. Likewise, regulatory B cells (Breg) confer neuroprotection through an IL-10-dependent mechanism, but do not enter the ischemic brain [5].

IMMUNE CELL ENTRY INTO THE ISCHEMIC BRAIN

The forebrain with exception of the olfactory bulb is covered by a continuous layer of astrocyte cell bodies in rodents or by cytoplasmic processes of marginal astrocytes in primates known as the glia limitans superficialis. The astrocytes are in close contact with the basal lamina attached to the pia mater, the innermost leptomeningeal membrane. This structure forms a tight barrier to the subarachnoid space and limits the exchange of solutes and cells between the brain parenchyma and the cerebrospinal fluid (CSF). The glia limitans extends along penetrating blood vessels and together with pericytes, the basement membrane and interendothelial tight junctions forms the BBB of the parenchymal capillary bed that renders the blood vessel of the cerebral vasculature impermeable to macromolecules and circulating cells. In the course of inflammation these barriers are altered and leukocyte penetrate into the CNS via multiple routes including (1) blood-to-parenchyma through a weakened BBB, (2) blood-to-CSF via the choroid plexus, (3) blood-to-CSF via leptomeningeal vessel, and (4) CSF-to-parenchyma across the glia limitans [6].

Parenchymal Entry

In the course of cerebral ischemia the BBB undergoes several changes that are reversible at first but are more permanent at later time points. The first opening is due to alterations of transendothelial vesicular transport and loss of endothelial tight junctions with concomitant transient downregulation or redistribution of junction proteins such as occludins, claudin, and zonula occludens-1. Although the impairment of the BBB caused by endothelial tight-junction dysfunction occurs within hours after cerebral ischemia, a second opening of the BBB occurs in later phases of the ischemic insult and is characterized by loss of vascular cells, including endothelial cells and pericytes, degradation of the basal membrane, and retraction of astrocytic end feet from the gliovascular unit. It is during this phase that substantial infiltration of peripheral immune cells is observed in the ischemic territory. Within the reperfused territory leukocytes adhere to the activated endothelium of arterioles and postcapillary venules via endothelial upregulation of adhesion molecules (Fig. 28.1). This interaction does not result in leukocyte extravasation and endothelial adherence is a relatively short-lived event that, depending on the experimental model, reverts back to preischemia levels within hours after reperfusion. The reasons for this remain elusive, but the existence of a double barrier—endothelial and glial basement membrane—could be a contributing factor. As a result, blood-borne immune cells, specifically neutrophils, are often found in the ischemic territory forming cuffs that surround the vessel, without entering the neuropil. Still, perivascular immune cells may contribute to tissue injury by releasing diffusible neurotoxic factors, ROS, RNS, and extracellular proteases in particular. Overall, although immune cell entry from the vasculature within the ischemic parenchyma is likely, hard evidence for this route of entry is still missing.

Choroid Plexus

The choroid plexus is a highly vascularized structure that attaches to the ependymal cells lining all four cerebral ventricles and separates the ventricles from the subarachnoid space. The endothelium of the choroid plexus is fenestrated allowing solutes and intravascular cells to cross the endothelial cell layer, whereas the barrier function is upheld by the choroid epithelial cells that are interconnected by continuous tight junctions. The stroma of the choroid plexus contains a sizable number of resident macrophages and dendritic cells that express major histocompatibility complex (MHC) class II molecules and may present antigens to T cells entering the CSF. Under homeostasis the choroid plexus is the entry site for patrolling lymphocytes, mostly CD4$^+$ central memory T cells. To facilitate leukocyte trafficking, the choroid plexus epithelium expresses constitutively intercellular adhesion molecule 1 (ICAM-1) and vascular cell adhesion molecule 1 (VCAM-1), which together with mucosal vascular addressin cell adhesion molecule-1, are upregulated under inflammatory conditions. The choroid plexus epithelial cells also constitutively express the chemokine CCL20 (chemokine (C-C motif) ligand) that acts on the (C-C chemokine receptor) CCR6 receptor on the surface of IL-17-secreting lymphocytes to promote their CSF entry. Given that IL-17 γδT cells have been implicated in stroke pathophysiology, the selective recruitment of IL-17$^+$ T cells across the choroid plexus might be relevant for the outcome of ischemic brain injury.

Meninges

The brain and the spinal cord are enclosed by three membranes: the fibrous dura mater; the highly vascularized arachnoid, which with its trabecular structure spans the subarachnoid space filled with CSF; and the pia mater, which covers the brain parenchyma and is in close contact with astrocytes forming the glia limitans at the surface of the brain. Cerebral blood vessels are embedded within the arachnoid trabeculae before they enter the brain parenchyma. Similarly to the choroid plexus, the meninges are populated by bone marrow-derived perivascular myeloid cells that express the microglia/macrophage marker IbaI and MHC class II molecules, whereas a subpopulation of meningeal macrophages expresses the dendritic cell marker CD11c. Little is known about the physiological role of meningeal macrophages but given their proximity to blood vessels and their possible interaction with CSF lymphocytes, it has been hypothesized that these cells play a role in the CNS immune surveillance. The other constitutive meningeal immune cell population identified in humans and rodents are mast cells primarily located in the dura mater. Because mast cells contain granules with vasoactive molecules and proteases, they have been implicated in BBB disruption and in promoting neutrophil extravasation in the course of cerebral ischemia. Several studies have addressed the role of leptomeningeal vessel as a source of blood-borne immune cells after stroke. Consistent with a meningeal origin, neutrophils are found on the abluminal site of leptomeningeal vessel within hours after stroke in permanent and transient ischemia models in rodents. A strong association of neutrophils with leptomeningeal vessel has also been observed in tissue samples from human stroke victims. Whether neutrophils that extravasated to the subarachnoid space go on to infiltrate the ischemic territory remains to be established, but the fact that accumulation in the meninges precedes the appearance of neutrophils in the brain parenchyma supports such a scenario.

ALTERATIONS OF THE PERIPHERAL IMMUNE SYSTEM AFTER STROKE

Although peripheral immune cells contribute to ischemic brain injury, the ischemic brain in turn has a profound effect on the composition and behavior of the peripheral immune system. This effect is mediated by the systemic release of DAMP molecules from the ischemic territory, humoral signals generated by hypothalamic stress response centers, and neural signals of the autonomous nervous system. The response is characterized by an early state of hyperinflammation, followed by a phase of immunosuppression with increased susceptibility to infection (Fig. 28.2). Finally, increased immune cell access to the injured brain and leakage of normally "hidden" CNS antigens to the periphery evokes an adaptive immune response that can contribute to the long-term outcome after stroke. Stroke severity is a main determiner for these effects and many of the changes to the peripheral immune system discussed later are only observed in stroke models that result in large ischemic injuries, a correlation also observed in human stroke. In addition, stroke-induced changes in the peripheral immune system show lateralization and the net immunomodulatory autonomic output after ischemia might depend on brain structures damaged.

Hyperinflammation: The Early Response to Ischemic Brain Injury

In experimental stroke, the earliest peripheral immune response to ischemic brain injury is characterized by elevated serum cytokine levels (IL-6, interferon-γ, (chemokine (C-X-C motif) ligand) CXCL1) and increased production of inflammatory mediators in circulating and splenic immune cells [tumor necrosis factor (TNF), IL-6, IL-2, CCL2, and CXCL2] within hours after ischemia [7]. The response is generally transient and most parameters return to baseline levels 24h after stroke. Comparable changes can be observed in human patients with stroke. TNF and IL-6 are increased in patients at stroke onset (<24h) and IL-6 serum levels are positively correlated with stroke severity and unfavorable outcome.

Stroke-Induced Immunodeficiency Syndrome

The early activation of the immune system is superseded by a state of systemic immunosuppression that predisposes to poststroke infections. Accordingly, complications from pulmonary or urinary tract infections have been observed in ~20% of patients with stroke [8]. Studies on the immune status of patients with stroke found prolonged peripheral lymphopenia and reduced T-cell responsiveness. Ischemic brain injury leads to sustained decrease in blood and splenic B, T, and NK

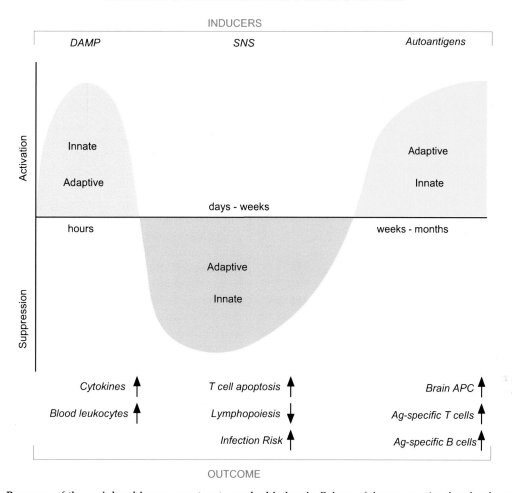

FIGURE 28.2 **Response of the peripheral immune system to cerebral ischemia.** Release of danger-associated molecular patterns (DAMP) from the ischemic territory into systemic circulation activates the innate immune system leading to increase in circulating inflammatory cytokines and circulating myeloid cells, particularly neutrophils. Some adaptive immune functions such as T-cell reactivity to mitogens are also increased. Activation of the sympathetic nervous system (SNS) is pivotal for the development of stroke-induced immunodeficiency syndrome, which predisposes patients after stroke to infections, by inducing T-cell apoptosis in secondary lymphoid organs and reducing lymphopoiesis in the bone marrow by skewing hematopoietic stem cells to the myeloid lineage. Some activities of innate immune cells, such as bacterial phagocytosis, are also suppressed. Autoantigens released from the ischemic territory and processed by antigen-presenting cells (APCs) can lead to expansion of reactive B and T cells in regional lymph nodes or directly in the affected brain region. This adaptive response might be inflammatory and deleterious (Th1/Th17) or tolerogenic and potentially beneficial (Th2).

cells in mice undergoing transient middle cerebral artery occlusion. These changes are observed 12h after ischemia and persist for several weeks. The decrease in lymphocytes is correlated with increased splenocyte apoptosis, spleen atrophy, and Treg expansion. The sympathetic nervous system is fundamentally involved in this response and inhibition of adrenergic signaling by the β-adrenergic receptor antagonist propranolol is sufficient to lower bacteremia and bacterial colonization of the lungs and significantly increases survival rates along with preservation of splenic and circulating lymphocyte populations in mice. Whether the splenic response to cerebral ischemia is induced by similar mechanisms in all stroke models and in humans remains to be determined. Studies in rats concluded that the loss of splenic lymphocytes was not due to increased apoptosis but due to activation of α-adrenergic receptors on trabecular and capsular smooth muscle cells that leads to spleen contraction and expulsion of immune cells into the circulation. Although spleen size was decreased 1 day after stroke, spleen volume was restored 3 days thereafter, arguing against long-lasting effects of cerebral ischemia on spleen physiology. As in rats, splenic size loss in humans is transient and there is a tendency of increased splenic volume 4 days after stroke. The importance of these species differences for the development of stroke-induced immunodeficiency syndrome has yet to be elucidated. The immunosuppressive effects of ischemic brain injury are not limited to the spleen. In the bone marrow, tyrosine hydroxylase and norepinephrine levels increase 1 day after transient middle cerebral artery occlusion in mice. This triggers a response in mesenchymal stromal cells, probably through activation of β3-adrenergic receptors, resulting in the reduction of homeostatic and cell retention factors IL-7, Cxcl12 (SDF-1), VCAM-1, stem cell factor, and angiopoietin-1. Downregulation of these factors increases hematopoietic stem cell proliferation. This proliferative response, however, does not profit all arms of blood cell lineages equally. The hematopoietic system becomes skewed toward the myeloid lineage, whereas the lymphoid lineage is suppressed [9]. Collectively the data suggest that the brain is a strong regulator of the peripheral immune system by regulating development and homeostasis of splenic and BM immune cell populations and that increased sympathetic output after stroke is the main efferent branch responsible for these effects.

Late-Phase Adaptive Immune Response

Disruption of the BBB during acute stroke releases novel CNS antigens that are normally sequestered in the brain and exposes them to the systemic immune system. Meningeal lymphatic vessels, which run along the venous sinus and terminate in the deep cervical lymph nodes, are likely to be involved in the cranial export of antigenic macromolecules and antigen-presenting cells (APCs) after ischemic brain injury. Evidence for antigen-specific T-cell reactivity has been found in animal models of stroke. There is an ever increasing list of CNS-derived peptides that can induce a peripheral T-cell response after stroke, including, but not limited to, peptides derived from myelin basic protein (MBP), neuron-specific enolase, proteolipid protein, N-methyl-D-aspartate receptor 2A, and microtubule-associated protein. Peptide-reactive B and T cells can be found in cervical lymph nodes and spleen as early as 4 days after transient focal ischemia in mice and T- and B-cell-rich infiltrates, which resemble tertiary lymphoid organs, are found in chronic stages of experimental stroke. In human stroke, APCs loaded with CNS-derived peptides have been found in T-cell zones of cervical lymph nodes and palate tonsils. Association studies of this small patient cohort indicated that increased reactivity to neuronal-derived antigens was correlated with smaller infarct size and better long-term outcome, whereas greater reactivity to MBP was correlated with larger infarcts and worse outcome. This dichotomy might be an indication that the adaptive immune response to ischemic brain injury can be skewed toward reactive (Th1/Th17) or tolerogenic (Th2) phenotypes. This interpretation is supported by studies in rodent stroke models that indicate a beneficial effect of a Th2 immune response on stroke outcome. Future studies will have to address whether a tolerogenic immune response is linked to favorable stroke outcome in humans [10].

CONCLUSION

A growing body of evidence suggests that ischemia-induced inflammation might play an important role in various stages of cerebral ischemic injury. Bidirectional interactions between the injured brain and the peripheral immune system not only are important for the development of the ischemic injury but also strongly affect the immune status of the organism as a whole. The initial immune response triggered by stroke is largely pro-inflammatory, whereas stroke-induced immunodeficiency syndrome, although being desirable for limiting the deleterious effects of postischemic inflammation, poses a severe risk for bacterial infections and unfavorable outcome. Because inflammation and peripheral immune cells that infiltrate the ischemic brain might also be involved in repair processes, it will be of importance to determine how the functionality of these systems is affected by stroke. The use of antiinflammatory strategies in ischemic stroke therapy is attractive because they have a wider therapeutic window than the now-predominant approaches based on reperfusion. However, clinical trials that utilize antileukocyte agents have failed to show benefits. Immunomodulation by enhancing the activity of reparatory neutrophils and macrophages or by skewing the adaptive immune system to a "tolerized" state might be a more promising approach than indiscriminate antiinflammatory therapies to limit the deleterious effects of poststroke inflammation. A comprehensive therapeutic approach based on antiinflammatory strategies will require a more complete understanding of the multifaceted effects of inflammation in the ischemic brain.

Acknowledgments

This work was supported by the National Institutes of Health grant NS081179.

References

[1] Iadecola C, Anrather J. The immunology of stroke: from mechanisms to translation. Nat Med 2011;17:796–808.
[2] Cuartero MI, Ballesteros I, Lizasoain I, Moro MA. Complexity of the cell–cell interactions in the innate immune response after cerebral ischemia. Brain Res 2015. http://dx.doi.org/10.1016/j.brainres.2015.04.047.
[3] Kawabori M, Yenari MA. The role of the microglia in acute CNS injury. Metab Brain Dis 2014. http://dx.doi.org/10.1007/s11011-014-9531-6.
[4] Gliem M, Schwaninger M, Jander S. Protective features of peripheral monocytes/macrophages in stroke. Biochim Biophys Acta 2015. http://dx.doi.org/10.1016/j.bbadis.2015.11.004.
[5] Liesz A, Hu X, Kleinschnitz C, Offner H. Functional role of regulatory lymphocytes in stroke: facts and controversies. Stroke 2015;46:1422–30.
[6] Engelhardt B, Ransohoff RM. Capture, crawl, cross: the T cell code to breach the blood–brain barriers. Trends Immunol 2012;33:579–89.

[7] Offner H, et al. Experimental stroke induces massive, rapid activation of the peripheral immune system. J Cereb Blood Flow Metab 2005;26:654–65.

[8] Prass K, et al. Stroke-induced immunodeficiency promotes spontaneous bacterial infections and is mediated by sympathetic activation reversal by poststroke T helper cell type 1-like immunostimulation. J Exp Med 2003;198:725–36.

[9] Courties G, et al. Ischemic stroke activates hematopoietic bone marrow stem cells. Circ Res 2015;116:407–17.

[10] Becker KJ. Activation of immune responses to brain antigens after stroke. J Neurochem 2012;123(Suppl. 2):148–55.

CHAPTER

29

Cytotoxic and Vasogenic Brain Edema

R.F. Keep, A.V. Andjelkovic, G. Xi

University of Michigan, Ann Arbor, MI, United States

INTRODUCTION

Cerebral edema is a buildup of fluid in the brain. It occurs after brain ischemia and different types of cerebral hemorrhage, as well as other conditions such as traumatic brain injury and brain neoplasms. It is a major clinical issue. Because of the encasing skull, cerebral edema can cause increased intracranial pressure (ICP), reduced cerebral blood flow, brain herniation and death. Swelling of parenchymal cells can also disrupt the physical architecture of the brain and reduce the size of the brain extracellular space. This review examines normal fluid flow in the brain, mechanisms that cause brain edema (focusing on cerebral ischemia and hemorrhage), current methods of treatment and future potential directions.

NORMAL FLUID MOVEMENT IN THE BRAIN

An understanding of brain edema requires knowledge of normal fluid movement in the brain to elucidate how it becomes deranged in disease states. Broadly, fluid movement in the body follows hydrostatic or osmotic gradients. Thus, for example, in peripheral capillaries, a pressure gradient between the vasculature and tissue interstitial fluid drives fluid into the tissue; an osmotic (oncotic) pressure, due to plasma proteins, moves water from tissue to blood; and the difference in fluid movement between these two processes is drained by the lymphatic system. In other tissues involved in fluid handling, ion transport drive water fluxes (e.g., kidney) by creating an osmotic gradient. Tissues that are involved in fluid handling also express aquaporins (AQPs), water channels that facilitate water movement down such osmotic gradients.

In brain, the general consensus has been that most of the bulk flow of fluid occurs at the choroid plexuses in the cerebral ventricles [cerebrospinal fluid (CSF) production], as the result of vectoral ion transport [1]. A portion of the fluid (~30%) production is thought to occur across the blood–brain barrier (BBB), at the cerebral microvasculature, also linked to ion transport [2]. Unlike peripheral capillaries, the cerebral endothelium forms a tight barrier as the cells are linked by tight junctions (TJs). This greatly limits hydrostatically driven water flow [2]. Some water is also generated in the brain as a by-product of metabolism. Fluid from the BBB/brain is thought to drain into the CSF system. CSF is absorbed into blood or the systemic lymphatic system from the subarachnoid CSF space at sites including the arachnoid villi, the cribriform plate, and the spinal nerve roots space, with the brain lacking an intrinsic lymphatic system [2].

Primer on Cerebrovascular Diseases, Second Edition
http://dx.doi.org/10.1016/B978-0-12-803058-5.00029-1

However, there are some current controversies [1,2]. Questions have been raised over whether the choroid plexuses are the main site of CSF secretion, although the preponderance of data favors that they are [1]. There have also been questions over whether the brain has a lymphatic system. It has been proposed that there are lymph vessels in the dura mater over the brain and that there is a brain glymphatic system [3]. In the latter, fluid enters the brain in the periarterial spaces, moves into brain parenchyma through AQP4 channels on astrocyte end feet, and exits the brain via the perivascular space around the venous system [3]. Such a glymphatic system would important implications (e.g., for the clearance of waste products from brain), but more work is needed to prove the hypothesis [2].

The glymphatic hypothesis places astrocytic end feet AQP4 with a central role in regulating brain fluid flow. In addition to being present on astrocyte end feet, AQP4 is highly expressed on the cells of the glia limitans and the ependyma, sites at the brain/CSF interface, and AQP4 may have a role in controlling fluid movements at those sites [4]. It should be noted that the presence of an AQP in a cell may alter the rate at which it comes in osmotic equilibrium after ion shifts. It may, therefore, regulate cell and extracellular space volume and thus have important consequences for interstitial fluid flow. Apart from AQP4, AQP1 is also present in brain at the apical membrane of the choroid plexus epithelium where it is thought to participate in CSF secretion [4].

Under normal conditions there is no net movement of fluid into and out of brain, although rapid fluctuations occur related to the cardiac cycle and respiration. In contrast, with ischemic and hemorrhagic stroke, fluid influx exceeds efflux and edema results.

CEREBRAL EDEMA

Edema Classifications

In neurological conditions associated with increased brain water content, that water is derived from the bloodstream. Although metabolic water generation could theoretically contribute, its rate of generation is much smaller than the rate of edema formation found in stroke. Thus for there to be a net increase in brain water, fluid influx from the blood is no longer equal to efflux. However, the underlying cause of that shift varies in different conditions. Although other classification systems have been proposed, the most commonly used is that proposed by Klatzo [5], who coined the terms vasogenic and cytotoxic edema for situations where the underlying causes are vascular or parenchymal injury.

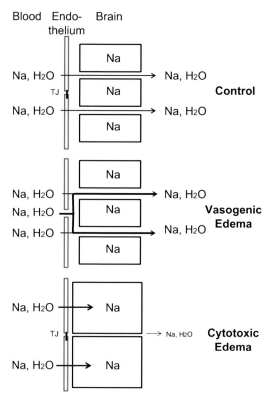

FIGURE 29.1   Under control conditions water/ion fluxes into brain match efflux. Paracellular movement between cerebral endothelial cells is greatly limited by the presence of tight junctions (TJ) and ion/water flow is probably transcellular. Inside the brain, there are marked differences in ionic composition between the intra- and extracellular spaces. For example, intracellular Na^+ concentration is much lower and K^+ higher because of Na^+/K^+-ATPase activity. In vasogenic edema, the cerebrovasculature is disrupted, here depicted as a loss of TJs. This enhances movement of fluid (water, ions, proteins) from blood to brain and can dilate the extracellular space. In cytotoxic edema, parenchyma cells swell and the size of the extracellular space is reduced. Thus in brain ischemia, reduced cerebral blood flow results in loss of cell Na^+/K^+-ATPase activity and an influx of Na^+ and water into the cell causing swelling. In addition, the influx of Na^+ into the cell reduces extracellular Na^+ and there is a movement of Na^+ from blood to brain that exceeds an efflux of potassium that also occurs. A net movement of ions from blood to brain results in a movement of water (edema). In addition, parenchymal cell swelling will compress the extracellular and perivascular spaces potentially limiting edema clearance. In mixed edema there is a combination of vascular and parenchymal damage leading to edema.

Vasogenic Edema

Vasogenic edema (Fig. 29.1) results from enhanced vascular permeability. There is a buildup of fluid, particularly in white matter, and an increased size of the brain extracellular space [5]. As noted earlier, under normal conditions, the tightness of the BBB limits hydrostatically driven fluid flux into brain, but that flux (water + solutes) is enhanced following BBB disruption. In addition, such disruption will allow entry of plasma proteins into brain dissipating an oncotic gradient that would draw water out of brain. Those proteins may also participate in stroke-induced brain injury, e.g., fibrinogen and prothrombin are converted in the brain to fibrin and thrombin, which are potent inflammatory agents.

Three underlying mechanisms may be involved in enhancing cerebrovascular permeability following stroke [6]. (1) There can be alterations in endothelial TJ function. At the later stages of stroke, this can involve the loss of TJ proteins (e.g., claudin-5, occludin, zonula occludens-1), but there can also be subtler changes that alter the interactions and cellular location of those proteins, resulting in barrier hyperpermeability. Progress is being made in delineating those mechanisms that may identify therapeutic targets to limit TJ disruption and edema formation. (2) Under normal conditions there is limited pinocytosis at the BBB. However, after stroke there is a pronounced increase in the number of endothelial vesicles and it is proposed that increased transcytosis contributes to BBB disruption and edema formation. (3) It has also been proposed that endothelial cell death is a major component of BBB disruption after stroke. Such death occurs, but its role in barrier disruption/edema formation is uncertain as the loss of the endothelium normally results in coagulation cascade activation limiting vessel perfusion.

Genetic deletion of AQP4 increases brain water content in models of vasogenic edema (e.g., persistent ischemia and brain tumors) [4]. This may reflect an effect on the clearance of vasogenic edema [4].

Cytotoxic Edema

In contrast to vasogenic edema, the underlying cause of cytotoxic edema (Fig. 29.1) is parenchymal. It is associated with parenchymal cell swelling and a reduced extracellular space [5]. Brain cells possess an array of ATP-dependent transporters (e.g., Na^+/K^+-ATPase), secondarily active transporters that rely on ion gradients generated by ATP-dependent transporters (e.g., the Na^+/H^+-antiporters and Na^+/Ca^{2+} exchangers), and ion channels, the permeability of which are directly or indirectly ATP dependent (e.g., ATP-, Ca^{2+}-, and voltage-dependent channels). In ischemia and other cytotoxic events, cellular ATP levels fall inhibiting ATP-dependent transporters and particularly Na^+/K^+-ATPase. The latter causes an influx of Na^+ into the cell that exceeds the loss of K^+ resulting in cell swelling. Further ion changes occur through effects on secondarily active transporters and ion channels [7]. For example, release of cellular glutamate during ischemia results in astrocyte swelling and activates N-methyl-D-aspartate channels that lead to an influx of Na^+ and Ca^{2+} into neurons. In addition to the effects through preexisting transporters and channels, transporter/channel expression is also changed in stroke. Thus the

nonselective cation channel sulfonylurea 1-transient receptor potential melastatin 4 (SUR1-TRPM4) is not normally expressed in brain but it is upregulated in multiple cell types after ischemia. That channel may play a prominent role in Na^+ movement into cells (and across the BBB) after stroke [7].

In contrast to vasogenic edema, AQP4 deletion reduces edema formation in situations of cytotoxic edema such as acute cerebral ischemia and traumatic brain injury [4]. AQP4 may facilitate the ion-driven movement of water into the intracellular space occurring in those conditions.

It should be noted that a movement of fluid from the extracellular to the intracellular space will, of itself, not cause an increase in overall brain water content. However, in cerebral ischemia, entry of Na^+ into parenchymal cells reduces the extracellular concentration of Na^+ providing an electrochemical gradient for movement of Na^+ into brain from blood. That Na^+ movement may be transporter and/or channel mediated [7]. Ischemia also causes a loss of cellular K^+, increased extracellular K^+, and a gradual loss of K^+ to blood. However, the gain in brain Na^+ outweighs the loss in K^+ and the resultant gain in brain cations (and associated anions) draws water into brain [8].

In addition, in cytotoxic edema, the movement of water from extracellular to intracellular space also reduces the size of the former. This can impede the normal flow of interstitial fluid through the brain, contributing to the buildup of edema and also delaying the resolution of that edema.

Mixed Edema

It should be noted that cerebral ischemia and intracerebral hemorrhage have mixed forms of edema, with both vasogenic and cytotoxic components. The relative importance of these components to brain edema formation varies with time and the nature of the stroke.

Edema Resolution

In rodent models of stroke, edema peaks at 3–7 days after ictus and then resolves. In humans, that process is generally longer. Resolution of edema involves multiple processes. In ischemic stroke, injured/dead cells are removed, the BBB reseals, and, in many cases, there is restoration of blood flow. In addition, in hemorrhagic stroke, the clot gradually resolves. These changes remove the underlying mechanisms that induce edema formation. In addition, these changes will enhance fluid outflow from the brain by reducing compression of the extracellular and perivascular spaces.

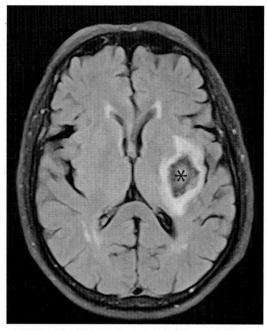

FIGURE 29.2 An example of brain edema surrounding an intracerebral hemorrhage (*) detected using FLAIR (fluid-attenuated inversion recovery) magnetic resonance imaging. The scan was from a patient 1 day post ictus.

IMAGING AND MONITORING BRAIN EDEMA

Multiple modalities can be used to detect clinical and experimental brain edema. These include magnetic resonance imaging (MRI) and X-ray computed topography scans, either examining the affected tissue directly (edema causes high signal on T2-weighted MRI and low signal on T1-weighted imaging; Fig. 29.2) or looking for other parameters such as a midline shift and ventricular ablation. Diffusion-weighted imaging and apparent diffusion coefficient sequences can aid in distinguishing cytotoxic edema, with restricted diffusion, from vasogenic edema, with normal or increased diffusion [9]. Measuring ICP also gives an indirect evaluation of brain edema formation.

TREATMENT

The treatments for cerebral edema have not changed greatly for decades. These include the use of osmotic agents such as mannitol and hypertonic saline to draw water out of the brain, diuretics or drains to impact CSF volume, hyperventilation to reduce cerebral blood volume, and decompressive craniectomy to allow the brain to swell [10]. Although steroids are beneficial in some forms of cerebral edema (e.g., that associated with brain tumors), they are generally not effective in cerebral ischemia or hemorrhage. It should also be noted that systemic osmotic agents rely on the establishment of an osmotic gradient between blood and brain. In situations of BBB compromise, the agent may enter the brain negating the gradient and also, potentially, causing a rebound effect when plasma osmolality falls.

The only current clinically effective therapy for cerebral ischemia is early reperfusion with either tissue plasminogen activator or thrombectomy. Early reperfusion can limit parenchymal cell injury and, thereby, cytotoxic edema. However, delayed reperfusion can exacerbate BBB damage and cause greater vasogenic edema, with hemorrhagic transformation being an ultimate example with blood rather than just fluid entering brain.

There are currently no therapies that prevent BBB disruption in patients with stroke. However, considerable progress is being made in understanding the BBB changes that occur in stroke, such as endothelial TJ modifications [6]. Hopefully such studies will lead to the identification of therapeutic targets that will limit BBB disruption and vasogenic edema.

As noted earlier, brain edema formation is associated with ion shifts between blood and brain and between the brain extracellular and intracellular spaces. A number of approaches have been tried to modulate those changes preclinically [7]. For example, inhibiting Na-K-Cl cotransport and Na/H exchange at the BBB reduces ischemic brain edema formation in rodents. Another target is the nonselective cation channel SUR1-TRPM4, which is inhibited by glyburide. A number of studies have shown the importance of this channel in both vasogenic and cytotoxic edema [7] and glyburide is currently in phase II clinical trial for malignant edema in stroke (NCT01794182). Although AQPs may also be a clinical target, there are as yet no available therapeutics that specifically inhibit AQPs.

CONCLUSION

Brain edema is a major cause of mortality and morbidity after ischemic and hemorrhagic stroke. Edema can result from events at the cerebrovasculature and/or in parenchymal cells. Although treatments have not changed substantially in decades, a greater understanding of the underlying pathology suggests several new promising targets.

References

[1] Spector R, Keep RF, Robert Snodgrass S, Smith QR, Johanson CE. A balanced view of choroid plexus structure and function: focus on adult humans. Exp Neurol 2015;267:78–86.
[2] Hladky SB, Barrand MA. Mechanisms of fluid movement into, through and out of the brain: evaluation of the evidence. Fluids Barriers CNS 2014;11:26.

[3] Simon MJ, Iliff JJ. Regulation of cerebrospinal fluid (CSF) flow in neurodegenerative, neurovascular and neuroinflammatory disease. Biochim Biophys Acta 2016;1862:442–451.

[4] Zador Z, Stiver S, Wang V, Manley GT. Role of aquaporin-4 in cerebral edema and stroke. Handb Exp Pharmacol 2009:159–70.

[5] Klatzo I. Presidental address. Neuropathological aspects of brain edema. J Neuropathol Exp Neurol 1967;26:1–14.

[6] Prakash R, Carmichael ST. Blood–brain barrier breakdown and neovascularization processes after stroke and traumatic brain injury. Curr Opin Neurol 2015;28:556–64.

[7] Khanna A, Kahle KT, Walcott BP, Gerzanich V, Simard JM. Disruption of ion homeostasis in the neurogliovascular unit underlies the pathogenesis of ischemic cerebral edema. Transl Stroke Res 2014;5:3–16.

[8] Betz AL, Keep RF, Beer ME, Ren XD. Blood–brain barrier permeability and brain concentration of sodium, potassium, and chloride during focal ischemia. J Cereb Blood Flow Metab 1994;14:29–37.

[9] Ho ML, Rojas R, Eisenberg RL. Cerebral edema. AJR Am J Roentgenol 2012;199:W258–73.

[10] Steiner T, Ringleb P, Hacke W. Treatment options for large hemispheric stroke. Neurology 2001;57:S61–8.

CHAPTER

30

Spreading Depolarizations

D.Y. Chung, C. Ayata

Harvard Medical School, Boston, MA, United States

Spreading depolarizations (SDs) are recurrent waves of intense neuronal and glial depolarization that develop in apparently spontaneous fashion in ischemic stroke, intracranial hemorrhage, and trauma. Although Leão first described spreading depression, an SD wave in normal (i.e., uninjured) brain, in 1944, it was not until the late 1970s that periinfarct spreading depression-like waves were detected in experimental focal cerebral ischemia [1]. Today we know that SDs occur after most types of injury in human brain, and worsen both tissue and functional neurological outcome [2].

BASIC CHARACTERISTICS

The electrophysiological events underlying SDs have been reviewed in detail [3]. SDs are associated with massive transmembrane K^+, Ca^{2+}, Na^+, and water shifts, as well as cell swelling. The depolarization and ion fluxes create a characteristic 20- to 30-mV extracellular negative slow potential shift (Fig. 30.1). In otherwise normal brain, all transmembrane ion and water balances are restored within less than a minute, which is even more metabolically costly than seizures. Once triggered, SDs slowly propagate within the gray matter at a rate of ~3–6 mm/min by way of contiguity. When SDs occur in or propagate into otherwise normal tissue, complete neuronal membrane depolarization precludes all spontaneous and evoked activity resulting in electrocorticogram (ECoG) depression. Because Leão discovered and described the basic features of SD in normal brain by using ECoG recordings with high-pass filtering, he only detected ECoG depression rather than the massive pan-depolarization, and termed the phenomenon spreading depression. But since the ECoG is already depressed in injured tissue (e.g., ischemic penumbra), SD events can not cause further ECoG depression in these regions. Therefore, SD (spreading *depolarization*) is a more accurate and inclusive term that has replaced spreading *depression*. Nevertheless, the term spreading depression has historical significance, and is well recognized and widely used by the scientific and clinical

Primer on Cerebrovascular Diseases, Second Edition
http://dx.doi.org/10.1016/B978-0-12-803058-5.00030-8

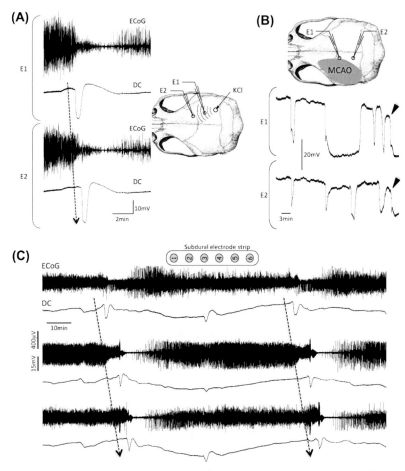

FIGURE 30.1   **Spreading depolarizations in normal and injured brain.** (A) Electrophysiological tracings of spreading depolarization (SD) show-ing electrocorticogram (ECoG) depression spatiotemporally coincident with a large negative extracellular direct-coupled (DC) potential shift that lasts up to a minute when recorded by two serial intracortical microelectrodes in the rat *(Modified from Ayata C. Spreading depression: from serendipity to targeted therapy in migraine prophylaxis. Cephalalgia 2009;29:1095–1114. With permission from Sage.).* (B) Electrophysiological tracings show-ing large and often prolonged negative extracellular DC potential shifts associated with periinfarct SDs recorded by two intracortical microelectrodes in the mouse. Note the terminal depolarization at the end of the recording in both electrodes sequentially *(arrowhead).* MCAO, ischemic territory after middle cerebral artery occlusion. (C) Sample ECoG and DC potential tracings from human brain showing SDs in the setting of traumatic brain injury recorded by subdural electrode strips. ECoG traces are bipolar between adjacent channels on the strip. DC traces are referential recordings from an extracranial reference. Note the sequential involvement of electrode pairs by SD (Courtesy of Dr. Jed Hartings.).

community, especially in the migraine field where SD occurs in normal brain to create the perception of aura.

IMPACT ON INJURY OUTCOME

SDs in injured brain often last longer than in healthy brain because blood flow-, O_2-, and glucose-dependent restoration of transmembrane ion gradients is often delayed, and because, in contrast to the large hyper-emic response they evoke in normal brain, SDs cause vasoconstriction and further reduce tissue perfusion and oxygenation in injured brain (Fig. 30.2) [3]. Under-standably, the vasoconstrictive effect, combined with the tremendous metabolic burden they impose on the tissue, severely worsens supply–demand mismatch, and is highly detrimental to the survival of already

metabolically compromised brain tissue. As a result, SDs expand the injury (e.g., ischemic core), and con-tribute to neurological worsening. Additional mecha-nisms by which SDs can be detrimental in injured brain are (1) disruption of the blood–brain barrier, potentially leading to malignant cerebral edema, mass effect, and herniation, and (2) triggering of epileptic seizures. Confirming the pathophysiological impor-tance of SDs, pharmacological SD inhibitors (e.g., NMDA receptor blockers) reduce infarct volumes and improve neurological outcomes in experimental mod-els of ischemic stroke [4]. As a cautionary note, how-ever, SDs may also have beneficial effects particularly at later stages of injury maturation and recovery when supply–demand mismatch is no longer a critical factor in tissue outcome. They have been shown to stimulate neurogenesis, and may even promote angiogenesis.

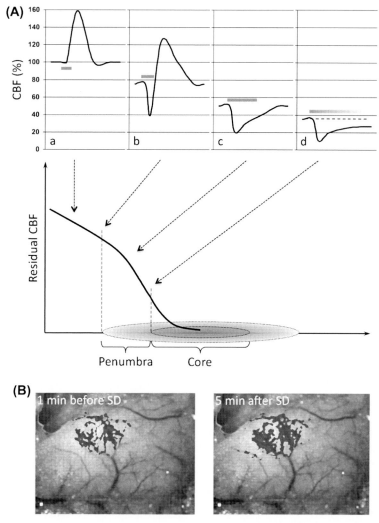

FIGURE 30.2 **CBF response to SDs in focal ischemic brain.** (A) Diagram illustrating the transformation of the CBF response to periinfarct SDs recorded in tissue (*upper graphs*) with increasing severity of ischemia as shown on the CBF profile across focal ischemic tissue (*lower graph*). In nonischemic tissue, SDs evoke a predominantly hyperemic response (a), whereas in mildly ischemic penumbra (b) a biphasic response is observed. In moderately ischemic penumbra (c), the response is mainly a monophasic hypoperfusion. In the severely ischemic core–penumbra junction (d), both the DC shift and the hypoperfusion may not recover completely after an SD. Horizontal bars represent the DC shift during SDs (*Modified from Ayata C, Lauritzen M. Spreading depression, spreading depolarizations, and the cerebral vasculature. Physiol Rev 2015;95: 953–993. With permission.*). (B) Speckle contrast images from a representative experiment showing the abrupt expansion (>100%) of severely hypoperfused cortex (i.e., ≤20% residual CBF; shown in black) by a single SD event after acute distal middle cerebral artery occlusion in the mouse. The imaging field covers the entire right hemisphere. (*Modified from Shin HK, Dunn AK, Jones PB, Boas DA, Moskowitz MA, Ayata C. Vasoconstrictive neurovascular coupling during focal ischemic depolarizations. J Cereb Blood Flow Metab 2006;26:1018–1030. With permission from Sage.*).

NATURAL HISTORY OF SDs IN INJURED BRAIN

Experimentally, SDs have been most frequently studied in the setting of ischemic stroke, where they are often called periinfarct SDs. This is in part because, unlike in human brain, animal models of other brain injury states (e.g., subarachnoid hemorrhage and trauma) do not consistently yield spontaneous SDs. During acute arterial occlusion, SDs originate within the partially ischemic but still viable periinfarct tissue, starting with the onset of anoxic depolarization in the core, and continuing to emerge for many hours and days even after successful reperfusion in focal cerebral ischemia.

SDs can originate from the same focus or from multiple different foci in each brain. Once triggered, they propagate through and beyond the ischemic penumbra in virtually all species studied to date, including human. Of note, SDs are obligate gray matter events and do not originate from or propagate into white matter. Moreover, in gyrencephalic brains, which are relatively resistant to SDs compared with lissencephalic brains, sulci and larger pial vessels can act as barriers to propagation [5]. The propagation pattern can evolve over time in an unpredictable fashion, and reentrant patterns can emerge causing a single SD wave to circle around the injured tissue numerous times, effectively multiplying its detrimental impact. As noted earlier, SDs can occur for up to a week or more after ischemic stroke, and even longer in other brain injury states such as subarachnoid hemorrhage, the latter likely upon emergence of focal ischemic zones. As such, SDs offer a wide potential therapeutic window after ischemic stroke, intracranial hemorrhage, and traumatic brain injury.

TRIGGER FACTORS AND MODULATORS

Data in ischemic stroke implicate focal supply–demand mismatch transients in metastable periinfarct hot zones as a trigger for SDs [6]. Either increased demand (e.g., functional activation) or reduced supply (e.g., episodic hypoxia or hypotension) can trigger SDs by worsening the supply–demand mismatch in critically ischemic periinfarct tissue. As such, the volume of metastable periinfarct tissue is a critical determinant of SD frequency. Larger infarcts or those with large volumes of penumbra in gray matter are more likely to develop SDs, whereas pure white matter infarcts do not develop SDs. The propensity to develop SDs is further enhanced by genetic (e.g., migraine mutations), systemic physiological (e.g., hypoglycemic, hypotensive, hypoxic transients), and environmental (e.g., sensory stimulation) factors. Conversely, normobaric hyperoxia and migraine prophylactic drugs can suppress the occurrence of SDs or limit their propagation to improve tissue and neurological outcome. Indeed, a large number of pharmacological targets can suppress the susceptibility to SDs, particularly NMDA receptor antagonists such as ketamine [7].

DETECTION OF SDs

The negative extracellular potential shift (~20–30 mV) and the attendant ECoG depression are the electrophysiological signatures of SDs. Two factors have made SDs notoriously difficult to detect in human brain using non-invasive methods. First, the negative potential shift is too slow to develop and recover (over many seconds), and can only be detected using unfiltered or direct-coupled (DC) recordings. Hence, the typical electrical noise (e.g., muscle artifacts, fields generated by the environment) associated with scalp electroencephalography (EEG), normally eliminated by the powerful high-pass filters of conventional EEG amplifiers, obscures the negative potential shift caused by SDs in DC recordings. Second, both the propagating depolarization and

ECoG depression are relatively focal at any point in time (i.e., millimeters of brain tissue) and their magnitude diminishes exponentially with increasing distance (e.g., between the brain tissue and the scalp electrode). Hence, they can only be detected reliably using intracortical or subdural electrodes.

These obstacles have been overcome by the remarkable work of the international multicenter Cooperative Studies on Brain Injury Depolarizations (COSBID) collaboration over the past 10 years [8–10]. COSBID implemented subdural electrode strips placed during open craniotomy procedures in patients with large ischemic stroke, subarachnoid or intracerebral hemorrhage, or traumatic brain injury. The data revealed an abundance of SDs in injured human brain, and provided a clinical proof-of-concept that SDs: (1) are associated with worse clinical outcomes, (2) continue to occur for many days after injury onset, and (3) can be suppressed by pharmacological agents.

THERAPEUTIC TARGETING OF SDs

Recognition that SDs contribute to lesion growth in acute and subacute stages of brain injury has renewed the interest in therapeutic targeting of SDs to prevent injury growth. However, the need for invasive monitoring (i.e., subdural or intracortical recordings) to detect SDs has thus far hampered efforts to design and launch large-scale multicenter clinical trials. There is considerable interest in noninvasive detection methods using surface electrophysiology with data processing, or optical tools such as near-infrared or diffuse correlation spectroscopy. On the other hand, because SDs have been detected in close to 100% of patients with ischemic stroke, and in up to 80% of patients with aneurysmal subarachnoid hemorrhage, the argument has been made to forego detection of SDs to obviate the need for a craniectomy, to simplify enrollment and trial execution, and to minimize potential complication rates. Another challenge facing clinical trials is that the most potent inhibitors of SDs also have potent neurological side effects such as depressed level of consciousness. Although self-limited, such side effects may interfere with clinical assessments, require endotracheal intubation, and obligate intensive care unit (ICU) stays. Nevertheless, technologically advanced ICU monitoring and management may in the near future allay such concerns and allow clinical trials to move forward. Novel pharmacological and device inhibitors of SDs may provide additional opportunities. Last but not least, an improved understanding of the systemic physiological and environmental factors that predispose to SDs or worsen their impact may alter the overall clinical management of patients with brain injury.

CONCLUSION

SDs occur frequently for days to weeks after primary brain injury in both experimental animals and human brain. There is strong associative evidence in humans, and compelling experimental evidence that SDs lead to extensive secondary brain injury. There is currently no clinically proven intervention to mitigate SDs and their impact on injury outcome. However, efforts for clinical translation have gained momentum thanks to technological advances in the field of neurocritical care.

References

[1] Branston NM, Strong AJ, Symon L. Extracellular potassium activity, evoked potential and tissue blood flow. Relationships during progressive ischaemia in baboon cerebral cortex. J Neurol Sci 1977;32:305–21.

[2] Lauritzen M, Dreier JP, Fabricius M, Hartings JA, Graf R, Strong AJ. Clinical relevance of cortical spreading depression in neurological disorders: migraine, malignant stroke, subarachnoid and intracranial hemorrhage, and traumatic brain injury. J Cereb Blood Flow Metab 2011;31:17–35.

[3] Ayata C, Lauritzen M. Spreading depression, spreading depolarizations, and the cerebral vasculature. Physiol Rev 2015;95:953–93.

[4] Shin HK, Dunn AK, Jones PB, Boas DA, Moskowitz MA, Ayata C. Vasoconstrictive neurovascular coupling during focal ischemic depolarizations. J Cereb Blood Flow Metab 2006;26:1018–30.

[5] Santos E, Scholl M, Sanchez-Porras R, et al. Radial, spiral and reverberating waves of spreading depolarization occur in the gyrencephalic brain. Neuroimage 2014;99:244–55.

[6] von Bornstadt D, Houben T, Seidel JL, et al. Supply-demand mismatch transients in susceptible peri-infarct hot zones explain the origins of spreading injury depolarizations. Neuron 2015;85:1117–31.

[7] Ayata C. Spreading depression: from serendipity to targeted therapy in migraine prophylaxis. Cephalalgia 2009;29:1095–114.

[8] Dohmen C, Sakowitz OW, Fabricius M, et al. Spreading depolarizations occur in human ischemic stroke with high incidence. Ann Neurol 2008;63:720–8.

[9] Bosche B, Graf R, Ernestus RI, et al. Recurrent spreading depolarizations after subarachnoid hemorrhage decreases oxygen availability in human cerebral cortex. Ann Neurol 2010;67:607–17.

[10] Hinzman JM, Andaluz N, Shutter LA, et al. Inverse neurovascular coupling to cortical spreading depolarizations in severe brain trauma. Brain 2014;137:2960–72.

CHAPTER

31

Hypertension

T.M. De Silva[1,2], F.M. Faraci[1]

[1]The University of Iowa Carver College of Medicine, Iowa City, IA, United States; [2]Monash University, Clayton, VIC, Australia

INTRODUCTION

Normal brain function critically depends on adequate levels of perfusion under baseline conditions and in the face of changing cellular demands, often in the presence of underlying vascular disease. Alterations in cerebral blood flow (CBF) are primarily determined by changes in diameter of resistance vessels, which includes both arteries and arterioles in brain. Vascular disease causes profound changes in these segments of the circulation. Consequences of cerebral vascular disease commonly

include hypoperfusion (low levels of baseline CBF), ischemic or hemorrhagic stroke, microbleeds, and cognitive deficits [1,2]. With some clinical conditions, increases in CBF or hyperperfusion can occur. Of the known risk factors for cerebrovascular disease, hypertension is associated with some of the most profound effects. Chronic hypertension is very common in adults and is currently the number one risk factor for overall disease burden and health loss worldwide.

In this chapter, we briefly highlight the effects of acute and chronic hypertension on large and small vessels in

Primer on Cerebrovascular Diseases, Second Edition
http://dx.doi.org/10.1016/B978-0-12-803058-5.00031-X

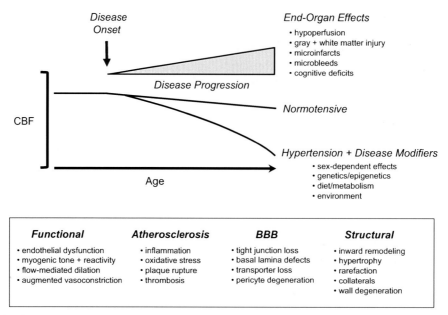

FIGURE 31.1 Vascular disease, its progression over time, and some key end-organ effects are shown schematically. Effects on relative cerebral blood flow (CBF) are illustrated for normotensive and hypertensive individuals. Major disease modifiers are also listed. These modifiers can increase or decrease the rate at which CBF progressively declines. The box in the lower portion of the figure lists examples of key vascular changes that occur in hypertension. Endothelial dysfunction refers to impairment of both endothelial nitric oxide synthase (eNOS)-dependent and eNOS-independent mechanisms as well as neurovascular coupling.

brain. As noted by others, the term small vessels refer to small arteries and arterioles within the pial circulation on the brain surface, along with arterioles, capillaries, and venules within the parenchyma [1,3]. We highlight changes in vascular function and structure that impact baseline CBF as well as adaptive changes in CBF (Fig. 31.1). An example of the latter is neurovascular coupling. This collective process normally occurs in a controlled spatial and temporal manner to match delivery of glucose, oxygen, and other nutrients with variations in cellular activity.

Mechanisms that produce hypertension include activation of the systemic or local renin–angiotensin–aldosterone systems (RAAS), activation of the sympathetic nervous system, renal changes that result in sodium and fluid retention, along with contributions by the immune system, and perhaps the microbiota of the gut. In addition, levels and progression of hypertension are modified by sex-dependent mechanisms, genetics and epigenetics, diet, metabolism, the environment, and aging (Fig. 31.1). In addition to insight obtained clinically, experimental models are being commonly used in an effort to better define these relationships and include various genetic, renal, and pharmacological models of hypertension.

VASCULAR FUNCTION

Endothelium-Dependent Effects

Vascular endothelial cells have many functions in health and disease, including exerting diverse effects on multiple cell types throughout the circulation. Endothelium affects vessel diameter through its influence on underlying smooth muscle via several endothelium-dependent mechanisms. Through these effects, endothelial cells influence resting CBF and mediate or contribute to varied adaptive vascular responses. The greatest known contributor in this regard is the vasodilator nitric oxide (NO), produced by endothelial NO synthase (eNOS). A tonic vasodilator influence of eNOS-dependent signaling is present in arteries and arterioles on the brain surface as well as arterioles within the parenchyma [4]. In parenchymal arterioles, both eNOS-dependent and eNOS-independent mechanisms have robust effects on microvascular tone [4,5], the latter may occur mainly through endothelium-dependent hyperpolarization of underlying vascular muscle. In this role, endothelial cells are critical mediators of vasodilator responses to physical forces (shear stress), neurotransmitters, metabolic factors, and therapeutic agents. In addition, endothelial cells control vascular permeability (see later discussion) and play a fundamental role in thromboresistance through antiplatelet, antithrombotic, and fibrinolytic mechanisms [6].

The loss of normal endothelium-dependent mechanisms is a key (often initiating) event in the pathogenesis of vascular disease [4,7]. The phrase endothelial dysfunction describes collective endothelial-based abnormalities that promote oxidative stress, low-grade inflammation, increased vascular tone and permeability, atherosclerosis, and thrombosis. For both basic and clinical-focused research, perhaps the most studied end point in this regard has been the endothelium-dependent control of vasomotor tone. It is particularly noteworthy that in addition to regulating local perfusion, changes in the endothelial control of vascular tone in disease are predictive of clinical events including stroke [4,7].

Hypertension affects each element of endothelial function in large arteries and microvessels. In relation to vascular function, chronic hypertension is associated with impairment of both basal and agonist-induced endothelium-dependent vasodilation [4]. Both low-grade inflammation and oxidative stress within the vessel wall are commonly seen in animal models and humans with hypertension. The progression of atherosclerosis is accelerated in both cerebral and peripheral blood vessels. Along with the loss of protective vasodilator mechanisms, endothelial cells during hypertension produce factors that increase vascular tone and promote abnormal vascular growth including prostanoids, angiotensin II, endothelins, and reactive oxygen species (ROS) [4].

Acute hypertension (or hypertension emergencies) presents additional cerebrovascular and neurological challenges. The presence of preexisting chronic hypertension is a key risk factor for acute hypertension. In addition, both normotensive and chronically hypertensive persons may exhibit periods of acute hypertension as a result of myocardial infarction, aortic dissection, drug abuse or misuse, head injury, and hypertensive disorders of pregnancy, among other causes.

Acute hypertension also affects endothelial integrity, vascular tone, and vascular permeability. Injury to endothelium following rapid increases in arterial pressure occurs throughout the cerebral circulation including in arterioles, capillaries, and venules. Large increases in CBF can occur during acute hypertension. In addition to loss of blood–brain barrier (BBB) integrity and the consequences of an influx of protein and other molecules, this form of endothelial damage promotes adhesion and entry of immune cells along with brain edema. Such a sequence of events is seen in hypertensive encephalopathy and posterior reversible encephalopathy syndrome.

Neurovascular Coupling

Neurovascular coupling is the term used to describe alterations in local perfusion that occur in response to changes in neuronal activity. Also known as functional hyperemia, neurovascular coupling involves the integration of multiple signaling molecules (or events) and cell types. Imaging approaches that are commonly used in basic and clinical neuroscience, such as functional MRI, are based on this local hemodynamic response [8].

Intact neurovascular coupling requires an integrated multicellular response to provide the perfusion needs that result from acute and often changing focal neuronal activation. In addition to local perfusion changes driven by neurons and astrocytes, dilation of arterioles and arteries upstream from the site of activation also occurs via endothelium-dependent mechanisms [8]. Thus the endothelial dysfunction noted earlier may also contribute to diminished neurovascular coupling. The identity and relative importance of various cell types and signaling molecules and their contribution to neurovascular coupling are subjects of debate, particularly in regions like the somatosensory cortex (where most studies have focused). Mechanisms that underlie neurovascular coupling also vary regionally and may be better defined in the cerebellum, basal forebrain, and hippocampus. Regardless of the underlying mechanism(s), a common experimental finding is impairment of neurovascular coupling in both experimental models (genetic and pharmacological) and humans with chronic hypertension. Along with endothelial dysfunction and hypoperfusion, impaired neurovascular coupling is thought to contribute to tissue injury, loss of function, and cellular degeneration over time (Fig. 31.1).

In relation to the fundamental vascular changes described earlier, some common underlying mechanisms may be involved [4,9]. Those mechanisms include oxidative stress driven by key sources of ROS (e.g., nicotinamide adenine dinucleotide phosphate oxidases and mitochondria) and their downstream effectors including peroxynitrite and poly ADP ribose polymerase. Additional interacting systems may include prostanoids and endothelins. The RAAS has been broadly implicated in hypertension and vascular disease based on multiple lines of evidence. Peptide and hormone effectors within this system (e.g., angiotensin II and aldosterone) promote vascular disease and dysfunction, even when the circulating levels of these molecules are insufficient to increase arterial pressure. Although their relative importance is not clear, both the circulating and the tissue RAAS may contribute to disease progression. Lastly, the immune system has emerged as a key contributor to mechanisms that produce hypertension. The transcription factor NF-κB, a central integrator of immune-related responses including in the vasculature, is an additional target of angiotensin II and ROS. Elements of this system are known to contribute to large-vessel disease (e.g., atherosclerosis and aortic stiffening) and may be important in small-vessel disease as well. Thus both circulating and local (e.g., microglia and perivascular macrophages) immune cells may affect various elements of vascular biology in brain.

Autoregulation

The term autoregulation describes the ability of the brain to maintain a relatively constant level of CBF over a range of perfusion pressures [10]. Among the factors that mediate autoregulation, myogenic mechanisms are generally considered to be the most important. Myogenic tone (tone generated by a vessel when pressurized) and myogenic responses (changes in tone that occur with changes in transmural pressure) are seen through a large portion of the brain circulation: cerebral arteries, and pial and parenchymal arterioles. Myogenic mechanisms contribute to autoregulation in that these vessels constrict with increases in pressure and dilate with reductions in transmural pressure [10]. Size-dependent responses to pressure have been described both in vivo and in vitro, with greater levels of myogenic tone being seen in pial and parenchymal arterioles than in large cerebral arteries [5].

Myogenic mechanisms have been studied widely using isolated vessels and it is well established that the cellular basis for the response resides in vascular muscle [10]. Despite these efforts, some key elements remain unclear or are at least debated. For example, what actually senses changes in pressure to then activate downstream pathways that alter vessel diameter remains unsettled. Key intermediates may include integrins, G-proteins, ion channels and other regulators of membrane potential, kinase-mediated events, and ultimately contractile proteins.

Myogenic mechanisms change during hypertension [10]. Myogenic tone is increased in isolated arteries and arterioles from hypertensive models. The relationship of CBF to arterial pressure, often presented as the autoregulatory curve, commonly shifts to the right (to higher levels of blood pressure). Thus a normal level of resting CBF can be present, but at higher arterial pressure. Functionally, this shift or resetting protects against uncontrolled increases in CBF if further moderate increases in arterial pressure occur. However, because the entire autoregulatory curve shifts, the brain is also at risk for a reduction in CBF with decreases in blood pressure. This is a well-described clinical problem in patients with essential or primary hypertension. The underlying causes for the shift are not entirely clear, but are thought to result collectively from both functional and structural changes that occur in the vasculature. With very large or severe increases in arterial pressure, the ability of the vasculature to autoregulate is overwhelmed and CBF can increase markedly. As noted earlier, these acute hemodynamic changes can result in BBB damage and cerebral edema.

BLOOD–BRAIN BARRIER

The BBB is a highly specialized structural and functional barrier formed by and between adjacent cerebral endothelial cells [11]. Although endothelial cells are the site of the barrier, its properties and function are influenced by other cell types including astrocytes and pericytes, additional cells in what is often referred to as the neurovascular unit. The BBB regulates the movement of

ions, molecules, and cells between the circulation and the brain, preventing most changes in the constituents of the circulation from reaching and affecting the brain. Key features of the normal BBB include: (1) adjacent endothelial cells are interconnected by tight and adherens junction molecules linked to the cytoskeleton, (2) limited transcytosis, and (3) the presence of an array of transporter proteins facilitating the inward movement of needed molecules (e.g., glucose) and the removal of unwanted ones.

There is considerable evidence that the BBB is affected by acute and chronic hypertension. Loss of BBB integrity is thought to be key element of cerebrovascular disease including small-vessel disease [2]. For example, a heterogenous loss of BBB integrity occurs in many regions including areas involved in memory and autonomic control. Both blood pressure–dependent and blood pressure–independent changes in BBB permeability can occur. At least some of the changes may be due to direct effects of angiotensin II on AT1 receptors, not effects of increased arterial pressure per se [12]. The implications of such changes are several fold. For example, loss of BBB integrity disrupts normal brain fluid balance resulting in edema formation and increases in intracranial pressure. Increased permeability of the BBB may allow angiotensin II or other circulating factors to gain access to central AT1 receptors that drive neurohumoral responses including activity of sympathetic nerves that further increase arterial pressure.

VASCULAR STRUCTURE

Some of the most prominent effects of hypertension on the cerebrovasculature can be seen when one examines changes in vascular structure (Fig. 31.1) [4,10]. First, increases in the cross-sectional area of the vessel wall (or hypertrophy), are commonly seen and represent an adaptive response that reduces wall stress during hypertension. Depending on specific features, hypertrophy may also encroach on the vessel lumen and thus increase vascular resistance. Increases in the cross-sectional area of the vessel wall occur in many models of hypertension and are present in hypertensive humans. In contrast, hypertension can also promote degeneration and thinning of the vessel wall over time contributing to leakage of molecules from the circulation along with microbleeds and intracerebral hemorrhage. Key events that underlie these changes include oxidative stress along with changes in activity of matrix metalloproteinases and endogenous tissue inhibitors of metalloproteinases in vascular cells and the extracellular matrix.

Second, a three-dimensional rearrangement of the vessel wall occurs known as inward remodeling. The result is a reduction in lumen diameter that is not due to changes in vascular tone or vascular mechanics. Inward vascular remodeling is commonly seen in arterioles and small arteries in some, but not all forms of hypertension. There is evidence that similar remodeling occurs in small cerebral arteries in humans. The RAAS and its downstream targets are key determinants of inward remodeling. Even nonpressor doses of angiotensin II, when administered chronically, produce inward remodeling of cerebral arterioles.

The two changes in vascular structure outlined earlier are both adaptive and maladaptive. Inward remodeling of arteries and arterioles protect downstream capillaries, venules, and the BBB from increases in microvascular pressure when systemic pressure is elevated. However, the reduction in lumen diameter also increases minimal vascular resistance, thus limiting vasodilator responses and collateral-dependent perfusion during ischemia.

Third, rarefaction has been described in arterioles and capillaries in experimental models and humans with hypertension. This loss of blood vessels is important for several reasons. In addition to potential effects on CBF, oxygen delivery, and local PO_2 levels at baseline, the integrity of collateral vessels in the pial circulation is a determinant of tissue loss after occlusion of larger arteries upstream (such as that which occurs during thrombotic occlusion of a large cerebral artery). Reductions in both collateral number and diameter have been described in models of hypertension [13]. Resistance of cerebral vessels is increased during chronic hypertension, including when the vasculature is maximally dilated. The inward remodeling and rarefaction described earlier are major contributors to this change, potentially affecting resting CBF, limiting vasodilator responses, and predisposing to greater injury in the face of ischemia.

CONSEQUENCES OF VASCULAR DISEASE DURING HYPERTENSION

The functional and structural changes outlined earlier impact CBF through effects on vascular resistance and vasodilator responses. These effects generally worsen with age (Fig. 31.1). The result can be hypoperfusion at rest combined with dysregulation in the moment-to-moment control of CBF. Vascular resistance is elevated so vasodilator reserve is reduced including in collateral vessels. In addition to these perfusion-related changes,

hypertension promotes atherosclerosis, thrombosis, loss of BBB integrity, and microhemorrhages. As vascular disease progresses, the process impacts other cell types in the brain including gray and white matter. Reductions in local perfusion and PO_2 levels promote white matter injury, an effect that is thought to be a key contributor to cognitive deficits. It is through such effects that hypertension is believed to promote dementias and other forms of neurological dysfunction.

Acknowledgments

Work summarized in this chapter was supported by research grants from the National Institute of Health (HL-62984, HL-113863, NS-09465), the Department of Veteran's Affairs (BX001399), the Fondation Leducq (Transatlantic Network of Excellence), and the National Health and Medical Research Council of Australia (1053786).

References

[1] Joutel A, Faraci FM. Cerebral small vessel disease: insights and opportunities from mouse models of collagen IV-related small vessel disease and cerebral autosomal dominant arteriopathy with subcortical infarcts and leukoencephalopathy. Stroke 2014;45:1215–21.

[2] Wardlaw JM, Smith C, Dichgans M. Mechanisms of sporadic cerebral small vessel disease: insights from neuroimaging. Lancet Neurol 2013;12:483–97.

[3] Pantoni L, Gorelick PB. Cerebral small vessel disease. Cambridge, England: Cambridge University Press; 2014.

[4] Faraci FM. Protecting against vascular disease in brain. Am J Physiol 2011;300:H1566–82.

[5] Cipolla MJ, Smith J, Kohlmeyer MM, Godfrey JA. SKCa and IKCa Channels, myogenic tone, and vasodilator responses in middle cerebral arteries and parenchymal arterioles: effect of ischemia and reperfusion. Stroke 2009;40:1451–7.

[6] Badimon L, Vilahur G. Thrombosis formation on atherosclerotic lesions and plaque rupture. J Intern Med 2014;276:618–32.

[7] De Silva TM, Faraci FM. Microvascular dysfunction and cognitive impairment. Cell Mol Neurobiol 2016;36:241–58.

[8] Hillman EM. Coupling mechanism and significance of the BOLD signal: a status report. Annu Rev Neurosci 2014;37:161–81.

[9] Iadecola C. The pathobiology of vascular dementia. Neuron 2013;80:844–66.

[10] Cipolla MJ. The cerebral circulation. San Rafael (CA): Morgan & Claypool Life Sciences; 2009. p. 1–59.

[11] Tietz S, Engelhardt B. Brain barriers: crosstalk between complex tight junctions and adherens junctions. J Cell Biol 2015;209:493–506.

[12] Biancardi VC, Stern JE. Compromised blood–brain barrier permeability: novel mechanism by which circulating angiotensin II signals sympathoexcitatory centers during hypertension. J Physiol 2016;594:1591–600.

[13] Moore SM, Zhang H, Maeda N, Doerschuk CM, Faber JE. Cardiovascular risk factors cause premature rarefaction of the collateral circulation and greater ischemic tissue injury. Angiogenesis 2015;18:265–81.

CHAPTER

32

Risk Factors: Diabetes

A. Ergul[1], S.C. Fagan[2]

[1]Augusta University, Augusta, GA, United States; [2]University of Georgia College of Pharmacy, Augusta, GA, United States

INTRODUCTION

Diabetes is an endocrine disease with devastating vascular consequences [1]. In the cerebrovasculature, accelerated atherosclerosis of the large vessels is believed to contribute to complications, such as stroke and transient ischemic attacks. There is growing body of evidence that microvascular disease also contributes to stroke and other neurological diseases including Alzheimer disease and vascular cognitive impairment [1]. This chapter will summarize the epidemiological data on diabetes and cerebrovascular diseases with a focus on stroke and cognitive impairment and discuss the impact of diabetes-mediated pathological changes in cerebrovascular function and structure on stroke and cognitive impairment in preclinical models.

DIABETES AND STROKE

Diabetic patients are at two- to sixfold increased risk of stroke, compared with their age-matched counterparts, and the highest relative risks occur in young patients with type 1 diabetes [2]. As reviewed in 2012, determining the effect of glycemic control on stroke prevention has been challenging but evidence from multiple trials suggests that tight glycemic control reduces the risk of cerebrovascular disease and it takes many years of follow-up to demonstrate this benefit in patients [1]. In addition to a higher risk, acute and long-term outcomes of stroke are also worse in patients with diabetes. Patients with diabetic ischemic stroke are at a twofold higher relative risk of 30-day mortality as well as long-term disability [3]. Although there is a lack of evidence regarding optimal acute management of hyperglycemia after stroke, the American Heart Association recommends managing hyperglycemia to a target glucose level of <300 mg/dL [4]

and trials are ongoing to test the impact of glucose control in the acute stroke period. There is clear evidence that diabetes decreases the thrombolytic efficacy while increasing the risk of hemorrhagic transformation in patients receiving tissue plasminogen activator (rtPA), the only Food and Drug Administration–approved treatment for acute ischemic stroke [1]. However, there are no guidelines with respect to the use of rtPA in diabetic patients.

The underlying reasons why diabetic patients suffer to a greater extent from acute ischemic stroke remain elusive. There is no clear evidence that diabetic patients present with larger cerebral infarctions. Indeed, the landmark Trial of Org 10172 in Acute Stroke Treatment (TOAST) showed that hyperglycemia worsened outcome in nonlacunar stroke but not in lacunar stroke [1]. Evidence also suggests that lacunar infarcts resulting from occlusion of small penetrating arterioles are more common in diabetic patients [1]. It also has to be noted that silent infarcts may be more common in diabetic patients, contributing to higher incidence of cognitive impairment [1]. It is also unclear whether the duration and severity of preexisting diabetes is critical for acute stroke injury and recovery. A meta-analysis suggested that patients who present with hyperglycemia and no history of diabetes suffer the most from acute ischemic stroke [3].

DIABETES AND COGNITIVE IMPAIRMENT

Diabetic individuals develop cognitive decline with a prevalence as high as 40% in poorly controlled and long-standing diabetes [5], yet cognitive impairment remains to be one of the less understood and less studied complications of diabetes. This is in part due to the complexity and wide spectrum of cognitive deficits observed. There are also differences between type 1 and type 2 diabetes.

Primer on Cerebrovascular Diseases, Second Edition
http://dx.doi.org/10.1016/B978-0-12-803058-5.00032-1

Patients with type 1 diabetes are younger and suffer from cognitive dysfunction in multiple areas including intelligence, information processing, psychomotor efficiency, memory, learning, problem solving, and executive functioning resulting in poor school and professional performance [6]. Lack of insulin, a known neuroprotective growth factor, is believed to be an important factor in the development of these symptoms. Confounding factors, such as dyslipidemia, insulin resistance, and hypertension may alter the severity and mechanisms of cognitive impairment in type 2 diabetes, which represents more than 90% of all diabetic cases in the United States [7]. These individuals are older and develop vascular cognitive impairment, which recently replaced the term vascular dementia. In most cases, there is no full blown dementia but most experience tragic loss of intellectual and physical abilities, quality of life, creativity, and productivity. Early in the disease, this form of cognitive decline presents with a different histopathology and lacks neurofibrillary tangles and amyloid deposition that are characteristics of Alzheimer disease. However, white matter lesions including astrogliosis, microbleeds, and demyelination in the periventricular area (leukoaraiosis) are often present.

Although there may be different forms of cognitive impairment and cognitive function can be influenced by many factors in type 1 and type 2 diabetes, common findings are the association with the microvascular complications, such as retinopathy and nephropathy, and changes in cerebral blood flow. Regulation of blood flow by myogenic, metabolic, and neuronal mechanisms is key to normal brain function. Clinical studies suggest there are regional changes in cerebral blood flow characterized by a mixture of hyperperfused and hypoperfused areas in diabetes [6]. Overall there seems to be hypoperfusion of the brain. Functional MRI data show that longstanding type 1 diabetic patients with background retinopathy fail to increase regional cerebral blood flow, known as functional hyperemia during cognitive tasks, an important indicator of neurovascular uncoupling. It has been reported that there are deficits in information processing, attention, and concentration in these patients. It is clear that in all forms of cognitive deficits, vascular disease plays an important role and as such American Heart Association recently issued a statement entitled "Vascular contributions to cognitive impairment and dementia."

CEREBROVASCULAR FUNCTION AND STRUCTURE IN DIABETES—LESSONS FROM PRECLINICAL MODELS

Cerebrovasculature is important not only for maintenance of blood flow to provide nutrients and remove metabolites from this highly metabolically active organ but also for structural and functional stability of the blood–brain barrier (BBB). Undoubtedly, there is a compelling need to further our understanding of micro- and macrovascular complications of diabetes with relevance to pathogenesis and management of stroke and cognitive impairment. Development of vasculoprotective therapeutic strategies may benefit both of these complications in diabetes. Preclinical models offer an opportunity to identify the mechanisms and consequences of cerebrovascular damage due to diabetes.

Cerebrovascular Structure in Diabetes

Studies, mostly using a streptozotocin (STZ)-induced type 1 model of diabetes, demonstrated ultrastructural changes with widening of the basement membrane due to collagen deposition and amorphous nodules described as "cotton tufts." Since pericytes and astrocytic end feet sit on the basement membrane and serve as a functional bridge between the vasculature and neuronal cells of the brain, these changes disrupt the integrity of the neurovascular unit [1]. In addition, there is significant vascular remodeling that can occur as early as 4 weeks after the STZ induction of diabetes (blood glucose levels over 300 mg/dL) when vessels become tortuous. Vascular remodeling (increased tortuosity, collateral numbers, and collateral size) indicative of arteriogenesis is also observed in the Goto-Kakizaki (GK) rat, a mild and lean model of type 2 diabetes [1,8]. As expected, these alterations are associated with increased BBB permeability. Although tight junction proteins, occludin, and zona occludens are decreased, matrix metalloproteases (MMPs), especially MMP-2 and MMP-9, are increased.

Not only the structure, but also the architecture of the vasculature can change with diabetes. Just like in diabetic retinopathy, there is increased aberrant neovascularization of the brain both in the cortex and striatum in multiple models of diabetes including GK rats and db/db mice [8]. This neovascularization response includes both arteriogenesis and angiogenesis, and appears to be unique to the brain and the retina. Vascular indices, such as vascular volume, surface area, lumen diameter of penetrating arterioles, and branch density are greater in the diabetic animals (Fig. 32.1). The ratio of unperfused capillaries to total vasculature is also increased. This augmented angiogenesis is associated with poor vessel wall maturity as these vessels are leaky and lack appropriate pericyte coverage. The pattern of the neovascularization resembles tumor angiogenesis and appears as if endothelial cells lost their guidance cues and ability to prune the newly formed vessels. Although there is an overall increase in vascularization of the brain in diabetic animals, the midsection of the brain shows the greatest extent [8]. This region is the area where most hemorrhagic transformation is seen if these animals are

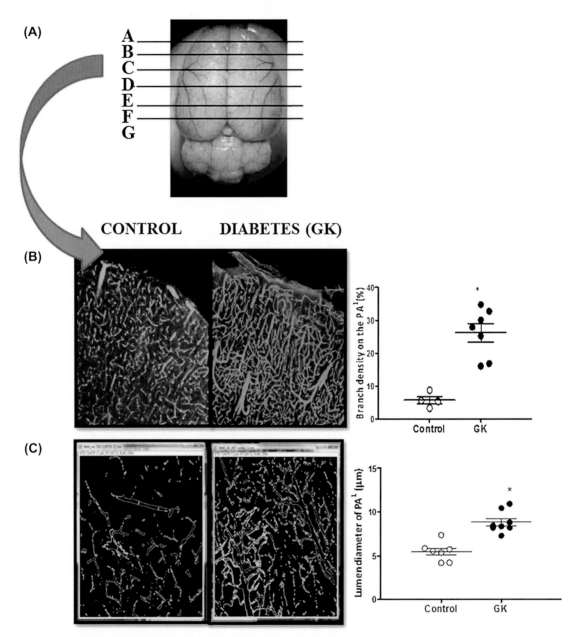

FIGURE 32.1 **Comparison of cerebral vascularization in control and diabetic Goto-Kakizaki (GK) rats.** (A) A representative image depicting the coronal sections processed using a brain matrix (sections A–H) for the measurement of neovascularization parameters. (B) Z-stack images of FITC-perfused cerebrovasculature were acquired from section C using confocal microscopy and reconstructed with the Volocity software. The analysis of the skeletonized images demonstrated increased branch density in diabetic animals. (C) Inner walls of penetrating arterioles (PA) and the immediate branched from the PA (PA[1], yellow) were outlined using the Fiji software. GK rats exhibited a significant increase in vessel diameter. *p < .01 vs. control. Mean ± SEM, n = 6–8. *Adapted from Prakash, et al., PLoS One 2013;8(2):e56264.*

subjected to ischemic stroke (see later discussion). Interestingly, vascularization and capillary branching in the dentate gyrus of the hippocampus, an area associated with memory and learning processes, is decreased in the same GK model, suggesting that there may be differences in the angiogenic response in very specialized areas of the brain. Increased vascular endothelial growth factor and reactive oxygen and nitrogen species signaling appears to be the underlying mechanism. Regulation of blood glucose by metformin prevents this dysfunctional

cerebral neovascularization in the GK model, suggesting that hyperglycemia is the upstream trigger [8].

Cerebrovascular Function in Diabetes

Cerebrovascular function is also affected negatively in diabetes. Endothelium-dependent relaxation is attenuated in part due to increased oxidative stress, disturbances in NO synthesis and production, and impairment of vascular smooth muscle ion channels. Diabetes also promotes hyperreactivity to vasoconstrictor agents, such as endothelin-1 and angiotensin II [1]. The duration of the disease influences vascular reactivity to vasoactive agents. An important property of the cerebral vasculature is its autoregulatory capacity, mostly due to myogenic behavior of the vascular smooth muscle in arterioles. Cerebral vessels constrict and dilate in response to increasing or decreasing pressure, respectively. This property is critical for maintaining a constant cerebral blood flow. Myogenic tone is increased in early diabetes and as animals age, they lose this ability, which limits the ability of the cerebrovasculature to regulate blood flow, rendering the brain more vulnerable to changes in pressure. Just like in patients with diabetes, cerebral blood flow is decreased and functional hyperemia, an indicator of neurovascular coupling, is blunted in the GK model of diabetes [9].

A UNIFYING HYPOTHESIS: RELEVANCE OF MICROCEREBROVASCULAR DISEASE TO ISCHEMIC STROKE AND COGNITIVE IMPAIRMENT

A critical review of the emerging literature in the last decade suggests that stroke outcomes in experimental models replicate clinical findings in diabetic animals showing worse functional outcomes after stroke. The use of different models of diabetes with varying disease severity also provided very valuable information with regard to infarct size, hemorrhagic transformation, and edema and their relevance to functional outcomes. For example, in STZ-induced type 1 diabetes and the db/db mouse model of type 2 diabetes, both of which present with severe hyperglycemia, infarct size and edema are greater than that observed in control animals [1]. There is also greater bleeding into the brain after stroke. GK rats, a mild and lean model as discussed earlier, present with smaller infarcts mainly localized to the striatum as compared with control animals 24 h after stroke. Again, there is more hemorrhage formation and edema than found in control animals. Despite the fact that these rats have smaller infarctions, their neurologic deficit is greater suggesting that the infarct size is not the sole determinant of poor outcomes in diabetes [1]. As discussed under "cerebrovascular structure" in the previous section, these animals display extensive cerebrovascular remodeling and neovascularization [1]. It is highly likely that newly formed immature vessels are susceptible to reperfusion injury and bleeding causing hemorrhagic transformation of the infarct. Prevention of vascular remodeling and aberrant neovascularization by chronic glycemic control with metformin before stroke decreases hemorrhage and improves functional

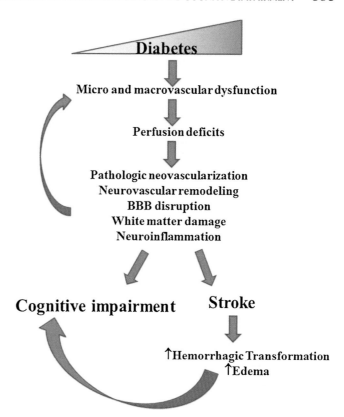

FIGURE 32.2 **Working model.** Diabetes mediates early vascular dysfunction triggering changes in cerebral blood flow, which in turn stimulates dysfunctional neovascularization and neurovascular remodeling. These pathological changes in the cerebrovasculature create a microenvironment that renders the brain more susceptible to ischemic stroke and cognitive impairment. *BBB*, blood–brain barrier.

outcomes, emphasizing the importance of vascular protection and glycemic control in diabetic stroke [1].

With respect to stroke recovery in diabetic models, our knowledge is limited. Studies showed that neuronal plasticity is greatly blunted in type 1 diabetes [10]. Interestingly, in the GK model of diabetes that presents with extensive immature vascularization at baseline and greater hemorrhagic transformation after stroke, there is significant vasoregression in the recovery period following ischemic injury. Furthermore, there is astrogliosis and swelling of astrocytic end feet. Coinciding with these changes, sensorimotor recovery as well as cognitive function is significantly impaired [11]. Metformin treatment started after stroke provides glycemic control and also prevents the massive loss of vascular tissue, in part by regulation of the redox signaling that promotes endothelial cell death.

Cognitive impairment has been demonstrated in both type 1 and type 2 models of diabetes [12]. Although most of these studies focused on neuronal changes, it is intriguing that the GK model, which presents with increased BBB permeability and cerebrovascular disease at both small and large vessels, also exhibits cognitive and memory deficits in multiple domains [11,12].

CONCLUSION

The cerebral circulation is an early target in diabetes. Endothelial dysfunction can occur very early in the disease and gradually lead to changes in cerebral blood flow that creates a hypoxic environment (Fig. 32.2), which in turn leads to pathological remodeling and neovascularization of the cerebrovasculature. Thus early interventions to improve cerebrovascular function have the potential to limit devastating cerebral complications of diabetes.

References

[1] Ergul A, Kelly-Cobbs A, Abdalla M, Fagan SC. Cerebrovascular complications of diabetes: focus on stroke. Endocr Metab Immune Disord Drug Targets 2012;12(2):148–58.

[2] Almdal T, Scharling H, Jensen JS, Vestergaard H. The independent effect of type 2 diabetes mellitus on ischemic heart disease, stroke, and death: a population-based study of 13,000 men and women with 20 years of follow-up. Arch Intern Med 2004;164(13):1422–6.

[3] Capes SE, Hunt D, Malmberg K, Pathak P, Gerstein HC. Stress hyperglycemia and prognosis of stroke in nondiabetic and diabetic patients: a systematic overview. Stroke 2001;32(10):2426–32.

[4] Jauch EC, Saver JL, Adams Jr HP, et al. Guidelines for the early management of patients with acute ischemic stroke: a guideline for healthcare professionals from the American Heart Association/American Stroke Association. Stroke 2013;44(3):870–947.

[5] Dejgaard A, Gade A, Larsson H, Balle V, Parving A, Parving HH. Evidence for diabetic encephalopathy. Diabet Med 1991;8(2):162–7.

[6] Kodl CT, Seaquist ER. Cognitive dysfunction and diabetes mellitus. Endocr Rev 2008;29(4):494–511.

[7] Sima AA. Encephalopathies: the emerging diabetic complications. Acta Diabetol 2010;47(4):279–93.

[8] Ergul A, Abdelsaid M, Fouda AY, Fagan SC. Cerebral neovascularization in diabetes: implications for stroke recovery and beyond. J Cereb Blood Flow Metab 2014;34(4):553–63.

[9] Kelly-Cobbs AI, Prakash R, Coucha M, et al. Cerebral myogenic reactivity and blood flow in type 2 diabetic rats: role of peroxynitrite in hypoxia-mediated loss of myogenic tone. J Pharmacol Exp Ther 2012;342(2):407–15.

[10] Sweetnam D, Holmes A, Tennant KA, Zamani A, Walle M, Jones P, et al. Diabetes impairs cortical plasticity and functional recovery following ischemic stroke. J Neurosci 2012;32(15):5132–43.

[11] Prakash R, Li W, Qu Z, Johnson MA, Fagan SC, Ergul A. Vascularization pattern after ischemic stroke is different in control versus diabetic rats: relevance to stroke recovery. Stroke 2013;44(10):2875–82.

[12] Duarte JM. Metabolic alterations associated to brain dysfunction in diabetes. Aging Dis 2015;6(5):304–21.

CHAPTER

33

Risk Factors: Aging

A.F. Logsdon, B.P. Lucke-Wold, C.L. Rosen, J.D. Huber

West Virginia University, Morgantown, WV, United States

INTRODUCTION

Aging is the primary, nonmodifiable risk factor for cerebrovascular disease (CVD). Two of the most detrimental diseases related to aged cerebrovascular dysfunction are stroke and vascular dementia. Aging is associated with changes to the immune system, neural networks, and vascular regulation. Vascular compliance is impaired with aging, which may contribute to vascular dementia. Aging can produce numerous detrimental outcomes to multiple organ systems. Damaged organs can release a host of inflammatory factors that compromise the cerebrovasculature. Comorbid diseases associated with peripheral organ dysfunction enhance the inflammatory response to brain damage. The risk of functional and cognitive deficits due to cerebrovascular inflammation complicates outcome in CVD. In essence, the age-related comorbidities can both increase susceptibility to CVD and worsen patient outcome.

Both stroke and vascular dementia limit the brain's blood supply. The brain requires 20% of the body's consumed energy resources, which are distributed by

Primer on Cerebrovascular Diseases, Second Edition
http://dx.doi.org/10.1016/B978-0-12-803058-5.00033-3

normal vascular perfusion. Age-related complications include atherosclerosis (AS), cerebral amyloid angiopathy, and small vessel disease. Age-related comorbidities, such as cardiovascular disease, respiratory obstructive disease, and diabetes mellitus type 2, also prevent the adequate distribution of nutrients and energy into the brain. These conditions are known to occlude vessels, disrupt vascular function, and can lead to acute areas of infarct. Most importantly, these problems reduce glucose distribution to the brain. Here, we discuss how aging is a risk factor for CVD. We examine how age-related complications disrupt vascular function and integrity, thereby increasing the risk of ischemic stroke and vascular dementia. We also highlight mechanistic changes caused by aging, and therapeutic options that may benefit the elderly clinical population.

AGED BLOOD–BRAIN BARRIER AND NEUROINFLAMMATION-AGING

The blood–brain barrier (BBB) maintains an optimal cellular milieu for normal brain function. In the aged human brain, BBB integrity is not lost but becomes more susceptible to immune and inflammatory mediators. Challenges to the BBB may account, in part, for vascular dementia often observed in the elderly [1]. The aged BBB is more permeable than a young BBB, due to changes in glial, mitochondrial, and tight junction protein function [2]. Aging also damages astrocyte podocytes and the basement membrane that comprise the neurovascular unit. Neurotoxic proteins and peripheral immune cells, typically excluded from the brain, in the aged brain, can now enter the parenchyma when BBB integrity is compromised (Fig. 33.1). These brain xenobiotics trigger neuroinflammatory cascades and generate reactive oxygen species, which both contribute to a basal level of inflammation and increased risk of neuronal cell death (Fig. 33.1). As such, it has been reported that aged animals exhibit worsened outcomes following experimental brain injury. Therefore, therapeutics designed to improve outcomes associated with CVD should focus on maintaining BBB integrity in aged animal models.

Neuroinflammation and oxidative stress are both enhanced in the aged brain. The elderly possess a chronic low-grade level of inflammation within the brain, termed *inflamm-aging* [3]. Not surprisingly, the aged rat also exhibits higher basal levels of circulating proinflammatory cytokines and other markers of oxidative stress. Not only does the aged brain have higher basal levels of inflammation, but it also loses its ability to cope with the challenges of vascular dysfunction and ischemic insults. The infiltration of toxic proteins and peripheral immune cells is exacerbated in the aged animal following neural injury possibly due to a loss in BBB integrity. The

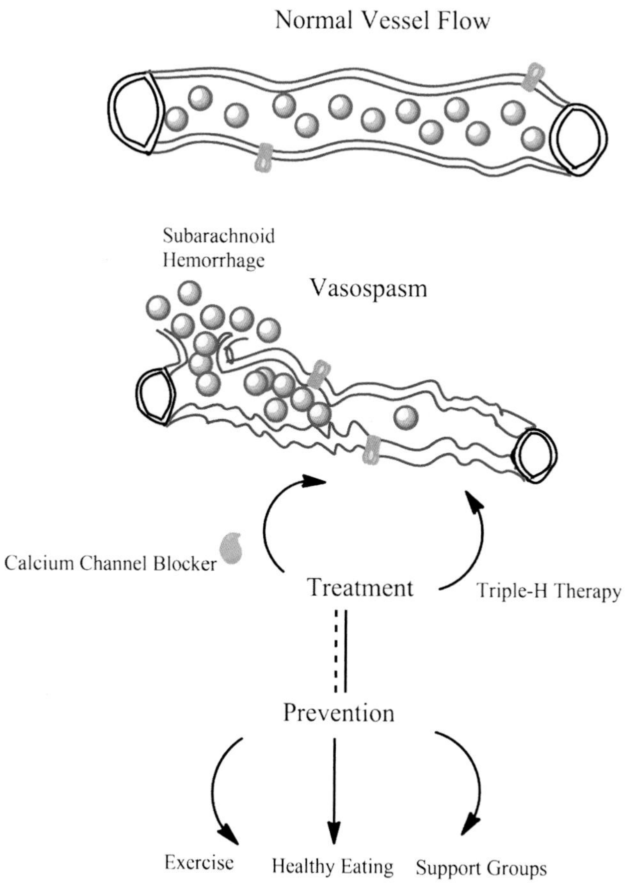

FIGURE 33.1 Schematic of young neurovascular unit (NVU), compared with an aged NVU. The young NVU is more intact and a normal extracellular milieu is maintained. The aged NVU has endothelial cells that are prone to amyloid beta (AB) deposition and atherosclerotic plaque (AP) development. Pericytes are lost and tight junction (TJ) proteins integrity is reduced. The extracellular milieu can have toxic protein (TP) and peripheral immune cell (PIC) infiltration, which exacerbates the inflammatory signals that triggers neuronal cell death. *Ast*, astrocyte; *BL*, basal lamina; *End*, Endothelial cell; *Mic*, Microglia; *Mit*, mitochondria; *Neu*, Neuron; *Nu*, nucleus; *Per*, pericyte; *RBC*, red blood cell; *VL*, vessel lumen.

infiltrating peripheral immune cells release proinflammatory signals into the brain parenchyma (Fig. 33.1). These warning signs can activate neighboring microglia, the brain's immune cell. Once a certain threshold is attained, microglia can aberrantly release additional inflammatory signals that damage neighboring neurons.

STROKE

Atherosclerosis

Atherosclerosis (AS) is common among the elderly and is known to contribute to cerebral infarction and hemorrhage. Atherosclerotic plaques build up in blood vessels over time and eventually break off and occlude smaller vessels in the brain (Fig. 33.2). Researchers are

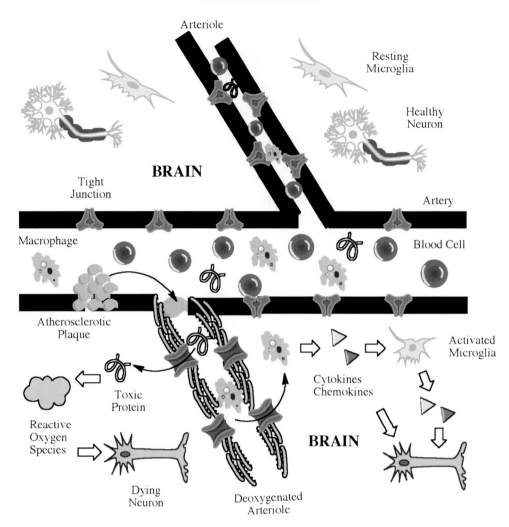

FIGURE 33.2   Schematic of cerebrovascular in the event of an atherosclerotic plaque occlusion. An occlusion would restrict blood flow from the main arteries into the smaller arterioles and into the capillaries. Restricted blood flow will reduce oxygen and nutrient delivery to important brain regions. The loss in blood flow will also reduce vascular integrity and compliance. The loss in integrity can lead to the infiltration of neurotoxin into the brain parenchyma, which can increase oxidative stress, neuroinflammation, and can cause neuronal cell death.

investigating how AS develops to reduce plaque development and possible occlusion. These occlusions restrict blood flow and oxygen delivery to important brain regions. The lack of oxygen and nutrient delivery can trigger many detrimental effects including leukocyte diapedesis, neuroinflammation, and ROS generation (Fig. 33.2). These effects are toxic to neuronal cells and may contribute to CVD. A limitation in stroke research is that aged rodents are highly resistant to AS development due to their limited exposure to a typical high-fat human diet [4]. This preclinical limitation makes it difficult to model human plaque development; however, new genetic manipulation techniques have enabled for better AS modeling.

Inflammation, oxidative stress, and immune cell accumulation are common factors that contribute to AS pathology in the brains of elderly patients [5]. Hypertension is another common risk factor for the

development of AS, and may play a crucial role in vessel wall damage. High blood pressure is a common symptom associated with AS and is known to contribute to BBB breakdown. BBB breakdown allows toxic proteins and peripheral immune cells to enter the brain parenchyma causing damage to neuronal cells (Fig. 33.2). Overall, AS treatments remain limited; however, a better understanding of AS development has helped to establish lifestyle changes as the most effective preventative strategy.

Cerebral Amyloid Angiopathy

It is well known that amyloid beta (Aβ) aggregation within the extracellular milieu is a hallmark feature of Alzheimer disease (AD). The prevailing theory is that an increase in Aβ production will increase Aβ deposition. In the elderly, Aβ can deposit into the blood vessel wall and cause endothelial cell dysfunction (Fig. 33.1). This pathology is known as cerebral amyloid angiopathy (CAA). CAA is a common disease pathology observed in the aged human brain [6]. Aβ deposition can contribute to hemorrhage, and poor clinical outcomes, with symptoms including headache, seizures, and vomiting. CAA has been associated with BBB breakdown, which is known to expose the brain to harmful toxins normally restricted to the cerebrovasculature. BBB integrity is lost over time and is associated with comorbid metabolic diseases observed in the aging population. No current treatment options exist to reduce CAA; however, more focus should be given to age-related comorbidities to combat CVD.

Lacunar Infarct

Over time brain areas with decreased perfusion can become ischemic and trigger a transient ischemic attack. Transient ischemic attack, or mini strokes, has been shown to contribute to vascular dementia over time. Approximately 10% of patients develop dementia after a mini stroke. Age-related comorbidities increase the risk for mini strokes and lacunar infarct (LI). LIs are small-vessel occlusions that often occur due to hypertension and can contribute to long-term cognitive deficits. The incidence of dementia is even higher among patients with LI than in patients experiencing a mini stroke. LIs have been shown to disrupt axonal transport and cause memory deficits [7]. BBB dysfunction is also thought to cause LI in white matter regions of the brain. Small-vessel disease can cause neurotoxic protein extravasation into the perivascular space and contribute to neuronal cell death. Unfortunately, small-vessel diseases cannot be treated with antithrombolytic therapy like large-vessel occlusions. Endothelin antagonists and neurotrophins are currently being investigated in preclinical models of small vessel diseases [8].

VASCULAR DEMENTIA

Vascular Compliance and Vasospasm

Central arterial compliance is an important indicator of whether a patient will go on to develop vascular dementia. Blood vessels of patients with vascular dementia have exhibited significantly reduced arterial and venous compliance [5]. When blood vessels struggle to respond to fluctuations in vascular load, the brain becomes at risk for transient ischemic attack. Neurodegenerative diseases, such as AD and Lewy body disease, have been shown to have vascular compliance issues related to pathologic protein accumulation. An emerging theory proposes that reduced vascular compliance can produce a vasospasm as blood enters the brain.

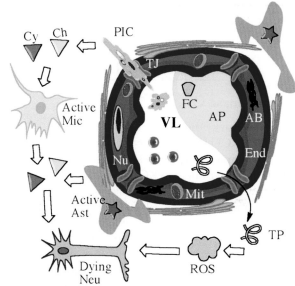

FIGURE 33.3 Schematic of vasospasm following cerebrovascular damage. Subarachnoid hemorrhage, for example, can lead to reduced vascular flow, which can trigger harmful vasospasms. Vasospasm can be treated with calcium channel blockers and triple-H therapy (hypertension, hypervolemia, and hemodilution therapy). Vasospasm can also be prevented with exercise, healthy eating, and support groups.

A vasospasm is a well-known, and sometimes life-threatening, outcome observed in patients with CVD (Fig. 33.3). A vasospasm can eventually damage brain tissue and possibly contribute to cognitive decline in the elderly clinical population. Vasospasm is frequently associated with age-related comorbidities, such as hypertension and cardiovascular disease. In patients suffering an ischemic stroke, a vasospasm can often occur after surgical intervention. The underlying triggers are still being investigated, but vasospasm may be a causative factor for vascular dementia. Calcium channel blockers, endothelin-1 antagonists, and simple lifestyle interventions all seem to be the most effective strategies to combat vasospasm in patients with CVD [9].

Cognitive Decline

A human's cognitive performance diminishes with age, and becomes even worse in patients with CVD [1]. The National Institutes of Health has now classified dementia into five groups: AD, dementia with Lewy bodies, frontotemporal dementia, vascular dementia, and mixed dementias. Among all classifications of human dementia, one thing remains the same—all dementias are correlated with CVD. Proteins associated with neurodegeneration, such as tau protein, Aβ, and Lewy bodies, often have perivascular distribution. Ongoing work investigates how vascular dysfunction contributes to pathologic changes associated with aging. Preclinical models of neural injury are now being used to elucidate the link between vascular damage and subsequent cognitive decline [10].

THERAPEUTIC TARGETS

Protein Kinase C

Aging changes the function and expression of protein kinase C (PKC) [10]. In particular, age-related decreases in hippocampal PKC expression have been linked to spatial memory deficits. PKC activation has been shown to improve cognitive performance in aged animals [11]. As such, modulators of PKC activity have become a viable target for improving outcomes in patients with stroke and vascular dementia. In the case of cerebrovascular function, PKC is known to translocate to the endothelial membrane to regulate vasoconstriction and vasospasm. PKC dysfunction may be attributed to age-related deficits in endothelial membrane stabilization mechanisms. Furthermore, age-related comorbidities are often associated with changes in PKC activity. Aberrant PKC has been shown to increase matrix metalloproteinases and reduce BBB integrity. The loss in BBB integrity can release neurotoxins into the brain parenchyma and cause neuronal cell death. We must determine how age-related comorbidities change PKC activity to develop novel therapeutics for age-related CVD.

Apolipoprotein E

The effects of apolipoprotein E (ApoE) are far reaching and extend beyond the metabolism of cholesterol. It has been proposed that ApoE may regulate smooth muscle cells. The ApoE receptor is expressed in endothelial cells surrounding AS plaques. In particular, the ApoE ε4 allele has been considered a genetic risk factor for BBB dysfunction and vascular dementia [12]. Patients with at least one ApoE ε4 allele often develop CVD and early-onset AD. In human patients with the ApoE ε4 allele a study observed accelerated pericyte loss and BBB breakdown [13]. Further work is needed to elucidate the role that ApoE ε4 and other genetic factors play in the development of CVD. Currently, no treatments are available that specifically target genetic risk factors for age-related diseases.

RESEARCH DISCONNECT

Over the past few decades, CVD has been primarily modeled in young animals. However, the number one risk factor for CVD is aging; begging the question: why study age-related diseases in young animal models? Many factors come into play as we age including comorbid diseases, reduced vascular compliance, and hypertension. Young animal models fail to exhibit these additional factors causing a translational disconnect between research and the clinic. Aged animals exhibit different biochemical and behavioral responses compared with young animals. As a result, clinical trials for stroke treatments have continued to fail. Moving forward research of CVD should be studied in aged rodent models or transgenically modified mice. To elucidate how age-related CVD contributes to morbidity, it is first essential to understand age-related morbidities in relevant models.

CONCLUSION

Stroke and vascular dementia are two extremely prevalent diseases affecting the elderly. These diseases are associated with extensive cerebrovascular compromise. In this chapter, we look at aging as the primary risk factor for CVD. BBB dysfunction and augmented neuroinflammation are two vital components contributing to CVD in the aging population. Age-related comorbidities also play a key role in CVD development. We discuss how stroke is an age-related disease and how aging contributes to AS, CAA, and LI. We also discussed how vascular dementia is an age-related disease and how aging contributes to vascular compliance and vasospasm. All of the age-related factors contribute to cognitive decline in some way.

Current treatment options seek to protect the brain from additive injury and aim to improve injury outcome. We lay out viable treatment options including endothelin-1 inhibition, manipulation of the ApoE ε4 allele, and modulation of PKC. Treatment options are not currently available for several of these factors and should be employed in conjunction with lifestyle change. Until these concerns are addressed, cognitive impairment will continue to burden elderly patients and clinical trials for CVD will continue to fail. Ongoing preclinical studies should use aged animal models or relevant transgenic models to truly evaluate age-related disease. This initial step will increase translatability of the preclinical findings and provide better therapeutic solutions for CVD.

References

[1] Montagne A, Barnes SR, Sweeney MD, Halliday MR, Sagare AP, Zhao Z, et al. Blood–brain barrier breakdown in the aging human hippocampus. Neuron 2015;85:296–302.

[2] Zlokovic BV. Neurovascular pathways to neurodegeneration in Alzheimer's disease and other disorders. Nat Rev Neurosci 2011;12:723–38.

[3] Franceschi C, Campisi J. Chronic inflammation (inflammaging) and its potential contribution to age-associated diseases. J Gerontol A Biol Sci Med Sci 2014;69(Suppl. 1):S4–9.

[4] Breslow JL. Mouse models of atherosclerosis. Science 1996;272:685–8.

[5] Grinberg LT, Thal DR. Vascular pathology in the aged human brain. Acta Neuropathol 2010;119:277–90.

[6] Jaunmuktane Z, Mead S, Ellis M, Wadsworth JD, Nicoll AJ, Kenny J, et al. Evidence for human transmission of amyloid-beta pathology and cerebral amyloid angiopathy. Nature 2015;525:247–50.

[7] Hinman JD, Lee MD, Tung S, Vinters HV, Carmichael ST. Molecular disorganization of axons adjacent to human lacunar infarcts. Brain 2015;138:736–45.

[8] Bath PM, Wardlaw JM. Pharmacological treatment and prevention of cerebral small vessel disease: a review of potential interventions. Int J Stroke 2015;10:469–78.

[9] Muroi C, Seule M, Mishima K, Keller E. Novel treatments for vasospasm after subarachnoid hemorrhage. Curr Opin Crit Care 2012;18:119–26.

[10] Lucke-Wold BP, Turner RC, Logsdon AF, Simpkins JW, Alkon DL, Smith KE, et al. Common mechanisms of Alzheimer's disease and ischemic stroke: the role of protein kinase C in the progression of age-related neurodegeneration. J Alzheimers Dis 2015;43:711–24.

[11] Hongpaisan J, Xu C, Sen A, Nelson TJ, Alkon DL. PKC activation during training restores mushroom spine synapses and memory in the aged rat. Neurobiol Dis 2013;55:44–62.

[12] Bell RD, Winkler EA, Singh I, Sagare AP, Deane R, Wu Z, et al. Apolipoprotein E controls cerebrovascular integrity via cyclophilin A. Nature 2012;485:512–6.

[13] Halliday MR, Rege SV, Ma Q, Zhao Z, Miller CA, Winkler EA, et al. Accelerated pericyte degeneration and blood–brain barrier breakdown in apolipoprotein E4 carriers with Alzheimer's disease. J Cereb Blood Flow Metab 2015;36:216–27.

CHAPTER

34

Risk Factors: Gender and Sex

F. Sohrabji[1], P.D. Hurn[2]

[1]Texas A&M Health Science Center, Bryan, TX, United States; [2]University of Michigan, School of Nursing, Ann Arbor, MI, United States

INTRODUCTION

Reductions in stroke incidence and mortality are ongoing in the United States; however, the burden of stroke remains high for women. Stroke is the third leading cause of death for women, and roughly 60% of stroke deaths in the United States are in women. In part, this is due to the fact that women are frequently older at the time of stroke, display more nonclassical stroke symptoms that may alter early stroke recognition, and tend to have more severe stroke damage and persistent disability. Since women live longer than men, stroke-related disability and institutionalization more strongly affects women than men throughout much of the developed world. Clinical and basic science research is beginning to unveil the many unique aspects of stroke pathobiology in women vs. men, and some of these findings may lead to new, gender-specific therapy in future.

This chapter focuses on risk rather than outcome and addresses ischemic stroke, except where indicated otherwise. Although traditional cardiovascular and stroke risk factors apply to women, there are also risk factors that are more prevalent in women than men (e.g., atrial fibrillation, diabetes mellitus, migraine with aura, and depression) and female-specific risk factors that bear consideration (e.g., pregnancy, menopausal hormone status and use of replacement therapies, preeclampsia, and oral contraceptive use). Greater detail on risk and risk reduction strategies is available in the American Heart Association/American Stroke Association Guidelines for the Prevention of Stroke in Women [1].

EXAMPLES OF TRADITIONAL, WELL-STUDIED RISK FACTORS: HYPERTENSION AND HYPERLIPIDEMIA

Hypertension is the best recognized and most common modifiable risk factor for stroke in both sexes. However, some studies suggest that women are more likely to be hypertensive than men and have a higher risk of first stroke in the presence of hypertension than men. Data from the National Health and Nutrition Evaluations Survey (NHANES) also emphasize that there are time-based sex differences in the prevalence of hypertension. Although men may be more likely to experience hypertension before 45 years of age, prevalence of hypertension in women is more common after 65 years.

There are also differences on average in lipid profiles for adult women vs. men; in general women have a more favorable lipid profile on average than men. For example, women have a higher prevalence of elevated total cholesterol (greater than 200 mg/dL) than do men, whereas men have a higher prevalence of low-density lipoprotein greater than 130 mg/dL and high density lipoprotein less than 40 mg/dL [2].

CLINICAL RISK FACTORS MORE PREVALENT IN WOMEN: ATRIAL FIBRILLATION, DIABETES MELLITUS, MIGRAINE WITH AURA, AND DEPRESSION

Atrial Fibrillation

Atrial fibrillation (AF) is the most common arrhythmia and a serious public health problem as its incidence is steadily increasing worldwide, particularly in the developed countries. AF is associated with a four- to fivefold increased risk of ischemic stroke and an increased association with increased death and disability from stroke. Most epidemiological series indicate that AF is diagnosed with equal frequency in both sexes and more frequently in men, but AF in women is diagnosed at a greater age. Sex differences in epidemiology and prognosis of AF are also beginning to unfold (e.g., the Global Burden of Disease 2010 study and the Framingham Heart Study). Some, but

not all, reports indicate that women with AF demonstrate a different prognosis as compared with men, suffering a greater incidence of stroke with higher mortality. Numerous studies of patients with AF have reported female sex as an independent risk factor for stroke (for review, see Ref. [3]). The causal links for this observation are not yet known; hypotheses center on differences in the degree of heart failure and hypercoagulability related to left atrial enlargement with blood stasis. Female sex has been incorporated into risk stratification tools that can be applied to decision making for AF anticoagulation prophylaxis, although the utility and appropriateness of sex-specific anticoagulation is largely untested [1,3].

Diabetes Mellitus

There is significant evidence that women with diabetes are at increased risk for stroke relative to their male counterparts. The biological mechanisms behind this sex difference are not well understood at present. Furthermore, part of the excess risk has been proposed to be related to disparities in diabetes management between men and women. Obesity is also strongly related to stroke risk, as measured by increased body mass index and waist circumference. Abdominal obesity is particularly common in postmenopausal women, and as discussed later, stroke risk rises during the menopausal years.

Migraine With Aura

The relationship between migraine with aura and stroke has been under investigation for decades, but only recently have the biological factors in this relationship begun to be elucidated, for example, the role of cortical spreading depression, endothelial dysfunction, and vasospasm. The absolute risk of stroke in migraineurs is relatively low, and factors, such as migraine frequency and association of aura with nausea and vomiting alter the risk profile, as do the use of oral contraceptives (OCs), smoking, and other intertwined stroke risk factors. Women are four times more likely than men to experience migraines, and the association between migraine aura and ischemic stroke is higher in women [1]. Women under 55 years of age who experience migraine with aura have increased risk of both ischemic stroke and intracerebral hemorrhage.

Depression

Much of the evidence around depression focuses on its role in retarding recovery after stroke, but it bears emphasis that depression is associated with increased risk for cerebral ischemic events in both sexes. Although numerous studies demonstrate that depression and psychological stressors are more common in women, there is not full agreement that these factors enhance risk for stroke in a sex-dependent manner. Further study in this area is essential, including the development of sex-specific stroke risk scoring.

FEMALE-SPECIFIC CLINICAL RISK FACTORS: MENOPAUSE, HORMONE REPLACEMENT THERAPY, ORAL CONTRACEPTIVES, AND PREGNANCY

Estradiol has been linked to stroke risk in three ways. First, there is a temporal link between the average age of menopause and the ages of 45–54 years, a time in which stroke risk doubles in women relative to men. This link has led to the view that loss of ovarian hormones contributes to stroke risk. Second, postmenopausal hormone replacement therapy (HRT) has been shown to contribute to stroke risk, although the mechanisms surrounding this connection are far from clear. Third, OCs, particularly those with high estrogen content, pose a stroke risk for young women.

Menopause and HRT

The average age of menopause in the United States is 52 years, and is usually preceded by a period of 5–7 years, called perimenopause, where women will experience menopause-related vasomotor symptoms and declining titers of ovarian hormones. In the Framingham Heart Study, where the average age of menopause was 49 years, stroke risk was compared in two age groups: women 42–54 years old and 55+ years old. Women who had a younger age at menopause had a higher risk for stroke as compared with women who experienced menopause at older ages. However, other studies have argued that this association does not hold after adjusting for age, hypertension, and physical activity. Furthermore, time since menopause was also not significantly associated with cerebral infarction or hemorrhagic stroke. In general, ethnic and geographic variables, age at menarche, diet, and other factors more likely to affect stroke risk than chronological age at menopause.

The connection between HRT and stroke incidence presents a more complicated picture. In some of the first retrospective case–control studies, neither estrogen therapy nor estrogen/progestin therapy increased stroke risk, although it was notable in several reports that ischemic or hemorrhagic stroke rates increased in women during the first 6 months of therapy. Better known is the Women's Health Initiative (WHI), a study of women aged 50–85 years, which showed HRT increased stroke risk. In this randomized, double-blind, placebo-controlled multicenter trial, stroke risk was increased in women with estrogen alone or complexed with progestins [4], particularly in older women (60–69 years of age). Observational trials have shown less agreement in the role of HRT as a risk factor for stroke. Some of the lack of agreement among the various studies may be due to

differences among the subjects' age and time lapse since menopause at which HRT was initiated. Furthermore, in some studies, subjects randomized to HRT differ in important physiological traits (blood pressure, lean body mass) from women in the general population. In general, there appears to be an interaction between HRT and hypertension in that increased stroke risk is more prominent among hypertensive women who use HRT, unlike normotensive women who use HRT with no increase in stroke risk. The route and dose of estrogen therapy may also influence stroke risk, for example, the use of high-dose transdermal patches has been found particularly to increase stroke risk. Lastly, the Heart and Estrogen-Progestin Replacement Study that evaluated HRT in women with known coronary artery disease showed no increase in the number of stroke, the number of fatal strokes, or transient ischemic attacks. In summary, the complexities of research in this area support the current view that HRT is an important factor to consider when appraising stroke risk in women. In addition, available evidence indicates that HRT should not be used for primary or secondary prevention of stroke in postmenopausal women.

Oral Contraceptives

Many studies have emphasized that among healthy, normotensive users, OC usage is not a major risk factor for women. However, a number of studies conclude that premenopausal women, typically at low risk for ischemic stroke, may be at increased risk with the use of OCs. An early meta-analysis of OC use and stroke risk [5] indicated that, although the overall risk for ischemic stroke is increased 2.75 times by OC use, the absolute risk was small, in part because stroke is relatively rare in this population and in part because of the overshadowing power of risk factors, such as hypertension, diabetes mellitus, hypercholesterolemia, and smoking. Most importantly, OC formulations have been optimized over many years by utilizing third-generation progestins and reducing estrogen dosage as a means of mitigating the risk of adverse cardiovascular events. For example, usage of high-dose, but not low-dose, estrogen formulations increase stroke risk substantially, as shown in World Health Organization studies.

Pregnancy

Pregnancy and stroke is a highly important area of women's health and deserving of more substantial study than can be found in this review (for comprehensive reviews, see Refs. [6,7]). It should also be emphasized that complications of pregnancy and/or the postpartum period, such as preeclampsia, gestation diabetes, and hypertension, are all relevant to stroke risk and are predictive of long-term risk of chronic disease as early as the year following delivery. Although pregnancy is widely considered to be sex-specific stroke risk factor (as much as a 13-fold increase for ischemic stroke in some studies), there is little consensus at present on the level of, or the mechanisms underlying, the increased risk. Our relatively recent understanding that the woman's risk is greatest during the postpartum period is important but poorly understood. Current hypotheses include associated hormonal changes over the course of the pregnancy, increases in clotting activity related to placental expulsion, and the natural elevation of estradiol that increases plasma coagulation factors, such as plasminogen activator inhibitor.

In general, stroke risk is increased with advanced maternal age (35+ years), ethnicity (African-American), and pregnancy-associated complications, such as preeclampsia and gestational diabetes. Preeclampsia and eclampsia are diagnosed in approximately one-third of stroke cases during pregnancy, associated with severe systolic hypertension, and particularly carry risk for hemorrhagic stroke. There is also some suggestion that this pathology may influence stroke risk years after pregnancy. For example, after multivariate adjustment for age, race, and number of pregnancies, women with a history of preeclampsia may be more likely to have a non–pregnancy-related ischemic stroke.

ROLE OF SEX IN FUNDAMENTAL STROKE PATHOBIOLOGY

Much clinical and animal-based research has demonstrated a sexually dimorphic epidemiology in ischemic stroke that persists well beyond the menopausal years and is evident in children even before puberty. However, once stroke occurs, outcomes in women are not favorable as compared with men. Women can sustain greater damage with high mortality, and considerable loss of quality of life. So much can be gained by understanding the details of sex-specific stroke pathobiology.

One research focus has been to understand the contribution of sex hormones to stroke outcomes, specifically estrogen, which has been largely touted to be neuroprotective. In older animals that best simulate clinical stroke circumstances, estrogen treatment does not reliably protect the brain from experimental stroke and may, in fact, exacerbate stroke damage [8,9], unlike the steroid's protective effect in the young. It has also become quite clear that sex hormones cannot fully account for the nuances of female vs. male cerebrovascular vulnerability. Accordingly, a second research focus has been on the role of genetic sex, i.e., the comparative vulnerability of XX vs. XY cells. Ongoing experimental results challenge the long-held assumption that the vulnerability and biological disease in cells are independent of the sex of those cells.

For example, at the bench, technical solutions now allow sex-stratified, in vitro cell systems to be explored. Findings suggest that male and female neurons can proceed to cell death via differing molecular pathways. Male neurons are more susceptible to glutamate- and oxidant-induced injury, whereas female neurons are more sensitive to triggers of cell death by apoptosis. Similarly, male vs. female astrocytic or cerebral microvascular endothelial cells show large differences in vulnerability to oxygen–glucose deprivation, a type of "in vitro" ischemia [10], emphasizing that sex, as well as gender, influences many aspects of stroke risk, vulnerability and outcome. Sex differences in the response to stroke are also reflected in processes as fundamental as the epigenome.

Not surprisingly, sex differences in cell death pathways foreshadow sex-specific responses to neuroprotectants. Thus in experimental studies, minocycline, an antiinflammatory, reduces infarct volume in males but not females, whereas inhibition of Let-7f, a small noncoding RNA and a new class of therapeutic molecules, is effective in females but not in males [9].

References

[1] Bushnell C, McCullough LD, Awad IA, Chireau MV, Fedder WN, Furie KL, et al. Guidelines for the prevention of stroke in women: a statement for healthcare professionals from the American Heart Association/American Stroke Association. Stroke 2014;45(5):1545–88.
[2] Go AS, Mozaffarian D, Roger VL, Benjamin EJ, Berry JD, Blaha MJ, et al. Heart disease and stroke statistics–2014 update: a report from the American Heart Association. Circulation 2014;129(3):e28–92. http://dx.doi.org/10.1161/01.cir.0000441139.02102.80.
[3] Poli D, Antonucci E. Epidemiology, diagnosis and management of atrial fibrillation in women. Int J Women's Health 2015;7:605–14.
[4] Wassertheil-Smoller S, Hendrix SL, Limacher M, Heiss G, Kooperberg C, Baird A, et al. Effect of estrogen plus progestin on stroke in postmenopausal women: the Women's Health Initiative: a randomized trial. JAMA 2003;289(20):2673–84. http://dx.doi.org/10.1001/jama.289.20.2673.
[5] Gillum LA, Mamidipudi SK, Johnston SC. Ischemic stroke risk with oral contraceptives: a meta-analysis. JAMA 2000;284(1):72–8.
[6] Gongora MC, Wenger NK. Cardiovascular complications of pregnancy. Int J Mol Sci 2015;16(10):23905–28. http://dx.doi.org/10.3390/ijms161023905.
[7] Grear KE, Bushnell CD. Stroke and pregnancy: clinical presentation, evaluation, treatment and epidemiology. Clin Obstet Gynecol 2013;56(2):350–9.
[8] Sohrabji F, Welsh CJ, Reddy DS. Sex differences in neurological diseases. In: Shansky R, Johnson Jr JE, editors. Sex differences in the central nervous system. London: Academic Press; 2015. p. 297–323.
[9] Selvamani A, Sathyan P, Miranda RC, Sohrabji F. An antagomir to microRNA Let7f promotes neuroprotection in an ischemic stroke model. PLoS One 2012;7(2):e32662. http://dx.doi.org/10.1371/journal.pone.0032662.
[10] Liu M, Hurn PD, Roselli CE, Alkayed NJ. Role of P450 aromatase in sex-specific astrocytic cell death. J Cereb Blood Flow Metab 2007;27(1):135–41.

C H A P T E R

35

Mechanisms of Stroke Recovery

K.L. Ng, S.T. Carmichael
University of California Los Angeles, Los Angeles, CA, United States

INTRODUCTION

The consequences of stroke can be chronically disabling and overall represents the leading cause of long-term disability [1]. A greater understanding of the mechanisms of stroke recovery will facilitate clinical, pharmacologic, and cell-based approaches to stroke recovery and rehabilitation. The infarct core at the center of the territory of occluded artery results in the cellular death of neurons and supporting cellular elements (glial cells) and thus, ultimately, the impairment of sensory and motor function. The surviving peri-infarct zone brain region adjacent to the infarct core and connected brain regions shows heightened neuroplasticity, allowing the remapping of sensory and motor functions from damaged areas and thus promoting recovery from stroke [1]. Mechanisms of recovery in these areas are multifaceted and hold to the inherent limitations of the endogenous cellular and molecular mechanisms involved in neural repair and to the time frame in which stroke recovery is possible. A useful framework to understand the mechanisms of stroke recovery consists of three fundamental themes: (1) timeline for stroke recovery, (2) excitatory neuronal signaling, and (3) promotion of growth programs, such as neurogenesis, oligodendrogenesis, and axonal sprouting. Here we discuss the mechanisms and time course of tissue repair after ischemic stroke.

TIMELINE FOR STROKE RECOVERY

Following ischemic stroke, the old adage "time heals all wounds" may not necessarily apply, but the machinery set forth by poststroke recovery events evolves with each phase of recovery. The normal course of stroke recovery can be divided into three overlapping phases: acute, subacute, and chronic.

The acute phase occurs during the initial minutes to days of the ischemic event. The initial clinical improvement occurs independent of patient behavior and physical or environmental stimulation and is attributed to the resolution of the effects of edema, diaschisis, and return of circulation within the penumbra. Edema surrounding the infarct may disrupt nearby neuronal functioning. As the edema subsides, these neurons may regain function, allowing for early recovery. The normalization of blood flow to the penumbra, the nonischemic area of low to moderate blood flow surrounding the infarct core, can facilitate early recovery by restoration of excitability and network responsiveness to previously nonfunctioning neurons. Diaschisis, the inhibition or suppression of surrounding cortical tissue or of cortical regions at a distance that are interconnected with the infarct core, may be partially resolved with reduction in edema, and neuronal function may return if the connected area of the brain is intact. Further improvement in diaschisis will continue into the subacute phase.

A few days to 3 months after the ischemic event is the subacute phase. During this period the endogenous mechanisms of neural repair are primed and engagement in rehabilitation is potentially most beneficial for stroke recovery. Patients who have had stroke exhibit most recovery from physical and sensorimotor impairments and near-plateaus on standardized scales of impairment, self-care, and mobility. Studies in humans and animal models show that most recovery from impairment during this period occurs as a result of both natural reorganization and increased responsiveness to enriched environments and training. Patients who engage in rehabilitation with repetitive practice have greater improvement in those skills (skilled use of the extremities) than in those who did not undergo practice. Beyond the 3-month period, the chronic phase, improvement on standardized impairment scale, such as the

Primer on Cerebrovascular Diseases, Second Edition
http://dx.doi.org/10.1016/B978-0-12-803058-5.00035-7

Fugl-Meyer scale, is substantially less than that seen during the subacute phase. Nevertheless, improvement can still be induced by rehabilitative therapies at this stage.

EXCITATORY NEURONAL SIGNALING

Poststroke neural repair involves remapping of functions within the peri-infarct cortex. Functional recovery in this peri-infarct tissue involves changes in neuronal excitability that alter the brain's representation of motor and sensory functions. The concentration of $GABA_A$ (γ-aminobutyric acid receptor) synaptic subunit receptors that mediate phasic GABA inhibition, which leads to rapid and transient hyperpolarization, is reduced in the peri-infarct cortex. In contrast to phasic GABA inhibition, tonic or extrasynaptic GABA inhibition, which controls the overall membrane potential of postsynaptic neurons and their propensity to fire in response to excitatory inputs, increases in the peri-infarct cortex. Immediately after ischemic stroke, there is an increase in GABAergic tonic inhibition in the peri-infract tissue, thus reducing excitability. In the acute period the increase in tonic inhibition represents an innate safety mechanism to minimize neuronal damage. However, the increase in tonic neuronal inhibition persists beyond the acute period and effectively hinders recovery [2,3]. Poststroke treatment with a subtype-selective inverse agonist of the $\alpha 5$-subunit of extrasynaptic $GABA_A$ receptors restores tonic inhibition to prestroke levels and leads to early and sustained improvement of motor function [2]. In line with the protective role of tonic inhibition immediately after a stroke, the treatment is only effective if delayed until 3 days after the stroke; initiating treatment before this period increases infarct size [2]. Evidence suggests that the overall effect of decreased phasic inhibition and increased tonic inhibition is a diminished neuronal excitability that, when reversed, promotes functional remapping and behavioral recovery.

It is significant and not surprising that an experimentally induced increase in glutamatergic excitability in the subacute phase of stroke was found to parallel recovery. Whereas early administration of an agonist for the glutamate α-amino-3-hydroxy-5-methyl-4-isoxazolepropionic acid (AMPA) receptor increases infarct size, a 5-day delay in treatment does not change infarct size but improves limb motor control in a dose-dependent way, suggesting a beneficial effect on recovering tissue [3,4]. Furthermore, blocking AMPA receptor signaling for 5 days immediately after the stroke induction impairs motor control. These studies establish a timeline from acute to subacute stroke for the manipulation of GABA and glutamate signaling, which enhances recovery if performed 3–5 days after the stroke but increases stroke damage if performed before this period. Interestingly,

AMPA receptor signaling in the peri-infarct cortex is enhanced by the local brain-derived neurotrophic factor (BDNF) release. BDNF is known to regulate synaptic transmission and activity-dependent plasticity, to promote survival and sprouting of dendrites and axons, and to promote synaptogenesis and angiogenesis. The recovering circuitry in the peri-infarct area may be primed for activity-dependent BDNF release and therefore neurologic recovery [3].

PROMOTION OF GROWTH PROGRAMS

Neurogenesis

In neurogenesis in adults, the neural progenitor cells (NPCs) reside in the mammalian brain and continuously generate new neurons. New neurons are generated by NPCs in restricted regions, namely, the subventricular zone (SVZ) of the lateral ventricle (LV) and the dentate gyrus of the hippocampus, and then migrate to their final destination, maturing into functional neurons. Ischemic stroke elicits a regenerative response in the peri-infarct area to increase the number of NPCs in SVZ and provides signals for migration of newborn immature neurons into the peri-infarct area. Although only a small number of newborn immature neurons reach maturation and integrate into the damage area, these cells exert a positive trophic influence on the ischemic microenvironment to promote recovery [5]. In rodent stroke models, there is an increase in mitotic activity within the SVZ that peaks between 7 and 10 days, subsequently decreases during 3–5 weeks after stroke, and thereafter continues at lower levels [6]. The stroke signals that initiate the neurogenic response are myriad and yet to be fully elucidated; however, there is a close association between poststroke neurogenesis and the vasculature. Newborn immature neurons are found to be associated with blood vessels after stroke [5]. Pharmacologic blockade of angiogenesis after stroke significantly reduces the number of immature neurons in the peri-infarct tissue [5]. Animal studies have identified specific ligand–receptor signaling systems that causally link neurogenesis and angiogenesis in peri-infarct tissue. Stromal-derived factor 1 (SDF-1) and angiopoietin 1 (Ang-1) that are expressed by endothelial cells and upregulated by hypoxia in angiogenic blood vessels in the peri-infarct cortex stimulate the migration of immature neurons from the SVZ through their receptors Tie2 and CXCR4 (C-X-C chemokine receptor type 4). The blockade of these receptors prevents the migration of immature neurons into the peri-infarct cortex [5]. Neurovascular coupling also exists with erythropoietin (EPO)—pharmacologic doses of EPO enhance cell proliferation in the SVZ and increase the appearance of immature neurons in peri-infarct tissue [7]. Conditional knockdown of the EPO receptor in the nervous system reduces the number of immature neurons in the peri-infarct cortex [8]. With neurogenesis the SVZ may serve as a constant reservoir of new neurons after stroke, with the potential to extend the window of opportunity for therapeutic treatments. EPO has not been proven as a therapeutic candidate for stroke recovery because of the side effects but it illustrates the utility of enhancing neural progenitor responses with a drug that activates the neurovascular unit after stroke [7].

Oligodendrogenesis

Mature oligodendrocytes are vulnerable to ischemic stroke and when injured are unable to generate myelin leading to impairment of axonal function, thus contributing to the neurologic deficits of stroke. In white matter ischemia, demyelination and axonal degradation occur rapidly in the infarct core; however, steady restoration of oligodendrocytes and remyelination has been observed in the peri-infarct area. Oligodendrocyte progenitor cells (OPCs), present in the adult mammalian brain, continuously differentiate into mature myelinating oligodendrocytes in the gray and white matter throughout life and are important in the process of neurologic recovery after stroke. Animal studies show that the proliferative response of OPCs peaks 5–7 days after stroke induction and declines 14–28 days later. Unfortunately, counter to this regenerative response is failure of these proliferating OPCs to differentiate into mature oligodendrocytes [9]. There is growing data in nonstroke models that suggest inducing neuronal activity via direct stimulation or through a learning paradigm to promote proliferation, survival of oligodendrocytes, and myelination. Currently there is no data to support activity-dependent oligodendrogenesis in ischemic stroke models, but it remains a potentially rewarding avenue of exploration. Further in-depth understanding of oligodendrogenesis will be important in remyelination, preventing axonal degeneration, and neurologic recovery from ischemic stroke.

Axonal Sprouting

After stroke within the peri-infarct tissue, existing neurons can sprout new connections allowing for a degree of regional cortical remapping that closely correlates with recovery after stroke in humans. In this process, surviving adult cortical neurons must engage in a neuronal growth program: develop a growth cone, extend an axon, and form new connections. Axonal sprouting is intricate and requires a complex of molecules that tie membrane-signaling events to growth cone phosphorylation cascades, cytoskeletal rearrangement, and new gene transcription. Gene expression occurs at specific time points in the axonal sprouting process after stroke, with an initial phase, the trigger, beginning 1–3 days after the ischemic event. The initiation and maintenance phases of the sprouting response occur 7 and 14 days after stroke, respectively. The maturation phase occurs at 28 days during which axonal sprouting has formed new patterns of connections that can be detected anatomically [1]. Neuronal-growth-promoting genes are induced in sequential waves after the stroke that correspond to these distinct time points in the sprouting response. The adult brain is normally inhibitory to axonal sprouting; however, stroke induces a time window allowing for a permissive environment for axonal sprouting. Animal studies support the notion of a 2–3-week poststroke window for axonal sprouting, allowing for the induction of neuronal-growth-promoting genes, the removal of growth inhibitory proteins, and a delay in the reintroduction of many growth inhibitory genes.

The time window illustrates one clear principle: axonal sprouting after stroke requires reduction in the growth inhibitory environment in addition to the induction of a growth-promoting program within the peri-infarct neurons.

Stroke induces the production of growth inhibitory proteins, such as neurite outgrowth inhibitor protein A (NogoA), chondroitin sulfate proteoglycans, the semaphorin IIIa receptor neuropilin 1, and myelin-associated glycoprotein and of growth-promoting proteins including growth associated protein 43 (GAP43), cytoskeleton-associated protein 23 (CAP23), myristoylated alanine-rich C-kinase substrate (MARCKS), and small proline repeat rich protein 1 (SPRR1). In humans and models of stroke, GAP43, a growth cone phosphoprotein, is highly linked to axonal sprouting. GAP43 messenger RNA and protein expression is increased from 3 days to 1 month after stroke [10]. This time-dependent graded expression correlates to a marker for synapses, suggesting a progression toward axonal sprouting in peri-infarct cortex and demonstrates a window for axonal sprouting. Growth inhibitory proteins, such as NogoA, are targets for therapy; counteracting their effects strongly promotes axonal sprouting in models of stroke and spinal cord injury [7]. Ephrin-A5 that is induced in reactive astrocytes within the peri-infarct area inhibits axonal sprouting thus limiting the motor recovery. Blocking ephrin-A5 signaling after stroke stimulates axonal sprouting and improves functional recovery. More interestingly, although behavioral activity alone increased axonal sprouting after stroke, blocking ephrin-A5 in combination with behavioral activity induced further axonal sprouting in cortically relevant areas [11]. The result suggests that although the poststroke environment may support axonal sprouting, there is a limit to cortical reorganization even in the presence of a relevant activity, e.g., poststroke rehabilitation, and the blockade of inhibitory cues in conjunction with an activity-based intervention has the potential to extend the window of neurologic recovery. In general, several candidate axonal growth inhibitors have been shown to play a role in blocking axonal sprouting and recovery in stroke, such as ephrin-A5 and NogoA, and efforts are underway to develop biologics or small molecules that block these growth inhibitors.

CONCLUSIONS

Stroke produces an abrupt change in the functional status of adjacent brain areas but mobilizes neural repair and reorganization in such areas. Tonic GABA and AMPA glutamate receptor signaling highlights the importance of timing in neuronal excitability after stroke. Acute GABA signaling counteracts the deleterious effects of early glutamate release, while continued signaling into the chronic phase prevents the brain from full recovery. There is a close association between neurogenesis and angiogenesis, which share similar molecular signaling systems during recovery. Although oligodendrogenesis is explored extensively in the development and other white matter injury models, it is an understudied topic in the setting of stroke. Conceivably the final outcome of axonal sprouting represents the greatest anatomic evidence of recovery after stroke, but it is clear that there are limitations to the endogenous machinery potentially lowering the ceiling of neurologic recovery. The molecular and cellular mechanisms of repair after stroke are complex. However, there is a window for optimal repair in the subacute phase of stroke, and during this period the interaction of environment/rehabilitation with pharmacologic therapies will positively affect the natural course of recovery after stroke.

References

[1] Carmichael ST. Cellular and molecular mechanisms of neural repair after stroke: making waves. Ann Neurol May 2006;59(5):735–42.

[2] Clarkson AN, Huang BS, Macisaac SE, Mody I, Carmichael ST. Reducing excessive GABA-mediated tonic inhibition promotes functional recovery after stroke. Nature November 11, 2010; 468(7321):305–9.

[3] Carmichael ST. Brain excitability in stroke: The yin and yang of stroke progression. Arch Neurol February 2012;69(2):161–7.

[4] Clarkson AN, Overman JJ, Zhong S, Mueller R, Lynch G, Carmichael ST. AMPA receptor-induced local brain-derived neurotrophic factor signaling mediates motor recovery after stroke. J Neurosci March 9, 2011;31(10):3766–75.

[5] Ohab JJ, Fleming S, Blesch A, Carmichael ST. A neurovascular niche for neurogenesis after stroke. J Neurosci December 13, 2006;26(50):13007–16.

[6] Thored P, Arvidsson A, Cacci E, et al. Persistent production of neurons from adult brain stem cells during recovery after stroke. Stem Cells March 2006;24(3):739–47.

[7] Carmichael ST. Translating the frontiers of brain repair to treatments: starting not to break the rules. Neurobiol Dis February 2010;37(2):237–42.

[8] Tsai PT, Ohab JJ, Kertesz N, et al. A critical role of erythropoietin receptor in neurogenesis and post-stroke recovery. J Neurosci January 25, 2006;26(4):1269–74.

[9] Rosenzweig S, Carmichael ST. The axon-glia unit in white matter stroke: Mechanisms of damage and recovery. Brain Res October 14, 2015;1623:123–34.

[10] Stroemer RP, Kent TA, Hulsebosch CE. Neocortical neural sprouting, synaptogenesis, and behavioral recovery after neocortical infarction in rats. Stroke November 1995;26(11):2135–44.

[11] Overman JJ, Clarkson AN, Wanner IB, et al. A role for ephrin-A5 in axonal sprouting, recovery, and activity-dependent plasticity after stroke. Proc Natl Acad Sci USA August 14, 2012;109(33):E2230–9.

NEUROPROTECTION

36

N-Methyl-D-Aspartate Receptors Remain Viable Therapeutic Targets for Stroke

C. Potey, Y.T. Wang

University of British Columbia, Vancouver, BC, Canada

INTRODUCTION

Stroke is a major cause of morbidity and mortality in North America and worldwide [1]. The mechanisms underlying neuronal injuries following stroke are still not fully understood, and involve multiple factors. One particularly well-characterized factor is excitotoxicity that results at least partly from the overactivation of N-methyl-D-aspartate subtype glutamate receptors (NMDARs) [2–4]. However, this idea has been challenged by several failed stroke clinical trials [5,6]. The reason underlying the discrepancy between basic research and clinical trials remains unknown. A few explanations have been suggested and these include, but are not limited to, (1) the inability to use NMDAR antagonists at protective doses due to the potential blocking of normal brain function and neuronal survival; (2) the inability to administer these drugs within the therapeutic window due to time required for patient transport and diagnosis; and (3) poor trial design and heterogeneity of the patient pool [6]. Although the last factor stresses the need for better designed clinical trials, advancements in our understanding of distinct intracellular pathways linking NMDAR activation to neuronal death through experimental studies may allow scientists to develop more promising NMDAR-based stroke treatments that target specific death signaling pathways without affecting the normal functions of the receptor. This increased specificity not only translates into reduced side effects but also extends the therapeutic window in which the drug may be efficaciously administered.

In this chapter, we will address the roles of NMDARs in regard to the pathophysiology of stroke, and discuss how they can be implicated in early and late consequences. We will also discuss the implications of some existing and future pharmacological agents aimed at

NMDARs as potential stroke therapeutics (see Fig. 36.1 for a graphical summary).

N-METHYL-D-ASPARTATE SUBTYPE GLUTAMATE RECEPTORS AT THE ACUTE PHASE OF STROKE

NMDARs are ionotropic glutamate receptors highly permeable to cations. They mediate rapid excitatory synaptic transmission. The activation of NMDARs requires two concomitant stimuli: binding of glutamate and of the coagonist glycine, and membrane depolarization to remove magnesium cations blocking the channel pore at resting potential. Many neurological processes depend on the physiological activation of NMDA receptors, including synaptic plasticity, learning, memory, and neuronal survival. The NMDAR is a heterotetramer containing subunits from three different families, GluN1, GluN2 (A-D), and GluN3 (A-B). Every tetramer is composed of two GluR1 subunits, and at least two GluN2 subunits. It is now known that the composition of the tetramer influences ion permeability, location of the receptor on the membrane, and downstream effectors coupling [7]. GluN2A-containing NMDARs are mostly located at the synapse, thus playing a physiological role in excitatory neurotransmission. Their activation is associated with many brain functions and neuronal survival. On the contrary, GluN2B-containing NMDARs are preferentially extrasynaptic and are implicated in mediating neuronal death during early stage of neurodevelopment and under pathological conditions.

The physiological activation of NMDARs is mandatory for basal brain functioning, whereas their pathological overactivation has been linked to a specific type of cell death, referred to as excitotoxicity. Excitotoxicity has been

Primer on Cerebrovascular Diseases, Second Edition
http://dx.doi.org/10.1016/B978-0-12-803058-5.00036-9

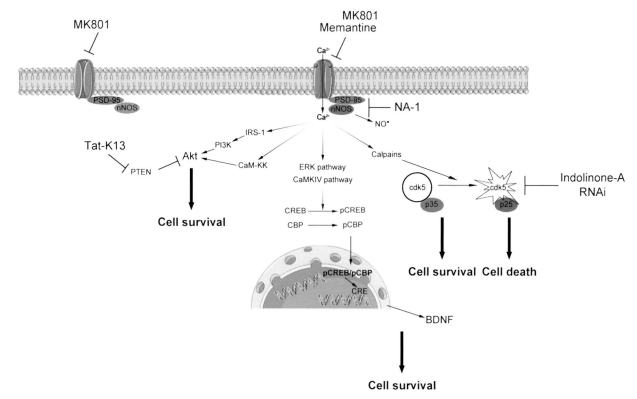

FIGURE 36.1 **Major pathways downstream of the N-methyl-D-aspartate receptor and their pharmacological interventions.** *BDNF,* brain-derived neurotrophic factor; *CaMKIV,* calcium/calmodulin-dependent protein kinase type IV; *CaM-KK,* calcium-calmodulin-dependent-protein-kinase kinase; *CBP,* CREB-binding protein; *cdk5,* cyclin-dependent kinase five; *CRE,* CREB response elements; *CREB,* cAMP response element-binding protein; *ERK,* extracellular signal-regulated kinases; *IRS-1,* insulin-receptor substrate-1; *MK-801,* dizocilpine; *NA-1,* commercial name of TAT-NR2B9c; *nNOS,* neuronal nitric oxide synthase; *pCBP,* phosphoCBP; *pCREB,* phosphoCREB; *PI3K,* phosphoinositide-3 kinase; *PSD-95,* postsynaptic density protein-95; *PTEN,* phosphatase and tensin homolog; *RNAi,* RNA interference. *Adapted from the Servier Medical Art database, accessible from* http://www.servier.com/Powerpoint-image-bank.

described in various neurological diseases; its role in ischemic stroke is prominent and constitutes one of the early steps of the pathophysiological cascade initiated following vascular occlusion. Indeed, downstream from the blocked artery, the drastic reduction in blood flow and subsequent reduction in oxygen and nutrient supplies puts a great stress on cells. One of the initial responses in neurons is an increased release of glutamate. This, associated with the compromised glutamate uptake due to the failure of the energy-dependent glutamate transporters, leads to extracellular glutamate accumulation. The accumulation results in a dual effect contributing to neuronal death: (1) a major intracellular influx of calcium responsible for the activation of proteases, nucleases, and other enzymes that will degrade substrates and produce metabolites in excess like NO· or reactive oxygen species; and (2) a strong and prolonged stimulation of receptors, including extrasynaptic NR2B-containing NMDARs, followed by the activation of various downstream effectors [8].

Given the magnitude of this phenomenon, NMDARs appear to be significant pharmacological targets for acute stroke. Dizocilpine (MK-801) is a noncompetitive antagonist of NMDARs. It was found to be neuroprotective when

administered preventively and up to 2h following experimental stroke. The demonstration of its neuroprotective potential in rodents led to the development of numerous antagonists with different pharmacological profiles: competitive at the glutamate-binding site (selfotel), competitive at the glycine-binding site (gavestinel), ion channel blockers (aptiganel), and NR2B selective (CP-101,606). Promising preclinical results led to several clinical trials in stroke and in traumatic brain injury; however, these trials proved to be disappointing, as no benefit was shown. Furthermore, it appeared that the widespread NMDAR blockade was responsible for serious psychotomimetic and cardiovascular side effects, undetected in rodents. Some of the trials (selfotel and aptiganel) were even stopped prematurely for safety reasons [9]. In addition to these concerns, the use of these molecules is limited by a very narrow time window, as NMDAR activation is a very early event and these interventions become useless once the downstream pathways coupled to the activated receptor are initiated. It is not surprising in that context that the trials of nonspecific NMDAR antagonists were then dropped [10].

However, one antagonist, memantine, currently used to treat patients with Alzheimer disease (AD), still seems of interest because of its distinct pharmacodynamic profile. Memantine is an NMDAR antagonist. As such, it binds only to the activated receptor, preventing the ion flux to go through the channel, and also dissociates rapidly from the closed channel. Under physiological conditions, the synaptic NMDAR ion channel opens only transiently by the presynaptically released glutamate, and hence is not substantially affected by memantine. Under excitotoxic conditions, elevated extracellular glutamate concentrations keep the NMDA channel open, and so accessible to memantine. Thus, memantine interferes with NMDARs in case of glutamate overload under pathological conditions and does not impair their physiologic activation. The available experimental data are in favor of a neuroprotective effect of memantine at the acute phase of stroke [10]. At the beginning of 2016, there were two clinical trials evaluating memantine in stroke recovery in the clinicaltrials.gov database (NCT02144584; NCT02535611).

With the exception of memantine, several large-scale stroke clinical trials have failed to confirm the efficacy of NMDAR antagonist therapies [6]. Research for effective NMDAR-based stroke therapies has taken a new direction. To be safe, efficient, and usable within a time window allowing potential clinical use, perhaps the target should not be the receptor itself, but the latter steps of its downstream pathways. This new strategy sparked a lot of new research interest in trying to decipher the underlying mechanisms involved and to design new therapeutic approaches either promoting NMDAR-dependent neuronal survival or halting death pathways.

Pathways Leading to Cell Death [4,7]

The Postsynaptic Density Protein-95/Neuronal Nitric Oxide Synthase Pathway

Many intracytoplasmic proteins establish interactions with the cytoplasmic tail of NMDARs. Among them, the scaffolding protein postsynaptic density protein-95 (PSD-95) binds to GluN2 subunits, and interacts with the neuronal isoform of the nitric oxide synthase (nNOS). After the NMDAR activation, PSD-95 undergoes a conformational change that brings nNOS closer to the channel pore through which the calcium flux is traveling, leading to the calcium-dependent activation of nNOS.

This interaction is important for intracellular signaling under physiological conditions, wherein NO· is produced at low levels as a fast-acting messenger. In contrast, under ischemic conditions, NO· is produced in great quantities and exerts a deleterious role contributing to excitotoxic neuronal damage. The GluN2/PSD-95/nNOS interaction has been demonstrated to be increased under ischemic conditions. The inhibition of nNOS by genetic deletion or pharmacological antagonism is neuroprotective in stroke; however, it completely suppresses its enzymatic activity and is deleterious in the long run

due to its physiological implications. Knocking out PSD-95 to abolish the GluN2/PSD-95/nNOS interaction seems a more promising and safer strategy. It is linked to a reduction of the NMDAR-dependent production of NO· and to reduced excitotoxic stress.

It is in this perspective that the Tat-NR2B9c peptide was designed. Tat-NR2B9c is a small interference peptide that mimics the last nine amino acids of the carboxy-terminal extremity of GluN2B, and is able to disrupt the GluNR2/PSD-95/nNOS complex through dissociating PSD-95 from GluN2. It has few effects on the physiological function of the NMDAR, as it does not affect NMDAR currents, NMDAR-dependent synaptic plasticity, and function of nNOS. Indeed, under ischemic conditions, Tat-NR2B9c (commercial name NA-1) has been shown to be protective against excitotoxicity and reduces nNOS activation. It is neuroprotective in experimental stroke in rodents and nonhuman primates.

These promising results led to a phase 2 study in testing its therapeutic potential in reducing the procedure-induced ischemic brain damage in a subpopulation of patients undergoing endovascular repair of aneurysms (Evaluating Neuroprotection in Aneurysm Coiling Therapy, NCT00728182). Following this successful clinical trial, a phase 3 study (Field Randomization of NA-1 Therapy in Early Responders, NCT02315443) was designed to assess the efficacy of NA-1 as a neuroprotectant in patients with stroke and it commenced in early 2015.

Calpains

Calpains are calcium-dependent endopeptidases. Among the different identified isotypes, calpain-I is activated after NMDAR stimulation. This leads to major proteolysis, resulting in brain damage. Calpain activation is NMDAR-dependent as it requires calcium influx through the channel. It, however, may also lead to intracellular calcium overload by degrading the membrane calcium–sodium exchanger, NCX3. This phenomenon is NR2B dependent, as it is blocked by NR2B-selective antagonists, but not by NR2A-selective inhibitors.

The selective inhibition of calpains is neuroprotective in vitro within a broader time frame than NMDAR blockade and with no effect on NMDAR channel function, which is indicative of a downstream function for calpains. It is interesting to note that NMDAR subunits GluN1, GluN2A, and GluN2B are calpain substrates, and that NMDAR activation leads to calpain cleavage of the cytoplasmic tail of these subunits. The downstream signaling is then disrupted and results in altered receptor function. Similar to nNOS, as mentioned earlier, calpains have important physiological functions by themselves and are required for the maintenance of cerebral homeostasis. Direct blockade of their enzymatic activity may potentially be responsible for side effects. The identification of further downstream effectors is required to avoid this pitfall.

Among these effectors, p35 seems to be of great interest. p35 is a brain-specific regulatory activator of the cyclin-dependent kinase 5 (cdk5); the complex formed by p35 and cdk5 is important for intracellular signaling and neuronal survival. p35 is a calpain substrate and is cleaved into p25. This truncated form of p35 also has activating properties on cdk5, but their association leads to a deregulated activation of cdk5, and a change in substrate specificity. The levels of p25 are increased in brain tissue following experimental stroke and in postmortem brain tissue from patients with AD. The pharmacological inhibition, or knocking out, of cdk5 prevents p25-mediated tissue damage in experimental stroke. As for today, pharmacological inhibitors such as indolinone-A, or RNAi, are neuroprotective in rodents. But further research is needed to characterize their efficacy and safety profile in clinical use.

Pathways Associated With Cell Survival [4,7]

The Akt/Glycogen-Synthase Kinase-3/Phosphatase and Tensin Homolog Pathway

Akt is a serine/threonine kinase that is critically required for neuronal survival. Its activity depends on synaptic activity. The entry of calcium following the opening of NMDAR channel pore triggers this cell survival pathway via potential mechanisms. It leads to the phosphorylation of the insulin-receptor substrate-1 (IRS1), phosphorylated IRS1 then binds and activates phosphoinositide-3 kinase (PI3K), and PI3K in turn activates Akt. The calcium influx may also activate the Akt pathway via the calcium-calmodulin-dependent-protein-kinase kinase, of which Akt is a substrate. Akt will in turn phosphorylate various substrates, thereby either activating cell survival signaling or inhibiting cell death signaling molecules. One of the substrates is the glycogen-synthase kinase-3 (GSK3), involved in cell death signaling. Akt phosphorylation reduces GSK3 activity, thus inhibiting its downstream cell death pathway. The PI3K/Akt pathway plays a major role in cell survival, and is the direct or indirect target of many neuroprotective strategies in stroke. In vitro and in vivo data support a protective role for Akt in stroke. Ischemia-reperfusion triggers the activation of mitogen-activated protein kinases that are part of apoptotic pathways, like JNK. The activation of the Akt pathway can exert its neuroprotective actions by negatively regulating JNK signaling.

The phosphatase and tensin homolog (PTEN) is a lipid/protease phosphatase and a negative regulator of the PI3K/Akt pathway. It opposes PI3K/Akt signaling by dephosphorylating and inactivating the PI3K/Akt cascade. In particular, PTEN forms complexes with GluN2B-containing NMDARs via interactions with the cytoplasmic domain of GluN1. Activation of the NMDARs also induces the interaction between PTEN and PSD-95. This interaction has been shown to strengthen extrasynaptic NMDAR currents, thereby contributing to cell death. Supporting its anti-cell survival roles by opposing PI3K/Akt pathway, evidence indicates that low levels of cytoplasmic PTEN are associated with the activation of the Akt pathway and neuroprotection. In addition, PTEN also has a nuclear location where it functions as a tumor suppressant and participates in the regulation of cell division. Its translocation from the nuclear to the cytosol compartment is thought to be a major mechanism in promoting tumorigenesis. Evidence accumulated suggests that there is a post-stroke increase in PTEN nuclear translocation through a GluN2B-containing NMDAR-dependent death pathway.

Consistent with a critical role of stroke-related PTEN nuclear translocation in mediating ischemic/excitotoxic neuronal damage, an interference peptide, Tat-K13, which inhibits the PTEN nuclear translocation without interfering with PTEN cytoplasm-located functions, was developed. The K13 peptide has the potential of being a novel therapeutic agent following stroke. It protects cells against excitotoxic stress in vitro, and is neuroprotective in vivo when administered up to 6h after stroke onset in animal models.

The cAMP Response Element Binding Protein/Brain-Derived Neurotrophic Factor Pathway

The cAMP response element-binding protein (CREB) is a key transcription factor for neuronal survival. Once activated, CREB binds to specific DNA sequences called CREB response elements, and to its coactivator CREB-binding protein (CBP). This dimer then induces or represses the expression of target genes including c-fos, brain-derived neurotrophic factor (BDNF), neuroenkephalines, and so on.

The phosphorylation and activation of both CREB and CBP is NMDAR dependent, mostly via two pathways: the fast-acting calcium/calmodulin-dependent protein kinase type IV pathway, and the delayed extracellular signal-regulated kinases (ERK) pathway. NMDARs, therefore, are implicated in the up- or down-regulation of these target genes. Synaptic NMDARs, particularly those containing GluN2A subunits, are implicated in the expression of BDNF in response to neuronal activity. Contrarily, the activation of extrasynaptic, mostly GluN2B-containing, NMDARs inhibit BDNF expression. The upregulation of the expression of BDNF is also self-maintained. Once released, BDNF will bind to its Trk receptors and set off the ERK-dependent signaling cascade, leading to a sustained BDNF expression.

BDNF is a major growth factor known to be involved in neuronal and synaptic growth and differentiation. It is a major player in poststroke tissue repair. It has been shown to be neuroprotective in vitro and in vivo. Physiologically, its production is enhanced after stroke via an NMDAR-dependent manner. Many neuroprotective strategies exert their effects, directly or indirectly, by activating the CREB/BDNF pathway. However, to date, there has been no glutamatergic therapeutic agent promoting this pathway. Developing pharmacological agents that may positively modulate synaptic receptors, mostly GluN2A containing, and/or negatively modulate extrasynaptic receptors, largely GluN2B containing, may be a promising strategy.

N-METHYL-D-ASPARTATE SUBTYPE GLUTAMATE RECEPTORS AND THE LATE CONSEQUENCES OF STROKE

Glutamate, present in excess at the acute phase of stroke, exerts deleterious effects on the cerebral parenchyma. However, the normal functioning of the glutamatergic system is also critically important during later recovery processes following stroke. Glutamate signaling is involved in poststroke brain plasticity. During stroke recovery, glutamate signaling is increased in the periinfarct cortical areas where it is involved in cortical remapping. This effect appears to be α-amino-3-hydroxy-5-methyl-4-isoxazolepropionic acid receptor (AMPAR) dependent, and is mandatory for sensorimotor recovery [11]. NMDARs, on the other hand, based on their well-documented role in learning and memory, could play a role in mediating the cognitive consequences of stroke.

Involvement of N-Methyl-D-Aspartate Subtype Glutamate Receptors in Learning and Memory

The pivotal role of NMDARs in various memory processes, especially spatial learning, working memory, object recognition, and olfactory memory, has been established by numerous experimental studies using a combination of behavioral and electrophysiological methods.

NMDARs appear to mediate certain forms of synaptic plasticity necessary for learning and memory, particularly long-term potentiation (LTP) and long-term depression (LTD), occurring at the glutamatergic synapses in the hippocampus and amygdala. LTP and LTD, respectively, represent a strengthening and a weakening of synaptic transmission, and are thought to be mechanistic substrates for many aspects of learning and memory. Genetic or pharmacological reduction in NMDAR functions in CA1 results in impaired hippocampal LTP and spatial learning deficits. On the contrary, the enhancement of NMDAR-dependent synaptic transmission results in improved learning and better mnesic performances. Interfering with NMDARs after the learning step has occurred does not impair memory restitution. Similar results supporting such a role for NMDARs in learning but not in consolidating memory have been reported in rodents, nonhuman primates, and humans [12].

LTP and LTD are expressed, in response to cerebral activity, as an alteration in neurotransmitter release at the presynaptic level and/or by a change in the density of AMPARs on the postsynaptic membrane. The interplay between NMDARs and AMPARs is of the highest importance. After glutamate release, the influx of sodium through the ion channel coupled to AMPARs induces membrane depolarization. The depolarization and simultaneous glutamate binding lead to NMDAR activation. AMPAR activation is necessary for NMDAR activation. The calcium influx crossing through the activated NMDARs will activate intracellular cascades leading to persistent alterations in function and/or trafficking (increase in exocytosis for LTP and endocytosis for LTD) of AMPARs, leading to the respective expression of LTP and LTD [13].

NMDARs have been suggested to have a direct impact on the direction of the resulting activity-dependent plasticity—activation of synaptic, largely GluN2A-containing NMDARs promotes the induction of LTP, whereas activation of extrasynaptic, predominantly GluN2B-containing NMDARs favors the induction of LTD [14]. Although the underlying mechanisms are still to be precisely identified, the subunit specificity may be in part attributed to the difference of these subtype receptors in coupling with distinct intracellular cellular signaling molecules.

Poststroke Cognitive Impairment and Dementia

Cognitive impairment and dementia are a frequent yet overlooked consequence of stroke. They form a complex entity as this pathology regroups dementias from various etiologies: vascular, neurodegenerative, or from a mixed origin. It is estimated that one of ten patients with stroke will develop dementia after a first stroke and one of three after a multiple occurrence of stroke. Clinical studies report a prevalence ranging from 7% to 41%, based on the stroke type and on the population included [15]. Its clinical expression can be delayed after stroke, although the actual onset is thought to be precocious. The severity and time course of the pathology depend on the stroke type, the location of the lesion, the presence of vascular risk factors, and the presence of prestroke dementia. One hypothesis is that stroke primes neurodegenerative mechanisms leading ultimately to cognitive impairment and dementia.

Although the pathophysiology is still unclear, accumulating experimental data, although still scarce, provide some insights into the mechanisms underlying poststroke cognitive impairment. Models of global cerebral ischemia mimic the blood flow reduction and subsequent cerebral ischemia occurring during cardiac arrest. These models induce direct hippocampal lesions, entailing impairment of NMDAR-dependent LTP and deficits in spatial learning and memory. Animals subjected to transient focal ischemia, as in the middle cerebral artery occlusion model, also develop spatial memory deficits, but in a delayed fashion. The impairment of hippocampal NMDAR-dependent LTP is seen as early as 24h after stroke, evidenced by the absence of cell death and before any cognitive deficit is observed. Typically this occurs 4–5weeks after the procedure [16]. The precise mechanisms involved in the poststroke impairment of LTP remain unknown. Whether it directly relies on NMDARs or it is an indirect consequence of other disturbances, like abnormal gamma-aminobutyric acid inhibition, has yet to be determined.

There is still a lack of human data to corroborate the hypothesis of an NMDAR-dependent dysfunction in mediating the poststroke cognitive impairment. A postmortem histological study reported a loss of glutamatergic synapses in the temporal cortex of patients suffering from poststroke vascular dementia, and another revealed neuronal atrophy in the hippocampus of patients suffering from delayed poststroke dementia. Although no other lesions usually associated with AD, such as amyloid plaques or neurofibril tangles, were found [17], it has been suggested that poststroke dementia shares some features with AD. In AD and other chronic neurodegenerative diseases, glutamate receptors are activated by low but sustained concentrations of glutamate. This phenomenon, named slow excitotoxicity, leads to a delayed excitotoxic cell death. It is thought to be a consequence of energy deficit to and/or of an abnormal functioning of the NMDAR [10]. Whether it is implicated in the cerebral dysfunction associated with poststroke dementia has yet to be investigated.

Pharmacological Interventions for Poststroke Dementia

As the pathophysiology is being deciphered, there is still no treatment specifically designed for poststroke dementia. Since NMDARs are involved in both acute neuronal damage following stroke and in maintaining normal synaptic plasticity and function of learning/memory, they constitute interesting targets upon which novel therapeutic agents can be developed for treating poststroke dementia. To this end, it is interesting to note that memantine has been approved for use in patients suffering from mild to moderate AD, and exerts beneficial effects in patients with vascular dementia [15]. In addition, novel approaches targeting downstream steps of NMDAR pathway could also be of interest, but would require a proper preclinical assessment first.

CONCLUSION

NMDARs play a major part in the pathophysiology of ischemic stroke at both acute and delayed phases. Their roles at the acute phase is well known and was the starting point for essential research on developing NMDAR-based neuroprotectants for stroke, with promising therapeutic candidates. Data on their implications in the cognitive consequences of stroke at the later phase are still scarce. Based on their physiological importance in normal learning and memory, and their pathological involvement in many forms of chronic neurodegeneration and dementia, they also appear to be promising therapeutic targets.

Acknowledgments

CP is the recipient of a postdoctoral fellowship granted by the regional council of Nord-Pas-De-Calais, France; and YTW is the holder of the Heart Stroke Foundation of British Columbia and Yukon Chair in Stroke Research. The authors also thank Ms. Rebecca Wiens for her excellent editorial assistance.

References

[1] Mozaffarian D, Benjamin EJ, Go AS, et al. Heart disease and stroke Statistics—2016 update a report from the American heart association. Circulation 2016;133:e38–60.

[2] Hardingham GE, Fukunaga Y, Bading H. Extrasynaptic NMDARs oppose synaptic NMDARs by triggering CREB shut-off and cell death pathways. Nat Neurosci 2002;5:405–14.

[3] Aarts MM, Arundine M, Tymianski M. Novel concepts in excitotoxic neurodegeneration after stroke. Expert Rev Mol Med 2003;5:1–22. http://dx.doi.org/10.1017/S1462399403007087.

[4] Lai TW, Zhang S, Wang YT. Excitotoxicity and stroke: identifying novel targets for neuroprotection. Prog Neurobiol 2014;115:157–88.

[5] Gladstone DJ, Black SE, Hakim AM, and the Heart and Stroke Foundation of Ontario Centre of Excellence in Stroke Recovery. Toward wisdom from failure: lessons from neuroprotective stroke trials and new therapeutic directions. Stroke 2002;33:2123–36.

[6] Tymianski M. Stroke in 2013: disappointments and advances in acute stroke intervention. Nat Rev Neurol 2014;10:66–8.

[7] Lai TW, Shyu W-C, Wang YT. Stroke intervention pathways: NMDA receptors and beyond. Trends Mol Med 2011;17:266–75.

[8] Moskowitz MA, Lo EH, Iadecola C. The science of stroke: mechanisms in search of treatments. Neuron 2010;67:181–98. http://dx.doi.org/10.1016/j.neuron.2010.07.002.

[9] Muir KW. Glutamate-based therapeutic approaches: clinical trials with NMDA antagonists. Curr Opin Pharmacol 2006;6:53–60.

[10] Hoque A, Hossain MI, Ameen SS, et al. A beacon of hope in stroke therapy—blockade of pathologically activated cellular events in excitotoxic neuronal death as potential neuroprotective strategies. Pharmacol Ther 2016;160:159–79.

[11] Carmichael S. Brain excitability in stroke: the yin and yang of stroke progression. Arch Neurol 2012;69:161–7. http://dx.doi.org/10.1001/archneurol.2011.1175.

[12] Rezvani AH. Involvement of the NMDA system in learning and memory. In: Animal models of cognitive impairment. 1st ed. Frontiers in neuroscienceBoca Raton: CRC Press/Taylor & Francis; 2006.

[13] Takeuchi T, Duszkiewicz AJ, Morris RGM. The synaptic plasticity and memory hypothesis: encoding, storage and persistence. Phil Trans R Soc Lond B Biol Sci 2014;369:20130288.

[14] Shipton OA, Paulsen O. GluN2A and GluN2B subunit-containing NMDA receptors in hippocampal plasticity. Phil Trans R Soc Lond B Biol Sci 2014;369:20130163.

[15] Ihara M, Kalaria RN. Understanding and preventing the development of post-stroke dementia. Expert Rev Neurother 2014;14:1067–77. http://dx.doi.org/10.1586/14737175.2014.947276.

[16] Li W, Huang R, Shetty RA, et al. Transient focal cerebral ischemia induces long-term cognitive function deficit in an experimental ischemic stroke model. Neurobiol Dis 2013;59:18–25. http://dx.doi.org/10.1016/j.nbd.2013.06.014.

[17] Kalaria RN. Cerebrovascular disease and mechanisms of cognitive impairment evidence from clinicopathological studies in humans. Stroke 2012;43:2526–34.

CHAPTER

37

Neuroprotectants: Reactive Oxygen Species (ROS) Based

E.D. Hall

University of Kentucky College of Medicine, Lexington, KY, United States

INTRODUCTION

Oxygen-free-radical-induced membrane lipid peroxidation (LP) is a highly validated secondary injury mechanism that occurs in focal and global cerebral ischemia and subarachnoid hemorrhage (SAH). It has been firmly established as a major contributor to multiple aspects of ischemic, postischemic, and posthemorrhagic pathophysiologic conditions caused by the oxidative damage to lipids and proteins of the neural cell membrane.

SOURCES OF REACTIVE OXYGEN SPECIES IN CEREBRAL ISCHEMIA AND SAH

Superoxide Radical

The primordial radical formed in most biologic processes is superoxide radical ($O_2 \cdot^-$). Within the ischemic nervous tissue, a number of sources of superoxide radical are operative within the first minutes and hours of the onset of ischemia, particularly after postischemic reperfusion. These sources include the arachidonic acid cascade (i.e., prostaglandin synthase and 5-lipoxygenase activity), enzymatic oxidation (i.e., monoamine oxidase) or autoxidation of biogenic amine neurotransmitters (e.g., dopamine), mitochondrial leak, xanthine oxidase activity, and the oxidation of extravasated hemoglobin. Over the first few postischemic hours and days, activated microglia and infiltrating neutrophils and macrophages provide additional sources of $O_2 \cdot^-$.

Superoxide, which is formed by the single electron reduction of oxygen, may act as either an oxidant (electron acceptor) or reductant (electron donor). Although superoxide itself is reactive, its direct reactivity toward biologic substrates in aqueous environments is questioned. Moreover, once formed, superoxide undergoes spontaneous dismutation to form hydrogen peroxide (H_2O_2) in a reaction that is markedly accelerated by the enzyme superoxide dismutase (SOD) ($O_2 \cdot^- + O_2 \cdot^- + 2H^+ \rightarrow H_2O_2 + O_2$).

In solution, superoxide actually exists in equilibrium with the hydroperoxyl radical formed by the protonation of superoxide ($O_2 \cdot^- + H^+ \rightarrow HO_2 \cdot$). The pKa of this reaction is 4.8 and the relative concentrations of $O_2 \cdot^-$ and $HO_2 \cdot$ depend on the H^+ concentration. Therefore, at a pH around 6.8 the ratio of $O_2 \cdot^- / HO_2 \cdot$ is 100/1, whereas at a pH of 5.8 the ratio is only 10/1. Thus under conditions of tissue acidosis of a magnitude known to occur within the ischemic nervous system, a significant amount of the $O_2 \cdot^-$ formed will exist as $HO_2 \cdot$. Furthermore, compared with $O_2 \cdot^-$, $HO_2 \cdot$ is considerably more lipid soluble and is a far more powerful oxidizing agent. Therefore, as the pH of a solution falls, and the equilibrium between $O_2 \cdot^-$ and $HO_2 \cdot$ shifts toward greater formation of $HO_2 \cdot$, superoxide becomes more reactive, particularly toward lipids. In addition, whereas the dismutation of $O_2 \cdot^-$ to H_2O_2 is exceedingly slow at neutral pH in the absence of SOD, $HO_2 \cdot$ will dismutate to H_2O_2 far more readily at acidic pH values because the rate constant for $HO_2 \cdot$ dismutation is on the order of 10^8 times greater than that for $O_2 \cdot^-$. Thus in an acidic environment, $O_2 \cdot^-$ is converted to the more reactive, more lipid-soluble $HO_2 \cdot$ and its rate of dismutation to H_2O_2 is greatly increased.

Iron and the Formation of Hydroxyl Radical and Iron–Oxygen Complexes

The central nervous system (CNS) is an extremely rich source of iron. Under normal circumstances, low-molecular-weight forms of redox-active iron in the brain are

Primer on Cerebrovascular Diseases, Second Edition
http://dx.doi.org/10.1016/B978-0-12-803058-5.00037-0

maintained at extremely low levels. Extracellularly, the iron transport protein transferrin tightly binds iron in the Fe^{3+} form. Intracellularly, Fe^{3+} is sequestered by the iron storage protein ferritin. Although both ferritin and transferrin have very high affinity for iron at neutral pH and effectively maintain iron in a noncatalytic state, both proteins readily give up their iron at pH values of 6.0 or less. The iron in ferritin can also be released by reductive mobilization by $O_2 \cdot^-$. Therefore, within the ischemic CNS environment where pH is typically lowered and several sources of superoxide are active, conditions are favorable for the potential release of iron from storage proteins. Once iron is released from ferritin or transferrin, it can actively catalyze oxygen radical reactions.

The second source of catalytically active iron is hemoglobin. SAH or intracerebral hemorrhage (ICH) places hemoglobin in contact with nervous tissue. Although hemoglobin itself can stimulate oxygen radical reactions, it is more likely that iron released from hemoglobin is responsible for hemoglobin-mediated LP. Iron is released from hemoglobin by either H_2O_2 or lipid hydroperoxides (LOOHs), and this release is further enhanced as the pH falls to 6.5 or below. Therefore, hemoglobin may catalyze oxygen radical formation and LP either directly or through the release of iron by H_2O_2, LOOH, and/or acidic pH.

Free iron or iron chelates participate in further free radical production at two levels: (1) the autoxidation of Fe^{2+} provides an additional source of $O_2 \cdot^-$ ($Fe^{2+} + O_2 \rightarrow Fe^{3+} + O_2 \cdot^-$) and (2) Fe^{2+} is oxidized in the presence of H_2O_2 to form hydroxyl radical (Fenton reaction, $Fe^{2+} + H_2O_2 \rightarrow Fe^{3+} + \cdot OH + OH^-$) or perhaps a ferryl ion ($Fe^{2+} + H_2O_2 \rightarrow Fe^{3+}OH + OH^-$). Both $\cdot OH$ or $Fe^{3+}OH$ are extraordinarily potent initiators of LP.

Peroxynitrite-Mediated Free Radical Formation

Another source of $\cdot OH$ formation is the peroxynitrite pathway. This involves neurons, endothelial cells, neutrophils, macrophages, and microglia that can produce two radicals, namely, $O_2 \cdot^-$ and nitric oxide ($\cdot NO$), the latter from nitric oxide synthetase. The two radical species can combine to form peroxynitrite anion ($OONO^-$) that, at physiologic pH, largely undergoes protonation (pKa = 6.8) to become peroxynitrous acid (ONOOH). However, ONOOH is an unstable acid that readily decomposes to give two potent oxidizing radicals, $\cdot OH$ and nitrogen dioxide ($O_2 \cdot^- + \cdot NO \rightarrow ONOO^- + H^+ \rightarrow ONOOH \rightarrow \cdot NO_2 + \cdot OH$). Alternatively, $ONOO^-$ can combine with carbon dioxide (CO_2) to form nitrosoperoxocarbonate, which can then decompose into nitrogen dioxide and carbonate radical ($ONOO^- + CO_2 \rightarrow ONOOCO_3^- \rightarrow \cdot NO_2 + \cdot CO_3$). Thus, this mechanism provides another source of the highly reactive free radicals that are operative within the ischemic (or postischemically reperfused) CNS. Moreover, $\cdot NO_2$ may also initiate LP or nitrate (i.e., inactivate) cellular proteins. A particularly attractive aspect of this scenario is that each of the peroxynitrite forms ($ONOO^-$, ONOOH, or $ONOOCO_3^-$) have relatively long half-lives (~1 s) and consequently are potentially more diffusible [i.e., they have a larger diffusion radius than their highly reactive diffusion-rate-limited radicals ($\cdot OH$, $\cdot NO_2$, or $\cdot CO_3$) that react instantaneously with the nearest susceptible polyunsaturated fatty acid they come into contact with, as explained in the next section]. Therefore, peroxynitrite offers a mechanism by which free radical damage may occur at a site remote from the actual location of ROS formation.

BASICS OF FREE-RADICAL-INDUCED LIPID PEROXIDATION

"Initiation" of LP occurs when a radical species ($R\cdot$) attacks and removes an allylic hydrogen from a polyunsaturated fatty acid (LH) resulting in a radical chain reaction ($LH + R\cdot \rightarrow L\cdot + RH$). In the process the initiating radical is quenched by the receipt of the hydrogen electron from the polyunsaturated fatty acid (e.g., arachidonic acid). This, however, converts the latter into a lipid or alkyl radical ($L\cdot$). This step sets the stage for a series of "propagation" reactions that begin when the alkyl radical reacts with a mole of O_2 forming a lipid peroxyl radical ($LOO\cdot$) ($L\cdot + O_2 \rightarrow LOO\cdot$). The $LOO\cdot$ then reacts with a neighboring LH within the membrane and steals its electron, forming a lipid hydroperoxide (LOOH) and a second alkyl radical ($LOO\cdot + LH \rightarrow LOOH + L\cdot$). Furthermore, the LOOH can be decomposed by interaction with iron. When the LOOH reacts with iron in the ferrous form (Fe^{2+}), an alkoxyl radical ($LO\cdot$) is formed, and when it reacts with iron in the ferric form (Fe^{3+}), $LOO\cdot$ is formed; both of which can contribute to further propagation reactions among the free-radical-susceptible polyunsaturated fatty acids that are the major constituents of cell membranes.

Ultimately, the LP process leads to "fragmentation" or "scission" reactions in which the peroxidized polyunsaturated fatty acid breaks down to give rise to the neurotoxic aldehydic end products **4-hydroxynonenal (4-HNE)** and **2-propenal (acrolein)**. These end products contain reactive carbonyl moieties that can covalently bind to basic amino acids, such as lysine or histidine as well as sulfhydryl-containing cysteine residues in cell proteins, thus impairing protein functions. Fig. 37.1 shows the chemistry of peroxidation of the polyunsaturated arachidonic acid ending with the generation of the neurotoxic carbonyl 4-HNE. Fig. 37.2 illustrates the covalent chemical reaction of the carbonyl-containing

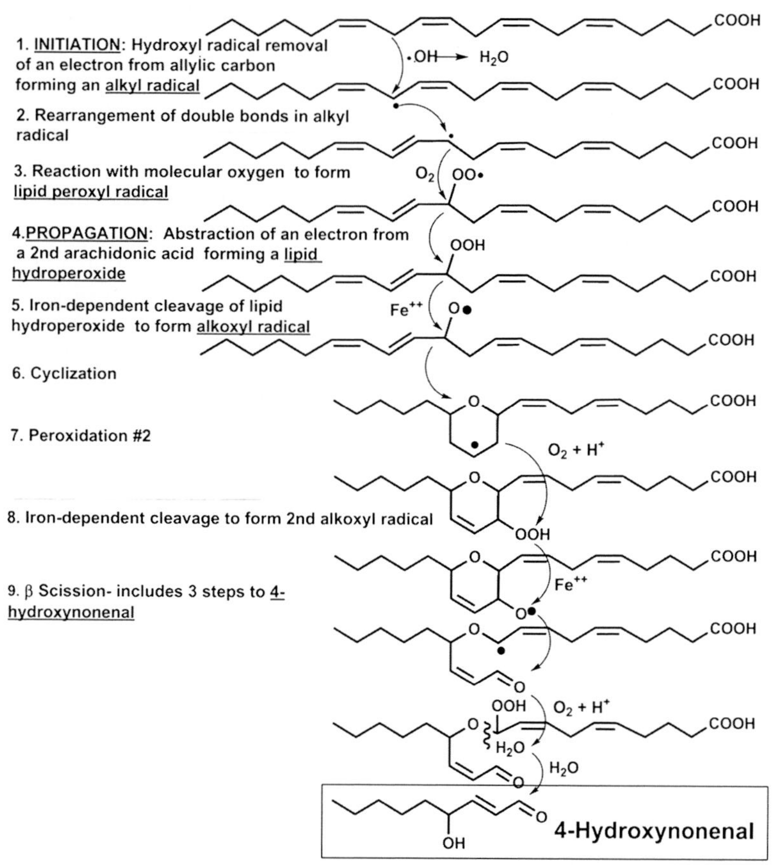

FIGURE 37.1 Peroxidation of the polyunsaturated arachidonic acid resulting in the generation of the neurotoxic aldehyde 4-hydroxynonenal (4-HNE).

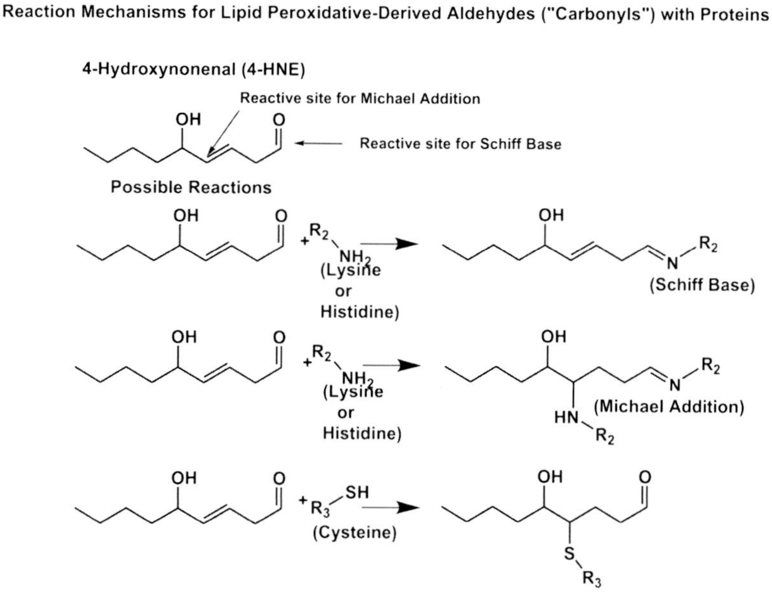

FIGURE 37.2 Reaction mechanisms for 4-hydroxynonenal (4-HNE) modification of proteins.

III. NEUROPROTECTION

4-HNE with cellular proteins that occurs via either Schiff base formation or the Michael addition reactions.

Another aldehydic product formed as a by-product of LP is malondialdehyde (MDA), which has been used extensively as an experimental biomarker for postischemic or posttraumatic LP. Although MDA can bind to cellular proteins, it does not share the neurotoxic effects of 4-HNE or acrolein. Furthermore, it is also formed during enzymatic LP (e.g., prostaglandin synthase or 5-lipoxygenase activity) and is therefore not specific for free-radical-induced LP.

In any event, a large body of literature has established the role of ROS-induced LP in cerebral ischemic pathophysiologic conditions including cellular and vasogenic edema, compromise of ionic homeostasis, disruption of mitochondrial function, enhancement of glutamate-mediated excitotoxicity, and induction of cerebral vasospasm.

MECHANISMS FOR PHARMACOLOGIC INHIBITION OF OXIDATIVE DAMAGE IN ACUTE ISCHEMIC OR HEMORRHAGIC STROKE

Based on the aforementioned outline of the steps involved in oxygen-radical-induced oxidative damage, and LP in particular, a number of potential mechanisms for its pharmacologic inhibition are apparent.

Prevention of the ROS Formation

The first category includes compounds that inhibit the initiation of LP and other forms of oxidative damage by preventing the formation of inorganic ROS. For instance, nitric oxide synthesis inhibitors, by limiting $\cdot NO$ production, can limit peroxynitrite formation. However, they also have the potential to interfere with the physiologic roles that $\cdot NO$ is responsible for, including antioxidant effects that have an important role as a scavenger of lipid peroxyl radicals (e.g., $LOO \cdot + \cdot NO \rightarrow LOONO$). Another approach to block postischemic radical formation is the inhibition of the arachidonic acid cascade during which the formation of $O_2 \cdot^-$ is produced as a by-product.

Chemical Scavenging of ROS

The second indirect LP inhibitory approach involves chemically scavenging the inorganic and LP-derived radical species (e.g., $O_2 \cdot^-$, $\cdot OH$, $\cdot NO_2$, $\cdot CO_3$, $LOO \cdot$, $LO \cdot$) before they have a chance to steal an electron from a polyunsaturated fatty acid and initiate or propagate LP chain reactions. An example of this type of compound is the nitrone spin-trapping radical scavenger NXY-059 shown in Fig. 37.3 [8], which was shown to reduce infarct volume in a wide variety of preclinical focal ischemic stroke paradigms with a therapeutic window as long as 4 h. However, in two phase 3 clinical trials carried out with a 6-h treatment window, no significant improvement in poststroke outcomes was observed [8].

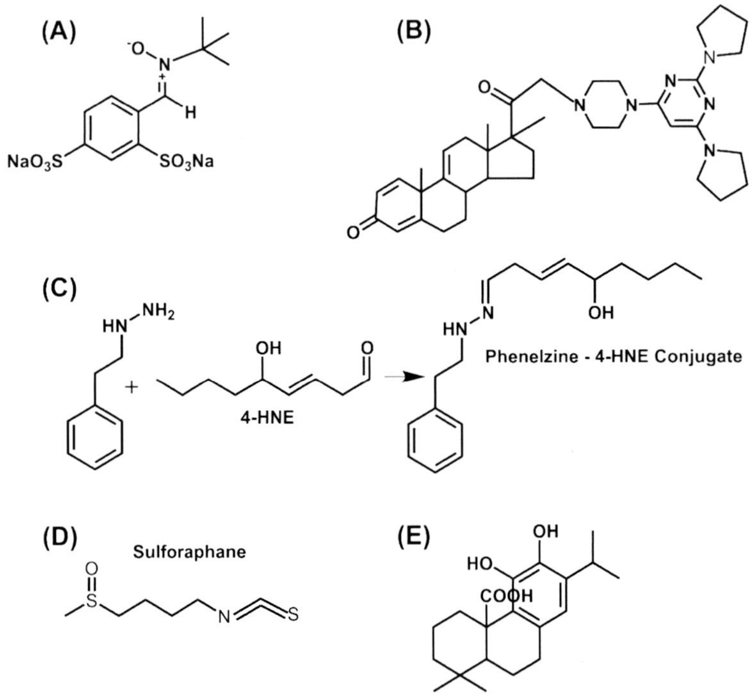

FIGURE 37.3　Chemical structures of antioxidant compounds shown to be neuroprotective in stroke models. (A) Free radical spin trap: NXY-059, (B) lipid peroxidation inhibitor: tirilazad, (C) carbonyl scavenger: phenelzine, (D) Nrf2/ARE activators: sulforaphane, and (E) carnosic acid.

A general limitation of these first two approaches is that they would be expected to have a short therapeutic window such that the inhibiting drugs would have to be administered rapidly to have a chance to interfere with the initial burst of free radical production and LP that has been documented in stroke models, particularly those that include a phase of reperfusion (i.e., reoxygenation). Although it is believed that ROS production persists for minutes and even some hours after reperfusion, the major portion of inorganic radical formation is an early event (often referred to as free radical burst) that may peak too quickly to pharmacologically inhibit, unless the antioxidant compound was already present when the ischemic event occurred or if it was available for administration immediately thereafter.

Lipid Peroxidation Inhibition

In contrast to the indirect-acting antioxidant mechanisms discussed earlier, the third category involves stopping the "chain reaction" propagation of LP once it has begun. The most frequently demonstrated way to accomplish this is by scavenging lipid peroxyl (LOO·) radicals. The endogenous prototypical scavenger of these lipid radicals is alpha-tocopherol (vitamin E or Vit E), which can donate an electron from its phenolic hydroxyl (OH) moiety to quench the LOO· radicals, in particular. However, this scavenging process is only stoichiometric one Vit E can only quench one LOO· and in the process vitamin E loses its antioxidant efficacy and becomes vitamin E radical (LOO· + Vit E → LOOH + Vit E·). Although Vit E· is relatively harmless, it also cannot scavenge another LOO· until it is reduced back to its active form by receiving an electron from another endogenous reducing agent, such as ascorbic acid (vitamin C) or glutathione (GSH). Although this tripartite LOO· antioxidant defense system (Vit E, Vit C, GSH) works fairly effectively in the absence of a major ischemic event, numerous studies have shown that each of these antioxidants is rapidly consumed during the early minutes and hours of neural insults. Thus it has long been recognized that more effective pharmacologic LOO· and LO· scavengers are needed. Furthermore, it is expected that compounds that could interrupt the LP process after it has begun would be able to exert a more practical neuroprotective effect (i.e., possess longer antioxidant therapeutic window).

LP-inhibiting antioxidants have been shown to be neuroprotective in focal, global, and SAH stroke models via their ability to limit propagation reactions by scavenging LOO· or LO· or by decreasing the lipid membrane fluidity, which decreases the rate at which lipid radicals can interact with nearby polyunsaturated fatty acids. An example is the LP inhibitor tirilazad (Fig. 37.3) whose mechanism includes the ability to react with lipid radicals along with a decrease in membrane phospholipid fluidity, which produces a dual mechanism attenuation of LP propagation. Tirilazad has been documented to produce neuroprotective effects in a variety of preclinical stroke, SAH, and neurotrauma models [3,4].

However, an inherent problem with inhibiting the propagation phase of LP as an antioxidant neuroprotective strategy is that it is associated with a clinically practical, but limited, therapeutic efficacy window. Indeed, phase 3 clinical trials of tirilazad in patients who have had focal stroke and enrolled within 6h of the onset of ischemic symptoms showed no improved outcome, and the later trials with increased tirilazad doses actually showed worsened outcome [1]. In contrast, the final phase 3 stroke trial involving initiation of tirilazad administration within a shorter treatment initiation window of 4h appeared to show some improvement in stroke outcome. However, that trial was stopped after enrollment of only 111 patients because of concern over the poorer outcomes in tirilazad-treated patients in the aforementioned stroke trials with a 6-h window [2]. This suggests that the neuroprotective efficacy window for LP inhibition as an antioxidant strategy may be limited to the first 4h after ischemic stroke onset.

In contrast, phase 3 trials of tirilazad in patients with aneurysmal SAH produced more encouraging results than those conducted in patients who have had focal ischemic stroke. Initiation of tirilazad administration within the first 48h after SAH (along with the calcium channel blocker nimodipine, which was already FDA-approved for aneurysmal SAH) was shown to significantly improve both 3-month recovery and survival of the patients [5]. Although a significant effect was observed in males and females analyzed together, examination of the effect of tirilazad in each gender separately revealed that the improved outcome was almost entirely attributable to male patients with SAH.

Similarly, in a phase 3 trial of tirilazad use in moderate and severe traumatic brain injury (TBI), in which treatment begun within the first 4h after TBI, tirilazad was shown to decrease 6-month mortality in the subset of moderately injured male patients with traumatic SAH, which was more than 50% of the enrolled subjects. Mortality was reduced from 24% in the male patients who received placebo to only 6% in the tirilazad-treated patients ($p < .042$) and in the severely injured male patients from 43% to 34% ($p < .071$) [6]. However, no mortality reduction was seen in females with TBI and no overall improvement in neurologic recovery was apparent in the total population of male and female patients with TBI with mixed traumatic pathologic conditions [6]. Although tirilazad achieved regulatory approval for aneurysmal SAH in several European countries and

Australia and New Zealand, the US FDA approval was never obtained and further development was stopped by the sponsor Pharmacia & Upjohn Company LLC.

NEW ANTIOXIDANT APPROACHES

Carbonyl Scavenging

During the past decade, investigators have identified a novel antioxidant mechanistic approach that may be more neuroprotective, has a longer therapeutic window than LOO· scavenging, and can possibly reverse some aspects of LP damage. This new mechanistic strategy involves the chemical scavenging of the neurotoxic LP-derived aldehydic (carbonyl-containing) end products 4-HNE and acrolein, which can induce additional oxidative damage. The pharmacologic scavenging approach involving interception of these highly reactive products is referred to as "carbonyl scavenging." The most studied compounds contain hydrazine moieties that can covalently react with carbonyl groups via a Schiff base-type reaction forming unreactive hydrazones. An example of a clinically used hydrazine-containing drug is phenelzine (Fig. 37.3), an antidepressant that has been reported to be neuroprotective in focal and global ischemia-reperfusion stroke models in concert with a reduction in "aldehyde load" [9].

Based on studies to date on prototype carbonyl scavengers, it is theoretically possible that this approach may have greater efficacy and a longer therapeutic time window because of the longer half-life of 4-HNE and acrolein than the approach of attempting to scavenge rapidly reactive free radicals that initiate LP and protein oxidation almost instantaneously after their formation (i.e., with a diffusion rate-limited rate constant). Furthermore, carbonyl scavenging may be able to reverse LP-generated reactive aldehyde impairment of cellular proteins after they have reacted.

Nrf2-Antioxidant Response Element (ARE) Activation

Another promising ROS-based means to achieve neuroprotective effects involves pharmacologic activation of the endogenous antioxidant defense system, which is largely regulated by the nuclear factor E2-related factor 2/antioxidant response element (Nrf2/ARE) pathway. Nrf2 activation and the multimechanistic upregulation of antioxidant and anti-inflammatory genes has shown promise in acute ischemic stroke, ICH, and TBI models. Postischemic treatment with either of the natural products, sulforaphane, found in high concentrations in broccoli, and

carnosic acid (Fig. 37.3), found in the herb rosemary, have been demonstrated to reduce focal stroke infarct size together with an upregulation of Nrf2/ARE associated genes [7,10]. Additional ongoing studies have demonstrated that these compounds are neuroprotective in ICH and TBI models.

CONCLUSION

Despite the failure of the previous phase 3 clinical trial attempts to demonstrate neuroprotective effects of the once-promising ROS-based neuroprotectants tirilazad and NXY-059 in patients who have had acute stroke, the important involvement of free-radical-induced oxidative damage remains firmly established as a neuroprotective target. New antioxidant approaches that may possess a longer therapeutic efficacy window, such as carbonyl scavenging and Nrf2/ARE activation, are actively being studied in acute neurologic injury models. In addition, the potential combinations of antioxidants that possess complementary neuroprotective mechanisms are also being explored.

References

[1] Committee TIS. Tirilazad mesylate in acute ischemic stroke: a systematic review. Stroke 2000;32:9.
[2] Haley Jr EC. High-dose tirilazad for acute stroke (RANTTAS II). RANTTAS II Investigators. Stroke 1998;29:1256–7.
[3] Hall ED, McCall JM, Means ED. Therapeutic potential of the lazaroids (21-aminosteroids) in acute central nervous system trauma, ischemia and subarachnoid hemorrhage. Adv Pharmacol 1994;28:221–68.
[4] Hall ED, Vaishnav RA, Mustafa AG. Antioxidant therapies for traumatic brain injury. Neurotherapeutics 2010;7:51–61.
[5] Kassell NF, Haley Jr EC, Apperson-Hansen C, Alves WM. Randomized, double-blind, vehicle-controlled trial of tirilazad mesylate in patients with aneurysmal subarachnoid hemorrhage: a cooperative study in Europe, Australia, and New Zealand. J Neurosurg 1996;84:221–8.
[6] Marshall LF, Maas AI, Marshall SB, Bricolo A, Fearnside M, Iannotti F, et al. A multicenter trial on the efficacy of using tirilazad mesylate in cases of head injury. J Neurosurg 1998;89:519–25.
[7] Satoh T, Kosaka K, Itoh K, Kobayashi A, Yamamoto M, Shimojo Y, et al. Carnosic acid, a catechol-type electrophilic compound, protects neurons both in vitro and in vivo through activation of the Keap1/Nrf2 pathway via S-alkylation of targeted cysteines on Keap1. J Neurochem 2008;104:1116–31.
[8] Sutherland BA, Minnerup J, Balami JS, Arba F, Buchan AM, Kleinschnitz C. Neuroprotection for ischaemic stroke: translation from the bench to the bedside. Int J Stroke 2012;7:407–18.
[9] Wood PL, Khan MA, Moskal JR, Todd KG, Tanay VA, Baker G. Aldehyde load in ischemia-reperfusion brain injury: neuroprotection by neutralization of reactive aldehydes with phenelzine. Brain Res 2006;1122:184–90.
[10] Zhao J, Kobori N, Aronowski J, Dash PK. Sulforaphane reduces infarct volume following focal cerebral ischemia in rodents. Neurosci Lett 2006;393:108–12.

CHAPTER

38

Neuroprotectants: Cell-Death Based

S. Namura

Morehouse School of Medicine, Atlanta, GA, United States

INTRODUCTION

Research in the 1990s provided fundamental discoveries of the molecular mechanisms of apoptosis, one type of cell death. In the nematode *Caenorhabditis elegans*, the genetic control of the programmed cell death during development was elucidated [1]. Cloning of the cell death–related genes was completed. Among them, *ced-3* and *ced-4* are death-promoting genes and *ced-9* is death inhibiting. The identification of their mammalian homologues suggested that the cell death mechanism might be conserved from nematode to humans. In addition, there was the speculation that the death mechanism might be shared by various types of cells and by various types of death stimuli. A great deal of interest was generated in translating the knowledge to human disease conditions, including cerebrovascular diseases, with a hope to identify new therapeutic targets. Moreover, the recognition of apoptosis, in turn, promoted mechanistic studies concerning necrosis, the predominant form of ischemic cell death. Necrosis had been thought to be merely an unregulated cellular crisis. A group of proteases called caspase are homologous to the *ced-3* gene product. Apaf-1 family and Bcl-2 family members are mammalian homologues of the *ced-4* and *ced-9* gene products, respectively. In this chapter, the molecular mechanisms of cell death will be summarized (Fig. 38.1). Experimental studies reporting neuroprotection by targeting these death mechanisms will also be described.

CASPASE

Caspase is the family of cysteine proteases that are Ced-3 homologues. In mammals, more than 10 caspases have been identified. Caspases are made as inactive proform. After being cleaved, they become active. Among the caspases, caspase-3 has the highest homology to Ced-3. It cleaves multiple apoptosis substrates, being considered as the main effector in apoptosis. Pro-caspase-3 is activated by being cleaved by initiator caspases, caspase-9 and caspase-8, which are, respectively, activated by the mitochondrial pathway and the extrinsic pathway, as described later. There is evidence suggesting that caspases are activated in the neuronal injury caused by brain ischemia [2,3]. In addition, intracerebroventricular injection of caspase inhibitors, such as z-VAD-fmk and z-DEVD-fmk, has been shown to attenuate DNA laddering in the brain after mild ischemia [4].

MITOCHONDRIAL PATHWAY

Mitochondria play a major role in cell death signal transduction in many types of cells, including neurons. Death signals are initiated by various conditions including DNA damage, changes in intracellular calcium level, free radical formation, and serum deprivation. The signals converge to the mitochondria. Upon the arrival of the death signal, the permeability transition pore (PTP), which is localized on the mitochondrial membrane, opens. Through PTP, cytochrome c is released from the mitochondria into the cytosolic space. Under the presence of dATP/ATP, cytochrome c binds Apaf-1, forming a complex called apoptosome. This apoptosome recruits pro-caspase-9, resulting in caspase-9 activation, which further activates effector caspases, such as caspase-3 and caspase-7. An immunosuppressant cyclosporine A prevents PTP from opening, which has been studied as a potential neuroprotectant targeting the mitochondrial pathway [5].

BCL-2 FAMILY

Bcl-2 family members are Ced-9 homologues. They control cell death by regulating the cytochrome c release

Primer on Cerebrovascular Diseases, Second Edition
http://dx.doi.org/10.1016/B978-0-12-803058-5.00038-2

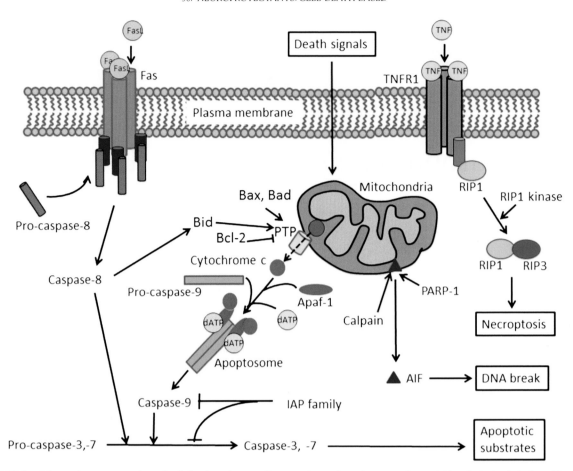

FIGURE 38.1   Schematic representation of cell death pathways. Caspase-dependent cleavage of apoptotic substrates is induced by the mitochondrial pathway in which cytochrome c release from the mitochondria plays a pivotal role in initiation. In some cell types, the death receptor-caspase-8 axis also activates caspase-3 and -7. Caspase-independent pathways, either by the tumor necrosis factor (TNF) receptor (TNFR)-receptor interacting protein (RIP) axis or the poly-ADP-ribose polymerase (PARP)-1-apoptosis-inducing factor (AIF) axis, lead to energy crisis and DNA break, resulting in programmed-type necrosis. Arrows indicate activation of the targets, whereas lines with blunt ends indicate inactivation. *IAP*, inhibitors of apoptosis protein; *PTP*, permeability transition pore.

from the mitochondria. Among them, Bcl-2 and Bcl-xL inhibit cell death. In contrast, there are members (Bax, Bak, Bid, and Bad) that promote cell death. Bax and Bad are thought to open PTP by directly acting on PTP. The inhibitory Bcl-2 members inhibit cytochrome c release by interacting with Bax and Bad so that the cytochrome c release is inhibited. The activity of some of the Bcl-2 family members is controlled by phosphorylation and proteolysis. Transgenic mice overexpressing Bcl-2 are more resistant against brain ischemia [6].

EXTRINSIC PATHWAY

In immune cells, death signal is transmitted via death receptors (DRs): Fas and tumor necrosis factor (TNF) receptor-1 (TNFR1). When DRs are activated, pro-caspase-8 is recruited to the intracellular domain of DR, forming death-inducing signaling complex. By so doing, caspase-8 is cleaved and activated. The activated

caspase-8 in turn cleaves and activates the downstream effector caspases, caspase-3 and caspase-7. Cross talk between the mitochondrial pathway and DR-caspase-8 pathway has been suggested. In some cell types, activated caspase-8 cleaves Bid. As described earlier, Bid belongs to the Bcl-2 family. The cleaved C-terminal product of Bid translocates to the mitochondria, opening PTP. Consequently, cytochrome c is released from the mitochondria, initiating the mitochondrial pathway. Blocking Fas and TNF could be potential interventions against stroke. Neutralizing antibodies against Fas ligand and TNF demonstrated brain protection against stroke in mice [7].

INHIBITOR OF APOPTOSIS

Inhibitors of apoptosis proteins (IAPs) are endogenous inhibitors of apoptosis. IAP was first identified in baculoviruses. Later, it was shown that IAPs are conserved across the species. Among IAPs, X-linked IAP and neuronal apoptosis inhibitory protein (NAIP) are extensively studied. They directly inhibit caspase-3, -7, and -9. NAIP1-deficient mice are more vulnerable to kainic acid toxicity [8]. The delivery of adenovirus vector overexpressing NAIP attenuated hippocampal injury in rats after global ischemia [9]. Thus, IAPs may be potential targets for neuroprotection against stroke.

CASPASE-INDEPENDENT CELL DEATH

Since caspase activation through the mitochondrial pathway requires dATP/ATP, and since ischemia disturbs energy production, typical apoptosis is rather rare in ischemic neuronal death. Apoptosis after ischemia is more apparent in neonatal brains than in adult ones. Which (apoptosis or necrosis) is more prominent depends on the severity of ischemia, cell type, brain region, and age. Typical necrosis under extremely severe conditions has been considered as unregulated events that lack intracellular signaling. In contrast, programmed types of necrosis have been reported. In the presence of caspase inhibitors, TNF-α receptor activation results in the interactions between receptor interacting protein (RIP)-1 and RIP-3. RIP-3 activation stimulates enzymes that are involved in energy metabolism. The stimulation of the energy metabolism-related enzymes leads to necrosis. This programmed necrosis is termed necroptosis [10]. RIP-1 inhibition by necrostatin-1 has been shown to protect the brain against ischemia in mice.

Other players in caspase-independent cell death include poly-ADP-ribose polymerase-1 (PARP-1) and apoptosis-inducing factor (AIF). PARP-1 is an enzyme that is activated by damaged DNA, transferring ADP-ribose from β-nicotinamide adenine dinucleotide (NAD^+) to acceptor molecules. PARP-1 promotes DNA repair. However, under severe stress, such as ischemia, overly activated PARP-1 depletes NAD^+ and ATP, causing an energy crisis. PARP inhibitors have shown to protect against ischemia in animals [11]. PARP-1 also leads to cell death via another pathway. The produced poly-ADP-ribose polymer translocates to the mitochondria, resulting in the release of AIF from the mitochondria [12]. The released AIF translocates to the nucleus, contributing to DNA break and chromatin alterations. Calpain I, a calcium-dependent protease that cleaves various structural proteins, has been shown to cleave the N-terminal portion of AIF, liberating AIF from the mitochondria [13]. Overexpression of calpastatin, a calpain inhibitor, provided neuroprotection against ischemia. Another type of proteases cathepsins have been implicated in degradation of intracellular organelles including plasmalemma and lysosomal membrane. These proteases may also be potential targets for neuroprotection.

CONCLUSION

Experimental studies during the past two decades have provided substantial evidence that multiple cell death pathways are turned on in the brain after stroke. Animal studies using transgenic approach and pharmacological intervention have suggested that some of these pathways play roles in neuronal injury caused by stroke. For predicting whether those pathways, alone or in combination, could be clinically feasible therapeutic targets for neuroprotection, more questions remain to be answered. Which pathway is more important? Do those death pathways also play roles in human brains? To answer these questions, specific drugs with appropriate bioavailability are required. Such drugs must be delivered into ischemic brain area where blood flow is limited. They must cross the blood–brain barrier and further plasma membrane, reaching the intracellular target molecules. Developing a safe and clinically testable neuroprotectant targeting these cell death pathways is a huge challenge. To circumvent this challenge, there are efforts to identify off-target efficacies of drugs that have been clinically used for purposes other than neuroprotection. Minocycline, a semisynthetic tetracycline, is such an example [14]. Besides its antibiotic effect, minocycline has multiple actions including inhibition of mitochondrial PTP, PARP-1, and matrix metalloproteinase. The Minocycline to Improve Neurologic Outcome in Stroke (MINOS) study (ClinicalTrials.gov Identifier: NCT00630396) reported encouraging results in dosing and safety. A large randomized controlled trial remains to be conducted to assess its effect on neurological outcome. A phase 2 clinical trial for aneurysmal subarachnoid hemorrhage is ongoing (ClinicalTrials.gov Identifier: NCT02113176).

References

[1] Metzstein MM, Stanfield GM, Horvitz HR. Genetics of programmed cell death in C. elegans: past, present and future. Trends Genet 1998;14:410–6.

[2] Chen J, Nagayama T, Jin K, Stetler RA, Zhu RL, Graham SH, et al. Induction of caspase-3-like protease may mediate delayed neuronal death in the hippocampus after transient cerebral ischemia. J Neurosci 1998;18:4914–28.

[3] Namura S, Zhu J, Fink K, Endres M, Srinivasan A, Tomaselli KJ, et al. Activation and cleavage of caspase-3 in apoptosis induced by experimental cerebral ischemia. J Neurosci 1998;18:3659–68.

[4] Endres M, Namura S, Shimizu-Sasamata M, Waeber C, Zhang L, Gómez-Isla T, et al. Attenuation of delayed neuronal death after mild focal ischemia in mice by inhibition of the caspase family. J Cereb Blood Flow Metab 1998;18:238–47.

[5] Friberg H, Wieloch T. Mitochondrial permeability transition in acute neurodegeneration. Biochimie 2002;84:241–50.

[6] Martinou JC, Dubois-Dauphin M, Staple JK, Rodriguez I, Frankowski H, Missotten M, et al. Overexpression of BCL-2 in transgenic mice protects neurons from naturally occurring cell death and experimental ischemia. Neuron 1994;13:1017–30.

[7] Martin-Villalba A, Hahne M, Kleber S, Vogel J, Falk W, Schenkel J, et al. Therapeutic neutralization of CD95-ligand and TNF attenuates brain damage in stroke. Cell Death Differ 2001;8:679–86.

[8] Holcik M, Thompson CS, Yaraghi Z, Lefebvre CA, MacKenzie AE, Korneluk RG. The hippocampal neurons of neuronal apoptosis inhibitory protein 1 (NAIP1)-deleted mice display increased vulnerability to kainic acid-induced injury. Proc Natl Acad Sci USA 2000;97:2286–90.

[9] Xu DG, Crocker SJ, Doucet JP, St-Jean M, Tamai K, Hakim AM, et al. Elevation of neuronal expression of NAIP reduces ischemic damage in the rat hippocampus. Nat Med 1997;3:997–1004.

[10] Degterev A, Huang Z, Boyce M, Li Y, Jagtap P, Mizushima N, et al. Chemical inhibitor of nonapoptotic cell death with therapeutic potential for ischemic brain injury. Nat Chem Biol 2005;1:112–9.

[11] Curtin NJ, Szabo C. Therapeutic applications of PARP inhibitors: anticancer therapy and beyond. Mol Aspects Med 2013;34:1217–56.

[12] Yu SW, Andrabi SA, Wang H, Kim NS, Poirier GG, Dawson TM, et al. Apoptosis-inducing factor mediates poly(ADP-ribose) (PAR) polymer-induced cell death. Proc Natl Acad Sci USA 2006;103:18314–9.

[13] Cao G, Xing J, Xiao X, Liou AK, Gao Y, Yin XM, et al. Critical role of calpain I in mitochondrial release of apoptosis-inducing factor in ischemic neuronal injury. J Neurosci 2007;27:9278–93.

[14] Hess DC, Fagan SC. Repurposing an old drug to improve the use and safety of tissue plasminogen activator for acute ischemic stroke: minocycline. Pharmacotherapy 2010;30:55S–61S.

CHAPTER

39

Comprehensive Concept of Regenerative Medicine for Ischemic Stroke With Bone Marrow Stromal Cells

S. Kuroda[1,2]

[1]University of Toyama, Toyama, Japan; [2]Hokkaido University Graduate School of Medicine, Sapporo, Japan

INTRODUCTION

Few drugs have been developed to effectively rescue the patients with ischemic stroke in spite of the huge efforts to develop them for longer than 50 years [1]. As alternative approach, cell therapy has recently been expected as one of the promising strategies to enhance functional recovery after ischemic stroke. A variety of cells have been studied as the candidate donor cells for this purpose. These include embryonic stem (ES) cells, neural stem cells, inducible pluripotent stem (iPS) cells, and bone marrow stromal cells (BMSCs). Of these, the BMSCs may have the most enormous therapeutic potential among them, because they can be obtained from the patients themselves and easily expanded without posing any ethical and immunological problems. The BMSCs are non-hematopoietic cells

in the bone marrow and regulate the proliferation and differentiation of hematopoietic cells. The transplanted BMSCs significantly enhance functional recovery after the insults in animal models of ischemic stroke. On the other hand, studies during 2012 to 2013 showed that the adult stem cells, including BMSCs, are observed in the peripheral blood and play an important role in repairing the injured tissues [2,3].

Based on these observations, some of preliminary clinical testing has already been conducted to evaluate the safety and therapeutic effects of BMSC transplantation for the patients with both acute and chronic neurological disorders [2,3]. However, it should be reminded that a variety of questions or problems still remains to be solved in order to establish BMSC transplantation as scientifically proven entity in clinical situation [2]. This chapter reviews recent knowledge

Primer on Cerebrovascular Diseases, Second Edition
http://dx.doi.org/10.1016/B978-0-12-803058-5.00039-4

on therapeutic impacts of BMSC transplantation on ischemic stroke.

BMSC TRANSPLANTATION FOR ISCHEMIC STROKE

Recent studies have shed light on the mechanisms through which the transplanted BMSCs enhance functional recovery after cerebral infarct. They aggressively migrate toward the damaged lesions through some chemokines. In 2007, CXCR4, a specific receptor for stromal cell–derived factor (SDF)-1α, is believed to play an important role in their migration in the CNS [4]. There are few studies on whether the engrafted BMSCs retain their proliferative activity in the host brain or not. Yano et al. (2005) labeled the GFP-expressing BMSCs with a superparamagnetic iron oxide (SPIO) agent and transplanted into the ipsilateral striatum of the mice infarct brain. As the results, they found that the BMSCs actively proliferate, migrate toward the lesion, and partially express the neuronal phenotype in the host brain after transplantation [5].

Nowadays, the BMSCs are known to produce some neuroprotective or neurotrophic factors and support the survival of the host neural cells. This hypothesis is readily reasonable because the BMSCs per se support the homing, proliferation, and differentiation of the hematopoietic cells in the bone marrow by producing a variety of cytokines. They release soluble neuroprotective factors, including nerve growth factor (NGF), hepatocyte growth factor (HGF) and brain-derived neurotrophic factor (BDNF), and significantly ameliorate glutamate-induced damage of neurons [6]. The BMSCs markedly promote the neurite extension from the neurons in the organotypic slice of the brain and spinal cord, and also protect the neurovascular integrity between basement membrane and astrocyte end-feet and ameliorate brain damage in stroke-prone spontaneous hypertensive rats [7]. Shichinohe et al. [8] demonstrated that the BMSCs serve the "nursing effect" to the damaged neurons and activate the neural stem cells in the host brain by producing BDNF. Therefore, the transplanted BMSCs trigger endogenous signaling pathways of survival and repair in neurons by secreting soluble neurotrophic factors.

Both neutrophils and macrophages are well known to play an important role in the early inflammation after cerebral infarct. Indeed, their inflammatory response may be an essential process to clear cellular debris and initiate the healing pathways. Simultaneously, however, these inflammatory reactions may also give rise to cytotoxic damage to the surviving neurons, astrocytes, and endothelial cells in the peri-infarct area. The BMSCs have currently been investigated as donor cells for novel cell therapy to prevent and to treat clinical disease associated with aberrant immune response. In the host, the BMSCs may attenuate pro-inflammatory cytokine and chemokine induction, reduce pro-inflammatory cell migration into sites of injury and infection [9]. Therefore, the transplanted BMSCs may prevent excessive inflammatory response and prevent further tissue damage in the peri-infarct area.

The BMSCs are believed to differentiate into neural cells in the host's brain. This theory is based on the findings that BMSCs simulate neuronal morphology and express the proteins specific for neurons [10]. Studies in 2006 showed that the BMSCs can alter their gene expression profile in response to exogenous stimuli and increase the genes related to the neural cells [11]. Sanchez-Ramos et al. [10] showed that a small fraction of BMSCs cultured in epidermal growth factor (EGF) or retinoic acid/BDNF expressed nestin, NeuN, or GFAP, and that the proportion of NeuN-expressing cells increased when BMSC were co-cultured with fetal mouse midbrain neurons. Wislet-Gendebien et al. [12] co-cultured the BMSCs with cerebellar granule cells and assessed their fates. They found that the nestin-expressing BMSCs express other neuronal markers and that BMSC-derived neuron-like cells fire single-action potentials in response to neurotransmitters such as glutamate. Hokari et al. [6] also demonstrated that a certain subpopulation of the BMSCs morphologically simulated the neuron and expressed the neuron-specific proteins without any evidence of cell fusion, when co-cultured with the neurons. These findings strongly suggest that at least a certain subpopulation of the BMSCs have the potential to alter their gene expression profile and to differentiate into the neural cells in response to the surrounding environment. More importantly, the findings indicate that only the subgroup of BMSCs with potential of neural differentiation can survive in the host brain for a long time (>4 weeks).

Based on these observations, the exogenous transplantation of BMSCs is now believed to enhance functional recovery through multiple mechanisms, including nursing effect, antiinflammatory action, and neural cell differentiation, in patients with ischemic stroke.

MUSE CELL TRANSPLANTATION FOR ISCHEMIC STROKE

Very recently, Dezawa and coworkers successfully isolated stress-tolerant adult human stem cells from cultured skin fibroblasts and BMSCs. These cells can self-renew, express a set of genes associated with pluripotency, and differentiate into endodermal, ectodermal, and mesodermal cells both in vitro and in vivo. When transplanted into immunodeficient mice by local or intravenous injection, they were integrated into damaged skin, muscle, or liver and differentiated into cytokeratin 14-, dystrophin-, or albumin-positive cells in the respective tissues. Furthermore, they can be efficiently isolated as SSEA-3-positive cells. Unlike authentic ES cells, their proliferation activity is not very high, and they do not form teratoma in immunodeficient mouse testes. The findings are quite attractive, because non-tumorigenic stem cells with the ability to generate the multiple cell types of the three germ layers can be obtained through easily accessible adult human mesenchymal cells without introducing exogenous genes [13]. These cells were named as multilineage-differentiating stress-enduring (Muse) cells. Furthermore, they have proven that Muse cells are a primary source of induced pluripotent stem (iPS) cells in human fibroblasts [14]. The results strongly suggest that a certain subpopulation of BMSCs may have the biological properties of neural differentiation and contribute to regenerate the infarct brain. When Muse cells are directly engrafted into the murine model of ischemic stroke, they can survive in the infarct brain, differentiate into the neurons, and promote the recovery of motor function [15].

CONCLUSION

Direct transplantation of BMSCs/Muse cells may be one of promising strategies to promote functional recovery in patients with ischemic stroke in very near future. Further translational approaches would accelerate clinical application of cell therapy for ischemic stroke, using these bone marrow–derived cells.

References

[1] Savitz SI, Fisher M. Future of neuroprotection for acute stroke: in the aftermath of the SAINT trials. Ann Neurol May 2007;61(5):396–402.

[2] Abe K, Yamashita T, Takizawa S, Kuroda S, Kinouchi H, Kawahara N. Stem cell therapy for cerebral ischemia: from basic science to clinical applications. J Cereb Blood Flow Metab July 2012;32(7):1317–31.

[3] Kuroda S. Bone marrow stromal cell transplantation for ischemic stroke – its multi-functional feature. Acta Neurobiol Exp (Wars) 2013;73(1):57–65.

[4] Shichinohe H, Kuroda S, Yano S, Hida K, Iwasaki Y. Role of SDF-1/CXCR4 system in survival and migration of bone marrow stromal cells after transplantation into mice cerebral infarct. Brain Res September 22, 2007;1183:138–47.

[5] Yano S, Kuroda S, Shichinohe H, Hida K, Iwasaki Y. Do bone marrow stromal cells proliferate after transplantation into mice cerebral infarct?—a double labeling study. Brain Res December 14, 2005;1065(1–2):60–7.

[6] Hokari M, Kuroda S, Shichinohe H, Yano S, Hida K, Iwasaki Y. Bone marrow stromal cells protect and repair damaged neurons through multiple mechanisms. J Neurosci Res April 2008;86(5):1024–35.

[7] Ito M, Kuroda S, Sugiyama T, et al. Transplanted bone marrow stromal cells protect neurovascular units and ameliorate brain damage in stroke-prone spontaneously hypertensive rats. Neuropathology October 2012;32(5):522–33.

[8] Shichinohe H, Ishihara T, Takahashi K, et al. Bone marrow stromal cells rescue ischemic brain by trophic effects and phenotypic change toward neural cells. Neurorehabil Neural Repair January 2015;29(1):80–9.

[9] Auletta JJ, Deans RJ, Bartholomew AM. Emerging roles for multipotent, bone marrow-derived stromal cells in host defense. Blood February 23, 2012;119(8):1801–9.

[10] Sanchez-Ramos J, Song S, Cardozo-Pelaez F, et al. Adult bone marrow stromal cells differentiate into neural cells in vitro. Exp Neurol August 2000;164(2):247–56.

[11] Yamaguchi S, Kuroda S, Kobayashi H, et al. The effects of neuronal induction on gene expression profile in bone marrow stromal cells (BMSC)—a preliminary study using microarray analysis. Brain Res May 4, 2006;1087(1):15–27.

[12] Wislet-Gendebien S, Hans G, Leprince P, Rigo JM, Moonen G, Rogister B. Plasticity of cultured mesenchymal stem cells: switch from nestin-positive to excitable neuron-like phenotype. Stem Cells March 2005;23(3):392–402.

[13] Kuroda Y, Kitada M, Wakao S, et al. Unique multipotent cells in adult human mesenchymal cell populations. Proc Natl Acad Sci USA May 11, 2010;107(19):8639–43.

[14] Wakao S, Kitada M, Kuroda Y, et al. Multilineage-differentiating stress-enduring (Muse) cells are a primary source of induced pluripotent stem cells in human fibroblasts. Proc Natl Acad Sci USA June 14, 2011;108(24):9875–80.

[15] Yamauchi T, Kuroda Y, Morita T, et al. Therapeutic effects of human multilineage-differentiating stress enduring (MUSE) cell transplantation into infarct brain of mice. PLoS One 2015;10(3):e0116009.

CHAPTER

40

Neuroprotectants: Temperature

A.C. Klahr, F. Colbourne
University of Alberta, Edmonton, AB, Canada

INTRODUCTION

The history of therapeutic hypothermia (HYPO) as a treatment for brain injury is quite a fascinating one. For millennia, cooling has been used to reduce edema and inflammation. Hippocrates recommended using ice packs on wounded soldiers. However, it was Napoleon's surgeon Dominique-Jean Larrey (1766–1842) who noted the benefit of HYPO after he observed that wounded soldiers who stayed closer to the fire were likely to die earlier [1]. Experimental studies on HYPO began in the late 1800s and early 1900s when it was noted by Stefani, Deganello, and Trendelenburg that cooling reduces brain activity and metabolism [2]. In the 1940s and 1950s, small clinical trials and animal experiments indicated that HYPO improved outcome, although later unsuccessful studies tempered enthusiasm for this treatment for decades. In the late 1980s and early 1990s, meticulous animal research confirmed that HYPO is a powerful neuroprotectant against ischemia [3,4]. Indeed, of over one thousand treatments developed in animal models, HYPO is the only one that has been repeatedly successful in clinical trials, although there is still controversy.

A key challenge in HYPO research was, and still is, to find optimal cooling parameters (depth, duration, and delay) that cause the fewest complications while providing significant benefit. To answer this, one must consider how insult severity, comorbidities, cooling method, and other factors influence treatment efficacy and choice of treatment parameters. Here we discuss animal research that has sought to define cooling protocols in an effort to successfully translate this treatment to the ischemic and hemorrhagic stroke patient population.

THE RAPEUTIC HYPOTHERMIA FOR ISCHEMIA

The pathological processes underlying ischemic injury are temperature dependent. Cooling applied during and after ischemia reduces metabolic demand, oxidative stress, excitotoxicity, inflammation, edema, blood–brain barrier breakdown, and so on. Intuitively, lower temperatures should confer better neuroprotection. However, earlier studies showed that moderate HYPO (29°C) led to mortality, especially during rewarming. Later, it was found that mild drops in temperature (34°C) during ischemia prevented CA1 cell death [5]. Experiments using several neuroprotective drugs were also discovered to be effective through drug-induced HYPO, often of only a few degrees Celsius. When these treated animals were kept normothermic there was no or little benefit of the drug itself. Altogether, these studies established that mild HYPO (e.g., 32–25°C) was safe and effective [3–6]. The optimal depth of cooling, however, is still controversial as some argue that milder temperatures are equally or more protective with less risk of complications.

The durability of protection with mild HYPO depends on timing and treatment parameters. Brief cooling is highly protective when applied during the insult, but often not when started afterward. Here protracted HYPO (e.g., 24 h) is needed for a long-lasting reduction in CA1 injury, especially when delayed for hours after global ischemia [6]. Similar results have been found in animal models of focal ischemia. In this setting, a meta-analysis revealed that HYPO reduced infarction by up to 44% after focal ischemia in animal models, but the issue of treatment duration was less clear [4]. Individual studies, however, show that prolonged cooling provides better protection in focal ischemia. To optimize protection, one must vary

Primer on Cerebrovascular Diseases, Second Edition
http://dx.doi.org/10.1016/B978-0-12-803058-5.00040-0

HYPO parameters to take into account the type and severity of brain injury. For example, longer cooling is needed to attain enduring benefit against more severe (longer) insults.

Currently, HYPO is used to treat some types of ischemia based on successful clinical trials. For instance, mild HYPO for 12–24 h reduced morbidity and mortality in patients with cardiac arrest. Interestingly, data suggest that maintenance of normothermia is equally as effective [7], but additional work is needed to prove this. Cooling also improved outcome in neonates with hypoxic-ischemic injury [8]. Phase II clinical trials have confirmed the safety of this treatment for patients with focal ischemia, but studies testing the effectiveness of HYPO are currently underway [9]. Here, converging evidence from animal studies as well as the positive correlation between fever and poor outcome after ischemic stroke strongly suggests that fever prevention and induced cooling will be promising treatments for focal ischemia.

Given that longer durations of HYPO are often needed to maximize protection, one must question whether HYPO harms recovery processes after ischemia. This is possible as many mechanisms of injury and repair overlap spatially, temporally, and mechanistically (e.g., glutamate neurotransmission in the penumbra in the days and weeks after stroke). Fortunately, animal studies show that HYPO does not negatively impact neuronal plasticity. For example, 21 days of localized HYPO did not affect behavior, dendritic shape, or cause brain damage, and this cooling had no meaningful impact on physiological variables (e.g., heart rate) in rats without brain injury [10]. In some cases, HYPO actually improves recovery by augmenting neurogenesis, angiogenesis, and synaptogenesis after brain injury [11]. Although prolonged postischemic HYPO is not thought to harm plasticity, there is at some point diminishing returns and other risks emerge (e.g., systemic side effects, interference with assessment, and delaying rehabilitation).

In addition to HYPO depth, duration, and delay, there are other parameters that determine efficacy (e.g., rewarming rate and method of cooling) along with numerous other considerations (e.g., insult severity, age, and use of drug cotreatments). Notably, animal studies have used brain selective (focal) and whole-body (systemic) HYPO approaches. Both methods have also been used in the clinic, and they have their advantages and disadvantages [2,3,12]. For instance, systemic cooling, which is the most commonly used in rodent experiments, induces shivering that may increase metabolic activity in the brain. Shivering is normally mitigated in patients with meperidine and other drugs. These appear to be effective and safe, but they should be more thoroughly assessed to determine their impact on HYPO's

neuroprotective properties. Likewise, other agents may interact with cooling. Perhaps the most important one is tissue plasminogen activator (tPA), a clot-busting drug given to patients with ischemic stroke. The combination appears to be safe in that cooling does not seem to diminish the effectiveness of tPA, which also does not hinder HYPO neuroprotective properties, and HYPO appears to reduce the risk of hemorrhagic transformation from giving tPA [13,14].

THERAPEUTIC HYPOTHERMIA FOR INTRACEREBRAL HEMORRHAGE

Many cell death mechanisms overlap between ischemia and intracerebral hemorrhage (ICH), such as excitotoxicity, inflammation, oxidative stress, and edema. Therefore, one would expect hyperthermia and cooling to have similar effects in ischemia and ICH. This is only partly true, likely because some mechanisms of injury vary considerably between ischemic and hemorrhagic stroke [3]. In an ICH, primary damage largely results from mechanical trauma of bleed dissecting through the brain parenchyma, which is not directly temperature sensitive. There is possibly some ischemia surrounding the area where the vessel ruptured [3]. Secondary injury occurs over time from several factors, such as inflammation, blood–brain barrier disruption, edema, and raised intracranial pressure, which can be attenuated by HYPO. As to functional and histological efficacy, however, cooling has not been consistently beneficial despite using treatment protocols that work well in ischemia (e.g., 1–7 days at 33°C). This could be partly due to the greater role blood neurotoxicity plays in ICH, and the fact that our rat studies show that cooling does not mitigate thrombin or iron-induced secondary injury. Prolonged HYPO (10 days) has been tested in a small clinical trial where cooling appears to reduce edema while improving outcome without an effect on hematoma size [15]. Larger randomized controlled trials are underway in patients with ICH, but preclinical rat studies raise concerns. Specifically, administering HYPO soon after ICH can aggravate bleeding either by increasing blood pressure or causing coagulopathy. Such findings emphasize how the effectiveness of HYPO varies with the type the brain insult (e.g., hemorrhage vs. ischemia), and it brings us back to the early worry with cooling—safety. With regard to hyperthermia/ fever, both clinical and animal findings are presently unclear. For instance, our work shows that mild hyperthermia appears to have no impact in rat models of ICH. Thus, at this time the optimal temperature management parameters are not known, but in some ways they are likely to overlap with those used in ischemia assuming safety issues can be managed.

CONCLUSION

Hypothermia is a robust neuroprotectant against global and focal ischemia, whereas hyperthermia is harmful. The ability of cooling to mitigate injury after ICH is more controversial but warrants further study. The translation of HYPO has been hindered by safety concerns, inadequate knowledge of how cooling provides neuroprotection, the use of less-than-ideal treatment parameters, and difficulty with regulating temperature. This has been somewhat resolved through the work of many laboratories that have focused on: comprehensive assessment of efficacy (e.g., long-term outcome measures), use of clinically relevant protocols (e.g., prolonged mild cooling), use of better cooling methods (e.g., hypothermia drugs) and cotherapies (e.g., drugs to stop shivering), and mechanistic work. Although our knowledge on the mechanisms of neuroprotection and potential side effects has grown substantially, much remains to be done to maximize efficacy while avoiding complications. A better appreciation for how HYPO (and hyperthermia) impacts brain injury will also improve neuroprotection drug studies, which have been frequently compromised by inadequately measuring and regulating temperature during and after surgery.

References

[1] Remba SJ, Varon J, Rivera A, Sternbach GL. Dominique-Jean Larrey: the effects of therapeutic hypothermia and the first ambulance. Resuscitation March 2009;81(3):268–71.
[2] Rothman SM. The therapeutic potential of focal cooling for neocortical epilepsy. Neurotherapeutics April 2009;6(2):251–7.
[3] MacLellan CL, Clark DL, Silasi G, Colbourne F. Use of prolonged hypothermia to treat ischemic and hemorrhagic stroke. J Neurotrauma February 13, 2009;26:313–23.
[4] van der Worp HB, Sena ES, Donnan GA, Howells DW, Macleod MR. Hypothermia in animal models of acute ischaemic stroke: a systematic review and meta-analysis. Brain December 2007;130(Pt. 12):3063–74.
[5] Busto R, Dietrich W, Globus M-T, Valdés I, Scheinberg P, Ginsberg M. Small differences in intraischemic brain temperature critically determine the extent of ischemic neuronal injury. J Cereb Blood Flow Metab 1987;7:729–38.
[6] Colbourne F, Corbett D. Delayed and prolonged post-ischemic hypothermia is neuroprotective in the gerbil. Brain Res 1994;654:265–7.
[7] Nielsen N, Wetterslev J, Cronberg T, et al. Targeted temperature management at 33°C versus 36°C after cardiac arrest. N Engl J Med December 5, 2013;369(23):2197–206.
[8] Shankaran S, Laptook AR, Ehrenkranz RA, et al. Whole-body hypothermia for neonates with hypoxic-ischemic encephalopathy. N Engl J Med October 13, 2005;353(15):1574–84.
[9] Lyden PD, Hemmen TM, Grotta J, Rapp K, Raman R. Endovascular therapeutic hypothermia for acute ischemic stroke: ICTuS 2/3 protocol. Int J Stroke January 2014;9(1):117–25.
[10] Auriat A, Klahr A, Silasi G, et al. Prolonged hypothermia in rat: a safety study using brain-selective and systemic treatments. Ther Hypothermia Temp Manag 2012;2:37–43.
[11] Yenari MA, Han HS. Influence of therapeutic hypothermia on regeneration after cerebral ischemia. Front Neurol Neurosci 2013;32:122–8.
[12] Gluckman PD, Wyatt JS, Azzopardi D, et al. Selective head cooling with mild systemic hypothermia after neonatal encephalopathy: multicentre randomised trial. Lancet February 19–25, 2005;365(9460):663–70.
[13] Sena ES, Jeffreys AL, Cox SF, et al. The benefit of hypothermia in experimental ischemic stroke is not affected by pethidine. Int J Stroke April 2012;8(3):180–5.
[14] Tang XN, Liu L, Koike MA, Yenari MA. Mild hypothermia reduces tissue plasminogen activator-related hemorrhage and blood–brain barrier disruption after experimental stroke. Ther Hypothermia Temp Manag June 2013;3(2):74–83.
[15] Staykov D, Wagner I, Volbers B, Doerfler A, Schwab S, Kollmar R. Mild prolonged hypothermia for large intracerebral hemorrhage. Neurocrit Care April 2013;18(2):178–83.

CHAPTER

41

Drug Delivery to the Central Nervous System

S.K. Sriraman[1],[a], G. Salzano[1],[a], V. Torchilin[1],[2]

[1]Northeastern University, Boston, MA, United States; [2]King Abdulaziz University, Jeddah, Saudi Arabia

INTRODUCTION

The effective delivery of pharmaceutical agents into the central nervous system (CNS) still represents a significant challenge to modern drug delivery. Much of this can be attributed to the presence of the blood–brain barrier (BBB), which consists of a network of tight junctions formed between cerebral capillary endothelial cells along with the surrounding basal lamina, astrocytes, pericytes, and microglia. The integrity of these tight junctions is mediated by a number of important proteins such occludins, claudins, cadherins, and a variety of cell adhesion molecules. This dynamic environment is a critical regulator of brain homeostasis and tightly regulates the entry of agents into the CNS. This barrier is thought to allow diffusion of gas molecules, such as O_2 and CO_2 and lipophilic small molecules (less than 400 Da) while still being largely impermeable to hydrophilic and macromolecular drugs. However, most lipophilic drugs are plagued by poor pharmacokinetic properties in systemic circulation. Although the use of polymer and lipid-based nanoparticles initially served as an attractive approach to increase the systemic circulation of these drugs, they still remain largely impermeable to the BBB [1].

In order for therapeutic drugs to be able to affect the CNS, they need to traverse this barrier, and it should also be ensured that there is a sufficient concentration of drug reaching the intended targets. Therefore, a number of interesting strategies have been employed to traverse the BBB by specifically targeting its components, or by simply circumventing the BBB altogether by employing nonsystemic routes, allowing for direct drug distribution into the CNS. These strategies have been reviewed in greater detail in the following discussion.

CROSSING THE BBB

As mentioned earlier, the effective treatment of brain diseases represents a significant challenge to pharmaceutical research due to their delicate environment. Many approaches aimed at overcoming the BBB to promote drug delivery for the treatment of brain diseases, including neurodegenerative diseases and brain cancer, have been proposed. Some of these approaches involve directly affecting BBB function either due to its temporary disruption by physical means, or by specifically targeting one or more of its biological components.

Disruption of the BBB

Focused ultrasound (FUS) is a noninvasive strategy where ultrasound mediates the temporary disruption of the BBB, resulting in an efficient and targeted delivery of active molecules from the blood circulation to the desired areas of the brain. FUS, in particular low-power ultrasound, is often combined with the intravenous administration of microbubbles containing a contrast agent [2]. The microbubbles act to concentrate the ultrasound energy, resulting in an improvement of both safety and reproducibility of the FUS. Once the microbubbles reach the ultrasound field, they undergo a stable expansion and contraction causing a mechanical stimulation of the blood vessels, thereby allowing therapeutic agents temporarily to move into the brain [2]. Multifunctional microbubbles containing a conventional chemotherapeutic agent, doxorubicin, and conjugated with superparamagnetic iron oxide nanoparticles were designed to release both agents upon FUS exposure [3]. In an animal model of brain tumor, the concurrent BBB opening and drug delivery were achieved during the FUS exposure. The use of FUS has also shown to reduce

---

[a] Both these authors contributed equally to this work.

Primer on Cerebrovascular Diseases, Second Edition
http://dx.doi.org/10.1016/B978-0-12-803058-5.00041-2

plaque pathology in a model of Alzheimer's disease in the absence of drug delivery, suggesting the broad range of applications of FUS [2]. The feasibility of a future clinical translation of FUS has been proven by the successful demonstration of its safety in nonhuman primates [2]. Transient BBB disruption can be also achieved via intra-arterial administration of osmotic agents, most commonly mannitol. The administration of the hyperosmolar agent mannitol induced endothelial cell dehydration and subsequent constriction thereby opening the tight junctions of the BBB for up to 4h. The mannitol-based BBB transient disruption followed by intra-arterial administration of chemotherapeutic agents is estimated to increase drug delivery by 10–100 times in comparison with the delivery of drugs alone [4]. However, it must be mentioned that despite its first use in patients in the 1970s, the disruption of the BBB by osmotic agents is not part of standard practice due to the low compliance from patients and the increased risk of seizures and strokes. Another approach for BBB disruption is the non-thermal irreversible electroporation (NTIRE) method. NTIRE is a soft-tissue ablation strategy, which uses a pulsed electric current to increase the permeability of the cell membrane by causing micro- and nanopores in the cell membrane. The advantage is that the effects of the electric current only affect the cell membrane and not the surrounding areas, including blood vessels. Although its clinical safety still needs to be assessed further, use of NTIRE has been shown to significantly reduce the tumor volume in a model of glioma [5].

Modulation of Drug Efflux Pumps at the BBB Interface

Inhibition of Drug Efflux Pumps

As discussed earlier, although the BBB serves as a tight anatomical barrier that regulates the entry of drugs, it can be temporarily disrupted to force the entry of pharmaceutical agents. However, even after this "forced" entry, the BBB also contains a number of drug efflux pumps that can extrude these external agents from the brain thereby reducing their therapeutic efficacy. One of the most prominent efflux proteins at the BBB is the P-glycoprotein (P-gp), which belongs to the ATP-binding cassette family of proteins. It is encoded by the multidrug resistance gene (mdr1/abcb1) and is a 170-kDa transmembrane protein that is responsible for the disposition of a wide array of substrates. This is a major impediment to the CNS delivery of drugs for the treatment of neural diseases as well as brain cancer chemotherapy. Therefore the inhibition of these efflux mechanisms represents a more specific and less disruptive approach to enhance the delivery of drugs to the CNS.

Initially, the use of small-molecule inhibitors of P-gp showed limited efficacy and safety in clinical settings.

However, the third generation of P-gp inhibitors, such as tariquidar, elacridar, zosuquidar, and valspodar has demonstrated significant therapeutic potency and safety. A comprehensive study carried out by Kemper et al. investigated the penetration of paclitaxel (PCT), an anticancer drug that is a known substrate of P-gp, into the brain [6]. The study investigated the relative therapeutic efficiencies of a range of P-gp inhibitors such as cyclosporine A and elacridar. In a murine model, the inhibitors were administered orally (p.o.) before PCT administration. It was found that, although cyclosporine A was able to mediate three-fold higher brain penetration of PCT, elacridar allowed for a five-fold increase in brain PCT levels. On comparison with a P-gp knockout mouse model (reference control), an optimized dosing regimen of elacridar allowed for PCT brain levels of 80–90% further highlighting the therapeutic potential of this combination regimen.

In a separate study, sphingolipid-based drugs targeting the sphingosine-1-phosphate receptor also inhibited basal P-gp activity thereby allowing for a five-fold increase in PCT levels [7]. Similarly, intracerebroventricular (i.c.v.) injection of vascular endothelial growth factor also mediated acute reduction of P-gp activity allowing for the increased brain distribution of morphine as well as verapamil [8]. These studies thus highlight the increased therapeutic potential of CNS drugs by targeting efflux pumps at the BBB.

Targeting to Cross the BBB by Transcytosis

Another valid approach is the conjugation of drugs, or the functionalization of nanocarriers incorporating a drug, with specific peptides or proteins capable of transporting the so-called Trojan horse complex, across the BBB via receptor-mediated transcytosis. In particular, by using this strategy, the Trojan horse complex from the blood is transported transcellularly across the brain endothelium via a native receptor with consequent release of the drug at a desired site of the brain (Fig. 41.1). Transferrin (Tf), insulin, and low-density lipoproteins are some examples of appropriate ligands used to transport therapeutics to the brain endothelium through receptor-mediated transcytosis. The interaction of Tf with the relative abundance of Tf receptors (TfR) on the BBB endothelial brain cells rationalized the wide use of Tf as a specific ligand on the surface of nanoparticles for brain targeting. Tf-conjugated self-assembly nanoparticles encapsulating zoledronic acid (NPs-ZOL-Tf) showed a significant higher antitumor efficacy when compared with NPs not functionalized with Tf, in an orthotropic glioblastoma model of immunosuppressed mice [9]. This effect was correlated with a higher intratumor localization of the NPs-ZOL-Tf in intracranial xenografts suggesting that the presence of Tf on the surface of NPs increases the antitumor efficacy

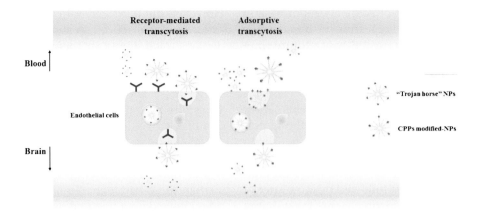

FIGURE 41.1   Transcytosis of nanoparticles across the BBB. *CPP*, cell-penetrating peptide; *NP*, nanoparticle.

of ZOL in glioblastoma due to its increased ability to cross the BBB [9].

Liposomes coated with glutathione-conjugated polyethylene glycol (G-Technology) is another example of a receptor-mediated transcytosis approach that has entered the clinical stage of evaluation. The glutathione receptor is highly expressed in the BBB. After assessment of the clinical safety and the enhanced delivery of drugs into the brain by using G-Technology, the efficacy of doxorubicin formulated with glutathione-conjugated polyethylene glycol is currently under evaluation in patients with brain metastasis.

Adsorptive transcytosis is a second type of transcytosis. Specifically, it is a charge-dependent endocytic internalization event that occurs with cationized molecules, including cell-penetrating peptides (CPPs). CPPs, such as the well-known TAT peptide (transduction domain of human immunodeficiency virus type-1), are short cationic peptides that are able to facilitate the internalization of nanocarriers across cell membranes by absorbing tightly to the negatively charged cell surface. Yao Qin et al. covalently conjugated TAT with cholesterol for preparing doxorubicin-loaded liposomes for glioma therapy [10]. The results from biodistribution studies indicated an evident preferential distribution of the TAT-modified liposomes in the brain tissues in glioma-bearing rats. The survival time of animals treated with TAT-modified liposomes was significantly longer than in the other groups. Interestingly, the treatment with TAT-modified liposomes showed the best anti-tumor effects, with a long-term remission in 40% of the animals.

CIRCUMVENTING THE BBB

Bypassing the BBB represents a safer alternative to directly altering BBB function. These approaches include but are not limited to direct injection of the drug into the cerebrospinal fluid, brain parenchyma, and intranasal administration.

I.C.V. injection of drugs represents a more direct means to ensure brain distribution of drugs by avoiding the major rate-limiting steps such as pharmacokinetics in circulation and diffusion across the BBB. For example, i.c.v. infusion of valproic acid in amygdala-kindled rats increased its therapeutic efficacy by suppressing the onset of seizures while avoiding potential systemic toxicity [11].

Similarly, intrathecal injection of drugs directly into the cerebrospinal fluid also represents an attractive approach while being more patient compliant. Gray and colleagues had demonstrated transgene delivery using adeno-associated virus in nonhuman primates by intrathecal administration [12]. Interestingly at the administered dose, around 2% of the entire brain was transduced covering most regions of the CNS.

Alternatively, implants or reservoirs may also be introduced surgically to allow for controlled release from a drug depot [11]. It must also be noted that, although parenchymal administration into cerebrospinal fluid (CSF) allows us to circumvent the BBB, the distribution of drugs from the CSF to the brain tissue may be slow and limited primarily by diffusion [1].

Although the above-mentioned approaches involve direct intra-cerebral and intrathecal application of drugs, these methods are highly invasive and are not a practical means for administering on a repeated or chronic basis. Intra-nasal administration therefore represents a very noninvasive and highly patient-compliant method for the delivery of drugs to the CNS. The nasal submucosal space is enriched by CSF flow tracts leading to the olfactory lobe of the brain. The drugs are therefore thought to move via nerve sheaths or by axonal transport along the neurons. This therefore allows for direct CNS entry of drugs bypassing the BBB. However, as is the case with the BBB, hyrdrophilic molecules as well as macromolecules do not permeate through to the brain via the intranasal routes as easily as hydrophobic molecules. To overcome this limitation, peptide drugs may be co-administered with CPPs to facilitate delivery to the brain parenchyma. As was demonstrated by Kamei and Takeda-Morishita, peptide insulin co-administered with penetratin was able to reach the cerebellum, cerebral cortex, and the brain stem following their intranasal administration in mice [13]. Furthermore, the role of the BBB in the intranasal route was also determined. Application of sodium caprate, a known BBB paracellular absorption enhancer, confirmed that the BBB played a very minor role in the brain distribution of insulin when co-administered intranasally with penetratin. Moreover, this result indicated the possibility that penetratin could mediate transport of insulin via the trigeminal nerve thus shedding more light on the mechanism of transport from the nasal cavity.

CONCLUSION

It is evident that significant strategies have been developed to overcome some of the challenges associated with delivering drugs to the CNS. Some of these approaches involve directly traversing the BBB or simply circumventing the BBB. Strategies to traverse the BBB involve its temporary disruption by physical means such as ultrasound or by specifically targeting some of its components like drug efflux proteins. Similarly, targeting drugs with ligands to facilitate transcytosis across the BBB as well as the use of alternative delivery routes such as intracerebral, intranasal, etc. have also been employed. It is also important to keep in mind the importance of patient compliance. Although some strategies allow for enhanced drug deposition in the brain, these are not very patient compliant. Therefore the development of novel patient-compliant approaches such as intranasal delivery systems and FUS represent a very promising direction for the delivery of CNS agents.

References

[1] Pardridge WM. Drug transport across the blood–brain barrier. J Cereb Blood Flow Metab 2012;32(11):1959–72.

[2] Burgess A, Shah K, Hough O, Hynynen K. Focused ultrasound-mediated drug delivery through the blood–brain barrier. Expert Rev Neurother 2015;15(5):477–91.

[3] Fan C-H, Ting C-Y, Lin H-J, et al. SPIO-conjugated, doxorubicin-loaded microbubbles for concurrent MRI and focused-ultrasound enhanced brain-tumor drug delivery. Biomaterials 2013;34(14):3706–15.

[4] Miller G. Breaking down barriers. Science 2002;297(5584):1116–8.

[5] Garcia P, Pancotto T, Rossmeisl J, et al. Non-thermal irreversible electroporation (N-TIRE) and adjuvant fractionated radiotherapeutic multimodal therapy for intracranial malignant glioma in a canine patient. Technol Cancer Res Treat 2011;10(1):73–83.

[6] Kemper EM, van Zandbergen AE, Cleypool C, et al. Increased penetration of paclitaxel into the brain by inhibition of P-glycoprotein. Clin Cancer Res 2003;9(7):2849–55.

[7] Cannon RE, Peart JC, Hawkins BT, Campos CR, Miller DS. Targeting blood–brain barrier sphingolipid signaling reduces basal P-glycoprotein activity and improves drug delivery to the brain. Proc Natl Acad Sci 2012;109(39):15930–5.

[8] Hawkins BT, Sykes DB, Miller DS. Rapid, reversible modulation of blood–brain barrier P-glycoprotein transport activity by vascular endothelial growth factor. J Neurosci 2010;30(4):1417–25.

[9] Porru M, Zappavigna S, Salzano G, et al. Medical treatment of orthotopic glioblastoma with transferrin-conjugated nanoparticles encapsulating zoledronic acid. Oncotarget 2014;5(21):10446.

[10] Qin Y, Chen H, Zhang Q, et al. Liposome formulated with TAT-modified cholesterol for improving brain delivery and therapeutic efficacy on brain glioma in animals. Int J Pharm 2011;420(2):304–12.

[11] Serralta A, Barcia JA, Ortiz P, Durán C, Hernández ME, Alós M. Effect of intracerebroventricular continuous infusion of valproic acid versus single i.p. and i.c.v. injections in the amygdala kindling epilepsy model. Epilepsy Res 2006;70(1):15–26.

[12] Gray SJ, Kalburgi SN, McCown TJ, Samulski RJ. Global CNS gene delivery and evasion of anti-AAV-neutralizing antibodies by intrathecal AAV administration in non-human primates. Gene Ther 2013;20(4):450–9.

[13] Kamei N, Takeda-Morishita M. Brain delivery of insulin boosted by intranasal coadministration with cell-penetrating peptides. J Control Release 2015;197:105–10.

CHAPTER

42

Ischemic Tolerance: In Situ and Remote Pre- and Postconditioning

H. Zhao
Stanford University, Stanford, CA, United States

INTRODUCTION

"Thus, when Heaven is about to confer a great office on any man, it first exercises his mind with suffering ... it exposes his body to hunger ... it stimulates his mind, hardens his nature, and supplies his incompetencies." More than 2000 years ago, the Chinese philosopher Mencius believed that a person who experiences prior suffering is able to take on more responsibility throughout life. Friedrich Nietzsche, the German philosopher, restates this as "That which does not kill us makes us stronger." In the 1880s, the German pharmacologist Hugo Schulz described that the growth of yeast could be stimulated by small doses of poisons. In 1943, the term "hormesis" was coined, indicating that a low dose of poison is stimulating, whereas a high dose is toxic, an observation in almost all biological systems [1]. This dose-dependent effect of hormesis, when applied to stroke research, is similar to the concept of ischemic tolerance. Ischemic tolerance occurs when a brief sublethal ischemic insult is applied before a subsequent injurious insult. The term ischemic tolerance is often interchangeable with the term ischemic preconditioning (PreC) but, more specifically, ischemic PreC refers to the brief, subinjurious ischemia conducted before a subsequent injurious ischemic insult, whereas ischemic tolerance refers to the increased ability of the ischemic organ to resist injury after the brief ischemic event. Ischemic tolerance or PreC is protective against ischemia in many organs, including the heart and brain. In addition, ischemic tolerance or PreC can be mimicked by cross-tolerance, which is induced by heterogeneous stimuli or stressors other than ischemia [2]. The stimuli for PreC can also be applied after reperfusion, which is defined as postconditioning (PostC). In stroke, ischemic conditioning (both PreC and PostC) is performed in situ in the brain;

thus it is defined here as in situ ischemic conditioning. In contrast to in situ ischemic conditioning, remote ischemic conditioning refers to ischemia conducted in an organ other than the brain, which can also attenuate brain injury after stroke.

IN SITU PreC AND PostC

The first study addressing in situ ischemic PreC against brain injury in 1964 described a prior anoxic insult that was found to protect against hippocampal CA1 neuronal death [2]. The concept of ischemic PreC was formally defined in 1986 by Murry et al. in a myocardial ischemia model in dog [3], and it was replicated in the brain by Kitagawa et al. [4] in a forebrain ischemic model in gerbil. In situ ischemic PreC in brain research has two therapeutic time windows—rapid and delayed. Rapid PreC is induced within a few hours, whereas delayed PreC is induced from 24h to a few days, before the subsequent injurious cerebral ischemia or stroke. Ischemic PreC performed between these two time windows does not generate protective effects against brain injury. The protective effects of PreC have been proven in both in vivo and in vitro models. In vivo models include forebrain ischemia, focal cerebral ischemia, and hypoxia/ischemia, and in vitro models include neuronal cultures and organotypic brain slice cultures.

More recently, in situ ischemic PostC has also been found to protect against brain injury after cerebral ischemia or stroke [5]. As mentioned earlier, PreC is a transient subinjurious ischemia conducted before ischemia onset, whereas PostC is performed after postischemic reperfusion. Typically, ischemic PostC refers to the interruption of reperfusion by using a mechanical method, which causes brief and repeated occlusions of

Primer on Cerebrovascular Diseases, Second Edition
http://dx.doi.org/10.1016/B978-0-12-803058-5.00042-4

the cerebral blood vessels after reperfusion. Nevertheless, ischemic PostC can also be induced by a single, brief period of ischemia or anoxia, and in vitro models of hypoxic PostC have also been established by using the oxygen-glucose deprivation model [6]. In contrast to ischemic PreC, three therapeutic time windows for ischemic PostC have been arbitrarily defined—rapid, intermediate, and delayed. Rapid PostC is conducted immediately or within a few minutes after reperfusion; intermediate PostC is performed from a few hours to 12 h after reperfusion; and delayed PostC can be induced from 24 h to a few days after reperfusion [6]. Rapid and intermediate PostC robustly reduce infarct sizes as measured 2–3 days after stroke, and were found to offer long-term protection for up to 2 months. Like rapid PostC, delayed PostC that is conducted 2 days after forebrain ischemia can attenuate hippocampal neuronal death. Nevertheless, delayed hypoxic PostC performed on day 5 after stroke does not reduce infarction, but it can attenuate delayed thalamic atrophy [6].

CROSS-TOLERANCE INDUCED BY HETEROLOGOUS CONDITIONING

Past studies have revealed the extensive pathological mechanisms of brain injury after stroke. Brain ischemia results in ATP depletion, which leads to ion disruption and imbalance in the distribution of Na^+/K^+ across cellular membranes. This redistribution of ions results in anoxic or ischemic depolarization, and leads to glutamate release and increased intracellular Ca^{2+}, which then initiates various cascades in the cell signaling pathways of both apoptosis and necrosis. In general, any hazard stimuli that can induce one of these pathological cascades can be used as a stimulus for PreC. Accordingly, PreC induced by a stimulus or stressor other than ischemia is called heterologous conditioning, whereas ischemic PreC is defined as homologous PreC. The protective effects induced by heterologous conditioning are referred to as cross-tolerance.

In summary, in addition to ischemic homologous PreC, ischemic tolerance can be also induced by many stimuli, including anoxia/hypoxia, hyperoxia, glutamate/N-methyl-D-aspartate, cerebral spreading depression, anesthesia, hypothermia/hyperthermia, inflammation (lipopolysaccharide), free radicals, and metabolic inhibition. Like ischemic PreC, ischemic PostC can also be mimicked by a broad range of stimuli and stressors [2]. Until recently, these types of conditioning were conducted in the brain as in situ conditioning. Now ischemic tolerance can also be induced remotely, as described in the following section.

REMOTE CONDITIONING

Transient ischemia conducted in a peripheral organ to protect against brain injury after stroke is defined as remote conditioning [7]. Remote conditioning includes PreC, perconditioning (PerC, applied during ischemia) and PostC [8]. In animal models, most remote conditioning is conducted in a hind limb by occluding the femoral artery for 5–10 min, followed by reperfusion for 5–10 min, and repeated for three to five cycles. Like in situ PreC, remote PreC occurs before stroke onset. As discussed earlier, in situ ischemic PreC has a two-phase therapeutic time window and does not offer protection when conducted between these two windows. In contrast, remote PreC appears to generate protection at three therapeutic time windows, as robust neuroprotection was found when limb ischemia was conducted immediately, 12 and 24 h before stroke onset [7].

Remote ischemic PostC often uses the same parameters as remote PreC, except the limb ischemia is performed after postischemic reperfusion [8]. When limb ischemia is induced immediately after reperfusion, or even a few hours after reperfusion, it protects against brain injury. In addition, limb ischemia is performed after stroke onset while the brain blood vessel is still occluded, which is termed remote ischemic PerC, and is neuroprotective. Remote PerC may be better suited for clinical translation, as reperfusion is not a prerequisite. With advancements of basic research in the laboratory, a number of medical centers in the world have conducted pilot clinical trials for remote PreC, which have proved its safety, feasibility, and potential efficacy for improving brain functions after stroke [9].

THE UNDERLYING PROTECTIVE MECHANISMS OF IN SITU AND REMOTE CONDITIONING

It seems that the protective effects of different patterns of conditioning are comparable, including in situ PreC and PostC, with homologous and heterologous stimuli, as well as remote PreC and PostC. Nevertheless, their underlying protective mechanisms must differ to some extent, at least their mechanisms must differ at the triggering stages. In situ PreC mainly stimulates and adapts the brain to resist the subsequent injurious ischemia, but in situ PostC mainly interrupts reperfusion or interferes with the pathological events that occur after reperfusion. In addition, remote PreC or PostC protects against brain injury via protective signaling transferred from the remote ischemic organs. Despite these differences, their protective mechanisms may eventually converge on common cellular and molecular pathways at the later stages of brain injury and recovery.

IN SITU PreC

The protective mechanisms of in situ PreC have been most extensively studied, and it is now known that the underlying mechanisms differ between rapid and delayed PreC [2]. Rapid PreC, which is induced from within a few minutes to 2–3h before stroke onset, involves changes in ion channel permeability and post-translational modifications of proteins, whereas delayed PreC involves gene activation and de novo protein synthesis. It is likely that these distinct mechanisms are responsible for differing protective effects; for example, rapid PreC, unlike delayed PreC, does not offer permanent protection.

In general, in situ PreC directly stimulates the brain and adapts the brain to resist subsequent injurious ischemia [2]. There are many cell signaling pathways involved in this process [10]. First, PreC improves energy preservation and slows down metabolism, which renders the ischemic brain more resistant when ATP is depleted after stroke. Second, mitochondrial functions are preserved, including mitochondrial membrane potentials, opening of ATP-sensitive K^+ channels, and antioxidative systems. Third, promotion of antiapoptotic proteins and inhibition of proapoptotic proteins of the Bcl-2 family leads to inhibition of apoptosis and enhanced neuronal survival. Fourth, the balance of excitatory and inhibitive systems shifts toward inhibition of neurotoxicity, thus toward neuronal survival. These effects include delaying the onset of anoxic/ischemic depolarization, inhibiting the amount of glutamate release, attenuating the sensitivity of glutamate receptors, and promoting gamma-aminobutyric acid receptor activity. Fifth, various cell survival signaling pathways are activated, whereas cell death pathways are inhibited. These include promotion of heat shock protein expression, reductions in the release of cytochrome C and apoptosis inducing factor release, inhibition of caspase activity, and increases in Akt and ERK (extracellular-signal regulated kinase) phosphorylation. Sixth, proinflammatory responses are inhibited, including inhibition of inflammatory transcriptional factors, inflammatory cell infiltrations, and toll-like receptor activity [10].

IN SITU PostC

As discussed in the section "Cross-tolerance induced by heterologous conditioning," a broad range of stimuli can induce in situ PostC, including mechanical interruption of reperfusion, a single period of brief ischemia or anoxia, anesthesia, or neurotoxic agents. Classically, ischemic PostC is induced by the mechanical interruption of reperfusion, which consists of a few cycles of 10- to 30-s occlusions and reperfusions [6]. When PostC interrupts reperfusion, it inhibits reperfusion injury that is caused by the abrupt supply of blood flow to the ischemic brain, which leads to large amounts of reactive oxygen species (ROS) production. Therefore, ischemic PostC robustly reduces ROS production by attenuating or interrupting reperfusion [6]. Nevertheless, little is known whether other patterns of PostC, such as pharmacological PostC, also results in reduced ROS production. In addition, little is known about the protective mechanisms of intermediate and delayed ischemic PostC. Despite the obvious temporal differences between ischemic PreC and PostC, it seems that both protect against brain injury by converging on some common protective mechanisms. These common mechanisms include improved metabolism and glucose uptake, promoted opening of the mitochondrial KATP channels, enhanced ε-protein kinase C (PKC) but inhibited δPKC activity, and inhibited apoptosis and neuroinflammation [6]. Thus, ischemic PostC and PreC may share some common protective mechanisms at the late stage of brain injury and recovery.

REMOTE PreC AND PostC

It is not surprising that remote conditioning protects against brain injury, as many types of heterologous conditioning offer protection against brain injury, and remote conditioning may generate various factors that are similar to some stressors involved in heterologous conditioning. An additional key reason that remote conditioning can exert its effects on the ischemic brain is the broad cross talk between the brain and peripheral organs [11]. For example, brain injury after stroke confers injury to multiple peripheral organs, including skeletal muscle atrophy, myocardial injury, immunodepression, and spleen atrophy, as well as lung injury due to pneumonia. Conversely, peripheral organs exert their effects on the brain, and alterations of the functions of peripheral organs may promote or inhibit brain injury. For instance, promoting or inhibiting T cells, macrophages, and neutrophils exacerbates or attenuates brain injury, respectively; immune cells released from the spleen robustly increase, whereas splenectomy robustly decreases brain infarction. With this concept of bidirectional communication between the brain and peripheral organs, it is not difficult to appreciate that remote conditioning can protect against brain injury after stroke, as limb ischemia for remote conditioning may generate a broad range of stressors that are able to reach the brain to effect brain injury. The brain interacts with peripheral organs through at least three pathways: the immune system, the nervous system, and the circulatory system. Remote conditioning is often induced by a series of brief transient ischemic insult applied to a limb. Transient ischemia results in the release of free radicals, adenosine, ATP, and other inflammatory cytokines, which may directly stimulate the afferent nerves that transduce signaling to the brain. Indeed, inhibition of the afferent nerve pathway in rats abolished the protective effect of remote PreC. In addition, released factors may circulate in the blood and reach the brain to stimulate endothelial cells in the periventricular regions, which indirectly convey protective signaling to the ischemic brain. Furthermore, limb ischemia may also alter the function of circulating immune cells, which may pass protective signaling from the ischemic limb to the brain. Through these communication pathways, limb ischemia may attenuate brain edema, BBB opening, and inhibit the inflammatory response in the ischemic brain. Eventually, the protective mechanisms of remote conditioning may converge on similar mechanisms as other patterns of conditioning, including in situ PreC and PostC, as discussed earlier.

CONCLUSION

In situ and remote conditioning, including both PreC and PostC, robustly attenuate brain injury after stroke. Ischemic homologous conditioning can be mimicked by a broad range of heterologous stressors. Both PreC and PostC protect against brain injury at rapid and delayed time windows, and often offer long-term protective effects. Although their protective mechanisms may differ at the early stages due to the nature of the stressors and the different onset of time, their mechanisms seem to converge on some common cellular and molecular

survival and apoptotic pathways. These studies of PreC and PostC may open novel avenues for clinical translation and stroke treatment.

References

[1] Calabrese EJ. Hormesis: why it is important to toxicology and toxicologists. Environ Toxicol Chem 2008;27(7):1451–74.

[2] Dirnagl U, Simon RP, Hallenbeck JM. Ischemic tolerance and endogenous neuroprotection. Trends Neurosci 2003;26(5):248–54.

[3] Murry CE, Jennings RB, Reimer KA. Preconditioning with ischemia: a delay of lethal cell injury in ischemic myocardium. Circulation 1986;74(5):1124–36.

[4] Kitagawa K, et al. 'Ischemic tolerance' phenomenon found in the brain. Brain Res 1990;528(1):21–4.

[5] Zhao H, Sapolsky RM, Steinberg GK. Interrupting reperfusion as a stroke therapy: ischemic postconditioning reduces infarct size after focal ischemia in rats. J Cereb Blood Flow Metab 2006;26(9):1114–21.

[6] Zhao H. Ischemic postconditioning as a novel avenue to protect against brain injury after stroke. J Cereb Blood Flow Metab 2009;29(5):873–85.

[7] Ren C, et al. Limb remote-preconditioning protects against focal ischemia in rats and contradicts the dogma of therapeutic time windows for preconditioning. Neuroscience 2008;151(4):1099–103.

[8] Wang Y, et al. Ischemic conditioning-induced endogenous brain protection: applications pre-, per- or post-stroke. Exp Neurol 2015;272:26–40.

[9] Nikkola E, et al. Remote ischemic conditioning alters methylation and expression of cell cycle genes in aneurysmal subarachnoid hemorrhage. Stroke 2015;46(9):2445–51.

[10] Stetler RA, et al. Preconditioning provides neuroprotection in models of CNS disease: paradigms and clinical significance. Prog Neurobiol 2014;114:58–83.

[11] Zhao H, et al. From rapid to delayed and remote postconditioning: the evolving concept of ischemic postconditioning in brain ischemia. Curr Drug Targets 2012;13(2):173–87.

MOLECULAR MECHANISMS

43

Mechanisms of Neuron Death (Necrosis, Apoptosis, Autophagy) After Brain Ischemia

B.R. Hu, C.L. Liu

Shock Trauma and Anesthesiology Research Center, University of Maryland School of Medicine, Baltimore, MD, United States

INTRODUCTION

Brain ischemia refers to a neurological condition that brain blood flow is insufficient to meet metabolic demand. There are two kinds of brain ischemia: (1) focal ischemia in which ischemia is confined to a specific region of the brain and (2) global ischemia, which affects the entire area of brain or forebrain tissue. Focal brain ischemia is a subtype of stroke along with subarachnoid hemorrhage and intracerebral hemorrhage. Global brain ischemia may occur in many pathological conditions, such as cardiorespiratory arrest.

Necrosis

Necrotic cell death or necrosis is an accidental type of cell death in living tissue and always caused by pathological factors, such as energy failure after brain ischemia. Cell death due to necrosis is always passive, and thus does not need activation of a particular cellular signaling pathway. A typical necrotic process due to brain ischemia encompasses swelling of the cell and subcellular organelles, followed by multiple organelle damage, the loss of cell membrane integrity, and an uncontrolled release of cellular contents into the surrounding extracellular space (Fig. 43.1). As a result, necrosis usually initiates an inflammatory response in the surrounding areas of cell injury, which may be contained locally. However, severe necrosis may also lead to systemic inflammation in the remote organs, such as thymus, spleen, and small and large intestines. The systemic inflammatory response may be mediated by the circulating inflammatory signals, such as damage-associated molecular pattern molecules released from necrotic cell or tissue. Untreated necrosis can lead to a buildup of decomposing dead tissue and cell debris at or near the site of the cell death, resulting in tissue infarction.

Apoptosis

Apoptotic cell death or apoptosis is an active cellular process, can occur naturally, and requires activation of particular apoptotic pathways in multicellular organisms. In contrast to necrosis, which is a form of accidental cell death, apoptosis is physiologically required for tissue or organ formation during development or lifecycle. Therefore, the naturally occurring apoptosis is evolutionarily designed to provide beneficial effects to the organism by eliminating unwanted cells. The morphological characteristics of apoptosis include cell membrane blebbing, cell shrinkage, chromatin condensation, and formation of apoptotic body (Fig. 43.2). Unlike necrosis, apoptosis usually does not initiate the inflammatory responses as apoptotic bodies can be quickly removed by phagocytes [1]. Apoptosis occurs naturally and extensively during normal brain development, which is critical for the establishment of a definitive pattern of neuronal connections. However, some pathological conditions may accidentally initiate the same apoptotic pathways to induce unwanted cell death. For example, in immature or developing neurons, an episode of brain hypoxia-ischemia (HI) turns on the apoptotic pathways to induce neuronal death [2].

Apoptotic pathways can be divided into caspase-dependent or caspase-independent. The caspase-dependent pathway can also be classified into the extrinsic and intrinsic pathways. The extrinsic pathway is mediated by the so-called death receptors including tumor necrosis factor (TNF) receptors, Fas, and TNF-related apoptosis-inducing ligand (TRAIL) receptor. The specific binding

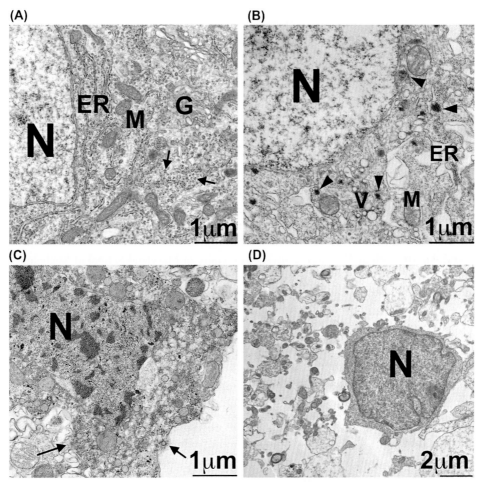

FIGURE 43.1   Electron micrographs of cortical neurons from a sham-operated rat (A) and rats subjected to 120 min of focal ischemia followed by 1 h (B) and 24 h (C) of reperfusion (Hu et al., 2001), and a postnatal 26-day rat subjected to 30 min of hypoxia-ischemia followed by 48 h of recovery (D) [5]. (A) In sham, ribosomal rosettes (*arrows*), the endoplasmic reticulum (ER), mitochondria (M), nucleus (N), and Golgi (G) are normally distributed; (B) at 1 h of reperfusion, ribosomes are clumped into aggregates (*arrowheads*). Golgi apparatus disappears to form vacuoles (V). The ER and mitochondria (M) are severely swollen. The nucleus (N) seems not changed; (C) At 24 h of reperfusion, a necrotic neuron shows membrane damage (*arrows*), shrunken nucleus with clumped tigroid chromatin, irregular amorphous organelles, and vesicular structures and vacuoles; (D) an acute necrotic neuron shows "burst" to release its entire content.

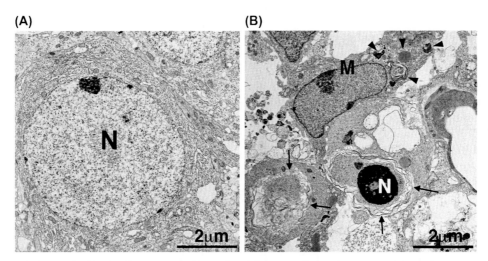

FIGURE 43.2   Electron micrographs of dentate gyrus (DG) neurons from a postnatal 7-day rat pup subjected to 60 min of hypoxia-ischemia (HI) followed by 48 h of recovery [5]. (A) A normal DG neuron from the contralateral hemisphere after HI and (B) apoptotic DG neurons from the ipsilateral hemisphere after HI. The apoptotic neurons have classical apoptotic ultrastructural features including condensed nucleus as a dark mass ball (N), formation of filament bundles (*arrows*), and a phagocyte, probably microglia (M), attaching to the apoptotic neuron. The phagocyte contains both primary and secondary lysosomes (*arrowheads*). Scale bar = 2 μm.

of Fas-ligand, TNF-α, or TRAIL to their corresponding receptors leads to recruitment of the adaptor-protein Fas-associated death domain or complex-I/II to activate caspase-8, followed by activation of caspase-3. In the intrinsic apoptotic pathway, release of cytochrome c from mitochondria leads to activation of caspase-9. Activated caspase-9 works together with apoptotic factor-1 (Apaf-1) to activate caspase-3. Activated caspase-3 executes apoptosis via cleaving specific substrates, such as caspase-activated DNase (CAD) and poly(ADP-ribose) polymerase (PARP) to induce apoptosis. The caspase-independent apoptotic pathway is mediated via apoptosis-inducing factor (AIF). AIF normally resides just behind the mitochondrial outer membrane. When a mitochondrion is damaged, AIF may be released to the cytoplasm and then translocate to the nucleus to induce DNA fragmentation and chromatin condensation, resulting in apoptosis. In different cell types and conditions, increasing number of other factors, such as B-cell lymphoma 2 (Bcl-2), Bcl2-associated X protein (BAX), Bax-like BH3 protein (BID), Bcl-2 homologous antagonist/killer (BAK), or Bcl-2-associated death promoter (BAD), B-cell lymphoma-extra large (Bcl-Xl), are also involved in the apoptotic pathways, suggesting that the apoptotic processes are considerably more complicated and diversified than the simplified caspase-dependent and caspase-independent pathways described earlier.

Despite the widespread use of apoptosis or necrosis, in some cases, dying cells may have molecular and morphological overlap features of necrosis and apoptosis, which has occasionally been referred to as "necroptosis" and "apoptosis-like cell death."

Autophagy

Autophagy is the cellular machinery of late endosome/lysosomal degradation of unnecessary or dysfunctional cellular components. There are three basic types of autophagy: macroautophagy, microautophagy, and chaperone-mediated autophagy. Macroautophagy is the process in which aberrant cellular components or organelles are sequestered by double-layer membranes to form autophagosomes (APs), and then the APs fuse with lysosomes to degrade the AP's cargo. Microautophagy is mediated by direct lysosomal engulfment of aberrant cytoplasmic contents. Chaperone-mediated autophagy refers to the chaperone-dependent selection and delivery of a particular group of aberrant cytosolic proteins to lysosomes for degradation.

Macroautophagy is the major type of autophagy, thus commonly referred to as autophagy (hereafter). Fig. 43.3 shows simplified basic steps of the (macro) autophagy pathway, consisting of: (1) AP formation; (2) AP-to-late endosome/lysosome fusion to form autophagolysosome or simply autolysosome (AL); and (3) lysosomal degradation of AP and its cargo by hydrolases. The biochemical cascade for AP formation begins with ATG1 (ATG = autophagic gene-related protein). Activated ATG1 facilitates incorporation of two key ATG complexes into a double-membrane cistern known as phagophore: (1) ATG12-ATG5 complex and (2) ATG8-phosphatidylethanolamine (PE) complex. In mammalian cells, ATG8 is also known as microtubule-associated protein 1A/1B-light chain 3-I (LC3-I), whereas ATG8-PE

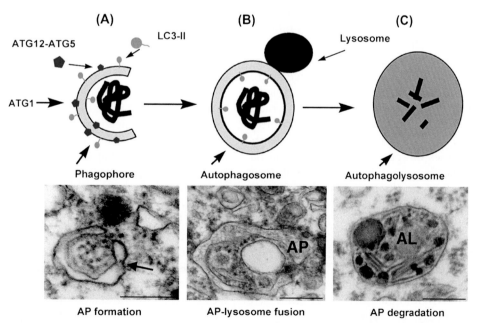

FIGURE 43.3 (Top) Simplified basic steps of the autophagy pathway consisting of autophagosome (AP) formation, AP-to-lysosome fusion, and AP and its cargo degradation. (Bottom) Electron micrographs of autophagic ultrastructures of hippocampal neurons from rats subjected to 15 min of ischemia followed by 4 h of reperfusion. (A) The arrow denotes a double membrane cistern or phagophore. (B) An AP containing dilated endoplasmic reticulum with ribosomes. (C) An autophagolysosome (AL) with partially digested cellular structures. ATG, autophagic gene-related protein; LC3, light chain 3. Scale bar = 0.2 μm.

is the active form and referred to as LC3-II. After incorporation of ATG12-ATG5 and LC3-II complexes, the phagophore is able to grow to envelope aberrant cellular components and organelles, and eventually becomes an AP. Therefore, the electron microscopy manifestation of double membrane structures containing cellular components and organelles represents the gold standard for quantification of cellular APs [3]. An AP is then merged with a late endosome/lysosome to form an AL for bulk degradation of both AP and its cargo content. Autophagy is a very sophisticated nonstop life-sustaining process to main cell homeostasis, which is active under normal conditions and further enhanced in response to cellular stress [4].

APOPTOSIS, NECROSIS, AND AUTOPHAGY AFTER BRAIN ISCHEMIA

Neuronal apoptosis has been repeatedly and consistently observed in neonatal HI models [1,5]. However, it is still a subject of debate about the role of apoptosis in the etiology of neuronal death in adult brain after ischemia.

It may be important to understand neuronal death modes after brain ischemia as therapeutic intervention may target key molecular events involved in the particular cell death process. It is generally held that persistent or permanent brain ischemia will lead to typical necrotic cell death, followed by inflammatory responses. Neuronal necrosis often occurs in the core area after focal brain ischemia. However, after a brief episode of global cerebral ischemia followed by reperfusion and in the penumbra after focal ischemia, some neurons may die selectively, whereas the others, as well as glial and vascular cells may be preserved, i.e., the selective neuronal death. It remains controversial whether selective neuronal death in the penumbra after focal ischemia, or some population of neurons (e.g., hippocampal CA1 pyramidal neurons) after transient global ischemia, is apoptotic or necrotic, or a combination of both types of cell death.

Since the discoveries of apoptotic genes and mechanisms about two decades ago, selective neuronal death after brain ischemia has usually been deciphered as the apoptotic type. Many studies have reported that proapoptotic genes, such as caspases, BAX, BID, BAK, and BAD, and antiapoptotic genes, such as Bcl-Xl and Bcl-2, are up- or downregulated in brain tissue samples after ischemia. Release of cytochrome c from mitochondria, upregulation of Apaf-1, activation of CAD and PARP, and translocation of AIF into the nuclei after brain ischemia have also repeatedly been reported.

However, the hypothesis of apoptotic neuronal death after brain ischemia has constantly been challenged by a considerable number of observations. First, it has been consistently and repeatedly observed that the morphological feature of selective neuronal death in mature brain is not apoptotic, but rather necrotic (Fig. 43.1) [2,5–7]. Second, assays of internucleosomal DNA damage with TUNEL (transferase-mediated dUTP nick end labeling) or electrophoretic DNA ladder have very frequently been used to identify apoptosis in many brain ischemia studies, but both assays are not apoptosis specific [8,9]. This is because both TUNEL and DNA ladder assays reflect DNA damage that occurs also in necrotic neurons. As a result, both TUNEL staining or electrophoretic DNA ladder are also positive in necrotic cells. Therefore, TUNEL positive and appearance of the DNA ladder may be insufficient evidence to conclude that selective neuronal death is apoptotic.

The active caspase-3 assays may be relatively specific for caspase-dependent apoptosis. Several studies show that, in neonatal brain, almost all dead neurons after HI are positive for active caspase-3 immunostaining, but it is extremely hard to find active caspase-3-positive dead neurons in the ipsilateral hippocampus of mature brain after HI (Fig. 43.4A). As controls, active caspase-3 is negative in the contralateral hippocampus [2,5]. These results are consistent with the fact that typical morphological features of apoptosis can be seen in immature neonatal neurons after HI (Fig. 43.2), but only necrotic morphological features can be observed in mature neurons after ischemia (Fig. 43.1). Furthermore, along with brain maturation, the mRNA of caspases-3 is gradually and dramatically reduced in brain samples, suggesting that caspase-3 may not even be expressed in neurons when brain is matured (Fig. 43.4A,B). Other apoptosis-related mRNAs, such as caspase-2, Bcl-2, and BAX, are also markedly reduced during brain maturation, whereas Fas, Fas-ligand (FASL), and Bcl-x L/S are not obviously changed (Fig. 43.4B). Two housekeeping genes L32 and GADPH are unchanged (Fig. 43.4B). These studies suggest that the caspase-3-dependent apoptotic machinery may be faded out or significantly diminished in neurons during brain maturation.

Several factors may be accountable for the apoptotic and necrotic neuronal death discrepancy in the literature. For example, many studies were carried out using neuronal culture ischemia-like models, such as hypoxia, hypoglycemia, oxygen glucose deprivation, or oxidative stress. However, primary cultured neurons may not represent mature, rather than developing neurons, as primary neuronal cultures are mostly derived either from rodent E18 embryos or from newborns. This is consistent with the fact that active caspase-3 is highly positive in 7-day primary neuronal cultures after exposure to oxidative stress inducer tert-butyl hydroperoxide (Fig. 43.4C), whereas active caspase-3-positive neurons are extremely hard to find in P60 mature neurons after HI (see Fig. 43.3A). Furthermore, ischemic brain injury is not limited

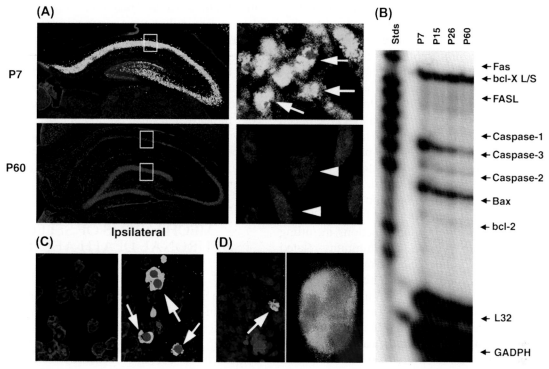

FIGURE 43.4   (A) Confocal microscopic images of active caspase-3 (*green*) and propidium iodide or PI (*red*) double staining. Brain sections were obtained from postnatal 7-day (P7) rat pups and a postnatal 60-day (P60) mature rat subjected to HI followed by 48h of recovery [5]. (Left) Hippocampal sections. (Right) Higher magnification of the CA1 inset on the left. Arrows denote active caspase-3-positive dead neurons. Arrowheads denote caspase-3-negative dead neurons. PI stains nucleic acids in both normal and dead neurons of paraformaldehyde-fixed tissue sections, as the fixation leads to PI permeability into cells. PI-stained dead neurons in P60 CA1 region become polygonal in shape (*arrowheads*) [5]. (B) Ribonuclease protection assay (RPA) of apoptotic mRNAs. Brain cortical tissue samples were from postnatal 7-, 15-, 26-, and 60-day rats. RPA was carried out according to the manufacturer instruction (Thermo Fisher Scientific, USA). The RNA probe standards (Stds) were on the first lane on the left. The housekeeping gene mRNAs L32 and GADPH were unchanged. *L32*, large ribosomal 32 mRNA; *GADPH*, glyceraldehyde-3-phosphate dehydrogenase. (C) Active caspase-3 (*green*) and PI (*red*) double staining of neuronal cultures treated with vesicle (*left*) or tert-butyl hydroperoxide (*right*). Arrows denote active caspase-3-positive dead neurons. (D) A higher magnification of the inset of the P60 DG area in (a-left). Arrow denotes an active caspase-3-positive dead neuron. The right image is a higher magnification of the left.

to neurons, but may also occur in astrocytes, oligodendroglia, microglia, and invading leukocytes (e.g., macrophages, neutrophils, and lymphocytes). These nonneuronal cells should have strong intrinsic apoptotic machinery and may undergo apoptotic cell death after brain ischemia. Moreover, adult brain contains some progenitor cells and immature neurons mostly located in the subventricular zone and the subgranular zone of dentate gyrus. These immature cells in adult brain should still have built-in apoptotic machinery and thus may undergo apoptosis after brain ischemia. This view is supported by the observation that caspase-3-positive cells may very occasionally be found in the subventricular and subgranular zones of adult brain after global brain ischemia or HI (Fig. 43.4D).

The Prosurvival and Prodeath Role of Autophagy After Brain Ischemia

Evolutionarily, autophagy is a prosurvival mechanism for cleaning up cellular "garbage" to maintain cell homeostasis. Therefore, autophagy is constitutively required for neuronal survival. This is supported by key autophagy component knockout mouse studies, in which autophagy deficiency leads to accumulation of APs and neurodegeneration [10]. Numerous compelling pharmacological studies have also indicated that impairment of the autophagy pathway results in accumulation of APs, aberrant organelles, and cell death [11]. These studies support the prosurvival role of autophagy in pathological conditions.

Although controversial, it has repeatedly been suggested that the excessive autophagy activity may contribute to selective neuronal death after brain ischemia, the so-called autophagic cell death (ACD). The term "autophagic cell death or ACD" was originally coined in the 1970s in a purely morphological context to refer to dying cells containing numerous ALs, and was, in fact, not to suggest autophagy inducing cell death [12]. The prodeath hypothesis suggests that excessive activation of autophagy might induce cell death via overdigestion of cellular contents or by activation of apoptotic enzymes. An argument against

ACD may be that the large-scale accumulation of APs in dying cells may not be the cause of cell death, but rather reflects a cellular attempt to remove aberrant components and organelles or interruption of the late steps of the autophagy pathway after brain ischemia [13]. Morphological and histochemical studies so far did not prove a causative relationship between the autophagic process and cell death. ACD may also be defined as a type of cell death that can be suppressed by specific inhibition of the autophagy pathway. Although some pharmacological studies have shown that inhibition of autophagy reduces cell death, none of the presently available pharmacological agents are exclusively for inhibition of autophagy. Even some key autophagic proteins like LC3, ATG5, and ATG12 have autophagy-independent functions, including the role in cell death, endocytosis, and immunity-related GTPase trafficking [14,15]. For these reasons, the presence of ACD in pathological conditions remains elusive.

The inconsistent conclusions drawn from previous studies about the role of autophagy in selective neuronal death after brain ischemia may also be owing to misunderstanding of the experimental results. First, increases in AP numbers or the level of LC3-II protein are often misinterpreted as upregulation of autophagy in pathological conditions. In the field of tissue ischemia and reperfusion, several studies show that an increase in LC3-II protein level after brain ischemia is not because of upregulation of autophagy activity, but rather a result of impairment of the later lysosomal degradation step of the autophagy pathway or disruption of autophagic flux [11]. Autophagic flux refers to the complete dynamic process of the entire autophagy pathway including AP formation, AP fusion with late endosome/lysosome, and late endosome/lysosomal degradation of AP and its cargo. An increase only in the LC3-II level is insufficient to suggest upregulation of autophagy because it is only an upstream component of the entire autophagy pathway. In many cases, an increase in the LC3-II level is caused by impairment of lysosomal degradation of AP. For example, overproduction of aberrant cellular substances and damaged organelles after brain ischemia may overwhelm the downstream lysosomal degradation of AP, resulting in an increase in the LC3-II level in brain tissue samples [13]. Accordingly, the use of the LC3-II level as an autophagic activity marker must be complemented by assays to determine the overall autophagic flux, to permit a correct interpretation of the results [4]. Second, some studies use nonspecific autophagic agents or methods, and thus these studies may be inconclusive. Chemical inhibitors for autophagy are widely used for studying autophagy, but none of the current chemical agents are specific for autophagy [14]. Third, various cell types, such as neurons, astrocytes, oligodendroglia, microglia, and invading leukocytes (e.g., macrophages, neutrophils, and lymphocytes) may contribute to the changes in the tissue autophagic activity

after brain ischemia. Therefore, change in the tissue LC3-II level may reflect a net result of mixed changes among diverse cell types after brain ischemia. Autophagy is likely to be activated in microglia and invading inflammatory cells to digest engulfed contents. Inhibition of autophagy might reduce the activity of phagocytes attacking ischemic tissues. Therefore, studying of the entire autophagy process, rather than a small portion of it (e.g., increase in LC3-II level), and dissecting the autophagy changes in different brain cell types may be critical to understanding the real role of autophagy in selective neuronal death after brain ischemia.

MECHANISMS OF SELECTIVE NEURONAL DEATH AFTER BRAIN ISCHEMIA

Much brain ischemia research has been focused on the studies of the selective neuronal death, probably because selective neuronal death is more likely rescued or prevented in the ischemic penumbra than the pan-necrosis in the ischemic core after focal brain ischemia. Significant progress has been made toward our understanding of the underlying mechanisms of selective neuronal death after brain ischemia. In addition to the apoptotic cell death pathways and autophagy, many hypotheses have been postulated, including excitotoxicity, calcium influx, acidosis, inhibition of protein synthesis, excessive production of reactive oxygen species, subcellular organelle damage or dysfunction, endoplasmic reticulum stress, stress response, protein misfolding and aggregation, and so forth. Earlier electron microscopic studies show that electron-dense deposits are accumulated in living CA1 pyramidal neurons destined to undergo selective neuronal death after transient cerebral ischemia [6,16]. At that time, the identity and molecular composition of these deposits were unknown. Hu and colleagues have carried out a series of morphological, biochemical, and molecular studies demonstrating that these electron-dense deposits are protein aggregates made of misfolded proteins [11]. Protein aggregates are clumped together with organelle membranes, which become the foremost subcellular structures in neurons before selective neuronal death occurs after brain ischemia. Therefore, protein aggregation and multiple organelle failure may contribute to selective neuronal death after brain ischemia [11].

In summary, ischemic stroke leads to necrotic neuronal death in the ischemic core area and selective neuronal death in ischemic penumbra. It remains controversial whether selective neuronal death after a brief episode of global cerebral ischemia and in the ischemic penumbra after focal ischemia is apoptotic or necrotic, or a combination of both types of cell death. Autophagy in neurons is likely an indispensable step for removing toxic substances and damaged organelles after brain ischemia. Autophagy in inflammatory cells, however, might be a double-edged sword; overactivation might make inflammatory cells capable of attacking repairable brain tissue especially at the early stage after ischemic stroke.

References

[1] Kerr JF, Wyllie AH, Currie AR. Apoptosis: a basic biological phenomenon with wide-ranging implications in tissue kinetics. Br J Cancer 1972;26:239–57.

[2] Hu BR, Liu CL, Ouyang Y, Blomgren K, Siesjo BK. Involvement of caspase-3 in cell death after hypoxia-ischemia declines during brain maturation. J Cereb Blood Flow Metab 2000;20:1294–300.

[3] Eskelinen EL. Maturation of autophagic vacuoles in mammalian cells. Autophagy 2005;1:1–10.

[4] Klionsky DJ, et al. Guidelines for the use and interpretation of assays for monitoring autophagy. Autophagy 2012;8:445–544.

[5] Liu CL, Siesjo BK, Hu BR. Pathogenesis of hippocampal neuronal death after hypoxia-ischemia changes during brain development. Neuroscience 2004;129:113–23.

[6] Deshpande J, Bergstedt K, Linden T, Kalimo H, Wieloch T. Ultrastructural changes in the hippocampal CA1 region following transient cerebral ischemia: evidence against programmed cell death. Exp Brain Res 1992;88:91–105.

[7] Colbourne F, Sutherland GR, Auer RN. Electron microscopic evidence against apoptosis as the mechanism of neuronal death in global ischemia. J Neurosci 1999;19:4200–10.

[8] Enright H, Nath KA, Hebbel RP. Internucleosomal cleavage of DNA is insufficient evidence to conclude that cell death is apoptotic. Blood 1994;83:2005–7.

[9] Grasl-Kraupp B, Ruttkay-Nedecky B, Koudelka H, Bukowska K, Bursch W, Schulte-Hermann R. In situ detection of fragmented DNA (TUNEL assay) fails to discriminate among apoptosis, necrosis, and autolytic cell death: a cautionary note. Hepatology 1995;21:1465–8.

[10] Hara T, Nakamura K, Matsui M, et al. Suppression of basal autophagy in neural cells causes neurodegenerative disease in mice. Nature 2006;441:885–9.

[11] Luo T, Park Y, Sun X, Liu C, Hu BR. Protein misfolding, aggregation, and autophagy after brain ischemia. Transl Stroke Res 2013;4:581–8.

[12] Puyal J, Ginet V, Clarke PG. Multiple interacting cell death mechanisms in the mediation of excitotoxicity and ischemic brain damage: a challenge for neuroprotection. Prog Neurobiol 2013;105:24–48.

[13] Liu CL, Gao Y, Barrett J, Hu BR. Protein aggregation and autophagy after brain ischemia. J Neurochem 2010;115:68–78.

[14] Mizushima N, Yoshimori T, Levine B. Methods in mammalian autophagy research. Cell 2010;140:313–26.

[15] Boya P, Reggiori F, Codogno P. Emerging regulation and functions of autophagy. Nat Cell Biol 2013;15:713–20.

[16] Kirino T, Tamura A, Sano K. Delayed neuronal death in the rat hippocampus following transient forebrain ischemia. Acta Neuropathol 1984;64:139–47.

C H A P T E R

44

Mechanisms of Glial Death and Protection

C. Matute[1,2,3]

[1]Achucarro Basque Center for Neuroscience, Zamudio, Spain; [2]CIBERNED, Madrid, Spain; [3]Universidad del País Vasco-UPV/EHU, Leioa, Spain

INTRODUCTION

Cerebrovascular diseases cause tissue damage to both gray and white matter, which contribute about half of the CNS volume and differ in structure and cellular composition. White matter exclusively contains axons and their glial cell partners including fibrous astrocytes, oligodendrocytes (myelinating and nonmyelinating), and microglia. Gray matter harbors neurons and is rich in protoplasmic astrocytes, which shape synaptic transmission as they partner with nerve endings and postsynaptic elements to form the tripartite synapse.

Pharmacological developments of potential treatments for stroke have failed in clinical trials because they typically aimed at protecting neurons from postischemic damage and neglected glial cells, especially

Primer on Cerebrovascular Diseases, Second Edition
http://dx.doi.org/10.1016/B978-0-12-803058-5.00044-8

oligodendrocytes, which are highly vulnerable to shortage of oxygen and nutrients. Oligodendrocytes are most abundant in white matter whose damage is a major cause of functional disability in cerebrovascular disease and the majority of ischemic strokes.

Early animal studies indicated that oligodendrocytes can be damaged by even brief focal ischemia [1], preceding by several hours the appearance of necrotic neurons in ischemic regions. In addition, pathological changes after ischemic insults include segmental swelling of myelinated axons and the formation of spaces or vacuoles between the myelin sheath and axolemma [1,2]. These observations confirm that oligodendrocytes and myelin are vulnerable to ischemia and that their damage proceeds independently from neuronal injury.

Stroke, therefore, produces disability not only as a result of dysfunction of neurons and synapses, but also by primary or secondary damage to oligodendrocytes and other glial cells. This chapter summarizes current knowledge of the molecular mechanisms of ischemic injury to glia and discusses its translational implications for the treatment of stroke (Table 44.1).

GLIA METABOLISM

Glucose is the primary energy source in the adult brain. Glucose transporter proteins on endothelial cells, glial cells, neurons, and axons are necessary for glucose uptake from the circulation and into cells. Astrocytes take up glucose in their end feet surrounding the capillaries and store glucose residues as glycogen. In addition to glucose, lactate can also support brain energy metabolism and function. Thus astrocyte glycogen is quickly mobilized to produce lactate that can be delivered to neurons and axons ensuring function during high activity or when glucose supply is limited. Lactate is impermeable and is transported across cell membranes by monocarboxylate transporters present in neurons and glia. Lactate enters neurons and sustains their function by producing ATP via oxidative phosphorylation. Lactate is also taken up by oligodendrocytes and their myelin sheath via MCT1 transporters, and utilized for lipid metabolism and myelin synthesis. In vitro evidence suggests that oligodendrocytes consume lactate at a higher rate than neurons or astrocytes, apparently to support the

TABLE 44.1   Glial Cell Damage and Protection in Ischemia

Glial Cell Type	Model and/or Preparation	Target	Protecting Agent
Oligodendrocytes	OGD in dissociated cultures	AMPA/kainate receptors	CNQX
	OGD in dissociated cultures	Glutamate uptake	Dihydrokainic acid
	Chemical ischemia in slices and optic nerve	NMDA receptors	MK-801, 7-chlorokynurenic acid, D-AP5
	OGD in optic nerve	AMPA/kainate receptors	NBQX
	Perinatal ischemia	NMDA receptors	Memantine
	Hypoxia-ischemia	P2X7 receptors	Brilliant Blue G
	OGD in dissociated cultures	P2X7 receptors/pannexin-1	Brilliant Blue G/mefloquine
	MCAO	A2A receptors	SCH58261
Myelin	Chemical ischemia in rat optic nerve	NMDA receptors	7-Chlorokynurenic acid
	Chemical ischemia in rat optic nerve	P2X7 receptors	Brilliant Blue G
	Perinatal ischemia	NMDA receptors	Memantine
Oligodendrocyte progenitors	Perinatal hypoxia-ischemia	Microglia	Minocycline
Oligodendrocytes and astrocytes	Optic nerve	Adrenoreceptors and nAChR	Propofol, mecamylamine
Astrocytes	Ligation of carotids	S-100 protein	Arundic acid
		Oxidative stress	Melatonin
		Blood supply	Adrenomedullin

AMPA, α-Amino-3-hydroxy-5-methyl-4-isoxazolepropionic acid; CNQX, 6-cyano-7-nitroquinoxaline-2,3-dione; D-AP5, D(−)-2-amino-5-phosphonovaleric acid; MCAO, middle cerebral artery occlusion; nAChR, nicotinic acetylcholine receptor; NBQX, 2,3-dioxo-6-nitro-7-sulfamoilbenzo(f)quinoxalina; NMDA, N-methyl-D-aspartic acid; OGD, oxygen-glucose deprivation.

high lipid demand associated with myelin manufacture, and myelination is rescued during hypoglycemia when exogenous lactate is supplied.

During partial ischemia, when glucose would still be present, although reduced, increased glycolysis in astrocytes, and possibly in oligodendrocytes, can contribute usable energy substrate to neurons and axons, although the mechanism(s) that signals axon metabolic need and mediates glial substrate production is still unknown. An attractive possibility is that neurotransmitters (i.e., glutamate and ATP) released from discharging axons signal surrounding astrocytes, and possibly oligodendrocytes, to release fuel in the form of lactate, delivered via cytoplasmic compartments within the myelin sheath [3], which can be quickly used by the axons.

OLIGODENDROCYTES ARE VULNERABLE TO EXCITOTOXICITY FOLLOWING ISCHEMIA

Cells of the oligodendroglial lineage express receptors to excitatory neurotransmitters including glutamate and ATP that are stimulated under physiological and pathological conditions.

Oligodendrocytes express functional α-amino-3-hydroxy-5-methyl-4-isoxazolepropionic acid (AMPA), kainate, and N-methyl-D-aspartic acid (NMDA) receptors, which can be activated during ischemic injury. Glutamate signaling in oligodendrocytes is relevant to myelination since action potentials travelling along axons can release glutamate that promotes the local synthesis of major myelin proteins [4]. In turn, glutamate homeostasis is controlled by Na^+-dependent glutamate transporters expressed mainly in astrocytes, and also in oligodendrocytes. Glutamate transporters are necessary to maintain very low basal levels of extracellular glutamate (the range is mid-nanomoles to low micromoles), and transporter blockade is sufficient to induce excitotoxic damage to oligodendrocytes. Under ischemic conditions, however, cells may depolarize and accumulate intracellular Na^+ leading to reversal of Na^+-dependent glutamate transport and toxic glutamate release. Thus collapse of ionic gradients during ischemia, especially the transmembrane Na^+ gradient, causes glutamate efflux that can be blocked by glutamate transport inhibitors. Astrocytes may predominate in this process, or merely contribute along with oligodendrocytes and axons. On the other hand, microglia express the cystine/glutamate antiporter, which can release glutamate in response to oxidative stress, and thus constitute another source of toxic glutamate release.

ATP signaling in oligodendrocytes occurs through P2X and P2Y receptors. P2X receptors are highly Ca^{2+} permeable, and P2Y receptors mobilize Ca^{2+} from intracellular stores. In particular, the low-affinity P2X7 receptor is expressed at relatively high levels in oligodendrocytes and myelin and could be activated by ATP release in pathological states. In contrast, P2X receptors with higher affinity might be activated by ATP released during axonal electrical activity and/or from astrocytes. Microglia express several P2X and P2Y receptors that act as sensors of damage and trigger a potent microglial inflammatory reaction. Postischemic anoxic depolarization triggers the release of ATP, in addition to glutamate. The mechanisms underlying ATP release include opening of pannexin hemichannels, activation of calcium homeostasis modulator channels (CalMH), and exit of the transmitter through pore-forming P2X7 receptor itself. Regardless of the source, excess of extracellular ATP causes oligodendrocyte excitotoxicity and demyelination.

MECHANISMS OF OLIGODENDROCYTE DAMAGE, PROTECTION, AND REPAIR IN STROKE

Table 44.1 summarizes major molecular pathways of glial cell damage after ischemia and promising agents that act on these pathways in a protective fashion.

Energy shortage during ischemia results in failure of ion pumps. In turn, it promotes the destructive accumulation of intracellular Ca^{2+}, primarily as a result of reverse Na^+/Ca^{2+} exchange, abetted by L-type Ca^{2+} channel activation. The loss of transmembrane Na^+ gradients also causes slowed or reversed Na^+-dependent glutamate uptake, and glutamate slowly accumulates in the extracellular space. High extracellular glutamate activates a complex sequence of pathological events in oligodendrocytes that are reminiscent of neuronal excitotoxicity.

Energy deprivation activates protective as well as destructive mechanisms in oligodendrocytes [5,6]. Both adenosine and gamma-aminobutyric acid seem to be produced or released, respectively, at low concentrations after energy deprivation and operate synergistically to improve functional recovery. However, adenosine acting through A_{2A} receptors on oligodendrocytes could also have deleterious effects. Paradoxically, both agonists and antagonists of α2 noradrenergic receptors improve recovery after ischemic/reperfusion injury, apparently by reducing axonal Na^+ and Ca^{2+} accumulation.

Among glial cells, oligodendrocytes are the primary victims of excitotoxicity. Overactivation of oligodendrocyte glutamate receptors causes Ca^{2+} overload of the cytosol leading to endoplasmic reticulum (ER) stress, mitochondrial depolarization, oxidative stress, and Bax-mediated, caspase-dependent, and caspase-independent cell death and myelin destruction. Thus oligodendrocytes can be partially protected from irreversible ischemic injury, including perinatal ischemia, by glutamate receptor antagonists and glutamate uptake inhibitors (Table 44.1). Ischemia induces an inward current in "young" oligodendrocytes that is mediated, in part, by NMDA and AMPA/kainate receptors, and increases Ca^{2+} levels in myelin itself (an effect that is abolished by NMDA receptor antagonists) causing ultrastructural damage to myelin, and perhaps secondarily to axon cylinders as well [6].

However, oligodendrocytes in adult or old rodents behave very differently during ischemia. Although oligodendrocytes express NMDA receptors throughout life, these receptors do not participate in ischemic injury in fully mature tissue; blocking these receptors actually worsens ischemic damage [7]. In adult and old animals, dysregulation of intracellular $[Ca^{2+}]$ remains a crucial feature of irreversible ischemic injury, but the dramatic benefit of Ca^{2+}-free extracellular fluid is lost for unclear reasons. It is possible that Ca^{2+} release from intracellular Ca^{2+} stores becomes more critical during ischemia in oligodendrocytes from older animals. Although NMDA receptors are not involved in ischemic injury in older animals, glutamate excitotoxicity is enhanced, at least partly due to increased glutamate release during ischemia. These age differences in the pathophysiology of oligodendrocyte ischemic damage highlight the need for age-specific stroke therapies.

Intracellular levels of ATP decline and extracellular ATP is elevated during cerebral ischemia, coincident with secondary anoxic depolarization. The rise in extracellular ATP during ischemia is sufficient to activate $P2X_7$ receptors and kill neurons and oligodendrocytes; blocking $P2X_7$ receptors is protective. Ischemia causes ATP release by opening of oligodendrocyte pannexin channels, which leads to mitochondrial depolarization and oxidative stress culminating in oligodendrocyte death and myelin destruction. These pathological events are attenuated by $P2X_7$ receptor antagonists, by the ATP-degrading enzyme apyrase, and by blockers of pannexin hemichannels [8].

Antioxidants may attenuate poststroke glial cell damage. Oligodendrocytes are very susceptible to oxidative stress for two reasons: they lack a high-potency antioxidant system and they have high iron content. When exposed to hypoxia or ischemia, these cells exhibit robust production of superoxide radical, lipid peroxidation, and conversion of iron stores to the oxidizing agent, ferrous ion [2]. The antioxidant ebselen significantly reduces axonal and oligodendrocyte damage as well as the neurological deficit associated with transient ischemia when administered 2h after the onset of stroke (Table 44.1). More antioxidants should be tested in models of ischemia to gain deeper insight into the therapeutic potential of this class of compounds in attenuating oligodendrocyte death.

Other agents that may ameliorate oligodendrocyte ischemic damage include minocycline, citicoline, and arundic acid (Table 44.1). Minocycline is a potent inhibitor of microglia and a neuroprotective agent of oligodendrocyte damage after hypoxia-ischemia in neonatal animal models. Daily postinsult treatment with minocycline abolished neuroinflammation and attenuated damage of oligodendrocyte precursors. Its efficacy in adults is untested.

Oligodendrocyte precursor cells (OPCs) in the adult brain contribute to replenish the mature oligodendrocyte population. After damage to the later caused by prolonged cerebral hypoperfusion, OPCs compensate for oligodendrocyte loss by differentiating into mature oligodendrocytes via mechanisms that are only partially understood. Thus astrocytes support the maturation of OPCs by secreting brain-derived neurotrophic factor (BDNF) [9]. As hypoperfusion may damage oligodendroglia in stroke and in vascular dementia, regulating astrocytic BDNF expression may provide a broad therapeutic approach for cerebrovascular disorders.

MECHANISMS OF ASTROCYTE DAMAGE AND PROTECTION IN STROKE

Astrocytes are the most abundant cell type within the central nervous system. They play essential roles in maintaining normal brain function, as they are a critical structural and functional part of the tripartite synapses and the neurovascular unit, and communicate with neurons, oligodendrocytes, and endothelial cells. After an ischemic stroke, astrocytes perform multiple functions both detrimental and beneficial, for neuronal survival during the acute phase [10]. At later stages after injury, astrocytes also contribute to angiogenesis, neurogenesis, synaptogenesis, and axonal remodeling, and thereby promote neurological recovery. Thus the pivotal involvement of astrocytes in normal brain function and responses to an ischemic lesion designates them as excellent therapeutic targets to improve functional outcome following stroke.

Astrocytes are generally more resistant than neurons and oligodendrocytes to ischemia [10]. Thus astroctyes are better preserved in the penumbra of the infarct, and a subpopulation of them within the ischemic core remains viable and metabolically active after the onset of reperfusion. Oxidative stress is a major mechanism leading to astrocyte demise in ischemia and, accordingly, melatonin enhances astrocyte survival during reperfusion. In addition, vasodilation with adrenomedullin gene delivery also improves the outcome of the astrocyte population following ischemic insults.

On the other hand, arundic acid interferes with astrocyte activation during injury and controls the expression of S100 Ca^{2+}-binding protein that is primarily expressed in astrocytes. The levels of S100 correlate with the volume of the cerebral infarct and treatment with arundic acid before and after ischemia greatly reduced the levels of S100 in astrocytes (Table 44.1). The mechanism of protection by arundic acid may be downregulation in astrocytes of inducible nitric oxide synthase, with decreased nitric oxide production and, consequently, less toxicity to neighboring cells including oligodendrocytes. Finally, citicoline has neuroprotective effects in a model of chronic hypoperfusion, although the mechanism of action is not clear. It also promotes neurogenesis, which may contribute to repair after ischemic damage.

Astrocytes play an essential role in the induction of brain ischemic tolerance without producing any noticeable brain damage [11]. Astrocytic activation correlates with ischemic tolerance and is accompanied by P2X7 receptor upregulation in activated astrocytes. Importantly, induction of ischemic tolerance with a sublethal ischemic insult (preconditioning) is abolished in P2X7 receptor knockout mice. Thus astrocytes play indispensable roles in inducing ischemic tolerance, and upregulation of P2X7 receptors in astrocytes is essential. In contrast, P2X7 receptors play a deleterious role after ischemia in the absence of preconditioning, as they mediate damage to neurons and oligodendrocytes and P2X7 receptor antagonists are strongly protective [6].

CONCLUSION

Astrocytes and oligodendrocytes are highly vulnerable to stroke. Ischemia causes these glial cells to lose ion homeostasis, due to loss of ATP, which results in Ca^{2+} overload. This process is accelerated by glutamate- and ATP-mediated overactivation of ionotropic receptors. Based on recent experimental work, several strategies for therapeutic intervention in glial cells after stroke seem promising.

References

[1] Pantoni L, Garcia JH, Gutierrez JA. Cerebral white matter is highly vulnerable to ischemia. Stroke 1996;27:1641–6. discussion 1647.
[2] Shereen A, Nemkul N, Yang D, Adhami F, Dunn RS, Hazen ML, et al. Ex vivo diffusion tensor imaging and neuropathological correlation in a murine model of hypoxia-ischemia-induced thrombotic stroke. J Cereb Blood Flow Metab 2011;31:1155–69.
[3] Fünfschilling U, Supplie LM, Mahad D, Boretius S, Saab AS, Edgar J, et al. Glycolytic oligodendrocytes maintain myelin and long-term axonal integrity. Nature 2012;485:517–21.
[4] Wake H, Lee PR, Fields RD. Control of local protein synthesis and initial events in myelination by action potentials. Science 2011;333:1647–51.
[5] Fern RF, Matute C, Stys PK. White matter injury: ischemic and nonischemic. Glia 2014;62:1780–9.
[6] Matute C, Domercq M, Pérez-Samartín A, Ransom BR. Protecting white matter from stroke injury. Stroke 2013;44:1204–11.
[7] Baltan S, Besancon EF, Mbow B, Ye Z, Hamner MA, Ransom BR. White matter vulnerability to ischemic injury increases with age because of enhanced excitotoxicity. J Neurosci 2008;28:1479–89.
[8] Domercq M, Perez-Samartin A, Aparicio D, Alberdi E, Pampliega O, Matute C. P2x7 receptors mediate ischemic damage to oligodendrocytes. Glia 2010;58:730–40.
[9] Miyamoto N, Maki T, Shindo A, Liang AC, Maeda M, Egawa N, et al. Astrocytes promote oligodendrogenesis after white matter damage via brain-derived neurotrophic factor. J Neurosci 2015;35:14002–8.
[10] Liu Z, Chopp M. Astrocytes, therapeutic targets for neuroprotection and neurorestoration in ischemic stroke. Prog Neurobiol October 9, 2015. http://dx.doi.org/10.1016/j.pneurobio.2015.09.008, pii: S0301-0082(15)30001-0, [Epub ahead of print].
[11] Hirayama Y, Ikeda-Matsuo Y, Notomi S, Enaida H, Kinouchi H, Koizumi S. Astrocyte-mediated ischemic tolerance. J Neurosci 2015;35:3794–805.

CHAPTER

45

Mechanisms of Endothelial Injury and Blood–Brain Barrier Dysfunction in Stroke

P.T. Ronaldson, T.P. Davis

University of Arizona, Tucson, AZ, United States

INTRODUCTION

The blood–brain barrier (BBB) is an essential physical and biochemical barrier that separates the CNS from the systemic circulation. It is formed by a monolayer of capillary endothelial cells that interact with each other as well as with other components of the neurovascular unit (i.e., astrocytes, microglia, neurons, pericytes, extracellular matrix) to precisely maintain CNS homeostasis. In addition, the BBB restricts brain permeation of xenobiotics in an effort to reduce the probability of CNS toxicity. During ischemic stroke, brain microvascular endothelial cells are triggered by pathophysiological stimuli [i.e., reactive oxygen species (ROS), inflammatory mediators], leading to disruption of tight junction (TJ) protein complexes and subsequent barrier opening. This process is clinically significant because BBB dysfunction is well known to contribute to deleterious events following stroke, such as intracerebral hemorrhage and vasogenic edema. Understanding cellular responses in the setting of stroke provides an opportunity to develop novel pharmacological approaches that promote BBB repair by modulating endothelial injury.

DISRUPTION OF THE BLOOD–BRAIN BARRIER IN ISCHEMIC STROKE

Considerable research has demonstrated that BBB dysfunction in stroke is biphasic in nature. To this end, acute breakdown of the BBB is demarcated by an immediate enhancement in solute permeability observed at approximately 4–6h after an ischemic insult followed by a second barrier opening that occurs 2–3 days after stroke [1]. The onset of BBB dysfunction directly corresponds to the severity of ischemia [1]. Acute BBB

breakdown involves four distinct processes: (1) endothelial swelling, (2) endothelial membrane disruption, (3) disruption of TJ protein complexes between adjacent endothelial cells, and (4) total vascular dysfunction [1]. Perturbation of extracellular matrix (i.e., type IV collagen, heparan sulfate proteoglycan, laminin, fibronectin, perlecan) is prominently involved in this acute BBB dysfunction. Activation of proteinases, including matrix metalloproteinases (MMPs), is a critical component of early breakdown of extracellular matrix and subsequent BBB dysfunction in stroke [2]. Involvement of MMPs in BBB dysfunction has been demonstrated in in vivo experimental stroke models. Furthermore, elevation of MMP9 levels has been observed in patients diagnosed with acute ischemic stroke [2]. MMPs degrade the extracellular matrix that comprises the basal lamina and directly compromises the BBB by degradation of TJ constituent proteins (i.e., claudin-5, occludin). In addition, MMP-mediated opening of the BBB in ischemic stroke is associated with activation of endothelial nitric oxide synthase (eNOS) and inducible NOS (iNOS), leading to an enhancement in nitric oxide (NO) signaling.

Experimental models of focal cerebral ischemia have provided essential information on solute leak across the BBB in the context of stroke. Using the transient middle cerebral artery occlusion (MCAO) rodent model, increased leak of sucrose, a vascular marker that does not typically cross the BBB, was demonstrated in the ischemic hemisphere but not in the contralateral hemisphere [2]. BBB disruption following an ischemic insult is significant and can allow blood-to-brain leak of large molecules, such as Evans blue-albumin [2]. Evans blue dye, when unconjugated to plasma proteins, is a relatively small molecule with a molecular weight of 960.8 Da. In vivo, Evans blue dye irreversibly binds to

Primer on Cerebrovascular Diseases, Second Edition
http://dx.doi.org/10.1016/B978-0-12-803058-5.00045-X

serum albumin causing formation of a very large, solute–protein complex (i.e., in excess of 60,000 Da) that can only traverse the BBB under significant pathological stress as is known to occur following focal ischemia. Of particular note, Evan's blue-albumin leak following experimental ischemic stroke was directly correlated with redistribution of critical TJ proteins occludin, claudin-5, and zonula occludens (ZO)-1 [2].

Reorganization of TJ protein complexes and associated leak across the BBB enables vascular fluid (i.e., water) to readily move across the microvascular endothelium, which leads to the development of vasogenic edema. MCAO studies have shown that enhanced blood-to-brain movement of sodium exacerbates water movement across the BBB. This change in the sodium gradient across the microvascular endothelium is facilitated by increased functional expression of the Na–K–Cl cotransporter as well as Na–H exchangers NHE1 and/or NHE2 [2]. Disruption of sodium gradients across the BBB during ischemic stroke can also involve upregulation of sodium-dependent glucose transporters, such as sodium-glucose cotransporter (SGLT). Specifically, pharmacological inhibition of SGLT in MCAO rats significantly reduced infarct and edema ratios, which implies that this transporter may be a critical determinant of stroke outcome [2].

PATHOLOGICAL MECHANISMS OF ENDOTHELIAL INJURY IN ISCHEMIC STROKE

Oxidative Stress

Functional BBB integrity is disrupted by production of ROS and subsequent oxidative stress (Fig. 45.1). Superoxide anion, a potent ROS generated when molecular oxygen is reduced by only one electron, is a known mediator of stroke pathology. Superoxide dismutase (SOD) enzymes precisely regulate biological activity of superoxide anion, a by-product of normal physiological processes. Under oxidative stress conditions, superoxide is produced at high levels that overwhelm the metabolic capacity of SOD and contribute to endothelial dysfunction at the BBB [3,4]. This endothelial injury is intensified by conjugation of superoxide and NO to form peroxynitrite, a cytotoxic and proinflammatory molecule. Peroxynitrite causes significant damage to cerebral microvessels through lipid peroxidation, protein nitrosylation, consumption of endogenous antioxidants [i.e., reduced levels of glutathione (GSH) at the brain microvascular endothelium], and mitochondrial failure. Peroxynitrite formation in BBB endothelial cells becomes more likely with activation of eNOS and iNOS

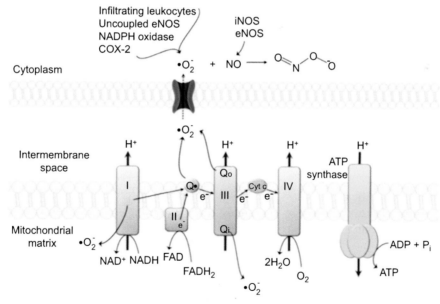

FIGURE 45.1 **Generation of reactive oxygen species (ROS) in brain microvascular endothelial cells.** During disease, mitochondrial superoxide levels increase via nitric oxide (NO) inhibition of cytochrome complexes and oxidation of reducing equivalents in the electron transport chain. Complex I and both sides of complex III (i.e., Qi and Qo sites) are the most common sources of mitochondrial superoxide. Superoxide generated within the intermembrane space of mitochondria can reach the cytosol through voltage-dependent mitochondrial anion channels. Superoxide levels further increase via cyclooxygenase-2, NADPH oxidase, uncoupled endothelial nitric oxide synthase (eNOS), and infiltrating leukocytes. The resulting high levels of superoxide coupled with the activation of NO-producing eNOS and inducible NOS (iNOS) increase the probability of peroxynitrite formation. Peroxynitrite-induced cellular damage includes protein oxidation, tyrosine nitration, DNA damage, poly(ADP-ribose) polymerase activation, lipid peroxidation, and mitochondrial dysfunction. *Adapted from Thompson BJ, Ronaldson PT. Drug delivery to the ischemic brain. Adv Pharmacol 2014;71:165–202.*

because NO diffuses easily through membranes and readily reacts with superoxide anion. Using a proteomics approach, cultured human brain endothelial cells were shown to produce a series of reactive/inflammatory proteins in response to oxidative stress conditions including thrombospondin-1 (TSP-1), chemokine ligand-1, heat shock protein-1, serum amyloid-A1, and annexin-5 [5]. Of particular note, both TSP-1 and chemokine ligand-1 were significantly increased in brain tissue obtained from patients with stroke, suggesting that these proteins may have utility as biomarkers for acute ischemic stroke [5].

Oxidative stress contributes to disruption of endothelial cell–cell interactions and BBB injury by promoting redistribution and/or downregulation of critical TJ proteins, such as claudin-5, occludin, ZO-1, and junctional adhesion molecule 1 [4]. For example, decreased expression of occludin is associated with increased BBB permeability as shown in an in vivo model of hypoxia/reoxygenation (H/R) stress, an established component of ischemic stroke [4]. In addition, H/R stress causes trafficking of occludin away from BBB TJ protein complexes, a mechanism that contributes to solute leak across the brain microvascular endothelium [4]. In the context of experimental stroke, reorganization of TJ complexes also contributes to movement of vascular fluid across the microvascular endothelium (i.e., leak) and development of vasogenic edema [4].

The increase in brain microvascular permeability observed in response to ROS involves changes to biochemical components of BBB, such as transcellular transport pathways. For example, the endogenous BBB transporter Oatp1a4 is upregulated following global cerebral hypoxia [3,6]. This is particularly noteworthy because Oatp1a4 is capable of promoting blood-to-brain transport of therapeutic drugs relevant to stroke treatment, such as 3-hydroxy-3-methylglutaryl coenzyme A reductase inhibitors (i.e., statins) [6].

Inflammation

Inflammatory stimuli are critical mediators of BBB dysfunction in the setting of stroke. Previous research has demonstrated that inflammatory responses in focal cerebral ischemia are primarily mediated via production of proinflammatory cytokines [i.e., tumor necrosis factor (TNF)-α, interleukin (IL)-1β], which appear within a few hours following ischemic insult [7]. Signaling mediated by proinflammatory cytokines leads to an increase in microvascular expression of adhesion molecules, which is associated with transmigration of activated neutrophils, lymphocytes, or monocytes into brain parenchyma. Under normal conditions, expression of vascular adhesion molecules, such as ICAM-1 and VCAM-1, is barely detectable in brain microvessel endothelial cells; however, their expression is dramatically increased in response to cytokines associated with ischemic stroke, such as

TNF-α [8] and IL-1β [9]. In addition, the brain microvascular endothelium itself is involved in the production and secretion of proinflammatory mediators. For example, brain endothelial production of TNF-α during ischemia is triggered by early release of inflammatory factors, such as high-mobility group box-1 (HMGB1) from neurons [10]. Similar to ROS, proinflammatory mediators can also modulate biochemical characteristics of the BBB, such as P-glycoprotein (P-gp) and breast cancer resistance protein [2]. Proinflammatory factors can also regulate the physical barrier characteristics of the brain microvascular endothelium via altering expression of TJ proteins occludin and ZO-1 [2]. Taken together, these observations imply that inflammation plays a critical role in exacerbating BBB dysfunction in the setting of ischemic stroke.

THERAPEUTIC OPPORTUNITIES

Targeting BBB Tight Junctions

Since oxidative stress is known to alter BBB expression of TJ protein complexes and increase paracellular solute leak, antioxidant drugs represent a potential pharmacological approach to protect BBB integrity in stroke [4]. For example, 4-hydroxy-2,2,6,6-tetramethylpiperidine-N-oxyl (TEMPOL) shows SOD-like activity toward the superoxide anion as well as reactivity with hydroxyl radicals, nitrogen dioxide, and the carbonate radical [4]. In vivo, TEMPOL preserves occludin localization at the TJ and attenuates BBB sucrose leak following global cerebral hypoxia (Fig. 45.2) [4]. TEMPOL inhibits breakage of disulfide bonds on occludin monomers and thus prevents breakdown of occludin oligomeric assemblies and subsequent blood-to-brain leak of circulating solutes [4]. Restoration of BBB functional integrity following TEMPOL administration coincided with a decrease in nuclear translocation of HIF-1α and a decrease in microvascular expression of the cellular stress marker heat shock protein 70 in rats subjected to H/R stress [4]. Taken together, these data provide evidence that the TJ can be targeted pharmacologically for the purpose of reducing both oxidative stress–associated injury to the brain microvascular endothelium and blood-to-brain solute leak.

Targeting Endogenous BBB Transporters

The ability of a drug to elicit a pharmacological effect at the level of the BBB requires the achievement of efficacious concentrations within the endothelial cell cytoplasm. This therapeutic objective depends on multiple mechanisms of transport that may include uptake into the cell by an influx transporter and/or extrusion from the endothelium by an efflux transporter. For many drugs, it is this discrete balance between influx and

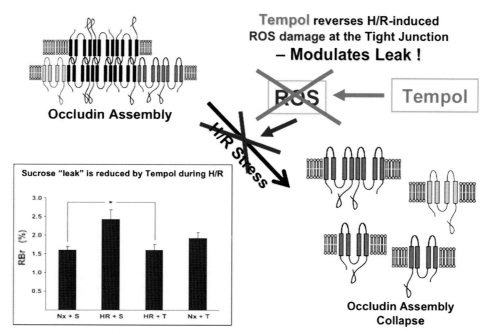

FIGURE 45.2   **Effect of 4-hydroxy-2,2,6,6-tetramethylpiperidine-*N*-oxyl (TEMPOL) on hypoxia/reoxygenation (H/R)-mediated disruption of the tight junction.** Reactive oxygen species (ROS) and subsequent oxidative stress are known to disrupt assembly of critical tight junction (TJ) proteins such as occludin. Our results show that administration of TEMPOL, by scavenging ROS, prevents disruption of occludin oligomeric assemblies. Furthermore, TEMPOL attenuates the increase in sucrose leak across the blood–brain barrier (BBB) observed in animals subjected to H/R stress. Taken together, our studies with TEMPOL demonstrate that the TJ can be targeted pharmacologically in an effort to preserve BBB functional integrity during ischemic stroke. *Adapted from Ronaldson PT, Davis TP. Blood–brain barrier integrity and glial support: mechanisms that can be targeted for novel therapeutic approaches in stroke. Curr Pharm Des 2012;18(25):3624–44.*

efflux that determines if a therapeutic agent will be able to protect the BBB. The complexity of drug transporter biology at the BBB is further underscored by the observation that functional expression of transporters can be dramatically altered by oxidative stress and inflammation [2–4]. A thorough understanding of the regulation and functional expression of endogenous BBB transporters in stroke is essential for the provision of effective pharmacotherapy. Furthermore, such information will enable effective targeting of transporters and/or transporter regulatory mechanisms, thus allowing endogenous BBB transport systems to be exploited for purposes of conferring BBB protection and/or repair.

Considerable research has focused on studying mechanisms that limit endothelial transport by studying P-gp-mediated transport activity [4]; however, clinical trials targeting P-gp with small-molecule inhibitors have been unsuccessful in improving pharmacotherapy due to inhibitor toxicity and/or enhanced drug distribution to peripheral tissues. An alternative approach for optimizing delivery of drugs across the endothelial plasma membrane is to focus on BBB transporters that are involved in cellular uptake of drugs [6]. One potential candidate is Oatp1a4, which is known to transport statins. Evidence suggests that statins can act as protective agents at the BBB, an effect that is independent of the well-documented effects on cholesterol biosynthesis for these drugs [6]. For example, atorvastatin administration during the acute

phase of cerebral ischemia prevented the increase in BBB permeability that is commonly associated with this disease [4]. Studies in an in vivo model of global cerebral hypoxia have shown that Oatp1a4 can effectively transport statins at the BBB [6], suggesting that these drugs can be delivered to the endothelial cell cytoplasm by targeted Oatp1a4-mediated delivery.

Although pathophysiological stressors can modulate BBB transporters, such changes must be controlled to provide optimal delivery of drugs. For example, increased functional expression of Oatp1a4 only occurs after 1h hypoxia followed by up to 1h reoxygenation [6]. If Oatp1a4 is to facilitate effective delivery of drugs (i.e., statins), its functional expression must be precisely controlled over a more desirable time course than is possible by relying on pathophysiological changes. This objective can be accomplished by pharmacological targeting of Oatp regulatory pathways such as the transforming growth factor-β (TGF-β) system [4,6]. TGF-βs are cytokines that signal by binding to a heterotetrameric complex of type I and type II receptors. The type I receptors, also known as activin receptor-like kinases (ALKs) propagate intracellular signals through phosphorylation of receptor-specific Smad proteins [i.e., (R)-Smads] and subsequent modulation of gene transcription. At the BBB, only two ALK receptors (ALK1, ALK5) have been identified [4]. Pharmacological inhibition of TGF-β/ALK5 signaling can increase Oatp1a4 functional expression [6]. This suggests that targeting of

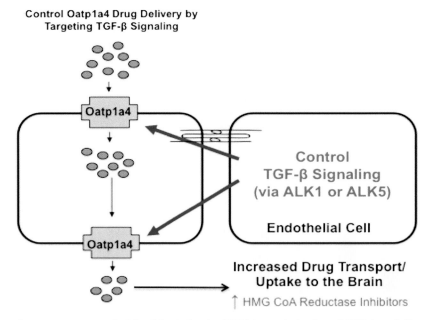

FIGURE 45.3   **Targeting Oatp transporters at the blood–brain barrier (BBB) for optimization of CNS drug delivery.** Results from our studies demonstrate that targeting Oatp transporters during pathophysiological stress can modify CNS drug delivery. Oatp1a4 facilitates brain delivery of drugs that may exhibit efficacy in treatment of peripheral inflammatory pain or cerebral hypoxia such as statins and opioid peptide analgesics. The transforming growth factor (TGF)-β signaling pathway enables control Oatp isoforms by targeting TGF-β receptors [i.e., activin receptor-like kinase (ALK)1, ALK5] with small-molecule therapeutics. *HMG CoA*, 3-hydroxy-3-methylglutaryl coenzyme A.

the TGF-β/ALK5 pathway may enable improved control of BBB Oatp1a4 expression and/or activity, thereby providing novel strategies for improved BBB protection in stroke (Fig. 45.3).

Optimization of drug delivery is not the only benefit that can be achieved from targeting transporters. BBB transporters mediate flux of endogenous substrates, many of which are essential to the cellular response to pathological insult. One such substance is the endogenous antioxidant GSH. During oxidative stress, GSH is rapidly oxidized to glutathione disulfide (GSSG). Therefore, the redox state of a cell is represented by the ratio of GSH to GSSG. In vitro studies using human and rodent brain microvascular endothelial cells have demonstrated that hypoxia reduces intracellular GSH levels and decreases the GSH:GSSG ratio, suggesting significant oxidative stress occurs at the level of the BBB in stroke [4]. Using an in vivo model, oxidative stress was shown to cause BBB disruption characterized by altered expression/assembly of TJ proteins [4]. These TJ modifications correlated with increased BBB permeability to sucrose, an established vascular marker, and dextrans [2,4]. Such an enhancement in BBB permeability can result in leak of neurotoxic substances from blood into brain and/or contribute to vasogenic edema. BBB protection and/or repair in the context of stroke are paramount to protecting the brain from neurological damage. One approach that can accomplish this therapeutic objective is to prevent cellular loss of GSH from endothelial

cells (i.e., preservation of the endothelial antioxidant defense system) by targeting endogenous BBB transporters. BBB transporters that can transport GSH and GSSG include MRPs/Mrps. Both GSH and GSSG are effluxed from brain microvascular endothelial cells by multidrug resistance protein (MRP) isoforms such as MRP1/Mrp1, MRP2/Mrp2, and MRP4/Mrp4 [4]. It is well known that increased cellular concentrations of GSH are cytoprotective, whereas processes that promote GSH loss from cells are damaging. Therefore, it stands to reason that pharmacological targeting of Mrps during oxidative stress may have profound therapeutic benefits.

Previous studies have shown that Mrp expression and/or activity can change in response to pathophysiological stressors [4]. Altered BBB expression of Mrps may prevent endothelial cells from retaining effective GSH concentrations. A thorough understanding of signaling pathways involved in Mrp regulation during oxidative stress will enable development of pharmacological approaches to target Mrp-mediated efflux (i.e., GSH transport) for the purpose of preventing BBB dysfunction in diseases with an oxidative stress component. One intriguing pathway is signaling mediated by nuclear factor E2-related factor-2 (Nrf2), a sensor of oxidative stress. In the presence of ROS, the cytosolic Nrf2 repressor Kelch-like ECH-associated protein 1 (Keap1) undergoes structural alterations that cause dissociation from the Nrf2-Keap1 complex. This enables Nrf2 to translocate to the nucleus and induce transcription

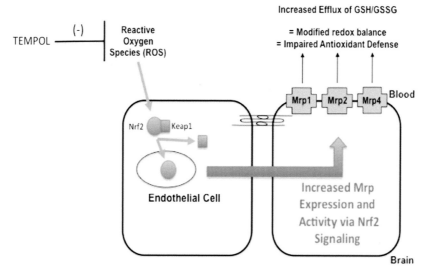

FIGURE 45.4 **Prevention of blood–brain barrier (BBB) dysfunction by targeting Mrp isoforms in cerebral endothelial cells.** Results from our laboratory demonstrate increased expression of Mrp1, Mrp2, and Mrp4 at the BBB following a hypoxia/reoxygenation (H/R) insult. Furthermore, H/R stress is known to suppress glutathione (GSH) levels and increase glutathione disulfide (GSSG) concentrations in the brain. We propose that changes in GSH/GSSG transport occur during H/R as a result of altered functional expression of at least one Mrp isoform. Since Nrf2, a reactive oxygen species (ROS) sensitive transcription factor, is known to regulate Mrps, we hypothesize that this pathway is a critical regulatory mechanism for Mrps at the BBB. 4-Hydroxy-2,2,6,6-tetramethylpiperidine-N-oxyl (TEMPOL), an ROS-scavenging antioxidant, is a pharmacological tool that can be utilized to understand how targeting activation of the Nrf2 pathway can control Mrp expression/activity.

of genes that possess an antioxidant response element at their promoter. It has been demonstrated that activation of Nrf2 signaling induces expression of Mrp1, Mrp2, and Mrp4 [4]. An emerging concept is that Nrf2 acts as a double-edged sword [11]: on the one hand, Nrf2 is required for protecting tissues from oxidative stress; on the other, its activation can lead to deleterious effects. Therefore, an alteration in the balance of Mrp isoforms via activation of Nrf2 signaling may adversely affect redox balance and antioxidant defense at the brain microvascular endothelium. Indeed, this points toward a need for rigorous study of pharmacological approaches (i.e., use of antioxidant drugs such as TEMPOL) that can modulate Nrf2 signaling and control expression of Mrp isoforms and/or GSH transport at the BBB (Fig. 45.4).

SUMMARY

The field of BBB biology, particularly the study of TJ protein complexes and endogenous transport systems, has rapidly advanced over the past decade. It is now established that TJs are dynamic in nature and can organize and reorganize in response to oxidative stress and inflammation, mechanisms that are involved in endothelial injury and BBB dysfunction in the setting of stroke. These changes can lead to increased BBB permeability to endogenous and exogenous solutes. Transporters (i.e., Oatps, Mrps) provide a considerable opportunity to protect the BBB and/or promote BBB repair by either

facilitating endothelial uptake of drugs with cytoprotective/antioxidant properties or by preventing cellular loss of critical endogenous substances (i.e., GSH). Furthermore, molecular mechanisms involved in regulating endogenous BBB transport systems, such as TGF-β/ALK5 signaling and the Nrf2 pathway are just now being identified and characterized. These discoveries have identified discrete molecular targets that can be exploited for control of BBB xenobiotic transport. Perhaps targeting of novel drugs to influx transporters such as Oatp1a4 or to efflux systems, such as Mrps will lead to significant advances in the treatment of diseases such as ischemic stroke. Characterization of intracellular signaling pathways that can regulate functional expression of BBB transporters provides another approach for pharmacological control of transport systems in an effort to prevent BBB dysfunction. Future work will continue to provide more information on the interplay of cell–cell interactions, transporters, and signaling pathways at the BBB endothelium and how these systems can be effectively targeted. Ultimately, data derived from these studies will enable achievement of more precise and more efficient drug concentrations within brain microvessel endothelial cells and improved BBB protection and/or repair following ischemic stroke.

Acknowledgments

This work was supported by grants from the National Institutes of Health to P.T.R. (R01 NS084941) and to T.P.D. (R01 NS42652 and R01 DA11271).

References

[1] Prakash R, Carmichael ST. Blood–brain barrier breakdown and neovascularization processes after stroke and traumatic brain injury. Curr Opin Neurol 2015;28(6):556–64. Epub Sept 19, 2015.
[2] Ronaldson PT, Davis TP. Blood–brain barrier integrity and glial support: mechanisms that can be targeted for novel therapeutic approaches in stroke. Curr Pharm Des 2012;18(25):3624–44.
[3] Thompson BJ, Ronaldson PT. Drug delivery to the ischemic brain. Adv Pharmacol 2014;71:165–202.
[4] Ronaldson PT, Davis TP. Targeting transporters: promoting blood–brain barrier repair in response to oxidative stress injury. Brain Res 2015;1623:39–52.
[5] Ning M, Sarracino DA, Kho AT, Guo S, Lee SR, Krastins B, et al. Proteomic temporal profile of human brain endothelium after oxidative stress. Stroke 2011;42(1):37–43.
[6] Thompson BJ, Sanchez-Covarrubias L, Slosky LM, Zhang Y, Laracuente ML, Ronaldson PT. Hypoxia/reoxygenation stress signals an increase in organic anion transporting polypeptide 1a4 (Oatp1a4) at the blood–brain barrier: relevance to CNS drug delivery. J Cereb Blood Flow Metab 2014;34(4):699–707.
[7] Teresaki Y, Liu Y, Hayakawa K, Pham LD, Lo EH, Arai K. Mechanisms of neurovascular dysfunction in acute ischemic brain. Curr Med Chem 2014;21(18):2035–42.
[8] Guo S, Stins M, Ning M, Lo EH. Amelioration of inflammation and cytotoxicity by dipyridamole in brain endothelial cells. Cerebrovasc Dis 2010;30(3):290–6.
[9] Wu L, Walas S, Leung W, Sykes DB, Lo EH, Lok J. Neuregulin1-β decreases IL-1β-induced neutrophil adhesion to human brain microvascular endothelial cells. Transl Stroke Res 2015;6(2):116–24.
[10] Qiu J, Nishimura M, Wang Y, Sims JR, Qiu S, Savitz SI, et al. Early release of HMGB-1 from neurons after the onset of brain ischemia. J Cereb Blood Flow Metab 2008;28(5):927–38.
[11] Ma Q. Role of Nrf2 in oxidative stress and toxicity. Annu Rev Pharmacol Toxicol 2013;53:401–26.

CHAPTER

46

The Neurovascular Unit

E.H. Lo
Harvard Medical School, Boston, MA, United States

INTRODUCTION

The concept of the "neurovascular unit" arose during a 2001 workshop convened by National Institute of Neurological Disorders and Stroke to address barriers in the pursuit of therapeutics and diagnostics for stroke [1]. The development of this concept was driven in part by accumulating failures in translating neuroprotective compounds from the laboratory into positive results in clinical trials. Fundamentally, this idea is based on the logic that protecting neurons alone from cell death is not enough. Stroke therapies should rescue homeostatic signaling between all cell types in the brain comprising neuronal, glial, and vascular compartments.

THE NEUROVASCULAR UNIT IN CNS FUNCTION

The importance of cell–cell signaling in the neurovascular unit can be first observed during the development of the CNS. Cross talk between vascular and neuronal progenitor cells augments neurogenesis and neuronal maturation. Cross talk between radial glia and newborn neurons allow the patterned construction of cortical

Primer on Cerebrovascular Diseases, Second Edition
http://dx.doi.org/10.1016/B978-0-12-803058-5.00046-1

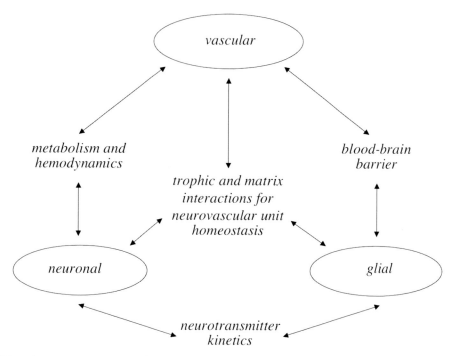

FIGURE 46.1 CNS function requires cross talk between multiple cell types from neuronal, glial, and vascular compartments. Neurotransmission is predicated on precisely coordinated release–reuptake kinetics. The blood–brain barrier is based on cell–cell signaling between the endothelium, astrocytes, and pericytes. The coupling between cerebral blood flow and metabolism requires cross talk between all neuronal, glial, and vascular cells. Trophic and matrix signaling is necessary for homeostasis and coordination between all cell types. *Adapted from Lo et al., Stroke 2004.*

layers. Signals from the growing vascular tree guide the migration of oligodendrocyte precursor cells in white matter.

In adult brain, cell–cell signaling within the neurovascular unit underlies almost all of normal CNS function (Fig. 46.1). The basic phenomenon of neurotransmission requires coordination between neuronal release and astrocytic reuptake for a wide range of neurotransmitters. Maintenance of synaptic plasticity is now thought to involve ongoing monitor-and-adjust pruning functions of microglia. Proper functioning of the blood–brain barrier requires cross talk between cerebral endothelium and adjacent pericytes and astrocyte end feet. Homeostasis in white matter is mediated by trophic signaling between axons, myelin sheaths, and the renewing pool of oligodendrocyte precursor cells. Even the revolutionary method of functional MRI itself is predicated on the spatial and temporal coupling between metabolic demands of neuronal depolarization and the hemodynamic response in arterioles and capillaries. Ultimately, the CNS works not just because neurons are firing action potentials, but also because cells from multiple neuronal, glial, and vascular compartments are all working in coordination [2]. Hence, the pathophysiology of stroke

cannot be investigated without taking into account the underlying mechanisms of function and dysfunction within the entire neurovascular unit.

THE NEUROVASCULAR UNIT IN ACUTE STROKE

The initial impetus for neuroprotection after stroke was triggered in part by the discovery of excitotoxicity. After cerebral ischemia, energetic perturbations disrupt glutamate transmitter homeostasis, and overactivation of N-methyl-D-aspartate (NMDA), α-amino-3-hydroxy-5-methyl-4-isoxazolepropionic acid (AMPA), and metabotropic glutamate receptors ultimately lead to ionic imbalance and neuronal death. Furthermore, ischemic injury also perturbs mitochondrial function, leading to the generation of reactive oxygen and nitrogen species, all of which underlie oxidative stress and neuronal dysfunction. Finally, at least in experimental model systems, cerebral ischemia may also upregulate programmed cell death pathways, comprising apoptosis, autophagy, and necroptosis. In cell culture models of oxygen–glucose deprivation and animal models of focal cerebral ischemia, blockade of these various excitotoxic, oxidative stress, and cell death pathways appears to significantly reduce neuronal injury (please see relevant chapters in this Primer for detailed discussion of these neuronal death mechanisms). However, none of these neuroprotective strategies have been proved effective in clinical trials so far.

The lack of effective neuroprotection in human stroke has been widely analyzed and debated. Perhaps, the treatment time windows were overoptimistic in early clinical trials. Or patient populations were not optimized in many of these prior "take-all-comers" design. Perhaps, the standard dichotomized chi-squared modified Rankin score was not sensitive enough. Or maybe some of these neuroprotective compounds did not effectively penetrate the blood–brain barrier. And of course, cell culture and animal models may not accurately capture all aspects of clinical stroke in humans. There will surely be many reasons involved. But an important one may involve the fundamental idea that neuroprotection requires the protection of not only neurons but the entire neurovascular unit.

Translation in stroke research is challenging because the mechanisms of stroke pathophysiology are complex and remain ill-defined [3]. For ischemic stroke, reversing the initial insult by restoring blood flow is an "upstream" approach with large effects in selected patients. But "downstream" neuroprotection may be more subtle and difficult to achieve. This may be where the neurovascular unit comes into play. Disruption of glutamate release–reuptake cross talk between neurons and astrocytes worsen excitotoxicity [4]. Downregulation of antioxidant genes in glial cells render neurons more vulnerable to ischemic oxidative stress [4]. Alterations in cell–cell and cell–matrix signaling at the gliovascular interface, for example, those pathways involving matrix metalloproteinases, promote blood–brain barrier damage [5]. Disruption of cross talk between endothelium and smooth muscle cells may alter hemodynamic function [6]. Damaged cerebral endothelium may produce lower levels of various neurotrophic factors thus rendering neuronal populations more susceptible to cell death [7]. In damaged white matter, loss of signaling between oligodendrocyte precursor cells, endothelium, and astrocytes may worsen demyelination. Hence, therapeutic approaches for acute stroke must do more than just save neurons. Rescuing cell function and cell–cell signaling in all cell types throughout the entire neurovascular unit should be an important goal [8].

THE NEUROVASCULAR UNIT IN STROKE RECOVERY

After acute ischemic stroke, cells in the severely ischemic core will die rapidly. However, penumbral areas surrounding this core may be transiently sustained by collaterals, so some brain cells should temporarily survive. Hence,

the penumbra may represent a metastable state where the process of recovery versus cell death is in a dynamic and precarious balance; the penumbra is not only dying but it is also actively trying to recover [9].

Many mediators in stroke pathophysiology come from the larger families of damage-associated-molecular-pattern and pathogen-associated-molecular-pattern signals, and these mediators are known to possess biphasic properties. Therefore, many of the therapeutic targets for stroke may not be static over time. For example, overactivation of NMDA receptors will cause excitotoxicity and neuronal death in the acute phase, but without proper NMDA signaling, recovery cannot take place when patients move into rehabilitation units [10]. Uncontrolled oxidative stress is damaging but homeostatic levels of free radicals are required for angiogenesis and endothelial progenitor cell function [11]. Matrix metalloproteinases disrupt the microvessel matrix and blood–brain barrier during acute stroke, but these proteases are necessary for neurovascular remodeling during stroke recovery [12]. Poststroke inflammation itself may also be nuanced; M1-like macrophages/microglia are deleterious, whereas M2-like cells may promote tissue repair [13]. Hence, targeting the penumbra in stroke, and perhaps more broadly treatment of any CNS disorder, may require careful attention to the signals involved as the brain transitions from initial injury into endogenous modes of remodeling and recovery [9].

SUMMARY

Targeting the entire neurovascular unit is required for stroke. However, the mechanisms of cell–cell signaling within the neurovascular unit are multifactorial and complex (Fig. 46.2). There are multiple signals in multiple cells. Mediators may play biphasic roles as the stroke-damaged brain transitions from initial injury into repair. Furthermore, all of these signals and substrates are affected by a whole host of risk factors comprising genetic, physiologic, and lifestyle-induced changes. Finally, the neurovascular unit is not isolated but interacts with peripheral responses occurring in cardiac, immune, and hormonal systems. Because these mechanisms are complex, a systems biology approach may eventually be required. Mapping gene response networks in recovering neurons may give insight into mechanisms of how the stroke-damaged brain attempts to rewire [14]. Mapping the astrocyte transcriptome may allow one to probe the role of reactive gliosis after stroke [15]. A brain vasculome has been proposed that may provide a way to correlate experimental findings with human (genome-wide association studies) GWAS genes as well as a database for biomarker mining [16]. For investigating stroke recovery, the concept of help-me signaling has been proposed as a way to dissect

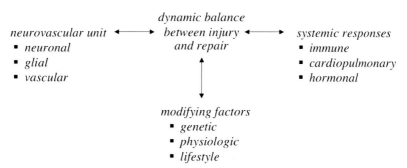

FIGURE 46.2 The neurovascular unit represents a conceptual framework that seeks to capture the multicellular signaling that occurs in stroke pathophysiology. Within the CNS itself, cross talk between neuronal, glial, and vascular cells mediate the central response to injury. Once the damaged CNS begins to respond, interactions are initiated between central and peripheral systems comprising hormonal, cardiac, and immune pathways. Over time, these signals and substrates demonstrate biphasic properties as the stroke-damaged brain begins to transition from initial injury into subsequent repair. And ultimately, the entire network of interactive and recursive signaling loops are modified by "risk factors" coming from genetic, physiologic, and lifestyle categories. *Adapted from Lo, Stroke 2013.*

the signals and substrates of non-cell-autonomous mechanisms within the neurovascular unit as it attempts to remodel and repair itself after injury and [17]. Ultimately, the neurovascular unit may provide a powerful conceptual framework to investigate causal mechanisms that can be targeted for stroke therapeutics.

Acknowledgments

The neurovascular unit concept was developed by the entire research community, not just any one single scientist. The author apologizes to many colleagues who could not be included as coauthors and whose work could not be cited, due to space limitations.

References

[1] Lo EH, Dalkara T, Moskowitz MA. Mechanisms, challenges and opportunities in stroke. Nat Rev Neurosci 2003;4:399–415.

[2] Lok J, Gupta P, Guo S, Kim WJ, Whalen MJ, van Leyen K, et al. Cell–cell signaling in the neurovascular unit. Neurochem Res 2007;32:2032–45.

[3] Lo EH. 2013 Thomas Willis award lecture: causation and collaboration for stroke research. Stroke 2014;45:305–8.

[4] Barreto G, White RE, Ouyang Y, Xu L, Giffard RG. Astrocytes: targets for neuroprotection in stroke. Cent Nerv Syst Agents Med Chem 2011;11:164–73.

[5] Yang Y, Rosenberg GA. Matrix metalloproteinases as therapeutic targets for stroke. Brain Res 2015;1623:30–8.

[6] Zhang JH, Badaut J, Tang J, Obenaus A, Hartman R, Pearce WJ. The vascular neural network—a new paradigm in stroke pathophysiology. Nat Rev Neurol 2012;8:711–6.

[7] Guo S, Kim WJ, Lok J, Lee SR, Besancon E, Luo BH, et al. Neuroprotection via matrix-trophic coupling between cerebral endothelial cells and neurons. Proc Natl Acad Sci USA 2008;105:7582–7.

[8] Moskowitz MA, Lo EH, Iadecola C. The science of stroke: mechanisms in search of treatments. Neuron 2010;67:181–98.

[9] Lo EH. A new penumbra: transitioning from injury into repair after stroke. Nat Med 2008;14:497–500.

[10] Wahlgren NG, Martinsson L. New concepts for drug therapy after stroke. Can we enhance recovery? Cerebrovasc Dis 1998;8(Suppl. 5):33–8.

[11] Ushio-Fukai M, Urao N. Novel role of NADPH oxidase in angiogenesis and stem/progenitor cell function. Antioxid Redox Signal 2009;11:2517–33.

[12] Zhao BQ, Wang S, Kim HY, Storrie H, Rosen BR, Mooney DJ, et al. Role of matrix metalloproteinases in delayed cortical responses after stroke. Nat Med 2006;12:441–5.

[13] Hu X, Leak RK, Shi Y, Suenaga J, Gao Y, Zheng P, et al. Microglial and macrophage polarization—new prospects for brain repair. Nat Rev Neurol 2015;11:56–64.

[14] Li S, Nie EH, Yin Y, Benowitz LI, Tung S, Vinters HV, et al. GDF10 is a signal for axonal sprouting and functional recovery after stroke. Nat Neurosci 2015;18:1737–45.

[15] Zamanian JL, Xu L, Foo LC, Nouri N, Zhou L, Giffard RG, et al. Genomic analysis of reactive astrogliosis. J Neurosci 2012;32:6391–410.

[16] Guo S, Zhou Y, Xing C, Lok J, Som AT, Ning M, et al. The vasculome of the mouse brain. PLoS One 2012;7:e52665.

[17] Xing C, Lo EH. Help-me signaling: non-cell autonomous mechanisms of neuroprotection and neurorecovery. Prog Neurobiol 2016. http://dx.doi.org/10.1016/j.pneurobio.2016.04.004.

47

Mitochondrial Mechanisms During Ischemia and Reperfusion

Z.Z. Wei[1], L. Wei[2]

[1]Emory University School of Medicine and Atlanta Veterans Affairs Medical Center, Decatur, GA, United States;
[2]Emory University School of Medicine, Atlanta, GA, United States

INTRODUCTION

The mitochondrion keeps a continuous workflow for oxidation of energy substances that are obtained from the blood to produce adenosine triphosphate (ATP) for cellular functions. Neural cells consume this energy derived from the ATP hydrolysis for maintaining ionic gradients across plasma membranes. During ischemia, oxygen supply and ATP production are rapidly impaired, followed by exhaustion of energy substances. This condition blocks mitochondrial respiration and glycolytic activity, two major energy metabolism events. Restoration of the local blood flow after an ischemic event for short term immediately reinitiates the energy production when ATP and other energy substances/metabolites recover mostly in parallel. Although recovery of energy demand prevails in all the postischemic tissues regardless of the regional sensitivity difference to ischemic insults, complete restoration of the ATP content may be delayed to several hours later.

During ischemic reperfusion, alterations in mitochondrial dynamics occur and lead to acute cell death. This disruption of mitochondrial Ca^{2+} homeostasis may impair ATP-producing mitochondrial respiration and cause energy deficiency in postischemic cells. Regulation of the phosphate receptors, voltage-dependent anion channels, Na^+/Ca^{2+} exchanger, and mitochondrial permeability transition pore (mPTP), which were precisely controlled, plays significant and dual roles in reperfusion injury.

Calcium is involved in the maintenance of electrical potential of inner mitochondrial membrane. In hypoxic and ischemic conditions, the inefficient Na^+/K^+ ATPase activity causes accumulation of sodium in the cytoplasm, which drives Ca^{2+} influx through the Na^+/Ca^{2+} exchanger contributing to the accumulation of intracellular Ca^{2+}. The intracellular concentration affects mitochondrial calcium dynamics. Mitochondrial respiratory activity is restored during early reperfusion when intracellular calcium is still increased. Mitochondrial calcium increase is then reversed within the first hour after reperfusion. A delayed local increase may be restricted in some ischemia-susceptible regions at as long as 24 h after reperfusion. This event can lead to ATP loss and energy deficiency, contributing to the progression of ischemic reperfusion injury. Characterization of the molecular basis for reduced mitochondrial activity and its involvement in ischemia-induced changes has helped to understand the mitochondrial mechanisms during ischemia and reperfusion.

ADENOSINE TRIPHOSPHATE AND ENERGY METABOLISMS

Mitochondria from ischemic brains show a decreased ability for ATP generation. This is correlated to a reduced capacity for respiratory activity and cell death in the acute phase of ischemia. In the adult brain after ischemia, glycogen is rapidly consumed by aerobic glycolysis, generating ATP and pyruvate. Resynthesizing of ATP in the reperfusion period required a certain time in all the affected tissues, although there may be differences between an ischemia-susceptible region and an ischemia-resistant region. The ability and maintenance for relatively high ATP concentrations in cells has indicated integrity of tissue after ischemia and an operative intracellular control against ischemic cell death and brain injury. The late reductions of ATP in ischemia-susceptible regions may indicate the disrupted mitochondrial functions (Fig. 47.1).

Primer on Cerebrovascular Diseases, Second Edition
http://dx.doi.org/10.1016/B978-0-12-803058-5.00047-3

Besides, measurements of other energy-related metabolites such as glucose provide evidence for mitochondrial activity and energy metabolism during blood flow recirculation. Reduction of glucose oxidation in the ischemic brain compared with normal brain persists within the first several hours. Evidences show that an increased energy requirement in the brain tissue can result in elevated glucose utilization. However, the coupling of blood flow to glucose utilization can be significantly varied among different brain areas in different physical and pathologic conditions. In postischemic brains, the generalized reductions of glucose metabolism are associated with the low local energy requirements in short term, and do not seem to result from a loss of metabolic capacity due to changes of blood flow in the affected tissue. Studies also suggest that the continuous reductions in the requirement for ATP are associated with long-term changes in the brain that has been subjected to ischemic insults.

Within the first few hours of reperfusion, pyruvate is reduced, suggesting a reduced aerobic glycolytic activity. This activity catalyzed by the pyruvate dehydrogenase complex, was inhibited to limit glucose oxidation after ischemia, accounting for the reduction in energy-producing mitochondrial function during the first few hours of reperfusion. Through anaerobic glycolysis, pyruvate is catalyzed to lactate, with generation of NAD^+. The lactate level is of significant prognostic value for outcome after neonatal hypoxic-ischemic injury. Moreover, the inability to survive the accumulated oxidative stress in postischemic tissue may prolong the exposure of some neurons to oxidative damage as the reperfusion period progresses.

However, during the first few hours of reperfusion, decreased energy demands prevail in the ischemic brain, indicated by an oxidative metabolism reduction. Tissue slices reveal an attenuated metabolic response to chemical depolarization and an impairment of mitochondrial function. Although at this time the brain tissue tends to recover normal function, the mitochondria-associated problem fails to respond to increases in energy requirements. Mitochondrial deficiency due to impaired mitochondrial function leads to further deleterious changes and potentially compromises secondary injury of ischemia-susceptible neurons in the long term.

ION CHANNEL REGULATION OF MITOCHONDRIAL CALCIUM

Calcium influx and efflux regulates the mitochondrial responses to ischemia. Mitochondrial calcium levels and biochemistry is fueled by both the extracellular calcium and calcium released from the endoplasmic reticulum. An increase in mitochondrial calcium augments the activity of mitochondrial matrix enzymes and eventually increases ATP production. However, the entry of excess calcium in response to hyperactivity of neurons and astrocytic mitochondrial dysfunction during ischemic reperfusion triggers necrosis (Fig. 47.2).

Endoplasmic-reticulum-to-mitochondria calcium mobility is pushed through specialized contact sites called mitochondrial-associated endoplasmic reticulum membranes. Inositol 1,4,5-triphosphate (IP_3) receptors, voltage-dependent anion channels in the outer mitochondrial membrane, and the uniporter complex promote this calcium mobility into mitochondrial matrix. In physical conditions, mitochondrial Na^+/Ca^{2+} exchanger is finally responsive for calcium removal when intracellular sodium concentration is beyond a certain threshold.

A family of large-conductance voltage- and calcium-activated potassium channels has been suggested as a promising neuroprotective target. These channel proteins expressed in plasma membrane and mitochondria, allow for attenuation of Ca^{2+} influx and reactive oxygen species (ROS) production in mitochondria upon

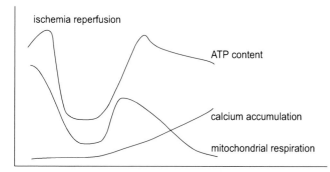

FIGURE 47.1 Relationships between mitochondrial respiration, ATP production, and mitochondrial calcium after ischemia and reperfusion. Hypothetical curves are plotted.

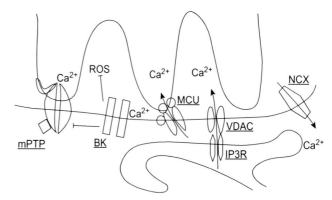

FIGURE 47.2 Regulation of mitochondrial calcium after ischemic reperfusion. The arrows indicate the flow of calcium ions from cytoplasm and endoplasmic reticulum into mitochondrial matrix. *BK,* (big potassium) channel; *IP3R,* inositol trisphosphate receptor; *mPTP,* mitochondrial permeability transition pore; *MCU,* mitochondrial calcium uniporter; *NCX,* sodium, calcium exchanger; *ROS,* reactive oxygen species; *VDAC,* voltage-dependent anion channels.

membrane depolarization and/or intracellular Ca^{2+} ion accumulation [1].

CELL DEATH AND MITOPHAGY

The mPTP regulates mitochondrial Ca^{2+} homeostasis as well and its prolonged opening can result in necrotic cell death after ischemia [2]. In addition, the release of cytochrome c, endonuclease G, Smac/DIABLO (direct inhibitor of apoptosis-binding protein with low pI), and apoptosis-inducing factor into the cytoplasm activates apoptotic cascades. For population of neurons particularly sensitive to short-term ischemia, for example, in the hippocampus CA1 area, these cell death signaling molecules are sufficient to cause neuronal loss. Overloaded Ca^{2+}, ROS, and hydrogen ions and the interactions with polyphosphate (polyP) in the inner mitochondrial membrane can induce mPTP opening and membrane potential dissipation, leading to impaired respiratory chain, halted ATP synthesis, organelle swelling, and outer membrane rupture [3]. PolyP potentially interacts with many mitochondrial proteins, by which polyP and mPTP could be regulated through direct protein-binding activities [4]. In vivo studies have showed that release of polyP from astrocytes evokes calcium signals and mitochondrial deletion of polyP reduces mitochondrial membrane potential and calcium-induced mitochondrial permeability transition [5]. Depolarized mitochondrial membrane potentials and delayed opening of mPTPs have been considered as a neuroprotective strategy. For example, dl-3-n-butylphthalide protects against ischemic damage via blocking release through mPTP.

Permeabilization of the outer mitochondrial membrane via Bcl-2-associated X protein/Bcl-2 homologous antagonist killer and mPTP mechanisms activates phagophore formation leading to autophagy or initiates irreversible cell death pathways [6]. Mitochondrial autophagy, or mitophagy, promotes turnover of mitochondria through clearance of dysfunctional mitochondria [7]. After hypoxia and ischemia, many signaling molecules including 5′ adenosine monophosphate-activated protein kinase–mammalian target of rapamycin, Bcl-2/adenovirus E1B 19 kDa protein-interacting protein 3, forkhead box O3, hypoxia-inducible factor-1α, nuclear factor kappa-light-chain-enhancer of activated B cells, and signals of endoplasmic reticulum stress and oxidative stress have been identified as triggers to mitophagy [8]. An elimination of affected mitochondria can occur selectively when fission complex forms to degrade the dysfunctional fragment. PINK1 and Parkin in between mitochondrial membranes are natural regulators of mitophagy [9]. Interestingly, hypothermia significantly reduced Parkin-mediated mitophagy and autophagic cell death (Fig. 47.3) [10].

MITOCHONDRIAL TRANSFER AND BIOGENESIS

Stem cell transplantation holds great promise for stroke. Transplantation of bone marrow–derived mesenchymal stem cells in the acute phase exerts neuroprotective benefits after ischemia [11]. Importantly, a mitochondrial transfer through connexin-43 (Cx-43)-containing gap junctional channels has been demonstrated in an acute lung injury model [12]. This highlights a therapeutic mechanism of cell transplantation therapy by correcting energy downshift during acute ischemia.

Mitochondrial biogenesis following ischemia improved energy production capacity and thus is considered as a clinically relevant endogenous neuroprotective response. However, ischemic insult has been shown to inhibit the mitochondrial biogenesis. Mitochondrial fission and fusion processes, which are considered as central events of mitochondrial biogenesis, are controlled by nuclear transcriptional factors. Peroxisome proliferator-activated receptor (PPAR) gamma-1α (PGC-1α) and nuclear respiratory factor 1 (NRF-1), are two key nuclear factors that regulate mitochondrial biogenesis (Fig. 47.4) [13].

OVERVIEW

Evidence is provided that ischemia and reperfusion cause parallel changes in mitochondrial activity and ATP production. The initial reperfusion recovers energy-related metabolites and mitochondrial function. Calcium mobility from cytoplasm and endoplasmic reticulum into mitochondria and dynamic alteration of oxidative glucose metabolism restrict the respiratory capacity and lead to delayed cell death. Mitochondrial dysfunction contributes to neurodegenerative processes after ischemic stroke, making mitochondrial homeostasis a potential target for pharmacological and cell-based therapies.

Abbreviations

AIF Apoptosis-inducing factor
AMPK 5′ Adenosine monophosphate-activated protein kinase
BAK Bcl-2 Homologous antagonist killer
BAX Bcl-2-Associated X protein
Bcl-2 B-Cell lymphoma 2
BNIP3 Bcl-2/adenovirus E1B 19 kDa protein-interacting protein 3
Cx-43 Connexin 43
FoxO3 Forkhead box O3

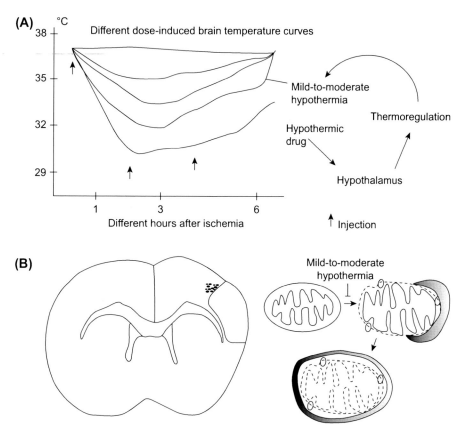

FIGURE 47.3 Mitophagy-related neuroprotective strategy by neurotensin receptor agonists. Targeting neurotensin receptor can activate the hypothalamus temperature regulation and induce a mild-to-moderate hypothermia (A). Mild-to-moderate hypothermia has been shown to be effective in preventing mitophagy and autophagic cell death (B).

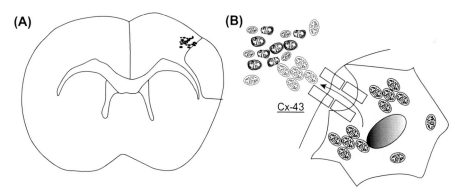

FIGURE 47.4 Schematic graph indicating the mitochondrial transfer hypothesis by cellular therapy. In a focal ischemic model, detrimental responses such as mitophagy leading to the degradation of mitochondria happened in penumbra regions (A). Cellular therapy by intracranial transplantation of cells potentially provides the metabolic support via mitochondrial transfer mechanisms (B). Cx-43, connexin 43.

HIF-1α Hypoxia-inducible factor 1
IP$_3$ Inositol trisphosphate
mPTP Mitochondrial permeability transition pore
mTOR Mammalian target of rapamycin
NF-κB Nuclear factor kappa-light-chain-enhancer of activated B cells
NRF-1 Nuclear respiratory factor 1
polyP Polyphosphate
PGC-1α Peroxisome proliferator-activated receptor gamma coactivator 1-alpha
ROS Reactive oxygen species

References

[1] Liao Y, Kristiansen AM, Oksvold CP, Tuvnes FA, Gu N, Runden-Pran E, et al. Neuronal Ca²⁺-activated K⁺ channels limit brain infarction and promote survival. PLoS One 2010;5(12):e15601.
[2] Kwong JQ, Molkentin JD. Physiological and pathological roles of the mitochondrial permeability transition pore in the heart. Cell Metab 2015;21(2):206–14.
[3] Seidlmayer LK, Juettner VV, Kettlewell S, Pavlov EV, Blatter LA, Dedkova EN. Distinct mPTP activation mechanisms in ischaemia–reperfusion: contributions of Ca²⁺, ROS, pH, and inorganic polyphosphate. Cardiovasc Res 2015;106(2):237–48.

[4] Wei ZZ, Vatcher G, Tin AH, Teng JL, Wang J, Cui QH, et al. Positively-charged semi-tunnel is a structural and surface characteristic of polyphosphate-binding proteins: an in-silico study. PLoS One 2015b;10(4):e0123713.

[5] Abramov AY, Fraley C, Diao CT, Winkfein R, Colicos MA, Duchen MR, et al. Targeted polyphosphatase expression alters mitochondrial metabolism and inhibits calcium-dependent cell death. Proc Natl Acad Sci USA 2007;104(46):18091–6.

[6] Kubli DA, Gustafsson AB. Mitochondria and mitophagy: the yin and yang of cell death control. Circ Res 2012;111(9):1208–21.

[7] Youle RJ, Narendra DP. Mechanisms of mitophagy. Nat Rev Mol Cell Biol 2011;12(1):9–14.

[8] Heo JM, Ordureau A, Paulo JA, Rinehart J, Harper JW. The PINK1-PARKIN mitochondrial ubiquitylation pathway drives a program of OPTN/NDP52 recruitment and TBK1 activation to promote mitophagy. Mol Cell 2015;60(1):7–20.

[9] Lazarou M, Sliter DA, Kane LA, Sarraf SA, Wang C, Burman JL, et al. The ubiquitin kinase PINK1 recruits autophagy receptors to induce mitophagy. Nature 2015;524(7565):309–14.

[10] Choi KE, Hall CL, Sun JM, Wei L, Mohamad O, Dix TA, et al. A novel stroke therapy of pharmacologically induced hypothermia after focal cerebral ischemia in mice. FASEB J 2012;26(7):2799–810.

[11] Wei ZZ, Gu X, Ferdinand A, Lee JH, Ji X, Ji XM, et al. Intranasal delivery of bone marrow mesenchymal stem cells improved neurovascular regeneration and rescued neuropsychiatric deficits after neonatal stroke in rats. Cell Transplant 2015a;24(3):391–402.

[12] Islam MN, Das SR, Emin MT, Wei M, Sun L, Westphalen K, et al. Mitochondrial transfer from bone-marrow-derived stromal cells to pulmonary alveoli protects against acute lung injury. Nat Med 2012;18(5):759–65.

[13] Vosler PS, Graham SH, Wechsler LR, Chen J. Mitochondrial targets for stroke: focusing basic science research toward development of clinically translatable therapeutics. Stroke 2009;40(9):3149–55.

CHAPTER

48

Excitotoxicity and Stroke

D.W. Choi[1,2]

[1]State University of New York at Stony Brook, Stony Brook, NY, United States; [2]Korea Institute of Science and Technology, Seoul, South Korea

INTRODUCTION

A long-standing goal of stroke research has been to elucidate the mechanisms responsible for the high vulnerability of brain tissue to hypoxic-ischemic insults. Understanding these mechanisms will aid the development of specific countermeasures that could be employed in the acute treatment of stroke victims.

One such mechanism, "excitotoxicity," was implicated in the 1980s as a prominent contributor to neuronal death after transient or permanent focal brain ischemia, and became the focus of many laboratory and clinical studies during the 1990s. However, the failure of several novel antiexcitotoxic therapies to show efficacy in clinical trials diminished interest. This chapter will briefly summarize the excitotoxic concept, and outline how it might be more effectively countered to reduce brain damage during stroke.

WHAT IS EXCITOTOXICITY?

In keeping with its widespread role in central fast excitatory neurotransmission [1], glutamate is present at millimolar levels throughout CNS gray matter. Under normal conditions, most of this glutamate is highly concentrated within nerve terminals, and energy-dependent cellular uptake rapidly clears synaptically released glutamate from the extracellular space. Thus neurons are exposed only briefly and focally to glutamate in the course of excitatory neurotransmission. This normal synaptic glutamate exposure is not injurious.

Primer on Cerebrovascular Diseases, Second Edition
http://dx.doi.org/10.1016/B978-0-12-803058-5.00048-5

However, under certain disease conditions, for example, when tissue energy stores are compromised by hypoxia-ischemia, increased glutamate efflux from depolarized neurons and astrocytes together with reduced cellular glutamate uptake can produce an abnormally large and sustained buildup of extracellular glutamate that is no longer restricted to synaptic zones [2]. In addition, energy depletion reduces the ability of neurons to correct or manage the cellular changes induced by glutamate exposure, especially increases in intracellular calcium. Sustained exposure to high concentrations of extracellular glutamate, with or without energy depletion can quickly kill neurons, a phenomenon named "excitotoxicity" by John Olney [3], and primarily triggered by ion channels directly gated by glutamate (see later discussion).

Abnormal activation of other neurotransmitter pathways, or ionic/metabolic derangements mediated by other events, can also promote neuronal injury by mechanisms convergent with those underlying excitotoxicity. Although vulnerability to excitotoxicity was originally thought to be an exclusive property of neurons, mature oligodendrocytes [4] can also be damaged by glutamate receptor overactivation, perhaps contributing to ischemic white matter damage. Even certain cells outside the brain, such as renal cells, can be injured by glutamate receptor overactivation [5].

HOW EXCITOTOXICITY KILLS NEURONS

Almost half a century ago, Olney observed that high systemic doses of glutamate produced characteristic pathological changes in the circumventricular regions of young rodent or monkey brain, for example, the arcuate nucleus of the hypothalamus, lacking a full blood–brain barrier [3]. Within 30 min of glutamate administration, electron microscopy revealed that neuronal cell bodies and dendrites had developed massive acute swelling followed by degeneration of intracellular organelles and nuclear pyknosis; axons were relatively spared. Over the next few hours, neurons degenerated and underwent phagocytosis by macrophages.

Glutamate activates three major families of ionophore-linked (ionotropic) receptors, classified by their preferred agonists: N-methyl-D-aspartate (NMDA), a-amino-3-hydroxy-5- methyl-4-isoxazolepropionic acid (AMPA), and kainate; multiple functional receptor subunits from each family have been cloned [6]. The channels gated by NMDA, AMPA, and kainate receptors are permeable to both sodium and potassium. Channels gated by NMDA receptors, but only a small subset of channels gated by AMPA or kainate receptors (see later discussion), additionally possess high permeability to calcium. These different glutamate receptor subtypes do not participate

equally in excitotoxicity [7]. After brief (minutes) intense glutamate exposure, all ionotropic glutamate receptors contribute to neuronal depolarization and acute toxic swelling, but NMDA receptors dominantly mediate subsequent neuronal death. Although brief exposure to high concentrations of kainate or AMPA will not kill most forebrain neurons, more extended exposure (hours) to even low concentrations of AMPA or kainate will produce widespread neuronal death.

Rapidly triggered, NMDA receptor-mediated toxicity critically depends on the presence of extracellular calcium, consistent with the idea that this toxicity is initiated by excessive calcium influx through the calcium-permeable NMDA receptor-gated channel. As the molecular biology of NMDA receptors has unfolded, it has become clear that different NMDA receptor subtypes and subcellular localizations can contribute differentially to excitotoxicity. In particular, extrasynaptic NMDA receptors located outside of synapses mediate most of the toxicity induced by broad cellular exposure to glutamate, whereas synaptic NMDA receptors and subsequent nuclear calcium signaling play a larger role in antiapoptotic effects [8] (see later discussion). In part reflecting this dominant participation of extrasynaptic NMDA receptors, activation of the NR2B subtype of NMDA receptors may be more lethal than the activation of the NR2A subtype [9]. Key interacting molecular partners of NR2B receptors in mediating excitotoxic death have been identified, including the postsynaptic density protein PSD-95, which couples the receptor to nitric oxide synthase [10] (see later discussion), as well as the death associated protein kinase 1 (DAPK1), which phosphorylates the receptor and upregulates current through its channel in an amplifying positive-feedback loop [11].

Slowly triggered, AMPA/kainate receptor-mediated excitotoxicity also may be initiated by excessive calcium influx. Most channels gated by AMPA or kainate receptors have limited calcium permeability, but calcium influx can also occur indirectly via other pathways activated in part downstream of ionotropic glutamate receptors, including voltage-gated calcium channels, transient receptor potential channels, acid-sensing ion channels, stretch-activated channels, and the membrane sodium/calcium exchanger [7,10,12].

In addition, some AMPA or kainate receptors may gate channels permeable to calcium (in the case of AMPA receptors, due to an RNA-edited form of the GluR2 subunit) [13]. A small number of cortical neurons (in particular, a subset of GABAergic inhibitory neurons) may express large numbers of calcium-permeable AMPA receptors and as a result may be especially vulnerable to damage mediated by AMPA receptor overactivation. On these neurons, AMPA receptors may essentially function like NMDA receptors on other neurons, mediating large calcium currents and triggering a relatively rapid form of excitotoxicity. Expression of calcium-permeable AMPA receptors and consequent heightened vulnerability to glutamate-induced death may not be a static property, but rather be modulated on certain neurons by prior ischemic insults [14].

Glutamate also activates a family of metabotropic receptors (mGluRs) that activate second messenger systems rather than directly gating ion channels [15]. For the most part, mGluRs do not directly mediate excitotoxic injury but rather modify it by modulating circuit excitability (and hence glutamate release) as well as downstream injury pathways, including apoptosis. Actions are complex, for example, although activation of group I mGluRs (mGluR1 and mGluR5) generally produces excitatory, proexcitotoxic effects, these receptors may also directly inhibit NMDA receptor activation [16].

Excessive transmembrane calcium influx combines with excessive release from intracellular stores, including release from endoplasmic reticulum mediated by ryanodine and inositol trisphosphate receptors, to elevate intracellular-free calcium and initiate multiple feed-forward cytotoxic consequences [7,8,10]. Several catabolic enzymes are calcium activated, including proteases capable of breaking down structural proteins, phospholipases capable of breaking down cell membranes (and liberating arachidonic acid), and endonucleases capable of breaking down genomic DNA. Excess intracellular-free calcium is taken up by mitochondria, resulting in electron transport disturbances, reduced energy production, generation of reactive oxygen species, cytochrome c release, and caspase-3 activation, the latter events initiating programmed cell death signaling leading to apoptosis. Toxic-free radicals can also form as a consequence of arachidonic acid metabolism as well as nitric oxide synthase activity, which generates nitric oxide [17]. Excessive nitric oxide overactivates the nuclear enzyme poly(ADP-ribose)polymerase, leading to NAD^+ depletion and further mitochondrial failure [18,19]. Many of these downstream events, in particular oxidative damage to cellular constituents, have been implicated in the pathogenesis of ischemic brain injury independent of considering excitotoxicity. Although excitotoxicity is not exclusively responsible for these events, the idea that excitotoxicity is a key driver provides valuable convergence between different lines of investigation, and supports the notion that ischemic cell death is mediated by a defined array of final common pathways.

Besides derangements in intracellular calcium homeostasis, the opening of glutamate receptor-gated channels directly alters the homeostasis of several other ions that can further damage cells. As noted earlier, sodium entry depolarizes cells, promoting calcium entry through voltage-gated calcium channels and the sodium–calcium exchanger; it also drives chloride and water entry, triggering cellular swelling and tissue edema. The potassium permeability of glutamate receptor-gated channels permits potassium efflux, which lowers intracellular potassium concentrations and facilitates programmed cell death [20], in part likely through increasing caspase-3 activation. Finally, intracellular-free zinc concentrations rise due to transmembrane zinc influx through multiple calcium channels and exchangers [21], as well as release from intracellular metalloproteins driven by oxidative stress [22]. This increased intracellular zinc also has cytotoxic consequences, including damage to mitochondria [23], and may synergize with elevated intracellular calcium in enhancing delayed rectifier potassium currents, lowering intracellular potassium and thus facilitating apoptosis [22].

Intense excitotoxic insults induce neuronal cell swelling and an unspecific cell death (necrosis) attributable to myriad parallel lethal derangements. However, since glutamate receptor overstimulation capably activates programmed cell death pathways, more moderate excitotoxic insults (for example, occurring in the penumbra of a stroke) may be able to progress more slowly to apoptosis.

ISCHEMIA INITIATES EXCITOTOXICITY

Although all mammalian cells will eventually die if deprived sufficiently of oxygen, central neurons are far more vulnerable to hypoxia-ischemia than most other cells in the body. Additional to the high energy requirements of neurons, this vulnerability likely substantially reflects the impact of excitotoxicity, which might be considered a "ride-along" cost in the emergence of advantageous fast glutamatergic central nervous signaling during evolution.

Microdialysis measurements have indicated that extracellular glutamate rises sharply during hypoxia-ischemia, and likely easily exceeds neurotoxic levels [2,7]. Other animal experiments have established that excitatory pathway deafferentation can reduce hypoxic-ischemic neuronal injury, that the pathologies of hypoxic-ischemic injury and excitotoxic injury are similar, and that neuronal calcium overload and oxidative stress occur prominently in both conditions [24]. The strongest evidence implicating excitotoxicity in the pathogenesis of hypoxic-ischemic neuronal death is provided by studies from many laboratories showing that experimental interventions capable of attenuating excitotoxicity can reduce this death, both in vitro (e.g., cultured neurons exposed to oxygen-glucose deprivation) and in vivo, in animal models of stroke or cardiac arrest/resuscitation [3,7]. Such neuroprotective effects have been observed with a wide range of pharmacological, genetic, or even physical (hypothermia) approaches, capable of reducing excitotoxicity in various ways: including reduction of glutamate release, blockade of NMDA or AMPA/kainate receptors, or interference with downstream cytotoxic events [25].

CLINICAL DISAPPOINTMENT—AND POSSIBLE EXPLANATIONS

Several companies developed NMDA receptor antagonist drugs and brought them forward into phase II or phase III stroke clinical trials during the 1990s, but these programs failed, in two cases (CNS-1102, CGS 19755) showing worrisome trends toward worse outcome in the treated group [25]. It has been pointed out that these efforts fell short of fully testing the NMDA receptor/excitotoxicity hypothesis, because doses were too low (limited by mechanism-associated behavioral side effects, such as hallucinations) or too late (6h after stroke onset, limited by the practicality of enrolling sufficient patients at the earlier time points demonstrated to work in animal models). A smaller number of pilot studies with AMPA antagonists was also performed without encouraging results, in one case (ZK20075) unsurprisingly depressing consciousness [26].

It is possible that the contribution of excitotoxicity to focal ischemic injury is highly species dependent, prominent only in rodents and other animal models but not in humans. But this scenario would be at odds with the highly conserved nature of the glutamate transmitter system across vertebrates, the face validity of animal models of stroke (compare with animal models of more complex insults, such as Alzheimer disease), and the quantitative constancy of relationships between ischemia (degree, time) and resultant brain tissue injury across many animals. A more likely explanation lies in the limitations of past clinical trials— a trial combining adequate dose and immediate (<2h) treatment has not yet been done.

Furthermore, evidence has emerged implicating programmed cell death as an important contributor to ischemic brain damage in addition to excitotoxic necrosis, perhaps especially in areas of brain exposed to threshold insult levels and contributing to a delayed progression of tissue infarction over days [27]. Although as noted earlier excitotoxicity may participate in activating programmed cell death pathways, once these pathways are activated, moderate levels of raised intracellular-free calcium can be neuroprotective, staving off apoptosis [28]. Thus aggressively lowering calcium entry through the sustained application of NMDA antagonists might not only fail to rescue neurons undergoing programmed cell death, but might even promote this event. It seems plausible that natural human stroke, with a sometimes stuttering

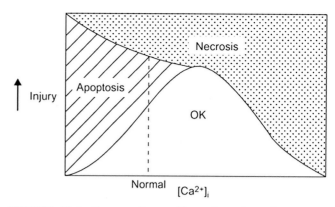

FIGURE 48.1 Concept diagram illustrating relationships among insult severity, intracellular-free calcium levels, and a survival-apoptosis-necrosis cell fate continuum. A single insult might lead either to apoptosis or to necrosis, depending on insult severity and intracellular calcium, with low calcium or milder insults favoring apoptosis. *Reprinted from Choi DW. Calcium: still center-stage in hypoxic-ischemic neuronal death. Trends Neurosci 1995;18:58–60.*

onset of ischemia and potentially large penumbral areas across complex gyri, might exhibit a greater contribution from ischemic apoptosis relative to fulminant excitotoxic necrosis, in comparison with a precisely induced experimental stroke in a lissencephalic laboratory rat (Fig. 48.1).

A prediction of this calcium duality scenario is that a controlled increase in calcium influx and intracellular-free calcium elevation can reduce ischemic apoptosis; this was borne out in cultured neurons [29]. Furthermore, administration of the partial NMDA agonist, D-cycloserine, improved functional outcome in rats subjected to transient middle artery occlusion [30].

POSSIBLE PATHS FORWARD—MAYBE STILL STARTING WITH NMDA ANTAGONISTS

As noted earlier, there are potentially many ways to interfere with ischemia-induced excitotoxicity, including reduction of glutamate efflux, enhanced clearance from extracellular space, blockade of ionotropic receptors, modulation of interacting membrane receptors or transmembrane ion fluxes, or amelioration of downstream injury effectors, especially oxidative stress and programmed cell death.

NMDA antagonist drugs may yet work, but in light of recent understandings the approach should be adjusted to improve therapeutic index: blocking excitotoxic necrosis with less neurobehavioral disturbances or enhancement of ischemic apoptosis. This might be accomplished with existing subtype-unspecific agents by refining the timing of application, perhaps utilizing neuroimaging techniques to help stage pathophysiology in individual patients. The

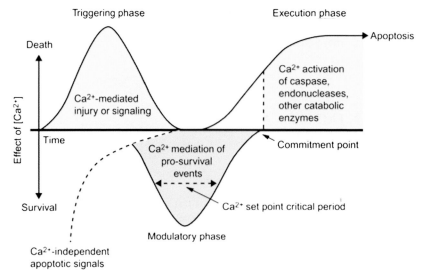

FIGURE 48.2 Concept diagram illustrating a possible time- and intracellular-free calcium dependence of ischemic excitotoxicity. Glutamate receptor overactivation leads to initial calcium overload, which if severe drives early necrosis; more moderate overactivation allows programmed cell death signaling to proceed to a modulatory phase where elevated calcium is neuroprotective and low calcium facilitates progression to apoptosis. In the final execution phase of apoptosis, calcium activation of catabolic enzymes contributes to cell death. *Reprinted from Yu SP, Canzoniero LMT, Choi DW. Ion homeostasis and apoptosis. Curr Opin Cell Biol 2001;13:405–11. The importance of calcium entry location, synaptic versus extrasynaptic [8,10], is not incorporated into this diagram.*

goal would be to block early intense receptor activation, but ease up on blockade later when NMDA receptor activation is lower and useful in inhibiting apoptosis. Spatial control of application, for example, by delivering agent through an endovascular device catheter, might also be useful in this regard, achieving higher drug concentrations in central injury zones with more intense receptor overactivation than in the penumbral areas where apoptosis might be more prominent (Fig. 48.2).

Another way to improving the neuroprotective effectiveness of unspecific NMDA antagonists would be to combine these drugs with other agents attacking other points in excitotoxic cascades or additional injury processes, with heightened attention to temporal sequences and potentially complex opposing actions. For example, although loss of zinc homeostasis and elevation of intracellular zinc likely participates in ischemic injury, completely preventing zinc elevations might increase early excitotoxic NMDA receptor activation, due to loss of zinc inhibition of NMDA receptor activity, and also preclude a late zinc-induced phosphorylation and upregulation of NMDA receptors potentially beneficial in limiting ischemic apoptosis [31]. A combination that might be tested early on is the combination of an NMDA antagonist with another drug designed to reduce ischemic apoptosis.

In the longer term, adjusting the targeting of candidate NMDA antagonists will likely be desirable. Several small companies are now working on the development of NR2B-preferring antagonists, at least in one case combining this with an additional auspicious property of pH

dependence: producing greater receptor blockade at the lower pH levels induced by ischemia, and less blockade at normal tissue pH. Use dependency (greater receptor antagonism at higher levels of receptor activation) and partial antagonism (less likely than full blockade to promote calcium starvation and apoptosis) are additional attributes that might be dialed into future drug candidates. When the specific membrane milieu of extrasynaptic NMDA receptors becomes better defined, it might prove possible to block these preferentially, independent of the receptor subtypes involved. Selective decoupling of extrasynaptic NMDA receptors from intracellular postsynaptic density proteins, such as PSD-95, may be a way to reduce downstream toxic events with relative sparing of immediate synaptic signaling [9,10].

References

[1] Watkins JC. L-Glutamate as a central neurotransmitter: looking back. Biochem Soc Trans 2000;28:297–309.
[2] Benveniste H, Drejer J, Schousboe A, Diemer NH. Elevation of the extracellular concentrations of glutamate and aspartate in rat hippocampus during transient cerebral ischemia monitored by intracerebral microdialysis. J Neurochem 1984;43:1369–74.
[3] Rothman SM, Olney JW. Glutamate and the pathophysiology of hypoxic–ischemic brain damage. Ann Neurol 1986;19:105–11.
[4] Mcdonald JW, Althomsons SP, Hyrc KL, Choi DW, Goldberg MP. Oligodendrocytes from forebrain are highly vulnerable to AMPA/kainate receptor-mediated excitotoxicity. Nat Med 1998;4:291–7.
[5] Bozic M, Valdivielso JM. The potential of targeting NMDA receptors outside the CNS. Expert Opin Ther Targets 2015;19:399–413.

[6] Hollmann M, Heinemann S. Cloned glutamate receptors. Ann Rev Neurosci 1994;17:31–108.

[7] Choi DW. Glutamate neurotoxicity and diseases of the nervous system. Neuron 1988;1:623–34.

[8] Hardingham GE, Bading H. Synaptic versus extrasynaptic NMDA receptor signalling: implications for neurodegenerative disorders. Nat Rev Neurosci 2010;11:682–96.

[9] Sun Y, Zhang L, Chen Y, Zhan L, Gao Z. Therapeutic targets for cerebral ischemia based on the signaling pathways of the GluN2B C terminus. Stroke 2015;46:2347–53.

[10] Szydlowska K, Tymianski M. Calcium, ischemia and excitotoxicity. Cell Calcium 2010;47:122–9.

[11] Tu W, Xu X, Peng L, Zhong X, Zhang W, Soundarapandian MM, et al. DAPK1 interaction with NMDA receptor NR2B subunits mediates brain damage in stroke. Cell 2010;140:222–34.

[12] Choi DW. Calcium: still center-stage in hypoxic-ischemic neuronal death. Trends Neurosci 1995;18:58–60.

[13] Higuchi M, Single F, Köhler M, Sommer B, Sprengel R, Seeburg P. RNA editing of AMPA receptor subunit GluR-B: a base-paired intron-exon structure determines position and efficiency. Cell 1993;75:1361–70.

[14] Pellegrini-Giampietro DE, Zukin RS, Bennett MV, Cho S, Pulsinelli WA. Switch in glutamate receptor subunit gene expression in CA1 subfield of hippocampus following global ischemia in rats. Proc Natl Acad Sci USA 1992;89:10499–503.

[15] Nicoletti F, Bockaert J, Collingridge GL, Conn PJ, Ferraguti F, Schoepp DD, et al. Metabotropic glutamate receptors: from the workbench to the bedside. Neuropharmacology 2011;60:1017–41.

[16] Yu SP, Sensi SL, Canzoniero LMT, Buisson A, Choi DW. Membrane-delimited modulation of NMDA currents by metabotropic glutamate receptor subtypes 1/5 in cultured mouse cortical neurons. J Physiol 1997;499:721–32.

[17] Dawson VL, Dawson TM, London ED, Bredt DS, Snyder SH. Nitric oxide mediates glutamate neurotoxicity in primary cortical cultures. Proc Natl Acad Sci USA 1991;88:6368–71.

[18] Mandir A, Poitras M, Berliner A, Herring WJ, Guastella DB, Feldman A, et al. NMDA but not non-NMDA excitotoxicity is mediated by poly(ADP-ribose) polymerase. J Neurosci 2000;20:8005–11.

[19] Alano CC, Garnier P, Ying W, Higashi Y, Kaupinnen TM, Swanson RA. NAD$^+$ depletion is necessary and sufficient for poly(ADP-ribose) polymerase-1-mediated neuronal death. J Neurosci 2010;30:2967–78.

[20] Yu SP, Canzoniero LMT, Choi DW. Ion homeostasis and apoptosis. Curr Opin Cell Biol 2001;13:405–11.

[21] Choi DW, Koh JY. Zinc and brain injury. Annu Rev Neurosci 1998;21:347–75.

[22] McCord MC, Aizenman E. The role of intracellular zinc release in aging, oxidative stress, and Alzheimer's disease. Front Aging Neurosci 2014;6:1–16.

[23] Medvedeva YV, Weiss JH. Intramitochondrial Zn^{2+} accumulation via the Ca^{2+} uniporter contributes to acute ischemic neurodegeneration. Neurobiol Dis 2014;68:137–44.

[24] Choi DW, Rothman SM. The role of glutamate neurotoxicity in hypoxic-ischemic neuronal death. Annu Rev Neurosci 1990;13:171–82.

[25] Lai TW, Zhang S, Wang YT. Excitotoxicity and stroke: identifying novel targets for neuroprotection. Prog Neurobiol 2014;115:157–88.

[26] Ginsberg MD. Neuroprotection for ischemic stroke: past, present and future. Neuropharmacology 2008;55:363–89.

[27] Du C, Hu R, Csernansky CA, Hsu CY, Choi DW. Very delayed infarction after mild focal cerebral ischemia: a role for apoptosis? J Cereb Blood Flow Metab 1996;16:195–201.

[28] Lee JM, Zipfel GJ, Choi DW. The changing landscape of ischaemic brain injury mechanisms. Nature 1999;399:A7–14.

[29] Canzoniero LMT, Babcock DJ, Gottron FJ, Grabb MC, Manzerra P, Snider BJ, et al. Raising intracellular calcium attenuates neuronal apoptosis triggered by staurosporine or oxygen-glucose deprivation in the presence of glutamate receptor blockade. Neurobiol Dis 2004;15:520–8.

[30] Dhawan J, Benveniste H, Luo Z, Nawrocky M, Smith SD, Biegon A. A new look at glutamate and ischemia: NMDA agonist improves long-term functional outcome in a rat model of stroke. Future Neurol 2011;6:823–34.

[31] Manzerra P, Behrens MM, Canzoniero LMT, Wang XQ, Heidinger V, Ichinose T, et al. Zinc induces a Src family kinase-mediated up-regulation of NMDA receptor activity and excitotoxicity. Proc Natl Acad Sci USA 2001;98:11055–61.

CHAPTER

49

Oxidative and Nitrosative Stress

J.M. Modak[1], L.D. McCullough[2]

[1]Beth Israel Deaconess Medical Center, Boston, MA, United States; [2]The University of Texas Health Science Center at Houston, Houston, TX, United States

Oxidative stress plays a critical role in cerebral ischemic injury and occurs when there is an overproduction of free radicals and reactive oxygen species (ROS) beyond the ability of a biologic system to neutralize their adverse effects. ROS include oxygen ions, free radicals, and peroxides, and are products of cellular metabolism. Iron and its metabolites are crucial in the formation as well as destruction of ROS [1]. These molecules are involved in normal physiologic processes, such as synaptic transmission, cell signaling, induction of mitogenic responses, and immune defense. Under physiologic conditions the cerebral vascular tone is controlled in part by endothelium-derived nitric oxide (NO) and other ROS. However, under periods of ischemic stress leading to mitochondrial uncoupling, ROS are generated at higher rates than the system's ability to clear them, leading to oxidative damage.

ROS are also produced as part of the natural immune response. A prime example is the inflammatory burst produced by neutrophils that protects the host against pathogens. However, when inflammation is unchecked, ROS are produced in pathological levels leading to damage not just to the invading organism but also to the host. Although stroke is a sterile injury, cellular damage leads to immune activation via the activation of danger associated molecular patterns both in the periphery and within the brain itself. The induction of poststroke inflammation induces expression of adhesion molecules and proinflammatory cytokines post stroke [2]. Furthermore, leukocyte adhesion and aggregation lead to the release of additional free radicals, further propagating secondary injury. Leukocyte adhesion and aggregation also lead to vascular stasis and occlusion, which further limits tissue perfusion and destabilizes the blood–brain barrier.

Inflammation, reperfusion injury, excitotoxicity, calcium imbalance, and respiratory inhibition all contribute to the generation of free radicals including superoxide ions, hydroxyl radicals, and NO-like reactive nitrogen species (RNS), such as the potent oxidant peroxynitrite (ONOO⁻). These molecules react with molecular targets including proteins, lipids, and/or nucleic acids contributing to cellular dysfunction or death.

Neurons have high oxygen consumption, relatively low antioxidants (i.e. catalase and glutathione), and are exquisitely vulnerable to oxidative stress. In addition, neuronal cell membranes are rich in polyunsaturated fatty acids, and hence prone to ROS damage. Brain cells also have higher levels of iron, which acts as a prooxidant under pathologic conditions. ROS lead to lipid peroxidation, further producing aldehydes, alkanes, and dienals, which are toxic to neurons and white matter, and induce apoptosis. Studies have revealed variable susceptibility of specific brain regions to oxidative injury. Dopaminergic areas seem to be specifically sensitive to ROS/RNS-induced injury.

Cell death can be seen as a spectrum comprising apoptosis, necrosis, necroptosis, and parthanatos. Necroptosis (a form of regulated necrosis) involves proteins such as receptor-interacting serine/threonine-protein kinase 3 and Mixed Lineage Kinase domain like (protein), which are induced by death receptors, interferons, toll-like receptors, and other mediators. Enhanced mitochondrial ROS production was found to be the terminal event in experimental mouse cells undergoing necroptosis. Ferroptosis is a type of regulated necrosis that is characterized by an iron-dependent production of ROS [2–4]. Cell death involving poly-ADP ribose (PAR) polymerase formation is termed as parthanatos.

OXIDATIVE STRESS IN ISCHEMIC STROKE

Arterial blockade by a thrombus or an embolus results in impaired cerebral perfusion. A complex biochemical cascade is triggered after an ischemic injury to the brain. Oxidative stress plays a major role in the series

Primer on Cerebrovascular Diseases, Second Edition
http://dx.doi.org/10.1016/B978-0-12-803058-5.00049-7

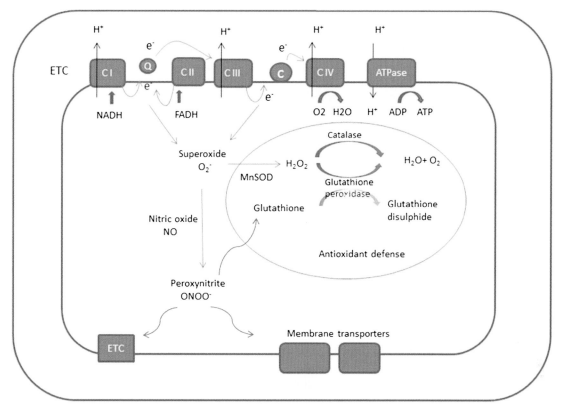

FIGURE 49.1 **Oxidative stress and mitochondrial dysfunction.** Electrons that leak from the electron transport chain (ETC) react with molecular oxygen to form the superoxide radical O_2^-. This reacts with NO to form peroxynitrite $ONOO^-$, which further disrupts the ETC and membrane transport proteins, and depletes antioxidants. *SOD*, superoxide dismutase.

of events. Depletion of ATP inhibits the Na^+/K^+-ATPase pump, subsequently leading to electrolyte imbalance, which further results in anoxic depolarization of the membrane and promotes membrane instability. This usually takes place at the core of the ischemic stroke, and further expansion of the core depends on this biochemical cascade. The deprivation of oxygen and glucose and increased intracellular Ca^{2+}, Na^+, and ADP leads to increased mitochondrial production of ROS. Normal cell homeostasis depends on effective mitochondrial calcium buffering. Increased calcium levels induce opening of the mitochondrial permeability transition pore, which results in diffusion of molecules including ROS from the mitochondria to the cytoplasm, further exacerbating mitochondrial dysfunction. Along with leakage from the mitochondrial electron transport chain, free radicals are also generated via action of NADPH oxide synthases, xanthine oxidase, and cyclooxygenase (Fig. 49.1).

THE SUPEROXIDE RADICAL

The superoxide radical (O_2^-) is the primary ROS. O_2^- has both direct and indirect effects on the vascular smooth muscle. It can inactivate NO, causing impaired vasodilatation. Xathine oxidase can also generate O_2^- in the vessel wall.

O_2^- affects endothelium-dependent relaxation by opening calcium-dependent potassium channels, and lowers the threshold of platelet aggregation in the presence of thrombin, collagen, and ADP. This is a likely effect of impaired availability of NO, which acts as a potent inhibitor of platelet aggregation. Induction and promotion of platelet aggregability has been uniquely linked to O_2^-, and these effects are not observed with OH^- or H_2O_2 [4]. Apart from the effects on the vessel walls, superoxide has been shown to inactivate mitochondrial enzymes including aconitase and complex I and II of the electron transport chain.

HYDROGEN PEROXIDE: A SUPEROXIDE DISMUTANT

Accumulation of lactic acid due to anaerobic glycolysis, is a result of energy depletion, and leads to cellular acidosis. An increase in the H^+ ion concentration further enhances the rate of conversion of superoxide anion (O_2^-) to H_2O_2 or a hydroperoxy radical (HO_2). Superoxide dismutase [SOD; the enzyme that alternately catalyzes the dismutation of O_2^- into either molecular oxygen (O_2) or hydrogen peroxide (H_2O_2)], is also important in limiting the deleterious effects of ROS in ischemic tissue. Of the three forms of SODs, copper–zinc SOD and manganese

SOD are abundant in neural tissues. H_2O_2 is chemically more stable than superoxide, and diffuses more easily across the cell membranes. It can act both as a vasodilator or a vasoconstrictor. High concentrations of H_2O_2 can produce vasoconstriction, followed by vasodilatation. H_2O_2 also promotes lipid peroxidation, which is important in the lipid-rich brain. Uncontrolled lipid peroxidation has shown to trigger nonapoptotic cell death pathways via iron-dependent enzymatic degradation.

NITRIC OXIDE

NO is a major molecule involved in oxidative stress. It can exert protective as well as deleterious effects. This depends on the source of NO, its redox state as well as the local tissue concentration. A combined effect of glutamate efflux, Ca^{2+} influx, and N-methyl-D-aspartate (NMDA) receptor activation leads to an increase in the activity of nitric oxide synthase (NOS). This further acts on cellular oxygen (O_2) and L-arginine leading to generation of NO. Synthesis of NO can take place in glial cells, endothelium, macrophages, and leukocytes, in addition to neurons. Experimental evidence from animal models demonstrates significant increases in NO, from 10 nM to 2.2 μM, in the cerebral cortex after ischemic stroke.

NO release from endothelial NOS (eNOS, type III) results in vasodilation and inhibits microvascular adhesion and hence is protective. NO derived from eNOS diffuses into the smooth muscle, and activates guanyl cyclase, which leads to an increase in intracellular cGMP and subsequent vascular relaxation. Neuronal NOS (nNOS, type I), which is released by neurons subjected to ischemia, and inducible NOS (iNOS, type II) released by the glial cells contribute to brain injury. Experimental studies have also shown that endothelial dysfunction is closely related to a reduced NO bioavailability. Knockout mice lacking nNOS and iNOS have smaller infarcts after experimental stroke; in contrast, mice deficient in eNOS have increased infarct. Ischemia leads to an initial rise in the constitutive NOS activity (the eNOS and nNOS isoforms) due to increased intracellular Ca^{2+}. The initial rise in NO may be neuroprotective owing to its vasodilator effect and a resultant improvement in blood flow. However, during the later phases of ischemia, a sustained increase from iNOS derived from microglia and other inflammatory cells is seen. The inhibition of iNOS minimizes leukocyte attachment to the endothelium, and may serve as a potential target against endothelial injury, but much more work is needed to understand the complex secondary signaling induced by NO.

The generation of NO is directly proportional to neurologic deterioration and brain injury. NO inhibits cytochrome c oxidase, and modulates mitochondrial oxygen consumption. NO competes with oxygen at complex IV of the electron transport chain and reversibly inhibits the rate of oxidative phosphorylation. Cytochrome c and apoptosis-inducing factors (AIFs) are released from the mitochondria, which further promote cell death. NO can further inhibit other mitochondrial enzymes including complexes I, II via formation of iron–NO complexes. It can also lead to protein sulfhydryl oxidation. NO has a short half-life and it is usually oxidized into the nitrite or the nitrate form (NOx).

Studies have indicated that endothelium-derived NO is reduced in individuals with stroke risk factors, such as smoking, hypertension, and hyperlipidemia. Oxidative modification of low-density lipoprotein present in the intima can be induced by ROS, and this is a key step in initiating atherosclerosis. Plasma NOx levels (nitrate and nitrite combined) were 22% lower in patients with ischemic stroke compared with healthy controls. Patients with stroke had significantly lower plasma amino acid levels (L-arginine, L-citrulline, serine, and ornithine), and higher plasma cyclic GMP levels than controls. Furthermore, lower NOx levels were associated with severe stroke and poor outcomes [5].

PEROXYNITRITE

Highly reactive NO and superoxide can combine to form peroxynitrite ($ONOO^-$), a toxic anion, and may further produce a nitrosonium cation (NO^+) or a nitroxyl anion (NO^-). Experimental studies have revealed that $ONOO^-$ in low concentrations can cause vasoconstriction, but in higher concentration causes vasodilatation. Peroxynitrite specifically targets electron-rich moieties like thiols, iron–sulfur substrates, thus affecting cell cycle and signal transduction proteins. Higher generation of $ONOO^-$ further leads to DNA damage and activation of poly-ADP-ribose polymerase (PARP)-1 ultimately leading to caspase-independent apoptosis.

Hypertension is a major stroke risk factor. Angiotensin II increases ROS formation in the brain, specifically peroxynitrite, by the reaction of NO (involving iNOS) with NOX_2-derived superoxide. Increased production of RNS that nitrosylate and alter protein structure and function is known as "nitrosative" stress. Under some conditions, NOS may produce large amounts of ROS, instead of NO; this uncoupling may be instrumental in brain injury during ischemic stroke.

IRON AND OXIDATIVE STRESS

Iron also plays a critical role in ROS metabolism. Various enzymes including those of the mitochondrial electron transport chain, xanthine oxidase, NADPH oxidase, and cytochrome P450 enzymes depend on iron for the production of superoxide and other free radicals.

The reaction of peroxides with Fe^{2+} to yield soluble hydroxyl (HO·) or lipid alkoxy (RO·) radicals is called the Fenton reaction. Redox active iron pools are found in the cytoplasm as well as in organelles like mitochondria and lysosomes. Thus both iron-dependent enzymes and labile iron contributes to ROS-mediated cell damage and death. Studies have hence proposed iron chelators such as deferiprone or deferoxamine to treat stroke, and more specifically intracerebral hemorrhage (where there is considerable iron derived from the increase in local heme) to reduce ROS-induced injury [1]. Clinical trials have been initiated in intracerebral hemorrhage, but large efficacy studies are needed.

REACTIVE OXYGEN SPECIES AND CELL SIGNALING PATHWAYS

ROS can indirectly trigger the release of other proteins like cytochrome c. Cytochrome c, after its release from the mitochondria, binds to deoxy-ATP and apoptotic protease activating factor (Apaf-1). This further activates caspase 9 and downstream caspases (cysteine aspartic proteases), leading to an increase in the oxidative DNA lesions. In addition, ROS can activate caspase-activated DNAase (CAD), which can further lead to apoptosis.

The caspase cascade can be triggered either by extrinsic or intrinsic mechanisms. In the extrinsic pathway, apoptosis is initiated when stimuli such as ischemia activate members of the tumor necrosis factor-alpha receptor superfamily death domain receptors. Ischemia can cause translocation of the Bcl-2 (B-cell lymphoma 2) associated death promoter (Bad) from the cytosol to the mitochondria. Furthermore, ROS also affects p53 signaling pathways. The p53 upregulated modulator of apoptosis interacts with bcl2 or Bax, promotes cytochrome c, and plays a role in neuronal death [5]. Cytochrome c translocation from mitochondria to the cytosol leads to apoptosome formation, caspase cleavage, and complement activation. Proapoptotic proteins like Bax, Bid, and Bim may disrupt the mitochondrial membrane potential and promote caspase release. Caspase 3 cleavage and release of CAD initiates DNA fragmentation. This constitutes the intrinsic or the mitochondrial pathway of caspase activation wherein release of cytochrome c into the cytoplasm promotes apoptosis as described earlier.

Caspases have thus been implicated in causing DNA damage, pathologic programmed cell death. Caspase cascades usually occur in the ischemic stroke penumbra; these processes are energy dependent, and hence occur in areas with some residual blood flow. Studies have shown that TAT-BH4, an inhibitor of apoptosis, protects cells from ROS-induced injury.

Other signaling pathways are induced by ischemia including caspase-independent PARP-1 activation. DNA oxidation induced by ischemia activates repair enzymes including PARP. Overactivation of PARP-1 leads to depletion of NAD+ levels, impairing glycolysis and mitochondrial respiration. This results in ATP depletion, impaired cellular homeostasis, and neuronal death, as well as induction of mitochondrial to nuclear translocation of AIF and the production of PAR polymers to toxic levels. In cases of ischemic stroke, PARP activation has been noted in two phases, initially in the neurons, and a few days later in infiltrating inflammatory cells [5].

GLUTAMATE EXCITOTOXICITY

Extracellular glutamate levels increase after ischemia. Raised intracellular calcium levels, apart from free radicals, may promote glutamate release. Furthermore, inhibition of glutamate uptake by free radicals contributes to glutamate-related excitotoxicity leading to extended neuronal depolarization and energy failure. Stimulation of α-amino-3-hydroxy-5-methyl-4-isoxazolepropionic acid (AMPA) receptors generates O_2^-, which reacts with NO to produce $ONOO^-$. AMPA inhibition is being studied as a potential therapeutic strategy in patients with stroke. Other extrasynaptic glutamate receptors include NMDA and kainate receptors, which when pathologically activated allow for the influx of Ca^{2+} and promote cell death.

Cell death can also occur via direct activation of channels, such as the transient receptor potential (TRP) channel or acid-sensing ion channels (ASIC). The melastatin TRP (TRPM) channels TRPM 2 and TRPM 7 have been the most closely implicated in cell death associated with anoxia-induced ROS. TRPM2 channels are activated by hydrogen peroxide and gated by ADP ribose. ASIC is also implicated in inducing oxidative cell death [6].

SEX DIFFERENCES IN OXIDATIVE STRESS AND STROKE

Current evidence suggests sexual differences in the mitochondrial mechanism and cell signaling pathways implicated in oxidative stress during ischemia/reperfusion. Within the mitochondria, males predominantly use proteins as a fuel source, and females utilize lipids. Furthermore, studies have revealed that isolated male mitochondria have a greater calcium uptake capacity than females. This has been related to activation of TRPM2 channels following injury. Thus females have more favorable mitochondrial responses in the setting of ischemia, and may be relatively resilient to oxidative or nitrosative injury.

Following ischemic injury, a higher degree of nNOS induction is observed in males. Furthermore, in vitro studies have revealed that male hippocampal tissue subjected to NMDA produces greater amount of nitratres

TABLE 49.1   Male and Female Susceptibility to Ischemia/Reperfusion Injury

Cell signaling pathway	Males	Females
Oxidative and nitrosative stress	• Higher production of ROS/RNS	• Higher endogenous antioxidants
Excitotoxicity	• Susceptible to glutamate excitotoxicity	• Relatively low intracellular and mitochondrial calcium uptake
Autophagy	• Utilize proteins as biofuels under stress	• Utilize lipids as biofuels under stress
Cell death pathways	• Caspase-independent pathway	• Caspase-dependent pathway

RNS, reactive nitrogen species; *ROS*, reactive oxygen species.
Adapted from Demarest TG, McCarthy MM. Sex differences in mitochondrial (dys) function: implications for neuroprotection. J Bioenerg Biomembr 2015 47:173–88.

and has a higher susceptibility to death, in comparison with females. Experimental studies in animals also suggest that caspase-dependent cell death predominates in females, whereas caspase-independent pathways predominate in males.

Male animals with targeted deletions of nNOS, PARP, or AIF have smaller infarcts after experimental middle cerebral artery occlusion. However, these manipulations do not have a significant effect in females. Inhibitors of nNOS, such as 7-NI (nitroimidazole) or PARP inhibitors like PJ-34 have been found to have similar effects on the infarct size in males [7,8]. (Fig. 49.3, Table 49.1). This may have major ramifications on therapeutic drug development.

OXIDATIVE STRESS DURING VARIOUS STAGES OF ISCHEMIA

High concentrations of ROS are known to cause lipid peroxidation and protein denaturation, ultimately leading to cytoskeletal damage and death during acute ischemic injury (Fig. 49.2).

Interestingly, apart from its role in overt ischemia, ROS also contribute to ischemic preconditioning, in which a sublethal insult protects against a more a subsequent more severe injury. Animal studies have revealed that NO and superoxide are involved in lipopolysaccharide (LPS)-induced tolerance to brain ischemia. The vasoprotective effects of LPS involve low levels of peroxynitrite. However, more severe injury leads to high levels of peroxynitrite, nitrosylation, and cell death.

Oxidative stress during reperfusion is also a major cause of brain injury. Xanthine dehydrogenase oxidizes to form xanthine oxidase (XO) and generates superoxide ions despite low oxygen tension. Hypoxia further

enhances the formation of superoxide radicals and hydrogen radicals by activating phospholipase C, and increasing the levels of xanthine oxidase in the blood [9]. Studies have shown than XO inhibitors like oxypurinol block the production of ROS after ischemia and may have some potential to reduce injury. Trials examining the administration of the antioxidant uric acid in patients with stroke showed reduced infarct growth in women and in patients with hyperglycemia, perhaps reflecting the increased oxidative burden or reduced antioxidant capacity in these groups [10,11].

Another enzyme complex, NADPH oxidase, can amplify ischemia reperfusion injury. Among its five isoforms, NOX2 and NOX4 are located in the hippocampus and the cerebral cortex. Mice deficient in the NOX2 subunit and mice treated with the NOX inhibitor apocynin had significantly less infarct damage compared with the wild-type mice. In another murine study, NOX2 and O_2−were elevated for up to 7 days, and NOX4 from 7 to 28 days post stroke in the ipsilateral cortex and striatum [12]. Animal studies in rats subjected to transient ischemia/reperfusion have demonstrated a direct correlation between oxidative/nitrosative biomarkers such as NO and malondialdehyde with neurologic deficits [12].

Therefore in ischemic stroke mitochondrial production of ROS is seen in the initial stages, which later diminishes after membrane depolarization. Subsequent ROS production occurs secondary to XO activation and then reperfusion leads to further ROS production via NADPH signaling.

ANTIOXIDANTS: DEFENSE AGAINST REACTIVE OXYGEN SPECIES

The cellular redox homeostasis is balanced by an endogenous antioxidant system. Some major cellular antioxidants include SOD, catalase, glutathione, and glutathione peroxidase. Glutathione acts as a major antioxidant, and it functions in neutralization of ROS and other free radicals. It is also instrumental in the regulation of the NO cycle. Glutathione exists in reduced and oxidized forms; the relative proportion determines the level of oxidative stress. Antioxidant enzymes are essential for normal mitochondrial homeostatic functions. The reduced form of glutathione helps maintain the integrity of mitochondrial complex I. SODs block the release of cytochrome c, caspase induction, and modulate the DNA repair enzymes, thus preventing apoptosis. Catalase helps in breakdown of H_2O_2 into gaseous oxygen and water.

Paroxynase 2 (PON2) is an antioxidant enzyme that is found in relatively high levels in astrocytes, compared with neurons, and the highest levels are seen in dopaminergic regions (which are known to be susceptible to oxidative damage). Another key molecule regulating cellular defense mechanisms is the nuclear factor erythroid 2 related factor

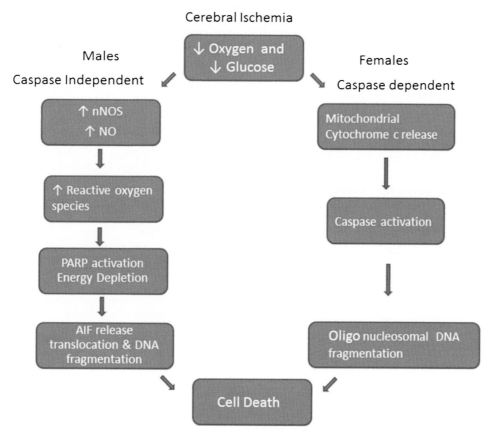

FIGURE 49.2 **Cell death after cerebral ischemia: Sexually dimorphic pathways.** Males are sensitive to caspase-independent poly-ADP-ribose polymerase (PARP)/nitric oxide (NO)-mediated cell death, whereas females appear to be exquisitely sensitive to caspase-dependent cell death signaling. *AIF*, apoptosis-inducing factor; *nNOS*, neuronal nitric oxide synthase.

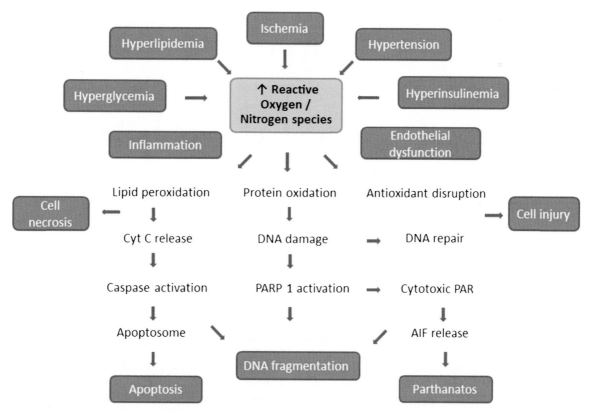

FIGURE 49.3 **Reactive oxygen species (ROS) and cell death.** Schematic of stimuli and consequence of ROS-mediated cell death. *Cyt C*, cytochrome c; *AIF*, apoptosis-inducing factor; *PAR*, poly-ADP-ribose; *PARP*, poly-ADP-ribose polymerase.

2 (Nrf2), a leucine-zipper transcription factor that binds to the antioxidant response element and modulates genes involved in cellular antioxidant and inflammatory defense. Many of these agents are being studied as potential therapeutic targets against oxidative injury.

In summary, oxidative stress and the generation of ROS is a major contributor to ischemic injury. Oxidant-antioxidant-targeted therapies may serve as future avenues for stroke therapy. Additional benefits of antioxidant therapy may be seen in specific groups such as diabetic patients and women. Clinical trials need to analyze these groups carefully to ensure that any potential benefit of antioxidant therapies is not lost.

References

[1] Dixon SJ, Stockwell BR. The role of iron and reactive oxygen species in cell death. Nat Chem Biol 2014;10:9–17.

[2] Tobin MK, Bonds JA, Minshall RD, et al. Neurogenesis and inflammation after ischemic stroke: what is known and where we go from here. J Cereb Blood Flow Metab 2014;34:1573–84.

[3] Pasparakis M, Vandenabeele P. Necroptosis and its role in inflammation. Nature 2015;517:311–20.

[4] Allen CL, Bayraktutan U. Oxidative stress and its role in the pathogenesis of ischaemic stroke. Int J Stroke 2009;4:461–70.

[5] Rashid PA, Whitehurst A, Lawson N, et al. Plasma nitric oxide (nitrate/nitrite) levels in acute stroke and their relationship with severity and outcome. J Stroke Cerebrovasc Dis 2003;12:82–7.

[6] Baron A, Lingueglia E. Pharmacology of acid-sensing ion channels – physiological and therapeutical perspectives. Neuropharmacology 2015;94:19–35.

[7] Manwani B, McCullough LD. Sexual dimorphism in ischemic stroke: lessons from the laboratory. Womens Health (Lond) 2011;7:319–39.

[8] Demarest TG, McCarthy MM. Sex differences in mitochondrial (dys)function: implications for neuroprotection. J Bioenerg Biomembr 2015;47:173–88.

[9] Awooda HA, Lutfi MF, Sharara GG, et al. Oxidative/nitrosative stress in rats subjected to focal cerebral ischemia/reperfusion. Int J Health Sci (Qassim) 2015;9:17–24.

[10] Amaro S, Llull L, Renú A, Laredo C, Perez B, Vila E, et al. Uric acid improves glucose-driven oxidative stress in human ischemic stroke. Ann Neurol 2015;77:775–832.

[11] Llull L, Laredo C, Renú A, Pérez B, Vila E, Obach V, et al. Uric acid therapy improves clinical outcome in women with acute ischemic stroke. Stroke 2015;46:2162–7.

[12] Taylor CJ, Weston RM, Dusting GJ, et al. NADPH oxidase and angiogenesis following endothelin-1 induced stroke in rats: role for nox2 in brain repair. Brain Sci 2013;3:294–317.

CHAPTER

50

Protein Kinases in Cerebral Ischemia

A.P. Raval, M.A. Perez-Pinzon, K.R. Dave

University of Miami, Miami, Florida, United States

INTRODUCTION

Protein kinases are key regulators of cell processes. Protein kinases are defined as enzymes that transfer a phosphate group onto an acceptor amino acid in a substrate protein. This process is called phosphorylation and can be reversed by phosphatases, enzymes that remove phosphoryl moieties from target proteins. Protein phosphorylation is an essential mechanism by which intracellular and extracellular signals are transmitted throughout the cell and into the nucleus. It has also been shown that protein phosphorylation regulates various aspects of cellular function, such as division, metabolism, movement, and apoptosis. Because protein kinases play crucial roles in cellular signaling and are highly expressed in the CNS, any disruption of their action can alter cellular function(s) of the neurovascular unit and cause cell death and ultimately, cerebrovascular disease.

The current chapter describes the role of three important protein kinases, i.e., protein kinase A (PKA), protein kinase B (PKB), and protein kinase C (PKC), on the progression of post-ischemic brain injury and their associated cell signaling events. Although there are substantial differences among kinases, their responses to ischemic stress show many similarities.

Protein Kinase A

PKA, also known as cyclic AMP (cAMP)-dependent protein kinase, is a family of enzymes whose activity depends on cellular levels of cAMP. In an unstimulated cell, PKA exists as a tetramer consisting of two regulatory subunits, each containing two cAMP binding sites and two catalytic subunits. In the absence of cAMP, these subunits bind to each other and the activity of the complex is low. Following cell stimulation by a signaling molecule, an increase in cAMP levels results in binding of cAMP to the regulatory subunit and promotes the dissociation of the tetramer. Dissociation of the tetramer sets the catalytic domains free, which then translocate to the nucleus and activate cyclic-AMP response binding protein (CREB). Once activated, it binds to a cyclic-AMP response elements (CRE) region of DNA, and is then bound by CREB-binding protein (CBP). The subsequent binding of these proteins ultimately allows this process to switch certain genes on or off [12].

CREB is a transcription factor that is known to play a role in neuronal responses to ischemia [12]. Studies from various laboratories have demonstrated that after an episode of global ischemia the expression of phosphorylated CREB and CBP is induced in CA1 hippocampal neurons. A variety of protein kinases use CBP as a substrate for posttranslational modification, and it has been implied that postischemic calcium influx leads to the activation and recruitment of CBP. In fact, several stimuli associated with cerebral ischemia could account for the phosphorylation-mediated activation of CREB [1]. These include Ca^{2+} influx through L-type voltage-gated channels and NMDA receptor-gated Ca^{2+} channels or the release of neurotrophic factors, such as nerve growth factor and brain-derived neurotrophic factor (BDNF). In general, target genes of CREB include encodings for metabolic enzymes, transcription factors, neurotransmitters, cell cycle–related proteins, cell survival–related molecules, growth factors, immune regulatory proteins, and structural proteins. Among these, genes targeted by CREB activation include those that induced the production of molecules that have protective functions (such as PGC1α, Bcl-2, and BDNF) in neurons as well as inflammatory functions (such as interleukin-6 and cyclooxygenase 2). Furthermore, studies have suggested that CREB may also be involved in the acquisition of ischemic tolerance, a phenomenon that occurs after sublethal ischemic stress, and in survival of newborn neurons in the dentate gyrus of the hippocampus after ischemia. Overall, the functions of CREB and CRE-mediated systems are crucial for the recovery of neural function, and are important in synaptic plasticity, neurogenesis, and axon growth.

Protein Kinase B

PKB is also known as Akt. Akt is a serine/threonine-specific protein kinase that plays a critical role in controlling the balance between survival and death pathways in cells. The Akt kinase family comprises three highly homologous isoforms: Akt1 (PKBα), Akt2 (PKBβ), and Akt3 (PKBγ) [4,14]. Akt isoform knockout mice studies demonstrated that the functions of the different Akt kinases are not completely overlapping and the diverse activities of Akt are due to isoform-specific signaling [4]. It has also been demonstrated that Akt1 and Akt3 are considered to be crucial to brain function and are expressed abundantly in the brain, whereas Akt2 is expressed predominantly in the heart and brown adipose tissue. Genetic knockouts of various Akt isoforms have shown that both Akt1 and Akt3 deficiency lead to smaller brain sizes, and the loss of Akt3 is associated with only mild neurological deficits. One study has also demonstrated that ischemia causes differential expression of Akt isoforms. Specifically, it has been shown that Akt1 and Akt3 degrade rapidly along with delayed reductions in Akt2. This study also suggested that Akt1 and Akt3 promote neuronal survival after focal ischemia through the mediation of multiple signaling pathways, such as enhancing the downstream factor mammalian target of rapamycin (mTOR) [14]. Importantly, this study concluded that compared with Akt1, Akt3 could be a more important Akt isoform for protection against poststroke injury [14].

In general, Akt is activated by insulin, various growth factors, and survival factors, and Akt functions in a wortmannin-sensitive pathway involving phosphatidylinositol 3-kinase (PI3K) [5]. In response to the activation of PI3Ks, Akt binds to phosphorylated lipids in the plasma membrane with an amino-terminal pleckstrin homology domain. Once recruited to the plasma membrane, Akt is phosphorylated at two sites: within the T-loop of the catalytic domain (Thr308, Akt1 residue) and within the carboxyl terminal hydrophobic domain (Ser473, Akt1 residue) by phosphoinositide-dependent kinase 1 (PDK1) and mTOR complex 2 (mTORC2), respectively. Interestingly, the targeting of Akt to the plasma membrane, independent of external stimuli (or PI3K activity), results in Akt activation, which strongly suggests that activation is limited predominantly by recruitment to the plasma membrane rather than the direct modulation of PDK1 and/or mTORC2 activity. Akt is transiently localized to the plasma membrane during activation. Following activation it phosphorylates substrates distributed throughout the cell to regulate multiple cellular functions (Fig. 50.1).

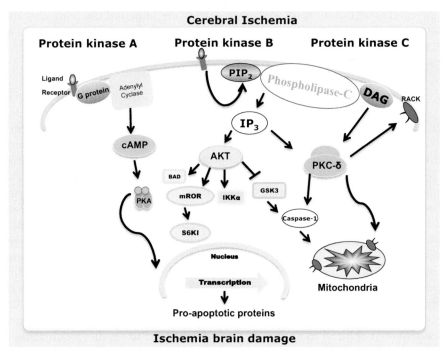

FIGURE 50.1   The figure describes the role of three important protein kinases [i.e. protein kinase A (PKA), protein kinase B, and protein kinase C (PKC)] on the progression of postischemic brain injury and their associated cell signaling events. *BAD*, Bcl-2-associated death promoter; *cAMP*, cyclic adenosine monophosphate; *DAG*, diacylglycerol; *GSK-3*, glycogen synthase kinase three; *IKK alpha*, I kappaB kinase subunit alpha; *IP3*, inositol triphosphate; *mTOR*, mechanistic target of rapamycin; *PIP2*, phosphatidylinositol biphosphate; *PKA*, protein kinase A; *PKC*, protein kinase C; *PLC*, phospholipase C; *RACK*, receptor for activated C-kinase; *S6K1*, protein S6 kinase one.

It has been shown that cerebral ischemia alters levels of phosphorylated Akt in the penumbra. Inhibition of Akt phosphorylation by an intracerebroventricular injection of a PI3K inhibitor (LY294002) exacerbated postischemic neuronal damage [7]. Phosphorylated Akt promotes cell survival and prevents apoptosis by inactivating several targets, including Bcl-2-associated death promoter (BAD), glycogen synthase kinase-3 (GSK3), fork head transcription factors, and caspase-9 [7]. BAD is one of the downstream factors of the Akt pathway, and a proapoptotic member of the Bcl-2 family that can displace Bax from binding to Bcl-2 and Bcl-xL, resulting in cell death [7]. Although Akt has been extensively studied and is considered to be neuroprotective in ischemic stroke, there is some controversy in the field. Stroke in Akt1 knockout and Akt transgenic animals resulted in lower cerebral ischemic damage compared with the wild-type controls, potentially due to developmental and global effects of traditional transgenic or gene knockout mouse models [7,14].

One study demonstrated that dietary omega-3 polyunsaturated fatty acid (n-3 PUFA) supplementation reduces ischemic brain damage by promoting Akt signaling in neonatal rats. This study demonstrated that n-3 PUFA enhances the biosynthesis of phosphatidylserine in neurons. Phosphatidylserine on the neuronal membrane facilitates membrane translocation of Akt in a phosphatidylinositol trisphosphate–dependent

manner and enhances the subsequent phosphorylation and activation of Akt. In addition, this study also suggested that phosphatidylserine may also directly interact with Akt and enhance Akt phosphorylation/activation. Activated Akt may then inactivate the proapoptotic protein GSK-3β and, perhaps, other substrates, resulting in suppression of neuronal cell death after ischemia [15]. The dietary supplement-mediated activation of Akt signaling can have broad implication in the prevention of ischemic brain damage.

Protein Kinase C

PKC is a group of multifunctional protein kinases that phosphorylate serine and threonine residues in many target proteins. PKC represents a structurally homologous group of proteins of which, at present, 12 isozymes have been identified and can be split into three broad categories with multiple cellular roles: conventional (α, βI, βII, and γ), novel (δ, ε, η, and θ), and atypical, (ζ, λ, μ, and ι) [2]. A hallmark of PKC activation is by translocation of the enzyme from the cytosolic (soluble) fraction to the cell particulate fraction, which includes the plasma and mitochondrial and nuclear membrane. Apart from translocation, binding of the aforementioned stimulators (hormones, growth factors, and neurotransmitters, etc.) to their receptors activates phospholipase C and generates diacylglycerol, a secondary messenger [2]. In turn, the diacylglycerol activates only novel PKCs, whereas conventional PKCs require calcium for their activation. In addition to secondary messenger–mediated PKC activation, posttranslational modifications of PKC were also shown to cause the activation of selected PKC isozymes in both normal and diseased states [2]. These posttranslational modifications include activation by proteolysis between the regulatory and catalytic domains, which was noted to occur for PKC-delta (PKCδ) in the hippocampal neurons owing to an ischemic episode. Apart from proteolysis, translocation also activates PKCδ. It is now known that PKCδ is specifically upregulated and rapidly activated in response to reperfusion-induced intracellular signaling; PKCδ mRNA is shown to be upregulated within 24h after both focal and global cerebral ischemia in respective ischemia-sensitive brain regions [2,10]. PKCδ is responsive to reactive oxygen species and other apoptotic mediators including caspase, which causes cell apoptosis in a stimulus- and tissue-specific manner. PKCδ regulates the expression and function of apoptosis-related proteins [2,10].

Another novel PKC isozyme, PKC epsilon (PKCε), was shown to play a pivotal role in protection of brain from ischemic injury following ischemic tolerance [9]. Different signal transduction pathways that are downstream of PKCε have been described, including the following: the Src family of protein tyrosine kinases, the mitogen-activated protein kinase (MAPK) p38, the MAPK/extracellular signal-regulated kinase (ERK) kinase MEK1/2, and the serine/threonine kinase Akt (suggesting cross talk between two groups of protein kinases). These and other yet-to-be-identified signaling pathways lead to posttranslational modifications of proteins as well as transcriptional activation and de novo protein synthesis leading to neuroprotection [3,8,13].

PKCε appears to participate in several steps in the process of ischemia tolerance following ischemic preconditioning (IPC). In an earlier study we observed that IPC activates PKCε early via the NMDA-Ca^{2+}-diacylglycerol pathway [9]. This early activation of PKCε, which in turn activates cell survival pathways, suggests its participation in the induction/initiation/early phase of IPC. As mentioned earlier, PKCs translocate to the particulate/membrane fraction following activation. We observed that PKCε is translocated to mitochondria following IPC [11] where it activates the mitochondrial K$^+$-ATP channel. It is plausible that activation of the mitochondrial K$^+$-ATP channel further magnifies PKC activation via the reactive oxygen species–phospholipase C pathway. Post-IPC increase in levels of PKCε in synaptosomes indicates that it preserves mitochondrial and synaptic functions during and following an otherwise lethal episode of ischemia [3]. PKCε also increases expression of BDNF, which via electrophysiological modifications participates in IPC-induced ischemia tolerance [6]. We also observed that anoxic depolarization was delayed in hippocampal slices harvested from PKCε activator-treated rats compared with control peptide-treated slices subjected to oxygen/glucose deprivation [6]. These results suggest that activated PKCε is able to maintain membrane potential for longer durations following preconditioning. Overall, these studies indicate that PKCε plays an important role in the induction of ischemic tolerance following IPC.

CONCLUSION

In summary, all three protein kinases (PKA, PKB, and PKC) appear to act as mediators in either induction of ischemia tolerance, or in cell-damaging pathways following cerebral ischemia. Considering the wide array of signals that can affect their activities and that their activities can in turn regulate several downstream pathways, these protein kinases appear to be potential therapeutic targets either to induce ischemia tolerance in patients at risk of cerebral ischemia, or to suppress cell death machinery poststroke. Preclinical studies evaluating their therapeutic potential are thus needed.

References

[1] Aronowski J, Grotta JC, Waxham MN. Ischemia-induced translocation of Ca^{2+}/calmodulin-dependent protein kinase II: potential role in neuronal damage. J Neurochem 1992;58(5):1743–53.

[2] Bright R, Mochly-Rosen D. The role of protein kinase C in cerebral ischemic and reperfusion injury. Stroke 2005;36(12):2781–90. http://dx.doi.org/10.1161/01.STR.0000189996.71237.f7.

[3] Dave KR, DeFazio RA, Raval AP, Torraco A, Saul I, Barrientos A, et al. Ischemic preconditioning targets the respiration of synaptic mitochondria via protein kinase C epsilon. J Neurosci 2008;28(16):4172–82. http://dx.doi.org/10.1523/JNEUROSCI.5471-07.2008.

[4] Gonzalez E, McGraw TE. The Akt kinases: isoform specificity in metabolism and cancer. Cell Cycle 2009;8(16):2502–8.

[5] Hanada M, Feng J, Hemmings BA. Structure, regulation and function of PKB/AKT–a major therapeutic target. Biochim Biophys Acta 2004;1697(1–2):3–16. http://dx.doi.org/10.1016/j.bbapap.2003.11.009.

[6] Neumann JT, Thompson JW, Raval AP, Cohan CH, Koronowski KB, Perez-Pinzon MA. Increased BDNF protein expression after ischemic or PKC epsilon preconditioning promotes electrophysiologic changes that lead to neuroprotection. J Cereb Blood Flow Metab 2015;35(1):121–30. http://dx.doi.org/10.1038/jcbfm.2014.185.

[7] Noshita N, Lewen A, Sugawara T, Chan PH. Evidence of phosphorylation of Akt and neuronal survival after transient focal cerebral ischemia in mice. J Cereb Blood Flow Metab 2001;21(12):1442–50. http://dx.doi.org/10.1097/00004647-200112000-00009.

[8] Perez-Pinzon MA, Stetler RA, Fiskum G. Novel mitochondrial targets for neuroprotection. J Cereb Blood Flow Metab 2012;32(7):1362–76. http://dx.doi.org/10.1038/jcbfm.2012.32.

[9] Raval AP, Dave KR, Mochly-Rosen D, Sick TJ, Perez-Pinzon MA. Epsilon PKC is required for the induction of tolerance by ischemic and NMDA-mediated preconditioning in the organotypic hippocampal slice. J Neurosci 2003;23(2):384–91.

[10] Raval AP, Dave KR, Prado R, Katz LM, Busto R, Sick TJ, et al. Protein kinase C delta cleavage initiates an aberrant signal transduction pathway after cardiac arrest and oxygen glucose deprivation. J Cereb Blood Flow Metab 2005;25(6):730–41. http://dx.doi.org/10.1038/sj.jcbfm.9600071.

[11] Raval AP, Dave KR, DeFazio RA, Perez-Pinzon MA. epsilonPKC phosphorylates the mitochondrial K(+) (ATP) channel during induction of ischemic preconditioning in the rat hippocampus. Brain Res 2007;1184:345–53. http://dx.doi.org/10.1016/j.brainres.2007.09.073.

[12] Tanaka K. Alteration of second messengers during acute cerebral ischemia - adenylate cyclase, cyclic AMP-dependent protein kinase, and cyclic AMP response element binding protein. Prog Neurobiol 2001;65(2):173–207.

[13] Thompson JW, Narayanan SV, Perez-Pinzon MA. Redox signaling pathways involved in neuronal ischemic preconditioning. Curr Neuropharmacol 2012;10(4):354–69. http://dx.doi.org/10.2174/157015912804143577.

[14] Xie R, Cheng M, Li M, Xiong X, Daadi M, Sapolsky RM, et al. Akt isoforms differentially protect against stroke-induced neuronal injury by regulating mTOR activities. J Cereb Blood Flow Metab 2013;33(12):1875–85. http://dx.doi.org/10.1038/jcbfm.2013.132.

[15] Zhang W, Liu J, Hu X, Li P, Leak RK, Gao Y, et al. n-3 polyunsaturated fatty acids reduce neonatal hypoxic/ischemic brain injury by promoting phosphatidylserine formation and Akt signaling. Stroke 2015;46(10):2943–50. http://dx.doi.org/10.1161/STROKEAHA.115.010815.

CHAPTER

51

Ischemia Regulated Transcription Factors: Hypoxia Inducible Factor 1 and Activating Transcription Factor 4

I. Alim[1,2], S.S. Karuppagounder[1,2], R.R. Ratan[1,2]

[1]Burke Medical Research Institute, White Plains, NY, United States; [2]Weill Medical College of Cornell University, New York, NY, United states

INTRODUCTION

Loss of blood flow during stroke creates significant cellular stress via a host of stresses including hypoxia, glucose deprivation, and oxidative stress. Preconditioning studies teach us that homeostatic programs to cell stress are rapidly engaged via activation of preexisting proteins and sustained via adaptive gene expression [1,2]. Whether activated prior or after a stroke, if these adaptive mechanisms fail to restore homeostasis at a single cell level, then cell death mechanisms are engaged. For postmitotic neurons, deciding to die is a critical, irreversible decision, so transcription of death proteins adds another layer of regulation. Reversing hypoxia, glucose deprivation, or oxidative stress individually via small molecules has not been effective clinically. Failure is likely because each stress creates parallel and overlapping pathways to trigger cell death. Accordingly, a successful therapeutic will target all three pathways by restoring homeostasis. Modulating transcriptional programs that affect a large cassette of genes at cellular, local, and systemic levels provides a strategy for targeting many parallel and interacting cell stress pathways. A number of studies have focused on identifying downstream transcriptional changes that occur before neurons under stress commit to a cell death pathway [1–4]. In this chapter, we discuss two known stroke-activated transcription factors, hypoxia inducible factor (HIF) and activating transcription factor 4 (ATF4) and potential therapeutic strategies targeting these transcription factors to prevent neuronal death.

Primer on Cerebrovascular Diseases, Second Edition
http://dx.doi.org/10.1016/B978-0-12-803058-5.00051-5

HYPOXIA-MEDIATED TRANSCRIPTION CHANGES DURING STROKE

Ischemic stroke leads to a decrease in essential cellular fuels, such as oxygen (hypoxia) and glucose (hypoglycemia), required for cell metabolism and survival. Because of the high-energy demands of synaptic activity, neurons critically depend on oxygen and have evolved mechanisms to detect and deal with hypoxia. These adaptive mechanisms involve changes in gene expression that would favor nonaerobic energy generation and increase vascularly derived oxygen delivery, thus counteracting the deleterious effects of hypoxia on energy production. To facilitate recovery of damaged neurons and prevent neuronal death, many have attempted to identify adaptive transcriptional programs and determine how these can be augmented for therapeutic advantage. The HIF transcriptional system has emerged as a key regulatory system that responds to hypoxia at both a local and systemic level [5] and aspects of hypoxia signaling have emerged as key targets for therapeutic intervention.

HYPOXIA REGULATES HYPOXIA-INDUCIBLE FACTOR SYSTEM

Reductions in oxygen levels below a critical threshold can trigger a series of elegantly evolved biochemical events that lead to the stabilization of the transcription factor HIF-1α and its isoforms (HIF-2α, HIF-3α). Hypoxia-stabilized HIF-1α translocates into the nucleus, forms a heterodimer transcription factor complex with constitutively synthesized HIF-1β, and recruits p300/CBP for transactivation of target genes [6,7]. This HIF transcriptional factor complex binds specifically to a conserved promoter sequence known as the hypoxia response element (HRE; 5′-RCGTG-3′) in the 3′ enhancer region of HIF target genes [5]. A number of genes enhanced during hypoxia by HIF-1α are known to have protective and reparative properties [e.g., erythropoietin (EPO), vascular endothelial growth factor, and glycolytic enzymes], whereas others trigger cell death or mitophagy [e.g., BCL2/adenovirus E1B 19kDa protein-interacting protein 3(BNIP3)] [5,8,9]. The structure of HIF transcriptional proteins is divided into four domains: an N-terminus with a basic helix-loop-helix structure for DNA binding, a central domain with a per-ARNT-sim domain for heterodimerization or ligand activation, an oxygen-dependent (ODD) domain that confers oxygen-dependent stability, and C and N terminal transactivation domains that regulate the basal transcriptional machinery.

Preconditioning with a sublethal dose of hypoxia in neonatal rats increases HIF-1α and HIF-2α protein expression and upregulates mRNA for HIF target genes [1,2]. Pretreatment of neonatal animals with known "hypoxia mimics," such as cobalt chloride and deferoxamine (DFO) provided neuroprotection during hypoxia [1]. These studies initially suggested that pharmacological induction of the hypoxic adaptive response via small molecules could be neuroprotective.

REGULATION OF ADAPTIVE TRANSCRIPTION BY HIF–PROLYL HYDROXYLASES

Increased stability of HIF-1α (and its isoforms) in response to hypoxia depends on the inhibition of post-translational hydroxylation of prolines in the HIF ODD domain. During normoxia, hydroxylation of HIF-1α is mediated by a group of enzymes known as HIF prolyl hydroxylases (HIF-PHDs). HIF-PHDs belong to a super-family of 2-oxoglutarate (2OG)-dependent hydroxylases and are expressed in three isoforms (HIF-PHD1–3) in humans. All isoforms have a common structure of eight double stranded β-helix core fold. These enzymes use oxygen, iron, and 2OG to hydroxylate proline residues (pro-402 and 546) in HIF-1α, which signals for the recruitment of the von Hippel-Landau (pVHL)-E3 ubiquitin ligase complex and subsequently leads to proteasomal degradation of HIF-1α (Fig. 51.1). Graded hypoxia in vitro progressively increases HIF-1α stability and transactivation of target genes [10], suggesting that HIF-PHDs are highly sensitive oxygen sensors that regulate HIF-dependent hypoxia response. Using this model, studies have used pharmacological inhibitors of HIF-PHDs to stabilize HIF-1α as a therapeutic strategy to prevent hypoxic/ischemic damage in a number of organs including the brain [10–13].

INHIBITING HIF-PHDs AS A THERAPEUTIC STRATEGY

Since HIF-PHD function is sensitive and depends on the availability of its cofactors, the most common pharmacological approaches attempt to limit these cofactors, such as using iron chelation [DFO and ciclopirox (CPO)] and 2OG analogs (3,4, dihydroxybenzoic acid) [1,4,12,14]. HIF-PHD inhibitors have been shown to be neuroprotective in in vitro models of oxidative stress [4] and excitotoxicity [15]; and animal models of ischemic stroke [12]. Although the use of iron chelators (DFO and CPO) are promising in providing neuroprotection following stroke, a challenge faced when using this strategy is to reduce iron to inhibit the HIF-PHDs without interfering with physiological iron-dependent functions. Iron is essential for a number of processes including mitochondrial function, cell signaling, and

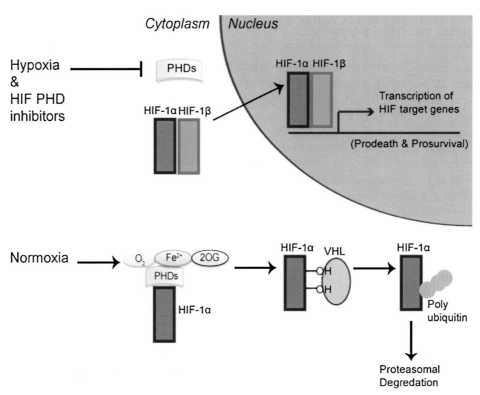

FIGURE 51.1 **Mechanism for hypoxia-dependent hypoxia inducible factor (HIF)-1α regulation.** Hypoxia/HIF–prolyl hydroxylase (PHD) inhibitors (top half) inactivate HIF-PHDs, and stabilize HIF-1α. HIF-1α dimerizes with HIF-1β and translocates to the nucleus to transactivate HIF target genes. Under normoxic conditions (bottom half), HIF-PHDs hydroxylate prolines in HIF-1α. Hydroxylation recruits von Hippel-Landau (pVHL), leading to polyubiquitination and proteasomal degradation.

myelination [16,17]. The potential side effects of using nonselective inhibitors of HIF-PHDs have led to the development of strategies to identify specific small-molecule inhibitors [18].

HIF-PHD INHIBITION TO PREVENT OXIDATIVE STRESS

Oxidative stress–dependent neuronal death can be prevented in vitro by the use of iron chelators [4,12]. Iron is generally thought to play a role in oxidative stress by catalyzing the production of reactive hydroxyls [a reactive oxygen species (ROS)] via Fenton chemistry. In theory hydroxyl radicals cause oxidative damage in proteins, lipids, or DNA, which is thought to be the primary mediator for neuronal death [19]. Iron chelators have been suggested to inhibit this process, but this model remains controversial. Other studies suggest that iron chelators also prevent oxidative-dependent cell death by inhibiting HIF-PHDs [12,20]. The neuroprotection by iron chelators during oxidative stress is associated with increased HIF-1α binding to HRE and increased transactivation of HIF-dependent genes

(EPO) [4]. Structurally diverse and iron chelation-independent HIF-PHD inhibitors show neuroprotection following oxidative stress [12]. In addition, knockdown of HIF-PHD1 also afforded neuroprotection [20]. Together these studies show that HIF-PHD activity is an alternative iron-dependent mechanism of oxidative-dependent cell death and that inhibition of HIF-PHDs is sufficient for neuroprotection.

Neuroprotection by HIF-PHD inhibition during oxidative stress was originally thought to be mediated by increased HIF-1α and HIF-2α stability, and transactivation of antioxidant enzymes (superoxide dismutase) [21]. Contrary to this model, studies using molecular knockdowns of HIF-1α, HIF-2α, or both show that HIF is not critical for adaptive mechanisms following oxidative stress and ischemic stroke [9,20,22], likely because HIF is also involved in the regulation of prodeath genes. This leads to the question of how HIF-PHD inhibition provides neuroprotection following oxidative stress. In addition to HIFs, HIF-PHDs can hydroxylate and regulate a number of different proteins, such as iron regulatory protein-2, RNA polymerase 2, mitogen-activated protein kinase organizer-1, and ATF4 [23,20].

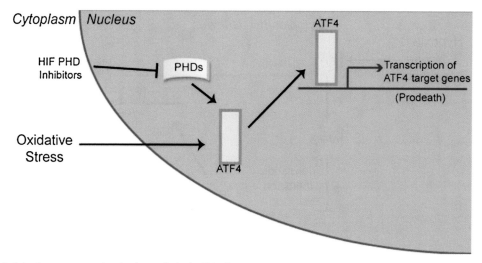

FIGURE 51.2 **Model of neuroprotection by hypoxia inducible factor (HIF)–prolyl hydroxylase (PHD) inhibitors during oxidative stress.** Oxidative stress increases activating transcription factor 4 (ATF4) activity leading to transcription of prodeath ATF4 target genes. HIF-PHD inhibitors protect during oxidative stress by inhibiting PHD-dependent ATF4 function.

ATF4 IS AN OXIDATIVE STRESS–INDUCIBLE TRANSCRIPTION FACTOR REGULATED BY HIF-PHDs

Of the potential targets of HIF-PHD, ATF4 is a known transcription factor that is induced by oxidative stress in neurons [3,23]. Originally named cAMP-response element binding protein 2, ATF4 is a member of the family of basic leucine zipper proteins. In addition to oxidative stress, other cellular stresses, such as viral infection, endoplasmic reticulum stress, and translation inhibition also activates ATF4 [3,24,25]. Under these stress conditions, neurons attempt to protect themselves by inhibiting protein synthesis through the phosphorylation and inhibition of the translation initiation protein eukaryotic initiation factor 2 alpha (eIF2α) [24]. Paradoxically, during this inhibition of protein synthesis there is an increase in the expression of ATF4. This is due to eIF2α-dependent upstream open reading frames (uORFs) located in the 5′ untranslated region of the ATF4 ORF. The uORFs are preferentially used as translation start sites by eIF2α, thus skipping the ATF4 ORF. When eIF2α is inhibited, this preference for translation at uORF is lost and ATF4 can be translated. This mechanism is known as the unfolded protein response (UPR).

A number of target genes of ATF4 are involved in the synthesis of the antioxidant glutathione, such as the X_C^- transporter, and glycine transporter [24], and this led many to initially conclude that ATF4 was involved in prosurvival mechanisms during oxidative stress. However, a number of studies have demonstrated that ATF4 mediates oxidative stress–dependent cell death [3,26]. In addition to potential antioxidant genes, ATF4

also upregulates a cassette of genes associated with cell death, such as tribbles 3 homolog, C/EBP homologous protein (CHOP), and activating transcription factor 3 [3,26,27]. ATF4 target gene CHOP is a transcription factor that forms heterodimers with ATF4 and transactivates growth arrest and DNA damage gene 34 (GADD34). GADD34 can prematurely initiate protein synthesis during UPR [27], which in turn enhances ROS generation [26].

HIF-PHDs have been shown to regulate ATF4 stability through the hydroxylation of five prolines in the ATF4 protein sequence [23]. The mechanism of how hydroxylation regulates ATF4 is not fully understood as unlike HIF-1α, its stability is independent of pVHL. However, mutation of the five prolines in ATF4 has been shown to affect the stability ATF4 [23]. This would suggest that neuroprotection by HIF-PHD inhibitors following oxidative stress is through PHD-dependent posttranslational modifications modulation of ATF4 (Fig. 51.2).

GLUCOSE DEPRIVATION AND HIF ACTIVITY

A number of studies have shown that glucose is required for the hypoxia-induced stabilization of HIF-1α in a number of cell types including neurons [14,28,29]. During oxygen-glucose deprivation (OGD) there is a decrease in HIF-1α accumulation due to a decrease in the translation of HIF-1α [14]. HIF-1α is preferentially translated during global translation suppression by hypoxia and the addition of glucose deprivation during hypoxia likely adds additional stress to neurons. This inhibition

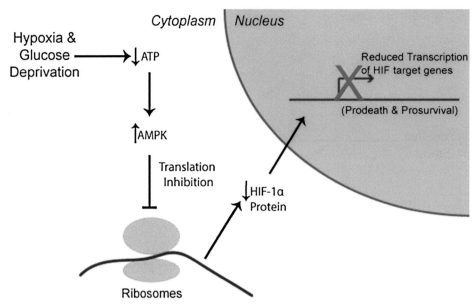

FIGURE 51.3 **Proposed model for hypoxia inducible factor (HIF)-1α reduction during oxygen and glucose deprivation.** Combination of hypoxia and glucose deprivation decreases ATP levels in neurons, leading to an increase in AMP-activated protein kinase (AMPK), which in turn inhibits global protein synthesis. Decreased synthesis of HIF-1α leads to a reduction in transactivation of HIF response genes.

of HIF-1α translation is associated with reduced adenosine triphosphate (ATP) levels within neurons, which is not observed during hypoxia [14]. A potential model explaining how OGD exacerbates cell death is that during ATP scarcity, neurons activate AMP-activated protein kinase (AMPK) to maintain metabolic homeostasis. To reduce the demand for ATP in cells, AMPK activity suppresses protein synthesis, including HIF-1α (Fig. 51.3) [30]. Since glucose depletion during stroke modulates the stress response by inhibiting synthesis of HIF, the efficacy of therapeutic strategies using HIF-PHD inhibitors would also be affected by glucose deprivation.

CONCLUSION

In this chapter, we briefly discuss how hypoxia, oxidative stress, and hypoglycemia can contribute to determining survival or death in neurons. Although it has been traditionally thought that neuronal death is primarily through physical damage of neuronal compartments, we discuss here an alternative model, where transcriptionally dependent prosurvival and prodeath genes are integrated to lead to damage or recovery. Specifically, HIF-PHD–dependent regulation of ATF4 rather than HIF transcription factors may be critical to neuronal survival during each of these conditions. Finally, we suggest that inhibition of HIF-PHDs could be a potential therapeutic strategy to stroke as it regulates adaptive responses for a number of stresses rather than each individually. The full panoply of these adaptive responses triggered by HIF-PHD inhibition is only beginning to be explored.

References

[1] Bergeron M, Gidday JM, Yu AY, Semenza GL, Ferriero DM, Sharp FR. Role of hypoxia-inducible factor-1 in hypoxia-induced ischemic tolerance in neonatal rat brain. Ann Neurol September 2000;48(3):285–96.

[2] Mu D, Jiang X, Sheldon RA, et al. Regulation of hypoxia-inducible factor 1alpha and induction of vascular endothelial growth factor in a rat neonatal stroke model. Neurobiol Dis December 2003;14(3):524–34.

[3] Lange PS, Chavez JC, Pinto JT, et al. ATF4 is an oxidative stress-inducible, prodeath transcription factor in neurons in vitro and in vivo. J Exp Med May 12, 2008;205(5):1227–42.

[4] Zaman K, Ryu H, Hall D, et al. Protection from oxidative stress-induced apoptosis in cortical neuronal cultures by iron chelators is associated with enhanced DNA binding of hypoxia-inducible factor-1 and ATF-1/CREB and increased expression of glycolytic enzymes, p21(waf1/cip1), and erythropoietin. J Neurosci 1999;19(22):9821–30.

[5] Wang GL, Semenza GL. Characterization of hypoxia-inducible factor 1 and regulation of DNA binding activity by hypoxia. J Biol Chem October 15, 1993;268(29):21513–8.

[6] Jiang BH, Rue E, Wang GL, Roe R, Semenza GL. Dimerization, DNA binding, and transactivation properties of hypoxia-inducible factor 1. J Biol Chem July 26, 1996;271(30):17771–8.

[7] Chan DA, Sutphin PD, Yen SE, Giaccia AJ. Coordinate regulation of the oxygen-dependent degradation domains of hypoxia-inducible factor 1 alpha. Mol Cell Biol August 2005;25(15):6415–26.

[8] Semenza GL, Roth PH, Fang HM, Wang GL. Transcriptional regulation of genes encoding glycolytic enzymes by hypoxia-inducible factor 1. J Biol Chem September 23, 1994;269(38):23757–63.

[9] Barteczek P, Li L, Ernst AS, Bohler LI, Marti HH, Kunze R. Neuronal HIF-1alpha and HIF-2alpha deficiency improves neuronal survival and sensorimotor function in the early acute phase after ischemic stroke. J Cereb Blood Flow Metab January 8, 2016, pii: 0271678X15624933, [Epub ahead of print].

[10] Epstein AC, Gleadle JM, McNeill LA, et al. C. elegans EGL-9 and mammalian homologs define a family of dioxygenases that regulate HIF by prolyl hydroxylation. Cell October 5, 2001;107(1):43–54.

[11] Ivan M, Haberberger T, Gervasi DC, et al. Biochemical purification and pharmacological inhibition of a mammalian prolyl hydroxylase acting on hypoxia-inducible factor. Proc Natl Acad Sci USA October 15, 2002;99(21):13459–64.

[12] Siddiq A, Ayoub IA, Chavez JC, et al. Hypoxia-inducible factor prolyl 4-hydroxylase inhibition. A target for neuroprotection in the central nervous system. J Biol Chem 2005;280(50):41732–43.

[13] Aminova LR, Chavez JC, Lee J, et al. Prosurvival and prodeath effects of hypoxia-inducible factor-1alpha stabilization in a murine hippocampal cell line. J Biol Chem February 4, 2005;280(5):3996–4003.

[14] Karuppagounder SS, Basso M, Sleiman SF, et al. In vitro ischemia suppresses hypoxic induction of hypoxia-inducible factor-1alpha by inhibition of synthesis and not enhanced degradation. J Neurosci Res August 2013;91(8):1066–75.

[15] Li D, Bai T, Brorson JR. Adaptation to moderate hypoxia protects cortical neurons against ischemia-reperfusion injury and excitotoxicity independently of HIF-1alpha. Exp Neurol August 2011;230(2):302–10.

[16] Lee DL, Strathmann FG, Gelein R, Walton J, Mayer-Proschel M. Iron deficiency disrupts axon maturation of the developing auditory nerve. J Neurosci April 4, 2012;32(14):5010–5.

[17] Robbins E, Pederson T. Iron: its intracellular localization and possible role in cell division. Proc Natl Acad Sci USA August 1970;66(4):1244–51.

[18] Smirnova NA, Rakhman I, Moroz N, et al. Utilization of an in vivo reporter for high throughput identification of branched small molecule regulators of hypoxic adaptation. Chem Biol April 23, 2010;17(4):380–91.

[19] Winterbourn CC. Toxicity of iron and hydrogen peroxide: the Fenton reaction. Toxicol Lett December 1995;82–83:969–74.

[20] Siddiq A, Aminova LR, Troy CM, et al. Selective inhibition of hypoxia-inducible factor (HIF) prolyl-hydroxylase 1 mediates neuroprotection against normoxic oxidative death via HIF- and CREB-independent pathways. J Neurosci 2009;29(27):8828–38.

[21] Peng YJ, Nanduri J, Khan SA, et al. Hypoxia-inducible factor 2alpha (HIF-2alpha) heterozygous-null mice exhibit exaggerated carotid body sensitivity to hypoxia, breathing instability, and hypertension. Proc Natl Acad Sci USA February 15, 2011;108(7):3065–70.

[22] Chen C, Hu Q, Yan J, et al. Early inhibition of HIF-1alpha with small interfering RNA reduces ischemic-reperfused brain injury in rats. Neurobiol Dis March 2009;33(3):509–17.

[23] Koditz J, Nesper J, Wottawa M, et al. Oxygen-dependent ATF-4 stability is mediated by the PHD3 oxygen sensor. Blood November 15, 2007;110(10):3610–7.

[24] Harding HP, Zhang Y, Zeng H, et al. An integrated stress response regulates amino acid metabolism and resistance to oxidative stress. Mol Cell March 2003;11(3):619–33.

[25] Luo S, Baumeister P, Yang S, Abcouwer SF, Lee AS. Induction of Grp78/BiP by translational block: activation of the Grp78 promoter by ATF4 through and upstream ATF/CRE site independent of the endoplasmic reticulum stress elements. J Biol Chem September 26, 2003;278(39):37375–85.

[26] Han J, Back SH, Hur J, et al. ER-stress-induced transcriptional regulation increases protein synthesis leading to cell death. Nat Cel Biol May 2013;15(5):481–90.

[27] Novoa I, Zeng H, Harding HP, Ron D. Feedback inhibition of the unfolded protein response by GADD34-mediated dephosphorylation of eIF2alpha. J Cel Biol May 28, 2001;153(5):1011–22.

[28] Sudarshan S, Sourbier C, Kong HS, et al. Fumarate hydratase deficiency in renal cancer induces glycolytic addiction and hypoxia-inducible transcription factor 1alpha stabilization by glucose-dependent generation of reactive oxygen species. Mol Cell Biol August 2009;29(15):4080–90.

[29] Osada-Oka M, Hashiba Y, Akiba S, Imaoka S, Sato T. Glucose is necessary for stabilization of hypoxia-inducible factor-1alpha under hypoxia: contribution of the pentose phosphate pathway to this stabilization. FEBS Lett July 16, 2010;584(14):3073–9.

[30] Liu L, Cash TP, Jones RG, Keith B, Thompson CB, Simon MC. Hypoxia-induced energy stress regulates mRNA translation and cell growth. Mol Cel February 17, 2006;21(4):521–31.

IV. MOLECULAR MECHANISMS

C H A P T E R

52

Lipid Mediators

L. Belayev, R.S. Freitas, S.J. Marcell, N.G. Bazan

Louisiana State University Health New Orleans, New Orleans, LA, United States

PLATELET-ACTIVATING FACTOR AND OTHER BIOACTIVE LIPIDS

A target for cerebral ischemia is phospholipids from plasma membranes of neural cells. Phospholipid molecules of membranes from neurons, glial cells, and other neural cells store a variety of lipid messengers. Receptor-mediated events, or changes in intracellular events $[Ca^{2+}]$, such as those that occur during excitatory neurotransmission in activity-dependent synaptic plasticity, uncompensated oxidative stress, and other disruptors of homeostasis, activate phospholipases that catalyze the release of precursors of bioactive mediators from phospholipids. These messengers then participate in intracellular and/or intercellular signaling pathways. Accordingly, contemporary research into bioactive lipids has focused on their neurobiological significance and role in diseases.

Cerebral ischemia unsettles the tightly regulated events that control the production and accumulation of lipid messengers and their precursors, such as docosahexaenoic acid (DHA), free arachidonic acid (AA), diacylglycerol, and platelet-activating factor (PAF, 1-O-alkyl-2-acyl-*sn*-3-phosphocholine), under physiological conditions. Rapid activation of phospholipases, particularly of phospholipase A_2 (PLA_2), occurs at the onset of cerebral ischemia [1]. There are a wide variety of PLA_2s and current investigations aim to define those affected by ischemia. For example, in addition to the role(s) of intracellular PLA_2s in lipid messenger formation, a low-molecular-weight secretory PLA_2 synergizes glutamate-induced neuronal damage. Pathways leading to PLA_2 activation/release are part of normal neuronal function, whereas ischemia-reperfusion enhances these events, overproducing PLA_2-derived lipid messengers (e.g., enzymatically produced AA or DHA oxygenation derivatives, nonenzymatically generated lipid peroxidation products, and other reactive oxygen species)

involved in neuronal damage. Among the consequences of PLA_2 activation by ischemia are alterations in mitochondrial function by the rapid increase in the brain free fatty acid pool size (e.g., uncoupling of oxidative phosphorylation from respiratory chain) and the generation of lipid messengers.

PAF is a very potent and short-lived lipid messenger. It is known to have a wide range of actions: as a mediator of inflammatory and immune responses, as a second messenger, and as a potent inducer of gene expression in neural systems. Thus, in addition to its acute roles, PAF can potentially mediate longer-term effects on cellular physiology and brain functions.

PLATELET-ACTIVATING FACTOR CONTRIBUTES TO EXCITOTOXICITY BY ENHANCING GLUTAMATE RELEASE

PAF accumulates during cerebral ischemia, and inhibition of this process plays a critical role in neuronal survival. PAF is a potent, short-lived phospholipid mediator of leukocyte functions, platelet aggregation, and proinflammatory signaling [1]. PAF accumulates in the brain after cerebral ischemia, and, in conjunction with glutamate release and glutamate reuptake inhibition, leads to intracellular Ca^{2+} overload, mitochondrial dysfunction, generation of reactive oxygen species, and inflammation-mediated excitotoxicity [2]. Furthermore, the brain is endowed with a variety of degradative enzymes that rapidly convert PAF to biologically inactive lyso-PAF [1]. Taking these findings together, PAF, when overproduced at the synapse during ischemia, will promote enhanced glutamate release that in turn, through the activation of postsynaptic receptors, will contribute to excitotoxicity. Fig. 52.1 outlines the role of PAF as a presynaptic messenger.

We demonstrated that LAU-0901, a novel PAF receptor antagonist (Figs. 52.1 and 52.2), is neuroprotective in

Primer on Cerebrovascular Diseases, Second Edition
http://dx.doi.org/10.1016/B978-0-12-803058-5.00052-7

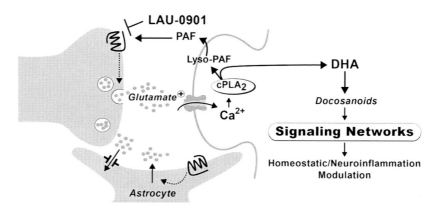

FIGURE 52.1 Simplified cartoon of phospholipases A_2 (PLA$_2$) at the synapse. A depolarizing stimulus at the presynaptic terminal triggers glutamate release. Glutamate binds to the *N*-methyl-D-aspartate receptor and as a consequence an influx of calcium in the postsynaptic neuron occurs. Calcium-mediated activation of the cytoplasmic PLA$_2$ results in the release of arachidonic acid, docosahexaenoic acid (DHA), and lyso-platelet-activating factor (PAF), the PAF precursor. During synaptic plasticity events, PAF may also activate gene expression that in turn is probably involved in long-term alterations of synaptic function (not shown in this diagram). Cell surface PAF-receptor antagonists confer neuroprotection during ischemia-reperfusion and inhibit PAF-induced glutamate release from hippocampal neurons and CA1 long-term potentiation (LTP) formation, presumably through the same mechanism. The inhibitory effects of this antagonist on glutamate release could account in part for its neuroprotection in ischemia-reperfusion.

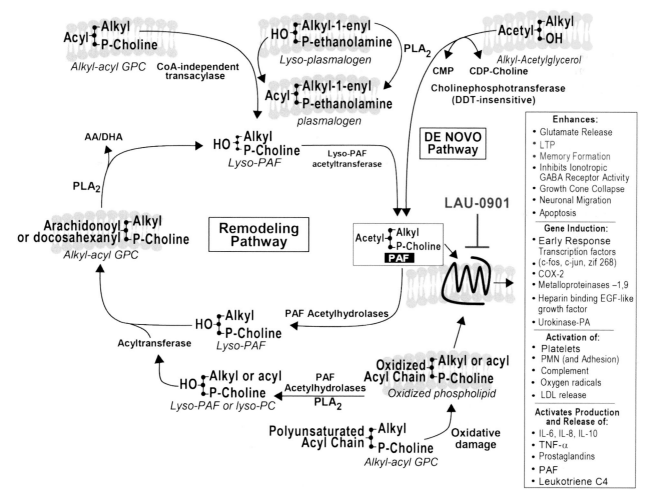

FIGURE 52.2 Platelet-activating factor (PAF) synthesis and degradation. The remodeling pathway or "PAF cycle" from the membrane phospholipid PAF precursor, alkyl-acyl-glycerophosphorylcholine (GPC; left) to the biologically active PAF (right). The remodeling route includes the production of lyso-PAF, which is generated from the PAF precursor alkyl-acyl-GPC either directly by the action of phospholipase A_2 (PLA$_2$) or by the transfer of the *sn*-2 acyl moiety to a "donor" lyso-plasmalogen (not shown), which is mobilized from membrane plasmalogen by phospholipase action. The de novo route of PAF synthesis involves the direct transfer of a choline moiety to alkyl-acetylglycerol. Note that PAF-AH inactivates all PAF molecules, regardless of their biosynthetic route, and also inactivates oxidatively damaged phospholipids (shorter peroxidated acyl group at C2) that possess biological activity at the PAF receptor. *AA*, arachidonic acid; *DHA*, docosahexaenoic acid; *GABA*, gamma-aminobutyric acid; *CDP*, cytidine diphosphate; *COX*, cyclooxygenase; *DDT*, 1,1-bis-(p-Chlorophenyl)-2,2,2-trichloroethane; *EGF*, epidermal growth factor; *LTP*, long-term potentiation; *LDL*, low-density lipoprotein; *IL*, interleukin; *PA*, plasminogen activator; *PC*, phosphatidylcholines; *PMN*, polymorphonuclear; *TNF*, tumor necrosis factor.

experimental stroke [3]. Rats were treated with LAU-0901 (60mg/kg) or vehicle 2h after the onset of stroke. The behavioral deficit was improved by LAU-0901 treatment on days 1, 3, and 7, as compared with vehicle treatment. Total lesion volumes computed from T2-weighted images were significantly reduced by LAU-0901 on days 1, 3, and 7 (by 83%, 90%, and 96%, respectively), which was consistent with decreased edema formation. Histopathology revealed that LAU-0901 treatment reduced cortical and subcortical infarct volumes, attenuated microglial infiltration, and promoted astrocytic and neuronal survival. In addition, we established that LAU-0901 confers enduring ischemic neuroprotection when animals were allowed to survive for 1month after focal ischemic insult [3]. LAU-0901 improved behavior, beginning on day 1 (by 29%), and this improved behavior persisted throughout the 30-day survival period (42%). There was also an increased volume of noninfarcted brain tissue loss relative to the unlesioned hemisphere, as compared with the vehicle group. These findings suggest LAU-0901 is a promising neuroprotectant, thus providing the basis for future therapeutics for ischemic stroke.

In addition to its modulatory effect on synaptic transmission and neural plasticity, PAF activates receptor-mediated immediate early gene expression. Since PAF is a phospholipid and can pass through membranes, it is rapidly taken up by cells. An intracellular binding site with very high affinity, yet pharmacologically distinct from the presynaptic site, was found in the brain [4]. Prostaglandin-endoperoxide synthase 2 [prostaglandin G/H synthase and cyclooxygenase (COX)], also known as PGS-2, COX-2, and TIS-10, catalyzes the cyclooxygenation and peroxidation of AA into PGH_2, the precursor of biologically active prostaglandins, thromboxanes, and prostacyclin. PGS-1 also catalyzes the same first committed step of the AA cascade. PGS-2, however, is expressed in response to mitogenic and inflammatory stimuli and is encoded by an early response gene. In contrast, PGS-1 expression is not subject to short-term regulation. Neurons in the hippocampus, as well as in a few other brain regions, are unlike other cells in that they display basal levels of PGS-2 expression [5]. This expression is modulated by synaptic activity, long-term potentiation, and involves the N-methyl-D-aspartate class of glutamate receptors [4]. PAF is a transcriptional activator of PGS-2, since PAF induces mouse PGS-2 promoter-driven luciferase activity transfected in neuroblastoma cells (NG108-15 or SHSY5Y) and NIH 3T3 cells.

OMEGA-3 POLYUNSATURATED FATTY ACIDS

Omega-3 polyunsaturated fatty acids (PUFAs) (also known as n-3 fatty acids, found abundantly in fish oil) are essential nutrients because they must be obtained from dietary sources since the body cannot synthesize them [6]. The PUFAs also include the n-6 (or omega-6) types; linoleic and arachidonic acids belong to this family. The major types of omega-3 PUFAs are eicosapentaenoic acid (EPA) and DHA, and their precursor, alpha-linolenic acid (ALA). EPA and DHA are found primarily in fatty fish, such as salmon, and in fish-oil supplements. Sources of ALA include flax seed, canola, soybean, walnuts, and green leafy vegetables. Fish-oil supplements are among the most widely used dietary supplements. The main n-3 PUFA in the brain is DHA, comprising 10–20% of the total fatty acid composition, whereas the n-3 PUFAs ALA, EPA, and docosapentaenoic acid comprise only 1% of total brain fatty acid composition [2,6]. DHA is also associated with neuronal membrane stability and the functions of serotonin and dopamine-mediated neurotransmission, which might connect to the etiology of mood and cognitive manifestations of depression. In contrast, EPA is important in the balance of the immune and neuronal functions in that it antagonizes membrane AA (an n-6 PUFA) and reduces prostaglandin E2 synthesis [5]. Omega-3 PUFAs may provide a range of neurobiological activities via modulation of neurotransmitters, inflammation, oxidation, and neuroplasticity [6], which could contribute to their psychotropic effects. Dietary omega-3 fatty acids are required to maintain cellular functional integrity, and overall they are necessary to human health. Numerous reports demonstrate the beneficial effects of fish oil on human diseases, such as arthritis, Alzheimer's disease, lung fibrosis, inflammatory bowel diseases, cerebral ischemic injury, atherosclerosis, asthma, cardiovascular diseases, cancer, and depression [2], as well as in preventing sudden death after myocardial infarction.

DOCOSAHEXAENOIC ACID IS NEUROPROTECTIVE IN CEREBRAL ISCHEMIA

Cerebral ischemia is characterized by a rapid accumulation (within minutes) of free fatty acids, including DHA (22:6, n-3) and AA (20:4, n-6), due to increases in intracellular calcium and activation of phospholipases [1]. This free pool of DHA and AA is then converted via enzymatic processes, as well as free radical–mediated lipid peroxidation, into a cascade of activated pro- and anti–inflammatory mediators, the makeup of which ultimately drives the cell toward survival or programmed cell death (Fig. 52.2) [2].

DHA is concentrated in cellular membranes of the central nervous system and is vital for proper brain function [1]. It is also necessary for the development of the nervous system, including vision [6,7]. DHA is a natural element of our diet and is found in cold water fatty fish, including salmon, tuna, mackerel, sardines, shellfish, and herring. Several epidemiological studies indicate that consumption of fish is associated with reduced risk of ischemic stroke and coronary heart disease. DHA deficiency, as observed with aging and dementia, impairs memory and learning and promotes age-related neurodegenerative diseases. Vertebrates do not have adequate metabolic capacity to biosynthesize DHA; thus, vertebrates, including humans, depend on the diet to supply this fatty acid [2,6]. DHA is involved in excitable membrane function and neuronal signaling and has been implicated in neuroprotection [8].

DHA is well known as a robust neuroprotectant against experimental stroke [8]. We demonstrated that DHA improves behavioral function, decreases infarct volume, and promotes cell survival in the ischemic penumbra as well as resolution of cerebral edema during 1 week of survival after focal cerebral ischemia in rats [8]. In addition, the therapeutic window shows that DHA is neuroprotective when administered up to 5 h after experimental stroke during a 7-day survival period [8]. DHA in phospholipids is also a reservoir for biologically active mediators, and under conditions that disrupt homeostasis it is transformed into the neuroprotective docosanoids. DHA is the precursor of potent bioactive structures responsible for the many properties attributed to DHA, which is important in regulating biological systems [1].

DOCOSANOIDS: NEUROPROTECTIN D1

Neuroprotectin D1 (NPD1; 10R, 17S-dihydroxy-docosa-4Z, 7Z, 11E, 15E, 19Z hexaenoic acid), the first identified member of this group of mediators (the docosanoids), inhibits oxidative stress–induced proinflammatory gene expression and promotes cell survival both in vitro and in vivo [6,9]. The identification of the protective docosanoid NPD1 in brain ischemia-reperfusion [9], as well as in neural cells exposed to oxidative stress, has uncovered a key survival signaling event leading to neuroprotection. NPD1 is a pleiotropic modulator of inflammation resolution [10]. DHA is converted to the bioactive mediator NPD1 through a pathway initiated by 15-lipoxygenase-1 (15-LOX-1), followed by epoxidation and hydrolysis reactions [9]. NPD1 synthesis is triggered by oxidative stress and/or neurotrophins. The evolving concept is that DHA-derived docosanoids set in motion endogenous signaling to sustain homeostatic synaptic and circuit integrity [1]. NPD1 elicits neuroprotective activity in brain ischemia-reperfusion and in oxidative-stressed retinal cells. When NPD1 is infused during ischemia-reperfusion or added to retinal pigment epithelial (RPE) cells during oxidative stress, apoptotic DNA damage is downregulated [7,9]. NPD1 also upregulates the antiapoptotic Bcl-2 proteins Bcl-2 and Bcl-xL and decreases proapoptotic Bax and Bad expression [2]. Moreover, NPD1 inhibits oxidative stress–induced caspase-3 activation. NPD1 also inhibits interleukin-1β-stimulated

expression of COX-2. Overall, NPD1 protects cells from oxidative stress–induced apoptosis [7,9]. Because photoreceptors are progressively impaired after RPE cell damage in retinal degenerative diseases, understanding how these signals contribute to retinal cell survival may lead to the development of new therapeutic strategies [1,7]. NPD1 also exhibits potent inflammatory-resolving bioactivity by inhibiting leukocyte infiltration, COX-2 expression, and interleukin-1β-induced nuclear factor κB activation in experimental ischemic stroke [9]. NPD1 also counteracts apoptosis in RPE cells [1,7]. Moreover, NPD1 bioactivity demonstrates that DHA is not only a target of lipid peroxidation, but also is the precursor to a neuroprotective signaling response to ischemia-reperfusion, thus opening up newer avenues of therapeutic exploration in stroke, neurotrauma, spinal cord injury, and neurodegenerative diseases, such as Alzheimer's disease, aiming to upregulate this novel cell-survival signaling.

CONCLUDING REMARKS

1. **Significance of lipid mediators in intracellular signaling pathway and apoptosis:** In cerebrovascular diseases, the significance of the PLA_2-related signaling triggered by ischemia-reperfusion may be part of events finely balanced between neuroprotection and neuronal cell death. The precise events that would tilt this balance toward the latter are currently being explored. It is interesting to note that PAF, being short-lived and rapidly degraded by PAF acetylhydrolase, is a long-term signal with consequences to neurons through COX-2-sustained expression. COX-2 is localized in the nuclear envelope and perinuclear endoplasmic reticulum. The overexpression of hippocampal COX-2 during cerebral ischemia and seizures may in turn lead to the formation of neurotoxic metabolites (e.g., superoxide). Current investigations aim to determine whether or not other messengers cooperate to enhance neuronal damage (e.g., nitric oxide) and the possible involvement of astrocytes and microglial cells. Further understanding of these potentially neurotoxic events involving lipid messengers and COX-2 will permit the identification of new strategies and define therapeutic windows for the management of cerebrovascular diseases.

2. **Significance of DHA and NPD1 in neuroprotection:** DHA, the precursor of NPD1, is an essential omega-3 fatty acid concentrated in the central nervous system. After being systemically administered, free DHA is carried through the blood stream and concentrated in synapses and cellular membranes in the brain and retina. Under uncompensated oxidative stress like the condition generated by ischemia-reperfusion, DHA is converted to NPD1 through 15-LOX-1 and released from the cell membranes. NPD1 promotes

cell survival and inhibits apoptosis, and it attenuates leukocyte infiltration and inflammatory gene expression. Administration of DHA and NPD1 improves behavioral function, decreases infarct volume, and stimulates cell survival in the ischemic penumbra, as well as resolution of cerebral edema in 1 week of survival after focal cerebral ischemia in rats. In addition, the therapeutic window shows that DHA is neuroprotective when administered up to 5 h after experimental stroke.

Acknowledgment

This work was supported by NIGMS grant GM103340 and NINDS grant NS046741 (NGB).

References

[1] Bazan NG. Lipid signaling in neural plasticity, brain repair, and neuroprotection. Mol Neurobiol 2005;32:89–103.

[2] Bazan NG, Molina MF, Gordon WC. Docosahexaenoic acid signalolipidomics in nutrition: significance in aging, neuroinflammation, macular degeneration, Alzheimer's, and other neurodegenerative diseases. Annu Rev Nutr 2011;31:321–51.

[3] Belayev L, Eady T, Khoutorova L, Atkins K, Obenaus A, Cordoba M, et al. Superior neuroprotective efficacy of LAU-0901, a novel platelet-activating factor antagonist, in experimental stroke. Transl Stroke Res 2012;3:154–63.

[4] Eady TN, Khoutorova L, Anzola DV, Hong SH, Obenaus A, Mohd-Yusof A, et al. Acute treatment with docosahexaenoic acid complexed to albumin reduces injury after a permanent focal cerebral ischemia in rats. PLoS One 2013;8:e77237.

[5] Takemiya T, Yamagata K. Intercellular signaling pathway among endothelia, astrocytes, and neurons in excitatory neuronal damage. Int J Mol Sci 2013;14:8345–57.

[6] Bazan NG, Calandria JM, Gordon WC. Docosahexaenoic acid and its derivative neuroprotectin D1 display neuroprotective properties in the retina, brain, and central nervous system. Nestle Nutr Inst Workshop Ser 2013;77:121–31.

[7] Mukherjee P, Marcheselli V, de Rivero Vaccari J, Gordon W, Jackson F, Bazan N. Photoreceptor outer segment phagocytosis attenuates oxidative stress-induced apoptosis with concomitant neuroprotectin D1 synthesis. Proc Natl Acad Sci USA 2007;104:13158–63.

[8] Belayev L, Khoutorova L, Atkins K, Eady T, Hong S, Lu Y, et al. Docasahexaenoic acid therapy of experimental ischemic stroke. Transl Stroke Res 2011;2:33–41.

[9] Marcheselli V, Hong S, Lukiw W, Tian X, Gronert K, Musto A, et al. Novel docosanoids inhibit brain ischemia-reperfusion-mediated leukocyte infiltration and pro-inflammatory gene expression. J Biol Chem 2003;278:43807–17.

[10] Serhan C, Fredman G, Yang R, Karamnov S, Belayev L, Bazan N, et al. Novel proresolving aspirin-triggered DHA pathway. Chem Biol 2011;18:976–87.

CHAPTER

53

Mitogen-Activated Protein Kinase Signaling in Cerebrovascular Disease

F.C. Barone

SUNY Downstate Medical Center, New York, NY, United States

MITOGEN-ACTIVATED PROTEIN KINASE SIGNALING

Cells respond to their extracellular environment via common intracellular signaling systems.

Mitogen-activated protein kinases (MAPKs) are serine–threonine kinases that mediate intracellular signaling involved in regulating protein and cell functioning related to membrane, intra- and intercellular processes and transformation, proliferation/growth, differentiation, survival, and death. The mammalian MAPK family consists of an extracellular signal–regulated kinase (ERK) signaling arm and the stress-activated protein kinases (SAPK) consisting of a p38 arm and c-Jun NH2-terminal kinase (JNK) arm, with ERK, p38, and JNK each existing in several isoforms. As shown in Fig. 53.1, the cascade of MAPK signaling consists of a MAPK kinase kinase (MAP3K), a MAPK kinase (MAP2K), and a MAPK (left side). MAP3Ks phosphorylate and activate MAP2Ks, which in turn phosphorylate and activate MAPKs. The phosphorylated MAPK then interacts with cellular substrates and/or translocates to the nucleus to modulate diverse biological responses. Thus, the activated MAPKs ERK, p38, and JNK (right side) phosphorylate various target/substrate proteins including transcription factors (e.g., Elk-1, ATF2, and c-Jun, respectively) that modify cellular function and phenotype. Growth factor stimulation of the ERK pathway modulates the activity of many cellular proteins and transcriptional factors modifying cell functioning, proliferation, differentiation, migration, and survival. The SAPK p38 and JNK pathways are weakly activated by growth factors, but are strongly activated by stress stimuli such as ischemia and/or hemorrhage, oxygen free radicals, and inflammatory cytokines such as interleukin-1 beta (IL-1β) and tumor necrosis factor alpha (TNFα) resulting in altered transcription, translation, and activation of factors involved in cell proliferation, differentiation, inflammatory cytokines and inflammation, and cellular apoptosis/death.

Although initially identified as independent arms of intracellular signaling pathways activated in response to distinct extracellular stimuli, it is now understood that there is significant cross talk/interaction between these three MAPK signaling arms. This suggests that these pathways can also be considered as components of much larger signaling networks that still needs to be understood. As examples, both ERK and p38 pathways activate common transcription factors such as Elk-1, Sap-1a, and cyclic-AMP response element–binding protein (CREB). Some stimuli activate CREB through p38 and ERK, whereas others can activate ERK through activation of the p38 pathway. In addition, ERK and p38 pathways share common intracellular kinase substrates and can mediate growth factor and stress-induced activation

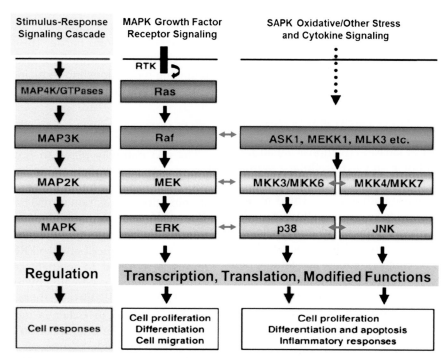

Mitogen-Activated Protein Kinase (MAPK) Signaling

FIGURE 53.1 MAPK pathways mediate an intracellular signaling cascade (indicated by *vertical unidirectional arrows*) that is initiated by extracellular stimuli and/or intracellular changes. MAP3Ks are activated by MAP4Ks or GTPases and phosphorylate (activate) MAP2Ks that phosphorylate and activate MAPKs. Activated MAPKs phosphorylate substrate proteins (namely, transcription factors and other proteins) to modify/regulate cellular functioning such as proliferation, differentiation, migration, and inflammation, membrane channels and/or receptors, and cell survival or death). The MAPK family includes ERK, p38, and JNK. ERK MAPK pathway signaling is initiated by receptor tyrosine kinase (RTK) stimulation–induced GTPase activation of RAS to ERK. Stress-activated protein kinase (SAPK) pathway signaling is initiated by extracellular or intracellular stress (e.g., induced by ischemia, hemorrhage, oxygen free radicals, inflammation, and inflammatory cytokines) that induces activation of ASK1, MEKK1, or MLK3 to p38 or JNK. Significant interactions that can occur between the ERK/MAPK and the SAPK/p38 with or without JNK signaling arms of MAPK signaling are indicated by the horizontal bidirectional arrows.

of CREB. It is clear that these signaling systems are far from being fully understood and new data continue to emphasize the interactions or cross talk between MAPK signaling in cellular changes. Alternative signaling pathways will emerge as our knowledge of kinase biology increases via the evaluation of transgenic animals and the development of new, selective kinase inhibitors. Thus, modulation of a single kinase may alter signaling through cross talk or via alternative/to be discovered pathway interactions. All these can depend on signal strength, duration, and location, and the precise control of kinase activation, including the detailed temporal activations in development or in disease pathology, and the specific scaffolding proteins that tether MAPK signaling components into discrete signaling cascades. Thus, the complexity of kinase signaling changes and their complex interactions must be considered with outcome changes in animal models of cerebrovascular disease using selective small-molecule inhibitors as is discussed in the following sections. The reference list provided [1–10], although a very short list, does provide the primary connection to the highly summarized information contained in this chapter.

ISCHEMIC STROKE

Ischemic stroke results in significant brain injury and disability. Cerebral ischemia produces many pathological changes that results in impaired cellular regulation, intercommunication and cell loss/brain infarction. Following ischemia many factors are released, including growth factors, cytokines, glutamate, and oxygen free radicals that can stimulate the activation of MAPK pathways. The consequence of kinase activation is dependent on changes in intracellular and extracellular environment (i.e., influences from other cell changes), the cell type, the number of kinase pathways activated at any given time, and the duration of kinase activation. The complex balance between ischemic cell survival and death-mediated changes in cerebral ischemia has been emphasized by many researchers. Kinase activation after ischemia can be related to other cell changes or be central to cell death. Future optimum therapeutic intervention to improve outcome might require selective or a specific profile of kinase inhibitors as addressed in the following section.

ERK/MAPK

Consistent with this complexity, ERK activation has been shown to be critical to neurotrophic survival signaling, but is also important in cell death. ERK pathway signaling increases in isolated neuronal and mixed-cellular brain cultures exposed to ischemia or stressful-toxic stimuli, and in the brain of animal experimental stroke models, suggesting its activation is central to cell death. In experimental stroke, ERK phosphorylation (pERK) and activation increases within one to several hours in neurons and glia in the ischemic "penumbral" regions (i.e., those areas involved in infarct expansion). More importantly, selective MEK inhibitors reduce cytotoxic/ischemic cell death and infarct size, and is accompanied by an inhibition of ERK activation, suggesting the involvement of this pathway in ischemic cell death. However, since downstream changes from ERK are not always changed as predicted, other arms of MAPK signaling might be involved (i.e., the protection might alter signaling through alternative pathway signaling arms as cautioned earlier). One must also consider vascular factors. After ischemia, neuroprotection can be directly in brain cells, or can be due to increasing blood flow to these brain cells and/or by blocking vascular leakiness and edema that are also important factors in ischemic brain injury (e.g., see section Hemorrhagic Stroke). Importantly, the Raf/MEK/ERK pathway is activated and appears to be a key mechanism involved in vasoconstriction produced by many potent vasoconstrictors (e.g., endothelin, angiotensin II, and noradrenaline) and in vascular leakiness produced by inflammatory mediators and matrix-degrading proteins in ischemia. Thus, since Raf/MEK/ERK activation is key to vasoconstriction and vascular/blood–brain barrier (BBB) leaks, inhibition of this signaling arm might provide neuroprotection by enhancing blood flow and reducing vascular permeability/edema as well as directly blocking cell death.

p38/SAPK

The p38 pathway plays an important role in transducing signals involved in cell survival, cell death/apoptosis, and inflammatory cytokine production and inflammation. Activation of p38 has been shown repeatedly to be associated with neuronal death/apoptosis, and selective p38 inhibition promotes cell survival. Apparently, a balance between ERK and stress-activated kinases is required to mediate cell survival. The p38 pathway is strongly activated by the inflammatory cytokines TNFα and IL-1β (i.e., these increase after stroke and are involved in ischemia-induced cell death and neuro- and vascular inflammation). Ischemia stroke activates p38 in microglia, astrocytes, and neurons, including downstream signaling to ATF2 within the first and for several hours up to at least 24h (i.e., when infarcts have fully evolved). The majority of studies investigating p38 activity following cerebral ischemia describe maintained activation in brain regions destined to die after the cessation of blood flow. An earlier activation of p38 compared to ERK is suggested. Again, the fate of ischemic cells depends on the balance between the activation of ERK survival and SAPK (also see later for JNK) cell death pathways, but continued, detailed, parallel studies are required to fully understand the specific timing of ERK and p38 activations. Selective p38 inhibitors reduce ischemic or stress-induced neuronal apoptotic cell death (in vivo and in vitro) and stroke-induced brain infarctions and post-stroke neurological deficits. The p38 MAPK pathway contributes significantly to inflammatory cytokine production and to brain neuroinflammation mediated by glial cells (e.g., microglia and astrocytes). Since cytokine, inflammation and apoptotic pathways converge, p38 inhibition can be key to protection against both neurovascular inflammation and delayed cell death that occurs poststroke. Blocking p38 directly and indirectly protects brain cells, reduces vascular and brain inflammation and edema, and improves blood flow during ischemia. Clearly, the necessary pharmacokinetic profile and brain penetration and selectivity for current and future p38 inhibitors that might be useful alone or in combination with other interventions still require more ischemic stroke research.

JNK/SAPK

Signaling via the JNK/SAPK pathway occurs in response to similar stress stimuli that activate p38 and, like p38, is associated with the induction of apoptosis, proliferation, and differentiation. JNK isoform knockouts are either embryonic lethal (attributed to increased brain apoptosis) or exhibit decreased sensitivity to glutaminergic excitotoxicity. Increased hippocampus JNK activation following global forebrain ischemia is variable from several hours to several days and is associated with increased p–c-Jun. Similarly, following focal ischemia–reperfusion, JNK and c-Jun activation occurs within and/or around the evolving infarction. However, there is significant variability between different studies compared with data from ERK and p38. However, this occurs earlier and is more consistent for hemorrhagic stroke (see the following section). The early and sustained activation of JNK associated with vulnerable cells and infarct evolution suggests that it plays a significant role in cell death. Like p38 activation, JNK activation is also involved in free radical–mediated cell death. Molecules that inhibit the JNK pathway without inhibiting the p38 pathway also reduce apoptotic cell loss. A selective inhibitor when administered after reperfusion stroke significantly reduces infarction, cognitive deficits, and post-stroke apoptotic cells.

HEMORRHAGIC STROKE

Hemorrhagic stroke results in significant brain injury and morbidity/mortality. Similar to ischemic stroke there are no brain protective therapies approved/available for hemorrhagic stroke. Vascular leak—BBB permeability and brain edema, elevated intracranial pressure, and compromised microcirculation contribute to brain injury resulting in death, morbidity, and disability after experimental subarachnoid hemorrhage (SAH). SAH induces cellular stress responses and results in the activation of all arms of MAPK signaling. MAPK activation occurs via SAH-induced growth factors, oxidative stress, and inflammatory cytokines. Vascular endothelial growth factor (VEGF) increases after SAH increasing BBB permeability, edema, acute arterial spasm, and infarction. VEGF binds with its RTK (tyrosine kinase receptor) and activates MAPK signaling pathways that promote other molecular changes, especially related to apoptotic cell death and later vasoconstriction and additional ischemic damage. SAH activated ERK, p38, and JNK are specifically involved in cell death. All three arms of MAPK signaling induce apoptosis in the brain and in cerebral vessels. Inhibition of the ERK signaling pathway reduces cerebral vasospasm, the degree of vascular contractile receptor upregulation, the vascular contraction to potent vasoconstrictors, the upregulation of inflammatory cytokines and matrix-degrading matrix metalloproteinase, and the reduced blood flow that occurs following experimental SAH. p38 pathway inhibition reduces the vasospasms, the increased vascular and CSF levels of TNFα, mitochondrial dysfunction, and the vascular and brain cell apoptosis that occurs post-SAH. JNK is activated by cerebral ischemia, as stated earlier, but also by blood components early after SAH. Significant research using selective JNK inhibitors/inhibition demonstrates improved outcome in experimental SAH. After experimental SAH, JNK inhibition prevents c-Jun phosphorylation and suppresses most SAH induced responses including the increased VEGF, the activation of apoptosis and cell death, the increased neuronal and BBB injury, the brain swelling/edema, the cerebral vasospasm and vascular injury, and the mortality and neurological deficits. The powerful protective effects of JNK inhibition might be due to its combined actions on the severe SAH pathology of combined ischemia, blood stimulation and vasoconstriction, and the consecutive neurovascular inflammation. As expected for direct MAPK inhibition, JNK inhibition (i.e., as also typically seen for direct p38 inhibition) does not affect the phosphorylation status of JNK but does reduce phosphorylation of its downstream targets. It is important to mention that JNK pathway inhibition that induces strong protection in ischemic stroke and SAH described earlier, also improves outcome in experimental intracerebral hemorrhage (i.e., the other major type of hemorrhagic stroke). Delayed post-intracranial hemorrhage intervention with JNK inhibition reduces brain infarct, swelling/edema, and neurological deficits.

SUMMARY

There is now significant evidence that ERK, p38, and JNK pathways activation occurs, and that inhibition of this activation exhibits significant efficacy to improve outcome in animal models of cerebrovascular disease as discussed earlier for ischemic and hemorrhagic stroke. It is important to note that the results of countless studies (not mentioned in the preceding sections) demonstrate that the mechanism of action of many protective interventions clearly involves the reduction of MAPK activation induced in experimental stroke. There are also several cases where dephosphorylation of activated MAPK also provides brain-protective effects. Additionally, there is significant involvement of MAPK signaling in ischemic conditioning—brain-protection paradigms of all types suggesting new opportunities for brain protection in the future. Finally, significant evidence is accumulating to supporting MAPK activation in the pathology mechanisms of the major neurodegenerative diseases (i.e., these also have significant cerebrovascular components as well). For stroke, the potential to intervene early, without imaging delays to differentiate ischemia from hemorrhage can be a very important future opportunity for MAPK in the future. There is close homology between kinases, with 400–500 of them predicted to be encoded by the human genome. Therefore, there is a great potential for compounds to provide an optimum inhibition profile for significant outcome improvement in the future (i.e., rather than specificity for a single target that might not be as effective). Inhibition of a kinase toward the top of its signaling cascade can have a much more widespread effect as compared with inhibition of a single downstream kinase. For example, targeting ERK directly will result in a more selective inhibition than the upstream inhibition described earlier, and therefore, the biological consequences can be very different and certainly need to be understood. It is confirmed that MAPK pathways involve a highly complex network of signaling cascades, many of which are activated following the onset of cerebrovascular ischemia or hemorrhage. A greater understanding of kinase signaling and their functional consequences on multiple measures (e.g., apoptosis, mediators, blood flow, and metabolism in relation to brain injury, function, and more detailed effects at the multiple MAPK kinase signaling and target proteins) will be essential.

References

[1] Barone FC, Feuerstein GZ. Inflammatory mediators and stroke: new opportunities for novel therapeutics. J Cereb Blood Flow Metab 1999;19(8):819–34. Review.

[2] Barone FC, Irving EA, Ray AM, Lee JC, Kassis S, Kumar S, et al. Inhibition of p38 mitogen-activated protein kinase provides neuroprotection in cerebral focal ischemia. Med Res Rev. 2001;21(2):129–45. Review.

[3] Force T, Kuida K, Namchuk M, Parang K, Kyriakis JM. Inhibitors of protein kinase signaling pathways: emerging therapies for cardiovascular disease. Circulation 2004;109(10):1196–205.

[4] Irving EA, Bamford M. Role of mitogen- and stress-activated kinases in ischemic injury. J Cereb Blood Flow Metab 2002;22(6):631–47. Review.

[5] Kaminska B, Gozdz A, Zawadzka M, Ellert-Miklaszewska A, Lipko M. MAPK signal transduction underlying brain inflammation and gliosis as therapeutic target. Anat Rec (Hoboken) 2009;292(12):1902–13.

[6] Kim EK, Choi EJ. Compromised MAPK signaling in human diseases: an update. Arch Toxicol 2015;89(6):867–82.

[7] Michel-Monigadon D, Bonny C, Hirt L. c-Jun N-terminal kinase pathway inhibition in intracerebral hemorrhage. Cerebrovasc Dis 2010;29(6):564–70.

[8] Peti W, Page R. Molecular basis of MAP kinase regulation. Protein Sci 2013;22(12):1698–710.

[9] Suzuki H, Hasegawa Y, Kanamaru K, Zhang JH. Mitogen-activated protein kinases in cerebral vasospasm after subarachnoid hemorrhage: a review. Acta Neurochir Suppl 2011;110(Pt 1): 133–9.

[10] Uehling DE, Harris PA. Recent progress on MAP kinase pathway inhibitors. Bioorg Med Chem Lett 2015;25(19):4047–56.

CHAPTER

54

Rho-Associated Kinases in Cerebrovascular Disease

H.-H. Kim, C. Ayata

Harvard Medical School, Boston, MA, United States

INTRODUCTION

More than two decades after their discovery, Rho-associated kinase (ROCK) is at the center stage of therapeutic development in many diverse diseases ranging from cardiovascular to neurological, metabolic, and ocular disorders. Because ROCK activity is pivotal in the regulation of numerous processes in virtually all cell types, the therapeutic indications have been wide ranging and growing. It is well established that upregulation of ROCK activity is a central mechanism contributing to vascular dysfunction in chronic hypertension, hyperlipidemia, type 1 and 2 diabetes, and atherosclerosis, as chronic vascular risk factors. ROCK inhibition lowers the arterial blood pressure, prevents cardiac hypertrophy in disease models, and inhibits the development of atherosclerosis. As such, it has found a broad range of potential cardiovascular indications for development. ROCK inhibition has a multitude of pleiotropic effects converging to provide additive or synergistic benefits, including antiinflammatory, antiplatelet, antiatherosclerotic, vasodilator, antiapoptotic, and axon growth promoting effects, all of which are predicted to be beneficial in cerebrovascular diseases [1]. Indeed, ROCK inhibition is in part responsible for the beneficial effects of statins, which inhibit the production of isoprenoid building blocks of cholesterol that are also critical for posttranslational modification, subcellular localization, and function of Rho subfamily of small-molecular-weight monomeric GTPases, the immediate upstream regulators of ROCK function.

RHO-ASSOCIATED KINASES

The Rho-associated kinases 1 (aka ROKβ, p160ROCK, ROCK1, ROCK I) and 2 (aka ROKα, ROCK2, ROCK II) are ~160 kDa serine/threonine kinases that are among the downstream effectors of the small GTPase RhoA. The Rho/ROCK pathway acts as a molecular switch. Activated when bound to GTP-Rho, ROCK regulates numerous processes involving cytoskeletal rearrangement, such as activation, adhesion and aggregation of platelets, leukocyte migration, cell motility and proliferation, and smooth muscle contraction. The two ROCK isoforms show 65% sequence homology (58% in the Rho-binding domain and 92% in the kinase domain). Both ROCK1 and ROCK2 are expressed ubiquitously throughout the tissues from early embryonic development to adulthood. However, their subcellular localizations, upstream regulators, and tissue expression patterns differ. ROCK2 is expressed highest in the brain, muscle, and heart, whereas ROCK1 is abundantly expressed in the lung, liver, thymus, kidney, testis, spleen, and leukocytes [2]. ROCK1 is expressed at low levels in the brain, predominantly in glial cells and not in neurons. Moreover, there is evidence suggesting that ROCK1 and ROCK2 serve different functions in different cell types and molecular pathways. Unfortunately, genetic global homozygous deletion of either ROCK isoform is embryonic or perinatal lethal albeit via different mechanisms [3], but they can to some extent also compensate for loss of each other's activity.

RHO-ASSOCIATED KINASE INHIBITORS

Currently available ROCK inhibitors compete for the ATP-binding site of the kinase. Since the availability of fasudil and Y-27632, the two originally described isoform-nonselective ROCK inhibitors that have relatively low potency (K_i in the high-nanomolar range) and poor selectivity over other kinases, numerous other ROCK inhibitors have been developed with higher potency (sub- to low-nanomolar range) and ROCK selectivity, and one with 200-fold selectivity for ROCK2 (KD-025, formerly SLX2119; IC_{50} [The drug concentration that achieves 50% inhibition of enzyme activity in any given assay.] $0.11 \mu M$) over ROCK1 ($IC_{50} > 20 \mu M$), and its derivatives in development, as reviewed in detail in Refs. [2,4]. Given the differential tissue expression patterns and functions of the two isoforms, further development of isoform-selective ROCK inhibitors is likely to open new indications and minimize side effects.

RHO-ASSOCIATED KINASE INHIBITION IN EXPERIMENTAL MODELS OF CEREBROVASCULAR DISEASES

ROCK inhibitors have multiple distinct but complementary and converging mechanisms of action on various different cell types (Table 54.1) that altogether are predicted to be beneficial in the prevention, acute treatment, recovery, and rehabilitation of cerebrovascular diseases (Fig. 54.1) [1,3]. Presumably owing to these pleiotropic effects, ROCK inhibitors have been uniformly efficacious in a wide range of experimental models including focal and global ischemia, and subarachnoid hemorrhage [1]. In a meta-analysis of focal cerebral ischemia, inhibition of the Rho/ROCK pathway was efficacious in reducing infarct volume (by ~30% after accounting for publication bias) and neurological deficits in both sexes, particularly in transient and thrombotic models, when administered up to 6h after occlusion onset, in a relatively dose-dependent manner [5]. The primary mechanism of action in the hyperacute stage appeared to be improved collateral flow in an endothelial nitric oxide synthase–dependent fashion [6]. Efficacy was retained in animal models of comorbidities, such as hyperlipidemia and diabetes, and safety and efficacy in combination with statins (overlapping mechanisms of action) has been shown [7,8]. More importantly, ROCK2-selective inhibitor KD-025 has also provided lasting efficacy comparable with isoform-nonselective inhibitors in acute focal cerebral ischemia, with a dose–response relationship and a therapeutic window of 3h, and without exacerbating hemorrhagic transformation within the infarct or causing severe hypotension [8]. The latter is an important consideration in focal ischemic stroke as hypotension has the potential to diminish the perfusion in ischemic brain, especially from extracranial to intracranial collaterals. Another potential adverse effect of ROCK inhibition in ischemic stroke, particularly upon chronic use, may be the suppression of neovascularization and vascular hypertrophy. Therefore, it remains to be seen whether chronic ROCK inhibition decreases cerebral collateral formation in patients with intracranial stenoses.

ROCK inhibition has also been beneficial in ameliorating tissue injury and neurological deficits after global

TABLE 54.1 Pleiotropic Effects of ROCK Inhibition Relevant for Cerebrovascular Diseases

Endothelium	↑ eNOS expression/activity ↓ Migration ↓ BBB disruption and permeability ↓ Actin contractility and stiffness of the endothelium ↑ Fibrinolytic activity
Vascular smooth muscle	↓ Myosin light chain phosphorylation ↓ Tone and vasospasm ↓ Mechanotransduction ↓ Proliferation and intimal thickness
Platelet	↓ Adhesion and aggregation ↓ Bone marrow production ↓ Adhesion to fibrinogen ↓ Thrombus contraction ↑ Bleeding times
Neuron	↓ Growth cone collapse and neurite retraction ↑ Axonal sprouting and regeneration ↑ Microtubule assembly ↓ Intermediate and neurofilament depolymerization ↑ Proliferation, mobilization and neuronal differentiation of stem cells ↓ NMDA receptor activity ↓ Subacute apoptotic death after injury
Oligodendrocyte precursor cells	↑ Process extension, survival, proliferation, differentiation, and myelination
Astrocyte	↓ Repulsive axon guidance molecules ↑ Prosurvival genes (e.g., BDNF, GCSF) ↓ Genes promoting cell motility and migration ↑ EAAT2 expression and glutamate transport ↑ GFAP, stellate shape, spindly processes ↓ Actin stress fibers
Microglia	↓ Chemokine release and chemotaxis ↓ Hypoxia-induced proinflammatory phenotype
Leukocytes	↓ Leukocyte-endothelial interactions ↓ Infiltration and migration ↓ Proinflammatory adhesion molecule expression ↓ MMP9 expression

BBB, blood–brain barrier; *BDNF*, brain derived neurotrophic factor; *EAAT*, excitatory amino acid transporter; *eNOS*, endothelial nitric oxide synthase; *GCSF*, granulocyte colony stimulating factor; *GFAP*, glial fibrillary acidic protein; *NMDA*, N-methyl-D-aspartate.

cerebral ischemia, including brief but severe models such as gerbil bilateral common carotid artery occlusion, and chronic but mild models such as rat bilateral common carotid artery occlusion or severe stenosis models [1]. Similarly, ROCK inhibition has been efficacious in animal models of subarachnoid hemorrhage, and even diminished cerebral aneurysm formation and growth [9]. It should be noted that enhanced endothelial thrombolytic activity and reduced platelet activation pose theoretical concerns for

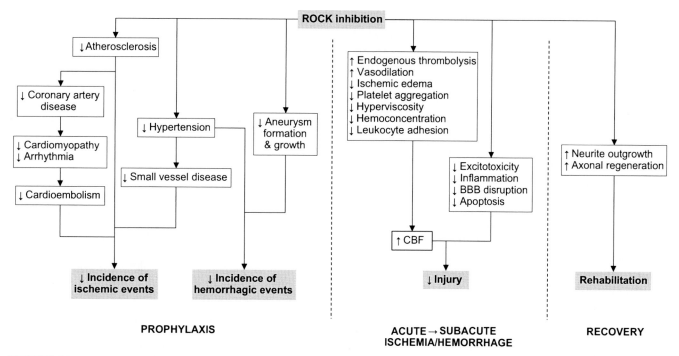

FIGURE 54.1   Selected mechanisms of action of Rho-associated kinase (ROCK) inhibitors in the prophylaxis, acute treatment, and rehabilitation of ischemic or hemorrhagic stroke. By ameliorating the endothelial dysfunction and atherosclerosis associated with vascular risk factors, chronic ROCK inhibition has the potential to reduce the risk for the most common stroke subtypes (i.e., cardioembolic, large-artery, and small-vessel strokes). The endothelial nitric oxide synthase (eNOS)-dependent acute hemodynamic improvement in focal ischemia may be complemented by antithrombotic, antiinflammatory, and antiedema effects; a direct neuroprotective mechanism has also been implicated. After the subacute stage, ROCK inhibition may improve functional rehabilitation via enhanced axonal regeneration, as shown in spinal cord injury. Although these mechanisms remain largely speculative, existence of multiple potential pathways for a beneficial effect makes ROCK a favorable target. *BBB*, blood–brain barrier; *CBF*, cerebral blood flow. *Modified from Shin HK, Salomone S, Ayata C. Targeting cerebrovascular Rho-kinase in stroke. Expert Opin Ther Targets 2008;12(12):1547–64.*

ROCK inhibition in the setting of intracerebral or subarachnoid hemorrhage. Animal models of subarachnoid hemorrhage in which ROCK inhibition was tested and found safe and efficacious have so far been direct cisternal blood injection models, which, in contrast to endovascular perforation or elastase-induced spontaneous aneurysmal rupture models in rats and mice, does not pose a risk for continuing or recurrent hemorrhage. Therefore, safety of ROCK inhibition remains to be determined in models mimicking the aneurismal subarachnoid hemorrhage stage before securing the aneurysm, or intracerebral hemorrhage models, such as collagenase injection.

Neurological recovery and rehabilitation has been another area in which ROCK inhibitors have fairly consistently shown benefit. Neurological injury creates an environment where molecular factors (e.g., Nogo-A) form a barrier to neuroplasticity, triggering axonal growth cone collapse, retraction bulb formation, and neurite withdrawal, and inhibiting oligodendrocyte precursor cell differentiation and process extension. These growth inhibitors all converge to activate RhoA/ROCK pathway as their downstream target in neurons, oligodendrocyte precursor cells, and astrocytes, initiating multiple deleterious signaling pathways causing secondary injury and inhibiting plasticity and recovery.

Ischemic or traumatic injury upregulates brain Rho/ROCK expression and activity in neurons, endothelial cells, and astrocytes within hours after stroke onset, which persists for weeks. There is ample preclinical data showing that ROCK inhibitors enhance neurological function after traumatic spinal cord as well as peripheral and optic nerve injury, and in a couple of studies after ischemic stroke [10]. Therefore, a comprehensive testing of ROCK inhibition in various models of ischemic gray and/or white matter injury will pave the way for its translation in stroke recovery.

RHO-ASSOCIATED KINASE INHIBITORS IN CLINICAL USE FOR CEREBROVASCULAR DISEASES

Despite intense efforts to develop and clinically test ROCK inhibitors for various indications (e.g., atherosclerosis, chronic renal failure, diabetic retinopathy, glaucoma, psoriasis, and spinal cord injury), only two (fasudil and ripasudil) of more than 170 ROCK inhibitors have been approved for clinical use to date. In Japan, fasudil has been in clinical use against cerebral vasospasm after aneurysmal subarachnoid hemorrhage since 1995, and its efficacy has been confirmed in a meta-analysis [11]. In one pilot trial of ischemic stroke, fasudil was found efficacious, although trial design and patient selection have made it difficult to interpret and extrapolate the result to the general stroke population. Nevertheless, clinical experience from already completed trials in pulmonary hypertension, heart failure, and stable angina will undoubtedly facilitate the translation of ROCK inhibitors in cerebrovascular diseases.

Interestingly, leukocyte ROCK activity has been shown to increase and peak 48h after ischemic stroke [12]. Indeed, ROCK activity in circulating leukocytes is a reliable biomarker for metabolic syndrome, hyperlipidemia, hypertension, and heart failure, to assess disease severity and response to therapies, and is an independent predictor of cardiovascular risk including risk for recurrent stroke [3].

CONCLUSION

The Rho/ROCK pathway is a critical molecular switch in virtually all cell types that turns on numerous potentially deleterious downstream effects, rendering ROCK a relevant therapeutic target in the prevention, acute treatment, and rehabilitation of many diverse cerebrovascular diseases. As a result, ROCK inhibitors have multiple potentially synergistic mechanisms of action that are beneficial in ischemic as well as hemorrhagic stroke. Although such pleiotropic mechanisms are expected to afford better efficacy in cerebrovascular diseases, they may also increase unwanted adverse effects particularly upon prolonged use. Efforts are under way to develop isoform-selective pharmacological inhibitors of ROCK with favorable drug-like properties and bring them to clinical trials.

References

[1] Shin HK, Salomone S, Ayata C. Targeting cerebrovascular Rho-kinase in stroke. Expert Opin Ther Targets 2008;12(12):1547–64.

[2] Loirand G. Rho kinases in health and disease: from basic science to translational research. Pharmacol Rev 2015;67(4):1074–95.

[3] Shimokawa H, Satoh K. 2015 *ATVB* Plenary Lecture: translational research on rho-kinase in cardiovascular medicine. Arterioscler Thromb Vasc Biol 2015;35(8):1756–69.

[4] Feng Y, LoGrasso PV, Defert O, Li R. Rho kinase (ROCK) inhibitors and their therapeutic potential. J Med Chem 2015;56.

[5] Vesterinen HM, Currie GL, Carter S, et al. Systematic review and stratified meta-analysis of the efficacy of RhoA and Rho kinase inhibitors in animal models of ischaemic stroke. Syst Rev 2013;2:33.

[6] Shin HK, Salomone S, Potts EM, et al. Rho-kinase inhibition acutely augments blood flow in focal cerebral ischemia via endothelial mechanisms. J Cereb Blood Flow Metab 2007;27(5):998–1009.

[7] Shin HK, Huang PL, Ayata C. Rho-kinase inhibition improves ischemic perfusion deficit in hyperlipidemic mice. J Cereb Blood Flow Metab 2014;34(2):284–7.

[8] Lee JH, Zheng Y, von Bornstadt D, et al. Selective ROCK2 inhibition in focal cerebral ischemia. Ann Clin Transl Neurol 2014;1(1):2–14.

[9] Zoerle T, Ilodigwe DC, Wan H, et al. Pharmacologic reduction of angiographic vasospasm in experimental subarachnoid hemorrhage: systematic review and meta-analysis. J Cereb Blood Flow Metab 2012;32(9):1645–58.

[10] Watzlawick R, Sena ES, Dirnagl U, et al. Effect and reporting bias of RhoA/ROCK-blockade intervention on locomotor recovery after spinal cord injury: a systematic review and meta-analysis. JAMA Neurol 2014;71(1):91–9.

[11] Liu GJ, Wang ZJ, Wang YF, et al. Systematic assessment and meta-analysis of the efficacy and safety of fasudil in the treatment of cerebral vasospasm in patients with subarachnoid hemorrhage. Eur J Clin Pharmacol 2012;68(2):131–9.

[12] Feske SK, Sorond FA, Henderson GV, et al. Increased leukocyte ROCK activity in patients after acute ischemic stroke. Brain Res 2009;1257:89–93.

CHAPTER

55

Akt-GSK3β Pro-survival Signaling Pathway in Cerebral Ischemic Injury

W. Zhang[1,2], R.A. Stetler[1,2], J. Chen[1,2]

[1]University of Pittsburgh, Pittsburgh, PA, United States; [2]Fudan University, Shanghai, China

INTRODUCTION

The balance between apoptotic signals and survival signals determines the fate of neurons after cerebral ischemia. Unobstructed activation of caspases in pathological conditions will lead to cell death; however, activation of phosphoinositide 3-kinase (PI3K)/Akt signaling pathway may counteract the apoptotic process and support neuron survival and recovery from ischemic insults. Akt, also called protein kinase B (PKB), is a 57kD serine/threonine protein kinase that is composed of an N-terminal pleckstrin homology (PH) domain, a central catalytic domain, and a C-terminal regulatory domain.

AKT ACTIVATION AND REGULATION

Full activation of Akt occurs by successive phosphorylation of Akt by serine/threonine kinases at Thr-308 within the catalytic domain and at Ser-473 within the C-terminus. This requires the upstream spatial coordination of kinases with Akt, and is largely mediated by the availability of the membrane phospholipid phosphoinositol-(3,4,5)-phosphate (PIP_3). By binding Akt via its PH domain, PIP_3 recruits both Akt and upstream kinases, including the serine/threonine kinase 3-phosphoinositide-dependent kinase-1 (PDK1), to plasma membrane. Once in close proximity to Akt, PDK1 can directly phosphorylate Akt on Thr308. In addition to the dominant role of PDK1, Akt phosphorylation on Thr308 may also be mediated by unidentified non-PDK1 kinases or autophosphorylation [1]. Phosphorylation of Akt at Ser-473 is largely dependent on PI3K family activity, and may serve as a regulatory step to Thr308 phosphorylation. Importantly, and as mentioned later, phosphorylation of Akt at Ser-473 does not always correlate with kinase activity.

Phosphatidylserine (PS) is a membrane phospholipid that regulates the Akt phosphorylation process by interacting with both Akt and PDK1 at the cytoplasmic membrane. PS, the major acidic phospholipid class in the cerebral cortex, is exclusively localized in the inner leaflet of membrane where proteins—including Akt—dock for activation. Akt contains the basic residues R15 and K20 that bind to PS in anionic domains of the cytoplasmic leaflet, and appear to be required for Akt membrane binding and conformational changes that promote activation. PS is also specifically required for the association of PDK1 with the plasma membrane and its subsequent activation of Akt. Disruption of the PS-PDK1 interaction blocks membrane localization of PDK1 and leads to diminished phosphorylation of Akt at Thr308. In addition to PDK1 phosphorylation of Akt at Thr308, the binding between the basic residues in the regulatory domain of Akt and PS facilitates phosphorylation of Ser473 by mTOR. Although phosphorylation of only Ser473 is not sufficient to stimulate its cell survival activity, full Akt activity is classically regulated by both phosphorylation sites (Thr308 and Ser473). Externalization of PS induced by cell death stimuli, including cerebral ischemia, impedes the activation of Akt and stimulates microglia phagocytosis of apoptotic cells.

Regulation of Akt activation is also highly controlled by the bioavailability of PIP_3. PIP_3 is generated by phosphorylation of phosphatidylinositol phosphate 2 (PIP_2) via PI3K. PI3K is normally present in cytosol and can be activated directly by recruitment to an activated Trk receptor. Active PI3K then phosphorylates phosphatidylinositol phosphate 2 (PIP_2) to PIP_3, which in turn can bind to Akt. Counter to the activity of PI3K, phosphatase and tensin homolog (PTEN) dephosphorylates PIP_3 back to PIP_2, leading to the subsequent inactivation

Primer on Cerebrovascular Diseases, Second Edition
http://dx.doi.org/10.1016/B978-0-12-803058-5.00055-2

of Akt. Phosphorylation of PTEN blocks its ability to interact with PIP, and thus is permissive to Akt activation. In addition to the regulation of PTEN by phosphorylation, S-nitrosylation of PTEN at Cys83 inhibits the phosphatase activity of PTEN, thereby preventing dephosphorylation of PIP$_3$ and allowing the recruitment and activation of Akt [2].

ACTIVATION AND REGULATION OF AKT IN CEREBRAL ISCHEMIC INJURY

Numerous studies suggest that cerebral ischemic insults can alter Akt phosphorylation states. In transient focal ischemic models, phosphorylation of Akt at Thr308 decreases at a slightly delayed timeframe (6 h following reperfusion), whereas phosphorylation of Akt at Ser473 transiently increases within the first few hours of reperfusion, then decreases (9–24 h following reperfusion) [1,3]. During the course of reperfusion, differences in p-Akt(Ser473) are observed in the penumbra versus the ischemic core. In the penumbral regions, transient increase of p-Akt (Ser473) occurs within the first few hours following reperfusion, but subsides by 9 h after reperfusion [1,3]. In the ischemic core, p-Akt (Ser473) decreases and remains at lower levels after reperfusion [1,3]. In a model of permanent focal ischemia, phosphorylation at both Ser473 and Thr308 are suppressed at extended time points following permanent MCAO [4]. Akt phosphorylation at Ser473 increases at acute time points following global brain ischemia in the CA1 of ischemic rats, but the duration of the increase is highly variable across the literature [5–7]. For example, many groups describe a robust upregulation of p-Akt (Ser473) within the first 24 h that subsequently subsides in rat global ischemic models [6–8]. However, Yano et al. noted an overall decrease in p-Akt (Ser473) in the gerbil model of global ischemia. These differences are likely due to subtleties between global ischemia models and species.

It is important to note that Akt phosphorylation at Ser473 is associated with, but not necessarily reflective of, full catalytic activity. Indeed, despite detecting increased p-Akt (Ser473) levels at early time points following focal ischemia, Zhao et al. [1] found *decreased* Akt activity using an in vitro kinase activity assay with lysates obtained from focal ischemic brain (also at early time points following focal ischemia). Likewise, the robust early rise in p-Akt (Ser473) following global ischemia was not associated with any change in Akt activity using lysates obtained from CA1 of global ischemic rats [8]. Thus, although early phosphorylation of Akt at Ser473 occurs, this may represent an early response to injury that is muted before robust activation of the pro-survival signaling cascade in vulnerable ischemic regions.

Significant data suggest an inverse correlation between phosphorylation or activity of Akt and cell injury in both focal and global ischemia models [3,7,9]. The presence of phosphorylated Akt or activated downstream components of the Akt signaling pathway does not overlap temporally with markers of apoptotic cell death. Following global ischemic injury in Wistar rats, the early activation of p-Akt (Ser473) subsides by 36 h after reperfusion in the CA1 region of the hippocampus, at which point cytosolic cytochrome c and caspase-3 activity begin to be detected [6]. Using a marker of Akt activity (phosphorylation of GSK3β at Ser9), no significant colocalization occurs within neurons between Akt activity and TUNEL labeling for cell death [7]. Likewise, p-Akt(Ser473) expression does not overlap with TUNEL-positive neurons in a mouse model of transient focal ischemia [3]. Conditions that decrease ischemic neuronal death, such as transgenic overexpression of Cu/ZnSOD, estradiol administration, hypothermia, and ischemic pre- or postconditioning, are also associated with the promotion of Akt phosphorylation and/or activity [1,4,5,7,9]. Thus, cells with activated Akt likely have not (yet) fully initiated cell death signaling, and may be still at a point of rescue or recovery (Fig. 55.1).

The inhibition of PTEN phosphatase activity is associated with increased Akt activity and neuroprotection. Following transient focal cerebral models, phosphorylated PTEN (i.e., inactive PTEN) immediately decreases during early time points, and either returns to normal or even to slightly increased levels 24 h postischemia/reperfusion or remains suppressed [1,9]. PTEN negatively regulates Akt, and thus the patterns observed following cerebral ischemia suggest that delayed PTEN may participate in the Akt inactivation (mentioned in the preceding paragraphs) after transient cerebral ischemia in vulnerable cellular populations. Similarly, an acute transient decrease of PTEN phosphorylation (thus an increase in PTEN activity) occurs rapidly in CA1 following global ischemia that returns to baseline by 12 h after reperfusion [8]. This is consistent with the lack of Akt activity despite increased phosphorylation at Ser473. Likewise, S-nitrosylation of Cys83 occurs following focal ischemia [2].

As an interesting note, the relationship between loss of PTEN and cellular survival is primarily described in neurons. However, in ischemic astrocytes, loss of PTEN exacerbates ischemic damage and stimulates astrogliosis, which is detrimental for the recovery of neurological functions. The difference in PTEN activity between cell types remains an open area of investigation.

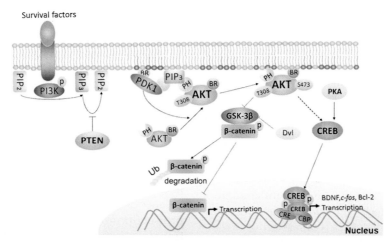

FIGURE 55.1 **Akt signaling leading to transcriptional regulation supporting cell survival after cerebral ischemic injury.** In cells protected from ischemic cell death, Akt is activated through the coordination of several upstream kinases and scaffolding proteins, including PDK1 and PI3K–PIP3. The activation of Akt leads to the suppression GSK3β, and alteration of downstream signaling pathways, including β-catenin, creating a cellular environment associated with cell survival. Inhibition of GSK3β by active Akt allows β-catenin to remain unphosphorylated and transcriptionally active. CREB is another downstream target of Akt that promotes cell survival that is activated in cellular protection against ischemic injury. However, CREB activation can be stimulated by upstream kinases other than Akt, and the role of Akt in promoting CREB activity is not well established in models of cerebral ischemic injury and ischemic neuroprotection.

AKT REGULATES TRANSCRIPTIONAL MACHINERY LEADING TO AN ENVIRONMENT SUPPORTING CELL SURVIVAL

Upon phosphorylation, Akt plays a pivotal role in mediating survival signaling by both impacting the cytoplasmic cell death machinery and by regulating expression of nuclear proteins. Akt regulates transcription by phosphorylation of critical components that alter the balance of pro-survival gene transcription, including glycogen synthase kinase 3β (GSK3β) and cyclic AMP-responsive element-binding protein (CREB), which influence expression of an array of pro-survival genes.

GLYCOGEN SYNTHASE KINASE 3β

GSK3 is a serine/threonine kinase that is a major downstream target of Akt. Of the two isoforms of GSK3, GSK3β is highly expressed in the central nervous system and is particularly abundant in neurons. Overexpression of GSK3β or increased GSK3β activity is often associated with apoptosis. Through an unknown mechanism, phosphorylation at Tyr216 of GSK3β occurs and is associated with the promotion of its activity and perpetuation of cellular injury. Active pTyr216–GSK3β was transiently observed within the first 6h after reperfusion following focal cerebral ischemia, but not in global cerebral ischemia [7]. Direct inhibition of GSK3β activity decreased injury in both ischemic cultured neurons as well as infarct volume following focal ischemia. Conversely,

Akt-mediated phosphorylation of GSK3β at Ser9 leads to suppression of GSK3β kinase activity, and is associated with signaling permissive of neuronal protection. Indeed, increased presence of pSer9–GSK3β is associated with neuroprotective paradigms against some models of cerebral ischemia [10], but not others [9].

The exact downstream effects of GSK3β inhibition that contribute to ischemic neuroprotection are still not well defined, but several downstream transcriptional pathways affected by GSK3β have been identified as potential modes to confer ischemic cerebral protection, including β-catenin and inflammatory pathways. β-catenin is a major transcription coactivator of gene products that can control cell survival following cerebral ischemia. Under conditions where GSK3β is active (i.e., phosphorylated at Tyr216), GSK3β phosphorylates β-catenin at Ser or Thr residues of the N-terminal region. Phosphorylated β-catenin is highly unstable and quickly targeted toward degradation, and thus suppresses the expression of cell survival gene products. In cerebral ischemic settings, decrease of β-catenin protein is associated with vulnerable ischemic regions. Inactivation of GSK3β by Akt-mediated phosphorylation at Ser9 leads to the stabilization of β-catenin, thus allowing β-catenin to translocate to nucleus and promote gene expression of antiapoptotic proteins. The Wnt pathway may also be relevant for linking activated Akt to GSK3β, and thus promoting GSK3β inhibition. By binding to its transmembrane receptor Frizzled and the co-receptor lipoprotein related protein 5 and 6 (LRP5/6), Wnt recruits the protein disheveled (Dvl), which coordinates PI3K/Akt to inactivate GSK3β and thus promote β-catenin-mediated gene expression. Together, these actions could promote neuron survival after ischemic brain damage. Proof of principle for the involvement of Wnt signaling in ischemic neuronal survival was provided by the observation that inhibition of Dickkopf-1 (Dkk-1, a negative regulator of Wnt signaling) increased neuronal survival following global ischemic injury.

Activation of GSK3β may also impact inflammatory responses in the brain, which is relevant to ischemic settings. These effects appear to be cell type- and context-specific, and may explain why the activity of GSK3β is variable between different cerebral ischemic models. Much of the focus of the role of GSK3β in inflammatory signaling centers on its effects on NF-κB, a major transcriptional activator that regulates the expression of dozens of inflammation-related gene products. Overexpression of GSK3β or a mutant that is resistant to Akt phosphorylation suppressed nuclear factor kappa B (NF-κB) activity in cultured astrocytes, resulting in spontaneous cell death. In cultured microglia stimulated with the proinflammatory molecule lipopolysaccharide (LPS), inhibition of GSK3β suppressed the transactivation activity of the NF-κB subunit p65, thereby suppressing Tnf-alpha expression. In this way, GSK3β activation in microglia is associated with induction of the inflammation cascade. Finally, in a model of permanent focal ischemia, inhibition of GSK3β suppressed the inflammatory response.

OTHER AKT TARGETS IMPACTING TRANSCRIPTION

In addition to modulation of the cellular environment by negatively regulating GSK3β transcription, Akt is capable of promoting cell survival by phosphorylating several other downstream targets, leading to transcriptional modifications. However, Akt-mediated control of these targets has not been established in the context of cerebral ischemia. For example, although Akt can phosphorylate the transcriptional factor cyclic AMP-responsive element-binding protein (CREB) on Ser133, which promotes transcription of genes related to cell survival, it appears as though protein kinase A, Ca^{2+}/calmodulin-dependent kinase, and mitogen-activated protein kinase may be the primary upstream kinases responsible for CREB activation in neuroprotection against cerebral ischemia.

Akt also phosphorylates Forkhead box transcription factor class Os (FOXOs). In its dephosphorylated state, FOXOs lead to the transactivation of many deleterious genes. Phosphorylation by Akt suppresses the transcription factor activity of FOXO, and thus promotes cell survival. A correlation exists following cerebral ischemia wherein Akt activity decreases concomitant to increased FOXO dephosphorylation and activity. In hypoxic preconditioning, inhibition of PI3K led to dephosphorylation of FOXO1,

implicating Akt in the phosphorylation and suppression of FOXO-targeted gene expression. More expansive studies to examine the role of Akt and FOXOs following ischemia are needed to further understand the impact of Akt on the Forkhead box branch of transcription-mediated cell death signaling.

SUMMARY

Overall, phosphorylation of Akt appears to be highly involved in cellular recovery after cerebral ischemia. Balance between PI3K and PTEN tightly regulates the production of PIP_3, which determines the membrane translocation of Akt and subsequent activation. Pro-survival effects of Akt following ischemic injury depend at least in part on the downstream substrates that target transcription, including GSK3β and FOXO, promoting a variety of pro-survival gene expression and suppressing deleterious gene expression to counteract cerebral ischemia-induced neuronal death.

References

[1] Zhao H, Shimohata T, Wang JQ, et al. Akt contributes to neuroprotection by hypothermia against cerebral ischemia in rats. J Neurosci 2005;25(42):9794–806.

[2] Numajiri N, Takasawa K, Nishiya T, et al. On-off system for PI3-kinase-Akt signaling through S-nitrosylation of phosphatase with sequence homology to tensin (PTEN). Proc Natl Acad Sci USA 2011;108(25):10349–54.

[3] Noshita N, Lewen A, Sugawara T, Chan PH. Evidence of phosphorylation of Akt and neuronal survival after transient focal cerebral ischemia in mice. J Cereb Blood Flow Metab 2001;21(12):1442–50.

[4] Perez-Alvarez MJ, Maza Mdel C, Anton M, Ordonez L, Wandosell F. Post-ischemic estradiol treatment reduced glial response and triggers distinct cortical and hippocampal signaling in a rat model of cerebral ischemia. J Neuroinflammation 2012;9:157.

[5] Kawano T, Fukunaga K, Takeuchi Y, et al. Neuroprotective effect of sodium orthovanadate on delayed neuronal death after transient forebrain ischemia in gerbil hippocampus. J Cereb Blood Flow Metab 2001;21(11):1268–80.

[6] Ouyang YB, Tan Y, Comb M, et al. Survival- and death-promoting events after transient cerebral ischemia: phosphorylation of Akt, release of cytochrome C and Activation of caspase-like proteases. J Cereb Blood Flow Metab 1999;19(10):1126–35.

[7] Endo H, Nito C, Kamada H, Nishi T, Chan PH. Activation of the Akt/GSK3beta signaling pathway mediates survival of vulnerable hippocampal neurons after transient global cerebral ischemia in rats. J Cereb Blood Flow Metab 2006;26(12):1479–89.

[8] Miyawaki T, Ofengeim D, Noh KM, et al. The endogenous inhibitor of Akt, CTMP, is critical to ischemia-induced neuronal death. Nat Neurosci 2009;12(5):618–26.

[9] Gao X, Zhang H, Takahashi T, et al. The Akt signaling pathway contributes to postconditioning's protection against stroke; the protection is associated with the MAPK and PKC pathways. J Neurochem 2008;105(3):943–55.

[10] Zhang F, Signore AP, Zhou Z, Wang S, Cao G, Chen J. Erythropoietin protects CA1 neurons against global cerebral ischemia in rat: potential signaling mechanisms. J Neurosci Res 2006;83(7):1241–51.

CHAPTER

56

Heat Shock Proteins and the Stress Response

J.Y. Kim, M. Yenari

University of California, San Francisco and the San Francisco Veterans Affairs Medical Center, San Francisco, CA, United States

INTRODUCTION

In response to ischemia, cellular responses lead to the induction of a variety of stress proteins, included among them are the heat shock proteins (Hsps) and immediate early genes (IEGs). Hsps are a family of stress proteins thought to be involved in chaperone functions, such as protein folding, trafficking, and repair. Their expression can be constitutive or inducible depending on the family member. Constitutively expressed members exist in all cell compartments, and appear essential for development and cellular function. Inducible forms can be induced following a variety of external stress including ischemia, but were originally described following heat stress [1]. Their induction appears to be part of an orchestrated stress response. Work over the past two decades has also established that some Hsps also function as cytoprotectants. Hsps have long been known to serve as protein chaperones in the sense that they assist in protein folding and the correct attainment of functional three-dimensional configuration, while preventing incorrect folding and protein aggregation [2]. They have also been shown to affect cellular signaling [2], and have been extensively studied in the setting of cerebral ischemia and demonstrated to provide protection against both global and focal cerebral ischemia.

HEAT SHOCK PROTEINS

Hsps are classified according to their molecular mass, and include Hsp100, Hsp90, Hsp70, Hsp60, Hsp40, and the small Hsp families. The best-studied class is Hsp70, the 70-kDa class that includes an inducible form also known as Hsp72, Hsp70i, or simply Hsp70. Hsp70 interacts with hydrophobic peptide segments of unstructured proteins in an ATP-dependent manner. Hsp70 also contains a C-terminal substrate-binding domain that identifies unstructured polypeptide segments, and an N-terminal ATPase domain that assists in protein folding, alternating between an ATP-bound, open state with low substrate affinity and an ADP-bound closed state with high substrate affinity [2]. In studies of cerebral ischemia, Hsp70 was observed to be induced in brain regions that were relatively resistant to ischemic insults. Hence, the notion of a "molecular penumbra" was introduced, and raised questions as to whether this expression was an epiphenomenon of the injury, or an active participant in cell survival [3]. Subsequent studies using strategies to increase or inhibit Hsp70 expression have consistently shown that Hsp70 protects the brain and brain cells against experimental cerebral ischemia, neurodegenerative disease models, epilepsy, and trauma. Through its chaperone properties, it has been shown to reduce protein aggregates and intracellular inclusions [4]. Two other stress proteins studied in brain ischemia include Hsp27 and Hsp32 (also known as heme oxygenase, or HO-1) [5]. In addition to their function in protein processing, Hsps appear to protect the brain by affecting several cell death and immune response pathways [1].

THE HEAT SHOCK RESPONSE

The molecular mechanism for regulation of Hsps expression depends on the activity of a unique transcription factor-heat shock factor 1 (HSF1) that can bind to the 5'promoter regions of all Hsp genes and trigger transcription [6]. Under homeostatic conditions, Hsps are located intracellularly and are bound to HSF1 [1]. An appropriate stress, such as heat, ischemia, and other causes of accumulation of unfolded proteins leads to the dissociation of Hsps from HSF, leaving Hsps free to bind target proteins. In the stressed cell, dissociated HSF is transported to the nucleus where it is phosphorylated,

Primer on Cerebrovascular Diseases, Second Edition
http://dx.doi.org/10.1016/B978-0-12-803058-5.00056-4

possibly by protein kinase C, to form activated trimers. These trimers bind to highly conserved regulatory sequences on the heat shock gene known as heat shock elements (HSEs). Once bound to HSEs, HSFs bind to the promoter region of Hsp genes, leading to more Hsp generation [6]. Newly generated Hsps can then bind denatured proteins and act as a molecular chaperone by contributing to repair, refolding, and trafficking of damaged proteins within the cell. Hsp90 can also influence Hsp70, since Hsp90 is bound to HSF1. When Hsp90 dissociates from HSF1, HSF1 is liberated to bind HSEs, leading to more Hsp70 induction.

HSP70 IN ISCHEMIC STROKE

Hsp70 in Experimental Cerebral Ischemia

Overexpression of Hsp70 in experimental models leads to protection against a variety of acute insults, such as cerebral ischemia, trauma, and hemorrhage [6]. During homeostatic conditions, inducible Hsp70 levels are low; however, its expression is significantly increased following injury. In experimental cerebral ischemia, Hsp70 has been shown to lead to neuroprotection [1]. Viral vector–mediated Hsp70 overexpression has been shown to improve survival of neurons and astrocytes from ischemic and ischemia-like insults, including oxygen glucose deprivation and focal and global cerebral ischemia [4]. Similarly, transgenic mice that overexpress Hsp70 are protected from these ischemic insults, whereas their deficiency exacerbates outcome [7,8].

Pharmacological induction of Hsp70 is also possible, and has been shown to protect the brain in experimental models. The best studied Hsp70 inducers include the ansamycins, geldanamycin, and 17-(allylamino) geldanamycin (17AAG). These compounds induce Hsp70 through their ability to inhibit Hsp90, and have been shown to protect the brain from experimental stroke and traumatic brain injury when given exogenously [6]. Further, geldanamycin and 17-AAG have been tested in humans, albeit for other indications. Although the safety profile of the ansamycins has not been ideal, other Hsp70-inducing compounds include radicicol, BIIB021, and geranylgeranylacetone, some of which have been or are currently under study at the clinical level for cancer.

Mechanisms of Hsp70 Protection

Hsp70 has been assumed to protect the cell via its chaperone functions. Indeed, overexpression of Hsp70 appears to prevent protein aggregation and redistribution of ubiquitin [2,4]. However, other studies have shown that Hsp70 can interfere with apoptosis at multiple levels. Hsp70 has been shown to interfere with the

recruitment of procaspase-9 into the apoptosome, and the activity of apoptosis protease activating factor-1, which is required for the formation of the apoptosome, which leads to the cleavage of procaspase-9 into its active form, caspase-9 [2]. Hsp70 also prevents apoptosis inducing factor (AIF) from translocating from the mitochondria to the nucleus where it can induce caspase-independent cell death. Overexpressing Hsp70 has also been shown to lead to increased levels of Bcl-2 [9], and Bcl-2 is an antiapoptotic protein which acts to block mitochondrial release of cytochrome c and AIF. Hsp70 overexpression also appears to inhibit mitochondrial release of the pro-apoptotic Bcl-2 family member Bax [10], and directly inhibits the effector caspase, caspase-3 [9].

Hsp70 is also known to play a role in modulating inflammation caused by cerebral ischemia. As an anti-inflammatory molecule, Hsp70 inhibited production of proinflammatory cytokines in cultured microglia and macrophages and in stroke models [6]. Hsp70 also suppresses the release of other proinflammatory factor molecules, such as matrix metalloproteinases (MMPs), inducible nitric oxide synthase (iNOS), and reactive oxygen species. Hsp70 also decreases NADPH oxidase activity in neutrophils while increasing superoxide dismutase in phagocytes [2]. Hsp70 also inhibits activation of the inflammatory transcription factor, nuclear factor kappa B (NF-kB). Hsp70 was also found to bind to and block NF-kB and its inhibitor protein (IkB), and disrupt function of the kinase of IkB [2]. In addition to NF-kB inhibition, Hsp70 overexpression in astrocytes inhibited JNK and p38, effectively downregulating the expression of proinflammatory genes [11]. Inducing Hsp70 in cultured astrocytes reduced expression and activation of MMP-9 in oxygen–glucose deprivation [6].

Although much of the work in brain ischemia surrounding the role of Hsp70 in modulating inflammation has focused on its role in intracellular signaling, work in related areas indicates that Hsp70 plays a different role extracellularly [12]. Hsp70 can be secreted by astrocytes and Schwann cells, or released by dying cells [2]. In these settings, Hsp70 may act as a danger signal, by acting upon receptors, such as Toll-like receptor-4 and -2, and leading to activation of immune cells and elaboration of proimmune molecules, thus potentially contributing to worsened damage.

OTHER HEAT SHOCK PROTEINS

Hsp27

Hsp27 possesses many similarities to Hsp70, with the exception that it does not require ATP for its actions [5]. Animals overexpressing Hsp27 were protected against excitotoxin exposure and experimental stroke

[5,13]. Mechanisms of this protective effect have largely been attributed to the ability of Hsp27 to interfere with apoptosis. Like Hsp70, Hsp27 has been shown to prevent mitochondrial release of cytochrome c and prevent formation of the apoptosome. Hsp27 may also directly interact with procaspase-3 and prevent Bax translocation to the mitochondria.

Hsp32 (Heme Oxygenase)

Hsp32, more commonly known as heme oxygenase-1 (HO-1), is an enzyme involved in heme catabolism. HO-1 is inducible, and is considered a member of the stress protein family, because it contains an HSF in its promoter. It converts heme into biliverdin, carbon dioxide, and ferrous iron. It is induced by similar factors as other Hsps [5]; however, the literature surrounding HO-1 in brain ischemia and related conditions is conflicting. Studies of HO-1-deficient animals have shown that HO-1 knockout mice have worsened outcomes in ischemic stroke models, but improved outcome in models of brain hemorrhage. Reasons for these discrepancies may be due to the differential effects of its metabolites and the setting and location where HO-1 is active.

Immediate Early Genes

Like the heat shock proteins, IEGs are rapidly upregulated in response to cell stress including ischemia and trauma. They are thought to play roles in signaling cascades involved cell growth and differentiation under noninjury conditions. In comparison with Hsps, IEGs are observed within minutes to hours of ischemia onset, and before the emergence of Hsps. The two most widely studied IEGs in brain ischemia are c-fos and c-jun, and their respective proteins, Fos and Jun. The expression of IEG proteins in brain ischemia models has been observed in brain regions correlated to delayed neuronal death, leading some to speculate that these genes may participate in cell death mechanisms evolving over days, such as apoptosis. Fos and Jun proteins form a heterodimer, AP-1, which acts as a transcription factor [14]. AP-1 leads to the upregulation of several molecules known to promote death, such as CD95 (also known as Fas) and iNOS [3]. Consistent with this, gene knockdown of c-fos with subsequent reduction of AP-1 activity appears to improve outcome from experimental stroke, thus supporting a death-promoting property of AP-1 and its Fos and Jun subunits [15].

CONCLUSION

Many studies support the neuroprotective effect of Hsps in cerebral ischemia. The most widely studied is Hsp70, which has shown consistent neuroprotective effects in different injury models. It appears to have multiple protective mechanisms, and can be pharmacologically induced with agents that have been tested in humans for other indications. Other stress genes and their respective proteins also hold promise as beneficial endogenous protectants, but may be less developed in terms of how this property may be applied clinically. IEGs, in contrast, have been studied less, but accumulating data suggest that they may be involved in promoting cell death in response to stress.

References

[1] Kelly S, Yenari MA. Neuroprotection: heat shock proteins. Curr Med Res Opin 2002;18(Suppl. 2):s55–60.
[2] Giffard RG, Han RQ, Emery JF, Duan M, Pittet JF. Regulation of apoptotic and inflammatory cell signaling in cerebral ischemia: the complex roles of heat shock protein 70. Anesthesiology August 2008;109(2):339–48.
[3] Sharp FR, Lu A, Tang Y, Millhorn DE. Multiple molecular penumbras after focal cerebral ischemia. J Cereb Blood Flow Metab July 2000;20(7):1011–32.
[4] Giffard RG, Yenari MA. Many mechanisms for hsp70 protection from cerebral ischemia. J Neurosurg Anesthesiol January 2004;16(1):53–61.
[5] Sharp FR, Zhan X, Liu DZ. Heat shock proteins in the brain: role of Hsp70, Hsp 27, and HO-1 (Hsp32) and their therapeutic potential. Transl Stroke Res December 2013;4(6):685–92.
[6] Kim N, Kim JY, Yenari MA. Anti-inflammatory properties and pharmacological induction of Hsp70 after brain injury. Inflammopharmacology June 2012;20(3):177–85.
[7] Lee SH, Kim M, Yoon BW, et al. Targeted hsp70.1 disruption increases infarction volume after focal cerebral ischemia in mice. Stroke December 1, 2001;32(12):2905–12.
[8] Kim JY, Kim N, Zheng Z, Lee JE, Yenari MA. The 70kDa heat shock protein protects against experimental traumatic brain injury. Neurobiol Dis October 2013;58:289–95.
[9] Yenari MA, Liu J, Zheng Z, Vexler ZS, Lee JE, Giffard RG. Antiapoptotic and anti-inflammatory mechanisms of heat-shock protein protection. Ann N Y Acad Sci August 2005;1053:74–83.
[10] Stankiewicz AR, Lachapelle G, Foo CP, Radicioni SM, Mosser DD. Hsp70 inhibits heat-induced apoptosis upstream of mitochondria by preventing Bax translocation. J Biol Chem November 18 2005;280(46):38729–39.
[11] Kim JY, Yenari MA, Lee JE. Regulation of inflammatory transcription factors by heat shock protein 70 in primary cultured astrocytes exposed to oxygen-glucose deprivation. Neuroscience Feburary 12 2015;286:272–80.
[12] Giffard RG, Lee JE, Yenari MA. Hsp70 in stroke-brain protection, intracellular and extracellular roles. In: Morel ECV, editor. Heat-shock proteins: New Research. Hauppauge. Nova Science Publishers, Inc.; 2008. p. 243–63.
[13] Latchman DS. Protective effect of heat shock proteins in the nervous system. Curr Neurovascular Res January 2004;1(1):21–7.
[14] Raivich G, Behrens A. Role of the AP-1 transcription factor c-Jun in developing, adult and injured brain. Prog Neurobiol April 2006;78(6):347–63.
[15] Liu PK, Salminen A, He YY, et al. Suppression of ischemia-induced fos expression and AP-1 activity by an antisense oligodeoxynucleotide to c-fos mRNA. Ann Neurol October 1994;36(4):566–76.

CHAPTER

57

Noncoding RNAs and Stroke

A. Dharap[1], R. Vemuganti[2]

[1]JFK Medical Center, Edison, NJ, United States; [2]University of Wisconsin, Madison, WI,
United States

INTRODUCTION

Stroke (cerebral ischemia)-induced brain damage is a complex injury that begins with loss of oxygen and nutrient flow to the brain due to blockage of one or more cerebral arteries by plaque formation leading to embolism. This acute shortage of essential elements triggers a plethora of molecular events that ultimately result in widespread inflammation, oxidative stress, edema, apoptotic and necrotic cell death, and ultimately scarring of the tissue. This in turn results in loss of function in the affected regions that induces serious disabilities in the stroke survivors. Although the molecular and cellular mechanisms underlying the progression of ischemic pathophysiology have been well studied, a majority of the studies were focused on proteins and protein-coding genes. Advances in the discovery and characterization of various classes of noncoding RNAs (ncRNAs) revealed a novel layer of regulatory mechanisms that is important for the modulation of transcription and translation, and normal cellular physiology. In this chapter, we discuss the studies that evaluated the significance of ncRNAs in promoting brain damage and/or plasticity after stroke.

NONCODING RNAs

Early studies on the mechanisms of polypeptide synthesis identified two distinct types of RNAs within the translational machinery that did not encode peptides. These include the ribosomal RNAs (rRNA), which interact with a variety of proteins to form ribonucleoprotein complexes that catalyze peptide bond formation to assemble amino acids into polypeptide chains, and the transfer RNAs (tRNAs), which recognize, bind, and transport individual amino acids to the rRNA complexes to be incorporated into the growing polypeptide chains. Together, these two classes of ncRNAs are indispensable to protein translation. Over the past 20 years, many studies including the Human Genome Project and the Encyclopedia of DNA Elements (ENCODE) indicated that the majority of the RNAs transcribed by the mammalian genome are ncRNAs. To date, 20 subtypes of ncRNAs have been conclusively identified that differ from each other in their biogenesis, size, subcellular localization, and function [1]. The three subtypes that have been studied widely in the mammalian brain are microRNAs (miRNAs), piwi-interacting RNAs (piRNAs), and long noncoding RNAs (lncRNAs).

MicroRNAs

The miRNAs are a highly conserved class of small ncRNAs that are 18–25 nucleotides (nt) long. Mechanistically miRNAs target mRNA transcripts via complementary base pairing in the 3′untranslated regions to prevent their translation in the cytosol. However, studies also showed the presence of mature miRNAs in the nucleus where they target the DNA or nascent antisense RNAs to either repress or induce transcription. The miRNAs are abundant in the mammalian CNS, and were shown to play major roles during brain development, maturation of dendrites and spines, and in synaptic function and plasticity. The miRNAs are known to be packaged in exosomes, which can be transferred between cells to facilitate cell-to-cell communication throughout the CNS [2]. Due to their indispensable physiological roles, miRNA dysregulation has been implicated in the progression of brain damage in acute and chronic conditions such as stroke, traumatic brain injury, epilepsy, tumors, and neurodegenerative and psychiatric disorders. As a

Primer on Cerebrovascular Diseases, Second Edition
http://dx.doi.org/10.1016/B978-0-12-803058-5.00057-6

result, miRNAs are prime candidates for the development of blood-based biomarkers and therapies to limit secondary brain damage.

Long Noncoding RNAs

The lncRNAs are a class of large ncRNAs that are >200 nt in length and are highly enriched in the CNS [3]. Some of them can be as large as 17,000 nucleotides long. They are predominantly transcribed from intergenic regions (genomic stretches between protein-coding sequences), but may also be generated from antisense to protein-coding genes; via bidirectional transcription (head-to-head or tail-to-tail with respect to protein-coding genes); in the sense direction overlapping protein-coding genes; or via splicing of introns. LncRNAs share several features with protein-coding transcripts, such as splicing and polyadenylation; however, a majority of them lack phylogenetic sequence conservation and are devoid of open reading frames (and therefore translation potential). LncRNAs exhibit the most diverse functions in the ncRNA family. At the transcriptional level, they are implicated in cis- or trans-regulation via chromatin landscape modifications, scaffolding of macromolecular protein complexes, recruitment of transcription factors, and chromosome looping. At the posttranscription level, they play a role in mRNA stability and splicing, and miRNA binding [3]. LncRNAs are expressed in all cells and tissue types, but usually at levels that are magnitudes of order lower than that of mRNAs. Although they generally exhibit poor evolutionary sequence conservation, hundreds of lncRNAs are highly conserved among primates suggesting a role in higher order brain functions.

Piwi-Interacting RNAs

The piRNAs are a distinct class of small ncRNAs that are ~30 nt in length and interact with the piwi family of proteins to form functional complexes that inhibit transposons. They are present as clusters throughout the genome (1–100 kb in length), and each cluster can contain from a dozen to up to a few thousand piRNAs. Like miRNAs, piRNAs may originate from the noncoding regions of the genome or within protein-coding stretches. They predominantly target RNAs derived from transposons (primarily retrotransposons) to cleave the transcripts and prevent them from getting reverse-transcribed and incorporated into DNA. They are therefore referred to as the guardians of the genome that prevent excessive mutations. Early studies in germ cells showed that piRNAs are important for spermatogenesis and embryonic development. Further studies showed that piRNAs are also robustly expressed in the CNS and other somatic tissues suggesting a diverse array of functional roles in mammals.

miRNAs AND ISCHEMIC BRAIN DAMAGE

Temporal changes in the cerebral miRNA expression profiles after stroke have been evaluated in humans and rodents using microarrays and high-throughput sequencing [1,4]. These studies have encompassed various types of ischemia (transient vs. permanent and focal vs. global) and have established that miRNA expression in the brain alters swiftly and robustly as early as a few minutes and lasts up to several days after ischemia. In addition to the CNS, blood miRNAs were also reported to be altered in response to stroke, and miRNAs such as miR-298, miR-155, and miR-362-3p show potential to be used as biomarkers for stroke progression. Importantly, many of the miRNAs altered after stroke were implicated in vascular function, angiogenesis, and neural function.

Various factors, such as diet, environment, sex, and health can influence gene expression in mammals. Due to hormonal differences, susceptibility to stroke varies significantly between males and females. Accordingly, post-stroke gene expression and physiological changes also vary between the sexes [5]. Young females are more resistant than young males to stroke. This heterogeneity can influence miRNA expression profiles of patients with stroke and thereby influence downstream applications, such as the development of biomarkers or therapies. An experimental stroke study showed that males and females share a common set of ischemia-responsive miRNAs in addition to sex-specific changes [6]. Expression of several miRNAs was significantly altered exclusively in female rats following transient focal ischemia and, importantly, circulating blood miRNAs were altered differentially at 2–5 days after stroke as a function of sex [6]. Infarct volume and sensory-motor deficits also demonstrated sexually dimorphic differences with significantly worse outcomes in young adult males, middle-aged males, and middle-aged females as compared to young adult females. One study showed that downregulation of miR-23a led to the de-repression of its target X-linked inhibitor of apoptosis (XIAP) in males after stroke [7]. As XIAP binds and sequesters caspases and thus inhibits the proapoptotic caspase-3, downregulation of miR-23a resulted in reduced post-stroke apoptosis in males. Conversely, miR-23a increased in females after stroke leading to decreased XIAP and hence increased apoptosis after stroke. Thus, there is a functional consequence of the sexual dimorphism in miRNA responses after stroke.

Numerous studies have been conducted to understand the importance of miRNAs that are altered after stroke. For example, miRNAs miR-15b, miR-29, and miR-497 induced after stroke have been implicated in modulating Bcl-2 expression and thus promoting neuronal death in the post-ischemic brain. Concurrently, miRNAs, such as miR-324 and miR-155 enhance the inflammatory response, and miR-145 represses neuroprotective superoxide dismutase-2 leading to oxidative stress and neuronal death. Other miRNAs, such as miR-134 induce post-stroke neuronal death by modulating the expression of the cyclic AMP response element-binding protein, and miR-29c by de-repression of DNA methyltransferase 3a. The miRNAs are also thought to regulate chaperones and astrocyte functions to alter the post-stroke outcome. Details on these and several other functional studies on ischemia-relevant miRNAs have been reviewed elsewhere [1,4,8].

Many miRNAs are also known to play important roles in neuroprotection and post-stroke recovery. For example, miR-126 modulates vascular integrity and vessel growth, whereas miR-9 promotes smooth muscle cell proliferation during hypoxia by interacting with hypoxia-inducible factor1-α. miR-181c and miR-21 minimize neuronal apoptosis after ischemia by suppressing the proinflammatory response of tumor necrosis factor-α and FasL in activated microglia. miR-146a induced after ischemia is thought to potentiate Toll-like receptor signaling and thus neuroprotection. miR-21 is thought to promote cell survival in the ischemic penumbra by repressing the proapoptotic FAS ligand, and miR-223 protects hippocampal neurons during global cerebral ischemia by targeting the glutamate receptors GluR2 and NR2B. The miR-124 and miR-17-92 clusters are implicated in post-stroke neurogenesis via modulation of the Notch and Sonic hedgehog signaling pathways. The functional roles of miRNAs in post-ischemic neuroprotection have been reviewed in detail previously [1,4,8]. A short-duration ischemic attack that does not induce brain damage is known to induce tolerance to subsequent long-duration ischemic attacks. The mechanisms that modulate this phenomenon known as ischemic preconditioning are not completely understood. However, many miRNAs induced during the acquisition of ischemic tolerance are thought to play a central role in preventing post-stroke brain damage [9]. Collectively, these studies demonstrate the involvement of miRNAs in regulating a variety of post-stroke physiological pathways that promote either cell death or survival, and modulating individual miRNAs may lead to therapeutic benefits. However, due to their functional redundancy and overlapping spatiotemporal expression, further experimentation is needed to design miRNA-based stroke therapeutics.

LncRNAs IN THE POST-ISCHEMIC BRAIN

Compared to miRNAs, expression and functions of lncRNAs in stroke pathophysiology have been less studied. A recent study from our lab evaluated the expression profiles of lncRNAs as a function of reperfusion time after transient focal cerebral ischemia in adult rats [10]. This study showed that ~5% of the lncRNAs expressed in the brain are altered significantly as early as 3h after stroke and many of the changes sustain to at least 1day after stroke. Bioinformatics analysis showed that many of the lncRNAs altered after stroke exhibit >90% homology with exons of protein-coding transcripts, but the potential open reading frames are truncated in them compared to their viable protein-coding counterparts. Many stroke-responsive lncRNAs also failed to yield any proteins/peptides when subjected to in vitro translation. Interestingly, several stroke-responsive lncRNAs were observed to bind to the chromatin-modifying proteins Sin3A and coREST, which are cofactors of the RE-1 silencing transcription factor (REST) [11]. This indicates that lncRNAs play a role in the epigenetic control of gene expression. REST is known to mediate post-ischemic neuronal death and lncRNAs might play an important role in scaffolding the essential cofactors like coREST and Sin3A to facilitate REST-induced gene suppression. Although most stroke-responsive lncRNAs were observed to be intergenic, some were intragenic to gene loci that are important for the stroke outcome, such as the transcription factor Fos, Dclk1, and the astrocytic activation marker GFAP. Further experimentation using gain- and loss-of-function strategies is necessary to evaluate the functional significance of these and other lncRNAs in post-stroke brain damage.

piRNAs IN THE POST-ISCHEMIC BRAIN

The piRNAs are known to inhibit retrotransposon activity and thus maintain fitness of the genome. A study from our laboratory showed that stroke significantly alters the expression of many piRNAs in the cerebral cortex of adult rats [12]. Interestingly, the transcription factor families that have binding sites on the putative gene promoters of stroke-responsive piRNAs were observed to be redundant indicating that specific transcription factors might be responsible for the changes in the piRNA transcriptome after stroke. Furthermore, piRNAs altered after stroke were observed to target a specific set of retrotransposons in a redundant manner. As transposon-induced mutagenesis disables protein-coding genes and therefore translation, altered piRNA levels after stroke might be important for maintaining genome integrity in the post-stroke brain by eliminating these retrotransposons. Further studies are needed to verify the roles of piRNAs in post-stroke brain damage.

CONCLUDING REMARKS

Although studies conducted to date show that stroke rapidly alters the expression and actions of various classes of ncRNAs in the brain, the functional significance of these changes for post-stroke epigenetics, transcription and translation leading to secondary brain damage and motor dysfunction needs to be evaluated. The ncRNAs may serve as future biomarkers and therapeutic targets to minimize the post-stroke pathophysiology.

Acknowledgments

This work was supported by NIH grant NS079585, VA merit review grant 1I01BX002985, and American Heart Association grant 15IRG23050015.

References

[1] Vemuganti R. All's well that transcribes well: non-coding RNAs and post-stroke brain damage. Neurochem Int 2013;63(5):438–49.

[2] Xin H, Li Y, Chopp M. Exosomes/miRNAs as mediating cell-based therapy of stroke. Front Cell Neurosci 2014;8:377.

[3] Baker M. Long noncoding RNAs: the search for function. Nat Methods 2011;8(5):379–83.

[4] Saugstad JA. Non-coding RNAs in stroke and neuroprotection. Front Neurol 2015;6:50.

[5] Selvamani A, Sathyan P, Miranda RC, Sohrabji F. An antagomir to microRNA Let7f promotes neuroprotection in an ischemic stroke model. PLoS One 2012;7(2):e32662.

[6] Murphy SJ, Lusardi TA, Phillips JI, Saugstad JA. Sex differences in microRNA expression during development in rat cortex. Neurochem Int 2014;77:24–32.

[7] Siegel C, Li J, Liu F, Benashski SE, McCullough LD. miR-23a regulation of X-linked inhibitor of apoptosis (XIAP) contributes to sex differences in the response to cerebral ischemia. Proc Natl Acad Sci USA 2011;108(28):11662–7.

[8] Ouyang YB, Stary CM, Yang GY, Giffard R. microRNAs: innovative targets for cerebral ischemia and stroke. Curr Drug Targets 2013;14(1):90–101.

[9] Dharap A, Vemuganti R. Ischemic pre-conditioning alters cerebral microRNAs that are upstream to neuroprotective signaling pathways. J Neurochem 2010;113(6):1685–91.

[10] Dharap A, Nakka VP, Vemuganti R. Effect of focal ischemia on long noncoding RNAs. Stroke 2012;43(10):2800–2.

[11] Dharap A, Pokrzywa C, Vemuganti R. Increased binding of stroke-induced long non-coding RNAs to the transcriptional corepressors Sin3A and coREST. ASN Neuro 2013;5(4):283–9.

[12] Dharap A, Nakka VP, Vemuganti R. Altered expression of PIWI RNA in the rat brain after transient focal ischemia. Stroke 2011;42(4):1105–9.

CHAPTER

58

Cytokines and Chemokines in Stroke

I. Ballesteros[1,3], M.I. Cuartero[1,2], J.M. Pradillo[1,2], M.A. Moro[1,2], I. Lizasoain[1,2]

[1]Universidad Complutense, Madrid, Spain; [2]Instituto de Investigación Hospital 12 de Octubre (i+12), Madrid, Spain; [3]Memorial Sloan-Kettering Cancer Center, New York, NY, United States

INTRODUCTION

Cells involved in the immune system need to be tightly interconnected to elaborate a collaborative and well-organized response against noxious stimuli. With this aim, cells communicate either by direct contact mediated by different plasma membrane molecules or by the synthesis and release of small proteins called *cytokines*. These cytokines are produced in the earliest stages of cellular activation, alerting other cells of the existence of an on-going immune response. In addition, at later phases, they may participate in resolution and repair of the inflamed tissue. In some instances they exert a local effect, whereas in others they may also act on distant targets. Their functions include the regulation of innate immunity and inflammation including the recruitment of leukocytes to the conflictive region (mediated by specific cytokines known as *chemokines*), the control of adaptive immunity, and the generation of newborn cells derived from hematopoietic precursors, among others. In the context of stroke, we will focus on the role of cytokines and chemokines on innate immune response and inflammation including the mechanisms for its resolution, and on some of the active reparative processes that take place in the injured brain.

CYTOKINES AS SIGNALING MOLECULES IN THE INNATE IMMUNE RESPONSE OF THE ISCHEMIC BRAIN

Innate immunity is an evolutionary old defense system that generates a fast inflammatory response that is generic to all types of pathogens or tissue damages and does not confer immune memory to the host.

The innate immune system is activated after stroke as a consequence of hypoxia-mediated cell death, which leads to the release of tissue "danger signals" known as damage-associated molecular patterns (DAMPs), like fibrinogen, heat shock proteins, or extracellular ATP. Recognition of DAMPs by pattern recognition receptors, such as Toll-like receptors (TLRs) or NOD-like receptors (NLRs), activates downstream signaling pathways in immune cells, including nuclear factor-κB, mitogen-activated protein kinases (MAPK), and type-I interferon pathways. This culminates in the upregulation of proinflammatory mediators, including cytokines and chemokines that, in turn, activate resident glial cells, mainly microglia, and recruit blood-borne cells to the ischemic tissue [1]. Accumulating evidence shows that postischemic inflammation plays critical roles in stroke outcome and that innate immunity can be associated with not only damage to the host tissue but also resolution of inflammation and neurorepair.

CYTOKINES AND NEUROINFLAMMATION

The inflammatory response to ischemic brain injury is orchestrated by a variety of cytokines released by brain cells, such as neurons and glia, and by blood-borne infiltrating cells. Among them, tumor necrosis factor (TNF), interleukin (IL)-1, and IL-6 attract considerable interest as putative biomarkers of stroke outcome severity and potential therapeutic targets because they strongly affect the development of ischemic brain damage in animal models and their levels are increased in blood and in cerebrospinal fluid (CSF) of patients with stroke. In contrast, other cytokines, such as IL-4, IL-10, and transforming growth factor (TGF)-β exert immunoregulatory and antiinflammatory effects and promote monocyte/macrophage polarization toward an M2 phenotype [2].

Primer on Cerebrovascular Diseases, Second Edition
http://dx.doi.org/10.1016/B978-0-12-803058-5.00058-8

TNF-α is a cytokine strongly involved in the post-stroke inflammatory damage. A rapid upregulation/release of TNF-α occurs following stroke both experimentally and clinically. Its expression is elevated in neurons during the first hours after the insult, whereas it is increased later in glia and blood-borne immune cells. This is accompanied by an increased expression of its converting enzyme, TACE/ADAM17, which produces the soluble form of this cytokine by proteolytic cleavage. TNF-α-mediated activation of its receptors (TNFR1 and 2) has neurotoxic repercussions, leading to cell necrosis, activation of caspases and other apoptotic factors, and induction of inflammatory enzymes and metalloproteinases. However, the role of TNF-α in stroke is controversial due to its ability to activate also neuroprotective pathways. In fact, TNF-α is involved in the phenomenon of ischemic tolerance both experimentally and clinically. In this setting, this cytokine reduces the rise in intracellular Ca^{2+}, promotes antioxidant defenses, and increases glutamate clearance from the extracellular space [3].

IL-1 is another key cytokine involved in ischemic brain damage. Cerebral levels of both IL-1α and IL-1β increase within hours after brain ischemia; IL-1α is mainly induced in microglia and IL-1β can be released by all the elements of the neurovascular unit. In astrocytes and microglia, ischemia-induced IL-1β production involves activation of TLR4 and p38 MAPK. Both IL-1 forms are initially produced as precursors and require proteolytic cleavage by multiprotein complexes called inflammasomes (including the NLR subset of NLRPs) to yield a mature, secreted, biologically active molecule. Once IL-1 is secreted, its binding to the IL-1 receptor 1 produces gliosis and endothelial activation, leading to an increase in cytokines/chemokines and MMP-9 production, blood–brain barrier (BBB) disruption, and immune cell infiltration. Despite the established detrimental roles of IL-1, this cytokine also participates in the induction of ischemic tolerance by regulation of its production to promote a shift toward an antiinflammatory state that contributes to neuroprotection [4].

Stroke-induced TLR4 activation and TNF-α/IL-1β release also increase the production of IL-6. This pleiotropic cytokine is released from all components of the neurovascular unit and blood-borne macrophages, and high IL-6 plasma levels correlate with stroke severity and poor clinical outcome. The deleterious actions in of IL-6 in cerebral ischemia are it produces gliosis, activates endothelial cells, and increases BBB damage and the synthesis and release of different chemokines that induce blood immune cell infiltration into the brain parenchyma. By contrast, IL-6 also appears to play a neuroprotective role, stimulating BBB recovery, angiogenesis, and neurogenesis.

Although other inflammatory cytokines act during this inflammatory response, such as IL-17, IL-23, or the interferon family, there is also an increase in the expression of several cytokines with antiinflammatory properties and associated with tissue repair.

One of them is IL-10, whose levels are associated with better outcome in patients. After cerebral ischemia, IL-10 is produced mainly by immune cells, such as Treg lymphocytes, monocytes/macrophages, and microglia. This cytokine has remarkable suppressive effects on the production of proinflammatory cytokines, and also modulates neuronal vulnerability to excitotoxic ischemic damage.

TGF-β is another pleiotropic immunoregulatory cytokine with a crucial role in the development of an antiinflammatory milieu. TGF-β is elevated in the blood of patients 1 day after stroke, and in the rodent ischemic brain for at least 1 week, primarily in microglia/macrophages. TGF-β has been associated with inhibition of MCP-1 and macrophage inflammatory protein (MIP)-1α production. It also induces the formation of the glial scar, inhibits Th1 and Th2 responses, and promotes Treg cell development.

IL-4 is another classical antiinflammatory cytokine. After cerebral ischemia, it is produced by CD4 Th cells and also by astrocytes and neurons, whereas IL-4 receptors are found mainly on microglia/macrophages. Among its actions, this cytokine downregulates astroglial activation, exerts neurotrophic effects by inducing NGF secretion, and has important roles in the resolution of inflammation by promoting a microglial resting phenotype and M2 activation of monocyte/macrophages [5].

THE ROLE OF CHEMOKINES AFTER STROKE

Chemokines are a family of chemoattractant cytokines that play a pivotal role in leukocyte migration. They have been classified into four main subfamilies: CXC, CC, CX3C, and XC, all of them exerting their biological effects by interacting with G-protein-coupled receptors known as *chemokine receptors*. Chemokines work through concentration gradients mediating cell migration toward areas of high chemokine concentrations. In addition to their role in immune system communication, chemokines are also expressed by other cell types. For example, in the CNS, neurons constitutively express CX3CL1 (also known as fractalkine) to control microglia activation through the CX3CL1/CX3CR1 axis. Chemokines expression is induced after stroke by the action of proinflammatory cytokines, such as IL-1, TNFα, and IL-6 on resident and infiltrated cells of the ischemic tissue. Two main chemokine networks are thought to be the recruiters of neutrophils and monocytes into the ischemic tissue: the CXCL8 (IL-8)/CXCR2 and the CCL2/CCR2 axis. In addition, the CX3CL1/CX3CR1 axis and SDF-1 (CXCL12) have also an important role in stroke outcome and in neurorepair.

CXCL8 (IL-8) mediates neutrophil transmigration and activation in humans. In rodents, CXCL1 (also known as keratinocyte-derived chemokine, KC) and CXCL2 (MIP-2) are assumed to mediate neutrophil migration instead of IL-8/CXCL8, as the rodent homolog for CXCL8 has not been found. Several studies have detected an increase in circulating CXCL1 and CXCL8 after stroke. CXCL8 binds to its receptor CXCR2 to mediate neutrophil recruitment. Human and murine CXCR2 share a 75% homology and neither human CXCL8 nor murine CXCL1 or CXCL2 can induce neutrophil accumulation in Cxcr2$^{-/-}$ mice, confirming that CXCR2 is the primary receptor for these ligands. As expected, Cxcr2$^{-/-}$ mice also exhibit a significant decrease in neutrophil infiltration and infarct size in the ischemic tissue [6]. However, other molecules apart from CXCL2, such as leukotriene B4, can also exert potent chemoattractant properties in neutrophils.

CCL2 (also known as MCP-1) is the most potent activator of CCR2 signaling, leading to monocyte transmigration. The implication of the CCL2/CCR2 axis in mediating the monocyte infiltration into the inflamed tissue has also been shown in other neuroinflammatory conditions, such as EAE or traumatic brain injury. CCR2 is mainly expressed on Ly6C^{hi} inflammatory monocytes and Ccr2$^{-/-}$ mice display an increased number of monocytes in the bone marrow and a decreased number of circulating monocytes, consistent with an important role of CCR2 in monocyte egress from bone marrow into the peripheral circulation. CCL2 has been detected in both serum and CSF of patients with stroke. Higher CCL2 blood levels are associated with poor long-term outcome and recurrence; however, higher CCL2 levels 3 days after stroke are associated with neurological improvement. Both, *Ccr2* and *Ccl2* deficiency have been shown to be neuroprotective in animal models of ischemic stroke. CCR2 also plays a positive role in preventing hemorrhagic transformation and maintaining neurovascular unit integrity after cerebral ischemia.

As previously indicated, CX3CL1 (fractalkine) is constitutively expressed by neurons and binds to its microglial receptor CX3CR1 leading to an inhibition of microglial activation. The role of constitutive CX3CL1-CX3CR1 has important implications in stroke [7]. As a consequence of neuronal injury, loss of CX3CL1-CX3CR1 signaling results in enhanced microglial activation, significantly reducing ischemic damage and inflammation. In addition, stroke induces upregulation and cleavage of neuronal CX3CL1, which is thought to mediate microglia recruitment to the sites of inflammation, where it becomes activated and promotes functional recovery of the CNS.

SDF-1 (CXCL12) has been reported to play a critical role in neuroprotection after stroke by mediating the migration of neural progenitor cells (NPCs). SDF-1 expression increases in the infarct core and penumbra in animal models of cerebral ischemia. However, SDF-1 plasma levels in humans after stroke did not show a clear increase when compared with control plasma levels. SDF-1 is the CXCR4 receptor ligand, which is found at the surface of neurons, astrocytes, microglia, bone marrow–derived cells, and other progenitor cells. Experimental studies have shown that SDF-1 mediates the recruitment of bone marrow–derived cells and NPCs to the site of injury.

IMPLICATION OF CHEMOKINES AND CYTOKINES IN NEUROREPAIR MECHANISMS AFTER STROKE: FOCUS ON NEUROGENESIS AND ANGIOGENESIS

During the chronic phase, patients with stroke undergo a spontaneous recovery due to, among other mechanisms, enhanced neuroplasticity. This process includes the generation of new neurons (neurogenesis) and blood vessels (angiogenesis), axonal sprouting, and even the modulation of synapses. Importantly, cytokines and chemokines that modulate the acute phase of stroke may also modulate some of these injury-induced neurorepair mechanisms.

Cerebral ischemia promotes the proliferation of neural stem cells (NSC) located in the subventricular zone (SVZ) and the migration of newly generated neuroblasts toward the ischemic boundary region, where they can differentiate into mature neurons to replace lost neurons or to release neurotrophic factors that mediate tissue repair. Under physiological conditions, a proper balance between pro- and antiinflammatory cytokines and chemokines results in appropriate levels of adult neurogenesis but the disturbance of this balance after ischemia may alter the neurogenic process. In this sense, the number of neuroblasts recruited to the damaged areas is correlated with the extent of the inflammatory process; however, the survival of newly generated neurons is also affected by the ischemic microenvironment, as most of the new neurons are unable to integrate into the synaptic circuits. The increased levels of pro- and antiinflammatory cytokines including IL-6, TNFα, IL-10, and TGF-β found after stroke alter both proliferation and migration of SVZ progenitors and therefore may directly affect this neurorepair process [8]. For example, after cerebral ischemia, SVZ NPCs express TNF-R1 and its activation through TNF-α negatively regulates stroke-induced SVZ progenitor proliferation. By contrast, IL-6 and IL-10 might act as positive regulators of stroke-induced neurogenesis increasing NPCs proliferation and neuroblast generation in the ipsilateral SVZ [9,10].

As previously described for leukocyte migration to the injury site, chemokines also play an important role for driving NPCs toward sites of neuroinflammation

[11]. One of the best characterized is the CXCL12 (SDF-1)-CXCR4 axis, which modulates the activation, maturation, function, and recruitment of progenitor cells toward the ischemic lesion. Under physiological conditions, SDF-1 from ependymal cells promotes NSC quiescence; however after ischemia, SDF-1, secreted predominantly by astrocytes, microglia, and endothelial cells, increases the activation of type C and type B NSCs. SDF-1 not only modulates NPC activation but also drives CXCR4-expressing NPCs and neuroblasts toward the injured tissue, this being correlated with improved behavioral outcome after stroke. Supporting this, SDF-1 inhibition with a CXCR4 antibody or the intracerebroventricular administration of SDF-1 in mice decreases or promotes, respectively, neuroblast migration to the injured area after stroke. Another important chemotactic factor for NPCs is CCL2 (MCP-1). Migrating neuroblasts toward the ischemic tissue express the MCP-1 receptor CCR2, an expression consistent with the results observed in Ccl2$^{-/-}$ and Ccr2$^{-/-}$ mice, which display decreased neuroblast migration after ischemia, corroborating the important role of the MCP-1/CCR2 chemokine axis for driving neuroblast migration after damage.

In addition to the previously described direct effects of cytokines/chemokines on NPCs and neuroblasts, several evidences indicate that these factors may also exert their pro- or antineurogenic effects in an indirect way. In this sense, under physiological conditions, microglia are crucial for the clearance of apoptotic newborn neurons by phagocytosis, which in turn mediates the final neurogenic balance. In the stroke context, proneurogenic microglia expressing IGF-1 accumulate in the SVZ, and their accumulation mitigates apoptosis and promotes the proliferation and differentiation of NSCs [12]. This indicates that, depending on the surrounding milieu, microglia may present different phenotypes able to modulate neurogenic responses after stroke.

The formation of new vessels after cerebral ischemia is another critical mechanism that drives long-term recovery and repair. Angiogenesis requires the mobilization of endothelial progenitor cells (EPCs) from the bone marrow to the ischemic tissue as well as the reorganization and removal of damaged tissue. All these processes are modulated by the inflammatory microenvironment presenting both beneficial and detrimental roles. As discussed, despite the ambivalent effects ascribed to IL-6 during the acute phase of cerebral ischemia, IL-6 presents a clear beneficial role during the chronic phase. In fact, IL-6 released from resident brain cells promotes angiogenesis in the delayed phases of stroke. Different evidences also point toward a role for TNF-α in poststroke angiogenesis by modulating erythropoietin levels in cerebral endothelial cells. The role of SDF-1 in recruiting EPCs toward the injured tissue has been extensively explored. The CXCR4 receptor on the surface of EPCs plays a fundamental role in mobilization and cell migration toward the levels of SDF-1 found at the injury site and, in turn, promoting angiogenesis. Finally, fractalkine, which is expressed constitutively in neurons and mediate the recruitment of immune cells after stroke, is also implicated in stroke-induced angiogenesis. Exogenous administration of fractalkine enhances endothelial cell proliferation and EPCs migration and, consequently, promotes neovascularization and attenuates ischemic damage.

With this background, it is clear that both cytokines and chemokines can modulate neurorepair processes after stroke and affect long-term outcome. It is interesting to note that neuroblasts migrate in association with blood vessels toward the injured tissue. Given this close link between neurogenesis and angiogenesis in both physiological and pathological conditions it is not to wonder that several of the previous described factors, for instance, SDF-1, modulate the recruitment of both NPCs and EPCs when administered early after ischemia and lead to enhanced recovery.

THERAPEUTIC APPROACHES AND FUTURE DIRECTIONS

Given the role of cytokines and chemokines in both ischemic damage and mechanisms for resolution and repair, their pharmacological manipulation could be beneficial for stroke treatment. It is the case of the naturally occurring selective antagonist of IL-1, interleukin-1 receptor antagonist (IL-1ra), which has neuroprotective properties in experimental models of brain injury including animals with comorbidities, following STAIR committee recommendations, and which has been shown to decrease inflammatory markers and to improve outcome in patients with stroke [4].

Future studies and clinical trials are needed to elucidate the utility of the intervention at these levels to decrease lesion size in the acute phase or to foster neurorepair at the chronic phase of stroke.

References

[1] Iadecola C, Anrather J. The immunology of stroke: from mechanisms to translation. Nat Med 2011;17:796–808.
[2] Gordon S, Martinez FO. Alternative activation of macrophages: mechanism and functions. Immunity 2010;32:593–604.
[3] Watters O, O'Connor JJ. A role for tumor necrosis factor-alpha in ischemia and ischemic preconditioning. J Neuroinflammation 2011;8:87.
[4] Galea J, Brough D. The role of inflammation and interleukin-1 in acute cerebrovascular disease. J Inflamm Res 2013;6:121–8.
[5] Amantea D, Micieli G, Tassorelli C, Cuartero MI, Ballesteros I, Certo M, Moro MA, Lizasoain I, Bagetta G. Rational modulation of the innate immune system for neuroprotection in ischemic stroke. Front Neurosci 2015;9:147.

[6] Gelderblom M, Weymar A, Bernreuther C, Velden J, Arunachalam P, Steinbach K, Orthey E, Arumugam TV, Leypoldt F, Simova O, Thom V, Friese MA, Prinz I, Hölscher C, Glatzel M, Korn T, Gerloff C, Tolosa E, Magnus T. Neutralization of the IL-17 axis diminishes neutrophil invasion and protects from ischemic stroke. Blood November 1, 2012;120(18):3793–802.

[7] Soriano SG, Amaravadi LS, Wang YF, Zhou H, Yu GX, Tonra JR, Fairchild Huntress V, Fang Q, Dunmore JH, Huszar D, Pan Y. Mice deficient in fractalkine are less susceptible to cerebral ischemia-reperfusion injury. J Neuroimmunol 2002;125:59–65.

[8] Christie KJ, Turnley AM. Regulation of endogenous neural stem/progenitor cells for neural repair-factors that promote neurogenesis and gliogenesis in the normal and damaged brain. Front Cell Neurosci January 18, 2013;6:70.

[9] Wang J, Xie L, Yang C, Ren C, Zhou K, Wang B, Zhang Z, Wang Y, Jin K, Yang GY. Activated regulatory T cell regulates neural stem cell proliferation in the subventricular zone of normal and ischemic mouse brain through interleukin 10. Front Cell Neurosci September 14, 2015;9:361.

[10] Meng C, Zhang JC, Shi RL, Zhang SH, Yuan SY. Inhibition of interleukin-6 abolishes the promoting effects of pair housing on post-stroke neurogenesis. Neuroscience October 29, 2015;307:160–70.

[11] Ruan L, Wang B, ZhuGe Q, Jin K. Coupling of neurogenesis and angiogenesis after ischemic stroke. Brain Res October 14, 2015;1623:166–73.

[12] Thored P, Heldmann U, Gomes-Leal W, Gisler R, Darsalia V, Taneera J, Nygren JM, Jacobsen SE, Ekdahl CT, Kokaia Z, Lindvall O. Long-term accumulation of microglia with pro-neurogenic phenotype concomitant with persistent neurogenesis in adult subventricular zone after stroke. Glia June 2009;57(8):835–49.

CHAPTER

59

Growth Factors and Cerebrovascular Diseases

T. Lippert, H. Nguyen, Q. Colburn, C.V. Borlongan

University of South Florida Morsani College of Medicine, Tampa, FL, United States

Cerebrovascular diseases, such as stroke and traumatic brain injury, pose as significant unmet clinical needs, as there are limited therapeutic options for many of these disorders. Because a major pathologic hallmark of the progression of these diseases is characterized by a massive inflammatory response partially caused by depletion of growth factors in the brain [1,2], replenishment of these soluble therapeutic molecules may offer a regenerative process for these central nervous system (CNS) disorders. That these growth factors may modulate brain homeostasis via regulation of inflammation, among other cell survival processes, such as neurogenesis, angiogenesis, and antiapoptosis, highlights their role in abrogating cell death in response to a cerebrovascular injurious event [3,4]. Here we briefly discuss specific therapeutic molecules shown to exert functional benefits in cerebrovascular diseases.

Many of the secreted growth factors that are upregulated after a brain insult may elicit a ligand–receptor mechanism of therapeutic action [5,6] (Table 59.1). Glial cell line-derived neurotrophic factor (GDNF), initially shown to exert beneficial effects in Parkinson disease and thereafter in stroke and traumatic brain injury (TBI), acts on the CNS via a transmembrane receptor tyrosine kinase and the ligand-binding component GDNF family receptor α1 (GFR-α1) [2,3]. Another growth factor that affords therapeutic effects against stroke and TBI is the vascular endothelial growth factor (VEGF), with its signaling pathway mainly mediated by VEGFR1 and VEGFR2 receptors. Hepatocyte growth factor (HGF) is another growth factor with robust regenerative action against cerebrovascular diseases that is achieved by partly binding with IgG to promote tyrosine kinase receptor (RYK) phosphorylation coupled with epidermal growth factor

Primer on Cerebrovascular Diseases, Second Edition
http://dx.doi.org/10.1016/B978-0-12-803058-5.00059-X

TABLE 59.1   Summary of Growth Factors and Their Anti-Inflammatory Effects Against Cerebrovascular Diseases

Growth Factor	Cerebrovascular Disease Indication	Upstream Mechanism of Action (Target Receptor)	Anti-inflammation as Downstream Mechanism of Action
GDNF	Stroke and TBI	Tyrosine kinase receptor and ligand-biding component of GFR-α1	Anti-inflammation
VEGF	Stroke and TBI	VEGFR1 and VEGFR2 receptors	Anti-inflammation
HGF	Stroke and TBI	Binds partially with IgG; promotes RYK phosphorylation	Anti-inflammation
BDNF	Stroke and TBI	TrkB, p75NTR	Anti-inflammation
SCF	Stroke	SCF/c-kit pathway; Bcl-2 upregulation through PI3k/Akt signaling mechanism	Anti-inflammation
SDF-1α	Stroke and TBI	CXCR4	Anti-inflammation
NT4/5	Stroke	TrkB, p75NTR	Anti-inflammation
NGF	Stroke and TBI	TrkB, p75NTR	Anti-inflammation
bFGF	Stroke and TBI	bFGF receptor	Anti-inflammation
IGF	Stroke and TBI	IGF receptor	Anti-inflammation

bFGF, basic fibroblast growth factor; NT4/5, neurotrophin 4/5.

receptor kinase activity. Similarly, the brain-derived neurotrophic factor (BDNF) is another well-characterized trophic factor that is capable of producing therapeutic effects against cerebrovascular diseases, with high BDNF expression through TrkB (tropomyosin related kinase B) the p75(NTR) (p75 neurotrophin receptor) signaling pathway detected in the ischemic penumbra after stroke or in periimpact and remote areas (i.e., hippocampus) after TBI associated with reduced pathologic symptoms. By playing the role of a c-kit receptor ligand, the stem cell factor (SCF) has also been implicated in inducing progenitor cell recruitment to specific areas of the injured CNS, and such SCF/c-kit pathway may also confer antiapoptotic effects linked to Bcl-2 (B-cell lymphoma 2) upregulation through phosphatidylinositol 3-kinase (PI3k)/Akt signaling mechanisms. Another relevant therapeutic molecule against stroke and TBI is the stromal-cell-derived factor 1α (SDF-1α), which possesses chemotactic and chemokine properties linked to the CXC chemokine receptor 4 (CXCR4) ligand. Similar overexpression of other growth factors, such as neurotrophin 4/5, nerve growth factor, basic fibroblast growth factor, and insulinlike growth factor 1 (IGF-1), with corresponding increased upregulation in their receptors [p75NTR, TrkB, basic fibroblast growth factor receptor, and IGF receptor (IGF-R)] in stroke and TBI, point to the ligand–receptor mechanism of therapeutic action by these molecules.

Treatment with these growth factors consistently results in sequestration of secondary cell death, especially in the neighboring ischemic penumbra and periimpact area in stroke and TBI, respectively.

Interestingly, the ligand–receptor action appears to be mostly localized to blood vessels, implicating that the therapeutic action of these growth factors may be related to the protection of the neurovascular unit [7,8]. Moreover, because cerebrovascular diseases have been shown to manifest when the neurovasculature is compromised, targeting the blood vessels for protection appears to be a logical treatment strategy for stroke and TBI. Indeed, plaque buildup has been associated with vascular vulnerability to inflammatory response, further suggesting the intimate role between inflammation and neurovascular compromise in cerebrovascular diseases and that enhancing the therapeutic action of growth factor signaling within the blood vessels may improve therapeutic outcome of growth factor therapy. Not only facilitating growth factor targeting of the neurovascular unit but also improving the delivery of the therapeutic molecules across the blood–brain barrier (BBB), with emphasis on abrogating inflammation, may render better treatment effects. Growth factors, such as GDNF, VEGF, HGF, and, in particular, BDNF acting via SDF-1α/CXCR4 binding, provide neuroprotection against inflammatory processes that are rampant in cerebrovascular diseases.

Inflammation is a potential treatment for cerebrovascular disease and this is dependent on the basic pathologic condition of cerebrovascular disorders, in that these pathologic conditions become evident when abnormal blood circulation to the brain is recognized, resulting in limited or no blood flow to the affected areas of the brain. To this end, atherosclerosis was established as a primary condition that can cause cerebrovascular disease and has been shown to be associated with plaque-filled blood vessels displaying aberrant high cholesterol levels concomitant with increased inflammation. Such plaque obtrusion of blood vessels can result in reduced or a blockade of blood flow to the brain, causing stroke, and can complicate subsequent brain insults such as TBI [9,10]. Similarly, defective arterial branching as in the case of aneurysm represents another form of cerebrovascular disease, which can also arise from abnormal cholesterol deposition and can be exacerbated by an inflammatory response, with a consequential ruptured aneurysm implicated in hemorrhagic stroke or intracranial aneurysm coincident with TBI [11,12]. Accordingly, maintenance of blood vessel integrity, including preservation of BBB permeability following an ischemic or traumatic event, may sequester the inflammation-plagued secondary cell death and render neuroprotection against cerebrovascular diseases.

Despite this knowledge of therapeutic effects of growth factors and their abrogation of inflammation through ligand–receptor mechanism, equally compelling evidence indicates adverse side effects when targeting such growth factor signaling pathways. For example, BDNF overexpression in the hippocampus may influence neurogenesis but it is also associated with subsequent development of epilepsy. Moreover, following experimental TBI, acute BDNF administration has not been efficacious, and clinically, higher cerebrospinal fluid (CSF) BDNF concentration after TBI may be detrimental because of the injury and age-related increase in proapoptotic BDNF target receptors. It is noteworthy that such detrimental outcomes may be related to postinjury transient BBB alterations, suggesting that the initiation, and the acute or chronic regimen of growth factor administration, will require cautious consideration to avoid the deleterious enhancement of brain damage after stroke and TBI. Further studies optimizing the treatment regimen of growth factors are needed, as the relationship between their therapeutic properties and pathologic effects remains poorly understood. An in-depth understanding of the expression of these growth factors and their receptors in the onset and progression of injury may offer an insight into their beneficial or deleterious effects on cerebrovascular diseases.

To enhance the potential of growth factor treatment for CNS disorders, novel biomedically engineered devices have been explored in aiding growth factor therapy, especially in facilitating the delivery of secreted therapeutic molecules to the injured brain. The low permeability of the BBB to many large molecules represents an obstacle for the peripheral administration of growth factors to the CNS. Local administration and disruption of the BBB are needed to achieve improved bioavailability of growth factors in the brain. Increasing growth factor bioavailability in the CNS

with a minimally invasive method has demonstrated the utility of innovative approaches, such as encapsulated cell therapy, viral vectors, and genetically engineered molecular Trojan horses [13]. Recent findings indicate that a hyaluronic acid–based hydrogel scaffold with (poly)lactic-co-glycolic acid microspheres containing angiopoietin 1 and VEGF, enhances the beneficial effects of VEGF by employing a controlled vessel permeability approach. Such scaffolding technique designed to improve growth factor bioavailability in the CNS similarly has been attempted with encouraging results using GDNF, HGF, BDNF, and SCF. However, similar to the stand-alone growth factor treatment, such fabrication has been documented to be accompanied by adverse side effects, in that the scaffolds may increase the risk of brain edema.

As an alternative to the scaffolding approach, there is increasing evidence demonstrating therapeutic effects of secreted RNAs or exosomes of growth factors after a cerebrovascular injury, which offers a novel line of investigation in growth factor treatment. Originally, long noncoding RNA or microRNA was found to be secreted by tumor cells in the CNS. Recently, laboratory results have shown that this molecule is capable of reaching injured or degenerating cells and may be able to change their phenotypic profile. Transplanted stem cells secrete several microRNAs, such as metastasis-associated lung adenocarcinoma transcript 1 (MALAT1) and nuclear enriched abundant transcript 1 (NEAT1), and the increased expression of NEAT1 and MALAT1 has been achieved through an intravenous transplant of human adipose stem cells, with beneficial results in a TBI model [14]. Interestingly, the therapeutic action of exosomes primarily entails abrogation of peripheral inflammation, notably dampening of splenic inflammatory cells [14], suggesting that although traditionally the entry of growth factors (albeit, their exosomes) into the brain may be a prerequisite for neuroprotection, targeting the peripheral source of inflammation (i.e., spleen) may equally produce beneficial effects against cerebrovascular diseases.

Growth factor treatment represents a promising strategy for treating cerebrovascular diseases. Over the past two decades, the therapeutic use of growth factors for brain injury has provided encouraging scientific basis for their use in the hospitals. The advent of novel biomedically engineered tools may allow growth factors to better infiltrate the brain by circumventing the BBB, facilitating the delivery of growth factors from the periphery to treat the injured brain. In the end, the exploitation of these growth factors, trophic factors, and RNAs, either as stand-alone treatments or in conjunction with biomedically engineered devices, is likely to improve the therapeutic outcome of growth factor therapy for cerebrovascular diseases.

Acknowledgments

Cesario V. Borlongan (CVB) is supported by the National Institutes of Health, National Institute of Neurological Disorders and Stroke 1R01NS071956, 1R01NS090962, and 1R21NS089851, Department of Defense W81XWH-11-1-0634, James and Esther King Foundation for Biomedical Research Program, SanBio Inc, KM Pharmaceuticals, NeuralStem Inc, and Karyopharm Inc. The content is solely the responsibility of the authors and does not necessarily represent the official views of the sponsors.

Author Contributions

Conceived the topics of this paper: CVB. Wrote the first draft of the manuscript: CVB. Contributed to the writing of the manuscript: CVB. Agreed with manuscript hypotheses: CVB. Jointly developed the structure and arguments for the paper: CVB. Made critical revisions and approved final version: CVB. All authors reviewed and approved the final manuscript.

Disclosures and Ethics

The authors have read and confirmed their agreement with the ICMJE authorship. All authors declare no conflicts of interest. The authors have also confirmed that this chapter is unique and not under consideration or published in any other publication.

References

[1] Roth TL, Nayak D, Atanasijevic T, Koretsky AP, Latour LL, McGavern DB. Transcranial amelioration of inflammation and cell death after brain injury. Nature 2014;505(7482):223–8.

[2] Garcia-Bonilla L, Racchumi G, Murphy M, Anrather J, Iadecola C. Endothelial CD36 Contributes to postischemic brain injury by promoting neutrophil activation via CSF3. J Neurosci 2015;35(44):14783–93.

[3] Acosta SA, Tajiri N, Shinozuka K, Ishikawa H, Sanberg PR, Sanchez-Ramos J, et al. Combination therapy of human umbilical cord blood cells and granulocyte colony stimulating factor reduces histopathological and motor impairments in an experimental model of chronic traumatic brain injury. PLoS One 2014;9(3):e90953.

[4] Jickling GC, Liu D, Stamova B, Ander BP, Zhan X, Lu A, et al. Hemorrhagic transformation after ischemic stroke in animals and humans. J Cereb Blood Flow Metab 2014;34(2):185–99.

[5] Kaneko Y, Tajiri N, Su TP, Wang Y, Borlongan CV. Combination treatment of hypothermia and mesenchymal stromal cells amplifies neuroprotection in primary rat neurons exposed to hypoxic-ischemic-like injury in vitro: role of the opioid system. PLoS One 2012;7(10):e47583.

[6] Meeker RB, Poulton W, Clary G, Schriver M, Longo FM. Novel p75 neurotrophin receptor ligand stabilizes neuronal calcium, preserves mitochondrial movement and protects against HIV associated neuropathogenesis. Exp Neurol 2015;275(P3):182–98.

[7] Guo S, Kim WJ, Lok J, Lee SR, Besancon E, Luo BH, et al. Neuroprotection via matrix-trophic coupling between cerebral endothelial cells and neurons. Proc Natl Acad Sci USA 2008;105(21):7582–7.

[8] Liu Z, Chopp M. Astrocytes, therapeutic targets for neuroprotection and neurorestoration in ischemic stroke. Prog Neurobiol 2015. http://dx.doi.org/10.1016/j.pneurobio.2015.09.008.

[9] Rom S, Dykstra H, Zuluaga-Ramirez V, Reichenbach NL, Persidsky Y. miR-98 and let-7g* protect the blood–brain barrier under neuroinflammatory conditions. J Cereb Blood Flow Metab 2015;35(12):1957–65. http://dx.doi.org/10.1038/jcbfm.2015.154.

[10] Xu L, Guo ZN, Yang Y, Xu J, Burchell SR, Tang J, et al. Angiopoietin-like 4: a double-edged sword in atherosclerosis and ischemic stroke? Exp Neurol 2015;272:61–6.

[11] Ibrahim GM, Morgan BR, Macdonald RL. Patient phenotypes associated with outcomes after aneurysmal subarachnoid hemorrhage: a principal component analysis. Stroke 2014;45(3):670–6.

[12] Tajiri N, Lau T, Glover LE, Shinozuka K, Kaneko Y, van Loveren H, et al. Cerebral aneurysm as an exacerbating factor in stroke pathology and a therapeutic target for neuroprotection. Curr Pharm Des 2012;18(25):3663–9.

[13] Emerich DF, Orive G, Thanos C, Tornoe J, Wahlberg LU. Encapsulated cell therapy for neurodegenerative diseases: from promise to product. Adv Drug Deliv Rev 2014;67-68:131–41.

[14] Tajiri N, Acosta SA, Shahaduzzaman M, Ishikawa H, Shinozuka K, Pabon M, et al. Intravenous transplants of human adipose-derived stem cell protect the brain from traumatic brain injury-induced neurodegeneration and motor and cognitive impairments: cell graft biodistribution and soluble factors in young and aged rats. J Neurosci 2014;34(1):313–26.

CHAPTER

60

Tissue Plasminogen Activator Signaling in the Normal and Diseased Brain

R. Bronstein, S.E. Tsirka

Stony Brook University, Stony Brook, NY, United States

Tissue-type plasminogen activator (tPA) is currently the only U.S. Food and Drug Administration–approved therapy for the acute treatment of ischemic stroke [1]. The most extensively studied function of tPA is its primary activity, namely, the proteolytic conversion of the zymogen plasminogen (plg) into the active protease plasmin, which in turn is essential for the lysis of blood clots [2]. Human tPA is a serine protease composed of 527 residues with four functional domains on its A chain [finger, epidermal growth factor (EGF), kringle] and one (protease) on the B chain [1]. In the central nervous system (CNS) tPA is expressed in neurons and glial cells and released in an activity-dependent manner via exocytosis [3]. Its activity is regulated through specific protein inhibitors, plasminogen activator inhibitor 1, and neuroseprin. It does not have one specific receptor, but can function on and modulate other receptors and components of the extracellular matrix (ECM). In addition to serving as a critical hemolytic node during the fibrinolysis cascade, tPA is involved in a number of other important functions in the brain and spinal cord. In the brain, these divergent roles broadly impact the normal as well as ischemic cerebral vasculature and parenchymal structures. The preponderance of this primer will focus on the critical role that neuronal and glial tPA signaling in the normal brain, and how this signaling is perturbed in the ischemic cerebrum.

SIGNALING IN THE NORMAL CNS

Signaling Through N-Methyl-D-Aspartate Receptors

A role for tPA in either normal synaptic function (neuroprotection) or exaggerated neuronal stimulation (excitotoxicity) through glutamate receptors has been an area of persistent investigation [1]. Endogenously, tPA expressed in hippocampal neurons is synthesized in the synaptodendritic compartment and is rapidly upregulated upon metabotropic glutamate receptor activation in a mechanism that involves regulated cytoplasmic polyadenylation [4].

N-methyl-D-aspartate receptors (NMDARs) are members of a large family of ionotropic glutamate receptors, which mediate fast synaptic transmission in the CNS. These receptors are obligatory heterotetramers made up of eight alternatively spliced isoforms (GluN1, GluN2, and/or GluN3), with the first and third binding glycine and second binding L-glutamate. They are essential components of the synaptic cleft, involved in calcium-mediated glutamatergic neurotransmission essential for processes as diverse as movement and memory consolidation. They can also be found extrasynaptically, a localization thought to primarily underpin their function in excitotoxic neuronal death [1]. Recombinant tPA administration to cultured hippocampal neurons affects calcium flux. Following stimulation of glutamate release presynaptically, recombinant tPA was reported to inhibit the resultant synchronous spontaneous calcium oscillations [5]. The proteolytic activity of tPA was critical for NMDAR-mediated calcium currents, as the enzymatically inactive tPA mutant, with the active site Ser478 residue mutated to alanine, no longer had any effect [5]. To control for active plasmin being the causal agent of changes in calcium flux, cultures were further incubated with α_2-antiplasmin, which ablated tPA-mediated calcium signaling [5]. In the totality of the literature, the interactions between NMDARs and tPA have been most heavily scrutinized given the explicit need to understand physiological signaling, which is eventually perturbed in various disease states including ischemia.

Primer on Cerebrovascular Diseases, Second Edition
http://dx.doi.org/10.1016/B978-0-12-803058-5.00060-6

Signaling Through Proteolytic Cleavage of ECM

tPA is secreted in a single-chain form (sc-tPA) from neurons and glial cells and processed into a two-chain form (tc-tPA) by plasmin or kallikreins. Proteolytic activation of tPA does not preclude sc-tPA from being proteolytically active, as the single-chain form can act as an effector of epidermal growth factor receptor (EGFR) and N-methyl-D-aspartate (NMDA) signaling. Both sc-tPA and tc-tPA modulate the cross talk between EGFR and NMDA receptors on neurons: tPA-mediated EGFR activation leads to a downregulation of NMDAR function. Low levels of sc-tPA and tc-tPA are thought to mediate this cross talk, which clearly points to an antiexcitotoxic effect of tPA on neurons, as opposed to that seen with high-level administration of tPA.

Another pointed role for tPA involves its effect on neurite outgrowth either during development and/or neurogenesis, or following injury [6]. tPA has been shown to degrade ECM components through the generation of plasmin, allowing the extension of neurites under normal developmental cues or following temporal lobe epilepsy (TLE) [6]. Some of these pathways have implicated phospholipase-D1 (Pld1) as the driver of tPA secretion from the growth cone in an excitation-dependent manner, thus regulating the neurite outgrowth necessary during normal hippocampal development, as well as aberrant growth due to excitotoxic insult [6]. The mechanism of how Pld1 might promote vesicular tPA release is unclear, likely involving the activation of protein kinase C (PKC) through generation of phosphatidic acid, and possibly diacylglycerides, both of which can stimulate PKC at the cell membrane [6]. These results clearly outline a role for tPA exocytosis from neurons in the normal growth and sprouting of neurites as well as that potentially caused by TLE.

Other Interactions

Outside of its well-established interaction with NMDARs, tPA maintains other prominent roles in the CNS. In addition to the roles outlined earlier, which mostly focus on the adult animal, tPA has been shown to be an important player during development. The growth and organization of the cerebellar nuclei are critical for innate and learned motor behavior, and several groups have described a role for tPA in this process [7]. The primary output of the cerebellar cortex comes from the actions of Purkinje neurons (PNs), sculpted by recurrent inhibitory networks within the cerebellum [7]. As mentioned previously, PKC and its several isoforms that are expressed in the CNS are thought to be stimulated by the local levels and activity of tPA [6,7]. PKCγ activity rather than its levels are reported to be modulated by endogenous or exogenous tPA leading to the suppression of PN dendritic development [7].

The other predominant cell type within the cerebellum, responsible for modulating the excitatory tone onto the PNs is the granule cell, employing the parallel fibers (PFs), which synapse on the dendritic spines of PNs [7]. The structure of these very important PF–PN synapses is thought to be structurally altered by brain-derived neurotrophic factor (BDNF) levels. BDNF production is tightly controlled by tPA within the granule cells of the cerebellum [7]. A causal role for tPA in cerebellar development and the pruning of immature PNs has yet to be established; however, there is increasing evidence that endogenous levels of tPA play an important role through signaling effectors in this system [7]. Mature BDNF levels are increased by the proteolytic action of tPA to plasmin, which cleaves proBDNF to BDNF, an increase abrogated by the administration of NMDAR inhibitor MK801, which is then associated with TrkB activation.

SIGNALING IN THE DISEASED CNS

Signaling Through NMDA Receptors

Excitotoxic neuronal injury overloads the neurons and drives them to apoptotic death [8]. A role for tPA in NMDR function has long been postulated starting with a 2001 study indicating that NMDAR-linked excitotoxic cell death is enhanced by the presence of ~280 nM of tPA acting through the proteolytic cleavage of the NMDA NR1 subunit in a plasminogen-independent fashion [1]. Cerebral ischemia is a prime culprit in the generation of large excitotoxic zones, and therefore the NR1 subunit has been considered as a potential therapeutic target in the treatment of stroke and other brain pathologies [1]. More recent in vivo findings, however, demonstrated that at the more realistic, low concentrations found in the brain, tPA actually protects neurons against excitotoxin-induced cell death [9]. Additional totally divergent roles for the interaction between tPA and NMDARs have been proposed. For example, one group demonstrated that the interaction between tPA and NMDARs is critical in neurovascular coupling, a process by which the activation of NMDARs through neuronal activity leads to the release of nitric oxide, an essential vasodilator in the CNS [3]. These results demonstrate that it is potentially tPA binding to different sites and/or subunit configurations of the NMDAR complex that underpins its specific role in the healthy or diseased brain.

Signaling Through EGF Receptors

Most studies of the contribution of tPA to excitotoxicity have focused on extracellular tPA at the behest of cytosolic tPA. Intracellular tPA may serve as a stand-alone signaling substrate and, although tPA has been shown to

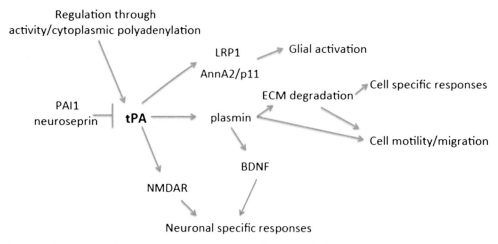

FIGURE 60.1 **Diverse pathways engaging tissue-type plasminogen activator (tPA).** tPA, acting either as a protease or nonproteolytically, can regulate neuronal activity and glial cell activation and motility. Its own expression is increased with neuronal activity both transcriptionally and posttranscriptionally within the cell, and through the action of protease inhibitors [plasminogen activator inhibitor 1 (PAI1) and neuroseprin]. *BDNF*, Brain-derived neurotrophic factor; *ECM*, extracellular matrix; *LRP*, lipoprotein-receptor related protein; *NMDAR*, N-methyl-D-aspartate receptor.

play a role in NMDAR signaling, the intracellular component has not been adequately looked at. Some evidence has emerged that EGFR plays an important role within neuronal somata. One study has demonstrated that cultured hippocampal neurons exposed to oxygen/glucose deprivation, a prominent model of ischemic injury, have varied reactions to the absence of normal endogenous levels of tPA. This effect seems to be specific to EGFRs, as perturbations of NMDARs and lipoprotein-receptor related proteins (LRPs) during titration of endogenous tPA levels did not have an effect on the level of neuroprotection. However, mitigating the interaction between endogenous tPA and EGFRs predisposed these cells to excitotoxic injury and cell death. Overall it appears that intracellular and extracellular tPAs have different substrates as well as functions, potentially contributing to excitotoxic insult outside the cell while being neuroprotective from within.

Glial Activation and tPA

In addition, cytokine-like behavior of tPA has been described in the CNS. This involves the enhancement of microglial activation through nonproteolytic interactions with the heterotetramer Annexin A2/p11 complex, which result in activation of the integrin-linked kinase pathway. tPA's interaction with LRP1, a member of the LDL receptor family, has been shown to induce the activation of matrix metalloproteinase 9 through activation of Mek1 and (extracellular-signal-regulated kinases) ERK1/2, and inflammatory nuclear factor κB signaling. Association of tPA with LRP1 was also described as a facilitator of transcytosis across the blood–brain barrier and into the CNS.

tPA signaling through LRP and other effectors has also been shown to regulate macrophage and microglial

cell migration, and the activation of the Rac1 pathway [10]. This migration has been described mainly in pathological settings, such as hemorrhagic stroke and epilepsy. The engagement of chemokines and their receptors (e.g., CCL2/CCR2) is mostly responsible for the cell motility and migration.

CONCLUSION

We have put forth evidence for tPA participating in a diverse array of physiologically and pathologically relevant neural mechanisms (Fig. 60.1). Classically this critical component of the normal clotting cascade has been implicated in functions as divergent as ECM clearance through the generation of plasmin, to cytosolic sc-tPA signaling leading to activation of the EGF receptor family, or nonproteolytic, cytokine functions. These varied roles point to tPA as a highly modular, versatile protein capable of great subunit specificity when it comes to carrying out its job in the physiological and injured CNS. The multitude of roles and functions in different settings also suggests that a single approach toward its exogenous administration or its inhibition, needs to be designed very carefully, as it may result in consequences irrelevant to its intended action.

References

[1] Yepes M, Roussel BD, Ali C, Vivien D. Tissue-type plasminogen activator in the ischemic brain: more than a thrombolytic. Trends Neurosci 2009;32:48–55.
[2] Nolin WB, Emmetsberger J, Bukhari N, Zhang Y, Levine JM, Tsirka SE. tPA-mediated generation of plasmin is catalyzed by the proteoglycan NG2. Glia 2008;56:177–89.

[3] Park L, Gallo EF, Anrather J, Wang G, Norris EH, Paul J, et al. Key role of tissue plasminogen activator in neurovascular coupling. Proc Natl Acad Sci USA 2008;105:1073–8.

[4] Shin CY, Kundel M, Wells DG. Rapid, activity-induced increase in tissue plasminogen activator is mediated by metabotropic glutamate receptor-dependent mRNA translation. J Neurosci 2004;24:9425–33.

[5] Robinson SD, Lee TW, Christie DL, Birch NP. Tissue plasminogen activator inhibits NMDA-receptor-mediated increases in calcium levels in cultured hippocampal neurons. Front Cell Neurosci 2015;9:404.

[6] Zhang Y, Kanaho Y, Frohman M, Tsirka S. Regulated secretion of tissue plasminogen activator during neurite outgrowth. J Neurosci 2005;25:1797–805.

[7] Li J, Yu L, Gu X, Ma Y, Pasqualini R, Arap W, et al. Tissue plasminogen activator regulates Purkinje neuron development and survival. Proc Natl Acad Sci USA 2013;110:E2410–9.

[8] Tsirka S, Gualandris A, Amaral D, Strickland S. Excitotoxin induced neuronal degeneration and seizure are mediated by tissue-type plasminogen activator. Nature 1995;377:340–4.

[9] Yepes M. Tissue-type plasminogen activator is a neuroprotectant in the central nervous system. Front Cell Neurosci 2015;9:304.

[10] Lin L, Hu K. Tissue plasminogen activator and inflammation: from phenotype to signaling mechanisms. Am J Clin Exp Immunol 2014;3:30–6.

CHAPTER

61

Matrix Metalloproteinases and Extracellular Matrix in the Central Nervous System

G.A. Rosenberg

The University of New Mexico, Albuquerque, NM, United States

INTRODUCTION

Brain cells are surrounded by extracellular matrix (ECM) made up of large protein molecules. Interstitial fluid (ISF) moves between the cells and along perivascular spaces, mixing with the cerebrospinal fluid (CSF) to act as the lymph of the brain. Early investigators realized that a mechanism to remove waste products of metabolism and to deliver nutrients to the cells was essential, and they postulated the ISF/CSF, acting as a third circulation, performed this essential function. However, they had no idea of the size of the space and no way to visualize it until early electron micrographs showed a space between the cells, using a method of freeze substitution that preserved the water in the brain, demonstrating an appreciable extracellular space (ECS) [1].

Studies have revealed that the ECM is an essential component of the central nervous system (CNS). Estimated to comprise 15–20% of the brain tissue, its complex role in brain development and injury is beginning to emerge. ECM in the adult CNS is localized to three principal compartments: the basal lamina, the perineuronal nets (PNNs), and the neural interstitial matrix in the parenchyma (http://www.nature.com/nrn/journal/v14/n10/full/nrn3550.html) (Fig. 61.1) [2]. In the CNS, the basal lamina separates endothelial cells from parenchymal tissue and surrounds the pial surface; it is made up of collagen, laminin–nidogen (also known as entactin) complexes, fibronectin, dystroglycan, and perlecan. The basal lamina is a major site of action of the matrix-degrading metalloproteinases (MMPs), which act on the proteins in the basal lamina to disrupt the blood–brain barrier (BBB). The third component of the ECM is the interstitial matrix consisting of a dense network of proteoglycans, hyaluronan, tenascins, and link proteins.

PNNs are a layer of lattice-like matrix that enwraps the surface of the soma and dendrites; they are mainly composed of hyaluronan, chondroitin sulfate proteoglycans (CSPG), link proteins, and tenascin R, and play a direct role in the control of CNS plasticity. Their removal is one way in which plasticity can be reactivated in the adult CNS [3].

Primer on Cerebrovascular Diseases, Second Edition
http://dx.doi.org/10.1016/B978-0-12-803058-5.00061-8

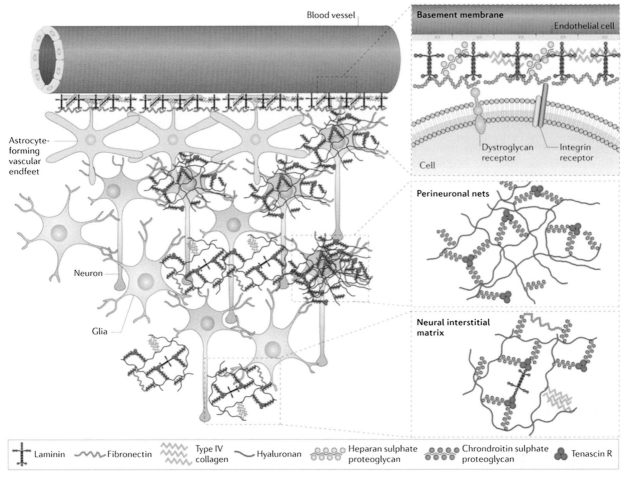

FIGURE 61.1   The three major compartments of the extracellular matrix (ECM) in the central nervous system (CNS). ECM components are arranged into basement membranes that lie outside cerebral vessels, condensed as perineuronal nets around the cell bodies and dendrites of neurons or diffusely distributed as the neural interstitial matrix between cells of the CNS parenchyma. The pink glial cells depict astrocytes, oligodendrocytes, or microglia [2].

BASAL LAMINA AND ENDOTHELIAL CELLS

The CNS basal lamina is a continuous ECM structure that lines the parenchymal side of cerebral microvessels, connecting the endothelial cell compartment to astrocytic end feet (Table 61.1). Collagen type IV in the basal lamina is a site of action of MMPs, the major enzymes involved in the maintenance of the integrity of the BBB. In cerebral ischemia, MMP-2 and MMP-9 are overexpressed, causing disruption of the BBB with leakage of fluids from blood to brain [4]. The resulting leakiness of the BBB contributes to brain damage by facilitating hemorrhage and vasogenic edema, or by preventing sufficient reperfusion.

The basal lamina around vessels forms a sheet-like structure and supports epithelia and vascular endothelial cells; it consists of structural elements, including type IV collagens and elastin; specialized proteins, such as laminins, entactin/nidogen, fibronectin, and vitronectin; and the proteoglycans, heparan sulfate, perlecan, and agrin; assembly and organization of the basal lamina in general involve polymerization of laminins and collagens, linkage by nidogens and a critical regulatory role of laminin [5].

The neurovascular unit (NVU), which is a more correct term for the BBB, plays a critical role in preserving normal brain function. Molecules attempting to enter the brain encounter a series of structures that impede their passage. There are specialized proteins in the clefts between endothelial cells that create tight junctions, forming the first layer of the BBB; the tight junction proteins maintain a high electrical resistance. The basal lamina is the next layer of the NVU, which slows the passage of proteins, but does not prevent them from entering the brain. As in other organs, the basal lamina may have a filtering

TABLE 61.1 Components of the Extracellular Matrix of the Central Nervous System (CNS)

Component	Function	
Basal lamina	Major component of the neurovascular unit	
Perineuronal nets (PNNs)	Surrounds cells and synapses	PNNs are mainly composed of hyaluronan, chondroitin sulfate proteoglycans (CSPG), link proteins, and tenascin R. PNNs play a direct role in the control of CNS plasticity, and their removal is one way in which plasticity can be reactivated in the adult CNS
Neural interstitial matrix	Fills the space between cells	Complex glycoproteins

TABLE 61.2 Endothelial Basal Lamina Proteins

Proteins
Collagen type IV
Laminin-nidrogen (entactin)
Fibronectin
Dystroglycan
Perlecan
Heparan sulfate proteoglycans

function based on molecular weight. Embedded within the basal lamina layer are the pericytes, which are smooth muscle-like cells that also act as macrophages. Astrocyte end feet form the next layer, and have an important role in the water management in the brain, containing the water channel, aquaporin, and MMP-2 and membrane type 1 metalloproteinase (MT1-MMP or MMP-14).

Both capillaries and postcapillary venules are involved in cellular transport with the postcapillary venule playing a significant role in inflammation. Electron microscopy reveals that there are two basal laminas that are apposed to each other and function to control leukocyte cellular transport: a vascular wall/endothelial basal lamina and a parenchymal basal lamina of the glia limitans of astrocytes [6]. The major components of the basal lamina are shown in Table 61.2. Assembly and organization of basal laminas in general involve polymerization of laminins and collagens, linkage by nidogens, and a critical regulatory role of laminin [5]. Under normal conditions, the two basal laminas are in close contact and appear as one under light microscopy. The parenchymal basal lamina is formed by astrocytes and its ECM composition differs somewhat from other basal lamina, containing mainly laminin isoforms.

Laminins are glycoproteins with several isoforms, comprising a combination of alpha, beta, and gamma chains [7]. Collagen IV and fibronectin are secreted by endothelial cells, astrocytes and pericytes; they function in basal lamina assembly with collagen IV stabilizing the basal lamina by retaining laminin, nidogen, and perlecan. Collagen IV and fibronectin increased electrical

resistance in blood vessels. Nidogen (enactin) links laminin and collagen IV and binds to many other ECM proteins. Agrin and perlecan are heparan sulfate proteoglycans found in the basal lamina [8].

COMMON FEATURES OF THE MMPs IN NERVOUS TISSUES AND CEREBRAL BLOOD VESSELS

The dynamic nature of the ECM depends on constitutive remodeling by several protease families, including the MMPs. There are 24 mammalian MMPs, and collectively they degrade all ECM proteins. Undesirable actions of abnormally upregulated MMPs include cytotoxicity, disruption of the BBB, and promotion of neuroinflammation. MMPs, however, are known to have reparative functions in the postacute phase of CNS injury. In mouse models of spinal cord injury, increased MMP-2 immunoreactivity is linked to greater sparing of white matter, reduced glial scarring, and improved locomotor recovery. In spinal cord injured rats, the local delivery of fibroblasts overexpressing CD147 (also known as basigin and extracellular MMP inducer), which promote the expression of multiple MMPs, improved locomotor recovery associated with axonal regeneration. In peripheral nerve injury, MMP-2 protein localized at the axon tip may play a part in nerve regeneration by inactivating inhibitory CSPGs, thus deinhibiting permissive laminin and increasing nerve regeneration to muscle targets.

The major MMPs involved in brain are MMP-2, MMP-3, MMP-9, MMP-10, and MMP-14. Although others, such as MMP-8 and MMP-13, are found in brain, they appear to play limited roles. MMPs are secreted as latent enzymes that require activation by either cleavage of a portion of the molecule or rearrangement without cleavage; this provides a protective mechanism to control proteolysis and prevent unwanted tissue damage. To maintain latency a complex series of interactions are required for activation, including the intervention of other enzymes and free radicals. Several MMPs are constitutively expressed and found normally in brain and CSF. The two main constitutive enzymes are MMP-2 and MMP-14, which act close to the cell membranes, and are most likely involved in continual remodeling of the perivascular ECM proteins. Latent MMP-2 is found in astrocytes that are an essential component of the NVU. Astrocytes have processes that cerebral blood vessels; the close proximity of the foot processes to the basal lamina is an important feature of the control of permeability of the by the MMPs [9]. Activation of proMMP-2 occurs through the action of a trimolecular complex formed by proMMP-2, tissue inhibitor of metalloprotienases-2 (TIMP-2), and MMP-14. This unique arrangement causes the complex to be tethered to the ECS between the foot processes and the basal lamina, limiting the action of the active MMP-2 to the space next to the blood vessels. Furin is an activator of MMP-14, which in turn activates proMMP-2 to a 62-kDa active form. MMPs can be toxic if released untethered into the brain, and a mechanism exists that constrains the activity of the active MMP-2 to localized regions of the brain. TIMP-2 is one of four TIMPs, and the one more closely associated with MMP-2 since it joins with MMP-14 to bind with MMP-2 providing the mechanism of activation. In this manner, the active form of MMP-2 is contained in the contiguous space next to the basal lamina of the cerebral blood vessel. Plasmin is another enzyme involved in the activation of MMP-2, and plasmin is released from plasminogen by the action of urokinase plasminogen activator (PA) and tissue PA. Latent MMP-9 is activated by MMP-3 and free radicals.

Astrocytes mainly produce MMP-2, which is found in the end feet that surround the blood vessels. Astrocyte end feet have MMP-14. MMP-9 is found in the endothelial cells. Pericytes produce MMP-3; microglia contain MMP-9 and MMP-3. During hypoxic injury, hypoxia inducible factor-1α is induced, leading to the transcription of a large number of genes that aid in the removal of damaged tissues and initiate the repair process [10]. One of the genes activated is *fur*, which leads to the production of furin, an activator of MMP-3 [11]. Thus, furin links hypoxia with the activation of MMP-9 through MMP-3 and furin. More damage is done to the cells of the brain by the inducible enzymes MMP-3 and MMP-9 than by MMP-2 since they are released into the ECS without the constraint of being attached to the membranes.

ENZYMES AND ECM

The dynamic nature of the ECM depends on remodeling by proteases. Many MMPs are upregulated in pathological conditions, such as stroke, multiple sclerosis, bacterial meningitis, and brain and spinal cord injury. Since MMPs are involved in both tissue injury and repair, the timing of upregulation is important. When the MMPs are induced during an injury, they participate in cytotoxicity, disruption of the

BBB, and promotion of neuroinflammation. On the other hand, during repair, the MMPs have reparative functions in the postacute phase of CNS injury, and facilitate angiogenesis and neurogenesis. Blocking MMP-9, which breaks down the matrix to allow for the growth, impairs axonal repair. MMPs participate in myelinogenesis as shown by the finding of deficient myelination in the corpus callosum of MMP-9 and/or MMP-12 null mice from postnatal days 7 to 14 compared with that of wild-type mice; insulin-like growth factor-1 (IGF-1) was involved in the deficient myelination in MMP null mice; the addition of IGF-1 normalized the lack of maturation of oligodendrocytes that occurred in cultures from MMP-12 null mice [12]. In mouse models of spinal cord injury, increased MMP-2 immunoreactivity is linked to greater sparing of white matter, reduced glial scarring, and improved locomotor recovery [13].

In the acutely injured spinal cord, MMP-9 contributes to disruption of the blood–spinal cord barrier, and the influx of leukocytes into the injured cord, as well as apoptosis; early pharmacologic inhibition of MMP-2 and MMP-9 results in an improvement in long-term neurological recovery and is associated with reduced glial scarring and neuropathic pain. During wound healing, MMP-2 plays a critical role in limiting the formation of an inhibitory glial scar, and mice that are genetically deficient in this protease showed impaired recovery [14].

Another enzymatic approach is the use of chondroitinase ABC to remove the glycosaminoglycan chains from CSPG. Chondroitinase ABC has been repeatedly successful in promoting axonal regeneration and behavioral recovery after traumatic spinal cord injury in rats; it may be equally useful in fostering remyelination, although the remaining intact core protein of CSPGs may still have the capacity to inhibit oligodendrocyte precursor cells. In vitro, chondroitinase ABC prevented the CSPG-mediated inhibition of the morphological maturation of human oligodendrocytes.

Overall, the administration of proteases is one approach to remove inhibitory ECM components, such as CSPGs and hyaluronan. Although local enzyme delivery to a focal lesion would be the likely route of administration, and is a reasonable strategy in localized insults such as traumatic spinal cord injury, this is problematic in conditions in which there are multifocal lesions such as those found in multiple sclerosis.

CONCLUSION

Highly complex molecules are found in the ECS of the CNS. A basal lamina surrounds cerebral blood vessels, impeding, but not preventing, molecular transport. PNNs surround neurons and synapses, and large glycoproteins form the interstitial space. Proteases attack the components of the ECM, remodeling the matrix during repair, but disrupting the matrix in injury. Enzymes that breakdown the matrix may be beneficial in tissue repair.

References

[1] Van Harreveld A, Collewijn H, Malhotra SK. Water, electrolytes, and extracellular space in hydrated and dehydrated brains. Am J Physiol 1966;210(2):251–6.

[2] Lau LW, Cua R, Keough MB, et al. Pathophysiology of the brain extracellular matrix: a new target for remyelination. Nat Rev 2013;14(10):722–9.

[3] Kwok JC, Dick G, Wang D, et al. Extracellular matrix and perineuronal nets in CNS repair. Dev Neurobiol 2011;71(11):1073–89.

[4] Rosenberg GA. Neurological diseases in relation to the blood–brain barrier. J Cereb Blood Flow Metab 2012;32(7):1139–51.

[5] Yurchenco PD, Patton BL. Developmental and pathogenic mechanisms of basement membrane assembly. Curr Pharm Des 2009;15(12):1277–94.

[6] Owens T, Bechmann I, Engelhardt B. Perivascular spaces and the two steps to neuroinflammation. J Neuropathol Exp Neurol 2008;67(12):1113–21.

[7] Hallmann R, Horn N, Selg M, et al. Expression and function of laminins in the embryonic and mature vasculature. Physiol Rev 2005;85(3):979–1000.

[8] Wolburg H, Noell S, Wolburg-Buchholz K, et al. Agrin, aquaporin-4, and astrocyte polarity as an important feature of the blood–brain barrier. Neuroscientist 2009;15(2):180–93.

[9] Rosenberg GA, Cunningham LA, Wallace J, et al. Immunohistochemistry of matrix metalloproteinases in reperfusion injury to rat brain: activation of MMP-9 linked to stromelysin-1 and microglia in cell cultures. Brain Res 2001;893(1–2):104–12.

[10] Semenza GL. Oxygen sensing, hypoxia-inducible factors, and disease pathophysiology. Annu Rev Pathol 2014;9:47–71.

[11] Remacle AG, Shiryaev SA, Oh ES, et al. Substrate cleavage analysis of furin and related proprotein convertases. A comparative study. J Biol Chem 2008;283(30):20897–906.

[12] Larsen PH, DaSilva AG, Conant K, et al. Myelin formation during development of the CNS is delayed in matrix metalloproteinase-9 and -12 null mice. J Neurosci 2006;26(8):2207–14.

[13] Levine JM, Cohen ND, Heller M, et al. Efficacy of a metalloproteinase inhibitor in spinal cord injured dogs. PLoS One 2014;9(5):e96408.

[14] Zhang H, Chang M, Hansen CN, et al. Role of matrix metalloproteinases and therapeutic benefits of their inhibition in spinal cord injury. Neurotherapeutics 2011;8(2):206–20.

MODELS AND METHODS

62

Animal Models of Focal Ischemia

M.J. Cipolla

University of Vermont, Burlington, VT, United States

Ischemic stroke is a highly heterogeneous disease with variation in size, location, and cause of occlusion. In addition, differences in underlying comorbidities, including hypertension, diabetes, and aging, and sex-specific sensitivity to ischemia make replicating this condition in one animal model impossible. Yet animal models have and will continue to be an integral and important part of understanding stroke pathophysiology and in developing treatments. Understanding the strengths and limitations of the different models of focal ischemia as well as experimental and physiological variables that influence stroke outcome are critical to successful use of these models.

APPROACHES TO INDUCING FOCAL ISCHEMIA

The most common subtype of human stroke is occlusion of the middle cerebral artery (MCA) or its distal branches and thus the most relevant and widely used animal models involve different means of MCA occlusion (MCAO). The most common and reproducible approaches to induce MCAO are either through a small craniotomy to expose distal branches of the MCA (distal MCAO) that are ligated, clipped, or cauterized, or via insertion of a filament intraarterially into the internal carotid artery that is advanced to the bifurcation of the MCA (filament MCAO). Although distal MCAO is often used as a permanent ischemia model, reperfusion can be induced by temporary ligatures or thrombolysis in an embolic model. The disadvantage of the distal MCAO model is the need for craniotomy that can damage structures including the eye, temporalis muscle, and zygomatic arch. The filament MCAO model avoids opening the skull, but is not without complications. Subarachnoid hemorrhage can be easily induced if filaments are too sharp or advanced too far. Filaments coated with silicone

help to avoid this complication and also allow for different filament dimensions that are helpful to adjust to the size or strain of animal. The advantage of the filament model is that restoration of blood flow is readily achieved by removing the filament. This is particularly relevant to modeling endovascular therapy with stent retrievers, now a standard of care for certain patients with large-vessel occlusions.

Local injection of the vasoconstrictor endothelin has also been used to induce focal ischemia but the injection requires a craniotomy for placement of a small cannula. However, the cannula can be left in place and ischemia induced after anesthesia is withdrawn. The disadvantage of using endothelin is that the sensitivity to vasoconstriction varies depending on species, strain, sex, and comorbidity, as well as the presence or absence of anesthesia, making control of ischemia and timing of reperfusion difficult and variable.

Several embolic and thrombotic models have been developed to better mimic a common mechanism of occlusion and/or to study thrombolysis. Embolic models typically involve either injection of a suspension of small clots or careful intraarterial placement of a large clot into the MCA. These embolic models are more difficult to control ischemia and have high variability in infarct size. Thromboembolic models can induce direct thrombus formation and involve injection of thrombin or ferric chloride, or use a photothrombotic dye that is activated by certain wavelength light through a thinned skull or craniotomy into either proximal MCA or distal MCA branches. Although thromboembolic models better mimic the mechanism of human stroke, the timing of reperfusion can be variable and uncertain. Although thromboembolic models are useful for studying the effects of thrombolysis and tissue plasminogen activator (tPA) as a stroke treatment, it should be noted that different species have different fibrinolytic systems and thus may not be as relevant for human stroke [1].

Primer on Cerebrovascular Diseases, Second Edition
http://dx.doi.org/10.1016/B978-0-12-803058-5.00062-X

In general, the approaches listed here can and have been used in different species, large and small, including mice, rats, cats, dogs, pigs, and nonhuman primates. The benefit of small rodents is that they are readily available and cost-effective, with more available reagents than larger species, and easier to use especially if a craniotomy is needed. The disadvantage of using rodents is that their lissencephalic brain is not as complex as the gyrencephalic brain of humans and primates and, therefore, the complexity of stroke pathology is different. Other differences in vascular anatomy and cellular responses likely confound translation from rodents to humans. For example, the anatomy of leptomeningeal collaterals has been shown to vary over ~15 different mouse strains that impacts the size of infarction [2]. However, the advantages of using rodents in most studies generally outweigh the disadvantages.

INDUCTION OF MIDDLE CEREBRAL ARTERY OCCLUSION: CONSIDERATIONS FOR ANESTHESIA AND MONITORING PHYSIOLOGICAL VARIABLES

In general, a surgical approach is needed to induce MCAO. Thus experience with appropriate surgical techniques, including sterile conditions, use of anesthesia, and controlling and monitoring physiological variables, is critical to reproducibility and reliability of results.

Although a few models induce focal ischemia without anesthesia (e.g., endothelin occlusion, clot injection), it is generally impractical to perform these models on unanesthetized animals. Thus understanding the effects of anesthesia on factors that influence the ischemic cascade or induction of ischemia is important. Ketamine and barbiturates are intrinsically neuroprotective and are not recommended for use in stroke models. Chloral hydrate is an injectable anesthesia that does not appreciably affect blood flow or depress neuronal function [3], making it the preferred choice for studies involving measuring seizure or periinfarct depolarizations. The disadvantage to chloral hydrate is it can lower blood pressure, even in hypertensive animals, and requires an intravenous (IV) line. Inhalational anesthetics such as isoflurane are preferred in stroke models if not measuring brain excitability because the depth of anesthesia is easily controlled and animals recover quickly. However, isoflurane is a cerebral vasodilator that can make inducing ischemia difficult if the level is not kept at <2%. Inhalation anesthetics are also toxic and thus should be contained through ventilation or trapping. Laboratory personnel should be monitored for exposure. A combination of anesthesia can also be used that involves initial use of isoflurane for instrumentation (intubation,

placement of flow probes, catheters, EEG leads, etc.) followed by tapering off isoflurane and onto an injectable anesthesia, such as chloral hydrate. Care should be taken to not overanesthetize during the change of anesthesia, but this approach is relatively simple since isoflurane is rapidly eliminated. Animals should be continuously monitored to maintain appropriate surgical anesthesia including periodic testing for lack of eye blink or toe pinch reflex, slow deep breathing, and increased blood pressure indicating pain.

Physiological parameters including blood gases (Pao_2 and $Paco_2$), pH, blood pressure, and temperature are important to monitor and potentially control during MCAO surgeries because they affect stroke outcome. CO_2 levels affect cerebral blood flow (CBF) through vasodilation or vasoconstriction and thus maintaining $Paco_2$ within the physiological range of 35–45 mm Hg will eliminate this as a variable to outcome. Similarly, O_2 levels can influence blood flow and the pathophysiology of stroke and thus should be ~90–100 mm Hg. Blood gases can be monitored through periodic arterial blood sampling (suggest at least one before induction of MCAO and once during MCAO) in rats; however, mice are a challenge because of their low blood volume. Pulse oximetry and CO_2 exhalation can also be used if blood samples are not feasible. Focal ischemia also causes significant fluctuations in ions, including Na^+, K^+, and H^+ due to ion channel inactivation during hypoxia/ischemia that can affect outcome. Thus controlling arterial pH to ~7.4 can eliminate this as a variable. If surgeries are prolonged (>1 h), intubation and mechanical ventilation to control blood gases and pH are recommended. Arterial blood pressure is also known to influence stroke outcome. If blood pressure falls too low, stroke outcome is worsened. Because anesthesia can lower blood pressure, it should be continuously monitored through an arterial catheter and adjusted with phenylephrine or norepinephrine if needed. This is particularly relevant when using hypertensive animals that may be normotensive during MCAO surgery with the use of certain anesthesia. Temperature also affects stroke outcome. Brain hypothermia is highly neuroprotective and since anesthesia can decrease body temperature, it is important to monitor and maintain core body temperature with a heating pad or lamp. However, there are cases in which monitoring physiological parameter without adjusting them is appropriate. For example, some neuroprotective pharmacologic agents are effective at reducing stroke severity due to promoting hypothermia (e.g., NBQX (2,3-dihydroxy-6-nitro-7-sulfamoylbenzo[f]quinoxaline-2,3-dione) and $MgSO_4$) [4]. In contrast, some agents may cause hyperthermia and would, therefore, be contraindicated for stroke therapy.

MONITORING CEREBRAL BLOOD FLOW DURING MIDDLE CEREBRAL ARTERY OCCLUSION AND REPERFUSION

An important variable in focal stroke is reduction in blood flow and the extent of reperfusion (if applicable). Thus monitoring changes in blood flow is also important and can be used for both appropriate modeling and outcome measures. Imaging (e.g., high-resolution computed tomography or MRI) is not an option for many laboratories and even if available, protocols need to be optimized for its use in stroke modeling. Changes in blood flow can be monitored noninvasively by laser Doppler flowmetry or laser speckle. The most common method used during MCAO to assess changes in CBF is laser Doppler flowmetry; however, care should be taken for appropriate probe placement over the central MCA territory (e.g., bregma −2 mm anterior, +4 mm lateral). If two probes are used, pial collateral flow can be measured but again care should be taken for appropriate probe placement to measure retrograde flow (e.g., bregma +2 mm anterior, +2 mm lateral). In most cases, it is important to produce an appropriate drop in CBF (e.g., >70% from baseline at central MCA territory) to induce infarction; however, the sensitivity to ischemia varies with age, sex, and comorbid state. Thus a drop of 70% from baseline will produce variable outcomes depending on those states.

Invasive measures of CBF can be used as well to give absolute measures of blood flow (in milliliters per 100 g tissue per minute) and include hydrogen clearance, microspheres or iodo[14C]antipyrine. Hydrogen clearance has the advantage that repeated measurements can be made in the same area, but the disadvantage that a craniotomy is necessary for probe placement (even though probes can be made as small as <20 μm). Microspheres and iodo[14C]antipyrine can only obtain one measurement per time point but have the advantage of measuring multiple brain regions.

COMMON OUTCOME MEASURES

The most common outcome measures when performing stroke models are infarct size, edema formation, and neurological or behavioral assessments. For most studies, randomization of groups is important and can be easily done using online randomization tools. In addition, blinding of investigators to treatment groups at some stage of the experiment is also important to prevent bias in the outcome, but may not be possible or necessary at every stage. For example, if pregnant animals are used or the compound to be tested emits an odor it is difficult for the experimenter to be completely blinded. At a minimum, the investigator performing the assessment of outcome (e.g., infarct, edema, or neurological score) should be blinded to group.

Infarct size can be measured histologically using tetrazolium salts (TTC), hematoxylin and eosin (H&E), or cresyl violet (CV) staining. TTC is a vital dye that is excluded by dead cells that appear bright white compared with live tissue that is dark red. The advantage of TTC staining is that it requires less preparation and is, therefore, faster than other histological methods including H&E and CV that require either frozen sections or paraffin-embedding of tissue. However, because it is a vital dye, it may not be visible at early times of MCAO when damage is minimal, especially in normal healthy animals. Conversely, TTC staining for infarct determination is not recommended for use at long time periods after MCAO (e.g., 7 days) when invading cells will also stain red and be indistinguishable from neuronal cell death. Because the side of infarction can be swollen due to edema formation, it is appropriate to determine infarct by first subtracting the ipsilateral area from the contralateral area (edema correction) to avoid underestimation of infarct. The difference in size between the hemispheres can also be used as a gross estimation of edema formation. If edema formation is an important outcome measure, the difference between wet and dry brain weights is a more accurate measure [5]. This is accomplished by weighing the brain, usually divided ipsilateral and contralateral just after it is removed from the skull, then dried overnight at 90–100°C and weighed again dry. The difference between wet and dry weights is the water content.

Diffusion-weighted MRI can also be used as a noninvasive measure of infarction that appears hyperintense. The advantage of imaging infarction is that its evolution can be followed over time. In addition, therapeutic intervention can also be assessed temporally and because it is noninvasive MRI can be used in combination with other measures, such as immunohistochemistry or Western blot.

Behavioral assessments for stroke outcome are an important outcome measure stemming from the clinical assessment of neurological deficit scoring. However, unlike measures in human stroke, behavioral measurements in rodents are more complex, especially after surgery. A range of behavioral tests have been developed to quantify sensorimotor deficit and recovery as well as tests to determine cognitive deficits (e.g., Morris water maze). General neurological scoring systems have been developed as well (e.g., score range from 0 to 3) to quantify forelimb flexion, circling, and lateral push [6]. More sensitive measures (e.g., 21 point scoring system) that assess both motor and sensory deficits are also established [8]. Care should be taken to not perform assessments too soon after stroke surgery as deficits may reflect the response to anesthesia or postsurgical pain and discomfort.

USE OF STROKE MODELS FOR UNDERSTANDING PATHOPHYSIOLOGY VERSUS TREATMENTS

Numerous tools are available to help unravel the complexities of the pathophysiology of ischemic stroke, including models of focal ischemia combined with genetic mouse models and pharmacologic agents. The design of mechanistic, hypothesis-driven studies is different from those focused on obtaining preclinical data for translation to human studies or clinical trials. For mechanistic studies, the dosing, timing, and route of administering agents can largely be ignored from a clinical standpoint because the goal is not treatment but understanding a pathway or mechanism. Thus it is appropriate to pretreat animals or treat in ways that are not clinically relevant (e.g., intracerebral injection) to inhibit or downregulate a protein, cell process, or ion channel. In fact, using transgenic or knockout animals does not allow for relevant timing of treatment as all animals are inherently "pretreated." In contrast, studies that are designed to obtain preclinical data need to have treatments that are clinically relevant (e.g., IV), be performed in a relevant time frame (e.g., delayed treatment or at time of reperfusion), and take into consideration the hemodynamics of stroke (e.g., that only the penumbra is being treated). Preclinical studies should also be done in the presence of clinically relevant doses of tPA (0.9 mg/kg) to assure that there is not an interaction between the treatment and thrombolysis that will make either less effective or induce more damage. Lastly, guidelines developed by the Stroke Therapy Academic Industry Roundtable (STAIR) should be followed for preclinical studies [7]. These guidelines were developed to help provide consistency and reliability in therapeutic testing of neuroprotective compounds between models and laboratories. These guidelines highlight the need for preclinical studies to take into consideration species, sex, comorbidities, dose–response relationships, randomization and blinding, and physiological monitoring, and include long- and short-term end points [7].

COMORBIDITIES AND SEX DIFFERENCES

Although stroke occurs in all demographics, most patients with stroke have some comorbidity, such as hypertension or diabetes, or are aged. The presence of these comorbidities changes the sensitivity to ischemia and risk of adverse events and is an important consideration when assessing therapies or treatments that are meant for the general stroke population. In addition, half the stroke population is female, yet very few stroke studies include female animals, presumably because estrogen is neuroprotective and levels change with the estrous cycle. A relevant model for studying females is ovariectomized, aged female animals to mimic the postmenopausal woman who is most likely to have stroke compared with younger females. The use of hypertensive, diabetic, or aged animals in stroke studies is not trivial. They are costly and have high mortality usually due to larger infarction, greater edema, and the propensity for hemorrhagic transformation. Thus, a reasonable approach is to determine if a treatment is efficacious in young, healthy animals initially. If it is not, then the study can stop without a large investment. If the treatment works in younger healthy animals, then it is important to test its efficacy in other models that better mimic the stroke population to determine not only efficacy but also adverse events. For example, hypertensive animals and humans have greater core infarction with little salvageable tissue (small penumbra) [8]. If a treatment is to be used in hypertensive humans, it seems reasonable to determine if the treatment will work in these animals. Commonly used models of hypertension include the spontaneously hypertensive rats (SHR) and SHR stroke prone; however, concerns exist that these models have specific features related to their inbred genetics. It is, therefore, useful to compare outcomes with another model of hypertension such as a renal hypertensive model. Acute hyperglycemia occurs in ~40% of patients with stroke independent of diabetes that is known to worsen stroke outcome [9]. However, injection of glucose into animals to cause hyperglycemia before stroke can decrease CBF due to osmotic redistribution of fluid into the peritoneum and worsen stroke outcome unrelated to the effects of hyperglycemia [10]. Several models of type 1 and type 2 diabetes exist that can be used to model some of the effects of these states during MCAO, but long-term studies are difficult because of their high morbidity and mortality. If mortality is high in animals with comorbidities, the ischemic duration can be shortened provided it is kept consistent between groups.

References

[1] Karges HE, Funk KA, Ronneberger H. Activity of coagulation and fibrinolysis parameters in animals. Arzneimittelforschung 1994;44(6):793–7.
[2] Zhang H, Prabhakar P, Sealock R, Faber JE. Wide genetic variation in the native pial collateral circulation is a major determinant of variation in severity of stroke. J Cereb Blood Flow Metab 2010;30(5):923–34.
[3] Olson DM, Sheehan MG, Thompson W, Hall PT, Hahn J. Sedation of children for electroencephalograms. Pediatrics 2001;108(1):163–5.
[4] Nurse S, Corbett D. Neuroprotection after several days of mild, drug-induced hypothermia. J Cereb Blood Flow Metab 1996;16:474–80.
[5] Keep RF, Hua Y, Xi G. Brain water content. A misunderstood measurement? Transl Stroke Res 2012;3(2):263–5.

[6] Macrae IM. Preclinical stroke research–advantages and disadvantages of the most common rodent models of focal ischaemia. Br J Pharmacol 2011;164(4):1062–78.

[7] Stroke Therapy Academic Industry Roundtable. Recommendations for standards regarding preclinical neuroprotective and restorative drug development. Stroke 1999;30:2752–8.

[8] Letourneur A, Roussel S, Toutain J, Bernaudin M, Touzani O. Impact of genetic and renovascular chronic arterial hypertension on the acute spatiotemporal evolution of the ischemic penumbra: a sequential study with MRI in the rat. J Cereb Blood Flow Metab 2011;31(2):504–13.

[9] Mandava P, Martini SR, Munoz M, Dalmeida W, Sarma AK, Anderson JA, et al. Hyperglycemia worsens outcome after rt-PA primarily in the large-vessel occlusive stroke subtype. Transl Stroke Res 2014;5(4):519–25.

[10] Duckrow RB. Decreased cerebral blood flow during acute hyperglycemia. Brain Res 1995;703(1–2):145–50.

C H A P T E R

63

Animal Models: Global Ischemia

K. van Leyen[1,2], K. Yigitkanli[3]

[1]Massachusetts General Hospital, Charlestown, MA, United States; [2]Harvard Medical School, Boston, MA, United States; [3]Polatli Government Hospital, Ankara, Turkey

INTRODUCTION

Global cerebral ischemia (GCI) is a major factor in the high mortality following cardiac arrest, which causes around 300,000 fatalities per year in the United States. Among the survivors, most have lifelong cognitive deficits. To model GCI in animals, a number of global ischemia models were established in the 1970s. An initial focus lay on large animals, such as cats, dogs, and monkeys. With increasing costs and regulatory difficulties for large animal studies, currently rodent models of GCI are more common. Especially mice are used frequently, because of the availability of genetic knockouts. To reflect the clinical situation more closely, models of cardiac arrest with cardiopulmonary resuscitation (CA/CPR) have also been established, which will only briefly be discussed here.

RODENT MODELS: MICE, RATS, GERBILS, AND THE ARCTIC GROUND SQUIRREL

From the 1960s on, decapitation has been used as a very simple way to induce GCI, and it is still sometimes used today. Although this model is helpful in following neurodegeneration, the main drawback is that it is irreversible and does not allow for resuscitation and studying treatment effects. The most commonly used rodent models involve occlusion of either both common carotid arteries in mice [two-vessel occlusion (2-VO)], or the carotid arteries plus prior cauterization of the vertebral arteries in rats [four-vessel occlusion (4-VO)]. Both have been used extensively and modified over the years [1].

In mice, the 2-VO model involving bilateral common carotid artery occlusion (BCCAO) (Fig. 63.1) is usually maintained for 10–25 min. The challenge is to achieve a consistent level of injury, without causing excessive mortality. At the lower end of ischemic duration, brain injury is highly variable and can be completely absent. At 25 min and above, mortality increases rapidly. In our hands, 12 min of GCI in CD1 mice produced variable levels of injury, but 20 min GCI consistently resulted in widespread injury to cortex, striatum, and hippocampus, with around 30% mortality. This applied to both CD1 mice and C57Bl6 mice, which is important because many gene knockouts are available in this strain. Many groups have successfully used this relatively simple model, for example, 22 min of GCI led to consistent striatal injury along with damage to the hippocampus [2]. Approaches

Primer on Cerebrovascular Diseases, Second Edition
http://dx.doi.org/10.1016/B978-0-12-803058-5.00063-1

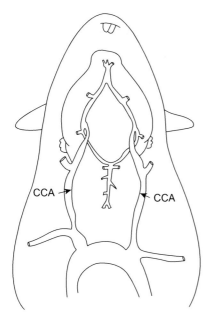

FIGURE 63.1 **Schematic drawing of the two-vessel occlusion model in rodents.** Global ischemia is induced by transient occlusion of both common carotid arteries (CCA). This basic model can be modified in several different ways, for example, by prior permanent blockade of the vertebral arteries creating a four-vessel occlusion model, or by including hypotension as component of the model. *Produced by Yigitkanli A, Yigitkanli K (unpublished).*

to improve consistency typically combine carotid artery occlusion with hypotension. This blood pressure reduction can be achieved either through vasodilatory drugs, or through bleeding via internal jugular vein or via the femoral artery [3]. Other attempts to standardize involve analyzing mice for the presence of a complete posterior communicating artery, which can confound the model by providing a source of collateral blood flow.

In rats, typically the 4-VO model is used to induce GCI. Twenty-four hours before BCCAO, the vertebral vessels are cauterized, and the subsequent carotid artery occlusion leads to successful GCI in around 50–75% of rats. Similar to the 2-VO model in mice, injury occurs in hippocampus, striatum, and cortex [4].

One surprising aspect of many GCI studies is the almost exclusive focus on injury to the hippocampus. Although especially the hippocampal CA1 region is known to be highly vulnerable, as stated earlier, other brain regions including cortex and striatum are also heavily affected. One reason for this neglect of damage to those regions may be the more diffuse injury encountered. This makes a simple histological determination more challenging than in the CA1 formation, where around 90% of all neurons may be damaged following prolonged GCI. These analytical difficulties can be overcome by using stains, such as FluoroJade-B or FluoroJade-C, which specifically label damaged neurons. The brightly stained cells can easily be counted, giving a semiquantitative measure of

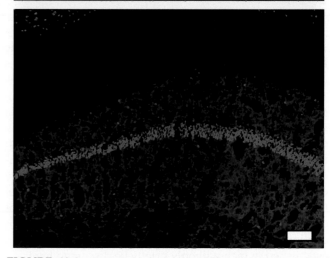

FIGURE 63.2 **Extensive neural injury following 20 min of global ischemia in mice.** Staining with FluoroJade-B 72 h after transient global cerebral ischemia shows neuronal injury in cortex (top), striatum (middle), and the CA1 region of the hippocampus (bottom). Scale bar, 100 μm.

injury. It should be noted that FluoroJade staining does not quantify injury to other cell types (Fig. 63.2).

In all of these models, temperature control is extremely important. Hypothermia is highly neuroprotective, and indeed currently the preferred treatment for patients with cardiac arrest. The original findings of the benefits of hypothermia came from animal models, emphasizing their value in testing therapeutic approaches. Lack of temperature control consequently leads to high variability of the model. One approach to stabilize the cerebral temperature during GCI involves warming the head with simultaneous body cooling [5].

Besides rats and mice, other species of the order Rodentia have been used for GCI studies. Global forebrain ischemia in gerbils was traditionally used as a GCI model, because gerbils do not have a functional posterior communicating artery and thus have less confounding collateral flow. Many of the important earlier findings in the global ischemia field were made in gerbils [6]. Nonetheless, many gerbils do not develop an infarction following forebrain ischemia. Furthermore, they are prone to epileptic seizures, and thus gerbils are not as commonly used nowadays. A very interesting model is the arctic ground squirrel [7]. During hibernation, it lowers its core body temperature to close to freezing. Because of its known resistance to ischemic injury, studying protective mechanisms in GCI in arctic ground squirrels may give novel insights into GCI pathology.

LARGER ANIMALS: DOGS, CATS, AND MONKEYS

In larger animals, early models of global ischemia typically involved the use of an inflatable neck cuff to block circulation to the brain (reviewed in Ref. [1]). The vertebral arteries need to be separately occluded in this model, because they are not compressed by the neck cuff. In monkeys and cats, the neck cuff technique was combined with hypotension to produce an adequate reduction in blood supply to the brain. Overall, cardiac arrest models are more commonly used in larger animals than global ischemia models, in part due to the historical use of dogs and swine in cardiovascular research.

COMPARING GLOBAL CEREBRAL ISCHEMIA WITH CARDIAC ARREST

The models described earlier produce global ischemia to the brain, but generally spare the rest of the organism. This is desired when specifically intracerebral injury mechanisms are to be studied, but the clinical situation reflected in global ischemia usually involves cardiac arrest, followed by cardiopulmonary resuscitation (CA/CPR). Consequently, cardiac arrest models have been developed in several animal species, aimed at mimicking the systemic effects of CA/CPR. Return of spontaneous circulation typically requires resuscitation by chest compression

and administration of epinephrine. These models require intense monitoring of physiological parameters, which is relatively easier in large animals. Swine subjected to ventricular fibrillation induced via electroshock have long been used in cardiovascular research, in part because their cardiovascular physiology is similar to humans. In young piglets, cardiac arrest can be induced by hypoxia-ischemia. Controlled mild asphyxia has been used in rats [8], whereas potassium injection is typically used to induce cardiac arrest in mice, followed by chest compression and ventilation to achieve cardiopulmonary resuscitation [9]. Although CPR is usually successful even after prolonged cardiac arrest in these models, overall survival is usually limited to a few days. This is not much different from human out of hospital cardiac arrest, which has strikingly low long-term survival rates.

In comparing GCI with CA/CPR for the study of neurological injury, several factors need to be considered. GCI models are typically simpler to apply, but collateral blood supply can be a confounding factor. CA/CPR models are more reflective of clinical cardiac arrest, but are also more technically challenging. They represent a systemic global ischemia, so physiological monitoring becomes much more important. Injury may also occur to other organs, especially the liver, leading to complex interactions of diverse injury pathways. From a practical point of view, CA needs to be extended to at least 7 min, otherwise no intracerebral injury is encountered. Despite these differences, many neuropathological consequences are common to both GCI and CA/CPR models. A striking finding typical for both GCI and CA/CPR studies is the time delay with which injury arises. Although focal ischemia models reflecting experimental stroke are typically fully developed after 24 h, injury in GCI and CA/CPR mainly increases from 1 to 3 days after the ischemic event. In both cases, hippocampus, striatum, and cortex can all be affected, although especially the CA1 region of the hippocampus is particularly vulnerable. Although many described injury pathways are common to both models, and in quite a few cases overlap with findings in focal ischemia, there are also exceptions. For a complete understanding and to support the robustness of mechanistic observations, it is therefore important to study injury pathways across injury models and organisms.

CONCLUSION AND OUTLOOK

A variety of animal models of global ischemia have been developed to mimic the brain injury in cardiac arrest pathology. Each model has its advantages and limitations. Choosing the appropriate animal model is crucial for better evaluation and testing of novel neuroprotective strategies, fostering successful future translation to the clinic.

References

[1] Traystman RJ. Animal models of focal and global cerebral ischemia. ILAR J 2003;44(2):85–95.

[2] Yoshioka H, Niizuma K, Katsu M, Sakata H, Okami N, Chan PH. Consistent injury to medium spiny neurons and white matter in the mouse striatum after prolonged transient global cerebral ischemia. J Neurotrauma 2011;28(4):649–60. http://dx.doi.org/10.1089/neu.2010.1662.

[3] Kristian T, Hu B. Guidelines for using mouse global cerebral ischemia models. Transl Stroke Res 2013;4(3):343–50. http://dx.doi.org/10.1007/s12975-012-0236-z.

[4] Pulsinelli WA, Brierley JB, Plum F. Temporal profile of neuronal damage in a model of transient forebrain ischemia. Ann Neurol 1982;11(5):491–8. http://dx.doi.org/10.1002/ana.410110509.

[5] Kofler J, Hurn PD, Traystman RJ. SOD1 overexpression and female sex exhibit region-specific neuroprotection after global cerebral ischemia due to cardiac arrest. J Cereb Blood Flow Metab 2005;25(9):1130–7. http://dx.doi.org/10.1038/sj.jcbfm.9600119. pii:9600119.

[6] Moskowitz MA, Kiwak KJ, Hekimian K, Levine L. Synthesis of compounds with properties of leukotrienes C4 and D4 in gerbil brains after ischemia and reperfusion. Science 1984;224(4651):886–9.

[7] Dave KR, Anthony Defazio R, Raval AP, Dashkin O, Saul I, Iceman KE, et al. Protein kinase C epsilon activation delays neuronal depolarization during cardiac arrest in the euthermic arctic ground squirrel. J Neurochem 2009;110(4):1170–9. http://dx.doi.org/10.1111/j.1471-4159.2009.06196.x. pii:JNC6196.

[8] Katz L, Ebmeyer U, Safar P, Radovsky A, Neumar R. Outcome model of asphyxial cardiac arrest in rats. J Cereb Blood Flow Metab 1995;15(6):1032–9. http://dx.doi.org/10.1038/jcbfm.1995.129.

[9] Minamishima S, Kida K, Tokuda K, Wang H, Sips PY, Kosugi S, et al. Inhaled nitric oxide improves outcomes after successful cardiopulmonary resuscitation in mice. Circulation 2011;124(15):1645–53. http://dx.doi.org/10.1161/CIRCULATIONAHA.111.025395.

CHAPTER

64

Animal Models: Cerebral Hemorrhage

Q. Li, J. Wang

The Johns Hopkins University School of Medicine, Baltimore, MD, United States

INTRODUCTION

Worldwide, stroke is the second leading cause of death, affecting approximately 6 million people in 2012. Intracerebral hemorrhage (ICH) accounts for 13% of all stroke cases and is associated with high mortality and morbidity. As many as 50% of patients die within 1 month after ICH onset, and only 20% of patients have functional independence at 6 months. However, preclinical research is less focused on ICH than on ischemic stroke, and only a few phase 3 clinical trials have been completed. To better mimic the clinical pathological progress of ICH, researchers have developed several preclinical animal models in pigs, dogs, rabbits, cats, and rodents, and in different brain regions, including striatum, cerebroventricles, cortex, hippocampus, and hypothalamus. Two well-established and commonly used preclinical ICH models are the whole blood model and the collagenase model (Fig. 64.1). Besides these two, researchers also use the microballoon insertion model to mimic the mass effect of ICH, and hypertensive mice for induction of spontaneous ICH.

BLOOD INJECTION MODEL

The blood injection model was first used in rats and was later adapted for use in mice [1]. In this model, whole blood is drawn from the animal itself or from a donor animal and injected directly into the brain. Fresh or heparinized blood can be used, as can blood

B-ICH   C-ICH

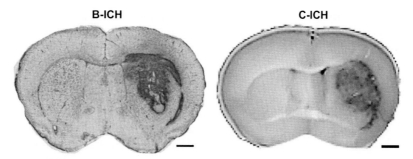

FIGURE 64.1   Representative images of unstained mouse brain sections obtained 72 h after 10 μL of autologous whole blood (B-ICH) or 0.075 U of collagenase (0.5 μL; C-ICH) was injected into the striatum of C57BL/6 mice. Blood was observed primarily in the striatum, with some blood pooled within the corpus callosum. Scale bar = 100 μm. *ICH*, Intracerebral hemorrhage. *Modified from Wang J, Fields J, Doré S. Brain Res. 2008;1222:214–21. Image of the C-ICH brain section was added.*

components or metabolites such as platelets, thrombin, hemin (oxidized heme), or iron. The volume of the blood or blood metabolite used varies among studies and species. In most studies with mice, researchers have injected 10–40 μL of whole blood directly into striatum (x: 0.8; y: 2.0; z: 2.8) at 0.5–2 μL/min. In rats, up to 280 μL is injected at 10 μL/min (3.5 mm to the right of and anteroposterior to the bregma; z: 6 mm). However, because only a limited volume of blood can be infused into the mouse striatum before it begins to flow back along the needle track, a double-injection method was devised. The investigator injects a small amount of blood (one-third of the total volume), waits for it to clot (5–10 min), and then pushes the needle deeper and injects the remaining blood [1,2]. This method largely prevents the blood backflow and makes the size of the hematoma more reproducible. Even a triple-injection model has been used [2].

Blood metabolite injection can be used to study the effects of a single element on brain damage, whereas the whole blood model mimics almost every aspect of ICH pathology. The use of autologous whole blood mimics ICH better than donor blood because the donor blood causes more severe brain edema and induces more inflammatory responses [1]. However, donor blood or blood components can be drawn from fluorescently labeled animals, enabling researchers to observe blood cell diffusion, absorption, and even engulfment by phagocytes.

The blood injection model is easy to perform and reproducible and permits control of hematoma size, but the time required to complete an injection in one animal is three times that of the collagenase model. Additionally, the blood model does not fully represent human ICH pathology because it lacks the underlying vascular pathology and rupture; the hematoma size does not continue to increase in the early hours, whereas most patients with ICH have continuous bleeding; and the injected blood can accumulate under the corpus callosum [1,2]. However, it simplifies the

disease in a way that allows investigators to better study the effects of blood itself and may be a good model for studying white matter injury because extravasated blood travels along white matter fiber tracks.

COLLAGENASE INJECTION MODEL

In the second commonly used ICH model, bacterial enzyme collagenase is injected into the striatum. Injection of 0.0345–2 U/0.5 μL collagenase causes a breakdown of the blood–brain barrier (BBB) and results in active breeding for 6 h or longer during the acute phase of ICH [1]. The procedure is very easy and quick and does not require as much technical expertise as the blood injection model. An experienced investigator can generate 16 animal models in one day. The hematoma location is very reproducible, but its size may vary because different animals react differently to collagenase. Some researchers believe that bacterial collagenase may cause more severe inflammation than blood injection, although our group and others have not observed activation of microglia in cultures treated with collagenase in vitro [1]. Interestingly, Kleinig et al. [3] reported that hemoglobin crystallization is more prominent in the autologous blood injection model than in the collagenase injection model, and may exaggerate associated inflammatory responses.

The collagenase model is relevant to the clinical condition because bleeding continues in 14–20% of all patients with ICH and lasts for over 6 h in 17% of cases [1]. As a result of BBB breakdown, many blood cells infiltrate the hematoma (B cells, T cells, monocytes, macrophages, neutrophils, etc.). Thus the collagenase model provides a good platform to study blood cell infiltration and the role of various blood cells after ICH (inflammation and brain repair). Of course, this model does not mimic clinical ICH in every respect. Collagenase induces rupture and bleeding of many small vessels, whereas a deep small artery rupture is the main cause of ICH in humans. Additionally, the dosage of collagenase injected into the brain is much greater than the level of endogenous collagenase in human ICH [1].

COMPARISON OF BLOOD AND COLLAGENASE MODELS

Hematoma/Injury Volume and Clearance

MacLellan et al. [4] performed a direct comparison between the collagenase model and blood injection model in rats by injecting 100 μL of autologous whole blood (drawn from the tail) or 0.2 U of collagenase into rat striatum. The two methods induced similar sized hematomas after 1 h of infusion. The hematoma remained stable for the first 4 h in the blood injection model but continued to expand significantly in the collagenase injection model, consistent with findings from other studies [1]. At 4–6 weeks, the injury volume was significantly greater in the collagenase model than in the blood injection model, as measured by magnetic resonance imaging (MRI), indicating that collagenase-induced tissue loss continues or that hematoma clearance is faster in the blood model [4].

As we and others have noted, collagenase induces a large, round hematoma in the striatum with little blood in the white matter, whereas blood injection produces a narrow, slit-like or umbrella-shaped lesion with more blood beneath the corpus callosum (Figs. 64.1 and 64.2). In mice, 12 μL of autologous whole blood and 0.0345 U of collagenase produced similar lesion volumes at 24 h, although the hemoglobin content was greater in the blood injection model [5]. Furthermore, hematoma clearance takes approximately 21 days in the blood injection model (15 μL) and approximately 14 days in the collagenase model (0.1 U collagenase, our unpublished data).

Brain Edema

In patients with ICH, larger hemorrhage leads to more perihematomal edema, which is detectable on day 1 and peaks on days 7–11 after the initial bleed. The increased perihematomal edema could be due to red blood cell lysis and hemoglobin- or thrombin-induced toxicity. No direct comparison of edema has been made between the collagenase and blood models in rats, but the time course has been shown in each model. In the blood injection model, brain water content (an indicator of brain edema) begins to increase at 2 h, reaches a maximum at 48 h, and remains

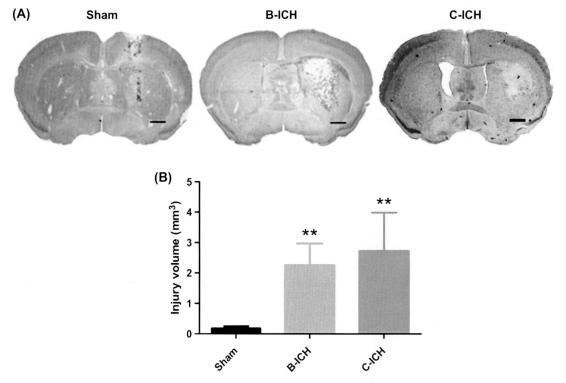

FIGURE 64.2   (A) Representative brain sections from C57BL/6 mice obtained 72 h after the sham procedure (Sham), intrastriatal infusion of 10 μL autologous whole blood (B-ICH), or intrastriatal infusion of 0.075 U collagenase (0.5 μL; C-ICH). Brain sections were stained with Luxol fast blue/cresyl violet. Intrastriatal infusion of 10 μL blood and 0.075 U collagenase produced extensive brain damage, whereas the sham procedure produced very limited needle track damage. Scale bar = 100 μm. (B) Quantification shows significantly larger brain injury volume in the intracerebral hemorrhage (ICH) groups than in the sham group at 72 h ($n = 7$/group, **$p < .01$). *Modified from Wang J, Fields J, Doré S. Brain Res. 2008;1222:214–21. Image and injury volume data from the C-ICH model were added.*

elevated for at least 5 days [6]. In the collagenase model, brain water content begins to increase at 3h, peaks at 24–48h, and remains elevated for at least 3 days. In mice, a direct comparison of edema has been made between injections of 30 μL blood, 0.0375 U collagenase (low dose), and 0.075 U collagenase (high dose) [7]. The three methods induced comparable brain edema in cortex (24 and 72h), but both blood and high-dose collagenase induced more severe brain edema in basal ganglia than did low-dose collagenase at both time points [7]. Another study reported that 0.0383 U collagenase increased the water content significantly more than did 20 μL blood injection [5].

Brain Atrophy/White Matter Injury

Barratt et al. [5] compared changes in striatal and ventricle size in the two mouse ICH models. Striatal shrinkage was significantly greater at 9 weeks than at 5 weeks post-ICH but did not differ significantly between the blood and collagenase models. In contrast, the collagenase injection (0.0383 U) model, but not the blood injection (20 μL) model, caused an increase in ipsilateral ventricle volume at 9 weeks after ICH. The same group also assessed axonal connections with the axonal tracer cholera toxin subunit B and found that collagenase injection caused significant loss of cortical projections from both ipsilateral and contralateral cortex; blood injection caused the loss of fewer projections and only from contralateral cortex [5]. MacLellan et al. [4] used MRI to calculate the volume of corpus callosum in each hemisphere and evaluate white matter injury and/or brain atrophy in rats. The ipsilateral corpus callosum volume was reduced in both models, but the loss was more severe in the collagenase model. Although slight changes were noted, the contralateral corpus callosum volume was unaffected by ICH at 6 weeks [4]. Data from the mouse blood injection model are limited, but we were able to detect gray and white matter damage and recovery in the collagenase injection model in mice by MRI/diffusion tensor imaging [8].

Functional Outcomes

MacLellan et al. [4] found that the collagenase injection model induced more severe neurologic deficits than did the blood injection model in rats from day 1 to day 28 post-ICH. Animals subjected to the blood model were fully recovered at day 21 after ICH, whereas those subjected to the collagenase model lacked spontaneous recovery and had long-term neurologic deficits up to 28 days [2,4]. Moreover, several sensorimotor behavioral tests have been combined to compare the functional outcome (sensorimotor deficits) between the two models. Krafft et al. [7] found no significant differences between mice evaluated with the Garcia neuroscore, the corner turn test, the forelimb placing test, the wire hanging test, and the beam walking test after injection with blood (30 μL) or collagenase (0.075 U). Additionally, the performance on the latter two tests did not correlate with the extent of brain edema at 24 or 72h after ICH. They also reported that the blood injection model failed to mimic the persistent disability of patients with ICH and may not be an appropriate model for investigating the long-term functional recovery of ICH or assessing long-term efficacy of novel compounds.

Cell Death

In rodents, the major forms of cell death after ICH are necrosis and apoptosis. Xue and Del Bigio [9], who investigated cell death in rats at 1h, 4h, and 1, 2, 3, 7, and 21 days after ICH, reported that dying neurons were characterized by cytoplasmic pyknosis and eosinophilia in both models. Peak neuronal death occurred on day 2 after ICH and then diminished on day 7 in the blood model, but was still observed on day 21 in the collagenase model. Also using rats, MacLellan et al. [4] looked at longer time points and found that, compared with the blood (100 μL) injection model, the collagenase (0.2 U) model induced much more neuronal loss in the ipsilateral striatum and, surprisingly, in the area outside of the hematoma in both hemispheres at 6 weeks after ICH. Propidium iodide and TUNEL (terminal deoxynucleotidyl transferase dUTP nick end labeling) staining have been used to evaluate the total cell death in mice after both models of ICH. Barratt et al. [5] observed a similar number of TUNEL-positive cells in both models at 7 and 21 days after ICH, but the collagenase (0.0383 U) model induced significantly more cell death than did the blood injection model (20 μL) at 1 day after ICH.

BBB Breakdown

MRI has revealed that the BBB is permeable at 30 min after collagenase injection and remains so for 7 days, whereas in the blood model, the BBB exhibits small increases in permeability in the surrounding rim of the hematoma 1 day after whole blood injection and remains slightly permeable through day 14 [2]. Disruption of the BBB plays an important role in hematoma expansion, edema formation, and inflammatory cell infiltration from blood. Researchers have also examined the effect of ICH on matrix metalloproteinases (MMPs), a class of protein that plays a vital role in BBB dysfunction. MMP-9 expression level and enzyme activity are elevated after ICH in both the blood and collagenase injection models. Moreover, mice treated with MMP inhibitors and MMP-9 knockout mice have a better outcome than controls after experimental ICH [1].

Inflammation: Neutrophil Infiltration

Because of the increased BBB permeability after ICH, neutrophils and other immune cells are able to infiltrate into the brain. Neutrophils can be identified by immunohistochemical analysis with anti-myeloperoxidase antibody. In the blood injection model, neutrophil infiltration was observed around the hematoma at 1 day, was diminished at day 3, but was still present for up to 10 days [1,2]. In the collagenase model, infiltrating neutrophils appeared in and around the hematoma of mice at 5 h and peaked at 3 days [1]. Direct comparison indicated that neutrophil infiltration was greater in the blood model than in the collagenase model at 24 h after ICH and that most neutrophils were clustered around the hemoglobin crystals [3].

Inflammation: Microglia/Macrophage Activation

After insult, brain resident microglia and infiltrating macrophages become activated and contribute to neuroinflammation and brain repair. Studies have shown that microglia/macrophage activation can be seen as early as the first 1–4 h in both ICH models [1]. In the collagenase model, the activation peaks at 7 days and returns to normal by 21 days, whereas in the blood injection model, the activation peaks at 3–7 days and persists for 4 weeks [1]. Barratt et al. [5] demonstrated that Iba-1 immunoreactivity was much higher in the blood injection (20 μL) model than in the collagenase injection (0.0383 U) model at both 3 and 7 days after ICH.

Inflammation: Cytokine Secretion and Changes of Adhesion Molecules

Liesz et al. [10] compared pro- and antiinflammatory factors and adhesion molecule expression in the two ICH models using real-time polymerase chain reaction during the first week post-ICH. They found that mRNA expression levels of proinflammatory factors [interleukin (IL)-1β, tumor necrosis factor-α, IL-6, and interferon-γ], antiinflammatory factors (IL-10), and adhesion molecules (VCAM-1 (vascular cell adhesion molecule-1) and ICAM-1 (intercellular adhesion molecule-1)) were increased in both models; however, the changes were more pronounced with blood injection than with collagenase injection. Moreover, this difference was associated with worse outcome after blood injection (30 μL) than after collagenase injection (0.045 U), despite equal ICH volumes.

CONCLUSION

Animal models have greatly improved our understanding of the pathophysiology of ICH and have provided a valuable platform for testing potential therapeutic strategies. However, as with other preclinical disease models, no single animal model can mimic all clinical features of ICH. To contend with this drawback, investigators who perform translational studies increasingly use more than one ICH model in their work [8]. The most commonly used strategy is to use a combination of the collagenase and the blood injection models to confirm the major findings.

In our efforts to develop effective therapies for ICH that can be verified in preclinical studies as well as in clinical trials, we must be aware of the pitfalls of ICH animal models. (1) Most clinical ICH occurs in middle-aged and aged individuals and in those with hypertension. Therefore using middle-aged or aged animals or spontaneous hypertensive animals provides a more clinically relevant model, but most translational studies are still carried out in young, healthy animals. (2) Considering the spontaneous nature of ICH, a better model for mimicking clinical ICH would entail spontaneous arterial rupture or bleeding and rebleeding. (3) Anesthesia, surgery, needle insertion, and type of collagenase may all affect the final results. (4) No standard histologic test indices correlate with functional outcomes such as behavioral tests. Therefore it is very difficult to evaluate the full therapeutic effect and side effects of a potential drug. (5) The recovery of patients after ICH may take several months, but most preclinical studies focus only on the acute phase. More preclinical studies should examine the later stages or the recovery stage of ICH to gather information on long-term histologic and functional outcomes.

To make progress in ICH treatment, new animal models are needed that better mirror human ICH pathology. To this end, researchers should utilize different ages, genders, or species of animals and test different injection materials in preclinical studies. Additionally, more communication among basic researchers, translational researchers, and physicians can help to select/generate the best ICH models, improve surgical procedures, and optimize outcomes.

References

[1] Wang J. Preclinical and clinical research on inflammation after intracerebral hemorrhage. Prog Neurobiol 2010;92:463–77.
[2] Manaenko A, Chen H, Zhang JH, Tang J. Comparison of different preclinical models of intracerebral hemorrhage. Acta Neurochir Suppl 2011;111:9–14.
[3] Kleinig TJ, Helps SC, Ghabriel MN, Manavis J, Leigh C, Blumbergs PC, et al. Hemoglobin crystals: a pro-inflammatory potential confounder of rat experimental intracerebral hemorrhage. Brain Res 2009;1287:164–72.
[4] MacLellan CL, Silasi G, Poon CC, Edmundson CL, Buist R, Peeling J, et al. Intracerebral hemorrhage models in rat: comparing collagenase to blood infusion. J Cereb Blood Flow Metab March 2008;28:516–25.
[5] Barratt HE, Lanman TA, Carmichael ST. Mouse intracerebral hemorrhage models produce different degrees of initial and delayed damage, axonal sprouting, and recovery. J Cereb Blood Flow Metab 2014;34:1463–71.

[6] Wu H, Zhang Z, Li Y, Zhao R, Li H, Song Y, et al. Time course of upregulation of inflammatory mediators in the hemorrhagic brain in rats: correlation with brain edema. Neurochem Int 2010;57:248–53.

[7] Krafft PR, McBride DW, Lekic T, Rolland WB, Mansell CE, Ma Q, et al. Correlation between subacute sensorimotor deficits and brain edema in two mouse models of intracerebral hemorrhage. Behav Brain Res 2014;264:151–60.

[8] Zhao X, Wu T, Chang CF, Wu H, Han X, Li Q, et al. Toxic role of prostaglandin E2 receptor EP1 after intracerebral hemorrhage in mice. Brain Behav Immun 2015;46:293–310.

[9] Xue M, Del Bigio MR. Intracerebral injection of autologous whole blood in rats: time course of inflammation and cell death. Neurosci Lett 2000;283:230–2.

[10] Liesz A, Middelhoff M, Zhou W, Karcher S, Illanes S, Veltkamp R. Comparison of humoral neuroinflammation and adhesion molecule expression in two models of experimental intracerebral hemorrhage. Exp Transl Stroke Med 2011;3:11.

CHAPTER

65

Animal Models of Neonatal Stroke/Ischemia

Z.S. Vexler

University of California, San Francisco, San Francisco, CA, United States

INTRODUCTION

Development of age-appropriate models of cerebral ischemia and/or hypoxia has allowed knowledge that brain immaturity at the time of injury plays a key role in the pattern of brain damage. Furthermore, studies of ischemia in immature animals of differing ages have made apparent how rapidly changing the injury response is within the "immaturity spectrum," in part due to different developmental steps of individual brain regions at any given time and the existence of cell type–specific susceptibility to injury. We will summarize available models of at-term ischemic brain injury in rodents and in larger species and discuss how the maturational stage of the brain at the time of an insult affects the underlying injury components with regard to the mechanisms of hypoxic-ischemic encephalopathy (HIE) and stroke in human infants.

DEVELOPMENTAL PATTERNS OF ISCHEMIA-ASSOCIATED BRAIN DAMAGE IN PRETERM AND TERM INFANTS

Hemorrhage and periventricular white matter injury, including periventricular leukomalacia, are the most common types of ischemia-related injuries in preterm human babies. These injury patterns predominate largely due to a weak germinal matrix and the immaturity of oligodendrocyte progenitors. Unlike in preterm, at term ischemia-related injury is no longer diffuse and is manifested focally in gray matter regions, most commonly in the striatum, thalamus, and cortex. Perinatal arterial ischemic stroke is frequent, up to 1 in 2300 live infant births, and produces significant morbidity and severe long-term neurological and cognitive deficits, including cerebral palsy, neurodevelopmental disabilities, and impaired vision [1]. Although injury types are different in preterm and at term, infection and inflammation are the major predisposing and/or modulatory factors in ischemic injury in both age groups [2].

LARGE ANIMAL MODELS OF ISCHEMIA-RELATED INJURY

Various models of hypoxia, hypoxia-ischemia (H-I), and focal stroke have been developed in rodents and in larger species to mimic an array of injuries seen in the human infant. Sheep, pigs, and rabbits are the most commonly used nonrodent mammal species to

Primer on Cerebrovascular Diseases, Second Edition
http://dx.doi.org/10.1016/B978-0-12-803058-5.00065-5

induce H-I in the immature brain. The advantages of these models are that these species have a white/gray matter ratio closer to the human brain than rodents. Cerebral H-I models in fetal sheep induced by bilateral transient occlusion of the carotid arteries during midgestation and in late gestation demonstrated predominant white matter lesions and deep gray matter injury following H-I during midgestation but cortical laminar necrosis in the late-gestation fetus, vulnerability patterns that correlate well with human pathology. At the same time, findings in sheep are relevant to preterm stages in the human because sheep are precocial species. Nevertheless, studies in sheep and in newborn piglets have helped in the development of therapeutic hypothermia for infants with HIE. In rabbits, intrauterine ischemia induced by the Tan lab around 22 days gestation mimicked injury in preterm, whereas intrauterine ischemia at the end of gestation, at 29 days, mimicked injury at term, demonstrating predictive value of animal modeling.

RODENT MODELS OF AT-TERM/ NEONATAL FOCAL STROKE AND HYPOXIA-ISCHEMIA

Principally, two types of models have been developed, an H-I model and focal arterial stroke model. An H-I model in postnatal day 7 (P7)–P9 rats and mice was essentially the only available model related to cerebral ischemia in neonatal rodents for many years. This model consists of a combined ligation of the common carotid artery of one hemisphere and systemic hypoxia for various periods of time. Due to the presence of a systemic hypoxic exposure, the H-I model more closely mimics HIE in human infant than focal arterial stroke (reviewed in Refs. [2, 3]).

Several factors needed to be considered for developing neonatal HIE or stroke models. First, rodent age is of crucial importance. Given that individual regions of rodent brain mature at a different pace, choosing a single postnatal day as comprehensive reflection of at-term human brain was not trivial. Multiple parameters taken into consideration, including neuroanatomy, the dynamics of neurogenesis, synaptogenesis, gliogenesis, and myelination, have demonstrated that brains of P7–P10 rodents most closely resemble brain development at 36–40 weeks of gestation in humans [4]. Second, genetic background markedly affects susceptibility to ischemia-related injury in the neonate [5]. Comparisons of mice of three strains commonly used to generate transgenic strains (C57Bl/6, 129Sv, and CD1), as well as three hybrids of these strains (C57Bl/6x129Sv, CD1xC57Bl/6, and CD1x129Sv) subjected to H-I at P7, showed that CD1 strains are particularly susceptible to brain damage

in this model, whereas 129Sv strains are resistant. The reasons for differing susceptibility are not well understood, but lack of caspase-1 and caspase-11-mediated inflammasome signaling in 129Sv points to the role of inflammatory mechanisms in H-I injury. Third, species-dependent susceptibility to ischemic injury and the response to treatments may also be critical but direct comparisons of H-I outcomes in neonatal rats and mice are scarce. Finally, animal sex is also of critical importance. Stroke, cerebral palsy, and related developmental disorders are more common in males than in females, but the reasons for these differences are not sufficiently studied. The existence of intrinsic sex-specific differences in cell death pathways in the neonatal period has been demonstrated following neonatal H-I [6] and focal ischemia [7]. There is also increasing evidence that the response to therapy may be sex dependent, as has been shown for protection of female but not male neonates against H-I by an iNOS (inducible nitric oxide synthase) inhibitor 2-iminobiotin (reviewed in Refs. [2, 3]).

To better understand mechanisms of perinatal arterial stroke, age-appropriate animal stroke models have been developed by modifying the monofilament middle cerebral artery occlusion (MCAO) model established by Longa et al. in 1989 for studying ischemic brain injury in adult rats. Ashwal et al. first developed a model of transient MCAO (tMCAO) in juvenile, P14–P18 spontaneously hypertensive rats. We developed a monofilament tMCAO model in normotensive P7 rat pups (Derugin et al.[8]). The tMCAO model produces severe cerebral blood flow (CBF) disruption in the region of the occluded middle cerebral artery (MCA) and confirmed partial reperfusion upon suture removal, as is evident from contrast-enhanced MRI studies. Developing a model of transient suture occlusion was important because reperfusion is a common situation seen in human neonates with arterial stroke. Another age-appropriate rat stroke model was developed by Renolleau et al. [9] with the use of a combined permanent MCAO and transient occlusion of the common carotid artery in P7 rats. The monofilament tMCAO model in P7 rat proved useful for short-term stroke studies [8, 10] but was associated with limited ability of neonates to thrive over longer postreperfusion periods. To make the model suitable for long-term studies, we modified the surgical procedure to produce less invasive surgical procedure [11] that did not require the use of vessel electrocoagulation and use of aneurysm clips. We then developed tMCAO model in P10 rats [12] and tMCAO model in P9 mice [13]. Varying the duration of tMCAO has allowed for induction of injuries of differing severity [13] and MRI-based confirmation of recirculation following suture retraction in the tMCAO model has allowed for studies of reperfusion.

METHOD FOR INDUCTION OF tMCAO IN P7 RATS

Animals

The described experimental procedures were performed in accordance with National Institutes of Health guidelines for humane handling of animals with prior approval from the Committee of Animal Research at University of California San Francisco. Female Sprague Dawley rats with a 6-day-old litter (10–11 pups per litter) were obtained from Simonson Laboratories (Gilroy, CA). The mother and her litter were given food and water ad libitum and housed in a temperature/light controlled animal care facility until the pups are 7 days old.

Suture Preparation

Proper suture coating is critical to achieve CBF blockage to the MCA and to prevent bleeding. In earlier studies [8, 11], we were preparing coated sutures in the laboratory. A 2-mm end of a 12-mm-long 6-0 Dermalon monofilament suture (United States Surgical, Norwalk, CT) was lightly scratched on all sides with fine, 400 silicone carbide paper, a silicone-based polysiloxane impression material (CutterSil, Miles Inc. Dental Products, South Bend, IN) was prepared by mixing 0.6 mL of the base with one drop of the liquid activator for 20–30 s, and the 2-mm scratched end of the suture was dipped in the silicone, lightly wiped against paper to remove excess of silicone and immediately taped upside down. Each suture was examined under a microscope before use and the diameter measured with a micrometer. The targeted dimensions were defined as a distal width of 0.18 mm, tapering over a 2-mm segment to 0.1 mm. Sutures that did not have a smooth continuous taper or those with a distal tip width of >0.2 mm or <0.14 mm were rejected. Recently, we began using commercially available 6-0 precoated filaments (Catalog# 602123PK10, Doccol Corp., Sharon, MA), which produce similar incidence and injury extent (unpublished).

Animal Surgery

The P7 pups are anesthetized utilizing 3% isoflurane in a mixture of 70% N_2O/30% O_2, and then the isoflurane is reduced to 1–1.5%, as needed. Throughout the 15–20 min procedure, the animal temperature is maintained at 38°C by a heating pad and an overhead lamp. The surgical incision is made directly over the right common carotid (CC), external carotid (EC), and internal carotid (IC) bifurcation. Beginning at its origin, a 1-mm segment of IC is carefully dissected and tied off with a single strand of silk at its origin [11]. A second silk suture is looped around the IC, just above the pterygopalatine artery, and gently pulled laterally

to prevent retrograde blood flow. A small incision is made in the proximal isolated IC vessel segment. The coated 6-0 occluding suture is inserted and advanced 7.5–9.0 mm, depending on the pup weight. In smaller pups, 12–14 g, the suture is advanced 7.5–8.5 mm; pups over 15 g have the suture filament advanced 8.5–9 mm. The lower knot is gently tightened, securing the suture, and the skin incision is closed with three interrupted 7-0 silk sutures. The pups are transferred in a container placed on a heating pad under the lamp until animals are recovered from anesthesia; hyperthermia is avoided. The pups are returned to dams after they are fully recovered from anesthesia.

For reperfusion, following anesthesia, the middle skin suture is removed, the knot securing the occluding suture is loosened, and the suture filament together with the coating is gently pulled out and the knot retightened. Typically, there is no blood loss during either occlusion or reperfusion. The pups are transferred in a container placed on a heating pad under the lamp until animals are recovered from anesthesia. Animal weights are recorded daily. If no weight gain is seen within 2 days after tMCAO, animals are gavage fed with 2% condensed milk until signs of weight gain are evident.

Caveats, Limitations, and Concerns

A number of factors can affect the successful blockade of the MCA. Blockade of the MCA may be inadequate. The weight of animals can vary substantially even within the same litter thus affecting the length of the suture advancement and the thickness of the suture needed for proper placement of the tip of the suture in proximity to the MCA. The use of a shorter than necessary suture will not result in blockage of the MCA, whereas the use of a longer than needed suture may produce only partial disruption of CBF through the MCA as the thickest portion of the suture, its coated tip, will be far from the MCA.

Another potential concern is that suture coating may be insufficient. Uncoated suture may perforate the vessel, whereas thick or uneven coating will make the suture more rigid, making it more difficult to advance or retract. Excessive coating may affect the ability to reperfuse due to vasospasm and affected responsiveness of the vessel. Reuse of the coated suture may affect consistency of results, as the suture will become more rigid and/or coating may slide off. Lack/delay in weight gain due to reduced nursing may adversely affect longer-term studies. Gavage feeding with milk may be necessary if weight gain of injured pups is diminished [11].

It is also important to acknowledge that intralitter variations in the incidence may occur and that variable extent of recovery in individual pups over time may follow nearly identical spatial injury pattern within 24 h after tMCAO [11].

METHOD FOR INDUCTION OF tMCAO IN P10 RATS

We used a similar approach for inducing tMCAO model in P10 rats, insertion of the filament into the external carotid artery lumen and gently advancing suture through the internal carotid artery following electrocoagulation of pterygopalatine, occipital, superior thyroid, and lingual arteries [12]. A 5-0 nylon monofilament suture was used in these larger rat pups [12]. Reperfusion was achieved by reanesthetizing the animal and removing the suture 1.5 or 3 h following MCAO. Histological assessment demonstrated large and consistent injury regions 24 h after reperfusion following both short and long occlusion periods. No assessments of completeness of reperfusion were performed.

METHOD FOR INDUCTION OF tMCAO IN P9–P10 MICE

Transient 1.5- or 3-h MCAO is induced in unsexed P9 mice in a manner similar to that detailed in P7 rat [8], with modifications [11]. Following induction of anesthesia with 2% isoflurane in O_2/N_2O, isoflurane is reduced to 1.25% and using aseptic techniques, a midline cervical incision is made and the CCA exposed. The skin and muscle are gently retracted with home built microretractors, exposing the ICA. A single thread from a 7-0 silk suture is used to tie a temporary loose knot at the origin of the ICA, which is pulled medially, and the long end clipped to the skin. A second strand of 7-0 suture is looped 1 mm below the origin of the ICA, and gently retracted laterally to prevent retrograde blood flow. An arteriotomy is then made. A coated 8-0 nylon suture is inserted and advanced 4–5 mm, depending on animal weight. The lower 7-0 suture is then gently tied with a releasable knot to keep the filament in place, and prevent retrograde bleeding from the arteriotomy. The skin is closed and the pups are allowed to recover. An overhead heat lamp is used to maintain temperature of the pups. Reperfusion is achieved by reanesthetizing the pups, removing a single skin suture to expose the cervical vessels, untying both 7-0 suture strands, removing the occluding 8-0 filament, and applying a small amount of gel foam and pressure to the arteriotomy for 30 s, after which the skin is resutured. We recently began using commercially available 7-0 precoated filaments (Catalog # 701712PK5Re, Doccol Corp., Sharon, MA) and obtained similar injury extent and incidence (unpublished).

MODEL-SPECIFIC OXIDATIVE AND INFLAMMATORY MECHANISMS OF ISCHEMIC INJURY IN NEONATES

The developing brain is very sensitive to oxidative injury, inflammation, and excitotoxicity following H-I or stroke (reviewed in Ref. [3]). From translational perspective, it would be important to compare whether the contributions of oxidative and inflammatory mechanisms in the pathophysiology of H-I and stroke models are similar. However, with very few exceptions, there are no studies that in detail compared how the model (tMCAO vs. H-I) or the maturation stage of the brain at the time of an insult (P7 vs. P9) affects injury induction, evolution, and/or repair. For example, blood–brain barrier (BBB) leakage has been demonstrated to be low within 24 h after reperfusion in P7 rats subjected to a 3-h tMCAO [14], whereas a transient region-specific increase in BBB leakage to small and large molecules was shown in the H-I model in P9 mice within 24 h [15]. Although data in both neonatal models are consistent in demonstrating a markedly more limited extent of BBB disruption than in adult rats and mice following tMCAO, it is unknown if the differing extent of BBB integrity patterns between tMCAO in neonatal rats and H-I in neonatal mice is due to differing model or differing species. Studies of angiogenesis following tMCAO in P7 and in P10 rats demonstrated the presence of a subtle and delayed angiogenic response irrespective of whether stroke was induced at P7 or P10 (reviewed in Ref. [3]), but it is unknown whether this range of ages may be a factor in myelination after tMCAO.

Neuroinflammation is a characteristic feature of stroke progression in the adult and in the neonate and is a major injury modifier. Although the critical role of brain immaturity for the pattern of neuroinflammatory responses has been firmly established [2, 3], it remains largely unknown if the behavior of glial cells and peripheral leukocytes is affected by the presence of systemic hypoxia in the H-I model compared with effects following ischemia-reperfusion.

To summarize, ischemia-related injury to early postnatal brain impacts many key neurodevelopmental processes that undergo maturation changes during these time frames, leading to various abnormalities later in life. There is no single animal model that recapitulates the complexity of the human condition but utilization of several ischemia-related age-specific models in rodents and in larger species, and of both sexes, enables the enhanced understanding of brain pathology and should help development of novel therapies for the immature brain.

Sources of Funding

NINDS_NS80015, NINDS_NS44025, NINDS_NS76726, The Leducq Foundation DSRR_P34404.

References

[1] Nelson KB. Perinatal ischemic stroke. Stroke; a journal of cerebral circulation February 2007;38(Suppl. 2):742–5.

[2] Hagberg H, Mallard C, Ferriero D, et al. The role of inflammation in perinatal brain injury. Nat Rev Neurol April 2015;11(4):192–208.

[3] Fernandez-Lopez D, Natarajan N, Ashwal S, Vexler ZS. Mechanisms of perinatal arterial ischemic stroke. J Cereb Blood Flow Metab June 2014;34(6):921–32.

[4] Semple BD, Blomgren K, Gimlin K, Ferriero DM, Noble-Haeusslein LJ. Brain development in rodents and humans: Identifying benchmarks of maturation and vulnerability to injury across species. Prog Neurobiol 2013 Jul-Aug;106-107:1–16.

[5] Sheldon RA, Sedik C, Ferriero DM. Strain-related brain injury in neonatal mice subjected to hypoxia-ischemia. Brain Res 1998;810(1–2):114–22.

[6] Hagberg H, Wilson MA, Matsushita H, et al. PARP-1 gene disruption in mice preferentially protects males from perinatal brain injury. J Neurochem September 2004;90(5):1068–75.

[7] Renolleau S, Fau S, Goyenvalle C, Charriaut-Marlangue C. 'Sex, neuroprotection, and neonatal ischemia'. Dev Med Child Neurol June 2007;49(6):477. author reply–8.

[8] Derugin N, Ferriero DM, Vexler ZS. Neonatal reversible focal cerebral ischemia: a new model. Neurosci Res 1998;32(4):349–53.

[9] Renolleau S, Aggoun-Zouaoui D, Ben-Ari Y, Charriaut-Marlangue C. A model of transient unilateral focal ischemia with reperfusion in the P7 neonatal rat: morphological changes indicative of apoptosis. Stroke; a journal of cerebral circulation 1998;29(7):1454–60. discussion 61.

[10] Manabat C, Han BH, Wendland M, et al. Reperfusion differentially induces caspase-3 activation in ischemic core and penumbra after stroke in immature brain. Stroke; a journal of cerebral circulation January 2003;34(1):207–13.

[11] Derugin N, Dingman A, Wendland M, Fox C, Vexler ZS. Magnetic resonance imaging as a surrogate measure for histological subchronic endpoint in a neonatal rat stroke model. Brain Res 2005;1066:49–56.

[12] Mu D, Jiang X, Sheldon RA, et al. Regulation of hypoxia-inducible factor 1alpha and induction of vascular endothelial growth factor in a rat neonatal stroke model. Neurobiol Dis December 2003;14(3):524–34.

[13] Woo MS, Wang X, Faustino J, et al. Genetic deletion of CD36 enhances injury after acute neonatal stroke. Ann Neurol 2012;72:961–70.

[14] Fernandez-Lopez D, Faustino J, Daneman R, et al. Blood–brain barrier permeability is increased after acute adult stroke but not neonatal stroke in the rat. The Journal of neuroscience : the official journal of the Society for Neuroscience July 11, 2012;32(28):9588–600.

[15] Ek CJ, D'Angelo B, Baburamani AA, et al. Brain barrier properties and cerebral blood flow in neonatal mice exposed to cerebral hypoxia-ischemia. J Cereb Blood Flow Metab January 28, 2015;35(5):818–27.

CHAPTER

66

Animal Models: Vascular Models of Cognitive Dysfunction

E. Hamel

McGill University, Montréal, QC, Canada

INTRODUCTION

Convincing evidence supports a link between vascular disease, cognitive impairment and dementia, including Alzheimer disease (AD) [1]. A compromised cerebral vasculature with structural alterations, impaired dilatory capacity, and failure to maintain brain perfusion and adjust flow in response to increased neuronal activity may result in cerebral dysfunctions that translate into cognitive failure. In human, a vasculopathology is a key factor in vascular cognitive impairment (VCI) and vascular dementia (VaD), the second most common form of dementia after

Primer on Cerebrovascular Diseases, Second Edition
http://dx.doi.org/10.1016/B978-0-12-803058-5.00066-7

AD, notwithstanding the fact that VaD and AD often coexist [1]. VaD can be caused by diverse pathological conditions that become manifest with aging and impede on the cerebral vasculature. Therefore, the multifaceted origin of VaD is difficult to faithfully recapitulate in animal models. We will review selected animal models in which a chronic cerebral hypoperfusion appears as the main culprit of the cognitive decline, admitting that there is no perfect model. We will attempt to identify some commonality between the models that may help better understand the human disease and identify possible target pathways that may lead to new treatment options.

COGNITIVE DYSFUNCTION OF VASCULAR ORIGIN IN HUMAN

In humans, cognitive impairment or dementia resulting from vascular dysfunctions can be due to causes as diverse as cardiovascular diseases, metabolic diseases, and genetic or sporadic small-vessel diseases (SVDs), age being generally an underlying key factor [1]. For the most part, patients with VCI and VaD, as well as patients with AD with vascular pathology, are characterized by discrete infarcts, chronic vasculopathy, and diffuse lesions in deep gray nuclei and white matter (WM). The vascular pathology is not identical depending on the disease etiology, but its main landmarks are a lack of integrity of the blood vessel (thickened walls; fibrosis; pericyte, endothelial, or smooth muscle cell degeneration), deficits in baseline perfusion and cerebral blood flow (CBF) evoked by increased neuronal activity (functional hyperemia or neurovascular coupling), reduced vascular compliance, cerebral autoregulation and endothelial function, and a compromised blood–brain barrier (BBB). In addition, abnormalities in neuronal circuits, glial cells, and WM (such as WM hyperintensities or leukoaraiosis seen on imaging modalities) are likely to be implicated in the cascade of events that leads to deterioration of cognitive function. Accordingly, it has been hypothesized that chronic brain hypoperfusion is a key event in triggering cognitive failure. It is important that animal models recapitulate as many aspects of the human disease as possible, and particularly the chronic cerebral hypoperfusion, to get insight into the underlying pathogenic cascade and help advance treatment options. The animal models selected in this short review all meet this prerequisite (Table 66.1).

TABLE 66.1 Vascular Models of Cognitive Dysfunction

Model	Main Species Used	Cerebrovascular Pathology[a]	Cerebrovascular Hypoperfusion	Cognitive Deficit	Main References
OCCLUSION MODELS					
BCAO	Rat	No, these models	Very effective in	Yes, of variable	2–4
BCAS	Mouse	do not induce a	reducing brain	severity	
Single vessel occlusion	Rat	vascular pathology	perfusion		
Others and variants	Rat/mouse				
VASCULAR RISK FACTORS					
Hypertension	Rat/mouse	Yes	Yes	Yes	2–5
Atherosclerosis	Rat/mouse	Yes	Yes	Yes	
Chronic heart failure	Rat/mouse	No	Yes	Yes	
Myocardial infarction	Rat/mouse	No	Yes	Yes/No	
Diabetes	Rat/mouse	Yes	Yes	Yes	
GENETIC FORMS OF SVD					
CAA	Mouse	Yes	Yes	Yes (?)	6, 7
Collagen type IV-SVD	Mouse	Yes	Yes	NR	
CADASIL	Mouse	Yes	Yes	NR	
RVCL	NA	–	–	–	
CARASIL	NA	–	–	–	
CEREBROVASCULAR DISEASE					
TGF-β1	Mouse	Yes	Yes	Yes (can be very subtle)	8, 9

[a]Refers to a loss of integrity of the cerebral blood vessels that can correspond to structural alterations, microvascular degeneration, thickened basement membrane, damaged vessel walls, fibrosis, arteriosclerosis, etc. Abbreviations: ?, unknown whether the deficit is directly imputed to CAA or other deleterious effects of soluble Aβ species; BCAO, bilateral carotid artery occlusion; BCAS, bilateral carotid artery stenosis; CAA, cerebral amyloid angiopathy; CADASIL, cerebral autosomal-dominant arteriopathy with subcortical infarcts and leukoencephalopathy; CARASIL, cerebral autosomal recessive arteriopathy with subcortical infarcts and leukoencephalopathy; NA, no validated model currently available; NR, no available reports of the cognitive deficits in these mouse models; RVCL, retinal vasculopathy with cerebral leukodystropy; SVD, small-vessel disease; TGF, transforming growth factor.

ANIMAL MODELS: ADVANTAGES AND LIMITATIONS

Occlusion Models

Several models have been developed in rat and mouse that induce global chronic cerebral hypoperfusion. The most popular models are those of bilateral common carotid occlusion (BCAO) in rat and bilateral common carotid artery stenosis (BCAS) in mouse. These models have been extensively used to induce chronic cerebral hypoperfusion accompanied by cognitive deficits. They recapitulate several aspects of VaD, such as reduced brain perfusion, WM damage, BBB disruption, inflammation, infarcts, and cognitive deficits in several tests related to spatial (rat) and working and reference memory (mouse). They are very effective at reducing brain perfusion. However, they bear the risk of inducing neuronal damage, mainly in CA1 (and CA3) neurons of the hippocampus that are important in mnemonic processes, and which are highly susceptible to ischemia as can be induced by the initial drop in CBF at the time of occlusion. The possibility of using gradual occlusion models in the rat (BCAGO or two-vessel gradual occlusion) with cuffs that swell gradually may be preferred to avoid the risk a cofounding neuronal damage. In the mouse BCAS model, the diameter of the coils will determine the severity of the cognitive impairment and coils need to be carefully selected [2–4]. Reports of somatosensory impairment, motor deficits, and damage to visual pathway raise caution in the control of these animal models for reliable interpretation of possible cognitive deficits. Multiple variants of the occlusion models exist [4], including refined models of microinfarcts at the level of a single penetrating artery that lead to cognitive impairment.

Overall, occlusion models have been widely used and characterized; they can be combined with other risk factors to precipitate or increase the severity of the cognitive impairment. However, although they effectively reduce perfusion to the brain, they do not mimic a vascular pathology as seen in the human disease.

Vascular Risk Factor Models

Cardiovascular disease (CVD) and, mainly, hypertension are the primary risk factors for both VaD and AD in the aging population [1]. The vascular pathology of CVD may result in structural changes of brain vessels, reduced CBF, altered autoregulation, BBB alterations, microvascular changes, WM lesions, microbleads, brain atrophy, and cerebral inflammation, which will lead to cognitive deficits. Different models of CVDs exist.

Hypertension: Several models of hypertension have been developed and found to be associated with cognitive dysfunction and dementia [5]. Among the most readily used models of hypertension, the spontaneous hypertensive rat (SHR) and the stroke-prone SHR models have been widely reviewed for their ability to recapitulate vascular fibrosis, BBB damage, and susceptibility to stroke events when combined with specific diets [3–5]. When using these models, animals that display motor deficits, visual lesions, and mild coordination deficits should be eliminated when assessing cognitive function [4].

Atherosclerosis: Models of atherosclerosis are based on alterations of ApoE, ApoB, or low-density lipoprotein receptor genes with different degrees of atherosclerosis, a pathology that can be accelerated by factors, such as a high-fat or high-cholesterol diet [2]. Animal models of atherosclerosis show cognitive deficits in several behavioral tests measuring spatial or episodic memory. In some models, atherosclerotic lesions in brain vessels are limited, which complicates definite conclusions on the link between atherosclerosis of brain vessels and cognitive deficits.

Chronic heart failure and myocardial infarction: Models of chronic heart failure (transverse aortic constriction) or myocardial infarction (permanent ligation of the left anterior descending coronary artery) have been extensively described and reviewed [2]. They result in cerebral hypoperfusion that may be accompanied or not by cognitive impairments depending on the model.

Diabetes: The number of animal models of type 1 and type 2 diabetes is increasing. The most commonly used mouse or rat models have been genetically selected, genetically modified, or pharmacologically induced (streptozotocin, alloxan). Some models have been shown to develop long-term cognitive deficits that have been attributed to alterations of the brain microvasculature, BBB disruption, and other neuronal and glial dysfunction [4]. However, more information is needed. For instance, better discriminating the effects of high-fat diet, often used in combination with diabetes and associated with atherosclerosis and dementia, and obesity from those of diabetes on the cognitive decline would be useful [4].

Additional work is needed to better decipher the mechanisms through which CVD and metabolic diseases lead to cognitive dysfunction. Since in a large proportion of cases, vascular risk factors can be controlled or prevented, adequate pharmacotherapy or change in lifestyle habits may have significant impact in the outcome on cognitive failure in these patients.

Genetic Forms of SVD Models

Most SVDs are sporadic and result from chronic cardiovascular diseases or other risk factors combined with aging, whereas familiar forms of SVD with earlier age of onset have been identified, and the related genetic mutations have been associated with specific facets of the disease [6, 7]. Similar to sporadic SVD, familial SVD affects the structure and function of brain vessels, can remain silent for years before the clinical portrait of WM hyperintensities, ischemic stroke, hemorrhagic events, leukoaraiosis, and cognitive decline becomes manifest. Genetic forms of SVD encompass hereditary cerebral amyloid angiopathy (CAA, mutations of the amyloid precursor protein, APP), collagen type IV-SVD (mutations of the collagen type IV alpha 1 (Col4A1), or alpha 2 (col4A2) gene), cerebral autosomal-dominant arteriopathy with subcortical infarcts and leukoencephalopathy (CADASIL, mutations in the extracellular domain of NOTCH3, Notch3^{ECD}), cerebral autosomal recessive arteriopathy with subcortical infarcts and leukoencephalopathy (CARASIL, mutations of the high temperature requirement A serine peptidase, HTRA1), and retinal vasculopathy with cerebral leukodystropy (RVCL, mutations of the three prime repair exonuclease 1, TREX1). The former account for most forms of familial SVD, whereas the two latter are more rare. Yet, more CARASIL cases are being identified since the genes have been characterized [2, 6].

Animal models of CAA: Although the Dutch mutation of the human APP may represent the archetype of familial CAA, transgenic APP mouse models will not be covered here. Indeed, literally all APP mutations will result in increased levels of soluble amyloid-β (Aβ) peptide that affect both cerebrovascular and cognitive function, in addition to inducing various degrees of CAA depending on the mutations. It is thus very difficult to discriminate the role of CAA from that of soluble Aβ species in the cognitive deficits of APP mice.

Collagen type IV-SVD models: Col4a1 and Col4a2 mutants have been generated and found highly relevant to cerebrovascular changes seen in patients with SVD, such as recurrent hemorrhagic strokes, alterations in the vascular basement membrane, damaged blood vessel walls, and age-related angiopathy [6]. However, no information could be found on cognitive deficits in these genetically modified mice, which would appear as an essential requirement to establish a link of causality between the cerebrovascular pathology and the cognitive dysfunctions.

CADASIL models: Different knock-in and transgenic mutant NOTCH3 mice have been generated, some of which replicate the most characteristic hallmarks of the disease, such as NOTCH^{ECD} aggregates and extracellular deposits of granular osmophilic material in brain and peripheral blood vessels [6, 7]. Particularly, transgenic PAC-Notch3^{R169C} mutant mice develop cerebrovascular dysfunctions corresponding to decreased myogenic tone, cerebral autoregulation, neurovascular coupling, and mild hypoperfusion and, with aging, astrogliosis and diffuse WM lesions but no stenosis of the cerebral blood vessels, lacunar infarcts, or BBB lesions. Notably,

patients with CADASIL have an increased incidence of migraine aura, and another transgenic Notch3 mutant has a reduced threshold for cortical spreading depression, the electrophysiological substrate for migraine aura [6]. PAC-Notch3^{R169C} mutant mice are considered the best model of CADASIL, and are thought to recapitulate the presymptomatic stage rather than the full spectrum of the human disease. However, no information is available on learning, memory, attention, or other types of cognitive performance in CADASIL mouse models, which limits any conclusion on the direct relationship between cerebrovascular pathology and cognitive impairment.

CARASIL and RVCL models: There is no established model of these two SVDs. CARASIL-associated HTRA1 mutations correspond to a loss of function as a repressor of transforming growth factor-β (TGF-β) family signaling, which is a key regulator in the synthesis of extracellular matrix proteins [7]. Accordingly, patients with CARASIL show increased levels of TGF-β1 in brain vessels with arteriosclerosis associated with thickened blood vessel walls; increased levels of collagen, fibronectin, and other basement membrane proteins; and degeneration of smooth muscle cells. Animal models of CARASIL-related HTRA1 mutations will be greatly useful to further decipher the relationship between the cerebrovascular pathology and the cognitive dysfunction seen in this form of familial SVD.

Cerebrovascular Pathology Models

Transgenic TGF-β1 mice (TGF mice) that overexpress a constitutively active form of TGF-β1 in brain astrocytes were originally developed as a model of the AD cerebrovascular pathology [8]. However, they could be highly relevant to CARASIL based on the HTRA1 mutation translating into increased TGF-β1 signaling. TGF mice display structural changes of the brain vasculature with increased basement membrane proteins (collagen IV, fibronectin, laminin, perlecan), thickened vessel walls, microhemorrhages, string vessel pathology (degenerating capillaries), and perivascular astrogliosis. There is no CAA pathology as originally thought since the accumulated proteins (detected with Congo red or thioflavin-S) do not correspond to Aβ peptide. TFG mice feature cerebrovascular dysfunctions characterized by reduced resting CBF, neurovascular uncoupling, and impaired cerebrovascular dilatory and contractive function [9]. Reduced levels of synaptophysin and neurogenesis have been reported in the hippocampus of TGF mice. Cognitive deficits in spatial (Morris water maze) and recognition (novel object recognition) memory develop in TGF mice with increasing age, but their severity varies depending on the study, as highlighted in Ref. [9]. Overall, TGF mice appear as a highly relevant model of

VaD with a primary cerebrovascular pathology and an age effect on the severity of the cerebrovascular and cognitive deficits. Studies are needed to better validate this model for additional features of VaD such as leukoaraiosis or other types of WM lesions.

Other Animal Models

Other models of VaD or SVD, and variants of the models described succinctly here, exist. They have not been covered here due to space limitation. However, excellent reviews can be consulted for more detail [2–4, 6].

COMMONALITIES AND PATHOPHYSIOLOGY

The vascular injuries that primarily characterize VaD and SVD may lead to decreased resting brain perfusion, neurovascular uncoupling, microinfarcts, microbleeds, BBB breakdown, and, ultimately, failure of neuronal function and cognitive impairment. Oxidative stress and neuroinflammatory processes are commonly reported in the brain of patients with SVD and VaD, but they are rather nonspecific markers of several neurological diseases. In contrast, the increased fibrotic process of brain arterioles and capillaries, characterized by accumulation of collagen and other proteins in the extracellular matrix, has been highlighted as a shared feature of SVD, VaD, AD, and several cardiovascular or metabolic diseases associated with cognitive impairment [10]. This fibrotic process, of which collagen IV is a core component, is largely regulated by TGF-β1 signaling in blood vessels. Interestingly, increased TGF-β1 levels have been reported in brain tissue or brain vessels in patients with AD, CADASIL, CARASIL, and, possibly, col4-SVD [6-8, 10]. Considering the phenotype of transgenic TGF mice (see earlier discussion), there is a need for further investigating pathologies of the vascular basement membrane and their consequences on cerebrovascular and neuronal function at the cellular and molecular levels.

CONCLUSION

The availability of neuroimaging techniques for imaging progressive changes in cerebrovascular alterations, such as vessel morphology, hemodynamic responses, and vascular integrity in animal models (including in mice), concurrently with behavioral testing should allow unprecedented progress in our ability to decipher the relationship between cerebrovascular pathology and cognitive decline. Developing new or refining existing animal models of vascular pathology and testing their impact on cognitive function and their response to therapy should help better understand the human disease. Such studies could lead to new treatment opportunities notwithstanding the possibility to revisit treatments already used for vascular disease.

References

[1] Iadecola C. The pathobiology of vascular dementia. Neuron 2013;80(4):844–66.

[2] Bink DI, Ritz K, Aronica E, van der Weerd L, Daemen MJ. Mouse models to study the effect of cardiovascular risk factors on brain structure and cognition. J Cereb Blood Flow Metab 2013;33(11):1666–84.

[3] Hainsworth AH, Markus HS. Do in vivo experimental models reflect human cerebral small vessel disease? A systematic review. J Cereb Blood Flow Metab 2008;28(12):1877–91.

[4] Venkat P, Chopp M, Chen J. Models and mechanisms of vascular dementia. Exp Neurol 2015;272:97–108.

[5] Obari D, Ozturk Ozcelik S, Girouard H, Hamel E. Cognitive dysfunction and dementia in animal models of hypertension. In: Girouard H, editor. Hypertension and the Brain as an End-Organ target. Springer; 2016. p. 71–97 [chapter 5].

[6] Joutel A, Faraci FM. Cerebral small vessel disease: insights and opportunities from mouse models of collagen IV-related small vessel disease and cerebral autosomal dominant arteriopathy with subcortical infarcts and leukoencephalopathy. Stroke 2014;45(4):1215–21.

[7] Yamamoto Y, Craggs L, Baumann M, Kalimo H, Kalaria RN. Review: molecular genetics and pathology of hereditary small vessel diseases of the brain. Neuropathol Appl Neurobiol 2011;37(1):94–113.

[8] Wyss-Coray T, Lin C, Sanan DA, Mucke L, Masliah E. Chronic overproduction of transforming growth factor-β1 by astrocytes promotes Alzheimer's disease-like microvascular degeneration in transgenic mice. Am J Pathol 2000;156:139–50.

[9] Tong XK, Hamel E. Simvastatin restored vascular reactivity, endothelial function and reduced string vessel pathology in a mouse model of cerebrovascular disease. J Cereb Blood Flow Metab 2015;35(3):512–20.

[10] Thompson CS, Hakim AM. Living beyond our physiological means: small vessel disease of the brain is an expression of a systemic failure in arteriolar function: a unifying hypothesis. Stroke 2009;40(5):e322–30.

CHAPTER

67

Animal Models: Nonhuman Primates

M. Tymianski[1,2,3]

[1]Krembil Research Institute, Toronto, ON, Canada; [2]University of Toronto, Toronto, ON, Canada; [3]University Health Network, Toronto, ON, Canada

INTRODUCTION

The notion that stroke studies in nonhuman primates (NHPs) will enable the translation of a promising therapy to humans is theoretically sound, but unproven in the absence of a positive human study. Moreover, tissue plasminogen activator (the only widely approved acute ischemic stroke therapy) was translated to human usefulness with rabbit models of thromboembolic stroke, and not primates [1]. Conversely, clinical trials with the neuroprotectant NXY-059 were conducted after promising preclinical results in NHPs, but they resulted in a failed phase III human trial [2]. Thus factors other than the animal model may need to be accounted for in the effort to translate promising neuroprotectants to clinical usefulness. Nonetheless, NHPs may provide an opportunity to further validate promising therapies, especially neuroprotectants. If so, then the choice of species, the method of inducing the stroke, and the choice of outcome measures may be critical in ensuring that a primate study is predictive of human results.

THE CHOICE OF SPECIES

The New World primates are native to South, Central, and Southern North America and include: *Callitrichidae, Cebidae, Aotidae, Pitheciidae,* and *Atelidae.* They are generally smaller than Old World primates and exhibit less, lissencephalic, or near-lissencephalic brains. *Callithrix jacchus* (the common marmoset) and *Saimiri sciureus* (the squirrel monkey) are most widely used [3]. Lissencephalic species may offer an advantage over gyrencephalic species in studies requiring consistent, focal strokes because motor, sensory, and visual representation is on the cortical surface, which allows precise functional mapping with minimal brain

exposure before ischemia. By contrast, in gyrencephalic species, mapping of specific regions may not be as precise when motor representation lies on cortex within a sulcus [4]. Lissencephalic species are also smaller making them easier to house, manage, and conduct experiments. However, a potential disadvantage is that they exhibit fewer similarities to the human brain as compared with gyrencephalic primates. Although unproven, this difference may limit the capacity of stroke experiments conducted in lissencephalic species to predict the effects of a promising therapy in humans. An example of this has been the apparent ineffectiveness of studies in marmosets to predict success with the neuroprotectant NXY-059 [2].

Old World primates are native to the African and Asian continents and include the family *Cercopithecidae.* They are gyrencephalic, possessing larger brains with a more complex cortical and subcortical organization and a vascular anatomy that resemble that of the human brain [3]. Due to their size, Old World species require specialized housing facilities with larger cages and more environmental enrichment than the smaller New World species. In general they also require greater resources and technical expertise for anesthesia, surgery, and recovery. Historically, the most common species used in stroke experiments has been the *Papio anubis* (baboon). Symon and colleagues used this species to characterize the ischemic penumbra, specifically the relationships between time, blood flow, and the likelihood of cerebral infarction after middle cerebral artery occlusion (MCAO) [5]. Such research provided the foundation for stroke research to come by establishing the concept that therapeutic intervention could reduce infarction in brain-at-risk. However, baboons are large, requiring significant resources to house, and are highly aggressive, which limits the extent of clinical neurological evaluation that can be practically conducted.

Primer on Cerebrovascular Diseases, Second Edition
http://dx.doi.org/10.1016/B978-0-12-803058-5.00067-9

Macaque monkeys (genus *Macaca*) have a gyrence-phalic brain with cortical and subcortical anatomy similar to human. *Macaca fascicularis* (cynomolgus macaque) and *Macaca mulatta* (rhesus macaque) have been used in permanent and transient MCAO models. The vascular anatomy of the macaque closely resembles the human with less collateralization than the baboon [6]. Among the macaques the cynomolgus may be more similar to humans in that it may have less collateralization than the rhesus as demonstrated by the need to occlude the ACA (anterior cerebral artery) along with the distal middle cerebral artery (MCA) to produce a cortical stroke in the rhesus [3] versus clipping only the first segment of the MCA in the cynomolgus to produce an equivalent stroke [3]. Macaque monkeys are highly social and more easily trained than baboons, enabling researchers to conduct neurological evaluations before and after stroke more readily.

METHODS OF STROKE CREATION

Since a primary purpose of NHP stroke research is to predict the human situation, it is critical that the strokes produced be conducted in a manner that simulates the strokes under study in humans. The method of stroke creation is thus a factor that must be considered [7]. Focal ischemia related to MCA stroke is the most common syndrome of human acute ischemic stroke and, as such, has been the most commonly tested in rodent and NHP stroke models. Strokes in the MCA distribution affect motor, sensory, and cognitive function and offer several targets for functional assessment. However, these strokes have inherent variability in anatomical localization based on variations in collateral blood flow between subject animals and increased variability in neurobehavioral outcomes based on the mixed nature of deficits in a given stroke distribution. This has led some investigators to pursue models that produce more restricted strokes than those covering the entire MCA distribution to control variability. For example, we have described an experimental approach to preclinical testing of neuroprotectants [8] that predicted efficacy in a corresponding clinical trial [9]. This was achieved by matching the strokes produced in humans undergoing endovascular aneurysm repair [9] to a preclinical model in primates in which similar embolic strokes were produced by the intracarotid injection of emboli, resulting in identical strokes in the primates that could then be subjected to the therapeutic intervention before, or in parallel with, the clinical study [8].

Various other models for producing restricted ischemic lesions are described in specific motor MCA territories such as those controlling the hand region to increase the sensitivity of motor assessments [4]. Focal strokes of end-artery distributions like the

lenticulostriate arteries have also been used [3]. Strokes in other anterior and posterior cerebrovascular distributions have not been widely pursued in preclinical studies.

The method of vessel occlusion must also be considered. An open approach to the MCA through a pterional craniotomy or transorbital route is a commonly employed technique, using either a surgical clip or suture ligation to achieve vessel occlusion [3,10]. Although the transorbital method provides relatively consistent strokes as evaluated by imaging and histological analysis, the technique requires enucleation of the eye and prevents behavioral testing that requires binocular vision. The craniotomy approach results in a potentially painful incision site with decreased function of the temporalis muscle as it is dissected; however, in skilled hands the surgery is generally well tolerated and the animal is prepared for a full battery of behavioral tests when it recovers. To balance issues associated with surgical approaches, endovascular techniques have been developed to create a temporary occlusion of the MCA by wedging a microcatheter tip into the MCA origin. This method also requires an occlusion of the posterior cerebral artery to decrease collateral flow. Other endovascular models employ autologous blood clots to achieve vessel occlusion. These models have the advantage of more closely modeling the pathophysiology underlying thromboembolic stroke in humans with the opportunity to achieve reperfusion with standard thrombolytic therapies; however, the variability in stroke distribution, volume, and neurological outcomes in endovascular models is high. Finally, photothrombotic methods have been used in limited numbers to achieve cerebral vessel occlusion in NHPs. This model can be undertaken with a minimal surgical opening and can be combined with thrombolytic therapies to achieve reperfusion. The strokes produced are generally smaller, are located in more distal vascular distributions and are variable in anatomic localization.

The stroke model in baboons requires enucleation of the eye followed by occlusion of the MCA and both ACAs to create a reproducible region of ischemia in the frontotemporal cortex, deep gray nuclei, and white matter of the MCA distribution. This ischemic area can be measured using various imaging modalities such as T2 and diffusion-weighted magnetic resonance imaging (MRI). Moreover, it can be defined using anatomical techniques that are commonly used in stroke research such as staining with the mitochondrial dye 2,3,5-triphenyl tetrazolium chloride. A permanent MCAO using this approach produces marked edema requiring prolonged intensive care of the animal and is associated with a mortality of about 30%. Therefore a transient ischemia model has been developed to facilitate recovery experiments; however, this produces stroke primarily in subcortical nuclei, sparing the cortex [3]. Thus the baboon provides a reasonable representation of human cortical anatomy with a gyrencephalic brain, similar subcortical anatomy, and equivalent white matter tracts. However, the baboon cerebral vasculature is critically different from that of humans due to a rich collateralization. Due to this, the obtaining of a cortical stroke necessitates the occlusion of the MCA and bilateral A1 segments of the ACA to achieve a stroke in the MCA distribution, and then this stroke is quite large and potentially fatal to the animal. This limits the ability to conduct long-term survival studies. Neurobehavioral assessments in this model are limited to observational scoring due to the aggressive nature of the baboon. Enucleation of the eye may obscure detailed cognitive, sensory, and motor neurobehavioral tasks requiring intact binocular vision. The inability to consistently produce cortical strokes with a temporary MCAO precludes, at least in our minds, using the baboon for stroke studies intended to model the human situation given that humans who suffer strokes, even with reperfusion, often sustain cortical damage.

Cynomolgus macaques afford the opportunity for producing a consistent MCAO, evaluation by MRI and/or histological techniques, and for extensive neurobehavioral assessments. Additionally, the rate of stroke progression (the extent of the ischemic penumbra) is controllable based on the location of the MCAO in relation to the orbitofrontal branch and origin of lenticulostriate arteries [10]. Thus it is possible to model both rapidly progressive strokes, as well as strokes that maintain a sizable ischemic penumbra for longer periods, such as those that are exhibited by patients who benefit from endovascular reperfusion or delayed thrombolytic therapy. Thus the flexibility of the cynomolgus macaque models, the ability to produce strokes that mimic various clinical situations, and the opportunity for neurobehavioral assessment makes this species our first choice for NHP stroke studies if resources and expertise allow.

ASSESSMENT OF FUNCTIONAL OUTCOME

Neurobehavioral tests to assess motor function in the upper extremities and to test for sensory neglect are highly pertinent following MCA stroke because stroke in this districation affects primary motor cortex, premotor regions, and sensory cortex related to arm and hand function. To undertake these types of tasks, animals must be trained to achieve minimum baseline scores before stroke onset. For example, the two tube choice test is a test of hand preference and hemispatial neglect [3,8,10]. The hill and valley staircase task requires the monkey to reach through vertical slots in an acrylic panel attached to the front of the cage to retrieve food rewards from a five-step staircase. This task separates hemiparesis from hemisensory neglect by testing motor function in each arm in the isolated hemisensory field. Tasks based on a six well Kluver board where animals must use fine pincer grasp to retrieve multiple items from wells within a plate have been used to test motor control and planning in NHPs [3]. Normal animals complete the task bimanually with pincer grip, by picking marshmallow pieces from two columns of three wells moving away from the cage on the left and right of center. Animals with hemisensory neglect will leave the marshmallows in the column on the affected side, whereas animals with hemiplegia will slowly gather rewards from both sides with the unaffected hand and mildly hemiparetic animals will lose pincer grasp in the affected hand but cup the hand and scoop marshmallow pieces out of the wells with the affected hand. With progressive motor impairment there is increased latency to complete the task.

Cognitive tasks play an important role in assessing models of global cerebral ischemia and in follow-up of animals in chronic focal cerebral ischemia studies as memory and learning are impaired in both scenarios and can potentially affect human measures of stroke outcome. The simplest test of working memory in primates is the Delayed Response test in which the monkey is presented with an uncovered pair or matrix of wells with a food reward in a single well and is then required to recall the location of a food reward in a covered matrix of wells after a delay in time [3]. Visual discrimination can be added to this task by presenting monkeys with a shape or color on a card and then after a delay presenting the monkey with a matrix of covered wells with patterns or colors on the covers where a food reward is baited in the well matching the stimulus. The Object Retrieval Detour Task is designed to evaluate the cognitive elements of motor planning in reaching movements and has been used in cynomolgus macaques with chronic stroke. Higher cognitive function is assessed with a battery of three tests including the Delayed Nonmatching to Sample Task, the Delayed Recognition Span Task, and the Conceptual Set Shifting Task that have been used in studies of dementia in aged rhesus macaques. These tests have also been used to study hypertensive leukoencephalopathy in cynomolgus macaques. They evaluate recognition memory, pattern recognition, visuospatial memory, and executive functions. Although there exists a range of individual tasks to enable the assessment of motor, sensory, and cognitive deficits following strokes in NHPs, a battery of several functional outcome measures may provide more information and match, as closely as possible, the outcome measures that will be included in clinical trials.

SURROGATE OUTCOME MEASURES

Measures of tissue death, metabolic dysfunction or failure, and improvements in essential cellular homeostatic function have been used as surrogate measures of tissue integrity to detect tissue salvage in animal models and human clinical trials. However, the validity of these surrogate measures in estimating functional outcome has been questioned. Surrogate measures have not been consistently linked to functional neurological outcome. Tissue salvage is commonly defined as a reduction in necrotic tissue on histological sections. Tissue integrity is also frequently measured using MRI diffusion-weighted and T2-weighted measures as these correlate well with histology [3,8,10]. Metabolic measures that correlate with tissue death have also been used in stroke studies. Of these the most common is the vital stain 2, 3, 5-triphenyltetrazolium chloride that delineates regions with functional cellular respiration from those without, with close correlation to histological measures of tissue necrosis. Positron emission tomography of metabolic activity where thresholds of activity correlate to irreversible tissue damage have been employed to measure tissue integrity in NHPs and predict resultant regions of tissue necrosis. Cellular homeostatic function is disturbed after stroke with both irreversible and reversible dysfunction; however, measures of some cellular functions that correlate strongly with tissue death have the advantage of defining tissue loss based on early loss of vital intracellular processes that may precede cell death. For instance, global protein expression and gene transcription have been used in limited studies to measure cellular function after stroke [10]. Expression of the cytoskeletal protein microtubule associated protein-2 has been used as a marker to define regions of brain that have lost essential cellular function after stroke in NHPs.

CONCLUSION

Research to date has not resolved which primate model of stroke most closely models the human condition, or whether studies of neuroprotectants in NHPs suffice to predict the results of a human trial. The chief hypothesis in primate stroke studies is that using high-order species may allow researchers to simulate the human situation as closely as possible in the preclinical setting. However, it is also plausible that the cellular and molecular biology of the human brain is not so substantially different from that of rodents, and the failures of translation seen to date are due to other causes. Chief among these may be flawed clinical trial design or execution, which may not have taken into account lessons already learned in lower-order species. Therefore until the translational road map is clearer, stroke research in NHPs may be best used to better understand any potential genetic, biological, or anatomical differences between primates and lower species, or as a final preclinical step before embarking on a clinical trial.

Acknowledgment

Supported by grants to MT from the Canadian Stroke Networks and the Heart and Stroke Foundation of Canada. MT is a Canada Research Chair (Tier 1) in Translational Stroke Research.

References

[1] Zivin JA, Fisher M, DeGirolami U, Hemenway CC, Stashak JA. Tissue plasminogen activator reduces neurological damage after cerebral embolism. Science 1985;230(4731):1289–92.

[2] Shuaib A, Lees KR, Lyden P, Grotta J, Davalos A, Davis SM, et al. NXY-059 for the treatment of acute ischemic stroke. N Eng J Med 2007;357(6):562–71.

[3] Cook DJ, Tymianski M. Nonhuman primate models of stroke for translational neuroprotection research. Neurotherapeutics 2012;9(2):371–9.

[4] Nudo RJ, Larson D, Plautz EJ, Friel KM, Barbay S, Frost SB. A squirrel monkey model of poststroke motor recovery. ILAR J 2003;44(2):161–74.

[5] Astrup J, Siesjo BK, Symon L. Thresholds in cerebral ischemia— the ischemic penumbra. Stroke 1981;12:723–5.

[6] de Crespigny AJ, D'Arceuil HE, Maynard KI, He J, McAuliffe D, Norbash A. Acute studies of a new primate model of reversible middle cerebral artery occlusion. J Stroke Cerebrovasc Dis 2005;14(2):80–7.

[7] Howells DW, Porritt MJ, Rewell SS, O'Collins V, Sena ES, van der Worp HB, et al. Different strokes for different folks: the rich diversity of animal models of focal cerebral ischemia. J Cereb Blood Flow Metab 2010;30(8):1412–31.

[8] Cook DJ, Teves L, Tymianski M. A translational paradigm for the preclinical evaluation of the stroke neuroprotectant Tat-NR2B9c in gyrencephalic nonhuman primates. Sci Transl Med 2012;4(154):154ra133.

[9] Hill MD, Martin RH, Mikulis D, Wong JH, Silver FL, Terbrugge KG, et al. Safety and efficacy of NA-1 in patients with iatrogenic stroke after endovascular aneurysm repair (ENACT): a phase 2, randomised, double-blind, placebo-controlled trial. Lancet Neurol 2012;11(11):942–50.

[10] Cook DJ, Teves L, Tymianski M. Treatment of stroke with a PSD-95 inhibitor in the gyrencephalic primate brain. Nature 2012;483(7388):213–7.

CHAPTER

68

Cerebral Blood Flow Methods

C. Waeber

University College Cork, Cork, Ireland

INTRODUCTION

Although the brain represents only 2% of the total body mass in humans, it accounts for a fifth of the body's basal O_2 consumption and a quarter of its glucose use. A continuous supply of blood and nutrients is essential for its functioning and cerebral blood flow (CBF) must constantly match the demands of brain activity. Because it lacks energy stores, the brain is particularly vulnerable not only to acute ischemic events, but also to chronic impairment in the coupling between CBF and neuronal activity. It is therefore essential to have accurate methods to determine absolute CBF, or at least to monitor changes in CBF following various interventions. Many methods have been developed to monitor CBF clinically as well as in experimental animal models. Because of space constraints, this review will focus on the latter. Methods aimed at monitoring CBF in patients have been reviewed elsewhere [1].

Owing to their small size, rodents (the most commonly used animals in cerebrovascular research) present unique technical challenges when compared with humans. For instance, a 1-mm spatial resolution that would be considered excellent in humans does not resolve many mouse brain structures, and only a limited volume of blood can be sampled from small animals to measure the arterial input function, which is essential for quantitative CBF analysis using tracers. In addition, many biomedical research laboratories do not have access to technologies used routinely in a clinical setting. The emphasis of this review will therefore be on CBF methods that are both easier and relatively less costly to implement.

METHODS REQUIRING BRAIN PENETRATION

These methods require the implantation of probes in one site (or at most a few discrete sites) and are therefore ill-suited for a regional assessment of CBF changes in laboratory animals.

Thermal Clearance or Thermal Diffusion Flowmetry

Since Gibbs' demonstration in 1933 that changes in CBF could be detected with the use of a heated thermocouple, many investigators have developed thermal systems to measure CBF [2]. Although they initially only provided an assessment of flow in arbitrary units and, hence, of relative changes, they can now achieve reliable CBF measurements in absolute terms. Thermal clearance flowmetry relies on the continuous measurement of the changes in brain thermal conductivity caused by CBF changes. Two thermistors are implanted, one of which is heated to approximately 2°C above the tissue temperature. The proximal thermosensor is located several millimeters away from the distal sensor, outside of its thermal field, thereby allowing continuous monitoring of tissue temperature and compensation of baseline fluctuations. The power dissipated by the heated element to keep itself at a constant elevated temperature provides a direct measure of the brain's ability to transport heat. This thermal transfer includes not only convective effects induced by blood flow, but also the intrinsic conductive properties of the tissue. These two components must therefore be separated to measure CBF. This was initially achieved by euthanizing the animal to determine the zero flow current. Modern systems enable quantification of tissue perfusion in absolute units by determining the conductive properties of the tissue from the initial rate of propagation of the thermal field and by subtracting this component from the total heat transfer as the determinant of the thermal convection component, making no-flow calibration unnecessary. The geometric arrangement of the two thermistors required

Primer on Cerebrovascular Diseases, Second Edition
http://dx.doi.org/10.1016/B978-0-12-803058-5.00068-0

for determination of thermal conductivity and convection makes this technique impractical to measure CBF in rodents. But progresses in microfabrication methods have led to the development of thermal flow microsensors small enough to be implanted in rat cortex that can yield continuous, real-time, and quantitative CBF measurement with high long-term accuracy and high temporal resolution [3].

Hydrogen Clearance

Kety and Schmidt were the first to attempt to measure global CBF in 1945 [2]. They used nitrous oxide to determine the global CBF based on Fick's principle, which relies on the premise that the total uptake of a tracer by an organ is equal to the product of the blood flow to the organ and the arterial-venous concentration difference of the tracer. The hydrogen clearance is a variation of Kety and Schmidt's technique that uses H_2 to assess the CBF. H_2 is not normally present in the body, is metabolically inert, dissolves readily into lipids, diffuses rapidly in tissues, such as brain, and is rapidly eliminated by the pulmonary circulation, thereby fulfilling the key criteria for tracer clearance studies developed by Kety and Schmidt. To measure H_2 clearance, one or several electrodes are inserted into the brain (polarized to +400 mV with respect to a subcutaneous reference Ag/AgCl electrode), H_2 is administered either by respiration or intraarterially, it is allowed to be cleared from arterial blood, and the exponential clearance rate of H_2 from the tissue is monitored [4]. This method can be used in both anesthetized and conscious animals to simultaneously measure local CBF at multiple sites, and measurements can be repeated many times over periods ranging from hours to days. It is important to note that absolute CBF values measured with this technique are often significantly lower than those found with other methods. This is likely to reflect actual CBF alterations that occur as a result of electrode insertion, rather than an artifact of the technique. Indeed, a transient flow reduction in the whole cerebral hemisphere has been linked to the induction of spreading depressions after electrode insertion [5]. These spreading depressions, and their effect on measured CBF, can be minimized by using thinner electrodes (50 μm) and/or by implanting the electrodes 1–2 weeks ahead of the experiments.

Helium Clearance

The principle of this method is identical to that of the H_2 clearance method. But critically, helium levels must be measured by aspirating tissue gases via a relatively large gas-sampling cannula (0.7-mm outer diameter) implanted into the brain, and analyzing the gas samples by mass spectrometry. This makes this method technically more challenging and invasive, but gives it an advantage over polarographic H_2 detection in that mass spectrometry enables the simultaneous monitoring of tissue Pco_2 and Po_2 at the site of CBF measurement [6].

METHODS REQUIRING THE INJECTION OF A RADIOACTIVE TRACER

CBF measurements with inert radioactive substances are another implementation of the Fick principle. They enable the simultaneous quantitative measurement of blood flow in all brain regions, with a high spatial resolution and without the need for expensive neuroimaging infrastructure. Autoradiographic analysis with three different tracers can be used to simultaneously measure CBF, glucose use, and protein synthesis. A main drawback is that autoradiographic tracer quantification provides only a "snapshot" of CBF at a single time point. It is possible to use extracranial detectors to monitor the brain levels of krypton-85 and xenon-133 injected into the carotid artery (because these tracers are rapidly eliminated by lungs, multiple measures can be obtained in the same animal) [2]. But this temporal resolution comes at the cost of spatial resolution in rodents, since their brain is small with respect to the size of the detector. Radiographic or tomographic imaging methods, such as positron emission tomography, computed tomography scanning, nuclear magnetic resonance, or single photon emission computed tomography for the regional detection of their respective tracers are more relevant clinically [1,2] and not discussed here.

Autoradiographic techniques provide a powerful way to quantify CBF with a high level of spatial resolution. The first tracer used for this purpose, [131I]trifluoroiodomethane, had numerous drawbacks, and was eventually replaced with [14C]iodoantipyrine, which is still commonly used today. There are two ways to use quantitative autoradiographic detection of a diffusible tracer, or indicator, to measure regional CBF: the indicator fractionation and tissue equilibration methods.

Tissue Equilibration

In this method, brain tissue is assumed to be in equilibrium with venous blood (i.e., the indicator is freely and rapidly exchanged between plasma, red cells, and a single, well-mixed brain tissue compartment). Practically, [14C]iodoantipyrine (the most commonly used tracer) is infused intravenously at a constant or an increasing rate, to obtain a monotonic rise in the plasma indicator concentration. Accurately timed arterial blood samples are collected (at approximately 5-s intervals) from the time the infusion begins until the experimental animal is decapitated (after 30–60 s). The brain is removed as quickly as possible to minimize tracer diffusion, and all samples are prepared for radioactivity measurement. Tracer concentrations in various brain regions are best quantified by exposing X-ray-sensitive films to 20-μm-thick cryostat brain sections along with calibrated radioactivity standards, but it is also possible to roughly dissect out various brain regions for liquid scintillation counting. CBF is calculated from the tracer concentration in the tissue at the end of the experimental period, the time course of the arterial tracer concentration, and the tissue/blood partition coefficient of the tracer (0.8 for [14C]iodoantipyrine in brain). Corrections for lag time and dead space washout in the arterial collection catheter must also be applied when measuring CBF in small animals such as mice [7].

Indicator Fractionation

This method assumes complete brain extraction of the tracer and negligible return to the blood compartment. It is simpler to implement than the tissue equilibration technique: arterial blood sampling is begun by withdrawing blood into an arterial catheter and syringe system at a constant rate; the indicator is then rapidly injected intravenously as a bolus. A short time after injection (usually 10 s), the experimental animal is decapitated, and arterial blood withdrawal is stopped. Radioactivity levels in blood and tissue samples are measured as described for the tissue equilibration method. A variation of the indicator fractionation technique uses labeled microspheres. Although these were originally radiolabeled, the fact that they can also be fluorescently labeled offers practical advantages (fluorescence is safer and more convenient to use, it is easier to detect several fluorescent tags than to monitor different isotopes, and working with fluorescent molecules does not require specific permits).

OPTICAL TECHNIQUES

Optical techniques based on dynamic light scattering, such as laser Doppler flowmetry (LDF) or laser speckle contrast imaging (LSCI), are increasingly used to monitor CBF. Although these two techniques vary in measurement geometry and analysis, both use intrinsic signals from brain tissue interactions and therefore do not require radioactive tracers or exogenous dyes.

Laser Doppler Flowmetry

LDF is commonly used to monitor relative CBF in vivo at one (possibly two or three) location(s) with high temporal resolution. Some of its features are therefore similar to those of thermal and H_2 clearance: excellent time resolution, but lack of spatial information. The latter can be alleviated by having the laser perform a two-dimensional scan of the brain surface, at the cost of decreasing the temporal resolution. An advantage of LDF is that it does not require brain tissue penetration, but an important drawback with respect to thermal and H_2 clearance is that it does not afford an absolute measurement of CBF.

Laser Speckle Contrast Imaging

An alternative to LDF is LSCI, which enables quantification of CBF over a wide area of cerebral cortex. The physics behind LDF and LSCI is similar: the former uses the Doppler effect to detect the frequency shift of laser light as it encounters a moving red blood cell, enabling local relative measurements of blood velocity, whereas in LSCI, a camera records the pattern of speckle that occurs when a coherent radiation interacts with a medium containing randomly distributed particles (e.g., brain tissue), producing random interference patterns of scattered light [8]. The changes in speckle pattern with time contain information about the speed of the moving particles encountered in the tissue. Because LSCI uses a camera rather than a scanning system to collect laser light after it is scattered off of the tissue surface, its spatial resolution does not come at the cost of scanning time. In LSCI, the camera exposure time is in the millisecond range to provide an integration period slightly longer than the motion dynamics of the specimen; this results in a blurring of the raw speckle image. In areas containing faster moving particles there is more blurring of the speckles during the exposure, lowering the spatial contrast of the speckles. The spatial blurring is quantified by computing at each point in the image the speckle contrast value K from the surrounding 5 × 5-or 7 × 7-pixel square area; K can theoretically vary between 0 (the scattering particles move so fast that the speckles average out) and 1 (no motion and therefore no blurring of the speckle pattern). Yet, although the basic principle of LSCI is simple, the relationship between K and the underlying CBF is a complex function that continues to be the subject of many theoretical and applied papers [8].

LSCI requires optical access to the brain surface (e.g., by implanting a cranial window, thinning the skull bone, or possibly by imaging through the skull in the case of mice), but no direct contact with the brain. It is therefore an ideal noninvasive tool to obtain detailed CBF measurements without disturbing brain physiology. However, a significant limitation to LSCI is its shallow penetration depth, the top 700 μm of tissue accounting for more than 95% of the detected signal. Although different wavelengths lead to slightly different sampling depths, these differences are relatively minor. Most lasers used for CBF monitoring operate in red to near-infrared wavelengths (600–800 nm), where the light has low absorption with high scattering coefficients, allowing for deeper light penetration.

Other limitations of LSCI are its narrow dynamic range and its inability to measure absolute CBF. These drawbacks are negligible when studying small changes in CBF dynamics in an acute setting. However, they severely limit the ability to compare CBF changes in a chronic setting, as well as large changes (>50%) in an acute setting, as each single exposure used for LSCI has a limited CBF range in which the measurements are accurate (for example, a longer exposure is necessary to capture the slower speckle fluctuation in regions with lower flow). This has led to the development of multiexposure speckle imaging (MESI) [8] that can accurately monitor CBF changes over periods of several weeks, even in the presence of severely altered CBF, as seen following experimental stroke. Light scattered from static elements, such as a thinned or an intact skull, also limits the accuracy of the blood flow changes measured with LSCI; another advantage of MESI is that it enables accurate determination of flow changes even in the presence of such stationary elements.

Because MESI can quantify in individual animals large regional perfusion deficits and subsequent CBF recovery over several weeks or even months, it is a uniquely useful tool to study functional plasticity and vascular remodeling associated with tissue repair following brain injury. A novel way of analyzing speckle variation in frequency domain, termed frequency-domain laser speckle imaging, has been used to quantify absolute CBF, with minimal impact from changes in illumination conditions [9]; this approach might further expand the applications of LSCI to long-term or longitudinal studies and enable the quantitative comparison of CBF from different experiments or subjects. Real-time CBF imaging has even been achieved in conscious and freely moving rats [10], further increasing the practical applications of the technique and demonstrating its ability to accommodate study designs aimed at correlating brain physiology with behavioral assessment.

CONCLUSION

As mentioned in the introduction, the main emphasis of this short review was on techniques that are relatively easy to implement and do not require costly imaging infrastructure. Although it tried to highlight the pros and cons of each approach, it should be clear that none of these techniques is ideal, since there is currently no cost-effective, noninvasive method to quantify absolute CBF in the whole brain, with a wide dynamic range and a high degree of temporal and spatial resolution.

References

[1] Le Roux PD, Lam AM. Monitoring cerebral blood flow and metabolism. In: Ruskin KJ, Rosenbaum SH, Rampil IJ, editors. Fundamentals of Neuroanesthesia. Oxford: Oxford University Press; 2014. p. 25–49.

[2] Zauner A, Muizelaar JP. Measuring cerebral blood flow and metabolism. In: Reilly P, Bullock R, editors. Head Injury. Physiology and Management of Severe Closed Injury. London: Chapman & Hall; 1997. p. 217–27.

[3] Li C, Wu PM, Wu Z, et al. Highly accurate thermal flow microsensor for continuous and quantitative measurement of cerebral blood flow. Biomed Microdevices October 2015;17(5):87.

[4] Young W. H2 clearance measurement of blood flow: a review of technique and polarographic principles. Stroke Sep–Oct 1980;11(5):552–64.

[5] Verhaegen MJ, Todd MM, Warner DS, James B, Weeks JB. The role of electrode size on the incidence of spreading depression and on cortical cerebral blood flow as measured by H2 clearance. J Cereb Blood Flow Metab March 1992;12(2):230–7.

[6] Seylaz J, Pinard E, Meric P, Correze JL. Local cerebral PO2, PCO2, and blood flow measurements by mass spectrometry. Am J Physiol. September 1983;245(3):H513–8.

[7] Jay TM, Lucignani G, Crane AM, Jehle J, Sokoloff L. Measurement of local cerebral blood flow with [14C] iodoantipyrine in the mouse. J Cereb Blood Flow Metab February 1988;8(1):121–9.

[8] Kazmi SM, Richards LM, Schrandt CJ, Davis MA, Dunn AK. Expanding applications, accuracy, and interpretation of laser speckle contrast imaging of cerebral blood flow. J Cereb Blood Flow Metab July 2015;35(7):1076–84.

[9] Li H, Liu Q, Lu H, Li Y, Zhang HF, Tong S. Directly measuring absolute flow speed by frequency-domain laser speckle imaging. Optics Express August 25, 2014;22(17):21079–87.

[10] Miao P, Lu H, Liu Q, Li Y, Tong S. Laser speckle contrast imaging of cerebral blood flow in freely moving animals. J Biomed Optics September 2011;16(9):090502.

CHAPTER

69

Magnetic Resonance Imaging of Stroke

M.J.R.J. Bouts[1,2,3,4], O. Wu[2], R.M. Dijkhuizen[1]

[1]University Medical Center Utrecht, Utrecht, The Netherlands; [2]Massachusetts General Hospital, Charlestown, MA, United States; [3]Leiden University, Leiden, The Netherlands; [4]Leiden University Medical Center, Leiden, The Netherlands

INTRODUCTION

Neuroimaging methods have become indispensable tools for diagnosis of tissue status and aid in treatment decision making after stroke. Especially magnetic resonance imaging (MRI), which allows noninvasive and longitudinal measurement of multiple (early) biomarkers of brain tissue injury, can provide important insights in stroke lesion development. In addition to its diagnostic potential in the clinic, MRI offers a valuable method for in vivo studies on stroke pathophysiology, treatment, and recovery in experimental animal models.

For the theory and principles behind MRI, we refer to a large variety of available textbooks and websites that describe (the basics of) MR physics, contrast mechanisms, and pulse sequences. In brief, MRI takes advantage of the natural abundance of protons and their intrinsic magnetic properties for the generation of signals that reflect the properties and state of underlying tissue(s); signals that can be employed for the construction of images that contain anatomical or functional information. To that aim, MRI protocols can be sensitized to proton density, MR relaxation times T_1, T_2, and T_2^*, diffusion, perfusion, and flow of water molecules in tissues. In the following paragraphs, we will describe in brief how different MRI protocols can be applied to assess brain injury after stroke. Due to space limitations we can only refer to a few review papers, but these include references to relevant original publications.

T_1-, T_2-, AND T_2^*-WEIGHTED MRI

Brain tissue damage after stroke may be identified with standard proton density MRI and T_1-, T_2-, or T_2^*-weighted MRI protocols [1,2]. Higher proton density, and T_1 and T_2 values after stroke are typically caused by increased interstitial water or vasogenic edema, which usually occurs at subacute to chronic stages of infarct progression. A clinically frequently applied sensitive method for detection of cerebral ischemic lesions is fluid-attenuated inversion recovery (FLAIR) MRI, which depicts tissue areas with prolonged T_2 while suppressing MRI signal from cerebrospinal fluid [1].

Hemorrhages, in which deposits of hemosiderin cause local magnetic field distortions leading to T_2^* shortening, are detectable (as hypointensities) with gradient-echo T_2^*-weighted MRI or susceptibility-weighted MRI [1,2]. Bleedings are usually preceded by blood–brain barrier (BBB) disruption, which may be detected with T_1-weighted MRI after intravenous injection of paramagnetic gadolinium (Gd)-containing contrast agent [2,3]. Tissue accumulation of paramagnetic contrast agent after leakage across the BBB leads to local T_1 shortening, giving rise to an increase in T_1-weighted signal intensity. BBB permeability can be quantified with dynamic contrast-enhanced (DCE) MRI and tracer kinetic analysis, which enables calculation of a blood-to-brain transfer constant (K_{trans} or K_i) [3]. Accumulation of intravenously injected contrast agent in brain parenchyma early after stroke has been found to be predictive of subsequent hemorrhagic transformation, conceivably as a result of (thrombolysis-induced) reperfusion injury [2].

Dynamic acquisition of $T_2^{(*)}$-weighted MRI sensitized to paramagnetic deoxygenated hemoglobin, i.e., blood oxygenation-level dependent (BOLD) MRI, allows measurement of hemodynamic activity, for example, linked to cerebrovascular reactivity or neuronal activity—the latter forms the basis of functional MRI (fMRI) [1,2]. Cerebral autoregulation can be assessed with BOLD MRI in combination with a vasodilatory challenge, induced by injection of the vasodilator acetazolamide or by CO_2

inhalation. This normally leads to increased BOLD MR signal intensity, which may be reduced or absent in tissue affected by stroke [2]. With BOLD fMRI of the hemodynamic response to task-related neuronal activation (resulting in increase in local blood oxygenation) or "resting-state" neuronal activity one can obtain information on the poststroke brain's functional state and connectivity [4] (although the BOLD response may be significantly affected by altered vascular reactivity or neurovascular coupling after stroke). The application of fMRI to assess brain (re)organization, by measuring (changes in) spatiotemporal patterns of neuronal signals/responses, in relation to stroke recovery is a growing area of research.

DIFFUSION-WEIGHTED MRI

Diffusion-weighted MRI exploits the inherent sensitivity of MRI to motion. Pulsed magnetic field gradients sensitize the MRI signal to the self-diffusion of water molecules, and resultant signal changes allow calculation of the apparent diffusion coefficient (ADC) or mean diffusivity of tissue water [5]. By measuring the diffusion of water in at least six different directions, additional information can be obtained on diffusion directionality, for example, expressed by the fractional anisotropy, based on the diffusion tensor [5]. Diffusion tensor imaging (DTI) is not typically employed in acute clinical stroke diagnosis because of relatively long acquisition times. Nevertheless, because of its sensitivity to diffusion anisotropy in oriented neuronal fiber tracts, DTI is suitable for evaluating effects on white matter structure, such as axon/myelin degeneration or remodeling, which are usually observed at chronic stages after stroke [5].

Since its first application in the early 1990s, diffusion-weighted imaging (DWI) has developed into a distinctively sensitive method for the detection of acute ischemic lesions after stroke [1,2,5]. It is assumed that early diffusion changes associated with cytotoxic edema (cellular swelling resulting from net displacement of tissue water from extracellular to intracellular compartments due to ionic pump failure) give rise to a reduction of the ADC, expressed by hyperintensity on diffusion-weighted images [1,2,5]. This is considered as one of the earliest hallmarks of acute ischemic tissue injury, and therefore an important feature for the identification of cerebral ischemic lesions. At later stages the ADC gradually increases from pseudonormal to elevated levels as a result of several pathophysiological processes, such as resolution of cytotoxic edema, loss of cell membrane integrity, vasogenic edema formation, and tissue cavitation [1,2].

Preclinical and clinical stroke studies have demonstrated a close relationship between acutely reduced tissue water diffusion and infarcted tissue at follow-up [1,2]. However, diffusion changes may normalize after reperfusion, which can be permanent (i.e., tissue salvage) or followed by secondary ADC reduction (i.e., delayed injury) [1,2]. This constrains the specificity of hyperintensity on diffusion-weighted images as a marker for tissue infarction, despite its high sensitivity.

MR ANGIOGRAPHY AND PERFUSION MRI

Large brain arteries can be visualized with magnetic resonance angiography (MRA) [1,2]. A frequently applied method, called time-of-flight MRA, measures inflow of fresh blood with a fully relaxed and magnetized, i.e., unsaturated, MR signal into an imaging plane in which the MR signal has been saturated from recurrent radiofrequency pulse excitations. Stenosis or occlusions, partial or complete, in carotid arteries can then be identified as signal voids. MRA may play a critical role in the identification and monitoring of patients who could benefit from endovascular thrombectomy, which has been recognized in several randomized clinical trials (MR CLEAN (multicenter randomized clinical trial of endovascular treatment for acute ischemic stroke in the Netherlands), ESCAPE (endovascular treatment for small core and anterior circulation proximal occlusion with emphasis on minimizing CT to recanalization times), EXTEND-IA (extending the time for thrombolysis in emergency neurological deficits - intra-arterial), SWIFT-PRIME (solitaire with the intention for thrombectomy as primary endovascular treatment), and REVASCAT (randomized trial of revascularization with solitaire FR device versus best medical therapy in the treatment of acute stroke due to anterior circulation large vessel occlusion presenting within eight hours of symptom onset)) as an effective intervention to improve stroke outcome through recanalization of thrombus-occluded intracranial arteries.

Where MRA depicts flow in the macroscopic vessels, perfusion MRI is sensitized to microvasular hemodynamics, thereby enabling measurement of parenchymal (micro)perfusion. Perfusion MRI measures the passage of intravascular endogenous or exogenous contrast agent through the imaging plane(s) [1–3]. Endogenous labeling of arterial blood [i.e., arterial spin labeling (ASL)] can be accomplished with directed radiofrequency pulses that magnetically label arterial blood before it enters the imaging plane [1–3]. Subsequently, cerebral blood flow (CBF) can be calculated from the arterial label-induced signal changes on T_1-weighted MRI, arising from exchange of magnetically labeled arterial water with brain tissue water. However, especially for perfusion imaging acutely after stroke, ASL is hampered by lack of sensitivity in low-flow areas, sensitivity to patient motion, relatively low signal-to-noise ratios, and long acquisition times. Therefore, another MRI approach involving dynamic tracking of the first passage of an intravenously injected bolus of paramagnetic Gd chelate, i.e., dynamic susceptibility contrast-enhanced (DSC) MRI [1–3], is usually preferred for perfusion imaging after acute clinical stroke. Passage of the paramagnetic contrast agent induces magnetic susceptibility effects that can be measured with T_2- (spin echo-based) or T_2^*-weighted (gradient echo-based) MRI. Similar to DCE MRI for quantification of BBB permeability (see earlier discussion), DSC MRI employs tracer kinetic theory to describe the hemodynamic condition of the underlying tissue. Contrast agent concentration-time curves can be generated based on the assumed linear relationship between contrast agent concentration and induced changes in transverse relaxation rates (ΔR_2 or ΔR_2^*) [1–3]. These concentration-time curves can be defined as a convolution of the arterial input function with a residue function (i.e., the tracer fraction in perfused tissue) scaled by the rate at which the tracer travels, i.e., the CBF. Different modeling approaches have been developed and applied to estimate the residue function enabling the calculation of hemodynamic parameters such as CBF, cerebral blood volume (CBV), and mean transit time (MTT) [3]. Despite the potential to characterize ischemic tissue status based on hemodynamic parameters (e.g., by CBF thresholding), consensus on thresholds to delineate hypoperfused areas that predict infarct progression or response to thrombolytic therapy has so far not been achieved [6].

Fig. 69.1 shows different MR images obtained from a rat after right-sided middle cerebral artery (MCA) occlusion. The MR angiogram and quantitative parametric maps of the T_2, ADC, K_i, CBF, CBV, and MTT, obtained within 2 h after stroke onset, reveal reduced flow through the right internal carotid artery with distinct occlusion of the MCA, resulting in an acute ischemic tissue lesion (low ADC) with significant hypoperfusion (low CBF, long MTT) in the MCA territory. After thrombolytic treatment with tissue plasminogen activator at 2 h poststroke, follow-up MRI at day 7 revealed tissue infarction (long T_2), BBB leakage (high K_i), and hemorrhagic transformation (T_2^*-weighted hypointensity).

PERFUSION-DIFFUSION MISMATCH

Estimating tissue salvageability is a key aspect in the prediction of tissue outcome after stroke. The amount of salvageable tissue determines treatment potential. By combining the sensitivity of DWI to detect acute ischemic tissue injury with the ability of perfusion MRI to delineate hypoperfused tissue, the acute ischemic lesion size can be directly compared with the total area of hypoperfusion. This has led to the concept of the perfusion-diffusion mismatch, i.e., hypoperfused tissue without diffusion abnormalities, as a marker of potentially salvageable tissue at risk (or in other words, the ischemic penumbra) [6,7]. Some phase II clinical trials have shown that patients with a perfusion-diffusion mismatch profit more from thrombolytic therapy at 3–6 h after stroke onset than those without,

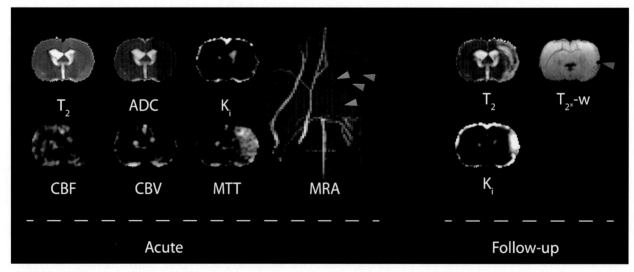

FIGURE 69.1   MRI-based parametric images of a coronal rat brain slice after right-sided middle cerebral artery occlusion. The acute (2 h post-stroke) unilateral ischemic lesion is most clearly depicted by reduced apparent diffusion coefficient (ADC), reduced cerebral blood flow (CBF), and prolonged mean transit time (MTT). The MR angiogram shows loss of flow in the ipsilateral internal (*green arrows*) and middle cerebral arteries (*blue arrows*). The animal was subsequently treated with recombinant tissue plasminogen activator (rt-PA) at 2 h poststroke, and follow-up imaging at 7 days poststroke revealed further tissue damage as indicated by T_2 prolongation, K_i (blood-to-brain transfer coefficient) increase, and T_2^*-weighted hypointensity, reflective of vasogenic edema, blood–brain barrier disruption, and hemorrhagic transformation, respectively.

whereas other trials that explicitly used the perfusion-diffusion mismatch for patient recruitment failed to observe a clear beneficial effect [6]. This discrepancy can be due to differences in thresholding standards or mismatch definitions. A particular drawback of the perfusion-diffusion mismatch concept is the possibility of recoverable diffusion abnormality (i.e., reversible tissue injury) and overestimation of perfusion deficiency (i.e., benign oligemia) [6]. Furthermore, the perfusion-diffusion mismatch concept does not consider other critical aspects that determine outcome and treatment response, such as occlusion site, lesion location, and collateral flow. Alternative mismatch concepts, such as the diffusion-FLAIR mismatch[8]—detection of an acute ischemic lesion with DWI but not with FLAIR—and the angiography-diffusion mismatch [9,10]—detection of intracranial vessel occlusion or stenosis with MRA, and a relatively small lesion detected with DWI—have also been introduced to identify potentially salvageable tissue at risk of infarction.

TISSUE CLASSIFICATION AND PREDICTION

Although estimation of tissue at risk of infarction by means of perfusion-diffusion mismatch analysis may aid in clinical decision making, quantitative appraisal may further extend utility in both clinical and preclinical stroke assessments. The definition of proper thresholds for the determination of tissue at risk of infarction has proved to be challenging due to a complex interplay of different hemodynamic and pathophysiological processes [a value of more than 5 or 6 s for the time to peak for tracer passage (T_{max}) is currently typically used as threshold for perfusion abnormality [6]]. It has been speculated that this complex relationship could be better apprehended by discerning their spatiotemporal correlations, which may be captured with multiple MRI indices acquired within a single scanning session. Such multiparametric voxel-wise methods allow classification of different tissue types by integrating various aspects of tissue viability, including time after stroke, hemodynamics, tissue condition, and possible intervention, into a single quantitative index [7]. These methods combine values of multiple MRI indices in a single vector, and relate these on a voxel-wise basis to a specific tissue condition (e.g., ischemic core) based on supervised or unsupervised learning algorithms. The resulting tissue theme maps provide a simplified representation of the underlying tissue status.

Unsupervised clustering approaches, such as iterative self-organizing data analysis (ISODATA), offer an unbiased means for tissue classification and may more reliably identify ischemic tissue based on T_2, diffusion, and perfusion data together, as compared with thresholding of these MRI parameters separately [7]. Alternatively, supervised learning algorithms can be employed to predict tissue outcome on a voxel-wise basis, based on prior knowledge of the underlying pathophysiology. Predictive algorithms can estimate the likelihood or probability of voxels belonging to a specific tissue class, thereby offering a single quantitative index that can be easily interpreted irrespective of the involved MRI parameters [7]. Different MRI-based predictive algorithms have been proposed for calculation of infarction risk. For example, with a generalized linear model (GLM) that combines T_2, diffusion, and perfusion MRI indices, heterogeneity in infarct probabilities can be detected within affected brain tissue (Fig. 69.2), which

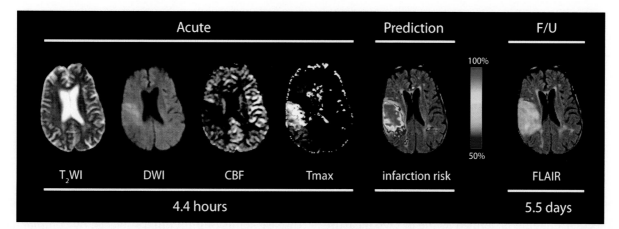

FIGURE 69.2 T_2-weighted (T_2WI), diffusion-weighted (DWI) and perfusion-weighted (CBF, T_{max}) brain images of a 57-year-old male acquired approximately 4.4 h after stroke onset. The area(s) with perfusion deficits (reduced CBF, prolonged T_{max}) and diffusion abnormality (DWI hyperintensity) mark acutely affected tissue that may progress to infarction. These acute MRI data were combined in a generalized linear model (GLM), which was developed from prior MRI data sets from patients with stroke with known outcome, to compute an infarction risk map. The tissue area with GLM-based prediction of infarction probability above 50% corresponded well with the area that eventually infarcted, as depicted as hyperintensity on the follow-up (F/U) fluid-attenuated inversion recovery (FLAIR) image.

may aid in improved identification of regions that are likely to favorably respond to thrombolytic therapy [7].

To conclude, MRI offers a versatile tool that holds unique potential for the characterization and monitoring of affected brain tissue after stroke. Combined assessment of vessel occlusion, tissue structure, perfusion, and functional responses provides a plethora of opportunities for stroke imaging in clinical as well as experimental settings. The use of concepts such as the perfusion-diffusion mismatch has opened up opportunities for individualized treatment planning in patients with acute stroke. Particularly the development of predictive voxelwise approaches, which allow for the integration of different aspects of stroke pathophysiology within a single index, may alleviate the necessity for clear thresholding standards, and in the end offer better support in individual clinical decision making or treatment efficacy monitoring.

References

[1] Baird AE, Warach S. Magnetic resonance imaging of acute stroke. J Cereb Blood Flow Metab 1998;18:583–609.

[2] Dijkhuizen RM, Nicolay K. Magnetic resonance imaging in experimental models of brain disorders. J Cereb Blood Flow Metab 2003;23:1383–402.

[3] Zaharchuk G. Theoretical basis of hemodynamic MR imaging techniques to measure cerebral blood volume, cerebral blood flow, and permeability. AJNR Am J Neuroradiol 2007;28:1850–8.

[4] Grefkes C, Fink GR. Connectivity-based approaches in stroke and recovery of function. Lancet Neurol 2014;13:206–16.

[5] Le Bihan D, Johansen-Berg H. Diffusion MRI at 25: exploring brain tissue structure and function. NeuroImage 2012;61:324–41.

[6] Fisher M, Albers GW. Advanced imaging to extend the therapeutic time window of acute ischemic stroke. Ann Neurol 2013;73:4–9.

[7] Wu O, Dijkhuizen RM, Sorensen AG. Multiparametric magnetic resonance imaging of brain disorders. Top Magn Reson Imaging 2010;21:129–38.

[8] Thomalla G, Cheng B, Ebinger M, Hao Q, Tourdias T, Wu O, et al. DWI-FLAIR mismatch for the identification of patients with acute ischaemic stroke within 4·5 h of symptom onset (PRE-FLAIR): a multicentre observational study. Lancet Neurol 2011;10:978–86.

[9] Lansberg MG, Thijs VN, Bammer R, Olivot J-M, Marks MP, Wechsler LR, et al. The MRA-DWI mismatch identifies patients with stroke who are likely to benefit from reperfusion. Stroke 2008;39:2491–6.

[10] Bouts MJ, Tiebosch IA, van der Toorn A, Hendrikse J, Dijkhuizen RM. Lesion development and reperfusion benefit in relation to vascular occlusion patterns after embolic stroke in rats. J Cereb Blood Flow Metab 2014;34:332–8.

CHAPTER

70

Principles and Methods of Molecular Imaging in Stroke

D.E. Kim

Dongguk University, Goyang, Republic of Korea

Molecular imaging is defined as the in vivo measurement of biological processes at the cellular and molecular levels [1]. The technique visualizes pathophysiologic processes noninvasively in real time, with the potential for serial monitoring, and provides information regarding specific molecular alterations underlying the disease status of individual subjects. By complementing conventional "anatomical or physiological" imaging, molecular imaging enables early detection of disease, staging of disease, and quantitative assessment of therapeutic response. Molecular imaging also contributes to the understanding of stroke pathophysiology in living animals and humans. This chapter describes the principles and methods of molecular imaging techniques, while providing a translational perspective in stroke.

IMAGING AGENTS AND MODALITIES

Various modalities of molecular imaging, such as optical imaging, computed tomography (CT), magnetic resonance imaging (MRI), and radionuclide imaging (single photon emission tomography, positron emission tomography, i.e., PET), have their own specific advantages and disadvantages [1–3]. Optical imaging has a high sensitivity and high spatial resolution, but it has poor depth penetration. CT has the advantage of short scan times, but disadvantages include the exposure to radiation. MRI has an excellent soft-tissue contrast and a high spatial resolution, and also allows for physiological and biochemical imaging. Unlike fluorescence imaging with femtomolar or picomolar detection limits, MRI and CT suffer from relatively poor sensitivity. Radionuclide imaging is inherently molecular, and has the potential for true signal quantification as well as a high sensitivity. Micro-PET, micro-MRI, and micro-CT are increasingly popular tools dedicated to molecular imaging of small animals; they provide higher spatial resolution, compared with clinical scanners [2].

For CT-based molecular imaging, gold nanoparticles (AuNPs) are frequently used. For MRI, superparamagnetic iron oxide (SPIO) and ultrasmall SPIO (USPIO) nanoparticles are widely used. Optical imaging largely depends on fluorescent proteins, fluorochromes, or quantum dots. Multimodal imaging probes report the signal through more than one imaging modality, thereby complementing each other in terms of spatiotemporal resolution, sensitivity, or depth penetration. Many studies have shown that MRI combined with ex vivo reflectance fluorescence imaging or in vivo fluorescence tomography is very useful for small-animal research. Dual modality magnetic resonance (MR)–PET scanners have begun to show promise as a powerful platform for molecular imaging in animals and humans [2,3].

Targeted imaging probes are synthesized through attachment to the nanoparticle of a ligand directed against a known target on the cell surface or modification of the nanoparticle surface with small molecules to modulate its uptake. High-throughput screening of phage and chemical libraries are useful for the identification of peptides specific for targets, leading to the discovery of novel probes [1–3]. Activatable imaging probes are silent at baseline and turned on after responding to specific biomolecular or environmental changes in real time, thereby amplifying output signal while reducing background noise [1–4].

IMAGING PLAQUE VULNERABILITY

Some atherosclerotic plaques are prone to rupture and thromboembolism, whereas others are clinically silent. Thus there is an unmet clinical need for tools to localize rupture-prone vulnerable plaques, so that risk-altering treatments can be offered to prevent stroke. These high-risk plaques are not well identified by current imaging techniques that measure stenosis and depict anatomic features, such as: large lipid-rich necrotic core, thin fibrous cap, intraplaque hemorrhage or thrombi, neovascularization, and microcalcification. As a molecular hallmark of vulnerable atherosclerotic plaques, researchers have focused on inflammatory activity by intraplaque monocytes/macrophages that secrete cathepsins and matrix metalloproteinases (MMPs), disorganizing the extracellular matrix and thus rendering the plaque susceptible to rupture [3,4].

Magnetic Resonance Nanoparticle Imaging of Inflammatory Cells

USPIO nanoparticles, which consist of ferromagnetic iron oxide particles with an overall size of 30 nm, are taken up by monocytes/macrophages that infiltrate atherosclerotic plaques [3]. Kooi et al. showed that these nanoparticles could produce a T_2^*-weighted MR susceptibility effect, visualizing carotid inflammation in humans after intravenous infusion (2.6 mg/kg suspended in normal saline) over 30 min [5].

Positron Emission Tomography Imaging of Inflammatory Cells

[18]F-fluorodeoxy glucose (FDG) is well known to accumulate in macrophage-rich areas of plaques. Carotid arteries harboring inflammatory atheromas exhibit a higher FDG PET signal than unaffected vessels, suggesting that the nuclear molecular imaging tool has the potential to detect vulnerable plaques [3].

Optical Imaging of Matrix-Disorganizing Protease Activity

Unlike "always-on" probes, such as targeted fluorescent probes relying on affinity ligands, protease-sensing fluorescent imaging agents developed by the Weissleder group [1] are "optically silent at injection" because of intramolecular autoquenching between closely spaced fluorochromes. After protease-mediated cleavage of the substrate sequence (e.g., GGPRQITAG for MMP-2/9), fluorochromes are spatially separated, dequenched, and become brightly fluorescent. Animal studies using cathepsin-B or MMP-2/9 activatable probes have shown that in vivo and ex vivo fluorescent

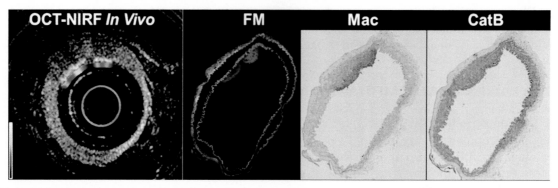

FIGURE 70.1 Dual-modality optical coherence tomography (OCT)/near-infrared fluorescence (NIRF) catheter imaging of microstructural morphology/inflammatory cathepsin-B (CatB) activity (color-coded) in rabbit iliac atheroma in vivo. In vivo imaging is closely matched by ex vivo fluorescence microscopy (FM) and immunohistochemistry for macrophage (Mac) and CatB. *Courtesy of Pf. Jin Won Kim, Korea University Guro Hospital.*

imaging could visualize the protease activity by monocyte/macrophages in the aortic or carotid atheromata of ApoE knockout mice or in human carotid endarterectomy specimens [3,4].

The clinical use of optical imaging is limited due to poor tissue penetration capability [1,3,4]. The attenuation of light by tissue is lowest in the near-infrared (NIR, 700–900 nm) region, with less photon absorption by blood hemoglobin, lipid, and water. Thus NIR fluorescent (NIRF) molecular imaging allows light to penetrate several centimeters into the body and offers substantially reduced tissue autofluorescence. Furthermore, this technology can be combined with a noninvasive optical tomography system or a fluorescence-sensing catheter-based system (Fig. 70.1), as shown by Jaffer's group [3,4].

IMAGING THROMBUS

Imaging the location, size, and physicochemical characteristics of thromboemboli will improve the safety and efficacy of chemical thrombolysis and mechanical thrombectomy in patients with hyperacute ischemic stroke [4,6,7]. Direct thrombus imaging may enable clinicians to estimate the dose of tissue plasminogen activator (tPA) or preferred methods for thrombus retrieval, and to predict the therapeutic response for the purpose of triaging patients. Furthermore, the molecular imaging technique is likely to help prevent recurrent stroke by aiding the determination of the stroke etiology and visualizing residual thrombi.

Magnetic Resonance-Based Direct Thrombus Imaging

Clinical trials showed that platelet-targeted or fibrin-targeted gadolinium (Gd)-based probes could enhance thrombi in the heart chambers, carotid arteries, and aortic arch [3]. Uppal et al. used a fibrin-specific molecular MRI agent, comprising afibrin-targeting peptide with two Gd-DOTA (1,4,7,10-tetraazacyclododecane-1,4,7,10-tetraacetic acid) chelates on each of the C and N-termini of the peptide (four Gd in total), to visualize occlusive thrombi in a rat model of embolic stroke [6]. They demonstrated positive contrast detection of the culprit intracranial thrombi, and reported that the concentration of Gd in the thrombus was 18 times that in the blood pool.

Computed Tomography-Based Direct Thrombus Imaging

CT remains the most widely used imaging technique in the diagnosis of hyperacute stroke. After introducing the first CT-based direct thrombus imaging technique using nontargeted glycol-chitosan AuNP (GC-AuNP), Kim's group developed fibrin-targeted GC-AuNP (fib-GC-AuNP) for high-resolution in vivo micro-CT imaging of cerebral thromboembolism as well as in situ carotid thrombosis in mice in a sensitive and quantitative manner (Fig. 70.2) [7]. Direct cerebral thrombus imaging enabled the prompt detection and quantification of millimeter-sized cerebral thrombi. The technique also enabled monitoring of tPA-mediated thrombolytic effect, which reflected histological stroke outcome. Moreover, recurrent thrombosis could be diagnosed up to 3 weeks without additional AuNP administration.

Ex Vivo Optical Direct Thrombus Imaging

Kim's group labeled thrombi with a Cy5.5 NIRF probe C15 [7] that was covalently linked to fibrin by factor-XIIIa coagulation enzyme. The fluorescently labeled thrombus was then injected into the left distal internal carotid artery of mice, near the middle cerebral artery (MCA)–anterior cerebral artery (ACA) bifurcation area. Ex vivo NIRF thrombus imaging at 24 h showed that fluorescent thrombus signal was

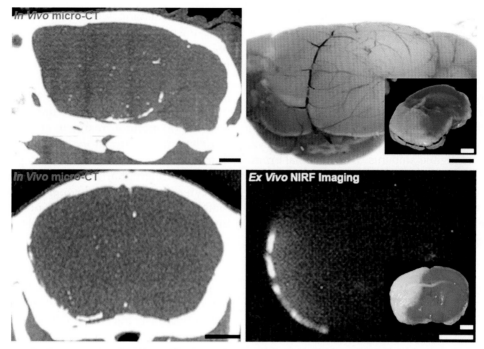

FIGURE 70.2 Dual-modality micro-computed tomography (CT)/near-infrared fluorescence (NIRF) imaging of cerebral thromboemboli (hyperdense/high-signal lesions) in vivo/ex vivo in a mouse model of embolic infarction, delineated by 2,3,5-triphenyl-tetrazolium chloride staining (insets). Intravenous injection of fibrin-targeted gold nanoparticles allows for direct clot visualization with in vivo micro-CT. Note that the in vivo micro-CT images correlate closely with the postmortem NIRF images of the removed brain, showing the thrombi prelabeled with a Cy5.5 probe before being placed in the cerebral artery.

located mostly in the left MCA–ACA bifurcation area. However, scattered emboli were also frequently observed in the adjacent or remote cerebral arteries (Fig. 70.2), reflecting the heterogeneous nature of the embolic stroke model with infarcts of varying size and location.

IMAGING APOPTONECROSIS

Recanalization therapy is often associated with neurological deterioration due to treatment-related hemorrhagic complications. Quantitative imaging assessment of the "salvageable" ischemic penumbra, which is the brain region at "risk" of infarction due to reduced blood flow, will contribute to avoiding futile or hazardous recanalization. Current MRI methods, relying on the well-known concept of diffusion–perfusion mismatch, hold promise as a faithful penumbral imaging method to select patients to be treated with recanalization therapy. However, clinical trials have failed to confirm the benefit. There is a need for more sophisticated molecular imaging markers of "reversible versus irreversible" cell injury along with physiologic neuroimaging modalities that can assess collateral circulation and blood–brain barrier (BBB) permeability to estimate the response and risk associated with recanalization therapy.

Necrosis Imaging

Based on a membrane impermeable vital dye (TO-PRO-1), Sosnovik's group developed a DNA-binding Gd-chelate (Gd-TO) to image necrotic cell death in vivo using MRI [8]. In a mouse model of myocardial infarction (MI), they observed hyperintense T1-weighted MR signal in the myocardium, where Gd-TO uptake was increased due to cell rupture, within 2h of MI. The signal peaked 9–18h after the onset of MI. This time course is consistent with histological data from previous immunohistochemical studies. A significant increase in the longitudinal relaxation rate (R1) in the infarct, which appears bright on T1-weighted MRI, was seen in mice injected with Gd-TO within 48h of MI, but not in those injected more than 72h after MI. Unlike Gd-TO, a conventional contrast agent (gadopentetic acid, Gd-DTPA) was completely washed out of acute infarcts within 2h of injection.

Apoptosis Imaging

Cerebral ischemia initiates apoptosis-related complex signaling pathways, involving the activation of caspase-3 and apoptosis-inducing factor. After a focal ischemic stroke, the irreversible necrotic core is surrounded by potentially salvageable penumbra, which is under threat of infarction. In this periinfarct zone, neuronal or nonneuronal cells may undergo apoptosis. Quantitative imaging and pharmacological blockade of apoptosis are important unmet goals for limiting the final infarct volume in clinics.

Apoptotic cells rapidly redistribute phosphatidylserine from the inner to the outer leaflet of the plasma membrane lipid bilayer. Annexin-V (36 kDa) has high affinity for phosphatidylserine-expressing cells, and lacks immunogenicity and toxicity [9]. Radiolabeled annexin-V, such as ^{99m}Tc-annexin-V, is accumulated mainly in the kidneys, and taken up in negligible amounts by the normal brain. Thus this probe has served as an ideal tracer for imaging apoptosis in animal models of ischemic stroke. In a study on 12 patients with acute ischemic stroke, Lorberboym et al. showed that ^{99m}Tc-HYNIC-annexin-V imaging was able to detect regions of ischemic neuronal injury; abnormal annexin uptake was present in all seven patients with cortical strokes and in one of five patients with subcortical strokes [10]. However, it is notable that annexin-V, which detects apoptotic cells with preserved membrane integrity, also labels necrotic cells by reaching and binding to intracellular phosphatidylserine through disrupted membranes [9].

Low-molecular-weight compounds have various advantages over large protein-based probes [9]. These small-molecule compounds target various other steps of the intrinsic and extrinsic apoptotic pathways rather than direct phosphatidylserine binding, such as caspase activation (e.g., isatin compounds to inhibit caspase), collapse of mitochondrial membrane potential, and the apoptotic membrane imprint that distinguishes apoptotic cells from their viable or necrotic counterparts.

In a rat model of permanent cerebral ischemia and in patients with acute ischemic stroke, in vivo PET imaging with ^{18}F-labeled 5-fluoropentyl-2-methyl-malonic acid (^{18}F-ML-10; molecular weight, 206) could successfully visualize apoptosis in the brain [11]. The radiolabeled small molecule responds to alterations in plasma membrane potential and phospholipid scrambling. The PET signal resulting from the uptake and accumulation of the tracer by apoptotic cells is lost upon membrane rupture. Thus the imaging probe distinguishes between apoptotic and necrotic cells.

IMAGING NEUROINFLAMMATION

Stroke-mediated inflammatory brain damage progresses up to several days, which is far beyond the 4.5–6-h time window for acute recanalization therapy. As Lo pointed out when describing the "new penumbra" concept, poststroke inflammation in the brain can not only exacerbate ischemic tissue damage but also support repair and regeneration of the damaged tissue, depending on the ever-changing balance between pro-versus antiinflammatory mediators. Thus, serial in vivo

molecular imaging would be a preferred method for studying or modulating the dynamic interplay of a plethora of inflammatory cells and mediators. Furthermore, imaging inflammation may guide clinicians to (1) select and risk stratify vulnerable patients with stroke who can benefit from antiinflammatory therapy, (2) determine the optimal time window for immunomodulatory therapy, and (3) monitor treatment efficacy [12].

Relevant inflammatory molecular targets include damage-associated molecular patterns, Toll-like receptors, microglia, neutrophils, macrophages, lymphocytes, astrocytes, myeloperoxidase (MPO), matrix-metalloproteinases, and BBB. Of note, a novel molecular imaging tool for stroke pathophysiology, including poststroke inflammation, needs to demonstrate target specificity with adjustments for confounders, such as BBB dysfunction-related signal due to the alteration of probe delivery and accumulation in the infarcted parenchyma.

Intravenously injected USPIOs and SPIOs are likely to be taken up by phagocytic cells, migrating and infiltrating into the inflammatory brain region associated with cerebral infarction [12]. These iron oxide–laden phagocytic cells perturb the nearby magnetic field, resulting in a local signal loss (i.e., negative contrast) on T2*-weighted MRI of the infarcted brain, as demonstrated in experimental ischemia and human stroke. The recruitment of hematogenous macrophages was shown to occur within a narrow time interval between days 3 and 6 after ischemia.

Chen's group was able to track the in vivo inflammatory response in a mouse model of transient MCA occlusion by using an activatable MRI agent that is aggregated after sensing MPO [13]. As part of the host defense system, MPO is a key inflammatory enzyme secreted by activated neutrophils and macrophages/microglia, generating highly reactive oxygen species to cause additional damage to the brain. In the presence of MPO, the 5-hydroxytryptamide moiety of bis-5-hydroxytryptamide-diethylenetriamine-pentatacetate gadolinium (MPO-Gd) is oxidized and radicalized. The radicalized MPO-Gd molecule can increase the T1-weighted signal by reacting with another radicalized MPO-Gd molecule, thereby forming a polymer, and also by binding to proteins, trapping the agent at the site of inflammation (Fig. 70.3). In vivo MPO imaging showed that the peak level of brain MPO activity occurred on day 3 after stroke.

IMAGING STEM CELLS

Stem cell therapy is a promising therapy for the stroke-damaged brain. However, it still faces major challenges in translating advances in cell-based therapies into clinical practice. For clinical translation of the potential novel

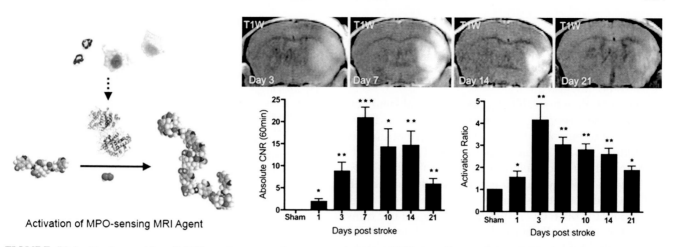

Activation of MPO-sensing MRI Agent

FIGURE 70.3 Myeloperoxidase (MPO)-sensing magnetic resonance imaging (MRI) that allows serial tracking of inflammation in stroke in vivo. MPO oxidizes the 5-hydroxytryptamide moiety of MPO-gadolinium that leads to oligomerization, and the activated agent can further bind to proteins. This leads to an increase in T1-weighted (T1W) signal. In a mouse model of focal ischemic stroke, MPO imaging over 3 weeks demonstrates that the absolute contrast-to-noise ratio (CNR) enhancement (left graph), which represents both BBB breakdown and MPO activation of the agent, peaks on day 7 ($P < .05$ on all days compared with the sham-operated animals). Activation ratio of the MPO agent (right graph) reveals that the highest MPO activity occurs on day 3 after stroke and remains elevated on day 21 ($P < .05$ on all days compared with the sham operated animals). *, $P < .05$; **, $P < .01$; ***, $P < .001$. *Reproduced from Breckwoldt et al. [13], with permission of Pf. John Chen, Harvard Medical School.*

therapy, the timing, route of delivery, implantation site, and dosage need to be optimized. In vivo molecular imaging will allow quantitative assessment of the spatiotemporal dynamics of cell survival and the functional state after transplantation [14].

Magnetic Resonance-Based Cell Tracking

de Vries et al. demonstrated that in vivo MR tracking of magnetically labeled cells was feasible in humans [15]. Dendritic cells were labeled by coculturing them with 200 mg/mL SPIO nanoparticles. After intranodal injection of the autologous cells, serial gradient-echo transversal MRI allowed in vivo visualization of cell migration to sentinel lymph nodes and assessment of the accuracy of the cell delivery. The limit of detection was 1.5×10^5 cells.

Compared with the contrast generated by ^{1}H-MRI detecting SPIO-labeled cells, particularly in nonhomogenous background tissue, such as infarcted lesions with hemorrhagic transformation, perfluorocarbon (PFC)-based cell tracking could enable more specific cell detection and more accurate cell quantification [16]. PFC-labeled cells were detectable using fluorine magnetic resonance imaging (^{19}F-MRI), and ^{19}F existed in biological tissue at an extremely low concentration. Therefore false-positive cell detection by ^{19}F-MRI is unlikely; there was no background signal from the host's tissues [16].

It is notable that both SPIO- and PFC-based MRI do not provide information on cell viability, graft functionality, or cell–cell interactions [14–16]. Moreover, cell division and subsequent dilution of intracellular nanoparticles can limit long-term visualization of cells.

Dead or phagocytosed cells can also cause false-positive contrast.

Luciferase Imaging-Based Cell Tracking

Luciferases are oxidative enzymes used in bioluminescence. Firefly luciferase is widely used as a bioluminescent reporter in many areas of biotechnology [1]. Firefly luciferase generates bioluminescence by catalyzing the oxidative decarboxylation of its substrate luciferin in the presence of O_2 to yield photon emission (light). After intravenous or intraperitoneal injection of luciferin in living mice, luciferase-expressing cells can be tracked noninvasively by using a bioluminescence imager. Unlike fluorescence imaging, there is no endogenous background signal.

Studies on living mice demonstrated that the recruitment of luciferase-expressing stem cells to cerebral infarcts can be visualized and assessed in a quantitative manner by serial in vivo bioluminescent imaging. These studies showed that bioluminescent imaging is a sensitive imaging modality, although optical imaging is affected by photon attenuation in intervening tissues, such as skin, skull, or hair.

Vandeputte et al. reported on bioluminescence imaging of stroke-induced endogenous neural stem cell (eNSC) response [17]. They injected conditional (Cre-mediated flip-excision) lentiviral vectors encoding firefly luciferase in the subventricular zone of Nestin-Cre transgenic mice. Increased eNSC proliferation caused a significant increase in the bioluminescent imaging signal 2–14 days after focal ischemic stroke. The signal decreased after 3 months. The bioluminescent imaging signal relocalized from the subventricular zone toward the infarct region during the 2 weeks following the stroke. These imaging findings indicate that nestin[+] eNSCs originating from the subventricular zone respond to stroke injury by increased proliferation and migration toward the infarct region.

Daadi et al. genetically engineered neural stem cells with a triple fusion reporter gene to express red fluorescence protein and herpes simplex virus-truncated thymidine kinase, which enabled PET imaging by trapping the radiolabeled reporter probe ^{18}F-fluoro-hydroxymethylbutyl-guanine transported into the cells [18]. In addition, these transduced NSCs were labeled with SPIO for MRI. Over 3 weeks after stereotaxic transplantation of the cells in rats with focal ischemic stroke, T2-weighted MRI showed the NSC grafts as hypointense areas in the striatum and infarcts as hyperintense areas. Furthermore, PET imaging revealed increased metabolic activity in the stroke region and visualized functioning grafted cells in vivo.

CONCLUSION

Molecular imaging complements traditional imaging by providing a fundamentally new type of information that allows for a timely and precise assessment of pathophysiologic states at the cellular and molecular levels. It facilitates the early detection of neurovascular disease, refines risk assessment, aids in the selection of individualized therapies, and monitors the efficacy of such therapies in a sensitive and quantitative manner. As the era of personalized medicine is on the horizon, molecular imaging and its theranostic application will likely play a pivotal role in both basic research and clinical practice for the improved treatment and prevention of stroke.

References

[1] Weissleder R, Mahmood U. Molecular imaging. Radiology 2001;219:316–33.

[2] de Kemp RA, Epstein FH, Catana C, Tsui BM, Ritman EL. Small-animal molecular imaging methods. J Nucl Med 2010;51(Suppl. 1): 18S–32S.

[3] Chen IY, Wu JC. Cardiovascular molecular imaging focus on clinical translation. Circulation 2011;123:425–43.

[4] Lee DK, Nahrendorf M, Schellingerhout D, Kim DE. Will molecular optical imaging have clinically important roles in stroke management, and how? J Clin Neurol 2010;6:10–8.

[5] Kooi ME, Cappendijk VC, Cleutjens KB, Kessels AG, Kitslaar PJ, Borgers M, et al. Accumulation of ultrasmall superparamagnetic particles of iron oxide in human atherosclerotic plaques can be detected by in vivo magnetic resonance imaging. Circulation 2003;107:2453–8.

[6] Uppal R, Ay I, Dai G, Kim RO, Sorensen AG, Caravan P. Molecular MRI of intracranial thrombus in a rat ischemic stroke model. Stroke 2010;41:1271–7.

[7] Kim JY, Ryu JH, Schellingerhout D, Sun IC, Lee SK, Jeon S, et al. Direct image of cerebral thromboemboli using computed tomography and fibrin-targeted gold nanoparticles. Theranotics 2015;5:1098–114.

[8] Huang S, Chen HH, Yuan H, Dai G, Schuhle DT, Mekkaoui C, et al. Molecular MRI of acute necrosis with a novel DNA-binding gadolinium chelate kinetics of cell death and clearance in infarcted myocardium. Circ Cardiovasc Imaging 2011;4:729–37.

[9] Niu G, Chen X. Apoptosis imaging: beyond annexin V. J Nucl Med 2010;51:1659–62.

[10] Lorberboym M, Blankenberg FG, Sadeh M, Lampl Y. In vivo imaging of apoptosis in patients with acute stroke: correlation with blood–brain barrier permeability. Brain Res 2006;1103:13–9.

[11] Reshef A, Shirvan A, Waterhouse RN, Grimberg H, Levin G, Cohen A, et al. Molecular imaging of neurovascular cell death in experimental cerebral stroke by PET. J Nucl Med 2008;49:1520–8.

[12] Jander S, Schroeter M, Saleh A. Imaging inflammation in acute brain ischemia. Stroke 2007;38:642–5.

[13] Breckwoldt MO, Chen JW, Stangenberg L, Aikawa E, Rodriguez E, Qiu S, et al. Tracking the inflammatory response in stroke in vivo by sensing the enzyme myeloperoxidase. Proc Natl Acad Sci USA 2008;105:18584–9.

[14] Chao F, Shen Y, Zhang H, Tian M. Multimodality molecular imaging of stem cells therapy for stroke. Biomed Res Int 2013;2013:849819.

[15] de Vries IJ, Lesterhuis WJ, Barentsz JO, Verdijk P, van Krieken JH, Boerman OC, et al. Magnetic resonance tracking of dendritic cells in melanoma patients for monitoring of cellular therapy. Nat Biotechnol 2005;23:1407–13.

[16] Ahrens ET, Bulte JWM. Tracking immune cells in vivo using magnetic resonance imaging. Nat Rev Immunol 2013;13:755–63.

[17] Vandeputte C, Reumers V, Aelvoet SA, Thiry I, Swaef SD, Van den Haute C, et al. Bioluminescence imaging of stroke-induced endogenous neural stem cell response. Neurobiol Dis 2014;69:144–55.

[18] Daadi MM, Klausner J, Li Z, Sofilos M, Sun G, Wu JC, et al. Imaging neural stem cell graft-induced structural repair in stroke. Cell Transplant 2013;22:881–92.

CHAPTER

71

Experimental Methods for Measuring Blood Flow in Brain Capillaries

N. Nishimura

Cornell University, Ithaca, NY, United States

Blood reaches the cortex through larger arterioles, then disperses through the bulk of the tissue through capillaries, and then coalesces back into venules to exit the brain tissue. Much of the gas and nutrient exchange occurs at the capillary level, so that understanding blood flow at this scale is important for both normal and disease physiology. Advancements in optical microscopy now enable researchers to assess flow in even the tiniest vessels of the cerebrovasculature. The smallest capillaries are just large enough to let red blood cells squeeze through (about 3 μm diameter in mice) so that they can be visualized easily by optical methods. In brain, however, most of the vessels are in the depth of the tissue, so it was not until the development of two-photon microscopy that studies of blood flow in the capillaries became common [3,7]. Since then, several other methods have been developed that can image with sufficient depth and spatial resolution to be useful in studies of capillary structure and function. Two-photon microscopy (also known as multiphoton microscopy, two-photon excited fluorescence microscopy, and two-photon laser scanning microscopy) has become a powerful experimental tool for in vivo studies of not only blood flow, but also many other cellular measurements, and hence it is commonly available. This chapter will describe some of the experimental and measurement techniques used to measure capillary blood flow, focusing on two-photon approaches.

TWO-PHOTON MICROSCOPY OF FLUORESCENT SIGNALS

In models including mice and rats, two-photon microscopy enables imaging of fluorescent labels with better than 1-μm spatial resolution deep in semiopaque tissues such as the brain. For in vivo studies, fluorescent labels can be endogenous or exogenous dyes, but many studies take advantage of the specificity of using genetic strategies to label particular cells or structures. For brain, a window of either glass or merely thinned bone is produced above the region of interest. The easiest way to label the blood vessels is an intravenous injection (usually in rodents via the tail or saphenous vein or retro-orbitally) of dextran-conjugated dyes (Fig. 71.1A) [4]. The length of time the dye stays in the circulation depends on the size of the dextran; 50–100 μL of 5% w/v dye conjugated to a dextran of 70 kDa dissolved in saline lasts several hours in mice. Dextran-conjugated dyes are available in many colors, so it is possible to choose a color that does not overlap other indicators in the experiment. In adult brain, normal blood vessels, including individual capillaries, are anatomically stable for many months. Because the pattern of branches is unique, the vessels provide a convenient way of repeatedly finding the same capillaries, neurons, or other cells. One can start with the large vessels, which are easy to see on the brain surface, and follow the branches down to specific capillaries reliably across weeks to months of imaging (analogous to following a city map to find a specific address).

Circulating cells, including red and white blood cells, do not take up the dextran-conjugated dye, so they appear as dark patches within the vessels (Fig. 71.1B). Previous studies have also used red blood cells that were extracted, labeled in vitro, and reinjected, but this complicated procedure is not necessary with two-photon imaging. Since red blood cells are the vast majority of cells in circulation, most blood flow speed measurements described here rely on the red blood cells. In fact, the occasional white blood cell often temporarily decreases the blood flow speed in a capillary, but no extra label needs to be added to see this effect because these cells also show up as dark regions against the fluorescent dye.

Primer on Cerebrovascular Diseases, Second Edition
http://dx.doi.org/10.1016/B978-0-12-803058-5.00071-0

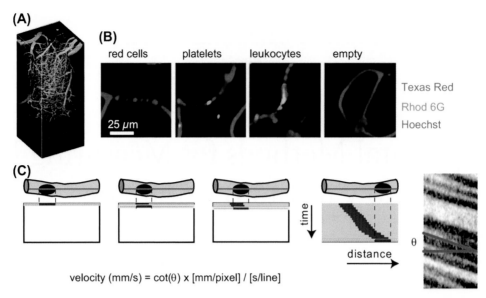

$$velocity\ (mm/s) = cot(\theta) \times [mm/pixel] / [s/line]$$

FIGURE 71.1   (A) Three-dimensional rendering of cortical blood vessels in mouse brain imaged with two-photon microscopy. Vessels were visualized by injecting dextran-conjugated fluorescent dye. *Adapted from Kobat D, Durst ME, Nishimura N, Wong AW, Schaffer CB, Xu C. Deep tissue multiphoton microscopy using longer wavelength excitation. Opt Express 2009;17:13354–64.* (B) Fluorescent labels used to provide cellular information. Blood cells show up as dark areas when the blood plasma is labeled with intravenous dye injections (*red*, Texas Red dextran). Red blood cells, nucleated white blood cells, and platelets can be distinguished with injections of rhodamine 6G (*green*) and Hoechst (*blue*). *Adapted from Santisakultarm TP, et al. Stalled cerebral capillary blood flow in mouse models of essential thrombocythemia and polycythemia vera revealed by in vivo two-photon imaging. J Thromb Haemost 2014;12: 2120–30. http://dx.doi.org/10.1111/Jth.12738.* (C) A schematic and example data showing line-scans. The excitation laser is repeatedly scanned along a vessel. The motion of cells produces streaks in the line-scans. The angle of the streaks is related to the speed of the cell motion.

Other fluorescent labeling techniques can also be used to label additional types of cells. Transgenic mice may have monocytes or other cells labeled by fluorescent protein expression. In some cases, labeled cells are taken from a donor animal and injected into a recipient in adoptive transfer. Several exogenous indicators also provide convenient ways to label particular cell types (Fig. 71.1B). An intravenous injection of rhodamine 6G (0.05 mL of 0.1% w/v per mouse) labels several types of white blood cells, including monocytes, neutrophils, and some lymphocytes, as well as platelets [10]. The DNA-binding dye Hoechst 33342 (0.05 mL of 0.5% w/v per mouse) can also be injected in the circulation with little effect in healthy animals. It will not label cells in the healthy brain because it does not cross the blood–brain barrier, but the nuclei in circulating cells such as white blood cells will fluoresce at blue wavelength. Since platelets do not have nuclei, Hoechst will not label platelets and provides a convenient way to distinguish thrombi in capillaries from leukocyte plugs when using rhodamine 6G [10].

LINE-SCANS OVERCOME LIMITATIONS OF MEASUREMENT SPEED

Two-photon microscopy is a scanning microscopy meaning that the image is formed point by point, by moving the focus of the laser through a sample and measuring the fluorescence generated at each spot. Confocal microscopy works in a similar way and has also been used to image capillaries in the brain, but cannot image as deep as two-photon microscopy because of the effect of light scattering in the tissue. The scanning approach to image formation limits the speed of events that can be captured. The majority of two-photon microscopes scan across the image region one line at time and require at least 0.5 ms per line. Most red blood cells in capillaries are moving fast enough that full frame images will not capture the same blood cells in successive frames, making it impossible to quantify blood flow speed. To increase time resolution the laser focus is repeatedly scanned along the length of a vessel segment, called a "line-scan," instead of scanning a full image plane, essentially capturing a very thin image in the middle of the blood vessel (Fig. 71.1C). Scan speeds are generally fast enough that they will easily capture the same red blood cells in multiple successive line-scans, thus enabling the speed of the cells to be measured. This data is often displayed by aligning the sequential line-scans so that time is along the vertical axis and path is along the vessel on the horizontal axis, also known as a kymograph, or space–time image. A stationary object, which is always in the same part of the vessel, will result in a straight, vertical line. Red blood cells moving along the length of the vessel generate diagonal streaks. The slower the cell moves, the more vertical the streak. The

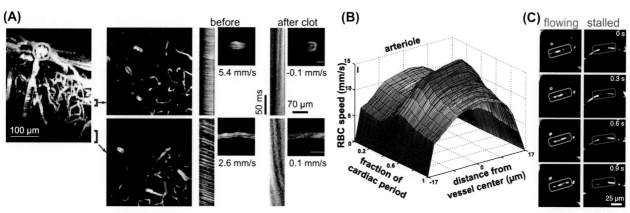

FIGURE 71.2 Line-scans were used to measure red blood cell (RBC) speed before and after the occlusion of a penetrating arteriole. (A) Coronal and tangential projections of selected image planes (depth indicated by brackets) of rat microvasculature after occlusion of a penetrating arteriole (*yellow ellipse*). Specific capillary vessel segments (*red*) were imaged and measured with line-scans taken both before and after the induction of clots in the penetrating arteriole. (B) Line-scans and images of the capillaries taken both before and after the induction of clots. *Adapted from Nishimura N, Schaffer CB, Friedman B, Lyden PD, Kleinfeld D. Penetrating arterioles are a bottleneck in the perfusion of neocortex. Proc Natl Acad Sci USA 2007;104:365–70.* (C) RBC speed varies with both the radial position across an arteriole and time across the cardiac cycle. *Adapted from Santisakultarm TP, et al. In vivo two-photon excited fluorescence microscopy reveals cardiac- and respiration-dependent pulsatile blood flow in cortical blood vessels in mice. Am J Physiol Heart Circulatory Physiol 2012;302:H1367–77. http://dx.doi.org/10.1152/ajpheart.00417.2011.* (D) Images or image stacks can be used to distinguish capillaries that are stalled for at least several seconds. In flowing vessels, the pattern of dark and light intensities will change from frame to frame, whereas in vessels without flow, the dark and bright regions stay in the same place when compared across image frames. *Adapted from Santisakultarm TP, et al. Stalled cerebral capillary blood flow in mouse models of essential thrombocythemia and polycythemia vera revealed by in vivo two-photon imaging. J Thromb Haemost 2014;12:2120–30. http://dx.doi.org/10.1111/Jth.12738.*

faster the cell moves, the more horizontal the streak. The speed of the red blood cell along the vessel is proportional to the inverse of the slope of the streak, or the cotangent of the angle made with the horizontal axis. By measuring the angle of these streaks, we can measure the speed (Fig. 71.2A).

In capillaries, the red blood cells must go in a single file through the narrow vessel. Therefore, in such "plug" flow, the speed of the cells is the same throughout the whole vessel. Note that this measurement reports only the speed of the red blood cells and the fluid of the blood plasma might be moving at a different rate around the red blood cell. In larger vessels, the red blood cell flow varies across the profile of the vessel (Fig. 71.2B). Line-scans can still be used to measure the blood flow speeds at the center (or any other location) of the vessel if the flow is laminar and the cells are moving in the same direction. Because the red blood cell flow speeds in most venules and arterioles follow a smooth, approximately parabolic profile across the vessel when averaged over fluctuations caused by heartbeat, the volume of flow (in units of liquid volume per time) scales as the product of the cross-sectional area of the vessel and the centerline blood cell speed. This makes area × speed a convenient metric to use when monitoring changes in perfusion in venules and arterioles [6]. Because the blood plasma does not move at the same speed as the red blood cells, this measurement is best used to compare changes in volumetric flow caused by diameter and red blood cell speed alterations in the same vessel.

In capillaries, in which blood cells pass through one at a time, it is possible to measure the flux, that is, the number of red blood cells passing through the vessel per time [3]. In practice, this is difficult in vessels with high flow speed in which the shape of each red blood cell becomes less clear, so it may be necessary to consider any bias in the selection of vessels with low flow speed for such measurements. Tube hematocrit, the fraction of volume within a vessels taken up by red blood cells, can be estimated if one assumes a fixed volume for each red blood cell and then calculates the volume of the vessel lumen from the fluorescent plasma [9]. In larger-diameter vessels, the number of red blood cells can vary with radial position, so any estimate of hematocrit would need to account for this effect.

ANALYSIS METHODS

One method of finding blood flow speeds is to measure the angle of the streaks in the line-scan images. Manually performing this with a program such as ImageJ is tedious (although manually measuring some of the data is always a good check of any automated analysis), so several automated image analysis methods have been developed. Generally, the line-scan data is broken into segments of short periods (often 10–50 m) and the algorithms calculate the predominant angles of the streaks within that period. This is done both for computation efficacy and to capture variations in flow speed over time. In almost all brain blood vessels, the speed of red blood cells will vary with the heart rate and other physiologic processes such as breathing and vasomotion [9]. For steady-state measurements, it is good to average a measurement over many heart-beat and breathing cycles. Although vasomotion is not always present in all preparations (not always in mice), it is advisable to sample enough time to cover several cycles of vasomotion (~40 s is typical). Fast speed variations, such as heart rate, result in a change of angle within an analysis segment. In such data sets, different algorithms may have different sensitivity levels to the curvature of the streaks in the line-scan image.

The basis for algorithms for calculating the angles include the Radon transform [1,9], which relies on the variation of intensities created by red blood cells and the fluorescent plasma, and singular value decomposition [3], which can be used in conjunction with rotating an image to detect the dominant angles. Vessels with very high flow speed, such as arteries, may not generate obvious stripes in a line-scan image. In laminar flow, a series of red blood cells produces a particular pattern of dark and light fluorescence that maintains itself as it shifts down the vessel. Algorithms based on the idea that the line-scan is fast enough to capture a particular pattern of red blood cells several times before they move out of the field of view have been developed, and can calculate velocities in vessels with very high flow speed [2]. For example, red blood cell speeds as fast as 84 mm/s was measured using algorithms based on correlation methods [2]. Most of the authors of the studies referred to in this chapter have made their analysis algorithms freely available, although many require access to MATLAB software.

VIEWING STALLED VESSELS

In some cases, especially with faster scanners such as resonant scanners, the images may be fast enough that the motion of red blood cells can be tracked individually from image frame to frame. However, even with slower galvanometer scanners, the flow in individual capillaries can be distinguished as moving or stalled on the timescale of several seconds. In some systems, this can be an interesting measurement in itself [10]. Three-dimensional data is often acquired in stacks in which sequential images are taken spaced by a fixed distance axially, often about 1 μm (Fig. 71.1A). In the brain, the vast majority of vessels span multiple image planes meaning that a single capillary can be imaged for several seconds in time, even in an image stack. Because capillaries only fit red blood cells in single file, a red blood cell not moving in any portion of the vessel indicates that the whole capillary has no red blood cell motion. With this

method, flowing vessels appear to have flickering dark and light patterns in the stack, but stalled vessels show stationary dark spots (Fig. 71.2B). This method was used to show that in mouse models of polycythemia vera or essential thrombocythemia, in which there are abnormal numbers of red blood cells or platelets, respectively, stacks of images revealed a large fraction of capillaries in the brain were not flowing and had blood cells that remained stationary over the course of several seconds [10]. In healthy mouse cortex, very few capillaries are not flowing at any time.

OTHER METHODS AND FUTURE OUTLOOK

Recent developments in optical coherence tomography suggest that this may be an excellent way to evaluate capillary blood flow or flux [5,11]. Because these measurements are based on changes in signal caused by motion similar to the Doppler effect, it is easier to measure fast-flowing rather than slow-flowing vessels. The approach is label free and can be faster in evaluating a volume of tissue than point-scanning techniques such as two-photon microscopy. The elimination of labels is convenient for clinical translation, although labels such as promoter-driven fluorescent protein expression can be useful in experiments to distinguish different types of cells. Although the interpretation of such signals is not as straightforward as with fluorescent labeling, this method has generated nice maps of blood flow in single vessels. Improvements in photoacoustic imaging and several other methods, even magnetic resonance imaging, are almost close to achieving measurements in small vessels and could be excellent candidates for the future. For now, as many institutions acquire tools such as two-photon microscopes, these optical methods may be the method of choice for many investigations. A major advantage of using multiphoton microscopy is that it can also be used for many other types of measurements such as neural activity, oxygenation, or chemical concentration, enabling multimodal studies on the relationship between cerebrovasculature and its function [8].

References

[1] Drew PJ, Blinder P, Cauwenberghs G, Shih AY, Kleinfeld D. Rapid determination of particle velocity from space-time images using the radon transform. J Comput Neurosci 2009;29(1–2):5–11.

[2] Kim TN, Goodwill PW, Chen Y, Conolly SM, Schaffer CB, Liepmann D, et al. Line-scanning particle image velocimetry: an optical approach for quantifying a wide range of blood flow speeds in live animals. PLoS One 2012;7:e38590. http://dx.doi.org/10.1371/journal.pone.0038590.

[3] Kleinfeld D, Mitra PP, Helmchen F, Denk W. Fluctuations and stimulus-induced changes in blood flow observed in individual capillaries in layers 2 through 4 of rat neocortex. Proc Natl Acad Sci USA 1998;95:15741–6.

[4] Kobat D, Durst ME, Nishimura N, Wong AW, Schaffer CB, Xu C. Deep tissue multiphoton microscopy using longer wavelength excitation. Opt Express 2009;17:13354–64.

[5] Lee J, Jiang JY, Wu W, Lesage F, Boas DA. Statistical intensity variation analysis for rapid volumetric imaging of capillary network flux. Biomed Opt Express 2014;5:1160–72. http://dx.doi.org/10.1364/BOE.5.001160.

[6] Nishimura N, Rosidi NL, Iadecola C, Schaffer CB. Limitations of collateral flow after occlusion of a single cortical penetrating arteriole. J Cereb Blood Flow Metab 2010;30:1914–27.

[7] Nishimura N, Schaffer CB, Friedman B, Lyden PD, Kleinfeld D. Penetrating arterioles are a bottleneck in the perfusion of neocortex. Proc Natl Acad Sci USA 2007;104:365–70.

[8] Sakadzic S, Lee J, Boas DA, Ayata C. High-resolution in vivo optical imaging of stroke injury and repair. Brain Res 2015;1623:174–92. http://dx.doi.org/10.1016/j.brainres.2015.04.044.

[9] Santisakultarm TP, et al. In vivo two-photon excited fluorescence microscopy reveals cardiac- and respiration-dependent pulsatile blood flow in cortical blood vessels in mice. Am J Physiol Heart Circulatory Physiol 2012;302:H1367–77. http://dx.doi.org/10.1152/ajpheart.00417.2011.

[10] Santisakultarm TP, et al. Stalled cerebral capillary blood flow in mouse models of essential thrombocythemia and polycythemia vera revealed by in vivo two-photon imaging. J Thromb Haemost 2014;12:2120–30. http://dx.doi.org/10.1111/Jth.12738.

[11] Srinivasan VJ, et al. Micro-heterogeneity of flow in a mouse model of chronic cerebral hypoperfusion revealed by longitudinal Doppler optical coherence tomography and angiography. J Cereb Blood Flow Metab 2015;35:1552–60. http://dx.doi.org/10.1038/jcbfm.2015.175.

C H A P T E R

72

Genomic Tools

F.R. Sharp, G.C. Jickling, B. Stamova
University of California at Davis, Sacramento, CA, United States

INTRODUCTION

Although an enormous amount of work has been performed trying to understand the molecular events associated with stroke in animal models, much less has been done assessing the molecular biology of stroke in humans. A number of human studies have examined clotting pathways and some inflammatory pathways; most of the studies have focused on candidate molecules or pathways that are thought to be important in human stroke or hemorrhage. We took a different approach by examining all of the messenger RNA (mRNA) expressed in the blood of patients with strokes, transient ischemic attacks (TIAs), and brain hemorrhage. Since all of the mRNAs are known, this approach will capture all of the molecules and pathways in blood associated with the stroke, TIA, hemorrhage, or other cerebrovascular event.

In humans, the only viable avenue at present is to study the blood. However, blood is a particularly relevant organ for stroke since ischemic stroke due to cardioembolism, large-vessel atherosclerosis, and lacunar disease involves clotting, which includes blood proteins, blood platelets, and blood leukocytes [1]. Evidence for clotting as being the primary event for all causes of stroke includes the fact that tissue plasminogen activator (tPA) is effective for all ischemic strokes including those due to cardioembolism, large-vessel disease, and lacunar disease. Thus, by examining the molecular biology of peripheral blood leukocytes and platelets using our genomics approach, one can examine factors that potentially promote or prevent clotting and thus play key roles in ischemic stroke and in brain hemorrhage. In addition, peripheral blood leukocytes and platelets interact with brain endothelium and atherosclerotic plaques, which changes their gene expression and function. By assessing the whole genome of blood leukocytes and platelets, we are able to assess the biology of stroke and intracerebral hemorrhage.

Primer on Cerebrovascular Diseases, Second Edition
http://dx.doi.org/10.1016/B978-0-12-803058-5.00072-2

ISCHEMIC STROKE VERSUS CONTROLS AND INTRACEREBRAL HEMORRHAGE

We took blood from animals 24 h after they had experimental ischemic strokes, brain hemorrhage, status epilepticus, hypoxia, and hypoglycemia [2]. The RNA was isolated and processed on whole genome microarrays. Although no single gene could distinguish these conditions, groups of genes we termed "gene expression profiles" did distinguish ischemic stroke from intracerebral hemorrhage. Indeed, there was a unique expression profile for each condition in peripheral blood [2].

Human studies demonstrated similar findings. There were unique gene expression profiles in whole blood of patients who had ischemic strokes compared with matched controls without strokes [3]. Gene expression changed within 3 h of a stroke with hundreds of genes with altered expression by 24 h after a stroke [3]. These results were confirmed in a larger follow-up study that showed that a panel of less than 20 genes could distinguish strokes from controls with greater than 85% accuracy [4]. Studies using RNAseq to measure alternative splicing of mRNA transcripts in blood showed that there are specific spliced gene profiles that can distinguish patients with ischemic strokes from patients with intracerebral hemorrhages and control patients [5].

MRNA PROFILES FOR CAUSES OF STROKE

Since different causes of stroke have a different pathogenesis, we postulated that this would be reflected in different gene expression profiles in blood. Indeed, several studies have shown that there are specific mRNA profiles in blood of patients with ischemic strokes caused by cardioembolism compared with large-vessel and lacunar strokes [6]. These profiles suggest unique pathogenesis for each stroke cause. Moreover, we propose that the profiles will potentially be useful for predicting the causes of cryptogenic strokes where the causes are unknown or at least uncertain based upon current methodology [6]. Predicting causes of cryptogenic strokes is important since our data suggest as many as 60% of cryptogenic strokes are cardioembolic, which would mean they should be treated with anticoagulants rather than antiplatelets to prevent future strokes [6].

TRANSIENT ISCHEMIC ATTACKS

The diagnosis of transient neurological events as true brain vascular TIAs can be quite challenging particularly in emergency room settings. We postulated that even brief cerebral ischemia would change gene expression in peripheral leukocytes and showed this in a rat study where 5 and 10 min of focal ischemia produced changes of gene expression 24 h later in adult rat blood [7]. These results were confirmed in patients with TIAs, where two gene expression profiles were associated with TIAs. A follow-up study demonstrated that mRNA profiles in peripheral blood could distinguish transient neurological events of probable ischemic etiology from those of nonischemic etiology (e.g., migraine and seizures) [8]. Future studies will examine the question of whether gene expression in peripheral blood leukocytes and platelets can predict which patients with TIA will go on to have strokes.

HEMORRHAGIC TRANSFORMATION

The most feared complication of administering tPA to patients with acute stroke is producing hemorrhagic transformation or frank hemorrhages, which increases morbidity and mortality. We tested the hypothesis that gene expression in peripheral blood leukocytes and platelets within 3 h of stroke and before tPA therapy would predict which patients would develop hemorrhagic transformation. Indeed, a panel of six genes predicted hemorrhagic transformation with 80% sensitivity [9]. Several of these genes were involved in clotting or associated with transforming growth factor beta signaling [9].

MICRO RNA REGULATION OF mRNA

Most of the aforementioned work was focused on whether specific cerebrovascular disorders led to specific changes of mRNA in peripheral blood leukocytes and platelets. Until recently, we had paid much less attention to the potential mechanisms by which the mRNA might be regulated. To address this issue, we turned our attention to relatively recently discovered microRNA (miRNA). miRNA are small noncoding RNA approximately 20 nucleotides long that bind to complementary sequences in target mRNA and either promote mRNA degradation or decrease translation of the mRNA to protein. There are now known to be over 1000 human miRNA and most importantly, they are believed to directly regulate >60% of all known mRNA and probably indirectly regulate virtually all mRNA in cells.

Based upon these facts, we examined miRNA in blood and brain of animals following ischemic stroke, intracerebral hemorrhage, and status epilepticus. Dozens of miRNA were regulated for each condition, and most were specific for each condition [10]. Of interest, the miRNA regulated in brain were generally different from those regulated in blood [10]. Some of the miRNA regulated in brain were found to be also regulated in blood, and these were interpreted to be miRNA released by injured brain that entered the blood [10]. We have shown that miRNAs are also regulated in the blood of humans following ischemic stroke [11]. Ongoing studies are designed to confirm these results and also will describe the miRNA–mRNA networks in peripheral leukocytes and platelets following ischemic stroke that are likely crucial in stroke pathogenesis.

ALTERNATIVE SPLICING OF mRNA AND THE FUTURE

The measurement of mRNA is being revolutionized by the development of RNAseq. This is a method by which mRNAs are converted to cDNA, and then these are sequenced and measured. The advantages of RNAseq over arrays are that new RNA can be discovered, but more importantly, all of the alternatively spliced mRNA in a cell can theoretically be identified and measured quantitatively. Alternative splicing is the process by which a single gene codes for multiple proteins. For this to occur, exons of a gene are either included or excluded from the final mRNA. Thus, a single gene produces multiple alternatively spliced mRNA transcripts that produce different proteins that may have similar or different functions. The importance of alternative splicing is that it is the process by which ~20,000 genes give rise to between >200,000 and 1 million proteins, and helps explain why the genome of every cell is identical but the phenotype and proteome of different cells is different. The relevance of alternative splicing to all of our previous studies is that every cerebrovascular disease, each of its causes, and differences of outcomes are all likely a function of differential alternative splicing. Thus, we were among the first to demonstrate that alternative splicing is unique for each cause of ischemic stroke and differs from intracerebral hemorrhage [5]. It is likely that alternative splicing is cell and disease specific, and that all of the previous studies mentioned earlier will require restudy using RNAseq methodology.

The studies to date represent just the beginning stages of the attempts at unraveling the molecular biology of stroke. As techniques become more available to isolate single cell types, and as RNAseq becomes more familiar to a new generation of stroke scientists, great advances in the field are expected.

References

[1] Jickling GC, Liu D, Ander BP, Stamova B, Zhan X, Sharp FR. Targeting neutrophils in ischemic stroke: translational insights from experimental studies. J Cereb Blood Flow Metab 2015;35(6):888–901.

[2] Tang Y, Lu A, Aronow BJ, Sharp FR. Blood genomic responses differ after stroke, seizures, hypoglycemia, and hypoxia: blood genomic fingerprints of disease. Ann Neurol 2001;50(6):699–707.

[3] Tang Y, Xu H, Du X, et al. Gene expression in blood changes rapidly in neutrophils and monocytes after ischemic stroke in humans: a microarray study. J Cereb Blood Flow Metab 2006;26(8):1089–102.

[4] Stamova B, Xu H, Jickling G, et al. Gene expression profiling of blood for the prediction of ischemic stroke. Stroke 2010;41(10):2171–7.

[5] Dykstra-Aiello C, Jickling GC, Ander BP, et al. Intracerebral hemorrhage and ischemic stroke of different etiologies have distinct alternatively spliced mRNA profiles in the blood: a pilot RNA-seq study. Transl Stroke Res 2015;6(4):284–9.

[6] Jickling GC, Stamova B, Ander BP, et al. Prediction of cardioembolic, arterial, and lacunar causes of cryptogenic stroke by gene expression and infarct location. Stroke 2012;43(8):2036–41.

[7] Sharp FR, Jickling GC, Stamova B, et al. Molecular markers and mechanisms of stroke: RNA studies of blood in animals and humans. J Cereb Blood Flow Metab 2011;31(7):1513–31.

[8] Jickling GC, Zhan X, Stamova B, et al. Ischemic transient neurological events identified by immune response to cerebral ischemia. Stroke 2012;43(4):1006–12.

[9] Jickling GC, Ander BP, Stamova B, et al. RNA in blood is altered prior to hemorrhagic transformation in ischemic stroke. Ann Neurol 2013;74(2):232–40.

[10] Liu DZ, Tian Y, Ander BP, et al. Brain and blood microRNA expression profiling of ischemic stroke, intracerebral hemorrhage, and kainate seizures. J Cereb Blood Flow Metab 2010;30(1):92–101.

[11] Jickling GC, Ander BP, Zhan X, Noblett D, Stamova B, Liu D. microRNA expression in peripheral blood cells following acute ischemic stroke and their predicted gene targets. PloS One 2014;9(6):e99283.

C H A P T E R

73

Proteomes and Biomarkers of the Neurovascular Unit

A.S. Haqqani, D.B. Stanimirovic
National Research Council of Canada, Ottawa, ON, Canada

INTRODUCTION

Brain endothelial cells (BECs) that form the blood–brain barrier (BBB) exhibit a unique polarized phenotype characterized by firmly sealed tight junctions, thick and elaborate luminal glycocalyx, and a multitude of transporters and ion pumps responsible for controlled exchange of water and nutrients between blood and brain compartments. Adjoining pericytes embedded in the vascular basement membrane, astrocytes projecting end feet that envelope capillaries and microvessels, and microvessel-innervating neurons form the neurovascular unit (NVU). Both the specialized phenotype of the BEC and functional regulation of the NVU are dependent on integrated molecular interactions of this cellular network [1].

The NVU is an important player in brain pathologies spanning brain tumors and cerebrovascular, neuroinflammatory, and neurodegenerative diseases [1]. Brain vessels undergo a profound molecular and functional remodeling in response to pathology; the NVU proteome changes associated with this remodeling can be analyzed using a variety of advanced mass spectrometry (MS) techniques. The key techniques include isotope labeling–based or label-free proteomics methods, glycoproteomics, and targeted quantitative MS. It should be noted that the application of these techniques in both research and clinical environments typically requires "core facilities" grouping together expensive equipment and specialized expertise.

The technical objective of discovery proteomics is the broadest possible protein coverage that includes low abundant and rare proteins. In contrast, targeted proteomics focuses on a limited number of proteins, usually selected from discovery approaches, that can be monitored and quantified (in absolute concentrations) in multiplexed fashion over a time course or in patient's samples. Various physiological and pathological aspects of the NVU have been dissected using these MS-based approaches (Table 73.1).

The generic workflows used for MS-based analyses of the NVU are shown in Fig. 73.1. Isolation of various NVU components that have been subjected to various conditions or pathology is followed by sample preparation to extract proteins. For labeling-based MS methods, paired samples are isotopically labeled followed by mixing of the pairs for differential protein expression analysis. Protein solubilization and proteolytic digestion into MS-analyzable/quantifiable peptides is followed by liquid chromatography (LC)-coupled MS analysis (tandem MS for discovery proteomics or selected-reaction monitoring for targeted quantification). Finally, protein identification/quantification, data analysis, and integration are performed using various bioinformatics tools.

The advances in applying MS-based approaches to interrogate NVU are summarized in this chapter, as both a resource and inspiration for future creative application of these techniques in the cerebrovascular research field.

NEUROVASCULAR UNIT "CARTA"

Discovery proteomic analyses of the NVU in various experimental paradigms are being assembled currently into a large repository called the NVU "Carta" (Fig. 73.2). The NVU "Carta" aims to: (1) catalog and map protein composition of BBB endothelial cells to facilitate understanding of their unique phenotype including tight junctions and polarized expression of transporters and receptors, important for physiological homeostasis and drug delivery; (2) identify protein–protein interactions (interactome) among different constituents of the NVU involved in BEC phenotype induction, maintenance, and modulation; (3) identify NVU-specific biomarkers and therapeutic targets for cerebrovascular diseases.

Primer on Cerebrovascular Diseases, Second Edition
http://dx.doi.org/10.1016/B978-0-12-803058-5.00073-4

TABLE 73.1   Examples of Proteomes Generated From Brain Endothelial Cells and Brain Vessels and Biomarker Studies

NVU Component	Fraction/Isolation	MS Method	Paradigm/Pathology	References
BEC (human, rat, mouse)	Whole cell	Proteomics, glycoproteomics Quantitative targeted proteomics	Hypoxia/ischemia; inflammatory stimulation	[3–5,16]
	Luminal/abluminal membranes	Proteomics, glycoproteomics Quantitative targeted proteomics	TNF, IFN	[2,17]
	Lipid rafts	Proteomics		[6,7]
	Microvesicles/exosomes	Proteomics, glycoproteomics	Transcytosis trigger (FC5)	[13]
	Glycocalyx	Proteomics, glycoproteomics	TNF, IFN	[2,5]
Isolated vessels	Gradient/filtration	Multidimensional protein identification technology		[10]
	Laser capture microdissection	ICAT proteomics	Global cerebral ischemia	[11]
	Dissection	Proteomics	Pial vessels, arteries	[9]
Biomarkers	Serum	Proteomics (ICAT proteomics; degradomics)	TBI; stroke	[14,15]

BEC, Brain endothelial cell; *ICAT*, isotope-coded affinity tag; *IFN*, interferon; *MS*, mass spectrometry; *NVU*, neurovascular unit; *TBI*, traumatic brain injury; *TNF*, tumor necrosis factor.

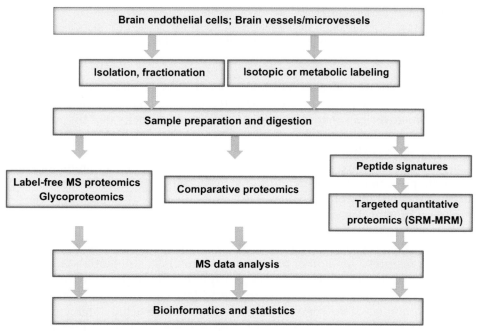

FIGURE 73.1   A schematic of generic protein identification and quantification workflows to interrogate neurovascular unit. *MRM*, Multiple-reaction monitoring; *MS*, mass spectrometry; *SRM*, selected-reaction monitoring.

To achieve these goals, three general research paradigms have been applied: (1) assembly of proteomes of isolated pure cellular and subcellular components of the NVU followed by their in silico reconstruction into models/networks that illustrate NVU functions; (2) proteomic analyses of whole brain vascular tissues (consisting of mixtures of various cellular/acellular NVU components), often from brains affected by pathology; (3) profiling of brain endothelium–specific biomarkers in accessible body fluids as descriptors of cerebrovascular pathologies using both discovery and targeted proteomic approaches. We will briefly illustrate advances in NVU analyses using each of these approaches.

BRAIN ENDOTHELIAL CELL "CARTA"

BECs have been extensively studied using proteomic approaches. Comprehensive proteome repositories of cultured human, rat, and mouse BECs have been built over the past decade. Our (National Research Council of Canada) databases (BEC "Carta")

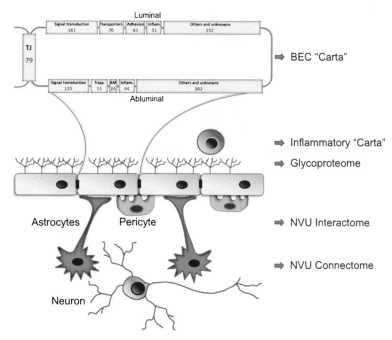

FIGURE 73.2 Neurovascular unit (NVU) proteome "Carta" is a comprehensive data base/atlas of protein composition of the brain endothelial cell (BEC Carta) membranes, intracellular compartments, secreted proteome, and surface glycoproteome, as well as those of interacting cells, including inflammatory cells, astrocytes, and neurons. This database is a resource to build NVU "interactomes" and "connectomes" that could serve as a source of biomarkers and targets to modulate NVU function or pathology. *BEC*, Brain endothelial cell; *NVU*, neurovascular unit; *TJ*, tight junction.

include proteomes and glycoproteomes of luminal and abluminal membranes, lipid rafts, intracellular sorting compartments, microvesicles, and exosomes, and secreted proteomes of primary and immortalized human (HBEC-D3) and rat (SV-ARBEC) BECs under control, hypoxic, and inflammatory conditions [2–4]. A schematic of the BEC proteome map depicting the number and relative abundance of proteins belonging to functionally important protein classes, such as junctional proteins, influx and efflux transporters, adhesion molecules, receptors involved in endocytosis and receptor-mediated transcytosis, and other protein classes is shown in Fig. 73.2, upper panel. This BEC proteome "Carta" has become an important resource to unravel molecular underpinnings of the complex BEC physiology including polarization of drug transport and transcytosis pathways across the BBB. Differential protein expression analyses of BEC exposed to "simulated" pathological conditions, including inflammation [2], cellular hypoxia [3], and oxidative stress [5] facilitated the identification of novel therapeutic targets for brain diseases. For example, proteome analyses of lipid rafts of human BECs identified adhesion molecules ALCAM (CD166) and MCAM (CD146) as pathogenic mediators of lymphocyte recruitment in experimental and clinical multiple sclerosis [6,7]. To further facilitate the discovery of "druggable" interactions between subsets of T cells and BECs we have built an intercellular protein–protein interaction network of "contact" and "communication" interactomes between patient-derived Th17 lymphocytes and human BECs activated by inflammatory insults [2,8]. Among more than 4500 unique proteins and glycoproteins identified in the human BEC, 25–30% responded significantly to the inflammatory stimulation; in addition, 2850 cellular, 1875 membrane, and 450 secreted proteins and glycoproteins were identified in the interleukin-23-activated Th17 cells [8]. More than 180 interacting protein pairs were found between Th17 membrane proteins and human BEC glycocalyx proteins as potential "accessible" targets for disrupting Th17 adhesion and transmigration across the BBB [8]. This example demonstrates how similar in silico-reconstructed interactomes of the NVU could be exploited to develop therapeutic strategies for cerebrovascular and neuroinflammatory diseases.

PROTEOME OF CEREBRAL VESSELS

The alternative to examining the proteome of a specific, pure cellular component of the NVU is the analysis of intact vessels isolated from normal or pathology-affected brains. Although the exact cellular origin of identified proteins cannot be fully discerned using this approach, the advantages include better preservation of in vivo phenotype and dissecting the effects of disease on the NVU in situ. The microvessels are usually separated from the brain tissue using gradient centrifugation methods, whereas larger vessels/arteries could be obtained by dissection. For example, one study [9] published detailed proteome of arteries dissected from the circle of Willis of 6-month-old mice using two proteomics approaches, gel-free nanoLC-MS/MS and gel-based GelLC-MS/MS, to maximize the protein coverage. This provided an opportunity to compare and contrast the protein "makeup" of pial/extracerebral arteries with the previously published proteome of intraparenchymal brain vessels [10]. Comparatively, the most striking difference between the arterial and microvascular data sets was the approximately ninefold increase in the number of unique BBB-specific (e.g., "TJ and adhesion proteins" and "membrane transporter and channel proteins") proteins in the later, suggesting an important difference in the BBB "specialization" between cerebral arteries and intracortical microvessels.

Brain vessels in situ can also be collected from brain sections using rapid staining- [11] or immuno-laser-capture microdissection (LCM) protocols [12]. When coupled to appropriate sample preparation and sensitive MS methods adjusted to small samples, this approach provides a unique glimpse into in vivo changes of the NVU proteome, avoiding artifacts associated with vessel isolation from brain homogenates. For example, isotope labeling–based proteomic analyses of LCM-captured brain vessels in the animal model of global cerebral ischemia/reperfusion identified seven dynamic temporal patterns of vascular protein changes that participated in events leading to the BBB disruption, inflammatory endothelial activation, and initiation of cell cycle and angiogenesis [11].

TRANSLATIONAL PROTEOMICS: BIOMARKERS FOR CARDIOVASCULAR DISEASES

The cerebrovascular surface area (>20 m^2 in human) is exquisitely enriched in thick glycocalyx and unique proteins providing an ample source of potential circulatory brain-vessel specific biomarkers measurable by MS techniques. The molecular nature of these biomarkers ranges from soluble extracellular domains and glycans shed from uniquely glycosylated BBB proteins to microvesicles/exosomes released from BECs (Fig. 73.3). The circulatory exosomes have drawn much attention as potential tissue- and disease-specific microcontainers of both proteomic and genomic, notably microRNA, biomarkers. BEC-derived exosomes contain more than 500 unique proteins not cataloged in other tissues [13] (Fig. 73.3). BEC exosomes could also "externalize" brain-specific biomarkers into the

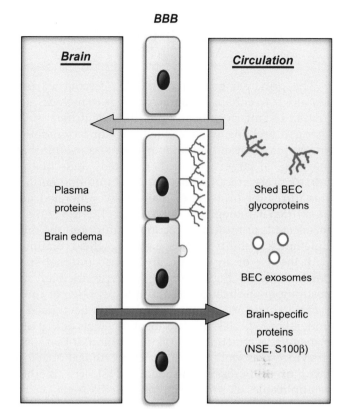

FIGURE 73.3 Neurovascular unit (NVU)-specific peripheral biomarkers of disease. The expansive surface area of brain vessels exposed to both brain and circulatory compartments is an exquisite source of circulatory biomarkers that could be used to evaluate clinical disease. Pathological processes in majority of brain diseases either affect or originate in brain vessels, leading to changes in molecular and functional attributes of the NVU. The blood–brain barrier (BBB) is often disrupted leading to plasma protein and water extravasation and brain edema, as well as efflux of brain-specific proteins into the blood stream. Pathology-induced modifications of the NVU could be monitored by analyzing brain endothelial cell (BEC)-specific shed glycoproteins or extracellular microvessicles and exosomes in accessible body fluids, a rich source of disease biomarkers. BEC exosomes have been shown to express over 500 unique protein biomarkers not present in exosomes of other cells/tissues.

blood stream during pathological conditions. They also play roles in transcytosis of blood-borne molecules into the brain and in cell–cell communication within the NVU.

Beyond BEC-specific biomarkers, brain pathologies, notably ischemic stroke or traumatic brain injury (TBI), often disrupt the BBB sufficiently to allow the egress of brain-specific proteins, such as neuron-specific enolase (NSE) or S100B, into circulatory compartment where they can be monitored by targeted proteomics as biomarkers of BBB disruption and repair. Biomarker responses to therapeutic intervention could be used as "surrogates" for clinical outcomes [14] to support clinical decisions and personalize therapeutic treatments [15].

QUANTITATIVE TARGETED PROTEOMICS: FROM DISCOVERY TO APPLICATION

The principal goal of the NVU "discovery" proteomics is to enhance protein coverage to include rare, low abundant, specifically modified or differentially changed proteins, whereas the transition to validation and application of identified proteins as useful pharmacological targets requires "targeted" and "quantitative" proteomics. In this approach, unique peptide signatures for proteins of interest are generated and quantified in biological samples using LC–MS "tuned" to selected- or multiple-reaction monitoring; the multiplexing capability of this method, when combined with the use of an isotopically labeled internal standard, achieves highly sensitive, reproducible, and absolute quantification of multiple peptides of interest in a single MS run. This method has been used to understand species-specific absolute levels of key BBB transporters affecting central disposition of therapeutics [16,17], as well as to support preclinical development of BBB-crossing antibodies [18]. The method is applicable in preclinical pharmacokinetic, target exposure, and protein stability studies, as well as in the clinical monitoring of peptide/protein biomarker sets. Expansion of targeted LC–MS/MS applications into clinical laboratories will likely grow in scope and importance exponentially.

CONCLUSION

MS techniques, from "discovery" to "application" mode, have found a substantial utility in studies of the NVU and cardiovascular diseases. The principal outcomes have been increased understanding of the molecular and functional complexity of the brain vascular bed, identification of targets for developing imaging or interventional approaches, and discovery of brain vessel–specific biomarkers deployable in pharmacoproteomic studies and clinical trials. The "bedside" application of large data sets generated by whole-proteome interrogations of the NVU in cellular, preclinical, and patient studies requires reduction, rationalization, and confirmation of clinical utility.

References

[1] Stanimirovic DB, Friedman A. Pathophysiology of the neurovascular unit: disease cause or consequence? J Cereb Blood Flow Metab 2012;32:1207–21.
[2] Haqqani AS, Stanimirovic DB. Intercellular interactomics of human brain endothelial cells and Th17 lymphocytes: a novel strategy for identifying therapeutic targets of CNS inflammation. Cardiovasc Psychiatry Neurol 2011;2011:175364.
[3] Haqqani AS, Kelly J, Baumann E, Haseloff RF, Blasig IE, Stanimirovic DB. Protein markers of ischemic insult in brain endothelial cells identified using 2D gel electrophoresis and ICAT-based quantitative proteomics. J Proteome Res 2007;6:226–39.
[4] Haqqani AS, Hill JJ, Mullen J, Stanimirovic DB. Methods to study glycoproteins at the blood–brain barrier using mass spectrometry. Methods Mol Biol 2011;686:337–53.
[5] Ning M, Sarracino DA, Kho AT, Guo S, Lee S-R, Krastins B, et al. Proteomic temporal profile of human brain endothelium after oxidative stress. Stroke 2011;42:37–43.
[6] Cayrol R, Wosik K, Berard JL, Dodelet-Devillers A, Ifergan I, Kebir H, et al. Activated leukocyte cell adhesion molecule promotes leukocyte trafficking into the central nervous system. Nat Immunol 2008;9:137–45.
[7] Larochelle C, Cayrol R, Kebir H, Alvarez JI, Lécuyer M-A, Ifergan I, et al. Melanoma cell adhesion molecule identifies encephalitogenic T lymphocytes and promotes their recruitment to the central nervous system. Brain 2012;135:2906–24.
[8] Haqqani AS, Stanimirovic DB. Prioritization of therapeutic targets of inflammation using proteomics, bioinformatics, and in silico cell–cell interactomics. Methods Mol Biol 2013;1061:345–60.
[9] Badhwar A, Stanimirovic DB, Hamel E, Haqqani AS. The proteome of mouse cerebral arteries. J Cereb Blood Flow Metab 2014;34:1033–46.
[10] Chun HB, Scott M, Niessen S, Hoover H, Baird A, Yates J, et al. The proteome of mouse brain microvessel membranes and basal lamina. J Cereb Blood Flow Metab 2011;31:2267–81.
[11] Haqqani AS, Nesic M, Preston E, Baumann E, Kelly J, Stanimirovic D. Characterization of vascular protein expression patterns in cerebral ischemia/reperfusion using laser capture microdissection and ICAT-nanoLC-MS/MS. FASEB J 2005;19:1809–21.
[12] Lu Q, Murugesan N, Macdonald JA, Wu S-L, Pachter JS, Hancock WS. Analysis of mouse brain microvascular endothelium using immuno-laser capture microdissection coupled to a hybrid linear ion trap with Fourier transform-mass spectrometry proteomics platform. Electrophoresis 2008;29:2689–95.
[13] Haqqani AS, Delaney CE, Tremblay T-L, Sodja C, Sandhu JK, Stanimirovic DB. Method for isolation and molecular characterization of extracellular microvesicles released from brain endothelial cells. Fluids Barriers CNS 2013;10:4.
[14] Haqqani AS, Hutchison JS, Ward R, Stanimirovic DB. Biomarkers and diagnosis; protein biomarkers in serum of pediatric patients with severe traumatic brain injury identified by ICAT-LC-MS/MS. J Neurotrauma 2007;24:54–74.
[15] Ning MM, Lopez M, Sarracino D, Cao J, Karchin M, McMullin D, et al. Pharmaco-proteomics opportunities for individualizing neurovascular treatment. Neurol Res 2013;35:448–56.
[16] Ohtsuki S, Ikeda C, Uchida Y, Sakamoto Y, Miller F, Glacial F, et al. Quantitative targeted absolute proteomic analysis of transporters, receptors and junction proteins for validation of human cerebral microvascular endothelial cell line hCMEC/D3 as a human blood–brain barrier model. Mol Pharm 2013;10:289–96.
[17] Kubo Y, Ohtsuki S, Uchida Y, Terasaki T. Quantitative determination of luminal and abluminal membrane distributions of transporters in porcine brain capillaries by plasma membrane fractionation and quantitative targeted proteomics. J Pharm Sci 2015;104:3060–8.
[18] Farrington GK, Caram-Salas N, Haqqani AS, Brunette E, Eldredge J, Pepinsky B, et al. A novel platform for engineering blood–brain barrier-crossing bispecific biologics. FASEB J 2014;28:4764–78.

CHAPTER

74

Blood Biomarkers in Acute Stroke

C. Foerch

Goethe University, Frankfurt am Main, Germany

INTRODUCTION

Blood biomarkers may improve the diagnostic workup and may facilitate therapeutic decision making in acute stroke patients. This includes the differentiation of "true" stroke patients from stroke mimics, the discrimination between intracerebral hemorrhage (ICH) and ischemic stroke, the prediction of complications of thrombolysis, the etiologic classification of ischemic stroke, and the prediction of recurrent events in patients having a first transitory ischemic attack (TIA). Since 1990s, significant progress has been achieved in biomarker research in these potential fields of application. Here, an overview on relevant studies is provided.

METHODOLOGICAL ASPECTS

Most research on blood biomarkers in stroke has been conducted using three different methodological strategies [1]. The first is to search for a single protein biomarker or a protein biomarker panel to be used as a diagnostic test. This hypothesis-driven approach includes proteins associated with glial or neuronal tissue injury (such as glial fibrillary acidic protein (GFAP), S100B, and neurofilament protein), blood–brain barrier damage (matrix metalloprotease 9 (MMP-9)), inflammation (C-reactive protein, CRP, interleukin 6, IL-6), and coagulation (fibrinogen, D-dimer). They are typically measured on standard enzyme-linked immunosorbent assay (ELISA) or multiplex ELISA platforms. The second strategy is based on proteomics technology, which allows a hypothesis-free approach for protein biomarker research. Analyses are typically performed using two-dimensional gel electrophoresis and mass spectrometry. Several promising candidate biomarkers have been identified by this approach, but prospective validation in independent cohorts of stroke patients was not always successful. The third approach is to use modern high-throughput microarray technologies to study RNA expression profiles in whole blood. Interesting findings from explorative studies have been published, but translation into clinical routine is challenging, mostly due to the high methodological efforts associated with the use of these techniques.

IDENTIFICATION OF PATIENTS WITH ISCHEMIC STROKE

The identification of "true" stroke patients in a collective of patients with symptoms suggestive of acute stroke has always been a prioritized field in blood biomarker research. As ischemic stroke is a heterogeneous disease, it was speculated that a biomarker panel rather than a single biomarker might be able to differentiate between ischemic stroke patients and stroke mimics. Pilot studies identified several protein biomarkers that were found elevated in the blood of patients with ischemic stroke, but not in healthy controls. However, a prospective validation study on more than 1000 patients suspected of having acute stroke revealed disappointing diagnostic accuracy values for a four-marker panel including S100B, MMP, D-dimer, and brain natriuretic peptide (BNP) for differentiating "true" stroke patients from stroke mimics [2]. It is likely that the various stroke mimic conditions including migraine, brain tumors, and epileptic seizures also cause alterations in biomarker levels, which reduce the diagnostic potential of the measure. In 2011, caspase-3 and D-dimer were described as candidate proteins for a panel differentiating between ischemic stroke patients and stroke mimics. Other biomarker panels included eotaxin, epidermal growth factor receptor, and prolactin. However, prospective independent validation of these markers is still missing.

Primer on Cerebrovascular Diseases, Second Edition
http://dx.doi.org/10.1016/B978-0-12-803058-5.00074-6

RNA expression profiles derived from whole blood were also investigated in this context [1]. A 22-gene panel reached a 78% sensitivity and a 80% specificity for differentiating ischemic stroke from controls. Later on, a study found 1335 genes with a different expression pattern between ischemic stroke and controls. A panel consisting of 18 of these 1335 genes differentiated ischemic stroke patients from controls. A critical point is to compare RNA expression between ischemic stroke patients and patients with stroke mimicking conditions (rather than healthy controls). In this context, a study reported that a 97-gene profile may differentiate ischemic stroke patients from patients with myocardial infarction and patients with vascular risk factors.

Potential Applications

A biomarker test that could reliably differentiate between a "true" ischemic stroke patient and a stroke mimic patient will likely improve the triage of acute stroke patients. As of 2016, about 20–30% of all patients treated on a stroke unit are patients admitted under the suspicion of an acute stroke but who turned out to have alternative diagnoses. However, due to the heterogeneity of the disease and the interference of mimicking conditions on the biomarker levels, this field of research is extremely challenging, and no sufficient biomarker test has been established so far for translation into routine clinical practice.

DIFFERENTIATION BETWEEN INTRACEREBRAL HEMORRHAGE AND ISCHEMIC STROKE

In recent years, progress has been made regarding the differentiation of ischemic stroke and ICH by means of serum biomarkers. Pathophysiological studies provided insights into the different kinetics of necrosis. In ICH, hematoma expansion causes rapid cellular destruction and cytolysis, whereas in ischemic stroke necrosis occurs not before 6–12h after symptom onset. Thus, within the first 6h after symptom onset, biomarkers indicating necrosis are supposed to be elevated in ICH and low in ischemic stroke. GFAP as an abundantly present protein in astrocytes has been extensively studied in this context. Pilot studies revealed elevated GFAP values in the serum of ICH patients as compared to patients with ischemic stroke. Later on, a prospective multicenter trial has been performed in order to determine the diagnostic accuracy of such a GFAP test [3]; 205 patients have been included. Median (interquartile) serum GFAP values were 1.91 µg/L (0.41–17.66) in ICH patients and 0.08 µg/L (0.02–0.14) in ischemic stroke patients. A cut-off point of 0.29 µg/L revealed a sensitivity of 84% and a specificity of 96% for differentiating ICH from ischemic stroke. In ischemic stroke, few false-positive GFAP determinations were obtained, likely resulting from falsely reported time windows or an additional brain trauma during stroke occurrence (e.g., a fall). On the other hand, a plausible explanation for the reduced sensitivity (false-negative testing in case of ICH) has not been found so far. One reason might be ICH that predominantly enters the ventricular system, thereby leading to less parenchymal damage. As of 2016, a GFAP point of care test is not available, and studies that determined GFAP in the prehospital phase have not been performed.

The hypothesis that ICH and ischemic stroke can be differentiated by means of GFAP is also supported by studies on other astroglial markers such as S100B (a calcium-regulating protein). It was demonstrated that S100B is slowly released in ischemic stroke, but is rapidly elevated in acute ICH. Further evidence was derived from studies on traumatic brain injury suggesting that GFAP is increased in patients with focal mass lesions (i.e., cerebral hemorrhage) as compared to patients having sheer lesions only. A broad analysis of GFAP in manifold neurological diseases revealed that only diseases with acute cell damage (such as ICH and traumatic brain injury) lead to detectable GFAP release. In contrast, in "pure" subarachnoid hemorrhage (i.e., without parenchymal damage), GFAP was not found elevated in the acute phase. However, GFAP may be used as a prognostic parameter in subarachnoid hemorrhage patients, as it was shown to increase in patients developing vasospasm-associated ischemic strokes.

Biomarker panels were also investigated to differentiate between ICH and ischemic stroke. Apolipoprotein CI and apolipoprotein CIII differentiated ICH from ischemic stroke with a sensitivity of 94% and a specificity of 87%. Recent panels included retinol-binding protein 4 and GFAP, UCH-L1 and GFAP, and biomarkers of the S100B/RAGE pathway.

Potential Applications

A point of care test that reliably differentiates between ischemic stroke and ICH could improve the triage of acute stroke patients. While patients suspected for having ICH might be delivered to larger hospitals with neurosurgical facilities, patients suspected of having ischemic stroke might be referred to nearby stroke units. These aspects are particularly of importance in rural areas. Furthermore, a point of care test may allow rapid (i.e., prehospital) treatment. In particular, blood pressure lowering in patients with suspected ICH may be an easy treatment strategy. In anticoagulated patients who develop an acute stroke syndrome, a positive GFAP testing suspected for ICH may allow rapid reversal of anticoagulation using specific antidots. In contrast, as no biomarker test has been identified so far that is able to rule out ICH, it is not possible to date to treat patients suspected of having ischemic stroke with tissue plasminogen activator (tPA) solely based on a biomarker test, even if the time gain may balance additional risks.

PREDICTING BLEEDING COMPLICATIONS AFTER THROMBOLYSIS

Several studies have addressed the question whether blood biomarkers can predict hemorrhagic complications after thrombolysis. Blood biomarkers may provide unique molecular information on the status of the blood–brain barrier. The MMP-9 has been extensively studied in this context. In general, in stroke patients, higher MMP-9 values are considered to be associated with larger infarct volumes and a worse functional outcome. More specifically, imaging studies demonstrated that elevated MMP-9 values are associated with blood–brain barrier disruption in acute stroke. An association between elevated MMP-9 values and hemorrhagic transformation was first demonstrated in untreated acute stroke patients. Montaner et al. later on investigated whether MMP-9 can be used as a predictor of hemorrhagic complications after thrombolysis in acute stroke. Higher baseline (pretreatment) MMP-9 serum levels were found associated with more severe bleeding complications (i.e., parchenymal hemorrhage) [4].

Cellular fibronectin (c-Fn) is a component of the extracellular matrix and was also studied in this context. A prospective study in tPA-treated stroke patients demonstrated a gradual association between elevated c-Fn levels and hemorrhagic complications after thrombolysis. A similar result was obtained for MMP-9 in this study. Later on, a larger validation study was performed including 134 thrombolyzed acute stroke patients. Both pretreatment serum levels of c-Fn and MMP-9 were higher in patients who later on developed hemorrhagic complications. Sensitivity values for the two markers were high (92–100%), but specificity was not optimal (60–74%) [5].

Potential Applications

A biomarker that predict hemorrhagic complications after thrombolysis would be highly desirable, in order to increase the safety of the therapy. However, considering that the risk of ICH per se is low after thrombolysis and specificity measures of the tests mentioned earlier are not optimal, there might be a risk of inadvertently withholding tPA to stroke patients who would have been otherwise treated. The development of point of care tests for determining these biomarkers in a short period of time appears mandatory. Further developments of this potential field of application will include markers that predict vessel recanalization and infarct growth.

IDENTIFYING ISCHEMIC STROKE ETIOLOGY

Significant progress was achieved regarding the use of biomarkers for identifying stroke etiology. A study that included 707 patients with acute ischemic stroke reported a panel consisting of BNP, soluble receptor for advanced glycation end products, and D-dimer to be associated with cardioembolic stroke [6]. BNP has been further investigated in this context. The cardiac hormone is released from myocytes in case of atrial dilatation, and elevated BNP values were found associated with atrial fibrillation. In 2015, a pooled data metaanalysis was published that confirmed the association between elevated BNP values and cardioembolic strokes and described predictive models with high diagnostic accuracy [7].

A study was published investigating whether gene expression profiles in the peripheral blood can differentiate between cardioembolic and large-vessel causes of stroke, and whether stroke etiology can be predicted in a group of cryptogenic stroke patients [8]. The authors identified a 40-gene profile that differentiated cardioembolic stroke from large-vessel stroke with >95% sensitivity and specificity. Furthermore, a 37-gene profile was found that distinguished cardioembolic stroke due to atrial fibrillation from nonatrial fibrillation causes with >90% sensitivity and specificity. Further studies demonstrated that gene expression profiling can also be used to differentiate between lacunar and nonlacunar strokes and that stroke etiology in patients with cryptogenic strokes can be determined in conjunction with neuroimaging data.

Potential Applications

The availability of anticoagulants to prevent stroke originating from atrial fibrillation stresses the need for tools that reliably identify the presence of atrial fibrillation during the first days after stroke onset. Among strategies that expand the monitoring of heart rhythms, biomarker may play a role in this context. Patients having elevated markers associated with cardioembolic stroke may undergo a more intense screening procedure.

PREDICTING RECURRENT-EVENTS IN PATIENTS WITH TRANSITORY ISCHEMIC ATTACK

Another interesting field of application of blood biomarkers in stroke is the prediction of recurrent-events after a first TIA. Biomarkers could add to existing predictive models that are based on clinical, brain imaging, and carotid imaging data. A 2014 study included 1292 patients with TIA and stroke and evaluated the predictive potential of 14 blood biomarkers [9]. Primary end point was the 90-day risk of recurrent stroke. Overall, only weak associations between the biomarkers and recurrent stroke risk were found, and the predictive potential was considered to be limited.

Copeptin, a hypothalamic stress hormone, has also been evaluated in this context. Higher copeptin levels were found in TIA patients who later on developed a cerebrovascular re-event as compared to patients without re-events. A prospective multicenter study was performed including 302 patients with TIA. Twenty-eight patients had recurrent stroke or TIA within 3 months [10]. The association of copeptin with all recurrent cerebrovascular events was not significant, but copeptin was associated with recurrent strokes. Copeptin improved the prognostic value of the ABCD2 score for the prediction of stroke.

Potential Applications

Similar to what was described earlier for the diagnosis of ischemic stroke, the prediction of re-events after a first stroke or TIA based on biomarkers is challenged by the heterogeneous pathophysiology. However, biomarkers may well add to established prediction models. Large-scaled validation studies are needed.

SUMMARY

In summary, while earlier biomarker studies in stroke rather focused on correlating biomarker levels with infarct volume and prognosis, the field has recently moved to investigate clinically relevant questions. Unfortunately, a "global" stroke biomarker has not been identified. Furthermore, due to the heterogeneity of the disease, a marker or a marker panel for ischemic stroke appear to be difficult to identify. Interesting candidate biomarkers for ICH and a cardioembolic stroke origin have been described, but large-scaled validation studies are missing.

References

[1] Jickling GC, Sharp FR. Blood biomarkers of ischemic stroke. Neurotherapeutics 2011;8:349–60.
[2] Laskowitz DT, Kasner SE, Saver J, Remmel KS, Jauch EC, Group BS. Clinical usefulness of a biomarker-based diagnostic test for acute stroke: The biomarker rapid assessment in ischemic injury (brain) study. Stroke 2009;40:77–85.
[3] Foerch C, Niessner M, Back T, Bauerle M, De Marchis GM, Ferbert A, et al. Diagnostic accuracy of plasma glial fibrillary acidic protein for differentiating intracerebral hemorrhage and cerebral ischemia in patients with symptoms of acute stroke. Clin Chem 2012;58:237–45.
[4] Montaner J, Molina CA, Monasterio J, Abilleira S, Arenillas JF, Ribo M, et al. Matrix metalloproteinase-9 pretreatment level predicts intracranial hemorrhagic complications after thrombolysis in human stroke. Circulation 2003;107:598–603.
[5] Castellanos M, Sobrino T, Millan M, Garcia M, Arenillas J, Nombela F, et al. Serum cellular fibronectin and matrix metalloproteinase-9 as screening biomarkers for the prediction of parenchymal hematoma after thrombolytic therapy in acute ischemic stroke: a multicenter confirmatory study. Stroke 2007;38:1855–9.
[6] Montaner J, Perea-Gainza M, Delgado P, Ribo M, Chacon P, Rosell A, et al. Etiologic diagnosis of ischemic stroke subtypes with plasma biomarkers. Stroke 2008;39:2280–7.
[7] Llombart V, Antolin-Fontes A, Bustamante A, Giralt D, Rost NS, Furie K, et al. B-type natriuretic peptides help in cardioembolic stroke diagnosis: pooled data meta-analysis. Stroke 2015;46:1187–95.
[8] Jickling GC, Xu H, Stamova B, Ander BP, Zhan X, Tian Y, et al. Signatures of cardioembolic and large-vessel ischemic stroke. Ann Neurol 2010;68:681–92.
[9] Segal HC, Burgess AI, Poole DL, Mehta Z, Silver LE, Rothwell PM. Population-based study of blood biomarkers in prediction of subacute recurrent stroke. Stroke 2014;45:2912–7.
[10] De Marchis GM, Weck A, Audebert H, Benik S, Foerch C, Buhl D, et al. Copeptin for the prediction of recurrent cerebrovascular events after transient ischemic attack: results from the coRisk study. Stroke 2014;45:2918–23.

CHAPTER

75

Rodent Behavioral Tests Sensitive to Functional Recovery and Compensation

T. Schallert[1], D.L. Adkins[2]

[1]The University of Texas at Austin, Austin, TX, United States; [2]Medical University of South Carolina, Charleston, SC, United States

INTRODUCTION

Research scientists rely upon rodent models to help us better understand and test potential treatments for stroke survivors. Although there are limitations in the use of rodents to study human neurological disorders, it is possible to identify and focus on the functional overlaps and similarities between species' brains and behaviors when designing studies investigating stroke prevention, acute neuroprotection, and chronic poststroke interventions. There is overwhelming evidence that the fundamental organization of the rat brain, for instance, is not that different from humans in that there is much overlap in sensory and motor systems, and cognitive, emotion, and attention processes. Researchers can utilize both species-typical and cross-species behavioral patterns to sensitively probe the impact of stroke and determine the efficacy of possible clinically relevant treatments through careful selection and analysis of tests of rodents' behaviors and by making careful comparisons to similar human behavior or cognitive processes.

Common rodent models of focal ischemic or hemorrhagic stroke cause damage to the sensorimotor cortex and/or striatum, producing sensory and motor impairments primarily in forelimb or hind limb function, and in more severe models, experimental stroke can damage the hippocampus and related brain structures, producing impairments in cognitive domains. The severity of injury and the resolution of impairments through interventions or treatments can be assessed utilizing well-established rodent behavioral tests in these stroke models (see Refs. [1–4]). However, one major criticism of using rodent models for stroke research has been that

often researchers report that rodents quickly "recover" from stroke-induced impairments, whereas humans do not. In most cases, this recovery is not true brain repair or functional recovery, but actually is a demonstration of the ability of the animal to learn new compensatory behaviors that mask ongoing impairments. Rodents, as is also true with humans, are really good at compensating for impairments, and thus many studies have reported "recovery" on tasks over time when in fact with careful methods, ongoing impairments are unmasked and compensatory behaviors can be revealed [1,2]. Through careful selection of behavioral tests, usually using both quantitative and qualitative methods of analysis, it is possible to more sensitively distinguish between behavioral patterns used to compensate for impairments versus true brain repair or remodeling. There is no agreed-upon battery of neurological tests that are routinely used to screen for clinically effective treatments. However, there are a number of well-established behavioral tests commonly used that are sensitive to lesion severity or lesion location. Some of these tests are more suited for acute studies, whereas others can reveal chronic ongoing impairments in function. There are also varying degrees to which these tests can be used to distinguish between recovery of function and compensatory behaviors.

In this brief review, we will describe in more detail a few commonly used tests of rodent sensory, motor, and cognitive function. These specific behavioral tests have been chosen for further discussion because they can provide both quantitative and qualitative data or methods to distinguish compensatory behaviors from true recovery of function. Additionally, we have compiled a more extensive list, although not exhaustive, of common

Primer on Cerebrovascular Diseases, Second Edition
http://dx.doi.org/10.1016/B978-0-12-803058-5.00075-8

poststroke rodent behavioral assays and indicate some of their limitations and strengths and whether they are best for rats, mice or both (Table 75.1; adapted and expanded from Ref. [4]).

FORELIMB FUNCTION TASKS

Hemiparesis of the hand and arm are the most prevalent, enduring, and disabling impairments following stroke in humans and there are many well-established tests developed to assess upper-extremity impairments in rodents following different unilateral stroke models ([2]; Table 75.1). One highly reliable task is the single-pellet reaching task. Although it is more time consuming and requires prior training, the test is sensitive to chronic upper-extremity impairments, including digit function, and can be used to reveal compensatory behaviors and recovery of normalized movement patterns.

Single-Pellet Reaching Task

Although there are several variations, most single-pellet reaching tasks for mice and rats are fairly similar in their general methods for administering the task, collecting data, and analyzing the data (e.g., Refs. [5,6]). Because this is a food-motivated task, animals are first food restricted (~85–90% initial starting weight), acclimated to a clear-reaching chamber and then are trained to reach through a narrow window to retrieve food pellets (or seeds for mice). Pellets are placed in a shallow well on a shelf placed outside the window at a distance between ~1 and 2 cm. Once animals show a limb preference, researchers can encourage the use of just that one limb by inserting a wall in the cage ipsilateral to the reaching limb or placing a bracelet or cast on the non-reaching limb to inhibit that limb's ability to reach the pellet. Over several days to weeks, animals are trained to reach through the window, grasp the pellet, withdraw their limb and place the pellet in their mouths. This is defined as a "Success." Animals are trained to an asymptotic level of ~40–60% success rate, maintained over two to three consecutive days. A success rate is calculated as the percentage of successful reaches of the total number of reach attempts [(total successes/total reach attempts) ×100]. Quantitative data can also be collected which includes the number of "Failures" and "Drops". A Failure is when an animal either reaches the upper limit of times without grasping the pellet or reaches for the pellet and knocks it out of the well. A Drop is when an animal grasps the pellet, retracts the arm into the cage but fails to place the pellet into its mouth. Drops may also denote a sensory impairment because the animal fails to recognize that a pellet is in its paw.

When assessing limb impairments and recovery, a minimum of 10–30 trials should be used. The specific number of trials depends upon the frequency of test administration and study design. When testing the efficacy of an intervention or treatment, the assignment of animals to groups should be counterbalanced to ensure reaching performance is matched at baseline (before stroke induction), poststroke, and prior to treatment (when possible). This task is highly sensitive to distal forelimb impairments for up to 11 months poststroke, even following fairly small unilateral strokes in rats [7]. There are several variations of skilled reaching tasks, and other reach-to-grasp tasks, that can also be used to evaluate forelimb dexterity [2].

Single-Pellet Reaching Movement Analysis

The single-pellet reaching task not only provides robust data on the rat's ability to perform the skilled forelimb task successfully, but also allows researchers to perform qualitative assessment of how normal or abnormal components of reaching movement are. Rodents, as with humans and nonhuman primates, have a fairly standard pattern of movements that make up a successful reach. Following injury, many rodents (and patients) compensate for impairments by altering standard movement patterns and employing compensatory behaviors. Using a movement rating scale, developed by Whishaw and colleagues [5], researchers can sensitively detect forelimb movements that reveal enduring impairments or compensatory strategies in reaching and grasping motor action patterns following brain injury by analyzing reaching patterns via slow-motion video playback.

Reaching movement analysis is rated through slow-motion playback of successful reaches. Data should be collected from baseline and then at appropriate poststroke and posttreatment time points. On average, 3–10 successful reach attempts are needed per time point to formulate an average abnormality score for each movement component that makes up a successful reach. Although there is variation in the number of components assessed in the literature, most studies break each reach into eight movement components: (1) Aim: The elbow is adducted while the digits are aligned with the midline of the reaching window and are oriented toward the food pellet. (2) Advance: The limb is advanced directly through the reaching window, initially above and beyond the food pellet. (3) Digits open: As the limb is advanced, the digits open and extend toward the pellet (4) Pronation: The wrist pronates over the pellet. (5) Grasp: The pads of the palm or the digits touch the food and the food is grasped by closure of the digits around the pellet. (6) Supination 1: The paw is dorsiflexed and supinated 90 degrees as the limb is withdrawn through

TABLE 75.1   Tests of Experimental Stroke Impairments

Tests	Symptoms Evaluated	Sensitive Acutely/ Chronically	Species	Disadvantages	Advantages
SENSORIMOTOR TESTS					
Adhesive removal test	Sensory	+/+	R & M	Training needed; less sensitive in models without a cortical injury	High sensitivity long after ischemic damage, including MCAO models with small cortical damage
Beam walking tests	Motor coordination	+/+	R & M	Training needed; compensatory bias in beams without a ledge	Good sensitivity in rats and mice long after ischemic damage; addition of a ledge increases sensitivity
Corner test	Whiskers sensitivity	+/+	M	Less sensitive in models with cortical lesions; low evidence in rats	Simple and fast; no training needed; evaluates long-term dysfunctions in striatal infarcts
Rotarod test	Motor coordination and balance	+/+	R & M	Low sensitivity in mice 72 h after MCAO; confounding factors; training needed	Good sensitivity and reliability in rats
Bilateral tactile stimulation test	Tactile discrimination	+/−	R & M	Task can be time consuming	Magnitude of sensory asymmetry is sensitive to minor impairments, long-lasting deficits in MCAO, and focal ischemia
Open field test	Locomotor activity	+/−	R & M	Low sensitivity with small infarcts; only useful in short-term protocols	Assesses anhedonia and stress
Cylinder test	Motor coordination	+/−	R & M	Decreased sensitivity in mice with small infarcts	Easy and fast to apply; it is sensitive 1 month after ischemia
Forelimb placing	Vibrissa evoked limb placement	−/+	R	It is subjective (double blind); sensitivity decreases due to a high spontaneous recovery	Simple and fast; detects neurological impairments in models of striatal and/or cortical lesions
Swim task	Forelimb inhibition	+/+	R		Little training needed; resistant to recovery
Staircase/ Montoya test	Fine motor coordination and sensory neglect	+/+	R & M	Long training needed; Large number of excluded animals	Good sensitivity and reliability in MCAO models with small infarcts, and long after the surgery
Single-pellet	Fine motor coordination, sensory neglect, forelimb movement abnormalities	+/+	R & M	Long training times; time-consuming video analysis	Great sensitivity and reliability up to at least 11 months after focal damage
Pasta handling	Manual dexterity; oral motor deficits	+/+	R & M	Must acclimate animals to being filmed while eating; can be difficult to quantify	Simple to administer; no shaping or training
Foot fault	Motor coordination	+/−	R & M	Poor sensitivity in mild ischemia	Very sensitive in severe ischemia models
Ladder task	Forelimb and hind limb function; motor coordination and paw grip	+/+	R & M	Animals learn to compensate unless ladder rung spacing is varied	Minimal training; qualitative and quantitative data; sensitive chronically
Elevated body swing test	Muscle strength	+/+	R & M	Less experience in mice	Simple and fast; no training needed; sensitive even 30 days after MCAO

Continued

TABLE 75.1 Tests of Experimental Stroke Impairments—cont'd

Tests	Symptoms Evaluated	Sensitive Acutely/ Chronically	Species	Disadvantages	Advantages
COGNITIVE TESTS					
Morris water maze	Spatial learning and memory	+/+	R & M	Less sensitive in mice 72h after MCAO; training needed	Very good sensitivity in rats long after ischemia
Passive avoidance tests	Avoidance learning	+/+	R & M	Expensive equipment; training needed	Very high sensitivity long after ischemia in rats and mice; easy to perform
NEUROLOGICAL SCALES					
Modified neurological severity score	General neurological assessment	+/−	R & M	Low sensitivity at late times after ischemia; subjective	Gives an overall degree of ischemic injury; easy and fast

Tests of experimental stroke impairments. Summary of several commonly used behavioral tests used to assess neurological damage following experimental stroke in rats (R) and mice (M), including tests that have shown sensitivity (+) during the acute and chronic periods after stroke.
MCAO, Middle cerebral artery occlusion.
Adapted from Zarruk JG, Garcia-Yebenes I, Romera VG, et al. Neurological tests for functional outcome assessment in rodent models of ischaemic stroke. Rev Neurol 2011;53(10):607–18.

the reaching window. (7) Supination 2: The paw is supinated again by approximately 45 degrees to bring the pellet to the mouth. (8) Release: The digits are opened and the pellet is released into the mouth. Each movement is then rated with a score of 0 (normal), 0.5 (slightly abnormal), and 1 (absent or highly abnormal). Data then can be analyzed as the total abnormality score across all the reaching components (averaged over trials, usually five successful reaches).

The Cylinder Task

The cylinder task examines the relative symmetry of forelimb use for postural support in rodents during exploration within a tall cylindrical testing apparatus. This task does not require prior training. Briefly, rats and mice are placed in a transparent cylinder placed either over a mirror or on a flat surface. Animals are then allowed to explore the environment for several minutes (on average 2 min or 20 wall touches) while being videotaped. Through slow-motion video replay, the researcher records the number of times each forelimb is used together or individually for weight support during rears and lands and weight support or shifting with the forelimbs against the cylinder wall during vertical exploration. Before injury, rodents use each limb fairly evenly, but after damage to the motor cortex and/or striatum, rats and mice begin relying upon the contralesional limb for weight support. This test is sensitive to long-term deficits, small unilateral strokes, and scores do not change with repeated testing [1,2,6,7]. Animals will often compensate for contralesional limb impairments by relying upon both limbs together, especially for wall support. Thus to account for this compensatory behavior, limb use can be calculated as

follows: [(ipsi-limb + ½ both limbs)/(ipsi-limb + contra-limb)] × 100]. Factoring in the compensatory behavior, the sensitivity of the task to longer-term or subtle impairments can be recorded.

HIND LIMB FUNCTION TASKS

An estimated one of three stroke survivors are unable to walk independently and up to two-thirds may experience critical limitations in functional walking and be at risk for additional declines in physical mobility and independent walking. There are several tests available to test hind limb function and locomotion following experimental strokes in rodents. Although rodents are quadrupeds and have a lower center of gravity, likely altering the neuromechanics of walking between species, stroked-induced changes in the basic functions of the hind limbs and species-specific locomotion can be assessed and used to determine efficacy of treatments.

Foot Fault or Ladder Rung Task

There are several forms of foot fault tasks that effectively assess forelimb and hind limb function during locomotion. Generally, while animals are being videotaped, they are encouraged to walk across an elevated grid or ladder that requires them to carefully place their paws onto metal or plastic grid edges or ladder rungs. Through the use of slow-motion video replay, the number of steps taken and the number of slips with any of the four paws can be collected.

The ladder rung walking task also allows sensitive discrimination between subtle disturbances of motor function by combining qualitative and quantitative analyses of skilled walking. Limb placement is scored by assessing the paw position on the ladder rung on a 0–6 scale, as follows: (0) Total miss. Zero points are given when a limb completely misses a rung and the limb falls through the rungs. (1) Deep slip. One point is given if the limb is placed on a rung and then slips off when weight bearing and causes a fall. (2) Slight slip. Two points are scored when a limb is placed on a rung, slips off when weight bearing, but does not result in a fall. (3) Replacement. Three points are given when the limb is placed on a rung, but before weight bearing, it is quickly lifted and placed on another rung. (4) Correction. Four points are scored when the limb is aimed at one rung, but then is placed on a different rung without touching the first one or four points are given if a limb is placed on a rung and is quickly repositioned while on the same rung. (5) Partial placement. Five points are recorded if the heel or toes of the hind limb (or wrist or digits of the forelimb) are placed on the rung. (6) Correct placement. Six points are assigned when the midportion of the palm of a limb is placed on the rung with full weight support.

By altering the spacing of the rungs, researchers can challenge animals, reducing learned compensatory strategies, such as body support with the tail and nonimpaired limb. The test can also reveal compensation through nonimpaired foot placement errors. Thus, varied rung placement improves the sensitivity of the task for chronic assessment of hind and forelimb function, coordination, and paw/foot impairments [8].

Ledge Beam Test

Hind limb use can be assessed by encouraging rats to walk down a narrow (1.5 cm wide) elevated beam toward their home cage or a darkened goal cage. The elevated beam can be fitted with a step-down ledge to serve as a "crutch" that the animal can then use to catch a slipping limb (2 cm wide and 2 cm below main beam). The animal's reliance upon this crutch discourages the development of other compensatory behaviors, i.e., tail or ipsilateral limb.

Training an animal to walk the beam begins by placing the rat near the goal and then over trials increasing the distance between the goal and the animal's starting place. Noninjured animals will keep all four limbs on the main beam surface while they walk; however, injured animals will chronically use the ledge to support their impaired hind limb. The number of foot faults per step are scored for each hind limb. Although there can be some recovery over the first couple of weeks after stroke, use of the ledge does continue in most models of injury. However, researchers can remove the ledge and the deficits usually reappear until the animal begins using compensatory behaviors (i.e., tail positioning) to mask foot

faults. After compensatory behaviors appear, the ledge can be replaced and the reliance with the impaired hind limb reappears. Thus, it is possible that by testing with and without the ledge, true brain recovery versus compensatory motor learning may be distinguished [1].

COGNITIVE TASKS OVERVIEW

Strokes in humans commonly result in considerable impairments in learning, memory, mental flexibility (perseveration), and inhibitory control. These cognitive dysfunctions are also common in animal models of stroke that causes hippocampal and/or frontal cortical damage. The Morris water maze [9] is a commonly used test in mice and rats and provides both quantitative and qualitative data that can reveal impairment severity, compensatory strategies, and location of damage. In brief, rats or mice are placed in a tank of water and are required to learn and remember cues (outside or inside the tank) to find an escape platform submerged in the water at a fixed location. When first placed in the tank, most animals will naturally swim toward a visible structure that might provide an escape route, thus most animals first search for an escape by swimming along the wall of the tank (thigmotaxis). When this strategy fails, they begin to search the interior of the tank, using several different search strategies [1]. Over many trials, the animal will become quicker at finding the hidden platform and eventually learns to associate cues in the room with the location of the platform. To test for spatial memory, the platform is removed and the amount of time the animal spends in the quadrant in which the platform had been is measured. Animals with damage to the hippocampus or related areas often spend less time in the correct quadrant. Moving the platform within one quadrant or to other quadrants can provide valuable data on perseveration. Often, when the platform is first moved, rodents will return to the thigmotaxic search strategy briefly before searching the pool. Hippocampus damage alters how quickly and efficiently rodents switch between search strategies [1]. Thus, qualitative analysis of an animal's search strategy can reveal impairments such as perseveration.

CONCLUSION

The tasks discussed in this chapter highlight tests that provide sensitive quantitative and qualitative data on fore and hind limb function, locomotion, and cognition. Although this is not a comprehensive discussion of all common behavioral tests used in stroke research, we did highlight those tests that provide sensitive measures of both functional recovery and compensation. In Table 75.1, we include a larger list of tasks used to assess poststroke impairments in rodent models and include their strengths and limitations and if they are sensitive for acute and/or chronic studies. Although there will always be limitations in using quadrapedal animals to model human neurological disorders, we can improve our preclinical models by understanding the strengths and weaknesses of the tasks and build into our tasks a way to measure compensatory behaviors that often mask enduring impairments.

References

[1] Schallert T. Behavioral tests for preclinical intervention assessment. NeuroRx 2006;3(4):497–504.

[2] Kleim JA, Boychuk JA, Adkins DL. Rat models of upper extremity impairment in stroke. Ilar J 2007;48(4):374–84.

[3] Schaar KL, Brenneman MM, Savitz SI. Functional assessments in the rodent stroke model. Exp Transl Stroke Med 2010;2(1):13.

[4] Zarruk JG, Garcia-Yebenes I, Romera VG, et al. Neurological tests for functional outcome assessment in rodent models of ischaemic stroke. Rev Neurol 2011;53(10):607–18.

[5] Whishaw IQ, Pellis SM, Gorny BP, Pellis VC. The impairments in reaching and the movements of compensation in rats with motor cortex lesions: an endpoint, videorecording, and movement notation analysis. Behav Brain Res 1991;42(1):77–91.

[6] Adkins DL, Hsu JE, Jones TA. Motor cortical stimulation promotes synaptic plasticity and behavioral improvements following sensorimotor cortex lesions. Exp Neurol 2008;212(1):14–28.

[7] O'Bryant AJ, Adkins DL, Sitko AA, Combs HL, Nordquist SK, Jones TA. Enduring poststroke motor functional improvements by a well-timed combination of motor rehabilitative training and cortical stimulation in rats. Neurorehabil Neural Repair 2014:30(2);143–54.

[8] Metz GA, Whishaw IQ. The ladder rung walking task: a scoring system and its practical application. J Vis Exp 2009;(28). http://dx.doi.org/10.3791/1204. [pii: 1204].

[9] Bromley-Brits K, Deng Y, Song W. Morris water maze test for learning and memory deficits in Alzheimer's disease model mice. J Vis Exp 2011;(53). http://dx.doi.org/10.3791/2920. [pii: 2920].

CLINICAL CHAPTERS

CLINICAL ASPECTS: MEDICAL AND SURGICAL

76

Transient Focal Neurological Events

C. Stephen[1], L.R. Caplan[2]

[1]Massachusetts General Hospital, Boston, MA, United States; [2]Harvard University, Beth Israel Deaconess Medical Center, Boston, MA, United States

The most common temporary focal neurological events are transient ischemic attacks (TIAs), migrainous accompaniments, seizures, and functional psychogenic episodes [1].

TRANSIENT ISCHEMIC ATTACKS

A TIA is defined as a transient episode of neurological dysfunction caused by focal brain, spinal cord, or retinal ischemia, without acute infarction [2,3]. In past years TIAs were arbitrarily characterized as lasting less than 24h. Modern brain and vascular imaging showed that ischemic attacks that lasted longer than 1h were most often associated with brain infarction. Most TIAs last less than 1h.

Symptoms

TIAs are caused by decreased blood flow to a local portion of the brain. Decreased perfusion can be due to diminished blood flow to a brain region because of in situ occlusion of a supply artery, or from embolism to that artery. Symptoms are focal, suggesting that they are attributable to dysfunction of a localized area of the brain. Symptoms are transient when the arterial blockage passes (e.g., following dissolution or distal passage of an embolus) or when the collateral circulation is able to restore adequate perfusion to the ischemic region. The symptoms and signs can fluctuate depending upon the adequacy of perfusion, which is in turn related to systemic factors (blood volume, cardiac output, blood pressure, blood viscosity) and local factors (e.g., propagation and embolization of clot, development of collateral circulation).

The symptoms of TIAs depend upon the underlying vascular cause. When ischemia is caused by penetrating artery disease, for example, the symptoms are usually stereotyped (e.g., numbness of the face, arm, and leg on one side of the body, or hemiparesis). In contrast, attacks due to large artery occlusive disease have different symptoms in different attacks:

- When the carotid artery is narrowed or occluded, patients may have monocular blindness on the side of the occlusive lesion, weakness of the contralateral hand, or aphasia.
- Occlusive lesions of the internal carotid artery in the neck or head, or of its major intracranial branches, the middle and anterior cerebral arteries, cause loss of cerebral hemisphere functions.
- Occlusive lesions of the vertebral arteries in the head and neck and of the basilar artery cause brainstem and cerebellar symptoms and deficits.
- Occlusive disease of the posterior cerebral arteries (PCAs) causes visual and somatosensory symptoms due to loss of function in the lateral thalamus and occipital lobe regions supplied by the PCAs.

Some TIAs, such as those causing transient monocular blindness, diplopia, and aphasia, are very specific for one vascular territory, whereas others, such as limb weakness or numbness, are compatible with a number of different territories.

The occurrence of a TIA is not helpful in predicting the presence of brain damage, the cause and mechanism of the ischemia, or the prognosis [4]. Transient brain ischemia can be caused by a variety of different conditions:

- Brain embolism arising from the heart, aorta, or proximal arterial vessels.
- Occlusive lesions of either large extracranial or intracranial arteries. These large artery lesions are associated with intermittent hypoperfusion of the symptomatic regions of the brain, or with embolization of fibrin-platelet (white thrombi) or

erythrocyte-fibrin (red thrombi) clots into the distal brain circulation. Emboli often break up or pass through the vasculature, explaining the transiency of the neurological symptoms.

- Occlusive lesions of small microscopic-sized intracranial arteries and arterioles.
- Hypercoagulability and diseases that cause thrombosis of small vessels (e.g., thrombotic thrombocytopenic purpura).

An important consideration in patients who present with TIA-like symptoms is whether these symptoms are due to cerebrovascular disease or another condition, a "stroke mimic" [5–8]. Stroke mimics account for up to 30% of code stroke presentations, and up to 60% of potential TIAs. The most common stroke mimics are seizures, migrainous auras, and functional neurological disorders [1,5]. Making a clinical diagnosis based on the history and examination is important as imaging cannot definitely rule out stroke.

SEIZURES

Seizures account for up to 38% of TIA-stroke mimics [1,5,9]. Risk factors for seizures include prior strokes, small vessel disease, head injuries, brain tumors, subdural hematomas (SDHs), vascular malformations, and space-occupying lesions. Patients can present in any stage of the seizure or postictal period; initial ictal symptoms or signs may be unwitnessed or subtle and can go unrecognized by the patient (particularly in those who have cognitive impairment) and medical professionals. Presentation is highly variable depending on the seizure focus, both with regard to ictal and postictal symptoms as well as the duration of symptoms.

During the ictal phase, spontaneous neuronal discharge most often leads to positive phenomena including focal or generalized limb jerking or myoclonic jerks and positive sensory symptoms such as paresthesias or simple positive visual phenomena such as bright colors or flashes of light. Positive phenomena are less common in patients who have TIAs. Frequent epileptiform discharges can also interrupt speech and thought processes resulting in brief periods of speech arrest or aphasia at times with florid paraphasic errors. Postictally, there is cortical suppression resulting in negative symptoms and signs such as weakness, numbness, or speech difficulty. The classical Todd's paresis, which often presents with hemiparesis, occurs commonly after generalized tonic-clonic or focal motor seizures [10]. The duration of postictal dysfunction may last for days or longer. A clue to a convulsive etiology is the tempo of onset, which evolves rapidly over several seconds with spreading focal involvement or secondary generalization. This allows differentiation from the much slower progression seen in migraine or the classically abrupt onset seen in stroke. A reduced level of consciousness is also more common in seizures than in TIAs. Electroencephalography can be used to identify epileptiform discharges. When attacks similar to the presenting event have occurred for years, seizures and migraine are more likely than stroke. Hyperperfusion seen in patients with seizures on computed tomography (CT) or MRI-perfusion imaging may be helpful and different from the pattern found in stroke [11].

MIGRAINE

Migraine is a very common TIA mimic. The incidence of migraine declines with age, but it is still relatively frequent in the older population [12]. A thorough history is important to detail the progression of symptoms and neurological deficits as well as the co-occurrence of a migrainous sounding headache. In comparison with the sudden onset symptoms in stroke and very rapid spread in seizures, migraine auras have a characteristic, slow spread during several minutes due to cortical spreading depression [13]. The sensory symptoms typically begin with positive signs that can involve different sensory modalities. Patients may have headache before, during, or after the episode. Often, the first symptoms are a positive visual aura, a sensation of fluttering, bright flashes of light, scintillations, shimmering, or a change in perspective. A migrainous visual aura is often characterized by gradual movement across the visual field leaving behind in its trail negative phenomenon in the form of an area of decreased vision that slowly resolves. In patients with stroke, there is no clear spread and involvement is in only one visual field as opposed to bilaterally in migraines and does not spread. The "scintillating scotoma," a combination of both negative and positive phenomena seen in migraine, is not seen in other conditions. When the visual symptoms clear, another positive phenomenon in the form of a tactile sensory aura may begin. Paresthesias most often involve the hand and face with legs and torso much less commonly affected. The sensation can be described as tingling, prickling, or burning sensations. The abnormal sensation starts focally, whether in a finger or leg, or on the face and will slowly evolve and may even spread to the contralateral side. This nonanatomic distribution also provides an additional clue that the symptoms do not respect the focal anatomic confines seen in strokes. The sensory aura resolves slowly, with the first affected area clearing last and often leaves behind a feeling of decreased sensation in its wake. As these positive symptoms recede, patients may then develop other "negative" symptoms such as weakness, aphasia, or other problems. Aphasic auras generally involve paraphasic errors in addition to word-finding difficulties. Associated weakness is generally slight in comparison with cerebrovascular causes and has a rostrocaudal typical sensory march. Patients with a clear history of migraine or prior migrainous sounding headaches with aura have an increased likelihood of an atypical migrainous aura particularly if there is a history of previous complex features.

Migrainous aura without headache, now termed acephalgic migraine, was first described in Miler Fisher's classic paper where he detailed migraine accompaniments in individuals above age 45 years, most often in the absence of headache [14]. Migraine accompaniments involving sensory, motor, and aphasic symptoms almost always co-occur with a visual aura. Table 76.1 details features distinguishing migraine accompaniments from TIA.

TABLE 76.1 Features Differentiating Migraine Accompaniments From Cerebral Ischemia

Symptom Type	Migraine Aura	Transient Ischemic Attack
Visual	Positive visual phenomena (flashes of light)	Negative symptoms (loss of vision)
	Involvement of both visual fields	Unilateral
	Slowly moving across visual field	Static
	Average duration 15–60 min	Average duration 3–10 min
		Amaurosis fugax
Sensory	Gradual build-up/ evolution	Abrupt onset
	Positive phenomena (paresthesia)	Negative symptoms (numbness)
	Cheiro-oral distribution (hand and face)	Unilateral weakness or sensory loss without slow spread
	Repetitive attacks of identical nature	Variable attacks
	Sequential progression from one modality to another (visual, sensory, speech)	Simultaneous appearance of symptom modalities (e.g., motor and sensory) and body parts (face and arm at the same time)
	Sequential progression from one body part another	Symptoms appear and disappear simultaneously
	Area involved first resolves last	Flurry of stereotyped attacks uncommon
	Flurry of attacks	Average duration 5–10 min
	Average duration 20–30 min	

A particularly challenging mimic at any age is migraine with brainstem aura [15]. Onset is most common in youth but there are late-onset cases. The onset is most often visual that can have both positive or negative features and other symptoms localizable to the brainstem including slurred speech, vertigo, tinnitus, hypacusis, diplopia and oscillopsia, and gait and appendicular ataxia. Patients may also report paresthesias in the perioral region or in the limbs and some may also have decreased consciousness. Examination is highly variable and may reveal ophthalmoplegia, ataxia, dysarthria, or other cranial nerve abnormalities.

TRANSIENT GLOBAL AMNESIA

Transient global amnesia (TGA) usually occurs in older age and is characterized by the sudden onset of profound anterograde amnesia with more variable retrograde memory loss, followed by a gradual recovery, which may last several hours and almost always less than 24 h [17,18]. Most cases are single events, although some patients have recurrent attacks. Patients appear confused, and invariably ask repetitive questions about why they are here and sometimes where they are. They are unable to retain information given to them. Patients may be anxious and can be agitated, although some become quiet and withdrawn. Despite the dramatic memory loss and disorientation to place and time, patients have otherwise intact cognitive function and no other neurological symptoms. They may be able to drive safely, calculate, play a musical instrument, or even deliver a speech. Following resolution of symptoms, there is permanent memory loss for the events during the attack. Caplan produced a set of criteria for TGA [19,34] that was later refined by Hodges and Warlow (Table 76.2) [20]:

TGA is associated with physical, sudden, or unexpected psychological stress as well as positive emotional responses; funerals as well as celebrations have been

TABLE 76.2 Hodges and Warlow Diagnostic Criteria for TGA [20]

1. Attacks must be witnessed and information available from a capable observer who was present for most of the attack.
2. There must be a clear-cut anterograde amnesia during the attack.
3. Clouding of consciousness and loss of personal identity must be absent, and the cognitive impairment limited to amnesia (i.e., no aphasia, apraxia).
4. There should be no accompanying focal neurological symptoms during the attack and no significant neurological signs afterward.
5. Epileptic features must be absent.
6. Attacks must resolve within 24 h.
7. Patients with recent head injury or active epilepsy (i.e., remaining on medication or one seizure in the past 2 years) are excluded.

temporally related. Most patients endorse precipitating events that include recent sexual activity, physical exertion, swimming in cold water ("amnesia by the sea"), a hot bath, or shower [17,18]. These triggering events are similar to those of migraineurs. Patients may have a past history of migraine [21].

VERTIGO

Vertigo is a common reason for presentation to emergency departments. It is often difficult to differentiate peripheral vestibulopathy from cerebrovascular TIAs affecting the vestibulocerebellar pathways [22,23]. There are different causes of acute dizziness including those of peripheral origin (vestibular neuritis, labyrinthitis, benign postural positional vertigo, and Meniere disease) and central origin (mainly brain infarction and TIA and migraine with brainstem aura).

The acute vestibular syndrome is characterized by persistent vertigo or dizziness and can be accompanied by nausea or vomiting, worsening with head movement, unsteady gait, and nystagmus [24]. Patients with posterior circulation ischemic disease including TIAs very rarely have isolated attacks of vertigo without other accompanying symptoms attributable to brainstem or cerebellar dysfunction. In peripheral causes, vertigo is often triggered by sudden movements and positional changes. Assessing features such as timing and triggers has significant clinical value [22]. CT is particularly unreliable in diagnosing acute posterior circulation infarction [24].

Bedside examination presents a way to differentiate stroke from peripheral causes of vertigo. The mnemonic H.I.N.T.S. (detailing the steps of eye examination) to I.N.F.A.R.C.T. (the examination features favoring a stroke—Impulse Normal, Fast-phase Alternating or Refixation on Cover-Test) is useful in differentiating peripheral from a central cause (see Table 76.3) [25]. Features suggestive of a peripheral etiology include a unilaterally abnormal head impulse test, direction-fixed, horizontal, or horizontal more than torsional nystagmus in addition to absent skew deviation by the alternate cover test. In testing nystagmus, the fast phase beats away from the abnormal side. In comparison, a central pattern may include any of the following: a normal head impulse test, bilateral direction-changing nystagmus (or mainly vertical or torsional), and the presence of skew deviation. The HINTS approach was found to have a HINTS sensitivity of 96.5% and specificity 84.4% and was superior to standard risk factor analysis in differentiating a central from peripheral cause of vertigo [25].

TABLE 76.3   H.I.N.T.S. to I.N.F.A.R.C.T.

"H.I.N.T.S." Eye Examination	Stroke Findings: "I.N.F.A.R.C.T." (Any of These)
Head Impulse (right- and leftward)	Impulse Normal (bilaterally normal)
Nystagmus type (gaze testing)	Fast-phase Alternating (direction-changing nystagmus)
Test of Skew (alternate cover test)	Refixation on Cover Test (looking for skew deviation)
H.I.N.T.S. PLUS	
Presence of hearing loss. Generally on the side of abnormal head impulse test (side opposite fast phase of nystagmus)	Unilateral hearing loss generally indicates an ischemic cause (labyrinthine/lateral pontine infarction) as opposed to viral labyrinthitis

Central patterns: (1) bilaterally normal head impulse test with spontaneous or gaze-evoked nystagmus; (2) bilateral, direction-changing, horizontal gaze-evoked nystagmus (or predominantly vertical or torsional nystagmus); (3) skew deviation.

Adapted from Kerber KA, Newman-Toker DE. Misdiagnosing dizzy patients–common pitfalls in clinical practice. Neurol Clin 2015;33:565–75; Kerber KA, Meurer WJ, West BT, Fendrick AM. Dizziness presentations in U.S. emergency departments, 1995–2004. Acad Emerg Med 2008;15:744–50.

HYPOGLYCEMIA

Metabolic abnormalities can result in symptoms that can be focal or global in distribution. These metabolic derangements as well as toxic consequences of drugs are increasingly prevalent in the era of significant polypharmacy in older age. Hypoglycemia, in the diabetic patient taking oral hypoglycemic medications or insulin can present in a manner similar to a stroke. Hypoglycemia causing hemiplegia was soon recognized after the introduction of insulin [26]. The ubiquitous use of bedside fingerstick blood glucose testing now makes this readily detectable. There is mandatory checking of blood glucose levels before recombinant tissue plasminogen activator administration, which should be routine in any potential stroke evaluation. Focal deficits often resolve quickly after the administration of intravenous glucose but in rare cases can take several hours [26].

FUNCTIONAL NEUROLOGICAL DISORDERS

Although once considered rare in older age, functional neurological disorders (formerly psychogenic or conversion disorder) are increasingly recognized in older individuals. They are one of the most frequent stroke mimics, and account for up to 40% of nonstroke cases [1,5,27]. Most patients are younger than the usual stroke cohort. They may have repeat presentations with similar symptoms [28]. They may be given thrombolytics because of the time constraints to adequately assess these patients given the detailed history and clinical examination needed for diagnosis. There is a strong female preponderance [29]. Risk factors include trauma (physical or psychological), psychiatric disorders, adverse life events, prior somatization, and a past history of functional neurological symptoms or other functional syndrome [28]. Often patients do not have an identifiable trigger or psychiatric comorbidity. Functional weakness is characterized by inconsistent features on examination such as variable or intermittent effort, "give-way," or collapsing weakness, a seemingly limp arm avoiding the face when dropped from a height or a dragging monoplegic gait. Weakness is more commonly hemiparetic than monoparetic and a positive Hoover sign has high diagnostic sensitivity [30]. Other features include midline splitting sensory loss, "astasia abasia" uneconomical gait, or distractible functional ataxia.

NEUROMUSCULAR CONDITIONS

Myasthenia

Myasthenia gravis due to nonselective weakness of eye, face, or bulbar muscles may be confused with a posterior circulation stroke [31]. A clue is that weakness is typically generalized and fatigable. Complex ophthalmoplegia either in isolation or in association with bulbar dysfunction and dysarthria can also imitate a brainstem stroke; however, the variability, lack of adherence to one cranial nerve or vascular territory, and the diurnal variation (worse at the end of the day) and lack of generally acute onset differentiate these conditions from an acute vascular event.

Peripheral Neuropathy

Focal neuropathy or plexopathy can also be confused with stroke. The limitation to a single peripheral nerve territory of common compressive neuropathies is the key factor as well as the mechanism of action of compression by prolonged abnormal posture (often postsurgery) by pressure by an object or sleeping position. "Saturday night palsy," where the radial nerve is compressed at the spiral groove, often over a bench or the armrest of a chair, is particularly prevalent in alcoholics or drug addicts as a result of sleeping in an unusual position due to somnolence caused by the substance. Clinically, there is wrist drop with finger drop and radial distribution sensory loss. The clinical examination is complicated in this wrist dropped position. When the hand is propped up, there is normal median and ulnar nerve function pointing to a focal neuropathy as opposed to a stroke.

BRAIN TUMORS, SUBDURAL HEMATOMAS, AND SPACE-OCCUPYING LESIONS

SDHs and brain tumors can also cause acute neurological symptoms and may be focal, depending on their location and nature [32]. A history of recent falls or anticoagulation administration suggests SDH, which can be differentiated from stroke by CT and symptoms may be transient [33]. A history of current or past cancer and a progressive course suggest a brain tumor. Tumors can also be a frequent radiological differential for stroke on a CT given hypodensity of cytotoxic edema often appearing similar to the vasogenic edema seen in tumors, which does not respect vascular territories. Tumors may also represent another differential for TIA. So-called transient tumor attacks involve transient neurological symptoms not attributable to another cause in the setting of brain tumors. Accounts have included attacks in the presence of meningiomas and other tumor types [35]. As part of large multicenter TIA study, Coleman et al. identified the following features as being more consistent with a tumor than a vascular etiology: focal jerking or shaking, pure (particularly positive) sensory symptoms, loss of consciousness, and isolated aphasia or speech arrest [36].

SYNCOPE AND VERTEBROBASILAR INSUFFICIENCY

In one study, syncope accounted for 13% of stroke mimics in an old-age population [9]. In patients with recurrent syncope or presyncope, there is concern particularly in older patients that this may be the result of vertebrobasilar insufficiency. Posterior circulation TIAs only very rarely cause temporary loss of consciousness and transient loss of consciousness is predominantly found only in extracranial disease of all four vessels and has been termed "brain claudication." In practice, basilar occlusive disease invariably presents with additional motor and brainstem signs.

PRIMARY OPHTHALMOLOGIC CONDITIONS

Aside from the visual aura seen in migraine, transient visual loss, double vision, visual obscurations, and blurring can be caused by a number of ophthalmological conditions that are prevalent in older persons and can be confused with stroke or TIA. Diplopia is often a sudden-onset phenomenon and as such, nonischemic causes may often be confused with stroke. Causes are best differentiated by whether there are normal eye movements or not.

If eye movements are entirely normal, the most common cause is a decompensated phoria and if abnormal include the gamut of causes of a cranial nerve palsy affecting eye movements (III, IV, and VI and internuclear opthalmoplegia being most common) or a neuromuscular junction problem. A decompensated phoria can be elucidated by history, with a description of episodic diplopia worse in the evenings when it is dark, when the patient is tired or after alcohol. Although eye movements are normal, eye deviation can be induced on alternate cover testing. Decompensated phoria is also the most common cause of postoperative diplopia, although care must be taken to differentiate it from myasthenia and other serious causes [37]. Regarding visual loss, one can differentiate between positive symptoms (with flashes of light or other scintillations such as seen in migraine and retinal or vitreous detachment)

or negative symptoms, which include amaurosis fugax or a cerebral artery occlusion causing a TIA or other primary ophthalmological causes as detailed in Table 76.4. Retinal detachment may mimic amaurosis fugax and is more common in older persons and in individuals following cataract surgery. Vitreous hemorrhage and acute cystic maculopathy are the most common causes of visual loss after cataract surgery. Glaucoma can also cause acute visual blurring lasting minutes to hours and should be investigated in the absence of other causes. Although it is usually associated by orbital pain, it can be painless. Cataracts can also produce visual obscurations, although this is rarely acute in nature. Any older patient with transient visual loss should also be evaluated for giant cell arteritis with erythrocyte sedimentation rate and C-reactive protein and further workup depending on the degree of suspicion.

TABLE 76.4   Presentation and Causes of Sudden Visual Loss

Duration	Potential Causes	Associated Findings
Seconds	Dry eye or other tear film abnormalities	Foreign body sensation, lacrimation
	Papilledema	Headache, tinnitus, bilateral disc edema
	Compressive optic neuropathy (orbit)	Vision loss in certain positions of gaze, extraocular muscle restriction, proptosis
	Orthostatic hypotension	Near syncope
1–10 min	Amaurosis fugax	Unilateral, cardiovascular risk factors
	Other transient ischemic attack	Homonymous hemianopia or other visual field loss, extremity numbness or weakness, abnormality of speech, ataxia, cranial nerve abnormalities. Cardiovascular and stroke risk factors
	Ocular stroke*—central retinal artery occlusion, ophthalmic artery occlusion, branch retinal artery occlusion, cilioretinal artery occlusion	Painless, variable loss of vision, cardiovascular and stroke risk factors
	Central retinal vein occlusion*	Painless vision loss, retinal hemorrhages, venous engorgement and tortuosity, cotton wool spots, optic disc edema
	Giant cell arteritis (arteritic ischemic optic neuropathy)*	Variable visual loss, headache, jaw claudication, temporal artery tenderness, myalgia, weight loss
10–60 min	Migrainous aura	Scintillating scotoma, other positive or negative visual phenomena, associated headache with photophobia, phonophobia, nausea/vomiting
Minutes to hours	Transient angle-closure glaucoma	Halos around lights, nausea, eye or head pain, narrow anterior chamber angle, conjunctival injection, corneal edema, high intraocular pressure
Days or longer	Vitreous hemorrhage	Painless vision loss, floaters, hazy vision, obscured view of the retina
	Acute maculopathy	Metamorphopsia (distorted vision in which a grid of straight lines appears wavy)
	Retinal detachment	Increase in floaters, photopsias (flashes of light), visual field defect
	Optic neuritis	Pain with eye movement, variable visual loss, later relative afferent pupillary defect and abnormal color vision, vision starts to improve by 2–4 weeks
	Nonarteritic ischemic optic neuropathy	Painless visual loss, disc address, cardiovascular risk factors

*Some causes of sudden visual loss can mimic stroke.

DEMYELINATING CONDITIONS

Demyelinating disorders are less common but do occur in old age. Although they usually do not have the acute onset suggestive of a stroke, they can rarely present with sudden-onset deficits and with brief repeated transient episodes, most often of ataxia with dysarthria [38,39].

References

[1] Caplan LR, Stephen C. Transient focal neurological events in stroke in the older person. Oxford: Oxford University Press; 2016. [in press].

[2] Albers GW, Caplan LR, Easton JD, et al. Transient ischemic attack–proposal for a new definition. N Engl J Med 2002;347:1713.

[3] Easton JD, Saver JL, Albers GW, et al. Definition and evaluation of transient ischemic attack. A scientific statement for healthcare professionals from the American Heart Association/American Stroke Association Stroke Council; Council on Cardiovascular Surgery and Anesthesia; Council on Cardiovascular Radiology and Intervention; Council on Cardiovascular Nursing; and the Interdisciplinary Council on Peripheral Vascular Disease. Stroke 2009;40:2276–93.

[4] Caplan LR. Caplan's stroke, a clinical approach. 4th ed. Philadelphia: Elsevier/Saunders; 2009.

[5] Lioutas VA, Sonni S, Caplan LR. Diagnosis and misdiagnosis of cerebrovascular disease. Curr Treat Options Cardiovasc Med June 2013;15(3):276–87.

[6] Libman RB, Wirkowski E, Alvir J, et al. Conditions that mimic stroke in the emergency department. Implications for acute stroke trials. Arch Neurol 1995;52:1119–22.

[7] Merino JG, Luby M, Benson RT, Davis LA, Hsia AW, et al. Predictors of acute stroke mimics in 8187 patients referred to a stroke service. J Stroke Cerebrovasc Dis 2013;(8):e397–403.

[8] Hand PJ, Kwan J, Lindley RI, et al. Distinguishing between stroke and mimic at the bedside: the brain attack study. Stroke 2006;37:769–75.

[9] Kose A, Inal T, Armagan E, Kıyak R, Demir AB. Conditions that mimic stroke in elderly patients admitted to the emergency department. J Stroke Cerebrovasc Dis 2013;22(8):e522–527.

[10] Todd RB. Clinical lectures on paralysis, certain diseases of the brain, and other affections of the nervous system. 1854. p. 284–307. London: John Churchill.

[11] Hedna VS, Shukla PP, Waters MF. Seizure mimicking stroke: role of CT perfusion. J Clin Imaging Sci 2012;2:32.

[12] Prencipe M, Casini AR, Ferretti C, et al. Prevalence of headache in an elderly population: attack frequency, disability, and use of medication. J Neurol Neurosurg Psychiatry 2001;70:377–81.

[13] Pietrobon D, Moskowitz MA. Pathophysiology of migraine. Annu Rev Physiol 2013;75:365–91.

[14] Fisher CM. Late-life migraine accompaniments – further experience. Stroke 1986;17(5):1033–42.

[15] Kirchmann M, Thomsen LL, Olesen J. Basilar-type migraine: clinical, epidemiologic, and genetic features. Neurology 2006;66(6):880–6.

[16] Vongvaivanich K, Lertakyamanee P, Silberstein SD, Dodick DW. Late-life migraine accompaniments: A narrative review. Cephalalgia 2015;35(10):894–911.

[17] Caplan LR. Transient global amnesia in Handbook of Clinical Neurology (revised series). In: Vinken PJ, Bruyn GW, Klawans H, editors. Clinical neuropsychology, ;45. Amsterdam: Elsevier Science Publishing; 1985. p. 205–18. Frederiks JAM.

[18] Caplan LR. In: Markowitsch HJ, editor. Transient global amnesia: characteristic features and overview in transient global amnesia and related disorders. Toronto: Hogrgrefe & Huber Publishing; 1990. p. 15–27.

[19] Caplan LR. Transient global amnesia: criteria and classification. Neurology 1986;36:441.

[20] Hodges JR, Warlow CP. The aetiology of transient global amnesia. A case-control study of 114 cases with prospective follow-up. Brain 1990;113:639–57.

[21] Caplan LR, Chedru F, Lhermitte F, Mayman C. Transient global amnesia and migraine. Neurology 1981;31:1167–70.

[22] Kerber KA, Newman-Toker DE. Misdiagnosing dizzy patients – common pitfalls in clinical practice. Neurol Clin 2015;33:565–75.

[23] Kerber KA, Meurer WJ, West BT, Fendrick AM. Dizziness presentations in U.S. emergency departments, 1995–2004. Acad Emerg Med 2008;15:744–50.

[24] Tarnutzer AA, Berkowitz AL, Robinson KA, Hsieh YH, Newman-Toker DE. Does my dizzy patient have a stroke? A systematic review of bedside diagnosis in acute vestibular syndrome. CMAJ 2011;183:E571–92.

[25] Hwang DY, Silva GS, Furie KL, Greer DM. Comparative sensitivity of computed tomography vs. magnetic resonance imaging for detecting acute posterior fossa infarct. J Emerg Med 2012;42:559–65.

[26] Wallis WE, Donaldson I, Scott RS, Wilson J. Hypoglycemia masquerading as cerebrovascular disease (hypoglycemic hemiplegia). Ann Neurol 1985;18(4):510–2.

[27] Vroomen PC, Buddingh MK, Luijckx GJ, De Keyser J. The incidence of stroke mimics among stroke department admissions in relation to age group. J Stroke Cerebrovasc Dis 2008;17(6):418–22.

[28] Stone J, Warlow C, Sharpe M. The symptom of functional weakness: a controlled study of 107 patients. Brain 2010;133:1537–51.

[29] Parry AM, Murray B, Hart Y, Bass C. Audit of resource use in patients with non-organic disorders admitted to a UK neurology unit. J Neurol Neurosurg Psychiatry 2006;77:1200–1.

[30] McWhirter L, Stone J, Sandercock P, Whiteley W. Hoover's sign for the diagnosis of functional weakness: a prospective unblinded cohort study in patients with suspected stroke. J Psychosom Res 2011;71:384–6.

[31] Libman R, Benson R, Einberg K. Myasthenia mimicking vertebrobasilar stroke. J Neurol 2002;249(11):1512–4.

[32] Snyder H, Robinson K, Shah D, Brennan R, Handrigan M. Signs and symptoms of patients with brain tumors presenting to the emergency department. J Emerg Med 1993;11(3):253–8.

[33] Moster ML, Johnston DE, Reinmuth OM. Chronic subdural hematoma with transient neurological deficits: a review of 15 cases. Ann Neurol 1983;14(5):539–42.

[34] Daly DD, Svein HJ, Yoss RE. Intermittent cerebral symptoms with meningiomas. Arch Neurol 1961;5:287–93.

[35] Ross RT. Transient tumor attacks. Arch Neurol 1983;40:633–6.

[36] Coleman RJ, Bamford JM, Warlow CP. The UK TIA Study Group. Intracranial tumours that mimic transient cerebral ischaemia: lessons from a large multicentre trial. J Neurol Neurosurg Psychiatry 1993;56:563–6.

[37] Wray SH. Eye movement disorders in clinical practice: signs and syndromes. New York: Oxford University Press; 2014. p. 142–6.

[38] Delgado MG, Santamarta E, Sáiz A, Larrosa D, García R, Oliva P. Fluctuating neurological symptoms in demyelinating disease mimicking an acute ischaemic stroke. BMJ Case Rep April 2, 2012.

[39] Marcel C, Anheim M, Flamand-Rouviére C, et al. Symptomatic paroxysmal dysarthria-ataxia in demyelinating diseases. J Neurol 2010;257(8):1369–72.

CHAPTER

77

Types of Stroke and Their Differential Diagnosis

C.-P. Chung

Taipei Veterans General Hospital, National Yang Ming University, Taipei, Taiwan

INTRODUCTION

Stroke is typically defined as the sudden onset of a neurologic deficit caused by vascular abnormalities. There are two main categories: (1) ischemia (infarction)—restricted blood flow due to vascular stenosis/occlusion and the consequent insufficient oxygen and glucose supply to an area of the central nervous system (CNS) and (2) hemorrhage—bleeding in or around the brain. Ischemic stroke is more common than hemorrhagic stroke (70% vs. 30% globally); a higher incidence of hemorrhagic stroke was noted in the Asian, Hispanic, and African–American populations [1].

Ischemic stroke can be further characterized into subtypes based on the different involved vascular territories, mechanisms, and causes. This chapter will introduce each subtype of stroke classified by each of these different methods. Each stroke subtype's pathophysiology, clinical characteristics, and imaging will be discussed. Venous diseases, silent brain infarcts/hemorrhage, and retina or spinal strokes are not discussed in this chapter.

ISCHEMIC STROKE

Classification Based on Involved Vascular Territory

The brain tissue is perfused and given nutrition by two vascular systems, the anterior and posterior circulation. Based on the infarct location, ischemic stroke can be classified as anterior circulation ischemic stroke (ACIS) or posterior circulation ischemic stroke (PCIS). There are several differences between ACIS and PCIS; for example, atrial fibrillation is noted more frequently in ACIS in most studies [2] and the rate of intracranial hemorrhage after intravenous thrombolysis is significantly less common in PCIS [3]. Therefore, their evaluation and management

strategies should be different. Patients with ACIS and PCIS often present their own localization-specific neurologic deficits. Magnetic resonance imaging with diffusion-weighted imaging sequences will often be needed to accurately differentiate these two categories of stroke.

Anterior Circulation Ischemic Stroke

The anterior circulation of the brain is supplied by the internal carotid arteries (ICAs) and their branches including large intracranial vessels, middle cerebral arteries (MCAs) and anterior cerebral arteries (ACAs), and small vessels (perforating arteries from intracranial ICAs, MCAs, and ACAs). The territory comprises most of the cerebrum except the occipital and medial temporal lobes. Ischemic stroke occurring in the anterior circulation accounts for 70–80% of all ischemic strokes.

The clinical features can present as MCA syndrome, ACA syndrome, or lacunar syndrome (LS). The common clinical manifestations are dysarthria, central facial palsy, and motor and/or sensory involvement of the contralateral extremities. The typical presentations of MCA territory infarction, the most common form of ACIS, are dysarthria, contralateral central facial palsy, ipsilateral deviation of eyes, contralateral hemianopsia, contralateral extremities weakness and sensory deficits, and cortical signs including aphasia (left MCA) or hemineglect (right MCA).

The three main causes of ACIS are large artery atherosclerosis (LAA), cardiac embolism (CE), and small vessel occlusion (SVO). The major mechanisms are (1) embolism: emboli can arise either from proximal atherosclerotic large artery lesions, mainly the extracranial and intracranial ICA and the mainstem MCA (artery-to-artery emboli), or from thrombi and other sources within the heart or the aorta (cardiogenic emboli); (2) thrombosis: in situ thrombosis at the site of atherosclerotic plaque in a large brain-supplying artery (usually over the vascular bifurcation region); and (3) hemodynamic compromise: a combination of large artery atherosclerotic severe stenosis/occlusion and generalized hypoperfusion caused by systemic conditions such as dehydration, blood loss, or cardiac failure. These cerebral infarct patterns can be single, multiple, or scattered; can involve the territory or subterritory; and can be isolated subcortical or cortical.

Lacunar infarction (small vessel occlusion) in the anterior circulation is caused by single perforating artery occlusion, and the associated small vascular pathologic conditions include lipohyalinosis or microatheroma. Perforating artery blood flow can also be blocked by large artery atheroma at the orifice of the branch (atheromatous branch occlusive disease).

Posterior Circulation Ischemic Stroke

Posterior circulation of the brain is the territory supplied by the vertebrobasilar arteries and their branches including the posteroinferior cerebellar arteries, anteroinferior cerebellar arteries, superior cerebellar arteries, posterior cerebral arteries, and perforating arteries. Brainstem, cerebellum, occipital lobe, medial temporal lobe, and thalamus are supplied by the posterior circulation.

The most common clinical features of PCIS are homolateral weakness and sensory deficits, which are hardly used for differentiating it from ACIS. Unique presentations of PCIS occur in less than 10% of patients, including crossed weakness or sensory deficits, hemianopia or quadrantanopsia, Horner syndrome, and cranial nerve signs, particularly oculomotor nerve palsy [4]. Compared with ACIS, onset of stuttering neurologic deficits and nonfocal symptoms such as dizziness, vertigo, nausea/vomiting, and unsteadiness are more common in PCIS. These facts may result in delayed identification of PCIS in the clinical setting. Progressive stroke and precedent transient ischemic attack (TIA) symptoms have also been found more frequently in PCIS than in ACIS. The longest progression time being reported in PCIS is 14 days after the onset [5].

Besides LAA, CE, and SVO, there are two other causes that are reported more frequently in PCIS than in AICS. One is basilar artery atheromatous branch occlusive disease (BABO) with perforating artery occlusion at the openings. BABO results in paramedian brainstem (most commonly pontine) infarction, which mimics lacunar infarction (SVO) radiologically and clinically, although the former infarction is originated from large artery atherosclerotic disease and the latter is from small vessel disease [6]. The other cause occurring more frequently in PCIS is arterial dissection, usually in younger individuals, than the other causes. Arterial dissections causing PCIS are most commonly initiated in the vertebral artery (VA). As in large artery atherosclerotic disease, intracranial arterial dissections [V4 part of VA and basilar artery (BA)] are involved more frequently in the Asian population with arterial dissection-related PCIS than in the Western population [6]. Both BABO and arterial dissection need extensive or high-resolution imaging investigations of large artery vessel walls.

Classification Based on Mechanism

Ischemic stroke is caused by restrictive blood flow related to different mechanisms. Acute management and prevention strategy should be tailored accordingly.

Hemodynamic Compromise

Ischemic stroke attributable to hemodynamic compromise is often called low-flow infarction. This mechanism explains or contributes to about 10% of brain infarction [7]. This type of ischemic stroke results from critically reduced cerebral perfusion pressure (CPP) by two key elements: (1) high-grade stenosis or occlusion of artery, usually upstream from the circle of Willis, e.g., extracranial carotid bifurcation, intracranial ICA (siphon region), or proximal MCA and (2) compromised circulatory compensation. The artery disease of this type is often atherosclerosis, a slowly progressive process that allows for circulatory compensation. Moyamoya disease and dissection are less common. The cascade of compensatory mechanisms is switched on in response to reduced CPP: (1) cerebral vasomotor reactivity dilating the cerebral microvessels and (2) establishment of collateral circulation including the circle of Willis, the reversed ophthalmic artery flow from the external carotid artery system, and leptomeningeal anastomoses [7]. When the extent of reduced CPP is beyond the ability of these compensations, brain infarction follows. Rupture of an atherosclerotic plaque and thrombosis of a large artery may accelerate the process of hemodynamic compromise.

Ischemic stroke of hemodynamic compromise may be precipitated by a sudden drop in blood pressure (BP), which exacerbates the CPP reduction. These conditions include orthostatic hypotension, dehydration, myocardial ischemia, cardiac arrhythmias, inappropriate treatment of hypertension, and perioperative complications. Patients may have syncope, fainting, pallor, or diaphoresis before the stroke onset.

The other clinical characteristics of low-flow infarction include stuttering onset, fluctuating neurologic deficits, and frequently preceding TIA. Several neurologic syndromic deficits can develop, among them, slowly progressive visual loss in one eye and intermittent arm or leg rhythmic movements are pathognomonic symptom/signs for this type of ischemic stroke.

The chainlike or rosarylike arrangement of infarct lesions over the centrum semiovale with or without the cortical watershed area involvement is the characteristic

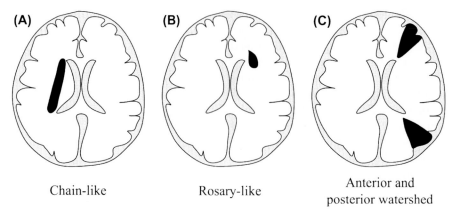

FIGURE 77.1 Characteristic lesional patterns of stroke caused by hemodynamic compromise: (A) chainlike, (B) rosarylike, and (C) anterior and posterior watershed.

lesional pattern (Fig. 77.1). These areas are zones of marginal irrigation or of lowest perfusion pressure that are susceptible to ischemia.

Embolism

Embolism is the most common mechanism of ischemic strokes. A blood clot formed elsewhere travels via bloodstream to intracranial arteries, acutely blocks the blood flow, and leads to infarction over the supplied brain territory. The commonest origins of embolism are from the heart (CE), aorta, and neck or intracranial arteries (artery-to-artery embolism).

The frequently associated cardiac abnormalities causing thrombus formation in the heart and the consequent cardiogenic embolic stroke are atrial fibrillation, valvular disease, valve replacement, atrial flutter, sick sinus syndrome, recent myocardial infarction, dilated cardiomyopathy, atrial myxoma, and infective or nonbacterial endocarditis.

Artery-to-artery embolism is usually caused by ruptured plaque or thrombosis formation within atherosclerotic or dissected arteries. Recurrent embolic stroke in the same vascular territory supplied by a large artery with more than 50% atherosclerotic stenosis or with an unstable plaque (ulcerated) falls into this category. Besides atherosclerosis, cervicocephalic artery dissection is another example of artery-to-artery embolic stroke, in which embolism is from the intramural thrombus.

Hypercoagulable states also predispose to the development of thrombosis in the vessels, including veins and arteries. Many diseases are associated with a hypercoagulable state, such as malignancy, infection, autoimmune diseases (e.g., antiphospholipid syndrome), hematologic diseases (e.g., deficiency of certain endogenous anticoagulants, thrombotic thrombocytopenic purpura), or hyperhomocysteinemia.

Emboli traveling through the bloodstream usually lodge in more distal arteries, although a main artery trunk can also be obstructed by a large embolus. The arteries are occluded so rapidly that the compensatory system is unable to be establish in time. The clinical characteristics of embolic stroke, compared with hemodynamic stroke, are sudden onset, maximal neurologic deficits at onset, and commoner cortical signs with or without seizure attacks. Embolism, especially cardiogenic embolism, can be washed out by blood flow soon after onset (reperfusion); in this case, abrupt neurologic deficits will be followed by dramatic improvement or disappearance of symptoms. The characteristic infarct patterns on imaging of embolic stroke are lesions that involve more than one vascular territory, scattered cortical lesions, and/or hemorrhagic transformation (caused by delayed reperfusion after embolism washout).

Thrombosis

This mechanism refers to the obstruction of the vessels caused by in situ thrombosis and the consequent brain infarction of the territory supplied by the vessels. Combined embolic stroke is not uncommon.

The rate of vascular obstruction is more rapid than that of stroke of hemodynamic compromise. The common scenarios are ruptured plaque with in situ thrombosis in an atherosclerotic large artery stenosis or SVO by a vascular disease (e.g., lipohyalinosis) or microatheroma. The other causes include arterial dissection, vasculitis, and hypercoagulable states caused by malignancy, infection, autoimmune disease, or hematologic abnormalities.

Classification Based on Etiology

The cause of ischemic stroke is revealed by clinical manifestations and investigatory studies including brain imaging, vascular and cardiac evaluations, and other laboratory studies. The three most common causes are LAA, CE, and SVO, totally contributing 60–80% of all ischemic strokes. The cause of stroke determines the outcomes and different acute management and secondary preventive strategy should be followed accordingly;

for example, antiplatelet therapy for secondary prevention of LAA-related ischemic stroke and anticoagulant therapy for CE ischemic stroke.

Large Artery Atherosclerosis

Large arteries refer to extracranial arteries (common carotid artery, ICA, VA) and intracranial arteries (ICA, MCA, ACA, posterior cerebral artery, and vertebrobasilar artery). Atherosclerosis is often generalized; involvement is usually most severe at arterial bifurcations (Fig. 77.2), such as the carotid bifurcation and the ICA–MCA junction, where turbulence flow is prominent. The cause of ischemic stroke is classified as LAA if the infarction territory is supplied by an atherosclerotic large artery with >50% stenosis (mechanisms can be hemodynamic, thromboembolic, or both) or <50% stenosis with occlusion of perforating arteries at their opening (atheromatous branch occlusive disease) (Fig. 77.3). Clinically, atheromatous branch occlusive disease mimics SVO (lacunar infarction), but unlike the latter, the former originates from large artery disease (atherosclerotic plaques that obstruct the ostia of perforating arteries that branch from the parent large artery). This condition is seen in deep brain infarction with proximal MCA atherosclerotic stenosis or small pontine infarction with or without BA atherosclerotic stenosis.

Vascular duplex ultrasonography, computed tomographic angiography, and/or magnetic resonance angiography are the investigatory tools for vascular pathogenesis study.

Cardiac Embolism

CE ischemic stroke occurs when patients have high-risk cardiac source of embolism such as atrial fibrillation, valvular disease, valve replacement, atrial flutter, sick sinus syndrome, recent myocardial infarction, dilated cardiomyopathy, atrial myxoma, or infective endocarditis. In addition, the infarction-associated large artery has no or little extent of atherosclerosis.

Echocardiography and 24-h Holter monitor are needed to investigate these CE risk factors.

Small Vessel Occlusion

Ischemic stroke of SVO is defined as a small infarction (<1.5 cm) in cerebral striatocapsular regions, pons, or thalamus region caused by the obstruction of a single perforating artery. Perforating arteries arise from the distal VA, BA, proximal MCA, and the arteries forming the circle of Willis. The associated small vessel pathologic conditions are, most frequently, lipohyalinosis (a destructive vessel lesion characterized by the loss of normal arterial architecture with medial hypertrophy with mural foam cells and fibrinoid necrosis) and, less commonly, microatheroma at the orifice of the branch [8]. Aging and hypertension are the most important risk factors. This type of stroke may be accompanied by other manifestations of small vessel diseases such as

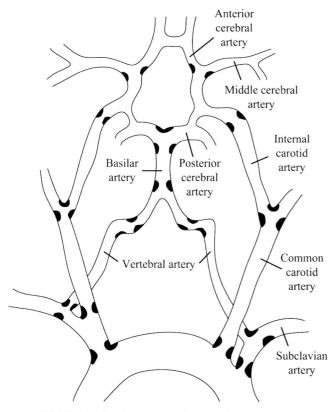

FIGURE 77.2   Typical sites of atherosclerotic plaques.

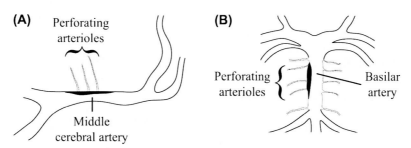

FIGURE 77.3   (A, B) Illustration of atheromatous branch occlusive disease.

age-related white matter lesions and microbleeds. The risks of LAA and CE need to be excluded in this category of ischemic stroke.

The most common LS is pure motor LS: complete or incomplete paralysis of the face, arm, and leg on one side, not accompanied by sensory deficits; visual field deficits; and cortical signs. The other less-recognized types of LS are pure sensory syndrome, dysarthria-clumsy hand syndrome, and ataxic hemiparesis.

Other Determined Causes

The other determined causes of ischemic stroke include arterial dissection, moyamoya disease, radiation-accelerated atherosclerosis, vasoconstriction, vasculitis, and hypercoagulable states caused by malignancy, infection, autoimmune disease, or hematologic abnormalities.

Undetermined Cause

Patients with incomplete investigations, with conflicting study results (e.g., both LAA and CE are potentially involved), or negative studies (cryptogenic stroke) are classified into this category.

The cause of ischemic stroke remains unknown despite a complete evaluation in about 20–40% of patients (cryptogenic ischemic stroke) [9]. Occult embolism from the heart, aorta, or other origins has been deemed as its pathophysiology. In one longitudinal and randomized study, atrial fibrillation was detected in 12.4% of patients with cryptogenic ischemic stroke by using an insertable cardiac monitor within 12 months of stroke [9].

HEMORRHAGIC STROKE

Hemorrhagic stroke is mainly composed of intracerebral hemorrhage (ICH) and subarachnoid hemorrhage (SAH); the incidence rate is about 3:1. Arterial aneurysms and the other vascular malformations are the major causes of SAH, which is not discussed further in this chapter.

Intracerebral Hemorrhage

The causes of ICH include hypertensive arteriosclerosis, amyloid angiopathy, vascular malformations, intracranial tumor, bleeding disorders, vasculitis, drugs (sympathomimetic agents), and trauma.

Hypertensive ICH is the most common hemorrhagic stroke, which is caused by an acute rise in BP or chronic hypertensive vasculopathy. The typical locations are the basal ganglia (35–44%), thalamus (10–25%), cerebellum (5–10%), and pons (5–9%). The severity of stroke and outcomes are greatly determined by the size of the hematoma; a hematoma of 60 mL in volume and coma (Glasgow Coma Score less than 9) are associated with greater than 90% mortality [10]. Compared with ischemic stroke, ICH more often results in increased intracranial pressure signs (headache, consciousness change, nausea, vomiting, etc.) early after the onset of the neurologic deficits. Enlargement of hematoma with neurologic deterioration is common within the first 24 h of onset [10].

Cerebral amyloid angiopathy (CAA) refers to amyloid accumulation in the vessel walls of small and medium-sized arteries and veins of the cerebral cortex and adjacent leptomeninges. Rupture of the amyloid-laden vessels results in hemorrhages that are typically located in the superficial lobar region (lobar ICH). CAA increases exponentially with age; it is found in about 5–8% of the population in their seventh decade versus 58% of those older than 90 years [10]. CAA also has the tendency to recur. Recurrent ICHs of the lobar region in the elderly are the distinguishing feature of CAA when compared with hypertensive ICH.

Acknowledgment

Thanks to Li-An Ho for the illustrations.

References

[1] Krishnamurthi RV, Feigin VL, Forouzanfar MH, Mensah GA, Connor M, Bennett DA, et al. Global burden of diseases, injuries, risk factors study 2010 (GBD 2010); GBD stroke experts group. Global and regional burden of first-ever ischaemic and haemorrhagic stroke during 1990–2010: findings from the global burden of disease Study 2010. Lancet Glob Health 2013;1:e259–281.

[2] De Marchis GM, Kohler A, Renz N, Arnold M, Mono ML, Jung S, et al. Posterior versus anterior circulation strokes: comparison of clinical, radiological and outcome characteristics. J Neurol Neurosurg Psychiatry 2011;82:33–7.

[3] Sarikaya H, Arnold M, Engelter ST, Lyrer PA, Mattle HP, Georgiadis D, et al. Outcomes of intravenous thrombolysis in posterior versus anterior circulation stroke. Stroke 2011;42:2498–502.

[4] Tao WD, Liu M, Fisher M, Wang DR, Li J, Furie KL, et al. Posterior versus anterior circulation infarction: how different are the neurological deficits? Stroke 2012;43:2060–5.

[5] Vuilleumier P, Bogousslavsky J, Regli F. Infarction of the lower brainstem. Clinical, aetiological and MRI-topographical correlations. Brain 1995;118:1013–25.

[6] Chung CP, Yong CS, Chang FC, Sheng WY, Huang HC, Tsai JY, et al. Stroke etiology is associated with outcome in posterior circulation stroke. Ann Clin Transl Neurol 2015;2:510–7.

[7] Ringelstein EB, Dittrich R, Stogbauer F. Borderzone infarcts. In: Caplan LR, Gijn JV, editors. Stroke Syndromes. 3rd ed. Cambridge (UK): Cambridge University Press; 2012. p. 480–500.

[8] Lammie GA. Pathology of small vessel stroke. Br Med Bull 2000;56:296–306.

[9] Sanna T, Diener HC, Passman RS, Di Lazzaro V, Bernstein RA, Morillo CA, et al. CRYSTAL AF Investigators. Cryptogenic stroke and underlying atrial fibrillation. N Engl J Med 2014;370:2478–86.

[10] Sutherland GR, Auer RN. Primary intracerebral hemorrhage. J Clin Neurosci 2006;13:511–7.

CHAPTER

78

Small Artery Occlusive Diseases

Y. Yamamoto

Kyoto Katsura Hospital, Kyoto, Japan

INTRODUCTION

Two Different Subtypes of Small Artery Occlusive Diseases

Small artery occlusive diseases, known as lacunar stroke, constitute approximately 25% of ischemic strokes and are particularly frequent among Hispanics and Asians. C. Miller Fisher performed detailed clinico-pathological examinations and described two different subtypes of lacunar strokes: those caused by an intrinsic disease of the small arteries (40–200 μm diameter) categorized as small artery disease such as arteriolosclerosis, lipohyalinosis, or fibrinoid necrosis, and those caused by proximal atherosclerotic disease of the larger-caliber perforating arteries (200–850 μm diameter) [1].

Small Multiple Lacunar Infarctions

Intrinsic disease of the small arteries tends to occur as small multiple infarcts with leukoaraiosis and usually asymptomatic lacunar infarcts. Prognosis for mortality, recurrent stroke, cognitive impairment, and overall functional outcome in patients with lacunar stroke with multiple silent lacunar lesions is known to be more unfavorable than in patients with single lacunar infarction [2].

Branch Atheromatous Disease and Progressive Motor Deficits

Conversely, a single lacunar infarction associated with proximal atherosclerotic disease tends to show progressive motor deficits during the acute stage. Single perforating artery territory infarction caused by focal atherosclerotic disease of the termed branch atheromatous disease (BAD) [3]. For example, single infarction in the territory of the lenticulostriate artery (LSA) with BAD mechanism may be located adjacent to the middle cerebral artery (MCA) and visible in multiple axial slices on MRI. Unilateral infarction in the territory of the paramedian pontine arteries with BAD mechanism may extend to the surface of the pontine base. Studies reported BAD was more often associated with atherosclerotic markers such as coronary artery disease and intracranial atherosclerosis but had a lower prevalence of small artery disease markers such as leukoaraiosis, multiple small lacunar infarcts, and microbleeds [4]. In our study of 394 consecutive patients with penetrating artery territory infarcts, 95 patients in the territory of LSA (36.1%) and 78 patients in the territory of paramedian pontine artery (59.5%) were classified as BAD. Among BAD type, progressive motor deficits were found in 40% in the LSA and 41% in the paramedian pontine artery [5]. A single lacunar infarct associated with atherosclerosis should be treated so as not to lead to progressive motor deficits in the acute stage, whereas small multiple infarcts should be carefully tracked to avoid recurrent stroke and progressive cognitive decline.

CHARACTERISTICS OF SMALL ARTERY OCCLUSIVE DISEASES ON SPECIFIC ARTERIES

Anterior Choroidal Arteries

Vascular Supply

The anterior choroidal artery (AChA) is a small artery that originates from the internal carotid artery (ICA) 2–5 mm distal to the posterior communicant artery (PComA) in diameter between 90 and 600 μm. The AChA supplies the zone between the striatum anterolaterally and the thalamus posteromedially. The AChAs initially course posteromedially within

Primer on Cerebrovascular Diseases, Second Edition
http://dx.doi.org/10.1016/B978-0-12-803058-5.00078-3

the carotid cistern and run posterolaterally within the crural cistern and terminate at the inferior limit of the choroidal fissure to supply the choroid plexus. The very proximal perforators supply the optic tract, medial segment of the globus pallidus, and the genu of the internal capsule. The next branches supply the temporal structures, such as the uncus, piriform cortex, amygdala, and head of the hippocampus. The third branches supply the middle third of the cerebral peduncle, substantia nigra, and red nucleus. The most distal perforating arteries vary in size from 200 to 610 μm, supply the posterior two-thirds of the posterior limb of the internal capsule, the tail of the caudate nucleus, the lateral geniculate body, the retrolenticular part including the origin of the geniculocalcarine tract, and the auditory radiation and lateral thalamic nuclei [6].

The posterior limb of the internal capsule, optic tract, lateral geniculate body, medial temporal lobe, and medial part of the pallidum are universally recognized as AChA supplied areas [7]. Whether or not the vascular supply to the anterior limb of the internal capsule, genu of the internal capsule, posterior part of the putamen, substantia nigra, red nucleus, and posterior corona radiata is derived from the AChA remains a source of controversy.

Clinical Symptoms

The clinical symptoms related to AChA territory infarcts are variable because the AChA supplies various anatomical areas. The most common findings are contralateral hemiparesis, contralateral hemihypesthesia, and dysarthria. Sohn et al. studied 127 patients with AChA infarction and found dysarthria in 108 patients (85.0%), hemiparesis in 103 patients (81.1%), and facial palsy in 93 patients (73.2%) [8]. In our study of 192 consecutive patients with AChA territory infarction, progressive motor deficits were found in 31 patients (16.3%), whereas progressive motor deficits were noted in 27.4% in 431 patients with lenticulostriate artery territory infarction during the same period. Contralateral sensory abnormalities are variable and mostly identified as sensorimotor syndrome. Ipsilateral visual field defect (homonymous hemianopia/quadrantanopia/sectoranopia) is variable depending on the involvement of the lateral geniculate body and the geniculocalcarine tract. Although the AChA supply the hilum and the anterolateral portion of the lateral geniculate body, the lateral choroidal arteries from the posterior cerebral artery (PCA) supply the medial and posterior portions. The most common visual field defects are a homonymous hemianopia, an upper quadrantanopia, and a quadruple sectoranopia.

Stroke Mechanism

The most commonly presumed stroke mechanism is intrinsic small artery disease. Large artery disease

represented by a stenosis or occlusion of the ICA is less common. Cardioembolism is not as frequent as generally emphasized. In the series of Sohn et al., 90/127 patients (70.9%) were classified as small artery disease, 26/127 (20.5%) as large artery disease, and 6/127 patients (4.7%) as cardioembolism [8]. Although small artery disease tends to cause single infarcts in the territory of the posterior limb of the internal capsule and medial part of the pallidum, large artery disease or cardioembolism tends to extend to the lateral geniculate body or medial temporal lobe.

Lenticulostriate Arteries

Vascular Supply

The LSAs are the major microvessels irrigating the basal ganglia and corticospinal tracts, brain regions particularly susceptible to stroke. The LSAs range in number from 2 to 12 and in size between 80 and 900 μm, mostly originated from the M1 segment of the MCA, either as singular vessels or by their own common trunks. The majority of the perforators course medially and then abruptly change their direction passing laterally and dorsally just before entering the anterior perforated substance. The LSAs supply the superior part of the head of the caudate nucleus and the anterior limb of the internal capsule, the putamen, most of the lateral segment of the globus pallidus, the superior part of the genu of the internal capsule, and the rostral part of the posterior limb of the internal capsule and the corona radiata [9]. Heubner's artery arising from the anterior cerebral artery supplies the inferior part of the head of the caudate nucleus.

Atherosclerotic Disease of the Middle Cerebral Arteries

Intracranial atherosclerotic disease has been emphasized as playing an important role in the development of LSA infarction. High-resolution MRI can show small atherosclerotic plaques on the wall of the MCA with or without stenosis on magnetic resonance angiography in patients with LSA territory infarction (Fig. 78.1). Such small atherosclerotic plaques may cause ischemic lesions via occlusion of the origin of the LSA leading to BAD of the LSA. LSA territory infarction sometimes becomes extensive producing severe motor deficits. Half of the LSAs in some patients branch from a single trunk from the MCA [10]. Occlusion of such a common trunk leads to ischemia in its entire region of the LSA territory (Fig. 78.2). Occlusion of an entire individual lateral perforating artery can result in an infarct measuring 41.6 × 15.5 mm on average, up to a maximum of 53 × 41 mm in case of large common stem, which exceeds the maximum of 20 mm usually cited as

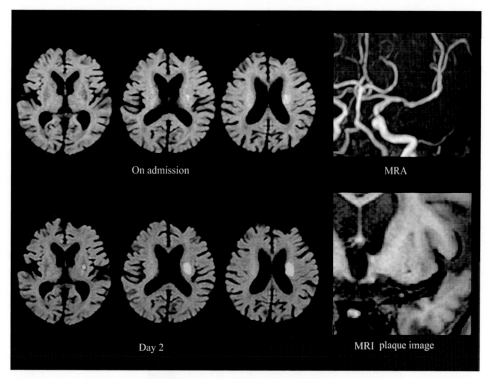

FIGURE 78.1   A 74-year-old man who developed progressive right hemiparesis. Mild stenosis in the MCA on magnetic resonance angiography and high intensity in the wall of the MCA on plaque image.

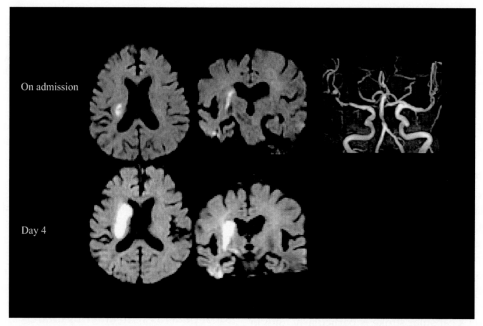

FIGURE 78.2   An 87-year-old woman who developed progressive left hemiparesis and neglect. A huge infarct in the territory of lenticulostriate arteries.

the upper limit of size for penetrating artery territory infarcts. Since atherosclerotic plaques usually develop in the perforating arteries ranging from 400 to 900 µm in size located in the main stems, BAD can cause large subcortical infarction in the territory of the LSA [9]. The corticospinal tracts cross the LSA territory at the posterosuperior quadrant [11], so patients with posteriorly located infarcts in the LSA territory may have severe motor deficits accompanied with progressive motor deficits probably due to stepwise involvement of corticospinal tracts by infarcts. Combined treatment of antiplatelet agents is often recommended as

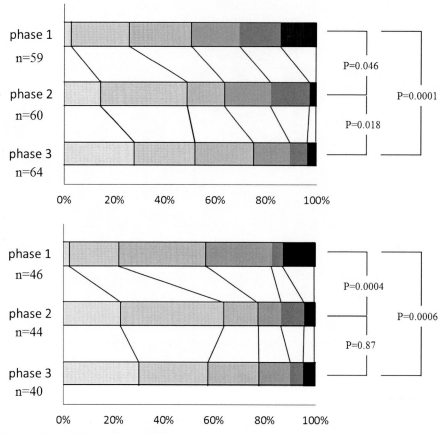

FIGURE 78.3   The efficacy of a combined use of antiplatelet agents was tested. During 12 years, 313 consecutive patients with intracranial branch atheromatous disease (BAD) located within the territories of lenticulostriate arteries (top) and anterior pontine arteries (bottom) were prospectively collected. Phase 1 (2001–05, $n=105$), the medical treatment that was considered best; phase 2 (2005–09, $n=104$), a combined treatment of argatroban, cilostazol, and edaravone; and phase 3 (2009–12, $n=104$), additional clopidogrel on top of the phase 2 protocol. Functional outcome was assessed by the modified Rankin Scale at 1 month after stroke onset.

treatment for progressive motor deficits in the LSA (Fig. 78.3) [12].

Capsular Warning Syndrome

A "capsular warning syndrome" characterized by recurrent stereotyped transient ischemic episodes often precedes infarction in the corresponding anatomical location, usually in the LSA territory [13]. The mechanism is believed to be small single perforating artery disease. Hemodynamic ischemic process in the territory of the penetrating arteries caused by high-grade stenoses of proximal penetrating arteries due to microatherosclerosis has been posited. The risk of developing a capsular stroke is particularly high in patients with capsular warning syndrome.

Thalamic Arteries

The blood supply to the thalamus originates from the PCAs or the PComAs. They are divided into four classical territories consisting of the thalamic–subthalamic, tuberothalamic, thalamogeniculate, and posterior choroidal arteries [14,15]. The thalamic arteries vary between individuals with respect to the parent artery from which each branch arises, and the number and position of the arteries.

Tuberothalamic Arteries

The tuberothalamic arteries originate from the middle third of the posterior communicating artery. In one-third of the population, it is absent and the paramedian artery assumes that territory. The tuberothalamic arteries irrigate the anterior nuclei, which are connected to the mamillothalamic tract, the anterior limbic system, and the medial and prefrontal cortex. Most patients with anterior infarcts show unique neuropsychological deficits known as "anterior behavioral syndrome," characterized by fluctuating levels of consciousness, personality changes, and impairment of recent memory and new learning. Language disturbances occur in left lesions and visual spatial deficits occur in right lesions.

Paramedian Arteries

The paramedian arteries arise from the P1 segment of the PCA. They are also known as the "mesencephalic artery" or the "posterior thalamosubthalamic paramedian artery." The paramedian arteries sometimes originate directly from the proximal segment of one of the PCAs. A single arterial trunk may branch from the P1 segment of one of the PCAs and then divide to supply both thalami, which is known as "the artery of Percheron." The paramedian artery principally supplies the dorsomedial nucleus, internal medullary lamina, and intralaminar nuclei. Unilateral thalamic infarction in the territory of the paramedian artery causes neuropsychological disturbances, such as a decreased level of consciousness, vertical gaze paresis, cognitive impairment, and personality changes. Confusion, agitation, aggression, and apathy are often observed. Amnesia frequently occurs and the role of the intralaminar and dorsomedial nuclei is posited. Bilateral infarction in the paramedian artery territory can result in a devastating outcome. Disorientation, confusion, hypersomnolence, coma, and severe memory impairment often occur accompanied by vertical eye movement abnormalities.

Inferolateral Arteries

The inferolateral thalamogeniculate arteries arise from the P2 segment of the PCA and consist of 5–10 arteries. The inferolateral arteries supply the ventrolateral nucleus that has connections to the cerebellum and the motor and prefrontal cortex and ventroposterior nuclei that receive inputs from the medial lemniscal and spinothalamic pathways and ventromedian nucleus. The ventrolateral nucleus receives inputs from the trigeminothalamic pathway. The most common clinical pattern after inferolateral infarct is the thalamic syndrome described by Dejerine and Roussy, i.e., sensory loss to a variable extent, with impaired extremity movement, sometimes with thalamic pain. Executive dysfunction, cognitive signs, and aphasia can occur probably related to the impairment of the thalamocortical projections from the posterolateral nuclei of thalamus.

Posterior Choroidal Arteries

The posterior choroidal arteries originate from the P2 segment of the PCA comprising a number of branches. These supply the subthalamic nucleus and midbrain, the medial half of the medial geniculate nucleus, the posterior parts of the intralaminar nuclei including central lateral and centromedian nuclei, and the pulvinar nuclei. The most frequent signs are hypoesthesia and homonymous horizontal sectoranopsia when the lateral geniculate body is involved.

Superficial Penetrating Arteries

The Superficial Penetrating Arteries Are Susceptible to Embolic Stroke Mechanisms Than the Deep Penetrating Arteries

The superficial penetrating arteries known as the white matter medullary arteries arise from the cortical branches of the MCA and irrigate the centrum semiovale, whereas the deep penetrating arteries stem from the MCA trunk and irrigate the basal ganglia and corona radiata. Although both the deep and superficial penetrating arteries originate from the MCA, the stroke mechanism of these two arterial systems differs. Yonemura et al. studied 582 consecutive patients with acute ischemic stroke including 38 patients with small infarcts in the territory of the superficial penetrating arteries and 68 patients in the territory of the deep penetrating arteries [16]. Patients in the superficial penetrating artery group had more frequently an abrupt onset of symptoms (63% versus 26%; $P = .0002$), emboligenic heart diseases (34% versus 12%; $P = .0054$), and occlusive carotid and/or MCA diseases (53% versus 19%; $P = .0004$) compared with patients in the deep penetrating artery group. Such findings suggest infarcts in the superficial penetrating artery showed a higher prevalence of embolic mechanism including cardioembolism or artery-to-artery embolism. In our study of 74 consecutive patients with medullary artery infarcts compared with 71 patients with small infarcts of the LSA, potential causes of embolism including cardiogenic and proximal large artery disease were significantly higher in the medullary artery group (30 patients, 40%) than in the small LSA group (14 patients, 19%) ($P = .0071$).

Long Insular Artery

The long insular artery (LIA) is one of the superficial penetrating arteries that arises from the insular segment of the MCA and supplies the insular cortex, extreme capsule, claustrum, and external capsule, and often extends to the corona radiata (Fig. 78.4). Tamura et al. reported eight patients with infarction in the territory of this artery and compared the findings with 50 patients who had LSA territory infarction [17]. They found abrupt onset was more common in the LIA group (five, 62.5%) than in the LSA group (six, 12.0%) ($P = .04$) and the combined prevalence of embolic high-risk sources and moderate-risk sources was higher in the LIA group (three, 37.5%) than in the LSA group (four, 8.0%) ($P = .048$). As the long insular artery runs close to the lenticulostriate arteries, an infarction in the territory of the long insular artery is often difficult to discriminate from that of the lenticulostriate arteries. As embolic mechanisms are a more common etiology in the LIA group than in the LSA group, adding a coronal view of MRI may be recommended for a corona radiata infarction.

White Matter Lesions or Leukoaraiosis

Vascular Supply and Vascular Pathology

White matter lesions (WMLs) or leukoaraiosis indicate small vessel vascular brain disease as well as degenerative or inflammatory processes. WMLs appear

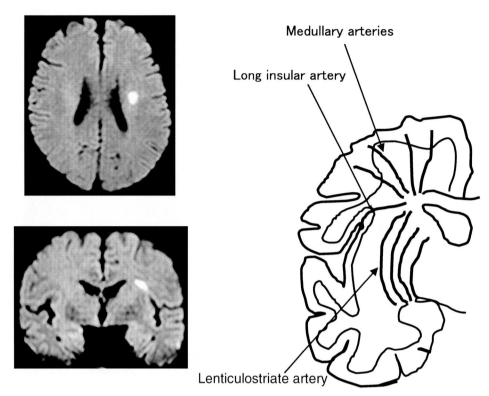

FIGURE 78.4 The infarction of the long insular artery that is one of the superficial penetrating arteries arising from the insular segment of the middle cerebral artery.

as hyperintense periventricular or subcortical patchy or confluent areas on T2 or fluid-attenuated inversion recovery MRI sequence. The pathology of WML includes a triad of demyelination, axonal loss, and lacunar infarcts in the periventricular and subcortical white matter. Cerebral white matter receives its blood supply through long medullary arteries, which have an average diameter of 100–200 μm, originating from the pial arteries. Arteriosclerosis, hyalinosis, and focal fibrinoid necrosis are observed in the long medullary arteries. Although lacunar infarcts are produced when the ischemic damage is focal and severe enough to result in a small area of necrosis especially in the basal ganglion, diffuse WMLs are caused by leakage of fluid from abnormal arterioles and capillaries with resultant gliosis. Some WMLs are caused by diffuse ischemia related to reduced blood flow.

Clinical Manifestations

Patients with extensive WMLs are clinically characterized by executive dysfunction, abnormal gait, urinary incontinence, pseudobulbar palsy, mood disturbances, and dementia. Extensive WML is also associated with a higher incidence of primary stroke, recurrent stroke, progressive cognitive impairment, poor functional outcome, and death. The severity of leukoaraiosis was independently associated with larger infarct volume, increased risk of hemorrhagic transformation,

and parenchymal hematoma following thrombolysis or intraarterial thrombectomy for treatment of acute ischemic stroke.

Blood Pressure as Major Risk Factor

Hypertension has consistently been associated with cross-sectional WML volume and longitudinal measured WML progression. In longitudinal studies, progression of white matter hyperintensities is associated with uncontrolled hypertension especially with high nighttime blood pressure (Fig. 78.5) [18,19]. Nondipper, especially riser status, in which nocturnal blood pressure decline on 24-h blood pressure monitoring is absent, is strongly associated with extensive WML and a progression of poststroke vascular dementia. Even though daytime blood pressure appears to be normal, it is better to monitor nighttime blood pressure in patients with extensive WML. Moreover, chronic kidney diseases are also independently associated with extensive WML [20]. Aggressive blood pressure control considered to protect renal function may reduce progression of cognitive impairment and onset of dementia associated with stroke, although the optimal level of blood pressure for older patients is still being debated. As nondipper status is considered to be associated with salt-sensitive hypertension, use of diuretics may be effective.

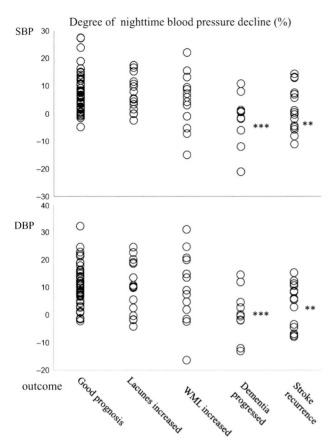

FIGURE 78.5 One hundred and five patients with symptomatic lacunar infarcts were tracked for 3.2 years applying 24-h blood pressure (BP) monitoring. A reduced nocturnal BP dip had an adverse effect on the development of dementia and symptomatic stroke attack.

References

[1] Fisher CM. Lacunar infarcts: a review. Cerebrovasc Dis 1991;1:311–20.

[2] Boiten J, Lodder J, Kessels F. Two clinically distinct lacunar infarct entities? A hypothesis. Stroke 1993;24:652–6.

[3] Caplan LR. Intracranial branch atheromatous disease: a neglected, understudied, and underused concept. Neurology 1989;39:1246–50.

[4] Nah HW, Kang DW, Kwon SU, Kim JS. Diversity of single small subcortical infarctions according to infarct location and parent artery disease: analysis of indicators for small vessel disease and atherosclerosis. Stroke 2010;41:2822–7.

[5] Yamamoto Y, Ohara T, Hamanaka M, Hosomi A, Tamura A, Akiguchi I. Characteristics of intracranial branch atheromatous disease and its association with progressive motor deficits. J Neurol Sci 2011;304:78–82.

[6] Tanriover N, Kucukyuruk B, Ulu MO, Isler C, Sam B, Abuzayed B, et al. Microsurgical anatomy of the cisternal anterior choroidal artery with special emphasis on the preoptic and postoptic subdivisions. J Neurosurg 2014;120:1217–28.

[7] Hamoir XL, Grandin CB, Peeters A, Robert A, Cosnard G, Duprez T. MRI of hyperacute stroke in the AChA territory. Eur Radiol 2004;14:417–24.

[8] Sohn H, Kang DW, Kwon SU, Kim JS. Anterior choroidal artery territory infarction: lesions confined to versus beyond the internal capsule. Cerebrovasc Dis 2013;35:228–34.

[9] Marinkovic SV, Milisavljevic MM, Kovacevic MS, Stevic ZD. Perforating branches of the middle cerebral artery. Microanatomy and clinical significance of their intracerebral segments. Stroke 1985;16:1022–9.

[10] Cho ZH, Kang CK, Han JY, Kim SH, Kim KN, Hong SM, et al. Observation of the lenticulostriate arteries in the human brain in vivo using 7.0T MR angiography. Stroke 2008;39:1604–6.

[11] Konishi J, Yamada K, Kizu O, Ito H, Sugimura K, Yoshikawa K, et al. MR tractography for the evaluation of functional recovery from lenticulostriate infarcts. Neurology 2005;64:108–13.

[12] Yamamoto Y, Nagakane Y, Makino M, Ohara T, Koizumi T, Makita N, et al. Aggressive antiplatelet treatment for acute branch atheromatous disease type infarcts: a 12-year prospective study. Int J Stroke 2014;9:E8.

[13] Makita N, Yamamoto Y, Nagakane Y, Ashida S, Mizuno T. Very prolonged capsular warning syndrome. J Neurol Sci 2015;352:115–6.

[14] Schmahmann JD. Vascular syndromes of the thalamus. Stroke 2003;34:2264–78.

[15] Carrera E1, Bogousslavsky J. The thalamus and behavior: effects of anatomically distinct strokes. Neurology 2006;66:1817–23.

[16] Yonemura K1, Kimura K, Minematsu K, Uchino M, Yamaguchi T. Small centrum ovale infarcts on diffusion-weighted magnetic resonance imaging. Stroke 2002;33:1541–4.

[17] Tamura A, Kasai T, Akazawa K, Nagakane Y, Yoshida T, Fujiwara Y, et al. Long insular artery infarction: characteristics of a previously unrecognized entity. Am J Neuroradiol 2014;35:466–71.

[18] Yamamoto Y, Akiguchi I, Oiwa K, Hayashi M, Kimura J. Adverse effect of nighttime blood pressure on the outcome of lacunar infarct patients. Stroke 1998;29:570–6.

[19] Yamamoto Y, Akiguchi I, Oiwa K, Hayashi M, Ohara T, Ozasa K. The relationship between 24-hour blood pressure readings, subcortical ischemic lesions and vascular dementia. Cerebrovasc Dis 2005;19:302–8.

[20] Yamamoto Y, Ohara T, Nagakane Y, Tanaka E, Morii F, Koizumi T, et al. Chronic kidney disease, 24-h blood pressure and small vessel diseases are independently associated with cognitive impairment in lacunar infarct patients. Hypertens Res 2011;34:1276–82.

CHAPTER

79

Anterior Circulation: Large Artery Occlusive Disease and Embolism

M. Mehdiratta[1,2], Y. Perez[1,2]

[1]Trillium Health Partners, Mississauga, ON, Canada; [2]University of Toronto, Toronto, ON, Canada

INTRODUCTION

Ischemic brain infarctions involving the anterior circulation are by far the most common type of ischemic stroke due to the relative amount of blood that flows to this area as opposed to the posterior circulation. Mechanisms causing anterior circulation cerebral infarction can be separated into three broad categories: cardiogenic embolism, artery-to-artery embolism, and hemodynamic insufficiency due to low perfusion.

Cardiogenic embolism accounts for about 25% of all ischemic strokes and is usually due to atrial fibrillation (either paroxysmal or sustained). Other mechanisms postulated in formation of cardiogenic emboli include left atrial appendage thrombus, paradoxical embolism via a patent foramen ovale, and aortic arch plaque embolism. Emboli tend to involve large- and medium-sized anterior circulation arteries and collect at bifurcations and trifurcations (e.g., middle cerebral artery (MCA) bifurcation/trifurcation, top of the basilar, posterior cerebral artery (PCA)). Emboli may also spontaneously thrombolyze themselves due to natural intrinsic thrombolytics processes resulting in emboli to smaller penetrating arteries.

In up to 25% of cases of suspected brain embolism, a cause cannot be found giving way to a new term in stroke neurology, Embolic Stroke of Undetermined Source (ESUS). Currently clinical trials are underway to determine optimal treatment for patients with ESUS.

Artery-to-artery embolism can also occur. The source is usually the carotid artery bifurcation where atherosclerotic disease tends to occur resulting in plaque formation and subsequent emboli. Intracranial stenosis that usually occurs in the MCA can result in thrombosis and embolism as well giving rise to different stroke syndromes.

VASCULAR ANATOMY AND PATHOLOGY

Cerebral Vasculature Anatomy

The aortic arch divides into three branches: the left common carotid artery, the left subclavian artery, and the brachiocephalic trunk which gives rise to the right subclavian and right common carotid artery.

"The internal carotid arteries arise from the common carotid artery at the carotid bifurcation. The internal carotid artery (ICA) and its main branches supply the cerebral hemisphere and part of the deep, subcortical areas. The origin of the ICA in the neck at approximately C4–C5 makes it readily accessible to examination by auscultation (for bruit) and noninvasive vascular studies. Atherosclerosis is the predominant condition affecting the extracranial ICA and is restricted to the first 2 cm of its origin. Fibromuscular dysplasia and dissection of the carotid arteries are other causes of artery-to-artery embolism emanating from the carotid artery.

Obstructive lesions at the site may cause hemodynamic changes in the distal regions of the hemispheric blood supply, in the so-called border zones between the major cerebral circulations [anterior cerebral artery (ACA), middle cerebral artery (MCA), and posterior cerebral artery (PCA)], or may serve as a source of intra-arterial embolism obstructing the intracranial circulation" (from previous edition).

The ICA passes through the petrous temporal bone and the cavernous sinus and reaches the subarachnoid space at the base the brain. It divides into the middle and anterior cerebral arteries but before this bifurcation it gives rise to the anterior choroidal artery and the PCA.

The anterior choroidal artery is long and thin and supplies the optic tract, parts of the choroid plexus and cerebral peduncle, and deeper areas such as parts of the internal capsule and hippocampus. The posterior communicating artery joins the PCA (which is part of the vetebrobasilar system).

The ACA supplies the medial aspect of frontal and parietal lobes.

The large MCA supplies the lateral aspect of the cerebral hemispheres including the precentral gyrus, postcentral gyrus, and insula. It is divided into an M1 and M2 branch followed by deeper branches.

Large arteries such as the MCAs can be prone to emboli being lodged at the M1 or proximal MCA or at the bifurcation between the M1 and M2 arteries. Atherosclerosis can also occur in the proximal MCA. Large emboli to the M1 branch are being removed using embolectomy given recent data showing the efficacy of this type of procedure.

There are small perforating arteries supplying the deep cerebral structures. These smaller arteries are subject to atherosclerosis in the form of lipohyalinosis that causes smaller deep infarcts. These so-called "lacunar" infarcts can result in pure motor or pure sensory syndromes if the anterior circulation perforators are involved.

Clinical Manifestations of Anterior Circulation Stroke Syndromes

Even with the development of sophisticated neuroimaging techniques, lesion localization in stroke neurology has remained central to the diagnosis and management of stroke syndromes. In the acute phase, localization is essential to the accurate diagnosis of cerebral dysfunction in a vascular territory as well as to rarer manifestations of stroke allowing the planning of acute interventions.

Ischemia Related to Extracranial Internal Carotid Artery Disease

Transient ischemic attacks (TIAs) and ischemic strokes are commonly associated with stenosis or occlusion of the proximal extracranial ICA causing embolism or less commonly hypoperfusion. While embolism will result in occlusion of a more distal artery intracranially, occlusion can be asymptomatic or result in a critical reduction in blood flow, especially when collateral circulation is lacking. TIAs are the main warning symptom of proximal extracranial ICA disease and typically involve the retina or hemispheres [1]. The clinical presentation of TIAs is heterogeneous and depends on the territory of the occluded vessel.

Transient Monocular Blindness

Classically described as a curtain coming down into the field of vision of one eye, this type of altitudinal loss of vision is infrequent. More commonly the vision loss is described as a blackout, graying, or dimming of vision that can last seconds to several minutes or longer. This is usually the result of an embolism to the ophthalmic artery or its branches [1].

An ischemic ophthalmopathy can also result from hypoperfusion of the retina. Monocular vision loss in this case is usually triggered by bright lights and described as bleaching of objects followed by brief vision loss. This is a form of retinal claudication due to hypoperfusion related to proximal ICA occlusion or severe stenosis. It is associated with a progressive loss of visual acuity and neovascularization of the retina [2].

Limb-Shaking Transient Ischemic Attacks

This type of TIA presents as a short period of involuntary jerking movements of the arm and/or leg lasting minutes, mimicking a focal seizure. It is associated with hemodynamically significant ICA stenosis or occlusion [3]. The transient ischemia in this type of TIA is thought to be due to transient hypoperfusion in the context of reduced flow to the hemisphere when cerebral vascular reactivity is impaired as vascular beds become maximally dilated and the hemispheric hemodynamic reserve has been exhausted.

Unilateral Watershed Territory Ischemic Syndromes

Watershed infarcts occur at the border zone between two arterial systems [2]. Anteriorly, they involve the junction between MCA and ACA territories. Posteriorly, they involve the junction between MCA and PCA. They are best recognized radiologically as a wedge area of infarction in the frontal or parieto-occipital cortex. There is also a subcortical border zone between the lenticulostriate perforating branches and the cortical branches of the MCA. Radiologically, these watershed infarcts are located along the centrum semiovale and classically appear as a "string of pearls" (Fig. 79.1).

While bilateral watershed infarcts are associated with severe systemic hypotension, unilateral watershed infarcts can be seen in association with severe stenosis or occlusion of the ICA. The mechanism of infarction is thought to be hemodynamic, embolic, or a combination of both. The clinical presentation of unilateral watershed infarcts is variable, depending on its location (Fig. 79.2).

Ischemia Related to Intracranial Carotid Artery Disease

Occlusion of the distal carotid artery near its termination can occur as a result of atherosclerosis or embolism from a more proximal source. This usually results in a severe stroke with features of both ACA and MCA territory ischemia as well as some impairment in level of consciousness. These are discussed under their respective headings.

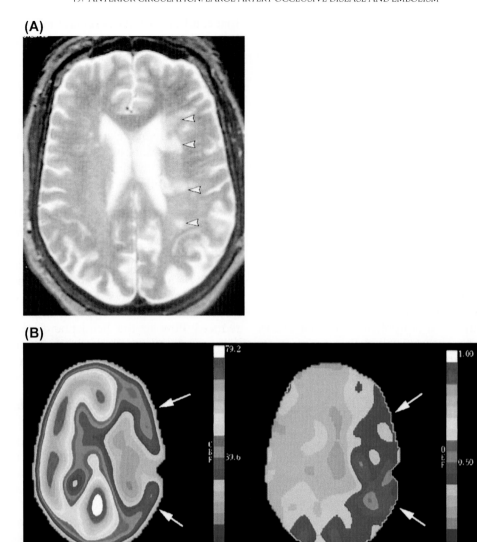

FIGURE 79.1   Linear pattern of deep white matter infarction in the corona radiata and increased OEF in a 72-year-old woman presenting with aphasia 104 days before MR and PET. Her symptoms had improved but not resolved at the time of these examinations. Angiograms (not shown) obtained on the day of symptom onset demonstrated complete occlusion of the left carotid artery, with collateral flow to the left hemisphere provided by the anterior communicating artery and retrograde flow through the left ophthalmic artery. MR image depicted frontal lobe infarction (not shown) correlating with the patient's ischemic symptoms. (A) T2- weighted spin-echo MR image (2500/90 [repetition time msec/echo time msec]; one signal acquired) through the corona radiata shows several white matter infarctions (arrowheads) parallel to the long axis of the lateral ventricle. (B) PET image shows reduced CBF (*arrows*, left image) and increased OEF (*arrows*, right image) in the hemisphere ipsilateral to the occluded carotid artery. *Reprinted, with permission, from Radiology 2001;220:195-201. ©RSNA.*

Clinical Features of Middle Cerebral Artery Territory Ischemia

The syndrome of complete MCA territory infarction presents as hemiplegia and hemisensory deficits involving the contralateral face, arm, and leg; hemianopia and contralateral gaze; and head deviation. Global aphasia is seen with dominant hemisphere lesions while hemineglect and anosagnosia can be seen with nondominant hemisphere lesions. This is usually associated with a proximal M1 branch occlusion, either from embolism or focal atherosclerosis, and portends an unfavorable prognosis [4]. Malignant brain edema can also develop in the infarcted region leading to increased intracranial pressure and life-threatening herniation.

Occlusion of the M1 distally sparing the lenticulostriate perforators will lead to a more superficial infarct with sparing of the deeper external capsule. This results clinically in a lesser involvement of the motor and sensory functions of the legs than of the face and arms.

In the usual anatomical configuration, the M1 or main trunk of the MCA divides into the superior and inferior M2 divisions. Embolism is the major cause of M2 occlusive disease.

Infarction in the territory of the superior M2 division results in predominantly motor and language deficits with relative sparing of sensory and visual functions. The most common syndrome involves weakness of the face and arm more than leg along with a Broca's aphasia with dominant hemisphere lesions or hemineglect with nondominant hemisphere lesions.

With infarction in the territory of the inferior M2 division, sensory and visual deficits predominate. The clinical manifestation includes sensory deficits in the contralateral face and arm more than leg with mild motor deficits, hemianopsia, Wernicke's aphasia with dominant hemisphere lesions, hemineglect, and behavioral changes with nondominant hemisphere lesions.

Clinical Features of Anterior Cerebral Artery Territory Ischemia

The common clinical presentation of ACA territory infarcts involves weakness and sensory deficits predominantly in the contralateral leg and foot. This is usually associated with an ACA branch occlusion from embolism or less commonly focal atherosclerosis. Other features can include mutism or abulia, transcortical motor aphasia with dominant hemisphere involvement, and behavioral disturbances. Unilateral left-sided apraxia can occur as a disconnection syndrome when there is involvement of the corpus callosum [5]. Less commonly urinary incontinence can occur.

Clinical Features of Anterior Choroidal Artery Ischemia

The anterior choroidal artery supplies the inferior part of the posterior limb of the internal capsule including retrolenticular projections, the optic tract, lateral geniculate body, and optic radiations. The classic syndrome involves contralateral hemiparesis and hemisensory deficit, with a homonymous hemianopia with sparing of the horizontal meridian. More commonly however, anterior choroidal infarcts present as a pure motor, sensorimotor, or ataxic–hemiparesis syndrome [6] (see Fig. 79.3).

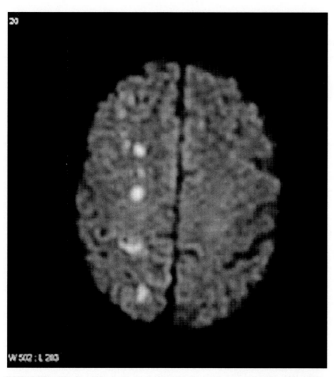

FIGURE 79.2 Watershed infarct: deep (internal) border zones infarct. ≥3 lesions, each ≥3 mm in diameter in a linear fashion parallel to the lateral ventricles in the centrum semiovale or corona radiata, that sometimes become more confluent and band-like.

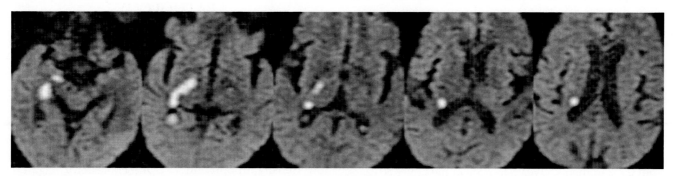

FIGURE 79.3 An associated infarct of choroid plexus of adjacent right lateral ventricle with increased signal on diffusion. http://www.neuroradiologycases.com/2012/03/anterior-choroidal-artery-territory.html.

References

[1] North American Symptomatic Carotid Endarterectomy Trial Collaborators. Beneficial effect of carotid endarterectomy in symptomatic patients with high-grade carotid stenosis. N Engl J Med 1991;325:445–53.

[2] Ringelstein ER, Stogbauer F. Border zone infarcts. In: Bogousslavsky J, Caplan L, editors. Stroke syndromes 2nd edition. Cambridge University Press; 2001. p. 564–83.

[3] Fisher CM. Concerning recurrent transient cerebral ischemic attacks. CMAJ 1962;86:1091–9.

[4] Mohr JP, Kejda-Scharler J. Middle cerebral artery territory syndromes. In: Caplan L, Van Gijn J, editors. Stroke syndromes 3rd edition. Cambridge University Press; 2012. p. 344–63.

[5] Bogousslavsky J, Regli F. Anterior cerebral artery infarction in the Lausanne stroke registry. Arch Neurol 1990;47(2):144–50.

[6] Brazis PW, Masdeu JC, Biller J. Vascular syndromes of the brainstem forebrain and cerebellum. In: Localization in clinical neurology 7th edition. Lippincott, Williams and Wilkins; 2017.

CHAPTER

80

Carotid Artery Disease

P. Bhattacharya[1], S. Chaturvedi[2]

[1]Saint Joseph Mercy Oakland, Pontiac, MI, United States; [2]University of Miami Miller School of Medicine, Miami, FL, United States

INTRODUCTION

Numerous pathological processes affect the cervical and intracranial carotid artery. The most common and widely studied is atherosclerosis of the cervical carotid artery. Atherosclerosis of the intracranial carotid arteries has the same pathophysiological basis and can present with similar symptoms. It has a higher prevalence among African-American and Asian populations. Some other diseases affecting the carotid artery that are occasionally encountered in practice include carotid artery dissections, Moyamoya disease, and fibromuscular dysplasia. This chapter primarily focuses on cervical carotid artery atherosclerotic disease.

Cervical carotid artery atherosclerosis is well established as a risk factor for ischemic stroke. It is responsible for about 7% of all ischemic strokes. However, the prevalence of asymptomatic carotid disease is much higher. The prevalence of moderate (50–69%) carotid stenosis varies from 0.2% in men <50 years to 7.5% in men over 80 years. Among women, prevalence ranges from 0% under 50 years to 5.0% over 80 years. Among men, prevalence of severe stenosis ranges from 0.1% under 50 years to 3.1% over 80 years. Among women, the prevalence ranges from 0% under 50 years to 0.9% over 80 years [1]. Traditional atherosclerotic risk factors such as arterial hypertension, diabetes mellitus, dyslipidemia, cigarette smoking, low physical activity, and older age increase the risk of carotid stenosis. With increasing numbers of risk factors in an individual, the risk of carotid stenosis progressively increases.

PATHOLOGY

Carotid atherosclerosis is an inflammatory disease. Inflammation plays a pivotal role in the initiation, propagation, and eventual rupture of atherosclerotic plaque [2]. Normal endothelium is in a quiescent, antiinflammatory state, with excess production of nitric oxide—regarded as protective for the endothelium. Vascular risk factors such as arterial hypertension, diabetes mellitus, oxidized low-density lipoprotein (LDL) cholesterol, very low–density lipoprotein (VLDL) cholesterol, cigarette smoking, hyperhomocysteinemia, certain infections (e.g., *Chlamydia pneumoniae*), and mechanical shear stresses in the region of the carotid bulb convert the endothelium into a pro-inflammatory state [2]. Oxygen-derived free radicals are produced, which oxidize LDL cholesterol. Oxidized LDL in turn activates other endothelial cells, attracts monocytes to the area, and activates monocytes to macrophages, that engulf the oxidized LDL resulting in foamy macrophages [2]. Oxidative stress also activates nuclear factor-kappa B (NF-kB). This molecule attracts inflammatory cells to the area, but also increases the production of matrix metalloproteinase 9 (MMP9). MMP9 causes degradation of extracellular matrix, that weakens the fibrin cap of a fully formed plaque, resulting in rupture [2].

The American Heart Association (AHA) has described various stages of plaque evolution [3]. The initial lesion involves activated macrophages. This progresses to a fatty streak where foamy macrophages are formed. Next, intermediate lesions are formed, that involve increase in the numbers of foamy macrophages. Some of these macrophages die, leaving extracellular lipid collection. Up to this stage, plaques are asymptomatic [3]. With increasing extracellular lipid, the plaque is now called an atheroma. This progresses to a fibroatheroma with a defined lipid necrotic core and a fibrous cap [3]. When sufficient amount of necrotic lipid accumulates, it may crystallize to form cholesterol crystals. The jagged crystals cause a rupture of the fibrous cap, or may rupture the vasa vasorum of the artery resulting in intraplaque hemorrhage [4]. Such a complicated lesion is the perfect setting for initiation of thrombosis. Once the necrotic lipid core begins to form, the patient is prone to develop clinical symptoms.

MECHANISMS OF SYMPTOMS FROM CAROTID OCCLUSIVE DISEASE

Carotid artery stenosis or occlusion results in symptoms in the territories of supply downstream to the affected artery: the ipsilateral hemisphere or the ipsilateral retina. This could involve two different mechanisms. (1) Low hemodynamic pressure in the internal carotid system will result in episodes of reduction in blood flow. This usually leads to transient retinal ischemia or decreased perfusion to the hemisphere. The hemispheric ischemia more commonly affects watershed territories between major arteries such as the middle cerebral (MCA) and anterior cerebral artery (ACA), or the MCA and posterior cerebral arteries (PCA). (2) With slow progression of carotid atherosclerotic stenosis, collateral circulation may develop through the external carotid system or from the posterior circulation. Such a collateral network may protect against ischemia to the hemisphere or the retina. Emboli (cholesterol crystals, platelet plugs, thrombi forming in situ) in these cases can propagate

downstream to the ophthalmic artery or its branches such as the central retinal artery or the posterior ciliary artery; or the intracranial carotid artery or its branches such as the MCA or the ACA and produce symptoms corresponding to the respective territory.

CEREBRAL HEMISPHERIC SYNDROMES

In its most devastating form, an artery-to-artery embolus from a carotid plaque can result in infarction of the entire MCA territory ipsilaterally. The patient presents with severe hemiplegia and hemisensory loss. Infarction of the frontal eye fields results in conjugate gaze deviation toward the side of the infarction (away from the side with hemiplegia). Involvement of the dominant hemisphere results in global aphasia. Infarction of the nondominant MCA territory results in asomatognosia, anosognosia, motor impersistence, contralateral neglect, dressing and constructional apraxia, and aprosodia. With involvement of both superficial and deep regions of the hemisphere the patient may become progressively somnolent or comatose. Such patients are at risk of developing malignant cerebral edema [5]. If edema progresses, signs of subfalcine herniation (that may compromise the ACA blood supply, increasing the area of infarction) or transtentorial (uncal) herniation may appear. Transtentorial herniation can result in PCA territory infarction as well. Without decompressive hemicraniectomy, the mortality rate in such infarctions is around 80%.

Infarction involving the superior division of the MCA results in partial features of the syndrome described earlier. The contralateral hemiparesis tends to be mostly faciobrachial, with relatively mild lower limb weakness. Eye deviation is less common. Involvement of the dominant hemisphere produces features of a global aphasia initially, which recovers over days to a predominant expressive aphasia. Infarction in the nondominant hemisphere produces the features described earlier to varying degrees. Poststroke depression may develop after hemispheric strokes; seen more commonly following left hemispheric strokes. However, the precise localization is uncertain [5].

Infarction involving the inferior division of the MCA results in a syndrome with predominant visual field defects and cognitive abnormalities. Hemiparesis (usually face and arm) and hemisensory impairment are mild and sometimes transient. Depending on the extent of infarction, the visual deficit may be a quadrantanopia or a hemianopia. Infarction in the dominant hemisphere results in a receptive aphasia. If features of global aphasia are present, they recover quickly to a receptive aphasia. Visual and sensory hemineglect are common with nondominant hemisphere infarctions. An acute confusional state with agitation may commonly occur. Branch infarcts within the superior or inferior divisions produce limited versions of the syndromes described earlier [5].

Hemodynamic stressors in the setting of severe carotid artery stenosis or occlusion can result in infarction in the watershed territories between the ACA and the MCA or the MCA and the PCA. Examples of such stressors are patients with active heart disease with hypotension and syncope, elevated hematocrit secondary to cigarette smoking, blood loss, and postprandial hypotension. Weakness typically affects the proximal extremity more than the distal extremity. In cases with bilateral carotid occlusive disease, infarctions in the bilateral ACA–MCA watershed areas can result in bilateral proximal arm weakness, classically described as the "man in the barrel" syndrome. Ischemic attacks in bilateral carotid occlusive disease may resemble vertebrobasilar ischemia. A transcortical motor aphasia may occur in dominant hemisphere infarcts.

Patients with carotid artery occlusion or severe stenosis may present with intermittent episodes of involuntary jerking movements of the contralateral arm and leg, often associated with variable duration and degree of weakness of the limbs. These are referred to as limb-shaking transient ischemic attacks. They typically last less than 5 min. They are generally precipitated by exercise, rising from the sitting position, or coughing. Ischemia in the watershed areas is believed to be the underlying basis. Electroencephalogram recordings in these patients demonstrate focal slowing in these watershed areas.

The most common pattern of weakness in infarction of the ACA territory involves severe leg weakness and to a lesser extent proximal arm weakness. The distal hand and the face are less commonly affected. Involvement of bilateral ACAs or of a single dominant ACA can result in infarction of bilateral ACA territories, resulting in paraparesis. Hemisensory loss is also most prominent in the contralateral lower extremity. Infarction of the medial frontal lobe results in akinetic mutism or abulia or an amotivational syndrome. Dominant hemisphere ACA territory infarcts can result in varying degrees of language dysfunction including comprehension difficulty, anomia, alexia, and paraphasias.

Infarction of the anterior choroidal artery produces a classic syndrome of hemiplegia, hemisensory loss, and homonymous hemianopia. Rarely, one of the fetal communications between the carotid circulation and the posterior circulation persist, such as a persistent trigeminal artery. In such instances, embolus from a carotid plaque may travel into the PCA territory and cause infarction, leading to prominent visual field defects and neuropsychological abnormalities depending on the extent of infarction. Finally, acute focal neurological deficits can occur in other noncerebrovascular conditions that need

to be differentiated from infarcts. Examples of these conditions include and are not limited to seizures, neoplasms, infections, hypoglycemia, and other metabolic abnormalities. Conversion disorders should also be considered in the category of stroke mimics.

RETINAL ISCHEMIA

Transient ischemia of the retina is a classic symptom of ipsilateral carotid occlusive disease. Referred to as transient monocular blindness (TMB), the patient presents with single or often recurrent stereotypical episodes of complete or partial, graying or obscuration of vision in one eye, lasting seconds or more commonly minutes, followed by complete recovery of vision. The episodes may recur over hours, days, or weeks, and sometimes over months. Patients describe the visual phenomenon as an ascending or descending curtain or blinds moving sideways. Rarely, it is described as moving tracks of light. An ipsilateral headache can sometimes occur. In patients with critical carotid stenosis, or impending central retinal artery occlusion, the episodes may occur frequently over a period of hours, a phenomenon referred to as crescendo TMBs.

Ophthalmological evaluation may reveal cholesterol emboli as bright shiny structures blocking the angles of the vessels. Sometimes, they may be seen only with orbital pressure. These emboli, referred to as Hollenhorst plaques may be present in asymptomatic patients also and predict an increased future risk of systemic vascular events and ischemic stroke. Other findings that are less frequently noted are platelet fibrin emboli that appear as grayish white columns within blood vessels and focal cotton wool spots.

Central retinal artery occlusion presents with a sudden, unremitting, painless loss of vision in one eye. On examination of the affected eye, there is no direct light response, but the indirect/consensual light response and near reflex is preserved. If some degree of vision is preserved, an afferent pupillary defect (APD) may be seen. On ophthalmoscopic examination, classic findings may take at least an hour to develop. The ischemic retina has a pale ground glass appearance to it. At the macula, the normal red color of the choroid shows through as a cherry red spot. These findings persist for the next few days. Eventually they resolve, and the patient is left with optic atrophy.

THE CAROTID BRUIT

Asymptomatic carotid stenosis is sometimes detected on physical examination by detection of a carotid bruit. The presence of a carotid bruit increases the risk of stroke and transient ischemic attacks. In a large multiethnic study of asymptomatic subjects bruits were present in 4.1% of the population. Sensitivity of the bruit to detect carotid artery stenosis was 56%, specificity was 98%; positive predictive value was only 25% and the negative predictive value was 99% [6]. Thus, absence of a bruit does not exclude carotid stenosis. Also, in today's era of intensive medical management, the benefit of detection of asymptomatic carotid stenosis does not outweigh the risk of early revascularization for stenosis. Therefore, referring every patient with a bruit for a carotid artery evaluation may not be an effective practice [7].

OTHER LESS COMMON CLINICAL FEATURES OF CAROTID ARTERY OCCLUSIVE DISEASE

A gradual and progressive syndrome of cognitive dysfunction may develop in patients with hemodynamically significant carotid stenosis or carotid occlusion, even without any evidence of infarction. The mechanism is believed to be a hemodynamic failure resulting in chronically decreased cerebral perfusion. This is supported by evidence that patients with increased oxygen extraction fraction on positron emission tomography (PET) imaging have significantly worse scores on cognitive tests. Revascularization procedures such as EC–IC bypass surgery do not improve cognitive status beyond what is achieved by medical management alone [8].

Some physical examination findings may less commonly be noted in carotid occlusive disease. Facial pulsations are sometimes felt due to increased blood flow through the collateral circulation of the external carotid artery. These collaterals open up over time and feed the orbit following chronic occlusion or progressive severe stenosis of the cervical carotid circulation. Patients sometimes develop a partial Horner's syndrome (miosis and ptosis) due to involvement of the autonomic fibers in the carotid sheath around the internal carotid artery. This is more commonly associated with dissections. Sweating is spared as the autonomic fibers to the sweat glands travel along the external carotid artery. Rarely iris ischemia can result in a dilated pupil.

CAROTID DISSECTION AND MOYAMOYA DISEASE

Carotid artery dissections involve the development of a hematoma within the wall of the carotid artery. It typically occurs about 2 cm distal to the origin of the internal carotid artery in the neck. It is a common cause of ischemic stroke in young and middle-aged adults. The most common etiology is trauma to the carotid artery. However, spontaneous dissections also occur. These are believed to originate from congenital areas of weakness in the vessel wall with a superimposed inciting factor such as an innocuous trauma or infection. Dissections can also occur in vasculopathies such as fibromuscular dysplasia and connective tissue disorders. Ischemic strokes present with the same hemispheric syndromes and retinal ischemic syndromes described earlier. Patients can have concomitant pain (or isolated pain) in the neck or head. They can acutely develop a Horner's syndrome due to involvement of cervical sympathetic fibers. Carotid dissection should be suspected in patients who present with a painful Horner's syndrome. Compression by the enlarged arterial wall can result in hypoglossal nerve (cranial nerve (CN) XII) palsy and less frequently palsies of CNs IX–X and XI. Therefore, lower CN involvement along with pain should raise concern for dissection. Occasionally the dissection segment extends intracranially, in which case the arterial wall balloons out as a pseudoaneurysm that can rupture causing subarachnoid hemorrhage [9].

Moyamoya disease is a chronically progressive, occlusive disease involving the terminal portions of the internal carotid arteries bilaterally. An abnormal vascular network of fragile collaterals develops in the region of the stenosis, supplying the deep portions of the cerebral hemispheres. The cause of Moyamoya disease is unknown. A similar vascular pattern can develop secondary to other diseases such as atherosclerosis, autoimmune disease (Graves' disease), meningitis, neurofibromatosis type 1, sickle cell disease, and cranial irradiation. The incidence is about 10 times higher in east Asian countries compared to the Western hemisphere, and twice as common in women compared to men. The disease presents with ischemic strokes in children and ischemic events or intracerebral hemorrhages in adults. Ischemic events tend to be multiple and recurrent. Patients commonly report migraine headaches due to stimulation of dural pain receptors by the collateral circulation. Cortical ischemia can result in development of epilepsy among children. Basal ganglia injury can result in movement disorders such as chorea [10].

References

[1] de Weerd M, Greving JP, Hedblad B, Lorenz MW, Mathiesen EB, O'Leary DH, et al. Prevalence of asymptomatic carotid artery stenosis in the general population: an individual participant data meta-analysis. Stroke 2010;41(6):1294–7.

[2] Roquer J, Segura T, Serena J, Castillo J. Endothelial dysfunction, vascular disease and stroke: the ARTICO study. Cerebrovasc Dis (Basel, Switz) 2009;27(Suppl. 1):25–37.

[3] Stary HC, Chandler AB, Dinsmore RE, Fuster V, Glagov S, Insull Jr W, et al. A definition of advanced types of atherosclerotic lesions and a histological classification of atherosclerosis. A report from the Committee on Vascular Lesions of the Council on Arteriosclerosis, American Heart Association. Circulation 1995;92(5):1355–74.

[4] Mughal MM, Khan MK, DeMarco JK, Majid A, Shamoun F, Abela GS. Symptomatic and asymptomatic carotid artery plaque. Expert Rev Cardiovas Ther 2011;9(10):1315–30.

[5] Mohr JP, Kejda-Scharler J. Middle cerebral artery territory syndromes. In: Caplan LR, van Gijn J, editors. Stroke syndromes. Cambridge University Press; 2012. p. 344–63.

[6] Ratchford EV, Jin Z, Di Tullio MR, Salameh MJ, Homma S, Gan R, et al. Carotid bruit for detection of hemodynamically significant carotid stenosis: the Northern Manhattan Study. Neurol Res 2009;31(7):748–52.

[7] Jonas DE, Feltner C, Amick HR, Sheridan S, Zheng ZJ, Watford DJ, et al. Screening for asymptomatic carotid artery stenosis: a systematic review and meta-analysis for the U.S. Preventive Services Task Force. Ann Intern Med 2014;161(5):336–46.

[8] Marshall RS, Festa JR, Cheung YK, Pavol MA, Derdeyn CP, Clarke WR, et al. Randomized Evaluation of Carotid Occlusion and Neurocognition (RECON) trial: main results. Neurology 2014;82(9):744–51.

[9] Debette S, Leys D. Cervical-artery dissections: predisposing factors, diagnosis, and outcome. Lancet Neurol 2009;8(7):668–78.

[10] Guey S, Tournier-Lasserve E, Herve D, Kossorotoff M. Moyamoya disease and syndromes: from genetics to clinical management. Appl Clin Genet 2015;8:49–68.

CHAPTER

81

Posterior Circulation: Large Artery Occlusive Disease and Embolism

R.J. Wityk

The Johns Hopkins University School of Medicine, Baltimore, MD, United States

INTRODUCTION

In many respects, strokes involving the posterior circulation parallel those of the anterior circulation. Ischemic infarcts may be due to atherothrombosis of large arteries, lacunes from occlusion of small penetrating arteries, or embolism from either the heart or proximal vascular sites. Unlike those of the anterior circulation, however, the vessels of the posterior circulation are more difficult to study by noninvasive means. Carotid stenosis in the neck, for example, can be easily assessed by duplex ultrasound, but similar disease in the extracranial vertebral artery may require angiography for reliable detection. The complexity of structures in the brain stem also makes localization of infarction more difficult. Nevertheless, certain clinical patterns help localize the lesions and suggest which vessels are involved [1]. These clinical findings aid the physician in determining the most likely mechanism of stroke and the most appropriate therapy.

VASCULAR ANATOMY OF THE POSTERIOR CIRCULATION

Vessels comprising the posterior circulation include the vertebral arteries arising from the subclavian arteries, three paired cerebellar arteries, the basilar artery, and two posterior cerebral arteries (PCAs) arising from the top of the basilar artery (Fig. 81.1). The structures supplied by these vessels include the brain stem, cerebellum, thalamus, medial temporal lobes, and occipital lobes. There are two types of vessels that supply the brain stem and cerebellum. Long circumferential arteries such as the cerebellar arteries run superficially and supply the lateral brain stem and the cerebellar hemispheres. Smaller penetrating arteries arise from the dorsal surface of the basilar artery and supply the basis pontis and medial portions of the brain stem. These smaller arteries do not have collaterals and occlusions of these vessels result in brain stem lacunes.

Primer on Cerebrovascular Diseases, Second Edition
http://dx.doi.org/10.1016/B978-0-12-803058-5.00081-3

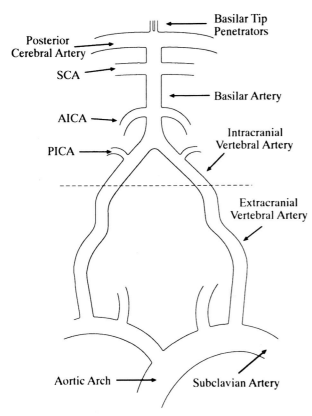

FIGURE 81.1 Vascular anatomy. Schematic representation of the arteries of the posterior circulation. The *dotted line* represents the division between the intracranial and extracranial vessels. *AICA,* anterior inferior cerebellar artery; *PICA,* posterior inferior cerebellar artery; *SCA,* superior cerebellar artery.

Posterior circulation ischemia may cause slurred speech, motor weakness, sensory change, and visual loss, similar to that found with anterior circulation stroke. Clinical features pointing to brain stem or cerebellar ischemia, however, are vertigo, nystagmus, gait or limb ataxia, double vision, bilateral motor weakness, and coma (Table 81.1). Findings such as aphasia, hemineglect, and higher-order sensory loss (astereognosis, agraphesthesia) are more commonly seen in anterior circulation ischemia.

SUBCLAVIAN ARTERY DISEASE

Atherosclerosis of the subclavian artery also commonly involves the origin of the vertebral artery. "Subclavian steal" refers to a condition in which there is severe stenosis or occlusion of the subclavian artery proximal to the origin of the vertebral artery. Low pressure in the subclavian artery distal to the lesion leads to "stealing" of blood from the vertebral artery [2]. Angiography in this situation shows normal antegrade blood flow in the right vertebral artery, but reversed flow down the left vertebral artery into the subclavian artery.

TABLE 81.1 Clinical Features That Suggest Posterior Circulation Ischemia

Coma
Dizziness, vertigo
Nystagmus, ocular bobbing
Diplopia, oculomotor palsy, skew deviation
Horners syndrome
Acute unilateral hearing loss
Lower motor neuron pattern of facial weakness
Bilateral perioral or tongue paresthesias
Bilateral motor weakness, bilateral Babinski signs
Ataxic gait with wide base and unsteadiness on turns
Limb ataxia, past-pointing on finger-to-nose testing, "cerebellar" tremor
Crossed sensory loss (e.g., loss of sensation on the right face and left arm/leg)
Severe dysarthria and dysphagia
Higher-order visual loss (prosopagnosia, color agnosia, visual agnosia)

On examination of a patient with left subclavian steal, for example, one may find a left supraclavicular bruit and decreased amplitude or delay of the left radial pulse. Occasionally patients have symptoms such as dizziness, parethesias, blurred vision, and ataxia, but many have no symptoms at all. Exercise of the affected arm is said to precipitate symptoms (by drawing off more blood into the arm), but this is actually a rare finding. Although patients may have transient neurological symptoms, the risk of stroke is small. Many patients with subclavian stenosis are asymptomatic, and the condition is an incidental finding on examination.

VERTEBRAL ARTERY ORIGIN DISEASE

Atherosclerotic disease of the vertebral artery origin (VAO) is common in patients with peripheral vascular disease, but because of difficulty in noninvasive detection of these lesions, the natural history of VAO stenosis is uncertain. If both vertebral arteries are of good caliber, then occlusion of one extracranial vertebral artery can be asymptomatic. On the other hand, in some patients one vertebral artery is either very small, occluded, or ends before connecting with the basilar artery (an anatomical variant). In these individuals with "basilarization" of the remaining vertebral artery, disease in this artery can produce symptoms similar to basilar artery stenosis.

Atherosclerosis of the VAO can lead to transient ischemic attacks (TIAs) or stroke in a manner parallel to a carotid artery stenosis in the neck [3]. Spells may be hemodynamic in nature (particularly if the other vertebral artery is small or diseased) or due to platelet fibrin emboli from the site of atherosclerosis. Patients have TIAs with dizziness, ataxic gait, slurred speech, transient weakness or numbness of the extremities, and diplopia. If the extracranial vertebral artery thromboses, a clot may break off and embolize into the intracranial circulation, causing stroke. An embolus typically lodges in the intracranial vertebral artery (ICVA), at the apex of the basilar artery, or in a PCA.

Patients with VAO stenosis and TIAs often respond to antiplatelet therapy (e.g., aspirin or clopidogrel). Those with continued spells on antiplatelet agents can be treated with anticoagulation with warfarin, but the benefit of long-term anticoagulation in this setting is uncertain. Anticoagulants are often used in patients with acute vertebral artery thrombosis to prevent propagation of the clot or distal embolism. These patients can be treated with warfarin for several months and then switched to aspirin, because by that time the thrombus has probably organized and poses less of a threat for embolism. For patients with vertebral origin stenosis who have continued TIAs despite medical treatment, the stenosis can be treated by endovascular means, typically with a stent placement [4]. Case series suggest that this is a relatively safe procedure with resolution of TIAs, but no large, randomized clinical trial has been performed to compare the long-term outcome of stenting with medical therapy alone.

EXTRACRANIAL VERTEBRAL ARTERY DISSECTION

Dissection of the extracranial carotid or vertebral arteries occurs more commonly than previously thought [5]. Patients tend to be younger than those with atherosclerotic stroke, with the most common age being the late 30s or early 40s. Dissection may occur spontaneously or be associated with a trivial trauma. Lacerations or dissections of the vertebral artery can occur with severe neck trauma, such as with a motor vehicle accident or gunshot wound. Patients with fibromuscular dysplasia are at higher risk for spontaneous dissection. The extracranial vertebral artery can also be injured by vigorous neck manipulation during chiropractic maneuvers or by iatrogenic trauma during central line placement.

Patients typically present with pain in the posterior neck which radiates to the occiput. Because of disruption of the vascular endothelium, the site of dissection is a potential source for thrombus formation. A common story is that patients have neck pain for several days and then seek medical attention when they develop

acute posterior circulation ischemia. A vertebral artery dissection can occasionally be seen on a magnetic resonance imaging (MRI) scan. The mural hematoma has certain signal characteristics that can be identified on axial images. Magnetic resonance angiography (MRA) shows narrowing or absence of the vessel, but the findings are nonspecific. Computed tomographical angiography (CTA) may demonstrate the vascular irregularities of dissection better than MRA. Conventional angiography, however, shows either occlusion or tapered narrowing of the artery and can be diagnostic. Short-term anticoagulation is often used to prevent artery-to-artery embolism, but there is no clinical trial evidence to show that this is superior to antiplatelet therapy alone. The dissection often resolves in 2–4 months. Noninvasive imaging such as MRA is very useful in following these patients to determine when vessel patency is restored.

INTRACRANIAL VERTEBRAL OCCLUSION

The most common stroke with occlusion of the ICVA is the lateral medullary stroke (Wallenberg syndrome). ICVA occlusion is often accompanied by occlusion of the posterior inferior cerebellar artery (PICA), which typically arises from this segment (Fig. 81.1). Table 81.2 lists the signs and symptoms of the complete syndrome. If the ICVA is occluded, the empirical use of intravenous heparin can be considered to prevent propagation of the clot. However, subadventitial dissection of the ICVA can lead to rupture and subarachnoid hemorrhage (a known

TABLE 81.2   Clinical Features of the Lateral Medullary Syndrome

Dizziness, vertigo
"Lateropulsion" toward the side of the lesion when sitting
Gait ataxia with tendency to fall toward the side of the lesion
Ipsilateral limb ataxia
Double vision, skew deviation
Homers syndrome (ipsilateral)
Nystagmus, dysmetria of eye movements
Ipsilateral facial numbness, facial pain
Contralateral loss of pinprick on the arm and leg
Dysarthria, dysphagia
Nausea and vomiting
Posterior neck or occipital pain
Hiccups
Autonomic dysfunction, fluctuating blood pressure

cause of high spinal subarachnoid hemorrhage in the absence of an aneurysm or arteriovenous malformation (AVM)).

It is critically important to look for an associated cerebellar stroke if the PICA is occluded. A large PICA territory cerebellar stroke can swell over 24–48h and either compress the lower brain stem or block the fourth ventricle, causing obstructive hydrocephalus (Fig. 81.2). Patients need to be followed closely in an intensive care unit for signs of decreased alertness or progressive brain stem compromise. The treatment is urgent surgical decompression of the edematous infarct. Some patients may do well with temporary ventriculostomy for hydrocephalus alone. When the condition is diagnosed and treated early enough, the outcome is generally excellent, even after surgical resection of half the cerebellum.

The ICVA is a common site for atherosclerosis. An orthostatic trigger to TIAs suggests the contralateral vertebral artery is either hypoplastic or is narrowed by disease as well. In the past, it was common to use anticoagulants for patients with vertebrobasilar ischemic symptoms, but the WASID trial [6a] showed no advantage of dose-adjusted anticoagulation with warfarin over aspirin alone for secondary prevention. Endovascular treatment of intracranial stenosis, including

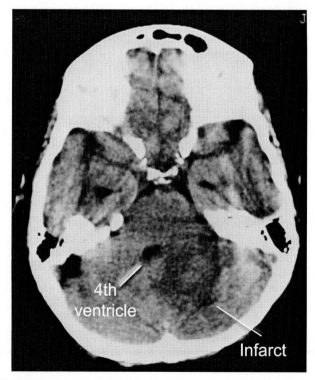

FIGURE 81.2 Cerebellar stroke. Computed tomographical scan of a patient with a left lateral medullary stroke also showing an infarction in the territory of the left posterior inferior cerebellar artery (PICA). The cerebellar infarct is edematous and displaces the fourth ventricle to the right. Continued swelling of the infarct can lead to brain stem compression and/or obstructive hydrocephalus.

ICVA and basilar artery lesions, is technically feasible, but is much riskier than other cerebrovascular procedures (e.g., VAO stenting). The only clinical trial comparing stenting to medical therapy found that patients in the medical arm did slightly better than those in the stenting arm [6b]. The current treatment recommendation is the use of antiplatelet agents and aggressive risk factor control with use of statin agents.

BASILAR ARTERY DISEASE

Patients with stenosis of the basilar artery frequently have TIAs with dizziness, slurred speech, ataxia, double vision, and unilateral or bilateral motor weakness. Occasionally patients present with syncope, but when they wake up, typically they have signs and symptoms of posterior circulation ischemia. An elderly patient with recurrent spells of dizziness or lightheadedness presents a diagnostic dilemma, because these symptoms can also be due to a variety of other causes, including inner ear disease, orthostatic hypotension, cardiac arrhythmia, vestibular migraine, and medication side effect. These patients need to be questioned carefully as to whether any of their spells were accompanied by other symptoms, such as subtle paresthesias around the mouth, slurred speech, or diplopia (not just "blurred vision").

Patients with basilar artery thrombosis often have stuttering progression of deficits over hours to days. Patients who present in coma or with bilateral motor weakness have a poor prognosis. Intubation is usually required to provide adequate respiration and protect the airway from aspiration. Hemodynamic instability is common, with labile hypertension and sometimes arrhythmias. Anticoagulation with intravenous heparin is often used, particularly if there is fluctuation in the neurological examination, but there has been no rigorous assessment of this therapy.

Because of the poor prognosis with basilar occlusion, endovascular procedures such as intra-arterial thrombolysis, mechanical clot retrieval, and angioplasty/stenting can be used within the first 12h after stroke onset to restore vessel patency and perhaps improve outcome. There has been no large, randomized clinical trial of endovascular therapy versus medical therapy alone, but there are a number of case series showing technical ability to open the basilar artery in a timely fashion, and anecdotal cases of remarkable neurological improvement after aggressive treatment. The ENDOSTROKE study reported a series of 148 patients with basilar artery occlusion treated with endovascular means (the most common being the Solitaire stent-retriever device), in whom the basilar artery could be recanalized in about 80% [7].

Disease of penetrating branches of the basilar artery causes lacunar stroke of the brain stem. The most common lacunar syndrome is pure motor hemiparesis, with weakness of the face, arm, and leg in the absence of objective sensory loss or higher-order cortical deficit. The infarct involves one side of the basis pontis and is usually evident on MRI of the brain stem. The clinical picture is identical to that for lacunar stroke involving the posterior limb of the internal capsule. Other less common lacunar syndromes include clumsy-hand dysarthria and ataxic hemiparesis. Patients with lacunar stroke in the absence of large vessel occlusive disease are treated with antiplatelet agents.

CEREBELLAR INFARCTION

PICA stroke was described earlier. Typical clinical features of cerebellar stroke are dysarthria, dizziness, nystagmus, vomiting, and ataxia of the limbs and gait. Limb ataxia manifests as dysmetria or past-pointing on finger-to-nose and heel-to-shin testing. The hemiataxia is ipsilateral to the infarct. Gait ataxia manifests as a broad-based, staggering gait, as if intoxicated. Patients tend to fall toward the side of the lesion. Sometimes the ataxia is subtle and seen only with tandem gait or when walking in a circle. Infarcts involving the midline of the cerebellum can produce little or no limb ataxia and be difficult to distinguish from inner ear disease if gait is not tested. With MRI, the vascular distribution of cerebellar infarction can now be well defined [8]. Strokes involving the inferior cerebellum (PICA territory) are due to either atherothrombosis or embolism, whereas strokes in the superior cerebellum (superior cerebellar artery territory) are more commonly due to embolism.

THALAMIC INFARCTION

The thalamus is supplied by penetrating arteries from the basilar artery apex and the proximal portions of the PCAs. Unilateral thalamic infarction results in a hemisensory loss which typically splits the midline and involves face, arm, leg, chest, and abdomen ("pure sensory stroke") [9]. With the acute infarct, there is loss of multiple sensory modalities, but during recovery, some patients develop paresthesias or an unpleasant, burning sensation in the affected areas (postthalamic pain syndrome of Dejerine–Roussy). Because proprioceptive fibers also run through the thalamus, patients with thalamic infarction may have hemiataxia on finger-to-nose and heel-to-shin testing. The thalamus also plays a role in cognition and language. Medial thalamic infarcts can cause cognitive disturbance, memory loss, agitation, excessive sleepiness, and alteration of day–night cycles. Dominant hemisphere thalamic infarction can produce aphasia in some patients.

POSTERIOR CEREBRAL ARTERY TERRITORY INFARCTION

The PCA territory includes the medial temporal lobes and the occipital lobes. Most PCA strokes are embolic and typically involve only the distal territory in the occipital lobes [10]. Infarction of one occipital lobe results in a homonymous hemianopia. Patients are usually aware of the deficit and can compensate for the visual loss by turning their head. Infarction of the inframedial temporal and occipital regions (particularly, if bilateral) may produce higher-order visual difficulties such as inability to recognize faces (prosopagnosia), difficulty recognizing objects (visual agnosia), and trouble identifying colors (color agnosia). Stroke involving the left occipital lobe can result in the syndrome of alexia without agraphia, in which patients are able to write, but are unable to read, even words they had previously written.

POSTERIOR CIRCULATION EMBOLISM

The posterior circulation receives about 20% of the cerebral blood flow, and it is not surprising that about 10–20% of cardioembolic strokes occur in this region. Emboli to the vertebrobasilar system tend to lodge either in the ICVA, the apex of the basilar artery, or in the distal territory of the PCAs. An embolus that lodges at the top of the basilar artery presents with a typical constellation of findings (Table 81.3) [11]. The clot obstructs the PCAs as well as penetrators from the basilar apex, which supply parts of the midbrain and medial thalami. Because the neurological findings are subtle, patients are often admitted to the hospital with a diagnosis of metabolic delirium before the true cause is found.

IMAGING STUDIES

MRI has significantly improved our ability to detect infarcts in the brain stem and cerebellum. CT tends to be limited due to bony artifact in the posterior fossa, but is sensitive to bleeds. Very small strokes such as lateral medullary infarction can now be detected on routine MRI using diffusion-weighted imaging. These small lesions may be overlooked by the radiologist unless the

TABLE 81.3 Clinical Features of "Top of the Basilar" Stroke

Sleepiness, altered sleep–wake cycle

Impairment of upward gaze, diplopia

Small, irregular or poorly reactive pupils

Confusion, memory loss, delirium

Visual field defects

MRA is an excellent noninvasive test for the posterior circulation and can diagnose basilar or vertebral artery stenosis with an accuracy approaching 80–90% (Fig. 81.3) [12]. Often, however, a vertebral artery is very small or not seen, and one cannot tell if this is a pathological condition or an anatomical variant by MRA alone. In addition, MRA tends to overestimate the degree of stenosis. The combined use of TCD and MRA may yield the best accuracy if the two studies show concordant findings. CTA is also a good way to study the posterior circulation, but requires a detailed review of all the axial images, since bone is present and cannot be eliminated easily as with projection MRA images.

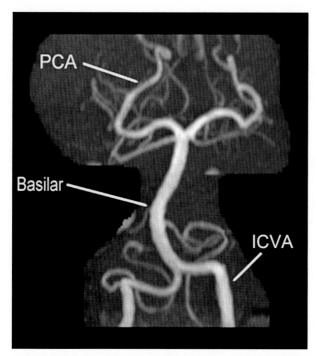

FIGURE 81.3 Magnetic resonance angiography. This anterior–posterior view shows the intracranial vertebral and the basilar arteries, with smaller branches.

clinician supplies information about the neurological findings and suspected diagnosis.

Cerebral angiography is the gold standard for detecting posterior circulation occlusive disease. Angiography of the vertebrobasilar system is not riskier than carotid angiography and should not be underutilized. Useful noninvasive studies include duplex imaging of the extracranial vertebral arteries and transcranial Doppler studies of the intracranial segments. The vertebral arteries in the neck can be partially assessed by duplex ultrasound, which can image a short segment of the vertebral artery and determine the direction of blood flow. The intracranial vertebral and basilar arteries can be assessed by transcranial Doppler (TCD) ultrasound through the foramen magnum. Stenosis in a large intracranial vessel causes high-velocity blood flow, which can be quantified by TCD.

References

[1] Caplan LR. Posterior circulation disease: clinical findings, diagnosis, and management. Cambridge (MA): Blackwell Science; 1996.

[2] Hennerici M, Klemm C, Rautenberg W. The subclavian steal syndrome: a common vascular disorder with rare neurologic deficits. Neurology 1988;38:669–73.

[3] Caplan LR, Amarenco P, Rosengart A, et al. Embolism from vertebral artery origin occlusive disease. Neurology 1992;42(8):1505–12.

[4] Piotin M, Spelle L, Martin J, et al. Percutaneous transluminal angioplasty and stenting of the proximal vertebral artery for symptomatic stenosis. AJNR Am J Neuroradiol 2000;21(4):727–31.

[5] Mokri B, House OW, Sandok BA, Piepgras DG. Spontaneous dissections of the vertebral arteries. Neurology 1988;38:880–5.

[6] [a] Chimowitz MI, Kokkinos J, Strong J, et al. The Warfarin-Aspirin Symptomatic Intracranial Disease Study. Neurology 1995;45:1488–93.
[b] Chimowitz MI, Lynn MJ, Derdeyn CP, et al. Stenting versus aggressive medical therapy for intracranial arterial stenosis. N Engl J Med 2011;365(11):993–1003.

[7] Singer OC, Berkefeld J, Nolte CH, et al. Mechanical recanalization in basilar artery occlusion: the ENDOSTROKE study. Ann Neurol 2015;77(3):415–24.

[8] Amarenco P. The spectrum of cerebellar infarctions. Neurology 1991;41(7):973.

[9] Bogousslavsky J, Regli F, Uske A. Thalamic infarcts clinical syndromes, etiology, and prognosis. Neurology 1988;38(6):837.

[10] Pessin MS, Lathi ES, Cohen MB, Kwan ES, Hedges TR, Caplan LR. Clinical features and mechanism of occipital infarction. Ann Neurol 1987;21(3):290–9.

[11] Caplan LR. "Top of the basilar" syndrome. Neurology 1980;30:72–9.

[12] Rother J, Wentz KU, Rautenberg W, Schwartz A, Hennerici M. Magnetic resonance angiography in vertebrobasilar ischemia. Stroke 1993;24(9):1310–5.

CHAPTER

82

Primer in Cerebrovascular Disease: Innominate and Subclavian Disease

S.L. Zettervall, F. Shuja, M.L. Schermerhorn

Beth Israel Deaconess Medical Center, Boston, MA, United States

INTRODUCTION

Diseases of the subclavian and innominate artery are rare entities that impact less than 5% of the population. Limited data are available to describe their prevalence and natural history, and current knowledge is derived predominantly from case reports and small case series.

EVALUATION AND WORKUP

Diseases involving the subclavian and innominate arteries are largely asymptomatic and discovered incidentally on radiographic imaging performed for other reasons. Any patient suspected of, or diagnosed with, subclavian or innominate disease, should undergo a complete history and physical examination including a thorough cardiac, neurological, and vascular evaluation. Vascular examination should include bilateral upper extremity pulses, individual upper extremity blood pressure measurements, auscultation for bruits, evaluation for pulsatile masses, documentation of upper extremity claudication or rest pain, and evidence of upper extremity embolization. A thorough neurological examination should evaluate both anterior and posterior circulation. Finally, patients should be examined for symptoms of compression of the central structures surrounding the aortic arch as well as more distal subclavian arteries. These symptoms may include nerve palsy resulting from compression of the brachial plexus, hoarseness from compression of the recurrent laryngeal nerve, and tracheal compression/deviation.

Imaging of the aortic arch has evolved rapidly since mid-1990s. Historically, aortography was considered the gold standard for diagnosis of brachiocephalic disease; however, it has been replaced by computerized tomography (CT) and magnetic resonance (MR) angiography both for the purpose of diagnosis and operative planning. CT or MR of the brain should also be performed before any planned revascularization. Presence of recent infarcts may affect the timing of an intervention due to the associated risk of reperfusion injury. Knowledge of the circle of Willis patency along with concomitant carotid bifurcation disease can aid in operative planning. Duplex ultrasound can provide indirect evidence of proximal subclavian or innominate stenosis, such as reversal of flow in vertebral arteries, or reduced flow velocities in right common carotid and subclavian arteries. However, its value can be limited due to overlying bony structures.

Prior to intervention, a thorough preoperative risk assessment is necessary for all patients. The incidence of concomitant coronary atherosclerosis approaches 40% [1], and cardiac evaluation is recommended particularly if open surgery is being considered. This assessment should comprise of a transthoracic echocardiogram and 12-lead electrocardiogram. Patients with a low ejection fraction (<50%) or ischemic changes on their electrocardiogram should be referred for additional workup with a stress test or coronary angiography.

SUBCLAVIAN AND INNOMINATE STENOSIS

Etiology and Clinical Presentation

Stenosis of the subclavian or innominate artery is a rare condition identified in less than 5% of patients undergoing aortic arch angiography [2]. The most common etiology is atherosclerosis; however, other causes include vasculitis (e.g., Takayasu's), radiation exposure, and dissection. A severe lesion is defined as stenosis greater than 75% of the vessel diameter, a deep ulcerated plaque, or thrombus within the arterial lumen.

Most patients are asymptomatic, and referred by primary care physicians after they find asymmetry in the upper extremity blood pressures. Clinical manifestations vary based on etiology, presence of single or multivessel disease, and the anatomical location of the disease. When present, symptoms can be classified as "embolic," or "low-flow" phenomenon. Clinical presentation may include stroke, transient ischemic attacks, or upper extremity ischemia (claudication, rest pain, or digital embolization). Stenosis of the proximal subclavian artery can cause subclavian steal syndrome. This occurs due to reversal of flow in the ipsilateral vertebral artery to provide blood supply to the arm, resulting in vertebro-basilar insufficiency with resultant vertigo, nausea, and imbalance. While vertebral flow reversal with a proximal subclavian/innominate stenosis may be seen commonly, steal symptoms are, in fact, rare. Occlusion of the innominate artery can cause subclavian–carotid steal causing anterior cerebral symptoms (aphasia, hemiparesis) in the ipsilateral hemisphere. In patients who have had a prior coronary artery bypass with the internal mammary artery, a subclavian–coronary steal may cause angina secondary to flow reversal in the bypass.

Management

All symptomatic patients merit consideration of revascularization, particularly those with evidence of embolization. Patients with low-flow symptoms particularly those affecting the upper extremities, can often be observed. There is no universal consensus, however, about management of asymptomatic patients. Some authors advocate revascularization for severe (>75%), asymptomatic disease in patients with reasonable operative risk, while others propose a conservative approach with medical optimization of cardiovascular risk factors [3]. There is agreement, however, that revascularization of severe, asymptomatic stenosis of subclavian artery is warranted prior to coronary artery bypass if the ipsilateral internal mammary artery is to be used as a conduit. Reversal of vertebral flow without symptoms of vertebro-basilar insufficiency can be safely observed [4]. All patients, regardless of intervention, should be treated with aspirin and a statin.

Several factors influence the optimal treatment strategy for subclavian stenosis. These include disease location, extent, calcification, its relationship to the ipsilateral vertebral artery, and the overall operative risk of the patient. Most surgeons employ an "endovascular-first" strategy for focal disease, not involving the vertebral origin. Brachial access is often preferred over femoral access, especially for flush occlusions. Both balloon and self-expanding stents are used in the subclavian artery. Patients with failed endovascular intervention, long segment disease, lesions that risk coverage of the vertebral artery and those without an ostial stump may be considered for open reconstruction. A number of extra-anatomical bypass options exist for these patients. The carotid–subclavian bypass is the most common open surgical procedure. It involves a supraclavicular incision lateral to the sternocleidomastoid muscle, limited dissection of the subclavian artery, and a conduit (usually prosthetic) to perform a bypass between the common carotid artery and the subclavian artery. A carotid–subclavian transposition is another option that carries the advantage of avoiding prosthetic material, but requires more extensive dissection of the proximal subclavian artery and the internal mammary arteries. Furthermore, a proximal vertebral artery origin occasionally precludes this procedure. Other extra-anatomical options include axillo-axillary and subclavian–subclavian bypass.

Unlike subclavian stenosis, innominate artery stenosis is treated almost exclusively with endovascular means. This can be done percutaneously via a transfemoral or transbrachial approach. Alternatively, many surgeons now approach innominate lesions via a cervical common carotid cut-down, affording easy access to the lesion while maximizing embolic protection by carotid clamping and flushing prior to the restoration of antegrade carotid flow. Balloon expandable stents are preferred due to their more precise deployment and greater outward radial force. Distal embolic protection devices are utilized in most instances during femoral approaches to innominate disease. Retrograde vertebral flow typically persists for 10–15 s after revascularization which minimizes the risk of posterior circulation embolization. Among patients in whom endovascular intervention is unsuccessful or not appropriate, open surgical interventions may include a transthoracic bypass (ascending aorta to innominate, or to common carotid and subclavian artery), innominate endarterectomy, or an extra-anatomical cervical bypass (left-to-right carotid bypass). These are associated with a higher risk of cardiopulmonary complications (Fig. 82.1).

Outcomes

No prospective randomized trials have been performed to compare endovascular and open surgical treatment strategies. Few studies have directly compared the mortality, patency, and re-intervention following endovascular and open intervention for subclavian stenosis [5,6]. AbuRahma et al. compared carotid–subclavian bypass versus stenting for isolated subclavian

(A)   **(B)**

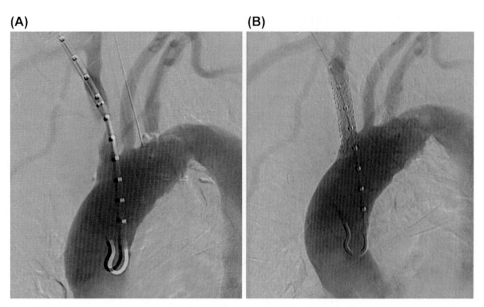

FIGURE 82.1   **Stenosis of the innominate artery.** (A) Angiogram demonstrating stenosis of the innominate artery. (B) Angiogram of innominate artery following stent placement for stenosis of the innominate artery.

artery disease in 121 patients treated with endovascular techniques and 51 patients treated with carotid–subclavian bypass. There were no perioperative strokes in the study cohort, and 30-day mortality was 0% among bypass patients and 0.8% among stent patients. Early 30-day primary patency rate was 100% for bypass and 99% for the stent group. The 5-year patency rate was 96% for the bypass versus 70% for the endovascular group [6]. Song et al. studied 252 patients, who underwent intervention over a 21-year period (endovascular = 148, bypass = 104), with no differences in perioperative mortality or major neurological complications identified. However, 10-year patency was 89% in the bypass group as compared to 49% in the stent group, and re-intervention was more common in the endovascular cohort (47% vs. 7%) [7]. The Texas Heart Institute published their experience with 391 brachiocephalic revascularizations (endovascular = 162, bypass = 229). Surgical bypass was superior to endovascular treatment in terms of freedom from graft intervention or failure at 5 years (93% vs. 84%). There was no difference between the two arms in terms of patient satisfaction surveys, with both groups exceeding 95% satisfaction [8].

Similar to subclavian artery revascularization, data regarding mortality, perioperative complications, and patency following innominate revascularization are limited and rely on case reports and small series. Following open repair, historical studies have identified mortality rates of 3–15%, perioperative neurological complication rates of 0–4%, and primary patency of 97% [9–11]. In one of the few case series of endovascular intervention, 72 patients underwent angioplasty or stenting, with no perioperative deaths or major neurological events. No

strokes were identified; however, 2.6% of patients experienced postoperative transient ischemic attacks. Primary patency was noted to be 100% at 12 months and 82% at 8 years with no differences between the angioplasty and stent groups [12].

While long-term patency data favor open revascularization techniques, it is reasonable to apply endovascular approaches first. They are associated with low risk at the time of intervention, shorter hospitalization, early resumption of normal activities, and preservation of surgical options if needed.

Postoperative Care and Surveillance

Recurrent stenosis of the subclavian and innominate artery can result in severe cardiac and neurological consequences including myocardial infarction and stroke. There are few data regarding the use of antiplatelet therapies for intervention in the supra-aortic trunk. We recommend the use of periprocedural dual antiplatelet therapy for all patients undergoing endovascular arch intervention for stenosis. While this treatment is not widely studied in this population due to the low frequency of disease intervention, data from carotid stenting have shown proven survival benefits, and the risk of perioperative stent thrombosis or embolization outweighs the lack of evidence in this population.

There are no clear guidelines for postintervention surveillance in patients treated for subclavian or innominate stenosis. However, due to early risk of restenosis, a duplex of the extra-cranial carotids and graft should be performed every 6 months for the first year following open intervention and annually thereafter. Among patients treated with endovascular intervention, duplex imaging is recommended every 6 months for the first 2 years following intervention and annually thereafter. If duplex imaging is unable to provide adequate visualization of the subclavian or innominate arteries, CT angiography can be completed at the same intervals. Patients should also have bilateral arm pressures measured routinely. A difference of 15–20 mmHg is considered significant and warrants radiographic imaging.

SUBCLAVIAN AND INNOMINATE ANEURYSMAL DISEASE

Etiology and Clinical Presentation

Aneurysms of the subclavian and innominate artery are rare accounting for less than 5% of all aneurysms. Most commonly, aneurysmal disease is degenerative; however, aneurysms of the aortic arch can result from connective tissue disorders, fibromuscular dysplasia, trauma, syphilis, or vasculitis. About 30–50% of all patients diagnosed with aneurysmal disease of the aortic arch are found to have concurrent aneurysms elsewhere, making complete physical examination and evaluation for other aneurysmal disease pertinent [13].

Subclavian aneurysms are the most common arch aneurysms accounting for about 50% of all supra-aortic aneurysms [14]. They can occur proximally, most often as a result of degenerative disease, or distally due to thoracic outlet obstruction including bony obstruction. Innominate aneurysms are much less common accounting for 3% of supra-aortic vessel aneurysms and result from similar disease processes as subclavian aneurysms [14].

Most arch vessel aneurysms are discovered incidentally with only 25% demonstrating symptoms at the time of diagnosis [14]. Presenting symptoms are similar to other subclavian/innominate pathologies and are covered in earlier sections of this chapter.

Management

The natural history of arch vessel aneurysms is the same as aneurysms elsewhere: slow growth with risks of rupture or thromboembolism that increases with larger diameter. To avoid these devastating consequences, symptomatic and large subclavian and innominate aneurysms should be repaired once diagnosed. However, it should be noted that no guidelines clearly identify size recommendations for intervention. Traditionally, open repair has been considered the standard of care for both innominate and subclavian aneurysms in patients of reasonable risk profile without contraindications.

These contraindications include poor pulmonary function, hemodynamic instability due to trauma, and prior sternotomy or left thoracotomy. Contemporary surgical repair involves resection or endoaneurysmorrhaphy and reestablishment of arterial continuity with either an end-to-end anastomosis or, more commonly, an interposition arterial prosthetic graft. Despite durable long-term results with open repair, there can be considerable morbidity and mortality. In one of the largest series of open repair of innominate aneurysms, elective open repair was associated with an in-hospital mortality of 4.3%, while the mortality of emergency repair of ruptured aneurysms was 50% [15].

Since 1990s, endovascular treatment with stent grafts has expanded leading to a dramatic reduction in procedural morbidity and mortality [16–18]. This has led to treatment of higher-risk patients with suitable anatomy, to avoid the morbidity of open surgery. There are several anatomical considerations in this location. The midportion of the subclavian artery is most amenable to endograft placement. There must be an adequate proximal and distal landing zones, and vertebral artery coverage should be avoided. The carotid artery origin may be covered if a carotid–carotid bypass is performed, or a Chimney stent is placed via a carotid artery cut down. Potential problems with endograft repair include extrinsic compression between the clavicle and first rib and stroke from embolic debris dislodged into the right common carotid artery [19]. Historically, endovascular intervention was reserved for patients with connective tissue disorders as well as coagulopathy or hemodynamic instability because it allows for more rapid vascular control through the use of proximal balloon occlusion [20]. However, with latest advances in endovascular techniques, it has become the first-line treatment for all patients with suitable anatomy. Finally, various hybrid procedures involving branched endografts, or coil embolization of the proximal subclavian artery aneurysm combined with carotid–subclavian bypass or transposition are also becoming increasingly widespread due to the combination of long-term outcomes offered from open repair, while avoiding the risk of cardiopulmonary bypass necessary for many open operations. However, long-term outcome data are limited due to the novelty of this approach.

Outcomes

Data assessing the mortality, stroke rate, and patency among patients with subclavian and innominate aneurysms are limited by the rarity of these conditions. As of 2016, small series and case reports provide basic parameters with significant limitations due to sample size, variable patient populations, and a wide disease extent.

Among patients with untreated supra-aortic aneurysms, mortality has been reported to be 16% with risk of stroke to be 33% over a mean follow-up of 14 months [14]. Only one small series retrospectively compared endovascular and open treatment among 21 patients with subclavian aneurysms with no 30-day mortality in the open or endovascular treatment groups. This study found patency rates of 90% at 3 years following endovascular repair and 91% at 7 years after open repair [21]. This study in conjunction with prior small case series have found excellent long-term durability with open repair with no data currently available to assess long-term patency of endovascular intervention. The operative mortality rates found in this study, however, are significantly lower than prior reports and metaanalysis, which have identified perioperative mortality rates of 3% following elective open repair and 4% following elective endovascular repair [22].

Conclusion

Subclavian and innominate diseases are rare entities with no current practice guidelines to direct overall management or operative approach. However, certain recommendations are agreed upon: (1) essentially all symptomatic lesions merit intervention; (2) all patients with occlusive disease should be treated with aspirin and statin therapy; and (3) endovascular treatment should be the first-line therapy for patients with suitable anatomy. There is no consensus for invasive treatment of asymptomatic occlusive or aneurysmal disease. As the use of endovascular therapy continues to expand, a better understanding of the long-term outcomes following endovascular intervention is likely.

References

[1] Kieffer E, Sabatier J, Koskas F, Bahnini A. Atherosclerotic innominate artery occlusive disease: early and long-term results of surgical reconstruction. J Vasc Surg 1995;21(2):326–36. Discussion 336–327.
[2] English JL, Carell ES, Guidera SA, Tripp HF. Angiographic prevalence and clinical predictors of left subclavian stenosis in patients undergoing diagnostic cardiac catheterization. Catheter Cardiovasc Interv 2001;54(1):8–11.
[3] Cherry KJ. Direct reconstruction of the innominate artery. Cardiovasc Surg 2002;10(4):383–8.
[4] Takach TJ, Reul GJ, Gregoric I, et al. Concomitant subclavian and coronary artery disease. Ann Thorac Surg 2001;71(1):187–9.
[5] Modarai B, Ali T, Dourado R, Reidy JF, Taylor PR, Burnand KG. Comparison of extra-anatomic bypass grafting with angioplasty for atherosclerotic disease of the supra-aortic trunks. Br J Surg 2004;91(11):1453–7.
[6] AbuRahma AF, Bates MC, Stone PA, et al. Angioplasty and stenting versus carotid-subclavian bypass for the treatment of isolated subclavian artery disease. J Endovasc Ther 2007;14(5):698–704.
[7] Song L, Zhang J, Li J, et al. Endovascular stenting vs. extrathoracic surgical bypass for symptomatic subclavian steal syndrome. J Endovasc Ther 2012;19(1):44–51.
[8] Takach TJ, Duncan JM, Livesay JJ, et al. Brachiocephalic reconstruction II: operative and endovascular management of single-vessel disease. J Vasc Surg 2005;42(1):55–61.
[9] Cherry Jr KJ, McCullough JL, Hallett Jr JW, Pairolero PC, Gloviczki P. Technical principles of direct innominate artery revascularization: a comparison of endarterectomy and bypass grafts. J Vasc Surg 1989;9(5):718–24.
[10] Vogt DP, Hertzer NR, O'Hara PJ, Beven EG. Brachiocephalic arterial reconstruction. Ann Surg 1982;196(5):541–52.
[11] Azakie A, McElhinney DB, Higashima R, Messina LM, Stoney RJ. Innominate artery reconstruction: over 3 decades of experience. Ann Surg 1998;228(3):402–10.
[12] Paukovits TM, Lukács L, Bérczi V, Hirschberg K, Nemes B, Hüttl K. Percutaneous endovascular treatment of innominate artery lesions: a single-centre experience on 77 lesions. Eur J Vasc Endovasc Surg 2010;40(1):35–43.
[13] Pairolero PC, Walls JT, Payne WS, Hollier LH, Fairbairn 2nd JF. Subclavian-axillary artery aneurysms. Surgery 1981;90(4):757–63.
[14] Cury M, Greenberg RK, Morales JP, Mohabbat W, Hernandez AV. Supra-aortic vessels aneurysms: diagnosis and prompt intervention. J Vasc Surg 2009;49(1):4–10.
[15] Kieffer E, Chiche L, Koskas F, Bahnini A. Aneurysms of the innominate artery: Surgical treatment of 27 patients. J Vasc Surg 2001;34(2):222–8.
[16] Criado E, Marston WA, Ligush J, Mauro MA, Keagy BA. Endovascular repair of peripheral aneurysms, pseudoaneurysms, and arteriovenous fistulas. Ann Vasc Surg 1997;11(3):256–63.
[17] May J, White G, Waugh R, Yu W, Harris J. Transluminal placement of a prosthetic graft-stent device for treatment of subclavian artery aneurysm. J Vasc Surg 1993;18(6):1056–9.
[18] Davidian M, Kee ST, Kato N, et al. Aneurysm of an aberrant right subclavian artery: treatment with PTFE covered stentgraft. J Vasc Surg 1998;28(2):335–9.
[19] Schoder M, Cejna M, Holzenbein T, et al. Elective and emergent endovascular treatment of subclavian artery aneurysms and injuries. J Endovasc Ther 2003;10(1):58–65.
[20] Becker GJ, Benenati JF, Zemel G, et al. Percutaneous placement of a balloon-expandable intraluminal graft for life-threatening subclavian arterial hemorrhage. J Vasc Interv Radiol 1991;2(2):225–9.
[21] Zehm S, Chemelli A, Jaschke W, Fraedrich G, Rantner B. Long-term outcome after surgical and endovascular management of true and false subclavian artery aneurysms. Vascular 2014;22(3):161–6.
[22] Vierhout BP, Zeebregts CJ, van den Dungen JJ, Reijnen MM. Changing profiles of diagnostic and treatment options in subclavian artery aneurysms. J Endovasc Ther 2010;40(1):27–34.

CHAPTER

83

Vertebral Artery Disease

L. Catanese, L.R. Caplan

Harvard University, Beth Israel Deaconess Medical Center, Boston, MA, United States

VASCULAR ANATOMY

The vertebral arteries (VAs) originate from the most proximal portion of the subclavian arteries (SAs) and constitute the birthplace of the so-called posterior circulation. An important feature of the SA system is that the right artery originates from the innominate artery, whereas the left arises as the last brachiocephalic branch of the aortic arch, providing different sources of blood supply to the vertebrobasilar system. After branching from the proximal portions of the SAs, the VAs ascend taking a posterior route within the neck on their way to the foramen magnum. While in the neck, they travel within the transverse foramina of the cervical vertebral bodies (C5–6 to C1) to finally exit at the level of the atlas before piercing the dura and entering the foramen magnum [1]. This pathway represents the extracranial portion of the VA [extracranial VA (ECVA)]. After entering the posterior fossa, the intracranial VA (ICVA) gives off penetrating arteries to the medulla: a pair of posteroinferior cerebellar arteries (PICAs) and the anterior and posterior spinal arteries. After branching off, the ICVA courses from a lateral to a medial position joining in the midline at the level of the medullopontine junction to form the basilar artery (BA). The VA is commonly divided into four portions: V1 runs from the origin to the transverse foramen of C5–6, V2 is encompassed within the transverse foramina (C5–6 to C2), V3 runs from C2 to the dura, and V4 runs from the dura to its confluence with the BA. The V4 segment represents the only intradural segment of this artery (Fig. 83.1). There is considerable variability of vascular supply within the posterior circulation, such as a unilateral hypoplastic vessel or an ICVA ending in a PICA, which are among the most commonly encountered asymmetric cerebral vessels.

EXTRACRANIAL VERTEBRAL ARTERY

The first three segments (V1–3) that form the ECVA will be considered in separate groups (V1 and V2–3) because of their differences in disease nature and treatment.

Epidemiology and Pathophysiology

The most common vascular disease affecting the proximal portion of the VA is atherosclerotic plaque formation. Often stenosis begins in the SA and spreads into the orifice of the V1 segment, although in situ plaque formation is also common (Fig. 83.2). Proximal ECVA disease is often multiple and accompanied by occlusive disease of the carotid artery as well as the brachiocephalic, coronary, or peripheral limb vessels. Even when V1 stenosis is severe, it rarely leads to hemodynamically mediated brain infarction because of the high potential for collateral formation with the surrounding muscular branches of the ECVA, in addition to the thyrocervical trunk and external carotid artery branches. Embolism from ECVA stenosis is an important cause of brainstem and cerebellar infarction and is the mechanism through which ECVA-origin disease causes strokes. Hemodynamically mediated transient ischemic attacks (TIAs) are common but brief and usually abate with time [2]. Bilateral proximal stenosis or occlusion carries a higher risk of stroke but is almost never seen in isolation and, in general, severe deficits or death are attributable to severe intracranial occlusive disease that accounts for the strokes [3]. Several risk factors have been identified as contributors to ECVA atherosclerotic disease, such as concomitant peripheral, aortic, and coronary artery diseases; hypertension; hypercholesterolemia; smoking; and being a white male.

Primer on Cerebrovascular Diseases, Second Edition
http://dx.doi.org/10.1016/B978-0-12-803058-5.00083-7

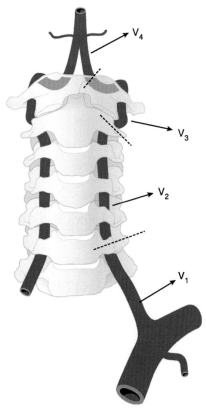

FIGURE 83.1　The different segments of the vertebral artery and its relationship with the vertebral bone landmarks. *From Caplan LR. Caplan's stroke: a clinical approach. 5th ed. UK: Cambridge University Press; 2016.*

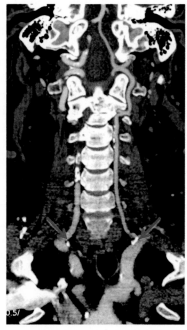

FIGURE 83.2　A 70-year-old man with history of hyperlipidemia, diabetes, and hypertension presented with dizziness. A CT angiogram of the neck showed bilateral ECVA origin stenosis (*red arrows*) secondary to atherosclerotic disease.

Other conditions that can lead to stroke originating in this portion of the VA include arterial dissections and aneurysms that have been shown to harbor clots with the potential to embolize distally. The distal ECVA segments (V2 and V3) rarely contain significant atherosclerotic lesions. Disease in these segments most often involves mechanical and physical perturbation of the vessel, such as arterial dissections, trauma, mechanical compression, and stretching related to head position [4]. Dissection is a common entity that affects the mobile segments of the vessels causing a tear and subsequent bleeding within the media of the arterial wall. The intramural hematoma can dissect longitudinally and can potentially tear through the intima exposing the hematoma to the arterial lumen. This situation can lead to activation of the coagulation cascade with subsequent distal embolization in the posterior circulation. The intramural hematoma also has the potential to expand, occluding the lumen and leading to hypoperfusion and brain infarction. Most often, the ECVA segment affected is the distal atlas loop portion of the vessel before it pierces the dura. A number of activities have been associated with the increased risk of dissection, which include but are not limited to trauma, forced prolonged neck postures such as those sustained in surgeries or resuscitation, and neck manipulation by chiropractors. However, at times, seemingly benign movements can function as triggers. Finally, VA injury within the intervertebral foramina has been well characterized in situations such as traumatic accidents causing facet joint dislocations. Proximal retrograde propagation of the clot from the site of the vessel injury has been documented in these cases. Atlantoaxial dislocation with secondary vessel compression due to motor vehicle accidents or conditions such as rheumatoid arthritis and Klippel–Feil syndrome as well as suicidal hangings and cervical osteophyte formation have been reported as more rare causes of V3 compression.

Clinical Presentation

Although often asymptomatic, ECVA disease can cause self-limited TIAs. Several reports describe dizziness, blurred or double vision, dysarthria, veering, and feeling off balance as being some of the most frequent posterior circulation symptoms, usually attributed to ischemia of the lateral medulla or cerebellum. If dissection is present, headache and/or neck pain is an almost invariable symptom. Pain often involves the trapezius, posterior neck, or occiput and precedes the neurologic symptoms by hours to weeks. Occasionally, ECVA dissections can present with cervical root pain if compressive aneurysmal dilations within the V2 segment develop, or even spinal cord infarction from occlusion of VA branches to the cervical spinal cord.

Diagnosis

Evaluation for ECVA disease should start with a detailed general physical examination including checking bilateral upper extremity pulses for differences in volume and timing of the pulse wave, auscultation of supraclavicular and posterior neck for bruits, checking blood pressure in both arms, and searching the distal areas for signs of distal embolization, such as digital ischemia. Pulse differences between the arms has been reported to be as sensitive as blood pressure checks in detecting SA stenosis [4]. V1 lesion bruits (posterior neck or mastoid process) are usually found contralateral to the site of narrowing, as they represent the compensatory turbulent flow in the opposite vessel. Diagnostic modalities for ECVA disease include ultrasonography, an excellent screening tool for occlusive disease; CT angiography (CTA), effective in detecting ECVA atherosclerotic ostial lesions and dissections; MR angiography (MRA), most sensitive for detecting arterial dissections with the fat-saturated T2-weighted sequence; and cerebral angiography, the gold standard for diagnosis of vascular lesions within the arterial bed. When an ECVA segment is not adequately visualized, as in the case of V2–3, cross-sectional MRI or MRA can add diagnostic value.

Treatment

Although prospective data is lacking, antiplatelet therapy such as using aspirin (81–325 mg/d) has been shown to be effective in nonocclusive disease. Anticoagulant agents, such as warfarin and heparin, are efficient in preventing red clots and have been safely used in the treatment of acute thrombosis and symptomatic stenotic ECVA disease. Anticoagulation and intravenous administration of thrombolytic agents appear to be safe in patients with acute dissections. The natural history of cervical arterial dissections indicates that strokes occur early after the initial episode of brain ischemia, often within the first 7 days, and their incidence decrease thereafter [5]. Experienced stroke clinicians often perform anticoagulation for the first few weeks to months after the diagnosis of cervical arterial dissection, followed by antiplatelet therapy for 6 months if the vessel remains occluded at the 3-month mark. Finally, when proximal ECVA surgery is performed, most often, VA transposition and bypasses have been used rather than endarterectomy. Experience with ECVA angioplasty/stenting shows not only a high success rate but also a high rate of restenosis preventing the routine use of this technique until more data clarifying patient selection and outcomes are available. Overall, the lack of prospective randomized data limits the prediction of long-term outcomes after surgical intervention in patients with significant ECVA disease [6,7].

INTRACRANIAL VERTEBRAL ARTERY

When discussing the ICVA, it is important to remember its highly variable anatomy. Often a contralateral ICVA is hypoplastic or ends in PICA, making the dominant VA, also known as *basilarized* ICVA, the sole blood flow distributor to the middle and distal portions of the posterior circulation. A *basilarized* ICVA occlusion can lead to devastating brainstem strokes.

Epidemiology and Pathophysiology

Most commonly, lesions affecting the ICVA are bilateral and atherosclerotic in nature. Other sources of ICVA stenosis or occlusion include proximal cardiac or arterial embolisms, dissection, dolichoectasia, and, in some instances, retrograde extension of a BA thrombus. Based on the New England Medical Center (NEMC) Posterior Circulation Registry, the largest posterior circulation stroke database to date, atherosclerotic intracranial disease is most commonly seen in the African-American and Asian population and tends to affect the distal third of the ICVA at or near the vertebrobasilar junction [3]. Vascular cholesterol deposition becomes evident during the third decade of life with the presence of fatty streaks. The initial cholesterol deposition tends to increase over years to form fibrous plaques within the arterial intima, which can lead to vessel stenosis, occlusion, or, if plaque rupture occurs, thromboembolic phenomena (Fig. 83.3). According to the NEMC Registry, risk factors for ICVA occlusive disease in the order of frequency include hypertension, hypercholesterolemia, diabetes, and smoking.

Embolism into the ICVA has been described from either cardiac sources or proximal vessels such as the aorta, SA, and ECVA. Arterial dissections affecting the V4 segment can lead to strokes in various ways including distal embolization of intramural clot material into the posterior circulation, intramural clot expansion leading to ICVA occlusion, aneurysmal dilation of the dissection site causing mass effect, and SAH or intracerebral hemorrhage if vessel rupture occurs (Fig. 83.4). Dissection has often extended into the BA causing fatal infarctions. In some occasions, chronic aneurysms can harbor thrombi and function as a source of distal intra-arterial embolization.

Clinical Presentation

The NEMC Posterior Circulation Registry showed that the clinical course and presentation of patients with ICVA disease varied considerably. About one-third had strokes without preceding TIAs, and one-fourth had either TIAs preceding strokes or isolated TIAs. The most common stroke within the ICVA circulation is the lateral medullary or Wallenberg syndrome.

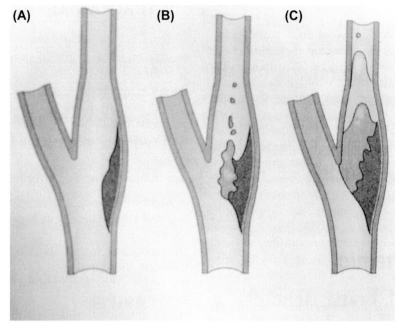

FIGURE 83.3   (A) Plaque; (B) plaque with platelet-fibrin emboli; (C) plaque with occlusive thrombus. *Adapted with permission from Elsevier from Caplan's stroke: a clinical approach. 4th ed.*

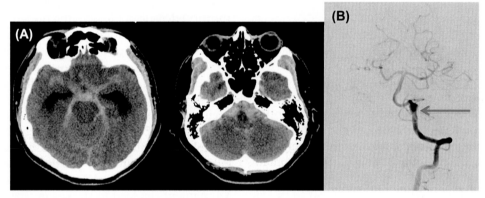

FIGURE 83.4   (A) Two sections of a head CT showing a large subarachnoid hemorrhage in a 41-year-old patient who presented to the emergency department with thunderclap headache. (B) The patient's cerebral angiogram showed aneurysmal dilation of the V4 segment (*red arrow*) of the left vertebral artery. The patient had a V4 dissection complicated by aneurysmal dilation and subarachnoid hemorrhage.

The syndrome's rich and assorted clinical features illustrate the proximity of the different structures within the lateral medulla. Some of the most common signs and symptoms reflect dysfunction of specific systems and structures, such as vestibulocerebellar, sensory, sympathetic, visual, cranial nerves (CNs), respiratory, whereas others are less specific such as headaches, nausea, vomiting, and hiccups. Vestibulocerebellar dysfunction is found in approximately 90% of patients and is related to the involvement of the vestibular nuclei or their connections. Often patients report dizziness; pure vertigo; a sensation of being off balance and pulled or falling toward one side; or just feeling seasick. Blurred vision, diplopia, and oscillopsia have been reported and represent vestibuloocular reflex dysfunction. Nystagmus is

a sign that nearly always accompanies these symptoms and is frequently horizontal and rotational [8]. The fast phase of nystagmus localizes to the side of the lesion as it attempts to return the eyes to the dysfunctional side. Ataxia, or lack of coordination of fine and voluntary movements, can affect the limbs and the trunk leading to significant impairment in walking. Although gait abnormality is one of the most persistent features, it tends to improve after 3–6 months. Sensory complaints are very common and tend to be crossed, affecting the ipsilateral face and contralateral body. Pain, unpleasant dysesthetic feelings, and numbness in the face are often reported and represent the involvement of the nucleus of the spinal tract of CN V. In other patients, loss of pain and temperature sensation is limited to the

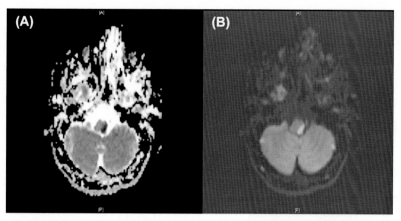

FIGURE 83.5 (A) ADC and (B) DWI MRI sequences showing an acute (black on ADC and bright on DWI) left lateral medullary infarct in a 52-year-old patient who presented with left-sided Horner syndrome and facial paresis and right-sided hemiparesis.

contralateral limbs and contralateral body, occasionally presenting clinically with a sensory level on the trunk. An absent corneal reflex may accompany the facial sensory disturbances. Although generally found on physical examination rather than by report, loss of pain and temperature sensations in the contralateral body is seen as a result of the involvement of the lateral spinothalamic tract. Sympathetic dysfunction due to ischemia of the descending sympathetic nervous system fibers manifests as an ipsilateral Horner syndrome that seldom includes hemianhidrosis. Involvement of the nucleus ambiguous leads to dysarthria, hoarseness, and dysphagia. Although dysphagia can be severe enough to cause aspiration, it often improves within weeks and months after the stroke. The control of inspiratory and expiratory automaticity lies within the lateral medulla and its impairment can lead to the so-called Ondine curse. Patients with this syndrome lack respiratory automaticity and therefore develop apnea during sleep. Death from nocturnal apnea can occur if prompt diagnosis is not made. Headaches can precede brain ischemia and tend to be localized to the ipsilateral occipital and mastoid regions. Hiccups are relatively common but the mechanism is not clear. Nausea and vomiting can occur in response to intense vertigo or due to involvement of the structures close to the solitary tract and its nucleus. Although rare, abnormalities of thermal regulation, labile blood pressure and heart rate, gastrointestinal autonomic dysfunction, and abnormal sweating have also been described.

In some patients, medial medullary infarcts accompany lateral medullary infarction as part of a hemimedullary syndrome. In contrast to the lateral medullary syndrome, the most consistent feature of the medial medullary syndrome is contralateral hemiparesis secondary to the involvement of the pyramids. Other findings for this syndrome include contralateral paresthesias and loss of posterior column sensory modalities (proprioception and vibration) due to ischemia in the medial lemniscus and ipsilateral tongue paresis from intraparenchymal 12th nerve fiber involvement.

Finally, PICA territory infarction can occur from either embolism to the ICVA or in situ ICVA atherosclerosis with superimposed thrombus. Infarction within this territory will be discussed separately.

Diagnosis

For evaluation of patients with medullary infarction, MRI has proved to be significantly superior to CT because of the extensive bone artifact within the posterior fossa encountered in the latter (Fig. 83.5). On the other hand, CT scans can be useful in assessing patients with cerebellar infarcts for any signs of herniation or hydrocephalus. Patients with PICA territory cerebellar infarcts, especially if young, can develop mass effect on the fourth ventricle leading to hydrocephalus. This condition is a neurologic emergency and decompressive surgery may be required as a lifesaving measure. CTA and MRA are sensitive modalities to detect vascular lesions within this territory (Fig. 83.6). Cerebral angiography is seldom performed unless for therapeutic purposes such as clot retrieval or when the diagnosis remains uncertain. Transcranial Doppler (TCD) is a useful noninvasive tool for assessing the ICVA.

Treatment

Antithrombotic therapy is an accepted first-line therapy in patients with chronic occlusive or nonocclusive ICVA disease. In those with acute occlusive ICVA disease or dissections, anticoagulants have been prescribed for a few weeks or months and deemed safe by experienced stroke clinicians, although there is

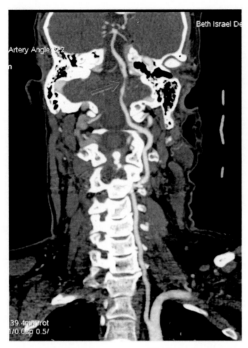

FIGURE 83.6   Coronal view of a head and neck CTA showing a left ICVA dissection (*red arrow*).

insufficient data to date assessing their utility. Finally, stenting is seldom recommended after the results from

the Stenting and Aggressive Medical Management for Preventing Recurrent Stroke in Intracranial Stenosis (SAMMPRIS) trial were published, which showed that patients treated with maximal medical therapy had better outcomes when compared with those who underwent stent placement. More studies that evaluate treatment modalities within this population are needed.

References

[1] Caplan LR. Caplan's stroke: a clinical approach. 5th ed. UK: Cambridge University Press; 2016.
[2] Fisher CM. Occlusion of the vertebral arteries. Arch Neurol 1970;22:13–9.
[3] Caplan LR, Witlyk RJ, Glass TA, et al. New England Medical Center posterior circulation stroke registry. Ann Neurol 2004;56(3):389–98.
[4] Caplan LR. Vertebrobasilar system syndromes. 2nd ed. UK: Cambridge University Press.
[5] Caplan LR. Dissection of brain-supplying arteries. Nat Clin Pract Neurol 2008;4:34–42.
[6] Grotta JC, Albers GW, Broderick JP, Kasner S, Lo EH, Mendelow D, et al. Stroke: pathophysiology, diagnosis and management. 6th ed. Elsevier.
[7] Caplan LR. Atherosclerotic vertebral artery disease in the neck. Curr Treat Opt Cardiovasc Med July 2003;5(3):251–6.
[8] Searls DE, et al. Symptoms and signs of posterior circulation ischemia in the New England medical Center posterior circulation registry. Arch Neurol 2012;69(3):346–51.

CHAPTER

84

Basilar Artery Disease

J. Tarsia[1], L.R. Caplan[2]

[1]Ochsner Health Systems, New Orleans, LA, United States; [2]Harvard University, Beth Israel Deaconess Medical Center, Boston, MA, United States

ANATOMY

The two vertebral arteries merge at the pontomedullary junction to form the basilar artery (BA). This artery lies within the prepontine cistern and is the main stem of the posterior circulation. It directly supplies a large territory of vital brain tissue including the brainstem and cerebellum and provides the main conduit for blood flow to the thalami and medial temporal and parietal lobes. The BA gives rise to several groups of penetrating branches that directly supply the brainstem. The major branches are the caudally located anteroinferior cerebellar arteries (AICAs) and

Primer on Cerebrovascular Diseases, Second Edition
http://dx.doi.org/10.1016/B978-0-12-803058-5.00084-9

rostrally located superior cerebellar arteries (SCAs). The rostral portion of the BA divides into two posterior cerebral arteries (PCAs) at the pontomesencephalic junction.

There are three major groups of small penetrating arteries that leave the BA and directly supply the brainstem. The first and shortest group is the *anteromedial group* (i.e., paramedian perforating arteries) that penetrates the basilar sulcus. The most caudal arteries of this group at the vertebrobasilar junction are the penetrating arteries of the foramen cecum of the medulla. Rostral to these arteries are the main perforators that supply the anteromedial territory of the pons from the ventral surface to the tegmentum. The most rostral of this group are the arteries penetrating the interpeduncular fossa (i.e., inferior rami of the interpeduncular fossa) [1].

The second group is the *anterolateral group* (i.e., short circumferential perforating arteries). Arteries of this group mainly supply the anterolateral territory of the pons and do not extend to the tegmentum. The third and longest group of penetrating arteries is the *lateral group* (i.e., long circumferential arteries). In conjunction with small penetrators from the proximal SCA (rostrally) and proximal AICA (caudally), the long circumferential arteries supply the lateral pontine territory, extending from the ventrolateral surface to the pontine tegmentum. The most rostrolateral pontine tegmentum is supplied by the SCA alone [1].

The AICA is the largest circumferential branch of the BA. It usually arises from the proximal or middle segment. It first contributes penetrators (i.e., recurrent penetrating arteries of AICA) to the caudal lateral pontine territory including the exiting roots of cranial nerve (CN) VII and CN VIII and the lateral tegmentum of the pontomedullary junction. It also supplies the middle cerebral peduncle. The AICA gives off an internal auditory artery that supplies the vestibular (i.e., the vestibular artery) and cochlear (i.e., the common cochlear artery) structures of the inner ear. The AICA terminally supplies the anteroinferior cerebellum and flocculus [1–3].

The SCAs arise a few millimeters proximal to the bifurcation of the BA into the PCAs, separated by the exit of CN III, at the pontomesencephalic junction. They circle the rostral pons or caudal midbrain and divide into medial (mSCA) and lateral (lSCA) branches. The lSCA initially gives off penetrators to the rostrolateral pons and midbrain tegmentum, the superior cerebellar peduncle, dentate nuclei, and eventually the lateral portion of the superior cerebellar hemisphere. The mSCA courses around the midbrain and with the collicular artery, a branch of the PCA, supplies the colliculi and the posterior territory of the midbrain. The mSCAs further supply the superior vermis, dentate nucleus, and medial portion of the superior surface of the cerebellar hemisphere. It is important to point out that the mSCA and lSCA anastomose with the vermian and hemispheric branches of the

posteroinferior cerebellar arteries (PICAs), respectively. This forms an important network of collateral blood supply that could bypass an occlusion of the BA proximal to the origin of the SCA [1,3,4].

PATHOLOGY AND PATHOPHYSIOLOGY

Atherosclerosis is the major contributor to BA pathologic conditions. Elevated and thickened fibrous plaques encroach upon the lumen leading to stenosis. Cracks in the plaque may lead to superimposition of red thrombi, which may occlude the artery at the point of stenosis or embolize distally. Atherosclerotic plaques may also be eccentric, representing arterial wall thickening and positive remodeling without significant compromise of the vessel lumen. These eccentric lesions can still lead to thin fibrous cap rupture, exposure of lipid core, and formation of superimposed thrombus [5]. Negative remodeling may also occur in which arterial thickening leads to concentric narrowing or collapse of the vessel lumen [6].

Emboli to the BA arise from the heart, aorta (e.g., atrial fibrillation, atherosclerotic plaque of the aortic arch), or the vertebral arteries in the neck and head (atherosclerotic plaques or dissections). It may be difficult to clinically distinguish in situ thrombosis from embolism. Embolism presents suddenly without warning signs, whereas patients with in situ disease of the BA commonly present with warning signs or transient ischemic attacks (TIAs).

The New England Medical Center Posterior Circulation Registry (NEMC-PCR) prospectively studied 407 patients with posterior circulation ischemia from 1988 through 1996 [7]. Moderate to severe BA occlusive disease was present in 109 (27%) of these patients. Three separate subgroups were identified within this population: those with isolated intrinsic BA atherostenosis (45%), those with BA disease as part of multiple artery occlusive disease found elsewhere (41%), and those with embolism to the BA without intrinsic basilar disease (14%). All groups had a high frequency of multiple vascular risk factors such as hypertension, hyperlipidemia, diabetes, smoking history, peripheral vascular disease, and coronary artery disease [7]. The Basilar Artery International Cooperation Study (BASICS) Registry prospectively studied 519 patients with BA occlusions who received treatment in 48 centers from 2002 to 2007. Embolism was the cause of occlusion in 215 of 592 patients (36%) [8]. This number coincides with the proportion of embolism in the NEMC-PCR when considering only complete occlusions ($n = 12/32$, 38%). The NEMC-PCR also found embolism to be more common in the rostral portion, but this was limited to 12 patients [7].

Dissections involving the BA are relatively rare. They compose about 5–10% of vertebrobasilar intracranial dissections in several databases [9,10]. The BA is more often involved as distal extensions of the intracranial V4 segment of the vertebral artery. Isolated dissections are most common in the middle or distal segment of the vessel. These dissections can extend into the AICA or SCA branches or disrupt the flow of penetrating arteries, leading to ischemia. Intracranial dissections are also prone to aneurysmal dilation and/or rupture because of the relatively poor media and adventitia compared to that of the extracranial arteries. BA dissections mostly present with ischemia rather than subarachnoid hemorrhage [10]. Thrombosis at the site of the dissection or within the aneurysmal dilation can cause distal embolization. Often patients have multiple attacks or progressive dysfunction of the brainstem. If the aneurysmal dilation is large enough, it may also directly compress the brainstem, leading to local symptoms.

The BA may also be subject to dilatative arteriopathy, also known as dolichoectasia. The underlying abnormal connective tissue leads to the formation of vertebrobasilar nonsaccular intracranial aneurysms. Fusiform lesions are characterized as focal dilation along an isolated segment of the BA, whereas dolichoectatic BAs involve uniform dilation along the entire course of the artery. Transitional lesions are a combination of both underlying dolichoectatic changes (i.e., elongation, dilation, and tortuosity) and superimposed focal fusiform dilation [11]. These large perturbations of BA structure lead commonly to compressive symptoms or thrombosis with ischemic symptoms. Less often they may rupture resulting in hemorrhage. Recurrence rates of ischemic stroke are high and mortality is increased except in case of strokes that are purely dolichoectatic [11,12].

Other Less Common Pathophysiologic Findings of the Basilar Artery [3]

Migraine	Occlusions and persistent narrowing
Hyperparathyoidism with hypercalcemia	Reversible narrowing due to vasconstriction
Congenital anomalies (i.e., duplications, fenestrations)	Predisposition to thrombus formation
Infectious causes (syphilis, HZV infection, bacterial meningitis, Lyme meningitis, TB, fungal infections, AIDS)	Inflammation leading to thrombus formation and possible aneurysm formation
Inflammatory connective tissue disease (SLE, PAN, Granulomatosis with polyangiitis)	Inflammation, hypercoagulability, and vasculitic changes
Trauma with clivus fracture	Compression

HZV, Herpes zoster virus; PAN, polyarteritis nodosa; SLE, systemic lupus erythematosus; TB, tuberculosis.

CLINICAL SYNDROMES OF BASILAR ARTERY ISCHEMIA

The location of BA occlusion plays a critical role in clinical presentation. Occlusions of the proximal and middle segments of the artery lead to a distinct clinical syndrome when compared to occlusions that occur at the distal portion of the artery. The two very distinct syndromes caused by such occlusions and the possible signs and symptoms of each are detailed in the following sections.

Middle (Pontine) Brainstem Ischemia

Occlusions of the proximal and middle BA lead to primarily pontine ischemia involving the bulk of the perforator territory and the AICA. Collateral blood flow from the vertebral artery branches that course laterally around the brainstem and from the perforators that arise from the distal BA allow for retained perfusion of the lateral portions of the base and tegmentum of the pons. As a result, the lesions are usually paramedian and result in motor and oculomotor symptoms and signs. Sensory and vestibular functions are usually spared. Lesions of the basis pontis, where descending corticospinal tract fibers are located, lead to predominantly motor symptoms. Given the central and singular position of the BA, bilateral motor symptoms are almost always present. Although bilateral, the symptoms are often asymmetric, appearing as a hemiplegia or hemiparesis with only very subtle signs of upper motor neuron disease (i.e., Babinski sign) on the contralateral side. In addition to motor weakness, the patient may also have repetitive limb twitching movements. These can be quite variable, at times unilateral and contralateral to the predominant weakness, and may represent fractions of decerebrate posturing with rigidity and stiffness. The term locked-in syndrome refers to the most severe form of motor paralysis in which the patient remains conscious but is quadriplegic with loss of all voluntary movements other than vertical eye movements and blinks.

Ataxia and limb incoordination can also occur secondary to the involvement of the corticopontocerebellar fibers, which cross through the pons to reach the cerebellum. Involvement of the middle cerebellar peduncle supplied by the AICA can also contribute to ataxia. Bulbar weakness with voluntary loss of movement of the jaw, tongue, and face muscles leads to several symptoms including slurred speech, difficulty swallowing, and poor control of saliva. Infarcts that involve the dorsal surface and descending corticofacial projections may impair volitional facial weakness but spare the fibers that control facial movements during emotion (i.e., laughter), which pass from the thalamus and supplementary motor area in the tegmentum. Similar lesions may lead to pseudobulbar release phenomena, with exaggerated crying and laughing and hypersensitivity to emotional stimulation. There may also be exaggerated pharyngeal weakness and clamping of the jaw. The loss of volitional control may also extend to one's own control of regulating the depth and rate of breathing.

Oculomotor signs are the second most common findings in patients with pontine ischemia secondary to proximal or middle BA occlusion. The effect on horizontal gaze centers in the pons leads to specific findings that can help localization. Depending on which structures are involved [i.e., nucleus and rootlets of CN VI, paramedian pontine reticular formation (PPRF), medial longitudinal fasciculus (MLF)], several different horizontal gaze disturbances are found. These include unilateral or bilateral horizontal conjugate gaze palsy, internuclear ophthalmoplegia (INO), and one-and-a-half syndrome. Vertical oculomotor symptoms such as vertical nystagmus, skew deviation, and ocular bobbing are also found. Horizontal gaze paretic nystagmus, dissociated horizontal nystagmus, and rhythmic vertical nystagmus are sometimes present. Ptosis of the upper eyelids caused by disruption of the descending sympathetic fibers in the lateral tegmentum may occur. This finding is typically bilateral but usually asymmetric, often more severe on the side of profound motor weakness and usually spares pupillary involvement, contrary to other lesions leading to Horner syndrome. Pontine lesions, in general, may spare pupillary changes, although some patients have small pupils that retain their reaction to light.

Sensory symptoms are not common in BA occlusion at this level because many of the ascending sensory tracts course too lateral or dorsal to be routinely affected. If present, sensory symptoms can present as paresthesias or hemihypesthesia due to involvement of the medial lemniscus in the dorsomedial portion of the basis pontis. Intermittent burning pain in the midline of the face has also been reported and may be secondary to decussating fibers from the nucleus of the spinal tract of CN V. More commonly, patients may present with perioral paresthesias. Tinnitus and hearing loss may also occur, either secondary to involvement of brainstem auditory nuclei and tracts or due to ischemia of CN VIII or inner ear structures. The internal auditory artery that supplies the inner ear structures is a branch of AICA.

Coma and reduced level of consciousness in BA thrombosis is due to ischemia of the reticular activating system located in the paramedian tegmentum of the pons and midbrain. The level of the rostral tegmental pons is the territory of maximal overlap for lesions that lead to coma [13]. This area of the brain is supplied by the anteromedial group of pontine perforators directly from the BA as well as lesions of the SCA in more rostral occlusions [1].

Top of the Basilar Syndrome (Upper Brainstem Ischemia)

Occlusion of the rostral portion of the BA can lead to ischemia of the midbrain, thalami, and temporal and occipital lobes. Because the BA narrows as it courses the cephalad, most smaller emboli reach the terminal portion of the artery. The great majority of the top of the basilar occlusions are embolic. The rostral BA contributes its supply to the midbrain via perforators of the basilar apex, the SCAs, and the immediate segments of the PCAs and their respective branches (i.e., collicular arteries, posteromedial choroidal arteries). The constellation of symptoms from BA occlusion in this segment is referred to as top of the basilar syndrome [14].

As opposed to the oculomotor findings discussed earlier in pontine infarcts, midbrain and diencephalic infarcts lead to oculomotor dysfunction primarily in the vertical plane as well as convergence. Bilateral cortical projections descend on the rostral interstitial nucleus of the MLF, the vertical gaze center located in the midbrain tegmentum. Lesions here result in voluntary upgaze palsy, more commonly bilateral rather than unilateral. Contraversive (ipsilateral eye up) skew deviation may also be present due to lesions in this area or the rostral interstitial nucleus of Cajal.

Ocular convergence is also controlled in the medial rostral midbrain tegmentum. Lesions here may cause one or both eyes to rest in or down and in at resting position. This hyperconvergence may manifest as adductor contractions of both eyes on attempted upgaze. On attempted horizontal gaze, there may be convergence of the abducting eye, commonly referred to as a pseudo-sixth palsy. Collier sign or retraction of the upper eyelid may also be present.

More caudal lesions in the midbrain may cause CN III palsy secondary to nuclear or fascicular damage. CN IV involvement is less common, given its more dorsal location and short course of fascicles as they exit the dorsal surface of the brainstem. Rostral brainstem lesions may also lead to pupillary changes due to disruption of the pupillary light reflex arc. Pupil response may be slow or absent or fixed and dilated if the Edinger–Westphal nuclei are affected. The pupil may also change shape and become more oval (corectopia).

A wide range of perturbations in alertness and behavior characterize these lesions. Hypersomnolence, as opposed to coma, is common. Patients may be apathetic or may develop mutism. They have an altered sleep–wake dreaming cycle and often have difficulty distinguishing dreams from reality. They may answer inappropriately to questions that were never asked or have no bearing in reality and make errors in the personal time dimension or current location. These symptoms may be due to the close anatomical relationship between the reticular activating system and the centers of control of the sleep–wake cycle. Peduncular hallucinosis describes patients who have vivid visual hallucinations, but tactile and auditory components may also be present. These patients have often suffered a lesion in the pars reticulata of the substantia nigra.

CLINICAL PRESENTATION AND OUTCOMES FROM THE REGISTRY DATA: NEMC-PCR AND BASICS

TIAs were present in 66% of patients in the NEMC-PCR study. These were often multiple and spanned over the course of several months, commonly with a crescendo effect near the time of presentation. This presentation more commonly reflected underlying intrinsic basilar atherosclerotic disease. Those with embolism often did not have a TIA, presented more acutely with progression over the course of hours, and more often, the rostral portion of the BA was affected.

The most common clinical presentations for those with intrinsic BA atherosclerotic disease were bulbar and pseudobulbar signs (82%) followed by hemiparesis or hemiplegia (54%), sensory symptoms (49%), cerebellar signs (49%), and vertigo (48%). For those with diffuse atherosclerotic disease of the intracranial vasculature, bulbar or pseudobulbar signs (75%), vertigo (69%), and cerebellar signs (61%) were the most prominent findings, with similar rates of hemiparesis, oculomotor signs, and sensory symptoms as that of atherosclerosis confined to the BA. In contrast, those with embolism to the BA rarely presented with cerebellar, sensory, or vertiginous findings but had a higher rate of decreased level of consciousness and tetraplegia than the other groups. Of note, the embolic subgroup only contained 12 subjects.

The NEMC-PCR also suggested that outcome of BA occlusive disease was not as dismal as historically viewed. Only 75% of patients had slight disability or absence of deficits and the mortality rate was 3%. The outcome was worse in those with embolic cause. The data regarding prognosis in the BASICS registry was in stark contrast to that found in the NEMC-PCR. At 1 month, 30% had severe disability requiring significant or complete assistance with activities of daily living and ambulation [modified Rankin Scale (mRS) = 4–5]. At this point, 39% of patients had died. The differences in mortality were likely due to the nature of those patients who were enrolled and studied—hyperacute, complete occlusions undergoing treatment with severe strokes averaging an NIHSS (National Institutes of Health Stroke Scale) of 22 points.

References

[1] Tatu L, Moulin T, Bogousslavsky J, Duvernoy H. Arterial territories of human brain: brainstem and cerebellum. Neurology 1996;47(5):1125–35.

[2] Lee H. Audiovestibular loss in anterior inferior cerebellar artery territory infarction: a window to early detection? J Neurol Sci 2012;313(1–2):153–9.

[3] Caplan LR. Vertebrobasilar ischemia and hemorrhage: clinical findings, diagnosis, and management of posterior circulation disease. 2nd ed. United Kingdom: Cambridge University Press; 2015.

[4] Amarenco P, Roullet E, Goujon C, Chéron F, Hauw JJ, Bousser MG. Infarction in the anterior rostral cerebellum (the territory of the lateral branch of the superior cerebellar artery). Neurology 1991;41(2 (Pt 1)):253–8.

[5] Vergouwen MD, Silver FL, Mandell DM, Mikulis DJ, Krings T, Swartz RH. Fibrous cap enhancement in symptomatic atherosclerotic basilar artery stenosis. Arch Neurol 2011;68(5):676.

[6] Ma N, Jiang WJ, Lou X, et al. Arterial remodeling of advanced basilar atherosclerosis: a 3-tesla MRI study. Neurology 2010;75(3):253–8.

[7] Voetsch B, DeWitt LD, Pessin MS, Caplan LR. Basilar artery occlusive disease in the New England medical center posterior circulation registry. Arch Neurol 2004;61(4):496–504.

[8] Schonewille WJ, Wijman CA, Michel P, et al. Treatment and outcomes of acute basilar artery occlusion in the Basilar Artery International Cooperation Study (BASICS): a prospective registry study. Lancet Neurol 2009;8(8):724–30.

[9] Debette S, Compter A, Labeyrie MA, et al. Epidemiology, pathophysiology, diagnosis, and management of intracranial artery dissection. Lancet Neurol 2015;14(6):640–54.

[10] Kim BM, Kim SH, Kim DI, et al. Outcomes and prognostic factors of intracranial unruptured vertebrobasilar artery dissection. Neurology 2011;76(20):1735–41.

[11] Flemming KD, Wiebers DO, Brown RD, et al. The natural history of radiographically defined vertebrobasilar nonsaccular intracranial aneurysms. Cerebrovasc Dis 2005;20(4):270–9.

[12] Shapiro M, Becske T, Riina HA, Raz E, Zumofen D, Nelson PK. Non-saccular vertebrobasilar aneurysms and dolichoectasia: a systematic literature review. J Neurointerv Surg 2014;6(5):389–93.

[13] Parvizi J, Damasio AR. Neuroanatomical correlates of brainstem coma. Brain 2003;126(Pt 7):1524–36.

[14] Caplan LR. "Top of the basilar" syndrome. Neurology 1980;30(1):72–9.

CHAPTER

85

The Heart and Stroke

M.C. Leary[1,2], H.A. Yacoub[1]

[1]Lehigh Valley Hospital and Health Network, Allentown, PA, United States; [2]University of South Florida, Tampa, FL, United States

OVERVIEW

Globally, stroke is the fifth leading cause of death in Western countries after cancer and heart disease. There are several ways in which stroke and heart disease interrelate [1]. Cardiogenic stroke occurs when (1) clots embolize from the heart and reach the brain via the arterial circulation, or (2) as a result of severe heart failure and cerebral hypoperfusion. Medications as well as certain cardiac procedures performed to treat heart disease can lead to adverse neurological complications such as ischemic stroke. Many vascular diseases are also known to affect both the heart and the brain. For example, there is a strong correlation between cardiac disease and cerebral atherosclerosis, which escalates the risk of thrombotic stroke. Lastly, central nervous system (CNS) diseases can influence the heart and its function [1].

STROKE CLASSIFICATION

It is estimated that about 87% of strokes are ischemic and the remaining 13% are hemorrhagic. Using the Trial of Org 10172 in Acute Stroke Treatment (TOAST) criteria, ischemic strokes may be further subdivided into the following types [2,3]:

1. Thrombosis or embolism associated with large vessel atherosclerosis.
2. Embolism of cardiac origin (cardioembolic stroke).
3. Small vessel occlusion (lacunar stroke).
4. Other determined cause (including hypoperfusion stroke).
5. Undetermined (cryptogenic) cause (no cause identified, more than one cause, or incomplete investigation).

The incidence of each cause of ischemic stroke is variable and dependent upon patient's age, sex, race, geographical location, risk factors, clinical history, physical findings, and the results of various tests [1,2].

DIRECT CARDIOGENIC BRAIN EMBOLISM

Cardiogenic cerebral embolism is responsible for about 20% of all ischemic strokes [1]. Cardiac sources of embolism include blood clots, tumor fragments, infected and noninfected vegetations, calcified particles, and atherosclerotic debris [2]. As more advanced diagnostic techniques have been developed, more causative cardiac abnormalities and their association with stroke have been recognized [1]. Conditions that are known to lead to systemic embolization may be subdivided into a high-risk and a low-risk groups based on their embolic potential (Table 85.1) [2].

Etiology

As seen in Table 85.1, certain cardiac sources have much higher rates of initial and recurrent embolism. The more common sources are reviewed in the following. The risk of embolism also varies within individual cardiac abnormalities depending on many factors. For example, in patients with atrial fibrillation, the presence of associated heart disease, patient's age, duration, chronic versus intermittent fibrillation, and atrial size, all influence embolic risk. Additionally, it is important to remember that the presence of a potential cardiac source of embolism does not definitely indicate that the stroke was caused by an embolus from the heart. Coexistent occlusive cerebrovascular disease is relatively common. In the Lausanne Stroke Registry, among patients with potential cardiac embolic sources, 11% of patients had severe cervicocranial vascular occlusive disease (>75% stenosis) and 40% had mild to moderate stenosis proximal to brain infarcts [4].

Primer on Cerebrovascular Diseases, Second Edition
http://dx.doi.org/10.1016/B978-0-12-803058-5.00085-0

Common Sources of Cardiac Emboli

Intracardiac Thrombi

Atrial Fibrillation (AF): Persistent and paroxysmal AF is a potent predictor of first and recurrent stroke, affecting more than 2.7 million Americans. The principal adverse consequence of AF is ischemic stroke, with AF patients being 3–5 times more likely to have a stroke than those without AF. In patients with brain emboli caused by a cardiac source, there is a history of nonvalvular AF in about one-half of all cases, left ventricular thrombus in almost one third, and valvular heart disease in one fourth [1,5]. Patients with AF typically have more severe strokes as well as longer transient ischemic attacks (TIAs) than patients with ischemic stroke from other causes. Stroke prevention

TABLE 85.1   Classification of Cardiac Causes of Embolism

High Potential for Embolism

1. Intracardiac thrombi
   a. Atrial arrhythmia
      i. Valvular atrial fibrillation
      ii. Nonvalvular atrial fibrillation
      iii. Atrial flutter
   b. Ischemic heart disease
      i. Recent myocardial infarction
      ii. Chronic myocardial infarction, especially with left ventricular aneurysm
   c. Nonischemic cardiomyopathy
   d. Prosthetic valves and devices
2. Intracardiac vegetations
   a. Native valve endocarditis
   b. Prosthetic valve endocarditis
   c. Nonbacterial thrombotic endocarditis
3. Intracardiac tumors
   a. Atrial myxoma
   b. Papillary fibroelastoma
   c. Other tumors
4. Aortic atheroma
   a. Thromboembolism
   b. Cholesterol crystal emboli

Low Potential for Embolism

1. Potential precursors of intracardiac thrombi
   a. Spontaneous echo contrast/"smoke"
   (in the absence of atrial fibrillation)
   b. Left ventricular aneurysm without a clot
   c. Mitral valve prolapse
2. Intracardiac calcifications
   a. Mitral annular calcification
   b. Calcific aortic stenosis
3. Valvular anomaly
   a. Fibrin stands
   b. Giant Lambl's excrescences
4. Septal defects and anomalies
   a. Patent foramen ovale
   b. Atrial septal aneurysm
   c. Atrial septal defect

in patients with AF and other heart diseases is discussed in a separate chapter.

Ischemic Heart Disease with Intracavitary Thrombus: Patients with large anterior myocardial infarctions (MI) associated with a left ventricular (LV) ejection fraction of <40% and antero-apical wall-motion abnormalities are at increased risk for developing mural thrombus due to stasis of blood in the ventricular cavity as well as endocardial injury with associated inflammation [5,6]. Ventricular thrombi can also occur in patients with chronic ventricular dysfunction caused by coronary artery disease, hypertension, and dilated cardiomyopathy. Stroke is less common among uncomplicated MI patients but can occur in up to 12% of patients with acute MI complicated by a left ventricular thrombus. The rate of stroke is higher in patients with anterior rather than inferior MI and reaches up to 20% in those with large anteroseptal involvement. The incidence of embolism is highest during the period of active thrombus formation during the first 1–3 months, with substantial risk remaining even beyond the acute phase in patients with persistent myocardial dysfunction, congestive heart failure, or AF [1,5].

Congestive Heart Failure: Current data indicates that congestive heart failure affects an estimated 5.1 million Americans. Patients with ischemic and nonischemic dilated cardiomyopathy have a similarly increased stroke risk by a factor of 2 to 3, accounting for an estimated 10% of ischemic strokes [1,5,7]. Stroke rates may be higher in certain subgroups, including patients with prior stroke or TIA, lower ejection fraction, LV noncompaction, peripartum cardiomyopathy, and Chagas heart disease [8]. The 5-year recurrent stroke rate in patients with cardiac failure has been reported to be as high as 45% [1,5,8].

Prosthetic Valves and Devices: The magnitude of risk for brain embolism from a diseased heart valve depends on the nature and severity of the disease, as well as the surgical procedure performed. Notably, atrial fibrillation often coexists with valve disease.

Intracardiac Vegetation

Emboli, including ischemic strokes, are common in patients with infective endocarditis [1]. Specific vegetation characteristics, such as large size, mobility, and location on the mitral valve, as well as vegetations related to a *Staphylococcus aureus* infection, are associated with an increased risk of symptomatic embolism [9,10]. Cerebral complications are the most severe and frequent extracardiac sequelae of infective endocarditis, occurring in an estimated 15–20% of patients. These complications include ischemic and hemorrhagic stroke (preceding the diagnosis of infective endocarditis in 60% of patients), TIA, silent cerebral embolism, mycotic aneurysm, brain abscess, and meningitis [1,9,10]. Unfortunately, these patients can develop multiple neurological complications (Figs. 85.1–85.3). Intracranial hemorrhage (ICH) occurs in about 5% of patients with infective endocarditis [1]. The ICH is typically attributed to ruptured mycotic aneurysms, which can cause fatal subarachnoid bleeding. However, ICH in these patients can also result from septic erosion of the arterial wall causing rupture, without the presence of a well-delineated aneurysm. An infected embolus can also cause a hemorrhagic infarction with massive bleeding in patients who are on anticoagulation at the time of the event [1,9,10]. Embolization usually stops when the infection is controlled. Warfarin does not prevent embolization and is contraindicated in patients with endocarditis and known cerebral embolism, unless indicated in patients with prosthetic valves or pulmonary embolism [1].

Intracardiac Tumors

Primary cardiac tumors, which can be associated with cerebral embolism and stroke, are extremely rare with atrial myxomas constituting over half of the cases. Papillary fibroelastoma is another histologically benign tumor associated with embolism. Embolic events have long been thought to occur in patients with cardiac tumors secondary to embolization of tumor fragments [1,11]. Aortic valve and left atrial tumors pose the greatest anatomical risk for embolism. Furthermore, patients with smaller tumors, minimal symptomatology, and no evidence of mitral regurgitation also have a high risk of embolism. The presence of mitral regurgitation and decreased functional status (New York Heart Association III/IV) are actually protective against the occurrence of embolism in patients with intracardiac masses. Cardiac tumors can be resected with low early mortality, and late survival after operation in the context of an embolic stroke is similar to patients with cardiac tumors who undergo resection for other indications [1,11].

Aortic Atheroma

Emboli can also arise from the aorta [1]. Ulcerated atheromatous plaques are often found at necropsy in patients with ischemic strokes, particularly in patient with cryptogenic stroke [1,12]. Transesophageal echocardiography (TEE) often shows aortic atheromas, but technical factors can limit visualization of the entire arch. Large (>4 mm), protruding mobile aortic atheromas are

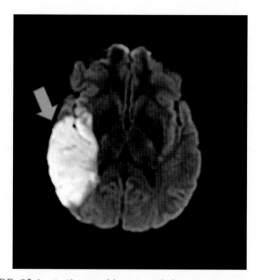

FIGURE 85.1 A 42-year-old man with known intravenous drug abuse initially presented with left-sided weakness and neglect due to an acute ischemic stroke in the right middle cerebral artery territory, as seen with diffusion-weighted magnetic resonance imaging (DWI-MRI).

Left **Right**

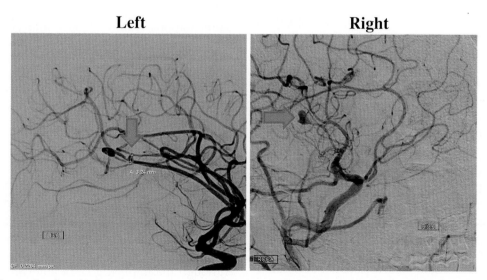

FIGURE 85.2 Conventional cerebral angiography identified bilateral peripheral middle cerebral artery pseudoaneurysms (*red arrows*) with occluded outflow of the right middle cerebral artery superior division branch and diseased outflow segments of the left middle cerebral artery in a 42-year-old man with infective endocarditis.

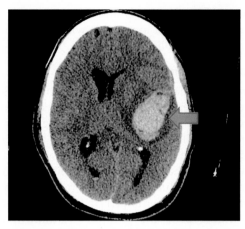

FIGURE 85.3 Despite antibiotic treatment, the patient developed an acute intraparenchymal hemorrhage (*arrow*) from the left middle cerebral artery mycotic aneurysm 2 weeks later.

especially likely to cause embolic strokes and are associated with a higher rate of stroke recurrence [1,13]. For this subgroup of patients, the optimal treatment–antiplatelet monotherapy versus anticoagulation for the prevention of ischemic stroke and TIA due to aortic disease is not clear. Other recommended therapies include lipid-lowering therapy (e.g., statins), blood pressure control, smoking cessation, and, in patients with diabetes, glycemic control. In the setting of nonmobile protruding aortic arch atheroma, antiplatelet agents are the treatment regimen of choice for stroke prevention in patients with aortic arch disease [1,14].

BRAIN HYPOPERFUSION DUE TO CARDIAC PUMP FAILURE

After cardiac arrest or prolonged hypotension, brain imaging and necropsy can show infarction in the border zones between the major brain-supplying intracranial arteries [1,15,16]. Cerebral watershed or border zone infarcts involve the junction of the distal fields of two nonanastomosing arterial systems, which are hemodynamic risk zones, and account for up to 10% of all strokes [16,19]. There are two types of watershed infarcts:

1. Cortical: occurring at the junction between cortical territories of the anterior, middle, and posterior cerebral arteries. These are typically located between the anterior cerebral artery (ACA) and middle cerebral artery (MCA) regions or between the MCA and posterior cerebral artery (PCA) territories.
2. Subcortical: occurring in the white matter between the deep and superficial perfusion systems of the middle cerebral artery [16,17].

Bilateral border-zone distribution strokes are traditionally attributed to systemic hypoperfusion. The weight of data suggests that most instances of hypoperfusion caused by cardiac pump failure cause either no symptoms or attacks of transient brain ischemia without resulting in major brain infarction. Cardiac arrest and systemic hypoperfusion even for 10–15 min are rarely associated with large border-zone infarcts even when neuronal ischemia in vulnerable areas has developed. Stroke related to cerebral hypoperfusion can also develop postcardiac surgical procedures or in other settings associated with protracted hypotension [1,15–17].

Brain perfusion during cardiac surgery is an important factor in predicting the presence and severity of ischemic brain damage after surgery. In one randomized study, cardiac surgery patients whose mean arterial pressure was maintained in the 50–60 mmHg range had a stroke frequency of 7.2% compared to 2.4% in patients whose mean arterial pressures were maintained in the 80–100 mmHg range [16,18]. The clinical picture of watershed infarctions is often progressive or fluctuating. Symptoms may be mild and underestimated. One characteristic feature is the occurrence of early-onset partial seizures, which occur more often with cortical watershed infarcts than in other forms of stroke [16,20].

NEUROLOGICAL EFFECTS FROM CARDIAC DRUGS

Drugs given to patients with cardiac disease often have neurological adverse effects that can be confused with symptoms of stroke. Digoxin can cause visual hallucinations, yellow vision, and general confusion. Patients may become acutely comatose after receiving very large doses of intravenous lidocaine, with adverse reactions of sedation, irritability, twitching, and seizures. Amiodarone can cause ataxia, weakness, tremors, paresthesias, visual symptoms, a parkinsonian-like syndrome, and occasionally delirium, even at normal dosages [1].

NEUROLOGICAL COMPLICATION FROM ENDOVASCULAR CARDIAC PROCEDURE

Patients with heart disease are diagnosed, treated, and even cured with a variety of cardiac procedures. Although the implicit goal of any cardiac intervention, whether diagnostic or therapeutic, is to improve the patient's quality of life, there are several risks associated with these procedures.

Cardiac Catheterization, Coronary Artery Angioplasty, and Stenting

Strokes and TIAs are known complications of heart catheterization, with the cerebrovascular impairment usually occurring during or within several minutes post procedure. Both ischemic stroke and intracerebral hemorrhage have been reported [1]. Potential mechanisms for ischemic cerebrovascular events during cardiac angiography include cerebral embolism from a local clot, catheter tip thromboembolism, atherosclerotic plaque or cholesterol embolism, air emboli, arterial vasospasm, and/or hypotension [1,21]. Mechanism of intracranial hemorrhage is suspected to be related to intraprocedural anticoagulation. The rate of stroke in patients undergoing percutaneous coronary interventions for both stable and unstable coronary artery disease including angioplasty for acute MI is estimated to be 0–4% [1].

Electrophysiologic Procedures and Electrical Cardioversion

Thromboembolic stroke can be a complication of cardiac electrophysiologic procedures, including radiofrequency catheter ablation of arrhythmia. Stroke risk appears to be <2% [1,21], but multicenter data is limited. Additionally, electrical cardioversion may be used in the treatment of AF and atrial flutter. Stroke caused by cardioversion has been estimated to occur in 1.3% of patients. Anticoagulation before and after cardioversion lowers the risk of embolism [1,21].

Intra-aortic Balloon Pump

Intra-aortic balloon pumps (IABPs) are used in patients with severe left ventricular failure or cardiogenic shock. The IABP is inserted into the patient's midthoracic aorta to maintain adequate perfusion. Spinal cord infarcts can occur in patients with IABPs caused by local thromboembolism, aortic dissection, aortic atherosclerotic plaque rupture, or local hypoperfusion [1,21].

Transcatheter Aortic Valve Replacement (TAVR) and Cardiac Surgery

This is discussed in Chapter 126.

CARDIAC COMPLICATIONS ASSOCIATED WITH BRAIN LESIONS

Cardiac muscle changes, known as myocytolysis, arrhythmias, pulmonary edema, electrocardiogram (ECG) changes, and sudden death, are all associated with stroke and other brain disease [1,22,23].

Cardiac Lesions

The two most common lesions found in the hearts of patients dying with acute CNS lesions are patchy regions of myocardial necrosis and subendocardial hemorrhage known as myocytolysis [1,22,23]. One study found a high incidence of myocardial abnormalities in patients dying of an ICH or other brain lesions that increase intracranial pressure rapidly [1,24]. Stress-related release of catecholamines and possibly corticosteroids may be responsible, in part, for the cardiac lesions found in patients with CNS lesions [1,24].

Electrocardiographic and Enzyme Changes

Stroke patients, especially those with subarachnoid hemorrhage (SAH), can have various arrhythmias, ECG changes, and cardiac enzyme abnormalities including an elevated MB isoenzyme of creatine kinase (MB-CK) [1]. ECG changes may include a prolonged QT interval, depressed ST segments, flat or inverted T waves, and U waves [1,22]. Less often, tall, peaked T waves and elevated ST segments are noted. The ECG changes and arrhythmias correlate significantly with increased levels of MB-CK in stroke patients [1,25]. During days 4 to 7 after stroke, there is usually a slow increase and later a decrease in serum MB-CK levels, a pattern that is different from that associated with acute MI and thought to be due to smoldering low-grade necrosis such as patchy, focal myocytolysis [1,25]. Myocardial enzyme release and echocardiographic regional wall-motion abnormalities are also associated with impaired left ventricular performance after SAH. In severely affected patients, reduction of cardiac output may elevate the risk of vasospasm-induced cerebral ischemia [1,26]. It is therefore reasonable to consider obtaining a transthoracic echocardiogram in patients with SAH.

Arrhythmias

Various cardiac arrhythmias have been found in stroke patients, most frequently sinus bradycardia, sinus tachycardia, and premature ventricular contractions [1,22]. Some arrhythmias are manifestations of primary cardiac problems, but others are undoubtedly secondary to brain lesions. The incidence of sinus tachycardia and bradycardia is maximal on the first day after intracerebral hemorrhage [27]. Ventricular bigeminy, atrioventricular dissociation and block, ventricular tachycardia, AF, and bundle-branch blocks are found less often [28]. Arrhythmias are more common in patients who have primary brainstem lesions or brainstem compression [1].

Echocardiographic Changes

Takotsubo cardiomyopathy is also known as *broken heart syndrome* or *transient left ventricular apical ballooning*. The name *takotsubo* cardiomyopathy was initially coined because the shape of the end-systolic left ventriculogram was thought to resemble an octopus catcher used in Japan [29]. It has been reported in patients with severe emotional stress, but also has been identified in both SAH and ischemic stroke patients [29]. In ischemic stroke patients, takotsubo cardiomyopathy occurs soon after stroke onset, is commonly asymptomatic, and is associated with insular damage [30]. Prognosis is generally very good with full recovery in most patients, but there is increased morbidity in patients with SAH [31].

Neurogenic Pulmonary Edema

Acute pulmonary edema may complicate strokes, especially SAH, subdural hemorrhage, primary spinal cord hemorrhage, and posterior circulation ischemia and/or hemorrhage [1,25]. Pulmonary edema has been found in 70% of patients with fatal SAH and correlates with the development of increased intracranial pressure [32]. The pulmonary edema can develop despite normal cardiac function [25]. The most relevant imaging method is the chest X-ray, where diffuse hyperintensive infiltrates in both lungs are apparent.

Sudden Death

Sudden death associated with stressful situations, including so-called *voodoo death*, typically involves CNS mechanisms [1]. Establishing a cause–effect relationship between sudden death and stroke is complicated, because these patients usually have risk factors for coronary disease as well [1]. However, it has been observed that patients with lateral medullary and pontine infarcts die unexpectedly and also have a high incidence of autonomic dysregulation (e.g., labile blood pressure and tachycardia). Postmortem analysis of stroke patients who died suddenly without any evidence of coronary disease often shows myocytolysis and myofibrillar degeneration [33].

COEXISTENT VASCULAR DISEASES AFFECTING BOTH HEART AND BRAIN

Atherosclerosis

The most common and important vascular disease that affects both the brain and the heart is atherosclerosis. The most frequent cause of death in stroke patients is coronary artery disease, and extra- and intracranial arterial atherosclerosis is common in patients with coronary artery disease [1]. In patients considered for cardiac surgery who have symptoms of brain ischemia, it is important to define the extent of cerebrovascular disease preoperatively by noninvasive means, ultrasound and/or magnetic resonance angiography, as well as to define cardiac and coronary artery anatomy and function. In some patients with excessive surgical risks, anticoagulation may represent an alternative treatment. Clearly, optimal medical therapy should be instituted preoperatively and continued after surgery [1].

Systemic Arterial Hypertension

Acute and chronic high blood pressure damages deep, penetrating small intracranial arteries, accelerates the development of atherosclerosis in the extracranial and large intracranial arteries, and results in ischemic syndromes of lacunar infarction, diffuse white matter ischemic changes, and intracerebral hemorrhage. Hypertension is also frequent in patients with aneurysmal SAH and may contribute to enlargement and rupture of aneurysms [1].

Arterial Dissection

Aortic dissection involving the innominate or common carotid arteries is a well-known cause of stroke and TIA. Transient or permanent neurological symptoms at onset of aortic dissection are not only frequent (17–40% of the patients), but often dramatic and may mask the underlying condition. Disorders of consciousness, syncope, and seizures frequently occur at the onset of aortic dissection. Aortic dissections can affect the outflow of supra-aortal, spinal, as well as extremity arteries, leading to a variety of neurological symptoms including disturbances of brain, spinal cord, or peripheral nervous system. Symptoms of acute ischemic stroke are the most common initial neurological finding. Spinal cord ischemia on the basis of aortic dissection is a much rarer syndrome and more common with distal aortic dissections. Involvement of the peripheral nerves can occur as ischemic neuropathy, ischemic plexopathy, or due to the direct compression of a nerve by the enlarging false arterial lumen. The latter comprises Horner's syndrome and hoarseness of voice caused by vocal cord paralysis due to compression of the left recurrent laryngeal nerve, which is also known as cardiovocal syndrome or Ortner's syndrome. It is important to recognize aortic dissection in ischemic stroke patients in particular. Thrombolysis as an emergency stroke therapy may be life threatening for these patients, because of the high risk of fatal rupture of the ascending aorta or the aortic arch or of dissection into the pericardium. Routine chest X-ray and being alert to physical examination findings such as hypotension, asymmetrical pulses, or cardiac murmur may reduce the risk of delayed diagnosis or misdiagnosis.

Neurological symptoms at onset or in the postoperative course of aortic dissection are not necessarily associated with increased mortality [34].

References

[1] Leary MC, Caplan LR. Cerebrovascular disease and neurologic manifestations of heart disease. In: O'Rourke RA, Walsh RA, Fuster V, editors. HURST's The Heart. 43th ed. New York (NY): McGraw-Hill; 2016. Chapter 90.

[2] Saric M, Armour AC, Arnaout MS, et al. Guidelines for the use of echocardiography in the evaluation of a cardiac source of embolism. J Am Soc Echocardiogr 2016;29:1–42.

[3] Adams Jr HP, Bendixen BH, Kappelle LJ, Biller J, Love BB, Gordon DL, et al. Classification of subtype of acute ischemic stroke. Definitions for use in a multicenter clinical trial. TOAST. Trial of Org 10172 in Acute Stroke Treatment. Stroke 1993;24:35–41.

[4] Bogousslavsky J, Cachin C, Regli F, et al. Cardiac sources of embolism and cerebral infarction: clinical consequences and vascular concomitants. The Lausanne Stroke Registry. Neurology 1991;41:855–9.

[5] Kernan WN, Ovbiagele B, Black HR, et al. Guidelines for the prevention of stroke in patients with stroke and transient ischemic attack. Stroke 2014;45(7):2160–236.

[6] Fuster V, Halperin JL. Left ventricular thrombi and cerebral embolism. N Engl J Med 1989;320:392–4.

[7] Go AS, Mozaffarian D, Roger VL, Benjamin EJ, Berry JD, et al. Heart disease and stroke statistics—2013 update: a report from the American Heart Association. Circulation 2013;127:e6–245.

[8] Sacco RL, Shi T, Zamanillo MC, et al. Predictors of mortality and recurrence after hospitalized cerebral infarction in an urban community. The Northern Manhattan Stroke Study. Neurology 1994;44:626–34.

[9] Hoen B, Duval X. Infective endocarditis. NEJM 2013;368:1425–33.

[10] Hart RG, Kagan-Hallet K, Joerns SE. Mechanisms of intracranial hemorrhage in infective endocarditis. Stroke 1987;18:1048–56.

[11] ElBardissi AW, Dearani JA, Daly RC, et al. Embolic potential of cardiac tumors and outcome after resection. Stroke 2009;40:156–62.

[12] Amarenco P, Duyckaerts C, Tzourio C, et al. The prevalence of ulcerated plaques in the aortic arch in patients with stroke. N Engl J Med 1992;326:221–5.

[13] The French Study of Aortic Plaques in Stroke Group. Atherosclerotic disease of the aortic arch as a risk factor for recurrent ischemic stroke. N Engl J Med 1996;334:1216–21.

[14] Ferrari E, Vidal R, Chevallier T, et al. Atherosclerosis of the thoracic aorta and systemic emboli: efficacy of anticoagulation and influence of plaque morphology on recurrent stroke. J Am Cardiol 1999;33:1317–22.

[15] Caplan LR, Wong KS, Gao S, Hennerici MG. Is hypoperfusion an important cause of strokes? If so, how? Cerebrovasc Dis 2006;21:145–53.

[16] Juergenson I, Mazzucco S, Tinazzi M. A typical example of cerebral watershed infarct. Clin Pract 2011;1:e114. http://dx.doi.org/10.4081/cp.2011.e114. Published online 2011 Nov 18.

[17] Momjian-Mayor I, Baron JC. The pathophysiology of watershed infarction in internal carotid artery disease (review of cerebral perfusion studies). Stroke 2005;36:567–77.

[18] Gold JP, Charlson ME, Williams-Russo P, et al. Improvement of outcomes after coronary artery bypass: a randomized trial comparing intraoperative high vs. low mean arterial pressure. J Thorac Cardiovasc Surg 1995;110:1302–14.

[19] Jorgensen L, Torvik A. Ischemic cerebrovascular diseases in an autopsy series. Prevalence, location, pathogenesis, and clinical course of cerebral infarcts. J Neurol Sci 1969;9:285–320.

[20] Denier C, Masnou P, Yacouba M, et al. Watershed infarctions are more prone than other cortical infarcts to cause early-onset seizures. Arch Neurol 2010;67:1219–23.

[21] Adams HP. Neurologic complications of cardiovascular procedures. In: Biller J, editor. Iatrogenic Neurology. Boston (MA): Butterworth-Heinemann; 1998. p. 51–61.

[22] Caplan LR, Hurst JW. Cardiac and cardiovascular findings in patients with nervous system disease: strokes. In: Caplan LR, Hurst JW, Chimowitz MI, editors. Clinical Neurocardiology. New York (NY): Marcel Dekker; 1999. p. 303–12.

[23] Schlesinger MJ, Reiner L. Focal myocytolysis of heart. Am J Pathol 1955;31:443–59.

[24] Kolin A, Norris JW. Myocardial damage from acute cerebral lesions. Stroke 1984;15:990–3.

[25] Norris JW, Hachinski V. Cardiac dysfunction following stroke. In: Furlan AJ, editor. The Heart and Stroke. London (United Kingdom): Springer-Verlag; 1987. p. 171–83.

[26] Mayer SA, Homma S, Lennihan L, et al. Myocardial injury and left ventricular performance after subarachnoid hemorrhage. Stroke 1999;30:780–6.

[27] Myers MG, Norris JW, Hachinsky VC, et al. Cardiac sequelae of acute strokes. Stroke 1982;13:838–42.

[28] Stober T, Sen S, Anstatt T, Bette L. Correlation of cardiac arrhythmias with brainstem compression in patients with intracerebral hemorrhage. Stroke 1988;19:688–92.

[29] Ako J, Sudhir K, Farouque O, et al. Transient left ventricular dysfunction under severe stress: brain-heart relationship revisited. Am J Med 2006;119:10–7.

[30] Yoshimura S, Toyoda K, Nagasawa H, et al. Takotsubo cardiomyopathy in acute ischemic stroke. Ann Neurol 2008;64:547–54.

[31] Hakeem A, Marks AD, Bhatti S, Chang SM. When the worst headache becomes the worst heartache!. Stroke 2007;38:3293–5.

[32] Wier BK. Pulmonary edema following fatal aneurysmal rupture. J Neurosurg 1978;49:502–7.

[33] Baranchuk A, Nault MA, Morillo CA. The central nervous system and sudden cardiac death: what should we know? Cardiol J 2009;16:105–12.

[34] Gaul C, Dietrich W, Erbguth FJ. Neurological symptoms in aortic dissection: A challenge for neurologists. Cerebrovasc Dis 2008;26: 1–8.

CHAPTER

86

Aortic Arch Artherosclerotic Disease

J.A. Abbatemarco[1], M.C. Leary[1,2], H.A. Yacoub[1]

[1]Lehigh Valley Hospital and Health Network, Allentown, PA, United States; [2]University of South Florida, Tampa, FL, United States

INTRODUCTION

Ischemic stroke etiologies can be classified into thrombotic, embolic, or cerebral infarcts secondary to systemic hypoperfusion. Aortic arch disease has been a subject of study for many years as a potential source of embolic stroke and is common in individuals over 60 years of age. Complex, ulcerated plaques are common, and mural thrombi can superimpose upon ulcers. Mural thrombi can occasionally be loosely adherent and mobile, increasing the risk of embolization to the brain, retina, or peripheral organs. Large plaques, defined as those ≥4 mm in size, have been shown to further increase the risk of stroke [1,2] and have been linked to a 2.5- to 9-fold increase in stroke risk in case–control studies [3]. The presence of complex features such as ulceration has also been suggested to further increase the risk of stroke in individuals with aortic arch disease [4]. Atherosclerotic disease of the aortic arch has thus been considered a potential source of cerebral ischemia in patients with cryptogenic stroke.

Varying degrees of aortic arch disease are found in about 1 in 4 patients with an embolic event [5]. The advanced use of transesophageal echocardiography (TEE)—the gold standard for evaluating aortic arch disease—has enhanced the ability to detect and characterize aortic arch plaques. Atherosclerosis burden of the thoracic aorta, particularly in lesions with complex characteristics and size of >4 mm, has been considered a major risk factor for both cerebral ischemia and peripheral embolization [6]. Interestingly, lesions in the descending thoracic aorta, located distal to the anatomical branch points of the great vessels, have been considered a potential source of cerebral embolism through retrograde aortic flow [7,8]. Retrograde flow-related emboli have been suggested as a potential cause of retinal or cerebral infarcts in 24% of patients with cryptogenic stroke [9]. Stroke prevention in patients with atherosclerotic burden of the aortic arch has been controversial, and there

is no known optimal stroke prevention strategy. In this chapter, we will summarize some seminal studies along with some of the most recent studies investigating aortic arch disease as a source of ischemic stroke. Stroke treatment and prevention strategies will also be discussed.

Aortic Arch Plaques as a Risk Factor for Ischemic Stroke

Aortic arch plaques >4 mm in thickness are found in one-third of patients with stroke of an unknown source, which accounts for about one-third of the total ischemic stroke population over the age of 60 years [10]. Amarenco et al. [11] found that the presence of ulcerated plaques in the aortic arch is associated with stroke of an unknown cause. It has also been reported that the association between aortic arch atheroma and stroke is stronger when the size of plaque is ≥4 mm [10]. Moreover, Jones et al. [12] demonstrated that the presence of aortic arch atheromas with a size of ≥5 mm or with mobile elements is an independent risk factor for ischemic stroke. Louis R. Caplan [13] reports that large protruding plaques in the ascending aorta and transverse arch are a major source of aortoembolic stroke, and the risk increases with mobile plaques.

Disparities concerning the underlying relationship between aortic arch disease and stroke have also been described. Petty et al. [14] conducted a population-based study of 1135 subjects and found that complex aortic atherosclerotic plaques of ≥4 mm in thickness, with or without mobile debris, was not a significant risk factor for stroke of unknown source or transient ischemic attack. They did, however, report an association between complex aortic plaques and non-cryptogenic stroke and, in general, concluded that the presence of aortic arch debris is an indicator for generalized atherosclerosis and increases the risk of cerebrovascular disease [14].

Prospective Studies

In 1996, the French Study of Aortic Plaques in Stroke Group [2], demonstrated an annual risk of recurrent stroke in 11.9 per 100 person-years in 45 patients with aortic arch plaques ≥4mm in size as compared with 3.5 per 100 person-years in 143 patients with plaques from 1 to 3.9mm and 2.8 in 143 patients with no plaque. This seminal work was a prospective study of patients consecutively admitted for cerebral infarcts over a period of 2–4years. Additionally, the investigators looked at overall vascular risk of stroke, myocardial infarction, peripheral emboli, and vascular death for the same aortic arch plaque variables. They noted an overall vascular risk of stroke, myocardial infarction, peripheral emboli, and vascular death in 26 per 100 person-years in patients with aortic arch plaques ≥4mm in size as compared with 9.1 per 100 person-years in patients with plaques from 1 to 3.9 mm and 5.9 in patients with no plaque. In patients with cryptogenic stroke at entry and with plaques ≥4mm in thickness, the event rates at 1year were 16% for recurrent cerebral infarct and 26% for all cardiovascular events. Multivariate analysis showed that aortic arch plaques ≥4mm in thickness were significant predictors of new brain infarcts, independent of the presence of carotid artery stenosis, atrial fibrillation, or peripheral artery disease, with a relative risk of 3.8. The study further provided a strong argument for a causality link between plaques ≥4mm in thickness and cerebral infarcts. Plaques ≥4mm in thickness in the aortic arch were also found to be strong and independent predictors of all vascular events (relative risk = 3.5) [2].

In a later study, Fujimoto et al. investigated the relationship between stroke recurrence and the characteristics of aortic arch atherosclerosis lesions, particularly the size [15]. Included in this study were 283 patients with brain embolism and no significant occlusive lesions in their cerebral arteries. All patients underwent TEE, at which time the intima–media thickness and extension and mobility of the aortic arch atherosclerotic lesions were measured. In investigating the relationship between stroke recurrence and the various characteristics of the lesions seen during the TEE, the investigators found an intima–media thickness of ≥4mm in 67 patients. In 51 of those patients, the lesion extended to the origin of the arch branch points of the great vessels (brachiocephalic trunk, left subclavian, and common carotid arteries). Recurrence of ischemic stroke was seen in 32 of the 283 total followed patients. Aortic atheroma ≥4mm, atheroma extending to the arch branches, or both, were seen more frequently in patients with recurrent strokes, a difference that reached statistical significance. After adjusting for other variables such as age and hypertension, an atheroma with a size of ≥4mm and extending to the arch branches was an independent predictor of stroke recurrence [15].

COMPLEX ATHEROMATOUS PLAQUES IN THE DESCENDING AORTA AND STROKE

Plaques in the proximal aortic arch have been established as an important source of cerebral and peripheral embolization. As mentioned earlier, plaques in the descending thoracic aorta have been considered a potential source of embolization. These plaques are found distal to the arch branch points of the great vessels, and cerebral embolization thus occurs through retrograde flow. Katsanos et al. [16] conducted a systemic review and meta-analysis of 11 prospective observational studies reporting the prevalence of complex atheromas in the descending aorta in patients with stroke and in unselected populations. The aim was to evaluate the potential association of this particular type of atherosclerotic burden, characterized by TEE, with ischemic stroke. The prevalence of complex atheromas in the descending aorta was higher in patients with stroke compared with unselected individuals: patients with stroke had about a fourfold higher prevalence of complex atheromatous plaques in the descending aorta when compared with unselected individuals. However, there was no significant difference in the prevalence of complex atheromatous plaques in the descending thoracic aorta between patients with stroke of an unknown source and unclassified cerebral ischemia. Therefore, based on this systematic review, there is no evidence for a direct causal relationship between descending aortic atherosclerotic disease and cerebral embolization.

Stroke prevention in patients with atherosclerotic disease of the descending aortic arch has not been studied specifically. In our practice, strategies of stroke prevention in these patients are similar to those with aortic arch disease in general.

STROKE PREVENTION IN PATIENTS WITH AORTIC ARCH ATHEROSCLEROSIS

The optimal antithrombotic strategy for stroke prevention in patients with atherosclerotic disease of the aortic arch is not well studied. There are no trials that provide an ideal stroke-preventive therapy in this patient population. Although stroke mechanism is likely thromboembolic, the efficacy of antithrombotic or anticoagulation therapy has not been well-investigated in a trial. In the majority of patients, antiplatelet therapy, along with a statin, is utilized. Anticoagulation should be considered in patients with severe atherosclerosis burden and recurrent cerebral or retinal ischemia despite the use of antiplatelet therapy. The use of the novel anticoagulants, direct prothrombin inhibitors, and factor Xa inhibitors, has not been investigated.

A Prospective Study

Di Tullio et al. [17] studied the relationship between aortic arch plaques and recurrent stroke in 516 patients. This was part of the Patent Foramen Ovale in Cryptogenic Stroke Study (PICSS) in which patients were double-blindly randomized to antiplatelet versus anticoagulation therapy. The study evaluated stroke recurrence and death rate over a 2-year follow-up period. Large aortic plaques, particularly those with complex morphology, were associated with a significantly higher risk of recurrent stroke and/or death. Interestingly, event rates were similar in the warfarin and aspirin groups, supporting the notion that large, complex, aortic plaques are associated with an increased risk of recurrent stroke and death at 2 years in spite of the choice of stroke-preventive strategy.

A Prospective Randomized, Controlled Trial

A prospective randomized, controlled, open-labeled trial with blinded end-point evaluation tested the superiority of aspirin combined with clopidogrel over warfarin in patients with ischemic stroke, transient ischemic attack, or peripheral embolism with plaques in the thoracic aorta ≥4 mm in size and no other apparent sources of stroke [18]. The primary end points included cerebral infarct, myocardial infarction, peripheral embolism, vascular death, or intracranial hemorrhage. The trial, interrupted after 349 patients were randomized over a period of 8 years and 3 months, demonstrated occurrence of the primary end point in 7.6% of patients on aspirin plus clopidogrel and in 11.3% in those on warfarin. Major hemorrhages, including intracranial hemorrhage, were seen in four patients on aspirin plus clopidogrel and six patients on warfarin. Vascular deaths occurred in six patients randomized to warfarin and in none of the patients on the combined antiplatelet therapy. Although the combination of aspirin and clopidogrel significantly reduced the rate of vascular deaths, this outcome could have been obtained by chance due to the lack of power in the study. Therefore, the results of this study should be interpreted with caution.

In our practice, we use aspirin and a high-dose statin in the majority of patients with stroke and aortic arch disease for stroke prevention. We advocate the use of warfarin in patients with large plaques of ≥4 mm in size, especially if ulcerated with mobile debris, on the basis of the evidence that these lesions can extend and propagate.

CONCLUSION

Aortic arch disease appears to be a major risk factor for ischemic stroke. The presence of ulcerations or mobile components seems to further increase this risk.

Large plaques (≥4 mm) are strong predictors of recurrent stroke and other vascular events, and remain associated with a higher risk of recurrent stroke and death despite medical therapy. Extension of the atheroma to the arch branches also seems to be an independent risk factor for stroke recurrence. The optimal strategy of stroke prevention in patients with aortic arch atherosclerosis remains unclear, and more studies are warranted. The majority of stroke centers utilize aspirin, a statin, and risk factor modifications strategies as appropriate.

References

[1] Tunick PA, Rosenzweig BP, Katz ES, Freedberg RS, Perez JL, Kronzon I. High risk for vascular events in patients with protruding aortic atheromas: a prospective study. J Am Coll Cardiol 1994;23(5):1085–90.

[2] The French Study of Aortic Plaques in Stroke Group. Atherosclerotic disease of the aortic arch as a risk factor for recurrent ischemic stroke. N Engl J Med 1996;334(19):1216–21.

[3] Di Tullio MR, Sacco RL, Gersony D, Nayak H, Weslow RG, Karqman DE, et al. Aortic atheromas and acute ischemic stroke: a transesophageal echocardiographic study in an ethnically mixed population. Neurology 1996;46(6):1560–6.

[4] Di Tullio MR, Sacco RL, Savoia MT, Sciacca RR, Homma S. Aortic atheroma morphology and the risk of ischemic stroke in a multiethnic population. Am Heart J 2000;139(2 Pt 1):329–36.

[5] Sheikhzadeh A, Ehlermann P. Atheromatous disease of the thoracic aorta and systemic embolism. Clinical picture and therapeutic challenge. Z Kardiol 2004;93(1):10–7.

[6] Tunick PA, Kronzon I. Atheromas of the thoracic aorta: clinical and therapeutic update. J Am Coll Cardiol 2000;35(3):545–54.

[7] Sharma U, Tak T. Aortic atheromas: current knowledge and controversies: a brief review of the literature. Echocardiography 2011;28(10):1157–63.

[8] Chhabra L, Niroula R, Phadke J, Spodick DH. Retrograde embolism from the descending thoracic aorta causing stroke: an underappreciated clinical condition. Indian Heart J 2013; 65(3):319–22.

[9] Tenenbaum A, Motro M, Shapira I, Feinberg MS, Schwammenthal E, Tanne D, et al. Retrograde embolism and atherosclerosis development in the human thoracic aorta: are the fluid dynamics explanations valid? Med Hypotheses 2001;57(5):642–7.

[10] Amarenco P, Cohen A, Tzourio C, Bertrand B, Hommel M, Besson G, et al. Atherosclerotic disease of the aortic arch and the risk of ischemic stroke. N Engl J Med 1994;331(22):1474–9.

[11] Amarenco P, Duyckaerts C, Tzourio C, Hénin D, Bousser MG, Hauww JJ. The prevalence of ulcerated plaques in the aortic arch in patients with stroke. N Engl J Med 1992;326(4):221–5.

[12] Jones EF, Kalman JM, Calafiore P, Tonkin AM, Donnan GA. Proximal aortic atheroma. An independent risk factor for cerebral ischemia. Stroke 1995;26(2):218–24.

[13] Caplan LR. The aorta as a donor source of brain embolism. In: Caplan LR, Manning WJ, editors. Brain embolism. New York: Informa Healthcare; 2006. p. 187.

[14] Petty GW, Khandheria BK, Meissner I, Whisnant JP, Rocca WA, Sicks JD, et al. Population-based study of the relationship between atherosclerotic aortic debris and cerebrovascular ischemic events. Mayo Clin Proc 2006;81(5):609–14.

[15] Fujimoto S, Yasaka M, Otsubo R, Oe H, Nagatsuka K, Minematsu K. Aortic arch atherosclerotic lesions and the recurrence of ischemic stroke. Stroke 2004;35(6):1426–9.

[16] Katsanos AH, Giannopoulos S, Kosmidou M, Voumvourakis K, Parissis JT, Kyritsis AP, et al. Complex atheromatous plaques in the descending aorta and the risk of stroke: a systemic review and meta-analysis. Stroke 2014;45(6):1764–70.

[17] DiTullio MR, Russo C, Jin Z, Sacco RL, Mohr JP, Patent Foramen ovale in Cryptogenic Stroke Study Investigators, et al. Aortic arch plaques and risk of recurrent stroke and death. Circulation 2009;119(17):2376–82.

[18] Amarenco P, Davis S, Jones EF, Cohen AA, Heiss WD, Aortic Arch Related Cerebral Hazard Trial Investigators, et al. Clopidogrel plus aspirin versus warfarin in patients with stroke and aortic arch plaques. Stroke 2014;45(5):1248–57.

C H A P T E R

87

Brain Injury From Cerebral Hypoperfusion

M. De Georgia, B. Miller

University Hospitals Case Medical Center, Neurological Institute, Cleveland, OH, United States

INTRODUCTION

There is a spectrum of brain injury from cerebral hypoperfusion ranging from transient loss of consciousness to devastating anoxic neuronal damage. At the root of this injury is the fact that with limited stores of high-energy phosphate compounds and a high metabolic demand, the brain is critically dependent on cerebral blood flow (CBF) to continuously supply oxygen and glucose. CBF in turn is governed by the Hagen–Poiseuille equation, correlating directly with cerebral perfusion pressure (CPP) (which is the difference between mean arterial pressure and intracranial pressure) and vessel radius and inversely with blood viscosity and vessel length. Blood pressure is the product of cardiac output and systemic vascular resistance. Cardiac output also correlates directly with CBF, independent of blood pressure. A decrease in blood pressure and/or cardiac output results in cerebral hypoperfusion.

When mild, cerebral hypoperfusion can lead to syncope, a temporary loss of consciousness and motor control characterized by a relatively rapid onset, brief duration, and spontaneous and full recovery. More prolonged hypoperfusion, together with extracranial vascular disease, can cause selective damage to vulnerable areas of the cerebral cortex that lie in between vascular territories and lead to the so-called watershed territory infarcts. In cardiac arrest, it is a sudden ventricular arrhythmia or asystole that leads to the immediate loss of cardiac output. This is accompanied by respiratory failure and reduced partial pressure of oxygen (hypoxemia). The combination of the two conditions culminates in the dual nature of hypoxic-ischemic brain injury.

SYNCOPE

Syncope is common, and it is estimated that 40% of the population has experienced syncope at some point in their lives. With a bimodal age distribution, there is a higher incidence between the ages of 10 and 30 years and in those older than 65 years. The frequency continues to rise as patients age; the 10-year cumulative incidence of syncope is approximately 11% for those who are 70–79 years old and 18% for those older than 80 years [1]. Although the cause of syncope is diverse, the clinical picture is remarkably uniform; patients feel light-headed and experience a frightening sensation of impending faint, confusion, and unsteadiness accompanied by pallor and diaphoresis. If they are able to lie down immediately upon losing consciousness, they generally awaken rapidly. If they do not or cannot, continued hypoperfusion may produce seizure-like tonic limb and trunk rigidity or clonic spasms as

Primer on Cerebrovascular Diseases, Second Edition
http://dx.doi.org/10.1016/B978-0-12-803058-5.00087-4

they collapse to the ground. Depending on the cause and duration of hypoperfusion, patients regain consciousness in several seconds to less than a minute. Establishing the cause is often difficult, as the event is transient and most patients have normal results on physical examination. In general, the causes of syncope can be grouped into three broad categories: reflex mediated, orthostatic hypotension, and cardiac causes.

Reflex-mediated syncope is the most common type identified in approximately two-thirds of cases. It is characterized by increased vagal tone plus peripheral sympathetic inhibition resulting in bradycardia or asystole and hypotension. The most common subtype, vasovagal syncope, usually occurs after prolonged standing. Typically, patients experience a sensation of warmth, nausea, and light-headedness just before losing consciousness. Although reflex syncope is the most frequent cause, asystolic pauses caused by sinus arrest or atrioventricular block is the most frequent direct mechanism in more than half the cases. This distinction is important because mechanism-specific treatment has been shown to be more effective than cause-specific treatment. Other patients have situational syncope that follows specific activities, such as coughing, swallowing, micturition, or defecation. Possible mechanisms in such patients include increased intrathoracic pressure resulting in diminished venous return and cardiac output along with a vagal surge, bradycardia, and vasodilation resulting in hypotension. These patients should be counseled on avoiding dehydration and maintaining adequate salt intake. Carotid sinus syncope mostly occurs in the elderly (and primarily in men) and when carotid massage causes severe bradycardia and/or hypotension leading to syncope. Medical therapy for reflex-mediated syncope includes the administration mineralocorticoids [e.g., fludrocortisone (Florinef)], vasoactive agents (e.g., midodrine), acetylcholinesterase inhibitors, and serotonin reuptake inhibitors. Beta-blockers have not been shown to be effective in randomized trials and are no longer recommended [2].

Orthostatic hypotension occurs when a patient's blood pressure decreases after moving to an upright posture. Usually this is because of a reduction in intravascular volume or impairment in the baroreflex, the latter is due to autonomic dysfunction or medication effects. Treatment centers on volume expansion. Compression stocking can be used to minimize venous pooling and Florinef and midodrine can also be used in refractory cases.

Cardiac syncope includes syncope from arrhythmias or structural heart disease. Arrhythmias are by far the most common cause of cardiac syncope, with bradycardia being more frequent than tachycardia. However, syncope caused by ventricular tachyarrhythmias in patients with poor left ventricular function or channelopathies (i.e., long QT syndrome, Brugada syndrome) is predictive of sudden cardiac death, so when identified, these patients need urgent cardiac electrophysiologic testing and often an implantable cardioverter defibrillator (ICD). In general, features of cardiac syncope include an abnormal electrocardiogram (ECG), palpitations before syncope, syncope during effort or in the supine position, and absence precipitating factors [3]. Although less common, structural heart disease (e.g., severe aortic stenosis, mitral stenosis, hypertrophic cardiomyopathy, atrial myxoma) can also cause syncope.

The prognosis of syncope depends on its cause. Noncardiac syncope generally has an excellent prognosis. For example, adolescents or young adults may have attacks of vasovagal syncope for many years without harm (other than the possible injury that comes from falling). Cardiac syncope can be more serious. In the Evaluation of Guidelines in Syncope Study 2 (EGSYS 2) death from any cause occurred in 9.2% of patients during a mean follow-up period of 614 days after syncope. Among those who died, 82% had an abnormal ECG and/or heart disease. In patients without an abnormal ECG and/or heart disease, only 3% died [4].

WATERSHED TERRITORY INFARCTS

Areas in between vascular territories tend to be especially vulnerable to hypoperfusion because of their distance from the main arterial supply. There are three such watershed areas in the brain: (1) the *anterior watershed*, around the superior frontal sulcus, lies in between the territories of the anterior cerebral artery (ACA) and middle cerebral artery (MCA); (2) the *posterior watershed*, the parietotemporal or parieto-occipital area, lies in between the MCA and posterior cerebral artery (PCA); and (3) the *internal watershed*, the white matter of the centrum semiovale and corona radiata, lies in between the deep and superficial perforators of the MCA (Fig. 87.1). Although hypoperfusion is the main cause of watershed territory infarcts, there may also be a synergistic interaction among hypoperfusion, intra-arterial embolism, and large artery stenosis such that reduced cerebral perfusion from proximal vessel stenosis limits the washout of distal microemboli that then end up lodging in the anatomic watershed areas [5].

Watershed territory infarcts from hypoperfusion account for approximately 2–5% of acute ischemic strokes [6]. Not surprisingly, they tend to occur more commonly in the elderly because of the greater prevalence of heart disease and hypotension combined with extracranial large artery disease in them. The onset and progression of these kinds of strokes is unique. Unlike embolic strokes that occur suddenly or atherothrombotic and lacunar strokes that often occur in a stuttering

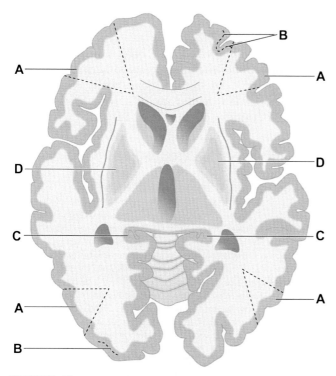

FIGURE 87.1 Drawing of a horizontal section of the cerebrum showing common patterns of hypoxic-ischemic brain damage. (A) Border zone infarct between the anterior and middle cerebral arteries and between the middle and posterior cerebral arteries. (B) Zones of laminar necrosis within the cerebral cortex. (C) Hippocampal necrosis. (D) Necrosis of nerve cells within the globus pallidus and putamen. *Adapted from Caplan LR, Hurst JW, Chimowitz MI. Clinical neurocardiology. New York: Marcel Dekker; 1999.*

pattern of discrete steps, watershed territory strokes tend to gradually worsen over many hours.

Patient symptoms reflect the anatomy of the border zones. Lesions in the anterior watershed area between the ACA and the MCA produce motor transcortical aphasia in the language-dominant hemisphere. Similarly, dominant posterior infarcts between the MCA and the PCA produce sensory transcortical aphasia. Bilateral anterior cortical infarcts can produce bibrachial paresis, also known as the man-in-the-barrel syndrome. Such bilateral infarcts can mimic posterior circulation territory ischemia by producing a confusing array of bilateral motor, sensory, and visual symptoms (although dysarthria, dysphagia, and diplopia are usually absent). Bilateral posterior cortical watershed infarcts that disconnect the visual cortex can produce syndromes of visual agnosia, such as the Balint syndrome characterized by asimultanagnosia (perception of parts of objects but not the whole), optic ataxia (poor hand-eye coordination), and gaze apraxia (inability to gaze in the desired direction). Syndromes of pure visual agnosia and cortical blindness can also occur.

The radiographic findings in watershed territory infarcts are sensitive but not specific; that is, infarcts in the border zones, although usually reflect a hemodynamic mechanism, may also be due to emboli or atherothrombosis. Similar-appearing white matter infarcts may result from lacunar mechanisms and occlusions of proximal branches of the MCA.

Watershed territory strokes can be prevented by treatment of the predisposing causes, particularly hypotension, heart failure, and arrhythmias. Hypertensive patients with extracranial carotid or vertebral arterial stenosis should have their antihypertensive medications monitored carefully to avoid overtreatment and hypotension.

The prognosis of watershed territory stroke also depends on the location and extent of the injury but is generally unfavorable. The mortality rate after a stroke from a hemodynamic mechanism has been estimated to be 9–10% per year, in part because of the high likelihood of accompanying coronary artery disease and heart failure in many patients [6].

HYPOXIC-ISCHEMIC ENCEPHALOPATHY

Out-of-hospital cardiac arrest affects up to 325,000 people in the United States each year. An additional 200,000 have cardiac arrest in the hospital [7]. The resulting brain injury is a major cause of mortality and morbidity. Neurons in the cerebral cortex, hippocampus, thalamus, cerebellum, and basal ganglia are the most sensitive to hypoxic-ischemic insults and may be selectively damaged. Clinically, immediately after cardiac arrest, all patients are initially comatose. The degree of brain damage then determines their clinical course and ultimate outcome. Patients with diffuse and severe hypoxic-ischemic neuronal damage may worsen to the point that the brain, as a whole, ceases to function. When careful bedside examination and confirmatory tests reveal the permanent cessation of all clinical function of the brain, in the absence of any potentially reversible metabolic or toxic confounders, the patient is considered brain dead. The three cardinal findings in brain death are coma or unresponsiveness, absence of brainstem reflexes, and apnea. Brain death is considered the legal determination of death in every state in the United States and in many other Western countries. Seizures or multifocal myoclonus are frequently observed in comatose survivors of cardiac arrest. Myoclonic status epilepticus is a particularly grave complication and is often resistant to treatment with anticonvulsants. Treated or untreated, nearly all patients with myoclonic status epilepticus have dismal outcomes. The electroencephalograms of these patients usually show a burst suppression pattern indicating profound, diffuse cortical neuronal dysfunction.

Patients with time will often ascend through stages of increasing awareness, beginning with spontaneous eye opening, then a vegetative state in which they are poorly responsive to environmental stimuli, and finally increasing voluntary behavior and a minimally conscious state. They can plateau at each stage or can continue on to full recovery depending on the extent of their injury. Those in a vegetative state [referred to as permanent vegetative state (PVS) when the state has lasted longer than 30 days] typically have the unusual appearance of wakefulness without awareness. They retain sleep-wake cycles, lie with their eyes open, blink, swallow, cough, sneeze, and retain normal brainstem-mediated reflexes. Despite this wakefulness, they have no measurable evidence of self-awareness or their environment. With adequate medical and nursing care, many such patients can survive for years in this state. Finally, some patients can have a good recovery but still be plagued with patchy retrograde amnesia mimicking Korsakoff syndrome. Some patients can perform their activities of daily living, but on more thorough neuropsychologic testing, they were found have severe deficits of cognition, memory, praxis, and motivation.

Brain computed tomographic (CT) scans are usually deceptively normal immediately after cardiac arrest, but after 48 h, loss of distinction between gray and white matter on brain CT scans has been reported to be predictive of death. Magnetic resonance imaging (MRI) is more sensitive than CT for anoxic-ischemic injury, and several studies have demonstrated that severe cortical changes on diffusion-weighted imaging (DWI) or fluid-attenuated inversion recovery (FLAIR) are strongly associated with poor outcome [8].

The prognosis of hypoxic-ischemic encephalopathy after cardiac arrest has been extensively studied. Predictors of poor neurologic outcome include absent brainstem reflexes or absent or extensor motor responses at 72 h, early myoclonus status epilepticus, and absent bilateral cortical N20 responses by somatosensory evoked potential (SSEP) after 24 h [9]. In patients who have undergone therapeutic hypothermia, clinicians are advised to delay examination until at least 72 h after rewarming (to ensure that administration of all sedatives has been discontinued for at least 12 h).

References

[1] Soteriades ES, Evans JC, Larson MG, et al. Incidence and prognosis of syncope. N Engl J Med 2002;347(12):878–85.

[2] Task Force for the Diagonsis and Management of Syncope; European Society of Cardiology, et al. Guidelines for the diagnosis and management of syncope (version 2009). Eur Heart J 2009;30(21):2631–71.

[3] Del Rosso A, Ungar A, Maggi R, et al. Clinical predictors of cardiac syncope at initial evaluation in patients referred urgently to a general hospital: the EGSYS score. Heart 2008;94(12):1620–6.

[4] Ungar A, Del Rosso A, Giada F, et al. Early and late outcome of treated patients referred for syncope to emergency department: the EGSYS 2 follow-up study. Eur Heart J 2010;31(16):2021–6.

[5] Caplan LR, Hennerici M. Impaired clearance of emboli (washout) is an important link between hypoperfusion, embolism, and ischemic stroke. Arch Neurol 1998;55(11):1475–82.

[6] Gottesman RF, Sherman PM, Grega MA, et al. Watershed strokes after cardiac surgery: diagnosis, etiology, and outcome. Stroke 2006;37(9):2306–11.

[7] Morrison LJ, Neumar RW, Zimmerman JL, et al. Strategies for improving survival after in-hospital cardiac arrest in the United States: 2013 consensus recommendations: a consensus statement from the American Heart Association. Circulation 2013;127(14):1538–63.

[8] Mlynash M, Campbell DM, Leproust EM, et al. Temporal and spatial profile of brain diffusion-weighted MRI after cardiac arrest. Stroke 2010;41(8):1665–72.

[9] Wijdicks EF, Hijdra A, Young GB, Bassetti CL, Wiebe S. Practice parameter: prediction of outcome in comatose survivors after cardiopulmonary resuscitation (an evidence-based review): report of the Quality Standards Subcommittee of the American Academy of Neurology. Neurology 2006;67(2):203–10.

CHAPTER

88

Stroke and Eye Findings

V. Biousse, N.J. Newman

Emory University School of Medicine, Atlanta, GA, United States

INTRODUCTION

Stroke patients often have visual symptoms and signs, the characteristics of which vary mostly depending on the type of vessel involved (arteries versus veins), the type of stroke (ischemic or hemorrhagic), and the size of the arteries involved (large versus small artery disease). Because the blood supply to the eye is mostly provided by branches of the ophthalmic artery, the first intracranial branch of the internal carotid artery (ICA), patients with anterior circulation cerebral ischemia may present with ipsilateral visual changes. Binocular visual loss or abnormal extraocular movements are common with vascular diseases affecting the posterior circulation, which provides the blood supply to the occipital lobes and posterior fossa. In addition, a number of ocular vascular disorders are associated with numerous systemic diseases affecting the eyes and the brain simultaneously [1,2].

Ocular Symptoms and Signs Associated With Carotid Artery Disease and Anterior Circulation Ischemia

Monocular (ipsilateral to the affected ICA) transient and permanent visual symptoms and signs may develop in patients with carotid artery disease or those with emboli from the aortic arch or the heart. Contralateral homonymous visual field defects, bitemporal visual field defects, and bilateral simultaneous visual loss may also result from anterior circulation ischemia, particularly when the disease is bilateral [1,2].

Transient Monocular Visual Loss

Although transient monocular visual loss may result from retinal transient ischemic attack (TIA), it is important to emphasize that there are numerous nonvascular causes of transient monocular visual loss, which require an emergent detailed ocular examination; therefore, an ophthalmic consultation should be obtained before obtaining an extensive workup for presumed vascular transient monocular visual loss (so-called "amaurosis fugax"). Retinal TIAs classically present as sudden, complete blackout of vision in one eye, lasting few minutes without residual visual loss. Ocular examination may show retinal emboli or may be normal. The workup is similar to that of a cerebral TIA, including stroke protocol brain imaging, since about 25% of patients with acute retinal ischemia have concomitant acute cerebral ischemia [3]. Multiple episodes of transient visual loss or painful transient visual loss in the elderly should raise the possibility of giant cell arteritis [2].

Permanent Monocular Visual Loss

Partial or complete monocular loss of vision may occur in patients with ipsilateral carotid artery disease, or emboli in the anterior circulation. This results most often from a central retinal artery occlusion (CRAO) (Fig. 88.1) or from one or multiple branch retinal artery occlusions (BRAOs) (Fig. 88.2). In such cases, emboli may be seen in the affected vessels. The most common emboli that occlude retinal arterioles are made of cholesterol, fibrin platelets, and calcium fragments (Table 88.1). Acute retinal ischemia should be evaluated emergently, in a similar manner as with acute cerebral infarctions [2].

Venous stasis retinopathy and ocular ischemic syndrome are caused by severe carotid obstructive disease and poor collateral circulation. The retinopathy is characterized by insidious onset, dilated and tortuous retinal veins, peripheral microaneurysms, and dot-blot hemorrhages in the mid-peripheral retina, associated with neovascularization when ocular ischemic syndrome develops (Fig. 88.3). Patients with this condition develop progressive and often irreversible visual loss, sometimes

Primer on Cerebrovascular Diseases, Second Edition
http://dx.doi.org/10.1016/B978-0-12-803058-5.00088-6

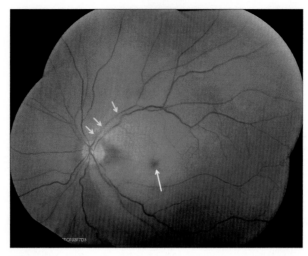

FIGURE 88.1   **Acute central retinal artery occlusion.** Funduscopic photograph showing an acute central retinal artery occlusion in the left eye. Note the attenuated central retinal artery with segmental arterial narrowing (*yellow arrows*). The ischemic retina is edematous and appears whitish, and there is a cherry red spot (*white arrow*). Within few weeks, the retinal vessels often recanalize and the retina has an almost normal appearance; however, the optic disk eventually becomes pale.

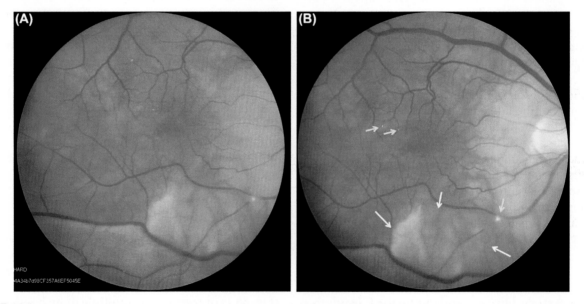

FIGURE 88.2   **Retinal emboli.** Funduscopic photograph showing refractile cholesterol retinal emboli (*yellow arrows*) associated with inferior retinal ischemia (whitish discoloration of the retina; *white arrow*) in a patient with aortic arch atheroma (A). One week later, repeat funduscopic photographs showed new emboli superiorly (*yellow arrows*) (B).

associated with cerebral hypoperfusion, and the prognosis is poor [1,4].

Ischemic optic neuropathy, anterior or posterior, results from small vessel disease and is usually not associated with large artery disease, explaining why carotid artery evaluation is not usually indicated in patients with classic anterior ischemic optic neuropathy [5].

Permanent Binocular Visual Loss

Partial or complete contralateral homonymous hemianopic visual field defects may result from cerebral ischemia in the anterior circulation. This is most often caused by occlusion of branches of the middle cerebral artery, but it may result from occlusion of the anterior choroidal artery or some of its branches to the optic tract and lateral geniculate body [1,6].

Horner Syndrome

Horner syndrome is a classic sign of carotid dissection and may also occur in patients with atherosclerotic carotid artery disease. The Horner syndrome may be isolated (often with ipsilateral facial pain) or may be associated with other neurological symptoms and signs of carotid artery disease [1].

TABLE 88.1   Most Common Types of Retinal Emboli

Type of Retinal Emboli	Source of Emboli	Funduscopic Appearance
Cholesterol (Hollenhorst plaque)	Ipsilateral internal carotid artery Aortic arch	Yellow, refractile
		Multiple in 70% of cases
		Appear wider than the arteriole
		Often at an arteriole bifurcation
Platelet fibrin	Carotid thrombus Aortic arch thrombus Cardiac thrombus Cardiac prosthesis	White–gray, pale, not refractile
		Often multiple
		Within small retinal arterioles
Calcium	Calcified atheromatous plaque Calcified cardiac valve	White and large
		Usually isolated
		In the proximal segment of the central retinal artery or its branches
Infectious	Infective endocarditis	White spots (Roth's spots)
	Candidemia	Multiple
Talc	Intravenous drugs	Yellow, refractile
		Multiple
Neoplasm	Cardiac myxoma	White, gray
		Often multiple
Fat	Fat emboli in the setting of leg facture	Whitish spots with hemorrhages and cotton wool spots
		Multiple

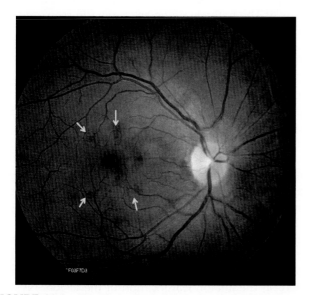

FIGURE 88.3   **Chronic ocular ischemia.** Funduscopic photograph showing multiple dot-blot retinal hemorrhages (*yellow arrows*) in the right eye of a patient with right internal carotid occlusion.

Referred Pain

Isolated facial pain (often periocular) may be a symptom of carotid occlusive disease, even without other symptoms and signs of vascular disease. It is usually a referred pain resulting from ischemia or compression of the trigeminal branches. It may also be part of the ocular ischemic syndrome [1].

Ocular Motor Nerve Paresis

Exceptionally, patients with acute occlusion or severe ICA stenosis may develop one or more ocular motor nerve pareses on the side of the occlusion, either in isolation or with signs of ocular ischemia [1].

Visual Symptoms and Signs Associated With Posterior Circulation Disease

The vertebrobasilar system supplies the entire brain stem ocular motor system as well as the posterior visual sensory pathways and visual cortex. For this reason, ocular motor and visual symptoms and signs play a major role in the diagnosis of vascular disease in the vertebrobasilar system [1,7].

Transient Binocular Visual Loss

Episodes of transient binocular visual loss are common in vertebrobasilar ischemia (Table 88.2). The visual loss is always bilateral, with both eyes being affected simultaneously and symmetrically. The change in vision may be described as a sudden "gray-out of vision," "a sensation of looking through fog or smoke," or "the feeling that someone has turned down the lights," classically

TABLE 88.2   Differential Diagnosis of Cerebral Visual Loss Classified by Mechanism

Mechanism	Cause of Vision Loss
Vascular	Vertebrobasilar ischemia (posterior cerebral artery territory)
	Cerebral anoxia
	Cerebral venous thrombosis (superior sagittal sinus)
	Hypertensive encephalopathy (posterior reversible encephalopathy syndrome)
	Eclampsia
Head trauma	Occipital lobe injury
Occipital mass	Tumor
	Abscess
	Vascular (aneurysm, arteriovenous malformation)
	Hemorrhage
Demyelinating disease	Multiple sclerosis
Infection	Occipital abscess
	Meningitis
	Progressive multifocal leukoencephalopathy
	Creutzfeld–Jacob disease
Toxic	Cyclosporine
	Tacrolimus
Metabolic	Hypoglycemia
	Porphyria
	Hepatic encephalopathy
Migraine	Migrainous visual aura
Seizure	Occipital lobe seizures
Degenerative	Alzheimer disease/posterior cortical atrophy

lasting no more than few minutes. Transient, complete 90–180 degree inversion of the visual image may rarely occur [7].

Homonymous Visual Field Defects

An isolated homonymous visual field defect of sudden onset is the hallmark of a vascular lesion in the occipital lobe in the territory supplied by the posterior cerebral artery (PCA). The visual field defect may be complete or incomplete, but when it is incomplete or scotomatous, it is usually congruous. When there is a complete homonymous hemianopia, macular sparing is common, and the occipital pole is usually spared. Patients with a homonymous visual field defect caused by ischemia in the PCA territory have normal visual

acuity. When both occipital lobes are infarcted, visual acuity is usually severely impaired, but the amount of visual loss is symmetric in both eyes. In some cases of occipital lobe infarction, the anterior portion of the lobe is unaffected, resulting in sparing of part or all of the peripheral 30 degree of the contralateral, monocular temporal field—the temporal crescent. Some patients initially experience complete blindness, with vision returning in the ipsilateral homonymous visual field within minutes. Pain in the ipsilateral eye or over the ipsilateral brow (contralateral to the hemianopia) is a classic symptom in such patients. This pain is referred from the tentorial branches of the trigeminal nerve [6].

Disorders of Higher Cortical Function

A number of syndromes described by patients as "difficulty seeing" or "difficulty reading" may result from posterior circulation cerebral ischemia, including alexia without agraphia, Gerstmann syndrome, associative visual agnosia, prosopagnosia, visual neglect, cerebral blindness, Balint's syndrome, simultagnosia, and achromatopsia [1,6].

Visual Hallucinations

Formed visual hallucinations may be caused by vertebrobasilar ischemia. These hallucinations, which may last 30 min or more, may be associated with decreased consciousness, but they usually occur in an otherwise alert patient who is aware that the visual images are not real. The hallucinations are generally restricted to a hemianopic field, and they are often complex [1,6].

Diplopia

Transient binocular horizontal or vertical diplopia is a common manifestation of vertebrobasilar ischemia. The diplopia may result from transient ischemia of the ocular motor nerves or their nuclei (ocular motor nerve paresis) or from transient ischemia to supranuclear or internuclear ocular motor pathways (skew deviation, internuclear ophthalmoplegia, gaze paresis). In most cases, the diplopia is not isolated, and the patient has other neurological symptoms suggesting vertebrobasilar ischemia [1,6,7].

Persistent disturbances of eye movements are extremely common in patients with vertebrobasilar ischemia. Ocular motor nerve paresis, internuclear ophthalmoplegia, supranuclear deficits, and nystagmus develop based on the anatomical location of the lesion. Nystagmus produces oscillopsia, which is often described by the patients as "jumping of vision" [1,6,7].

Ocular Symptoms and Signs Associated With Cerebral Vasculitis

Both infectious and noninfectious inflammation affecting the central nervous system can produce visual symptoms. In some vasculitides with a predilection for

TABLE 88.3   Characteristics of the Three Autosomal Dominant Cerebro-Ocular Vasculopathies Mapped to Chromosome 3

Name	Ocular Findings	Neurological Findings	Other Findings
Hereditary Vascular Retinopathy (HVR)	Microangiopathy Retinal periphery and posterior pole	Multiple small lesions in gray and white matter Headaches	Raynaud phenomenon
Cerebro-Retinal Vasculopathy (CRV)	Microangiopathy Posterior pole	Cerebral pseudotumors Extensive white matter lesions Dementia Headaches Death <55 years	
Hereditary Endotheliopathy, Retinopathy, Nephropathy, and Stroke (HERNS)	Microangiopathy Posterior pole	Cerebral pseudotumors Extensive white matter lesions Dementia Headaches Stroke Death <55 years	Renal involvement

large arteries, such as Takayasu's arteritis, ocular ischemia is common. Most often, vasculitis produces retinal vasculitis with retinal vascular occlusions and visual loss, often associated with ocular inflammation (uveitis).

Giant cell arteritis should be considered in all patients over 50 years presenting with acute optic nerve or retinal ischemia. Permanent visual loss is common in giant cell arteritis, and is often preceded by transient monocular visual loss [1].

Ocular Symptoms and Signs Associated With Noninflammatory Cerebral Vasculopathies

Susac Syndrome

Susac syndrome describes young patients with multiple bilateral branch retinal arterial occlusions, hearing loss, and neurological symptoms suggestive of a brain microangiopathy. It classically occurs in young women but can affect men. Affected patients have recurrent multiple BRAOs that are most often bilateral, progressive hearing loss (low and medium frequencies, and best demonstrated by audiogram), and various neurological presentations including psychiatric changes and encephalopathy (which varies in severity from mild memory loss and personality changes to severe cognitive dysfunction, confusion, psychiatric disorders, seizures, and focal neurological symptoms and signs). The disease usually has a chronic relapsing course punctuated by frequent remissions and exacerbations.

Retinal fluorescein angiography classically shows retinal arterial wall hyperfluorescence, which is indicative of disease activity. Brain MRI typically shows multiple enhancing small lesions in both the white and gray matter, with classic involvement of the corpus callosum. The MRA is normal, but catheter angiography has shown evidence of vasculopathy in some patients. In few cases, brain biopsy was performed and showed microinfarcts with some minimal perivascular lymphocytic infiltration, but no true vasculitis [8].

Hereditary Retinopathies

A number of rare hereditary retinopathies are associated with central nervous system abnormalities. Genetic characterization of three of these syndromes has shown that there is an overlap among these entities (Table 88.3) [9].

Aneurysms, Fistulas, and Vascular Malformations

Aneurysms

The most common neuro-ophthalmic manifestations of intracranial aneurysms are secondary to local mass effect on adjacent cranial nerves. The pulsatile process of the aneurysm may be as important as the direct mass effect. Aneurysms arising from the ophthalmic artery, the cavernous carotid artery, the anterior communicating artery, or the ICA can result in unilateral optic neuropathy or chiasmal visual field defects. Rarely, a homonymous hemianopia can result from compression of the optic tract. Multiple ocular motor nerve palsies can result from an aneurysm involving the cavernous carotid artery. Posterior

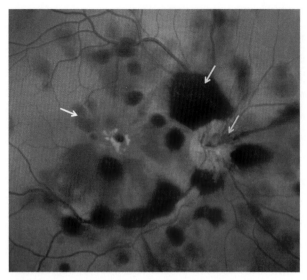

FIGURE 88.4 **Right Terson syndrome.** Funduscopic photograph showing multiple intraocular hemorrhages in a patient with acute aneurysmal subarachnoid hemorrhage. Note the intraretinal hemorrhages (*white arrow*), preretinal hemorrhages (*yellow arrow*), and disk edema (*blue arrow*). There was a dense vitreous hemorrhage in the left eye and the funduscopy could not be visualized.

communicating artery aneurysms classically produce an isolated third nerve palsy, whereas isolated trochlear or abducens nerve palsies only rarely result from aneurysmal compression.

Aneurysmal rupture producing a subarachnoid hemorrhage may be associated with intraocular hemorrhages (so-called "Terson syndrome") (Fig. 88.4). The presumed mechanism is that of acute raised intracranial pressure with sudden elevation of ocular central venous pressure. When an aneurysm arises from the cavernous carotid artery or from its branches in the cavernous sinus, rupture results in carotid–cavernous fistula, not subarachnoid hemorrhage [1,10].

Carotid–Cavernous Sinus Fistula

Direct carotid–cavernous fistula resulting from direct communication of the cavernous carotid artery and the cavernous sinus with resultant high velocity of blood flow most often results from trauma or from rupture of a preexisting aneurysm of the cavernous carotid artery. Ocular manifestations are usually obvious and include proptosis, periorbital swelling, chemosis, dilation of the episcleral vessels, orbital bruit, ophthalmoplegia, elevated intraocular pressure, dilation or occlusion of the retinal veins, and optic disc edema.

Dural carotid–cavernous sinus fistulas result from indirect communications between branches of the internal and external carotid arteries and the cavernous sinus. They are more common in middle-aged and elderly women. Neuro-ophthalmic manifestations depend on the blood flow velocity and vary from the full classical

appearance of direct fistulas to isolated bruit, cranial nerve palsies, or simply a "red eye."

Orbital imaging with CT or MRI classically shows dilation of the superior ophthalmic veins, enlargement of the extra-ocular muscles, or enlargement of the cavernous sinus. Catheter angiography is still the best way to confirm a carotid–cavernous fistula which can often be treated with an endovascular approach at the same time [1,10].

Arteriovenous Malformations

Intracranial arteriovenous malformations may result in intraparenchymal hemorrhage, subarachnoid hemorrhage, seizures, or mass effect. Occipital arteriovenous malformations can cause episodic visual symptoms mimicking the visual aura of migraine. Homonymous hemianopia is common in occipital lesions, most often as the result of bleeding or as a complication of treatment. Posterior fossa arteriovenous malformations usually cause intermittent or permanent diplopia, often associated with other neurological signs. Orbital arteriovenous malformations produce an acute or subacute orbital syndrome, including proptosis, chemosis, ophthalmoplegia, visual loss, and elevation of the intraocular pressure.

Retinal arteriovenous malformations are rare. They are sometimes isolated or are associated with intracranial or facial arteriovenous malformations [1,10].

Cavernous Hemangioma

Cavernous hemangiomas are most common in the posterior fossa and usually produce diplopia when they bleed. Familial cavernous hemangiomas are often multiple and may be associated with grape-like appearing small retinal hemangiomas that are usually asymptomatic [1,10].

Venous Disease

Central and Branch Retinal Vein Occlusions

Central retinal vein occlusion (CRVO) and branch retinal vein occlusions (BRVO) are common ocular disorders in elderly patients often associated with atheromatous vascular risk factors. Occlusion of the central retinal vein is presumed to result from compression by the central retinal artery within the optic nerve. Patients with CRVO complain of blurry vision, the severity of which depends on the severity of the CRVO and associated retinal ischemia. Funduscopic examination reveals diffuse retinal hemorrhages, retinal and macular edema, dilated and tortuous veins, and optic disk edema. This disorder is primarily managed by the ophthalmologist, and the goals of the treatment are to treat macular edema and prevent or treat neovascularization and its complications.

Cerebral Venous Thrombosis

Cerebral infarction related to cerebral venous thrombosis may present with acute focal neurological signs, including homonymous hemianopia or cranial nerve palsies, usually in the setting of headaches or altered mental status. Most neuro-ophthalmic manifestations of cerebral venous thrombosis, however, are related to increased intracranial pressure, and include papilledema and diplopia from uni- or bilateral sixth nerve palsies.

Cavernous sinus thrombosis is extremely rare and produces acute painful proptosis with chemosis, ophthalmoplegia, venous stasis retinopathy, and visual loss.

CONCLUSION

Cerebrovascular diseases are commonly associated with neuro-ophthalmological symptoms and signs, which mostly depend on the type, the size and the location of the vessels involved, and the mechanism of the vascular lesion.

References

[1] Biousse V, Newman NJ. Neuro-ophthalmology illustrated. 2nd ed. New-York (NY): Thieme; 2016.

[2] Biousse V, Newman N. Retinal and optic nerve ischemia. Continuum (Minneap Minn) 2014;20(4 Neuro-ophthalmology): 838–56.

[3] Biousse V. Acute retinal arterial ischemia: an emergency often ignored. Am J Ophthalmol 2014;157:1119–21.

[4] Mendrinos E, Machinis TG, Pournaras CJ. Ocular ischemic syndrome. Surv Ophthalmol 2010;55:2–34.

[5] Biousse V, Newman NJ. Ischemic optic neuropathies. N Engl J Med 2015;372:2428–36.

[6] Caplan L. Posterior circulation ischemia: then, now, and tomorrow. The Thomas Willis Lecture-2000. Stroke 2000;31:2011–23.

[7] Moncayo J, Bogousslavsky J. Vertebro-basilar syndromes causing oculo-motor disorders. Curr Opin Neurol 2003;16:45–50.

[8] Rennebohm R, Susac JO, Egan RA, Daroff RB. Susac's syndrome–update. J Neurol Sci 2010;299:86–91.

[9] Kolar GR, Kothari PH, Khanlou N, Jen JC, Schmidt RE, Vinters HV. Neuropathology and genetics of cerebroretinal vasculopathies. Brain Pathol 2014;24:510–8.

[10] Biousse V, Newman NJ. Intracranial vascular abnormalities. Ophthalmol Clin North Am 2001;14:243–64.

CHAPTER

89

Spinal Cord Strokes

R. Bhole[1], L.R. Caplan[2]

[1]University of Tennessee Health Science Center, Memphis, TN, United States; [2]Harvard University, Beth Israel Deaconess Medical Center, Boston, MA, United States

Spinal cord strokes are a rare but important differential consideration in central nervous system vascular disease. As in the brain, spinal cord strokes can be divided into two large groups—ischemic and hemorrhagic. They account for about 1% of all strokes and about 5–8% of acute myelopathies [1,2].

SPINAL CORD VASCULAR SUPPLY

The arterial supply of the spinal cord is unique. A large single anterior spinal artery runs in the ventral midline from the spinomedullary junction at the foramen magnum to the filum terminale. In contrast paired smaller posterior spinal arteries are located on the dorsal surface, which often form a rich plexus of small vessels (Fig. 89.1) [3]. The central portion between the two zones of supply has often been called the border zone or watershed region of the spinal cord.

The anterior spinal artery originates from the two vertebral arteries at the base of the skull, feeds the medulla, and descends in the midline through the foramen magnum to supply the cervical spinal cord. The artery is fed by a series of 5–10 unpaired radicular arteries that originate from the vertebral arteries and the aorta and its branches. The blood supply is most marginal in the upper thoracic

Primer on Cerebrovascular Diseases, Second Edition
http://dx.doi.org/10.1016/B978-0-12-803058-5.00089-8

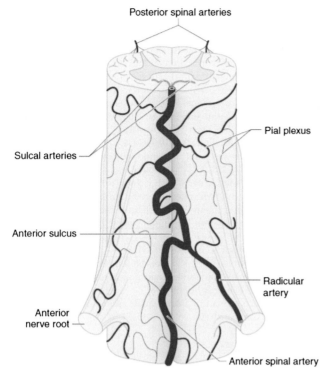

FIGURE 89.1 Cross-section of spinal cord, showing arterial patterns of supply. The anterior spinal artery is a single midline artery that courses in the anterior fissure. This artery divides into left and right sulcal arteries that supply the anterior horns and white matter. There are two posterior spinal arteries, one on each side, which form an anastomotic rete from which branches emerge to supply the posterior gray horns and the posterior columns. *With permission from Caplan LR. Spinal cord vascular disease in Caplan's stroke a clinical approach. 4th ed. Philadelphia: Elsevier; 2009 [chapter 15].*

region (T2–T4) which has been referred to as the longitudinal watershed region of the spinal cord. The largest radicular artery, the artery of Adamkiewicz, arises from the aorta most often between the T9 and T12 regions.

The paired posterior spinal arteries are fed by smaller radicular arteries at nearly every spinal level. These supply the dorsal columns and posterior gray matter. The blood supply of the anterior portion of the cord is much more vulnerable than that of the posterior portion and can be decompensated by occlusion of a large radicular branch or lesions of the aorta.

SPINAL CORD ISCHEMIA AND INFARCTION

The understanding of spinal cord infarction parallels the evolution of knowledge about the mechanism of brain infarcts.

Spinal cord infarctions can be subdivided as follows [4]:

- Bilateral, predominantly anterior (Fig. 89.2A). These patients have bilateral motor and spinothalamic-type sensory deficits. Posterior column sensory functions (vibration and position sense) are spared.

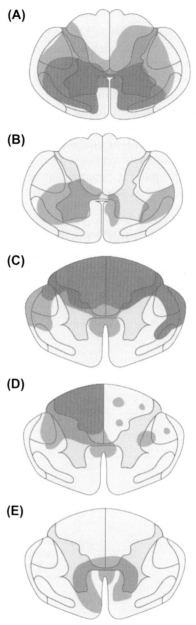

FIGURE 89.2 Cartoon of patterns of spinal cord infarction. *Dark gray* indicates usual extent, *while light gray* indicates potentially larger area of ischemia. (A) Anterior bilateral infarction. (B) Anterior predominantly unilateral infarction. (C) Posterior bilateral infarction. (D) Posterior predominantly unilateral infarction. (E) Central spinal cord. *With permission from Caplan LR. Spinal cord vascular disease in Caplan's stroke a clinical approach. 4th ed. Philadelphia: Elsevier; 2009 [chapter 15].*

- Unilateral, predominantly anterior (Fig. 89.2B). The motor deficit is a hemiparesis below the lesion and contralateral spinothalamic tract sensory loss—a Brown–Séquard syndrome.
- Bilateral, predominantly posterior (Fig. 89.2C). Posterior column type sensory loss below the lesion with variably severe bilateral pyramidal tract signs.
- Unilateral, mostly posterior (Fig. 89.2D). Ipsilateral hemiparesis and posterior column sensory loss.

- Central (Fig. 89.2E). Bilateral pain and temperature loss with spared posterior column and motor functions. Similar to the deficits produced by a syrinx.
- Transverse. Loss of motor, sensory, and sphincter functions below the level of the lesion. Anterior patterns are more common than posterior especially after aortic surgery.

CAUSES OF SPINAL ISCHEMIA

The main causes of ischemia to the spinal cord are listed in Table 89.1 and discussed in the following paragraphs.

1. Disease of the aorta is undoubtedly the most commonly recognized cause of spinal cord infarction [5]. Most often, paraplegia is recognized after repair of thoracic and abdominal aortic aneurysms. During repair, flow through radicular arteries to the anterior spinal artery is compromised.

Similar findings are noted in unruptured aneurysms of the aorta, dissections of the aorta, traumatic rupture of the aorta, thromboembolic aortic occlusions, and ulcerative aortic plaque disease. Thrombi and plaques can obstruct the orifices of the radicular arteries. Dissections can tear or interrupt the orifices of the spinal cord feeding arteries, sometimes over a long area. Cholesterol crystals and other plaque components can embolize to spinal arteries and block branches. In some patients the spinal ischemia will develop insidiously and can be misdiagnosed as motor neuron disease or diabetic amyotrophy because of selective ischemia involving anterior horn neurons and, sometimes, the pyramidal tracts [6].

2. Embolism. Infective endocarditis is a recognized cause of brain and spinal cord embolic infarcts. Atrial myxoma and nonbacterial thrombotic (marantic) endocarditis are other disorders in which small particles embolize to the spinal cord fibrocartilaginous embolism presumably arises from disk herniations, predominantly affects young

TABLE 89.1 Causes of Spinal Cord Ischemia/Infarction

Arterial	Aortic Disease	Aortic Aneurysms	
		Aortic dissection	
		Traumatic rupture	
		Aortic surgery	Aortic aneurysm repair
	Embolic disease	Infective endocarditis	
		Atrial myxoma	
		Nonbacterial thrombotic endocarditis	
		Fibrocartilaginous	Disk herniation, pregnancy, puerperium
	Dissections	Vertebral artery dissections	Cervical cord infarction
		Aortic dissections	Block orifices of spinal radicular arteries
	Infections and inflammations	Tuberculosis, syphilis	Spinal arteritis
		Fungal (cryptococcosis, coccidiomycosis),	
		Lyme disease	
		Schistosomiasis,	Parasite invading blood vessel or granulomatous inflammation
	Toxic	Heroin injection	Prolonged vasoconstriction
		Cocaine inhalation	
	Global ischemia	Severe hypotension Shock Cardiac arrest	T4–T8 vulnerable segment (longitudinal watershed zone)
	Iatrogenic	Endovascular procedures involving aorta	Plaque dislodged that embolize to spinal radicular arteries
Venous	Spinal AVM or dural fistulas		Venous thrombosis precipitating spinal cord infarction
	Spinal cavernous malformations		

VI. CLINICAL ASPECTS: MEDICAL AND SURGICAL

women, and nearly always involves the cervical spinal cord [7]. Some patients have been pregnant, puerperal, or on oral contraceptives. Minor trauma, sudden neck motion, or lifting is often mentioned as an immediate precipitant.

3. Dissections. Vertebral artery dissection with cervical cord infarction has been reported. Aortic dissections can block the orifices of the spinal radicular arteries leading to spinal cord infarction.

4. Infections and inflammation. Inflammation of the meningeal coverings of the spinal cord can spread to the spinal arteries, causing acute spinal cord infarcts. This phenomenon is similar to Heubner's arteritis found in the brain in patients who have tuberculosis and syphilis. Chronic adhesive arachnoiditis from any cause can also lead to scarring and obliteration of the spinal penetrating arteries and ischemia of the central spinal cord.

5. Drug abuse. Spinal cord ischemia has been reported after heroin injection and cocaine inhalation. The most likely mechanism of spinal cord damage after drug abuse is prolonged vasoconstriction.

6. Global ischemia. Spinal cord ischemia may also develop during severe hypotension, clinical shock, and cardiac arrest. The damage is most common in the thoracic spinal cord between T4 and T8 segments, which is the most vulnerable region of the spinal cord being the longitudinal watershed region [8].

7. Iatrogenic. Endovascular procedures and arteriography involving the aorta can cause spinal cord infarction by dislodging plaque components that embolize to the spinal radicular arteries.

8. Venous Infarcts. Venous spinal cord infarctions occur but are probably less common than venous brain infarcts. Thrombosis of spinal veins or increased pressure in the venous system can cause cord infarction. Venous hypertension is an important contributor to spinal cord infarction in patients with spinal dural fistulas. In some patients venous occlusion is due to a hypercoagulable state, for example, related to visceral cancer, septic thrombophlebitis, or spinal tumors with venous compression. Venous infarction is most often caused by mechanical compression, infection, and inflammation, which obliterate the veins.

SPINAL VASCULAR MALFORMATIONS

Like brain malformations, spinal vascular malformations include telangiectasias, venous malformations (angiomas), varices, cavernous malformations, and arteriovenous malformations (AVMs). Most common are the AVMs or fistulas on the surface of the cord [9].

a. Arteriovenous Malformations or Fistulas

Spinal vascular malformations often present with ischemia rather than hemorrhage and some can cause bleeding and ischemia. Veins draining an AVM can become occluded and draining veins can cause thrombose, precipitating spinal infarction.

Spinal malformations can be divided into two large groups: dural (type I) and intradural (type II). Dural malformations are more common than those that are entirely intradural. Spinal dural fistulas occur predominantly in men (4:1 ratio) between 40 and 70 years of age, and involve mostly the lower thoracic and lumbosacral segments and usually have one arterial feeder. The most frequent presentation of spinal dural fistulas is that of progressive neurologic worsening, often with acute deteriorations. Less common is an acute apoplectic onset.

Dural fistulas usually do not cause subarachnoid bleeding, except when the lesions are cervical [10]. The cervical fistulas that cause subarachnoid hemorrhage (SAH) are fed by the vertebral arteries and are often located near the cervicomedullary junction. About 10% of patients with spinal AVMs have SAH [11]. Thoracic, lumbar, and sacral fistulas rarely present with epidural hemorrhage and rarely SAH.

Intradural AVMs can be located in the space outside of the spinal cord or can be both extra and intramedullary. Usually there are multiple dilated veins and venous pressure is high.

Spinal TIAs are more frequent in patients with spinal dural fistulas than with other spinal vascular lesions. These lesions cause symptoms because of venous hypertension and occasional thrombosis of the venous drainage system.

DIAGNOSTIC EVALUATION OF PATIENTS PRESENTING WITH SPINAL CORD ISCHEMIA

Diffusion weighted MRI scans often show acute spinal cord infarction. Early MRI scans can be negative and in an appropriate clinical setting a repeat scan is often mandated. Recent ischemia causes edema with increased signal on T2-weighted scans and older infarcts also have abnormal increased T2 signal.

The key finding that suggests the presence of a dural fistula is serpiginous dilated veins on the surface of the spinal cord. These are often seen on T2-weighted and gadolinium-enhanced images and on contrast enhanced MR angiography. Using MR angiography techniques, phase display after contrast injection can show the direction of flow within the epidural veins to indicate the likely location of the arterial feeders. The coiled serpiginous veins are usually visible along the dorsal cord surface during myelography which still has a

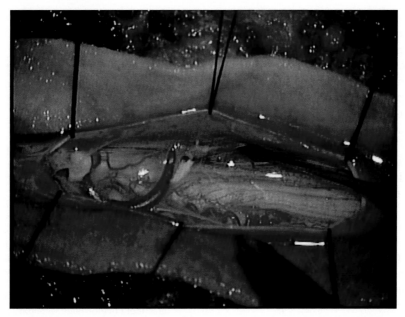

FIGURE 89.3 Intraoperative view of the surface of the spinal cord after the dura has been opened. Tortuous dilated veins can be seen along the surface of the spinal cord. *Courtesy of Adam S. Arthur MD, Department of Neurosurgery, Semmes–Murphey Brain and Spine Institute and University of Tennessee Health Science Center, Memphis, TN, USA.*

place in diagnosis. Fig. 89.3 is an intraoperative photo that shows these enlarged veins on the surface of the spinal cord. Myelographic films should be taken with the patients lying supine to show the abnormal veins. Intradural AVMs are referred to as low-flow because angiography results in slow, low-volume filling of the lesions. Selective spinal digital subtraction angiograms (DSA) in expert hands often shows the feeding arteries, but occasionally the feeding arteries do not get opacified. Fig. 89.4 is a DSA of a spinal AVM. Surgical exploration is urged when clinical findings are typical and abnormal veins are clearly present on myelography or angiographic examinations. Ligation of arterial feeders usually prevents worsening, so that the venous structures need not be removed [12].

In most patients in whom dural fistulas have been treated effectively, muscle strength and gait improve, although sacral spinal cord dysfunction (urination, defecation, sexual function) often remains unchanged. Surgical treatment is not as successful for intradural AVMs as for dural ones, but cavernous angiomas can occasionally be removed.

SPINAL HEMORRHAGES

Hematomyelia describes bleeding into the substance of the spinal cord parenchyma. The most common cause is trauma. Onset can be immediate or delayed. Usually the area around the central canal and gray matter are involved, most often in the cervical region. The usual signs are neck pain, weakness, and areflexia in the arms, associated with a cape-like distribution of pain and temperature loss. Other causes include AVMs,

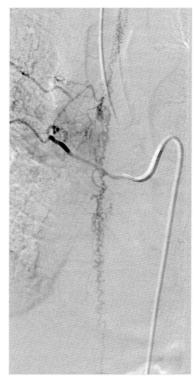

FIGURE 89.4 Digital subtraction angiogram of spinal cord showing tortuous dilated veins along surface of the spinal cord. *Courtesy of Adam S. Arthur MD, Department of Neurosurgery, Semmes–Murphey Brain and Spine Institute and University of Tennessee Health Science Center, Memphis, TN, USA.*

anticoagulation, hemophilia and other bleeding disorders, and hemorrhage into spinal cord tumors (ependymoma, melanoma, angioblastoma), as well as (rarely) syrinxes. The MRI shows a swollen blood-filled cord.

Spinal epidural hematomas are approximately four times more common than spinal subdural hematomas. Each most often occurs in patients who are on anticoagulants. Some patients have had liver disease and portal hypertension. Lumbar punctures have been known to precipitate these bleeds in patients on anticoagulants. The earliest symptom is pain in the back, usually in the neck. This is followed by radicular pain, usually in one or both arms. The earliest symptoms closely mimic disk herniation syndromes. Within hours, or rarely, days, sensory and motor signs develop in the legs, bowel, and bladder, and sexual dysfunction ensues. Usually, weakness and sensory loss are symmetric, but a Brown–Séquard distribution may be found.

MRI is superior to CT scanning in defining the location and extent of hematomas. Signs are almost invariably progressive, unless the lesions are decompressed. Anticoagulation should be reversed using vitamin K, fresh frozen plasma, or prothrombin complex concentrates particularly those enriched with factor VII, factor X, and prothrombin.

Decompression is urgent and should be pursued as soon as feasible considering the INR findings. Outcome depends on the severity of the deficit before surgery, the duration of spinal cord compression, and the rapidity of onset of the paraplegia.

References

[1] Buchan AM, Barnett HJM. Infarction of the spinal cord. In: Barnett HJM, Mohr JP, Stern B, Yatsu F, editors. Stroke: pathophysiology, diagnosis and management. New York: Churchill Livingstone; 1986. p. 707–19.

[2] De Sèze J, Stojkovic T, Breteau G, et al. Acute myelopathies. Clinical, laboratory and outcome profiles in 79 cases. Brain 2001;124:1509–21.

[3] Gillilan L. The arterial blood supply of the human spinal cord. J Comp Neurol 1958;110:75–103.

[4] Novy J, Carruzzo A, Maeder P, Bogousslavsky J. Spinal cord ischemia. Clinical and imaging patterns, pathogenesis, and outcomes in 27 patients. Arch Neurol 2006;63:1113–20.

[5] Picone AL, Green RM, Ricotta JR, et al. Spinal cord ischemia following operations on the abdominal aorta. J Vasc Surg 1986;3:94–103.

[6] Herrick MK, Mills PE. Infarction of spinal cord: two cases of selective grey matter involvement secondary to asymptomatic aortic disease. Arch Neurol 1971;24:228–41.

[7] Bots GT, Wattendorff AR, Buruma OJ, Roos RA, Endtz LJ. Acute myelopathy caused by fibrocartilaginous emboli. Neurology 1981;31:1250–6.

[8] Cheshire WP, Santos CC, Massey EW, Howard Jr JF. Spinal cord infarction: etiology and outcome. Neurology 1996;47:321–30.

[9] Tobin WD, Layton DD. The diagnosis and natural history of spinal cord arteriovenous malformations. Mayo Clin Proc 1976;51:637–46.

[10] Hemphill III JC, Smith WS, Halbach VV. Neurologic manifestations of spinal epidural arteriovenous malformations. Neurology 1998;50:817–9.

[11] Caroscio JT, Brannan T, Budabin M, Huang YP, Yahr MD. Subarachnoid hemorrhage secondary to spinal arteriovenous malformation and aneurysm: report of a case and review of the literature. Arch Neurol 1980;37:101–3.

[12] Heros R. Arteriovenous malformations of the spinal cord. In: Ojemann RG, Heros RC, Crowell RM, editors. Surgical management of cerebrovascular disease. 2nd ed Baltimore: Williams & Wilkins; 1988. p. 451–66.

CHAPTER

90

Unique Features of Aneurysms by Location

B.J. Cord, K.R. Bulsara

Yale University School of Medicine/Yale New Haven Hospital, New Haven, CT, United States

INTRODUCTION

The rupture of intracranial aneurysms is associated with severe morbidity and mortality. Prehospital mortality estimates range from 10% to 15% [1]. The rate of rerupture if untreated is 40% over 4 weeks with much of that risk early on; rerupture carries an 80% mortality underscoring the importance of early treatment for ruptured aneurysms. In patients who survive the initial hemorrhage long enough to undergo treatment, there is an additional 30% of 1-month mortality [2], and more than 50% of long-term survivors have difficulty in activities of daily living [3]. The incidence of aneurysmal rupture is about 10 per 100,000 persons per year, with an average prevalence of aneurysms (ruptured or nonruptured) of 2–5% [4]. With such devastating consequences, accurate assessment of the natural history of incidentally found, unruptured aneurysms and preventative treatment is necessary. Aneurysm characteristics and behavior can vary drastically according to location, thus affecting their management (see Table 90.1).

The overall frequency of intracranial aneurysms ranges from 2% to 5% [4]. Their site distribution is nonrandom, with unruptured aneurysms occurring in the internal carotid artery (ICA, 28%), anterior cerebral artery (ACA, 27%), middle cerebral artery (MCA, 36%), and posterior circulation (8%). ICA aneurysms include cavernous, ophthalmic, anterior choroidal, and posterior communicating artery (PCOM) aneurysms. ACA aneurysms are predominantly composed of anterior communicating artery (ACOM) aneurysms. Posterior circulation aneurysms include those of the posterior cerebral artery (PCA), vertebrobasilar system, and cerebellar arteries. The site distribution of aneurysms that present with subarachnoid hemorrhage (SAH) is discordant (see Table 90.2). This evidence suggests a differential susceptibility for aneurysm formation and rupture according to location. Indeed, several prospective studies document

differential risk of rupture based on location and aneurysm size [5,6]. Particularly, they ascribe an extremely small risk of rupture to small anterior circulation aneurysms, which paradoxically, make up the majority of ruptured aneurysms [7–10]. An increasing number of intracranial aneurysms are incidentally found with noninvasive advanced imaging techniques during the routine workup of other pathology and increasing resolution allows for the routine detection of increasingly smaller aneurysms [11,12].

Aneurysm pathogenesis is intimately related to the interplay of structural and hemodynamic effects. Most aneurysms arise at the intersection of two vessels with a subset arising at nonbifurcation points along vessels with significant curvature; in both situations the configuration gives rise to altered flow dynamics. Endothelial cells respond to alterations in shear stress and flow patterns with subsequent effects on the structure of the vascular cell wall composition. As an aneurysm develops, its geometry (i.e., size, shape, and relative configuration to the parent vessel) instantaneously dictates hemodynamic forces. The resulting wall shear stress is again sensed by

TABLE 90.1 Characteristics of Aneurysms That Can Vary by Aneurysm Location

HEMODYNAMIC
Intravascular pressure wave
Branch angle
Turbulent flow

GENETIC
Acquired, site-specific mutations during development

STRUCTURAL
Adjacent structures (nerves, eloquent structures)
Perforating vessels
Native vessel thickness
Developmental variations (i.e., azygous anterior cerebral artery)

Primer on Cerebrovascular Diseases, Second Edition
http://dx.doi.org/10.1016/B978-0-12-803058-5.00090-4

TABLE 90.2   Prevalence of Aneurysms by Location in Autopsy and Clinical Series

Study	Type	ICA (%)	ACA (%)	MCA (%)	Posterior (%)	Overall Prevalence (%)	Size
Inagawa and Hirano (1990) [19]	Autopsy	28	27	36	8	0.8	89% <10 mm
Agarwal et al. (2014) [11]	Incidental CTA	33	26	19	21	1.8	76% <10 mm
Igase et al. (2012) [12]	Incidental MRA	48	17	27	9	8.4	48% <3 mm, 92% <7 mm
Lee et al. (2015) [7]	Ruptured presentation	34	39	19	10	–	46% <5 mm, 93% <10 mm

the endothelium, and the tensile pressure is sensed by vascular mural cells; both factors are then transmitted into biological signals to the vessel wall affecting subsequent remodeling [13]. In many instances, vessel wall remodeling is able to achieve a new equilibrium with hemodynamic stresses, and aneurysmal stability is achieved; in other instances the vessel wall remodeling is insufficient and hemodynamic forces overcome wall strength resulting in rupture. As hemodynamic forces are a fundamental driver of aneurysm formation and growth, the local vascular configuration (i.e., branching pattern and angles, flow speed, and wall thickness) is likely related to regional differences in aneurysm behavior.

It is reasonable to postulate that there may be location-specific differences in gene expression independent of hemodynamic properties, although this has not specifically been tested. The cerebral vascular development is a sequential and highly orchestrated process where different segments arise at different times, rendering the opportunity for segment-specific patterning or even mutagenesis to occur. Indeed, this may underlie the phenomenon of mirror aneurysms seen in 12% of patients [14,15] most commonly in the MCA bifurcation. The distinct evolution of different segments of the cerebral vasculature also provides an opportunity for divergence of base components of the vessels. These events may give rise to a "positional identity" and differential susceptibility of aneurysmal formation to cells comprising the components of the vascular wall in different locations. Indeed, aneurysm location is a major factor in current models of future rupture risk.

Aneurysm location can also affect presentation. Mass effect or thromboembolic events into a specific vascular distribution can occur. Additionally, rupture with associated hematoma formation can also imbue symptoms referable to specific location. The risks of associated aneurysm treatment also vary with location due to accessibility and local perforators both from an open surgical standpoint and from and endovascular perspective.

Multiple aneurysms are present in about 20% of cases, which can make determination of the ruptured aneurysm difficult. Multiple aneurysms are present in about 20% of cases that necessitates accurately distinguishing the ruptured aneurysm versus nonruptured aneurysm(s), so that

the at risk, ruptured aneurysm can be treated. Traditionally, the identification of the aneurysm that ruptured in these is imprecise and relies on assessment of the pattern of SAH (which varies by location in classical aneurysms) and evaluation of the relative risk of rupture for each aneurysm taking into account the aneurysms location, size, and morphology. This method is likely to result in misidentification of the ruptured aneurysm and inadequate hemorrhage source protection in a subset of cases. More recently, advanced imaging of vessel wall inflammation using new MRI sequences has been shown to accurately distinguish ruptured and nonruptured aneurysms in patients with SAH and multiple aneurysms [16,17].

Factors independent of location can precipitate aneurysm formation and subsequent rupture. Hypertension, smoking, female, increasing age, dysmorphic morphology, and size have all been associated with increasing risk of rupture [4]. There is a congenital predisposition to aneurysm formation in polycystic kidney disease, and first-degree relatives of patients with SAH or unruptured aneurysms [18].

INTERNAL CAROTID ARTERY ANEURYSMS

ICA aneurysms account for about 30% of all unruptured intracranial aneurysms [8,19], and 35% of ruptured aneurysms in clinical series [7,8]. Cavernous segment, ophthalmic, anterior choroidal, and PCOM aneurysms are typically included in this category. However, these numbers may shift in accordance with newer imaging that allows for the detection of smaller aneurysms and aneurysms in more difficult locations such as the ICA [12].

Cavernous segment aneurysms comprise 2–9% of all aneurysms and follow a relatively benign course [20]. The risk of SAH is exceedingly low due to their extradural location within the cavernous sinus (0.2–0.4% yearly), although rupture can cause the formation of carotid–cavernous fistulae. Additional symptoms may stem from local mass effect on the cranial nerves (CNs) of the cavernous sinus (CN IV, VI, II, and V_1) or ischemic stroke; 61% present with CN palsies, 23% with retro-orbital pain, and 10% with trigeminal pain [21]. As the risk of SAH is low these aneurysms are prone to becoming quite large prior to becoming symptomatic adding to treatment morbidity and mortality. Newer treatment strategies with flow diverters may prove to improve treatment outcomes.

It is important to properly distinguish cavernous segment aneurysms from other intradural paraclinoid aneurysms, as intradural aneurysms carry a higher risk of devastating SAH. The optic strut is typically used as the defining landmark with those aneurysms distal to this landmark being considered intradural, and those proximal considered intradural. In 86% of cases, the ophthalmic artery arises intradurally, it arises extradurally in 6.7%, and arises from between the two dural rings in the remaining instances [22]. Clinical application of this can be difficult, and it may be more prudent to consider aneurysms at or near the optic strut to be "junctional," with the possibility of being intradural [23]. It should also be noted that some larger cavernous segment artery aneurysms which originate extradurally, may extend upward with a portion of their dome traversing the dural barrier giving them the ability to rupture into the subarachnoid space [23].

Blister aneurysms are classically described as non-branch point aneurysms arising from the dorsal and anterior wall of the ICA. They represent a minority of ICA aneurysms (0.9–6.5%) most often presenting ruptured (likely due to the difficulty in discerning an aneurysm that small on routine imaging). Believed to arise from a dissection, the arterial wall is extremely fragile and repeated imaging often demonstrates progression to a more saccular shape over a span of days. Re-hemorrhage is the norm, with surgical treatment rendering high morbidity and mortality [24]. Flow diverters or even single traditional stents may improve outcomes [25–27].

Unruptured PCOM aneurysms can be associated with a CN III palsy depending on their orientation. The third nerve begins its course traveling between the PCA and subclavian artery (SCA) before traveling along the PCOM, hence aneurysms of the basilar, PCA, and SCA can also compress the nerve [18]. Third nerve compression from an aneurysm typically involves the pupil on clinical examination, whereas a pupil-sparing third nerve palsy is more likely to result from microvascular ischemia. Most PCOM aneurysms causing isolated CN III palsy are 4mm or larger, but smaller aneurysms have also been reported [28,29]. Retro-orbital pain is a common symptom, preceding CN compromise by 10days. Full or partial recovery can be achieved with treatment, and is correlated with the duration of CN palsy. It often proceeds with initial improvement in the levator palpebrae, and then the medial rectus muscle [29]. Notably, decompression of the nerve by open resection of the aneurysm is not always necessary for return of CN function, as complete recovery has also been seen

with endovascular treatment that presumably does not engender a rapid decrease in the aneurysmal mass effect but would be expected to decrease pulsatility [30].

SAH from ICA aneurysms typically fills the basal cisterns with the classic starburst appearance and can be somewhat asymmetric in density toward the side of the ruptured aneurysm; 20% of intracerebral hemorrhage (ICH) associated with aneurysmal rupture are from ICA aneurysms [31,32]. While aneurysmal subdural hemorrhage (SDH) is more rare (0.9–5.8%), ruptured PCOM aneurysm is an associated risk factor for SDH [33]. The occurrence of associated SDH predicts a worse outcome [34].

ANTERIOR COMMUNICATING ARTERY ANEURYSMS

While the ACA is not the most common site for aneurysm development, it is the most common site of rupture [7–10]. Within the ACA circulation, the ACOM complex is the most frequent site for aneurysm formation and rupture. Incidentally found anterior circulation aneurysms in general have a lower risk of rupture than posterior circulation aneurysms when accounting for known risk factors such as size [5]. However, many studies have demonstrated a relatively overrepresentation of small-ruptured ACOM aneurysms [7–10]. It is difficult to reconcile this conflicting data to arrive at the best management strategy for aneurysms at this location.

Common configurations of the ACOM complex associated with aneurysm morphology underscore the association of hemodynamics and aneurysm formation. Patients with ACOM aneurysms are more likely to have nonsymmetric A1 segments [35]; furthermore, the A1 segment morphology correlates with the direction of projection of the aneurysm [36], likely due to the direction of the jet stream. The incidence of an azygous ACA is low in the general population (<1%); however, the incidence of an azygous ACA in patients with pericallosal artery aneurysms is high (17%), and the incidence of aneurysms in patients with azygous ACA is also high (33%) [37]. Most distal artery aneurysms are also found along the ACA/pericallosal arteries [38], shearing against the falx may be a contributing factor to their pathogenesis.

Ruptured ACOM aneurysms classically produce anterior interhemispheric SAH, which can help with initial localization of the ruptured aneurysm. Intraventricular hemorrhage (IVH) is also common. Additionally, ICH and SDH can also occur. In SAH with frontal ICH location, a ruptured ACOM aneurysm is usually the culprit (52.9%) [31]. The ICH size from ruptured ACOM aneurysms are more often <50mL in volume [32]. SDH associated with ACA aneurysms are typically inferiorly directed ACOM aneurysms. Due to its location, the mass effect of a growing ACOM aneurysm can cause compression of the optic nerve or chiasm resulting in visual symptoms. Rupture of an ACA aneurysm may result in frontal lobe damage resulting in decreases in consciousness, motivation, memory, and akinetic mutism. ACA vasospasm following SAH can manifest as confusion, drowsiness, leg weakness, and decreased speech or abulia.

The variability in ACOM region vascular anatomy, especially in relation to smaller perforating vessels that may not be visible on cross-sectional imaging, or even fully apparent on digital subtraction angiography (DSA) can make the treatment of such aneurysms challenging. Local perforators supply the optic chiasm, anterior commissure, anterior hypothalamus, infundibulum, the genu of the internal capsule and the anterior part of the globus pallidus. The recurrent artery of Heubner often arises from the ACOM–A2 junction and supplies the head of caudate. It is at risk for accidental occlusion during surgical procedures and can result in significant morbidity.

MIDDLE CEREBRAL ARTERY ANEURYSMS

Unruptured MCA aneurysms are common and account for about one-third of all unruptured aneurysms. In patients presenting with SAH, ruptured MCA aneurysms account for about 20% (see Table 90.1). About 80% of MCA aneurysms arise from the bifurcation and commonly project along the axis of the main MCA trunk [39]. Proximal MCA aneurysms account for 16% and distal MCA aneurysms account for 4%, aneurysms at both of these sites tend to be smaller in size than MCA bifurcation aneurysms [39].

The pattern of SAH is typically asymmetric, weighted more heavily on the side of the aneurysm and extending into the Sylvain fissure. While 20–30% of patients with SAH will present with ICH, MCA aneurysms are more commonly associated with ICH than aneurysms at other locations, with about 40–50% of SAH-associated hematomas caused by MCA aneurysms [31,32,39]. Temporal and perisylvian ICH are predominantly caused by MCA aneurysms [31], and are more likely to be large in size (>50 mL). Increasing distal location of MCA aneurysm yields increased risk of ICH, as does lateral projection of bifurcation aneurysms [39]. Basal ganglia hemorrhages are occasionally associated with distal lenticulostriate artery aneurysms [40]. ICH is a predictor for poor outcome and may necessitate operative decompression on initial presentation.

Ruptured MCA aneurysms are associated with higher rates of severe, persistent neurological deficits including visual field deficits, delayed epilepsy, and severe hemiparesis than ruptured aneurysms in other locations [39,41]. Visual field deficits occur in 20% of patients. Delayed epilepsy occurred in 20%, but the rate was significantly higher when temporal ICH was present.

Severe hemiparesis was associated with more proximal MCA aneurysm location [39]. Mycotic and neoplastic aneurysms most commonly arise in the MCA territory, likely stemming from hemodynamic effects. MCA vasospasm can result in hemiparesis, and if on the side of the dominant lobe, aphasia.

Surgical clipping of MCA bifurcation aneurysms is still often the treatment of choice as their endovascular treatment can be difficult; however, with newer endovascular devices more of these may be amenable to this treatment modality in the coming years.

VERTEBROBASILAR AND CEREBELLAR ARTERY ANEURYSMS

Unruptured posterior circulation saccular aneurysms have a lower prevalence than anterior circulation aneurysms, with an estimated prevalence of 8%; however, their natural history often appears to be less favorable; 10% of all aneurysms presenting as first-time hemorrhage are posterior circulation aneurysms. Natural history studies of unruptured aneurysms ascribe a higher risk of rupture for posterior circulation aneurysm compared to similarly sized anterior circulation aneurysms suggesting differences in the anterior and posterior circulation. Similar to ICA terminus aneurysms, basilar tip aneurysms are subjected to a strong perpendicular flow jet with the aneurysm commonly pointing superiorly. They are located at the bifurcation and often involve the bilateral PCA origins, and commonly the origin of the SCA given its close proximity. The complex anatomy of the posterior fossa complicates treatment: the vessels are in close proximity and send perforators to the brain stem, the lower CNs are also intertwined within the vessels, the vascular anatomy can be highly variable.

The pattern of SAH seen on noncontrast head CT in ruptured posterior circulation aneurysms typically includes posterior fossa SAH, often with IVH. Supratentorial SAH is also not infrequent. Isolated IVH is often a result of ruptured posterior inferior cerebellar artery (PICA) aneurysms with hydrocephalus on presentation [42]. ICH in the posterior circulation occurs with relatively lower frequency as compared to anterior circulation aneurysms, and also typically smaller in volume (7.2% <50 mL, 2.2% >50 mL) [32].

Similarly to PCOM aneurysms, unruptured basilar tip, basilar–SCA junction, and PCA aneurysms may also present with CN III compression due to the anatomical location in relation to the nerve [18]. Ruptured PICA aneurysms are more likely to affect the lower CNs and require percutaneous gastrostomy and endotracheal tube placement, likely contributing to their overall worse clinical outcomes than ruptured aneurysms in other locations (63% vs. 32% modified Rankin Score >2 at 3 years) [43].

A large percentage of cerebellar aneurysms are distal (28.6%), with distal PICA aneurysms making up the majority of those (~73%). These aneurysms most commonly present with hemorrhage [38]. Surgical management of these aneurysms often requires trap and bypass techniques or distal vessel sacrifice. This can be complicated by brain stem infarcts (Wallenberg syndrome) from critical perforators. Posterior perforator aneurysms (17 in the literature) present exclusively with rupture, and are often difficult to visualize and diagnose. Their location on critical perforators place them at increased risk of intervention, as unintended parent vessel sacrifice will cause stroke in critical brain structures. Almost 90% of posterior circulation perforator aneurysms reported occurred off basilar artery perforators. Their visualization was more prominent in the capillary and venous phases of the angiogram due to slow filling of small vessels. A significant number autothrombosed presumably due to this slow filling. Those that were treated, may have faired better with surgical rather than endovascular approaches. Re-hemorrhage risk is unclear [44].

The natural history of vertebrobasilar dolichoectasia and nonsaccular aneurysms is not well delineated. They include fusiform dilatation of once normal vessels (considered aneurysms with separate in-flow and out-flow) and variably include dissecting aneurysms. The incidence of vertebrobasilar fusiform aneurysms is generally low (0.05% of population), but likely increased in cases of stroke [45]. They present incidentally (~30%), with mass effect composed of CN or brain stem compression and possible hydrocephalus (~30%), ischemic stroke (~30%), and SAH or ICH (7%). Overall mortality is 43% over 7.1 years with slightly less than half of those due to nonneurological causes. Those who present asymptomatically and do not progress to growth have a relatively benign course with 0.4% annual mortality; in contrast, those who present with stroke, hemorrhage, brain stem compression, or develop unstable growth go on to have poor outcomes. Hemorrhagic risk is about 1.1% yearly but increases to 6% yearly if growth is observed. Prior stroke strongly predicts recurrent stroke (5.6% yearly). Brain stem compression predicts an 11% yearly mortality, and aneurysmal growth a 6.1% yearly mortality [46]. The distribution of stroke is most often in the pons or thalamus [45]. Treatment is complex including surgical/endovascular flow reduction or reversal and flow diversion, often with less than satisfactory results [45,46].

The symptoms of vertebrobasilar vasospasm after SAH is less specific than ACA or MCA vasospasm and manifests as generalized neurological deterioration and decreased level of consciousness. Posterior circulation aneurysms in general have a worse prognosis than anterior circulation aneurysms, coupled with their more challenging surgical treatment, there has been an increasing trend toward endovascular treatment which is also often more complex than similar treatments in the anterior circulation.

CONCLUSIONS

The pathogenesis of aneurysms and their progression to rupture depend on a complex interaction of hemodynamic on the vessel wall as dictated by local vascular anatomy and underlying inherent properties of the vessel wall constituents. The aneurysm's location influences its presentation, severity of morbidity in the event of rupture, and difficulty of treatment making classification by location of paramount importance in disease management.

References

[1] Huang J, van Gelder JM. The probability of sudden death from rupture of intracranial aneurysms: a meta-analysis. Neurosurgery 2002;51(5):1101–5. [discussion 1105-1107].

[2] van Asch CJ, Luitse MJ, Rinkel GJ, van der Tweel I, Algra A, Klijn CJ. Incidence, case fatality, and functional outcome of intracerebral haemorrhage over time, according to age, sex, and ethnic origin: a systematic review and meta-analysis. Lancet Neurol 2010;9(2):167–76.

[3] van Gijn J, Kerr RS, Rinkel GJ. Subarachnoid haemorrhage. Lancet 2007;369(9558):306–18.

[4] Weir B. Unruptured intracranial aneurysms: a review. J Neurosurg 2002;96(1):3–42.

[5] Wiebers DO, Whisnant JP, Huston 3rd J, Meissner I, Brown Jr RD, Piepgras DG, et al. Unruptured intracranial aneurysms: natural history, clinical outcome, and risks of surgical and endovascular treatment. Lancet 2003;362(9378):103–10.

[6] Investigators UJ, Morita A, Kirino T, Hashi K, Aoki N, Fukuhara S, et al. The natural course of unruptured cerebral aneurysms in a Japanese cohort. N Engl J Med 2012;366(26):2474–82.

[7] Lee GJ, Eom KS, Lee C, Kim DW, Kang SD. Rupture of very small intracranial aneurysms: incidence and clinical characteristics. J Cerebrovasc Endovasc Neurosurg 2015;17(3):217–22.

[8] Carter BS, Sheth S, Chang E, Sethl M, Ogilvy CS. Epidemiology of the size distribution of intracranial bifurcation aneurysms: smaller size of distal aneurysms and increasing size of unruptured aneurysms with age. Neurosurgery 2006;58(2):217–23. [discussion 217-223].

[9] Forget Jr TR, Benitez R, Veznedaroglu E, Sharan A, Mitchell W, Silva M, et al. A review of size and location of ruptured intracranial aneurysms. Neurosurgery 2001;49(6):1322–5. [discussion 1325-1326].

[10] Joo SW, Lee SI, Noh SJ, Jeong YG, Kim MS, Jeong YT. What is the significance of a large number of ruptured aneurysms smaller than 7 mm in diameter? J Korean Neurosurg Soc 2009;45(2):85–9.

[11] Agarwal N, Gala NB, Choudhry OJ, Assina R, Prestigiacomo CJ, Duffis EJ, et al. Prevalence of asymptomatic incidental aneurysms: a review of 2,685 computed tomographic angiograms. World Neurosurg 2014;82(6):1086–90.

[12] Igase K, Matsubara I, Igase M, Miyazaki H, Sadamoto K. Initial experience in evaluating the prevalence of unruptured intracranial aneurysms detected on 3-tesla MRI. Cerebrovasc Dis 2012;33(4):348–53.

[13] Meng H, Tutino VM, Xiang J, Siddiqui A. High WSS or low WSS? Complex interactions of hemodynamics with intracranial aneurysm initiation, growth, and rupture: toward a unifying hypothesis. AJNR Am J Neuroradiol 2014;35(7):1254–62.

[14] Meissner I, Torner J, Huston 3rd J, Rajput ML, Wiebers DO, Jones Jr LK, et al. International study of unruptured intracranial aneurysms I: mirror aneurysms: a reflection on natural history. J Neurosurg 2012;116(6):1238–41.

[15] Lasjaunias PL. Segmental identity and vulnerability in cerebral arteries. Interv Neuroradiol 2000;6(2):113–24.

[16] Matouk CC, Mandell DM, Gunel M, Bulsara KR, Malhotra A, Hebert R, et al. Vessel wall magnetic resonance imaging identifies the site of rupture in patients with multiple intracranial aneurysms: proof of principle. Neurosurgery 2013;72(3):492–6. [discussion 496].

[17] Chalouhi N, Hoh BL, Hasan D. Review of cerebral aneurysm formation, growth, and rupture. Stroke 2013;44(12):3613–22.

[18] Cianfoni A, Pravata E, De Blasi R, Tschuor CS, Bonaldi G. Clinical presentation of cerebral aneurysms. Eur J Radiol 2013;82(10):1618–22.

[19] Inagawa T, Hirano A. Autopsy study of unruptured incidental intracranial aneurysms. Surg Neurol 1990;34(6):361–5.

[20] Ambekar S, Madhugiri V, Sharma M, Cuellar H, Nanda A. Evolution of management strategies for cavernous carotid aneurysms: a review. World Neurosurg 2014;82(6):1077–85.

[21] Goldenberg-Cohen N, Curry C, Miller NR, Tamargo RJ, Murphy KP. Long term visual and neurological prognosis in patients with treated and untreated cavernous sinus aneurysms. J Neurol Neurosurg Psychiatry 2004;75(6):863–7.

[22] Javalkar V, Banerjee AD, Nanda A. Paraclinoid carotid aneurysms. J Clin Neurosci 2011;18(1):13–22.

[23] Carlson AP, Loveren HR, Youssef AS, Agazzi S. Junctional internal carotid artery aneurysms: the Schrodinger's cat of vascular neurosurgery. J Neurol Surg B Skull Base 2015;76(2):150–6.

[24] Gonzalez AM, Narata AP, Yilmaz H, Bijlenga P, Radovanovic I, Schaller K, et al. Blood blister-like aneurysms: single center experience and systematic literature review. Eur J Radiol 2014;83(1):197–205.

[25] Grant RA, Quon JL, Bulsara KR. Oversized self-expanding stents as an alternative to flow-diverters for blister-like aneurysms. Neurol Res 2014;36(4):351–5.

[26] Bulsara KR, Kuzmik GA, Hebert R, Cheung V, Matouk CC, Jabbour P, et al. Stenting as monotherapy for uncoilable intracranial aneurysms. Neurosurgery 2013;73(1 Suppl. Operative). [discussion 80-85].

[27] Ediriwickrema A, Williamson T, Hebert R, Matouk C, Johnson MH, Bulsara KR. Intracranial stenting as monotherapy in subarachnoid hemorrhage and sickle cell disease. J Neurointerv Surg 2013;5(2):e4.

[28] Elmalem VI, Hudgins PA, Bruce BB, Newman NJ, Biousse V. Underdiagnosis of posterior communicating artery aneurysm in noninvasive brain vascular studies. J Neuroophthalmol 2011;31(2):103–9.

[29] Yanaka K, Matsumaru Y, Mashiko R, Hyodo A, Sugimoto K, Nose T. Small unruptured cerebral aneurysms presenting with oculomotor nerve palsy. Neurosurgery 2003;52(3):553–7. [discussion 556-557].

[30] Birchall D, Khangure MS, McAuliffe W. Resolution of third nerve paresis after endovascular management of aneurysms of the posterior communicating artery. AJNR Am J Neuroradiol 1999;20(3):411–3.

[31] Bruder M, Schuss P, Berkefeld J, Wagner M, Vatter H, Seifert V, et al. Subarachnoid hemorrhage and intracerebral hematoma caused by aneurysms of the anterior circulation: influence of hematoma localization on outcome. Neurosurg Rev 2014;37(4):653–9.

[32] Guresir E, Beck J, Vatter H, Setzer M, Gerlach R, Seifert V, et al. Subarachnoid hemorrhage and intracerebral hematoma: incidence, prognostic factors, and outcome. Neurosurgery 2008;63(6):1088–93. [discussion 1093-1084].

[33] Biesbroek JM, Rinkel GJ, Algra A, van der Sprenkel JW. Risk factors for acute subdural hematoma from intracranial aneurysm rupture. Neurosurgery 2012;71(2):264–8. [discussion 268-269].

[34] Biesbroek JM, van der Sprenkel JW, Algra A, Rinkel GJ. Prognosis of acute subdural haematoma from intracranial aneurysm rupture. J Neurol Neurosurg Psychiatry 2013;84(3):254–7.

[35] Krasny A, Nensa F, Sandalcioglu IE, Goricke SL, Wanke I, Gramsch C, et al. Association of aneurysms and variation of the A1 segment. J Neurointerv Surg 2014;6(3):178–83.

[36] Feng W, Zhang L, Li W, Zhang G, He X, Wang G, et al. Relationship between the morphology of A-1 segment of anterior cerebral artery and anterior communicating artery aneurysms. Afr Health Sci 2014;14(1):83–8.

[37] Katz RW, Horoupian DS, Zingesser L. Aneurysm of azygous anterior cerebral artery. A case report. J Neurosurg 1978;48(5):804–8.

[38] Rodriguez-Hernandez A, Zador Z, Rodriguez-Mena R, Lawton MT. Distal aneurysms of intracranial arteries: application of numerical nomenclature, predilection for cerebellar arteries, and results of surgical management. World Neurosurg 2013;80(1–2):103–12.

[39] Rinne J, Hernesniemi J, Niskanen M, Vapalahti M. Analysis of 561 patients with 690 middle cerebral artery aneurysms: anatomic and clinical features as correlated to management outcome. Neurosurgery 1996;38(1):2–11.

[40] Choo YS, Kim YB, Shin YS, Joo JY. Deep intracerebral hemorrhage caused by rupture of distal lenticulostriate artery aneurysm: a report of two cases and a literature review. J Korean Neurosurg Soc 2015;58(5):471–5.

[41] Koskela E, Setala K, Kivisaari R, Hernesniemi J, Laakso A. Visual field findings after a ruptured intracranial aneurysm. Acta Neurochir 2014;156(7):1273–9.

[42] Kleinpeter G. Why are aneurysms of the posterior inferior cerebellar artery so unique? Clinical experience and review of the literature. Minim Invasive Neurosurg 2004;47(2):93–101.

[43] Williamson RW, Wilson DA, Abla AA, McDougall CG, Nakaji P, Albuquerque FC, et al. Clinical characteristics and long-term outcomes in patients with ruptured posterior inferior cerebellar artery aneurysms: a comparative analysis. J Neurosurg 2015;123(2):441–5.

[44] Ding D, Starke RM, Jensen ME, Evans AJ, Kassell NF, Liu KC. Perforator aneurysms of the posterior circulation: case series and review of the literature. J Neurointerv Surg 2013;5(6):546–51.

[45] Serrone JC, Gozal YM, Grossman AW, Andaluz N, Abruzzo T, Zuccarello M, et al. Vertebrobasilar fusiform aneurysms. Neurosurg Clin N Am 2014;25(3):471–84.

[46] Shapiro M, Becske T, Riina HA, Raz E, Zumofen D, Nelson PK. Non-saccular vertebrobasilar aneurysms and dolichoectasia: a systematic literature review. J Neurointerv Surg 2014;6(5):389–93.

CHAPTER

91

Clinical Aspects of Subarachnoid Hemorrhage

M.J. Koch, B. Choi, C.J. Stapleton, A.B. Patel

Massachusetts General Hospital and Harvard Medical School, Boston, MA, United States

INTRODUCTION

Aneurysmal rupture leading to subarachnoid hemorrhage (SAH) represents a dire clinical entity with profound neurologic and systemic manifestations. Despite several advances in microsurgical and endovascular technology, mortality rates remain high, and survivors often suffer substantial neurologic morbidity; of all patients with aneurysm rupture, 25% die within the first 24h and nearly half die within 30 days [1]. As such, proper diagnostic evaluation and early clinical management of patients with ruptured aneurysms is essential, and delays in this process have been shown to significantly worsen outcome [2]. Here we provide a brief review of the epidemiology and natural history of aneurysmal SAH as well as recommendations regarding the present state of diagnosis and management of this clinical entity.

EPIDEMIOLOGY

In the United States, incidence of aneurysmal SAH is 2 per 100,000 population—approximate number of cases annually [3,4]. In Japan and the Netherlands, incident rates reach as high as 22 per 100,000, suggesting a genetic and/or environmental component to their incidence. The average age at presentation is 55 years with a female predilection, which is thought to be related to hormonal influences and estrogen deficiency in postmenopausal women [5].

Cerebral aneurysms occur mainly as saccular or fusiform lesions. Saccular aneurysms are the most common cause of aneurysmal SAH and are thought to occur because of persistent pressure on a branch point or turning point, or as a result of a diseased vessel in the setting of underlying vascular disease. As with systemic vascular disease, several modifiable risk factors such as hypertension, smoking, and cocaine use are associated with a higher incidence of these lesions [6,7]. Fusiform lesions are thought to represent the end product of a prior dissection, atherosclerotic changes, or defects in collagen synthesis (e.g., Ehlers–Danlos type IV syndrome, Marfan syndrome, autosomal dominant polycystic kidney disease, pseudoxanthoma elasticum) [7]. Histologically, this manifests as a vessel outpouching with vessel intima and adventitia without the presence of arterial media, accounting for the fragility of these lesions and their tendency to rupture.

As demonstrated in the Nordic twin cohort, most aneurysms in practice are only modestly influenced by genetic predisposition when compared to environmental risk factors [8]. However, this should be weighed against the fact that a family history of aneurysmal SAH has been shown to be among the strongest independent predictors [6]. Even when excluding the aforementioned rare inherited connective tissue disorders, first-degree relatives of patients with SAH are at significantly increased risk when compared to the general population [7].

Most unruptured aneurysms do not rupture [9–11]. The greatest risk of rupture in incidentally discovered lesions is a history of SAH. Further risk stratification to determine the need for prophylactic surgical or endovascular obliteration is performed based on other known modifiers such as aneurysm size, interval growth of an aneurysm, aneurysm location, and irregular morphology [10,12].

DIAGNOSIS

Clinically, the primary symptom of rupture is acute onset headache, traditionally described as a thunderclap headache or the "worst headache of life" in reference to its severity and rapid onset [13]. Studies have

Primer on Cerebrovascular Diseases, Second Edition
http://dx.doi.org/10.1016/B978-0-12-803058-5.00091-6

demonstrated that up to 20% of patients who initially present with this symptom alone will have an SAH [14]. The sudden and severe headache may also be accompanied by some degree of altered mental status, nausea or vomiting, and meningismus. Although seizures are less common, they have been shown to portend a poor prognosis [15]. Patients may also report a history of headache that precedes a major SAH by several weeks; these sentinel headaches are thought to reflect a minor hemorrhage or "warning leak." [16,17].

Non–contrast-enhanced computed tomography (CT) represents the primary diagnostic tool for SAH, with a sensitivity approaching 100% within the first 6 h of peak headache onset [18,19]. However, because the sensitivity of CT decreases with time from onset, most guidelines recommend a lumbar puncture (LP) when initial imaging is negative in the appropriate clinical setting. Classically, LP for SAH will reveal both red blood cells that do not clear from tube one to four and xanthochromia. Although it may be absent within the first hours of the ictus, the presence of xanthochromia in the delayed setting has a specificity for cerebral aneurysm that reaches approximately 95% [19]. Delayed or subacute presentation can be further diagnosed by magnetic resonance imaging (MRI). While < 12 h CT may provide superior resolution, MRI has also been shown to reasonably evaluate for SAH [20].

Appropriate clinical suspicion and diagnosis are essential for appropriate treatment and patient outcome, as ruptured aneurysms not treated within 24 h of presentation are associated with an almost 15% risk of rerupture [16,17,21]. Furthermore, early diagnosis can facilitate the transfer of patients to those centers specializing in the care of patients with SAH (i.e., centers treating > 10 SAH cases a year demonstrate improved outcomes when compared with centers that treat less number of cases) [22].

CLINICAL COURSE

Patients with SAH can present with a range of neurologic symptoms, with the clinical severity ranging from a moderate headache to a nonresponsive and comatose state. The Hunt–Hess score or the World Federation of Neurosurgical Societies scale is used to stratify patients depending on their clinical severity. Regardless, advanced cardiac life support (ACLS) and advanced trauma life support (ATLS) strategies are primary in initial clinical management, as SAH is diagnosed often after immediate cardiopulmonary stabilization [23,24].

Acute management of SAH rests on two principles: (1) preventing rerupture and (2) mitigating and treating the effects of intraparenchymal, subarachnoid, and intraventricular blood. Physiologically, clot formation around the dome of a ruptured aneurysm prevents continuous hemorrhage of arterial blood into the subarachnoid space. Initial medical management focuses on the prevention of increase in arterial pressure and maintenance of that clot. Maintaining the blood pressure below a systolic blood pressure (SBP) of 140 mm Hg or a mean arterial pressure (MAP) of less than 90 mm Hg minimizes the risk of rerupture and is a means of temporizing the patient until definitive surgical or endovascular intervention [25]. Clinically, rerupture manifests as a worsened headache, seizure, or decreased level of consciousness. It is often associated with an acute spike in blood pressure and intracranial pressure. Maintenance of the clot integrity is optimized with reversal of antiplatelet and anticoagulant therapy. Patients with therapeutic or supratherapeutic international normalized ratio (INR) will usually receive emergent reversal with four-factor prothrombin complex concentrate (PCC). Those on aspirin or clopidogrel (Plavix) will receive platelets with an overall platelet goal of > 100,000.

Intraparenchymal, subarachnoid, and intraventricular blood manifests both intracranially and systemically. The most concerning feature of intracranial blood is the increase in intracranial pressure and local mass effect in the case of intraparenchymal clot. Clinically, the elevation in intracranial pressure and the cortical and meningeal irritation by blood products lead to the "worst headache of life," nuchal pain/rigidity, nausea or vomiting, confusion/agitation, decreased level of consciousness, and seizures. Immediately, patients are given antiepileptic drugs (AEDs), 1 g of levetiracetam or 20 mg/kg of phosphenytoin [17,26,16]. Analgesia with acetaminophen and administration of opiates are initiated to minimize pain and the associated blood pressure elevation, while maintaining the level of consciousness for neurologic examinations.

Mass effect from intraparenchymal clot can represent an especially grave circumstance necessitating emergent decompression to prevent uncal herniation and death. These lesions are more often associated with middle cerebral artery aneurysms and manifest with decrease in consciousness and, at extreme stages, third-nerve compression with pupillary dilatation. Radiographically, an acute hemorrhage in the middle fossa measuring greater than 15 cc may require emergent embolization/clipping and decompression. Both open surgical evacuation of the hemorrhage and clipping of the lesion or endovascular embolization with subsequent decompression are the current treatment options, with neither treatment showing significant difference in outcomes.

Communicating, or noncommunicating, hydrocephalus may occur as a result of subarachnoid blood. Acutely progressive hydrocephalus and ventricular enlargement represent a clinical emergency necessitating emergent diversion to prevent transtentorial herniation [27]. Hydrocephalus is managed with an external ventricular drain placed into the lateral ventricle.

VASOSPASM

The development of cerebral vasospasm typically peaks at 7–10 days following hemorrhage, with risk spanning approximately 2 weeks from the original ictus. The degree and density of SAH have been shown to correlate with the probability of developing vasospasm [28,29]. Although cerebral vasospasm may be asymptomatic at first, it has the potential to lead to symptomatic delayed cerebral ischemia, often involving watershed territories or other vascular areas distal to the aneurysm [30]. As such, several measures may be taken to reduce the incidence of vasospasm and its downstream effects. Such measures include the prophylactic use of nimodipine, which has been shown to improve outcomes [31]. Additionally, many clinicians use the "Triple-H" (hypertension, hemodilution, hypervolemia) therapy to treat and minimize the risk of vasospasm; however, there is equivocal clinical evidence supporting its use [32,33]. In case of failure of medical therapy, endovascular strategies such as intra-arterial pharmacotherapy with verapamil may be used [34].

TREATMENT OF CEREBRAL ANEURYSMS

A cerebral aneurysm can be diagnosed either as a result of aneurysmal rupture and/or progressive neurologic deficit or incidentally on cranial imaging for an unrelated pathologic condition. Incidentally discovered lesions are further evaluated with dedicated noninvasive vascular imaging techniques, such as CT angiogram and magnetic resonance angiography (MRA), and are then further stratified requiring the gold standard of evaluation, cerebral angiogram. Treatment of these lesions is not without risks; hence, several prospective trials have evaluated aneurysmal characteristics associated with an increased risk of rupture and the need for prophylactic treatment (i.e., lesion growth, daughter sacs, prior aneurysmal rupture). This information is used in combination with particular technical factors of aneurysmal treatment via open surgical [35] or endovascular intervention, and following the patient's preferences to determine the strategy of intervention or observation is warranted. A multidisciplinary expert discussion ultimately continues to play a central role in determining the most appropriate treatment for any given lesion. Lesions are typically addressed on a case-by-case basis, with anatomic and morphologic variables combined with patient characteristics to dictate an optimal care plan.

CONCLUSION

Aneurysmal SAH is an acute medical emergency requiring prompt recognition and medical/surgical evaluation.

References

[1] Broderick JP, Brott TG, Duldner JE, Tomsick T, Leach A. Initial and recurrent bleeding are the major causes of death following subarachnoid hemorrhage. Stroke July 1994;25(7):1342–7.

[2] Kassell NF, Kongable GL, Torner JC, Adams Jr HP, Mazuz H. Delay in referral of patients with ruptured aneurysms to neurosurgical attention. Stroke July–August 1985;16(4):587–90.

[3] Ingall T, Asplund K, Mahonen M, Bonita R. A multinational comparison of subarachnoid hemorrhage epidemiology in the WHO MONICA stroke study. Stroke May 2000;31(5):1054–61.

[4] Labovitz DL, Halim AX, Brent B, Boden-Albala B, Hauser WA, Sacco RL. Subarachnoid hemorrhage incidence among Whites, Blacks and Caribbean Hispanics: the Northern Manhattan Study. Neuroepidemiology 2006;26(3):147–50.

[5] Longstreth WT, Nelson LM, Koepsell TD, van Belle G. Subarachnoid hemorrhage and hormonal factors in women. A population-based case-control study. Ann Intern Med August 1, 1994;121(3): 168–73.

[6] Vlak MH, Rinkel GJ, Greebe P, Greving JP, Algra A. Lifetime risks for aneurysmal subarachnoid haemorrhage: multivariable risk stratification. J Neurol Neurosurg Psychiatry June 2013;84(6):619–23.

[7] Broderick JP, Brown Jr RD, Sauerbeck L, et al. Greater rupture risk for familial as compared to sporadic unruptured intracranial aneurysms. Stroke June 2009;40(6):1952–7.

[8] Korja M, Silventoinen K, McCarron P, et al. Genetic epidemiology of spontaneous subarachnoid hemorrhage: Nordic Twin Study. Stroke November 2010;41(11):2458–62.

[9] Morita A, Kirino T, Hashi K, et al. The natural course of unruptured cerebral aneurysms in a Japanese cohort. N Engl J Med June 2012;366(26):2474–82.

[10] Wiebers DO, Whisnant JP, Huston J, et al. Unruptured intracranial aneurysms: natural history, clinical outcome, and risks of surgical and endovascular treatment. Lancet July 2003;362(9378): 103–10.

[11] International Study of Unruptured Intracranial Aneurysms Investigators. Unruptured intracranial aneurysms–risk of rupture and risks of surgical intervention. N Engl J Med December 1998;339(24): 1725–33.

[12] Lindner SH, Bor AS, Rinkel GJ. Differences in risk factors according to the site of intracranial aneurysms. J Neurol Neurosurg Psychiatry January 2010;81(1):116–8.

[13] Gorelick PB, Hier DB, Caplan LR, Langenberg P. Headache in acute cerebrovascular disease. Neurology November 1986;36(11):1445–50.

[14] Morgenstern LB, Luna-Gonzales H, Huber Jr JC, et al. Worst headache and subarachnoid hemorrhage: prospective, modern computed tomography and spinal fluid analysis. Ann Emerg Med September 1998;32(3 Pt 1):297–304.

[15] Butzkueven H, Evans AH, Pitman A, et al. Onset seizures independently predict poor outcome after subarachnoid hemorrhage. Neurology November 14, 2000;55(9):1315–20.

[16] Bassi P, Bandera R, Loiero M, Tognoni G, Mangoni A. Warning signs in subarachnoid hemorrhage: a cooperative study. Acta Neurol Scand October 1991;84(4):277–81.

[17] Kowalski RG, Claassen J, Kreiter KT, et al. Initial misdiagnosis and outcome after subarachnoid hemorrhage. JAMA February 18, 2004;291(7):866–9.

[18] Perry JJ, Stiell IG, Sivilotti ML, et al. Sensitivity of computed tomography performed within six hours of onset of headache for diagnosis of subarachnoid haemorrhage: prospective cohort study. BMJ 2011;343:d4277.

[19] Cortnum S, Sorensen P, Jorgensen J. Determining the sensitivity of computed tomography scanning in early detection of subarachnoid hemorrhage. Neurosurgery May 2010;66(5):900–2. discussion 903.

[20] Fiebach JB, Schellinger PD, Geletneky K, et al. MRI in acute subarachnoid haemorrhage; findings with a standardised stroke protocol. Neuroradiology January 2004;46(1):44–8.

[21] Naidech AM, Janjua N, Kreiter KT, et al. Predictors and impact of aneurysm rebleeding after subarachnoid hemorrhage. Arch Neurol March 2005;62(3):410–6.

[22] Andaluz N, Zuccarello M. Recent trends in the treatment of cerebral aneurysms: analysis of a nationwide inpatient database. J Neurosurg June 2008;108(6):1163–9.

[23] Hunt WE, Hess RM. Surgical risk as related to time of intervention in the repair of intracranial aneurysms. J Neurosurg January 1968;28(1):14–20.

[24] Report of World Federation of Neurological Surgeons Committee on a Universal Subarachnoid Hemorrhage Grading Scale. J Neurosurg June 1988;68(6):985–6.

[25] Liu-Deryke X, Janisse J, Coplin WM, Parker Jr D, Norris G, Rhoney DH. A comparison of nicardipine and labetalol for acute hypertension management following stroke. Neurocrit Care 2008;9(2):167–76.

[26] Hart RG, Byer JA, Slaughter JR, Hewett JE, Easton JD. Occurrence and implications of seizures in subarachnoid hemorrhage due to ruptured intracranial aneurysms. Neurosurgery April 1981;8(4):417–21.

[27] Rajshekhar V, Harbaugh RE. Results of routine ventriculostomy with external ventricular drainage for acute hydrocephalus following subarachnoid haemorrhage. Acta Neurochir (Wien) 1992;115(1–2):8–14.

[28] Harrod CG, Bendok BR, Batjer HH. Prediction of cerebral vasospasm in patients presenting with aneurysmal subarachnoid hemorrhage: a review. Neurosurgery April 2005;56(4):633–54. discussion 633–654.

[29] Abla AA, Wilson DA, Williamson RW, et al. The relationship between ruptured aneurysm location, subarachnoid hemorrhage clot thickness, and incidence of radiographic or symptomatic vasospasm in patients enrolled in a prospective randomized controlled trial. J Neurosurg February 2014;120(2):391–7.

[30] Rabinstein AA, Friedman JA, Nichols DA, et al. Predictors of outcome after endovascular treatment of cerebral vasospasm. AJNR Am J Neuroradiol November–December 2004;25(10):1778–82.

[31] Barker FG, Ogilvy CS. Efficacy of prophylactic nimodipine for delayed ischemic deficit after subarachnoid hemorrhage: a meta-analysis. J Neurosurg March 1996;84(3):405–14.

[32] Sen J, Belli A, Albon H, Morgan L, Petzold A, Kitchen N. Triple-H therapy in the management of aneurysmal subarachnoid haemorrhage. The Lancet Neurol October 2003;2(10):614–21.

[33] Rinkel GJ, Feigin VL, Algra A, van Gijn J. Circulatory volume expansion therapy for aneurysmal subarachnoid haemorrhage. The Cochrane Database Syst Rev 2004;(4):CD000483.

[34] Feng L, Fitzsimmons BF, Young WL, et al. Intraarterially administered verapamil as adjunct therapy for cerebral vasospasm: safety and 2-year experience. AJNR Am J Neuroradiol September 2002;23(8):1284–90.

[35] Brilstra EH, Algra A, Rinkel GJ, Tulleken CA, van Gijn J. Effectiveness of neurosurgical clip application in patients with aneurysmal subarachnoid hemorrhage. J Neurosurg November 2002;97(5):1036–41.

CHAPTER

92

Clinical Aspects of Intracerebral Hemorrhage

K.C. Wu[1], A.P. See[1,2], M.A. Aziz-Sultan[1]

[1]Brigham and Women's Hospital, Harvard Medical School, Boston, MA, United States; [2]Boston Children's Hospital/Harvard Medical School, Boston, MA, United States

INTRODUCTION

Stroke is the second most frequent cause of death in the world, with 6.7 million globally documented cases in 2012 alone. This chapter focuses on intracerebral hemorrhage (ICH), the least common type of stroke, accounting for about 13% of all stroke victims, occurring at an annual rate of about 12–15 cases per 100,000 people. Despite having the lowest incidence of any type of stroke, the mortality rate exceeds 30% and only one-fifth

Primer on Cerebrovascular Diseases, Second Edition
http://dx.doi.org/10.1016/B978-0-12-803058-5.00092-8

of patients make a meaningful recovery to independence [1], making this an incredibly morbid, costly, and high impact disease. In this chapter we review the basic classification, clinical presentation, radiographic findings, and management of patients with ICH.

CLASSIFICATION

Strokes are divided into two main categories: ischemic and hemorrhagic. In ischemic stroke, diminished blood flow leads to inadequate oxygenation of brain parenchyma and neuronal infarction. Hemorrhagic stroke occurs most often due to vascular injury resulting in hematoma formation and local mass effect on surrounding neural tissue.

ICH is made up of a range of pathologies with different natural courses, workup, and management (Table 92.1). It is important for clinicians to recognize the major types of ICH in order to choose the appropriate diagnostic studies and treatment options.

ICH is categorized as primary or secondary depending on etiology. Primary ICH has a higher incidence and

TABLE 92.1 Causes of Intracerebral Hemorrhage

Primary Causes of ICH
Chronic hypertension
Amyloid angiopathy
Secondary Causes of ICH
Trauma
Vascular (e.g., arteriovenous malformation, aneurysms, cavernous malformations, dural arteriovenous fistulas, and dural venous sinus thrombosis)
Vasculopathies (e.g., Moyamoya disease, granulomatosis with polyangiitis, polyarteritis nodosa, Takayasu's arteritis, and giant cell/temporal arteritis)
Neoplastic (e.g., metastases, glioblastoma, oligodendroglioma, lymphoma, and pituitary adenoma)
Infections (e.g., herpes simplex virus, cerebral abscess, and infectious aneurysm)
Coagulopathy (e.g., liver disease, renal failure, thrombocytopenia, platelet dysfunction, various chemotherapy agents, aspirin, plavix, warfarin, heparin, and factor X inhibitors)
Hemorrhagic conversion of ischemic stroke
Drugs (e.g., EtOH, cocaine, amphetamines/sympathomimetics, and tobacco abuse)
Post-procedural (e.g., following: tPA administration, carotid endarterectomy, craniotomy for tumor, or vascular malformation resection)
IVH of prematurity/germinal matrix hemorrhage
Idiopathic

consists of two major diseases: cerebral amyloid angiopathy (CAA) and hypertensive hemorrhage. Together these two entities are responsible for greater than 80% of ICH, with hypertensive hemorrhage implicated in about 55% of total ICH cases.

Hypertensive hemorrhage occurs most often in the deep regions of the brain such as the basal ganglia, thalamus, pons, and cerebellum (Fig. 92.1). Chronic hypertension causes lipohyalinosis of small arteries and arterioles. Hyaline, collagen, protein, and fat replace arterial smooth muscle, thin the adventitia, and eventually form atheromas. The stiffened and narrowed lumen results in small areas of brain ischemia followed by encephalomalacia known as lacunes. This term is derived from the Latin word "lacunae," which appropriately means "lake." In addition to an increased risk of ischemic events, weakening of the tunica media at vessel bifurcations leads to a 2% annual risk of hemorrhage in patients with chronic hypertension [2]. Furthermore, the weakened muscular layer is unable to vasoconstrict in response to local hemorrhage, resulting in continued bleeding and ICH progression. Previously it was thought that weakened arterial walls form "Charcot–Bouchard microaneurysms" that rupture leading to hemorrhage. However, it has since been shown by electron microscopy that these "microaneurysms" are actually miniature hematomas located in the subadventitial or extravascular spaces.

CAA results from deposition of beta-amyloid proteins into the media of small meningeal and cortical vessels. These proteins exhibit "apple-green birefringence," upon exposure to Congo red stain under polarized light. Cerebral amyloid deposition occurs over time and has been found in about half of all patients over 70 years of age [3]. Over time, the amyloid proteins replace the smooth muscle cells found in arteries leading to decreased compliance and an increased bleeding risk. The annual recurrent hemorrhage rate is as high as 10.5% [4]. Apoprotein E2 and E4 are two alleles that have been implicated in predisposition to amyloid angiopathy, with the E4 allele

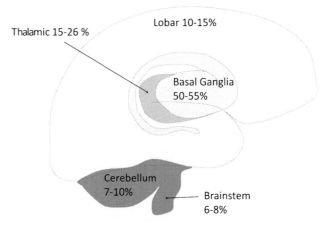

FIGURE 92.1 Anatomic distribution of intracerebral hemorrhage.

being associated with increased mortality and a younger age of initial ICH.

Secondary ICH encompasses a wide range of pathologies, as reviewed in Table 92.1. Although it encompasses many more diseases, the overall incidence of secondary ICH is much lower than primary ICH. Each of these diseases has a unique presentation, workup, and management.

Traumatic ICH or "hemorrhagic cerebral contusion" occurs most commonly in the frontal, temporal, or occipital poles due to acceleration/deceleration injuries that cause these regions to strike against the bony prominences of the cranium. In patients with severe contusions, there is a risk of delayed "blossoming" of the ICH, which may occur within 72h of the injury, leading to mass effect, edema, and rapid neurological deterioration due to elevated intracranial pressure (ICP).

The presence of a tumor can drastically alter diagnosis and management of ICH. The frequency of spontaneous ICH from a cerebral neoplasm averages 2–3%; however, this can be much higher with certain types of tumors. Intraparenchymal hemorrhage from vascular lesions is most often caused by arteriovenous malformations (AVMs) or cavernous malformations. The annual hemorrhage risk of an un-ruptured AVM is about 2–4%, and as high as 7.45% in patients who have bled previously [5]. Cavernous malformations have a lower propensity for bleeding with annual rates of <1% for incidentally found lesions, 2% for symptomatic lesions not related to hemorrhage, and 4.5% for lesions with symptomatic hemorrhage [6].

Anticoagulation is often used to reduce venous and arterial thromboembolic risk in patients with certain cardiac, stroke, or malignancy histories. Common conditions include atrial fibrillation, implanted vascular stents, prior strokes, deep vein thrombosis, pulmonary embolism, and inherited or acquired coagulopathic states. Oral anticoagulants have been associated with a 10-fold increased risk of ICH in patients over 50 years [7], of which 60% are fatal [8]. These hematomas have a larger volume and risk of continued expansion when compared with other causes of ICH (Fig. 92.2). Novel oral anticoagulants appear to have a more favorable profile of reduced hemorrhagic risk [9,10], but have yet to be as frequently prescribed as traditional vitamin K antagonists.

CLINICAL PRESENTATION

It is important to remember that no single presentation encompasses all patients with ICH, as there are various causes of this disease. As with all patients with neurological disease, the location of the injury is often the most important factor in determining presentation (Table 92.2). However, the classic description of ICH is that it occurs

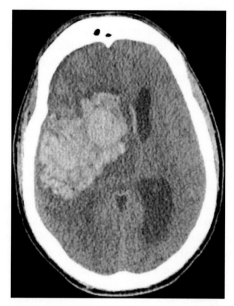

FIGURE 92.2 A gentleman in his fifth decade of life with a history of deep vein thrombosis (DVT)/pulmonary embolism (PE) taking Coumadin presented with progressive lethargy and decreased responsiveness. Head CT revealed a large right-sided frontal–temporal–parietal hematoma with associated hydrocephalus.

TABLE 92.2 Locations and Presentations of Intracerebral Hemorrhage

Location	Classic Presentation
Putamen	Progressive, contralateral hemiparesis/hemiplegia, coma, death. Headache in 14%.
Thalamus	Contralateral hemisensory loss, can extend to hemiparesis when internal capsule is involved. Vertical gaze palsy, retraction nystagmus, skew deviation, loss of convergence, ptosis + miosis + fixed and dilated pupils if spreads to midbrain. Potential for hydrocephalus. Headache in 30%.
Frontal lobe	Contralateral hemiparesis, usually mild leg/facial weakness. Frontal headache in 50%.
Parietal lobe	Contralateral hemisensory loss and possible hemiparesis. Contralateral inferior quadrantopsia.
Occipital lobe	Ipsilateral eye pain, contralateral homonymous hemianopsia.
Temporal lobe	Neglect, aphasia, intact repetition. Contralateral superior quadrantopsia.
Cerebellar	Hydrocephalus/herniation, lethargy, nausea/vomiting, headache. Hypertension + bradycardia + irregular respirations due to compression of fourth ventricle or cerebral aqueduct/intraventricular involvement. Dysmetria or bulbar dysnfunction.
Intraventricular	Severe headache, photophobia, nausea, vomiting, neck stiffness, hydrocephalus (lethargy, altered mental status, bradycardia, respiratory failure).

during the daytime with a smooth and progressive onset as the volume of the hematoma expands. This is in contrast to an acute ischemic stroke with symptoms that present as both sudden and maximum at onset due to vessel occlusion. ICH is associated with severe headache (40%) and vomiting (50%), eventually leading to worsening focal deficits related to location of the hemorrhage, and ultimately can result in cerebral herniation, coma, and death. Hypertension is noted in up to 90% of ICH patients as part of a classical triad seen in elevated ICP. Seizures indicated by a sudden change in examination with stable cranial imaging might also occur following ICH. Late deterioration may occur days to weeks later due to late re-bleeding or delayed hydrocephalus, which is more common in cases of ICH with associated intraventricular hemorrhage (IVH), due to impaired cerebrospinal fluid reabsorption.

RADIOLOGIC FINDINGS

The first-line imaging obtained for patients with ICH is head computed tomography (CT). It is both fast and ideal for showing acute hemorrhage as a hyperdense mass relative to brain parenchyma. This allows the clinician to immediately differentiate hemorrhage from ischemic stroke. It also shows acute structural complications such as herniation, IVH, or hydrocephalus that may require emergent treatment. As ICH patients are often acutely ill and at risk of rapid decompensation, a head CT allows for quick assessment and decision making regarding further diagnostic course, prognosis related to clot burden, and

subsequent treatment decisions. The rate of re-bleed within 24 h has been shown to steadily decline with time, starting as high as 30% in the first 1–3 h [11], and dropping as low as 14% within the 24 h period between initial and follow-up scans [12]. At our institution, we repeat a head CT at 6 h from initial imaging in nonoperative patients to ensure hematoma stability. We follow the same imaging protocol for patients with traumatic cerebral contusions as well, although data by Anandalwar et al. 2016 suggest that in certain cases of minimal head injury with ICH, observation is sufficient in lieu of repeat imaging [13].

Hypertensive ICHs occur most commonly in areas where the arteries receive excess shear force due to high-velocity blood flow, namely at early branching points of the proximal arterial system just beyond the circle of Willis or the basilar artery. This leads to hemorrhage in the deep structures in the brain (Figs. 92.3 and 92.4). Unfortunately, given the fascicular organization of the brain, the highly condensed "information highways" that carry the majority of information are located in these deep structures, making hemorrhage in these areas particularly devastating.

CAA should be suspected in older patients with recurrent lobar hemorrhages. Magnetic resonance imaging (MRI) often demonstrates multiple subcortical micro-hemorrhages with dropout regions of blooming on sagittal weighted imaging (SWI) (Figs. 92.5 and 92.6), highly suggestive of amyloid angiopathy. However, CAA may be difficult to see on standard T1 or T2 MRI imaging.

ICH patients suspected of having an underlying vascular lesion should undergo CT angiography (CTA), MR

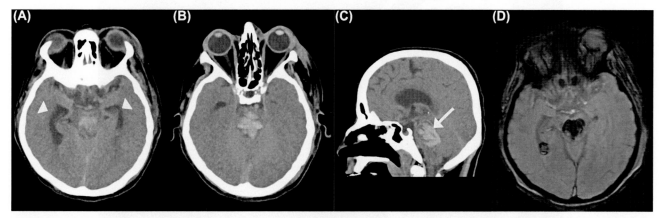

FIGURE 92.3 A woman in the sixth decade of life with a history of chronic hypertension, diabetes mellitus, coronary artery disease status post-coronary artery bypass graft, and on lifetime aspirin 325 mg daily, was found unresponsive with a systolic BP of 220 by first responders. (A) Axial plane noncontrast CT head through the temporal lobes demonstrates pontine hyperdensity consistent with pontine ICH. Prominent bilateral temporal horns (*arrowheads*) are concerning for hydrocephalus. (B) Axial-plane noncontrast CT head through the upper pons better demonstrates the extent of the pontine ICH. (C) Sagittal-plane noncontrast CT head through the midline demonstrates the pontine ICH with obstruction of the fourth ventricle (*arrow*) consistent with the previously observed prominent temporal horns and suspected obstructed hydrocephalus. (D) Axial plane gradient echo MR brain demonstrates signal dropout blooming artifact consistent with pontine ICH as well as right temporal horn blooming artifact suggesting intraventricular extension of hemorrhage.

angiography (MRA), digital subtraction cerebral angiography, or some combination of these studies. Halpin et al. examined the role of angiography in 102 patients presenting with ICH on head CT [14]. CT findings suspicious for an underlying vascular lesion included subarachnoid or intraventricular hemorrhage, abnormal calcification or

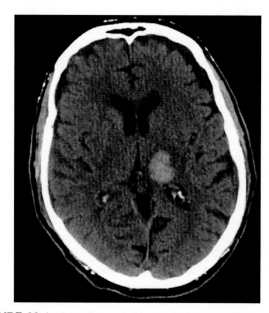

FIGURE 92.4 A gentleman in his eighth decade of life with a history of chronic hypertension presents with progressive lethargy, dysarthria, right-sided weakness and right facial droop. Blood pressure was found to be 190/90 upon arrival to the emergency room. His initial head CT revealed a large left-sided thalamic hemorrhage, likely of hypertensive etiology.

location of hemorrhage, and prominent vascular structures. In 84% of these patients an underlying AVM or aneurysm was demonstrated by angiography. Developmental venous anomalies can occasionally be seen on head CT, or more definitively on CTA, and are associated with cavernous malformations in about 20–25% of cases. MRI/MRA is useful in determining the age of blood products and can detect cavernous malformations, which are occult on cerebral angiography. MRA is also more sensitive than CTA at detecting dural arteriovenous fistulas, due to signal saturation of the skull at the dural surface in CT. If no lesion is seen on the initial round of imaging workup, a delayed contrast MRI is often performed 6–8 weeks following the initial hemorrhage so that after hematoma reabsorption, vascular lesions may be more readily identified. Fig. 92.7 depicts a ruptured mycotic aneurysm, which classically occurs in those with a pathological heart valve and bacteremia. An echocardiogram may show vegetation on the abnormal valve. Mycotic aneurysms are typically located in the distal arterial branches, rather than the proximal vessels seen most often in classic cerebral aneurysms.

A tumor can sometimes be seen on CT scan as a non-hemorrhagic hypodense lesion within the hyperdense ICH. Occasionally, surrounding edema is noted as well. These findings, unusual hemorrhagic locations, or a suspicious history (e.g., a cancer history), warrant an MRI with contrast to rule out tumor as the possible cause of hemorrhage (Fig. 92.8). If no lesion is identified on initial MRI but there remains suspicion for a neoplastic etiology, a repeat MRI should be performed in 6–8 weeks, as acute blood products can obscure an underlying tumor.

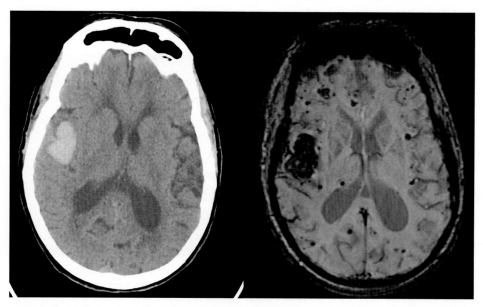

FIGURE 92.5 A woman in the ninth decade of life with a history of atrial fibrillation and deep venous thrombosis treated with Coumadin presented with right facial weakness and slurred speech. (Left) Axial noncontrast CT head demonstrating right sylvian fissure and superior temporal gyrus ICH. There is also demonstration of prominent ventricles and atrophy within the left hemisphere. (Right) Axial gradient echo sequence of brain MRI demonstrating signal dropout blooming within the right temporal lobe consistent with the ICH observed on the CT, but also multifocal disseminated micro-hemorrhages of variable chronicity not visible on CT, consistent with cerebral amyloid angiopathy.

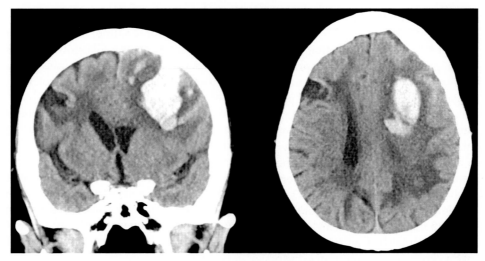

FIGURE 92.6   A woman in the ninth decade of life presents a year after right frontal intraparenchymal hemorrhage with new weakness and aphasia. (Left) Coronal-plane CT head demonstrates left frontal hyperdensity of acute ICH and right frontal hypodensity suggestive of atrophy 12 months after prior ICH. (Right) Axial-plane CT head re-demonstrates bilateral, serial, superficial, unprovoked ICH which should prompt high clinical suspicion for cerebral amyloid angiopathy.

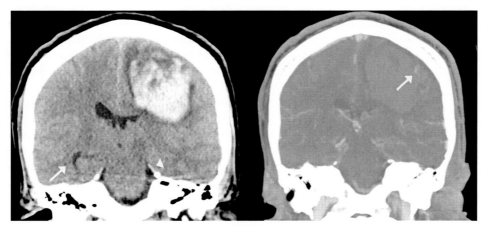

FIGURE 92.7   A gentleman in the fifth decade of life was found unconscious. He has a history of intravenous drug use complicated by endocarditis involving a mechanical aortic valve replacement and indefinite Coumadin anticoagulation for his mechanical valve. (Left) Coronal plane noncontrast CT.

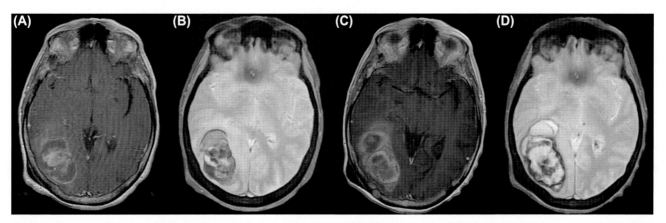

FIGURE 92.8   A woman in the 7th decade of life has a history of non-small cell lung cancer with known metastases to the brain and vertebral column who has undergone whole brain radiation therapy presented with left hand and leg weakness. (A) Axial plane post-contrast T1 MR brain during staging demonstrates a right occipito-parietal rim-enhancing lesion. (B) Axial plane gradient echo MR brain during staging demonstrates intralesional signal drop-out blooming artifact consistent with hemorrhagic neoplasm. (C) Axial plane post-contrast T1 MR during presentation with worsening of symptoms demonstrates mixed T1 signal and increase in volume of the lesion consistent with recurrent hemorrhage of the tumor. (D) Axial plane gradient echo MR brain during worsened symptoms corroborates the T1 MR findings with mixed signal-dropout suggesting blood product of variable chronicity.

MANAGEMENT

Management of ICH is a nuanced topic that requires a complete understanding of the patient's health prior to the inciting event, as well as their prognosis, and desired quality of life. One must gather as much information as possible from the patient, close contacts, neurological examination, laboratory data, and imaging findings when deciding upon management. Unfortunately, ICH is a highly morbid disease with a paucity of strong data supporting current medical and surgical treatment strategies. Hemphill et al. introduced an ICH scoring system in 2001 to predict 30-day mortality (Table 92.3), which can help guide clinical decision making. In 2015 the American Heart Association and the American Stroke published an up-to-date treatment guideline [15] for clinicians. We summarize the major concepts in the following paragraphs.

The initial goal of treatment is to reduce the risk of acute re-bleed and treat any pathology leading to immediate mortality. Blood pressure control and reversal of coagulopathy are priorities in acute hemorrhage. In parallel with these goals, emergent management of ICP is a priority and may include the use of intravenous mannitol or hypertonic saline. In patients with depressed consciousness, such as a Glasgow Coma Scale (GCS) of less than 8, rapid intubation provides both airway protection and a means of respiratory control. With secured mechanical ventilation the patient can be hyperventilated, causing cerebral arterial constriction thereby reducing ICP, and also safely receive sedating and paralyzing medications to decrease cerebral metabolic demand. A bedside external ventricular drain may be necessary in patients with acute hydrocephalus. Decompressive operations are an option in select patients who are acutely decompensating, and can sometimes be combined with clot evacuation. Upon stabilization, admission to a neurological ICU for close observation and medical optimization is warranted.

To reduce the risk of re-bleeding, strict blood pressure parameters are maintained with attention to premorbid perfusion dependency. Maximizing hemostatic ability including optimization of coagulation factors, and platelet count and function are equally important. In patients with thrombocytopenia, or using antiplatelet medications, platelet transfusion helps to form the initial platelet plug that acts as a scaffold for clot formation. Following formation of a platelet plug, coagulation factors have a major role in clot stability. Patients with coagulopathies such as those with liver disease or on anticoagulation may be given exogenous factors through transfusions of fresh frozen plasma, prothrombin concentrate complex, or even fast-acting antibodies that disable the latest generation of anticoagulants. Vitamin K should be given to those with functional livers to maintain coagulation factor production over a sustained time period.

Patients with supratentorial ICH are at increased risk of seizures and some centers will pretreat with anticonvulsant prophylaxis of set duration [16]. In cases of an unexplainable worsening neurological examination, an EEG can be used to diagnose the possibility of seizures. Untreated seizures can lead to increased intracranial pressure, re-rupture of vascular pathologies, or accelerated herniation in mass-occupying lesions.

Surgical management of ICH is a controversial topic, yet the goals of surgery in the acute setting of decompensated ICH are straightforward: decompression, clot removal, and treatment of hydrocephalus if present. During workup of a stable patient, if an underlying lesion such as a vascular malformation or tumor is noted in the region of hemorrhage, the patient may elect to undergo surgery, with the goal being to removal to reduce risk of future hemorrhage. Aneurysmal sources represent the lesions with greatest incidence of acute re-hemorrhage. In contrast, other vascular malformations have only a slight increase in short-term hemorrhage rate and intervention is planned according to the patient's ability to safely undergo surgery.

TABLE 92.3   ICH Score System and Prognosis

Presentation	Value	Score
GCS score	3–4	2
	5–12	1
	13–15	0
Age (years)	≥80	1
	<80	0
Location	Infratentorial	1
	Supratentorial	0
ICH volume (cc)	≥30	1
	<30	0
Intraventricular blood	Present	1
	Absent	0
ICH score = Total # of points	Total points = 0–6	

PROGNOSIS BASED ON ICH SCORE

ICH Score	30-Day Mortality (%)
0	0
1	13
2	26
3	72
4	97
5	100
6	100[a]

[a]*No patients obtained an ICH score of 6 in this study.*
Adapted from Hemphill JC, Bonovich DC, Besmertis L, Manley GT, Johnston SC. The ICH score: a simple, reliable grading scale for intracerebral hemorrhage. Stroke 2001;32(4):891–97.

Although there is limited data on surgical management of ICH, most surgeons prefer conservative management for deep hemorrhages, hemorrhages within vital structures (e.g., brain stem, deep white matter tracts, or basal nuclei), low-volume (<10 mL) hemorrhages, or for patients with minor deficits and no apparent underlying lesion. Patients with supratentorial hemorrhage and an exceedingly poor examination, such as a GCS motor score of 4 or less, are also not thought to be reasonable surgical candidates given a low likelihood of meaningful recovery [21].

There are two major groups of patients who warrant surgical management. Emergent surgery is indicated as a life-saving measure in cases of significant mass effect leading to herniation. Cerebellar hemorrhage >3 cm in diameter with brain stem compression or evolving hydrocephalus should be considered for emergent surgery due to the small volume of the posterior fossa and high risk of tonsillar and/or upward cerebellar herniation with hematoma expansion. Patients with large volume supratentorial hemorrhage causing shift and subsequent herniation at the level of the falx, tentorium, or foramen magnum should also be considered for emergent clot evacuation.

Another group of patients who require surgery are those with accessible lesions who develop delayed neurological deterioration due to hematoma growth. To preserve function, a decompressive operation should be highly considered. For example, a middle-aged patient with putamenal hemorrhage and progressive weakness due to an enlarging hematoma compressing the internal capsule would benefit from decompression. Removal of the hematoma gives an improved likelihood of restoring function.

STITCH I and II are the largest randomized studies on surgical versus medical management of supratentorial ICH [17,18]. The STITCH I trial is a prospective parallel-group trial that compared early craniotomy for hematoma evacuation (<30 h after symptoms) to medical management in 1033 patients. Follow-up focused on mortality, as well as disability based on the GCS. This 5-year trial showed no benefit from early surgery when compared to conservative management. However, there was a trend toward better outcomes in those with superficial hemorrhage. This trial enrolled all patients, including those with devastating hemorrhages (e.g., patients with IVH and hydrocephalus, deep hemorrhage, or massive hemorrhage) that surgeons would usually not operate on due to predictably of poor outcomes. There were also a high percentage of participant crossovers from the medical to the surgical arm upon rapid neurological decline. Due to these criticisms, STITCH II was initiated, almost a decade later, in a select patient group that was more likely to show benefit from surgery. Patients who were enrolled had small, superficial hemorrhage with GCS > 8 and no hydrocephalus or IVH. It was hypothesized that patients with these criteria would have better surgical outcomes; however, the results were once again equivocal. It is likely because the natural course of these selected patients was relatively benign, and these patients would have shown significant improvement with time regardless of management strategy. There is likely an as yet unidentified subgroup of patients who fall between those in the two STITCH trials who would benefit from surgical intervention.

As we continue our attempts to define which patients would benefit from current methods of open surgical intervention of ICH, innovative technologies are simultaneously being developed that aim to make hematoma evacuation more expeditious and less invasive. Endoscopic clot removal devices [19–22], local tissue plasminogen activator (tPA) administration [23], and stereotactic aspiration [24] are some of the novel methods under investigation. With the development of this new frontier of minimally invasive techniques, we hope to decrease surgical morbidity, safely gain access to deeper lesions, and improve patient outcomes.

CONCLUSION

ICH is a vast and nuanced topic with a significant impact on public health due to the associated high morbidity and mortality. It is important to have an in-depth understanding of the presentation and natural course of ICH. With improvements in preventative care and new diagnostic tests, we are reducing the incidence and time to diagnosis in cases of ICH. However, our medical and surgical treatment strategies have not improved significantly, and stubbornly high rates of morbidity and mortality persist in these patients [25]. Continued advancements in our understanding and treatments for ICH are needed to preserve the lives and neurological function of patients who suffer from intracranial hemorrhage.

References

[1] Counsell C, Boonyakarnkul S, Dennis M, et al. Primary intracerebral haemorrhage in the Oxfordshire Community Stroke Project. Cerebrovasc Dis 1995;5(1):26–34.

[2] Furlan AJ, Whisnant JP, Elveback LR. The decreasing incidence of primary intracerebral hemorrhage: a population study. Ann Neurol 1979;5(4):367–73. http://dx.doi.org/10.1002/ana.

[3] Vinters HV, Gilbert JJ. Cerebral amyloid angiopathy: incidence and complications in the aging brain. II. The distribution of amyloid vascular changes. Stroke 1983;14(6):924–8.

[4] O'Donnell HC, Rosand J, Knudsen KA, et al. Apolipoprotein E genotype and the risk of recurrent lobar intracerebral hemorrhage. N Engl J Med 2000;342(4):240–5.

[5] Pollock BE, Flickinger JC, Lunsford LD, Bissonette DJ, Kondziolka D. Factors that predict the bleeding risk of cerebral arteriovenous malformations. Stroke 1996;27(1):1–6.

[6] Kondziolka D, Monaco EA, Lunsford LD. Cavernous malformations and hemorrhage risk. Prog Neurol Surg 2013;27:141–6.

[7] Franke CL, de Jonge J, van Swieten JC, Op de Coul AA, van Gijn J. Intracerebral hematomas during anticoagulant treatment. Stroke 1990;21(5):726–30.

[8] Hart RG, Boop BS, Anderson DC. Oral anticoagulants and intracranial hemorrhage. Facts and hypotheses. Stroke 1995;26(8):1471–7.

[9] Connolly SJ, Ezekowitz MD, Yusuf S, et al. Dabigatran versus warfarin in patients with atrial fibrillation. N Engl J Med 2009;361(12):1139–51. http://dx.doi.org/10.1056/NEJMoa0905561.

[10] Granger CB, Alexander JH, McMurray JJV, et al. Apixaban versus warfarin in patients with atrial fibrillation. N Engl J Med 2011;365(11):981–92.

[11] Brott T, Broderick J, Kothari R, et al. Early hemorrhage growth in patients with intracerebral hemorrhage. Stroke 1997;28(1):1–5.

[12] Fujii Y, Tanaka R, Takeuchi S, Koike T, Minakawa T, Sasaki O. Hematoma enlargement in spontaneous intracerebral hemorrhage. J Neurosurg 1994;80(1):51–7.

[13] Anandalwar SP, Mau CY, Gordhan CG, et al. Eliminating unnecessary routine head CT scanning in neurologically intact mild traumatic brain injury patients: implementation and evaluation of a new protocol. J Neurosurg January 2016:1–7.

[14] Halpin SF, Britton JA, Byrne JV, Clifton A, Hart G, Moore A. Prospective evaluation of cerebral angiography and computed tomography in cerebral haematoma. J Neurol Neurosurg Psychiatry 1994;57(10):1180–6.

[15] Hemphill JC, Greenberg SM, Anderson CS, et al. Guidelines for the Management of Spontaneous Intracerebral Hemorrhage: A Guideline for Healthcare Professionals From the American Heart Association/American Stroke Association. Stroke 2015;46(7):2032–60.

[16] Neshige S, Kuriyama M, Yoshimoto T, et al. Seizures after intracerebral hemorrhage; risk factor, recurrence, efficacy of antiepileptic drug. J Neurol Sci 2015;359(1–2):318–22.

[17] Mendelow AD, Gregson BA, Fernandes HM, et al. Early surgery versus initial conservative treatment in patients with spontaneous supratentorial intracerebral haematomas in the International Surgical Trial in Intracerebral Haemorrhage (STICH): a randomised trial. Lancet (London, Engl) 2005;365(9457):387–97.

[18] Mendelow AD, Gregson BA, Rowan EN, Murray GD, Gholkar A, Mitchell PM. Early surgery versus initial conservative treatment in patients with spontaneous supratentorial lobar intracerebral haematomas (STICH II): a randomised trial. Lancet (London, Engl) 2013;382(9890):397–408.

[19] Tan LA, Lopes DK, Munoz LF, et al. Minimally invasive evacuation of intraventricular hemorrhage with the Apollo vibration/suction device. J Clin Neurosci January 2016.

[20] Spiotta AM, Fiorella D, Vargas J, et al. Initial multicenter technical experience with the Apollo device for minimally invasive intracerebral hematoma evacuation. Neurosurgery 2015;11(Suppl. 2):243–51. discussion 251.

[21] Przybylowski CJ, Ding D, Starke RM, Webster Crowley R, Liu KC. Endoport-assisted surgery for the management of spontaneous intracerebral hemorrhage. J Clin Neurosci 2015;22(11):1727–32.

[22] Zhou X, Chen J, Li Q, Ren G, Yao G, Liu M, et al. Minimally invasive surgery for spontaneous supratentorial intracerebral hemorrhage: a meta-analysis of randomized controlled trials. Stroke 2012;43:2923–30.

[23] Morgan T, Zuccarello M, Narayan R, Keyl P, Lane K, Hanley D. Preliminary findings of the minimally-invasive surgery plus rtPA for intracerebral hemorrhage evacuation (MISTIE) clinical trial. Acta Neurochir Suppl 2008;105:147–51.

[24] Akhigbe T, Okafor U, Sattar T, Rawluk D, Fahey T. Stereotactic-guided evacuation of spontaneous supratentorial intracerebral hemorrhage: systematic review and meta-analysis. World Neurosurg 2015;84(2):451–60.

[25] Zahuranec DB, Lisabeth LD, Sánchez BN, et al. Intracerebral hemorrhage mortality is not changing despite declining incidence. Neurology 2014;82(24):2180–6.

CHAPTER

93

Clinical Aspects of Intraventricular Hemorrhage

B. Daou[1], D. Hasan[2], P. Jabbour[1]

[1]Thomas Jefferson University and Jefferson Hospital for Neuroscience, Philadelphia, PA, United States; [2]University of Iowa Hospital and Clinics, Iowa City, IA, United States

INTRODUCTION

Intraventricular hemorrhage (IVH) is the presence of blood within the ventricular system including the lateral, third and fourth ventricles. Primary IVH refers to bleeding directly into the ventricular system within the brain, from an intraventricular source or a lesion contiguous to the ventricles. Primary IVH is uncommon. Secondary IVH occurs more frequently (70% of IVH patients) and refers to bleeding extending from the parenchyma or subarachnoid space into the ventricular chambers. Once blood is inside the ventricles, it mixes with the cerebrospinal fluid (CSF) and circulates toward the subarachnoid space, which may result in obstructive hydrocephalus and increased intracranial pressure (ICP) that may be associated with significant morbidity or mortality.

ETIOLOGY OF INTRAVENTRICULAR HEMORRHAGE IN INFANTS

Premature infants, especially those born before 32 weeks of gestation and neonates with low birth weight (<1500 g) are at an increased risk of developing IVH compared to full-term infants. The reported rate of IVH is about 20–45% in infant with birth weight less than 1500 g [1–3]. In premature infants, bleeding occurs in small blood vessels in the subependymal or germinal matrix [4]. The pathogenesis of neonatal germinal matrix IVH is thought to result from hemodynamic changes and alterations in cerebral blood flow coupled with impaired autoregulation resulting in disturbance of perfusion to the delicate cellular structures of the germinal matrix located periventricularly and injury of the fragile blood vessels of the germinal matrix [5]. These changes lead to potential hypoxic ischemic encephalopathy, cell death, and subsequent hemorrhage within this injured brain tissue.

IVH in term infants, toddlers, or children is more frequently associated with trauma. It can also be associated with inherited coagulopathies and vascular malformations.

Several additional risk factors may contribute to an increased risk of IVH including: maternal hypertension during pregnancy, maternal chorioamnionitis, preeclampsia, breech presentation, intrapartum asphyxia, malformed or weak blood vessels in the brain, respiratory distress, mechanical ventilation, blood-clotting abnormalities, shaken baby syndrome, and head injury.

CLINICAL PRESENTATION OF INTRAVENTRICULAR HEMORRHAGE IN INFANTS

Neonatal IVH usually occurs in the first 72 h after birth. There are several nonspecific signs and symptoms that suggest IVH in infants, full-term infants, toddlers, or children including lethargy and excessive sleep, decreased feeding, nausea, vomiting, weak suckling, apnea, abnormal eye movement, abnormal and persistent crying, hypotonia, and decreased reflexes. More severe presentations may include seizures, cranial nerve abnormalities, bulging fontanelles, hemodynamic instability, and decreased blood count. On the other hand, IVH in infants can be clinically silent, detected only on routine screening.

Primer on Cerebrovascular Diseases, Second Edition
http://dx.doi.org/10.1016/B978-0-12-803058-5.00093-X

TABLE 93.1   Grading of IVH in Premature Infants [6]

Grade I	Bleeding occurs just in the germinal matrix
Grade II	Bleeding also occurs inside the ventricles without significant enlargement
Grade III	Ventricles are enlarged by the accumulated blood
Grade IV	Bleeding extends into the adjacent/periventricular brain tissue

GRADING OF INTRAVENTRICULAR HEMORRHAGE IN INFANTS

In the preterm infants, IVH is often classified into four grades depending on whether the bleeding is confined to the germinal matrix region or if it extends into the ventricles or brain parenchyma (Table 93.1) [6]. Grades I (mild) and II (moderate) are the most common and are often associated with minimal clinical complications. Grades III and IV IVH have a higher-risk morbidity and mortality, including long-term brain injury and neurodevelopmental impairment [6]. One of the main sequelae of grade III or IV IVH is obstructive hydrocephalus requiring intervention.

DIAGNOSIS OF INTRAVENTRICULAR HEMORRHAGE IN INFANTS

Cranial ultrasound is the method of choice for screening, diagnosis, and follow-up of IVH in infants. These infants require repeated imaging to monitor the progression of IVH and potential hydrocephalus.

In older children, noncontrast head CT is used more frequently to diagnose IVH because of the marked limitation of cranial ultrasounds to give detailed imaging of the brain due to closure of skull sutures and thickening of the skull.

TREATMENT OF INTRAVENTRICULAR HEMORRHAGE IN INFANTS

Several methods have been shown to help in preventing the development of neonatal IVH in preterm infants including antenatal corticosteroids [7], delayed clamping of the umbilical cord [8], and prompt resuscitation to avoid hemodynamic instability and metabolic abnormalities that may impair cerebrovascular autoregulation.

Interventions that could decrease the morbidity and mortality of obstructive hydrocephalus associated with IVH include diuretic therapy [9], repeated lumbar or intraventricular puncture [10], injection of intraventricular thrombolytics [11], or a combination of interventions including drainage using external ventricular drain (EVD), irrigation, and fibrinolytic therapy (DRIFT) [12].

In premature babies with low birth weight (<1500 g), repeated lumbar or mainly ventricular punctures can be employed to decrease the effects of increased ICP secondary to obstructive hydrocephalus. Ventriculoperitoneal shunting is the permanent solution for persistent hydrocephalus secondary to IVH. Shunting can be performed when the infant reaches a reasonable weight (>1.5 kg) [13].

ETIOLOGY OF INTRAVENTRICULAR HEMORRHAGE IN ADULTS

Spontaneous Primary Intraventricular Hemorrhage

Primary IVH confined to the ventricular system accounts for about 30% of all IVH cases [14], and only about 3% of all spontaneous intracerebral hemorrhages (ICHs) [15]. Primary IVH is typically caused by intraventricular aneurysm rupture, small ependymal, subependymal, or choroid plexus vascular malformations (Figs. 93.1 and 93.2) or intra/periventricular neoplasms (Table 93.2).

Intraventricular Hemorrhage Secondary to Intracerebral Hemorrhage

IVH secondary to ICH results from expansion of an existing intraparenchymal hemorrhage into the ventricular system. This is most commonly due to hypertensive hemorrhage, especially basal ganglia hemorrhage and lobar hemorrhage [14]. Deep, subcortical structures tend to be most at risk for IVH, including caudate, putamen, thalamus, pons, and cerebellum locations [16].

Arteriovenous malformations (AVM) (Fig. 93.3) and dural arteriovenous fistulas (dAVFs) (Fig. 93.4) of any size and location with deep cerebral venous drainage can rupture into both adjacent parenchyma and ventricular systems. Less commonly, periventricular AVMs that reach the ventricular wall can rupture primarily into the ventricle. Extension of ICH into the ventricles has been consistently demonstrated as an independent predictor of poor outcome [17].

Intraventricular Hemorrhage Secondary to Aneurysmal Subarachnoid Hemorrhage

In patients with aneurysmal subarachnoid hemorrhage (SAH), the frequency of IVH is reported to be between 28% and 50%, especially with Hunt and Hess grades 3–5 (Fig. 93.5) [18]. IVH is an independent predictor of death and poor outcome after aneurysmal

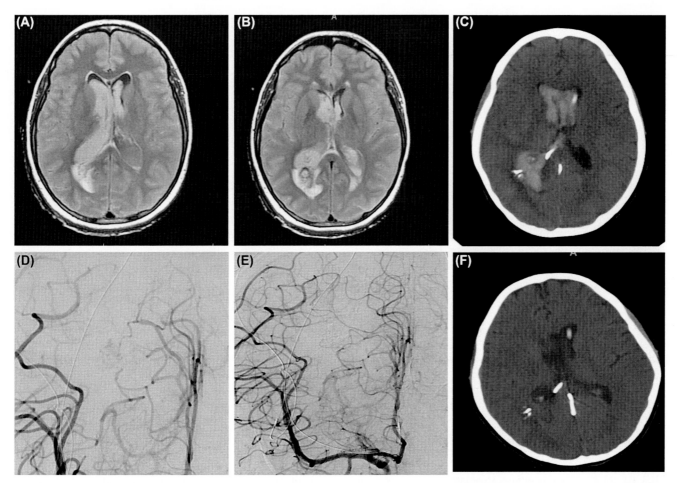

FIGURE 93.1   MRI showing a large amount of intraventricular hemorrhage primarily in the right lateral ventricle, within the third ventricle, fourth ventricle and within the frontal and occipital horns of the left lateral ventricle, with associated subarachnoid hemorrhage (A, B). There is some enlargement of the ventricular system compatible with hydrocephalus. Head CT shows the IVH (C). A left frontal ventriculostomy catheter was placed. Angiogram shows a ventricular AVM fed by distal right PCA feeder and draining into the straight sinus (D). Successful embolization with Onyx was performed (E). Follow-up CT shows that there is improvement in IVH (F).

SAH, with high mortality in these patients [17]. IVH has been associated with secondary complications including hydrocephalus, cerebral ischemia, clinical vasospasm, and poor long-term outcomes [18].

Posterior circulation aneurysms are the most common aneurysm type that results in concomitant IVH due to their anatomical proximity to the fourth ventricular foramina (Fig. 93.6). A large amount of blood in the fourth ventricle with relatively less blood in the suprasellar basal cisterns is especially suggestive of a posterior circulation saccular aneurysm rupture (e.g., posterior inferior cerebellar artery aneurysm), whereas rupture of anterior and posterior communicating artery aneurysms tends to fill the lateral and third ventricles.

Traumatic Intraventricular Hemorrhage

The prevalence of traumatic IVH among patients with blunt head trauma is low, between 0.4% and 4% [19]. However, IVH has been found to occur in 10–35% of moderate to severe traumatic brain injuries (e.g., penetrating trauma such as gunshot wound or severe blunt trauma) [20]. Outcomes of patients with traumatic IVH are poor. In one study, only 30% of patients made a functional recovery [19], and in another study less than half the patients were independent at a 6-month follow-up [21]. Patients with a lower Glasgow Coma Scale (GCS) score on presentation, older patients with traumatic IVH, patients with higher volume of blood and patients with hemorrhage in the third or fourth ventricles tend to have worse outcomes [19,21]. When blood involves the lateral ventricles only, a better outcome is expected. Isolated traumatic IVH is rare. Traumatic IVH usually results from the spread of an adjacent intracerebral hematoma, may be secondary to hypoxia, secondary to coagulopathy, related to stretch- or shear-related tears of ependymal vessels or choroid plexus, or related to diffuse axonal injury along with hemorrhage in the corpus callosum and dorsolateral brain stem [21]. Outcomes in patients with isolated traumatic IVH are better than patients with traumatic IVH and associated brain injury. This suggests

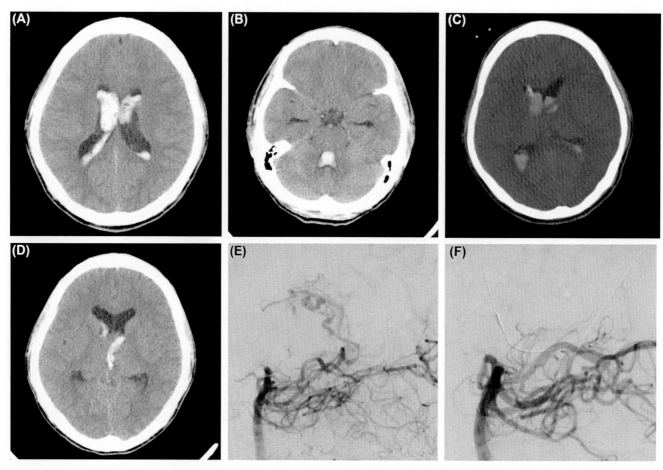

FIGURE 93.2   CT shows diffuse, acute intraventricular hemorrhage within the lateral, third and fourth ventricles and mild hydrocephalus (A, B, C) and a left thalamic subependymal intracerebral hemorrhage (D). A right frontal ventriculostomy catheter was placed. Angiogram shows a tuft of vessels in the ventricular surface, consistent with anterior ventral thalamic/intraventricular AVM with deep venous drainage into the vein of Galen (E). The AVM was embolized using NBCA followed by radiosurgery (F).

TABLE 93.2   Etiology of Intraventricular Hemorrhage

Primary IVH	Secondary IVH
Intraventricular/periventricular tumors Choroid plexus tumors Ependymoma Intraventricular metastases Adjacent parenchymal tumors such as glioblastoma multiforme (GBM) or metastases	Hypertensive intracerebral hemorrhage
Intraventricular/periventricular vascular malformations: Intraventricular aneurysms Arteriovenous malformations Dural arteriovenous fistulas Cavernous malformations especially the ones with subependymal location	*Vascular malformations:* Aneurysms Arteriovenous malformations Dural arteriovenous fistulas Cavernous malformations
Trauma	Trauma
Coagulopathy	Tumors
Moyamoya disease	Coagulopathy
Fibromuscular dysplasia	Vasculitis
Neonatal germinal matrix IVH	Pituitary apoplexy
Idiopathic	Hemorrhagic infarction

that the associated brain injury in these patients is accountable for the poor outcomes rather than the traumatic ventricular hemorrhage. The hemorrhage usually does not occur without extensive associated damage, and so the outcome is rarely good.

SYMPTOMS OF INTRAVENTRICULAR HEMORRHAGE IN ADULTS

Patients presenting with primary IVH often complain of a sudden onset of headache, nausea, and vomiting and occasionally altered mental status and/or level of consciousness. Occasionally, patients may have progressive or fluctuating symptoms. Focal neurological signs are either minimal or absent with primary IVH, and most commonly involve cranial nerve abnormalities [22,23]. Focal or generalized seizures are not typical but may occur. As clinical symptoms and signs of primary IVH are related to a sudden increase in ICP, symptoms can sometimes be minimal if the ventricles have not become distended, and there is no increase in ICP [22,23].

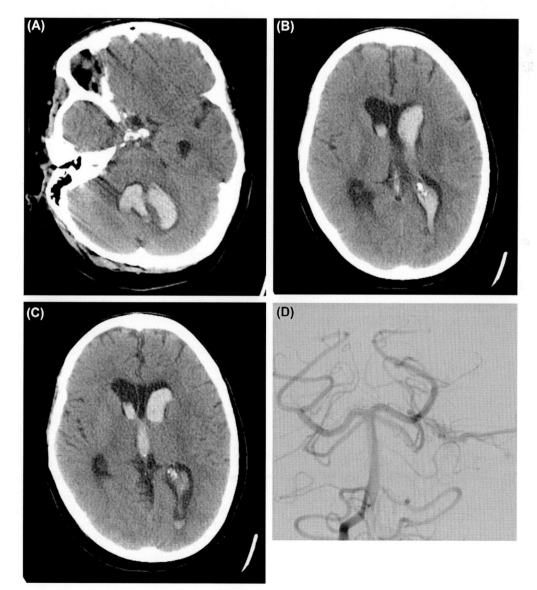

FIGURE 93.3   A 59-year-old female presented with acute onset of headache, nausea, vomiting, and altered mental status. She was found to have an intraventricular hemorrhage with casting of the ventricles associated with mild dilatation of ventricular system and large cerebellar hemorrhage (A, B, C). We placed an emergent ventriculostomy. An angiogram showed an AVM arising from the right posterior inferior cerebellar artery. Patient had a suboccipital craniectomy and resection of the AVM and evacuation of the hemorrhage. The patient had a poor outcome.

FIGURE 93.4   CT showing acute hemorrhage in the deep left cerebellum (A) with intraventricular extension of hemorrhage within the lateral, third, and fourth ventricles and mild hydrocephalus (B, C). A right frontal ventriculostomy catheter was placed. An angiogram was performed and showed a left tentorial dural AVF fed by distal branches of the superior cerebellar artery, and the tentorial artery (D). Successful embolization using Onyx embolic agent was performed. The patient made a tremendous neurological recovery.

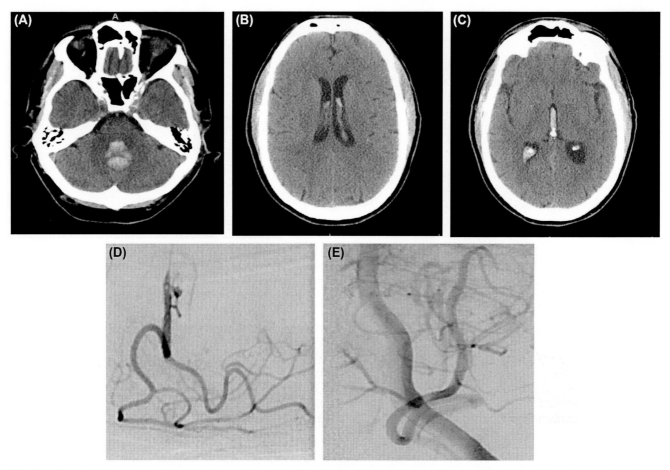

FIGURE 93.5 CT shows a posterior fossa hemorrhage including the fourth ventricle, extending into the third and lateral ventricles. There is hyperdense clot in the fourth ventricle, extending through the cerebral aqueduct into the third ventricle and bilateral lateral ventricles (A, B, C). We performed an angiogram that showed a distal left PICA aneurysm as the cause of hemorrhage (D). The aneurysm was successfully embolized using Onyx (E).

In patients with secondary IVH, neurological symptoms and signs are related to ICH (hemiplegia, hemisensory loss, visual defects, stupor, and coma) or SAH (worst headache of life, nausea, vomiting, and meningismus) and will vary depending upon the location and size of the hemorrhage. Blood clots may obstruct CSF drainage resulting in acute obstructive hydrocephalus, especially with blood in the third or fourth ventricle. Patients may also develop communicating hydrocephalus [24]. One-half to two-thirds of patients with IVH have some degree of hydrocephalus on the initial computerized tomography (CT) scan [22–24]. Symptomatic cerebral vasospasm may occur as well in unusual cases of primary IVH [25].

DIAGNOSIS OF INTRAVENTRICULAR HEMORRHAGE IN ADULTS

Diagnosis of IVH with or without associated ICH, SAH, or hydrocephalus can be confirmed by a noncontrast head CT scan showing presence of blood inside the ventricles. On noncontrast head CT, blood appears hyperdense in the ventricles and tends to pool dependently, best seen in the occipital horns when a small amount of blood is present within the ventricles. Occasionally, the amount of blood in the ventricles is significant with ventricular blood clots resulting in formation of casts of the entire ventricular chambers.

Magnetic resonance imaging (MRI) further helps in identifying the etiology of IVH and is more sensitive than CT in identifying very small amounts of blood in the ventricles. T2 weighted imaging, fluid-attenuated inversion recovery (FLAIR) imaging, and susceptibility weighted imaging (SWI) sequences specifically can be used to diagnose small amount of IVH. Within 48 h from IVH onset, blood will appear as hyperintense relative to the CSF signal on MRI. As time progress, the signal attenuates and blood become more hypointense. If no etiology is identified, and there is no obvious trauma or coagulopathy, catheter angiography can aid in identifying vascular malformations, dAVFs, and aneurysms.

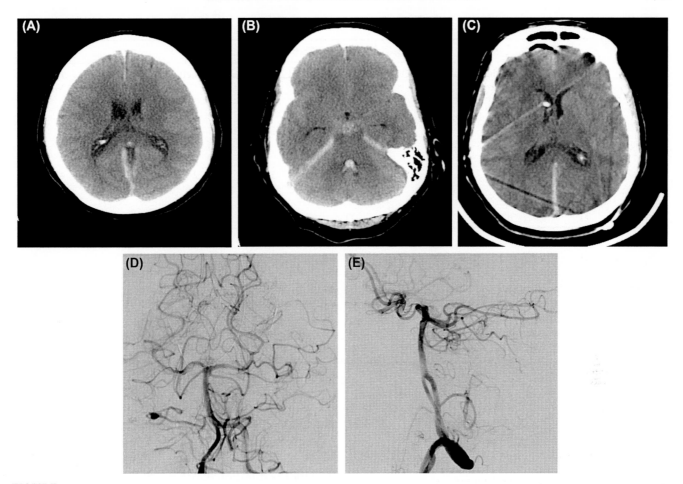

FIGURE 93.6 CT shows intraventricular hemorrhage in all of the ventricles, with the greatest component in the fourth ventricle (A, B, C). Additionally there is a thin, hypodense subdural collection along the left aspect of the falx. There is a cisternal subarachnoid hemorrhage. A right frontal ventriculostomy catheter was placed. Angiogram shows a ruptured distal right anterior inferior cerebellar artery aneurysm (D). The aneurysm was successfully embolized using Onyx (E).

GRADING OF INTRAVENTRICULAR HEMORRHAGE IN ADULTS

Several scoring systems based on CT scans have been implemented to estimate the amount and severity of IVH. The most widely used grading system is the "Graeb score" that is calculated based on the amount of blood in each ventricle separately (Table 93.3) [26]. The maximum Graeb score is 12: grade 1–4 is referred to as mild, 5–8 as moderate, and 9–12 as severe [26]. The Graeb score can be used for predicting patient outcomes following IVH. One study reported that in patients with an ICH volume greater than 60 cm [3], a Graeb score ≥6 was significantly associated with acute hydrocephalus, whereas a score ≤5 was associated with a GCS score ≥12 on admission [27].

The expanded or modified Graeb score was developed to differentiate specific regions of the ventricular system and give a more accurate measure of change in IVH volume over time (Table 93.4) [14,28]. It divides each of the lateral ventricles into compartments: temporal tip, lateral body, and occipital horn in addition

TABLE 93.3 Graeb Score [26]

Lateral ventricles	1. Point for trace blood
	2. Points for less than half the ventricle filled with blood
	3. Points for more than half the ventricle filled with blood
	4. Points for ventricle filled with blood and expanded
Third and fourth ventricles	1. Point for blood present with normal size ventricle
	2. Points for ventricle filled with blood and expanded

to third and fourth ventricles. It also takes into account the estimated IVH volume in each compartment. Extra points are given for expansion of the separate ventricular compartments, thus the maximum score is 32.

Hallevi et al. proposed another IVH scoring system, the IVH score (IVHS) that also assigns a number to each lateral ventricle based on the extent of bleeding (0–3), then adds one point for the presence of hydrocephalus (Table 93.5). The third and fourth ventricles each are

TABLE 93.4   Modified Graeb Score [14]

Blood %	Score for Each Ventricle							
	Right Temporal Tip	Right Lateral Ventricle	Right Occipital Horn	Left Temporal Tip	Left Lateral Ventricle	Left Occipital Horn	Third Ventricle	Fourth Ventricle
None	0	0	0	0	0	0	0	0
≤25	1	1	1	1	1	1	2	2
>25 to ≤50	1	2	1	1	2	1	2	2
>50 to ≤75	2	3	2	2	3	2	4	4
>75 to 100	2	4	2	2	4	2	4	4
Expanded	+1	+1	+1	+1	+1	+1	+1	+1

TABLE 93.5   IVH Score [29]

Lateral ventricles	1. Point for blood filling up to one-third of the ventricle
	2. Points for blood filling one-third to two-thirds of the ventricle
	3. Points for blood filling more than two-thirds of the ventricles
Third and fourth ventricles	0 No blood
	1 Partial or complete filling with blood
Presence of hydrocephalus	0 No
	1 Yes
Formula to calculate score	$IVHS = 3 \times (RV + LV) + III + IV + 3 \times H$
Convert to IVH volume	[IVH volume (mL) $= e^{IVHS/5}$]

III, third ventricle score; *IV*, fourth ventricle score; *H*, hydrocephalus; *IVHS*, intraventricular hemorrhage score; *LV*, left lateral ventricle score; *RV*, right lateral ventricle score.

assigned a score of 0 for no blood and 1 for partial or complete filling with blood. The IVHS ranges from 0 (no IVH) to 23 (all ventricles filled with blood and hydrocephalus present) [29].

TREATMENT OF INTRAVENTRICULAR HEMORRHAGE IN ADULTS

Placement of External Ventricular Drain for Intraventricular Hemorrhage and Obstructive Hydrocephalus

Acute obstructive hydrocephalus with signs and symptoms of increased ICP (altered mental status, loss of consciousness, nausea, and vomiting) following IVH is a life-threatening condition, and requires immediate management. Even though ICP can be managed medically with sedation and osmotic diuretics, medical therapy is often insufficient necessitating the need for EVD placement for drainage of CSF and ICP monitoring.

Usually a right frontal EVD at the Kocher's point is preferred unless specific contraindications are present. Generally, EVD placement is well tolerated, with a relatively low incidence of complications [17]. The most frequent complications are intraparenchymal hemorrhage, catheter-related infection, and obstruction requiring reinsertion [17,30]. The incidence of ventriculostomy-related infections has been reported to be between 0% and 22% [17]. Infection necessitates replacement of the EVD and may prolong hospital stay, antibiotic-associated cost and morbidities, and occasionally life-threatening sequelae [17]. Protocols of prophylactic antibiotic treatment have varied widely, including perioperative use, or continued use for the duration of drainage.

EVD is associated with a 0–33% risk of hemorrhagic complication [17,30]. It is usually associated with multiple passes, use of drill twister, concomitant use of anticoagulation and/or antiplatelet therapy, and bleeding disorders.

Inadequate placement of the EVD catheter tip is not infrequent. It could be as high as 12.3% as determined by postprocedural CT scans [31]. Several techniques could minimize complications including the use of the Ghajar guide, stealth navigating system, and using the appropriate skull landmarks.

EVD catheter occlusion secondary to blood clots usually requires either repeated flushing of the catheter to clear the tubing or EVD replacement using a larger lumen catheter. In a prospective case–control study, 59% of IVH with an EVD required catheter replacement [14,32]. To minimize the risk of catheter obstruction, a second EVD may be placed in the contralateral ventricle or the EVD can be placed contralateral to the ventricle with greatest IVH volume.

In the setting of very large IVH (>40 mL) with casting and mass effect, a single catheter may be ineffective in removing substantial hematoma from the contralateral side, especially if the foramina of Munro is obstructed; in this case, bilateral simultaneous EVD catheters may increase clot resolution [14,33].

If the patient is responsive, follows commands, and head CT shows mild hydrocephalus, the patient can be monitored closely with frequent neurological examinations and head CT to monitor the size of the ventricles [34]. Occasionally, permanent CSF diversion using ventriculoperitoneal shunting is required. It is important to note that placement of an EVD does not eliminate the morbidity and mortality of IVH that is related to underlying damage from the toxic effects of ventricular blood on adjacent periventricular brain tissue [17].

Use of Intraventricular Thrombolysis in Conjunction With External Ventricular Drain

Given the modest effect of EVD on mortality and poor outcomes associated with IVH, more effective strategies are required. In preclinical studies using animal models of IVH, larger volume of blood clot injected into the ventricles and prolonged exposure of the ventricles to blood were associated with a higher risk of pathological changes including subarachnoid fibrosis, extensive ependymal cell loss, and subependymal glial proliferation on the lateral ventricle walls and mortality. These changes were minimized or reversed when lytic therapy (urokinase or tissue plasminogen activator (tPA)) was injected into the ventricles [35–38]. Observational studies that followed, showed that intraventricular thrombolysis was effective in improving IVH clearance, reducing ICP, and decreasing mortality when used in conjunction with EVD [39–41]. A systematic review showed that the fatality rate for conservative treatment in the setting of severe IVH due to SAH or ICH was 78%, 58% with EVD, and 6% with EVD and fibrinolytic therapy [42]. Furthermore, the poor outcome rate for conservative treatment was 90%, 89% with EVD, and 34% with EVD and fibrinolytic agents [42]. Based on these smaller case series and animal models [43], the safety and feasibility of intraventricular thrombolysis was tested in a Phase II prospective randomized clinical trial, which showed that intraventricular thrombolysis with urokinase demonstrated earlier resolution of intraventricular blood clots when compared with EVD treatment alone [44]. However, urokinase was withdrawn from the market because of safety concerns, and recombinant tissue plasminogen activator (rt-PA) became the thrombolytic of choice for IVH. The Clot Lysis Evaluating Accelerated Resolution of Intraventricular Hemorrhage (CLEAR-IVH) trial was a phase 2 randomized clinical trial that included 48 patients treated with 3 mg rt-PA or placebo [45]. Patients treated with rt-PA had a more rapid clearance of IVH and improved outcome as compared to placebo [45]. Blood clot resolution in CSF follows first-order kinetics, and the injection of thrombolytic agents into the ventricular space could increase the rate of clot resolution [44]. The Clot Lysis: Evaluating Accelerated Resolution of Intraventricular Hemorrhage Phase III (CLEAR III), a randomized clinical trial testing the benefit of clot removal for IVH with EVD and rt-PA versus EVD and placebo just closed in 2014 [46]. Preliminary analyses showed that low doses rt-PA (1.0 mg) is safe in patients with stable IVH clots and may increase lysis rates [14].

Open Craniotomy and Surgical Evacuation of Intraventricular Hemorrhage

There have been mixed results with surgical removal of IVH using an open craniotomy with direct surgical evacuation and utilization of minimally invasive instruments (e.g., neuroendoscope) for drainage of IVH [14,47,48]. Zhang et al. compared 22 patients with IVH and less than 30 mL ICH who underwent neuroendoscopic aspiration of IVH within 48 h to a control group of 20 patients with IVH treated with EVD and intraventricular urokinase [49]. More patients in the neuroendoscopy group showed good recovery after 2 months of surgery, whereas the difference in mortality rate between the two groups was not statistically significant [49]. The literature on patients undergoing open craniotomy with direct surgical evacuation of IVH secondary to SAH is mixed [50,51]. The use of neuro-endovascular approaches to extract intraventricular blood has also been described in some cases [52]. The management strategy in cases with secondary IVH should include treatment of both the primary cause of IVH (e.g., aneurysm and AVM) and the potential secondary effect of obstructive hydrocephalus.

PROGNOSIS

Prognosis of patients who develop IVH is variable, extending from a good functional recovery to very poor outcomes and death, depending on the etiology of hemorrhage. Secondary IVH, for example, due to a large hypertensive ICH carries a higher risk of death than primary IVH (up to 80%) [14]. Advanced age, the extent of IVH (high Graeb score), GCS score ≤8, development of obstructive hydrocephalus are also associated with a higher risk of poor outcomes [23,24].

References

[1] Jain NJ, Kruse LK, Demissie K, Khandelwal M. Impact of mode of delivery on neonatal complications: trends between 1997 and 2005. Journal Matern Fetal Neonatal Med June 2009;22(6):491–500.

[2] Wilson-Costello D, Friedman H, Minich N, Fanaroff AA, Hack M. Improved survival rates with increased neurodevelopmental disability for extremely low birth weight infants in the 1990s. Pediatrics April 2005;115(4):997–1003.

[3] Stoll BJ, Hansen NI, Bell EF, et al. Neonatal outcomes of extremely preterm infants from the NICHD Neonatal Research Network. Pediatrics September 2010;126(3):443–56.

[4] Roland EH, Hill A. Germinal matrix-intraventricular hemorrhage in the premature newborn: management and outcome. Neurol Clin November 2003;21(4):833–51. vi–vii.

[5] Soul JS, Hammer PE, Tsuji M, et al. Fluctuating pressure-passivity is common in the cerebral circulation of sick premature infants. Pediatr Res April 2007;61(4):467–73.

[6] Papile LA, Burstein J, Burstein R, Koffler H. Incidence and evolution of subependymal and intraventricular hemorrhage: a study of infants with birth weights less than 1,500gm. J Pediatr April 1978;92(4):529–34.

[7] Crowley P. Prophylactic corticosteroids for preterm birth. Cochrane Database Syst Rev 2000;(2):Cd000065.

[8] Committee Opinion No.543: Timing of umbilical cord clamping after birth. Obstet Gynecol December 2012;120(6):1522–6.

[9] Whitelaw A, Kennedy CR, Brion LP. Diuretic therapy for newborn infants with posthemorrhagic ventricular dilatation. Cochrane Database Syst Rev 2001;(2):Cd002270.

[10] Whitelaw A. Repeated lumbar or ventricular punctures in newborns with intraventricular hemorrhage. Cochrane Database Syst Rev 2001;(1):Cd000216.

[11] Whitelaw A. Intraventricular streptokinase after intraventricular hemorrhage in newborn infants. Cochrane Database Syst Rev 2000;(2):Cd000498.

[12] Whitelaw A, Jary S, Kmita G, et al. Randomized trial of drainage, irrigation and fibrinolytic therapy for premature infants with posthemorrhagic ventricular dilatation: developmental outcome at 2 years. Pediatrics April 2010;125(4):e852–858.

[13] Volpe JJ. Intraventricular hemorrhage in the premature infant–current concepts. Part I. Ann Neurol January 1989;25(1):3–11.

[14] Hinson HE, Hanley DF, Ziai WC. Management of intraventricular hemorrhage. Curr Neurol Neurosci Rep March 2010;10(2):73–82.

[15] Darby DG, Donnan GA, Saling MA, Walsh KW, Bladin PF. Primary intraventricular hemorrhage: clinical and neuropsychological findings in a prospective stroke series. Neurology January 1988;38(1):68–75.

[16] Hallevi H, Albright KC, Aronowski J, et al. Intraventricular hemorrhage: Anatomic relationships and clinical implications. Neurology March 11, 2008;70(11):848–52.

[17] Dey M, Jaffe J, Stadnik A, Awad IA. External ventricular drainage for intraventricular hemorrhage. Curr Neurol Neurosci Rep February 2012;12(1):24–33.

[18] Rosen DS, Macdonald RL, Huo D, et al. Intraventricular hemorrhage from ruptured aneurysm: clinical characteristics, complications, and outcomes in a large, prospective, multicenter study population. J Neurosurg August 2007;107(2):261–5.

[19] Atzema C, Mower WR, Hoffman JR, Holmes JF, Killian AJ, Wolfson AB. Prevalence and prognosis of traumatic intraventricular hemorrhage in patients with blunt head trauma. J Trauma May 2006;60(5):1010–7. discussion 1017.

[20] Barkley J, Morales D, Hayman L, Diaz-Marchan P. Static neuroimaging in the evaluation of TBI. In: Zasler N, Katz D, Zafonte R, editors. Brain injury medicine: principles and practice. Demos Medical Publishing; 2006. p. 140–3.

[21] LeRoux PD, Haglund MM, Newell DW, Grady MS, Winn HR. Intraventricular hemorrhage in blunt head trauma: an analysis of 43 cases. Neurosurgery October 1992;31(4):678–84. discussion 684–675.

[22] Marti-Fabregas J, Piles S, Guardia E, Marti-Vilalta JL. Spontaneous primary intraventricular hemorrhage: clinical data, etiology and outcome. J Neurol April 1999;246(4):287–91.

[23] Flint AC, Roebken A, Singh V. Primary intraventricular hemorrhage: yield of diagnostic angiography and clinical outcome. Neurocrit Care 2008;8(3):330–6.

[24] Passero S, Ulivelli M, Reale F. Primary intraventricular haemorrhage in adults. Acta neurologica Scandinavica February 2002;105(2):115–9.

[25] Gerard E, Frontera JA, Wright CB. Vasospasm and cerebral infarction following isolated intraventricular hemorrhage. Neurocrit Care 2007;7(3):257–9.

[26] Graeb DA, Robertson WD, Lapointe JS, Nugent RA, Harrison PB. Computed tomographic diagnosis of intraventricular hemorrhage. Etiology and prognosis. Radiology April 1982;143(1):91–6.

[27] Nishikawa T, Ueba T, Kajiwara M, Miyamatsu N, Yamashita K. A priority treatment of the intraventricular hemorrhage (IVH) should be performed in the patients suffering intracerebral hemorrhage with large IVH. Clin Neurol Neurosurg June 2009;111(5):450–3.

[28] Morgan TC, Dawson J, Spengler D, et al. The Modified Graeb Score: an enhanced tool for intraventricular hemorrhage measurement and prediction of functional outcome. Stroke March 2013;44(3):635–41.

[29] Hallevi H, Dar NS, Barreto AD, et al. The IVH score: a novel tool for estimating intraventricular hemorrhage volume: clinical and research implications. Crit Care Med March 2009;37(3):969–74. e961.

[30] Dey M, Stadnik A, Riad F, et al. Bleeding and infection with external ventricular drainage: a systematic review in comparison with adjudicated adverse events in the ongoing Clot Lysis Evaluating Accelerated Resolution of Intraventricular Hemorrhage Phase III (CLEAR-III IHV) trial. Neurosurgery March 2015;76(3):291–300. discussion 301.

[31] Saladino A, White JB, Wijdicks EF, Lanzino G. Malplacement of ventricular catheters by neurosurgeons: a single institution experience. Neurocrit Care 2009;10(2):248–52.

[32] Huttner HB, Staykov D, Bardutzky J, et al. Treatment of intraventricular hemorrhage and hydrocephalus. Der Nervenarzt December 2008;79(12):1369–70. 1372–1364, 1376.

[33] Hinson HE, Melnychuk E, Muschelli J, Hanley DF, Awad IA, Ziai WC. Drainage efficiency with dual versus single catheters in severe intraventricular hemorrhage. Neurocrit Care June 2012;16(3):399–405.

[34] Tuhrim S, Horowitz DR, Sacher M, Godbold JH. Volume of ventricular blood is an important determinant of outcome in supratentorial intracerebral hemorrhage. Crit Care Med March 1999;27(3):617–21.

[35] Pang D, Sclabassi RJ, Horton JA. Lysis of intraventricular blood clot with urokinase in a canine model: Part 1. Canine intraventricular blood cast model. Neurosurgery October 1986;19(4):540–6.

[36] Mayfrank L, Kissler J, Raoofi R, et al. Ventricular dilatation in experimental intraventricular hemorrhage in pigs. Characterization of cerebrospinal fluid dynamics and the effects of fibrinolytic treatment. Stroke January 1997;28(1):141–8.

[37] Wagner KR, Xi G, Hua Y, et al. Ultra-early clot aspiration after lysis with tissue plasminogen activator in a porcine model of intracerebral hemorrhage: edema reduction and blood–brain barrier protection. J Neurosurg March 1999;90(3):491–8.

[38] Narayan RK, Narayan TM, Katz DA, Kornblith PL, Murano G. Lysis of intracranial hematomas with urokinase in a rabbit model. J Neurosurg April 1985;62(4):580–6.

[39] Coplin WM, Vinas FC, Agris JM, et al. A cohort study of the safety and feasibility of intraventricular urokinase for nonaneurysmal spontaneous intraventricular hemorrhage. Stroke August 1998;29(8):1573–9.

[40] Findlay JM, Grace MG, Weir BK. Treatment of intraventricular hemorrhage with tissue plasminogen activator. Neurosurgery June 1993;32(6):941–7. discussion 947.

[41] Naff NJ, Carhuapoma JR, Williams MA, et al. Treatment of intraventricular hemorrhage with urokinase: effects on 30-day survival. Stroke April 2000;31(4):841–7.

[42] Nieuwkamp DJ, de Gans K, Rinkel GJ, Algra A. Treatment and outcome of severe intraventricular extension in patients with subarachnoid or intracerebral hemorrhage: a systematic review of the literature. J Neurol February 2000;247(2):117–21.

[43] Dey M, Stadnik A, Awad IA. Thrombolytic evacuation of intracerebral and intraventricular hemorrhage. Curr Cardiol Rep December 2012;14(6):754–60.

[44] Naff NJ, Hanley DF, Keyl PM, et al. Intraventricular thrombolysis speeds blood clot resolution: results of a pilot, prospective, randomized, double-blind, controlled trial. Neurosurgery March 2004;54(3):577–83. discussion 583–574.

[45] Naff N, Williams MA, Keyl PM, et al. Low-dose recombinant tissue-type plasminogen activator enhances clot resolution in brain hemorrhage: the intraventricular hemorrhage thrombolysis trial. Stroke November 2011;42(11):3009–16.

[46] Ziai WC, Tuhrim S, Lane K, et al. A multicenter, randomized, double-blinded, placebo-controlled phase III study of Clot Lysis Evaluation of Accelerated Resolution of Intraventricular Hemorrhage (CLEAR III). Int J Stroke June 2014;9(4):536–42.

[47] Longatti PL, Martinuzzi A, Fiorindi A, Maistrello L, Carteri A. Neuroendoscopic management of intraventricular hemorrhage. Stroke February 2004;35(2):e35–38.

[48] Li Y, Zhang H, Wang X, et al. Neuroendoscopic surgery versus external ventricular drainage alone or with intraventricular fibrinolysis for intraventricular hemorrhage secondary to spontaneous supratentorial hemorrhage: a systematic review and meta-analysis. PLoS One 2013;8(11):e80599.

[49] Zhang Z, Li X, Liu Y, Shao Y, Xu S, Yang Y. Application of neuroendoscopy in the treatment of intraventricular hemorrhage. Cerebrovasc Dis 2007;24(1):91–6.

[50] Lagares A, Putman CM, Ogilvy CS. Posterior fossa decompression and clot evacuation for fourth ventricle hemorrhage after aneurysmal rupture: case report. Neurosurgery July 2001;49(1):208–11.

[51] Shimoda M, Oda S, Shibata M, Tominaga J, Kittaka M, Tsugane R. Results of early surgical evacuation of packed intraventricular hemorrhage from aneurysm rupture in patients with poor-grade subarachnoid hemorrhage. J Neurosurg September 1999;91(3):408–14.

[52] Longatti P, Fiorindi A, Di Paola F, Curtolo S, Basaldella L, Martinuzzi A. Coiling and neuroendoscopy: a new perspective in the treatment of intraventricular haemorrhages due to bleeding aneurysms. J Neurol Neurosurg Psychiatry December 2006;77(12):1354–8.

CHAPTER

94

Clinical Aspects of Subdural Hemorrhage (SDH)

A. Ozpinar, B. Jankowitz

University of Pittsburgh Medical Center, Pittsburgh, PA, United States

ACUTE SUBDURAL HEMATOMA

Epidemiology

Acute subdural hematoma (ASDH) is an intracranial space-occupying lesion that often occurs because of the tearing of bridging or cortical surface veins, secondary to a physical head trauma. ASDHs are often accompanied by cortical contusions, parenchymal hematomas, or global shearing injury such as diffuse axonal injury (DAI). Approximately 10–20% of all patients admitted with a traumatic brain injury (TBI)

have an ASDH. The incidence increases to 60% for patients with a Glasgow Coma Scale (GCS) score of 8 or less. Approximately 70% of such patients are older than 45 years and 40% are older than 65 years, with a male:female ratio of 3:1. As life expectancy rises, the incidence of ASDH is also expected to rise. The Traumatic Coma Data Bank reports that most ASDHs are caused by motor vehicle accidents (MVAs) and falls. MVAs are more frequent in the younger population (15–30 years) and falls are more frequent in the age group of 45–80 years [1,2].

Primer on Cerebrovascular Diseases, Second Edition
http://dx.doi.org/10.1016/B978-0-12-803058-5.00094-1

Physiologic Changes Associated With ASDH

Patients with an ASDH often have markedly decreased cerebral blood flow (CBF) immediately after injury, despite having normal blood pressure and arterial oxygenation. CBF is thought to decrease because of an increase in intracranial pressure (ICP) and a subsequent decrease in cerebral perfusion pressure (CPP). Other local phenomena such as dysautoregulation, cerebral vasospasm, and reduced metabolic demand contribute to decreased CBF and the associated morbidity [1].

Clinical Findings

Symptoms of ASDH can vary depending on the size and associated cerebral injuries. Small ASDHs can present with headache and meningismus. Larger ASDHs often present with altered consciousness, pupillary asymmetry, or hemiparesis. A dilated pupil can be seen on the ipsilateral side of the hematoma as the lesion causes indirect compression of the oculomotor nerve [cranial nerve (CN) III]. Motor findings are usually contralateral but can also be ipsilateral because of the Kernohan notch phenomena. If the hematoma is compressing the ipsilateral cerebral peduncle, motor weakness is typically contralateral. However, if the hematoma compresses the contralateral cerebral peduncle against the tentorium, weakness is on the ipsilateral side. This can be seen radiographically as an indentation in the contralateral cerebral crus by the tentorium. Careful attention should be paid to the elderly patients on anticoagulants, as even a mild head trauma can cause severe delayed ASDHs. Thus, some have suggested that this patient population should be admitted for close monitoring.

Diagnostic Evaluation

The imaging modality of choice for evaluating ASDHs is noncontrast computed tomography (CT) (Fig. 94.1A,B). ASDH has a characteristic hyperdense, crescentic appearance along the convexities of the brain, which can also track along the falx and tentorium. CT can also visualize cerebral edema, intracerebral and subarachnoid blood, midline shift, and the basal cisterns. As seen in parenchymal intracerebral hemorrhage (ICH), studies have demonstrated that contrast extravasation during CT angiography (CTA) is able to stratify patients with ASDH into those at high risk and those at low risk for hematoma expansion and in-hospital mortality [3]. Thus, CTA could be of use when determining surgical candidacy and medical management in an otherwise minimally symptomatic patient. There is no role for MRI in evaluating ASDH.

Treatment and Outcomes

Mortality in all patients with ASDH ranges from 50% to 90%, with the majority of deaths directly attributable to the underlying brain injury. The morbidity associated with ASDH varies drastically with the presenting GCS score, anticoagulant status, and other prognostic indicators such as alcohol use, hypoxia, or hypotension, or difficulty managing ICP [1].

The goal of surgery in ASDH is to remove the mass effect of the blood and prevent secondary injury to the brain (Fig. 94.1D,E). Initial determination of surgical necessity should be driven by evaluation of a patient's GCS score, pupillary examination, and CT results. The Congress of Neurological Surgeons Guidelines for the Surgical Management of Traumatic Brain Injury indicate that any ASDH more than 10 mm in thickness or causing greater than 5 mm of midline shift should be evacuated regardless of the GCS score. Patients with a GCS score of 8 or less should undergo clot removal if the GCS score decreases by 2 or more points by the time of admission, they present with asymmetric or fixed and dilated pupils, or the ICP exceeds 20 mm Hg. Early surgical intervention (within 2–4 h) improves outcome [4].

Poor outcomes following severe head injury have been strongly correlated with patient age, GCS score, and pupillary examination. These associations often deter physicians from more aggressive management of severe injury. For example, patients older than 70 years with GCS score <9 have a 15% chance of a good outcome. These studies also demonstrate that patients older than 75 years who exhibit extensor posturing or unilateral/bilateral fixed and dilated pupils [Glasgow outcome scale (GOS) 3–5] have no significant recovery [5–7].

CHRONIC SUBDURAL HEMATOMAS

Epidemiology

Chronic subdural hematoma (CSDH) is a common neurosurgical disorder and typically result from asymptomatic subdural hematomas that incompletely resorb and slowly grow. They are classically described as being at least 3 weeks old, although the exact age is often difficult to determine. Because of the increased population of elderly individuals and the use of antithrombotic therapies, the incidence of CSDH has been on the rise and is estimated to be 14.0 per 100,000 person-years. History of a head trauma is present in less than 50% of patients, and some risk factors predisposing patients to CSDH include alcohol abuse, epilepsy, and coagulopathies (including anticoagulant therapies). In the elderly population, the mortality rates are comparable to those of hip fractures, with in-house mortality rate being 16.7% and

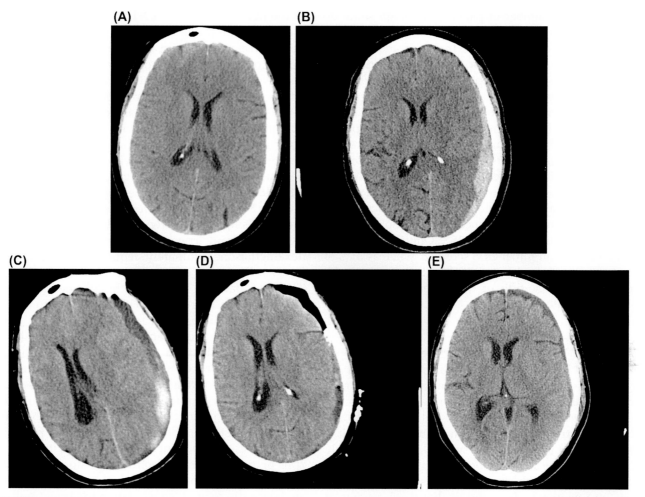

(A) **(B)**

(C) **(D)** **(E)**

FIGURE 94.1 Brain CT images showing evolution of subdural hematoma in a 70-year-old male. (A) One year before fall. (B) Image taken when patient presented with headache after fall from standing. The left acute panhemispheric subdural hematoma (SDH) is causing minimal midline shift. Effacement of the cortical sulci is seen. This SDH was managed conservatively. (C) The patient returns 11 days later with altered mental status. There is now a 1.3 cm ASDH or CSDH causing >1 cm of midline shift. The patient was taken urgently for surgical evacuation with a craniotomy. (D) Immediate postsurgical imaging shows complete resolution of the SDH with lessening of the midline shift. (E) Image taken 6 days after operation. Note the further reexpansion of the brain with visible cortical sulci and near-complete resolution of the midline shift. There is still some mixed-density subdural fluid remaining. Patient was discharged to rehabilitation with baseline mentation and motor strength.

the 6-month and 1-year mortality rates being 26.3% and 32%, respectively [8,9].

Pathogenesis

CSDH occurs in the dural border cell layer, which is a loose cellular layer located between the dura and the arachnoid matter. Multiple processes likely contribute to the development of CSDH. After an initial tear of the subdural bridging vein, a small hematoma forms that dissects into the dural border cell layer to form a subdural hematoma. One mechanism of growth involves the osmotic effect of blood products drawing fluid into the subdural space. Another cause is the repeated minor traumas leading to tearing of additional bridging veins and rebleeding into the same space. A key feature of CSDHs is the formation of vascularized membranes by fibroblasts around the hematoma. These membranes have fragile vessels that are also prone to tearing in response to minor trauma.

Cerebral atrophy and increased venous fragility are often key predisposing factors for CSDH. As the brain ages, its volume decreases and the space between the brain and the dura increases. Consequently, the brain moves more during minor traumas (mild, asymptomatic falls) and the bridging veins in the subdural space are exposed to greater stretching forces.

Clinical Findings

The clinical signs and symptoms of CSDH are often nonspecific. A history of head trauma may be absent. The diagnosis of CSDH is easily missed if there is no high index of suspicion. CSDHs remain asymptomatic in a significant number of patients, although some do present with high ICP, mental status changes, and even coma. In between these two extremes, some common symptoms are sensorimotor, neuropsychiatric, and mood disturbances and simple refractory headaches. In a randomized controlled trial (RCT), patients frequently presented with more than one symptom. Gait difficulties and falls were the most frequent presenting symptoms (57%) followed by extremity weakness (35%), decreased cognition (35%), and confusion (33%). Other presenting symptoms include headache, drowsiness or coma, speech impairment, seizure, incontinence, visual disturbance, and vomiting. Of all the symptoms mentioned, headache is a surprisingly uncommon symptom in the elderly, with an incidence of 12–16%. Only 7% of patients presented with an initial GCS score ≤8, whereas 81% of patients presented with an initial GCS score above 12. Approximately one-fifth of the patients had bilateral subdural hematomas [10].

Diagnostic Evaluation

The most important feature in diagnosing CSDH is a high index of suspicion. This diagnosis should be considered in any patient with a history of head trauma, a change in mental status, and/or focal neurologic deficits. Noncontrast CT is the primary method for diagnosis of CSDH (Fig. 94.1C). A sub-ASDH can be nearly isodense with the adjacent cortex and can be difficult to differentiate from the normal brain tissue. Certain features on imaging such as the medial displacement of the gray matter–white matter interface and failure of the cortical sulci to reach the inner skull table can point toward the presence of an isodense subdural hematoma. CSDHs more than 3 weeks old appear as a hypodense collection along the convexities and are often crescent shaped. They will often have isodense and hyperdense fragments visualized on presentation because of the recurrent hemorrhage from the membranes and bridging veins. As the capsule of a CSDH undergoes neoangiogenesis, it enhances with contrast administration. A mixed-density pattern or the fluid level may indicate rebleeding into a preexisting CSDH or hygroma. MRI is further helpful at discerning a CSDH from a subdural hygroma [11].

Treatment

The goal of treatment in patients with a CSDH is to prevent neurologic decline and avoid recurrent bleeding. Management consists of nonsurgical and surgical approaches. Steroids, antiepileptic drugs (AEDs), and angiotensin-converting enzyme inhibitors (ACEIs) have been the mainstay of nonsurgical treatment options. Corticosteroids are increasingly being used

in the management of CSDH. A study by Glover et al. found that the use of corticosteroids in animal models demonstrated a reduced rate of membrane and vessel formation. Thus, the rationale for corticosteroid use has been due to its anti-inflammatory antiangiogenic effects that will prevent hematoma growth and recurrence. A study by Lopez et al. reported good clinical outcomes in patients treated with steroid monotherapy. They grouped patients into those with a Markwalder Grading Score (MGS) of 1–2 who were assigned to the dexamethasone protocol (4 mg every 8 h, reevaluation after 48–72 h, then slow tapering of dose) and those with an MGS of 3–4 who were assigned to the surgical protocol (single frontal twist drill drainage to a closed system, without irrigation). A total of 22 patients on dexamethasone ultimately required a surgical drain (21.8%). Favorable outcome was obtained in 96% and 93.9% of patients treated with dexamethasone and surgical drain, respectively [12]. However, to date, no randomized studies have been done and the use of corticosteroid monotherapy remains controversial and surgeon dependent. In several studies, prophylactic treatment of seizures with AEDs after evacuation has been shown to decrease the development of seizures. However, other studies have shown no difference in seizure occurrence in control patients and in those who were administered AEDs. It is our practice not to use AEDs unless the patient experiences a witnessed seizure or has abnormal electroencephalographic (EEG) activity. As new pathophysiologic aspects of CSDH are being discovered, conservative treatment will also most certainly evolve. In particular, the identification of angiogenic cytokines that lead to the formation of leaky vessels within the outer membrane of the hematoma may offer a therapeutic target. In limited retrospective studies, the antiangiogenic properties of ACEIs have been shown to decrease the recurrence of CSDH from 18% to 5% and to decrease the levels of vascular endothelial growth factor within the hematoma [13].

Although controversy remains with nonoperative treatment options, there is consensus on the necessity of surgery in patients who present with focal deficits and altered mental status or with a hematoma that is more than 1 cm in thickness. Surgical treatment options include craniotomy, burr hole craniostomy (BHC), and a bedside procedure known as subdural evacuating port system (SEPS) [14].

Craniotomy requires the creation of a larger hole in the skull, typically 3–6 cm in diameter, followed by irrigation and drainage of the hematoma (Fig. 94.1D,E). Indications for a craniotomy include more complicated hematomas such as those with calcification or extensive fibrinous membrane organization. However, hematoma tends to occur more frequently with craniotomy than with BHC [8].

A BHC involves the creation of one or two small holes in the bone, typically 1 cm in diameter, followed by evacuation and irrigation of the hematoma. Currently, BHC is the most widely used technique for evacuating a CSDH. A large RCT found that the efficacy of BHC can be enhanced with the postoperative insertion of a subdural drain for 2 days. The RCT demonstrated a decrease in hematoma recurrence from 24% to 9% and decrease in mortality in the first 6 months following surgery (18–10%) [10,11].

SEPS is a new and less invasive surgical option that can be performed under local anesthesia at the bedside. It effectively involves a small bolt or port screwed into the skull over the largest width of the CSDH, which facilitates gradual drainage. SEPS has been shown to have a lower risk of infection and decreased need for reoperation than other methods [15].

Outcomes

The morbidity and mortality in patients with CSDH range from 0% to 32% and 0% to 25%, respectively. A retrospective study demonstrated that in patients aged 55–74 years, the mortality risk is 17 times higher after developing a CSDH. Age, GCS score at presentation, medical comorbidities, and coagulopathy have all been identified as poor prognostic factors. Morbidity in these patients is related to general and central nervous system complications. The central nervous system complications can take the form of seizures, acute subdural hemorrhage and ICH, tension pneumocephalus, subdural empyema, and wound infection, and some common general complications are pneumonia and venous thromboembolism. Recurrence of CSDH is not a trivial problem, and the recurrence rate ranges from 5% to 30%. It is usually defined as clinical deterioration or failure to improve after primary intervention along with consistent radiologic findings. Placement of a drain and cessation of antithrombotic medication help reduce recurrence.

The question of when to restart anticoagulation after the diagnosis or treatment of a CSDH is a challenging issue. One needs to balance the risk of recurrence with the complications of thromboembolism. Individual patient factors need to be considered, such as indications for anticoagulation, patient's age, postoperative brain expansion, and the likelihood of repeated hematoma. Typically, administration of anticoagulants is withheld for at least a week and restarted only after obtaining radiographic evidence of stability or nonrecurrence.

Overall, appropriate surgical management of CSDHs can significantly improve a patient's neurologic status and likelihood of a favorable long-term outcome.

References

[1] Karibe H, Hayashi T, Hirano T, Kameyama M, Nakagawa A, Tominaga T. Surgical management of traumatic acute subdural hematoma in adults: a review. Neurol Med Chir 2014;54:887–94.

[2] Seeling JM, Becker DP, Miller JD, Greenberg RP, Ward JD, Choi SC. Traumatic acute subdural hematoma: major mortality reduction in comatose patients treated within four hours. N Engl J Med 1981;304:1511–8.

[3] Romero JM, Kelly HR, Delgado Almandoz JE, Hernandez-Siman J, Passanese JC, Lev MH, et al. Contrast extravasation on CT angiography predicts hematoma expansion and mortality in acute traumatic subdural hemorrhage. AJNR Am J Neuroradiol August 2013;34(8):1528–34.

[4] Bullock MR, Chesnut R, Ghajar J, Gordon D, Hartl R, Newell DW, et al. Surgical management of traumatic brain injury author group: surgical management of acute subdural hematomas. Neurosurgery 2006;58:S16–24.

[5] Chesnut RM, Marshall LF, Klauber MR, Blunt BA, Bladwin N, Eisenberg HM, Jane JA, et al. The role of secondary brain injury in determining outcome from severe head injury. J Trauma 1993;34:216–22.

[6] Signorini DF, Andrews PJ, Jones PA, Wardlaw JM, Miller JD. Predicting survival using simple clinical variables. A case study in traumatic brain injury. J Neurol Neurosurg Psychiatr 1999;66:20–5.

[7] Jamjoom A. Justification for evacuating acute subdural hematomas in patients above the age of 75 years. Injury 1992;23:518–20.

[8] Kolias AG, Chari A, Santariusa T, Hutchinson PJ. Chronic subdural haematoma: modern management and emerging therapies. Nat Rev Neurol 2014;10:570–8.

[9] Miranda LB, Braxton E, Hobbs J, Quigley MR. Chronic subdural hematoma in the elderly: not a benign disease. J Neurosurg 2011;114:72–6.

[10] Santarius T, et al. Use of drains versus no drains after burr-hole evacuation of chronic subdural haematoma: a randomised controlled trial. Lancet 2009;374:1067–73.

[11] Santarius T, Kolias AG, Hutchinson PJ. In: Quinones-Hinojosa A, editor. Schmidek and Sweet: operative neurosurgical techniques: indications, methods and results. Saunders; 2012. p. 1573–8.

[12] Delgado-Lopez PD, et al. Dexamethasone treatment in chronic subdural haematoma. Neurocirugia (Astur) 2009;20:346–59.

[13] Weigel R, Hohenstein A, Schlickum L, Weiss C, Schilling L. Angiotensin converting enzyme inhibition for arterial hypertension reduces the risk of recurrence in patients with chronic subdural hematoma possibly by an antiangiogenic mechanism. Neurosurgery 2007;61:788–92.

[14] Ducruet AF, et al. The surgical management of chronic subdural hematoma. Neurosurg Rev 2012;35:155–69.

[15] Balser D, Rodgers SD, Johnson B, Shi C, Tabak E, Samadani U. Evolving management of symptomatic chronic subdural hematoma: experience of a single institution and review of the literature. Neurol Res 2013;35(3):233–42.

CHAPTER

95

Cerebral Venous Thrombosis

P. Canhão, J.M. Ferro

University of Lisbon, Lisbon, Portugal

INTRODUCTION

Cerebral venous thrombosis (CVT) accounts for 0.5–1% of all strokes. CVT is more frequent in children and young adults and affects women three times more than males. Its clinical presentation is diverse and diagnosis is often challenging.

CVT may arise in several clinical settings, such as pregnancy or puerperium, malignancy, head or neck infections, inflammatory disorders, and after some diagnostic and therapeutic procedures.

CVT is now recognized with higher frequency because of increased awareness among neurologists and other physicians and a wider access to magnetic resonance imaging (MRI) or computed tomography (CT) venography, which are required to establish the diagnosis of CVT.

Most patients have a good prognosis; however, about 5% may die in the acute phase, and on long-term follow-up, 15% may be left dependent or may die. Some clinical and imaging features have been described to identify patients at risk of a poor prognosis, who deserve more intensive surveillance.

EPIDEMIOLOGY

In a nationwide hospital-based series in Portugal the incidence of CVT was estimated to be 0.22/100,000 per year [1]. In a cross-sectional study in two provinces in the Netherlands, the overall incidence of CVT was 1.32/100,000 per year [2]. The incidence of CVT is higher in developing countries and in children. In the Canadian registry of CVT, in infants and children younger than 18 years, the incidence of CVT was 0.67/100,000 [3].

Among children, CVT is more common in neonates than in older children. In adults, CVT affects predominantly young female patients, and the incidence apparently decreases in older subjects. In the International Study on Cerebral Vein and Dural Sinus Thrombosis (ISCVT), the largest international cohort of adult patients with CVT, the median age of the patients studied was 37 years and less than 10% of the patients were older than 65 years [4].

ETIOLOGY

CVT may occur in the course of many diseases or may be precipitated by a plethora of conditions (Table 95.1). At least one risk factor for CVT can be identified in more than 85% of adult patients, and multiple factors may be found in half [4].

Prothrombotic conditions, inherited or acquired, were found in 34% of patients in the ISCVT. Inherited factors, such as factor V Leiden, prothrombin G20210A polymorphism, and protein S, protein C, and antithrombin deficiencies were detected in up to 22% of patients [4]. They may be the single cause of CVT or may increase the individual risk of the patient with other predisposing conditions. In pediatric patients, prothrombotic disorders, in many cases caused by an acquired disorder, are also common and were found in 32% of patients in the Canadian registry of CVT [3].

Systemic and nervous system infections are responsible for about 10% of CVT [4] and may be more significant in developing countries [5]. The most common precipitating infections are otitis, mastoiditis, and sinusitis. Facial cutaneous infections may cause cavernous sinus thrombosis. Head and neck infections were a risk factor for CVT in 18% of children and were predominant in preschool children [3].

Malignancy accounted for 7.4% of CVT in the ISCVT. Sinus or vein thrombosis can result from local compression or invasion by a tumor, can be caused

TABLE 95.1   Risk Factors Associated With Cerebral Venous Thrombosis

PERMANENT RISK FACTORS

Thrombophilia conditions

  Antithrombin, protein C, and protein S deficiency

  Factor V Leiden; prothrombin G20210A mutation

  Antiphospholipid antibody syndrome

  Homocystinuria, hyperhomocysteinemia

  Nephrotic syndrome

Neoplasia

  Central nervous system (e.g., meningioma)

  Solid tumor outside the central nervous system

  Hematologic (leukemia, lymphoma)

Hematologic disorders

  Paroxysmal nocturnal hemoglobinuria

  Sickle cell disease

  Severe anemia

  Polycythemia (primary or secondary)

  Thrombocythemia (primary or secondary)

Inflammatory diseases

  Behçet disease

  Systemic lupus erythematosus

  Inflammatory bowel disease

  Primary vasculitis

  Sarcoidosis

Other

  Dural fistula

  Congenital heart disease

  Thyroid disease (hyper- or hypothyroidism)

TRANSIENT RISK FACTORS

Pregnancy and puerperium

Infection

  Central nervous system (meningitis, abscess, empyema)

  Ear, sinus, mouth, face, and neck

  Systemic (sepsis, endocarditis, tuberculosis, HIV)

Drugs

  Oral contraceptives, hormone replacement therapy

  L-asparaginase, cyclosporine, tamoxifen, medroxyprogesterone acetate, anabolic steroids, glucocorticoids, thalidomide, epoetin alfa therapy, IV immunoglobulin

TABLE 95.1   Risk Factors Associated With Cerebral Venous Thrombosis—cont'd

TRANSIENT RISK FACTORS

Mechanical precipitants

  Head trauma

  Lumbar puncture, spinal-epidural anesthesia, myelography

  Neurosurgical procedures

  Jugular vein catheter

Other disorders

  Dehydration

  Diabetic ketoacidosis

by a prothrombotic state underlying cancer, can be drug related (e.g., L-asparaginase, steroids, hormonal treatment, angiogenesis inhibitors), can be associated with interventions or surgical procedures (e.g., jugular catheters, irradiation, neurosurgery), and, less commonly, may be associated with local or systemic infections. Cancer may also be a risk factor for CVT in children, particularly in non-neonates, and was described in 13% of patients [3].

In neonates, acute systemic illnesses are the age-dependent cause of CVT. They were present in 84% of neonates; the most frequent illnesses were perinatal complications (in 51%), of which hypoxic encephalopathy was most common, and dehydration (in 30%).

A gender-specific risk factor was present in 65% of women in the ISCVT. In fertile women, the most significant risk factors were oral contraceptives, in 54%; pregnancy, in 6%; and puerperium, in 14%.

Despite a comprehensive investigation, in almost 15% of adult patients, a cause could not be identified. Those patients need to be followed up because CVT can be the first manifestation of antiphospholipid antibody syndrome, Behçet disease, myeloproliferative disorders, intestinal inflammatory disease, and cancer.

CLINICAL PRESENTATION

CVT presents with a wide spectrum of symptoms and modes of onset. Over half of the patients have a subacute onset, and a few patients have a chronic protracted course until the diagnosis. An acute onset, the hallmark of arterial stroke, is found in only one-third of the patients. The most common symptoms and signs are headache, papilledema, seizures, focal neurologic deficits, and altered mental state and decreased alertness [6].

Headache occurs in almost 90% of patients. It may be localized or diffuse and usually has a gradual onset, increasing over several days. CVT may lead to increased intracranial pressure (ICP), and in those cases, headache is typically severe, dull, generalized, and exacerbates with Valsalva maneuvers and with decubitus. Visual obscurations may occur, coinciding with bouts of increased headache intensity, and patients may complain of visual loss or diplopia, nausea, or vomiting. Papilledema may be present. A few patients with CVT have a sudden explosive onset of severe head pain (i.e., thunderclap headache) that mimics subarachnoid hemorrhage. In a few other patients, headache may resemble migraine with aura. Typically, headache coexists with other symptoms and signs but may be a sole symptom in up to 9% of patients with CVT.

Motor weakness with monoparesis or hemiparesis, sometimes bilateral, is the most frequent focal sign in CVT, occurring in over one-third of patients. Aphasia occurs in almost 20% of patients, whereas sensory symptoms and visual field defects are less frequent. Seizures are more common in CVT than in other types of strokes, affecting almost 40% of patients. Severe cases of CVT may present with dysexecutive syndrome, delirium, decreased alertness, or coma, which may result from extensive dural sinus thromboses or thrombosis of the deep venous system.

CVT symptoms and signs can be combined in three major presentation syndromes: (1) isolated intracranial hypertension syndrome, (2) focal syndrome (focal deficits, seizures, or both), and (3) encephalopathy (bilateral or multifocal signs, mental status changes, stupor, or coma). Rarely, CVT presents as multiple cranial nerve palsies. Examples are cavernous sinus thrombosis (featuring oculomotor palsies, orbital pain, proptosis, and chemosis) and the Collet-Sicard syndrome with multiple lower cranial palsies in lateral sinus or jugular vein thrombosis.

DIAGNOSIS

The confirmation of the diagnosis of CVT relies on the detection of thrombi in the dural sinuses or cerebral veins.

CT scan is usually the first examination performed, essentially to exclude other diseases, such as brain tumor, infection, or subdural hematoma. CT scan results are normal in up to one-third of patients with CVT, and most of the findings are nonspecific. Direct signs of CVT, present in about one-third of cases, correspond to the visualization of the thrombus (Fig. 95.1): the cord sign (thrombosed vein); the dense triangle sign (thrombus in the sinus), and the empty delta sign (the nonopacified thrombus surrounded by the contrast-enhanced collateral veins of the sinus wall). Brain lesions can be visualized in 60–80% of patients and include hypodense lesions not respecting an arterial territory, hemorrhagic infarctions, or diffuse brain edema (Fig. 95.2). Some patterns may be evocative of CVT, such as multiple bilateral lesions, bilateral thalamic edema, small juxtacortical hemorrhages, and convexity subarachnoid hemorrhage.

MRI in combination with MR venography or CT in combination with CT venography are reliable alternatives to arterial angiography (IA) angiography to confirm the diagnosis of CVT. Cerebral IA angiography is rarely required nowadays, as it is used in the diagnosis of cases

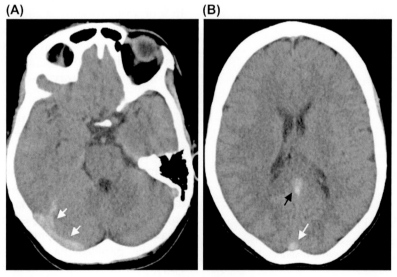

(A)   **(B)**

FIGURE 95.1   Noncontrast head scan CT shows hyperdensity of (A) the right transverse sinus (*white arrows*) and (B) the superior sagittal sinus (*white arrow*) and straight sinus (*black arrow*).

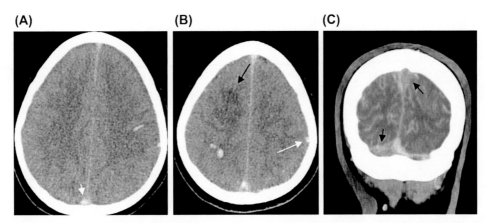

FIGURE 95.2   (A,B) Noncontrast head CT scan shows bilateral small juxtacortical hemorrhages, a decreased attenuation at the right frontal lobe (*black arrow*), the dense triangle sign (*small white arrow*), and the cord sign (*white arrow*). (C) CT venogram showing low-density contrast (*arrows*) in nonperfusing thrombosed segments.

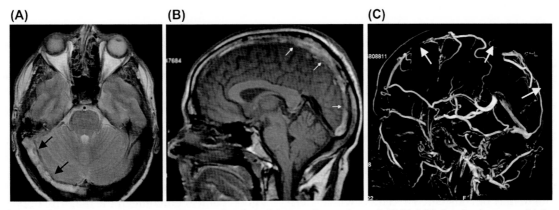

FIGURE 95.3   (A) T2 MRI showing hyperintensity signal at the right sigmoid sinus (*arrows*). (B) T1 MRI showing hypersensitivity signal at the superior sagittal sinus. (C) Magnetic resonance venogram showing thrombosis of the superior sagittal sinus (*arrows*).

where other imaging methods are doubtful, in severe cases when urgent endovascular treatment is planned, and to exclude dural sinus fistulae as the cause of CVT.

The combination of an abnormal MR signal in a sinus and a corresponding absence of flow on MR venography support the diagnosis of CVT (Fig. 95.3). T2*-weighted MRI is the sequence of choice to diagnose thrombosis of a cortical vein, where the clot can be visualized as strongly hypointense in the affected vein (Fig. 95.4). MRI is also useful to depict the extension of parenchymal lesions: nonhemorrhagic (venous infarcts and brain edema) and hemorrhagic lesions. In contrast to arterial stroke, venous strokes appear larger in fluid-attenuated inversion recovery (FLAIR) than in diffusion-weighted imaging (DWI) sequences (Fig. 95.4).

MR venography and CT venography have several limitations and potential pitfalls in the diagnosis of cortical vein thrombosis and partial sinus occlusion and in the distinction between dural sinus hypoplasia and thrombosis. Clinical judgment and the combination of several imaging methods are often required to establish the diagnosis.

D-dimer values are increased in many patients with CVT. However, D-dimer values may be normal in patients with isolated headache, thrombosis of a single sinus or vein, or longer duration of symptoms [7,8].

TREATMENT

Besides treatment of the cause of CVT, treatment of CVT includes antithrombotic treatment and symptomatic treatment.

Antithrombotic Treatment

Anticoagulation is the mainstay therapy for patients with recent CVT and should be started as soon as the diagnosis is confirmed. Two randomized control trials and a meta-analysis of these trials found that when compared with placebo, anticoagulant treatment was associated with a pooled relative risk of death of 0.33 [95% confidence interval (CI), 0.08–1.21] and a risk of death or dependency of 0.46 (95% CI, 0.16–1.31). Although these trials did not achieve statistical significance, there is a broad consensus concerning the use of anticoagulation. Guidelines recommend patients with CVT, with no

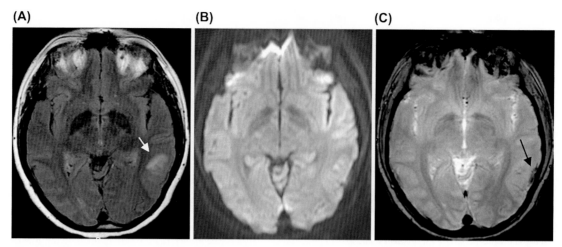

FIGURE 95.4 (A) MRI FLAIR sequence showing high-intensity venous infarct in the left temporal lobe (*arrow*). (B) The diffusion-weighted imaging sequence does not show changes in signal intensity in the corresponding area. (C) T2*-weighted MRI showing a linear hypointense signal in a cortical thrombosed vein (*arrow*).

contraindications to anticoagulation, to be treated either with intravenous (IV) heparin adjusted for activated partial thromboplastin time (aPTT) or with body-weight-adjusted low-molecular-weight heparin (LMWH), even in the presence of cerebral hemorrhagic venous infarcts [7,8]. Subcutaneous LMWH in therapeutic dosage has a better safety profile and is preferable to aPTT-adjusted IV heparin.

For patients who develop progressive neurologic worsening despite adequate anticoagulation, endovascular treatment may be considered at centers with experience with this therapy [7,8]. The efficacy of endovascular treatment, with mechanical and/or pharmacologic thrombolysis, is only based on small case series. Patients with large infarcts and impending herniation may not benefit with local thrombolysis. They may instead need decompressive hemicraniectomy.

Once the patient is stable, oral anticoagulant treatment can be started. The optimal duration of anticoagulation after CVT is unknown. The European Federation of Neurological Societies (EFNS) and the American Heart Association/American Stroke Association (AHA/ASA) guidelines suggest that when CVT is related to a transient risk factor, anticoagulants may be used for a short period (3–6 months). In patients with idiopathic CVT or CVT associated with mild thrombophilia, the period of anticoagulation must be longer (6–12 months). Patients with combined or severe thrombophilia (e.g., two or more prothrombotic abnormalities; antithrombin, protein C or S deficiency, antiphospholipid syndrome, cancer), or recurrent venous thrombosis, need permanent anticoagulation.

Symptomatic Treatment

Treatment of Intracranial Hypertension

In the acute phase, single or multiple large infarcts or hemorrhagic lesions, or even diffuse brain edema, may cause a severe increase in ICP. General recommendations to control acutely increased ICP include elevating the head of the bed, using osmotic diuretics, such as mannitol, intensive care unit admission with sedation, mechanical ventilation, and ICP monitoring. Patients whose condition deteriorates because of large space-occupying venous infarcts have a high risk of dying from cerebral herniation. In such cases an emergent decompressive hemicraniectomy may be lifesaving [7,8]. Experience from different centers with this procedure shows a good clinical outcome in the majority of patients.

Patients who develop chronic intracranial hypertension should be treated to improve headache and prevent visual loss. Although its efficacy is not proved, administration of acetazolamide or furosemide is a therapeutic option. If severe headaches continue or if vision is threatened, repeated lumbar punctures may be required, and in refractory cases, cerebrospinal fluid (CSF) shunting or fenestration of the optic nerve sheath can be considered.

Treatment and Prevention of Seizures

Antiepileptic medication may be prescribed in patients presenting with seizures and focal cerebral supratentorial lesions on admission CT/MRI because recurrent seizures are more likely to develop in such patients. The optimal duration of antiepileptic drug treatment after CVT is unknown, but it may be reasonable to continue antiepileptic drug treatment for 1 year in patients with early seizures and hemorrhagic lesions on admission brain scan [7].

PROGNOSIS

Several prospective series demonstrated that the prognosis of CVT is much better than other types of stroke. In the ISCVT, 79% of patients recovered completely. The pooled rate of death or dependency at the end of follow-up in prospective studies was 15% [4]. In a meta-analysis of prospective and retrospective studies, the overall rate of complete recovery was 88%; acute death, 5.6%; and death at the end of follow-up, 9.4% [9].

The clinical course of CVT is often unpredictable. In the ISCVT, about one-fourth of the patients deteriorated after admission, sometimes several days after the diagnosis. Worsening could include decreased alertness, mental state disturbance, new seizures, worsening of a previous symptom or a new focal deficit, increase in headache severity, or visual loss.

Several predictors of poor long-term prognosis were derived from the ISCVT cohort: age (older than 37 years), male gender, mental status disorder, coma, brain hemorrhage on admission CT/MRI, deep cerebral venous system thrombosis, CNS infection, and malignancy [4]. About 4% of patients died in the acute phase. The main cause of acute death is transtentorial herniation secondary to a large hemorrhagic lesion. Other causes of death are herniation caused by multiple lesions or diffuse brain edema, status epilepticus, medical complications, and pulmonary embolism.

Patients surviving the acute phase of CVT may present late complications. Headaches distress 14% of patients severely enough to require bed rest or hospital admission. MRI and MR venography may be needed to exclude recurrent CVT. Seizures may occur in up to 11% of patients, being more frequent in patients who had acute-phase seizures or acute-phase motor deficits, and who presented with supratentorial hemorrhagic brain lesion on the admission CT/MRI. Thrombotic events, specifically deep venous thrombosis of the limbs or pelvis and pulmonary embolism, occur in about 5% of patients. Recurrent CVT is rare, and sometimes difficult to document. It is useful to perform an MRI/MR angiography 3–6 months after CVT occurrence to document the extent of recanalization. If new symptoms suggesting CVT recurrence occur, MRI and MR venography should be performed and the images compared with previous ones. Severe visual loss caused by intracranial hypertension is very rare, but patients need careful visual monitoring to prevent it. About half of the survivors of CVT feel depressed or anxious and demonstrate minor cognitive or language deficits, which may prevent resumption of previous levels of professional activity.

References

[1] Ferro JM, Correia M, Pontes C, Baptista MV, Pita F, Cerebral Venous Thrombosis Portuguese Collaborative Study Group (Venoport). Cerebral vein and dural sinus thrombosis in Portugal: 1980-1998. Cerebrovasc Dis 2001;11:177–82.

[2] Coutinho JM, Zuurbier SM, Aramideh M, Stam J. The incidence of cerebral venous thrombosis: a cross-sectional study. Stroke 2012;43:3375–7.

[3] deVeber G, Andrew M, Adams C, Bjornson B, Booth F, Buckley DJ, et al. Cerebral sinovenous thrombosis in children. N Eng J Med 2001;345:417–23.

[4] Ferro JM, Canhão P, Stam J, Bousser MG, Barinagarrementeria F, ISCVT Investigators. Prognosis of cerebral vein and dural sinus thrombosis: results of the international study on cerebral vein and dural sinus thrombosis (ISCVT). Stroke 2004;35:664–70.

[5] Khealani BA, Wasay M, Saadah M, Sultana E, Mustafa S, Khan FS, et al. Cerebral venous thrombosis: a descriptive multicenter study of patients in Pakistan and Middle East. Stroke 2008;39:2707–11.

[6] Bousser MG, Ferro JM. Cerebral venous thrombosis: an update. Lancet Neurol 2007;6:162–70.

[7] Einhäupl K, Stam J, Bousser MG, De Bruijn SF, Ferro JM, Martinelli I, et al. EFNS guideline on the treatment of cerebral venous and sinus thrombosis in adult patients. Eur J Neurol 2010;17:1229–35.

[8] Saposnik G, Barinagarrementeria F, Brown Jr RD, Bushnell CD, Cucchiara B, Cushman M, et al. Diagnosis and management of cerebral venous thrombosis: a statement for healthcare professionals from the American Heart Association/American Stroke Association. Stroke 2011;42:1158–92.

[9] Dentali F, Gianni M, Crowther MA, Ageno W. Natural history of cerebral vein thrombosis: a systematic review. Blood 2006;108:1129–34.

CHAPTER

96

Developmental Venous Anomalies

A.P. See[1], M.A. Silva[2], M.A. Aziz-Sultan[2]

[1]Brigham and Women's Hospital; Boston Children's Hospital/Harvard Medical School, Boston, MA, United States;
[2]Harvard Medical School, Boston, MA, United States

INTRODUCTION

Developmental venous anomalies (DVAs) are rare, usually asymptomatic congenital cerebrovascular malformations consisting of a radial network of medullary veins draining into a dilated central channel. Originally called venous angiomas, DVAs are functional variants of normal venous anatomy that drain brain parenchyma, which is better described by the DVA nomenclature [1,2]. DVAs have a 2.5% incidence in postmortem studies and are the most common developmental vascular malformation [3,4]. There are no known genetic risk factors, but DVAs are associated with cavernous malformations [5–7]. Up to 33% of patients with DVAs are comorbid for cavernous malformations [8]. If a DVA possesses features of other vascular anomalies, it is considered a "mixed" vascular malformation [5].

DVAs are thought to be congenital remnants of fetal venous vasculature. They likely arise due to an arrest of venous development and persistence of fetal medullary veins and the large collecting trunk into which they drain during 6–12 weeks of gestation [5,9]. Venous drainage of the brain parenchyma can be conceptualized as two primary cortical periependymal routes consisting of superficial pial veins and deep medullary veins [1]. DVAs serve as the sole venous drainage for regions of brain tissue that lack pial or medullary venous drainage, draining into normal collecting veins outside of the parenchyma [1,4,10]. It has been proposed that recurrent subclinical microhemorrhages from a DVA can promote angiogenesis and predispose the patient to cavernous malformations [6,11].

RADIOPATHOLOGICAL CHARACTERISTICS

The characteristic *caput medusae* (so called because of its resemblance to the head of the Greek god,

Medusa) appearance of a DVA can be seen using contrast angiography (Fig. 96.1A–B) [12]. DVAs consist of thin-walled, dilated medullary veins draining into a large central trunk lacking a smooth muscle or elastic layer and without direct arterial supply [13,14]. The collecting vein can be identified by contrast-enhanced computed tomography (CT) (Fig. 96 1C–D), but is best visualized using gadolinium-enhanced magnetic resonance imaging (MRI) (Fig. 96.1E–H), which would demonstrate trans-cerebral linear flow voids with radial extensions on T2 sequences [8–10,12]. Magnetic resonance (MR) angiography is particularly useful for identifying associated malformations, such as cavernous malformations, which typically appear in close proximity to the DVA [9]. Diagnostic cerebral angiography is typically not applied for further characterization of lesions identified on CT or MR, especially since the associated cavernoma is angiographically occult.

CLASSIFICATION AND KEY DESCRIPTIVE ATTRIBUTES

Because DVAs are normal anatomical variants that are functionally required by the parenchymal tissue they drain, it is important to classify them based on their drainage territory. Most DVAs are supratentorial, appearing most often in the frontal lobe, but one-third are seen in the cerebellum and brainstem (Fig. 96.2) [1,8,10]. Supratentorial DVAs can either drain paraventricularly into superficial cortical veins or subcortically into deep subependymal veins [1,9]. The corpus callosum is the most common site of supratentorial DVAs [1]. A sinus pericranii may also be thought of as an extradural variant of a DVA that drains into the midline dural sinus [1]. DVAs can range from small, focal lesions to large malformations that cover an entire hemisphere [9].

Primer on Cerebrovascular Diseases, Second Edition
http://dx.doi.org/10.1016/B978-0-12-803058-5.00096-5

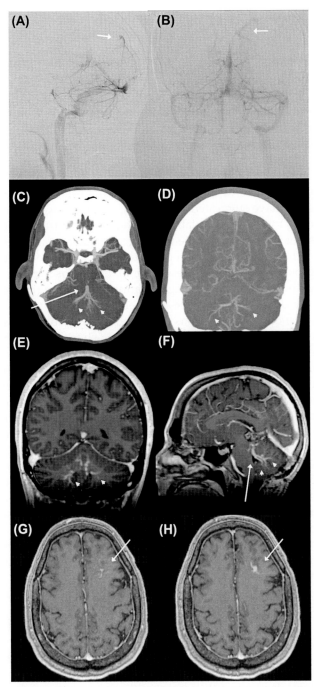

FIGURE 96.1 (A) Lateral projection and (B) anterior–posterior projection of digital subtraction angiography (DSA) of a left parietal DVA (*arrow*) with several feeders draining into a central vein as in a caput medusa. (C) Axial and (D) coronal head computed tomography angiography (CTA) of a DVA in the posterior fossa with dem*onstration of the multiple feeders (arrowheads) draining into a precerebellar vein (arrow).* (E) Coronal and (F) sagittal slices of gadolinium-enhanced magnetic resonance imaging (MRI) demonstrating the same precerebellar DVA. (G) and (H) Serial axial slices demonstrating multiple venous feeders supplying a central draining vein in a left frontal DVA (*arrow*).

While the medusa-like appearance of DVAs is typically conserved across patients, so-called mixed vascular malformations exhibit features of other vascular deformities and must be managed accordingly [5]. These

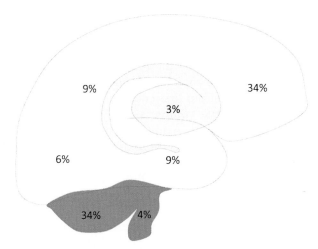

FIGURE 96.2 Distribution of DVAs in the frontal (34%), parietal (9%), occipital (6%), and temporal (9%) lobes, basal ganglia (3%), cerebellum (34%), and brainstem (4%).

mixed lesions often contain microshunts that drain into the DVA, and are clinically more aggressive with a higher risk of hemorrhage [10]. DVAs are most commonly associated with cavernous malformations [8].

PRESENTATION AND NATURAL HISTORY

The vast majority of DVAs are asymptomatic and discovered incidentally by CT or MRI in patients presenting with symptoms unrelated to the DVA [8,15]. While cerebral hemorrhaging was a common concern for patients with DVAs, it has been shown that the risk of hemorrhage is exceedingly low and almost exclusively observed in cases of comorbid cavernomas, which have a bleeding risk of 1% per year [8]. However, cavernous malformation hemorrhage is typically benign relative to other intracranial hemorrhages, such as subarachnoid hemorrhage; it may present as headache or subclinically, with many cavernous malformations presenting as seizure foci not acutely related to hemorrhage. Hemorrhaging is most often seen in close proximity with the cavernous malformation and is rarely associated with the location of the DVA [8].

Seizure and headache are the most common clinical presentations leading to diagnosis of DVAs [4,8]. However, there is little evidence to suggest a positive correlation between DVA location and seizure or headache location [8,12]. While seizures are not locally associated with DVAs, the abnormal development of the venous vasculature is thought to suggest aberrant neuronal migration that may underlie epilepsy [8]. In addition to seizures and headaches, cerebellar DVAs are often diagnosed during imaging to evaluate gait ataxia, diplopia, and dizziness [12].

The pathophysiology underlying symptomatic DVAs includes mass effect and hemodynamic changes. In 20% of symptomatic presentations, the mass effect of the DVA, especially the dilated collecting channel, can compress adjacent intracranial structures and cause hydrocephalus (from mesencephalic aqueduct compression) or nerve compression symptoms [4,10,13]. Trigeminal neuralgia, hemifacial spasm, and tinnitus are the most commonly reported consequences of DVA mass effect [4].

Local flow through the DVA can be disrupted by increased inflow or decreased outflow in 70% of cases [4,10]. These hemodynamic changes increase pressure within the DVA and can cause headaches, seizures, and increased risk of hemorrhage [4,15]. Increased inflow is most commonly a result of microshunting and arterial drainage into the DVA [4]. Diminished outflow can arise through a variety of mechanisms. Locally, thrombosis of DVA vessels and stenosis of the collecting channel (especially at the dural opening) can impede outflow and cause venous ischemia and infarction [4,9,15]. Thrombosis of the venous drain can also increase the risk of secondary hemorrhage [13,16,17]. The abnormal architecture of the dilated veins that lack smooth muscle or elastic layers predisposes these vessels to hemodynamic irregularity and thrombosis [15]. Diffuse venous hypertension from arteriovenous shunting can also impede outflow [4]. Collectively, these hemodynamic changes can increase DVA pressure and exceed the capacity of the compromised DVA vessels that lack smooth muscle or elastic layers, leading to headaches and increased risk of hemorrhage [4].

While the mass effect and hemodynamic changes associated with DVAs can be symptomatic, these malformations do not grow, and there is no evidence to suggest that DVAs have the potential for vascular proliferation [4,13]. Asymptomatic DVAs do not require treatment or follow-up.

TREATMENT OPTIONS

Although DVAs are typically thought of as benign lesions, and in the absence of symptomatology, they are not intervened upon and typically not followed. However, when they present in association with other cerebrovascular abnormalities, most commonly cavernous malformations, but occasionally also arteriovenous malformations, there has been a reported incidence of presentation with symptoms resulting from venous congestion or hemorrhage. Historically, resection of DVAs was associated with severe venous infarct because the region does not contain any normal draining veins and depends on the DVA for venous outflow. Furthermore, given a benign natural history, asymptomatic lesions are not treated [18], even during treatment of

associated cavernomas or arteriovenous malformations [4]. Although cases can present with venous congestion, venous infarct, or hemorrhage, these phenomena are attributed to the arteriovenous shunting pathology or the cavernous malformation, and not the DVA. Therefore, although the AVMs are treated with radiosurgery, embolization, or resection (not reported), and cavernous malformations are treated with resection, the DVA itself is not intervened upon [4,19,20].

DVAs may also occur due to insufficient venous outflow due to venous congestion or thrombosis. The management of thrombosis is not well established, but some have suggested minimizing and treating systemic risks for hypercoagulability, such as oral contraceptives, smoking, and inherited hypercoagulable states [16]. Although they did not report duration of total treatment, groups have reported resolution of thrombus in patients without underlying coagulopathy at first imaging follow-up, usually within 4–12 weeks [21].

Uncommonly, lesions may become symptomatic due to local mass effect or hemodynamic changes but these are managed with medical interventions for thrombosis or surgical interventions targeting the symptoms. For example, in cases presenting with hydrocephalus from compression of the foramen of Monro or the cerebral aqueduct, shunting or third ventriculostomy has been a successful mode of treatment [4]. On the other hand, half of patients with compressive neuropathy underwent surgical decompression , while the other half were clinically observed [22]. Again, resection of the DVA itself may be complicated by venous infarction and associated morbidity, including reports of death [23]. In cases presenting with severe hemorrhage, some patients may require operative intervention, such as an external ventricular catheter or a hemicraniectomy, for temporary management of increased intracranial pressure.

CONSIDERATIONS FOR MANAGEMENT

It is critical to distinguish whether the DVA is the source of symptomatology, since cavernous malformations also frequently present with recurrent hemorrhage and seizures, while AVMs also frequently present with hemorrhage and significant arteriovenous shunting phenomena. These are more often source of the symptoms and are typically treated without manipulation of the DVA.

The second critical consideration pertains to the pathophysiology of the symptoms, which can be local mass effect, or flow related. Symptomatic DVAs are managed with treatments targeting the symptomatology without interrupting the venous drainage, which is thought to be a critical venous pathway for the associated parenchymal territory. There is a relatively high rate of venous infarction following sacrifice of

DVAs. Reported local mass effects include hydrocephalus and compressive neuropathy. These are treated with shunting and microvascular decompressions, respectively.

A final consideration pertains to the nature of altered flow in the DVA. While altered drainage and shunting can increase flow through the DVA relative to other veins, thrombosis and stenosis may reduce the capacity for outflow from the DVA. Thrombosis may lead to venous infarction and hemorrhage in the territory of a DVA. It has been suggested that focal stenosis near the drainage into a central sinus may be a nucleus for thrombus formation [9,24]. Cross-sectional imaging suggesting thrombosis or stenosis should be further evaluated with cerebral angiography to evaluate the nature of flow through the DVA. Subtotal thrombosis may prompt consideration of anticoagulation. Furthermore, arterialization of the DVA is better defined on cerebral angiography. In the absence of direct arterial supply, it has been suggested that there is no increase in hemorrhage rate [25]. Without arterial supply, these lesions are also less amenable to endovascular approaches. However, identification of direct arterial supply or an associated AVM is generally associated with higher risk of hemorrhage or seizure [25]. These have been successfully approached with embolization.

Core principles in defining the risk and management plan of a DVA are: attribution of presenting symptoms to DVA or another source, identifying mechanism of DVA symptomatology for indirect treatment of these symptoms as an alternative to the DVA, and defining the DVA flow pattern.

References

[1] Lasjaunias P, Burrows P, Planet C. Developmental venous anomalies (DVA): the so-called venous angioma. Neurosurg Rev 1986;9(3):233–42.

[2] Toulgoat F, Lasjaunias P. Vascular malformations of the brain. Handbook Clin Neurol 2013;112:1043–51.

[3] Sarwar M, McCormick WF. Intracerebral venous angioma. Case report and review. Arch Neurol May 1978;35(5):323–5.

[4] Pereira VM, Geibprasert S, Krings T, et al. Pathomechanisms of symptomatic developmental venous anomalies. Stroke December 2008;39(12):3201–15.

[5] Awad IA, Robinson Jr JR, Mohanty S, Estes ML. Mixed vascular malformations of the brain: clinical and pathogenetic considerations. discussion 188 Neurosurgery August 1993;33(2):179–88.

[6] Chalouhi N, Dumont AS, Randazzo C, et al. Management of incidentally discovered intracranial vascular abnormalities. Neurosurg Focus December 2011;31(6):E1.

[7] Rigamonti D, Spetzler RF. The association of venous and cavernous malformations. Report of four cases and discussion of the pathophysiological, diagnostic, and therapeutic implications. Acta Neurochir 1988;92(1–4):100–5.

[8] Topper R, Jurgens E, Reul J, Thron A. Clinical significance of intracranial developmental venous anomalies. J Neurol Neurosurg Psychiatry August 1999;67(2):234–8.

[9] Truwit CL. Venous angioma of the brain: history, significance, and imaging findings. Am J Roentgenology December 1992;159(6):1299–307.

[10] Cohen JE, Boitsova S, Moscovici S, Itshayek E. Concepts and controversies in the management of cerebral developmental venous anomalies. Isr Med Assoc J November 2010;12(11):703–6.

[11] Perrini P, Lanzino G. The association of venous developmental anomalies and cavernous malformations: pathophysiological, diagnostic, and surgical considerations. Neurosurg Focus 2006;21(1):e5.

[12] Rigamonti D, Spetzler RF, Medina M, Rigamonti K, Geckle DS, Pappas C. Cerebral venous malformations. J Neurosurg October 1990;73(4):560–4.

[13] Kiroglu Y, Oran I, Dalbasti T, Karabulut N, Calli C. Thrombosis of a drainage vein in developmental venous anomaly (DVA) leading venous infarction: a case report and review of the literature. J Neuroimaging April 2011;21(2):197–201.

[14] Abe M, Hagihara N, Tabuchi K, Uchino A, Miyasaka Y. Histologically classified venous angiomas of the brain: a controversy. discussion 11 Neurologia Medico-Chirurgica January 2003;43(1):1–10.

[15] Masson C, Godefroy O, Leclerc X, Colombani JM, Leys D. Cerebral venous infarction following thrombosis of the draining vein of a venous angioma (developmental abnormality). Cerebrovasc Dis May-Jun 2000;10(3):235–8.

[16] Hammoud D, Beauchamp N, Wityk R, Yousem D. Ischemic complication of a cerebral developmental venous anomaly: case report and review of the literature. J Comput Assist Tomography Jul-Aug 2002;26(4):633–6.

[17] Amemiya S, Aoki S, Takao H. Venous congestion associated with developmental venous anomaly: findings on susceptibility weighted imaging. J Magn Reson Imaging December 2008;28(6):1506–9.

[18] Abla AA, Lekovic GP, Turner JD, de Oliveira JG, Porter R, Spetzler RF. Advances in the treatment and outcome of brainstem cavernous malformation surgery: a single-center case series of 300 surgically treated patients. discussion 414–405 Neurosurgery February 2011;68(2):403–14.

[19] Geibprasert S, Krings T, Pereira V, Lasjaunias P. Infantile dural arteriovenous shunt draining into a developmental venous anomaly. Interv Neuroradiology March 2007;13(1):67–74.

[20] Kurita H, Sasaki T, Tago M, Kaneko Y, Kirino T. Successful radiosurgical treatment of arteriovenous malformation accompanied by venous malformation. Am J Neuroradiol March 1999;20(3):482–5.

[21] Merten CL, Knitelius HO, Hedde JP, Assheuer J, Bewermeyer H. Intracerebral haemorrhage from a venous angioma following thrombosis of a draining vein. Neuroradiology January 1998;40(1):15–8.

[22] Korinth MC, Moller-Hartmann W, Gilsbach JM. Microvascular decompression of a developmental venous anomaly in the cerebellopontine angle causing trigeminal neuralgia. Br J Neurosurg February 2002;16(1):52–5.

[23] Mori K, Seike M, Kurisaka M, Kamimura Y, Morimoto M. Venous malformation in the posterior fossa: guidelines for treatment. Acta Neurochir 1994;126(2–4):107–12.

[24] Konan AV, Raymond J, Bourgouin P, Lesage J, Milot G, Roy D. Cerebellar infarct caused by spontaneous thrombosis of a developmental venous anomaly of the posterior fossa. Am J Neuroradiology February 1999;20(2):256–8.

[25] Ruiz DS, Yilmaz H, Gailloud P. Cerebral developmental venous anomalies: current concepts. Ann Neurol September 2009;66(3):271–83.

CHAPTER

97

Vein of Galen Arteriovenous Malformations

A.P. See[1,2], D.B. Orbach[2]

[1]Brigham and Women's Hospital; [2]Boston Children's Hospital/Harvard Medical School, Boston, MA, United States

Vein of Galen arteriovenous malformations (VOGMs) are a rare congenital vascular anomaly in which an arteriovenous shunt is established to the median prosencephalic vein of Markowski (MPV), a fetal vein typically present between weeks 7 and 12 of gestation [1]. Although true incidence of the condition is difficult to ascertain, VOGMs represent the earliest-appearing pediatric cerebrovascular malformation [2]. Other than a minority of patients, in whom a mutation in the RASA1 gene can be demonstrated, there is no demonstrated genetic risk factor for the development of VOGM [3]. The two pathophysiological challenges faced early on by patients are increased cardiac output and increased cerebral venous pressure. As a result, common presentations of VOGM include high-output right- and left-sided cardiac failure (including pulmonary hypertension), macrocrania with hydrocephalus, neurodevelopmental delays or reversals, and seizures [4]. Since the mid-1980s, management strategies have evolved away from surgical approaches toward the use of a combination of medical and endovascular approaches, timed to balance the risks of the malformation and the intervention, and fostering normal development. Although definitive anatomic cure remains an objective, even in cases where only partial closure of the malformation can be safely achieved, patients may be entirely neurodevelopmentally intact.

EMBRYOLOGY

After closure of the two ends of the neural tube in the fourth week of development, the amniotic fluid no longer supplies the developing central nervous system [1]. During the choroidal stage of embryological development, the MPV provides venous drainage of the anterior cerebral and choroidal arteries during the 6th through 11th weeks of gestation [5]. During continued development, the venous drainage of the choroid plexus is assumed by the paired internal cerebral veins, leading to regression of the MPV in an anterior-to-posterior fashion, with only the most caudal segment of the MPV contributing to the origin of the vein of Galen (VOG) [1].

In patients with VOGM, the MPV does not undergo its normal regression, with concomitant development of the deep venous system; thus, formation of the normal internal cerebral veins, basal veins, VOG, and straight sinus often does not occur [1]. Rather, persistent fetal venous drainage patterns are often seen [6], including infratentorial drainage via anterior and lateral pontomesencephalic veins, persistent falcine sinuses, and persistent occipital sinuses. The venous outflow often eventually joins the normal transverse sinus pathway leading to distal venous drainage [7]. Thus embryologically speaking, a more appropriate name for VOGM would have been median vein malformation.

ANATOMY AND CLASSIFICATION

VOGM, which involves a fetal vein normally absent from postnatal anatomy, as described earlier, should be distinguished from aneurysmal dilatation of the VOG. The latter consists of an anatomically normally developed VOG draining a deep brain arteriovenous malformation (AVM), resulting in enlargement of the vein, as can occur in any brain AVM.

Additionally, the persistent MPV occupies a potential space absent in the normal anatomy, bounded from above by the inferior pial surface of the fornix, and from below by the roof of the third ventricle [1]. Therefore VOGM represents an anatomically extrapial abnormality, as opposed to the pial compartment occupied by brain AVMs, including those involving the mature VOG.

The two most commonly used classification schemes for VOGM are by Lasjaunias and Yasargil.

The Lasjaunias classification is morphological, dividing VOGM into choroidal and mural subtypes. The choroidal subtype has supply from multiple choroidal arterial feeders that converge in a complex network of dysplastic vessels that resembles an AVM nidus, onto the MPV, typically along its anterior-superior wall. In contrast, the mural subtype has a direct fistula in the wall of the MPV, potentially as morphologically straightforward as a single-hole fistula.

The Yasargil classification scheme includes four subtypes, although only three of the four correspond to true VOGM. Yasargil type I has a direct arteriovenous shunt between the arterial feeders and the MPV; this corresponds to the Lasjaunias mural subtype. Yasargil type II has arterial feeders converging onto a nidus prior to draining into the MPV; this corresponds to the Lasjaunias choroidal subtype. Yasargil type III is a hybrid of types I and II with both direct fistulous inflow and interspersed nidal vascular anatomy. Type IV is not a VOGM, but rather a pial AVM with venous drainage into a dilated but mature VOG.

Ultimately, these classification schemes provide a framework to assess each individual case. In considering the anatomy of a VOGM, there should be a thorough description of the arterial supply, the location and structure of the arteriovenous connection, the pathologic venous anatomy (caliber and morphology of the MPV), and the drainage pathway of the deep and infratentorial brain. All these factors should then be taken into consideration, in the context of the clinical presentation and prognosis, to construct a treatment plan.

Arterial supply is most often derived from the choroidal supply, such as the medial and lateral posterior choroidal arteries and occasionally the anterior choroidal artery and the pericallosal artery [1,7]. Prenatally, the choroidal arterial supply is intimately related to the diencephalic and mesencephalic arteries, and this may be an explanation for the common involvement in VOGM of the thalamostriate perforators from the basilar bifurcation and the proximal segments of the posterior cerebral arteries [7].

The presence of a VOGM and the embryologic MPV, in most cases, precludes the presence of normal deep venous drainage anatomy [7,8]. However, it is certainly possible for the normal deep venous system to develop and be utilized for drainage, even in the presence of a VOGM, although angiographic visualization of this normal venous drainage pattern may be obscured by the prominent arteriovenous shunting [9]. Deep venous drainage patterns figure prominently in constructing a safe treatment strategy.

In addition to the course of the venous drainage, the presence of venous outflow restriction is a second important consideration in prognostication and treatment planning. Venous outflow restriction can reduce the quantity of arteriovenous shunting, but may also result in intracranial venous hypertension. The abnormal anatomy eventually will flow into a point of normal venous anatomy (such as the torcular or sigmoid sinus), and venous outflow restriction balances with arteriovenous shunting to create a pattern of clinical presentation. Thus, for example, progressive stenosis at the sigmoid–jugular transition point may herald irreversible global neurological injury and functional decline due to venous hypertension, despite the resulting decreased flow through the malformation.

CLINICAL PRESENTATION

In 1964 Gold [10] described three classical clinical presentations of VOGM: neonatal high-output cardiac failure, pediatric hydrocephalus with developmental delay, and pediatric and adult neurologic and hemorrhagic complications.

Antenatal diagnosis is possible during the third trimester using ultrasound and color Doppler, with the latter demonstrating high-velocity flow through the AVM. This finding should be supplemented by fetal MRI, which offers a detailed assessment of the state of the brain parenchyma, as well as the morphology of the malformation [7]. In utero echocardiography can also demonstrate cardiac hypertrophy [4]. An early series of VOGM diagnosed antenatally, published in 1994, found the majority (94%, 16/17) to rapidly develop cardiac failure in the first few hours of life [4]. But with the proliferation of fetal MRI, prenatally identified cases more evenly split between those that remain asymptomatic in the neonatal period and those that cause early heart failure.

Neonatal (0–2 months) presentation manifests as failure to thrive with workup demonstrating high-output cardiac failure, the presenting condition in 94% of all neonatal presentations [4]. These neonatal patients are often first evaluated for congenital heart disease and may undergo nondiagnostic invasive cardiac imaging, which typically shows no structural cardiac anomalies. Furthermore, other signs, such as facial venous prominence, may alert the astute clinician to a noncardiac etiology [7,8]. Low-resistance arteriovenous shunting and pulmonary hypertension themselves are associated with tachycardia, cardiomegaly, and failure to thrive due to difficulty feeding [11] and may lead to hepatic and or renal dysfunction, as captured in the Bicêtre neonatal evaluation score for neonates with VOGM [12].

Increasing age is associated with lower chance of severe cardiac failure; after the first month, VOGMs typically manifest with macrocephaly, hydrocephalus, or cognitive developmental delay [8,11]. These symptoms are associated with global intracranial venous hypertension, which may impair cerebrospinal fluid (CSF) reabsorption in the absence of a ventricular obstruction [5]. Furthermore, diversion of arterial blood through the malformation and away from the normal parenchyma may lead to ischemic neurologic injury [13]. In some cases, local mass effect of the malformation may result in compression of the aqueduct or the third ventricle and obstructive ventriculomegaly, although surgical drainage of CSF in the absence of prior treatment of the underlying VOGM is associated with high morbidity from intracranial hemorrhage [14]. Nonneurologic symptoms in this stage also include external jugular venous congestion, which may manifest with epistaxis, mild proptosis, and dilated cutaneous veins of the head and neck [13].

Finally, there are case reports of older patients with previously unrecognized VOGM, who may present with headaches and occasionally with hemorrhage related to arteriovenous shunting and venous reflux [11].

PROGNOSIS

Patients not receiving treatment for VOGM have an overall mortality rate of 77–90% [11]. Historical studies from before the development of endovascular treatment reported mortality of 55–90% in malformations presenting with severe congestive heart failure, whereas the application of modern endovascular techniques has decreased mortality to 10–20% [8,11,12].

Endovascular advances and improved understanding of treatment timing has significantly improved outcomes. In the largest cohort study to date, published in 2006, the mortality rate was reported 11% with moderate or severe neurologic morbidity in 26% [12,15]. Although in earlier studies the definition of "favorable outcome" centered on angiographic obliteration, recent studies include neurocognitive outcomes, in which normal or mild development delay is considered the goal of treatment and the focus of a favorable outcome. Modern endovascular therapy achieves favorable outcome in an estimated 61% of patients, and moderate to severe delay in 23% [11].

Several interrelated factors have prognostic value. The age of presentation often correlates with clinical severity and morphology of the lesion: as Yasargil noted, the choroidal type is associated with earlier clinical presentation and a worse outcome, although the presence of venous outflow restriction may somewhat mitigate this [12]. Patients younger than 1 month presenting with clinical symptoms have a 33% chance of good outcome and a 36% mortality rate, those presenting between 1 month and 2 years have a 75% chance of good outcome and 7% mortality, and those presenting after 2 years have a mortality rate of 3.2% [11].

Presentation with cardiac failure is associated with a poor prognosis, with high-output left-sided cardiac failure and refractory pulmonary hypertension being the leading causes of mortality. Severe arteriovenous shunting is associated with hepatic and renal hypoperfusion during diastolic aortic flow reversal. In fact, cardiac function and the presence of heart failure are useful surrogates for the overall risk of mortality in VOGM: in an antenatal diagnosis case series, five cases had cardiac hypertrophy on in utero echocardiography. Four of these (80%) had irreversible heart failure and parenchymal brain injury [4], whereas the remainder of the cases were first managed medically with successful control of heart failure within the first week. Two of 16 (12%) underwent embolization at 2 months for progression of heart failure.

Neurologic morbidity can occur early, with parenchymal loss and calcification during the antenatal stage. In early infancy, progressive global parenchymal loss can occur, manifesting as a "melting brain" [12,16]. Venous congestion may lead to intracranial venous hypertension, hydrocephalus, and cognitive developmental delay [7]. In particular, absence of venous sinus drainage is associated with the severe parenchymal injury due to pial venous reflux [12]. Untreated or undertreated arteriovenous shunting may lead to pronounced developmental delay and seizures in the long term [12].

GOALS OF TREATMENT

The overarching goals of treatment are threefold: prevention of morbidity and mortality from cardiac failure, normal neurologic development, and anatomic cure of the lesion. Treatment decision making is driven by the risk benefit profile at each stage of consideration. In neonates, endovascular intervention carries a higher intrinsic risk and is technically challenging due to the size and fragility of the involved vessels [7,8]. Technical considerations of size severely restrict the usable range of catheter and device choices in the neonatal period, and trans-umbilical artery access may be very helpful [17]. This is further complicated by the risk of anesthesia faced by neonates with cardiac insufficiency.

The treatment goal in neonates suffering from medically refractory life-threatening cardiac failure is to adequately decrease arteriovenous shunting. An approximately 30% reduction of the lesion is reported to provide adequate hemodynamic change for immediate cardiac symptom relief [8], whereas complete embolization may be associated with significant increase in cardiac afterload. Therefore in cases of complete embolization, potential inotropic vasopressor support remains under consideration during monitoring in the intensive care unit [8].

For patients with medically manageable cardiac insufficiency or no cardiac issues, serial follow-up with head ultrasound, MRI, and neurodevelopmental examinations is used until 3–6 months, when embolization is typically undertaken [5,8]. It is believed that this age represents a therapeutic window during which the patient is developed enough to tolerate serial interventions with low morbidity, and before the onset of permanent neurocognitive injury or permanently deranged CSF and venous dynamics [7].

Endovascular treatment options are not without complication, including the risk of obstructing normal deep venous drainage, with resultant venous infarct and hemorrhage [7]; as mentioned earlier, although it has often been assumed that the deep venous system does not drain into the MPV [8], there have been several reports demonstrating use of the normal deep venous drainage pathway, despite the presence of the VOGM [9]. An additional potential risk involves diversion of arterial blood away from the malformation in the setting of chronic venous congestion and dysfunction of cerebral autoregulation, posing a risk of hemorrhage in the surrounding parenchyma [18]. Delivery of the embolic agent through the high-flow lesion into the pulmonary arterial system, causing pulmonary embolus, is another recognized possible complication of endovascular therapy.

TREATMENT OPTIONS

Until the mid-1980s, VOGMs were associated with significant mortality and morbidity despite attempts at surgical anatomic cure. In the current era, medical management (diuretics, inotropic medications, mechanical ventilation) is used to temporize cardiovascular morbidity while awaiting an optimum endovascular therapeutic window [11]. It has been proposed

that prenatal patients can be managed in utero with maternal digoxin when cardiomegaly (cross-sectional area) exceeds half of the chest area [19]. Pretreatment evaluation includes clinical examination, echocardiography, and cerebral anatomic imaging: MRI with diffusion-weighted imaging and magnetic resonance angiography and venography demonstrate the baseline brain parenchyma, as well as the arterial and venous abnormality [19].

Both transarterial and transvenous approaches have been used for endovascular treatment. Although more technically challenging, transarterial embolization is associated with a lower morbidity and a higher success rate and is the preferred approach at most high-volume centers. Even without direct treatment of the venous drainage, transarterial embolization may reduce the mass effect of the dilated venous anatomy (Figs. 97.1 and 97.2), with more successful preservation of the normal deep venous drainage [7]. However, in particular cases, the transvenous approach may provide better access to the lesion, and may be a necessary alternative once the transarterial embolization options are exhausted [7,12]. Transvenous access is itself complicated by the risk of venous perforation, and relatively high incidence of venous stroke and hemorrhage, in particular if the deep venous system drains via the MPV [17]. A more recent proposal involves retrograde transvenous catheterization of the arterial feeding pedicles, allowing for a more controlled embolization.

Transarterial approaches include the trans-umbilical artery approach in neonates, whereby the umbilical artery is catheterized before the fifth postnatal day, before it stenoses to become the medial umbilical ligaments [17]. Transfemoral access is the standard approach in postneonatal infants and children. Transvenous approaches typically involve femoral venipuncture, but may also involve jugular vein or even trans-torcular access.

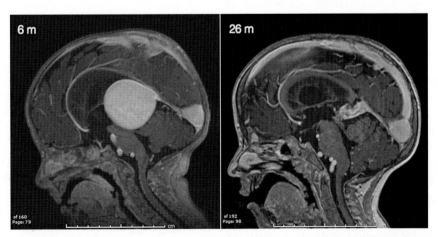

FIGURE 97.1 Postcontrast sagittal T1-weighted magnetic resonance images of the same patient; on the left, pretreatment, at 6 months, and on the right, posttreatment, at 26 months. Note the dramatic reduction in size of the massive median prosencephalic vein of Markowski, despite all embolization having been within the feeding arterial pedicle.

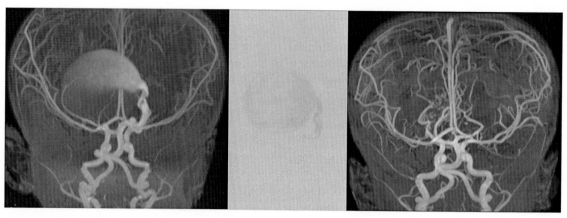

FIGURE 97.2 Magnetic resonance angiography (MRA) maximal intensity projection images in the same patient as seen in Fig. 97.1; on the left, pretreatment, at 6 months, and on the right, posttreatment, at 26 months. The pretreatment MRA shows direct fistulous inflow from a prominent arterial pedicle into the massively enlarged median prosencephalic vein of Markowski (MPV). The middle image shows a frontal angiographic view of microcatheterization of the feeding arterial pedicle just before embolization, with a hole in the left side of the MPV directly accepting the high-speed arterial inflow.

Although the data are confounded by case complexity, reviews of transarterial and transvenous approaches have found good angiographic and neurologic outcomes in 77% of patients undergoing transarterial approaches, 62% of patients undergoing combined transarterial and transvenous approaches, but close to zero good outcomes in patients undergoing purely transvenous approaches [11].

Transarterial embolization is typically accomplished via off-label use of liquid embolic agents (approved for brain AVM embolization): N-butyl-cyanoacrylate or ethylene vinyl alcohol (Onyx), whereas transvenous embolization is most frequently accomplished using detachable coils [20]. Although associated with a slightly higher rate of vessel rupture, detachable coils may sometimes provide embolic support when used in conjunction with liquid embolics [7].

Treatment may be accomplished in a single session, or more commonly in a staged fashion. Staging is favored to minimize hemodynamic change following any individual procedure; given the extent of arteriovenous shunting, normal perfusion pressure breakthrough and venous thrombosis are potential complications of single-stage complete embolizations [5,7]. A common treatment plan in a cardiovascularly uncomplicated case might involve transarterial embolization starting at 4–5 months staged over two to three sessions spaced by 3–8 weeks.

ONGOING CHALLENGES

Despite the well-documented natural history of VOGM and the significant decreases in mortality achieved by endovascular approaches, there remains a subset of patients who present with severe parenchymal brain injury from the outset, for whom the prognosis remains dismal. There is another subset of patients who suffer moderate to severe neurologic cognitive developmental delay despite successful endovascular treatment. Both of these patient groups clearly need better early diagnosis and more effective treatment, representing potential areas for marked improvement in the management of VOGM. Additionally, the degree of embolization of the arteriovenous shunting needed to achieve good long-term risk reduction remains unknown, so that in particular cases with complex lesions, it is unclear how many successive stages of embolization are appropriate. Finally, the risk factors, etiology, and very long-term prognosis of this developmental abnormality all remain unknown.

References

[1] Raybaud CA, Strother CM, Hald JK. Aneurysms of the vein of Galen: embryonic considerations and anatomical features relating to the pathogenesis of the malformation. Neuroradiology 1989;31(2):109–28. http://dx.doi.org/10.1007/BF00698838.

[2] Long DM, Seljeskog EL, Chou SN, French LA. Giant arteriovenous malformations of infancy and childhood. J Neurosurg 1974. http://dx.doi.org/10.3171/jns.1974.40.3.0304.

[3] Revencu N, Boon LM, Mulliken JB, et al. Parkes Weber syndrome, vein of Galen aneurysmal malformation, and other fast-flow vascular anomalies are caused by RASA1 mutations. Hum Mutat 2008;29(7):959–65. http://dx.doi.org/10.1002/humu.20746.

[4] Rodesch G, Hui F, Alvarez H, Tanaka A, Lasjaunias P. Prognosis of antenatally diagnosed vein of Galen aneurysmal malformations. Childs Nerv Syst 1994;10(2):79–83. http://dx.doi.org/10.1007/BF00302765.

[5] Gailloud P, O'Riordan DP, Burger I, et al. Diagnosis and management of vein of galen aneurysmal malformations. J Perinatol 2005;25(8):542–51. http://dx.doi.org/10.1038/sj.jp.7211349.

[6] Lasjaunias P, Garcia-Monaco R, Rodesch G, Terbrugge K. Deep venous drainage in great cerebral vein (vein of Galen) absence and malformations. Neuroradiology 1991;33(3):234–8.

[7] Pearl M, Gomez J, Gregg L, Gailloud P. Endovascular management of vein of Galen aneurysmal malformations. Influence of the normal venous drainage on the choice of a treatment strategy. Childs Nerv Syst 2010;26(10):1367–79. http://dx.doi.org/10.1007/s00381-010-1257-0.

[8] Garcia-Monaco R, De Victor D, Mann C, Hannedouche A, Terbrugge K, Lasjaunias P. Congestive cardiac manifestations from cerebrocranial arteriovenous shunts. Endovascular management in 30 children. Childs Nerv Syst 1991;7(1):48–52.

[9] Levrier O, Gailloud PH, Souei M, Manera L, Brunel H, Raybaud C. Normal galenic drainage of the deep cerebral venous system in two cases of vein of Galen aneurysmal malformation. Childs Nerv Syst 2004;20(2):91–7. http://dx.doi.org/10.1007/s00381-003-0841-y. discussion 98-99.

[10] Gold A, Ransohoff J, Carter S. Vein of Galen malformation. Acta Neurol Scand Suppl 1964;40(Suppl. 11):1–31.

[11] Khullar D, Andeejani AMI, Bulsara KR. Evolution of treatment options for vein of Galen malformations. J Neurosurg Pediatr 2010;6(5):444–51. http://dx.doi.org/10.3171/2010.8.PEDS10231.

[12] Lasjaunias PL, Chng SM, Sachet M, Alvarez H, Rodesch G, Garcia-Monaco R. The management of vein of Galen aneurysmal malformations. Neurosurgery 2006;59(Suppl. 5):184–94. http://dx.doi.org/10.1227/01.NEU.0000237445.39514.16.

[13] Hoang S, Choudhri O, Edwards M, Guzman R. Vein of Galen malformation. Neurosurg Focus 2009;27(5):E8. http://dx.doi.org/10.3171/2009.8.FOCUS09168.

[14] Jea A, Bradshaw TJ, Whitehead WE, Curry DJ, Dauser RC, Luerssen TG. The high risks of ventriculoperitoneal shunt procedures for hydrocephalus associated with vein of Galen malformations in childhood: case report and literature review. Pediatr Neurosurg 2010;46(2):141–5. http://dx.doi.org/10.1159/000319399.

[15] Li A-H, Armstrong D, terBrugge KG. Endovascular treatment of vein of Galen aneurysmal malformation: management strategy and 21-year experience in Toronto. J Neurosurg Pediatr 2011;7(1):3–10. http://dx.doi.org/10.3171/2010.9.PEDS0956.

[16] Alvarez H, Garcia Monaco R, Rodesch G, Sachet M, Krings T, Lasjaunias P. Vein of Galen aneurysmal malformations. Neuroimaging Clin N Am 2007;17(2):189–206. http://dx.doi.org/10.1016/j.nic.2007.02.005.

[17] Berenstein A, Masters LT, Nelson PK, Setton A, Verma R. Transumbilical catheterization of cerebral arteries. Neurosurgery 1997;41(4):846–50. http://dx.doi.org/10.1097/00006123-199710000-00014.

[18] Rangel-Castilla L, Spetzler RF, Nakaji P. Normal perfusion pressure breakthrough theory: a reappraisal after 35 years. Neurosurg Rev 2015;38(3):399–404. http://dx.doi.org/10.1007/s10143-014-0600-4. discussion 404-405.

[19] Berenstein A, Fifi JT, Niimi Y, et al. Vein of Galen malformations in neonates: New management paradigms for improving outcomes. Neurosurgery 2012;70(5):1207–13. http://dx.doi.org/10.1227/NEU.0b013e3182417be3.

[20] Thiex R, Williams A, Smith E, Scott RM, Orbach DB. The use of Onyx for embolization of central nervous system arteriovenous lesions in pediatric patients. Am J Neuroradiol 2010;31(1):112–20. http://dx.doi.org/10.3174/ajnr.A1786.

CHAPTER

98

Carotid Cavernous and Other Dural Arteriovenous Fistulas

C.J. Griessenauer[1], L. He[2]

[1]Harvard Medical School, Boston, MA, United States; [2]Vanderbilt University, Nashville, TN, United States

INTRODUCTION

Intracranial dural arteriovenous fistulas (dAVFs) are acquired vascular lesions that usually involve the intracranial dural sinuses and comprise less than 10% of all intracranial vascular lesions [1,2]. They are commonly divided into carotid cavernous fistulas (CCFs) and other dAVFs. Usually branches of the external carotid artery (ECA), internal carotid artery (ICA), or vertebral artery or a combination thereof form a direct connection with dural

Primer on Cerebrovascular Diseases, Second Edition
http://dx.doi.org/10.1016/B978-0-12-803058-5.00098-9

sinus and/or intracranial veins resulting in an arteriovenous shunt. Management is directed by symptomatology, location, and angioarchitecture of the lesion. Arterialization of intracranial veins through retrograde venous flow is classically associated with increased risk of cerebral hemorrhage.

CAROTID CAVERNOUS FISTULAS

CCFs are abnormal communications between the ICA or ECA and their branches and the cavernous sinus and comprise approximately 35% of all dAVFs [2].

Classification

Numerous classification systems have been applied to CCFs. The simplest classification divides CCFs into direct and indirect fistulas. Direct fistulas result from a defect in the ICA wall, from trauma or rupture of a cavernous ICA aneurysm, and are usually high-flow fistulas. Indirect fistulas are low-flow fistulas and the equivalent of a dAVF of the cavernous sinus and comprise the majority of CCFs encountered in clinical practice. The most widely adopted system to classify CCFs is the Barrow classification where the angioarchitecture of the arterial side of the fistula determines type [3]. However, the Barrow classification is not very practical from a clinical and therapeutic point as symptomatology and current treatment approach are influenced largely by venous drainage. In addition, most CCFs are indirect fistula and fall under Barrow type D since there is always some supply from meningeal branches of both ICA and ECA. One of the authors have proposed an updated five-tier classification system utilizing venous drainage, which captures symptomatology, endovascular treatment approach, and outcome (Fig. 98.1; Table 98.1) [4].

Symptomatology

Symptomatology results from venous congestion, hypertension, thrombosis, hemorrhage, neural compression, and/or ischemia from vascular steal [5]. With posterior venous drainage, increased pressure in the cavernous sinus and vascular steal can result in cranial nerve deficits manifesting as ophthalmoplegia, diplopia, ptosis, or anisocoria (types I and II). With anterior venous drainage (types II and III), increased venous pressure results in a rise in intraocular pressure leading to loss of vision, ocular pain, glaucoma, and retinal hemorrhage. Manifestations in the orbit include chemosis, exophthalmos, periorbital pain, and blepharedema. Most daunting is cortical symptomatology such as hemorrhage and seizures resulting from retrograde drainage

into the superficial middle cerebral vein or the posterior fossa via the petrosal vein (types IV and V) [4].

Management

Treatment of CCFs is indicated for symptomatic lesions presenting with visual impairment or disfigurement due to ocular or orbital involvement, cranial nerve deficits, cortical venous drainage, and hemorrhage [5–7]. A portion of CCFs will regress spontaneously due to stasis and thrombus formation with elevation of venous pressure beyond a critical point. Treatment of CCFs is accomplished using manual compression or endovascular embolization. Manual compression has a low cure rate and rarely provides definitive treatment [2]. Endovascular transvenous embolization is the most effective technique [2]. Types 1 and 2 are usually treated via the posterior transvenous approach through the inferior petrosal sinus, whereas type 3 is treated via the anterior transvenous approach through the superior ophthalmic vein. Type 5 lesions require transarterial treatment and ICA sacrifice [4]. Endovascular transarterial embolization is rarely curative and may be used for palliation in selected cases. Radiosurgery has shown some promising results but is not widely utilized for this indication [2].

OTHER DURAL ARTERIOVENOUS FISTULAS

dAVFs are abnormal arteriovenous communications between branches of the ECA, ICA, or the vertebral artery and venous structures within the meninges or cortical veins without intervening nidus or capillary bed. When first described, they were initially thought to be a subset of arteriovenous malformations (AVMs); however, now they are recognized as a separate entity and thought to generally be acquired. Although classification of dAVF is generally based upon venous drainage patterns, they can also be categorized depending on their anatomical location.

Pathogenesis

In adults, dAVFs are generally thought to be acquired. Commonly there is a history of trauma, craniotomy, or venous sinus thrombosis. Two hypotheses for dAVF formation have been proposed [8,9]. One involves enlargement of existing physiologic arteriovenous shunts secondary to elevated local venous pressure from trauma or craniotomy leading to dAVF formation [8]. The second involves outflow obstruction of a normal sinus due to thrombosis leading to venous hypertension and neoangiogenesis resulting in formation of a dAVF [8].

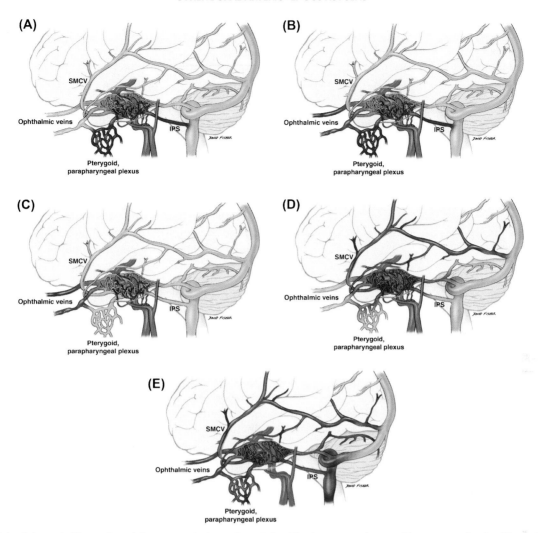

FIGURE 98.1   Schematic illustration of the venous drainage–based classification system for carotid cavernous fistulas. Normal venous anatomy is colored light blue. Preferential drainage of the individual fistula types are colored dark red. (A–E) Types I–V, respectively. *IPS*, inferior petrosal sinus; *SMCV*, superficial middle cerebral vein. *Reprinted from Thomas AJ, Chua M, Fusco M, Ogilvy CS, Tubbs RS, Harrigan MR, et al. Proposal of venous drainage-based classification system for carotid cavernous fistulae with validity assessment in a multicenter cohort. Neurosurgery 2015;77(3):380–5 with kind permission from Neurosurgery Wolters Kluwer Health.*

TABLE 98.1   Proposed Venous Drainage–Based Classification System for Carotid Cavernous Fistulas

Type	Venous Drainage
Type I	Posterior/inferior drainage only
Type II	Posterior/inferior and anterior drainage
Type III	Anterior drainage only
Type IV	Retrograde drainage into cortical veins ± other routes of venous drainage
Type V	High flow direct shunt between cavernous internal carotid artery and cavernous sinus (Barrow type A) ± multiple routes of venous drainage

Posterior/inferior drainage: inferior petrosal sinuses, superior petrosal sinus, pterygoid, and parapharyngeal plexus. Anterior drainage: superior and inferior ophthalmic veins. Cortical drainage: superficial middle cerebral veins, perimesencephalic, and cerebellar venous system. *Reproduced with permission from Thomas A, Chua M, Fusco M, et al. Neurosurgery. © 2015 Wolters Kluwer Health, Inc.*

Classification

Two main classification systems are commonly used for dAVFs and based on the venous drainage pattern (Table 98.2) [10,11]. In both, drainage through cortical veins with evidence of cortical venous reflux place the patient at higher risk for hemorrhagic complications.

Anatomical Location

The most common location for dAVF is at the junction of the transverse-sigmoid sinus. This location is also the least likely to be associated with retrograde venous drainage, thus the likelihood for hemorrhagic complications is low. dAVFs involving the tentorial-incisura region are prone to cortical venous reflux and frequently present with posterior fossa subarachnoid hemorrhage

TABLE 98.2   Two Main Classification Systems Commonly Used for Dural Arteriovenous Fistulas

Cognard	Borden	Risk of Hemorrhage
Type I: Anterograde drainage into dural sinus/meningeal vein	Type I: Anterograde drainage into dural sinus/meningeal vein	Low
Type IIa: Retrograde drainage into dural sinus/meningeal vein without cortical reflux		Low
Type IIb: Anterograde drainage into dural sinus/meningeal vein with retrograde drainage into cortical vein	Type II: Anterograde drainage into dural sinus and retrograde into cortical veins	Moderate
Type IIa + b: Retrograde drainage into dural sinus/meningeal vein and retrograde cortical venous drainage		Moderate
Type III: Retrograde cortical venous drainage without venous ectasia	Type III: Isolated retrograde drainage into cortical veins, trapped segment of sinus with reflux into cortical veins, venous varix/dural lake with reflux into cortical veins	High
Type IV: Retrograde cortical venous drainage with venous ectasia		High
Type V: Spinal venous drainage		High

or intraparenchymal hemorrhage secondary drainage into perimesencephalic or cerebellar veins. Within the anterior cranial fossa, dAVFs are supplied by meningeal arteries of the ECA or ICA (Fig. 98.2). In this location there are higher rates of hemorrhagic complications secondary to a higher frequency of drainage into cortical veins and retrograde reflux [12,13]. Other less frequent locations for dAVF include the superior sagittal sinus and deep venous structures (straight sinus, torcula, and galenic complex). Pial drainage of dAVFs involving these deep venous structures is associated with increased risk of hemorrhage. Some posterior fossa dAVFs drain into the venous plexus associated with the spinal cord. In these cases, myelopathy or spinal subarachnoid hemorrhage may be the presenting symptomatology.

Symptomatology

Tinnitus is the most common presenting symptom of dAVF with the majority of patients presenting in the fifth or sixth decade of life and usually correlates with a more benign venous drainage pattern (Cognard and Borden type I) [9]. With retrograde venous drainage and/or cortical venous reflux, patients may also exhibit symptoms related to increased venous hypertension including headache, seizures, hemorrhage (including subdural, intraparenchymal, and/or subarachnoid), increased intracranial pressure, cranial neuropathy, dementia, or other focal neurologic deficits [10]. In these higher cases (Cognard types IIb–V, Borden types II–III), the risk is higher and annual mortality rates of 10.4% and annual intracranial hemorrhage rates of 6.9% have been reported [9–11]. Any changes in the patients presenting symptoms should trigger further imaging and

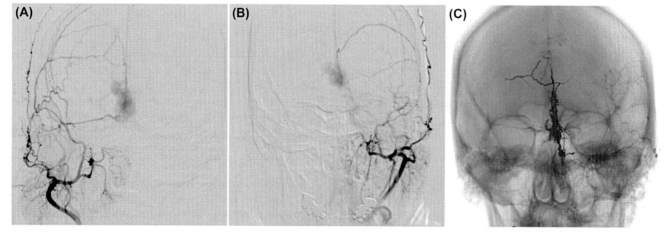

FIGURE 98.2   (A, B) Anteroposterior views of digital subtraction angiography of right and left external carotid artery injections, respectively. There is filling of the fistulous region by various branches of the external carotid artery including the middle meningeal, occipital, ascending pharyngeal, and posterior auricular arteries. (C) Onyx cast after transarterial embolization through the left middle meningeal artery alone. Due to properties of the embolic material, the entirety of the fistulous site was occluded with embolization of a single pedicle.

additional workup as this may represent progression of originally benign lesions [14].

Management

Treatment of dAVF is generally considered for patients with evidence of cortical venous drainage (Cognard IIb–V, Borden types II and III) as these lesions are thought to have higher risk of hemorrhage [9,12]. Treatment modalities include endovascular embolization, open surgical ligation, and/or radiosurgery.

Endovascular treatment is accomplished via transarterial or transvenous routes. The goal of any endovascular procedure is to effectively seal the fistulous region itself, usually with liquid embolic argents such as ethylene-vinyl alcohol copolymer [EVOH; Onyx (ev3 Endovascular, Inc., Plymouth, MN)] or N-butyl cyanoacrylate (n-BCA). Transarterial delivery of these agents occurs after distal catheterization of an arterial feeding pedicle that allows for optimal penetration of the fistulous connection [9]. It is usually not necessary to embolize through every single pedicle as the embolic material will penetrate other pedicles from the fistula site. Short-term follow-up data indicate initial cure rates of 95% with Onyx transarterial embolization [15]. Consideration should be given to potential dimethyl sulfoxide toxicity, entrapment of the microcatheter, prolonged fluoroscopy times, and potential unintended embolization as neither Onyx nor n-BCA can necessarily be controlled.

Transvenous embolization occurs through catheterization of the involved dural sinus or, in some cases, a cortical vein, with deposition of coils and/or liquid embolic agents [9]. Obliteration of the fistula is accomplished safely as long as the affected sinus is not involved in drainage of normal parenchyma. Transvenous embolization dAVF is especially appealing if there is a large fistulous region or if there are many tortuous arterial feeding pedicles difficult to access via the transarterial route.

Surgical ligation of the fistulous site is indicated for endovascular treatment failure (e.g., posterior fossa dAVF) or in cases where endovascular therapy carries significant morbidity (e.g., blindness in anterior fossa dAVF secondary to meningeal feeders from the ophthalmic artery). In these situations, identification of the fistulous connection with either clip ligation or bipolar cautery is required. In general, postoperative catheter angiography should be performed to confirm complete obliteration of the dAVF.

Radiosurgery for dAVF has generally been reserved for lesions that have failed all other treatment modalities. As with radiosurgery for AVM, there is a latency period between treatment and resolution during which there may be additional risk of hemorrhage [16]. Furthermore, long-term follow-up data on dAVF treated with radiosurgery are limited [9].

With respect to Cognard type I–IIa or Borden type I dAVF, it is reasonable to pursue conservative management with observation given that the long-term progression rates for these lesions are low at 3.2% [14]. Risk factors for progression of dAVF include multiple dAVF, evidence of venous hypertension, and younger age at presentation [14]. Changes in the patients presenting symptoms and/or new neurologic changes should prompt further imaging and workup to assess for progression.

References

[1] Al-Shahi R, Bhattacharya JJ, Currie DG, Papanastassiou V, Ritchie V, Roberts RC, et al. Prospective, population-based detection of intracranial vascular malformations in adults: the Scottish Intracranial Vascular Malformation Study (SIVMS). Stroke 2003;34(5):1163–9. http://dx.doi.org/10.1161/01.STR.0000069018.90456.C9.

[2] Harrigan M, Deveikis J. Dural arteriovenous fistulas. In: Handbook of cerebrovascular disease and neurointerventional technique second edition. New York: Humana Press, Springer; 2013. p. 603.

[3] Barrow DL, Spector RH, Braun IF, Landman JA, Tindall SC, Tindall GT. Classification and treatment of spontaneous carotid-cavernous sinus fistulas. J Neurosurg 1985;62(2):248–56. http://dx.doi.org/10.3171/jns.1985.62.2.0248.

[4] Thomas AJ, Chua M, Fusco M, Ogilvy CS, Tubbs RS, Harrigan MR, et al. Proposal of venous drainage-based classification system for carotid cavernous fistulae with validity assessment in a multicenter cohort. Neurosurgery 2015;77(3):380–5. http://dx.doi.org/10.1227/NEU.0000000000000829.

[5] Jung K-H, Kwon BJ, Chu K, Noh Y, Lee S-T, Cho Y-D, et al. Clinical and angiographic factors related to the prognosis of cavernous sinus dural arteriovenous fistula. Neuroradiology 2011;53(12):983–92. http://dx.doi.org/10.1007/s00234-010-0805-3.

[6] Halbach VV, Hieshima GB, Higashida RT, Reicher M. Carotid cavernous fistulae: indications for urgent treatment. AJR Am J Roentgenol 1987;149(3):587–93. http://dx.doi.org/10.2214/ajr.149.3.587.

[7] Kim DJ, Kim DI, Suh SH, Kim J, Lee SK, Kim EY, et al. Results of transvenous embolization of cavernous dural arteriovenous fistula: a single-center experience with emphasis on complications and management. AJNR Am J Neuroradiol 2006;27(10):2078–82.

[8] Chung SJ, Kim JS, Kim JC, Lee SK, Kwon SU, Lee MC, et al. Intracranial dural arteriovenous fistulas: analysis of 60 patients. Cerebrovasc Dis 2002;13(2):79–88.

[9] Gandhi D, Chen J, Pearl M, Huang J, Gemmete JJ, Kathuria S. Intracranial dural arteriovenous fistulas: classification, imaging findings, and treatment. AJNR Am J Neuroradiol 2012;33(6):1007–13. http://dx.doi.org/10.3174/ajnr.A2798.

[10] Borden JA, Wu JK, Shucart WA. A proposed classification for spinal and cranial dural arteriovenous fistulous malformations and implications for treatment. J Neurosurg 1995;82(2):166–79. http://dx.doi.org/10.3171/jns.1995.82.2.0166.

[11] Cognard C, Gobin YP, Pierot L, Bailly AL, Houdart E, Casasco A, et al. Cerebral dural arteriovenous fistulas: clinical and angiographic correlation with a revised classification of venous drainage. Radiology 1995;194(3):671–80. http://dx.doi.org/10.1148/radiology.194.3.7862961.

[12] Awad IA, Little JR. Dural arteriovenous malformations. In: Barrow DL, editor. Intracranial vascular malformations. American Association of Neurological Surgeons; 1990. p. 219–26.

[13] Awad IA, Little JR, Akrawi WP, Ahl J. Intracranial dural arterio-venous malformations: factors predisposing to an aggressive neu-rological course. J Neurosurg 1990;72(6):839–50. http://dx.doi.org/10.3171/jns.1990.72.6.0839.

[14] Hetts SW, Tsai T, Cooke DL, Amans MR, Settecase F, Moftakhar P, et al. Progressive versus nonprogressive intracranial dural arte-riovenous fistulas: characteristics and outcomes. AJNR Am J Neu-roradiol July 2015. http://dx.doi.org/10.3174/ajnr.A4391.

[15] Chandra RV, Leslie-Mazwi TM, Mehta BP, Yoo AJ, Rabinov JD, Pryor JC, et al. Transarterial onyx embolization of cranial dural arteriovenous fistulas: long-term follow-up. AJNR Am J Neurora-diol 2014;35(9):1793–7. http://dx.doi.org/10.3174/ajnr.A3938.

[16] Guo WY, Pan DH, Wu HM, Chung WY, Shiau CY, Wang LW, et al. Radiosurgery as a treatment alternative for dural arterio-venous fistulas of the cavernous sinus. AJNR Am J Neuroradiol 1998;19(6):1081–7.

CHAPTER

99

Cavernous Malformations

G. Lanzino, K.D. Flemming

Mayo Clinic, Rochester, MN, United States

INTRODUCTION

Cavernous malformations (CMs) are the second most common form of vascular malformation of the central nervous system after developmental venous anomalies (DVAs). The incidence of CMs is 0.34–0.53% in autopsy series [1,2] and 0.39–0.47% in MRI series [3,4]. The over-all male to female ratio is 1:1 and the mean age at pre-sentation is 30.6 years [5]. CMs can occur sporadically or in a familial form. In the familial form, the mode of inheritance is autosomal dominant and more than 50% of patients affected have multiple lesions [6].

CMs are often associated with other vascular malfor-mations with DVAs and capillary telangiectasias being the most common ones. Development of CM adjacent to a preexisting DVA has been well documented and this has led to speculations regarding a potential role of DVAs in the genesis of CMs. CMs can develop as a result of previous radiation treatment.

The distribution of CMs follows the relative mass of brain parenchyma and about 80% involve the supraten-torial compartment. CMs occur within the spinal cord and have also been observed within cranial nerves. Although histologically similar, cavernous angiomas involving the dural sinuses and especially the cavern-ous sinus exhibit specific clinical features and should not be considered part of the same disease process. Because blood flows through CMs very slowly, these lesions are not apparent on catheter angiography and have been, in the past, considered within the group of "angiographi-cally occult" vascular malformations. For these same reasons, their diagnosis was difficult and elusive before the widespread introduction of MRI.

GENETICS

Three genes have been linked to the familial form of CMs: CCM1 (KRIT1), CCM2 (macalvernin), and CCM3 (PDCD10). Each displays an autosomal dominant inheritance with incomplete penetrance and variable expression. KRIT-1 knockout mice show marked vascu-lar defects. However, CCM1 gene knockout alone does not result in CM formation, rather a second gene defect may be required (two-hit hypothesis) [7]. Each locus is involved in signaling pathways related to angiogenesis, endothelial tight junction stability, and cell integrity. It has been found that CCM1 and 2 affect RhoA (a GTPase protein) activity. When RhoA activity is increased due to lack of CCM1 or 2, vascular instability results. This is an important target for potential medications to reduce leakiness from CM.

PATHOLOGY

CMs are formed by thin-walled cavities ("caverns") covered with endothelium filled with blood products of different ages. These caverns are contained in a vascular stroma. Unlike arteriovenous malformations, there is no intervening brain parenchyma within the lesions. The endothelial cells lining the caverns forming the CM do not have tight junctions. The lack of tight junctions may be responsible for the leakage of hemosiderin, which stains the surrounding brain parenchyma even in absence of frank hemorrhage and creates a typical MRI appearance. Dystrophic microcalcifications are common in chronic established lesions and a peripheral rim of hemosiderin-storing macrophages is present in the adjacent gliotic brain.

CLINICAL PRESENTATION

CMs of the brain can present with seizures, can grow through multiple episodes of microhemorrhage with formation of multiple microcaverns with resultant mass effect or can present with frank hemorrhage. In the past, hemorrhage from a CM has been variably defined. However, a consensus statement has identified precise criteria for the definition of hemorrhage in patients with CMs: presence of acute or subacute onset of symptoms (headache, seizures, decreased consciousness, or new/worsened neurological deficit related to the anatomic location of the CM) associated with radiological or pathological evidence of recent hemorrhage within the lesion or in the parenchyma around it [8]. An increase in size of the CM without other evidence of recent hemorrhage or the mere presence of a hemosiderin ring around the lesion cannot be considered a surrogate for hemorrhage in absence of any of the clinical and radiological features mentioned before [8].

NATURAL HISTORY

Similar to other vascular lesions, the natural history of CMs is dependent upon the mode of presentation. The overall annual rates of hemorrhage for CMs presenting with hemorrhage, nonhemorrhagic symptoms, or for incidental lesions are 6.2%, 2.2%, and 0.33%, respectively [9]. In a population-based cohort of 134 patients the risk of hemorrhage was also higher for CMs presenting with hemorrhage [10]. The median interval before a recurrent bleed in patients presenting with hemorrhage is 8 months [9]. Clinical impairment after the first hemorrhage is usually mild with the majority of patients fully recovering within the first month [11]. Case-fatality rate is extremely low. In

patients who have suffered a first bleed, risk of rehemorrhage is higher shortly after the presenting bleed as there is evidence of a clustering of hemorrhages over the first 1–2 years after the first bleed and, in absence of recurrent hemorrhage, the risk decreases over time [9,10,12]. Based on these observations, the time interval elapsed from the last bleeding episode becomes an important factor in deciding whether invasive treatment is indicated in a patient with symptomatic hemorrhage.

There is no evidence that pregnancy, delivery, or the postpartum period are associated with an increased risk of bleeding from a CM; thus pregnancy is not contraindicated in patients with known diagnosis of CM [13]. In pregnant women with CMs, there is no contraindication to a vaginal delivery. Patients who need to take antiplatelet medications or to be on anticoagulation can take these medications as indicated since these drugs do not increase the risk of bleeding from a CM [14].

The natural history of incidental and asymptomatic CMs is benign; consequently there is no reason to consider invasive treatment in such cases. In an unselected series of 107 patients with incidental CMs, only one hemorrhage related to the CM occurred during 1311 patient-years of follow-up, corresponding to a bleeding rate of 0.08%/year [15]. Similarly, in a large group of pediatric patients with incidental CMs the hemorrhage rate at follow-up was exceedingly low (0.2%/year) [16].

DIAGNOSIS

In the absence of recent hemorrhage, nonenhanced computed tomography (CT) scan may not show small CMs. If the lesion is large enough, it may appear as an area of hyperdensity on CT occasionally with small scattered intralesional calcifications. On MRI, CMs often appear as heterogeneous, "popcorn" like well-circumscribed lesions with mixed signal intensity related to intralesional blood products in different stages of evolution. Although most "established" CMs share this "classic" MRI appearance, the MRI appearance of CMs is not always typical. After an acute bleed, the boundaries of an underlying CM can be difficult to distinguish early on and become clearer over time.

Functional MRI is helpful in patients with lesions located in or adjacent to eloquent areas to estimate the risk of surgical treatment and for surgical planning. T1 and fluid attenuated inversion recovery images identify the boundaries of the CM and are helpful in understanding how close the lesion is to the pial or ependymal surface (a factor very important in assessing surgical risk). To assess how close a CM is to the surface, T2 sequences should not be utilized because of the "blooming" artifact of the peripheral hemosiderin content, which creates a false assessment of how close the CM is to the surface. Advanced MRI with tractography is a valid tool to plan the surgical approach and avoid critical tracts. Contrast MRI is helpful in delineating the presence and location of an associated DVA.

Catheter angiography is not helpful in the diagnosis of CMs, although it may be indicated in the acute phase after a bleed to rule out other vascular malformations since the diagnosis of CM is not always immediately apparent on axial imaging in such cases.

INDICATIONS FOR TREATMENT

Patient selection, timing of surgery, and surgical planning are the most important steps in the evaluation of surgical candidates. Surgery is usually indicated after a symptomatic bleed to prevent further hemorrhage. In patients with CMs in highly eloquent areas such as the brainstem and the thalamus, the decision to proceed with surgical intervention must be carefully weighed against the risk of surgery. In several such cases, we opt for no intervention even after a symptomatic bleed if patients have recovered completely or exhibit only minor residual symptoms.

The role of surgery for patients with seizures related to CMs will be discussed later. Surgery is not recommended in patients with incidental lesions because of the very benign natural history of such incidental lesions. CMs that show progressive enlargement over time may be considered for surgical treatment. Patients with multiple lesions, usually encountered in the familial form, are a challenge. In general, surgery is indicated in these cases for the symptomatic lesion only, whereas associated CMs are treated only if adjacent to the target symptomatic one and their resection adds only very low additional risk.

Surgery is contraindicated for CMs truly buried within eloquent areas such as the thalamus, internal capsule, and the brainstem without coming close to a pial or ependymal surface; for elderly patients and in general when the risks of surgery outweigh the potential benefits.

TIMING OF SURGICAL TREATMENT

Surgical resection of a ruptured CM is easier after a recent hemorrhage. It is our preference to perform surgery in the 2–3 weeks following a bleeding episode before reabsorption of fresh blood takes place. Fresh hemorrhage facilitates exposure of the plane around the CM and creates a buffer against normal brain. More "established" and chronic CMs are more difficult and traumatic to remove than those with fresh extracapsular hemorrhage since

they are much more adherent to the surrounding gliotic brain and this characterization must be considered when considering surgery and assessing perioperative risk.

CHOICE OF OPERATIVE APPROACH

Goals of surgery for CMs differ in relation to the mode of presentation and include prevention of recurrent seizures, improvement of symptoms of mass effect, and prevention of hemorrhages. Removal of the CM eliminates the risk of future hemorrhage and, in patients with seizures, surgery can result in long-term complete seizure control [17].

The ideal surgical trajectory to the lesion avoids violation of normal brain parenchyma while minimizing the need for brain retraction. Some lesions that may not come to the surface may reach or approach the surface of brain hidden within a sulcus. DVAs of variable size are often associated with CMs [18,19]. Damage of the main collectors of a DVA can result in major surgical morbidity as they participate to the drainage of normal brain. Thus, when a DVA is present in association with a CM, the position of the DVA influences the choice of the surgical approach.

For supratentorial lesions not immediately visible on the surface two basic approaches can be used: a transulcal or a transgyral approach. It is not clear whether disruption of the U-fibers connecting adjacent gyri is worse than disruption of vertical tracts seen with a transgyral approach [20]. We prefer a transulcal approach because often CMs not immediately visible on the convexity come close to a surface in the depth of a sulcus. To avoid retraction and damage to sulcal vessels it is important to open the arachnoidal planes of the sulcus widely and under the highest magnification.

RADIOSURGERY

The role of radiosurgery for CMs continues to be controversial and we reserve it for the rare truly unresectable CMs that demonstrate aggressive clinical behavior.

PRINCIPLES OF SURGICAL TREATMENT

Surgical technique depends on the location of the lesion, presence of an associated venous anomaly, and presence of hemorrhage. Once the CM is exposed, there are basically two microsurgical techniques that depend on location (superficial or deep, eloquent versus noneloquent brain) and size (small or large).

For superficial small lesions in eloquent and noneloquent areas and for lesions buried in noneloquent areas without extracapsular hemorrhage, a circumferential dissection is preferred utilizing the hemosiderin-stained gliotic plane, which usually surrounds the lesion to develop a cleavage plane all around the CM. DVAs are frequently associated with CMs and also participate to the drainage of normal brain. The CM must be carefully separated from the main trunks of the DVA. Inadvertent sacrifice of these channels may result in venous congestion, edema, and hemorrhage. Smaller radicles of the associated DVA can be identified at the periphery of the lesion and, if necessary, can be gently coagulated and divided but the larger collectors must be respected.

After resection of the more obvious caverns, the surgeon is faced with a decision regarding the surrounding gliotic brain. In eloquent areas, this portion of the brain must be left alone. In other areas, judicious removal of hemosiderin-stained surrounding parenchyma should be considered to improve extent of resection, potentially decrease the potential for epileptogenesis, and facilitate interpretation of postoperative MRI.

CMs in eloquent areas and those that do not reach the surface may often be removed through an incision smaller than the CM itself with a piecemeal resection. Internal debulking, at first, targets larger caverns with liquefied blood. After the CM is partially emptied, a cleavage plane between the CM and the surrounding gliotic, hemosiderin-stained brain parenchyma is carefully developed.

CAVERNOUS MALFORMATIONS ASSOCIATED WITH SEIZURES

CMs present with seizures in approximately 50% of the time [21]. The typical seizure associated with a CM is one of focal onset, which may secondarily generalize. The mechanism by which CMs cause seizures is unclear, but is probably related to accumulation of epileptogenic blood products and iron deposition in the brain surrounding the lesion.

The risk of developing epilepsy after a first seizure as presentation of a CM is 94%. Thus, these patients should be started on anticonvulsants and the initial management should include a detailed history and a wake and sleep electroencephalography (EEG) [22]. Routine use of anticonvulsants is not recommended for patients with CMs found incidentally (4%) or after hemorrhage (6%) because of the very low 5-year risk of seizures (4% and 6%, respectively) [21,22].

If a patient has breakthrough seizures despite the use of two anticonvulsants it is quite unlikely that addition of a third medication will be of benefit. Patients with pharmacoresistant epilepsy should undergo evaluation in a specialized center. Although medical management is a reasonable first option in patients with a first ever seizure related to a CM [6], we feel that in patients with CMs in noneloquent areas, surgical resection should be considered even after a first ever seizure as surgery increases the likelihood that patients can be seizure-free long term. In patients with seizures resistant to medical therapy, the likelihood of being seizure-free after surgery is negatively correlated with increased time of onset. Therefore, there is agreement that it is not necessary in these patients to wait until the classic criteria for truly medically refractory epilepsy are met and presurgical evaluation should be considered after failure of a single drug. In patients with long-standing seizures or frequent seizures the outcome after surgery is less favorable than in patients with short seizure history or rare seizures.

In the situation where a patient presents with seizures and multiple cavernomas, the burden is to determine which cavernoma is the source of the epilepsy. Typically the cavernoma that is the largest and hence the most hemorrhagic is the culprit. However, a collaborative effort with a neuroepileptologist is essential in these patients to achieve the best long-term outcome. In patients with multiple CMs and seizures, video-EEG monitoring is mandatory.

Treatment strategies for patients with epilepsy include lesionectomy alone, lesionectomy with resection of the surrounding gliosis, and electrocorticography-tailored resection. In patients with seizures, if neurologically permissible, resection of the surrounding adjacent hemosiderin-stained brain is preferred to increase the likelihood of seizure control.

FUTURE DIRECTIONS

At present, there are no modifiable risk factors for preventing hemorrhage from CM. Ongoing studies are evaluating the role of select biomarkers as potential modifiable risk factors. In addition, drug therapy targeted at RhoA activity is being investigated as a possible alternative to surgery, especially in those patients where surgery is deemed high risk.

References

[1] McCormick WF. Pathology of vascular malformations of the brain. In: Wilson CB, Stein BM, editors. Intracranial arteriovenous malformations. Baltimore: Williams & Wilkins; 1984.

[2] Otten P, Pizzolato GP, Rilliet B, Berney J. 131 cases of cavernous angiomas (cavernomas) of the CNS, discovered by retrospective analysis of 24,535 autopsies. Neurochirurgie 1989;35:128–31.

[3] Robinson JR, Awad IA, Little JR. Natural history of the cavernous angioma. J Neurosurg 1991;75:709–14.

[4] Del Curling Jr O, Kelly DL, Elster AD, Craven TE. An analysis of the natural history of cavernous angiomas. J Neurosurg 1991;75:702–8.

[5] Gross BA, Lin N, Du R, Day AL. The natural history of intracranial cavernous malformations. Neurosurg Focus 2011;30(6):E24.

[6] Batra S, Lin D, Recinos PF, Zhang J, Rigamonti D. Cavernous malformations: Natural history, diagnosis and treatment. Nat Rev Neurol 2009;5:659–70.

[7] Revencu N, Vikkula M. Cerebral cavernous malformation: new molecular and clinical insights. J Med Genet 2006;43:716–21.

[8] Al-Shahi Salman R, Berg MJ, Morrison L, Awad IA. Angioma Alliance Scientific Advisory Board. Hemorrhage from cavernous malformations of the brain: definition and reporting standards. Stroke 2008;39(12):3222–30.

[9] Flemming KD, Link MJ, Christianson TJ, Brown Jr RD. Prospective hemorrhage risk of intracerebral hemorrhage. Neurology 2012;78:632–6.

[10] Al-Shahi Salman R, Hall JM, Horne MA, Moultrie F, Josephson CB, Bhattacharya JJ, et al. Scottish Audit of Intracranial Vascular Malformations (SAIVMs) collaborators. Untreated clinical course of cerebral cavernous malformations: a prospective, population-based cohort study. Lancet Neurol 2012;11:217–24.

[11] Cordonnier C, Al-Shahi S, Bhattacharya JJ, Counsell CE, Papanastassiou V, Ritchie V, et al. Differences between intracranial vascular malformation types in the characteristics of their presenting hemorrhages: prospective, population-based study. J Neurol Neurosurg Psych 2008;79:47–51.

[12] Barker FG, Amin-Hanjani S, Butler WE, Lyons S, Ojemann RG, Chapman PH, et al. Temporal clustering of hemorrhages from untreated cavernous malformations of the central nervous system. Neurosurgery 2001;49:15–25.

[13] Witiw CD, Abou-Hamden A, Kulkarni AV, Silvaggio JA, Schneider C, Wallace MC. Cerebral cavernous malformations and pregnancy: Hemorrhage risk and influence on obstetric management. Neurosurgery 2012;71:626–31.

[14] Flemming KD, Link MJ, Christianson TJ, Brown Jr RD. Use of antithrombotic agents in patients with intracerebral cavernous malformations. J Neurosurg 2013;118:43–6.

[15] Moore SA, Brown Jr RD, Christianson TJ, Flemming KD. Long-term natural history of incidentally discovered cavernous malformations in a single-center cohort. J Neurosurg 2014;120:1188–92.

[16] Al-Holou WN, O'Lynnger TM, Pandey AS, Gemmete JJ, Thompson BG, Muraszko KM, et al. Natural history and imaging prevalence of cavernous malformations in children and young adults. J Neurosurg Ped 2012;9:198–205.

[17] Dodick DW, Cascino GD, Meyer FB. Vascular malformations and intractable epilepsy: outcome after surgical treatment. Mayo Clin Proc 1994;69:741–5.

[18] Perrini P, Lanzino G. The association of venous developmental anomalies and cavernous malformations: pathophysiological, diagnostic, and surgical consideration. Neurosurg Focus 2006;21(1):e5.

[19] Rammos SK, Maina R, Lanzino G. Developmental venous anomalies: current concepts and implications for management. Neurosurgery 2009;65:20–30.

[20] Bortolotti C, Nannavecchia B, Lanzino G, Perrini P, Andreoli A. Supratentorial cavernous malformations. In: Lanzino G, Spetzler RF, editors. Cavernous malformations of the brain and spinal cord. New York: Thieme; 2008. p. 65–70.

[21] Josephson C, Leach J, Duncan R, Roberts RC, Counsell CE, Al-Shahi Salman R. Scottish Audit of Intracranial Vascular Malformations (SAIVMs) steering committee and collaborators. Seizure risk from cavernous or arteriovenous malformations: prospective population-based study. Neurology 2011;76:1548–54.

[22] Rosenow F, Alonso-Vanegas MA, Baumgartner C, Blümcke I, Carreño M, Gizewski ER, et al. Surgical Task Force, Commission on Therapeutic Strategies of the ILAE. Cavernoma-related epilepsy: review and recommendations for management – report of the Surgical Task Force of the ILAE Commission on Therapeutic Strategies. Epilepsia 2013;54:2025–35.

CHAPTER

100

Spinal Vascular Malformations

B.M. Howard, D.L. Barrow
Emory University School of Medicine, Atlanta, GA, United States

Vascular malformations of the spinal cord are a heterogeneous group of anomalies ranging from simple, dural arteriovenous fistulae to complex arteriovenous malformations that may involve the spinal cord parenchyma as well as adjacent spinal and extraspinal structures. Although spinal vascular malformations are uncommon, they represent a potentially curable cause of progressive myelopathy. As with most uncommon disorders, a high index of suspicion is necessary to make the diagnosis.

Primer on Cerebrovascular Diseases, Second Edition
http://dx.doi.org/10.1016/B978-0-12-803058-5.00100-4

SPINAL VASCULAR ANATOMY

To understand the pathophysiology and treatment of spinal cord vascular malformations requires knowledge of spinal vascular anatomy [2]. The spinal cord receives arterial blood supply through two distinct arterial systems, a single anterior spinal artery and paired posterior spinal arteries (Fig. 100.1). The anterior spinal artery is formed by the convergence of two branches of the distal vertebral arteries that join anterior to the cervicomedullary junction and descend in the anterior median sulcus. The anterior spinal artery receives segmental blood supply from radiculomedullary arteries arising from the vertebral arteries, costocervical or thyrocervical trunks, and intercostal, lumbar, and sacral arteries. Typically, two large radiculomedullary branches supplement the anterior spinal artery. A large radicular feeder usually arises at the C5 or C6 level from the vertebral or ascending cervical artery and is called the artery of cervical enlargement. The largest of the radiculomedullary branches is known as the artery of Adamkiewicz, which usually arises from a lower intercostal artery and enters the spinal canal on the left side at a variable level between T8 and L4. The artery of Adamkiewicz ascends anterolateral to the cord and takes a hairpin turn to descend to the conus medullaris (CM). As radiculomedullary arteries enter the dura at the nerve root sleeve, small branches supply the adjacent dura.

The paired posterior spinal arteries run longitudinally along the dorsolateral aspect of the spinal cord behind the dorsal nerve roots. Similar to the anterior spinal artery, the posterior spinal arteries receive radiculomedullary feeders from the vertebral, intercostal, and lumbar arteries. The anterior spinal artery supplies the anterior two-thirds of the spinal cord, and the posterior spinal arteries supply the posterior columns and portions of the lateral columns. The posterior spinal arteries join with the distal anterior spinal artery at the CM to form the cruciate anastomosis.

The venous drainage of the spinal cord is through small radial veins that course through the spinal cord parenchyma into the coronal venous plexus along the surface of the spinal cord. The convergence of the coronal venous plexus on the surface of the spinal cord forms the medullary veins that exit at the nerve root sleeves.

CLASSIFICATION AND EPIDEMIOLOGY

Most spinal vascular malformations can be classified into one of four types that differ in etiology, anatomy, pathophysiology, radiological appearance, and treatment. Additionally, cavernous malformations occur in the spinal cord.

The type I spinal vascular malformation is most common and consists of a single arteriovenous fistula imbedded within the dura around the proximal nerve root sleeve and adjacent spinal dura. The feeding artery is a dural branch off the radiculomedullary ramus of an intercostal or lumbar segmental artery. From the dural nidus, outflow of the fistula is intradural through the medullary vein into the dorsal venous plexus along the surface of the spinal cord (Fig. 100.2 top left). These lesions typically occur in the lower thoracic region in males, most commonly in the fifth to sixth decades of life.

Less commonly, type I spinal AVMs have two or more arterial feeders that enter at separate segmental levels.

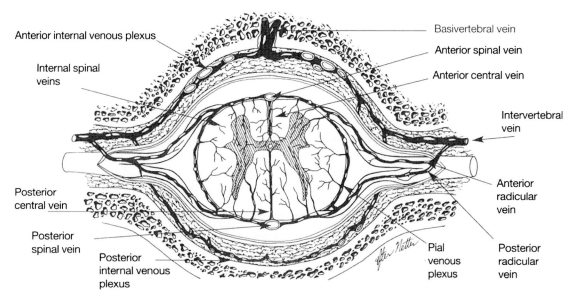

FIGURE 100.1 Diagrammatic illustration of normal spinal cord anatomy. *Reprinted with permission of American Association of Neurological Surgeons.*

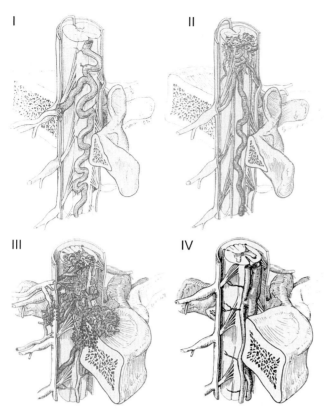

FIGURE 100.2 Diagrammatic illustration of the four types of spinal vascular malformations.

The additional dural branches travel within the dura to the fistula nidus where they converge to communicate with the efferent vein. Type I AVMs are subdivided into types Ia and Ib based on the presence of one or multiple arterial feeders, respectively [2].

Type II spinal cord AVMs, or glomus AVMs, are intramedullary vascular malformations that have a true, compact nidus within the parenchyma of the spinal cord (Fig. 100.2 top right). Type II AVMs are less common than type I AVMs, occur most frequently in the cervical spine with equal sex incidence, and present in a younger population. The nidus is often fed by the anterior spinal artery, but may have multiple feeders.

Type III spinal AVMs, also known as juvenile AVMs, are quite rare, occur with equal sex incidence and usually present in adolescents and young adults. These complex, extensive lesions have both intramedullary and extramedullary components and often extend into extraspinal tissues as well (Fig. 100.2 bottom left).

Type IV spinal AVMs, or perimedullary arteriovenous fistulae, occur more commonly in males and are intermediate in incidence between the more common type I and less common types II and III spinal AVMs [3,4]. The intradural extramedullary fistula drains directly into an enlarged venous outflow of varying size. They

present through a broader age range than other spinal AVMs. These lesions are most commonly supplied by the anterior spinal artery and are located anterior to the spinal cord, most often at the CM (Fig. 100.2 bottom right).

Spetzler et al. previously published a modified classification of spinal vascular abnormalities that includes separate categories for aneurysms and neoplasms of a vascular nature in addition to the four types of malformations described [9]. This expanded classification scheme includes lesions outside the narrow focus of this chapter; however, the authors describe CM AVMs as a subtype of spinal vascular anomaly distinct from the types I–IV as aforementioned. While anecdotal, our single center experience has lead us to believe that the CM AVM deserves consideration as a unique entity with idiosyncratic pathophysiology and treatment considerations.

PATHOPHYSIOLOGY, HEMODYNAMICS, AND NATURAL HISTORY

Type I Spinal AVMs

Although type I spinal AVMs are believed to be acquired lesions, the mechanism for their development is not understood. These low-flow, high-pressure arteriovenous shunts convey increased pressure through the venous outflow tract from the spinal cord veins into the valveless coronal venous plexus; thereby transmitting increased pressure to the spinal cord parenchyma [2]. Type I spinal AVMs rarely, if ever, present with hemorrhage. Patients typically present with painful, progressive myeloradiculopathy as a result of venous hypertension. The most common symptom associated with type I spinal AVMs is radicular pain. At the time of diagnosis, however, most patients have experienced lower extremity weakness and sensory disturbances and may have bowel or bladder dysfunction. About 10–15% of patients experience the acute onset of symptoms. Publications report a delay between symptom onset and diagnosis of a year or more for the majority of patients [8,12]. In 1926, Foix and Alajouanine reported acute necrotizing myelitis due to a dural spinal arteriovenous malformation [6]. This is an extremely rare cause of acute onset of neurological symptoms and is believed to be due to venous thrombosis within the draining intramedullary veins of the lesion.

In patients with type I spinal AVMs, symptoms are frequently exacerbated by particular activities or positions. This may well be related to the rostrally directed venous outflow, which is impeded when the patient in the upright position or with certain activities.

The prognosis for patients with type I spinal AVMs is clearly related to the condition of the patient at the time of treatment [1]. Therefore, a high index of suspicion and early diagnosis is of paramount importance to assure the patient a good recovery. Patients who present for treatment with severe neurological deficits are unlikely to recover fully.

Type II Spinal AVMs

Type II AVMs are congenital, high-flow, and high-pressure lesions characterized by multiple feeding arteries and venous drainage that is both rostral and caudal to the malformation. They may cause symptoms either by hemorrhage or vascular steal phenomenon. Patients experiencing hemorrhage from a type II spinal AVM typically present with the acute onset, severe myelopathy often associated with localized pain. The neurological deficit usually reaches its maximum shortly after the hemorrhage and, unless complete, gradually improves over the ensuing days to weeks.

Type II spinal AVMs may present with a slowly progressive neurological deficit due to the shunting of blood away from the normal spinal cord parenchyma into the lower resistance AVM. The prognosis for patients with type II spinal AVMs is dependent upon the neurological condition of the patient at the time of presentation, particularly if due to intraparenchymal hemorrhage. Furthermore, the anatomy of the lesion determines the difficulty of surgical or endovascular treatment, and, thus, has a significant impact on the ultimate prognosis.

Type III Spinal AVMs

Type III spinal AVMs are congenital lesions that have high-flow, high-pressure characteristics. These anomalies are extraordinarily difficult to treat due to involvement of not only the spinal cord parenchyma, but also extramedullary and extraspinal tissues. They may present with hemorrhage or a progressive neurological deficit from steal and are associated with a very poor prognosis.

Type IV Spinal AVMs

Type IV spinal AVMs, or intradural perimedullary arteriovenous fistulae, typically present with a progressive neurological deficit due to venous hypertension, but, on occasion, may present with hemorrhage. This diverse group of lesions has variable hemodynamics ranging from a simple, single arteriovenous fistula to giant, multipediculated, high-flow lesions. They have been subclassified into three types, IVa, IVb, and IVc, depending upon their anatomical and hemodynamic

characteristics [2,4]. Type IVa spinal AVMs are simple, extramedullary fistulae that are fed by a single arterial branch, most often the anterior spinal artery, located in the lower thoracic region near the conus. Type IVb lesions are intermediate-sized fistulas, usually with more than one arterial feeder. Type IVc are giant, multipediculated lesions that are high-flow, high-pressure anomalies.

Type IV spinal AVMs may be congenital or acquired [3]. Indeed, we have seen clearly congenital lesions in newborns and have reported a case of a type IV fistula in a patient who had undergone resection of an intramedullary spinal cord ependymoma with normal preoperative spinal angiography. She later presented with a progressive myelopathy that proved to be due to an acquired arteriovenous fistula that was successfully treated. As with type I lesions, prognosis is related to the patient's neurological condition at the time of diagnosis.

Conus Medullaris AVMs

CM AVMs, typically, are low-flow and high-pressure with multiple points of arteriovenous shunting supplied both by the anterior and by the posterior spinal arteries. An associated nidus is often identified, most commonly in an extramedullary location and involving the pia. Less frequently, the nidus is found within the parenchyma of the conus medullaris. Both the feeding arteries and draining veins are intertwined with the nerve roots of the cauda equina, often extensively so [9]. Patients most often present with myelopathy from venous hypertension, but can present with radiculopathy if the nidus or engorged draining veins cause compression of nerve roots. Hemorrhage may also lead to acute onset symptoms [13]. The epidemiology and natural history of CM AVMs is not well studied, but is likely similar to that of type I AVMs.

RADIOLOGICAL FEATURES

Although proper classification and treatment planning of spinal AVMs depends upon high-quality spinal angiography, most spinal AVMs are identified on MRI.

Myelography

While heavily relied upon in the past, the advent of high-resolution and time-resolved magnetic resonance imaging (trMRI) has limited the role of myelography for the diagnosis of spinal vascular malformations. Myelography will often outline the shadow of the dilated venous structures associated with types I and IV spinal AVMs.

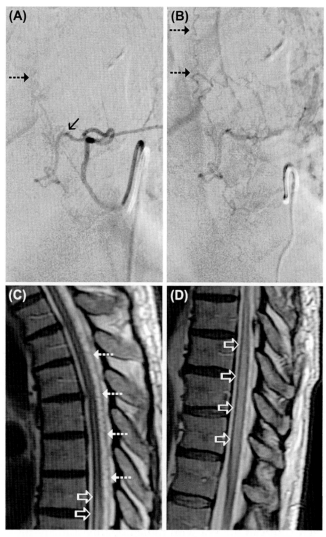

FIGURE 100.3 Type I spinal AVM. (A) Early and (B) late phase, AP spinal angiograms demonstrate an arteriovenous fistula within the dural nerve root sleeve fed by the left L2 segmental artery (*solid arrow*). Later film shows the dilated, arterialized draining vein on the dorsal surface of the spinal cord (*black, dashed arrows*). (C) Sagittal T-2 weighted MRI reveals multiple flow voids dorsal to the spinal cord rostral to the level of the fistula (*white, dashed arrows*). (D) Abnormal T-2 hyperintense signal is seen within an expanded spinal cord from the conus medullaris to the T9 level consistent with venous hyperemia secondary to fistula-dependent venous outflow obstruction (*open arrows*).

Magnetic Resonance Imaging

For type I spinal AVMs, MRI typically reveals the dilated, tortuous, draining veins which appear as prominent flow voids on the dorsal surface of the spinal cord. T2 hyperintensity within the spinal cord parenchyma is often seen, consistent with venous hyperemia (Fig. 100.3C and D). The draining vein may extend for long distances from the site of the actual fistula, and thus conventional MRI is of little benefit in localizing the site of the lesion. However, contrast-enhanced trMRI has been shown to be a useful initial study for suspected type I spinal

AVMs with high concordance with conventional angiography (Fig. 100.4). Pre-angiogram trMRI helps to localize the fistula to limit the number of catheterized vessels as well as contrast and radiation dose [7]. For types II and III spinal AVMs, MRI shows the intraparenchymal component of the lesion, its relationship to the surface of the spinal cord (Figs. 100.5 and 100.6, respectively) and associated intraparenchymal hemorrhage, which may influence surgical approaches. For type III spinal AVMs, MRI reveals additional spinal and extraspinal components with large flow voids associated with dilated vascular channels (Fig. 100.6). Like type I lesions, type IV spinal AVMs are seen on MRI primarily as a dilated, tortuous draining vein that emanates from the perimedullary arteriovenous fistula. There may also be flow voids within the parenchyma of the conus seen in patients with type IV AVMs. CM AVMs are characterized by flow voids, both of the nidus and the more serpiginous draining veins. MRI is imperative to identify intra- versus extramedullary nidus preoperatively.

Angiography

Spinal angiography is the definitive neuroimaging study required to accurately classify a spinal AVM and plan appropriate treatment, whether endovascular, surgical, or combined therapy. Complete spinal angiography may require bilateral vertebral angiography and selective catheterization and injection of branches of the subclavian, intercostal, lumbar, or sacral arteries. On rare occasions, intracranial dural AVMs can be associated with venous drainage that involves the spinal cord, and complete cerebral angiography is necessary to rule out such an anomaly presenting as a spinal vascular malformation.

For type I dural AVMs, individual feeders are selectively catheterized to demonstrate the precise location of the AVM (Fig. 100.3A). To identify type Ib lesions is important as failure to obliterate all feeding arteries at the time of surgery will fail to cure the lesion. Identification of the anterior spinal artery above and below the malformation is paramount to be sure the spinal cord does not share common blood supply with the AVM. While uncommon, this situation has important therapeutic implications. Venous drainage of most type I spinal AVMs is directed rostrally (Fig. 100.3B).

Types II (Fig. 100.5) and III (Fig. 100.6) spinal AVMs may be supplied by multiple feeding arteries, particularly type III lesions. Superselective catheterization of the feeding arteries of intraparenchymal lesions can be helpful to better understand the angio-architecture and is necessary for therapeutic embolization, whether preoperative or as definitive treatment (Fig. 100.5C).

Type IV spinal AVMs are typically fed by the anterior spinal artery, usually in the low thoracic or thoracolumbar region but occasionally may be fed by the posterior

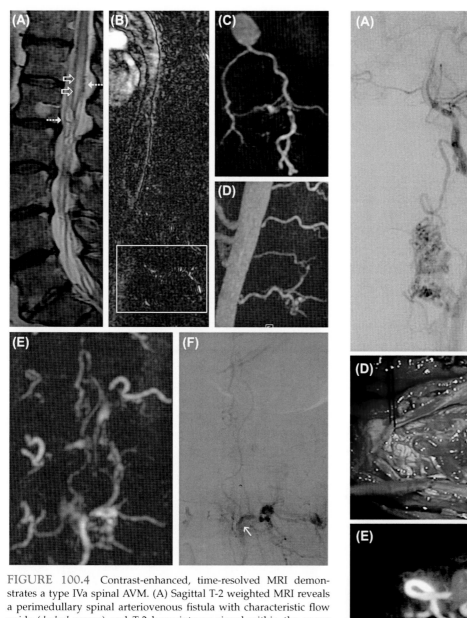

FIGURE 100.4 Contrast-enhanced, time-resolved MRI demonstrates a type IVa spinal AVM. (A) Sagittal T-2 weighted MRI reveals a perimedullary spinal arteriovenous fistula with characteristic flow voids (*dashed arrows*) and T-2 hyperintense signal within the conus medullaris (*open arrows*). (B) Sagittal, early phase, contrast-enhanced, trMRI reveals early opacification of the left L2 segmental artery and associated draining vein (*white box*). (C and D) Axial and sagittal maximum intensity projections (MIP) more clearly demonstrate the anatomy of the fistula. (E and F) Coronal MIP precisely recapitulates the anatomy of the malformation as demonstrated on AP angiography. The *white arrow* demarcates the feeding L2 segmental artery.

spinal artery and may have multiple feeders (Fig. 100.7). Although these lesions are readily seen on MRI, careful review of the spinal angiogram is necessary to appropriately classify the AVM and appreciate the precise site of the fistula. If one reviews only the mid-arterial phase, the anatomy of the lesion will not be appreciated, as the large, dilated draining veins will obscure what may be a simple fistula.

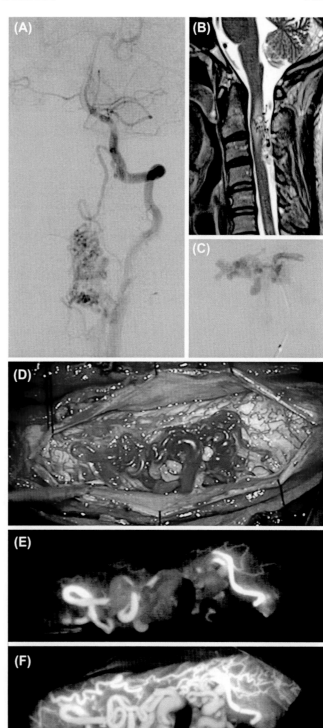

FIGURE 100.5 Type II spinal AVM. (A) AP spinal angiogram of an intramedullary AVM fed by the anterior spinal artery. (B) Sagittal T-2 weighted MRI reveals a compact nidus within the spinal cord with dorsally oriented arterial feeding vessels and draining veins. (C) Superselective angiogram of the anterior spinal artery following preoperative embolization of the AVM. (D) An intra-operative image demonstrates the angio-architecture of the AVM and the normal spinal cord parenchyma. (E and F) Intra-operative indocyanine green video angiography demonstrates the feeding arteries in the early phase (E) and the draining veins and normal spinal cord vessels in the late phase (F).

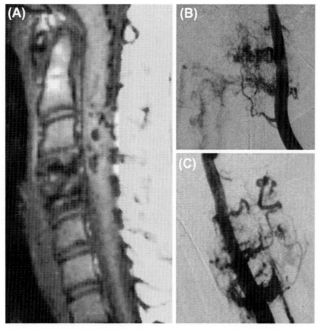

FIGURE 100.6   Type III spinal AVM. (A) Sagittal MRI of an extensive juvenile spinal AVM with involvement of the spinal cord parenchyma, spine, and extraspinal structures. (B and C) Left and right vertebral angiograms, respectively, demonstrating a portion of the extensive vascular supply to this complex vascular malformation.

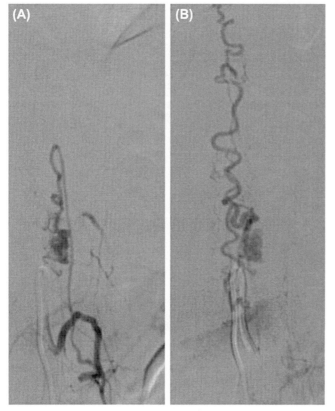

FIGURE 100.7   Type IV spinal AVM. (A) Early phase of AP spinal angiogram shows the anterior spinal artery with its characteristic hairpin turn and ending in a perimedullary arteriovenous fistula. (B) Later phase of the angiogram demonstrated the large, dilated draining veins.

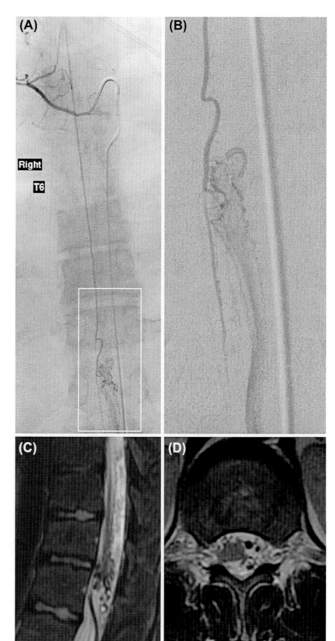

FIGURE 100.8   Conus medullaris AVM. (A) AP spinal angiogram reveals a conus medullaris AVM fed by the anterior spinal artery. (B) Blown up AP angiogram of the area of interest depicted by the white box in panel A, which demonstrates the angio-architecture of the malformation. (C and D) Sagittal and axial T2-weighted MRI sequences that show the nidus of the malformation is both intra- and extramedullary.

Both the anterior and posterior spinal arteries often supply CM AVMs and individual feeders may be superselected to clearly identify individual feeding vessels. While the angio-architecture of the nidus is better visualized on angiography, MRI is necessary preoperatively to determine the precise location of the nidus relative the parenchyma of the CM and the cauda equina (Fig. 100.8).

TABLE 100.1   Summary of Characteristics and Treatment of Spinal Arteriovenous Malformations

Type	Anatomy	Presentation	Epidemiology	Location	Treatment
I	dAVF at the nerve root sleeve	Progressive myelopathy from venous hypertension	Thought to be acquired; male predominance; age >40	Lower thoracic, upper lumbar	Surgery or embolization
II	Intramedullary, compact nidus, no intervening parenchyma	Acute, hemorrhage	Congenital, presents in children and young adults	No predilection	Combined embolization followed by surgical extirpation
III	Intramedullary nidus, extramedullary and extraspinal extension	Acute, hemorrhage	Adolescents and young adults	Cervical and thoracic	Combined embolization followed by surgical extirpation; worse outcomes than type II
IV	Intradural, extramedullary arteriovenous fistula Type A: Single feeding artery with small venous enlargement Type B: Multiple feeding arteries with venous enlargement Type C: Giant, multipediculated feedings arteries and large, engorged veins	Progressive myelopathy from venous hypertension	No age or sex predilection	Type A: Thoracolumbar Type C: Cervicothoracic	Type A: Surgery Type B: Embolization or surgery Type C: Embolization with subsequent surgery

dAVF, Dural arteriovenous fistula.
Modified from Schuette AJ, Cawley CM, Barrow DL. Spinal dural fistula. In: Bendok BR, Naidech AM, Walker MT, Batjer HH, editors. Hemorrhagic and ischemic stroke: Medical, imaging, surgical and interventional approaches. New York: Thieme; 2012. p. 344–55.

TREATMENT

Given the poor natural history of symptomatic spinal cord AVMs, most lesions can be treated with risk that is significantly less than that of the natural history. Treatment options for spinal AVMs include embolization, surgical excision, or a combination. Optimal treatment is best determined following complete angiography and classification of the malformation (Table 100.1).

Type I Spinal AVMs

Treatment of type I spinal AVMs by both surgical excision and embolization has been reported extensively [1,2,11]. The goal of either endovascular embolization or surgical treatment is to interrupt or disconnect the fistula by occluding the distal portion of the feeding vessel and the proximal portion of the efferent arterialized vein.

A variety of materials have been used to embolize type I AVMs, including polyvinyl alcohol (PVA) particles, cyanoacrylate polymers and onyx [8]. Although the use of these agents is associated with high initial occlusion rates, rate of recanalization of PVA is unacceptably high; glue and onyx may pass through the fistula into the draining vein, thus exacerbating the venous hypertension. Moreover, if the fistula is fed from the same radiculomedullary artery as the anterior spinal artery, embolization is contraindicated.

Surgical treatment of type I spinal AVMs has evolved significantly since the first successful treatment by Elsberg in 1916. Interestingly, he performed a rather limited resection of the dural fistula alone, consistent with modern technique [5]. As of 2016, we treat virtually all type I spinal AVMs by surgical excision (Fig. 100.9). Following precise localization with spinal angiography, a laminectomy is performed at the level of the fistula. The dural surface is carefully explored under the operating microscope to search for extradural evidence of the nidus, although this is typically not seen. The dura is opened and the intradural arterialized vein is located. The site of the fistula is identified under the microscope along the nerve root sleeve. An indocyanine green (ICG) video angiogram is then completed to confirm the site of the fistula. The efferent vein is coagulated with bipolar cautery and divided with microscissors. The inner surface of the dura at the site of the nidus is carefully inspected and coagulated with care taken to interrupt all feeding vessels. A final ICG angiogram is often completed to assure that the fistula has been disconnected. Gradually, the turgor of the dilated, arterialized dorsal vein decreases. If this is not observed, one should suspect the presence of a type Ib lesion and inspect for additional feeding arteries. The dura is then closed in a watertight fashion using running 4-0 Neurolon suture. With this limited approach, the risk of surgery is less than that of embolization, which is less effective than surgery to completely obliterate type I spinal AVMs for the majority of patients [8,11].

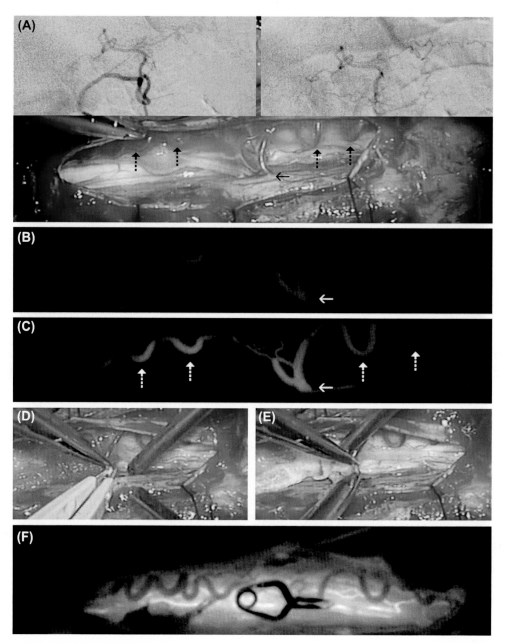

FIGURE 100.9   Intraoperative photographs that demonstrate anatomy of and surgical technique for disconnection of the type I AVM depicted in Fig. 100.3. (A) Insets: Early and late phase angiograms oriented caudal to rostral (left to right) in the orientation of the surgical field. The fistula is seen at the left L2 nerve root sleeve (*solid arrow*) with dilated, serpiginous draining vein both cranial and caudal to the fistula (*dashed arrows*). (B and C) Early and late phase indocyanine green (ICG) video angiograms, respectively, confirm the site of the fistula (*solid arrow*) and demonstrate the draining veins. (D and E) The fistula is cauterized and cut sharply with microscissors. (F) A final ICG angiogram reveals that the fistula has been disconnected and that the normal vessels of the spinal cord fill, but the vein associated with the fistula does not.

Type II Spinal AVMs

Type II lesions are formidable to treat. Oftentimes, a combination of embolization and surgical therapy is required for complete obliteration. Occasionally, those lesions supplied by a single arterial feeder can be completely obliterated by endovascular procedures. More commonly, embolization results in partial obliteration and surgical resection is required for a cure (Fig. 100.5C). The operative approach to type II spinal AVMs is dictated by their location within the spinal cord, anatomy of the arterial supply, and position of associated intraparenchymal hemorrhages. For those that approach the dorsal surface of the spinal cord, a laminectomy and midline myelotomy will often provide adequate exposure. Those that come closest to the ventral surface of the spinal cord can be exposed either by anterior approaches or, more commonly, a posterolateral approach (Fig. 100.10). We perform a posterolateral approach

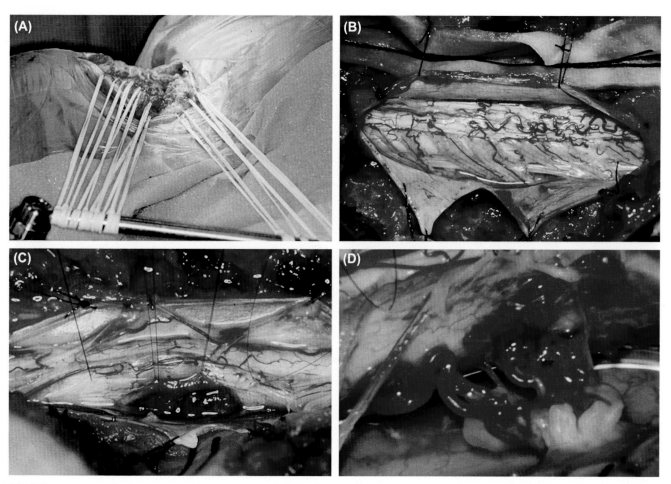

FIGURE 100.10 Posterolateral surgical approach to a type II AVM of the cervical spinal cord. (A) Fish hooks secured to a Leyla bar are used to hold the edges of the incision in the depressed position to aid in visualization of the surgical corridor. The cranio-caudad axis is oriented left to right. (B) The dura is opened sharply and retracted with suture. The exiting nerve routes are visualized. (C) After the dentate ligaments are cut, pial stitches are used to gently rotate the spinal cord to visualize the anterior surface and the AVM. (D) High-powered magnification of the AVM using the operating microscope.

to cervical lesions by maintaining the incision open with multiple fish hooks attached to a Leyla bar, which depresses the wound edges, rather than using self-retaining retractors that tend to lift the edges of the incision. A far-lateral bony exposure is accomplished by drilling the medial portion of the facet joints. After opening the dura, the dentate ligaments are divided and a view across the ventral surface of the spinal cord is achieved in much the same manner as the skull base surgeon resects bone from the skull to improve exposure and minimize retraction of neural tissue.

Once the malformation is exposed, resection is performed by careful bipolar cauterization along the edge of the malformation, coagulating and dividing feeding arteries as they enter the AVM, in much the same way a cerebral AVM is removed.

Type III Spinal AVMs

Type III spinal AVMs are extremely difficult to manage by any means. Successful treatment of a type III

spinal AVM is only achieved through staged embolization and surgical resection to gradually obliterate the various components of the malformation [10]. Treatment is associated with high risk and the prognosis is poor.

Type IV Spinal AVMs

The goal of treatment of type IV spinal AVMs is to completely occlude the arteriovenous fistula while preserving the arterial supply and venous drainage to the spinal cord. This can be accomplished by direct microsurgical disconnection or by endovascular occlusion [3,8,11]. As with type I lesions, the major problems with embolization are inability to catheterize all feeding vessels, recanalization of the fistula, and passage of the embolic material through the fistula into the draining vein and worsening venous hypertension. Surgery is the preferred treatment modality for most type IVa and IVb AVMs. Endovascular management of type IVa lesions is complicated by the difficulty of catheterizing the distal end of the long thin anterior spinal artery that usually supplies these fistulae. Embolization of

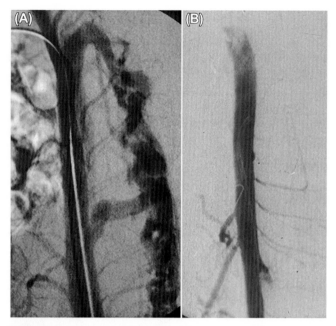

FIGURE 100.11 (A) AP Spinal angiogram of a type IVc, giant, multipediculated arteriovenous fistula in an infant. (B) Postembolization angiogram documents obliteration of the multiple fistulae and normal filling of the anterior spinal artery.

the multiple small feeders to type IVb fistulae may also be difficult or impossible. The large, high-flow, multipediculated type IVc lesions, however, are very difficult if not impossible to treat surgically and are readily managed by embolization due to the large size of the multiple feeding vessels [3] (Fig. 100.11).

Conus Medullaris AVMs

The primary objectives when treating CM AVMs are to disrupt the fistulae, similar to the treatment of types I and IV lesions and to resect any associated nidus while leaving the blood supply to the CM unperturbed. The practice in our center is to treat CM AVMs surgically, either with or without embolization. Although these lesions are challenging to treat, favorable long-term results can be obtained with this strategy [13].

CONCLUSION

Although relatively rare, spinal vascular malformations are an important group of lesions because they represent a potentially curable cause of progressive myelopathy. The physician must maintain a high index of suspicion to establish the diagnosis early in the course of the disorder if the patient is to achieve optimal benefit from treatment. Therefore, one should always suspect a spinal vascular malformation in the patient with an unexplained myelopathy or the individual carrying the diagnosis of demyelinating disease or transverse myelitis in whom the clinical picture and diagnostic tests are not entirely characteristic. MRI has made the diagnosis much simpler but high-quality spinal angiography is still necessary to confirm the diagnosis, classify the lesion, and plan appropriate treatment. Favorable outcomes can be achieved for patients treated with surgical or endovascular therapy if the intervention is appropriately completed in a timely fashion.

References

[1] Aminoff MJ, Logue V. The prognosis of patients with spinal vascular malformations. Brain J Neurol 1974;97:211–8.

[2] Anson JA, Spetzler RF. Spinal dural arteriovenous malformations. In: Awad IA, Barrow DL, editors. Dural Arteriovenous Malformations. Park Ridge, IL: American Association of Neurological Surgeons; 1993. p. 175–91.

[3] Barrow DL, Colohan AR, Dawson R. Intradural perimedullary arteriovenous fistulas (type IV spinal cord arteriovenous malformations). J Neurosurg 1994;81:221–9.

[4] Djindjian M, Djindjian R, Rey A, Hurth M, Houdart R. Intradural extramedullary spinal arterio-venous malformations fed by the anterior spinal artery. Surg Neurol 1977;8:85–93.

[5] Elsberg CA. Diagnosis and Treatment of Surgical Diseases of the Spinal Cord and Its Membrane. Philadelphia, PA: Saunders; 1916.

[6] Foix C, Alajouanine T. La myélite nécrotique subaigue. Rev Neurol 1926;2:1–42.

[7] Saindane AM, Boddu SR, Tong FC, Dehkharghani S, Dion JE. Contrast-enhanced time-resolved MRA for pre-angiographic evaluation of suspected spinal dural arterial venous fistulas. J Neurointerv Surg 2015;7:135–40.

[8] Schuette AJ, Cawley CM, Barrow DL. Spinal dural fistula. In: Bendok BR, Naidech AM, Walker MT, Batjer HH, editors. Hemorrhagic and ischemic stroke: medical, imaging, surgical and interventional approaches. New York: Thieme; 2012. p. 344–55.

[9] Spetzler RF, Detwiler PW, Riina HA, Porter RW. Modified classification of spinal cord vascular lesions. J Neurosurg 2002;96:145–56.

[10] Spetzler RF, Zabramski JM, Flom RA. Management of juvenile spinal AVM's by embolization and operative excision. Case report. J Neurosurg 1989;70:628–32.

[11] Steinmetz MP, Chow MM, Krishnaney AA, Andrews-Hinders D, Benzel EC, Masaryk TJ, et al. Outcome after the treatment of spinal dural arteriovenous fistulae: a contemporary single-institution series and meta-analysis. Neurosurgery 2004;55:77–87 [discussion 87–78].

[12] Symon L, Kuyama H, Kendall B. Dural arteriovenous malformations of the spine. Clinical features and surgical results in 55 cases. J Neurosurg 1984;60:238–47.

[13] Wilson DA, Abla AA, Uschold TD, McDougall CG, Albuquerque FC, Spetzler RF. Multimodality treatment of conus medullaris arteriovenous malformations: 2 decades of experience with combined endovascular and microsurgical treatments. Neurosurgery 2012;71:100–8.

CHAPTER

101

Reversible Cerebral Vasoconstriction Syndromes

A.B. Singhal[1], M.A. Topcuoglu[1,2]

[1]Massachusetts General Hospital and Harvard Medical School, Boston, MA, United States; [2]Hacettepe University Hospitals, Ankara, Turkey

INTRODUCTION

The term *"reversible cerebral vasoconstriction syndrome"* (RCVS) refers to a group of conditions characterized by segmental narrowing and dilatation of multiple intracerebral arteries lasting days to weeks, usually accompanied with recurrent thunderclap headaches [1–5]. Historically, patients with RCVS have been misinterpreted as having primary angiitis of the central nervous system (PACNS) or subarachnoid hemorrhage from ruptured brain aneurysms due to overlapping features such as headache, strokes, and cerebral angiographic narrowing [4,6]. Moreover, RCVS has remained under-recognized because it has been reported using variable terminology, including migrainous vasospasm, migraine angiitis, Call's or Call–Fleming syndrome, thunderclap headache–associated vasospasm, drug-induced cerebral arteritis, postpartum cerebral angiopathy, eclamptic vasospasm, benign angiopathy of the central nervous system (CNS), and CNS pseudovasculitis. The adoption of the broad term RCVS since 2007 [3] has encouraged relatively large retrospective and prospective studies that have helped to characterize the syndrome.

EPIDEMIOLOGY

Women are predominantly affected with a female:male (F:M) ratio ranging from 2:1 to 10:1. Larger studies have consistently reported a mean age of 42–43 years [2,4,5]. Cases have been reported in children as young as 8 years and in women up to age 65 years. The average age of affected women is higher than that of affected men. Women seem to have a higher frequency of underlying conditions such as migraine and depression, and develop more severe manifestations as compared to men. Reports have been published from numerous countries and regions, suggesting that all races and ethnic groups are affected. RCVS is being reported with increasing frequency either due to increased awareness of the clinical and imaging features, or improved detection of cerebral angiographic abnormalities due to advances in CT- and MR-angiography, or a truly higher incidence due to greater exposure to risk factors such as illicit drugs and vasoconstrictive medications.

RISK FACTORS AND PATHOPHYSIOLOGY

Most experts agree that the cerebral angiographic abnormalities result from an abnormality in the control of cerebrovascular tone. The mechanism underlying the recurrent thunderclap headaches that characterize this syndrome is not known. Table 101.1 shows a list of reported triggers and risk factors published in the literature. While a role for recall or publication bias cannot be excluded, many risk factors appear to implicate serotonergic or sympathomimetic pathways. The diverse range of risk factors suggests that additional pathophysiological mechanisms may be involved. The presence of reversible brain edema in patients with RCVS, and the high frequency of reversible cerebral angiographic abnormalities in patients with the posterior reversible leukoencephalopathy syndrome (PRES), suggests an overlapping pathophysiology between these syndromes [7]. At the molecular level, several nonhormonal molecular and biological factors (oxidative stress, prostaglandins, endothelial progenitor cells, endothelin, and others) have been implicated.

Primer on Cerebrovascular Diseases, Second Edition
http://dx.doi.org/10.1016/B978-0-12-803058-5.00101-6

TABLE 101.1　Risk Factors and Triggers

1. **Changing levels of female reproductive hormones:** childbirth (post-partum angiopathy) ovarian stimulation, oral contraceptive pills

2. **Headache disorders:** primary thunderclap headache, migraine, benign sexual headache, benign exertional headache

3. **Vasoconstrictive substances:** selective serotonin reuptake inhibitors (SSRIs), serotonin–noradrenaline reuptake inhibitors (SNRIs), triptans, isometheptene, ergotamine tartrate, methergine, cough and cold suppressants (phenylpropanolamine, pseudoephedrine), diet and energy pills (amphetamine derivatives, hydroxycut), epinephrine, bromocriptine, lisuride; illicit drugs (cocaine, ecstasy, marijuana, lysergic acid diethylamide); chemotherapeutic drugs (tacrolimus, cyclophosphamide); blood products (red blood cell transfusions, erythropoietin), intravenous immune globulin, interferon alpha, nicotine patches, licorice, ma huang, pheochromocytoma, carcinoid tumor

4. **Miscellaneous:** hypercalcemia, porphyria, unruptured saccular cerebral aneurysm, head trauma, spinal subdural hematoma, postcarotid endarterectomy, neurosurgical manipulation of intracerebral arteries, carotid glomus tumor, tonsillectomy, neck surgery, high altitude, swimming

As of 2016 there is no epidemiological evidence to support a causal relationship between vasoconstrictive drugs and RCVS; however, these drugs have been implicated due to the temporal relationship between exposure and onset, and their known vasoconstrictive effects. Several lines of evidence implicate imbalances in the estrogen–progesterone axis in triggering RCVS in women: female preponderance, onset after childbirth, and the association between RCVS and oral contraceptive pills or hormonal manipulation. Sexual activity is associated with RCVS onset in men. Some authors have suggested that the reversible angiographic abnormalities result from a mild, transient inflammatory process; however, autopsy studies, brain biopsy, arterial histopathology, serological tests, and cerebrospinal fluid (CSF) examination results have not shown any evidence for inflammation.

CLINICAL FEATURES

The clinical hallmark of RCVS is its dramatic and unforgettable onset with a sudden, excruciating headache that reaches peak intensity within seconds, meeting the definition for *thunderclap headache*. These headaches are usually diffuse or located at the vertex, can recur for days to weeks, and are different from migraine headache. Less than 10–15% of patients develop subacute headache or have onset without any headache. Generalized tonic-clonic seizures can develop at onset, however recurrent seizures or epilepsy is rare.

Single or multiple sudden, severe headaches can be the only clinical manifestation of RCVS. The severe head pain makes patients agitated, and may contribute to hypertension at the time of presentation. Cohort studies of admitted patients show that one-third to half go on to develop complications such as ischemic or hemorrhagic strokes or cerebral edema, with concomitant focal neurological deficits, over a span of 1–2 weeks. Visual symptoms and signs are common (scotomas, blurring, hemianopia, cortical blindness, partial or complete

Balint syndrome). Other neurological findings include hemiplegia, hyper-reflexia, ataxia, tremor, aphasia, and alterations in mental status as well as coma in patients with large strokes.

In general, headaches subside and clinical deficits improve over time. In the absence of large destructive brain lesions, complete or near-complete clinical recovery occurs within few weeks. Angiographic abnormalities may resolve earlier or later than the clinical deficits, but invariably resolve before the 3-month time point when follow-up vascular studies are typically obtained. Rare patients (2–3%) develop progressive angiographic narrowing that can result in massive strokes and fatal outcome.

LABORATORY TEST RESULTS

Routine blood tests and urinalysis are usually normal. Urine vanillylmandelic acid and 5-hydroxyindoleacetic acid may be considered to exclude vasoactive tumors such as carcinoid pheochromocytoma that have been associated with RCVS. Serum and urine toxicology screens are useful to investigate for triggers such as marijuana and cocaine. Erythrocyte sedimentation rate (ESR), C-reactive protein (CRP) level, infectious and rheumatological panel tests, and CSF examination are indicated to rule out mimics when the diagnosis remains insecure after a careful review of history and imaging findings (see section Diagnostic Approach). There is no role for brain biopsy or temporal artery biopsy other than to rule out cerebral vasculitis.

BRAIN IMAGING RESULTS

Urgent brain and vascular imaging is necessary to exclude secondary causes of thunderclap headache, such as ruptured brain aneurysms, cerebral artery dissection, embolic stroke, cerebral venous sinus thrombosis, pituitary apoplexy, intracranial hemorrhage, and meningitis.

Despite the presence of widespread and often severe cerebral vasoconstriction, about 30–70% of RCVS patients have normal parenchymal brain imaging upon admission. The combination of thunderclap headache, initially normal brain MRI findings, and multifocal cerebral artery vasoconstriction, is diagnostic for RCVS [1]. Inpatient cohort studies show that about 80% eventually develop abnormal brain imaging results from small convexity (non-aneurysmal) subarachnoid hemorrhages, ischemic strokes, parenchymal hemorrhages, reversible brain edema (PRES), and rarely, subdural hemorrhage. Any lesion combination can be present. Indeed, the *"normal-becoming-abnormal"* evolution of brain MRI is distinctive for RCVS [1,5]. The topography of ischemic strokes is somewhat unique to this syndrome in that they are typically bilateral, symmetric, initially localize to cortical–subcortical junctions, and involve the watershed regions of the anterior, middle, and posterior cerebral arteries or the superior and posterior inferior cerebellar arteries. Cerebral perfusion defects are similarly located, consistent with the distribution of arterial vasoconstriction. Recent studies have shown that RCVS is the most frequent cause of convexal subarachnoid hemorrhages in individuals below age 60 years. Parenchymal hemorrhages are typically lobar, rarely multiple in number, and their location is similar to that of infarcts, suggesting a mechanistic role for ischemia–reperfusion injury. Fluid-attenuated inversion recovery (FLAIR) MRI often shows dot- or linear-shaped hyperintensities within sulcal spaces, presumably from slow flow within dilated surface vessels.

The main diagnostic feature of RCVS is multifocal areas of smooth tapering cerebral artery narrowing and dilatation ("sausage on a string" appearance). Historically, this appearance has been attributed to PACNS; however, most patients with the latter have an irregular, notched appearance of cerebral arteries as can be expected from an inflammatory process. Angiographic abnormalities in RCVS can be documented using transfemoral, CT-, or MR-angiography [8]. Transcranial Doppler (TCD) ultrasound can show elevated blood flow velocities, but many patients have normal blood flow velocities despite the presence of diffuse vasoconstriction. Hence, this modality has greater utility in monitoring angiographic progression. TCD ultrasound studies have confirmed the presence of abnormal cerebral vasoreactivity (reduced breath-holding index), reflecting altered cerebral vascular tone in RCVS.

It is important to note that the cerebral vasoconstriction starts distally and progresses proximally; hence initial angiographic studies can be normal in patients with otherwise classic presentations of RCVS. In such cases a follow-up vascular imaging study may be justified after an interval of 3–5 days. Dissection of the carotid or vertebral arteries has been documented in many patients with RCVS and may be related to dynamic changes in arterial caliber or acute swings in blood pressure. A higher incidence of cervical artery dissection, unruptured cerebral aneurysms, and cavernous malformations suggests that a hitherto-unidentified subtle connective tissue disorder may be present in RCVS. Systemic arteries are typically spared. The time course of vasoconstriction is variable, but most patients show resolution within 3 months.

DIAGNOSTIC APPROACH

The diagnosis of RCVS is fairly straightforward and can be made at the bedside, upon initial clinical presentation, with nearly 100% accuracy by the experienced clinician [1]. The key clinical and angiographic features, summarized over a decade ago, include recurrent severe thunderclap headaches and multifocal intracranial arterial narrowing and dilatation, in the absence of a ruptured cerebral aneurysm and without evidence for mimics such as cerebral vasculitis [3]. However, since at least the 1960s, cases with RCVS have been misdiagnosed as having PACNS due to overlap in the clinical features (headache, focal deficits) and imaging features (stroke, arterial irregularities). Only recently has it been recognized that the characteristic of headaches is quite distinct (thunderclap-type versus chronic dull aches), and that the brain lesion patterns and angiographic features also differ substantially (Figs. 101.1–101.3). A recent study comparing the features of 159 patients with RCVS to 47 patients with PACNS [1] provides criteria that can diagnose RCVS and distinguish it from PACNS, namely (1) recurrent thunderclap headache (98% specificity, 99% positive predictive value); (2) single thunderclap headache combined with either normal neuroimaging, border zone infarcts, or vasogenic edema (100% specificity and 100% positive predictive value); (3) abnormal angiography, but no thunderclap headache and neuroimaging showing no brain lesions (compared with deep or brain stem infarcts in PACNS).

As stated in the preceding paragraphs, it is important to first exclude the more common causes of thunderclap headache with urgent brain and vascular imaging (aneurysmal subarachnoid hemorrhage, parenchymal hemorrhage, cerebral artery dissection, cerebral venous sinus thrombosis, middle and posterior cerebral artery embolism). If imaging excludes these etiologies and shows smooth segmental multifocal narrowing, the diagnosis of RCVS is secure. In isolation, the angiographic abnormalities can raise concern for intracranial atherosclerosis, infectious arteritis, vasculitis, moyamoya disease, fibromuscular dysplasia, and other cerebral arteriopathies; however, these entities can be excluded by a careful medical history. If vascular imaging proves to be normal, and the patient exhibits atypical symptoms (neck stiffness, confusion, fever, and so

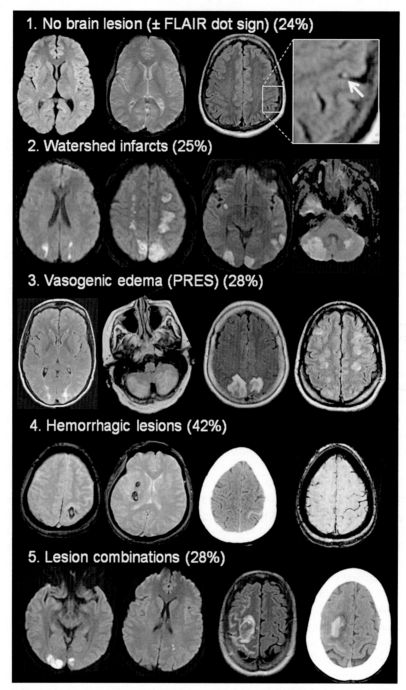

FIGURE 101.1 Brain lesions in reversible cerebral vasoconstriction syndrome (RCVS). Representative brain images from patients with RCVS are shown to highlight different lesion patterns. The numbers in parentheses show the percentages of the lesion patterns; totals exceed 100% due to lesion combinations. Pattern 1, no acute parenchymal lesion. Normal axial diffusion-weighted (DWI), gradient-echo (GRE) and fluid-attenuated inversion recovery (FLAIR) images. The hyperintense dot sign is present on FLAIR (far right, *arrow*). Pattern 2, borderzone/watershed infarcts. Far left, DWI showing typical symmetric, posterior infarcts that spare the cortical ribbon. Middle and far right, DWI shows widespread watershed infarcts. Pattern 3, vasogenic edema. Subcortical crescent-shaped T2-hyperintense lesions consistent with the posterior reversible encephalopathy syndrome (PRES) on FLAIR. Pattern 4, hemorrhagic lesions. The two left images (axial GRE) show simultaneous lobar and deep intraparenchymal hemorrhages. The two right images show convexal subarachnoid hemorrhages on CT and axial GRE. Pattern 5, lesion combinations. The two left images show bilateral watershed infarcts on DWI and the two right images show lobar as well as convexal subarachnoid hemorrhages on axial FLAIR and CT, all in the same patient. *Reprinted with permission from John Wiley & Sons, Inc., from Annals Neurol 2016;79(6):882–94.*

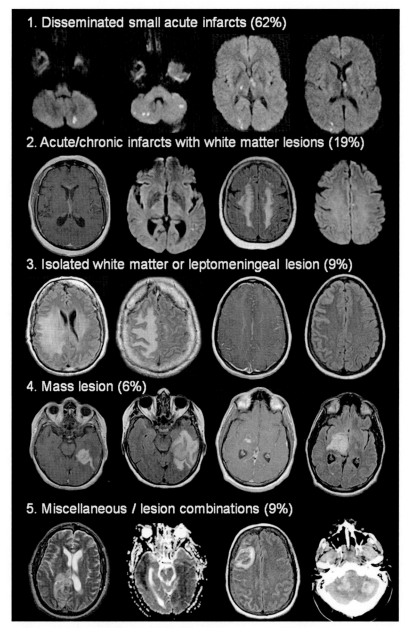

FIGURE 101.2 **Brain lesions in primary angiitis of the central nervous system (PACNS).** Representative brain images from patients with PACNS highlighting different lesion patterns. The numbers in parentheses show the percentages of the lesion patterns; totals exceed 100% due to lesion combinations. Pattern 1, disseminated small acute infarcts. Axial diffusion-weighted images (DWI) showing multiple infarcts in the same patient. Pattern 2, acute, subacute or chronic infarcts with underlying T2-hyperintense white matter lesions, often with contrast enhancement. Left, contrast-enhanced axial T1 image showing an enhancing corpus callosum lesion; middle, two DWI showing punctate infarcts; and right, axial fluid-attenuated inversion recovery (FLAIR) image showing diffuse white matter lesions, all in the same patient. Pattern 3, isolated white matter or leptomeningeal lesions without infarcts. The two left images show diffuse white matter hyperintense lesions on axial FLAIR in one patient. The two right images show right hemispheric leptomeningeal lesions on contrast-enhanced axial T1 and FLAIR in another patient. Pattern 4, mass lesion. The two left images show a mass lesion on contrast-enhanced axial T1 and FLAIR from a patient with biopsy-proven amyloid beta–related angiitis. The two right images show a mass lesion on contrast-enhanced axial T1 and FLAIR from a patient with biopsy-proved granulomatous angiitis. Pattern 5, miscellaneous. The two left images show large territorial infarcts on axial FLAIR and apparent diffusion coefficient images. The two right images show lobar hemorrhages (FLAIR) and simultaneous cerebellar hemorrhages (CT) in two separate patients with sympathomimetic drug-induced PACNS. *Reprinted with permission from John Wiley & Sons, Inc., from Annals Neurol 2016;79(6):882–94.*

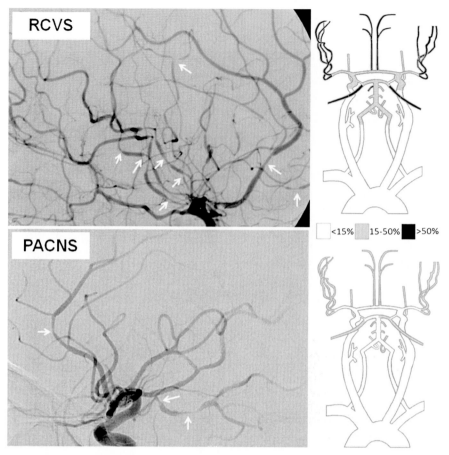

FIGURE 101.3 **Cerebral angiographic features of RCVS and PACNS.** Top, sagittal digital subtraction angiogram (DSA) image showing the typical widespread "sausage on a string" appearance in RCVS (*arrows*). Bottom, DSA showing the typical irregular notched appearance in PACNS (*arrows*). The corresponding figures on the right are schematic representations of the distribution of angiographic abnormalities in RCVS and PACNS. The angiographic abnormalities in RCVS are relatively symmetric, proximal, and more severe; the arterial caliber changes typically begin after dural penetration. *Reprinted with permission from John Wiley & Sons, Inc., from Annals Neurol 2016;79(6):882–94.*

on), then alternate diagnosis such as CT-negative subarachnoid hemorrhage and meningitis should be considered and ruled out by CSF examination. A common mistake is to attribute the severe onset headache to a severe migraine, and treat patients with antimigraine agents (which can precipitate further vasoconstriction).

MANAGEMENT

Appropriate management rests on establishing a prompt and accurate diagnosis, excluding mimics, removing the offending vasoconstrictive trigger (e.g., selective serotonin reuptake inhibitors, SSRIs), and offering symptomatic treatment for the severe head pain and agitation. Triptans and other vasoconstrictive antimigraine agents are contraindicated. Stool softeners and bed rest are advocated to minimize the Valsalva maneuver, which can trigger recurrent thunderclap headaches. Pharmacological blood pressure manipulation is best avoided since lowering blood pressure can compromise cerebral perfusion, and raising it can worsen the vasoconstriction or induce reperfusion injury. Acute seizures may warrant treatment; however, long-term seizure prophylaxis is probably unnecessary. Patients should be counseled to avoid physical exertion, the Valsalva maneuver, and other known triggers of recurrent headaches for a few weeks. Usual stroke preventive medications, such as antiplatelets, anticoagulants, and cholesterol-lowering agents, are not indicated.

Because rare cases can develop a progressive clinical course, it is reasonable to admit patients for observation for the first few days after symptom onset. Patients with early ischemic stroke, and those with prior hypertension, depression, and serotonergic antidepressant exposure, may be at higher risk for clinical worsening over the ensuing 10–14 days. On the other hand, those with convexal subarachnoid hemorrhage and normal parenchymal imaging on admission appear to have a lower risk and may not require intensive monitoring or prolonged observation. As the vast majority of patients recover spontaneously within days to weeks, specific drugs and interventions to address cerebral vasoconstriction should be avoided as far as possible unless the patient develops progressive neurological deficits. Calcium channel blockers such as nimodipine and verapamil do not relieve vasoconstriction but may be offered for a few weeks to relieve the intensity of headache. Glucocorticoids have been associated with worse outcome, and in fact may contribute to clinical worsening, so in our opinion they are contraindicated in RCVS [9]. Intra-arterial vasodilator therapy has equivocal effects: while clinical and angiographic improvement or stabilization has been documented, some reports show relentless progression and even rebound vasoconstriction or ischemic–reperfusion injury after intra-arterial vasodilator treatment. Substantial clinical experience and judgment are required to determine the appropriate threshold for intervention.

PROGNOSIS AND COUNSELING

The long-term outcome of RCVS is excellent. Some patients go on to have chronic headaches or depression [10]. Patients should be counseled to avoid future exposure to potential precipitants such as marijuana, diet pills, exercise stimulants, and potent serotonergic antidepressants. If clinically warranted to treat long-term depression, start with less vasoconstrictive antidepressants (amitriptyline, bupropion), although it should be noted that a small number of patients have been reexposed to more potent serotonergic antidepressants without recurrence of RCVS. Genetic counseling is not necessary, since there are no reports of family members being affected. The risk of RCVS or related syndromes (eclampsia/PRES) in future pregnancies is uncertain,

but it is prudent to closely monitor and promptly treat hypertension, proteinuria, and other signs of pregnancy-induced hypertension. Patients should be reassured that the risk of a single recurrent thunderclap headache is low, and that the risk of a second episode of RCVS with recurrent thunderclap headaches and complications such as stroke is negligible.

References

[1] Singhal AB, Topcuoglu MA, Fok JW, et al. Reversible cerebral vasoconstriction syndromes and primary angiitis of the central nervous system: clinical, imaging, and angiographic comparison. Ann Neurol 2016;79:882–94.

[2] Ducros A, Boukobza M, Porcher R, Sarov M, Valade D, Bousser MG. The clinical and radiological spectrum of reversible cerebral vasoconstriction syndrome. A prospective series of 67 patients. Brain 2007;130:3091–101.

[3] Calabrese LH, Dodick DW, Schwedt TJ, Singhal AB. Narrative review: reversible cerebral vasoconstriction syndromes. Ann Intern Med 2007;146:34–44.

[4] Muehlschlegel S, Kursun O, Topcuoglu MA, Fok J, Singhal AB. Differentiating reversible cerebral vasoconstriction syndrome with subarachnoid hemorrhage from other causes of subarachnoid hemorrhage. JAMA Neurol 2013;70:1254–60.

[5] Singhal AB, Hajj-Ali RA, Topcuoglu MA, et al. Reversible cerebral vasoconstriction syndromes: analysis of 139 cases. Arch Neurol 2011;68:1005–12.

[6] Hajj-Ali RA, Singhal AB, Benseler S, Molloy E, Calabrese LH. Primary angiitis of the CNS. Lancet Neurol 2011;10:561–72.

[7] Singhal AB. Postpartum angiopathy with reversible posterior leukoencephalopathy. Arch Neurol 2004;61:411–6.

[8] Chen SP, Fuh JL, Wang SJ, et al. Magnetic resonance angiography in reversible cerebral vasoconstriction syndromes. Ann Neurol 2010;67:648–56.

[9] Singhal AB, Topcuoglu MA. Glucocorticoid-associated worsening in reversible cerebral vasoconstriction syndrome. Neurology 2016. [in press].

[10] John S, Singhal AB, Calabrese L, et al. Long-term outcomes after reversible cerebral vasoconstriction syndrome. Cephalalgia 2016;36:387–94.

CHAPTER

102

Spontaneous Dissections of Cervicocephalic Arteries

E.C. Leira¹, J. Biller²

¹University of Iowa, Iowa City, IA, United States; ²Loyola University Chicago, Stritch School of Medicine, Maywood, IL, United States

ETIOLOGY AND PHYSIOPATHOLOGY

Dissection is a focal arteriopathy caused by a hematoma inside the arterial wall. It occurs most commonly in the extracranial and most mobile segments of the internal carotid and vertebral arteries. The specific vulnerability of this particular vascular segment to trauma is probably related to structural anatomical factors. The extracranial carotid segment is freely movable with a fixation at the entry into the carotid canal [1]. Similarly, the extracranial vertebral artery also has high mobility when passing through the transverse foramina of the cervical spine [1]. Not surprisingly, a history of cervical trauma in the prior month is strongly associated with a cervicocephalic artery dissection (CAD) [2]. But the types of trauma leading to wall hematomas are diverse. A gross traumatic event involving the head and neck, such as motor vehicle accidents, assault strangulation or hanging attempt, is sometimes recalled. However, the trauma is reported to be "mild" in nearly 90% of cases [2]. This includes trauma included in activities of daily living such as heavy lifting, spinal manipulative therapy [3], or practicing sports that involve rotation and accelerations. Nevertheless, the reporting of these relatively mild injuries in patients directly questioned about trauma raises natural concerns for recall biases so the concept of mechanical trigger event for CAD might be more appropriate [2]. Occasionally, the early symptoms of medullary ischemia, such as paroxysmal sneezing, can be mistaken as a trigger [4]. Also, the timing between the neck mechanical insult and neurological symptoms is quite variable, ranging from seconds to weeks, which further complicates the establishment of

a causal relationship [5]. When a mechanical trigger is minor or absent, the dissection can be considered spontaneous. While a predisposing arterial tendency to CAD is suspected in those cases, most patients do not have any identifiable disease or abnormal phenotype. A familial aggregation has been suggested, that would indicate a genetic predisposition [6]. A minority of CAD are secondary to vulnerability due to identified connective tissue disorders, such as Ehlers–Danlos syndrome Type IV, Marfan syndrome, or osteogenesis imperfecta Type I [7].

Once an intimal tear occurs in the wall of the artery, the blood enters between the layers of the arterial wall splitting the media forming a false lumen. An intimal flap often separates the true lumen from the false lumen, which might be connected. If the hematoma is formed between the intima and the media layers, a long segment of arterial stenosis in the true lumen or total occlusion will occur. Conversely, if the hematoma occurs between the media and the adventitia, the artery may undergo a pseudoaneurysmal dilatation [1]. Pathological changes in CAD occur at the media–adventitia border, and includes vacuolar degeneration, capillary neoangiogenesis, and erythrocyte extravasation [7]. Facial or neck pain can result from the compression of arterial nerve endings by the intramural hematoma. The arterial expansion of the mural hematoma in CAD can also compress the sympathetic fibers wrapped around the carotid artery and cause Horner syndrome, or lower cranial neuropathies (cranial nerve (CN) IX to XII; less commonly CN V and VII) that may be mistaken for infiltrating skull base tumors. Often CAD results in cerebral ischemia. The infarcts from CAD that are commonly attributed to artery–artery embolization from a secondary luminal thrombus over the injured vessel, and less likely to a hemodynamic (flow-reduction) mechanism [8]. Less commonly, the rupture of an intracranial pseudoaneurysm can result in a subarachnoid hemorrhage, typically with intracranial vertebral artery dissections (Fig. 102.3).

EPIDEMIOLOGY

Given that CAD can present with only headaches or even asymptomatic, and has a tendency to heal spontaneously, it is likely that the true incidence might be underestimated. The annual incidence of CAD was 2.6 per 100,000 in Olmsted County, Minnesota [9]. The mean age was 45.8 years, underscoring a higher incidence of CAD in younger patients [9]. Overall, CAD represents 25% of the strokes of the less common category ("other etiology") in the Trial of ORG 10172 in Acute Stroke Treatment (TOAST) classification [10]. But the prevalence of CAD in stroke patients is significantly higher (13%) when only patients younger than 45 years are considered [11]. Several risk factors/associations for nontraumatic CAD have been proposed and include cervical spine manipulation, recent upper respiratory infection, aortic root diameter >34 m/m, migraine, hyperhomocysteinemia, hypertension, hypercholesterolemia, α1-antitrypsin deficiency, and a relative diameter change (>11.8%) during the cardiac cycle of the common carotid artery [12,13] (Table 102.1).

The extracranial internal carotid artery is involved in about two-thirds of cases of dissection, the extracranial vertebral artery in about one-third, and both are involved in 5% of cases. Multivessel CADs and/or multivessel cervicocephalic and visceral arterial dissections should suggest an underlying arteriopathy even in the absence of angiographic abnormalities [14].

CLINICAL PRESENTATION

The clinical presentation of CAD is variable. It ranges from an asymptomatic incidental finding, headache or neck pain, pulsatile bruit or tinnitus, Horner syndrome, lower cranial nerve palsies, cerebral infarction (Fig. 102.1), retinal infarctions, subarachnoid hemorrhage (Fig. 102.3), or a combination of all these symptoms. The symptoms vary depending on whether the CAD is in the carotid or in the vertebral artery [15].

Isolated headache is not uncommon with CAD, and is the presenting symptom in half of the patients with carotid dissection, and one-third of the patients with vertebral artery dissection. A previous history of migraine is present in about 20% patients [16]. Still, the type of pain with CAD is usually different and more intense than any previous headaches. A favorable response of a CAD-induced headache to triptans can occur, and therefore this should not be used to exclude a CAD diagnosis [17]. Some patients report a "thunderclap" presentation, which is an unusually severe headache that builds up in less than 1 min to maximal intensity, particularly with vertebral artery dissections [15]. The headache is usually described as unilateral and ipsilateral to the side of the arterial lesion. In carotid dissections it typically involves the face, particularly around the eye, and anterior neck. This type of pain was classically known as carotidynia. Conversely, in vertebral artery dissections the pain tends to refer to the posterior part of the neck, occiput, and sometimes radiating to the front of the head.

An ipsilateral Horner syndrome might be present. This syndrome is characterized by ptosis, miosis, and anisocoria more noticeable in a dark environment. Anhydrosis is typically absent in internal carotid artery dissections because fibers that activate the sweating travel through the external carotid artery, but it might be present with the central Horner syndrome of vertebral artery dissections and lateral medullary infarctions. The Horner syndrome in carotid dissections can be transient, and subtle enough that might only be detected with topical pupillary pharmacological testing [18]. Compression of the lower cranial nerves can result in dysphonia and dysphagia. Tinnitus is a symptom seen more commonly with carotid dissection than with vertebral. The tinnitus tends to be ipsilateral to the lesion, and pulsatile.

Cerebral infarctions are more common with vertebral artery dissections (Fig. 102.2), but tend to be more severe with carotid artery lesions.

Typically a significant ipsilateral headache or neck pain precedes cerebral infarctions from CAD in days or weeks, a clinical feature that helps the diagnosis. This pattern is different than the most common atherothrombotic or embolic infarctions, which are not typically preceded pain, and whenever present is usually not of significant intensity. Carotid dissection-induced ischemia (Fig. 102.1) can range from retinal transient ischemic attacks (TIAs) and infarctions, TIAs and small scattered embolic lesions in the middle cerebral artery (MCA) territory or the border zones between the MCA and the anterior and posterior cerebral arteries (either from hypoperfusion or embolism), to a large (complete) MCA territory infarction. Dissection of the MCA itself (Fig. 102.4) can occur and produce similar symptoms of an ICA dissection except for the retinal symptoms and cranial nerve palsies. Patients with MCA dissection typically have a unilateral headache rather than neck pain. In addition to a postganglionic Horner syndrome, other neuro-ophthalmological manifestations of internal carotid artery dissections include central or branch retinal artery occlusions, ophthalmic artery occlusion, ischemic optic neuropathy (anterior or posterior), and homonymous hemianopia.

Infarctions with vertebral artery dissection typically involve the adjacent dorsolateral medulla (Wallenberg

TABLE 102.1   Factors Associated With Cervical Artery Dissection

Blunt head and neck trauma
Cervical manipulation
A variety of contact and noncontact sports
Fibromuscular dysplasia (FMD)
Marfan syndrome
Ehlers–Danlos syndrome Type IV
Pseudoxanthoma elasticum
Loeys–Dietz syndrome
Turner syndrome
William syndrome
Menkes disease
Osteogenesis imperfecta Type I
Hereditary hemochromatosis
Coarctation of the aorta
Autosomal dominant polycystic kidney disease (ADPKD)
α1-Antitrypsin deficiency
Lysyl oxidase deficiency
Ultrastructural connective tissue abnormalities
Mutations in the gene for type III and type I collagen
Irregular collagen fibrils and fragmentation of elastic fibers
Cystic medial necrosis
Reticular fiber deficiency
Accumulation of mucopolysaccharides
Elevated arterial elastase content
Atherosclerosis
Arterial hypertension
Widened aortic root
Moyamoya disease
Homocystinuria (cystathionine β-synthase deficiency)
Hyperhomocysteinemia
677T genotype methylenetetrahydrofolate reductase
Migraines
Recent infection (mainly respiratory tract infections; many of these patients may have a transient arteriopathy)
Luetic arteritis
Lentiginosis
Arterial redundancies (coils, kinks, and loops) especially if bilateral
Styloid process length
Sympathomimetic drug abuse
Current use of oral contraceptives

Adapted from Biller J, Sacco RL, Albuquerque FC, et al. Cervical arterial dissections and association with cervical manipulative therapy: a statement for healthcare professionals from the American Heart Association/American Stroke Association. Stroke 2014;45(10):3155–3174, on behalf of the American Heart Association Stroke Council.

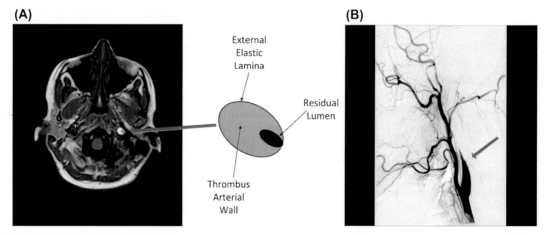

FIGURE 102.1   A 27-year-old woman complained left-sided headaches 1 week post-partum. Three days later, she had transient aphasia. (A) MRI shows a crescent sign in the left ICA. (B) Cerebral angiogram shows a tapering sign in the left ICA consistent with dissection.

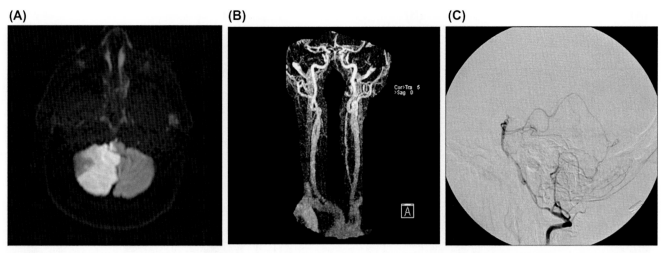

FIGURE 102.2   A 27-year-old man developed hiccups, vertigo, and imbalance shortly after a neck manipulation. (A) MRI DWI brain shows large right PICA territory and lateral medulla infarctions. (B) MRA neck shows a right vertebral artery dissection confirmed by (C) angiography.

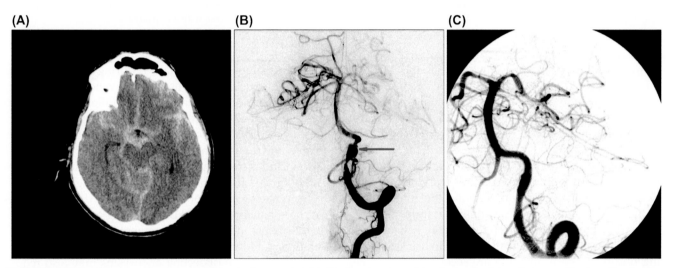

FIGURE 102.3   A 66-year-old man complained of unusual right-sided neck pain and headaches for 2 days before suddenly collapsing in a coma. (A) Noncontrast CT of head shows extensive subarachnoid hemorrhage and hydrocephalus. (B) Angiogram shows a left vertebral artery dissection with pseudoaneurysm (*arrow*). (C) Follow-up angiogram after vertebral artery reconstruction with a flow-diverting stent. *Courtesy of Prof. E. Samaniego.*

syndrome) and the posterior inferior cerebellar artery (Fig. 102.2). Embolism from a vertebral artery dissection may also cause medial medullary infarction, cerebellar infarction, and posterior cerebral artery distribution infarction. Less commonly, a vertebral artery dissection can result in a high cervical spinal cord infarction or cervical radiculopathy. Brain stem infarctions from embolism are probably less common given that the brain stem penetrator vessels are perpendicular to the flow in the basilar artery. Vertebral artery dissections can extend to the basilar artery, and in rare cases originate in the basilar artery itself. That can lead to significant brain stem ischemia from occluding multiple perforators. Subarachnoid hemorrhage can be a presentation of intracranial CAD, and is more common with vertebral artery dissections

[19] and can also occur with MCA pseudoaneurysms (Fig. 102.3). Subarachnoid hemorrhage has a poor prognosis with high likelihood of rebleeding and a poor neurological outcome.

DIAGNOSIS

The diagnosis of CAD is based on the history and imaging demonstration of the dissection. The differential diagnosis includes conditions that can result in unusually strong or apoplectic headaches with focal neurological symptoms, such as cerebral venous thrombosis, reversible cerebral vasoconstrictive syndrome, posterior reversible encephalopathy syndrome, or sacular brain

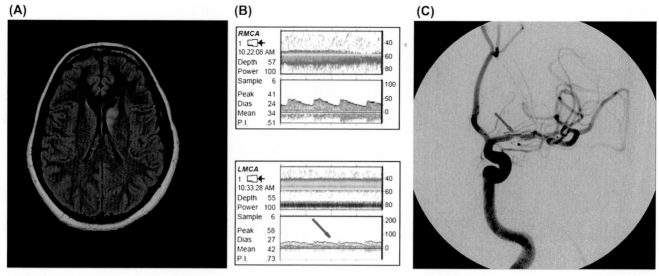

FIGURE 102.4   A 28-year-old woman complained of left-sided headaches shortly after starting an exercise program. Two weeks later she developed transient aphasia. (A) MRI shows residual infarction in the left caudate nucleus. (B) Transcranial Doppler shows marked alteration of the flow in the left MCA despite normal MRA. (C) Cerebral angiogram shows a left MCA dissection with a flap *(arrow)*.

aneurysms. Brain imaging, such as CT and MRI, can be used to diagnose cerebral infarction or bleeding, as well as to exclude mimics. While the diagnosis of carotid dissection can be done sometimes based on a "crescent sign" seen on axial MRI of the brain alone (Fig. 102.1), arterial neuroimaging is usually needed for diagnosis. This can be done by either magnetic resonance angiography (MRA), or computerized tomography angiography (CTA). Catheter angiography remains the gold standard but given its invasiveness is usually left for complicated cases, or when planning an intervention. Because CAD is a disease of the arterial wall, the luminal arterial imaging findings seen in imaging are secondary. The exception is carotid ultrasound or axial MRI where the wall hematoma can be visualized. The typical arterial imaging finding is a gradual tapering of the arterial lumen (Fig. 102.1), a finding that is not typical of atherosclerosis. A lumbar puncture may be necessary to diagnose small subarachnoid hemorrhage from pseudoaneurysm in certain cases. Pharmacological ocular testing might be used to confirm a Horner syndrome.

MANAGEMENT

The management of a CAD includes (1) acute stroke management (ischemic and subarachnoid hemorrhage); (2) strategies to prevent subsequent cerebral infarction or subarachnoid hemorrhage; and (3) management of pain and other complications. The acute management of a cerebral infarction caused by CAD typically involves the same procedure as any other ischemic stroke patient. After all, even if suspected, the diagnosis the CAD will not be confirmed

until arterial imaging is obtained during the stroke evaluation. This includes the use of intravenous recombinant tissue plasminogen activator (rtPA). While rtPA could theoretically worsen the arterial wall hemorrhagic process, the beneficial thrombolytic effect in the distal embolus predominates, and it has been deemed to be safe in practice [20]. It also includes mechanical thrombectomy in selected cases, although the dissection and false lumen might technically complicate the access of the device to the distal thrombus. In those patients with CAD presenting as a subarachnoid hemorrhage, the acute management is the same as in an aneurysmal subarachnoid hemorrhage. Like saccular brain aneurysms, those dissecting aneurysms may also require mechanical stenting or coiling.

The second strategy is to prevent further (or the first) ischemic cerebral events. The options are medical or endovascular treatment. Most patients are treated medically. The overall prognosis of CAD is considered to be good. Complete resolution of the arterial lesion was observed in about half of the cases [9]. The risk of stroke recurrence is lower than atherosclerosis and ranges 0.3–1%/year [21,22]. In this benign context, an invasive endovascular treatment of the CAD does not appear to be justified in most cases. The prevention of events is typically done with antithrombotic medications. While there is also a theoretical concern of exacerbating the arterial wall hemorrhagic process, the goal is to prevent luminal thrombus that can produce embolism. For decades, there has been a great deal of controversy of whether to use antiplatelet agents or anticoagulants. The practice was clearly divided, with little data to support either position [23]. In 2015, the results of the long-awaited Cervical Artery Dissection Stroke Study (CADISS) trial were published [21]. This was a prospective randomized open blinded end-point trial comparing the benefit of antiplatelets with anticoagulants in preventing further stroke in extracranial CAD. A total of 250 patients were randomized (118 carotid and 132 vertebral) in a course of 7 years, which underscores the difficulty in doing clinical trials in this condition. The antithrombotics were initiated within 7 days, mean time to randomization was 3.65 days, and were maintained for 3 months. The primary outcome was ipsilateral stroke or death in an intention to treat analysis. This end point was achieved by a similarly low number of patients in either group (2% antiplatelets and 1% anticoagulants), and there was no statistical difference (OR = 0.335, 0.006–4.23). A major criticism of this study was the relatively long window of enrollment which might have potentially excluded neurologically unstable patients [24], and the exclusion of intracranial dissections. Nevertheless, this study provides strong evidence of equipoise between antiplatelets and anticoagulants in preventing ischemic events in CAD, and is likely to shift practice in favor of the antiplatelets given the better safety record. Endovascular therapy should be considered for patients with unstable or progressive symptoms despite best medical therapy. Endovascular therapy is typically considered if a pseudoaneurysm develops, particularly in an intracranial location. This underscores the importance of performing follow-up imaging in these patients to assess for the development of this complication. The conversations about potential complications of CAD, including pseudoaneurysms and the need for endovascular treatment, are very distressing for patients and their families to the point of inducing posttraumatic stress disorder [25], so they must be conducted with caution. The neck pain and headache are symptoms that often require acute management with analgesics. Some patients have persistent pain beyond a few weeks and months that may require long-term strategies to treat neuropathic pain, such as antiepileptic drugs.

References

[1] Kim YK, Schulman S. Cervical artery dissection: pathology, epidemiology and management. Thromb Res 2009;123(6):810–21.

[2] Engelter ST, Grond-Ginsbach C, Metso TM, et al. Cervical artery dissection: trauma and other potential mechanical trigger events. Neurology 2013;80(21):1950–7.

[3] Smith WS, Johnston SC, Skalabrin EJ, et al. Spinal manipulative therapy is an independent risk factor for vertebral artery dissection. Neurology 2003;60(9):1424–8.

[4] Swenson AJ, Leira EC. Paroxysmal sneezing at the onset of lateral medullary syndrome: cause or consequence? Eur J Neurol 2007;14(4):461–3.

[5] Frisoni GB, Anzola GP. Vertebrobasilar ischemia after neck motion. Stroke 1991;22(11):1452–60.

[6] Grond-Ginsbach C, de Freitas GR, Campos CR, et al. Familial occurrence of cervical artery dissection–coincidence or sign of familial predisposition? Cerebrovasc Dis 2012;33(5):466–70.

[7] Debette S. Pathophysiology and risk factors of cervical artery dissection: what have we learnt from large hospital-based cohorts? Curr Opin Neurol 2014;27(1):20–8.

[8] Morel A, Naggara O, Touze E, et al. Mechanism of ischemic infarct in spontaneous cervical artery dissection. Stroke 2012;43(5):1354–61.

[9] Lee VH, Brown Jr RD, Mandrekar JN, Mokri B. Incidence and outcome of cervical artery dissection: a population-based study. Neurology 2006;67(10):1809–12.

[10] Kolominsky-Rabas PL, Wiedmann S, Weingartner M, et al. Time trends in incidence of pathological and etiological stroke subtypes during 16 years: the Erlangen Stroke Project. Neuroepidemiology 2015;44(1):24–9.

[11] Kristensen B, Malm J, Carlberg B, et al. Epidemiology and etiology of ischemic stroke in young adults aged 18 to 44 years in northern Sweden. Stroke 1997;28(9):1702–9.

[12] Rubinstein SM, Peerdeman SM, van Tulder MW, Riphagen I, Haldeman S. A systematic review of the risk factors for cervical artery dissection. Stroke 2005;36(7):1575–80.

[13] Micheli S, Paciaroni M, Corea F, Agnelli G, Zampolini M, Caso V. Cervical artery dissection: emerging risk factors. Open Neurol J 2010;4:50–5.

[14] Rodallec MH, Marteau V, Gerber S, Desmottes L, Zins M. Craniocervical arterial dissection: spectrum of imaging findings and differential diagnosis. Radiographics 2008;28(6):1711–28.

[15] von Babo M, De Marchis GM, Sarikaya H, et al. Differences and similarities between spontaneous dissections of the internal carotid artery and the vertebral artery. Stroke 2013;44(6):1537–42.

[16] Silbert PL, Mokri B, Schievink WI. Headache and neck pain in spontaneous internal carotid and vertebral artery dissections. Neurology 1995;45(8):1517–22.

[17] Leira EC, Cruz-Flores S, Leacock RO, Abdulrauf SI. Sumatriptan can alleviate headaches due to carotid artery dissection. Headache 2001;41(6):590–1.

[18] Leira EC, Bendixen BH, Kardon RH, Adams Jr HP. Brief, transient Horner's syndrome can be the hallmark of a carotid artery dissection. Neurology 1998;50(1):289–90.

[19] Yamada M, Kitahara T, Kurata A, Fujii K, Miyasaka Y. Intracranial vertebral artery dissection with subarachnoid hemorrhage: clinical characteristics and outcomes in conservatively treated patients. J Neurosurg 2004;101(1):25–30.

[20] Engelter ST, Dallongeville J, Kloss M, et al. Thrombolysis in cervical artery dissection–data from the Cervical Artery Dissection and Ischaemic Stroke Patients (CADISP) database. Eur J Neurol 2012;19(9):1199–206.

[21] Markus HS, Hayter E, Levi C, Feldman A, Venables G, Norris J. Antiplatelet treatment compared with anticoagulation treatment for cervical artery dissection (CADISS): a randomised trial. Lancet Neurol 2015;14(4):361–7.

[22] Touze E, Gauvrit JY, Moulin T, Meder JF, Bracard S, Mas JL. Risk of stroke and recurrent dissection after a cervical artery dissection: a multicenter study. Neurology 2003;61(10):1347–51.

[23] Engelter ST, Brandt T, Debette S, et al. Antiplatelets versus anticoagulation in cervical artery dissection. Stroke 2007;38(9):2605–11.

[24] Caplan LR. Antiplatelets vs anticoagulation for dissection: CADISS nonrandomized arm and meta-analysis. Neurology 2013;80(10):970–1.

[25] Speck V, Noble A, Kollmar R, Schenk T. Diagnosis of spontaneous cervical artery dissection may be associated with increased prevalence of posttraumatic stress disorder. J Stroke Cerebrovasc Dis 2014;23(2):335–42.

[26] Biller J, Sacco RL, Albuquerque FC, et al. Cervical arterial dissections and association with cervical manipulative therapy: a statement for healthcare professionals from the American Heart Association/American Stroke Association. Stroke 2014;45(10):3155–74.

CHAPTER

103

Stroke Secondary to Trauma

G.D. Reddy, S. Gopinath
Baylor College of Medicine, Houston, TX, United States

INTRODUCTION

Stroke is common following trauma. Indeed, direct cerebrovascular injury during a traumatic event can often lead to immediate ischemic and hemorrhagic complications. However, recent definitions for hemorrhagic stroke have, by policy, excluded etiologies that are secondary to trauma [1]. This was done to emphasize differences in presentation and outcomes between traumatic and nontraumatic etiologies, with nontraumatic etiologies including arterial hypertension, vascular malformation, underlying mass lesion, or coagulopathy.

In many clinical situations, this distinction becomes diagnostically difficult for two reasons. First it is often impossible to determine whether a nontraumatic event preceded a traumatic one. For example, in situations involving an intracerebral hemorrhage after a fall, patients can be hypertensive as part of the Cushing reflex. In such situations, it can be clinically unclear whether the instigating event was the hypertensive bleed or the fall. Second, hemorrhagic complications can develop in a delayed fashion following a traumatic event. Thus, although they are not directly caused by the trauma, they are a secondary consequence of it and, in many situations, due to the same etiologies as nontraumatic hemorrhagic stroke, including vascular dysfunction and coagulopathy.

However, to accommodate the new definition of stroke, in this chapter we will focus only on hemorrhagic complications that are not directly related to the traumatic event. This is in contrast to prior editions of this chapter in that primary injuries following trauma, such as immediate posttraumatic intracerebral hemorrhage and subarachnoid hemorrhage (SAH), will not be discussed as they no longer fall within the definition of stroke. Similarly, extraaxial hematomas such as subdural and epidural hematomas will also not be discussed,

even if they present in a delayed fashion, as they too no longer fall under the classification of stroke utilizing the new definitions. With these stipulations, there remain two broad categories for stroke following trauma, ischemic and hemorrhagic.

ISCHEMIC STROKE

Ischemic stroke following traumatic injury to the head is common and evidence of ischemic damage has been found in approximately 90% of deaths secondary to traumatic brain injury [2]. There are two methods in which this occurs. One is a primary injury that is directly from trauma to the brain parenchyma and local vasculature from the event itself. The other is delayed and is considered a secondary brain injury. Indeed, delayed ischemic development is considered one of the most common mechanisms of secondary brain injury after a traumatic event. Studies have shown that the time course of this ischemic development has two peaks [3]. The first occurs within the first 24 h and the second develops at approximately 2–3 days following injury.

DECREASED CEREBRAL PERFUSION PRESSURE

The instigating factor for the acute ischemic development is usually an elevated intracranial pressure (ICP) from either posttraumatic hemorrhage or edema. As the ICP rises in the closed cranial cavity, there is diminished cerebral perfusion pressure to the parenchyma, leading to the decreased oxygen delivery and resulting ischemia. This can be a local effect, with decreased perfusion occurring only near the area of injury, or a more global effect, with a decrease in perfusion occurring throughout multiple areas of the brain.

Primer on Cerebrovascular Diseases, Second Edition
http://dx.doi.org/10.1016/B978-0-12-803058-5.00103-X

A clarification needs to be made regarding the latter situation since global ischemic events are also no longer considered stroke according to recent definitions [1]. However, this distinction was made to remove cases of global hypoperfusion typically from causes outside the central nervous system, such as cardiogenic shock, hypovolemia, or hypoxia from drowning. The situation in posttraumatic circumstances, however, more resembles that following a malignant middle cerebral artery infarct where cerebral edema is leading to increased ICP and further diminishing cerebral blood flow. In the posttraumatic situation, compounding the effect of cerebral hypoperfusion is the precarious state of the traumatized brain. It has been shown in multiple studies that the injured brain is inherently more vulnerable to secondary ischemic insults. This primarily stems from the loss of several neuroprotective mechanisms, including autoregulation, which would otherwise serve to preserve cerebral blood flow in the presence of a decreased cerebral perfusion pressure.

Given the relative frequency of hypoperfusion-mediated ischemic events, cerebral blood flow monitoring has become routine in traumatic brain injury situations. Although the gold standard method for detecting cerebral blood flow to the brain, positron emission tomography, is often impractical in critically injured patients, bedside devices designed for detecting both global and local ischemia have been developed and are recommended in the guidelines for the management of traumatic brain injury set forth by the Brain Trauma Foundation [4]. These include an intraparenchymal brain oxygenation probe, which is great for monitoring local hypoperfusion and internal jugular monitors of venous oxygenation (SjvO2), which can assess for global cerebral hypoperfusion. Numerous studies have shown poor outcomes when values obtained from these monitors fall below certain thresholds. For SjvO2 monitors, this value is typically less than 50% oxygenation, although it is likely that an appreciable portion of the brain must become ischemic before this value is reached. Similarly, for intraparenchymal brain oxygenation monitors, the value utilized is usually less than a partial pressure of 20 mm Hg. Therapies guided by these instruments and aimed at promoting brain oxygenation by increasing oxygen delivery are suggested to improve outcomes, although no controlled clinical trial has yet been conducted [5].

When posttraumatic cerebral hypoperfusion is secondary to an elevated ICP, treatment options depend entirely on the etiology and are outlined in the traumatic brain injury guidelines [4]. If a surgically addressable mass lesion is present, surgical evacuation is usually the most effective course of action. Alternatively, if there is global cerebral edema present, initial treatment options include elevating the head, intubation with hyperventilation, CSF drainage via a ventriculostomy, decreasing cerebral metabolism with sedation and paralytics, and osmotic diuresis with mannitol. If these strategies prove ineffective, decompressive hemicraniectomy is often the next step to promote cerebral perfusion pressure. Initiation of a pentobarbital coma can also be utilized to decrease cerebral metabolism.

VASOSPASM

Aside from an increasing ICP, another common reason for the development of ischemia following traumatic brain injury is vasospasm, which is estimated to occur in up to 70% of patients following severe brain injury [6]. Although it is the most common etiology for the subacute development of ischemia following traumatic brain injury, most patients with posttraumatic vasospasm have no clinical symptoms, making its true incidence higher than reported. Despite its prevalence, the etiology underlying it remains poorly understood, although there is an association with the presence of SAH and a low presenting Glasgow Coma Scale.

It can present within the first 3 days and may also present for up to 10 days. In comparison with vasospasm from aneurysmal SAH (aSAH), this onset tends to be earlier and the duration shorter. Investigations can be initially done at bedside using transcranial Dopplers (TCDs) to assess the velocities in major intracranial vessels. Similar to aSAH, increased velocities are indicative of vasospasm. In contrast to aSAH, however, peak velocities as measured by TCDs tend to occur sooner and usually between 5 and 7 days. Definitive diagnosis requires further imaging with at least computed tomography (CT) angiography or magnetic resonance angiography and the gold standard remains a conventional angiogram. Perfusion scans are also helpful in identifying developing ischemic consequences secondary to decreased perfusion.

Unlike aSAH, no proven treatment strategies exist. Indeed, although the oral calcium channel blocker nimodipine has shown success in decreasing morbidity and poor outcome in aSAH, its benefit in posttraumatic vasospasm is less clear. Most studies have shown improvement in TCD velocities or radiographic vasospasm; however, some have shown improvement in clinical outcome and others have not. Triple H therapy, or hypertension, hypervolemia, and hemodilution, which is another treatment strategy employed for aSAH, is also usually not feasible in cases of traumatic vasospasm given other pathologies that are usually present. For example, hypertension is not advisable in traumatic cases of an intracranial bleed, which unlike cases of aSAH, cannot be secured. Similarly, hypervolemia and hemodilution are also not typically avoided given the presence of cerebral edema, which is also common following traumatic head injury. Newer treatment regimens, such as intraarterial papaverine or calcium channel blockers, intravenous or intraarterial milrinone, and balloon angioplasty, have also been reported with varying degrees of radiographic and clinical success in a few cases, although not enough data exist to make a strong recommendation.

VASCULAR INJURY

Outside direct trauma to the head, delayed strokes can also result from vascular injury to the vessels supplying the brain, most notably of either the carotid or vertebral arteries. Although cervical artery dissection accounts for only approximately 2% of all ischemic strokes, it is one of the most common causes in young patients, where it accounts for up to 25% of all ischemic strokes [7]. The etiology is believed to be secondary to emboli forming from the thrombus at the dissection site. Treatment for minor cases of dissection include either antiplatelet or anticoagulant medications for 3–6 months to prevent emboli formation and randomized control studies have shown no difference in efficacy between these two modalities [7]. Treatment for more significant injury in which there is significant compromise of the affected vessel or for medically refractory

injury requires either open repair or endovascular stenting of the injured vessel. In cases of complete cervical transection, treatment often requires emergent surgical decompression and repair of the injured vessel. Ischemic strokes in such situations can arise either immediately from direct hypoperfusion or in a delayed fashion from thromboembolic phenomenon (Fig. 103.1). Similarly, cervical artery aneurysm and pseudoaneurysms are uncommon, but are known pathologies that can occur following a trauma and result in thromboembolic ischemic stroke. Treatment for these entities is similar to dissections, but usually entails either endovascular or open surgical repair of the affected vessel.

HEMORRHAGIC STROKE

The second form of stroke that follows traumatic brain injury is hemorrhagic. As mentioned earlier, given the new definitions for stroke, we will not discuss the primary injuries from intracerebral contusions. However, outside of these immediate injures, there are also secondary injuries that are hemorrhagic in nature and include the delayed development of intracerebral hemorrhages as well as the hemorrhagic progression of an existing contusion. Although these terms have been utilized synonymously in the past, studies have begun to differentiate between them.

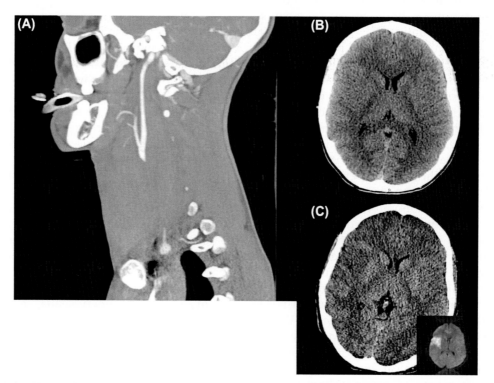

FIGURE 103.1   (A) A 20-year-old patient presented after a stab wound to the neck with complete transection of the common carotid artery and distal reconstitution of both the internal and external branches. (B) Initial computed tomography of the head (CTH) showed no stroke. (C) Follow-up CTH 24 h after the injury shows a left middle cerebral artery distribution stroke, confirmed by diffusion-weighted MRI (inset).

DELAYED TRAUMATIC INTRACEREBRAL HEMORRHAGE

Delayed traumatic intracerebral hemorrhage (DTICH), also known as "Spät-apoplexie" was first described by Bollinger in the late 1800s and refers to the development of intracerebral hemorrhage in areas of the brain that were previously normal. Most studies describing this etiology were conducted before the widespread use of CT scans, but more modern studies require an initial CT head be done within 6 h of injury before a follow-up CT showing an intracerebral hemorrhage can be classified as a delayed intracerebral hemorrhage. However, the degree of hemorrhage on the initial CT to qualify as only a delayed intracerebral hemorrhage is varied in the literature, with some studies requiring no evidence of hemorrhage and others only minimal hemorrhage. Within the flexibility of these variable definitions, the presence of delayed intracerebral hemorrhages is estimated to occur in up to 8% of patients with severe head injury [8]. Multiple factors have been postulated to contribute to this phenomenon, including dysfunction in autoregulation, the development of coagulation abnormalities, changes in pressure dynamics secondary to treatment implementations, the necrosis of blood vessels in injured brain areas, and cerebral venous sinus thrombosis. The overall prognosis from this condition is poor, with most studies reporting a mortality of 50% or higher [8], however, in younger patients, some studies have shown much better results with no mortality and overall good clinical outcomes [9]. Treatment strategies are similar to those for direct intracerebral hemorrhages with correction of underlying coagulopathies and surgical evacuation if there is significant mass effect.

HEMORRHAGIC PROGRESSION

Hemorrhagic progression of a contusion (HPC) refers to the continued expansion of a hemorrhagic lesion or the development of new noncontiguous lesions in the hours to early days following an injury to the head. It is often felt to be secondary to continued bleeding from injured vessels following a traumatic injury. However, it is variably present, with approximately 50% of contusions progressing in most studies. Studies have also shown that larger contusions tend to have a greater increase in size. This has given rise to new theories regarding its etiology, which include the development of a posttraumatic coagulopathy to trauma-induced changes in the microvasculature [10]. Using the hypothesis that an underlying coagulopathy contributes to HPC, studies evaluating the use of various hemostatic agents, including activated factor VII, have been performed, but none so far has demonstrated an improvement in mortality or disability in patients with traumatic brain injury. With the alternative hypothesis of posttrauma-induced microvascular changes, treatments aimed at inhibiting the molecular cascade that leads to these changes have had some success in animal models, but have not been investigated in humans. In general, current management of hemorrhagic progression mimics that of DTICH, with primarily symptomatic management, surgical evacuation for large lesions causing significant mass effect, and conservative therapy for smaller lesions.

	Onset	Treatment
Ischemic		
Decreased cerebral perfusion	Acute to subacute	Decrease intracranial pressure, increase cerebral oxygenation
Vasospasm	Subacute	Intraarterial therapy, intravenous milrinone, oral nimodipine[a]
Vascular injury	Acute to subacute	**Minor:** antiplatelet/anticoagulation **Severe:** surgery/endovascular therapy
Hemorrhagic		
DTICH/ hemorrhagic progression	Subacute	Correct coagulopathy Decrease mass effects

[a] No strategy had demonstrated clear effectiveness.

CONCLUSION

In conclusion, despite the restrictions placed by modern definitions, stroke remains a common phenomenon following traumatic brain injury and one of the most dreaded secondary complications. It can be classified as either ischemic or hemorrhagic. Ischemic causes are usually secondary to ICP-mediated hypoperfusion or vasospasm. Hemorrhagic causes on the other hand are usually multifactorial but are thought to be related to some degree of microvascular dysfunction and coagulopathy. Regardless of the classification or etiology, it remains a poor prognostic factor for functional recovery following traumatic brain injury.

References

[1] Sacco RL, et al. An updated definition of stroke for the 21st century: a statement for healthcare professionals from the American Heart Association/American Stroke Association. Stroke 2013;44:2064–89.
[2] Graham DI, et al. Ischaemic brain damage is still common in fatal nonmissile head injury. J Neurol Neurosurg Psychiatry 1989;52:346–50.
[3] Liu S, et al. Posttraumatic cerebral infarction in severe traumatic brain injury: characteristics, risk factors and potential mechanisms. Acta Neurochir (Wien) 2015;157:1697–704.

[4] Brain Trauma Foundation. American association of Neurological Surgeons & Congress of Neurological Surgeons. Guidelines for the management of severe traumatic brain injury. J Neurotrauma 2007;24(Suppl. 1):S1–106.

[5] Beynon C, Kiening KL, Orakcioglu B, Unterberg AW, Sakowitz OW. Brain tissue oxygen monitoring and hyperoxic treatment in patients with traumatic brain injury. J Neurotrauma 2012;29:2109–23.

[6] Kramer DR, Winer JL, Pease BAM, Amar AP, Mack WJ. Cerebral vasospasm in traumatic brain injury. Neurol Res Int 2013;2013:415813.

[7] CADISS trial investigators, et al. Antiplatelet treatment compared with anticoagulation treatment for cervical artery dissection (CADISS): a randomised trial. Lancet Neurol 2015;14:361–7.

[8] Cooper PR. Delayed traumatic intracerebral hemorrhage. Neurosurg Clin N Am 1992;3:659–65.

[9] Alvarez-Sabín J, Turon A, Lozano-Sánchez M, Vázquez J, Codina A. Delayed posttraumatic hemorrhage. 'Spät-apoplexie'. Stroke 1995;26:1531–5.

[10] Kurland D, Hong C, Aarabi B, Gerzanich V, Simard JM. Hemorrhagic progression of a contusion after traumatic brain injury: a review. J Neurotrauma 2012;29:19–31.

CHAPTER

104

Vascular Cognitive and Behavioral Disorders

Q. Hao, M. Johansen, R.J. Wityk

The Johns Hopkins University School of Medicine, Baltimore, MD, United States

INTRODUCTION

Ischemic infarcts in the brain can manifest with a variety of cognitive and behavioral disorders. For many years, use of autopsy and clinical–pathological correlation in stroke patients formed the basis for the anatomic localization of various cognitive functions. In addition to focal infarcts, diffuse cerebrovascular disease can also be associated with subcortical white matter injury and secondary vascular dementia. A full description of vascular dementia is beyond the scope of this chapter. We chose to focus on some of the interesting cognitive and behavioral disorders that occur with focal ischemic infarcts, both large and small.

Nondominant Hemispheric Stroke

Patients with right hemispheric stroke can have a variety of impairments, including left hemiparesis, left hemisensory loss, and left homonymous hemianopia, depending upon the location and size of the stroke. (For the purpose of discussion, we only consider here findings in right-handed individuals, who invariably have language function localized to the left hemisphere. Right hemispheric infarcts would not be expected to have any problems with aphasia).

Involvement of the right parietal lobe often produces left "neglect," defined as an impairment of attention or response to stimuli in the hemispace contralateral to the lesion, and not attributable to a primary sensory or motor deficit [1,2]. It can involve a variety of modalities, including visual, tactile, auditory and even olfactory (not common), and motor. "Extinction" is a form of neglect that occurs when the patient neglects left-sided stimuli only when bilateral stimuli are presented. For example, the patient may have normal visual fields when tested to single finger wiggle on the right and left sides independently (i.e., no hemianopia present), but when finger wiggle is presented in both fields simultaneously, the patient recognizes the stimulus only in the right visual field, but not in the left visual field (e.g., extinguishes on the left side).

Primer on Cerebrovascular Diseases, Second Edition
http://dx.doi.org/10.1016/B978-0-12-803058-5.00104-1

Some patients with acute right parietal stroke have profound inattention to the left environment that may take days to improve. Patients do not notice anything on their left side, shaving only half of the face, eating food on only half of the plate, and being unable to find items placed on their left side. On examination, they often have a right gaze preference, but with careful testing, there is no evidence for a gaze palsy or oculomotor abnormality. When asked to draw a clock or copy a picture of a house, they can draw the right half of the image well, but may completely leave out the left side of the image. A simple test is to draw a line on a piece of paper and ask the patient to make an "X" in the middle of the line. The patient with left hemineglect will put the "X" to the right of the midline. Sometimes it is difficult to differentiate severe hemineglect from homonymous hemianopia as they may overlap and require more specific assessment tools [1].

A right hemispheric parietal lobe lesion is well known to be the cause of neglect. In the past few years, detailed brain mapping studies have identified these more precise localizations: the inferior parietal sulcus, the temporo-parietal junction, the superior temporal gyrus, and the posterior intraparietal sulcus [2].

Anosognosia, Visual Anomia, and Prosopagnosia

Anosognosia was originally used by Babinski in 1914 to describe patients who lack awareness of the left-sided motor deficits due to a right hemisphere lesion [3]. As other patients were described with unawareness of other types of deficits, anosognosia now refers to the lack of subjective experience for a wide range of neurological and neuropsychological disturbances, including vision deficits, amnesia, and aphasia.

The patients with anosognosia of left-sided hemiplegia may deny their deficit and even deny that the left limbs belong to them when they are held up into their right visual field. They will repeatedly attempt to get up out of a chair and fall to the left, as if unaware of the deficits. In lesser degrees of neglect or in the recovery phase, patients may be aware of their paralysis, but appear unconcerned (termed "anisodiaphoria" by Babinski), often speaking of the weakness in an emotionless tone. Anton's syndrome is a type of anosognosia of visual deficit in patients with complete cortical blindness. They claim that they can see and often confabulate objects or scenes in front of them [4]. In practical terms, anosognosia negatively interferes with neurorehabilitation due to poor adherence to rehabilitation activities [4].

Anosognosia of left-sided deficits is associated with lesions in right hemisphere, specifically in the dorsolateral portions of frontal cortical area, parieto-temporal area, and

right insula area [5]. Nurmi and Jehkonen [5] report left hemispheric lesions can cause anosognosia for right-sided motor deficits. Patients with Anton's syndrome commonly have bilateral occipital–parietal lesions.

Visual anomia or optic aphasia was first described by Freund in 1889 and refers to being unable to name objects presented in visual modality, but able to name the same objects when perceived tactually or when given a verbal definition of their function and use [6]. It is a type of disconnection between the right occipital lobe visual cortex (visual encoding) and the left hemisphere language cortex (verbal semantics). Patients usually have normal fluency of speech with intact aural comprehension and repetition. The infarct is in the left occipito-temporal lobe and posterior corpus callosum in the vascular territory of left posterior cerebral artery (PCA). The right occipital lobe can recognize the object but cannot transmit the information via corpus callosum to the left hemisphere where the semantic system for naming is located (see section Disconnection Syndrome).

Prosopagnosia was first described by Quaglino and Borelli in 1867 and refers to inability to perceive and recognize facial identity [7]. The patients are unable to recognize the faces of close friends and relatives or pictures of famous people, but able to recognize familiar people by voice, mannerisms, and by posture or gait pattern. They may recognize some faces if there are other features, such as glasses, hairstyles, or scars, or if in a specific encounter (e.g., recognizing a physician in hospital but not on the street). It may not be restricted to faces, but also the other complex but familiar visual stimuli, such as architectural landmarks or pets. Facial recognition deficits may be an isolated symptom or may coexist with other symptoms, such as mild object agnosia, complex visual disturbances, or memory deficits. Patients with acquired prosopagnosia usually are aware of their social difficulties and feel distressed and dysphoric. Recent studies have shown that prosopagnosia can be caused by bilateral temporal–occipital lesions, but also less commonly with unilateral lesions (on the right side), involving the fusiform or temporo-occipital gyri and the right posterior temporal lobe, in the territories of the middle cerebral artery (MCA) and the PCA.

Disconnection Syndromes

One of the current tenets of behavioral neurology is that processing of complex cognitive functions is distributed over disparate sites in the cortex, connected by white matter tracts. This hypothesis leads to the possibility of an impairment of cognitive function by a restricted white matter lesion that disconnects two cortical sites. An example of this was discussed earlier with visual anomia.

One of the most dramatic examples is the syndrome of alexia without agraphia (also called pure alexia, pure word blindness, and letter-by-letter alexia) which was first described by Dejerine in 1892 [8]. The patient is able to read letters, but not the words. Writing is normal with proper grammar and meaning. The patient is able to write a sentence or two on a blank page, but when this page is taken away and given back to the patient later in the examination, they are unable to read their own words. Verbal fluency and auditory comprehension are normally intact. Many patients have associated color anomia, right hemianopia, and short-term memory loss.

The syndrome is usually caused by an infarct in the left PCA territory involving the left medial occipito-temporal cortex and extending into the splenium of the corpus callosum. The left occipital infarct causes a right hemianopia. Visual information from the left visual field (right occipital lobe) is unable to reach the left hemispheric language centers via the corpus callosum, hence the visual image of the word cannot be connected with the semantic meaning of the word.

Thalamic Infarcts

Although many of the preceding clinical syndromes are usually associated with moderate to large-sized infarcts, there are a number of well-described cognitive/behavioral changes caused by very small strokes. The thalamus serves as a relay center for information from the periphery to the brain. The most common stroke syndrome is that of hemisensory loss, when the relay center for sensory information on one side of the body (the ventral-posterolateral thalamus) is damaged, preventing projection of fibers to the parietal lobe. Detailed pathological studies have delineated the vascular supply and the effects of infarction in other areas of the thalamus, and high-resolution magnetic resonance imaging (MRI) studies have confirmed these findings in larger number of patients. The two arteries that are discussed are the polar artery ("tuberothalamic artery") and the paramedian artery ("thalamoperforating artery").

The polar artery usually originates from the middle third of the posterior communicating artery and predominantly supplies the ventral anterior (VA) nucleus, the rostral part of the ventral lateral (VL) nucleus, the ventral pole of the medial dorsal (MD) nucleus, and part of the anterior nuclear group [9]. Patients with strokes in this territory have prominent neurobehavioral deficits with apathy, withdrawal, and disorientation, but not with the same degree of drowsiness seen in infarcts in other thalamic territories. The majority of patients demonstrate severe deficits in laying down new memories.

The clinical presentation differs depending on infarct laterality. Those on the nondominant side present with prominent visual–spatial memory impairment with accompanying neglect. The dominant (normally left)-sided lesions impair formation of recent memories and are frequently accompanied by a transcortical motor aphasia. Some patients have "emotional" facial weakness contralateral to the lesion, described as notable facial asymmetry only during emotional displays, but retained motor strength during voluntary facial activation. The natural history of patients with polar artery infarcts appears to be one of gradual improvement over weeks to months. The marked cognitive and behavioral change despite a very small lesion as seen on MRI has led to the term "strategic infarct dementia." The presumed mechanism of an isolated lesion is small vessel disease, but polar infarcts may coexist with acute infarcts in the PCA territory due to other causes.

The paramedian artery arises from the proximal (P1) segment of the PCA and the rostral basilar artery (mesencephalic segment). The intralaminar nuclear group, part of the dorsomedial nucleus and the internal medullary lamina are the structures most commonly supplied, as well as the rostral midbrain and cerebral peduncle. A number of variations of normal anatomy of this artery have been described in autopsy studies and explain some unusual infarct patterns. Paramedian arteries can arise from the origins of the right and left PCA, each causing a unilateral medial thalamic infarct [10]. A common variation described by Percheron has the paramedian artery arising from one side, but crossing the midline to supply both sides of the medial thalamus [11].

Unilateral infarction occurring in the paramedian distribution is associated with impaired and fluctuating levels of arousal [9], memory impairment that occurs with both anterograde and retrograde memory loss, and often subtle oculomotor and pupillary changes due to rostral midbrain infarction. The terms thalamic dementia and strategic infarct dementia are often used to describe the persistent and diffuse cognitive impairment with paramedian thalamic strokes, particularly if bilateral. The concept has since been further defined and is used to refer to the strange constellation of amnesia with profound personality changes, disinhibition, and bizarre behavior similar to a primary psychiatric disorder [12]. Thalamic dementia is most apparent after improvement in the patient's depressed level of consciousness. Unilateral strokes have a variable degree of severity and may improve over days to weeks. A patient with the artery of Percheron variation, however, can have bilateral medial thalamic strokes from a single vessel occlusion, resulting in more severe memory loss and a depressed level of consciousness, including stupor and coma. A form of akinetic mutism can occur in which the person maintains consciousness, but is unresponsive [13]. Strokes in the paramedian artery can be isolated and presumed due to small vessel disease, but infarcts may also occur with disease of the proximal PCA and the rostral basilar artery.

Distal basilar artery strokes involving the rostral midbrain and medial thalamus manifest as the "top of the basilar syndrome," a unique presentation of retained motor function accompanied by somnolence, vivid hallucinations, and dream-like behavior [14]. Strokes extending into the rostral midbrain have oculomotor abnormalities, such as disorders of vertical gaze, convergence, skew deviation, and a pseudo-sixth nerve palsy. Memory impairment is also described as a temporary Korsakoff-like amnestic state. Peduncular hallucinosis, so named in reference to the cerebral peduncles, consists of vivid visual hallucinations often occurring at twilight while the patient is still awake and can occur in "top of the basilar" syndrome as well as posterior thalamic infarcts. Caplan also describes patients with "unusual reports" referring to bizarre, absurd responses to orientation questions with suggestibility from the environment [14]. The prominent neuropsychological symptoms resulting from paramedian artery infarcts must be recognized because delays in diagnosis can adversely impact correct medical treatment.

Frontal Lobe Infarction

Lesions in the frontal lobe can result in behavioral changes, such as prominent slowing, loss of initiation, and blunting of affect. Abulia, occurring as a result of medial frontal lobe strokes, is a phenomenon of inattention in a patient who otherwise appears awake. When talked to or given a command, the patient may actually respond, but after an extraordinarily long delay. However, they seem to respond more quickly to automatic behaviors, such as chewing food when it is placed in the mouth, or with answering a ringing telephone. The two structures most commonly implicated in abulia are the cingulate gyrus and the supplemental motor area. Both regions are involved in unilateral anterior cerebral artery infarctions [15]. The term akinetic mutism has been used to describe a state of reduced behavioral impetus with decreased activity, slowed responsiveness, and verbal inertia with the thought that akinetic mutism represents the more extreme form of the abulia spectrum [13]. Abulic patients retain the ability to perform reflexive actions and it is believed that damage in the supplementary motor area leads to a form of motor neglect with decreased desire to move the limb. Disruptions of frontal–subcortical circuits have been implicated in human behavioral disturbances to include a functional link with the thalamus [16]. It would naturally

follow that infarcts occurring in the cingulate gyrus, a structure essential in cortex connectivity, would manifest with behavioral impairment. The anterior cingulate syndrome has been described as patients presenting with akinetic mutism, aspontaneity, incontinence, and lack of emotion [13]. In humans, these symptoms must be carefully distinguished from depression which is known to be increased in poststroke patients.

References

[1] Ting DSJ, Pollock A, Dutton GN, et al. Visual neglect following stroke: current concepts and future focus. Surv Ophthalmol 2011;56(2):114–34.

[2] Maxton C, Dineen RA, Padamsey RC, Munshi SK. Don't neglect 'neglect'– an update on post stroke neglect. Int J Clin Pract 2013;67(4):369–78.

[3] Langer KG, Levine DN. Babinski, J. (1914). Contribution to the study of the mental disorders in hemiplegia of organic cerebral origin (Anosognosia). Translated by K.G. Langer & D.N. Levine: Translated from the original Contribution à l'Étude des Troubles Mentaux dans l'Hémiplégie Organique Cérébrale (Anosognosie). Cortex 2014;61:5–8.

[4] Prigatano GP. Anosognosia: clinical and ethical considerations. Curr Opin Neurol 2009;22(6):606–11.

[5] Nurmi ME, Jehkonen M. Assessing anosognosias after stroke: a review of the methods used and developed over the past 35 years. Cortex 2014;61:43–63.

[6] Ferreira CT, Giusiano B, Ceccaldi M, Poncet M. Optic aphasia: evidence of the contribution of different neural systems to object and action naming. Cortex 1997;33(3):499–513.

[7] Barton JJS. Disorders of face perception and recognition. Neurologic Clinics 2003;21(2):521–48.

[8] Dejerine J. Contribution a l'étude anatamo-pathologic et clinique des différentes variétes de cecite verbale. Mem Soc Biol 1892;4:61–90.

[9] Schmahmann JD. Vascular syndromes of the thalamus. Stroke 2003;34(9):2264–78.

[10] Tatemichi TK, Steinke W, Duncan C, et al. Paramedian thalamopeduncular infarction: clinical syndromes and magnetic resonance imaging. Ann Neurol 1992;32:162–71.

[11] Percheron G. Les artères du thalamus humain, II artères et territoires thalamique paramédians de l'artère basilar communicante. Rev Neurol 1976;132:309–24.

[12] Carrera E, Bogousslavsky J. The thalamus and behavior: effects of anatomically distinct strokes. Neurology 2006;66(12):1817–23.

[13] Nagaratnam N, Nagaratnam K, Ng K, Diu P. Akinetic mutism following stroke. Journal of Clinical Neuroscience 2004;11(1):25–30.

[14] Caplan LR. "Top of the basilar" syndrome. Neurology 1980;30:72–9.

[15] Bogousslavsky J, Regli F. Anterior cerebral artery territory infarction in the lausanne stroke registry: clinical and etiologic patterns. Arch Neurol 1990;47(2):144–50.

[16] Cummings JL. Frontal-subcortical circuits and human behavior. Arch Neurol 1993;50:873–80.

MEDICAL CONDITIONS AND STROKE

105

Stroke in Children

K. Sivakumar[1], K.R. Duncan[1], M.C. Leary[1,2]

[1]Lehigh Valley Hospital and Health Network, Allentown, PA, United States; [2]University of South Florida, Tampa, FL, United States

INTRODUCTION

There are various types of stroke in children, including arterial ischemic strokes, venous infarctions due to venous sinus thrombosis, intracerebral hemorrhages, and subarachnoid hemorrhages. Stroke is a relatively rare occurrence in children, but can lead to significant morbidity and mortality, and is one of the top 10 causes of death in children aged 1–18 years [1,2]. It is important to be aware of the fact that children with strokes present differently than adults, and that the clinical symptoms can vary, depending on the age of the child [3,4]. Children also have risk factors for stroke that are less common than in adults, thus the cause of stroke differs considerably with age. Two main age groups can be distinguished: perinatal/neonatal (week 22 of pregnancy to 1 month of life), and children aged 1 month to 18 years. Each of these groups has different frequencies of stroke etiologies. Additionally, the cause of stroke in these groups also does vary depending on geographical, economic, and environmental factors. For example, tuberculous meningitis is an important cause of pediatric stroke in India, but is unusual in the USA [4].

EPIDEMIOLOGY

A stroke in children is typically considered to be a relatively rare event. The reported incidence of combined ischemic and hemorrhagic pediatric stroke ranges from 1.2 to 13 cases per 100,000 children under 18 years of age [4]. However, pediatric stroke is likely more common than we may realize, as it is thought to be frequently undiagnosed or misdiagnosed. This may be due to a variety of factors including a low level of suspicion by the clinician or due to patients who present with subtle symptoms that mimic other diseases. This, in turn, can lead to a delay in the diagnosis [3]. Boys are more likely to have a stroke than girls, even after controlling for differences in frequency of causes such as trauma [1,5]. Additionally, African American children are more at risk compared to Caucasian and Asian children, and this difference remains true even after accounting for sickle cell disease (SCD) patients with stroke [1,3,6].

PERINATAL AND NEONATAL STROKE

As mentioned before, perinatal stroke is often undiagnosed or misdiagnosed due to the subtleness of signs and symptoms. As a result, the incidence of perinatal stroke is only an estimate. Ischemic stroke is anticipated to occur in about 1:3500 of live births, and hemorrhagic stroke in about 1:16,000 live births [2,7,8]. Of the identified perinatal strokes, about 80% are ischemic arterial strokes and the remaining 20% are due to either cerebral venous sinus thrombosis or primary brain hemorrhage [2,9,10]. The critical period in the pregnancy for stroke is the end of the second trimester and the entire third trimester [11]. Obvious signs of stroke may be observed within hours to days of birth. Acute symptoms of neonatal stroke can include seizures, periods of apnea with staring, coma/listlessness, focal weakness/hemiparesis, or other focal deficits [2,4,12]. Perinatal strokes are most likely to initially present with focal seizures or lethargy in the first few days after birth [3]. Seizures are quite common acutely, occurring in up to 75%. Typically the seizures are focal, mainly involving only one extremity [10].

Specific types of stroke will also present differently depending on the age of the child [3]. For example, cerebral venous sinus thrombosis can present in all ages with fever and lethargy, but young infants can also have decreased oral intake or respiratory distress as their initial symptoms. Physical examination of the infant may reveal dilated scalp veins, eyelid swelling, or a large anterior fontanelle, whereas an older child would likely present with slowly progressive signs, such as vomiting, headache, or other phenomena associated with increased intracranial pressure. A subarachnoid hemorrhage can also present as irritability and a bulging fontanelle in infants, whereas older children may instead complain of sudden acute headache, neck pain, or photophobia [3,13–15].

Unfortunately, especially in neonatal stroke, there can be a delay in recognizing the event. Subtle signs may not be observed in the newborn period, and symptoms may only be identified as the child grows and develops over the first year. One of the most common signs found in a child with a prior neonatal stroke is a hand preference, consistently reaching out for objects with only one hand before the age of 1 year. Missed developmental milestones, unilateral weakness/hemiparesis, later development of seizures, or the presence of another focal deficit can all be clues that a perinatal stroke occurred previously [2,12].

Etiology of stroke in neonates can be complex and may be due to a combination of factors, including cardiac disorders, coagulopathy, infection, trauma, maternal medications and toxins, maternal placenta disorders, and intrauterine or perinatal asphyxia [2,4,10]. Neonatal ischemia is often caused by cardiac disease, sepsis with vascular collapse, and hypertension. Heart abnormalities in the fetus, as well as hypoglycemia and twin-to-twin transfusion syndrome are conditions in which arterial strokes are more prominent. Additionally, hypoxic–ischemic injury is also relatively common in neonates, and can be caused by intrauterine asphyxia, birth-related problems, uterine and placental abruption, respiratory insufficiency after birth (such as with meconium aspiration), recurrent apnea, hyaline membrane disease in premature infants, and severe congenital heart disease [4]. The most vulnerable areas for hypoxic–ischemic injuries are the cerebral cortex, particularly the hippocampi, as well as the cerebellar cortex and pontine nuclei in the brainstem. On neuroimaging, observable lesions are most commonly noted in the parasagittal regions, deep periventricular white matter, and within the basal gangli/thalami [4].

The vast majority of lesions in premature infants are located within the periventricular white matter. Premature infants are also susceptible to developing brain hemorrhages, including germinal matrix hemorrhages that usually occur either in the periventricular region with spread into the adjacent ventricle or in the cerebellum. Most germinal matrix hemorrhages occur during the first 3 postnatal days, especially during the first hours [4,16]. Significant subdural and subarachnoid bleeding can also occur with more severe birth trauma or coagulopathy [4].

In most cases, the cause of perinatal stroke remains unknown. Parents should be counseled that the risk of having another child with perinatal stroke is extremely low, with recurrent stroke risk being <1% [2,11]. It is also very important for mothers to know that there is usually nothing they did or did not do during their pregnancy that caused their child's stroke [2,9].

STROKE IN CHILDREN

As with stroke in neonates, childhood stroke may be missed due to a lack of awareness that stroke can occur at any age, but it also can be misdiagnosed. Perhaps in part this is because stroke in children can present very differently depending upon the individual. Childhood stroke is estimated to occur in about 1.2–13 per 100,000 children per year [3,17]. For children who have a first stroke between 1 month and 18 years, the risk of recurrent stroke is 15–18% [11]. Children with cardiac disease can have up to a five-fold increased recurrence risk as compared to children without cardiac disease [18]. In children with stroke, 50% are ischemic, with hemorrhages comprising the remaining 50%, which is very different from neonates and from adults.

Children and adolescents with stroke may have atypical presentations when compared with adult patients [10]. Symptoms of childhood stroke include sudden or gradual focal numbness or weakness, sudden loss of vision or diplopia, sudden confusion or speech difficulty, new onset seizures, diminished level of consciousness, or sudden severe headache associated with vomiting or sleepiness [2,4,12]. Similar to other age groups, children often present with hemiplegia. However, unlike in adults, the hemiplegia usually resolves within a week with only minor motor deficits dependent on the severity of the stroke [4]. Sensory or cognitive abnormalities tend to be unusual in children unless the infarctions are bilateral [4,19]. The full neurological deficits may only emerge as the child develops, but can impact their life permanently. Of children who survive a stroke, about 60% will have permanent neurological deficits [1,2,11].

Hemorrhagic Stroke

Vascular malformations continue to be the most common cause of intracranial bleeding in this age group [4,10]. If all individuals under the age of 20 are included, however, then aneurysms are even more common than vascular malformations as a cause of CNS bleeding. In one series of 124 patients with subarachnoid hemorrhage (SAH), 50 patients had aneurysms and 33 had arteriovenous malformations (AVMs) [4,20]. Among three series, 36% of young patients had aneurysmal bleeding, whereas 27% bled from an AVM [20]. An aneurysm generally becomes symptomatic by the age of 2 or after the age of 10 years. These tend to be more common in children with other systemic diseases such as those with Marfan's, EDS, polycystic renal disease, and with coarctation of the aorta [4,10].

Intracranial malformations of the vasculature are usually present at birth, however, they tend not become symptomatic until the early to late adult years. Arteriovenous vascular malformations continue to be one of the more frequent causes of intracranial bleeding in the pediatric and adolescent demographic, however, studies have shown that the less than 10% of AVMs are diagnosed before age 10. Most do not become symptomatic until adulthood [4].

A rarity which is found solely in neonates and younger children is a vein of Galen malformation. In this condition, the vein of Galen is massive, forming a large varix, with the straight sinus being large and tortuous. A round hypodense mass behind the third ventricle, connected to a prominent torcula by the straight sinus, is a typical observation found on CT. The most common presenting syndrome during the neonatal period and infancy is high-output congestive heart failure as a result of the large volume of shunted blood. A loud cranial bruit is typically audible [4]. Younger children can present with IVH, SAH, or seizures. Additionally hydrocephalus and symptoms of hydrocephalus (headaches, double vision, poor balance, urinary incontinence, personality changes, and mental impairment) can occur. The distribution of other AVMs in children can be infratentorial, supratentorial, or involving the basal ganglia or thalamus. The recurrence of hemorrhages continues to be greatest in children who have had previous arteriovenous malformations [4].

Other conditions such as pheochromocytoma, cocaine and amphetamine use, or acute glomerulonephritis can also cause hemorrhaging. Infection also continues to be a cause in this age group with infective endocarditis predominating.

Ischemic Stroke

Pediatric brain ischemia has many etiologies, but there are four unusually common risk factors often found in children with ischemic stroke, including cardiac disease, infection, trauma, and migraine. Other causes are, however, possible (Table 105.1). Over 50% of children with ischemic stroke have a known risk factor at the time of infarct. In at least 2/3 of children who have stroke, more than one stroke risk factors can be identified after a thorough work-up [10]. In teenagers, drug abuse also can be a prominent risk. As with subarachnoid and intracerebral hemorrhages, the basal

TABLE 105.1 Causes of Pediatric Brain Ischemia: Age 1–18 years

Migraine

Trauma: Including dissection and other vascular injury; abusive head trauma, including choking and whiplash-shake injuries, oral foreign body trauma to the carotid artery.

Cardiac: Congenital heart disease with right-to-left shunt; Tetralogy of Fallot, transposition of the great vessels, tricuspid atresia, atrial septal defect, ventricular septal defect, cardiomyopathy, endocarditis, and pulmonary arteriovenous fistula; patent foramen ovale; rheumatic heart disease; prosthetic valve; aortic valve disease; myxoma; cardiomyopathy, acute myocardial infarction; chronic cardiac ischemia; arrhythmia; endocarditis.

Drugs: Including amphetamines, cocaine, and heroin.

Infection: Bacterial meningitis (especially *Haemophilus influenza*, pneumococci, and streptococci); facial, otic, and sinus infection; acquired immunodeficiency syndrome; dural sinus occlusion and infection; tuberculous meningitis; varicella zoster virus with post-varicella arteriopathy.

Genetic and metabolic: Neurofibromatosis; hereditary connective tissue disorders such as Marfan's and Ehlers–Danlos syndrome; pseudoxanthoma elasticum; homocystinuria; Menkes' kinky hair disease; hypoalphalipoproteinemia; familial hyperlipidemia; methylmalonic aciduria; MELAS syndrome (mitochondrial, encephalomyopathy, lactic acidosis and stroke-like episodes); CADASIL (cerebral autosomal-dominant arteriopathy with subcortical infarcts and leukoencephalopathy); CARASIL (cerebral autosomal-recessive arteriopathy with subcortical infarcts and leukoencephalopathy); Fabry's disease; HERNS (hereditary endotheliopathy with retinopathy, nephropathy, and stroke).

Hematological and neoplastic: Sickle cell anemia; purpuras; leukemia; L-asparaginase and aminocaproic acid therapy; radiation vasculopathy; antithrombin deficiency; protein C deficiency; protein S deficiency; factor V Leiden; prothrombin gene mutation; fibrinolytic system disorders; deficiency of plasminogen activator; antiphospholipid antibody syndrome; increased factor VIII; cancer; thrombocytosis; polycythemia; thrombotic thrombocytopenic purpura; disseminated intravascular coagulation.

Systemic disease: Rheumatological, gastrointestinal; renal; hepatic; pulmonary.

Others: Moyamoya syndrome; arteritis, collagen vascular disease; local infection; Takayasu's syndrome; Behcet's syndrome; venous sinus thrombosis; head and neck infections; dehydration; coagulopathy; paroxysmal nocturnal hemoglobinuria; puerperal or pregnancy-related; Sneddon's syndrome; fibromuscular dysplasia.

Premature atherosclerosis: Hyperlipidemias; hypertension; diabetes; smoking; homocystinuria.

Female hormone related: Oral contraceptive pills; pregnancy; puerperium; eclampsia; dural sinus occlusion; arterial and venous infarcts; peripartum cardiomyopathy.

Rheumatic and inflammatory: Systemic lupus erythematosus, rheumatoid arthritis; sarcoidosis; Sjogren's syndrome; scleroderma; polyarteritis nodosa; cryoglobulinemia; Crohn's disease; ulcerative colitis, collagen vascular disease; Behcet's syndrome; Cogan's disease.

ganglia, internal capsule, and thalamus continue to be more common locations for injury in ischemic strokes. The distribution of ischemia in these deeper structures usually coincides with the proximal portions of the internal carotid artery (ICA) and middle cerebral artery (MCA) being involved. The proximal vasculature is generally spared due to the presence of good collateral circulation and blood supply [4].

If the etiology of the stroke is not apparent, other causes in this age group should also be investigated, such as hematological and genetic disorders, depending on age and history. Hyperlipidemia, hypertension, diabetes and smoking are less likely associated with ischemic strokes in children however with the incidence of obesity increasing, these risk factors should be noted and screened for, particularly in teenagers.

WORK-UP

Comprehensive assessment of neurological status, clinical presentation, and radiological imaging is necessary in neonates, infants, and children of all ages [3,4,10]. In children, head CT will demonstrate hemorrhages well, but is generally considered inadequate to diagnose ischemic stroke, and MRI should be used for diagnosis. Brain MRI is more sensitive for acute ischemia than CT, particularly with use of diffusion weighted imaging in the hyperacute time period. In addition, brain MRI provides better visualization of the posterior fossa, and can also exclude stroke mimics. In addition to brain imaging, neurovascular evaluation should be undertaken with either magnetic resonance angiography (MRA) or computed tomography angiography (CTA) of the head and neck. Cranial ultrasound can be helpful in neonates. Initial work-up should include electrocardiogram and complete blood count including platelets, electrolytes, urea nitrogen, creatinine, serum glucose, prothrombin time (PT) and international normalized ratio (INR), activated partial thromboplastin time (aPTT), blood cultures, oxygen saturation, and transthoracic echocardiography (TTE) with agitated saline study to evaluate for possible cardiac source of embolism, including right-to-left shunt [21–23].

There is an additional battery of tests that needs to be considered if no cause or risk factor is immediately evident (Table 105.2) [21–23]. Identifying the cause of a stroke is a necessity in the attempt to prevent future events, but also provides some relief to the family and aids in answering questions about prognosis and recurrence. In many neonatal cases a definitive cause for the stroke is never found [10]. Additional risk factors need to be considered in teenagers, such as smoking, substance abuse, oral contraceptive pills, hypercholesterolemia, diabetes, and hypertension.

STROKE MIMICS

There are many diseases that may mimic a stroke [3]. Complicated migraines can cause focal neurological symptoms that typically resolve within 24h, and should be considered if there is a family history of migraine or hemiplegic migraine [3,24]. Focal seizures can result in subsequent transient postictal hemiparesis (Todd's paralysis), but stroke should be considered if the duration of the deficit is prolonged relative to the duration of the preceding seizure. It is also important to consider intracranial neoplasms and intracranial infections such as brain abscess, meningitis, and encephalitis [3,25]. Although rare, alternating hemiplegia of childhood is also

TABLE 105.2 Further Testing to Evaluate Cause of Stroke in Children

- Hypercoagulable evaluation: Protein C functional, protein S free and total or protein S functional, antithrombin activity, lipoprotein (a), Homocysteine, Prothrombin gene mutation, Factor V Leiden gene mutation, anticardiolipin antibodies (IgG and IgM), Beta2-glycoprotein I antibodies (IgG and IgM), lupus anticoagulant tests, including dilute Russell viper venom time and dilute activated PTT, Factor VIII activity, and D-dimer

- Vasculitis evaluation: Erythrocyte sedimentation rate, antinuclear antibody assay, C-reactive protein level, human immunodeficiency virus testing, rapid plasma reagin (RPR)

- Toxicology screen

- Blood alcohol level

- Liver function tests

- Lactate levels in serum and CSF

- Plasma and urine amino acids, methionine

- Pregnancy test in female children of childbearing potential

- Cardiac enzymes and troponin, if there is clinical suspicion of myocardial ischemia

- Hemoglobin electrophoresis in patients with possible sickle cell disease

- Molecular genetic testing

- Lumbar puncture, if there is clinical suspicion for subarachnoid hemorrhage and head CT scan is negative for blood, or if there is suspicion for an infectious or inflammatory etiology of stroke

- Transesophageal echocardiography (TEE) if TTE is nondiagnostic, or if there is a high index of suspicion for a cardioembolic source

- Transcranial Doppler when MRA or CTA are nondiagnostic and there is a high index of clinical suspicion for intracranial large artery disease

- Holter monitor if there is suspicion for cardiac arrhythmia, particularly atrial fibrillation

- Conventional cerebral angiography

- Muscle biopsy, particularly if MELAS is considered (Ebers, Jausch, Pediatric Stroke Working Group)

TABLE 105.3 Common Childhood Stroke Mimics

Benign	Not Benign
Migraine	Seizures/Post-ictal paralysis
Psychogenic	Tumor
Musculoskeletal abnormalities	Cerebellitis
Delirium	Drug toxicity
Episodic hypertensive episodes	Idiopathic intracranial hypertension
	Intracranial abscess
	Meningitis

a possibility, particularly if there is a distinct history of episodes of hemiplegia that last rarely longer than a day, alternate between sides, and present in a child with progressive developmental regression [26]. Hypoglycemia can also cause focal, stroke-like deficits [3,4] (Table 105.3).

TREATMENT

Neonatal and Perinatal

Short-term treatment first includes the early identification of the stroke and rapid evaluation by pediatric specialists. Supportive care is indicated for all types of perinatal stroke [2,10]. Attention to maintaining airway, breathing, and circulation with specific attention to oxygenation and ventilation is imperative. Hypoxia may be the cause of the stroke or may have a significant contribution in neonates and infants [10]. Rehabilitation is key to help with long-term outcome. Long-term treatment encompasses multiple therapies and pediatric subspecialists. It is important to note that the child's rehabilitation needs may change as they grow and develop [2,8].

Childhood

Treatment is also supportive, and as with neonatal stroke, rehabilitation is paramount in helping with long-term outcome. This should include speech therapy, occupational therapy, and physical therapy. Additionally, educational services to address educational and developmental needs of children can be provided at the local level with state supervision. Formal neuropsychology testing can also provide direction. Long-term treatment encompasses multiple therapies and pediatric subspecialists.

Ischemic Stroke: Secondary Prevention

Once the type of stroke is identified, treatment depends on the etiology. Anticoagulation is also often used in children with arterial dissection, dural sinus thrombosis, coagulation disorders, high risk of embolism, or

progressive deterioration during the initial evaluation of a new cerebral infarction [3]. Regarding secondary prevention in general for this population, Sträter et al. performed a small prospective multicenter follow-up study in 135 consecutively recruited children aged 6 months to 18 years with a first episode of ischemic stroke. Their data have provided evidence that low-dose low-molecular weight heparin is not superior to aspirin and vice versa in preventing recurrent stroke in white pediatric stroke patients [27]. Fullerton et al. also enrolled 355 children with acute ischemic stroke at 37 international centers during 2009–2014 and followed them prospectively for recurrent stroke. The data indicated that children with acute ischemic stroke, particularly those with arteriopathy, remain at high risk for recurrent stroke despite increased utilization of antithrombotic agents. The investigators concluded that therapies directed at the arteriopathies themselves are likely needed [28]. Future studies to determine the optimal acute treatment of childhood stroke and secondary prevention and risk factor modification will be critical [3].

Management of stroke in children with SCD should be mentioned. Ischemic strokes in this population require hydration and simple or partial exchange transfusion to achieve a hemoglobin SS fraction of less than 30% and a hemoglobin level not greater than 10 g/dL to avoid problems of hyperviscosity [3]. Evaluation for a structural vascular lesion in children with SCD and a hemorrhagic stroke is reasonable. This is because there is often an underlying aneurysm with potential for rebleeding in adolescents with SCD who present with an SAH [3,29].

Acute Ischemic Stroke Treatment

The clot-busting drug tissue plasminogen activator (tPA) is a key treatment of ischemic stroke in adults. However, no consensus exists about the use of IV tPA in older adolescents who otherwise meet the standard adult tPA eligibility criteria. IV tPA has been utilized on a case by case basis in children older than 2 years of age who presented within less than 4.5 h of initial symptom onset. The safety and efficacy of IV tPA in children is unknown and is generally not recommended as standard therapy outside of a clinical trial. Children presenting with possible acute ischemic stroke should undergo rapid evaluation in consultation with a pediatric stroke specialist, to determine management strategies and candidacy for acute intervention [3,10].

Acute Hemorrhages

Hemorrhagic strokes may require medical management beyond supportive measures. Prevention of rebleeding includes correction of coagulation defects and hematological disorders. Severe coagulation-factor deficiencies should receive replacement therapy. Surgical management of hemorrhagic strokes is controversial. There may be benefit of early surgical evacuation in patients with clinical deterioration due to mass effect. Some researchers argue that children may warrant more aggressive intervention given their lack of cerebral atrophy which, in older adults, could potentially accommodate some degree of hematoma expansion [3]. Vascular anomalies should be corrected as soon as is clinically feasible [2,10].

CONCLUSION

In sum, treatment of stroke in children is important, yet understudied in this patient population. It depends on the type of stroke and should be considered in both the short and long term. Management strategies in children are primarily extrapolated from adult studies, but with different considerations regarding short-term anticoagulation and guarded recommendations regarding thrombolytics [3,10]. Regardless of whether the stroke occurs in a neonate or child, rapid transfer to a tertiary pediatric medical center is warranted.

References

[1] American Heart Association. Facts knowing no bounds: Stroke in infants, children and youth. 2013. Retrieved from: https://www.heart.org/idc/groups/heart-public/@wcm/@adv/documents/downloadable/ucm_302255.pdf.

[2] Aucutt-Walter N, Burnett N, Delametter G, et al. An overview of pediatric stroke: Prenatal through teenager. Project for Expansion of Education in Pediatric Stroke (PEEPS) <https://www.med.unc.edu/neurology/divisions/stroke/pdfs/PEEPSProviderGuideFinal.pdf.

[3] Tsze DS, Valente JH. Pediatric stroke: A review. Emerg Med Int 2011. http://dx.doi.org/10.1155/2011/734506. Published online December 27, 2011.

[4] Caplan LR. Stroke in children and young adults. In: Caplan LR, editor. Caplan's Stroke. 4th ed. Philadelphia PA: Saunders-Elsevier; 2009.

[5] Golomb MR, et al. Male predominance in childhood ischemic stroke findings from the International Pediatric Stroke Study. Stroke 2009;40:52–7.

[6] Fullerton HJ, Wu YW, Zhao S, Johnston SC. Risk of stroke in children: ethnic and gender disparities. Neurology 2003;61(2):189–94. 158.

[7] Calgary Pediatric Stroke Program. Perinatal stroke: classification. 2015. Retrieved from: http://perinatalstroke.com/stroke classification.

[8] Calgary Pediatric Stroke Program. Frequently asked questions on perinatal stroke. 2015. Retrieved from: http://perinatalstroke.com/faq.

[9] Kirton A, deVeber G. Life after perinatal stroke. Stroke 2013;44:3265–71.

[10] Roach ES, Golomb MR, Adams R, Biller J, Daniels S, deVeber G, et al. Management of stroke in children. Stroke 2008;39:2644–91.

[11] International Alliance for Pediatric Stroke. Get the facts about pediatric stroke. 2014. Retrieved from: http://iapediatricstroke.org/about_pediatric_stroke.aspx.

[12] American Heart Association. Strokes can happen at any age. 2014. Retrieved from: http://www.strokeassociation.org/STROKEORG/AboutStroke/StrokeInChildren/What-is-Pediatric-StrokeInfographic_UCM_466477_SubHomePage.jsp.

[13] Carvalho KS, Bodensteiner JB, Connolly PJ, Garg BP. Cerebral venous thrombosis in children. J Child Neurol 2001;16:574–80.

[14] Imai WK, Everhart Jr FR, Sanders Jr JM. Cerebral venous sinus thrombosis: report of a case and review of the literature. Pediatrics 1982;70:965–70.

[15] Shevell MI, Silver K, O'Gorman AM, Watters GV, Montes JL. Neonatal dural sinus thrombosis. Pediatr Neurol 1989;5:161–5.

[16] Garcia JH, Pantoni L. Strokes in childhood. Semin Pediatr Neurol 1995;2:180–91.

[17] Jordan LC, Hillis AE. Challenges in the diagnosis and treatment of pediatric stroke. Nat Rev Neurol 2001;7:199–208.

[18] Brankovic-Sreckovic V, Milic-Rasic V, Jovic N, Milic N, Todorvic S. The recurrence risk of ischemic stroke in childhood. Med Princ Pract 2004;13:153.

[19] Brower MC, Rollins N, Roach ES. Basal ganglia and thalami infarction in children; Cause and clinical features. Arch Neurol 1996;53:1252–6.

[20] Sedzimir CB, Robinson J. Intracranial hemorrhages in children and adolescents. J Neurosurg 1973;38:269–81.

[21] Elbers J, Wainwright MS, Amlie-Lefond C. The pediatric stroke code: early management of the child with stroke. J Pediatr 2015;167:19.

[22] Jauch EC, Saver JL, Adams Jr HP, et al. Guidelines for the early management of patients with acute ischemic stroke: a guideline for healthcare professionals from the American Heart Association/American Stroke Association. Stroke 2013;44:870.

[23] Paediatric Stroke Working Group. Stroke in childhood: clinical guidelines for diagnosis, management and rehabilitation. November 2004. www.rcplondon.ac.uk/pubs/books/childstroke/.

[24] Eeg-Olofsson O, Ringheim Y. Stroke in children. Clinical characteristics and prognosis. Acta Paediatr Scand 1983;72(3):391–5.

[25] Riela AR, Roach ES. Etiology of stroke in children. J Child Neurol 1993;8(3):201–20.

[26] Younkin DP. Diagnosis and treatment of ischemic pediatric stroke. Current Neurol Neurosci Rep 2002;2(1):18–24.

[27] Sträter R, et al. Aspirin versus low-dose low-molecular-weight heparin: antithrombotic therapy in pediatric ischemic stroke patients a prospective follow-up study. Stroke 2001;32:2554–8.

[28] Fullerton HJ, Wintermark M, Hills NK2, Dowling MM, Tan M, Rafay MF, VIPS Investigators, et al. Risk of recurrent arterial ischemic stroke in childhood: a prospective international study. Stroke January 2016;47(1):53–9.

[29] Mallouh AA, Hamdan JA. Intracranial hemorrhage in patients with sickle cell disease. Am J Dis Child 1986;140:505–6.

106

Ischemic Stroke in the Young

F.D. Testai[1], J. Biller[2]

[1]University of Illinois at Chicago, Chicago, IL, United States; [2]Loyola University Chicago, Stritch School of Medicine, Maywood, IL, United States

INTRODUCTION

Age constitutes one of the most important nonmodifiable risk factors for cerebral infarction. About 15% of all the ischemic strokes occur in young adults. Since stroke predominately affects the elderly, most of the data gathered in pivotal clinical trials and population-based studies were obtained using cohorts of older patients with coexisting cardiac disease and atherosclerosis. In comparison to older individuals, however, patients aged 15–50 years have a lower prevalence of traditional vascular risk factors and a higher representation of uncommon causes of stroke; these differences necessitate a particularized, thoughtful approach and cause-specific treatments. This chapter hence focuses on the epidemiology, pathogenesis, and diagnostic investigations of ischemic stroke in the young.

EPIDEMIOLOGY OF STROKE IN THE YOUNG

The occurrence of stroke increases with age. In the Northern Manhattan Stroke (NOMAS) study, the reported annual *incidence* of stroke in individuals aged 20–44 years was 23 per 100,000 habitants [1]. Of these, 45% were ischemic strokes, 30% intracerebral hemorrhages, and 26% subarachnoid hemorrhages. In comparison, in individuals older than 45 years, 80% of the strokes were ischemic, 15% intracerebral hemorrhages, and 5% subarachnoid hemorrhages. The stroke incidence also has racial and ethnic variations. In the NOMAS study, for example, the relative risk of stroke was 2.4 times higher in blacks and 2.5 times higher in Hispanics than in whites. Similarly, in the Greater Cincinnati/Northern Kentucky Stroke Study, the risk of first stroke in blacks relative to whites was 2.2 for subjects ≤34 years of age, 5.0 for 35–44 years, and 2.6 for 45–54 years [2].

Because stroke is a preventable condition, its incidence is influenced by the implementation of effective prevention programs. Analysis of temporal trends has shown a declining incidence of ischemic stroke in the elderly. In comparison, the rate of stroke in subjects aged 20–54 years has increased over time, particularly for the year 2005 compared to earlier time periods [3]. Similarly, data from the US Nationwide Inpatient Sample show that the rate of stroke discharges for subjects of 15–44 years increased from 23% to 53% between 1995 and 2008 [4]. These trends were accompanied by an increased prevalence of traditional vascular risk factors in young individuals, including arterial hypertension, diabetes mellitus, dyslipidemia, obesity, and smoking.

The *prevalence* of stroke in the general population increases exponentially with age. Data from the 2009–2012 National Health and Nutrition Examination Survey show that prevalence of stroke is about 0.2% for men and 0.5% for women aged 20–39 years; in comparison, the prevalence of stroke in men and women aged 60–79 years are 6.1% and 5.2%, respectively (Table 106.1) [5]. A substantial number of strokes are asymptomatic. This has raised the concern that the prevalence of stroke reported in observational studies and national surveys may underestimate the burden of cerebrovascular diseases in the general population. In the NOMAS study, for example, the prevalence of asymptomatic stroke in relatively young adults aged ≤65 years of age was 9.5% [6]. Also, in a single-center retrospective study including young adults aged 18–50 years admitted with first-ever ischemic stroke, silent ischemic lesions were observed in about 28% of patients [7]. Moreover, in the Helsinki Young Stroke Registry, 13% of the patients had radiological evidence of previous silent cerebral ischemia which was associated with type 1 diabetes mellitus, obesity, smoking, and increasing age [8].

Primer on Cerebrovascular Diseases, Second Edition
http://dx.doi.org/10.1016/B978-0-12-803058-5.00106-5

TABLE 106.1　Prevalence of Stroke by Age and Sex (National Health and Nutrition Examination Survey: 2009–2012) [5]

Age (years)	Men (%)	Women (%)
20–39	0.2	0.7
40–59	1.9	2.2
60–79	6.1	5.2
>80	15.8	14.0

CAUSES OF ISCHEMIC STROKE IN THE YOUNG

Establishing the cause of stroke in the young is often challenging. While some causes of stroke in the young and the elderly are similar (e.g., carotid atherosclerosis and cardiac arrhythmia), the pathogenesis of cerebral ischemia in young individuals is more pleomorphic. Moreover, the cause of stroke remains uncertain in a substantially higher percentage of younger patients than older patients despite thorough diagnostic evaluations. From a practical standpoint, it is useful to categorize ischemic stroke in the young into the following major groups: cardioembolism; atherosclerotic large artery occlusive disease of the carotid or vertebrobasilar circulation; nonatherosclerotic large artery occlusive disease (e.g., cervical arterial dissection and moyamoya disease); occlusive disease of small penetrating arteries that arise from the major intracranial arteries (also termed small vessel disease); prothrombotic states; miscellaneous causes (e.g., migraine, oral contraceptives, substance abuse, and genetic); and stroke of undetermined cause (cryptogenic stroke). A detailed list of specific diseases within each of these major groups is shown in Table 106.2. In addition, there are numerous stroke mimics that should be considered in the differential diagnosis of stroke and transient ischemic attack (TIA) (Table 106.3).

The stroke etiology varies with the cohort investigated. In the Stroke Prevention in Young Adults Study, the most common cause of stroke in individuals ≤40 years was cardioembolism. In comparison, subjects aged 40–50 years were more likely to experience either lacunar or large vessel stroke [9]. In the Baltimore–Washington Stroke Cooperative study, 31% of the strokes in subjects of 15–45 years of age were caused by cardioembolism, 20% by coagulopathies, 20% by penetrating artery disease, 11% by nonatherosclerotic large artery occlusive disease, 9% by use of illicit drugs, and 4% by atherosclerotic larger artery occlusive disease [10]. Similarly, in a single-center retrospective study using the Get With The Guidelines—Stroke database, the most common cause of ischemic stroke in subjects of 18–45 years of age was cardioembolism (47%), followed by nonatherosclerotic large artery occlusive disease (21%), coagulopathy (4%), small vessel disease (7%), and atherosclerotic large artery occlusive disease (2%) [11]. In a 2015 prospective multicenter study of consecutive young patients aged 16–55 years with ischemic stroke, the most common etiology was cardioembolism (32%) followed by cervical dissection (17%), large artery atherosclerosis (11%), and small vessel disease (9%). Seventy-three percent of the subjects had hypercholesterolemia (47%), smoking (43%), or hypertension (35%). These findings highlight the role of modifiable vascular risk factors in the genesis of cerebral ischemia [12]. The percentages of cryptogenic stroke in these studies vary significantly (4–50%) and depend on the extent of the ancillary investigations and the population studied.

CLINICAL CLUES SUGGESTING SPECIFIC CAUSES OF STROKE

Although many heterogeneous disorders need to be considered when attempting to establish the cause of stroke in the young, a detailed history and thorough bedside examination are useful for narrowing the diagnostic possibilities. Atherosclerotic large artery occlusive disease should be strongly considered in patients with traditional vascular risk factors, coronary or peripheral vascular disease, family history of early atherosclerosis, history of multiple TIAs, and audible carotid or other cervical bruits. Prominent facial pain (usually periorbital) preceding or accompanying the stroke and the presence of an ipsilateral Horner syndrome should suggest carotid artery dissection. Prominent neck pain in a patient with signs and symptoms of stroke in the posterior fossa should suggest vertebral artery dissection. Frequently, dissections are preceded by trauma, neck manipulation, or vigorous exercise; however, spontaneous (nontraumatic) dissections are not infrequently encountered, particularly in patients with conditions associated with vascular fragility, such as fibromuscular dysplasia (FMD). History of atrial fibrillation (AF), cardiac valve replacement, cardiomyopathy, recent myocardial infarction, rheumatic mitral stenosis or the presence of palpitations, dyspnea, or lower extremity edema should raise concerns of cardioembolism. Stroke occurring during Valsalva maneuver (e.g., while straining) or after prolonged immobilization should suggest the possibility of paradoxical embolism through an interatrial septal defect. Lipohyalinosis or atherosclerosis of small penetrating arteries should be considered in patients with an established history of arterial hypertension or diabetes mellitus who present with a stroke syndrome and subcortical infarct of 1.5 cm in size. Prothrombotic states should be considered in patients with a personal or family history of venous or arterial thromboses

TABLE 106.2   Potential Causes of Ischemic Stroke in the Young

ATHEROSCLEROTIC LARGE ARTERY DISEASE

Predisposition to early atherosclerosis in patients with familial hyperlipidemia, onset of diabetes in childhood or adolescence, severe hypertension, homocystinuria (especially, homozygote state), and radiation therapy to neck or cranium

NONATHEROSCLEROTIC LARGE- OR MEDIUM-SIZED ARTERY DISEASE

Relatively Common	Rare
Carotid or vertebral artery dissection	Moyamoya pattern from neurofibromatosis type 1
Fibromuscular dysplasia	Sarcoid vasculopathy
Idiopathic moyamoya disease	Infectious
Sickle cell disease associated vasculopathy	Aspergillosis
	Mucormycosis
	Tuberculosis
	Herpes zoster vasculopathy
	Associated angiitis of the nervous system
	Polyarteritis nodosa
	Granulomatosis with polyangiitis
	Churg–Strauss syndrome (eosinophilic granulomatosis with polyangiitis)
	Takayasu disease
	Systemic lupus erythematosus
	Lymphomatoid granulomatosis
	Giant cell arteritis
	Primary angiitis of the central nervous system (PACNS)

PENETRATING ARTERY DISEASE

Relatively Common	Rare
Lipohyalinosis	Infectious causes (usually associated with meningeal infection)
Atherosclerosis	Syphilis
	Cryptococcus
	Tuberculosis
	Associated angiitis of the nervous system
	Behçet's disease
	Susac syndrome
	Granulomatosis with polyangiitis
	Hypersensitivity vasculitis
	Cogan's syndrome
	Eales disease
	Primary angiitis of the central nervous system (PACNS)

CARDIOEMBOLISM

Definite or Probable Association	Possible Association
Infective endocarditis	Right-to-left interatrial shunt without associated deep vein thrombosis or pulmonary embolus
Prosthetic heart valve	Interatrial septal aneurysm, especially the types associated with an interatrial shunt
Acute myocardial infarction	Akinetic or hypokinetic ventricular segment without associated thrombus
Right-to-left interatrial shunt with associated deep vein thrombosis or pulmonary embolus	Mitral annular calcification
Atrial fibrillation/atrial flutter	Spontaneous contrast (associated with cardiomyopathy, atrial fibrillation, or mitral stenosis)
Mitral stenosis	Bicuspid aortic valve, especially if calcified
Left atrial thrombus	Mitral valve prolapse
Nonbacterial thrombotic (marantic) endocarditis	
Intracardiac tumors (left atrial myxoma and papillary fibroelastoma)	

PROTHROMBOTIC STATES

Definite or Probable Association	Possible Association
Antiphospholipid antibody syndrome	Protein C deficiency
Sickle cell disease	Protein S deficiency
Cancer-related thrombosis Polycythemia vera	Antithrombin deficiency
Essential thrombocytosis	Dysfibrinogenemia
Thrombotic thrombocytopenic purpura	Disorders of fibrinolysis (e.g., dysplasminogenemia and elevated plasminogen activator inhibitor)
Disseminated intravascular coagulation	Factor V Leiden mutation
	Prothrombin gene mutation (G20210A)
	Methylenetetrahydrofolate reductase gene mutation
	Cancer

Continued

TABLE 106.2   Potential Causes of Ischemic Stroke in the Young—cont'd

MISCELLANEOUS

Substance abuse (especially with cocaine and amphetamines)
Migraine with aura
Oral contraception
Pregnancy, peri- and postpartum-associated stroke (e.g., cardiomyopathy, vasculopathy, prothrombotic state, and amniotic fluid embolism)
Mitochondrial encephalomyopathy, lactic acidosis, and stroke-like episodes syndrome (MELAS)
Hereditary endotheliopathy with retinopathy, nephropathy, and stroke (HERNS)
CADASIL and CARASIL (cerebral autosomal dominant/recessive arteriopathy with subcortical infarcts and leukoencephalopathy)
Reversible cerebral vasoconstriction syndrome (RCVS)
Sneddon syndrome
HaNDL syndrome (headache and neurological deficits with cerebrospinal fluid lymphocytosis)
Osler–Weber–Rendu syndrome or hereditary hemorrhagic telangiectasia (HHT)
SMART syndrome (stroke-like migraine Attacks after Radiation therapy)
CLIPPERS (chronic lymphocytic inflammation with pontocerebellar perivascular enhancement responsive to steroids)
Recurrent extracranial cervical carotid artery vasospasm
Fabry disease (also known as angiokeratoma corporis diffusum and alpha-galactosidase A deficiency)

STROKE OF UNDETERMINED CAUSE (CRYPTOGENIC STROKE)

TABLE 106.3   Stroke Mimics [22,23]

Psychogenic

Migraine

Musculoskeletal abnormalities

Delirium

Hypertensive emergency

Seizure disorder

Postictal paralysis

Tumor

Acute disseminated encephalomyelitis

Multiple sclerosis

Cerebellitis

Encephalitis

Meningitis

Drug toxicity

Idiopathic intracranial hypertension

Subdural empyema

Intracranial abscess

Reversible posterior leukoencephalopathy syndrome

Arteriovenous malformation

Metabolic causes

Peripheral vestibulopathy

Miller–Fisher syndrome

Pharyngeal–cervical–brachial variant of Guillain–Barré syndrome

Myasthenia gravis

Syncope

(especially, if the thrombosis is unprovoked or involve unusual sites such as the upper extremity), in patients with cancer, and in patients with an intraluminal carotid or vertebrobasilar clot without underlying occlusive disease. The presence of antiphospholipid antibodies (or antiphospholipid syndrome) contributing to the cause of stroke should be considered in patients with multiple miscarriages, recurrent thromboembolism, livedo reticularis, thrombocytopenia, or history of connective tissue diseases, such as systemic lupus erythematosus (SLE) or *Sjögren's* syndrome (Fig. 106.1). Stroke in patients with SLE can also be related to early large artery atherosclerosis, Libman–Sacks endocarditis, thrombotic thrombocytopenic purpura, and, rarely, from cerebral vasculitis. The presence of cancer raises the possibility of a prothrombotic state or nonbacterial thrombotic (marantic) endocarditis (NBTE). Livedo reticularis with or without antiphospholipid syndrome and no other stigmata of connective tissue disorders should raise the suspicion of *Sneddon* syndrome, which is a noninflammatory arteriopathy that may affect cerebral vessels of variable caliber.

Substance abuse raises the possibility of stroke related to vasospasm, vasculitis, or infective endocarditis (if intravenous drugs are used). This last condition can present with fevers, back pain, arthralgias, and stigmata of systemic microembolism. Cocaine, amphetamines, and marijuana are associated with the highest risk of stroke. Although the use of oral contraceptives has been associated with an increased risk of stroke in large epidemiological studies, the overall risk to an individual patient remains modest. Therefore, a comprehensive diagnostic evaluation should be performed in all young stroke patients taking oral contraceptives to rule out the possibility of another cause of stroke. While migraine

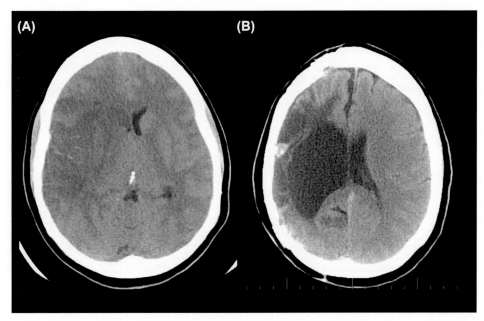

FIGURE 106.1 **Head computed tomography (HCT) images in a patient with antiphospholipid antibody syndrome (APLA).** HCT obtained in a 28 years old female with history of SLE, pulmonary embolism, APLA syndrome, and hereditary spherocytosis. The original HCT (A) shows a hypodense area in the territory of the right middle cerebral artery (MCA) consistent with embolism. There is, in addition, mass effect with efface-ment of the right lateral ventricle and mild right-to-left midline shift. The subsequent HCT, obtained 10 years after the index stroke (B), shows residual right MCA-territory encephalomalacia, dystrophic calcifications, and *ex vacuo* dilation of the right lateral ventricle.

has been reported to cause stroke, the diagnosis of migraine stroke should not be based solely on the pres-ence of unusual headaches associated with stroke. The presence of headache after stroke is common regardless of the cause of stroke. Migraine stroke should be diag-nosed only in patients with a history of migraines with aura who have a persistent neurological deficit after a typical migraine attack. Additionally, a thorough diag-nostic evaluation for other causes of stroke should be entirely nondiagnostic. The presence of thunderclap headaches during pregnancy or puerperium, in asso-ciation with vasoactive medications (e.g., selective sero-tonin reuptake inhibitors (SSRIs), migraine medications, and sympathomimetic decongestants), or during physi-cal activity should raise the possibility of reversible cere-bral vasoconstriction syndrome (RCVS) or *Call–Fleming* syndrome. Headache in the presence of encephalopathy, hearing loss, and visual changes suggests the diagnosis of retinocochleocerebral vasculopathy or *Susac syndrome*. Headache, cognitive decline, behavioral changes, and subcortical stroke at young age, in addition, may be seen in patients with cerebral autosomal dominant arteriopa-thy with subcortical infarcts and leukoencephalopathy (CADASIL) (Fig. 106.2).

Different central and systemic vasculitides may course with cerebral ischemia. Primary angiitis of the central nervous system (PACNS) presents with focal neurological deficits in association with headaches, sei-zures, and encephalopathy rather than a single stroke. They usually have inflammatory cerebrospinal fluid

(CSF), and angiography may show multiple areas of segmental narrowing of small- and medium-sized arteries, although larger arteries may occasionally be involved. PACNS is a disorder that is overdiagnosed in young patients with stroke. Since the angiographic findings are not specific for vasculitis (seen also with migraine, diffuse intracranial atherosclerosis, RCVS, and lymphomatous or infectious infiltration of arteries), brain biopsy may be required to establish the diagnosis. Reduced visual acuity, headache, polymyalgia rheu-matica, jaw claudication, and scalp tenderness suggest the diagnosis of giant cell arteritis. Vestibuloauditory dysfunction, interstitial keratitis, arthralgias, and skin rash may be seen in *Cogan* syndrome. Eales disease (also known as primary perivasculitis of the retina), in comparison, presents with prominent visual loss which may be related to retinal ischemia, vitreous hemor-rhage, or retinal detachment.

A significant family history of stroke at young age sug-gests a heritable predisposition. Genome-wide associa-tion studies have identified specific loci associated with cardioembolic stroke (4q25 and 16q22) and large artery atherosclerotic stroke (7p21 and 6p21). From the mecha-nistic standpoint, some of these loci may participate in the development of stroke risk factors; 4q25 and 16q22, for example, correlate with AF [13]. The study of genetic polymorphisms in the occurrence of ischemic stroke has been confounded by limited replicability. In addition, stroke subtypes considered to be genetically distinct (such as large artery atherosclerosis and small vessel

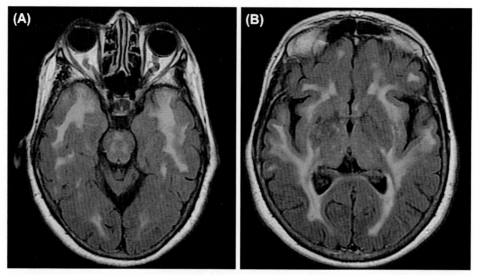

FIGURE 106.2 **Brain magnetic resonance imaging (MRI) obtained in a patient with CADASIL.** CADASIL is an autosomal dominant condition caused by a mutation in the *notch 3* gene. Patients present with white matter hyperintensities in T2-weighted and fluid-attenuated inversion recovery (FLAIR) images. Most commonly, these lesions involve the temporal pole (A) and external capsule (B).

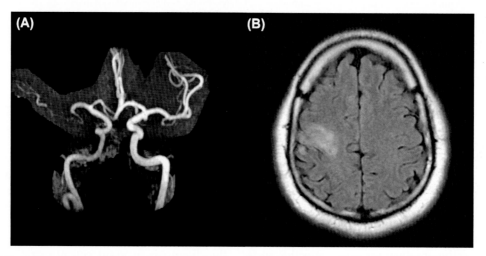

FIGURE 106.3 **Magnetic resonance (MR) images obtained in a patient with sickle cell disease (SCD).** SCD is an autosomal recessive hemoglobinopathy caused by a mutation in the β-globin gene. Patients with SCD might develop painful crises, large artery occlusive disease, and stroke at early age. The MR angiogram (A) shows an occlusion of the right middle cerebral artery. MR images show hyperintense signal in fluid-attenuated inversion recovery (FLAIR) sequence (B) denoting chronic ischemic changes in the MCA territory.

disease) seem to share genetic variants [14]. There are, however, physical findings suggestive of monogenic diseases that increase the risk of ischemic stroke, including sickle cell disease (anemia, painful crises, and hand–foot swelling) (Fig. 106.3); Fabry disease (angiokeratomas, acroparesthesias, kidney disease, and corneal opacities); Marfan syndrome (tall stature, arachnodactyly, and ectopia lentis); Ehlers–Danlos type IV (large eyes, micrognathia, small stature, translucent skin, ecchymoses, and hypermobility of small joints); neurofibromatosis (neurofibromas and café-au-lait spots); HERNS syndrome (hereditary endotheliopathy with retinopathy, nephropathy, and stroke); *Osler–Weber–Rendu* syndrome or hereditary hemorrhagic telangiectasia (HHT) which presents

with telangiectasias, recurrent thromboembolism, and hemorrhagic complications (Fig. 106.4); or mitochondrial encephalomyopathy, lactic acidosis, and stroke-like episodes syndrome (MELAS) which is characterized by muscle weakness and pain, recurrent headache, seizures, neuropsychiatric disorder, and lactic acidosis.

AN APPROACH TO DIAGNOSTIC TESTING IN YOUNG STROKE PATIENTS

The following tests are recommended routinely in all young (15–50 years) ischemic stroke patients: complete blood count, platelets, prothrombin time/

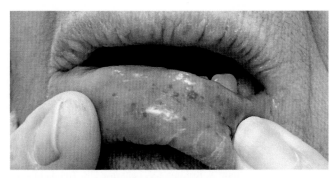

FIGURE 106.4 **Mucocutaneous telangiectasias observed in a patient with HHT or Osler–Weber–Rendu disease.** HHT is an autosomal dominant disease caused by genetic mutations (mainly endoglin (*ENG*) gene mutations) that result in tumor growth factor (TGF)-β superfamily receptor dysfunction and the development of vascular dysplasias. Patients present with skin and mucocutaneous telangiectasias, hemorrhagic complications, and hepatic or pulmonary arteriovenous malformations resulting in recurrent thromboembolism and cerebral abscess.

International Normalized Ratio (PT/INR), activated partial thromboplastin time (aPPT), cholesterol, high-density lipoproteins, low-density lipoproteins, triglycerides, fluorescent treponemal antibody-absorption (FTA-ABS) test, electrocardiography, brain imaging (MRI preferable to CT), and noninvasive cerebrovascular imaging (carotid Doppler, computed tomography angiography (CTA) or magnetic resonance angiography (MRA)). If noninvasive imaging is negative and the clinical suspicion of intracranial occlusive disease (e.g., intracranial atherosclerosis, moyamoya disease, and vasculitis) is low, digital subtraction angiography (DSA) is probably not necessary. If the clinical suspicion of intracranial occlusive disease is high, however (e.g., in the setting of multiple stereotypical TIAs and normal extracranial studies), or if noninvasive cerebrovascular imaging suggests intracranial occlusive disease, DSA is usually warranted. Blood cultures, transesophageal echocardiography (TEE), and CSF analysis are recommended for patients with fever. In African-American patients and in people with origins in equatorial Africa, the Mediterranean basin, or Saudi Arabia, a sickle cell screening (followed by hemoglobin electrophoresis, if the screening test is positive) should also be included in the routine studies to rule out sickle cell disease. If the diagnosis is established by these routine tests (e.g., angiography shows atherosclerotic carotid artery occlusion or a cervical artery dissection), the diagnostic evaluation is terminated. In patients with normal routine blood tests and cerebrovascular imaging, or in patients with an occluded intracranial artery without a proximal large artery source of embolus, echocardiography is recommended. This is particularly the case in patients with MR findings suggestive of an embolic source. Transesophageal echocardiography (TEE) rather than transthoracic echocardiography (TTE) is the preferred procedure because of its higher sensitivity and specificity for detecting cardiac abnormalities such as an interatrial septal aneurysm, an atrial septal defect, small vegetations, and left atrial spontaneous contrast ("smoke") possibly representing erythrocyte microaggregates that form when there is stasis of blood in the left atrium. Bubble contrast echocardiography is routinely performed in this setting to identify a right-to-left shunt through an interatrial septal defect. If a right-to-left shunt is found, venous imaging of the lower extremities and pelvis is performed to rule out deep venous thrombosis. Observational studies show that thrombophilias increase modestly to the risk of cerebral ischemia, particularly in young adults with cryptogenic stroke. Therefore, if the diagnostic studies outlined earlier are unrevealing, it is reasonable to measure antithrombin (AT) activity, protein C antigen and activity, protein S antigen and activity, D-dimer, fibrin split products, fibrinogen, serum homocysteine, and genotyping for methylenetetrahydrofolate reductase (MTHFR), factor V Leiden, and prothrombin G20210A mutation. Because low levels of protein C and protein S and high levels of fibrinogen may be acute-phase responses to stroke, it is advisable to repeat these tests at least 3 months after stroke to determine whether the abnormalities persist. Also, anticoagulation therapy influences the results of some of these tests (e.g., protein C, protein S, and AT); thus, testing should be performed when the patient is not on anticoagulant treatment. Protein C, protein S, and AT deficiencies are rare but increase the risk of venous thrombosis. Also, observational studies in young individuals have shown an odds ratio of myocardial infarction, ischemic stroke, and TIA of 2–4 for individuals carrying MTHFR C677T, factor V Leiden, or prothrombin G20210A mutations compared to wild type. Antiphospholipid antibodies, including lupus anticoagulant and antibodies against cardiolipin and β2-glycoprotein 1, increase the risk of both first and recurrent arterial and venous thrombosis. Elevated IgG titers have been detected in young individuals with cerebral infarction. In comparison, IgM antibodies are considered acute reactants and may increase in the course of infectious processes. The diagnosis of antiphospholipid syndrome requires persistent titers of antiphospholipid antibodies in the context of fetal loss and/or recurrent arterial or venous thrombosis.

TREATMENT AND PROGNOSIS IN YOUNG STROKE PATIENTS

Stroke in older patients is closely related to traditional vascular risk factors and the development of atherosclerotic disease. In comparison, the causes of stroke in the

TABLE 106.4 Functional Outcome of Young Adults With Ischemic Stroke After a Mean Followup of 11.7 years (n = 240) [19]

Outcome	%
Able to walk	95
Independent for daily activities	90
Barthel index = 100	74
Modified Rankin scale = 0	27
Able to return to work	53
Adjustment in occupation	23
Retirement due to illness	35
Recurrent stroke	25

young are more heterogeneous and can be uncommon. Thus, the treatment, rate of recurrence, and ultimately the outcome in this subgroup are closely related to the identification of the stroke etiology.

The treatment of ischemic stroke in young patients includes the use of antiplatelet agents and the aggressive correction of coexistent vascular risk factors as recommended by stroke-prevention guidelines [15]. The use of statins is reasonable in cases of atherosclerosis. Also, anticoagulation may be indicated in appropriate circumstances, including patients with AF or antiphospholipid syndrome. In addition, some conditions may require unique approaches, as is the case of immunomodulation in CNS vasculitides or enzyme replacement treatment in Fabry disease. The reader is referred to the individual chapters of each disease for additional information.

The prognosis of young patients with stroke is relatively good. Based on data obtained in different observational studies, the stroke fatality in patients aged 15–49 years is about 3% at 1 month, 5% at 1 year, and 11% at 5 years [16]. In general, the survival rate in young patients doubles that of older individuals. The 5-year rate of recurrent cerebral ischemia in patients aged <50 years is 9.4%, and this increases with advanced age, type 1 diabetes mellitus, heart failure, large artery atherosclerosis-related stroke subtype, and previous TIA [17]. From the functional standpoint, about 70–88% of young stroke victims achieve independence defined as a modified Rankin Scale score of ≤2 [12,18–20]. Both diabetes mellitus and stroke severity at admission are predictors of poor functional outcome. Despite this seemingly good outcome, a substantial number of stroke survivors suffer from residual neurocognitive disorders including depression (30–45%), chronic fatigue (54%), memory loss (41%), anxiety (19%), and sleep-related disorders (36%) [19,20] (Table 106.4). Also, young patients with stroke are left with significant socioeconomic sequelae with only 50–70% of them being able to return to work and 7% facing divorce [18–21].

References

[1] Jacobs BS, Boden-Albala B, Lin IF, Sacco RL. Stroke in the young in the Northern Manhattan stroke study. Stroke 2002;33(12):2789–93.

[2] Zakaria T, Lindsell CJ, Kleindorfer D, et al. Age accounts for racial differences in ischemic stroke volume in a population-based study. Cerebrovasc Dis 2008;26(4):376–80. http://dx.doi.org/10.1159/000151641.

[3] Kissela BM, Khoury JC, Alwell K, et al. Age at stroke: temporal trends in stroke incidence in a large, biracial population. Neurology 2012;79(17):1781–7. http://dx.doi.org/10.1212/WNL.0b013e318270401d.

[4] George MG, Tong X, Kuklina EV, Labarthe DR. Trends in stroke hospitalizations and associated risk factors among children and young adults, 1995–2008. Ann Neurol 2011;70(5):713–21. http://dx.doi.org/10.1002/ana.22539.

[5] Mozaffarian D, Benjamin EJ, Go AS, et al. Heart disease and stroke statistics–2015 update: a report from the American Heart Association. Circulation 2015;131(4):e29–322. http://dx.doi.org/10.1161/CIR.0000000000000152.

[6] Prabhakaran S, Wright CB, Yoshita M, et al. Prevalence and determinants of subclinical brain infarction: The Northern Manhattan study. Neurology 2008;70(6):425–30. http://dx.doi.org/10.1212/01.wnl.0000277521.66947.e5.

[7] Gioia LC, Tollard E, Dubuc V, et al. Silent ischemic lesions in young adults with first stroke are associated with recurrent stroke. Neurology 2012;79(12):1208–14. http://dx.doi.org/10.1212/WNL.0b013e31826aacac.

[8] Putaala J, Kurkinen M, Tarvos V, Salonen O, Kaste M, Tatlisumak T. Silent brain infarcts and leukoaraiosis in young adults with first-ever ischemic stroke. Neurology 2009;72(21):1823–9. http://dx.doi.org/10.1212/WNL.0b013e3181a711df.

[9] Trivedi MM, Ryan KA, Cole JW. Ethnic differences in ischemic stroke subtypes in young-onset stroke: The stroke prevention in young adults study. BMC Neurol 2015;15(1):221. http://dx.doi.org/10.1186/s12883-015-0461-7.

[10] Kittner SJ, Stern BJ, Wozniak M, et al. Cerebral infarction in young adults: The Baltimore-Washington cooperative young stroke study. Neurology 1998;50(4):890–4.

[11] Ji R, Schwamm LH, Pervez MA, Singhal AB. Ischemic stroke and transient ischemic attack in young adults: risk factors, diagnostic yield, neuroimaging, and thrombolysis. JAMA Neurol 2013;70(1):51–7. http://dx.doi.org/10.1001/jamaneurol.2013.575.

[12] Goeggel Simonetti B, Mono ML, Huynh-Do U, et al. Risk factors, aetiology and outcome of ischaemic stroke in young adults: The swiss young stroke study (SYSS). J Neurol 2015;262(9):2025–32. http://dx.doi.org/10.1007/s00415-015-7805-5.

[13] Cheng YC, Cole JW, Kittner SJ, Mitchell BD. Genetics of ischemic stroke in young adults. Circ Cardiovasc Genet 2014;7(3):383–92. http://dx.doi.org/10.1161/CIRCGENETICS.113.000390.

[14] Holliday EG, Traylor M, Malik R, et al. Genetic overlap between diagnostic subtypes of ischemic stroke. Stroke 2015;46(3):615–9. http://dx.doi.org/10.1161/STROKEAHA.114.007930.

[15] Kernan WN, Ovbiagele B, Black HR, et al. Guidelines for the prevention of stroke in patients with stroke and transient ischemic attack: a guideline for healthcare professionals from the American Heart Association/American Stroke Association. Stroke 2014;45(7):2160–236. http://dx.doi.org/10.1161/STR.0000000000000024.

[16] Putaala J, Curtze S, Hiltunen S, Tolppanen H, Kaste M, Tatlisumak T. Causes of death and predictors of 5-year mortality in young adults after first-ever ischemic stroke: The Helsinki Young Stroke Registry. Stroke 2009;40(8):2698–703. http://dx.doi.org/10.1161/STROKEAHA.109.554998.

[17] Putaala J, Haapaniemi E, Metso AJ, et al. Recurrent ischemic events in young adults after first-ever ischemic stroke. Ann Neurol 2010;68(5):661–71. http://dx.doi.org/10.1002/ana.22091.

[18] Leys D, Bandu L, Henon H, et al. Clinical outcome in 287 consecutive young adults (15 to 45 years) with ischemic stroke. Neurology 2002;59(1):26–33.

[19] Varona JF, Bermejo F, Guerra JM, Molina JA. Long-term prognosis of ischemic stroke in young adults. study of 272 cases. J Neurol 2004;251(12):1507–14. http://dx.doi.org/10.1007/s00415-004-0583-0.

[20] Waje-Andreassen U, Thomassen L, Jusufovic M, et al. Ischaemic stroke at a young age is a serious event—final results of a population-based long-term follow-up in Western Norway. Eur J Neurol 2013;20(5):818–23. http://dx.doi.org/10.1111/ene.12073.

[21] Synhaeve NE, Schaapsmeerders P, Arntz RM, et al. Cognitive performance and poor long-term functional outcome after young stroke. Neurology 2015;85(9):776–82. http://dx.doi.org/10.1212/WNL.0000000000001882.

[22] Kim BJ, Kang HG, Kim HJ, et al. Magnetic resonance imaging in acute ischemic stroke treatment. J Stroke 2014;16(3):131–45. http://dx.doi.org/10.5853/jos.2014.16.3.131.

[23] Shellhaas RA, Smith SE, O'Tool E, Licht DJ, Ichord RN. Mimics of childhood stroke: characteristics of a prospective cohort. Pediatrics 2006;118(2):704–9. http://dx.doi.org/10.1542/peds.2005-2676.

C H A P T E R

107

Rare Genetic Causes of Stroke

M.K. Tobin, F.D. Testai

University of Illinois at Chicago, Chicago, IL, United States

INTRODUCTION

Cerebrovascular diseases constitute the fourth leading cause of death in the United States annually [1] and the third leading cause of mortality in developed countries [2]. In addition, it is the number one cause of permanent disability globally [3,4] and the second most common cause of dementia [5,6]. Many risk factors for cerebrovascular diseases have been established including nonmodifiable factors such as age, gender, and race, as well as acquired risk factors such as hypertension, smoking, diabetes, and obesity. These factors, however, only account for a portion of the stroke risk [7] suggesting that other variables, including genetics, must be involved in the etiology of stroke.

The exact contribution of genetics to the incidence of stroke still remains largely unknown; however, it is clear that stroke can result from both monogenic and polygenic diseases. Common monogenic causes of stroke include cerebral autosomal dominant arteriopathy with subcortical infarcts and leukoencephalopathy (CADASIL) and its autosomal recessive form, CARASIL, as well as sickle cell disease, and Fabry disease. These diseases are covered elsewhere in this book. This chapter focuses on the rarer monogenic and polygenic causes of stroke, including mitochondrial encephalomyopathy, lactic acidosis, and stroke-like episodes (MELAS), hereditary endotheliopathy with retinopathy, nephropathy, and stroke (HERNS), homocystinuria, moyamoya disease, and inherited connective tissue disorders, including type IV collagen α1-chain gene (COL4A1) mutation, Marfan syndrome, and vascular Ehlers–Danlos syndrome (VEDS).

MONOGENIC DISEASES

About 5% of stroke cases result from monogenic disease; however, this number likely underestimates the true number of cases as there are upward of 50 single gene causes of stroke [8]. Some of these gene mutations result in stroke as a part of a systemic syndrome, while some others result in clinical manifestations limited to the central nervous system (CNS) [9]. The more common monogenic diseases associated with stroke are summarized in Table 107.1.

Primer on Cerebrovascular Diseases, Second Edition
http://dx.doi.org/10.1016/B978-0-12-803058-5.00107-7

TABLE 107.1 Summary of Monogenic Causes of Stroke

Disease	Genes Involved	Mechanism of Stroke	Age of Presentation
MELAS (mitochondrial encephalomyopathy with lactic acidosis and stroke-like episodes syndrome)	tRNA (Leu) A3243G [76] tRNA (Leu) T3271C [12] tRNA (Lys) A8344G [14] tRNA (Leu) A3260G [13] mtDNA deletion [15] ND4 A11084G [16] Cytochrome c oxidase subunit III T9957C [17] ND5 G13513A [18]	Unclear though metabolic failure has been suggested [19]	Variable
HERNS (hereditary endotheliopathy with retinopathy, nephropathy, and stroke)	TREX1 [29]	Blood–brain barrier dysfunction [30]	30–40 years
Homocystinuria	Cystathionine-β-synthase Has been linked to >130 genes	Small and large vessel disease, arterial dissection, cardioembolism	Variable
Inherited connective tissue disorders			
VEDS (vascular type of Ehlers–Danlos syndrome)	COL3A1 [69]	Arterial dissection	Childhood
Marfan syndrome	FBN1 [77] TGFBR1, TGFBR2, SMAD3, TGFB2, TGFB3, SKI, EFEMP2, COL3A1, FLNA ACTA2, MYH11, MYLK, and SLC2A10 [55]	Arterial dissection, cardioembolism	Childhood
Type IV collagen α1-chain gene mutation	COL4A1 [34]	Small vessel disease; blood–brain barrier dysfunction[34]	<50 years
Hereditary hemorrhagic telangiectasia	ENG [78,79] ALK1 [78,79]	Telangiectasias, arteriovenous malformations and carotid–cavernous fistulas	Childhood
Arterial tortuosity syndrome	SLC2A10 [80]	Elongation, tortuosity and aneurysm of the medium-sized and large arteries	<30 years
Autosomal dominant polycystic kidney disease	PKD1 and PKD2 [81]	Arterial dissection and intracranial aneurysms	<50 years
Osteogenesis imperfecta	COL1A1, COL1A2, LEPRE1, CRTAP, FKBP10, PPIB [82]	Aortic dissection, intracranial aneurysm, carotid–cavernous fistula	Childhood
Moyamoya disease	Linked to chromosomes 3p24.2–p26, 6q25, 8q23, 12p12, 17q25[10]	ICA stenosis with neovascularization	Juvenile (<5 years) Adulthood (30–50 years)
CADASIL[a]	NOTCH3 [83]	Small vessel disease	30–60 years
CARASIL[a]	HTRA1 [84]	Small vessel disease	25–35 years
Fabry disease[a]	GLA [85]	Small and large vessel disease	<40 years
Sickle cell disease[a]	HBB [86]	Small and large vessel disease	Childhood
Neurofibromatosis type 1	NF1 [87]	Small and large vessel disease	Childhood
Pseudoxanthoma elasticum	ABCC6 [88]	Small and large vessel disease	Childhood

[a]These diseases are covered elsewhere in this book but are included here for completeness. CADASIL, cerebral autosomal dominant arteriopathy with subcortical infarcts and leukoencephalopathy; CARASIL, cerebral autosomal recessive arteriopathy with subcortical infarcts and leukoencephalopathy.

MITOCHONDRIAL ENCEPHALOMYOPATHY, LACTIC ACIDOSIS, AND STROKE-LIKE EPISODES (MELAS)

Genetic Basis. The genetic basis for MELAS results from mutations in a number of different mitochondrial or nuclear genes associated with the respiratory chain. Despite several gene mutations being reported, around 80% of all MELAS patients have an A3243G point mutation in the leucine mitochondrial transfer RNAs (tRNAs) [10,11]. Additionally, there have been three other tRNA point mutations reported including two others in the leucine tRNA gene, the T3271C [12] and A3260G [13] mutations, and one in the lysine tRNA gene, A8344G [14]. Other mutations include a mitochondrial DNA deletion [15] and mutations to different respiratory chain genes including the *ND4* gene [16], the *cytochrome c oxidase subunit III* gene [17], and the *ND5* gene [18].

Clinical Findings. MELAS is a multiorgan disease. However, different cells have different amounts of mutated mitochondrial DNA (heteroplasmy) and tissue-specific metabolic requirements. In addition, MELAS has significant genotypic heterogeneity. These factors explain the variations in severity and organ involvement seen among patients. The most striking feature of MELAS are the stroke-like episodes where patients present with focal deficits, including hemianopia, aphasia, hemiparesis, and cortical blinding [19]. Further, vomiting, headache, and seizures can be seen during exacerbations. MELAS has been associated with another mitochondrial disease, MERRF, which is characterized by myoclonic epilepsy with ragged-red fibers [14,20]. Because of the disruption in the respiratory chain and mitochondrial dysfunction, blood lactate levels and lactate/pyruvate ratio are elevated. In some cases, lactic acid levels may be normal in the blood but increased in the CSF, particularly during stress. Additional clinical findings include muscle weakness, ptosis, pigmentary retinopathy, sensorineural hearing loss, neuropsychiatric disorders, cardiomyopathy, and diabetes [21]. Brain imaging studies reveal bilateral calcifications in the basal ganglia. Stroke-like episodes correlate with MRI/CT findings consistent with cerebral infraction. These lesions are typically asymmetric, localize to the parietal and/or occipital lobes, and do not follow a well-defined vascular territory [19]. Noninvasive angiographic studies are usually unrevealing. Proton magnetic resonance spectroscopy can identify regions with decreased *N*-acetyl-aspartate and increased lactate-to-creatine ratio consistent with decreased neuronal viability and anaerobic metabolism, respectively.

Disease Course. MELAS often has an onset as early as the teenage years with stroke-like episodes. The course of the disease, however, is highly variable ranging from asymptomatic with normal early development to insidious onset, rapid disease progression, and premature death as early as the fourth decade of life [22].

Diagnosis. MELAS patients usually have decreased respiratory chain enzyme activity in skeletal muscle. Mitochondrial DNA sequencing can be performed on blood, skeletal muscle, urinary sediment, hair follicles, and buccal mucosa. However, due to the process of heteroplasmy, the load of mutated DNA varies among tissues and may be, in some cases, undetectable. Therefore, a negative genetic screening test does not necessarily rule out the diagnosis of MELAS. The muscle biopsy may demonstrate ragged red fibers that are considered the histological hallmark of this condition. These fibers stain positive to Gomori trichrome and cytochrome oxidase. In addition, they may have areas of increased succinate dehydrogenase staining suggestive of mitochondrial dysfunction and compensatory proliferation. On electron microscopy, affected tissue shows increase in number and size of mitochondria with paracrystalline bodies.

Management. Unfortunately, there are no current successful treatment strategies for MELAS; however, numerous approaches have been tried including dietary supplementation with vitamins and coenzymes and administration of redox compounds. Most therapies that have been utilized aim at improving the function of the respiratory chain and have included coenzyme Q10 [23], nicotinamide [24], combination therapy of cytochrome c/vitamin B1/vitamin B2 [25], idebenone [26], and sodium dichloroacetate [27], among others. Nonetheless, these therapies have only demonstrated marginal benefit to these patients. More recently, the supplementation with arginine has been shown to decrease the severity and frequency of stroke-like episodes. A bolus of arginine 0.5 g/kg given within 3 h of symptom onset followed by 0.5 g/kg administered as a continuous infusion for 24 h for the next 3–5 days is recommended for MELAS patients presenting with signs or symptoms suggestive of metabolic stroke. In addition, long-term prophylaxis with a daily dose of arginine 0.15–0.30 g/kg administered orally in three divided doses is recommended for MELAS patients with history of stroke [28].

HEREDITARY ENDOTHELIOPATHY WITH RETINOPATHY, NEPHROPATHY, AND STROKE (HERNS)

Genetic Basis. HERNS is an autosomal dominant condition that has been attributed to a c-terminal frameshift mutation in the *TREX1* gene that encodes a 3′–5′exonuclease [29]. The truncated product preserves the exonuclease activity; however, the ability of the protein to maintain its perinuclear localization is lost. *TREX1* is located on chromosome 3p21.1–p21.3 which has also been linked to cerebroretinal vasculopathy (CRV) and hereditary vascular retinopathy (HVR), and these three diseases share some clinical presentations. These conditions have collectively been called retinal vasculopathy with cerebral leukodystrophy (RVCL) [29].

Clinical Findings. HERNS is systemic, noninflammatory, and progressive vasculopathy that affects, preponderantly, arterioles and capillaries of brain, retina, and kidneys. Clinical findings associated with HERNS include stroke, nephropathy, migraine, mood disorders, and cognitive decline. Additionally, HERNS patients may have a variety of ophthalmological findings including retinopathy, telangiectasia, macular edema, microaneurysms, and capillary obliteration [30]. Some patients, in addition, may present with Reynaud phenomenon. Brain imaging studies reveal vasogenic edema with contrast-enhancing white matter lesions suggesting blood–brain barrier dysfunction and increased capillary permeability [30].

Disease Course. Disease onset is typically during the third to fourth decade of life with patients most frequently presenting with retinopathy and progressive visual loss. In its early stage, the fundoscopic findings may be difficult to differentiate from hypertensive or diabetic retinopathy. Other pathognomonic features of this condition, including stroke, neurocognitive decline, migraine, and nephropathy, begin later with death generally occurring within 10 years of disease onset [31].

Diagnosis. The diagnosis of HERNS is based on genetic testing. The differential diagnosis includes CADASIL, CARASIL, amyloid angiopathy, as well as inflammatory and inherited connective tissue diseases, such as lupus erythematosus and type IV collagen α1-chain mutation.

Management. Presently, there are no effective or curative treatments for HERNS and most commonly patients undergo symptomatic treatment. Laser treatments have been performed for retinal changes, though largely unsuccessfully. Corticosteroid administration has been shown to decrease cerebral edema and is beneficial in those patients with large edematous CNS lesions. It has been demonstrated that CRV and radiation-induced cerebral necrosis share histological features suggesting that anticoagulation, which may be beneficial for radiation-induced necrosis, could also benefit CRV and HERNS patients [32,33]. However, the use of anticoagulation therapy has only been reported in a single HERNS patient, and it was demonstrated to be detrimental due to bleeding complications.

TYPE IV COLLAGEN α1-CHAIN MUTATION (COL4A1)

Genetic Basis. The *COL4A1* gene encodes for the highly conserved α1 chain of the type IV collagen. Type IV collagen is a heterotrimer of two α1 chains and one α2 chain. This protein is ubiquity expressed in the body and contributes to maintaining the stability and function of basal membranes. There have been 14 mutations described in the *COL4A1* gene, 12 of which are missense mutations involving highly conserved glycine residues in the helical domain of the α1 chain. These mutations compromise folding and impair normal collagen IV assembly [34–42]. The abnormal collagen IV affects vascular integrity and results in fragile vessels that are particularly vulnerable to acquired stressors.

Clinical Findings. Different phenotypes with overlapping characteristics have been reported in association with *COL4A1* gene mutations: cerebral small vessel disease with hemorrhages, autosomal dominant-type 1 porencephaly, and hereditary angiopathy with nephropathy, aneurysms, and muscle cramps (HANAC) syndrome [43]. The abnormal collagen IV results in a microvasculopathy that presents, clinically, with both ischemic and hemorrhagic stroke [43]. Additionally, patients can experience migraine as well as seizure disorders. Systemic findings of small vessel disease include ocular involvement with cataracts [41], retinal vessel tortuosity [39], glaucoma, microcornea, and retinal hemorrhage [39]. At the kidney level, *COL4A1* patients may have large renal cysts and hematuria [39]. Additional findings include muscle crumps, Raynaud's phenomenon, and cardiovascular abnormalities including mitral valve prolapse [44] and supraventricular tachycardia [39]. Brain imaging often reveals leukoencephalopathy with bilateral and symmetric white matter hyperintensities in the supratentorial periventricular areas [42], lacunar infarcts, and hemorrhage depending on the etiology of the stroke. In addition, single or multiple intracranial aneurysms, usually located in the terminal portion of the internal carotid artery (ICA), and porencephaly may be observed.

Disease Course. *COL4A1* gene mutation is an autosomal dominant condition with variable onset. Patients usually present with stroke during the third or fourth decade of life [34]. Other neurological manifestations, like seizures, can present much earlier in childhood in the teen years or younger [34,35,41].

Diagnosis. The diagnosis of *COL4A1* gene mutation is done by genetic testing that should include DNA coding and flanking regions. The differential diagnosis includes CADASIL, CARASIL, HERNS, and amyloid angiopathy.

Management. The management of this condition is typically supportive in nature treating symptoms as they occur with no specific therapy available. Vessels of

COL4A1 patients are particularly vulnerable to environmental stressors. Thus, screening and aggressive treatment of hypertension and other stroke risk factors are recommended.

HOMOCYSTINURIA

Genetic Basis. Homocystinuria is a heterogeneous group of autosomal recessive diseases characterized by the abnormal metabolism of methionine and the buildup of homocysteine in both the blood and urine [10]. The classic form of homocystinuria is related to a mutation in the *CBS* gene that encodes for *cystathionine-β-synthase* (CBS). CBS is a pyridoxine (vitamin B6)-dependent enzyme that catalyzes the conversion of homocysteine to cystathionine that is a precursor of cysteine. Less commonly, homocystinuria can be caused by mutations in genes that participate of the metabolism of folate (*MTHFR*), synthesis of methylcobalamin (*MMADHC*), or methylation of homocysteine to methionine (*MTR* and *MTRR*) (Fig. 107.1). Patients with homocystinuria suffer of early cardiovascular and cerebrovascular disease. Several mechanisms contribute to the pathogenesis of the disease, including endothelial damage, vascular smooth muscle cell proliferation, and increased lipid peroxidation. In addition, there is an upregulation of prothrombotic factors (XII and V) and a downregulation of nitric oxide [10].

Clinical Findings. Findings associated with elevated levels of homocysteine include thromboembolic events, seizures, mental retardation, skeletal deformation (marfanoid habitus), and downward dislocation of the lens (ectopia lentis) [10]. Half of all patients with CBS mutations experience some sort of thromboembolic event, with 32% of those events causing stroke [45].

Disease Course. Because of the increased tendency toward thrombosis and accelerated atherosclerosis, patients often present first with vascular problems in the form of stroke or myocardial infarction [46–48]. Two phenotypic variants of homocystinuria have been described: pyridoxine-responsive and pyridoxine-nonresponsive homocystinuria. The responsiveness to pyridoxine is determined by measuring the plasma level of homocysteine and methionine before and after an oral dose of 100 mg pyridoxine. In general, it is considered that the pyridoxine-responsive form is a milder form of homocystinuria.[49].

Diagnosis. The diagnosis of homocystinuria can be established by measuring amino acids in plasma and urine. The classic form presents with elevated homocysteine, total homocysteine, homocysteine–cysteine mixed

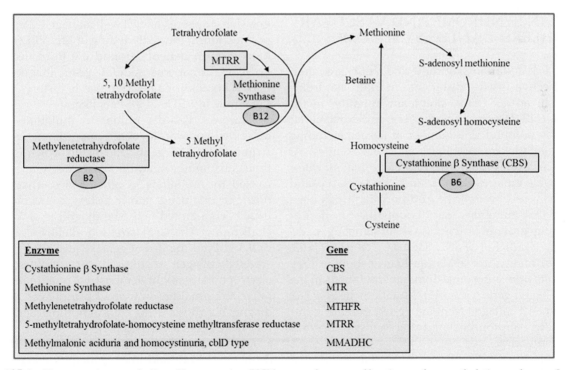

Enzyme	Gene
Cystathionine β Synthase	CBS
Methionine Synthase	MTR
Methylenetetrahydrofolate reductase	MTHFR
5-methyltetrahydrofolate-homocysteine methyltransferase reductase	MTRR
Methylmalonic aciduria and homocystinuria, cblD type	MMADHC

FIGURE 107.1 **Homocysteine metabolism.** Homocysteine (HCY) enters the transsulfuration or the remethylation pathways. In the transsulfuration pathway, HCY is condensed with serine by cystathionine-β-synthase, with vitamin B6 as a cofactor, to form cystathionine which is a precursor of cysteine. In the remethylation pathway, HCY is converted to methionine by methionine synthase, which requires vitamin B12. The mutation in MMADHC, an enzyme that participates in the synthesis of methylcobalamin, may affect the activity of the methionine synthase and increase the levels of homocysteine. After each cycle, the methionine synthase becomes inactive and is regenerated by MTRR. Betaine is a methyl donor that provides an alternative route to metabolize HCY to methionine.

disulfide, and methionine. The diagnosis can be confirmed by measuring CBS activity in fibroblasts or *CBS* gene sequencing.

Management. The main goal of the treatment is to normalize the levels of homocysteine and prevent thrombotic complications. Patients are placed on a methionine-restricted diet supplemented with the cofactors of CBS (vitamin B6) and methionine synthase (vitamin B12 and folate). In addition, adult patients are treated with oral betaine that provides an alternative pathway to convert the excess of homocysteine to methionine. Betaine is typically used orally at a dose of 6–9 g/day divided into two doses [49]. The VISP trial showed that while vitamin treatment decreased levels of homocysteine in the blood, it does not seem to decrease overall stroke risk [50,51]. However, the patients enrolled in this trial only had modestly elevated levels of blood homocysteine that might have influenced the sensitivity to detect a beneficial effect. Furthermore, the effect of vitamin therapy also seems to be affected by age and concurrent antithrombotic therapy [50–52]. Women with homocystinuria have a greater risk for thromboembolism than the average which peaks in the peripartum period [53]. Thus, prophylactic anticoagulation with low molecular-weight heparin has been recommended during the third trimester and the first 6 weeks post-partum [49].

MARFAN SYNDROME AND VASCULAR EHLERS–DANLOS SYNDROME

Genetic Basis. Marfan syndrome and VEDS are inherited connective tissue disorders. Marfan syndrome results from an autosomal dominant mutation in the fibrillin 1 (*FBN1*) gene. Fibrillin 1 is an extracellular matrix protein found in a number of tissues including the heart, elastic arteries, lung, and lens zonules. At the vascular level, mutations in *FBN1* ultimately cause vascular wall inflammation, intimal hyperplasia, and collapse of vessel walls [54]. Additionally, it has been suggested that mutations that disrupts the function of transforming growth factor-β (TGF-β) signaling pathways (e.g., *TGFBR1, TGFBR2, SMAD3, TGFB2, TGFB3,* and others) may cause a Marfan-like phenotype [55]. VEDS results from autosomal dominant mutations in the type III collagen gene (*COL3A1*). Mutations in *COL3A1* prevent type III collagen from integrated into the extracellular matrix appropriately leading to loss of connective tissue integrity.

Clinical Findings. Patients with Marfan syndrome present with cardiac abnormalities such as mitral valve prolapse, aortic valve incompetence, and aortic dissection, joint hypermobility, tall stature, arachnodactyly (long and thin fingers), dolichostenomelia (increased arm span), thoracolumbar scoliosis, ectopia lentis, and

pectus carinatum or excavatum [10,21]. Neurological complications in Marfan syndrome are rare with a prevalence in 3.5% of patients and include transient ischemic attack (TIA; 65% of patients), cerebral infarction (10% of patients), spinal cord infarction (10% of patients), subdural hematoma (10% of patients), and subarachnoid hemorrhage (5% of patients) [56]. Cardioembolism due to valvular heart disease or cardiac arrhythmia constitutes one of the most common neurovascular complications in Marfan syndrome. Ehlers–Danlos syndrome patients typically present with acrogeria (emaciated face with prominent facial bones and sunken cheeks), thin and hyperextensible skin, hypermobile joints, decreased muscle tone, joint deformities (per planus, hallus valgus, genu recurvatum, and kyphoscoliosis), ecchymoses, and hematomas. In VEDS, joint hyperextensibility may be absent and the skin is translucent but not necessarily extensible. Gastrointestinal perforation or organ rupture may be the presenting sign of VEDS [10,21]. Vascular complications include intracranial aneurysm, arterial dissection, and spontaneous rupture of medium and large arteries [10,21,57]. Dissections usually involve the proximal branches of the aortic arch, the descending aorta, the intracranial or extracranial segments of the ICA, and the vertebral arteries. In addition, carotid–cavernous fistulae have been described.

Disease Course. Marfan syndrome is typically diagnosed during childhood and has a relatively slow progression if patients have it well managed. Death typically occurs around the sixth decade of life. VEDS is typically diagnosed in patients around the third decade of life; however, in patients under 18 years, diagnosis is made in 60% because of a positive family history. Median survival time in VEDS is 50 years [58].

Diagnosis. Because numerous mutations have been described in Marfan syndrome, genetic testing may require full *FBN1* gene sequencing which is costly and time consuming. In addition, Marfan syndrome can be caused by mutations in other genes. Therefore, diagnostic criteria using family history and clinical findings have been adopted [59]. The diagnosis of VEDS is also challenging. The large size and allelic heterogeneity of *COL3A1* limit the use of gene sequencing. The study of secreted collagen in cultured skin fibroblasts using gel electrophoresis might demonstrate quantitative or qualitative abnormalities in type III procollagen. In addition, diagnostic criteria using clinical findings and family history have been developed [60].

Management. There is no cure for either Marfan syndrome or VEDS, and management is usually achieved through regular examination and supportive care. Patients are often advised to avoid highly strenuous activity to decrease stress to the cardiovascular system. Additionally, β-blockers have been demonstrated to be helpful in Marfan syndrome patients with aortic dilatation [61]. Because of the increased risk of dissections and vascular perforation, conventional arterial angiography should be discouraged in patients with these conditions.

MOYAMOYA DISEASE

Genetic Basis. While no single gene has been causally linked to moyamoya disease, familial forms of the disease exist. This and the much higher prevalence among Japanese people indicate a genetic component to the disease [62]. Furthermore, cases of moyamoya disease have been demonstrated in identical twins [63]. Moyamoya has been linked to chromosomes 3p24.2–p26, 6q25, 8q23, 12p12, and 17q25[10] as well as to HLA genes [64,65]. It has been proposed to be inherited as an autosomal dominant disease with incomplete penetrance [66].

Clinical Findings. Moyamoya disease is a noninflammatory and nonarteriosclerotic progressive stenosis of the distal ICA and its proximal branches. The reduced cerebral blood flow results in the expression of vascular growth factors that induce the development of weak collateral vessels. More rarely, moyamoya affects the posterior circulation. Moyamoya "disease" does not have recognized risk factors. Moyamoya "syndrome," in comparison, refers to a progressive vasculopathy—radiologically undistinguishable of moyamoya disease—that is associated with predisposing conditions such as cerebral radiation, Down's syndrome, sickle cell disease, tuberous sclerosis, and neurofibromatosis type 1. Conventional angiography reveals ICA stenosis with the new collateral vessels appearing as a "puff of smoke" around the stenotic segment. Clinical manifestations of moyamoya disease depend on the age. Young patients with early moyamoya present with ischemic stroke due to vascular stenosis. Is has been estimated that about 50–75% of the patients with moyamoya suffer of TIA or stroke. The cerebral ischemia may be precipitated by hyperventilation and the presentation varies depending on the location and severity of the initial event. In addition, patients may have refractory migraine-like headache. In comparison to young patients, older counterparts present with cerebral hemorrhage due to rupture of weak collateral vessels. Hemorrhage is a common cause of death in moyamoya and its prevalence is 10–40% [67].

Disease Course. Juvenile moyamoya disease usually present in patients less than 5 years old with the adult form appearing in individuals between 30 and 50 years [10]. Prognosis is highly variable and depends largely on the severity of the stenosis and progression of the disease. Prognosis is worse, however, in the juvenile form of the disease because of an increased risk for subsequent hemorrhage [68].

Diagnosis. The diagnosis of moyamoya is based on the following criteria: (a) stenosis in the distal ICA and/or proximal anterior cerebral artery or middle cerebral artery, (b) presence of neovascularization in the vicinity of the arterial narrowing, and (c) absence of predisposing factors. The definite diagnosis of moyamoya disease requires of (c) and either (a) or (b), bilaterally. The presence of the typical radiological findings unilaterally is categorized as probable moyamoya disease.

Management. While medical management with antiplatelet agents is often utilized, these strategies often have limited success because of the high rate of recurrent strokes and the increased risk of cerebral hemorrhage in older patients. Consequently, surgical intervention with direct, indirect, and combined bypass is used both in the juvenile and adult form of the disease. Direct bypass includes superficial temporal artery–middle cerebral artery bypass and occipital artery–middle cerebral artery bypass. Indirect approaches include encephaloduroarteriosynangiosis, encephalomyosynangiosis, encephalogaleosynagiosis, and encephaloduroarteriomyosynagiosis. Combined approaches combine both a direct and indirect bypass. These are performed either unilaterally or bilaterally depending on the severity of the ICA stenosis and the vessel involved [10,68]. Observational studies have shown that flow augmentation reduces the rate of recurrent stroke from 67% to 4%. In addition, the probability of remaining stroke-free over the 5 years post-surgery may be as high as 96% [67]. Calcium-channel blockers are frequently used in patients with intractable headaches or migraines. Since these agents may cause hypotension and compromise cerebral perfusion, they must be used with caution in these patients.

POLYGENIC DISEASES

Stroke represents a highly complex condition with single gene disorders only representing a small minority of stroke cases. Stroke is often multifactorial with contributions from both genetic and environmental factors. What is more likely the case with polygenic diseases is a contribution from many affected alleles which, individually, have low relative risks [69]. Previously reported candidate gene approaches have linked numerous different genes with increased stroke risk. Some of the more common identified genes include those associated with hemostasis and with homocysteine metabolism. Table 107.2 summarizes some of the genes identified as causing an increased risk of stroke.

Diseases of Hemostasis. Genes involved in maintaining hemostasis make for logical candidate genes for increasing the risk of stroke as mechanisms increasing the incidence of thrombotic events could lead to an increased incidence of stroke. While numerous mutations have been described

TABLE 107.2   Summary of Polymorphisms Associated With an Elevated Stroke Risk

Gene	Polymorphism	Function of Gene
Prothrombin	G20210A [89]	• Cleaved by factor X to thrombin • Mutation results in increased levels
Fibrinogen	G455A [71]	• Cleaved by thrombin to fibrin • High levels are associated with arterial thrombosis
Factor V Leiden	G1691A [71]	• Helps convert prothrombin to thrombin • mutations result in resistance to activated protein C which increases stroke risk
MTHFR	C677T [90]	• Catalyzes the conversion of 5,10-methylene-THF to 5-methyl-THF • Mutation leads to an increase of homocysteine in the blood

MTHFR, methylenetetrahydrofolate reductase; *THF*, tetrahydrofolate.

in a large number of coagulation factors, three such genes have demonstrated an increased relative risk for stroke occurrence. They are *prothrombin* [70], *fibrinogen* [71], and mutations to *factor V*, namely the factor V Leiden mutation [71]. Furthermore, mutations in *prothrombin* have also been associated with venous thrombosis [72–74].

Diseases of Homocysteine Metabolism. As discussed previously, increased blood levels of homocysteine are an independent risk factor for stroke. However, hyperhomocysteinemia must be distinguished from homocystinuria as the genetic cause for each are different, and the blood levels are much different with homocystinuria patients having blood levels of homocysteine >100 µmol/L versus 15–100 µmol/L in hyperhomocysteinemia. Additionally, while homocystinuria is definitively linked to many different single gene mutations, patients can present with hyperhomocysteinemia in the presence of vitamin deficiency including vitamin B12, vitamin B6, and folate deficiency [75]. The most common gene variant present in patients with mild hyperhomocysteinemia is point mutation in the *methylenetetrahydrofolate reductase (MTHFR)* gene causing a 50% reduction of enzymatic activity and a buildup of homocysteine [71].

CONCLUSIONS

Stroke is a devastating disease and a major cause of death or permanent disability in millions of people annually. Significant progress has been made to identify and characterize single causes of stroke. However, the genetics behind polygenic and multifactorial stroke remain largely uncharacterized. Furthermore, while candidate gene approaches have been successful in identifying novel mutations in stroke patients, especially in those patients with a single gene mutation, these types of techniques do not work as well for multifactorial stroke patients. Because of the necessary sample size required to adequately screen candidate genes, newer, more powerful technologies, including genome-wide association studies (GWAS) have been employed more recently to attempt to link

new polymorphisms to multifactorial stroke. Significant advances have been made with these newer technologies, and they offer promising new insight into both the diagnosis and management of these patients.

References

[1] Towfighi A, Saver JL. Stroke declines from third to fourth leading cause of death in the United States: historical perspective and challenges ahead. Stroke 2011;42(8):2351–5.

[2] Murray CJ, Lopez AD. Global mortality, disability, and the contribution of risk factors: Global Burden of Disease Study. Lancet 1997;349(9063):1436–42.

[3] Donnan GA, Fisher M, Macleod M, Davis SM. Stroke. Lancet 2008;371(9624):1612–23.

[4] Go AS, Mozaffarian D, Roger VL, et al. Heart disease and stroke statistics–2014 update: a report from the American Heart Association. Circulation 2014;129(3):e28–92.

[5] Fratiglioni L, Launer LJ, Andersen K, et al. Incidence of dementia and major subtypes in Europe: A collaborative study of population-based cohorts. Neurologic Diseases in the Elderly Research Group. Neurology 2000;54(11 Suppl. 5):S10–5.

[6] Lobo A, Launer LJ, Fratiglioni L, et al. Prevalence of dementia and major subtypes in Europe: A collaborative study of population-based cohorts. Neurologic Diseases in the Elderly Research Group. Neurology 2000;54(11 Suppl. 5):S4–9.

[7] Sacco RL, Ellenberg JH, Mohr JP, et al. Infarcts of undetermined cause: the NINCDS Stroke Data Bank. Ann Neurol 1989;25(4):382–90.

[8] Ferro JM, Massaro AR, Mas JL. Aetiological diagnosis of ischaemic stroke in young adults. Lancet Neurol 2010;9(11):1085–96.

[9] Hassan A, Markus HS. Genetics and ischaemic stroke. Brain 2000;123(Pt 9):1784–812.

[10] Francis J, Raghunathan S, Khanna P. The role of genetics in stroke. Postgrad Med J 2007;83(983):590–5.

[11] Enter C, Muller-Hocker J, Zierz S, et al. A specific point mutation in the mitochondrial genome of Caucasians with MELAS. Hum Genet 1991;88(2):233–6.

[12] Goto Y, Nonaka I, Horai S. A new mtDNA mutation associated with mitochondrial myopathy, encephalopathy, lactic acidosis and stroke-like episodes (MELAS). Biochim Biophys Acta 1991;1097(3):238–40.

[13] Nishino I, Komatsu M, Kodama S, Horai S, Nonaka I, Goto Y. The 3260 mutation in mitochondrial DNA can cause mitochondrial myopathy, encephalopathy, lactic acidosis, and strokelike episodes (MELAS). Muscle Nerve 1996;19(12):1603–4.

[14] Mancuso M, Orsucci D, Angelini C, et al. Phenotypic heterogeneity of the 8344A>G mtDNA "MERRF" mutation. Neurology 2013;80(22):2049–54.

[15] Zupanc ML, Moraes CT, Shanske S, Langman CB, Ciafaloni E, DiMauro S. Deletion of mitochondrial DNA in patients with combined features of Kearns-Sayre and MELAS syndromes. Ann Neurol 1991;29(6):680–3.

[16] Lertrit P, Noer AS, Jean-Francois MJ, et al. A new disease-related mutation for mitochondrial encephalopathy lactic acidosis and strokelike episodes (MELAS) syndrome affects the ND4 subunit of the respiratory complex I. Am J Hum Genet 1992;51(3):457–68.

[17] Manfredi G, Schon EA, Moraes CT, et al. A new mutation associated with MELAS is located in a mitochondrial DNA polypeptide-coding gene. Neuromuscul Disord 1995;5(5):391–8.

[18] Santorelli FM, Tanji K, Kulikova R, et al. Identification of a novel mutation in the mtDNA ND5 gene associated with MELAS. Biochem Biophys Res Commun 1997;238(2):326–8.

[19] Hirt L. MELAS and other mitochondrial disorders. In: Caplan LR, Bogousslavksy J, editors. Uncommon Causes of Stroke. Cambridge University Press; 2008. p. 149–53.

[20] Nakamura M, Nakano S, Goto Y, et al. A novel point mutation in the mitochondrial tRNA(Ser(UCN)) gene detected in a family with MERRF/MELAS overlap syndrome. Biochem Biophys Res Commun 1995;214(1):86–93.

[21] Terni E, Giannini N, Brondi M, Montano V, Bonuccelli U, Mancuso M. Genetics of ischaemic stroke in young adults. BBA Clin 2015;3:96–106.

[22] Scaglia F, Northrop JL. The mitochondrial myopathy encephalopathy, lactic acidosis with stroke-like episodes (MELAS) syndrome: a review of treatment options. CNS Drugs 2006;20(6):443–64.

[23] Abe K, Matsuo Y, Kadekawa J, Inoue S, Yanagihara T. Effect of coenzyme Q10 in patients with mitochondrial myopathy, encephalopathy, lactic acidosis, and stroke-like episodes (MELAS): evaluation by noninvasive tissue oximetry. J Neurol Sci 1999;162(1):65–8.

[24] Majamaa K, Rusanen H, Remes A, Hassinen IE. Metabolic interventions against complex I deficiency in MELAS syndrome. Mol Cell Biochem 1997;174(1–2):291–6.

[25] Tanaka J, Nagai T, Arai H, et al. Treatment of mitochondrial encephalomyopathy with a combination of cytochrome C and vitamins B1 and B2. Brain Dev 1997;19(4):262–7.

[26] Napolitano A, Salvetti S, Vista M, Lombardi V, Siciliano G, Giraldi C. Long-term treatment with idebenone and riboflavin in a patient with MELAS. Neurol Sci 2000;21(Suppl. 5):S981–2.

[27] Saitoh S, Momoi MY, Yamagata T, Mori Y, Imai M. Effects of dichloroacetate in three patients with MELAS. Neurology 1998;50(2):531–4.

[28] Koenig MK, Emrick L, Karaa A, et al. Recommendations for the management of strokelike episodes in patients with mitochondrial encephalomyopathy, lactic acidosis, and strokelike episodes. JAMA Neurol 2016;73(5):591–4.

[29] Richards A, van den Maagdenberg AM, Jen JC, et al. C-terminal truncations in human 3′-5′ DNA exonuclease TREX1 cause autosomal dominant retinal vasculopathy with cerebral leukodystrophy. Nat Genet 2007;39(9):1068–70.

[30] Jen JC, Baloh RW. Hereditary endotheliopathy with retinopathy, nephropathy, and stroke (HERNS). In: Caplan LR, Bogousslavksy J, editors. Uncommon Causes of Stroke. Cambridge University Press; 2008. p. 255–7.

[31] Ophoff RA, DeYoung J, Service SK, et al. Hereditary vascular retinopathy, cerebroretinal vasculopathy, and hereditary endotheliopathy with retinopathy, nephropathy, and stroke map to a single locus on chromosome 3p21.1-p21.3. Am J Hum Genet 2001;69(2):447–53.

[32] Grand MG, Kaine J, Fulling K, et al. Cerebroretinal vasculopathy. A new hereditary syndrome. Ophthalmology 1988;95(5):649–59.

[33] Niedermayer I, Graf N, Schmidbauer J, Reiche W. Cerebroretinal vasculopathy mimicking a brain tumor. Neurology 2000;54(9):1878–9.

[34] Lanfranconi S, Markus HS. *COL4A1* mutations as a monogenic cause of cerebral small vessel disease: a systematic review. Stroke 2010;41(8):e513–518.

[35] Breedveld G, de Coo IF, Lequin MH, et al. Novel mutations in three families confirm a major role of COL4A1 in hereditary porencephaly. J Med Genet 2006;43(6):490–5.

[36] de Vries LS, Koopman C, Groenendaal F, et al. *COL4A1* mutation in two preterm siblings with antenatal onset of parenchymal hemorrhage. Ann Neurol 2009;65(1):12–8.

[37] Gould DB, Phalan FC, Breedveld GJ, et al. Mutations in *COL4A1* cause perinatal cerebral hemorrhage and porencephaly. Science 2005;308(5725):1167–71.

[38] Gould DB, Phalan FC, van Mil SE, et al. Role of COL4A1 in small-vessel disease and hemorrhagic stroke. N Engl J Med 2006;354(14):1489–96.

[39] Plaisier E, Gribouval O, Alamowitch S, et al. *COL4A1* mutations and hereditary angiopathy, nephropathy, aneurysms, and muscle cramps. N Engl J Med 2007;357(26):2687–95.

[40] Shah S, Kumar Y, McLean B, et al. A dominantly inherited mutation in collagen IV A1 (COL4A1) causing childhood onset stroke without porencephaly. Eur J Paediatr Neurol 2010;14(2):182–7.

[41] Sibon I, Coupry I, Menegon P, et al. *COL4A1* mutation in Axenfeld-Rieger anomaly with leukoencephalopathy and stroke. Ann Neurol 2007;62(2):177–84.

[42] Vahedi K, Kubis N, Boukobza M, et al. *COL4A1* mutation in a patient with sporadic, recurrent intracerebral hemorrhage. Stroke 2007;38(5):1461–4.

[43] Alamowitch S, Plaisier E, Favrole P, et al. Cerebrovascular disease related to *COL4A1* mutations in HANAC syndrome. Neurology 2009;73(22):1873–82.

[44] Aguglia U, Gambardella A, Breedveld GJ, et al. Suggestive evidence for linkage to chromosome 13qter for autosomal dominant type 1 porencephaly. Neurology 2004;62(9):1613–5.

[45] Hassan A, Hunt BJ, O'Sullivan M, et al. Homocysteine is a risk factor for cerebral small vessel disease, acting via endothelial dysfunction. Brain 2004;127(Pt 1):212–9.

[46] Dafer RM, Love BB, Yilmaz EY, Biller J. Mitochondiral and metabolic causes of stroke. In: Caplan LR, Bogousslavksy J, editors. Uncommon Causes of Stroke. Cambridge University Press; 2008. p. 413–22.

[47] Clarke R, Lewington S. Homocysteine and coronary heart disease. Semin Vasc Med 2002;2(4):391–9.

[48] Sacco RL, Roberts JK, Jacobs BS. Homocysteine as a risk factor for ischemic stroke: an epidemiological story in evolution. Neuroepidemiology 1998;17(4):167–73.

[49] Yap S, Boers GH, Wilcken B, et al. Vascular outcome in patients with homocystinuria due to cystathionine beta-synthase deficiency treated chronically: a multicenter observational study. Arterioscler Thromb Vasc Biol 2001;21(12):2080–5.

[50] Toole JF, Malinow MR, Chambless LE, et al. Lowering homocysteine in patients with ischemic stroke to prevent recurrent stroke, myocardial infarction, and death: the Vitamin Intervention for Stroke Prevention (VISP) randomized controlled trial. JAMA 2004;291(5):565–75.

[51] Towfighi A, Arshi B, Markovic D, Ovbiagele B. Homocysteine-lowering therapy and risk of recurrent stroke, myocardial infarction and death: the impact of age in the VISP trial. Cerebrovasc Dis 2014;37(4):263–7.

[52] Arshi B, Ovbiagele B, Markovic D, Saposnik G, Towfighi A. Differential effect of B-vitamin therapy by antiplatelet use on risk of recurrent vascular events after stroke. Stroke 2015;46(3):870–3.

[53] Novy J, Ballhausen D, Bonafe L, et al. Recurrent postpartum cerebral sinus vein thrombosis as a presentation of cystathionine-beta-synthase deficiency. Thromb Haemost 2010;103(4):871–3.

[54] Judge DP, Dietz HC. Marfan's syndrome. Lancet 2005;366(9501):1965–76.

VII. MEDICAL CONDITIONS AND STROKE

[55] Morisaki T, Morisaki H. Genetics of hereditary large vessel diseases. J Hum Genet 2016;61(1):21–6.

[56] Wityk RJ, Zanferrari C, Oppenheimer S. Neurovascular complications of marfan syndrome: a retrospective, hospital-based study. Stroke 2002;33(3):680–4.

[57] Pepin M, Schwarze U, Superti-Furga A, Byers PH. Clinical and genetic features of Ehlers-Danlos syndrome type IV, the vascular type. N Engl J Med 2000;342(10):673–80.

[58] Pepin MG, Schwarze U, Rice KM, Liu M, Leistritz D, Byers PH. Survival is affected by mutation type and molecular mechanism in vascular Ehlers-Danlos syndrome (EDS type IV). Genet Med 2014;16(12):881–8.

[59] Faivre L, Collod-Beroud G, Ades L, et al. The new ghent criteria for marfan syndrome: what do they change? Clin Genet 2012;81(5):433–42.

[60] Beighton P, De Paepe A, Steinmann B, Tsipouras P, Wenstrup RJ. Ehlers-Danlos syndromes: revised nosology, Villefranche, 1997. Ehlers-Danlos National Foundation (USA) and Ehlers-Danlos Support Group (UK). Am J Med Genet 1998;77(1):31–7.

[61] Cunha L. Marfan's syndrome. In: Caplan LR, Bogousslavksy J, editors. Uncommon Causes of Stroke. Cambridge University Press; 2008. p. 131–4.

[62] Graham JF, Matoba A. A survey of moyamoya disease in Hawaii. Clin Neurol Neurosurg 1997;99(Suppl. 2):S31–5.

[63] Fukui M. Current state of study on moyamoya disease in Japan. Surg Neurol 1997;47(2):138–43.

[64] Han H, Pyo CW, Yoo DS, Huh PW, Cho KS, Kim DS. Associations of moyamoya patients with HLA class I and class II alleles in the Korean population. J Korean Med Sci 2003;18(6):876–80.

[65] Inoue TK, Ikezaki K, Sasazuki T, Matsushima T, Fukui M. Analysis of class II genes of human leukocyte antigen in patients with moyamoya disease. Clin Neurol Neurosurg 1997;99(Suppl. 2):S234–7.

[66] Mineharu Y, Takenaka K, Yamakawa H, et al. Inheritance pattern of familial moyamoya disease: autosomal dominant mode and genomic imprinting. J Neurol Neurosurg Psychiatry 2006;77(9):1025–9.

[67] Scott RM, Smith ER. Moyamoya disease and moyamoya syndrome. N Engl J Med 2009;360(12):1226–37.

[68] Adams Jr HP, Davis P, Hennerici M. Moya-moya syndrome. In: Caplan LR, Bogousslavksy J, editors. Uncommon Causes of Stroke. Cambridge University Press; 2008. p. 465–78.

[69] Dichgans M. Genetics of ischaemic stroke. Lancet Neurol 2007;6(2):149–61.

[70] Casas JP, Hingorani AD, Bautista LE, Sharma P. Meta-analysis of genetic studies in ischemic stroke: thirty-two genes involving approximately 18,000 cases and 58,000 controls. Arch Neurol 2004;61(11):1652–61.

[71] Bersano A, Ballabio E, Bresolin N, Candelise L. Genetic polymorphisms for the study of multifactorial stroke. Hum Mutat 2008;29(6):776–95.

[72] Sykes TC, Fegan C, Mosquera D. Thrombophilia, polymorphisms, and vascular disease. Mol Pathol 2000;53(6):300–6.

[73] Endler G, Mannhalter C. Polymorphisms in coagulation factor genes and their impact on arterial and venous thrombosis. Clin Chim Acta 2003;330(1–2):31–55.

[74] Girolami A, Simioni P, Scarano L, Carraro G. Prothrombin and the prothrombin 20210 G to A polymorphism: their relationship with hypercoagulability and thrombosis. Blood Rev 1999;13(4):205–10.

[75] Hankey GJ, Eikelboom JW. Homocysteine levels in patients with stroke: clinical relevance and therapeutic implications. CNS Drugs 2001;15(6):437–43.

[76] Goto Y, Nonaka I, Horai S. A mutation in the tRNA(Leu)(UUR) gene associated with the MELAS subgroup of mitochondrial encephalomyopathies. Nature 1990;348(6302):651–3.

[77] Romaniello F, Mazzaglia D, Pellegrino A, et al. Aortopathy in marfan syndrome: an update. Cardiovasc Pathol 2014;23(5):261–6.

[78] Lesca G, Plauchu H, Coulet F, et al. Molecular screening of ALK1/ACVRL1 and ENG genes in hereditary hemorrhagic telangiectasia in France. Hum Mutat 2004;23(4):289–99.

[79] Porteous ME, Curtis A, Williams O, Marchuk D, Bhattacharya SS, Burn J. Genetic heterogeneity in hereditary haemorrhagic telangiectasia. J Med Genet 1994;31(12):925–6.

[80] Coucke PJ, Willaert A, Wessels MW, et al. Mutations in the facilitative glucose transporter GLUT10 alter angiogenesis and cause arterial tortuosity syndrome. Nat Genet 2006;38(4):452–7.

[81] Torres VE, Harris PC, Pirson Y. Autosomal dominant polycystic kidney disease. Lancet 2007;369(9569):1287–301.

[82] Biggin A, Munns CF. Osteogenesis imperfecta: diagnosis and treatment. Curr Osteoporos Rep 2014;12(3):279–88.

[83] Joutel A, Corpechot C, Ducros A, et al. Notch3 mutations in CADASIL, a hereditary adult-onset condition causing stroke and dementia. Nature 1996;383(6602):707–10.

[84] Chabriat H, Joutel A, Dichgans M, Tournier-Lasserve E, Bousser MG. CADASIL. Lancet Neurol 2009;8(7):643–53.

[85] Linthorst GE, Bouwman MG, Wijburg FA, Aerts JM, Poorthuis BJ, Hollak CE. Screening for Fabry disease in high-risk populations: a systematic review. J Med Genet 2010;47(4):217–22.

[86] Razvi SS, Bone I. Single gene disorders causing ischaemic stroke. J Neurol 2006;253(6):685–700.

[87] Wallace MR, Marchuk DA, Andersen LB, et al. Type 1 neurofibromatosis gene: identification of a large transcript disrupted in three NF1 patients. Science 1990;249(4965):181–6.

[88] Chassaing N, Martin L, Calvas P, Le Bert M, Hovnanian A. Pseudoxanthoma elasticum: a clinical, pathophysiological and genetic update including 11 novel ABCC6 mutations. J Med Genet 2005;42(12):881–92.

[89] Poort SR, Rosendaal FR, Reitsma PH, Bertina RM. A common genetic variation in the 3'-untranslated region of the prothrombin gene is associated with elevated plasma prothrombin levels and an increase in venous thrombosis. Blood 1996;88(10):3698–703.

[90] Casas JP, Bautista LE, Smeeth L, Sharma P, Hingorani AD. Homocysteine and stroke: evidence on a causal link from mendelian randomisation. Lancet 2005;365(9455):224–32.

CHAPTER

108

Stroke in Fabry Disease

L.R. Caplan[1], F.D. Testai[2]

[1]Harvard University, Beth Israel Deaconess Medical Center, Boston, MA, United States; [2]University of Illinois at Chicago, Chicago, IL, United States

Fabry disease (FD), also known as *angiokeratoma corporis diffusum*, is an X-linked lysosomal storage disorder caused by deficiency of α-galactosidase A (α-gal). This defect causes the accumulation of glycosphingolipids in cells, including endothelial and vascular smooth muscle cells of small arterioles. These vessels become dysfunctional and undergo progressive stenosis resulting in ischemia and end-organ damage. Stroke constitutes one of the most devastating complications of FD.

EPIDEMIOLOGY

The prevalence of FD in the general population ranges from 1 in 117,000 to 1 in 476,000; in males, the range increases to 1 in 40,000 to 1 in 60,000 [1,2]. FD can have a subclinical course and remain unrecognized, suggesting that population studies may underestimate the true prevalence of this disease. Data from newborn screening programs seem to support this hypothesis (Table 108.1). A study performed in Italy (n = 37,104) showed that the frequency of FD in males may be as high as 1 in 3100 [3].

The frequency of FD in stroke patients is significantly higher than in the general population. In an early German study that included 721 adults aged 18–55 years with cryptogenic stroke, a biologically significant mutation in the *GAL* gene was found in 4.9% of men and 2.4% of women [5]. Based on these findings, it was hypothesized that FD was responsible for up to 1.2% of the strokes of unknown origin in young patients. Later studies done in other cohorts reported an incidence of FD in patients with cryptogenic stroke of 0–2.3% in men and 0–2.6% in women [6]. The Stroke in Young Fabry Patients (SIFAP-1) study was the largest initiative designed to prospectively investigate the incidence of FD in young stroke patients. Definite FD occurred in 0.5% of the 5023 patients (median age 46 years) and probable FD in 0.4%. About 74% of the qualifying events were related to brain ischemia and 20% were recurrent events. Silent strokes were observed in 20% of the patients who had first-ever strokes. The mean number of silent lesions was significantly higher in FD than in non-FD patients (1.7 ± 2.2 versus 1.0 ± 2.2; P = 0.025) [7].

PATHOPHYSIOLOGY

FD is caused by a mutation in the *GAL* gene, which is located in the long arm of the X-chromosome. *De novo* mutations occur so that the absence of a family history consistent with FD does not preclude the diagnosis. To date, more than 600 mutations have been described. The deficiency in α-gal leads to the accumulation of glycosphingolipids, particularly globotriaosylceramide (Gb3) and its deacylated derivative, lyso-globotriaosylceramide (lyso-Gb3). These molecules are stored in the cells of most organs, including the heart (cardiomyocytes and valvular cells), kidney (tubular and glomerular cells), neurons, and vascular cells (endothelial and smooth muscle cells).

The first manifestations become evident early in life. Depending on the residual enzyme activity, however, late presentations might occur. During embryogenesis, one copy of the X-chromosome in females undergoes irreversible inactivation or lyonization. Thus, females are mosaics of cell populations with different wild-type/mutated X-chromosome ratios and α-gal activities. The process of lyonization is genetically and epigenetically regulated and both random and skewed patterns of X-chromosome inactivation have been described in FD [14]. Heterozygous females with random X-chromosome inactivation have delayed manifestation of symptoms. In comparison, organ-specific skewed X-chromosome inactivation results in tissue-specific α-gal deficiency and, consequently, pleomorphic presentations. The occurrence

Primer on Cerebrovascular Diseases, Second Edition
http://dx.doi.org/10.1016/B978-0-12-803058-5.00108-9

of genotype–phenotype correlations has been posited. As such, genetic mutations resulting in renal or cardiac variants of FD have been described. It is posited that carriers of the mutation S126G and A143T may be particularly susceptible to stroke [7]. Familial studies show that phenotypic heterogeneity exists within family members carrying the same mutation, suggesting the importance of environmental or acquired factors [15].

CLINICAL FEATURES

FD usually presents during childhood with signs and symptoms resulting from small fiber neuropathy and autonomic dysfunction. Painful crises (or acroparesthesias) are reported by almost 60–80% of the males with classic FD. These events, also known as "Fabry crises," are described as burning pain that may be precipitated by fever, exercise, or stress. Patients also have nonspecific gastrointestinal symptoms, such as vomiting, diarrhea, or constipation, which are attributed to autonomic dysregulation or deposition of Gb3 in the mesenteric vessels. Young FD individuals may have anhydrosis or hypohydrosis leading to heat and physical activity intolerance (Table 108.1).

During adolescence, patients develop chronic fatigue and angiokeratomas. These small dark-red papules, located usually in the periumbilical area, buttocks, groin, and upper thighs, denote endothelial cell damage with vascular dilation in the dermis (Fig. 108.1). Patients might also present with vasculopathy of the retinal and conjunctival vessels and corneal opacities (also known as *cornea verticillata*), which represent glycosphingolipid deposition in the corneal epithelium and Bowman's membrane. These lesions are detectable by slit lamp examination and do not always compromise visual acuity. Visual impairment may occur due to central retinal artery occlusion or dense cataracts.

TABLE 108.1   Rate of Fabry Disease Obtained in Newborn Screening Studies

Population	N	Rate[a]
Austria [44]	34,736	1 in 3859
Italy [3]	37,104	1 in 3100
Japan [45]	21,170	1 in 3014
Taiwan [4]	110,027	1 in 1512
United States [46] [b]	47,701	1 in 2355

[a]*Diagnosis based on enzyme activity and confirmed by genetic analysis.*
[b]*Rates based on statewide newborn screening program performed in the state of Missouri.*

The progressive accumulation of glycosphingolipids eventually leads to life-threatening complications. In the kidneys, Gb3 deposits are observed in the glomerular endothelium, tubular epithelial cells, podocytes, and arterioles. The earliest manifestation of renal damage is microalbuminuria, which is usually detected in the second or third decade of life. The kidneys progressively lose the ability to concentrate urine and eventually evolve into end-stage renal disease. Cardiac involvement results from the deposition of Gb3 and autonomic dysregulation. About 40% of the FD patients develop left ventricular

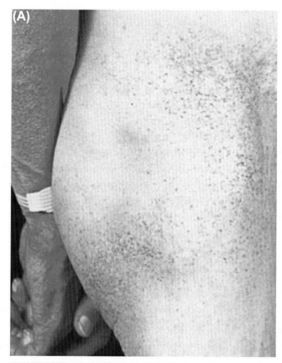

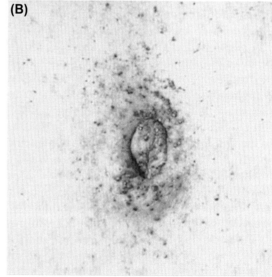

FIGURE 108.1   Angiokeratomas. (A) Scattered angiokeratomas on the thigh and buttocks. (B) A close up of the lesions.

hypertrophy, arrhythmia, angina, and dyspnea. Patients may also have myocardial ischemia due to Gb3 accumulation in coronary vessels or atherosclerosis [8].

A variety of peripheral and central nervous system complications have been described in FD (Table 108.2). Gb3 deposits are found in the peripheral nervous system and in neuronal populations associated with autonomic function, such as the hypothalamus, hippocampus, and brain stem [9,10]. In addition, the progressive deposition of Gb3 in endothelial and smooth muscle cells leads to cerebrovascular complications—brain ischemia, parenchymal hematomas, and subarachnoid hemorrhage [11].

TABLE 108.2 Clinical Manifestations of Fabry Disease

Organ	Sign/Symptom
Nervous system	Peripheral nervous system • Painful neuropathy (mainly small fiber) • Dysautonomia • Neuropathic pain (acroparesthesias) • Heat intolerance • Hypohydrosis or anhydrosis Central nervous system • Vestibular/cochlear dysfunction (tinnitus, vertigo, hearing loss) • Stroke (ischemic or hemorrhagic) • Psychiatric disorders • Cognitive impairment
Gastrointestinal tract	Nausea Vomiting Diarrhea or constipation Postprandial bloating Early satiety Weight loss
Skin	Angiokeratomas Hypohydrosis/anhydrosis Teleangiectasias
Eyes	Corneal opacities (cornea verticillata) Cataracts Retinal and conjunctival vasculopathy
Kidneys	Microalbuminuria and proteinuria Impaired concentration function (polyuria) Hyperfiltration Kidney failure
Heart	Arrhythmias Valvular insufficiency Left ventricular hypertrophy Cardiomyopathy Premature coronary artery disease Ischemic heart disease
Respiratory	Exertional dyspnea Wheezing Chronic cough
Skeletal	Decreased bone density

The incidence of stroke in FD ranges from 7% to 48% [12,13]. Fabry Registry data (n = 2446) show that the prevalence of stroke in FD is higher in males (6.9%) than in females (4.3%); about 87% of these are ischemic strokes and 13% are hemorrhagic strokes. In this cohort, the median age of stroke was 39 years for males and 46 years for females. Stroke was the first manifestation of FD in 50% of the males and 38% the females [11]. Compared to the general US population, FD patients have an increased burden of stroke at all age groups, particularly at young age when the risk of stroke can be 4 to 12 times higher than that observed in non-FD patients (Fig. 108.2).

STROKE PATHOGENESIS

The relentless accumulation of glycosphingolipids in endothelial and vascular smooth muscle cells leads to a microvasculopathy characterized by arteriolar remodeling, dysregulation, and stenosis. Gb3 reduces the expression of the $K_{Ca}3.1$ channel in endothelial cells and causes vascular dysfunction [16]. Aberrant vascular reactivity is further supported by positron emission tomography studies which show an increased cerebrovascular response to the infusion of acetazolamide [17]. Gb3 in endothelial cells also enhances the production of reactive oxygen species and the expression of adhesion molecules that facilitate the transmigration of immune cells through the vessel wall [18]. This immune response is associated with the secretion of matrix metalloproteinases, particularly matrix metalloproteinase-9, that degrade the lamina elastica and contribute to cardiovascular remodeling [19].

During late 2000s, lyso-Gb3 has been identified as a possible pathogenic mediator of vascular injury in FD. Gb3, lyso-Gb3, and circulating proangiogenic factors, such as sphingosine-1-phosphate, induce vascular smooth muscle cell proliferation [20,21]. This, along with the accumulation of glycosphingolipids in endothelial cells, leads to intima-medial thickening of the aorta, carotid, and brachial arteries. In addition, large- and middle-sized vessels, particularly those in the posterior circulation, dilate in response to vascular dysregulation and remodeling. The nature of this observation is unclear but it has been suggested that the vertebrobasilar system may be especially susceptible to oxidative stress. Gb3 accelerates the progression of atherosclerosis [22,23]. FD patients also have an increased prevalence of factors associated with thrombosis, including hyperhomocysteinemia, factor V Leiden, and protein Z gene polymorphism [24]. Fabry patients have an increased prevalence of vascular risk factors, including kidney disease, hypertension, cardiac arrhythmia, valvular disease, and cardiomyopathy.

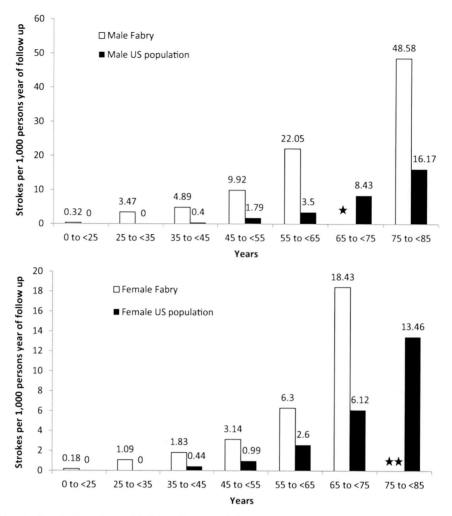

FIGURE 108.2   Incidence of stroke in patients with Fabry disease and the general US population by age. *Bars* represent the number of stroke per person-year of follow-up for male (upper panel) and female (lower panel) [11]. ★There were no males in this age group. ★★None of the women in this age group had stroke.

Young Fabry patients often have intracranial arterial dolichoectasia. Many strokes are due to penetrating artery disease or the dolichoectasia.

NEUROIMAGING

Microvascular damage caused by the progressive deposition of Gb3 leads to endothelial damage and the development of chronic white matter hyperintensities apparent in T2-weighted images. These lesions may be simple, multiple, or confluent, are usually symmetric, localize in the subcortical or periventricular area, and increase progressively with age. Also, bilateral increased posterior thalamus signal intensity may be observed on T1-weighted images (pulvinar sign) [25,26]. This finding represents microvascular calcification due to regional hypoperfusion [27]. Dolichoectasia and tortuosity of the basilar artery is commonly seen in FD patients. The

presence of this pulvinar sign and a widened basilar artery diameter should suggest FD [28–30].

DIAGNOSIS

In males the diagnosis is made by measuring α-gal activity in peripheral leukocytes or plasma. Females with FD may have normal peripheral α-gal activity. Heterozygous females and suspected cases of FD should undergo *GAL* genotyping. Elevated plasma and urine levels of α-gal substrates, Gb3 and lyso-Gb3, suggest the diagnosis of FD. Plasma Gb3 levels in females are usually lower than in males. Urine Gb3 levels may be normal in late-onset FB patients and spuriously elevated in coronary artery disease and other conditions. The measurement of glycosphingolipids, particularly urine Gb3, can be used to monitor the responsiveness to treatment in patients with abnormal baseline levels, but have limited diagnostic utility.

GAL gene sequencing constitutes the gold standard for the diagnosis of FD. Owing to the X-linked inheritance, there is no male-to-male transmission, and heterozygous females have a 50% risk of transmitting the mutated gene to their offspring.

TREATMENT

Many studies show and confirm the utility of enzyme replacement therapy (ERT) [31–41]. ERT with recombinant α-gal (agalsidase β; Fabrazyme, Genzyme Corp., Cambridge, MA) was approved by the US Food and Drug Administration in 2003. Two alternative recombinant α-gal forms (agalsidase α and agalsidase β) are available in Europe. The recommended dose for agalsidase α (Replagal, Shire, Human Genetic Therapies, Inc.) is 0.2 mg/kg every 2 weeks. Expert panels have developed guidelines for the assessment and management of FD patients [42,43]. ERT is currently recommended for all males and as soon as organ involvement is noted in females. Due to the potential development of blocking antibodies, close monitoring and delayed initiation of ERT may be considered in asymptomatic males. Aggressive control of modifiable vascular risk factors is recommended in all cases. Treatment with antithrombotics, either antiplatelet agents or oral anticoagulants depending on the circumstances should also be considered.

References

[1] Meikle PJ, Hopwood JJ, Clague AE, Carey WF. Prevalence of lysosomal storage disorders. JAMA 1999;281(3):249–54. [pii:joc80368].

[2] Poorthuis BJ, Wevers RA, Kleijer WJ, et al. The frequency of lysosomal storage diseases in the Netherlands. Hum Genet 1999;105(1–2):151–6.

[3] Spada M, Pagliardini S, Yasuda M, et al. High incidence of later-onset Fabry disease revealed by newborn screening. Am J Hum Genet 2006;79(1):31–40. [pii:S0002-9297(07)60021-4].

[4] Lin HY, Chong KW, Hsu JH, et al. High incidence of the cardiac variant of Fabry disease revealed by newborn screening in the Taiwan Chinese population. Circ Cardiovasc Genet 2009;2(5):450–6. http://dx.doi.org/10.1161/CIRCGENETICS.109.862920.

[5] Rolfs A, Bottcher T, Zschiesche M, et al. Prevalence of Fabry disease in patients with cryptogenic stroke: a prospective study. Lancet 2005;366(9499):1794–6. [pii:S0140-6736(05)67635-0].

[6] Marquardt L, Baker R, Segal H, et al. Fabry disease in unselected patients with TIA or stroke: population-based study. Eur J Neurol 2012;19(11):1427–32. http://dx.doi.org/10.1111/j.1468-1331.2012.03739.x.

[7] Rolfs A, Fazekas F, Grittner U, et al. Acute cerebrovascular disease in the young: the stroke in young Fabry patients study. Stroke 2013;44(2):340–9. http://dx.doi.org/10.1161/STROKEAHA.112.663708.

[8] Zarate YA, Hopkin RJ. Fabry's disease. Lancet 2008;372(9647):1427–35. http://dx.doi.org/10.1016/S0140-6736(08)61589-5.

[9] Kaye EM, Kolodny EH, Logigian EL, Ullman MD. Nervous system involvement in Fabry's disease: clinicopathological and biochemical correlation. Ann Neurol 1988;23(5):505–9. http://dx.doi.org/10.1002/ana.410230513.

[10] Rodrigues LG, Ferraz MJ, Rodrigues D, et al. Neurophysiological, behavioral and morphological abnormalities in the Fabry knockout mice. Neurobiol Dis 2009;33(1):48–56. http://dx.doi.org/10.1016/j.nbd.2008.09.001.

[11] Sims K, Politei J, Banikazemi M, Lee P. Stroke in Fabry disease frequently occurs before diagnosis and in the absence of other clinical events: natural history data from the Fabry registry. Stroke 2009;40(3):788–94. http://dx.doi.org/10.1161/STROKEAHA.108.526293.

[12] Gupta V, Bhinge KN, Hosain SB, et al. Ceramide glycosylation by glucosylceramide synthase selectively maintains the properties of breast cancer stem cells. J Biol Chem 2012;287(44):37195–205. http://dx.doi.org/10.1074/jbc.M112.396390.

[13] Vedder AC, Linthorst GE, van Breemen MJ, et al. The Dutch Fabry cohort: diversity of clinical manifestations and Gb3 levels. J Inherit Metab Dis 2007;30(1):68–78. http://dx.doi.org/10.1007/s10545-006-0484-8.

[14] Echevarria L, Benistan K, Toussaint A, et al. X-chromosome inactivation in female patients with Fabry disease. Clin Genet 2015. http://dx.doi.org/10.1111/cge.12613.

[15] Niemann M, Rolfs A, Stork S, et al. Gene mutations versus clinically relevant phenotypes: Lyso-Gb3 defines Fabry disease. Circ Cardiovasc Genet 2014;7(1):8–16. http://dx.doi.org/10.1161/CIRCGENETICS.113.000249.

[16] Park S, Kim JA, Joo KY, et al. Globotriaosylceramide leads to K(ca)3.1 channel dysfunction: a new insight into endothelial dysfunction in Fabry disease. Cardiovasc Res 2011;89(2):290–9. http://dx.doi.org/10.1093/cvr/cvq333.

[17] Moore DF, Altarescu G, Herscovitch P, Schiffmann R. Enzyme replacement reverses abnormal cerebrovascular responses in Fabry disease. BMC Neurol 2002;2:4.

[18] Shen JS, Meng XL, Moore DF, et al. Globotriaosylceramide induces oxidative stress and up-regulates cell adhesion molecule expression in Fabry disease endothelial cells. Mol Genet Metab 2008;95(3):163–8. http://dx.doi.org/10.1016/j.ymgme.2008.06.016.

[19] Shah S, Saver J, Kidwell C, et al. A multicenter pooled, patient-level data analysis of diffusion-weighted MRI in TIA patients. Stroke 2007;38(2):463–4.

[20] Brakch N, Dormond O, Bekri S, et al. Evidence for a role of sphingosine-1 phosphate in cardiovascular remodelling in Fabry disease. Eur Heart J 2010;31(1):67–76. http://dx.doi.org/10.1093/eurheartj/ehp387.

[21] Aerts JM, Groener JE, Kuiper S, et al. Elevated globotriaosylsphingosine is a hallmark of Fabry disease. Proc Natl Acad Sci USA 2008;105(8):2812–7. http://dx.doi.org/10.1073/pnas.0712309105.

[22] Altarescu G, Moore DF, Schiffmann R. Effect of genetic modifiers on cerebral lesions in Fabry disease. Neurology 2005;64(12):2148–50. [pii:64/12/2148].

[23] Bodary PF, Shen Y, Vargas FB, et al. Alpha-galactosidase A deficiency accelerates atherosclerosis in mice with apolipoprotein E deficiency. Circulation 2005;111(5):629–32. [pii:01.CIR.0000154550.15963.80].

[24] Lenders M, Karabul N, Duning T, et al. Thromboembolic events in Fabry disease and the impact of factor V leiden. Neurology 2015;84(10):1009–16. http://dx.doi.org/10.1212/WNL.0000000000001333.

[25] Ginsberg L. Nervous system manifestations of Fabry disease: data from FOS – the Fabry outcome survey. In: Mehta A, Beck M, Sunder-Plassmann G, editors. Fabry disease: perspectives from 5 years of FOS.. Oxford: Oxford PharmaGenesis; 2006. NBK11613 [bookaccession].

[26] Burlina AP, Manara R, Caillaud C, et al. The pulvinar sign: frequency and clinical correlations in Fabry disease. J Neurol 2008;255(5):738–44. http://dx.doi.org/10.1007/s00415-008-0786-x.

[27] Takanashi J, Barkovich AJ, Dillon WP, Sherr EH, Hart KA, Packman S. T1 hyperintensity in the pulvinar: key imaging feature for diagnosis of Fabry disease. AJNR Am J Neuroradiol 2003;24(5):916–21.

[28] Fellgiebel A, Keller I, Martus P, et al. Basilar artery diameter is a potential screening tool for Fabry disease in young stroke patients. Cerebrovasc Dis 2011;31(3):294–9. http://dx.doi.org/10.1159/000322558.

[29] Fellgiebel A, Keller I, Marin D, et al. Diagnostic utility of different MRI and MR angiography measures in Fabry disease. Neurology 2009;72(1):63–8. http://dx.doi.org/10.1212/01.wnl.0000338566.54190.8a.

[30] Fazekas F, Enzinger C, Schmidt R, et al. Brain magnetic resonance imaging findings fail to suspect Fabry disease in young patients with an acute cerebrovascular event. Stroke 2015;46(6):1548–53. http://dx.doi.org/10.1161/STROKEAHA.114.008548.

[31] Eng CM, Guffon N, Wilcox WR, et al. Safety and efficacy of recombinant human alpha-galactosidase A–replacement therapy in Fabry's disease. N Engl J Med 2001;345(1):9–16. http://dx.doi.org/10.1056/NEJM200107053450102.

[32] Schiffmann R, Ries M, Timmons M, Flaherty JT, Brady RO. Long-term therapy with agalsidase alfa for Fabry disease: safety and effects on renal function in a home infusion setting. Nephrol Dial Transplant 2006;21(2):345–54. gfi152.

[33] Imbriaco M, Pisani A, Spinelli L, et al. Effects of enzyme-replacement therapy in patients with Anderson-Fabry disease: a prospective long-term cardiac magnetic resonance imaging study. Heart 2009;95(13):1103–7. http://dx.doi.org/10.1136/hrt.2008.162800.

[34] Schiffmann R, Askari H, Timmons M, et al. Weekly enzyme replacement therapy may slow decline of renal function in patients with Fabry disease who are on long-term biweekly dosing. J Am Soc Nephrol 2007;18(5):1576–83. ASN.2006111263.

[35] Rombach SM, Smid BE, Bouwman MG, Linthorst GE, Dijkgraaf MG, Hollak CE. Long term enzyme replacement therapy for Fabry disease: effectiveness on kidney, heart and brain. Orphanet J Rare Dis 2013;8:47–1172. http://dx.doi.org/10.1186/1750-1172-8-47. 8–47.

[36] Anderson LJ, Wyatt KM, Henley W, et al. Long-term effectiveness of enzyme replacement therapy in Fabry disease: results from the NCS-LSD cohort study. J Inherit Metab Dis 2014;37(6):969–78. http://dx.doi.org/10.1007/s10545-014-9717-4.

[37] Rombach SM, Smid BE, Linthorst GE, Dijkgraaf MG, Hollak CE. Natural course of Fabry disease and the effectiveness of enzyme replacement therapy: a systematic review and meta-analysis: effectiveness of ERT in different disease stages. J Inherit Metab Dis 2014;37(3):341–52. http://dx.doi.org/10.1007/s10545-014-9677-8.

[38] Linthorst GE, Hollak CE, Donker-Koopman WE, Strijland A, Aerts JM. Enzyme therapy for Fabry disease: neutralizing antibodies toward agalsidase alpha and beta. Kidney Int 2004;66(4):1589–95. http://dx.doi.org/10.1111/j.1523-1755.2004.00924.x.

[39] Rombach SM, Aerts JM, Poorthuis BJ, et al. Long-term effect of antibodies against infused alpha-galactosidase A in Fabry disease on plasma and urinary (lyso)Gb3 reduction and treatment outcome. PLoS One 2012;7(10):e47805. http://dx.doi.org/10.1371/journal.pone.0047805.

[40] Giugliani R, Waldek S, Germain DP, et al. A phase 2 study of migalastat hydrochloride in females with Fabry disease: selection of population, safety and pharmacodynamic effects. Mol Genet Metab 2013;109(1):86–92. http://dx.doi.org/10.1016/j.ymgme.2013.01.009.

[41] Warnock DG, Bichet DG, Holida M, et al. Oral migalastat HCl leads to greater systemic exposure and tissue levels of active alpha-galactosidase A in Fabry patients when co-administered with infused agalsidase. PLoS One 2015;10(8):e0134341. http://dx.doi.org/10.1371/journal.pone.0134341.

[42] Eng CM, Germain DP, Banikazemi M, et al. Fabry disease: guidelines for the evaluation and management of multi-organ system involvement. Genet Med 2006;8(9):539–48. http://dx.doi.org/10.1097/01.gim.0000237866.70357.c6.

[43] Desnick RJ, Brady R, Barranger J, et al. Fabry disease, an under-recognized multisystemic disorder: expert recommendations for diagnosis, management, and enzyme replacement therapy. Ann Intern Med 2003;138(4):338–46. [pii:200302180-00014].

[44] Mechtler TP, Stary S, Metz TF, et al. Neonatal screening for lysosomal storage disorders: feasibility and incidence from a nationwide study in Austria. Lancet 2012;379(9813):335–41. http://dx.doi.org/10.1016/S0140-6736(11)61266-X.

[45] Inoue T, Hattori K, Ihara K, Ishii A, Nakamura K, Hirose S. Newborn screening for Fabry disease in Japan: prevalence and genotypes of Fabry disease in a pilot study. J Hum Genet 2013;58(8):548–52. http://dx.doi.org/10.1038/jhg.2013.48.

[46] Hopkins PV, Campbell C, Klug T, Rogers S, Raburn-Miller J, Kiesling J. Lysosomal storage disorder screening implementation: findings from the first six months of full population pilot testing in Missouri. J Pediatr 2015;166(1):172–7. http://dx.doi.org/10.1016/j.jpeds.2014.09.023.

CHAPTER

109

Moyamoya Disease and Syndrome

J.S. Kim

Asan Medical Center, University of Ulsan, Seoul, South Korea

INTRODUCTION

Moyamoya disease (MMD) is a chronic, occlusive cerebrovascular disease with an unknown etiology. It is characterized by steno-occlusive changes at the terminal portion of the internal carotid artery (ICA), and an abnormal vascular network at the base of the brain [1]. When there are causative diseases or associated conditions, the terms "moyamoya syndrome" and "angiographic moyamoya" are often used [2].

EPIDEMIOLOGY

MMD is relatively common in the East Asian countries [3]. The reported incidence and prevalence rate in Japan were 0.35 and 3.16/100,000, respectively, in 1995, which increased to 0.54 and 6.03/100,000, in 2003. An epidemiological study from Korea based on National Health Insurance (NHI) data reported that MMD prevalence increased from 6.3/100,000 in 2004 to 9.1/100,000 in 2008. From 2007 to 2011, the annual incidence increased from 1.7 to 2.3 per 100,000 persons. In a study based on the Taiwan NHI program, the annual incidence of MMD was 0.15/100,000, while a study conducted in the area of Nanjing, China showed a prevalence of 3.92/100, 000 during 2000–2007. National data on mainland China are not available. Both in Japan and Korea, the male-to-female ratio was about 1:2, and 10–15% of the patients had MMD family history. There were two age peaks, at ages 10–20 and 35–50 years (Fig 109.1).

Studies from outside of Asia are rare. In the states of Washington and California in the United States, the incidence of MMD was 0.086/100,000. The incidence was the highest among Asians, followed by African Americans, Whites, and Hispanics. However, a recent study based on the Nationwide Inpatient Sample (NIS) reported that MMD appears to be distributed among the races/ethnic groups based on their proportions of the total US population. Similar to East Asian patients, there was a bimodal age distribution, and female-to-male ratio was 2.2, but familial occurrence seems to be less common (about 2%).

There is an increasing incidence and prevalence of MMD over time worldwide. This finding may indicate an actual increase in patients with MMD. A more plausible explanation would be an increment in the number of newly diagnosed patients following the recent advent of noninvasive diagnostic tools, such as magnetic resonance angiography (MRA) and computed tomography angiography (CTA). Another reason is an increasing number of survivors due to improved management.

PATHOLOGY AND PATHOGENESIS

Pathology

The main pathological changes of the stenotic segment of MMD are fibrocellular thickening of the intima, irregular undulation of the internal elastic lamina, medial thinness (attenuation of media), and decrease in the outer diameter. Histopathological findings in the distal ICA have shown proliferation of smooth muscle cells or endothelium, and stenosis or occlusion associated with fibrocellular thickening of the intima [4]. The moyamoya vessels are the dilated perforating arteries associated with various histopathological changes including fibrin deposits in the wall, fragmented elastic lamina, attenuated media, and microaneurysms formation. Aside from the moyamoya vessels, another important finding of MMD is cortical microvascularization, which is characterized by increased microvascular density and vascular diameter.

Primer on Cerebrovascular Diseases, Second Edition
http://dx.doi.org/10.1016/B978-0-12-803058-5.00109-0

Genetics Underlying MMD

A 2011 genome-wide association study identified the ring finger protein 213 (*RNF213*) gene in the 17q25.3 region as a susceptibility gene for MMD among East Asian population [5]. A single nucleotide polymorphism (SNP) of c.14576G>A variant in *RNF213* was detected in 95% of familial MMD and 79% of sporadic cases among Japanese patients. Homozygous c.14576G>A variant of *RNF213* appears to predict early-onset and a severe form of MMD. In addition, novel variants in *RNF213* in non-c.14576G>A were recently found in Caucasian and Chinese MMD patients.

The exact function of *RNF213* and the role of SNP of *RNF213* in the pathogenesis of MMD are still undetermined. In vivo experiments reported in 2013 using genetically engineered mice (*RNF213*-deficient mice and *RNF213*-knock-in mice expressing a missense mutation in mouse *RNF213*, p.R4828K) failed to produce histological and angiographic findings of MMD under normal condition. On the contrary, post-ischemic angiogenesis was significantly enhanced in mice lacking *RNF213* after chronic hind-limb ischemia. These results, together with the low penetrance rate of *RNF213* polymorphisms, suggest the importance of environmental factors in addition to a genetic predisposition [6]. Autoimmunity, infection or chronic inflammatory conditions, and cranial irradiation have been suggested to be candidates for secondary insults in MMD patients.

Moyamoya Syndrome With Associated Conditions

MMD-like vasculopathy associated with other disease conditions is called the "moyamoya syndrome."

The disease conditions include: (1) *genetic, hereditary disorders*: neurofibromatosis type 1, Down syndrome, Noonan syndrome, and trisomy 12p syndrome; (2) *hematological disorders*: sickle cell disease, essential thrombocythemia, hereditary spherocytosis, protein C deficiency, and protein S deficiency; (3) *connective tissue diseases*: systemic lupus erythematosus, antiphospholipid syndrome, livedo reticularis; (4) *infectious or chronic inflammatory conditions*: pneumococcal meningitis tuberculous meningitis, HIV infection, leptospirosis, cat scratch disease, pulmonary sarcoidosis, Behçet's disease; (5) *metabolic diseases*: diabetes mellitus, thyrotoxicosis, and hyperhomocysteinemia; (6) *vascular injury*: radiation therapy; and (7) *others*: renovascular hypertension and oral contraceptive use, especially in cigarette smokers.

It remains unclear how these disorders are related to the moyamoya vasculopathy, but they may play the role of triggering factors or secondary insults in susceptible patients.

Biomarkers Underlying Vascular Stenosis and Aberrant Angiogenesis

Circulating Vascular Progenitor Cells

An increased level of circulating endothelial progenitor cells (EPCs), and defective angiogenic function of EPCs were reported in MMD patients, suggesting that abnormal angiogenesis is involved in the pathogenesis of MMD. In addition, the smooth muscle progenitor cells (SPCs) from the peripheral blood of MMD patients were found to make more irregularly arranged and thickened tubules, suggesting that a defect in the cell maturation process may occur in the SPCs from the MMD patients [4].

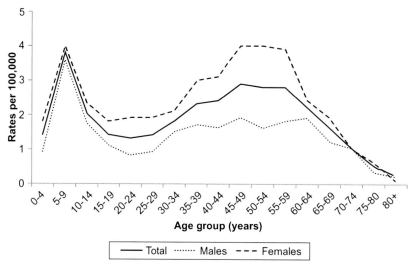

FIGURE 109.1 Incidence of moyamoya disease by sex and age in Korea, 2007–2011. *From Ahn IM, Park DH, Hann HJ, Kim KH, Kim HJ, Ahn HS. Incidence, prevalence, and survival of moyamoya disease in Korea: a nationwide, population-based study. Stroke 2014; 45:1090–95.*

Cytokine and Their Polymorphisms

Various cytokine levels and their polymorphisms have also been reported to be associated with MMD, including: (1) growth factors, such as vascular endothelial growth factor (VEGF), fibroblast growth factor (FGF), platelet-derived growth factor (PDGF), and hepatocyte growth factor (HGF); (2) cytokines related to vascular remodeling and angiogenesis, such as matrix metalloproteinases (MMPs) and their inhibitors, hypoxia-inducible factor-1α (HIF-1), and cellular retinoic–binding protein-1 (CRABP1); and (3) cytokines related to inflammation. However, changes in these factor levels may be secondary to ischemic brain damage rather than the cause of MMD [4].

CLINICAL FEATURES

The clinical presentations of MMD include transient ischemic attacks (TIA), ischemic strokes, hemorrhagic strokes, seizures, headache, and cognitive impairment. The incidence of each symptom varies according to patient age. In children, ischemic symptoms, especially TIAs, are predominant (70%). Intellectual decline, seizures, and involuntary movements are also common in this age group. On the other hand, adult patients present with intracranial hemorrhages more often than pediatric patients.

An ischemic event is the most important clinical manifestation of MMD. Cerebral hypoperfusion due to progressive major vessel occlusion results in repeated hemodynamic TIAs or ischemic strokes in children or young adults. The TIAs may be associated with episodes of hyperventilation, for example, running, crying, or eating hot noodles. Less often, patients develop territorial infarction due to embolism or thrombotic occlusion, in the distribution of the middle cerebral artery (MCA), anterior cerebral artery (ACA), posterior cerebral artery (PCA), or multiple territories (Fig 109.2). In one study of 410 pediatric MMD, ischemic symptoms were attributed to the MCA territory in 92, the ACA territory in 52, and the PCA territory in 10 of the cases [7]. Posterior circulation stroke is less common and tends to develop in the late stages of MMD. PCA involvement may be one of the factors related to a poor prognosis [7]. In MMD patients, the infarct topography does not often fit with the classical

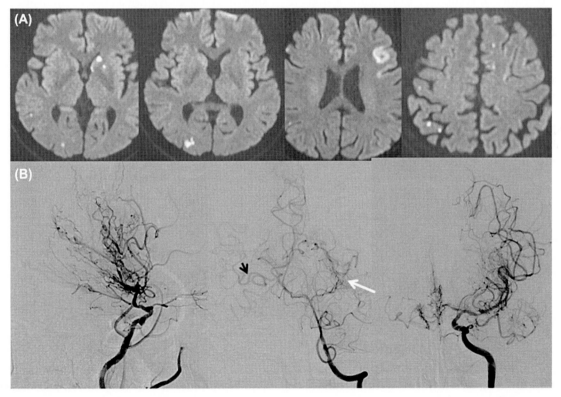

FIGURE 109.2 A 65-year-old woman suddenly developed dysarthria and speech disturbances. She had a history of intermittent, brief, left limb weakness whenever she got tired since 10 years before. Examination showed she had transcortical motor aphasia. Diffusion-weighted MRI showed multiple scattered infarction in the bilateral middle cerebral artery (MCA) and left anterior cerebral artery (ACA) territory (A). Transfemoral angiography showed left terminal internal carotid artery (ICA) occlusion with extensive basal collaterals (B, left image). Right posterior cerebral artery (PCA) was relatively intact and partly supplied blood flow to the MCA territory (black arrow), but there was involvement of left PCA with collateral formation (white arrow) (B, middle image). Left MCA showed stenosis and extensive collateral vessels. Basal collaterals were also observed around the occluded left ACA (B, right image).

vascular territories. This is probably due to altered vascular territories secondary to the long-standing major vessel occlusion, along with diversely developed collateral channels.About 30% of the MMD patients present with intracerebral hemorrhages (ICH), or less commonly, subarachnoid hemorrhage (SAH), secondary to the friable collateral vessels harboring microaneurysms or false aneurysms. One case–control study showed that the location of hemorrhage differed between primary ICH and ICH associated with MMD [8]. The orders of frequency were putamen (46.2%), thalamus (19.4%), and pons (14%) in cases of primary ICH, whereas they were intraventricular (37.6%), lobar (23.7%), and putamen (22.6%) in MMD patients. Rupture of focal microaneurysm formation in the abnormally dilated anterior choroidal artery branches may explain the high prevalence of intraventricular hemorrhages in MMD patients. In addition, it has been shown that gradient-echo T2*-weighted or susceptibility-weighted magnetic resonance imaging (MRI) identified cerebral microbleeds in 28–46% of MMD patients [3].

Significant brain hypoperfusion may lead to cognitive impairment or intellectual disability (intellectual developmental disorder) in children. Seizures occur in about 5% of patients, secondary to ischemic lesions or hypoperfusion; they usually develop early in childhood. Headache either presents as a symptom of MMD or develops following bypass surgery. About 22% of MMD children have prominent headaches. Although uncommon, MMD may present as involuntary movements, usually among children; in one study, 17 of 410 (4%) pediatric MMD patients developed involuntary movements [7], that include chorea, dystonia, and dyskinesia, presumably due to cerebral perfusion defect in the basal ganglia or cerebral cortical areas.

DIAGNOSIS

It was previously required that steno-occlusive changes at distal ICA should be present bilaterally for a definitive diagnosis of MMD (Fig 109.2). However, after a 2015 revision (Research Committee of MMD of the Japanese Ministry of Health, Labor, and Welfare, 2015), the diagnostic criteria of definitive MMD also included patients with unilateral terminal ICA steno-occlusion. The definitive diagnosis of MMD requires catheter cerebral angiography in unilateral cases, while bilateral cases can be promptly diagnosed by either catheter cerebral angiography or MRA.

Based on various angiographic findings, Suzuki and Takaku [1] proposed six stages of angiographic evolution: Stage 1. Segmental narrowing of the distal portion of the ICA. Stage 2. Initial appearance of basal moyamoya, and segmental narrowing of the proximal portion of the

ACA and MCA. Stage 3. The basal moyamoya becomes very prominent. The proximal portion of the ACA and MCA are no longer visualized, and their distal branches may be visualized via collaterals from branches of the PCA. Stage 4. The basal moyamoya begins to disappear. The proximal portion of the PCA becomes narrowed. Stage 5. The basal moyamoya becomes less apparent. All the major intracranial arteries are no longer visualized. Stage 6. The basal moyamoya is absent. Only meningeal–pial collaterals arising from branches of external carotid arteries supply the cerebral hemispheres. However, a stepwise progression from Stage 1 through Stage 6 is observed in a limited number of patients, and the practical value of the classification remains questionable.

MRI is the method of choice in detecting symptomatic or asymptomatic ischemic brain lesions. Moreover, MRI provides clinicians some clues to the suspicion of MMD: the absence of flow voids in the distal ICA and MCA, multiple signal voids in the basal ganglia and dilated leptomeningeal and cortical collateral vessels. However, for definite diagnosis, MRA or CTA should additionally be used that can noninvasively detect steno-occlusion of the distal ICA, MCA, and ACA. MRA and CTA are, however, less sensitive than catheter angiography in showing basal moyamoya vessels, small aneurysms, and assessing the status of the external carotid artery circulation. Catheter cerebral angiography is generally required before considering bypass surgery.

It is often challenging to diagnose MMD in patients with early stage of Suzuki's angiographic grading [1], when the abnormal vascular network is not yet evident. To improve the diagnostic capability, high-resolution vessel wall MRI has been used. A decrease in the outer diameter and concentric and weak enhancement in the bilateral terminus of the ICA, proximal ACA, or MCA are characteristics of MMD [9]. High-resolution MRI may provide the supportive information for the accurate diagnosis of MMD, especially in the early angiographic stage.

PROGNOSIS AND TREATMENT

Prognosis and Medical Treatment

Previous studies have shown that the prognosis is guarded for both pediatric and adult MMD patients. In one study that followed pediatric MMD patients for 4–15 years showed that they developed occasional TIA or headache in 33%, intellectual and/or motor impairment in 26%, and 11% required special school or care after reaching the teenage years. In medically treated adult symptomatic patients, the 5-year risk of recurrent ipsilateral stroke was as high as 65%. Patients with initial hemorrhagic presentation tended to develop hemorrhagic strokes while an ischemic presentation was usually followed by recurrent ischemic strokes. Overall, the annual stroke risk was 4.5% [10].

Although physicians frequently use antiplatelets especially in patients who present with ischemic stroke, there still are no randomized, controlled trials that appropriately examined the efficacy of antiplatelets in the prevention of stroke in MMD patients. There also are no proven therapies that can reverse or halt the progression of arterial steno-occlusion. Based on the nonrandomized study results that showed the relative reduction of ischemic strokes in patients in the surgically treated group, surgical therapy has become the mainstay of therapy in the treatment of MMD.

Surgical Treatment

The main goal of surgical revascularization is to prevent cerebral infarction by improving cerebral blood flow (CBF). Generally accepted indications for revascularization include recurrent clinical symptoms due to: (1) apparent cerebral ischemia or (2) decreased regional CBF, vascular response, and reserve in perfusion studies. However, the stroke incidence in MMD patients with relatively stable hemodynamic status seems to be high, and the factors predicting strokes in asymptomatic MMD have yet to be clarified.

Direct Revascularization for Ischemic MMD

Since the 1970s, direct bypass has been used in MMD patients (Fig 109.3). Following successful anastomosis between donor and recipient arteries, improvement in flow is achieved immediately. For the donor, the superficial temporal artery (STA) is selected in most of the cases. Craniotomy is performed after harvesting the STA from the scalp. Most often used recipients are cortical branches of the MCA. To reinforce PCA territory blood flow, the occipital artery (OA) is also used as a donor for anastomosis to the cortical branch of the PCA. Various indirect bypass methods can be combined. During the follow-up period, digital subtraction and quantitative MRA are used to evaluate the patency and the amount of bypass flow. After direct revascularization, the incidence of newly developed cerebral infarction is acceptable

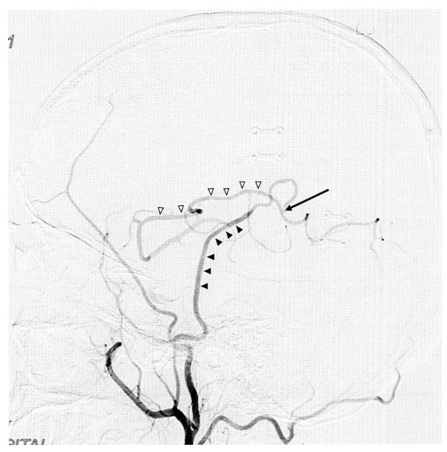

FIGURE 109.3 Postoperative angiography of the external carotid artery. *Black arrowheads* indicate the superficial temporal artery (donor artery). The black arrow indicates anastomosis site to the angular artery (recipient artery). The middle cerebral artery is supplied from the superficial temporal artery with reverse direction (*white arrowheads*). From Kim T, Oh CW, Bang JS, Kim JE, Cho WS. Moyamoya disease: treatment and outcomes, J Stroke 2016;18:21–30.

(0–3.8%), with an annual incidence of 0.2%. These favorable outcomes can be explained by improvement of CBF and vascular reservoir capacity. The relative risk reduction for cerebral infarction by direct revascularization was reported to be 70.7% [10].

Indirect Revascularization for Ischemic MMD

Direct bypass is a difficult procedure in pediatric patients or adult patients with advanced disease because of the small caliber of the recipient artery. In such cases, indirect bypass using various connective tissues has been used to promote leptomeningeal collateral formation. The surgical procedures can be classified according to the various tissues covering the brain: encephalo-duro-arterio synangiosis (EDAS), encephalo-myo synangiosis (EMS), encephalo-myo-arterio synangiosis (EMAS), encephalo-duro-arterio-myo synangiosis (EDAMS), encephalo-arterio synangiosis (EAS), and so on. Bi-frontal indirect bypass may be considered if the patient has frontal lobar hypoperfusion. Indirect bypass for PCA territories has also been described.

The incidence of newly developed cerebral infarction after indirect revascularization is relatively high as compared to that achieved through the direct method. In pediatric patients, 5.2–13.0% experienced postoperative infarction, whereas the incidence was 8.9–14.3% in adult patients. Five-year stroke incidence was 9% and 13% in pediatric and adult patients, respectively [10].

Revascularization Surgery for Hemorrhagic MMD

The benefit of surgery in patients with hemorrhagic presentation remains uncertain. Recent studies showed some evidence of effectiveness of surgical revascularization for hemorrhagic MMD patients. However, further studies are needed to prove the benefit of revascularization for hemorrhagic MMD patients.

Perioperative Complications

Postoperative stroke with permanent neurological deficits developed in 1.6–16.0%, and was more frequent in adults than in pediatric patients. Permanent neurological deficits developed in 0.9–8.0% of those with ischemic stroke. Postoperative diffusion-weighted imaging (DWI) defined lesions were observed in 13 (9.3%) of the 140 patients who underwent direct bypass operations, although only four lesions (2.9%) were associated with a permanent deficit. Hemorrhagic strokes leading to permanent disability, as a complication of revascularization for MMD, developed in 0.7–8.0% in previous studies. Postoperative epidural hematomas requiring surgical treatment developed in 4.8% of pediatric MMD patients. In addition, postoperative hyperperfusion syndrome occurs in 21.5–50.0% of the patients. Patients develop transient neurological deterioration with increased blood flow. Adult-onset MMD or MMD patients presenting with hemorrhagic strokes have a relatively high risk for symptomatic hyperperfusion. During surgery, the scalp arteries (i.e., STA and OA) are harvested and implanted into the intracranical space. This procedure may cause scalp ischemia after revascularization, which occurs in 17.6–21.4% of the patients [10].

References

[1] Suzuki J, Takaku A. Cerebrovascular "moyamoya" disease. Disease showing abnormal net-like vessels in base of brain. Arch Neurol 1969;20:288–99.

[2] Natori Y, Ikezaki K, Matsushima T, Fukui M. 'Angiographic moyamoya' its definition, classification, and therapy. Clinical Neurology and Neurosurgery 1997;99(Suppl 2):S168–72.

[3] Kim JS. Moyamoya disease: Epidemiology, clinical features, and diagnosis. J Stroke 2016;18:2–11.

[4] Bang OY, Fujimura M, Kim SK. The pathophysiology of moyamoya disease: An update. J Stroke 2016;18:12–20.

[5] Kamada F, Aoki Y, Narisawa A, Abe Y, Komatsuzaki S, Kikuchi A, et al. A genome-wide association study identifies RNF213 as the first moyamoya disease gene. Journal of human genetics 2011;56:34–40.

[6] Fujimura M, Sonobe S, Nishijima Y, Niizuma K, Sakata H, Kure S, et al. Genetics and biomarkers of moyamoya disease: Significance of RNF213 as a susceptibility gene. J Stroke 2014;16:65–72.

[7] Kim SK, Cho BK, Phi JH, Lee JY, Chae JH, Kim KJ, et al. Pediatric moyamoya disease: An analysis of 410 consecutive cases. Ann Neurol 2010;68:92–101.

[8] Nah HW, Kwon SU, Kang DW, Ahn JS, Kwun BD, Kim JS. Moyamoya disease-related versus primary intracerebral hemorrhage: [corrected] location and outcomes are different. Stroke 2012;43:1947–50.

[9] Ryoo S, Cha J, Kim SJ, Choi JW, Ki CS, Kim KH, et al. High-resolution magnetic resonance wall imaging findings of moyamoya disease. Stroke 2014;45:2457–60.

[10] Kim T, Oh CW, Bang JS, Kim JE, Cho WS. Moyamoya disease: Treatment and outcomes. J Stroke 2016;18:21–30.

CHAPTER

110

Varicella Zoster Virus and Stroke

M.A. Nagel

University of Colorado School of Medicine, Aurora, CO, United States

INTRODUCTION

Varicella zoster virus (VZV) is a ubiquitous, neurotropic alphaherpesvirus. Primary infection usually causes varicella (chickenpox), after which virus becomes latent in cranial nerve, dorsal root, and autonomic ganglionic neurons, including enteric ganglionic neurons and adrenal cells.

With a decline in VZV-specific cell-mediated immunity in elderly and immunocompromised individuals, virus reactivates from one or more ganglia and typically travels along nerve fibers peripherally to produce herpes zoster (shingles) in the corresponding dermatome(s), which can be complicated by postherpetic neuralgia. VZV can also travel centrally to produce ocular and neurological disease with or without rash, including stroke due to productive virus infection of cerebral arteries (VZV vasculopathy). Early cases of VZV vasculopathy were described as a contralateral hemiparesis associated with herpes zoster ophthalmicus or as a postvaricella angiopathy, typically presenting as a hemiparesis during or shortly after varicella. Today, the clinical spectrum of VZV vasculopathy has expanded to include transient ischemic attacks (TIAs), hemorrhagic or ischemic stroke, aneurysm, and sinus thrombosis, as well as giant cell arteritis (GCA)—all occurring with or without rash. Since VZV is latent in >90% of the world population and will reactivate in 50% of individuals by 85 years of age, and since our aging population continues to increase, VZV vasculopathy will continue to play a significant role in stroke. Herein, the epidemiology of stroke associated with zoster, the clinical features and management of VZV vasculopathy and its pathogenesis will be reviewed.

EPIDEMIOLOGY

Multiple population-based studies have shown that: (1) the incidence of stroke after zoster is greater than in age-matched control patients; (2) stroke risk is greater immediately following zoster and if zoster occurs in the ophthalmic distribution; and (3) antiviral treatment decreases stroke risk. Notable reports include a study from the UK Clinical Practice Research Datalink which showed a decreasing risk of stroke over time following zoster in all dermatomes, with a statistically significant age-adjusted incidence at 1–4 weeks after zoster (1.63), 5–12 weeks after zoster (1.42), and 13–26 weeks after zoster (1.23), but no increase at later times [1]. In patients with zoster in the ophthalmic distribution, the risk of stroke was increased threefold at 5–12 weeks after zoster. Importantly, among 55% of zoster patients who received oral antiviral therapy, the stroke risk was reduced compared to that in untreated zoster patients, indicating the value of antiviral treatment in reducing stroke incidence after zoster. In 2016, in the first US population-based study, the risk of stroke within 3 months of zoster was increased 1.53-fold [2]. In the pediatric population, about one-third of arterial ischemic stroke is associated with varicella [3]. Together, these studies suggest that zoster and varicella are risk factors for stroke and that antiviral therapy may decrease this risk.

CLINICAL FEATURES, LABORATORY ABNORMALITIES, AND DIAGNOSIS OF VZV VASCULOPATHY

Patients with VZV vasculopathy may present with mental status changes, headache, vision loss, and focal motor and/or sensory deficits with or without rash. The course is often protracted with waxing and waning features. In a study of 30 patients with virologically confirmed VZV vasculopathy [4], zoster or varicella rash was present in 63% of patients, with the average time from rash to neurological symptoms and signs of 4.1 months. CSF analysis revealed a pleocytosis in

Primer on Cerebrovascular Diseases, Second Edition
http://dx.doi.org/10.1016/B978-0-12-803058-5.00110-7

67% of patients, with the average time from neurological symptoms and signs to CSF analysis of 4.2 months. Brain MRI and CT abnormalities were present in 97%, typically seen as enhancing lesions at gray–white matter junctions. Of 23 patients analyzed by angiography, 50% had abnormalities predominantly in both large and small arteries, 37% in small arteries, and 13% in large arteries. In 2015, high-resolution MRI (HRMR) analysis of patients with virologically verified VZV vasculopathy showed various patterns of stenosis, vessel wall thickening, and enhancement, primarily in terminal internal carotid artery segments and the M1 segment of the middle cerebral arteries [5]. After antiviral treatment, HRMR revealed improvement of stenosis, as well as reduced vessel wall enhancement and thickening, suggesting the value of HRMR in the management of VZV vasculopathy.

Of the 30 patients with VZV vasculopathy, virological confirmation of VZV vasculopathy based on the detection of VZV DNA or anti-VZV antibody in CSF showed that only 30% of the CSF samples contained VZV DNA, whereas 93% had anti-VZV IgG antibody in CSF, including a reduced serum/CSF ratio of anti-VZV IgG that confirmed intrathecal synthesis of anti-VZV IgG antibody [4]. However, when patients were divided into immunocompromised and immunocompetent groups, there was significantly higher VZV DNA in CSF of immunocompromised versus immunocompetent individuals (54% versus 16%, respectively). Overall, while both PCR and detection of antibody to VZV in CSF are highly specific, detection of anti-VZV IgG antibody in CSF is the more reliable test to diagnose VZV vasculopathy due to the chronic nature of disease.

In children, post-varicella cerebral angiopathy is usually monophasic and presents as a hemiparesis with involvement of the middle cerebral artery. In a study of 24 children with post-varicella cerebral infarction, the time from varicella rash to arterial ischemic stroke was 4 months [6], similar to that seen in adults with zoster. All had infarctions in the distribution of the middle cerebral artery, mostly in the M1 segment with 42% having additional involvement of anterior and posterior cerebral arteries, the internal carotid artery (ICA), and terminal branches of the ICA and collaterals. All 14 children tested for the presence of serum anti-VZV IgG were positive and 6 children tested for serum anti-VZV IgM were negative. One of eight children had VZV DNA in CSF. One child with a T-cell immunodeficiency and nummular keratitis had VZV DNA in the vitreous humor. None of the children were treated with antivirals. After a median of 27 months, 25% had recurrent TIAs. Twenty-two of the 24 children were re-imaged and 50% had improvement of arterial disease, 18% were stable and 32% progressed with recurrent TIAs and infarcts. In 2014, the live-attenuated varicella vaccine strain was shown to cause VZV

vasculopathy in an immunodeficient child [7], pointing to the need for caution in vaccinating potentially immunocompromised children.

Less commonly, VZV vasculopathy can present as aneurysms with subarachnoid hemorrhage, as well as sinus thrombosis. A notable case was of a 41-year-old woman with systemic lupus erythematosus on methotrexate, who developed zoster in multiple dermatomes and severe headache. Two months later, four-vessel digital subtraction angiography revealed nine anterior circulation aneurysms. Antiviral treatment led to clinical improvement, reduction in size of most aneurysms, and complete resolution of the two largest aneurysms. Another notable case was of a 30-year-old man who developed varicella followed 1 week later with neurological deficits; MR venography demonstrated a transverse and sigmoid sinus thrombosis.

TREATMENT OF VZV VASCULOPATHY

Patients with virologically confirmed VZV vasculopathy are treated with intravenous acyclovir, 10–15 mg/kg every 8 h for 14 days. Since arterial inflammation is also present, prednisone 1 mg/kg daily for 5 days can be given concurrently. Administration of both antivirals and corticosteroids is supported by a study of 30 patients with virologically confirmed VZV vasculopathy, in which 75% of patients treated with both antivirals and steroids improved or stabilized compared to 66% treated with antivirals alone; however, statistical analysis could not be carried out due to variations in antiviral and steroid dose and duration [4]. Another course of antiviral therapy may be necessary when disease recurs, frequently in immunocompromised patients.

VZV IN GIANT CELL ARTERITIS

In the past few years, the spectrum of VZV vasculopathy has expanded to include not only the intracranial circulation, but the extracranial circulation including temporal arteries (TAs) from patients with GCA, the most common systemic vasculitis in the elderly which can be complicated by stroke. GCA is characterized by a constellation of symptoms including head/scalp pain, temporal artery tenderness, jaw claudication, polymyalgia rheumatica, fever, night sweats, weight loss, fatigue, and elevated inflammatory markers (erythrocyte sedimentation rate and C-reactive protein). TA biopsy reveals transmural inflammation and medial damage, with multinucleated giant/epithelioid cells in skip lesions (GCA-positive TAs). In many clinically suspect cases, TA biopsy may be pathologically negative (GCA-negative TAs), attributable to missed skip lesions. Since vision loss frequently occurs, patients are immediately treated with corticosteroids, with 50% of patients relapsing after discontinuation of therapy or progressing to stroke and vision loss despite therapy. The cause of GCA was unknown. However, since the pathology of VZV-infected cerebral arteries and TAs from patients with GCA are identical, multiple studies were done demonstrating VZV antigen in these TAs, which supported the notion that VZV arterial infection induces the immunopathology of GCA.

The most comprehensive, retrospective study supporting the notion that GCA is a VZV vasculopathy analyzed multiple formalin-fixed, paraffin-embedded sections from TAs of patients with clinically suspect GCA, as well as normal TA biopsies from subjects >50 years of age, for the presence of VZV antigen by immunohistochemistry. Overall, VZV antigen was found in 73/104 (70%) GCA positive and 58/100 (58%) GCA negative TAs compared to 11/61 (18%) normal TAs ($p < .0001$) [8]. VZV antigen was present predominantly in adventitia and present in skip areas. Of 58 GCA-positive, VZV antigen-positive TAs examined, all contained cellular DNA and 23 (40%) contained VZV DNA. Of 58 GCA-negative, VZV antigen-positive TAs examined, 51 contained cellular DNA, 9 (18%) of which contained VZV DNA. Nine of 11 normal VZV antigen-positive TAs contained cellular DNA, of which 3 (33%) contained VZV DNA. Adventitial inflammation was seen adjacent to viral antigen in 26 (52%) of 58 GCA-negative subjects whose TAs contained VZV antigen. No inflammation was seen in normal TAs containing VZV antigen. VZ virions were also found in a GCA-positive TA that contained VZV antigen indicating productive virus infection [8]. The finding of VZV antigen predominantly in adventitia of GCA-positive TAs, and VZV with inflammation in the adventitia of GCA-negative TAs supports the notion that after VZV reactivation from ganglia, virus spreads transaxonally to the arterial adventitia followed by transmural spread with accompanying inflammation.

PATHOGENESIS

Because no animal model exists for VZV vasculopathy, pathogenesis studies have been restricted to arteries from subjects with VZV vasculopathy, as well as primary human cerebrovascular cells infected with VZV. Immunohistochemical analysis of cerebral and temporal arteries from three patients with virologically confirmed VZV vasculopathy revealed VZV antigen in the arterial adventitia early in infection and in the media and intima later in the course of disease—consistent with VZV reactivation from ganglia followed by transaxonal spread to adventitia, transmural spread, and

accompanying inflammation [9]. VZV-infected arteries were characterized by a disrupted internal elastic lamina, a thickened intima composed of cells expressing alpha-smooth muscle actin, and a paucity of medial smooth muscle cells. VZV-infected arteries contained CD4+ and CD8+ T cells, CD68 + macrophages, and rare B cells expressing CD20 [10]. Arteries from early VZV vasculopathy contained abundant neutrophils in the adventitia, which are absent in late VZV vasculopathy. Inflammatory cells were absent in control arteries. Although the exact mechanisms of vascular remodeling triggered by VZV are unknown, soluble factors secreted by inflammatory cells and VZV-infected vascular cells play an important role. Specifically, in a VZV-infected artery, a thickened intima was seen overlying an arterial region with inflammation; another region without inflammation of the same artery showed normal intima [10]. Furthermore, compared to uninfected cells, VZV-infected human brain vascular adventitial fibroblasts secrete increased matrix metalloproteinases-1, -3, and -9 which have been implicated in vascular remodeling and aneurysm formation.

CONCLUSIONS

VZV vasculopathy produces TIAs, stroke, aneurysm, sinus thrombosis, and GCA with and without zoster or varicella rash due to productive virus infection, inflammation, and pathological vascular remodeling of arteries. Characteristic features include imaging and angiographic abnormalities involving both large and small vessels. The absence of rash, a CSF pleocytosis or VZV DNA in CSF does not exclude the diagnosis, and the best test for diagnosing VZV vasculopathy is detection of anti-VZV IgG in the CSF. Treatment consists of intravenous acyclovir with prednisone. Because >90% of individuals are infected with VZV and are at risk for reactivation, VZV vasculopathy should be considered in the evaluation of stroke patients, particularly in elderly and immunocompromised patients, as antiviral treatment can be effective.

Acknowledgments

Supported in part by National Institutes of Health grants AG032958 and NS094758.

References

[1] Langan SM, Minassian C, Smeeth L, Thomas SL. Risk of stroke following herpes zoster: a self-controlled case-series study. Clin Infect Dis 2014;58:1497–503.
[2] Yawn BP, Wollan PC, Nagel MA, Gilden D. Risk of stroke and myocardial infarction after herpes zoster in older adults in a US community population. Mayo Clin Proc 2016;91:33–44.

[3] Askalan R, Laughlin S, Mayank S, Chan A, MacGregor D, Andrew M, et al. Chickenpox and stroke in childhood: a study of frequency and causation. Stroke 2001;32:1257–62.

[4] Nagel MA, Cohrs RJ, Mahalingam R, Wellish MC, Forghani B, Schiller A, et al. The varicella zoster virus vasculopathies: clinical, CSF, imaging, and virologic features. Neurology 2008;70:853–60.

[5] Cheng-Ching E, Jones S, Hui FK, Man S, Gilden D, Bhimraj A, et al. High-resolution MRI vessel wall imaging in varicella zoster virus vasculopathy. J Neurol Sci 2015;351:168–73.

[6] Miravet E, Danchaivijitr N, Basu H, Saunders DE, Ganesan V. Clinical and radiological features of childhood cerebral infarction following varicella zoster virus infection. Dev Med Child Neurol 2007;49:417–22.

[7] Sabry A, Hauk PJ, Jing H, Su HC, Stence NV, Mirsky DM, et al. Vaccine strain varicella-zoster virus-induced central nervous system vasculopathy as the presenting feature of DOCK8 deficiency. J Allergy Clin Immunol 2014;133:1225–7.

[8] Gilden D, White T, Khmeleva N, Boyer PJ, Nagel MA. VZV in biopsy-positive and-negative giant cell arteritis: analysis of 100+ temporal arteries. Neuroimmunol Neuroinflamm 2016;3:e216.

[9] Nagel MA, Traktinskiy I, Azarkh Y, Kleinschmidt-DeMasters B, Hedley-Whyte T, Russman A, et al. Varicella zoster virus vasculopathy: analysis of virus-infected arteries. Neurology 2011;77:364–70.

[10] Nagel MA, Traktinskiy I, Stenmark KR, Frid MG, Choe A, Gilden D. Varicella-zoster virus vasculopathy: immune characteristics of virus-infected arteries. Neurology 2013;80:62–8.

C H A P T E R

111

Stroke and Migraine

A. Yakhkind[1], J. Castaldo[2], M.C. Leary[2,3]

[1]Brown University, Providence, RI, United States; [2]Lehigh Valley Hospital and Health Network, Allentown, PA, United States; [3]University of South Florida, Tampa, FL, United States

INTRODUCTION

Migraine headache is both a prevalent and potentially debilitating condition attributed to both environmental and hereditary factors. It is associated with significant comorbidities including depression, dysautonomia, cognitive dysfunction, vertigo, and stroke. Migraine accounts for approximately one-third of neurologic disease burden in daily adjusted life years and ranked in the top 30 (of 176) of the highest burdens of disease in most areas of the world [1]. Much has been learned about the pathophysiology of migraine in the last 20 years that can shed light on its association with stroke. In this chapter, we review the current data, summarize the proposed causality, and discuss the clinical pearls that both relate and distinguish migraine and stroke.

WHAT IS MIGRAINE?

Migraine headache is primarily a neuronal disorder with subsequent secondary vasomotor effects on intracranial and extracranial cerebral blood vessels. It may be acute and intermittent, or it may become a chronic, daily disabling condition. Migraine is three times more prevalent in women than in men and, in about a third of people, can be accompanied by a visual or sensory aura [2]. An episode, which may last typically from 4 up to 72 hours, can potentially be intractable and lead to hospitalization. A diagnosis of migraine is characterized by the presence of at least two of four characteristics, including (1) throbbing character, (2) unilaterality, (3) potential of becoming severe, and (4) worsening with physical exertion, and at least one of the two autonomic components including (1) nausea/vomiting/anorexia or (2) sensitivity to light, sound, touch, and smell.

Primer on Cerebrovascular Diseases, Second Edition
http://dx.doi.org/10.1016/B978-0-12-803058-5.00111-9

WHAT IS THE RELATIONSHIP OF MIGRAINE TO STROKE?

Epidemiology

Small-Vessel Ischemic Stroke

An association between migraine and increased risk of ischemic stroke, particularly among younger adults and women, has been widely discussed. On average, migraine with aura (MA) doubles the risk of ischemic stroke, particularly of the posterior circulation including the cerebellum and thalamus [2,3].

White matter abnormalities, thought to be small-vessel ischemic lesions, are four times more likely to be seen on MRI of the brain among patients who experience MA than those who do not [3]. Moreover, MRI findings have shown that white matter abnormalities occur in proportion to the frequency and severity of migraine attacks and increase the overall lifetime risk of symptomatic stroke in these patients.

Carotid and Vertebral Dissection

There is an association of migraine with intracranial and extracranial large-vessel dissection, which is poorly understood. People with cervical carotid or vertebral dissection are twice as likely to have a history of migraine (with or without aura) than the general population. This association with migraine increases with the number of dissections and is not found to depend on age or gender, possibly limited by the overall low prevalence of dissection [4].

Cardioembolism

Patent foramen ovale (PFO) occurs in approximately one-quarter of the population and is associated with an increased risk of cryptogenic stroke, particularly in patients younger than 55 years [5]. Although the causal relationship of PFO to stroke is unclear and the relationship of PFO with migraine provocative, the prevalence of PFO and cerebral ischemic events in patients with MA has been shown to be roughly double that of the general population in a number of observational studies [2]. Closure of PFO with mechanical devices has been shown to reduce the risk of recurrent stroke with 5-year or longer follow-up and therefore may reduce the frequency of migraine headache [6].

Hemorrhagic Stroke

The literature is sparse on the relationship between migraine and hemorrhagic stroke. Two case controlled studies and one prospective cohort trial have identified an odds ratio of 1.8–2.3 for migraine with hemorrhagic stroke, highest in women with MA who are on oral contraceptives or older than 55 years [3].

Transient Ischemic Attack

The aura of migraine may be prolonged and may mimic transient ischemic attack (TIA) or stroke, acutely complicating emergent treatment decisions. Some patients with migraine and unilateral motor weakness (MUMS), described in the diagnosis section, often carry the diagnosis of "multiple TIAs" because the symptoms of weakness are reversible and may be difficult to distinguish from ischemic stroke. There are some clinical features that are helpful in this situation. The aura of migraine typically begins with paresthesias that migrate slowly over 20–40 min from one part of the body to another, whereas stroke paresthesia and weakness are abrupt and involve face, arm, and leg simultaneously. The weakness of MUMS is sometimes described as "give way" weakness that is largely due to poor effort with no hard signs on examination or imaging of a central lesion [7].

Pathophysiology

Cortical spreading depression (CSD) corresponds to changes in vasomotor vasculature of the cortex. Typically there is 1–2 min of hyperemia followed by 1–2 h of a 20–30% reduction in blood flow. The oligemic phase of migraine has long been hypothesized as a potential trigger for ischemic stroke. At the same time, CSD in itself can be triggered by anything from hypoperfusion to hypoxia, ischemia, hemorrhage, cerebral venous thrombosis, or emboli. Because CSD decreases recovery from ischemia, it has been proposed that the phenomenon can decrease the threshold for ischemic events [2].

Genetics

Cerebral autosomal dominant arteriopathy with subcortical infarcts and leukoencephalopathy (CADASIL) is a rare hereditary condition of the small arteries characterized by frequent MA, recurrent strokes, and progressive white matter degeneration. CADASIL has been determined to arise from a mutation in the NOTCH3 gene on chromosome 19, which controls a receptor important in gene regulation of vascular smooth muscle. A deficiency in the NOTCH3 protein leads to early apoptosis of vascular smooth muscle and damage to small arteries in the brain. Forty percent of patients with CADASIL have MA, which always begins one to two decades before the onset of stroke [2]. This points to a vascular relationship between the two conditions despite inconsistent evidence linking peripheral vascular disease to MA [8].

DIAGNOSIS AND PREVENTION

Diagnosis

The algorithms for the diagnosis of migraine and stroke diverge early, but the connections between the two conditions are important to keep in mind.

Approximately 20% of ischemic strokes are associated with some form of headache [3]. Headache with ischemic stroke is more common with cardioembolism. Carotid and vertebral dissections are even more frequently associated with head and neck pain. Reversible cerebral arterial vasospasm often begins with recurrent thunderclap migrainous-like headaches and can also cause hemorrhagic or ischemic stroke. Both migraine and large-vessel cerebral arterial dissections occur more frequently in patients who present with this syndrome [9].

Migraine can sometimes be associated with hemiparesis or plegia. Familial hemiplegic migraine is a rare hereditary autosomal dominant condition that is believed to be caused by calcium channel disruption leading to hyperexcitability and the occurrence of CSD in the motor or sensory parts of the cortex. The resulting aura is characterized by weakness or paralysis on one side and sometimes numbness, confusion, and aphasia. These symptoms can last up to 24 h and are usually followed by a migraine-type headache that distinguishes it from a TIA. Brain imaging is typically normal in hemiplegic migraine [10].

A large population of patients with migraine describe an ill-defined sense of heaviness or weakness on limbs ipsilateral to the side of their migraine headache. This phenomenon is consistent with MUMS, which was first described and characterized by Young in 2007 and is believed to be vastly underrecognized [11].

Migrainous stroke, or stroke that presents as migraine, is also rare. The exact incidence is unknown, but the largest study to date, conducted in 1986 by Henrich and colleagues, found seven cases of migraine symptoms in 244 patients presenting with first-time stroke. Migrainous stroke was thought to comprise 0.5–1.5% of all ischemic infarcts and 10–14% of ischemic infarcts of the young in the studied population [12]. No pattern of stroke etiology has been found with these types of symptoms. Imaging has shown infarcts predominantly in the posterior circulation of the brain as well as vessel spasm, vessel wall hyperplasia, embolism, and dissection. Migranous infarcts have been reported in patients with migraine with and without aura [1].

If a patient presents with symptoms of migraine and stroke, even with a family history of one or the other, neither etiology should be ruled out until an accurate diagnosis is made in order to prevent complications.

Prevention

Migraine affects the quality of life of millions of Americans and often goes undiagnosed and undertreated. The increasing evidence of a link between migraine and stroke makes it all the more important to effectively diagnose and appropriately treat people with migraine to prevent future stroke.

Depending on the frequency and severity of attacks, migraine treatment starts with abortive nonsteroidal antiinflammatory drugs and 5-hydroxytryptamine 1A, 1B receptor agonists (triptans and ergots) and moves to a wide array of preventative therapies that include antiepileptic medications, antidepressants, minerals, and herbs. Ergots have been linked to white matter abnormalities at high doses, but no association with stroke has been found in patients who use ergots and triptans as prescribed [13]. That said, patients with cardiac risk factors should not be prescribed these medications as their effect on stroke in this context remains unclear.

Given the association between migraine and stroke, the early treatment and prevention of cardiac risk factors is imperative. Low-dose aspirin has been found to both mitigate migraine and prevent stroke, and remains a safe and viable option. As the common pathophysiological and genetic links between stroke and migraine are further elucidated, more effective treatment and prevention strategies will become available.

References

[1] Murray CJ, Vos T, Lozano R, Naghavi M. Disability-adjusted life years (DALYs) for 291 diseases and injuries in 21 regions, 1990-2010: a systematic analysis for the global burden of disease study 2010. Lancet 2012;380:2197–223. http://dx.doi.org/10.1016/S0140-6736(12)61689-4.

[2] T1 K, Chabriat H, Bousser MG. Migraine and stroke: a complex association with clinical implications. Lancet Neurol 2012;11:92–100. http://dx.doi.org/10.1016/S1474-4422(11)70266-6.

[3] Kurth T, Diener HC. Migraine and stroke: perspectives for stroke physicians. Stroke 2012;43:3421–6.

[4] Rist PM, Diener HC, Kurth T, Schurks M. Migraine, migraine aura, and cervical artery dissection: a systematic review and meta-analysis. Cephalalgia 2011;31:886–96.

[5] Furlan AJ, Reisman M, Massaro J, Mauri L, Adams H, Albers GW, et al. CLOSURE I Investigators. Closure or medical therapy for cryptogenic stroke with patent foramen ovale. N Engl J Med 2012;366:991–9. http://dx.doi.org/10.1056/NEJMoa1009639.

[6] Patti G, Pelliccia F, Gaudio C, Greco C, Hand PJ, Kwan J, et al. Meta-analysis of net long-term benefit of different therapeutic strategies in patients with cryptogenic stroke and patent foramen ovale. Am J Cardiol 2015;115:837–43. http://dx.doi.org/10.1016/j.amjcard.2014.12.051. Epub January 6, 2015.

[7] Hand PJ, Kwan J, Lindley RI, Dennis MS, Wardlaw JM. Distinguishing between stroke and mimic at the bedside: the brain attack study. Stroke 2006;37:769–75. Epub February 16, 2006.

[8] Sacco S, Ripa P, Grassi D, Pistoia F, Ornello R, Carolei A, et al. Peripheral vascular dysfunction in migraine: a review. J Headache Pain 2013;14:80. http://dx.doi.org/10.1186/1129-2377-14-80.

[9] Singhal AB, Hajj-Ali RA, Topcuoglu MA, Fok J, Bena J, Yang D, et al. Reversible cerebral vasoconstriction syndromes: analysis of 139 cases. Arch Neurol 2011;68:1005–12. http://dx.doi.org/10.1001/archneurol.2011.68. Epub April 11, 2011.

[10] Russell MB, Ducros A. Sporadic and familial hemiplegic migraine: pathophysiological mechanisms, clinical characteristics, diagnosis, and management. Lancet Neurol 2011;10:457–70.

[11] Young WB, Gangal KS, Aponte RJ, Kaiser RS. Migraine with unilateral motor symptoms: a case-control study. J Neurol Neurosurg Psychiatry 2007;78:600–4. Epub October 20, 2006.

[12] Henrich J, Sandercock P, Warlow C, Jones L. Stroke and migraine in the Oxfordshire community stroke project. J Neurol 1986;233:257–62.

[13] Roberto G, Raschi E, Piccinni C, Conti V, Vignatelli L, D'Alessandro R, et al. Adverse cardiovascular events associated with triptans and ergotamines for treatment of migraine: systematic review of observational studies. Cephalalgia 2015;35:118–31. http://dx.doi.org/10.1177/0333102414550416. Epub September 22, 2014.

CHAPTER

112

Cerebrovascular Complications of Cancer

D. Mehta[1], M. El-Hunjul[1], M.C. Leary[1,2]

[1]Lehigh Valley Hospital and Health Network, Allentown, PA, United States; [2]University of South Florida, Tampa, FL, United States

As cancer and stroke are the second and fifth leading causes of mortality in the United States, it is not surprising to encounter a patient with these concomitant diagnoses [1]. Various cerebrovascular disorders can occur within the oncological population, complicating the overall clinical course, treatment, and long-term outcome of cancer patients. Detailed investigation and precise diagnosis of cerebrovascular disorders in cancer patients is important for several reasons. Early recognition of acute stroke may allow the cancer patient access to interventional thrombolytic, surgical, and endovascular therapies, and improve overall patient outcome. Secondary stroke prevention therapies are also guided by the etiology of the particular cerebrovascular event. Additionally, diagnostic workups in young or cryptogenic stroke patients without overt cancer can lead to the first recognition of the underlying malignancy. This chapter presents an overview of cerebrovascular complications from cancer [2].

STROKE DIRECTLY DUE TO TUMOR

Intratumoral Parenchymal Hemorrhage

Intracranial hemorrhage into brain tumors is a relatively frequent occurrence, reported in 1.7–9.6% of all intracranial hemorrhages. Metastatic tumors are more often associated with hemorrhage than are primary tumors [2–6]. The most common primary central nervous system (CNS) malignancies associated with intratumoral hemorrhage are oligodendroglioma, glioblastoma, and germ-cell tumors. Regarding metastatic malignancies associated with hemorrhage, three of the most common culprits are melanoma, lung cancer, and choriocarcinoma [6,7]. The diagnosis of thyroid metastases is a suggested consideration in a patient with multiple hemorrhagic masses [8] (Fig. 112.1). Predisposing factors associated with intratumoral hemorrhage include head trauma, hypertension, coagulopathy, shunting procedures, and anticoagulation [4,7]. Histological factors associated with intratumoral hemorrhage include rapid tumor growth, tumor necrosis, vessel thrombosis, the presence of multiple thin-walled vessels, and tumor invasion of adjacent cerebral vessels/vessel wall degeneration [5,6]. Patients with parenchymal hemorrhage associated with brain metastases can benefit from steroids and external radiation to reduce cerebral edema. Surgical evacuation of a single hematoma also may be helpful.

Subdural Hemorrhage

Subdural hematomas have been reported in association with a wide variety of malignancies, but typically tend to occur with dural tumor metastases [2,3,9].

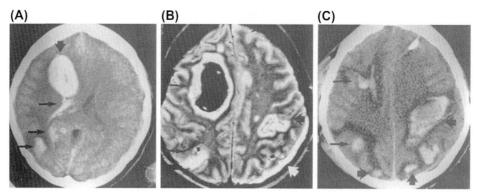

FIGURE 112.1   (A) (B) (C). Multiple hemorrhagic lesions (arrows) in a patient with underlying thyroid malignancy.

Neoplastic infiltration of the dura typically results either from hematogenous spread of tumor via dural vessels or from direct extension of skull metastases. Histologically, the tumors most frequently associated with subdural hematoma include gastric carcinoma, prostate carcinoma, breast cancer, leukemia, and lymphoma [2,3]. Histological examination of the dura with biopsy or cytology studies of the subdural fluid is necessary to confirm the tumoral origin of the subdural hematoma. Treatment of dural metastatic–associated hemorrhage is palliative, including drainage of subdural fluid and brain radiation therapy [10]. If a cancer patient with subdural hematoma undergoes surgical treatment, an adequate biopsy of the dural membrane should be obtained. Radiation therapy should then be administered once the diagnosis is confirmed [2].

Neoplastic Infiltration of Vessels

Venous Infiltration

Thrombosis of cerebral veins or dural sinuses is a rare event in any patient population, including the oncological population. When obstruction of cerebral venous drainage in cancer patients occurs, a frequent culprit is invasion or compression of cortical veins or dural sinuses by tumor [2,11]. The most common sinus affected by metastases is the superior sagittal sinus [10]. A variety of malignancies have been reported in association with sinus thrombosis, including leukemia, lymphoma, neuroblastoma, breast carcinoma, lung carcinoma, cervical carcinoma, gallbladder carcinoma, Ewing's sarcoma, and myeloma [2,3,10]. The proposed mechanism for dural sinus thrombosis is also similar to that of subdural hematoma: skull or dural metastases infiltrate or compress the sinus, producing stasis, thrombosis, and occlusion [10]. Heparin has been beneficial in reducing morbidity and mortality in patients with sinus thrombosis without cancer. Radiation therapy should be a consideration in patients with superior sagittal sinus occlusion due to tumor invasion [10].

Arterial Infiltration

Neoplastic infiltration of arterial vessels has been reported to cause both hemorrhagic and ischemic strokes. Commonly, neoplastic infiltration of arteries results first in aneurysm formation, with subsequent aneurysm rupture, causing intracerebral and/or subarachnoid hemorrhage. Less often, aggressively destructive tumoral invasion produces vessel breakdown and rupture without formation of a true aneurysm. Neoplastic aneurysms are typically small in size, and are often located in the distal cerebral arterial branches, in contrast to saccular aneurysms which typically arise in proximal cerebral arteries around the circle of Willis [12]. Thus, the location of these hemorrhages on imaging involves more peripheral cortical areas, rather than deeper subcortical structures. Neoplastic aneurysms most commonly arise in patients with choriocarcinoma, lung carcinoma, and cardiac myxoma [2,3,12–14]. These same tumors can also produce tumor emboli, which may directly occlude an arterial vessel, producing an ischemic stroke [2,3,12–14].

Ischemic stroke has also been associated with tumor infiltration of arteries. Autopsy reports in cases of cancer patients with leptomeningeal metastases have documented ischemic strokes due to neoplastic arterial wall infiltration [2,3,15,16]. Metastases to the leptomeninges is a relatively uncommon complication of systemic cancer [2], but can occur with breast cancer, lung cancer, melanoma, and hematological malignancies [16]. These patients present with abrupt, focal neurological deficits in addition to the typical features of leptomeningeal metastases such as meningeal irritation and nerve root lesions [2,15,16].

Tumor Embolus: Interestingly, although tumor embolus is generally thought to be one typical mechanism by which focal cerebral metastases result, ischemic stroke directly secondary to tumor embolism is rare. Indeed, most of the reported cases in the literature of ischemic stroke secondary to tumor emboli have been due to primary intracardiac tumors. In general, half of these tumors are myxomas, with the other 50% being papillary fibroelastomas (papillomas), hamartomas, primary malignant neoplasms, teratomas, and other uncommon intracardiac tumors [17]. The most common malignant neoplasms metastasizing to the heart include lung carcinoma and breast carcinoma, followed by melanoma, lymphoma, leukemia, and sarcoma [2,17,18]. Multiple microscopic tumor emboli to small vessels may clinically present as an encephalopathy. History and physical examination is helpful in identifying a potential cardiac tumor as a source, with dyspnea, peripheral edema, and precordial murmurs being the most common cardiac clinical manifestations [2]. The complication rate and rapidity of tumor progression are unpredictable in cardiac tumors. Thus, the treatment of choice in these patients is prompt surgical resection. Additionally, there is evidence to suggest that cerebral aneurysms may improve after resecting the primary cardiac tumor [2].

STROKE DUE TO REMOTE EFFECTS OF TUMOR: COAGULOPATHY

Coagulation dyscrasias resulting in thrombosis and/or hemorrhage can commonly complicate the natural history of cancer and its treatment. The coagulopathy can be preexisting or the neoplasm can indirectly induce a hypercoagulable state, complicating therapy and survival.

HYPERCOAGULABILITY AND THROMBOSIS

Coagulation abnormalities in cancer patients was first described by Trousseau in 1865, who described accelerated bleeding times and an associated thrombophlebitis, in greater than 60% of the oncological patients he observed [19]. Since then, a relationship between hypercoagulability and a variety of malignancies has been reported. The neoplasms most frequently associated with thrombosis include colon cancer, gallbladder cancer, gastric cancer, lung cancer, ovarian cancer, pancreatic cancer, paraprotein disorders, and myeloproliferative syndromes [20,21]. Both venous and arterial occlusions can occur as a result of an oncological-associated hypercoagulable state.

Neoplastic cells can activate the clotting system directly by generating thrombin or indirectly by stimulating mononuclear cells to synthesize and express procoagulants. Disseminated intravascular coagulation (DIC) and elevated D-dimer levels are more likely to be seen in stroke patients with cancer than in stroke patients without malignancy [22,23]. Acquired protein S deficiency causing venous sinus occlusion has been reported in association with leukemia, multiple myeloma, and pancreatic adenocarcinoma. Venous sinus thrombosis due to marked acquired protein C deficiency has also been reported in association with hepatocellular carcinoma, chronic lymphocytic leukemia, acute monocytic leukemia, and acute myeloid leukemia [2,3].

Mucin-producing adenocarcinoma: Mucin-producing adenocarcinomas—especially pancreatic, colon, breast, lung, prostate, and ovarian adenocarcinomas—have been associated with both venous and arterial thrombosis in the CNS. Adenocarcinomas are thought to potentiate thrombi via production of mucin, a high molecular weight "sticky" molecule that is secreted directly into the bloodstream by the tumor, precipitating a viscous hypercoagulable state. Mucin interacts with certain cell adhesion molecules on endothelial cells, platelets, and lymphocytes to induce formation of platelet-rich microthrombi [23,24]. Anticoagulation with low molecular weight (LMW) heparin is recommended in these patients.

Nonbacterial thrombotic endocarditis (NBTE): NBTE is the most common cause of ischemic stroke in the oncology population. In 1938, Sproul first described widespread venous thrombosis, multiple arterial infarcts, and NBTE in patients with pancreatic cancer [25]. Since then, several studies have linked NBTE with cancer and embolic phenomena, typically to the arterial circulation, although venous occlusions have been reported as well [2,26]. NBTE vegetations are most commonly located on the aortic and mitral valves, but also occur on the tricuspid and pulmonic valves. Dual valve involvement occurs in about 12% of patients [27]. Although NBTE is associated with a variety of neoplasms, studies suggest that it is most common in adenocarcinoma patients. Of the adenocarcinomas themselves, NBTE is strongly associated with pancreatic cancer compared with other forms of adenocarcinomas (10.34% risk vs. 1.55% risk) [28]. NBTE can result in either focal, diffuse (i.e., encephalopathy), or mixed neurological deficits when there is embolization from the affected cardiac valve to the brain. Spinal cord infarction has also been reported [2,3].

Treatment of NBTE should focus on the underlying etiology of the coagulation disorder, such as the neoplasm itself or an underlying sepsis/DIC picture indirectly due to the neoplasm or its treatment. Additionally, patients with NBTE are routinely anticoagulated provided there is no contraindication (e.g., CNS bleeding). The rationale for this approach is based upon clinical experience and retrospective studies, as well as the known fragile nature of vegetations and the high rates of recurrent and extensive embolization in this population. Therapeutic dose subcutaneous LMW heparin or intravenous unfractionated heparin should be used, rather than warfarin or oral direct thrombin or factor Xa inhibitors (e.g., dabigatran, apixaban, edoxaban, and rivaroxaban). Although no formal comparison

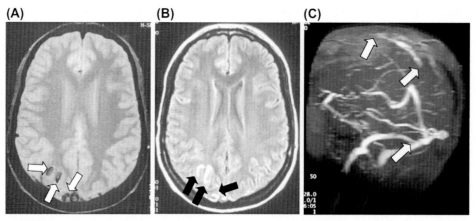

FIGURE 112.2   Sagittal sinus thrombosis in patient with non–small cell lung cancer as well as positive antiphospholipid IgG antibodies.

between heparin and warfarin has been reported in patients with NBTE, older studies suggest that warfarin is less effective than heparin in reducing the rate of recurrent embolization [2,3,26].

Antiphospholipid antibody syndrome: Acute ischemic stroke as well as venous infarcts associated with antiphospholipid antibodies can be the first manifestation of malignancy. Accordingly, some have suggested that Trousseau's syndrome and antiphospholipid syndrome due to malignancy overlap. In particular, solid tumors can be associated with the development of antiphospholipid antibodies, enhancing the thrombophilic risk [29]. (Fig. 112.2) In general, for patients with a malignancy, routine anticoagulant prophylaxis with LMW heparin in cancer patients with positive antiphospholipid antibodies is strongly indicated [30].

COMBINED HYPERCOAGUABILITY/ BLEEDING DIATHESIS

Normal physiological hemostasis involves a balance between coagulation/thrombus formation and fibrinolysis. DIC is characterized by poor coagulation factor control, which results in the acceleration of fibrin formation and subsequent lysis. Patients may have an excessive hypercoaguability/thrombotic picture, uncontrolled fibrinolysis and hemorrhage, or both processes simultaneously. DIC in cancer patients has certainly been associated with both ischemic stroke and intracerebral hemorrhage [2,3]. The DIC course, chronic versus acute, may be an important predisposing factor for the type of stroke that occurs in oncological patients [2,3]. In treating acute DIC in cancer patients, management is controversial, and the final decision, in part, must be based on the individual patient's situation. Certain patients will have temporization of their hemorrhage simply with infusion of platelets, fresh frozen plasma, and cryoprecipitate therapies. Specific

treatment for the underlying malignancy may be helpful, such as using estrogen or androgen deprivation therapy in patients with prostate cancer [2,3].

Bleeding Diathesis/Hemorrhage

Primary fibrinolysis: Both DIC and systemic primary fibrinolysis may coexist in patients with acute promyelocytic leukemia and in patients with prostate cancer, escalating risk of CNS hemorrhage. Primary fibrinolysis alone is also possible [2,3]. There are several potentially possible mechanisms for intracerebral hemorrhage due to primary fibrinolysis: direct tumor production of tissue plasminogen activator (t-PA), direct tumor production of urokinase plasminogen activator (u-PA), and direct protease digestion of fibrinogen. Laboratory findings suggesting the diagnosis of primary fibrinolysis may include a normal platelet count in the setting of hemorrhage associated with hypofibrinogenemia, elevated fibrinogen degradation products, and negative paracoagulation tests. Treatment consists of administering cryoprecipitate or fresh frozen plasma. If DIC is convincingly excluded, ε-aminocaproic acid, or tranexamic acid may also be given [31].

Hyperleukocytic syndrome: The primary neurological manifestation of the hyperleukocytic syndrome is intracranial hemorrhage [32]. The hyperleukocytic syndrome is a distinct entity typically affecting patients with acute myelogenous leukemia. The increased number of abnormal white blood cells, or myeloblasts, elevates the patient's white blood cell count, which can invade and directly damage blood vessels walls. Vessels most vulnerable to this type of invasion are vessels in the CNS and in the pulmonary circulation [32]. Effectiveness of varying treatments is unclear, especially in the setting of intracranial hemorrhage, although lowering the blast count is the general goal. Case reports suggest that if intracranial hemorrhage occurs, the survival rate in these patients is minimal [32].

Thrombocytopenia: Thrombocytopenia itself is not uncommon in the cancer population and poses risk for CNS hemorrhage [2,3,31]. Thrombocytopenia-associated cerebral hemorrhage in oncology patients may be secondary to extensive marrow infiltration by tumor, peripheral destruction of platelets due to tumor-associated hypersplenism, underproduction of platelets due to chemotherapy-induced toxicity, or a combination of the three [2,31]. Other causes of thrombocytopenia-associated stroke in the cancer population include thrombocytopenia due to DIC, autoimmune dysfunction, and microangiopathic hemolytic anemia (MAHA). Patients with isolated thrombocytopenia and acute CNS hemorrhage should receive platelet transfusions.

Microangiopathic hemolytic anemia: Neurological complications with MAHA are rare; however, there are case reports of MAHA-associated CNS hemorrhages [2,3].

Vitamin K deficiency: Deficiencies in vitamin K have been associated with intracerebral hemorrhages. Although typically hereditary, vitamin K deficiency can also occur in patients with poor dietary intake of vitamin K, as well as in patients receiving antibiotic therapy, which sterilizes the gut. Both of these situations are relatively common in oncology patients, and vitamin K deficiency can occur relatively rapidly in this population [2,3].

STROKE IN THE SETTING OF TUMOR, BUT NOT RELATED TO TUMOR

Patients with cancer may also have risk factors for stroke independent of their neoplasm, including hypertension, atrial fibrillation, hypercholesterolemia, diabetes, heart disease, and tobacco usage. It is certainly possible for an oncological patient to have an ischemic stroke secondary to nononcologically related factors, such as small and large vessel atherosclerosis, cardioembolism, paradoxical embolism, arterial dissection, hypercoagulable state, and migraine. Additionally, cancer patients may have a hemorrhage due to hypertension, amyloid angiopathy, or coagulopathy irrespective of their tumor [2].

SEQUELAE OF CANCER TREATMENT

Unfortunately for the oncology patient, the treatment of their particular neoplasm may be an additional risk for developing a hemorrhagic or ischemic stroke. Treatment-induced stroke in the cancer population may result secondary to diagnostic procedures, palliative therapies, chemotherapy, or due to complications of surgical resection/biopsy of the tumor [2,3].

Palliative Therapy

Radiation-induced atheromatous disease: The typical presentation of extensive radiation-induced atheromatous disease occurs in head and neck cancer patients. On physical examination, these patients often have extensive postradiation skin atrophy and fibrosis of the tissues overlying the diseased vascular segment in the neck [2,3]. Radiation-induced necrosis of the mandible may also be another strong clue of underlying radiation-induced vascular disease in patients with head and neck neoplasms [3]. With head and neck cancer patients, arterial injury due to radiation treatments may present in one of two ways: either as arterial rupture or as arterial stenosis/occlusion. Postradiation rupture of the carotid artery typically occurs within 2 to 16 weeks after radical neck surgery and radiation therapy [2,3]. From a neurological standpoint, carotid rupture clinically can present as a unilateral hemispheric watershed infarction, with the specific neurological symptoms correlating to affected areas of the brain. Oncology patients with radiation therapy-associated arterial stenosis/occlusion may have up to double the incidence of stroke and of significant hemodynamic lesions when compared to age-matched controls [33] (Fig. 112.3A and B). It has been recommended that any patient surviving 5 years beyond radiation therapy be followed with noninvasive screening tests such as Duplex ultrasound [33]. Depending on the specific patient and the anatomy/vasculature involved, antiplatelet agents, carotid endarterectomy, or carotid angioplasty/stenting are potential treatment options.

Surgery-Associated Stroke

There are a variety of direct mechanisms during oncological surgical procedures through which stroke may result. Pneumonectomy for pulmonary cancer has been associated with perioperative and postoperative stroke secondary to tumor emboli. Surgical manipulation of the lung can promote release of tumor emboli, especially in the setting of vascular tumor invasion in the lung [2,3,34]. Stroke has also resulted from head and neck cancer resections due to several etiologies. With regard to oncological neurosurgical procedures, venous infarction has been reported in association with craniotomies and craniectomies. Although rare, dural sinus thrombosis is a potentially life-threatening complication of these procedures. Clinically, headache, visual obscuration, and papilledema have been noted in these patients. Heparin may be necessary in this situation, despite recent surgery [35].

Chemotherapy

Multiple factors contribute to the increased risk of stroke in cancer patients, and unfortunately, chemotherapy is no exception to the rule. Potential causes of cerebral thrombosis due to antineoplastic therapy include

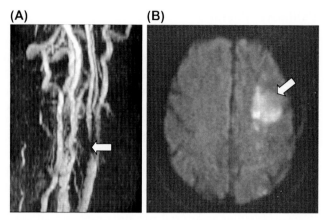

FIGURE 112.3 (A) Left internal carotid artery atherosclerotic disease due to radiation therapy from prior radiation treatment. (B) Acute ischemic stroke from the carotid atherosclerotic disease in the same patient, as seen on diffusion-weighted magnetic resonance imaging.

endothelial cell damage, vasospasm, and acquired coagulation abnormalities. Of all the chemotherapeutic agents associated with stroke, L-asparaginase is a well-known culprit and may present as cortical infarction, capsular infarction, intracerebral hemorrhage, hemorrhagic infarction, or dural sinus thrombosis. Venous thrombosis is very common. Hormone therapies and hematopoietic growth factors are additional precipitants of arterial and venous thromboses [10]. In addition to increasing the risk of thrombotic stroke, chemotherapy has also been linked with hematological factors predisposing cancer patients to hemorrhage. Thrombocytopenia, MAHA, DIC, and thrombotic thrombocytopenic purpura have all been linked to chemotherapeutic agents and pose risk for intracranial hemorrhage [2,3].

Aside from altering the coagulation cascade in cancer patients, chemotherapeutic agents can indirectly contribute to stroke through other mechanisms. Immunosuppressed leukemic patients have been reported to have cerebral infarcts secondary to vessel infiltration with fungi such as *Mucor*, *Aspergillus*, and *Candida* [2,3].

References

[1] http://www.cdc.gov/nchs/fastats/leading-causes-of-death.htm.
[2] Leary MC, Saver JL. Cerebrovascular complications of cancer. In: Wen P, Schiff D, editors. Cancer Neurology: A Guide to Clinical Management. Totowa, (NJ): Humana Press Inc.; 2002.
[3] Graus F, Rogers LR, Posner JB. Cerebrovascular complications in patients with cancer. Medicine 1985;64:16–35.
[4] Drake CG, McGee D. Apoplexy associated with brain tumors. Can Med Ass J 1961;84:303–5.
[5] Little JR, Dial B, Belanger G, Carpenter S. Brain hemorrhage from intracranial tumor. Stroke 1979;10:283–8.
[6] Iwama T, Ohkuma A, Miwa Y, Sugimoto S, Itoh T, Takada M, Tanabe Y, Funakoshi T, Sakai N, Yamada H. Brain tumors manifesting as intracranial hemorrhage. Neurol Med Chir 1992;32:130–5.
[7] Wakai S, Yamakawa K, Manaka S, Takakura K. Spontaneous intracranial hemorrhage caused by brain tumor: its incidence and clinical significance. Neurosurgery 1982;10:437–44.
[8] Isoda H, Takahashi M, Arai T, Ramsey RG, Yokoyama T, Mochizuki T, Yamamoto I, Kaneko M. Multiple haemorrhagic brain metastases from papillary thyroid cancer. Neuroradiology 1997;39:198–202.
[9] Minette SE, Kimmel DW. Subdural hematoma in patients with systemic cancer. Mayo Clin Proc 1989;64:637–42.
[10] Rogers LR. Cerebrovascular complications in cancer patients. Oncology 1994;8:23–30.
[11] Hickey WF, arnick MB, Henderson IC, Dawson DM. Primary cerebral venous thrombosis in patients with cancer – a rarely diagnosed paraneoplastic syndrome. Am J Med 1982;73:740–50.
[12] Helmer FA. Oncotic aneurysm. J Neurosurg 1976;73:740.
[13] Ho KL. Neoplastic aneurysm and intracranial hemorrhage. Cancer 1982;50:2935.
[14] Roeltgen DP, Weimer GR, Patterson LF. Delayed neurological complications of left atrial myxoma. Neurology 1981;31:8.
[15] Klein P, Haley C, Wooten GF, VandenBerg SR. Focal cerebral infarctions associated with perivascular tumor infiltrates in carcinomatous leptomeningeal metastases. Arch Neurol 1989;46:1149–52.
[16] Wasserstrom WR, Glass JP, Posner JB. Diagnosis and treatment of leptomeningeal metastases from solid tumors. Cancer 1982;49:759–72.
[17] Molina JE, Edwards JE, Ward HB. Primary cardiac tumors: experience at the University of Minnesota. Thorac Cardiovasc Surg 1990;38:183–91.
[18] Hanfling SM. Metastatic cancer to the heart, review of the literature and report of 127 cases. Circulation 1960;22:474–83.
[19] Trousseau A. Phlegmasia alba dolens. In: Clinique Medicale de L'Hotel Dieu de Paris 3. 2nd ed. Paris: Balliere; 1865.
[20] Bick RL. Coagulation abnormalities in malignancy: a review. Semin Thromb Hemost 1992;18:353–72.
[21] Amico L, Caplan LR, Thomas C. Cerebrovascular complications of mucinous cancers. Neurology 1989;39:522–6.
[22] Uemura J, Kimura K, Sibazaki K, et al. Acute stroke patients have occult malignancy more often than expected. Eur Neurol 2010;64:140–4.
[23] Dearborn JL, Urrutia VC, Zeiler SR. Stroke and cancer-a complicated relationship. J Neurol Transl Neurosci 2014;2:1039.
[24] Chaturvedi P, Singh AP, Batra SK. Structure, evolution, and biology of the MUC4 mucin. FASEB J 2008;22:966–81.
[25] Sproul EE. Carcinoma and venous thrombosis: the frequency of association of carcinoma in the body or tail of the pancreas with multiple venous thrombus. Am J Cancer 1938;34:566–85.
[26] el-Shami K, Griffiths E, Streiff M. Nonbacterial thrombotic endocarditis in cancer patients: pathogenesis, diagnosis, and treatment. Oncologist 2007;12:518.
[27] Biller J, Challa VR, Toole JF, Howard VJ. Nonbacterial thrombotic endocarditis. A neurologic perspective of clinicopathologic correlations of 99 patients. Arch Neurol 1982;39:95–8.
[28] Gonzalez Quintela A, Candela MJ, Vidal C, Roman J, Aramburo P. Non-bacterial thrombotic endocarditis in cancer patients. Acta Cardiol 1991;46:1–9.
[29] Miesbach W. Antiphospholipid antibodies and antiphospholipid syndrome in patients with malignancies: features, incidence, identification, and treatment. Semin Thromb Hemost 2008;34:282–5.
[30] Marshall AL, Connors JM. Anticoagulation for noncardiac indications in neurologic patients: comparative use of non-vitamin K oral anticoagulants, low-molecular-weight heparins, and warfarin. Curr Treat Options Neurol 2014;16:309.
[31] Rosen PJ. Bleeding problems in the cancer patient. Hematol/Oncol Clin North Am 1992;6:1315–28.

[32] Dabrow MB, Wilkins JC. Hematologic Emergencies. Management of hyperleukocytic syndrome, DIC, and thrombotic, thrombocytopenic purpura. 1993;5:193–202.

[33] Elerding SC, Fernandez RN, Grotta JC, Lindberg RD, Causay LC, McMurtrey MJ. Carotid artery disease following external cervical irradiation. Ann Surg 1981;194:609–15.

[34] Lefkovitz NW, Roessmann UR, Kori SH. Major cerebral infarction from tumor embolus. Stroke 1986;17:555–7.

[35] Keiper GL, Sherman JD, Tomsick TA, Tew JM. Dural sinus thrombosis and pseudotumor cerebri: unexpected complications of suboccitpital craniotomy and translabyrinthine craniectomy. J Neurosurg 1999;91:192–7.

CHAPTER

113

Radiation Vasculopathy

N.M. Rao, H.V. Vinters, J.L. Saver

David Geffen School of Medicine at University of California, Los Angeles, Los Angeles, CA, United States

INTRODUCTION

Radiation-induced arterial damage was described shortly after the advent of the X-ray by Roentgen in the late 1800s [1]. In the setting of radiation therapy for malignancy, this arterial change has been described in multiple vessels throughout the body, from the capillary bed to the aorta. Pertinent to stroke, radiation therapy to the neck may result in steno-occlusive disease of the carotid and vertebral arteries in patterns atypical for standard atherosclerosis. Furthermore, radiation delivered to cranial targets may result in moyamoya syndrome, cavernomas, arteriovenous malformations (AVMs), and hemorrhage. Radiation injury to cerebral vessels may also be pursued as a therapeutic intervention, for instance, to treat vascular malformations. This chapter discusses the pathophysiology, incidence, monitoring, and treatment of radiation vasculopathy relevant to cerebrovascular disease.

PATHOPHYSIOLOGY

Radiation-induced damage to cerebral blood vessels may be divided grossly into acute, subacute, and chronic time periods. Endothelial cells are particularly radiosensitive and are among the first to succumb to radiation damage. In a study of irradiated canine femoral arteries by Fonkalsrud et al., endothelial cell disruption and fibrin deposition was seen within 48 h of radiation exposure. After the 1st week, this initial luminal injury was followed by progressive cellularity, fibrosis, and necrosis of the media, along with hemorrhage and inflammation of the adventitia [2]. Other subacute changes include dilation of the blood vessel lumen, vessel wall thickening, endothelial cell nuclear enlargement, and astrocyte hypertrophy. Within the brain, these vascular changes lead to blood–brain barrier breakdown and accompany perivascular inflammation, edema, necrosis, and demyelination seen in the surrounding brain tissues [3]. Advances in radiation dosage optimization and cotreatment with steroids have reduced the incidence of these acute and subacute complications of radiation.

Of increasing prevalence are the chronic effects of radiation to large vessels, both intracranially and extracranially. Histological sections of chronic large vessel radiation vasculopathy show connective tissue proliferation, dense hyalinization of the vessel wall, including thickening of the intima, internal elastic lamina, and adventitia [4]. There are several pathological features that separate radiation vasculopathy from other forms of vascular disease (Fig. 113.1). For instance, in studies of coronary arteries affected by radiation vasculopathy, medial thinning, and adventitial fibrosis distinguished these vessels from standard coronary atherosclerosis [5]. Furthermore, the severe atherosclerotic change was limited to vessels within the field of irradiation.

Primer on Cerebrovascular Diseases, Second Edition
http://dx.doi.org/10.1016/B978-0-12-803058-5.00113-2

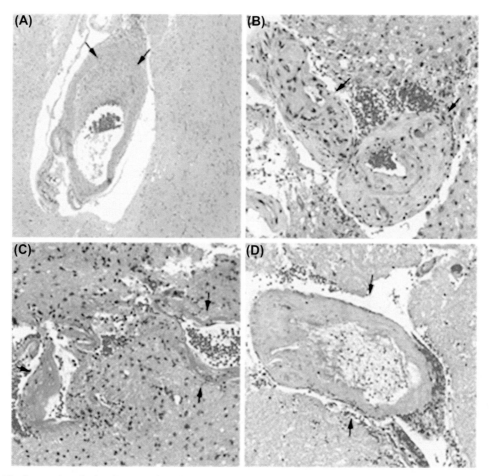

FIGURE 113.1   The "many faces" of radiation-induced vasculopathy. All panels are micrographs from portions of glioblastoma in two subjects who had previously been radiated, including one autopsy and one surgical specimen. (A) Meningeal artery, arrows indicate internal elastic lamina (IEL). Note pronounced intimal hyperplasia between IEL and lumen. (B) Arrows indicate two arteries with marked thickening of their walls resulting from proliferation of various cellular elements. Surrounding brain parenchyma shows pronounced reactive change. (C) Arrows indicate fibrinoid necrosis in arteries at the edge of glioblastoma. (D) Arrows indicate hyaline, non-amyloid thickening of an artery in a radiated glioblastoma. (All micrographs are from H&E-stained sections.)

Another distinguishing feature unique to radiation vasculopathy is the direct damage to the vascular endothelium. Endothelial cells are exquisitely sensitive to the effects of radiation, and this initial insult may provide the inflammatory nidus for further atherosclerosis. Additionally, endothelial cell loss in the vasa vasorum leads to progressive dysfunction of the vessel's own blood supply, gradually choking the arterial wall of circulation and leading to further stenosis [3]. The prominent damage to the vasa vasorum also separates radiation vasculopathy from traditional atherosclerosis. Interestingly, it is often noted that large vessels with significant stenosis due to radiation are associated with abundant collateral vasculature [6]. Formation of these collaterals is a testament to the slow progressive nature of radiation vasculopathy and may allude to the angiogenic nature of radiation as well.

As for additional similarities to traditional atherosclerosis, risk factors such as diabetes, smoking, hyperlipidemia, and hypertension have an additive effect on irradiated vessels. Although these comorbid genetic and environmental pressures may hasten the development of stenosis in irradiated vessels, the progression is still relatively slow. Studies of the prevalence of radiation vasculopathy have established that these long-term pathological changes occur over years to decades.

INCIDENCE

Chronic radiation vasculopathy was once thought to be a rare occurrence, as patients often expired from their malignancies long before complications of the treatment could develop. However, radiation therapy has since become an established, effective primary and adjuvant treatment for a multitude of cancers. In the case of head and neck cancers, radiation therapy alone can be curative. As a consequence of advances in treatment, survivors of previously deadly malignancies are increasingly experiencing the long-term sequelae of radiation injury.

Following radiotherapy to the neck, the incidence of carotid stenosis within the field of irradiation ranges from 20% to 50%, depending on the type of radiotherapy, age at onset, comorbid atherosclerotic risk factors, and time from radiotherapy to evaluation. For instance, in a prospective cohort of 44 patients with unilateral radiotherapy, Brown et al. showed an incidence of 0.6 ipsilateral and 0.6 contralateral carotid artery stenosis of >50% per 100 person-years when evaluated within the first decade after radiotherapy. However, at >15 years after radiotherapy, the incidence jumped to 21.3 and 5.3 per 100 person-years in the ipsilateral and contralateral carotid artery, respectively [7]. Other studies also support an elevated incidence of cervical artery stenosis after radiotherapy to the neck with a time course measured in years to decades after the initial therapy [6].

Radiotherapy directed at the skull and brain may lead to a more diverse range of sequelae, including cavernomas, AVMs, hemorrhages, and moyamoya syndrome. The exact incidence of these complications is not fully characterized. In a case series of 100 survivors of primary central nervous system tumors, Passos et al. showed an elevated incidence of microbleeds (29%) and cavernomas (19%) at a mean 16.7 years after radiotherapy [8]. In a case series of 345 young patients who received cranial radiation for brain tumors, 9.6% developed vascular abnormalities and 3.5% progressed to moyamoya syndrome [9]. As for the time frame of development, in a meta-analysis yielding 54 patients with cranial radiation therapy who developed moyamoya syndrome, the median latent period was 40 months, with half of the patients developing moyamoya syndrome within 4 years, and 95% within 12 years. Young patients (<5 years) who received radiotherapy to the parasellar region had the highest risk of moyamoya syndrome [10].

In addition to steno-occlusive vasculopathy, cerebral cavernous malformations after intracranial radiation have also been described. Although also seen in adults, this is particularly reported in the pediatric population, where multiple cavernomas are often found within the field of radiation. Similar to steno-occlusive lesions, the average time from radiation treatment to development of cavernous malformation is approximately a decade, with higher radiation dose associated with a shorter latency time [11].

RADIOLOGICAL MANIFESTATIONS

Patients with radiation vasculopathy tend to develop radiographic pathology atypical of standard atherosclerosis. The angiographic appearance of radiation vasculopathy may take the form of diffuse lesions with tandem stenosis, dissecting flaps, and ulcerative caps. In a study by Zou et al., when compared to standard atherosclerotic cervical artery disease, radiation vasculopathy was more frequently bilateral, and associated with multiple areas of stenosis or complete occlusions [6]. Additionally, steno-occlusive disease of the vertebral, common and external carotid arteries were more prevalent in irradiated vessels. Patients with radiation vasculopathy develop more robust collateral circulation, supporting the indolent progression of the disease and potential angiogenic effects of radiation as mentioned earlier [6].

Radiographic evaluation of the vessel wall has also revealed several distinguishing features of radiation vasculopathy. For instance, an ultrasound study by Cheng et al. found radiation-affected carotid arteries to have greater vessel wall thickness and smaller luminal diameter than standard atherosclerotic plaques, although the plaque morphology did not differ significantly between the two groups [12]. In an MRI comparison between radiation vasculopathy and moyamoya disease, prominent ring enhancement and thickening of the affected vessel walls were unique to irradiated patients, indicative of the differing etiologies of these disease processes [13]. Examples of the multivessel steno-occlusive disease and extensive collateralization of adjacent vessels is shown in Fig. 113.2.

MONITORING

The slow progressive nature of steno-occlusive disease after radiation generally allows adequate time for the generation of collateral circulation. Thus, it may progress unnoticed for decades, until a critical threshold is reached and cerebrovascular ischemia occurs. Once identified however, radiation vasculopathy may progress faster than atherosclerotic arteriopathy. Therefore, it is important for physicians to recognize the potential symptoms of radiation vasculopathy in at-risk patients. Primary care or oncology specialists who regularly follows these patients should be aware of any complaints concerning for transient ischemic events. Imaging and referral should be performed on such patients.

As of 2016, there are no detailed guidelines regarding screening with imaging studies for high-risk patients. As for patients who have received radiation to the head and neck, in 2007 the American Society of Neuroimaging recommended cervical carotid artery screening 10 years after treatment. Patients with history of radiation to the head and brain often have frequent brain parenchymal imaging performed to monitor for cancer recurrence, but these protocols may not include intracranial vessel imaging. Accordingly, evaluating physicians must still be vigilant of complaints that may represent transient neurological attacks and request vessel imaging for at-risk patients.

TREATMENT OPTIONS

The optimal treatment strategy for cervical radiation vasculopathy is still under investigation. With regard to procedural options for large vessel disease such as cervical carotid artery stenosis, both endarterectomy and stenting have been utilized to successfully treat radiation-related lesions; however, one is not clearly superior over the other in all cases [14]. Radiation to the neck poses several obstacles to treatment that must be considered. Due to fibrosis of the vessel wall, scar tissue obscuring anatomical landmarks, fragile connective tissue, and poor wound healing, radiation therapy may result in a "hostile neck" with regard to open surgery. Tandem and atypical lesions with a propensity for restenosis pose barriers to endovascular options. Traditionally, patients undergoing corrective procedures for large vessel disease were felt to be at high risk for complication, particularly with open surgery. However, in a meta-analysis by Fokkema et al., both carotid endarterectomy and stenting had equivalent perioperative cerebrovascular event rates (3.5 vs. 3.9% respectively, P=0.77). In this study, patients undergoing endarterectomy had higher rates of cranial nerve injury, whereas higher rates of late cerebrovascular events and restenosis were associated with stenting [14]. Given the heterogeneity of the lesions and potentially multiple comorbid variables, each patient should be evaluated on a case-by-case basis to determine optimal management.

Additional consideration must be taken for patients undergoing subsequent surgical procedures to the head and neck—ligation of what might be thought of as a minor branch of the external carotid or thyrocervical trunk may prove catastrophic in the case of patients who are relying on collateral vessels in lieu of occluded major cervical vessels for brain perfusion (Fig. 113.2).

For intracranial pathology such as moyamoya syndrome, intracranial revascularization procedures such as encephalo-duro-arterio-synangiosis as per idiopathic moyamoya have been utilized with good success [9]. For both intracranial and extracranial disease, traditional stroke risk factors such as smoking, hypertension, hyperlipidemia, diabetes, and obesity should be addressed as these act additively on vessels damaged by radiation.

CEREBROVASCULAR COMPLICATIONS OF CARDIAC RADIATION

Cardiovascular complications to thoracic radiation for treatment of conditions such as breast cancer and Hodgkin's lymphoma include vasculopathy of the aortic arch, cardiac valves, and coronary arteries. Furthermore, radiation damage to the heart itself may cause myocyte loss and fibrosis of the myocardium and pericardium, leading to restrictive cardiomyopathy, diastolic heart failure, and conduction abnormalities [15]. These complications may lead to stroke through cardioembolism or arterioembolism.

Similar to cerebrovascular radiation vasculopathy, radiation-induced cardiac disease poses unique challenges. Pericardial fibrosis and scar tissue formation are barriers to surgical procedures. Heavy calcifications of the aortic arch and coronary arteries may impede endovascular approaches. Corrective options should be

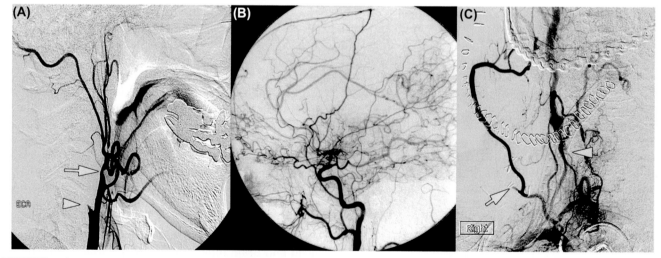

FIGURE 113.2 Angiograms of irradiated cervical and intracranial vessels. (A) Occluded left internal carotid artery (arrowhead) with multiple stenoses of the external carotid artery (arrow). (B) Moyamoya collaterals in a patient with history of radiation to the skull base. (C) 57-year-old man with a history of squamous cell carcinoma and radiation therapy to the neck 8 years prior to presentation. On the day after a mandibular reconstruction, he was found to have watershed infarctions of the right frontal and parietal lobes. The angiogram showed bilateral common carotid artery occlusions and filling of the internal carotid arteries (arrowhead on right internal carotid artery) from the costocervical trunk via the inferior thyroid arteries through the superior thyroid arteries. On the right, the transverse cervical artery was anastamosed to the graft feeding the reconstructed mandible (arrow). Reduction in collateral circulation may have precipitated the cerebral infarct.

weighed depending on each patient's unique anatomy and comorbid risk factors. Cardiovascular risk factors such as smoking, diabetes, and hypertension also have additive effects on the radiation damaged heart, and should be aggressively controlled. Akin to radiation vasculopathy, cardiac radiation damage may be more aggressive than traditional cardiac disease and frequent monitoring should be considered.

THERAPEUTIC EFFECTS OF RADIATION ON CEREBRAL BLOOD VESSELS

The sclerosing effect of radiation injury on blood vessels has been used advantageously in the treatment of cerebrovascular abnormalities such as AVMs and dural arteriovenous fistulas (AVF). In particular, radiation therapy is an attractive option in patients with vascular malformations not amenable to endovascular or surgical approaches. For AVMs, obliteration rates are 60–70% overall 5 years after two or more radiosurgery procedures. Patients with small (<3 cm^3) AVMs may reach 70–90% obliteration rates 3 years after a single radiosurgery procedure [16]. As for dural AVF, in a meta-analysis by Chen et al., 73% of cavernous sinus and 58% of noncavernous sinus dural AVF achieved complete obliteration with stereotactic radiosurgery [17]. However, given the prolonged time course of radiation injury, eventual obliteration of the malformation may take several years. During which time, the patient may be unprotected from hemorrhagic risk. Furthermore, the radiation itself may precipitate hemorrhage, cavernous malformations, cyst formation, or injury to the brain tissue resulting in neurological dysfunction or seizures.

FUTURE PROSPECTS

As increasing numbers of cancer survivors are outliving their malignancies thanks to advances in chemotherapy and radiation treatments, radiation vasculopathy will become all the more prevalent. Practitioners caring for these patients must be diligent in monitoring for the complications of radiation-induced vascular disease. In addition to ongoing research to further our understanding of both the angiodestructive and angiogenic effects of radiation, it may soon be possible to conduct large-scale randomized trials of therapies for patients with various forms of radiation vasculopathy and optimize treatment for patients with this increasingly common disease.

Acknowledgments

We acknowledge Dr. David Liebeskind for assistance with the angiographic images of radiation vasculopathy.

References

[1] Gassmann A. Zur histologie der roentgenulcera. Fortschr Geb Roentgenstr 1899;2:199–207.

[2] Fonkalsrud EW, Sanchez M, Zerubavel R, Mahoney A. Serial changes in arterial structure following radiation therapy. Surg Gynecol Obstet September 1977;145(3):395–400.

[3] Rahmathulla G, Marko NF, Weil RJ. Cerebral radiation necrosis: a review of the pathobiology, diagnosis and management considerations. J Clin Neurosci April 2013;20(4):485–502.

[4] Brant-Zawadzki M, Anderson M, DeArmond SJ, Conley FK, Jahnke RW. Radiation-induced large intracranial vessel occlusive vasculopathy. AJR Am J Roentgenol January 1980;134(1):51–5.

[5] Virmani R, Farb A, Carter AJ, Jones RM. Pathology of radiation-induced coronary artery disease in human and pig. Cardiovasc Radiat Med 1999 Jan-Mar;1(1):98–101.

[6] Zou WX, Leung TW, Yu SC, Wong EH, Leung SF, Soo YO, et al. Angiographic features, collaterals, and infarct topography of symptomatic occlusive radiation vasculopathy: a case-referent study. Stroke February 2013;44(2):401–6.

[7] Brown PD, Foote RL, McLaughlin MP, et al. A historical prospective cohort study of carotid artery stenosis after radiotherapy for head and neck malignancies. Int J Radiat Oncol Biol Phys 2005;63:1361–7.

[8] Passos J, Nzwalo H, Marques J, Azevedo A, Netto E, Nunes S, et al. Late cerebrovascular complications after radiotherapy for childhood primary central nervous system tumors. Pediatr Neurol September 2015;53(3):211–5.

[9] Ullrich NJ, Robertson R, Kinnamon DD, Scott RM, Kieran MW, Turner CD, et al. Moyamoya following cranial irradiation for primary brain tumors in children. Neurology March 20, 2007;68(12):932–8.

[10] Desai SS, Paulino AC, Mai WY, Teh BS. Radiation-induced moyamoya syndrome. Int J Radiat Oncol Biol Phys July 15, 2006;65(4):1222–7.

[11] Cutsforth-Gregory JK, Lanzino G, Link MJ, Brown RD, Flemming KD. Characterization of radiation-induced cavernous malformations and comparison with a nonradiation cavernous malformation cohort. J Neurosurg 2015;122(5):1214–22.

[12] Cheng SW, Ting AC, Wu LL. Ultrasonic analysis of plaque characteristics and intimal-medial thickness in radiation-induced atherosclerotic carotid arteries. Eur J Vasc Endovasc Surg December 2002;24(6):499–504.

[13] Aoki S, Hayashi N, Abe O, Shirouzu I, Ishigame K, Okubo T, et al. Radiation-induced arteritis: thickened wall with prominent enhancement on cranial MR images report of five cases and comparison with 18 cases of Moyamoya disease. Radiology June 2002;223(3):683–8.

[14] Fokkema M, et al. Stenting versus surgery in patients with carotid stenosis after Previous cervical radiation therapy: systematic review and meta-analysis. Stroke 2012;43:793–801.

[15] Groarke JD, Nguyen PL, Nohria A, Ferrari R, Cheng S, Moslehi J. Cardiovascular complications of radiation therapy for thoracic malignancies: the role for non-invasive imaging for detection of cardiovascular disease. Eur Heart J March 2014;35(10):612–23.

[16] Niranjan A, Lunsford LD. Stereotactic radiosurgery guideline for the management of patients with intracranial arteriovenous malformations. Prog Neurol Surg 2013;27:130–40.

[17] Chen CJ, Lee CC, Ding D, Starke RM, Chivukula S, Yen CP, et al. Stereotactic radiosurgery for intracranial dural arteriovenous fistulas: a systematic review. J Neurosurg February 2015;122(2):353–62.

CHAPTER

114

Aortic Dissection and Stroke

J.S. Veluz[1], C.M. Vial[2]

[1]St. Mary Medical Center, Langhorne, PA, United States; [2]Sutter Health/Palo Alto Medical Foundation, Palo Alto, CA, United States

INTRODUCTION

Aortic dissection occurs in about 3–4 per 100,000 persons per year. A common clinical error is to equate aortic dissection with aneurysm. This misconception unfortunately has been propagated since the earliest descriptions of these diseases, describing a "Dissecting Aneurysm of the Aorta" [1]. While some aneurysms can be complicated by, and predispose to, dissection, and dissections can become aneurysmal over time, the disease processes are relatively distinct.

Aortic aneurysm is an abnormal dilation of the diameter of the aorta usually caused by longstanding atherosclerotic disease, hypertension, and/or a history of smoking. Aortic dissection, in contrast, is the result of an intimal tear of the aorta. This tear is often rapidly propagated within the three layers of the aortic wall to various extent due to the forces exerted by the systemic blood pressure, which quite commonly is elevated. Thus blood rushes directly into the separated wall of the aorta creating a false passageway for blood to travel. This "false lumen" of blood flow may thrombose, rupture, or obstruct blood flow to branch vessels, including coronary and carotid arteries, visceral organs, or the extremities. Dissection may also extend to the aortic valve sinuses and cause acute aortic insufficiency. Risk factors for aortic dissection include hypertension, male gender, bicuspid aortic valve, connective tissue disorders such as Marfans and Ehlers–Danlos syndrome, and age greater than 60.

Several classifications exist for aortic dissection, but the most common and functionally useful is the Stanford classification [2]. Dissections that occur proximal to the left subclavian artery are termed Type A. Aortic dissection occurring distal to the left subclavian artery are described as Type B. This distinction of location of the dissection dictates natural history, management, and surgical approach, if necessary. Type A dissections represent surgical emergencies.

Left untreated, Type A dissections carry a mortality rate of 50% within 48 h, and 90% at 2 weeks. Death is usually due to tamponade or aortic rupture. In contrast, Type B aortic dissections are usually managed medically, and have an overall lower short-term mortality, but with a progressive long-term mortality. Surgery for Type B dissection, when necessary, is usually to address complications of malperfusion or later development of aneurysm.

Since brain perfusion arises normally from the aortic arch (carotid arteries) and its branches (vertebral arteries by way of the subclavian arteries), dissection involving this region can lead to acute neurological injury in the form of branch obstruction or embolism. Thus all organs are potentially at risk of ischemia from dissection, typically categorized as either involvement of the visceral, extremity, or neurological (cerebral or spinal) vessels. Most reports find that the innominate (brachiocephalic) artery is the most commonly involved with dissection, followed by the left carotid, and then left subclavian arteries. The vertebral arteries normally take their origin from the subclavian arteries, and can be secondarily affected by subclavian artery dissection.

Epidemiological studies report an incidence of aortic dissection of 5–30 cases/million people. About 17–40% of acute dissections present with evidence for neurological injury [3].

Neurological injury, when manifest, can present as focal neurological deficit due to brain injury, encephalopathy, or paralysis due to either spinal cord ischemia or acute limb ischemia. It is generally believed that aortic dissection presenting with neurological injury has a higher morbidity and mortality, but there is some conflicting data.

Type A Aortic Dissection

Type A dissection represents a surgical emergency. Mortality is related to acute rupture of the dissected aorta in the untreated patient. Surgical treatment, including

Primer on Cerebrovascular Diseases, Second Edition
http://dx.doi.org/10.1016/B978-0-12-803058-5.00114-4

the ancillary modalities used to achieve repair, is commonly an enigma for the noncardiac surgeon. However, it is vitally important for caregivers and consultants to understand the basic concepts of the surgical approach to render an informed and accurate opinion, especially with regard to neurological injury after such procedures, and their attendant prognoses.

Operative goals for Type A dissection are to replace the ascending aorta and usually part of the aortic arch with a synthetic tube graft usually made of woven polyester–derived fiber (Dacron). Concomitant procedures if necessary include aortic valve replacement or repair, replacement of the aortic sinuses of valsalva with coronary artery reimplantation (Bentall procedure), coronary artery bypass grafting, and/or replacement of the entire aortic arch with branch reimplantation.

Preoperative Injury

Neurological injury can involve the brain and/or spinal cord after dissection. The mechanism of injury is usually due to malperfusion and/or hypoperfusion due to occlusion of branch vessels due to an intimal dissection flap. Patients frequently have a pericardial effusion due to transudative effusion from the dissected aorta into the pericardium, leading to varying degrees of tamponade and hypotension. Stroke due to embolism is much less likely. Initial treatment is to complete the aortic repair/replacement with the anticipation that restoration of flow could improve overall cerebrovascular perfusion. It is extremely important to understand that branch vessels as well as the aorta distal to the repair, if already dissected at presentation, will likely remain dissected after surgical repair. More often than not, this residual dissection is usually well tolerated acutely as long as occlusion was not already present. Thrombolysis is contraindicated in acute aortic dissection due to the risk of hemorrhage, aortic rupture, and cardiac tamponade, as well as the need for emergent surgery. While chest pain and back pain are classic presenting signs of aortic dissection, it is important to note that 10–30% of acute Type A aortic dissections primarily manifest as neurological injury [4]. Thus, in the patient with unexplained acute neurological injury, an asymmetry of pulses on examination should prompt concern for aortic dissection, especially in the current era of tissue plasminogen activator (TPA) given for acute stroke. TPA given in the setting of aortic dissection is generally disastrous.

Intraoperative Injury

Central nervous system (CNS) injury can occur due to techniques used to repair the dissected aorta. Cardiopulmonary bypass (CPB) utilizing the heart–lung machine and its techniques itself can liberate debris/emboli into the systemic arterial circulation leading to stroke. General considerations for the noncardiac surgeon provider include the need for systemic heparinization (to an activated clotting time of greater than 460–480 seconds for several hours), exposure of the patient's blood to foreign surfaces (reservoirs, filters, tubing, and cannulae) and the attendant inflammatory response to such exposure, anesthetic medications that affect CNS function both positively and negatively, and the use of hypothermia, along with pH management.

In order to replace part or all of the aortic arch, a common technique utilizes deep hypothermic circulatory arrest (DHCA). Because operating on the aortic arch requires a period of no blood flow to allow for a bloodless operative field, carotid and vertebral artery flow (in addition to the rest of the body) is usually temporarily arrested by halting the CPB machine, thus creating a period of necessary, but controlled whole-body ischemia. Cerebral protection techniques during this time of arrested circulation include hypothermia (typically cooling the patient to 15–24°C and packing the head in ice), systemic steroids, systemic mannitol (to decrease cerebral edema), along with alternative cerebral perfusion strategies outlined later. Postoperatively, the lingering effects of DHCA along with the anesthetic may suppress normal brain function, and should not necessarily be equated with brain injury in the first several hours after surgery.

Surgically created debris may be liberated during this portion of the repair, as the aortic arch is often involved with atherosclerotic and calcific disease, predisposing to emboli. Retrograde cerebral perfusion (RCP) as outlined in the following paragraph, may help "washout" such debris in the cerebral vessels.

Alternative perfusion strategies to the brain during DHCA can involve the use of either antegrade cerebral perfusion (ACP) or RCP. *Antegrade*: Typically, after the patient has been cooled to the desired temperature and given protective therapies (steroids, mannitol, packing head in ice), the heart–lung machine is placed on standby. During this time, there is no circulation within the patient and the ventilator is quiescent. If the site of arterial cannulation (return of oxygenated blood to the body from the heart–lung machine) for CPB is the right subclavian/axillary artery, the heart–lung machine can be run at lower flow during DHCA, thus perfusing the right subclavian, innominate and right vertebral, right common carotid, and right internal carotid arteries with oxygenated blood. Blood will run down to the surgically opened aortic arch, which can be controlled by temporary clamps placed on the arch branches as needed to maintain a bloodless field. Depending on the degree of collateralization from the circle of Willis, left-sided cerebral perfusion can also be achieved to some extent. DHCA (without ACP or RCP) has been shown to be well tolerated up to 30–60 min with low risk of neurological injury, after which time, there is a significant increase in the risk as the arrest time increases. *Retrograde*: This strategy utilizes the already placed cannula in the superior vena cava (venous line for systemic drainage back to the heart–lung machine) to provide oxygenated blood under low pressure to the upper body and CNS *venous* system. Oxygenated blood is therefore delivered to the brain tissue by way of retrograde venous flow. Data regarding the efficacy of antegrade versus retrograde cerebral protection techniques are mixed [5].

Temperature and pH Management and Risk

By systemically cooling the patient, the acid–base physiology of the patient drifts to a more alkaline state. Depending on the operative center, two strategies are used to account for this metabolic derangement. An alpha-stat strategy does not correct for the alkaline shift due to temperature. This strategy is generally more common as it is simpler, and maintains cerebral autoregulation, which may result in less cerebral emboli, which has been shown to be decreased with this technique. A pH-stat strategy corrects for the alkaline change by adding carbon dioxide to the CPB circuit. While physiological pH is maintained with respect to temperature for pH-stat, some studies suggest a higher risk of cerebral embolism, the clinical significance of which is not completely known. There is believed to be loss of cerebral autoregulation with pH-stat management [6].

The speed at which the patient's body is cooled and rewarmed for DHCA can lead to neurological injury. Gas emboli may form from too rapid rewarming of the patient due to dissolved gas changing to a less soluble state. Studies have shown that faster rewarming results in a larger artery–jugular oxygen content difference and higher cognitive dysfunction in the elderly [7].

After repair of the aorta, systemic perfusion is restarted, the patient is warmed, the heart begins to beat, the ventilator is reinitiated, and the patient is ultimately weaned from CPB. Surgical hemorrhage at this time is common, exacerbated by coagulopathy from cooling, inflammation, and contact with the surfaces of the heart–lung machine. Transfusion of blood, platelets, plasma, and cryoprecipitate is common. In cases of severe operative bleeding, recombinant factor VII has also been used. Use of these hemostatic agents may contribute to the formation, or exacerbation of, micro-emboli, which may predispose to stroke.

Postoperative Care

Early postoperative encephalopathy and other neurological changes are extremely common after surgical repair, in part due to anesthesia, varying medications, and recent hypothermia. These changes include altered mental status, loss of pupillary reflex, occulocephalic reflex, and loss of corneal reflex that do not necessarily indicate permanent brain injury or brain death in the early postoperative state.

Malperfusion is difficult to assess in the operating theater. If organ malperfusion clinically exists postoperatively, imaging is recommended. Treatment depends on the type of malperfusion, but may involve catheter-based fenestration, angioplasty, and/or stenting.

Type B Aortic Dissection

Type B aortic dissection may occur de novo, or may exist as the untreated remnant of dissected aorta after a Type A repair. The majority of these patients are managed medically, particularly with regard to hypertension, which exists in about 67% of patients with acute Type B dissection. The combined 30-day neurological complication rate from Type B dissection is about 10%. Neurological injury may manifest as stroke, spinal cord ischemia, encephalopathy, hypoxia, or peripheral nerve ischemia.

Complicated Type B dissection usually results from malperfusion. In the case of neurological injury, the mechanism of injury needs to be investigated, as to whether it is embolic or perfusion related. Creating fenestrations in the dissected aorta, whether by surgical or catheter-based means, can help restore normal perfusion to the ischemic territory. The International Registry of Aortic Dissection had provided a database to better understand and treat this difficult disease. Studies in 2014 suggest that early endovascular stent grafting may result in better long-term outcomes, including survival [8].

References

[1] Jordan WM. Dissecting aneurysm of the Aorta. Br Med J 1954;2:131–2.

[2] Daily PO, Trueblood HW, Stinson EB, Wuerflein RD, Shumway NE. Management of acute aortic dissections. Ann Thorac Surg 1970;10(3):237–47.

[3] Meszaros I, Morocz J, Szlavi J, Schmidt J, Tornoci L, Nagy L, Szep L. Epidemiology and clinicopathology of aortic dissection. Chest 2000;117:1271–8.

[4] Gaul C, Dietrich W, Friedrich I, Sirch J, Erbguth FJ. Neurological symptoms in type A aortic dissections. Stroke 2007;38:292–7.

[5] Tokuda Y, Miyata H, Motomura N, Oshima H, Usui A, Takamoto S. Brain protection during ascending aortic repair for Stanford type A acute aortic dissection surgery—nationwide analysis in Japan. Circ J 2014;78:2431–8.

[6] Hoover LR, Dinavahi R, Cheng WP, Cooper JR, Marino MR, Spata TC, Daniels GL, Vaughn WK, Nussmeier NA. Jugular venous oxygenation during hypothermic cardiopulmonary bypass in patients at risk for abnormal cerebral autoregulation: influence of alpha-Stat versus pH-stat blood gas management. Anesth Analg 2009;108(5):1389–93.

[7] Mukherji J, Hood RR, Edelstein SB. Overcoming challenges in the management of critical events during Cardiopulmonary Bypass. Semin Cardiothorac Vasc Anesth 2014;18(2):190–207.

[8] Luebke T, Brunkwall J. Type B aortic dissection: a review of Prognostic factors and meta-analysis of treatment options. Aorta 2014;2(6):265–78.

CHAPTER

115

Coagulopathies and Ischemic Stroke

S.M. Koleilat, B.M. Coull

The University of Arizona College of Medicine, Tucson, AZ, United States

INTRODUCTION

In ischemic stroke coagulopathies are characterized by a condition in which blood is too quick to clot. Such coagulation abnormalities may be genetic or acquired. Coagulation abnormalities are identified as the cause of ischemic stroke in less than 1% of unselected series. Accordingly, in the vast majority of patients with ischemic stroke an extensive evaluation to identify a coagulopathy is not warranted. In few specific setting testing for a coagulopathy may provide diagnostic evidence for an otherwise unexplained stroke.

The common coagulopathies and estimates of their prevalence in a general population are listed in Table 115.1. These coagulopathies sometimes referred to as thrombophilias are mostly associated with venous thrombotic episodes (VTE) [1]. They can be associated with arterial occlusive events including stroke under circumstances of paradoxical embolization and in association with cerebral sinus and vein occlusions.

ANTITHROMBIN

Antithrombin inactivates thrombin as well as coagulation factors X, IX, XI, and XII. More than 250 loss-of-function mutations are described causing either decreased circulating levels of protein or decreased activity. Antithrombin deficiencies are a leading contributor to thrombophilia associated VTE but are not associated with stroke as a sole contributor.

PROTEIN C AND PROTEIN S

Protein C, which is activated by thrombin, complexes with endothelial protein C receptor and thrombomodulin and together with protein S forms the activated protein C complex that inactivates activated coagulation factors V and VIII. This is a pivotal step in regulation of the coagulation cascade. Protein C and protein S deficiencies are less common than antithrombin deficiency and either low levels or activity can give rise to VTE. Low protein C levels have been associated with silent strokes in adult population studies and apparent arterial strokes in children [2,3]. Venous thrombosis that include cerebral vein and intracranial sinus thrombosis is most associated with these deficiencies, but arterial thrombosis and stroke have been reported [4].

ACTIVATED PROTEIN C RESISTANCE

A point mutation in which adenine is substituted for guanine at nucleotide 1691 in the gene coding for factor mutation, often termed the "factor V Leiden mutation," is the most common inherited predisposition to thrombosis occurring in about 5% of Caucasians. The mutation alters the cleavage site of factor V making it more resistant to the anticoagulant effect of activated protein C (APC-R). A similar but less severe gain-of-function mutation is described in Asian populations (factor V Hong Kong). Heterozygotes are estimated to have a sevenfold increase in the risk of thrombosis, while homozygotes are estimated to have an 80-fold risk. In some cases, APC-R may not be associated with the factor V Leiden mutation. APC-R can also be acquired due to immune-mediated interference with the APC complex. Immune causes of APC-R be detected by functional clotting studies.

APC-R is strongly linked to venous thrombosis and has been found in 20–60% of young individuals with unexplained venous thrombosis. So far several meta-analyses have failed to show a consistent relationship between APC-R and stroke. In one small prospective

Primer on Cerebrovascular Diseases, Second Edition
http://dx.doi.org/10.1016/B978-0-12-803058-5.00115-6

TABLE 115.1 Common Coagulopathies and Estimates of Their Prevalence in a General Population

Coagulopathy Abnormality	Prevalence in General Population (%)
Antithrombin deficiency	0.02–0.2
Protein C deficiency	0.2–0.4
Protein S deficiency	0.03–0.1
Factor V Leiden (heterozygous)	5
Factor V Leiden (homozygous)	0.2
Prothrombin G20210A (heterozygous)	2
Prothrombin G20210A (homozygous)	0.02
Elevated levels of FVIII	3–5
Dysfibrinogenemia	0.001
Hyperhomocysteinemia	5–27
Non-O blood group	55–57

study of unusual site thromboses, factor V Leiden was the most identified cause of thrombophilia among individuals with cerebral venous occlusions at 16% [5].

FIBRINOGEN AND FIBRINOLYTIC ABNORMALITIES

An elevated plasma fibrinogen level is an independent risk factor for cardiovascular events including stroke [3]. After cerebral infarction, fibrinogen levels rise and may remain elevated for over 6 weeks or more. Although the fibrinogen level after stroke is correlated with recurrent stroke risk, as an acute phase reactant it does not have adequate sensitivity or specificity to be of use in most clinical situations. In very rare instances, dysfibrinogenemia may be associated with an increased risk of thrombosis. Some dysfibrinogenemias can be detected by abnormalities in the thrombin time determination.

A variety of abnormalities that impair fibrinolysis have been associated with arterial thromboses including stroke. Among these are plasminogen deficiency, plasminogen activator deficiency (plasminogen activator inhibitor 1, PAI-1), and thrombin-activatable fibrinolysis inhibitor (TAFI) [3]. Lipoprotein (a) also known as Lp(a) is a composite low-density serum lipoprotein linked to apolipoprotein(a) that is structurally similar to plasminogen and thereby competes with plasminogen for the fibrin-binding site. Elevated levels of Lp(a) are associated with increased risk of stroke in some studies. Factor XII/pre-kallikrein deficiency has also been associated with increased propensity for thrombosis. So far a consistent picture of

the role of these observations for assessing the risk of ischemic stroke has not emerged. At present, assays for t-PA and PAI-1 and TAFI are not useful for clinical monitoring in individual patients.

Prothrombin G20210A

This mutation results in about 30% increase in plasma levels of prothrombin (factor II). The mutation is the second most common thrombophilia in Caucasians with a heterozygosity prevalence of 2–3% and 6% in individuals with thrombosis. Both arterial and venous thrombosis are reported with the mutation including an increased risk of arterial stroke and cerebral vein occlusion.

HOMOCYST(E)INEMIA

Elevated plasma homocyst(e)ine levels have been identified as a risk factor for cardiovascular thrombotic events including ischemic stroke [3]. Elevated levels may be present in up to 30% of persons with ischemic stroke. Hyperhomocyst(e)inemia most often results from common mutations in either of two genes affecting the metabolism of homocyst(e)ine or as a consequence of dietary deficiencies, gastrointestinal (GI) malabsorption, or smoking-induced depletion of necessary cofactors for homocyst(e)ine metabolism. These cofactors include vitamins B6, B12, and folic acid. It remains uncertain the degree to which elevated homocysteine levels are a thrombophilia as opposed to a vasculopathy. The inborn error of metabolism that causes homocystinuria appears to contribute an increased risk for stroke at an early age via injury to the vascular wall. It is possible that both mechanisms are important.

Although elevated homocyst(e)ine levels are epidemiologically linked to stroke, the relationship in an individual patient is less certain. Plasma levels of homocyst(e)ine can be lowered with dietary supplementation with vitamin B12, folate, and pyridoxine but the benefits in doing so remain uncertain. One large randomized prospective study did show a 24% risk reduction in stroke with dietary vitamin supplementation [6].

OTHER COAGULATION ABNORMALITIES

Additional important coagulation abnormalities associated with stroke include antiphospholipid antibody syndrome, sickle cell anemia, platelet disorders, the coagulopathies of cancer, and those associated with inflammatory syndromes such as heparin-induced thrombocytopenia. Clotting disorders such as these may

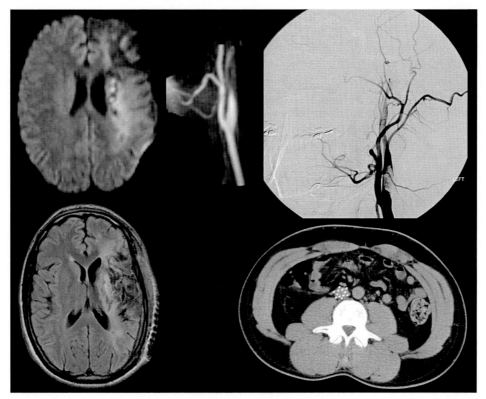

FIGURE 115.1 This 40-year-old man experienced a left hemispheric infarction associated with a left carotid artery dissection at an age of 31 years. He subsequently developed bilateral deep vein thrombosis in both legs extending to the vena cava. As shown (lower right) an inferior vena cava filter was deployed. He subsequently was found to have antiphospholipid antibodies, and he also experienced heparin-induced thrombocytopenia. He is free of recurrent thrombotic events since being placed on warfarin (target INR 2.0–3.0).

coalesce together in a given individual with consequent severe stroke as illustrated in Fig. 115.1. Each of these is covered in a separate chapter of this book. The special circumstances of stroke in the young is also covered in a separate chapter.

WHO SHOULD BE EVALUATED FOR COAGULOPATHIES?

There is no clear answer or current consensus to the question of which stroke patients should be screened for coagulation abnormalities or what specific tests should be ordered. There is general consensus that the yield from a thrombophilia evaluation is low and that the decision to embark on such investigations should be individualized until more large studies become available. Stroke patients in whom the yield of screening is likely to be highest are young patients, those with repeated unexplained strokes, and individuals with a prior history of thrombosis (particularly venous). Patients with unexplained cerebral venous thrombosis (cortical vein or intracranial sinus thrombosis) should be investigated for hypercoagulable conditions, especially APC-R. A suggested approach is summarized in Tables 115.2 and 115.3.

TABLE 115.2 Screening for Coagulopathies in Patients With Ischemic Stroke

Patients under age 30 with no obvious cause for stroke
Young patients with multiple unexplained strokes
Young patients with a prior history of venous thrombosis
Patients with a strong family history of thrombosis
Patients with abnormalities on routine screening coagulation tests

TABLE 115.3 Laboratory Screening Tests for Coagulopathies in Selected Patients

Protein C, protein S, and antithrombin levels by functional assay
Free protein S antigen
Prothrombin G20210A determination
Lipoprotein (a) level
Clotting factor VIII level
Fasting homocyst(e)ine level
Factor V Leiden by polymerase chain reaction and/or functional assay for activated protein C resistance
Thrombin time for dysfibrinogenemia

SUMMARY

Although thrombosis is the underlying pathophysiology of most ischemic strokes, a specific predisposing coagulopathy can only be identified in a small minority of patients. As our ability to evaluate the coagulation system grows more precise the percentage of patients with a definable coagulation abnormality or combined abnormalities will increase.

References

[1] Mannucci PM, Franchini M. Classic thrombophilic gene variants. Thromb Haemost 2015;114:885–9.
[2] Knuiman MW, Folsom AR, Chambles LE, Liao D, Wu KK. Association of hemostatic variables with MRI-detected cerebral abnormalities: the atherosclerosis risk in communities study. Neuroepidemiology 2001;20:96–104.
[3] de Lau LML, Leebeek FWG, de Maat MPM, Koudstaal PJ, Dippel DWJ. A review of hereditary and acquired coagulation disorders in the aetiology of ischaemic stroke. Int J Stroke 2010;5:385–94.
[4] Stam J. Thrombosis of the cerebral veins and sinuses. N Engl J Med 2005;352:1791–8.
[5] Ma K, Wells P, Guzman C, Anderson D, Blostein M, Hirsh A, et al. A multicenter prospective study of risk factors and treatment of unusual site thrombosis. Thromb Res 2016;144:100–5.
[6] Lonn E, Yusuf S, Arnold MJO, Sheridan P, McQueen MJ, Pogue J, et al. Homocysteine lowering with folic acid and B vitamins in vascular diseae. N Engl J Med 2006;354:1567–77.

C H A P T E R

116

Antiphospholipid Antibody Syndrome

C.H. Topel, R.L. Brey

University of Texas Health Science Center at San Antonio, San Antonio, TX, United States

INTRODUCTION

The antiphospholipid syndrome (APS) is defined as a constellation of clinical and laboratory features including thrombosis and/or obstetrical complications associated with the presence of the following antiphospholipid antibodies (aPL): anticardiolipin antibody (aCL), anti-β2-glycoprotein 1 antibody (a-β2-GP1) or the lupus anticoagulant (LA) [1]. The unfortunate misnomer used for the coagulation-based tests, "Lupus Anticoagulant," is a historical phenomenon. Phospholipid-dependent coagulation assays were found to be prolonged in some patients with systemic lupus erythematosus (SLE), and initially thought to be a risk for bleeding. Paradoxically, patients with a positive LA were found to have a higher frequency of thrombosis. This set the stage for the research that followed, leading to the recognition of APS as a clinical syndrome, studies of disease mechanism and pathogenesis, and treatment.

Unlike many hypercoagulable states, thrombosis can occur commonly in the arterial as well as the venous circulation. The most common vascular territory for arterial thrombosis is in the brain, causing stroke. Although many other clinical manifestations have been linked to aPL, there is insufficient evidence to warrant inclusion in the APS diagnostic criteria. It is important to recognize that a single aPL-positive laboratory test at low titer or present on only a single occasion (with or without clinical manifestations) is

Primer on Cerebrovascular Diseases, Second Edition
http://dx.doi.org/10.1016/B978-0-12-803058-5.00116-8

insufficient to make the diagnosis of APS. Treatment with long-term warfarin at an (international normalized ratio) INR-producing dose of 2.0–3.5 (target 2.5) is the current recommended treatment for patients with APS. Adjunctive treatments include hydroxychloroquine and statins. Search for vitamin D deficiency in patients with APS is important as vitamin D deficiency is both common in APS and has been associated with an increased frequency of thrombotic events. Newer immunomodulatory therapies and oral direct thrombin inhibitors are currently being studied as effective therapies for patients with APS [2].

WHAT ARE ANTIPHOSPHOLIPID ANTIBODIES?

Antiphospholipid antibodies are a heterogeneous group of autoantibodies, either inherited or acquired, associated with an increased risk for thrombosis and obstetrical complications. As the aPL field has evolved, the number of aPL with different specificities that have been associated with thrombosis and obstetric risk has expanded [1–5]. Commonly tested aPL include aCL (IgG, IgM, or IgA) antibodies, a-β2-GP1 (IgG, IgM, or IgA) antibodies, and LA (also called lupus inhibitor). The presence of these autoantibodies can lead to an autoimmune hypercoagulable state. Pathogenic aPL bind to phospholipid-binding proteins, which in turn activate cell surface receptors leading to changes in intracellular signaling pathways creating a proinflammatory or hypercoagulable environment. It should be stressed, however, that the mere presence of one or more of these antibodies in an asymptomatic individual does not necessarily increase the risk for either thrombosis or pregnancy morbidity in such individuals. It is possible that some aPL are not pathogenic or other factors in addition to aPL presence are needed for pathogenicity. The body of research that has emerged over the last decade suggests that treatment decisions cannot be made on the basis of a positive aPL test alone, and both laboratory data and clinical manifestations must be considered.

CLINICAL AND LABORATORY FEATURES OF ANTIPHOSPHOLIPID ANTIBODY SYNDROME

The diagnostic clinical features of APS include stroke, venous thrombosis, and obstetrical complications, but many other clinical features have been recognized as well (Table 116.1). These nondiagnostic clinical features can be commonly seen; however, there is insufficient

TABLE 116.1 Clinical Features Associated With Antiphospholipid Syndrome

DIAGNOSTIC (SEE TABLE 116.2 FOR SPECIFIC CRITERIA)
Arterial thrombosis—any organ
Venous thrombosis—any organ
Miscarriage

ASSOCIATED BUT NONDIAGNOSTIC
Livedo reticularis
Pulmonary hypertension
Left-sided cardiac valvular lesions
Libman–Sacks endocarditis
Myxomatous mitral valve degeneration
Intracardiac thrombi
Thrombocytopenia
Chorea
Transient focal neurologic events
Transverse myelitis

evidence to include them among the diagnostic clinical features. Nonetheless, the presence of these nondiagnostic features can be helpful in some cases to suggest the need for a full APS evaluation. Likewise, the presence of a prolonged activated partial thromboplastin time in patients not receiving heparin or a false-positive rapid plasma reagin can also be clues to check for the presence of aPL.

DIAGNOSIS OF ANTIPHOSPHOLIPID ANTIBODY SYNDROME

Table 116.2 lists the APS diagnostic clinical and laboratory features. The diagnosis of APS requires one of the diagnostic clinical manifestations with the presence of one or more of the laboratory criteria that are moderately positive and persistently present at least 12 weeks apart [1]. Low positive aPL values are excluded because they have not been associated with APS-associated clinical manifestations, and the variability of most assays is quite high in the low positive range, leading to a high false-positive rate. Thrombosis occurs most commonly in the venous circulation, but about 30% of APS-associated thromboses occur in the arterial circulation. When arterial thrombosis occurs, it is most commonly seen in the brain for reasons that are not clear. The panel of aPL that, by international consensus, is used to determine criteria for APS includes aCL, a-β2-GP1, and LA. Both aCL and a-β2-GP1 are measured directly using an enzyme-linked immunosorbent assay, whereas LA is detected indirectly by means of a

TABLE 116.2 Revised 2006 Classification Criteria for Antiphospholipid Syndrome [1]

CLINICAL CRITERIA

Vascular thrombosis—one or more clinical episodes of arterial, venous, or small-vessel thrombosis in any tissue or organ confirmed by objective validated criteria (e.g., imaging or histopathology)

Pregnancy morbidity (at least one must be present):

1. One or more unexplained deaths of a morphologically normal fetus at or beyond the 10th week gestation
2. One or more premature births or a morphologically normal neonate at or before the 34th week gestation because of eclampsia or recognized features of placental insufficiency
3. Three or more unexplained consecutive spontaneous abortions before the 10th week gestation with maternal anatomic or hormonal abnormalities and maternal and paternal chromosomal causes excluded

LABORATORY CRITERIA (AT LEAST ONE MUST BE PRESENT):

1. Lupus anticoagulant present in plasma on two or more occasions at least 12 weeks apart
2. Anticardiolipin antibody (IgG or IgM isotype) present in serum or plasma, present in medium to high titer on two or more occasions at least 12 weeks apart
3. Anti-β2-glycoprotein 1 antibody (IgG or IgM isotype) present in serum or plasma, present in medium to high titer on two or more occasions at least 12 weeks apart

phospholipid-dependent coagulation assay [1]. Less is known about thrombosis risk associated with antiphosphatidylserine (aPS), and other aPL, and thus these are not included in the panel used to diagnosis APS.

EPIDEMIOLOGY

The prevalence of aPL in the healthy general population is estimated to be 1–5% [1]. The prevalence in different disease categories, such as people with lupus, first ischemic stroke, deep vein thrombosis, or pregnancy morbidity, is quite variable. The APS Action Group published a review focused on the prevalence of aPL and estimated that aPL are positive in approximately 13% of patients with stroke, 11% with MI (myocardial infarction), 9.5% with DVT (deep venous thrombosis), and 6% with pregnancy morbidity. Recurring criticisms of the body of literature examining the role of aPL in clinical manifestations noted by this group and others are that many studies evaluated for only a few of the specific aPL now thought to be important in thrombosis, most did not perform the tests needed to evaluate for LA, many accepted low positive antibody titers, and many looked at aPL values at only one time point [5]. Thus many of these studies did not include subjects fulfilling the criteria for APS. In addition, some studies combined subjects with venous, arterial, and obstetric manifestations,

and there may be differences in pathogenic antibody specificities that could lead to the different types of thrombotic manifestations.

The evidence most clearly supports the high risk of recurrent thrombosis in patients fulfilling criteria for the APS, particularly in younger individuals (risk increased nearly 2.5-fold) [4]. However, predicting thrombosis risk in older individuals with APS, or in asymptomatic individuals with a persistently positive aPL test, is less clear.

WHEN TO SCREEN FOR aPL IN PATIENTS WITH CEREBROVASCULAR DISEASE

Given the evidence that aPL are most strongly associated with stroke in young people, it is important to consider and screen for all of the APS diagnostic aPL tests in these patients (aCL, a-β2-GP1, and LA). This approach is also suggested in patients regardless of age who have recurrent stroke and no other clear etiology for stroke. If some of the nondiagnostic clinical manifestations such as thrombocytopenia or livedo reticularis and others (see Table 116.1) are present, or if an incident stroke occurs in a patient of any age with a history of fetal loss or spontaneous venous thrombosis, screening is indicated. Finally, patients with SLE and other connective tissue diseases have a high prevalence of aPL, thus an incident stroke occurring in this population should be screened for aPL.

PREDICTING THROMBOSIS RISK IN APS AND aPL-POSITIVE PATIENTS

Much attention has been paid to identifying a specific aPL profile, along with any additive effect from other clinical variables such as traditional risk factors for atherosclerotic vascular disease on thrombosis risk prediction. In one study looking only at aPL profile on thrombosis risk in patients with SLE, multiple aPL, including two antigen specificities that are not included in the diagnostic APS criteria (aPS/prothrombin combination) were measured and a total score was assigned based on the number of positive results. The prevalence of APS clinical manifestations for scores in ascending quintiles from 0 to ≥60 was 10%, 26%, 29%, 56%, and 89%. The rate of incident thrombosis was 5.144/100 person-years for moderate to high scores versus 1.455/100 person-years in the group with a score of 0. Further, the prevalence of recurrent thrombosis on antithrombotic therapy was also higher among patients with a moderate to high score as compared with those of a lower score (odds ratio, 5.4; p = .00015) [6]. Another more simplified risk assessment model including these same antibodies in addition to hyperlipidemia and hypertension further refines risk prediction [7].

MECHANISMS OF ANTIPHOSPHOLIPID ANTIBODY-ASSOCIATED DISEASE

The pathophysiologic processes associated with thrombotic and pregnancy morbidity manifestations are varied, including oxidant-mediated endothelial cell injury, dysregulation of coagulation pathways, and complement activation. However, there is a limited understanding of the origin, specificities, and precise pathologic mechanisms of aPL. As mentioned earlier, a thrombosis risk prediction tool including vascular risk factors in addition to aPL-positive results was most robust. This suggests a "two-hit" hypothesis whereby the presence of aPL alone may not be sufficient to lead to clinical manifestations. Rather, their presence in concert with other factors that may damage vascular endothelium and platelets may be needed for thrombosis to occur [8]. These clinical observations coupled with in vitro work demonstrating endothelial cell activation by aPL bound to phospholipid-binding proteins in the setting of prior injury supports the hypothesis that aPL may act in concert with other vascular risk factors that damage cells leading to disease expression.

Other pathogenic mechanisms pertaining specifically to stroke include their effects on the heart [9]. A variety of cardiac valvular lesions have been associated with aPL antibodies making cardiac emboli a potential stroke mechanism in some patients. Echocardiography (primarily two dimensional, transthoracic) is abnormal in one-third of patients, typically demonstrating nonspecific left-sided valvular (predominantly mitral) lesions, characterized by valve thickening. These may represent a potential cardiac source of stroke.

TREATMENT

The current accepted treatment for APS is heparin followed by long-term anticoagulation with warfarin, with a target INR of 2.0–3.5 (target 2.5) [2]. Unfortunately, long-term warfarin treatment is difficult to manage due to numerous drug and food interactions, the need for frequent monitoring, bleeding risk, and the knowledge that warfarin therapy is not effective in all patients with APS. It is also crucial to vigorously treat patients with APS for all traditional vascular risk factors. In addition, vitamin D has important immunomodulatory functions, and vitamin D deficiency may occur in as many of 50% of patients with APS. Low vitamin D levels correlate with arterial and venous thrombosis. Thus it is important to search for and correct vitamin D deficiency in all patients with APS using guidelines for the general population [2].

There are no data on the use of oral direct thrombin inhibitors or factor Xa inhibitors in patients with APS. A trial is ongoing (Rivaroxaban in Antiphospholipid Syndrome), a prospective, open-label, noninferiority randomized controlled trial in patients with APS comparing rivaroxaban to warfarin. The addition of hydroxychloroquine to warfarin may be more effective in decreasing thrombosis risk and lowering antibody levels. Hydroxychloroquine is recommended for all patients with both APS and SLE, and as a possible adjunct therapy in patients with APS without SLE. Statins have been shown in animal models to modulate aPL-induced prothrombotic effects on endothelial cells. Current recommendations are to use statins for thrombosis prevention only in patients who have an otherwise accepted indication for their use or in patients with APS who have a recurrent thrombotic episode on warfarin [2].

Other treatment approaches based on newly understood mechanisms involving immunomodulatory and cell signaling pathways are also being studied. B-cell inhibition with rituximab and belimumab has been studied in patients with APS and SLE in small phase II studies, and is not recommended except for patients with APS who have failed other therapies. Complement inhibition and small peptide therapy are therapeutic strategies that are being considered [2].

The need and efficacy of primary prevention strategies in patients with positive aPL tests is not clear. Earlier studies did not find a benefit in treatment with low-dose aspirin in thrombosis prevention in these patients. A meta-analysis of 11 studies including 1208 aPL-positive patients and 139 thrombotic events revealed that long-term low-dose aspirin conveyed a 50% risk reduction for occurrence of a first thrombotic event compared with patients not receiving aspirin therapy [10]. Subgroup analyses were done for aPL-positive only patients, aPL-positive patients with SLE, and aPL-positive patients with obstetric complications only and no arterial or venous thrombosis. There was no difference in risk reduction among these groups. Nonetheless, the authors conclude that the use of low-dose aspirin as primary prevention remains controversial, because of the varied methodology used among the studies evaluated (number, type, and frequency of aPL testing) and that no significant risk reduction was seen when only prospective studies or those with the best methodological quality were considered [10]. As with patients with APS, vigorous attention to risk factor reduction is important to decrease thrombosis risk.

References

[1] Gomez-Puerta JA, Cervera R. Diagnosis and classification of the antiphospholipid syndrome. J Autoimmun 2014;48-49:20–5.
[2] Erkan D, Aguiar CL, Andreade D, Cohen H, Cuadrado MJ, Danowski A, et al. 14th International Congress on Antiphospholipid Antibodies: task force report on antiphospholipid syndrome treatment trends. Autoimmune Rev 2014;13(6):685–96.

[3] Pezzini A, Grassi M, Lodigiani C, Patella R, Gandolfo C, Zini A, et al. Predictors of long-term recurrent vascular events after ischemic stroke at young age. Circulation 2014;129(26):1668–76.

[4] Sciascia S, Sanna G, Khamashta MA, Cuadrado MJ, Erkan D, Andreoli L, et al. APS Action. The estimated frequency of antiphospholipid antibodies in young adults with cerebrovascular events: a systematic review. Ann Rheum Dis 2015;74(11):2028–33.

[5] Andreoli L, Chighizola CB, Banzato A, Pons-Estel GJ, Ramire de Jesus G, Erkan D. Estimated frequency of antiphospholipid antibodies in patients with pregnancy morbidity, stroke, Myocardial Infarction, and deep vein thrombosis: a critical review of the literature. Arthritis Care Res 2013;65(11):1869–73.

[6] Sciascia S, Murru V, Sanna G, Roccatello D, Khamashta MA, Bertolaccini ML. Clinical accuracy for diagnosis of antiphospholipid syndrome in systemic lupus erythematosus: evaluation of 23 possible combinations of antiphospholipid antibody specificities. J Thromb Haemost 2012;10(12):2512–8.

[7] Sciascia S, Sanna G, Murru V, Roccatello D, Khamashta MA, Bertolaccini ML. GAPSS: the global anti-phospholipid syndrome score. Rheumatology 2013;52(8):1397–403.

[8] Meroni PL, Borghi MO, Raschi E, Tedesco F. Pathogenesis of antiphospholipid syndrome: understanding the antibodies. Nat Rev Rheumatol 2011;7:330–9.

[9] Zuily S, Regnault V, Selton-Suty C, Eschwège V, Bruntz JF, Bode-Dotto E, et al. Increased risk for heart valve disease associated with antiphospholipid antibodies in patients with systemic lupus erythematosus: meta-analysis of echocardiographic studies. Circulation 2011;124(2):215–24.

[10] Arnaud L, Mathian A, Ruffatti A, Erkan D, Tektonidou M, Cervera R, et al. Efficacy of aspirin for the primary prevention of thrombosis in patients with antiphospholipid antibodies: an international and collaborative meta-analysis. Autoimmun Rev 2014;13(3):281–91.

CHAPTER

117

Primary Platelet Disorders

R.M. Dafer

Rush University Medical Center, Chicago, IL, United States

INTRODUCTION

Platelets are small anucleated discoid-shaped blood cells 2–3 μm in diameter, which are essential for regulating hemostasis. They are produced in the bone marrow by megakaryocytes, circulating in the blood stream for 5–10 days before being destroyed by phagocytosis in the spleen and liver. A normal platelet count is between 150,000 and 450,000/mm^3 of blood. Platelets provide hemostasis through adhesion, aggregation, and coagulation properties. In response to a vessel wall injury, platelets are instantly activated, adhering themselves to the exposed extracellular matrix [1]. This adhesive effect depends on the interaction between platelet plasma membrane glycoproteins (GPIb) and subendothelial adhesion molecules such as laminin, von Willebrand factor (vWF), fibronectin, and other types of collagen [1]. Platelet activation then leads to dense-granule secretion of calcium, thromboxane, and adenosine diphosphate, which help initiate platelet aggregation [2–3]. This subsequently triggers the release of platelet granule contents, accelerating platelet aggregation, a process mediated by glycoprotein IIb-IIIa. Aggregation requires the binding of fibrinogen to its receptor on activated platelets. The release of other substances (i.e. platelet factor 4, beta-thromboglobulin, and platelet-derived growth factor) augments aggregation and facilitates thrombus formation. More platelets are then incorporated in the growing thrombus by binding to fibrin, thus leading to clot retraction [2–3].

Platelet disorders including thrombocytopenia and thrombocytosis are associated with higher risk of ischemic and hemorrhagic strokes through several mechanisms including increased platelet production, dysfunctional platelets, or autoimmune conditions.

Evaluation of platelet disorders includes objective clinical assessment of bleeding or clotting history, physical examination, and quantifiable and qualitative laboratory investigations. Diagnostic testing includes measurement of platelet count, size, and morphology, platelet granulation, tests for platelet hemostatic function, vWF screening tests, platelet function analyzer (PFA-100) closure time, platelet flow cytometry, and specific assays of platelet aggregation and glycoprotein analysis. Genetic studies and bone marrow examination are essential to differentiate between various types of myeloproliferative disorders (MPDs).

THROMBOCYTOSIS AND ISCHEMIC STROKE

Thrombocytosis is defined as a platelet count greater than $450,000/mm^3$. The three major pathophysiological causes of thrombocytosis are (1) clonal, including essential thrombocythemia and other MPDs; (2) familial, including rare cases of nonclonal myeloproliferation due to thrombopoietin mutations; and (3) reactive, in which thrombocytosis occurs secondary to a variety of acute and chronic clinical conditions.

Essential Thrombocythemia

Essential (primary) thrombocythemia (ET) or clonal thrombocytosis is a rare nonreactive chronic MPD of unknown etiology characterized by increased platelet count and short platelet survival [4]. The disease affects adult patients of any age, with an incidence of 1.0 per 100,000 population [5]. ET is associated with increased risk of arterial and venous thrombotic events, in particular in older patients with known vascular risk factors for atherosclerosis. Clinical manifestations include headache, visual disturbances, dizziness, lightheadedness, ocular ischemia, and venous and arterial thrombosis. Transient ischemic attack (TIA) and arterial cerebral ischemia (CI) occur in about 25% of patients, and often account for approximately two-thirds of all vascular events. Cerebral vein thrombosis can complicate ET and can be the first manifestation of the disease [6]. Ischemic stroke is more likely to occur in patients with conventional vascular risk factors for atherosclerosis, in the absence of clear correlation between the platelet count and the stroke occurrence [7].

Diagnosis of ET is made when platelet count is sustained $>450,000/mm^3$, with specific morphology on bone marrow biopsy showing proliferation in the megakaryocytic lineage with increased numbers of large mature cells in subjects not meeting the criteria for other MFDs or myeloid neoplasms, and in the presence of Janus kinase-2 (*JAK2*) VF-mutation, less commonly *CALR* mutation or other clonal markers, when causes of reactive thrombocytosis have been excluded.

Cerebral ischemic changes usually occur predominantly in the periventricular or subcortical regions, although watershed ischemia in the absence of large-vessel arterial stenosis may occur (Fig. 117.1) [7–8].

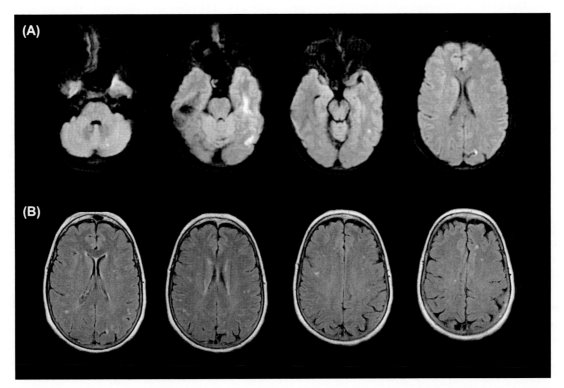

FIGURE 117.1   MRI brain in a patient with essential thrombocythemia. (A) Scattered areas of restricted diffusion consistent with acute infarctions supra- and infratentorially. (B) Diffuse subcortical white matter changes on fluid-attenuated inversion recovery.

Echocardiography may reveal aortic and mitral valvular lesions, including leaflet thickening and vegetations, similar to those described in nonbacterial thrombotic endocarditis [7].

Early management is necessary to reduce thrombotic complications. Strict control of vascular risk factors and smoking cessation may reduce the risk of thrombotic event [7–8]. Although stroke risk does not directly correlate with the severity of thrombocytosis, treatment aims at reducing platelet count to a target of $400,000/mm^3$. Aspirin is typically the antiaggregant of choice, and is usually recommended in patients with ischemic events and those with vascular risk factors [9–10]. Caution should be used in patients with platelet counts $1,000,000/mm^3$ in patients with associated von Willebrand disease due to compromised hemolysis and potential increased bleeding risk [11]. The prophylactic efficacy of antiplatelet agents in asymptomatic subjects with ET and low vascular risk has not been tested. Cytoreductive therapy is indicated in high-risk patients as first-line [10]. Other agents such as interferon alpha, busulfan, and anagrelide may effectively induce hematological remission in ET and should be used as second-line drugs of choice [10].

Familial Thrombocytosis

Familial thrombocytosis (FT) is a rare genetically heterogeneous disorder caused by thrombopoietin gene mutation. FT has been associated with thrombotic episodes including myocardial infarction and ischemic stroke [12]. Clinical manifestations and management are similar to ET.

Secondary Thrombocytosis

Secondary or reactive thrombocytosis (RT), defined as a platelet count greater than $400,000/mm^3$, is often a normal physiologic response. RT is by far the most common cause of thrombocytosis, observed in over 80% of people. RT may occur due to overproduction of thrombopoietin, interleukin-6, and other proinflammatory cytokines, in response to various stressors including infective state, chronic inflammatory conditions, bleeding disorders, surgical stress, trauma, hemolysis, iron deficiency anemia, post-splenectomy, and malignancy [1] (Table 117.1). RT is usually transient and benign. Arterial occlusive diseases and stroke have been reported in patients with iron deficiency anemia [13–17], secondary to cardiopulmonary bypass, and post-splenectomy [18–19]. Management includes the treatment of the underlying predisposing condition. Antiplatelet therapy is not usually indicated.

TABLE 117.1   Causes of Reactive Thrombocytosis

Chronic inflammatory disorders
Connective tissues disorders
Inflammatory bowel disease
Surgery
Cardiopulmonary bypass
Postsplenectomy
Trauma
Iron deficiency anemia
Hemolytic anemia
Infection
Malignancy
Drugs
Vincristine
Cytokines

THROMBOCYTOPENIA AND ISCHEMIC STROKE

Thrombotic Thrombocytopenic Purpura

Thrombotic thrombocytopenic purpura (TTP) is a multisystem thrombotic microangiopathy characterized by thrombocytopenia, microvascular thrombosis, microangiopathic hemolytic anemia (MAHA), and end-organ damage. The incidence of TTP is 3.7–10 per million, with greater occurrence in women and blacks [20–21]. TTP can be inherited, due to a deficiency in ADAMTS13 enzyme activity (<10%), or acquired/autoimmune, due to development of autoantibodies against ADAMTS13 autoantibodies. ADAMTS13 deficiency is a risk factor for the development of myocardial infarction, ischemic stroke, cerebral malaria, and preeclampsia [20,22]. TTP can also be associated with pregnancy, organ transplant, systemic lupus erythematosus, systemic infections, and various drugs either through a dose-related toxicity (e.g., mitomycin C, cyclosporine, and tacrolimus) or immune-mediated reaction (e.g., quinine, ticlopidine, clopidogrel, trimethoprim, and simvastatin) [23–25]. Clinical picture includes fever, fatigue, arthralgia, myalgia, bleeding complications, jaundice, chest discomfort, and renal impairment. Neurological manifestations range from altered mental status, headaches, and visual disturbances, to neurological deficits due to cerebral vascular events, seizures, and coma [23,24].

Diagnosis is made by clinical picture, the presence of MAHA, low platelet count $<50,000/mm^3$, and reduced ADAMTS13 activity. TTP is usually life-threatening, requiring urgent diagnosis and prompt therapeutic intervention. Plasma exchange (PEX) remains the mainstay of treatment, reducing mortality to less than 10%. Large-volume plasma infusion is indicated when delay

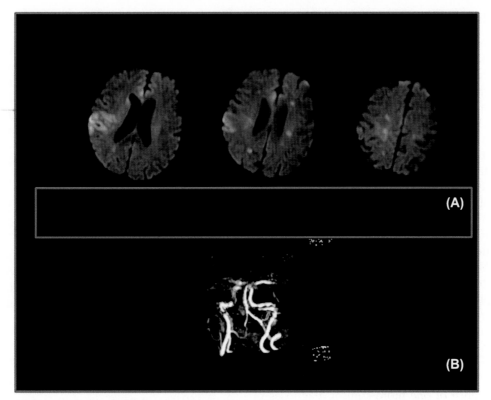

FIGURE 117.2   MRI brain in a patient with heparin-induced thrombocytopenia. (A) Diffusion-weighted imaging showing bilateral scattered infarctions. (B) MRA circle of Willis showing bilateral anterior cerebral arteries and middle cerebral arteries occlusion.

in arranging PEX is anticipated. In refractory cases, rituximab should be considered in conjunction with PEX and steroids. Novel therapeutics such as recombinant ADAMTS13 and gene therapy are under development [23,24,26]. Eculizumab, an anti-C5 monoclonal antibody, has been used in refractory TTP which failed conventional treatment with PEX, steroids, and rituximab [26].

Antiplatelet therapy with low doses aspirin is safe, and may be considered when platelet count is >50,000/mm³.

Immune-Mediated Heparin-Induced Thrombocytopenia

Heparin-induced thrombocytopenia (HIT) is a clinicopathological prothrombotic state caused by the binding of heparin-dependent IgG antibodies to heparin/platelet factor 4 complexes, thus causing platelet activation. HIT usually occurs within 5–10 days of exposure to heparin products, predominantly unfractionated heparin or low-molecular-weight heparin, and less commonly fondaparinux [27]. Rarely, HIT can occur in the absence of prior exposure to heparin. HIT occurs in 1 in 5000 hospitalized patients, with an incidence of 1–3% after cardiac surgery [28].

HIT is associated with a high incidence of paradoxical arterial and venous prothrombotic state. Deep vein thrombosis, pulmonary embolism, peripheral arterial thrombosis, warfarin necrosis, and CI may occur (Fig. 117.2) [28].

Diagnosis is based on reduction in platelet count by more than 50% (usually between 40,000 and 80,000/mm³), with thrombotic events occurring 5–10 days following exposure to heparin. The presence of heparin-dependent antibodies together with the clinical manifestations of arterial or venous thrombosis is necessary to establish the diagnosis. Anti-PF4-heparin enzyme-linked immunosorbent assay (ELISA) is very sensitive for HIT but with low positive predictive value. Functional washed-platelet activation assays (heparin-induced platelet activation test and serotonin release assay) are confirmatory, with a better specificity compared with ELISA [27,28].

Once HIT is suspected, heparin products should be immediately discontinued and alternative anticoagulants should be considered. In patients with thrombotic events, nonheparin anticoagulants such as argatroban or danaparoid are preferred over vitamin K antagonist (VKA) [29]. In patients with HIT without thrombosis, VKA may be considered [29]. VKA should be generally avoided until the platelets have reverted to a count above 150,000. New oral anticoagulants are preferred over VKA, and may be considered as alternative treatment options [27,29–31].

Antiphospholipid Antibodies

Antiphospholipid syndrome (APS) is an autoimmune prothrombotic state associated with arterial and venous thrombosis, recurrent fetal loss, and thrombocytopenia. The exact mechanism of thrombosis in APS remains unclear. Circulating pathogenic antiphospholipid (aPL) antibodies has been implicated in the pathogenesis of thrombosis in APS through binding to the platelet membrane, and initiating various inflammatory responses including resistance to natural anticoagulants, activation of endothelial cells, and dysregulated activation of platelets [32]. Thrombocytopenia may be present in up to 25% of patients with APS [33,34].

APS may affect any tissue or organ, with ischemic stroke being one of the most common complications of the disorder. Neurological manifestations of APS are broad: they include TIA, CI, cerebral dural venous sinus thrombosis, epilepsy, migraine-like headaches, movement disorders, psychiatric manifestations, and dementia. Despite the role for β2-glycoprotein I-dependent aPL in the development of APS, the pathogenic mechanism for thrombosis remains not well known [32].

Criteria for the diagnosis of APS-related stroke include the presence of one more clinical diagnoses of ischemic stroke, with persistent elevated titers of autoantibodies of the IgG or IgM isotype detected by ELISA for anti–β2-glycoprotein I or aPL, or by lupus anticoagulant assays, on two or more consecutive testings at least 12 weeks apart [32,36].

Management of APS includes the use of various antithrombotic therapies. The antiphospholipid antibody stroke study showed no difference between aspirin and low- to moderate-intensity anticoagulation with international normalized ratio of 1.8–2.4 with regards to reduction in thromboembolic events in patients with APS and ischemic stroke [35]. Alternative therapies such as low-molecular-weight heparin and new oral anticoagulants may be considered [37–39]. In catastrophic APS with multiple organ involvement, anticoagulation is the option of choice. Glucocorticoids, intravenous immunoglobulin (IVIG), and immunosuppressant agents such as cyclophosphamide and rituximab should be considered in conjunction with anticoagulation [32,40].

Hyperaggregable Platelets

Platelets play an important role in CI. During an acute ischemic event, platelets become activated, and elevated levels of platelet-release proteins and circulating platelet aggregates are observed. Whether platelet aggregability is related to early-stage arteriosclerosis or whether an increase in platelet count is an acute phase reactant in response to ischemia remains unclear.

Sticky platelet syndrome (SPS) is a congenital autosomal dominant thrombophilic thrombocytopathy, characterized by an increased in vitro platelet aggregation in response to low concentrations of adenosine diphosphate and/or epinephrine [41]. SPS is an important cause of life-threatening thrombotic complications including fetal loss, growth retardation, myocardial infarction, ischemic optic neuropathy, recurrent venous thromboembolism, TIA, and strokes [41–42]. Antiplatelets are the therapy of choice [43–44].

PLATELET DISORDERS AND HEMORRHAGIC STROKE

Various platelet disorders, hereditary, autoimmune, and acquired, have been associated with the increased risk of bleeding complications including spontaneous intracerebral hemorrhage (ICH) and subarachnoid hemorrhage (SAH) [45].

Thrombocytopenia

Thrombocytopenia is defined as platelet count less than 140,000/mm^3. Causes include decreased platelet production (e.g., in alcohol or MDS), increased splenic sequestration of platelets with normal platelet survival (e.g., in liver cirrhosis with congestive splenomegaly), increased platelet destruction or consumption (both immunologic and nonimmunologic causes), dilatational states (e.g., in red blood cell or exchange transfusion), or combined mechanisms (Table 117.2) [46]. ICH is one of the most serious bleeding complications of severe thrombocytopenia, usually occurring when the platelet count drops below 10,000–20,000/mm^3 [47]. Rarely, SAH may occur, in particular in the setting of underlying infection with human immunodeficiency virus or staphylococcal bacteremia, or in patients with superimposed hypercoagulability state such as in HIT [48–50].

Primary Immune Thrombocytopenia

Primary immune thrombocytopenia (PIT) or idiopathic thrombocytopenic purpura is an acquired autoimmune disorder characterized by an isolated decrease in circulating platelets in the absence of other conditions known to cause thrombocytopenia. In PIT, an abnormal autoantibody, usually immunoglobulin G with specificity for one or more platelet membrane glycoproteins, binds to circulating platelet membranes resulting in splenic sequestration and phagocytosis by mononuclear macrophages. The resulting shortened life span of platelets in the circulation, together with incomplete compensation by increased platelet production by bone marrow megakaryocytes, leads to decreased platelet count.

TABLE 117.2 Causes of Thrombocytopenia

Hereditary
   Glycoprotein abnormalities
      Bernard–Soulier disorder
      Glanzmann thrombasthenia
      Platelet-type von Willebrand disorder
      Wiskott–Aldrich syndrome
   Storage pool disease
      Gray platelet syndrome

Immune mediated
   Human immunodeficiency virus infection
   Connective tissues disorders
   Infections
   Inflammatory diseases
   Medications
   Posttransfusion

Platelet sequestration with congestive splenomegaly

Nonimmune platelet destruction
   Thrombotic thrombocytopenic purpura
   Disseminated intravascular coagulation
   Cardiopulmonary bypass
   Liver disease
   Infections
   Sepsis
   Uremia
   Pregnancy

Bone marrow dysfunction
   Infiltrative
   Myelodysplastic disorders
   Neoplastic
   Folic acid deficiency
   Iron deficiency
   Alcohol

Dilution
   Exchange transfusion
   Red blood cell transfusion

PIT is usually acute in children, and chronic in adult patients. Symptoms are mild and self-limiting, often restricted to purpura, petechia, minimal mucosal bleeding, or visceral hemorrhage [51]. ICH is an extremely rare cause of morbidity and mortality in PIT, usually occurring in children with platelet count below 10,000/mm³ [51–57]. Several cases of ischemic stroke in patients with PIT have been reported [58–61].

In the absence of specific laboratory testing, PIT is a diagnosis of exclusion after all other causes of thrombocytopenia are ruled out. Remission is usually spontaneous. In patients with bleeding complications, the first line of treatment includes administration of oral corticosteroids. IVIG, anti-D immune globulin, rituximab, and platelet transfusion may be used in conjunction with steroids in patients with severe bleeding or when a rapid increase in platelet count is necessary such as preoperatively. Splenectomy should be reserved to critical patients with severe thrombocytopenia or major bleeding, when other forms of therapeutic modalities have been ineffective [51].

Hereditary Platelet Dysfunction Disorders

Hereditary platelet dysfunction disorders (HPDDs) belong to a heterogeneous group of diseases associated with decreased platelet production, abnormal platelet morphology, and impaired platelet function. The degree of thrombocytopenia and functional abnormality of platelets determines the clinical manifestations [1]. Most HPDDs are associated with mild bleeding tendencies, usually mucocutaneous, and rarely visceral, hemorrhages. Severe bleeding complications are uncommon.

Glanzmann thrombasthenia is a rare hereditary platelet surface receptor disorder of GPIIb/IIIa, which results in faulty platelet aggregation and diminished clot retraction [62]. Spontaneous mucocutaneous hemorrhages are the main clinical presentation. Rare cases of ICH have been reported in the setting of traumatic injuries [63]. Recombinant activated factor VII appears to be safe and effective for the treatment of nonsurgical bleeds [62,64].

Bernard–Soulier syndrome (BSS) is a rare disorder characterized by thrombocytopenia <20,000/mm³, decreased platelet adhesion, abnormal prothrombin consumption, and low-surviving large platelets. It is associated with increasing risk of spontaneous severe bleeding complications, including ICH [1,65,66].

Platelet-type von Willebrand disorder is another platelet adhesion disorder characterized by enlarged platelets and defective platelet adhesion due to spontaneous binding of vWF to GPIbα and cleavage of vWF multimers [66,67].

Other rare platelets dysfunction conditions include storage pool disease, dense granule abnormalities, disorders of platelet secretion and signal transduction, defects in cytoskeletal regulation, and Myosin heavy chain 9-related diseases [1,66].

Diagnosis of HPDD is difficult, and specific tests are not readily available. Extensive and cumbersome diagnostic testing for platelet function and morphology is necessary to differentiate between various causes of platelets disorders. This includes platelet count and smear, full coagulation screening (prothrombin time, activated partial thromboplastin time, fibrinogen and thrombin time), bleeding time, PFA-100 measurements, platelets aggregation assays, platelet nucleotides content and release, platelet flow cytometry, and molecular analysis [1].

Management is directed at preventing bleeding complications. Caution should be used during surgical intervention and labor. The use of antithrombotic agents should be avoided. In most cases, bleeding is minor. Nasal packing is usually applied in patients with epistaxis. Platelet transfusions are indicated in patients with severe trauma-induced hemorrhage. Desmopressin may be used prophylactically to reduce bleeding tendency. Recombinant factor VIIa proved successful in stopping severe hemorrhage, especially in patients with GS and

BSS. Hematopoietic stem cell transplantation and gene therapy are promising investigational options [1].

Acquired Platelet Dysfunction

Platelet dysfunction commonly occurs in response to various stressors including surgical procedures, hepatic and renal insufficiency, systemic infections, underlying hematologic diseases, and following exposures to certain medications [68].

Drug-induced thrombocytopenia (DIT) is a relatively common condition occurring in approximately 20–25% of patients with systems illness (Table 117.3).

TABLE 117.3   Common Drugs Associated With Thrombocytopenia

Anesthetics

Antiplatelets
  Aspirin
  Thienopyridines
  GPIIb/IIa inhibitors
  Dipyridamole
  Cilostazol

Anticoagulants
  Direct thrombin inhibitors
  Factor Xa inhibitors
  Heparin
  Warfarin

Antihistamines

B-lactam antibiotics
  Cephalosporins
  Penicillins

Cardiovascular drugs
  Beta-blockers
  Calcium channel blockers

Chemotherapeutic agents

Herbal supplements
  Black dried fungus
  Feverfew
  Garlic
  Ginger
  Ginkgo biloba
  Ginseng
  Meadowsweet willow

Nonsteroidal antiinflammatory agents

Psychotropic drugs
  Phenothiazines
  Selective serotonin reuptake inhibitors
  Serotonin and norepinephrine reuptake inhibitors
  Tricyclic antidepressants

Steroids

Miscellaneous
  Alcohol
  Clofibrate
  Iodine contrast

A remarkable drop in platelet count usually occurs within 2–3 days of exposure to the agent [68,69]. DIT can occur as a consequence of bone marrow suppression and decreased platelet production, or accelerated platelet destruction via immune-mediated process [70]. The condition should be differentiated from other causes of thrombocytopenia, in particular PIT and HIT. DIT is reversible once the drug in question is discontinued. In severe bleeding, the administration of IVIG and PEX may be warranted. Platelet transfusion should be considered in patients with severe hemorrhage and organ failure [71–72].

CONCLUSION

Hereditary and acquired platelet disorders are associated with an increased risk of arterial and venous cerebral thrombosis and rarely ICH. Mechanisms of stroke in platelet disorders are diverse: they include inflammatory and autoimmune dysregulation, endothelial injury, and hypercoagulability state. Abnormal platelet count is common in the setting of various medical and surgical conditions. Careful monitoring and thought workup is necessary in patients with persistent platelet count abnormality in particular in the setting of bleeding and thromboembolic events.

References

[1] Michelson AD. Platelet function disorders. In: Arcuri R, Hann I, Owen S, editors. Pediatric hematology. New Jersey: Blackwell Publishing; 2007. p. 562–82.

[2] Andrews RK, Berndt MC. Platelet physiology and thrombosis. Thromb Res 2004;114:447–53.

[3] Nieswandt B, Pleines I, Bender M. Platelet adhesion and activation mechanisms in arterial thrombosis and ischaemic stroke. J Thromb Haemost 2011;9(Suppl. 1):92–104.

[4] Michiels JJ, Berneman Z, Van Bockstaele D, van der Planken M, De Raeve H, Schroyens W. Clinical and laboratory features, pathobiology of platelet-mediated thrombosis and bleeding complications, and the molecular etiology of essential thrombocythemia and polycythemia vera: therapeutic implications. Semin Thromb Hemost 2006;32:174–207.

[5] Griesshammer M, Gisslinger H, Heimpel H, Lengfelder E, Reiter A. Chronische myeloproliferative Erkrankungen. Leitlinien Deutsche Gesellschaft für Hämatologie und Onkologie (DGHO). 2006. www.dgho.de/cms/php?id=705.

[6] Artoni A, Bucciarelli P, Martinelli I. Cerebral thrombosis and myeloproliferative neoplasms. Curr Neurol Neurosci Rep 2014;14:496.

[7] Pósfai É, Marton I, Szőke A, Borbényi Z, Vécsei L, Csomor A, et al. Stroke in essential thrombocythemia. J Neurol Sci 2014;336:260–2.

[8] Kato Y, Hayashi T, Sehara Y, Deguchi I, Fukuoka T, Maruyama H, et al. Ischemic stroke with essential thrombocythemia: a case series. J Stroke Cerebrovasc Dis 2015;24:890–3.

[9] Landolfi R, Marchioli R, Kutti J, Gisslinger H, Tognoni G, Patrono C, et al. Efficacy and safety of low-dose aspirin in polycythemia vera. N Engl J Med 2004;350:114–24.

[10] Barbui T, Barosi G, Birgegård G, Cervantes F, Finazzi G, Griesshammer M. Philadelphia-negative classical myeloproliferative neoplasms: critical concepts and management recommendations from European LeukemiaNet. J Clin Oncol 2011;29:761–70.

[11] Budde U, van Genderen P. Acquired von Willebrand disease in patients with high platelet counts. Semin Thromb Hemost 1997;1997(23):425–31.

[12] Teofili L, Giona F, Torti L, Cenci T, Ricerca BM, Rumi C, et al. Hereditary thrombocytosis caused by MPLSer505Asn is associated with a high thrombotic risk, splenomegaly and progression to bone marrow fibrosis. Hematologica 2010; 95:65–70.

[13] Akins PT, Glenn S, Nemeth PM, Derdeyn CP. Carotid artery thrombus associated with severe iron-deficiency anemia and thrombocytosis. Stroke 1996;27:1002–5.

[14] Azab SF, Abdelsalam SM, Saleh SH, Elbehedy RM, Lotfy SM, Esh AM, et al. Iron deficiency anemia as a risk factor for cerebrovascular events in early childhood: a case-control study. Ann Hematol 2014;93:571–6.

[15] Belman AL, Roque CT, Ancona R, Anand AK, Davis RP. Cerebral venous thrombosis in a child with iron deficiency anemia and thrombocytosis. Stroke 1990;21:488–93.

[16] Scoditti U, Colonna F, Ludovico L, Trabattoni G. Mild thrombocytosis secondary to iron-deficiency anemia and stroke. Riv Neurol 1990;60:146–7.

[17] Naito H, Naka H, Kanaya Y, Yamazaki Y, Tokinobu H. Two cases of acute ischemic stroke associated with iron deficiency anemia due to bleeding from uterine fibroids in middle-aged women. Intern Med 2014;53:2533–7.

[18] Crowley JJ, Hannigan M, Daly K. Reactive thrombocytosis and stroke following cardiopulmonary bypass surgery: case report on three patients. Eur Heart J 1994;15:1144–6.

[19] Pommerening MJ, Rahbar E, Minei K, Holcomb JB, Wade CE, Schreiber MA, et al. Splenectomy is associated with hypercoagulable thrombelastography values and increased risk of thromboembolism. Surgery 2015;158:618–26.

[20] Terrell DR, Williams LA, Vesely SK, Lämmle B, Hovinga JA, George JN. The incidence of thrombotic thrombocytopenic purpura-hemolytic uremic syndrome: all patients, idiopathic patients, and patients with severe ADAMTS-13 deficiency. J Thromb Haemost 2005;3:1432–6.

[21] Török TJ, Holman RC, Chorba TL. Increasing mortality from thrombotic thrombocytopenic purpura in the United States – analysis of national mortality data, 1968–1991. Am J Hematol 1995;50:84–90.

[22] Zheng XL. ADAMTS13 and von Willebrand factor in thrombotic thrombocytopenic purpura. Annu Rev Med 2015;66:211–25.

[23] Ruggenenti P, Noris M, Remuzzi G. Thrombotic microangiopathy, hemolytic uremic syndrome, and thrombotic thrombocytopenic purpura. Kidney Int 2001;60:831–46.

[24] Scully M, Hunt BJ, Benjamin S, Liesner R, Rose P, Peyvandi F, et al. British Committee for Standards in Haematology. Guidelines on the diagnosis and management of thrombotic thrombocytopenic purpura and other thrombotic microangiopathies. Br J Haematol 2012;158:323–35.

[25] Khodor S, Castro M, McNamara C, Chaulagain CP. Clopidogrel-induced refractory thrombotic thrombocytopenic purpura successfully treated with rituximab. Hematol Oncol Stem Cell Ther 2015:S1658–3876.

[26] Mannucci PM, Cugno M. The complex differential diagnosis between thrombotic thrombocytopenic purpura and the atypical hemolytic uremic syndrome: laboratory weapons and their impact on treatment choice and monitoring. Thromb Res 2015;136:851–4.

[27] Greinacher A. Heparin-induced thrombocytopenia. N Engl J Med 2015;373:252–61.

[28] Bakchoul T, Greinacher A. Recent advances in the diagnosis and treatment of heparin-induced thrombocytopenia. Ther Adv Hematol 2012;3(4):237–51.

[29] Linkins LA, Dans AL, Moores LK, Bona R, Davidson BL, Schulman S, et al. Treatment and prevention of heparin-induced thrombocytopenia: antithrombotic therapy and prevention of thrombosis, 9th ed: American College of Chest Physicians Evidence-based Clinical Practice Guidelines. Chest 2012;141(2 Suppl):e495S–530S.

[30] Sharifi M, Bay C, Vajo Z, Freeman W, Sharifi M, Schwartz F. New oral anticoagulants in the treatment of heparin-induced thrombocytopenia. Throb Res 2015;135:607–9.

[31] Walenga JM, Prechel M, Jeske WP, Hoppensteadt D, Maddineni J, Iqbal O, et al. Rivaroxaban–an oral, direct Factor Xa inhibitor–has potential for the management of patients with heparin-induced thrombocytopenia. Br J Haematol 2008;143:92–9.

[32] Giannakopoulos B, Krilis SA. The pathogenesis of the antiphospholipid syndrome. N Engl J Med 2013;368:1033–44.

[33] Cuadrado MJ, Mujic F, Munoz E, Khamashta MA, Hughes GRV. Thrombocytopenia in the antiphospholipid syndrome. Ann Rheum Dis 1997;56:194–6.

[34] Feldmann E, Levine SR. Cerebrovascular disease with antiphospholipid antibodies: immune mechanisms, significance, and therapeutic options. Ann Neurol 1995;37(Suppl. 1):S114–30.

[35] Levine SR, Brey RL, Tilley BC, et al. Antiphospholipid antibodies and subsequent thrombo-occlusive events in patients with ischemic stroke. JAMA 2004;291:576–84.

[36] Miyakis S, Lockshin MD, Atsumi T, Branch DW, Brey RL, Cervera R, et al. International consensus statement on an update of the classification criteria for definite antiphospholipid syndrome (APS). J Thromb Haemost 2006;4:295–306.

[37] Cohen H, Doré CJ, Clawson S, Hunt BJ, Isenberg D, Khamashta M, et al. Rivaroxaban in antiphospholipid syndrome (RAPS) protocol: a prospective, randomized controlled phase II/III clinical trial of rivaroxaban versus warfarin in patients with thrombotic antiphospholipid syndrome, with or without SLE. Lupus 2015;24:1087–94.

[38] Betancur JF, Bonilla-Abadía F, Hormaza AA, Jaramillo FJ, Cañas CA, Tobón GJ. Direct oral anticoagulants in antiphospholipid syndrome: a real life case series. Lupus 2016;25. pii:0961203315624555. [Epub ahead of print].

[39] Woller SC, Stevens SM, Kaplan DA, Branch DW, Aston VT, Wilson EL, et al. Apixaban for the secondary prevention of thrombosis among patients with antiphospholipid syndrome: Study Rationale and Design (ASTRO-APS). Clin Appl Thromb Hemost 2015;22. [Epub ahead of print].

[40] Panichpisal K, Rozner E, Levine SR. The management of stroke in antiphospholipid syndrome. Curr Rheumatol Rep 2012;14:99–106.

[41] Holliday PL, Mammen E, Gilroy J. Sticky platelet syndrome and cerebral infarction in young adults. In: Presented at the Ninth International Joint Conference on Stroke and Cerebral. 1983. [Abstract] Phoenix, Arizona Circulation 1983 (Suppl.).

[42] Chaturvedi S, Dzieczkowski JS. Protein S deficiency, activated protein C resistance and sticky platelet syndrome in a young woman with bilateral strokes. Cerebrovasc Dis 1999;9:127–30.

[43] Kubisz P, Stasko J, Holly P. Sticky platelet syndrome. Semin Thromb Hemost 2013;39:674–83.

[44] Kubisz P, Stanciakova L, Stasko J, Dobrotova M, Skerenova M, Ivankova J, et al. Sticky platelet syndrome: an important cause of life-threatening thrombotic complications. Expert Rev Hematol 2016;9:21–35.

[45] Ziai WC, Torbey MT, Kickler TS, Oh S, Bhardwaj A, Wityk RJ. Platelet count and function in spontaneous intracerebral hemorrhage. J Stroke Cerebrovasc Dis 2003;12:201–6.

[46] Sekhon SS, Roy V. Thrombocytopenia in adults: a practical approach to evaluation and management. South Med J 2006; 99:491–8.

[47] Estcourt LJ, Stanworth SJ, Collett D, Murphy MF. Intracranial haemorrhage in thrombocytopenic haematology patients–a nested case-control study: the InCiTe study protocol. BMJ 2014;4:e004199.

[48] Landi PJ, Fernández Torrejón GA, Rausch S, Reunión Anatomoclínica. Subarachnoid hemorrhage and thrombocytopenia associated to staphylococcal bacteremia. Medicina (B Aires) 2014;74:497–501.

[49] Mehta BP, Sims JR, Baccin CE, Leslie-Mazwi TM, Ogilvy CS, Nogueira RG. Predictors and outcomes of suspected heparin-induced thrombocytopenia in subarachnoid hemorrhage patients. Interv Neurol 2014;2:160–8.

[50] Silvestrini M, Floris R, Tagliati M, Stanzione P, Sancesario G. Spontaneous subarachnoid hemorrhage in an HIV patient. Ital J Neurol Sci 1990;11:493–5.

[51] Cines DB, Blanchette VS. Immune thrombocytopenic purpura. N Engl J Med 2002;346:995–1008.

[52] Butros LJ, Bussel JB. Intracranial hemorrhage in immune thrombocytopenic purpura: a retrospective analysis. J Pediatr Hematol Oncol 2003;25:660–4.

[53] El Koraïchi A, Mounir K, El Haddoury M, El Kettani SE. Cerebral haemorrhage: an unusual complication of idiopathic thrombocytopenic purpura. Ann Fr Anesth Reanim 2011;30:92.

[54] Krivit W, Tate D, White JG, Robison LL. Idiopathic thrombocytopenic purpura and intracranial hemorrhage. Pediatrics 1981;67:570–1.

[55] Psaila B, Petrovic A, Page LK, Menell J, Schonholz M, Bussel JB. Intracranial hemorrhage (ICH) in children with immune thrombocytopenia (ITP): study of 40 cases. Blood 2009; 114:4777–83.

[56] Ranger A, Szymczak A, Fraser D, Salvadori M, Jardine L. Bilateral decompressive craniectomy for refractory intracranial hypertension in a child with severe ITP-related intracerebral haemorrhage. Pediatr Neurosurg 2009;45:390–5.

[57] Roganovic J, Kalinyak K. Intracranial hemorrhage in children with idiopathic thrombocytopenic purpura. Pediatr Int 2006; 48:517.

[58] Mahawish K, Pocock N, Mangarai S, Sharma A. Cerebral infarction in idiopathic thrombocytopenic purpura: a case report. BMJ Case Rep 2009;2009.

[59] Otsuki T, Funakawa T, Sugihara T, Kanzaki A, Wada H, Inoue T, et al. Multiple cerebral infarctions in a patient with refractory idiopathic thrombocytopenic purpura. J Intern Med 1997;241:249–52.

[60] Theeler BJ, Ney JP. A patient with idiopathic thrombocytopenic purpura presenting with an acute ischemic stroke. J Stroke Cerebrovasc Dis 2008;17:244–5.

[61] Rhee HY, Choi HY, Kim SB, Shin WC. Recurrent ischemic stroke in a patient with idiopathic thrombocytopenic purpura. J Thromb Thrombolysis 2010;30:229–326.

[62] Solh T, Botsford A, Solh M. Glanzmann's thrombasthenia: pathogenesis, diagnosis, and current and emerging treatment options. J Blood Med July 08, 2015;6:219–27.

[63] Vigren P, Ström JO, Petrini P, Callander M, Theodorsson A. Treatment of spontaneous intracerebral haemorrhage in Glanzmann's thrombasthenia. Haemophilia 2012;18:e381–3.

[64] Di Minno G, Zotz RB, d'Oiron R, Bindslev N, Di Minno MN, Poon MC, et al. The international, prospective Glanzmann Thrombasthenia Registry: treatment modalities and outcomes of non-surgical bleeding episodes in patients with Glanzmann thrombasthenia. Haematologica 2015;100:1031–7.

[65] Pham A, Wang J. Bernard-Soulier syndrome: an inherited platelet disorder. Arch Pathol Lab Med 2007;131:1834–6.

[66] Balduini CL, Savoia A. Genetics of familial forms of thrombocytopenia. Hum Genet 2012;131:1821–32.

[67] Othman M. Platelet-type von Willebrand disease: a rare, often misdiagnosed and underdiagnosed bleeding disorder. Semin Thromb Hemost 2011;37:464–9.

[68] Bonfigliio MF, Traeger SM, Kier KL, Martin BR, Hulisz DT, Verbeck SR. Thrombocytopenia in intensive care patients: a comprehensive analysis of risk factors in 314 patients. Ann Pharmacother 1995;29:835–40.

[69] Shalansky SJ, Verma AK, Levine M, Spinelli JJ, Dodek PM. Risk markers for thrombocytopenia in critically ill patients: a prospective analysis. Pharmacotherapy 2002;22:803–13.

[70] Visentin GP, Liu CY. Drug induced thrombocytopenia. Hematol Oncol Clin North Am 2007;21:685.

[71] Drews RE. Critical issues in hematology: anemia, thrombocytopenia, coagulopathy, and blood product transfusions in critically ill patients. Clin Chest Med 2003;24:607–22.

[72] Nguyen TC, Stegmayr B, Busund R, Bunchman TE, Carcillo JA. Plasma therapies in thrombotic syndromes. Int J Artif Organs 2005;28:459–65.

CHAPTER

118

Stroke and Sickle Cell Disease

A.P. Ostendorf[1], E.S. Roach[1], R.J. Adams[2]

[1]Ohio State College of Medicine, Columbus, OH, United States; [2]Medical University of South Carolina, Charleston, SC, United States

INTRODUCTION

Sickle cell disease (SCD) is a powerful risk factor for stroke. Individuals with SCD are at risk for hemolytic anemia, painful vaso-occlusive crises, infection, and cerebrovascular disorders. Those with sickle cell anemia (SCA), defined as homozygous hemoglobin S or hemoglobin S-beta thalassemia, have an even greater risk for stroke beginning in early childhood, as up to 10% will have stroke without primary prevention. Red blood cells sickle under conditions of low blood oxygen tension due to a missense mutation in the beta hemoglobin protein substituting valine in place of glutamic acid. This causes a decrease in hemoglobin solubility under low oxygen tension and subsequent crystallization, distorting the red blood cell membrane. The sickled cells increase blood viscosity, enhance cellular adhesion, and cause vascular endothelial injury. Therefore, individuals with SCD have increased risk for cerebrovascular injuries including silent cerebral infarctions, overt ischemic stroke, and intracranial hemorrhage among other neurological complications. Prompt identification of individuals at risk for cerebrovascular complications of SCD, diagnosis and treatment of acute stroke, and secondary prevention strategies are crucial to limiting morbidity and mortality in individuals with SCD.

SILENT CEREBRAL INFARCTIONS

Silent cerebral infarcts, those that are not associated with abnormalities on neurological exam, occur in 30–35% of children with SCA and are likely secondary to acute hemodynamic instability in the setting of chronic anemia (Fig. 118.1A). Children with SCA and silent infarcts have an increased risk for overt ischemic stroke and neurocognitive deficits. The Silent Cerebral Infarct Multi-Center Clinical Trial (SIT) was a large, randomized trial of regular blood transfusions in children with SCA and silent or overt stroke that demonstrated a strong protective effect (relative risk reduction 58%) in decreasing the incidence of overt stroke or enlarged silent cerebral infarct [1]. Although screening neuroimaging should not be routinely performed [2], regular transfusion therapy is beneficial after identifying a child with silent cerebral infarcts in order to decrease their future risk for overt cerebral infarction.

OVERT CEREBRAL ISCHEMIC INFARCTION

Prevalence and Pathophysiology

Cerebral infarction is the most recognizable neurological complication in SCD patients (Fig. 118.1B). Infarction accounts for 75% of the strokes in SCD and is the most common stroke type in children. The average age for the first recognized cerebral infarct is 7.7 years with most occurring between ages 5 and 15 years. Markers of risk for cerebrovascular disease have been difficult to determine and in the majority of cases, no predisposing factors are recognized.

Large vessel occlusive disease occurs in the proximal anterior cranial vessels, most commonly in the intracranial portion of the internal carotid artery (ICA) (Fig. 118.1C) beyond the origin of the ophthalmic artery. This vasculopathy may be due to repeated trauma from sickled cells in areas of flow turbulence. Sickled cells have a higher viscosity and an increased adherence to the vascular endothelium, causing rheological injury, direct mechanical trauma, and endothelial damage leading to thrombus formation. Reparative processes may also contribute to pathology through smooth muscle hypertrophy, vessel wall thickening, and stenosis.

Chronic anemia causes increased cerebral blood flow (CBF) due to arteriolar vasodilation. This dilation

Primer on Cerebrovascular Diseases, Second Edition
http://dx.doi.org/10.1016/B978-0-12-803058-5.00118-1

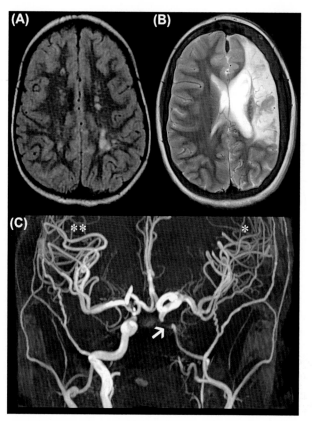

FIGURE 118.1 Brain MRI may reveal chronic silent cerebral infarcts (panel A; FLAIR sequence) or chronic overt cerebral ischemic infarcts (panel B; T2 sequence) in individuals with SCD. Large vessel occlusive disease (*arrow*) and subsequent development of moyamoya syndrome (* and **) may be visualized on MRA (panel C).

may alter the vasomotor regulatory responses to CBF decreases during arterial occlusion or decrease in hematocrit. The latter explains why an acute decrease in hematocrit or other exacerbations of SCD are at times associated with cerebral infarction.

Primary Prevention

Presymptomatic screening and treatment is successful in preventing stroke in individuals with SCD identified by transcranial Doppler (TCD) ultrasonography, a noninvasive method of measuring direction and velocity of blood flow in large intracranial vessels. Blood velocity is directly related to CBF and inversely to vessel diameter, thus blood velocity can be increased by 2 to 3 times normal in patients with severe arterial narrowing (Fig. 118.2). A pivotal study revealed that increased CBF velocities in large intracranial vessels measured with TCD is a reliable biomarker for elevated stroke risk in children with SCA, and intermittent screening is necessary to monitor those children who develop higher risk for ischemic infarct [3]. The validation of this noninvasive screening test for risk stratifying individuals with SCD facilitated the design and completion of several key clinical trials. The Stroke

Prevention Trial in Sickle Cell Anemia (STOP) demonstrated that ischemic infarcts can be prevented in children with increased CBF velocities on TCD through the use of intermittent blood transfusions (relative risk reduction 92%), although STOP2 demonstrated the protective effect is lost if transfusions are discontinued [4,5].

Based on these data and others, the 2014 National Heart, Lung, and Blood Institute (NHLBI) recommendations included annual TCD screening for children with SCA from age 2 years until age 16 and referral for chronic transfusion therapy in individuals with conditional (170–199 cm/s) or elevated (>200 cm/s) CBF velocities [2]. Chronic transfusion therapy results in lower risk for stroke, improved growth, fewer pain crises, and fewer episodes of acute chest syndrome [2]. However, this treatment has profound implications for individuals with SCD and elevated TCD velocities. It is resource-intensive and limited in availability. It also confers risk of adverse effects, such as alloimmunization, hemolytic transfusion reactions, and, most importantly, iron overload. Ferritin levels must be monitored and iron chelation with desferrioxamine or deferasirox instituted if iron overload occurs.

The burden of chronic transfusion therapy has prompted many researchers and clinicians to search for alternative strategies to decrease the rate of complications from SCD. Hydroxyurea, also called hydroxycarbamide, is a ribonucleotide reductase inhibitor that increases fetal hemoglobin (HbF) in individuals with SCD, inhibits several mechanisms of vaso-occlusion, and increases nitric oxide release. Guidelines from the NHBLI strongly recommend it in individuals with SCA and frequent pain crises, acute chest syndrome, and chronic anemia, and in infants and older children with SCA [2]. Data from the TCD With Transfusions Changing to Hydroxyurea (TWiTCH) noninferiority trial demonstrated children with elevated TCD velocities, and no evidence of severe vasculopathy on MRA (magnetic resonance angiography) can switch to hydroxyurea maintenance after 1 year of chronic transfusion therapy without elevated risk of stroke [6]. In summary, an evidence-based approach to primary stroke prevention in individuals with elevated TCD velocities should rely on chronic transfusion therapy for 1 year, followed by transition to hydroxyurea in the absence of significant intracranial vasculopathy.

Adults with SCD may benefit from general risk factor reduction similar to those strategies employed in the general population. These include management of hypertension, dyslipidemia, diabetes, obesity, physical inactivity, sleep apnea, nutrition, carotid disease, atrial fibrillation, nutrition, intracranial atherosclerosis, intracardiac thrombus, valvular heart disease, smoking cessation, or comorbid hypercoagulable state [7]. TCD screening and chronic transfusion therapy were performed in children and their efficacy in adults is uncertain.

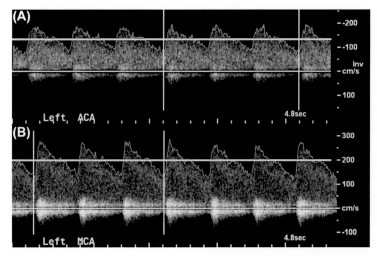

FIGURE 118.2 Transcranial Doppler flow velocities in the left ACA and left MCA in an individual with SCD. The yellow lines demarcate flow velocities of 135 cm/s (normal) in the ACA and 200 cm/s in the MCA (elevated).

Clinical Presentation

About 10% of individuals with SCA will have ischemic infarcts despite optimal primary prevention [2]. Individuals with ischemic infarcts may present with focal motor deficits, numbness, visual field deficits, dysarthria, aphasia, or ataxia. Sensory complaints are less prominent, probably due to the young age of the patients. Severe headache, generalized seizures, and loss of consciousness, when they occur, are more often due to intracerebral hemorrhage. Focal seizures occur in 10–33% of cases and are presumptive evidence of the onset of an ischemic event. Transient ischemic attacks (TIAs) preceding stroke have been recognized in less than 10% of cases. The majority (80%) of children with significant hemiparesis have large artery occlusive disease in either the internal carotid or the middle and anterior cerebral arteries.

Acute Evaluation

Prompt evaluation of cerebrovascular symptoms in a young individual with SCD presenting with neurological symptoms is critical to reduce associated morbidity and mortality. Meningitis, for which SCD children are at high risk, should also be considered. A noncontrast cranial computed tomography (CT) scan of the head is the first step and is an effective screening tool sensitive for ruling out subdural hematoma and intraparenchymal hemorrhage. However, acute ischemic infarct may not be associated with CT abnormalities. Consensus recommendations for the next step in urgent evaluation are magnetic resonance imaging (MRI) and MRA, including when subtle or mild signs or symptoms suggest TIA [2]. Some larger centers have developed urgent stroke imaging protocols with sequences specific for MRI and MRA

in individuals with SCD, thus head CT and associated radiation may be avoided if MR techniques are immediately available.

In some patients in whom complex vascular findings are suspected, conventional cerebral angiography may be necessary. Angiography carries a higher inherent risk in individuals with SCD. Nevertheless, experienced clinicians can safely perform angiography once the sickle hemoglobin (HbS) is lowered to less than 30% by hydration and transfusion.

The common infarction patterns seen in SCD are wedge-shaped infarcts in a large vessel territory (Fig. 118.3A and B) or watershed infarctions at the interface between the middle cerebral artery (MCA) and anterior cerebral artery (ACA). Angiography may show focal areas of stenosis or occlusion in the distal ICA and proximal middle (Fig. 118.3C) and anterior cerebral arteries. The basilar system is rarely involved and the extracranial carotid artery usually undergoes only secondary changes reflecting distal intracranial. Also seen are variable irregularities of the more distal vessels and recruitment of collateral pathways from the circle of Willis, leptomeningeal, and extracranial vessels. In up to 30% of SCD patients with abnormal angiograms, the collateral formation may be extensive enough to be defined as moyamoya syndrome.

Acute Treatment

Consensus guidelines for the acute treatment of ischemic stroke in children with SCA recommend exchange transfusion [2]. In individuals unable to receive immediate exchange transfusion, simple transfusion with hydration aimed at reducing the percentage of HbS to less than 30% may be considered. Treatment should be carried out in an intensive care setting while monitoring

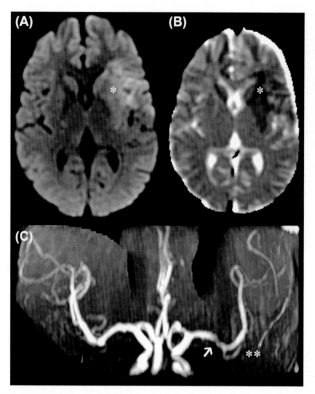

FIGURE 118.3 Acute overt cerebral ischemic infarct in an individual with SCD presenting with aphasia and right hemiparesis. Diffusion restriction is present in the territory of the superior division of the left MCA (*) as shown with diffusion weighted imaging (panel A) and apparent diffusion coefficient (panel B) sequences. Vascular imaging with MRA revealed partial narrowing in the proximal left MCA (*arrow*) and more severe occlusion in the M2 segment.

for increased intracranial pressure or stroke extension. The prognosis for survival can be good and survivors may recover with few motor deficits. Large or multiple infarcts, however, cause residual neurological impairment.

In adults with SCD, little evidence exists that is specific for adults with SCD outside of the general treatment strategies for tissue plasminogen activator (tPA) and supportive management discussed elsewhere in this text.

Secondary Prevention

Ischemic stroke recurs in 46–90% of children with SCD untreated with chronic transfusions or hematopoietic stem cell transplantation. Consensus guidelines for the secondary prevention of ischemic stroke in children or adults with SCA recommend monthly simple or exchange transfusions; or hydroxyurea therapy if a transfusion program is not possible [2]. The most effective long-term therapy is chronic transfusion therapy, which reduces recurrence risk to about 10%. Transient ischemic events while on transfusion therapy occur almost twice as commonly as stroke, but their relationship to the risk of eventual cerebral infarction remains

to be determined. Simple transfusions are likely less effective than exchange transfusion in secondary stroke prevention, and partial exchanges are preferable as they lessen the overall iron load. Transfusions and iron chelation with a target HbS of <30% is considered by most experts as the current goal of therapy [2]. If simple transfusions are given, the hematocrit should not be allowed to exceed 35mg/dL since increased blood viscosity could further impair CBF.

Observational cohort studies suggest that hydroxyurea used in addition to chronic transfusion therapy is likely more effective in reducing stroke recurrence. The Stroke With Transfusions Changing to Hydroxyurea (SWiTCH) study found hydroxyurea alone inferior to chronic transfusions in secondary stroke prevention [8]. However, hydroxyurea should be prescribed for individuals with a history of stroke who are unable to undergo chronic transfusion therapy. Recommended dosing for hydroxyurea in infants and children is 20 and 15mg/kg/day for adults [2]. Furthermore, adults with SCD and history of ischemic stroke likely benefit from secondary prevention strategies employed in those without SCD, such as risk factor management and antiplatelet therapy discussed elsewhere in this text [7].

Curative treatment for SCD is allogeneic bone marrow transplantation in children. Indications for transplantation include stroke and frequent, severe pain crises. Projected event-free survival of symptomatic patients after transplantation is as high as 86%, although transplanted individuals may have an increased risk for intracranial hemorrhage.

INTRACRANIAL HEMORRHAGE

Clinical Presentation and Pathophysiology

Intracranial hemorrhage occurs in children and young adults with SCD at a mean age of 25 years and is more frequent in individuals with a prior cerebral infarction. Hemorrhage typically presents with headache, seizures, altered mental status, and focal neurological signs. It may occur as a primary event due to rupture of a collateral vessel or aneurysm, or hemorrhage may develop as in a brain region previously injured with an ischemic infarct.

Individuals with SCD are at higher risk for subarachnoid hemorrhage (3–4%) than the general population (1–2%) and may also develop spontaneous epidural bleeding. Aneurysms are more likely to rupture at a younger age or smaller size, and multiple aneurysms are common in the same individual. This predisposition for vessel rupture and aneurysm formation is likely secondary to chronic arterial wall damage due to physical trauma from sickled cells. Chronic hemodynamic stresses may instigate degenerative processes and fractionation of the arterial wall elastic lamina and medial atrophy. Consequently, arterial wall dilation, aneurysm formation, or simply a break in the vessel wall may all lead to subarachnoid hemorrhage. Spontaneous epidural hematoma may rarely occur. While epidural hemorrhages typically occur in the setting of trauma, individuals with SCD are more likely to have a spontaneous event thought to be secondary to skull infarction or changes in the bone secondary to medullary hematopoiesis.

Acute Evaluation and Treatment

Urgent neuroimaging should be obtained when an individual with SCD presents with signs or symptoms consistent with intracranial hemorrhage as described earlier. The most commonly employed modality remains a noncontrast cranial CT. Some children's hospitals have developed rapid, tailored MRI protocols for individuals with SCD and suspected neurovascular injury that typically include gradient echo sequences and arterial angiography. If intracranial hemorrhage is present and vascular imaging has not been obtained or is nondiagnostic, conventional angiography should be performed to detect surgically correctable aneurysms. If neuroimaging reveals blood in the third ventricle, the patient should be observed for delayed deterioration secondary to acute hydrocephalus requiring prompt CSF diversion. Patients should also be evaluated for coagulopathy, illicit drug use, and other important differential diagnoses of stroke in the young, although such problems are no more common in individuals with SCD than the general population.

At the time of hemorrhage, patients should receive drugs to reduce cerebral vasospasm, and increased intracranial pressure must be managed. Individuals with SCD should be transfused to a HbS less than 30% prior to angiography or surgery due to an increased periprocedural complication risk. The effect of chronic transfusion therapy on risk for recurrent hemorrhage is unknown. We typically recommend that transfusion therapy be maintained if the angiogram shows aneurysms not amenable to repair, large artery stenosis, or moyamoya.

OTHER NEUROLOGICAL PROBLEMS

Recurrent headaches occur in over one-fourth of individuals with SCD without apparent cause. Severe headache with nausea and vomiting, especially including depression of consciousness, should raise concern for intracranial hemorrhage. The pathophysiology is likely related to increased CBF, partly due to the anemia. Individuals receiving chronic transfusion therapy may report their headaches are less frequent in the immediate posttransfusion period. Headaches are more frequent with lower hemoglobin or frequent pain events, but do not appear to be associated with silent cerebral infarcts [9]. It remains to be determined if mild to moderate headaches are associated with a long-term increased risk for overt stroke. Management should follow standard approaches to migraine or nonmigraine headache, and individuals may benefit from chronic headache prophylaxis.

Individuals with SCD and isolated silent cerebral infarcts or overt stroke may have significant cognitive impairment. Children with left hemisphere cerebral infarcts have global decline on both performance and verbal IQ. Children with right hemisphere infarcts show decline only in performance IQ. Adults with SCA have poorer cognitive performance associated with increasing degree of anemia and older age, although performance IQ was independent of previous infarcts [10]. Children with SCD and stroke or with unexplained learning problems should be evaluated with neuropsychological testing to assess their educational needs and vocational potential.

CONCLUSION

Individuals with SCD have an elevated risk for neurovascular injury, headaches, and neurocognitive deficits. Urgent cranial CT or MRI should be obtained in all patients with suspected stroke or hemorrhage. The mainstay of treatment for acute ischemic infarct is transfusion therapy, and patients with hemorrhage may require surgical treatment for aneurysm or thrombus. Stroke can be prevented in at-risk individuals through the use of chronic transfusion therapy transitioning to hydroxyurea therapy for selected individuals. Screening is age-dependent and typically relies on TCD, MRI, and MRA to identify SCD-associated vasculopathy. Knowledgeable practitioners can decrease the significant morbidity and mortality associated with SCD through prompt treatment of acute neurological injuries and techniques for both primary and secondary prevention. Finally, although SCD is a strong risk factor for stroke, neurologists should evaluate for other potential etiologies, especially in adults.

References

[1] DeBaun MR, Gordon M, McKinstry RC, Noetzel MJ, White DA, Sarnaik SA, et al. Controlled trial of transfusions for silent cerebral infarcts in sickle cell anemia. N Engl J Med 2014;371(8):699–710.

[2] U.S. Department of Health and Human Services, National Institutes of Health, National Heart, Lung, and Blood Institute. Evidence-based management of sickle cell disease: expert panel report, 2014. 2014.

[3] Adams R, McKie V, Nichols F, Carl E, Zhang DL, McKie K, et al. The use of transcranial ultrasonography to predict stroke in sickle cell disease. N Engl J Med 1992;326(9):605–10.

[4] Adams RJ, McKie VC, Hsu L, Files B, Vichinsky E, Pegelow C, et al. Prevention of a first stroke by transfusions in children with sickle cell anemia and abnormal results on transcranial Doppler ultrasonography. N Engl J Med 1998;339(1):5–11.

[5] Adams RJ, Brambilla D, Optimizing Primary Stroke Prevention in Sickle Cell Anemia (STOP 2) Trial Investigators. Discontinuing prophylactic transfusions used to prevent stroke in sickle cell disease. N Engl J Med 2005;353(26):2769–78.

[6] Ware RE, Davis BR, Schultz WH, Brown RC, Aygun B, Sarnaik S, et al. Hydroxycarbamide versus chronic transfusion for maintenance of transcranial doppler flow velocities in children with sickle cell anaemia-TCD with transfusions changing to hydroxyurea (TWiTCH): a multicentre, open-label, phase 3, non-inferiority trial. Lancet 2016;(10019):661–70.

[7] Kernan WN, Ovbiagele B, Black HR, Bravata DM, Chimowitz MI, Ezekowitz MD, et al. Guidelines for the prevention of stroke in patients with stroke and transient ischemic attack: a guideline for healthcare professionals from the American Heart Association/American Stroke Association. Stroke J Cereb Circ 2014;45(7):2160–236.

[8] Ware RE, Helms RW, SWiTCH Investigators. Stroke with transfusions changing to hydroxyurea (SWiTCH). Blood 2012;119(17):3925–32.

[9] Dowling MM, Noetzel MJ, Rodeghier MJ, Quinn CT, Hirtz DG, Ichord RN, et al. Headache and migraine in children with sickle cell disease are associated with lower hemoglobin and higher pain event rates but not silent cerebral infarction. J Pediatr 2014;164(5):1175–80.e1.

[10] Vichinsky EP, Neumayr LD, Gold JI, Weiner MW, Rule RR, Truran D, et al. Neuropsychological dysfunction and neuroimaging abnormalities in neurologically intact adults with sickle cell anemia. JAMA 2010;303(18):1823–31.

C H A P T E R

119

Pregnancy, Hormonal Contraception, and Postmenopausal Estrogen Replacement Therapy

C. Lamy, J.-L. Mas

Paris Descartes University, Paris, France

There is an increasing emphasis on gender-specific issues in stroke, such as the role of endogenous and exogenous female hormones. This chapter reviews available data on the risk of stroke related to pregnancy, hormonal contraception, and postmenopausal hormone replacement therapy (HRT).

STROKE IN PREGNANCY

Epidemiology

The reported incidence of stroke during pregnancy and the puerperium ranges from 4 to 34 per 100,000 deliveries or pregnancies [1–3]. The US Nationwide Inpatient Sample (NIS; a 20% stratified sample of all discharges in the United States) identified 2850 pregnancies complicated by stroke in the years 2000–01, for a rate of 34.2 per 100,000 deliveries [2]. The mortality rate was 1.4 per 100,000. Pregnancy-related intracerebral hemorrhage (ICH) was associated with the highest morbidity and mortality. Recent analysis of data from the NIS demonstrated that between 1994–95 and 2006–07, the rates of antenatal and postpartum hospitalizations for all types of strokes increased by 47% and 83%, respectively [4]. This increase was thought to be related to the increase in hypertensive disorders, obesity, and heart disease.

The 6 weeks postpartum and, particularly, the several days around delivery are times of increased risk for ischemic and hemorrhagic stroke as well as for cerebral venous thrombosis (CVT) [2,5]. A nationwide Swedish cohort study found that cerebral infarction was 33.8 times (95% CI: 10.5–84.0) more likely to develop in the 3 days surrounding delivery (defined as 1 day before and 2 days after delivery), and 8.3 times (95% CI: 4.4–14.8) more likely in the subsequent 6 weeks after delivery. Antenatally, the risk was negligible (OR, 2.2; 95% CI, 0.8–4) [6].

Primer on Cerebrovascular Diseases, Second Edition
http://dx.doi.org/10.1016/B978-0-12-803058-5.00119-3

Risk Factors

Pregnancy and delivery bring substantial changes in hemodynamic demands, hemostasis, vessel wall, and blood volume, which peak by the end of pregnancy and may take up to several weeks to resolve after delivery. Advanced maternal age and hypertension, whether preexisting, gestational, or secondary to preeclampsia/eclampsia, are the most important risk factors for stroke in pregnancy [2]. In the NIS sample analysis, the incidence of stroke increased from 35.3 per 100,000 in women aged 30–34 years to 58.1 per 100,000 in those aged 35–39 years [2]. Compared to women without hypertension, women with hypertension complicating pregnancy are six- to ninefold more likely to have a stroke. Several other risk factors have been described, including heart disease, obesity, preexisting vascular disease, collagen vascular disease, diabetes mellitus, renal disease, multiple pregnancy, higher parity, postpartum infection or hemorrhage and transfusion, fluid, electrolyte, and acid–base disorders [2]. Cesarean delivery has also been associated with postpartum stroke although a causal relationship has not been clearly demonstrated.

Diagnosis and Causes

The diagnostic approach to suspected stroke in a pregnant woman should not be significantly different than that for a nonpregnant young woman [1,3]. Pregnancy does not contraindicate radiological procedures; MRI is the imaging study of choice during pregnancy, but gadolinium contrast should be avoided because it crosses the placenta and has unknown effects on development. Head computed tomography (CT) and catheter cerebral angiography, if necessary, are considered to be reasonably safe in pregnancy if the uterus is shielded.

Strokes related to various etiologies have been reported in pregnancy and the puerperium, the most important causes being preeclampsia and eclampsia, cardioembolism, postpartum cerebral angiopathy, CVT, and rupture of a cerebrovascular malformation [1,3].

Preeclampsia/eclampsia is a complex multisystem disorder, characterized by widespread endothelial dysfunction and vasospasm. Preeclampsia complicates about 5% of pregnancies in the United States. It is defined as progressively worsening high blood pressure that occurs in the setting of proteinuria (≥300 mg of protein in a 24-h urine specimen). Hemolysis, elevated liver enzymes, and low platelets (HELLP) syndrome is possible. Preeclampsia usually occurs during the third trimester of pregnancy. Eclampsia is defined as the occurrence of seizures in a patient with preeclampsia. Postpartum eclampsia is rare and occurs mainly within 48 h after delivery, but delayed eclampsia (occurring within several weeks after delivery) is possible.

The typical presentation of eclampsia consists of headache, mental status changes, and seizures. Visual troubles are frequent. The most characteristic MRI pattern is the presence of relatively symmetric hypersignals on T2 and FLAIR sequences involving the subcortical white matter of posterior regions. Lesions are usually hypointense or isointense on diffusion-weighted (DW) sequences, with an increased ADC, indicating reversible vasogenic edema, corresponding to posterior reversible encephalopathy syndrome (PRES) [7] (Fig. 119 1). A reversible angiopathy with focal vasoconstriction, as well as even a string-of-beads appearance, consistent with vasospasm (see reversible cerebral vasoconstriction syndrome), has occasionally been demonstrated at catheter angiography, MR angiography, or transcranial Doppler (TCD). Eclamptic patients may also have ischemic stroke or intracranial hemorrhage. Intracranial hemorrhage is associated to a poor maternal and fetal prognosis.

In addition to pharmacological control of hypertension, the efficacy of magnesium sulfate for seizure prophylaxis has been well demonstrated in randomized trials to decrease the risk of stroke in women with eclampsia. A systematic study and meta-analysis has shown that women who have had preeclampsia have an increased risk of cardiovascular disease, including hypertension, ischemic heart disease, stroke, and venous thromboembolism in later life [6]. The mechanism underlying this association remains to be defined.

Reversible cerebral vasoconstriction syndrome (RCVS) has been associated with various conditions, such as postpartum or exposure to various vasoactive substances. RCVS is characterized by recurrent thunderclap headaches over a few days to 2 weeks after delivery, with or without additional neurological symptoms. Unlike eclampsia, postpartum RCVS is limited to the nervous system, and most patients have a history of an uncomplicated pregnancy and delivery. Cerebrospinal fluid (CSF) is normal or shows a moderate pleocytosis. Arterial imaging shows multiple segmental narrowing of medium-sized cerebral arteries, which resolves on subsequent examination within few weeks, suggesting transient vasospasm. Rapid spontaneous clinical recovery occurs in the majority of cases. However, the condition is not always benign. The potential complications of RCVS include cortical subarachnoid hemorrhage, intracerebral hemorrhage, posterior reversible encephalopathy syndrome, mainly during the first week postpartum, and ischemic strokes over watershed zones, mainly during the second week. In numerous reported cases, vasoactive drugs had been given in the preceding days or hours. The efficacy of nimodipine has been suggested in aborting thunderclap headache but not

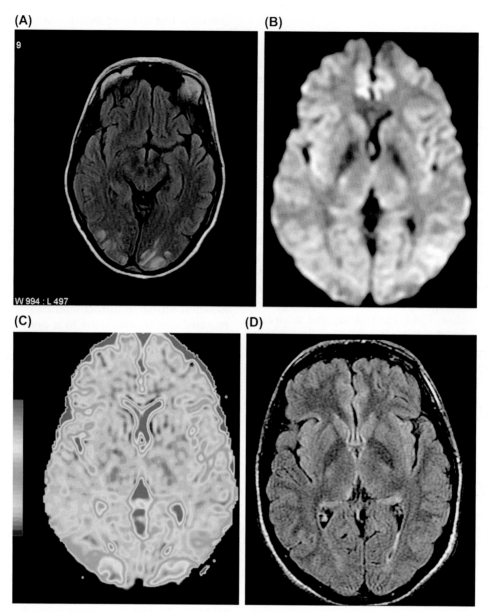

FIGURE 119.1   Eclampsia in a 32-year-old woman. Severe headaches, generalized seizures, and blurred vision. Complete clinical recovery after delivery. (A) MRI showing bilateral high signal in the occipital regions on axial FLAIR sequences. (B) MRI showing normal signal in the same regions on diffusion-weighted sequences. (C) ADC map showing an increased ADC in the posterior regions. (D) Two-weeks follow-up MRI. Complete regression of lesions.

been proven against the hemorrhagic and ischemic complications.

Cardioembolism is common in stroke during pregnancy and the puerperium, especially in developing countries. Women with heart disease, particularly those with mechanical valve prosthesis, have an increased risk of maternal and fetal complications. Peripartum cardiomyopathy is an uncommon cause of pregnancy-related stroke and is diagnosed by symptoms of heart failure occurring during the peripartum period in previously healthy women and the absence of a determined cause of cardiomyopathy.

Cerebral venous thrombosis: Pregnancy and the puerperium period are times of increased risk for venous thrombosis for women, including CVT. This higher risk is attributed to the prothrombotic state of pregnancy and the puerperium. The incidence of CVT during pregnancy and the puerperium is estimated at 1 in 2500 deliveries to 1 in 10,000 deliveries in western countries [1,3]. CVT often presents with a severe headache with symptoms of increased intracranial pressure. Most pregnancy-related CVT occurs in the third trimester or the first four postpartum weeks. Puerperal infections and dehydration may contribute to the high frequency of CVT in developing countries. Testing for prothrombotic conditions can be beneficial for the management of patients.

Intracranial hemorrhage is a serious complication during pregnancy and the puerperium and is associated with a substantial maternal mortality, contributing to more than 5–12% of all maternal deaths. The main causes are eclampsia and rupture of a cerebral vascular malformation [1,3]. Whether pregnancy increases the risk of rupture of existing cerebrovascular malformations is controversial. A population study revealed that the risk of aneurysmal subarachnoid hemorrhage is not increased during pregnancy, delivery, and the puerperium; the relative risk was 0.6 (95% CI: 0.3–1.3) and the standardized incidence ratio was 0.6 (95% CI: 0.2–1.1). A retrospective review of 979 Chinese women with intracranial arteriovenous malformation (AVM) also showed no increased risk of hemorrhage of AVM during pregnancy and the puerperium (OR for rupture of AVM during pregnancy compared to the control period: 0.71; 95% CI: 0.61–0.82) [8].

Management

Management of patients with pregnancy-related stroke is largely the same as that of nonpregnant patients, with more consideration of maternal and fetal risks [1,3]. Choice of therapy for secondary prevention depends on the etiology of the stroke. After the first trimester, there is substantial evidence that low-dose aspirin (50–150 mg/day) is safe for the fetus and newborn infant. Data on the safety of aspirin during the first trimester is more limited. Because aspirin crosses the placenta, its use during first-trimester organogenesis could increase the risk of birth defects, but potential benefits may warrant the use of the drug in pregnant women. Alternative antiplatelet agents have not been studied. For these reasons, the American recommendations suggest that low-dose aspirin, unfractionated heparin or low-weight-molecular heparin, or no treatment could be acceptable during the first trimester depending on the clinical context and the maternal attitude toward risk [9]. Anticoagulation during pregnancy is indicated for high-risk thromboembolic conditions, such as hypercoagulable state or mechanical heart valves, and for CVT. Heparin is the anticoagulant of choice because it does not cross the placenta. Anticoagulant treatment for CVT during pregnancy should be continued throughout pregnancy and at least 6 weeks postpartum (for a total minimum duration of treatment of 6 months). Data on thrombolytic therapy in pregnant women with acute ischemic stroke are scarce; they come from a limited number of case reports in the literature. Animal data suggest that recombinant tissue plasminogen activator is unlikely to have teratogenic effects. Data are even more limited for thrombectomy. The risks and benefits of thrombolytic therapy and/or thrombectomy for pregnant women and fetuses must be considered cautiously, on an individual basis [1,3].

Studies have shown that treatment of ruptured arterial aneurysms is beneficial to both mother and fetus. Treatment of a ruptured cerebral vascular malformation should proceed based solely on neurological criteria. While proper counseling is imperative, a history of pregnancy-related stroke or CVT should not be considered as a contraindication for subsequent pregnancy [10].

HORMONAL CONTRACEPTION

Numerous studies have shown that oral contraceptive pills (OCPs) with a high estrogen content greatly increase the risk of stroke (relative risk around 4), both ischemic and hemorrhagic [6]. The doses of estrogen in combined estrogen–progestin pills were subsequently lowered, and four meta-analyses summarizing over 30 years of studies concluded that the risk of stroke in estrogen–progestin pill users was about twice as that of nonusers. In one of these meta-analyses that separated case–control and cohort studies, an increased risk of stroke in estrogen–progestin pill users was only present in case–control studies (OR, 2.13; 95% CI, 1.59–2.86). The risk was increased for ischemic strokes but not for hemorrhagic strokes in the two cohort studies that reported stroke subtypes [6].

Recent data from a very large cohort of about 1.6 million Danish women, followed up over a 15-year period, showed that the relative risks of thrombotic stroke and myocardial infarction were increased by a factor of 1.3–2.3 among users of intermediate-dose (30–40 µg of ethinyl estradiol [EE]) OCPs, and by a factor of 0.9–1.7 among users of low doses of EE (20 µg), according to progestin type [11]. The overall absolute risk of stroke was low at approximately 2 events per 10,000 women per year. These risks are slightly lower than previously reported, and for an individual woman, the risk is quite small. Other routes of administration (vaginal ring, transdermal patches) were also associated with an increased risk of stroke and myocardial infarction (2.5 and 3.2, respectively). Duration of use did not change the risk estimates. In contrast, none of the progestin-only products, including levonorgestrel-releasing intrauterine devices and subcutaneous implants, significantly increased the risk of thrombotic stroke or myocardial infarction [11].

The risk for stroke among women using OCPs increases exponentially with older age (from 3.4 per 100,000 women aged 15–19 years to 64.4 per 100,000 women aged 45–49 years) [11]. The RATIO (Risk of Arterial Thrombosis in relation to Oral Contraceptives) study, based on a population-based cohort in the Netherlands, found that the risk of stroke is further increased in OCPs users who smoked, had hypertension,

TABLE 119.1   OR of Ischemic Stroke in Current Users of OCPs in the RATIO Study, According to Other Risk Factors [12]

	Patients/Control Women, n	OR (95% CI)
SMOKING		
No	36/183	2.8 (1.5–5.0)
Yes	66/165	4.4 (2.7–7.3)
HYPERTENSION		
No	86/327	2.7 (1.8–4.0)
Yes	16/19	7.6 (3.5–16.3)
HYPERCHOLESTEROLEMIA		
No	97/344	2.1 (1.5–3.1)
Yes	5/3	10.8 (2.3–49.9)
DIABETES		
No	99/345	2.3 (1.6–3.4)
Yes	2/2	5.3 (0.7–42.6)
OBESITY		
No	77/300	2.2 (1.5–3.0)
Yes	22/37	4.6 (2.4–8.9)

hypercholesterolemia, diabetes, or who were obese [12] (Table 119.1). Migraine, particularly migraine with aura, has also been associated with an increased risk of stroke in OCP users. Identifying women with such risk factors is therefore recommended in order to manage modifiable risk factors. The risk associated with OCPs can be minimized by abstinence from smoking and by checking blood pressure, with avoidance of hormonal contraceptive use if blood pressure is raised. A consensus statement from both headache and stroke experts suggests screening for and treatment of all traditional stroke risk factors in women with migraine but does not state that low-dose OCP use is contraindicated [6]. The RATIO study also found that many prothrombotic mutations and biomarkers of endothelial dysfunction, such as von Willebrand factor and ADAMTS13, increased the risk of ischemic stroke in women using OCPs.

With regard to venous thromboembolism, users of OCPs with desogestrel, gestodene, or drospirenone were at least twice at risk compared with users of OCPs with levonorgestrel. A similar increase was reported for CVT. Based on these findings, second-generation OCPs appear as a better first-line choice than third-generation OCPs [6]. Several studies have also shown that the use of OCPs and the existence of a genetic prothrombotic condition increase the risk of venous thrombosis in a multiplicative way; a similar effect has been observed in CVT. However, screening for prothrombotic mutations before starting oral contraceptive therapy is not recommended

because of their low prevalence in otherwise healthy women, especially in the absence of a positive family history [6].

HORMONE REPLACEMENT THERAPY

While women under age 50 years have generally a lower incidence and prevalence of stroke than men, the risk of stroke roughly doubles in women during the 10 years after menopause, emphasizing the need to screen for and manage risk factors that are also increasing during this time period [13]. Endogenous estrogen levels decline during the menopausal transition, leading to a relative androgen excess, which could contribute to the increased cardiovascular risk factors in women.

Previous observational cohort studies have shown that the risk of heart disease is about 50% lower in women who use estrogen than in nonusers. However, their results were conflicting in relation to stroke risk, with some showing decreased risk, some showing no effects, and some showing increased risk. Subsequent randomized trials were negative for both primary and secondary prevention of stroke (Table 119 2). In postmenopausal women with known coronary heart disease, the Heart and Estrogen/Progestin Replacement Study (HERS) found an early increased risk but overall no difference in coronary disease events with combination HRT (conjugated equine estrogen [CEE] and medroxyprogesterone [MP]) compared with placebo, as well as a nonsignificantly increased risk of stroke [14]. The Women's Estrogen for Stroke Trial (WEST) [15] found that, overall, estrogen alone in postmenopausal women with recent stroke or transient ischemic attack had no effect on recurrent stroke (fatal and nonfatal combined). However, there was an early increased risk of stroke in the first 6 months after randomization, and the estrogen group had nearly three times the rate of fatal strokes as those on placebo. In the Women's Health Initiative (WHI) [16] trial, a randomized multicenter trial involving 16,608 healthy postmenopausal women, those taking CEE plus MP had an approximately 31% increase in total stroke risk compared with those taking placebo. However, the absolute risk of stroke in the treatment groups was small, resulting in only 8 strokes per 10,000 women in each year of HRT use. This increased stroke risk was significant for ischemic but not for hemorrhagic stroke. In the WHI estrogen-alone study, among 10,739 postmenopausal women with a hysterectomy, use of CEE increased the risk of ischemic stroke by 55%, but had no effect on hemorrhagic stroke. Meta-analyses of existing trials have confirmed these findings, showing a 29% elevated total stroke risk associated with HRT use.

TABLE 119.2 Randomized Studies of HRT. Risk of Stroke

Study	Number of Subjects	Mean Age (years)	HRT	Primary Endpoint	(CI 95%)
HERS	2763	66.7	CEE/MP	Ischemic stroke	RR 1.18 (0.83–1.67)
				Fatal stroke	RR 1.61 (0.73–3.55)
				Nonfatal stroke	RR 1.18 (0.83–1.66)
WEST	664	71.0	17β-estradiol	Stroke or death	RR 1.10 (0.80–1.40)
				Nonfatal stroke	RR 1.00 (0.70–1.40)
				Ischemic stroke	RR 1.00 (0.60–1.40)
				Fatal stroke	RR 2.90 (0.90–9.00)
WHI	16,608	68.3	CEE/MP	Ischemic stroke	HR 1.44 (1.09–1.90)
				Fatal stroke	HR 1.20 (0.58–2.50)
WHI	10,739		CEE	Stroke	HR 1.39 (1.10–1.77)
				Nonfatal stroke	HR 1.39 (1.05–1.84)
				Fatal stroke	HR 1.13 (0.54–2.34)

HR, hazard ratio; *RR*, relative ratio.

Similar results have been shown in randomized trials for selective estrogen modulators (such as raloxifene and tibolone). The results of these studies had a great impact on decision-making by women and their health care providers and led to a marked reduction in HRT usage.

In addition to differences in study design, the divergent findings between studies may result from differences in route of administration and dose of hormone therapy. Oral estrogens seem to be responsible for deleterious proinflammatory and prothrombotic effects. Transdermal estradiol would limit these effects and may be a safer alternative than oral estrogens given the lack of exposure to first-pass liver metabolism and the lack of increase in clotting factors and inflammatory markers. However, there are only limited observational data exploring the different routes of administration of estrogen and stroke risk [6,13].

Differences in cardiovascular outcome may also be accounted for by the time from menopause until the start of HRT (the "timing hypothesis"). The vascular effect of HRT might not be deleterious if HRT is given early after menopause in women with less-advanced atherosclerosis, in whom endothelial dysfunction may be improved or reversed with physiological levels of estrogen replacement [13]. Therefore, there have been controversies about WHI's applicability to women just entering menopause. In a secondary analysis of the WHI study, women who were 10 years or less since menopause had a nonsignificant benefit from hormone therapy (HR, 0.76; 95% CI, 0.50–1.16), whereas there was an increased risk in women 20 years or more since menopause. However, this trend was not seen in women who had stroke events [6]. More research is needed to determine the safest and most effective formulation, dose, and duration of hormone therapy that will treat vasomotor symptoms without increasing the risk for stroke. Women who are experiencing menopausal symptoms should be individually evaluated for their baseline risk for developing breast cancer, venous thromboembolism, and coronary heart disease recurrence. After weighing the risks of these events against quality of life, the lowest effective estrogen dose should be used for the shortest possible time. In addition, for women at increased risk of coronary heart disease, transdermal hormone therapy should be the first choice over oral formulations [6].

References

[1] Grear KE, Bushnell CD. Stroke and pregnancy: clinical presentation, evaluation, treatment, and epidemiology. Clin Obstet Gynecol 2013;56:350–9.

[2] James AH, Bushnell CD, Jamison MG, Myers ER. Incidence and risk factors for stroke in pregnancy and the puerperium. Obstet Gynecol 2005;106:509–16.

[3] Tang SC, Jeng JS. Management of stroke in pregnancy and the puerperium. Expert Rev Neurother 2010;10:205–15.

[4] Kuklina EV, Tong X, Bansil P, George MG, Callaghan WM. Trends in pregnancy hospitalizations that included a stroke in the United States from 1994 to 2007: reasons for concern? Stroke 2011;42:2564–70.

[5] Sharshar T, Lamy C, Mas JL. Incidence and causes of strokes associated with pregnancy and puerperium; a study in public hospitals of Ile de France. Stroke 1995;26:930–6.

[6] Bushnell C, McCullough LD, Awad IA, et al. Guidelines for the prevention of stroke in women: a statement for healthcare professionals from the American Heart Association/American Stroke Association. Stroke 2014;45:1545–88.

[7] Lamy C, Oppenheim C, Mas JL. Posterior reversible encephalopathy syndrome. In: Biller J, Ferro JM, editors. Neurologic aspects of systemic disease, part III. Elsevier; 2014. p. 1687–701.

[8] Liu XJ, Wang S, Zhao YL, et al. Risk of cerebral arteriovenous malformation rupture during pregnancy and puerperium. Neurology 2014;82:1798–803.

[9] Kernan WN, Ovbiagele B, Black HR, et al. Guidelines for the prevention of stroke in patients with stroke and transient ischemic attack: a guideline for healthcare professionals from the American Heart Association/American Stroke Association. Stroke 2014;45:2160–236.

[10] Lamy C, Hamon JB, Coste J, Mas JL. Ischemic stroke in young women: risk of recurrence during subsequent pregnancies. French Study Group on Stroke in Pregnancy. Neurology 2000;55:269–74.

[11] Lidegaard O, Lokkegaard E, Jensen A, Skovlund CW, Keiding N. Thrombotic stroke and myocardial infarction with hormonal contraception. N Engl J Med 2012;366:2257–66.

[12] Kemmeren JM, Tanis BC, van den Bosch MA, et al. Risk of Arterial Thrombosis in Relation to Oral Contraceptives (RATIO) study: oral contraceptives and the risk of ischemic stroke. Stroke 2002;33:1202–8.

[13] Lisabeth L, Bushnell C. Stroke risk in women: the role of menopause and hormone therapy. Lancet Neurol 2012;11:82–91.

[14] Simon JA, Hsia J, Cauley JA, et al. Postmenopausal hormone therapy and risk of stroke: the heart and estrogen-progestin replacement study (HERS). Circulation 2001;103:638–42.

[15] Viscoli CM, Brass LM, Kernan WN, Sarrel PM, Suissa S, Horwitz RI. A clinical trial of estrogen-replacement therapy after ischemic stroke. N Engl J Med 2001;345:1243–9.

[16] Wassertheil-Smoller S, Hendrix SL, Limacher M, et al. Effect of estrogen plus progestin on stroke in postmenopausal women: the Women's Health Initiative: a randomized trial. JAMA 2003;289:2673–84.

CHAPTER

120

Toxicity/Substance Abuse

R. Sweis, J. Biller

Loyola University Chicago, Stritch School of Medicine, Maywood, IL, United States

INTRODUCTION

Stroke is the second leading cause of mortality and most common cause of disability worldwide. About 30% of stroke survivors are permanently disabled, and 20% require institutionalized care. In 2002, the cost of stroke was estimated at $49.4 billion [1]. Although stroke in adults younger than 55 years comprises only 10% of stroke, it remains a significant source of morbidity and mortality [2]. At a younger age, stroke is also associated with higher societal costs compared to older age groups [3]. In 2009, a national health survey reported an estimated 2.2 million people in the United States used illicit drugs. About 17,000 deaths were related to drug use in 2000, and over 1 million emergency department visits were attributed to drugs in 2007 [4]. The major causes of death related to drug use are acquired immune deficiency syndrome (AIDS), overdose, suicide, and accidents. However, cerebrovascular complications also pose a threat and remain a significant source of morbidity and mortality. Illicit drug use remains the most common predisposing factor for stroke in patients under 35 years, and those of age 15–44 years are 6.5 times more likely to suffer a stroke as compared to nondrug users [4]. Epidemiological data remain scarce with a need for more population-based studies assessing the link between drug use and ischemic stroke, intracerebral hemorrhage (ICH), and subarachnoid hemorrhage (SAH) [4].

Sloan and colleagues reported that 12.1% of 422 patients aged 15–44 years had recent drug use in the setting of acute ischemic stroke, and in 4.7%, drug use was the most likely cause of stroke [4]. In 2007, Westover et al. reported that drug use was associated with 14.4% of ICH and 14.4% of ischemic stroke in 1935 patients [4,5]. The most commonly associated drugs with stroke are cocaine, amphetamines, ecstasy, heroin, phencyclidine (PCP), lysergic acid diethylamide (LSD), cannabis, tobacco, and ethanol.

Primer on Cerebrovascular Diseases, Second Edition
http://dx.doi.org/10.1016/B978-0-12-803058-5.00120-X

CANNABIS

Marijuana is the most widely used illicit drug in the United States. Cannabis is consumed by about 181 million people worldwide, with 13.1 million dependent on it. About 7 million people in the United States use cannabis weekly [6]. Its use in older adults (50+ years) is expected to rise [7], but is greatest among adults aged 18–25 years [8]. In the United States, young adults aged 18–25 years who used cannabis rose from an incidence of 5.8% in 1965 to 50% in 2002.

The desired nonpermanent effects of cannabis are euphoria, self-confidence, and relaxation which start within minutes of use and last 2–3 h [9]. Adverse effects of cannabis include red eyes, possible weight loss, thunderclap headache, psychiatric symptoms, memory alterations, motor incoordination, poor executive functioning, sedation, and cardio- and cerebrovascular effects. Because of the widespread use of cannabis, it is difficult to establish a causal relationship [10,11].

Cannabinoids (CBs) consist of a wide group of compounds with various affinities to G-protein-coupled membrane-bound receptors classified according to their source. CB receptors are located in the brain, spleen, blood vessels, cells of the immune system, and heart [8]. Three of the major classes include the endogenous form, the phytocannabinoids derived from cannabis and include delta-9-tetrahydrocannabinol (delta-9-THC), cannabidiol (CBD), and cannabinol (CBN), and the synthetic CBs manufactured artificially in the laboratory [12].

Benefits of Cannabis

The CBs hold much promise in not only improving stroke but also traumatic and anoxic brain injury outcome. There is a rise in evidence that the endocannabinoids and botanical nonpsychoactive cannabidiol derivatives from the cannabis plant have a multitude of beneficial effects, particularly with ischemic stroke. Animal models of stroke have shown cannabidiol increases cerebral blood flow (CBF) mediated by 5-hydroxytryptamine (serotonin) receptor 1A (5HT1A) receptors and induces vasorelaxation [12].

A major goal of acute stroke care entails salvaging the penumbra and extra-penumbral regions of brain by preventing further growth of the infarct zone. Necrosis and apoptosis within the stroke core lead to free radical formation, glutamate release, and an inflammatory cascade leading to accumulation of intracellular calcium and cell death [12]. Endocannabinoids accumulate in ischemic tissue with CB 1 receptor (CB1 receptor) activation resulting in neuroprotective mechanisms including inhibition of glutamate release, decrease in intracellular calcium, hypothermia, decreased reactive oxygen species, and expression of brain-derived neurotrophic factor (BDNF). CB 2 receptor (CB2 receptor) activation leads to a decrease of leukocyte adhesion and cytokine release [12].

Synthetic CB1/CB2 receptor agonists are associated with a reduction of infarct size and hypothermia in animal models. CB1 receptor antagonists, however, have also shown a reduction of infarct size. The conflicting results prove the complexity of the ECS; others hypothesize that the CB receptor inhibitors may be targeting nonCB receptors but still producing neuroprotective effects. Studies are also investigating modulation of CB1/CB2 receptor activation with promising results on reducing infarct size with stimulation of CB2 and inhibition of CB1 [12]. CB may also play a beneficial role in post-stroke rehabilitation with studies reporting CB1 and CB2 receptor expression in neural stem and progenitor cells [12].

Cons of Cannabis

Delta-9-THC is the main active ingredient in cannabis and most often implicated in depression, psychosis, and anxiety [14,15]. Delta-9-THC is a psychoactive substance with vasoconstrictor effects. It has been associated with arteriopathies of small muscular arteries of the legs, coronary arteries, and cerebral circulation based on angiographic studies. A study by Ntholang and collaborators was the first study demonstrating histological evidence of hyperplastic tunica intima with associated luminal stenosis of cerebral arteries in young cannabis users [16].

Despite the known underreporting, the rate of cannabis-related cardiovascular complications, including peripheral, cerebral, and cardiac, has steadily risen over the past 5 years, with acute coronary syndromes and peripheral arteriopathies comprising the majority of complications [17]. Cannabis is associated with arterial disease resulting in stroke with a high rate of occurrence in the posterior circulation, myocardial infarction (MI), and limb arteritis. An incidence of 2–39% of cannabis-associated stroke has been reported [9]. The temporal relationship between symptom onset and resolution with starting and stopping cannabis, respectively, suggests cannabis is a trigger of these arterial disorders [14] (Matteo).

Cannabis-related ischemic strokes are limited to case reports and population-based studies. Cannabis-related hemorrhagic stroke is also limited to few case reports. Pathogenesis of hemorrhagic stroke may be related to transient cannabis-induced hypertension with altered autoregulation. Most reported cases of cannabis-related ischemic stroke are seen in patients younger than 50 years of age without traditional vascular risk factors [18]. Concomitant alcohol use and tobacco use have been suggested as risk factors. The pathogenesis of cannabis-related ischemic stroke is thought to be due to altered cerebral autoregulation and regional hypoperfusion, or an acute inflammatory cascade with thrombosis. It remains uncertain if platelet activation is involved in the pathogenesis of THC-triggered procoagulant properties. In vivo studies are necessary to evaluate the role on THC-platelet activation leading to a procoagulant effect [18].

Other mechanisms include arterial hypotension, cardioembolism, vasospasm, vasculitis, or reversible cerebral vasoconstriction syndrome (RCVS) [13,15,19]. Chronic cannabis is associated with increased cerebrovascular resistance [19]. Because delta-9-THC increases CBF, transcranial Doppler (TCD) was studied in marijuana users assessing blood flow velocity in the anterior and middle cerebral arteries (ACAs and MCAs, respectively) in control groups and participants using marijuana in different intensities. Blood flow velocity was also measured for 30 days of marijuana abstinence [20,40]. The pulsatile index, which is a measure of cerebrovascular resistance, and systolic velocity were elevated in marijuana users compared to control groups and persisted despite abstinence. This increase in cerebrovascular resistance may account for strokes as well as cognitive deficits seen in chronic marijuana users [20,40].

Laced marijuana poses a serious danger, increasing risk of stroke in users. Marijuana can be laced with cocaine, crack (also called bazooka), PCP, heroin, and even embalming fluid [22]. Although reports are infrequent, unregulated marijuana can easily be adulterated by dealers but most often occurs at the user level. Powdered cocaine or crack is sprinkled into a joint or blunt in order to combine the stimulant effects of cocaine with the depressant effects of marijuana. A joint can also be immersed in formaldehyde (a known carcinogen), mimicking the effects of PCP in which the user develops hallucinations, euphoria, panic, or rage. Laced marijuana undoubtedly poses cardio- and cerebrovascular complications and should be avoided [22].

AMPHETAMINES

Amphetamines have been used medically since the early 20th century and long been used to treat children and adolescents for attention deficit hyperactivity disorder (ADHD). Amphetamines are now increasingly prescribed to adults for maintenance therapy of ADHD and narcolepsy but remain the most commonly abused prescription drugs. In 2000, prescription amphetamines exceeded 8 million. The pharmacokinetics of amphetamines differ among children and adults, posing a cardiovascular risk in adults with prolonged amphetamine use [24].

Amphetamines produce their effect by increasing the synaptic levels of dopamine, biogenic amines, norepinephrine, and serotonin. They also disrupt vesicular storage of dopamine and inhibit its degradative enzymes (monoamine oxidases, MAO-A and MAO-B), promoting dopamine accumulation within the cytoplasm. The behavioral effects are mediated through dopamine modulation [24]. Amphetamines exist as two stereoisomers, the L-enantiomer (levoamphetamine) that produces more cardiovascular and peripheral effects as compared to the D-enantiomer (dextroamphetamine). Pharmaceutical amphetamines are either D-amphetamine or a mixture of the D- and L-amphetamine salts [24]. The most potent and most often abused amphetamine is methamphetamine (meth) which undergoes hepatic metabolism and has an active metabolite that functions as a hallucinogen. Users experience euphoria, increased motor movements, decreased appetite, and increased libido. Adverse effects include bad halitosis, nervousness, convulsions, irritability, paranoia, hypertension, coma, and death. Addiction and tolerance with chronic use are inevitable [4,11].

The effects of prolonged amphetamine use have not been explored and warrant further research. They are generally considered to have a safe drug profile, but studies reported during late 2010s and early 2010s have raised concern of their safety with reports of associated sudden death, MI, SAH, and ischemic and hemorrhagic stroke, especially in a younger population [25,26]. Amphetamines have also been associated with multiorgan injury including cardiomyopathy, renal and liver failure, respiratory compromise, memory changes, and psychiatric manifestations [4].

Amphetamines and ICH/SAH

ICH is a more common adverse effect than ischemic stroke after sympathomimetic use [27]. The association between amphetamines and ischemic or hemorrhagic stroke is based solely on case series [28]. Most epidemiological studies assessing this association have been conducted in underserved settings or largely based on the minority population, but small case–control studies have demonstrated a strong risk factor of stroke in the insured, urban populations with amphetamine use [28]. Most case series report that the risk of hemorrhagic stroke due to amphetamines is twice that of cocaine use [4]. Amphetamines and other related sympathomimetic drugs can result in ICH within minutes to hours after exposure. The acute rise of blood pressure leading to ICH has been observed even after first time use of amphetamines. Transient arterial hypertension after drug exposure is thought to cause ICH in locations typical of a hypertensive hemorrhage including the basal ganglia, thalamus, pons, and subcortical white matter. Hypertension induced by amphetamine use is also thought to contribute to aneurysm formation with resulting SAH [26]. Some studies have refuted this demonstrating that in patients presenting with amphetamine-related SAH, autopsy failed to show aneurysm or arteriovenous malformations [26]. Although rare, methamphetamines and synthetic cannabis have also been associated with spinal SAH due to ruptured thoracic radicular artery pseudoaneurysm rupture in a single case report [29].

Other mechanisms of ICH include altered cerebrovascular autoregulation, coagulopathy, vasculitis, and hemorrhagic transformation of ischemic stroke [30]. Cerebral vasoconstriction has been reported with cocaine, amphetamine, ephedrine, phenylephrine, LSD, and heroin, but it is difficult to distinguish between vasospasm and vasculitis. It is suspected that as vasospasm resolves and perfusion is restored, arterial rupture can occur [30].

Amphetamines and Ischemic Stroke

Cocaine and amphetamines have the strongest association with ischemic stroke [4]. Compared to nonusers, amphetamines increase the odds of suffering a stroke by 4 times [4]. Ischemic events occur due to a rise in catecholamine levels, vasoconstriction of extra- and intracerebral vasculature, and formation of oxygen-free radicals. Histological evidence demonstrates amphetamines produce vessel wall necrosis, atherosclerosis, and occlusion leading to infarcts [4]. Cerebral vasculitis reducing vessel caliber with resulting infarction has also been postulated as a possible mechanism of amphetamine-related ischemic stroke. One case series revealed the presence of fibrinoid necrosis, luminal thrombosis, and mixed inflammatory cells in small- and medium-sized arteries, identical to a necrotizing angiitis as seen with polyarteritis nodosa [26,31]. Other histopathological studies have countered these findings, showing no evidence of inflammation and varying extent of vessel wall medial necrosis [26]. Animal studies have shown a marked decrease in vessel caliber of the internal carotid (ICA), middle cerebral (MCA), and anterior cerebral arteries (ACAs) after only 1 week of methamphetamine use [26]. On autopsy, animals also had petechial and SAH, ischemia, and infarction. Some have postulated that atherosclerosis occurs with time as methamphetamines cause repeated episodes of hypertension with resulting microinfarction in the vaso vasorum and atherosclerosis [26].

COCAINE

Crack cocaine became available for recreational use in the United States more than two decades ago. More than 30 million Americans have tried cocaine once and 5 million report regular use [33]. The annual prevalence of cocaine use among adults in 2011 in Western and Central Europe was 1.2% and 1.5% in North America, and as high as 5.5–8.8% in young adults age 18–25 years. Cocaine users have a mortality rate 4–8 times higher than their age-sex peers in the general population [3]. Abuse has increased especially among African-Americans from low socioeconomic status in urban areas [34].

Cocaine is an alkaloid derived from the leaves of *Erythroxylum coca*. The illicit forms of cocaine are cocaine hydrochloride and cocaine alkaloid also known as crack cocaine. Crack cocaine or "crack" is the cocaine alkaloid "free base" which is a crystalline chemical that is heat stable and can be smoked and absorbed by the lungs. It is often the form smoked by addicts from low socioeconomic status and is very addictive [34].

Acute and chronic adverse effects include anxiety, intense high followed by dysphoria, sleeplessness, hypertension, addiction, loss of appetite, nasal septal perforation, ciliary madarosis, long pinky nail (cake nail), lung damage, and death from overdose [11]. Cocaine abuse is associated with both ischemic and hemorrhagic stroke. The majority of patients with cocaine intoxication present with ischemic rather than hemorrhagic stroke [35]. Larger studies are needed to quantify the risk and assess stroke type, concomitant amphetamine use, duration and frequency of cocaine use, and hypertension variation. The relationship between stroke and cocaine is influenced by other factors including demographics, lifestyle, patterns of cocaine use, and concomitant use of other drugs or medical problems.

Cocaine and ICH/SAH

SAH and ICH have been temporally related to cocaine abuse. ICH in patients with cocaine intoxication is associated with concomitant IVH and poorer prognosis compared to ICH due to other etiologies [35]. Multiple, bilateral, deep, and convexity cerebral hemorrhages are rare but have been associated with smoking crack cocaine and consuming large quantities of ethanol. A vascular lesion has been discovered in 78% of cocaine abusers with SAH and 48% of those with ICH. Cocaine induces acute arterial hypertension with rupture of an aneurysm or atriovenous malformation (AVM). In the absence of a vascular lesion, etiologies still include acute hypertension as well as hemorrhagic conversion following ischemia. No pathological evidence exists demonstrating an association between cocaine and cerebral vasculitis unlike the sympathomimetics [36].

Cocaine is reported in about 33% of aneurysmal SAH (aSAH). Cocaine users with aSAH are often younger with anterior circulation aneurysms [37]. The effect of cocaine on aSAH-associated outcome remains uncertain. A study published in *Stroke* in 2013 demonstrated acute cocaine use is associated with a higher risk of aneurysm re-rupture and hospital mortality due to cocaine-induced hypertension and vasospasm, respectively. Vasospasm is historically known to negatively impact outcome in aSAH, and aneurysm re-rupture is also linked to higher case-fatality rates. The association of cocaine use and delayed cerebral ischemia was not significant nor was the impact on functional outcome [37].

Cocaine and Ischemic Stroke

The pathophysiological mechanisms of cocaine-related ischemic stroke include vasospasm, microischemia, endothelial dysfunction with thrombus formation, and vasculitis [35]. Autopsies have also shown platelet-rich arterial thrombi suggesting platelet activation as the underlying pathophysiological mechanism [38]. Crack cocaine induces more rapid blood levels and a more rapid high hydrochloride compared to cocaine hydrochloride, which is snorted or administered intravenously. It is also associated with a higher incidence of ischemic strokes. Regardless of route of administration, strokes occur shortly after use with a predilection for the brain stem. Spinal cord infarctions have also been reported following cocaine use. Bowel ischemia, MI, and an eosinophilic myocarditis can also occur with cocaine use. Concomitant ethanol use carries a synergistic action with cocaine, likely potentiating its effects. In the presence of ethanol, cocaine is metabolized to cocaethylene which binds more powerfully to the monoamine transport proteins [41].

A double-blind randomized controlled trial of 24 healthy and neurologically normal men with mean age of 29 years and mean lifetime cocaine use of eight exposures studied the effect of low-dose intravenous cocaine administration on the cerebral vasculature. Cerebral magnetic resonance angiography (MRA) was performed at baseline and 20 min after infusion. In the absence of other stroke risk factors, cocaine administration induced dose-related cerebral vasoconstriction. Prior cocaine use revealed a statistically significant dose-related effect, suggesting greater lifetime use predisposed users to a higher likelihood of cerebral vasoconstriction [39].

Reduced CBF has been postulated as a cause of the neuropsychological deficits seen in cocaine users. Studies using positron emission tomography (PET)/single-photon emission computed tomography (SPECT) have indicated cerebral perfusion deficits in cocaine abusers [20,40]. Transcranial Doppler (TCD) sonography has been used to study cerebral hemodynamics including blood flow velocity and pulsatility. Blood flow velocity and pulsatility were measured in the anterior and middle cerebral arteries using TCD in cocaine users within 3 days of use and again 28 days after abstinence.

There was a significant difference in velocities and pulsatility between cocaine abusers and the control group, with increased cerebrovascular resistance that persisted despite abstinence of 1 month. Further studies are needed to assess if pharmacological intervention would impact cocaine-related cerebrovascular resistance and outcome [20,40].

The risk of hemorrhagic conversion after ischemic stroke increases in the setting of hyperglycemia, high National Institutes of Health Stroke Scale (NIHSS) score, use of antiplatelet agents, and leukoaraiosis. Leukoaraiosis may also reflect chronic cocaine use [35]. Despite small retrospective studies showing no increased risk of hemorrhage with thrombolysis in patients presenting with acute ischemic strokes in the setting of cocaine use, the actual risk of hemorrhagic conversion remains uncertain and mandates further research [35].

HEROIN

Heroin or diacetylmorphine is a semisynthetic derivative of morphine which is found in opium. Heroin crosses the blood–brain barrier where it binds to mu receptors and creates a feeling of analgesia, euphoria, miosis, dry mouth, chronic rhinorrhea, and respiratory depression [2,11]. Systemic symptoms of heroin include eosinophilia, elevated immune and gamma globulins, hemolysis with positive Coombs test, and lymph node hypertrophy [41]. Heroin is often injected intravenously, though smoking or inhaling heroin is also increasingly popular [2].

Heroin is more often associated with ischemic cerebral or spinal stroke. It often follows intravenous use after a period of abstinence but can occur immediately after use or delayed by 6–24 h [41]. Risk of stroke is elevated when using heroin and other illicits such as cocaine or LSD [42]. Other neurological complications reported with heroin include transverse myelopathy, septic embolism with abscess formation, and bilateral basal ganglia necrosis [43].

Heroin and Ischemic Stroke

Mechanisms of heroin-associated stroke include cardioembolism secondary to infective endocarditis, anoxic injury secondary to hypoxemia and hypotension due to heroin-induced shock, and infective vasculitis secondary to heroin adulterants [2]. Heroin is often adulterated with quinine, lactose, and other diluents triggering angiitis secondary to a hyperimmune response with eosinophilia that can play a role in stroke [41,42]. Further immunological studies are warranted to study heroin-associated stroke. This immune reaction can occur several weeks after heroin exposure or with chronic addicts who develop sensitivity after a period of abstinence [43].

Inhaling or sniffing heroin can lead to a risk of hypereosinophilic syndrome (HES). HES is a primary or secondary proliferation of eosinophils in the peripheral blood ranging >1.5 nL or within the bone marrow, lasting at least 6 months or requiring treatment lowering eosinophils [44]. Eosinophils wreak havoc by depositing within the heart, central nervous system (CNS), and lungs causing end-organ damage. Common neurological manifestations include mono- or poly-neuropathies, stroke, and psychiatric disorders. HES can lead to ischemic stroke in 12% of cases [44]. Strokes associated with HES are often multiple and occur in multiple vascular territories, suggesting a cardioembolic source. Eosinophilic myocarditis or Loeffler's fibroblastic endocarditis leads to endocardial destruction, myocardial fibrosis, and formation of intracardiac thrombi [44]. TCD studies have also detected microemboli supporting a cardioembolic source. Other mechanisms include deposition of eosinophilic proteins in the endothelium, causing damage in the small and larger arterioles. A component of the eosinophilic granule, the eosinophilic cationic protein, also increases blood viscosity, and with an elevated eosinophil count results in hypercoagulability and higher risk of stroke [44].

Heroin and ICH

Heroin-associated ICH is uncommon but is thought to occur due to mycotic aneurysm rupture in the setting of endocarditis, pyogenic arteritis, hemorrhagic transformation of an infarct, or due to hypertension, if other illicits are ingested [2]. Mechanisms of heroin-associated myelopathy include hypotension, vasculitis, hypersensitivity, or a direct toxic effect of heroin. Hypotension is an unlikely cause and no patterns have correlated with spinal cord watershed zones. With a hypersensitivity reaction, heroin undergoes haptenation with an in vivo protein. This may be a protein specific to the spinal cord with resulting local inflammation, ischemia, and tissue injury [45].

ALCOHOL

Excessive alcohol intoxication is the third leading cause of preventable death with binge drinking accounting for more than half these deaths [46]. There is class IA evidence that reducing alcohol intake also reduces blood pressure. It remains unclear if reducing alcohol intake will also reduce the incidence of stroke. There is a clear association between binge drinking and circadian blood pressure variation (class II, level B) and time of onset of ICH and SAH (class II, level B) [47].

An inverted J-shaped relationship has been demonstrated between alcohol intake and CBF in the absence of stroke using PET. Mean global CBF is greatest in populations with the least amount of consumption (<1 drink/week) and lowest in the heaviest drinkers (>15/week) [48]. A linear relationship exists between alcohol and hemorrhagic stroke [1,47].

"Heavy" drinking is defined as at least three standard-sized drinks per day. "Light–moderate" intake is less than three standard-sized drinks per day. Sex and age may lower the upper limit of light–moderate drinking or raise it for others. Standard-sized portions of wine, liquor, or beer contain about the same amount of alcohol [49].

Alcohol and Ischemic Stroke

Observational studies have demonstrated an association between light to moderate alcohol intake and decreased risk of ischemic stroke, thought to be due to atherosclerosis prevention, although no clear explanation has been found. A cross-sectional analysis has indicated light to moderate alcohol intake is associated with decreased atherosclerotic burden in the proximal aortic arch which could explain the lower incidence of ischemic stroke. Protective effects of alcohol on the vasculature include regulation of lipids and fibrinolysis, decreased platelet aggregation, coagulation factors, inflammation, and insulin resistance [50].

An increased risk of cardioembolism is expected in heavy drinkers with underlying risk factors such as atrial fibrillation or cardiomyopathy. Variability of blood pressure leading to arrhythmias can also precipitate stroke. Increased blood flow during early intoxication may cause artery-to-artery emboli in atherosclerotic vessels. Small vessel disease can occur via alcohol-induced hypertension [47].

Despite the decreased risk of ischemic stroke with mild to moderate alcohol consumption, this risk is modified in the setting of other concomitant risk factors that include apolipoprotein E (apoE) genotype. ApoE4 positivity with moderate alcohol consumption is associated with an increased risk of ischemic stroke [51].

Jones et al. [52] performed a study refuting that light to moderate drinking reduces risk of ischemic stroke; they examined drinkers in the Atherosclerosis Risk in Communities study and found no association between light–moderate drinking and decreased ischemic stroke risk. Heavy drinking increased risk for ischemic and hemorrhagic stroke while moderate drinking increased risk for ICH [23].

Alcohol and ICH/SAH

Alcohol-induced hypertension is a risk for ICH with severity of ICH greater in heavy drinkers compared to moderate drinkers. Moderate consumption is neutral or harmful for hemorrhagic stroke. Concomitant liver cirrhosis also raises this risk. The risk of alcohol on SAH remains unclear. The effects of smoking and drinking have not been separately studied in SAH but both increase blood pressure transiently which may pose risk for aneurysm rupture [49].

A multivariate regression analysis was performed in 1714 patients with hemorrhagic stroke. Independent risk factors for ICH included hypertension, diabetes, menopause, current cigarette smoking, alcoholic drinks ≥2/day, caffeinated drinks ≥5/day, and caffeinated drugs [53].

Chronic alcohol use has been shown to pose no increase in risk for symptomatic ICH post IV thrombolysis despite having a decreased likelihood of receiving IV tissue plasminogen activator (tPA). Chronic alcoholism is not an exclusion for IV tPA. Factors increasing risk of hemorrhage after IV thrombolysis for ischemic stroke include hyperglycemia, age, high blood pressure, and white matter intensities [54].

CIGARETTES

One in five US adults is a regular smoker with initiation occurring during teenage years. The economic cost of smoking on society has totaled nearly $193 billion per year [55]. Tobacco smoke contains over 4000 different chemicals including heavy metals and toxins that result in free radical formation and vascular endothelial injury leading to accelerated atherosclerosis. Cigarette smoking also causes a procoagulant state induced by a change in concentrations of hemostatic and inflammatory markers. Smoking also promotes fibrinogen formation, a decrease in fibrinolytic therapy, an increase in platelet aggregation, and polycythemia. A decrease in CBF is also a clear association with cigarette use [55].

Cigarettes and Ischemic Stroke

The Framingham study demonstrated that cigarette smoking increases risk of all types of stroke due to arterial thromboembolism, is dose related, and independent of history of hypertension [56,57]. Even in younger patients aged 15–45 years, a prospective study indicated smokers were 1.6 times more likely to have an ischemic stroke (95% CI, 1.07–2.42) with a dose-dependent effect with each additional pack year. This study reported no significant differences in the smoking status of all subjects with various stroke subtypes including atherosclerotic, vasculopathic, cardioembolic, or cryptogenic [56].

Cigarette use also increases the risk of silent cerebral infarction. Chronic smokers have a higher incidence

of small vessel ischemic disease [57]. The relative risk of ischemic stroke is maximal in middle age adults, declines with advanced years, and most toxic in females especially those taking oral contraceptive pills (OCPs) or who suffer of migraine with aura [58].

A dose–response relationship between smoking and stroke risk is well established for secondhand smoking. Nonsmokers have a higher prevalence of stroke that increases relative to their significant other's intensity and duration of cigarette use [55]. A causal relationship between secondhand smoking and stroke has been suggested even at low levels of exposure. Exposure to secondhand smoke causes platelet aggregation, thrombosis, endothelial dysfunction, and inflammation [59].

Two-thirds of young patients continue to smoke despite the effectiveness of smoking cessation in reducing the risk of recurrent stroke [58]. After quitting, within 1–2 months, smoking-related stroke risk secondary to hypercoagulability decreases to that of nonsmokers. At 5 years, the stroke risk is reduced to that of a nonsmoker. After 1 year, the risk of heart disease is reduced to half and by 15 years, the risk of heart disease is that of a nonsmoker [55].

Snuff is one of the most widespread forms of smokeless tobacco. The cardiovascular risk associated with snuff remains controversial. A case–control study in men has shown no increase in risk of stroke with snuff use compared to smoking cigarettes. They nonetheless had a higher risk of stroke compared to nonusers of tobacco [60].

Controversy exists regarding use of electronic cigarettes (e-cigarettes). The American Heart Association (AHA) supports electronic cigarette use in regions with smoke-free air laws. There is some exposure of nicotine, organic compounds, and fine particles with e-cigarettes, but there is insufficient evidence that the exhaled aerosol is dangerous to the public. Some argue that e-cigarette use promotes nicotine addiction and "renormalizes smoking behavior" [61].

Cigarettes and SAH

Important modifiable risk factors for aSAH in the younger population include hypertension, cocaine use, and cigarette smoking, with the latter being the most important modifiable risk factor in a case–control study of men and women aged 18–49 years. They also reported lean body mass, low educational achievement, and family history of hemorrhagic stroke as independent risk factors for aSAH [62]. The risk of SAH is heightened with hypertension and cigarette use, although the risk persists despite abstinence of cigarette smoking as an aneurysm may have already formed. Cigarettes degrade elastin in vessel walls with resulting dilation under higher pressures or turbulence. Cigarettes also reduce the activity of α-1-antitrypsin which is a proteolytic enzyme that prevents breakdown of collagen and elastin [33].

CONCLUSION

Stroke is the second leading cause of death in individuals older than 60 years [63]. A healthy lifestyle consisting of abstinence of smoking, healthy diet including low to moderate alcohol intake, exercise, and maintaining optimal body weight are effective in lowering risk of coronary artery disease, diabetes, and cancer than any other single factor. A healthy lifestyle has also been shown to significantly lower risk of ischemic stroke [64]. A 44–79% reduced risk of ischemic stroke is associated with a healthy lifestyle even in patients, specifically men, already at higher risk of stroke due to other cardiovascular risk factors (atrial fibrillation, heart failure, hypertension, hyperlipidemia, and diabetes) [63]. A healthy lifestyle also embodies avoidance of illicit drug use, specifically cocaine, heroin, and amphetamines, though profound use continues to rise and has become a major source of stroke-related morbidity and mortality in the younger population. The effect of cannabis on ischemic stroke, however, remains controversial. Given the perception that cannabis is harmless and legalization continues to be debated, health care professionals and the general public need to be made aware of the increased risk of cardio- and cerebrovascular disorders.

References

[1] Reynolds K, Lewis LB, Nolen JDL, et al. Alcohol consumption and risk of stroke. JAMA 2003;289(5):579–88.

[2] Kumar N, Bhalla MC, Frey JA, Southern A. Intraparenchymal hemorrhage after heroin use. Am J Emerg Med 2015;33:1109.e3–4.

[3] Sordo L, Indave BI, Degenhardt L, et al. Cocaine use and risk of stroke: a systematic review. Drug Alcohol Depend 2014;142:1–13.

[4] Esse K, Fossati-Bellani M, Traylor A, et al. Epidemic of illicit drug use, mechanisms of action/addiction and stroke as a health hazard. Brain Behav 2011:44–54.

[5] Westover AN, McBride S, Haley RW. Stroke in young adults who abuse amphetamines or cocaine. Arch Gen Psychiatry 2007;64:495–502.

[6] Singh NN, Pan Y, Muengtaweeponsa S, et al. Cannabis-related stroke: case series and review of literature. J Stroke Cerebrovasc Dis 2012;21(7):555–60.

[7] Hemanchandra D, McKetin R, Cerbuin N, et al. Heavy cannabis users at elevated risk of stroke: evidence from a general population survey. Aust N Z J Public Health 2015. http://dx.doi.org/10.1111/1753-6405.12477. Online.

[8] Mittleman MA, Lewis RA, Maclure M, et al. Triggering myocardial infarction by marijuana. Circulation 2001;103:2805–9.

[9] Inal T, Kose A, Koksal O, et al. Acute temporal lobe infarction in a young patient associated with marijuana abuse: an unusual cause of stroke. World J Emerg Med 2014;5(1):72–4.

[10] Santos AF, Rodrigues M, Mare R, et al. Recurrent stroke in a young cannabis user. J Neuropsychiatry Clin Neurosci 2014;26:1.

[11] Johnson MD, Heriza TJ, Dennis CS. How to spot illicit drug abuse in your patients. Postgrad Med 1999;106(4):199–218.

[12] Latorre JGS, Schmidt EB. Cannabis, cannabinoids, and cerebral metabolism: potential applications in stroke and disorders of the central nervous system. Curr Cardiol Rep 2015;17(72):1–7.

[13] Oyinloye O, Nzeh D, Yusuf A, et al. Ischemic stroke following abuse of marijuana in a Nigerian adult male. J Neurosci Rural Pract 2014;5(4):417–9.

[14] Desbois AC, Cacoub P. Cannabis-associated arterial disease. Ann Vasc Surg 2013;27:996–1005.

[15] Matteo I, Pinedo A, Gomez-Beldarrain M, et al. Recurrent stroke associated with cannabis use. J Neurol Neurosurg Psychiatry 2005;76:435–7.

[16] Ntholang O, McDonagh R, Nicholson S, et al. Is intimal hyperplasia associated with cranial arterial stenosis in cannabis-associated cerebral infarction? Vol. ;10. World Stroke Organization; 2015. p. E56–9.

[17] Jouanjus E, Lapeyre-Mestre M, Micallef J, et al. Cannabis use: signal of increasing risk of serious cardiovascular disorders. J Am Heart Assoc 2014;3:e000638.

[18] Ince B, Benbir G, Yuksel O, et al. Both hemorrhagic and ischemic stroke following high doses of cannabis consumption. Presse Med 2014. //dx.doi.org/10.1016/j.lpm.2014.05.022.

[19] Barber PA, Pridmore H, Krishnamurthy V, et al. Cannabis, ischemic stroke, and transient ischemic attack: a case-control study. Stroke 2013;44:2327–9.

[20] Herning RI, Better WE, Tate K, et al. Cerebrovascular perfusion in marijuana users during a month of monitored abstinence. Neurology 2005;64:488–93.

[21] Deleted in review.

[22] Musto DF. Opium, cocaine and marijuana in American history. Sci Am 1991;265(1):40–7.

[23] Jones RT. Cardiovascular system effects of marijuana. J Clin Pharmacol 2002;42(11S):58S–63S.

[24] Berman SM, Kuczenski R, McCracken JT, et al. Potential adverse effects of amphetamine treatment on brain and behavior: a review. Mol Psychiatry 2009;14:123–42.

[25] Cooper WO, Habel LA, Sox CM, et al. ADHD drugs and serious cardiovascular events in children and young adults. NEJM 2011;365:1896–904.

[26] Ho EL, Josephson SA, Lee HS, et al. Cerebrovascular complications of methamphetamine abuse. Neurocrit Care 2009;10:295–305.

[27] Bruno A, Nolte K, Chapin J. Stroke associated with ephedrine use. Neurology 1993;43:1313–6.

[28] Petitti DB, Stephen S, Quesenberry C, et al. Stroke and cocaine or amphetamine use. Epidemiology 1998;9(6):596–600.

[29] Ray WZ, Krisht K, Schabel A, et al. Subarachnoid hemorrhage from a thoracic radicular artery pseudoaneurysm after methamphetamine and synthetic cannabinoid abuse: case report. Global Spine J 2013:119–24.

[30] Renard D, Gaillard N. Brain hemorrhage and cerebral vasospasm associated with chronic use of cannabis and buprenorphine. Cerebrovasc Dis 2008;25:282–3.

[31] Citron BP, Halpern M, McCarron M, et al. Necrotizing angiitis associated with drug abuse. NEJM 1970;283:1003–11.

[32] Deleted in review.

[33] Qureshi AI, Suri F, Guterman L, et al. Cocaine use and the likelihood of nonfatal myocardial infarction and stroke: data from the Third National Health and Nutrition Examination Survey. Circulation 2001;103:502–6.

[34] Giraldo EA, Taqi MA, Vaidean GD. A case-control study of stroke risk factors and outcomes in African American stroke patients with and without crack-cocaine abuse. Neurocrit Care 2012;16:273–9.

[35] Baud MO, Brown EG, Singhal NS, et al. Immediate hemorrhagic transformation after intravenous tissue-type plasminogen activator injection in 2 cocaine users. Stroke 2015;46:e167–9.

[36] Green RM, Kelly KM, Gabrielsen T, et al. Multiple intracerebral hemorrhages after smoking "crack" cocaine. Stroke 1990;21:957–62.

[37] Chang TF, Kowalski RG, Caserta F, et al. Impact of acute cocaine use on aneurysmal subarachnoid hemorrhage. Stroke 2013;44:1825–9.

[38] Cagienard F, Schulzki T, Reinhart WH. Cocaine in high concentrations inhibits platelet aggregation in vitro. Clin Hemorheol Microcirc 2014;57:385–94.

[39] Kaufman MJ, Levin JM, Ross MH, et al. Cocaine-induced cerebral vasoconstriction detected in humans with magnetic resonance angiography. JAMA 1998;279:376–80.

[40] Herning RI, King DE, Better WE, et al. Neurovascular deficits in cocaine abusers. Neuropsychopharmacology 1999;21:110–8.

[41] Caplan LR, Biller J. Non-atherosclerotic vasculopathies. In: Caplan's stroke: a clinical approach. 5th ed. Cambridge University Press; 2016.

[42] Brust JC, Richter RW. Stroke associated with addiction to heroin. J Neurol Neurosurg Psychiatry 1976;39:194–9.

[43] Woods BT, Strewler GJ. Hemiparesis occurring six hours after intravenous heroin injection. Neurology 1972;22:863–6.

[44] Bolz J, Meves SH, Kara K, et al. Multiple cerebral infarctions in a young patient with heroin-induced hypereosinophilic syndrome. J Neurol Sci 2015;356:193–5.

[45] McCreary M, Emerman C, Hanna J, et al. Acute myelopathy following intranasal insufflation of heroin: a case report. Neurology 2000;55(2):316–7.

[46] Sull JW, Yi SW, Nam CM, et al. Binge drinking and hypertension on cardiovascular disease mortality in Korean men and women: a Kangwha cohort study. Stroke 2010;41:2157–62.

[47] Hillbom M, Saloheimo P, Juvela S. Alcohol consumption, blood pressure, and the risk of stroke. Curr Hypertens Rep 2011;13:208–13.

[48] Christie IS, Price J, Edwards L, et al. Alcohol consumption and cerebral blood flow among older adults. Alcohol 2008;42:269–75.

[49] Klatsky AL. Alcohol and cardiovascular diseases: where do we stand today. J Intern Med 2015;278:238–50.

[50] Kohsaka S, Jin Z, Rundek T, et al. Alcohol consumption and atherosclerotic burden in the proximal thoracic aorta. Atherosclerosis 2011;219:794–8.

[51] Mukamal KJ, Chung H, Jenny NS, et al. Alcohol use and risk of ischemic stroke among older adults: the Cardiovascular Health Study. Stroke 2005;36:1830–6.

[52] Jones SB, Loehr L, Avery CL, et al. Midlife alcohol consumption and the risk of stroke in the atherosclerosis risk in communities study. Stroke 2015;46:3124–30.

[53] Feldmann E, Broderick JP, Kernan WN, et al. Major risk factors for intracerebral hemorrhage in the young are modifiable. Stroke 2005;36:1881–5.

[54] Gattringer T, Enzinger C, Fischer R, et al. IV thrombolysis in patients with ischemic stroke and alcohol abuse. Neurology 2015;85:1592–7.

[55] Shah RS, Cole JW. Smoking and stroke: the more you smoke the more you stroke. Expert Rev Cardiovasc Ther 2010;8(7):917–32.

[56] Love BB, Biller J, Jones MP, et al. Cigarette smoking. Arch Neurol 1990;47:693–8.

[57] Hossain M, Mazzone P, Tierney W, et al. In vitro assessment of tobacco smoke toxicity at the BBB: do antioxidant supplements have a protective role? BMC Neurosci 2011;12(92):1–18.

[58] Girot M. Smoking and stroke. Presse Med 2009;38(7–8):1120–5.

[59] Oono IP, Mackay DF, Pell JP. Meta-analysis of the association between secondhand smoke exposure and stroke. J Public Health 2011;33(4):496–502.

[60] Asplund K, Nasic S, Janlert U, et al. Smokeless tobacco as a possible risk factor for stroke in men: a nested case-control study. Stroke 2003;34:1754–9.

[61] Bhatnagar A, Whitsel LP, Ribisl KM, et al. Electronic cigarettes: a Policy Statement from the American Heart Association. Circulation 2014;130:1418–36.

[62] Broderick JP, Viscoli CM, Brott T, et al. Major risk factors for aneurysmal subarachnoid hemorrhage in the young are modifiable. Stroke 2003;34:1375–81.

[63] Larsson SC, Akesson A, Wolk A. Primary prevention of stroke by a healthy lifestyle in a high-risk group. Neurology 2015;84:2224–8.

[64] Chiuve SE, Rexrode KM, Spiegelman D, et al. Primary prevention of stroke by healthy lifestyle. Circulation 2008;118:947–54.

CHAPTER

121

Functional Disorders Presenting to the Stroke Service

G. Nielsen[1], M.J. Edwards[2], J. Stone[3]

[1]UCL Institute of Neurology, London, United Kingdom; [2]St Georges University of London, London, United Kingdom; [3]University of Edinburgh, Edinburgh, United Kingdom

INTRODUCTION

About one-quarter of patients admitted to specialist stroke units will be discharged with a diagnosis other than stroke; these patients are commonly referred to as stroke mimics [1]. Functional neurological disorder (FND) is one of the commonest causes of stroke mimic, representing with a reported range of 7–45% [2–4]. FND can be broadly defined as a condition characterized by neurological symptoms that lack internal consistency, are genuine, but not explained by a defined disease process. Also called conversion disorder, or referred to as psychogenic, FND is common in neurological practice making up about 16% of new referrals to neurology outpatient clinics [5].

Within specialist stroke services, FND or functional stroke mimic (FSM) represents a reported range of 1.4–8.4% of all admissions [2,4,6–8] and 0.5–5% of all patients who receive intravenous thrombolysis for presumed stroke (23–47% of stroke mimics receiving thrombolysis) [3,9,10]. Large discrepancies in the numbers exist in the literature and are related to the difficulties of case ascertainment. Many studies are likely to underestimate the incidence of FND as these patients may be represented in other diagnostic categories such as migraine, hemiplegic migraine, seizure, vertigo, or may be discharged without a clear diagnosis.

ETIOLOGY

The etiology of FND is not fully understood and there is debate over the relative contribution of psychopathology. Disagreement here is represented in the different preferences of terminology. The terms conversion disorder and psychogenic presume an etiology based in psychopathology. Critics of these terms point out the absence of clear or substantial psychological factors in many patients; problems explaining symptom mechanism in purely psychological terms and the limitations to research and treatment imposed by a monothematic etiological model [11,12]. Attributing all FND to psychological causes is a bit like blaming all strokes on smoking. A full discussion of etiological theories is beyond the scope of this brief introduction to FND. In summary, evidence from clinical and laboratory studies support the concept of a biopsychosocial etiological model. Functional neuroimaging studies, while limited by low numbers and mixed results have shown hypoactivity in areas associated with action selection and abnormal connections between limbic structures and motor areas [13]. Clinical and laboratory studies of symptoms demonstrate the central importance of abnormally focused attention directed toward the body [13]. When this attention is directed away from the body, symptoms resolve to a greater or lesser extent. Beliefs and expectations about symptoms and an abnormal sense of agency have also been described as part of the mechanism driving FND. A biopsychosocial etiological model that is generally well accepted considers symptoms in terms of a heterogeneous mixture of predisposing, precipitating, and perpetuating factors [14]. We encourage clinicians to consider mechanism of the symptoms. For example, a panic attack (causing unilateral sensory symptoms), acute dissociation, migraine, physical injury, asymmetric pain, and fatigue are all physiological experiences that can form the basis for FND [15].

Primer on Cerebrovascular Diseases, Second Edition
http://dx.doi.org/10.1016/B978-0-12-803058-5.00121-1

DIAGNOSIS

Differentiating symptoms due to FND from stroke can be particularly challenging in the emergency department, when the patient may be anxious and confused, and there is pressure to make an early diagnosis within the window of opportunity to receive intravenous thrombolysis. There may be clues such as a gradual onset, a dissociative (nonepileptic) seizure or a history of multiple previous functional symptoms. Psychological problems such as anxiety, depression, and recent stress are more common in functional disorder but not universal and should not be used to make a diagnosis [16]. The key to a positive diagnosis is positive identification of internal inconsistency during the physical examination or incongruity with recognized neurological disease [17].

A number of clinical signs have been described to identify FND, with high specificity and sensitivity, Hoover's sign for functional lower limb weakness may be the most useful [18,19] (Fig. 121.1). This describes weakness of hip extension that returns to normal power with contralateral hip flexion against resistance. Pain on assessment or sensory inattention/neglect may lead to false positives on some tests. Parietal stroke is especially prone to lead to an apparent "functional" presentation. Nonetheless this is a sign that appears to perform reasonably well in an acute stroke setting. For example it had a sensitivity of 63% and a specificity of 100% in one study of 337 patients with suspected stroke [19].

Functional facial symptoms are common, usually due to muscle overactivity, typically orbicularis oculi, orbicularis oris, and platysma muscles with a depressed eyebrow or pulled down mouth. This can give the appearance of combined upper and lower facial weakness and is one of the few stroke mimics that may pass a FAST (Facial drooping, Arm weakness, Speech difficulties, and Time) test, but when combined with unilateral functional limb weakness, should be a red flag for a functional disorder [20] (Fig. 121.2). Examples of incongruity with recognized clinical disease include midline splitting of sensory disturbance, global pattern of limb weakness, and a tubular visual field (Table 121.1) [21]. In general, it is easier to base the diagnosis on motor symptoms, as they are more amenable to objective examination, whereas sensory symptoms are subjective experiences. Where sensory symptoms dominate the history, a thorough physical examination often reveals new or initially unobserved motor symptoms that can support the diagnosis [13].

Investigations

Functional disorder may coexist or be triggered by neurological disease. Be prepared to make a diagnosis of stroke *and* a functional disorder in some patients where the clinical picture clearly indicates this. It is important to conclude the period of investigation as quickly as possible to enable the patient to move toward treatment.

Clinical and Demographic Characteristics of Functional Stroke Mimic (FSM)

On average, patients with FSM tend to be younger with a greater proportion of women than patients with stroke and other (medical) stroke mimics. One study of 1165 consecutive admissions to a stroke unit found the average age of FSM was 49 years, stroke 71 years, and medical stroke mimic 63 years [4]. The proportion of

Test hip extension – it's weak   Test contralateral hip flexion against resistance – hip extension has become strong

FIGURE 121.1   Hoover's sign of functional leg weakness. Weak hip extension that returns to normal with contralateral hip flexion against resistance. *Reproduced courtesy BMJ publications; Stone J. The bare essentials: Functional symptoms in neurology. Pr Neurol 2009;9:179–89.*

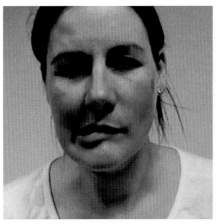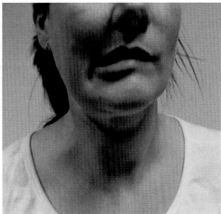

FIGURE 121.2 Unilateral platysmal contraction can produce facial asymmetry that may be misinterpreted as facial weakness. When present functional limb weakness is usually ipsilateral (reproduced). *Reproduced from Stone J. Functional neurological disorders: The neurological assessment as treatment. Neurophysiol Clin Neurophysiol 2014;44:363–7.*

TABLE 121.1   Clinical Signs of Functional Neurological Disorder

MOTOR SIGNS

Hoover's sign	Hip extension weakness that returns to normal when the contralateral hip is flexed against resistance.
Give-way weakness/collapsing weakness	Muscle power is initially generated on testing which quickly gives way or collapses.
Hip abductor sign	Hip abduction weakness that returns to normal with contralateral hip abduction against resistance.
Dragging leg gait	Gait pattern characterized by dragging a weak leg behind.
Clear signs of inconsistency	For example, weak ankle plantarflexion on testing but the patient is able to walk on their toes.
Hemifacial muscle overactivity presenting with unilateral limb symptoms	Overactivity of orbicularis oculus, orbicularis oris, and/or platysma giving the appearance of a facial droop.
Sternomastoid test	Weakness of head turning to affected arm and leg in functional hemiparesis.
Drift without pronation test	During a "pronator drift" test, the forearm may not pronate in a functional hemiparesis.
Global pattern of weakness	Flexors and extensors equally affected, for example, wrist flexion and wrist extension.

SENSORY SIGNS

Midline splitting of sensory loss	Sensory loss demarcated at the midline, in particular reduced vibration sense on one side of the frontal bone of the skull or sternum.
Tubular visual field	A visual field that when tested does not expand with distance from the patient.

INCONGRUITY WITH KNOWN NEUROLOGICAL DISEASE

Global distribution of weakness	Weakness affecting extensor and flexor muscle groups equally.
Dissociative nonepileptic attacks	Seizure-like episodes that have features that can distinguish them from epilepsy, such as ictal weeping, eyes closed during episodes, prolonged unresponsiveness lasting longer than 2 min, and normal EEG during episode.

women in each group was 63%, 46%, and 50%, respectively. In this study, the most common presentations in 98 consecutive FSMs were weakness or numbness of the arm (64%), leg (45%), and face (24%). Other common presenting symptoms were dysarthria (18%), dysphasia (14%), posterior circulation territory symptoms (14%), and visual symptoms (11%).

Functional neurological symptoms rarely occur in isolation and multiple comorbidity of other functional symptoms may point toward the diagnosis of FSM. Pain, fatigue, sleep disturbance, gastrointestinal symptoms, headache, memory/concentration problems, and bladder dysfunction are among the most frequently reported complaints in FND [22]. Headache at symptom onset appears to be reported with significantly higher frequency in functional stroke cases compared to ischemic stroke. Nazir et al. (2005) reported headache on presentation in 47% of 141 FSMs compared to 26% in stroke or TIA, with an adjusted odds ratio of 3.7 (95% CI 1.8–7.7) [6].

Misdiagnosis and Diagnostic Uncertainty

With limitations of diagnostic technology and the importance of making a quick diagnosis, it may be inevitable that a small proportion of patients with FSM will enter the stroke treatment pathway. This group is exposed to unnecessary risks, such as treatment with intravenous thrombolysis. The rate of symptomatic intracerebral hemorrhage in stroke mimics is relatively low at 0.5% [9]. This may lead the stroke clinician to err on the side of caution, reasoning that the odds of a poor outcome are higher when thrombolysis is withheld from a case of ischemic stroke than when administered to a case of functional stroke. However, outcome in FND in this situation can also be adversely affected by incorrect use of thrombolysis, by reinforcing erroneous illness beliefs, and delaying correct diagnosis and appropriate treatment.

Longitudinal follow-up studies have shown that rates of misdiagnosis in FND are relatively low at less than 4% [23]. However, it is rarely a simple and clear-cut diagnosis. Comorbid complaints in FND may require separate assessment leading to additional diagnoses requiring specific treatment (e.g., migraine, chronic low back pain, and diabetic neuropathy). Neurological disease, such as Parkinson's disease and multiple sclerosis, is a strong risk factor for developing functional neurological symptoms; therefore, the diagnosing clinician should be prepared to make two diagnoses or to recognize functional symptoms against a background of existing neurological disease [24,25]. FND may also exist as a prodromal state for neurological disease. For example, a functional gait disturbance and psychiatric symptoms may occur prior to the onset of firm signs of motor neuron disease, multiple system atrophy, or Parkinson's disease [16]. These examples illustrate the importance of being upfront at times about diagnostic uncertainty and providing neurological follow-up.

Another diagnostic gray area is the presence of new symptoms in chronic stroke without progression of clinical signs or evidence of further ischemia. This *decompensation* may be classified as a medical mimic when associated with a medical event such as infection or FND if associated with anxiety or belief that a new stroke has occurred [4]. Finally, some symptoms of rare neurological disease can appear functional in nature, such as alien hand phenomenon in corticobasal degeneration or the variable nature of stiff person syndrome [16].

Malingering and Factitious Disorders

Factitious disorder and malingering are common concerns of clinicians when a diagnosis of FND is being considered [14]. Malingering is where symptoms are feigned for material gain, and factitious disorder is where symptoms are feigned for the purpose of receiving medical attention. These are distinct categories separate from FND. Proving that symptoms are feigned requires evidence of a major discrepancy between reported and observed function (i.e., patients should be able to accurately report what they can and cannot do, even if they are not accurate about reporting how variable their symptoms are). This evidence is rarely obtained and in clinical practice, such patients are likely to be rare. Some clinicians may interpret symptom variability or exaggeration of symptoms as evidence of malingering. Symptom variability can be explained by the amount of attention invested in a movement or action. It may be true that some patients with FND exaggerate symptoms to convince of the need to be taken seriously, but this is different from exaggeration to deceive.

MANAGEMENT

A thorough assessment followed by effective communication of the diagnosis is an essential first step in the therapeutic management of the patient with FND [17,26]. In some cases this alone will facilitate resolution of symptoms. A useful starting point is to ask the patient to list their symptoms at the beginning of the consultation. This allows the patient to feel satisfied that these experiences have been taken into account when the diagnosis is communicated. Asking the patient about pain, fatigue, and dissociation or depersonalization can validate these experiences and open potential avenues of treatment. It is also important to clarify the patients understanding and expectations as this can highlight issues that need to be addressed in order to help the patient shift their focus from seeking assessment and investigations to treatment and symptom management.

Doctors are often uncomfortable delivering the diagnosis of FND, perhaps put off by the thought of discussing sensitive issues relating to mental health or to avoid conflict and patient dissatisfaction [26]. In our experience, the diagnosis is generally well accepted when it is explained in terms of the positive signs noted on examination. Clinical signs, such as Hoover's sign were previously used to "catch patients out," but in 2010s they were described as a useful part of the diagnostic explanation [27]. By showing the patient their clinical signs, they develop an understanding of how it is possible to experience involuntary motor symptoms in the absence of disease. Clinical signs demonstrate the potential for symptom reversibility and help explain the rationale for rehabilitation (see Table 121.2).

We generally go easy on questions and discussions of a psychological nature on initial assessment. Patients commonly interpret psychologically focused questioning and explanations as "not being believed," and they are not essential for diagnosis. We explain that the etiology is multifactorial and varies from person to person. Psychological factors can be an important part of the problem for many people but not necessarily everyone. It is helpful to state that you believe the patient has genuine symptoms that are outside of their control, that you do not think they are "going crazy" or that it is "all in their head." It is our experience that some patients are more willing to explore the relevance of psychological factors after their physical concerns have been addressed, perhaps after starting or completing a course of physical rehabilitation.

When considering treatment, we recommend a stepped care model, where the complexity of treatment can be escalated according to the patients need [28]. The model starts with assessment and therapeutic diagnostic explanation as described earlier. The next step is a referral for treatment in secondary care for psychological and or physical rehabilitation.

Physical Rehabilitation

For patients who have motor symptoms, physical rehabilitation is a logical starting point. There is growing evidence for physical-based rehabilitation designed specifically for FND. Outcomes from one randomized delayed-start controlled trial ($n = 60$) [29], and several large cohort studies ($n > 45$) are promising [30,31]. Small to moderate gains in measures of physical function and quality of life are reported that are maintained at follow-up beyond one year. The composition of these treatments is varied, but there appears to be several key ingredients that have been described elsewhere in detail. There is an interdisciplinary approach to treatment with a smooth transition from medical consultation to therapy. Treatment is based on a shared formulation for symptoms. For example, symptoms are considered as a disconnection between brain and body and physical rehabilitation aims to reestablish this connection [30]. Education and self-management is an important component of many programmes. Rehabilitation is directed toward task practice (e.g., transfers, walking, activities of daily living, and leisure) rather than being impairment based (i.e., strengthening exercises, balance exercises, and so on). Sequential motor learning is used, often with distraction and redirection of attention to normalize movement patterns. In general walking aids, wheelchairs, splints, orthotics, and adaptive aids should be

TABLE 121.2 Explaining the Diagnosis [26]

Component	Example
Explain what is wrong	You have functional neurological disorder/functional weakness.
Explain the mechanism	There is a problem with the way your brain is communicating with your body/controlling movement/perceiving sensory information.
Explain how you made the diagnosis	You have a positive Hoover's sign. You can see that you are unable to push your leg down when you try, but when you lift your other leg, the muscles turn on automatically.
Explain what this means	This means that the wiring is intact and the problem lies with the function of the brain and not structure of the brain, nerves of muscles.
Explain that it is common	This is a common and genuine cause of weakness.
Explain that you believe them	I do not think that you are making it up or imagining it.
Explain that it is reversible, but it can take time and work	Because this is a problem with abnormal functioning rather than structural damage, it is possible for it to improve. However, sometimes this takes a lot of time and effort.
Explain what they do not have	You have not had a stroke and you do not have other neurological disease.
Introduce the role of psychological factors	Panic, anxiety, or other psychological problems can be part of this problem for some people. When that is the case, assessment and treatment of these can help symptoms improve.
Explain the importance of self-help	There are some things that you can do which will help this problem improve…
Suggest a treatment plan and follow-up	I will refer you to a physiotherapist and/or psychologist who has experience with this problem and arrange a follow-up appointment in 3 months from now.
Provide written information	Send a clinic letter with information on the diagnosis. Refer the patient to written or online information, for example, www.neurosymptoms.org.

avoided in patients who are rehabilitation candidates, although they can improve quality of life in patients who have failed treatment. Chronic pain and fatigue are common features of FND. When present, successful rehabilitation needs to consider these in the treatment approach. In line with this approach to treatment, consensus recommendations for physiotherapy treatment have been published [32].

Psychological Treatment

Psychological treatments have had a long association with FND yet the evidence base to support its use in motor symptoms is limited. A randomized controlled trial of a cognitive behavioral therapy (CBT) based, guided self-help intervention for patients with a wide range of functional neurological symptoms, including motor symptoms, showed benefits in subjective health at 3 months and the physical function domains of the SF36 questionnaire at 3 and 6 months. The number needed to treat was 8 [33]. There is some evidence for psychodynamic interpersonal therapy combined with neurological consultation [34,35], although these studies have low numbers and other limitations due to pragmatic experimental designs. There is better evidence for psychological treatment of the specific presentation of dissociative (nonepileptic) attack disorder [36].

Multidisciplinary Rehabilitation Programs

There is nonrandomized evidence from specialist inpatient multidisciplinary rehabilitation involving psychiatry, neurology, physical and other therapists [37,38]. These studies report modest improvements in physical and psychological measures, with at least some benefit lasting at long-term follow-up. These programs are usually reserved for patients with more severe symptoms and comorbidities.

Novel Treatments

A number of novel treatments have been described in the literature. These include therapeutic sedation, transcranial magnetic stimulation, EMG biofeedback, and electrical stimulation [32,39]. These have not been tested under controlled conditions, but they are likely to be useful treatment adjuncts in selected patients. Hypnosis in addition to multidisciplinary rehabilitation has been tested in one randomized controlled trial but was found not to be of added benefit [40].

Suggestion, Deception, and Placebo

Projecting confidence in recovery is an important part of early diagnosis and treatment, but it is important to set realistic expectations. A common myth is that all patients with FND are vulnerable to suggestion and that this should be exploited in treatment by suggesting that symptoms will be short lived and resolution complete. This is an unrealistic expectation for all patients and failure to achieve this can lead to problems such as undermining confidence in the diagnosis, seeking further investigations, and leaving the patient vulnerable to exploitation. Deceptive treatment strategies are described in the literature as useful for patients who fail standard rehabilitation approaches [41]. There is no data on the longer-term outcomes of deceptive treatments, but it would seem likely that symptom relapse is more probable. This appears to be the case for placebo treatments with curative responses, such as botulinum toxin given with suggestion that effect will be immediate [42]. We personally think deceptive approaches are unethical.

PROGNOSIS

The prognosis of FND is generally considered poor, although individual outcomes vary widely. A systematic review of long-term follow-up studies found that at an average of 7.4 years, 40% of patients were the same or worse, and it seems the majority of patients remain symptomatic [43]. These data are based on a broad cross section of patients with FND, including those with chronic symptoms at baseline assessment. There are limited data on the prognosis of first episode acute onset FND, a group that may have a higher rate of symptom resolution. Gargalas et al. found that about 60% of patients admitted acutely to a specialist stroke unit with FND were referred to another clinical service in the year following discharge, suggesting these patients remained symptomatic [4]. A number of prognostic indicators have been identified, longer duration of symptoms prior to diagnosis and the presence of personality disorder are among the most powerful predictors of poor outcome, while high satisfaction with care has been shown to predict positive outcome [43].

CONCLUSION

FSMs are common and the acute stroke doctor needs to have experience in differentiating these symptoms from those caused by cerebrovascular disease and other conditions mimicking stroke. The diagnosis should be made on positive clinical signs, backed up by relevant investigations. An early diagnosis has been associated with better outcome; therefore, the acute stroke service has a unique opportunity to make a clear diagnosis and triage for appropriate treatment. Patients presenting with FND are a heterogeneous group, and treatment should reflect this. Comorbidity accounts for a significant proportion of the symptom burden, and specific assessment and treatment of this is an important part of management. We suggest a flexible approach to treatment with escalation in a stepped care model as necessary. Detailed evidence-based recommendations are available to guide physiotherapy in addition to psychological and other multidisciplinary treatment.

References

[1] Gibson LM, Whiteley W. The differential diagnosis of suspected stroke: a systematic review. J R Coll Physicians Edinb 2013;43:114–8.

[2] Tobin WO, Hentz JG, Bobrow BJ, Demaerschalk BM. Identification of stroke mimics in the emergency department setting. J Brain Dis 2009;1:19–22.

[3] Lewandowski C, Mays-Wilson K, Miller J, Penstone P, Miller DJ, Bakoulas K, et al. Safety and outcomes in stroke mimics after intravenous tissue plasminogen activator administration: a single-center experience. J Stroke Cerebrovasc Dis 2015;24:48–52.

[4] Gargalas S, Weeks R, Khan-Bourne N, Shotbolt P, Simblett S, Ashraf L, et al. Incidence and outcome of functional stroke mimics admitted to a hyperacute stroke unit. J Neurol Neurosurg Psychiatry 2015. http://dx.doi.org/10.1136/jnnp-2015-311114.

[5] Stone J, Carson A, Duncan R, Roberts R, Warlow C, Hibberd C, et al. Who is referred to neurology clinics?—The diagnoses made in 3781 new patients. Clin Neurol Neurosurg 2010;112:747–51.

[6] Nazir FS, Lees KR, Bone I. Clinical features associated with medically unexplained stroke-like symptoms presenting to an acute stroke unit. Eur J Neurol 2005;12:81–5.

[7] Vroomen PC, Buddingh MK, Luijckx GJ, De Keyser J. The incidence of stroke mimics among stroke department admissions in relation to age group. J Stroke Cerebrovasc Dis 2008;17:418–22.

[8] Shellhaas RA, Smith SE, O'Tool E, Licht DJ, Ichord RN. Mimics of childhood stroke: characteristics of a prospective cohort. Pediatrics 2006;118:704–9.

[9] Tsivgoulis G, Zand R, Katsanos AH, Goyal N, Uchino K, Chang J, et al. Safety of intravenous thrombolysis in stroke mimics: prospective 5-year study and comprehensive meta-analysis. Stroke 2015;46:1281–7.

[10] Zinkstok SM, Engelter ST, Gensicke H, Lyrer PA, Ringleb PA, Artto V, et al. Safety of thrombolysis in stroke mimics: results from a multicenter cohort study. Stroke 2013;44:1080–4.

[11] Kranick S, Ekanayake V, Martinez V, Ameli R, Hallett M, Voon V. Psychopathology and psychogenic movement disorders. Mov Disord 2011;26:1844–50.

[12] Stone J, Edwards MJ. How "psychogenic" are psychogenic movement disorders? Mov Disord 2011;26:1787–8.

[13] Edwards MJ, Fotopoulou A, Parees I. Neurobiology of functional (psychogenic) movement disorders. Curr Opin Neurol 2013;26:442–7.

[14] Stone J. The bare essentials: Functional symptoms in neurology. Pract Neurol 2009;9:179–89.

[15] Stone J, Warlow C, Sharpe M. Functional weakness: clues to mechanism from the nature of onset. J Neurol Neurosurg Psychiatry 2012;83:67–9.

[16] Stone J, Reuber M, Carson A. Functional symptoms in neurology: mimics and chameleons. Pract Neurol 2013;13:104–13.

[17] Edwards MJ, Bhatia KP. Functional (psychogenic) movement disorders: merging mind and brain. Lancet Neurol 2012;11:250–60.

[18] Daum C, Hubschmid M, Aybek S. The value of "positive" clinical signs for weakness, sensory and gait disorders in conversion disorder: a systematic and narrative review. J Neurol Neurosurg Psychiatry 2014;85:180–90.

[19] McWhirter L, Stone J, Sandercock P, Whiteley W. Hoover's sign for the diagnosis of functional weakness: a prospective unblinded cohort study in patients with suspected stroke. J Psychosom Res 2011;71:384–6.

[20] Kaski D, Bronstein AM, Edwards MJ, Stone J. Cranial functional (psychogenic) movement disorders. Lancet Neurol 2015;14:1196–205.

[21] Stone J, Carson A. Functional neurologic disorders. Contin (Minneap Minn) 2015;21:818–37.

[22] Stone J, Warlow C, Sharpe M. The symptom of functional weakness: a controlled study of 107 patients. Brain 2010;133:1537–51.

[23] Stone J, Smyth R, Carson A, Lewis S, Prescott R, Warlow C, et al. Systematic review of misdiagnosis of conversion symptoms and "hysteria". BMJ 2005;331:989.

[24] Parees I, Saifee TA, Kojovic M, Kassavetis P, Rubio-Agusti I, Sadnicka A, et al. Functional (psychogenic) symptoms in Parkinson's disease. Mov Disord 2013;28:1622–7.

[25] Stone J, Carson A, Duncan R, Roberts R, Coleman R, Warlow C, et al. Which neurological diseases are most likely to be associated with "symptoms unexplained by organic disease". J Neurol 2012;259:33–8.

[26] Stone J. Functional neurological disorders: the neurological assessment as treatment. Pract Neurol 2016;16:7–17.

[27] Stone J, Edwards M. Trick or treat? Showing patients with functional (psychogenic) motor symptoms their physical signs. Neurology 2012;79:282–4.

[28] Health Improvement Scotland. Stepped care for functional neurological symptoms.

[29] Jordbru AA, Smedstad LM, Klungsøyr O, Martinsen EW. Psychogenic gait disorder: A randomized controlled trial of physical rehabilitation with one-year follow-up. J Rehabil Med 2014;46:181–7.

[30] Czarnecki K, Thompson JM, Seime R, Geda YE, Duffy JR, Ahlskog JE. Functional movement disorders: successful treatment with a physical therapy rehabilitation protocol. Park Relat Disord 2012;18:247–51.

[31] Nielsen G, Ricciardi L, Demartini B, Hunter R, Joyce E, Edwards MJ. Outcomes of a 5-day physiotherapy programme for functional (psychogenic) motor disorders. J Neurol 2015;262:674–81.

[32] Nielsen G, Stone J, Matthews A, Brown M, Sparkes C, Farmer R, et al. Physiotherapy for functional motor disorders: a consensus recommendation. J Neurol Neurosurg Psychiatry 2015;86:1113–9.

[33] Sharpe M, Walker J, Williams C, Stone J, Cavanagh J, Murray G, et al. Guided self-help for functional (psychogenic) symptoms: a randomized controlled efficacy trial. Neurology 2011;77:564–72.

[34] Hubschmid M, Aybek S, Maccaferri GE, Chocron O, Gholamrezaee MM, Rossetti AO, et al. Efficacy of brief interdisciplinary psychotherapeutic intervention for motor conversion disorder and nonepileptic attacks. Gen Hosp Psychiatry 2015;37:448–55.

[35] Kompoliti K, Wilson B, Stebbins G, Bernard B, Hinson V. Immediate vs. delayed treatment of psychogenic movement disorders with short term psychodynamic psychotherapy: randomized clinical trial. Park Relat Disord 2014;20:60–3.

[36] Goldstein LH, Mellers JD, Landau S, Stone J, Carson A, Medford N, et al. Cognitive behavioural therapy vs standardised medical care for adults with Dissociative non-Epileptic Seizures (CODES): a multicentre randomised controlled trial protocol. BMC Neurol 2015;15:98.

[37] Demartini B, Batla A, Petrochilos P, Fisher L, Edwards MJ, Joyce E. Multidisciplinary treatment for functional neurological symptoms: a prospective study. J Neurol 2014;261:2370–7.

[38] McCormack R, Moriarty J, Mellers JD, Shotbolt P, Pastena R, Landes N, et al. Specialist inpatient treatment for severe motor conversion disorder: a retrospective comparative study. J Neurol Neurosurg Psychiatry 2013;85:895–900.

[39] Pollak TA, Nicholson TR, Edwards MJ, David AS. A systematic review of transcranial magnetic stimulation in the treatment of functional (conversion) neurological symptoms. J Neurol Neurosurg Psychiatry 2014;85:191–7.

[40] Moene FC, Spinhoven P, Hoogduin KAL, van Dyck R. A randomised controlled clinical trial on the additional effect of hypnosis in a comprehensive treatment programme for in-patients with conversion disorder of the motor type. Psychother Psychosom 2002;71:66–76.

[41] Shapiro AP, Teasell RW. Behavioural interventions in the rehabilitation of acute v. chronic non-organic (conversion/factitious) motor disorders. Br J Psychiatry 2004;185:140–6.

[42] Edwards MJ, Bhatia KP, Cordivari C. Immediate response to botulinum toxin injections in patients with fixed dystonia. Mov Disord 2011;26:917–8.

[43] Gelauff J, Stone J, Edwards M, Carson A. The prognosis of functional (psychogenic) motor symptoms: a systematic review. J Neurol Neurosurg Psychiatry 2014;85:220–6.

CHAPTER

122

CADASIL

H. Chabriat

Centre de référence pour les maladies rares des vaisseaux du cerveau et de l'œil (CERVCO), DHU-NeuroVasc and INSERM U1161, Université Denis Diderot, Paris, France

INTRODUCTION

The acronym CADASIL (Cerebral Autosomal Dominant Arteriopathy with Subcortical Infarcts and Leukoencephalopathy) is for an autosomal dominant inherited cerebral small-vessel disease responsible in adults for subcortical infarcts and leukoencephalopathy [8]. Mutations in NOTCH3 gene on chromosome 19 are responsible for the disease. CADASIL is observed in all countries worldwide. Hundreds of families have been reported in Europe. In a region of the West of Scotland of 1,418,990 individuals, 22 cases confirmed by genetic testing were detected that allows estimating the prevalence of the disease at 1.98 per 100,000 adults.

CLINICAL MANIFESTATIONS

The earliest clinical manifestations reported in nearly half of patients are attacks of migraine with aura that most often occur between 20 and 30 years. Ischemic events are reported by 70–80% of patients and usually occur around 50 years. Cognitive alterations of variable importance are present early during the course of the disease but they become evident between 50 and 60 years. Dementia is almost constant at the final stage of the disease and associated with severe gait and balance disturbances [1].

Unlike attacks of migraine without aura whose frequency in CADASIL is close to that observed in the general population, migraine attacks with aura are detected with a frequency more than five times higher than in the general population. Age at first attacks of migraine with aura is extremely variable from one individual to another; mean age is around 30 years and appears earlier in women (26 years) compared with men (35 years) [2]. The frequency of attacks is extremely variable from one subject to another, from two attacks per week to one attack every 4 years. The triggers are those usually reported during migraine attacks. In addition to migraine attacks with typical aura, more than half of patients report at least one atypical aura evoking basilar migraine, hemiplegic migraine, aura without headache, or aura of sudden onset. Some patients may even develop severe attacks leading to confusion, coma, hyperthermia, visual or motor deficit for several days, sometimes associated with cerebral edema, and intracranial hypertension.

Between 70% and 85% of patients have at least a transient ischemic attack or an ischemic stroke. The average age at onset of ischemic stroke is between 45 and 50 years (range 20–70 years) with no significant difference between women and men. Most manifestations are similar to those of typical lacunar syndromes. Some observations of territorial infarction have been rarely reported. The cerebral ischemic events can occur in the absence of any vascular risk factor. The real impact of vascular risk factors on the clinical phenotype of CADASIL is still poorly known. Active smoking was found to have a negative predictive value on the occurrence of ischemic events and clinical worsening. Brain hemorrhages are possible but exceptional. They have been observed in about a dozen of cases in the literature.

About 5–10% of patients can present with epileptic seizures, sometimes with partial, most commonly with generalized, seizures. The majority of epileptic individuals already had one or more cerebral infarction. Seizures can be observed in the absence of any visible cortical lesion on conventional MRI. They are usually well controlled by conventional treatments.

Mood disorders affect about 20% of patients; their frequency seems to vary within families. They can be inaugural. Some patients can present with episodes mimicking severe depression. Depressive symptoms may alternate with manic episodes as in bipolar disorder. Conventional antidepressants seem to be effective in the vast majority of cases.

Apathy defined as a lack of motivation responsible for a reduction in voluntary goal-directed activities is frequently detected during the progression of the disease (30–40% of cases). Apathetic patients seem to have a more severe clinical presentation than those free from apathy.

Decline in cognitive performances that can lead to dementia represents the second clinical manifestation of the disease after ischemic stroke. The profile of the initial cognitive deficit is usually heterogeneous. Executive dysfunction is the most frequent deficit. It can be detected even in young patients, decades before the onset of other manifestations of the disease. These cognitive changes frequently include alterations of attention and working memory. Cognitive decline increases over years and becomes progressively more diffuse and homogeneous and can include decrease of performances in all cognitive domains. Memory deficit of hippocampal type can even be observed in 20% of patients with CADASIL with memory alterations at the end stage of the disease. Cognitive decline may evolve acutely or increase progressively even in the absence of any ischemic event. Dementia is observed in one-third of symptomatic patients. Its frequency increases with age; it affects about 60% of patients over 60 years and is observed in 80% of subjects before death. When dementia is present, cognitive deficit is often extensive and involves executive function, attention, and memory, as well as reasoning and language. Severe

aphasia, apraxia, or agnosia is rarely observed. Semantic memory and recognition are preserved. Dementia can remain isolated in 10% of cases. It is frequently associated with gait disturbances, urinary incontinence, and pseudobulbar palsy. The final stage of the disease is characterized by progressive loss of independence with severe disability. Mean age at death is estimated around 64.6 years for men and 70.7 years for women.

IMAGING

MRI is always abnormal in symptomatic cases (after excluding migraine with aura). MRI signal abnormalities can also be detected during the presymptomatic phase of the disease from the age of 20 years. After 35 years, all carriers of the mutated gene usually have abnormal MRI.

Hyperintense lesions are observed on T2-weighted or fluid-attenuated inversion recovery (FLAIR) MRI (Fig. 122.1). They are more or less diffuse within the white matter, often punctate in basal ganglia, thalamus, and brainstem. The extent of T2/FLAIR hyperintensities increases with age, at onset signal abnormalities are usually punctate or nodular, of symmetrical distribution. They become gradually confluent and extend to the whole white matter. Periventricular occipital and frontal lesions are constant when MRI is abnormal. The frequency of signal abnormalities in the anterior pole of the temporal lobes has a crucial diagnostic value because of their specificity compared with signal anomalies detected in other areas as in other small-vessel diseases. In the brainstem, hyperintensities predominate in the central pons. An increase of diffusion and loss of anisotropy can be measured using diffusion tensor imaging techniques in all hyperintense areas as well as in the normal-appearing white matter.

On T1-weighted images, punctiform hypointense lesions corresponding to small cavities with cerebrospinal fluid (lacunes) are detected in 2/3 of symptomatic individuals. They are observed in the white matter, basal ganglia, or brainstem. The total load of lacunes is found highly correlated with the clinical severity. Dilated perivascular spaces (smaller hypointensities) are also frequent (Fig. 122.2). They are seen at the junction of the cortical ribbon and white matter in the temporal poles and external capsules T1 where they appear very specific to the disease.

Microbleeds are mainly observed on "gradient echo" or T2*-weighted images. They are detected in 25–69% of cases. Occurrence of microbleeds was found associated with blood pressure level, HbA1C, and extent of white matter hyperintensities [3]. The distribution of microbleeds differs from that of infarcts and white matter hyperintensities; they are often observed within the cerebellum.

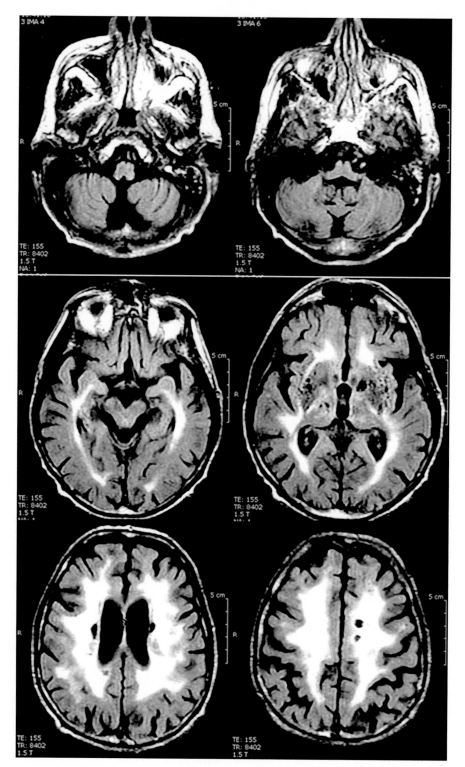

FIGURE 122.1   Fluid-attenuated inversion recovery (FLAIR) images showing the distribution of white matter hyperintensities on FLAIR images and the presence of lacunes in the left thalamus and centrum semiovale associated with multiple dilated perivascular spaces in basal ganglia in CADASIL.

Cortical or cerebellar lesions are exceptional. However, very small cortical infarcts have been already detected using 7-T MRI and confirmed postmortem [4]. Changes in the cortex thickness or of sulcal morphology were found related to the accumulation of subcortical lesions [5].

Cerebral angiography is usually normal in patients with CADASIL. Angiography can aggravate patients

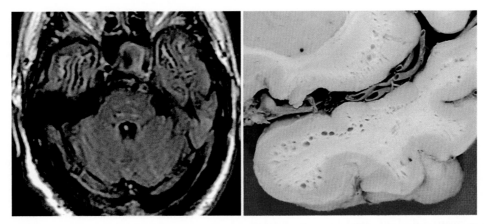

FIGURE 122.2   Dilated perivascular spaces and white matter hyperintensities in the temporal lobes. *Data obtained from Pr Fancoise Gray, Service d'anatomopathologie, Hopital Lariboisiere, Paris – Copyright Lancet Neurol, 2009. Data obtained from Chabriat H, Joutel A, Dichgans M, Tournier-Lasserve E, Bousser MG. Cadasil. Lancet Neurol 2009;8(7):643–53. .*

and lead to severe clinical manifestations and should be therefore avoided.

PATHOLOGICAL DATA

Macroscopic examination usually shows mild cerebral atrophy with no or very limited atherosclerotic lesions in the circle of Willis in CADASIL. The white matter appears gray, brown, or grainy with many cystic lesions. Lacunar infarcts are detected in the white matter and basal ganglia with numerous dilated perivascular spaces. Microscopic examination shows a diffuse pallor of myelin that usually spares U-fibers. White matter lesions predominate in periventricular areas and within the centrum semiovale. Small infarcts are located in the deep white matter. The media of cerebral and leptomeningeal arteries is thickened. Perforating arteries may appear stenotic and their wall with hyalinosis and adventitial fibrosis. The most specific aspect is the presence of granular deposits within the media corresponding to NOTCH3 extracellular products accumulation. On electron microscopy, this material appears granular and osmiophilic [granular osmiophilic material (GOM)] [6]. The presence of GOM is characteristic of the disease and is also detected in small vessels of other organs as well as in skin microvessels. NOTCH3 antibodies can be used to show the abnormal accumulation of the Notch3 protein in the vascular wall. The accumulation of the Notch3 protein within the vessel wall was shown to cause aggregation of other proteins of the extracellular matrix, particularly the TIMP3 protein and vitronectin altering the functional properties of the vessel wall [7].

DIAGNOSIS

The diagnosis of CADASIL is now easily confirmed by a genetic test. The NOTCH3 gene consists of 33 exons and encodes a protein of 2321 amino acids. This protein is expressed at the plasma membrane of vascular smooth muscle cells. It is a heterodimer consisting of an extracellular portion that contains 34 epidermal-like growth factor repeat units, each one containing six cysteine residues [8,9]. To date, more than a hundred mutations have been found causing CADASIL. They are, invariably, located within exons 2–24. All lead to an odd number of cysteine residues. Mutations in exon 4 are the most frequent mutations but the distribution of mutations varies by population. Approximately 95% of mutations are missense mutations but deletions and splice site mutations have also been reported. In most cases, heterozygous mutations are inherited from a patient having already a typical mutation. De novo mutations have also been described and homozygous cases are possible. The diagnosis of CADASIL is easily considered in a symptomatic case within a CADASIL family. The diagnosis can also be raised in patients with a personal or family history of migraine attacks with aura, ischemic stroke, dementia, and/or mood disorders and whose MRI reveals symmetrical and periventricular white matter lesions on T2-weighted MRI. The presence of MRI signal abnormalities affecting the subcortical white matter of the anterior portion of the temporal lobes and/or external capsules makes the diagnosis even more plausible.

Genetic counseling is essential in subjects at risk and is done through a multidisciplinary consultation before deciding genetic testing [10]. Electron microscopy study of skin biopsy or the use of anti-NOTCH3 antibodies on pathological examination can confirm the diagnosis in difficult cases.

TREATMENT

Aspirin is currently used for secondary prevention after stroke in CADASIL but the benefit of this treatment has not been shown. Anticoagulants are not recommended. To treat migraine attacks with aura, vasoconstrictors are not recommended in general because of the theoretical risk of reduced cerebral blood flow. Nonsteroidal antiinflammatory drugs and analgesics are therefore recommended as first-line treatments of migraine attacks. Acetazolamide, which significantly increases cerebral blood flow in patients with CADASIL, has been proposed as a potential preventive treatment of migraine attacks because of its strong efficacy already reported in few patients. Cholinesterase inhibitors are not of proven efficacy after a controlled clinical trial conducted in patients with CADASIL with executive dysfunction and/or mild dementia. Some improvement of performances in executive function was detected with donepezil. Antihypertensive drugs should be used with caution because of the risk of cerebral hypoperfusion in CADASIL. Rehabilitation techniques are needed to maintain as long as possible patient independence and motor performances. Finally, psychological support is absolutely crucial for the family and the patient.

References

[1] Chabriat H, Joutel A, Dichgans M, Tournier-Lasserve E, Bousser MG. Cadasil. Lancet Neurol 2009;8(7):643–53.

[2] Guey S, Mawet J, Herve D, Duering M, Godin O, Jouvent E, et al. Prevalence and characteristics of migraine in CADASIL. Cephalalgia 2015. pii:0333102415620909.

[3] Viswanathan A, Guichard JP, Gschwendtner A, Buffon F, Cumurcuic R, Boutron C, et al. Blood pressure and haemoglobin A1c are associated with microhaemorrhage in CADASIL: a two-centre cohort study. Brain 2006;129(Pt 9):2375–83.

[4] Jouvent E, Mangin JF, Duchesnay E, Porcher R, During M, Mewald Y, et al. Longitudinal changes of cortical morphology in CADASIL. Neurobiol Aging 2012;33(5):e29–36. 1002.

[5] Jouvent E, Poupon C, Gray F, Paquet C, Mangin JF, Le Bihan D, et al. Intracortical infarcts in small vessel disease: a combined 7-T postmortem MRI and neuropathological case study in cerebral autosomal-dominant arteriopathy with subcortical infarcts and leukoencephalopathy. Stroke 2011;42(3):e27–30.

[6] Ruchoux MM, Brulin P, Brillault J, Dehouck MP, Cecchelli R, Bataillard M. Lessons from CADASIL. Ann NY Acad Sci 2002;977:224–31.

[7] Monet-Lepretre M, Haddad I, Baron-Menguy C, Fouillot-Panchal M, Riani M, Domenga-Denier V, et al. Abnormal recruitment of extracellular matrix proteins by excess Notch3ECD: a new pathomechanism in CADASIL. Brain 2013;136(Pt 6):1830–45.

[8] Joutel A, Corpechot C, Ducros A, Vahedi K, Chabriat H, Mouton P, et al. Notch3 mutations in CADASIL, a hereditary adult-onset condition causing stroke and dementia. Nature 1996;383(6602):707–10.

[9] Joutel A, Vahedi K, Corpechot C, Troesch A, Chabriat H, Vayssiere C, et al. Strong clustering and stereotyped nature of Notch3 mutations in CADASIL patients. Lancet 1997;350(9090):1511–5.

[10] Reyes S, Kurtz A, Herve D, Tournier-Lasserve E, Chabriat H. Presymptomatic genetic testing in CADASIL. J Neurol 2012;259.

CHAPTER

123

Stroke and Infection: Tuberculosis, Brucellosis, Syphilis, Lyme Disease and Listeriosis

R. Sabzwari, D.H. Slade

Loyola University Medical Center and Edward Hines Jr. Veteran Administration Hospital, Hines, IL, United States

Neurological complications may occur as sequelae of systemic infection, but can be a presenting feature with certain pathogens. An association exists between ischemic strokes and infection, with a number of proposed mechanisms by which vascular disease develops into central nervous system (CNS) infections. The pathophysiological mechanisms may involve vasculitis, affecting primarily the vessels at the base of the brain in the setting of meningitis, an immune-mediated parainfectious process leading to vasospasm or thrombosis, or a hypercoagulable state in combination with endothelial dysfunction resulting from activation of inflammatory and procoagulant cascades [4].

Patients with neurotuberculosis, neurosyphilis, neurobrucellosis, and neuroinvasive listeriosis can initially have meningitis which may be asymptomatic, but can progress to a chronic form. Cerebrospinal fluid (CSF) analysis is notable for lymphocytic pleocytosis, high protein content, and low glucose levels, and hence is a more sensitive guide in diagnosis, treatment, and cure of the disease.

Stroke is a serious complication with CNS infections and is clinically indistinguishable from a noninfectious etiology. Proper management requires a high index of suspicion, prompt recognition and early initiation of appropriate therapy, given progression to devastating neurologic outcomes.

NEUROSYPHILIS

Neurosyphilis is a globally endemic disease caused by the spirochete *Treponema pallidum*. With the development of penicillin, there was a decline in the incidence of syphilis, followed by a resurgence since 2000. *T.pallidum* rapidly invades the CNS early in the course of infection, though most patients remain asymptomatic and the infection is cleared from the CNS. Symptomatic neurosyphilis may occur at any stage of disease.

Pathogenesis

T. pallidum invades, usually following sexual activity through small breaks in the skin or intact mucosa, manifesting later as a chancre at the site of infection. The bacteria then rapidly disseminate to virtually every organ system. CNS invasion occurs quickly, though the majority of patients remain asymptomatic. Infection produces a vasculitis, leading to endarteritis of the large- and medium-sized vessels. Obliterative endarteritis is characterized by fibroblast proliferation in the intima, a thinning of the media, and fibrosis of the adventitia. This inflammation leads to thrombosis and infarction. Infection with *T. pallidum* may also directly involve the brain parenchyma, causing frontal and temporal lobe atrophy [1].

Clinical Features

Early symptoms of neurosyphilis include headache, vertigo, and insomnia, typically with a sudden onset. Stroke symptoms are directly related to the areas of infarction, with the middle cerebral artery (MCA) and basilar artery (BA) commonly involved. Patients may develop hemiplegia, hemianesthesia, homonymous hemianopia, and aphasia. If the vessels of the spinal

Primer on Cerebrovascular Diseases, Second Edition
http://dx.doi.org/10.1016/B978-0-12-803058-5.00123-5

cord are involved, syphilitic meningomyelitis presents as spastic weakness of the legs, loss of sphincter control, and sensory loss. Parenchymal involvement in the brain manifests as general paresis with features of dementia and psychosis or in the spinal cord as tabes dorsalis with loss of proprioception and vibratory sense [1].

Diagnosis

Diagnosis of neurosyphilis is based on CSF analysis that shows an elevated white blood cell (WBC) count ≥5 cells/mL with lymphocyte predominance, elevated protein, and a reactive CSF Venereal Disease Research Laboratory (VDRL) test. The sensitivity of CSF VDRL has not been clearly established, and while a positive CSF VDRL confirms neurosyphilis, a negative test does not exclude it. There are no pathognomonic radiological findings, and imaging is of little diagnostic value.

Treatment

Penicillin is the treatment of choice for neurosyphilis. If the patient is penicillin allergic, desensitization should be performed. Treatment is with aqueous crystalline penicillin G 18–24 million units per day intravenously (IV), given either by continuous infusion or as 3–4 million units IV every 4h for 10–14 days. If penicillin cannot be used, then ceftriaxone 2g IV or intramuscularly (IM) every 24h may be used in nonpregnant patients for 10–14 days, although the data on efficacy are limited.

NEUROBRUCELLOSIS

Neurobrucellosis is rarely seen in most developed countries, but remains endemic in the Middle East, India, and Latin America. *Brucella* spp. are aerobic, intracellular gram-negative coccobacilli. Four pathogenic species, *Brucella abortus*, *Brucella suis*, *Brucella canis*, and *Brucella melitensis* infect humans. *B. melitensis* is the most pathogenic. Most infections occur secondary to direct exposure to animals or consumption of contaminated animal products including unpasteurized milk.

Pathogenesis

Brucellosis occurs following contact with infected animals, unpasteurized milk products, or by inhalation. The bacteria then migrate to lymph nodes and survive intracellularly within macrophages. CNS involvement is mediated both by the direct effect of the bacteria and the resulting immune response. CNS invasion occurs either via transport within macrophages or direct endothelial cell invasion. *Brucella* invades and replicates within

astrocytes and microglia. Activation of the innate immune system triggers an inflammatory response which leads to astrogliosis, a characteristic feature of neurobrucellosis. This inflammatory response is primarily responsible for the clinical features of neurobrucellosis. Ischemic stroke results from vascular and perivascular inflammation leading to lacunar infarcts, microhemorrhages, and venous thrombosis. A second mechanism by which stroke may occur is rupture of a mycotic aneurysm. Hydrocephalus may be seen if granulomatous inflammation within the CSF blocks reuptake by the arachnoid villi [2].

Clinical Features

Acute neurobrucellosis presents with high fever, malaise, headaches, and arthralgias. Meningeal signs, confusion, papilledema, hepatosplenomegaly, back pain, ataxia, paresthesias, paraplegia, and urinary or fecal incontinence may also be present. Patients may develop stroke, acute or chronic meningoencephalitis, myelitis, radiculitis, isolated cranial or peripheral nerve involvement, brain abscess, subarachnoid bleeding, or neuropsychiatric symptoms. While mortality is low at less than 1%, morbidity may be significant [2].

Diagnosis

The first step in the diagnosis of neurobrucellosis is recognition of neurological signs and symptoms and exclusion of other causes. CSF analysis shows increased WBC count with lymphocytic predominance, increased protein content, and a low glucose level. A positive CSF culture is confirmatory for neurobrucellosis, though sensitivity is low. Bone marrow cultures may aid in the diagnosis of chronic brucellosis. Testing should also be performed for *Brucella* antibodies in the CSF. Imaging studies including CT and MRI of the brain may show evidence of infarct, leptomeningeal enhancement, and nodular lesions (Figs. 123.1 and 123.2) [3].

Treatment

Recommended treatment for neurobrucellosis consists of 6 months of antibiotic therapy, which may be extended for a longer duration based on clinical response. Treatment is challenging due to the intracellular nature of the organism. Antibiotics with activity against *Brucella* include ceftriaxone, rifampin, doxycycline, and aminoglycosides (streptomycin and gentamicin). Regimens vary, but typically include multiple agents. The combination of doxycycline 100 mg IV or PO (orally) BID (twice daily), rifampin 600–900 mg PO daily, and ceftriaxone 2g IV every 12h is one example. Repeat CSF studies should be performed to confirm normalization as evidence of cure [3].

NEUROBORRELIOSIS

Lyme disease, one of the most common tickborne infections, is caused by *Borrelia burgdorferi* in the USA and other *Borrelia* species in Europe and Asia. In the USA, the incidence is highest in the North East, northern Midwest, and occasionally in northern California. *Ixodes scapularis*, the deer tick, serves as the vector, with the vast majority of cases occurring in summer months. The nymph stage of the tick is responsible for most human cases, but adult ticks can also transmit the infection. The tick must remain attached for 24–48 h to successfully infect humans.

Initial symptoms may include fever, headaches, fatigue, and a characteristic spreading rash called erythema migrans. If untreated, hematogenous dissemination of the spirochetes occurs and may involve the joints, heart, or CNS, resulting in Lyme neuroborreliosis (LNB).

Pathogenesis

After *B. burgdorferi* enters the host following an infected tick bite, several mechanisms in the pathogenesis of LNB have been proposed.

CNS invasion may occur through the bloodstream or along the peripheral nerves. Even though a definite entry of *Borrelia* has been proven by culture methods and polymerase chain reaction (PCR), the exact mechanism is not yet understood. Borrelial-induced chemokines can be found in high concentrations in the CSF, which may contribute to bacterial migration along the peripheral nerves or lymphatic vessels to the CNS. The spirochetes interact with receptors on glial cells, inducing the production of pro-inflammatory cytokines including interleukin-6 or tumor necrosis factor alpha (TNF-α). These cytokines are neurotoxic and may provoke an autoimmune reaction leading to LNB [5].

Clinical Features

After an incubation period of about 3–32 days, progressive infection from the early localized erythema migrans to late manifestations, including CNS involvement is seen in patients who have not received antimicrobial therapy.

The neurological abnormalities include lymphocytic meningitis, encephalitis, cranial nerve abnormalities (particularly peripheral facial nerve palsy), cerebellar ataxia, radiculoneuritis, and myelitis [6]. A small number of patients present with ischemic stroke secondary to cerebral vasculitis. Stroke can occur several weeks to months after a period of nonspecific symptoms such as headaches and malaise. LNB should be considered as a possible

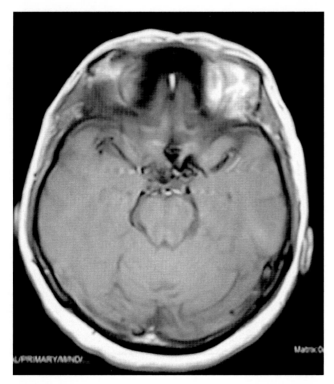

FIGURE 123.1 MRI images of a 51-year-old woman with neurobrucellosis showing a suprasellar cistern located extra axial lesion, hypointense on axial T2 weighted and T1 images. *Used with permission from Elsevier.*

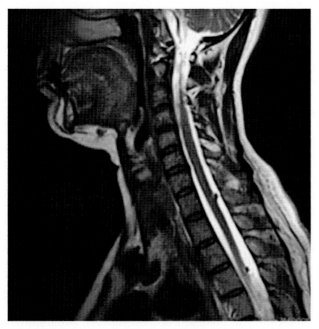

FIGURE 123.2 Sagittal T2 weighted image of MRI of the cervical region, showing intradural, extramedullary hypointense nodular lesions (consistent with granuloma formation) in a 21-year-old-male with neurobrucellosis and meningitis. *Used with permission from Elsevier.*

etiology of otherwise unexplained cerebral vasculitis and stroke, particularly when there is a history of tick bites, erythema migrans, facial palsy, or radicular symptoms.

Diagnosis

The diagnosis of LNB relies on the epidemiology, clinical history and examination, CSF analysis, and antibody studies of serum and CSF.

B. burgdorferi is a tissue-based organism and does not remain in blood or CSF for prolonged durations. Cultures are not readily available, but serological diagnosis is widely used. The Centers for Disease Control and Prevention (CDC) recommends a two-step process when testing blood for antibodies against this bacterium. ELISA (enzyme-linked immunosorbent assay) is the screening test and western blot (WB) is used for confirmation, using the same blood sample. By week 4, the test is positive in 70–80% of patients with Lyme infection. CSF analysis shows a lymphocytic pleocytosis, elevated protein content, and normal glucose levels. CSF immunoglobulins IgG, IgM, or IgA directed against *B. burgdorferi* may be seen. Lyme antibodies are more often present during acute, rather than chronic neuroborreliosis in the CSF.

Imaging with MRI may reveal meningeal enhancement in patients presenting with stroke, suggesting an underlying inflammatory disease.

Treatment

Patients with isolated peripheral facial nerve palsies can be treated with oral doxycycline (100 mg every 12 h). Amoxicillin (500 mg TID, three times daily) or cefuroxime (500 mg BID) are reasonable alternatives, if there is a contraindication to therapy with doxycycline (e.g., children less than 8 years of age, pregnant or breast-feeding women, and allergy to tetracycline).

Preferred parenteral regimens for patients with neurological abnormalities include ceftriaxone (2 g IV daily) or penicillin G (18–24 million units/day in divided doses).

A Jarisch–Herxheimer reaction may occur during treatment of any form of Lyme disease. Some patients may experience worsening of symptoms in the initial 24 h of therapy, resulting from the host's inflammatory reaction to dying spirochetes.

Duration of treatment is usually 14–28 days and results in significant clinical improvement even though it may take longer for complete normalization of CSF.

NEUROTUBERCULOSIS

Tuberculosis (TB) is one of the world's deadliest diseases, and it is estimated that one-third of the population is infected. The majority of the cases present as pulmonary disease, with CNS involvement accounting for about 1% of all cases.

It can manifest as tuberculous meningitis (TBM), spinal arachnoiditis, or a tuberculoma. Cerebrovascular complications, including stroke may occur after TBM in 15–57% of patients with advanced disease [8]. Early recognition and treatment initiation is key to improved outcomes.

Pathogenesis

Mycobacterium tuberculosis (MTB) is an aerobic, non-spore forming acid-fast bacterium.

The pathogenesis involves entry of MTB into the systemic circulation following an initial pulmonary infection. It subsequently invades the CNS in which it establishes a focus called Rich focus, either in the meninges, subpial, or subependymal region of the brain or spinal cord. This focus may then rupture into the subarachnoid space or ventricular system causing meningitis.

The likelihood of invasion into the CNS depends on the magnitude of bacteremia and the extent of the cell-mediated immune response to the infection [8].

Clinical Features

Patients with TBM may often have a subacute clinical presentation with nonspecific symptoms, such as low-grade fever, weight loss, malaise, and encephalopathy earlier in the course of disease. Subsequently, progressive symptoms include nausea, vomiting, lethargy, confusion, and cranial neuropathies. Stupor, coma, and seizures occur in advanced stages of the disease. Cranial nerve involvement in TBM occurs as a result of ischemia or entrapment of the nerves in basal exudates. Fibrotic changes in the later stages cause permanent loss of function in these nerves.

Stroke occurs as a complication of TBM. Patients often present with monoplegia, hemiplegia, or quadriplegia. Depending on the location of the infarct, manifestations may include ataxic gait, sensory impairment, seizures, cranial nerve palsies, movement disorders, apraxia, aphasia, and hypothalamic disturbances.

Diagnosis

A positive acid-fast smear or culture for MTB from a CSF sample is considered the gold standard for definitive diagnosis of neurotuberculosis. CSF analysis shows a lymphocytic pleocytosis (neutrophilic predominance can be seen early in infection), elevated protein level (>100 mg/dL) and low glucose levels. MTB PCR should be obtained on CSF sample if there is high suspicion. Because of widespread resistance, antibiotic susceptibility testing should be completed on all positive MTB cultures.

The prognostic value of TST (tuberculin skin test) and I GRA (interferon gamma release assay) in the diagnosis of CNS TB is limited.

MRI is the imaging modality of choice and is superior to CT in the evaluation of suspected TBM and its sequelae [7]. Findings include choroid plexus enhancement with ventricular enlargement or thick meningeal enhancement—either of which are highly suggestive of TBM. Cerebral infarcts can be seen in nearly 30% of cases (Figs. 123.3 and 123.4) [9].

Treatment

CNS TB is associated with significant mortality, and therefore prompt initiation of antimycobacterial therapy is key to survival. For patients, in whom there is strong clinical suspicion based on history, epidemiology, and suggestive CSF findings, treatment should be started while awaiting results of specific diagnostic tests. Rifampin/rifabutin, isoniazid (INH), pyrazinamide (PZA), and ethambutol constitute first-line therapy.

INH and PZA have excellent CNS penetration and should always be included as part of the antimycobacterial regimen. After the initial 2 months of combination therapy with four drugs, de-escalation to rifampin and INH for the remainder of the treatment course is recommended, once the susceptibility of the isolate to both of those drugs is confirmed.

Given the risk of optic neuritis with ethambutol, baseline ophthalmological examination should be completed

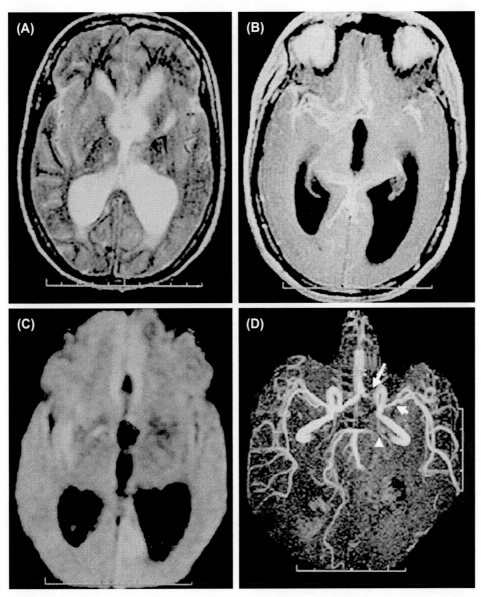

FIGURE 123.3 Cranial MRI and MR angiography of a 32-year-old woman with stage III tuberculous meningitis. (A) T2 sequence showing left anterior limb of internal capsule hyperintensity and hydrocephalous. (B) T1 contrast showing meningeal enhancement. (C) DWI showing acute infarction of the right genu of the internal capsule. (D) MR angiography showing narrowing of proximal MCA (*short arrow*) and anterior cerebral (*long arrow*) arteries and nonvisualization of posterior cerebral artery (*arrow head*). *Used with permission from Elsevier.*

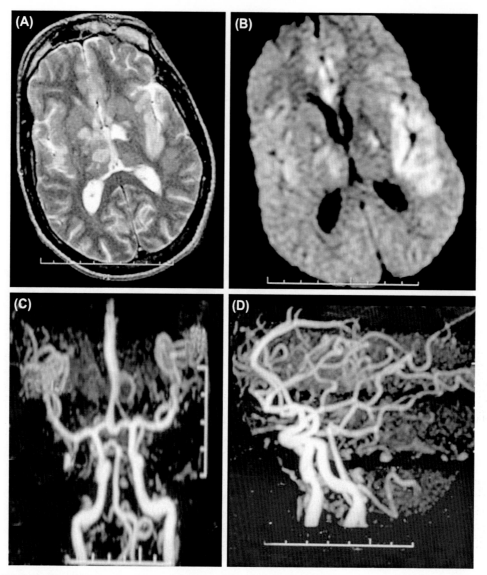

FIGURE 123.4   Cranial MRI and MR angiography of a 30-year-old man with stage III meningitis showing multiple infarcts involving both the basal ganglia and left cortical infarct on the middle cerebral arterial territory on both T2 (A) and DWI (B). Angiography however is normal (C and D). *Used with permission from Elsevier.*

for patients starting this drug. Vitamin B6 should be administered with INH to prevent peripheral neuropathy and liver function tests should be monitored.

Second-line agents include ethionamide, cycloserine, kanamycin, capreomycin, and para-amino salicylic acid, but are considered less efficacious and associated with greater toxicity.

High-dose corticosteroids should be used as adjunctive therapy and have shown benefit in treatment of CNS TB.

NEUROINVASIVE LISTERIOSIS

Listeriosis is a zoonotic infection caused by a gram-positive, facultatively anaerobic, intracellular bacterium called *Listeria monocytogenes*. Invasive disease including meningitis, encephalitis, or bacteremia may occur in the immunosuppressed host. The mean incubation period for neuroinvasive disease is about 31 days, and differs significantly for *L. monocytogenes*-induced gastroenteritis (about 24 h).

The most common CNS manifestation of *L. monocytogenes* is meningoencephalitis. Brain abscess, cerebritis, and rhomboencephalitis are rare but serious complications and associated with significant mortality. Ingestion of contaminated food appears to be the most common route of transmission of *L. monocytogenes*.

Pathogenesis

L. monocytogenes possesses an affinity to cause CNS infections, mainly involving the brain stem and meninges in persons with deficiencies in cell-mediated immunity. This, combined with the ability of the organism to penetrate the blood–brain barrier, accounts for its virulence. Invasion of the CNS may occur via a hematogenous route, within circulating leukocytes or via a neural route in which bacteria penetrate into oral tissue, are phagocytosed by tissue macrophages, and then invade the cranial nerves.

They reach the CNS by moving in retrograde direction through nerve axons, followed by their spread intracellularly to the brain parenchyma, resulting in clinical rhomboencephalitis.

Clinical Features

Patients with *L. monocytogenes* meningitis often have a subacute clinical presentation. Fever, nausea, headaches, and stiff neck occur in majority of the patients. Focal neurological deficits, fluctuating mental status, ataxia, or seizures are indicative of parenchymal involvement and subsequent meningoencephalitis.

Brain stem involvement with *L. monocytogenes* results in rhomboencephalitis. Patients initially present with nausea, vomiting, headaches, and fever with progressive development of signs and symptoms of stroke, including asymmetrical cranial nerve deficits, cerebellar signs, hemiparesis, or hemisensory deficits. Rhomboencephalitis due to *L. monocytogenes* is associated with high mortality and devastating neurological sequelae.

Diagnosis

Isolation of *L. monocytogenes* from blood, CSF, or an otherwise sterile site is required for diagnosis. CSF analysis reveals elevated WBC count and a neutrophilic pleocytosis. Protein levels are elevated and glucose level is often low.

Gram stain of the CSF may be negative in about two-thirds of patients and blood cultures are positive in only 50%, which makes it extremely challenging to make a definitive diagnosis upon initial presentation.

MRI of the brain is considered superior to CT scan, especially in the setting of brain stem involvement with rhomboencephalitis.

Treatment

Empiric therapy for meningitis in immunosuppressed patients should target *L. monocytogenes*, if the clinical suspicion is high.

Ampicillin and penicillin G are considered the drug of choice. Combination therapy with IV gentamicin to achieve synergy is often used. Adjuvant steroid therapy has failed to demonstrate favorable outcomes or improve mortality/morbidity [10].

Trimethoprim–sulfamethoxazole is considered an alternative in patients with severe penicillin allergy. Duration of treatment largely depends on the degree of CNS involvement. Meningitis should be treated for at least 21 days, and brain abscess and rhomboencephalitis should be treated for about 6 weeks, depending on the clinical and radiographic resolution of infection.

References

[1] Ghanem KG. Neurosyphilis: a historical perspective and review. CNS Neurosci Ther 2010;16:e157–68.

[2] Gul HC, Erdem H, Bek S. Overview of Neurobrucellosis: a pooled analysis of 187 cases. Int J Infect Dis 2009;13:e339–43.

[3] Guven T, Ugurlu K, Ergonul O, Celikbas AK, Gok SE, Selcuk C, et al. Neurobrucellosis: clinical and diagnostic features. Clin Infect Dis 2013;56:1407–12.

[4] Chow FC, Marra CM, Cho TA. Cerebrovascular disease in central nervous system infections. Semin Neurol 2011;31:286–306.

[5] Rupprecht TA, Koedel U, Fingerle V, Pfister H. The pathogenesis of lyme neuroborreliosis: from infection to inflammation. Mol Med 2008;14(3–4):205–12.

[6] Pachner AR, Steiner I. Lyme neuroborreliosis: infection, immunity, and inflammation. Lancet Neurol 2007;6:544–52.

[7] Patkar D, Narang J, Yanamandala R, Lawande M, Sha GV. Central nervous system tuberculosis, pathophysiology and imaging findings. Neuroimaging Clin N Am 2012;22:677–705.

[8] Misra UK, Kalita J, Maurya PK. Stroke in tuberculous meningitis. J Neurol Sci 2011;303:22–30.

[9] Cherian A, Thomas SV. Central nervous system tuberculosis. Afr Health Sci 2011;11:116–27.

[10] Arslan F, Meynet E, Sunbul M, Sipahi OR, Kurtaran B, Kaya S, et al. The clinical features, diagnosis, treatment, and prognosis of neuroinvasive listeriosis: a multinational study. Eur J Clin Microbiol Infect Dis 2015;34:1213–21.

CHAPTER

124

HIV and Stroke*

A. Bryer, K. Bateman
University of Cape Town, Cape Town, South Africa

INTRODUCTION

There is an elevated risk of ischemic stroke in human immunodeficiency virus (HIV)-infected compared to HIV-uninfected individuals, independent of age and traditional vascular risk factors. In individuals with untreated HIV infection, the odds of an ischemic stroke are more than five times greater than in uninfected individuals matched for age and sex [1]. The risk is highest in the setting of worse virological control, low CD4 cell counts, and, particularly, short duration on treatment. The increased risk of stroke persists, but appears to taper off over years, after combination antiretroviral therapy (cART) is initiated. This suggests that an immune reconstitution inflammatory syndrome may be a contributing factor. Large, prospective studies comparing HIV-infected individuals on cART with controls have shown that HIV infection is associated with a 17–40% overall increase in the risk of ischemic stroke, after matching for traditional vascular risk factors and demographics [2,3].

In addition, the use of certain antiretroviral therapies, such as older protease inhibitors and abacavir, may increase this risk; however, the effects of cART, both short and long term, on cerebrovascular risk are not yet clear.

CAUSES OF STROKE IN HIV INFECTION

The most common causes or mechanisms cited for stroke development in HIV-infected patients are:

- cerebral infections,
- cardiac embolism,
- prothrombotic states,
- substance abuse,
- HIV-associated vasculopathies,
- accelerated atherosclerosis.

CEREBRAL INFECTIONS

Cerebral infections are important causes of secondary infective vasculitis and may cause both cerebral infarcts and hemorrhages. Some infections, such as syphilis or tuberculosis, may cause cerebral infarctions even in immunocompetent individuals. However, immunosuppressed people with HIV are more at risk of opportunistic infections, and a wider range of organisms have been associated with stroke in this setting. Stroke occurring in the weeks to months following cART initiation may be due to the immune reconstitution inflammatory syndrome either mounting a "paradoxical" response to antigen from a recent past infection in the host, or autoimmune target, or causing the "unmasking" of an occult central nervous system (CNS) opportunistic infection. Intracerebral hemorrhage has been documented in patients with infectious or neoplastic intracerebral space-occupying lesions such as primary CNS lymphoma, metastatic Kaposi's sarcoma, and cerebral toxoplasmosis [4], and can mimic stroke. Selected cerebral infections that have been associated with stroke and vasculitis are listed in Table 124.1.

CARDIAC EMBOLISM

Cardiac sources of emboli have been well described. These include HIV-associated cardiac dysfunction (including cardiomyopathy), infective endocarditis, and nonbacterial thrombotic endocarditis [5].

PROTHROMBOTIC STATES

A coagulation abnormality is often found associated with HIV and stroke. Protein S deficiency and elevated antiphospholipid antibodies have been associated with

---

* This chapter has drawn heavily upon material from Kathleen Bateman and Alan Bryer, *Stroke in Persons Infected with HIV*, in Louis Caplan, José Biller (eds), Uncommon Causes of Stroke, 3rd Edition, (forthcoming, February 2018), © Cambridge University Press, reproduced with permission.

TABLE 124.1   Infections Associated With Cerebral Vasculitis

More Common

Varicella zoster virus infection
Syphilis
Tuberculosis
Cryptococcosis
Herpes simplex virus infection
Cytomegalovirus infection

Less Common

Aspergillosis
Candidiasis
Coccidioidomycosis
Mucormycosis
Toxoplasmosis
Trypanosomiasis

HIV and stroke but are probably epiphenomena, and not significant. Several procoagulant and inflammatory markers are increased in plasma samples of HIV-infected patients, such as fibrinogen, D-dimers, P-selectin, tissue factor, and von Willebrand factor [6,7]. A prothrombotic state characterized by elevated von Willebrand factor and low ADAMTS13 levels has been found in HIV-positive arterial young strokes when compared to age-matched HIV-negative young stroke controls and cART-naive HIV-positive nonstroke controls [8]. The contribution that these or other procoagulant factors make in the causation of vascular events is not yet clearly established.

SUBSTANCE ABUSE

Patients may be predisposed to developing stroke by virtue of falling into a specific risk group for HIV. In particular, this category includes all patients who misuse drugs intravenously and use substances that are known to cause stroke such as cocaine, heroin, and amphetamines.

HIV-ASSOCIATED VASCULOPATHY AND VASCULITIS

The co-occurrence of systemic vasculitis and infection with HIV was the subject of an early review [9]. Efforts to determine whether the coexisting diseases were causally or merely coincidentally associated were limited by the absence of controlled epidemiological investigations, leaving only clinical descriptions and pathological studies to address the issue that has not been adequately resolved. The issue is compounded by the fact that HIV infection is frequently associated with coexisting infections—such as Epstein–Barr virus (EBV), varicella zoster virus (VZV), cytomegalovirus (CMV), hepatitis B virus, and others—all of which have been linked causally to various vasculitic syndromes.

HIV-related cerebral vasculitis is often cited as a cause of stroke. This may be the result of the misinterpretation of perivascular inflammatory cells, which are common in the HIV setting, and the inclusion of cases in which either potential or more likely causes of vasculitis have not been excluded. The pathological diagnosis of a vasculitis should be reserved for cases in which there is infiltration of the vessel wall by inflammatory cells with subsequent vessel wall damage. Thus, the term vasculopathy is preferable as it encompasses a broader range of vessel dysfunction than vasculitis.

The primary vasculopathies of the CNS described in association with HIV include:

1. small-vessel vasculopathy,
2. large- and medium-vessel vasculopathy.

SMALL VESSEL HIV-ASSOCIATED VASCULOPATHY

This is largely an asymptomatic vasculopathy that has been described in patients who have died with AIDS and is characterized by hyaline small-vessel wall thickening, perivascular space dilatation, rarefaction and pigment deposition with vessel wall mineralization, and occasional perivascular inflammatory cell infiltrates, and is associated with microinfarcts.

LARGE- AND MEDIUM-VESSEL HIV-ASSOCIATED VASCULOPATHY

This was first described in children and then later in adults. The vasculopathy involves both extracranial arteries (including the carotids but also other large arteries) as well as medium-sized intracranial vessels, particularly those of the circle of Willis. The extracranial vasculopathy presents with occlusive disease or aneurysm formation and is most likely secondary to vessel wall damage after leukocytoclastic vasculitis. The intracranial vasculopathy is characterized as medium-vessel occlusion with or without fusiform dilatation, stenosis, and vessel caliber variation by angiography and can present as either ischemic stroke or intracranial hemorrhage. [10].

A number of possible mechanisms have been proposed as the pathogenesis of this condition, but the exact role of HIV in the etiology of the condition remains uncertain.

ACCELERATED ATHEROSCLEROSIS

The use of cART has dramatically increased the life span of people infected with HIV. As the HIV-infected population has aged, the spectrum of disease has changed with cardiovascular disease and non-AIDS malignancies becoming more prevalent. Controversy exists as to whether atherosclerosis in HIV-infected patients is accentuated by a greater prevalence of traditional vascular risk factors, or in fact accelerated by both traditional and additional novel vascular risk factors directly or indirectly related to HIV disease.

In differing populations around the world, the risk of cardiovascular disease appears to be increased in HIV infection particularly in individuals with higher viral loads and CD4 cell counts below 200 cells/mm^3, related to but also independent of traditional vascular risk factors [11,12]. There is conflicting evidence that treatment with older protease inhibitors or with abacavir (from the nucleoside reverse transcriptase inhibitor class) is associated with premature atherosclerotic vascular disease and elevated risk of cardiovascular disease [13–15]. The older protease inhibitors can induce a variety of metabolic abnormalities including hypercholesterolemia, hypertriglyceridemia, increased serum insulin and peptide C levels with insulin resistance, and lipodystrophy [16,17], but this does not seem to be a class effect, and there is evidence that the increased risk of cardiovascular disease may occur independent of lipid dysfunction [18].

Several mechanisms of increased cardiovascular risk may be at play in HIV infection, as soluble and cellular markers of inflammation (such as interleukin-6 or C-reactive protein), procoagulant factors, and markers of endothelial dysfunction are elevated in HIV. These mechanisms may be directly or indirectly related to the HIV itself, or mediated by other factors including treatment. Active inflammation may cause endothelial dysfunction which could potentially accelerate atherosclerosis. There is a relative lack of data using cerebrovascular disease as the primary end point, as most studies have explored risk of myocardial infarction. From the existing literature, however, there is some evidence to suggest that current cardiovascular risk profile tools underestimate the risk of cardiovascular events in HIV-infected individuals. In addition, studies have shown that HIV-infected patients may be overlooked for primary preventive therapies such as aspirin despite being eligible according to existing guidelines.

APPROACH TO STROKE THERAPY IN THE HIV-POSITIVE PATIENT

The management of the HIV-infected patient presenting with a stroke should be directed primarily toward determining the underlying cause of the stroke.

The occurrence of stroke in the younger patient who is HIV positive should not preclude comprehensive evaluation, as HIV status may be incidental particularly in a population with a high HIV-seropositive prevalence in the general population. A thorough history and clinical examination frequently reveal a likely cause for stroke. This should include enquiry concerning drug misuse and symptoms and signs of opportunistic infection such as tuberculosis and prior VZV.

Acquired immune deficiency syndrome (AIDS) is frequently found in developing countries and in medical environments in which a rational and cost-effective approach is recommended. A CT scan is essential in any young patient with a stroke who is HIV positive in order to establish whether there has been a hemorrhage or an infarct and to exclude other conditions that often mimic stroke in the HIV-positive patient such as toxoplasmosis, tuberculoma, or lymphoma. If the CT scan does not reveal evidence of a space-occupying lesion, a lumbar puncture is usually indicated in order to exclude chronic meningitis (tuberculosis, syphilis, fungal infection, zoster). Given the high prevalence of cardioembolic causes of stroke in this group, careful cardiac examination, electrocardiogram, and often echocardiography is required. Full blood count, serology for syphilis, and chest X-ray are routinely done. A CD4+ cell count (where available) provides information about the degree of immunosuppression in the HIV-infected person. This may help narrow the differential diagnosis as well as assist in the timing of cART initiation in untreated patients.

Other investigations including angiography (whether CT or magnetic resonance angiography, or digital subtraction angiography) should be individualized and should follow routine stroke management guidelines. Regardless of angiographic appearance, stroke should not be attributed to an HIV-associated vasculopathy unless thorough evaluation has excluded infective, such as VZV, and other causes.

The patient should ideally be treated in a stroke unit, and long-term care plans would need to be tailored to the patient's specific needs. The severity of the primary disease and the association of secondary opportunistic infections are of major importance in the management of these patients. The rehabilitation of the HIV-infected patient, particularly with a stroke, should not be neglected. With the predicted future increased use of cART in patients diagnosed with HIV infection, attention will need to be directed to the potential metabolic abnormalities (including the lipid abnormalities and atherogenic potential related to protease inhibitors) that may develop on treatment. It is unclear to what extent aggressive treatment of modifiable traditional vascular risk factors will benefit HIV-infected individuals and studies are underway to evaluate this. However, until improved risk prediction and optimal treatment strategies are known, careful assessment of cardiovascular risk using existing frameworks and appropriate primary and secondary prevention should be emphasized in the management of HIV-infected individuals, taking into account potential drug interactions (e.g., with statins) that may occur with concomitant cART use.

References

[1] Benjamin LA, Corbett EL, Connor MD, et al. HIV, antiretroviral treatment, hypertension and stroke in Malawian adults: a case-control study. Neurology 2016;86:1–10.

[2] Marcus JL, Leyden WA, Chao CR, et al. HIV infection and incidence of ischemic stroke. AIDS 2014;28(13):1911–9. http://dx.doi.org/10.1097/QAD.0000000000000352.

[3] Sico JJ, Chang CH, So-armah K, et al. HIV status and the risk of ischemic stroke among men. Neurology 2015;84:1933–40.

[4] Pinto AN. AIDS and cerebrovascular disease. Stroke 1996; 27:538–43.

[5] Berger JR, Harris JO, Gregarios J, Norenberg M. Cerebrovascular disease in AIDS: a case control study. AIDS 1990;4:239–44.

[6] Goeijenbier M, Van Wissen M, Van der Weg C, et al. Review: viral Infections and mechanisms of thrombosis and bleeding. J Med Virol 2012;84:1680–96.

[7] van den Dries LWJ, Gruters RA, Hovels-van der Borden SBC, et al. von Willebrand Factor is elevated in HIV patients with a history of thrombosis. Front Microbiol March 2015;6:1–8. http://dx.doi.org/10.3389/fmicb.2015.00180.

[8] Allie S, Stanley A, Bryer A, Meiring M, Combrinck MI. High levels of von Willebrand factor and low levels of its cleaving protease, ADAMTS13, are associated with stroke in young HIV-infected patients. Int J Stroke June 2015. http://dx.doi.org/10.1111/ijs.12550.

[9] Calabrese LH. Vasculitis and infection with the human immunodeficiency virus. Rheum Dis Clin North Am 1991;17:131–47.

[10] Tipping B, de Villiers L, Wainwright H, et al. Stroke in patients with human immunodeficiency virus infection. J Neurol Neurosurg Psychiatry 2007;78(12):1320–4. http://dx.doi.org/10.1136/jnnp.2007.116103.

[11] Freiberg MS, Chang C-CH, Kuller LH, et al. HIV infection and the risk of acute myocardial infarction. JAMA Intern Med 2013;173(8):614–22. http://dx.doi.org/10.1001/jamainternmed.2013.3728.

[12] Shahbaz S. Cardiovascular disease in human immunodeficiency virus infected patients: a true or perceived risk? World J Cardiol 2015;7(10):633. http://dx.doi.org/10.4330/wjc.v7.i10.633.

[13] Choi AI, Vittinghoff E, Deeks SG, Weekley CC, Li Y, Shlipak MG. Cardiovascular risks associated with abacavir and tenofovir exposure in HIV-infected persons. AIDS 2011;25(10):1289–98. http://dx.doi.org/10.1097/QAD.0b013e328347fa16.

[14] Young J, Xiao Y, Moodie EEM, et al. The effect of cumulating exposure to abacavir on the risk of cardiovascular disease events in patients from the Swiss HIV Cohort Study. J Acquir Immune Defic Syndr 2015;69(4). http://dx.doi.org/10.1097/QAI.0000000000000662. [Epub ahead of print].

[15] Cruciani M, Zanichelli V, Serpelloni G, et al. Abacavir use and cardiovascular disease events: a meta-analysis of published and unpublished data. AIDS 2011;25(16):1993–2004. http://dx.doi.org/10.1097/QAD.0b013e328349c6ee.

[16] Rabinstein AA. Stroke In HIV-infected patients: a clinical perspective. Cerebrovasc Dis 2003;15:37–44.

[17] Sklar P, Masur H. HIV infection and cardiovascular disease–is there really a link? N Engl J Med 2003;349:2065–7.

[18] Worm SW, Sabin C, Weber R, et al. Risk of myocardial infarction in patients with HIV infection exposed to specific individual antiretroviral drugs from the 3 major drug classes: the data collection on adverse events of anti-HIV drugs (D: A:D) study. J Infect Dis 2010;201(3):318–30. http://dx.doi.org/10.1086/649897.

CHAPTER

125

Stroke-Induced Immunodepression and Clinical Consequences

A. Meisel, S. Hoffmann

Charité – Universitätsmedizin Berlin, Berlin, Germany

INTRODUCTION

Infections are common complications in the acute phase after stroke affecting about one-third of all patients with stroke [1,2]. The most frequent poststroke infections are urinary tract infections and stroke-associated pneumonia (SAP), the latter showing consistent association with poor outcome [1,2]. Initial concepts on the etiology of SAP focused on stroke-facilitated aspiration. In fact, dysphagic

Primer on Cerebrovascular Diseases, Second Edition
http://dx.doi.org/10.1016/B978-0-12-803058-5.00125-9

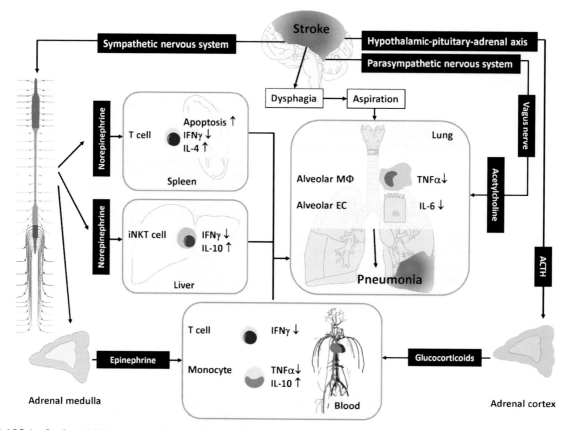

FIGURE 125.1   Stroke might impair swallowing (dysphagia) and thus cause aspiration. Moreover, stroke might activate the hypothalamic-pituitary-adrenal axis, the sympathetic nervous system including the adrenal-medullary axis, and cholinergic signaling via the vagus nerve (para-sympathetic nervous system). Subsequently, released glucocorticoids, epinephrine, norepinephrine, and acetylcholine bind cells of the adaptive and innate immunity via their cognate receptors inducing an immunodepressive state. Stroke-induced immunodepression syndrome (SIDS) is characterized by a rapid but temporary numerical decrease in peripheral blood lymphocyte subpopulations caused by apoptosis and lympho-cytic dysfunction in secondary lymphoid organs and blood switching from a pro- to an antiinflammatory mode of action. Moreover, monocytes as well as lung macrophages and epithelial cells are suppressed. Thereby, SIDS impairs the antibacterial host defense facilitating the develop-ment of pneumonia after aspiration of an otherwise harmless bacterial colonization. *ACTH*, adrenocorticotropic hormone; *EC*, epithelial cells; *IFN-γ*, interferon-γ; *IL*, interleukin; *iNKT*, invariant natural killer cells; *MΦ*, macrophages; *TNF-α*, tumor necrosis factor-α.

patients have an increased risk for SAP. However, dys-phagia alone is not sufficient for the development of SAP [3] and therapeutic measures to prevent aspira-tion do not effectively protect against SAP. Moreover, even silent cerebral infarctions are related to a higher incidence of SAP [2]. Accumulating experimental and clinical evidence suggests that stroke induces a rapid, temporary immunodepression mainly mediated by the autonomic nervous system [3]. Stroke-induced immunodepression syndrome (SIDS) is characterized by a downregulation of systemic cellular immune responses, i.e., rapid numerical decrease in peripheral blood lymphocyte subpopulations and functional shift or deactivation of lymphocyte subpopulations [1] as well as monocytes, lung macrophages, and epithelial cells resulting in an increased susceptibility for SAP [2,4,5] (Fig. 125.1). Alterations in markers for SIDS develop within 24 h after stroke onset and thereby pre-cede SAP, which usually occurs between day 2–7 after

stroke onset [1]. Within weeks, signs of SIDS as well as risk of pneumonia abate.

PATHWAYS OF STROKE-INDUCED IMMUNODEPRESSION SYNDROME

The nervous system and the immune system are engaged in intense bidirectional communication via complex humoral and neural pathways. Stroke dis-turbs this normally well-balanced interplay leading to brain inflammation and immunodepression [2]. Three major pathways of the autonomic nervous sys-tem (ANS) are identified for being involved in the mediation of neuroimmunomodulation (Fig. 125.1): the hypothalamic-pituitary-adrenal (HPA) axis, the sympathetic-adrenal-medullary axis, and the para-sympathetic cholinergic nervous system (mainly the vagus nerve) [3]. Overactivation of these signaling cascades might be caused directly by damaging or affecting the control centers of the ANS due to stroke. Furthermore, cerebral ischemia activates inflammatory pathways in the brain parenchyma causing the release of proinflammatory cytokines, which might indirectly activate control centers of the ANS [1]. For example, clinical and experimental evidence indicate that proinflammatory cytokines produced by damaged brain tissue might activate the HPA axis [2]. Glucocorticoids produced as a result of HPA axis activation have well known antiinflammatory and immunosuppressive effects. They prevent inflammation by suppressing the production of many proinflammatory mediators including cytokines, prostaglandins, and nitric oxide and by enhancing the release of antiinflammatory mediators such as interleukin (IL)-10 and transforming growth factor [3]. Moreover, glucocorticoids decrease the capacity of monocytes/macrophages and dendritic cells to present antigen to T lymphocytes and to elicit immune responses by downregulating the expression of major histocompatibility complex (MHC) class II and costimulatory molecules [3]. The hypothalamus is functionally linked with other autonomic centers allowing for a synchronization of glucocorticoid responses with the cholinergic pathway [1]. The vagal "cholinergic antiinflammatory pathway" might reduce the production of proinflammatory cytokines such as tumor necrosis factor (TNF)-α, IL-1β, and IL-18 by activated macrophages in inflamed peripheral organs and tissues [3]. The sympathetic nervous system innervates primary and secondary lymphoid organs and thereby regulates immunity. Hence, overactivation of cholinergic as well as sympathetic pathways might cause immune dysfunction [4,5]. For example, the cholinergic pathway suppresses pulmonary macrophages and epithelial cells of the innate immunity thereby facilitating pneumonia after stroke [5]. Moreover, secondary lymphatic organs like spleen and thymus exhibit an atrophy after focal cerebral ischemia [4]. The extensive sympathetic innervation of immune organs and the presence of adrenergic receptors on almost all leukocytes indicate the strong influence of sympathetic activity on immune function [3]. Stroke induces an overactivation of the adrenergic system causing the release of catecholamines from sympathetic nerve terminals and the adrenal medulla [3]. This results in a pronounced antiinflammatory phenotype in lymphocytes, monocytes, and macrophages [1]. Interestingly, sympathetic activation can also result in gastrointestinal dysmotility which increases the risk of aspiration pneumonia [1] thereby increasing the susceptibility for SAP via facilitating the two currently pathophysiologically postulated concepts on the development of SAP, aspiration pneumonia and SIDS.

TABLE 125.1 Summary of Blood-Based Immune and Stress Markers for Predicting Poststroke Infections in Mice and Humans

Biomarker Day 1	Murine	Human
Norepinephrine↑, cortisol↑	✔	✔
(CD4-T-) lymphocytes↓	✔	✔
Lymphocyte IFN-γ↓	✔	✔
Monocyte TNF-α↓	✔	✔
Monocyte HLA-DR expression↓	?	✔

IFN, interferon; *TNF*, tumor necrosis factor.

LABORATORY MARKERS FOR STROKE-INDUCED IMMUNODEPRESSION

Considering the high incidence of SAP even in specialized stroke units, an identification of patients at high risk for SAP is desirable and might help to justify increased monitoring and tailored prophylactic measures in these patients. Clinical parameters such as stroke severity, older age, impaired level of consciousness, and decreased bulbar reflexes are known clinical risk factors for SAP, supporting the concept of aspiration pneumonia. However, a more refined patient identification including biomarkers could be helpful. Table 125.1 summarizes biomarkers identified as predictors for SAP from experimental and clinical studies. The overactivation of the ANS results in increased levels of cortisol and norepinephrine after stroke with higher concentrations being found in patients with infections compared with patients without infections [3]. Overall, the mediated changes in neuroimmunomodulation result in a switch from a proinflammatory Th1 response to an antiinflammatory Th2 response, a rapid numerical decrease in peripheral blood lymphocyte subpopulations, changes in cytokine levels, and functional deactivation of monocytes. In detail, a catecholamine-mediated reduced interferon-γ (IFN-γ)-secretion by T and natural killer (NK) cells appears to be an essential cause in the impaired antibacterial defense after brain ischemia [4]. IFN-γ normally enhances MHCII expression on monocytes. Consistently, a decreased monocytic HLA-DR (human leukocyte antigen - antigen D related) expression has been identified as a strong predictor for SAP in various studies [3]. Impaired monocyte function results in insufficient antigen presentation and may, thus, contribute to reduced lymphocyte responses. Furthermore, monocytes swiftly decrease the production of proinflammatory cytokines such as TNF-α in response to acetylcholine and glucocorticoids [1,3]. An overall dramatic loss of T-lymphocytes can be found in patients with stroke. However, analysis of lymphocyte subsets showed a consistent association between a numerical decrease of CD4+ T helper cell counts with SAP [4].

SENSE AND NONSENSE OF STROKE-INDUCED IMMUNODEPRESSION

Considering the harmful effects of SIDS by facilitating stroke-associated infections, the question arises whether SIDS is a mere immunological blunder ("paralysis of immunity") or whether it may harbor beneficial effects. The inflammatory cascade induced by acute stroke is usually linked to the progression of brain damage [2]. Thus, SIDS might be an adaptive response protecting the nervous tissue against excessive inflammation. The disruption of the blood–brain barrier after stroke allows for an encounter of normally hidden central nervous system (CNS) antigens by the systemic immune system [3]. This encounter can occur in the injured brain itself as well as in the periphery and might promote priming and activation of lymphocytes that are reactive to CNS antigens [1,6]. In fact, cellular immune responses to myelin-associated antigens and other brain antigens are seen in stroke survivors and seem to be more robust than in patients with multiple sclerosis [6]. One study demonstrated that autoimmune responses to myelin basic protein in patients with stroke are associated with a decreased likelihood of good outcome, even after adjusting for baseline stroke severity and patient age [6]. Further data supporting the hypothesis of SIDS being an adaptive response to limit ischemia-induced inflammation in the brain come from experimental studies. For example, depletion of CD4$^+$ T cells (see earlier discussion) reduced infarct volume in experimental stroke [1]. Moreover, infiltrating T cells are the main source of IFN-γ production, which has been shown to mediate neurotoxic effects in ischemic brain tissue [1]. Therefore depletion of circulating T-cell populations and suppression of IFN-γ secretion as key mechanisms of SIDS might counteract the inflammatory brain after stroke [2]. Teleologically, SIDS might thus represent an adaptive immune response to limit ischemia-induced inflammation and autoreactive responses in the brain at the price of an increased susceptibility to poststroke infections owing to reduced inflammatory drive [1].

THERAPEUTIC IMPLICATIONS

Immunomodulation in the Prevention of SAP

Given the experimental and clinical evidence supporting the concept of SIDS and its negative impact on poststroke infections and stroke outcome, immunomodulatory therapies to reconstitute antimicrobial host immune defenses after stroke may provide a new therapeutic approach. Blocking the mediation of SIDS seems intriguing due to the possibility of positively influencing overall immunological changes after stroke but harbors the potentially hazardous effect of boosting ischemia-induced inflammation and autoreactive responses in the brain. In fact, in experimental stroke, blocking SIDS promoted autoreactive CNS antigen-specific T-cell responses in the brain [4,7], which might cause worse functional outcome in patients with stroke [6]. Another therapeutic approach in the treatment of SIDS are selective immunomodulatory strategies. For example, administration of α-galactosylceramide, an activator of invariant NK T cells, increased systemic IFN-γ levels and reduced bacterial infections after experimental stroke [1]. Likewise, immune reconstitution by adoptive transfer of IFN-γ producing T and NK cells from wild-type mice greatly diminished bacterial burden [4]. Consistently, systemic application of IFN-γ prevented poststroke infections in an experimental stroke model [4]. However, given that brain immune interactions after stroke might have protective as well as destructive effects in the brain itself and on the human organism as a whole, development of immunomodulatory strategies is not straightforward [2]. Further research on this topic is urgently needed since immunomodulation after stroke may not only allow for a prevention of poststroke infections but also might harbor further beneficial effects. One study demonstrated that B-lymphocyte responses to stroke occur in the brain of mice that subsequently developed cognitive deficits as well as in human subjects with stroke and dementia. More importantly, pharmacologic ablation of B lymphocytes with an anti-CD20 antibody prevented the appearance of delayed cognitive deficits after experimental stroke [8].

Preventive Antibacterial Treatment in Acute Stroke

Since immunomodulation currently is not a viable therapeutic option, preventive antibiotic treatment raised hope for a more effective prevention of SAP and, thus, improvement of stroke outcome. In an animal model of stroke, preventive antibacterial treatment dramatically improved mortality and reduced infarct sizes [2]. A Cochrane systematic review on clinical trials assessing the efficacy and safety of antibiotics in the prevention of poststroke infections concluded that preventive antibiotic treatment reduces the risk of poststroke infections [1]. Two large randomized-controlled trials investigated whether or not prophylactic antibiotic therapy prevents poststroke infections and improves long-term outcome in acute stroke. In The Preventive Antibiotics in Stroke Study (PASS) [9], preventive antibiotic therapy did not improve long-term outcome after stroke but reduced the overall infection rate. However, incidence of SAP was not significantly reduced. In the STROKE-INF study [10], antibiotic prophylaxis also failed in reducing the frequency of SAP and did not improve 3-month outcome in dysphagic patients with stroke. These two trials are of considerable interest because they provide class 1 evidence recommending against preventive antibiotic treatment after stroke. Thus, therapeutic measures for the prevention of SAP remain limited to standard measures such as dysphagia screening, swallowing and breathing exercises, early mobilization, upper body elevation, as well as avoidance of acid-suppressive medication and prophylactic enteral tube feeding. Despite the negative findings, PASS and STROKE-INF might stimulate research to further improve the pathophysiological understanding and diagnostic criteria of SAP as well as approaches to prevent and treat SAP. For example, more sophisticated approaches to prevent SAP might be developed based on a refined SAP prediction using biomarkers and immune modulating strategies starting already in the hyperacute phase of stroke.

CONCLUSIONS

The occurrence of SAP is associated with poor outcome even in patients treated in specialized stroke units receiving treatment in compliance with current guidelines. Growing evidence supports that SIDS is an essential cause for poststroke infections. Blood-based immune and stress markers have been demonstrated to identify patients at high risk for poststroke infections as well as patients with unfavorable outcome. In the light of two recent large randomized clinical trials recommending against preventive antibiotic treatment in patients with stroke, biomarker guidance might contribute to a more refined SAP prediction in patients at high risk for poststroke infections and stimulate research on immune modulating strategies. However, further studies are needed to address the apparently heterogeneous clinical consequences of SIDS with its beneficial and harmful effects for the brain following stroke.

References

[1] Chamorro A, Meisel A, Planas AM, Urra X, van de Beek D, Veltkamp R. The immunology of acute stroke. Nat Rev Neurol July 2012;8(7):401–10.

[2] Dirnagl U, Klehmet J, Braun JS, et al. Stroke-induced immunodepression: experimental evidence and clinical relevance. Stroke February 2007;38(Suppl. 2):770–3.

[3] Meisel C, Schwab JM, Prass K, Meisel A, Dirnagl U. Central nervous system injury-induced immune deficiency syndrome. Nat Rev Neurosci October 2005;6(10):775–86.

[4] Prass K, Meisel C, Hoflich C, et al. Stroke-induced immunodeficiency promotes spontaneous bacterial infections and is mediated by sympathetic activation reversal by poststroke T helper cell type 1-like immunostimulation. J Exp Med September 1, 2003;198(5):725–36.

[5] Engel O, Akyuz L, da Costa Goncalves AC, et al. Cholinergic pathway suppresses pulmonary innate immunity facilitating pneumonia after stroke. Stroke November 2015;46(11):3232–40.

[6] Becker KJ, Kalil AJ, Tanzi P, et al. Autoimmune responses to the brain after stroke are associated with worse outcome. Stroke October 2011;42(10):2763–9.

[7] Romer C, Engel O, Winek K, et al. Blocking stroke-induced immunodeficiency increases CNS antigen-specific autoreactivity but does not worsen functional outcome after experimental stroke. J Neurosci May 20, 2015;35(20):7777–94.

[8] Doyle KP, Quach LN, Sole M, et al. B-lymphocyte-mediated delayed cognitive impairment following stroke. J Neurosci February 4, 2015;35(5):2133–45.

[9] Westendorp WF, Vermeij JD, Zock E, et al. The Preventive Antibiotics in Stroke Study (PASS): a pragmatic randomised open-label masked endpoint clinical trial. Lancet April 18, 2015;385(9977):1519–26.

[10] Kalra L, Irshad S, Hodsoll J, et al. Prophylactic antibiotics after acute stroke for reducing pneumonia in patients with dysphagia (STROKE-INF): a prospective, cluster-randomised, open-label, masked endpoint, controlled clinical trial. Lancet November 7, 2015;386(10006):1835–44.

CHAPTER

126

Cerebrovascular Complications of Cardiac Surgery

J.S. Veluz[1], M.C. Leary[2,3]

[1]St. Mary Medical Center, Langhorne, PA, United States; [2]Lehigh Valley Hospital and Health Network, Allentown, PA, United States; [3]University of South Florida, Tampa, FL, United States

BACKGROUND

Cardiac surgery has undergone a rapid and extraordinary development since mid-1940s. Many operations that were once considered experimental are now commonplace, with thousands of open heart procedures performed annually in the United States. At present, an estimated 1 million patients undergo cardiac surgery throughout the world every year. Neurological impairment is a well-known complication of cardiac surgery, resulting in longer hospitalizations, increased costs, and an escalation in morbidity and mortality. The types of neurological complications vary and can include peripheral neuropathy, encephalopathy, cognitive impairment, and stroke. Overall, in prospective studies, transient neurological complications have been noted in 61% of cardiac surgery patients [1,2]. Although refinements in surgical and anesthesia techniques have improved neurological outcomes, the number of elderly patients undergoing cardiac surgery has also increased, and thus cerebrovascular complications in particular continue to occur.

One prospective study of 16,184 consecutive patients indicates that the specific stroke risk depends on the type of cardiac surgical procedure performed [3]. While the overall incidence of stroke in cardiac surgery is generally estimated to be 4.6%, the risk of ischemic stroke varies depending on the actual surgical procedure undertaken: coronary artery bypass grafting (CABG), 3.8%; beating-heart CABG, 1.9%; aortic valve surgery, 4.8%; mitral valve surgery, 8.8%; double or triple valve surgery, 9.7%; and CABG and valve surgery, 7.4% [1,3–5]. Regarding combined CABG and a left-sided cardiac procedure (such as aortic or mitral valve replacement,

AVR or MVR, respectively), it has been estimated that 15.8% of patients have neurological complications: 8.5% with stroke or transient ischemic attack (TIA) and 7.3% with new intellectual deterioration [6]. Thus, a combined procedure appears to carry a higher cerebrovascular risk than CABG performed in isolation.

CARDIAC SURGICAL PROCEDURES

Coronary Artery Bypass Graft Surgery

CABG is the most common major cardiovascular operation performed. Preoperative factors that elevate perioperative CABG stroke risk include the diagnoses of diabetes, prior stroke, older age, female gender, smoking, hypertension, left main coronary disease, mild renal impairment, and elevated high-sensitivity preoperative C-reactive protein. Additionally, both preoperative stroke and TIA are also risk factors for in-hospital mortality [1].

Early studies suggested a decrease in postoperative stroke rates in patients undergoing off-pump CABG compared to patients undergoing the traditional on-pump operation. This was thought to be due to less aortic manipulation during off-pump surgery, but conflicts in the literature exist [1]. A single-center study of 2516 consecutive patients noted that off-pump CABG reduced the incidence of early postoperative stroke (symptoms noted just after emergence of anesthesia). However, the risk of delayed stroke (normal neurologically emerging from anesthesia, but symptoms presenting within 30 days after surgery) was no different between the on- and off-pump CABG patients [1,7]. Similarly, in 2013, the CORONARY (CABG Off or On Pump Revascularization Study) trial

Primer on Cerebrovascular Diseases, Second Edition
http://dx.doi.org/10.1016/B978-0-12-803058-5.00126-0

enrolled 4752 patients, randomly assigning them to on-pump versus off-pump CABG. Stroke rates, as well as quality of life and cognitive function did not differ significantly between the two groups at 30 days and 1 year [8]. The potential mechanisms of cerebral infarction in the cardiac surgery population is explored in a later section.

Valve Surgery: Aortic Valve Replacement and Mitral Valve Replacement

Aortic Valve Replacement

Patients aged 50–69 years who received mechanical valves had better long-term survival after AVR than those with bioprosthetic valves. The risk of stroke was similar; however, patients with bioprostheses had a higher risk of aortic valve reoperation and a lower risk of major bleeding. The specific risk of ischemic stroke during AVR is estimated to be 2.3–4.8% [3,9,10].

Mitral Valve Replacement

A systematic review and meta-analysis of data on octogenarians who underwent MVR or mitral valve repair (MVRpr) yielded 16 retrospective studies. The pooled proportion of postoperative stroke were 4% for patients undergoing MVR (6 studies, 2945 patients, 95% CI: 3–7%) and 3% in those undergoing MVRpr (3 studies, 348 patients, 95% CI: 1–8%) [11].

MINIMALLY INVASIVE CARDIAC SURGICAL PROCEDURES

Minimally Invasive Direct Coronary Artery Bypass

It is unclear whether minimally invasive direct coronary artery bypass (MIDCAB) procedures have a lessened stroke risk because further data are needed. One study suggests that minimally invasive CABG may lessen the risk of major adverse neurological and cardiac events compared with traditional off-pump CABG [12]. This operation has generally fallen out of favor in lieu of off-pump techniques or a robotic approach.

Minimally Invasive Mitral Valve Surgery

The data are conflicting regarding minimally invasive mitral valve surgery (MIMVS) and stroke. One center's result with direct-access, MIMVS demonstrated a low rate of stroke and TIA, with a total of 0.28% of patients having either stroke or TIA [13]. However, a meta-analysis of MIMVS by Modi et al. of six eligible studies found no difference in neurological events between sternotomy and MIMVS groups [14,15]. Additionally, this past year, an analysis of the Society of Thoracic Surgeons (STS) Database demonstrated that stroke was more common among less-invasive mitral valve surgery patients, with an odds ratio of about 2 as compared with traditional sternotomy [14,16]. This increased risk of stroke was attributed to potentially inadequate de-airing, fibrillating-heart techniques, and prolonged CPB and cross-clamp times [14].

Minimally Invasive Aortic Valve Replacement

Most studies have demonstrated no difference in morbidity or mortality between minimally invasive AVR (MIAVR) and conventional AVR [14].

PERCUTANEOUS CARDIAC SURGICAL PROCEDURES

Thoracic Endovascular Aortic Repair

Thoracic endovascular aortic repair (TEVAR) is an evolving minimally invasive aortic therapy developed during the early 1990s. Since then, there has been a growing comfort and a growing popularity for this intervention, especially to treat descending thoracic aorta pathology. Meta-analysis of TEVAR therapy during early 2010 in the descending thoracic aorta demonstrated lower 30-day mortality and paraplegia when compared with traditional open repair [14,17]. Rates of stroke and mortality beyond 1 year were all similar. The use of TEVAR in ascending aorta repairs is also advancing, although repairs involving the aortic arch require branched grafts and greater technical skill and expertise to ensure proper perfusion of the head vessels [14]. Exact stroke rates in this population are not clear, as further data are needed.

Transcatheter Aortic Valve Replacement

Transcatheter AVR (TAVR) is a minimally invasive catheter-based surgical procedure for patients with severe symptomatic aortic stenosis, who are considered to be either high-risk or inoperable candidates for traditional cardiac surgery. The 2012 Placement of Aortic Transcatheter Valves (PARTNER) B trial found the rate of symptomatic stroke at 1 year among TAVR patients to be double that of patients assigned to medical therapy alone [1,18]. Initial presentation of these data caused concern, however follow-up data from the same trial have demonstrated that the risk of stroke or TIA is similar between those randomized to TAVR versus surgical AVR: 15.9% in the TAVR group compared to 14% in the surgical AVR group [1,19]. Notably, the incidence of symptomatic stroke post-TAVR has significantly improved over time, presumably as experience with this procedure has grown and delivery technology/valve design has been optimized.

A 2013 meta-analysis of 14 studies, including 2 randomized controlled trials and 11 observational studies found no significant differences between TAVR and AVR in terms of all-cause and cardiovascular-related mortality, including stroke. The incidence of stroke was not significantly different between TAVR and AVR during the periprocedural period (2.6% vs. 2.3%). A subgroup analysis of just randomized controlled trials however, identified a higher combined incidence of stroke or TIAs in the TAVR group compared to the AVR group [1,10].

Data collected from the 2013 Society of Thoracic Surgeons/American College of Cardiology Transcatheter Valve Therapy (STS/ACC TVT) registry of 7710 patients who underwent TAVR showed an average TAVR stroke rate of about 2% [20]. For comparison, the 2013 STS surgical valve replacement stroke rates were between 1.1% and 7% [21,22]. Another large registry published in 2015 found the stroke rate at 1 year to be 4.1% [23].

Mechanism of TAVR-Induced Stroke

The valves being replaced in TAVR procedures tend to be heavily calcified and sometimes require balloon valvuloplasty prior to the placement of the device. Placement of the Edwards Lifesciences Sapien valve requires a balloon inflation to expand the stent support structure of the valve. Medtronic's Corevalve is self-expanding and is deployed by releasing it out of a catheter into position. Irrespective of the specific valve used, both the valvuloplasty and valve deployment/implantation are blamed for causing a shower of atheroemboli, which can flow directly up the carotid arteries to the brain, resulting in stroke [24]. Stroke rates for TAVR in trials using Corevalve range from 2% to 5.8% [22,25]. In trials for Edwards Lifesciences Sapien valve, the stroke rate ranges from 5% to 7.8% [21]. The Corevalve US Pivotal Trial High Risk Study paid close attention to stroke detection and identified a post-procedural stroke rate of 3.9% for Corevalve and 3.1% for surgical repair. These stroke rates rose to 5.8% and 7% by one year, respectively [22]. While higher than reported rates in some earlier trials, the stroke percentages were statistically comparable between surgical and TAVR arms of this trial. Both transcranial Doppler and MRI data suggest that almost all TAVR patients experience silent new onset silent cerebral ischemia. Despite this concerning data, however, the embolic events do not clearly seem to correlate with clinical outcomes.

TAVR Embolic Protection Systems

Decreasing the TAVR risk of stroke in the future may include improving risk prediction, valve system technology and technique, periprocedural medical therapy, and the use of embolic protection devices [24]. There are several embolic protection systems in development and in clinical trials, including Claret's Sentinel (Montage in Europe) dual filter Cerebral Protection System, Keystone Heart's Triguard device, the Edwards Lifesciences Embrella embolic deflector, and the Emboline Inc. Prosheath embolic protection system. The Emboline Inc. Prosheath embolic protection system is the first device that is theoretically capable of capturing all emboli released during TAVR procedures, and initiation of a clinical trial assessing this device is anticipated shortly. Data on the Embrella's recent PROTAVI-C Pilot Study of 53 patients were presented at the 2014 Transcatheter Cardiovascular Therapeutics (TCT) meeting. The Embrella embolic deflection device for cerebral protection during TAVR was noted to actually increase the number of silent cerebral ischemic lesions on magnetic resonance imaging (MRI), but there was also a significant reduction in single lesion volume as well as an absence of large total infarct volumes. Whether this reduction in lesion size correlates to improved neurological outcomes is uncertain. All of these defects, evaluated via MRI, were small, clinically silent, and resolved within few weeks [24,26,27]. Also reported at TCT 2014 was preliminary data from the CLEAN-TAVI trial, which examined the impact of Claret's Sentinel Cerebral Protection System device in patients with severe aortic stenosis who were at increased surgical risk. The use of this cerebral protection device during TAVR significantly reduced the number of cerebral lesions as well as the volume of cerebral lesions in the protected brain regions as determined by MRI [27].

As of 2016 the clinical significance of cerebral emboli detected during MRI and/or ultrasound are not fully understood. Prospective studies specifically evaluating neurological outcomes and interventions are necessary, as well as long-term clinical outcomes.

MECHANISM OF STROKE IN CARDIAC SURGERY

Atherothrombotic, Hemodynamically Mediated Brain Infarcts: Hypoperfusion Stroke

Carotid artery disease causes about a third of post-CABG stroke [1,4]. The major concern regarding cardiac surgery in patients with carotid disease is whether the hemodynamic stress of heart surgery leads to underperfusion of areas supplied by already stenotic or occluded arteries, causing ischemic stroke. This concern underlies neck auscultation for bruits, ultrasound carotid artery testing, and other preoperative testing prior to CABG. The chance of perioperative hypoperfusion stroke does escalate with increasing severity of carotid disease [1,4]. CABG patients with no significant carotid stenosis have a perioperative stroke risk of <2%, while the risk increases to 3% with a unilateral 50–99% stenosis and to 5% with bilateral 50–99% stenosis. Carotid occlusion has the highest perioperative stroke risk at 7–11% [4].

The best approach to the management of concomitant severe carotid and coronary artery disease remains unanswered. Despite the lack of randomized data, the American College of Cardiology/American Heart Association (ACC/AHA) guidelines recommend carotid endarterectomy (CEA) in patients with symptomatic carotid stenosis or in asymptomatic patients with unilateral or bilateral internal carotid artery stenosis of 80% or greater either prior to, or combined with, CABG. As of 2016, there is no consensus as to which surgical approach is superior [1,28]. Carotid artery stenting (CAS) has also been recently introduced as an alternative revascularization modality in high-risk patients. Whether CAS should be performed prior to CABG is still a point of debate at this time.

Brain Embolism

An argument against a hypoperfusion as a potential etiology for stroke in cardiac surgery patients is the timing of the infarct itself. It appears that strokes occur more frequently *after* recovery from anesthesia [1]. If the mechanism of stroke were hemodynamic, the major circulatory stress would be intraoperative, and patients would awaken from anesthesia with the deficit. This scenario is not typically the case. Postoperative emboli may arise from preexisting cardiac abnormalities (such as hypofunctioning ventricles, dilated atria, and aortic atheromas) or from postoperative arrhythmias such as atrial fibrillation. Additionally, postoperative activation of coagulation factors in cardiac surgery patients can promote hypercoagulability, precipitating occlusive thrombosis in atherostenotic arteries, and causing intra-arterial embolism. Thromboembolic infarction often occurs in the days following surgery when cessation of anticoagulation is necessary. Evidence also links intraoperative and postoperative embolism to aortic ulcerative atherosclerotic lesions [1].

As far as potential causes for intra-operative embolism, cross-clamping of the ascending aorta and aortotomy can liberate cholesterol or calcific plaque debris [1]. Given that atherosclerosis of the ascending aorta is a significant risk for perioperative stroke, it was postulated that avoiding direct manipulation of this area may improve neurological outcome postoperatively. Epiaortic ultrasound scanning is thought to be superior to both manual palpation of the ascending aorta and transesophageal echocardiography (TEE) in detecting atherosclerosis, particularly noncalcified plaque. It has led to adjustments in surgical management in patients undergoing CABG, such as modification of cannulation, clamping, anastomotic technique, and temperature management, as well as consideration for alternative technique (off-pump CABG) [1,29]. The overall stroke risk was lower in patients who had intraoperative epiaortic ultrasound, compared with all patients undergoing cardiac surgical procedures [1,29,30].

ACUTE ISCHEMIC STROKE TREATMENT IN CARDIAC SURGERY PATIENTS

With regard to interventional treatment of cardiac surgery patients with an acute embolic stroke, unfortunately, given their recent surgery, these patients are not candidates to receive intravenous tissue plasminogen activator (t-PA). However, these patients may still be candidates for mechanical endovascular acute stroke treatment [1].

OTHER CEREBROVASCULAR COMPLICATIONS AND STROKE MIMICS

Postoperative Encephalopathy: Microemboli and Other Causes

Many patients are hard to rouse postoperatively. Delirium, agitation, confusion, and other alterations in mental status are common, and typically referred to as encephalopathy. Clinical and imaging studies usually do not show important focal neurological signs or large focal infarcts. The incidence of encephalopathy varies. In the Cleveland Clinic prospective series, 11.6% of cardiac surgery patients were encephalopathic on the 4th postoperative day [31]. Encephalopathy has multiple causes including iatrogenically generated cerebral microemboli, diffuse hypoxic–ischemic insults from hypotension, as well as toxic–metabolic factors such as infection or medications. Particularly relevant in encephalopathy are haloperidol, narcotics, and sedatives.

Postoperative Intracranial Hemorrhage

Intracerebral or subarachnoid hemorrhages have occasionally been reported after cardiac surgery, most commonly in children who had repair of congenital heart disease [32] or in cardiac transplantation patients [33]. The postulated mechanism involves an abrupt increase in brain blood flow with rupture of small intracranial arteries unprepared for the new load. Usually, there is a prolonged period when cardiac output is low, and this output is suddenly increased by the surgery. Abrupt increases in brain blood flow or pressure in other situations have also been associated with intracerebral hemorrhage [1].

Stroke Mimics: Postoperative Peripheral Nerve Complications and Other Issues

Brachial plexus and peripheral nerve lesions frequently develop after cardiac surgery and can be confused with central nervous system (CNS) complications. The most common deficit is a unilateral brachial plexopathy characterized by shoulder pain and usually weakness and numbness of one hand. Brachial plexopathy is probably caused by either sternal retraction or positioning of the arm and shoulders during surgery, with traction on the lower trunk of the brachial plexus. Ulnar, peroneal, and saphenous nerve injuries are also common and are also related to positioning. Diaphragmatic and vocal cord paralysis are likely related to local effects of the cardiac surgery on the recurrent laryngeal and phrenic nerves. Postoperative Horner syndrome may be caused by manipulation of the sympathetic chain, but carotid dissection (particularly in surgical patients undergoing aortic dissection repair) should be excluded [1]. Finally, serotonin syndrome and neuroleptic malignant syndrome have been reported after cardiac surgery, and may be related to methylene blue administration in heart surgery patients who have been exposed to certain antidepressants.

References

[1] Leary MC, Caplan LR. Cerebrovascular disease and neurologic manifestations of heart disease. Chapter 90 In: O'Rourke RA, Walsh RA, Fuster V, editors. HURST's The Heart. 43rd ed. New York, NY: McGraw-Hill 2017 [2017 in press].

[2] Shaw PJ, Bates D, Cartlidge NEF, et al. Early neurological complications of coronary artery bypass surgery. Br Med J 1985;291:1384–7.

[3] Bucerius J, Gummert JF, Borger MA, et al. Stroke after cardiac surgery: a risk factor analysis of 16,184 consecutive adult patients. Ann Thorac Surg 2003;75:472–8.

[4] Naylor AR, Mehta Z, Rothwell PM, et al. Carotid artery disease and stroke during coronary artery bypass: a critical review of the literature. Eur J Vasc Endovasc Surg 2002;23:283–94.

[5] Ricotta JJ, Char DJ, Cuadra SA, et al. Modeling stroke risk after coronary artery bypass and combined coronary artery bypass and carotid endarterectomy. Stroke 2003;34:1212–7.

[6] Gansera B, Angelis I, Weingartner J, et al. Simultaneous carotid endarterectomy and cardiac surgery: additional risk or safety procedure? J Thorac Cardiovasc Surg 2003;51:22–7.

[7] Nishiyama K, Horiguchi M, Shizuta S, et al. Temporal pattern of strokes after on-pump and off-pump coronary artery bypass graft surgery. Ann Thorac Surg 2009;87:1839–44.

[8] Lamy A, Devereaux PJ, Prabhakaran D, et al. Effects of off-pump and on-pump coronary-artery bypass grafting at 1 year. NEJM 2013;368:1179.

[9] Glaser N, Jackson V, Holzmann MJ, Franco-Cereceda A, Sartipy U. Aortic valve replacement with mechanical vs. biological prostheses in patients aged 50-69 years. Eur Heart J November 11, 2015:ehv580.

[10] Cao C, Ang SC, Indraratna P, et al. Systematic review and meta-analysis of transcatheter aortic valve implantation versus surgical aortic valve replacement for severe aortic stenosis. Ann Cardiothorac Surg January 2013;2(1):10–23. http://dx.doi.org/10.3978/j.issn.2225-319X.2012.11.09.

[11] Andalib A, Mamane S, Schiller I, et al. A systematic review and meta-analysis of surgical outcomes following mitral valve surgery in octogenarians: implications for transcatheter mitral valve interventions. EuroIntervention February 2014;9(10):1225–34. http://dx.doi.org/10.4244/EIJV9I10A205.

[12] Poston RS, Tran R, Collins M, et al. Comparison of economic and patient outcomes with minimally invasive versus traditional off-pump coronary artery bypass grafting techniques. Ann Surg 2008;248:638–46.

[13] Sharony R, Grossi EA, Saunders PC, et al. Minimally invasive reoperative isolated valve surgery: early and mid-term results. J Card Surg 2006;21:240–4.

[14] Iribarne A, Easterwood R, Chan EYH, et al. The golden age of minimally invasive cardiothoracic surgery: current and future perspectives 2011;7:333–46.

[15] Modi P, Hassan A, Chitwood WR. Minimally invasive mitral valve surgery: a systematic review and meta-analysis. Eur J Cardiothorac Surg 2008;34:943–52.

[16] Gammie JS, Zhao Y, Peterson ED, O'Brien SM, Rankin JS, Griffith BP. Less-invasive mitral valve operations trends and outcomes from the society of thoracic surgeons adult cardiac surgery database. Ann Thorac Surg 2010;90:1401–10.

[17] Cheng D, Martin J, Shennib H, et al. Endovascular aortic repair versus open surgical repair for descending thoracic aortic disease a systematic review and meta-analysis of comparative studies. J Am Coll Cardiol 2010;55(10):986–1001.

[18] Makkar RR, Fontana GP, Jilaihawi H, PARTNER Trial Investigators, et al. Transcatheter aortic valve replacement for inoperable severe aortic stenosis. N Engl J Med 2012;366:1696–704.

[19] Mack MJ, Leon MB, Smith CR, PARTNER Trial 1 Investigators, et al. 5-year outcomes of transcatheter aortic valve replacement or surgical aortic valves replacement for high surgical risk patients with aortic stenosis: a randomized controlled trial. Lancet 2015;385:2477–84.

[20] Mack M, Brennan M, Brindis R, et al. Outcomes following transcatheter aortic valve replacement in the United States. JAMA 2013;310(19):2069–77. http://dx.doi.org/10.1001/jama.2013.282043.

[21] Leon M, Smith C, Mack M, et al. Transcatheter aortic-valve implantation for aortic stenosis in patients who cannot undergo surgery. N Engl J Med October 21, 2010;2010(363):1597–607. http://dx.doi.org/10.1056/NEJMoa1008232.

[22] Adams D, Popma J, Reardon M, et al. Transcatheter aortic-valve replacement with a self-expanding prosthesis. N Engl J Med May 8, 2014;2014(370):1790–8. http://dx.doi.org/10.1056/NEJMoa1400590.

[23] Holmes Jr DR, Brennan JM, Rumsfeld JS, STS/ACC TVT Registry., et al. Clinical outcomes at 1 year following transcatheter aortic valve replacement. JAMA 2015;313:1019–28.

[24] Wimmer NJ, Williams DO. Transcatheter aortic valve replacement and stroke. Circ Cardiovasc Interv June 2015;8(6):e002801.

[25] Popma JJ, et al. Transcatheter aortic valve replacement using a self-expanding bioprosthesis in patients with severe aortic stenosis at extreme risk for surgery. JACC 2014. http://dx.doi.org/10.1016/j.jacc.2014.02.556.

[26] Samim M, Agostoni P, Hendrikse J, et al. Embrella embolic deflection device for cerebral protection during transcatheter aortic valve replacement. J Thorac Cardiovasc Surg 2015;149:799–805.

[27] <http://www.tctconference.com>.

[28] Brott TG, Halperin JL, Abbara S, et al. 2011 ASA/ACCF/AHA/AANN/AANS/ACR/ASNR/CNS/SAIP/SCAI/SIR/SNIS/SVM/SVS Guideline on the management of patients with extracranial carotid and vertebral artery disease. J Am Coll Cardiol 2011;57(8):e16–94.

[29] Yamaguchi A, Adachi H, Tanaka M, et al. Efficacy of intraoperative epiaortic ultrasound scanning for preventing stroke after coronary artery bypass surgery. Ann Thorac Cardiac Surg 2009;15:98–104.

[30] Rosenberger P, Shernan SK, Löffler M, et al. The influence of epiaortic ultrasonography on intraoperative surgical management in 6051 surgical patients. Ann Thorac Surg 2008;85:548–53.

[31] Breur AC, Furlan AJ, Hanson MR, et al. Central nervous system complications of coronary artery bypass graft surgery: prospective analysis of 421 patients. Stroke 14:682–7.

[32] Humphreys RP, Hoffman JH, Mustard WT, et al. Cerebral hemorrhage following heart surgery. J Neurosurg 1975;43:671–5.

[33] Sila CA. Spectrum of neurologic events following cardiac transplantation. Stroke 1989;20:1586–9.

DIAGNOSTIC TESTING

127

Clinical Stroke Diagnosis

L.R. Caplan

Harvard University, Beth Israel Deaconess Medical Center, Boston, MA, United States

INFORMATION USED FOR STROKE DIAGNOSIS [1,2]

Clinicians must first decide on the key questions to be asked. Answers are difficult unless the questions are clearly framed. In neurology, two diagnostic questions always require an answer: (1) *what* is the disease mechanism—the pathology and pathophysiology? and (2) *where* is the lesion(s)—the anatomy of the disorder? In stroke patients, the "what" question concerns which of the five stroke mechanisms (hemorrhage—subarachnoid or intracerebral; ischemia—thrombotic, embolic, or decreased global perfusion) is present. Before distinguishing among stroke mechanisms, clinicians should first ask whether the findings could be caused by a nonvascular process, such as a brain tumor, metabolic abnormality, infection, intoxication, seizure disorder, or traumatic injury that mimics stroke (Chapter 127 is devoted to stroke mimics). The *where* question concerns the anatomic location of the condition, in the brain and in the vascular system that supplies and drains the brain.

Different data are used to answer these two different questions. In determining stroke mechanism, the *what* question, the following clinical bedside data are most helpful:

1. Ecology—the past and present personal and family illnesses
2. Presence and nature of past strokes or transient ischemic attacks (TIAs)
3. Activity at the onset of the stroke, such as physical effort
4. Temporal course and progression of the findings (Was the stroke onset sudden with the deficit maximal at onset? Did the deficit improve, worsen, or remain the same after onset? If it worsened, did this occur in a stepwise, remitting, or gradually progressive fashion? Were there fluctuations between normal and abnormal?)
5. Accompanying symptoms such as headache, vomiting, seizures, and decreased level of consciousness

Information about these items can all be gleaned from a thorough and thoughtful history from the patient, a review of physicians' and medical records, and data collected from observers, family members, and friends. The general physical examination, which uncovers disorders not known from the history, adds to the data used for diagnosing the stroke mechanism. Elevated blood pressure, cardiac enlargement, murmurs, arrhythmia, and vascular bruits are examples of physical findings that influence identification of the stroke mechanism.

Diagnosis of stroke location—the *where* question—is made using very different information:

1. Analysis of the neurological symptoms and their distribution
2. Findings on neurological examination
3. Findings from brain and vascular imaging

Mechanism and anatomic diagnoses are not absolute. More realistic are estimates of probabilities. In one patient, intracerebral hemorrhage may be by far the most likely diagnosis, but embolism and thrombosis are also possible and should not be eliminated from consideration. In another patient, there might be a toss-up between thrombosis and embolism.

The process of diagnosis involves two basic techniques: (1) hypothesis generation and testing and (2) pattern matching.

THE INDUCTIVE METHOD— SEQUENTIAL HYPOTHESIS GENERATION AND TESTING AND PATTERN MATCHING [1,2]

Hypothesis generation should begin as soon as the first information about the patient becomes available. As the patient or another individual relates the history, the clinician should be thinking of possible diagnoses. The overview given by the patient or other historian generates hypotheses and queries. The clinician then asks the patient and available other questions whose answers should help confirm or refute the hypotheses about the two questions that should be answered, "What" and "Where." Anatomic hypotheses are also generated. A left hemiparesis raises the possibility of a right cerebral or brain stem lesion. Ask about accompanying visual, sensory, or brain stem symptoms that would help generate a more specific anatomic localization.

The other technique used by clinicians is pattern matching. Clinicians identify a constellation of findings that match their mental images of patterns of stroke mechanisms and pathology and anatomy.

After the history, the clinician should plan the examination (general and neurological) to test the hypotheses generated from the history. After the examinations the hypotheses and probabilities are revised. The laboratory and imaging testing are planned to further test the hypotheses and to arrive at workable diagnoses.

IMPACT OF VARIOUS FINDINGS ON THE "WHAT" DIAGNOSIS

Ecology

Included within ecology are prior medical diseases and demographic data that might predispose the patient to have one or more of the various stroke mechanisms. The presence of diabetes and coronary artery disease strongly favors a diagnosis of associated atherosclerosis of the extracranial cervical arteries and a thrombotic (or artery-to-artery embolus) mechanism of stroke. The presence of prior heart disease raises the possibility of arrhythmia, mural thrombosis, ventricular aneurysm, and valvular heart disease, all potential sources of brain embolism. The presence of hypertension increases the probability of intracerebral hemorrhage (ICH), especially if the hypertension is severe. Anticoagulation increases the risk of intracranial hemorrhage. Neck and/or face pain in young physically active individuals raises the possibility of arterial dissection. A stroke that develops rapidly in a young person while defecating or during sex raises the possibility of a patent foramen ovale (PFO) or other cardiac shunt causing a brain embolus [3].

Prior Cerebrovascular Symptoms Especially TIAs

Prior cerebrovascular events should be given considerable importance. Recent TIAs in the same vascular territory are frequent precursors of thrombotic stroke, so their presence, especially when multiple, is virtually diagnostic of that stroke mechanism. Prior events in different vascular territories raise the possibility of brain embolism or

hypercoagulability. The presence, nature, and duration of TIAs are important. Information about the presence of TIAs must be vigorously and repeatedly sought. Many patients are quite naive about the functions of the body, especially the nervous system; they attribute their weakness, lack of feeling, and visual deficits to the local limbs or to the eyes; they often do not understand that the central nervous system control of these functions has been damaged. There were multiple attacks, when was the first and when was the last? Are they getting more frequent or is the interval between attacks becoming longer? Are the episodes stereotyped and nearly identical in all attacks? How long do the TIAs last? What is the shortest, longest, and average duration? Are attacks becoming longer or shorter? Are TIAs provoked by standing or activity?

Activity at Onset

Traditional teaching is that most thrombotic strokes occur when the circulation is least active and most sluggish (e.g., during the night or during a nap, with the deficit usually noticed on arising). Embolism and hemorrhage, in contrast, are more likely to develop when the circulation is more active or when blood pressure rises. Data now show that most ischemic and hemorrhagic strokes actually occur during the morning hours, especially between 10 a.m. and noon after the patient has awakened and begun daily activities [10]. A significant number of hemorrhages do occur at night, and thrombotic deficits can occur during activity but it is unusual for a thrombotic stroke or a lacunar infarct to develop during vigorous physical activity or during sex. A particularly common time for embolism to occur is on arising at night to urinate, the so-called matudinal (morning) embolus. Certain physical activities and situations are related to particular stroke subtypes. Valsalva maneuver immediately before stroke symptoms should suggest paradoxical embolism related to an increase in right atrial pressure in patients with a PFO. Coughing or a vigorous sneeze can also shake loose an embolic particle, resulting in brain embolism. Physical efforts that involve neck trauma or sudden neck movements and stroke after neck manipulations should raise suspicion of arterial dissection [4]. Arterial dissections have also been described after labor and during the postpartum period and after weight lifting [4,5].

Early Course of Development of the Deficit

The early course gives important information about the stroke mechanism. I encourage clinicians should construct "course of illness" graphs that show the temporal pattern of the findings [6]. Fig. 127.1 shows the course of illness in a patient with the following history: a previously hypertensive man suddenly became aphasic and

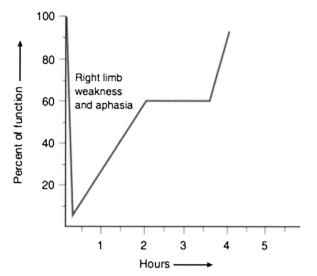

FIGURE 127.1   Course of illness graph in a patient described in the text.

LOCALIZATION AND DETECTION OF THE VASCULAR LESION

Having pursued the history as thoroughly as possible, the clinician should be ready to perform a general and neurological examination. While proceeding, the principal aims should be kept in mind. They are (1) to detect vascular and cardiac abnormalities that aid in determining stroke mechanism and localization of vascular lesions and (2) to localize the process within the central nervous system. Once the clinician knows where the lesion is in the brain, knowledge about the anatomy of the vascular supply, about the risk factors in the patient, and about the results of the vascular examination help the clinician predict the most likely vascular location and process in that individual patient.

Findings From Examination of the Heart

The diagnosis of cardiogenic embolism is important because its evaluation and treatment differ from intrinsic disease of the neck and intracranial arteries. A detailed history of possible cardiac symptoms, angina, myocardial infarction, palpitations or arrhythmia, congestive heart failure, and rheumatic heart disease is as important as the neurological history. The heart should be examined thoroughly, taking time to estimate size, character, and quality of heart sounds and gallops; listening for murmurs is not enough.

Findings From Examination of the Vascular System and the Eyes

Examination of the available systemic and neck arteries can yield clues to the presence of atherosclerosis or diminished flow not detectable by history. The radial pulse should be palpated for at least a minute, seeking any irregularities. Feel the radial pulse simultaneously, looking for a significant difference in the strength of the pulses or a delay on one side. The femoral and foot pulses should be palpated. Listening for a bruit over the carotid arteries in the neck and the vertebral arteries in the supraclavicular region and over the mastoid slope in the back of the neck is important. In patients suspected of having atrial fibrillation, the cardiac rate should be compared with the radial pulse rate to look for dropped pulses.

The eyes provide a window into the body's vascular system and can yield clues concerning stroke mechanism. Subhyaloid hemorrhages, large round hemorrhages with a fluid level, represent sudden bleeding below the retina and almost always reflect a sudden change in intracranial pressure. They are often seen in patients with SAH and also occur in acutely developing large ICHs. The severity of hypertensive retinopathy and arteriosclerotic

hemiplegic while eating lunch with his family. When initially examined, he could not speak and had a severe right hemiplegia. Two hours later, he was much improved and could lift his right leg and say a few words. Four hours after the symptoms began, he had returned to normal except for minor weakness of the right hand and arm. The improvement soon after onset of the deficit argues strongly against an ICH. The deficit, which was maximal at onset and was unassociated with headache, is most compatible with an embolic mechanism.

The process of eliciting the historical details is called "walking through" the course of illness with the patient. Most patients have difficulty quantifying their deficits and estimating the course of their illness. When patients are asked to describe their activities, an alert observer can often better gauge the course of development of the deficit.

Accompanying Symptoms

Headache, especially if sudden and severe at onset, is an invariable symptom of aneurysmal subarachnoid hemorrhage (SAH). Sudden release of blood into the subarachnoid space increases intracranial pressure and usually leads to severe headache, vomiting, and a decrease in the level of consciousness. In ICH, the focal deficit usually develops progressively, and only later, when there has been enlargement of the hematoma, do headache, vomiting, and decreased consciousness develop. Loss of consciousness is common in SAH and is rare in ischemic stroke unless the ischemia involves the brain stem bilaterally. Occasionally, transient loss of consciousness occurs in patients with ischemic strokes particularly those due to embolism. Seizures are rare in the early period after stroke onset; their presence argues for embolic stroke or ICH.

changes are important to note. Examination of the retina can also yield signs of embolism (cholesterol crystals, fibrin-platelet, and calcific emboli), most often from the carotid artery but sometimes from the heart and its valves or from the aorta.

STROKE LOCALIZATION

Clinical localization of the brain lesion is primarily from the patient's description of their neurological symptoms and the findings on neurological examination [1,2]. It would be impossible to review here the full details of the neurological examination. The neurological findings do not have much impact on the diagnosis of stroke mechanism, although they do help with the anatomic location of the lesion and the nature of the neurological deficits that require attention. Detailed discussions of the neurological examination are published elsewhere [2,7]. The most important and most frequently missed signs of brain dysfunction involve abnormalities of (1) higher cortical function, (2) level of alertness, (3) the visual and oculomotor systems, and (4) gait. These are parts of the examination most often overlooked by nonneurologists that provide key clues to anatomic localization. Cognitive tests should assess memory, language function (spoken and written), and visual-spatial abilities [2,7–9] (drawing and copying). Examining eye movements and the visual fields is essential in all patients. Motor strength, coordination, deep tendon reflexes, sensation, and gait are always examined.

Be sure to test the strength and coordination in each limb proximally and distally. In central lesions, the most important weakness is usually in the shoulder abductors, arm extensors, finger extensors and abductors, thigh flexors, leg flexors, and foot and toe dorsiflexors and everters. Check for drift of the outstretched hands. Try to estimate the relative motor strength in face, arms, hands, and legs. In hemiparetic patients, are any of these regions disproportionately affected or preserved? Test coordination of each limb by the finger-nose, toe-object maneuvers. It is informative to elicit the Babinski responses. In patients with cerebral lesions, higher sensory functions, such as position sense, object recognition, and extinction, are more often affected than elementary pin or touch perception.

Some patients with cerebellar lesions have a normal examination when recumbent or seated but cannot walk. Observation of gait also gives a great deal of information about motor function and its symmetry. Is there dragging of one foot, delay in hip flexion on one side, or less arm swing on one side? Are tremors or odd posturing of a limb seen as the patient walks?

Common Localization Patterns

Neurological signs most often fall into recognizable patterns that predict the likely anatomical localization of the brain lesion. The neurological symptoms and signs can usually be placed in one of seven general categories. The process is simply one of pattern recognition, that is, matching the patient's clinical deficit with that of patients with known lesions in one of the following regions. Also, are there expected findings that are absent or unexpected added findings beyond those described among these patterns?

1. Left hemisphere lesion [in the anterior hemisphere in the territory of the internal carotid artery (ICA) and its middle cerebral artery (MCA) and anterior cerebral artery (ACA) branches]: Aphasia, right limb weakness, right limb sensory loss, right visual field defect, reduced right conjugate gaze, difficulty reading, writing, and calculating
2. Right hemisphere lesion (in ICA-ACA-MCA distribution): Neglect of the left visual space, difficulty drawing and copying, left visual field defect, left limb motor weakness, left limb sensory loss, reduced left conjugate gaze, extinction of the left stimulus of two simultaneously given visual or tactile stimuli
3. Left posterior cerebral artery (PCA) lesion: Right visual field defect, difficulty reading with retained writing ability, difficulty naming colors and objects presented visually, normal repetition of spoken language, numbness and sensory loss in the right limbs
4. Right PCA lesion: Left visual field defect, often with neglect, left limb numbness and sensory loss
5. Vertebrobasilar territory infarction [10,11]: Spinning dizziness; diplopia; weakness or numbness of all four limbs or bilateral regions; crossed motor or sensory findings (e.g., numbness or weakness of one side of the face and the opposite side of the body); ataxia; vomiting; headache in the occiput, mastoid, or neck; bilateral blindness or dim vision; on examination, nystagmus or dysconjugate gaze, gait or limb ataxia out of proportion to weakness, bilateral recently acquired weakness or numbness (i.e., one side not due to an old stroke or other defect), crossed signs, bilateral visual-field defects, amnesia

IMAGING AND LABORATORY EVALUATION

After the clinician has tabulated in the mind the neurological abnormalities, they should step back from the bedside and think. Where is the lesion likely to be? If

TABLE 127.1 Suggestions for Blood Testing in Patients With Brain Ischemia (Stroke or Transient Ischemic Attack)

Hemoglobin and hematocrit

White blood cell count (and differential if too high or too low)

Platelet count

Activated partial thromboplastin time

Prothrombin time—international normalized ratio

Serum fibrinogen level

Blood sugar

Serum calcium

Total cholesterol and high-density lipoprotein and low-density lipoprotein cholesterol

Blood urea nitrogen and electrolytes (sodium, chloride, potassium, and carbon dioxide)

Homocysteine

C-reactive protein and erythrocyte sedimentation rate

there is more than one possible or probable location, they might think of further bedside testing that could distinguish among these possibilities. The imaging and laboratory evaluation should be planned with the likeliest brain and vascular lesions in mind. In general, a brain [computed tomography (CT) or MRI] image and vascular imaging of the cervicocranial arteries and veins computed tomography angiography, magnetic resonance angiography, or neck and transcranial ultrasound should be ordered. These studies are discussed in separate chapters in this Primer. In many patients the heart and aorta should be studied by electrocardiography and echocardiography, and often rhythm monitoring. The blood constituents (red and white cells and platelets) should be quantified. Coagulation functions are also screened. There are chapters in this Primer on coagulopathies. I list in Table 127.1 the recommended blood testing in patients with suspected brain ischemia.

References

[1] Caplan's stroke, a clinical approach. 5th ed. Cambridge: Cambridge University Press; 2016.

[2] Caplan LR, Hollander J. The effective clinical neurologist. 3rd ed. Shelton (CT, USA): PMPH; 2011.

[3] Caplan LR. Clinical diagnosis of brain embolism. Cerebrovasc Dis 1995;5:79–88.

[4] Caplan LR. Dissections of brain-supplying arteries. Nat Clin Pract Neurol 2008;4:34–42.

[5] Baffour FI, Kirchoff-Torres KF, Einstein FH, Karakash S, Miller TS. Bilateral internal carotid artery dissection in the postpartum period. Obstet Gynecol 2012;119:489–92.

[6] Caplan LR. Course-of-illness graphs. Hosp Pract 1985;20:125–36.

[7] Caplan LR. In: Fisher M, Bogousslavsky J, editors. The neurological examination in textbook of neurology. Boston: Butterworth-Heinemann; 1998. p. 3–18.

[8] Hier D, Mondlock D, Caplan L. Behavioral deficits after right hemisphere stroke. Neurology 1983;33:337–44.

[9] Caplan LR, Bogousslavsky J. In: Bogousslavsky J, Caplan LR, editors. Abnormalities of the right cerebral hemisphere in stroke syndromes. Cambridge: Cambridge University Press; 1995. p. 162–8.

[10] Savitz S, Caplan LR. Current concepts: vertebrobasilar disease. N Engl J Med 2005;352:2618–26.

[11] Caplan LR. Vertebrobasilar ischemia and hemorrhage: clinical findings, diagnosis, and management of posterior circulation disease. Cambridge (UK): Cambridge University Press; 2014.

CHAPTER

128

Cardiac Electrophysiology in Stroke Investigation: Holter, Event Monitor, and Long-Term Monitoring

P.A. Santucci

Loyola University Chicago, Stritch School of Medicine, Maywood, IL, United States

Identifying the etiology of ischemic stroke is critical to providing optimum therapy and preventing future events. Unfortunately, it is not always a simple task to establish the source, which all too often results in the diagnosis of "cryptogenic" stroke. Since cardiac sources of emboli are a well-known and significant cause of ischemic stroke, and may require different patient management, substantial effort has been made into the investigation for cardiac sources. The most important cardiac etiology for embolus is atrial fibrillation (AF), which may account for approximately 15–20% of cases of ischemic stroke. Atrial flutter, although less common, also can result in stroke, and its detection is also important.

However, not all patients with AF as the cause of stroke are known to have the diagnosis before the embolic event, nor do they necessarily present with AF at the time of the stroke. Many patients with AF have no cardiac symptoms, and the diagnosis is made only incidentally or after varying levels of search. Paroxysmal and persistent AF are considered to have a similar risk of stroke; therefore a presentation in sinus rhythm does not exclude AF as the cause of stroke. In fact, it is often in the period after the spontaneous conversion of fibrillation that embolization occurs.

BACKGROUND

Prevalence

After initial investigation, up to 40% of ischemic strokes are classified as cryptogenic stroke (CS) [1].

However, subsequent evaluation has shown that a significant portion of these patients will manifest previously undiagnosed AF. The frequency of newly diagnosed AF varies but has been found in 12–30% or more of patients with CS using long-term monitoring [1–4]. A number of factors influence the percentage in which AF is diagnosed in various studies, including the specific population evaluated; the extent of initial evaluation leading to the diagnosis of CS; the method, duration, and frequency of rhythm monitoring; and the duration of AF defined as relevant.

Value of Cardiac Monitoring

Standard practice dictates some form of initial cardiac rhythm assessment in cases of ischemic stroke of uncertain etiology. In particular, an assessment for AF is appropriate. However, due to the paroxysmal nature of many cases of AF, initial electrocardiography (ECG) alone will not detect all patients with AF. Studies looking at more extended monitoring have shown that sensitivity increases with more prolonged or repeated investigation. The optimal method and duration of monitoring remains uncertain, with several considerations influencing clinical decisions discussed later. The detection of occult AF frequently affects therapy, with anticoagulation resulting in a significantly decreased risk of recurrent stroke in these patients as compared with antiplatelet therapy alone. Extended monitoring has led to significantly increased usage of guideline-directed anticoagulation in these patients.

Primer on Cerebrovascular Diseases, Second Edition
http://dx.doi.org/10.1016/B978-0-12-803058-5.00128-4

OPTIMIZING DETECTION

Type of Monitoring

Historically, an initial ECG following stroke was often the only rhythm assessment performed. Multiple options are available for additional cardiac monitoring, including serial ECGs, Holter monitors, inpatient continuous telemetry, event monitors, outpatient continuous telemetry, and implantable cardiac monitors. External event monitors and outpatient telemetry monitors exist in various forms including standard wired two- or three-channel recorders, or more recently patch recorders without wires. Monitors that do not have full continuous recording should have the capability of automated detection of arrhythmias due to the frequently asymptomatic nature of many arrhythmias in this population. Those that rely on patient activation are not suitable for this purpose. External monitors are typically placed for 7–30 days with better yields seen with longer durations.

Implantable cardiac monitors are becoming increasingly used due to accumulating evidence for additional diagnostic yield for more prolonged monitoring beyond 30 days. These devices are leadless devices placed subcutaneously typically in the left parasternal region. Downsizing of these devices with advancements in technology now allows for injection of these devices in a brief outpatient procedure with minimal discomfort (Fig. 128.1). These devices allow for continuous monitoring for up to 3 years or more. Studies show additional yield in the first 6–12 months, and beyond [2,4,5]. Although the causal relationships between the index stroke and detections of AF many months afterward are unclear, the potential need for anticoagulation in these patients may still reduce the incidences of future ischemic stroke.

In addition, patients with preexisting implantable cardiac rhythm devices (i.e., pacemakers or implantable defibrillators) usually have robust device-acquired diagnostic data that detect the presence and details of AF episodes (Fig. 128.2). Capabilities vary by device and manufacturer, and to a large extent depend on the presence or absence of an atrial lead. However, in addition to detecting fibrillation of flutter, these devices can supply a plethora of long-term data, including dates of occurrence, arrhythmia durations, heart rates, and overall AF burden (Fig. 128.3). Many devices will record intracardiac atrial electrograms so that the accuracy of the diagnosis of AF can be confirmed (Fig. 128.4). These devices have been utilized in patients with stroke to newly detect (or exclude) AF. Furthermore, these devices may in some cases be used to guide therapy, by potentially allowing a more accurate evaluation of the benefits and risks of long-term anticoagulation in patients undergoing rhythm control therapy either medically, or by catheter ablation or surgical procedures, in which AF is thought to have been potentially effectively suppressed

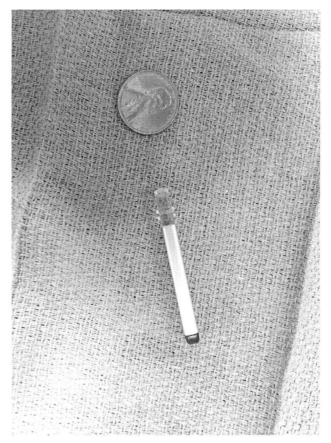

FIGURE 128.1 Implantable cardiac monitor, which can be injected subcutaneously and monitor the cardiac rhythm for approximately 3 years.

or eliminated. In addition, patients without a history of stroke, but with subclinical detections of AF by these devices have been shown to have subsequent increased risk of stroke, thus potentially allowing for prevention of even an initial ischemic stroke [6].

As of this time, there is no proven role for invasive intracardiac diagnostic electrophysiologic testing for the evaluation of CS, and there is no indication for such testing.

Duration and Frequency of Monitoring

Beyond a baseline ECG, initial 24-h monitoring can be accomplished with inpatient telemetry or by Holter monitoring. Holter monitoring has detected AF in 1–6% of patients with stroke of uncertain etiology at an initial evaluation [1]. Although earlier guidelines recommended a minimum 24-h period of rhythm monitoring, it has become common to monitor for significantly longer periods to improve the detection of occult AF. The baseline ECG and initial 24-h Holter monitoring have limited sensitivity in detecting new AF and studies examining various forms of additional monitoring show significant stepwise improvement with more prolonged monitoring [3,7]. One approach has been intermittent reevaluation with periodic ECGs or Holter monitoring, but this too

Brady/CRT Counters	Reset Before Last 17 Jun 2015	Since Last Reset 13 Oct 2015
Counters		
% A Paced	56	43
% RV Paced	100	100
% LV Paced	100	100
Intrinsic Promotion		
Rate Hysteresis		
% Successful	0	0
Atrial Burden		
Atrial Burden %	1	1
Episodes by Duration		
< 1 minute	824	442
1 min - < 1 hr	67	42
1 hr - < 24 hr	11	9
24 hr - < 48 hr	0	0
> 48 hr	0	0
Total PACs	839.9K	1.1M
Ventricular Counters		
Total PVCs	31.9K	26.9K
Three or More PVCs	0	0

FIGURE 128.2 Diagnostic information stored in an implanted pacemaker, showing 1% burden of atrial fibrillation and durations.

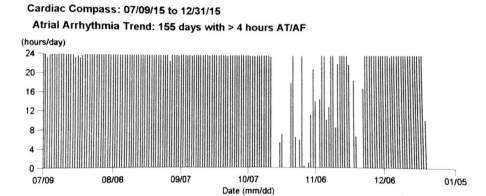

FIGURE 128.3 Other diagnostic information stored in an implanted pacemaker, showing timing and burden of atrial fibrillation (AF) episodes.

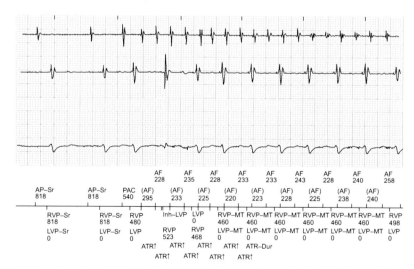

FIGURE 128.4 Stored electrograms in an implanted device. Atrial electrogram shows the detailed intracardiac recording of the onset of atrial fibrillation (AF).

has limited sensitivity. Current guidelines note the value of more prolonged continuous monitoring to 30 days or more to increase the detection rates [8]. In 30-day cardiac event monitor belt for recording atrial fibrillation after a cerebral ischemic event (EMBRACE) trial, AF detection was seen in 16.1% using a 30-day event-triggered recorder [3]. There has been accumulating evidence that prolonged monitoring up to 3 years with the use of implantable cardiac monitors will result in increased diagnoses of AF [4,5]. In the cryptogenic stroke and underlying AF (CRYSTAL-AF) trial, extended monitoring using an implanted cardiac monitor revealed AF in 8.9%, 12.4%, and 30% at 6, 12, and 30 months versus only 1.4% in the control group [2,4]. However, the costs rise with the duration and frequency of monitoring.

Predictors of Atrial Fibrillation

Beyond the aforementioned considerations is whether the optimal form of cardiac monitoring is uniform among all patients with CS, or should vary based on patient characteristics. The likelihood of finding an arrhythmia relevant to stroke management depends not only on the monitoring methods, but also on the pretest probability of arrhythmias. This can be influenced by the patient's age, comorbidities, and other risk factors for cardiac versus noncardiac causes of stroke. In particular, the presence of preexisting heart disease significantly increases the likelihood of AF and should indicate the need for a more thorough exclusion of arrhythmias [9]. There is also some evidence that the likelihood of AF detection also correlates with CHA2DS2VASc [CHA2DS2VASc is a commonly used stroke risk estimator for atrial fibrillation, which stands for Congestive heart failure, Hypertension, Age (>75 = 2 points), Diabetes, and Stroke/TIA (2 points); Vascular disease (1 point for peripheral arterial disease, previous MI, or aortic atheroma), Age (65–75, 1 point) and Sc-Sex category: female gender (1 point).] scores [10]. In addition, details of any preexisting or discovered heart disease may also indicate a risk for cardiac sources of emboli unrelated to rhythm.

Furthermore, specific characteristics of the stroke itself, such as the location, size, or other characteristics might be relevant to the source of stroke and may affect who should receive a more intensive search for an arrhythmic cause. Ischemic strokes associated with AF tend to be associated with larger infarct size, more severe neurologic consequences, and anterior circulation territory infarctions [7]. Similarly, if characteristics can be identified that indicate a very low yield from extensive monitoring, a reduction of patient burden and cost savings might be realized. For example, lacunar infarcts appear to have a low association with AF [7]. At this time, this is an area of ongoing investigation, and there are no definitive answers to determine specific monitoring protocols based on specific stroke characteristics.

RELEVANCE OF MONITOR FINDINGS

In addition to the issue of limited sensitivity of initial 24-h monitoring, there may also be a concern regarding specificity. A question not yet fully answered is whether the early period after a stroke results in episodes of new poststroke AF that are not relevant to the etiology of stroke, and may falsely suggest the existence of AF before the stroke. Some preliminary data have suggested that poststroke AF may be an epiphenomenon rather than the cause of the stroke in some cases. Additional data will be needed to clarify whether and how often this may be the case.

Duration and Burden of Atrial Fibrillation

Uncertainty remains on the duration and burden of AF that is clinically relevant. Although historically AF of less than 24-h duration is considered unlikely to result in cardiac thrombus, it has become apparent that patients with findings of much shorter duration AF have increased risk of stroke. Although transient AF lasting <30 s has generally not been accepted as clinically important, durations as short as 6 min down to 30 s have been used in more recent studies as enough to warrant consideration of preventive therapy. Many patients with initial detections in these shorter ranges have significantly longer duration episodes found at other times with more extended evaluation.

Temporal Relationships and Competing Causes of Stroke

Another consideration in clinical decision making involves the sometimes inconsistent temporal relationship between AF and stroke. Investigation of patient with long-term monitoring has shown stroke occurrence in patients with AF at times when there have been no recent episodes of AF [11]. This calls into question not only whether AF is at all responsible for the stroke, but even whether it is cardioembolic in nature at all. Complicating matters further is that some patients with potential cardiac sources of emboli might also have arterial sources and some patients with nonembolic stroke might also have occult AF. The presence of coronary arterial disease not only might increase the risk of a cardiac cause, but also may suggest a higher risk of extracardiac arterial disease. This issue of potentially "competing causes" of ischemic stroke may influence the need for additional complete investigation even in some patients who do not carry a diagnosis of CS [8]. Regardless of whether AF was the cause of the prior stroke, these patients generally meet criteria for anticoagulation to prevent future strokes related to AF. These findings may indicate that treatment of either the AF alone (with anticoagulation) or an arterial source alone (with antiplatelet therapy) may not be fully protective. However, treating both possible sources can become difficult, since combining both antiplatelet therapy and anticoagulation can significantly increase the risk of bleeding. Best practice still requires keen judgment by the practitioner on not only whether these patients have more than one source of stroke that should also be considered, but also which potential sources create enough future risk to be treated without creating excessive added risk.

EVALUATION

In addition to an ECG, the initial evaluation should include a careful history and examination to assess for cardiac abnormalities that might indicate a high probability of arrhythmic or nonarrhythmic cardiac sources of embolism. If no prior cardiac evaluation has been performed, a screening echocardiogram should be considered to assess for underlying structural heart disease. Since AF is often clinically silent and often undetected at the initial evaluation following ischemic stroke, additional monitoring for the detection of AF beyond the initial ECG has become standard in the evaluation of CS. At a minimum, 24–72 h of initial inpatient or outpatient monitoring is reasonable. If no AF is seen, consideration should be given to more prolonged monitoring up to 30 days with external monitoring, or beyond via a preexisting cardiac rhythm device or by an implantable cardiac monitor. At this time, whether all or most patients should receive long-term monitoring is undecided. Generally, the extent of monitoring should probably be tailored to the patient. Those in whom there is a reasonable suspicion or probability of finding AF, and in whom therapy may be determined by the finding, are good candidates for more long-term monitoring, including the use of implantable cardiac monitors. Patient preferences and cost considerations should also be taken into account.

SUMMARY

As cardiac sources of ischemic stroke often require anticoagulation as opposed to the antiplatelet therapy typically used for arterial sources, the diagnosis of AF or atrial flutter is critical to appropriate management. Missed cases can lead to significantly higher risks of recurrent strokes, and thus the investigation into causes of CS should involve an adequate evaluation for occult cases of arrhythmic causes.

References

[1] Glotzer TV, Ziegler PD. Cryptogenic stroke: is silent atrial fibrillation the culprit? Heart Rhythm 2015;12:234–41.
[2] Sanna T, Diener HC, Passman RS, Di Lazzaro V, Bernstein RA, Morillo CA, et al. Cryptogenic stroke and underlying atrial fibrillation. N Engl J Med 2014;370:2478–86. http://dx.doi.org/10.1056/NEJMoa1313600.
[3] Gladstone DJ, Spring M, Dorian P, Panzov V, Thorpe KE, Hall J, et al. Atrial fibrillation in patients with cryptogenic stroke. N Engl J Med 2014;370:2467–77. http://dx.doi.org/10.1056/NEJMoa1311376.
[4] Brachmann J, Morillo CA, Sanna T, Di Lazzaro V, Diener HC, Bernstein RA, et al. Uncovering atrial fibrillation beyond short-term monitoring in cryptogenic stroke patients: three-year results from the cryptogenic stroke and underlying atrial fibrillation trial. Circ Arrhythm Electrophysiol January 2016;9(1):e003333. http://dx.doi.org/10.1161/CIRCEP.115.003333.
[5] Cotter PE, Martin PJ, Ring L, Warburton EA, Belham M, Pugh PJ. Incidence of atrial fibrillation detected by implantable loop recorders in unexplained stroke. Neurology 2013;80:1546–50.
[6] Healey JS, Connolly SJ, Gold MR, Israel CW, Van Gelder IC, Capucci A, et al. Subclinical atrial fibrillation and the risk of stroke. N Engl J Med 2012;366:120–9.
[7] Jabaudon D, Sztajzel J, Sievert K, Landis T, Sztajzel R. Usefulness of ambulatory 7-day ECG monitoring for the detection of atrial fibrillation and flutter after acute stroke and transient ischemic attack. Stroke 2004;35:1647–51.
[8] Kernan WN, Ovbiagele B, Black HR, Bravata DM, Chimowitz MI, Ezekowitz MD, et al. Guidelines for the prevention of stroke in patients with stroke and transient ischemic attack: a guideline for healthcare professionals from the American heart association/American stroke association. Stroke 2014;45:2160–236.
[9] Rizos T, Horstmann S, Dittgen F, Täger T, Jenetzky E, Heuschmann P, et al. Preexisting heart disease underlies newly diagnosed atrial fibrillation after acute ischemic stroke. Stroke 2016;47:336–41. http://dx.doi.org/10.1161/STROKEAHA.115.011465.
[10] Friberg L, Rosenqvist M, Lindgren A, Terént A, Norrving B, Asplund K. High prevalence of atrial fibrillation among patients with ischemic stroke. Stroke September 2014;45(9):2599–605. http://dx.doi.org/10.1161/STROKEAHA.114.006070. Epub 2014 Jul 17.
[11] Daoud E, Glotzer T, Wyse DG, Ezekowitz M, Hilker C, Koehler J, et al. Temporal relationship of atrial tachyarrhythmias, cerebrovascular events, and systemic emboli based on stored device data: a subgroup analysis of TRENDS. Heart Rhythm April 25, 2011;8(9):1416–23.

CHAPTER

129

Cardiac Ultrasound in Stroke Investigation

W.J. Manning

Harvard Medical School, Boston, MA, United States

Abbreviations

AF atrial fibrillation
ASA atrial septal aneurysm
INR international normalized ratio
LA left atrium, left atrial
LAA left atrial appendage
LV left ventricle, left ventricular
MAC mitral annular calcification
MRI magnetic resonance imaging
PFO patent foramen ovale
TTE transthoracic echocardiography
TEE transesophageal echocardiography

Cardiac sources of stroke may account for 20–30% of the near 800,000 strokes that occur annually in the United States. Conventional transthoracic echocardiography (TTE) remains the cornerstone of noninvasive cardiac imaging for this disorder, although moderately invasive transesophageal echocardiography (TEE) is superior for the identification of most cardiac sources of emboli. Clinical paradigms in the use of TTE and TEE continue to evolve, with Appropriate Use Criteria classifying their use as appropriate [1]. In addition, newer imaging methods, such as cardiac magnetic resonance imaging (MRI) may also have an important role in identifying cardiogenic sources of embolism [2]. The clinical integration of these newer imaging methods on treatment and prognosis remains to be defined.

MAJOR SOURCES OF EMBOLI

Potential cardiac sources of thromboembolism are summarized in Table 129.1 and can be broadly categorized into masses (e.g., intracardiac thrombi/tumors, aortic atheroma, and vegetations) that migrate/embolize; an increased propensity for thrombus formation [e.g., left atrial (LA) spontaneous echo contrast, prominent mitral annular calcification (MAC), and left

ventricular (LV) dysfunction/aneurysm]; and "passageways" for paradoxical venous thromboembolism [e.g., patent foramen ovale (PFO) and atrial septal defects].

Left atrial anatomy and imaging: The body of the left atrium (LA) is a thin-walled, ovoid chamber that lies immediately anterior to the descending thoracic aorta and has a smooth endocardial surface. The body of the LA is well

TABLE 129.1   Sources of Cardioembolic Thromboembolism

MASSES

- Thrombi
  - Left atrium appendage
    - Rheumatic mitral stenosis
    - Atrial fibrillation
  - Prosthetic valve
    - Mechanical & biologic
    - Mitral & aortic
  - Left ventricle
    - Dyskinesis/aneurysm
    - Anteroapical infarction

- Tumors
  - Myxoma
  - Fibroelastoma

- Aortic atherosclerosis
  - Hypertension
  - Hypercholesterolemia
  - Takayashu arteritis

- Valvular heart disease
  - Vegetations/endocarditis
  - Calcific aortic stenosis

PROPENSITY FOR THROMBUS FORMATION

- Spontaneous echo contrast
- Mitral annular calcification
- Left ventricular cavity enlargement/apical systolic dysfunction
- Interatrial septal aneurysm

"PASSAGEWAYS" FOR PARADOXICAL EMBOLISM

- Patent foramen ovale
- Atrial septal defect

Primer on Cerebrovascular Diseases, Second Edition
http://dx.doi.org/10.1016/B978-0-12-803058-5.00129-6

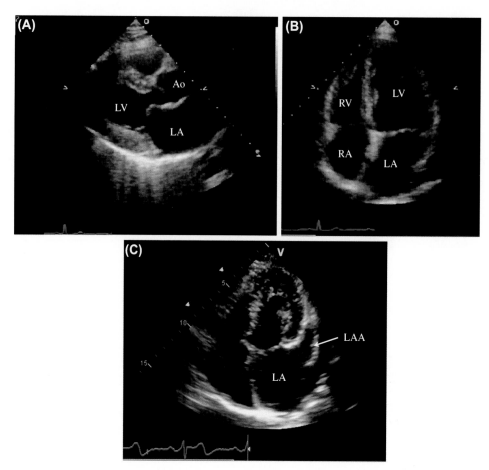

FIGURE 129.1 Transthoracic echocardiogram (TTE). (A) Parasternal long-axis view demonstrating the ascending aorta (Ao), left ventricle (LV), and left atrium (LA). (B) Apical four-chamber view. (C) Off-axis apical four-chamber view demonstrating the location of the left atrial appendage (LAA).

visualized and characterized on TTE (Fig. 129.1) and gender-specific body-size normograms for transthoracic LA size (often based on biplane measurements) demonstrated that increased LA size correlates with cardiovascular disease. The left atrial appendage (LAA) is a highly trabeculated, often multilobulated cul de sac arising from the midlateral LA near the entrance of the left upper pulmonary vein (Fig. 129.2). The LAA is poorly visualized on TTE, but the close proximity of the esophagus to the posterior LA makes TEE the ideal imaging tool for LA and especially LAA thrombus assessment. For patients not candidates for TEE (e.g., esophageal stricture and high sedation risk), long inversion time late gadolinium enhancement cardiac MRI is an excellent alternative [2].

Left atrial thrombi: LA thrombi are almost exclusively seen in association with atrial fibrillation (AF) (or flutter), prosthetic mitral valves with stenosis or inadequate anticoagulation, or rheumatic mitral stenosis. Left atrial thrombi represent nearly half of all cardiogenic thromboemboli. The vast majority of thrombi in AF and native mitral valve disease are found in the LAA. In the absence of a history of AF or mitral valve disease, the incidental finding of an LA/LAA thrombus is very rare [3] even in the setting of embolic

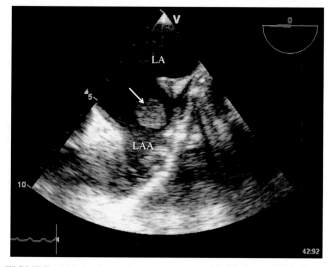

FIGURE 129.2 Transesophageal echocardiogram (TEE) demonstrating the left atrial appendage (LAA) with a thrombus (*arrow*).

stroke or transient cerebral ischemia (although detection of AF may require prolonged monitoring [4]).

Mitral stenosis: Native valve mitral stenosis is almost exclusively found in rheumatic heart disease with stroke recognized with both sinus rhythm and particularly with AF. The normal adult mitral valve area is 4–6 cm^2. In sedentary individuals, clinical symptoms may be lacking until the valve area is <1.5 cm^2. Progressive valve stenosis leads to increased LA pressure, chamber dilation, and endocardial wall fibrosis—all conditions that promote thrombus formation. Large thrombi within the body of the LA, especially large freely mobile "ball" thrombi, are almost exclusively seen with severe rheumatic (or prosthetic) mitral stenosis.

Atrial fibrillation: AF afflicts over 4 million people in the United States alone. With AF, there is loss of organized atrial electrical and mechanical activity leading to symptoms of palpitations and fatigue as well as stagnation/stasis of blood within the body of the LA and especially within the LAA (Fig. 129.3). Spontaneous echo contrast ("smoke-like" echoes), an imaging marker of stasis within the LAA and/or body of the LA is seen in 60% of AF patients and >85% of those with AF and LA thrombi. As the AF population greatly exceeds those with rheumatic mitral stenosis, the clinical association of atrial thrombus and AF is far more common, with stroke rates as high as 20%/year among very high-risk patients. In the setting of nonvalvular AF, atrial thrombi are most commonly located within or involve the LAA (Fig. 129.2) with a prevalence of 13% with new-onset AF [5], increasing to 27% with permanent AF, and 45% in those with AF and recent thromboembolism [6]. For the latter population, the frequency of LAA thrombus represents *residual* thrombus. TEE risk factors for thromboembolism include thrombus size and mobility.

While AF patients with mitral stenosis remain at highest risk for clinical thromboembolism (estimated at up to 20%/year), clinical risk factors for stroke in *nonvalvular* AF (no mitral valve stenosis and no prosthetic valves) disease include Congestive heart failure, Hypertension (systolic blood pressure >160 mmHg), Age >74 years, Diabetes, and prior Stroke/thromboembolism. These are best summarized by the CHADS$_2$ score (Table 129.2). The addition of Vascular disease, Age 65–74 years, and female Sex constitute the CHA$_2$DS$_2$-VAS score [7] and the current standard by which prophylactic anticoagulation is prescribed. In addition to these clinical risk factors, TTE and TEE imaging risk factors for thromboembolism include increased left ventricular systolic dysfunction, more pronounced spontaneous echo contrast, reduced LAA ejection velocity, LA/LAA thrombus, and aortic atheroma. Interestingly, moderate or great mitral regurgitation is not "protective" for LA/LAA thrombus in new-onset AF [5], but is protective against clinical stroke. This difference may represent pathophysiological differences between thrombus formation and thrombus migration. Absent from echocardiographic risk factors for stroke in AF is LA size. This is likely because of the confounding impact of LA dilation

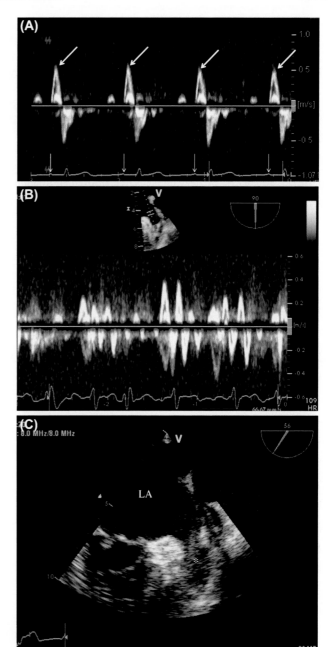

FIGURE 129.3 TEE with pulsed Doppler derived flow at the mouth of the left atrial appendage in a patient with (A) sinus rhythm. Note the rhythmic atrial ejection velocities (*heavy arrows*) occurring immediately after the p-wave on the electrocardiogram (*small vertical arrows*). (B) Similar Doppler flow spectra in a patient with atrial fibrillation. Note the irregular/chaotic LAA ejection velocities of variable intensities and without correspondence to the electrocardiogram. (C) TEE of LAA in a patient with severe spontaneous echo contrast (*) in the LAA.

promoting stasis and the "protective" benefit of mitral regurgitation. Among patients with atrial thrombi, larger size and increased mobility are risk factors for clinical thromboembolism. Even among patients without a clinical history of stroke, studies demonstrate subclinical stroke on head CT among 15% of patients with AF [8].

Systemic thromboembolism occurs in patients with dilated cardiomyopathy [see left ventricular (LV) thrombi] at an annual rate of about 4%. While LV thrombi are presumed to be the source of emboli, LAA thrombi may also be a culprit in this population.

Left ventricular thrombi: LV thrombi are most commonly associated with acute ST elevation anterior

TABLE 129.2 CHADS$_2$ and CHA$_2$DS$_2$-VAS Score Details

CHADS$_2$ Stroke Risk Strategy for Nonvalvular Atrial Fibrillation
C=Congestive heart failure or left ventricular systolic dysfunction
H=Hypertension, systolic blood pressure >160mmHg
A=Age >74 years
D=Diabetes
S$_2$=Stroke or transient neurological event (2 points)
Minimum score=0
Maximum score=6 (a stroke or transient event gets 2 points)
CHA$_2$DS$_2$-Vas Stroke Risk Strategy for Nonvalvular Atrial Fibrillation
C=Congestive heart failure or left ventricular systolic dysfunction
H=Hypertension, systolic blood pressure >160mmHg
A$_2$=Age >74 years (2 points)
D=Diabetes
S$_2$=Stroke or transient neurological event (2 points)
V=Vascular
A=Age 65–74 years
S=Sex, female gender
Minimum score=0
Maximum score=10 (a stroke or transient neurologic event and age >74 gets 2 points)

myocardial infarctions as well as dilated ischemic and nonischemic cardiomyopathies. LV thrombi are detected with TTE often with an endocardial definition echo contrast (Fig. 129.4). TEE is less optimal due to the frequent inability to visualize the true LV apex. Importantly, data suggest that long inversion time late gadolinium enhancement cardiac MRI is *superior* to both TTE and TEE for identifying LV thrombi (Fig. 129.5) [9], even in patients after a negative contrast echo study.

Myocardial infarction: Among patients with acute infarction, the risk of clinical stroke is up to 3.6% with a particular predilection for ST elevation/large anterior infarctions. The majority of thrombi develop within the first 2 weeks after infarction, and thrombus mobility and protrusion are associated with increased risk of stroke.

Dilated cardiomyopathy: Even in the absence of acute myocardial infarction, patients with LV systolic dysfunction and heart failure are at increased risk for stroke with the stroke risk related to the LV dysfunction severity. Like LV thrombus with myocardial infarction, the two most important risk factors for stroke are thrombus size and mobility. The recurrence rate may reach 20% in the first year.

Prosthetic valve thrombi: Thrombus formation and subsequent embolization may be seen in patients with prosthetic valves, especially mechanical mitral (and tricuspid) valves with suboptimal international normalized ratio (INR).

Due to acoustic shadowing on TTE, TEE is greatly preferred for evaluation of prosthetic valves, especially mitral prostheses. It is often assumed that patients with stroke and mechanical prostheses have prosthetic valve thrombi, especially if there is no other obvious cause and/or the INR is suboptimal. Tissue characteristics may help distinguish thrombus from pannus ingrowth.

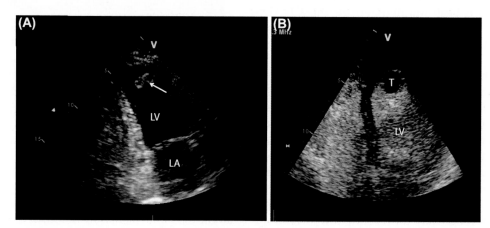

FIGURE 129.4 TTE in the apical view of a patient with an apical left ventricular thrombus (a) without echo contrast; (b) the thrombus (T) demonstrated with echo contrast.

Aortic atherosclerosis: Aortic atheroma are readily visualized with TEE and are commonly seen in patients with transient ischemic attack, stroke, or peripheral embolism. There is a higher incidence of thromboembolism when plaque is pedunculated and mobile, especially if located within the ascending aorta. Patients with complex (mobile or >4 mm) aortic atheroma (Fig. 129.6) and AF may be at particularly high risk for thromboembolism. Management usually includes statins. The benefit of antiplatelet versus anticoagulants is uncertain.

Endocarditis: After heart failure due to valvular regurgitation, arterial embolism is the most common life-threatening complication of endocarditis. Embolization during the 1st week of antibiotic treatment is most common with an increased risk of thromboembolism with *Staphylococcus aureus* and fungal endocarditis due to their relatively large vegetation size. TTE has a relatively low sensitivity for the detection of both native (Fig. 129.7) and prosthetic valve vegetations and is very poor for detection of abscesses. Thus, if the clinical suspicion for endocarditis is moderate or high, a TEE should be considered as the initial imaging modality [1]. Proceeding to initial TEE should also be considered if a "negative" TTE would lead to a TEE, such as patients with prosthetic valves [1]. Larger (>10 mm) and mobile vegetations are at higher risk for thromboembolism, especially in the setting of severe valvular regurgitation—making early surgery recommended [10].

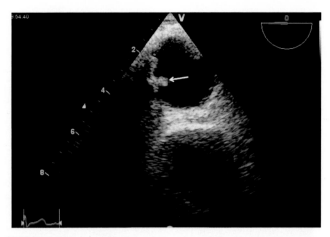

FIGURE 129.6 TEE of the descending thoracic aorta in a patient with complex atheroma (*arrow*).

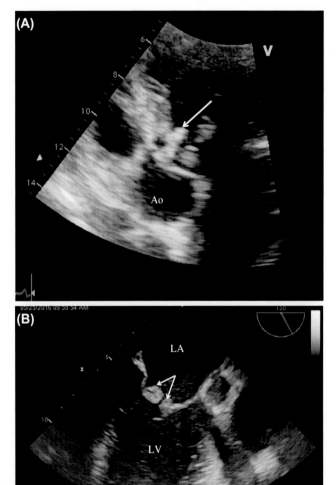

FIGURE 129.7 (A) TTE in a patient with a vegetation (*arrow*) on the aortic valve. (B) TEE in a patient with vegetations involving both leaflets of the mitral valve. Characteristically, vegetations are on the "upstream" side of the valve surfaces.

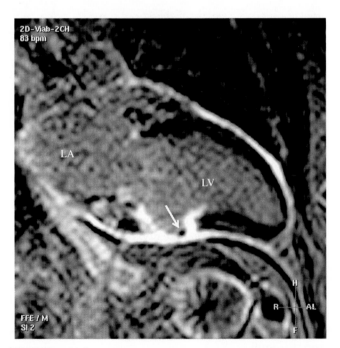

FIGURE 129.5 Cardiac magnetic resonance imaging. Late gadolinium enhancement technique. Note the small left ventricular (LV) thrombus (*arrow*) along the basal inferior wall surrounded by "bright" myocardium/scar.

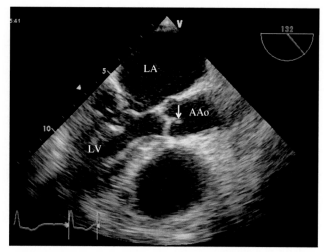

FIGURE 129.8   TEE in a patient with a Lambl's excrescent on the aortic valve (*arrow*).

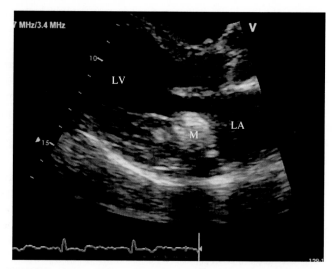

FIGURE 129.9   TTE in the parasternal orientation in a patient with severe mitral annular calcification.

Aortic valve calcification: Valvular calcification may involve a bicuspid aortic valve, a rheumatically deformed valve, and senile degenerative aortic stenosis. Uncommonly, the calcified material may spontaneously embolize. Thus, other causes for thromboembolism should be sought. For patients with aortic stenosis undergoing catheterization, catheter crossing of the aortic valve should be avoided to minimize stroke risk.

Valve excrescences: Valve excrescences, often referred to as Lambl's excrescence, are thin/linear, elongated, mobile structures seen near the aortic or mitral valve leaflet line of closure, especially in the elderly (Fig. 129.8). They do not appear to be a primary source of cardioembolism.

Mitral annular calcification: The most common TTE finding among older patients with a suspected cardiac source of embolism is mitral annular calcification (MAC), a highly echo-reflective area in the posterior portion of the mitral annulus (Fig. 129.9). The incidence of stroke and MAC severity appears to be linear [11], and there is an association of MAC with AF, hypertension, and aortic atherosclerosis. The association of MAC with stroke may be related to overlying thrombus or its association with aortic atheroma.

Intracardiac tumors: Primary cardiac tumors are extremely rare with myxomas constituting over half of these lesions. Myxomas are histologically benign tumors that most commonly are found attached to the midportion of the interatrial septum in the area of the foramen (Fig. 129.10). Myxomas are more common among women and over half may present with thromboembolism due to migration of tumor or overlying thrombi. Multiple myxoma may be present in the uncommon familial myxoma syndrome.

Another histologically benign tumor associated with embolism is papillary fibroelastoma. These tumors are usually highly mobile, pedunculated and most commonly

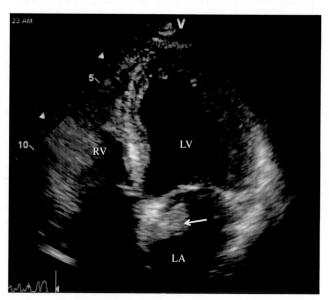

FIGURE 129.10   TTE in a patient with a myxoma (*arrow*) arising from the basal interatrial septum.

located on the aortic or mitral valves, but rarely cause valve dysfunction. Less commonly, they may be present on endocardial surfaces. Like myxomas, systemic embolization is due to migration of overlying thrombus or the tumor itself.

Both TTE and TEE are highly sensitive in detecting myxomas and papillary fibroelastomas, although TEE may provide more accurate anatomical details, such as the site of attachment, and, may help differentiate tumors from thrombi.

Left atrial spontaneous echo contrast: Spontaneous echo contrast or "smoke-like" echoes within the LA and/or LAA (Fig. 129.3) represent erythrocyte aggregation in low shear rate conditions. A majority of AF patients have at least mild spontaneous echo contrast, increasing to

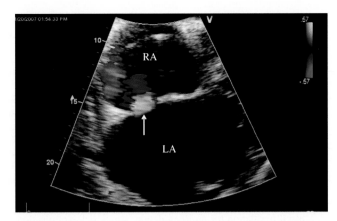

FIGURE 129.11 TTE with color flow Doppler in the subcostal orientation in a patient with a secundum-type atrial septal defect (*arrow*). There is left-to-right flow.

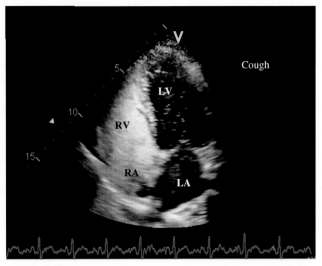

FIGURE 129.12 TTE after intravenous administration of agitated saline. There is premature (within three beats) appearance of microbubbles in the left heart with cough consistent with a patent foramen ovale.

nearly all AF patients with LAA thrombi. Among AF patients, spontaneous echo contrast is associated with reduced peak LAA ejection velocity and is an independent predictor of thromboembolic risk.

From a mechanistic perspective, the previously mentioned abnormalities provide the necessary pathophysiological substrate for thromboembolism. In the normal circulatory system, the pulmonary arteriolar and capillary beds provide a "filtering system" preventing venous thrombi from gaining access to the arterial system. For some patients, an abnormal connection between the right and left heart provide a conduit through which venous thrombi can bypass the pulmonary bed, and thereby cause "paradoxical" thromboembolism. The most common sites for "crossing" of these thrombi are at the atrial level.

Abnormalities of the interatrial septum: Abnormalities of the interatrial septum are associated with thromboembolism via two pathophysiological mechanisms: right-to-left shunting such as through a secundum atrial septal defect (Fig. 129.11) and interatrial septal aneurysm (inter-ASA). Patients with an atrial septal defect or PFO may have intermittent flow of blood from the right atrium to the LA. This can be induced on a daily basis by maneuvers which transiently increase right atrial pressure such as coughing or Valsalva during a bowel movement (Fig. 129.12).

A PFO during fetal development shunts blood from the right atrium to the LA, thereby bypassing the high resistance pulmonary circuit. At birth or shortly thereafter, the septum primum and the septum secundum fuse, closing the interatrial septum. A PFO occurs when this fusion is not complete. Ostium secundum atrial septal defects occur when there is excess resorption of the septum primum or inadequate formation of the septum secundum. Less common are sinus venosus and primum atrial septal defects which also promote right-to-left shunting.

PFO prevalence varies by definition. Autopsy studies demonstrated a PFO in nearly a 25% of all individuals, somewhat declining with age. A PFO may be noninvasively identified with TTE (and/or TEE) and intravenous agitated saline contrast with the premature appearance of LA microbubbles within three beats of full opacification of the right atrium. Imaging is usually performed at rest and with cough and Valsalva maneuvers designed to transiently increase right atrial pressure so as to promote right-to-left shunting. Valsalva maneuver provides the highest sensitivity. Shunt size is semiqualitatively graded as "trivial" (<10 bubbles), "small" (10–30 bubbles), and "large" (>30 bubbles) with this relative size corresponding to stroke risk. TEE is more sensitive for a PFO than TTE, so long as the patient is not overly sedated and is able to cooperate with maneuvers. Administration of agitated saline from the groin increases sensitivity, but is rarely performed due to logistical reasons. Other novel methods to detect PFO include intravenous saline injection with transcranial Doppler of the middle cerebral artery.

The mechanism of stroke in PFO is felt to be due to paradoxical venous embolism with the risk being highest in younger (<55 years) patients and minimal after age 65.

An ASA is far less common than a PFO and is a >10–15 mm excursion of the septum in the area of the fossa due to redundant atrial septal tissue. An ASA is commonly associated with a PFO and the mechanism for stroke in ASA has been ascribed to coexistent PFO (found in >50%), propensity for atrial arrhythmias, and direct thrombus formation in the neck of the aneurysm.

IMAGING PROCEDURES

TTE Versus TEE: TTE performed with agitated saline contrast is a minimally invasive procedure, while TEE is a moderately invasive procedure employing both sedation and poster-pharyngeal anesthesia. The close proximity of the esophagus to the posterior heart, the lack of intervening lung and bone, and the use of higher-frequency imaging transducers results in enhanced visualization of many of the mentioned cardiac "sources," particularly when a history suggestive of thromboembolism, including identification of intra-atrial thrombi and tumors, PFO, vegetations, aortic atheroma, and spontaneous echo contrast. TTE remains preferred for identification of LV regional systolic function and apical LV thrombi.

References

[1] Douglas PS, Garcia MJ, Haines DE, Lai WW, Manning WJ, Patel AR, et al. ACCF/ASE/ACCP/AHA/ASNC/HFSA/HRS/SCAI/SCCT/SCMR 2010 Appropriate use criteria for echocardiography. J Am Coll Cardiol 2011;57:1126–66.

[2] Kitkungvan D, Nabi F, Ghosn MG, Dave AS, Quinones M, Zoghbi WA, et al. Detection of LA and LAA Thrombus by CMR in patients referred for pulmonary vein isolation. JACC Cardiovasc Imaging. 2016; 7:809–18.

[3] Omran H, Rang B, Schmidt H, et al. Incidence of left atrial thrombi in patients in sinus rhythm and with a recent neurologic deficit. Am Heart J 2000;140:658–62.

[4] Sanna T, Diener HC, Passman RS, DiLazzaro V, Bernstein RA, Marillo CA, et al. Crytogenic stroke and underlying atrial fibrillation. N Engl J Med 2014;370:2478–86.

[5] Weigner MJ, Thomas LR, Patel U, Schwartz JG, Burger AJ, Douglas PS, et al. Transesophageal-echocardiography-facilitated early cardioversion from atrial fibrillation: short-term safety and impact on maintenance of sinus rhythm at 1 year. Am J Med 2001;110:694–702.

[6] Manning WJ, Silverman DI, Waksmonski CA, Oettgen P, Douglas PS. Prevalence of residual left atrial thrombi in patients presenting with acute thromboembolism and newly recognized atrial fibrillation. Arch Intern Med 1995;155:2193–7.

[7] Gage BF, van Walraven C, Pearce L, et al. Selecting patients with atrial fibrillation for anticoagulation: stroke risk stratification in patients taking aspirin. Circulation 2004;110(16):2287–92.

[8] Ezekowitz MD, James KE, Nazarian SM, et al. Silent cerebral infarction in patients with nonrheumatic atrial fibrillation. The veterans affairs stroke prevention in nonrheumatic atrial fibrillation investigators. Circulation 1995;92:2178–82.

[9] Weinsaft JW, Kim HW, Shah DJ, Klem I, Crowley AL, Brosnan R, et al. Detection of left ventricular thrombus by delayed-enhancement cardiovascular magnetic resonance prevalence and markers in patients with systolic dysfunction. J Am Coll Cardiol 2008;52:148–57.

[10] Kang DH, Kim YJ, Kim SH, Sun BJ, Kim DH, Yun SC, et al. Early surgery versus conventional treatment for infective endocarditis. N Engl J Med 2012;366:2466–73.

[11] Benjamin EJ, Plehn JF, D'Agostino RB, et al. Mitral annular calcification and the risk of stroke in an elderly cohort. N Engl J Med 1992;327:374–9.

CHAPTER

130

Overview of Neuroimaging of Stroke

K. Malhotra, D.S. Liebeskind

University of California, Los Angeles, Los Angeles, CA, United States

INTRODUCTION

Stroke incurs ischemic or hemorrhagic injury in the brain, typically associated with alterations in the cerebral circulation, including a variety of disorders that affect arteries, capillaries, and venous components. Imaging provides detailed information on brain tissue and vascular anatomy for clinicians to corroborate their clinical history and examination findings in acute stroke cases. The significance of imaging in stroke depends on unraveling the pathophysiology of acute ischemia and plays a pivotal role during the acute phase to formalize therapeutic plans. This chapter provides an overview on the role of neuroimaging in stroke, including computed tomography (CT), magnetic

Primer on Cerebrovascular Diseases, Second Edition
http://dx.doi.org/10.1016/B978-0-12-803058-5.00130-2

resonance imaging (MRI), and cerebral angiography with further elaboration in subsequent chapters. Across a variety of practice scenarios, clinicians often have a choice of various imaging modalities to yield further information on the care of patients with stroke. Imaging data serve as an extension of the clinical examination that routinely enhances the clinical evaluation and neurological localization in a patient with stroke. The following discussion delineates the key goals of neuroimaging in stroke to augment clinical decision making.

Overwhelmingly, the primary goal of imaging using noncontrast CT or MRI has been to differentiate hemorrhage from ischemia for consideration of intravenous (IV) thrombolysis. There has been a trend toward identifying early ischemic changes along with attendant early vessel signs on CT or MRI. "Time is brain" has been the mantra for decades, whereas recent advanced imaging modalities have enticed clinicians to modify it to "tissue is brain," largely disclosed by imaging. It has been shown that time alone is rudimentary and neuroimaging complements decision making in acute stroke cases.

In recent years, more comprehensive imaging techniques such as multimodal CT or MRI that encompasses angiography and perfusion images (further details in following subsections), have emerged that provide a detailed perspective of cerebrovascular disease pathogenesis, vascular anatomy, and the functional correlates of perfusion abnormalities. These imaging strategies have been used on a regular basis as they portray a snapshot of brain parenchyma and guide further therapeutic decisions. Various key elements of pathophysiology portrayed with imaging include large vessel occlusion (LVO), collateral blood flow patterns, resultant hemodynamic modification in blood flow, and the tissue injury that ensues.

COMMONLY USED IMAGING MODALITIES

CT comprised the predominant imaging modality to assess patients with stroke from the acute to chronic phases in previous eras as compared with the vast neuroimaging data available via multimodal CT/MRI more recently. Since the advent of MRI and concomitant newly refined multimodal techniques of vascular neuroimaging, the understanding of stroke has expanded in numerous directions. For acute stroke cases, multimodal CT typically includes noncontrast CT, CT angiography of head and neck (CTA), as well as CT perfusion (CTP). Multimodal MRI includes various sequences, such as diffusion-weighted imaging (DWI), apparent diffusion coefficient (ADC), fluid-attenuated inversion recovery (FLAIR), gradient recalled echo (GRE), and perfusion-weighted imaging in addition to MR angiography of head and neck (MRA) (Fig. 130.1).

CT has been the standard for initial and subsequent comparisons of imaging lesion evolution, yet limitations include failure to reliably discern lesions that are very early from stroke symptom onset and those located in specific locations such as the posterior fossa [1]. MRI with its numerous imaging sequences including DWI provides readily available data to prioritize and make decisions for thrombolysis or endovascular therapy, if applicable. The decision between CT and MRI is often based on an individual case approach or the availability of either technique in each stroke center. CT is the optimal imaging modality for MRI-incompatible patients due to specific pacemaker devices and implantable metallic objects or those patients who are hemodynamically unstable to remain supine for an extended period of time. Image acquisition using MRI may impose greater logistical challenges compared with CT, although it provides greater details and dimensions of brain tissue and corresponding vascular changes. Various institutions have developed multimodal CT or MRI protocols with demonstrated feasibility and utility in the early management of patients with stroke across acute settings.

ISCHEMIC STROKE

Ischemic stroke is a dynamic process that evolves from acute through subacute and chronic phases. Imaging in stroke focuses to determine the diagnosis and etiology, lesion localization, extent of ischemic evolution, therapeutic implications, and expected prognosis. Imaging evaluation in acute ischemic stroke (AIS) often revolves around the key parameters of core infarct volume, presence and extent of perfusion-diffusion mismatch, and collateral flow profiles or grades that refine therapeutic decisions. CT provides a solid platform for physicians to not only exclude hemorrhage, but also estimate the extent of infarct evolution. Rather than focusing solely on ruling out massive infarcts, the trend has shifted to evaluate subtle signs of ischemia including hypoattenuation or obscuration of the lentiform nuclei, loss of the insular ribbon, sulcal effacement, cortical hypodensity, and the presence of various hyperdense vessel findings. Alberta Stroke Program Early CT score (ASPECTS) is a topographic scoring system to assess early ischemic parenchymal changes on initial CT in patients with AIS of anterior circulation. Interobserver discrepancy has emerged as a major concern even among experienced stroke neurologists, to interpret early signs of ischemia using initial CT scans [2].

CT has been the primary neuroimaging modality with rapid and widespread availability, although increasing number of institutions are resorting to MRI-based technique that confers extensive data on cerebrovascular events. MRI involves multiple sequences that provide snapshots of brain tissue including prior silent

(A) (B) (C)

(D) (E) (F)

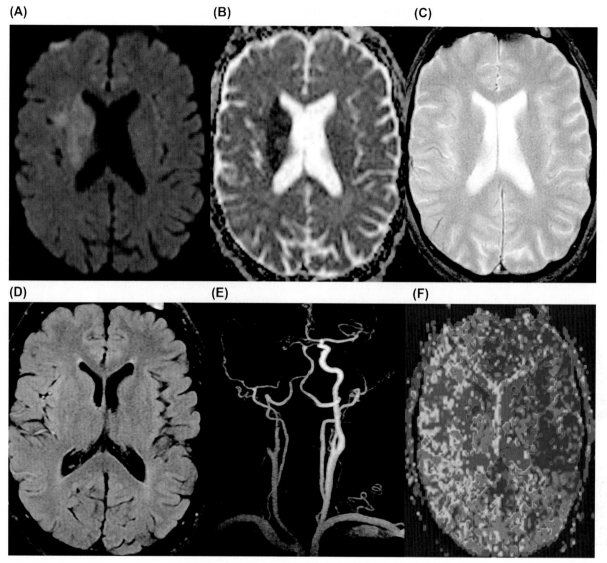

FIGURE 130.1 Demonstration of various MRI sequences as multimodal acquisition in a 70-year-old patient who presented with acute onset of left hemiplegia and right gaze preference: (A) DWI, (B) ADC, (C) GRE, (D) FLAIR, (E) MRA of the head and neck, and (F) perfusion-weighted imaging.

ischemia and other cerebrovascular pathology, in exquisite detail. DWI is the primary sequence used in patients with AIS that reveals restricted Brownian motion of water molecules secondary to cytotoxic edema. DWI has a sensitivity of 99% and a specificity of 92% in detecting ischemic changes as compared with CT [2]. DWI patterns of ischemia are especially pathognomonic for disease mechanisms pertaining to LVO, thromboembolism, small-vessel disease, hypoperfusion or border zone mechanisms [3] (Fig. 130.2). FLAIR sequences on MRI may reveal hypointense signal for space-occupying lesions with surrounding vasogenic edema, whereas old infarcts are hyperintense due to increased water content in brain tissue. Periventricular FLAIR hyperintensities portray old small-vessel lacunar infarcts and further

guide clinicians in formulating therapeutic plans for risk management of their patients. FLAIR vascular hyperintensities may also offer subtle clues about the presence of slow flow in collateral vessels downstream of an arterial occlusion.

Therapeutic decision making in AIS primarily involves DWI/ADC and FLAIR sequences once hemorrhage has been excluded on GRE. DWI hyperintensities with ADC hypointensity indicate cytotoxic edema associated with acute to subacute infarcts. Furthermore, lesions with mismatch between DWI and FLAIR sequences depict temporal aspects and the duration of ischemia [4]. Neuroimaging usually compliments the examination findings obtained by clinicians, whereas lesion patterns are particularly helpful in stroke. Distally

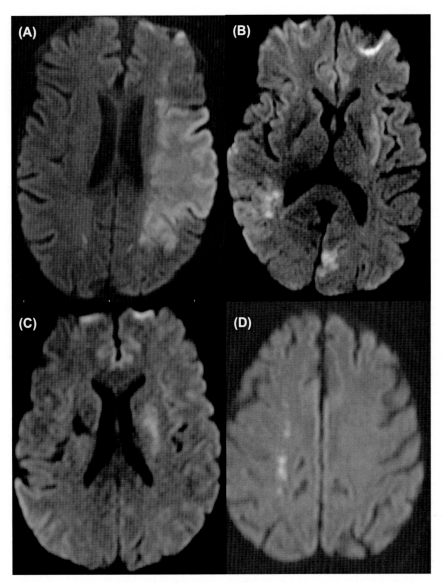

FIGURE 130.2 DWI sequences showing ischemic infarcts with different radiographic lesion patterns due to various etiologies: (A) large MCA territory infarction, (B) bihemispheric embolic pattern, (C) lacune, and (D) border zone infarct.

located cortical lesions likely may infer thromboembolic phenomena. The proximal nidus or embolic source may be artery–artery or from the heart, prompting further diagnostic testing. Subcortical, especially basal ganglia and deep white matter, lesions are more typically due to microvascular in situ disease or small-vessel ischemia, although proximal subocclusive lesions involving lenticulostriate branches of the middle cerebral artery (MCA) can also produce similar lesions. Border zone lesions of ischemia between the principal arterial territories in the brain may manifest as a string or archipelago of discrete cortical and subcortical lesions that implicate a proximal stenotic or occlusive lesion causing hypoperfusion.

Contrast enhancement of ischemic lesions is helpful in the age determination of any vascular lesion. Parenchymal enhancement is associated with disruption of the blood–brain barrier that usually presents in a gyriform or ring-like shape, located peripheral to the central ischemic lesion. It usually becomes visible around 4–7 days after the ischemic event, tends to resolve within 8 weeks, and may persist beyond 3 months in isolated cases. Serial imaging of infarct evolution over ensuing days or weeks illustrates the dynamic nature involved in cerebral ischemia and various stages may be encountered, including "fogging" or the transient disappearance of subacute lesions before chronic scar formation. Fogging may be seen with either CT or MRI and refers to gradual fading of lesion on follow-up serial scans usually 2 weeks after the onset of ischemia.

ANGIOGRAPHIC ASSESSMENT OF CEREBRAL VASCULATURE

Vascular obstruction can be due to either luminal narrowing (stenosis) or complete blockage limiting antegrade or forward blood flow (occlusion). Atherosclerotic plaques are the most important culprits leading to stenosis or distal thromboembolic phenomenon due to plaque rupture. Carotid ultrasound has been used for decades to assess intimal thickness and calculate focal stenosis in extracranial circulation. Calcified atheromas identified on CTA are less likely to embolize and thus have reduced risk of ischemic stroke [5]. The most commonly used grading technique to measure stenosis and reduction of luminal caliber is the North American Symptomatic Carotid Endarterectomy Trial method in extracranial carotid artery and the Warfarin-Aspirin Symptomatic Intracranial Disease (WASID) method in the intracranial circulation [6].

CTA is the most commonly and easily available technique to study intra- and extracranial cerebral vessels. It uses extended acquisition after intravenous contrast and provides crisp anatomic details including vessel lumen. MRA, especially time-of-flight (TOF) sequence, produces images due to blood flow in vessel lumen based on radiofrequency pulsing of tissue. As TOF mainly depends on the features of blood flow, it accentuates luminal flow and may not be as specific in providing a clear anatomical picture as provided by CTA. In a nutshell, CTA accentuates vascular anatomy, whereas MRA and ultrasound being flow-dependent imaging techniques provide a snapshot of dynamic blood flow.

Cerebral ischemia from vessel occlusion is the prime target for all current acute stroke interventions. Proximal occlusion can be visualized easily with dedicated angiographic studies as abrupt cutoff of vascular flow. Hyperdense vessel signs have been used for decades as a marker of MCA occlusion (Fig. 130.3), whereas only recently has this been extended to include occlusions of the internal carotid artery (ICA) or posterior cerebral artery (PCA). Various factors pertaining to clot visualization include location, composition, age, density, and regional flow dynamics that influence the appearance of hyperdense vessel sign. Thrombus of an intracranial artery is visible as hyperdensity on CT, or as blooming artifact on GRE sequences of MRI. Proximal arterial occlusion leads to sluggish blood flow downstream and if collateral supply is available, FLAIR vascular hyperintensities may be evident.

IMAGING UTILIZED FOR ENDOVASCULAR INTERVENTION

The safety and efficacy of mechanical thrombectomy with superiority over standard medical care including IV thrombolysis have been proved through

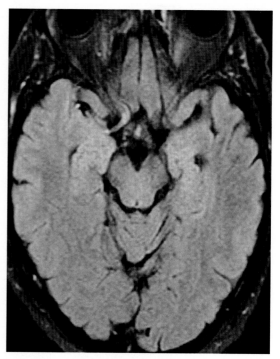

FIGURE 130.3   FLAIR sequence of MRI showing hyperintense right MCA vessel.

a litany of endovascular trials. The rapid evaluation of patients with AIS with readily accessible multimodal imaging, robust collateral flow, and refined revascularization techniques including mechanical stent retriever devices, has been colossal contributors toward the success of recent endovascular trials. Multimodal imaging techniques have clearly eased the patient selection and evaluation of potential endovascular intervention in the acute phase of ischemic stroke (Fig. 130.4).

Noninvasive imaging including CT with ASPECTS >6 or 7, LVO on vessel imaging, and good collateral supply on multiphase CTA provides a cogent algorithm to follow, especially in patients with AIS with hemispheric syndromes. Multiphase CTA has emerged as a relatively novel method to study collateral blood flow while capturing snapshots throughout the arterial, venous, and delayed venous phases. With total scan time of ~5 min, limited artifact susceptibility, and widely accessible approaches, multiphase CTA may further improve patient selection for endovascular cases.

Revascularization techniques aim to preserve salvageable brain tissue (i.e., penumbra) and prevent further progression of infarct core. Postprocedure CT or MRI is usually performed to evaluate for any reperfusion injury, including hemorrhage. Hemorrhagic transformation may be difficult to distinguish, especially when contrast staining is present on CT within the first 12–24 h (Fig. 130.5).

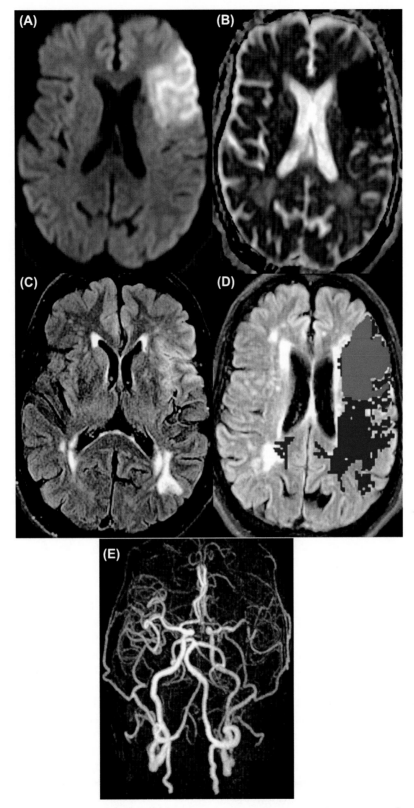

FIGURE 130.4 (A and B) MRI brain DWI sequence shows hyperintensity in left anterior MCA territory with corresponding hypointensity on apparent diffusion coefficient sequence consistent with restricted diffusion. (C) FLAIR sequence shows serpiginous hyperintense vessels corresponding to slow flow or collateral supply. (D) MR perfusion shows delayed time to peak in large left MCA territory suggesting mismatch perfusion defect. (E) MRA brain shows hypointense left ICA due to sluggish flow and narrowing in the proximal segment of superior left M2 MCA branch likely representing thrombus. (F) Diagnostic cerebral angiogram demonstrates significant stenosis in proximal segment of left ICA distal to bifurcation. (G) Successful angioplasty and stenting of left ICA with no residual stenosis. (H and I) Cerebral angiogram showing proximal left M2 occlusion with good flow post mechanical thrombectomy.

VIII. DIAGNOSTIC TESTING

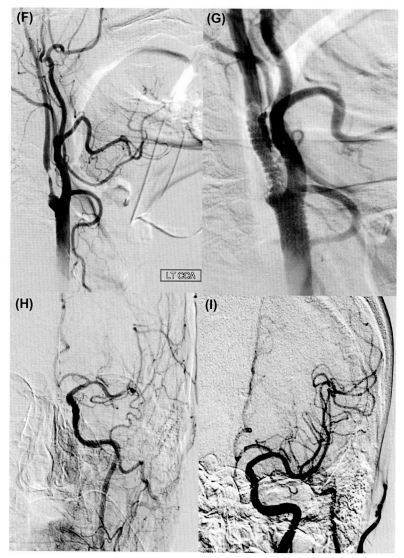

FIGURE 130.4 cont'd.

IMAGING MODALITIES TO ASSESS COLLATERAL CIRCULATION

Collateral blood flow refers to the potential vigorous network of vessels that play a major role in patients with ischemic stroke. It provides perfusion to the penumbral region reducing the progression of infarct size and thus improves clinical outcome with reperfusion therapies. Noninvasive imaging techniques provide real-time evaluation of various aspects of this robust collateral arterial supply. CTA is a more commonly used technique than MRA to assess the collateral status by direct comparison with the contralateral hemisphere or thorough visual inspection of the number of MCA subdivisions enhanced with intravenous contrast. Multiphase CTA has emerged as an important imaging technique to investigate collateral status in patients who are perfusion dependent.

Similar to CTA, MRA also evaluates collateral status using noninvasive vessel analysis and provides detailed data on collateral flow. As mentioned earlier, FLAIR vascular hyperintensities reflect slow retrograde collateral flow distal to the arterial occlusion, providing information on the extent of leptomeningeal collaterals. Flow diversion into the ipsilateral anterior cerebral artery (ACA) or PCA on MRA may also be indirect markers of collateral support to the ischemic MCA territory. Such presence of adequate collateral flow at initial presentation provides detailed information on auxiliary vascular flow to the ischemic area and further guides clinicians for therapeutic decisions including endovascular procedures.

NEUROIMAGING TO ASSESS CEREBRAL PERFUSION

Various experts initially argued against the use of perfusion while accepting "time is brain" as the sole rescue theory available for AIS management. This precluded the

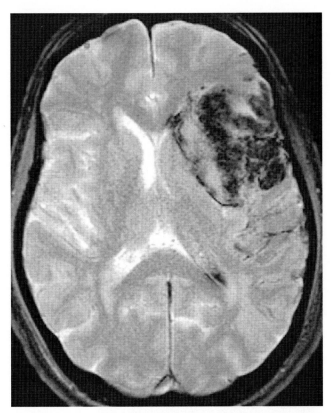

FIGURE 130.5 Repeat MRI with GRE sequence showing left frontal lobe parenchymal hemorrhage especially in the infarct zone with mild mass effect on left lateral ventricle. Also noted is extension of hemorrhage into intraventricular and left frontoparietal subarachnoid space.

critical aspect of assessing collateral supply as a part of cerebral perfusion. The current AIS treatment approach includes "tissue is brain" with the success of endovascular trials, although it continues to follow "time is brain" with the utmost priority.

With the advent of new multimodal imaging techniques, either CT or MR perfusion has become a more regular element of multimodal stroke imaging. Perfusion imaging delineates cerebral ischemia and illustrates severely compromised vascular territory and the infarct core. Perfusion mismatch models have proved their efficacy in depicting the extent of ischemic territory and have been incorporated into thrombolysis and endovascular therapeutic decision making. MR perfusion uses perfusion-diffusion mismatch to compare hypoperfusion with diffusion abnormality, whereas CTP correlates mean transit time with cerebral blood flow and cerebral blood volume. Gross visual comparison and estimation of perfusion defects helps to determine the mismatch between cerebral hemispheres. Various automated software have emerged for both CT and MR perfusion that have simplified and eased the acquisition process for colored predictive maps, although accuracy of such automated software remains yet to be confirmed.

INTRACEREBRAL HEMORRHAGE

Intracranial or intracerebral hemorrhage is defined as the presence of blood in either epidural, subdural, subarachnoid, or intraparenchymal spaces, although neurologists frequently encounter intraparenchymal hemorrhages on a routine basis. Various neuroimaging techniques have been used to detect and differentiate hemorrhage from ischemia in acute and chronic settings. Hemorrhagic lesions are defined based on type, location, and size of the hemorrhagic event. Primary intraparenchymal hemorrhage includes chronic hypertension and cerebral amyloid angiopathy, as the most common etiologies whereas secondary hemorrhage could be due to a variety of causes such as hemorrhagic transformation with trauma as a common cause. CT is routinely performed as the initial imaging technique to exclude intracranial hemorrhage in most primary and comprehensive stroke centers (Fig. 130.6A). Few comprehensive stroke centers have shifted their imaging paradigm toward MRI even for initial evaluation (Fig. 130.6B). GRE sequences of MRI detect blood products due to susceptibility artifacts and have been shown to be equivalent to CT [7]. GRE is routinely utilized to detect microbleeds and hemorrhagic changes in patients with acute stroke (Fig. 130.7).

New techniques have emerged to explore the potential flow dynamics and expansion of cerebral hemorrhage. The CTA "spot sign" refers to foci of enhancement within the hematoma that marks active bleeding and likely hematoma expansion [8]. Perihematomal edema and hematoma expansion can produce ischemia at the hematoma boundary due to mechanical compression or neurotoxicity from blood breakdown products, leading to poor outcome. Clinicians acquire serial neuroimaging using either CT or MRI to compare mass effect from edema or hematoma expansion and corroborate imaging with the clinical examination. CT is the preferred modality for initial and follow-up evaluation, although GRE sequence can also be followed if available looking for blooming artifact as described earlier. Follow-up imaging during the subacute phase is also routinely performed to consider restarting antithrombotic therapy in select cases if required and also to preclude any masked vascular lesions obscured in the hematomal bed. This information is critical for clinicians to make their therapeutic plans for patient as a whole rather than focusing on the hemorrhagic event in isolation.

DISTINCT CASES ASSOCIATED WITH ISCHEMIA/INFARCTION

Various mechanisms and causes for ischemia stand out with characteristic imaging patterns that further help in exploring the etiology of cerebrovascular event.

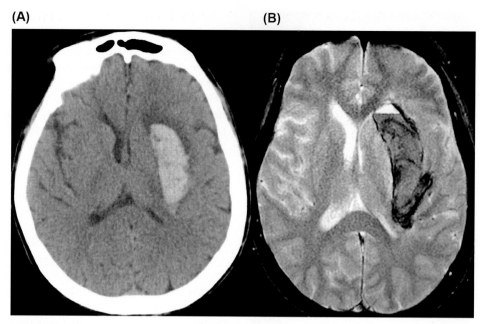

FIGURE 130.6   (A) CT head without contrast showing acute left basal ganglia hematoma with mild mass effect on left lateral ventricle. (B) MRI brain GRE sequence showing hypointensity in left basal ganglia.

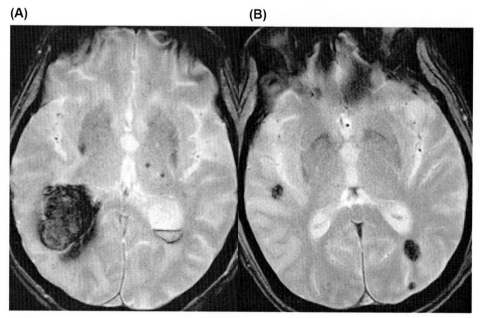

FIGURE 130.7   (A and B) MRI brain GRE sequence showing multiple microbleeds with right parietal lobar parenchymal hematoma with intraventricular extension.

Cerebral ischemia from vasospasm associated with subarachnoid hemorrhage is a well-known entity that may progress to infarction, especially with a severe degree of vasospasm. Transcranial Doppler (TCD) is a noninvasive method that utilizes pulsed wave Doppler ultrasonography to assess arterial flow velocities at bedside [9]. Serial TCD studies are performed in these patients with increased flow velocities in the major branches of circle of Willis correlating with the degree of vasospasm.

Various unusual causes of cerebral ischemia have been the focus of attention in the young patient population including cerebral vasculitis, moyamoya syndrome, fibromuscular dysplasia, and other congenital vascular anomalies. Venous flow may be altered due to various causes including mechanical trauma, infections, and hypercoagulable conditions such as pregnancy, malignancy, disseminated intravascular coagulation, or polycythemia. Dural sinus thrombosis or occlusion can produce hemorrhagic venous infarcts located in deep white matter that are distinct when compared with infarcts due to arterial occlusions. Superior or transverse sinus thrombosis presents with abnormality of flow voids on CT or MR venograms, whereas the torcula or confluence of sinuses may reveal an "empty delta sign" due to lack of contrast from occlusion or thrombus.

Neuroimaging in stroke provides essential data pertaining to etiology and pathophysiologic mechanisms, providing a basis for rational algorithms of decision making by clinicians. Multimodal CT/MRI provides detailed information on ischemic stroke evolution. The described imaging modalities not only influence therapeutic decision making regarding acute stroke, but may also drive subacute management in the intensive care unit and early implementation of stroke preventive measures. The rational use of neuroimaging modalities described in further detail in the following chapters is a basic component in stroke care ranging from acute interventions to early rehabilitation and effective implementation of stroke prevention.

References

[1] Mullins ME, Schaefer PW, Sorensen AG, Halpern EF, Ay H, He J, et al. CT and conventional and diffusion-weighted MR imaging in acute stroke: study in 691 patients at presentation to the emergency department. Radiology 2002;224:353–60.

[2] Malhotra K, Liebeskind DS. Imaging in endovascular stroke trials. J Neuroimaging 2015;25:517–27.

[3] Caplan LR, Wong KS, Gao S, Hennerici MG. Is hypoperfusion an important cause of strokes? If so, how? Cerebrovasc Dis 2006;21(3):145–53.

[4] Thomalla G, Cheng B, Ebinger M, Hao Q, Tourdias T, Wu O, et al. DWI-FLAIR mismatch for the identification of patients with acute ischaemic stroke within 4·5 h of symptom onset (PRE-FLAIR): a multicenter observational study. Lancet Neurol November 2011;10(11):978–86.

[5] Uwatoko T, Toyoda K, Inoue T, Yasumori K, Hirai Y, Makihara N, et al. Carotid artery calcification on multislice detector-row computed tomography. Cerebrovasc Dis 2007;24(1):20–6.

[6] Liebeskind DS. Imaging the future of stroke: I. Ischemia. Ann Neurol November 2009;66(5):574–90.

[7] Kidwell CS, Chalela JA, Saver JL, Starkman S, Hill MD, Demchuk AM, et al. Comparison of MRI and CT for detection of acute intracerebral hemorrhage. JAMA 2004;292:1823–30.

[8] Thompson AL, Kosior JC, Gladstone DJ, Hopyan JJ, Symons SP, Romero F, et al. Defining the CT angiography 'spot sign' in primary intracerebral hemorrhage. Can J Neurol Sci July 2009;36(4):456–61.

[9] Malhotra K, Conners JJ, Lee VH, Prabhakaran S. Relative changes in transcranial Doppler velocities are inferior to absolute thresholds in prediction of symptomatic vasospasm after subarachnoid hemorrhage. J Stroke Cerebrovasc Dis January 2014;23(1):31–6.

CHAPTER

131

CT, CT-Angiography, and Perfusion-CT Evaluation of Stroke

C. Federau, M. Wintermark

Stanford University, Stanford, CA, United States

INTRODUCTION

A stroke is a sudden loss of neurological function due to brain parenchyma damage caused by a cerebrovascular disorder. It can be caused either by the sudden loss or significant decrease of blood flow to a specific brain region (ischemic stroke), or by the rupture of blood vessels in the brain (hemorrhagic stroke). Together, stroke is the second most common cause of death worldwide, accounting for about 5.9 million deaths in 2010, almost equally distributed between ischemic and hemorrhagic strokes [1], but the latter accounts only for about 15% of all strokes. Stroke is also the leading cause of acquired disability in adults, and the third cause of loss of years of life. Ischemic stroke has a 6-month mortality rate of about 20%, and about 30% of patients who survive are handicapped at 6 months [2]. Hemorrhagic stroke is an even more devastating disease, with a 30-day mortality rate of almost 50%, and only 20% of survivors being functionally independent at 6 months [3].

Patients with acute stroke usually present with a focal neurological deficit, such as a hemisyndrome if the stroke involves the motor cortex, or a loss of speech if the stroke involves the Broca area. The patients can also present nonspecific signs, such as headache, nausea, and seizures. In general, symptoms are not specific to the etiology of the stroke, and a large variety of diseases, such as an intracranial tumor, epilepsy, or posterior reversible encephalopathy syndrome, can mimic stroke clinically. Imaging is therefore used in the acute setting to differentiate between the various possible etiologies of the symptoms, which then help with the choice of the appropriate therapeutic approach.

Computed tomography (CT) remains the preferred imaging modality for the early evaluation of a patient suspected of acute stroke at most institutions, due to its broad availability, speed, relative lack of contraindications (except for contraindications for iodinated contrast, mainly allergy and renal insufficiency), and relative low cost. Its main drawbacks are the ionizing radiation used to make the images, and the reduced brain tissue contrast compared to magnetic resonance imaging (MRI).

In the setting of acute stroke, CT imaging has two main goals: (1) confirm the diagnosis of hemorrhagic or ischemic stroke, and exclude other possible causes of the clinical symptoms and (2) evaluate the risk:benefit ratio for each individual patient to triage for the optimal therapy.

HEMORRHAGIC STROKE

Etiology and Therapy

The most common etiology for hemorrhagic stroke are trauma and hypertensive intracranial hemorrhage, followed by cerebral amyloid angiopathy, but a large number of other entities can cause intracerebral hemorrhage, including a ruptured aneurysm, an arteriovenous malformation, a hemorrhagic brain tumor, or a metastasis. Independent of the etiology of the hemorrhage, if the hemorrhage is large, the therapy usually consists of neurosurgical decompressive cranioectomy, to avoid a potentially fatal brainstem compression due to brain herniation caused by increased intracranial pressure.

Acute Imaging

Intracerebral hemorrhage is typically diagnosed on a noncontrast head CT (Fig. 131.1). When a large intracerebral hemorrhage is visualized, particular attention is required to identify signs of increased intracranial pressure: mass effect, displacement of the falx and the tentorium cerebelli, and the loss of visualization of the basal cisterns (Fig. 131.2). CT angiography (CTA) and postcontrast CT are used to detect active extravasation of contrast and to exclude other causes of hemorrhage, such as an aneurysm, vessel anomalies, or blood–brain barrier breakdown in a glioblastoma.

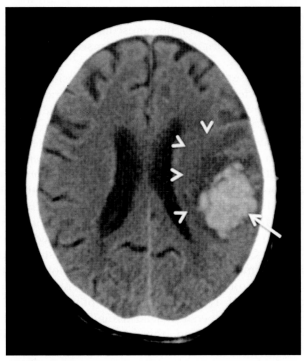

FIGURE 131.1 Acute bleeding on nonenhanced CT, visible as an area of hyperdensity (*arrow*). The surrounding area of hypodensity (*arrowhead*) corresponds to edema.

Follow-Up Imaging

Immediate follow-up of patients with intracerebral hemorrhage, especially if their clinical condition deteriorates, is usually achieved using noncontrast head CT, to see if the hematoma has grown due to rebleeding, and if the edema has increased in size and might necessitate surgical decompression. Because the hematoma can obscure the cause of the bleeding, a follow-up MRI examination after hematoma resolution is often necessary.

ARTERIO-OCCLUSIVE ISCHEMIC STROKE

Etiology and Therapy

The most common etiology for ischemic stroke is arterial occlusion, either due to the formation of a local thrombus, or through embolization of a thrombus typically arising from the heart or from an atherosclerotic plaque of the carotid bifurcation. Since 1995, intravenous administration of recombinant tissue plasminogen activator (rt-PA) was considered the standard of care for patients with acute ischemic stroke, if it could be administered within 3h after symptom onset [4], later extended to a 4.5h window [5]. Intraarterial administration of rt-PA has been shown to be more effective than intravenous administration, with reperfusion of around 66% [6]. In 2015, five large randomized clinical trials [7–11] showed in a spectacular way the efficacy of endovascular treatment up to 6h after symptom onset, and should now be considered the new standard of care for proximal middle cerebral artery (MCA) or internal carotid artery occlusion in this time window. The safety

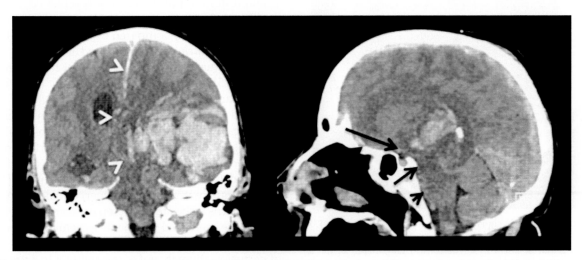

FIGURE 131.2 Massive intracerebral hemorrhage: Displacement of the falx and other midline structures (*arrowheads*) are visible on the coronal nonenhanced CT image. On the sagittal image, the suprasellar (*long arrow*), interpeduncular (*arrow*), and prepontine (*short arrow*) cisterns are effaced as a result of intracranial hypertension.

and efficacy of endovascular thrombectomy in carefully selected patients with acute ischemic anterior circulation stroke in an extended time window (6–16 h) is currently under investigation (DEFUSE 3).

Acute Imaging

Acute ischemic stroke due to arterial occlusion on noncontrast head CT typically features the loss of gray matter–white matter differentiation shortly followed by a confluent area of hypodense brain parenchyma respecting the vascular territory of an artery. In the early phase of a MCA stroke, this can usually be observed first at the level of the insula, and this is called the *loss of the insular ribbon sign* (Fig. 131.3). The blood clot can often be observed on noncontrast CT, as it is hyperdense, a sign called the *dense artery sign* (Fig. 131.3). Noncontrast head CT is also used to assess the so-called Alberta Stroke Program Early CT Score (ASPECTS), which is calculated by evaluating ten regions of brain parenchyma in two standardized levels of the MCA territory (Fig. 131.4), giving 1 point if the region is considered normal, and 0 if abnormal (i.e., hypodense). A normal ASPECTS is 10, and the portion of patients with bad outcome is increasing with decreasing ASPECTS [12]. Finally, noncontrast head CT is used to exclude intracranial hemorrhage, which is an absolute contraindication to intravenous rt-PA or endovascular recanalization.

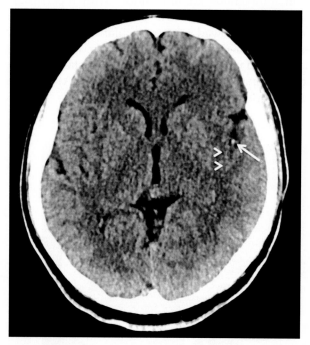

FIGURE 131.3 Early signs of stroke on noncontrast head CT. An M2 branch of the middle cerebral artery contains a blood clot (*arrow*), representing a *dense artery sign*. The cortical gray matter of the left insula is not as well differentiated as the one on the right side, especially in the posterior part (*arrow head*). This is called the *loss of the insular ribbon sign*.

Intracranial-CTA is used to confirm the presence and determine the exact site of the arterial occlusion. Maximum intensity projections (Fig. 131.5) and volume rendering (Fig. 131.6) can often help to quickly detect the arterial occlusion. CTA is also used to evaluate the quality of the collateral flow, which is an important prognostic factor for good outcome [13], and can be very variable from patient to patient. Cervical CTA is usually acquired together with the intracranial CTA to plan for the endovascular therapy, to evaluate the anatomy and the state of the arteries (atherosclerotic plaque, carotid or aortic dissection (Fig. 131.7), fibromuscular dysplasia), and to determine the origin of the stroke for secondary prevention purposes.

In perfusion-CT (PCT), different perfusion maps are reconstructed from multiple serial acquisitions during the wash-in and wash-out of a bolus of iodine-based contrast, using an *arterial input function* to deconvolute the tissue signal (Fig. 131.8), and when absolute quantification of *cerebral blood volume* (CBV) and *cerebral blood flow* (CBF) are attempted, a *venous output function* to correct for partial volume effect. Those maps are the CBF, the CBV, the *time to maximum* (Tmax), and the *mean transit time* (MTT) (Fig. 131.9). PCT has a higher sensitivity to detect acute ischemic stroke compared to noncontrast CT and can be used to distinguish stroke from stroke mimics, such as seizures. A number of investigators have proposed PCT as a tool for patient selection for treatment [14]. The underlying concept is to use PCT to differentiate irreversibly dead tissue from ischemic but salvageable tissue and from normally perfused brain. In this regard, the following definition have been established: the *ischemic core* is defined as the area of brain parenchyma that is considered dead, and the *penumbra* is defined as the area that is considered critically hypoperfused, and at risk of dying but still salvageable if appropriate restoration of blood flow is achieved; a *malignant profile* is defined as a patient with a poorer expected outcome with therapy compared to no therapy at all; finally, a *target mismatch profile* is defined if a patient presents with a mismatch profile but does not have a malignant profile [15], and should consequently be offered the revascularization therapy (Fig. 131.10).

As mentioned earlier, seizures are often accompanied with focal perfusion abnormalities and can mimic stroke. In the setting of epilepsy, focal hypoperfusion can be observed, with prolonged MTT and Tmax, and decreased CBF and CBV, but also focal hyperperfusion [16], also called compensatory, luxury hyperperfusion (Fig. 131.11). PCT patterns associated with seizures can be differentiated from those associated with acute ischemic stroke using the following criteria: the lack of vessel occlusion, the territory of perfusion anomaly not corresponding to a vascular territory, and the clinical history. The second important pitfall in PCT interpretation is the delay in blood flow arrival due to chronic precerebral cardiac and vascular disease. Low cardiac output can result in global increase in MTT and Tmax, and precerebral carotid stenosis, and aortic or carotid dissection can result in asymmetric increase in MTT and Tmax, which can be confused with an acute stroke. Therefore, cervical vessel imaging and evaluation is essential for correct interpretation of PCT imaging.

There is currently no broad consensus on the best method and modality to assess the ischemic penumbra and the malignant profile. At our institution, on CT, the *infarct core* is defined as region with CBF <30% compared to the nonischemic hemisphere, or alternatively as the hypodense area on noncontrast CT, and the *penumbra* as the region with Tmax >6 s. A *mismatch profile*

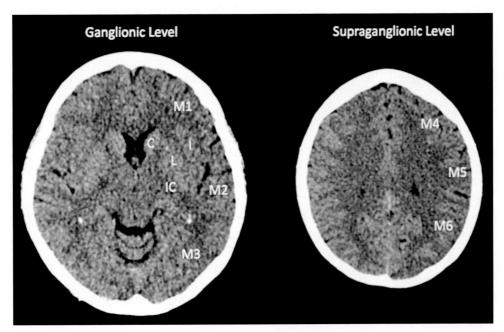

FIGURE 131.4   Alberta Stroke Program Early CT Score (ASPECTS) scheme. At the ganglionic level, M1–M3, insula (I), the lentiform nucleus (L), the caudate nucleus (C), and the posterior limb of the internal capsule (IC) are evaluated. At the supraganglionic level (defined as above the caudate nucleus), M4–M6 are evaluated. A point is subtracted from 10 for each additional region involved by the stroke.

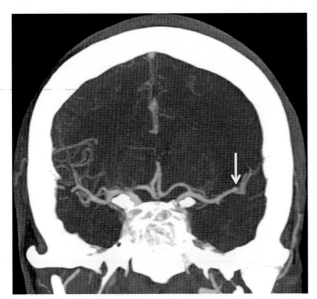

FIGURE 131.5   Maximum intensity projection in the coronal plane of a CT-angiography, showing the site of occlusion (*arrow*) at the superior division of the left middle cerebral artery. Note the difference in overall vessel density distally of the occlusion site, compared to the right side.

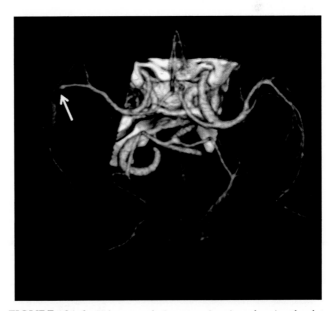

FIGURE 131.6   Volume rendering, superior view, showing the clot-related filling defect (*arrow*) at the bifurcation of the left middle cerebral artery.

is considered to be present when the difference (Tmax >6 s) – (core >15 mL) and the ratio (Tmax >6 s): core >1.8. A *malignant profile* is defined as an infarct core (CBF <30%) >70 mL or a Tmax >10 s lesion >100 mL. Those definitions are adapted for CT from the initial definitions made for MR in the DEFUSE study [15]. The various thresholds used in those assessments are dependent on the scanner and the postprocessing

software used, and efforts are currently made to standardize those values to cross-platform values. While no randomized clinical trial has definitely demonstrated the benefits on outcome of patient selection with PCT [14], perfusion imaging is part of the standard stroke protocol at our institution and many others.

Follow-Up Imaging

In the subacute phase, imaging is helpful to assess for a possible hemorrhagic transformation post revascularization therapy, which is especially feared after rt-PA treatment, and edema development, which might get life threatening and might necessitate surgical decompression. In the chronic phase, parenchymal atrophy and gliosis is observed in the stroke area, and is characterized by a loss of parenchyma, a widening of adjacent sulci, and a frank hypodensity on CT (Fig. 131.12).

VENO-OCCLUSIVE ISCHEMIC STROKE

Etiology and Therapy

Veno-occlusive ischemic stroke, caused by cerebral venous thrombosis (CVT), is a rare condition, which accounts for 0.5–1% of all strokes, and affects females three times more often than males. Most patients have a predisposing condition, such as prothrombotic coagulopathies (hereditary thrombophilias, hyperhomocysteinemia), elevated estrogen levels (oral contraceptive, hormone replacement therapy, pregnancy), dehydration, local infection, sepsis, inflammatory diseases, and malignancy. The treatment for CVT, even in the presence of hemorrhagic infarction, involves anticoagulation, appropriate use of antiepileptic drugs, and the treatment

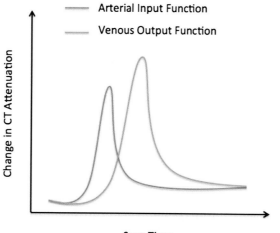

FIGURE 131.7 An internal carotid artery dissection on the left side in this young patient, which was later diagnosed with fibromuscular dysplasia. A crescent-shaped contrast filling (*short arrow*) is visible in the true lumen, which is compressed by the thrombus that has formed in the false lumen (*long arrow*). Note for comparison the patent lumen of the right internal carotid artery (*arrow head*).

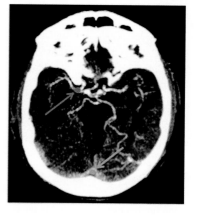

FIGURE 131.8 Arterial input function and venous output function are obtained (either automatically or drawn manually) in a large cerebral artery (typically the middle cerebral artery) and vein (typically the superior sagittal sinus) and are used to compute the perfusion-CT maps.

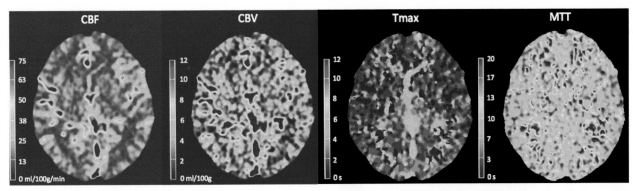

FIGURE 131.9 Normal-looking perfusion-CT images: cerebral blood flow (CBF, in mL/100 g/min), cerebral blood volume (CBV, in mL/100 g), time to maximum (Tmax, in s), and mean transit time (MTT, in s) maps.

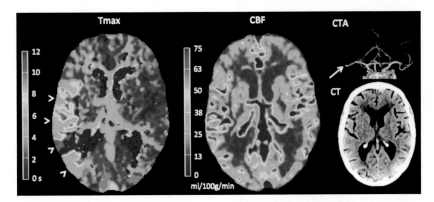

FIGURE 131.10 A region of increased Tmax is visible in the posterior territory of the right middle cerebral artery (MCA). There is no asymmetry visible on the CBF maps, as well as no anomalous hypodense area on noncontrast CT. The occlusion of the M1 segment of the right MCA is clearly visible on CTA. Therefore, with a large penumbra, no infarct core, and a well-accessible thrombus, this patient is an ideal candidate for endovascular thrombectomy, which was performed.

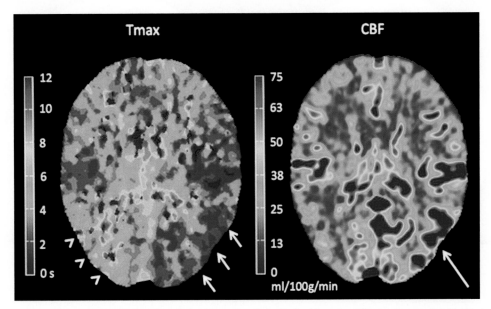

FIGURE 131.11 Epilepsy can sometimes be confounded with stroke clinically. In this patient, Tmax seems to be slightly increased in the right parietal region (*arrowheads*) compared to the contralateral side, but Tmax is in fact decreased (*short arrows*) and CBF (*long arrow*) is increased in the left parietal region, representing hyperperfusion related to a seizure.

of the underlying risk factors. Thrombolysis (catheter directed or systemic) or thrombectomy is used as a last resort, when deterioration continues despite adequate anticoagulation.

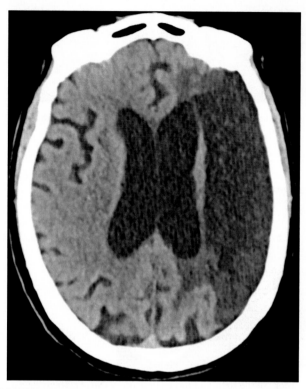

FIGURE 131.12 Chronic infarct of the left middle cerebral artery territory, showing a large hypodense territory of encephalomalacia.

Acute Imaging

Acute ischemia due to CVT involving venous sinuses usually forms a confluent area of hypodense brain parenchyma that does not respect an arterial territory, often interspersed with blood products (Fig. 131.13), but CVT limited to cortical veins can also occur without parenchymal infarction. The venous thrombus itself is hyperdense on noncontrast head CT (Fig. 131.14), and is often located at the level of the sagittal sinus, transverse sinus, or sigmoid sinus. A filling defect can be seen after contrast injection, and is called the *empty delta sign* if involving a venous sinus (Fig. 131.15). Thrombotic filling defects should not be confused with Pacchioni's granulations, which are extensions of the arachnoid membrane into the dural sinuses, have usually a lobulated, well-circumscribed aspect, and are more focal than a thrombus (Fig. 131.16). The role of CTA and PCT in the context of CVT is mostly the exclusion of an arterial occlusion.

Follow-Up Imaging

Repeated imaging is advised in cases of neurological deterioration, to assess mass effect following edema formation or intraparenchymal bleeding. In the subacute setting, follow-up imaging is indicated in selected cases to search for a possible dural arteriovenous fistula, which is a rare complication of CVT.

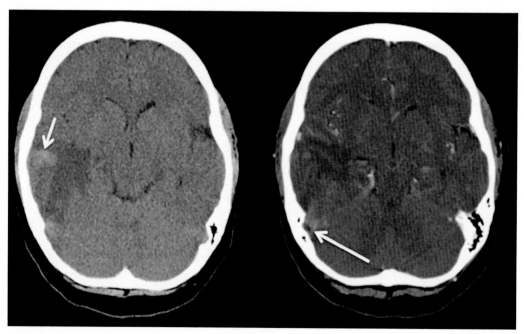

FIGURE 131.13 Cerebral venous thrombosis: venous infarction of brain parenchyma, which does not correspond to an arterial territory. It is interspersed with blood products (*short arrow*), visible on noncontrast head CT (left). The venous thrombus is visible as a filling defect on the postcontrast head CT (right) (*long arrow*).

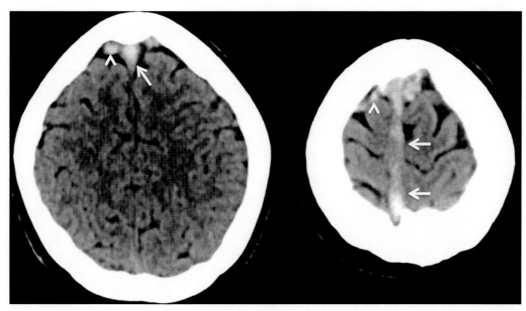

FIGURE 131.14 Cerebral venous thrombosis: a hyperdense thrombus is seen in the superior sagittal sinus (*arrows*) as well as in adjacent cortical veins on this noncontrast axial CT slices.

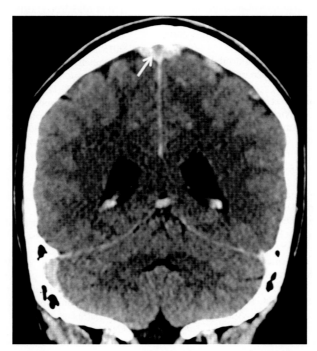

FIGURE 131.15 *Empty delta sign*: filling defect on this superior sagittal sinus on this contrast-enhanced coronal CT slice, corresponding to a venous thrombus.

VIII. DIAGNOSTIC TESTING

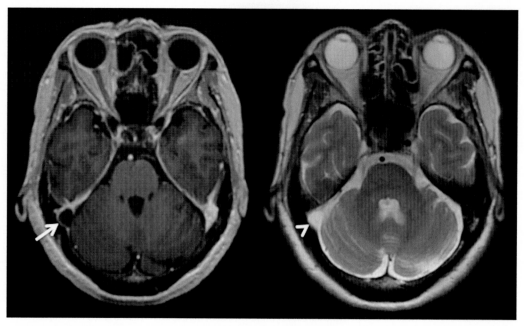

FIGURE 131.16   A Pacchioni's granulation seen on MR: Well-circumscribed filling defect in the right sigmoid sinus (*arrow*), which could be misinterpreted as a subocclusive short thrombus on this postcontrast gradient echo MR image. On the T2 images, the bright signal (*arrowhead*) proves that this is a Pacchioni's granulation, as those are filled with cerebrospinal fluid.

References

[1] Lozano R, Naghavi M, Foreman K, et al. Global and regional mortality from 235 causes of death for 20 age groups in 1990 and 2010: a systematic analysis for the Global Burden of Disease Study 2010. Lancet 2012;380(9859):2095–128.

[2] Slot KB, Berge E, Sandercock P, et al. Causes of death by level of dependency at 6 months after ischemic stroke in 3 large cohorts. Stroke J Cereb Circ 2009;40(5):1585–9.

[3] Broderick J, Connolly S, Feldmann E, et al. Guidelines for the management of spontaneous intracerebral hemorrhage in adults: 2007 update: a guideline from the American heart Association/American stroke association stroke Council, High blood pressure Research Council, and the quality of care and outcomes in Research Interdisciplinary Working Group. Stroke J Cereb Circ 2007;38(6):2001–23.

[4] Tissue plasminogen activator for acute ischemic stroke. The National Institute of Neurological Disorders and Stroke rt-PA Stroke Study Group. N Engl J Med 1995;333(24):1581–7.

[5] Emberson J, Lees KR, Lyden P, et al. Effect of treatment delay, age, and stroke severity on the effects of intravenous thrombolysis with alteplase for acute ischaemic stroke: a meta-analysis of individual patient data from randomised trials. Lancet 2014;384(9958):1929–35.

[6] Furlan A, Higashida R, Wechsler L, et al. Intra-arterial prourokinase for acute ischemic stroke. The PROACT II study: a randomized controlled trial. Prolyse in Acute Cerebral Thromboembolism. JAMA 1999;282(21):2003–11.

[7] Berkhemer OA, Fransen PS, Beumer D, et al. A randomized trial of intraarterial treatment for acute ischemic stroke. N Engl J Med 2015;372(1):11–20.

[8] Goyal M, Demchuk AM, Menon BK, et al. Randomized assessment of rapid endovascular treatment of ischemic stroke. N Engl J Med 2015;372(11):1019–30.

[9] Saver JL, Goyal M, Bonafe A, et al. Stent-retriever thrombectomy after intravenous t-PA vs. t-PA alone in stroke. N Engl J Med 2015;372(24):2285–95.

[10] Campbell BC, Mitchell PJ, Kleinig TJ, et al. Endovascular therapy for ischemic stroke with perfusion-imaging selection. N Engl J Med 2015;372(11):1009–18.

[11] Jovin TG, Chamorro A, Cobo E, et al. Thrombectomy within 8 hours after symptom onset in ischemic stroke. N Engl J Med 2015;372(24):2296–306.

[12] Puetz V, Dzialowski I, Hill MD, Demchuk AM. The Alberta Stroke Program Early CT Score in clinical practice: what have we learned? Int J Stroke 2009;4(5):354–64.

[13] Bang OY, Saver JL, Buck BH, et al. Impact of collateral flow on tissue fate in acute ischaemic stroke. J Neurol Neurosurg Psychiatry 2008;79(6):625–9.

[14] Heit JJ, Wintermark M. Imaging selection for reperfusion therapy in acute ischemic stroke. Curr Treat Options Neurol 2015;17(2):332.

[15] Albers GW, Thijs VN, Wechsler L, et al. Magnetic resonance imaging profiles predict clinical response to early reperfusion: the diffusion and perfusion imaging evaluation for understanding stroke evolution (DEFUSE) study. Ann Neurol 2006;60(5):508–17.

[16] Gelfand JM, Wintermark M, Josephson SA. Cerebral perfusion-CT patterns following seizure. Eur J Neurol 2010;17(4):594–601.

CHAPTER

132

MRI and MRA of Ischemic Stroke

R.G. González

Harvard Medical School, Boston, MA, United States

INTRODUCTION

We image patients with stroke syndromes to establish a diagnosis and to guide patient management, especially when there is a reasonable likelihood of intervention improving outcomes. Computed tomography (CT)- and magnetic resonance imaging (MRI)-based imaging, often in combination, are commonly used. With recent major advances in the treatment of the severe ischemic stroke caused by occlusions of the major cerebral arteries, a critical assessment is underway on how to optimize triage decisions to deliver the best care. The most important clinical and physiological determinants of patient outcomes include: severity of the neurological deficit, site of arterial occlusion, size of the infarct core as a function of the collateral circulation, and the success of recanalization. Imaging provides essential physiological information, but time is required. Thus a tension exists between obtaining most precise physiological information and the need to restore cerebral perfusion as soon as possible. This chapter focuses on how the use of MR results in optimal outcomes in patients undergoing endovascular intervention even outside the traditional time windows. Of course, MR is just as valuable in evaluating patients with all types of ischemic strokes and TIAs, because it is superior to all other imaging methods in revealing stroke physiology. While the focus here is on imaging the acute stroke patient, the basic MRI principles are applicable to all stroke and transient ischemic attack (TIA) patients. These basic principles are described here, without detailed documentation. The reader interested in the documentation is directed to the bibliography at the end, particularly the relevant chapters in the book [9] from which this contribution is in part derived.

MRI OF BRAIN ISCHEMIA

MR has greatly enhanced our understanding of ischemic stroke. It is a widely available and practical clinical tool, and is commonly employed to diagnose and guide the treatment of acute stroke patients. It is particularly useful in patients with major stroke syndromes caused by the occlusions of major arteries of the anterior circulation, because it is the most precise method to select patients that are most likely to benefit from endovascular thrombectomy. What follows is an overview of the MR methods most useful in evaluating the patient with ischemic stroke.

DIFFUSION MRI

Diffusion MRI, commonly designated DWI (diffusion-weighted imaging), is highly sensitive and specific in the detection of acute ischemic stroke at early time points when CT and conventional MR sequences are unreliable [5]. The initial DWI lesion is a marker of severe ischemia and progresses to infarction unless there is early reperfusion. The initial diffusion lesion volume correlates highly with final infarct volume as well as with acute and chronic neurologic assessment. Conventional CT and MRI cannot reliably detect infarction at early time points. Sensitivities are less than 50% for these modalities within 6 h of ictus. The sensitivity and specificity for DWI are in the upper 90% range [10].

Contrast mechanisms. Acute ischemia results in reduction in tissue water diffusivity, and is the principal source of lesion detection by DWI. The biophysical basis for this change is complex and not fully understood. Mechanisms include: failure of ionic pumps with net transfer of water

Primer on Cerebrovascular Diseases, Second Edition
http://dx.doi.org/10.1016/B978-0-12-803058-5.00132-6

from the extracellular to the intracellular compartment; reduced extracellular space volume and increased tortuosity of extracellular space pathways; increased intracellular viscosity from microtubule dissociation and fragmentation of other cellular components; increased intracellular space tortuosity; decreased cytoplasmic mobility; temperature decrease; and increased cell membrane permeability.

Time course. In rodent studies, experimental middle cerebral artery (MCA) occlusion is followed by a decline in water diffusivity as early as 10 min after occlusion. Diffusion coefficients pseudonormalize at about 48 h, and are elevated thereafter. In humans, decreased diffusion in ischemic brain tissue has been reported as early as 11 min after vascular occlusion. The apparent diffusion coefficient (ADC) continues to decrease with maximum reduction at 1–4 days. This decreased diffusion appears hyperintense on DWI (which combines T2 and diffusion weighting) and hypointense on ADC images. The ADC returns to baseline at 1–2 weeks. At this point, the infarct appears mildly hyperintense on the DWI images and isointense on the ADC images. Thereafter, while the DWI can be variable (slight hypointensity, isointensity, or hyperintensity, depending on the relative strength of the T2 versus diffusion components), the ADC is elevated because of increased water content of encephalomalacic tissue. The time course is influenced by a number of factors including infarct type and size, as well as patient age [1].

Reversibility. In the vast majority of stroke patients, the DWI abnormality represents tissue that is destined to infarct. The final infarct volume usually includes the initial DWI lesion and other surrounding tissue into which the infarct extends. However, at least partial DWI reversal following early reperfusion is observed. Studies in nonhuman primates have shown that temporary occlusions of cerebral arteries of less than an hour commonly result in true DWI reversal, while occlusions of 3 h or more followed by reperfusion produces a temporary pseudonormalization. Reversibility is variable with temporary occlusions of between 1 and 3 h. This is similar to the experience in human stroke. After 3 h of cerebral artery occlusion, the DWI abnormality represents irreversibly injured tissue regardless of the pseudonormalization that may be seen after reperfusion.

MAGNETIC SUSCEPTIBILITY IMAGING

Sequences that are sensitive to local disturbances in the magnetic field permit identification of blood products including thrombus that causes vascular occlusion, parenchymal hematoma, and chronic microhemorrhages. Such blood products produce hypointensities on T2*-weighted images created using gradient recalled echo sequences or can detect intraluminal thrombus (due to its high content of paramagnetic deoxyhemoglobin) in hyperacute infarcts as a linear low signal region of magnetic susceptibility. Susceptibility-weighted imaging (SWI) is a type of T2*-weighted imaging that utilizes both phase and magnitude information obtained from high-resolution 3D gradient echo-based sequences, which produces images that accentuate differences in magnetic susceptibility. SWI also demonstrates decreased signal in veins in regions with reduced perfusion due to the increase in intravascular deoxyhemoglobin.

FLAIR/T2- AND T1-WEIGHTED IMAGING

Immediately after occlusion, there is loss of the signal flow void of the occluded artery on T2-weighted images. On T2 FLAIR, high intravascular signal against the surrounding low-signal subarachnoid space is commonly observed in the region of impending infarction. This is thought to be due to slow flow within veins. The parenchyma exhibits detectable T2/FLAIR hyperintensity and T1 hypointensity because of an increase in tissue water content. This is due primarily to vasogenic edema. While the primary contrast mechanism identifying infarction is the same (increased water content), the conspicuity of the lesion is not the same, with the FLAIR abnormality more readily identified than T2, both of these are substantially more conspicuous than on T1-weighted images. Importantly, increased water content also results in sulcal effacement, which helps in identifying the infarction. None of the conventional MRI sequences are reliable before 6 h after ictus, but are highly reliable after 24 h.

In the subacute phase of cerebral infarction (1 day to 2 weeks), the progressive increase in edema results in increased T2 and FLAIR hyperintensity, more T1 hypointensity, better definition of the infarction and swelling along with gyral thickening, effacement of sulci and cisterns, effacement of adjacent ventricles, midline shift, and brain herniation. The swelling usually peaks at about 3 days and resolves by 7–10 days. There is increased T2 and FLAIR signal within the first week that usually persists, but there may be "MR fogging," another form of pseudonormalization. MR fogging occurs when the infarcted tissue signal intensity is similar to that of normal tissue. This is thought to be due to infiltration of inflammatory cells and other mechanisms, which produces a local water content that is similar to normal tissue.

After 2 weeks, the mass effect decreases, and the parenchyma develops tissue loss and gliosis. During the chronic stage of infarction (>6 weeks), necrotic tissue and edema are resorbed, the gliotic reaction is complete, and the blood–brain barrier is repaired. There is no longer parenchymal, meningeal, or vascular enhancement. There is tissue loss and there may be ex vacuo enlargement of ventricles, sulci, and cisterns. There is increased FLAIR/T2-hyperintensity and T1-hypointensity due to increased water content associated with encephalomalacia. Wallerian degeneration is characterized by T2-hyperintensity and tissue loss of the ipsilateral cortical spinal tract after MCA territory infarctions. Chronic infarcts can demonstrate gyriform T1 high signal most likely due to mineral deposition that occurs after laminar necrosis.

CONTRAST-ENHANCED T1-WEIGHTED IMAGING

Occasionally, contrast enhancement on T1-weighted images may be observed in the earliest stages of ischemic infarction. This occurs with severe ischemia and is a result of blood–brain barrier disruption of small pial arteries. The contrast is leaked into the subarachnoid spaces where it may persist for many hours. This effect is observed sporadically and is reversible. More typically, gyriform enhancement of the cerebral cortex peaks at 2–6 days. Parenchymal enhancement may be visible at 2–3 days but is consistently present at 6 days and may persist for 6–8 weeks.

MR ANGIOGRAPHY

Identification of a treatable vascular occlusion is paramount and must be done with alacrity. While many MR pulse sequences have been developed to assess the vessels of the head and neck, this is best accomplished with contrast-enhanced magnetic resonance angiography (ceMRA). It is performed with a short TR (10 ms) gradient echo sequence following an intravenous bolus of gadolinium. The gadolinium shortens blood T1 making it hyperintense on this sequence. Signal from all other tissues are suppressed. The MRA data must be obtained during peak arterial enhancement with the timing set by a test bolus or by automatic bolus detection. A variant is time resolved ceMRA, which depicts flow dynamics. We obtain ceMRA from the arch to a level above the temporal lobes in the coronal plane with a first-pass technique (Fig. 132.1). For an acquisition period of less than 2 min the following information is gained that is important in the planning of an endovascular procedure: the status of the aortic arch; status of the great arteries of the neck including their origins and the carotid bifurcation;

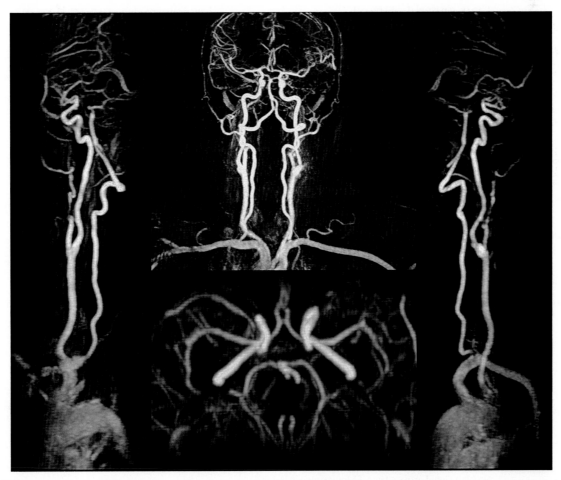

FIGURE 132.1 Frontal view of the complete MR angiogram extending from the aortic arch almost to the vertex (top center). Axial restricted MIPS of the head showing to best advantage the arteries of the circle of Willis (bottom center). Lateral oblique restricted MIPS demonstrating the right and left great arteries of the neck from their origin into the head.

status of the internal carotid and vertebral arteries at the skull base; status of the intracranial ICAs and M1 segments of the MCAs; and the status of the intracranial vertebral arteries and the basilar artery.

The major advantages of ceMRA over phase-contrast (PC) and time-of-flight (TOF) MRA are: it covers a much larger area of anatomy; it has a much shorter acquisition time, making it less susceptible to patient motion; and it has greater signal-to-noise ratio, less dephasing from turbulence, and does not suffer signal loss from saturation effects. CeMRA produces more anatomic images while noncontrast MRA reflects physiology more, and anatomy must be inferred from blood flow. This can be misleading when a vessel is not seen on TOF techniques due to very slow flow or flow reversal. The disadvantages of ceMRA are a narrow time window for data acquisition and that it cannot be repeated until the intravascular gadolinium agent is cleared. Moreover, it has a lower spatial resolution.

TIME-OF-FLIGHT AND PHASE-CONTRAST MRA

The short acquisition times and large coverage volumes make CE-MRA most suitable for evaluating patients being considered for endovascular embolectomy. For other stroke patients, TOF and PC MRA in their 2D or 3D forms may reveal clinically relevant information on the cerebrovascular system. Table 132.1 lists the advantages and disadvantages of each of these methods.

TABLE 132.1   Advantages and Disadvantages of Noncontrast MR Angiography Sequences

MRA Type	Advantages	Disadvantages
2D time-of-flight	Noninvasive	Overestimates vessel stenosis
	Can image slow flow	In-plane signal loss
	Can image large volume of tissue	Signal loss from turbulence
	Can be repeated if suboptimal	Low spatial resolution
		Artifact from T1 hyperintense lesions
3D time-of-flight	Noninvasive	Only small volumes due to marked saturation effects
	High spatial resolution	Cannot image slow flow because of saturation effects
	Shows complex vascular flow	Time consuming, susceptible to patient motion
	Less susceptible to intravoxel dephasing	Artifact from T1 hyperintense lesions
	Can be repeated if suboptimal	
	Can be obtained after contrast	
2D phase contrast	Can show direction and magnitude of flow	Low spatial resolution
	No artifact from T1 hyperintense lesions	More susceptible to turbulent dephasing than TOF MRA
	Can show very slow moving blood	Can have aliasing artifact
	Can be obtained after contrast	2D PC of neck is no longer than 2D TOF of neck
	Can be repeated if suboptimal	
3D phase contrast	High spatial resolution	Time consuming; 3D TOF MRA is faster, with similar resolution
	Does not show high signal artifact from T1 hyperintense lesions	Lower spatial resolution than 3D TOF MRA
	Fast, minimizing patient motion artifact	Occasionally underestimates stenosis
	Less susceptible to signal loss from flow turbulence	Must be obtained in arterial phase
	No saturation artifact	Cannot be repeated until IV contrast has cleared
	Good signal to noise	Requires rapid power injection of contrast
	Images large volume of tissue	
	No high signal artifact from T1 hyperintense lesions	
	Accurate in estimating stenoses	
	Helps differentiate occlusion from near-occlusion	

PERFUSION MRI

Imaging of cerebral hemodynamics can be accomplished with MR and may be useful in certain cases. A full discussion of this topic is outside the scope of this chapter. It is noted here that it is not necessary to make a decision on whether to triage a patient with a terminal ICA or proximal MCA occlusion to endovascular thrombectomy, because if the core is less than 100 mL, there will always be a large perfusion abnormality and the diffusion/perfusion mismatch will always be greater than 1.5 when the perfusion maps, such as *time to maximum* or *mean transit time*, are used [7]. This follows from the physiological fact that the MCA territory is greater than 200–250 mL. We do use perfusion MR on occasion, usually to confirm that a stroke syndrome is due to hemodynamic factors rather than a stroke mimic. This most typically occurs in patients without an identifiable DWI lesion.

MR IN ACUTE ISCHEMIC STROKE TRIAGE

Infarct growth rates. Time is a major determinant on how a patient with a major ischemic stroke syndrome is evaluated and treated. In most institutions, CT imaging is favored over MRI because it is inherently faster and logistically easier. The price for the advantage in time is a less-reliable measurement of the ischemic core. At the MGH, we typically perform CT/CT angiography (CTA) followed by DWI. The addition of the MR sequence adds 15–30 min resulting in continued core infarct growth before perfusion is restored. This begs the questions of how much infarct growth occurs during this period and is it clinically significant? There are data that provide insights into this issue.

Empirically we have observed no relationship between the size of a core infarct and the time after stroke onset in patients with documented terminal ICA or proximal MCA occlusion (Fig. 132.2) [6]. In a prospective study, patients with similar occlusions were scanned serially, and we observed that infarct growth followed a logarithmic pattern (Fig. 132.3) [4], which is similar to what is observed in experimental animal stroke studies. These observations are explained by a high variability in the collateral circulation: one patient with an MCA main stem occlusion may have a core infarct of less than 50 mL at 24 h while another with the *same occlusion* will have a core of greater than 150 mL at 3 h. The difference is a result of a robust collateral circulation in the former and a poor one in the latter [3].

A highly variable, logarithmic ischemic core growth pattern is fully consistent with the published literature. If one assumes that there is a threshold core infarct volume above which intervention does not

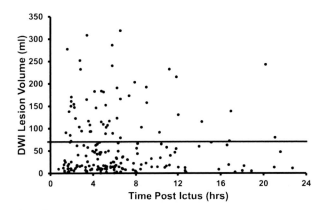

DWI Lesion Volumes in ICA/MCA Occlusions

FIGURE 132.2 DWI lesion volume versus estimated time since stroke onset. The scatter plot includes data from 186 acute ischemic stroke patients with occlusions of the ICA or proximal MCA. Each point depicts the DWI lesion volume from an MRI study performed at an estimated time after stroke onset. There is no correlation between size of the ischemic core and time of stroke onset. The horizontal line is set at about 70 mL. Note that at all times, the number of data points below this line exceeds the number above the line.

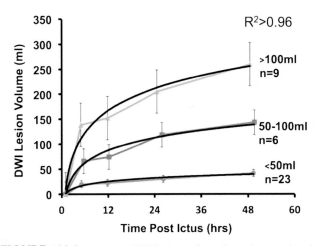

Infarct Growth Grouped by Baseline DWI Volume

FIGURE 132.3 Average DWI lesion volume in major anterior circulation strokes grouped by baseline DWI volume. Patients with baseline volumes of < 50 mL are represented in *blue circles*, *red squares* are patients with 50–100 mL DWI lesion volumes, and the green triangles represent patients with initial volumes > 100 mL. The horizontal axis is time from the stroke onset. Error bars represent the standard error. Logarithmic curves (*black lines*) correlated very highly with the experimental data.

result in favorable patient outcomes and may cause harm, then it is clear that with the passage of time fewer patients will be below this threshold. At some point, the odds ratio of a favorable outcome will be less than 1. A mathematical model using these data and 100 mL as the threshold infarct volume suggests this occurs at about 6 h.

However, if imaging is used to select for treatment only those patients below a threshold, then the effect of time is decoupled from outcomes and the odds ratio for a favorable outcome remains stable, above 1 even well beyond 6h post ictus. The clinical implications of these data are the following. Virtually all patients have small infarct cores early (<3h), and high-precision neuroimaging adds little additional information prior to intervention. Moreover, in this early time points, advanced neuroimaging is unreliable—DWI lesions are reversible, for example. Patients who may be treated later should have precise core imaging using DWI; suitable patients will be on the flat part of the logarithmic growth curve and there is a small price to pay for the large benefit of proper selection.

THE MGH STROKE IMAGING ALGORITHM

The imaging algorithm that we have adopted to evaluate the patient with acute stroke is shown in Fig. 132.4. It is based on the reliability of imaging to measure critical physiological factors, and was developed using an experienced and evidence-based approach [2]. The algorithm is applicable to all patients with stroke. The imaging algorithm answers three key pathophysiological questions: Is there a hemorrhage? Is there an occlusion that is treatable by endovascular embolectomy? Is the early core infarct small enough that endovascular embolectomy has a reasonable probability of helping the patient achieve a favorable outcome? Typically, a patient undergoes CT to assess for hemorrhage, immediately followed by CTA to identify a treatable embolus. The most precise way to assess the size of the early infarct core is with diffusion MRI. These questions may be answered with MRI using magnetic susceptibility imaging, MRA, and diffusion MRI. CT and CTA alone are relied upon when a patient cannot undergo MRI, but the results are not as robust. After over a decade of experience with CT perfusion, we found that it was unreliable for the selection of individual patients for endovascular embolectomy [11]. Fig. 132.5 is an example of a patient with a severe left MCA syndrome who arrived outside the standard time windows and who was evaluated solely with magnetic resonance sequences and had an excellent outcome after endovascular intervention.

We performed a prospective, observational clinical study to evaluate the MGH acute stroke imaging algorithm [8]. A total of 72 patients with MCA or terminal ICA occlusion identified by CTA had core infarct volume determination by DWI, and underwent endovascular thrombectomy. They were prospectively classified as Likely to Benefit (LTB) or Uncertain to Benefit (UTB) using DWI lesion volume and clinical

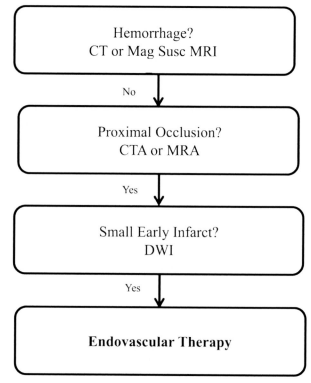

FIGURE 132.4   MGH acute stroke imaging algorithm. The imaging algorithm answers three key pathophysiological questions: Is there a hemorrhage? Is there an occlusion that is treatable by IAT? Is early core infarct small enough that IAT has a reasonable probability of helping the patient achieve a favorable outcome? Typically, a patient undergoes CT to assess for hemorrhage, immediately followed by CT angiography (CTA) to identify a treatable embolus. However, these may be substituted with magnetic susceptibility–weighted MRI and MRA. The most precise way to assess the size of the early infarct core is with diffusion MRI, an exam that takes less than 2min. CT and CTA alone are used when a patient cannot undergo MRI, but the results are not as robust.

criteria (age, NIHSS, time from onset, baseline mRS, life expectancy). An additional 31 patients were not imaged with MRI, and underwent endovascular thrombectomy after CT/CTA alone. The main outcome measure was a 90-day modified Rankin score (mRS), with favorable defined as 90-day mRS of ≤2 indicating that the patient was able to live independently. Forty patients were prospectively classified as LTB, and 32 as UTB. Reperfusion and prospective categorization as LTB were powerfully associated with favorable outcomes ($p < 0.001$ and $p < 0.005$, respectively). Reperfusion was achieved in 68% of the LTB patients and 75% of UTB patients (nonsignificant). Favorable outcomes were obtained in 53% and 25% of LTB and UTB patients that were treated, respectively ($p = 0.016$). Favorable outcomes were observed in 74% of LTB patients that had successful reperfusion compared to 33% of successfully reperfused UTB patients ($p = 0.004$). In the CT only group, favorable outcomes occurred in 29% which was significantly less than the

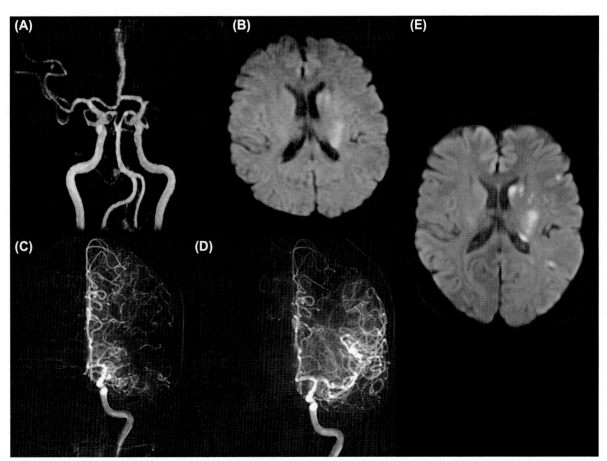

FIGURE 132.5 Successful embolectomy treated outside standard time window. The patient was a 62-year-old woman with a history of hypertension and hypothyroidism. She was last seen well going to bed at 23.00 h. She awoke at 07.00 h with right upper extremity numbness. At 08.00 h she arose, noted RUE and facial weakness, but remained ambulatory. She presented to an outside hospital where she underwent Telestroke evaluation by MGH neurologist and was noted to have a left MCA syndrome with an NIHSS score of 10. The unenhanced CT was notable for dense left MCA sign but no obvious signs of infarct. She was outside IV tPA window and she was medflighted to the MGH for urgent evaluation for endovascular thrombectomy. She arrived at 11.30 h (12.5 h from last seen well) with a left MCA syndrome and an NIHSS score of 14. She was taken directly to MR scanner where 3D TOF MRA demonstrated occlusion of the left MCA main stem (A). DWI showed a lesion in the left MCA territory that measured less than 30 mL (B). There was no evidence of hemorrhage. Initial cerebral angiogram confirmed a left M1 occlusion (C). A stent retriever device successfully recanalized this artery after a few minutes with antegrade excellent parenchymal perfusion (D). Follow-up DWI showed the initial ischemic lesions in addition to a few new scattered lesions involving the left MCA territory (E). The patient recovered nearly completely and was living independently at 3 months. *Images courtesy of B. Leslie-Mazwa.*

LTB MRI group, but similar to the UTB MRI group. The ratio of treated to screened patients was 1:3. MRI selection compares favorably to selection using CT techniques with the distinction that a higher proportion of screened patients were treated. This is demonstrated in Table 132.2, which compares the results of the MGH prospective observational study with results from five successful prospective randomized trials.

CONCLUSIONS

MR is the most effective imaging method for the evaluation of the patient with acute ischemic stroke. It is widely available and the time needed to perform it is similar to CT methods. Diffusion MRI is the most precise

TABLE 132.2 Comparison of Outcomes and Proportion of Patients Treated in MGH and Other Recently Published Trials

Study	Patients Treated	mRS 0–2 (%) Favorable Outcome	Proportion Treated
MGH-LTB	40	53	1:3
MGH-UTB	32	25	
MGH-CT	31	29	
EXTEND-IA	35	71	1:14
SWIFT PRIME	98	60	1:7.5
MR CLEAN	233	32	Unknown
ESCAPE	165	53	Unknown
REVASCAT	103	44	Unknown

method available to establish the ischemic core size and to infer the strength of the collateral circulation, which is critical for estimating the probability of success of endovascular thrombectomy in a patient with an occlusion of a major cerebral artery.

References

[1] Copen WA, Schwamm LH, Gonzalez RG, et al. Ischemic stroke: effects of etiology and patient age on the time course of the core apparent diffusion coefficient. Radiology 2001;221:27–34.

[2] Gonzalez RG, Copen WA, Schaefer PW, et al. The Massachusetts General Hospital acute stroke imaging algorithm: an experience and evidence based approach. J Neurointerv Surg 2013; 5(Suppl. 1):i7–12.

[3] Gonzalez RG, Hakimelahi R, Schaefer PW, et al. Stability of large diffusion/perfusion mismatch in anterior circulation strokes for 4 or more hours. BMC Neurol 2010;10:13.

[4] Gonzalez RG, Hakimelahi R, Schaefer PW, et al. Highly variable logarithmic growth of ischemic core in major anterior circulation ischemic strokes. (submitted for publication).

[5] Gonzalez RG, Schaefer PW, Buonanno FS, et al. Diffusion-weighted MR imaging: diagnostic accuracy in patients imaged within 6 hours of stroke symptom onset. Radiology 1999;210:155–62.

[6] Hakimelahi R, Vachha BA, Copen WA, et al. Time and diffusion lesion size in major anterior circulation ischemic strokes. Stroke 2014;45:2936–41.

[7] Hakimelahi R, Yoo AJ, He J, et al. Rapid identification of a major diffusion/perfusion mismatch in distal internal carotid artery or middle cerebral artery ischemic stroke. BMC Neurol 2012;12:132.

[8] Leslie-Mazwi TM, Hirsch JA, Falcone GJ, et al. Endovascular stroke treatment outcomes after patient selection based on magnetic resonance imaging and clinical criteria. JAMA Neurol 2016;73:43–9.

[9] Gilberto González JAH R, Lev MH, Schaefer PW, Schwamm LH. Acute ischemic stroke. Heidelberg: Berlin, Springer; 2011.

[10] Schaefer PW, Grant PE, Gonzalez RG. Diffusion-weighted MR imaging of the brain. Radiology 2000;217:331–45.

[11] Schaefer PW, Souza L, Kamalian S, et al. Limited reliability of computed tomographic perfusion acute infarct volume measurements compared with diffusion-weighted imaging in anterior circulation stroke. Stroke 2015;46:419–24.

CHAPTER

133

Transcranial and Cervical Ultrasound in Stroke

M.R. Vosko[1], D.W. Newell[2], A.V. Alexandrov[3]

[1]Kepler Universitätsklinikum, Linz, Austria; [2]Seattle Neuroscience Institute, Seattle, WA, United States; [3]The University of Tennessee Health Science Center, Memphis, TN, United States

INTRODUCTION

Transcranial Doppler ultrasound (TCD), or transcranial color-coded sonography (TCCS) and duplex ultrasound of extracranial brain supplying vessels are noninvasive methods that can be used at bedside for neurovascular examination [1] (Fig. 133.1). Ultrasound is the safest vessel imaging method that also allows repeated assessments and real-time monitoring. Brightness-mode (B-mode) display has been in clinical use for over four decades to evaluate the vessel wall and plaques in the extracranial arteries. In addition to color flow imaging, spectral Doppler depicts a waveform containing information about velocity, resistance, flow direction vasomotor responses, and presence of emboli [1].

In many countries, early examination of intracranial arteries is part of benchmarking parameters of an effective acute stroke treatment. With the increasing use of mechanical thrombectomy, there is a trend to perform imaging of cerebral arteries in most acute stroke patients using computed tomography angiography (CTA) or magnetic resonance angiography (MRA). These two modalities of vessel imaging are not often available worldwide on a 24/7 basis. CTA and MRA do not provide information about lesion morphology

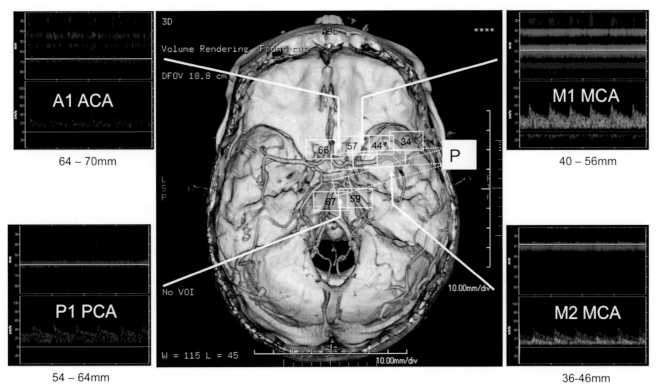

FIGURE 133.1 Bedside neurovascular ultrasound examination with portable equipment (transcranial and cervical ultrasound) and monitoring set for intracranial vasculature.

and real-time hemodynamic information. Thus, the role of ultrasound in stroke patients remains complementary to angiography providing additional important information.

TRANSCRANIAL DOPPLER ULTRASONOGRAPHY

TCD [2] uses a 2-MHz probe through the thin portions of the skull (trans-temporal, trans-nuchal (suboccipital) or trans-orbital windows) and can identify proximal basal intracranial arteries that often harbor emergent large vessel occlusions (ELVO) amenable to mechanical thrombectomy. The trans-temporal window is used to examine the M1 and proximal M2 segments of the middle cerebral artery (MCA), A1 segment of the anterior cerebral artery (ACA), terminal internal carotid artery (TICA), and P1/P2 segments of the posterior cerebral artery (PCA) (Fig. 133.2). Trans-orbitally we examine the ophthalmic artery (OA) and ICA siphon. The trans-occipital window (Fig. 133.3) is used to examine both terminal vertebral arteries (VA) and the entire course of the basilar artery (BA) [3]. Cerebellar arteries can also be detected but are not examined due to tortuosity and less predictable course. Key applications of TCD are summarized in Table 133.1.

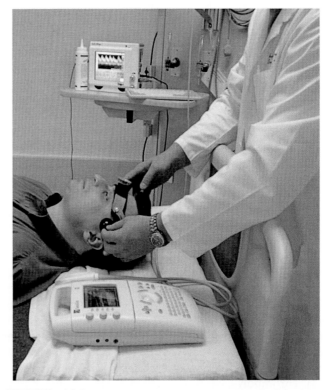

FIGURE 133.2 Power-motion Doppler examination of the intracranial vessels through trans-temporal approach. Depth and vessel localization are shown relative to CTA. *Images courtesy of Dr. Kristian Barlinn.*

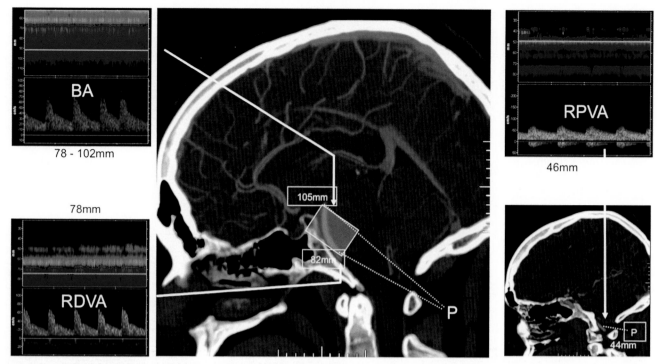

FIGURE 133.3   Power-motion Doppler examination of the terminal vertebral and basilar arteries vessels through suboccipital approach. Depth and vessel localization are shown relative to CTA. *Images courtesy of Dr. Kristian Barlinn.*

TABLE 133.1   Applications of Transcranial Doppler (TCD) or Transcranial Color-Coded Sonography (TCCS) in Patients With Acute Ischemic Stroke

1. Bedside confirmation of vascular origin of the presenting symptoms and determination of underlying stroke mechanism
2. Fast detection and localization of occlusion/stenosis
3. Detection of intracranial collaterals
4. Detection of cerebral embolism in real time and quantification of right-to-left shunt
5. Real-time monitoring of recanalization in patients treated with systemic thrombolysis
6. Determination of stroke pathogenic mechanism through real-time findings complementary to angiography
7. Detection of re-occlusion, air-embolism, hyperperfusion syndrome
8. Identification of patients at high risk of stroke, stroke recurrence, or deterioration

Acute Ischemic Stroke

Several studies provide evidence that TCD can offer useful information in acute ischemic stroke patients. TCD is capable to diagnose intracranial arterial steno-occlusive disease, mostly caused by atherosclerosis or thromboembolic occlusions, moyamoya disease, or sickle cell disease. TCD has >90% specificity in demonstrating MCA occlusions in patients with acute stroke [4]. Thrombolysis in Brain Ischemia (TIBI) scale for TCD can be used for diagnosis of an occlusion and recanalization [5]. TIBI criteria have excellent agreement with simultaneously performed catheter angiography [6].

Brunser et al. showed that TCD provided additional information to a multimodal radiological imaging in over half of hyperacute patients, and in 17% TCD findings changed management [7].

Detection of intracranial stenosis is relevant for therapeutic decision making. Current therapeutic strategies for intracranial stenosis favor dual antiplatelet therapy combined with a statin. Early detection of an intracranial stenosis could prevent recurrent strokes or transient ischemic attacks (TIAs). TCD can detect 50% or greater stenoses (based upon elevated blood flow velocities) [8]. Local validation of published velocity criteria is mandatory by the Intersocietal Accreditation Commission (www.iac.org/vascular).

Vasospasm in Subarachnoid Hemorrhage

Cerebral vasospasm after subarachnoid hemorrhage (SAH) is a preventable and reversible life-threatening condition. TCD can identify and monitor vasospasm, which is characterized by a segmental velocity increase in comparison to a focal increase by stenosis [9]. Aaslid et al. in 1984 classified "spastic" MCA on digital subtraction angiography (DSA) with mean flow velocities (MFVs) between 120/230cm/s on TCD [9]. Vasospasm must be discriminated from elevated velocity by hyperemia, where flow velocities in both ICA and MCA are elevated, while a greater increase in intracranial MCA than extracranial

ICA suggests MCA vasospasm (intracranial to extracranial MFV ratio of ≥3 and 6, or so-called Lindegaard ratio [10,11]). The original velocity criteria described by pioneers of TCD [9, 10] need to be reevaluated since neurocritical care advances in management of SAH likely changed cerebral hemodynamics compared to management paradigms in 1980s when these data were initially obtained.

Embolic Stroke

TCD is the gold standard for detection, localization, and quantification of cerebral embolization in real time. The ability of TCD to detect gaseous or solid microemboli was previously recognized during monitoring of carotid endarterectomy and cardiac surgery [12–14]. Intracranial microemboli can be detected in patients with atrial fibrillation, prosthetic heart valves, carotid artery stenosis, fibromuscular dysplasia, arterial dissection, and intracranial stenosis, as well as during invasive vascular procedures [14]. Monitoring for intracranial air microemboli after venous injection is useful in identifying patients with cardiac right–left shunts [15,16]. Emboli detection together with real-time assessment of vasomotor activity make TCD a unique noninvasive test that helps to determine stroke pathogenic mechanism and explain neurological deficit fluctuation in patients with stroke [1]. Mastering TCD should be essential for training vascular neurologists to work at a comprehensive stroke center level [1].

Intracranial Hemodynamics

Impaired intracranial hemodynamics on TCD identifies patients with cerebrovascular occlusive disease who are at higher risk for stroke recurrence or progression [17,18]. The use of TCD to evaluate vasomotor reserve (VMR) is based on the principle that changes in velocity are proportional to changes in flow through a vessel if the vessel diameter is constant. Measurement of the velocity of blood flow through the MCA can be used to detect relative changes in its blood flow due to changes in CO_2 concentration with voluntary breath-holding [17], hypoventilation, or apnea. With these principles, the functional capacity of the distal regulating vessels in the cerebral circulation can be assessed with TCD. Patients with diminished or exhausted vasomotor reactivity or arterial blood flow steal represent the high-risk groups [17,18].

CERVICAL DUPLEX ULTRASONOGRAPHY

Cervical duplex ultrasonography (CDU) is a noninvasive modality for the evaluation of both vessel wall features and blood flow parameters in the cervical arteries. CDU provides real-time information about the presence of an atherosclerotic plaque, its length, composition, protrusion and surface, range of resulting vascular stenosis or occlusion, presence of a dissection and other significant imaging findings in patients with stroke [1,14] (Table 133.2). CDU can directly visualize atherosclerotic plaque composition that can be classified based on their echogenicity. Uniformly hyperechoic carotid artery plaques are mainly comprised of fibrotic tissue needed for plaque stability. In contrast, heterogeneous (and predominantly hypoechoic) plaques consisting of matrix deposition, cholesterol accumulation, necrosis, calcification, and intraplaque hemorrhage are considered unstable being a source of artery-to-artery embolic strokes [19]. CDU reports should provide information on plaque presence, location, extent, texture, and surface: these data can help to prognosticate the risk of stroke beyond percent stenosis [1,19].

The mainstay of CDU is to screen patients with stroke or TIA for carotid artery stenosis and CDU reports should include prediction of the North American Symptomatic Carotid Stenosis Endarterectomy Trial (NASCET) strata, that is, normal, 0–49%, 50–69%, 70–99%, near occlusion, and total occlusion (Fig. 133.4). Peak systolic velocity (PSV), end-diastolic velocity (EDV), and the peak systolic internal carotid artery/common carotid artery (ICA/CCA) velocity ratio are essential ultrasound parameters for grading percent stenosis. PSV and EDV must be assessed in the pre-stenotic, stenotic, and post-stenotic segments of the vessel. The Society of Radiologists in Ultrasound Consensus Conference reached multidisciplinary agreement on criteria to predict clinically relevant strata of NASCET ICA stenosis [20]. Briefly, PSV of 125 cm/s and ICA/CCA ratio of 2 identify patients with 50% stenosis while PSV of 230 cm/s, ICA/CCA ratio of 4, and EDV of 100 cm/s identify patients with

TABLE 133.2   Applications of Carotid Duplex Ultrasonography (CDU) in Patients With Acute Ischemic Stroke

1. Information about plaque composition and surface (lipid, hemorrhage, fibrous content, ulceration, superimposed thrombus)
2. Diagnosis of degree of carotid artery stenosis (<50%, 50–69%, >70%, near occlusion, occlusion)
3. Diagnosis of degree of vertebral artery stenosis (<50%, >50%, occlusion)
4. Detection of intraluminal thrombus in cervical vessels
5. Diagnosis of subclavian steal syndrome
6. Diagnosis/suspicion of uncommon causes of stroke (cervical artery dissection, fibromuscular dysplasia, aortic arch dissection)
7. Complementary information in the diagnosis of temporal arteritis
8. Indirect information for distal intracranial vessel occlusion
9. Indirect information for heart rate and heart valves
10. Diagnosis of other conditions not related to acute stroke (e.g., carotid body tumor)

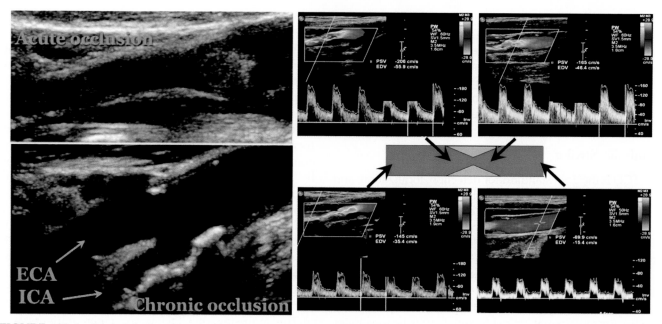

FIGURE 133.4 Cervical duplex ultrasound images of an acute thromboembolic total ICA occlusion (left inset, upper right image), chronic complete ICA occlusion with vessel collapse and fibrosis (left inset, lower image), and atherosclerotic 50–69% NASCET stenosis on B-mode, PSVs, and ICA/CCA ratios derived from angle-corrected velocity images (right inset).

70% stenosis. Like with other diagnostic ultrasound criteria, these SRU criteria have to be locally validated by each laboratory against their own angiographic studies/NASCET measurements as the gold standard.

CDU allows rapid detection of ICA thrombosis, differentiation with chronic ICA occlusion with or without preexisting atheromatous stenosis (Fig. 133.4, left panel). These acute findings may help in selecting endovascular reperfusion strategy or prompt consideration of carotid endarterectomy [1].

Ultrasound diagnosis and classification of VA stenosis is more demanding, as VA asymmetry is common and hypoplastic, and moreover, the VA may terminate in the posterior inferior cerebellar artery (PICA). Anatomic variants and abnormalities (stenosis, occlusion) of the contralateral VA also influence VA flow. Severe stenosis in the carotid arteries may also affect blood flow in the VA due to recruitment of collaterals. VA stenoses are most commonly located in the origin from the subclavian artery (V0 segment) followed by the atlas loop/intracranial segments with intertransverse segments less commonly affected. Criteria for VA stenoses are not based on strict PSV cutoff but on focal and significant PSV increase since tortuosity of the proximal VA segment, ICA lesions prompting compensatory flow changes and VA asymmetry may result in relatively high velocities. The velocity increase should be found over a relatively short segment of the VA with normal or decreased pre- and post-stenotic velocities [21]. Elongated and multiple stenoses in the VA may not produce focal velocity elevations that could be a source of false-negative CDU studies [1,14].

SUMMARY

TCD and CDU offer insights into the mechanisms and process of acute stroke and provide real-time information for clinical decision making in hyperacute stroke patients, selection for and monitoring effects of therapeutic interventions. Both tests can be repeatedly performed at bedside. Main limitations include operator dependency of ultrasound testing technique and absence of trans-temporal window particularly in older patients. Nonetheless, TCD and CDU are applicable to the vast majority of stroke patients, and successful practice of transcranial and cervical ultrasound requires in-depth knowledge of anatomy, physiology, scanning techniques, and interpretation principles that have to be taught during neurology residency and vascular neurology fellowship training.

References

[1] Alexandrov AV. Neurovascular Examination: Rapid Evaluation of Stroke Patients with Ultrasound Waveform Interpretation. Oxford: Wiley Publishers; 2013. ISBN: 978-1-4051-8530-1.

[2] Aaslid R, Markwalder TM, Nornes H. Noninvasive transcranial Doppler ultrasound recording of flow velocity in basal cerebral arteries. J Neurosurg 1982;57:769–74.

[3] Alexandrov AV, Sloan MA, Tegeler CH, Newell DN, Lumsden A, Garami Z, American Society of Neuroimaging Practice Guidelines Committee, et al. Practice standards for transcranial Doppler (TCD) ultrasound. Part II. Clinical indications and expected outcomes. J Neuroimaging 2012;22:215–24.

[4] Burgin WS, Malkoff M, Felberg RA, Demchuk AM, Christou I, Grotta JC, et al. Transcranial Doppler ultrasound criteria for recanalization after thrombolysis for middle cerebral artery stroke. Stroke 2000;31:1128–32.

[5] Demchuk AM, Burgin WS, Christou I, Felberg RA, Barber PA, Hill MD, et al. Thrombolysis in Brain Ischemia (TIBI) transcranial Doppler flow grades predict clinical severity, early recovery, and mortality in patients treated with tissue plasminogen activator. Stroke 2001;32:89–93.

[6] Tsivgoulis G, Ribo M, Rubiera M, Vasdekis SN, Barlinn K, Athanasiadis D, et al. Real-time validation of transcranial Doppler criteria in assessing recanalization during intra-arterial procedures for acute ischemic stroke: an international, multicenter study. Stroke 2013;44:394–400.

[7] Brunser A, Mansilla E, Hoppe A, Olavarría V, Sujima E, Lavados PM. The role of TCD in the evaluation of acute stroke. J Neuroimaging 2016;26:420–5.

[8] Zhao L, Barlinn K, Sharma VK, Tsivgoulis G, Cava LF, Vasdekis SN, et al. Velocity criteria for intracranial stenosis revisited: an international multicenter study of transcranial Doppler and digital subtraction angiography. Stroke 2011;42:3429–34.

[9] Aaslid R, Huber P, Nornes H. Evaluation of cerebrovascular spasm with transcranial Doppler ultrasound. J Neurosurg 1984;60:37–41.

[10] Lindegaard KF, Nornes H, Bakke SJ, Sorteberg W, Nakstad P. Cerebral vasospasm after subarachnoid haemorrhage investigated by means of transcranial Doppler ultrasound. Acta Neurochir (Wien) 1988;42(Suppl):81–4.

[11] Lindegaard KF. The role of transcranial Doppler in the management of patients with subarachnoid haemorrhage: a review. Acta Neurochir (Suppl) 1999;72:59–71.

[12] Padayachee TS, Bishop CCR, Gosling RG, et al. Monitoring middle cerebral artery blood flow velocity during carotid endarterectomy. Br J Surg 1986;73:98–100.

[13] Spencer MP, Thomas GI, Nicholls SC, Sauvage LR. Detection of middle cerebral artery emboli during carotid endarterectomy using transcranial Doppler ultrasonography. Stroke 1990;21:415–23.

[14] Alexandrov AV. Cerebrovascular Ultrasound in Stroke Prevention and Treatment. 2nd edition Oxford: Wiley-Blackwell Publishers; 2011. 9781405195768.

[15] Jauss M, Zanette E. Detection of right-to-left shunt with ultrasound contrast agent and transcranial Doppler sonography. Cerebrovasc Dis 2000;10:490–6.

[16] Spencer MP, Moehring MA, Jesurum J, Gray WA, Olsen JV, Reisman M. Power m-mode transcranial Doppler for diagnosis of patent foramen ovale and assessing transcatheter closure. J Neuroimaging 2004;14:342–9.

[17] Silverstrini M, Vernieri F, Pasqualetti P, et al. Impaired vasosmotor reactivity and risk of stroke in patients with asymptomatic carotid artery stenosis. JAMA 2000;283:2122–7.

[18] Palazzo P, Balucani C, Barlinn K, Tsivgoulis G, Zhang Y, Zhao L, et al. Association of reversed Robin Hood syndrome with risk of stroke recurrence. Neurology 2010;75:2003–8.

[19] Rundek T. Beyond percent stenosis: carotid plaque surface irregularity and risk of stroke. Int J Stroke 2007;2:169–71.

[20] Grant EG, Benson CB, Moneta GL, et al. Carotid artery stenosis: gray-scale and Doppler US diagnosis–Society of radiologists in ultrasound consensus conference. Radiology 2003;229:340–6.

[21] Bartels E, Fuchs HH, Flugel KA. Duplex ultrasonography of vertebral arteries: examination, technique, normal values, and clinical applications. Angiology 1992;43:169–80.

C H A P T E R

134

Conventional Cerebral Arteriography

D.C. Haussen, R.G. Nogueira

Emory University School of Medicine, Atlanta, GA, United States

INTRODUCTION

Cerebral angiography was introduced and expanded by the Portuguese neurologist Egas Moniz close to a century ago [1]. Significant morbidity and mortality in the first procedures lead to a transition to iodinated solution as contrast media, leading to satisfactory technical results. Despite the original "distrust" and controversy regarding this method and its future [2], cerebral angiography not only remains the gold standard for the diagnosis of vascular conditions affecting the craniocervical and cerebral vessels, but has developed itself into a very sophisticated interventional field. Even though neuroangiography is suited for the

Primer on Cerebrovascular Diseases, Second Edition
http://dx.doi.org/10.1016/B978-0-12-803058-5.00134-X

evaluation and treatment of craniocervical arteries, capillaries, and veins, this chapter specifically focuses on cerebral arterial angiography.

INDICATIONS

Conventional catheter-based arterial cerebral angiography constituted the only available cerebrovascular diagnostic test for decades, and has more recently shared indications with noninvasive imaging studies. CT angiogram, MR angiogram, cervical duplex, transcranial Doppler (TCD), and even SPECT/PET scan technology have evolved significantly and are now, often times, first-line modalities in the evaluation of the cerebral vasculature. Angiography is typically reserved for situations in which noninvasive imaging investigation is inconclusive or when interventional plans exist. The advantages of catheter angiography rely on its better spatial resolution (submilimetric), dynamic evaluation flow patterns of each individual artery, and high-temporal resolution (up to 30 frames per second).

A discussion regarding the indications and diagnostic accuracy of noninvasive imaging studies is beyond the scope of this chapter. The indications for diagnostic catheter angiography include accurate characterization of aneurysm angioarchitecture, allowing for optimal definition of shape, size, and location. Although microsurgical treatment may be performed based on CTA or MRA, conventional angiography prior to treatment of aneurysms revealed by noninvasive imaging studies is performed in most institutions (Fig. 134.1). In cases of arteriovenous malformations, conventional angiography is typically indicated for all patients. In individuals with suspicion for dural arteriovenous fistulas (including carotid-cavernous fistulas), an angiogram is indicated for all patients for the analysis of potential dangerous features, such as cortical venous drainage and venous ectasia. Subjects with intracerebral hemorrhages in atypical settings (such as lack of risk factors for hypertensive hemorrhage, atypical locations, hemorrhage in young age, and questionable findings on MR or CT) may benefit from more careful investigation with conventional angiography. We reserve conventional angiography in patients with dissections for when there is potential hypoperfusion or repeated embolic phenomena despite aggressive medical management, or when there is controversy regarding its diagnosis. The use of conventional angiography for the evaluation of carotid atherosclerotic disease has declined substantially, given the relatively high risk of navigation/catheterization of the aortic arch and craniocervical vessels in this specific patient population, and the relatively good accuracy and safety of noninvasive imaging for this condition. Typically, catheter angiography is reserved for cases

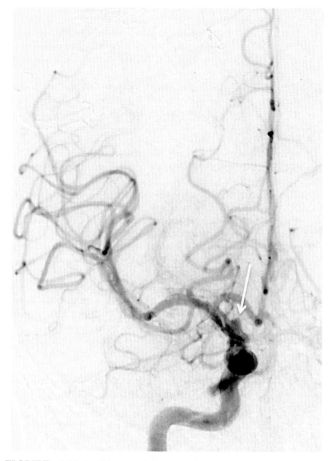

FIGURE 134.1 **Diagnostic conventional angiography: spatial resolution.** Arterial phase/frame of a frontal angiogram of the right internal carotid artery of a patient with subarachnoid hemorrhage and negative CT angiography. Conventional angiogram clearly demonstrated a nonbifurcating anterior carotid wall wide-necked aneurysm ("blister aneurysm") with <2mm height (*arrow*).

determined to be suitable candidates for stent revascularization or when doubt exists regarding etiology, lesion site (cervical and/or intracranial carotid disease), degree of stenosis, or when perfusion/flow patterns are unclear despite noninvasive imaging (Fig. 134.2). Intracranial atherosclerotic disease is a controversial topic, however, we may consider conventional angiography for patients with high-grade stenosis from atherosclerotic disease that have failed maximal medical therapy for risk-stratification. Although vasculopathies encompass a multitude of different etiologies and phenotypes (e.g., systemic or CNS vasculitis, reversible cerebral vasoconstrictive syndrome, intravascular lymphoma, infectious endarteritis, moyamoya changes, and cerebral vasospasm), we favor a cerebral angiogram in cases of unclear diagnosis. Although venous conditions, such as cortical vein or sinus thrombosis, can be identified on venous phases of arterial angiograms, the diagnosis and follow-up rely strongly on noninvasive imaging; angiogram can be performed in cases of equivocal diagnosis,

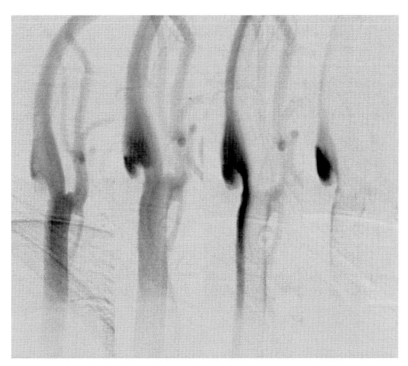

FIGURE 134.2 **Diagnostic conventional angiography: temporal and spatial resolution.** Lateral angiogram of the carotid artery of a patient with previous diagnosis of carotid dissection and recurrent strokes despite medical therapy. The dynamic evaluation revealed an atypical fibromuscular dysplasia shelf/web with significant stagnation of flow in the venous phase within the lesion.

or when flow patterns must be better depicted, or when manometric measures are required (such as suspicion for congestion secondary to venous outflow limitation). Finally, diagnostic conventional angiography is often required for neurosurgical or endovascular planning (such as balloon-test occlusions, Wada testing, petrosal sinus sampling, aneurysm treatment, and preoperative embolization of vascular lesions or tumors).

From the interventional perspective, modern sheaths, guide and distal access catheters, as well as microcatheters and microwires allow for unprecedented navigability and safety. The field of endovascular treatment of acute ischemic stroke has directly benefited from these advances and has dramatically evolved, with a multitude of different available devices and techniques. The technological development has led to a wide arsenal of coil variations, leading to the ability of providing remarkable coil/aneurysm conformation. Stents for coil-assisted embolization have also evolved considerably and now allow great deliverability, conformability, and support. The flow-diversion technology (with lower-porosity stents) revolutionized the aneurysm treatment arena, allowing the treatment of previously inoperable large and wide-necked aneurysms. Intra-arterial chemotherapy for retinoblastomas, lacrimal gland tumors, and gliomas are emerging as a potential oncological adjuvant treatment. The management of craniocervical blowout from tumor invasion or lacerations with covered stents is effective and safe. Embolization of tumors of the head

and neck are critical for bleeding control or as a preoperative risk reduction maneuver. Arteriovenous malformations, dural fistulas, and intracranial tumors can also be treated via embolization. Angioplasty/stenting can be performed in carotids and vertebrals (atherosclerosis, fibromuscular dysplasia, Takayasu's arteritis, dissections), selected intracranial steno-occlusive lesions, as well as in certain venous steno-occlusive conditions. The diagnosis and treatment of vasospasm from subarachnoid hemorrhage and even of reversible vasoconstriction syndrome can also be pursued.

BASIC CONCEPTS

Preprocedure

A diagnostic cerebral angiogram is typically an outpatient procedure scheduled after a clinical and basic laboratorial preprocedural evaluation. There are no absolute contraindications to a cerebral conventional angiogram, although there are certain patient characteristics that may veer unfavorably the risk–benefit ratio. Some operators prefer to discontinue anticoagulation prior to the procedure; in our center, we routinely perform angiography in anticoagulated patients. Platelet counts <75.000/mL are a relative contraindication [3]. Diabetics should only take half of the insulin morning dose due to the NPO status. Although controversial, metformin should be held for 48h post procedure as per the risk of lactic acidosis with contrast administration. Catheter-based angiography has been widely utilized in adults and in the elderly, and its safety has been reported in the pediatric population with the use of contemporary technology/techniques and procedural planning [4]. Pregnant patients would ideally undergo noninvasive imaging studies, however, if clinically indicated, it can be performed safely with appropriate measures.

Procedure

A 6-h fasting is normally recommended and peripheral IVs are obtained. Conscious sedation (typically with fentanyl and midazolam) is typically utilized for cooperative adults. The local anesthesia and pressure cause minimal discomfort, and the contrast injections in the craniocervical vasculature may generate transient and mild sensation of warmth, scotomas, transient nausea, dizziness, or neck pain. Monitored anesthesia care and general anesthesia are normally reserved for interventional procedures. Patients are characteristically placed supine in the angiography table and the area for access prepared in sterile fashion.

Many centers have been performing arterial access with a micropunture kit (21G needle) after local anesthesia is administered. Ultrasound guidance is commonly utilized. The most frequently accessed route is the common femoral artery. The upper extremity (radial or brachial artery) may constitute a useful option, with the advantages of eliminating the possibility of retroperitoneal hemorrhage and prolonged bed rest post procedure, and is particularly useful in elongated arches. Very rarely, direct carotid access may

be utilized in emergent cases or in cases with very difficult arch anatomy. Arterial sheaths may be used, and this is followed by the introduction and navigation of a catheter over a wire. Wires and most catheters and microcatheters utilized in neurointervention have hydrophilic coating (hydrophilic biomaterials are used for lubricity by reducing friction within the system and due to the hypothrombogenic potential compared to noncovered catheters/wires). The system (catheters and wires) is radiopaque and is navigated under fluoroscopic (X-ray) guidance across the aortic arch and maneuvered into the craniocervical vessels. A wide range of options exists in terms of catheter sizes and shapes, and the choice is tailored according to the predicted/encountered angioarchitecture. The use of continuous infusion and/or flushes with heparinized saline solution for prevention of thrombotic events is widespread. In our center, a partial bolus of intravenous heparin is typically administered for diagnostic angiograms.

The selection of the arteries to be catheterized/imaged is done according to the clinical indication (e.g., selective internal carotid artery angiography for middle cerebral artery stenosis, 3–4-vessel angiogram for aneurysmal subarachnoid hemorrhage, and a full angiogram (≥6 vessels) for dural arteriovenous fistulas or cryptogenic hemorrhages). Injection of nonionic, water-soluble, iodinated, low-osmolality contrast (safer and hypoallergenic) [5] is performed with standard high-resolution biplane digital subtraction angiography of the vessels of interest. The two planes allow the simultaneous imaging of two projections, which lead to less contrast use, shorter procedural times, and overall safer interventional cases (Fig. 134.3). A typical 3-vessel angiogram may require 60–100mL of contrast; contrast-sparing procedures may

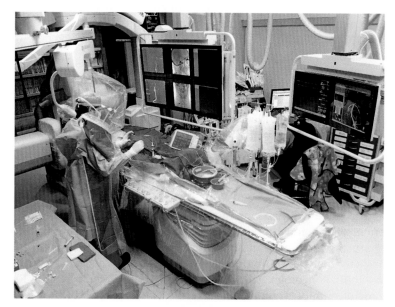

FIGURE 134.3 **Neurointerventional suite.** Neurointerventional room in use. The biplane angiography machine can be visualized, as well as the screen interface to the operator, and the equipment for radiation protection.

require as low as ≤20 mL of contrast solution. Gadolinium could be considered in cases of contraindication for iodinated contrast. After completion, either manual pressure or closure devices are used for hemostasis.

Post Procedure

Arterial access monitoring is routinely performed, with assessment of puncture site, pulse, temperature, visual inspection, neurological checks, and vital checks for a few hours (according to the puncture site, sheath size and use of closure device, required monitoring may be as low as 2 h). Recovery from conscious sedation is routinely performed in a recovery unit or step-down area, while post general-anesthesia/intervention patients are recovered in the (neuro)intensive care unit.

SAFETY

Neurological Complications

Neurological complications are the most feared complications of conventional angiography, and typically relates to embolism due to disruption of atherosclerotic plaques (thrombus and/or plaque fragments), dissections (thrombus), or iatrogenic flushing of air via catheters and wires. The largest reported series encompassed 2899 diagnostic cerebral angiograms (performed by attendings and by trainees from 1996 to 2001) and reported 0.5% permanent neurological complications and 0.9% transient/reversible changes [6]. The risks relate in great part to age and vascular risk factors [6]. Therefore, the indication and the procedural course may be adjusted (such as more aggressive heparinization and no protamine reversal in atherosclerotic patients, and selection of specific catheters and maneuvers).

Nonneurological Complications

Iodinated contrast media can generate various side effects. Anaphylactic reactions to intra-arterial contrast administration are less common than with intravenous use and are overall extremely uncommon. Less-significant reactions are more frequent and can usually be minimized or prevented with steroid and antihistaminic premedication regimens [3]. Nausea and vomiting are uncommon and readily improve with symptomatic medications. Although multiple definitions have been previously proposed, contrast-induced nephropathy is defined by the elevation of creatinine of ≥0.3 mg/dL or ≥50% relative increase, or urine output ≤0.5 mL/kg/h for ≥6 h within a narrow time window after contrast administration [7]. The expected renal complication is expected to happen in ≤0.2% [8]. The most significant risk factors for contrast-induced nephropathy include underlying kidney insufficiency, dehydration, diabetes, and congestive heart failure [9]. Although prophylactic intravenous hydration is commonly used and recommended, no preventative measures can be strongly recommended (e.g., bicarbonate and n-acetylcysteine) [10]. The rate of hematoma requiring transfusion or evacuation should be <0.5%, femoral artery pseudoaneurysm/fistula <0.2%, and femoral artery occlusion <0.2% [8]. Other complications are extremely rare, such as transient contrast neurotoxicity (scotomas, blindness, and amnesia).

PRESENT AND FUTURE

The advances in the neurointerventional suites have been fast paced. The angiography machines have evolved and now commonly incorporate flat panel technology, the ability to perform intraoperative CT scans (CT angiograms and even perfusion scans), and may be built in a "hybrid" design (within an operating room, or containing a CT or MRI scanner). New generation equipment has added safety and speed to the performance of angiography. Software development includes 3-dimensional roadmapping, rotational angiography (Fig. 134.4), and image noise reduction with resultant minimization of patient and operator radiation entrance doses [11,12]. The standard radiation protection strategy is the use of a conventional lead apron, thyroid shield, and lead eyeglasses. However, notwithstanding the supplementary use of table lead skirts and ceiling-suspended shields, the operator is typically exposed to substantial ionizing radiation. The aprons have been reported to cause significant musculoskeletal disorders/occupational hazards in the operators [13]. A system composed of an overhead suspended lead suit with a curved lead–acrylic head shield based on a ceiling-mounted monorail is now commercially available and provides enhanced radiation protection while eliminating the weight from the operator's body [11]. Architectural and logistical features are becoming more clearly important in the neurointerventional field. Our center was designed including the neurointerventional suite inside of the neuro-intensive care unit and adjacent to the CT scanner room. Neurointerventional suites can be constructed next to ORs, anesthesia recovery units, or emergency department [13]. The field of endovascular surgical neuroradiology is diverse and encompasses neurologists, neurosurgeons, and radiologists as operators. The multidisciplinary aspect has been critical for collaboration and field advancement. Finally, the most impressive and exciting variable is the continued exponential growth and refinement of neurointerventional tools. Navigability, trackability, and visualizability of sheaths,

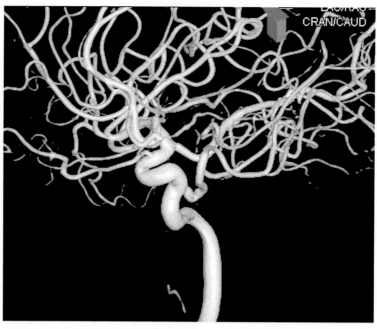

FIGURE 134.4   **3-dimensional angiography.** Rotational angiography allows for precise 3-dimensional reconstructions that optimize diagnostic accuracy and interventional procedural planning.

guide/distal access catheters, microcatheters, wires, and microwires have tremendously improved secondary to technological advances. From the device perspective, stent-retriever and thromboaspiration catheters have revolutionized the stroke field, and there are multiple other retriever devices under investigation. Flow diverting stents and other systems for wide-necked aneurysm treatment (such as the WEB device and pCON system) are allowing the treatment of aneurysms with very challenging angioarchitecture. Liquid embolization with solvent-suspended polymers has enhanced the ability to perform controlled head and neck and cerebral embolizations. Newer micromesh stent technology and embolic protection devices are promising technologies in order to perform safer craniocervical revascularization.

References

[1] Antunes JL. Egas Moniz and cerebral angiography. J Neurosurg April 1974;40(4):427–32.
[2] Moniz E. Cerebral angiography with thorotrast. Arch Neurol Psych 1933;29:1318–23.
[3] Harrigan MDJ. Handbook of cerebrovascular and neurointerventional technique. 2nd ed. Humana Press; 2013.
[4] Burger IM, Murphy KJ, Jordan LC, Tamargo RJ, Gailloud P. Safety of cerebral digital subtraction angiography in children: complication rate analysis in 241 consecutive diagnostic angiograms. Stroke October 2006;37(10):2535–9.
[5] Barrett BJ, Carlisle EJ. Metaanalysis of the relative nephrotoxicity of high- and low-osmolality iodinated contrast media. Radiology July 1993;188(1):171–8.
[6] Willinsky RA, Taylor SM, TerBrugge K, Farb RI, Tomlinson G, Montanera W. Neurologic complications of cerebral angiography: prospective analysis of 2,899 procedures and review of the literature. Radiology May 2003;227(2):522–8.
[7] Lakhal K, Ehrmann S, Chaari A, et al. Acute Kidney Injury Network definition of contrast-induced nephropathy in the critically ill: incidence and outcome. J Crit Care December 2011;26(6):593–9.
[8] Citron SJ, Wallace RC, Lewis CA, et al. Quality improvement guidelines for adult diagnostic neuroangiography. Cooperative study between ASITN, ASNR, and SIR. J Vasc Interv Radiol September 2003;14(9 Pt. 2):S257–62.
[9] McDonald JS, McDonald RJ, Comin J, et al. Frequency of acute kidney injury following intravenous contrast medium administration: a systematic review and meta-analysis. Radiology April 2013;267(1):119–28.
[10] Wichmann JL, Katzberg RW, Litwin SE, et al. Contrast-induced nephropathy. Circulation November 17, 2015;132(20):1931–6.
[11] Haussen DC, Van Der Bom IM, Nogueira RG. A prospective case control comparison of the ZeroGravity system versus a standard lead apron as radiation protection strategy in neuroendovascular procedures. J Neurointerv Surg October 21, 2015.
[12] Soderman M, Mauti M, Boon S, et al. Radiation dose in neuroangiography using image noise reduction technology: a population study based on 614 patients. Neuroradiology November 2013;55(11):1365–72.
[13] Norbash A, Klein LW, Goldstein J, et al. The neurointerventional procedure room of the future: predicting likely innovations in design and function. J Neurointerv Surg September 2011;3(3):266–71.

CHAPTER

135

Ultrasound Examination of the Aortic Arch in Stroke

M.R. Di Tullio
Columbia University, New York, NY, United States

INTRODUCTION

The presence of protruding or ulcerated atheromas in the proximal segment of the aorta is associated with an increased frequency of peripheral or cerebral embolic events [1–4]. The association, initially recognized in pathology studies [1], was confirmed in vivo with the introduction of transesophageal echocardiography (TEE), which rapidly became the diagnostic test of choice for detecting proximal aortic atheromas. TEE has allowed an accurate visualization of the proximal portion of the aorta and provided further evidence linking aortic atheromas with embolic events in numerous case–control, cross-sectional, and prospective cohort studies. This chapter discusses the technical aspects of the use of TEE for the detection of proximal aortic atheromas, a brief overview of the role of these atheromas as a risk factor for ischemic stroke, and some considerations regarding the measures to prevent recurrent embolic events.

TRANSESOPHAGEAL ECHOCARDIOGRAPHY

Technical Aspects

The transesophageal approach provides unique advantages in the ultrasound examination of the aorta. The proximity of the esophagus to the aorta and the absence of interposed structures along most of their course allow for the acquisition of high quality images that cannot be achieved with a transthoracic approach. Transducers with higher frequencies (usually 5–7.5 MHz) than those used in transthoracic imaging can be used, resulting in higher image resolution.

Moreover, modern transducers allow the examination to be performed along multiple planes for a more complete visualization of the vessel, and also allow the three-dimensional (3D) imaging and reconstruction of some segments of the vessel, which may add to diagnostic accuracy.

Examination of the Aorta in Stroke Patients

In the diagnostic work up of cerebral ischemia, the portion of the aorta proximal to the takeoff of the carotid arteries represents the focus of the examination, although retrograde embolization from slightly distal aortic segments has occasionally been described. The examination is accomplished by rotating and slowly withdrawing the ultrasound transducer from the level used to image the heart, and manipulating it until an optimal visualization of the aortic walls and lumen is obtained. The initial portion of the vessel (ascending aorta) can be accurately visualized from the valve level to the initial curvature of the arch (Fig. 135.1). The mid- and distal portions of the arch are also well visualized in all patients. A small portion of the vessel (proximal arch) cannot be visualized due to the interposition of the trachea between the esophagus and the aorta, and may therefore represent a "blind spot" of the examination; however, modern multiplane transducers allow a more complete visualization of the vessel in most patients. The high resolution of TEE allows the assessment of the presence and size of any intimal atheromas (Fig. 135.2), and provides important information on the presence of irregularities or ulcerations of the luminal surface of the atheroma, or the presence of superimposed thrombotic components (Fig. 135.3), which increase the embolic risk. The thickness of the atheroma can be assessed for risk stratification by obtaining measurements perpendicular to the major axis of the vessel (also Fig. 135.2), using the most advanced lesion in cases of multiple atheromas. TEE has high sensitivity and specificity for the detection of aortic atheromas. Its diagnostic accuracy for presence of superimposed thrombus is also high (sensitivity 91%, specificity 90%), while its sensitivity for detecting small surface ulcerations, which may carry additional embolic risk, is about 75%. Reproducibility of TEE measurements of aortic plaque thickness is good, with agreement of 84–88% for the diagnosis of large (≥4mm) plaque.

In the past decade, 3D imaging and reconstruction of ultrasound images have allowed a more complete assessment of the morphology of an aortic atheroma.

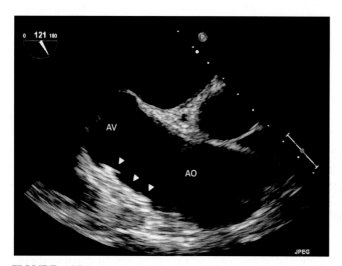

FIGURE 135.1 **Longitudinal view of the ascending aorta.** The entire ascending aorta (AO) is visualized from the aortic valve (AV) to the initial curvature of the arch. Small calcified atheromas (*arrows*) are visualized.

Fig. 135.4 shows an example of a large protruding atheroma by 3D TEE, which allows a thorough understanding of the actual extension into the aortic lumen, and of the atheroma morphology and associated embolic potential.

Safety and Tolerability in Stroke Patients

While a semi-invasive test that requires conscious sedation, TEE is a remarkably safe diagnostic test. Major complications are uncommon and mainly related to unsuspected preexisting esophageal or gastric disease. In over 10,000 patients, one death was observed. In an additional 2.7%, the test was unsuccessful because of impossible intubation (1.9%) or patient intolerance (0.8%) [5]. In another 15,381 consecutive patients, 2 deaths (0.01%) occurred, with an overall incidence of complications of 1.7% [6]. In 901 patients, intubation was unsuccessful in 1.2% of cases, with low incidence of major complications (0.6%), and no deaths. Our experience in stroke patients has shown no increased frequency of patient discomfort, unsuccessful intubation, or significant complications compared with stroke-free patients of similar age. Moreover, the test can be safely performed even in most patients of very advanced age.

PROXIMAL AORTIC ATHEROMAS AND ISCHEMIC STROKE

In the early 1990s, a large autopsy case–control study demonstrated a much greater prevalence of ulcerated aortic atheromas in patients over 60 years of age who

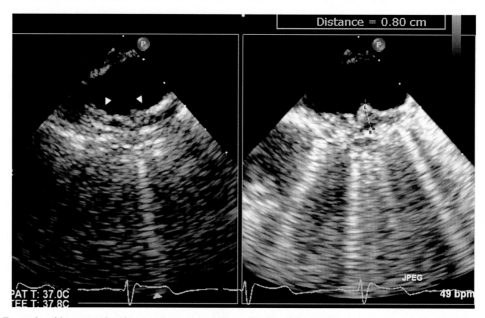

FIGURE 135.2 **Example of large aortic atheroma by transesophageal echocardiography.** Simultaneously displayed orthogonal views (left and right panels) show a diffusely atherosclerotic arch (*arrows*). Measurement of the atheroma maximal thickness (perpendicular to the major aortic axis) is shown in the right panel (0.8 cm).

had died from stroke than in patients who had died from other neurological diseases (26% vs. 5%; age-adjusted odds ratio 4.0; 95% confidence intervals 2.1–7.8) [1]. Moreover, the highest frequency of ulcerated atheromas was found in patients with unexplained (cryptogenic) stroke, suggesting that aortic atheromas could explain some of them, even because the increased risk was independent of the presence of known stroke determinants such as carotid stenosis and atrial fibrillation. Only 3% of all ulcerated atheromas were found in patients below the age of 60, underlining that this new stroke source was relevant in the elderly subgroup

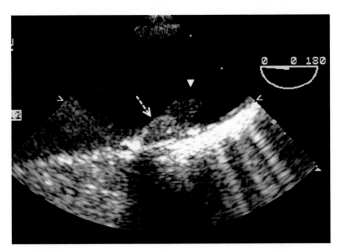

FIGURE 135.3 **Complex aortic atheroma by transesophageal echocardiography.** A large atheroma is seen at mid-arch (*large arrow*); a faintly echogenic component is attached to it (*small arrow*), which was highly mobile in real time imaging and represents superimposed thrombus.

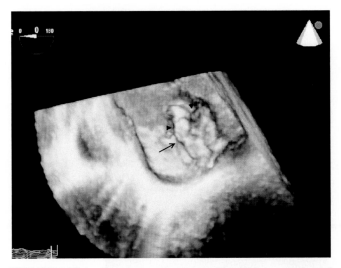

FIGURE 135.4 **Complex aortic atheroma by 3D transesophageal echocardiography.** 3D imaging and reconstruction has allowed the display of the entire atheroma (*large arrow*); the total volume of the atheroma is appreciated, along with protruding and ulcerated components (*small arrows*).

of the general population. TEE then played a crucial role in the evaluation of the embolic potential of aortic atheromas in vivo. Several case–control studies confirmed the association between aortic atheromas and ischemic stroke (Table 135.1). Although with different study populations and atheroma thickness cutoffs for risk stratification, all studies confirmed the role of arch atheromas as an independent stroke risk factor for stroke in the elderly. Subsequently, prospective TEE studies conducted in patients who had experienced a first ischemic stroke reported on an increased risk of recurrent events in patients with large or complex atheromas (Table 135.2). The presence of these atheromas was associated with a two- to almost fourfold increase in the risk of recurrent stroke in different studies.

Therefore, the cumulative evidence clearly shows that aortic arch atheromas are an independent risk factor for ischemic stroke and stroke recurrence in patients over 60 years of age. The importance of aortic atheromas could be even greater in elderly patients whose brain infarcts have no other obvious explanation.

ARCH ATHEROMAS AND STROKE AFTER CARDIAC SURGERY OR AORTIC PROCEDURES

The unsuspected presence of large proximal aortic atheromas may lead to severe thromboembolic complications, and especially stroke, after cardiac surgery requiring cannulation of the proximal aorta for extracorporeal circulation. In 921 consecutive patients undergoing cardiac surgery, the incidence of postoperative stroke was 8.7% in patients with atherosclerotic disease of the ascending aorta and 1.8% in patients without it (p < .0001), and aortic atherosclerosis was the strongest predictor of perioperative stroke [7]. When the presence of severe atherosclerosis is known, modification of the surgical technique may be adopted to decrease the stroke risk. Similarly, a high risk of stroke is present during or after cardiac catheterization of patients with severely atherosclerotic aortas. In 1000 patients undergoing percutaneous coronary interventions, placement of guiding catheter was associated with scraping debris from the aorta in more than 50% [8].

Before both cardiac surgery and transcatheter aortic procedures, testing for aortic atherosclerosis appears indicated in patients at higher risk for it, especially elderly patients with multiple atherosclerotic risk factors. While TEE may not be appropriate for this type of screening, transthoracic imaging from a suprasternal window can be performed (Fig. 135.5), which may help select patients in whom the more detailed TEE examination may be necessary.

TABLE 135.1   Proximal Aortic Atheromas and Ischemic Stroke Risk: Case–Control Studies

Author, Year	Stroke Patients/ Controls (N)	Age (Years)	Type of Atheroma	Controls (%)	Stroke Patients (%)	Adjusted Odds Ratio[a] (95% CI)
Amarenco et al., 1994 [1]	250/250	≥60	1–3.9 mm	22	46	4.4 (2.8–6.8)
			≥4 mm	2	14	9.1 (3.3–25.2)
Jones et al., 1995 [11]	215/202	≥60	<5 mm, smooth	22	33	2.3 (1.2–4.2)
			≥5 mm, complex	4	22	7.1 (2.7–18.4)
Di Tullio et al., 1996 [3]	106/114	≥40	≥5 mm	13	26	2.6 (1.1–5.9)
	30/36	<60		3	3	1.2 (0.7–20.2)
	76/78	≥60		18	36	2.4 (1.1–5.7)
Di Tullio et al., 2008 [4]	255/209	≥55	≥4 mm	24	49	2.4 (1.3–4.6)

CI, confidence interval [4].

[a]Adjusted for conventional stroke risk factors.

TABLE 135.2   Risk of Recurrent Stroke Associated With Proximal Aortic Atheromas: Prospective TEE Studies

Author, Year	Aortic Atheroma Present/Absent (N)	Follow-up (Months)	Type of Atheroma	Stroke Aortic Atheroma Absent (%)	Incidence Aortic Atheroma Present (%)	Adjusted[a] Relative Risk (95% CI)
FAPS[b], 1996 [2]	45/143	24–48	≥4 mm	2.8/y	11.9/y	3.8 (1.8–7.8)
Tanaka et al., 2006 [12]	97/139	42	≥3.5 mm	Not reported	Not reported	2.1 (1.2–3.7)
Fujimoto et al., 2004 [13]	51/232	40	≥4 mm, extending to branches	2.9/y	9.8/y	2.4 (1.1–5.2)
Di Tullio et al., 2009[c] [9]	101/415	24	≥4 mm	5.1/y	13.4/y	2.1 (1.04–4.3)
	44/415	24	≥4 mm, complex	5.1/y	13.7/y	2.6 (1.1–5.9)

[a]Adjusted for conventional stroke risk factors (also see text).
[b]French study of aortic plaque in stroke.
[c]Study conducted in patient with congestive heart failure; stroke may represent first episode or recurrence.

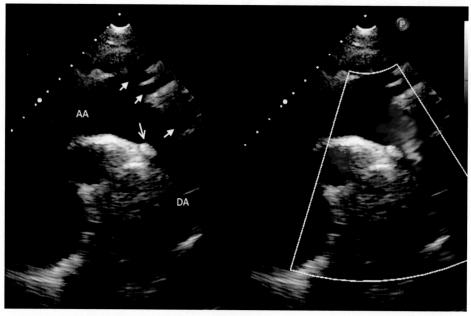

FIGURE 135.5   **Transthoracic aortic imaging from a suprasternal view.** *Left panel*: The aortic arch (AA) and descending aorta (DA) are visualized. A large atheroma (*large arrow*) is seen in the distal arch. The ostia of the head vessels (*small arrows*) are also visualized. *Right panel*: Color flow Doppler shows unobstructed laminar blood flow (blue color) in the distal arch and proximal descending aorta.

CONSIDERATIONS ON THERAPY AND PREVENTION OF RECURRENT EVENTS

As with all atherosclerotic diseases, the use of statins to lower serum cholesterol and possibly stabilize the lesion appears indicated in patients with large aortic atheromas, even in the absence of data from randomized clinical trials on the risk of recurrent stroke. Such trials cannot be performed given the widespread indications of the use of statins in patients surviving first stroke, and the consequent unethical nature of placebo-controlled studies.

Since more complications of arch atheromas are thromboembolic in nature, and because of the frequently detected activation of the coagulation system in patients with acute stroke and large arch atheromas [4], antithrombotic therapy has been attempted as a means to reduce the frequency of recurrent events, often with questionable results. While systemic anticoagulation appears indicated in the relatively infrequent patients with superimposed thrombus, its use in the more numerous patients with large but nonmobile atheromas is debatable and usually not recommended. In 516 patients with acute ischemic stroke treated with aspirin or warfarin, antithrombotic treatment did not affect the incidence of recurrent stroke and death over a follow-up of 2 years; large atheromas (≥4 mm) remained associated with an over twofold increase in risk of events (adjusted hazard ratio 2.12, 95% CI 1.04–4.32), especially among cryptogenic stroke patients (adjusted hazard ratio 6.42, CI 1.62–25.46) [9]. The combination of aspirin and clopidogrel was tested against warfarin in stroke patients with large (≥4 mm) atheromas in the Aortic arch Related Cerebral Hazard (ARCH) trial, but insufficient enrollment and lower-than-expected event rates resulted in the termination of the trial and consequent inconclusive results. On the available data, over a median follow-up of 3.4 years, the primary endpoint (stroke, myocardial infarction, peripheral embolism, vascular death, or intracranial hemorrhage) occurred in 7.6% of patients on combination therapy versus 11.3% of those on warfarin therapy (log-rank, $p = .2$; adjusted HR 0.76, 95% CI, 0.36–1.61, $p = .5$) [10].

Surgical endarterectomy for severely atherosclerotic aortas has been associated with considerable stroke risk, and should be reserved to carefully selected cases.

References

[1] Amarenco P, Duyckaerts C, Tzourio C, Henin D, Bousser MG, Hauw JJ. The prevalence of ulcerated plaques in the aortic arch in patients with stroke. N Engl J Med 1992;326:221–5.

[2] Atherosclerotic disease of the aortic arch as a risk factor for recurrent ischemic stroke. The French Study of Aortic Plaques in Stroke Group. N Engl J Med 1996;334:1216–21.

[3] Di Tullio MR, Sacco RL, Gersony D, Nayak H, Weslow RG, Kargman DE, Homma S. Aortic atheromas and acute ischemic stroke: a transesophageal echocardiographic study in an ethnically mixed population. Neurology 1996;46:1560–6.

[4] Di Tullio MR, Homma S, Jin Z, Sacco RL. Aortic atherosclerosis, hypercoagulability, and stroke the APRIS (Aortic Plaque and Risk of Ischemic Stroke) study. J Am Coll Cardiol 2008;52:855–61.

[5] Daniel WG, Erbel R, Kasper W, Visser CA, Engberding R, Sutherland GR, Grube E, Hanrath P, Maisch B, Dennig K. Safety of transesophageal echocardiography. A multicenter survey of 10,419 examinations. Circulation 1991;83:817–21.

[6] Oh JK, Seward JB, Tajik AJ. Transesophageal echocardiography. The echo manual. 2nd ed. Lippincott Williams & Wilkins; 1999. p. 23–36.

[7] van der Linden J, Hadjinikolaou L, Bergman P, Lindblom D. Postoperative stroke in cardiac surgery is related to the location and extent of atherosclerotic disease in the ascending aorta. J Am Coll Cardiol 2001;38:131–5.

[8] Keeley EC, Grines CL. Scraping of aortic debris by coronary guiding catheters: a prospective evaluation of 1,000 cases. J Am Coll Cardiol 1998;32:1861–5.

[9] Di Tullio MR, Russo C, Jin Z, Sacco RL, Mohr JP, Homma S. Aortic arch plaques and risk of recurrent stroke and death. Circulation 2009;119:2376–82.

[10] Amarenco P, Davis S, Jones EF, Cohen AA, Heiss WD, Kaste M, Laouenan C, Young D, Macleod M, Donnan GA. Clopidogrel plus aspirin versus warfarin in patients with stroke and aortic arch plaques. Stroke 2014;45:1248–57.

[11] Jones EF, Kalman JM, Calafiore P, Tonkin AM, Donnan GA. Proximal aortic atheroma. An independent risk factor for cerebral ischemia. Stroke February 1995;26(2):218–24.

[12] Tanaka M, Yasaka M, Nagano K, Otsubo R, Oe H, Naritomi H. Moderate atheroma of the aortic arch and the risk of stroke. Cerebrovasc Dis 2006;21(1-2):26–31.

[13] Fujimoto S, Yasaka M, Otsubo R, Oe H, Nagatsuka K, Minematsu K. Aortic arch atherosclerotic lesions and the recurrence of ischemic stroke. Stroke June 2004;35(6):1426–9.

SECTION IX

MANAGEMENT

136

Thrombolysis for Acute Ischemic Stroke

R. Gupta

Wellstar Health System, Marietta, GA, United States

In 1996, the FDA approved the use of intravenous alteplase for the treatment of acute ischemic stroke after publication of the National Institute of Neurological Disorders and Stroke (NINDS) trial. There were five randomized controlled trials that laid the groundwork for this initial approval and paved the path for thrombolytic therapy for acute stroke [1–5]. Over the next 20 years, significant advances have occurred in the treatment of acute stroke patients. The focus has shifted away from the efficacy and safety of the medication to the development of pathways to more rapidly identify patients who will benefit from early treatment. Additionally, there have been rapid advances in catheter-based approaches leading to six randomized controlled trials that have shown substantial benefit to patients suffering from strokes secondary to large vessel occlusions [6–11].

INTRAVENOUS TISSUE PLASMINOGEN ACTIVATOR

Intravenous tissue plasminogen activator (t-PA) was tested in five randomized trials in the early 1990s (Table 136.1) [1–5]. The NINDS trial demonstrated that patients treated under 3 h from onset had a 30% higher probability of an excellent recovery compared to the placebo group [1]. The other four trials (ATLANTIS A and B and ECASS I and II) failed to demonstrate this result and showed higher rates of intracranial hemorrhage [2–5]. The conflicting results were felt to be secondary to the expanded time window in the four trials as well as the use of a higher dose of intravenous t-PA of 1.1 mg/kg in ECASS I. The adoption rate for delivery of the medication was not only slow due to patients not presenting in time but also due to reluctance of clinicians to administer the medication because of the concerns of hemorrhagic risks. It is estimated that less than 2% of patients were receiving the drug after the several years after approval of the drug [12,13]. Moreover, community hospitals were not replicating results of the NINDS trial with higher

complication rates due to protocol violations further compounding the concerns surrounding hemorrhagic risks [12].

EXTENDED TIME WINDOW TO 4.5 H

A pooled analysis of patients from the five randomized trials demonstrated a benefit of intravenous t-PA to patients treated up to 4.5 h with an odds ratio of 1.4 [14]. This finding laid the foundation for the ECASS III study to assess if the drug was efficacious in the 3–4.5 h time window [15]. This study randomized 821 patients who were under the age of 80, presenting National Institutes of Health Stroke Scale (NIHSS) <25 and absence of prior stroke and diabetes mellitus, and presented between 3 and 4.5 h from last known normal. The study demonstrated a moderate benefit of administration of the drug in this population with 52.4% of patients receiving t-PA having a modified Rankin Scales (mRS) of 0–1% versus 45.2% in the placebo arm ($p < .04$, OR 1.34 [1.02–1.76]) [15]. The investigators found a 7.9% rate of symptomatic hemorrhage when using the NINDS definition in the treatment group. Despite the hemorrhages, patients benefited from the treatment thus prompting the 2013 American Heart Association guidelines to recommend treatment with intravenous t-PA up to 4.5 h if the patient meets the inclusion and exclusion criterion of ECASS III [16].

TIME TO TREATMENT EFFECTS

The pooled analysis by Hacke et al. demonstrated a crucial finding of the importance of time to treatment and its impact on outcomes. Table 136.2 demonstrates the decay in odds ratios for an excellent neurological outcome and increasing rate of symptomatic hemorrhage by each 90 min quartile. The number needed to treat for an excellent outcome in the 0–90 min quartile

Primer on Cerebrovascular Diseases, Second Edition
http://dx.doi.org/10.1016/B978-0-12-803058-5.00136-3

TABLE 136.1   Summary of Randomized Clinical Trials Used to Determine Time Curve for t-PA Administration

Study	Year Started	Time Window (h)	Number of Patients	Median NIHSS t-PA Group	Symptomatic ICH Rate (%)
NINDS Part 1	1991	0–3	291	14	5.6
NINDS Part 2	1991	0–3	333	14	7.1
ECASS I	1992	0–6	620	12	20
ECASS II	1996	3–6	635	11	8.3
ATLANTIS A	1991	0–6	142	10	11.3
ATLANTIS B	1993	0–5	613	10	7.6

TABLE 136.2   Summary of Outcomes and Hemorrhage Rates in Meta-Analysis [14]

Time (min)	Intravenous t-PA, mRS 0–1, N (%)	Placebo, mRS 0–1, N (%)	OR (95% CI) for Favorable Outcome	Symptomatic Hemorrhage (%)
0–90	161 (41)	150 (29)	2.81 (1.75–4.5)	3.1
91–180	302 (43)	315 (30)	1.55 (1.12–2.15)	5.6
181–270	390 (37)	411 (32)	1.40 (1.05–1.85)	5.9
271–360	538 (37)	508 (36)	1.15 (0.90–1.47)	6.9

is 1 in 3 patients. These findings were important as it stressed that delivery of the medication quicker had a higher safety profile and also offered the patient a much higher probability of an excellent recovery. The focus shifted away from if the treatment was efficacious, but more toward how to set up systems of care to rapidly identify patients and provide treatment quickly. The American Heart Association placed a focus on the door to needle time of 60 min as a key metric for stroke centers to achieve. One group demonstrated that from 2003 to 2011 they were able to reduce times for delivery of t-PA from 60 to 20 min by reducing the number of steps and developing a prehospital notification system [17]. The Get With The Guidelines database demonstrated that reducing door to needle times for delivery of t-PA in the hospital was associated with improved outcomes, lower mortality, and lower hemorrhagic complications [18].

LARGE VESSEL OCCLUSION STROKES

Patients suffering from strokes secondary to an occlusion of the middle cerebral artery stem, carotid artery, or basilar artery portend to have high rates of mortality and morbidity. Although natural history studies have been limited in patients without reperfusion, retrospective analysis have suggested poor outcomes for this population. One group showed that of 120 patients with documented large vessel occlusion of the anterior circulation only 25% had a good outcome. The patients

who had a carotid terminus occlusion or proximal M1 segment middle cerebral artery occlusion had an 8% and 13% chance of a good outcome, respectively [19]. Given the abysmal outcomes of this population and prior work showing low recanalization rates with intravenous alteplase in this population, catheter-based treatments or mechanical thrombectomy appeared to be the natural progression to treat this condition.

Unfortunately, the trajectory to show benefit of the treatment took 15 years. In 1999, when PROACT II was published there appeared to be promise of catheter-based approaches for middle cerebral artery occlusion. PROACT II was a randomized controlled study of patients with middle cerebral artery occlusions who were treated with 9 mg of pro-urokinase intra-arterially or intravenous heparin. A total of 180 patients were randomized (121 in the pro-urokinase arm and 59 in the heparin arm). The study showed patients treated with intra-arterial thrombolytics had a significantly higher probability of achieving a Rankin of 0–2 (40% vs. 25%, $p < .04$) with a number needed to treat of 1 in 7 [20]. These positive findings were present despite a 10% symptomatic hemorrhage rate in the treatment group. The FDA did not approve pro-urokinase for this indication despite the trial and the medication was not commercially available in the United States. The American Heart Association guidelines at the time states that it was reasonable to consider intra-arterial t-PA for patients with middle cerebral artery occlusions, but given that the dosage and pharmacological agent used in PROACT II was not available stronger evidence could not be given.

TABLE 136.3   Summary of Positive Randomized Controlled Trials Favoring Thrombectomy for Large Vessel Occlusion

Trial	Number of Patients	Time Window (h)	Good Outcome Thrombectomy Arm %	Good Outcome Medical Arm %	Number Needed to Treat
MR CLEAN	500	0–6	32.6	19.1	1 in 7
ESCAPE	316	0–12	53	29.3	1 in 4
SWIFT PRIME	196	0–6	60	35	1 in 4
EXTEND IA	70	0–6	71	40	1 in 3
THRACE	414	0–6	53	42	1 in 9
REVASCAT	206	0–8	43.7	28.2	1 in 6

MECHANICAL THROMBECTOMY

Given the poor natural history of patients with large vessel occlusion if left untreated and early evidence that recanalization seemed to improve outcomes, the foundation for mechanical devices was set. There have been several iterative changes in technology that allowed for the development of the stent retriever which was used in the six positive randomized controlled studies favoring thrombectomy over medical management. Table 136.3 summarizes the studies and the 1702 patients randomized in these trials [6–11]. All of the trials except EXTEND-IA used noncontrast CT imaging to assess for inclusion into the trial along with CT angiography to demonstrate a vascular occlusion. EXTEND-IA employed perfusion imaging to assess the impact of radiographic patient selection.

The number needed to treat to achieve a good neurological recovery at 90 days (mRS 0–2) ranged from 1 in 3 patients to 1 in 9 patients. Additionally, the symptomatic hemorrhage rate ranged from 0% to 3.6% in five of the trials with one outlier of 7.7% in MR CLEAN. These rates were substantially lower than the earlier published literature and likely reflect advances in operator technique and device iteration. Lastly, successful reperfusion rates defined as >50% of the target territory being perfused ranged from 59% to 88%.

The American Heart Association Guidelines now reflect a level IA evidence for mechanical thrombectomy in patients with large vessel occlusion under 6h from onset. Further trials are ongoing to assess the efficacy of the treatment beyond 6h.

CONCLUSIONS

Intravenous t-PA should be given to patients expeditiously when they present under 3h from symptom onset and meet the inclusion and exclusion criterion. Patients under the age of 80 years presenting in the 3–4.5h time window who also have not had a prior stroke and are not diabetic will have a modest benefit from intravenous t-PA. Mechanical thrombectomy should be offered to patients presenting with a large vessel occlusion of the middle cerebral artery or internal carotid artery territory under 6h from symptom onset. Outcomes for thrombolysis and mechanical thrombectomy are dependent on time and thus strategies to reduce times to treatment are required to ensure the highest rates of good outcomes.

References

[1] Tissue plasminogen activator for acute ischemic stroke. The National Institute of Neurological Disorders and Stroke rt-PA Stroke Study Group. N Engl J Med 1995;333:1581–7.

[2] Hacke W, Kaste M, Fieschi C, Toni D, Lesaffre E, von Kummer R, et al. Intravenous thrombolysis with recombinant tissue plasminogen activator for acute hemispheric stroke. The European Cooperative Acute Stroke Study (ECASS). JAMA 1995;274:1017–25.

[3] Hacke W, Kaste M, Fieschi C, von Kummer R, Davalos A, Meier D, et al. Randomised double-blind placebo-controlled trial of thrombolytic therapy with intravenous alteplase in acute ischaemic stroke (ECASS II). Second European-Australasian Acute Stroke Study Investigators. Lancet 1998;352:1245–51.

[4] Clark WM, Albers GW, Madden KP, Hamilton S. The rtPA (alteplase) 0- to 6-hour acute stroke trial, part A (A0276g): results of a double-blind, placebo-controlled, multicenter study. Thromblytic therapy in acute ischemic stroke study investigators. Stroke 2000;31:811–6.

[5] Clark WM, Wissman S, Albers GW, Jhamandas JH, Madden KP, Hamilton S. Recombinant tissue-type plasminogen activator (alteplase) for ischemic stroke 3 to 5hours after symptom onset: the ATLANTIS study: a randomized controlled trial: Alteplase Thrombolysis for Acute Noninterventional Therapy in Ischemic Stroke. JAMA 1999;282:2019–26.

[6] Berkhemer OA, Fransen PS, Beumer D, van den Berg LA, Lingsma HF, Yoo AJ, et al. A randomized trial of intraarterial treatment for acute ischemic stroke. N Engl J Med 2015;372:11–20.

[7] Goyal M, Demchuk AM, Menon BK, Eesa M, Rempel JL, Thornton J, et al. Randomized assessment of rapid endovascular treatment of ischemic stroke. N Engl J Med 2015;372:1019–30.

[8] Saver JL, Goyal M, Bonafe A, Diener HC, Levy EI, Pereira VM, et al. Stent-retriever thrombectomy after intravenous t-PA vs. t-PA alone in stroke. N Engl J Med 2015;372:2285–95.

[9] Campbell BC, Mitchell PJ, Kleinig TJ, Dewey HM, Churilov L, Yassi N, et al. Endovascular therapy for ischemic stroke with perfusion-imaging selection. N Engl J Med 2015;372:1009–18.

[10] Bracard S, Ducrocq X, Mas JL, Soudant M, Oppenheim C, Moulin T, et al. Mechanical thrombectomy after intravenous alteplase versus alteplase alone after stroke (THRACE): a randomised controlled trial. Lancet Neurol 2016. Epub ahead of print.

[11] Jovin TG, Chamorro A, Cobo E, de Miquel MA, Molina CA, Rovira A, et al. Thrombectomy within 8 hours after symptom onset in ischemic stroke. N Engl J Med 2015;372:2296–306.

[12] Katzan IL, Furlan AJ, Lloyd LE, Frank JI, Harper DL, Hinchey JA, et al. Use of tissue-type plasminogen activator for acute ischemic stroke: the Cleveland area experience. JAMA 2000;283:1151–8.

[13] Grond M, Stenzel C, Schmülling S, Rudolf J, Neveling M, Lechleuthner A, et al. Early intravenous thrombolysis for acute ischemic stroke in a community-based approach. Stroke 1998;29:1544–9.

[14] Hacke W, Donnan G, Fieschi C, Kaste M, von Kummer R, Broderick JP, et al. Association of outcome with early stroke treatment: pooled analysis of ATLANTIS, ECASS, and NINDS rt-PA stroke trials. Lancet 2004;363:768–74.

[15] Hacke W, Kaste M, Bluhmki E, Brozman M, Dávalos A, Guidetti D, et al. Thrombolysis with alteplase 3 to 4.5 hours after acute ischemic stroke. N Engl J Med 2008;359:1317–29.

[16] Jauch EC, Saver JL, Adams Jr HP, Bruno A, Connors JJ, Demaerschalk BM, et al. Guidelines for the early management of patients with acute ischemic stroke: a guideline for healthcare professionals from the American Heart Association/American Stroke Association. Stroke 2013;44:870–947.

[17] Meretoja A, Strbian D, Mustanoja S, Tatlisumak T, Lindsberg PJ, Kaste M. Reducing in-hospital delay to 20 minutes in stroke thrombolysis. Neurology 2012;79:306–13.

[18] Fonarow GC, Smith EE, Saver JL, Reeves MJ, Bhatt DL, Grau-Sepulveda MV, et al. Timeliness of tissue-type plasminogen activator therapy in acute ischemic stroke: patient characteristics, hospital factors, and outcomes associated with door-to-needle times within 60 minutes. Circulation 2011;123:750–8.

[19] Hernández-Pérez M, Pérez de la Ossa N, Aleu A, Millán M, Gomis M, Dorado L, et al. Natural history of acute stroke due to occlusion of the middle cerebral artery and intracranial internal carotid artery. J Neuroimaging 2014;24:354–8.

[20] Furlan A, Higashida R, Wechsler L, Gent M, Rowley H, Kase C, et al. Intra-arterial prourokinase for acute ischemic stroke. The PROACT II study: a randomized controlled trial. Prolyse in acute cerebral thromboembolism. JAMA 1999;282:2003–11.IX. MANAGEMENTIX. MANAGEMENTIX. MANAGEMENT

C H A P T E R

137

General Treatment of Stroke in Intensive Care Setting

A.M. Alkhachroum, M. De Georgia

University Hospitals Case Medical Center, Neurological Institute, Cleveland, OH, United States

INTRODUCTION

Meticulous critical care is key to achieve the best possible outcome after stroke. There has been considerable evolution in our approach to critically ill patients over the recent years, including the establishment of standardized protocols that cover both general critical care and neurocritical care. This chapter provides an overview of the treatment of critically ill patients with ischemic and hemorrhagic stroke in the intensive care unit (ICU).

GENERAL CRITICAL CARE

Blood Pressure Management

Elevated blood pressure (BP) is common in patients presenting with ischemic stroke; about 85% will be hypertensive and almost half will have a BP greater than 160 mm Hg [1]. The underlying mechanism of hypertension in stroke is thought to stem from the cerebral ischemic response, a derivative of the diving reflex. Many studies have demonstrated a U-shaped curve when it

Primer on Cerebrovascular Diseases, Second Edition
http://dx.doi.org/10.1016/B978-0-12-803058-5.00137-5

comes to BP—patients with low pressure, a systolic BP less than 140 mm Hg, have a worse outcome as do those with high pressure, a systolic BP more than 180 mm Hg; however, the exact reasons for this are complex and poorly understood. In patients who have received intravenous tissue plasminogen activator (t-PA), it is clear that the risk of hemorrhagic transformation correlates with high BP. In these patients, the target BP is less than 185/110 mm Hg before t-PA can be administered and less than 180/105 mm Hg for the next 24 h after administration [2]. For patients who have not received thrombolysis, the conventional thinking has been to not treat hypertension unless the BP is very high. In theory, patients with persistent vessel occlusions may be pressure dependent and their condition may worsen if the BP is lowered too much. Recent randomized trials have challenged this view [3]; however, current guidelines recommend only treating hypertension when the systolic BP is greater than 220 mm Hg. In the setting of intra-arterial intervention, there are no guidelines as to what the optimal BP target should be. In general, before revascularization, and in the setting of poor collaterals, maintaining a systolic BP greater than 150 mm Hg to optimize perfusion is reasonable. After revascularization, the goal is to maintain the systolic BP close to normal and consider targeting an even lower pressure (systolic BP less than 140 mm Hg) if risk factors for hemorrhage are present, including diabetes mellitus, prolonged time to puncture, use of intravenous or intra-arterial t-PA, atrial fibrillation, and high NIHSS (National Institutes of Health Stroke Score) [4].

BP management in patients with intracerebral hemorrhage (ICH) has also been controversial. As arterial pressure may be a driving force in hematoma expansion, lowering the BP acutely would seem beneficial. Lowering it too much, however, has raised concerns that patients' condition may worsen from reduced cerebral perfusion, especially in the setting of elevated intracranial pressure (ICP). A recent randomized control trial did demonstrate a benefit from targeting a systolic BP of less than 140 mm Hg [5]. More studies are needed but current guidelines suggest that this approach is safe and can be effective for improving functional outcomes [2].

Respiratory Care

Many patients with severe stroke require intubation because of respiratory failure, mainly because of poor airway control. Impaired oxygenation is also common from aspiration, atelectasis, or pulmonary edema. The FiO_2 should be adjusted to maintain an oxyhemoglobin saturation greater than 94%. In general, a protective ventilator strategy, extrapolated from patients with acute respiratory distress syndrome, is used targeting tidal volumes <6 mL/kg and sufficient positive end-expiratory pressure (PEEP) to prevent alveolar collapse while maintaining low plateau pressures. Excessively high PEEP levels should be avoided, as they may decrease cerebral venous outflow, cardiac output, and BP and result in impaired cerebral perfusion. Patients should be ventilated to maintain a normal $Paco_2$; hyperventilation should be avoided because it can result in cerebral vasoconstriction and reduced cerebral blood flow (CBF).

Sedation and Analgesia

Every attempt should be made to minimize sedation. Minimizing sedation, including daily sedation interruption trials, has been associated with a shorter duration of mechanical ventilation and a shorter ICU length of stay. When sedation is needed, nonbenzodiazepine sedatives (such as propofol or dexmedetomidine) are recommended. Deep sedation with benzodiazepines (such as midazolam or lorazepam) has been linked with increased delirium and worse long-term cognitive outcomes [6]. Narcotics, such as morphine, fentanyl, and remifentanil, should also be considered for analgesia.

Coagulopathy Reversal

In patients with ICH secondary to warfarin-related coagulopathy, rapid correction of the international normalized ratio (INR) is needed. Traditionally, fresh frozen plasma and vitamin K have been used. Recently, prothrombin complex concentrate (PCC), activated PCC FEIBA (factor VIII inhibitor bypassing activity), and recombinant activator factor VIIIa have emerged and may be faster and more effective. For example, PCC does not require crossmatching and it can be reconstituted quickly and administered rapidly in a small volume (20–40 mL), making it attractive for patients with heart failure and volume overload. For patients with hemorrhages secondary to dabigatran, idarucizumab, a monoclonal antibody, is now available. Hemodialysis has also been shown to be effective for removing dabigatran, but less so for rivaroxaban or apixaban. For patients taking antiplatelet agents, although intuitive, there is little evidence that platelet transfusions prevent hemorrhage growth or improve outcome. Finally, for ischemic stroke patients with thrombolysis-related hemorrhagic transformation, administration of cryoprecipitate is recommended to restore diminished fibrinogen levels.

Nutritional Support

Early nutritional support is important in critically ill patients but recent randomized controlled trials have not shown an unequivocal benefit with full-replacement feeding. Rather, it is recommended to initiate hypocaloric gastric feeding, along with providing micronutrients during the first week of critical illness. Managing micronutrition is important to prevent refeeding syndromes in the setting of deficient elements such as thiamine, potassium, and phosphate. Other elements (such as selenium, copper, manganese, zinc, and iron) and vitamins (E and C), and beta-carotene may reduce oxidative cellular damage and organ failure in critically ill patients. Electrolytes must be monitored closely and a hypoosmolar state, which can contribute to cerebral edema in patients with brain injury, avoided. When the stroke is severe and long-term enteral feeding is anticipated, percutaneous endoscopic gastrostomy tube placement should be considered and discussed early in the hospital course.

Fever Control

About a third of patients admitted with ischemic stroke will be febrile within the first 24 h of admission. Fever is also common after ICH, especially in patients with intraventricular hemorrhage. Fever worsens outcome in experimental models of brain injury and has similarly been associated with worse clinical outcomes in both ischemic and hemorrhagic stroke. Therefore, the goal should be maintenance of strict normothermia throughout the ICU stay. Methods of fever control include the liberal use of antipyretic medications (i.e., acetaminophen 650 mg every 4 h) and, if needed, the use of a cooling blanket or endovascular cooling device.

Glycemic Control

Hyperglycemia during the first 24 h is also associated with worse outcomes in both ischemic and hemorrhagic strokes. It is recommended to target blood glucose levels of 140–180 mg/dL [7]. Note that in the setting of traumatic brain injury, very tight glycemic control (less than 110 mg/dL) has been shown to precipitate brain energy crisis, namely, elevated lactate/pyruvate ratios and glutamate levels measured by intracerebral microdialysis. This has been linked with increased mortality. In critically ill patients who have had stroke, caution should be exercised regarding tight glycemic control.

Deep Venous Thrombosis Prophylaxis

Pulmonary embolism (PE) accounts for 10% of deaths after stroke. Emboli generally arise from deep venous thrombi (DVTs) that develop in the paralyzed lower limbs or pelvis of critically ill patients who have had stroke. Use of intermittent pneumatic compression devices and subcutaneous administration of unfractionated or low-molecular-weight heparin prevent DVT and PE. In patients with ICH, it is recommended to start intermittent pneumatic compression on admission and subcutaneous heparin after 24–48 h (once stability of the hematoma has been demonstrated by neuroimaging). When a DVT or PE has occurred, patients are usually made to undergo full anticoagulation with heparin followed by warfarin. Patients in whom anticoagulation is contraindicated may require placement of an inferior vena cava (IVC) filter. There are few guidelines regarding whether to anticoagulate or place an IVC filter in the patient with ICH in whom a DVT or PE has occurred, but considerations include timing (how far out from the hemorrhage the DVT/PE has occurred), cause of the hemorrhage, and documentation of stable hematoma size by neuroimaging.

Stress Ulcer Prophylaxis

Stress-related mucosal damage begins in the proximal stomach within hours of a critical illness. Prophylaxis with either histamine-2 receptor antagonists or proton pump inhibitors is recommended for the duration of the ICU stay; the goal is to target a gastric pH greater than 4. The cytoprotective agent sucralfate is more effective than placebo in reducing bleeding but has been shown to be inferior to histamine-2 receptor antagonists and proton pump inhibitors and is rarely used nowadays. Antacids are not recommended.

Decubitus Ulcer Prophylaxis

Critically ill patients who have had stroke are at high risk to develop decubitus pressure ulcers caused by immobilization, malnutrition, hemodynamic instability, and the use of vasoactive medications. Skin inspection should occur on each shift or more often in patients at high risk. ICU protocols for the prevention of pressure ulcer development should include pressure relief, moisture management, and nutrition support.

NEUROCRITICAL CARE

Intracranial Pressure Management

Patients with large ischemic strokes and ICHs often develop cerebral edema. Cerebral edema increases the intracranial volume. Because the cranium is essentially a fixed vault, any increase in intracranial volume results in an increase in ICP. The pressure–volume relationship approximates an exponential curve with the inflection point ranging from 20 to 25 mm Hg. This forms the theoretical basis for keeping the ICP below 20 mm Hg. Several studies (mainly on head trauma) have validated this threshold by demonstrating that patients with ICP values persistently greater than 20 mm Hg have a higher mortality rate. The singular goal of maintaining the ICP less than 20 mm Hg, however, does not take into consideration the effect of ICP on the cerebral perfusion pressure (CPP). The CPP is the difference between systemic mean arterial pressure and ICP and represents the net pressure gradient across the cerebral microvascular bed. Poor outcome associated with globally increased ICP may, in part, reflect the detrimental effects of low CPP and low CBF to the injured brain resulting in secondary ischemic brain injury. Thus, current guidelines for the management of increased ICP recommend maintaining an ICP <20 mm Hg as well as maintaining a CPP between 50 and 70 mm Hg depending on the status of cerebral pressure autoregulation [8]. Guidelines for ischemic stroke and ICH have generally followed these same thresholds despite less disease-specific data. Finally, both animal and clinical studies have demonstrated that elevations in ICP secondary to mass lesions may not be equally distributed throughout the cranium. Pressure gradients can develop and cause horizontal brain tissue shift and mechanical damage to the diencephalon long before globally increased ICP causes downward vertical shift. Thus, a single ICP number measured ipsilaterally may significantly underestimate the complexity of the intracranial dynamics within the cranium.

In general, patients with increased ICP should be mechanically ventilated with adequate sedation and analgesia. Ventilator settings should be adjusted to maintain normocapnia; prophylactic hyperventilation is not recommended, although brief periods of mild hyperventilation ($Paco_2$ 30–35 mm Hg) can be used to treat acute neurological deterioration caused by increased ICP. The core temperature should be maintained at normothermia to slight hypothermia (36.5–37°C). The mainstay of medical management of increased ICP is osmotic therapy, including administration of mannitol and hypertonic saline. The primary mode of action is thought to be the creation of an osmotic gradient between the systemic circulation and brain compartments across an intact blood–brain barrier, which leads to the efflux of water from the brain into the vascular compartment. This leads to a reduction in intracranial volume and decrease in ICP. Initially this also results in an increase in CPP, which, when autoregulation is intact, contributes to the lowering of ICP through a compensatory decrease in vessel radius and cerebral blood volume. Mannitol 20% is given as an intravenous bolus of 0.25–1 g/kg. This can be repeated as needed for sustained elevations in ICP if serum osmolality is less than 330 mOsm. Mannitol must be used judiciously, however, because the subsequent osmotic diuresis can lead to hypovolemia, decreased cardiac preload, and hypotension, which could compromise CPP, lead to vasodilation, and increase ICP again. In addition, as hyperosmolarity is maintained

over a prolonged period, its benefit decreases with time. When brain tissue is exposed to a serum osmolality of 315–330 mOsm, neurons and glia compensate by accumulating osmotic equivalents or idiogenic osmoles (mainly the amino acids glycine and taurine). Finally, the osmotic agents may accumulate in brain areas where the blood–brain barrier is disrupted. If systemic osmolality decreases, water may reaccumulate in the brain and potentially worsen the ICP. Hypertonic saline may be used instead of mannitol. Dosing includes a 30-mL bolus of 23.4% solution via a central line. Alternatively, a 250-mL bolus of 3% solution (also through a central line) or 2% solution (through a peripheral line) can be given followed by a continuous infusion at the rate of 30–150 mL/h. The target serum sodium is between 150 and 160 mEq/L.

In case of traumatic brain injury, these first-tier therapies are always exhausted before moving onto to second-tier therapies such as decompressive craniectomy and barbiturate therapy. However, in patients with stroke, such as those with malignant middle cerebral artery (MCA) territory infarction and in whom ICP gradients develop, decompressive craniectomy, given its immediate effectiveness, has moved up to be a first-tier therapy (see the following section). Similarly, barbiturate therapy is rarely used nowadays to treat stroke.

DECOMPRESSIVE CRANIECTOMY

Complete MCA territory infarction (the so-called "malignant infarction") is associated with a high mortality rate, approximately 80%, as a result of the development of cerebral edema, ICP gradients, brain tissue shift, and herniation. Decompressive craniectomy, by allowing the brain to swell outwards and equalizing the pressure gradients, can significantly reduce the mortality and improve the functional outcome in survivors. Specifically, according to three randomized trials, compared with maximal medical therapy alone, surgery was associated with a 50% absolute risk reduction (ARR) of death, with more patients surviving with a slight to moderate disability (ARR 23%) or a slight to moderately severe disability (ARR 51%) [9]. The criteria for performing decompressive craniectomy include evidence of a massive MCA territory infarction (>2/3 MCA territory), high NIHSS, reduced level of consciousness (somnolence), generally younger age (60 years or younger), and, perhaps more importantly, good prestroke functional status. Both dominant and nondominant hemispheric strokes should be considered. Surgery should be performed within 48 h of stroke onset.

Surgical decompression also plays an important role in posterior fossa infarcts and hemorrhages because of the risk of hydrocephalus and brainstem compression. The use of serial neurologic examinations and CT scanning enables the clinician to detect mass effect from progressive infarct edema, hematoma expansion, or hydrocephalus. In cerebellar strokes, surgery has a well-defined role. Prompt surgical decompression of the posterior fossa and placement of an intraventricular catheter if hydrocephalus is present can be lifesaving. In patients with supratentorial hemorrhage, conventional surgical clot removal alone has failed to demonstrate a clear benefit over medical management [10]. There is evidence that minimally invasive surgery with imaging-guided catheter insertion and intraclot thrombolysis may be beneficial but further studies are needed.

Seizures

Seizures after ischemic stroke are uncommon (<10%), although they may be more frequent in patients with hemorrhagic transformation. In ICH, on the other hand, the frequency of seizures is as high as 16%, especially in those with cortical hemorrhages, and studies of continuous electroencephalography (cEEG) report electrographic seizures in about 30% of patients with ICH. Nevertheless, prophylactic use of antiseizure medications is not recommended. Clinicians should be vigilant when looking for seizures by cEEG monitoring. Clinical seizures or electrographic seizures should be treated with antiseizure medications.

Goals of Care and Code Status

The goals of care should be determined as soon as possible on arrival to the ICU. Some patients may have clear wishes about goals of care, having expressed them in a living will or verbally to their family or friends. Regardless, it is important to provide every patient with the best possible care and to avoid early prognostication. Clinicians are especially cautioned about the early use of do-not-attempt-resuscitation (DNAR) orders, as studies have shown them to be independent predictors of outcome possibly because such orders may lead to less aggressive care and the creation a self-fulfilling prophecy.

References

[1] Qureshi AI, Ezzeddine MA, Nasar A, et al. Prevalence of elevated blood pressure in 563,704 adult patients with stroke presenting to the ED in the United States. Am J Emerg Med 2007;25(1):32–8.

[2] Jauch EC, Saver JL, Adams Jr HP, et al. Guidelines for the early management of patients with acute ischemic stroke: a guideline for healthcare professionals from the American Heart Association/American Stroke Association. Stroke 2013;44(3):870–947.

[3] Potter JF, Robinson TG, Ford GA, et al. Controlling hypertension and hypotension immediately post-stroke (CHHIPS): a randomised, placebo-controlled, double-blind pilot trial. Lancet Neurol 2009;8(1):48–56.

[4] Nogueira RG, Gupta R, Jovin TG, et al. Predictors and clinical relevance of hemorrhagic transformation after endovascular therapy for anterior circulation large vessel occlusion strokes: a multicenter retrospective analysis of 1122 patients. J Neurointerv Surg 2015;7(1):16–21.

[5] Anderson CS, Heeley E, Huang Y, et al. Rapid blood-pressure lowering in patients with acute intracerebral hemorrhage. N Engl J Med 2013;368(25):2355–65.

[6] Ruokonen E, Parviainen I, Jakob SM, et al. Dexmedetomidine versus propofol/midazolam for long-term sedation during mechanical ventilation. Intensive Care Med 2009;35(2):282–90.

[7] Hemphill 3rd JC, Greenberg SM, Anderson CS, et al. Guidelines for the management of spontaneous intracerebral hemorrhage: a guideline for healthcare professionals from the American Heart Association/American Stroke Association. Stroke 2015;46(7):2032–60.

[8] Bratton SL, Chestnut RM, Ghajar J, et al. Guidelines for the management of severe traumatic brain injury. VIII. Intracranial pressure thresholds. J Neurotrauma 2007;24(Suppl. 1):S55–8.

[9] Vahedi K, Hofmeijer J, Juettler E, et al. Early decompressive surgery in malignant infarction of the middle cerebral artery: a pooled analysis of three randomised controlled trials. Lancet Neurol 2007;6(3):215–22.

[10] Mendelow AD, Gregson BA, Rowan EN, Murray GD, Gholkar A, Mitchell PM. Early surgery versus initial conservative treatment in patients with spontaneous supratentorial lobar intracerebral haematomas (STICH II): a randomised trial. Lancet 2013;382(9890):397–408.

CHAPTER

138

Physiological Monitoring of Stroke in the Intensive Care Setting

R. Sweis, J. Biller

Loyola University Chicago, Stritch School of Medicine, Maywood, IL, United States

Physiological monitoring is the cornerstone of critical care; it allows for identifying hemodynamic instability and assessing response to therapy. Utility of most hemodynamic monitoring remains unproven and rather serves as a trigger for detection of cardiorespiratory instability. Some have argued that utility can only be proven if linked to a treatment protocol improving outcome [1]. Noninvasive physiological monitoring includes electrocardiogram (ECG), pulse oximetry, and arterial blood pressure measurement. Invasive monitoring includes arterial catheterization, central venous catheterization, pulmonary artery catheterization, intracranial pressure (ICP) monitoring, and esophageal Doppler echocardiography [1].

Management of acute ischemic stroke has evolved since 1990s with advances in reducing neuronal injury, post-stroke disability, and death. The advent of dedicated stroke units and neuro-intensive care settings has demonstrated a clear benefit with reduction of mortality independent of age, sex, or stroke severity [2]. Close monitoring of various physiological parameters within these stroke units permits early detection of cerebral edema, blood pressure derangements, pyrexia, hyperglycemia, and oxygenation with expedited care, therefore preventing secondary brain injury. Physiological monitoring of stroke in an intensive care setting is therefore crucial for neuroprotection and decreasing dependency and early mortality [2].

NONINVASIVE MONITORING

Electrocardiogram

Ischemic stroke, specifically of the insular cortex, has been associated with ECG abnormalities. Due to its diffuse interconnections with subcortical autonomic centers, the limbic system, and thalamus, the insular cortex plays a crucial role in autonomic function [3]. Insular infarction and its associated laterality can therefore result in autonomic derangements, specifically cardiac arrhythmias. Insular involvement is reported to be independently associated with heart rate ($\leq$64 beats/min), abnormal repolarization, atrial fibrillation (AF), and ventricular and supraventricular ectopic beats [3]. Right insular involvement has been found to result in ECG abnormalities associated with poor prognosis. When adjusted for age, sex, cardiovascular history, and handicap at admission, as well as lesion side, prolonged QTc interval and left bundle branch block were independent predictors of all-cause and vascular mortality at 2 years in right insular infarctions [3]. Right insular lesions are also significantly associated with 2-year all-cause (hazard ratio, 2.11; 95% confidence interval, 1.27–3.52) and vascular (hazard ratio, 2.00; 95% confidence interval, 1.00–3.93) death when adjusted for age, sex, cardiovascular history, and handicap at admission [3]. Insular involvement or elevated high-sensitivity cardiac troponin-T are also associated with a higher risk of paroxysmal AF [4]. These markers as well as older age (median age of 73 years), history of hypertension, and longer duration of ECG monitoring (3 days) increase the detection rate of unknown AF and may warrant a longer duration of ECG monitoring [4]. Acute insular infarction, rather than atherosclerosis at the carotid bifurcation and aortic arch, also contributes to impaired baroreceptor reflex sensitivity. The left insula predominates in causing baroreflex derangements, presumed to be secondary to parasympathetic outflow modulation [5].

OXIMETRY

Pulse oximetry is one of the most common modalities used for monitoring in the critical care setting. Continuous oximetry monitoring is required in patients

Primer on Cerebrovascular Diseases, Second Edition
http://dx.doi.org/10.1016/B978-0-12-803058-5.00138-7

with ischemic or hemorrhagic strokes in order to prevent secondary brain injury due to hypoxemia. This subset of patients is at risk of hypoxemia due to aspiration, hypoventilation, Cheyne–Stokes respiration, atelectasis, or pulmonary embolism [6,7]. Oxygen content of the brain depends on cerebral blood flow (CBF) and arterial oxygen content. If CBF falls below 10 mL/100 g/min, irreversible damage occurs due to lactic acid production, adenosine triphosphate (ATP) depletion, glutamate release, and sodium and calcium entry into cells progressing to cytotoxic edema and mitochondrial death [6]. Hypoxia also results in anaerobic metabolism, further worsening brain injury [6]. Continuous oximetry targeting safe and acceptable oxygen content is therefore a parameter that can improve outcome in stroke patients. Targeting oxygen saturation >95% has been suggested for patients with stroke [8]. Use of supplemental oxygen for nonhypoxic patients remains controversial. Hyperoxia has been associated with cerebral vasoconstriction and reduced flow, free radical formation during reperfusion, and increased mortality [6,9].

Pulse oximeters have many limitations including insensitivity in detecting hypoxemia in patients with elevated baseline levels of P_aO_2 or in the presence of carboxyhemoglobin or methemoglobin [12]. Intravenous dyes like methylene blue and blue, green, or black nail polish can cause falsely low S_pO_2 levels, whereas falsely elevated levels can be seen with fluorescent and xenon surgical lamps [13]. Motion artifact by far creates the most significant number of falsely low S_pO_2 readings. Motion causes equal absorption of both wavelengths resulting in a saturation of 85% [11]. Low perfusion states during systolic heart failure, vasoconstriction, or hypothermia also creates inaccuracy in S_pO_2 readings as the oximeter sensor is unable to detect a true signal [10].

END-TIDAL CO₂ MONITORING

Although initially used for anesthesia monitoring, capnography is now used to confirm airway patency and lung ventilation in the critical care setting, field resuscitation, conscious sedation, and emergency medicine. It can be very useful in patients with large hemispheric strokes who are intubated or nonintubated. A change in end-tidal CO_2 (etCO_2) reflects a range of pathophysiological states. An increase in etCO_2 can be associated with sepsis, malignant hyperthermia, decreased ventilation, carbon monoxide poisoning, return of circulation following cardiac arrest, and chronic obstructive pulmonary disease [14]. A drop in etCO_2 indicates pulmonary embolus, cardiac arrest, hypothermia, hypometabolic state, low cardiac output, esophageal intubation, and a disconnected ventilator [14]. Capnometry measures fractionated CO_2 (FCO_2) in tidal gas at the airway opening using

infrared absorption or mass spectrometry. Capnography graphically displays FCO_2 versus time [14]. The two main capnometers used in clinical practice are side stream or mainstream. Mainstream capnometers are comprised of an airway adaptor cuvette attached inline to the endotracheal tube in which infrared light source and sensor detect CO_2 absorption and measure PCO_2 [14]. Mainstream analysis is rapid though limited in nonintubated patients. With a side stream capnometer, a sampling line attaches to a T-piece adaptor at the airway opening for nonintubated patients where it analyzes CO_2 [14].

BLOOD PRESSURE

Post stroke, elevated blood pressure may occur due to physiological compensation for cerebral ischemia, pain, elevated ICP, or due to long-standing hypertension [15]. Cerebral autoregulation is impaired such that CBF is dependent on systemic blood pressure. Avoiding hypotension and falls in cerebral perfusion pressure (CPP) is therefore key in preventing infarction of an ischemic penumbra. Normal CBF in an adult is 55–60 mL/100 g/min, but with drops as low as 12 mL/100 g/min, neuronal death occurs [6]. Patients with chronic hypertension develop impaired autoregulation that occurs at higher systemic blood pressures of 120–160 mmHg rather than normal autoregulatory systemic pressures of 60–125 mmHg [6]. At mean arterial pressures (MAPs) less than 70 mmHg, this subset of patients develops brain hypoxia. General consensus recommends withholding antihypertensive medications in the acute stroke period. The American Stroke Association (ASA) recommends treating blood pressure exceeding 220/120 mmHg or MAP > 130 mmHg and 180/105 mmHg if thrombolytics have been administered to avoid hemorrhagic conversion [16]. Presence of hypertensive encephalopathy if discernible from the stroke syndrome, aortic dissection, acute renal failure, acute pulmonary edema, or acute myocardial infarction should also prompt treatment. Alpha and beta adrenergic blockers (labetalol), calcium channel blockers (nicardipine), and angiotensin converting enzyme (ACE) inhibitors (enaliprilat) are preferred over agents such as nitroprusside that can cause cerebral vasodilation [15].

TEMPERATURE MANAGEMENT

Temperature of >37°C has been associated with increased stroke mortality and morbidity. Hyperthermia is often due to a central mechanism in the setting of brain injury or in 20% of the time, due to an infectious complication of stroke such as pneumonia, urinary tract infection, or bloodstream infection. Fever occurs in 25–50% of patients presenting with acute ischemic stroke, and most often occurs with concomitant cerebral edema, midline shift, or hemorrhagic transformation [16]. It is recommended to maintain normothermia with temperature of 36–37°C using antipyretics, antimicrobials, and surface or intravascular cooling devices [17].

WATER HEMOSTASIS

A hyperosmolar hypovolemic state is often present in patients with ischemic strokes due to dysphagia, somnolence, decreased thirst, or concomitant infection resulting in hemoconcentration, increased viscosity, and hypotension, further causing stroke extension or recurrence [19,20]. Because elevated plasma osmolality is associated with worse stroke survival at 3 months, patients should be started on a continuous intravenous saline infusion, with more conservative infusion rates for patients with congestive heart failure [6].

GLYCEMIC CONTROL

Maintaining normoglycemia is crucial in preventing secondary brain injury. It is well validated that diabetic patients suffer of slow recovery and poor outcome post stroke. Hyperglycemia in nondiabetic patients is also associated with poor outcome [21,22]. Diabetic and hyperglycemic patients have chronic impairment of CBF and autoregulation with increased endothelial cell activation, reduced leukocyte, and erythrocyte deformability leading to thrombosis and stroke [6]. Lactic acid production within the ischemic penumbra due to decreased oxidative glucose metabolism causes neuronal, glial, and endothelial cell damage as well as mitochondrial dysfunction, causing edema and impairing microcirculation [23,24]. The recommendation as of 2016 is to avoid hypoglycemia (serum glucose less than 80 mg/dL) and avoid a glucose level >180 mg/dL [16].

INVASIVE MONITORING

Arterial Catheterization

Indwelling arterial catheterization for continuous hemodynamic monitoring has widespread use in the critical care setting. It is an easy and convenient bedside procedure that allows for serial blood gas analysis and routine blood sampling if necessary. The most commonly catheterized artery for hemodynamic monitoring is the radial artery, because it is associated with low complication rates and is easily accessible. Common complications include temporary arterial occlusion with a reported incidence of about 20% [25]. Other

major complications include pseudoaneurysm and sepsis occurring in 0.09% and 0.13% of cases, respectively [25]. The femoral artery is the second most commonly cannulated artery. It has a lower incidence of occlusion due to larger vessel diameter. Pseudoaneurysm and sepsis incidence rates were 0.3% and 0.44%, respectively [25]. Blood pressure curve from femoral arterial catheters is considered more accurate and reflects an estimate of aortic blood pressure. The axillary artery is rarely catheterized due to proximity to the internal carotid artery and concern of thromboembolism to the brain. Other arteries cannulated include the dorsalis pedis and brachial arteries with minimal complications, though few studies are available to report an incidence rate of complications [25].

PULMONARY ARTERY CATHETERIZATION

Until 2014, Swan–Ganz or pulmonary arterial catheters (PACs) were considered the gold standard of measuring central hemodynamics in critically ill and high-risk surgical patients worldwide. PACs have been criticized since 2015 due to their invasive nature, technical artifacts, and misinterpretation of PAC values especially in the setting of shunt or valvular disorders. Misinterpretation of PAC tracings by experienced clinicians remains very common [26]. Transesophageal echocardiography (TEE) is progressively being used as a noninvasive alternative to PACs in critically ill patients with hemodynamic instability. Similarly to PACs, TEE also provides measurements of central venous pressure, pulmonary arterial pressure, pulmonary capillary wedge pressure, pulmonary vascular resistance, stroke volume, cardiac output, and systemic vascular resistance [27].

CEREBROVASCULAR PHYSIOLOGY AND INTRACRANIAL PRESSURE MONITORING

ICP monitoring is common in patients with severe brain injury including traumatic brain injury (TBI), intracerebral hemorrhage (ICH), and subarachnoid hemorrhage (SAH), since persistently elevated ICP has been associated with poor clinical outcome. Despite increased use of ICP monitoring in patients with ischemic stroke, its value remains uncertain [28]. ICP is defined by the Monroe–Kellie doctrine stating that ICP is the sum of the pressures exerted by blood, tissue, and CSF in the intracranial vault. The cranium can accommodate up to an additional 150 mL of volume before ICP begins to rise. Compensatory mechanisms such as collapse of low-pressure veins and CSF egress from the cranial to the spinal subarachnoid space allow for a logarithmic compliance curve [29]. Elastance is the reciprocal of cerebral compliance. Once the compensatory mechanisms are exhausted, additional volume results in sharp rises in ICP [29].

CBF is a crucial physiological variable in brain injury patients as neuronal survival depends on metabolic demand and substrate. CBF depends on CPP defined as MAP minus ICP. CBF in adults is normally 50 mL/100 g/min and determined by blood viscosity, CPP, and vessel radius. CBF is tightly coupled to metabolism termed metabolic-flow coupling. Lowering metabolism with mechanisms such as sedation and hypothermia decreases CBF and therefore ICP [30]. Cerebral autoregulation is the ability of the cerebral circulation to maintain constant blood flow in the setting of changing perfusion pressures. Below the autoregulatory plateau of 60–160 mmHg, vasodilation is insufficient and ischemia ensues [31]. Above this plateau, increased intraluminal pressure dilates arterioles causing luxury perfusion with resulting disruption of the blood–brain barrier and development of edema. Autoregulation, which can be global, focal, or multifocal is impaired by hypoxia, ischemia, TBI, ICH, and SAH and causes CBF to passively follow arterial pressure [31]. CPP should be maintained between 50 mmHg and 70 mmHg to avoid cerebral ischemia and secondary brain injury [30].

CONCLUSION

Stroke is the second most common cause of mortality in the developed world with 5.5 million deaths worldwide [32]. The rise of an older population will result in an increase in stroke prevalence and mortality. Interventions for acute ischemic stroke or primary ICH halt stroke progression, decrease disability, and increase survival. Targeting the physiological monitoring of blood pressure, oxygen saturation, blood glucose, body temperature, and ICP are key elements in decreasing stroke-associated mortality. Continuous physiological monitoring allows for prompt medical interventions by nursing and physicians, improving stroke outcome.

References

[1] Pinsky MR. Hemodynamic evaluation and monitoring in the ICU. Chest 2007;132:2020–9.
[2] Stroke Unit Trialists' Collaboration. Organized inpatient (stroke unit) care for stroke. Cochrane Database Syst Rev 2007;(4):CD000197.
[3] Abboud H, Berroir S, Labreuche J, et al. Insular involvement in brain infarction increases risk for cardiac arrhythmia and death. Ann Neurol 2006;59:691–9.
[4] Scheitz JF, Erdur H, Haeusler KG, et al. Insular cortex lesions, cardiac troponin, and detection of previously unknown atrial fibrillation in acute ischemic stroke: insights from the troponin elevation in acute ischemic stroke study. Stroke 2015;46:1196–201.

[5] Sykora M, Diedler J, Rupp A, et al. Impaired baroreceptor reflex sensitivity in acute stroke is associated with insular involvement, but not with carotid atherosclerosis. Stroke 2009;40:737–42.
[6] Bhalla A, Wolfe CDA, Rudd AG. Management of acute physiological parameters after stroke. Q J Med 2001;94:167–72.
[7] Walshaw MJ, Pearson MG. Hypoxia in patients with acute hemiplegia. Br Med J 1984;228:15–7.
[8] Williams AJ. Assessing and interpreting arterial blood gases and acid base balance. Br Med J 1998;317:1213–6.
[9] Kety SS, Schmidt CF. The effects of altered tensions of carbon dioxide and oxygen on cerebral blood flow and cerebral oxygen consumption of normal young men. J Clin Invest 1984;27:484–92.
[10] Jubran A. Pulse oximetry. Crit Care 1999;3(2):11–7.
[11] Terry JB, Hanley DF. Physiologic monitoring of stroke in intensive care settings. Primer Cerebrovasc Dis 1997;166:678–84.
[12] Barker SJ, Tremper KK. Pulse oximetry: applications and limitations. Int Anesthesiol Clin 1987;25:155–75.
[13] Amar D, Neidzwski J, Wald A, et al. Fluorescent light interferes with pulse oximetry. J Clin Monit 1989;5:135–6.
[14] Anderson CT, Breen PH. Carbon dioxide kinetics and capnography during critical care. Crit Care 2000;4(4):207–15.
[15] Caulfield AF, Wijman C. Critical care of acute ischemic stroke. Crit Care Clin 2007;22:581–606.
[16] Coplin WM. Critical care management of acute ischemic stroke. Continuum 2012;18(3):547–59.
[17] De Keyser J. Antipyretics in acute ischemic stroke. Lancet 1998;352:6–7.
[18] Deleted in review.
[19] Harrison MJG. The influence of hematocrit in the cerebral circulation. Cerebrovasc Brain Metab Rev 1989;1:55–67.
[20] Yamaguchi T, Minematsu K. General care in acute stroke. Cerebrovasc Dis 1997;7(Suppl. 3):12–7.
[21] Fuller JH, Shipley MJ, Rose G, et al. Mortality from coronary heart disease and stroke in relation to degree of glycemia: the Whitehall study. Br Med J 1983;28:867–70.
[22] Asplund K, Hagg E, Helmers C, et al. The natural history of stroke in diabetic patients. Acta Med Scand 1980;207:267–75.
[23] Dietrich WD, Alonso O, Busto R. Moderate hyperglycemia worsens acute blood–brain barrier injury after forebrain ischemia in rats. Stroke 1993;24:111–6.
[24] Sulter G, Elting JW, De Keyser J. Increased serum neuron specific enolase concentrations in patients with hyperglycemic cortical ischemic stroke. Neurosci Lett 1998;253:71–3.
[25] Scheer BV, Perel A, Pfeiffer UJ. Clinical review: complications and risk factors of peripheral arterial catheters used for hemodynamic monitoring in anesthesia and intensive care medicine. Crit Care 2002;6(3):1–7.
[26] Petters Y, Bernards J, Mekeirele M, et al. Hemodynamic monitoring: to calibrate or not to calibrate? Part 1-calibrated techniques. Anesthesiol Intensive Ther 2015;47(5):487–500.
[27] Meersch M, Schmidt C, Zarbock A. Echophysiology: the transesophageal echo probe as a noninvasive Swan-Ganz catheter. Curr Opin Anesthesiol 2016;29:36–45.
[28] Schwab S, Aschoff A, Spranger M, et al. The value of intracranial pressure monitoring in acute hemispheric stroke. Neurology 1996;47(2):393–8.
[29] Levine J, Kumar M. Traumatic brain injury. Pract Neurocrit Care 2015:137–62. [Chapter 7].
[30] Mangat HS. Severe traumatic brain injury. Contin: Crit Care Neurol 2012:532–46.
[31] Torbey MT. Neurocritical care. New York: Cambridge University Press; 2010.
[32] Jones SP, Leathley MJ, McAdam JJ, et al. Physiological monitoring in acute stroke: a literature review. J Adv Nurse 2007;60(6):577–94.

CHAPTER

139

Hypertensive Encephalopathy

B. Bar

Loyola University Chicago, Stritch School of Medicine, Maywood, IL, United States

INTRODUCTION

Hypertensive encephalopathy is an acute syndrome characterized by acute hypertension (HTN) associated with various neurological symptoms and characteristic neuroimaging findings. The common clinical signs and symptoms of the condition include encephalopathy, seizures, headache, visual disturbances, and focal neurological deficits [1]. Hypertensive encephalopathy is an important clinical syndrome to recognize since early recognition and treatment can lead to good clinical outcomes. We now have an ever-increasing choice of antihypertensive medications that are easily titratable at our disposal. It is difficult to discuss hypertensive encephalopathy without discussing posterior reversible encephalopathy syndrome (PRES). PRES is a syndrome characterized by acute neurological symptoms associated with typically posterior vasogenic edema that is often reversible if the underlying etiology is treated rapidly. As will become clearer later on in this chapter it is now abundantly clear that PRES is not always posterior and also not always reversible, but for now this remains the most commonly used terminology to describe this syndrome. Since the publication of the prior issue of this book there has been increasing recognition that hypertensive encephalopathy represents one of the many manifestations of PRES. I therefore focus on important aspects regarding pathophysiology, imaging characteristics, and treatment as it relates to PRES and hypertensive encephalopathy.

PATHOPHYSIOLOGY

A brief overview of normal cerebral autoregulation is warranted prior to a discussion of pathophysiology. The human brain is extremely sensitive to decreases in cerebral blood flow (CBF) and therefore multiple mechanisms exist to ensure adequate CBF during varying physiological conditions. This has been termed cerebral autoregulation and occurs via changes in the diameter of the cerebral arteriolar wall. In response to increased cerebral perfusion pressure (CPP) the arterioles constrict to maintain a steady CBF and the converse is true in conditions of decreased CPP. In humans, the lower limit of autoregulation is about 40–60 mmHg mean arterial pressure (MAP) with an upper limit of 150–160 mmHg MAP [2]. It has been demonstrated that the autoregulation-induced alterations in CBF are mediated and modulated by several mechanisms such as cerebral myogenic vasomotor responses, arterial carbon dioxide tension, arterial oxygen tension, cerebral metabolism, and neurogenic control [3]. Directly relevant to hypertensive encephalopathy, both direct innervation of the cerebral arteries from cervical ganglia and stimulation of adrenergic receptors by circulating sympathomimetics prevent sudden increases of CBF associated with HTN and hypercapnia [3].

Our understanding of the exact pathophysiological changes that occur during PRES is incomplete. One theory is that sudden increases in arterial blood pressure overwhelm the compensatory autoregulatory mechanisms that exist. When the pressure rise is rapid and severe, the autoregulatory response might be insufficient, hyperperfusion can occur, and the blood–brain barrier breaks down, allowing the interstitial extravasation of plasma and macromolecules [4]. It is speculated that the reason for the posterior predominance of PRES is secondary to decreased sympathetic innervation of the cerebral vasculature in the posterior fossa. This theory is supported by the observation that 80–85% of patients have elevated blood pressure at the time of symptom onset [5]. However,

Primer on Cerebrovascular Diseases, Second Edition
http://dx.doi.org/10.1016/B978-0-12-803058-5.00139-9

PRES cannot be explained solely on the basis of severe acute HTN or sudden BP surges, and endothelial and blood–brain barrier dysfunction are likely key to the development of PRES [5]. An earlier alternative theory posits that vasoconstriction producing hypoperfusion could be responsible for the development of PRES. According to this theory, vasoconstriction secondary to evolving HTN and autoregulatory compensation leads to reduced brain perfusion, ischemia, and subsequent vasogenic edema [6]. Evidence to support this theory includes multiple studies showing that about 20–30% of patients who develop PRES are normotensive [6]. A variety of host factors including sepsis, transplantation, toxemia of pregnancy, autoimmune disease, chemotherapy, to name a few are beyond the scope of this chapter and have been found to be associated with the development of PRES. A commonality among these various conditions is immune activation and endothelial injury [6]. The clinical characteristics and imaging findings in these normotensive patients mirror the findings in patients who have hypertensive encephalopathy and PRES. Treatment of the underlying inciting factor in these patients also leads to clinical and radiological resolution in these patients. Ultimately, activation and dysfunction of the vascular endothelium underlie PRES, irrespective of whether the trigger is abrupt HTN of circulating chemical factors released during systemic illness [4].

IMAGING CHARACTERISTICS

It is important to recognize the typical imaging characteristics of hypertensive encephalopathy while also acknowledging that there are patients who present with hypertensive encephalopathy without radiological abnormalities. Brain CT can show areas of vasogenic edema, however MRI is much more sensitive for the typical findings observed in hypertensive encephalopathy. The typical MRI findings are areas of vasogenic edema predominantly in the parieto-occipital regions of both hemispheres [4]. The subcortical white matter is always involved and the best imaging sequence to observe these findings is on axial T2 fluid-attenuated inversion recovery (FLAIR) sequences. Besides the dominant parieto-occipital pattern, two other common patterns have been described as shown in Fig.139.1 [4]. Hemorrhage is a known complication of PRES and occurs in about 15% of patients [7]. Patients who develop hemorrhage due to hypertensive encephalopathy are an example of a group of patients who do not have complete resolution of their clinical and radiological abnormalities. One pattern of hemorrhage is a sulcal subarachnoid hemorrhage pattern which is seen in Fig.139.2 [7]. Restricted

diffusion has also been found in patients with PRES which interestingly resolves on follow-up imaging (Fig.139.3) [8]. PRES has been found to involve the brain stem, and multiple cases of PRES with involvement of the spinal cord [9] have been reported as well. Therefore, PRES ultimately is a clinical diagnosis that can be supported by imaging findings. Awareness of the various atypical radiological findings that can occur with PRES can reassure the clinician when the clinical picture and course is consistent with hypertensive encephalopathy.

MANAGEMENT OF HYPERTENSIVE ENCEPHALOPATHY

As of 2016 there are no randomized clinical trials to help guide the clinician in the management of hypertensive encephalopathy, and it is unlikely that any will be conducted. The initial goal in treating patients is to reduce blood pressure by 25% within the first few hours [10]. It is suggested that patients have close hemodynamic monitoring preferably with an arterial line to allow careful titration of antihypertensive medications. There are multiple titratable antihypertensive medications that can be given intravenously via a pump including labetolol, esmolol, nicardipine, fenoldopam, and clevidipine. These medications allow for a more controlled lowering of blood pressure thus decreasing the possibility of pronounced fluctuations of blood pressure, which should be avoided. The management of seizures with antiepileptic medications is done in a fashion that is similar to other disorders. Patients with PRES who are treated aggressively and early have a good prognosis with about 75–90% having complete recovery [4]. Patients who develop complications including hemorrhage, infarction, and significant cerebral edema can have permanent neurological deficits.

CONCLUSION

Hypertensive encephalopathy is characterized by acute HTN leading to a clinical syndrome with characteristic radiological findings. It is commonly recognized that hypertensive encephalopathy represents one of the etiologies of PRES. It is critical to recognize this syndrome early since this allows for rapid treatment and a good overall prognosis. The underlying pathophysiology leading to PRES is incompletely understood. Fortunately, the treatment of hypertensive encephalopathy is relatively straightforward and most patients make a complete recovery.

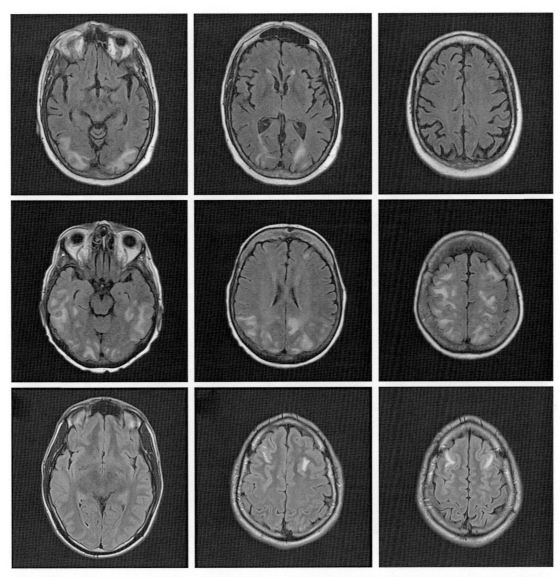

FIGURE 139.1 **Three radiological patterns of posterior reversible encephalopathy syndrome.** The dominant parieto-occipital pattern describes the involvement of only posterior brain (A–C). The holohemispheric watershed pattern (D–F) describes the predominant involvement of the border zone between the anterior and middle cerebral artery territories in a linear pattern spanning the frontal, parietal, and occipital lobes. The superior frontal sulcus pattern (G–I) manifests as more isolated involvement of the superior frontal ulcus without extension into the frontal pole.

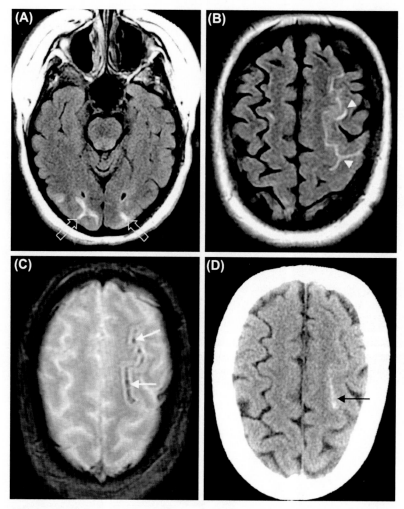

FIGURE 139.2 A 50-year-old woman with fever and severe hypertension. (A and B) FLAIR MR image demonstrates sulcal signal abnormality and PRES vasogenic edema in the left frontal lobe (*arrowheads*) and edema in the occipital lobes bilaterally (*open arrows*). (C) Gradient MR image demonstrates linear low signal intensity consistent with sulcal subarachnoid hemorrhage (*arrows*). (D) CT image demonstrates high attenuation consistent with the MR imaging appearance, further confirming the sulcal subarachnoid hemorrhage (*arrow*).

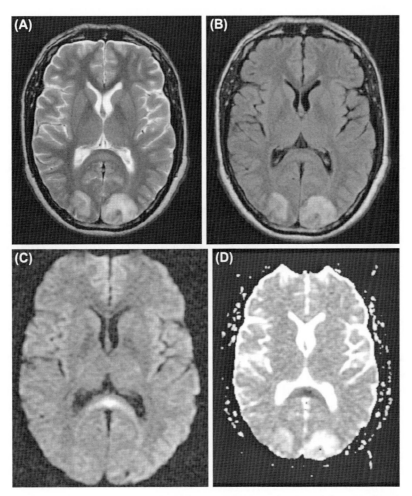

FIGURE 139.3   A 30-year-old male with uncontrolled hypertension presenting with a 2-week history of paroxysmal headaches and visual changes. His blood pressure was 198/120 mmHg on presentation. Brain MRI with (A) T2 and (B) fluid-attenuated inversion recovery sequence demonstrated white matter vasogenic edema in the occipital lobes. (C) Diffusion-weighted imaging and (D) apparent diffusion coefficient map demonstrated a transient splenial lesion, a region of reversible restricted diffusion in the splenium of the corpus callosum. This resolved on subsequent imaging (not shown).

References

[1] Li Y, Gor D, Walicki D, et al. Spectrum and potential pathogenesis of reversible posterior leukoencephalopathy syndrome. J Stroke Cerebrovasc Dis 2012;21(8):873–82.

[2] Kontos HA, Wei EP, Navari RM, Levasseur JE, Rosenblum WI, Patterson Jr JL. Responses of cerebral arteries and arterioles to acute hypotension and hypertension. Am J Physiol 1978;234(4):H371–83.

[3] ter Laan M, van Dijk JM, Elting JW, Staal MJ, Absalom AR. Sympathetic regulation of cerebral blood flow in humans: a review. Br J Anaesth 2013;111(3):361–7.

[4] Fugate JE, Rabinstein AA. Posterior reversible encephalopathy syndrome: clinical and radiological manifestations, pathophysiology, and outstanding questions. Lancet Neurol 2015;14(9):914–25.

[5] Rabinstein AA, Mandrekar J, Merrell R, Kozak OS, Durosaro O, Fugate JE. Blood pressure fluctuations in posterior reversible encephalopathy syndrome. J Stroke Cerebrovasc Dis 2012;21(4):254–8.

[6] Bartynski WS. Posterior reversible encephalopathy syndrome, part 2: controversies surrounding pathophysiology of vasogenic edema. AJNR Am J Neuroradiol 2008;29(6):1043–9.

[7] Hefzy HM, Bartynski WS, Boardman JF, Lacomis D. Hemorrhage in posterior reversible encephalopathy syndrome: imaging and clinical features. AJNR Am J Neuroradiol 2009;30(7):1371–9.

[8] Stevens CJ, Heran MK. The many faces of posterior reversible encephalopathy syndrome. Br J Radiol 2012;85(1020):1566–75.

[9] de Havenon A, Joos Z, Longenecker L, Shah L, Ansari S, Digre K. Posterior reversible encephalopathy syndrome with spinal cord involvement. Neurology 2014;83(22):2002–6.

[10] Hypertension EETFftMoA. 2013 Practice guidelines for the management of arterial hypertension of the European Society of Hypertension (ESH) and the European Society of Cardiology (ESC): ESH/ESC Task Force for the Management of Arterial Hypertension. J Hypertens 2013;31(10):1925–38.

CHAPTER

140

Management of Cerebral Edema/Intracranial Pressure in Ischemic Stroke

M.M. McDowell, A.F. Ducruet, R.M. Friedlander

University of Pittsburgh, Pittsburgh, PA, United States

INTRODUCTION

Stroke is the fifth leading cause of death in the United States, resulting in greater than 125,000 deaths annually [1,1a]. Large brain infarctions are associated with edema and, in severe cases, shift of intracranial contents and elevation of intracranial pressure (ICP). Depression of consciousness and other signs of brainstem compression represent symptoms concerning for severe cerebral edema and can be correlated to imaging findings of midline shift and herniation. High ICP from excessive cerebral edema, if untreated, may result in severe morbidity or mortality. In this chapter, we discuss the pathophysiology, assessment, and management of cerebral edema in ischemic stroke.

PATHOPHYSIOLOGY

Ischemic infarction results in the accumulation of toxic metabolites in the setting of breakdown of neural tissue regulation and loss of structural integrity in the surrounding architecture due to depletion of energy sources. The inability to power ATP sodium-potassium pumps results in neuronal depolarization, increase in intracellular calcium levels, the generation of free radicals, and cell death. The failure to reduce intracellular sodium concentrations results in the osmotic accumulation of water. Apoptosis and necrosis ensue hours after ischemic onset and progress over the course of days, resulting in neuronal injury extending into the subacute period. Apoptosis occurs predominantly in the periphery of ischemia known as the penumbra through two major pathways, namely, the intrinsic and extrinsic apoptotic pathways, mediated by cellular and extracellular signals of irreversible injury, respectively. Both pathways serve

to damage mitochondrial membranes and activate caspases, the proteins responsible for autolysis via promoting DNA cleavage [2]. Irreversible ischemic injury in the core area of infarction is mediated by necrosis caused by the inability to maintain even the basic regulatory mechanisms necessary for apoptosis due to energy depletion. Depletion of oxygen, glucose, and other energy substrates results in rapid transition to anaerobic glycolysis and acidosis from lactate production. Acidosis enhances free radical development and intracellular protein injury crucial to cellular homeostasis and mitochondrial integrity. Mitochondrial membrane breakdown results in further free radical formation. Free radicals react with phospholipid membranes while simultaneously degrading DNA bonds and preventing cellular repair and regulation. High concentrations of free radicals result in a localized inflammatory response and the activation of the innate immune system through the release of adhesion molecules. Neutrophils, macrophages, and monocytes accumulate in the ischemic area, worsening local ischemia by simultaneously increasing local energy requirements and by obstructing blood flow. Cells not succumbing to necrosis are thus subjected to a proapoptotic environment beyond the hyperacute period [3].

In a process distinct from, but parallel to, neuronal cell death, blood–brain barrier (BBB) breakdown causes the accumulation of cellular and plasma proteins in the interstitial tissues. The basal membrane of the BBB, which is composed of collagen type IV, heparin sulfate proteoglycan, fibronectin, laminin, and other proteins, is cleaved by the matrix metalloproteinases [3]. In the setting of postischemic reperfusion, hydrostatic pressure gradients further encourage water migration into the extravascular space, contributing to vasogenic edema in tissues already undergoing cytotoxic edema. As edema progresses, macroscopic swelling appears, compressing the adjacent structures until the volume of the brain tissue exceeds the intracranial volume, thus increasing ICP. Given the typically unilateral injury in acute stroke, this may additionally cause midline shift into the territory of the contralateral brain, herniation of critical tissues, compression of blood vessels, and ultimately additional infarction [4].

The most frequent occluded vessel that results in malignant intracranial hypertension is the middle cerebral artery (MCA). As the largest branch of the internal carotid artery (ICA), the MCA is frequently occluded by emboli from either carotid plaque or cardiac thromboembolism because of its higher rate of flow and ability to accommodate thrombi too large to enter the anterior cerebral artery (ACA). The MCA supplies the largest cerebral territory and thus its occlusion can result in a large infarction and the development of edema over a larger territory, which is directly related to the risk of progression to intracranial hypertension [5]. ICA occlusion likewise poses an increased risk of malignant edema. ACA occlusion may also occur in isolation or in the setting of terminal ICA occlusion; however, contralateral blood supply from the contralateral ACA often preserves tissue from as great a risk of infarction [5].

ASSESSMENT OF CEREBRAL EDEMA

Clinical Features

The most concerning feature of cerebral edema progression after a supratentorial stroke is a decline in the level of consciousness and ultimately uncal herniation causing cranial neuropathies and depression of respiration. Depression of consciousness is typically attributable to the compression of the thalamus and brainstem, where the centers of arousal are located [5]. Combined MCA and ACA infarction may reduce a patient's apparent level of consciousness because of apraxia of eyelid opening and aphasia, so radiographic correlation is recommended to determine if the patient is truly symptomatic from mass effect. Cerebellar infarctions with worsening edema may be identified by worsening cerebellar signs such as dizziness, vertigo, vomiting, and the decreasing ability to coordinate speech, gait, and eye movements in contrast to the frequent motor deficits seen in supratentorial injury. The proximity to the brainstem and the fourth ventricular outflow tract for cerebrospinal fluid necessitate vigilance when observing for decreased arousal, development of cranial nerve neuropathies such as ocular palsy, decreased respiratory drive, and signs of acute hydrocephalus. All these signs may present rapidly and, if not detected early, can rapidly cause permanent injury. Neurological deterioration

attributable to worsening cerebral edema in any territory and increased ICP typically occurs within 72–96 h, but can occur up to 10 days postictus [6].

Imaging

Early imaging in acute stroke relies heavily on head computed tomographic (CT) scanning and brain magnetic resonance imaging (MRI). MRI is capable of detecting early cytotoxic edema approximately 30 min postinsult using diffusion weighted imaging (DWI) sequences (Fig. 140.1). Early edema formation can be detected by decreased free motion in the extracellular space, a direct result of sodium-potassium pump dysfunction and a net flow of fluid from the extracellular interstitial space into the ischemic cells [4]. This earliest stage of infarction rarely manifests in severe ICP changes but can be predictive of future progression. After 3–6 h, DWI infarct volume greater than 82 cc is predictive of malignant edema requiring operative management because of the lack of response to medical treatment with a specificity of 87% and sensitivity of 91% [7].

CT scans allow for rapid determination of mass effect progression from cerebral edema, such as midline shift greater than or equal to 5 mm or frank uncal, tonsillar herniation, or subfalcine herniation (Fig. 140.2). Serial CT scans are recommended in the early phase of treatment of patients with large territory infarctions [5].

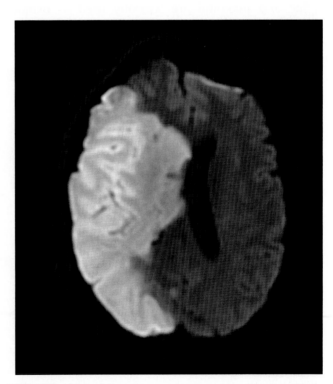

FIGURE 140.1 Magnetic resonance imaging of brain: diffusion weighted imaging sequence demonstrates a large right middle cerebral artery territory infarct.

Noninterventional Management of Intracranial Pressure

As discussed in the Clinical Features section, the most reliable clinical sign of worsening cerebral edema and developing ICP crisis is depression of consciousness. With diminished arousal there is a decrease in the patients' ability to protect their airway and maintain adequate levels of respiration. Rapid sequence intubation with the goal of maintaining normal levels of PaO_2 and $PaCO_2$ is frequently required because of the potential for rapid decline and the fact that persistent hypoxia may place further brain tissue at risk for infarction.

According to the Monro–Kellie hypothesis, the intracranial volume is fixed, and thus a change in the volume of parenchymal tissue results in an inverse change in blood volume or cerebrospinal fluid volume, or both [8]. A decrease in oxygenated blood flow may result if cerebral edema causes sufficient compression of vascular structures, which increases the resistance to blood flow as the diameter of the vessels decreases. Maintaining an adequate arterial pressure is thus necessary to prevent further ischemia, particularly as ICP from cerebral edema rises. Intravenous crystalloids and colloids are the first step to maintain cerebral perfusion by increasing intravascular volume [5]. Excessive hypertension should be avoided due to the risk of hemorrhage into infarcted tissue [5]. Hypertonic saline and mannitol are typically used as osmotic agents to reduce cellular fluid content in patients with acute clinical symptoms concerning for progression of cerebral edema or, at some centers, in patients who are at high risk of progressing to malignant intracranial hypertension. Empiric use of osmotic agents in the absence of symptoms is not recommended. Patients in critical condition due to ICP crisis or impending herniation are often treated with a mannitol bolus (1 g/kg), but this may only serve to delay herniation without surgical intervention. Hypertonic boluses ranging from 3% to 23% are also frequently used. Other measures include elevation of the head to 30 degrees and attention to optimizing venous outflow through careful head positioning. A barbiturate-induced coma may also be attempted, but evidence is lacking and inducing a reduction in consciousness diminishes the ability to monitor a patient's clinical state [5].

Surgical Management of Elevated Intracranial Pressure and Intracranial Monitoring

The possibility of clinical decline unresponsive to medical management of cerebral edema requires careful discussion with the patient and/or the power of attorney regarding surgical options. Placement of intracranial monitoring devices and external ventricular drains (EVDs) allow for direct measurement of ICP and drainage of cerebrospinal fluid so that there is a greater potential volume for expansion of edematous tissue [8]. Early placement of monitoring devices or EVDs is not recommended in the absence of clinical deterioration even

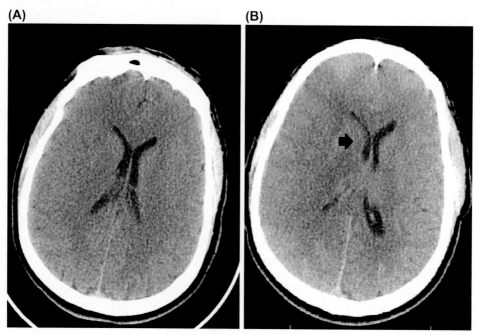

FIGURE 140.2 (A) Computed tomography of the head demonstrating subtle right middle cerebral artery territory infarct in early stroke period. (B) Interval scan 6 h later demonstrating worsening sulcal effacement and right to left shift of right basal ganglia resulting in effacement of lateral ventricle and shift of midline structures (*arrow*).

with radiographic evidence of cerebral edema and mass effect, as a benefit of routine monitoring in outcome is controversial and these surgical procedures carry an intrinsic risk of hemorrhage and postprocedural infection [9]. Patients with cerebellar infarct may develop hydrocephalus prior to brainstem compression and may benefit from ventriculostomy in the absence of symptoms attributable to brainstem compression [5]. In general, though, concomitant suboccipital decompression is recommended to avoid the risk of upward transtentorial herniation.

Decompressive hemicraniectomy for cerebral infarctions or suboccipital decompression for cerebellar infarctions is indicated in select cases with malignant brain edema, particularly in patients with a high-level functional independence prior to presentation. Both the surgical procedures are performed by removal of a portion of the skull, opening of the dura, and in some cases resection of infarcted brain tissue followed by closing of the skin without bone flap replacement in the acute phase to increase the volume of space available for parenchymal swelling. It is important to maximize the bony and dural opening and provide an expansile duraplasty to maximize the benefit of this procedure. Because of the surgical nature of the intervention, the data available are heterogeneous as to the exact indications and concomitant therapy that is ideal for a given patient, but as a whole, a clear reduction in mortality exists in patients whose condition deteriorates despite maximum medical management. A crucial issue to emphasize to the patient and the family is that functional independence may not be achieved in many cases and decompression is likely most beneficial in patients younger than 60 years [5]. However, recent data from the DESTINY II trial suggests that patients aged from 61 to 82 years can benefit to an extent from decompression in the setting of severe cerebral edema. In this study, 38% of patients who underwent hemicraniectomy within 48h recovered independence in some activities of daily living compared to 18% of patients who were solely medically managed, but no patients in either group survived with the ability to live independently in all activities of daily living [10]. A detailed discussion regarding a patient's beliefs as to what level of independence would result in a meaningful quality of life is indicated. The ideal time to perform decompression is unclear, but decompression before clinical signs of brainstem compression appear is generally recommended [5]. Some centers have suggested even more aggressive measures such as prophylactic decompression in young patients with very large infarct territories or even prophylactic resection of infarcted tissues [11]. After cerebral edema has resolved, the bone is replaced in a delayed fashion.

Mechanical Thrombectomy for Large-Vessel Occlusive Stroke

Mechanical thrombectomy in the setting of large-vessel occlusion has been conclusively demonstrated to improve outcome. Select patients undergoing effective mechanical revascularization demonstrate significant reduction in ischemic injury and the resulting cerebral edema, who are otherwise destined for malignant infarction. The use of endovascular clot retrieval has been demonstrated to be effective alone and in combination with tissue plasminogen activator (t-PA) in the recent MR CLEAN (Multicenter Randomized Clinical trial of Endovascular treatment for Acute ischemic stroke in the Netherlands), EXTEND-IA (Extending the Time for Thrombolysis in Emergency Neurological Deficits - Intra-Arterial), and ESCAPE (Emphasis on Minimizing CT to Recanalization Times) trials for MCA and distal ICA occlusions [12]. Patients most likely to benefit from thrombectomy include those who receive treatment within 6h, but a benefit is also seen beyond 6h in patients with small infarct cores and significant tissue at risk. Patients with high baseline functional independence and a large-volume penumbra should be aggressively considered for both treatments.

CONCLUSION

The morbidity associated with ischemic stroke extends beyond the initial insult because of the risk of generating clinically worrisome cerebral edema capable of raising ICP and causing compression of adjacent noninfarcted tissue. Careful monitoring for evidence of hydrocephalus, herniation, and worsening midline shift in conjunction with evaluation for worsening neurological deficits is critical in the early postinfarct period. Decline in consciousness is the most common sign of worsening cerebral edema. Medical management to reduce the volume of cerebral edema is the recommended initial intervention, but patients refractory to medical treatment should be considered for surgical decompression. Routine placement of intracranial monitoring devices or EVDs before the onset of severe cerebral edema is not recommended. Early treatment with thrombolytics or endovascular thrombectomy may reduce the risk of severe cerebral edema formation by reducing infarct volume.

References

[1] Heron M. Deaths: leading causes for 2011. Natl Vital Stat Rep July 27, 2015;64(7):1–96.
[1a] https://www.google.com/amp/www.medicalnewstoday.com/articles/amp/282929?client=safari#amph=1.
[2] Friedlander RM. Apoptosis and caspases in neurodegenerative diseases. N Engl J Med 2003;348(14):1365–75.

[3] Durukan A, Tatlisumak T. Acute ischemic stroke: overview of major experimental rodent models, pathophysiology, and therapy of focal cerebral ischemia. Pharmacol Biochem Behav 2007;87(1):179–97.

[4] Kanekar SG, Zacharia T, Roller R. Imaging of stroke: part 2, pathophysiology at the molecular and cellular levels and corresponding imaging changes. AJR Am J Roentgenol 2012;198(1):63–74.

[5] Wijdicks EF, Sheth KN, Carter BS, et al. Recommendations for the management of cerebral and cerebellar infarction with swelling: a statement for healthcare professionals from the American Heart Association/American Stroke Association. Stroke 2014;45(4):1222–38.

[6] Qureshi AI, Suarez JI, Yahia AM, et al. Timing of neurologic deterioration in massive middle cerebral artery infarction: a multicenter review. Crit Care Med 2003;31(1):272–7.

[7] Thomalla GJ, Kucinski T, Schoder V, et al. Prediction of malignant middle cerebral artery infarction by early perfusion- and diffusion-weighted magnetic resonance imaging. Stroke 2003;34(8):1892–9.

[8] Mokri B. The Monro–Kellie hypothesis: applications in CSF volume depletion. Neurology 2001;56(12):1746–8.

[9] Forsyth RJ, Raper J, Todhunter E. Routine intracranial pressure monitoring in acute coma. Cochrane Database Syst Rev 2015;11:CD002043.

[10] Juttler E, Unterberg A, Woitzik J, et al. Hemicraniectomy in older patients with extensive middle-cerebral-artery stroke. N Engl J Med 2014;370(12):1091–100.

[11] Curry Jr WT, Sethi MK, Ogilvy CS, Carter BS. Factors associated with outcome after hemicraniectomy for large middle cerebral artery territory infarction. Neurosurgery 2005;56(4):681–92. discussion 681–92.

[12] McDowell MM, Ducruet AF. Time is brain: a critical analysis of the EXTEND-IA and ESCAPE trials. World Neurosurg 2015;83(6):949–51.

CHAPTER

141

Surgery for Ischemic Strokes

J. Messegee, H. Yonas

University of New Mexico, Albuquerque, NM, United States

INTRODUCTION

Before the age of computed tomography (CT) and magnetic resonance imaging (MRI), urgent hemicraniectomy was not routinely performed for ischemic stroke, but rather for presumed massive edema related to tumors of the brain. In the modern era, the cause of the swelling is seldom in question. Screening CT imaging is the gold standard for identifying ischemic versus hemorrhagic causes of sudden neurologic decline and ultimately determines the immediate management. Ischemic changes due to cytotoxic edema in the ischemic core are evident on noncontrast CT within about 3–4h, and the later changes can be identified on serial imaging. Swelling usually peaks 3–5 days after the ischemic insult without reperfusion to the territory. Earlier reperfusion has been associated with earlier peak swelling due to more rapid edema formation and an increased incidence of hemorrhagic transformation [1].

The morbidity and mortality of large ischemic infarcts are very high. Without surgical decompression of a large middle cerebral artery (MCA) infarct accompanied by a decreasing level of consciousness, there is a historic 80% mortality rate despite best medical management [2]. Significant improvements in first-line stroke therapy over the years have changed this outcome dramatically. Intravenous tissue plasminogen activator (t-PA) administration within 3h of stroke onset has been the standard of care for ischemic stroke since 1995, but intravenous t-PA has been relatively ineffective for large-vessel occlusion [2a]. Recently, multiple randomized controlled trials have demonstrated significant benefit from endovascular mechanical thrombectomy for select patients with large-vessel occlusions [3–6]. These vessels primarily include the internal carotid artery (ICA) and MCA. If performed within 6h of the onset of stroke symptoms with or without t-PA administration, there is now level 1 evidence demonstrating improved efficacy,

Primer on Cerebrovascular Diseases, Second Edition
http://dx.doi.org/10.1016/B978-0-12-803058-5.00141-7

with approximately 80% of patients having functional independence at 90 days. As the medical community accepts this new standard of care for large-vessel occlusion, we can expect an increase in the number of patients being referred for stroke therapy. Despite the increased efficacy of treatment, a significant population of patients will either not benefit from thrombectomy or present outside the therapeutic window. Many of these patients will require consideration of surgical intervention for life-threatening malignant edema and/or hemorrhage.

The risk of hemorrhagic transformation of large ischemic infarcts increases with revascularization, and this is proportional to the duration and depth of ischemia and inversely proportional to the collateral circulation to that territory. These factors can be difficult to assess. Late or severe infarcts with poor collaterals will be evident as areas of hypodensity on CT imaging within as early as 3 h and profoundly apparent by 24 h. There is often a core of severely ischemic and permanently injured brain tissue surrounded by a penumbra of at-risk tissue that will likely benefit from revascularization. Only two of the aforementioned thrombectomy trials used perfusion imaging data in their study design.

The Alberta Stroke Program Early CT score (ASPECTS) helps evaluate the extent of permanently injured tissue by measuring ischemic changes in the anterior circulation on early CT and plays a role in the decision making for emergent thrombectomy [6a]. Some stroke centers have a standardized cutoff for candidates undergoing thrombectomy, where candidates with poor scores (often less than 6, with no ischemic changes scoring 10) are being excluded from thrombectomy because of the documented poor outcomes in this group of patients, despite successful revascularization.

As noted, large-vessel occlusion must be documented by CT angiography or MR angiography before considering endovascular intervention. Multiphase CT angiography using contrast can also play a role in evaluating collateral circulation. A delayed CT after the initial arterial phase CT, performed seconds or minutes after the first scan, may demonstrate late filling of regions at risk by retrograde or collateral pathways. These patients may benefit from revascularization, even outside the 6 h window. Likewise, a significant lack of collateralization on delayed arterial imaging may indicate a poor outcome in this patient population despite being within the time window. Further studies are ongoing and, if shown to be beneficial, may ultimately become the standard of care in determining candidacy for thrombectomy.

In addition to vascular imaging, perfusion studies using CT or MRI can also be performed. These are qualitative in their assessment of cerebral blood flow after processing contrast arrival and washout times. They are based on selecting a normal vessel on the image, such as the anterior cerebral artery (ACA) or contralateral MCA. This leads to a wide variation in the data, and interpretation is therefore useful only for symmetry assessment. Xenon-CT imaging provided quantitative blood flow data, but has not yet been approved by the FDA for use in diagnostics. Current guidelines do not require perfusion information to guide therapy, but will likely also play a role in the future.

Medical therapies include blood pressure control, end-tidal carbon dioxide (CO_2) management, temperature control, and fluid-electrolyte balance. Based on the stroke guidelines, permissive hypertension is the standard of care, but care must be taken to avoid hypotension, as this could lead to further secondary ischemic injury to the penumbra tissue. Excessive hypertension increases the risk of hemorrhagic transformation [7]. Osmotic therapies have limited utility in ischemic infarcts, as the dead tissue does not respond to diuresis, whereas normal brain parenchyma does, which can theoretically worsen shift and herniation. Current standards aim at maintaining sodium levels in the high normal range, thus avoiding hyponatremia. Normothermia is recommended to minimize secondary injury, but there is no evidence to suggest that therapeutic cooling in ischemic stroke improves overall outcome. Cerebral metabolism increases with fever, requiring more blood flow to the already hypoperfused tissue, thus hyperthermia must be avoided to minimize secondary ischemic injury. In patients who are ventilated, avoiding both hyper- and hypocarbia helps control intracranial pressure and does not cause additional injury by CO_2-induced cerebral vasoconstriction.

Patients who develop large hemispheric infarctions despite best medical interventions may progress to develop significant edema, as the dead tissue accumulates cytotoxic fluid. In the elderly population, this can often be tolerated because of prior diffuse cerebral volume loss. However, in younger patients, edema can lead to severe hemispheric swelling with transtentorial or uncal herniation, followed by brainstem compression, coma, and death. Evidence suggests that early decompression with hemicraniectomy with or without removing some of the necrotic tissue leads to improved outcomes. Often surgeons will remove the anterior temporal lobe to reduce lateral brainstem compression. If clinical deterioration occurs, there is an increase in morbidity and mortality. This can include infarction of additional brain tissue, which results from occlusion of the posterior cerebral artery or ACA. Recent randomized controlled trials performed across Europe demonstrated markedly improved survival after hemicraniectomy in this population [8]. Long-term outcome is, however, often accompanied by severe morbidity that often requires long-term care. Early discussion with the patient after stroke onset or with the family regarding the patient's wishes is imperative, preferably well before any consideration of decompression. Patients older than 60 years have had even worse outcomes in one randomized controlled trial, with 93% of the surviving patients requiring significant or complete care with daily living needs [9].

Patients with large cerebellar ischemic insults with little brainstem involvement are good surgical candidates for decompression and removal of ischemic tissue to prevent imminent direct brainstem compression and the potential for rapid deterioration. Presenting symptoms often include nausea, vomiting, vertigo, and ataxia. Facial and sixth nerve palsies can be early signs of significant brainstem compression that may rapidly be followed by a loss of brainstem function. Placement of an external ventricular drain to treat acute hydrocephalus combined with posterior fossa craniectomy often results in rapid return to baseline neurological function. Performing a ventriculostomy without decompression risks upward herniation, a potentially devastating secondary insult. Over time, cerebellar symptoms often improve significantly and most patients can live longer with minimal support.

In summary, young patients who present with large-territory infarctions, but good mental status and an acceptable amount of morbidity to the patient and family, should be considered for decompressive hemicraniectomy, as many of these patients will do well long term after prolonged rehabilitation. Early decision making and discussion with the family is necessary to prevent delays in treatment, which lead to herniation. Patients with large cerebellar infarcts without significant brainstem involvement should likewise be considered for decompression, as they historically return to independent function, even if they are older.

References

[1] Strbian D1, Sairanen T, Meretoja A, Pitkäniemi J, Putaala J, Salonen O, et al. Patient outcomes from symptomatic intracerebral hemorrhage after stroke thrombolysis. Neurology July 26, 2011;77(4):341–8.
[2] Shaw CM, Alvord EC, Berry RG. Swelling of the brain following ischemic infarction with arterial occlusion. Arch Neurol 1959;1:161–77.
[a] Tissue plasminogen activator for acute ischemic stroke. The National Institute of Neurological Disorders and Stroke rt-PA Stroke Study Group. N Engl J Med December 14, 1995;333(24):1581–7.
[3] Fransen PS, Berkhemer OA, Lingsma HF, Beumer D, van den Berg LA, Yoo AJ, et al. Time to Reperfusion and Treatment Effect for Acute Ischemic Stroke: A Randomized Clinical Trial. JAMA Neurol December 21, 2015:1–7.
[4] Campbell BC, Mitchell PJ, Kleinig TJ, Dewey HM, Churilov L, Yassi N, et al. Endovascular therapy for ischemic stroke with perfusion-imaging selection. N Engl J Med March 12, 2015;372(11):1009–18.
[5] Goyal M, Demchuk AM, Menon BK, Eesa M, Rempel JL, Thornton J, et al. Randomized assessment of rapid endovascular treatment of ischemic stroke. N Engl J Med March 12, 2015;372(11):1019–30.
[6] Saver JL, Goyal M, Bonafe A, Diener HC, Levy EI, Pereira VM, et al. Stent-retriever thrombectomy after intravenous t-PA vs. t-PA alone in stroke. N Engl J Med June 11, 2015;372(24):2285–95.
[a] Barber PA, Demchuk AM, Zhang J, Buchan AM. Validity and reliability of a quantitative computed tomography score in predicting outcome of hyperacute stroke before thrombolytic therapy. ASPECTS Study Group. Alberta Stroke Programme Early CT Score. Lancet May 13, 2000;355(9216):1670–4.
[7] Powers WJ, Derdeyn CP, Biller J, Coffey CS, Hoh BL, Jauch EC, et al. 2015 American heart association/American stroke association focused update of the 2013 guidelines for the early management of patients with acute ischemic stroke regarding endovascular treatment: a guideline for healthcare professionals from the American heart association/American stroke association. Stroke October 2015;46(10):3020–35.
[8] Vahedi K, Hofmeijer J, Juettler E, Vicaut E, George B, Algra A, et al. Early decompressive surgery in malignant infarction of the middle cerebral artery: a pooled analysis of three randomised controlled trials. Lancet Neurol March 2007;6(3):215–22.
[9] Juttler E, Unterberg A, Woitzik J, et al. Hemicraniectomy in older patients with extensive middle-cerebral-artery stroke. N Engl J Med 2014;370:1091–100.

CHAPTER

142

Clinical Status of Neuroprotection in Cerebral Ischemia

B.P. Walcott[1], C.J. Stapleton[2]

[1]University of California, San Francisco, San Francisco, CA, United States; [2]Massachusetts General Hospital and Harvard Medical School, Boston, MA, United States

INTRODUCTION

Cerebral ischemia, resulting in neuronal death, is seen in a wide range of conditions. Most often, it results from arterial occlusion secondary to thrombo-embolic or atherosclerotic diseases. Trauma is also a common cause of ischemic disease. In other instances, it can be iatrogenic in the setting of interventional treatments, such as those for carotid artery stenosis or cerebral aneurysms. Neuroprotective strategies in clinical practice fall into one of the three categories: to prevent disruption of ion homeostasis in the neurovascular unit before reaching ischemic threshold, to salvage at-risk territory (ischemic penumbra), and to prevent secondary injury in normal brain tissue near ischemia-related cerebral edema. In this chapter, we review these categories and their roles in the current clinical practice.

PRIMARY STROKE PREVENTION DURING INTERVENTIONS

Although stroke (cerebral ischemia) is most commonly a pathological and unexpected event, there are certain instances when certain therapeutic interventions temporarily limit blood flow to brain tissue, thereby placing the involved territory at risk for injury. An example of such instance is during carotid endarterectomy, when clamping of the common carotid artery temporarily arrests blood flow to the distal internal carotid artery. Another example is during brain aneurysm surgery, when on occasion, a segment of the parent vessel may be temporarily occluded to facilitate final dissection and clipping of the aneurysm.

During any of these instances, the amount of time the blood vessels can be occluded without resulting in stroke is highly variable. It appears to be dependent on many patient-specific factors, such as collateral circulation from other vascular territories, although the variability in maximally tolerated ischemia time from patient to patient is not well understood. Patients under general anesthesia are often closely monitored with somatosensory evoked potentials and motor evoked potentials to detect the onset of early ischemic changes. Measures to augment blood flow through collateral circulation, including systemic hypertension, are thought to be helpful and are routinely used in clinical practice [1].

Alternative strategies to maximize the ischemia interval are centered on decreasing cerebral metabolic demand. Studies have examined the use of mild hypothermia [2], although this was not found to be beneficial in patients undergoing surgery for ruptured brain aneurysms. Its potential benefit in patients undergoing elective procedures, or more specifically in those undergoing temporary occlusion of cerebral arteries, is not well studied. Another technique frequently used to decrease cerebral metabolic demand is anesthetic-induced electroencephalographic (EEG) burst suppression. Burst suppression is an EEG pattern in which high-voltage activity alternates with isoelectric quiescence. It is characteristic of an inactivated brain, is seen with deep levels of general anesthesia, and is thought be associated with a decrease in cerebral metabolic rate coupled with the stabilizing properties of ATP-gated potassium channels [3]. In clinical practice, to achieve 100% intensity burst suppression, the rate of infusion of an anesthetic (typically propofol) can be manipulated based on real-time feedback from EEG

Primer on Cerebrovascular Diseases, Second Edition
http://dx.doi.org/10.1016/B978-0-12-803058-5.00142-9

monitoring. The neuroprotective benefit of this technique to extend the ischemic window is a subject of ongoing investigation.

PENUMBRA SALVAGE

One of the greatest advancements in neuroprotective strategies for cerebral ischemia is the development of effective endovascular techniques to establish rapid reperfusion following the acute onset of thrombus/embolism in the proximal large cerebral vessels. While these patients can often present with profound neurological dysfunction, the amount of brain tissue affected (and resulting in neurological dysfunction) is often a

mix of areas of permanent tissue injury and hypoperfused territory. Magnetic resonance-based diffusion- and perfusion-weighted imaging (DWI and PWI) are widely utilized modalities to aid in treatment selection. While DWI bright regions typically indicate cytotoxic edema as a surrogate for permanent tissue injury, delayed PWI regions correspond to tissue with compromised hemodynamics. The ischemic penumbra is defined as tissue that is hypoperfused to such an extent that focal neurological symptoms arise, but where neurological function can be restored and tissue survival ensured by early reperfusion (Fig. 142.1). Randomized trials of stent-retriever devices have shown that thrombus removal and reperfusion is effective in improving outcomes by targeting restoration of blood flow to the

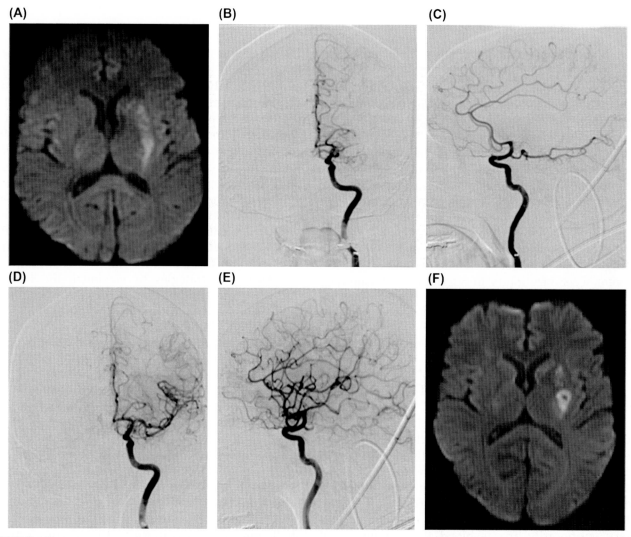

FIGURE 142.1   A 65-year-old male presented with right-sided hemiparesis and difficulty speaking. Computed tomographic angiogram demonstrated a left M1 segment middle cerebral artery occlusion (not shown). (A) diffusion-weighted imaging revealed a small bright area out of proportion to the patient's symptoms. He was brought for emergent thrombectomy, where an M1 segment occlusion was confirmed [(B) anteroposterior and (C) lateral]. After one pass with a stent retriever [(D) anteroposterior and (E) lateral], reperfusion in the distal middle cerebral artery territory was established. (F) The final area of ischemia was demonstrated on magnetic resonance imaging the following day.

ischemic penumbra [4,5]. This represents a paradigm shift from medical to procedural interventions as a main focus of neuroprotective strategies in cerebral ischemia for eligible patients.

PREVENTION OF SECONDARY BRAIN INJURY

Historically, experimental pharmacologic treatments targeted at a multitude of intracellular molecular pathways have failed to demonstrate clinical benefit when evaluated in randomized clinical trials. In theory, inhibition of inflammatory cytokines, prevention of excitotoxicity, reduction of apoptosis, and mitigation of the development of vasogenic edema should improve cellular survival in the area of ischemia. The purported neuroprotective effects of these therapies, although observed in preclinical animal models, have not been generally reproducible in human subjects. Among the hundreds of drugs evaluated, the vast majority have been proven ineffective in preventing further brain injury. Erythropoietin [6] and progesterone [7] have failed to demonstrate any superiority over placebo medication as neuroprotective agents.

A notable exception to these failures is a sulfonylurea receptor 1 (SUR1) antagonist [8,9]. In a phase II randomized clinical trial, inhibition of this receptor was shown to reduce the development of vasogenic edema following ischemic stroke, a key mediator of secondary brain injury. The pathogenesis of ischemic edema is thought to involve a stepwise progression through phases of cytotoxic, ionic, and vasogenic edema. This stepwise progression is driven by pathological changes in the transmembrane permeability of neurons, glia, and vascular endothelial cells composing the neurogliovascular unit. In the first phase (cytotoxic edema), hypoxia-induced failure of energy-dependent mechanisms [primarily the Na^+,K^+-adenosine triphosphatase (ATPase)] that maintain normal physiological ionic gradients across cell membranes causes thermodynamically driven abnormal ionic flux through channels and secondary active transporters. In particular, extracellular Na^+ flows down its concentration gradient into the intracellular compartment. This movement generates oncotic pressure that drives water into cells through aquaporins and other pathways, resulting in swelling and membrane blebbing of neurons, glia, and endothelial cells. Clinically, this is thought to represent the DWI bright area on MRI. Then, as ionic flux into cells depletes Na^+ from the extracellular space, an Na^+ gradient is established between the intravascular and extracellular spaces. This gradient drives transcapillary Na^+ movement across the blood–brain barrier (BBB) through a host of preexisting and newly expressed channels and transporters. The Na^+ flux simultaneously provides the electrochemical drive for Cl^- and oncotic drive for water to flow into the extravascular space also, resulting in an expansion of total extravascular brain volume, known as ionic edema. During cytotoxic and ionic edema, swelling of endothelial cells causes cytoskeletal rearrangements, hypoxia activates a transcriptional program that increases endovascular permeability, and necrotic cell death occurs in the ischemic core, which together contribute to further disruption of the BBB. Together, these processes result in vasogenic edema, in which capillaries become fenestrated, tight junctions are disrupted, and reverse pinocytosis occurs. This results in the leakage of macromolecules, ions, and water into the brain parenchyma. This can result in localized cerebral edema and brain compression, which clinically is associated with a high rate of morbidity and mortality.

A number of preclinical studies have shown that ionic movement through the SUR1-TRPM4 (transient receptor potential melastatin 4) nonselective cation channel, which is upregulated in the setting of ischemia, is a major component of the pathological ionic flux in cytotoxic and ionic edema. In animal studies, it has been demonstrated that this channel can be effectively blocked with a medication approved by the FDA for diabetes treatment, namely, glyburide (Cirara). This has formed the basis for a multicenter, double-blind, phase II clinical trial to determine the efficacy of SUR1 inhibition in the prevention of malignant edema following severe anterior circulation ischemic stroke (ClincalTrials.gov, identifier NCT01794182). Results of this trial demonstrated that the medication Cirara (Remedy Pharmaceuticals, Inc., New York, United States) resulted in a 53% mortality reduction when compared to placebo and the amount of brain swelling (measured by midline shift) was also reduced by half compared to placebo. This evidence, along with an excellent safety profile, suggests that Cirara has the potential to dramatically change the treatment of ischemic stroke, and supports further investigation in a phase III trial.

Another strategy to minimize secondary injury following ischemic stroke by reducing/mitigating cerebral edema is the practice of targeted hypernatremia. When administered in a bolus form, hyperosmolar agents such as mannitol and hypertonic saline have been shown to reduce total brain water content and decrease intracranial pressure. However, surprisingly little is known about the increasingly common clinical practice of inducing a state of sustained hypernatremia. The effects of a pharmacologically sustained state of hypernatremia are likely distinct from bolus-type therapy. In bolus therapy, water is acutely drawn out of the intracellular and extracellular interstitial spaces via a strong transient osmotic gradient originating from within the cerebral vasculature. It is not known whether this osmotic gradient continuously exists in a chronic state of hypernatremia or whether osmotic particles eventually traverse into the interstitial space. Insufficient evidence exists to recommend pharmacologic induction of hypernatremia as a treatment for cerebral edema. The strategy of vigilant avoidance of hyponatremia is currently a safer potentially more efficacious paradigm.

NEUROPROTECTION FROM VASOSPASM-RELATED ISCHEMIA

Aneurysmal subarachnoid hemorrhage is associated with a high mortality rate, often attributed to the initial hemorrhage event or rehemorrhage before aneurysm obliteration. For those who survive these early events, cerebral vasospasm can result in impaired cerebral blood flow and is the major cause of delayed ischemia and stroke. The mechanism is not completely understood, but vasoconstriction is thought to be due in part to an increase in expression of the potent vasoconstrictor, endothelin 1 (ET-1), and a decrease in that of the vasodilator, nitric oxide, in the vascular endothelium.

Although there is evidence that ET-1 has a basal level of constitutive activity, levels of ET-1 and its receptors, namely, ETrA (endothelin receptor A) and ETrB (endothelin receptor B), increase in endothelial cells, pericytes, and astrocytes following injury for more than 24 h. ETrA is thought to stimulate pericyte contraction via signaling through the $PLC\beta/IP_3/Ca^{2+}$ (phospholipase $C\beta$/inositol triphosphate) pathway. Although blockade of ET-1 action at ETrA improves perfusion in preclinical models, the use of endothelin antagonists, such as clazosentan (Actelion Pharmaceuticals Ltd., Allschwil, Switzerland), has not been shown to reduce vasospasm-related morbidity (ischemia) in a phase III randomized trial of patients with subarachnoid hemorrhage [10]. Further study to better understand the mechanisms of subarachnoid hemorrhage-induced vasospasm and its relationship with delayed ischemia is ongoing.

CONCLUSION

Neuroprotective strategies in cerebral ischemia have broadened beyond the search for a single pharmacologic agent. Manipulation of collateral blood supply and cerebral metabolic demand is now a routine measure during therapeutic procedures that temporarily block arterial blood flow to prevent ischemia. New technology, such as retrievable stent thrombectomy, has revolutionized the concept of neuroprotection, as at-risk brain tissue can now be rescued, given the high rates of rapid revascularization. For completed large-volume ischemic stroke, early inhibition of the SUR1-TRPM4 cation channel has been shown in randomized clinical trial to decrease morbidity and mortality by limiting secondary injury associated with massive brain swelling.

References

[1] Shuaib A, Butcher K, Mohammad AA, Saqqur M, Liebeskind DS. Collateral blood vessels in acute ischaemic stroke: a potential therapeutic target. Lancet Neurol 2011;10(10):909–21.

[2] Todd MM, Hindman BJ, Clarke WR, Torner JC. Mild intraoperative hypothermia during surgery for intracranial aneurysm. N Engl J Med 2005;352(2):135–45.

[3] Ching S, Purdon PL, Vijayan S, Kopell NJ, Brown EN. A neurophysiological–metabolic model for burst suppression. Proc Natl Acad Sci 2012;109(8):3095–100.

[4] Berkhemer OA, Fransen PS, Beumer D, et al. A randomized trial of intraarterial treatment for acute ischemic stroke. N Engl J Med 2015;372(1):11–20.

[5] Campbell BC, Mitchell PJ, Kleinig TJ, et al. Endovascular therapy for ischemic stroke with perfusion-imaging selection. N Engl J Med 2015;372(11):1009–18.

[6] Nichol A, French C, Little L, et al. Erythropoietin in traumatic brain injury (EPO-TBI): a double-blind randomised controlled trial. Lancet 2016;386(10012):2499–506.

[7] Skolnick BE, Maas AI, Narayan RK, et al. A clinical trial of progesterone for severe traumatic brain injury. N Engl J Med 2014;371(26):2467–76.

[8] Khanna A, Walcott BP, Kahle KT, Simard JM. Effect of glibenclamide on the prevention of secondary brain injury following ischemic stroke in humans. Neurosurg Focus 2014;36(1):E11.

[9] Simard JM, Chen M, Tarasov KV, et al. Newly expressed SUR1-regulated NC(Ca-ATP) channel mediates cerebral edema after ischemic stroke. Nat Med 2006;12(4):433–40.

[10] Macdonald RL, Higashida RT, Keller E, et al. Clazosentan, an endothelin receptor antagonist, in patients with aneurysmal subarachnoid haemorrhage undergoing surgical clipping: a randomised, double-blind, placebo-controlled phase 3 trial (CONSCIOUS-2). Lancet Neurol 2011;10(7):618–25.

CHAPTER

143

Cardiac Complications and ECG Abnormalities After Stroke

M.J. Schneck

Loyola University Chicago, Stritch School of Medicine, Maywood, IL, United States

INTRODUCTION

A significant overlap exists between cerebrovascular disease and cardiac disease. Cardiac disease is frequent in stroke patients, and cardiac abnormalities are also common following stroke. The most serious events include acute myocardial infarction (MI), heart failure, arrhythmias such as ventricular tachycardia, ventricular fibrillation, or atrial fibrillation, and cardiac arrest. Patients with cerebral atherosclerosis often have coronary artery disease (CAD) or peripheral vascular disease (PVD). Conversely, patients with CAD or PAD are at greater risk of stroke [1]. A number of studies have demonstrated that patients with cerebrovascular disease are at significant risk for subsequent MI or vascular death in the years following a stroke, with cardiac disease as the most likely cause of death in stroke patients over time [1–4]. In one inpatient acute stroke series, there was a cardiac cause of death in 35/846 patients (4%) with other serious cardiac adverse events in the first three months post stroke occurring in 161/84 patients (19%). Factors increasing the risk of serious cardiac events included severe clinical stroke, cardiac failure history, and renal dysfunction (creatinine > 1.3 mg/dl). ECG findings associated with increased risk included extra ventricular beats and prolonged QTc [2].

It is common for a cardiac cause of stroke to be identified only after the initial cerebrovascular event. It thus becomes critical to delineate whether those problems are secondary to stroke, coincidental, or the direct cause of the stroke.

ISCHEMIA AND HEART FAILURE

Concomitant stroke and MI is a well-described phenomenon [1–3]. The annual risk of MI in stroke patients is estimated at 2% and is cumulative over time [3]. Furthermore, acute coronary syndromes (ACSs) including angina, MI, and cardiac ischemia, occur within 3 months in about 6% of acute ischemic stroke patients [3]. Stroke, like diabetes mellitus (DM) is considered a coronary heart disease (CHD) risk equivalent in various models of cardiac risk. Therefore, patients with stroke without known CHD or MI may have a risk of subsequent cardiac events similar to that of patients with known CHD. As such, American Heart Association (AHA) guidelines suggest cardiac risk assessment in all stroke patients with noninvasive cardiac stress testing for stroke patients with high cardiovascular risk profiles including large vessel cervico-cerebral atherosclerosis [4]. These guidelines state that "all patients with ischemia stroke or TIA should undergo a comprehensive assessment of cardiovascular disease … to identify those with the highest likelihood of morbidity and mortality from unrecognized coronary heart disease" [4].

Troponin elevations in stroke patients are often lower than in patients with acute MI suggesting that the mechanism of troponin elevation is not due to CAD-related myocardial ischemia but may be the result of stroke-induced sympathetic stressors as so-called "demand ischemia" [4–6]. Troponin elevations have also been described in SAH patients, as well, with troponin elevations associated with higher risk of death and disability at discharge [7]. In the subarachnoid hemorrhage (SAH) patients, however, elevated troponin was not associated with a higher risk of significant disability or death beyond 3 months postdischarge.

A 2012 joint European and American guideline highlighted an MI definition based on elevated cardiac biomarkers and clinical evidence for ischemia [8]. The guidelines stated that the term "MI" should be used when "there is evidence of myocardial necrosis in a

Primer on Cerebrovascular Diseases, Second Edition
http://dx.doi.org/10.1016/B978-0-12-803058-5.00143-0

clinical setting consistent with myocardial ischemia." The guideline further states that MI is best diagnosed utilizing troponin as the preferred biomarker along with symptomatic ECG, or echocardiographic evidence of myocardial ischemia. This becomes important in the assessment of possible post-stroke coronary ischemia as blood troponin elevations in stroke patients may not always be the result of an ACS event. Troponin release is usually thought to reflect damage to cardiac tissue due to sustained cardiac ischemia [8]. In the absence of ACS, acute ischemic or hemorrhagic events, of themselves, are associated with elevated troponins, however, reflecting the phenomenon of cardiac demand ischemia. Additionally, other non-ACS etiologies of troponin elevation such as hypervolemia or heart failure, tachyarrhythmia or bradyarrhythmia, heart block, atrial fibrillation, heart failure, venous thromboembolism (VTE), myocarditis, myocardial contusion, aortic dissection, sepsis, respiratory failure and renal failure, autoimmune or inflammatory conditions, certain drug toxicities, and other non-CHD-related cardiac problems including hypertrophic cardiomyopathy and valvular heart disease, often occur in stroke patients. Troponin elevations, as a marker of ACS, are mainly useful in patients with high pretest probabilities of CHD and are less diagnostic in stroke patients without known CHD. In patients without known CHD, a troponin elevation may potentially lead to missed diagnosis of those other stroke complications such as sepsis or VTE.

Regardless of the underlying etiology, an elevated troponin is a significant cardiac biomarker in stroke patients as a number of series have found that elevated cardiac troponins are associated with an increased risk of cardiovascular and all-cause mortality regardless of the presence (or absence) of intrinsic CAD [3].

The presumed mechanism of myocardial injury manifesting as an elevated troponin is usually attributed to myocardial ischemia. Cardiac troponins can be released into the circulation, however, without myocyte injury in the context of increased cardiac muscle cell membrane permeability. The mechanism of this nonischemic, or "neurogenic stress cardiomyopathy" (NSC), is probably due to a phenomenon of "myocardial stunning," or catecholamine-induced myocardial injury [6]. Takotsubo cardiomyopathy is one form of NSC that appears more frequently in subarachnoid patients but can also occur in patients with acute ischemic strokes.

First described in Japan, takotsubo cardiomyopathy is a transient cardiac syndrome involving left ventricular apical akinesia, including ST segment elevations and elevated troponin levels, that can mimic ACS [6]. Coronary angiography in these patients shows no significant CAD, however. The term derives from the purported similarity of the akinetic left ventricle to a Japanese octopus fishing pot (Fig. 143.1). The mechanism for this cardiomyopathy is unclear but seems to be partly related to endogenous catecholamine–induced stunning of the myocardium. Estimates are that 20–30% of SAH patients may have a secondary cardiomyopathy, or some other regional ventricular wall motion abnormalities that are typically reversible in the absence of CAD. Stroke patients may also have a less specific hypokinesis of the basal and mid-left ventricular segments that is even more common than the apical or global left ventricular hypokinesis seen in takotsubo cardiomyopathy. Predictors of NSC after SAH include severity of neurological injury,

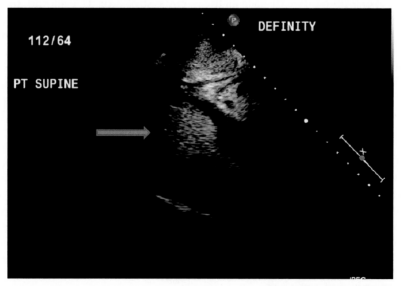

FIGURE 143.1   Echocardiographic lateral view with Definity contrast showing increased left ventricular volume in the context of an apical aneurysm with septal akinesis (that markedly improved 3–4 days later), 1 day following acute subarachnoid hemorrhage. Note the gourd-like appearance of the left ventricle.

troponin elevation, elevated brain natriuretic peptide, and female gender [6]. Elevation of cardiac troponins in SAH has 100% sensitivity and 91% specificity for SAH-related cardiac dysfunction. NSC is more likely in SAH patients where the wall motion abnormalities are not attributable to a specific coronary artery distribution and where the troponin release is relatively "minor" compared with the degree of left ventricular dysfunction. Creatine kinase MB (CK-MB) biomarkers may also be minimally elevated.

Of note, patients with NSC are at increased risk of cerebral vasospasm but are also at risk for ventricular arrhythmias especially in those patients with right insular lesions [6,9]. Strokes that encompass the insular cortex may also be associated with a number of cardiac adverse events including repolarization abnormalities, arrhythmias, neurogenic cardiac damage, heart failure, and sudden death, and the mechanism appears to be related to the autonomic control of cardiovascular function that appears to be centered in the insular cortex [6,9,10].

The management of stroke patients with elevated troponins requires a full diagnostic evaluation with treatment of any possible underlying disorder. Management of patients with NSC-induced cardiomyopathy includes diuretics for volume overload, and angiotensin converting enzyme inhibitors (ACE-I) or angiotensinogen receptor blockers (ARB) at least until left ventricular dysfunction has resolved. In the context of stroke, identification of patients with possible CAD and possible ACS as the cause for an elevated troponin requires a thorough cardiac investigation. In addition to echocardiography, some patients may require coronary angiography (either catheter-based or noninvasive angiography, as available) to distinguish between NSC- and CAD-related troponin elevations. For patients with ACS and large ischemic strokes or hemorrhagic stroke (intracerebral hemorrhage (ICH) or SAH), cardiac stenting may not immediately be possible, because of the risk of worsening hemorrhage with anticoagulation or combination antiplatelet therapies. In those instances, aggressive risk factor control with statins and beta blockade should still be implemented. Additionally, while in the past thrombolytic therapy for patients presenting simultaneously with acute MI and acute ischemic stroke was deemed absolutely contraindicated, intravenous and/or intra-arterial thrombolytic therapy may be an option in carefully selected patients [11].

ECG ABNORMALITIES AND ARRHYTHMIAS

Stroke patients frequently have cardiac arrhythmia and repolarization changes and are at higher risk of sudden death and fatal arrhythmias account for 6% of post-stroke death [6,8,10]. ECG changes may be mediated through abnormal discharges leading to increased norepinephrine changes [6,9]. Estimates are that 75–92% of patients with acute ischemic stroke develop new ECG abnormalities. Cardiac arrhythmias may then occur in upward of 60–70% of all stroke patients [6,8]. The most common abnormalities appear to be QT interval prolongations (45%), ST segment depressions (35%), and U-waves (28%). QT prolongation puts patients at risk for torsade de pointe and may be potentiated by hypokalemia; 50% of SAH patients may have hypokalemia at presentation and QT prolongation is more common in SAH than other acute strokes [12]. QT prolongation, is found in up to 71% of patients with SAH, 64% of patients with intraparenchymal hemorrhage, and 38% of patients with ischemic stroke and may be associated with elevated systolic blood pressures at time of stroke presentation [6].

ST segment elevations are more common in stroke patients with apical and mid-ventricular wall motion abnormalities, and those patients with possible concomitant CAD should undergo a full diagnostic evaluation [3,4]. Patients with NSC may also develop large symmetric T-wave inversions also known as cerebral T-waves (Fig. 143.2) [6]. Again, these ECG changes are thought due to catecholamine surges with resultant sympathomimetic-induced myocyte transcellular perturbations and/or myocardial necrosis and, as previously noted, are more common in right insular strokes presumably due to the autonomic control exerted by the right insula [8,9]. The incidence of serious arrhythmias is highest in first 24 h post admission with up to 25% of patients sustaining a serious arrhythmia in the first 3 days post stroke [10]. Only about 25% of patients have clinical symptoms related to these arrhythmias, however, so continued cardiac telemetry monitoring of all stroke patients while in hospital is essential.

Cardioembolic strokes comprise 14–30% of all ischemic strokes and atrial fibrillation may represent 50% of these cardioembolic strokes [13]. In general, cardioembolic strokes have a high in-hospital mortality rate. Serious arrhythmias, however, may only be detected after stroke occurrence and may be directly related to the size or location of stroke. Serious tachyarrhythmia due to atrial fibrillation are particularly frequent in acute ischemic stroke (Fig. 143.3). New onset paroxysmal or persistent atrial fibrillation is also more common in ischemic stroke, as compared with SAH or ICH. The extent to which some of these arrhythmias represent a preexisting or otherwise undetected cardiac condition is unknown, but development of arrhythmias should prompt a diagnostic evaluation for underlying cardiac dysfunction. In particular, depolarization abnormalities are of particular concern as they are associated with a greater risk of ventricular tachycardia or fibrillation.

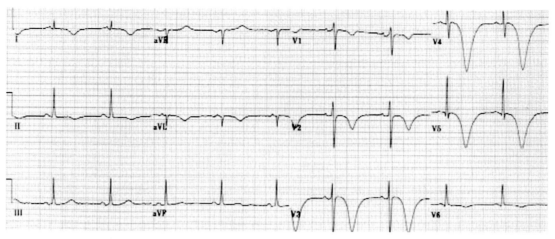

FIGURE 143.2 Prolonged QT segment and cerebral T-waves in the anterior ECG leads in a patient with subarachnoid hemorrhage.

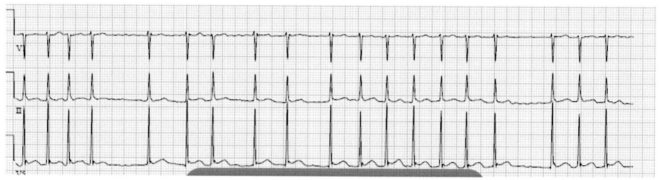

FIGURE 143.3 Atrial fibrillation with rapid ventricular response (heart rate of 115 beats per minute) in a patient with a large left middle cerebral artery ischemic stroke.

The cause of cardiac dysrhythmias likely reflects increased sympathetic tone and alterations in blood pressure and heart rate variability that are common post stroke [6,8–10]. Underlying CHD may also predispose stroke patients to cardiac arrhythmias. Many stroke patients who develop cardiac arrhythmias have normal cardiac function, however, and a central nervous system mechanism has been suggested for many of these arrhythmias [6,9,10]. Evidence suggests that right insula control of sinoatrial function is also the basis for an increased risk of bradycardia and hypotension that has been observed with right insular strokes. Conversely, tachycardia and arterial hypertension seem to be more common in patients with left insular lesions, though supraventricular arrhythmias usually occur more frequently in right hemispheric strokes possibly due to loss of parasympathetic control from the right insula [6,9].

Regardless, the evaluation and management of post-stroke arrhythmia depends on the type of arrhythmia and whether the patient is asymptomatic or having signs of hemodynamic instability. A neurogenic etiology remains a diagnosis of exclusion and a search for an underlying intrinsic cardiac etiology should be sought including structural cardiac abnormalities. Other arrhythmogenic causes that should be explored include electrolyte abnormalities, medications (especially drugs that prolong the QT interval), and noncardiac causes of arrhythmia such as VTE or sepsis. Treatment should focus on controlling the arrhythmia, as necessary, and treating the underlying cause of the arrhythmia. In particular, correction of electrolytes abnormalities with maintenance of normal serum electrolyte levels is essential to prevent acute stroke patients from progressing to a potentially fatal arrhythmia.

SUMMARY

Significant cardiac events post stroke may be the result of acute MI, heart failure, and ventricular and atrial arrhythmias. Concomitant stroke and cardiac events are common. Elevations in cardiac enzymes after acute ischemic stroke may not be indicative of myocardial injury, but an ACS should be always considered in those circumstances, and NSC represents a diagnosis of exclusion. All stroke patients should also undergo cardiac telemetry during the acute hospitalization since cardiac arrhythmias are common, especially in the first few days after stroke symptom onset.

References

[1] Alberts MJ, Bhatt DL, Mas JL, et al. Three-year follow-up and event rates in the international Reduction of Atherothrombosis for Continued Health Registry. Eur Heart J 2009;30(19):2318–26.

[2] Prosser J, MacGregor L, Lees KR, et al. Predictors of early cardiac morbidity and mortality after ischemic stroke. Stroke 2007;38:2295–302.

[3] Touze E, Varenne O, Chatellier G, et al. Risk of myocardial infarction and vascular death after transient ischemia attack and ischemic stroke: as systematic review and meta-analysis. Stroke 2005;36:2748–55.

[4] Adams RJ, Chimowitz MI, Alpers JS, et al. Coronary risk evaluation in patients with transient ischemic attack and ischemic stroke: a scientific statement for healthcare professionals form the stroke council and the council on clinical cardiology of the American Heart Association/American stroke association. Circulation 2003;34:2310–22.

[5] Naidech AM, Kreiter KT, Janua N, et al. Cardiac troponin elevation, cardiovascular morbidity, and outcome after subarachnoid hemorrhage. Circulation 2005;112(18):2851–6.

[6] Bybee KA, Prasad A. Stress-related cardiomyopathy syndromes. Circulation 2008;118:397–409.

[7] Thygesen K, Alpert JS, Jaffe AS, on behalf of the Joint ESC/ACCF/AHA/WHF Task Force for the Universal Definition of Myocardial Infarction, et al. Third universal definition of myocardial infarction. Circulation 2012;126. http://dx.doi.org/10.1161/CIR.0b013e31826e1058.

[8] Khechinasvili G, Asplund K. Electrographic changes in patients with acute stroke. A systematic review. Cerebovasc Dis 2002;14:65–6.

[9] Oppenheimer S. Cerebrogenic cardiac arrhythmias: cortical lateralization and clinical significance. Clin Auto Res 2008;16(1):6–11.

[10] Kallmunzer B, Breuer L, Kahl N, et al. Serious cardiac arrhythmias after stroke. Incidence, time course, and predictors—a systematic prospective analysis. Stroke 2012;43:2892–7.

[11] Fugate JE, Rabinstein AA. Absolute and relative contraindications to IV rt-PA for acute ischemic stroke. Neurohospitalist July 2015;5(3):110–21.

[12] Fukui S, Katoh H, Tsuzuki N, et al. Multivariate analysis of risk factors for QT prolongation following subarachnoid hemorrhage. Crit Care 2003;7(3):7–12.

[13] Arboix A, Alio J. Cardioembolic stroke: clinical features, specific cardiac disorders and prognosis. Curr Cardiol Rev 2010;6(3):150–61.

CHAPTER

144

Management of Hypertension in Stroke

J.A. Abbatemarco[1], K.R. Duncan[1], P. Varade[1,2]

[1]Lehigh Valley Hospital and Health Network, Allentown, PA, United States; [2]University of South Florida, Tampa, FL, United States

INTRODUCTION

Chronic hypertension is a known risk factor for acute ischemic as well as hemorrhagic strokes. There is an independent, graded relationship between the two, and hypertension is therefore aggressively treated in the outpatient environment in order to reduce future stroke risks [1]. Ischemic strokes comprise the majority of all strokes. Although they can present in a myriad of manifestations, patients will usually have a focal onset of neurological symptoms at admission hinting toward this diagnosis.

Cerebral autoregulation is integral to the delicate process of maintaining stable cerebral perfusion and brain tissue oxygenation against changes in the arterial blood pressure (BP). The classic description of cerebral autoregulation is that cerebral blood flow is maintained at a constant level across a wide range of mean arterial BP (60–150 mmHg) [2]. Autoregulation occurs through changes in cerebral vascular resistance leading to vasodilation when peripheral BP is reduced and vasoconstriction when BP is elevated. In the face of longstanding hypertension, which leads to morphological changes in the vessel walls, this

Primer on Cerebrovascular Diseases, Second Edition
http://dx.doi.org/10.1016/B978-0-12-803058-5.00144-2

"autoregulatory window" is shifted toward higher BP such that reductions in cerebral blood flow can occur when BP is lowered within the 60–150 mmHg range [3].

MANAGEMENT OF HYPERTENSION IN THE ACUTE STROKE PERIOD

Brain ischemia results from decreased cerebral blood supply leading to neuronal dysfunction. Stroke treatment should be initiated as early as possible in this acute period. It is not well established how long the acute stroke period lasts; however depending upon the pathology, this may be anywhere from 24 h to 7 days during which there may be a loss of cerebral autoregulation following the acute stroke. Subsequently, most of these patients with an ischemic stroke or transient ischemic attack (TIA) will have an acute increase in BP upon presentation to the hospital [4].

Elevated BP is common during acute ischemic strokes and often higher in patients with a history of hypertension than in those without premorbid hypertension. The BP typically decreases spontaneously during the acute phase of ischemic stroke, starting within 90 min after onset of stroke symptoms [5]. Extreme arterial hypertension is detrimental because it leads to encephalopathy, cardiac complications, and renal insufficiency. Extreme arterial hypotension, on the other hand, is also clearly detrimental by decreasing perfusion to multiple organs, especially the ischemic brain, exacerbating the ischemic injury. This is further associated with an increased tendency for hemorrhagic transformation of an ischemic lesion, especially in the face of pharmacological agents such as antifibrinolytics, anticoagulants, or antiplatelet therapy. Moderate arterial hypertension then, during the acute phase, may be advantageous by improving cerebral perfusion of the ischemic tissue without exacerbating edema or leading to hemorrhagic transformation [6].

BP management in acute ischemic strokes is therefore a delicate balance. There is a U-shaped curve for mortality and morbidity when an acute stroke patient is hypo- or hypertensive [7]. One acute ischemic stroke treatment trial, the Intravenous Nimodipine West European Stroke Trial (INWEST), found complications associated with BP lowering after using intravenous (IV) nimodipine therapy and worse clinical outcome at 21 days [8]. Another trial, the Continue Or Stop post-Stroke Antihypertensives Collaborative Study (COSSACS), compared the continuation of antihypertensive therapy to stopping preexisting antihypertensive drugs during acute hospitalization for ischemic stroke. This study was terminated prematurely; however, continuation of the antihypertensives did not reduce 2-week mortality or morbidity and was not associated with an increase in adverse events [9]. It is reasonable to discontinue or reduce premorbid antihypertensive

regimens at the onset of acute ischemic stroke as swallowing is often impaired, and responses to these medications may be less predictable during the acute phase. If needed, controlled BP lowering during an acute stroke can best be achieved with IV antihypertensive therapies. Despite multiple studies investigating various BP parameters during admission for acute ischemic stroke and clinical outcomes, it remains unclear what the risk–benefit ratio is for lowering or raising the BP during an acute ischemic stroke. At this time, the American Heart Association (AHA) recommends not to lower the BP during the initial 24 h of acute ischemic stroke unless the BP is >220/120 mmHg or there is a concomitant specific medical condition that would benefit from BP lowering [6]. It is then reasonable to initiate gradual long-term antihypertensive therapy after the initial 24 h from stroke onset in most patients.

Specific BP management recommendations have been established for acute ischemic stroke patients being considered for fibrinolytic therapy, which include a gentle approach to bringing the pressure below 185/110 mmHg to qualify for fibrinolytic therapy with IV tissue plasminogen activator (tPA). Once IV tPA is given, the BP must be maintained below 180/105 mmHg to limit the risk of intracerebral hemorrhage (ICH). Higher BP during the initial 24-h after IV tPA are also associated with a greater risk of spontaneous ICH in a linear fashion [10].

Elevated BP is also very common in hemorrhagic strokes because of a variety of factors, including stress, pain, increased ICP, and premorbid acute or persistent elevations in BP. High systolic BP (SBP) in hemorrhagic strokes is associated with greater hematoma expansion, neurological deterioration, and death and dependency after ICH [11]. The largest randomized clinical trial evaluating the efficacy of intensive BP lowering is INTERACT2, undertaken in 2839 patients which randomized patients receiving intensive BP-lowering treatment to <140 mmHg within 1 h versus standard treatment of SBP < 180 mmHg. Overall, the study concluded that early intensive BP lowering is safe and feasible and that surviving patients show modestly better functional recovery, with a favorable trend seen toward a reduction in the conventional clinical end point of death and major disability [12]. It is therefore reasonable for ICH patients to receive early treatment targeted to an SBP level <140 mmHg to improve their chances of achieving better functional recovery should they survive the condition.

The Seventh Report of the Joint National Committee (JNC7) does not make any specific recommendations regarding acute ischemic or hemorrhagic stroke patients. For those with BP > 180/120 mmHg complicated by evidence of impending or progressive target-organ damage, the general goals for BP control include reducing MAP by no more than 25% within minutes to 1 h. If the BP remains stable, reduce to 160/100 within next 2–6 h. It also recommends avoiding excessive falls that may precipitate renal, cerebral, or coronary ischemia. Further gradual reductions toward a normal BP can be implemented over the next 48 h [13]. Although these are not recommended for stroke patients, the general approach of a gradual decrease in the BP over the course of hours to days should be implemented in stroke patients as well. Furthermore, JNC7 stresses the importance of lifestyle modifications in the overall management of hypertension.

MANAGEMENT OF CHRONIC HYPERTENSION IN PATIENTS WITH CEREBROVASCULAR DISEASE

Treatment of hypertension is possibly the most important intervention for primary and secondary prevention of ischemic stroke. The prevalence of hypertension among patients with a recent ischemic stroke is about 70% [14]. The first major trial to demonstrate the effectiveness of hypertension treatment for secondary prevention of stroke was the Post-stroke Antihypertensive Treatment Study (PATS), which followed 5665 patients with a recent TIA or stroke for 24 months after randomization with antihypertensive (indapamide) versus a placebo. Recurrent stroke was observed in 44.1% of patients assigned to placebo (vs. 30.9% of those treated with an antihypertensive) [15]. These results of secondary prevention were subsequently confirmed in the Perindopril Protection Against Recurrent Stroke Study (PROGRESS), which randomized 6105 patients with a history of TIA or stroke to active treatment with perindopril-based regimen or placebo. Active therapy reduced the primary end point of stroke by 28% [16]. Another metaanalysis of 10 randomized trials concluded that treatment with antihypertensive drugs was associated with a significant reduction in recurrent strokes [17]. Most patients with cerebrovascular disease tolerate gradual reductions in BP to the desired <140 mmHg SBP and <90 mmHg DBP (diastolic) target range without difficulty and will further benefit from reductions in cardiovascular morbidity and mortality.

There are limited data that specifically assess the optimal BP target for secondary stroke prevention. In 2013, the results of the Secondary Prevention of Small Subcortical Strokes (SPS3) trial were presented. This study enrolled 3020 patients with small vessel disease and randomized them to two different target levels of SBP control, <150 mmHg versus <130 mmHg. There was no difference between the target groups with regard to the composite outcome of stroke, myocardial infarction, and vascular death [18]. The greatest benefit of more intensive BP lowering is on the prevention of ICH in patients with established small vessel stroke disease. The AHA/ASA guidelines recommend to start therapy at SBP > 140 mmHg or DBP > 90 mmHg for all adults with a history of stroke or

TIA. In those patients who are considered high risk (diabetes mellitus, heart failure, or chronic kidney disease) or with a history of ICH, the target BP should be lowered to <130 mmHg SBP and <80 mmHg DBP. Several lifestyle and diet modifications have been associated with BP reductions and are a reasonable part of a comprehensive antihypertensive therapy, including salt restriction, weight loss, consumption of a diet rich in fruits, vegetables, and low-fat dairy products and regular aerobic physical activity. The optimal drug regimen to achieve the recommended level of reductions is uncertain due to limited direct comparisons; however, available data suggest diuretics or a combination or diuretics with angiotensin-converting enzyme inhibitor may be useful [19].

References

[1] Meschia JF, Bushnell C, Boden-Albala B, Braun LT, Bravata DM, Chaturvedi S, et al. Guidelines for the primary prevention of stroke: a statement for healthcare professionals from the American Heart Association/American Stroke Association. Stroke 2014;45(12):3754–832.

[2] Lassen NA. Cerebral blood flow and oxygen consumption in man. Physiol Rev 1959;39(2):183–238.

[3] Strandgaard S. Autoregulation of cerebral blood flow in hypertensive patients. The modifying influence of prolonged antihypertensive treatment on the tolerance to acute, drug-induced hypotension. Circulation 1976;53(4):720–7.

[4] Harper G, Castleden CM, Potter JF. Factors affecting changes in blood pressure after acute stroke. Stroke 1994;25(9):1726–9.

[5] Broderick J, Brott T, Barsan W, Haley EC, Levy D, Marter J, et al. Blood pressure during the first minutes of focal cerebral ischemia. Ann Emerg Med 1993;22(9):1438–43.

[6] Jauch EC, Saver JL, Adams Jr HP, Bruno A, Demaerschalk BM, Khatri P, et al. Guidelines for the early management of patients with acute ischemic stroke: a guideline for healthcare professionals from the American Heart Association/American Stroke Association. Stroke 2013;44(3):870–947.

[7] Castillo J, Leira R, Garcia MM, Serena J, Blanco M, Davalos A. Blood pressure decrease during the acute phase of ischemic stroke is associated with brain injury and poor stroke outcome. Stroke 2004;35(2):520–6.

[8] Ahmed N, Nasman P, Wahlgren NG. Effect of intravenous nimodipine on blood pressure and outcome after acute stroke. Stroke 2000;31(6):1250–5.

[9] Robinson TG, Potter JF, Ford GA, Bulpitt CJ, Chernova J, Jagger C, et al. Effects of antihypertensive treatment after acute stroke in the Continue or Stop Post-Stroke Antihypertensives Collaborative Study (COSSACS): a prospective, randomised, open, blinded-endpoint trial. Lancet Neurol 2010;9(8):767–75.

[10] Ahmed N, Wahlgren N, Brainin M, Castillo J, Ford GA, Kaste M, et al. SITS investigators. Relationship of blood pressure, antihypertensive therapy, and outcome in ischemic stroke treated with intravenous thrombolysis: retrospective analysis from Safe Implementation of Thrombolysis in Stroke-International Stroke Thrombolysis Register (SITS-ISTR). Stroke 2009;40(7):2442–9.

[11] Hemphill 3rd JC, Greenberg SM, Anderson CS, Becker K, Bendok BR, Cushman M, et al. Guidelines for the management of spontaneous intracerebral hemorrhage: a guideline for healthcare professionals from the American Heart Association/American Stroke Association. Stroke 2015;46(7):2032–60.

[12] Anderson CS, Heeley E, Huang Y, Wang J, Stapf C, Delcourt C, et al. Rapid blood-pressure lowering in patients with acute intracerebral hemorrhage. New Engl J Med 2013;368(25):2355–65.

[13] National High Blood Pressure Education Program. The Seventh Report of the Joint National Committee on Prevention, Detection, Evaluation, and Treatment of High Blood Pressure. Bethesda (MD): National Heart, Lung, and Blood Institute; 2004.

[14] Lovett JK, Coull AJ, Rothwell PM. Early risk of recurrence by subtype of ischemic stroke in population-based incidence studies. Neurology 2004;62(4):569–73.

[15] PATS Collaborating Group. Post-stroke antihypertensive treatment study. A preliminary result. Chin Med J 1995;108(9):710–7.

[16] PROGRESS Collaborative Group. Randomised trial of a perindopril-based blood-pressure-lowering regimen among 6,105 individuals with previous stroke or transient ischaemic attack. Lancet 2001;358(9287):1033–41.

[17] Liu L, Wang Z, Gong L, Zhang Y, Thijs L, Staessen JA, et al. Blood pressure reduction for the secondary prevention of stroke: a Chinese trial and a systematic review of the literature. Hypertens Res 2009;32(11):1032–40.

[18] SPS Study Group. Blood-pressure targets in patients with recent lacunar stroke: the SPS3 randomised trial. Lancet 2013;382(9891):507–15.

[19] Kernan WN, Ovbiagele B, Black HR, Bravata DM, Chimowitz MI, Ezekowitz MD, et al. Guidelines for the prevention of stroke in patients with stroke and transient ischemic attack: a guideline for healthcare professionals from the American Heart Association/American Stroke Association. Stroke 2014;45(7):2160–236.

C H A P T E R

145

Management of Diabetes in Stroke

M. Emanuele[1], N. Emanuele[2]

[1]Loyola University Chicago, Stritch School of Medicine, Maywood, IL, United States; [2]Hines VA Medical Center, Hines, IL, United States

INTRODUCTION

Diabetes mellitus is a global health problem. About 350 million people worldwide are affected and this number is expected to increase by a staggering 50% in the next decade because of increases in the prevalence of obesity and inactivity [1]. Type 1 diabetes mellitus, which accounts for about 10% of all cases, is an autoimmune disease where there is immune destruction of the insulin-producing pancreatic β-cells [2]. Type 2 diabetes mellitus, representing 90% of people with diabetes, is caused by progressive pancreatic β-cell failure in the context of increased insulin resistance [2].

DIABETES AS A RISK FACTOR FOR STROKE

Diabetes is a well-recognized risk factor for stroke and there may be a sexual dimorphism in the stroke risk imposed by diabetes. A recent systematic review and meta-analysis of 64 cohorts, including 775,385 individuals and 12,539 strokes, is instructive in this regard [1]. The duration of follow-up was between 5 and 32 years and included data on both fatal and nonfatal strokes. Compared with individuals without diabetes, those with diabetes, not surprisingly, had higher systolic blood pressure, total cholesterol levels, body mass index, and waist circumference and lower high-density lipoprotein cholesterol fraction. The overall pooled relative risk (RR) for fatal and nonfatal strokes associated with diabetes was 2.28 [95% confidence interval (CI), 1.93–2.69] in women and 1.83 (1.60–2.08) in men. Adjustment for other cardiovascular risk factors had little effect on the RR for stroke, which remained higher in diabetic people but higher in diabetic women than in diabetic men. The RR for fatal stroke was 2.29 (1.73–3.04) in women and 1.74 (1.45–2.08) in men. Sensitivity analyses showed that the increased stroke risk did not vary by region, age, duration of follow-up study, smoking

Primer on Cerebrovascular Diseases, Second Edition
http://dx.doi.org/10.1016/B978-0-12-803058-5.00145-4

status, year of study baseline, stroke subtype, method of diabetes ascertainment, or data source. Possible reasons for the gender differences are beyond the scope of this chapter and discussed in reference [1]. The point is that diabetes imposes a large excess risk for stroke.

PATHOPHYSIOLOGY OF STROKE IN DIABETES

The heightened risk of stroke in people with diabetes mellitus is not surprising, given the major risk factors for stroke are largely present in patients with diabetes [3]. These include hyperglycemia, hypertension, dyslipidemia, obesity, and atrial fibrillation. Individuals with diabetes also exhibit a hypercoagulable state evidenced by increased levels of clotting factors and fibrinogen and increased levels of plasminogen activator inhibitor 1 and hyperreactive platelets, with enhanced adhesion and aggregation. Diabetes is a state with increased inflammatory markers as well.

DIABETES AND STROKE OUTCOMES

Individuals with diabetes not only are at higher risk for stroke but also have poorer outcomes after a stroke [4]. This is illustrated in a report from the South London Stroke Register on 3730 patients who had stroke, with events occurring between January 1995 and December 2011. Among them, 3177 (85.2%) had ischemic strokes and 553 (14.8%) had intracerebral hemorrhage. Outcome was measured by the Barthel index as either a good outcome (Barthel index ≥ 15) or a poor outcome (death or dependency, Barthel index 0–14). After 1, 5, and 10 years of an ischemic stroke event, diabetes was associated with poor outcome, with statistically significant odds ratios varying from about 1.5 to 1.7. There was no association of diabetes with intracerebral hemorrhage.

In the short-term setting, observational studies in humans have demonstrated that acute ischemic stroke with concomitant hyperglycemia increases stroke severity and functional impairment compared to those with normoglycemia. Individuals with hyperglycemia had a two-fold higher mortality at 90 days than those with normal glucose levels [5]. In acute stroke the initial ischemia is rarely complete due to collateral blood supply. The ischemic area of hypoperfused, but still viable, tissue surrounding the ischemic region may respond to tight glycemic control with a reduction ultimate brain injury [6]. Neuroimaging has demonstrated that acute hyperglycemia can increase final infarct size [7] and hyperglycemia is associated with reduced benefit from recanalization with thrombolytic therapy [8].

MANAGEMENT

Acute Management

Data indicate that diabetes in general and hyperglycemia in particular are potential risk factors for stroke and, by inference, that meticulous blood glucose control should be beneficial in the clinical setting of acute stroke. Despite these findings, a comprehensive Cochrane review [9] found that the intravenous administration of insulin targeting serum glucose levels within a specific range of 72–135 mg/dL in the first 24 h of symptom onset of an acute ischemic stroke did not provide benefit when compared with less intensive glycemic control in terms of functional outcome, death, or improvement in final neurological deficit. It is of note that the number of hypoglycemic episodes was significant. In fact, individuals with glucose levels maintained within a tighter range by intravenous insulin administration experienced a greater risk of symptomatic and asymptomatic hypoglycemia than those in the control group. As there is abundant data to support that hypoglycemia is detrimental to the cardiovascular system, it is prudent to avert it as much as possible. Hypoglycemia is associated with tachycardia, an increase in systolic blood pressure and decrease in diastolic blood pressure, and an increase in cardiac output and myocardial contractility. Moreover, hypoglycemia is reported to cause electrocardiographic changes of T-wave flattening or inversion and ST-segment depression consistent with ischemia and QT prolongation that can be associated with arrhythmias. Additionally, hypoglycemia has been noted to result in increased number of RBCs leading to increased blood viscosity; enhanced platelet aggregation; increased levels of platelet factor 4, thromboglobulin, coagulation factor VIII, and von Willebrand factor; and increased thrombin generation, all creating a hypercoagulable milieu and increasing stroke risk [10].

To avert hypoglycemia without compromising glycemic control, our institution has adopted protocols that are specific for patients with neurological compromise. While the hospital glycemic goal is 120–160 mg/dL, we have modified our hospital glycemic target for patients with head injury, status post craniotomy, stroke, or aneurysm to 120–180 mg/dL. In intensive care units, intravenous insulin infusions are managed with the assistance of a computer-generated program. This has decreased the incidence of hypoglycemia and reduced glucose excursions. Subcutaneous insulin protocols are used outside intensive care units. In general, a body-weight-based insulin calculation is used. The recommended total daily dose of insulin is 0.6 units/kg for type 2 diabetic patients, where 50% of the dose was given as basal insulin glargine and the remaining 50% as mealtime rapid-acting insulin divided equally among each meal. For individuals with type 1 diabetes, the calculation is 0.4 units/kg total insulin dose with half basal dose and half bolus. A finger-stick glucose test is performed before breakfast, lunch, and dinner and correction rapid-acting insulin is added to the mealtime dose if indicated. The correction insulin dose is based on the individual's total daily insulin dose [11].

Chronic Management

At the turn of the 21st century, there was data showing that good glycemic control prevented or delayed the microvascular complications of diabetes, such as retinopathy, nephropathy, and neuropathy. Although epidemiologic data showed that good blood glucose control was associated with better macrovascular outcomes (including stroke), there was need for prospective studies examining the impact of long-term glycemic management on macrovascular outcomes. Now a somewhat clearer picture has emerged with the publications of the Diabetes Control and Complications Trial (DCCT), the United Kingdom Prospective Diabetes Study (UKPDS), the Veterans Affairs Diabetes Trial (VADT), the Action to Control Cardiovascular Risk in Diabetes (ACCORD), and the Action in Diabetes and Vascular Disease: Preterax and Diamicron Modified Release Controlled Evaluation.

In the DCCT, a randomized controlled trial of intensive versus standard glycemic control in people with type 1 diabetes, there was a beneficial microvascular, but not macrovascular, effect during the interventional phase of the trial. The average glycated hemoglobin concentration in the standard group was 9% compared with 7% in the intensively treated participants. In a follow-up study, namely, the Diabetes Control and Complications Trial/Epidemiology of Diabetes Interventions and Complications (DCCT/EDIC), a beneficial macrovascular effect emerged at 17 years of follow-up [12]. There was a marked decrease in the occurrence of cardiovascular events in those who had been assigned to intensive glycemic treatment compared with those who had been assigned to standard glycemic therapy, even though the difference in glycated hemoglobin concentration between the study groups was no longer evident. Pertinent to this discussion, the risk of the first occurrence of nonfatal myocardial infarction, stroke, or death from cardiovascular disease was reduced to 57% with intensive treatment, as compared with conventional treatment (95% CI, 12–79; $P = .02$).

The UKPDS was a study on with newly diagnosed type 2 diabetes. In secondary randomization, metformin treatment in a subgroup of overweight individuals reduced the risk for stroke when compared with those randomized to chlorpropamide, glibenclamide, or insulin ($P = .032$) [13].

The VADT, in contrast to the UKPDS, enrolled individuals with long-term (diabetes duration was 11.5 years at entry) type 2 diabetes. These patients entered with a substantial cardiovascular burden. During the interventional phase of the trial, intensive glycemic control did not result in macrovascular benefit. The average glycated hemoglobin concentration in the standard group was 8.4% compared to 6.9% in the intensively treated participants. However, similar to the DCCT/EDIC, in a follow-up study, a beneficial macrovascular effect emerged at 10 years of follow-up [14]. There was a decrease in the occurrence of cardiovascular events in those who had been assigned to intensive glycemic treatment compared with those who had been assigned to standard glycemic therapy, even though the difference in glycated hemoglobin concentration between the study groups was no longer evident. There was a significant 17% reduced risk for overall cardiovascular events. This was driven largely by a decrease in the occurrence of nonfatal myocardial infarctions. There appeared to be no stroke benefit, but the number was small.

There is some emerging evidence that chronic maintenance of good blood glucose control may have beneficial effects on stroke reduction in both type 1 and type 2 diabetes, although, admittedly, it is not entirely as consistent and robust as expected. However, data from the ACCORD warrant caution on the aggressiveness of the targeted level of glucose control [15]. The ACCORD was a study on people with long-term type 2 diabetes. The average hemoglobin concentration in the standard group was 7.5% compared to 6.4% in the intensively treated participants. So the people in the intensive glycemic therapy group in ACCORD achieved substantially lower hemoglobin concentrations than the group in the VADT (6.9%). The ACCORD had to be stopped earlier than planned because those in the intensive group experienced increased all-cause mortality [hazard ratio (HR) 1.22 (1.01–1.46), $P = .04$)] and increased cardiovascular mortality [HR 1.35 (1.04–1.76), $P = .02$] when compared with those in the standard group. Thus there does not seem to be any prospective trial data to support the occurrence of hemoglobin concentration less than approximately 7%. Indeed some circumstances may dictate less aggressive targets. These include those at high risk for hypoglycemia with longer disease duration, important comorbidities, and insufficient resources and support systems.

There is no data to support one form of treatment over another. Management must be on an individualized basis with education, healthy diet, whatever exercise the patient can undertake, smoking cessation, and combinations of the vast array of oral and parenteral medications that are now available.

References

[1] Peters S, Huxley R, Woodward M. Diabetes as a risk factor for stroke in women compared to men: a systematic review and meta-analysis of 64 cohorts, including 775385 individuals and 12539 strokes. Lancet 2014;383:1973–80.
[2] American Diabetes Association. Classification and diagnosis of diabetes. Diabetes Care 2015;38(Suppl. 1):S8–16.

[3] Sander D, Kearney M. Reducing the risk of stroke in type 2 diabetes: pathophysiological and therapeutic perspectives. J Neurol 2009;256:1603–19.

[4] Bhalla A, Wang Y, Rudd A, Wolfe C. Differences in outcomes and predictors between ischemic and intracerebral hemorrhage. The South London stroke register. Stroke 2013;44:2174–81.

[5] Stead L, Gilmore R, Bellolio M, Mishra S, Bhagra A, Vaidyanathan L, et al. Hyperglycemia as an independent predictor of worse outcome in non-diabetic patients presenting with acute ischemic stroke. Neurocrit Care 2009;10:181–6.

[6] Bruno A, Williams L, Kent T. How important is hyperglycemia during acute brain infarction? Neurologist 2004;10:195–200.

[7] Bruno A, Levine S, Frankel M, Brott T, Lin Y, Tilley B, et al. Admission glucose level and clinical outcomes in the NINDS rt-PA Stroke Trial. Neurology 2002;59:669–74.

[8] Parsons M, Barber P, Chalk J, Darby D, Rose S, Desmond P, et al. Diffusion-and perfusion-weighted MRI response to thrombolysis in stroke. Ann Neurol 2002;51:28–37.

[9] Bellolio M, Gilmore R, Ganti L. Insulin for glycemic control in acute ischemic stroke (review). Cochrane Database Syst Rev 2014;(1). http://dx.doi.org/10.1002/14651858.CD005346.pub4. Art. No.: CD005346.

[10] Wright R, Frier B. Vascular disease and diabetes: is hypoglycaemia an aggravating factor? Diabetes Metab Res Rev 2008;24(5):353–63.

[11] O'Malley C, Emanuele M, Halasyamani L, Amin N. Bridge over troubled waters: safe and effective transitions of the inpatient with hyperglycemia. J Hosp Med 2008;3(S5):55–65.

[12] The Diabetes Control and Complications Trial/Epidemiology of Diabetes Interventions and Complications (DCCT/EDIC) Study research Group. Intensive diabetes treatment and cardiovascular disease in patients with type 1 diabetes. New Engl J Med 2005;353:2643–53.

[13] UK Prospective Diabetes Study (UKPDS) Group. Effect of Intensive Blood-glucose control with metformin on complications in overweight patients with type 2 diabetes (UKPDS 34). Lancet 1998;352:854–65.

[14] Hayward R, Reaven P, Wiitala W, Bahn G, Reda D, Ge L, et al. Emanuele N for the VADT Investigators. Follow-up of glycemic control and cardiovascular outcomes in type 2 diabetes. New Engl J Med 2015;372:2197–206.

[15] The Action to Control Cardiovascular Risk in Diabetes Study Group. Effects of Intensive glucose lowering in type 2 diabetes. New Engl J Med 2008;358:2545–59.

CHAPTER

146

Management of Atrial Fibrillation

K.A. Groshans[1], M.C. Leary[2,3]

[1]Walter Reed National Military Medical Center, Bethesda, MD, United States; [2]Lehigh Valley Hospital and Health Network, Allentown, PA, United States; [3]University of South Florida, Tampa, FL, United States

FREQUENCY AND PATHOGENESIS

Nonvalvular atrial fibrillation (AF) is a potent predictor of first and recurrent stroke, affecting more than 2.7 million Americans. An estimated 14–38% of patients with ischemic stroke have AF [1]. The frequency of ischemic stroke events in patients with AF is directly proportional to the mean age of populations studied. Of note, this data may be an underestimation due to the difficulty of detecting asymptomatic or paroxysmal AF [2,3]. The presence of AF greatly increases the risk of ischemic stroke, and this effect is seen more prominently as age increases and in female populations [1,2]. The prevalence of ischemic stroke associated with AF is less than 10% in patients under the age of 50, but rises to over 50% in patients above 90 years of age. In general, ischemic stroke patients with AF tend to be older, female, and have more comorbidities than patients with ischemic stroke without AF [2].

Patients with AF typically have more severe strokes as well as longer transient ischemic attacks (TIAs) than patients with ischemic stroke from other causes. Thus, strokes secondary to AF are generally associated with greater morbidity and mortality than strokes caused by other risk factors. Ischemic stroke caused by AF and cardioembolism carries a high mortality of 27.3%, compared with 21.7% for atherothrombotic strokes and 0.8% for lacunar infarcts. The higher impact of stroke associated with AF is believed to be related to cardioembolism as

Primer on Cerebrovascular Diseases, Second Edition
http://dx.doi.org/10.1016/B978-0-12-803058-5.00146-6

the primary mechanism [2–5]. In general, emboli formed in cardiac chambers are larger than those produced by atherosclerotic plaques or other sources. The majority of ischemic strokes in patients with AF are due to embolism of thrombi that form within the left atrium, particularly the left atrial appendage (LAA). LAA thrombi in AF are precipitated by sluggish flow from ineffective atrial contraction. However, other factors also contribute to LAA thrombus formation, including associated cardiovascular disease, age, and hematologic factors [3,6]. Further, an echocardiographic study indicated that LAA dysfunction can occur independently of the whole of the left atrium, suggesting a mechanism for the increased risk of ischemic stroke in paroxysmal AF [6,7].

It should be noted that cardioembolism does not account for all strokes in patients with AF. One study demonstrated that an estimated 70% of all strokes in patients with AF are secondary to cardioembolism, while the remaining 30% are composed of atherothrombotic and small vessel ischemic infarcts. This study highlights the fact that patients with AF often have several common comorbidities that may additionally contribute to the risk for stroke, such as chronic hypertension (HTN). HTN independently increases the risk for both ischemic and hemorrhage stroke. Ischemic strokes secondary to cardioembolism are also associated with a 71% rate of hemorrhagic conversion [3]. With the variation in comorbidities in each patient, evaluating risk involves clinical judgment and referring to the current evidence and classification schemes. This is critical, as estimating an individual AF patient's stroke risk is a key factor in determining an appropriate medication to prevent future stroke. As an example, a younger patient with lone AF does not have the same risk for future ischemic stroke or the same risk for developing bleeding complications as an elderly patient with several other comorbidities. Thus the management for these patients will differ.

THE ROLE OF ECHOCARDIOGRAPHY IN ESTIMATING STROKE RISK

Echocardiography is helpful in assessing risk of stroke in patients with AF. Left ventricular dysfunction seen on echocardiography significantly increases ischemic stroke risk in both low-risk patients (0.4–9.3% per year) and high-risk patients (4.4–15% per year). Other important echocardiographic findings associated with an elevated thromboembolic risk include left atrial thrombus, LAA size, LAA peak velocity, and spontaneous echocardiographic contrast. Of these factors, left ventricular dysfunction and left atrial size appear to have the greatest predictive value for thromboembolism. Therefore, these characteristic echocardiographic findings may be helpful in stratifying patients and guiding management for stroke prevention [6,8,9].

Transesophageal echocardiography (TEE) may also be helpful in the evaluation of patients after ischemic stroke. In patients with comorbid atherosclerotic disease and AF, echocardiography combined with electrocardiography may be able to elucidate the specific source of ischemic stroke in such patients. In approximately 45% of patients, residual LAA thrombus may be seen. TEE may also identify a source in patients with suspected cardioembolic stroke without a definitive history of AF. TEE is more sensitive than transthoracic echocardiography (TTE) for detecting spontaneous echodensities and atrial appendage thrombi. If cardioversion is anticipated, TEE is essential to exclude the presence of thrombi prior to cardioversion [6,8,9].

ISCHEMIC STROKE RISK FACTORS AND BLEEDING RISK

Among patients with nonvalvular AF, the annual ischemic stroke risk averages 3–4%. The absolute risk in an individual patient varies greatly based on the presence of risk factors. Although AF is considered an independent risk factor for stroke, this risk is relatively small when no comorbidities are present [1–5]. However, several risk factors have been shown to escalate ischemic stroke risk in conjunction with AF (Table 146.1). These risk factors include left ventricular systolic dysfunction or congestive heart failure, HTN, age >75 years, diabetes mellitus, previous stroke or TIA, vascular disease, and female sex. Of these factors, prior stroke or TIA and increasing age over 75 appear to be the most important. Vascular disease confers the smallest risk, with some studies not establishing a definitive correlation. Based on these factors, several risk stratification schemes have been developed. Two popular risk stratification models are the $CHADS_2$ and the CHA_2DS_2-VASc. The $CHADS_2$ does not include female sex and vascular disease. Both models try to take into account that the known risk factors for ischemic stroke in AF do not confer the same level of risk. Both $CHADS_2$ and CHA_2DS_2-VASc assign two points to prior stroke/TIA, however CHA_2DS_2-VASc additionally takes into account increasing age, which also confers greater risk. The CHA_2DS_2-VASc tool is the currently recommended model for estimating risk in a patient with AF [9–13] (Table 146.2).

Risk stratification can guide clinical decision making and help determine the risk and benefit of treating a particular patient. Patients with CHA_2DS_2-VASc scores of 0 are considered to be low risk with an ischemic stroke risk of 0.2% per year. A CHA_2DS_2-VASc score of 1 is considered intermediate risk with an annual rate of 0.6%. CHA_2DS_2-VASc scores of 2 or greater are considered to be high risk with annual risk of ischemic stroke ranging from 2.2% to 12.2% [9,11–13] (Table 146.3).

TABLE 146.1   Risk Stratification for Ischemic Stroke in Nonvalvular Atrial Fibrillation[a]

Risk Factor	No. of Ischemic Events	Univariable[b]		Multivariable[c]		CHA$_2$DS$_2$-VASc Score
		Hazard Ratio	95% CI	Hazard Ratio	95% CI	
Congestive heart failure	1905	1.28	1.21–1.35	0.98	0.92–1.04	1
Hypertension	2724	1.51	1.43–1.59	1.19	1.12–1.25	1
Age ≥75 years	4665	8.32	7.04–9.83	5.49	4.63–6.52	2
Diabetes mellitus	1070	1.34	1.25–1.43	1.19	1.11–1.27	1
Stroke/TIA						2
Ischemic stroke	2076	4.00	3.78–4.22	3.13	2.96–3.32	
Unspecified stroke	276	2.27	2.01–2.56	1.79	1.58–2.02	
TIA	546	2.05	1.88–2.24	1.59	1.45–1.73	
Vascular disease						1
MI	1261	1.24	1.17–1.33	1.05	0.98–1.12	
Peripheral arterial disease	366	1.37	1.23–1.52	1.18	1.05–1.31	
Vascular disease	1489	1.27	1.20–1.35	1.07	1.01–1.14	
Age 65–74 years	522	3.95	3.28–4.75	3.07	2.55–3.71	1
Female sex	3226	1.51	1.43–1.60	1.21	1.14–1.28	1

[a]All values are compared to the reference of patients <65 years old.
[b]Simple statistical analysis of the data.
[c]Multivariable Cox regression analysis of the study data.
Data derived from Friberg L, Rosenqvist M, Lip GYH. Evaluation of risk stratification schemes for ischaemic stroke and bleeding in 182 678 patients with atrial fibrillation: the Swedish atrial fibrillation cohort study. Eur Heart J 2012;33:1500–10.

TABLE 146.2   The CHA$_2$DS$_2$-VASc Tool

	CHA$_2$DS$_2$-VASc	Points
C	Congestive heart failure or left ventricular systolic dysfunction	1
H	Hypertension: blood pressure consistently above 140/90 mmHg (or treated hypertension on medication)	1
A2	Age ≥75 years	2
	Age 65–74	1
D	Diabetes mellitus	1
S2	Prior stroke or TIA or thromboembolism	2
V	Vascular disease (such as peripheral artery disease, myocardial infarction, aortic plaque)	1
A	Age 65–74	1
Sc	Sex category (i.e., female sex)	1
Your score		**Sum**

TABLE 146.3   CHA$_2$DS$_2$-VASc Score and Clinical Stroke Risk Estimation in Patients Not on Warfarin

CHA$_2$DS$_2$-VASc Score	Adjusted Stroke Rate (% Per Year)
0	0
1	1.3
2	2.2
3	3.2
4	4.0
5	6.7
6	9.8
7	9.6
8	6.7
9	15.2

Data derived from Lip GY, Nieuwlaat R, Pisters R, Lane DA, Crijns HJ. Refining clinical risk stratification for predicting stroke and thromboembolism in atrial fibrillation using a novel risk factor-based approach: the euro heart survey on atrial fibrillation. Chest 2010;137:263–72.

ANTITHROMBOTIC STRATEGIES FOR STROKE PREVENTION

Making the Decision to Anticoagulate

Before starting antithrombotic therapy in any AF patient, the benefit of anticoagulation versus the potential for serious bleeding complications must be considered. Based on ischemic stroke risk stratification, anticoagulation is recommended in patients with CHA_2DS_2-VASc scores of 2 or greater. In patients with CHA_2DS_2-VASc scores of 0 or 1, individual factors must be considered and risks and benefits of therapy must be weighed with the patient before proceeding with medication choice [9,11–13]. Anticoagulation has been shown to reduce the risk of stroke by about two-thirds in patients with AF. The benefits of antithrombotic therapy as stroke prevention in AF have been confirmed by numerous randomized control trials and meta-analyses. In the absence of significant contraindications, oral anticoagulant therapy is preferred for patients with moderate to high-risk AF (CHA_2DS_2-VASc scores of 2 or greater).

Stroke risk is also closely related to bleeding risk, and the benefit from stroke prevention needs to be balanced against the risk of bleeding. HAS-BLED is a scoring system developed to assess 1-year risk of major bleeding (intracranial bleeding, hospitalization, hemoglobin decrease $>2\,g/L$, and transfusion) in patients with atrial fibrillation (Table 146.4). A high HAS-BLED score (≥ 3) does not necessarily preclude anticoagulation. An elevated HAS-BLED score identifies the need for close clinical follow-up after the initiation of oral anticoagulation, but should not be used per se as a reason to avoid oral anticoagulation. It is important to assess risk on an individual basis, using CHA_2DS_2-VASc as well as HAS-BLED scores to guide shared decision-making with patients [9,14].

Treatment Options

The gold standard of therapy has always been the oral vitamin-K antagonist, warfarin. It has the most published data and most well-established profile of efficacy and safety [15,16]. Recently, novel oral anticoagulants (NOACs), including the direct factor Xa and direct thrombin inhibitors, have become FDA approved and available for the use in patients with AF [17–21]. Antiplatelet agents, including aspirin and clopidogrel, have not been shown to be as beneficial as the oral anticoagulants in the prevention of ischemic stroke in patients with AF [22–24].

Warfarin

Therapy with warfarin has been shown to have benefit in decreasing ischemic stroke relative to placebo with

TABLE 146.4   HAS-BLED Scoring System

HAS-BLED	Points
Hypertension	1
Abnormal liver function	1
Abnormal renal function	1
Stroke	1
Bleeding	1
Labile INRs	1
Elderly (age >65)	1
Drugs	1
Alcohol	1
Your score (maximum = 9)	**Sum**

acceptable rates of bleeding complications. In patients without valvular disease or prostheses, the target international normalized ratio (INR) is 2.0–3.0. This range produces the greatest therapeutic benefit with the fewest bleeding complications. A range of 2.5–3.5 is recommended in those with mechanical valves or valvular disease. Therapeutic levels of warfarin have been found to be associated with an acceptable increase in intracranial hemorrhage (ICH) versus those patients not on warfarin (0.46 vs. 0.23, respectively). Warfarin is also an option for patients with chronic kidney disease (a hypercoagulable state) and has been safely used in dialysis patients [15,16,24]. However, the narrow therapeutic margin of warfarin, in addition to associated food and drug interactions, requires frequent INR testing and dosage adjustments. These liabilities likely contribute to underuse of warfarin, and alternative therapies are needed.

Novel Oral Anticoagulants (NOACs)

There are two types of NOACs:

1. Oral direct thrombin inhibitors (DTIs): dabigatran etexilate.
2. Oral direct factor Xa inhibitors: apixaban, rivaroxaban, otamixaban, betrixaban, and edoxaban.

DTIs prevent thrombin from cleaving fibrinogen into fibrin. They bind to thrombin directly, rather than by enhancing the activity of antithrombin, as is done by heparin. The only oral DTI available for clinical use is dabigatran etexilate. Another oral agent, ximelagatran, was tested but development was discontinued in 2006 due to hepatotoxicity and cardiovascular events. Dabigatran was compared with warfarin in the landmark RE-LY trial, which was an open, prospective, randomized study, with blinded adjudication of events. In this trial, 18,113 patients with AF were assigned to dabigatran 110 mg twice a day, dabigatran 150 mg twice a day, or adjusted-dose warfarin. Dabigatran 110 mg twice a day performed similarly to warfarin in terms of efficacy, but had fewer bleeding complications. Dabigatran given at 150 mg twice a day had a lower stroke rate compared with warfarin, with less mortality but similar major bleeding rates. Of note, dyspepsia, dysmotility, and gastrointestinal reflux were twice as common in those who received dabigatran [17].

Direct factor Xa inhibitors prevent factor Xa from cleaving prothrombin to thrombin. They bind directly to factor Xa, rather than enhancing the activity of antithrombin III. Examples of oral direct factor Xa inhibitors include apixaban, rivaroxaban, otamixaban, betrixaban, and edoxaban. In recent years, several direct factor Xa inhibitors have also gone through randomized prospective clinical trials and been compared in terms of efficacy and safety to traditional warfarin therapy (Table 146.3). Thoestrial results have been favorable [17–21,25] (Table 146.5). Apixaban was associated with greater efficacy, decreased bleeding, and overall lower mortality than warfarin [18]. Rivaroxaban was deemed noninferior to warfarin in terms of efficacy as well as bleeding risks. Rivaroxaban had fewer fatal bleeding events than warfarin [19].

Additional meta-analyses have pooled the results from the RE-LY, ARISTOTLE, ROCKET AF, and ENGAGE AF-TIMI 48 trial and come to similar conclusions [17–21,25–27]. More recently, a 2014 Cochrane review compared the following factor Xa inhibitors to warfarin in patients with AF: apixaban, betrixaban, darexaban, edoxaban, idraparinux, and rivaroxaban. A lower rate of stroke and systemic embolic events as well a lower rate of death and ICH were found with the NOACs [28]. A second 2014 Cochrane review evaluated studies that compared direct thrombin inhibitors to warfarin and found no significant difference in the odds of vascular death and ischemic events [29]. Fatal and nonfatal major bleeding events, including hemorrhagic strokes, were less frequent with these agents. These meta-analyses support the broad concept that NOAC agents (direct thrombin and factor Xa inhibitors) are preferable to warfarin use in patients with AF. However, they cannot directly compare the relative advantages and disadvantages of the individual agents nor can they demonstrate whether the agents are equal in safety and efficacy. Also, the NOACs were studied in nonvalvular AF, and as a result, they are only approved for use in nonvalvular AF.

Important additional advantages of the NOAC agents are convenience (no requirement for routine testing of the INR), a high relative but small absolute reduction in the risk of ICH, lack of susceptibility to dietary interactions, and reduced susceptibility to drug interactions. Disadvantages include lack of efficacy and safety data in patients with chronic severe kidney disease, lack of easily available monitoring of blood levels and compliance, higher cost, variable availability of agents that can reverse NOAC anticoagulation, as well as the potential for unanticipated side effects that will subsequently become evident [17–19,21].

Antiplatelets

The Atrial Fibrillation Clopidogrel Trial With Irbesartan for Prevention of Vascular Events (ACTIVE) evaluated the safety and efficacy of the combination of aspirin plus clopidogrel in AF patients who were unsuitable candidates for vitamin K antagonist therapy. In those patients, the addition of clopidogrel to aspirin did reduce the risk of major vascular events, especially stroke, but also increased the risk of major hemorrhage. Antiplatelet monotherapy has not been shown to be effective in the prevention of stroke in patients with AF. However, antiplatelet therapy may be considered in patients with contraindications to oral anticoagulant therapy [22,23]. The 2014 American Heart Association guidelines for AF are recommended for choosing appropriate therapy [9] (Tables 146.4 and 146.5).

Assessing a particular patient's risk of ischemic stroke with CHA2DS-VASc scoring can guide the treatment decision between antiplatelet, warfarin, and an NOAC. Although treatment does need to be individualized, in general, the 2014 American Heart Association Guidelines recommended the management options (Table 146.6).

EVALUATION AND MANAGEMENT OF PATIENTS WITH ISCHEMIC STROKE

Following an ischemic stroke, about 11% of patients may be found to have new AF on cardiac telemetry. Many other patients may not be captured by telemetry alone. Therefore, workup for potential AF or a cardioembolic source may be warranted following an ischemic stroke event. It is recommended that patients undergo at least 24 h of cardiac monitoring following an ischemic stroke to rule out the presence of AF. In some patients, including those with cryptogenic stroke or TIA, prolonged ambulatory monitoring may help to identify paroxysmal AF. The ideal monitoring method has not been elucidated, however it is generally accepted that longer durations of monitoring are best for determining the presence of paroxysmal AF [2]. TTE is also recommended in patients with ischemic stroke who do not have any known risk factors. While TEE is more sensitive, it is less invasive and is a recommended starting point for evaluation of potential causes of emboli [8]. Laboratory studies including complete blood count, prothrombin time (PT)/INR, and lipid panels are generally indicated in all patients with ischemic stroke. A thrombin time may be included for patients using direct factor Xa inhibitors. The PT/INR is especially valuable in determining whether the

TABLE 146.5　Antithrombotic Therapy Efficacy and Safety Data

Event	Comparison	NOAC			Warfarin			Relative Risk (95% CI)	p Value[a]
		N	No of Patients With Event	Event Rate (%/yr)	N	No of Patients With Event	Event Rate (%/year)		
Stroke or systemic embolism	Dabigatran 110 mg versus warfarin	6015	182	1.53	6022	199	1.69	0.91 (0.74–1.11)	.34
	Dabigatran 150 mg versus warfarin	6076	134	1.11	6022	2	1.69	0.66 (0.53–0.82)	<.001
	Apixaban versus warfarin	9120	212	1.27	9081	265	1.60	0.79 (0.66–0.95)	.01
	Rivaroxaban versus warfarin	6958	188	1.70	7004	241	2.20	0.79 (0.66–0.96)	<.001
Major bleeding	Dabigatran 110 mg versus warfarin	6015	322	2.71	6022	397	3.36	0.80 (0.69–0.93)	.003
	Dabigatran 150 mg versus warfarin	6076	375	3.11	6022	397	3.36	0.93 (0.81–1.07)	.31
	Apixaban versus warfarin	9120	613	4.07	9081	877	6.01	0.68 (0.61–0.75)	<.001
	Rivaroxaban versus warfarin	7111	395	3.60	7125	386	3.40	1.03 (0.96–1.11)	.44
Intracranial hemorrhage	Dabigatran 110 mg versus warfarin	6015	27	0.23	6022	87	0.74	0.31 (0.20–0.47)	<.001
	Dabigatran 150 mg versus warfarin	6076	36	0.30	6022	87	0.74	0.40 (0.27–0.60)	<.001
	Apixaban versus warfarin	9120	52	0.33	9081	122	0.80	0.42 (0.30–0.58)	<.001
	Rivaroxaban versus warfarin	7111	55	0.50	7125	84	0.70	0.67 (0.47–0.93)	.02

[a]p values of less than .001 were considered the threshold for noninferiority to warfarin in these studies.

Data extracted from randomized control trial data Connolly SJ, Ezekowitz MD, Yusuf S, et al. Dabigatran versus warfarin in patients with atrial fibrillation. N Engl J Med 2009;361:1139–51; Granger CB, Alexander JH, McMurray JJV, et al. Apixaban versus warfarin in patients with atrial fibrillation. N Engl J Med 2011;365:981–92; Patel MR, Mahaffey KW, Garg J, et al. Rivaroxaban versus warfarin in nonvalvular atrial fibrillation. N Engl J Med 2011;365:883–91.

TABLE 146.6 Management Options[a]

Risk Group	Recommendation	Target INR for Warfarin
CHA$_2$DS$_2$-VASc ≥2	Warfarin or NOAC anticoagulation	2.0–3.0
CHA$_2$DS$_2$-VASc 1	Consider warfarin or NOAC if benefit outweighs risk	2.0–3.0
CHA$_2$DS$_2$-VASc 0	Shared decision-making	2.0–3.0
Prior stroke/ TIA	Warfarin or NOAC anticoagulation	2.0–3.0
Chronic kidney disease	Warfarin anticoagulation or NOAC with dose adjusted to renal function	2.0–3.0
End-stage chronic kidney disease	Warfarin anticoagulation, NOACs not recommended	2.0–3.0
Valvular disease	Warfarin anticoagulation	2.5–3.5
Mechanical prosthesis[b]	Warfarin anticoagulation	2.5–3.5

[a]Recommendations based on the 2014 AHA guidelines for the management of nonvalvular AF.
[b]Dabigatran is contraindicated in patients with AF and a mechanical heart valve.

patient was subtherapeutic on warfarin, a common cause of ischemic stroke in patients with AF [30].

In patients with AF who suffer an ischemic stroke, it is essential to restart oral anticoagulation as soon as possible. However, the timing of therapy is critical to avoid hemorrhagic conversion. In patients with small to moderate sized ischemic infarcts, warfarin (without loading) can usually be safely started after 24 h. Initiation of NOACs is generally recommended at the 48 h mark due to more rapid onset of anticoagulation than warfarin. With larger ischemic strokes or confirmed hemorrhage, withholding anticoagulation for 2 weeks is recommended. Evidence indicates that the use of aspirin may be beneficial before anticoagulation may be restarted [31].

Patients who suffer an ischemic stroke on warfarin or NOAC therapy must be evaluated for potential causes. In patients receiving warfarin, the most common cause of stroke is subtherapeutic INR or compliance with treatment regimen [9,15]. Most strokes occurring in patients on warfarin occur when the INR is below the therapeutic level of 2.0. If the INR was therapeutic, the etiology of stroke is usually due to small vessel disease, such as lacunar stroke, and in this situation, the authors suggest continuing warfarin, while further optimizing lacunar stroke risk factor (such as hypercholesterolemia or diabetes control). Others suggest increasing intensity of warfarin therapy to a range of 2.5–3.5 in this situation

[9,15]. The addition of antiplatelet therapy is generally not recommended due to the significant associated increase in bleeding complications [22,23]. In patients taking NOACs, compliance with treatment regimen is the most common cause of treatment failure [17–19]. It is important to determine whether patients on NOAC therapy have any issues taking their medications. In some patients, it may be helpful to switch from a twice-a-day drug, such as dabigatran or apixaban, to a once-a-day drug, such as rivaroxaban.

SUMMARY

There is no single perfect medication choice for patients with atrial fibrillation. The authors assert that a variety of factors should influence a physician's selection of therapy for their patient. Assessing a particular patient's risk of ischemic stroke with CHA2DS-VASc scoring can assist in making a treatment decision between antiplatelet, warfarin, and an NOAC. Additionally, assessing the patient's risk of significant bleeding with HAS-BLED scoring can also influence medication choice. We recommend that therapy should be tailored to the individual patient and their specific needs, including making an assessment of other comorbid medical issues, cost of the particular agent selected, history of patient compliance with medications, as well as patient preference.

References

[1] Wolf P, Abbott R, Kannel W. Atrial fibrillation as an independent risk factor for stroke: the Framingham study. Stroke 1991;22:983–8.
[2] Friberg L, Rosenqvist M, Lindgren A, et al. High prevalence of atrial fibrillation among patients with ischemic stroke. Stroke 2014;45:2599–605.
[3] Arboix A, Alioc J. Cardioembolic stroke: clinical features, specific cardiac disorders and prognosis. Curr Cardiol Rev 2010;6:150–61.
[4] Jorgensen H, Nakayama H, Reith J, Raaschou H, Olsen T. Acute stroke with atrial fibrillation: the Copenhagen stroke study. Stroke 1996;27:1765–9.
[5] Lin H, Wolf P, Kelly-Hayes M, Beiser A, et al. Stroke severity in atrial fibrillation: the Framingham study. Stroke 1996;27:1760–4.
[6] Hart RG. Management of atrial fibrillation. In: Welch KMA, Caplan LR, Reis DJ, Siesjo BK, Weir B, editors. Primer on cerebrovascular diseases. San Diego (CA): Academic Press; 1997. p. 786–90.
[7] Warraich H, Gandhavadi M, Manning W. Mechanical discordance of the left atrium and left atrial appendage: a novel mechanism to explain stroke in paroxysmal atrial fibrillation. J Am Coll Cardiol 2014;45:A2157.
[8] Mcnamara RL, Tamariz LJ, Segal JB, Bass EB. Management of atrial fibrillation: review of the evidence for the role of pharmacologic therapy, electrical cardioversion, and echocardiography. Ann Intern Med 2003;139:1018–33.
[9] January CT, Wann LS, Alpert JS, et al. 2014 AHA/ACC/HRS Guideline for the management of patients with atrial fibrillation: a report of the American College of Cardiology/American Heart Association Task Force on Practice Guidelines and the Heart Rhythm Society. Circulation 2014;130:E199–267.

[10] Hart R, Pearce L. Current status of stroke risk stratification in patients with atrial fibrillation. Stroke 2009;40:2607–10.

[11] Friberg L, Rosenqvist M, Lip GYH. Evaluation of risk stratification schemes for ischaemic stroke and bleeding in 182 678 patients with atrial fibrillation: the Swedish atrial fibrillation cohort study. Eur Heart J 2012;33:1500–10.

[12] Lip GY, Frison L, Halperin JL, Lane DA. Identifying patients at high risk for stroke despite anticoagulation: a comparison of contemporary stroke risk stratification schemes in an anticoagulated atrial fibrillation cohort. Stroke 2010;41:2731–8.

[13] Lip GY, Nieuwlaat R, Pisters R, Lane DA, Crijns HJ. Refining clinical risk stratification for predicting stroke and thromboembolism in atrial fibrillation using a novel risk factor-based approach: the euro heart survey on atrial fibrillation. Chest 2010;137:263–72.

[14] Pisters R, Lane DA, Nieuwlaat R, et al. A novel user-friendly score (HAS-BLED) to assess one year risk of major bleeding in atrial fibrillation patients: the Euro Heart Survey. Chest 2010;138:1093–100.

[15] Go AS, Hylek EM, Chang Y, et al. Anticoagulation therapy for stroke prevention in atrial fibrillation: How well do randomized trials translate into clinical practice? JAMA 2003;290:2685–92.

[16] Hart RG, Pearce LA, Aguilar MI. Meta-analysis: antithrombotic therapy to prevent stroke in patients who have nonvalvular atrial fibrillation. Ann Intern Med 2007;146:857–67.

[17] Connolly SJ, Ezekowitz MD, Yusuf S, et al. Dabigatran versus warfarin in patients with atrial fibrillation. N Engl J Med 2009; 361:1139–51.

[18] Granger CB, Alexander JH, McMurray JJV, et al. Apixaban versus warfarin in patients with atrial fibrillation. N Engl J Med 2011;365:981–92.

[19] Patel MR, Mahaffey KW, Garg J, et al. Rivaroxaban versus warfarin in nonvalvular atrial fibrillation. N Engl J Med 2011;365:883–91.

[20] Dentali F, Riva N, Crowther M, et al. Efficacy and safety of the novel oral anticoagulants in atrial fibrillation: a systematic review and meta-analysis of the literature. Circulation 2012;126:2381–91.

[21] Ruff CT, Giugliano RP, Braunwald E, et al. Comparison of the efficacy and safety of new oral anticoagulants with warfarin in patients with atrial fibrillation: a meta-analysis of randomized trials. Lancet 2014;383:955–62.

[22] Van Walraven C, Hart RG, Singer DE, et al. Oral anticoagulants vs aspirin in nonvalvular atrial fibrillation: an individual patient meta-analysis. JAMA 2002;288:2441–8.

[23] Connolly SJ, Eikelboom JW, Ng J, et al. Net clinical benefit of adding clopidogrel to aspirin therapy in patients with atrial fibrillation for whom vitamin K antagonists are unsuitable. Ann Intern Med 2011;155:579–86.

[24] Singer D, Chang Y, Fang M, et al. Should patient characteristics influence target anticoagulation intensity for stroke prevention in nonvalvular atrial fibrillation?: The ATRIA study. Circ Cardiovasc Qual Outcomes 2009;2:297–304.

[25] Chatterjee S, Sardar P, Biondi-Zoccai G, Kumbhani DJ. New oral anticoagulants and the risk of intracranial hemorrhage: traditional and Bayesian meta-analysis and mixed treatment comparison of randomized trials of new oral anticoagulants in atrial fibrillation. JAMA Neurol 2013;70:1486–90.

[26] Adam SS, McDuffie JR, Ortel TL, Williams Jr JW. .Comparative effectiveness of warfarin and new oral anticoagulants for the management of atrial fibrillation and venous thromboembolism: a systematic review. Ann Intern Med 2012;157:796–807.

[27] Ntaios G, Papavasileiou V, Diener HC, et al. Nonvitamin-K-antagonist oral anticoagulants in patients with atrial fibrillation and previous stroke or transient ischemic attack: a systematic review and meta-analysis of randomized controlled trials. Stroke 2012;43:3298–304.

[28] Bruins Slot KM, Berge E. Factor Xa inhibitors versus vitamin K antagonists for preventing cerebral or systemic embolism in patients with atrial fibrillation. Cochrane Database Syst Rev 2013;8:CD008980.

[29] Salazar CA, del Aguila D, Cordova EG. Direct thrombin inhibitors versus vitamin K antagonists for preventing cerebral or systemic embolism in people with non-valvular atrial fibrillation. Cochrane Database Syst Rev 2014;3:CD009893.

[30] Jauch EC, Saver JL, Adams HP, et al. Guidelines for the early management of patients with acute ischemic stroke: a guideline for healthcare professionals from the American Heart Association/ American Stroke Association. Stroke 2013;44:870–947.

[31] Chen ZM, Sandercock P, Pan HC, et al. Indications for early aspirin use in acute ischemic stroke: a combined analysis of 40 000 randomized patients from the Chinese Acute Stroke Trial and the International Stroke Trial. Stroke 2000;31:1240–9.

CHAPTER

147

Depression, Psychosis, and Agitation in Stroke

R. Tarawneh[1,3], J.L. Cummings[2,3]

[1]Cleveland Clinic, Cleveland, OH, United States; [2]Cleveland Clinic Las Vegas, NV, United States; [3]Cleveland Clinic Lerner College of Medicine of Case Western Reserve University, Cleveland, OH, United States

INTRODUCTION

Every year, stroke affects more than 15 million patients worldwide, resulting in death in 5.7 million patients and disability in another 5 million [1]. Neuropsychiatric disorders are common after stroke, including depression, anxiety, agitation, mania, apathy, emotional lability, psychosis, and fatigue (Table 147.1). Noncognitive neuropsychiatric disorders are distressing to patients, families, and caregivers in the poststroke period, and can significantly interfere with rehabilitation efforts following a stroke [1]. Moreover, these disorders have been shown to be associated with increased morbidity in the poststroke period. Despite their high prevalence and association with increased morbidity, these disorders remain underrecognized by health care professionals and may not always be adequately treated. We present the definition, clinical diagnosis, causes, and treatment of three of the most severe neuropsychiatric complications of stroke, i.e., depression, psychosis, and agitation. Common medications used for the treatment of the neuropsychiatric complications of stroke are summarized in Table 147.2.

POSTSTROKE DEPRESSION

The fifth edition of the US Diagnostic and Statistical Manual of Mental Disorders (DSM-5) [2] defines depression as depressed mood or anhedonia (loss of interest or pleasure) for at least 2 weeks, in addition to at least four of the following symptoms when they are persistent and cause impairment in social, occupational, or other important areas of functioning: substantial weight loss or gain, altered sleep pattern (insomnia or hypersomnia), psychomotor agitation or retardation, fatigue or loss of energy, feelings of worthlessness or inappropriate guilt, reduced concentration or indecisiveness, and recurrent thoughts of death or suicidal thoughts and plans.

Depression is the most common neuropsychiatric complication of ischemic stroke, with a prevalence of 25–47% in practice- and population-based studies [3]. Estimates from large systematic reviews suggest that 30–35% of patients who have had a stroke have depression up to 1 year following stroke onset. In one study of 301 patients who have had a stroke, major depression was present in 20% of the patients and minor depression (defined by the presence of more than two but less than five depressive symptoms) occurred in another 20% of patients [4]. In other studies, the prevalence of major depression (meeting *DSM-IIIR* or *DSM-IV* criteria) was estimated to be 2–31% at the time of admission, 31% at 3 months, 16% at 1 year, 19% at 2 years, and 29% at 3 years following stroke onset. Studies suggest that 15–30% of patients who have had a stroke with no symptoms of depression on admission develop depression during the first year following the stroke.

The course of poststroke depression is variable. Many patients with poststroke depression show remission within the first 3 months, whereas an estimated 13–52% continue to experience depressive symptoms after a 1-year follow-up [1], and a proportion may continue to experience depression for more than 3 years. Studies suggest that the mean duration of poststroke depression is 9 months, with cortical lesions showing a longer duration of poststroke depression than subcortical and brainstem lesions.

Poststroke depression is associated with worse functional outcomes and increased mortality after a stroke, even after correcting for other variables such as age, gender, and stroke location [3]. Patients who have had a stroke with post-stroke depression have greater impairments in activities of daily living at various time points following the stroke and are more likely

Primer on Cerebrovascular Diseases, Second Edition
http://dx.doi.org/10.1016/B978-0-12-803058-5.00147-8

TABLE 147.1   Neuropsychiatric Complications of Stroke

Neuropsychiatric Syndrome	Prevalence (%)	Putative Neuroanatomical Correlates
Depression	30–35% at admission	No clear association established
	19–52% at 2 years	Major depression: left-frontal and left-basal-ganglia lesions
Mania	≤2%	Right-hemisphere, bifrontal, basotemporal, subcortical, midline, right-basal-ganglia, and right thalamus lesions
Bipolar Disorder	Rare	Right subcortical , right basal ganglia, and right thalamus lesions
Anxiety	25–40% with depression	No clear association established
	6% without depression	With depression: left frontal cortical lesions
		Without depression: right parietal lesions
Emotional Lability	8–32%	Frontal lesions
Agitation with Aggression	6%	Left-anterior-lobe lesions
Fatigue	23–75%	No clear associations established
Apathy	35%	No clear associations established: possible associations with cingulate and subcortical structures (more common with bilateral lesions or unilateral lesions in the presence of a prior contralateral frontal lesion)
		With depression: left frontal and basal ganglia lesions
		Without depression: posterior internal capsule lesions
Personality Disorders	<1%	No clear association established
Psychosis and Psychotic Symptoms	<1–10%	No clear associations established, possibly more common with right-parietotemporal-occipital lesions
Pathological Laughing and Crying (Pseudobulbar Affect)	20%	Bilateral hemisphere lesions
Catastrophic Reaction	20%	Anterior cortical lesions
Anosognosia	24–43%	Right-hemisphere lesions

to experience ongoing functional decline even after resolution of the depressive symptoms [3]. Patients with minor or major poststroke depression have significantly less recovery in motor and language functions at 2 years compared with patients who have had a stroke with no post-stroke depression, and who had comparable stroke care. Cognitive impairment is more common in stroke patients with depression, particularly in association with left-hemisphere lesions. Stroke patients with depression were found to have a 3.4 times higher risk of mortality during a 10-year follow-up compared with those without depression in the poststroke period.

Given the high prevalence of stroke and its association with increased morbidity and mortality, current guidelines recommend that all patients be screened for mood disorders, including depression, within 6 weeks of a stroke [1]. In patients whose strokes result in significant motor or language deficits, a distinction between a normal adjustment reaction and a major depressive disorder should also be made. Persistent dysphoric mood, anhedonia, vegetative symptoms (e.g., insomnia or anorexia), and poor participation in rehabilitation point to a likely diagnosis of a major depressive disorder.

Several brief screening tools for depression have been validated in stroke cohorts and are available for use in clinical settings. These include the Beck Depression Inventory, the nine-item Patient Health Questionnaire, the Center for Epidemiologic Studies Depression Scale, the Hospital Anxiety and Depression Scale, and the Hamilton Depression Rating Scale. These tools are helpful in identifying patients who would benefit from more detailed neuropsychiatric evaluations and follow-up. Fatigue, psychomotor retardation, sleep, and appetite disturbance may represent somatic rather than neuropsychiatric complications of stroke; therefore, clinical judgment is often needed to assess whether these symptoms are indicative of poststroke depression. Furthermore, assessment of depression may be particularly challenging in the presence of aphasia, loss of vocal inflection, impaired attention, or altered facial expression patterns that may mask or mimic the symptoms of depression. Observer-rated screening tools, such as the Stroke Aphasia Depression Questionnaire and the Depression Intensity Scale Circles, may be particularly helpful in these cases.

Several studies have attempted to identify patient or stroke characteristics that are predictive of poststroke depression. Stroke severity, physical disability, cognitive impairment, social isolation, and a history of depression before the stroke have all been shown to be associated with a higher incidence of poststroke depression. Conversely, no clear associations have been established between poststroke depression and

TABLE 141.2 Psychotropic Medications for the Neuropsychiatric Complications of Stroke

Neuropsychiatric Syndrome	Psychotropic Medication	Daily Dose (mg)	Initial Dose (mg/d)	Dose Adjustment in Elderly or in Those with Medical Illness	Side Effects
Depression	Sertraline	50–200	50	Lower dose or less frequent dosing in hepatic impairment; elderly prone to hyponatremia with SSRIs; start at 12.5 mg/d in patients with Alzheimer disease	Diarrhea, nausea, headache, insomnia, sexual dysfunction, dizziness, dry mouth, fatigue, drowsiness
	Citalopram	40	20	Daily dose should not exceed 20 mg in patients with hepatic impairment, in poor CYP2C19 metabolizers, or when coadministered with CYP2C19 inhibitors (e.g., cimetidine, fluconazole, omeprazole) because of the risk of prolonged QT interval Concomitant administration with linezolid and IV methylene blue is contraindicated because of the risk for serotonin syndrome; elderly prone to hyponatremia and long QT interval	Dry mouth, nausea, sedation, insomnia Increased sweating Prolonged QT interval in higher doses
	Mirtazapine	15–45	15	Initial dose 7.5 mg in the elderly Monitor closely in patients with renal and hepatic impairment	Sedation, weight gain, increased appetite, constipation Agranulocytosis in 0.1% of patients
Bipolar Disorder Mania	Lamotrigine	100–400 (target dose for bipolar disorder is 200 mg without valproic acid use and 100 mg with valproic acid use)	Slow titration: 25 mg po qd for 2 weeks, then 50 mg po qd for 2 weeks, and finally 100 mg po qd for 1 week; start at 25 mg every other day when used with valproic acid	Dose reduction by 25% in patients with hepatic impairment without ascites and 50% in those with ascites; dose reduction in cases of severe renal impairment	Dizziness, diplopia, headache, ataxia, blurred vision, sedation, rhinitis, and Stevens-Johnson syndrome in 0.3–0.8% of children and 0.08–0.3% of adults when used with valproic acid
	Carbamazepine	400–1600 (divided into bid or tid dosing)	400 (divided into bid dosing)	Dose should be reduced by 25% in patients who are on hemodialysis or with GFR <10 mL/min; caution is necessary in patients with hepatic impairment	Ataxia, dizziness, drowsiness, nausea/vomiting, SIADH, Stevens-Johnson syndrome
	Valproic acid	750–1500 (maximum dose 60 mg/kg/d)	750 mg/d in divided doses (25 mg/kg/d for extended release)	Lower dose and slower titration in elderly and in patients with hepatic impairment, contraindicated in cases of severe hepatic impairment	Nausea/vomiting, headache, low platelet count, tremor, alopecia, sedation, ataxia, nystagmus, weight gain, acne, Stevens-Johnson syndrome, high ammonia levels, pancreatitis, increased levels of liver transaminases
	Lithium	900–2400 (divided into tid or qid dosing)	900 mg (divided into tid or qid dosing)	Lower doses in elderly	Leukocytosis, polyuria/polydipsia, hand tremor, ataxia, hypothyroidism (1–4%), confusion, memory impairment, headache
Anxiety	Sertraline	25–200	25	Lower dose or less frequent dosing in patients with hepatic impairment; elderly prone to hyponatremia with SSRIs; start at 12.5 mg/d in patients with Alzheimer disease	Diarrhea, nausea, headache, insomnia, sexual dysfunction, dizziness, dry mouth, fatigue, drowsiness
	Paroxetine	10–60	10	10 mg/d not to exceed 40 mg/d in cases of severe renal impairment (CrCl <30 mL/min); sedation and anticholinergic effects in elderly; elderly prone to hyponatremia	Nausea, insomnia, dry mouth, headache, asthenia, constipation, diarrhea, dizziness, sexual dysfunction, tremor

Continued

IX. MANAGEMENT

TABLE 147.2   Psychotropic Medications for the Neuropsychiatric Complications of Stroke,—cont'd

Neuropsychiatric Syndrome	Psychotropic Medication	Daily Dose (mg)	Initial Dose (mg/d)	Dose Adjustment in Elderly or in Those with Medical Illness	Side Effects
Fatigue	Buspirone	10–60	10–15 (divided into bid or tid dosing)	Not recommended in cases of severe renal or hepatic impairment	Dizziness
	Amantadine	200–400 (divided into bid, tid, or qid dosing)	200 (divided into bid dosing)	Dose adjustment in renal disease: 100mg every day if CrCl is 30–50mL/min, 100mg every other day if CrCl is 15–29mL/min, and 200mg every week if CrCl <15mL/min	Anxiety, anorexia, constipation, depression, peripheral edema, livedo reticularis, dry mouth, insomnia with bedtime dosing
Apathy	Methylphenidate	20–60	Immediate release form: 20–30 mg divided into bid or tid dosing	Contraindicated with glaucoma or within 2 weeks of use of MAO inhibitors, use cautiously with hypertension, heart failure, drug dependence, and alcoholism	Headache, hypertension, nausea, nervousness, psychosis, seizures, tachycardia, arrhythmias, risk of drug abuse and dependence, withdrawal associated with depression
Psychosis Agitation with Aggression Catastrophic Reaction	Quetiapine	150–750 (immediate release) 400–800 (extended release)	50	25–200mg in elderly (initially 25–50 mg/d); Dose adjustment needed in patients with hepatic impairment	Extrapyramidal syndrome/tardive dyskinesia (+/−), sedation (++), anticholinergic effects (dry mouth, constipation, urine retention) (++), orthostasis (++), prolonged QT interval (+), weight gain, hyperglycemia, hyperlipidemia, increased diastolic blood pressure, headache
	Olanzapine	10–20	5–10	Initially 1.25–2.5 mg/d; typical maintenance dose is 5 mg/d; maximum 10mg/d	Extrapyramidal syndrome/tardive dyskinesia (+), sedation (++), anticholinergic effects (dry mouth, constipation, urine retention) (++), orthostasis (+), prolonged QT interval (+), weight gain, hyperglycemia, hyperlipidemia, increased ALT, high prolactin
	Risperidone	2–6	1–2	Initially 0.25–0.5 mg/d; typical maintenance 1 mg/d; maximum 2 mg/d; dose adjustments are needed in patients with renal and hepatic impairment	Extrapyramidal syndrome/tardive dyskinesia (+++), sedation (+), anticholinergic effects [dry mouth, constipation, urine retention] (+), orthostasis (+), prolonged QT interval (+), weight gain, hyperglycemia, hyperlipidemia, increased ALT levels, high prolactin levels
Pseudobulbar Affect	Dextromethorphan/ quinidine	One capsule (20/10mg) po bid	One capsule (20/10mg) qd for 1 week	Safety not established in patients with severe renal or hepatic impairment, no dose adjustment needed for patients with mild or moderate renal or hepatic impairment	Diarrhea, dizziness, vomiting, cough, peripheral edema, increased GGT levels
	Nortriptyline	75–150	75–100 (divided into tid or qid dosing)	Initial dose is 30–50 mg/d and a maximum dose of 100mg/d in elderly (given once a day or in divided doses)	Fatigue, sedation, dry mouth, constipation, orthostasis, tachycardia, seizures, SIADH

ALT, alanine aminotransferase; *bid*, twice a day; *CrCl*, creatinine clearance; *CYP2C19*, cytochrome P450 2C19; *GFR*, glomerular filtration rate; *GGT*, gamma-glutamyl transferase; *IV*, intravenous; *MAO*, monoamine oxidase; *po*, orally; *qd*, every day; *qid*, four times a day; *SIADH*, syndrome of inappropriate secretion of antidiuretic hormone; *SSRIs*, selective serotonin reuptake inhibitors; *tid*, three times a day; *qid*, four times a day; (+/−), absent or negligible; (+), infrequent; (++), moderately frequent; (+++), frequent.

aCommonly used medications are listed in the table; the table is not intended to be inclusive of all medication options.

age, gender, level of education, stroke subtype, diabetes, or a history of prior stroke. While some studies suggest that strokes in certain anatomical locations, such as the left hemisphere, left basal ganglia, and lesions near the left frontal lobe, are associated with a higher incidence of depression, a consensus has not yet emerged on these associations [1]. As population and methodological differences between studies limit the generalizability of these findings, screening for poststroke depression is recommended in all patients.

The exact etiology and pathophysiologic mechanisms of poststroke depression remain to be elucidated; however, several hypotheses regarding the biological substrates of poststroke depression have been proposed. The biogenic amine hypothesis suggests that ischemic brain injury interrupts ascending axonal projections that contain the biogenic amines, serotonin and norepinephrine, from the brainstem to the cortex, thus resulting in reduced overall levels of serotonin and norepinephrine in axonal regions distant from the stroke location, and subsequent dysfunction of limbic structures in the basal ganglia, frontal lobe, and temporal lobe. Supporting this hypothesis is the observation that stroke patients with depression have significantly lower levels of the serotonin metabolite 5-hydroxyindoleacetic acid (5-HIAA) compared with nondepressed stroke patients with similar age, gender, and hemispheric stroke location. Another hypothesis suggests that increased production of pro-inflammatory cytokines such as interleukin(IL)-1β, tumor necrosis factor α (TNF-α), or IL-18 as a result of vascular injury results in widespread activation of the indoleamine 2,3-dioxygenase enzyme and subsequent reduction of serotonin levels in paralimbic structures such as the ventral lateral frontal cortex, polar temporal cortex, and the basal ganglia. Cytokine activation may be associated with increased activity in the hypothalamic–pituitary axis or alterations in the serotonergic system by reducing the availability of tryptophan, an essential precursor in the production of serotonin.

While some studies suggest a benefit with the use of antidepressants in the treatment of poststroke depression, other studies have failed to show significant benefits over placebo [1]. The interpretation of these results is confounded by the high rate of spontaneous resolution of poststroke depression, the small number of studies investigating antidepressants in poststroke depression, and methodological limitations of these studies. Results from large systematic reviews of pharmacologic interventions for poststroke depression suggest that antidepressant drugs are minimally effective in treating mild and moderate poststroke depression, with an increase in central nervous system and gastrointestinal side effects, but are of substantial benefit for severe poststroke depression. Evidence for the benefits of treatment of depression in the general population supports extrapolation to patients with poststroke depression.

Current guidelines [5] recommend pharmaceutical treatment (e.g., use of selective serotonin reuptake inhibitors such as citalopram, fluoxetine, or sertraline or, less commonly, heterocyclic antidepressants such as nortriptyline) for patients with poststroke depression. No controlled trials have been conducted on other classes of antidepressants (e.g., bupropion, venlafaxine, and mirtazapine) for poststroke depression treatment, although these are commonly used in clinical practice. There is no evidence of differences in efficacy between these drugs, and therefore, the choice of antidepressant should be guided by its side effect profile and patient comorbidities (e.g., antidepressants with sedative effects may be useful for patients with impaired sleep or increased anxiety). Trazodone has been associated with improved recovery in activities of daily living in patients with poststroke depression in whom cortisol suppression failed after dexamethasone administration. Retrospective studies suggest the use of stimulants (e.g., methylphenidate) in select cases, and electroconvulsive treatment in refractory cases that do not respond to medical treatment.

Treatment, when considered, should be started early in the poststroke period to improve functional outcomes and allow for maximum benefit from rehabilitation services. Symptoms such as emotional lability, anger outbursts, and executive dysfunction due to depression seem more likely to respond to treatment than other symptoms such as fatigue. Antidepressant therapy should be continued for at least 6 months after initial recovery, and follow-up is needed to monitor patient tolerance and drug efficacy [1]. Psychosocial support should be included in therapeutic planning with emphasis on increasing social interactions and physical activity. Cognitive-behavioral therapy may be considered in some patients as an alternative to antidepressant therapy, but there is still no clear evidence of its use in poststroke depression.

There is limited evidence to support the routine use of antidepressants for the prevention of poststroke depression. Although some reports suggest potential intermediate and long-term benefits of early antidepressant therapy in reducing the incidence of poststroke depression, the evidence is not adequate to support the routine use of antidepressants in the poststroke period, particularly given the high rate of spontaneous remission of early depressive symptoms and the relatively small size of these study cohorts.

POSTSTROKE PSYCHOSIS

Poststroke psychosis is a rare but serious complication of stroke, which may develop acutely following a stroke and last for days to months [1]. Psychosis refers to significant distortion of thought content. Psychotic symptoms include delusions, hallucinations, and ideas of reference. Delusions are fixed false beliefs that are resistant to change despite conflicting evidence (including delusions of persecution, theft, or infidelity). Hallucinations are abnormal perceptions in the absence of a corresponding sensory stimulus, which are not experienced by others. Hallucinations may occur in any sensory modality, with visual hallucinations being the most common in the post-stroke period.

It is estimated that 0.4–3.1% of patients who have had a stroke have a psychotic disorder [6], with a median interval of 6.1 months following a stroke. The incidence of a single episode of psychotic symptoms that does not meet the criteria for a psychotic disorder is likely to be higher. In one study, delusions were reported in 3–10% and hallucinations in 4% of patients who have had a stroke [1].

Several clinical tools are available for the evaluation of psychotic symptoms following a stroke. Most researchers use the Neuropsychiatric Inventory [7], which assesses the frequency and severity of any type of delusions or hallucinations, to evaluate poststroke psychosis. The other scales used to validate poststroke psychosis are the Positive and Negative Symptom Scale and the Brief Psychotic Rating Scale.

A higher incidence of poststroke psychosis has been reported in the presence of significant cortical or subcortical brain atrophy at the time of the stroke. No associations between psychotic symptoms and stroke locations have been clearly established in large studies; however, reports suggest that visual hallucinations may be more common following occipital strokes, and auditory hallucinations are more frequently associated with subcortical and brainstem strokes than with cortical strokes. Other studies report a higher incidence of poststroke psychosis with right-sided lesions. In a cross-sectional study of 641 patients who have had a stroke, four patients developed auditory hallucinations in the setting of right temporal strokes [8]. Risk factors for poststroke psychosis identified in other studies include right-hemisphere lesions and seizures, which may occur either before or shortly after the onset of psychosis.

Identifying whether psychotic features in a patient who have had a stroke represent emergence of prestroke psychiatric disease, complications of poststroke depression, or delirium in the setting of acute illness and hospitalization presents a diagnostic challenge for even the most experienced physicians [1]. Poststroke psychotic disorders may be particularly difficult to differentiate from poststroke depression with psychotic features. Given the frequent co-occurrence of poststroke psychosis and seizures in most studies, it is important to exclude seizures as an underlying cause of psychotic symptoms in the poststroke period.

There are limited data regarding the course or outcome of poststroke psychosis and the best modalities for treatment or prevention of psychotic symptoms in the poststroke period. Consultation with a psychiatrist for diagnostic and therapeutic planning is recommended. Two therapeutic approaches have been suggested for poststroke psychosis: treatment with antiepileptic drugs and treatment with antipsychotic medications. Antiepileptic drugs include valproate, carbamazepine, oxcarbazepine, lamotrigine, and phenytoin. Some antiepileptic drugs such as levetiracetam may worsen psychosis and behavioral abnormalities, particularly in the presence of an underlying psychiatric disease.

Studies suggest a higher mortality rate associated with the use of antipsychotics in elderly patients with dementia, and the risks and benefits of pharmacologic treatment with antipsychotics must be carefully weighed [9]. If stroke-related psychotic symptoms are mild and not distressing or harmful to the patient or caregivers, drug treatment may not be necessary. When pharmacologic treatment is recommended, atypical antipsychotics such as risperidone (0.5–1 mg given orally twice a day; liquid risperidone can be used as needed to facilitate administration), quetiapine (25–50 mg given twice daily orally), and olanzapine (2.5 mg by the oral or sublingual route) are commonly used. Comparative efficacy data are not available. Patients should be closely monitored to assess response to therapy and the dose should be titrated as needed. The underlying premorbid psychiatric disease should be vigorously sought. Cognitive assessments and screening of underlying depression are also recommended. In the absence of premorbid cognitive impairment, most poststroke psychotic symptoms resolve spontaneously in the weeks to months following the stroke, so long-term treatment with antipsychotics is often not required.

One retrospective study suggested that psychosis is associated with significantly increased mortality following a stroke; however, the exact mechanism underlying this association remains to be determined.

POSTSTROKE ANXIETY AND AGITATION

Anxiety symptoms include feeling wound up, tense, and restless, difficulty concentrating, irritability, increased muscle tension and sweating, fatigue, and sleep impairment [1]. Poststroke anxiety symptoms may also include posttraumatic stress symptoms, anticipatory anxiety about the risk of recurrence, and somatization. Agitation is defined as excessive motor activity, physical aggression, or verbal aggression in a patient with observed or inferred evidence of emotional distress [10].

Anxiety is an underreported complication of stroke that may significantly interfere with poststroke rehabilitation. Poststroke anxiety occurs often in the form of a generalized anxiety, rather than as discrete panic attacks, and may occur alone or in association with poststroke depression. Anxiety symptoms that are out of proportion to any external threat and are present for more than 6 months constitute the diagnosis of a generalized anxiety disorder (GAD). The prevalence of poststroke GAD is estimated to be 4–28%, with a higher prevalence of poststroke generalized anxiety symptoms that do not meet the criteria for GAD. Estimates suggest that 26% of men and 39% of women who have had a stroke develop anxiety in the poststroke period [11].

No clear association between poststroke anxiety and age, social class, family history of psychiatric disease, or stroke location has been established. There is some evidence to suggest that anxiety with depression may be more common in females than in males and anxiety without depression may be related to alcohol use. Some reports suggest that anxiety in association with comorbid depression is more common following left-cortical strokes, whereas isolated depression without anxiety is associated with left-subcortical strokes, and isolated anxiety in the absence of depression occurs more commonly following right-hemisphere locations. Severe anxiety that is associated with verbally or physically aggressive behavior in the poststroke period appears to be more frequent following left-anterior-hemisphere strokes.

There is a paucity of randomized clinical trials to guide the treatment of poststroke anxiety. The choice of agents is often influenced by the presence or absence of comorbid depression. Most anxiety symptoms respond well to antidepressants, particularly those with sedating properties. However, anxiety may be worsened by stimulants used in the treatment of depression. In the absence of response to antidepressants alone, other anxiolytics such as buspirone, alone or in combination with antidepressants, may be considered, given their low addictive potential and tolerable side effect profile. Long-term treatment with benzodiazepines should generally be avoided because these agents may worsen poststroke cognitive impairment, and even short-term treatment may cause paradoxical agitation in elderly patients.

Agitation is most commonly treated with antipsychotics as described earlier in the treatment of psychosis. Anecdotal evidence supports the use of beta-blockers, lithium, and carbamazepine in the treatment of severe poststroke agitation with aggressive behavior.

SUMMARY

Depression, psychosis, and agitation are serious neuropsychiatric complications of stroke that are associated with increased morbidity and mortality in the poststroke period [12]. These disorders are often underrecognized and undertreated in the poststroke period. Early diagnosis of these disorders will allow for more effective treatments and improvements in patient outcomes. Major depression may be severe and last for several years following a stroke. Poststroke depression is more frequent in the presence of functional and cognitive impairment, poor social support, and a history of depression. Certain stroke locations such as the left frontal and left basal ganglia may be associated with a higher incidence of poststroke depression. Poststroke depression can be effectively treated, which will likely result in reduced morbidity and mortality in the poststroke period. Psychosis is a relatively rare complication of stroke that is often seen in association with underlying cortical atrophy, premorbid psychiatric disease, and right parietotemporal strokes. Poststroke psychosis may be a manifestation of underlying poststroke epilepsy, which needs to be aggressively sought and treated. Given their association with increased mortality, particularly in elderly patients, antipsychotics should be used judiciously to control psychotic symptoms in the poststroke period. Agitation is associated with increased motor activity and aggression. Poststroke anxiety occurs often in association with poststroke depression and responds well to antidepressant treatment. Long-term treatment with benzodiazepines should be avoided, given their association with cognitive impairment. Further studies are needed to better understand the underlying pathophysiologic mechanisms of poststroke neuropsychiatric syndromes, which would allow for more targeted and individualized therapies.

References

[1] Hackett ML, Kohler S, O'Brien JT, Mead GE. Neuropsychiatric outcomes of stroke. Lancet Neurol 2014;13(5):525–34.
[2] American Psyhciatric Association. Diagnostic and statistical manual of mental disorders. 5th ed. Arlington: American Psychiatric Publishing; 2013.
[3] Williams LS. Depression and stroke: cause or consequence? Semin Neurol 2005;25(4):396–409.
[4] Kishi Y, Robinson RG, Kosier JT. The validity of observed depression as a criteria for mood disorders in patients with acute stroke. J Affect Disord 1996;40(1–2):53–60.
[5] Miller EL, Murray L, Richards L, et al. Comprehensive overview of nursing and interdisciplinary rehabilitation care of the stroke patient: a scientific statement from the American Heart Association. Stroke 2010;41(10):2402–48.
[6] Rabins PV, Starkstein SE, Robinson RG. Risk factors for developing atypical (schizophreniform) psychosis following stroke. J Neuropsychiatry Clin Neurosci 1991;3(1):6–9.

[7] Cummings JL. The Neuropsychiatric Inventory: assessing psychopathology in dementia patients. Neurology 1997;48(5 Suppl. 6):S10–6.

[8] Lampl Y, Lorberboym M, Gilad R, Boaz M, Sadeh M. Auditory hallucinations in acute stroke. Behav Neurol 2005;16(4):211–6.

[9] Corbett A, Ballard C. Antipsychotics and mortality in dementia. Am J Psychiatry 2012;169(1):7–9.

[10] Cummings J, Mintzer J, Brodaty H, et al. Agitation in cognitive disorders: International Psychogeriatric Association provisional consensus clinical and research definition. Int Psychogeriatr 2015;27(1):7–17.

[11] Burvill P, Johnson G, Jamrozik K, Anderson C, Stewart-Wynne E. Risk factors for post-stroke depression. Int J Geriatr Psychiatry 1997;12(2):219–26.

[12] Robinson RG, Paradiso S. Depression, psychosis, and agitation in stroke. Primer on cerebrovascular disease. Academic Press; 1997.

CHAPTER

148

Treatment of Vasculitis

R. Tehrani[1], R. Hariman[2]

[1]Loyola University Chicago, Stritch School of Medicine, Maywood, IL, United States; [2]Medical College of Wisconsin, Milwaukee, WI, United States

INTRODUCTION

Vasculitis, in general, is the inflammation of blood vessels. However, not all cases of vasculitis are identical. Vasculitis can vary based on a number of characteristics such as size of blood vessel involved, target organs, etiology, and the underlying pathogenesis which can range from infectious, malignant to connective tissue disease. When involving the central nervous or peripheral nervous system, diagnosis and treatment can be difficult due to variable presentations, lack of highly sensitive and specific diagnostic tests, and vasculitis mimics. A prompt diagnosis with treatment is essential for the prevention of morbidity and mortality (Table 148.1). We use categories from the 2012 International Chapel Hill Consensus Conference on the Nomenclature of Vasculitides to subgroup a number of vasculitides that can affect the central and peripheral nervous system, with focus on treatments [1].

LARGE-VESSEL VASCULITIS

Giant cell arteritis (GCA) is an immune-mediated vasculitis of medium- to large-sized arteries, which commonly involves cranial arteries, but may affect the aorta as well [2]. Clinical manifestations include headaches, fever, visual loss, limb claudication, fatigue, weight loss, and polymyalgia rheumatica symptoms. Inflammatory markers including ESR and CRP should be elevated. Before a diagnosis of GCA is confirmed with a temporal artery biopsy, treatment with high-dose corticosteroids should be initiated once clinical suspicion is present that the patient has this disease [2]. Typical starting doses are 60–80 mg/day until symptoms disappear and inflammatory markers normalize, typically 2–4 weeks after initiation. Typical tapers involve reducing the dose of prednisone by a maximum of 10% of the total daily dose every 1–2 weeks after reaching 40 mg/day. Flares, indicated by a return of symptoms or elevation of

TABLE 148.1   Vasculitis Treatment Options

Vasculitis		Treatment
Large-vessel vasculitis	Giant cell arteritis	High-dose glucocorticoids, tapered over 1–2 years
	Takayasu arteritis	Glucocorticoids initially, steroid-sparing agent
Medium-vessel vasculitis	Polyarteritis nodosa	Glucocorticoids plus cyclophosphamide for severe disease, methotrexate or azathioprine for mild disease
Small-vessel vasculitis (ANCA associated)	Granulomatosis with polyangiitis (Wegener's)	Cyclophosphamide or rituximab for severe disease along with glucocorticoids Glucocorticoids plus azathioprine or methotrexate for mild disease
	Microscopic polyangiitis	Cyclophosphamide or rituximab for severe disease along with glucocorticoids Glucocorticoids plus azathioprine or methotrexate for mild disease
	Eosinophilic granulomatosis with polyangiitis (Churg–Strauss)	Cyclophosphamide or rituximab for severe disease along with glucocorticoids Glucocorticoids plus azathioprine or methotrexate for mild disease
Variable vessel vasculitis	Behcet's	Glucocorticoids plus cyclophosphamide, azathioprine or tumor necrosis factor inhibitors
Single-organ vasculitis	Primary angiitis of the central nervous system	Glucocorticoids plus cyclophosphamide or rituximab
Vasculitis associated with systemic disease	Systemic lupus erythematosus	Glucocorticoids plus cyclophosphamide or rituximab
	Sjogren's syndrome	Further studies required

inflammatory markers, may require a lengthened taper. Patients generally will be on corticosteroid therapy for 1–2 years. Differing results have been seen in randomized controlled trials utilizing IV and oral steroids, but there has been no consensus that IV steroids have shown greater efficacy.

Given the average length of time utilizing corticosteroid treatment, various steroid-sparing agents have been investigated. Methotrexate (MTX) has been evaluated in three randomized controlled trials of GCA with divergent results. The trials suggest that MTX is, at best, a moderately effective steroid-sparing agent. Recently, case reports have suggested that the use of tocilizumab may be effective in patients with GCA in whom it has been difficult to taper glucocorticoids. There is evidence that IL-6 may be important in the pathogenesis of GCA. In one of these studies, 22 GCA patients with refractory disease were treated with tocilizumab (8 mg/kg once monthly) [3]. After a median follow-up of 9 months, 17 patients achieved remission, and the median prednisone dose was tapered from 19 mg to 5 mg daily. Tocilizumab remains a promising steroid-sparing agent, but more studies need to be performed.

Takayasu arteritis is a chronic inflammatory arteritis affecting large vessels, predominantly the aorta and its main branches. Vessel inflammation leads to wall thickening, fibrosis, stenosis, and thrombus formation [4]. Acute inflammation can destroy the arterial media and lead to aneurysm formation. A two-stage process has been suggested with a "prepulseless" phase characterized by nonspecific inflammatory features, followed by a chronic phase with development of vascular insufficiency. The disease commonly presents in the second or third decade of life. Symptoms consist of fever, malaise, weight loss, arthralgias, myalgias, diminished or absent pulses, limb claudication, and vascular bruits. Neurological features secondary to hypertension and ischemia include postural dizziness, seizures, headaches, and amaurosis. Treatments should aim to control disease activity and preserve vascular competence, while minimizing long-term side effects.[4] Steroids remain the mainstay or treatment for Takayasu arteritis, but response may be related to the stage of disease at which this treatment is introduced. Glucocorticoid-sparing agents such as azathioprine, mycophenolate, MTX, tocilizmab, leflunomide, anti-TNF alpha inhibitors, and cyclophosphamide have been tried, but studies have not shown one medication to be clearly superior to others. Providers may initiate one medication for 4–6 months, and switch immunosuppressants if there is lack of efficacy.

MEDIUM-VESSEL VASCULITIS

Polyarteritis nodosa (PAN) is a necrotizing angiitis involving the medium-sized arteries characterized by predominant skin, muscle, kidney, and gastrointestinal tract involvement. Peripheral neuropathies may be a presenting manifestation of PAN and can occur in 60% of patients. Central nervous system abnormalities develop in 40% of patients and include encephalopathies, subarachnoid hemorrhage, seizures, strokes, and cranial neuropathies. Given the strong association with hepatitis (especially, hepatitis B), hepatitis serologies should be checked when evaluating a patient with PAN.

Patients with concurrent PAN and hepatitis virus infection benefit from treatment with antivirals, but the timing of therapy relative to the use of immunosuppressive agents depends upon the severity of the vasculitis. In moderate to severe PAN, treatment typically entails both high-dose glucocorticoids and a second immunosuppressive drug, such as cyclophosphamide, followed by azathioprine or MTX for remission maintenance [5]. In patients with moderate to severe disease, but without organ or life-threatening involvement or progressive mononeuritis multiplex, oral glucocorticoids (prednisone 1 mg/kg) should be initiated with a gradual taper over 6–8 months. In severe disease with organ or life-threatening disease, glucocorticoid therapy such as intravenous methylprednisolone 500–1000 mg once daily for 3 days, followed by oral glucocorticoid therapy provides more rapid control of disease, although there are no randomized trials comparing these regimens in this patient population [5].

The optimal route, dose, or duration of treatment with cyclophosphamide for PAN is uncertain, and there are no randomized trials comparing intravenous and oral routes of therapy for PAN. Despite this, traditionally intravenous cyclophosphamide has been used in this disorder. Cyclophosphamide is generally continued for 4–6 months and then an alternative maintenance immunosuppressant such as azathioprine or MTX is often used for 18 months or longer depending on symptoms. Alternative immunosuppressants such as mycophenolate or rituximab have been used in resistant disease.

SMALL-VESSEL VASCULITIS

Granulomatosis with polyangiitis (Wegener's), microscopic polyangiitis (MPA), and eosinophilic granulomatosis with polyangiitis (Churg–Strauss) are diseases with similar clinical features which are classified under ANCA-associated vasculitidies. ANCA-associated vasculitides affect small- to medium-sized blood vessels, with a predilection for the respiratory tract and kidneys [6]. Neurological features may be due to contiguous granulomas from primary sites in the nasopharynx causing cranial neuropathies, cavernous sinus compression, diabetes insipidus, and vasculitis causing peripheral neuropathies, mononeuritis multiplex, encephalopathies, and stroke.

Cyclophosphamide and glucocorticoids have been standard therapy for ANCA-associated vasculitis remission induction since the early 1970s [6]. This regimen transformed the usual treatment outcome of death to a strong likelihood of disease control and temporary remission. Given the side effects of cyclophosphamide including infertility, cytopenias, infections, bladder injury, and malignancy, 2010 studies including the RAVE trial have demonstrated noninferiority with the use of rituximab for induction [6]. Initial therapy will depend on the severity of the disease and the organ systems involved. For mild disease (no signs of active glomerulonephritis, no organ-threating manifestation), treatment typically consists of a combination of glucocorticoids with MTX or azathioprine. For moderate to severe disease (organ-threatening or life-threatening manifestations), opinions differ on the initial induction immunosuppressive regimen. Glucocorticoids (oral or intravenous pulse dose) along with a combination of either cyclophosphamide (oral or intravenous) or rituximab is typically initiated. The RAVE trial suggested that severe ANCA-associated vasculitis treated with rituximab produced similar results compared to cyclophosphamide treatment. Limitations of this trial include excluding patients with alveolar hemorrhage severe enough to require mechanical ventilation and dose with advanced renal dysfunction (serum creatinine level >4.0 mg/dL) [6]. Studies have also tended to suggest that oral cyclophosphamide may have more proven efficacy compared to intravenous cyclophosphamide. Plasma exchange can be considered in severe renal dysfunction or diffuse alveolar hemorrhage as well.

VARIABLE VESSEL VASCULITIS

Behcet's disease can affect multiple organs, with central nervous system involvement not being uncommon. This is in contrast to the peripheral nervous system where involvement is rare and isolated to case reports. Neuro-Behcet's syndrome occurs in both acute and chronic progressive forms. The acute form typically presents as an acute meningoencephalitis that is self-limiting and responsive to steroids. This is unlike the chronic form which is progressive and involves symptoms such as ataxia and dementia. The chronic progressive form has been resistant to treatments to date (2016), but they are still used in management. Treatment consists of intravenous steroids followed by a taper of oral steroids in conjunction with an immunosuppressive such as cyclophosphamide, azathioprine, or tumor necrosis factor inhibitors even though there is no evidence to their efficacy in studies [7,8].

SINGLE-ORGAN VASCULITIS

Primary angiitis of the central nervous system is rare. Controversy exists in the literature on the best method for diagnosis ranging from biopsy to cerebral angiographic findings. Biopsy will histologically help exclude mimics. A number of presentations have been noted from focal or diffuse neurological symptoms to cerebral vascular events and cognitive dysfunction. No prospective studies to date have been done in regards to treatment of primary angiitis of the central nervous system. Current regimens have been based on the regimens of other systemic vasculitides. High-dose glucocorticoids are used for initial treatment in addition to oral or intravenous cyclophosphamide. Reports of the successful use of rituximab have been described, but further studies are needed [9].

VASCULITIS ASSOCIATED WITH SYSTEMIC DISEASE

Systemic lupus erythematosus is an autoimmune disease that can affect a number of organ systems including the central and peripheral nervous systems via antibody mediated or thrombotic disease. The clinical course can be variable. Central nervous system manifestations can range from headaches, seizures, demyelinating syndromes, and myelopathy, to cognitive dysfunction. Involvement of the peripheral nervous system is uncommon and typically related to vasculitis. It can present with autonomic disorders, cranial or peripheral neuropathies, and acute inflammatory demyelinating polyneuropathy. Unfortunately, there is a lack of controlled randomized clinical trials for the involvement of the nervous system. A number of treatments ranging from steroids and immunosuppressive therapies such as rituximab and cyclophosphamide are noted in the literature, but require further studies for recommendations [8].

Like systemic lupus erythematosus, Primary Sjogren's syndrome can also affect the central and peripheral nervous system in a variable clinical course. Central nervous system manifestations can range from encephalopathy to cerebral white matter and spinal cord lesions. Peripheral nervous system manifestations can include mononeuropathies, cranial neuropathies, polyneuropathies, and autonomic neuropathies. Several mechanisms of action have been hypothesized for the mechanism of neurological involvement, including vascular injury caused by anti-Ro and antineuronal antibodies [8]. Unfortunately, treatments have not been established and are based on empirical and clinical experience [10].

References

[1] Jennette JC, Falk RJ, Bacon PA, Basu N, Cid MC, Ferrario F, et al. 2012 Revised International Chapel Hill Consensus Conference nomenclature of vasculitides. Arthritis Rheum January 2013;65(1):1–11.

[2] Tehrani R, Ostrowski RA, Hariman R, Jay WM. Giant cell arteritis. Semin Ophthalmol 2008;23:99–110.

[3] Loricera J, Blanco R, Hernández JL, Castañeda S, Mera A, Pérez-Pampín E, et al. Tocilizumab in giant cell arteritis: Multicenter open-label study of 22 patients. Semin Arthritis Rheum June 2015;44(6):717–23.

[4] Johnston SL, Lock RJ, Gompels MM. Takayasu arteritis: a review. J Clin Pathol 2002;55:481–6.

[5] Guillevin L, Cohen P, Mahr A, Arène JP, Mouthon L, Puéchal X, et al. Treatment of polyarteritis nodosa and microscopic polyangiitis with poor prognosis factors: a prospective trial comparing glucocorticoids and six or twelve cyclophosphamide pulses in sixty-five patients. Arthritis Rheum 2003;49(1):93.

[6] Stone JH, Merkel PA, Spiera R, et al. Rituximab versus cyclophosphamide for ANCA-associated vasculitis. N Engl J Med 2010;363:221–32.

[7] Saip S, Akman-Demir G, Silva A. Neuro-Behcet syndrome. Handb Clin Neurol 2014;121:1703–23.

[8] Bougea A, Anagnostou E, Spandideas N, Triantafyllou N, Kararizou E. An update of neurological manifestations of vasculitides and connective tissue diseases: a literature review. Einstein December 2015;13(4):627–35.

[9] De Boysson H, Arquizan C, Guillevin L, Pagnoux C. Rituximab for primary angiitis of the central nervous system: report of 2 patients from the French COVA cohort and review of the literature. J Rheumatol December 2013;40(12):2102–3.

[10] Gono T, Kawaguchi Y, Katsumata Y, Takagi K, Tochimoto A, Baba S, et al. Clinical manifestations of neurological involvement in Primary Sjogren's syndrome. Clin Rheumatol April 2011;30(4):485–90.

CHAPTER

149

Endovascular Therapy for the Treatment of Cerebrovascular Disease

L. Rangel-Castilla, H.J. Shakir, A.H. Siddiqui

University at Buffalo, State University of New York, Buffalo, NY, United States

Abbreviations

3D three dimensional
ACoA anterior communicating artery
ADAPT a direct aspiration first-pass technique
AP anteroposterior
AVM arteriovenous malformation
CT computed tomographic
CTA CT angiography
DMSO dimethyl sulfoxide
DSA digital subtraction angiography
FRED Flow redirection endoluminal device
IA intracranial aneurysm
ISAT International Subarachnoid Aneurysm Trial
ISUIA International Study of Unruptured Intracranial Aneurysms
IVH intraventricular hemorrhage
LVIS Low-profile visualized intraluminal support
MCA middle cerebral artery
NBCA N-butylcyanoacrylate
NIHSS National Institutes of Health Stroke Scale
PED pipeline embolization device
SWIFT SOLITAIRE FR with the intention for thrombectomy
TICI thrombolysis in cerebral infarction
tPA tissue plasminogen activator
TREVO 2 Trevo versus Merci Retrievers for Thrombectomy Revascularization of Large Vessel Occlusions in Acute Ischemic Stroke
VA vertebral artery

INTRODUCTION

Endovascular approaches to the central nervous system have evolved tremendously over the last decade. It is one of the most rapidly growing specialties in neurosurgery and in medicine. The early 1980s were marked by primitive technologies associated with long and risky procedures and high rates of complications, often resulting in major catastrophes. As of 2016, neuroendovascular technology and procedures have evolved to the point that nowadays these interventions are performed routinely and safely at most major medical centers. Some of the applications of neuroendovascular technology in the treatment of cerebrovascular diseases include ischemic stroke, intracranial aneurysms (IAs), intracranial arteriovenous malformations (AVMs) and fistulas, and extracranial vascular diseases (e.g., carotid artery stenosis). In this chapter, we present the current status of neuroendovascular management of these cerebrovascular diseases.

STROKE

Stroke is the leading cause of long-term disabilities in America and the second most common cause of mortality worldwide. Ischemic stroke is the prevalent stroke type in 87% of patients. Less than 4% of them are treated with intravenous thrombolysis with recombinant tissue plasminogen activator (tPA) [1]. Most patients do not qualify for tPA therapy due to delayed presentation or multiple exclusion criteria. Endovascular management is an alternative for some of these patients. Intraarterial revascularization is indicated in patients with National Institutes of Health Stroke Scale (NIHSS) scores of ≥8, which may indicate a large-vessel occlusion. Endovascular revascularization targets patients with large-vessel occlusion. Exceptions include patients with severe visual deficit or isolated severe aphasia for whom endovascular therapy could be indicated even if the NIHSS score is <8. Endovascular technology for stroke has also evolved over the last few years [1,2]. Most recent trials have reported successful recanalization in approximately 85% of the cases compared to 50% with earlier technologies [e.g., pharmacological intraarterial thrombolysis or mechanical thrombectomy with

Primer on Cerebrovascular Diseases, Second Edition
http://dx.doi.org/10.1016/B978-0-12-803058-5.00149-1

the Merci retriever (Stryker Neurovascular)] [3]. With the publication of five randomized controlled trials in 2015 [4–8], mechanical thrombectomy, when used in combination with intravenous tPA, has demonstrated a significant radiographic and clinical benefit over traditional strategies with intravenous tPA alone. These results have placed endovascular therapy at the forefront of stroke treatment, redefining the standard of care (Fig. 149.1) [4,8,9]. Currently, stent retriever thrombectomy and primary aspiration thrombectomy are the most commonly used endovascular techniques in stroke intervention [2,10].

Stent Retriever Thrombectomy

Stents initially originated as an adjunct to treatment for IAs. Self-expanding stents were originally used to prevent coils from herniating into parent vessel during aneurysm embolization. However, they were noted to be effective anecdotally in retrieval of coils or clot debris prior to detachment [11]. These initial reports resulted in rapid development of a whole class of stent retrievers to capture intravascular thrombus and restore blood flow immediately to the affected brain territory by displacing the clot against the artery wall. The mechanism

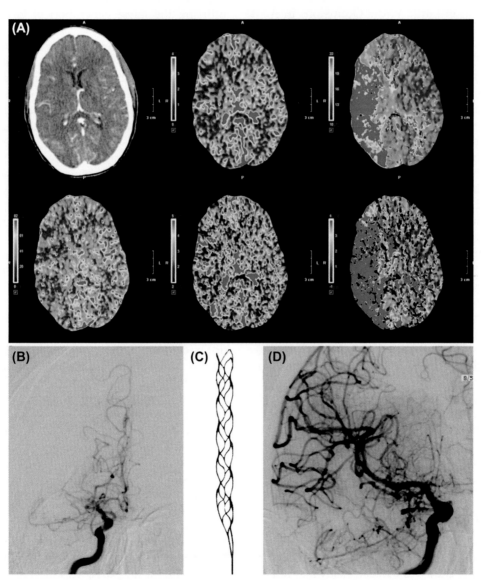

FIGURE 149.1 Endovascular thrombectomy. A 55-year-old man with a history of atrial fibrillation and recent surgery presented to the emergency department with acute stroke symptoms with an NIHSS score of 15. The patient was not a candidate for tPA therapy. (A) Computed tomographic (CT) perfusion image demonstrates increased time to peak with preserved volume in the right middle cerebral artery (MCA) territory. The patient was taken for emergent endovascular thrombectomy with stentriever. (B) Anteroposterior (AP) angiogram shows occlusion of the proximal segment of the MCA (TICI 0). (C) Illustration of a stentriever device (Trevo, Stryker). (D) AP angiogram demonstrates complete revascularization (TICI 3) after one stentriever pass.

is as follows: once deployed, the stentriever within the occluded segment is left for several minutes (typically 5 min); the clot becomes engaged and entrapped within the struts of the device; the clot and stent are retrieved together into a distal cervical guide (including balloon guides) while manually aspirating through the guide catheter (Fig. 149.1). The clot length determines the size of the stent retriever that is used. Two important trials demonstrated the superiority of stentrievers over first-generation mechanical thrombectomy devices (e.g., Merci) [12,13], the SOLITAIRE FR With the Intention for Thrombectomy (SWIFT) trial and the Thrombectomy Revascularization of large Vessel Occlusions (Trevo) versus Merci Retrievers for Thrombectomy Revascularization of Large Vessel Occlusions in Acute Ischemic Stroke (TREVO 2) trial [12,13]. Results of these trials led to Food and Drug Administration approval of the SOLITAIRE FR (ev3-Covidien) and Trevo (Stryker Neurovascular) stentrievers for treatment of acute large-vessel occlusion. As experience with stent retrievers in the treatment of acute large-vessel occlusion increased over time, new problems were encountered. One of these new problems was fragmentation and loss of clot particles into the same or different vascular territory, which could lead to partial recovery, neurological deterioration, perioperative morbidity, distal infarctions, and perioperative morbidity and ultimately result in poor outcomes or death.

Some of the strategies to reduce embolic complications include proximal aspiration with a balloon guide catheter, distal aspiration with an intermediate catheter, or the novel secure approach to prevent clot fragmentation: the Lazarus Cover (Medtronic). The Lazarus Cover device envelops a stent retriever and protects clot captured to maximize clot removal while eliminating migration of emboli to distal territory. In in vitro stroke models in which the stent retriever in conjunction with the Cover device was compared with a conventional guide catheter, stent retriever, and balloon guide catheter, application of the Cover device resulted in higher successful recanalization rates with no embolic events [14]. Hence, utilization of the Cover device proved to be more effective than use of the conventional guide catheter or balloon guide catheter.

Primary Aspiration Thrombectomy

Advances in endovascular technology have included the development of very large, easily trackable, aspiration thrombectomy catheters that can easily and reliably navigate into large intracranial arterial vessels. One of these advances includes a technique known as a direct aspiration first-pass technique (ADAPT), which involves application of the newest generation of large-bore aspiration catheters as a first approach for thrombectomy

and has demonstrated very promising results. In a multicenter prospective trial that included 98 acute ischemic stroke patients with 100 occluded large cerebral vessels, 78% achieved thrombolysis in cerebral infarction (TICI) 2b/3 revascularization [15]. The addition of a stentriever improved the TICI 2b/3 revascularization rate to 95%. The average time from groin puncture to at least TICI 2b recanalization was 37 min. Patients improved from a median admission National Institutes of Health Stroke Scale (NIHSS) score of 17 (12–21) to a 7.3 (1–11) at discharge. There were no symptomatic intracerebral hemorrhages. The authors concluded that ADAPT is a fast, safe, simple, and effective method that facilitates the approach to acute ischemic stroke thrombectomy by utilizing the latest generation of large-bore aspiration catheters.

Although stent retrievers were utilized in the vast majority of cases in the recently concluded trials of mechanical thrombectomy for acute ischemic stroke from large-vessel occlusion, there is expectation that mechanical approaches will continue to rapidly evolve and the tools we currently utilize will change. There is controversy at present as to what is the best technique for acute ischemic stroke, and trials are under way to compare ADAPT versus stent-retriever-based methodologies.

INTRACRANIAL ANEURYSMS

The management of IAs has changed dramatically as endovascular technology continues to evolve over time. The overall prevalence of unruptured IA is about 2–5%. Patient-related risk factors for IA rupture include older age, female sex, smoking, and hypertension. Aneurysm-related risk factors for IA rupture include size, posterior circulation location, irregular aneurysm shape, and certain locations in the anterior circulation (e.g., anterior communicating artery, ACoA). The International Study of Unruptured Intracranial Aneurysms (ISUIA-I) investigators found that the accumulative rupture rate of aneurysms <10 mm in diameter was <0.05% per year and aneurysms of the posterior circulation had a higher risk of rupture than those of the anterior circulation [16]. The results of ISUIA-II showed that aneurysms <7 mm with no previous history of SAH have almost 0% risk of rupture [17]; this is still the subject of significant controversy because most ruptured aneurysms are 7 mm or smaller. Aneurysms sized 7–12 mm, 13–24 mm, and 25 mm or more were found to have annual rupture rates of 2.6%, 14.5%, and 40%, respectively.

The goal of preventing hemorrhage or recurrent hemorrhage can only be achieved by excluding the aneurysm from the cerebral circulation, either by microsurgical or endovascular means. The decision of how, when, and whether to treat an aneurysm, should be an evidence-based decision centered on supporting safety and efficacy in combination with the patient's life expectancy and personal wishes. Aneurysm factors such as size, location, configuration, patient's anatomy, and surgeon's experience also play a role in the selection of treatment method. The safety of endovascular coiling compared with microsurgical clipping was evaluated and validated by the results of the International Subarachnoid Aneurysm Trial (ISAT), which demonstrated that for aneurysms considered to be amenable to either approach coiling is safer than microsurgical clipping in relatively healthy patients [18]. However, despite the promising results of reducing on unfavorable 1-year outcome by 7–10% after coiling, these significant differences were lost in the most recent follow-up of some of these studies because of delayed recurrence and potential rerupture risk following endovascular therapy. When comparing endovascular treatment with surgical occlusion, the efficacy of endovascular aneurysm occlusion appears less optimal. Many factors play a role in the decision-making of IA treatment modality. Patient factors include age, comorbidities, family history, and rupture scale (Hunt and Hess). Endovascular coiling has been associated with less significant adverse effects due to reduced invasiveness. Aneurysm factors also play an important role when deciding treatment modality; these include size, location, and configuration.

Endovascular Embolization of Intracranial Aneurysms

Endovascular techniques can be divided into the following categories: parent artery reconstruction with coil deposition (primary coiling, balloon-assisted coiling, and stent-assisted coiling); other new techniques (neck reconstruction devices and intraluminal occlusion devices); reconstruction with flow diversion; and deconstructive techniques involving parent artery sacrifice with or without arterial bypass. The techniques are briefly described in the following.

Primary coiling: A complete three-dimensional (3D) appreciation of the anatomy of the aneurysm, parent artery, and related branches is paramount for procedural planning. Primary coiling is the most common endovascular technique used in ruptured IAs. Careful attention should be given to sizing and distribution of the coils in the aneurysm. Homogeneous coil distribution within the aneurysm is the ideal goal, as this will decrease the recanalization rate due to compaction. Coils differ in sizes, length, shape, and configuration. For example, some coils are designed with a complex 3D shape as opposed to the helical coils. They are made to closely mirror the shape of the aneurysm. The 3D coils can be used to "frame" the outside of the aneurysm and create a basket to later place "filling" coils.

Coating coils are additions as opposed to bare platinum. They aid in filling the aneurysm by expanding material (hydrogel) or by preventing clot recanalization. The packing density of the aneurysm usually ranges from 20 to 40% for complete treatment, with the remaining volume of the aneurysm filled with thrombus. The goal is for the aneurysm parent vessel to endothelialize prior to the breakdown of the thrombus. If the thrombus dissolves prior to endothelialization, the aneurysm may recanalize and recur.

Newer strategies for primary coiling include next-generation coils (Athena, Medtronic), which are ribbon-like instead of platinum wires, and endosaccular devices designed to fit the body of aneurysms (Luna, Medtronic; and Web, Sequent Medical), with the goal of better obliteration of the aneurysm body and creation of optimal neck coverage for endothelialization.

Balloon-assisted coiling: This technique can be quite effective for wide-necked or bifurcation aneurysms. Balloon "remodeling" techniques help with neck remodeling, protection of a side branch, and stabilization of the microcatheter within the aneurysm. Balloon-assisted techniques use compliant balloons. The most commonly used devices are the Transform (Stryker Neurovascular), HyperForm (ev3-Covidien), HyperGlide (ev3-Covidien), and Scepter (MicroVention) balloons. The HyperForm balloon is highly conformable, allowing herniation of part of the balloon into the neck of the aneurysm. This is ideal when an artery is in close proximity to or involved in the aneurysm neck, promoting coil deflection away from the endangered arterial branch and neck of the aneurysm [19].

Stent-assisted coiling: Key uses of this technique include: (1) wide-necked aneurysms unable to hold the coil mass without risk of protrusion into the parent vessel; (2) in cases of residual coil mass extruded from the parent vessel, a stent can trap the coil mass; (3) to aid in coiling of giant aneurysms; and (4) in dissecting or fusiform aneurysms where there is no discernible neck for coil placement. The stent can be deployed before ("through stent technique") or after ("jailing technique") a microcatheter is placed in the aneurysm. The "jailing" technique stabilizes the microcatheter in the aneurysm before and during stent deployment (Fig. 149.2). Recently developed stents [e.g., low-profile visualized intraluminal support (LVIS; MicroVention) device, LVIS Jr. (MicroVention), and ATLAS (Stryker)] are easier to deploy through smaller microcatheters in comparison to earlier-generation stents [e.g., Neuroform (Stryker) and Enterprise (Codman)] [20]. Closed-cell (Enterprise) and braided (LVIS) stents can additionally be resheathed and therefore repositioned prior to deployment.

Most recently, stents have been introduced that are specifically designed to deal with bifurcations and the need for protection of both the branches and parent

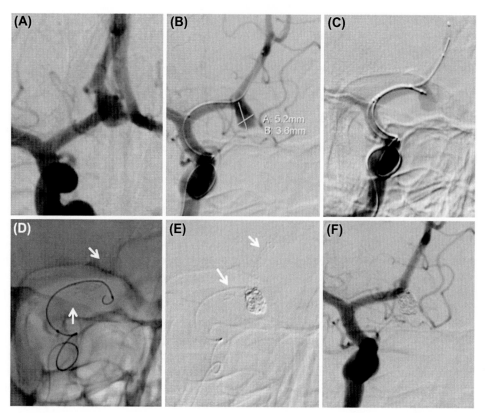

FIGURE 149.2   ACoA aneurysm. (A) AP view digital subtraction angiogram (DSA) of the anterior communicating complex showing an ACoA aneurysm predominantly filling from the right anterior cerebral artery. (B) AP view DSA showing the stent microcatheter at the ipsilateral distal A2 segment. (C) AP view DSA showing both delivery microcatheters (stent and coils) in preparation for "jailing" technique. (D) AP X-ray demonstrates the stent (*arrows*) deployed at the A1-A2 junction and the coil being deployed in the aneurysm; in this case, an LVIS Jr. stent (MicroVention) was used. (E and F) AP views of DSA after completion of stent-assisted coiling procedure demonstrate successful aneurysm obliteration. The *arrows* in (E) point to the stent.

vessel of wide-necked aneurysms. These devices include the Barrel (Medtronic) and PulseRider (Pulsar), and they are designed to obviate the need for Y-stenting, which was performed using two of the previously noted self-expanding stents, of such aneurysms.

Flow diversion: These devices disrupt aneurysmal inflow and outflow jets and cause stasis and thrombosis inside the aneurysm sac, even in the absence of coils. They induce neointimal proliferation in the arterial wall overlaying the stent struts, potentially leading to complete integration of the device within the intima (Fig. 149.3). Flow diverters include the pipeline embolization device [PED (ev3-Covidien)], silk flow diverter (Balt Extrusion), flow re-direction endoluminal device [FRED (MicroVention)], and the surpass flow diverter [Stryker], as well as many others available mainly outside the USA [21–23]. As of 2016, the PED is the most commonly used of all flow diverters available worldwide and the only flow diverter approved by the FDA for use in the USA. The PED is a self-expanding braided platinum and nickel–cobalt–chromium alloy stent. Compared to other nonflow diverting stents, the cells of the PED are smaller and more densely packed (low porosity), resulting in metal surface area exceeding 30–35% as compared to previously mentioned stents with metal surface area ranging from 6 to 20%. Flow diversion induces aneurysm thrombosis in a more delayed fashion (Fig. 149.3). Flow diversion may incur a higher risk of thromboembolic complications than other intracranial stents; adherence to a strict dual antiplatelet regimen is critical [24].

Vessel deconstruction: Parent vessel sacrifice remains an excellent option, typically for more proximally located large or dissecting aneurysms whereby the natural circle of Willis collaterals may continue to fill the distal vascular territories. Vessel deconstruction is typically performed after testing the vessel occlusion through temporary balloon catheter-mediated flow arrest using either advanced imaging or hypotension in conjunction with a clinical examination or electrophysiological monitoring. Vessel deconstruction (using coils, liquid embolics, or other vessel occlusion devices) can then be safely performed in patients who successfully tolerate balloon test occlusion.

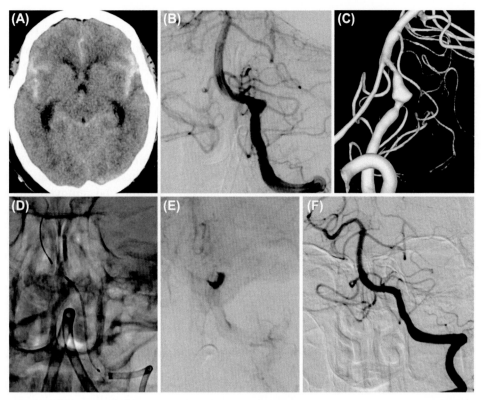

FIGURE 149.3 Ruptured vertebral artery (VA) aneurysm. (A) CT scan of the head of a 65-year-old woman who presented with aneurysmal subarachnoid hemorrhage. (B) AP, and (C) 3D, views of the dissecting VA aneurysm. Due to the configuration and anatomic characteristics, we decided to treat the aneurysm with flow diversion. (D) AP X-ray demonstrates flow diverter (PED, ev3-Covidien) deployment. (E) Immediate postoperative AP view DSA shows intra-aneurysmal flow stasis. (F) 3-month follow-up DSA demonstrates complete vessel reconstruction.

INTRACRANIAL ARTERIOVENOUS MALFORMATIONS

Most AVMs are sporadic and thought to be congenital. The occurrence rate for symptomatic disease is estimated at 0.94 per 10,000 person-years, with a prevalence of <10.3 per 100,000 [25]. Hemorrhage is the most common clinical presentation (Fig. 149.4), followed by seizures. These lesions cause neurological injury through two mechanisms: hemorrhage and/or vascular steal syndrome from the high-flow state leading to ischemia or seizures. Aneurysms that are associated with cerebral AVMs have a 7% rate of hemorrhage. The definitive treatment of an AVM by any modality requires complete obliteration of the nidus. The Spetzler–Martin grading system is based on AVM size (<3 cm = 1 point, 3–6 cm = 2 points, >6 cm = 3 points), pattern of venous drainage (superficial = 0 points, deep = 1 point), and eloquence of adjacent brain (noneloquent = 0 points, eloquent = 1 point) relative to the AVM; it correlates to clinical outcome after microsurgical resection with a higher grade corresponding to a poor surgical prognosis (Fig. 149.4) [26]. The Buffalo score has been proposed as a grading system to estimate complication rate for the endovascular treatment of AVMs [27]. The system accounts for features that make an angiographic approach more difficult including the number of arterial pedicles, diameter of pedicles, and the eloquence of brain at the AVM nidus [27].

Endovascular Embolization of Arteriovenous Malformations

Significant improvements in endovascular technology since the first description of embolization of cranial AVMs by Luessenhop and Spence [28] have increased the role of embolization to include palliative therapy, adjunct to microsurgical resection, adjunct to stereotactic radiosurgery, or, in a minority of cases, curative therapy. The dynamic nature of cerebral AVMs makes their treatment more unpredictable and difficult than that for other cerebral vascular diseases. Multiple arterial feeders may exist with unique anatomy (variable diameter and caliber, associated aneurysms, and difficult angles). Embolization of a single feeder may alter the flow and wall stress, possibly leading to rupture. A small caliber vessel makes cannulation difficult and increases the chances of embolic material reflux. Unless all feeders and the nidus are completely obliterated, the AVM has the potential to recur. Traditional endovascular treatment uses embolic material via a transarterial route;

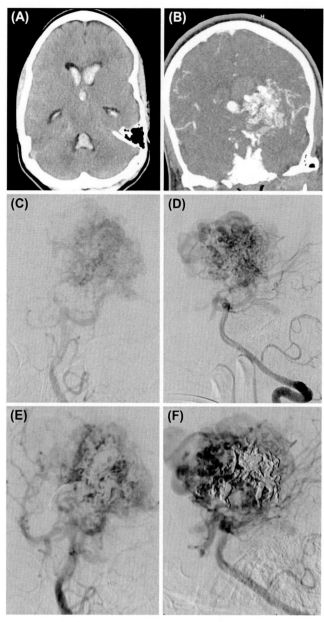

FIGURE 149.4 Ruptured AVM. An 18-year-old girl presented with IVH. (A) CT scan of the head shows acute IVH and hydrocephalus. (B) CT angiogram reveals a large left basal ganglia AVM (Spetzler-Martin Grade V). (C) and (D) Initial AP (C) and lateral (D) angiograms demonstrate the large AVM with arterial feeders predominantly from both posterior cerebral arteries. (E and F) AP (E) and lateral (F) angiograms show partial AVM embolization with onyx (ev3-Covidien). The plan is to continue with staged embolizations until radiographic cure is achieved or the size of the AVM is appropriate for radiosurgery.

however, transvenous and combined approaches have recently been described.

Endovascular embolization utilizes liquid embolics such as N-butylcyanoacrylate [NBCA (Trufill, Codman)], ethylene-vinyl alcohol [EVOH (Onyx, ev3-Covidien)], and coils [29]. Coils may be deployed as an adjunct for aneurysms associated with AVMs or to prevent

distal embolization of Onyx or NBCA during deployment while the embolic material is still in liquid form. NBCA is essentially glue mixed with ethiodized oil to make it radiopaque and decrease the polymerization time. NBCA hardens in an exothermic reaction when in contact with an alkalotic ionic material, such as blood. The injection must be preceded by the administration of an acidic nonionic solution, such as dextrose in water, to flush the catheter. The concentration of NBCA varies by adding more ethiodized oil, with lower concentrations allowing more distal migration of the solution. The microcatheter must be removed quickly after injection to avoid gluing it in place. Alternatively, detachable tip microcatheters have been developed to reduce this complication (e.g., Apollo microcatheter). Onyx is a mixture of EVOH and tantalum dissolved in dimethyl sulfoxide (DMSO) that is paste-like in consistency. After injection, the DMSO dissipates, hardening the paste. Multiple concentrations of Onyx are available: Onyx 18, Onyx 20 (none in the USA), Onyx 34, and Onyx 500. Several procedural considerations must be taken into account, including the time required to prepare the Onyx and the use of DMSO-compatible syringes and microcatheters. The injection of Onyx can be challenging; it must be done in a slow controlled manner, while monitoring for distal embolization or reflux (Fig. 149.4).

CONCLUSION

Endovascular therapies for cerebrovascular diseases are rapidly becoming the mainstay of interventional management for these diseases as compared to open microsurgical approaches. This is principally because of a markedly improved understanding of underlying pathologies as well as remarkable improvement in tools and technologies available for endovascular therapy. Endovascular therapy is a rapidly advancing field with multiple new techniques and strategies introduced each year for each category of disease. With increasing experience and improved tools, procedural success continues to get better while the inherent complications continue to decline. We expect this trend to continue for the foreseeable future.

References

[1] Mokin M, Snyder KV, Siddiqui AH, Hopkins LN, Levy EI. Endovascular management and treatment of acute ischemic stroke. Neurosurg Clin N Am July 2014;25(3):583–92.
[2] Mokin M, Khalessi AA, Mocco J, et al. Endovascular treatment of acute ischemic stroke: the end or just the beginning? Neurosurg Focus January 2014;36(1):E5.
[3] Deshaies EM, Singla A, Villwock MR, Padalino DJ, Sharma S, Swarnkar A. Early experience with stent retrievers and comparison with previous-generation mechanical thrombectomy devices for acute ischemic stroke. J Neurosurg July 2014;121(1):12–7.

[4] Berkhemer OA, Fransen PS, Beumer D, et al. A randomized trial of intraarterial treatment for acute ischemic stroke. N Engl J Med January 1, 2015;372(1):11–20.

[5] Campbell BC, Mitchell PJ, Kleinig TJ, et al. for the EXTEND-IA Investigators. Endovascular therapy for ischemic stroke with perfusion-imaging selection. N Engl J Med March 12, 2015;372(11):1009–18.

[6] Goyal M, Demchuk AM, Menon BK, Escape Trial Investigators, et al. Randomized assessment of rapid endovascular treatment of ischemic stroke. N Engl J Med March 12, 2015;372(11):1019–30.

[7] Jovin TG, Chamorro A, Cobo E, REVASCAT Trial Investigators, et al. Thrombectomy within 8 hours after symptom onset in ischemic stroke. N Engl J Med June 11, 2015;372(24):2296–306.

[8] Saver JL, Goyal M, Bonafe A, et al. Stent-retriever thrombectomy after intravenous t-PA vs. t-PA alone in stroke. N Engl J Med June 11, 2015;372(24):2285–95.

[9] Munich SA, Mokin M, Snyder KV, Siddiqui AH, Hopkins LN, Levy EI. Guest editorial: an update on stroke intervention. Neurosurgery September 2015;77(3):313–20.

[10] Mehta BP, Leslie-Mazwi TM, Simonsen CZ, et al. ADAPT FAST study: third-generation stroke thrombectomy devices place renewed focus on the elusive relationship between revascularization and good outcomes. J Neurointerv Surg 2016;8:321–3.

[11] Castano C, Dorado L, Guerrero C, et al. Mechanical thrombectomy with the Solitaire AB device in large artery occlusions of the anterior circulation: a pilot study. Stroke August 2010;41(8):1836–40.

[12] Nogueira RG, Lutsep HL, Gupta R, et al. Trevo versus Merci retrievers for thrombectomy revascularisation of large vessel occlusions in acute ischaemic stroke (TREVO 2): a randomised trial. Lancet October 6, 2012;380(9849):1231–40.

[13] Saver JL, Jahan R, Levy EI, et al. Solitaire flow restoration device versus the Merci retriever in patients with acute ischaemic stroke (SWIFT): a randomised, parallel-group, non-inferiority trial. Lancet October 6, 2012;380(9849):1241–9.

[14] Mokin M, Setlur Nagesh SV, Ionita CN, Mocco J, Siddiqui AH. Stent retriever thrombectomy with the Cover accessory device versus proximal protection with a balloon guide catheter: in vitro stroke model comparison. J Neurointerv Surg February 12, 2015.

[15] Turk AS, Frei D, Fiorella D, et al. ADAPT FAST study: a direct aspiration first pass technique for acute stroke thrombectomy. J Neurointerv Surg May 2014;6(4):260–4.

[16] International Study of Unruptured Intracranial Aneurysms Investigators. Unruptured intracranial aneurysms–risk of rupture and risks of surgical intervention. N Engl J Med December 10, 1998;339(24):1725–33.

[17] Wiebers DO, Whisnant JP, Huston 3rd J, International Study of Unruptured Intracranial Aneurysms, Investigators, et al. Unruptured intracranial aneurysms: natural history, clinical outcome, and risks of surgical and endovascular treatment. Lancet July 12, 2003;362(9378):103–10.

[18] Molyneux A, Kerr R, Stratton I, et al. International Subarachnoid Aneurysm Trial (ISAT) of neurosurgical clipping versus endovascular coiling in 2143 patients with ruptured intracranial aneurysms: a randomised trial. Lancet October 26, 2002;360(9342):1267–74.

[19] Moon K, Albuquerque FC, Ducruet AF, Crowley RW, McDougall CG. Balloon remodeling of complex anterior communicating artery aneurysms: technical considerations and complications. J Neurointerv Surg June 2015;7(6):418–24.

[20] Poncyljusz W, Bilinski P, Safranow K, et al. The LVIS/LVIS Jr. stents in the treatment of wide-neck intracranial aneurysms: multicentre registry. J Neurointerv Surg July 2015;7(7):524–9.

[21] Mohlenbruch MA, Herweh C, Jestaedt L, et al. The FRED flow-diverter stent for intracranial aneurysms: clinical study to assess safety and efficacy. AJNR Am J Neuroradiol June 2015;36(6):1155–61.

[22] Colby GP, Lin LM, Caplan JM, et al. Flow diversion of large internal carotid artery aneurysms with the surpass device: impressions and technical nuance from the initial North American experience. J Neurointerv Surg 2016;8:279–85.

[23] Briganti F, Leone G, Marseglia M, et al. Endovascular treatment of cerebral aneurysms using flow-diverter devices: a systematic review. Neuroradiol J 2016;28:365–75.

[24] Brasiliense LB, Hanel RA. Pipeline embolization device: lessons learned after 1000 aneurysms. World Neurosurg September–October 2014;82(3–4):248–50.

[25] Berman MF, Sciacca RR, Pile-Spellman J, et al. The epidemiology of brain arteriovenous malformations. Neurosurgery August 2000;47(2):389–96. discussion 397.

[26] Spetzler RF, Martin NA. A proposed grading system for arteriovenous malformations. J Neurosurg October 1986;65(4):476–83.

[27] Dumont TM, Kan P, Snyder KV, Hopkins LN, Siddiqui AH, Levy EI. A proposed grading system for endovascular treatment of cerebral arteriovenous malformations: Buffalo score. Surg Neurol Int 2015;6:3.

[28] Luessenhop AJ, Spence WT. Artificial embolization of cerebral arteries. Report of use in a case of arteriovenous malformation. J Am Med Assoc March 12, 1960;172:1153–5.

[29] Elsenousi A, Aletich VA, Alaraj A. Neurological outcomes and cure rates of embolization of brain arteriovenous malformations with n-butyl cyanoacrylate or Onyx: a meta-analysis. J Neurointerv Surg 2016;8:265–72.

CHAPTER

150

Surgery for Ischemic Infarcts

Y. Laviv, E.M. Kasper
Harvard Medical School, Boston, MA, United States

INTRODUCTION

The first surgical procedures on patients with massive infarctions were performed to alleviate severe brain swelling often erroneously attributed to a tumor. With the common use of computed tomography (CT) and magnetic resonance imaging (MRI), the pathophysiologic cause of severe brain swelling is now rarely in question. Although anatomic imaging studies may not show signs of massive hemispheric or cerebellar infarction for the first few hours, the abrupt onset of any neurologic deficit is highly suggestive of a stroke or transient ischemic attack [1]. The brain edema (usually a combination of cytotoxic and vasogenic edema) associated with an infarct increases usually 3–5 days after the insult, often resulting in massive swelling of the affected region and a dramatic change in tissue density. Depending on the location and size of the infarcted area, such massive swelling may cause life-threatening brain shift causing herniation and brainstem compression. Knowing the extent of the insult is of great importance—researchers who obtained quantitative tomographic measurements of cerebral blood flow within 1–2h of the onset of symptoms that were related to a massive middle cerebral artery (MCA) stroke (malignant MCA stroke) observed an absence of perfusion throughout the entire territory of the MCA [2].

Despite enormous progress in the field of reperfusion therapy for ischemic strokes (medical and interventional therapies alike), the reestablishment of perfusion is unwarranted and potentially harmful for this kind of MCA infarction. Clinical studies indicate that reperfusion of a region that has already become irreversibly ischemic can accelerate vasogenic edema and convert ischemic infarcts to hemorrhagic ones [3]. Medical therapies without the goal of reperfusion for massive infarction of the cerebral hemispheres or the cerebellum are directed toward preventing additional ischemic injuries

and controlling brain edema. Depending on the cause of vascular occlusion, anticoagulation is often indicated to prevent additional embolic events or thrombus propagation. A moderately elevated perfusion pressure must be maintained while avoiding both hypotension and excessive hypertension. Ideally, at least in the initial days, this should take place in a neurointensive care unit. Traditionally, osmotic diuretics administration and intubation combined with hyperventilation have been used when the mass effect is accompanied by a decreased mental status. However, controlled hyperventilation and/or administration of mannitol or hypertonic saline is of little value if a decision against decompressive surgery has already been made, as these measures show only a transient effect in most cases and is possibly associated with a detrimental rebound phenomenon after discontinuation and thus result in increased intracranial pressure (ICP) [4]. In case of concomitant hydrocephalus, placement of external ventricular drains with ICP monitors should be considered; however, shunt placement without performing a craniectomy is most likely not a successful strategy [4].

Despite such therapeutic efforts, 20–30% of massive infarctions resulted in irreversible secondary brainstem injuries caused by herniation and mass effect [5]. In some cases, and to prevent these secondary brainstem injuries, decompressive surgery is suitable. In decompressive craniectomy, part of the skull is removed to create more space for the swelling brain, thus immediately reducing increased ICP and alleviating the mass effect. Usually this intervention includes opening and expanding the underlining dura. Rarely additional space is created by resection of infarcted brain tissue (e.g., temporal lobectomy). The bone flap is banked in a freezer or implanted in the patient's abdominal wall, and patients wear helmets to protect the brain until the autologous bone or allograft prosthesis can be reimplanted in the second-stage procedure called cranioplasty. The optimal time to perform cranioplasty after decompressive craniectomy for strokes remains unknown, but the complication rate (e.g., hydrocephalus, infection) was slightly higher in case of early cranioplasty (within 10 weeks of craniectomy) and in patients with a ventriculoperitoneal shunt [6]. On the other hand, if bone flap replacement is delayed further, communicating hydrocephalus may develop requiring shunt placement [7].

Decompressive craniectomy is typically done for two types of strokes: malignant MCA strokes and cerebellar strokes.

MALIGNANT MCA STROKE

Territorial MCA infarcts are among the most devastating forms of ischemic insults and are associated with mortality rates as high as 80% [8]. Predictors of significant brain edema after MCA strokes are (1) NIHSS (National Institutes of Health Stroke Scale) scores of > 20 [left MCA (L-MCA)] or > 15 [right MCA (R-MCA)], (2) nausea/vomiting, (3) hypodensity in more than 50% of the MCA territory on initial head CT, (4) involvement of additional vascular territories, and (5) early large abnormalities on diffusion-weighted MRI [4]. The most commonly described signs of deterioration from hemispheric supratentorial infarction are ipsilateral pupillary dysfunction, varying degrees of mydriasis, and ocular adduction paralysis. Limb strength can worsen, progressing to extensor posturing. A false-localizing Babinski sign as a result of brainstem compression against the contralateral tentorium (Kernohan notch) can also occur. Abnormal respiratory patterns reflecting brainstem dysfunction typically occur late in the course; these include central neurogenic hyperventilation or ataxic respiratory patterns and periodic breathing [9].

The efficacy of hemicraniectomy in this context is now documented by several randomized controlled trials (RCTs) [10]. One of the studies is a pooled analysis of 1-year outcome data from three European multicenter RCTs: the French DECIMAL (decompressive craniectomy in malignant MCA infarcts) trial [11], the German DESTINY (decompressive surgery for the treatment of malignant infarction of the MCA) trial [12], and the Dutch HAMLET (hemicraniectomy after MCA infarction with life-threatening edema trial) [13]. Although the inclusion and exclusion criteria for these three trials differed somewhat, clinical results from 93 patients fulfilling a set of the following criteria were pooled: age 18–60 years, NIHSS > 15, infarct in > 50% of the MCA territory (or approximately > 145 cc), and randomization within 48h of symptoms [14]. Of note, patients with significant preexisting morbidities were excluded from this analysis (e.g., life expectancy < 3 years). In this pooled analysis, hemicraniectomy was associated with a

statistically significant reduction of mortality, from 75% to 24%. The modified Rankin Scale (mRS) was used to assess functional outcome at 1 year. Of the surviving surgical patients, 5% had an mRS of 5, 40% had 4, 38% had 3, and 17% had 2. Statistical analysis revealed that the proportion of patients with mRS ≤3 (i.e., good level of independency) was significantly high in the surgery group (43%) relative to the group receiving medical management only (21%) [14].

Two other RCTs studied the role of decompressive hemicraniectomy in the elderly (older than 60 years). The first study was reported by Zhao et al. [15]. This is a Chinese RCT that randomized 47 patients (ages 18–80 years) with infarcts in more than two-thirds of the MCA territory (approximately 200 cc) and with Glasgow Coma Scale ≤9. In this study, > 50% of the patients were older than 60 years. Patients with significant preexisting morbidities were excluded from the trial (e.g., baseline mRS >2). Surgical intervention was performed within 48h of symptom onset. Hemicraniectomy was found to be associated with a statistically significant reduction in mortality, from 60.9% to 12.5% ($P = 001$). Similar to the observations made in the HAMLET study [13], the proportion of patients with mRS <3 was not significantly different between the surgery group and the medically managed group (9% vs. 21%). At 1-year follow-up, 15% of the surviving surgical patients had an mRS of 5, 15% had 4, 60% had 3, and 10% had 2. The 1-year mRS for the surgical patients older than 60 years was 5 in 15% of the patients, 4 in 70%, and 3 in 15%.

The German DESTINY II trial [16] prospectively randomized (surgery vs. best medical management) 112 patients older than 60 years (median age, 70 years) with an infarct in more than two-thirds of the MCA territory, and NIHSS > 14 (for nondominant hemisphere infarcts) or > 19 (for dominant hemisphere infarcts). Surgical intervention was performed within 48h of symptom onset. As for younger patients, hemicraniectomy was associated with a significant reduction in mortality (from 76% to 43%). However, the functional outcome results were far less encouraging. At 1-year follow-up, < 10% of the surviving patients in either group had an mRS <3. Moreover, half of the surviving patients in both treatment groups continued to have an mRS of 4 (unable to walk without assistance and unable to attend to own bodily needs without assistance) and an additional one-third of the patients in both groups had a score of 5 (bedridden, incontinent, and requiring constant nursing care and attention). Severe depression was present in 80–100% of the survivors, irrespective of intervention. This has led some providers to decide not to perform surgery in this patient population if the dominant hemisphere is affected.

To conclude, irrespective of the patients' age, all published RCTs have consistently demonstrated that decompressive hemicraniectomy decreases the mortality associated with malignant MCA infarction from 46%–75% to 12%–27%, with absolute mortality risk reduction to about one-third to a half. Also, additional reasonable interpretation of the available data sets is that surgery may reduce morbidity in select patient populations (younger than 60 years) when performed within 48 h. The only good reason to question hemicraniectomy as a suitable treatment for malignant MCA infarction is the increased proportion of patients with an mRS of 4 post procedure, especially in older individuals. On the other hand, surgical treatment increased the probability of being fully independent from 2% to 14% and that of having no/minimal dependency from 21% to 43%. Even if hemicraniectomy for large MCA infarcts still leaves unanswered questions (e.g., timing surgery 48 h after symptom onset, resection of ischemic tissue), it should be contemplated as a key step in the treatment of territorial MCA infarction.

CEREBELLAR STROKE

Space-occupying edema is also a common complication in 17%–54% of patients with significant cerebellar infarction and may cause life-threatening conditions such as (1) obstructive hydrocephalus attributable to the mass effect in the posterior fossa resulting in blockage of the fourth ventricle, (2) direct compression of the midbrain and pons, (3) upward herniation of the superior cerebellum through the tentorial incisura, and (4) downward herniation of the tonsils through the foramen magnum [17].

Besides typical signs of cerebellar dysfunction, the most concerning clinical symptom of swelling is decreased level of consciousness. In addition, pontine compression may lead to ophthalmoparesis, breathing irregularities, and cardiac dysrhythmias [9]. However, deterioration from swelling or extension of the infarct into the brainstem cannot always be clinically distinguished. Accurate diagnosis requires the use of various radiologic modalities.

In patients who initially present with a massive cerebellar infarction and minimal brainstem dysfunction, timely surgical decompression can be lifesaving and can result in good functional outcome and independence [18].

Decompression for cerebellar strokes is by far less controversial than for malignant MCA strokes leading to the fact that RCTs have never been done for suboccipital craniectomies for cerebellar strokes.

Risk factors associated with development of brain edema after cerebellar stroke include impairment of consciousness, hypodensity in more than two-thirds of the cerebellar hemisphere, compression/displacement of the fourth ventricle, displacement of the brainstem,

obstructive hydrocephalus, compression of basal cisterns, and hemorrhagic transformation of cerebellar ischemic infarct [13].

Close neurologic and cardiovascular monitoring in an intermediate or intensive care stroke unit in patients with territorial cerebellar infarctions for up to 5 days is recommended, even if the patient seems to be stable [13].

To conclude, for cerebellar strokes as well as for malignant MCA strokes, timely surgical intervention can prevent secondary injuries and lead to improved survival quality, i.e., independent and productive individuals. On one hand, the only alternative to surgery is death in some of these critically ill patients. On the other hand, the alternate outcome to death can also be lifelong dependency. We therefore want to emphasize that the decision to operate or not should be based on both known prognostic factors and the patient's individual wishes.

References

[1] Mohr JP, Biller J, Hilal SK, et al. Magnetic resonance versus computed tomographic imaging in acute stroke. Stroke 1995;26(5):807–12.

[2] Kalia KK, Yonas H. An aggressive approach to massive middle cerebral artery infarction. Arch Neurol 1993;50(12):1293–7.

[3] Hacke W, Kaste M, Fieschi C, et al. Intravenous thrombolysis with recombinant tissue plasminogen activator for acute hemispheric stroke. The European Cooperative Acute Stroke Study (ECASS). JAMA 1995;274(13):1017–25.

[4] Michel P, Arnold M, Hungerbuhler HJ, et al. Decompressive craniectomy for space occupying hemispheric and cerebellar ischemic strokes: Swiss recommendations. Int J Stroke 2009;4(3):218–23.

[5] Ng LK, Nimmannitya J. Massive cerebral infarction with severe brain swelling: a clinicopathological study. Stroke 1970;1(3):158–63.

[6] Piedra MP, Ragel BT, Dogan A, Coppa ND, Delashaw JB. Timing of cranioplasty after decompressive craniectomy for ischemic or hemorrhagic stroke. J Neurosurg 2013;118(1):109–14.

[7] Waziri A, Fusco D, Mayer SA, McKhann 2nd GM, Connolly Jr ES. Postoperative hydrocephalus in patients undergoing decompressive hemicraniectomy for ischemic or hemorrhagic stroke. Neurosurgery 2007;61(3):489–93. discussion 493–484.

[8] Berrouschot J, Sterker M, Bettin S, Koster J, Schneider D. Mortality of space-occupying ('malignant') middle cerebral artery infarction under conservative intensive care. Intensive Care Med 1998;24(6):620–3.

[9] Wijdicks EF, Sheth KN, Carter BS, et al. Recommendations for the management of cerebral and cerebellar infarction with swelling: a statement for healthcare professionals from the American Heart Association/American Stroke Association. Stroke 2014;45(4):1222–38.

[10] Hatefi D, Hirshman B, Leys D, et al. Hemicraniectomy in the management of malignant middle cerebral artery infarction: lessons from randomized, controlled trials. Surg Neurol Int 2014;5:72.

[11] Vahedi K, Vicaut E, Mateo J, et al. Sequential-design, multicenter, randomized, controlled trial of early decompressive craniectomy in malignant middle cerebral artery infarction (DECIMAL Trial). Stroke 2007;38(9):2506–17.

[12] Juttler E, Schwab S, Schmiedek P, et al. Decompressive surgery for the treatment of malignant infarction of the middle cerebral artery (DESTINY): a randomized, controlled trial. Stroke September 2007;38(9):2518–25.

[13] Hofmeijer J, Kappelle LJ, Algra A, Amelink GJ, van Gijn J, van der Worp HB. Surgical decompression for space-occupying cerebral infarction (the Hemicraniectomy After Middle Cerebral Artery infarction with Life-Threatening Edema Trial [HAMLET]): a multicentre, open, randomised trial. Lancet Neurol 2009;8(4):326–33.

[14] Vahedi K, Hofmeijer J, Juettler E, et al. Early decompressive surgery in malignant infarction of the middle cerebral artery: a pooled analysis of three randomised controlled trials. Lancet Neurol 2007;6(3):215–22.

[15] Zhao J, Su YY, Zhang Y, et al. Decompressive hemicraniectomy in malignant middle cerebral artery infarct: a randomized controlled trial enrolling patients up to 80 years old. Neurocrit Care 2012;17(2):161–71.

[16] Juttler E, Unterberg A, Woitzik J, et al. Hemicraniectomy in older patients with extensive middle-cerebral-artery stroke. N Engl J Med Mar 20, 2014;370(12):1091–100.

[17] Juttler E, Schweickert S, Ringleb PA, Huttner HB, Kohrmann M, Aschoff A. Long-term outcome after surgical treatment for space-occupying cerebellar infarction: experience in 56 patients. Stroke 2009;40(9):3060–6.

[18] Sypert GW, A. Lvord ECJ. Cerebellar infarction. A clinicopathological study. Arch Neurol 1975;32(6):357–63.

C H A P T E R

151

Carotid Endarterectomy and Carotid Angioplasty and Stenting

C.J. Griessenauer, C.S. Ogilvy

Harvard Medical School, Boston, MA, United States

INTRODUCTION

Large-vessel arteriosclerotic disease and stenosis, most commonly of the extracranial carotid artery bifurcation and simply referred to as carotid stenosis, accounts for 15–20% of cerebral ischemic events. The two main mechanisms for these events are hemodynamic compromise, as the internal carotid artery (ICA) diameter is reduced by at least 70%, and artery-to-artery embolization from an ulcerated plaque. The carotid bifurcation is prone to arteriosclerotic plaque formation because of its geometric configuration resulting in reduced wall shear stress. Common risk factors for carotid stenosis include increasing age, smoking, hypertension, diabetes mellitus, coronary artery disease, peripheral vascular disease, chronic renal failure, and other metabolic abnormalities [1]. Both carotid endarterectomy (CEA) and carotid angioplasty and stenting (CAS) are established interventions for select patients with carotid stenosis (Table 151.1).

CAROTID ENDARTERECTOMY

In 2002, approximately 134,000 CEAs were performed in the United States for both symptomatic and asymptomatic carotid stenoses [1]. Three major randomized controlled trials compared CEA with the best medical therapy for symptomatic carotid stenosis.

North American Symptomatic Endarterectomy Trail

A total of 2885 patients with a TIA (transient ischemic attack) or minor stroke within 120 days and ipsilateral ICA stenosis of >30% were randomized to either the best medical therapy (including aspirin) or the best medical therapy and CEA [2]. The study was terminated early because of a considerable advantage for CEA. The ipsilateral stroke rate at 2 years in patients with ≥70% carotid stenosis was 26% in the medical group and 9% in the CEA group. For moderate stenosis (50–69%),

Primer on Cerebrovascular Diseases, Second Edition
http://dx.doi.org/10.1016/B978-0-12-803058-5.00151-X

TABLE 151.1   Summary of Notable Clinical Trials

Carotid Endarterectomy	Carotid Angioplasty and Stenting
Symptomatic	SAPPHIRE [12]
NASCET [2]	CREST [13]
ECAS [3]	SPACE and SPACE II [14]
VACS [4]	EVA-3S [15]
Asymptomatic	ICSS [16,17]
ACAS [6]	
Veterans Administration Cooperative Asymptomatic Trial [7]	
CASANOVA [8]	
Mayo Asymptomatic Carotid Endarterectomy Trial [9]	
ACST [10]	

the 5-year ipsilateral stroke rates were 22% and 15%, respectively. In the North American Symptomatic Carotid Endarterectomy Trial (NASCET) the degree of stenosis was determined using angiography and calculated using the following formula: % stenosis = 1 − [residual lumen diameter of the most stenotic portion of the ICA (mm)/lumen diameter of normal ICA distal to stenosis (mm)] × 100. In North America and other parts of the world the method has been widely adopted since it was developed.

European Carotid Surgery Trial

A total of 3024 patients with TIA, retinal infarction, or minor stroke within the previous 6 months and ipsilateral ICA stenosis were randomized to either the best medical therapy (aspirin was an option, but not required) or the best medical therapy and CEA [3]. In the European Carotid Surgery, the degree of stenosis was determined on angiography by comparing the diameter of the lumen at the most stenotic segment of the ICA with the assumed original diameter of the lumen at that site resulting in higher degrees of stenosis than those in the NASCET. The 3-year risk rates of major stroke or death in patients with ≥60% carotid stenosis, corresponding to approximately ≥80% in the NASCET, was 26% in the medical group and 14% in the CEA group.

Veterans Affairs Cooperative Study on Symptomatic Stenosis

This study enrolled 197 men with symptomatic carotid stenosis and randomized between the best

medical therapy (including aspirin) and the best medical therapy and CEA [4]. The trial was prematurely terminated as the NASCET and ECAS trial data were published. At an average follow-up of approximately 1 year the stroke rate in patients with >50% stenosis was 19% in the medical group and 7% in the CEA group.

Data from a pooled analysis including over 6000 patients from the NASCET, ECAS trial, and Veterans Affairs Cooperative Study on Symptomatic Stenosis (VACS) trial found an increased risk for stroke, with CEA for carotid stenosis of <30%, no effect for stenosis of 30–49%, a marginal benefit (absolute risk reduction of 4%) for stenosis of 50–69%, and absolute benefit (absolute risk reduction of 16%) for stenosis of ≥70% [5]. The benefit of CEA strongly depends on the perioperative complication rate. For patients with stenosis of ≥70% a complication rate of greater than 6% will offset the benefits of CEA, making the American Heart Association to recommend that CEA can only be performed by surgeons with a complication rate below that. Further analysis of the NASCET has offered insights into various aspects concerning the management of carotid stenosis. Patients with ulcerated plaques in the medical group had dramatically higher stroke rates and those in the CEA group had higher rates of perioperative complications. Patients with hemispheric ischemia in the medical group were at a higher risk for stroke than patients with retinal ischemia. Although there was direct correlation between stroke risk and the degree of stenosis in the medical group, patients with near occlusion had a lower risk of stroke compared to patients with 70–89% stenosis. Contralateral occlusion was associated with increased stroke risk in the medical and CEA groups. Intraluminal thrombus conferred increased risk in the medical and CEA groups. There was a greater net benefit with CEA in older patients with ≥50% stenosis.

Carotid stenosis is common and affects approximately 7% of all females and >12% of all males older than 70 years. Five major randomized controlled trials compared CEA for asymptomatic carotid stenosis.

Asymptomatic Carotid Arteriosclerosis Study

In the Asymptomatic Carotid Arteriosclerosis Study (ACAS) a total of 1662 patients with ≥60% stenosis were randomized to the best medical therapy (including aspirin) or the best medical therapy and CEA [6]. The trial was stopped prematurely because CEA was beneficial over medical therapy alone. The aggregate risk for ipsilateral stroke, any perioperative stroke, or death at 5 years was 11% in the medical group and 5% in the CEA group. Patients with a contralateral occlusion in the medical arm were less likely to have a stroke and CEA in these patients may in fact be harmful.

Veterans Administration Cooperative Asymptomatic Trial

A total of 440 men with ≥50% stenosis were randomized [7]. At 4 years, the rate of ipsilateral stroke was 20% in the medical group (including aspirin) and 8% in the CEA group. An overall mortality rate of 33% in the study limits the ability to broadly apply the results.

Carotid Artery Stenosis With Asymptomatic Narrowing: Operation Versus Aspirin

The Carotid Artery Stenosis with Asymptomatic Narrowing: Operation Versus Aspirin (CASANOVA) study enrolled 410 patients with 50–90% stenosis and did not identify a difference in the stroke rate between the medical and the CEA groups [8]. The study has been criticized for a significant number of patients crossing over.

Mayo Asymptomatic Carotid Endarterectomy Trial

In this study, 158 patients were randomized to receive either the best medical therapy with aspirin or CEA without aspirin [9]. Although there was no incidence of strokes or deaths in either group, there was a significantly higher rate of myocardial infarctions in the surgical group underlining the importance of antiplatelet therapy in these patients.

Asymptomatic Carotid Surgery Trial

A total of 3120 patients with carotid stenosis of 60–99% on duplex ultrasonography were randomized to immediate or deferred CEA [10]. The 5-year risk rate for stroke was 6% in the immediate group and 11% in the deferred group. The risk of stroke or death within 30 days of CEA was 3%.

The two largest clinical trials, ACAS and Asymptomatic Carotid Surgery Trial (ACST), on CEA for asymptomatic carotid stenosis found an annual stroke rate of approximately 2% in patients with ≥60% stenosis. However, due to advances in medical therapy, such as the use of antiplatelet agents and lipid-lowering agents, the annual stroke risk is likely lower, thus rendering the trials outdated in terms of stroke risk with medical therapy. Guidelines from the American Heart Association indicate that CEA for asymptomatic carotid stenosis of ≥60% angiographic stenosis may be beneficial when performed with a complication rate of <3%.

CAROTID ANGIOPLASTY AND STENTING

Since the late 1990s, CAS has become a popular alternative to CEA. In 2004, the FDA approved the procedure initially in patients with CEA and at high risk for adverse events and expanded the approval in 2011 to patients with CEA and at regular risk. In Medicare beneficiaries, there was a fourfold increase in the number of CAS claims from 1998 to 2007, while the number of CEAs performed decreased by about one-third during the same period [11]. Five randomized controlled trials compared CAS with CEA.

Stenting and Angioplasty With Protection in Patients at High Risk for Endarterectomy

A total of 334 patients considered high risk for CEA with symptomatic carotid stenosis of ≥50% or asymptomatic stenosis of ≥80% were enrolled in the Stenting and Angioplasty with Protection in Patients at High Risk for Endarterectomy (SAPPHIRE) study [12]. High-risk criteria included clinically significant cardiac disease, severe pulmonary disease, contralateral carotid occlusion, contralateral laryngeal nerve palsy, previous neck surgery or neck radiation, recurrent stenosis after CEA, and age above 80 years. The study was terminated early due to low recruitment. There was a trend toward a lower incidence of stroke, death, or myocardial infarction within 30 days of the procedure and of ipsilateral stroke or death from 31 days to 1 year in the CAS group (9%) compared with the CEA group (4%). At 3 years the results for both groups were essentially identical [1].

Carotid Revascularization Endarterectomy Versus Stent Trial

A total of 2502 patients with symptomatic carotid stenosis of ≥50% or asymptomatic stenosis ≥60% were randomized between CAS and CEA [13]. Primary endpoints were stroke, myocardial infarction, or death within 30 days or ipsilateral stroke within 4 years. The RX Acculink Carotid Stent System (Abbott Vascular, Santa Clara, California, United States) was used for CAS and combined with the Accunet embolic protection device in 96% of cases. The primary endpoint did not differ between the two groups. During the periprocedural, period there were more myocardial infarctions in the CEA group and more strokes in the CAS group. Patients older than 70 years fared slightly better with CEA, and periprocedural complications were more common in patients with symptomatic carotid stenosis when treated with CAS than with CEA. Advantages of the Carotid Revascularization Endarterectomy Versus Stent Trial (CREST) include the rigorous training of the operators and the use of only one type of stent with a high rate of distal embolic protection. Disadvantages include the enrollment of symptomatic and asymptomatic patients, a low threshold for diagnosis of myocardial infarction, and lack of screening for silent ischemic events [1].

Stent-Supported Percutaneous Angioplasty of the Carotid Versus Endarterectomy

A total of 1214 patients with symptomatic carotid stenosis of ≥50% were randomized to CAS or CEA [14]. The trial was stopped early due to shortage in funding and low enrollment. The study endpoints did not differ between both groups. The risk of ipsilateral stroke or death increased with age in the CAS group. The statistical analysis chosen in the Stent-Supported Percutaneous Angioplasty of the Carotid Artery Versus Endarterectomy (SPACE) trial failed to demonstrate noninferiority of CAS to CEA for periprocedural complications. Interpretation of data on the use of embolic protection devices is controversial. An embolic protection device was only used in 27% of CAS cases. The rate of periprocedural complications, however, did not differ based on the use of an embolic protection device.

The SPACE II trial is an ongoing trial on patients with asymptomatic stenosis ≥70% randomized to CAS, CEA, or the best medical therapy [1].

Endarterectomy Versus Angioplasty in Patients With Symptomatic Severe Carotid Stenosis

A total of 527 patients were enrolled with symptomatic carotid stenosis of ≥60% and randomized to CAS or CEA [15]. Enrollment was terminated early due to safety and futility reasons. The stroke or death rate within 30 days was significantly higher in the CAS group (9%) than that in the CEA group (3%). This high periprocedural complication rate with CAS significantly influenced outcomes at 4 years. In patients with angulations of ≥60% from the common carotid artery to the ICA and in whom embolic protection devices were not used, a higher complication rate with CAS was observed. The requirements for operators to participate in the trial were modest compared to those in the CREST, a wide variety of stents were used, and not all patients who underwent CAS received dual antiplatelet therapy.

International Carotid Stenting Study

A total of 1713 patients with symptomatic carotid stenosis of >50% were enrolled and randomized to CAS or CEA [16,17]. The number of fatal or disabling strokes and cumulative 5-year risk rates did not differ significantly between the CAS group (6%) and the CEA group (6%). Mainly, nondisabling strokes were more common in the CAS group. A substudy of the International Carotid Stenting Study (ICSS) found a significantly higher rate of ischemic lesions on MRI with CAS than with CEA.

Recurrent stenosis rates of >50% at 5 years after CAS are approximately 3% [18]. In the Endarterectomy Versus Angioplasty in Patients with Symptomatic Severe Carotid Stenosis (EVA-3S) trial, the restenosis rate of ≥50% after CAS was significantly higher than that after CEA at a mean follow-up period of 2.1 years. Restenosis, however, was not a significant risk factor for recurrent stroke or TIA [19].

Although there has not been a randomized controlled trial comparing CAS with or without embolic protection device, data indicate that periprocedural stroke rates are significantly lower with the use of such devices [20].

Current FDA Approval for Carotid Stenting

In 2004, the FDA approved the RX Acculink Carotid Stent System for patients at high risk with CEA, with symptomatic carotid stenosis of ≥50% or asymptomatic carotid stenosis of ≥80%, and with a reference vessel diameter of 4–9 mm. The following comorbidities are considered high risk for CEA: congestive heart failure classes III and IV, left ejection fraction <30%, unstable angina, contralateral carotid occlusion, recent myocardial infarction, previous CEA with recurrent stenosis, prior radiation treatment to the neck, and other conditions used to determine patients at high risk in clinical trials.

CAROTID ENDARTERECTOMY OR CAROTID ANGIOPLASTY AND STENTING

Abundant data from clinical trials help guide the decision-making process in patients with carotid stenosis and have led to the following conclusions. Asymptomatic carotid stenosis is fairly benign and risk factor modification and medical management are good initial treatment options. Duplex ultrasonography may be helpful in identifying patients with asymptomatic stenosis at risk for stroke. Patients younger than 70 years do better with CAS than older patients. Carotid stenting has a higher risk for stroke than CEA within 30 days of the procedure, whereas the stroke risk is similar for both procedures thereafter. CEA remains the standard of care for patients with symptomatic carotid stenosis, and CAS should be reserved for patients with contraindications for CEA.

References

[1] Harrigan M, Deveikis J. Extracranial cerebrovascular occlusive disease. In: handbook of cerebrovascular disease and neurointerventional technique second edition. New York: Humana Press, Springer; 2013. p. 737.

[2] North American Symptomatic Carotid Endarterectomy Trial Collaborators. Beneficial effect of carotid endarterectomy in symptomatic patients with high-grade carotid stenosis. N Engl J Med 1991;325(7):445–53. http://dx.doi.org/10.1056/NEJM199108153250701.

[3] Randomised trial of endarterectomy for recently symptomatic carotid stenosis: final results of the MRC European Carotid Surgery Trial (ECST). Lancet 1998;351(9113):1379–87.

[4] Mayberg MR, Wilson SE, Yatsu F, Weiss DG, Messina L, Hershey LA, et al. Carotid endarterectomy and prevention of cerebral ischemia in symptomatic carotid stenosis. Veterans Affairs Cooperative Studies Program 309 Trialist Group. JAMA 1991;266(23):3289–94.

[5] Rothwell PM, Eliasziw M, Gutnikov SA, Fox AJ, Taylor DW, Mayberg MR, Carotid Endarterectomy Trialists' Collaboration, et al. Analysis of pooled data from the randomised controlled trials of endarterectomy for symptomatic carotid stenosis. Lancet 2003;361(9352):107–16.

[6] Endarterectomy for asymptomatic carotid artery stenosis. Executive Committee for the Asymptomatic Carotid Atherosclerosis Study. JAMA 1995;273(18):1421–8.

[7] Hobson RW, Weiss DG, Fields WS, Goldstone J, Moore WS, Towne JB, et al. Efficacy of carotid endarterectomy for asymptomatic carotid stenosis. The Veterans Affairs Cooperative Study Group. N Engl J Med 1993;328(4):221–7. http://dx.doi.org/10.1056/NEJM199301283280401.

[8] Carotid surgery versus medical therapy in asymptomatic carotid stenosis. The CASANOVA Study Group. Stroke 1991; 22(10):1229–35.

[9] Results of a randomized controlled trial of carotid endarterectomy for asymptomatic carotid stenosis. Mayo Asymptomatic Carotid Endarterectomy Study Group. Mayo Clin Proc 1992;67(6):513–8.

[10] Halliday A, Mansfield A, Marro J, Peto C, Peto R, Potter J, MRC Asymptomatic Carotid Surgery Trial (ACST) Collaborative Group, et al. Prevention of disabling and fatal strokes by successful carotid endarterectomy in patients without recent neurological symptoms: randomised controlled trial. Lancet 2004;363(9420):1491–502. http://dx.doi.org/10.1016/S0140-6736(04)16146-1.

[11] Goodney PP, Travis LL, Malenka D, Bronner KK, Lucas FL, Cronenwett JL, et al. Regional variation in carotid artery stenting and endarterectomy in the Medicare population. Circ Cardiovasc Qual Outcomes 2010;3(1):15–24. http://dx.doi.org/10.1161/CIRCOUTCOMES.109.864736.

[12] Yadav JS, Wholey MH, Kuntz RE, Fayad P, Katzen BT, Mishkel GJ, Stenting and angioplasty with protection in patients at high risk for endarterectomy investigators, et al. Protected carotid-artery stenting versus endarterectomy in high-risk patients. N Engl J Med 2004;351(15):1493–501. http://dx.doi.org/10.1056/NEJMoa040127.

[13] Brott TG, Hobson RW, Howard G, Roubin GS, Clark WM, Brooks W, CREST Investigators, et al. Stenting versus endarterectomy for treatment of carotid-artery stenosis. N Engl J Med 2010;363(1):11–23. http://dx.doi.org/10.1056/NEJMoa0912321.

[14] SPACE Collaborative Group, Ringleb PA, Allenberg J, Brückmann H, Eckstein H-H, Fraedrich G, et al. 30 day results from the SPACE trial of stent-protected angioplasty versus carotid endarterectomy in symptomatic patients: a randomised non-inferiority trial. Lancet 2006;368(9543):1239–47. http://dx.doi.org/10.1016/S0140-6736(06)69122-8.

[15] Mas J-L, Trinquart L, Leys D, Albucher J-F, Rousseau H, Viguier A, EVA-3S investigators, et al. Endarterectomy Versus Angioplasty in Patients with Symptomatic Severe Carotid Stenosis (EVA-3S) trial: results up to 4 years from a randomised, multicentre trial. Lancet Neurol 2008;7(10):885–92. http://dx.doi.org/10.1016/S1474-4422(08)70195-9.

[16] International Carotid Stenting Study investigators, Ederle J, Dobson J, Featherstone RL, Bonati LH, van der Worp HB, et al. Carotid artery stenting compared with endarterectomy in patients with symptomatic carotid stenosis (International Carotid Stenting Study): an interim analysis of a randomised controlled trial. Lancet 2010;375(9719):985–97. http://dx.doi.org/10.1016/S0140-6736(10)60239-5.

[17] Bonati LH, Dobson J, Featherstone RL, Ederle J, van der Worp HB, de Borst GJ, International Carotid Stenting Study investigators, et al. Long-term outcomes after stenting versus endarterectomy for treatment of symptomatic carotid stenosis: the International Carotid Stenting Study (ICSS) randomised trial. Lancet 2015;385(9967):529–38. http://dx.doi.org/10.1016/S0140-6736(14)61184-3.

[18] Bosiers M, Peeters P, Deloose K, Verbist J, Sievert H, Sugita J, et al. Does carotid artery stenting work on the long run: 5-year results in high-volume centers (ELOCAS Registry). J Cardiovasc Surg 2005;46(3):241–7.

[19] Arquizan C, Trinquart L, Touboul P-J, Long A, Feasson S, Terriat B, EVA-3S Investigators, et al. Restenosis is more frequent after carotid stenting than after endarterectomy: the EVA-3S study. Stroke 2011;42(4):1015–20. http://dx.doi.org/10.1161/STROKEAHA.110.589309.

[20] Kastrup A, Gröschel K, Krapf H, Brehm BR, Dichgans J, Schulz JB. Early outcome of carotid angioplasty and stenting with and without cerebral protection devices: a systematic review of the literature. Stroke 2003;34(3):813–9. http://dx.doi.org/10.1161/01.STR.0000058160.53040.5F.

CHAPTER

152

Management of Asymptomatic Extracranial Carotid Artery Disease

V.A. Lioutas

Beth Israel Deaconess Medical Center, Boston, MA, United States

INTRODUCTION

The stroke risk and management principles for symptomatic internal carotid artery (ICA) stenosis are relatively well established and have a wide consensus among practicing clinicians, discussed in Chapter 80. The same, however, cannot be said for management of asymptomatic carotid stenosis (ACS), which is a more complex and challenging issue. An extensive in-depth analysis exceeds the scope of this chapter, but we will present the salient points of our understanding of ACS, as it has evolved over the years.

EPIDEMIOLOGY

Extracranial ICA atherosclerosis is an important potentially treatable cause of ischemic stroke. Its prevalence in the population increases with age: in men younger than 50 years the prevalence of moderate stenosis is 0.2% and of severe stenosis is 0.1%, whereas in women of the same age the prevalence is very close to 0% [1]. In contrast, moderate stenosis (50–74%) was found in ~7% of men and ~5% of women older than 65 years [2]. The percentages for severe stenosis (75–100%) were 2.3% and 1.1%, respectively [2]. In population-wide scale, approximately 7–10% of all first ischemic strokes is associated with extracranial carotid stenosis of >60% [3]. Its relative contribution to ischemic stroke is considerably higher in the black people (attributable incidence of 17 in 100,000) than in the Hispanic population (9 in 100,000) and the white people (5 in 100,000) [3,4].

REVASCULARIZATION

In cases of recently symptomatic hemodynamically significant ICA stenosis, the risk of recurrent stroke increases in proportion to the degree of vessel stenosis, and in general, CEA is recommended for patients with stenosis of 70% or more. The role of revascularization in ACS is less straightforward.

Carotid Endarterectomy

The first evidence of the benefit of using carotid endarterectomy (CEA) in addition to medical management in ACS was provided by a clinical trial of 444 patients recruited from 10 US Veterans Affairs medical centers [5], and it was followed by two large-scale studies in the United States and Europe:

The Asymptomatic Carotid Atherosclerosis Study (ACAS) [6] was conducted between 1987 and 1993 in the United States and Canada. A total of 1662 patients with 60–99% carotid stenosis and no clinical symptoms suggestive of TIA or stroke were randomized to receive maximum medical management versus maximal medical management plus CEA. The study reached the significance boundary after 2.7 years of median follow-up when only 9% of the patients had completed the 5-year follow-up. The imputed 5-year risk of ipsilateral stroke or death in the medical arm was 11% versus 5.1% in the CEA arm [relative risk reduction (RRR), 53%; $p = .004$; confidence interval (CI), 22–72%]. The 30-day perioperative risk rate of death and stroke in those who underwent CEA was 2.3%.

The international Asymptomatic Carotid Surgery Trial (ACST) [7] was conducted between 1993 and

Primer on Cerebrovascular Diseases, Second Edition
http://dx.doi.org/10.1016/B978-0-12-803058-5.00152-1

2003. The investigators randomized 3120 patients with ACS of 60–90% to either immediate CEA or deferral of any CEA in a 1:1 ratio. The results essentially mirrored those of the ACAS study: the 5-year risk rate of any type of stroke or perioperative mortality was 6.4% in the CEA group compared to 11.8% in the CEA deferral group ($p < 0.0001$; 95% CI, 2.96–7.75). An increase in early mortality in the CEA group caused by perioperative risk was counterbalanced and eventually overturned by a higher, but more evenly distributed, number of endpoint events in the medical arm. This statistically significant net reduction of stroke and death of ~5% was maintained in the 10-year follow-up [8] which suggests that the benefit from CEA was achieved in the early to mid-term follow-up period as the rate of endpoint events remained comparable in the 5- to 10-year follow-up period. The 30-day risk rate of stroke or death was 3.1%.

Carotid Artery Stenting

No study has specifically explored the efficacy of carotid artery stenting (CAS) versus medical management in carotid stenosis, and no study with published results was devoted to asymptomatic patients only. However, patients with ACS were included in most CAS trials. The early carotid angioplasty and stenting trials were discouraging, with relatively high periprocedural risk rates of stroke and mortality. In the Carotid Revascularization Endarterectomy Versus Stenting Trial (CREST) [9], the 4-year risk rate of stroke among patients with ACS undergoing CEA was 2.7% versus 4.5% with a periprocedural risk rate of 1.4% versus 2.5%. Therefore CAS is used as an alternative to CEA in select cases, especially in cases with challenging anatomy and vasculature.

Ongoing Trials

A large phase 3 multicenter trial in Europe, Africa, Asia, and North America is currently underway (ACST-2, ISRCTN 21144362), comparing CEA and CAS for long-term stroke prevention in patients with ACS of ≥70%. The study aims to recruit 3600 patients by 2019 and follow them for 5–10 years.

The CREST-2 trial (NCT02089217) consists of two parallel multicenter randomized trials conducted in the United States and Canada with an enrolment target of 2480 patients. One trial is comparing intensive medical management alone with intensive medical management plus CEA, whereas the parallel arm is comparing medical management alone with CAS plus medical management.

OFFICIAL GUIDELINES AND CURRENT STATUS OF PRACTICE

Based on the ACAS and ACST findings, revascularization has been recommended for moderate and severe ACS in select cases [10] and this has led to practice changes, especially in the United States [11]. A caveat is that selection criteria are not specified, which has allowed a rather loose interpretation and implementation of the guidelines. As a result, a large number of CEAs have been and continue to be performed on the basis of angiographic findings only, without further stratification of stroke risk.

However, as we will see, a closer look at the specific details of these trials, along with other factors, challenges a simplistic, "one-size-fits-all," mass intervention stance.

SUBGROUP ANALYSES, PITFALLS, AND OTHER LIMITATIONS OF PUBLISHED RANDOMIZED CEA TRIALS IN ACS

Gender

The efficacy and safety of CEA in ACS were superior in males, mirroring the results of the symptomatic carotid surgery trials. In ACAS, the RRR for men was 66% (85% CI, 36–82%) and 17% (95% CI, −96 to 65%) for women. Perioperative complications occurred in 3.6% of women and in 1.7% of men, although this was not statistically significant [6]. Similar trends were seen in the ACST: a definitive benefit was noticed in men while in women the absolute risk reduction was 4%, but with wide confidence intervals (0.74–7.41)[7]. It should be noted that fewer women than men were recruited to these studies and the studies were not powered to address this particular issue, but this discrepancy between females and males was maintained when pooling the results of both trials [11,12].

Stenosis Gradation

The symptomatic ICA surgical trials revealed a clear increase in recurrent stroke risk proportional to diameter reduction. On the contrary, in ACAS, no significant difference in the 5-year benefit in stroke and mortality reduction across different stenosis gradations was noted [6]. A similar result was observed in the ACST, although both studies were underpowered to address this specific issue.

Surgeon Selection and Applicability to General Practice

The process of surgeon selection in the two studies was highly selective, in particular for ACAS: 40% of the applicants were rejected and additional surgeons were removed during the trial on the basis of adverse events, resulting in excellent surgical outcomes (0.14% surgical mortality). Concerns that these results could not be replicated in routine clinical practice were confirmed by a meta-analysis of 46 surgical case series that showed an eight-fold risk of operative mortality and approximately three-fold risk of stroke and death compared with ACAS results [11,13]. Taking into account that the net effect of the intervention is partially determined by the perioperative complications, such a distinct difference in surgical harm raises concern for the generalizability of the ACAS results to the general population.

Medical Management

A major point of criticism in both the ACAS and ACST trials is that medical management has significantly evolved since they were conducted. As an example, the medical arm of the ACAS received 325 mg of aspirin plus reduction of blood pressure, cholesterol levels, blood glucose levels, and tobacco/ethanol use, all of which were rather loosely specified. It is noteworthy that the first commercially available statin, lovastatin, was released in 1987, the year that ACAS commenced.

It has been demonstrated robustly that the annual stroke risk rate in medically managed patients with ACS has been steadily declining between 1990 and 2010 [14–16]. It matched the annual risk rate of stroke or transient ischemic attack in surgically managed patients in the ACAS and ACST in 2005 and 2007, respectively [14].

The modern intensive medical therapy includes dual antiplatelet therapy with aspirin and clopidogrel, blood pressure control, and lowering the lipid levels [17]. Particular emphasis is placed on lifestyle modifications (healthy diet, frequent exercise, smoking cessation). It should be emphasized that the lipid-level-lowering regimen should include high-dose statin targeted at halting atherosclerotic plaque progression and should not be limited to merely achieving target levels of cholesterol and low-density lipoprotein [17].

Utility, Efficacy, and Cost-Effectiveness of Mass Interventions and Medical Management

It has been calculated that for 5 years, ~60 strokes can be prevented per 1000 CEAs performed for ACS, which means that hypothetically if the results of the ACAS and ACST are applied to all patients with ACS of >60%, 94% of the procedures would have been unnecessary. Even in the extreme scenario of hypothetical 0% procedural risk, ~80 strokes per 1000 CEAs performed would be prevented in 5 years [16].

The cost-effectiveness of modern intensive medical management is superior—even with conservative calculations, medical management alone is four to eight times cheaper than CEA as observed in the landmark surgical trials [14].

RISK STRATIFICATION: IDENTIFYING PATIENTS AT HIGH RISK

The advances of stroke prevention in ACS by medical management alone have led research efforts to identify patients at high risk, who are more likely to benefit from prophylactic CEA [18].

Microemboli Detection on Transcranial Doppler

Monitoring the middle and anterior cerebral arteries ipsilateral to a stenotic ICA for the presence of microembolic signals (MESs) has yielded promising results that have been replicated in prospective observational studies. In the largest of those studies, namely, the Asymptomatic Carotid Emboli Study, the odds ratio (OR) for future ipsilateral stroke was 5.35 in the presence of MESs [19]. A meta-analysis of all relevant observational studies [20] revealed a more than six-fold increase in the risk of ipsilateral stroke in the presence of MESs [16]. Another prospective study confirmed the reduction of MESs from 12.3% to 3.7% in a population of patients with ACS treated with aggressive medical management [21], suggesting that it could also be useful in monitoring the efficacy of medical therapy. Limitations of this method include the need for trained technicians to perform the study and time constraints (patients should be monitored for at least 30 min, ideally for 1 h).

Carotid Ultrasound Plaque Characteristics

The largest amount of data regarding carotid plaque characteristics on ultrasound has been obtained from the Asymptomatic Carotid Stenosis and Risk of Stroke (ACSRS) study, a prospective study of 1121 patients with ACS on medical management followed up for a mean period of 48 months [22].

Plaque echolucency was associated with increased annual stroke risk rate (3%), whereas mainly echogenic plaques had a substantially lower risk rate (0.4%) [22]. The ACSRS study investigators developed a computerized measurement of all plaque pixels after image normalization, known as the Gray-Scale Median (GSM), which, in effect, quantifies echolucency, i.e., low GSM scores correspond with echolucent plaques and high GSM scores with echogenic ones [23]. In the ACSRS study, low GSM scores were associated with a significantly higher annual stroke risk rate.

Juxtaluminal black area (JBA) was another feature identified as contributing a high risk for ipsilateral strokes. It is thought that it represents a necrotic lipid-rich core or an intraplaque hemorrhage [24]. Patients with a JBA>8mm² had an annual stroke risk rate of 4.8% versus 0.4% in those with JBA<4mm². The overall plaque area was also predictive of stroke: the annual stroke risk rate ranged from 1% in those with plaque area<40mm² to 4.6% in those with plaque area>80mm² [25].

By combining more than one of the above-computerized plaque analysis methods with baseline characteristics, the ACSRS study investigators developed an individualized stroke risk stratification schema with predicted annual stroke risk rates ranging from 0.2% to 10% [22].

Intraplaque Hemorrhage in MRI

Advances in carotid MRI have offered promising results in identification of asymptomatic patients at high risk [26]. The presence of intraplaque hemorrhage is a marker of plaque instability and has been linked to the occurrence of future ipsilateral stroke, with reported OR ranging from 2 to 10 in various studies [27,28]. Although carotid MRI is promising, its value has not been validated prospectively, it is not widely available in clinical practice, and its cost is not covered by insurers.

Silent Infarcts Identified in Brain CT or MRI Images

Clinically silent infarctions were present in 18% of the ACSRS subjects at the time of enrolment. The silent cerebral infarcts were associated with a three-fold increase in stroke risk [29]. Although this is an easy way to identify individuals at high risk, the major counterargument is that clinically silent but radiographically manifesting infarcts denote symptomatic and not asymptomatic stenosis.

Progression of Stenosis Severity

Progression of ICA stenosis occurred in 19.8% of the ACSRS participants, while the plaque remained unchanged in 76.4% and regressed in 3.8%. Ipsilateral stroke occurred in 16% of those with progression and in 9% of those with stable degree of stenosis and no strokes were seen in the patients with plaque regression (relative risk, 1.92; 95% CI, 1.14–3.25; $P=0.05$) [30]. The incidence of stenosis progression was inversely proportional to the degree of stenosis. Similar results were reported in other studies and support the proposition that patients with ACS should be monitored and revascularization offered if the progression of stenosis is not halted by medical management. However, the optimal frequency of sequential imaging and the threshold of progression severity above which intervention is advised remain unanswered.

CONCLUSION

Our understanding of ACS is evolving. The currently available data suggest that most patients are best managed medically. Revascularization with CEA or CAS certainly has a role in the management of ACS, although its benefit is maximized when used in a subset of individuals at high risk. Individual patient and plaque characteristics can help stratify patients and aid clinicians in the appropriate use of surgical interventions. Although promising, many of the plaque imaging methods need to be validated in large prospective patient cohorts.

LIST OF COMMONLY USED ABBREVIATIONS

ACS	Asymptomatic carotid stenosis
ICA	Internal carotid artery
ACAS	Asymptomatic Carotid Atherosclerosis Study
ACST	Asymptomatic Carotid Surgery Trial
CEA	Carotid endarterectomy
CAS	Carotid artery stenting
ACSRS	Asymptomatic Carotid Stenosis and Risk of Stroke

References

[1] de Weerd M, Greving JP, Hedblad B, et al. Prevalence of asymptomatic carotid artery stenosis in the general population: an individual participant data meta-analysis. Stroke 2010;41:1294–7.
[2] Fine-Edelstein JS, Wolf PA, O'Leary DH, et al. Precursors of extracranial carotid atherosclerosis in the Framingham Study. Neurology 1994;44:1046–50.
[3] O'Leary DH, Polak JF, Kronmal RA, et al. Distribution and correlates of sonographically detected carotid artery disease in the cardiovascular health study. The CHS collaborative research group. Stroke 1992;23:1752–60.
[4] Sacco RL, Roberts JK, Boden-Albala B, et al. Race-ethnicity and determinants of carotid atherosclerosis in a multiethnic population. The Northern Manhattan Stroke Study. Stroke 1997;28:929–35.
[5] Hobson 2nd RW, Weiss DG, Fields WS, et al. Efficacy of carotid endarterectomy for asymptomatic carotid stenosis. The veterans affairs cooperative study group. N Engl J Med 1993;328:221–7.
[6] Endarterectomy for asymptomatic carotid artery stenosis. Executive committee for the asymptomatic carotid atherosclerosis study. JAMA 1995;273:1421–8.

[7] Halliday A, Mansfield A, Marro J, et al. Prevention of disabling and fatal strokes by successful carotid endarterectomy in patients without recent neurological symptoms: randomised controlled trial. Lancet 2004;363:1491–502.

[8] Halliday A, Harrison M, Hayter E, et al. 10-year stroke prevention after successful carotid endarterectomy for asymptomatic stenosis (ACST-1): a multicentre randomised trial. Lancet 2010;376:1074–84.

[9] Brott TG, Hobson 2nd RW, Howard G, et al. Stenting versus endarterectomy for treatment of carotid-artery stenosis. N Engl J Med 2010;363:11–23.

[10] Brott TG, Halperin JL, Abbara S, et al. 2011 ASA/ACCF/ AHA/AANN/AANS/ACR/ASNR/CNS/SAIP/SCAI/SIR/ SNIS/SVM/SVS guideline on the management of patients with extracranial carotid and vertebral artery disease. Stroke 2011;42:e464–540.

[11] Rothwell PM, Goldstein LB. Carotid endarterectomy for asymptomatic carotid stenosis: asymptomatic carotid surgery trial. Stroke 2004;35:2425–7.

[12] Rothwell PM. ACST: which subgroups will benefit most from carotid endarterectomy?. Lancet 2004;364:1122–3. author reply 5–6.

[13] Bond R, Rerkasem K, Rothwell PM. Systematic review of the risks of carotid endarterectomy in relation to the clinical indication for and timing of surgery. Stroke 2003;34:2290–301.

[14] Abbott AL. Medical (nonsurgical) intervention alone is now best for prevention of stroke associated with asymptomatic severe carotid stenosis: results of a systematic review and analysis. Stroke 2009;40:e573–83.

[15] Marquardt L, Geraghty OC, Mehta Z, Rothwell PM. Low risk of ipsilateral stroke in patients with asymptomatic carotid stenosis on best medical treatment: a prospective, population-based study. Stroke 2010;41:e11–7.

[16] Naylor AR. Time to rethink management strategies in asymptomatic carotid artery disease. Nat Rev Cardiol 2012;9:116–24.

[17] Spence JD. Intensive risk factor control in stroke prevention. F1000Prime Rep 2013;5:42.

[18] Paraskevas KI, Spence JD, Veith FJ, Nicolaides AN. Identifying which patients with asymptomatic carotid stenosis could benefit from intervention. Stroke 2014;45:3720–4.

[19] Markus HS, King A, Shipley M, et al. Asymptomatic embolisation for prediction of stroke in the asymptomatic carotid emboli study (ACES): a prospective observational study. Lancet Neurol 2010;9:663–71.

[20] Spence JD, Tamayo A, Lownie SP, Ng WP, Ferguson GG. Absence of microemboli on transcranial doppler identifies low-risk patients with asymptomatic carotid stenosis. Stroke 2005;36:2373–8.

[21] Spence JD, Coates V, Li H, et al. Effects of intensive medical therapy on microemboli and cardiovascular risk in asymptomatic carotid stenosis. Arch Neurol 2010;67:180–6.

[22] Nicolaides AN, Kakkos SK, Kyriacou E, et al. Asymptomatic internal carotid artery stenosis and cerebrovascular risk stratification. J Vasc Surg 2010;52:1486–96. e1–e5.

[23] Nicolaides AN, Kakkos SK, Griffin M, et al. Effect of image normalization on carotid plaque classification and the risk of ipsilateral hemispheric ischemic events: results from the asymptomatic carotid stenosis and risk of stroke study. Vascular 2005;13:211–21.

[24] Griffin MB, Kyriacou E, Pattichis C, et al. Juxtaluminal hypoechoic area in ultrasonic images of carotid plaques and hemispheric symptoms. J Vasc Surg 2010;52:69–76.

[25] Kakkos SK, Griffin MB, Nicolaides AN, et al. The size of juxtaluminal hypoechoic area in ultrasound images of asymptomatic carotid plaques predicts the occurrence of stroke. J Vasc Surg 2013;57:609–18 e1. discussion 17–18.

[26] Hellings WE, Peeters W, Moll FL, et al. Composition of carotid atherosclerotic plaque is associated with cardiovascular outcome: a prognostic study. Circulation 2010;121:1941–50.

[27] Singh N, Moody AR, Gladstone DJ, et al. Moderate carotid artery stenosis: MR imaging-depicted intraplaque hemorrhage predicts risk of cerebrovascular ischemic events in asymptomatic men. Radiology 2009;252:502–8.

[28] Takaya N, Yuan C, Chu B, et al. Association between carotid plaque characteristics and subsequent ischemic cerebrovascular events: a prospective assessment with MRI–initial results. Stroke 2006;37:818–23.

[29] Kakkos SK, Sabetai M, Tegos T, et al. Silent embolic infarcts on computed tomography brain scans and risk of ipsilateral hemispheric events in patients with asymptomatic internal carotid artery stenosis. J Vasc Surg 2009;49:902–9.

[30] Kakkos SK, Nicolaides AN, Charalambous I, et al. Predictors and clinical significance of progression or regression of asymptomatic carotid stenosis. J Vasc Surg 2014;59:956–67. e1.

CHAPTER

153

Management of Cerebellar Hematomas and Infarcts

D.M. Heiferman, C.M. Loftus

Loyola University Chicago, Stritch School of Medicine, Maywood, IL, United States

Cerebellar hemorrhage and cerebellar infarction are two clinical entities with entirely different pathophysiologies, although the clinical syndromes are quite similar and the surgical management and operative considerations are nearly identical. For this reason, surgical treatment of either cerebellar hemorrhage or cerebellar infarction should be listed under the rather broader category of management of vascular-related mass effect in the posterior fossa.

CEREBELLAR HEMORRHAGE

Cerebellar hemorrhage is most often seen in known hypertensive patients, and the most common site of bleeding is in the area of the dentate nucleus, presumably from rupture of a branch of the superior cerebellar artery [1]. The hematoma thus begins in the cerebellar hemisphere, although it may extend medially to involve the vermis or the contralateral hemisphere, and therefore may dissect into the fourth ventricle, resulting in intraventricular hemorrhage [2,3]. Direct involvement of the brain stem is unusual, but brain stem compression from mass effect, as is seen in both cerebellar hemorrhage and infarction, is not only common, but is the major cause of morbidity and mortality from these conditions.

Besides hypertensive hemorrhage, other etiologies of a cerebellar hematoma need to be ruled out. Patients with hereditary, acquired, or iatrogenic coagulopathies are at significantly increased risk for cerebellar hemorrhage. Underlying structural causes of hemorrhage may be present in any patient, but are most common in younger patients, in nonhypertensive patients, or in patients with hematoma localized to the cerebellar

vermis (Fig. 153.1). Structural causes of hemorrhage would include primary or metastatic brain tumors, vascular anomalies, sterile or septic thromboemboli and, related to the above, hemorrhagic transformation of a cerebellar infarction. A careful search for an underlying structural cause needs to be made in all patients who present atypically, according to the criteria mentioned. In our experience, a traumatic etiology for cerebellar hemorrhage (as opposed to supratentorial traumatic hemorrhage) is extremely rare and has been seen only in association with major head trauma (Fig. 153.2). The exception to this rule would be patients with coagulopathies who may have either spontaneous cerebellar hemorrhage or cerebellar hemorrhage associated with very minor head trauma.

The signs and symptoms of cerebellar hemorrhage are very similar to those for cerebellar infarction and, once again, relate to the resultant mass effect in the posterior fossa. Heros has outlined the following three major stages that form the hallmarks of the clinical presentation of posterior fossa mass effect [1]. In the early stage, symptoms are related to destruction of cerebellar tissue and/or extension of hematoma into the subarachnoid space. These symptoms include dizziness, nausea, vomiting, headache, balance difficulties, and gait ataxia. These symptoms are more likely to be of an abrupt onset in cerebellar hemorrhage, and are more likely to be of a stuttering nature or insidiously progressive in cerebellar infarction. Patients in the early stage of the posterior fossa compression process are still relatively awake and alert.

The intermediate stage of posterior fossa compression ensues when mass effect increases and/or hydrocephalus results from either direct compression of the fourth ventricle or from extension of blood into the

Primer on Cerebrovascular Diseases, Second Edition
http://dx.doi.org/10.1016/B978-0-12-803058-5.00153-3

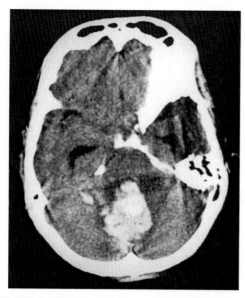

FIGURE 153.1 CT scan, axial section, without contrast. A large hemorrhage is evident in the cerebellar vermis. This patient was a young nonhypertensive woman who had become somnolent and ataxic 48h following open mitral valve commissurotomy. Although never proven pathologically, it was our feeling that this was most likely a hemorrhage from a cardiac origin embolus or a hemorrhagic infarct. Centrally located lesions of this kind are also more likely to be associated with underlying vascular anomalies or hemorrhage into a tumor bed.

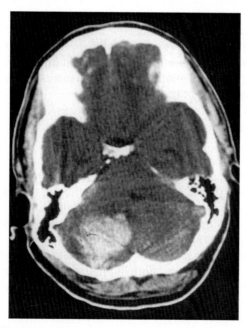

FIGURE 153.2 CT scan, axial section, without contrast. A large hemorrhage is evident in the cerebellar hemisphere with compression and deviation of the fourth ventricle. This particular case was a traumatic cerebellar hemorrhage, which, as mentioned in the text, is extremely rare in our experience. Surgical evacuation was carried out immediately because of the posterior fossa mass effect and the size of the lesion.

ventricular system with blockage of the cerebrospinal fluid pathways. This stage is marked by a depression of the level of consciousness and other neurological findings referable to increased intracranial pressure, including abducens nerve palsies. A mild hemiparesis and peripheral facial nerve deficit, from compression of the facial colliculus, as well as Babinski's sign, may be detectable at this time.

In the late stage of posterior fossa compression, coma ensues. At this point, brain stem compression is maximal and cardiovascular abnormalities are seen. Patients may progress rapidly through the three stages of posterior fossa compression particularly when significant hydrocephalus or large masses are present. This compounds the precarious nature of their medical management and has led to a great interest in optimizing treatment for both cerebellar hemorrhage and cerebellar infarction with early surgical intervention in appropriate cases.

DIAGNOSIS

The cornerstone of diagnosis for any patient suspected of having mass effect in the posterior fossa is the axial computed tomography (CT) scan. Many cases of cerebellar hemorrhage or infarction were missed, misdiagnosed, or identified only on autopsy prior to the advent of the CT scan given the often nonfocal clinical picture. A noncontrast head CT is indicated for any patient in whom there is clinical suspicion for posterior fossa compression and should be done as early as possible to facilitate rapid intervention. During evaluation of the posterior fossa on a CT scan, low-density areas of cerebellar parenchyma consistent with evolving infarction or high-density lesions, which may be indicative of a hematoma, should be identified, and the presence of associated intraventricular hemorrhage and/or hydrocephalus must be noted. Magnetic resonance imaging (MRI) is another imaging modality that is useful in understanding cerebellar pathology. On MRI, the extent of infarction can be evaluated on diffusion-weighted sequences, which can help to predict impending edema, and to further characterize structural lesions that may be underlying a cerebellar hemorrhage. Whether or not further diagnostic procedures are required will depend on the nature of the lesion identified on initial imaging. Angiography is no longer necessary in most cases. Any patient suspected of harboring a structural cause for cerebellar hemorrhage is studied with CT angiography (CTA) and/or magnetic resonance angiography (MRA), especially in nonhypertensive patients, patients with lesions in the vermis, or other unusual manifestations of cerebellar hemorrhage. This is with the provision that patients who are at risk for incipient worsening are managed with immediate surgical intervention. In our experience, the most common associated structural abnormality is not a vascular malformation, but rather hemorrhage into a primary or metastatic tumor.

TREATMENT

Prior to the availability of CT scans, the diagnosis of cerebellar hemorrhage was based on clinical suspicion or on postmortem findings. The early series' of McKissock and Fisher demonstrated that surgical evacuation was associated with the most favorable outcome, and highlights that the only good outcomes without surgery were in patients who remained alert throughout their course [3,4]. Patients with cerebellar hemorrhage often progress and deteriorate rapidly, again underscoring the necessity of an aggressive surgical approach [5].

The ability to diagnose and measure hematoma size on CT has changed management strategies in some cases, possibly due to our enhanced ability to identify a subset of patients with smaller lesions who may do well without surgery, but who might not have come to clinical attention in the pre-CT era. Numerous authors have attempted to develop criteria for surgery, mainly based on hematoma size, patient level of consciousness, presence of hydrocephalus, and/or presence of brain stem compression, either clinically or on imaging. Little et al. divided 10 patients with cerebellar hematomas into two groups based on initial clinical presentation; Group 1 had early impaired levels of consciousness, which was considered to be an early sign of brain stem dysfunction, and Group 2 had more benign neurological findings, often only with cerebellar findings. Five of six patients in Group 1 underwent surgery, with one being determined to be nonsurvivable, while patients in Group 2 were managed conservatively. Of the 10 total patients, there were two mortalities in Group 1 and all other patients recovered with minimal residual deficits [6]. Van der Hoop et al. reported good outcomes in 13 patients treated conservatively; in 4 of these the hematomas were >3 cm, but all 4 had Glasgow Coma Scores of 11 points or more, suggesting that level of consciousness may be an equally important guide to treatment [7]. Additional studies, using hematoma size, hemorrhage acuity, quadrigeminal cistern effacement, and fourth ventricular effacement as independent metrics for surgical decision making, corroborated the findings that those with large acute hematomas causing mass effect, although without poor neurological examinations, benefit most

from surgical decompression [8–12]. The American Heart Association/American Stroke Association 2015 Guidelines [13] recommend that "patients with cerebellar hemorrhage who are deteriorating neurologically or who have brain stem compression and/or hydrocephalus from ventricular obstruction should undergo surgical removal of the hemorrhage as soon as possible," and "initial treatment of these patients with ventricular drainage alone rather than surgical evacuation is not recommended."

Open suboccipital craniectomy, with either a paramedian or midline approach with hematoma evacuation is the standard surgical treatment for cerebellar hemorrhage. Stereotactic aspiration procedures have been described, but are currently less widely accepted. These techniques involve aspiration of as much clot as possible with postoperative delivery of fibrinolytic agents through an indwelling catheter to dissolve and aspirate the remaining hematoma [14,15]. Although hydrocephalus has been reported to be present in up to 75% of patients with cerebellar hemorrhage, ventricular drainage alone is felt by most to be inappropriate, with a high risk of upward herniation from posterior fossa mass effect [4,16,17].

The possibility of an error of omission outweighs that of an error of commission in these cases. The senior author's (C.M.L.) policy for cerebellar hemorrhage is straightforward. This pathology is managed conservatively only in alert patients with small hemorrhages. Any patient with a depressed level of consciousness, a large hemorrhage, or cisternal compression on CT undergoes evacuation. We do not quantify hemorrhages volumetrically. In most cases, especially the unusual ones, such as vermian or nonhypertensive bleeds, investigation for structural lesions with CTA and/or MRA is undertaken, if time permits, prior to surgery.

CEREBELLAR INFARCTION

Surgical intervention in cases of cerebellar infarction with consequent posterior fossa swelling and brain stem compression were first described in 1956 by Fairburne et al. and in a separate report by Lindgren [18,19]. Sporadic reports of clinical successes with various treatments for cerebellar infarction appeared in the literature prior to the advent of CT scanning, but the diagnostic capabilities prior to CT and MRI were quite limited, as previously discussed. Sypert and Alvord described extensively the pre-CT manifestations of cerebellar infarction, including the low percentage of accurate diagnoses [20]. Until CT was widely available, diagnosis relied upon both clinical suspicion and the angiographic appearance of an avascular posterior fossa mass with arterial occlusion. All too often in these patients, the diagnosis was unfortunately made post-mortem. Now, patients with early signs of posterior fossa compression can have immediate confirmation of cerebellar hemorrhage, cerebellar infarction, or other space-occupying process in the posterior fossa, as well as the presence of associated hydrocephalus, and can be treated accordingly.

Several treatment strategies have been proposed for cerebellar infarction with posterior fossa mass effect. Khan et al. reported good results in 10 of 11 patients treated either conservatively or with ventricular drainage only and attributed their single poor result to brain stem infarction at the time of vertebral artery occlusion [21]. Horwitz and Ludolph proposed a combined medical and surgical scheme in which patients were first intubated and received dexamethasone and mannitol. A ventriculostomy was then placed if the state of consciousness did not improve with medical treatment, and likewise, suboccipital craniectomy was proposed if the level of consciousness did not improve within few hours of ventriculostomy [22]. As in cerebellar hematomas, most literature now stresses that the surgical procedures proposed are relatively benign compared to the malignant natural history of the disease and the rather fulminant course that can occur with compression in the posterior fossa. Thus, both Heros and Chen et al. recommend combined ventricular drainage and suboccipital decompression with resection of infarcted tissue as primary treatment [23,24]. With this strategy, Chen et al. reported independent survival in 8 of 11 patients so treated [23]. Additionally, in a review of their series on surgical management of bilateral cerebellar infarction, Tsitsopoulos et al. demonstrated meaningful recovery when patients were without brain stem infarction [25].

Published in 1999, the German–Austrian Cerebellar Infarction Study (GASCIS), which was a multicenter cohort study of 84 patients, is the only prospective study of the treatment of cerebellar infarction. Treatment protocols, including medical management, ventriculostomy, and craniotomy, were per the discretion of the treating physician. It was found that the patient's state of consciousness was the strongest predictor of clinical outcome; the more awake a patient was throughout their hospitalization, the higher chance of a "good outcome" (which they noted to be a Rankin score ≤2 at 3 months post-stroke). Also to this point, Jauss et al. stated that among the awake and stuporous patients, there was no difference in outcome with any particular treatment. This did not apply to the comatose patients, as they all underwent

procedural management, with 16 of 19 undergoing surgical decompression. While these 19 patients' level of consciousness deteriorated to a comatose state, half of those that received a surgical decompression improved to a good outcome at 30 days, further giving credence to the importance of decompression and which may speak to the potential benefit of preemptive decompression, although the nonrandomized nonblinded design of this study clearly imparted significant selection bias. From this study, it appears that when significant cerebellar edema and neurological decompensation is present or expected, suboccipital decompression is a common practice [26].

The surgical procedures are straightforward and within the province of any neurosurgeon. A simple suboccipital craniectomy is performed with opening of the dura and resection of infarcted tissue followed by meticulous hemostasis. As would be suspected from the patterns of arterial occlusion, almost all such operations will be unilateral. Several authors stress that even patients in poor clinical condition (i.e., severely depressed level of consciousness) can do well, because of the physiological mechanism of brain stem compression accounting for their deficits [24,27]. It must be understood, however, that vertebral artery infarcts can also involve the brain stem primarily and, of course, the prognosis would be expected to be much poorer in patients of this kind (Fig. 153.3).

Ventriculostomy is a treatment option often reserved for scenarios of hydrocephalus without brain stem compression, with a concern for upward herniation when not paired with decompression of the posterior fossa. The limited literature available frequently favors decompression with or without ventriculostomy over ventriculostomy alone, given higher rates of good clinical outcomes in patients who underwent suboccipital craniectomy. When appropriate, the ventriculostomy is performed according to standard landmarks and long-term conversion to a shunt may be necessary based on the guidance of follow-up imaging [26,28,29].

The clinical syndrome of posterior circulation arterial occlusion and cerebellar infarction has been increasingly recognized since the advent of CT. An aggressive surgical approach is warranted since brain stem compression is potentially reversible in many of these patients. At present, surgical decompression of the posterior fossa and resection of infarcted tissue with/without ventricular drainage for concomitant hydrocephalus are recommended with the understanding that patients with infarction of the brain stem may represent a subgroup who are more likely to have a poor outcome.

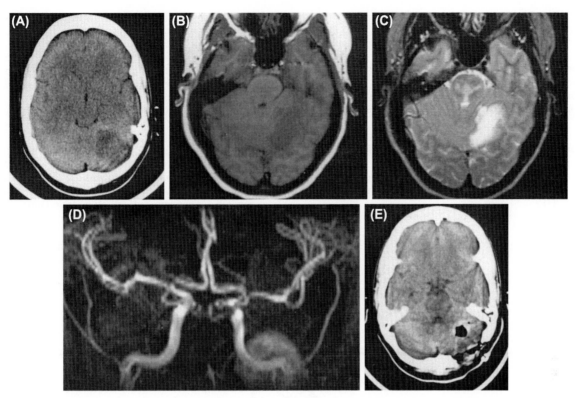

FIGURE 153.3 CT scans (A, E), MRI scans (Tl-weighted (B), T2-weighted (C)), and MR angiogram (D) of a 25-year-old woman who had become unconscious while being treated for a severe asthmatic attack. Enormous quantities of epinephrine had been administered in an outside emergency room. The axial CT demonstrated a large cerebellar infarction with significant mass effect (A). Axial MRI scans (B, C) confirmed the infarction and also showed significant brain stem involvement (C), whether from compression or ischemia. No angiogram was performed, but the vertebral-basilar system did not fill on MRA (D). At the family's request a posterior fossa decompression was performed (postoperative CT (E)), but the patient was left in what appeared to be a locked-in state and was allowed to expire with supportive care. Clearly the brain stem injury was too profound to be compatible with neurological salvage.

References

[1] Heros RC. Cerebellar hemorrhage and infarction. Stroke 1982;13(1):106–9.

[2] Freeman RE, Onofrio BM, Okazaki H, Dinapoli RP. Spontaneous intracerebral hemorrhage. Diagnosis and surgical treatment. Neurology 1973;23(1):84–90.

[3] Fisher CM, Picard EH, Polak A, Dalal P, Pojemann RG. Acute hypertensive cerebellar hemorrhage: diagnosis and surgical treatment. J Nerv Ment Dis 1965;140:38–57.

[4] McKissock W, Richardson A, Walsh L. Spontaneous cerebellar haemorrhage. Brain 1960;83(1):1–9.

[5] Ott KH, Kase CS, Ojemann RG, Mohr JP. Cerebellar hemorrhage: diagnosis and treatment. A review of 56 cases. Arch Neurol 1974;31(3):160–7.

[6] Little JR, Tubman DE, Ethier R. Cerebellar hemorrhage in adults. Diagnosis by computerized tomography. J Neurosurg 1978;48(4):575–9.

[7] van der Hoop RG, Vermeulen M, van Gijn J. Cerebellar hemorrhage: diagnosis and treatment. Surg Neurol 1988;29(1):6–10.

[8] Zieger A, Vonofakos D, Steudel WI, Dusterbehn G. Nontraumatic intracerebellar hematomas: prognostic value of volumetric evaluation by computed tomography. Surg Neurol 1984;22(5):491–4.

[9] Taneda M, Hayakawa T, Mogami H. Primary cerebellar hemorrhage. Quadrigeminal cistern obliteration on CT scans as a predictor of outcome. J Neurosurg 1987;67(4):545–52.

[10] Kobayashi S, Sato A, Kageyama Y, Nakamura H, Watanabe Y, Yamaura A. Treatment of hypertensive cerebellar hemorrhage–surgical or conservative management?. Neurosurgery 1994;34(2):246–50. discussion 250-241.

[11] Chin D, Carney P. Acute cerebellar hemorrhage with brainstem compression in contrast with benign cerebellar hemorrhage. Surg Neurol 1983;19(5):406–9.

[12] Kirollos RW, Tyagi AK, Ross SA, van Hille PT, Marks PV. Management of spontaneous cerebellar hematomas: a prospective treatment protocol. Neurosurgery 2001;49(6):1378–86. discussion 1386-1377.

[13] Hemphill 3rd JC, Greenberg SM, Anderson CS, Becker K, Bendok BR, Cushman M, et al. Guidelines for the management of spontaneous intracerebral hemorrhage: a guideline for healthcare professionals from the American Heart Association/American Stroke Association. Stroke July 2015;46(7):2032–60.

[14] Niizuma H, Suzuki J. Computed tomography-guided stereotactic aspiration of posterior fossa hematomas: a supine lateral retromastoid approach. Neurosurgery 1987;21(3):422–7.

[15] Mohadjer M, Eggert R, May J, Mayfrank L. CT-guided stereotactic fibrinolysis of spontaneous and hypertensive cerebellar hemorrhage: long-term results. J Neurosurg 1990;73(2):217–22.

[16] Auer LM, Auer T, Sayama I. Indications for surgical treatment of cerebellar haemorrhage and infarction. Acta neurochir 1986;79(2–4):74–9.

[17] Lui TN, Fairholm DJ, Shu TF, Chang CN, Lee ST, Chen HR. Surgical treatment of spontaneous cerebellar hemorrhage. Surg Neurol 1985;23(6):555–8.

[18] Fairburn B, Oliver LC. Cerebellar softening; a surgical emergency. Br Med J 1956;1(4979):1335–6.

[19] Lindgren SO. Infarctions simulating brain tumours in the posterior fossa. J Neurosurg 1956;13(6):575–81.

[20] Sypert GW, A. Lvord ECJ. Cerebellar infarction. A clinicopathological study. Arch Neurol 1975;32(6):357–63.

[21] Khan M, Polyzoidis KS, Adegbite AB, McQueen JD. Massive cerebellar infarction: "conservative" management. Stroke 1983;14(5):745–51.

[22] Horwitz NH, Ludolph C. Acute obstructive hydrocephalus caused by cerebellar infarction. Treatment alternatives. Surg Neurol 1983;20(1):13–9.

[23] Chen HJ, Lee TC, Wei CP. Treatment of cerebellar infarction by decompressive suboccipital craniectomy. Stroke 1992;23(7):957–61.

[24] Heros RC. Surgical treatment of cerebellar infarction. Stroke 1992;23(7):937–8.

[25] Tsitsopoulos PP, Tobieson L, Enblad P, Marklund N. Clinical outcome following surgical treatment for bilateral cerebellar infarction. Acta Neurol Scand 2011;123(5):345–51.

[26] Jauss M, Krieger D, Hornig C, Schramm J, Busse O. Surgical and medical management of patients with massive cerebellar infarctions: results of the German-Austrian cerebellar infarction study. J Neurol 1999;246(4):257–64.

[27] Laun A, Busse O, Calatayud V, Klug N. Cerebellar infarcts in the area of the supply of the PICA and their surgical treatment. Acta Neurochir 1984;71(3–4):295–306.

[28] Kudo H, Kawaguchi T, Minami H, Kuwamura K, Miyata M, Kohmura E. Controversy of surgical treatment for severe cerebellar infarction. J Stroke Cerebrovas Dis 2007;16(6):259–62.

[29] Juttler E, Schweickert S, Ringleb PA, Huttner HB, Kohrmann M, Aschoff A. Long-term outcome after surgical treatment for space-occupying cerebellar infarction: experience in 56 patients. Stroke 2009;40(9):3060–6.

CHAPTER

154

Surgical and Endovascular Management of Ruptured Posterior Circulation Aneurysms

R. Tahir, M. Kole

Henry Ford Health System, Detroit, MI, United States

INTRODUCTION

Nearly 30,000 Americans suffer from aneurysmal subarachnoid hemorrhage (SAH) every year, and the overall mortality rate is 40% [1]. About 5–15% of all intracranial aneurysms are located in the posterior circulation [2]. The posterior circulation provides the blood supply to the medulla, pons, midbrain, cerebellum, occipital lobes, and posterior parietal and posteroinferior temporal watershed zones. A comprehensive review of the posterior circulation vasculature is beyond the scope of this chapter, but an understanding of the anatomy allows for proper pattern recognition of the characteristic aneurysmal locations and stereotypical presentations, thus facilitating successful management. This chapter will review the endovascular and surgical treatment of ruptured posterior circulation aneurysms.

PRESENTATION AND DIAGNOSIS

The usual presentation of a ruptured posterior circulation aneurysm does not differ significantly from that of SAH caused by aneurysm rupture in other locations. Most frequently the patient presents with the "worst headache of my life." Other associated symptoms include nausea/vomiting (77%), loss of consciousness (53%), and nuchal rigidity (35%) [1].

The initial diagnostic test of choice is noncontrast computed tomographic (CT) scan (Class I evidence) [1]. Lumbar puncture is strongly recommended in cases in which the clinical suspicion for SAH is high but the CT scan result is negative. After confirmation of SAH, a noninvasive CT angiogram (CTA) of the head and neck can be rapidly performed to define the vascular anatomy (Class IIb evidence) [1]. Axial, coronal,

Primer on Cerebrovascular Diseases, Second Edition
http://dx.doi.org/10.1016/B978-0-12-803058-5.00154-5

sagittal, and three-dimensional (3D) reconstructions are extremely helpful to define anatomic relationships, aneurysmal geometry, and direct therapeutic decisions. More rapid, less invasive techniques such as CTA and magnetic resonance angiography (MRA) have shown comparable sensitivities in detecting aneurysms, but selective catheter cerebral angiography still remains the gold standard for diagnosing cerebral aneurysms in the setting of SAH (Class I evidence) [1]. It provides dynamic information on intracranial transit time and collateral circulation. It also has better spatial resolution to detect small aneurysms and adjacent perforating vessels. Rotational angiography allows for 3D reconstructions that are invaluable to the clinician in treatment planning. These advancements in diagnostic imaging have dramatically improved the understanding of the anatomy, thus facilitating successful treatment.

NATURAL HISTORY

The International Study of Unruptured Intracranial Aneurysms (ISUIA) demonstrated 5-year rupture rates of aneurysms based on their size and location [3]. Posterior communicating and posterior circulation aneurysms, especially basilar tip aneurysms, demonstrated the highest risk of rupture. The higher risk natural history was also associated with higher treatment risk. Variables associated with poor surgical and endovascular outcomes in the treatment of unruptured aneurysms included aneurysm diameter >12 mm and its location in the posterior circulation, particularly the basilar tip [3].

TREATMENT OPTIONS

There is no consensus on the best technique, endovascular versus microsurgical, for securing all the different types of ruptured posterior circulation aneurysms. The treatment modality is generally selected based on the risk-benefit ratio—the likelihood of the most definitive obliteration of the aneurysm with the least risk to the patient. Patient-specific considerations such as clinical status, anatomy, aneurysm location, aneurysm projection, and aneurysm geometry are important determinants of treatment approach and operator experience.

After aneurysmal rupture, the main objective is to secure the aneurysm and minimize secondary injury caused by mass effect, edema, hydrocephalus, and ischemia. The surgical exposure of posterior circulation aneurysms is more challenging and has higher risk than that of anterior circulation aneurysms [4]. Posterior circulation aneurysms require more involved skull base approaches necessitating dissection between cranial nerves, deep brainstem nuclei, and critical, tiny perforating brainstem vessels. These exposures frequently offer limited opportunity for proximal control of the aneurysm. For these reasons, at most centers, endovascular techniques are the preferred first treatment method for posterior circulation aneurysms. We will briefly review surgical and endovascular techniques in the following sections.

SURGICAL APPROACHES

The basilar bifurcation aneurysm is the most common aneurysm of the posterior circulation and accounts for 5–8% of all intracranial aneurysms (Fig. 154.1). The two main surgical methods to expose the basilar bifurcation are the subtemporal approach and the transsylvian approach. The third option is a combination of the two (a half-and-half approach). Each method has its advantages, disadvantages, proponents, and opponents. Generally, the selection of approach is based on

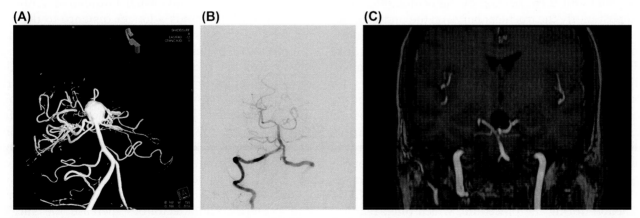

FIGURE 154.1 (A) A three-dimensional reconstruction of a right vertebral artery rotational angiogram demonstrates a 14-mm wide-necked basilar bifurcation aneurysm, and (B) immediate posttreatment right vertebral artery injection following selective coiling. (C) Coronal gadolinium-enhanced magnetic resonance angiogram shows stable occlusion of the aneurysm 3 years after treatment.

(A)	**(B)**

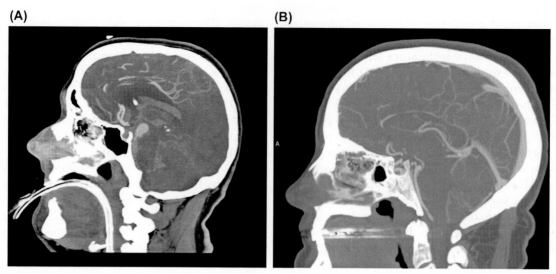

FIGURE 154.2 Computed tomographic angiogram of a sagittal reconstruction showing a ruptured basilar apex aneurysm in relationship to the posterior clinoid. (A) The aneurysm dome projects above the posterior clinoid. (B) A smaller lower-lying posteriorly directed basilar apex aneurysm in relationship to the posterior clinoid.

the aneurysm's location in relation to the posterior clinoid. The transsylvian approach is better suited for high-riding aneurysms, whereas the subtemporal approach is more appropriate for aneurysms at or slightly below the posterior clinoid (Fig. 154.2).

Subtemporal Approach

The subtemporal approach was popularized by Dr. Charles Drake [5]. The basilar bifurcation can be visualized perpendicularly and laterally through a 4-cm temporal craniotomy down to the level of the zygomatic root, flush with the middle cranial floor. After gentle retraction of the temporal lobe, the free edge of the tentorium is exposed. Retracting the tentorium several millimeters with a suture allows visualization of the arachnoid and identification of the oculomotor nerve and posteriorly the trochlear nerve on the undersurface of the tentorium. Dissection is continued through the arachnoid membrane and the membrane of Liliequist following the superior cerebellar artery (SCA) medially to the basilar artery. This approach allows for good proximal control and better access to the posteriorly projecting basilar tip aneurysms as well as the SCA aneurysms (Fig. 154.3). Although this approach offers a shorter distance to the basilar apex, the surgical window provides limited visualization of the contralateral P1 segment and contralateral thalamoperforators. In addition, this approach requires temporal lobe retraction during the procedure, which often requires cerebrospinal fluid (CSF) diversion. Modifications of

this method can be performed to gain lower exposure by incising the tentorium or combining with anterior petrosectomy or zygomatic osteotomy.

Frontotemporal Transsylvian Approach

The transsylvian approach was classically described by Dr. Gazi Yasargil for basilar apex aneurysms [6]. An oblique view of the basilar bifurcation can be obtained with a more familiar frontotemporal craniotomy. The sylvian fissure is split and dissection is extended to the optic and carotid cisterns. Through various anatomic corridors between the internal carotid artery and the optic nerve and between the internal carotid artery and the posterior communicating artery and oculomotor nerve, the Liliequist membrane can be dissected to gain exposure to the basilar bifurcation. Although this approach demands a longer distance to the basilar apex than the subtemporal approach (Fig. 154.4), it allows for a better view of anteriorly projecting and high-riding aneurysms of the basilar artery. In addition, the approach provides ideal access to anterior circulation aneurysms, if needed, for surgical clipping during the same procedure. Posteriorly projecting perforators are often obscured in this approach and proximal control can be difficult. For access to high-riding aneurysms an orbitozygomatic approach can be added. This involves removal of the superior and lateral walls of the orbit and zygomatic arch. Anterior and posterior clinoidectomies increase the surgical window for low-lying basilar aneurysms.

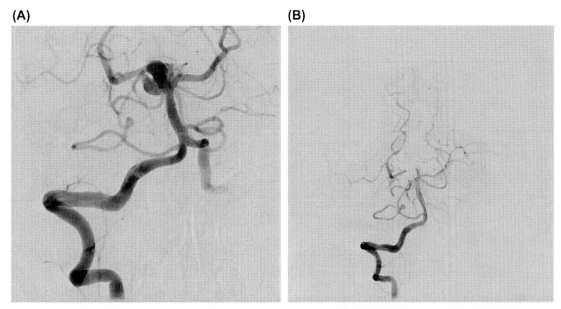

FIGURE 154.3 (A) A right vertebral artery angiogram shows a complex ruptured aneurysm of the right superior cerebellar artery. The neck of the aneurysm is located on the basilar artery between the P1 posterior cerebral artery and the superior cerebellar artery. Notice the cleft in the aneurysm at the level of the cranial nerve (CN) III. (B) Cerebral angiogram shows near-complete obliteration of the aneurysm after a stent-assisted technique.

Half-and-Half Approach

The half-and-half approach, temporopolar or pretemporal approach, utilizes a combination of transsylvian and subtemporal exposure. The sylvian fissure is widely dissected and the arachnoid connections between the frontal and temporal lobes are divided, allowing for mobilization of the temporal pole posterior to the middle cranial fossa floor exposing the tentorial edge.

Approaches for Midbasilar, Vertebrobasilar Junction, Vertebral, and PICA Aneurysms

Transpetrosal approaches including the retrolabyrinthine, translabyrinthine, and transcochlear approaches as well as the combined supratentorial and infratentorial approaches are generally reserved for large and giant midbasilar aneurysms [7]. Retrosigmoid and far lateral approaches are used for vertebrobasilar, posteroinferior cerebellar artery (PICA), and vertebral aneurysms.

Far Lateral Approach

A retrosigmoid craniotomy is performed down to the lateral edge of the foramen magnum with optional resection of the posterior arch of C1. Both the anterior brainstem and vertebral artery are easily visualized in this approach, thus facilitating proximal control of the ipsilateral vertebral artery. This approach allows for

surgical treatment of PICA and vertebral artery aneurysms. Several complications may arise from direct injury to the vertebral artery, PICA, and lower cranial nerves, or there may be atlanto-occipital instability because of aggressive removal of the arch of C1 or occipital condyle.

ENDOVASCULAR TREATMENTS

Since the advent of the Guglielmi detachable coil (GDC) [8], there has been tremendous innovation, evolution, and enthusiasm for endovascular techniques in the treatment of intracranial aneurysms. Endovascular treatment of ruptured posterior circulation aneurysms generally carries a lower risk than open surgery. However serious complications of the procedure can occur and include groin complications, contrast allergy, contrast nephropathy, vessel occlusion, stroke, intraprocedural rupture, incomplete obliteration, aneurysm recanalization, and coil migration.

Patient selection is critical to the success of endovascular procedures. Not all intracranial aneurysms can be treated with endovascular strategies as seen by a significant number of patients being excluded or crossing over from endovascular to surgical arms in the randomized controlled trials. A fundamental prerequisite for endovascular treatment is a thorough angiographic evaluation of the anatomy, with special attention to the neck of the aneurysm and its

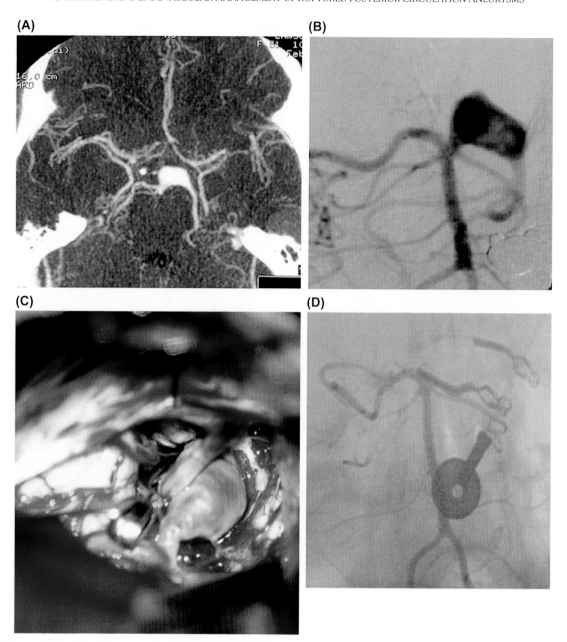

FIGURE 154.4   (A) Axial computed tomographic angiogram (CTA) and (B) right vertebral artery angiogram in Waters projection showing 16-mm left P1 dissecting type aneurysm. Aneurysms of the posterior cerebral artery (PCA) comprise 1% of all intracranial aneurysms, with the P1 segment being the most common site. Aneurysms in this location are frequently fusiform/dissecting type. Patients tend to tolerate proximal occlusion of the PCA without major deficits as long as the perforating and choroidal vessels are not occluded. (C) Intraoperative photograph from a left subtemporal approach illustrating proximal P1 clipping with a bayoneted straight Sugita clip. Cranial nerve (CN) III is easily seen in the way. (D) Postoperative angiogram demonstrating proximal occlusion of the P1 aneurysm.

relationship to the parent vessel and normal branches. If this evaluation cannot be performed, endovascular techniques are not recommended. Aneurysms with a wide neck (>5 mm) or with an unfavorable dome-to-neck ratio have proven to be less amenable to a definitive coiling. Small aneurysms (less than 2 mm) also pose a challenge for coiling, and large aneurysms have a higher rate of incomplete obliteration and subsequent recanalization.

ENDOVASCULAR TECHNIQUE

Most centers perform endovascular procedures for ruptured posterior circulation aneurysms under general anesthesia. Access is generally obtained from the common femoral artery. A 5F to 6F guide catheter is placed in the dominant or feeding vertebral artery. Using a coaxial technique, a microcatheter (1.7F–2.3F catheter distal tip) is placed within the aneurysm with the aid of a microwire (0.014 inch in diameter). If anatomically feasible, selective coiling of the aneurysm is performed by placing an appropriately sized framing coil. This is followed by additional packing coils, successively smaller and softer, until the aneurysm lumen is obliterated or no longer opacifies with contrast.

STENT-ASSISTED COILING

Stent-assisted coiling is an option for those aneurysms with a wide neck or unfavorable dome-to-neck ratios (Figs. 154.5A,B and 154.6A,B). Deploying a stent across the neck of the aneurysm allows the detachable coils to remain within the aneurysm and promotes flow remodeling from the aneurysm lumen. Stents can be placed across the neck of the aneurysm and bifurcating branches in different configurations. Stent deployment in small intracranial vessels has the inherent risk of in-stent thrombosis and stroke. Therefore patients must be appropriately loaded on dual antiplatelet therapy preferably before stent deployment and following the procedure for at least 6 weeks. Special considerations must be taken in those patients presenting with ruptured aneurysms, those with large intracerebral clots, and those who may require subsequent ventricular catheter placement and shunting.

BALLOON REMODELING

Similar to stent-assisted coiling, balloon remodeling allows for detachable coils to be placed in aneurysms with wide necks and unfavorable dome-to-neck ratios. A small compliant balloon is inflated across the neck of the aneurysm to protect the parent vessel during each coil deployment. If the coil placement is stable the balloon is deflated and the coil is detached.

FLOW DIVERSION

High-porosity flow-diverting stents, such as the Pipeline Embolization Device, were designed for untreatable wide-necked, giant, and fusiform aneurysms. The stent can be placed across the neck of the aneurysm to divert flow from the aneurysm. It alters the hemodynamic properties of the inflow and outflow pathways of the aneurysm, resulting in gradual thrombosis. Dual antiplatelet therapy is mandatory. The risks of flow-diverting stents are in-stent thrombosis, stroke, perforator occlusions, and distant and delayed hemorrhages. Although ruptured posterior circulation aneurysms currently represent an "off label" use of flow-diverting stents, this is an area of active investigation.

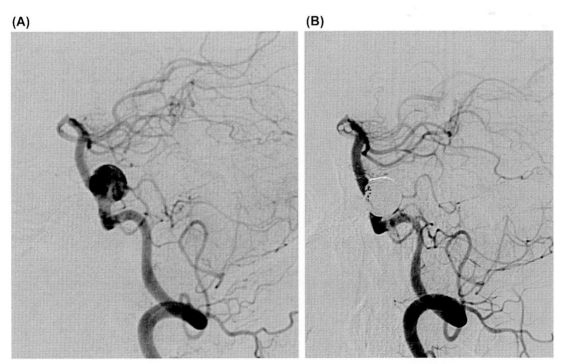

(A)

(B)

FIGURE 154.5   A left vertebral artery angiogram in lateral projection illustrating a wide-necked midbasilar artery aneurysm (A) before and (B) after stent-assisted coiling.

(A)   **(B)**

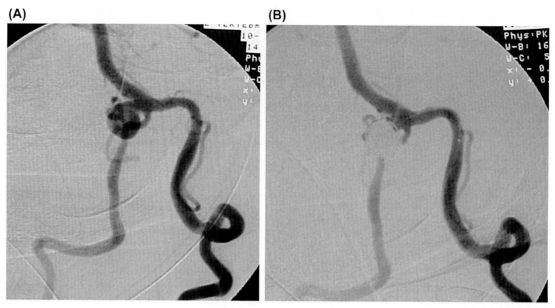

FIGURE 154.6   A cerebral angiogram in Waters projection of the left vertebral artery showing a 10-mm spherical vertebrobasilar junction (VBJ) aneurysm (A) before and (B) after selective coiling. VBJ aneurysms are located where the vertebral arteries converge to form the basilar artery. These aneurysms are often associated with proximal basilar artery fenestrations.

MULTIDISCIPLINARY APPROACHES

The treatment of ruptured posterior circulation aneurysms requires a multidisciplinary team with microsurgical and endovascular experience for treatment selection, performing the procedures, and managing vasospasm and subsequent complications. For complex posterior intracranial aneurysms or giant aneurysms, combined open and endovascular approaches are often applied.

CLIP VERSUS COILING

Several randomized prospective studies on the treatment of ruptured aneurysms have been published, highlighting the advantages and disadvantages of surgical versus endovascular therapy [9–13]. While the International Subarachnoid Aneurysm Trial (ISAT) demonstrated lower rates of disability in the coiling group for up to 7 years following rupture as well as a lower risk of death at 5 years, the extrapolation of these data to ruptured posterior circulation aneurysms is tenuous [10]. Posterior circulation aneurysms represented only 2.7% of the aneurysms in the patients enrolled in the ISAT. The ISAT investigators stated,

"Most participating centers considered endovascular treatment the favored option for posterior circulation aneurysms particularly aneurysms arising from the basilar artery because of the high surgical risk" [9]. The Barrow Ruptured Aneurysm Trial (BRAT) contained a slightly larger cohort of patients with posterior circulation aneurysms (13.7%) and demonstrated better outcome based on modified Rankin Scale (mRS) in the endovascular arm [11]. In the post hoc subgroup analyses, if the patients with posterior circulation aneurysms were excluded, there appeared to be only a marginal difference in outcome between clipping and coiling observed over a 6-year period [13]. This was not the case for aneurysms in the posterior circulation, where there appeared to be a sustained benefit of coil embolization over surgical clipping (Figs. 154.7 and 154.8) [13]. Despite the higher rates of incomplete occlusion, need for retreatment, and risk of rebleeding demonstrated in both the ISAT and the BRAT for patients with ruptured aneurysms, coil embolization seems to provide a sustained advantage over clipping for the treatment of posterior circulation aneurysms. The benefit of coiling is largely due to the reduced surgical morbidity and mortality rates rather than the completeness of occlusion or durability of treatment.

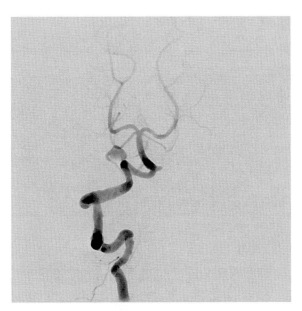

FIGURE 154.7   Right vertebral artery angiogram showing a distal posteroinferior cerebellar artery (PICA) aneurysm. Initial imaging of a ruptured PICA aneurysm may reveal perimedullary blood or preponderance of blood near the foramen magnum and sometimes may only reveal isolated fourth ventricular hemorrhage. PICAs comprise about 3% of all intracranial aneurysms. They may form at the vertebral artery (VA)-PICA junction or distally along the course of the PICA and can be saccular or fusiform in morphology. Greater incidence of fusiform geometry makes treatment challenging and less amenable to selective coiling. The small-vessel caliber and acute angulation of a PICA makes a stent-assisted or balloon remodeling technique less favorable. Direct surgical clipping, clip reconstruction, and/or trapping with or without bypass may be required. Because of the difficulty in accessing PICAs amidst the lower cranial nerves and their fusiform tendencies, the BRAT investigators hypothesize that patients with ruptured PICAs have worse outcomes than patients with ruptured aneurysms in other locations [13].

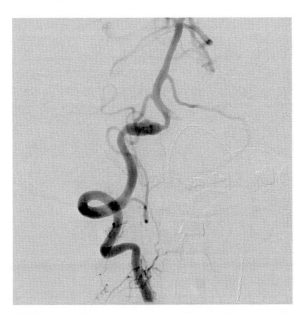

FIGURE 154.8   Right vertebral artery angiogram of a vertebral dissecting type aneurysm in the V4 segment presenting with subarachnoid hemorrhage. Aneurysm trapping or coil occlusion with vessel sacrifice is often required for treatment. Therefore, a thorough knowledge of the PICA, anterior spinal artery, and the contralateral vertebral artery filling is necessary.

References

[1] Bederson JB, Connolly Jr ES, Batjer HH, Dacey RG, Dion JE, Diringer MN, Duldner Jr JE, Harbaugh RE, Patel AB, Rosenwasser RH. American Heart Association. Guidelines for the management of aneurysmal subarachnoid hemorrhage. A statement for healthcare professionals from a special writing group of the Stroke Council, American Heart Association. Stroke 2009;40:994–1025.

[2] Welch KMA, Caplan LR, Reis DJ, Siesjo BK, Weir B. Primer on cerebrovascular diseases, vol. 121. San Diego: Academic Press; 1997. p. 455–62.

[3] Wiebers DO, Whisnant JP, Huston III J, Meissner I, Brown Jr RD, Piepgras DG, Forbes GS, Thielen K, Nichols D, O'Fallon WM, Peacock J, Jaeger L, Kassell NF, Kongable-Beckman GL, Turner JC. International Study of Unruptured Intracranial Aneurysms Investigators. Unruptured intracranial aneurysms: natural history, clinical outcome, and risks of surgical and endovascular treatment. Lancet 2003;362:103–10.

[4] Batjer HH, Samson DS. Causes of morbidity and mortality from surgery of aneurysms of the distal basilar artery. Neurosurgery 1989;25:904–15. Comment 915–916.

[5] Drake CG. Further experience with the surgical treatment of aneurysms of the basilar artery. J Neurosurg 1968;29:372–92.

[6] Yasargil MG, Antic J, Laciga R, de Preux J, Fideler RW, Boone SC. Microsurgical peritoneal approaches to aneurysm of the basilar bifurcation. Surg Neurol 1976;6:83–91.

[7] Gonzalez LF, Amin-Hanjani S, Bambakidas NC, Spetzler RF. Skull base approaches to the basilar artery. Neurosurg Focus 2005;19:E3.

[8] Guglielmi G, Vinuela F, Sepetka I, Macellari V. Electrothrombosis of saccular aneurysms via endovascular approach. Part 1: Electrochemical basis, technique and experimental results. J Neurosurg 1991;75:1–7.

[9] International Subarachnoid Aneurysm Trial (ISAT) Collaborative Group. International subarachnoid aneurysm trial (ISAT) of neurosurgical clipping versus endovascular coiling in 2143 patients with ruptured intracranial aneurysms: a randomized trial. Lancet 2002;360:1267–74.

[10] Molyneux AJ, Kerr RSC, Yu LM, Clarke M, Sneade M, Yarnold JA, Sandercock P, International Subarachnoid Aneurysm Trial (ISAT) Collaborative Group. International subarachnoid aneurysm trial (ISAT) of neurosurgical clipping versus endovascular coiling in 2143 patients with ruptured intracranial aneurysms: a randomised comparison of effects on survival, dependency, seizures, rebleeding, subgroups, and aneurysm occlusion. Lancet 2005;366:809–17.

[11] McDougall CG, Spetzler RF, Zabramski JM, Partovi S, Hills NK, Nakaji P, Albuquerque FC. The Barrow ruptured aneurysm trial. Clinical article J Neurosurg 2012;116:135–44.

[12] Spetzler RF, McDougall CG, Albuquerque FC, Zabramski JM, Hills NK, Partovi S, Nakaji P, Wallace RC. The Barrow ruptured aneurysm trial: 3-year results. J Neurosurg 2013;119:146–57.

[13] Spetzler RF, McDougall CG, Zambramski JM, Albuquerque FC, Hills NK, Russin JJ, Partovi S, Nakaji P, Wallace RC. The Barrow ruptured aneurysm trial: 6-year results. J Neurosurg 2015;123:609–17.

CHAPTER

155

Surgical and Endovascular Management of Patients With Unruptured Aneurysm

B. Daou, P. Jabbour

Thomas Jefferson University and Jefferson Hospital for Neuroscience, Philadelphia, PA, United States

INTRODUCTION

Intracranial aneurysms are abnormal focal dilations of blood vessels in the brain that result in weakening of the vessel wall making it more prone to rupture. They are common acquired lesions that are usually saccular. Rupture of an intracranial aneurysm results in subarachnoid hemorrhage (SAH), which can have devastating effects with high morbidity and mortality.

With the advances in imaging techniques, a large number of intracranial aneurysms are being detected before they rupture. There are several treatment strategies that can be employed to manage unruptured intracranial aneurysms (UIAs) to prevent the deadly outcomes associated with hemorrhage. Several factors have to be taken into consideration to decide the optimal treatment approach, including the natural history of cerebral aneurysms and risk factors for aneurysm formation, growth, and rupture. This chapter describes the different approaches in managing UIAs and the factors to consider in deciding the treatment strategy.

EPIDEMIOLOGY

The reported prevalence of intracranial aneurysms has varied greatly between different studies depending on the study design, the population studied, and the diagnostic tool used (angiography vs autopsy). In 2011, in a systematic review that included 68 studies, 94,912 patients from 21 countries reported the overall prevalence of UIAs to be 3.2% [1]. Most UIAs are located in the anterior circulation, predominantly in the proximal arterial bifurcations on the circle of Willis arising from the internal carotid artery, the middle cerebral artery, and the

anterior cerebral artery [1]. Other common sites include the junction of the anterior communicating artery and the anterior cerebral artery, the junction of the posterior communicating artery and the internal carotid artery, the middle cerebral artery branch points, and the internal carotid artery bifurcation. Posterior circulation aneurysms are less common, with most aneurysms occurring on the tip of the basilar artery, the junction of the superior cerebellar and anteroinferior cerebellar arteries and the basilar artery, and the junction of the vertebral artery and the posteroinferior cerebellar artery [2].

The prevalence of UIAs increases with age; is higher in women than in men, especially in women older than 50 years; and increases with certain disorders including autosomal dominant polycystic kidney disease (ADPKD), coarctation of the aorta, fibromuscular dysplasia, Ehlers-Danlos syndrome type IV, Marfan syndrome, multiple endocrine neoplasia I, neurofibromatosis 1, moyamoya disease, and brain arteriovenous malformations [2,3]. Prevalence of UIAs is higher in patients with a family history of SAH (two or more affected first-degree relatives), so it is recommended to screen for cerebral aneurysms in such patients. Approximately 15–30% of patients have multiple aneurysms [3].

CLINICAL PRESENTATION

Small aneurysms (<7 mm in size) do not usually cause symptoms and are frequently detected incidentally after neuroimaging for other reasons (e.g., headache, ischemic cerebrovascular disease, or transient ischemic attack) [3]. UIAs are often identified during evaluation of hemorrhage from another aneurysm. They may also be discovered when they present with mass effect, cranial nerve deficits (most commonly a third nerve palsy), seizures, or motor and sensory deficits [3]. Petrous ICA segment aneurysms are uncommon but may result in abducens nerve or trigeminal nerve deficits. Cavernous ICA segment aneurysms may result in compression of cranial nerves within the cavernous sinus and can cause cavernous sinus syndrome with ophthalmoplegia, chemosis, proptosis, Horner syndrome, or trigeminal sensory loss as they grow [2]. Expansion of clinoidal ICA segment aneurysms may result in retro-orbital headaches, facial pain, or visual loss from optic nerve compression. Ophthalmic ICA aneurysms may present with visual field defects and visual loss. Anterior communicating artery aneurysms are rarely symptomatic. Middle cerebral artery aneurysms may result in hemiparesis, hemisensory loss, visual field defect, or seizure [2]. Posterior communicating artery or basilar artery aneurysms may result in third cranial nerve palsy. Vertebrobasilar distribution aneurysms may result in bulbar dysfunction and signs and symptoms related to compression of brainstem structures.

IMAGING

Initial imaging should provide a complete evaluation of the anatomy of the aneurysm, with characterization of the neck size, neck-to-dome ratio, and the relationship to the surrounding vessels. The techniques of aneurysm imaging have expanded greatly with the advent of magnetic resonance angiography (MRA), computed tomography angiography (CTA), and digital subtraction angiography (DSA). These imaging techniques allow for aneurysm detection with high sensitivity and specificity [3]. DSA remains the gold standard for diagnosing cerebral aneurysm with the highest sensitivity and is the best modality to diagnose aneurysms smaller than 3 mm, especially when combined with three-dimensional (3D) rotational angiography that allows for more detailed imaging.

DSA is associated with procedural risks related to endovascular access. Although complications are uncommon (0.07%) [3], these include contrast-related events, access-related complications, vessel injury, and aneurysm rupture. CTA and MRA are noninvasive imaging techniques and can be used for detection and follow-up of UIAs.

RISK FACTORS FOR RUPTURE

The majority of UIAs will never rupture. The overall annual incidence of cerebral aneurysm rupture ranges from 0.05% per year to 5% per year, with most studies reporting an overall annual rupture risk rate around 1%

and an incidence of SAH around 9 per 100,000 population [4–7].

The risk of rupture of intracranial aneurysms may be affected by several factors. These include both patient and aneurysm characteristics. Patient characteristics that may increase the risk of aneurysm rupture include female gender, old age, family history of intracranial aneurysms, personal history of SAH from a different aneurysm, and a history of hypertension and ADPKD [3,5,7]. Furthermore, cigarette smoking is a powerful risk factor for aneurysm rupture and SAH [3]. Patients harboring unruptured cerebral aneurysms should be strongly counseled to quit smoking.

Aneurysm characteristics that affect rupture risk include:

1. aneurysm size: the first phase of the International Study of Unruptured Intracranial Aneurysms (ISUIA) reported that aneurysms <10 mm were much less likely to rupture than those that measured 10–24 mm (RR = 11.6) or ≥25 mm (RR = 59) [4] and the second ISUIA similarly reported that a larger aneurysm size increases the risk of aneurysm rupture but used a lower size cutoff (>7 mm) [5];
2. aneurysm growth during follow-up;
3. aneurysm location: the risk of rupture is higher for aneurysms located in the posterior circulation, then by aneurysms in the anterior communicating artery and posterior communicating artery, and finally by anterior cerebral artery aneurysms, middle cerebral artery UIAs, and internal carotid artery aneurysms [4–6];
4. aneurysm shape: risk of rupture is increased with multilobulated aneurysms, aneurysms with daughter sacs, and irregularity of the aneurysm [6];
5. multiplicity of aneurysms;
6. symptomatic status of the aneurysm [3]: sudden onset of symptoms can be the result of aneurysm expansion and is concerning for imminent rupture and should prompt for early evaluation and treatment.

MANAGEMENT

Decision to Treat

The treatment strategies for UIAs include conservative management, endovascular intervention, or surgical treatment. Optimal management of UIAs should be considered on a case-by-case basis depending on patient characteristics, risk factors, and procedural risks. Patients with UIAs who are considered for treatment should be fully informed about the potential risks and benefits of all treatment possibilities. Surgical experience and hospital volume are also factors that should be considered in deciding the treatment modality.

Young patients with a family history of SAH, a history of smoking, a symptomatic aneurysm, an aneurysm of size ≥7 mm, aneurysms with a change in the size or configuration during follow-up, aneurysms with multiple daughter sacs, and aneurysms located in the posterior circulation or anterior and postero-anterior communicating arteries should be strongly considered to undergo treatment because of increased susceptibility to rupture.

Patients with small aneurysms (<7 mm), without a family or personal history of SAH, and with an asymptomatic UIA and low hemorrhage risk by location, size, and morphology can be managed with periodic follow-up imaging. The treatment risk of UIAs increases in older patients (>65 years) and patients with significant medical comorbidities, so observation is a reasonable alternative in this age group.

Microsurgical Clipping

The main surgical treatment option is direct clip placement at the neck to isolate the aneurysm from the parent blood vessel (Fig. 155.1). Other surgical options include aneurysm wrapping or extracranial-intracranial bypass with parent vessel occlusion. Microsurgical clipping is performed under general endotracheal anesthesia. Neurophysiologic monitoring (somatosensory and motor evoked potentials, continuous electroencephalography,

and brainstem auditory evoked potentials with posterior circulation aneurysms) is commonly used intraoperatively and can play an important role in preventing complications by allowing adjustments of surgical and anesthetic techniques.

Several surgical approaches can be performed based on the location of the lesion. In general, a standard pterional craniotomy is adequate for managing aneurysms located in the anterior circulation. Skull base techniques can be used for large, complex aneurysms and for posterior circulation aneurysms. Temporary clip placement can be used to soften the aneurysm and allows for manipulation of the aneurysm without intraoperative rupture, and can facilitate permanent clip placement.

Surgical clipping results in complete aneurysm occlusion in greater than 90% of cases [8,9]. The reported morbidity and mortality rates have varied widely. A meta-analysis of 2460 patients reported a mortality of 2.6% and morbidity of 10.9% [9]. The prospective arm of the ISUIA followed up 1917 patients after clipping for UIA and reported an overall mortality of 2.3% and morbidity of 12.1% at 1 year after treatment [5]. Most common operative complications include intracerebral hemorrhage, seizure, postoperative stroke, incomplete occlusion, and recurrence.

Several factors should be considered before proceeding with surgical clipping for UIAs. The size and location

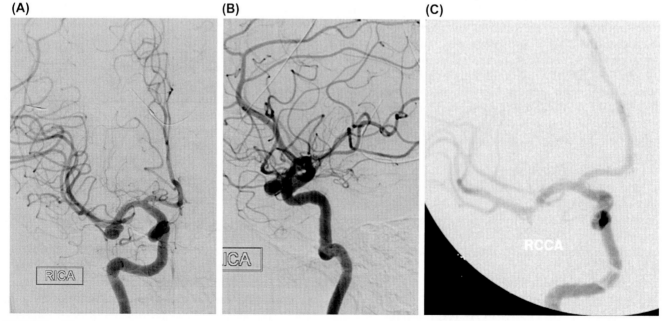

(A)   **(B)**   **(C)**

FIGURE 155.1 (A) A 33-year-old woman with a history of hypertension and smoking and a family history of intracranial aneurysms had an incidental finding of an intracranial aneurysm. An angiogram was performed. Frontal view of right internal carotid artery injection shows a right-sided middle cerebral artery aneurysm with a wide neck and vessel coming off the neck of the aneurysm that measured 6.3 × 8.1 mm. (B) Cerebral angiogram of right internal carotid artery injection, with the lateral view showing the aneurysm. (C) Because of the patient's young age and risk factors and the aneurysm's characteristics and location, the patient was treated with microsurgical clipping. An intraoperative angiogram performed after clipping shows complete occlusion of the aneurysm. *RCCA*, right common carotid artery; *RICA*, right internal carotid artery.

of aneurysms are highly associated with surgical risk. Larger aneurysms (>12 mm) and posterior circulation location are significant predictors of poor outcome [3]. Old age and medical comorbidities negatively influence outcome as well [3].

Endovascular Treatment

Endovascular management emerged as a treatment for intracranial aneurysms in the 1990s and has been used with increasing popularity since. Coiling represents the most commonly employed endovascular strategy where platinum coils are introduced into the aneurysm, causing local thrombosis and isolation of the aneurysm from the parent artery (Fig. 155.2). The use of coil embolization increased substantially after publication of the results of the International Subarachnoid Aneurysm Trial (ISAT) [10]. The ideal candidates for the use of coils are those with aneurysms with a narrow neck (<4 mm) and low dome-to-neck ratio (<2). In older patients, the benefit of endovascular treatment is greater than that of surgery, especially because the rate of surgical complications is higher in this age group. Endovascular management is associated with a reduction in procedural morbidity, length of hospital stay,

and mortality as compared to surgical clipping in select cases, but endovascular treatment has an overall higher risk of recurrence than microsurgical treatment and the latter confers more durable protection against aneurysm regrowth [3].

For endovascular management, the risk of unfavorable outcomes is approximately 4–5%, with a mortality risk of 1–2% according to data from the meta-analysis of existing literature [3]. The ISUIA included 451 patients who underwent endovascular coiling and reported that the 1-year morbidity rate was 6.4% and the mortality rate was 3.1% [5].

Complete aneurysm occlusion is achieved in 86.1% of coiled aneurysms based on postprocedure imaging, with recurrence reported in 24.4% and need for retreatment in 9.1% [11]. Large and giant wide-necked aneurysms with a high dome-to-neck ratio tend to have lower angiographic occlusion and higher recurrence and retreatment rates after coiling. However, with the continuous advancement in endovascular techniques, several options are available to target these lesions, including balloon inflation or stent placement, stent-assisted coiling (Fig. 155.3), liquid embolic agents, and flow diversion. Recanalization and retreatment rates are lower with these techniques. Flow diversion devices

(A)   **(B)**   **(C)**

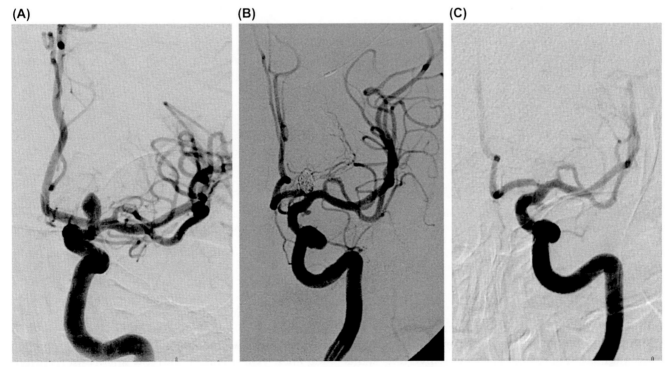

FIGURE 155.2   (A) A 40-year-old woman with a history of smoking was found to have an aneurysm during workup for vertigo. An angiogram was performed. Frontal view of the left internal carotid artery injection shows a 7 × 4.7 mm left internal carotid artery bifurcation aneurysm projecting superiorly. (B) Because of the patient's young age and smoking status, the decision was made to treat the aneurysm. The patient was treated with coil embolization, and complete aneurysm occlusion was achieved as shown in the postembolization cerebral angiogram. (C) A 6-month follow-up angiogram was performed to check for any residue or for recurrence of the lesion. Frontal view of the left internal carotid artery injection shows complete occlusion of the aneurysm.

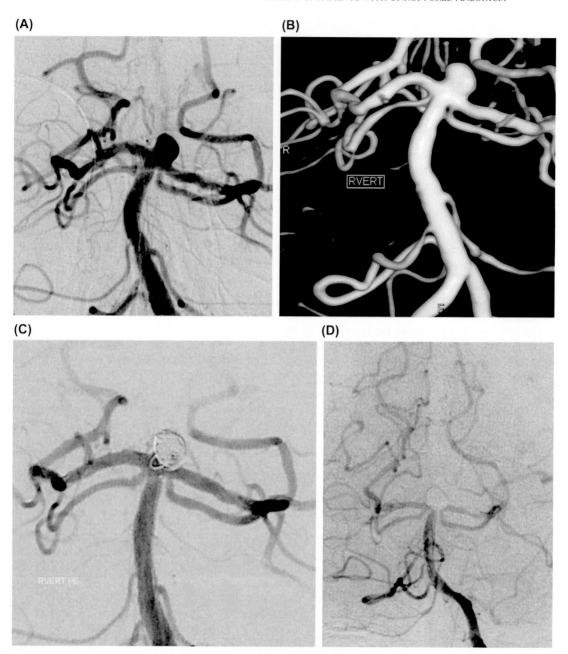

(A)

(B)

(C)

(D)

FIGURE 155.3   (A) A 55-year-old female with a history of hypertension and smoking was found to have an aneurysm during workup for a transient ischemic attack. Angiogram of the right vertebral artery (RVERT) injection shows a wide-necked basilar tip aneurysm measuring 5.7 × 4.4 mm. (B) Three-dimensional reconstruction showing the basilar tip aneurysm. (C) Because of the patient risk factors for aneurysm rupture, and the aneurysm location and wide neck, the patient was treated using stent-assisted coiling. Complete aneurysm occlusion was achieved. (D) A 6-month follow-up angiogram shows complete occlusion of the aneurysm.

including the Pipeline Embolization Device (PED) act by causing endoluminal reconstruction of the affected parent artery, thus promoting thrombosis within the aneurysm sac (Fig. 155.4) [12]. Flow-diverting stents are indicated for large and giant wide-necked aneurysms and unruptured saccular or fusiform intracranial aneurysms (>10 mm) in the internal carotid artery, from the petrous segment to the superior hypophyseal segment [12]. The PED has gained popularity because

it provides a more physiologic and durable treatment strategy than other endovascular interventions, and studies have confirmed its high success rate in achieving aneurysm occlusion and low aneurysm recurrence and retreatment rates [12].

Complications related to endovascular treatment of intracranial aneurysms occur in 5–10% of cases [3]. The main complications include thromboembolism, arterial dissection, aneurysm rupture, intracerebral hemorrhage,

(A) (B) (C) (D)

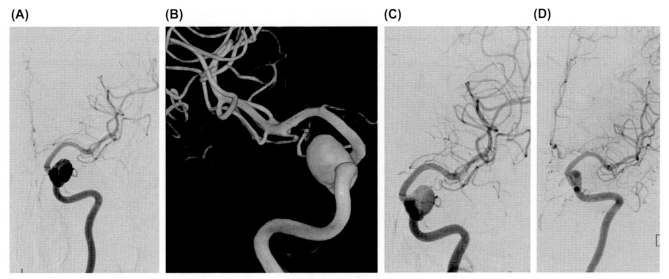

FIGURE 155.4   (A) A 65-year-old female with a history of smoking and a known cavernous internal carotid artery aneurysm that has been growing on follow-up imaging presented with worsening diplopia and facial pain. Cerebral angiogram of the left internal carotid artery injection shows a 15×12mm left cavernous internal carotid artery aneurysm. (B) Three-dimensional reconstruction showing the aneurysm. (C) Because the aneurysm was enlarging during follow-up and was causing cranial nerve deficits, the decision was made to treat the aneurysm. The Pipeline Embolization Device was used to treat the aneurysm because the aneurysm was large and wide necked. Angiogram after deployment of the flow diverter stent shows contrast stasis within the aneurysm. (D) A 1-year follow-up angiogram shows complete occlusion of the aneurysm.

and access-related complications including groin hematomas, infection, pseudoaneurysms, and contrast-related events.

FOLLOW-UP

Patients with UIAs who are managed conservatively without surgical or endovascular intervention should have an initial follow-up imaging 6–12months after diagnosis. Follow-up imaging should then be performed with MRA or CTA at regular intervals, every 1–2years, with subsequent decrease in intervals if stable. Strong consideration for treatment should be given to any aneurysm that grows over the follow-up period.

In patients with UIAs that are adequately clipped, no other follow-up imaging is obtained. Intraoperative angiography can be obtained to document complete aneurysm occlusion and verify the patency of branch vessels. Intraoperative Doppler sonography and intravenous indocyanine green videoangiography can also be used after surgical intervention. Long-term follow-up imaging should be obtained in aneurysms that are incompletely obliterated after initial clipping.

For endovascularly treated aneurysms, follow-up imaging with DSA or noninvasive methods is usually performed at 6months to 1year after treatment because of the higher potential for residual or recurrent aneurysms that can still result in hemorrhage. Later follow-up imaging can be obtained depending on the occlusion status of the aneurysm. MRA is a suitable option, as CT imaging may

be limited by artifacts from coil or stent placement. Large aneurysms, wide-necked aneurysms, and aneurysms with partial treatment have a greater risk for recurrence and should be evaluated with regular follow-up. Growing residuals or recurrences during follow-up should be considered for evaluation with an angiogram and retreatment.

References

[1] Vlak MH, Algra A, Brandenburg R, Rinkel GJ. Prevalence of unruptured intracranial aneurysms, with emphasis on sex, age, comorbidity, country, and time period: a systematic review and meta-analysis. Lancet Neurol 2011;10(7):626–36.

[2] Brown Jr RD, Broderick JP. Unruptured intracranial aneurysms: epidemiology, natural history, management options, and familial screening. Lancet Neurol 2014;13(4):393–404.

[3] Thompson BG, Brown Jr RD, Amin-Hanjani S, et al. Guidelines for the management of patients with unruptured intracranial aneurysms: a guideline for healthcare professionals from the American Heart Association/American Stroke Association. Stroke 2015;46(8):2368–400.

[4] Unruptured intracranial aneurysms–risk of rupture and risks of surgical intervention. International Study of Unruptured Intracranial Aneurysms Investigators. N Engl J Med 1998;339(24):1725–33.

[5] Wiebers DO, Whisnant JP, Huston 3rd J, et al. Unruptured intracranial aneurysms: natural history, clinical outcome, and risks of surgical and endovascular treatment. Lancet 2003;362(9378):103–10.

[6] Morita A, Kirino T, Hashi K, et al. The natural course of unruptured cerebral aneurysms in a Japanese cohort. N Engl J Med 2012;366(26):2474–82.

[7] Wermer MJ, van der Schaaf IC, Algra A, Rinkel GJ. Risk of rupture of unruptured intracranial aneurysms in relation to patient and aneurysm characteristics: an updated meta-analysis. Stroke 2007;38(4):1404–10.

[8] Kotowski M, Naggara O, Darsaut TE, et al. Safety and occlusion rates of surgical treatment of unruptured intracranial aneurysms: a systematic review and meta-analysis of the literature from 1990 to 2011. J Neurol Neurosurg Psychiatry 2013;84(1):42–8.

[9] Raaymakers TW, Rinkel GJ, Limburg M, Algra A. Mortality and morbidity of surgery for unruptured intracranial aneurysms: a meta-analysis. Stroke 1998;29(8):1531–8.

[10] Molyneux A, Kerr R, Stratton I, et al. International subarachnoid aneurysm trial (ISAT) of neurosurgical clipping versus endovascular coiling in 2143 patients with ruptured intracranial aneurysms: a randomised trial. Lancet 2002;360(9342):1267–74.

[11] Naggara ON, White PM, Guilbert F, Roy D, Weill A, Raymond J. Endovascular treatment of intracranial unruptured aneurysms: systematic review and meta-analysis of the literature on safety and efficacy. Radiology 2010;256(3):887–97.

[12] Becske T, Kallmes DF, Saatci I, et al. Pipeline for uncoilable or failed aneurysms: results from a multicenter clinical trial. Radiology 2013;267(3):858–68.

CHAPTER

156

Surgical and Endovascular Management of Patients With Giant Cerebral Aneurysms

J.S. Beecher, B.G. Welch

UT Southwestern Medical Center, Dallas, TX, United States

INTRODUCTION

Giant cerebral aneurysms (GCAs) have been defined by convention as aneurysms that are 2.5 cm or greater in diameter. These lesions constitute approximately 5% of intracranial aneurysms and are the result of progressive enlargement of a small aneurysm [19]. Descriptions of aneurysms as large as 8.5 cm in diameter exist in the literature [16]. Despite advances in both surgical and endovascular techniques, the GCA continues to be a formidable pathologic condition to manage primarily because of dysplasia of the aneurysm wall (calcification or random thickening) and the involvement of the surrounding structures through compression or adhesion.

The most common presentation of a GCA is an aneurysm mass that results in compression of nearby structures. That such growth and displacement can occur before presentation is evidence to the resiliency of the optic apparatus that abuts the ophthalmic and anterior communicating segments where these lesions appear more frequently. Transient ischemic attacks (TIAs) or strokes from distal embolization of intraluminal thrombus is a less common presentation. Seizures can result from cortical compression or deposition of hemosiderin

around the GCA [20,21] which is often the presentation of GCAs of the middle cerebral artery.

Historically these lesions were thought to have a benign course, but now it is well established that GCAs have a malignant clinical course and should be treated with the goal of complete obliteration, reduction of the mass effect, and reconstruction of the cerebral vasculature [12]. This chapter will focus on conventional microsurgical and endovascular management of GCAs. Regardless of the method of treatment, it is imperative that the cerebrovascular neurosurgeon remembers what Dr. Charles Drake observed in his study of giant aneurysms: "Every giant aneurysm is a unique entity and must be treated as such [9,17]."

MICROSURGERY

The basis for microsurgery of these formidable lesions has its foundation in three principles that apply to all GCAs in any location:

1. preparation and adequate exposure,
2. complete vascular control,
3. aneurysm decompression before clip placement.

Primer on Cerebrovascular Diseases, Second Edition
http://dx.doi.org/10.1016/B978-0-12-803058-5.00156-9

In addition, each individual locus in the cerebral vasculature has its own subset of additional necessities for a successful surgery and a good patient outcome. During surgical reconstruction of a GCA, one must always entertain multiple treatment possibilities before committing to a surgical solution. Although clip reconstruction is a desired outcome, calcified atheroma and thrombosis may render these solutions impossible without parent vessel occlusion (PVO). For this reason, revascularization or trapping strategy should always be a potential component of the surgical plan for the GCA. Although it has been routine to preserve the superficial temporal artery during the treatment of these aneurysms, improvements in imaging and endovascular techniques

have allowed for some refinement. The use of selective trial balloon occlusion (TBO) in the preoperative evaluation of the patient can provide improved knowledge of the necessary conduit for bypass and can also approximate a time window for temporary occlusion. Skull base approaches are invaluable in maximizing the workspace for the surgeon and should be carefully considered to allow for full GCA visualization, minimal manipulation of surrounding structures, and a relaxed and direct entry of the bypass conduit. Table 156.1 delineates useful techniques (by location) that should be implemented when treating GCAs.

The basic tenant of proximal and distal control is especially important in surgery for GCAs. It is mandatory to

TABLE 156.1 Summary of Surgical Strategies for Giant Cerebral Aneurysms (GCAs)

Circulation		Preparation	Identify	Structures to Resect	Reconstructive/ Deconstructive Considerations
Anterior	Ophthalmic	Cervical carotid exposure	Proximal ICA and OphA	Anterior clinoid, optic strut, and roof of optic canal; consider OZO	Sacrifice/trapping (suggest TBO)
Anterior	Superior hypophyseal	Cervical carotid exposure	Proximal ICA and OphA	Anterior clinoid	Sacrifice/trapping (suggest TBO)
Anterior	Posterior carotid wall		AChor, fetal or dominant PComm		Sacrifice/trapping (suggest TBO)
Anterior	ICA bifurcation		Medial and lateral lenticulostriate arteries and AChor	Frontal lobe around the aneurysm	Clip including the M1 or A1 vessels with an MCA bypass or patent AComm, respectively
Anterior	MCA bifurcation	Preservation of STA	Medial and lateral lenticulostriate arteries		STA-MCA bypass or other interposition graft to MCA
Anterior	AComm and ACAs	Consider effect of bypass on positioning	Rec. Heub. and perforating arteries	Gyrus rectus	A2 to A2 side-to-side, interposition, or A2 reimplantation bypass techniques
Posterior	Basilar apex	Consider bypass conduit	Thalamoperforators, bilateral PCAs and SCAs	Posterior clinoid	Exclusion of a P1 with a robust ipsilateral PComm
					OA to PCA bypass
Posterior	Basilar trunk/VB junction	Consider skull base exposure	Contralateral vertebral arteries, basilar perforators		Proximal ligation of one or both vertebral arteries or the basilar trunk itself after trial occlusion
Posterior	PICA	Consider bypass conduit	Distal PICA		OA to PICA bypass
					PICA-PICA bypass

Anterior and posterior aneurysms present different surgical challenges that require various solutions. Some of the strategies are higher risk or more technically demanding than others, and it is imperative that these possibilities are entertained, if not routinely performed.
Green, routinely performed and accepted as a surgical option; yellow, increased risk or challenging surgical technique; red, high surgical morbidity/mortality and should be considered as the last option or not at all.
ACA, anterior cerebral artery; *AChor*, anterior choroidal artery; *AComm*, anterior communicating artery; *ICA*, internal carotid artery; *MCA*, middle cerebral artery; *OA*, occipital artery; *OphA*, ophthalmic artery; *OZO*, orbitozygomatic osteotomy; *P1*, most proximal portion of PCA; *PCA*, posterior cerebral artery; *PComm*, posterior communicating artery; *rec. Heub.*, recurrent artery of Heubner; *SCA*, superior cerebellar artery; *STA*, superficial temporal artery; *TBO*, trial balloon occlusion; *VB*, vertebrobasilar.

have access to all afferent and efferent vessels involved in the aneurysmal segment, and the consideration of cervical carotid exposure is important.

After exposure and circumferential dissection of the GCA, decompression of the aneurysm before the final clip placement is crucial. This frequently requires temporary artery occlusion/trapping, opening of the aneurysm dome, evacuation of intra-aneurysmal contents, and deflation to complete the reconstruction of the parent vessel. Without this maneuver, it is unlikely that the clip will fully occlude the orifice of the aneurysm entrance. Without complete closure the clip may also be driven on to the parent vessel, resulting in stenosis or occlusion. It is important to realize that the first clip is not the final clip.

The use of an ultrasonic aspiration device is useful in rapidly achieving removal of the aneurysm contents (clot or atheroma) that are preventing clip application. It is highly important that the contents are removed expediently, such that the focus of evacuation is on the aneurysm neck and not the fundus. Clearing the fundus does not facilitate clip placement and takes away precious temporary artery occlusion time. At times, complete dome resection may be necessary leaving a "tuft" of soft flexible neck tissue for clip reconstruction. This eliminates the "umbrella" effect in which a nonpliable, calcific, thrombotic, or previously coiled dome prevents an aneurysm clip from closing by rigidly holding the neck tissue open. This may be seen in lesions in which the aneurysm dome adheres to the overlying brain tissue.

Despite advances in clip technology, there may be persistent bleeding from the aneurysm after initial clip placement. Although direct observation may suggest adequate clip length across the neck, the possibility of redundant aneurysmal sac tissue in the clip blades should be considered. A fenestrated clip to "jump" the redundancy and to apply pressure directly on the redundant region may supply additional closing strength; this is a variant of a booster clip.

In lesions that do not contain partial thrombus, decompression of the intrasaccular contents for proximal carotid GCAs often requires the suction/decompression technique that was described initially at our institution. The proximal carotid artery being a common location, it is imperative that anyone treating these aneurysms be experienced in this technique or a variation of it. The neurosurgeon must remember that for these proximal lesions the control is in the neck. In conjunction with a distal temporary clip in place the aneurysm will collapse and permit clipping or emptying of the aneurysm contents as necessary. The contribution of the ophthalmic artery is often underappreciated. Failure to control "the ophthalmic" during suction/decompression may allow for reversal of flow during temporary clipping and filling of the aneurysm or bleeding during aneurysm repair.

Initially suction and decompression was described as a direct carotid puncture with a 16-gauge angiocatheter connected to suction to reverse flow in the internal carotid artery (ICA) (Fig. 156.1).

With more institutions building "hybrid suites" that have biplane angiography in an operating room, it has been increasingly common to obtain proximal control utilizing endovascular techniques (Fig. 156.1). Although this replaces the neck dissection and carotid puncture, the use of anticoagulants and the intraluminal injury to the carotid artery are new risks introduced by these techniques. The comfort with either technique may vary by facility. There is no definitive evidence supporting one over the other.

Microsurgery Results

Microsurgical clipping or trapping, with or without revascularization, is the benchmark for the management of GCAs. These techniques have been used for decades with success and, as surgical methods have matured, the literature has demonstrated improvement in patient outcomes. In our own surgical series, we found that 92% of patients younger than 50 years with large or giant aneurysms had a Glasgow Outcome Score (GOS) of 4 or 5. However, in patients older than 50 years the outcome decreased dramatically to 51% with microsurgery for aneurysms of the anterior circulation [5]. Another large series that was heavily reliant on revascularization techniques demonstrated that 81% of 114 patients had a GOS of 4 or 5 with surgical mortality of 13% [10]. These results are from large institutions that are referral centers for this pathologic condition, and it is our opinion that any new technology or technique must meet or surpass these results.

ENDOVASCULAR SURGERY

Cerebrovascular disease management has been revolutionized by the advent of endovascular therapies. With the fundamentals of these techniques being discussed later, it is still relevant to discuss the progression of the techniques as they evolved to treat GCAs.

Parent Vessel Occlusion

Endovascular methods of PVO were an early installment in the less invasive management of GCAs. In 2009, van Rooij and Sluzewski reported a 14-year study in which 76 PVOs were performed for large or giant aneurysms with no early or late permanent neurologic ischemic deficits [4]. Although 70–75% of patients may tolerate PVO, this management strategy is highly dependent on the results of TBO before performing the definitive procedure. The TBO is performed in different methods across institutions and carries an isolated risk

(A) Angiocatheter   **(B)** Balloon catheter

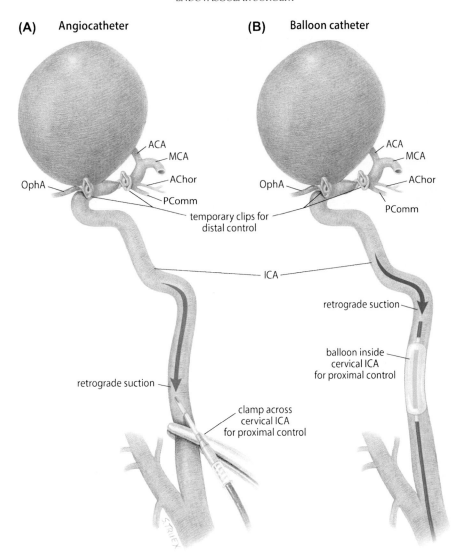

FIGURE 156.1   (A) Angiocatheter: a direct carotid puncture with proximal ligation through neck dissection allows for retrograde suction and decompression, with the benefit of direct visualization and access to the carotid artery. This avoids any concerns of balloon deflation while performing the procedure and has a lower risk of postoperative ischemic events. (B) Balloon catheter: endovascular balloon occlusion with retrograde suction to assist in decompression of a GCA. An 8F or 9F balloon catheter is placed in the cervical carotid. We recommend manual retrograde suction with careful attention to maintain a steady return of blood. This will prevent the carotid artery from collapsing and obstructing the catheter. The possibility of dissection depends on the mode of access. *ACA*, anterior cerebral artery; *AChor*, anterior choroidal artery; *ICA*, internal carotid artery; *MCA*, middle cerebral artery; *OphA*, ophthalmic artery; *PComm*, posterior communicating artery.

of 3% in some reports; [7] this additive risk and institutional experience with TBO should be considered when PVO is suggested as a treatment option. Nowadays the use of PVO as a primary endovascular therapy is indicated in older patients, patients with bleeding diathesis, and those with surgical comorbidities that would preclude more modern techniques.

Coil Embolization Without or With Stent Assistance

Coil embolization with or without a stent has been successfully used as a treatment modality for occlusion of GCAs; however, alternative techniques are often sought because of its high rates of recanalization and the unreliable resolution of symptoms related to mass effect. Compression of the surrounding neural structures is found to be the mechanism of neurologic deficit in 39–75% of patients with GCA; [11] therefore, the ideal treatment modality must be able to eliminate the mass of the aneurysm to allow for neurologic improvement.

One study that treated 501 aneurysms of all sizes with coiling alone reported 33.6% angiographic recurrence, with 20.7% being "sizable" recurrences and a greater than two-fold increase in the likelihood of recurrence if the aneurysm was 10 mm or larger [8]. In a review of the endovascular literature from 1994 to 2003, Parkinson et al. found the combined use of available endovascular

(A)   **(B)**

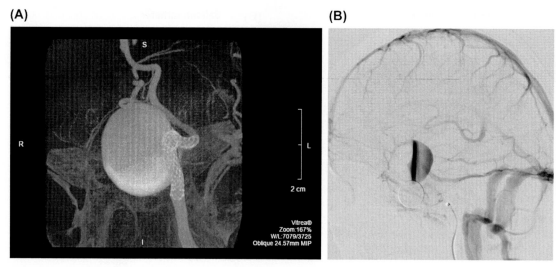

FIGURE 156.2   (A) Cone beam CT after placement of a flow-diverting device without the use of coils for a giant left superior hypophyseal aneurysm. (B) Lateral view of a left ICA injection demonstrating the "*eclipse* sign" after flow diversion in the venous phase, which corresponds to significant flow stagnation in the aneurysm.

techniques (coils, stent and coil, PVO, and embolic "glues") to treat giant intracranial aneurysms resulted in a 26.9% rate of recanalization [9]. Their review of 28 publications that included the treatment of 316 GCAs revealed a morbidity and mortality of 17.2% and 7.7%, respectively [9]. Another review by Jahshan et al. suggested that the rate of total aneurysm occlusion on the final follow-up angiogram was between 47.9% and 76.3% when embolization was performed primarily or with stent assistance [3]. Although stent-assisted coiling showed a higher percentage of persistent obliteration than coil embolization alone, the former's application to GCAs has not been found to be a permanent treatment solution and it rarely prevented mass effects. Furthermore, it is important for the practitioner and the patient to discuss that treatment of the GCA with coils or stent-assisted coiling is accompanied by an elevated rate of recanalization that may require further intervention.

Flow Diversion

Flow diversion is the result of attempts to augment stent-assisted therapies using intraluminal devices that improved on the initial "coil scaffold" provided by first-generation intracranial stents. By developing a device with decreased porosity and increased pore density producing endosaccular stagnation of flow, which promotes thrombosis and subsequent aneurysm occlusion [1,13]. Like traditional stents, flow diverters undergo the process of epithelialization and currently require the use of dual antiplatelet therapy to prevent distal emboli or thrombosis within the stent. Whereas the previous stents were used with coils, with flow diversion the use of coils remains a topic of discussion. Fig. 156.2A demonstrates stereotypic utilization of flow diversion without coils of

a proximal ICA GCA, and Fig. 156.2B is a classic example of the "eclipse sign" seen during flow diversion because of contrast stagnation in the aneurysm. Although a transient increase in mass effect has been observed, remodeling of the vessel wall and regression of the aneurysm frequently occur. This is similar to the apoptosis found at the dome of a coiled aneurysm approximately 1 year after embolization [14]. This process ultimately decreases the mass effect of the aneurysm and may reverse the symptoms of mass effect that are the presenting complaint in many patients.

Currently, the most commonly used flow diversion device is the Pipeline Embolization Device (PED; Covidien, Irvine, California, United States). The current iteration of the device (Pipeline FLEX) is a retrievable 48 strand braiding of cobalt chromium and platinum with 35% coverage when fully expanded [1]. This device was initially FDA approved in 2011 for the treatment of large or giant wide-necked aneurysms of the ICA from the petrous to the superior hypophyseal segments [1]. It has since demonstrated off-label utility in various studies in treating other aneurysms distal to the initially indicated segments.

The Flow Re-Direction Endoluminal Device (FRED) has the unique characteristic of a dual-layer design with a low-porosity inner mesh combined with a high-porosity outer stent. The benefits are easier deployment due to the outer stent's lower friction from being more porous and improved reliability of stent deployment due to increased radial force from having the two stents together. The dual layer unique to FRED may provide improved flow diversion when compared to the PED because of the combination of 16 wires from the high-porosity stent and the 48 wires from the low-porosity stent. The current trials and their data, along with long-term results, will determine the success of this device [15].

The SCENT (Safety and Effectiveness of an Intracranial Aneurysm Embolization System for Treating Large and Giant Wide Neck Aneurysms) trial was designed to evaluate the utility and safety of the Surpass stent (Stryker, Kalamazoo, Michigan, United States). This device has an increased coverage provided by 96 wires for increased lamination of flow. The results of this trial must be compared to those that demonstrated the safety and efficacy of the PED.

Flow diversion has been received as a revolutionary modality for treating GCAs. Providing the ability to recreate the vessel lumen without coils, thereby reducing symptoms of mass effect as the aneurysm thrombosis and regresses has demonstrated great promise. These devices are often reserved for treating GCAs that have evidence of calcium at the aneurysm neck, making clip reconstruction increasingly challenging; GCAs that have solitary afferent and efferent vessels; fusiform aneurysms; and older patients who cannot tolerate an elaborate complicated surgical intervention. It must be disclosed that using flow diversion in the distal ICA and beyond is off-label practice and is a burgeoning technique for management of intracranial GCA. Flow diversion has also been used with coils in the hopes to promote thrombosis of the aneurysm, and this technique is not well understood. There have been cases of delayed rupture after treatment with flow diversion, and there are many theories beyond the scope of this chapter as to why this may occur.

CONCLUSION

GCAs are one of the most challenging pathologic conditions encountered by the cerebrovascular practitioner. Because of GCAs' higher surgical morbidity than smaller aneurysms and their tendency to be resilient to treatment with conventional endovascular techniques, flow diversion has become an increasingly utilized, but frequently off-label, treatment alternative. In this discussion on the management of these complex lesions, it is crucial to understand that each GCA has individual characteristics that may suggest a more favorable outcome with surgical, endovascular, or combined strategies. We would emphasize that each case is worthy of review by a team of cerebrovascular specialists (neurosurgeons, interventional neuroradiologists, and interventional neurologists). Any institution that intends to manage these lesions must embrace a multidisciplinary approach that appropriately applies all treatment modalities to patient management.

References

[1] Eller JL, Dumont TM, Sorkin GC, et al. The Pipeline embolization device for treatment of intracranial aneurysms. Expert Rev Med Devices 2014;11(2):137–50.

[2] Deleted in review.

[3] Jahshan S, Abla AA, Natarajan SK, et al. Results of stent-assisted vs non-stent-assisted endovascular therapies in 489 cerebral aneurysms: single-center experience. Neurosurgery 2013;72(2):232–9.

[4] Van Rooij WJ, Sluzewski M. Endovascular Treatment of large and giant aneurysms. AJNR 2009;30:12–8.

[5] Hauck EF, Wohlfeld B, Welch BG, et al. Clipping of very large and giant unruptured intracranial aneurysms in the anterior circulation: an outcome study. J Neurosurg 2008;109:1012–8.

[6] Deleted in review.

[7] Mathis JM, Barr JD, Jungreis CA, et al. Temporary balloon test occlusion of the internal carotid artery: experience in 500 cases. AJNR 1998;16:749–54.

[8] Raymond J, Guilbert F, Weill A, et al. Long-term angiographic recurrences after selective endovascular treatment of aneurysms with detachable coils. Stroke 2003;34:1398–403.

[9] Parkinson RJ, Eddleman CS, Batjer HH, Bendok BR. Giant intracranial aneurysms: endovascular challenges. Neurosurgery 2006;59(Suppl. 3):103–12.

[10] Sughrue ME, Saloner D, Rayz VL, Lawton MT. Giant Intracranial aneurysms: evolution of management in a contemporary surgical series. Neurosurgery 2011;69:1261–70.

[11] Ha SW, Jang SJ. Clinical analysis of giant intracranial aneurysms with endovascular embolization. J Cerebrovasc Endovasc Neurosurg 2012;14(1):22–8.

[12] Nanda A, Sonig A, Banerjee AD, Javalkar VK. Microsurgical management of giant intracranial aneurysms: a single surgeon experience from Louisiana State University, Shreveport. World Neurosurgery 2012;81(5/6):752–64.

[13] Becske T, Kallmes DF, Saatci I, et al. Pipeline for Uncoilable or failed aneurysms: results from a multicenter clinical trial. Radiology 2013;267(3):858–68.

[14] Brinjikji W, Kallmes DF, Kadirvel R. Mechanisms of healing in coiled intracranial aneurysms: a review of the literature. AJNR 2015;36:1216–22.

[15] Möhlenbruch MA, Herweh C, Jestaedt L, et al. The FRED flow-diverter stent for intracranial aneurysms: clinical study to assess safety and efficacy. AJNR 2015;36:1155–61.

[16] Sadik AR, Budzilovich GN, Shulman K. Giant aneurysm of the middle cerebral artery. J. Neurosurg 1965;22:177–81.

[17] Drake CG. Giant intracranial aneurysms: experience in 174 patients. Clin. Neurosurg 1979;26:12–95.

[18] Deleted in review.

[19] Onuma T, Suzuki J. Surgical treatment of giant intracranial aneurysms. J Neurosurg 1979;51:33–6.

[20] Mehdorn HM, Chater NL, Townsend JJ, et al. Giant aneurysm and cerebral ischemia. Surg Neurol 1980;13:49–57.

[21] Whittle I, Allsop JL, Halmagyi GM. Focal seizures: An unusual presentation of giant intracranial aneurysms. Surg Neurol 1985;24:533–40.

CHAPTER

157

Surgical and Endovascular Management of Aneurysms in the Pediatric Age Group

E.R. Smith

Harvard Medical School, Boston, MA, United States

INTRODUCTION

Aneurysms are one of the most common vascular anomalies of the central nervous system but are far less common in children than in adults. These structurally abnormal areas of the arterial wall can cause bleeding, compression of adjacent structures, and concomitant loss of neurologic function. The epidemiology, pathophysiology, presentation, and treatment of pediatric intracranial aneurysms have features distinct from those of adults.

EPIDEMIOLOGY AND PATHOPHYSIOLOGY

The prevalence of unruptured, asymptomatic intracranial aneurysms (across all ages) is estimated at 3.2%, with only 0.5–4.6% of these present in children [1–6]. Hemorrhage from aneurysms in the pediatric population is also rare, with only 0.6% of all aneurysmal subarachnoid hemorrhage (SAH) occurring in patients younger than 19 years [3,6,7]. With a reported total of ~18,300 SAH cases annually in the United States, this means that there are only about 100 aneurysmal SAH cases each year in children [8,9]. Intracranial aneurysms are more common in males than females (especially in prepubertal children) in the ratio 2.7:1, with a shift to a female preponderance (similar to adults) in the ratio 3–5:1 in the postpubertal population [6,10–12].

The size and location of aneurysms in children differ when compared to those in adults. Pediatric patients are more likely to harbor aneurysms in the posterior circulation (25% in children vs. 8% in adults) and less likely to have anterior cerebral artery aneurysms (5–10% in children vs. 34% in adults), but the occurrence of internal carotid artery and middle cerebral artery lesions is roughly the same in both children and adults [3,6,7,11,13]. Children are less likely than adults to have multiple aneurysms and are two to four times more likely to have giant (>2.5 cm) aneurysms [5,6,10,13].

The pathophysiology of aneurysms in the pediatric population changes with increasing age. Younger children, particularly those younger than 5 years, predominantly have dissecting, fusiform aneurysms, whereas older children have a majority of saccular aneurysms [6,10,13]. Hereditary aneurysms are very rare, accounting for 5–20% of all reported cases in children and young adults, but less than 5% of prepubertal cases [14,15]. The causes of pediatric intracranial aneurysm are summarized in Table 157.1.

PRESENTATION AND EVALUATION

Most aneurysms are asymptomatic and often never detected. There are no formal screening guidelines for children with affected family members, with the exception of some rare genetic disorders, including hereditary hemorrhagic telangiectasia (HHT) [16,17]. Generally, only family members of sibling pairs or three first-degree relatives who harbor known intracranial aneurysms are recommended for screening, usually with magnetic resonance imaging (MRI)/magnetic resonance angiography (MRA) [14–16,18].

Table 157.2 outlines the common presentations of symptomatic aneurysms. In one series, over half of the pediatric patients with aneurysm presented with SAH—33% presented with mass effect symptoms and in 11% the aneurysm was found after trauma [6]. Children with SAH from aneurysm often present with a lower Hunt–Hess grade than adults, usually 1–3 [6,10,19,20] (Table 157.3).

Primer on Cerebrovascular Diseases, Second Edition
http://dx.doi.org/10.1016/B978-0-12-803058-5.00157-0

TABLE 157.1 Causes of Pediatric Intracranial Aneurysms [5,6,10,11,13,18,26,39]

Saccular	46–70%	More common in older children
Traumatic	5–40%	Closed head injury, postsurgical
Infectious	5–15%	Usually bacterial, often on more distal branches of the arterial tree because of the embolic nature of formation (*Staphylococcus*, cardiac disease)
Other		*Dissection* (spontaneous and posttraumatic)
		Genetic (polycystic kidney disease, fibromuscular dysplasia, Marfan syndrome, Ehlers-Danlos type IV, hereditary hemorrhagic telangiectasia, Klippel-Trénaunay syndrome)
		Hemodynamic (flow related from AVM or moyamoya disease)
Multiple	~5%	Up to 15% of older children with aneurysms have multiple lesions

TABLE 157.2 Presentation of Pediatric Intracranial Aneurysms [5,6,10,11]

Headache	~80%
Loss of consciousness	~25%
Seizure	~20%
Focal deficit	~20%
Vision changes	~10%

TABLE 157.3 The Hunt–Hess SAH Classification [19,20]

0	Unruptured
1	Asymptomatic/mild headache
2	Cranial nerve palsy, severe headache, nuchal rigidity
3	Mild focal deficit, lethargy or confusion
4	Stupor, hemiparesis, decerebrate
5	Coma, moribund appearance

Evaluation of a patient with a suspected SAH from an aneurysm includes history taking, neurologic and physical examination, and performing radiographic studies to define the anatomy of the lesion. Up to 90% of all nontraumatic SAH in children will result from a structural lesion [21]. These patients should be screened with computed tomographic (CT) scan and if a vascular lesion is suspected, a CT arteriogram (CTA) can be a means of identifying the presence of an aneurysm [16]. Although CTA lacks the details provided by a catheter-based arteriogram, it provides an immediate image outlining the key anatomy of a lesion, which is an information of critical importance in the setting of an acutely ill child.

TABLE 157.4 Radiographic Findings in Pediatric Intracranial Aneurysms [5,6,10,11]

SAH	~60%
IVH	~10–15%
ICH	~10–15%
Subdural	~1–3%

ICH, intracerebral hemorrhage; *IVH*, intraventricular hemorrhage.

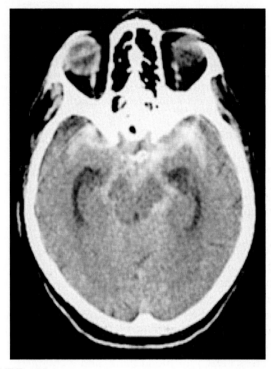

FIGURE 157.1 Axial noncontrast head CT image of aneurysmal subarachnoid hemorrhage. Note early ventricular dilatation in temporal horns.

Patterns of SAH in pediatric aneurysm CT studies are summarized in Table 157.4 (Fig. 157.1).

MRI is also useful in the diagnosis and delineation of the three-dimensional anatomy of an aneurysm, particularly with MRA. Overall, MRI/MRA was able to identify the source of SAH correctly in 66% of cases [21]. In contrast, use of catheter-based digital subtraction angiography (DSA)—the gold standard for imaging in aneurysm—was able to identify lesions in 97% of patients, versus 80% of the time without DSA [22]. Angiography generally includes bilateral injection of both the internal and external carotid arteries and the vertebral arteries to visualize all the vessels. Three-dimensional angiography with computer-generated reconstruction is increasingly used to depict the anatomy of lesions (Fig. 157.2). With catheter angiography, an analysis of 241 consecutive pediatric patients revealed a 0% complication rate during the procedure

(A)　　　　　　**(B)**　　　　　　**(C)**

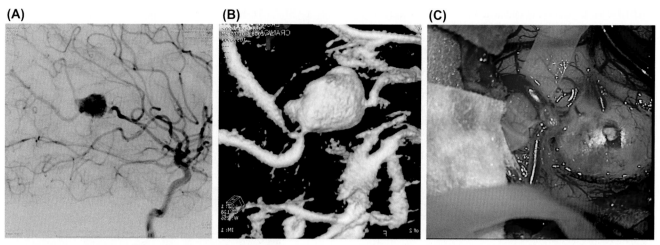

FIGURE 157.2　(A) Appearance of aneurysm on conventional digital subtraction angiography (DSA), with distal middle cerebral dissecting lesion. (B) The same lesion is visualized with three-dimensional reconstructions from the arteriogram. (C) An intraoperative photograph, as preparation is being made for a distal bypass and concomitant trapping of the aneurysm.

and a 0.4% postprocedural complication rate [23]. Evaluation should look for the following:

- lesion size, orientation, and location
- neck anatomy
- daughter blisters
- other aneurysms (especially important with infectious lesions).

If the aneurysm is fusiform the segment of vessel involved, delineation of normal borders, and involvement of perforators are important.

Standard preoperative laboratory studies, such as complete blood count (CBC), clotting times [prothrombin time (PT)/partial thromboplastin time (PTT)], type and cross (T&C) for blood bank, and a chemistry panel (Chem 7), should be considered as pretreatment studies for aneurysms.

TREATMENT AND SURGICAL/ ENDOVASCULAR INDICATIONS

The decision to treat or observe a given aneurysm can be complex and is best made by a multidisciplinary team including neurosurgeons, endovascular specialists, and neurologists. If treatment is planned the lesion can be successfully removed from circulation while maintaining normal blood flow to the brain. This can be achieved by open surgery, endovascular techniques, or a combination of approaches (Table 157.5).

The advent of endovascular therapy has revolutionized the treatment of pediatric aneurysms, providing options unavailable to earlier generations of physicians. The relatively noninvasive nature of the approach is particularly attractive in children. In experienced hands, the overall procedural risk can be extremely low, with excellent obliteration rates

TABLE 157.5　Treatment Approaches for Pediatric Intracranial Aneurysms

Surgical
Clipping (obliteration of lesion)
Clip reconstruction (rebuild arterial wall/lumen)
Trapping of lesion (with or without revascularization with bypass)
Endovascular
Embolization (coil/glue)
Stenting (with or without embolization)
Parent artery occlusion

after treatment [24]. There is a great deal of evolution within this field, limiting long-term outcome data. Consequently, it is important, whenever possible, to have cases reviewed at institutions that offer both endovascular and open techniques to provide a balanced approach to formulate care plans [25,26].

The rarity of pediatric aneurysms preclude firm evidence-based guidelines for treatment indications [16]. In general, aneurysms that have ruptured, aneurysms that enlarge over time, and symptomatic lesions should be considered for treatment. Depending on the location and patient status, it may be reasonable to treat aneurysms greater than 3 mm in size, particularly given the long expected life span of children. Mycotic aneurysms sometimes will regress with effective antibiotic therapy, obviating the need for other interventions. In addition, flow-related aneurysms located proximal to a lesion, such as an arteriovenous malformation (AVM), may regress following definitive treatment of the primary lesion (such as resection of the AVM), demonstrating another scenario when direct aneurysm treatment might not be required [27,28].

Although debatable, lesions 2 mm or smaller in size and those located outside the subarachnoid space are sometimes followed expectantly with serial imaging. Much controversy, on both sides, surrounds the large study of unruptured aneurysms published in 1998, but it is important to recognize that these were predominantly adult patients, with questionable relation to the pediatric population, given the differences in life span, aneurysm cause, and risk factors [1,7]. In addition, an international study on endovascular treatment versus surgical treatment of aneurysms, in adults, has added substantially to the evidence supporting endovascular therapies as the first-line treatment in many aneurysm cases [29,30].

Overall a review of the literature demonstrated that 53% of pediatric aneurysms were treated with open surgery and 35% were treated endovascularly, with the remainder undergoing mixed treatments, including bypass, combined procedures, and observation [31]. The percentage of cases treated solely with endovascular therapies has increased from 25% in 2000 to 50% in 2015.

INITIAL MANAGEMENT OF RUPTURED ANEURYSM CASES (SURGICAL AND ENDOVASCULAR)

The variability of aneurysm size, location, and presentation makes each case unique. There is an obvious difference between a critically ill child with an unexpected hemorrhage and an asymptomatic lesion. However, there are general principles applicable to all cases [32]. A ventriculostomy [(placement of an external ventricular drain (EVD)] can sometimes help with the control of elevated intracranial pressure and hydrocephalus from hemorrhage, but it is critical to recognize that overdrainage of cerebrospinal fluid (CSF) can result in hemorrhage from sudden shifts in aneurysmal transmural pressure. Appropriate patient preparation, with large-bore intravenous access, arterial line monitoring, and other adjuncts, such as a bladder catheter, is important. Controlling blood pressure and administration of antibiotics (through an EVD) or antiepileptic medications need to be considered. If surgical trapping is an option, the availability of a bypass graft (scalp artery, radial artery, etc.) with access needs to be planned in advance.

Pretreatment communication with nursing and anesthesia before the therapy (both operative and endovascular) will streamline care. The anesthesia team should be prepared with multiple large-bore intravenous catheters and should have blood products in the procedure room. Availability of intraoperative angiography and/or indocyanine green (ICG) on the microscope to confirm real-time aneurysm obliteration

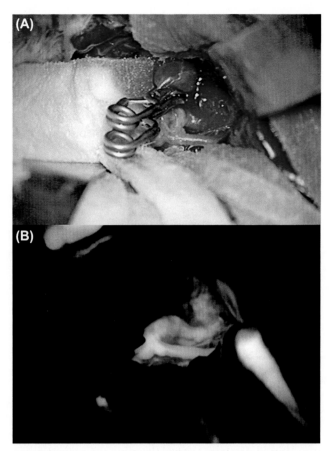

FIGURE 157.3 Intraoperative image of aneurysm clipping with (A) conventional microscopy and (B) indocyanine green (ICG) dye. The ICG image demonstrates patency of parent vessel (below clips) with complete obliteration of aneurysm (no filling with dye).

with parent vessel patency is helpful for open surgical cases (Fig. 157.3). Agents to manage temporary clipping (such as barbiturates or propofol) and hypothermia and vasoactive medications are important. Adenosine can be used to briefly reduce blood flow so that the anatomy of an unexpected rupture is better visualized.

After treatment, evaluation of parent vessel patency and aneurysm occlusion can be performed with intraoperative Doppler and ICG administration (for open surgical cases) and/or catheter angiography (in both open and endovascular cases).

COMPLICATIONS

- Bleeding from intraoperative rupture is the most immediate complication of surgery, and risks are magnified in smaller children, who have little reserve. The loss of a quarter of blood volume can induce shock and there may be rapid decompensation in children, which mandates careful monitoring and replacement of blood products by the operative team.

- Stroke may occur due to perforator/parent vessel occlusion or emboli/dissection caused by injury to the arterial tree during manipulation (reported in 6–8% cases) [10,26].
- Hydrocephalus can occur because of blood in the subarachnoid space or ventricles (reported in 14% cases) [10].
- Vasospasm can occur following SAH (reported in 21% of pediatric cases) [10].
- Cerebral salt wasting (CSW) is an often underrecognized consequence of SAH and should be considered in the posthemorrhage period if major fluid or sodium shifts are observed.

The overall morbidity and mortality vary widely dependent on age, aneurysm type, and presentation. Series have described average mortality rates of 1–3% and morbidity rates of 8–14% [26,33].

POSTTREATMENT CARE AND FOLLOW-UP

Immediate posttreatment care centers on confirming that the aneurysm has been fully obliterated (with imaging) and maintaining hemodynamic stability with good blood pressure control (keeping the patient normotensive to sometimes slightly hypertensive, in order to avoid vasospasm). In general most patients will spend a period in the intensive care unit and then often mobilize to the wards. In addition to immediate operative/endovascular complications, there are problems specific to SAH, including hydrocephalus, vasospasm, and CSW. For patients with SAH, the first week or so after ictus is the key period for developing these conditions. Frequent neurologic examinations should be supplemented with imaging (such as CT/MRI for hydrocephalus and CTA/MRA/transcranial Doppler/angiography for vasospasm). Some patients may benefit from the so-called "triple-H" therapy—hypertension, hypervolemia, and hemodilution—to reduce the risk of vasospasm (only after aneurysm treatment) [34,35]. The use of calcium channel blocking agents, such as nimodipine, is unclear in children.

Follow-up will frequently consist of an office visit about 1 month after treatment, and annually thereafter. In addition to the perioperative angiography to confirm obliteration of the aneurysm, an MRI/MRA at 6 months may be helpful as a baseline study to compare with subsequent annual MRI/MRA. Imaging is performed annually for 5 years, if feasible, with some centers suggesting lifetime imaging every 3–5 years thereafter [36]. An angiogram (DSA) is often performed 1 year after operation to confirm durable cure.

OUTCOMES

There is a wide range of reported outcomes for pediatric aneurysms, with "good" posttreatment outcomes ranging from 13% to 95% and treatment-related mortality ranging from 3% to 100% [37]. Shunting will be required for 14% of patients with hydrocephalus after SAH [10]. Overall, from a clinical perspective, from the patients who survive treatment, 91% (with a mean of 25 years of follow-up) go on to enjoy independent living, with high rates of university graduation and employment [12]. One study radiographically reviewed a group of 59 patients with aneurysms who were treated when they were children, with an average of 34 years of follow-up [38]. In this series, 41% developed recurrent or de novo aneurysms after treatment. The annual rate of hemorrhage was 0.4%, with four deaths. The only identified risk factor adding to recurrent or de novo aneurysm development was smoking [38]. Although these data are obviously limited in part by the absence of current imaging technologies and surgical techniques at the time of patient treatment, they, nonetheless, underscore the need for continued follow-up in pediatric patients with intracranial aneurysm.

The evidence relevant to long-term follow-up in endovascular treatment is far more limited because of the shorter period that this therapeutic modality has been available (and the evolution of device technology). Important for children, with their expected longer life span relative to adults, is the data from the International Subarachnoid Aneurysm Trial (ISAT) that suggests a higher rate of rebleeding over time caused by endovascular treatments [29,31]. The trial data suggest that long-term follow-up is particularly important in children treated with endovascular therapies.

COMMON CLINICAL QUESTIONS

A number of questions are commonly asked by patients and families when diagnosed with an intracranial aneurysm. These include the following:

1. What are some modifiable risk factors that contribute to aneurysm formation in children?
   Although adult intracranial aneurysms are often associated with modifiable risk factors, including hypertension, obesity, high cholesterol levels, diabetes, and alcohol and drug abuse, the only shared major risk seems to be smoking. It makes sense for children to control all the other risk factors as well, but smoking appears to have a major contribution particularly to the development of recurrent or growing residual aneurysms [6,38].

2. If an aneurysm is found in a child, what are the screening recommendations for siblings and other relatives?

There are no fixed screening recommendations for family members of children with intracranial aneurysms. Although the risk factors for acquired aneurysms (posttraumatic, postinfectious, postsurgical, etc.) in related family members is not genetic, the evidence for or against screening relatives of children with spontaneous aneurysms is limited. Current data suggests increased risk with a sibling pair or three first-degree relatives with known aneurysms [14,15,18].

3. Where are the most common locations of traumatic aneurysms?

Traumatic intracranial aneurysms can occur after trauma or surgery. They are most commonly found in the anterior circulation, especially around the cavernous sinus and pericallosal artery [18,35].

CONCLUSIONS

Pediatric intracranial aneurysms differ markedly from those found in adults in size, location, and pathogenesis. Treatment should be considered in patients with aneurysms that have ruptured, are growing, or have characteristics suggesting increased risk of rupture in the near future. Specific treatment modalities are case specific, although a general trend toward the increasing use of endovascular therapy has been apparent over the years. The evolution of surgical and endovascular technology has increased the scope of lesions that can be treated and has improved outcomes, but the rarity of pediatric aneurysms and the limited amount of long-term follow-up means that ongoing study is warranted.

References

[1] Vlak MH, Algra A, Brandenburg R, Rinkel GJ. Prevalence of unruptured intracranial aneurysms, with emphasis on sex, age, comorbidity, country, and time period: a systematic review and meta-analysis. Lancet Neurol 2011;10(7):626–36.

[2] Sedzimir CB, Robinson J. Intracranial hemorrhage in children and adolescents. J Neurosurg 1973;38(3):269–81.

[3] Locksley HB, Sahs AL, Knowler L. Report on the cooperative study of intracranial aneurysms and subarachnoid hemorrhage. Section II. General survey of cases in the central registry and characteristics of the sample population. J Neurosurg 1966;24(5):922–32.

[4] Roche JL, Choux M, Czorny A, et al. Intracranial arterial aneurysm in children. A cooperative study. Apropos of 43 cases. Neurochirurgie 1988;34(4):243–51.

[5] Gerosa M, Licata C, Fiore DL, Iraci G. Intracranial aneurysms of childhood. Child's Brain 1980;6(6):295–302.

[6] Gemmete JJ, Toma AK, Davagnanam I, Robertson F, Brew S. Pediatric cerebral aneurysms. Neuroimaging Clin North America 2013;23(4):771–9.

[7] Unruptured intracranial aneurysms–risk of rupture and risks of surgical intervention. International Study of Unruptured Intracranial Aneurysms Investigators. N Engl J Med 1998;339(24):1725–33.

[8] Roger VL, Go AS, Lloyd-Jones DM, et al. Executive summary: heart disease and stroke statistics–2012 update: a report from the American Heart Association. Circulation 2012;125(1):188–97.

[9] Go AS, Mozaffarian D, Roger VL, et al. Heart disease and stroke statistics–2013 update: a report from the American Heart Association. Circulation 2013;127(1):e6–245.

[10] Garg K, Singh PK, Sharma BS, et al. Pediatric intracranial aneurysms-our experience and review of literature. Childs Nerv Syst 2013.

[11] Lasjaunias P, Wuppalapati S, Alvarez H, Rodesch G, Ozanne A. Intracranial aneurysms in children aged under 15 years: review of 59 consecutive children with 75 aneurysms. Childs Nerv Syst 2005;21(6):437–50.

[12] Koroknay-Pal P, Lehto H, Niemela M, Kivisaari R, Hernesniemi J. Long-term outcome of 114 children with cerebral aneurysms. J Neurosurg Pediatr 2012;9(6):636–45.

[13] Allison JW, Davis PC, Sato Y, et al. Intracranial aneurysms in infants and children. Pediatr Radiol 1998;28(4):223–9.

[14] Broderick JP, Sauerbeck LR, Foroud T, et al. The familial intracranial aneurysm (FIA) study protocol. BMC Med Genet 2005;6:17.

[15] Brown Jr RD, Huston J, Hornung R, et al. Screening for brain aneurysm in the Familial Intracranial Aneurysm study: frequency and predictors of lesion detection. J Neurosurg 2008;108(6):1132–8.

[16] Roach ES, Golomb MR, Adams R, et al. Management of stroke in infants and children: a scientific statement from a special writing group of the American heart association stroke council and the council on cardiovascular disease in the young. Stroke; 2008;39(9):2644–91.

[17] Woodall MN, McGettigan M, Figueroa R, Gossage JR, Alleyne Jr CH. Cerebral vascular malformations in hereditary hemorrhagic telangiectasia. J Neurosurg 2014;120(1):87–92.

[18] Aeron G, Abruzzo TA, Jones BV. Clinical and imaging features of intracranial arterial aneurysms in the pediatric population. Radiographics 2012;32(3):667–81.

[19] Hunt WE, Hess RM. Surgical risk as related to time of intervention in the repair of intracranial aneurysms. J Neurosurg 1968;28(1):14–20.

[20] Hunt WE, Kosnik EJ. Timing and perioperative care in intracranial aneurysm surgery. Clin Neurosurg 1974;21:79–89.

[21] Beslow LA, Jordan LC. Pediatric stroke: the importance of cerebral arteriopathy and vascular malformations. Childs Nerv Syst 2010;26(10):1263–73.

[22] Al-Jarallah A, Al-Rifai MT, Riela AR, Roach ES. Nontraumatic brain hemorrhage in children: etiology and presentation. J Child Neurol 2000;15(5):284–9.

[23] Burger IM, Murphy KJ, Jordan LC, Tamargo RJ, Gailloud P. Safety of cerebral digital subtraction angiography in children: complication rate analysis in 241 consecutive diagnostic angiograms. Stroke 2006;37(10):2535–9.

[24] Lin N, Smith ER, Scott RM, Orbach DB. Safety of neuroangiography and embolization in children: complication analysis of 697 consecutive procedures in 394 patients. J Neurosurg Pediatr 2015:1–7.

[25] terBrugge KG. Neurointerventional procedures in the pediatric age group. Childs Nerv Syst 1999;15(11–12):751–4.

[26] Hetts SW, Narvid J, Sanai N, et al. Intracranial aneurysms in childhood: 27-year single-institution experience. AJNR Am J Neuroradiol 2009;30(7):1315–24.

[27] Hayashi S, Arimoto T, Itakura T, Fujii T, Nishiguchi T, Komai N. The association of intracranial aneurysms and arteriovenous malformation of the brain. Case report. J Neurosurg 1981;55(6):971–5.

[28] Watanabe H, Nakamura H, Matsuo Y, et al. [Spontaneous regression of cerebral arterio-venous malformation following major artery thrombosis proximal to dominant feeders: a case report]. No Shinkei Geka 1995;23(4):371–6.

[29] Molyneux A, Kerr R, Stratton I, et al. International Subarachnoid Aneurysm Trial (ISAT) of neurosurgical clipping versus endovascular coiling in 2143 patients with ruptured intracranial aneurysms: a randomised trial. Lancet 2002;360(9342):1267–74.

[30] Molyneux AJ, Kerr RS, Yu LM, et al. International subarachnoid aneurysm trial (ISAT) of neurosurgical clipping versus endovascular coiling in 2143 patients with ruptured intracranial aneurysms: a randomised comparison of effects on survival, dependency, seizures, rebleeding, subgroups, and aneurysm occlusion. Lancet 2005;366(9488):809–17.

[31] Beez T, Steiger HJ, Hanggi D. Evolution of management of intracranial aneurysms in children: a systematic review of the modern literature. J Child Neurol 2015.

[32] Sanai N, Auguste KI, Lawton MT. Microsurgical management of pediatric intracranial aneurysms. Childs Nerv Syst 2010;26(10):1319–27.

[33] Kakarla UK, Beres EJ, Ponce FA, et al. Microsurgical treatment of pediatric intracranial aneurysms: long-term angiographic and clinical outcomes. NeurosurgeryV 2010;67(2):237–49. discussion 250.

[34] Origitano TC, Wascher TM, Reichman OH, Anderson DE. Sustained increased cerebral blood flow with prophylactic hypertensive hypervolemic hemodilution ("triple-H" therapy) after subarachnoid hemorrhage. Neurosurgery 1990;27(5):729–39. discussion 739–740.

[35] Nahed BV, Ferreira M, Naunheim MR, Kahle KT, Proctor MR, Smith ER. Intracranial vasospasm with subsequent stroke after traumatic subarachnoid hemorrhage in a 22-month-old child. J Neurosurg Pediatr 2009;3(4):311–5.

[36] Tonn J, Hoffmann O, Hofmann E, Schlake HP, Sorensen N, Roosen K. "De novo" formation of intracranial aneurysms: who is at risk? Neuroradiology 1999;41(9):674–9.

[37] Huang J, McGirt MJ, Gailloud P, Tamargo RJ. Intracranial aneurysms in the pediatric population: case series and literature review. Surg Neurol 2005;63(5):424–32. discussion 432–423.

[38] Koroknay-Pal P, Niemela M, Lehto H, et al. De novo and recurrent aneurysms in pediatric patients with cerebral aneurysms. Stroke 2013;44(5):1436–9.

[39] Dunn IF, Woodworth GF, Siddiqui AH, et al. Traumatic pericallosal artery aneurysm: a rare complication of transcallosal surgery. Case report. J Neurosurg 2007;106(Suppl. 2):153–7.

C H A P T E R

158

Surgical Management of Pituitary Apoplexy

A.O. Jamshidi, D.M. Prevedello, A. Beer-Furlan, R.A. Hachem, B. Otto, R. Carrau

Wexner Medical Center, The Ohio State University, Columbus, OH, United States

INTRODUCTION

The term "pituitary apoplexy" (PA) was first described in context of five autopsies in which hemorrhagic necrosis was noted in pituitary adenomas; the studied patients had a history of clinical dyspituitarism and sudden death [1]. This condition is exceptionally rare relative to the frequency of pituitary adenomas; 0.6–9% of adenomas undergo clinical apoplexy while up to 20–25% of the general population may have pituitary tumors or cysts of variable size [2,3]. Although patients have been known to present up to 14 days after the onset of apoplectic symptoms, their clinical deterioration may

have begun prior to this event [2,4]. Unfortunately, most patients have pituitary tumors unbeknown to them because they do not have access to primary care physicians due to the lack of health insurance [4]. Therefore, many of these patients could avoid catastrophic neurological and endocrinological decline.

PATHOPHYSIOLOGY

Radiographically, hemorrhage within the sella is not specific to pituitary apoplexy. Hemorrhage in this location can be seen in other pathological conditions, such as Sheehan's syndrome, metastases, trauma, and coagulopathic states. When apoplexy occurs in the setting of a pituitary adenoma, both ischemia and hemorrhage are usually observed [5,6]. Typically, an enlarging adenoma with increased metabolic requirement can outstrip its arterial supply leading to necrosis or the tumor can compress its draining veins causing a venous infarct [7]; owing to this latter theory, engorgement of sphenoid sinus mucosa due to venous hypertension has been radiographically appreciated prior to apoplexy [2]. Others have hypothesized that the hemorrhagic event is associated with the intrinsic friability of tumor vessels with incomplete basal membranes or with atherosclerotic emboli [7,8]. Based on the authors' experience, the majority of the surgical cases seem to have a direct association to decreased venous drainage into the cavernous sinus with subsequent venous congestion followed by venous infarct. A hemorrhagic component may or may not be associated to the event and most likely depends on the severity of the venous congestion. Needless to say, there is disagreement regarding the etiology of pituitary apoplexy.

The general risk factors for apoplexy include acute increase or decrease in hypophyseal blood flow, hormone modulation from exogenous or endogenous sources, surgery (particularly cardiac), and anticoagulation [9]. Hypertension has been shown to be associated with apoplectic events and has been seen in about one-quarter of patients [6]. Alterations in intracranial pressure or systemic hypotension, which can be induced during cardiac surgery, can also cause changes in the portal circulation [10]. An increase in circulating hormones from stress or the use of dopamine agonists can lead to the same outcome [11,12]. However, most cases of apoplexy occur without any documented, precipitating factor [13,14]. Most often, this condition is seen in adults with null-cell adenomas and 80% of patients have no documented history of an adenoma upon presentation [2].

CLINICAL PRESENTATION

Headache and meningismus are the most common presenting signs of apoplexy. These symptoms are related to stretching of the hypophyseal capsule, which surround the anterior pituitary lobe, and/or to spilling of blood products in the subarachnoid space. Although hydrocephalus is not a common associated cause of headache, giant adenomas can cause obstruction of cerebrospinal fluid at the level of the third ventricle. Visual deterioration varies from decreased visual acuity to bitemporal hemianopsia to blindness and is due to mass effect on the optic apparatus; visual obscurations occur in 43–61% of patients [2]. Ophthalmoplegia also occurs

due to expansion of the adenoma outside of the confines of the sella.

The oculomotor nerve is most often effected followed by the abducens and the trochlear nerves [15]. The oculomotor nerve can be compressed in the oculomotor triangle by the tumor placing force on the nerve against the interclinoid ligament; distally it can be injured due to mass effect on the lateral wall of the cavernous sinus or direct tumor invasion into it. Trochlear, abducens, and trigeminal nerve palsies most often occur in conjunction with a third nerve palsy and the former usually recover before the oculomotor nerve. An isolated abducens nerve palsy suggests mass effect on the pons due to erosion of the dorsum sella or tumor invasion into Dorello's canal. Trochlear nerve palsy is usually not seen in isolation while the trigeminal nerve is typically quite resilient. Sensory changes of the face, for example, usually indicate that the adenoma has expanded beyond the lateral wall of the cavernous sinus and potentially in a posterior–inferior projection toward Meckel's cave [16].

Although these possible neurological deficits are significant, adrenal insufficiency causing cardiovascular shock is the most deadly, albeit rare, complication [2]. Recognition of this condition and resuscitation with exogenous corticosteroids are critical in this patient population. Permanent hypopituitarism after apoplexy can be as high as 70–80% of cases and even 100% in some series [17].

MANAGEMENT

Up to 73% of patients have evidence of hypopituitarism upon presentation; therefore, initial management includes the administration of stress dose steroids, electrolyte correction and resuscitation, particularly in the setting of hemodynamic distress. Ideally, the patient would have serum cortisol checked prior to the administration of corticosteroids; because the mineralocorticoid effect of dexamethasone and methylprednisolone is negligible, hydrocortisone (1–2 mg/kg) is the preferred agent in the setting of vascular collapse. Patients can also have dilutional or hypervolemic hyponatremia from a decrease in circulating cortisol because of excess antidiuretic hormone secretion or sodium–potassium pump malfunction. Exogenous steroids and an infusion of normal saline most often correct this condition; sodium should be frequently checked with the goal of normonatremia. Diabetes insipidus is also a potential finding in these patients and should be treated and managed accordingly.

In addition to cortisol, patients should have a full panel of endocrine hormones tested in order to assess for the degree of hypopituitarism and for the presence of a functional or nonfunctional adenoma. If the endocrine laboratories indicate the presence of a prolactinoma (generally greater than 200 µg/L), conservative management with dopamine agonists is an option. However, if the patient has visual deterioration and/or a cystic prolactinoma, transsphenoidal surgery is a reasonable option.

Most patients with acute vision loss and/or worsening visual function are taken to surgery for urgent decompression of the optic apparatus. Carefully selected patients with improving or no vision loss who present subsequent to the initial ictus or those with improving ophthalmoparesis may avoid surgery with the expectation that the extraocular palsy will resolve within several months (Fig. 158.1). Furthermore, patients with severe comorbidities and/or advanced age may be managed conservatively even in the presence of a neurological deficit since most often the blood products and necrotic tumor regress over time.

Because the goal of surgery is to decompress critical neural structures, the initial imaging obtained upon presentation can help determine the timing of surgical intervention. If the imaging reveals compression of the optic apparatus, the third ventricle (with resultant hydrocephalus) or the brain stem causing long-tract signs, then urgent surgical decompression is mandated (Figs. 158.2 and 158.3). Magnetic resonance imaging (MRI) is the most ideal study because it can show the relationship of the tumor with surrounding structures, indicate the location of the normal pituitary gland and suggest the consistency and quality of the tumor. A noninvasive angiographic study, such as a computed topography (CT) angiogram or magnetic resonance (MR) angiogram, can also be helpful for preoperative planning in order to understand the relationship of the tumor to the bony architecture of the skull base as well as to the arterial anatomy of the brain.

SURGICAL TECHNIQUE

The authors favor an endoscopic transsphenoidal approach for the surgical management of these tumors. The patient is positioned supine and secured in a three-pin head fixation system if the tumor involves cavernous sinus (cranial nerves three, four, or six) (Fig. 158.4). The authors advocate a four-handed, two-surgeon technique and do not routinely elevate a vascularized nasoseptal flap. A unilateral middle turbinate is removed during the exposure and a sphenoidotomy and ethmoidectomy are completed as needed in order to adequately visualize the tuberculum sella, sellar face, and clival recess in the midline as well as the carotid protuberance, optic impression, and the opticocarotid recesses laterally (Fig. 158.5). Once the dura is exposed and incised, typically a double suction technique can be used in order to efficiently aspirate necrotic tumor. Initially, the sellar floor is cleared

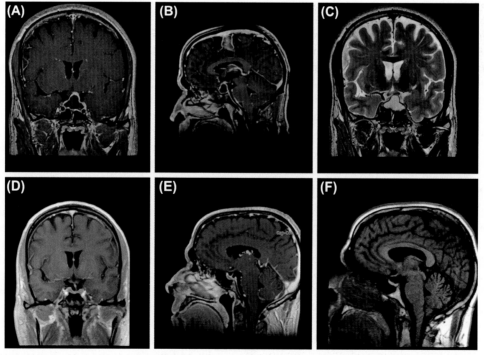

FIGURE 158.1   The patient presented to an outside facility with clinical apoplexy confirmed by MRI and endocrine evaluation. The tumor expanded the gland causing compression of the optic apparatus as seen in the contrast-enhanced T1 (A and B) and coronal T2 (C). The patient was given exogenous steroids and because his symptoms improved over 2 weeks when he presented to the authors, another MRI was ordered. The T1 images with (D and E) and without (F) contrast show resolution of his apoplectic tumor and he was managed conservatively.

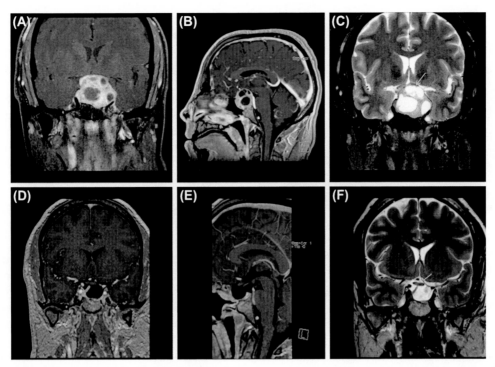

FIGURE 158.2   An ophthalmologist ordered an MRI for this patient due to complaints of deteriorating visual function and clinical pituitary apoplexy. Preoperative MRI revealed a large heterogeneous mass of the sella compressing the suprasellar cistern (A and B) and invading the bilateral cavernous sinuses (C). The tumor compressed the optic apparatus and flattened the left optic nerve (C, *yellow arrow*). The patient underwent an urgent EEA within 24 h of presentation to the authors. Postoperatively MRI showed that the gland and stalk were enhancing and shifted to the right (D and E); the bilateral cavernous sinuses and the optic apparatus were decompressed (F, *yellow arrow*).

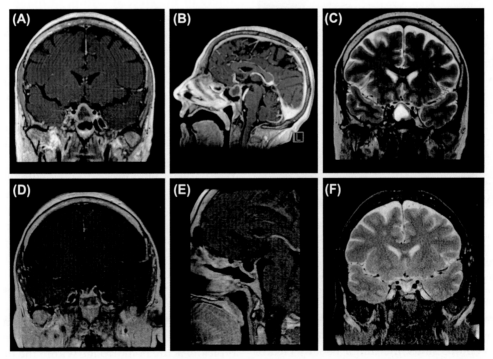

FIGURE 158.3   The patient presented with sudden vision decline and endocrinopathy with preoperative imaging that corroborated the diagnosis of pituitary apoplexy. Coronal and sagittal MRI with contrast (A, B) and coronal T2 (C) revealed a cystic sellar tumor with surpasellar extension that mildly displaced the optic apparatus and invaded the medial compartment of the cavernous sinus on the left (C). The patient underwent an EEA and postoperatively had resolution of preoperative visual symptoms due to the decompression of the optic nerves as seen in the postoperative MRI coronal and sagittal with contrast (D, E) and coronal T2 (F).

FIGURE 158.4 The patient is positioned supine and secured in a three-pin head fixation system because of the need for cranial nerve monitoring. The authors routinely monitor the cavernous sinus nerves when indicated by tumor expansion outside the sella; electromyography leads can be seen entering the orbits. The head is rotated to the right, and slightly tilted toward the contralateral shoulder, which favors right-handed surgeons. The abdomen is kept exposed and is prepped using standard sterile technique before draping in case of the need for a fat graft as part of the skull base reconstruction. The thigh is also exposed and prepped in case of a ICA rupture and the need for rapid access muscle grafting.

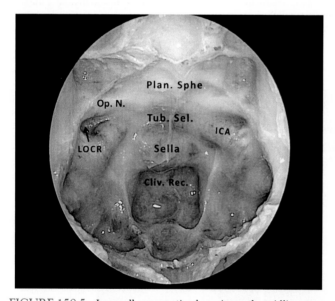

FIGURE 158.5 In a well-pneumatized specimen, the midline structures of the posterior, middle, and anterior cranial fossa can be seen as the clival recess (Cliv. Rec.), sella (notice the superior and inferior intercavernous sinus above and below the sella), tuberculum sella (Tub. Sel.), and planum sphenoidale (Plan. Sphe.), respectively. The optic nerve (Op. N.) can also identified relative to the paraclinoid carotid (ICA) and the lateral optico-carotid recess (LOCR).

of tumor followed by the lateral walls. If the mass has violated the cavernous sinus, the tumor usually creates a corridor that leads the surgeon laterally into this structure, which is critical if the patient suffers any degree of ophthalmoparesis. Attention is then brought to the superior component of the tumor and debulking continues. If

the mass is fibrotic and cannot be aspirated, microdissectors are used to establish a plane between the tumor and the pituitary gland. At the conclusion of the procedure, arachnoid may herniate down with the very thin layer of pituitary gland, which is known as the herniation of the diaphragma sellae indicating that the suprasellar cistern has been decompressed. The authors often place an inlay dura substitute matrix into the resection cavity; a mucoperichondrial flap elevated from the middle turbinate is then positioned over the skull base opening.

POSTOPERATIVE MANAGEMENT AND OUTCOME

Postoperatively, the patient should be closely monitored with frequent neurological checks and laboratories. A step-down floor is appropriate and if unavailable, admission to an intensive care unit is reasonable. Although there is not a universal protocol for postoperative imaging, the authors favor obtaining a head CT scan in order to assess for any immediate complications. Because of the risk of diabetes insipidus, serum sodium, urine specific gravity, and urine osmolality are analyzed every 6 h; in addition, nursing staff must record the hourly fluid intake and output. For patients with Cushing's disease, cortisol is measured frequently in anticipation of a "cortisol crash," which refers to the clinical setting of hypocortisolism and a serum cortisol of less than 2 µg/deciliter. If the patient has a nonfunctional pituitary adenoma, then cortisol and prolactin are measured the morning following surgery to ensure that the patient has pituitary function. Although uncommon, patients are routinely evaluated for a cerebrospinal fluid fistula in the postoperative phase as well.

The majority of patients (53–89%) experience improvement in their vision after transsphenoidal surgery [2]. Timing of surgery is controversial; the authors believe that patients who present with acute vision loss should undergo urgent decompression. Patients with a shorter duration of symptoms have the highest chance of recovering function with surgical management. Rarely, vision deteriorates postoperatively; the surgeon should be very cautious of this potential complication when a significant volume of suprasellar tumor is not removed via the transsphenoidal route. Although uncommon, residual tumor can suffer further apoplexy because its venous drainage is disrupted during surgery. Increased venous congestion then leads to swelling, venous infarction, and typically acute hemorrhage within the tumor; vision loss occurs subsequently from mass effect. This clinical situation most often warrants transcranial decompression.

Although cranial neuropathies typically improve after surgery, endocrine function can reverse in about one-quarter of apoplexy patients. Up to 80% of patients have long-term hypopituitarism and require hormone replacement. In addition to follow-up for hormone therapy with an endocrinologist, the authors believe that these patients require lifelong radiographic follow-up. Postoperatively, recurrence can occur in about 10% of patients within 10 years of diagnosis of a pituitary adenoma with a Ki-67 (marker of proliferation) less than 3%.

CONCLUSION

Albeit rare, the rate of pituitary apoplexy is variable in the literature because of an inconsistent definition of this diagnosis. This condition is a based on a clinical presentation that is supported by radiographic evidence of tumor with or without hemorrhage and endocrine collapse. Therefore, its diagnosis requires the scrutiny of a discerning clinician. Once discovered, these patients benefit primarily from hormone replacement and supportive care; surgery is secondary and must be considered within context of the patient's neurological complaints and overall medical condition.

References

[1] Brougham M, Heusner AP, Adams RD. Acute degenerative changes in adenomas of the pituitary body–with special reference to pituitary apoplexy. J Neurosurg 1950;7(5):421–39.

[2] Bi WL, Dunn IF, Laws Jr ER. Pituitary apoplexy. Endocrine 2015;48(1):69–75.

[3] Rajasekaran S, Vanderpump M, Baldeweg S, et al. UK guidelines for the management of pituitary apoplexy. Clin Endocrinol (Oxf) 2011;74(1):9–20.

[4] Jahangiri A, Clark AJ, Han SJ, Kunwar S, Blevins Jr LS, Aghi MK. Socioeconomic factors associated with pituitary apoplexy. J Neurosurg 2013;119(6):1432–6.

[5] Mohr G, Hardy J. Hemorrhage, necrosis, and apoplexy in pituitary adenomas. Surg Neurol 1982;18(3):181–9.

[6] Randeva HS, Schoebel J, Byrne J, Esiri M, Adams CB, Wass JA. Classical pituitary apoplexy: clinical features, management and outcome. Clin Endocrinol (Oxf) 1999;51(2):181–8.

[7] Biousse V, Newman NJ, Oyesiku NM. Precipitating factors in pituitary apoplexy. J Neurol Neurosurg Psychiatry 2001;71(4):542–5.

[8] Semple PL, Jane JA, Lopes MB, Laws ER. Pituitary apoplexy: correlation between magnetic resonance imaging and histopathological results. J Neurosurg 2008;108(5):909–15.

[9] Semple PL, Jane Jr JA, Laws Jr ER. Clinical relevance of precipitating factors in pituitary apoplexy. Neurosurgery 2007;61(5):956–61. [discussion 961–952].

[10] Yakupoglu H, Onal MB, Civelek E, Kircelli A, Celasun B. Pituitary apoplexy after cardiac surgery in a patient with subclinical pituitary adenoma: case report with review of literature. Neurol Neurochir Pol 2010;44(5):520–5.

[11] Chng E, Dalan R. Pituitary apoplexy associated with cabergoline therapy. J Clin Neurosci 2013;20(12):1637–43.

[12] Singh P, Singh M, Cugati G, Singh AK. Bromocriptine or cabergoline induced pituitary apoplexy: rare but life-threatening catastrophe. J Hum Reprod Sci 2011;4(1):59.

[13] Onesti ST, Wisniewski T, Post KD. Clinical versus subclinical pituitary apoplexy: presentation, surgical management, and outcome in 21 patients. Neurosurgery 1990;26(6):980–6.

[14] Semple PL, Webb MK, de Villiers JC, Laws Jr ER. Pituitary apoplexy. Neurosurgery 2005;56(1):65–72. [discussion 72–63].

[15] Kim SH, Lee KC, Kim SH. Cranial nerve palsies accompanying pituitary tumour. J Clin Neurosci 2007;14(12):1158–62.

[16] Jefferson G. Extrasellar Extensions of Pituitary Adenomas: (Section of Neurology). Proc R Soc Med 1940;33(7):433–58.

[17] Ayuk J, McGregor EJ, Mitchell RD, Gittoes NJ. Acute management of pituitary apoplexy–surgery or conservative management? Clin Endocrinol (Oxf) 2004;61(6):747–52.

CHAPTER

159

Surgical and Endovascular Management of Acutely Ruptured Arteriovenous Malformations

W.C. Rutledge, M.T. Lawton

University of California, San Francisco, San Francisco, CA, United States

INTRODUCTION

A brain arteriovenous malformation (AVM) is a tangle of dysplastic blood vessels characterized by abnormal connections between arteries and veins. Dilated arteries and a nidus drained by arterialized veins without intervening capillaries form a high-flow, low-resistance shunt between the arterial and venous systems. High flow through the feeding arteries, nidus, and draining veins may lead to rupture and intracerebral hemorrhage. AVMs are a leading cause of intracerebral hemorrhage in young adults. Patients with AVM rupture and intracerebral hemorrhage may have better outcomes than patients with intracerebral hemorrhage from other causes, but AVM rupture is still associated with significant morbidity and mortality and long-term disability. While an increasing number of AVMs are now discovered incidentally due to widespread brain imaging, patients frequently present with hemorrhage, seizure, or progressive neurological deficits. Diagnosis mainly involves CT, MRI, and angiography.

Most AVMs are solitary and occur sporadically, while multiple AVMs are associated with syndromes, such as hereditary hemorrhagic telangiectasia (HHT). AVMs have long been considered congenital; however, their pathogenesis is still not well understood, and a growing body of evidence suggests that AVMs can form de novo after a variety of insults associated with trauma, treatment of dural arteriovenous fistulas, tumors, cavernous malformations, and other AVMs. The overall incidence of ruptured and unruptured AVMs is about 1 per 100,000 person years.

There is controversy whether patients with unruptured AVMs should be treated or observed [1,2]. There is no primary medical therapy. The overall risk of hemorrhage is about 1–2% per year, however the risk of additional hemorrhage increases significantly once rupture occurs [3,4]. Deep locations, deep venous drainage, increasing age, and the presence of flow-related aneurysms may also increase the risk of additional hemorrhage. Seizures or progressive neurological deficits are other important indications for consideration of treatment, as they may result from silent AVM microhemorrhages associated with increased risk of more significant hemorrhage and frank rupture.

Acutely ruptured AVMs often present with an intracerebral hematoma, mass effect, and increased intracranial pressure causing neurological deterioration, and are operated upon immediately to relieve mass effect. The hematoma creates a corridor of exposure to the AVM and facilitates resection.

Selection of AVM patients for treatment requires balancing risk of treatment complications against the risk of hemorrhage in the natural history course. Modern treatment of AVMs is multimodal and includes microsurgical resection along with endovascular embolization and stereotactic radiosurgery, either as surgical adjuncts or as alternatives. Microsurgical resection has superior cure rates compared to endovascular embolization and stereotactic radiosurgery, however not all patients are good candidates for surgery. Resection of large and complex AVMs in eloquent brain is associated with poor neurological outcomes. Grading systems are used to predict neurological outcomes after AVM surgery and develop management plans. Patients with high surgical risk are considered for endovascular embolization, embolization of flow-related aneurysms, or palliative embolization to eliminate functional steal. Single-session stereotactic radiosurgery or volume-staged radiosurgery can also be used to "downgrade" an AVM, or reduce its volume and make it more favorable for embolization and microsurgical resection.

Primer on Cerebrovascular Diseases, Second Edition
http://dx.doi.org/10.1016/B978-0-12-803058-5.00159-4

AVM GRADING SYSTEMS

Grading systems are used to describe AVMs as well as develop management plans. The Spetzler–Martin and Lawton–Young supplementary grading systems were developed to predict neurological outcomes after AVM surgery. The Spetzler–Martin scale is the predominant classification scheme, and includes AVM size, eloquence of surrounding brain, and venous drainage patterns [5]. The Lawton–Young supplementary grading system supplements the traditional Spetzler–Martin system by incorporating additional factors important to surgical selection and outcome, including patient age, hemorrhagic presentation, and compactness [6,7]. A patient with a supplemented grade ≤6 is a viable candidate for surgery, while patients with grades >6 have a high risk for surgical complications and poor outcomes. Patients with grades >6 are then considered for radiosurgery, or observed. Incompletely obliterated AVMs are reconsidered for surgery when downgraded to a more manageable size. Judicious patient selection is essential to avoid complications and poor neurological outcomes.

AVMs are also classified by their location in the brain: frontal, temporal, and parieto-occipital lobes, ventricles, deep central core, brainstem, and cerebellum. Frontal AVMs are the most common and include the lateral frontal (Fig. 159.1), medial frontal, paramedian frontal, basal frontal, and Sylvian frontal subtypes. Temporal AVM subtypes include one for each surface of the temporal lobe: lateral temporal (Fig. 159.2), basal temporal, Sylvian temporal, and medial temporal. Parieto-occipital AVMs often have a more robust arterial supply than frontal and temporal AVMs and include four subtypes: lateral parieto-occipital (Fig. 159.3), medial parieto-occipital, paramedian parieto-occipital, and basal occipital. Ventricular and periventricular AVMs "float" in cerebrospinal fluid (CSF) and are easier to circumdissect than parenchymal and deep AVMs. Ventricular and periventricular subtypes include callosal (Fig. 159.4), ventricular body, atrial, and temporal horn AVMs. Deep AVMs require a high selection threshold, but favorable results can be expected with surgery. Subtypes include the pure Sylvian, insular, basal ganglial, and thalamic AVMs (Fig. 159.5). Similarly to deep AVMs, brainstem AVMs are often considered inoperable, but maybe located entirely on the pial surface, have higher hemorrhage rate than supratentorial AVMs, and may not be amenable to radiosurgery or embolization. Lateral brainstem AVMs in particular have adequate surgical exposures and are often surgically resectable. Subtypes include anterior midbrain, posterior midbrain, anterior pontine, lateral pontine (Fig. 159.6), anterior medullary, and lateral medullary. Finally, the cerebellar AVMs include five subtypes: suboccipital, tentorial, vermian (Fig. 159.7), tonsillar, and petrosal. Cerebellar AVMs are more likely to present with hemorrhage than supratentorial AVMs. Each subtype is characterized by unique arterial supply, draining veins, eloquent surrounding structures, surgical approach, and management strategy.

MICROSURGICAL RESECTION

Microsurgical resection remains the mainstay of treatment for ruptured and unruptured AVMs. Preoperative adjunctive embolization is routinely done except with AVMs with poor endovascular access or those arising from lenticulostriate arteries, thalamoperforators, or brainstem perforators where there is risk of retrograde escape of the embolic agent. In the acutely ruptured AVM, intracerebral hemorrhage causing mass effect, increased intracranial pressure, and neurological deterioration may preclude preoperative embolization. Immediate surgery allows for hematoma evacuation and relief of mass effect. Hemorrhage may facilitate surgery by separating the AVM from adjacent brain and creating working space around the AVM that minimizes transgression of normal brain. Hemorrhage may also obliterate some of the AVM's blood supply and reduce its blood flow.

Outcomes after AVM surgery in patients with favorable grades are excellent [8]. In two 2010 series, morbidity and mortality after microsurgery ranged from only 0.7% to 2.4% and 0% to 0.5%, respectively, with cure rates of 97–98% [8,9].

AVM surgery follows sequential steps that include exposure, subarachnoid dissection, defining the draining vein, defining the feeding arteries, pial dissection, parenchymal dissection, ependymal or deep dissection, and finally resection. Preservation of the draining vein until the end of the resection is critical to prevent intraoperative rupture. Occluding venous outflow prematurely causes increased intranidal pressure and distension of the malformation resulting in rupture and bleeding in the surgical field. Tamponade and suction are often not effective to control the bleeding and clear the field. Feeding arteries should be disconnected as close as possible to the AVM to prevent sacrifice of normal branches and infarcts in adjacent normal brain. As the AVM is circumferentially dissected and feeding arteries disconnected, the draining vein changes color from red to blue, indicating complete dearterialization of nidus and resectability.

ENDOVASCULAR EMBOLIZATION

Endovascular embolization is an important adjunct to microsurgery. Preoperative embolization is used to obliterate bleeding, flow-related aneurysms, but also to

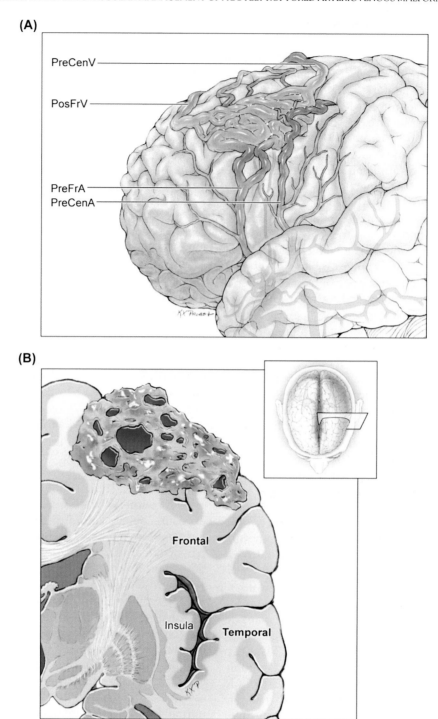

FIGURE 159.1   The lateral frontal AVM: (A) lateral and (B) coronal cross-sectional views. The frontal AVM subtypes also include the medial frontal, paramedian frontal, basal frontal, and Sylvian frontal.

eliminate surgically inaccessible, deep feeding arteries reducing blood flow to the AVM, and minimizing blood loss during surgery and operative times [10]. Embolized feeding arteries are also easier to coagulate and differentiate from en passage arteries that must be preserved.

In select patients, curative embolization is possible. Transarterial embolization with liquid embolic agents, such as Onyx, is associated with higher obliteration rates,

but increased risk of morbidity and mortality. In a large series of 350 patients treated with prolonged intranidal Onyx injection, complete obliteration was achieved in 51% with a mortality rate of only 1.4% and a morbidity rate of 7.1% [11].

An endovascular grading scale incorporating the number of feeding arteries, eloquence, and presence of an arteriovenous fistula component was described.

Fewer numbers of feeding arteries, noneloquence, and absence of an arteriovenous fistula were associated with endovascular cure and fewer complications [12]. Improvements in endovascular therapy will decrease its complication rate and increase the rate of complete obliteration and cure.

STEREOTACTIC RADIOSURGERY

Stereotactic radiosurgery is a viable treatment alternative to AVM surgery. While outcomes for small AVMs are excellent, radiosurgery has some risks, including hemorrhage during the latency period and damage to

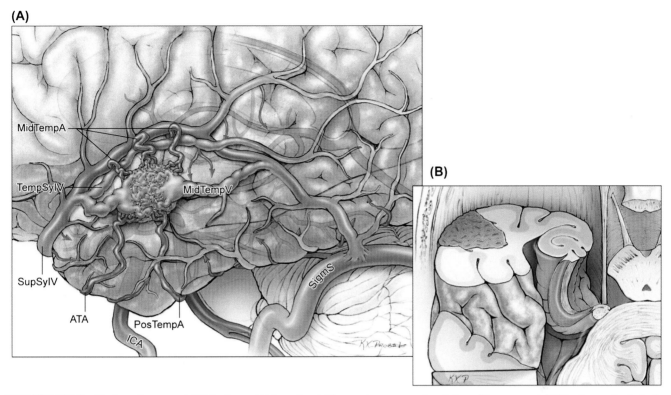

FIGURE 159.2 The lateral temporal AVM subtype (A) lateral and (B) superior cross-sectional views. The temporal AVM subtype also includes basal temporal, Sylvian temporal, and medial temporal.

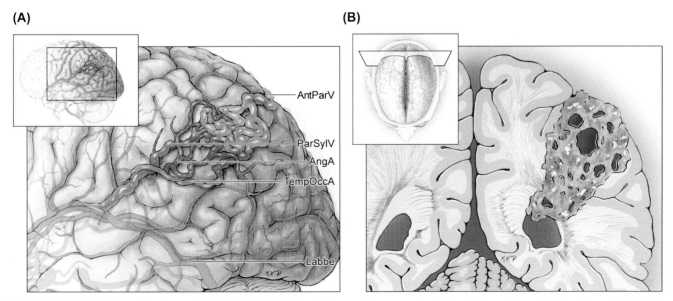

FIGURE 159.3 The lateral parieto-occipital AVM: (A) lateral and (B) coronal cross-sectional views. The parieto-occipital subtype also includes medial parieto-occipital, paramedian parieto-occipital, and basal occipital.

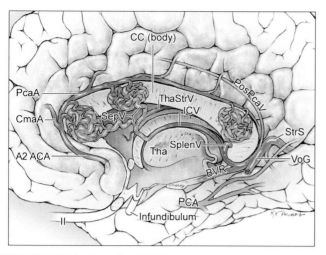

FIGURE 159.4 The callosal AVM, medial view. The ventricular and periventricular subtype also includes ventricular body, atrial, and temporal horn.

surrounding tissue. The likelihood of complete obliteration depends on the size of the lesion and the amount of radiation delivered. Higher doses are associated with morbidity from damage to surrounding structures. Large and complex AVMs in surgically inaccessible locations or with high surgical risk are candidates for volume-staged radiosurgery. Incompletely obliterated AVMs are reconsidered for AVM surgery.

CONCLUSIONS

Management of unruptured AVMs is controversial as the risk of treatment-associated morbidity and mortality must be weighed against the risk of spontaneous hemorrhage causing death, neurological deficits, and long-term disability. Once ruptured, however, the risk of further hemorrhage increases significantly and

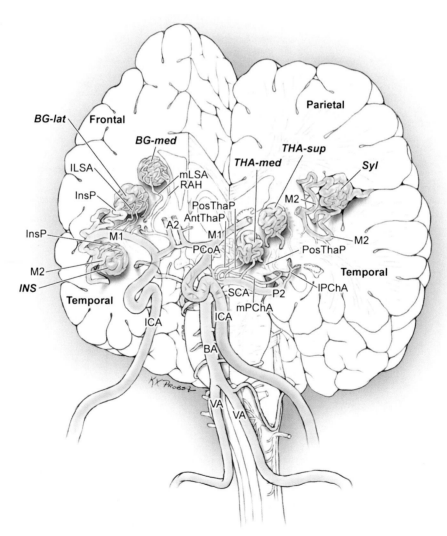

FIGURE 159.5 Overview of deep AVM subtypes: pure sylvian (SYL), insular (INS), basal ganglial (BG), and thalamic (THA) as seen in an anterior oblique, coronal cross-sectional view.

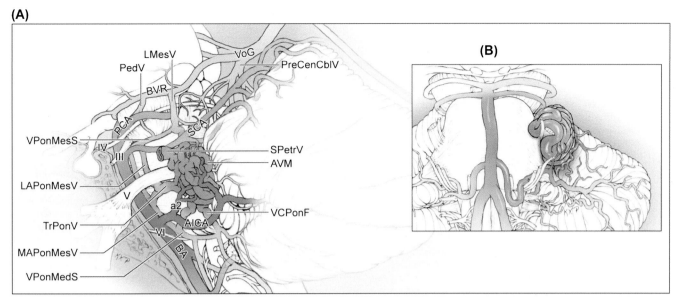

FIGURE 159.6 The lateral pontine AVM: (A) lateral and (B) anterior views. The brainstem subtype also includes anterior midbrain, posterior midbrain, anterior pontine, anterior medullary, and lateral medullary.

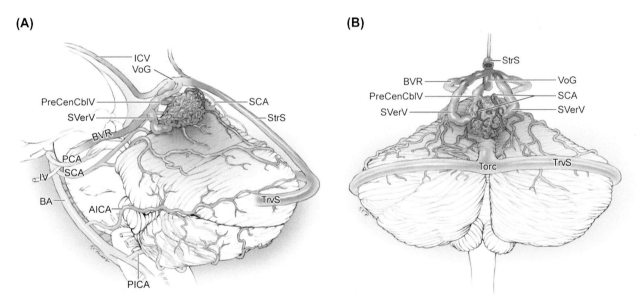

FIGURE 159.7 The vermian cerebellar AVM: (A) lateral and (B) posterior views. Vermian AVMs are the most common cerebellar AVMs. The cerebellar subtype also includes suboccipital, tentorial, tonsillar, and petrosal.

treatment is usually recommended. The supplemented Spetzler–Martin grading system is used to predict neurological outcomes after AVM surgery and to develop management plans. Microsurgical resection remains the mainstay of treatment due to its superior cure rates compared to endovascular embolization and stereotactic radiosurgery. Modern management of AVMs is multimodal and also includes endovascular embolization and stereotactic radiosurgery. Surgical outcomes in appropriately selected patients are excellent with high cure rates and little morbidity and mortality. Patients with a supplemented Spetzler–Martin grade ≤6 are considered first for surgery, but are also candidates for radiosurgery. When possible, endovascular embolization is performed as an adjunct to surgery to eliminate bleeding aneurysms or deep, surgically inaccessible feeding arteries. Additionally, newer embolic agents and delivery systems have improved obliteration rates with fewer complications, but curative embolization is only possible in select patients. Regardless, the goal of treatment in all cases remains complete AVM obliteration while preserving neurological function.

References

[1] Mohr JP, et al. Medical management with or without interventional therapy for unruptured brain arteriovenous malformations (ARUBA): a multicentre, non-blinded, randomised trial. Lancet 2014;383:614–21. http://dx.doi.org/10.1016/s0140-6736(13)62302-8.

[2] Rutledge WC, et al. Treatment and outcomes of ARUBA-eligible patients with unruptured brain arteriovenous malformations at a single institution. Neurosurg Focus 2014;37:E8. http://dx.doi.org/10.3171/2014.7.focus14242.

[3] Ondra SL, Troupp H, George ED, Schwab K. The natural history of symptomatic arteriovenous malformations of the brain: a 24-year follow-up assessment. J Neurosurg 1990;73:387–91. http://dx.doi.org/10.3171/jns.1990.73.3.0387.

[4] Kim H, Al-Shahi Salman R, McCulloch CE, Stapf C, Young WL. Neurology, vol. 83. American Academy of Neurology; 2014. p. 590–7.

[5] Spetzler RF, Martin NA. A proposed grading system for arteriovenous malformations. J Neurosurg 1986;65:476–83. http://dx.doi.org/10.3171/jns.1986.65.4.0476.

[6] Lawton MT, Kim H, McCulloch CE, Mikhak B, Young WL. A supplementary grading scale for selecting patients with brain arteriovenous malformations for surgery. Neurosurgery 2010;66:702–13. http://dx.doi.org/10.1227/01.neu.0000367555.16733.e1. [discussion 713].

[7] Kim H, et al. Validation of the supplemented Spetzler-Martin grading system for brain arteriovenous malformations in a multicenter cohort of 1009 surgical patients. Neurosurgery 2015;76:25–33. http://dx.doi.org/10.1227/neu.0000000000000556.

[8] Potts MB, et al. Current surgical results with low-grade brain arteriovenous malformations. J Neurosurg 2015:1–9. http://dx.doi.org/10.3171/2014.12.jns14938.

[9] Davidson AS, Morgan MK. How safe is arteriovenous malformation surgery? A prospective, observational study of surgery as first-line treatment for brain arteriovenous malformations. Neurosurgery 2010;66:498–504. http://dx.doi.org/10.1227/01.neu.0000365518.47684.98. [discussion 504-495].

[10] Vinuela F, et al. Combined endovascular embolization and surgery in the management of cerebral arteriovenous malformations: experience with 101 cases. J Neurosurg 1991;75:856–64. http://dx.doi.org/10.3171/jns.1991.75.6.0856.

[11] Saatci I, Geyik S, Yavuz K, Cekirge HS. Endovascular treatment of brain arteriovenous malformations with prolonged intranidal Onyx injection technique: long-term results in 350 consecutive patients with completed endovascular treatment course. J Neurosurg 2011;115:78–88. http://dx.doi.org/10.3171/2011.2.jns09830.

[12] Bell DL, et al. Application of a novel brain arteriovenous malformation endovascular grading scale for transarterial embolization. AJNR Am J Neuroradiol 2015;36:1303–9. http://dx.doi.org/10.3174/ajnr.A4286.

CHAPTER

160

Surgical and Endovascular Management of Unruptured Arteriovenous Malformations

C.J. Griessenauer, C.S. Ogilvy

Harvard Medical School, Boston, MA, United States

INTRODUCTION

Brain arteriovenous malformations (AVMs) are clusters of direct connections of arteries to draining veins without intervening capillary bed [1]. The three main components of an AVM are one or more feeding arteries, the nidus at the site of the arteriovenous shunt, and the draining venous structures. AVMs are high-flow, low-resistant shunts due to a significant pressure difference between the arterial and venous sides. The pressure gradient and resultant high flow trigger remodeling of both arteries and draining

Primer on Cerebrovascular Diseases, Second Edition
http://dx.doi.org/10.1016/B978-0-12-803058-5.00160-0

veins. Arteries may be dilated and thin walled due to degeneration of the media and elastic lamina or thickened from endothelial proliferation, hypertrophy of the media, and changes in the basal lamina. Remodeling of the venous system is referred to as arterialization and includes thickening of the wall due to cellular proliferation without an organized elastic lamina [1,2]. The draining veins commonly coalesce and form a major draining vein that eventually drains into a dural venous sinus.

AVMs are equally distributed between both hemispheres and are most commonly seen in the frontal and parietal lobes. Typically, AVMs are pyramid-shaped lesions with the base oriented toward the meninges and the apex pointing to the ventricle or deep into the brain. Three types of feeding arteries have been described and include terminal, pseudo-terminal, and indirect, or en passage, feeders [2]. The surrounding parenchyma may be stained from previous hemorrhage and exhibits edema, necrosis, and gliosis as a result of ischemic injury related to vascular steal and venous hypertension. The overlying meninges tend to have a thickened, fibrotic appearance. While there is usually no functional brain tissue within the AVM, vessels may be separated by normal brain in diffuse lesions [2]. These diffuse lesions with normal brain tissue intermingled between vessels are termed cerebral proliferative angiopathy and may be regarded as separate from classic AVMs [3]. Other distinguishing features include lobar or hemispheric involvement, the absence of dominant feeders or flow-related aneurysms, transdural supply, proximal stenosis of feeding arteries, and the absence of large, early draining veins [3,4].

The pathogenesis of the AVM has not been fully elucidated. The predominant theory is that AVMs are congenital in nature and result from incomplete or abnormal resolution of a primitive vascular plexus that occurs during early embryogenesis [2]. One explanation for why they are rarely detected in utero or in infants is that they first appear in utero but then continue to grow after birth. There is also growing evidences for postnatal de novo formation of these lesions [5–7]. Recent studies have identified some of the factors that may be involved in the formation of AVMs. One of them is endothelin-1, found throughout the normal cerebral vasculature, and a potent vasoconstrictor that plays a role in vascular cell growth. Local repression of endothelin-1 within the AVM has been implicated in the pathophysiology underlying AVMs [8]. Endothelial cell–specific tyrosine kinases that are normally found in developing embryonic blood vessels and vascular endothelial growth factor have been shown to be increased in association with AVMs [9].

EPIDEMIOLOGY AND NATURAL HISTORY

Estimates of the prevalence of AVMs range from 0.005% to 0.6% in the general population and are believed to be about one-tenth as common as intracranial aneurysms [2]. The incidence of first ever AVM hemorrhage is 0.51 per 100.000 person-years. AVMs are slightly more common in males and diagnosed at a mean age of 31.2 years [2]. Up to 9% of patients have multiple AVMs and most of these patients have an associated vascular syndrome, such as hereditary hemorrhagic telangiectasia, Wyburn–Mason syndrome, or Sturge–Weber syndrome. While the majority of AVMs are sporadic, familial intracranial AVMs have also been reported [2].

Hemorrhage was the most common manifestation prior to noninvasive imaging and is most commonly located in the brain parenchyma often with intraventricular extension. Isolated intraventricular hemorrhage and subarachnoid hemorrhage may also occur [10]. Cortically based AVMs are more likely to cause subarachnoid hemorrhage [11]. The initial hemorrhage or hemorrhage during follow-up appears to carry a lower morbidity than intracranial hemorrhage from other causes [12,13]. After hemorrhage, seizures are the second most common presenting symptom in about 20–25% of patients [2]. In unruptured AVMs, specific angioarchitectural characteristics, such as location, fistulous component in the nidus, venous outflow stenosis, and the presence of a long pial course of the draining vein, are the strongest predictors of seizures [14].

Since 2000s the detection rates of unruptured AVMs have doubled due to availability and advances in imaging [15,16]. Early retrospective data including previously ruptured AVMs estimated the annual rupture rate at about 4% [17]. A more recent prospective study estimated rupture rates as low as 0.9% per year for unruptured AVMs [18]. On the other hand, annual rupture rates may be as high as 34.4% for AVMs that have ruptured previously and are located deep in the brain with deep venous drainage [18]. A meta-analysis estimated the overall annual rupture rate at 3% with a rate of rupture of 2.2% and 4.5% for unruptured and previously ruptured AVMs, respectively [19]. The risk of rerupture is greatest in the first year after the initial hemorrhage at about 7% [20].

Features that pose increased risk of rupture include previous hemorrhage, particularly within the first year, deep location, deep venous drainage, associated aneurysms along the feeding vessels or within the nidus, location in the posterior fossa or intra- and periventricular, and venous outflow obstruction [18,19,21,22]. The effects of AVM size on hemorrhage risks are controversial [2]. Overall, an annual rupture risk of 2–4% is frequently cited for unruptured AVMs [2].

MANAGEMENT

Management options for patients with unruptured AVM include expectant management, surgery, radiosurgery, endovascular embolization, or a combination thereof [2,13,23]. Surgery and radiosurgery comprise the mainstay of treatment with embolization as a useful preparatory step for either of the two treatment options. Expectant management is indicated for large lesions that are difficult to treat and associated with significant morbidity and mortality.

Surgery

Surgical resection is the gold standard for small, accessible AVMs, and the decision-making process begins with stratification according to the Spetzler–Martin grading system. There is a good correlation of increasing morbidity and mortality with higher Spetzler–Martin grades [2,24]. Surgery accomplishes obliteration rates of 94–100% for Spetzler–Martin grade I to III AVMs, which account for about 60–80% of all AVMs. Surgical resection of small, asymptomatic AVMs, assuming a risk of major neurological morbidity and mortality <6.8%, offers the greatest overall quality of life over time compared to observation or radiosurgery [25]. The overall rate of postoperative mortality and permanent morbidity has been quoted to be 3.3% and 8.6%, respectively. For Spetzler–Martin grade I to III AVMs, mortality and permanent morbidity range from 0% to 3.9% and 0–5%, respectively [4]. It is considerably higher than that for Spetzler–Martin grade IV and V AVMs [4]. There is a paucity of data on obliteration rates for surgery on Spetzler–Martin grade IV and V AVMs alone as most of these lesions are subject to multimodality treatment. Seizure-free rates postsurgical resection range from 43.6% to 81% [4].

Endovascular Embolization

Endovascular embolization as the sole treatment for a cure of an AVM is uncommon and accomplished in only 5–10% of cases as most AVMs have multiple feeders and not all of them can be safely catheterized. In selected cases, palliative embolization may be considered to alleviate symptomatology secondary to vascular steal phenomenon. Embolization is currently applied mostly presurgically or preradiosurgically and is used to occlude deep feeders. Preoperative embolization decreases the morbidity associated with surgery for higher Spetzler–Martin grades to the level of lower grade lesions that are not embolized preoperatively [26]. Some argue that at least 50% of the AVM nidus has to be obliterated with embolization to accomplish a definitive gain during surgical resection [27]. Permanent

morbidity related to presurgical embolization varied from 4% to 8.9% [28]. Embolization prior to radiosurgery is controversial as it may result in decreased effectiveness of radiation. Possible mechanisms include reduced delivery of the radiation dose from radiopaque embolic material, increased angiogenesis due to hypoxia after embolization, and recanalization after embolization with nonadhesive embolic agents [4]. The primary goal is to decrease the size of the nidus, thus reducing the radiation dose necessary [2].

Medical Management With or Without Interventional Therapy for Unruptured Brain Arteriovenous Malformations Trial

The ARUBA trial was a prospective, nonblinded, randomized trial that enrolled 223 adult patients with unruptured AVMs at 39 clinical sites in nine countries and compared interventions (using any treatment modality alone or in combination) to medical management. The trial was stopped prematurely after the primary endpoint of stroke or death was reached and found to be significantly higher in the interventional compared to the observational arm (hazard ratio 0·27, 95% CI 0·14–0·54) [29]. The ARUBA trial, or the "elephant in the room of AVM management" as it has been referred to [30], has sparked animated discussions among all disciplines involved in the care of patients with AVMs. The enrolled cohort had a heterogenous group of unruptured AVMs where >10% of lesions were Spetzler–Martin grade IV AVMs and >25% were grade III AVMs, namely those known to carry higher risks with treatment [31]. Any comparison of an intervention with expectant management is initially in favor of the expectant management as any intervention comes with an upfront risk that may be offset over time due to the natural history of the disease. This is of particular importance for a lifelong disease, such as an untreated AVM. Furthermore, there was a high rate of embolization as the sole treatment (32% of patients), a treatment modality known to be associated with a low rate of obliteration. Only 5% of patients were treated with surgery, the only imminent cure for AVMs, despite two-thirds of the patients harboring Spetzler–Martin grade I and II AVMs, which were amendable to surgical treatment. Neither accurately reflect current AVM management in the United States. Likewise, some well-known and recognized clinical centers have published their experience with ARUBA-eligible patients and found excellent outcomes with surgery in patients with AVMs of Spetzler–Martin grades I and II [32]. One of the studies concluded that the results in ARUBA-eligible patients managed outside the trial led to an entirely different conclusion about AVM intervention, due to the primary role of surgery, judicious surgical selection with established outcome predictors, and technical expertise developed at high-volume AVM centers [33]. Lastly, Clinical equipoise is a necessary element in ensuring unbiased enrollment of patients into any trials. Concerns for lack of equipoise and resultant selective enrollment were the primary objection to participation in the trial and an explanation for the low number of clinical sites in the United States involved [31].

CONCLUSIONS

Given the complexity of decision-making in the management of brain AVMs, it is often of benefit to have a multidisciplinary team for the detailed discussions of risks and benefits of treatment of each modality and of projected combined treatment. In addition, the patient's age and overall medical condition is extremely important in balancing the risks of observation with no treatment compared to the risks of treatment. A team should consist of the surgeon experienced in the removal of brain AVMs, endovascular experts who routinely treat AVMs, and individuals who perform radiosurgery on a regular basis. The ARUBA data presented earlier has not significantly changed our management strategies for patients with brain AVMs. Young patients with low-grade lesions are often treated with surgical excision or embolization followed by surgical excision (Fig. 160.1). Older patients with higher Spetzler–Martin grade lesions are typically followed clinically without intervention.

One major advantage in the treatment of brain AVMs is to have individuals who perform endovascular treatments and open surgical resections. There is a greater appreciation for the detailed angioarchitecture acquired during embolization procedures. Armed

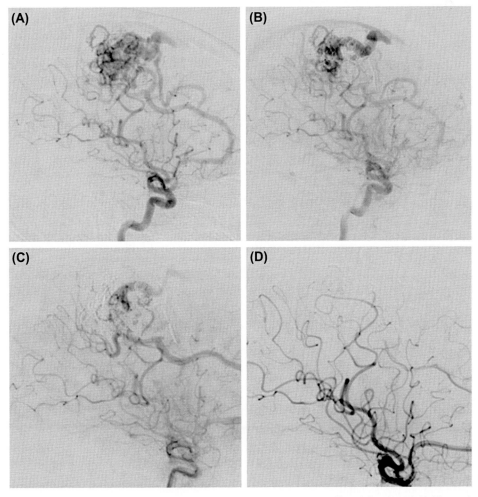

FIGURE 160.1   48-year-old female with right frontal Spetzler–Martin II arteriovenous malformation (panel A). The patient underwent preoperative embolization with ethylene–vinyl alcohol copolymer (Onyx, Covidien–Ev3, Plymouth, MN) at two separate occasions (panels B and C) and resection of the AVM. Postoperative angiogram shows complete resection of the AVM (panel D).

with this information, the surgeon has a much better sense of the anatomy and physiology of the AVM on the day of surgical excision. By having been the individual who treated the patient with embolization on anywhere from one to three previous treatments, the surgeon had time to digest the remaining details of the AVM for surgical excision. While it is difficult to prove how this helps statistically, individuals who perform embolization and then the surgery can attest to the increased safety and efficacy of this type of approach.

If indeed the person performing embolization and the one performing surgical resection are different individuals, there has to be excellent and detailed communication between the two regarding the anatomy and physiology of the AVM. Review session should be undertaken to salute all parties involved to understand the different components of the AVM, where embolization has been performed, and what remains for subsequent surgical or radiosurgical obliteration. Working in a center that manages a high volume of AVM patients facilitates an increased understanding of the pathophysiology and management of brain AVMs.

References

[1] Challa VR, Moody DM, Brown WR. Vascular malformations of the central nervous system. J Neuropathol Exp Neurol 1995; 54(5):609–21.

[2] Harrigan M, Deveikis J. Arteriovenous malformations. In: Handbook of Cerebrovascular Disease and Neurointerventional Technique. New York: Humana Press, Springer; 2009 . p. 511–27.

[3] Lasjaunias PL, Landrieu P, Rodesch G, Alvarez H, Ozanne A, Holmin S, et al. Cerebral proliferative angiopathy: clinical and angiographic description of an entity different from cerebral AVMs. Stroke 2008;39(3):878–85. http://dx.doi.org/10.1161/STROKEAHA.107.493080.

[4] Harrigan M, Deveikis J. Arteriovenous malformations. In: Handbook of Cerebrovascular Disease and Neurointerventional Technique. 2nd ed. New York: Humana Press, Springer; 2013. p. 571–602.

[5] Fujimura M, Kimura N, Ezura M, Niizuma K, Uenohara H, Tominaga T. Development of a de novo arteriovenous malformation after bilateral revascularization surgery in a child with moyamoya disease. J Neurosurg Pediatr 2014;13(6):647–9. http://dx.doi.org/10.3171/2014.3.PEDS13610.

[6] Miller BA, Bass DI, Chern JJ. Development of a de novo arteriovenous malformation after severe traumatic brain injury. J Neurosurg Pediatr 2014;14(4):418–20. http://dx.doi.org/10.3171/2014.7.PEDS1431.

[7] O'shaughnessy BA, Getch CC, Bendok BR, Batjer HH. Microsurgical resection of infratentorial arteriovenous malformations. Neurosurg Focus 2005;19(2):E5.

[8] Rhoten RL, Comair YG, Shedid D, Chyatte D, Simonson MS. Specific repression of the preproendothelin-1 gene in intracranial arteriovenous malformations. J Neurosurg 1997;86(1):101–8. http://dx.doi.org/10.3171/jns.1997.86.1.0101.

[9] Hatva E, Jääskeläinen J, Hirvonen H, Alitalo K, Haltia M. Tie endothelial cell-specific receptor tyrosine kinase is upregulated in the vasculature of arteriovenous malformations. J Neuropathol Exp Neurol 1996;55(11):1124–33.

[10] LeBlanc R, Ethier R, Little JR. Computerized tomography findings in arteriovenous malformations of the brain. J Neurosurg 1979;51(6):765–72. http://dx.doi.org/10.3171/jns.1979.51.6.0765.

[11] Perret G, Nishioka H. Report on the cooperative study of intracranial aneurysms and subarachnoid hemorrhage. Section VI. Arteriovenous malformations. An analysis of 545 cases of cranio-cerebral arteriovenous malformations and fistulae reported to the cooperative study. J Neurosurg 1966;25(4):467–90. http://dx.doi.org/10.3171/jns.1966.25.4.0467.

[12] Choi BS, Park JW, Kim JL, Kim SY, Park YS, Kwon H-J, et al. Treatment Strategy based on multimodal management outcome of Cavernous Sinus Dural Arteriovenous Fistula (CSDAVF). Neurointervention 2011;6(1):6–12. http://dx.doi.org/10.5469/neuroint.2011.6.1.6.

[13] van Beijnum J, Lovelock CE, Cordonnier C, Rothwell PM, Klijn CJM, Al-Shahi Salman R. SIVMS Steering Committee and the Oxford Vascular Study. Outcome after spontaneous and arteriovenous malformation-related intracerebral haemorrhage: population-based studies. Brain J Neurol 2009;132(Pt 2):537–43. http://dx.doi.org/10.1093/brain/awn318.

[14] Shankar JJS, Menezes RJ, Pohlmann-Eden B, Wallace C, terBrugge K, Krings T. Angioarchitecture of brain AVM determines the presentation with seizures: proposed scoring system. AJNR Am J Neuroradiol 2013;34(5):1028–34. http://dx.doi.org/10.3174/ajnr.A3361.

[15] Al-Shahi R, Bhattacharya JJ, Currie DG, Papanastassiou V, Ritchie V, Roberts RC, et al. Scottish intracranial vascular malformation study Collaborators. Prospective, population-based detection of intracranial vascular malformations in adults: the Scottish Intracranial Vascular Malformation Study (SIVMS). Stroke 2003;34(5):1163–9. http://dx.doi.org/10.1161/01.STR.0000069018.90456.C9.

[16] Stapf C, Mast H, Sciacca RR, Berenstein A, Nelson PK, Gobin YP, et al. New York Islands AVM Study Collaborators. The New York Islands AVM Study: design, study progress, and initial results. Stroke 2003;34(5):e29–33. http://dx.doi.org/10.1161/01.STR.0000068784.36838.19.

[17] Ondra SL, Troupp H, George ED, Schwab K. The natural history of symptomatic arteriovenous malformations of the brain: a 24-year follow-up assessment. J Neurosurg 1990;73(3):387–91. http://dx.doi.org/10.3171/jns.1990.73.3.0387.

[18] Stapf C, Mast H, Sciacca RR, Choi JH, Khaw AV, Connolly ES, et al. Predictors of hemorrhage in patients with untreated brain arteriovenous malformation. Neurology 2006;66(9):1350–5. http://dx.doi.org/10.1212/01.wnl.0000210524.68507.87.

[19] Gross BA, Du R. Natural history of cerebral arteriovenous malformations: a meta-analysis. J Neurosurg 2013;118(2):437–43. http://dx.doi.org/10.3171/2012.10.JNS121280.

[20] Halim AX, Johnston SC, Singh V, McCulloch CE, Bennett JP, Achrol AS, et al. Longitudinal risk of intracranial hemorrhage in patients with arteriovenous malformation of the brain within a defined population. Stroke 2004;35(7):1697–702. http://dx.doi.org/10.1161/01.STR.0000130988.44824.29.

[21] da Costa L, Wallace MC, Ter Brugge KG, O'Kelly C, Willinsky RA, Tymianski M. The natural history and predictive features of hemorrhage from brain arteriovenous malformations. Stroke 2009;40(1):100–5. http://dx.doi.org/10.1161/STROKEAHA.108.524678.

[22] Hernesniemi JA, Dashti R, Juvela S, Väärt K, Niemelä M, Laakso A. Natural history of brain arteriovenous malformations: a long-term follow-up study of risk of hemorrhage in 238 patients. Neurosurgery 2008;63(5):823–9. http://dx.doi.org/10.1227/01.NEU.0000330401.82582.5E. discussion 829–831.

[23] Hartmann A, Mast H, Mohr JP, Pile-Spellman J, Connolly ES, Sciacca RR, et al. Determinants of staged endovascular and surgical treatment outcome of brain arteriovenous malformations. Stroke 2005;36(11):2431–5. http://dx.doi.org/10.1161/01.STR.0000185723.98111.75.

[24] Spetzler RF, Martin NA. A proposed grading system for arterio-venous malformations. J Neurosurg 1986;65(4):476–83. http://dx.doi.org/10.3171/jns.1986.65.4.0476.

[25] McInerney J, Gould DA, Birkmeyer JD, Harbaugh RE. Decision analysis for small, asymptomatic intracranial arteriovenous malformations. Neurosurg Focus 2001;11(5):e7.

[26] Jafar JJ, Davis AJ, Berenstein A, Choi IS, Kupersmith MJ. The effect of embolization with N-butyl cyanoacrylate prior to surgical resection of cerebral arteriovenous malformations. J Neurosurg 1993;78(1):60–9. http://dx.doi.org/10.3171/jns.1993.78.1.0060.

[27] Viñuela F, Dion JE, Duckwiler G, Martin NA, Lylyk P, Fox A, et al. Combined endovascular embolization and surgery in the management of cerebral arteriovenous malformations: experience with 101 cases. J Neurosurg 1991;75(6):856–64. http://dx.doi.org/10.3171/jns.1991.75.6.0856.

[28] Castel JP, Kantor G. [Postoperative morbidity and mortality after microsurgical exclusion of cerebral arteriovenous malformations. Current data and analysis of recent literature]. Neurochirurgie 2001;47(2–3 Pt 2):369–83.

[29] Mohr JP, Parides MK, Stapf C, Moquete E, Moy CS, Overbey JR, et al. Medical management with or without interventional therapy for unruptured brain arteriovenous malformations (ARUBA): a multicentre, non-blinded, randomised trial. Lancet 2014;383(9917):614–21. http://dx.doi.org/10.1016/S0140-6736(13)62302-8.

[30] Van Loveren H. In: Current Treatments of AVM's (66th Southern Neurosurgical Society Annual Meeting). March 2015.

[31] Amin-Hanjani S. Aruba results are not applicable to all patients with arteriovenous malformation. Stroke J Cereb Circ 2014;45(5):1539–40. http://dx.doi.org/10.1161/STROKEAHA.113.002696.

[32] Nerva JD, Mantovani A, Barber J, Kim LJ, Rockhill JK, Hallam DK, et al. Treatment outcomes of unruptured arteriovenous malformations with a subgroup analysis of ARUBA (A Randomized Trial of Unruptured Brain Arteriovenous Malformations)-eligible patients. Neurosurgery 2015;76(5):563–70. http://dx.doi.org/10.1227/NEU.0000000000000663. discussion570; quiz 570.

[33] Rutledge WC, Abla AA, Nelson J, Halbach VV, Kim H, Lawton MT. Treatment and outcomes of ARUBA-eligible patients with unruptured brain arteriovenous malformations at a single institution. Neurosurg Focus 2014;37(3):E8. http://dx.doi.org/10.3171/2014.7.FOCUS14242.

CHAPTER

161

Surgical and Endovascular Management of Vertebrobasilar Atherosclerosis

D.M. Panczykowski, T.G. Jovin

University of Pittsburgh Medical Center, Pittsburgh, PA, United States

INTRODUCTION

Posterior circulation stroke and transient ischemic attack (TIA) constitute 20% of all cerebrovascular ischemic events, with vertebrobasilar atherosclerosis (VBA), the primary etiology in up to 35% [1]. Aggressive medical therapy may reduce the risk of recurrence; however, 30-day and 1-year recurrent stroke rates may be higher than 30%, and nearly 60% in the setting of hemodynamic insufficiency [2]. Although use of antiplatelet agents and strict control of atherosclerotic risk factors are the mainstay of treatment for VBA, some patients remain refractory to aggressive medical therapy necessitating revascularization. Careful analysis of lesion morphology and hemodynamics are necessary for both effective treatment selection and the provision of technically safe procedures that significantly lower the patient's risk of future stroke.

GENERAL FEATURES OF VERTEBROBASILAR ATHEROSCLEROSIS

Epidemiology and Natural History

Atherosclerotic disease of the vertebrobasilar system accounts for 25–35% of ischemic events in the posterior circulation [3]. Patients presenting with posterior

Primer on Cerebrovascular Diseases, Second Edition
http://dx.doi.org/10.1016/B978-0-12-803058-5.00161-2

circulation TIA or stroke in the presence VBA are at significantly greater risk of developing recurrent events than symptomatic lesions affecting the anterior lesion [3]. A review of patients suffering VBA found 46% of patients presenting with stroke or TIA suffered a recurrent event during the following 23 months, while a systematic review of symptomatic VBA found recurrent stroke risk to be as high as 33% within the first 90-days after initial event [3,4].

Anatomical Features of the Vertebrobasilar System

Each vertebral artery (VA) arises as the first branch from their respective subclavian artery. The extraosseous segment (V1) of the VA extends from the origin, behind the anterior scalene, to its entrance of the transverse foramen of the 6th cervical vertebra. The foraminal segment (V2) courses through the transverse foramina of C6 through C1, providing segmental branches to the spinal cord and musculature until it emerges beside the lateral mass of the atlas. The extraspinal segment (V3) then forms a posterior-medial loop around the atlantooccipital articulation, coursing on the superior surface of the arch of the atlas. The intracranial segment (V4) begins where the artery penetrates the dura, and fuses with the contralateral VA to form the basilar artery, ventral to the pontomedullary junction.

The normal diameter of the VA lumen is 3–5 mm, with unilateral hypoplasia of the right VA (diameter ≤2 mm) in 75–85% of cases where the vertebral arteries are asymmetric with respect to size [5]. The basilar artery is typically 3–4 mm in diameter and provides nearly 20 paramedian and circumflex perforating arteries to the pons and midbrain. The vertebrobasilar circulation irrigates the entire brain stem, cerebellum, occipital lobes, and part of the temporal lobes.

The intracranial arteries lack external elastic lamina, have thin adventitia, and are suspended within the leptomeninges, making them susceptible to injury from both endoluminal and microsurgical manipulation. Intracranial arteries also have well-developed internal elastic lamina and circularly arranged smooth muscle which can be damaged by over-dilation, possibly contributing to greater recoil following angioplasty. Numerous perforating arteries arise from the vertebrobasilar system that can be damaged and/or occluded by displacement of the atheromatous plaque during angioplasty and stent deployment or during microsurgical dissection and vessel manipulation.

There are several collateral pathways to the vertebrobasilar system that may compensate for flow-limiting disease, including leptomeningeal collaterals, persistent carotid-vertebrobasilar anastomoses, and external carotid artery (ECA)-segmental VA anastomoses.

However, the existence and hemodynamic capacity of these collaterals greatly varies between individuals. Anatomical variants that could potentially limit compensatory flow to the vertebrobasilar territory in the setting of VBA include hypoplasia or absence of the contralateral VA, posterior communicating arteries, and P1 segment of the PCAs [5].

Pathophysiological Features of Vertebrobasilar Atherosclerosis

VBA produces ischemic stroke through three primary mechanisms, (1) distal embolization, (2) occlusive disease at perforator ostia, and (3) hemodynamic failure [6]. Coexistence of embolism and hemodynamic failure portends a particularly poor prognosis [7,8]. Beyond conventional atherosclerotic risk factors, several pathophysiological characteristics contribute to the risk of stroke recurrence in VBA and have serious implications regarding tailoring treatment to the underlying disease.

The morphology of atherosclerotic lesions plays a significant role in both risk and timing of recurrent stroke. Thromboembolism from unstable atheroma rupture is considered to be an important mechanism of stroke in VBA [9]. Radiographic patterns suggestive of a thromboembolic mechanism include branch-vessel territory infarcts and/or perforator infarcts adjacent to the plaque. The existence of a period of thromboembolic lability in VBA has been supported by detection of emboli distal to the stenosis in patients presenting after stroke or TIA [9]. A systematic review investigating the timing of recurrent stroke in symptomatic VBA found the risk of recurrence within 30 days to be 33% suggesting a period of increased thromboembolic lability or hemodynamic instability in the weeks following plaque rupture [10]. Complicating intervention, unstable atherosclerotic lesions have also been cited as potential causes of the high periprocedural stroke and death rates that occur with early endovascular therapy for VBA [11].

Severity of stenosis is a traditional surrogate for hemodynamic status and has been associated with increased risk of stroke recurrence. The Oxford Vascular Study group found stenosis ≥50% in VBA to be associated with 3 times greater risk of recurrent stroke and/or TIA (up to 46% of patients) independent of cardiovascular risk factors, while the WASID study found twice the risk of stroke recurrence in those with >70% stenosis [4,12]. Conversely, other studies have found that stenosis >70% was not associated with risk of stroke recurrence, suggesting stenosis alone might be an insufficient surrogate of hemodynamic impairment [13]. The GESICA study analyzed patients with ≥50% stenosis and found that in patients suspected of having hemodynamic insufficiency based on symptomatology (multiple recurrent events, dependency of symptoms on body position, exercise, and so on) 61% developed recurrent TIA or stroke (vs. 32% of those without) [14]. The VERiTAS Study Group evaluated distal flow status in VBA using phase-contrast quantitative MRA (QMRA) and found that although parent vessel flow diminishes above 80% stenosis, hemodynamic insufficiency in the vascular territory supplied by the stenosed vessel was not well predicted by severity of stenosis [15]. These data support the idea that the hemodynamic consequences of high-grade intracranial stenosis are primarily determined by the presence and vigor of posterior circulation collaterals.

MANAGEMENT OF VERTEBROBASILAR ATHEROSCLEROSIS

Medical Management

Medical treatment goals for intracranial atherosclerosis have primarily been derived from the WASID, SAMMPRIS, and VISSIT trials [12,16,17]. Extrapolating from these, primary therapy for symptomatic VBA should include dual antiplatelet agents early after index event (e.g., aspirin and a thienopyridine or ticagrelor for 6 months), tight blood pressure control (≤130/80 mm Hg), LDL cholesterol control (≤70 mg/dL), excellent control of blood sugar, smoking cessation, and exercise. Despite institution of aggressive medical management demonstrating significant improvement in cardiovascular risk factors, the medical arms of SAMMPRIS and VISSIT trials still suffered a 12–15% annual risk of stroke in the first year after index event [16,17]. One reason for this is that medical management of VBA only addresses thromboembolic-mediated symptomatology, not the coexistent hemodynamic compromise that potentiates plaque remodeling and contributes to recurrent ischemia. As such, the hemodynamic status of the territory distal to the stenotic lesion should be evaluated in all patients with symptomatic VBA. Utilizing QMRA, Amin-Hanjani et al. reported that patients with diminished distal flow status demonstrated a significantly worse stroke-free survival at 24 months compared with the normal distal flow status group (71% vs. 100%, respectively) [2]. Distal flow status also remained an independent predictor of recurrent stroke regardless of stenosis severity and location of disease. Lesions that do not result in hemodynamic compromise are expected to benefit most from medical therapy, while those with stenosis of greater than 50%, presenting with recent symptom onset, and evidence of hemodynamic impairment are more likely to benefit from revascularization.

ENDOVASCULAR REVASCULARIZATION

Endoluminal revascularization addresses both hemodynamic- and thromboembolic-mediated disease processes in VBA. Angioplasty immediately improves regional blood flow and decreases in situ thrombosis and thromboembolism due to platelet activation from high shear stress. Long term, it promotes smooth neo-intimal growth preventing plaque ulceration and re-rupture. Despite these benefits, significant limitations exist including difficulties in device delivery with its inherent risks, vessel recoil or dissection, and acute thrombosis. As well, excessive neointimal hyperplasia with consequent risk of restenosis remains a significant long-term concern. However, with appropriate patient/lesion selection and technique, angioplasty with or without stenting may ultimately prove to be an effective treatment of VBA refractory to medical therapy.

Percutaneous Transluminal Angioplasty

Sundt et al. first reported percutaneous transluminal angioplasty (PTA) for intracranial atherosclerosis in 1980 [18]. Early PTA suffered from frequent technical complications including dissection, immediate elastic recoil, and delayed restenosis. Introduction of the concept of submaximal angioplasty led to a decrease in dissection rates from 75% to 14%, as well as reduced rates of residual stenosis and delayed restenosis [19]. Both shorter lesion length (<5 mm vs. >10 mm) and concentric plaque geometry (vs. eccentric or tortuous morphology) are important predictors of immediately successful PTA and risk of delayed restenosis [20]. Modern patient selection and angioplasty techniques have demonstrated more favorable perioperative stroke rates <8%, 1-year stroke rates of <6%, and long-term stroke rates of 2–3% at a mean follow-up of >40 months [21,22].

Concerns over dissection and restenosis along with mounting data from the coronary literature on improved outcome through the use of stents deflected interest in PTA alone in favor of PTA with stent placement; however, no randomized trial has yet compared these techniques. A retrospective comparison by Siddiq et al. demonstrated restenosis-free survival at 12 months of 68% in the PTA group and 64% for the stent-treated group with no differences in the rate of stroke or death at 2 years [23]. Discouraging results of recent trials evaluating intracranial stenting have led to renewed interest in PTA alone. Angioplasty alone remains an effective option in the setting of short, concentric atherosclerotic lesions, and may be the optimal choice for vasculature or lesions difficult to navigate larger devices across or when patients cannot tolerate dual antiplatelet therapy.

Self-Expanding Stents

At present, the only FDA-approved stent for the management of intracranial atherosclerosis is the self-expanding Wingspan stent (Stryker Neurovascular, Fremont, California, USA). Results from prospective registries were initially encouraging; however, the prospective, randomized SAMMPRIS trial demonstrated that aggressive medical therapy alone was superior to percutaneous transluminal angioplasty and stenting (PTAS) at 30 days and 1 year (rate of stroke or death ~6% vs. 15% and ~12% vs. 20%, respectively) [16]. This outcome discrepancy has been attributed to technical drawbacks resulting in a high rate of periprocedural complications. Additionally, patient selection (TIA or nondisabling stroke within 30 days of enrollment attributable to 70–99% stenosis) did not reflect the nuances of intracranial atherosclerosis evaluation and management. A large segment of the patient population presenting with symptomatic atherosclerotic disease were excluded from enrollment (i.e., patients presenting with disabling neurological deficit and/or unstable clinical course). Furthermore, significant time delays existed from index event to treatment (median time from qualifying event to randomization was 1 week, with revascularization being performed within 3 days of treatment assignment). Contrary to these study parameters, experience from the coronary literature suggests patients most likely to benefit from revascularization are those with severe symptoms and/or evidence of hemodynamic impairment in the affected vascular territory (evidenced by an unstable clinical course), especially when revascularized early after the index event [24]. Therefore, the efficacy of intracranial stenting remains to be truly elucidated. Despite these flaws, the SAMMPRIS trial did provide information regarding the efficacy of modern, aggressive medical therapy which produced a stroke rate nearly half of that seen in WASID (12% vs. 25%, respectively) [12,16]. Currently, the FDA only recommends Wingspan be used for patients meeting all of the following criteria:

- Age between 22 and 80 years.
- Intracranial stenosis of 70–99% due to atherosclerosis.
- Two or more strokes despite aggressive medical management.
- Most recent stroke occurring >7 days prior to planned stenting.

Alternative self-expanding stents have been evaluated since SAMMPRIS; with design differences being reduced radial force and improved device deliverability. A retrospective registry evaluating PTAS with the Enterprise stent (Cordis Neurovascular) demonstrated a periprocedural stroke rate of 7% with no further TIAs or strokes during a median follow-up of 26 months [25]. Retrospective review of the Solitaire self-expanding stent (ev3, Irvine, California, USA) found the risk of periprocedural stroke to be 9%, while recurrent ischemic events in the territory treated occurred in 5% during the mean follow-up period of 26 months [26].

Balloon-Mounted Stents

Balloon-mounted stents (BMSs) originally developed for cardiac interventions have also been applied with success to the treatment of VBA. This modality combines the advantages of both PTA and stenting, including immediate protection from recoil and dissection from PTA, as well as minimizing manipulation of multiple devices across the lesion (thought to be responsible for the high rate of periprocedural intracerebral hemorrhages noted in SAMMPRIS). BMSs have demonstrated less residual stenosis when compared to self-expanding stents (16% vs. 35%, respectively) [27,28]. Despite technical concerns regarding navigability, initial retrospective series have reported high technical success with comparable complication rates (0–26%) [29]. Specific to symptomatic VBA, Gomez et al. reviewed their use of the coronary BMS (Microstent II or GFX, Advanced Vascular Engineering, Inc. or Multilink Duet, Guidant, Inc.) [30]. Average stenosis was reduced from 71% to 10% without the occurrence of periprocedural stroke. Mazighi et al. reviewed consecutive symptomatic patients with intracranial atherostenotic lesions who were treated with BMS [31]. Their technical success rate was 99%, with a median reduction in stenosis from 85% to 0%. The 30-day stroke or death rate was 10%, and a TIA/stroke rate of <6% at 24 months. The first major prospective trial evaluating BMS specifically designed for intracranial circulation was the SSYLVIA study [32]. Technical success was achieved in 95% of cases, with a 30-day stroke rate of 7%; long-term stroke risk in the territory treated was 7% after 1 year. The VISSIT trial was a randomized trial comparing the BMS Vitesse against aggressive medical management alone for treatment of symptomatic intracranial stenosis >70% [17]. This trial was halted due to futility after interim analyses produced similar findings to those in SAMMPRIS. Like SAMMPRIS this study suffered from numerous methodological flaws, including a high crossover rate (17% of the medical arm suffered recurrent TIA or stroke and underwent revascularization), low technical success (<80%), and a lack of stratification for either lesion morphology or hemodynamic symptomatology.

Because restenosis remains a major concern, drug-eluting stents (DESs) have also been evaluated for treatment of intracranial atherostenotic lesions. In a case series that included 26 patients with intracranial lesions, Gupta et al. reported successful, balloon-mounted delivery of DES in all lesions, and a 6-month restenosis rate of 5% [33].

BMS remains a valid treatment option for patients suffering symptomatic VBA refractory to medical management with hemodynamic insufficiency and plaque morphology too complex for PTA alone. There is a continued need for technique as well as device improvement, with robust comparisons against current treatment modalities through prospective randomized trials.

Complications of Endovascular Treatment

Complications of angioplasty and stenting typically involve intraplaque dissection, intraplaque or distal branch wire perforation, plaque dislodgment, perforator occlusion, and restenosis due to vessel recoil and/or excessive in-stent neointimal growth. Periprocedural stroke is the most significant contributor to morbidity and mortality following endovascular VBA therapy, occurring in about 5–10% [13]. Periprocedural stroke typically occurs within 24 h of the procedure in the setting of unstable, ulcerated plaques and is most commonly due to occlusion of perforating arteries through compression of plaque into the arterial ostia ("snowplowing") and/or through embolization of plaque debris. An analysis by Levy et al. suggested that >90% stenosis, plaques >10 mm, unstable plaques undergoing intervention within 4 weeks of an ischemic event, tortuosity of vessels, thrombus within lesions, calcified plaques, perforator-rich zones, and other significant medical comorbidities all contribute to increased periprocedural stroke risk [34]. The presence of perforator infarcts on preprocedural MRI adjacent to the target lesion has also been associated with higher periprocedural stroke risk [35].

In-stent restenosis develops due to intimal hyperplasia and vessel remodeling and may occur in up to 40% of patients, reaching maximum severity between 3 and 6 months. Patient-related factors such as diabetes, as well as lesion-specific factors such as small vessel diameter and postprocedure, residual stenosis >30% have all been predictive of restenosis [34]. The use of DES may potentially improve rates of restenosis with long-term hemodynamically significant restenosis (≥50% luminal diameter) occurring in 0–5% at 5 years, as compared with up to 32% following placement of bare metal stents [33,34,36,37].

MICROSURGICAL REVASCULARIZATION

Indications for Microsurgical Revascularization

Despite advancements in medical and endovascular therapy for VBA, microsurgical revascularization remains a consideration for a subset of patients. Microsurgical revascularization for VBA may be performed through various procedures including in situ or graft interposition bypass, endarterectomy, or segment excision with re-anastomosis. Indications for microsurgical revascularization include symptomatic VBA refractory to maximal medical therapy due to hemodynamic phenomena that is not amenable to endovascular treatment as a result of access difficulties, severe vessel tortuosity, and/or vessel occlusion.

The success and durability of revascularization through bypass for stroke prevention in VBA has been evaluated in various retrospective series, with long-term patency rates of 78–100% and stroke-free survival rates of 92–98% (follow-up 9 months to 5 years) [38–40]. Although these series have shown improvement in symptoms and treatment durability, the morbidity and mortality associated with microsurgical revascularization (up to 20% and 8%, respectively) demands caution. Nonetheless, advances in preoperative evaluation, microsurgical technique, and perioperative neurointensive care since these early publications allow posterior circulation revascularization to be successfully undertaken in select patients without other options for treatment. However, the absence of any control patients in all case series reported thus far does not allow any conclusions to be drawn regarding superiority of this approach compared to aggressive medical therapy. Similar to endovascular treatment, this approach is reserved for patients who have repeatedly failed maximal medical therapy and for whom no other treatment options are thought to be available.

Bypass Options

Bypasses used for revascularization in symptomatic VBA can be categorized as in situ bypass or interposition graft bypass; the choice of which depends on the flow replacement needed and anatomy of the anastomotic site. Anastomotic target vessels in the treatment of VBA are typically one of four recipient arteries in the posterior circulation: posterior cerebral artery, superior cerebellar artery, anterior inferior cerebellar artery, and posterior inferior cerebellar artery. In situ bypasses are low-flow bypasses using pedicled arterial grafts as donor vessels such as the superficial temporal or occipital arteries. Interposition bypasses utilize radial artery or saphenous vein grafts (SVG), as high-flow conduits constructed between recipient vessels and either the ipsilateral VA or ECA. Phase-contrast quantitative magnetic resonance angiography can help preoperatively determine the perfusion replacement necessary, while intraoperative blood flow measurements made using a microvascular ultrasonic flow probe (Charbel Micro-Flow Probe, Transonic Systems, Inc.) and are helpful with confirming appropriate graft selection and predicting long-term graft patency.

CONCLUSION

VBA is a significant cause of posterior circulation stroke and carries serious risk for recurrent stroke despite aggressive medical therapy, the primary treatment approach to this disease. Atherosclerotic plaque morphology and hemodynamic features are significant determinants of risk and timing of stroke recurrence, as well as selection of the optimal treatment modality for medically refractory VBA. Microsurgical revascularization via bypass techniques is an alternative treatment option for lesions not amenable to endovascular intervention due to lesion complexity, tortuosity of vasculature, and/or vessel occlusion. Angioplasty alone is a viable treatment approach in the setting of short, concentric atherosclerotic lesions, gaining renewed interest following negative stenting trials. Stenting in addition to PTA permits treatment of longer, irregular atherosclerotic plaques at higher risk of dissection, restenosis, and/or those with unsatisfactory results following angioplasty. Robust trials are still needed to define the subgroups that respond best to each treatment modality, as well as refine current treatment techniques and develop more effective therapeutic modalities.

References

[1] Labropoulos N, Nandivada P, Bekelis K. Stroke of the posterior cerebral circulation. Int Angiol 2011;30(2):105–14.

[2] Amin-Hanjani S, Du X, Zhao M, Walsh K, Malisch TW, Charbel FT. Use of quantitative magnetic resonance angiography to stratify stroke risk in symptomatic vertebrobasilar disease. Stroke 2005;36(6):1140–5.

[3] Marquardt L, Kuker W, Chandratheva A, Geraghty O, Rothwell PM. Incidence and prognosis of > or =50% symptomatic vertebral or basilar artery stenosis: prospective population-based study. Brain 2009;132(Pt 4):982–8.

[4] Gulli G, Marquardt L, Rothwell PM, Markus HS. Stroke risk after posterior circulation stroke/transient ischemic attack and its relationship to site of vertebrobasilar stenosis: pooled data analysis from prospective studies. Stroke 2013;44(3):598–604.

[5] Thierfelder KM, Baumann AB, Sommer WH, et al. Vertebral artery hypoplasia: frequency and effect on cerebellar blood flow characteristics. Stroke 2014;45(5):1363–8.

[6] Flossmann E, Rothwell PM. Prognosis of vertebrobasilar transient ischaemic attack and minor stroke. Brain 2003;126(Pt 9):1940–54.

[7] Caplan LR. When should heparin be given to patients with atrial fibrillation-related embolic brain infarcts? Arch Neurol 1999;56(9):1059–60.

[8] Hen.erici M. Improving the outcome of acute stroke management. Hosp Med 1999;60(1):44–9.

[9] Markus HS, van der Worp HB, Rothwell PM. Posterior circulation ischaemic stroke and transient ischaemic attack: diagnosis, investigation, and secondary prevention. Lancet Neurol 2013;12(10):989–98.

[10] Gulli G, Khan S, Markus HS. Vertebrobasilar stenosis predicts high early recurrent stroke risk in posterior circulation stroke and TIA. Stroke 2009;40(8):2732–7.

[11] Alexander MD, Rebhun JM, Hetts SW, et al. Lesion location, stability, and pretreatment management: factors affecting outcomes of endovascular treatment for vertebrobasilar atherosclerosis. J Neurointerv Surg 2016;8(5):466–70. http://dx.doi.org/10.1136/neurintsurg-2014-011633, [Epub 2015 March 20].

[12] Chimowitz MI, Lynn MJ, Howlett-Smith H, et al. Comparison of warfarin and aspirin for symptomatic intracranial arterial stenosis. N Engl J Med 2005;352(13):1305–16.

[13] Abuzinadah AR, Alanazy MH, Almekhlafi MA, et al. Stroke recurrence rates among patients with symptomatic intracranial vertebrobasilar stenoses: systematic review and meta-analysis. J Neurointerv Surg 2016;8(2):112–6. http://dx.doi.org/10.1136/neurintsurg-2014-011458, [Epub 2015 December 11].

[14] Mazighi M, Tanasescu R, Ducrocq X, et al. Prospective study of symptomatic atherothrombotic intracranial stenoses: the GESICA study. Neurology 2006;66(8):1187–91.

[15] Amin-Hanjani S, Du X, Rose-Finnell L, et al. Hemodynamic features of symptomatic vertebrobasilar disease. Stroke 2015; 46(7):1850–6.

[16] Chimowitz MI, Lynn MJ, Derdeyn CP, et al. Stenting versus aggressive medical therapy for intracranial arterial stenosis. N Engl J Med 2011;365(11):993–1003.

[17] Zaidat OO, Fitzsimmons BF, Woodward BK, et al. Effect of a balloon-expandable intracranial stent vs medical therapy on risk of stroke in patients with symptomatic intracranial stenosis: the VISSIT randomized clinical trial. JAMA 2015;313(12):1240–8.

[18] Sundt Jr TM, Smith HC, Campbell JK, Vlietstra RE, Cucchiara RF, Stanson AW. Transluminal angioplasty for basilar artery stenosis. Mayo Clin Proc 1980;55(11):673–80.

[19] Connors 3rd JJ, Wojak JC. Percutaneous transluminal angioplasty for intracranial atherosclerotic lesions: evolution of technique and short-term results. J Neurosurg 1999;91(3):415–23.

[20] Mori T, Fukuoka M, Kazita K. Follow-up study after intracranial percutaneous transluminal cerebral balloon angioplasty. Interv Neuroradiol 2000;6(Suppl. 1):243–9.

[21] Cruz-Flores S, Diamond AL. Angioplasty for intracranial artery stenosis. Cochrane Database Syst Rev 2006;(3):CD004133.

[22] Marks MP, Marcellus ML, Do HM, et al. Intracranial angioplasty without stenting for symptomatic atherosclerotic stenosis: long-term follow-up. AJNR Am J Neuroradiol 2005;26(3): 525–30.

[23] Siddiq F, Vazquez G, Memon MZ, et al. Comparison of primary angioplasty with stent placement for treating symptomatic intracranial atherosclerotic diseases: a multicenter study. Stroke 2008;39(9):2505–10.

[24] Coronary Revascularization Writing G, Patel MR, Dehmer GJ, et al. ACCF/SCAI/STS/AATS/AHA/ASNC/HFSA/SCCT 2012 appropriate use criteria for coronary revascularization focused update: a report of the American College of Cardiology Foundation Appropriate Use Criteria Task Force, Society for Cardiovascular Angiography and Interventions, Society of Thoracic Surgeons, American Association for Thoracic Surgery, American Heart Association, American Society of Nuclear Cardiology, and the Society of Cardiovascular Computed Tomography. J Thorac Cardiovasc Surg 2012;143(4):780–803.

[25] Feng Z, Duan G, Zhang P, et al. Enterprise stent for the treatment of symptomatic intracranial atherosclerotic stenosis: an initial experience of 44 patients. BMC Neurol 2015;15(1):187.

[26] Duan G, Feng Z, Zhang L, et al. Solitaire stents for the treatment of complex symptomatic intracranial stenosis after antithrombotic failure: safety and efficacy evaluation. J Neurointerv Surg 2016;8(7):680–4.http://dx.doi.org/10.1136/neurintsurg-2015-011734, [Epub 2015 June 3].

[27] Miao ZR, Feng L, Li S, et al. Treatment of symptomatic middle cerebral artery stenosis with balloon-mounted stents: long-term follow-up at a single center. Neurosurgery 2009;64(1):79–84. [discussion 84–75].

[28] Kurre W, Berkefeld J, Sitzer M, Neumann-Haefelin T, du Mesnil de Rochemont R. Treatment of symptomatic high-grade intracranial stenoses with the balloon-expandable Pharos stent: initial experience. Neuroradiology 2008;50(8):701–8.

[29] Groschel K, Schnaudigel S, Pilgram SM, Wasser K, Kastrup A. A systematic review on outcome after stenting for intracranial atherosclerosis. Stroke 2009;40(5):e340–347.

[30] Gomez CR, Misra VK, Liu MW, et al. Elective stenting of symptomatic basilar artery stenosis. Stroke 2000;31(1):95–9.

[31] Mazighi M, Yadav JS, Abou-Chebl A. Durability of endovascular therapy for symptomatic intracranial atherosclerosis. Stroke 2008;39(6):1766–9.

[32] Investigators SS. Stenting of Symptomatic Atherosclerotic Lesions in the Vertebral or Intracranial Arteries (SSYLVIA): study results. Stroke 2004;35(6):1388–92.

[33] Gupta R, Al-Ali F, Thomas AJ, et al. Safety, feasibility, and short-term follow-up of drug-eluting stent placement in the intracranial and extracranial circulation. Stroke 2006;37(10):2562–6.

[34] Levy EI, Turk AS, Albuquerque FC, et al. Wingspan in-stent restenosis and thrombosis: incidence, clinical presentation, and management. Neurosurgery 2007;61(3):644–50. [discussion 650–641].

[35] Jiang WJ, Srivastava T, Gao F, Du B, Dong KH, Xu XT. Perforator stroke after elective stenting of symptomatic intracranial stenosis. Neurology 2006;66(12):1868–72.

[36] Gupta R, Schumacher HC, Mangla S, et al. Urgent endovascular revascularization for symptomatic intracranial atherosclerotic stenosis. Neurology 2003;61(12):1729–35.

[37] Park S, Lee DG, Chung WJ, Lee DH, Suh DC. Long-term outcomes of drug-eluting stents in symptomatic intracranial stenosis. Neurointervention 2013;8(1):9–14.

[38] Britz GW, Agarwal V, Mihlon F, et al. Radial artery bypass for intractable vertebrobasilar insufficiency: case series and review of the literature. World Neurosurg 2015;8:106–13. http://dx.doi.org/ 10.1016/j.wneu.2015.08.004, [Epub 2015 August 15].

[39] Ausman JI, Diaz FG, Vacca DF, Sadasivan B. Superficial temporal and occipital artery bypass pedicles to superior, anterior inferior, and posterior inferior cerebellar arteries for vertebrobasilar insufficiency. J Neurosurg 1990;72(4):554–8.

[40] Hopkins LN, Budny JL. Complications of intracranial bypass for vertebrobasilar insufficiency. J Neurosurg 1989;70(2):207–11.

CHAPTER

162

Extracranial–Intracranial Bypass Procedures

Z.A. Hage, S. Amin-Hanjani

University of Illinois at Chicago, Chicago, IL, United States

INTRODUCTION

Extracranial–intracranial (EC–IC) bypass procedures have played an important role in the treatment of cerebrovascular disease since their development in the 1960s. They serve two main purposes [1]: (1) flow replacement when managing challenging aneurysms or tumors requiring cerebral vessel sacrifice and (2) flow augmentation to treat cerebral ischemia mainly in the setting of atherosclerotic occlusive disease and moyamoya disease. In both settings, a new conduit for blood flow is established by suturing a graft vessel surgically to the intracranial circulation, thus providing revascularization to the brain. Over the years, the perfection of microsurgical anastomosis techniques in combination with significant advancements in imaging technology have molded the thought process involved in surgical cerebral revascularization using bypass. Moreover, important studies, such as the

1985 EC–IC bypass study as well as the 2011 Carotid Occlusion Surgery Study (COSS), have had significant impact on the use of EC–IC bypass for flow augmentation, particularly in the setting of carotid atheroocclusive disease. In this chapter, we briefly discuss indications, patient selection, hemodynamic assessment techniques, and bypass options pertaining to flow augmentation and flow replacement EC–IC bypass.

EC–IC BYPASS FOR FLOW REPLACEMENT

The concept of flow replacement is important when planning treatment of complex aneurysms where parent vessel sacrifice is a necessity, or when confronted with skull base tumors engulfing surrounding vasculature. When considering the carotid artery, for example, although a majority of patients may withstand occlusion, up to 30% may suffer a stroke if this vessel is

Primer on Cerebrovascular Diseases, Second Edition
http://dx.doi.org/10.1016/B978-0-12-803058-5.00162-4

sacrificed. This is primarily due to poor collateral vascular supply. Consequently, preoperative testing prior to carotid sacrifice is mandatory in order to identify those patients who would not safely tolerate carotid occlusion and would therefore need EC–IC bypass to preserve their cerebral blood flow (CBF). Patients at risk are identified using endovascular balloon occlusion testing (BOT), during which the patient's response to temporary carotid artery occlusion can be studied based on one or a combination of neurological, angiographic, electroencephalographic, and perfusion criteria. If the BOT is failed, flow replacement is needed and revascularization should be performed prior to vessel sacrifice. If the BOT is failed based on clinical criteria, a higher flow bypass is generally needed, while if it is failed based on subclinical perfusion deficits, a lower flow bypass may be adequate. A minority of specialists advocate revascularization in all cases regardless of BOT results, as they are concerned regarding the possibility of false-negative BOT or that younger patients may be at higher risk for development of contralateral aneurysms over time due to increased hemodynamic stress on collateral vessels. The sacrifice of more distal cerebral vessels, such as the middle cerebral artery or anterior cerebral artery branches, typically requires flow replacement prior to sacrifice, since these are end vessels and the collateral flow from the circle of Willis is lacking. Rarely, leptomeningeal collaterals may be adequate to avert major strokes, but it is unlikely that acute sacrifice of a major intracranial vessel can be tolerated without ischemia, and thus warrants bypass preemptively.

When the decision has been made to replace flow, different donor vessels with varying carrying capacities are available for bypass. Typically larger conduits can provide greater flows. Traditionally, when the need for high flow is anticipated, large conduits capable of greater carrying capacities, such as interposition vein grafts (saphenous vein), are employed. For intermediate flow bypass, the radial artery can be used as an interposition graft. For low flow bypass, the superficial temporal artery (STA) or occipital artery can be utilized as donor vessels, although the STA can provide flows as high as 100 mL/min. The use of the STA as a donor graft is most desirable as it requires only one anastomosis between the donor and the cerebral vessel recipient. Use of vein or radial artery grafts requires not only an intracranial anastomosis but also an extracranial anastomosis to the source of blood supply, typically the cervical carotid artery. Thus, additional surgery to harvest the graft and to expose the cervical carotid is needed, which adds complexity and time to the procedure. Additionally, STA grafts have a higher patency and longevity than interposition grafts. Ultimate determination of the choice of graft

can be tailored at the time of surgery by measuring the flow in the vessel to be sacrificed, and assessing the flow in the STA; if adequate, the STA can be used preferentially as the simpler option; otherwise, an interposition graft must be performed. Techniques for optimizing graft selection and technical success have been developed based on information from intraoperative blood flow measurements [2].

In certain situations, intracranial to intracranial bypass rather than extracranial to intracranial bypass can be performed if a suitable intracranial vessel segment is close enough to the vessel that needs to be revascularized, such as the branches of the MCA or distal anterior cerebral artery segments or posterior inferior cerebellar artery segments. In such instances, the vessels are anastomosed to each other in a side-to-side fashion maintaining the flow in the donor vessel while also providing blood flow to the distal territory of the recipient vessel. Such in situ bypasses obviate the need for EC–IC bypass grafts offering some advantages (i.e., no need to rely on the quality of the donor vessel and the additional morbidity of harvesting a graft) and disadvantages (i.e., putting another intracranial vascular territory at risk if the bypass fails).

With the advent of new endovascular techniques, the need for flow replacement bypass surgery to treat complex aneurysms that once were untreatable without that option is decreasing. Indeed, the development of devices such as flow diverters (endovascular stents that form a scaffold within the vessel in order to block flow into an aneurysm) now allow for endovascular options for treatment of some of the technically challenging large and giant aneurysms, which previously required surgery and EC–IC bypass. Nonetheless, EC–IC bypass remains an important strategy for select cases (Figs. 162.1–162.3).

EC–IC BYPASS FOR FLOW AUGMENTATION

Flow augmentation EC–IC bypasses are performed to treat chronic cerebral ischemic states caused by occlusive diseases of the cerebral vasculature. Assessment of cerebral hemodynamics is a key component of determining the need for a flow augmentation bypass. When brain perfusion is inadequate due to pathological states obstructing blood flow to the brain (e.g., atherosclerotic carotid occlusion), autoregulatory mechanisms lead to cerebrovascular vasodilation in an attempt to compensate and maintain CBF. The degree of blood flow reduction and the extent of these compensatory mechanisms can vary. If the compensatory autoregulatory vasodilation is able to maintain CBF, this is referred to as stage 1 hemodynamic compromise. However, when the mechanism of autoregulation is inadequate to maintain CBF, the brain

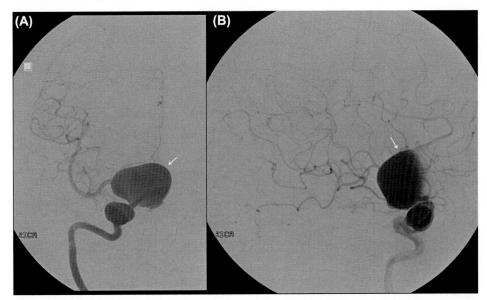

FIGURE 162.1   Digital subtraction angiogram of a right internal carotid artery (ICA) injection in AP (A) and lateral (B) projections showing a giant carotid ophthalmic artery aneurysm (*white arrows*). The second aneurysm seen in both projections is a cavernous segment ICA aneurysm. This 67-year-old woman presented with subarachnoid hemorrhage. Microsurgical treatment with trapping of both aneurysms and EC–IC bypass with a cadaver saphenous vein graft from the external carotid artery to the middle cerebral artery along with ligation of the cervical internal carotid artery was chosen, rather than endovascular stenting that has requirement for dual antiplatelets or anticoagulation and was not feasible in the setting of recent hemorrhage.

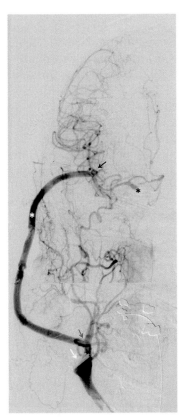

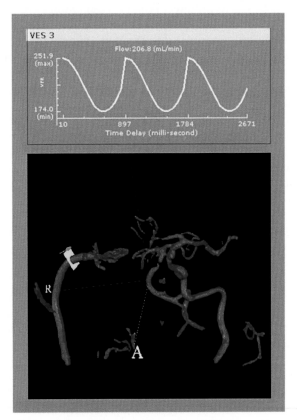

FIGURE 162.2   Digital subtraction angiogram of a right common carotid artery injection in AP projection for the patient depicted in Fig. 162.1, showing the completed EC–IC bypass with a cadaver saphenous vein graft (*white asterisk*) from the external carotid artery (*red arrow* pointing at anastomosis site) to the middle cerebral artery (*black arrow* pointing at anastomosis site) along with ligation of the cervical internal carotid artery (ICA) (*white arrow*). The bypass fills the cerebral vasculature retrograde all the way down to the supraclinoid segment of the right ICA (*black asterisk*).

FIGURE 162.3   Quantitative magnetic resonance angiography (MRA) 3D map using Noninvasive Optimal Vessel Analysis (NOVA) software, with flow measurement in the saphenous vein graft bypass at the site of the yellow cross section in the same patient measuring a mean of 206.8 mL/min. The *green arrow* adjacent to the yellow cross section indicates the direction of flow, which is from the neck toward the cranium depicted in Fig. 162.2.

IX. MANAGEMENT

tissue responds by an increase in oxygen extraction from the blood: this state of decreased CBF and compensatory increased oxygen extraction fraction (OEF) is referred to as "misery perfusion" or stage 2 hemodynamic failure. When the point of maximal oxygen extraction is reached, additional decrease in CBF will lead to brain infarction or stroke [3].

Based on these physiological compensatory processes, two general approaches can be utilized to assess the status of cerebral hemodynamics. One technique evaluates cerebrovascular reserve by first measuring blood flow at rest followed by a second measurement after introduction of a cerebral vasodilatory challenge (e.g., hypercapnia, acetazolamide, or a physiological stimulus, such as motor, visual, and auditory stimuli). In normal states, the vasodilatory challenge should induce a marked increase in CBF, which would be detected on the second measurement; if reduced or lacking, this signifies the lack of capacity for additional autoregulation due to preexisting vasodilation, and thus is an indication of underlying hemodynamic impairment. Imaging methods capable of measuring CBF quantitatively or qualitatively include computed tomography (CT) and magnetic resonance (MR)–based techniques. When a vasodilatory challenge is given, the autoregulation response can be classified into the following: (1) decreased augmentation in comparison with the contralateral hemisphere or previously determined normal values; (2) absence of augmentation; (3) paradoxical reduction also known as the "steal phenomenon" [3] indicative of progressive degrees of hemodynamic impairment and loss of cerebrovascular reserve. The second technique for hemodynamic assessment relies upon measurement of the OEF using oxygen-15 positron emission tomography (PET) to delineate patients with stage 2 hemodynamic failure based on the presence of increased OEF relative to absolute values or the contralateral normal hemisphere [3]. Patients with absent or paradoxical reduction response to vasodilatory stimuli or elevated OEF have been demonstrated to have an increased risk of recurrent cerebral ischemia in observational studies.

Keeping the previously mentioned concepts in mind, EC–IC bypass for flow augmentation can be entertained in patients with occlusive intracranial vascular disease resulting in the impairment of the blood flow to the brain and the risk of ischemic stroke. These conditions include moyamoya disease (a progressive vasculopathy resulting in narrowing and subsequent occlusion of the vessels at the base of the brain), and atherosclerotic stenosis or occlusion.

Moyamoya Disease

Moyamoya disease most frequently presents with findings of transient ischemic attacks (TIAs) or stroke:

this form of presentation is referred to as ischemic moyamoya disease and is more prevalent in the pediatric population [4]. Moyamoya can also present with intracranial hemorrhage; this type is known as hemorrhagic moyamoya disease and is more frequently evident in the adult population [4]. Hemorrhagic moyamoya is thought to be due to the hemodynamic stress on the fragile blood vessel collaterals that develop during disease progression when the brain is trying to form new channels for blood supply to counteract the natural course of the disease. These collateral vessels can develop aneurysms, which can be the source of hemorrhage, or more typically, can spontaneously rupture given their fragility and the underlying hemodynamic stress.

For ischemic moyamoya disease, the indications for bypass are the presence of ischemic symptoms; presence of hemodynamic impairment even in the absence of clinical symptoms can be considered for treatment, especially in the pediatric population where disease progression can be more aggressive and frequent. Testing for loss of cerebrovascular reserve can be performed using any of the techniques discussed earlier. Evidence for the effectiveness of EC–IC bypass is derived from multiple observational series [4]. Revascularization is performed using the standard STA–MCA bypass as a direct bypass graft, or via a procedure called indirect bypass, whereby surrounding tissues (such as the temporalis muscle, dura, periosteum) are laid over the brain surface to act as a source of new vessel ingrowth to the brain. The indirect bypass procedures, such as encephalo-duro-arterio-synangiosis (EDAS) or encephalo-myo-synangiosis (EMS) can be effective, particularly in pediatric patients, leading to the development of significant angiogenesis, presumably due to the milieu in moyamoya disease being conducive to collateral formation. The combination of direct and indirect bypasses offers the most robust revascularization strategy, particularly in adults.

For hemorrhagic moyamoya disease, recent evidence suggests that EC–IC bypass surgery using STA–MCA bypass can prevent rebleeding [5] and therefore revascularization is now considered an indicated procedure for the prevention of recurrent hemorrhage. This data comes from the results of the Japan Adult Moyamoya (JAM) trial published in 2014 [5]. This multicenter (22 Japanese institutes), prospective, randomized, controlled study enrolled 80 adult patients with moyamoya disease who had experienced intracranial hemorrhage within a year. The primary end point was defined as all adverse events, and the secondary end point was rehemorrhage. The patients were randomized to conservative treatment or bilateral EC–IC bypass and were followed for 5 years; 14.3% of patients in the surgical group suffered adverse events with significant morbidity versus 34.2% in the non-surgical group ($P = 0.057$). Kaplan–Meier cumulative curves for primary end points showed that the surgical group did significantly better, 3.2%/year for the surgical group versus 8.2%/year, $P = 0.048$. This was also observed with Kaplan–Meier cumulative curves for secondary end points, 2.7%/year for the surgical group versus 7.6%/year; $P = 0.042$. The conclusion of the study was that EC–IC bypass is effective in reducing rebleeding.

Atherosclerotic Disease

When considering patients with carotid occlusion or intracranial atherosclerosis, justifying flow augmentation EC–IC bypass becomes more challenging, and the patient selection process is more stringent. The reasons underlying this stem from the results of two major studies: the ECIC bypass study [6] and COSS [7].

In the 1960s and 1970s the volume of EC–IC bypass procedures was on the rise; nonetheless evidence supporting improved patient outcomes with bypass was lacking. Consequently, in 1977, a prospective randomized trial to study whether ECIC bypass (STA-MCA) with concomitant best medical therapy was superior to best medical therapy alone in patients with cerebrovascular ischemia was initiated; 1377 patients with ipsilateral ischemic stroke symptoms within 3 months prior to enrollment and proximal MCA stenosis or occlusion, distal internal carotid artery (ICA) stenosis (above C2 vertebral body), or extracranial ICA occlusion were included [6]. The results were as follows: no statistical difference in outcomes between the two groups within the 30 perioperative days and no benefit shown in reducing the long-term risk of stroke. Over an average 55.8-month follow-up, 31% had recurrent strokes in the surgical group versus 29% in the medical group, with no statistical difference ($P = 0.19$). These results led to a substantial decline in EC–IC bypass surgery for that indication. However, in the ensuing period after publication of the results, the study received considerable criticism [8]. The major concerns were that patients had not been selected based on hemodynamic criteria and that there was significant selection bias due to a large number of patients in whom bypass was performed outside of the trial at participating centers. This prompted the initiation of COSS [7] in 2002, which evaluated the usefulness of STA–MCA in reducing ipsilateral stroke in patients with complete ICA occlusion and elevated OEF in the cerebral hemisphere distal to the occlusion, in conjunction with best medical therapy. In this prospective trial, patients were randomized to surgery versus medical management following angiographic confirmation of ICA occlusion causing TIA or ischemic stroke within

120 days, in addition to increased OEF by PET scan (i.e., stage 2 hemodynamic failure). The results were published in 2011: despite excellent bypass graft patency rates as high as 96% at long-term follow-up, significant reduction in stroke incidence after the initial postoperative period, as well as improved cerebral hemodynamics with reduction in OEF on follow-up PET, the study was halted early due to futility analysis showing no overall benefit to surgery, with no significant difference in event rates between the medical and surgical groups. This was partly because medical management was more effective than anticipated in preventing recurrent strokes (22.7% in the medical group vs. 21 % in the surgical group, p = 0.78) and because of a relatively high 30-day perioperative event rate, 14.4%, in the surgical group. Nonetheless, it is noteworthy to mention that COSS did not address the subgroup of patients with recurrent ischemia and extensive cerebral hemodynamic disease who had already failed maximal medical therapy. For this reason, experts still believe that highly select patients with severe hemodynamic compromise, refractory to maximal medical therapy, may still benefit from revascularization (Figs. 162.4 and 162.5), if performed at centers where EC–IC bypass surgery can be done with sufficiently low perioperative morbidity [9]. Data indicate a favorable volume–outcome relationship in EC–IC bypass surgery [10] and thus support performance of these procedures at specialized centers.

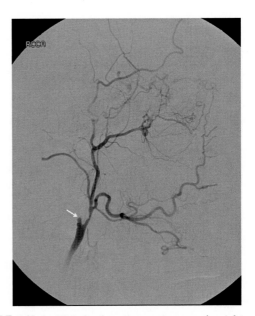

FIGURE 162.4 Digital subtraction angiogram of a right common carotid artery injection in lateral projection showcasing occlusion of the right cervical internal carotid artery slightly distal to its origin (*white arrow*) in a 40-year-old woman who presented with ipsilateral strokes and had lack of hemodynamic reserve. She failed maximal medical therapy and remained symptomatic. A right STA–MCA bypass using the parietal branch of the STA was performed for flow augmentation.

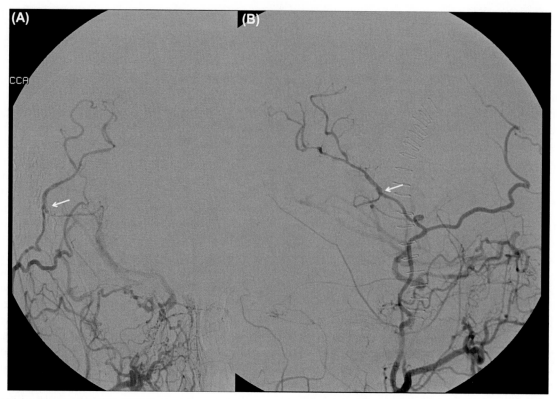

FIGURE 162.5 Digital subtraction angiogram of a right common carotid artery injection in AP (A) and lateral (B) projections showing the completed STA–MCA bypass (*white arrows* pointing at anastomosis site) in the same patient depicted in Fig. 162.4.

CONCLUSION

In conclusion, the role of EC–IC bypass procedures has significantly evolved over recent decades, particularly when considering treatment of cerebral ischemia due to athero-occlusive disease; only select patients remain potential candidates. Although the role of bypass for flow replacement in aneurysm surgery is more widely recognized, cases requiring this treatment option are relatively rare; new endovascular techniques have also been developed as alternate options, which have reduced the demand for direct surgical solutions. Nevertheless, EC–IC bypass remains an invaluable tool in the management of complex cases with no other viable options. Given the limited indications, the technically challenging nature of this surgery, and with better outcomes at high volume centers, patients requiring this treatment option are best managed in tertiary centers with adequate volumes and surgical expertise.

References

[1] Amin-Hanjani S. Cerebral revascularization: extracranial-intracranial bypass. J Neurosurg Sci 2011;55(2):107–16.

[2] Ashley WW, Amin-Hanjani S, Alaraj A, Shin JH, Charbel FT. Flow-assisted surgical cerebral revascularization. Neurosurg Focus 2008;24(2):E20.

[3] Derdeyn CP, Grubb Jr RL, Powers WJ. Cerebral hemodynamic impairment: methods of measurement and association with stroke risk. Neurology 1999;53(2):251–9.

[4] Kuroda S, Houkin K. Moyamoya disease: current concepts and future perspectives. Lancet Neurol 2008;7(11):1056–66.

[5] Miyamoto S, Yoshimoto T, Hashimoto N, et al. Effects of extracranial-intracranial bypass for patients with hemorrhagic moyamoya disease: results of the Japan Adult Moyamoya Trial. Stroke 2014;45(5):1415–21.

[6] Failure of extracranial-intracranial arterial bypass to reduce the risk of ischemic stroke. Results of an international randomized trial. The EC/IC Bypass Study Group. N Engl J Med 1985;313(19):1191–200.

[7] Powers WJ, Clarke WR, Grubb Jr RL, et al. Extracranial-intracranial bypass surgery for stroke prevention in hemodynamic cerebral ischemia: the Carotid Occlusion Surgery Study randomized trial. JAMA 2011;306(18):1983–92.

[8] Goldring S, Zervas N, Langfitt T. The extracranial-intracranial bypass study. A report of the committee appointed by the American Association of Neurological Surgeons to examine the study. N Engl J Med 1987;316(13):817–20.

[9] Amin-Hanjani S, Barker 2nd FG, Charbel FT, Connolly Jr ES, Morcos JJ, Thompson BG. Extracranial-intracranial bypass for stroke-is this the end of the line or a bump in the road? Neurosurgery 2012;71(3):557–61.

[10] Amin-Hanjani S, Butler WE, Ogilvy CS, Carter BS, Barker 2nd FG. Extracranial-intracranial bypass in the treatment of occlusive cerebrovascular disease and intracranial aneurysms in the United States between 1992 and 2001: a population-based study. J Neurosurg 2005;103(5):794–804.

CHAPTER

163

Surgical Aspects of Moyamoya Disease

M.K. Teo, V. Madhugiri, G.K. Steinberg
Stanford University School of Medicine, Stanford, CA, United States

Moyamoya disease (MMD) is a rare cerebrovascular disorder characterized by unilateral or bilateral steno-occlusion of the main trunks of the circle of Willis and the development of basal collateral channels, including hypertrophy of the lenticulostriate and thalamoperforating arteries, which results in the characteristic appearance of moyamoya vessels. Takeuchi first described the disease in 1957 as hypoplasia of the bilateral internal carotid arteries [1]. In 1969, Suzuki first employed the term "moyamoya" to describe the angiographic appearance of the enlarged collaterals, meaning "puff of smoke" in Japanese [2].

MMD is now known to affect people of diverse ethnicities, including Europeans, Americans, Japanese, Koreans, and Chinese. MMD is reported most often in Japan, with an estimated incidence of 0.54–0.94 per 100,000 and a prevalence of 6.03 per 100,000 population [3–5]. In Europe and North America, partly due to underdiagnosis or misdiagnosis, the true incidence is still unknown. A 2005 study estimated an incidence of 0.086 per 100,000 population in California and Washington [5a]. However, a 2012 study reported an incidence of 0.57/100,000 in the United States [6], which is closer to the Asian incidence. MMD has a bimodal age distribution with the first peak in the pediatric population around 5–9 years, and the second peak in mid-adulthood around 45–49 years. In the first decade of life, the diagnosis is equally common in males and females, but subsequently, females predominate with a ratio of 2–3:1 [7].

Familial occurrence has been reported in 5–10% of cases. Although genetic factors have a role in MMD, they remain to be fully elucidated. In Japan, it has been established that familial MMD is transmitted in an autosomal dominant fashion with incomplete penetrance. RNF213 was the first identified susceptibility gene for MMD, and was found to be mutated in 95% of Japanese patients with familial MMD, 73% of those with nonfamilial MMD, and only 1.4% of controls [8].

Histologically, the affected vessels show eccentric fibrocellular intimal thickening, smooth muscle cell proliferation in the media, and abnormalities of the internal elastic lamina, with no evidence of inflammation or atheroma [4]. The resulting steno-occlusion leads to hypoperfusion of cerebral vascular territories, and the watershed areas are particularly susceptible to ischemia. In response to hypoperfusion, thin-walled moyamoya collateral arteries (including the lenticulostriate) proliferate and enlarge. Moyamoya collaterals also lack smooth muscle cells, have incomplete internal elastic lamina, and may harbor aneurysmal dilatation, which could lead to hemorrhagic presentations (intracerebral, intraventricular, or subarachnoid hemorrhage).

In children, the most common presentation is cerebral ischemia. In a large Asian population study, 40% of those younger than 10 years of age presented with TIA and nearly 30% presented with cerebral infarction. Hemorrhagic presentation occurred in 3–9% of pediatric patients in both Asian and North American series. Other less common presentations include seizures, movement disorders, learning difficulties, and developmental delays. In adults, nearly 50% presented with intracranial hemorrhage in the Japanese series [9]. In the North American cohort, adult MMD still largely presented with cerebral ischemia (about 80%) [10].

SURGICAL MANAGEMENT

With a morbidity rate of over 65% in medically treated patients, surgical intervention has become the standard therapy for MMD patients. Emerging evidence has shown that surgical revascularization leads to a reduction in ischemic strokes in these patients compared with medical management only [11,12]. Until recently, the evidence suggesting that surgical revascularization improves outcomes for moyamoya patients presenting with hemorrhage was less compelling than for those presenting with ischemia. However, a landmark prospective, randomized study published in 2014 demonstrated that direct surgical revascularization significantly reduced subsequent rebleeding and improved outcome compared with conservative therapy in these patients as well [9].

Neurosurgical techniques for the treatment of MMD are divided into two main categories: direct and indirect revascularization. The principal difference between these two strategies lies in the method of cerebral reperfusion. Direct methods anastomose scalp arteries directly to intracranial arteries, immediately increasing perfusion to the affected cerebral territories. Additionally, over time, indirect neovascularization often occurs. Indirect methods aim to stimulate the development of a new vascular network. This can be achieved by using adjacent tissues (galea, muscle, scalp arteries, dura) or a distant graft (omentum) to cover the brain surface and to promote indirect collateralization. Indirect procedures include encephalomyosynangiosis (EMS), encephalogaleo[periosteal]synangiosis (EGPS), encephaloduroarteriosynangiosis (EDAS), encephaloduroarteriomyosynangiosis (EDAMS), pial synangiosis, multiple burr holes (MBH), and omental transposition (encephalo-omental synangiosis).

Currently there is no randomized controlled trial (RCT) that has determined whether direct or indirect revascularization is superior in the management of MMD. Some literature on this topic suggests that direct bypass is superior [13–15]. The main advantages of direct anastomosis are augmented flow immediately after surgery, a more consistent and higher extent of angiographic collateralization, superiority in restoring post-bypass cerebrovascular reserve capacity, more patients with symptomatic improvement, less recurrent ischemic risk, and more patients with stroke-free survival.

Endovascular treatment involving angioplasty with or without stenting of the stenotic vessels in MMD has been attempted without long-term success. Although there are no prospective trials examining this therapy, Khan et al. [16] reported failure of angioplasty or stenting in a series of five patients, all within just over a year of treatment. Due to the progressive steno-occlusive nature of the affected vessel, endoluminal therapy does not provide durable treatment.

OPERATIVE TECHNIQUE: DIRECT STA–MCA BYPASS

Preoperatively, patients undergo a thorough medical, cardiac, and anesthetic assessment with routine preoperative laboratory testing and relevant diagnostic imaging, which includes 6-vessel cerebral angiogram, MRI brain, and cerebral perfusion imaging with and without Diamox (PET, MR perfusion, TCDs). At our institution, we perform MR perfusion with and without Diamox,

and patients demonstrating poor cerebrovascular reserve or steal (indicating that the affected vascular territory is already maximally vasodilated to promote flow) are considered especially high risk for ongoing ischemia without treatment. These patients are also at higher risk for perioperative ischemic complications, thus particular care is taken to avoid hypotension perioperatively and in the recovery period. Intraoperatively, the patient's blood pressure is maintained at or above the preoperative baseline at all times.

The patient is positioned supine, and the head fixed in a Mayfield pin head holder with the operative side parallel to the floor (Fig. 163.1). A shoulder roll is used if cervical flexibility is restricted to avoid excessive rotational force on the cervical spine. From the preoperative angiogram, the most suitable superficial temporal artery (STA) branch (frontal vs. parietal branch) is chosen as the donor. Whenever possible, the parietal branch is used as it is behind the hairline, has a straighter course, and there is less risk of damage to the frontalis nerve branches during dissection. Using a handheld Doppler probe, the donor STA branch is transduced above the zygoma, followed to the convexity, and marked along its course. A superficial skin incision is made over the STA trunk just anterior to the tragus in the region of the zygoma and extended following the course of the parietal branch of the STA. In the case of harvesting the frontal STA branch, a curvilinear incision behind the hairline is performed and the STA is dissected from beneath the frontal skin flap. This is performed under microscopic magnification with the surgeon and assistant seated. Meticulous hemostasis is achieved, and careful dissection through the dermis and subcutaneous tissue is performed to isolate the STA vessel. Scalp veins are commonly encountered running parallel to the STA, and their identity is verified using a Doppler probe prior to coagulating. The goal is to dissect 9–10 cm of STA prior to performing the craniotomy. The nondonor branch of the STA (frontal branch if the parietal branch is used as the donor) is generally preserved for potential subsequent revascularization procedures; small branches can be coagulated and divided to mobilize the STA trunk. Papaverine is intermittently applied to the dissected STA to alleviate any spasm induced by mechanical manipulation of the vessel. A Doppler probe is also used to ensure that the flow is maintained during STA dissection.

After harvesting the STA vessel off the temporalis fascia, hook retractors are placed at the skin edges. The temporalis muscle is cut in an H-shaped fashion using cautery. The muscle is retracted anteriorly or posteriorly to expose the underlying skull. Burr holes are then placed at the inferior and superior limits of the exposure and a circular craniotomy (6 cm in diameter) is created. Hemostasis is achieved around the bone edges, and the dura is tacked up around the margin of the craniotomy to avoid epidural oozing. To prepare the

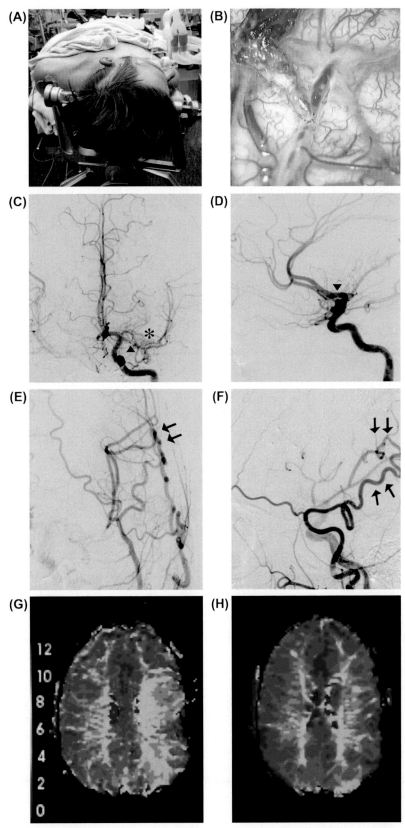

FIGURE 163.1 A 33-year-old man with intermittent right body numbness and speech impairment was diagnosed with left unilateral moy-amoya disease. He subsequently underwent left direct STA–MCA bypass. (A) Intraoperative positioning with head fixed in Mayfield pin, and scalp-marking of the parietal STA branch for the donor vessel. (B) Microscopic view after STA–MCA direct anastomosis using interrupted 10-0 suture. (C, D) Preoperative cerebral angiogram (AP, lateral views, ICA injection) showing left MCA occlusion (*arrowhead*) with moyamoya vessels formation (*asterisk*). (E, F) 6-months postoperative cerebral angiogram (AP, lateral views, ECA injection) showing widely patent direct STA–MCA bypass graft (*arrows*) using the parietal STA branch, now filling the MCA territories. (G) Preoperative MRI brain perfusion scan showing reduced perfusion in the left hemisphere, MCA region. (H) 6-months postoperative MRI perfusion showed improvement of left hemispheric perfusion.

donor artery, microscopic guidance is used to dissect 1–2 cm of the distal STA from the surrounding tissue using fine microscissors. Similarly, the proximal STA is dissected from its surrounding soft tissue cuff to create a site for placement of a proximal temporary clip. The ideal site for temporary clipping is distal to the takeoff of the unused frontal branch of STA; this allows continued flow through the STA into the unoccluded branch, reducing stagnant flow and the risk of thrombosis proximal to the temporary clip. Adequate mobilization of the STA is typically possible while preserving the unused frontal branch of STA. Furthermore, this preserved branch could also be used for future repeat revascularization should the need arise.

The dura is opened in cruciate fashion and tacked to the outer skin edges. With the cortex now exposed, the microscope is used to find a suitable recipient cortical middle cerebral artery (MCA) branch. The most important considerations are size of the donor and recipient vessels (0.9 mm or greater is optimal for both), location of the recipient M4 branch of MCA (away from the craniotomy edges is preferred), and orientation of the vessels (to avoid an acute angulation between the donor and recipient vessels).

The arachnoid over the potential recipient M4 artery is opened, and a 1 cm length of vessel is prepared for the anastomosis. Small perforators from the artery should be avoided; in some cases, they can be coagulated with bipolar cautery (at low power) so the artery can be elevated from the brain surface for placement of a high visibility background material underneath the vessel. Larger perforators can be spared by including their offtake from the MCA segment in the temporary vessel clips prior to the anastomosis. Before temporary clipping, blood flow is measured in the MCA branch and the cut STA using an ultrasonic flow probe (Charbel Micro-Flow-probe; Transonics Systems, Inc., Ithaca, NY). After the temporary clip is applied to the proximal STA, the cut STA segment is then flushed with heparinized saline. The STA–MCA anastomosis is performed in an end-to-side fashion. The STA is cut at a 45° angle, and a fish mouth is created to enlarge the opening.

After achieving mean arterial pressure (MAP) over 90 mmHg, hypothermia of 33°C, and burst suppression using propofol, specially designed Anspach-Lazic temporary clips (Peter Lazic GmbH, Tuttlingen, Germany) are placed proximally and distally on the recipient vessel. An arteriotomy is then made over the recipient MCA using microscissors, and the lumen is flushed with heparinized saline. Indigo-carmine is used to color the walls of the donor and recipient vessel to allow the lumen to be seen more easily and to facilitate microanastomosis. Monosof suture (10-0) (Covidien, Dublin, Ireland) is used to place anchoring sutures at the apices of the incision (toe stitch, followed by heel stitch). Sutures should be passed from outside the donor to inside the recipient, and then tied

on the outer surface of the anastomosis. Once the donor STA has been anchored, interrupted sutures are placed on each side of the anastomosis at close intervals. Great care must be taken not to catch the back wall of the vessel while suturing the anastomosis. Although not preferred by the senior author, a continuous suture technique can also be used, which involves leaving short loops of suture along the entire length of the vessel, which are tightened sequentially just prior to tying the suture.

Once the anastomosis is complete, the temporary clips on the recipient artery are released, and then the proximal clip on the STA is removed. Occasionally, additional sutures are needed to seal the anastomosis. Blood flow in the STA and MCA, proximal and distal to the anastomosis, is then measured with a quantitative, directional flow probe (Charbel Micro-Flow-probe; Transonics Systems, Inc., Ithaca, NY), and an intraoperative indocyanine green (ICG) angiogram is performed to confirm patency and function of the bypass graft.

During closure, the dura is loosely replaced over the cortex and graft with great care not to disturb flow, and the dural opening is covered with a piece of dural substitute graft. After enlarging the inferior burr hole to accommodate the entering STA graft, and drilling the underside of the bone flap to create room for the graft/anastomosis, the bone is replaced, avoiding any kinking or pressure on the vessel. The temporalis muscle is approximated, and the skin is closed with care to avoid injury to the proximal STA. The patency of the proximal STA graft and STA trunk are verified with a Doppler probe during the closure and at the end of the procedure.

INDIRECT BYPASS TECHNIQUE

In our practice, indirect procedures for MMD tend to be reserved for pediatric patients who have higher angioplasticity and better collateralization compared to adult patients. Furthermore, direct bypass is more difficult to perform in young children due to the extremely small size of the arteries. We also perform indirect procedures (usually EDAS) in adults who do not have complete occlusion of the symptomatic ICA or MCA and who have anterograde filling of the MCA distribution, where a direct STA–MCA graft might promote complete ICA or MCA occlusion. Repeat revascularization using the various indirect techniques is another indication for MMD patients who have ongoing symptoms due to inadequate revascularization from initial procedures.

For indirect bypass techniques using adjacent tissue, for example EDAS, the stages involve a scalp vessel harvest with soft tissue cuff, craniotomy, cruciate dural opening, multiple fenestrations of the pia-arachnoid layer, and scalp vessel placement over the exposed brain. The dura is then approximated over the vessel, the bone

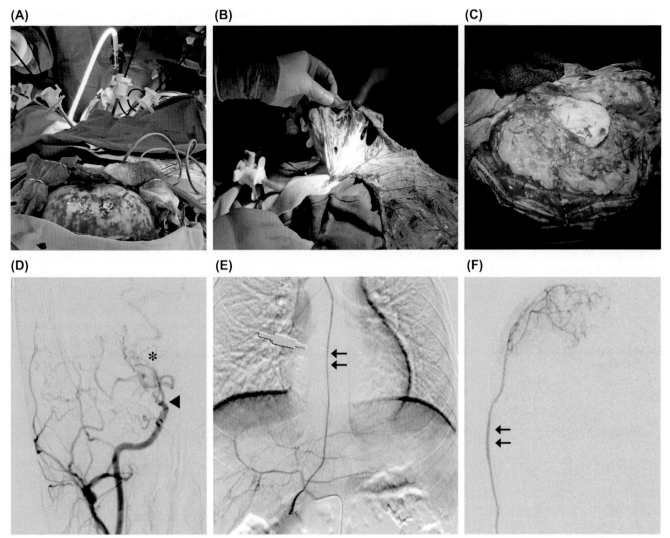

FIGURE 163.2 An 18-year-old girl with moyamoya disease had a bilateral indirect cranial bypass 3 years previously. She presented with left body numbness, and an angiogram showed right terminal ICA occlusion (*arrowhead*) and inadequate collateralization from previous indirect bypass grafts. She subsequently underwent encephalo-omental synangiosis to revascularize a large area of her right hemisphere. (A) Intraoperative view of exposed skull (note plating system in place due to previous craniotomy), and simultaneous endoscopic harvest of the greater omentum. (B) Omentum is now delivered out of peritoneal cavity and tunneled to the cranial compartment. (C) After dural opening, the omentum is placed over a large surface area of the exposed hemisphere. The skull is then thinned to avoid compression of the underlying brain prior to closure. (D) Preoperative cerebral angiogram (AP view, common carotid injection) showing left terminal ICA occlusion with moyamoya vessels, and reduced perfusion of a large part of the right cerebral hemisphere. (E, F) Postoperative celiac trunk injection showing the pedicle of the omental flap (gastroepiploic artery, *arrows*) along the thoracic cavity and cervical region to the cranial compartment.

flap is fashioned to allow the artery to enter and exit through the burr holes, and closure is performed, while ensuring the patency of the bypass graft.

In circumstances when no suitable adjacent tissue or scalp vessel is available for indirect bypass, or a large area of brain revascularization is needed, indirect bypass using distant tissue, for example, omental transposition, has become an attractive option. The omentum is harvested laparoscopically, while preserving the vascular pedicle supplied by the right gastroepiploic artery. It is then delivered to the cranial region, and placed over the hemisphere of interest for indirect collateralization (Fig. 163.2).

OPERATIVE RESULTS

At Stanford, we have performed over 1414 bypasses in more than 885 patients for moyamoya revascularization. We are strong advocates of direct bypass for MMD, as direct revascularization immediately increases blood flow to the ischemic brain, with visual proof of a good graft obtained intraoperatively using indocyanine green videoangiography. It is a more complex procedure to perform, however, especially in pediatric patients with small donor and recipient vessels. Our philosophy has been to attempt direct revascularization in all patients, except for pediatric patients < 4 years of age or patients with vessels less than 0.8 mm [10]. In our published series in 2009 with 557 surgeries for MMD (272 adult, 96 pediatric), 96.6% of adult and 67.2% of pediatric patients underwent direct bypass. Postoperative significant neurological deficit occurred in 15 patients undergoing 16 procedures (3.5% of procedures or 5.6% of patients). With a mean follow-up of 4.9 years, the overall cumulative risk of stroke or hemorrhage was 5.5%. With a mean angiographic follow-up of 1.5 years (range 0.5–9.4 years, median 0.6 years), 99% of the bypasses were patent. The few occluded direct STA grafts were noted to have formed indirect collaterals that were supplying the territory at risk.

Long-term follow-up also showed that patients had generally improved in terms of the level of disability following revascularization. The mean preoperative modified Rankin Score (mRS) was 1.62, whereas postoperatively it was 0.83 at a mean of 4.9 years. Only 5.2% of patients suffered a worsening of mRS. Of the patients who presented with TIAs, 82% were TIA-free at 1 month post operation and 92% at 1 year. The risk of strokes or intracranial hemorrhage beyond 30 days after surgery was very low at 0.8% [10].

The only RCT on MMD was recently conducted by 22 Japanese institutes, where 80 adult patients presenting with hemorrhagic MMD were randomized to conservative care or bilateral extracranial–intracranial direct bypass (42 surgical; 38 nonsurgical). Adverse events causing significant morbidity were observed in 6 patients in the surgical group (14.3%) and 13 patients in the nonsurgical group (34.2%), and Kaplan–Meier survival analysis revealed significant differences between the two groups (3.2%/years versus 8.2%/years; $P = 0.048$). Furthermore, rebleeding occurred in 5 patients in the surgical group (11.9%) and 12 in the nonsurgical group (31.6%), again significantly different in the Kaplan–Meier survival analysis (2.7%/years versus 7.6%/years; $P = 0.042$) [9].

CONCLUSION

Surgical revascularization for symptomatic MMD is recommended, and we advocate direct bypass in the majority of cases, as it also has the added benefit of indirect collateralization to the at risk territory after the initial increase in cerebral perfusion.

References

[1] Takeuchi K, Shimizu K. Hypoplasia of the bilateral internal carotid arteries. Brain Nerve (Tokyo) 1957;(9):37–43.
[2] Suzuki J, Takaku A. Cerebrovascular "moyamoya" disease. Disease showing abnormal net-like vessels in base of brain. Arch Neurol 1969;20(3):288–99.
[3] Baba T, Houkin K, Kuroda S. Novel epidemiological features of moyamoya disease. J Neurol Neurosurg Psychiatry 2008;79(8):900–4.
[4] Cook DJ, Mukerji N, Furtado SV, Steinberg GK. Moyamoya Disease. Berlin Heidelberg: Springer-Verlag; 2014.
[5] Kuriyama S, Kusaka Y, Fujimura M, et al. Prevalence and clinico-epidemiological features of moyamoya disease in Japan: findings from a nationwide epidemiological survey. Stroke 2008;39(1):42–7.
[5a] Uchino K, Johnston SC, Becker KJ, Tirschwell DL. Moyamoya disease in Washington State and California. Neurology September 27, 2005;65(6):956–8.
[6] Starke RM, Crowley RW, Maltenfort M, et al. Moyamoya disorder in the United States. Neurosurgery 2012;71(1):93–9.
[7] Gooderham PA, Steinberg GK. Intracranial-extracranial bypass surgery for moyamoya disease. In: Spetzler RF, Kalani Y, Nakaji P, editors. Neurovascular surgery. 2nd ed. Thieme; 2015.
[8] Kamada F, Aoki Y, Narisawa A, et al. A genome-wide association study identifies RNF213 as the first Moyamoya disease gene. J Hum Genet 2011;56(1):34–40.
[9] Miyamoto S, Yoshimoto T, Hashimoto N, et al. Effects of extracranial-intracranial bypass for patients with hemorrhagic moyamoya disease: results of the Japan Adult Moyamoya Trial. Stroke 2014;45(5):1415–21.
[10] Guzman R, Lee M, Achrol A, et al. Clinical outcome after 450 revascularization procedures for moyamoya disease. Clinical article. J Neurosurg 2009;111(5):927–35.
[11] Fung LW, Thompson D, Ganesan V. Revascularisation surgery for paediatric moyamoya: a review of the literature. Childs Nerv Syst 2005;21(5):358–64.
[12] Zipfel GJ, Fox Jr DJ, Rivet DJ. Moyamoya disease in adults: the role of cerebral revascularization. Skull Base 2005;15(1):27–41.
[13] Houkin K, Kamiyama H, Abe H, Takahashi A, Kuroda S. Surgical therapy for adult moyamoya disease. Can surgical revascularization prevent the recurrence of intracerebral hemorrhage? Stroke 1996;27(8):1342–6.
[14] Kazumata K, Ito M, Tokairin K, et al. The frequency of postoperative stroke in moyamoya disease following combined revascularization: a single-university series and systematic review. J Neurosurg 2014;121(2):432–40.
[15] Nakashima H, Meguro T, Kawada S, Hirotsune N, Ohmoto T. Long-term results of surgically treated moyamoya disease. Clin Neurol Neurosurg 1997;99(Suppl. 2):S156–61.
[16] Khan N, Dodd R, Marks MP, Bell-Stephens T, Vavao J, Steinberg GK. Failure of primary percutaneous angioplasty and stenting in the prevention of ischemia in Moyamoya angiopathy. Cerebrovasc Dis 2011;31(2):147–53.

CHAPTER

164

Prevention of Ischemic Stroke

C. Goshgarian[1], P.B. Gorelick[1,2]

[1]Mercy Health Hauenstein Neurosciences, Grand Rapids, MI, United States; [2]Michigan State University College of Human Medicine, East Lansing, MI, United States

INTRODUCTION

Stroke is well suited for prevention as it has a high prevalence, a number of modifiable factors proven to reduce stroke risk, and a high societal economic and disability burden [1]. An international observational study, INTERSTROKE, showed that 10 relatively common factors are associated with about 90% of stroke risk [2]. Hypertension accounted for almost 35% of the risk. Approaches to stroke prevention include the "high-risk" and "mass" approaches [1]. The latter approach aims to reduce risk in a population through health education, legislation, and economic measures to decrease exposure to risks in a broad manner. On the other hand, the high-risk approach is designed to screen for persons at high risk of stroke, for example, and then aggressively manage the risks which may require not only education and lifestyle modification, but also administration of medication [1]. As an example, the high-risk approach is applied by health care providers in office practice to select out persons for treatment with medication to lower risk. The mass and high-risk approaches are viewed as complementary means to reduce occurrence of stroke. Furthermore, we have learned over time with advancement and sophistication of clinical trials in stroke prevention that identification of stroke mechanism is important as a springboard to application of evidence-based efficacious and safe stroke prevention practices.

In this chapter we discuss ischemic stroke prevention in relation to first and recurrent stroke prevention. While worldwide there are a number of important guidelines for first and recurrent ischemic stroke prevention, based on limitations in the scope of this chapter we have chosen to review the most recent first and recurrent ischemic stroke prevention guidelines from the American Heart Association/American Stroke Association (AHA/ASA) [3, 4]. For the sake of simplicity and clarity of discussion,

AHA/ASA stroke prevention guidelines refer to three categories of stroke risk factors: generally nonmodifiable ones (e.g., age, low birth weight, race/ethnicity, and genetic factors); well-documented and modifiable risk factors (e.g., physical activity, dyslipidemia, diet and nutrition, raised blood pressure, obesity, diabetes mellitus, cigarette smoking, atrial fibrillation and other cardiac conditions, and carotid and intracranial artery stenosis); and less well-documented or potentially modifiable factors (e.g., migraine, metabolic syndrome, alcohol consumption, and sleep-disordered breathing) [3].

The topical discussions that follow are divided into the following key categories in relation to first and recurrent ischemic stroke prevention: (1) management of lifestyle and modifiable vascular risks; and (2) indications for aspirin and other antiplatelet medications, oral anticoagulants, closure of patent foramen ovale (PFO) and left atrial appendage device occlusion, carotid endarterectomy (CEA) and angioplasty/stenting, and extracranial to intracranial arterial bypass. Finally, determination of risk for atherosclerotic cardiovascular diseases (ASCVD), such as stroke, is deemed useful at the individual patient level. The reader is referred elsewhere for a more detailed discussion of risk prediction in ASCVD and stroke [5].

AHA/ASA GUIDELINE RECOMMENDATIONS FOR MANAGEMENT OF LIFESTYLE AND VASCULAR RISKS FOR ISCHEMIC STROKE PREVENTION

Based on similarities in management of lifestyle and vascular risks for first and recurrent ischemic stroke, we discuss AHA/ASA guideline recommendations primarily from the viewpoint of first stroke prevention [3].

Primer on Cerebrovascular Diseases, Second Edition
http://dx.doi.org/10.1016/B978-0-12-803058-5.00164-8

We provide additional comments when there is a difference in lifestyle or vascular risk management recommendations between first and recurrent ischemic stroke prevention [4].

Lifestyle Factors

For healthy adults moderate to vigorous-intensity aerobic *physical activity* at least 40 min per day, 3–4 days per week is recommended [3]. The recommendation is similar for recurrent ischemic stroke prevention; however, for those who cannot readily engage in physical activity due to stroke impairment or disability, a structured exercise program developed by a health care professional, such as a physiatrist or other rehabilitation specialist, may be helpful [4]. In relation to *diet and nutrition*, the Mediterranean diet supplemented with nuts may be considered [3], which is similar to guidance for recurrent ischemic stroke prevention [4]. In addition, reduction of intake in sodium to help lower blood pressure (e.g., 2.4 g/day or less [<1.5 g/day]) and a DASH diet are recommended as good options [3]. *Smoking cessation* is recommended as is avoidance of passive smoke. Counseling, nicotine replacement therapy, and other medications (e.g., bupropion and varenicline) also may be useful [3, 4]. *Weight reduction* for those who are overweight or obese is indicated to lower blood pressure, and therefore, reduce stroke risk [3]. Although screening for weight and determination of body mass index (BMI) is recommended, the usefulness of weight loss is uncertain in relation to recurrent stroke prevention [4]. For *alcohol consumption*, there is counseling to attempt to eliminate heavy drinking, and if one does drink, the aim is for ≤2 standard drinks/day for men and up to one standard drink/day for non-pregnant women [3,4].

Well- and Less Well-documented Vascular Risk Factors

Hypertension

As referred to earlier, *hypertension* is the most important modifiable risk factor for stroke. AHA/ASA stroke prevention guidance now calls for regular blood pressure (BP) screening and lifestyle modification and pharmacological management [3]. Prehypertensive persons (i.e., systolic BP (SBP) 120–139 mm Hg or diastolic BP (DBP) 80–89 mm Hg) should be screened annually. Currently, the BP target for treatment is <140/90 mm Hg, and BP lowering is considered more important than specific class of BP medication, though there may be compelling indications for administration of certain classes of BP-lowering medication [3]. Finally, self-measurement of BP is recommended to enhance BP control.

In relation to recurrent ischemic stroke prevention, the target is similar to that for primary stroke prevention (<140/90 mm Hg) with the exception of recent lacunar infarction whereby a reasonable target is a SBP <130 mm Hg [4].

Dyslipidemia

Based on national guidelines in the United States, statin therapy is indicated for primary prevention of ASCVD including stroke in persons with a 10-year risk ≥7.5% or for those with other indications (e.g., having clinical ASCVD, elevation of LDL-C ≥190 mg/dL, and persons aged 40–75 years with an LDL-C 70–189 mg/dL and a history of diabetes mellitus) [5]. Recommendations for recurrent stroke prevention state that those with ischemic stroke or transient ischemic attack (TIA) of atherosclerotic origin and an LDL-C level ≥100 mg/dL should receive a statin agent [4]. Furthermore, according to the primary stroke prevention guideline, potent statin-lowering drugs are indicated though a specific target LDL-C goal is not emphasized, whereas, in the recurrent stroke prevention guideline an LDL-C target is specified at a level of <100 mg/dL and optimally at 50–70 mg/dL or 50% of the pretreatment LDL-C value [3–5].

Diabetes Mellitus

The AHA/ASA prevention of a first stroke guideline highlights the importance of BP control and administration of a statin agent in persons with diabetes mellitus [3]. In addition, the uncertainty of the usefulness of aspirin as a primary prevention measure in persons with diabetes mellitus and a low 10-year risk of cardiovascular disease is discussed. The AHA/ASA recurrent ischemic stroke prevention guideline emphasizes the application of American Diabetes Association guidelines for management of glycemic control [4].

Migraine

For first stroke prevention, smoking cessation and alternatives to oral contraceptives (especially those containing estrogen) are recommended or considered, respectively [3]. Furthermore, treatment to reduce the frequency of migraine is considered reasonable, whereas closure of PFO in persons with migraine is not indicated [3].

Sleep-Disordered Breathing

Screening for sleep-disordered breathing and polysomnography should be considered [3]. In persons with sleep apnea, treatment with continuous positive airway pressure (CPAP) should be considered [4].

Homocysteinemia

Elevated blood homocysteine has been associated with increased risk for cardiovascular disease. However, in relation to recurrent ischemic stroke prevention, routine screening for hyperhomocysteinemia is not indicated. Although persons with mild to moderate elevations of homocysteine and who receive folate, vitamin B6, and vitamin B12 have lower homocysteine levels, this strategy has not been proven to reduce stroke risk [4].

AHA/ASA GUIDELINE RECOMMENDATIONS FOR MANAGEMENT OF ANTITHROMBOTICS, ORAL ANTICOAGULANTS, PFO CLOSURE AND RELATED SURGERY, CAROTID ENDARTERECTOMY, AND PREVENTION

We now discuss medical and surgical interventions for the prevention of first and recurrent stroke.

Atrial Fibrillation

Atrial fibrillation is an independent risk factor for first and recurrent ischemic stroke. Multiple risk stratification schemes are available to identify patients who may benefit from oral anticoagulation for stroke prevention. A commonly used strategy involves calculating the annual risk of ischemic stroke with the CHA2DS2-VASc score. AHA/ASA guidance recommends oral anticoagulation for nonvalvular atrial fibrillation (NVAF) patients with a CHA2DS2-VASc score of ≥2 and an acceptable risk of bleeding [3], whereas it is reasonable to omit treatment with oral anticoagulants if the CHA2DS2-VASc is 0 [3]. For patients with a CHA2DS2-VASc of 1, treatment with aspirin, an oral anticoagulant, or no antithrombotic treatment at all are all reasonable options [3].

Once it has been decided to pursue treatment with an oral anticoagulant, two treatment options exist: vitamin K antagonist therapy (e.g., warfarin) or one of the newer nonvitamin K antagonist oral anticoagulants (e.g., apixaban, dabigatran, edoxaban, and rivaroxaban) [3]. Traditionally, warfarin has been used as a mainstay treatment option. Warfarin is associated with a 64% relative risk reduction in stroke risk [6], whereas aspirin is associated with only a 39% risk reduction when compared to placebo in NVAF [6,7]. The novel oral anticoagulants (NOACs), including the direct thrombin inhibitor dabigatran and factor Xa inhibitors apixaban, rivaroxaban, and edoxaban, have provided clinicians and patients with alternative treatment options. NOAC have been proven to be at least as effective as warfarin in preventing ischemic strokes, and there is less intracranial hemorrhage. However, when deciding which agent to use, one has to consider several factors including cost, reversibility, monitoring, and side effect profile. For patients who cannot tolerate long-term oral anticoagulation, it is reasonable to consider left atrial appendage closure assuming the patient can tolerate anticoagulation for at least 45 days following the procedure [3]. This can be accomplished with the WATCHMAN device.

PFO and Atrial Septal Defect

PFO is a connection between the venous and arterial blood in the heart and acts as a potential source of paradoxical embolism in ischemic stroke. Atrial septal defects can be a nidus for clot formation. PFOs are common and may occur in 15–25% of the adult population [3], however, the importance of PFO in relationship to stroke causation is less clear. There is conflicting evidence whether PFO predisposes individuals to an increased risk of stroke. Due to this information and the fact that the PFO closure and pharmacological intervention come with risk, AHA/ASA guidance does not recommend routine intervention for PFO in primary stroke prevention [3]. Patients who have experienced a TIA or ischemic stroke and are found to have a PFO warrant additional investigations to search for lower extremity or pelvic deep vein thrombosis. AHA/ASA guidelines state that for patients with an ischemic stroke or TIA and both a PFO and a venous source of embolism, anticoagulation is indicated, depending on stroke characteristics [4]. An inferior vena cava filter is reasonable for patients with contraindications to anticoagulation (e.g., bleeding risk) [4]. AHA/ASA guidelines do not support routine PFO closure following a stroke or TIA, if no DVT is identified [4].

Asymptomatic Carotid Artery Stenosis

Asymptomatic extracranial internal carotid artery narrowing is a risk factor for ischemic stroke. Multiple imaging modalities can be used to detect carotid artery stenosis including cerebral angiography, CT angiography, carotid duplex, and MRA. Each test has advantages and disadvantages including sensitivity for detection, risks to the patient, and costs. The AHA/ASA does not provide specific recommendations as to which imaging tool to use. In patients with asymptomatic carotid artery stenosis a select subgroup of patients may benefit from prophylactic CEA or carotid artery stenting (CAS). This group may include those with high-grade carotid artery stenosis (e.g., >70%). Risks such as gender (women), age, and progression of stenosis may play an important role in determining whether to intervene surgically. AHA/ASA guidance indicates that it is reasonable to perform CEA in patients with asymptomatic carotid artery stenosis >70% if perioperative stroke, myocardial infarction (MI), and death risk is less than 3% [3]. Currently, debate exists as to the effectiveness of CEA and CAS in asymptomatic carotid artery stenosis as it is believed by some experts that modern medical management with statins, blood pressure–lowering agents, and lifestyle coaching may be adequate to prevent stroke in asymptomatic carotid artery stenosis. CREST 2 trial is an ongoing randomized controlled trial that compares the use of current best medical management with CEA and CAS for treatment of asymptomatic carotid artery stenosis and promises to clarify the debate over best medical versus surgical management in asymptomatic carotid artery stenosis.

Symptomatic Carotid Artery Stenosis

Three landmark studies demonstrated a statistical benefit of CEA compared to medical management in patients with high-grade carotid artery stenosis (70–99%). The studies included the European Carotid Surgery trial (ECST), the North American Symptomatic Carotid Endarterectomy Trial (NASCET), and the Veterans Affairs Cooperative Study Program (VACS). Therefore, AHA/ASA guidance recommends CEA for symptomatic carotid artery stenosis of 70–99% within the last 6 months, if perioperative risk is less than 6% [4]. There was no surgical benefit in patients with stenosis of less than 50% [4]. The benefit is less clear when there is 50–69% stenosis, and thus, proper patient selection is important [4]. For example, a subgroup analysis from the NASCET trial questioned the benefit of CEA in women due to less favorable outcomes, and those with higher surgical mortality risk and higher recurrent stenosis [8,9]. Also, the timing of CEA was an important factor in secondary stroke prevention. The attributable risk reduction from CEA decreased from 30% when the procedure was performed in the first two weeks to 18% at 2–4 weeks to 11% at 4–12 weeks [10]. Thus AHA/ASA guidance suggests performing CEA or CAS within 2 weeks of onset of the index stroke assuming there is no significant contraindication to early revascularization [4].

Carotid Artery Stenting

CAS is an alternative option to CEA in patients who are high-risk CEA candidates. Examples of patients who may benefit from CAS include those with stenosis above the level of the 2nd cervical vertebra, contralateral carotid artery stenosis, recurrent carotid artery stenosis following CEA, or history of neck radiation. Thus, AHA/ASA guidance recommends CAS as a reasonable alternative to CEA in patients with medical conditions that predispose them to high surgical risk or specific conditions that favor the use of CAS, such as radiation-induced stenosis or restenosis after CEA [4]. On the other hand, AHA/ASA guidelines state that CEA may be preferred when the patient is >70 years of age [4]. For younger patients,

CAS is equivalent to CEA in terms of risk for periprocedural complications (i.e., stroke, MI, or death) and long-term risk for ipsilateral stroke [4].

Antiplatelet Therapy for Primary Stroke Prevention

Aspirin is a cyclooxygenase inhibitor that ultimately leads to reduction in platelet aggregation and thus may prevent ischemic events. A potentially dangerous side effect of this medication is gastrointestinal ulcer and bleeding. Therefore, one must carefully weigh the risks and benefits when there is consideration to prescribe aspirin for first stroke prevention. AHA/ASA guidance suggests aspirin administration for the primary prevention of cardiovascular disease (including stroke) when the 10-year risk of ASCVD exceeds 10%, and the potential benefits of therapy are judged to outweigh the risks [3]. For lower risk individuals, aspirin is not recommended for first stroke prevention [3]. Outside of aspirin and cilostazol, other antiplatelet agents are not recommended for primary stroke prevention [3].

Antiplatelet Therapy for Recurrent Stroke Prevention

In the United States, aspirin (50–325 mg/day), clopidogrel (75 mg/day), or aspirin plus extended-release dipyridamole (25 mg/200 mg bid) may be used for recurrent ischemic stroke prevention [4]. With the exception for symptomatic intracranial occlusive disease, the combination of aspirin plus clopidogrel is not recommended for recurrent stroke prevention due to the increased risk of major and life-threatening bleeding.

Intracranial Atherosclerosis

Intracranial atherosclerosis is an important cause of stroke and is associated with a high rate of recurrent stroke. Stroke risk factors associated with decreased risk of recurrent stroke include an SBP <140 and an LDL-C ≤100 mg/dL [11]. Based on the results of the SAMMPRIS trial, best medical management is preferred over angioplasty and stenting with the Wingspan stent system for the treatment of symptomatic intracranial atherosclerosis. AHA/ASA guidance recommends that it may be reasonable to add clopidogrel 75 mg/day for 90 days to aspirin 325 mg/day for the treatment of TIA/stroke attributable to severe (70–99%) intracranial atherosclerosis [4]. Maintenance of SBP below 140 mm Hg and the addition of high potency statin therapy are recommended [4]. Use of the Wingspan stent is not recommended as the initial treatment for intracranial atherosclerosis [4]. For patients who have had recurrent ischemic stroke or TIA while on best medical management with aspirin, clopidogrel, SBP of less than 140 mm Hg, and high-dose statin therapy, the usefulness of angioplasty alone or placement of a Wingspan stent or other stent is unknown and is considered investigational [4].

CONCLUSION

Both first and recurrent ischemic stroke are highly preventable. Application of lifestyle, medical, and interventional therapies are indicated to reduce first and recurrent ischemic stroke risk. Over time, the benefits of aggressive lifestyle and medical management in stroke prevention have been emphasized. Targets for control of risk factors, such as LDL-cholesterol and blood pressure, will continue to undergo modification to further reduce cardiovascular risk. For example, in 2015, researchers from the SPRINT trial concluded that aggressive blood pressure management with a target SPB of less than 120 mm Hg resulted in lower rates of cardiovascular events and death compared to SPB of less than 140 mm Hg [12].

Stroke prevention is a dynamic and evolving area. Future stroke prevention recommendations will be based on new observational and randomized control trial results.

References

[1] Gorelick PB. The Future of Stroke Prevention by Risk Factor Modification. Stroke: Part III. In: Fisher M, editor. Handbook of Clinical Neurology, ;94. New York: Elsevier; 2009. p. 1261–76.

[2] O'Donnell MJ, Xavier D, Lui L, et al. Risk factors for ischemic stroke and interacerebral hemorrhagic stroke in 22 countries (the INTERSTROKE study): a case-control study. Lancet 2010;376:112–23.

[3] Meschia JF, Bushnell C, Boden-Abala B, et al. Guidelines for the primary prevention of stroke. A statement for healthcare professionals from the American Heart Association/American Stroke Association. Stroke 2014;45:3754–832.

[4] Kernan WN, Ovbiagele B, Black HR, et al. Guidelines for the prevention of stroke in patients with stroke and transient ischemic attack. A guideline for healthcare professionals from the American Heart Association/American Stroke Association. Stroke 2014;45:2160–236.

[5] Gorelick PB, Goldstein LB, Ovbiagele B. New guidelines to reduce risk of atherosclerotic cardiovascular disease. Implications for stroke prevention in 2014. Stroke 2014;45:945–7.

[6] Fang MC, Go AS, Chang Y, Borowsky LH, Pomernacki NK, Udaltsova N, et al. A new risk scheme to predict warfarin-associated hemorrhage: the ATRIA (Anticoagulation and Risk Factors in Atrial Fibrillation) Study. J Am Coll Cardiol 2011;58:395–401.

[7] Mant J, Hobbs FD, Fletcher K, Roalfe A, Fitzmaurice D, Lip GY, et al. BAFTA Investigators; Midland Research Practices Network (MidReC). Warfarin versus aspirin for stroke prevention in an elderly community population with atrial fibrillation (the Birmingham Atrial Fibrillation Treatment of the Aged Study, BAFTA): a randomized controlled trial. Lancet 2007;370:493–503.

[8] Barnett HJ, Taylor DW, Eliasziw M, Fox AJ, Ferguson GG, Haynes RB, et al. North American Symptomatic Carotid Endarterectomy Trial Collaborators. Benefit of carotid endarterectomy in patients with symptomatic moderate or severe stenosis. N Engl J Med 1998;339:1415–25.

[9] Ferguson GG, Eliasziw M, Barr HW, Clagett GP, Barnes RW, Wallace MC, et al. The North American Symptomatic Carotid Endarterectomy Trial: surgical results in 1415 patients. Stroke 1999;30:1751–8.

[10] Rothwell PM, Eliasziw M, Gutnikov SA, Warlow CP, Barnett HJ. Carotid Endarterectomy Trialists Collaboration. Endarterectomy for symptomatic carotid stenosis in relation to clinical subgroups and timing of surgery. Lancet 2004;363:915–24.

[11] Chaturvedi S, Turan TN, Lynn MJ, Kasner SE, Romano J, Cotsonis G, et al. WASID Study Group. Risk factor status and vascular events in patients with symptomatic intracranial stenosis. Neurology 2007;69:2063–8.

[12] SPRINT Research Group. A randomized trial of intensive versus standard blood-pressure control. N Engl J Med 2015;373:2103–16.

CHAPTER

165

Management of Lipid Metabolism

Y. Turan, A. Kozan, M.K. Başkaya

University of Wisconsin–Madison, Madison, WI, United States

INTRODUCTION

Recent studies have demonstrated that elevated low-density lipoprotein (LDL) cholesterol and total cholesterol (TC) levels and decreased levels of high-density lipoprotein (HDL) cholesterol are particularly significant risk factors for ischemic cerebrovascular disease—is the second leading cause of death after ischemic heart disease (IHD) [1]. The pathophysiological basis of both diseases refers to atherosclerosis.

Other determinants, such as diabetes mellitus (DM), arterial hypertension, cigarette smoking, and alcohol consumption, are considered as risk factors not only for IHD but also for ischemic cerebrovascular disease.

Atherosclerotic processes begin even in childhood and are affected over the time as a result of a genetic tendency and environmental and modifiable risk factors throughout the life. Therefore, it is recommended to begin preventative measures early in life. Maintaining a normal body weight, having a healthy diet, and regular physical activities are key points for prevention of atherosclerosis in childhood. On the other hand, dietary supplements, medical treatment, and surgical procedures, which include carotid endarterectomy (CEA), carotid artery stenting (CAS), and bypass surgery, are the treatment options for occlusive cerebrovascular diseases.

However, treatment recommendations have to be managed on a patient according to the current data and guidelines for restoring abnormal blood lipids to normal levels.

PATHOPHYSIOLOGY OF LIPID METABOLISM IN CEREBROVASCULAR DISEASE

Atherosclerosis overwhelmingly causes more morbidity and mortality in the world than any other disorder. Although epidemiological data related to atherosclerosis mortality typically reflect the death caused by IHD, carotid atherosclerotic disease is also associated with significantly high morbidity and mortality rates. It is characterized by intimal lesions called atheromas, which are the result of deterioration of lipid metabolism [2]. They eventually protrude into the vascular lumina resulting in cerebrovascular disease.

There are several risk factors, combination of which is the main indicator of the prevalence and severity of atherosclerosis. Some of them (age, gender, family history, and genetic abnormalities) are constitutional, therefore, are less modifiable, in contrast to other acquired risk factors (hyperlipidemia, arterial hypertension, DM, and cigarette smoking), which are more modifiable [2].

Several hypotheses have been proposed to understand the mechanism of its initiation and development [2]. The mainly supported one is the response-to-injury theory. According to this theory, the initial provocative reason for this disease to occur is an endothelial injury. The subsequent step is mainly dictated by distorted lipid profile and accumulation of modified lipoproteins in the vessel wall.

The first characteristic manifestation of lipid abnormality in cerebrovascular disease is the occurrence of fatty streaks [2]. Initially, these fatty streaks are not significantly raised and thus do not cause any disturbances in blood flow. These lesions usually begin in childhood and continue to increase both in size and number with advancing age. Fatty streaks initially appear as smooth areas of yellowish discoloration, flat spots that can coalesce into elongated streaks, 1cm long or longer, on the intimal surface of medium- to large-sized arteries. Histologically, they are composed of an accumulation of lipid-laden macrophages, known as foam cells, and extracellular matrix in the intima of vessels. With further evolution, fatty streaks will eventually transform to fibro-fatty atheromatous plaques consisting of proliferated smooth muscle cells, foam cells, extracellular lipid,

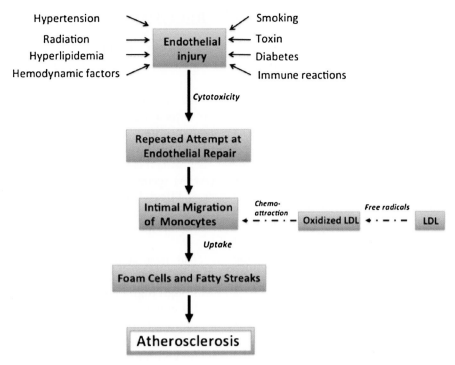

FIGURE 165.1 Pathogenesis of atherosclerosis.

and extracellular matrix (Fig. 165.1). These cells within the deep lamina of the lesions are sensitive to apoptosis, which is related to more macrophage accumulation and calcified vesicle formations resulting in the transformation of fatty streaks into atheromatous plaques. These atheromatous plaques growth and remodeling continue throughout the life span as a dynamic process. As the plaques increase in size, vessel stenosis occurs with a considerable increase in flow velocity and turbulence, although total flow is reduced. The irregular surface of the enlarging plaque can cause fragments to embolize.

Lipids are the most significant risk factors in the genesis of this atherosclerosis process. Hypercholesterolemia, more specifically, elevated LDL cholesterol and very low-density lipoprotein (VLDL) cholesterol levels, and decreased HDL cholesterol level are all major risk factors for atherosclerosis. Even in the absence of other risk factors, hypercholesterolemia is sufficient to stimulate lesion development.

The cornerstone aspect of response-to-injury hypothesis is chronic and repetitive endothelial injury [2], which is another risk factor in the development of atherosclerosis. Normally, endothelium functions as an active barrier between blood and other tissues. It is also responsible for inflammation and hemostasis in the blood circulation. The specific causes of endothelial dysfunction in early atherosclerosis are not completely understood. Etiological culprits include toxins from cigarette smoking, homocysteine, and even infectious agents. Inflammatory cytokines (e.g., interleukins 1 and 6 (IL-1 and IL-6), tumor necrosis factor alpha (TNF-α),

C-reactive protein (CRP), and phospholipase A2) can also stimulate the expression of proatherogenic genes in endothelial cells. Nevertheless, the two most important causes of endothelial dysfunction are hemodynamic disturbances and hypercholesterolemia. An increasing amount of adhesion molecules, cytokines, and free radicals derived from endothelial injury causes the migration and proliferation of more smooth muscle cells and macrophages in the vessel wall. Lipids, especially LDL, are taken up by these macrophages (foam cells) in this region. Because of the high chemoattractant property for monocyte-derived macrophages in plaque, oxidized LDL is picked up more rapidly. Oxidized LDL also leads to endothelial injury[3] and increases inflammatory and immune processes by cytokine secretion from lipid-enriched macrophages. On the other hand, HDL has antiatherogenic features that are responsible for reverse lipid transportation, maintenance of endothelial cell function, and prevention of thrombosis.

Besides these, high blood lipoprotein (a) levels are also risk factors for atherosclerosis and ischemic cerebrovascular disease, because they stimulate a release of a plasminogen activator inhibitor-1 (PAI-1), which retards the development of atherosclerosis.

Furthermore, loss of endothelium-derived nitric oxide causes endothelial vasodilator dysfunction that contributes to the atherosclerotic process. The atherosclerotic plaque process begins with a proliferation of smooth muscle cells, which synthesize connective tissues to yield an extended matrix for the plaque. Then, cholesterol accumulation in macrophages gives rise to mitochondrial dysfunction, increased cellular proteases, and cytokines' release, which results in apoptosis and necrosis. Necrosis of lipid-laden cells gives rise to a lipid core characteristic of onward lesions. After an endothelial injury occurs, reactive intercellular messengers and ingredients collect in the vessel wall leading to further destruction and plaque growth.

Lipid mediators are also closely engaged in tissue injury after the cerebrovascular event starts, which results in mediator genesis of inflammation, such as thromboxane A2, leukotrienes, and prostaglandins. These substances lead to accumulation of platelets and polymorphonuclear leukocytes. Leukotrienes also enhance vascular permeability, which causes the deterioration of the blood–brain barrier and brain edema. The interactions of these cause a tissue damage cascade [2].

LIPIDS AND ATHEROSCLEROSIS AND THEIR INTERACTION WITH OTHER RISK FACTORS

The other main treatable risk factors for atherosclerosis are arterial hypertension, DM, and tobacco use. Hypertension plays a significant role in the genesis of atherosclerosis in cerebrovascular disease. It enhances tension in the arterial wall, breaks repair processes, and leads to aneurysm formation. Hypertension is closely related to increased risk of subclinical or silent stroke occurrence in recent studies for cerebrovascular diseases [4].

DM is also another significant risk factor, increasing the stroke risk twice compared with nondiabetic patients [5]. This risk is higher in women than in men. Impaired glucose tolerance may also increase transient ischemic attack (TIA) and stroke risk. Endothelial dysfunction, dyslipidemia, and platelet and coagulation disorders are usually seen in diabetic patients, which result with atherosclerosis. A 2012 study showed that during the foam cell formation, the insulin triggers an increase in CD36 expression and a decrease in ABCA-1 expression, which leads to cholesterol accumulation in the macrophages [6]. In type 2 diabetic patients, elevated MCP-1 and IL-6 cytokine levels also contribute to atherosclerosis formation [7].

Cigarette smoking is also a known major risk factor that influences all phases from endothelial dysfunction to clinical cases in the formation of atherosclerosis. It decreases nitric oxide availability on endothelium and increases inflammatory markers, such as IL-1, IL-6, TNF-α, and CRP, which increase atherosclerotic processes. It also increases LDL oxidation and decreases paraoxonase enzyme activity in plasma that protects from LDL oxidation effects.

RECOMMENDATION FOR PREVENTION AND TREATMENT OF LIPID-ASSOCIATED CEREBROVASCULAR DISEASE

Atherosclerotic processes begin in childhood and are affected by genetic tendencies and environmental and modifiable risk factors throughout the course of life. Therefore, both the National Cholesterol Education Program (NCEP) and American Heart Association (AHA) recommend starting prevention at these early ages. Lifestyle modifications with average body weight, healthy diet, and regular physical activity are the cornerstones for prevention of atherosclerosis in childhood [8].

The first step of cerebrovascular disease prevention should be the reduction of high cholesterol level. Although high blood cholesterol and LDL levels have a well-known relationship with coronary artery disease, no similar results have been identified for cerebrovascular diseases. It is unlikely that hemorrhagic stroke is related with increased cholesterol levels and atherosclerosis pathophysiologically. However, ischemic stroke was found strongly associated with high total cholesterol and LDL levels in many studies [1]. Therefore, many guidelines recommend regular monitoring and lowered levels of high cholesterol for cerebrovascular diseases.

There are many ways to lower serum lipid levels [9]. A healthy diet and adequate nutrition education are important factors for improving blood lipids. Mediterranean-type diet is also recommended for the prevention of stroke, which includes vegetables, fruits, low-fat products, nontropical vegetable oils, and nuts. Calorie intake from saturated fatty foods also should be significantly reduced. Increased weight is a predictor of low HDL level in the body, so weight loss is very effective to achieve high blood HDL level. Lifestyle modifications, such as cigarette smoking cessation, limited alcohol consumption, and regular physical activities, are critical ways of reducing the risk of ischemic stroke and atherosclerosis. Cigarette smoking has a high risk in all types of stroke. Smokers have 2.58 times more stroke risk than nonsmokers. Other studies showed that the risk of stroke decreases after smoking cessation and it is eliminated within 5 years. Also excessive alcohol intake increases stroke risk. It also depends on the types of stroke and ethnicity. Physical activity and exercise are associated with low stroke rates. The underlying mechanism is not well understood. However, it has a dose–response relationship between the strength of exercise and stroke. Extremely vigorous exercises have more beneficial protective effects on cerebrovascular events [8]. Moderate to high exercise is suggested for at least 30–40 min most days of the week [9]. Moderate intensity results in breaking a sweat or elevating the heart rate. Specifically, regular exercise for longer than 12 weeks has also been shown to increase plasma HDL levels [8].

Dietary supplements are widely used in the United States [8]. About one-third to one-half of the population uses them regularly. There are many types of lipid-lowering supplements, such as vitamin E and C, garlic, fish oil, and soy products. It is hypothesized that use of vitamin E and C as major antioxidants can retard LDL oxidation processes for plaque formation. However, in the major clinical trials, the beneficial effects of antioxidants' use have not been shown to prevent stroke. One of the most practical ways to lower lipid levels in the blood is taking fish oil supplements. It contains omega-3 fatty acid, which inhibits the synthesis of VLDL and helps to reduce lipids. Garlic also reduces lipid and blood pressure levels and it is accepted as an antioxidant and anti-inflammatory substance. However, there is not any clear evidence for protective effects of garlic in atherosclerosis.

Nowadays, many useful drugs are used to lower blood lipid levels, but no magic single drug exists for all circumstances. Statins are the most likely group that improves lipid profile. These agents inhibit cholesterol synthesis and also have a role in the upregulation of LDL receptors. Combination therapy may be considered with a statin and niacin or fibrate in some high-risk patients. However, medical treatment needs patient education and systemic follow-up in management of lipid disorders.

The carotid artery bifurcation is the most affected region in carotid atherosclerosis. Expansion of plaque makes the lumen narrower, and ulceration accompanies with it. This process may result in TIAs or strokes. CEA, CAS, and medical management are conventional treatment options for carotid artery stenosis [9]. CEA is an efficient and safe treatment modality for both symptomatic and asymptomatic patients with carotid atherosclerosis and is suggested for patients with 70–99% carotid stenosis who have, at least, a 5-year life expectancy. The lesion also should be surgically accessible, and the patient ideally should have no prior endarterectomy, ipsilaterally or clinically, and cardiac, pulmonary, or other diseases that can increase the risks of anesthesia and surgery. CAS can be considered in these patients who are not appropriate for surgery. Patients, especially men, who have 50–69% stenosis and with new symptoms such as TIA and stroke are also recommended for CEA [9]. Bypass surgery can also be considered for symptomatic patients who have low cerebral blood flow due to atherosclerosis. However, additional data and a new consensus are required on behalf of this subject [10].

References

[1] Young AR, Ali C, Duretete A, Vivien D. Neuroprotection and stroke: time for a compromise. J Neurochem 2007;103(4):1302–9.

[2] Mitchell RN. Blood vessels. In: Kumar V, Abbas AK, Aster JC, editors. Robbins basic pathology. 9th ed. Philadelphia: Elsevier Saunders; 2013. p. 327–65.

[3] Vink H, Constantinescu AA, Spaan JA. Oxidized lipoproteins degrade the endothelial surface layer: implications for platelet-endothelial cell adhesion. Circulation 2000;101(13):1500–2.

[4] Vermeer SE, Longstreth Jr WT, Koudstaal PJ. Silent brain infarcts: a systematic review. Lancet. Neurol 2007;6(7):611–9.

[5] Luitse MJ, Biessels GJ, Rutten GE, Kappelle LJ. Diabetes, hyperglycaemia, and acute ischaemic stroke. Lancet. Neurol 2012;11(3):261–71.

[6] Park YM, Kashyap SR, Major JA, Silverstein RL. Insulin promotes macrophage foam cell formation: potential implications in diabetes-related atherosclerosis. Lab Invest 2012;92(8):1171–80.

[7] Fernandez-Real JM, Ricart W. Insulin resistance and chronic cardiovascular inflammatory syndrome. Endocr Rev 2003;24(3):278–301.

[8] Fletcher B, Berra K, Ades P, et al. Managing abnormal blood lipids: a collaborative approach. Circulation 2005;112(20):3184–209.

[9] Kernan WN, Ovbiagele B, Black HR, et al. Guidelines for the prevention of stroke in patients with stroke and transient ischemic attack: a guideline for healthcare professionals from the American Heart Association/American Stroke Association. Stroke 2014;45(7):2160–236.

[10] Rodriguez-Hernandez A, Josephson SA, Langer D, Lawton MT. Bypass for the prevention of ischemic stroke. World Neurosurg 2011;76(Suppl. 6):S72–9.

CHAPTER

166

Antiplatelet Agents: Mechanisms and Their Role in Stroke Prevention

M.K. Erdman[1], M.C. Leary[2,3]

[1]Los Angeles County Hospital and USC Medical Center, Los Angeles, CA, United States; [2]Lehigh Valley Hospital and Health Network, Allentown, PA, United States; [3]University of South Florida, Tampa, FL, United States

INTRODUCTION

Ischemic stroke presents several opportunities for medical intervention, distinguished by their relative timing:

1. primary prevention of a stroke;
2. control of tissue damage in the acute ischemic stroke setting;
3. secondary prevention following a transient ischemic attack (TIA) or stroke.

Considering the high morbidity and mortality associated with stroke, primary and secondary prevention are paramount. Traditionally, antiplatelets are used in this third setting, to prevent recurrent ischemic stroke. However, antiplatelet agents are not the drug of choice across the board to prevent all types of ischemic stroke. For example, an elderly woman with cardioembolic stroke due to atrial fibrillation would be better served with an anticoagulant to minimize risk of future cerebral infarction, rather than an antiplatelet agent. The decision whether a physician should utilize antiplatelet agents for secondary stroke prevention depends on the pathogenesis or cause of the initial cerebral infarct.

The mechanisms of action of antiplatelet agents present a compelling overlap with the pathogenic mechanisms that define brain ischemia; therefore, antiplatelet agents theoretically could have a role in all three stages of the stroke timeline. We herein review antiplatelet agent mechanisms in the context of stroke pathology for which their use is appropriate, outline the current index of antiplatelet agents, and highlight the practical concerns of these medications in their clinical use.

PATHOGENESIS OF STROKE AND CONTRIBUTING DISEASE STATES

The decision to use an antiplatelet agent versus an anticoagulant should be driven by the white-clot/red-clot paradigm [1,2]. There are significant differences in the manner in which white thrombi and red thrombi form, and the treatments used to prevent each type of clot vary (Table 166.1). Red erythrocyte–fibrin clots are treated with thrombolytic drugs and anticoagulants such as heparin, warfarin, and the newer novel oral anticoagulants. White platelet-fibrin clots do not contain red blood cells and are prevented better by antiplatelet agents. If a physician determines that their patient's stroke was likely due to an increased risk of forming white thrombi, then antiplatelet medications can rationally be selected for future stroke prevention [1,2].

White thrombi in general are both platelet and thrombin-rich, typically forming in fast-moving arteries where the endothelium has been damaged. Endothelial damage can initiate an intravascular process that ultimately disrupts blood flow to the brain (Fig. 166.1). The damage may occur with an apparent disruption of the endothelium, such as at the site of an atherosclerotic-ruptured plaque, as well as with more subtle irritations such as that seen with syphilitic vasculitis. The damaged barrier of endothelium invites platelet adhesion, a process mediated by GPIb/IX receptors on platelets and endothelial collagen-bound von Willebrand factor (vWF) multimers. The remainder of the process is a review of the normal cascade of platelet-mediated hemostasis [1,2].

Primer on Cerebrovascular Diseases, Second Edition
http://dx.doi.org/10.1016/B978-0-12-803058-5.00166-1

TABLE 166.1   Differences Between White and Red Thrombi

White Clots	Red Clots
Composed mainly of platelets and fibrin	Composed mainly of red blood cells and fibrin
Form within arteries	Form in arteries or veins
Form almost exclusively in areas that have an abnormal endothelial surface	Do not require an abnormal vessel wall or tissue thromboplastin
Do not form on the surface of myocardial infarcts	Form on the surface of myocardial infarcts
Often form on damaged heart valves	Often form on damaged heart valves
Develop in fast-moving bloodstreams	Develop in areas of slow blood flow
Prevented with antiplatelet agents	Prevented with anticoagulants

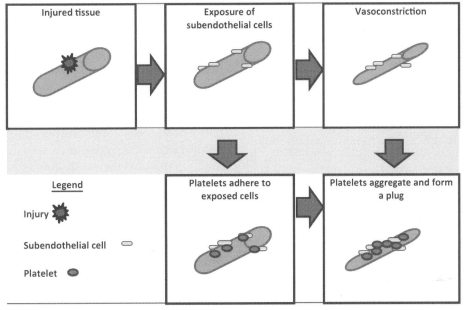

FIGURE 166.1   Formation of white thrombus.

Review of the Normal Cascade of Platelet-Mediated Hemostasis

Platelets are locally activated by thrombin, thromboxane, and adenosine diphosphate (ADP). Thrombin serves as the main bridge between platelet activation and the coagulation cascade. Thromboxane activates the GIIb/IIIa receptors on platelets and initiates platelet aggregation. ADP binds to the $P2Y_{12}$ G-protein-coupled receptor that, in turn, increases the platelet cytosolic calcium (Ca^{2+}) level and induces platelet activation. Additionally, the $P2Y_{12}$ receptor, also bound by ADP, causes a morphological change in the platelet to a more spherical shape and the development of pseudopodia. Following the thromboxane-induced activation of the GIIb/IIIa platelet-bound receptors, these receptors are then capable of binding either endothelial vWF or fibrinogen. Fibrinogen is coupled with another GIIb/IIIa receptor on a neighboring platelet. This process allows for the stabilization and propagation of the clot, respectively [2].

Multiple factors work to inhibit this process, which, when in a healthy state, provide the delicate balance required for hemostasis. Prostacyclin is released by the endothelium and counteracts the interaction of ADP and the $P2Y_{12}$ receptor. When prostacyclin binds to a separate platelet-bound G-protein-coupled receptor, it *decreases* the cytosolic Ca^{2+} and inhibits platelet activation. Prostacyclin has a synergistic antiaggregating effect with another vasodilator, nitric oxide (NO). NO is a gaseous signaling molecule synthesized in the vascular endothelium that crosses the cell membrane easily and prompts relaxation of vascular smooth muscle. The resultant vasodilation and increase in blood flow creates an unfavorable environment for thrombogenesis. NO also inhibits platelet adhesion by activating a molecular signaling pathway [2]. Each antiplatelet agent inhibits at least one of the processes in this cascade to a greater or lesser degree. Although the effects of the different antiplatelet agents are similar, their mechanisms can differ.

Stroke Risk Factors That Pose a Risk for White Thrombi

What risk factors are associated with platelet-dependent thrombus formation? The cascade of white clot formation can be triggered by a variety of diseases and conditions. Atherosclerotic disease can cause direct arterial endothelial insult and subsequent platelet fibrin-rich thrombus formation, particularly in areas of ulceration. Hypercholesterolemia is a risk factor for the development of atherosclerotic disease and also directly enhances platelet thrombus formation on injured arteries. Infections such as syphilis, bacterial and fungal meningitis, tuberculosis, vasculitis, sickle cell disease, and HIV all lead to an increase in stroke risk secondary to endothelial damage. Lastly, hyperhomocysteinemia, a result of either a vitamin deficiency (vitamin B_6, B_{12}, or folate) or inherited enzymatic defects, can also cause direct harm to the endothelium [1,2].

Smoking is a significant risk factor for stroke, by way of multiple mechanisms. As smoking predisposes to hypertension and hyperlipidemia by indirectly increasing the risk of stroke. However, smoking directly escalates the risk of white-thrombus formation by causing toxin-mediated damage to the vascular endothelium as well as causing platelets to be in a hyperaggregable state [2,3]. This phenomenon is also seen in diabetes mellitus, antiphospholipid antibody syndrome, and heparin-induced thrombocytopenia [2]. Collagen vascular diseases, such as systemic lupus erythematosus and scleroderma, are associated with anticardiolipin antibodies, and thus potential for platelet-dependent thrombus formation. These antibodies bind to the vascular endothelium and inhibit the release of prostacyclin, creating a prothrombotic state [4].

ANTIPLATELET AGENTS: MECHANISMS AND ROLE IN STROKE PREVENTION

In the United States, four antiplatelets agents have been approved by the FDA for prevention of vascular events in patients with prior stroke or TIA: aspirin, ticlopidine, clopidogrel, and combination aspirin/dipyridamole. In Asia and elsewhere cilostazol is also widely prescribed. In general, these agents reduce the relative risk of future ischemic stroke by 22%. There are important differences

between the various antiplatelet medications that have direct implications for therapeutic selection.

Aspirin

A medication so old that it dates back to Hippocrates, aspirin is a nonsteroidal antiinflammatory drug that works by irreversibly inhibiting platelet cyclooxygenase-1 causing a subsequent decrease in the production of prostacyclin and thromboxane. As detailed in the preceding section, prostacyclin functions as an inhibitor of platelet aggregation while thromboxane induces platelet aggregation and potentiates thrombin and ADP. The inhibition of thromboxane as an inducer of platelet aggregation, as well as proportionately smaller effect on thrombin and ADP is the basis for the rationale of multiple antiplatelet drug regimens. The prothrombotic effect of decreased prostacyclin is counterbalanced by the shunting of the arachidonic acid pathway to the production of lipoxygenase. Intermediates of this pathway, including hydroperoxyeicosatetraenoic acid and hydroxyl-eicosatetraenoic acid (15-HPETE and 15-HETE, respectively) inhibit platelet aggregation, meaning that a greater production of lipoxygenase discourages thrombosis [2,5].

Aspirin not only targets the cyclooxygenase enzyme in platelets, but also inhibits this enzyme in endothelial cells. The inhibition of prostacyclin in vascular endothelial cells imposes a prothrombotic risk; however, this is less relevant clinically than the impaired production of thromboxane. It is important to note that as nucleated cells, vascular endothelial cells have the potential to replete the cyclooxygenase-1 enzyme and restore function. This ability is not shared with platelets, in which aspirin irreversibly inhibits the entire store of cyclooxygenase-1 for the lifetime of the affected platelet (7–10 days) [2,6]. Improved endothelial function has been shown in the setting of atherosclerosis by modulating acetylcholine-induced vasodilation of peripheral arteries subsequent to an alteration in the driving endothelial cyclooxygenase pathway [2,7]. Aspirin may also impart an antiplatelet aggregating effect by an NO-mediated interaction with neutrophils, as well as protect the endothelium from oxidative stress caused by free radicals.

In a metaregression analysis of placebo-controlled trials of aspirin for secondary ischemic stroke prevention, the relative risk reduction for any type of stroke overall (hemorrhagic or ischemic) was estimated at 15%. The benefit of aspirin is relatively similar over a range of doses: from 50 to 1500 mg daily [8]. Some studies have shown variability in the effect of various doses on platelet function in vitro; some patients show more effective inhibition with higher aspirin dosage [9,10,11]. The side effects also vary by dose, including bleeding complications and gastrointestinal effects.

Thienopyridines

Ticlopidine

Thienopyridines, namely ticlopidine and clopidogrel, inhibit platelet function by working as noncompetitive antagonists to the $P2Y_{12}$ G-protein-coupled receptor found on the surface of platelets; this receptor typically binds ADP and activates platelet aggregation. More specifically, it is theorized that a metabolite of these drugs, possibly the same metabolite, is the active form. This was demonstrated when plasma from treated patients was shown to have an inhibitory effect on untreated platelets [12]. Ticlopidine has fallen out of favor in clinical practice due to a side effect profile that includes neutropenia, bone marrow aplasia, and thrombotic thrombocytopenic purpura [8].

Clopidogrel

Clopidogrel is an irreversible inhibitor of the $P2Y_{12}$ receptor, inhibiting the aggregation of a platelet for the entire lifespan of that platelet (7–10 days). Like aspirin it directly affects platelet metabolism and functions. It has a similar safety profile to aspirin, with a minor increase in the incidence of diarrhea and rare but serious instances of thrombotic thrombocytopenic purpura reported [8,13]. It has been shown to prevent both arterial and venous atherogenesis in animal models [14]. It is a once-daily medication, as compared to the twice-daily dosing regimen of ticlopidine. There have been several trials comparing clopidogrel to other antiplatelet agents (Table 166.2) [8,13,15–22]. Based on the result of the CHANCE trial, timing should be considered in the selection of this antiplatelet agent. The combination of aspirin and clopidogrel, initiated within 24 h of a minor ischemic stroke or TIA, may be effective in preventing recurrent stroke within the first 90 days. After those 90 days, based on other trial results, transition to a single antiplatelet agent would be appropriate. Observation from the survival curves of CAPRIE and PRoFESS indicate that clopidogrel is probably as effective as the combination of aspirin/dipyridamole and, by inference, aspirin.

There has been controversy over the effect of concomitant use of proton pump inhibitors (PPIs) and clopidogrel, with a decrease in the effectiveness of the latter drug. A 2015 study by Leonard et al. showed no significant difference in the hazard ratio for ischemic strokes in patients on clopidogrel and a PPI when compared to pantoprazole [23]. Inter-subject variability of efficacy has been partly attributed to alterations of the cytochrome P450 enzyme in some individuals.

Dipyridamole

Dipyridamole has a constellation of effects that are mediated by alterations in the molecular signaling pathways among vascular endothelium and platelets.

Essentially, dipyridamole inhibits phosphodiesterase; this leads to an increase in cGMP in vascular smooth muscle cells and, ultimately, vasodilation. Likewise, the levels of cAMP in platelets also increase, causing a blockade of platelet aggregation. This effect is enhanced by the prevention of platelet uptake of adenosine, and a stimulation of platelet adenylyl cyclase secondary to this increase in extracellular adenosine. Moreover, a variety of pleiotropic effects, including attenuation of the nuclear transcription factor nuclear factor kappa B (NF-κB) and a subsequent decrease in cytokine production, limit the oxidative stress and inflammation experienced by the vascular bed [24]. Its role in clinical practice is typically coupled with aspirin, and the combination has been shown to be effective for secondary stroke prevention (Table 166.3), although side effects can cause it to be less well tolerated [8,15,25–28]. Main side effects included headache and gastrointestinal symptoms.

Glycoprotein IIb/IIIa Inhibitors

Abciximab, eptifibatide, and tirofoban are three intravenous drugs that inhibit platelet aggregate by occupying the binding site of GIIb/IIIa. This prevents fibrinogen from binding, linking to a neighboring platelet, and propagating the clot. These drugs are typically used in coronary angioplasty, but have also been investigated for their effect in an acute ischemic stroke setting. Theoretically, they would increase the rate of recanalization, allow for improved microcirculation, and minimize tissue death secondary to ischemia. However, their efficacy is yet to be shown to outweigh the increased risk of bleeding; therefore, as of 2016, they are not recommended or FDA approved in the acute ischemic stroke setting [29].

Newer Antiplatelet Agents

Cilostazol is a reversible type III phosphodiesterase inhibitor that causes both vasodilation and inhibition of platelet aggregation by increasing the levels of cAMP. Moreover, there is some suggestion that it has an endothelial-protective mechanism and induces production of NO. It is the mainstay of treatment for intermittent claudication, but its potential as a secondary stroke preventative medication is starting to be explored. A noninferiority study comparing cilostazol to aspirin for the reduction of secondary strokes demonstrated no significant difference in the reduction of secondary stroke and a decreased rate of intracranial or systemic hemorrhage in the cilostazol group [30]. This study was done with an Asian population, so the findings may not be generalizable. A smaller study explored the therapeutic potential of cillostazol in the setting of acute small vessel infarct. The authors reported a reduced risk of early neurological

TABLE 166.2   Comparison of Clopidogrel With Other Antiplatelet Agents in Stroke Prevention

Study	N	Outcome
CAPRIE compared 1. aspirin to 2. clopidogrel in patients with prior stroke, MI, or peripheral vascular disease	19,000+	The annual rate of ischemic stroke, MI, or peripheral vascular disease was 5.32% in clopidogrel group and 5.83% of aspirin group. Study not designed to show superiority.
PRoFESS compared 1. aspirin/dipyridamole to 2. clopidogrel in patients with non-cardioembolic ischemic stroke	20,332	No difference in stroke rates over a mean of 2.5 years. Ischemic stroke occurred in 9.0% of the combination group compared with 8.8% on clopidogrel alone. The risk of gastrointestinal (GI) hemorrhages was much higher in the combination group.
MATCH compared 1. aspirin/clopidogrel to 2. clopidogrel in the prevention of vascular events among patients with a recent TIA or stroke	7599	No benefit of long-term combination therapy compared to clopidogrel alone over 3.5 years. The risk of major hemorrhages was significantly increased in the combination group compared to clopidogrel alone (1.3% absolute increase in life-threatening bleeding).
CHARISMA compared 1. aspirin/clopidogrel to 2. aspirin in patients with stroke or TIA within 5 years or multiple cerebrovascular risk factors	15,603	An analysis of the subgroup of patients that entered the study after having had a stroke showed an increased bleeding risk, but no benefit of long-term combination therapy compared to aspirin alone.
SPS3 compared 1. aspirin/clopidogrel to 2. aspirin in patients with MRI-confirmed lacunar stroke within 180 days	3026	The rate of recurrent ischemic or hemorrhagic stroke was 2.7%/year in the aspirin monotherapy group and 2.5% in the combination group. Ischemic stroke rate was lower in the combination group, but mortality and hemorrhage rate was higher for both intracerebral hemorrhage (ICH) and GI.
FASTER* compared 1. aspirin/clopidogrel to 2. aspirin in patients with TIA or prior stroke, patients were enrolled within 24 h of their prior stroke/TIA	392	There was a trend toward a reduced rate of ischemic events with combination therapy (7.1%) patients on combination therapy had a stroke within 90 days compared with 10.8% of patients on aspirin alone, with only a small 1% risk of symptomatic ICH. Study stopped early for low recruitment.
CHANCE* compared 1. aspirin/clopidogrel to 2. aspirin among patients with TIA or minor stroke treated within 24 h after the onset of symptoms	5170	Combination therapy was superior to aspirin alone in reducing stroke in the first 90 days and did not increase the risk of hemorrhage. Ischemic or hemorrhagic stroke was seen in 8.6% of combination therapy and in 11.7% of aspirin monotherapy patients. Subgroup analysis of patients with intracranial artery stenosis (ICAS) showed a higher rate of recurrent stroke than in those without ICAS, but no significant difference in the response of ICAS versus non-ICAS groups to the two antiplatelet therapies. In another substudy, clopidogrel plus aspirin compared to aspirin alone improved 90-day functional outcome as measured by modified Rankin scale.

deterioration, but insisted on a larger, randomized control trial [31]. Cilostazol is often used in patients with peripheral vascular occlusive disease. Until 2016 its effect on stroke prevention was mainly studied in Asians, and the patients studied had a high frequency of penetrating artery disease (lacunar strokes) and intracranial large artery disease. Its effect in Caucasian patients in the United States and Europe has not been studied in any depth.

Sarpogrelate has been used in Japan for the treatment of peripheral artery disease, and, like cilostazol, is being explored for its potential therapeutic benefit in stroke prevention. The mechanism of action of sarpogrelate is a selective 5-hydroxytryptamine (5-HT) receptor antagonist, inhibiting both platelet aggregation and vasoconstriction. In a randomized, double-blind, aspirin-controlled trial of sarpogrelate versus aspirin, aspirin was confirmed as the first-line treatment with regard to prevention of stroke recurrence. However, it is important to note that the sarpogrelate arm had fewer hemorrhagic events than aspirin [32].

Triflusal, a biochemical cousin to aspirin, irreversibly inhibits cyclooxygenase-1 in platelets, reducing thromboxane production, but does not impact endothelial arachidonic acid metabolism; it also increases the production

TABLE 166.3 Comparison of Aspirin/Dipyridamole to Other Antiplatelet Agents in Stroke Prevention

Study	N	Outcome
ESPS-1 compared 1. aspirin 325 mg daily plus dipyridamole 75 mg TID to 2. placebo in patients with prior TIA or stroke	2500	After 24 months, the rate of stroke or death was 16% among patients in the combination group compared with 25% among the patients on placebo (relative risk reduction 33%, $p < 001$).
ESPS-2 compared 1. aspirin 25 mg BID plus dipyridamole 200 mg BID 2. aspirin 25 mg BID 3. extended release dipyridamole 4. placebo in patients with prior TIA or stroke	6602	Compared with placebo, the risk of stroke on aspirin alone was reduced by 18% ($p = 013$), 16% with dipyridamole alone ($p = 039$), and 37% with the combination ($p < 001$). Compared to aspirin alone combination therapy reduced stroke risk by 23% ($p = 006$) and of stroke or death by 13% ($p = 056$). Bleeding was not significantly increased, but headache and GI symptoms were more common in the combination aspirin/dipyridamole group.
ESPRIT compared 1. aspirin (30–325 mg daily) 2. aspirin/dipyridamole in preventing stroke, MI, vascular death or major bleeding in patients with prior ischemic stroke or TIA within 6 months over 3.5 years	2739	The primary end point was observed in 13% of patients in the combination therapy group versus 16% in the aspirin alone group. There was a surprising reduced rate of major bleeding in the combination group, which may reflect event reporting bias.
EARLY compared 1. aspirin/dipyridamole to 2. aspirin 100 mg daily × 7 days then conversion to combination therapy in patients with ischemic stroke within 24 h of symptom onset in preserving neurological function	543	No difference in functional ability, as measured by modified Rankin scale at 90 days.
PRoFESS compared 1. aspirin/dipyridamole to 2. clopidogrel in patients with non-cardioembolic ischemic stroke	20,332	No difference in stroke rates over a mean of 2.5 years. Ischemic stroke occurred in 9.0% of the combination group compared with 8.8% on clopidogrel alone. The risk of GI hemorrhages was much higher in the combination group.

of NO, leading to vasodilation. It is most frequently used for cardiovascular indications, such as coronary angioplasty and peripheral artery disease. Considering the similar pharmacodynamics, it is not surprising that no study has demonstrated superiority to aspirin in reducing secondary stroke; however, triflusal has been associated with lower hemorrhagic complications [33].

Terutroban is a selective antagonist of thromboxane–prostaglandin receptors in both platelets and the vascular endothelium. Experimental studies showed some improved function in the endothelium and a decrease in the size of atherothrombotic plaques [34]. In a randomized clinical trial, however, the promise portended by laboratory work did not manifest clinically, and terutroban did not meet criteria for noninferiority compared to aspirin in the prevention of recurrent strokes [35].

Prasugrel is a P2Y12 platelet inhibitor indicated to reduce the rate of thrombotic cardiovascular events in patients with acute coronary syndrome. It has been shown to reduce the rate of a combined end point of cardiovascular death, myocardial infarction, or stroke compared to clopidogrel. However, difference between treatments was driven predominantly by MI, with no difference on strokes and little difference on CV death.

The risk of significant bleeding was substantial, and it is contraindicated in patients with active pathological bleeding or a history of TIA or stroke [36].

Vorapaxar is a reversible antagonist of the protease-activated receptor-1 expressed on platelets, but its long half-life makes it effectively irreversible. Vorapaxar inhibits thrombin-induced and thrombin receptor agonist peptide-induced platelet aggregation in in vitro studies. It has been shown to reduce the rate of a combined end point of cardiovascular death, MI, stroke, and urgent coronary revascularization. The use of vorapaxar is contraindicated in patients with a history of stroke, TIA, or ICH, because of an increased risk for ICH in this patient population [37].

Antiplatelet Resistance

A resistance to aspirin and clopidogrel is a phenomenon observed in a subset of patients and demonstrated by platelet function testing and attributed to a multifactorial cause. However, caution should be taken to resist the seemingly intuitive decision to switch to an alternative antiplatelet drug. There have been no clinical trials to indicate that switching antiplatelet agents reduces the risk for current events [8]. Additionally, an increased rate of death, bleeding, and ischemic events has been noted when drug therapy was changed in response to measured nonresponse. For this reason, routine testing of platelet function is not recommended by the American Heart Association/American Stroke Association Guidelines [8].

References

[1] Caplan LR. Anticoagulants to prevent stroke occurrence and worsening. Isr Med Assoc J 2006;8:773–8.

[2] Helgason C.M. Mechanisms of antiplatelet agents and the prevention of stroke. In"Primer on cerebrovascular diseases" (Welch K.M.A., Caplan L.R., Reis D.J., Siesjo B.K., Weir B., Eds.). Academic Press, San Diego.

[3] Celermajer DS, Sorensen KE, Georgakopoulos D, et al. Cigarette-smoking is associated with dose-related and potentially reversible impairment of endothelium-dependent dilation in healthy-young adults. Circulation 1993;88:2149–55.

[4] Escolar G, Font J, Reverter JC, et al. Plasma from systemic lupus-erythematosus patients with antiphospholipid antibodies promotes platelet-aggregation. Studies in a perfusion system. Arterioscler Thromb 1992;12:196–200.

[5] Awtry EH, Loscalzo J. Aspirin Circ 2000;101:1206–18.

[6] Jaffe EA, Weksler BB. Recovery of endothelial cell prostacyclin production after inhibition by low-doses of aspirin. J Clin Invest 1979;1979(63):532–5.

[7] Husain S, Andrews NP, Mulcahy D, Panza JA, Quyyumi AA. Aspirin improves endothelial dysfunction in atherosclerosis. Circulation 1998;97:716–20.

[8] Kernan WN, Ovbiagele B, Black HR, et al. Guidelines for the prevention of stroke in patients with stroke and transient ischemic attack: a guideline for healthcare professionals from the American Heart Association/American Stroke Association. Stroke 2014;45:2160–236.

[9] Alberts M, Bergman D, Molner E, Jovanovic BD, Ushiwata I, Teruya J. Antiplatelet effect of aspirin in patients with cerebrovascular disease. Stroke 2004;35:175–8.

[10] Helgason CM, Tortorice KL, Winkler SR, Penney DW, Schuler JJ, McClelland TJ, Brace LD. Aspirin response and failure in cerebral infarction. Stroke 1993;24:345–50.

[11] Coleman J, Alberts MJ. Effect of *Aspirin* dose, preparation, and withdrawal on platelet response in normal volunteers. Am J Cardiol 2006;98:838–41.

[12] Diminno G, Cerbone AM, Mattioli PL, Turco S, Iovine C, Mancini M. Functionally thrombasthenic state in normal platelets following the administration of ticlopidine. J Clin Invest 1985;75:328–38.

[13] Gent M, Beaumont D, Blanchard J, et al. A randomised, blinded, trial of clopidogrel versus aspirin in patients at risk of ischaemic events (CAPRIE). Lancet 1996;348:1329–39.

[14] Herbert JM, Frehel D, Vallee E, et al. Clopidogrel, a novel antiplatelet and antithrombotic agent. Cardiovasc Drug Rev 1993;11:180–98.

[15] Sacco RL, Diener H-C, Yusuf S, et al. Aspirin and extended-release dipyridamole versus clopidogrel for recurrent stroke. N Engl J Med September 18, 2008;359:1238–51.

[16] Diener HC, Bogousslavsky J, Brass LM, Match investigators, et al. Aspirin and clopidogrel compared with clopidogrel alone after recent ischaemic stroke or transient ischaemic attack in high-risk patients (MATCH): randomised, double-blind, placebo-controlled trial. Lancet 2004;364:331–7.

[17] Bhatt DL, Fox KA, Hacke W, Charisma Investigators, et al. Clopidogrel and aspirin versus aspirin alone for the prevention of atherothrombotic events. N Engl J Med April 20, 2006;354(16):1706–17.

[18] SPS3 Investigators, Benavente OR, Hart RG, McClure LA, Szychowski JM, Coffey CS, Pearce LA. Effects of clopidogrel added to aspirin in patients with recent lacunar stroke. N Engl J Med August 30, 2012;367(9):817–25.

[19] Kennedy J, Hill MD, Ryckborst KJ, Eliasziw M, Demchuk AM, Buchan AM, Faster Investigators. Fast assessment of stroke and transient ischaemic attack to prevent early recurrence (FASTER): a randomised controlled pilot trial. Lancet Neurol November 2007;6(11):961–9. Epub October 10, 2007.

[20] Wang Y, Wang Y, Zhao X, Chance Investigators, et al. Clopidogrel with aspirin in acute minor stroke or transient ischemic attack. N Engl J Med July 4, 2013;369(1):11–9.

[21] Liu L, Wong KS, Leng X, Chance Investigators, et al. Dual antiplatelet therapy in stroke and ICAS: subgroup analysis of CHANCE. Neurology September 29, 2015;85(13):1154–62.

[22] Wang X, Zhao X, Johnston SC, Chance Investigators, et al. Effect of clopidogrel with aspirin on functional outcome in TIA or minor stroke: CHANCE substudy. Neurology 2015;85:573–9.

[23] Leonard CE, Bilker WB, Brensinger CM, et al. Comparative risk of ischemic stroke among users of clopidogrel together with individual proton pump inhibitors. Stroke 2015;46:722–31.

[24] Balakumar P, Nyo YH, Renushia R, et al. Classical and pleiotropic actions of dipyridamole: not enough light to illuminate the dark tunnel? Pharmacol Res 2014;87:144–50.

[25] The ESPS Group. The European Stroke Prevention Study (ESPS). Principal end-points. Lancet 1987;2(8572):1351–4.

[26] Diener HC, Cunha L, Forbes C, Sivenius J, Smets P, Lowenthal A. European Stroke Prevention Study. 2. Dipyridamole and acetyl-salicylic acid in the secondary prevention of stroke. J Neurol Sci 1996;143:1–13.

[27] Halkes PHA, van Gijn J, Kappelle LJ, Koudstaal PJ, Algra A. Aspirin plus dipyridamole versus aspirin alone after cerebral ischaemia of arterial origin (ESPRIT): randomised controlled trial. Lancet 2006;367:1665–73.

[28] Dengler R, Diener HC, Schwartz A, Grond M, Schumacher H, Machnig T, Eschenfelder CC, Leonard J, Weissenborn K, Kastrup A, Haberl R, Early Investigators. Early treatment with aspirin plus extended-release dipyridamole for transient ischaemic attack or ischaemic stroke within 24 h of symptom onset (EARLY trial): a randomised, open-label, blinded-endpoint trial. Lancet Neurol 2010;9:159–66.

[29] Jauch EC, Saver JL, Adams HP, et al. Guidelines for the early management of patients with acute ischemic stroke: a guideline for healthcare professionals from the American Heart Association/American Stroke Association. Stroke March 2013;44:870–947.

[30] Shinohara Y, Katayama Y, Uchiyama S, et al. Cilostazol for prevention of secondary stroke (CSPS 2): an aspirin-controlled, double-blind, randomised non-inferiority trial. Lancet Neurol 2010;9:959–68.

[31] Nakase T, Sasaki M, Suzuki A. The effect of acute medication with cilostazol, an anti-platelet drug, on the outcome of small vessel brain infarction. J Stroke Cerebrovasc Dis 2014;23:1409–15.

[32] Shinohara Y, Nishimaru K, Sawada T, et al. Sarpogrelate-Aspirin Comparative Clinical Study for efficacy and safety in secondary prevention of cerebral infarction (S-ACCESS): a randomized, double-blind, aspirin-controlled trial. Stroke 2008;39:1827–33.

[33] Culebras A, Rotta-Escalante R, Vila J, et al. Triflusal vs aspirin for prevention of cerebral infarction: a randomized stroke study. Neurology 2004;62:1073–80.

[34] Viles-Gonzalez J.F., Fuster V., Corti R., Inventors, et al., Atherosclerosis regression and TP receptor inhibition: effect of S18886 on plaque.

[35] Bousser MG, Amarenco P, Chamorro A, et al. Terutroban versus aspirin in patients with cerebral ischaemic events (PERFORM): a randomised, double-blind, parallel-group trial. Lancet 2011;377:2013–22.

[36] Antman EM, Wiviott SD, Murphy SA, Voitk J, Hasin Y, Widimsky P, Chandna H, Macias W, McCabe CH, Braunwald E. Early and late benefits of prasugrel in patients with acute coronary syndromes undergoing percutaneous coronary intervention: a TRITON-TIMI 38 (TRial to Assess Improvement in Therapeutic Outcomes by Optimizing Platelet InhibitioN with Prasugrel-Thrombolysis In Myocardial Infarction) analysis. J Am Coll Cardiol 2008;51:2028–33.

[37] Morrow DA, Braunwald E, Bonaca MP, TRA 2P–TIMI 50 Steering Committee and Investigators, et al. Vorapaxar in the secondary prevention of atherothrombotic events. N Engl J Med 2012;366:1404–13.

CHAPTER

167

Anticoagulants in Stroke Treatment

B. Huisa[1], M. Fisher[2]

[1]University of California San Diego, San Diego, CA, United States; [2]University of California, Irvine, Irvine, CA, United States

INTRODUCTION

Anticoagulants continue to play a central role in the prevention of stroke. Unlike thrombolytics, anticoagulants do not degrade clot; rather, they prevent thrombus formation and propagation by reduction of fibrin formation. Anticoagulants act on different steps of the intrinsic and extrinsic coagulation pathways. Unfractionated heparin, low-molecular-weight heparins (LMWHs) and heparinoids are parenteral anticoagulants of rapid onset that act by indirect inhibition of thrombin and factor Xa via antithrombin. Warfarin (coumadin), the most commonly used oral vitamin K antagonist (VKA), inhibits the conversion of oxidized vitamin K epoxide into its reduced form, vitamin K. It diminishes the K-dependent γ-carboxylation of clotting factors II, VII, IX, and X, as well as the naturally occurring endogenous anticoagulant proteins C and S; the full antithrombotic effect of warfarin is not achieved for several days. During the past few years, a number of new oral anticoagulants (NOAC) have been added to the standard anticoagulant armamentarium. These novel anticoagulants selectively target thrombin (factor IIa) or factor Xa, have much more rapid onset (hours), better therapeutic window, and less drug and food interactions than VKAs (Fig. 167.1).

ANTICOAGULANTS FOR SECONDARY STROKE PREVENTION

Anticoagulation is typically indicated when stroke is produced by a cardiac source, with atrial fibrillation being the most common indication. Decision-making for use of anticoagulation in this setting is largely based on stroke risk stratification via the CHA2DS2-VASC (previously CHADS2) scoring system, combined with assessment of hemorrhagic risk. Atrial fibrillation and its treatments are discussed in more detail in Chapter 144. Another clinical indication for long-lasting anticoagulation is the presence of mechanical heart valve. In these patients, a more intense level of anticoagulation is needed to prevent stroke, and NOACs are not recommended. More intense anticoagulation can be achieved with warfarin by increasing the INR limit from 2.5 to 3.5. Other cardioembolic conditions in which anticoagulants have been used for secondary stroke prevention include cardiomyopathy, cardiac thrombi, and patent foramen ovale (PFO). The Warfarin versus Aspirin

Primer on Cerebrovascular Diseases, Second Edition
http://dx.doi.org/10.1016/B978-0-12-803058-5.00167-3

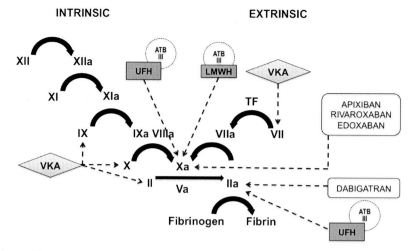

FIGURE 167.1   Unfractionated heparin (UFH) binds antithrombin III (ATB III), increasing its activity and causing indirect inhibition of mainly factor Xa and thrombin (IIa). Low-molecular-weight heparins (LMWHs) mainly produce indirect inhibition of factor Xa via ATB III. Vitamin K antagonists (VKAs) prevent thrombus formation by inhibition of factors II, VII, IX, and X. Dabigatran is a direct thrombin (IIa) inhibitor. Apixaban, rivaroxaban, and edoxaban act by direct inhibition of factor Xa.

in Reduced Cardiac Ejection Fraction (WARCEF) trial, although underpowered, did not show benefit for its composite endpoint of ischemic stroke, hemorrhagic stroke, or death in patients with cardiomyopathy in normal sinus rhythm [1]. In patients with PFO, platelet medications alone are generally the best option for stroke prevention. However, in the presence of deep venous thrombosis (DVT), 3–6 months of anticoagulation is indicated.

The use of anticoagulation to prevent stroke due to arterial sources is considered in special situations. Routine use of anticoagulation to prevent ischemic stroke in non-cardioembolic stroke has been largely eliminated following the disappointing results of the Warfarin–Aspirin Recurrent Stroke Study (WARSS) [2]. The long-term risk of intracranial hemorrhage (ICH) and life-threatening hemorrhage outweighed any reduction in thrombolytic events. However, high-risk patients who have failed platelet therapies or have serious contraindication to platelet medications may benefit from long-term anticoagulation.

Intra-arterial thrombosis is another indication to consider anticoagulation. Cervical arterial dissections are typically treated with anticoagulation for up to 6 months. However, due to the low risk of stroke recurrence with dissections, platelet medications are a reasonable alternative. Cerebral venous thrombosis, discussed in detail in Chapter 95, is typically treated with 3–6 months of anticoagulation.

Use of anticoagulants for stroke prevention in the setting of thrombophilic disorders is generally not required, and platelet medications are usually sufficient. Conversely, anticoagulation is a reasonable option for recurrent stroke in patients with antiphospholipid antibody syndrome and other thrombophilic disorders, who have failed antiplatelet therapy. Other chapters in this book discuss the use of anticoagulants to prevent stroke in different specific clinical settings (Table 167.1).

ISCHEMIC STROKE

Routine use of anticoagulation in acute ischemic stroke is not supported by current evidence. Many trials have tested different heparins for the treatment of acute ischemic stroke without proving benefit (Table 167.2). Recent Cochrane metaanalysis has confirmed that early heparin administration compared with no anticoagulation had a small reduction of recurrent ischemic stroke and venous thromboembolism at the cost of increased ICH, with no overall reduction in death or disability [3]. In order to identify high-risk stroke patients with low risk of hemorrhage who could benefit from acute anticoagulation, Whitely et al. used risk models based on simple clinical variables to stratify individual patient data from the five largest trials with heparins. The analysis did not provide support for targeting subgroups of patients who would benefit from treatment with early anticoagulant therapy for the prevention of thrombotic events or reduction of disability and death [4].

Clinicians may confront special situations in which anticoagulation should still be considered based on what is largely anecdotal evidence. These include presence of arterial lumen thrombosis, particularly in the posterior circulation, and subtotal carotid occlusion in the patient awaiting carotid endarterectomy. Moreover, the balance between risks and benefits of heparin remains unclear for patients with acute transient ischemic attack (TIA) or minor ischemic strokes, groups poorly represented in the randomized trials. These patients have very low risk of ischemic hemorrhagic transformation and initiation of early anticoagulation may be safer in these cases. Furthermore, the low frequency of ICH complications with the NOACs, and their rapid onset, make them promising candidates for acute stroke treatment.

TABLE 167.1 Chronic Anticoagulation for Prevention of Stroke

Condition	Evidence/ Effect[a]	Duration	Anticoagulant	Comments
A-Fib	Strong/beneficial	Indefinitely	Warfarin, NOA	INR 2–3. New oral anticoagulants are as effective as warfarin but less ICH complications.
Mechanical valve	Strong/beneficial	Indefinitely	Warfarin	INR 2.5–3.5 Dabigatran inferior to warfarin for stroke prevention.
Intracranial stenosis	Strong/no benefit	N/A	N/A	Not better than ASA and increase risk of bleeding as shown by WASID.
Small vessel	Limited/no benefit	N/A	N/A	Higher risk of ICH may be due to association with BBB disruption and microbleeds.
Aortic atheroma	Limited/unknown	Unknown	Warfarin	Consider in case of arterial thrombosis.
Extracranial dissection	Limited/unknown	3–6 months	Warfarin	Low risk of recurrent stroke.
Intracranial dissections	Limited/harm	N/A	N/A	Risk of SAH.
Cardiac thrombi	Limited/beneficial	6 months or resolution	Warfarin	Left atrial, left ventricular, or mural thrombi.
Acute anterior STEMI	Limited/reasonable	3 months	Warfarin	INR 2–3.
PFO + DVT	Limited/beneficial	6 months	Warfarin, NOA	In selected cases when anticoagulation is contraindicated, IVC filter or PFO closure might be considered. (Uncertain benefit).
Cardiomyopathy with EF < 35%	Limited/unknown	Indefinitely	Warfarin	Many patients might have cryptic atrial fibrillation or develop cardiac thrombi.
APAS	Limited/unknown	Indefinitely	Warfarin	Consider as alternative to platelet therapy. Some refractory patients might need INR levels of 2.5–3.5.
Sinus thrombosis	Limited/reasonable	3–6 months	Warfarin	INR 2–3. Controversial with ICH but recommended per guidelines.
Cryptogenic	Limited/unknown	Unknown	N/A	Paroxysmal atrial fibrillation and PFO are common causes of cryptogenic stroke. New anticoagulants are being tested.
Hypercoagulable	Limited/unknown	Indefinitely	Warfarin	Consider as alternate to platelet therapy.

APAS, Antiphospholipid antibody syndrome; *CHF*, congestive heart failure; *DVT*, deep venous thrombosis; *IVC*, inferior vena cava; *NOA*, new oral anticoagulants: apixaban, rivaroxaban, edoxaban and dabigatran; *PFO*, patent foramen ovale; *STEMI*, ST elevation myocardial infarction.
[a]*Levels of evidence by AHA guidelines [7,8].*

TABLE 167.2 Major Trials of Heparins in Acute Ischemic Stroke

Trial/Year	N	Drug	Comparison	Population	Window	Outcome	Comments
CESG (1983)	45	IVUFH	None	Cardioembolic Stroke	48 h	Death, ICH, recurrence	Non-blinded. No statistical difference.
Duke (1986)	225	IVUFH	Placebo	Non-cardioembolic	48 h	Death, functional outcome, ICH	Favors placebo control.
FISS (1995)	308	LMWH (fraxiparin)	Placebo	Ischemic stroke with motor deficit	24 h	Death, BI	Favors LMWH.
IST (1997)	19,435	UFH SQ 12500 BID, UFH SQ 5000 BID	ASA and placebo	All types	24 h	Death, recurrence	No difference between treatments.
FISS-bis (1998)	767	LMWH (fraxiparin)	Placebo	Ischemic stroke with motor deficit	24 h	Death, BI	No difference between treatments.
HAEST (2000)	449	LMWH (dalteparin)	ASA 160 mg		30 h	Recurrent Stroke	No difference between treatments.

Continued

TABLE 167.2   Major Trials of Heparins in Acute Ischemic Stroke—cont'd

Trial/Year	N	Drug	Comparison	Population	Window	Outcome	Comments
TOAST (1998)	281	LMWH (danaparoid)	Placebo	All types	24 h	GOS and BI	No difference between treatments.
TAIST (2001)	1486	LMWH (tinzaparin)	ASA	All types	48 h	mRS, BI, death	Two doses of tinzaparin. No difference between treatments.
TOPAS (2001)	404	LMWH (certoparin 8000 U BID, 5000 U BID, 3000 U BID)	Certoparin 3000 U QD	All types	12 h	BI	No difference between groups.
Camerlingo et al. (2005)	418	IVUFH	Saline	Non-lacunar	3 h	mRS	Favor to UFH.
FISS-tris (2007)	603	LMWH (nadroparin)	ASA	50% with ICS	48 h	mRS	No difference between treatments.

ASA, Aspirin; *GOS*, glasgow outcome scale; *ICH*, intracranial hemorrhage; *ICS*, intracranial stenosis; *IVUFH*, intravenous unfractionated heparin; *LMWH*, low-molecular-weight heparin; *mRS*, modified rankin score; *NS*, nonsignificant; *UFH SQ*, unfractionated heparin subcutaneous.

Intravenous tPA is contraindicated in patients taking direct factor X inhibitors and thrombin inhibitors, a topic under investigation. In patients using heparin, a pTT should be within normal limits. For patients using warfarin, INR should be less than or equal to 1.7. Intravenous thrombolysis beyond 3 h is contraindicated for patients taking any type of anticoagulant. In these cases, thrombectomy alone without tPA is a reasonable approach. Data have shown no increased risk of ICH or worse outcome in stroke patients on anticoagulation who underwent mechanical thombectomy [5].

DVT and pulmonary embolism (PE) are common complications in immobilized stroke patients. Low-dose LMWH has to be usually initiated within the 24 h of stroke to prevent venous thrombosis. Prevention of venous thromboembolism after acute ischemic stroke trial (PREVAIL) established superior clinical efficacy of LMWH over unfractionated heparin for DVT and PE prevention in immobilized stroke patients [6]. At prophylactic doses, anticoagulants entail very low risks of serious bleeding complications. For patients treated with intravenous t-PA (IV tPA), current guidelines indicate that administration of anticoagulants is contraindicated during the first 24 h after treatment. Intermittent venous compression stockings are an alternative in immobilized patients treated with IV tPA [7].

INITIATION AND DISCONTINUATION OF ANTICOAGULATION AFTER STROKE

Optimal timing for initiation of anticoagulation after ischemic stroke is controversial and clinically challenging. The risk of hemorrhagic transformation is a profound limitation on anticoagulant usage in this setting. Hemorrhagic transformation is the greatest concern following large cardiogenic infarctions, particularly during the first 4 days poststroke, and a delay of 4 to 14 days is commonly recommended [8].

Bridging warfarin with heparin has been common practice due to differential half-lives of vitamin K-dependent factors, which may result in a prothrombotic state during the first days after initiation of warfarin. However, there is no supporting evidence for this from clinical trials. For ischemic stroke, heparin bridging in clinical practice has not shown to prevent embolic events, but does appear to increase risk of bleeding. For brief interruption of VKA treatment for relatively low-risk patients (mean CHADS2 score 2), bridging appears to be counterproductive, that is, no benefit for protection against embolic events, but substantially higher risk of hemorrhage [9]. For higher risk patients having interruption of VKA treatment, use of bridging needs to be individualized.

ANTICOAGULATION: INTRACRANIAL HEMORRHAGE AND MICROBLEEDS

A difficult scenario for the clinician is to decide when to start or resume anticoagulation after ICH. The risk of rebleeding as well as the risk of ischemic stroke needs to be carefully evaluated. Specifically, patients with lobar location of ICH have a higher probability of rebleeding when compared to those with deep hemispheric bleeding. For patients with ICH who require urgent anticoagulation, current guidelines recommend waiting at least 1 week after the initial ICH [8]. Patients with cardiac thrombi, mechanical valves, and atrial fibrillation with high CHA2DS2-VASC scores generally have higher risk of cardioembolic stroke than risk of rebleeding.

Brain magnetic resonance imaging (MRI) routinely demonstrates cerebral microbleeds (CMB). CMB are the

MRI signature of microscopic hemorrhage in the brain and they can be easily demonstrated by using gradient echo and susceptibility-weighted imaging sequences. CMB prevalence is age-dependent, approaching 20% by age 60 and 40% by age 80, making them a common finding in patients in the typical age range for stroke.

The coexistence of ischemic and hemorrhagic stroke is termed "mixed cerebrovascular disease," and represents a troubling phenomenon for the stroke clinician. Most, if not all, stroke therapies involve a risk/benefit ratio that at least implicitly involves risks of thrombosis and hemorrhage. However, in mixed cerebrovascular disease, there is obvious presence of both kinds of phenomenon.

CMB can be characterized as a primary or secondary process, that is, whether the microscopic hemorrhage is due to a primary disruption of vascular integrity or is the consequence of ischemic injury ("hemorrhagic microinfarction") [10]. This distinction is crucial in stroke neurology, because presence of secondary CMB may indicate the need for more, not less, antithrombotic therapy. In contrast, presence of primary CMB, most commonly due to hypertensive vasculopathy or cerebral amyloid angiopathy, represents a warning flag for use of anticoagulants.

The distinction between primary and secondary CMB may be difficult. However, presence of CMB combined with the need for anticoagulation must trigger a careful assessment by the stroke clinician for the likely cause of the CMB phenomenon. [10] Extensive CMB (usually defined as five or more CMB) due to chronic hypertension represents a cautionary note for use of anticoagulants, and presence of cerebral amyloid angiopathy indicates that anticoagulation should probably be avoided. Much of the dilemma of anticoagulation in patients with CMB is likely to be resolved with use of the very promising mechanical occlusion devices directed at the left atrial appendage, the source of the vast majority of thrombi in atrial fibrillation. However, until such time as these devices become routine, anticoagulation in mixed cerebrovascular disease will remain a profound clinical challenge.

Funding Information

Supported by NIH NS20989.

References

[1] Homma S, Thompson JLP, Pullicino PM, et al. Warfarin and aspirin in patients with heart failure and sinus rhythm. N Engl J Med 2012;366(20):1859–69.

[2] Mohr JP, Thompson JL, Lazar RM, et al. A comparison of warfarin and aspirin for the prevention of recurrent ischemic stroke. N Engl J Med 2001;345(20):1444–51.

[3] Sandercock PA, Counsell C, Kane EJ. Anticoagulants for acute ischaemic stroke. Cochrane Database Syst Rev 2015;3:CD000024.

[4] Whiteley WN, Adams HP, Bath PM, et al. Targeted use of heparin, heparinoids, or low-molecular-weight heparin to improve outcome after acute ischaemic stroke: an individual patient data meta-analysis of randomised controlled trials. Lancet Neurol 2013;12(6):539–45.

[5] Rebello LC, Haussen DC, Belagaje S, Anderson A, Frankel M, Nogueira RG. Endovascular treatment for acute ischemic stroke in the setting of anticoagulation. Stroke 2015;46(12):3536–9.

[6] Sherman DG, Albers GW, Bladin C, et al. The efficacy and safety of enoxaparin versus unfractionated heparin for the prevention of venous thromboembolism after acute ischaemic stroke (PREVAIL Study): an open-label randomised comparison. Lancet 2007;369(9570):1347–55.

[7] Jauch EC, Saver JL, Adams HP, et al. Guidelines for the early management of patients with acute ischemic stroke: a guideline for healthcare professionals from the American Heart Association/American Stroke Association. Stroke 2013;44(3):870–947.

[8] Kernan WN, Ovbiagele B, Black HR, et al. Guidelines for the prevention of stroke in patients with stroke and transient ischemic attack: a guideline for healthcare professionals from the American Heart Association/American Stroke Association. Stroke 2014;45(7):2160–236.

[9] Douketis JD, Spyropoulos AC, Kaatz S, et al. Perioperative bridging anticoagulation in patients with atrial fibrillation. N Engl J Med 2015;373(9):823–33.

[10] Fisher M. Cerebral microbleeds: where are we now? Neurology 2014;83(15):1304–5.

CHAPTER

168

Rehabilitation for Disabling Stroke

B.H. Dobkin

University of California Los Angeles, Los Angeles, CA, United States

Rehabilitation after stroke concentrates on reducing physical and cognitive impairments and the disabilities they induce so as to return patients to more independently managed self-care, mobility, communication, and daily activities. Care goals include prevention of the complications of immobility, dysphagia, pain, bowel and bladder dysfunction, sleep and mood disorders, as well as treatment of risk factors for cardiovascular disease and stroke.

Optimal management to provide retraining strategies, assistive devices, and caregiver support requires an interactive team of therapists, nurses and aides, and physicians with neurological rehabilitation, primary care, and other expertise within a model of chronic care. Only 20% of patients who suffer a stroke are admitted for inpatient rehabilitation. They usually have a hemiparesis that prevents walking without human assistance on admission, as well as degraded independence for self-care. Fig. 168.1 is a typical decision tree for placement planning during the acute stroke hospitalization. Only 31% of stroke survivors receive any outpatient rehabilitation [1], significantly lower than expected if clinical guidelines were followed. One may assume that most of the 69% who had no formal rehabilitation received little or no education or support for activities of daily living (ADLs), mobility practice, and fitness exercise.

SCIENTIFIC BASES

Studies of the scientific bases for neurological rehabilitation after stroke have led to a general consensus: *Progressive practice of task-related skills assisted by a therapist in an adequate dose leads to improvement of motor, cognitive, or behavioral skills mostly restricted to what was trained, via mechanisms of activity-dependent* molecular, cellular, synaptic, and structural and physiological plasticity within spared neural ensembles and networks. This underlying adaptability or neuroplasticity, which is most highly manifest in animal models in the first month after onset and for 3 months in patients (see Chapter 35, Carmichael), enables reorganization and rewiring within residual perilesional, ipsilesional, and contralesional cortical and subcortical circuits, as well as the neural ensembles that represent, for example, a skilled movement [2]. Synaptic mechanisms of learning and memory also contribute to compensatory, adaptive strategies.

REHABILITATION OUTCOME MEASURES

The intention to address impairment, functional disability, activity, and participation requires a range of outcome measurement tools for clinical trials. There are many examples of each. This chapter includes studies with primary outcomes drawn from some of the most commonly used upper and lower extremity function tools that have high reliability [3]. Based on these measures, ADLs tend to reach a plateau of improvement by 12 weeks after onset of stroke. About 70% of gains above the baseline Fugl–Meyer Motor Assessment (FM, see later) score also evolve by three months after onset of mild–moderate hemiparesis [4]. Some patients with severe hemiparesis or hemiplegia in the first week (FM score <20) may improve their motor control if enough corticospinal tract fibers are spared. If no wrist or finger extension is present 2 weeks after an ischemic stroke, functional use of the hand is very unlikely. As many trials have shown, however, less robust but valued goal-specific improvements, such as faster walking speed or improved grasp and pinch with greater daily incorporation of the affected hand, can be made at any time after stroke, as long as enough motor control is latently available to enable progressive practice.

The FM of impairment has been used in the majority of upper extremity trials in hemiparetic persons.

Primer on Cerebrovascular Diseases, Second Edition
http://dx.doi.org/10.1016/B978-0-12-803058-5.00168-5

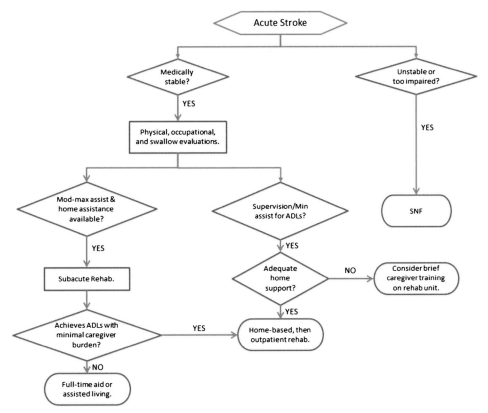

FIGURE 168.1   Decision tree for rehabilitation or other placement after acute stroke.

It grades increasingly complex multijoint movement tasks of the arms (top score 66) and legs (best score 34) from 0 to 3 (cannot perform and partially or fully performed motion), reflecting simple movements within a synergistic pattern to more selective motion. Maximal strength required, however, is only against gravity.

Casual or fastest walking speed over a 10- or 15-m flat surface and distance walked in 3 or 6 min are the most frequently employed ratio scale measures for changes in gait and mobility. Speed reflects many of the sensorimotor, balance, and cognitive components that go into safe walking. These tasks in a clinic setting may not reflect walking pace and endurance in the home or community. A commercial device with an accelerometer is often used for a few days during a clinical trial to reveal the number of steps taken daily outside of the clinic, but the accuracy for hemiplegic gait drops sharply as walking speed falls below 0.6 m/s.

Functional use of the affected upper extremity has been measured by ordinal scales, such as the Arm Research Action Test (ARAT), and timed tasks, such as the Wolf Motor Function Test (WMFT). The modified ARAT consists of four subtests including grasp, grip, pinch, and gross movement. All tasks are scored on a four-point scale from 0 to 3 where 0 reflects poor hand function and 3 reflects good hand function for a total possible score of 57. The WMFT uses 15 timed reach and grasp or pinch movements and two strength tasks. An unimpaired hand average score is about 5 s, but patients often average 15–45 s or get a 120 s score if unable to perform. The quality of the movement can also be scored on a 6-point ordinal scale.

The Functional Independence Measure (FIM, Table 168.1) is perhaps the most commonly employed American inpatient tool for assessing ADLs. The 0–7 scale reflects the burden of care on others to aid the disabled person. As a functional measurement, the person need not be able to use the affected arm at all to score as fully independent and only has to walk 150 feet at a speed as slow as 0.4 m/s to be graded independent. Scoring is much less sensitive to change three months after onset. The Stroke Impact Scale is a self-reported measure of domains relevant to activity and participation and it too may be less sensitive to change in patients with chronic stroke.

STAGES OF REHABILITATION

Acute Stroke Admission

Rehabilitation assessment and treatment can begin by two days after onset in medically stable patients (e.g., alert and interactive, afebrile, no orthostatic hypotension, and no side effects of new medications) who are not hemodynamically at risk for further hypoperfusion.

TABLE 168.1   Functional Independence Measure (FIM) Scores for Stroke Admissions and Discharges (National Uniform Data System for Medical Rehabilitation[a] Data for 101,885 Entries in 2014)

Items	Entry	Discharge
Self-care	18.9	29.0
Eating		
Grooming		
Bathing		
Dressing upper body		
Dressing lower body		
Toileting		
Sphincter control	7.4	9.9
Bladder management		
Bowel management		
Mobility and transfers	7.8	13.2
Bed-to-chair and wheelchair-to-chair transfer		
Toilet transfer		
Tub and shower transfer		
Locomotion	3.5	7.8
Walking or wheelchair use		
Climbing stairs		
Communication	8.0	10.2
Comprehension		
Expression		
Social cognition	11.1	14.5
Social interaction		
Problem solving		
Memory		
Total FIM rating (maximum score 126)	56.6	84.7

Burden of care rating for each sub item.
7 = Complete independence (timely, safely).
6 = Modified independence (device).
5 = Supervision.
4 = Minimal assistance (subject contributes 75% or more).
3 = Moderate assistance (subject contributes at least 50%).
2 = Maximal assistance (subject contributes at least 25%).
1 = Total assistance (subject contributes 0% up to 25%).
[a]UDSMR is a division of University of Buffalo Foundation Activities, Inc. Report generated 11/17/15.

The large multicenter, A Very Early Rehabilitation Trial (AVERT), compared mobilization within 24h after stroke onset to usual care to determine whether this inexpensive strategy could lead to better functional outcomes. The early group, however, was mobilized at a mean of 18h and only one-third by 12h, compared to 22h for the usual care group, which seems rather soon compared to most American stroke units. Patients with severe stroke and serious complications were excluded. The earlier group had somewhat worse outcomes on the modified Rankin Scale, but the findings do not support delaying mobilization unless medical contraindications require this.

Fig. 168.1 shows a typical decision tree for discharge placement after acute stroke. To admit a patient for subacute inpatient rehabilitation, the major decision-making components include the ability to learn, adequate motivation, adequate social support to avoid eventual placement in a skilled nursing facility, and capacity to participate in at least 3h of daily therapies.

Inpatient Subacute Rehabilitation

The primary goal is to enable patients to become independent enough to be manageable at home, with help by a trained caregiver or support services as needed. To best achieve this, transfers, walking, and toileting usually have to approach a minimally assisted level by FIM criteria. Although inpatient rehabilitation is called intensive, the reality across units internationally is for actual practice of walking to average about 17min a day [5]. By providing medical care and insight into the neural bases of manifest and subtle impairments and disabilities, the physician can lead the rehabilitation team toward problem-solving solutions. Much medical care usually accompanies inpatient rehabilitation, mostly related to premorbid diseases, further management of the etiology of the new stroke, adjustments of medications, bladder infections, pain, and mood and sleep disorders.

Tables 168.1 and 168.2 describe average admission and discharge data for a first stroke rehabilitation admission. In general, FIM scores <60 at discharge make it less likely that patients can return home without fulltime physical assistance. The rehabilitation team aims for the safest discharge setting and concentrates on the skills, assistive equipment, and home modifications necessary to return to the community. With lengths of stay averaging only 15days, home health rehabilitation services are usually needed to transfer what was learned as an inpatient to the home setting.

Outpatient Rehabilitation and Chronic Care

The goal for outpatient rehabilitation is to become as independent as feasible in usual home and community activities. A Medicare beneficiary's annual outpatient therapy expenditures, however, are capped at about $4000 for combined physical, occupational, and speech therapies. Careful documentation and a review process can allow some additional therapy. With potentially so

TABLE 168.2 Characteristics of Patients in UDSMR for Stroke in Table 168.1.

Age >64 years	66%
Gender: male	50.7%
Married	47%
Medicare payer	70%
Onset to rehabilitation transfer (days)	9.1
Length of stay (days)	14.7
LIVING SITE PREHOSPITAL	
Alone	26%
Family/relatives	68%
Friends	2%
Attendant	0.5%
LIVING SITE AT DISCHARGE	
Home/community	74%
Long-term care	17%
Acute care	9%

little formal therapy, physicians should encourage progressive practice, help patients set goals, and consider how to accomplish them during and after the period of outpatient therapies. Patients can be encouraged to practice more skillful, coordinated, and faster movements. They might aim to add skills to their repertoire (e.g., turning a key in a lock, setting a table, walking further at a higher cadence, or, if aphasic, practicing a group of single words or phrases to improve the ease of social communication).

Inpatient and outpatient rehabilitation services mostly promote independent activities of daily living, rather than a transition to a lifestyle of practice, exercise, and less-sedentary behavior. Across follow-up studies, persons with hemiplegic stroke rarely walk for more than 1-min intervals and most achieve little more than a 5–10 min continuous walk a day. They take about half the number of steps as their peers [6]. For stroke risk factor reduction and safer mobility, outpatients can be advised to gradually increase daily walking to achieve, for example, three continuous walks for at least 10 min each while reducing sedentary behavior, then aim for at least one 30-min walk. At any time after stroke, a pulse of therapy to improve a valued functional activity, such as walking or strengthening, and fitness should be considered. For example, up to 85% of hemiparetic persons are deconditioned by 6 months post stroke. The percentage of patients who were dependent in ADLs at 3 months after stroke (20–25%) may increase to 35–45% by 12 months [7].

SPECIFIC INTERVENTIONS

A variety of Cochrane Database System Reviews and other meta-analyses have examined specific rehabilitation interventions for a variety of outcomes. In general, no specific active intervention has been shown to improve outcomes better than another [8]. No high-quality evidence is available for any routine therapeutic approach in use. However, progressive, task-related practice in a high-enough dose (number of sessions per week, duration of active practice at each session, and feedback to optimize practice) does improve outcomes compared to nonspecific recommendations. At least 20 and perhaps 60 h of practice, with or without a therapist, are usually necessary to measure a significant change for the better, especially beyond the first 2–3 months after disabling stroke.

Mobility

Practice for walking starts with work on head and trunk control while seated on a mat, hip and knee flexion and extension in side-lying, standing in parallel bars, and gradually improv components of the gait cycle. Over-ground training emphasizes hip extension at the end of stance and clearance of the paretic foot to initiate leg swing, knee stability when loading the paretic leg in stance, and stepping with a more rhythmic and equal stance and swing times, as well as stride length, for each leg. An assistive device, such as a hemi-walker, quad cane or single-point cane, or ankle–foot orthosis to help clear the paretic foot and stabilize the knee, may enable compensation for leg weakness and add safety. A Cochrane review found positive correlations between the amount of over-ground training and small improvements in gait speed with no significant increase in the number of adverse events, such as falls [8].

Body weight–supported treadmill training (BWSTT) employs supervised, repetitive, task-related practice of walking. Patients wear a chest harness connected to an overhead lift and walk on a treadmill with some unloading of the legs and physical assistance as needed. Although this form of practice seems to meet the criteria for optimal practice conditions for motor learning, multiple randomized clinical trials (RCTs) revealed that the technique is no better than the same intensity of over-ground training or home-based exercise that did not include practice walking for the primary outcomes of 10-m walking speed and distance walked in 6 min [6]. BWSTT led to the notion of placing highly impaired patients in an electromechanically driven exoskeleton that helps control stepping on an integrated treadmill belt or over ground. To date, no trials have revealed any clear advantage compared to over-ground practice. Functional electrical stimulation of specific muscles for knee extension or foot dorsiflexion has grown rather sophisticated, but well-performed trials have not revealed any advantage over a molded ankle–foot orthosis [9].

One way to increase walking speed during inpatient therapy is to give feedback about walking speed a few times a week to the patient and the physical therapist. This simple strategy improved walking speed by over 20% compared to no feedback about performance [5,10].

Falls occur in 40% at least once in the first year after onset [6]. Safety precautions in the environment, elimination of sedatives, use of gait belts and assistive devices for stability, and use of a bedside urinal or a commode overnight instead of risking a walk to the toilet should be considered.

Functional Use of the Affected Upper Extremity

Therapy for the hemiparetic arm begins with attempts at movement of single joints, then gradually proceeds to incorporate more complex, multijoint actions, along with task-specific practice, such as reaching to grasp a coffee cup, a process known as behavioral shaping. Constraint-induced therapy (CIT) was developed to do this within the context of preventing nonuse of the affected arm and hand. A well-done trial tested a 10-day program of 6h a day of formal practice with a therapist and home-based practice while the unaffected hand was in a mitt to prevent its use [11]. The 60–80h of therapy gave far better results compared to no therapy in subacute patients. Efficacy trials, however, include only patients who have rather good baseline motor control, which means 5–10 degrees of wrist and finger extension. Variations such as only 2h of formal practice or not restraining the hand have also led to greater use of the affected hand in chronically impaired patients.

While these trials confirm the value of progressive, task-related practice, others do not. For example, the well-performed ICARE trial found no differences across three different intensities and types of therapy in moderately impaired participants starting 45days after stroke. All improved considerably on the WMFT, which was also the primary outcome for most CIT trials.

Imagining a movement or mental practice as an athlete might before performing a dive, as well as observation of an action, activate regions of the brain that are also activated by performing the same upper or lower extremity task. Virtual reality environment games that require reach, grasp, and pinch aim to better engage users and neural networks. All of these techniques could enhance practice and lead to greater improvements, but so far they have been tested mostly in subjects with mild–moderate impairments, leading to modest improvements. Repetitive bimanual practice aims to activate the bilateral motor cortices and enhance corticomotor

activation of the affected upper extremity. In small trials, bimanual practice has resulted in a similar degree of improvement as CIT at the same intensity. Functional electrical stimulation of a nerve/muscle group and robotic-assistive training may also lead to some gains in moderately impaired subjects, particularly in the FM score, but usually no greater than similar practice without either device. So far, highly impaired subjects in these trials have not improved more than conventionally treated subjects, including for a robotic-assistive training intervention [12].

Strategies to modulate the networks representing practiced tasks are evolving works-in-progress. Noninvasive brain stimulation by transcranial magnetic stimulation (TMS) or direct current stimulation (tDCS) by a variety of techniques may modestly improve hand function when performed in relation to practice and possibly when peripheral nerve stimulation is added. Use of a brain–computer interface to enhance neural feedback for learning during practice of a skill has also had some success in small studies and may be especially useful in highly impaired persons or in coordination with neural repair interventions.

Aphasia

Specific aspects of language, such as naming and word finding, can improve with targeted therapies and much practice, especially if extended by trained caregivers. For those who have adequate comprehension, successful training of social communication is most rewarding for patients and families.

Hemispatial neglect

Left-sided inattention or neglect of peripersonal space or limbs has been reduced in small trials employing visual cues and scanning, affected limb movement activation, wearing prisms in glasses that shift the visual field, sustained attention training, vestibular and optokinetic stimulation, mirror therapy that creates the visual illusion of movement of the affected hand, and various protocols for TMS, tDCS, and theta burst stimulation [13]. The improvements, however, do not usually extend to improved ADLs or walking.

Spasticity

Hypertonicity may evolve over several months post stroke, especially in those with the least motor control who had a bout of shoulder or wrist pain. Noxious input tends to promote greater tone along with shortening of muscle and tendon at a joint and changes in muscle fiber composition. Serotonergic medications may also enhance spinal cord mechanisms that increase hypertonicity and spasms. Immediate pain management with, for example, a nonsteroidal antiinflammatory drug or medication to lessen central pain if present may help prevent hypertonicity and contractures, as will ranging affected joints for a few minutes several times daily. Spasticity per se generally does not interfere with motor control. Rather, weakness, fatigability, and slow movements are the main clinically key problems in upper motor neuron dysfunction. If flexor postures of the elbow, wrist, or fingers induce pain or make daily care more difficult, focal injections of botulinum toxin may provide relief for up to three months and allow stretching exercises to better work.

Pharmacological augmentation

Drug trials to augment rehabilitation gains have a long history drawn from animal models of recovery, but even seemingly well-carried out trials are rarely positive or reproducible in human trials. The FLAME trial tested the serotonergic agent fluoxetine in combination with standard rehabilitative therapies and reported better FM scores for those patients who received the drug [14]. Noradrenergic stimulation with roboxetine has increased motor control for the affected hand, but trials are not large enough to draw any conclusion. Ropinirole and other dopaminergic drugs have not increased recovery, although amantadine in some studies had a positive effect [15]. Attempts to increase cerebral BDNF, NT3, and other neurotrophins to modulate plasticity have only been feasible in animal models so far due to poor crossing of the blood–brain barrier. The same holds for modulating tonic GABA inhibition of plasticity. The extrasynaptic NMDA receptor antagonist memantine improved spontaneous speech and naming skills in chronic stroke patients with aphasia [16]. Acupuncture is suspected to provide sensory input and activate opioids and perhaps other neural substances, but the better-designed, controlled trials of its use in addition to more conventional rehabilitation have not revealed efficacy.

New onset anxiety and depression after stroke are common, but often respond enough to medication and counseling to allow those affected to participate in therapy. The serotonergic reuptake inhibitors seem especially practical to use during subacute and chronic rehabilitation.

Exoskeletons, Robotic Aids, and Wearable Sensors

Technology may play a greater role in improving quality of life. Rather lightweight battery-driven lower extremity exoskeletons are commercially available to fully control or assist a slow stepping pattern. Designed initially for the military and paraplegic persons, the devices are being tested as training aids for hemiplegia. Robots and computer-controlled environmental aids that can assist disabled persons by monitoring their safety and fulfilling daily needs as "butlers" are becoming commercially available. These may enable more people to remain in their homes or travel (e.g., the computer-driven automobiles being tested).

Wearable motion sensors (accelerometers and gyroscopes) connected by Bluetooth to smartphone apps that can remotely monitor the type, quantity, and quality of upper and lower extremity movement are increasingly available [5,17]. They can be combined with sensors of vital signs and even body chemistry. Internet and smartphone access to patients combined with wearable sensors will enable precise feedback about skills training and exercise for fitness, making inexpensive telerehabilitation feasible.

Neural Repair

Cellular and other biological interventions to enhance recovery in highly impaired persons will have to be combined with applicable rehabilitation strategies to optimize their incorporation and action in neural networks. Animal models show promise. Several blockers of axonal regeneration and several cell types injected into the subcortical peri-infarct region are in human trials, but it is not yet clear that enough is understood about repair mechanisms to expect improved outcomes.

CONCLUSION

Subacute and chronic neurological rehabilitation after disabling stroke requires strategies to enhance mechanisms of recovery, along with attention to what patients need to increase their functional independence, daily activities, and health-related quality of life. Maintenance of gains over time requires motivation and self-management skills. An occasional pulse of goal-specific training with a therapist can lead to modest improvements with considerable practice. Exercise for strengthening and fitness is important for both secondary prevention and maintenance of ADLs.

References

[1] Centers for Disease Control. Outpatient rehabilitation among stroke survivors – 21 States and the District of Columbia, 2005. MMWR 2007;56:504–7.

[2] Cramer S, Sur M, Dobkin B, et al. Harnessing neuroplasticity for clinical applications. Brain 2011;134:1591–609.

[3] Bushnell C, Bettger J, Cockroft K, et al. Chronic stroke outcome measures for motor function intervention trials: Expert panel recommendations. Circ Cardiovasc Qual Outcomes 2015;8:S163–9.

[4] Krakauer J, Marshall R. The proportional recovery rule for stroke revisited. Ann Neurol 2015;78:845–7.

[5] Dorsch A, Thomas S, Xu X, Kaiser W, Dobkin B. SIRRACT: An international randomized clinical trial of activity feedback during inpatient stroke rehabilitation enabled by wireless sensing. Neurorehabil Neural Repair 2015;29:407–15.

[6] Duncan P, Sullivan K, Behrman A, et al. Body-weight-supported treadmill rehabilitation program after stroke. N Engl J Med 2011;364:2026–36.

[7] Ullberg T, Zia E, Petersson J, Norrving B. Changes in functional outcome over the first year after stroke: an observational study from the Swedish stroke register. Stroke 2015;46:389–94.

[8] Pollock A, Baer G, Campbell P, et al. Physical rehabilitation approaches for the recovery of function and mobility after stroke: major update. Cochrane Database Syst Rev 2014;4. [CD001920].

[9] Everaert D, Stein R, Abrams G, et al. Effect of a foot-drop stimulator and ankle-foot orthosis on walking performance after stroke: a multicenter randomized controlled trial. Neurorehabil Neural Repair 2013;27:579–91.

[10] Dobkin B, Plummer-D'Amato P, Elashoff R, Lee J, Group S. International randomized clinical trial, Stroke Inpatient Rehabilitation With Reinforcement of Walking Speed (SIRROWS) improves outcomes. Neurorehabil Neural Repair 2010;24:235–42.

[11] Wolf SL, Winstein CJ, Miller JP, et al. Effect of constraint-induced movement therapy on upper extremity function 3 to 9 months after stroke: the EXCITE randomized clinical trial. JAMA 2006;296:2095–104.

[12] Lo A, Guarino P, Richards L, et al. Robot-assisted therapy for long-term upper-limb impairment after stroke. N Engl J Med 2010;362:1772–83.

[13] Corbetta M. Hemispatial neglect: clinic, pathogenesis, and treatment. Semin Neurol 2014;34:514–23.

[14] Chollet F, Tardy J, Albucher J, et al. Fluoxetine for motor recovery after acute ischemic stroke (FLAME): a randomised placebo-controlled trial. Lancet Neurol 2011;10:123–30.

[15] Cramer S, BH D, Noser E, Rodriguez R, Enney L. Randomized, placebo-controlled, double-blind study of ropinirole in chronic stroke. Stroke 2009;40:3034–8.

[16] Barbancho M, Berthier M, Naval-Sanchez P, et al. Bilateral brain reorganization with memantine and constraint-induced aphasia therapy in chronic post-stroke aphasia: An ERP study. Brain Lang 2015;145–146:1–10.

[17] Dobkin B, Dorsch A. The promise of mHealth: daily activity monitoring and outcome assessments by wearable sensors. Neurorehabil Neural Repair 2011;25:788–98.

CHAPTER

169

Magnetic and Direct Current Stimulation for Stroke

S. Madhavan

University of Illinois Chicago, Chicago, IL, United States

INTRODUCTION

There has been a recent surge in literature regarding noninvasive techniques that stimulate the brain to better understand plastic changes following stroke, and to modulate neuroplasticity to enhance post-stroke motor recovery. Noninvasive brain stimulation (NIBS) can be broadly categorized into transcranial magnetic stimulation (TMS) and transcranial direct current stimulation (tDCS) (Table 169.1). The theory behind magnetic and direct current stimulation is that neuronal activity in cortical brain regions underlying the electrodes may be modulated by applied magnetic or electric currents. It has been proposed that modulation of cortical excitability induces neuroplasticity and thereby motor learning. Single- and paired-pulse magnetic stimulation are also used to study cortical excitability, cortical connectivity, and plasticity of the brain. This chapter aims to provide a broad overview of TMS and tDCS, discuss evidence of their effects, and provide considerations and recommendations for future clinical application.

Primer on Cerebrovascular Diseases, Second Edition
http://dx.doi.org/10.1016/B978-0-12-803058-5.00169-7

TABLE 169.1 Summary Table of NIBS Protocols

Type	Description	Use
Single-pulse TMS	0.25 Hz or greater	Measures cortical excitability
Paired-pulse TMS	2 pulses with ISI of 2–14 ms	Measures intracortical inhibition (2–5 ms ISI) or intracortical facilitation (7–14 ms ISI)
Low-frequency rTMS	≤1 Hz	Cortical inhibition
High-frequency rTMS	≥5 Hz	Cortical facilitation
Continuous TBS	3–5 pulses at 50 Hz, every 200 ms for 2 s with a 10 s off period (20 trains)	Cortical inhibition
Intermittent TBS	3–5 pulses at 50 Hz, every 200 ms for 40 s	Cortical facilitation
Anodal tDCS	0.5–2 mA of anodal current	Cortical facilitation
Cathodal tDCS	0.5–2 mA of cathodal current	Cortical inhibition

Hz, Hertz; *ISI*, interstimulus interval; *rTMS*, repetitive transcranial magnetic stimulation; *TBS*, theta burst stimulation; *tDCS*, transcranial direct current stimulation; *TMS*, transcranial magnetic stimulation.

HISTORICAL BACKGROUND

The concept of brain stimulation is not new and can be traced back to as early as the 1st century when live torpedo fish was used to deliver strong electric currents to patients suffering from migraines. "Therapeutic electricity" was still prevalent in the 18th century when electric currents were used to elicit different physiological effects, and identify cortical representations of limb movements. However, there was difficulty in focusing electricity to focal areas of the brain. As the skull is a poor conductor, high levels of electrical energy was needed causing widespread activation of the brain resulting in convulsions. Prior to the 1950s, brain stimulation required the electrodes to be directly placed on the exposed brain surface. In the 1960s, Bindman, Leopold, and Redfern performed experiments using low-level direct currents that led to long-lasting brain polarization accompanied by changes in sensory, motor, and emotional abilities. This led to a resurgence of studies exploring the use of direct current stimulation for brain polarization. In the 1980s, Barker, Jalinous, and Freeston first reported the use of magnetic stimulation over the motor cortex to elicit responses in the muscle. And TMS became an important tool in neurophysiology [1]. Meanwhile, the exploration of direct currents continued. In 1998, Priori and colleagues first demonstrated the effects of anodal currents on increased cortical excitability and coined the term "transcranial direct current stimulation." Subsequent experiments by Nitsche and Paulus demonstrated modulating effects of anodal (increases cortical excitability) and cathodal (decreases cortical excitability) tDCS on brain tissue in which the effects outlasted the duration of stimulation [2]. Since the introduction of single-pulse TMS in the 1980s and the revival of tDCS in the 1990s, there have been numerous forms of TMS and tDCS that have been developed and applied to a variety of neurological and psychiatric conditions.

BRAIN PLASTICITY IN STROKE

Before addressing the developments using NIBS to enhance post-stroke motor function, it is important to understand the changes in cortical excitability that accompany stroke recovery. Several longitudinal brain imaging (fMRI and PET) and TMS studies have explored neural correlates of stroke recovery. The general consensus is that there is a disruption of interhemispheric balance of cortical excitability after stroke. Particularly, there is decreased activity in the ipsilesional hemisphere and a corresponding increase in activity in the contralesional hemisphere. Many studies have shown that balance of between-hemisphere cortical excitability is associated with improved post-stroke functional recovery [3]. This forms the basis for neuromodulation using stimulation-based priming protocols in stroke. A balance of excitability can be achieved by either upregulating the ipsilesional hemisphere or downregulating the contralesional hemisphere using NIBS.

TRANSCRANIAL MAGNETIC STIMULATION

TMS has been widely used to study brain-related neurophysiology and modulate brain function using single, paired, or repetitive pulses of stimulation. TMS works on the principle of electromagnetic induction. A rapidly changing pulse of current is generated within the stimulator and is passed through a tightly wound copper wire (coil). This rapidly changing electrical field induces a rapidly changing magnetic field that noninvasively and painlessly penetrates the scalp and skull, and depolarizes neural membranes in the brain. When TMS is applied over the motor cortex, it induces efferent volleys along the descending corticomotor pathway. The summated action potential produces a visible muscle twitch, can be recorded using electromyography from the contralateral muscle, and is classically referred to as a "motor evoked potential"(MEP). Single-pulse and paired-pulse TMS have been used successfully to probe the neurophysiological mechanisms involved in motor planning, movement execution, cognitive aspects of motor behavior, as well as changes in corticomotor physiology following disease or trauma. In addition to the MEP characteristics (size, latency, and duration), other single-pulse TMS-related parameters used in probing neurophysiology include resting and active motor threshold, cortical silent period, input–output response curves, central conduction time, and cortical mapping. Paired-pulse TMS is a useful tool for examining inhibitory and excitatory intracortical circuits within the motor cortex. A subthreshold conditioning stimulus is applied prior to a suprathreshold stimulus. Depending on the duration of the interstimulus interval (ISI), the paired pulse can be used to study intracortical inhibition (2–5 m ISI) or intracortical facilitation (7–14 m).

TMS has been widely used to explore changes in cortical excitability after stroke along with the clinical correlates of recovery [4]. Presence of MEPs in the affected limb (via stimulation of the lesioned hemisphere) in the acute phase of stroke has been suggested to be a predictive factor of good recovery. Conversely, the absence of MEPs during the early phase is associated with subsequent poor recovery. Studies have also reported abnormal facilitation of MEPs from the contralesional hemisphere, which returned to normal or decreased in patients with good recovery. Intracortical inhibition in the contralesional hemisphere is reported to appear normal or better than in the ipsilesional hemisphere in patients with good recovery, whereas it might be persistently reduced in patients with poor recovery. Measures of cortical excitability from the ipsilesional and contralesional hemisphere have shown to correlate with degree of impairment or improvements in motor function after stroke.

REPETITIVE TMS

Since the development of TMS, a range of repetitive pulse TMS (rTMS) protocols have been developed to modulate the excitability of the motor cortex [5]. rTMS delivers repeated single magnetic pulses of the same intensity to a discrete brain area for a short interval of time (15–20 min). The rate or frequency of the magnetic pulses determines the direction of neuromodulation. A pulse frequency of 1 Hz (low frequency) suppresses cortical excitability, while pulses that are 5–20 Hz (high frequency) facilitate cortical excitability. Theta Burst Stimulation (TBS) is a more recent type of patterned rTMS paradigm consisting of short bursts of pulses at 50–100 Hz stimulation frequency that are repeated at 5 Hz ("theta frequency"). Depending on whether the pulses are continuous or intermittent, TBS has shown to inhibit or facilitate cortical excitability, respectively. Repetitive TMS-mediated changes in cortical excitability have been suggested to modulate synaptic efficiency and are mediated by LTP-like or LTD-like mechanisms that have been confirmed in animal rTMS studies.

The use of rTMS after stroke has been primarily to modulate the abnormal interhemispheric symmetry of cortical excitability either by downregulating excitability of the contralesional hemisphere with low-frequency rTMS or continuous TBS, or by upregulating excitability of the ipsilesional hemisphere using high-frequency or intermittent TBS. A large number of studies have examined the effects of rTMS on cortical excitability and inhibition and its association with function. There is promising evidence to suggest that priming the brain with rTMS is accompanied by clinically significant improvements in upper and lower limb motor function post stroke, which outlast the duration of stimulation [6]. Most of these studies involve repeated stimulation (~12 sessions) of the M1 area of the contralesional or ipsilesional hemisphere for about 15 min/session followed by motor therapy. Of interest, the magnitude of the effects accompanying the newer TBS protocols seem to outweigh the benefits of simple rTMS protocols.

TRANSCRANIAL DIRECT CURRENT STIMULATION

tDCS involves delivery of weak direct currents (0.5–2.0 mA) to the targeted cortical area using saline-soaked electrodes with a battery-powered generator. Depending on the polarity of stimulation, tDCS can upregulate or downregulate cortical excitability. Anodal tDCS facilitates cortical excitability while cathodal tDCS inhibits cortical excitability. The typical tDCS electrode montage for motor cortex stimulation involves the active electrode to be placed on the motor cortex (focally targeted using single-pulse TMS) and the reference electrode over the contralateral supraorbital region. Other bilateral montages, such as placement of two active electrodes over either motor cortex and two supraorbital reference electrodes, or one active electrode over the motor cortex and the reference electrode over the other motor cortex, have also been explored. The short-term mechanisms of tDCS are widely different from rTMS protocols in that they do not induce neuronal firing by suprathreshold neuronal membrane depolarization, but rather modulate spontaneous neuronal network activity by changing activity in the neuronal membrane potential. The aftereffects of tDCS, which outlasts the period of stimulation, similar to rTMS, may involve LTP- and LTD-like synaptic plasticity.

Numerous studies have reported the beneficial effects of anodal tDCS of the lesioned M1 or cathodal tDCS over the non-lesioned M1 in improving motor function of the paretic limb in stroke patients [7]. Typical dosage involves 10–15 min of 1–2 mA of anodal stimulation over the M1. Studies using tDCS over the upper or lower limb M1 have reported significant improvements in motor function after single or multiple sessions of tDCS. It is important to note that tDCS needs to be administered in combination or before therapy to maximize its potential.

SAFETY CONSIDERATIONS FOR NIBS

At present in the United States, the only neurological indication approved by the Food and Drug Administration for use of TMS is acute migraine. No major safety adverse events have been reported with TMS [8] or tDCS [9]. However, both modalities convey a few risks that can be reduced with proper patient selection. The induction of seizures is the most severe side effect of rTMS especially with high-frequency rTMS, and the occurrence rate is estimated at ≤1/10,000. Because risk is higher in people with previous seizures or brain lesions, or with use of medications that reduce the seizure threshold, these are considered relative contraindications to participate in TMS studies. Another potential adverse event could be heating or damaging metallic implants in or near the head. This can be managed by strictly excluding patients with such implants. Other transient side effects of rTMS include tingling, headache, dizziness, discomfort, hearing loss, and increase in anxiety. These events are reported to improve rapidly with multiple sessions and respond to over the counter treatments. As tDCS does not induce neuronal action potential, the safety of tDCS is higher compared to the TMS protocols. The most common side effects of tDCS includes tingling and itching at the site of stimulation. Mild headaches, nausea, and insomnia have also been infrequently reported. The presence of metal in the head, implanted devices, and severe scalp lesions are some of the contraindications that exclude patients from tDCS trials.

LIMITATIONS OF NIBS

Despite the encouraging results obtained with TMS and tDCS, there are numerous limitations and unresolved issues associated with NIBS [10]. Recent studies have reported high interindividual variability in the neuromodulation response to rTMS and tDCS. As in many stroke intervention studies, not all subjects show the expected improvement. This is in part due to lack of consideration of individual response patterns and individualization of stimulation parameters. The mechanisms of action are still a point of debate with a tendency to favor LTP-like enhancements of plasticity. It should be noted that the magnitude of improvement is varied among studies and may be dependent on the stimulation dosage, the type of patients recruited (acute, subacute, or chronic), lesion location, extent of structural and functional damage, outcome measures used, and the type of task performed in conjunction with NIBS.

FUTURE DIRECTIONS

NIBS represents a promising adjuvant for enhancing motor recovery after stroke. The application of NIBS to stroke patients is experimental and can be recommended only for scientific studies currently. Multicenter, large randomized controlled clinical trials are needed to address questions regarding underlying mechanisms, optimal stimulation parameters, and combination with other types of interventions to establish the role of NIBS in stroke recovery.

References

[1] George MS, Nahas Z, Kozel FA, et al. Mechanisms and state of the art of transcranial magnetic stimulation. J ECT 2002;18(4):170–81.

[2] Nitsche MA, Paulus W. Transcranial direct current stimulation–update 2011. Restor Neurol Neurosci 2011;29(6):463–92.

[3] Ward NS, Cohen LG. Mechanisms underlying recovery of motor function after stroke. Arch. Neurol 2004;61(12):1844–8.

[4] Dimyan MA, Cohen LG. Contribution of transcranial magnetic stimulation to the understanding of functional recovery mechanisms after stroke. Neurorehabil Neural Repair 2010;24(2):125–35.

[5] Tang A, Thickbroom G, Rodger J. Repetitive Transcranial Magnetic Stimulation of the Brain: Mechanisms from Animal and Experimental Models. Neuroscientist 2015.

[6] Le Q, Qu Y, Tao Y, Zhu S. Effects of repetitive transcranial magnetic stimulation on hand function recovery and excitability of the motor cortex after stroke: a meta-analysis. Am J Phys Med Rehabil 2014;93(5):422–30.

[7] Madhavan S, Shah B. Enhancing motor skill learning with transcranial direct current stimulation – a concise review with applications to stroke. Front Psychiatry 2012;3.

[8] Rossi S, Hallett M, Rossini PM, Pascual-Leone A. Safety, ethical considerations, and application guidelines for the use of transcranial magnetic stimulation in clinical practice and research. Clin Neurophysiol 2009;120(12):2008–39.

[9] Poreisz C, Boros K, Antal A, et al. Safety aspects of transcranial direct current stimulation concerning healthy subjects and patients. Brain Research Bulletin 2007;72(4–6):208–14.

[10] Hummel FC, Celnik P, Pascual-Leone A, et al. Controversy: noninvasive and invasive cortical stimulation show efficacy in treating stroke patients. Brain Stimul 2008;1(4):370–82.

CHAPTER

170

Management of Aneurysms and Vascular Malformations During Pregnancy and Puerperium

A.G. Larsen, A. Can, R. Du

Harvard Medical School, Boston, MA, United States

In this chapter, we review the management of aneurysms and vascular malformations in women during pregnancy and puerperium. The lesions discussed include moyamoya disease, dural arteriovenous fistula (dAVF), arteriovenous malformations, and cavernous malformations.

NEUROSURGICAL PROCEDURES DURING PREGNANCY

To determine recommendations for proper management of aneurysms and vascular malformations, we reviewed the literature of women during pregnancy and puerperium who were candidates for neurosurgical intervention.

In a single institutional retrospective study by Nossek et al.,[1] the records of 34 pregnant or early postpartum women who were candidates for neurosurgery were analyzed. Of those 34 women, 5 had vascular malformations categorized as cavernous malformation (CM) ($n=3$), moyamoya disease ($n=1$), and arteriovenous malformation (AVM) ($n=1$) [1]. Of the three patients with CM, one underwent a craniotomy and two were conservatively managed. One of the conservatively managed women developed diplopia. All three women delivered healthy babies at term with no obstetric complications. The woman with moyamoya disease presented hemorrhagically, and suffered a spontaneous abortion at 22 weeks. The woman with the AVM presented with an intracranial hemorrhage due to rupture and underwent endovascular embolization. She delivered at 36 weeks of gestation with no obstetric complications.

In the same study by Nossek et al.,[1] of the 16 women who underwent neurosurgery during pregnancy, 7 had no neurosurgical or obstetric complications, 5 had neurosurgical complications but all resolved, and 4 patients had preterm delivery with 3 out of 5 babies with Apgar scores less than 7 but had a favorable outcome. From the total series of 34, 12 patients delayed their surgery. Of the five patients who delayed their surgery until later in pregnancy, four developed neurosurgical complications. Of the seven patients who delayed their surgery until after delivery, four patients experienced complications: three with neurosurgical complications and one with obstetrical complications. This case series highlights the risks of delaying treatment in neurosurgical patients and the favorable outcomes when treatment was implemented in a timely fashion.

In a study by Cohen-Gadol et al.,[2] 19 of 34 patients who were candidates for neurosurgery underwent neurosurgical intervention during pregnancy. Twelve patients presented with vascular malformations including aneurysm ($n=5$), AVM ($n=5$), dAVF ($n=1$), CM ($n=1$). Of these 12 patients, clinical presentation included subarachnoid hemorrhage ($n=5$), intracerebral hemorrhage ($n=3$), and intraventricular hemorrhage ($n=1$). Six patients underwent immediate surgery for aneurysm clipping ($n=3$), AVM resection ($n=2$), and CM resection ($n=1$) after the patient rebled 1 week after initial presentation. Of the six other patients with cerebrovascular lesions, two patients (PICA aneurysm and AVM) underwent a Cesarean section and were conservatively managed, one patient received an urgent Cesarean section then

Primer on Cerebrovascular Diseases, Second Edition
http://dx.doi.org/10.1016/B978-0-12-803058-5.00170-3

aneurysm repair, one patient with dAVF underwent a therapeutic abortion followed by radiosurgery, one patient with an AVM was conservatively managed with radiosurgery delayed until after delivery, and one patient with an AVM underwent embolization complicated by a thalamic stroke. There was no fetal or maternal mortality or permanent morbidity associated with surgery. Overall, this study demonstrates the relative safety of neurosurgical intervention during pregnancy.

RADIATION EXPOSURE

The risks associated with radiation in diagnostic procedures should be considered when developing a neurosurgical treatment plan. The International Commission on Radiological Protection issued guidelines and summary recommendations for the use of medical radiation during pregnancy. The consequences of radiation differ among the stages of pregnancy, with the fetus most vulnerable to the effects of radiation during organogenesis weeks 3–8 of gestation and the early fetal period with central nervous system development weeks 8–25 of gestation [3]. Radiation absorbed by the fetus above the threshold of 100–200 mGy places the fetus at risk for death, nervous system abnormalities, growth retardation or malformation, and childhood cancer [3].

The most commonly used diagnostic imaging modalities in neurosurgery, including head CTs, plain X-rays, and cerebral angiograms expose the fetus to far less than 50 mGy of radiation if the abdomen is shielded [4]. During cerebral angiography, shielding the abdomen with lead results in exposure to the fetus of less than 1 mGy [5]. Therefore, according to the International Commission of Radiological Protection, the risk to the fetus of death, nervous system abnormalities, and growth retardation or malformation would be negligible since the neurosurgical diagnostic modalities have levels of fetal radiation absorption below the 100–200 mGy threshold [3].

RECOMMENDATIONS FOR THE MANAGEMENT OF VASCULAR MALFORMATIONS

Aneurysms and vascular malformations need to be carefully managed during pregnancy and puerperium to prevent fetal and maternal morbidity or mortality. This section discusses the risk of hemorrhage during pregnancy (Table 170.1) and the recommended practices for the treatment of these lesions (Table 170.2).

TABLE 170.1 Hemorrhage Risk During Pregnancy

	Baseline Annual Risk of Hemorrhage	Risk of Hemorrhage During Pregnancy[a]
Aneurysm	1.9% [7]	Unchanged: 1.4% [6,7]
Arteriovenous malformation	1.1% [13]	Increased: 8.1–8.3% [13,14]
Cavernous malformation	1.6–3.1% [15,16]; Rehemorrhage rate: 4.5–22.9% [16]	Unchanged: 3% [15,17]
Moyamoya	3.9% in women [23]	Unchanged: 2.8% [23]

[a]Risk of hemorrhage per pregnancy assuming each pregnancy is 9 months.

TABLE 170.2 Recommended delivery method

	Recommended Delivery Method
Aneurysm	Vaginal or Cesarean section [6,8]
AVM	Cesarean section [13]
CM	Vaginal or Cesarean section [15]
Moyamoya	Vaginal or Cesarean section [25–28]

Aneurysm

There have been varying reports on whether the risk of hemorrhage due to an aneurysm increases during pregnancy or puerperium. Kim et al. [6] analyzed the National Inpatient Sample (NIS) with about 20 million hospitalizations involving pregnancies on record and found 193 unruptured and 714 ruptured aneurysm cases in pregnancies without delivery, and 218 unruptured and 172 ruptured aneurysm cases in pregnancies with delivery. Assuming the prevalence of unruptured aneurysms among all women of pregnancy age was 1.8%, the estimated risk of aneurysm rupture was 1.4% during pregnancy and 0.05% during delivery [6]. In a meta-analysis by Rinkel et al., [7] of nine studies totaling 3907 patient-years, the annual risk of rupture was 1.9% (1.5–2.4%), which is similar to the risk during pregnancy reported by Kim et al.

In Kim et al.'s study, [6] the overall maternal mortality rate in the hospital for ruptured aneurysm with pregnancy was 9.5% and for pregnancy with delivery was 18%. The mortality rate in the pregnant population was greater for patients who did not undergo treatment with clipping or coiling for the ruptured aneurysm (10.2% without treatment versus 5.2% with treatment) [6]. The maternal mortality rates for those with ruptured aneurysms for pregnancy with delivery was more than three times higher for patients who did not undergo treatment with clipping or coiling (20.4% without treatment versus 6.7% with treatment) [6]. However, it is possible that this is confounded by those who were not treated due to the severity of their presentation. A study by Dias et al., [8] found that the maternal mortality rate from aneurysm rupture was 35% and the fetal mortality was 17%. There was no difference in outcome based on mode of delivery [6, 8]. Given the significant morbidity or mortality of not treating ruptured aneurysms, we recommend that the treatment of pregnant patients with ruptured aneurysms be based on neurosurgical rather than obstetric considerations.

Arteriovenous Malformation

Although data regarding risk of AVM hemorrhage during pregnancy are sparse and inconclusive, pregnancy is considered a significant risk factor for AVM hemorrhage [3]. During pregnancy, it is estimated that the plasma volume and cardiac output increases between 30% and 50% more than what is found in nonpregnant women [9]. The cerebral perfusion pressure of the middle cerebral artery was found to increase by 52% between 12 and 40 weeks of gestation [10]. Increased cerebral blood flow, cardiac output, and vascular volume during pregnancy are thought to contribute to the increased risk of AVM hemorrhage during pregnancy [11].

Dias and Sekhar [8] found in their analysis of 154 patients with spontaneous ICH during pregnancy that 23% of cases were due to ruptured AVMs. The overall maternal and fetal mortality rates from AVM-associated ICH were 28% and 14%, respectively [8]. For those who received surgical intervention, the maternal and fetal mortality rates were 23% and 0%, respectively. In contrast, among women who were treated conservatively with ruptured AVMs, maternal and fetal mortality were 32% and 23%, respectively. Almost 60% of the cohort with ICH due to AVM rupture presented to the hospital in an altered state of consciousness, which might increase the mortality rate even with surgical intervention [8]. In another report of 24 women with AVMs, hemorrhage was associated with a 49% fetal complication rate and a 26% fetal mortality rate [12].

In our retrospective study of 54 women with angiographic diagnosis of AVM, we found a hemorrhage rate of 8.1% per pregnancy, or an annual hemorrhage rate of 10.8% versus 1.1% in nonpregnant women, with a statistically significant hazard ratio of 7.91 for hemorrhage during pregnancy [13]. These findings are consistent with an 11.1% annual hemorrhage rate for women who become pregnant during the three-year latency interval between stereotactic radiosurgery and AVM obliteration [14].

Due to the increased risk of AVM hemorrhage during pregnancy, we recommend surgical resection if the aneurysm has hemorrhaged and if the resection can be performed safely from the neurosurgical standpoint. If an unruptured AVM is discovered during pregnancy, comprehensive, multidisciplinary counseling and evaluation of the risks of surgery versus continued conservative treatment are recommended.

Cavernous Malformation

There are varying reports on whether the risk of hemorrhage due to a cavernous malformation (CM) increases during pregnancy or puerperium. Kalani et al. [15] published a study of 168 pregnancies among 64 patients with cavernous malformations. From this cohort, the risk for symptomatic hemorrhage (e.g., change in neurological status or new-onset or exacerbation of seizure activity) was 3% for each pregnancy, and there was a higher risk of hemorrhage for pregnant women with a positive family history (3.6%) and a lower risk of hemorrhage for patients with isolated lesions and a negative family history (1.8%) [15]. As our group published in 2015, the overall annual hemorrhage rate for CMs ranged from 1.6% to 3.1% per patient-year across the literature in the general population [16]. The rate of hemorrhage was lower for previously unruptured CMs (0.3–0.6%) and for incidental CMs (0.08–0.2%). A study of 186 previously untreated women with cavernous malformations reported that the risk of hemorrhage from CMs did not statistically deviate from baseline during pregnancy, delivery, or the postpartum period [17]. This would suggest that asymptomatic cavernous malformations do not need to be treated during pregnancy.

As we reported in 2015, the increased rate of repeat hemorrhage reported in the literature ranges from 4.5% to 22.9% [16]. A case report by Flemming et al. [18] reported a successful brainstem cavernous malformation resection during pregnancy in a patient who had a repeat hemorrhage with worsening neurological deficits within the same week as the initial hemorrhagic presentation. Warner et al. [19] report a case in which surgery for a CM that had hemorrhaged was delayed due to the patient's pregnancy. The hemorrhage then recurred during pregnancy and the patient experienced remaining deficits. There was no difference in outcome based on mode of delivery [15]. We recommend that women who present with hemorrhage due to CM rupture during pregnancy should be considered for surgical intervention due to the increased risk of repeat hemorrhage and potential neurological deficits.

Moyamoya Disease

Moyamoya disease is characterized by progressive stenosis of the internal carotid arteries and the anterior and middle cerebral arteries [20]. The narrowing of the arteries results in hypoxia and formation of collateral blood supply. The delicate moyamoya collateral vessels pose a risk for hemorrhage [21]. Our retrospective study of 42 adult patients in North America with moyamoya disease reported the rate of hemorrhagic presentation of 17% [20]. The study by Ikezaki et al., [22] reported 69% of South Korean adult patients and 51% of Japanese patients with moyamoya disease presented with hemorrhage. The study found that adults were 26.7 times more likely to present with hemorrhage ($p < .0001$, OR = 26.7) and females were 2.34 times more likely to present with hemorrhage ($p = .008$, OR = 2.34).

Liu et al., [23] reported an annual hemorrhage rate of 3.9% for female patients with moyamoya disease, regardless of pregnancy status. In 144 pregnancies among 81 patients in this cohort, there were 4 hemorrhages resulting in a hemorrhage rate of 3.2% per year and 2.8% per pregnancy, which is similar to the overall rate. The authors concluded that pregnancy and puerperium are not risk factors for intracranial hemorrhage associated with moyamoya disease [23].

In a literature review of moyamoya disease and pregnancy by Komiyama et al., [24] there were 16 cases of cerebral hemorrhage during pregnancy or puerperium in 23 patients newly diagnosed during that period. Only one patient out of 30 with known moyamoya disease prior to pregnancy experienced hemorrhage during pregnancy, and this individual had not undergone a surgical bypass treatment for the disease. Thus, while hemorrhage during pregnancy and puerperium due to undiagnosed moyamoya disease lead to significant morbidity and mortality, those with known moyamoya disease can safely undergo pregnancy and delivery with proper monitoring of volume status and blood pressure. Good outcome is seen with both vaginal and Cesarean delivery modalities with careful monitoring [25–28].

CONCLUSION

Due to high mortality rates associated with intracranial hemorrhage, every necessary step should be taken to prevent the initial hemorrhage and potential rehemorrhage. The decision to treat a vascular malformation should be based on neurosurgical considerations, including hemodynamic changes, hemorrhage rate, rebleed rate, and complexity of the surgery.

References

[1] Nossek E, Ekstein M, Rimon E, Kupferminc MJ, Ram Z. Neurosurgery and pregnancy. Acta Neurochir (Wien) 2011;153(9):1727–35.

[2] Cohen-Gadol AA, Friedman JA, Friedman JD, Tubbs RS, Munis JR, Meyer FB. Neurosurgical management of intracranial lesions in the pregnant patient: a 36-year institutional experience and review of the literature. J Neurosurg 2009;111(6):1150–7.

[3] Pregnancy and medical radiation. Ann ICRP 2000;30(1):iii–viii. 1–43.

[4] Richard K, Osenbach LM. Neuroradiological diagnosis during pregnancy: risks and guidelines. In: Loftus CM, editor. Neurosurgical aspects of pregnancy. Thieme; 1995. p. 250.

[5] Harrigan MR. Cerebral Angiography. In: Lanzer P, editor. Catheter-based cardiovascular interventions: a knowledge-based approach. Springer; 2012.

[6] Kim YW, Neal D, Hoh BL. Cerebral aneurysms in pregnancy and delivery: pregnancy and delivery do not increase the risk of aneurysm rupture. Neurosurgery 2013;72(2):143–9. discussion 150.

[7] Rinkel GJ, Djibuti M, Algra A, van Gijn J. Prevalence and risk of rupture of intracranial aneurysms: a systematic review. Stroke 1998;29(1):251–6.

[8] Dias MS, Sekhar LN. Intracranial hemorrhage from aneurysms and arteriovenous malformations during pregnancy and the puerperium. Neurosurgery 1990;27(6):855–65. discussion 865–6.

[9] Capeless EL, Clapp JF. Cardiovascular changes in early phase of pregnancy. Am J Obstet Gynecol 1989;161(6 Pt 1):1449–53.

[10] Belfort MA, Tooke-Miller C, Allen JC, Saade GR, Dildy GA, Grunewald C, et al. Changes in flow velocity, resistance indices, and cerebral perfusion pressure in the maternal middle cerebral artery distribution during normal pregnancy. Acta Obstet Gynecol Scand 2001;80(2):104–12.

[11] Trivedi RA, Kirkpatrick PJ. Arteriovenous malformations of the cerebral circulation that rupture in pregnancy. J Obstet Gynaecol 2003;23(5):484–9.

[12] Robinson JL, Hall CS, Sedzimir CB. Arteriovenous malformations, aneurysms, and pregnancy. J Neurosurg 1974;41(1):63–70.

[13] Gross BA, Du R. Hemorrhage from arteriovenous malformations during pregnancy. Neurosurgery 2012;71(2):349–55. discussion 355-6.

[14] Tonetti D, Kano H, Bowden G, Flickinger JC, Lunsford LD. Hemorrhage during pregnancy in the latency interval after stereotactic radiosurgery for arteriovenous malformations. J Neurosurg 2014;(121 Suppl):226–31.

[15] Kalani MY, Zabramski JM. Risk for symptomatic hemorrhage of cerebral cavernous malformations during pregnancy. J Neurosurg 2013;118(1):50–5.

[16] Gross BA, Du R. Cerebral cavernous malformations: natural history and clinical management. Expert Rev Neurother 2015;15(7):771–7.

[17] Witiw CD, Abou-Hamden A, Kulkarni AV, Silvaggio JA, Schneider C, Wallace MC. Cerebral cavernous malformations and pregnancy: hemorrhage risk and influence on obstetrical management. Neurosurgery 2012;71(3):626–30. discussion 631.

[18] Flemming KD, Goodman BP, Meyer FB. Successful brainstem cavernous malformation resection after repeated hemorrhages during pregnancy. Surg Neurol 2003;60(6):545–7. discussion 547–8.

[19] Warner JE, Rizzo JF, Brown EW, Ogilvy CS. Recurrent chiasmal apoplexy due to cavernous malformation. J Neuroophthalmol 1996;16(2):99–106.

[20] Gross BA, Du R. The natural history of moyamoya in a North American adult cohort. J Clin Neurosci 2013;20(1):44–8.

[21] Scott RM, Smith ER. Moyamoya disease and moyamoya syndrome. N Engl J Med 2009;360(12):1226–37.

[22] Ikezaki K, Han DH, Kawano T, Kinukawa N, Fukui M. A clinical comparison of definite moyamoya disease between South Korea and Japan. Stroke 1997;28(12):2513–7.

[23] Liu XJ, Zhang D, Wang S, Zhao YL, Ye X, Rong W, et al. Intracranial hemorrhage from moyamoya disease during pregnancy and puerperium. Int J Gynaecol Obstet 2014;125(2):150–3.

[24] Komiyama M, Yasui T, Kitano S, Sakamoto H, Fujitani K, Matsuo S. Moyamoya disease and pregnancy: case report and review of the literature. Neurosurgery 1998;43(2):360–8. discussion 368-9.

[25] Sato K, Yamada M, Okutomi T, Kato R, Unno N, Fujii K, et al. Vaginal delivery under epidural analgesia in pregnant women with a diagnosis of Moyamoya disease. J Stroke Cerebrovasc Dis 2015;24(5):921–4.

[26] Sei K, Sasa H, Furuya K. Moyamoya disease and pregnancy: case reports and criteria for successful vaginal delivery. Clin Case Rep 2015;3(4):251–4.

[27] Takahashi JC, Ikeda T, Iihara K, Miyamoto S. Pregnancy and delivery in Moyamoya disease: results of a nationwide survey in Japan. Neurol Med Chir (Tokyo) 2012;52(5):304–10.

[28] Tanaka H, Katsuragi S, Tanaka K, Miyoshi T, Kamiya C, Iwanaga N, et al. Vaginal delivery in pregnancy with Moyamoya disease: experience at a single institute. J Obstet Gynaecol Res 2015;41(4):517–22.

Index

CPI Antony Rowe
Eastbourne, UK
January 04, 2018